Biomechanical Aspects of Soft Tissues

Biomechanical Aspects of Soft Tissues

Benjamin Loret

Fernando M. F. Simões

CRC Press
Taylor & Francis Group
Boca Raton London New York

CRC Press is an imprint of the
Taylor & Francis Group, an **informa** business

CRC Press
Taylor & Francis Group
6000 Broken Sound Parkway NW, Suite 300
Boca Raton, FL 33487-2742

First issued in paperback 2020

© 2017 by Taylor & Francis Group, LLC
CRC Press is an imprint of Taylor & Francis Group, an Informa business

No claim to original U.S. Government works

ISBN 13: 978-0-367-57434-5 (pbk)
ISBN 13: 978-1-4987-5239-8 (hbk)

Version Date: 20160609

Library of Congress Cataloging-in-Publication Data

Names: Loret, Benjamin, author. |Simões, Fernando Manuel Fernandes, author.
Title: Biomechanical aspects of soft tissues / Benjamin Loret and Fernando
Manuel Fernandes Simões.
Description: Boca Raton : Taylor & Francis, 2017. | Includes bibliographical
references and index.
Identifiers: LCCN 2016015588 | ISBN 9781498752398 (alk. paper)
Subjects: LCSH: Biomedical materials. | Tissue engineering.
Classification: LCC R857.M3 L67 2017 | DDC 610.28/4--dc23
LC record available at https://lccn.loc.gov/2016015588

Visit the Taylor & Francis Web site at
http://www.taylorandfrancis.com

and the CRC Press Web site at
http://www.crcpress.com

Crée comme Dieu
Ordonne comme un roi
Travaille comme un esclave
Constantin Brancusi

Contents

Preface

This book, devoted to the constitutive aspects of the biomechanics of soft tissues, is organized in three parts. Part I provides an in-depth description of the mechanics and thermodynamics that underlie the constitutive modeling of biological tissues in a continuum perspective. Although comments and examples are definitely aimed at soft tissues, the range of applications of this part is much larger and many of the tools that are presented apply to other fields, like mechanical engineering or civil engineering.

On the other hand, Parts II and III are solely devoted to the biomechanics of soft tissues. Part II presents a unified framework in which electro-chemo- mechanical couplings that are associated with the composition-induced function of three tissues with a fixed electric charge are exposed. As it is subjected during walking and jumping to a huge number of intense load cycles, as many as 1 million times a year, the articular cartilage that coats the knees and hips should be, mechanically speaking, endowed with a high endurance. Favored by overweight, osteoarthritis (OA) is the main pathology that affects articular cartilages. In the United States, OA affects 5% of the population and 70% of the population aged over 65, and costs \$8 billion annually. Osteoarthritic cartilage leads to decreased tensile and compressive stiffness, increased tissue hydration and permeability. The intervertebral disc provides a second instance of tissue whose composition and structure are best adapted to the physiological function, which is also mostly to sustain a load. The corneal stroma is another lamellar tissue, subjected to the much smaller ocular pressure, that is yet adequate to ensure transparency.

Part III addresses the growth of biological tissues. In fact, these three tis- sues are aneural, avascular and alymphatic in adults. In the case of genetic degeneration or accidental damage, their regeneration is practically nil. One method to treat damaged cartilage consists in removing the damaged part and in implanting biodegradable cell-polymer constructs. Once in place, artificial cartilage proliferates reasonably well. However, the fact that its mechanical properties are lower than those of native cartilage is an issue. Recent studies have shown that the mechanical conditions to which chondrocytes, the cartilage cells, are subjected in vivo and in vitro affect the synthesis and degradation of the extracellular matrix (ECM). The challenge of tissue engineering is to produce implantable tissue with physical properties similar to the native car- tilage. Articular cartilages are addressed in the book from the points of view of their normal functioning and of the production of substitutes through tissue engineering.

On the other hand, tumors take advantage of the vasculature of tissues. Mechanics is seen to play a very important role here as well and tumor pressure opposes drug delivery. Tumor tissue grows abnormally fast with respect to normal cells. The process from the alteration of the first cell to the formation of tumors is usually long. Unlike normal cells whose proliferation is regulated, tumor cells proliferate without physiological control. They tend to establish their own colony, damaging the lymphatic system, and recruiting vessels to their own benefit, in order to both attract nutrients and spread metastases. It has been recently discovered that the mechanical state that is dominating inside the tumor plays an important role in the overall process. In fact, fluid pressure is high and this has two diabolic consequences: (a) metastases easily spread out of the tumors and invade normal tissues; (b) drug delivery to these zones of high pressure is virtually impossible. Tricky methods to bypass the high fluid pressure barriers opposed to therapeutic agents have been proposed but have still to prove their efficiency.

In all examples that are addressed, the structural architecture, biochemical composition and, if necessary, molecular basis, of the tissues under consideration are provided in the running chapter. In the electronic version, references to the appropriate chapters of Part I are abundantly available via hyperlinks. The bibliography is also reachable through hyperlinks. During the exposition, a number of examples taken from the literature are reviewed and adapted to the framework. Some more comprehensive problems are treated in detail so as to highlight the concepts that have been presented, and also to familiarize the reader with the analytical/computational tools that are needed to fully benefit from the book. Along the way, some particular results are moved to the end of the chapter in a section devoted to exercises that are fully documented.

Authors

Dr. Benjamin Loret is professor of mechanics and civil engineering at the University of Grenoble, France. His research addresses the constitutive responses of engineering and biological materials to static and dynamic loadings. He has especially focused on the couplings of thermal, hydraulical, electrical, chemical and mechanical natures that are ubiquitous in fluid saturated porous media. Applications target innovative energy production systems and biomechanics of soft tissues.

Fernando M. F. Simões received his Ph.D. degree in civil engineering from the Instituto Superior Técnico of the Technical University of Lisboa, Portugal, in 1997. Presently, he is assistant professor at the Department of Civil Engineering, Architecture and GeoResources, University of Lisbon. His research interests include structural mechanics and biomechanics.

Chapter 1

Biomechanical topics in soft tissues

1.1 Introduction

Biomechanics applies the laws and the techniques of mechanics in the study of biological systems and phenomena. Indeed, mechanics provides the tools for:

- the construction of models of the musculo-skeletal system with applications in ergonomics, in sports, in the study of gait and in processes of rehabilitation, or in the development of orthoses and prostheses;
- the construction of physical and mathematical models to describe, simulate and predict the behavior of soft and hard tissues of the human body, like muscles, cartilages, skin, blood vessels, and bones, with applications in areas such as orthopedics, dental medicine, medical interventions in the cardio-vascular system or tissue engineering;
- the construction of models of fluid flows like blood circulation and breathing;
- the computational modeling of tissues.

Biomechanics aims at using these mathematical and computational tools to aid in the medical diagnosis, therapeutics and surgery planning. This work targets a few topics pertaining to the biomechanics of soft tissues. The subjects that have been addressed can be grouped in three main areas:

I.   Solids and multi-species mixtures as open systems: a continuum mechanics perspective;

II. Electro-chemomechanical couplings: tissues with a fixed electrical charge

III. Growth of biological tissues

Part I, including Chapters 2 to 11, presents the elements that are necessary to later consider the biomechanical issues in a porous media perspective. Basic notions of continuum mechanics that will be used throughout are collected in Chapter 2. Elements necessary to master the exchanges of mass, momentum and energy between constituents of an open mixture form the backbone of the exposition. Continuum thermodynamics of fluids and solids serves to guide the modeling attempts. The basic constitutive models that describe the mechanical response of soft tissues account for their peculiarities, namely anisotropy may be strong, deformations may be large, the solid constituents themselves may be endowed with a time dependent mechanical response. Moreover, emphasis is laid on the consequences of the interactions between the mechanical properties (elasticity, viscoelasticity), the transport properties (of biological fluids, macromolecules and ionic or neutral solutes), and the transfers of mass and energy.

Chapters 12 to 19, forming Part II, specialize in the electro-chemomechanical couplings in tissues with a fixed charge. Focus is on articular cartilages, on the corneal stroma and, in a minor mode, on the intervertebral disc. The fixed charge which is referred to sits on the solid parts of the tissues, in contrast to the mobile charges that are dissolved in the biological fluids. In that sense, the adjective "fixed" refers to space, or geometry, not to time. The change of the fixed charge associated with pH is considered at length in Chapter 17.

The presentation can be seen as macroscopic, in the sense that no formal change of scale is attempted. Still, a number of information from the lower scale, starting from the nanoscale, permeates at the scale of study. In that respect, we may mention the directional repartition of collagen fibers, the unbalanced charges in collagen and proteoglycans, the spatial organization of the lamellae in the corneal stroma and intervertebral disk.

Issues regarding growth of biological tissues are addressed in Part III, which includes Chapters 20 to 24. The perspective is on tissue engineering. Elements of biochemistry and mechanobiology are included in basic modeling attempts. In line with earlier chapters, the analysis considers tissues as saturated porous media where several solid constituents may grow due to exchanges of nutrients through the biological fluids. In fact, the last chapter which is devoted to the growth of solid tumors follows this same perspective.

Certain chapters cover in part or entirely purely mechanical issues. Chapter 5 addresses linear elasticity but lays emphasis on the anisotropic properties of soft tissues and on their conewise constitutive response, a large part of tissues being reinforced by collagen. This fiber-reinforced aspect is further emphasized in Chapter 12 where the concept of tissue fabric is introduced. The important notions of length scale and spatial heterogeneity are unfortunately not covered. On the other hand, intrinsic time dependence of the constitutive response is the subject of Chapter 8. Some purely mechanical aspects of the growth of tissues are covered by Chapter 22.

Thermal aspects in both solid and porous media contexts are exposed in Chapters 9 and 10. Chapter 9 is devoted to heat conduction in media that are in local thermal equilibria, Chapter 10 puts emphasis on mass and heat transfer in media that are not in instantaneous equilibrium, neither hydraulically nor thermally, at time scales of interest in soft tissues. Section 16.5 is devoted to generalized diffusion in hydrated tissues with thermal effects. While the subjects considered in these chapters are preliminary, the other chapters are all devoted to issues in a porous media context where mechanics and transport are addressed in the presence of couplings of chemical, electrical, bio-energetic and thermal natures.

Even if the developments of the models are self-content, we think that it is important to perform tests in an appropriate framework, which is referred to as initial and boundary value problem (IBVP). In order to help the reader, the responses to simple tests of several pertinent constitutive models gathered from the literature are re-phrased in that perspective. In fact, a correct specification of the boundary conditions is crucial, especially when couplings are present in the constitutive models.

In order to show how this work can be used, a few problems that are not treated per se in the text have been listed in this introductory chapter. For each of them, we provide a short description and we refer to the chapters, sections or pages of the work that contain elements that are of interest to begin discovering these problems. This approach both gives the opportunity to introduce to further adjacent topics of current interest in biomechanics and it also allows to discover the issues that are considered at length in the next chapters.

1.2   Diffusion, convection, osmosis: toward wearable artificial kidneys

An artificial kidney is an example of a device that makes use of countercurrents to enhance the transfer of small molecules (urea, creatinine) through a fixed membrane (Fig. 1.2.1). The leading governing phenomena are diffusion, convection and osmosis. The basics of diffusion and convection of solutes are exposed in Chapter 15, and the concepts

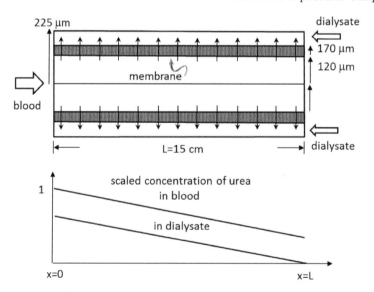

Fig. 1.2.1: Schematic of urea transfer from blood to dialysate in a hollow cylindrical fiber, according to Rambod et al. [2010].

of osmosis and semi-permeable membranes in Section 13.7. Problems of a similar nature are considered for the extravasation of drugs across the vascular wall in Section 15.8.3, for blood filtration and re-absorption by capillaries in Sections 15.9.3 and 24.8.1 and for coupled flows through electrically charged tissues in Chapter 16.

A wearable artificial kidney (WAK) is a device aimed at palliating dys- or non-functioning kidneys [Rambod et al. 2010]. It is lightweight, belt-type, battery driven, and it generates pulsatile flows of low rate since it operates around the clock. Dialysate regeneration is included.

Like a steady pump, the WAK pump pushes blood into the dialyzer but it pulls/sucks the dialysate out, which improves the filtration (a steady pump would push the dialysate out). The higher transmembranar pressure enhances filtration, convection and ultimately blood clearance. The two blood and dialysate inflows are pulsatile with the same frequency but out of phase.

The dialyzer consists of a number of long and thin hollow fibers, with a high specific surface (for filtration), where blood and the saline solution circulate in opposite directions and are separated by a membrane through which the solutes diffuse and are convected by the flow. The flow rates of blood and dialysate are controlled at the inlet and the pressures at the outlet. Maintaining a constant transmembranar pressure is crucial, given that the ultrafiltration rate depends on this pressure. This requirement is a technical challenge for conventional hemodialyzers that appears to be overcome by pulsatile flows [Rambod et al. 2010].

1.3 Issues in drug delivery

While a drug may be efficient once it has reached its target, its delivery remains a key issue. Oral and intravenous administrations of drugs are sometimes inefficient due to dilution and to physiological barriers. To reach a sufficient concentration at the targeted zone, such a high dose would need to be administered that it may induce toxic effects. When the barriers to the drugs are too tight, as in the case of blood–brain barriers, direct delivery through catheters may be the only viable issue.

Hydrogels as drug carriers and release controllers

Oral administration of a therapeutic agent is viable when it can diffuse to the targeted region. The technique consists in wrapping the active agent in a polymer network of dehydrated gel. Living bodies contain a substantial amount of water with dissolved ions. Once ingested, hydrogels[1.1] hydrate and swell. High hydration increases the diffusion coefficient of soluble drugs: water acts as a plasticizer that depresses the glass transition temperature, turning the polymer chains in the rubbery state where their mobility is higher. The active substance is released (due to concentration gradients) when the polymer network becomes sufficiently weak. Drug delivery depends on the diffusion and on the swelling of the coating polymer. The delivery process is controlled by the Deborah and swelling interface numbers which involve the characteristic time of polymer network relaxation, featured in Section 8.8, and the characteristic time of drug diffusion and liquid seepage, defined in Section 13.13.3. Hydration can be elicited by external stimuli, e.g., by an energy flux for thermosensitive polymers. Swelling of biological gels result from osmotic forces resisted by the elasticity of the network, a topic exposed in Section 13.7. In fact, each polymer presents its own peculiarities, of electrical nature (type of the electrical charges, ionic valences, electrical field, etc.), of chemical nature (ionic types, concentrations, pH, etc.) or of mechanical nature (deformations, stresses, pressures). Hydrophilic glycosaminoglycans drive swelling, and/or contraction, due to the presence of hydroxyl, carboxyl and sulfonic residues as exposed in Section 17.1. The nature of the polymer network and density of crosslinks[1.2] is an important ingredient. As a reference, we may contrast the tight organization of collagen fibers in articular cartilages aimed at minimizing strain and described in Chapter 14, from the weaker collagen interconnections in the cornea aimed at avoiding swelling and optimizing transparency reported in Section 19.1. Electro-chemomechanical couplings have been taken into account to describe the volume changes and stresses in articular cartilages and corneal stroma due to changes of the chemical composition, pH and mechanical conditions of the solution the tissues are bathed in.

The presence of crosslinks hinders dissolution of the therapeutic agent in the aqueous medium. The polymers should be able to deliver the therapeutic agent over the requested period of time. Indeed, delivery may need to be

---

[1.1] *Hydrogels* are crosslinked chains of swellable hydrophilic insoluble polymers. While water is the continuous phase in hydrogels, *gels* are semi-solid colloids of a solid and a liquid. In contrast to *solutions* where the solvent and the solute form a single phase, the continuous and dispersed components of a colloid form distinct phases.

[1.2] A crosslink is a covalent or ionic bond that links polymer chains. The chains of *chemical* hydrogels are crosslinked by covalent bonds while the crosslinks of *physical* hydrogels involve the much weaker hydrogen bonds or van der Waals interactions.

sustained over days/weeks/months or rest on a one-time basis. They should be biocompatible and biodegradable without toxicity. Their dissolution itself is negligible if it takes place after complete drug release.

When water seepage is slow with respect to polymer relaxation, drug diffusion obeys Fickian diffusion with a concentration depending on time t like t^α with $\alpha = 1/2$ as reported in Section 15.3.3. Semi-empirical expressions of the time dependence of drug concentration have been developed to cover more complex cases, e.g., for close characteristic times of polymer relaxation, drug diffusion and fluid seepage, for non-Fickian diffusion in the glassy state of the polymer network, or for burst release [Siepmann and Peppas 2001].

The presence of the polymer hinders diffusion of solvents and solutes: the effect is often embedded in a tortuosity factor that involves the volume fraction of the solid, an entity touched upon in Section 16.4.3. More elaborated models include some geometrical features of the diffusing species and of the polymer [Masaro and Zhu 1999]. Improving over the above models where the polymer network is seen as obstructing diffusion, hydrodynamic theories account for interactions between polymer network, solvent and solutes and are valid at higher concentrations: Sections 15.5 and 15.7.8 provide examples along this approach.

Drug delivery in the presence of strong barriers

The issue of drug delivery surfaces conspicuously for the treatments of neurodegenerative diseases and tumors of the central nervous system. Indeed, *the blood-brain barrier* in particular opposes the entrance of a number of macromolecules into the brain tissue. A new *invasive* technique termed "convection enhanced drug delivery" aims at alleviating this problem by bypassing the vasculature and introducing directly the drug into the brain tissue through an infusion catheter. With respect to the purely diffusive process described in Section 24.8.3, convection enhanced drug delivery increases the penetration depth [Linninger et al. 2008]. Infusion pressure, flow rate, catheter design, including port diameters, inter-port distance, and catheter position are unknowns to be optimized. While the hydraulic conductivity of the gray matter containing neuron cells is low and can be considered as isotropic, the white matter formed by long axons has directional physical properties. The introduction of directional properties in the permeability tensor is elaborated in Section 15.7. Some attempts to account for the heterogeneous substructure of the tissue are mentioned in Sections 15.7.8 and 19.4.5 for the stroma. In an alternative approach [Stöverud et al. 2012] use the self-diffusion tensor of water in the brain tissue obtained by magnetic resonance imaging (MRI) to infer the permeability and diffusion tensors of the tissue, the three entities being assumed to be coaxial. In fact, the brain surfaces, volumes and boundaries between regions and therefore the geometric reconstruction from MR images and the physical properties of the tissue are patient-dependent.

As the drug moves through the tissue, it can decompose via enzymatic reactions, be absorbed by cells and be released through the vasculature (blood clearance). Basics of enzyme kinetics are briefly exposed in Section 15.12. Interstitial pressure, membrane permeability and osmotic/oncotic pressures control the reaction (release/absorption) processes. These *reactive* processes are accounted for in the mass balances as thoroughly exposed in a general context in Sections 4.4, 10.2, and illustrated for fluid extravasation in solid tumors in Sections 15.8, 24.8-24.9, and for oxygen diffusion in Section 15.11.

As a routinely adopted property in this work, the brain tissue has to be considered deformable, and endowed either with an elastic or hyperelastic time-independent stress response as exposed in Chapters 5 and 6, or with a viscoelastic response, the subject of Chapter 8. In fact, changes of porosity may require a large deformation framework, a topic addressed in Chapter 6 and used for tissue growth in Chapters 22 and 23.

1.4 Macroscopic models of tissues, interstitium and membranes

Description of the components of tumor tissue in a drug delivery perspective is attempted through several porous media models in Section 24.6. Including tissue elasticity, the model in Section 24.8.2 accounts for two compartments, namely the interstitium and the capillaries, and three phases, the interstitium being composed of a solid and a pore fluid. It is used to describe the mass exchanges between the capillaries and the pores in the context of a systemic therapy.

For cryopreservation of whole organs, a cryoprotective agent CPA should be introduced in the intracellular space[1.3]. This procedure is realized by perfusion through the vascular system. A model should then include a finer description of the tissue constitution. In fact, the cell volume varies much from one tissue to the other, depending of course on the actual cellular activity: this activity is very small in adult articular cartilages, about 2 to 3% of

---

[1.3] A cryoprotective agent is a substance that prevents tissue damage by ice, increasing the solute concentrations in cells. Actually, the technique termed vitrification lowers the glass transition temperature and keeps the tissue in a glassy state.

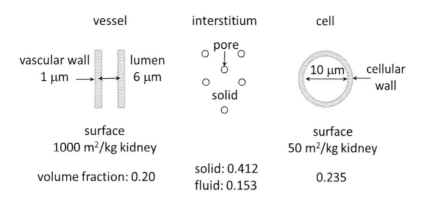

Fig. 1.4.1: Schematic of the porous media model with three compartments for rabbit kidney, each one containing species, namely CPA, ions, etc. with separate concentrations [Lachenbruch and Diller 1999].

total volume, but large in the cardiac muscle, and intermediate in kidneys, about 20% of tissue. In fact, the model of kidney considered by Lachenbruch and Diller [1999] involves the following items:

- the model includes three compartments, Fig. 1.4.1: (a) the vascular and lymphatic systems; (b) the interstitium; (c) the intracellular space. The concentrations of water, CPA and other solutes and the pressure are innate to each compartment;
- water and CPA move through the vascular membranes and cellular membranes. In developing the transport equations, attention is paid to the fact that the CPA concentration may be as high as 4 mole per liter, at variance with conventional formulations that consider dilute concentrations. As another consequence, the viscosity of the solution is expected to change with solute concentration;
- the transport of ions Na^+, Cl^-, Mg^{++}, Ca^{++}, is *passive* through the vascular membrane and interstitium as described in Chapters 15, 16 and Section 19.14. On the other hand, their transport through the cell membranes requires an energy source (*active transport*) as described for the corneal endothelium in Sections 19.3 and 19.15. At low temperature, the energy source is typically not available and transport through the membrane can take place only by diffusion and convection. Therefore, the membrane can be seen as practically impermeable to these solutes. It is also impermeable to large proteins by steric considerations;
- CPA and water diffuses through the cell membrane by coupled diffusion, a topic addressed in Chapters 15 and 16;
- the elastic response of the compartments and surrounding capsule is accounted for: Chapters 5 and 6 consider several aspects of elasticity.

Perfusion with glycerol, dimethyl sulfoxide and sucrose-based CPA results in swelling of the interstitium and shrinkage of the cells. Definite protocols for injection and removal of CPA are needed as the relative volume changes that cells can undergo without damage is known to be limited.

A large molar mass polypeptide that cannot cross the vascular membrane may be added to the perfusate to induce an osmotic pressure, see Section 15.8. The perfusate may also contain some agent that triggers active transport through the cellular membrane.

1.5 Energy couplings, passive and active transports

Energy couplings are ubiquitous in biomechanics. The basic idea of coupling requires an energy provider, an energy consumer and a device that is able to transfer the energy, with some overhead costs, between the two agents. The idea is tentatively illustrated in Fig. 4.8.2 in a restricted perspective where the provided and consumed energies are both of mechanical nature. In other examples developed in Section 4.8.3, the energy provided by a chemical reaction is used by another chemical reaction. The main biochemical energy source in living bodies is ATP. Its activation modifies the transport of solutes across cell membranes. The chemical equilibrium that would hold in absence of energy source is altered to "a dynamic equilibrium": examples described in Sections 19.3.3 and 20.2.2 show that the concentrations on the sides of the membranes depend explicitly on the energy source.

This phenomenon, termed active transport, is tied to the presence of the energy providers, the metabolic pumps, in the cell membranes. The corneal endothelium is a conspicuous instance of a cell membrane that features active transporters, refer to Section 19.3 for details. Active transport mechanisms in the corneal endothelium limits the infinite radial swelling, which might take place due to the weak radial crosslinking between lamellae. Swelling of the stroma would be due to the absorption of water from the aqueous humor through the leaky endothelium. The dynamic equilibrium that maintains the optimal hydration regarding transparency is evidenced by decreasing the temperature just above freezing so as to silence the metabolic pumps. Then the thickness of the cornea increases but, when the temperature is returned to a physiological value, the normal thickness is progressively retrieved after a few hours.

As just alluded to, energetic coupling alters a chemical equilibrium to a dynamic equilibrium. However, at variance with the example referred to so far, the energy source may not be of biochemical origin, but rather of mechanical origin, giving rise to a phenomenon that may be called *mechanobiological coupling*.

The idea is developed for the growth of cartilage tissues in Chapter 23. Tissue growth is an irreversible process, metabolism and catabolism being associated with distinct pathways. Energy of mechanical and biochemical natures is used to grow the mass and improve the mechanical properties of the tissue constituents. Heat dissipation is in some sense minimized: the tissue recovers as much energy as possible to drive the growth and stiffening/self-healing processes. Indeed in contrast to damage, stiffening contributes negatively to the dissipation inequality. In other words, the model developed in Section 23.1 is built so that the dissipation can only be positive or zero. The idea is that mechanisms that require energy (active processes) are scaled down if the energy provided by passive processes is not sufficient. The energy available to the active processes is limited by satisfaction of the dissipation inequality.

As another instance of mechanochemical coupling and active transport, Albro et al. [2008] provide experimental evidence that strain controlled cyclic loading may enhance reversibly the uptake of a macromolecule in an agarose hydrogel.

In certain engineering materials, a chemical energy source is dispatched to potentially improve their mechanical properties in case of damage: Section 23.8 presents a brief account of this subject. This process, referred to as "self-healing" to convey the idea that energy from the surroundings in the form of heat or light is not required to repair the material, features a *chemomechanical coupling*. Both mechanical and metabolic energy sources drive tissue growth: they generate mechanochemical and chemomechanical couplings, respectively.

1.6   More general direct and reverse couplings

While the muskulo-skeletal and cardiovascular systems are not addressed here, it is of interest to provide some brief overview as they display couplings of various physical natures with both similarities and differences with respect to articular cartilages.

As an organ, the heart receives from the body various types of information (electrical, chemical, mechanical) to which it reacts. As such, it is an object of study for automation specialists. Now clearly, this interaction depends on the behavior of the heart itself. Physiologically, the cardiac muscle is very far from cartilage. Articular cartilage is avascular, aneural and alymphatic, its adult cells occupy a small percentage of the medium, they are practically inactive, and the chemomechanical activity is extracellular. The cardiac muscle on the other hand is a syncitium: cells occupy a large portion of the space and they communicate quickly through gap junctions.

Although endowed with a certain autonomy by the sinoatrial node, a natural pacemaker, the heart is subjected to the injunctions of the central nervous system. The spatio-temporal diffusion of the electrical signal produces a contraction that is optimized due to the subtle conformation of the fibrous structure of the organ (the study of its morphogenesis touched upon in Section 21.7.1 is in its infancy). The electro-chemomechanical coupling is in fact mediated by chemistry, and, in the long range, by metabolism. The electrical signal elicits a number of electric currents through the cardiomyocyte membrane that give rise to calcium cycling and muscle contraction, an area termed electrophysiology. The membrane behaves as a capacitor of fixed capacitance as described in Section 13.6.3 and p. 481. There are three types of channels through the membrane, namely ligand-gated, voltage-gated and stretch-gated channels. For the myocardiac cells, the ligand is ATP and the channels are then called *ionic pumps*, in as far as ATP allows the transport of ions across these channels against a concentration gradient.

Coupling cell electrophysiology to bioenergetics aims at obtaining a realistic control of the metabolic activities of the cardiac cells by the ATP demand of crossbridges. Indeed, the parts of the cardiac metabolism (glycolysis, TCA cycle, β-oxidation, oxidative phosphorylation) are themselves described by complex models that involve enzymatic reactions defined by various kinetics, a subject touched upon in Sections 15.12 and 20.2. β-adrenergic factors reported in Section 20.2.4 are certainly required in the modeling of the cardiac function, and especially for the simulations of ischemia and hypoxia.

The inverse *mechanoelectric coupling* is also thought to be exhibited by stretch-dependent ionic channels and by commotio cordis, a phenomenon that gives rise to cardiac arrest without tissue damage in young subjects. At variance, cardiac arrest by hypoxia or anoxia is accompanied by local tissue damage.

At the scale of the body, it is also certain that lack of exercise leads to a decrease of the mass of skeletal muscles, tendon efficiency and to bone decalcification, a mechanobiological coupling. The issue is raised in a particularly acute way for deep space explorations due to the microgravity environment. Appropriate exercises to help maintain the integrity of the muskuloskeletal and cardiovascular systems of astronauts need to be devised [White et al. 2003].

1.7 Tissue engineering redirected to tumor tissue exploration

While Chapter 20 is devoted to decipher the effects of mechanics on tissue engineering, the mechanical environment induced in a tumor tissue is considered in Chapter 24. Knowledge from these two independent expositions may advantageously be combined in the future to gain better understanding of biological and biochemical effects on the mechanical properties and of the reciprocal mechanobiological effects in both domains.

Naturally derived and artificial 3D scaffolds provide a controllable environment in which the growth of tumor cells and their interactions with the extracellular matrix (ECM) can be examined. The physical properties, in particular the density, and the mechanical properties of the ECM are significantly modified in a tumor tissue, a point stressed in Section 24.1. The shear modulus of tumor tissues may be an order of magnitude higher than in normal tissues. Their architecture and composition are also altered, including abnormal collagen density and alignment, oxygen concentration, pH and concentration of growth factors. *In vitro* cultures on biodegradable scaffolds aim at deciphering the mechanotransductive agents that mediate the dynamics of these interactions.

In fact, biochemists have realized that traditional 2D cultures do not pay enough attention to the role of the ECM in cancer metastasis and that they may gain from techniques of tissue engineering. The proliferation and spreading of cell lines differ significantly according to the stiffness of the scaffold [Helmlinger et al. 1997]a. In tumor tissues, a failing lymphatic system disorganizes the circulation of fluids and hinders drug delivery. Whether the altered ECM properties are the cause or the consequence of the tumor cell development, these properties should be understood in view of a better assessment of the elements that participate in the resistance to drug delivery.

The scaffolds should provide a porous matrix, and be proteolytically degradable to permit drug migration and cell re-structuration. They are expected to be noncytotoxic, controllable in their mechanical and diffusion properties, and susceptible to a bioactivity comparable to natural tissues. Ideally, these three properties should not be interacting to allow a full *in vitro* exploration of their effects [Gill and West 2014]. Naturally derived scaffolds, using collagen and fibrin, are endowed with the three properties but modification of one of them inevitably reflects on one or the two others. Semi-synthetic hydrogels, obtained by modifying polysaccharide alginate and hyaluronic acid, and synthetic scaffolds, based on polylactic acid and polyglycolic acid, offer the advantages of being noncytotoxic while their stiffness may be controlled during the polymerization process, a point reported in Sections 20.4 and 20.5.

Bioactivity of polyethylene glycol (PEG) hydrogel matrices can be induced through incorporation of ligands, peptides and growth factors. They are fabricated by photopolymerization, a technique[1.4] overviewed in Section 20.8. The matrix stiffness may be controlled via the PEG concentration and the molar mass of the polymer chain.

1.8 From mechanics to biomechanics

Some living tissues are submitted to severe mechanical loading conditions. During daily life, walking and jumping, articular cartilages undergo repetitive stresses ranging from 3 to 18 MPa as many as 1 million times a year. Teeth are subjected to about 1 million loading cycles per year [Bayne 2005]. While ceramics mimic natural enamel, they often contain pores that act as stress concentrators and initiators of fractures. Switching from traditional synthetic restorative dental materials to truly biological biomaterials is expected to take place in 10 to 20 years and it will require significant progresses in a chain involving 3D scaffolds, stem cells and inductive growth factors. Longevity of the substitutes necessary requires a deeper understanding of the mechanobiological interactions that take place in a polycyclic context.

Therefore, substitutes should not only provide mechanical resistance in compression, tension and/or shear as well as biological function but also be compatible with the interactions between mechanics, biochemistry and biology. Elements of this vast perspective are exposed in this book.

[1.4] Use of light for cancer therapy is briefly addressed in Section 24.3.7.4.

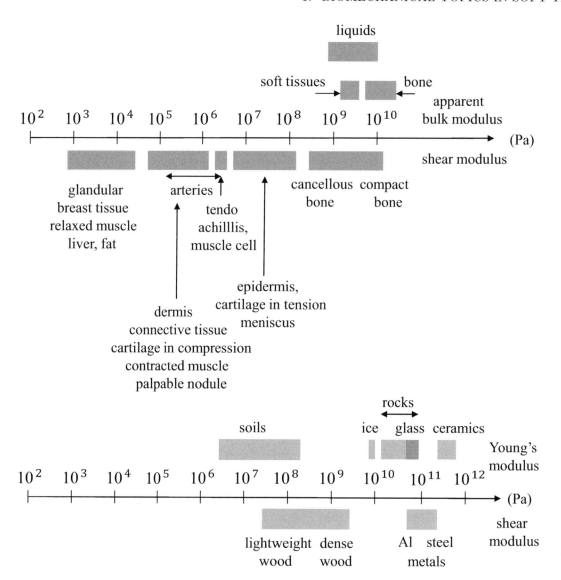

Fig. 1.8.1: Range of elastic moduli in biological tissues, compared with other engineering materials based on Sarvazyan et al. [1998], Wakatsuki et al. [2000], Gibson and Ashby [1988]. The values are to be seen as orders of magnitude, because, in fact, the moduli depend on temperature, state (glassy, transition, rubbery regimes for polymers), material density (woods, geomaterials), spatial loading direction (anisotropy), loading type (tension/compression), etc. Still, the bulk moduli are seen to range in a small interval while the shear moduli vary over seven orders of magnitude. As an implication, their pathological changes can be detected either through routine screening and more systematically and quantitatively through advanced techniques, like computer tomography, magnetic resonance elastography, etc.

That a biomechanical analysis cannot be content with a purely mechanical approach may be motivated in a number of domains, and in particular for biomaterials.

Biomaterials for orthopedic applications, namely for fixation of fractured bones, reconstruction of ligaments and tendons, and joint arthroplasty, are required to be endowed with good bio-compatibility, be corrosion-resistant and bio-degradable. Bioresorbable substitutes have attracted interest as they can be used instead of stainless steel or titanium alloys and can progressively be replaced by natural tissues.

Still, a number of difficulties have been reported when using polymer substitutes for tendon fixations, as they may be brittle, their degradation time is long, their osteo-integration is poor, and osteolysis and cyst formation may take place. The quest for biomaterials with a controllable degradation rate that allows a better tissue regeneration has attracted interest in magnesium based alloys [Farraro et al. 2014]. These biomaterials have a stiffness slightly lower than cortical bone, Fig. 1.8.1, which reduces stress shielding and therefore resorption, and their ductility reduces the risks of fracture. Their porous structure makes them usable as scaffolds for natural tissue growth and prone to cell infiltration, and favors their degradation. Special coatings have been developed to improve their cyto-compatibility and osteo-inductivity.

1.9 Mechanisms of injury of the knee, fracture mechanics

The subject of biomechanics in relation to car occupant crash protection grew rapidly in the 1950's and 1960's as the number of road traffic accidents surged. During frontal car collisions at moderate velocity (about 50 km per hour), the contact of the knee with the instrument panel takes place early while the vehicle is still moving ahead. Therefore, the velocity of the impact knee-panel is relatively low, about a third of the vehicle initial velocity according to Mackay [1992].

Knee injury results mostly in fracture of the patella (by impact on the instrument panel) and femoral condyles, and, to a lesser extent, in knee ligament and tendon damages.

In fact, the knee is stabilized by ligaments and tendons:
- ligaments connect bones together, in this case, femur and tibia. The medial collateral and lateral collateral ligaments minimize lateral motion. The anterior cruciate ligament (ACL) inserts in the posterior face of the femur and in the upper anterior face of the tibia and the posterior cruciate ligament (PCL) joints the posterior face of the tibia and the lower face of the femur. This inclination allows for some rotation of the bones in the sagittal plane but prevents forward and backward relative translation of the tibia and femur;
- tendons attach muscles to bones. The patellar tendon links the muscles of the thigh to the patella and inserts in the front of the tibia. Similarly, the hamstring muscles attach to the back of the tibia through tendons.

Collateral ligaments are injured by excessive varus or valgus flexion, Fig. 1.9.1. Condyles may be fractured and meniscus punctured if the axial load is large during flexion. The ACL may be ruptured by shocks from the rear hitting the tibia.

Still, osteoarthritis (OA) is the main pathology that affects articular cartilages. In the United States, OA affects 5% of the population and 70% of the population aged over 65. Osteoarthritis induces a decreased tensile and compressive cartilage stiffness, and an increased tissue hydration and permeability.

A possible cause of osteoarthritis is traced to the disruption of the crosslinks between collagen fibers. According to other interpretations, osteoarthritis may not be due to a self-degradation of cartilage components, but to proteolytic

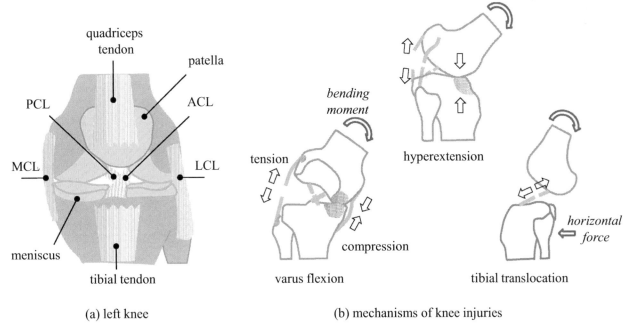

(a) left knee   (b) mechanisms of knee injuries

Fig. 1.9.1: (a) Sketch of ligaments in the left knee. ACL/PCL: anterior/posterior cruciate ligament; LCL/MCL: lateral/medial collateral ligament; (b) typical mechanisms of knee injury, based on Terezinsky [2005].

enzymes. These enzymes are generated by the chondrocytes themselves or diffuse into cartilage from an inflamed synovial liquid. Supra-physiological mechanical loads are also advocated to trigger osteoarthritis: as just suggested, these loads may take place during traumas due to traffic accidents or sport activities. Moreover, clinical and animal studies show that meniscectomy is often followed by osteoarthritic degeneration. In Chapter 14, which is devoted to the biomechanical analysis of articular cartilages, Section 14.3 contains a brief introduction to the subject.

Some specific sport activities contribute significantly to knee injuries, which amounts to 45% of all athletic injuries. The rupture of the ACL, medial meniscus and lateral meniscus contribute, respectively, to 20.3%, 10.8% and 3.7% [Majewski et al. 2006]. This study was performed in Switzerland and Germany and the sports leading to most injuries are skiing and soccer. In fact, the occurrence of the rupture of ACL is seven times higher for athletes. Analysis of video-taped handball matches has shown that ACL rupture was associated with a valgus knee flexion and tibia rotation. Knee braces constraining extension and modifying flexion at landing have been suggested to limit ACL rupture in skiing, basketball and handball [Yu et al. 2004].

1.10   Water and solid constituents of soft tissues

Since the chapters to follow adopt definitely the point of view that soft tissues should be regarded as mixtures of solid constituents bathed in biological fluids, it may be appropriate to provide some numbers that conform with this landscape.

For the carotid artery, water and proteoglycans contribute to 70% of wet weight [Humphrey and Rajagopal 2003]. The dry mass is composed at 50% by collagen types I, III, IV and V, at 30% by smooth muscle, at 20% by elastin, plus endothelial cells, fibroblasts, fibronectin and laminin.

Water contributes also to 60 to 70% of the weight of human skin. Collagen of type I primarily, as well as of type III, accounts for 70-80% of the dry weight, while the proportion of elastin is only about 0.5%, and that of proteoglycans 0.07-0.08% [Seehra and Silver 2006].

The proportion of water in corneal stroma ranges from 75 to 80%. The dry weight is contributed by collagen of types I and V (70%), proteoglycans and associated proteins (10%), and noncollagenous proteins (20%).

Water contributes to 60-80% of the weight of articular cartilages. Collagen of type II (50-75%) and proteoglycans (15-30%) constitute most of the dry weight. The composition evolves with age. For cartilages of distal bovine femur and humerus, the ratio of dry weight over wet weight increases progressively, 12% in third-trimester fetus, 20% in the 4-8 week old calf, 26% in the 1-2 year old animal to 32% in the 3-5 year old adult [Williamson et al. 2001].

Before entering detailed constitutive modelings, it might also be sound to keep in mind the ranges of moduli of soft and hard living tissues in comparison with common engineering materials, Fig. 1.8.1.

Part I

Solids and
multi-species mixtures
as open systems:
a continuum mechanics
perspective

Chapter 2

Elements of continuum mechanics

Continuum mechanics concepts are used throughout this textbook. To help the reader in concentrating on the biomechanical aspects of the presentation, a number of algebraic, tensor and differential relations have been gathered in this chapter. The exposition is necessarily compact. Broader and thorough introductions to the field and constitutive equations of continuum mechanics may be found in classical textbooks, e.g., Fung [1965][1977], Mandel [1966], Malvern [1969], Bowen [1989] *inter alii*.

2.1 Algebraic relations and algebraic operators

2.1.1 Notations and conventions for tensor calculus

Cartesian coordinates are used throughout, except in specified conditions. As a general guideline, the convention of summation over mute indices applies to indices which represent the components of vectors or tensors in cartesian bases, e.g., $a_i b_i$ has the meaning of the sum of the products $a_i b_i$, the subscript i varying from 1 to 3. To avoid ambiguity, a subscript bar prevents this summation, e.g., $a_{\underline{i}} b_{\underline{i}}$ refers to a single product.

On the other hand, summation is a priori not implied with indices that refer to the species of a mixture, usually denoted by the generic subscript or superscript k. The conventions are stated whenever necessary.

Let $\{\mathbf{e}_i^1\}$, $\{\mathbf{e}_i^2\}$ and $\{\mathbf{e}_i\}$, $i \in [1,3]$, be three cartesian bases. Here is a list of basic definitions that are used throughout:

- scalar product of two vectors: $\mathbf{a} = a_{1i}\,\mathbf{e}_i^1 = a_i\,\mathbf{e}_i$, $\mathbf{b} = b_{2j}\,\mathbf{e}_j^2 = b_j\,\mathbf{e}_j$:

$$\mathbf{a} \cdot \mathbf{b} = \begin{cases} a_{1i}\,b_{2j}\,\mathbf{e}_i^1 \cdot \mathbf{e}_j^2, \\ a_i\,b_i\,. \end{cases} \tag{2.1.1}$$

- dot product of two second order tensors $\mathbf{A} = A_{1ij}\mathbf{e}_i^1 \otimes \mathbf{e}_j^1$, $\mathbf{B} = B_{2kl}\mathbf{e}_k^2 \otimes \mathbf{e}_l^2$:

$$\mathbf{A} \cdot \mathbf{B} = A_{1ij}\,B_{2kl}\,(\mathbf{e}_j^1 \cdot \mathbf{e}_k^2)\,\mathbf{e}_i^1 \otimes \mathbf{e}_l^2\,. \tag{2.1.2}$$

- double dot product of two second order tensors:
  $\mathbf{A} = A_{1ij}\mathbf{e}_i^1 \otimes \mathbf{e}_j^1 = A_{ij}\mathbf{e}_i \otimes \mathbf{e}_j$, $\mathbf{B} = B_{2kl}\mathbf{e}_k^2 \otimes \mathbf{e}_l^2 = B_{kl}\mathbf{e}_k \otimes \mathbf{e}_l$:

$$\mathbf{A} : \mathbf{B} = \operatorname{tr}\mathbf{A}^{\mathrm{T}} \cdot \mathbf{B} = \operatorname{tr}\mathbf{A} \cdot \mathbf{B}^{\mathrm{T}} = \begin{cases} A_{1ij}\,B_{2kl}\,(\mathbf{e}_i^1 \cdot \mathbf{e}_k^2)\,(\mathbf{e}_j^1 \cdot \mathbf{e}_l^2), \\ A_{ij}\,B_{ij}\,, \end{cases} \tag{2.1.3}$$

and therefore the double product is commutative, $\mathbf{A} : \mathbf{B} = \mathbf{B} : \mathbf{A}$ and $\mathbf{A} : \mathbf{B} = \mathbf{A}^{\mathrm{T}} : \mathbf{B}^{\mathrm{T}}$. Note the following property that indicates that the double product actually decouples the symmetric and skew-symmetric parts,

$$\mathbf{A} : \mathbf{B} = \frac{\mathbf{A} + \mathbf{A}^{\mathrm{T}}}{2} : \frac{\mathbf{B} + \mathbf{B}^{\mathrm{T}}}{2} + \frac{\mathbf{A} - \mathbf{A}^{\mathrm{T}}}{2} : \frac{\mathbf{B} - \mathbf{B}^{\mathrm{T}}}{2}\,. \tag{2.1.4}$$

The double dot product of two dyads is just a particular case of the double dot product over second order tensors,

$$\left(\mathbf{a} \otimes \mathbf{b}\right) : \left(\mathbf{c} \otimes \mathbf{d}\right) = \left(\mathbf{a} \cdot \mathbf{c}\right) \left(\mathbf{b} \cdot \mathbf{d}\right). \tag{2.1.5}$$

- Euclidean norm of a vector $|\mathbf{v}| = [\mathbf{v} \cdot \mathbf{v}]^{\frac{1}{2}}$.
- Euclidean norm of a second order tensor $|\mathbf{A}| = [\mathbf{A} : \mathbf{A}]^{\frac{1}{2}}$.

The reader is invited to pay attention to the very definitions of the double product and of the Euclidean norm. The rules that a norm over a linear space should satisfy are exposed in textbooks of matrix analysis, e.g., Ortega [1987], Ciarlet [1989]. The norm of any entity, vector or tensor of any order, should be either positive or zero. It can be zero if and only if the entity itself vanishes. Consider some matrix products that can be obtained from a particular nonsymmetric 2×2 matrix $\mathbf{A}$,

$$\mathbf{A} = \begin{bmatrix} 0 & 1 \\ 0 & 0 \end{bmatrix}, \quad \mathbf{A}^2 = \begin{bmatrix} 0 & 0 \\ 0 & 0 \end{bmatrix}, \quad \mathbf{A} \cdot \mathbf{A}^{\mathrm{T}} = \begin{bmatrix} 1 & 0 \\ 0 & 0 \end{bmatrix}, \quad \mathbf{A}^{\mathrm{T}} \cdot \mathbf{A} = \begin{bmatrix} 0 & 0 \\ 0 & 1 \end{bmatrix}. \tag{2.1.6}$$

Consequently, the square of a matrix $\mathbf{A}$ cannot be used to build the norm in general. Still, it can if the matrix is symmetric, and then $\operatorname{tr} \mathbf{A}^2 = |\mathbf{A}|^2$. If the matrix is skew-symmetric, then $\operatorname{tr} \mathbf{A}^2 = -|\mathbf{A}|^2$.

As a side remark, a matrix for which there exists a positive integer p such that $\mathbf{A}^p$ vanishes is called nilpotent. A matrix is nilpotent if and only if all its eigenvalues vanish, e.g., Ortega [1987].

For a second order tensor $\mathbf{A}$ with components A_{ij} in a cartesian basis, its transpose $\mathbf{A}^{\mathrm{T}}$ has components A_{ji}. $\mathbf{A}$ is said to be symmetric if $\mathbf{A} = \mathbf{A}^{\mathrm{T}}$. The transpose operator satisfies the relations $(\mathbf{A}^{\mathrm{T}})^{\mathrm{T}} = \mathbf{A}$ and $(\mathbf{A} \cdot \mathbf{B})^{\mathrm{T}} = \mathbf{B}^{\mathrm{T}} \cdot \mathbf{A}^{\mathrm{T}}$. Note the trivial but useful equivalence,

$$\mathbf{A} + \mathbf{B} \quad \text{symmetric} \quad \Leftrightarrow \quad \mathbf{A} - \mathbf{B}^{\mathrm{T}} \quad \text{symmetric}. \tag{2.1.7}$$

If $\mathbf{A}$ is symmetric, its inverse is symmetric if it exists, and so is any power $\mathbf{A}^n$, $n \geq 1$. The converse implication does not hold: consider the counter-example below,

$$\mathbf{A} = \begin{bmatrix} 0 & 1 \\ -1 & 0 \end{bmatrix}, \quad \mathbf{A}^2 = \begin{bmatrix} -1 & 0 \\ 0 & -1 \end{bmatrix}. \tag{2.1.8}$$

Further striking differences in spectral properties between symmetric and nonsymmetric matrices are displayed in Exercise 9.1.

2.1.2 Algebraic operators

The identity over second order tensors is denoted by $\mathbf{I} = (I_{ij})$.

The permutation symbol $\mathbf{e}$, also called Levi-Civita tensor, is a third order tensor defined by

$$e_{ijk} = \begin{cases} 1 & \text{if } (i,j,k) \text{ is an even permutation of } (1,2,3), \\ -1 & \text{if } (i,j,k) \text{ is an odd permutation of } (1,2,3), \\ 0 & \text{otherwise}. \end{cases} \tag{2.1.9}$$

Some useful properties are collected below [Eringen 1967, p. 430],

$$\begin{aligned} e_{ijk}\, e_{mnp}\, A_{mnp} &= A_{ijk} - A_{ikj} + A_{jki} - A_{jik} + A_{kij} - A_{kji}, \\ e_{ijk}\, e_{kmn} &= I_{im}\, I_{jn} - I_{in}\, I_{jm}, \\ e_{ijk}\, e_{jkn} &= 2\, I_{in}, \\ e_{ijk}\, e_{ijk} &= 6. \end{aligned} \tag{2.1.10}$$

The definition of the permutation symbol can be extended to higher orders.

A one-to-one correspondence can be established between a second order skew-symmetric tensor $\mathbf{A}$, which is such that $\mathbf{A}^{\mathrm{T}} = -\mathbf{A}$ and a pseudo-vector $\vec{\mathbf{A}}$. In cartesian axes,

$$A_i = \frac{1}{2}\, e_{ijk}\, A_{jk}, \quad A_{ij} = e_{ijk}\, A_k\,, \tag{2.1.11}$$

or in matrix form,

$$\mathbf{A} = \begin{bmatrix} 0 & A_{12} & A_{13} \\ -A_{12} & 0 & A_{23} \\ -A_{13} & -A_{23} & 0 \end{bmatrix} = \begin{bmatrix} 0 & A_3 & -A_2 \\ -A_3 & 0 & A_1 \\ A_2 & -A_1 & 0 \end{bmatrix}, \quad |\mathbf{A}|^2 = 2\,|\vec{\mathbf{A}}|^2\,. \tag{2.1.12}$$

For any vector $\mathbf{b}$,

$$\mathbf{A}\cdot\mathbf{b} = -\vec{\mathbf{A}}\wedge\mathbf{b} = \mathbf{b}\wedge\vec{\mathbf{A}}\,, \tag{2.1.13}$$

where $\wedge$ is the vector product, or cross product, defined by Eqn (2.1.14). In fact $\vec{\mathbf{A}}$ is called a *pseudo*-vector because it does not follow tensor rules under change of reference frame as briefly indicated in Exercise 2.1.

The permutation symbol is instrumental to obtain the component form of a number of operators, e.g., the *vector product*,

$$(\mathbf{a}\wedge\mathbf{b})_i = e_{ijk}\, a_j\, b_k\,, \tag{2.1.14}$$

and the *determinant of a second order tensor* $\mathbf{A}$ in $\mathbb{R}^3$,

$$e_{mnp}\,\det\mathbf{A} = e_{ijk}\, A_{im}\, A_{jn}\, A_{kp} \quad\Rightarrow\quad \det\mathbf{A} = \begin{cases} e_{ijk}\, A_{i1}\, A_{j2}\, A_{k3}, \\[2mm] \frac{1}{6}\, e_{mnp}\, e_{ijk}\, A_{im}\, A_{jn}\, A_{kp}, \\[2mm] e_{mnp}\, A_{1m}\, A_{2n}\, A_{3p}\,, \end{cases} \tag{2.1.15}$$

namely, in explicit form,

$$\det\mathbf{A} = A_{11}\, A_{22}\, A_{33} - A_{11}\, A_{23}\, A_{32} - A_{22}\, A_{13}\, A_{31} - A_{33}\, A_{12}\, A_{21} + A_{12}\, A_{23}\, A_{31} + A_{13}\, A_{32}\, A_{21}\,. \tag{2.1.16}$$

These relations prove that the determinant of a second order tensor and of its transpose are the same, $\det\mathbf{A} = \det\mathbf{A}^{\mathrm{T}}$, $\forall\mathbf{A}$. Also the determinant of the product of two tensors is equal to the product of the determinants namely $\det\mathbf{A}\cdot\mathbf{B} = \det\mathbf{A}\,\det\mathbf{B}$. The proof is worthwhile just to manipulate the various pieces of Eqn (2.1.15):

$$\begin{aligned} \det\mathbf{A}\cdot\mathbf{B} &= e_{ijk}\, A_{im}\, B_{m1}\, A_{jn}\, B_{n2}\, A_{kp}\, B_{p3} \\[1mm] &= \left(e_{ijk}\, A_{im}\, A_{jn}\, A_{kp}\right)\left(B_{m1}\, B_{n2}\, B_{p3}\right) \\[1mm] &= e_{mnp}\,\det\mathbf{A}\,\left(B_{m1}\, B_{n2}\, B_{p3}\right) \\[1mm] &= \det\mathbf{A}\,\det\mathbf{B}\,. \end{aligned} \tag{2.1.17}$$

If a second order tensor $\mathbf{A}$ is invertible, its inverse is denoted by $\mathbf{A}^{-1}$. The transpose and inverse operators commute and therefore the shortcut notation $\mathbf{A}^{-\mathrm{T}} = (\mathbf{A}^{\mathrm{T}})^{-1} = (\mathbf{A}^{-1})^{\mathrm{T}}$ makes sense.

The following result is useful for a matrix whose components are in arbitrary order [Segel 1987, p. 19],

$$\det\begin{bmatrix} A_{ip} & A_{iq} & A_{ir} \\ A_{jp} & A_{jq} & A_{jr} \\ A_{kp} & A_{kq} & A_{kr} \end{bmatrix} = e_{ijk}\, e_{pqr}\,\det\mathbf{A}\,. \tag{2.1.18}$$

Note also the fundamental relations for the cartesian triad $(\mathbf{e}_1, \mathbf{e}_2, \mathbf{e}_3)$,

$$\mathbf{e}_i\otimes\mathbf{e}_j = e_{ijk}\,\mathbf{e}_k, \quad \det(\mathbf{e}_i, \mathbf{e}_j, \mathbf{e}_k) = e_{ijk}\,, \tag{2.1.19}$$

and the additional relations for $\mathbf{a}$, $\mathbf{b}$, $\mathbf{c}$ and $\mathbf{d}$ arbitrary vectors,

$$|\mathbf{a} \wedge \mathbf{b}|^2 = |\mathbf{a}|^2 |\mathbf{b}|^2 - |\mathbf{a} \cdot \mathbf{b}|^2,$$

$$(\mathbf{a} \wedge \mathbf{b}) \wedge \mathbf{c} = (\mathbf{a} \cdot \mathbf{c})\,\mathbf{b} - (\mathbf{b} \cdot \mathbf{c})\,\mathbf{a},$$

$$\mathbf{a} \wedge (\mathbf{b} \wedge \mathbf{c}) = (\mathbf{a} \cdot \mathbf{c})\,\mathbf{b} - (\mathbf{a} \cdot \mathbf{b})\,\mathbf{c},$$

$$(\mathbf{a} \wedge \mathbf{b}) \cdot \mathbf{c} = \mathbf{a} \cdot (\mathbf{b} \wedge \mathbf{c}) = \det(\mathbf{a},\,\mathbf{b},\,\mathbf{c}), \tag{2.1.20}$$

$$(\mathbf{a} \wedge \mathbf{b}) \cdot (\mathbf{c} \wedge \mathbf{d}) = (\mathbf{a} \cdot \mathbf{c})\,(\mathbf{b} \cdot \mathbf{d}) - (\mathbf{a} \cdot \mathbf{d})\,(\mathbf{b} \cdot \mathbf{c}),$$

$$(\mathbf{a} \wedge \mathbf{b}) \wedge (\mathbf{c} \wedge \mathbf{d}) = (\mathbf{a} \cdot \mathbf{c})\,(\mathbf{b} \wedge \mathbf{d}) - (\mathbf{a} \cdot \mathbf{d})\,(\mathbf{b} \wedge \mathbf{c}).$$

The fourth relation provides the algebraic volume defined by three vectors. While the vector product is anti-commutative, $\mathbf{a} \wedge \mathbf{b} = -\mathbf{b} \wedge \mathbf{a}$, the second and third relations show that it is not associative,

$$(\mathbf{a} \wedge \mathbf{b}) \wedge \mathbf{c} \neq \mathbf{a} \wedge (\mathbf{b} \wedge \mathbf{c}). \tag{2.1.21}$$

Another useful algebraic operator is the *trace* operator, e.g., of a second order tensor $\mathbf{A}$ expressed in cartesian axes,

$$\operatorname{tr} \mathbf{A} = A_{ii}. \tag{2.1.22}$$

A second order tensor decomposes uniquely into an isotropic or spherical part and a traceless or deviatoric part,

$$\mathbf{A} = \frac{1}{3}\,(\operatorname{tr}\mathbf{A})\,\mathbf{I} + \operatorname{dev}\mathbf{A}. \tag{2.1.23}$$

The following operator will be instrumental to manipulate the elasticity tensor,

$$(\mathbf{A}\,\overline{\underline{\otimes}}\,\mathbf{B})_{ijkl} = \frac{1}{2}\,(A_{ik}\,B_{jl} + A_{il}\,B_{jk}), \quad (\mathbf{A}\,\overline{\underline{\otimes}}\,\mathbf{B}):\mathbf{X} = \mathbf{A}\cdot\tfrac{1}{2}(\mathbf{X}+\mathbf{X}^{\mathrm{T}})\cdot\mathbf{B}^{\mathrm{T}}, \tag{2.1.24}$$

which, for $\mathbf{A}$, $\mathbf{B}$ and $\mathbf{X}$ symmetric second order tensors, simplifies to

$$(\mathbf{A}\,\overline{\underline{\otimes}}\,\mathbf{B}):\mathbf{X} = \mathbf{A}\cdot\mathbf{X}\cdot\mathbf{B}. \tag{2.1.25}$$

The component form,

$$(\mathbf{I}\,\overline{\underline{\otimes}}\,\mathbf{I})_{ijkl} = \frac{1}{2}(I_{ik}I_{jl} + I_{il}I_{jk}), \quad (\mathbf{I}\,\overline{\underline{\otimes}}\,\mathbf{I}):\mathbf{X} = \frac{1}{2}(\mathbf{X}+\mathbf{X}^{\mathrm{T}}), \tag{2.1.26}$$

shows that the operator $\overline{\underline{\otimes}}$ is the identity operator over symmetric second order tensors. Therefore $\mathbf{I}\,\overline{\underline{\otimes}}\,\mathbf{I}$ is the fourth order identity tensor $\mathbf{I}^{(4)}$ over symmetric second order tensors. Consequently, $\mathbf{I}\,\overline{\underline{\otimes}}\,\mathbf{I} - \frac{1}{3}\mathbf{I}\otimes\mathbf{I}$, with cartesian components $\frac{1}{2}(I_{ik}I_{jl} + I_{il}I_{jk}) - \frac{1}{3}I_{ij}I_{kl}$, is the fourth order deviator operator over symmetric second order tensors,

$$(\mathbf{I}\,\overline{\underline{\otimes}}\,\mathbf{I}):\mathbf{X} = \mathbf{X}, \quad \forall\,\mathbf{X} = \mathbf{X}^{\mathrm{T}},$$

$$(\mathbf{I}\,\overline{\underline{\otimes}}\,\mathbf{I} - \frac{1}{3}\mathbf{I}\otimes\mathbf{I}):\mathbf{X} = \operatorname{dev}\mathbf{X}, \quad \forall\,\mathbf{X} = \mathbf{X}^{\mathrm{T}}, \tag{2.1.27}$$

and

$$(\mathbf{I}\,\overline{\underline{\otimes}}\,\mathbf{I}):(\mathbf{I}\,\overline{\underline{\otimes}}\,\mathbf{I}) = \mathbf{I}\,\overline{\underline{\otimes}}\,\mathbf{I}, \quad (\mathbf{I}\,\overline{\underline{\otimes}}\,\mathbf{I}):(\mathbf{I}\,\overline{\underline{\otimes}}\,\mathbf{I} - \frac{1}{3}\mathbf{I}\otimes\mathbf{I}) = (\mathbf{I}\,\overline{\underline{\otimes}}\,\mathbf{I} - \frac{1}{3}\mathbf{I}\otimes\mathbf{I}),$$

$$(\mathbf{I}\,\overline{\underline{\otimes}}\,\mathbf{I} - \frac{1}{3}\mathbf{I}\otimes\mathbf{I}):(\mathbf{I}\,\overline{\underline{\otimes}}\,\mathbf{I} - \frac{1}{3}\mathbf{I}\otimes\mathbf{I}) = (\mathbf{I}\,\overline{\underline{\otimes}}\,\mathbf{I} - \frac{1}{3}\mathbf{I}\otimes\mathbf{I}). \tag{2.1.28}$$

Note the derivatives of an arbitrary nonzero symmetric second order tensor $\mathbf{X}$,

$$\frac{\partial|\mathbf{X}|}{\partial\mathbf{X}} = \frac{\mathbf{X}}{|\mathbf{X}|}, \quad \frac{\partial^2|\mathbf{X}|}{\partial\mathbf{X}^2} = \frac{1}{|\mathbf{X}|}\Big(\mathbf{I}\,\overline{\underline{\otimes}}\,\mathbf{I} - \frac{\mathbf{X}}{|\mathbf{X}|}\otimes\frac{\mathbf{X}}{|\mathbf{X}|}\Big), \quad \frac{\partial^2|\mathbf{X}|^2}{\partial\mathbf{X}^2} = 2\,\mathbf{I}\,\overline{\underline{\otimes}}\,\mathbf{I}. \tag{2.1.29}$$

Therefore the quadratic forms associated with the second derivatives are positive (semi-)definite, namely for an arbitrary nonzero symmetric second order tensor $\mathbf{Y}$,

$$\mathbf{Y}\cdot\frac{\partial^2|\mathbf{X}|}{\partial\mathbf{X}^2}\cdot\mathbf{Y} = \frac{|\mathbf{Y}|^2}{|\mathbf{X}|}\,\sin^2(\mathbf{X},\mathbf{Y}) \geq 0, \quad \mathbf{Y}\cdot\frac{\partial^2|\mathbf{X}|^2}{\partial\mathbf{X}^2}\cdot\mathbf{Y} = 2\,|\mathbf{Y}|^2 > 0. \tag{2.1.30}$$

Another partial differential relation is worth mentioning. For an invertible symmetric second order tensor $\mathbf{X}$,

$$\frac{\partial(\mathbf{X}^{-1})_{ij}}{\partial\mathbf{X}_{kl}} = -\frac{1}{2}\left((\mathbf{X}^{-1})_{ik}\,(\mathbf{X}^{-1})_{jl} + (\mathbf{X}^{-1})_{il}\,(\mathbf{X}^{-1})_{jk}\right). \tag{2.1.31}$$

Therefore, if $\mathbf{X}$ is symmetric,

$$\frac{\partial\mathbf{X}^{-1}}{\partial\mathbf{X}} = -\mathbf{X}^{-1}\,\overline{\underline{\otimes}}\,\mathbf{X}^{-1}. \tag{2.1.32}$$

2.2 Characteristic polynomial, eigenvalues and eigenvectors

2.2.1 Eigenvalues and eigenvectors

The *eigenvalues* $\lambda = \lambda(\mathbf{A})$ of the square matrix or tensor $\mathbf{A}$ of order n are the n real, or complex, roots of the *characteristic polynomial* $P_A(\lambda) = \det(\lambda\,\mathbf{I} - \mathbf{A})$. At least one nonzero (right) *eigenvector* $\mathbf{v}$, such that $(\mathbf{A} - \lambda\,\mathbf{I})\cdot\mathbf{v} = \mathbf{0}$ is associated with each eigenvalue λ. Given an eigenvalue λ, the set of vectors such that $(\mathbf{A} - \lambda\,\mathbf{I})\cdot\mathbf{v} = \mathbf{0}$ is a vector space termed eigenspace associated with λ. The dimension of this eigenspace is the *geometrical multiplicity* of λ, which might be different from its *algebraic multiplicity*. As examples, let us consider the matrices in $\mathbb{R}^2$,

$$\mathbf{A} = \begin{bmatrix} a & 0 \\ 0 & b \end{bmatrix}, \quad \mathbf{B} = \begin{bmatrix} a & 1 \\ 0 & b \end{bmatrix}. \tag{2.2.1}$$

The matrix $\mathbf{A}$ has the characteristic polynomial $P_A(\lambda) = (\lambda - a)\,(\lambda - b)$. If $a \neq b$, the eigenvalues $\lambda = a, b$ are both single, and the associated eigenspaces are spanned by the respective vectors $[1\ 0]^{\mathrm{T}}$ and $[0\ 1]^{\mathrm{T}}$ and have dimension 1. Therefore the algebraic and geometric multiplicities of the two eigenvalues are equal to 1. The eigenvalues of matrix $\mathbf{B}$ are also a and b and the associated eigenvectors are $[1\ 0]^{\mathrm{T}}$ and $[1\ b - a]^{\mathrm{T}}$.

When $a = b$, the eigenvalue $\lambda = a$ has algebraic multiplicity 2 for both matrices. Further, any vector is an eigenvector of the matrix $\mathbf{A}$ and the dimension of the eigenspace associated with the eigenvalue is equal to 2. On the other hand, the eigenspace of matrix $\mathbf{B}$ is then spanned by the single vector $\mathbf{a}_1 = [1\ 0]^{\mathrm{T}}$, so that the geometrical multiplicity of the eigenvalue is equal to 1. Such a matrix, for which the geometrical multiplicity of at least one eigenvalue is smaller than its algebraic multiplicity, is called *defective*. A generalized eigenvector $\mathbf{a}_2$ satisfies the relation $(\mathbf{B} - a\,\mathbf{I})\cdot\mathbf{a}_2 = \mathbf{a}_1$, yielding $\mathbf{a}_2 = [x\ 1]^{\mathrm{T}}$, with x arbitrary. The two vectors generate $\mathbb{R}^2$ and

$$\mathbf{P}^{-1}\cdot\mathbf{B}\cdot\mathbf{P} = \mathbf{B}, \quad \text{with} \quad \mathbf{P} = [\mathbf{a}_1\ \mathbf{a}_2]^{\mathrm{T}} = \begin{bmatrix} 1 & x \\ 0 & 1 \end{bmatrix}. \tag{2.2.2}$$

A real symmetric matrix in $\mathbb{R}^n$, $n \geq 1$, has real eigenvalues. Indeed, whatever the potential nature real or complex of the eigenvector $\mathbf{a}$ associated with the eigenvalue λ, the product $\overline{\mathbf{a}}\cdot\mathbf{A}\cdot\mathbf{a}$ is real, and therefore the eigenvalue λ which is equal to the Rayleigh quotient $\overline{\mathbf{a}}\cdot\mathbf{A}\cdot\mathbf{a}/(\overline{\mathbf{a}}\cdot\mathbf{a})$ is real. The eigenvectors of a real symmetric matrix form an orthonormal basis of $\mathbb{R}^n$ and, in this basis, the representation of the matrix, namely $\mathbf{diag}[\lambda_1\ \lambda_2\ \cdots]$, is diagonal, see, e.g., Ortega [1987]. Moreover, a positive definite symmetric matrix has strictly positive real eigenvalues.

Exercise 9.1 shows that the same properties do not hold for nonsymmetric matrices. In fact, for nonsymmetric matrices, the eigenvalues of the matrix and of its transpose are the same since $\det(\lambda\,\mathbf{I} - \mathbf{A}) = \det(\lambda\,\mathbf{I} - \mathbf{A}^{\mathrm{T}})$. On the other hand, at least one *right* eigenvector $(\mathbf{A} - \lambda\,\mathbf{I})\cdot\mathbf{v} = \mathbf{0}$ and one *left* eigenvector $\mathbf{w}\cdot(\mathbf{A} - \lambda\,\mathbf{I}) = \mathbf{0}$, which is also a right eigenvector associated with the transpose matrix, are associated with each eigenvalue λ. If $\mathbf{v}$ is a right eigenvector associated with a certain eigenvalue and $\mathbf{w}$ a left eigenvector associated with a different eigenvalue, then $\mathbf{v}\cdot\mathbf{w} = 0$ [Wilkinson 1965, Section 1.3].

The eigenvalues of (real) skew-symmetric tensors in $\mathbb{R}^3$ are all pure imaginary but one which is zero. For more details on skew-symmetric tensors in $\mathbb{R}^3$, see Section 2.5.5.1.

As for proper orthogonal tensors in $\mathbb{R}^3$, they have one eigenvalue equal to 1, and the other two are complex conjugate as indicated in Section 2.5.5.3.

Two remarkable algebraic properties that are illustrated by Figs. 2.2.1 and 2.2.2 are used to sort wave-speeds in saturated porous media in Section 11.5.4. The *separation theorem*, e.g. [Wilkinson 1965, Section 2.47], indicates that the $n - 1$ eigenvalues $\lambda'(\mathbf{A}')$ of the leading principal minor $\mathbf{A}' = \mathbf{A}_{[n-1,n-1]}$ of a symmetric matrix $\mathbf{A} = \mathbf{A}_{[n,n]}$,

$$\mathbf{A}_{[n,n]} = \begin{bmatrix} \mathbf{A}_{[n-1,n-1]} & \times \\ \times & \times \end{bmatrix}, \tag{2.2.3}$$

separate the n eigenvalues $\lambda(\mathbf{A})$ of the matrix $\mathbf{A}$, namely

$$\lambda_n(\mathbf{A}) \leq \lambda'_{n-1}(\mathbf{A}') \leq \lambda_{n-1}(\mathbf{A}) \leq \quad \cdots \quad \leq \lambda_2(\mathbf{A}) \leq \lambda'_1(\mathbf{A}') \leq \lambda_1(\mathbf{A}). \tag{2.2.4}$$

Let a, $\mathbf{a}$ and $\mathbf{A}$ be an arbitrary scalar, vector and symmetric matrix, respectively. The *interlacing theorem* states that the eigenvalues of $\mathbf{A}$ and of the rank one modification $\mathbf{A}' = \mathbf{A} + a\,\mathbf{a}\otimes\mathbf{a}$ interlace, namely for $a \geq 0$,

$$\lambda_n(\mathbf{A}) \leq \lambda'_n(\mathbf{A}') \leq \lambda_{n-1}(\mathbf{A}) \leq \lambda'_{n-1}(\mathbf{A}') \quad \cdots \quad \leq \lambda_1(\mathbf{A}) \leq \lambda'_1(\mathbf{A}'). \tag{2.2.5}$$

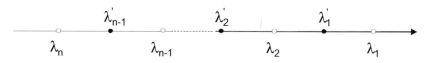

Fig. 2.2.1: Separated eigenvalues.

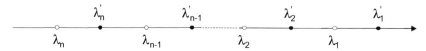

Fig. 2.2.2: Interlacing eigenvalues.

2.2.2   Characteristic polynomial of a second order tensor

For a second order tensor in $\mathbb{R}^3$, the characteristic polynomial has degree 3 in the eigenvalue λ,

$$P_A(\lambda) = \det(\lambda \, \mathbf{I} - \mathbf{A}) = \lambda^3 - I_1 \, \lambda^2 + I_2 \, \lambda - I_3 = 0 \,, \tag{2.2.6}$$

where I_1, I_2 and I_3 are the scalar invariants,

$$I_1(\mathbf{A}) = \text{tr } \mathbf{A}, \quad I_2(\mathbf{A}) = \frac{1}{2}\left((\text{tr } \mathbf{A})^2 - \text{tr } \mathbf{A}^2\right), \quad I_3(\mathbf{A}) = \det \mathbf{A} \,. \tag{2.2.7}$$

The second invariant $I_2(\mathbf{A})$ is also the trace of the matrix adjoint to $\mathbf{A}$, namely

$$I_2(\mathbf{A}) = \text{tr} \, (\det \mathbf{A} \, \mathbf{A}^{-1}) = \det \mathbf{A} \, \text{tr} \, \mathbf{A}^{-1} \,. \tag{2.2.8}$$

$I_3(\mathbf{A}) = \det \mathbf{A}$ is given in component form by Eqn (2.1.16), and,

$$I_1(\mathbf{A}) = A_{11} + A_{22} + A_{33}, \quad I_2(\mathbf{A}) = A_{11} \, A_{22} + A_{22} \, A_{33} + A_{33} \, A_{11} - A_{23} \, A_{32} - A_{13} \, A_{31} - A_{12} \, A_{21} \,. \tag{2.2.9}$$

The definition extends to tensors of higher order. Let $\mathbf{A}$ be an arbitrary tensor and $\mathbf{I}$ the identity tensor, both of order n. The characteristic polynomial $P_A(\lambda)$ associated with $\mathbf{A}$ is the polynomial of degree n defined by the relation, p. 440 in Eringen [1967],

$$P_A(\lambda) = \det(\lambda \, \mathbf{I} - \mathbf{A}) = \lambda^n + \alpha_1 \, \lambda^{n-1} + \alpha_2 \, \lambda^{n-2} + \cdots + \alpha_{n-1} \, \lambda^1 + \alpha_n \,, \tag{2.2.10}$$

where

$$\alpha_1 = -\text{tr } \mathbf{A},$$

$$\alpha_2 = -\frac{1}{2}\left(\alpha_1 \, \text{tr } \mathbf{A} + \text{tr } \mathbf{A}^2\right),$$

$$\alpha_3 = -\frac{1}{3}\left(\alpha_2 \, \text{tr } \mathbf{A} + \alpha_1 \, \text{tr } \mathbf{A}^2 + \text{tr } \mathbf{A}^3\right), \tag{2.2.11}$$

$$\vdots$$

$$\alpha_n = -\frac{1}{n}\left(\alpha_{n-1} \, \text{tr } \mathbf{A} + \alpha_{n-2} \, \text{tr } \mathbf{A}^2 + \cdots + \alpha_1 \, \text{tr } \mathbf{A}^{n-1} + \text{tr } \mathbf{A}^n\right).$$

In terms of the eigenvalues $\lambda_i(\mathbf{A})$, $i \in [1, n]$, of $\mathbf{A}$, the above formulas become

$$P_A(\lambda) = \prod_{i=1}^{n} (\lambda - \lambda_i) \,, \tag{2.2.12}$$

and, for $n = 3$,

$$I_1(\mathbf{A}) = -\alpha_1 = \lambda_1 + \lambda_2 + \lambda_3, \quad I_2(\mathbf{A}) = \alpha_2 = \lambda_1 \, \lambda_2 + \lambda_2 \, \lambda_3 + \lambda_3 \, \lambda_1, \quad I_3(\mathbf{A}) = -\alpha_3 = \lambda_1 \, \lambda_2 \, \lambda_3 \,. \tag{2.2.13}$$

The isotropic scalar invariants of second order tensors satisfy some interesting inequalities, as already touched upon in Section 2.5.7. For instance, let $\mathbf{A}$ be an arbitrary *second order tensor with real positive eigenvalues*. Then, by a direct calculation or using Eqn (2.2.8),

$$(I_1(\mathbf{A}))^2 \geq 3 \, I_2(\mathbf{A}), \quad I_1(\mathbf{A}) \, I_2(\mathbf{A}) - I_3(\mathbf{A}) = I_3(\mathbf{A}) \left(I_1(\mathbf{A}) \, I_1(\mathbf{A}^{-1}) - 1\right) > 0 \,, \tag{2.2.14}$$

since, according to Eqn (2.5.60), $I_1(\mathbf{A}) \, I_1(\mathbf{A}^{-1}) \geq 9$. In fact, according to Eqn (2.5.60) and Eqn (2.2.8),

$$\frac{1}{3} \, I_1(\mathbf{A}) \geq (I_3(\mathbf{A}))^{\frac{1}{3}} \geq \frac{3}{I_1(\mathbf{A}^{-1})} = 3 \, \frac{I_3(\mathbf{A})}{I_2(\mathbf{A})} \,, \tag{2.2.15}$$

whence, with equalities only if the three eigenvalues are equal,

$$I_1(\mathbf{A}) \geq 3 \, (I_3(\mathbf{A}))^{\frac{1}{3}}, \quad I_1(\mathbf{A}) \, I_2(\mathbf{A}) \geq 9 \, I_3(\mathbf{A}), \quad I_2(\mathbf{A}) \geq 3 \, (I_3(\mathbf{A}))^{\frac{2}{3}} \,. \tag{2.2.16}$$

2.2.3 Scalar invariants of the stress deviator

2.2.3.1 Spectral decomposition of a unit deviator

For the record, let us note the spectral decomposition of the unit deviator $\hat{\mathbf{S}}$,

$$\hat{\mathbf{S}} = \sum_{i=1}^{3} \hat{S}_i \, \mathbf{e}_i \otimes \mathbf{e}_i, \quad \hat{S}_1 \geq \hat{S}_2 \geq \hat{S}_3, \tag{2.2.17}$$

where the eigenvectors $\mathbf{e}_i$, $i \in [1,3]$, are of unit norm and the eigenvalues can be defined in terms of a single scalar $\ell \in [0, \pi/3]$, referred to as Lode angle,

$$\hat{S}_i = \sqrt{\frac{2}{3}} \cos \ell_i, \quad \ell_i = \ell - \frac{2\pi}{3}(i-1), \quad i \in [1,3]. \tag{2.2.18}$$

Note the properties of the unit deviator [Loret and Harireche 1991, Loret 1992],

$$\hat{S}_1 + \hat{S}_2 + \hat{S}_3 = 0, \quad \hat{S}_1^2 + \hat{S}_2^2 + \hat{S}_3^2 = 1 \quad \hat{S}_j \, \hat{S}_k = \hat{S}_i^2 - \tfrac{1}{2}, \quad i \neq j \neq k,$$

$$\hat{S}_1^3 + \hat{S}_2^3 + \hat{S}_3^3 = 3\,\hat{S}_1 \, \hat{S}_2 \, \hat{S}_3, \quad \hat{S}_1^4 + \hat{S}_2^4 + \hat{S}_3^4 = \tfrac{1}{2}. \tag{2.2.19}$$

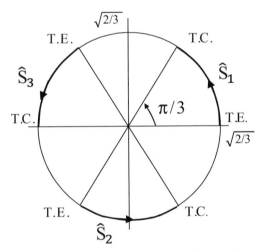

Fig. 2.2.3: Range of the eigenvalues of the unit deviator $\hat{\mathbf{S}}$.

For triaxial extension (T.E.), $\ell=0°$, and $\hat{S}_1 > \hat{S}_2 = \hat{S}_3$, while for triaxial compression (T.C.), $\ell=60°$, and $\hat{S}_1 = \hat{S}_2 > \hat{S}_3$. The three eigenvalues range in the following interval, see Fig. 2.2.3,

$$\hat{S}_1 \in \left[\frac{1}{2}\sqrt{\frac{2}{3}}, \sqrt{\frac{2}{3}}\right], \quad \hat{S}_2 \in \left[-\frac{1}{2}\sqrt{\frac{2}{3}}, \frac{1}{2}\sqrt{\frac{2}{3}}\right], \quad \hat{S}_3 \in \left[-\sqrt{\frac{2}{3}}, -\frac{1}{2}\sqrt{\frac{2}{3}}\right], \tag{2.2.20}$$

and satisfy the inequalities,

$$\hat{S}_1 + 2\,\hat{S}_3 \leq 0, \quad \hat{S}_3 + 2\,\hat{S}_1 \geq 0. \tag{2.2.21}$$

As an addition, two compact nonaligned perturbations with respect to a unit deviator are provided below. Indeed, consider the nonaligned unit deviators $\hat{\mathbf{S}}_P$ and $\hat{\mathbf{S}}_Q$. They can be cast in the following form [Loret 1992]:

$$\hat{\mathbf{S}}_P = \cos\alpha \, \hat{\mathbf{S}} - \sin\alpha \, \Delta\hat{\mathbf{S}}, \quad \hat{\mathbf{S}}_Q = \cos\alpha \, \hat{\mathbf{S}} + \sin\alpha \, \Delta\hat{\mathbf{S}}, \tag{2.2.22}$$

where we introduce the unit deviators $\hat{\mathbf{S}}$ and $\Delta\hat{\mathbf{S}}$ which are orthogonal and the angle $\alpha \in \,]\,0, \pi/2\,[$:

$$\hat{\mathbf{S}} = \frac{\hat{\mathbf{S}}_Q + \hat{\mathbf{S}}_P}{|\hat{\mathbf{S}}_Q + \hat{\mathbf{S}}_P|}, \quad \Delta\hat{\mathbf{S}} = \frac{\hat{\mathbf{S}}_Q - \hat{\mathbf{S}}_P}{|\hat{\mathbf{S}}_Q - \hat{\mathbf{S}}_P|}, \quad \tan\alpha = \frac{|\hat{\mathbf{S}}_Q - \hat{\mathbf{S}}_P|}{|\hat{\mathbf{S}}_Q + \hat{\mathbf{S}}_P|}, \tag{2.2.23}$$

and
$$|\hat{\mathbf{S}}| = 1, \quad |\Delta\hat{\mathbf{S}}| = 1, \quad \hat{\mathbf{S}} : \Delta\hat{\mathbf{S}} = 0 \,. \tag{2.2.24}$$
Due to the orthogonality condition, the unit deviator $\Delta\hat{\mathbf{S}}$ may be defined by the angle $\ell - \pi/2 + k\,\pi$, where k is an integer,
$$\Delta\hat{S}_i = (-1)^k \sqrt{\frac{2}{3}} \, \sin\left(\ell - \frac{2\pi}{3}(i-1)\right), \quad i \in [1,3] \,. \tag{2.2.25}$$
Consequently the non collinear unit deviators $\hat{\mathbf{S}}_P$ and $\hat{\mathbf{S}}_Q$ are defined by the Lode angles $\ell + (-1)^k\,\alpha$ and $\ell - (-1)^k\,\alpha$,
$$
\begin{aligned}
(\hat{S}_P)_i &= (-1)^k \sqrt{\frac{2}{3}} \, \cos\left(\ell + (-1)^k\,\alpha - \frac{2\pi}{3}(i-1)\right), \\
(\hat{S}_Q)_i &= (-1)^k \sqrt{\frac{2}{3}} \, \cos\left(\ell - (-1)^k\,\alpha - \frac{2\pi}{3}(i-1)\right), \quad i \in [1,3] \,.
\end{aligned}
\tag{2.2.26}
$$

2.2.3.2   The Lode angle: eigenvalues in terms of isotropic scalar invariants

Let $\mathbf{s}$ be the stress deviator. Its characteristic polynomial (2.2.6),
$$P_s(\lambda) = \det(\lambda\,\mathbf{I} - \mathbf{s}) = \lambda^3 + I_2\,\lambda - I_3 = 0 \,, \tag{2.2.27}$$
expresses in terms of the scalar invariants I_2 and I_3,
$$I_2 = -\frac{1}{2}\operatorname{tr}\mathbf{s}^2, \quad I_3 = \det\mathbf{s} = \frac{1}{3}\operatorname{tr}\mathbf{s}^3 \,. \tag{2.2.28}$$
The last relation is due to (2.2.33). The solution of this cubic equation for λ is obtained by introducing the Lode angle ℓ,
$$\lambda = \sqrt{\frac{2}{3}\operatorname{tr}\mathbf{s}^2}\,\cos\ell \,, \tag{2.2.29}$$
which transforms (2.2.27) to
$$4\cos^3\ell - 3\cos\ell = \cos 3\ell = \sqrt{6}\,\frac{\operatorname{tr}\mathbf{s}^3}{(\operatorname{tr}\mathbf{s}^2)^{\frac{3}{2}}} \,, \tag{2.2.30}$$
which has three independent solutions,
$$\ell_i = \ell - \frac{2\pi}{3}(i-1), \quad i = 1,2,3, \quad \text{with } \ell = \frac{1}{3}\cos^{-1}\left[\sqrt{6}\,\frac{\operatorname{tr}\mathbf{s}^3}{(\operatorname{tr}\mathbf{s}^2)^{\frac{3}{2}}}\right] \in [0, \frac{\pi}{3}] \,. \tag{2.2.31}$$
Incidentally, as a complement to the previous section, we have proved that the three eigenvalues of the stress can be expressed analytically as
$$\sigma_i = \frac{1}{3}\operatorname{tr}\boldsymbol{\sigma} + \sqrt{\frac{2}{3}\operatorname{tr}\mathbf{s}^2}\,\cos\ell_i, \quad i = 1,2,3 \,. \tag{2.2.32}$$

2.2.3.3   Some relations between isotropic scalar invariants

Let $\mathbf{A}$ and $\mathbf{B}$ be two arbitrary second order tensors. Here are some interesting relations involving the isotropic scalar invariants I_1, I_2 I_3 of a second order tensor, the isotropic scalar invariants of its deviatoric part, its determinant and the trace of its cubic power,
$$
\begin{aligned}
\mathbf{A} &= \operatorname{tr}\mathbf{A}\,\frac{\mathbf{I}}{3} + \operatorname{dev}\mathbf{A}, \quad \mathbf{B} = \operatorname{tr}\mathbf{B}\,\frac{\mathbf{I}}{3} + \operatorname{dev}\mathbf{B}, \\
\mathbf{A} : \mathbf{B} &= \frac{1}{3}\operatorname{tr}\mathbf{A}\operatorname{tr}\mathbf{B} + \operatorname{dev}\mathbf{A} : \operatorname{dev}\mathbf{B}, \\
I_2(\operatorname{dev}\mathbf{A}) &= I_2(\mathbf{A}) - \frac{1}{3}I_1(\mathbf{A})^2 = -\frac{1}{2}\operatorname{tr}(\operatorname{dev}\mathbf{A})^2 \,, \\
J_3(\mathbf{A}) &\equiv \operatorname{tr}\mathbf{A}^3 \,, \\
J_3(\operatorname{dev}\mathbf{A}) &= \operatorname{tr}(\operatorname{dev}\mathbf{A})^3 = \operatorname{tr}\mathbf{A}^3 - \operatorname{tr}\mathbf{A}^2\operatorname{tr}\mathbf{A} + \frac{2}{9}(\operatorname{tr}\mathbf{A})^3, \\
I_3(\mathbf{A}) &\equiv \det\mathbf{A} = \frac{1}{3}\operatorname{tr}\mathbf{A}^3 - \frac{1}{2}\operatorname{tr}\mathbf{A}\operatorname{tr}\mathbf{A}^2 + \frac{1}{6}(\operatorname{tr}\mathbf{A})^3, \\
I_3(\operatorname{dev}\mathbf{A}) &= \det(\operatorname{dev}\mathbf{A}) = \frac{1}{3}\operatorname{tr}(\operatorname{dev}\mathbf{A})^3 = \frac{1}{3}\operatorname{tr}\mathbf{A}^3 - \frac{1}{3}\operatorname{tr}\mathbf{A}^2\operatorname{tr}\mathbf{A} + \frac{2}{27}(\operatorname{tr}\mathbf{A})^3 \,.
\end{aligned}
\tag{2.2.33}
$$

Also, for $\mathbf{A}$ dimensionless, setting $\lambda = -1$ in the characteristic polynomial,

$$\det(\mathbf{I} + \mathbf{A}) = 1 + I_1(\mathbf{A}) + I_2(\mathbf{A}) + I_3(\mathbf{A}) \tag{2.2.34}$$

and more generally for any scalar a and second order tensor $\mathbf{A}$,

$$\begin{aligned}
I_1(a\,\mathbf{I} + \mathbf{A}) &= 3\,a + I_1(\mathbf{A}), \\
I_2(a\,\mathbf{I} + \mathbf{A}) &= 3\,a^2 + 2\,a\,I_1(\mathbf{A}) + I_2(\mathbf{A}) \\
I_3(a\,\mathbf{I} + \mathbf{A}) &= a^3 + a^2\,I_1(\mathbf{A}) + a\,I_2(\mathbf{A}) + I_3(\mathbf{A}).
\end{aligned} \tag{2.2.35}$$

2.2.4 Spectral analysis of a symmetrized dyadic product of two vectors

The symmetrized dyadic second order tensor $\boldsymbol{\epsilon}^{\otimes}$

$$\boldsymbol{\epsilon}^{\otimes} = \frac{1}{2}\,(\mathbf{n} \otimes \mathbf{g} + \mathbf{g} \otimes \mathbf{n}), \qquad \mathbf{n} \cdot \mathbf{n} = \mathbf{g} \cdot \mathbf{g} = 1 \tag{2.2.36}$$

has peculiar spectral properties [Loret and Rizzi 1997]. If $\mathbf{g} \neq \mathbf{n}$, the three eigenvalues ξ_i and orthogonal eigenvectors $\mathbf{v}_i$, $i \in [1,3]$, are

$$\begin{aligned}
\xi_1 &= \frac{1}{2}\,(\mathbf{g} \cdot \mathbf{n} - 1) < 0, & \mathbf{v}_1 &= \frac{1}{2\sqrt{-\xi_1}}\,(\mathbf{g} - \mathbf{n}); \\
\xi_3 &= \frac{1}{2}\,(\mathbf{g} \cdot \mathbf{n} + 1) > 0, & \mathbf{v}_3 &= \frac{1}{2\sqrt{\xi_3}}\,(\mathbf{g} + \mathbf{n}); \\
\xi_2 &= 0, & \mathbf{v}_2 &= \mathbf{v}_3 \wedge \mathbf{v}_1.
\end{aligned} \tag{2.2.37}$$

Conversely, for a symmetric second order tensor $\boldsymbol{\epsilon}$ to be of the dyadic form (2.2.36), its eigenvalues ξ_i, $i \in [1,3]$, $\xi_1 \leq \xi_2 \leq \xi_3$, must satisfy $\xi_1 \leq 0 = \xi_2 \leq \xi_3$. Moreover,
- if $\xi_1 = \xi_2 = 0$, one necessarily has $\mathbf{g} = \mathbf{n}$, and so $\boldsymbol{\epsilon} = \mathbf{n} \otimes \mathbf{n}$, $\xi_3 = 1$, $\mathbf{v}_3 = \mathbf{n}$, $\mathbf{v}_1 =$ any unit vector $\perp \mathbf{n}$;
- if $\xi_2 = \xi_3 = 0$, $\mathbf{g} = -\mathbf{n}$ and so $\boldsymbol{\epsilon} = -\mathbf{n} \otimes \mathbf{n}$, $\xi_1 = -1$, $\mathbf{v}_1 = \mathbf{n}$, $\mathbf{v}_3 =$ any unit vector $\perp \mathbf{n}$;
- if $\xi_1 < 0 = \xi_2 < \xi_3$, there exist two directions $\mathbf{n}$ and $\mathbf{g}$ such that $\boldsymbol{\epsilon}$ has the dyadic form $\boldsymbol{\epsilon}^{\otimes}$, namely

$$\frac{1}{\sqrt{\xi_3 - \xi_1}}\left(\sqrt{\xi_3}\,\mathbf{v}_3 + \sqrt{-\xi_1}\,\mathbf{v}_1\right) \quad \text{and} \quad \frac{1}{\sqrt{\xi_3 - \xi_1}}\left(\sqrt{\xi_3}\,\mathbf{v}_3 - \sqrt{-\xi_1}\,\mathbf{v}_1\right) \tag{2.2.38}$$

where the $\mathbf{v}_i$, $i=1,3$, are the unit eigenvectors of $\boldsymbol{\epsilon}$ associated with its eigenvalues ξ_i.

2.3 A few useful tensorial relations

2.3.1 Cayley-Hamilton theorem

The Cayley-Hamilton theorem states that a second order tensor satisfies its own characteristic polynomial. Indeed, for an arbitrary second order tensor $\mathbf{A}$ in $\mathbb{R}^3$,

$$\mathbf{CH}(\mathbf{A}) = \mathbf{A}^3 - I_1\,\mathbf{A}^2 + I_2\,\mathbf{A} - I_3\,\mathbf{I} = \mathbf{0}, \tag{2.3.1}$$

where, in accordance with (2.2.7) and (2.2.11) for $n = 3$, I_1, I_2 and I_3 are the scalar invariants of $\mathbf{A}$ defined by

$$I_1 = \operatorname{tr}\mathbf{A}, \quad I_2 = \frac{1}{2}\left((\operatorname{tr}\mathbf{A})^2 - \operatorname{tr}\mathbf{A}^2\right), \quad I_3 = \det\mathbf{A} = \frac{1}{3}\operatorname{tr}\mathbf{A}^3 - \frac{1}{2}\operatorname{tr}\mathbf{A}\operatorname{tr}\mathbf{A}^2 + \frac{1}{6}(\operatorname{tr}\mathbf{A})^3. \tag{2.3.2}$$

If $\mathbf{A}$ has three distinct eigenvalues, the proof is trivial: just write the characteristic polynomial for the three eigenvalues. For a general proof, see, e.g., Bowen [1989], p. 235.

As an application, let $\mathbf{A}$ be an invertible second order tensor, and $a \geq 0$ a positive scalar. The inverse of $\mathbf{A} + a\,\mathbf{I}$ is sought as a linear combination of $\mathbf{I}$, $\mathbf{A}$ and $\mathbf{A}^2$ and by use of Cayley-Hamilton theorem [Hoger and Carlson 1984]a,

$$\left(a\left(a(a + I_1) + I_2\right) + I_3\right)(\mathbf{A} + a\,\mathbf{I})^{-1} = \mathbf{A}^2 - (a + I_1)\,\mathbf{A} + \left(a(a + I_1) + I_2\right)\mathbf{I}. \tag{2.3.3}$$

Alternatively apply the Cayley-Hamilton theorem to $\mathbf{A} + a\,\mathbf{I}$ using (2.2.35).

2.3.2 Application to the polar decomposition $\mathbf{F} = \mathbf{Q} \cdot \mathbf{U}$

Consider the polar decomposition of the deformation gradient $\mathbf{F} = \mathbf{Q} \cdot \mathbf{U}$ detailed in Section 2.5.1. The scalar invariants of the associated right Cauchy-Green tensor $\mathbf{C} = \mathbf{F}^{\mathrm{T}} \cdot \mathbf{F} = \mathbf{U}^2$ express in terms of the scalar invariants of $\mathbf{U}$,

$$I_1(\mathbf{C}) = (I_1(\mathbf{U}))^2 - 2\,I_2(\mathbf{U}), \quad I_2(\mathbf{C}) = (I_2(\mathbf{U}))^2 - 2\,I_1(\mathbf{U})\,I_3(\mathbf{U}), \quad I_3(\mathbf{C}) = (I_3(\mathbf{U}))^2. \tag{2.3.4}$$

To prove these relations, just express the invariants in terms of the eigenvalues $\lambda_i(\mathbf{U})$ and $\lambda_i(\mathbf{C}) = (\lambda_i(\mathbf{U}))^2$, $i \in [1,3]$, as indicated by (2.2.13).

The converse relations which provide the scalar invariants of $\mathbf{U}$ in terms of the scalar invariants of $\mathbf{C}$ may be obtained through the following process. First obtain the eigenvalues $\lambda_i(\mathbf{C})$ of $\mathbf{C}$ using the procedure outlined in Section 2.2.3.2. Since these eigenvalues are positive, take their (positive) square-roots $\sqrt{\lambda_i(\mathbf{C})}$ to obtain the eigenvalues and invariants of $\mathbf{U}$. A different procedure suggested in Hoger and Carlson [1984]a is critized by Sawyers [1986].

As another application of the Cayley-Hamilton theorem, the pure deformation $\mathbf{U}$ and its inverse $\mathbf{U}^{-1}$ may be brought in the formats,

$$\mathbf{U} = I_1(\mathbf{U})\,(\mathbf{C} + I_2(\mathbf{U})\,\mathbf{I})^{-1} \cdot \left(\mathbf{C} + \frac{I_3(\mathbf{U})}{I_1(\mathbf{U})}\,\mathbf{I}\right), \quad \mathbf{U}^{-1} = \frac{1}{I_1(\mathbf{U})}\left(\mathbf{C} + \frac{I_3(\mathbf{U})}{I_1(\mathbf{U})}\,\mathbf{I}\right)^{-1} \cdot (\mathbf{C} + I_2(\mathbf{U})\,\mathbf{I}). \tag{2.3.5}$$

With help of (2.3.3), (2.3.4) and (2.3.5) and another use of Cayley-Hamilton theorem, $\mathbf{U}$ and $\mathbf{U}^{-1}$ can be expressed as linear combinations of $\mathbf{I}$, $\mathbf{C}$ and $\mathbf{C}^2$, with coefficients involving the invariants of $\mathbf{U}$ only,

$$\left(I_1(\mathbf{U})\,I_2(\mathbf{U}) - I_3(\mathbf{U})\right)\mathbf{U} = -\mathbf{C}^2 + \left(I_1^2(\mathbf{U}) - I_2(\mathbf{U})\right)\mathbf{C} + I_1(\mathbf{U})\,I_3(\mathbf{U})\,\mathbf{I}, \tag{2.3.6}$$

and

$$I_3(\mathbf{U})\left(I_1(\mathbf{U})\,I_2(\mathbf{U}) - I_3(\mathbf{U})\right)\mathbf{U}^{-1} = I_1(\mathbf{U})\,\mathbf{C}^2 - \left(I_1(\mathbf{U})\,(I_1^2(\mathbf{U}) - 2\,I_2(\mathbf{U})) + I_3(\mathbf{U})\right)\mathbf{C}$$
$$+ \left((I_1(\mathbf{U})\,I_2(\mathbf{U}) - I_3(\mathbf{U}))\,I_2(\mathbf{U}) - I_1^2(\mathbf{U})\,I_3(\mathbf{U})\right)\mathbf{I}. \tag{2.3.7}$$

Moreover, the invariants of $\mathbf{U}$ may be expressed in terms of the invariants of $\mathbf{C}$ via algebraic manipulations, as indicated just above. The results (2.3.6) and (2.3.7) are meaningful because $I_1(\mathbf{U})\,I_2(\mathbf{U}) - I_3(\mathbf{U})$ does not vanish, as shown by (2.2.14).

2.3.3 Generalized Cayley-Hamilton relation

Rivlin [1955] has generalized the Cayley-Hamilton theorem to three arbitrary second order tensors $\mathbf{A}$, $\mathbf{B}$, $\mathbf{C}$ in $\mathbb{R}^3$,

$$\mathbf{GCH}(\mathbf{A}, \mathbf{B}, \mathbf{C}) =$$
$$\mathbf{A} \cdot \mathbf{B} \cdot \mathbf{C} + \mathbf{A} \cdot \mathbf{C} \cdot \mathbf{B} + \mathbf{B} \cdot \mathbf{A} \cdot \mathbf{C} + \mathbf{B} \cdot \mathbf{C} \cdot \mathbf{A} + \mathbf{C} \cdot \mathbf{A} \cdot \mathbf{B} + \mathbf{C} \cdot \mathbf{B} \cdot \mathbf{A}$$
$$- \operatorname{tr}\mathbf{A}\,(\mathbf{B} \cdot \mathbf{C} + \mathbf{C} \cdot \mathbf{B}) - \operatorname{tr}\mathbf{B}\,(\mathbf{C} \cdot \mathbf{A} + \mathbf{A} \cdot \mathbf{C}) - \operatorname{tr}\mathbf{C}\,(\mathbf{A} \cdot \mathbf{B} + \mathbf{B} \cdot \mathbf{A}) \tag{2.3.8}$$
$$+ (\operatorname{tr}\mathbf{B}\operatorname{tr}\mathbf{C} - \operatorname{tr}\mathbf{B} \cdot \mathbf{C})\,\mathbf{A} + (\operatorname{tr}\mathbf{C}\operatorname{tr}\mathbf{A} - \operatorname{tr}\mathbf{C} \cdot \mathbf{A})\,\mathbf{B} + (\operatorname{tr}\mathbf{A}\operatorname{tr}\mathbf{B} - \operatorname{tr}\mathbf{A} \cdot \mathbf{B})\,\mathbf{C}$$
$$- (\operatorname{tr}\mathbf{A}\operatorname{tr}\mathbf{B}\operatorname{tr}\mathbf{C} - \operatorname{tr}\mathbf{A}\operatorname{tr}\mathbf{B} \cdot \mathbf{C} - \operatorname{tr}\mathbf{B}\operatorname{tr}\mathbf{A} \cdot \mathbf{C} - \operatorname{tr}\mathbf{C}\operatorname{tr}\mathbf{A} \cdot \mathbf{B} + \operatorname{tr}\mathbf{A} \cdot \mathbf{B} \cdot \mathbf{C} + \operatorname{tr}\mathbf{C} \cdot \mathbf{B} \cdot \mathbf{A})\,\mathbf{I} = \mathbf{0}.$$

A sketch of the proof is suggested in Exercise 2.5.

With $\mathbf{C} = \mathbf{A}$, the generalized Cayley-Hamilton theorem specializes to

$$\begin{aligned}\mathbf{SCH}(\mathbf{A}, \mathbf{B}) &= \mathbf{A}^2 \cdot \mathbf{B} + \mathbf{B} \cdot \mathbf{A}^2 + \mathbf{A} \cdot \mathbf{B} \cdot \mathbf{A}\\[4pt] &- \operatorname{tr}\mathbf{A}\,(\mathbf{A} \cdot \mathbf{B} + \mathbf{B} \cdot \mathbf{A}) - \operatorname{tr}\mathbf{B}\,\mathbf{A}^2\\[4pt] &- (\operatorname{tr}\mathbf{A} \cdot \mathbf{B} - \operatorname{tr}\mathbf{A}\operatorname{tr}\mathbf{B})\,\mathbf{A} + \tfrac{1}{2}((\operatorname{tr}\mathbf{A})^2 - \operatorname{tr}\mathbf{A}^2)\,\mathbf{B}\\[4pt] &- \left(\operatorname{tr}\mathbf{A}^2 \cdot \mathbf{B} - \operatorname{tr}\mathbf{A}\operatorname{tr}\mathbf{A} \cdot \mathbf{B} + \tfrac{1}{2}\operatorname{tr}\mathbf{B}\,((\operatorname{tr}\mathbf{A})^2 - \operatorname{tr}\mathbf{A}^2)\right)\mathbf{I} = \mathbf{0}.\end{aligned} \tag{2.3.9}$$

Two further algebraic manipulations provide two other useful linear dependence relations [Rivlin 1955]

$$\mathbf{A}^2 \cdot \mathbf{B} \cdot \mathbf{A} + \mathbf{A} \cdot \mathbf{B} \cdot \mathbf{A}^2 - \operatorname{tr}\mathbf{A}\,\mathbf{A} \cdot \mathbf{B} \cdot \mathbf{A} = (\operatorname{tr}\mathbf{A} \cdot \mathbf{B})\,\mathbf{A}^2 + \left(\operatorname{tr}\mathbf{A}^2 \cdot \mathbf{B} - \operatorname{tr}\mathbf{A}\operatorname{tr}\mathbf{A} \cdot \mathbf{B}\right)\mathbf{A}$$
$$- \det\mathbf{A}\,\mathbf{B} + \det\mathbf{A}\operatorname{tr}\mathbf{B}\,\mathbf{I}\,;$$

$$\mathbf{A}^2 \cdot \mathbf{B} \cdot \mathbf{A}^2 - \tfrac{1}{2}((\operatorname{tr}\mathbf{A})^2 - \operatorname{tr}\mathbf{A}^2)\,\mathbf{A} \cdot \mathbf{B} \cdot \mathbf{A} = -\det\mathbf{A}\,(\mathbf{A} \cdot \mathbf{B} + \mathbf{B} \cdot \mathbf{A}) + (\operatorname{tr}\mathbf{A}^2 \cdot \mathbf{B})\,\mathbf{A}^2 \tag{2.3.10}$$
$$+ \left((\det\mathbf{A}\operatorname{tr}\mathbf{B} - \tfrac{1}{2}((\operatorname{tr}\mathbf{A})^2 - \operatorname{tr}\mathbf{A}^2)\operatorname{tr}\mathbf{A} \cdot \mathbf{B}\right)\mathbf{A}$$
$$+ \det\mathbf{A}\,(\operatorname{tr}\mathbf{A} \cdot \mathbf{B})\,\mathbf{I}.$$

In view of future use, we record the special case where $\mathbf{A}$ is symmetric and $\mathbf{B}$ skew-symmetric,

$$\mathbf{A}^2 \cdot \mathbf{B} + \mathbf{B} \cdot \mathbf{A}^2 + \mathbf{A} \cdot \mathbf{B} \cdot \mathbf{A} = \operatorname{tr} \mathbf{A} \left(\mathbf{A} \cdot \mathbf{B} + \mathbf{B} \cdot \mathbf{A} \right) - \tfrac{1}{2} \left((\operatorname{tr} \mathbf{A})^2 - \operatorname{tr} \mathbf{A}^2 \right) \mathbf{B};$$

$$\mathbf{A}^2 \cdot \mathbf{B} \cdot \mathbf{A} + \mathbf{A} \cdot \mathbf{B} \cdot \mathbf{A}^2 - \operatorname{tr} \mathbf{A} \, \mathbf{A} \cdot \mathbf{B} \cdot \mathbf{A} = - \det \mathbf{A} \, \mathbf{B}; \qquad (2.3.11)$$

$$\mathbf{A}^2 \cdot \mathbf{B} \cdot \mathbf{A}^2 - \tfrac{1}{2} \left((\operatorname{tr} \mathbf{A})^2 - \operatorname{tr} \mathbf{A}^2 \right) \mathbf{A} \cdot \mathbf{B} \cdot \mathbf{A} = - \det \mathbf{A} \left(\mathbf{A} \cdot \mathbf{B} + \mathbf{B} \cdot \mathbf{A} \right).$$

Rivlin [1955] also provides a number of additional interesting relations for matrices, or tensors, in $\mathbb{R}^2$.

2.3.4   The tensor equation $\mathbf{X} \cdot \mathbf{A} + \mathbf{A} \cdot \mathbf{X} = 2\,\mathbf{S}$

Here are other instances where the use of the Cayley-Hamilton theorem combines with the theorems of representation of isotropic second order tensors.

Differentiating the polar decomposition yields the velocity gradient $\mathbf{L} = d\mathbf{R}/dt \cdot \mathbf{R}^{\mathrm{T}} + \mathbf{R} \cdot d\mathbf{U}/dt \cdot \mathbf{U}^{-1} \cdot \mathbf{R}^{\mathrm{T}}$. Taking the symmetric part provides an equation for the rate of the pure deformation

$$\frac{d\mathbf{U}}{dt} \cdot \mathbf{U}^{-1} + \mathbf{U}^{-1} \cdot \frac{d\mathbf{U}}{dt} = 2\,\mathbf{D}_* = 2\,\mathbf{R}^{\mathrm{T}} \cdot \mathbf{D} \cdot \mathbf{R}. \qquad (2.3.12)$$

An equation for the rotation,

$$\left(\frac{d\mathbf{R}}{dt} \cdot \mathbf{R}^{\mathrm{T}} \right) \cdot \mathbf{V} + \mathbf{V} \cdot \left(\frac{d\mathbf{R}}{dt} \cdot \mathbf{R}^{\mathrm{T}} \right) \cdot \mathbf{V} = \mathbf{L} \cdot \mathbf{V} - \mathbf{V} \cdot \mathbf{L}^{\mathrm{T}}, \qquad (2.3.13)$$

derives by forming the rhs of this expression and use of the polar decomposition in the form $\mathbf{F} = \mathbf{V} \cdot \mathbf{R}$.

The solution of these equations has been addressed in a number of papers, e.g., Sidoroff [1978], Guo [1984], Hoger and Carlson [1984]b. The presentation below follows their main idea, although it considers two sub-problems so as to minimize the algebras.

For $\mathbf{A}$ a positive definite second order tensor, and $\mathbf{S}$ and $\mathbf{T}$ two arbitrary second order (rate) tensors, we shall consider the two equations for the unknown second order tensors $\mathbf{X}$ and $\mathbf{Y}$,

$$\mathbf{X} \cdot \mathbf{A} + \mathbf{A} \cdot \mathbf{X} = 2\,\mathbf{S}, \quad \mathbf{Y} \cdot \mathbf{A}^{-1} + \mathbf{A}^{-1} \cdot \mathbf{Y} = 2\,\mathbf{T}. \qquad (2.3.14)$$

The idea is to view the solution $\mathbf{Y}$ (a rate) as an isotropic function of $\mathbf{A}$ and $\mathbf{T}$, linear in the rate $\mathbf{T}$. Section 2.11.3.5 provides generators of two symmetric second order tensors. The symmetry is not assumed here, but we proceed heuristically and use this basis,

$$\mathbf{I}, \ \mathbf{A}, \ \mathbf{A}^2, \ \mathbf{T}, \ \mathbf{T}^2, \ \mathbf{A} \cdot \mathbf{T} + \mathbf{T} \cdot \mathbf{A}, \ \mathbf{A} \cdot \mathbf{T}^2 + \mathbf{T}^2 \cdot \mathbf{A}, \ \mathbf{A}^2 \cdot \mathbf{T} + \mathbf{T} \cdot \mathbf{A}^2, \qquad (2.3.15)$$

from which we first eliminate the two terms which involve $\mathbf{T}^2$ and, second, with an eye toward (2.3.10) which holds for arbitrary tensors, we replace the three terms $\mathbf{I}$, $\mathbf{A}$, $\mathbf{A}^2$ by terms linear in $\mathbf{T}$ and build the solution on the set

$$\mathbf{A}^2 \cdot \mathbf{T} \cdot \mathbf{A}^2, \ \mathbf{A}^2 \cdot \mathbf{T} \cdot \mathbf{A} + \mathbf{A} \cdot \mathbf{T} \cdot \mathbf{A}^2, \ \mathbf{A}^2 \cdot \mathbf{T} + \mathbf{T} \cdot \mathbf{A}^2, \ \mathbf{A} \cdot \mathbf{T} \cdot \mathbf{A}, \ \mathbf{A} \cdot \mathbf{T} + \mathbf{T} \cdot \mathbf{A}, \ \mathbf{T}. \qquad (2.3.16)$$

Backsubstituting the linear combination in Eqn (2.3.14)$_2$ and using the Cayley-Hamilton theorem provides readily the coefficients, yielding

$$(I_1 I_2 - I_3)\,\mathbf{Y} = \qquad (2.3.17)$$

$$\mathbf{A}^2 \cdot \mathbf{T} \cdot \mathbf{A}^2 - I_1 \left(\mathbf{A}^2 \cdot \mathbf{T} \cdot \mathbf{A} + \mathbf{A} \cdot \mathbf{T} \cdot \mathbf{A}^2 \right) + (I_1^2 + I_2)\,\mathbf{A} \cdot \mathbf{T} \cdot \mathbf{A} - I_3 \left(\mathbf{T} \cdot \mathbf{A} + \mathbf{A} \cdot \mathbf{T} \right) + I_1 I_3\,\mathbf{T},$$

with I_1, I_2 and I_3 isotropic scalar invariants of $\mathbf{A}$. The solution $\mathbf{X}$ to Eqn (2.3.14)$_1$ deduces by setting $\mathbf{T} = \mathbf{A}^{-1} \cdot \mathbf{S} \cdot \mathbf{A}^{-1}$ and further use of the Cayley-Hamilton theorem,

$$(I_1 I_2 - I_3)\,I_3\,\mathbf{X} = \qquad (2.3.18)$$

$$I_1\,\mathbf{A}^2 \cdot \mathbf{S} \cdot \mathbf{A}^2 - I_1^2 \left(\mathbf{A}^2 \cdot \mathbf{S} \cdot \mathbf{A} + \mathbf{A} \cdot \mathbf{S} \cdot \mathbf{A}^2 \right) + (I_1 I_2 - I_3) \left(\mathbf{A}^2 \cdot \mathbf{S} + \mathbf{S} \cdot \mathbf{A}^2 \right) + (I_1^3 + I_3)\,\mathbf{A} \cdot \mathbf{S} \cdot \mathbf{A}$$

$$- I_1^2 I_2 \left(\mathbf{S} \cdot \mathbf{A} + \mathbf{A} \cdot \mathbf{S} \right) + \left((I_1^2 - I_2)\,I_3 + I_1 I_2^2 \right) \mathbf{S}.$$

Alternative, but equivalent, expressions may be obtained by use of the relations (2.3.10), e.g., Hoger and Carlson [1984]b. In the special case where $\mathbf{A}$ is symmetric and $\mathbf{S}$ skew-symmetric, we recover via (2.3.11) the solution of Guo [1984],

$$(I_1 I_2 - I_3)\,\mathbf{X} = -2 \left(\mathbf{A}^2 \cdot \mathbf{S} + \mathbf{S} \cdot \mathbf{A}^2 \right) + 2 \left(I_1^2 - I_2 \right) \mathbf{S}. \qquad (2.3.19)$$

Incidentally, this result indicates that the construction holds for both symmetric and skew-symmetric tensors, and therefore, by superposition, for arbitrary tensors $\mathbf{S}$ and $\mathbf{T}$.

So far, so good. But is the solution unique? Guo [1984] has proved that it is indeed unique if $\mathbf{A}$ is symmetric positive definite. Indeed then, he showed that the equation $\mathbf{X} \cdot \mathbf{A} + \mathbf{A} \cdot \mathbf{X} = \mathbf{0}$ has only the trivial solution $\mathbf{X} = \mathbf{0}$, by writing this equation in component form on the eigenaxes of $\mathbf{A}$. This result shows that the denominator $I_1 I_2 - I_3$ does not vanish, but this we knew already from (2.2.14).

2.3.5   Rank one and rank two modifications of the identity

Let $\mathbf{a}$ and $\mathbf{b}$ be two arbitrary vectors. If

$$\Delta = \det \left(\mathbf{I} + \mathbf{a} \otimes \mathbf{b} \right) = 1 + \mathbf{a} \cdot \mathbf{b} \neq 0 \,, \tag{2.3.20}$$

the inverse of this rank one modification of the identity keeps exactly its structure, namely

$$\text{Sherman} - \text{Morrison formula:} \quad \left(\mathbf{I} + \mathbf{a} \otimes \mathbf{b} \right)^{-1} = \mathbf{I} - \frac{\mathbf{a} \otimes \mathbf{b}}{1 + \mathbf{a} \cdot \mathbf{b}} \,. \tag{2.3.21}$$

Symmetric rank one modifications of the identity enjoy interesting properties. Let $\mathbf{n}$ be a unit vector, and x an arbitrary real number. The symmetric matrices,

$$\mathbf{I}(x) = \mathbf{I} + (x - 1)\,\mathbf{n} \otimes \mathbf{n} \,, \tag{2.3.22}$$

are form-invariant under multiplication, e.g., for x and y arbitrary real numbers,

$$\mathbf{I}(x) \cdot \mathbf{I}(y) = \mathbf{I}(y) \cdot \mathbf{I}(x) = \mathbf{I}(x\,y) = \mathbf{I} + (x\,y - 1)\,\mathbf{n} \otimes \mathbf{n} \,, \tag{2.3.23}$$

and, for $x > 0$, when raised to an arbitrary real exponent a,

$$\det \mathbf{I}(x) = x \,; \quad \mathbf{I}^{a}(x) = \mathbf{I}(x^{a}) = \mathbf{I} + (x^{a} - 1)\,\mathbf{n} \otimes \mathbf{n} \,. \tag{2.3.24}$$

They enjoy the following spectral properties:

$$\mathbf{I}(x) \cdot \mathbf{n} = x\,\mathbf{n}, \quad \mathbf{I}(x) \cdot \mathbf{n}^{\perp} = \mathbf{n}^{\perp}, \quad \text{with} \quad \mathbf{n}^{\perp} \cdot \mathbf{n} = 0 \,. \tag{2.3.25}$$

A useful extension of the Sherman-Morrison formula is obtained by replacing the identity $\mathbf{I}$ by an invertible second order tensor $\mathbf{A}$. Namely if

$$\det \left(\mathbf{A} + \mathbf{a} \otimes \mathbf{b} \right) = \det \mathbf{A} + \det \mathbf{A}\,(\mathbf{A}^{-1} \cdot \mathbf{a}) \cdot \mathbf{b} \neq 0 \,, \tag{2.3.26}$$

then

$$\left(\mathbf{A} + \mathbf{a} \otimes \mathbf{b} \right)^{-1} = \mathbf{A}^{-1} - \frac{\mathbf{A}^{-1} \cdot \mathbf{a} \otimes \mathbf{b} \cdot \mathbf{A}^{-1}}{1 + (\mathbf{A}^{-1} \cdot \mathbf{a}) \cdot \mathbf{b}} \,. \tag{2.3.27}$$

Another extension to rank two modifications of the identity can be recorded. Let $\mathbf{a}$, $\mathbf{b}$, $\mathbf{c}$ and $\mathbf{d}$ be four arbitrary vectors. If

$$\Delta = \det \left(\mathbf{I} + \mathbf{a} \otimes \mathbf{b} + \mathbf{c} \otimes \mathbf{d} \right) = (1 + \mathbf{a} \cdot \mathbf{b})\,(1 + \mathbf{c} \cdot \mathbf{d}) - (\mathbf{a} \cdot \mathbf{d})\,(\mathbf{b} \cdot \mathbf{c}) \neq 0 \,, \tag{2.3.28}$$

the inverse of this rank two modification of the identity keeps formally its structure, namely

$$\left(\mathbf{I} + \mathbf{a} \otimes \mathbf{b} + \mathbf{c} \otimes \mathbf{d} \right)^{-1} = \mathbf{I} + \mathbf{a} \otimes \mathbf{b}' + \mathbf{c} \otimes \mathbf{d}' \,, \tag{2.3.29}$$

with

$$\Delta\,\mathbf{b}' = -(1 + \mathbf{c} \cdot \mathbf{d})\,\mathbf{b} + (\mathbf{b} \cdot \mathbf{c})\,\mathbf{d}, \quad \Delta\,\mathbf{d}' = -(1 + \mathbf{a} \cdot \mathbf{b})\,\mathbf{d} + (\mathbf{a} \cdot \mathbf{d})\,\mathbf{b} \,. \tag{2.3.30}$$

The proof of these formulas proceeds by simple inspection, Exercise 2.6.

2.3.6   Square-root of a positive definite symmetric second order tensor

Let $\mathbf{B}$ be a positive definite symmetric second order tensor. If $\mathbf{B} = \sum_{i=1}^{3} b_i\,\mathbf{b}_i \otimes \mathbf{b}_i$ is its spectral decomposition, then of course its square root $\mathbf{B}^{\frac{1}{2}}$ is equal to $\sum_{i=1}^{3} b_i^{\frac{1}{2}}\,\mathbf{b}_i \otimes \mathbf{b}_i$. However this formula requires the eigenvectors. Morman [1986] has provided an expression of the square root that requires only the eigenvalues.

First, $\mathbf{B}$ satisfies its characteristic polynomial $\mathbf{B}^3 - I_1(\mathbf{B})\,\mathbf{B}^2 + I_2(\mathbf{B})\,\mathbf{B} - I_3(\mathbf{B})\,\mathbf{I} = \mathbf{0}$ where $I_1(\mathbf{B}) = \operatorname{tr} \mathbf{B}$, $I_2(\mathbf{B}) = (\operatorname{tr}^2 \mathbf{B} - \operatorname{tr} \mathbf{B}^2)/2$, $I_3(\mathbf{B}) = \det \mathbf{B}$ are the three isotropic scalar invariants of $\mathbf{B}$.

Second, the square root is an isotropic function of $\mathbf{B}$, Section 2.11.3.2, and, third, representation theorems indicate that an isotropic function of the second order tensor $\mathbf{B}$ can be expressed as a linear combination of $\mathbf{I}$, $\mathbf{B}$ and $\mathbf{B}^2$, namely $\mathbf{B}^{\frac{1}{2}} = \alpha_0\,\mathbf{I} + \alpha_1\,\mathbf{B} + \alpha_2\,\mathbf{B}^2$. The three coefficients α_i, $i \in [0,2]$, are calculated from the set of three equations (some details are provided in Exercise 2.10),

$$b_j^{\frac{1}{2}} = \alpha_0 + \alpha_1\,b_j + \alpha_2\,b_j^2\,, \quad j \in [1,3]\,. \tag{2.3.31}$$

The eigenvalues,

$$b_j = \frac{I_1}{3} + \frac{2}{3}\,(I_1^2 - 3\,I_2)^{\frac{1}{2}} \cos\ell_j, \quad j \in [1,3]\,, \tag{2.3.32}$$

can be expressed in terms of the Lode angles defined in (2.2.31) if the tensor $\mathbf{B}$ is not isotropic. Then, with help of (2.2.33), $I_2(\operatorname{dev}\mathbf{B}) = -\frac{1}{2}\operatorname{tr}(\operatorname{dev}\mathbf{B})^2$ is equal to $I_2(\mathbf{B}) - \frac{1}{3}I_1(\mathbf{B})^2$. The Lode angles may be calculated via the invariants of the tensor itself instead of the invariants of its deviator, namely

$$\ell_j = \ell - \frac{2\pi}{3}\,(j-1),\ j \in [1,3], \text{ with } \cos 3\,\ell = \frac{2\,I_1^3 - 9\,I_1\,I_2 + 27\,I_3}{2\,(I_1^2 - 3\,I_2)^{3/2}}\,. \tag{2.3.33}$$

Finally,

$$
\begin{aligned}
&1.\ b_1 \neq b_2 \neq b_3: \quad \mathbf{B}^{\frac{1}{2}} \;=\; && \sum_{i=1}^{3} b_i^{\frac{1}{2}}\,\frac{\mathbf{B}^2 - (I_1 - b_i)\,\mathbf{B} + I_3\,b_i^{-1}\mathbf{I}}{2\,b_i^2 - I_1\,b_i + I_3\,b_i^{-1}}; &&\\
&2.\ b_i = b_j \neq b_k: \quad \mathbf{B}^{\frac{1}{2}} \;=\; b_i^{\frac{1}{2}}\mathbf{I} \;+\; (b_k^{\frac{1}{2}} - b_i^{\frac{1}{2}})\,\frac{\mathbf{B}^2 - (I_1 - b_k)\,\mathbf{B} + I_3\,b_k^{-1}\mathbf{I}}{2\,b_k^2 - I_1\,b_k + I_3\,b_k^{-1}}; &&\\
&3.\ b_1 = b_2 = b_3: \quad \mathbf{B}^{\frac{1}{2}} \;=\; b_1^{\frac{1}{2}}\mathbf{I}\,. &&
\end{aligned}
\tag{2.3.34}
$$

2.4 The differential operators of continuum mechanics

Differential operators are needed to express the local forms of the balance equations. They are also used to define the kinematics of the continuum.

2.4.1 Definitions of the differential operators

We start by defining operators over scalars, vectors and tensors so as to be able to list a number of fundamental results that allow switching from volume to surface integrals and conversely. Next, we define these operators over tensors of various orders. We will interpret these first formulas as providing their left version in noncartesian axes and we will also define their right versions.

2.4.1.1 Standard differential operators

Let $\mathbf{x} = x_i\,\mathbf{e}_i$ be the coordinates of a point $\mathbf{x}$ in the cartesian axes $\mathbf{e}_i$, $i \in [1,3]$. In a first step, the basic operators *gradient* $\boldsymbol{\nabla}$, *divergence* denoted by div or $\boldsymbol{\nabla}\cdot$, and *rotational* denoted by curl or $\boldsymbol{\nabla}\wedge$ may be defined componentwise,

$$\boldsymbol{\nabla}\boldsymbol{\Phi} = \mathbf{e}_k\,\frac{\partial\boldsymbol{\Phi}}{\partial x_k}, \quad \operatorname{div}\boldsymbol{\Phi} = \boldsymbol{\nabla}\cdot\boldsymbol{\Phi} = \mathbf{e}_k\cdot\frac{\partial\boldsymbol{\Phi}}{\partial x_k}, \quad \operatorname{curl}\boldsymbol{\Phi} = \boldsymbol{\nabla}\wedge\boldsymbol{\Phi} = \mathbf{e}_k\wedge\frac{\partial\boldsymbol{\Phi}}{\partial x_k}\,, \tag{2.4.1}$$

for the smooth field $\boldsymbol{\Phi}$, whether vector or tensor. The first relation also defines the gradient of a scalar $\boldsymbol{\Phi}$. A field which is divergence free is said to be *solenoidal*. A field whose rotational vanishes is said to be *irrotational*.

It is useful to record the component forms in cartesian axes of the gradient, divergence and rotational of a vector,

$$(\boldsymbol{\nabla}\boldsymbol{\Phi})_{ij} = \frac{\partial\Phi_j}{\partial x_i}, \quad \boldsymbol{\nabla}\cdot\boldsymbol{\Phi} = \frac{\partial\Phi_j}{\partial x_j}, \quad (\boldsymbol{\nabla}\wedge\boldsymbol{\Phi})_i = e_{ijk}\,\frac{\partial\Phi_k}{\partial x_j}\,. \tag{2.4.2}$$

The divergence and the rotational of a second order tensor $\boldsymbol{\Phi}$ will be defined componentwise as

$$(\operatorname{div}\boldsymbol{\Phi})_i = \frac{\partial\Phi_{ji}}{\partial x_j}, \quad (\boldsymbol{\nabla}\wedge\boldsymbol{\Phi})_{ij} = e_{ikl}\,\frac{\partial\Phi_{lj}}{\partial x_k}\,. \tag{2.4.3}$$

The relations for $\boldsymbol{\Phi}$ scalar,

$$\boldsymbol{\nabla}\wedge(\boldsymbol{\nabla}\boldsymbol{\Phi}) = \mathbf{0}\,, \tag{2.4.4}$$

for $\boldsymbol{\Phi}$ vector,

$$\mathrm{div}\,(\boldsymbol{\nabla} \wedge \boldsymbol{\Phi}) = \mathbf{0},$$

$$\Delta\boldsymbol{\Phi} = \boldsymbol{\nabla}\mathrm{div}\,\boldsymbol{\Phi} - \boldsymbol{\nabla} \wedge (\boldsymbol{\nabla} \wedge \boldsymbol{\Phi}),\tag{2.4.5}$$

$$\boldsymbol{\Phi} \wedge (\boldsymbol{\nabla} \wedge \boldsymbol{\Phi}) = \boldsymbol{\nabla}(\tfrac{1}{2}\boldsymbol{\Phi}^2) - \boldsymbol{\Phi} \cdot \boldsymbol{\nabla}\boldsymbol{\Phi},$$

for Φ scalar and $\boldsymbol{\Psi}$ vector,

$$\boldsymbol{\nabla} \wedge (\Phi\,\boldsymbol{\Psi}) = \boldsymbol{\nabla}\Phi \wedge \boldsymbol{\Psi} + \Phi\,\boldsymbol{\nabla} \wedge \boldsymbol{\Psi},\tag{2.4.6}$$

and for $\boldsymbol{\Phi}$ and $\boldsymbol{\Psi}$ vectors,

$$\mathrm{div}\,(\boldsymbol{\Phi} \wedge \boldsymbol{\Psi}) = \boldsymbol{\Psi} \cdot \boldsymbol{\nabla} \wedge \boldsymbol{\Phi} - \boldsymbol{\Phi} \cdot \boldsymbol{\nabla} \wedge \boldsymbol{\Psi},\tag{2.4.7}$$

follow from (2.1.10) and the definitions of the divergence and rotational operators. Eqn (2.4.5) involves the Laplacian of a vector which is built from the gradients, namely $\Delta\boldsymbol{\Phi} = \boldsymbol{\nabla} \cdot (\boldsymbol{\nabla}\boldsymbol{\Phi})$.

- **Green's theorem** or **Gauss's theorem** or **divergence theorem**:

The divergence theorem transforms a surface integral into a volume integral or conversely. Let $\mathbf{n}$ be the unit outward normal to the surface $S = \partial V$ of the volume V which does not need to be simply connected. If the volume V is not simply connected, the surface S includes both the outer and the inner boundaries of V and $\mathbf{n}$ denotes always the outward normal. Therefore, the main condition on the domain V is that one should be able to define an interior and an exterior, excluding nonorientable surfaces like the Klein surface or the Möbius strip. In the formulas below, whatever its tensorial nature, the integrand considered is assumed sufficiently smooth so that there is no *discontinuity surface* of the gradient $\partial\boldsymbol{\Phi}/\partial\mathbf{x}$ crossing the volume. Otherwise, the formulas should be restricted to the volumes where the integrand is smooth: summation over these volumes will display the contributions due to the lack of smoothness, e.g., [Bowen 1989, p. 248].

For a proof of the divergence theorem, refer to a textbook of vector analysis.
For Φ scalar,

$$\int_{\partial V} n_k\,\Phi\,dS = \int_V \frac{\partial\Phi}{\partial x_k}\,dV, \quad i \in [1,3],\tag{2.4.8}$$

for $\boldsymbol{\Phi}$ vector,

$$\int_{\partial V} \mathbf{n} \cdot \boldsymbol{\Phi}\,dS = \int_V \mathrm{div}\,\boldsymbol{\Phi}\,dV, \quad \int_{\partial V} n_k\Phi_k\,dS = \int_V \frac{\partial\Phi_k}{\partial x_k}\,dV,\tag{2.4.9}$$

and

$$\int_{\partial V} \mathbf{n} \wedge \boldsymbol{\Phi}\,dS = \int_V \boldsymbol{\nabla} \wedge \boldsymbol{\Phi}\,dV, \quad \int_{\partial V} e_{ijk}\,n_j\Phi_k\,dS = \int_V e_{ijk}\frac{\partial\Phi_k}{\partial x_j}\,dV,\tag{2.4.10}$$

for $\boldsymbol{\Phi}$ tensor,

$$\int_{\partial V} \mathbf{n} \cdot \boldsymbol{\Phi}\,dS = \int_V \mathrm{div}\,\boldsymbol{\Phi}\,dV, \quad \int_{\partial V} n_k\,\Phi_{k\cdots}\,dS = \int_V \frac{\partial\Phi_{k\cdots}}{\partial x_k}\,dV.\tag{2.4.11}$$

The formula (2.4.10) deduces from (2.4.9) simply by use of the definition of the vector product (2.1.14) and of the rotational (2.4.2).

Analogous theorems transform an integral over a closed curve $\mathcal{C} = \partial S$ into an integral over the surface S it encloses and conversely, namely for Φ scalar,

$$\oint_{\mathcal{C}=\partial S} \mathbf{n}\,\Phi\,ds = \int_S \frac{\partial\Phi}{\partial\mathbf{x}}\,dS, \quad \oint_{\mathcal{C}=\partial S} n_k\,\Phi\,ds = \int_S \frac{\partial\Phi}{\partial x_k}\,dS,\tag{2.4.12}$$

and for $\boldsymbol{\Phi}$ vector or tensor of arbitrary order,

$$\oint_{\mathcal{C}=\partial S} \mathbf{n} \cdot \boldsymbol{\Phi}\,ds = \int_S \mathrm{div}\,\boldsymbol{\Phi}\,dS,$$

$$\oint_{\mathcal{C}=\partial S} \mathbf{t} \cdot \boldsymbol{\Phi}\,ds = \int_S \mathbf{e} \cdot (\boldsymbol{\nabla} \wedge \boldsymbol{\Phi})\,dS.\tag{2.4.13}$$

The last formula is known as **Stokes theorem** in $\mathbb{R}^3$ with $\mathbf{e}\,dS$ the oriented elementary surface. For a vector of the plane $(\mathbf{e}_1, \mathbf{e}_2)$ endowed with cartesian coordinates (x_1, x_2), the theorem reads (see [Mei 1995, p. 448], for a simple proof),

$$\oint_{\mathcal{C}=\partial S} \Phi_1\,dx_1 + \Phi_2\,dx_2 = \int_S \left(\frac{\partial\Phi_2}{\partial x_1} - \frac{\partial\Phi_1}{\partial x_2}\right) dx_1\,dx_2.\tag{2.4.14}$$

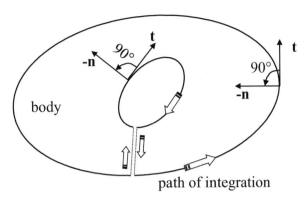

Fig. 2.4.1: Direct sense of integration for a body which is not simply connected. The inner and outer surfaces are connected by a branch cut, made by two arbitrarily close segments, which are run in the appropriate direction.

The curvilinear abscissa along the curve $\mathcal{C}$ is denoted by s. The symbol $\oint$ means integration over a closed curve in the *direct sense*, that is, the normal $-\mathbf{n}$ deduced from the tangent $\mathbf{t}$ by rotation of $+90°$ points to the interior of the surface. This definition serves also to define the sense of integration on inner boundaries if the surface is not simply connected, Fig. 2.4.1. Note that a similar notation might be used for the surface integration in the divergence theorem since the latter assumes the normal $\mathbf{n}$ to point to the exterior of the body. However the special integration symbol is used here only in the plane (x_1, x_2) where the tangent vector $\mathbf{t}\,ds$ and the normal $\mathbf{n}\,ds$ have components (dx_1, dx_2) and $(dx_2, -dx_1)$, respectively. These relations follow from (2.4.2) and (2.4.12).

- **Green's first and second identities**:

For Φ and Ψ twice differentiable scalars in the volume V, bounded by the surface $S = \partial V$ of incremental area $d\mathbf{S} = \mathbf{n}\,dS$,

$$\int_V \boldsymbol{\nabla}\Phi \cdot \boldsymbol{\nabla}\Psi \, dV = \int_{\partial V} \Phi\,\boldsymbol{\nabla}\Psi \cdot d\mathbf{S} - \int_V \Phi\,\Delta\Psi \, dV$$

$$\int_V \left(\Phi\,\Delta\Psi - \Psi\,\Delta\Phi \right) dV = \int_{\partial V} \left(\Phi\,\boldsymbol{\nabla}\Psi - \Psi\,\boldsymbol{\nabla}\Phi \right) \cdot d\mathbf{S} \,. \tag{2.4.15}$$

Green's first identity follows from the divergence theorem. The second identity is obtained by subtracting the first identities associated with Φ and Ψ and Ψ and Φ, respectively. Green's identities are key ingredients for the study of potential flows and constitute the starting point of boundary integral methods.

- **Representation theorems based on differential operators**:

We have seen in (2.4.4) and (2.4.5) that the rotational of a gradient and the divergence of a rotational vanish. The converses of these results are most useful in a number of applications:

Representation of a solenoidal displacement field:

There exists a twice differentiable solenoidal field $\boldsymbol{\psi}$, such that

$$\mathbf{u} = \boldsymbol{\nabla} \wedge \boldsymbol{\psi}, \quad u_i = e_{ijk} \frac{\partial \psi_k}{\partial x_j} \,, \tag{2.4.16}$$

if and only if $\operatorname{div} \mathbf{u} = \mathbf{0}$.

Representation of an irrotational displacement field:

In a simply connected domain, there exists a twice differentiable scalar field ϕ such that

$$\mathbf{u} = \boldsymbol{\nabla}\phi, \quad u_i = \frac{\partial \phi}{\partial x_i} \,, \tag{2.4.17}$$

if and only if the sufficiently smooth field $\mathbf{u}$ is irrotational, namely $\boldsymbol{\nabla} \wedge \mathbf{u} = \mathbf{0}$.

Helmholtz representation of a displacement field:

If $\mathbf{u}$ is a displacement field which is continuously differentiable, bounded, and bounded at infinity, there exist a differentiable scalar field ϕ, and a solenoidal field $\boldsymbol{\psi}$ such that

$$\mathbf{u} = \boldsymbol{\nabla}\phi + \boldsymbol{\nabla} \wedge \boldsymbol{\psi}, \quad u_i = \frac{\partial \phi}{\partial x_i} + e_{ijk} \frac{\partial \psi_k}{\partial x_j} \,. \tag{2.4.18}$$

Note that, due to (2.4.4) and (2.4.5), div $\mathbf{u} = \Delta\phi$ and $\boldsymbol{\nabla} \wedge \mathbf{u} = -\boldsymbol{\Delta\psi}$.

The condition that $\boldsymbol{\psi}$ should be divergence free is intended to reduce to three the number of independent scalar fields of the representation. The completeness of this representation is discussed in textbooks on dynamic elasticity where it is a most useful tool, e.g., Fung [1965], p. 184, or Achenbach [1975].

- **Conservative field**:

If $\int_C \boldsymbol{\Phi}(\mathbf{x}) \cdot d\mathbf{x}$ vanishes for any closed curve C in a domain, then there exists a differentiable field $\Phi = \Phi(\mathbf{x})$, that serves as a potential to $\boldsymbol{\Phi} = \boldsymbol{\Phi}(\mathbf{x})$ in this domain,

$$\boldsymbol{\Phi} = -\frac{\partial \Phi}{\partial \mathbf{x}}. \tag{2.4.19}$$

The theorem holds also when the integrand depends on variables that are not the space coordinates.

2.4.1.2 Scaling factors, left and right differential operators

The operators defined in the preceding section can be seen as *left* operators, in the sense that they operate on entities on their *right*. Their *right* counterparts which operate on entities on their *left* are provided here, in the context of curvilinear coordinates with which are associated the set of unit orthogonal vectors $\hat{\mathbf{e}}_i$ and the scaling factors h_i, $i \in [1,3]$. Arrows are superimposed over the operators to indicate if they operate to the right or to the left. If the coordinates are expressed in the format

$$\mathbf{x} = x_k \, \hat{\mathbf{e}}_k \,, \tag{2.4.20}$$

the scaling factors are defined by the relations

$$d\mathbf{x} = h_k \, \hat{\mathbf{e}}_k \, dx_k \,. \tag{2.4.21}$$

Starting from the left operators (2.4.1), their right counterparts follow through a *de visu* representation, for Φ scalar,

$$\overrightarrow{\boldsymbol{\nabla}}\Phi = \frac{\hat{\mathbf{e}}_k}{h_k}\frac{\overrightarrow{\partial}}{\partial x_k}\Phi, \quad \Phi\frac{\overleftarrow{\partial}}{\partial x_k}\frac{\hat{\mathbf{e}}_k}{h_k} = \Phi\overleftarrow{\boldsymbol{\nabla}}\,, \tag{2.4.22}$$

and for $\boldsymbol{\Phi}$ vector or tensor,

$$\overrightarrow{\boldsymbol{\nabla}}\boldsymbol{\Phi} = \frac{\hat{\mathbf{e}}_k}{h_k}\frac{\overrightarrow{\partial}}{\partial x_k}\boldsymbol{\Phi}, \quad \boldsymbol{\Phi}\frac{\overleftarrow{\partial}}{\partial x_k}\frac{\hat{\mathbf{e}}_k}{h_k} = \boldsymbol{\Phi}\overleftarrow{\boldsymbol{\nabla}}$$

$$\overrightarrow{\boldsymbol{\nabla}}\cdot\boldsymbol{\Phi} = \frac{\hat{\mathbf{e}}_k}{h_k}\cdot\frac{\overrightarrow{\partial}}{\partial x_k}\boldsymbol{\Phi}, \quad \boldsymbol{\Phi}\frac{\overleftarrow{\partial}}{\partial x_k}\cdot\frac{\hat{\mathbf{e}}_k}{h_k} = \boldsymbol{\Phi}\cdot\overleftarrow{\boldsymbol{\nabla}} \tag{2.4.23}$$

$$\overrightarrow{\boldsymbol{\nabla}}\wedge\boldsymbol{\Phi} = \frac{\hat{\mathbf{e}}_k}{h_k}\wedge\frac{\overrightarrow{\partial}}{\partial x_k}\boldsymbol{\Phi}, \quad \boldsymbol{\Phi}\frac{\overleftarrow{\partial}}{\partial x_k}\wedge\frac{\hat{\mathbf{e}}_k}{h_k} = \boldsymbol{\Phi}\wedge\overleftarrow{\boldsymbol{\nabla}} \,.$$

The definition of the gradient operators stems from the differentials,

$$d\Phi = d\mathbf{x} \cdot (\overrightarrow{\boldsymbol{\nabla}}\Phi) = (\Phi\overleftarrow{\boldsymbol{\nabla}}) \cdot d\mathbf{x}, \quad d\boldsymbol{\Phi} = d\mathbf{x} \cdot (\overrightarrow{\boldsymbol{\nabla}}\boldsymbol{\Phi}) = (\boldsymbol{\Phi}\overleftarrow{\boldsymbol{\nabla}}) \cdot d\mathbf{x}. \tag{2.4.24}$$

Therefore, the left and right gradients of a scalar are one and the same vector,

$$\overrightarrow{\boldsymbol{\nabla}}\Phi = \frac{\hat{\mathbf{e}}_k}{h_k}\frac{\overrightarrow{\partial}}{\partial x_k}\Phi = \Phi\overleftarrow{\boldsymbol{\nabla}} = \Phi\frac{\overleftarrow{\partial}}{\partial x_k}\frac{\hat{\mathbf{e}}_k}{h_k}\,. \tag{2.4.25}$$

The Laplacian of a scalar is built from the gradients,

$$\Delta\Phi = \overrightarrow{\boldsymbol{\nabla}}\cdot(\overrightarrow{\boldsymbol{\nabla}}\Phi) = \overrightarrow{\boldsymbol{\nabla}}\cdot(\Phi\overleftarrow{\boldsymbol{\nabla}}) = (\Phi\overleftarrow{\boldsymbol{\nabla}})\cdot\overleftarrow{\boldsymbol{\nabla}} = (\overrightarrow{\boldsymbol{\nabla}}\Phi)\cdot\overleftarrow{\boldsymbol{\nabla}}\,, \tag{2.4.26}$$

namely, in cartesian coordinates ($h_k = 1$, $k \in [1,3]$),

$$\Delta\Phi = \frac{\partial^2\Phi}{\partial x_k\partial x_k}\,. \tag{2.4.27}$$

The Laplacian of a vector $\mathbf{\Phi}$ is defined similarly, namely $\Delta\mathbf{\Phi} = \overrightarrow{\nabla}\cdot(\overrightarrow{\nabla}\mathbf{\Phi}) = (\mathbf{\Phi}\overleftarrow{\nabla})\cdot\overleftarrow{\nabla}$.

For a vector, the left and right gradients are the transpose of one another, the left and right divergences are one and the same scalar, and the left and right rotationals are opposite to one another,

$$\mathbf{\Phi}\overleftarrow{\nabla} = \left(\overrightarrow{\nabla}\mathbf{\Phi}\right)^{\mathrm{T}}, \quad \mathbf{\Phi}\cdot\overleftarrow{\nabla} = \overrightarrow{\nabla}\cdot\mathbf{\Phi}, \quad \mathbf{\Phi}\wedge\overleftarrow{\nabla} = -\overrightarrow{\nabla}\wedge\mathbf{\Phi}, \tag{2.4.28}$$

and, in cartesian coordinates,

$$(\overrightarrow{\nabla}\mathbf{\Phi})_{ij} = \frac{\partial\Phi_j}{\partial x_i}; \quad (\mathbf{\Phi}\overleftarrow{\nabla})_{ij} = \frac{\partial\Phi_i}{\partial x_j};$$

$$\mathbf{\Phi}\cdot\overleftarrow{\nabla} = \frac{\partial\Phi_k}{\partial x_k} = \overrightarrow{\nabla}\cdot\mathbf{\Phi}; \tag{2.4.29}$$

$$(\overrightarrow{\nabla}\wedge\mathbf{\Phi})_i = e_{ijk}\frac{\partial\Phi_k}{\partial x_j} = -(\mathbf{\Phi}\wedge\overleftarrow{\nabla})_i.$$

For a 2nd order tensor, the left and right divergences are the transpose of one another, and the left rotational is equal to the negative of the transpose of the right rotational of the transpose tensor, easier to write than to say,

$$\mathbf{\Phi}\cdot\overleftarrow{\nabla} = (\overrightarrow{\nabla}\cdot\mathbf{\Phi})^{\mathrm{T}}, \quad \mathbf{\Phi}\wedge\overleftarrow{\nabla} = -\left(\overrightarrow{\nabla}\wedge\mathbf{\Phi}^{\mathrm{T}}\right)^{\mathrm{T}}, \tag{2.4.30}$$

and, in cartesian coordinates,

$$(\overrightarrow{\nabla}\cdot\mathbf{\Phi})_i = \frac{\partial\Phi_{ji}}{\partial x_j}, \quad (\mathbf{\Phi}\cdot\overleftarrow{\nabla})_i = \frac{\partial\Phi_{ij}}{\partial x_j};$$

$$(\overrightarrow{\nabla}\wedge\mathbf{\Phi})_{ij} = e_{ikl}\frac{\partial\Phi_{lj}}{\partial x_k}, \quad (\mathbf{\Phi}\wedge\overleftarrow{\nabla})_{ij} = -e_{jkl}\frac{\partial\Phi_{il}}{\partial x_k}. \tag{2.4.31}$$

We may now rewrite two versions of the divergence theorem (2.4.11) for 2nd order tensors,

$$\int_{\partial V} \mathbf{n}\cdot\mathbf{\Phi}\, dS = \int_V \overrightarrow{\nabla}\cdot\mathbf{\Phi}\, dV, \quad \int_{\partial V} n_j\,\Phi_{ji}\, dS = \int_V \frac{\partial\Phi_{ji}}{\partial x_j}\, dV, \tag{2.4.32}$$

and

$$\int_{\partial V} \mathbf{\Phi}\cdot\mathbf{n}\, dS = \int_V \mathbf{\Phi}\cdot\overleftarrow{\nabla}\, dV, \quad \int_{\partial V} \Phi_{ij}\,n_j\, dS = \int_V \frac{\partial\Phi_{ij}}{\partial x_j}\, dV. \tag{2.4.33}$$

2.4.2 Application to kinematics

The motion which associates with a point $\underline{\mathbf{x}}$ another unique point $\mathbf{x} = \underline{\mathbf{x}} + \mathbf{u}$ is thus defined by the displacement vector $\mathbf{u}$. The latter can be seen as a function of time t and of the reference point $\underline{\mathbf{x}}$, or of the current point $\mathbf{x}$[2.1]. The gradients corresponding to these two points of view use the following notations:

$$\underline{\overleftarrow{\nabla}} = \frac{\overleftarrow{\partial}}{\partial\underline{\mathbf{x}}}, \quad \underline{\overrightarrow{\nabla}} = \frac{\overrightarrow{\partial}}{\partial\underline{\mathbf{x}}}, \quad \overleftarrow{\nabla} = \frac{\overleftarrow{\partial}}{\partial\mathbf{x}}, \quad \overrightarrow{\nabla} = \frac{\overrightarrow{\partial}}{\partial\mathbf{x}}. \tag{2.4.34}$$

Let us recall the component form of the left and right gradients with respect to the current configuration,

$$(\overrightarrow{\nabla}\mathbf{u})_{ij} = \frac{\partial u_j}{\partial x_i}, \quad (\mathbf{u}\overleftarrow{\nabla})_{ij} = \frac{\partial u_i}{\partial x_j}. \tag{2.4.35}$$

The deformation gradient $\mathbf{F}$ is defined through the relations,

$$\begin{aligned} d\mathbf{x} &= (\mathbf{x}\underline{\overleftarrow{\nabla}})\cdot d\underline{\mathbf{x}} = d\underline{\mathbf{x}}\cdot(\underline{\overrightarrow{\nabla}}\mathbf{x}) \\ &= \mathbf{F}\cdot d\underline{\mathbf{x}} = d\underline{\mathbf{x}}\cdot\mathbf{F}^{\mathrm{T}}, \end{aligned} \tag{2.4.36}$$

[2.1]To the exception of this chapter, which of the left or right differential operators is used is usually not indicated explicitly. To avoid ambiguity, the relations are provided in terms of operators and in component form.

and componentwise in terms of the scaling factors defined by (2.4.21) in the reference and current configurations,

$$F_{iJ} = \frac{h_i}{\underline{h}_J} \frac{\partial x_i}{\partial \underline{x}_J}, \quad (i, J) \in [1, 3] \text{ (no sum on i, J)}. \tag{2.4.37}$$

In cartesian axes,

$$d\mathbf{x} = dx_i \, \mathbf{e}_i = \mathbf{F} \cdot d\underline{\mathbf{x}} = (F_{iJ} \, \mathbf{e}_i \otimes \underline{\mathbf{e}}_J) \cdot (d\underline{x}_K \underline{\mathbf{e}}_K) = F_{iJ} \, \mathbf{e}_i \, d\underline{x}_J, \tag{2.4.38}$$

the component form of the deformation gradient simplifies to,

$$F_{iJ} = \frac{\partial x_i}{\partial \underline{x}_J}, \quad (i, J) \in [1, 3]. \tag{2.4.39}$$

The symmetric and skew-symmetric 2nd order tensors associated with the displacement vector,

$$\boldsymbol{\epsilon}(\mathbf{u}) = \tfrac{1}{2}\big(\mathbf{u}\overleftarrow{\boldsymbol{\nabla}} + (\mathbf{u}\overleftarrow{\boldsymbol{\nabla}})^{\mathrm{T}}\big) = \tfrac{1}{2}\big(\mathbf{u}\overleftarrow{\boldsymbol{\nabla}} + \overrightarrow{\boldsymbol{\nabla}}\mathbf{u}\big), \quad \epsilon_{ij} = \frac{1}{2}\big(\frac{\partial u_i}{\partial x_j} + \frac{\partial u_j}{\partial x_i}\big);$$

$$\boldsymbol{\omega}(\mathbf{u}) = \tfrac{1}{2}\big(\mathbf{u}\overleftarrow{\boldsymbol{\nabla}} - (\mathbf{u}\overleftarrow{\boldsymbol{\nabla}})^{\mathrm{T}}\big) = \tfrac{1}{2}\big(\mathbf{u}\overleftarrow{\boldsymbol{\nabla}} - \overrightarrow{\boldsymbol{\nabla}}\mathbf{u}\big), \quad \omega_{ij} = \frac{1}{2}\big(\frac{\partial u_i}{\partial x_j} - \frac{\partial u_j}{\partial x_i}\big), \tag{2.4.40}$$

are termed *infinitesimal strain* and *rotation tensor*, respectively (the matter is elaborated in Section 2.5.6). Equivalently, the displacement gradients decompose uniquely into a symmetric part and a skew-symmetric part,

$$\overrightarrow{\boldsymbol{\nabla}}\mathbf{u} = \boldsymbol{\epsilon} - \boldsymbol{\omega}, \quad \mathbf{u}\overleftarrow{\boldsymbol{\nabla}} = \boldsymbol{\epsilon} + \boldsymbol{\omega}. \tag{2.4.41}$$

The left rotational of a vector is defined componentwise in cartesian axes through the permutation symbol,

$$(\overrightarrow{\boldsymbol{\nabla}} \wedge \mathbf{u})_i = e_{ijk} \frac{\partial u_k}{\partial x_j}, \quad i \in [1, 3]. \tag{2.4.42}$$

Via (2.1.11), it is the pseudo-vector associated with the skew-symmetric tensor $-2\,\boldsymbol{\omega}(\mathbf{u})$. If $\vec{\omega}$ is the pseudo-vector associated with the skew-symmetric tensor $-\boldsymbol{\omega}$, then

$$\overrightarrow{\boldsymbol{\nabla}} \wedge \mathbf{u} = 2\,\vec{\omega}, \quad (\overrightarrow{\boldsymbol{\nabla}} \wedge \mathbf{u})_i = -e_{ijk}\,\omega_{jk} = 2\,\vec{\omega}_i, \quad i \in [1, 3]. \tag{2.4.43}$$

Consequently

$$\overrightarrow{\boldsymbol{\nabla}} \wedge \boldsymbol{\epsilon}(\mathbf{u}) = \frac{1}{2}\big(\overrightarrow{\boldsymbol{\nabla}} \wedge \mathbf{u}\big)\overleftarrow{\boldsymbol{\nabla}} = (\vec{\omega}(\mathbf{u}))\overleftarrow{\boldsymbol{\nabla}}. \tag{2.4.44}$$

Note also, with help of (2.1.10)$_2$,

$$\overrightarrow{\boldsymbol{\nabla}} \cdot \boldsymbol{\epsilon}(\mathbf{u}) = \overrightarrow{\boldsymbol{\nabla}}(\overrightarrow{\boldsymbol{\nabla}} \cdot \mathbf{u}) - \frac{1}{2}\,\overrightarrow{\boldsymbol{\nabla}} \wedge \overrightarrow{\boldsymbol{\nabla}} \wedge \mathbf{u}. \tag{2.4.45}$$

2.4.3 Spherical, cylindrical and polar coordinates

2.4.3.1 Spherical coordinates

The definitions of left and right differential operators stated in Section 2.4.1.2 are used to obtain the entities that characterize the kinematics in curvilinear coordinates.

Let $(\mathbf{e}_1, \mathbf{e}_2, \mathbf{e}_3)$ be a reference cartesian triad. The spherical triad of unit orthogonal vectors shown in Fig. 2.4.2,

$$\hat{\mathbf{e}}_r = \sin\theta\,\mathbf{e}_* + \cos\theta\,\mathbf{e}_3,$$

$$\hat{\mathbf{e}}_\theta = \cos\theta\,\mathbf{e}_* - \sin\theta\,\mathbf{e}_3,$$

$$\hat{\mathbf{e}}_\phi = \hat{\mathbf{e}}_r \wedge \hat{\mathbf{e}}_\theta = \mathbf{e}_*^\perp, \tag{2.4.46}$$

is built from the cartesian triad and from the two vectors,

$$\mathbf{e}_* = \cos\phi\,\mathbf{e}_1 + \sin\phi\,\mathbf{e}_2, \quad \mathbf{e}_*^\perp = -\sin\phi\,\mathbf{e}_1 + \cos\phi\,\mathbf{e}_2. \tag{2.4.47}$$

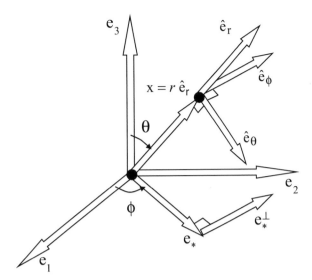

Fig. 2.4.2: Spherical coordinates. Starting from a cartesian triad $(\mathbf{e}_1, \mathbf{e}_2, \mathbf{e}_3)$, the spherical triad $(\hat{\mathbf{e}}_\theta, \hat{\mathbf{e}}_\phi, \hat{\mathbf{e}}_r)$ is deduced by applying first a rotation of angle ϕ about $\mathbf{e}_3$ and next a rotation of angle θ about $\mathbf{e}_2$.

Its differentials read

$$d\hat{\mathbf{e}}_r = d\theta\,\hat{\mathbf{e}}_\theta + \sin\theta\,d\phi\,\hat{\mathbf{e}}_\phi\,,$$

$$d\hat{\mathbf{e}}_\theta = -d\theta\,\hat{\mathbf{e}}_r + \cos\theta\,d\phi\,\hat{\mathbf{e}}_\phi\,, \qquad (2.4.48)$$

$$d\hat{\mathbf{e}}_\phi = -d\phi\,(\sin\theta\,\hat{\mathbf{e}}_r + \cos\theta\,\hat{\mathbf{e}}_\theta)\,.$$

The rotation matrix $\mathbf{R}$ that brings the cartesian axes $\mathbf{e}$ to the spherical axes $\hat{\mathbf{e}}$ has the form

$$\begin{bmatrix} \hat{\mathbf{e}}_\theta \\ \hat{\mathbf{e}}_\phi \\ \hat{\mathbf{e}}_r \end{bmatrix} = \mathbf{R}(\theta,\phi) \begin{bmatrix} \mathbf{e}_1 \\ \mathbf{e}_2 \\ \mathbf{e}_3 \end{bmatrix}, \quad \mathbf{R}(\theta,\phi) = \begin{bmatrix} \cos\theta\cos\phi & \cos\theta\sin\phi & -\sin\theta \\ -\sin\phi & \cos\phi & 0 \\ \sin\theta\cos\phi & \sin\theta\sin\phi & \cos\theta \end{bmatrix}. \qquad (2.4.49)$$

Let $\mathbf{R}(0,\phi)$ be the rotation about $\mathbf{e}_3$ and $\mathbf{R}(\theta,0)$ be the rotation about $\mathbf{e}_2$,

$$\mathbf{R}(0,\phi) = \begin{bmatrix} \cos\phi & \sin\phi & 0 \\ -\sin\phi & \cos\phi & 0 \\ 0 & 0 & 1 \end{bmatrix}, \quad \mathbf{R}(\theta,0) = \begin{bmatrix} \cos\theta & 0 & -\sin\theta \\ 0 & 1 & 0 \\ \sin\theta & 0 & \cos\theta \end{bmatrix}. \qquad (2.4.50)$$

Note that the rotation $\mathbf{R}(\theta,\phi)$ consists in applying first the rotation $\mathbf{R}(0,\phi)$ and next the rotation $\mathbf{R}(\theta,0)$. In fact, since the inverse and transpose of $\mathbf{R}(0,\phi)$ are equal to $\mathbf{R}(0,-\phi)$, and similarly the inverse and transpose of $\mathbf{R}(\theta,0)$ are equal to $\mathbf{R}(-\theta,0)$, one may check the relations,

$$\begin{aligned} \mathbf{R}(\theta,\phi) &= \mathbf{R}(\theta,0)\cdot\mathbf{R}(0,\phi), \\ \mathbf{R}^{-1}(\theta,\phi) &= \mathbf{R}^{\mathrm{T}}(\theta,\phi) = \mathbf{R}^{\mathrm{T}}(0,\phi)\cdot\mathbf{R}^{\mathrm{T}}(\theta,0) = \mathbf{R}(0,-\phi)\cdot\mathbf{R}(-\theta,0) \\ \mathbf{R}(0,\phi)\cdot\mathbf{R}(\theta,0) &= \mathbf{R}^{-1}(-\theta,-\phi)\,. \end{aligned} \qquad (2.4.51)$$

As expected, the rotation inverse to $\mathbf{R}(\theta,\phi)$ consists in applying first the rotation $\mathbf{R}(-\theta,0)$ followed by the rotation $\mathbf{R}(0,-\phi)$. The two rotations $\mathbf{R}(\theta,0)$ and $\mathbf{R}(0,\phi)$ do not commute since they involve distinct axes.

The motion associates the point $\mathbf{x}$ with a reference point $\underline{\mathbf{x}}$,

$$\underline{\mathbf{x}} = \underline{r}\,\hat{\underline{\mathbf{e}}}_r \longmapsto \mathbf{x} = r\,\hat{\mathbf{e}}_r\,, \qquad (2.4.52)$$

with differentials

$$d\underline{\mathbf{x}} = d\underline{r}\,\hat{\underline{\mathbf{e}}}_r + \underline{r}\,d\underline{\theta}\,\hat{\underline{\mathbf{e}}}_\theta + \underline{r}\,\sin\underline{\theta}\,d\underline{\phi}\,\hat{\underline{\mathbf{e}}}_\phi, \quad d\mathbf{x} = dr\,\hat{\mathbf{e}}_r + r\,d\theta\,\hat{\mathbf{e}}_\theta + r\,\sin\theta\,d\phi\,\hat{\mathbf{e}}_\phi. \tag{2.4.53}$$

Comparison of (2.4.53) and (2.4.21) yields the scaling factors in the reference and current configurations,

$$\begin{aligned}
\underline{h}_r &= 1, \quad \underline{h}_\theta = \underline{r}, \quad \underline{h}_\phi = \underline{r}\,\sin\underline{\theta}; \\
h_r &= 1, \quad h_\theta = r, \quad h_\phi = r\,\sin\theta.
\end{aligned} \tag{2.4.54}$$

Formally, the left and right gradients are defined through the relations, e.g., Malvern [1969], p. 652,

$$\overrightarrow{\boldsymbol{\nabla}} = \frac{\hat{\mathbf{e}}_k}{h_k}\,\frac{\overrightarrow{\partial}}{\partial x_k}, \quad \overleftarrow{\boldsymbol{\nabla}} = \frac{\overleftarrow{\partial}}{\partial x_k}\,\frac{\hat{\mathbf{e}}_k}{h_k}. \tag{2.4.55}$$

These operators can be used directly to obtain the gradient of a scalar u,

$$\overrightarrow{\boldsymbol{\nabla}}u = \left(\frac{\hat{\mathbf{e}}_k}{h_k}\,\frac{\overrightarrow{\partial}}{\partial x_k}\right)u = u\overleftarrow{\boldsymbol{\nabla}} = u\left(\frac{\overleftarrow{\partial}}{\partial x_k}\,\frac{\hat{\mathbf{e}}_k}{h_k}\right), \tag{2.4.56}$$

namely

$$\overrightarrow{\boldsymbol{\nabla}}u = u\overleftarrow{\boldsymbol{\nabla}} = \frac{\partial u}{\partial r}\,\hat{\mathbf{e}}_r + \frac{1}{r}\,\frac{\partial u}{\partial \theta}\,\hat{\mathbf{e}}_\theta + \frac{1}{r\,\sin\theta}\,\frac{\partial u}{\partial \phi}\,\hat{\mathbf{e}}_\phi, \tag{2.4.57}$$

the gradients of a vector $\mathbf{u}$,

$$\overrightarrow{\boldsymbol{\nabla}}\mathbf{u} = \left(\frac{\hat{\mathbf{e}}_k}{h_k}\,\frac{\overrightarrow{\partial}}{\partial x_k}\right)\mathbf{u}, \quad \mathbf{u}\overleftarrow{\boldsymbol{\nabla}} = \mathbf{u}\left(\frac{\overleftarrow{\partial}}{\partial x_k}\,\frac{\hat{\mathbf{e}}_k}{h_k}\right), \tag{2.4.58}$$

its rotationals,

$$\overrightarrow{\boldsymbol{\nabla}}\wedge\mathbf{u} = \left(\frac{\hat{\mathbf{e}}_k}{h_k}\,\frac{\overrightarrow{\partial}}{\partial x_k}\right)\wedge\mathbf{u} = -\mathbf{u}\wedge\overleftarrow{\boldsymbol{\nabla}}, \tag{2.4.59}$$

and its divergence $\operatorname{div}\mathbf{u} = \operatorname{tr}(\overrightarrow{\boldsymbol{\nabla}}\mathbf{u}) = \operatorname{tr}(\mathbf{u}\overleftarrow{\boldsymbol{\nabla}})$.

The right and left gradients of a vector $\mathbf{u}$, $\overrightarrow{\boldsymbol{\nabla}}\mathbf{u}$ and $\mathbf{u}\overleftarrow{\boldsymbol{\nabla}} = (\overrightarrow{\boldsymbol{\nabla}}\mathbf{u})^{\mathrm{T}}$, respectively, are also defined through the relation

$$d\mathbf{u} = (\mathbf{u}\overleftarrow{\boldsymbol{\nabla}})\cdot d\mathbf{x} = d\mathbf{x}\cdot\overrightarrow{\boldsymbol{\nabla}}\mathbf{u}. \tag{2.4.60}$$

In the axes $(\hat{\mathbf{e}}_r, \hat{\mathbf{e}}_\theta, \hat{\mathbf{e}}_\phi)$,

$$\mathbf{x} = r\,\hat{\mathbf{e}}_r, \quad \mathbf{u} = u_r\,\hat{\mathbf{e}}_r + u_\theta\,\hat{\mathbf{e}}_\theta + u_\phi\,\hat{\mathbf{e}}_\phi, \tag{2.4.61}$$

and the differentials $d\mathbf{u}$ and $d\mathbf{x}$ are obtained via (2.4.48) and (2.4.53). Hence

∘ the right gradient $\mathbf{u}\overleftarrow{\boldsymbol{\nabla}}$ of the vector $\mathbf{u}$ is read from the matrix format,

$$d\mathbf{u} = \begin{bmatrix} \dfrac{\partial u_r}{\partial r} & \dfrac{1}{r}\dfrac{\partial u_r}{\partial \theta} - \dfrac{u_\theta}{r} & \dfrac{1}{r\,\sin\theta}\dfrac{\partial u_r}{\partial \phi} - \dfrac{u_\phi}{r} \\[2ex] \dfrac{\partial u_\theta}{\partial r} & \dfrac{1}{r}\dfrac{\partial u_\theta}{\partial \theta} + \dfrac{u_r}{r} & \dfrac{1}{r\,\sin\theta}\dfrac{\partial u_\theta}{\partial \phi} - \dfrac{u_\phi}{r}\cot\theta \\[2ex] \dfrac{\partial u_\phi}{\partial r} & \dfrac{1}{r}\dfrac{\partial u_\phi}{\partial \theta} & \dfrac{1}{r\,\sin\theta}\dfrac{\partial u_\phi}{\partial \phi} + \dfrac{u_r}{r} + \dfrac{u_\theta}{r}\cot\theta \end{bmatrix} \begin{bmatrix} dr \\[2ex] r\,d\theta \\[2ex] r\,\sin\theta\,d\phi \end{bmatrix}. \tag{2.4.62}$$

For the special case of radial motions,

$$r = r(\underline{r}, t), \quad \theta = \underline{\theta}, \quad \phi = \underline{\phi}, \tag{2.4.63}$$

the only nonzero components of the deformation gradient (2.4.37) are

$$F_{rr} = \frac{dr}{d\underline{r}}, \quad F_{\theta\theta} = F_{\phi\phi} = \frac{r}{\underline{r}}. \tag{2.4.64}$$

2.4.3.3   Polar coordinates

Let $(\mathbf{e}_1, \mathbf{e}_2)$ be cartesian axes, and $\hat{\mathbf{e}}_r = \cos\theta\,\mathbf{e}_1 + \sin\theta\,\mathbf{e}_2$ and $\hat{\mathbf{e}}_\theta = -\sin\theta\,\mathbf{e}_1 + \cos\theta\,\mathbf{e}_2$ be the polar axes deduced by a direct rotation of angle θ as in (2.4.70).

Let us record the polar components of the displacement vector $\mathbf{u} = u_1\,\mathbf{e}_1 + u_2\,\mathbf{e}_2$, namely

$$u_r = \cos\theta\,u_1 + \sin\theta\,u_2,$$
$$u_\theta = -\sin\theta\,u_1 + \cos\theta\,u_2, \tag{2.4.82}$$

as well as the polar components of the symmetric stress $\boldsymbol{\sigma} = \sigma_{11}\,\mathbf{e}_1 \otimes \mathbf{e}_1 + \sigma_{22}\,\mathbf{e}_2 \otimes \mathbf{e}_2 + \sigma_{12}\,(\mathbf{e}_1 \otimes \mathbf{e}_2 + \mathbf{e}_2 \otimes \mathbf{e}_1)$, namely

$$\sigma_{rr} = \cos^2\theta\,\sigma_{11} + \sin^2\theta\,\sigma_{22} + 2\sin\theta\cos\theta\,\sigma_{12},$$
$$\sigma_{\theta\theta} = \sin^2\theta\,\sigma_{11} + \cos^2\theta\,\sigma_{22} - 2\sin\theta\cos\theta\,\sigma_{12}, \tag{2.4.83}$$
$$\sigma_{r\theta} = -\sin\theta\cos\theta\,(\sigma_{11} - \sigma_{22}) + (\cos^2\theta - \sin^2\theta)\,\sigma_{12}.$$

In the case where $\sigma_{zr} = 0$, $\sigma_{z\theta} = \sigma_{z\theta}(r)$ and $\sigma_{zz} = \sigma_{zz}(r, \theta)$, the component of the left divergence $\overrightarrow{\boldsymbol{\nabla}}\cdot\boldsymbol{\sigma}$ defined by (2.4.79) along z vanishes and the divergence adopts the format of polar coordinates,

$$[\overrightarrow{\boldsymbol{\nabla}}\cdot\boldsymbol{\sigma}] = \begin{bmatrix} \dfrac{\partial\sigma_{rr}}{\partial r} + \dfrac{1}{r}\dfrac{\partial\sigma_{\theta r}}{\partial\theta} + \dfrac{1}{r}(\sigma_{rr} - \sigma_{\theta\theta}) \\[2ex] \dfrac{\partial\sigma_{r\theta}}{\partial r} + \dfrac{1}{r}\dfrac{\partial\sigma_{\theta\theta}}{\partial\theta} + \dfrac{\sigma_{r\theta}}{r} + \dfrac{\sigma_{\theta r}}{r} \end{bmatrix}. \tag{2.4.84}$$

2.4.4   Convective acceleration

A particle located at position $\underline{\mathbf{x}}$ at a reference time, say $t = 0$, moves to the position $\mathbf{x} = \mathbf{x}(\underline{\mathbf{x}}, t)$ at time t. We may also take a picture of the particle positions at an arbitrary time t: this picture will not tell us where the particle located at position $\mathbf{x}$ was at time $t = 0$. To cover these different points of view, we adapt the arguments of the field Φ (although we should also change the notation for the field itself),

$$\Phi(\underline{\mathbf{x}}, t) = \Phi(\mathbf{x}, t) = \Phi(\mathbf{x}(\underline{\mathbf{x}}, t), t). \tag{2.4.85}$$

The derivatives with respect to time will reflect these differences,

$$\underbrace{\frac{\partial\Phi}{\partial t}_{|\underline{\mathbf{x}}}}_{\substack{\text{material} \\ \text{time derivative}}} = \underbrace{\frac{\partial\Phi}{\partial t}_{|\mathbf{x}}}_{\substack{\text{local time derivative} \\ \text{vanishes in steady flow}}} + \underbrace{\frac{\partial\Phi}{\partial\mathbf{x}}_{|\underline{\mathbf{x}}}\cdot\frac{\partial\mathbf{x}}{\partial t}}_{\substack{\text{convective rate of change} \\ \text{vanishes in spatially} \\ \text{uniform flow}}}. \tag{2.4.86}$$

The convective rate of change is due to the fact that the particle is moving in a landscape where the field Φ is not uniform.

In order to avoid the notation being more cumbersome than necessary, we will write

$$\begin{aligned} \frac{d\Phi}{dt} &= \frac{\partial\Phi}{\partial t} + \mathbf{v}\cdot\overrightarrow{\boldsymbol{\nabla}}\Phi \\[1ex] &= \frac{\partial\Phi}{\partial t} + \Phi\overleftarrow{\boldsymbol{\nabla}}\cdot\mathbf{v} \\[1ex] \frac{d\Phi...}{dt} &= \frac{\partial\Phi...}{\partial t} + v_l\frac{\partial\Phi...}{\partial x_l}, \end{aligned} \tag{2.4.87}$$

where $\mathbf{v}$ is the velocity,

$$\mathbf{v}(\underline{\mathbf{x}}, t) = \frac{\partial\mathbf{x}}{\partial t}_{|\underline{\mathbf{x}}} = \frac{\partial\mathbf{u}}{\partial t}_{|\underline{\mathbf{x}}}. \tag{2.4.88}$$

These relations apply to any field and to the position $\mathbf{x}$ and displacement $\mathbf{u} = \mathbf{x} - \underline{\mathbf{x}}$ in particular. Similarly the formulas apply to the velocity itself and the acceleration may be cast in two equivalent formats:

$$\mathbf{a} = \frac{d\mathbf{v}}{dt} = \frac{\partial \mathbf{v}}{\partial t} + \mathbf{v} \cdot \overrightarrow{\boldsymbol{\nabla}} \mathbf{v}$$

$$= \frac{\partial \mathbf{v}}{\partial t} + \mathbf{v} \overleftarrow{\boldsymbol{\nabla}} \cdot \mathbf{v} \tag{2.4.89}$$

$$a_k = \frac{dv_k}{dt} = \frac{\partial v_k}{\partial t} + v_l \frac{\partial v_k}{\partial x_l}.$$

The acceleration components follow, for example, from the right gradient (2.4.62) in spherical axes,

$$a_r = \frac{\partial v_r}{\partial t} + v_r \frac{\partial v_r}{\partial r} + \frac{v_\theta}{r} \frac{\partial v_r}{\partial \theta} + \frac{v_\phi}{r \sin\theta} \frac{\partial v_r}{\partial \phi} - \frac{1}{r}\left(v_\theta^2 + v_\phi^2\right)$$

$$a_\theta = \frac{\partial v_\theta}{\partial t} + v_r \frac{\partial v_\theta}{\partial r} + \frac{v_\theta}{r} \frac{\partial v_\theta}{\partial \theta} + \frac{v_\phi}{r \sin\theta} \frac{\partial v_\theta}{\partial \phi} + \frac{1}{r} v_r v_\theta - \frac{1}{r} v_\phi^2 \cot\theta \tag{2.4.90}$$

$$a_\phi = \frac{\partial v_\phi}{\partial t} + v_r \frac{\partial v_\phi}{\partial r} + \frac{v_\theta}{r} \frac{\partial v_\phi}{\partial \theta} + \frac{v_\phi}{r \sin\theta} \frac{\partial v_\phi}{\partial \phi} + \frac{1}{r} v_r v_\phi + \frac{1}{r} v_\theta v_\phi \cot\theta \,,$$

and from the right gradient (2.4.77) in cylindrical axes,

$$a_r = \frac{\partial v_r}{\partial t} + v_r \frac{\partial v_r}{\partial r} + \frac{v_\theta}{r} \frac{\partial v_r}{\partial \theta} + v_z \frac{\partial v_r}{\partial z} - \frac{1}{r} v_\theta^2$$

$$a_\theta = \frac{\partial v_\theta}{\partial t} + v_r \frac{\partial v_\theta}{\partial r} + \frac{v_\theta}{r} \frac{\partial v_\theta}{\partial \theta} + v_z \frac{\partial v_\theta}{\partial z} + \frac{1}{r} v_r v_\theta \tag{2.4.91}$$

$$a_z = \frac{\partial v_z}{\partial t} + v_r \frac{\partial v_z}{\partial r} + \frac{v_\theta}{r} \frac{\partial v_z}{\partial \theta} + v_z \frac{\partial v_z}{\partial z}\,.$$

A rewriting of the convective term of the acceleration using the identity $(2.4.5)_3$,

$$\frac{d\mathbf{v}}{dt} = \frac{\partial \mathbf{v}}{\partial t} + (\overrightarrow{\boldsymbol{\nabla}} \wedge \mathbf{v}) \wedge \mathbf{v} + \overrightarrow{\boldsymbol{\nabla}}(\tfrac{1}{2}\mathbf{v}^2)\,, \tag{2.4.92}$$

will be useful to derive a simple scalar condition for steady irrotational flow of inviscid fluids, termed Bernoulli equation. Still an alternative and equivalent form may be derived. According to (2.4.43), $\overrightarrow{\boldsymbol{\nabla}} \wedge \mathbf{v}$ is equal to $2\overrightarrow{\mathbf{W}}$ where $\overrightarrow{\mathbf{W}}$ is the pseudo-vector associated with the negative of the spin $\mathbf{W}(\mathbf{v})$ defined by (2.5.13). Then with the help of (2.1.13),

$$\frac{d\mathbf{v}}{dt} = \frac{\partial \mathbf{v}}{\partial t} + 2\overrightarrow{\mathbf{W}} \wedge \mathbf{v} + \overrightarrow{\boldsymbol{\nabla}}(\tfrac{1}{2}\mathbf{v}^2) = \frac{\partial \mathbf{v}}{\partial t} + 2\mathbf{W} \cdot \mathbf{v} + \overrightarrow{\boldsymbol{\nabla}}(\tfrac{1}{2}\mathbf{v}^2)\,. \tag{2.4.93}$$

2.5   Measures of strain

2.5.1   Polar decomposition of the deformation gradient

The invertible deformation gradient $\mathbf{F}$ can be decomposed uniquely into a product of a proper orthogonal tensor $\mathbf{Q}$ and of a pure deformation $\mathbf{U}$ or $\mathbf{V}$, namely

$$\mathbf{F} = \mathbf{Q} \cdot \mathbf{U} = \mathbf{V} \cdot \mathbf{Q}\,, \tag{2.5.1}$$

with $\mathbf{Q} \cdot \mathbf{Q}^{\mathrm{T}} = \mathbf{Q}^{\mathrm{T}} \cdot \mathbf{Q} = \mathbf{I}$, $\det \mathbf{Q} = 1$, and $\mathbf{U}$ and $\mathbf{V}$ symmetric definite positive tensors referred to as right and left stretch tensors, respectively. For a proof of the existence and uniqueness of the decomposition, see, e.g., Bowen [1989], p. 237. As an example, Exercise 2.3 proposes the decomposition of a motion describing a simple shear.

Incidentally note that the tensor $d\mathbf{Q}/dt \cdot \mathbf{Q}^{\mathrm{T}}$ is skew-symmetric: this result simply follows by derivation of the relation $\mathbf{Q} \cdot \mathbf{Q}^{\mathrm{T}} = \mathbf{I}$. Indeed

$$\mathbf{Q} \cdot \mathbf{Q}^{\mathrm{T}} = \mathbf{I} \Rightarrow \frac{d\mathbf{Q}}{dt} \cdot \mathbf{Q}^{\mathrm{T}} + \mathbf{Q} \cdot \frac{d\mathbf{Q}^{\mathrm{T}}}{dt} = \mathbf{0} \Rightarrow \frac{d\mathbf{Q}}{dt} \cdot \mathbf{Q}^{\mathrm{T}} = -\mathbf{Q} \cdot \left(\frac{d\mathbf{Q}}{dt}\right)^{\mathrm{T}} = -(\frac{d\mathbf{Q}}{dt} \cdot \mathbf{Q}^{\mathrm{T}})^{\mathrm{T}}. \tag{2.5.2}$$

The right and left Cauchy-Green tensors are defined, respectively, as

$$\mathbf{C} = \mathbf{F}^{\mathrm{T}} \cdot \mathbf{F} = \mathbf{U}^2, \quad \mathbf{B} = \mathbf{F} \cdot \mathbf{F}^{\mathrm{T}} = \mathbf{V}^2. \tag{2.5.3}$$

Thus $\mathbf{U} = \mathbf{C}^{\frac{1}{2}}$ and $\mathbf{V} = \mathbf{B}^{\frac{1}{2}}$.

Let $\mathbf{v}$ be an eigenvector of $\mathbf{C}$ associated with the eigenvalue λ, namely $\mathbf{C} \cdot \mathbf{v} = \lambda \mathbf{v}$. Multiplication of the two terms of the equation by $\mathbf{F}$ yields $(\mathbf{F} \cdot \mathbf{F}^{\mathrm{T}}) \cdot (\mathbf{F} \cdot \mathbf{v}) = \lambda (\mathbf{F} \cdot \mathbf{v})$. Consequently the eigenvalues of the left and right Cauchy-Green tensors are identical and, if $\mathbf{v}$ is an eigenvector of $\mathbf{C}$ associated with the eigenvalue λ, $\mathbf{F} \cdot \mathbf{v}$ is the eigenvector of $\mathbf{B}$ associated with the same eigenvalue.

Explicit expressions of the right and left stretches require taking the square-root of a symmetric second order tensor. One possibility is to use the spectral decomposition, including eigenvalues and eigenvectors. The approach of Section 2.3.6 makes use only of the eigenvalues. An alternative method based on repeated use of Cayley-Hamilton theorem has been devised by Hoger and Carlson [1984]a, Section 2.3.2.

2.5.2   Spectral decomposition of the right and left Cauchy-Green tensors

Since they are symmetric, the right and left stretches have a real spectrum[2.2]: the eigenvalues referred to as principal stretches are identical and remain strictly positive while the eigenvectors of the left stretch are rotated with respect to those of the right stretch through the rotation $\mathbf{Q}$:

$$\mathbf{U} = \sum_{i=1,3} \lambda_i \, \underline{\mathbf{e}}_i \otimes \underline{\mathbf{e}}_i, \quad \mathbf{V} = \sum_{i=1,3} \lambda_i \, \mathbf{e}_i \otimes \mathbf{e}_i, \quad \mathbf{Q} = \sum_{i=1,3} \mathbf{e}_i \otimes \underline{\mathbf{e}}_i, \tag{2.5.4}$$

with, for $i \in [1,3]$,

$$\mathbf{e}_i = \mathbf{Q} \cdot \underline{\mathbf{e}}_i = \sum_{j=1,3} Q_{ji} \, \underline{\mathbf{e}}_j, \quad \underline{\mathbf{e}}_i = \mathbf{Q}^{\mathrm{T}} \cdot \mathbf{e}_i = \sum_{j=1,3} Q_{ij} \, \mathbf{e}_j, \tag{2.5.5}$$

using the decomposition of $\mathbf{Q}$ on the two triads, $\mathbf{Q} = \sum_{i,j} Q_{ij} \, \mathbf{e}_i \otimes \mathbf{e}_j = \sum_{i,j} Q_{ij} \, \underline{\mathbf{e}}_i \otimes \underline{\mathbf{e}}_j$.

Consequently, the eigenvalues of the right and left Cauchy-Green tensors are identical, while their eigenspaces are equal to those of the right and left stretches, respectively,

$$\mathbf{C} = \mathbf{F}^{\mathrm{T}} \cdot \mathbf{F} = \mathbf{U}^2 = \sum_{i=1,3} \lambda_i^2 \, \underline{\mathbf{e}}_i \otimes \underline{\mathbf{e}}_i,$$

$$\mathbf{B} = \mathbf{F} \cdot \mathbf{F}^{\mathrm{T}} = \underbrace{\mathbf{V}^2}_{\mathbf{Q} \cdot \mathbf{U}^2 \cdot \mathbf{Q}^{\mathrm{T}}} = \sum_{i=1,3} \lambda_i^2 \, \mathbf{e}_i \otimes \mathbf{e}_i. \tag{2.5.6}$$

The spectral decomposition of the deformation gradient itself is implied by the spectral properties of the right and left stretches or of the Cauchy-Green tensors, namely,

$$\mathbf{F} = \sum_{i=1,3} \lambda_i \, \mathbf{e}_i \otimes \underline{\mathbf{e}}_i. \tag{2.5.7}$$

2.5.3   Lagrangian and Eulerian strain measures

A family of Lagrangian (right) and Eulerian (left) strain measures associated with the integer n and defined in terms of the right and left stretches is attributed to Seth [1964],

$$\mathbf{E}^{\mathrm{L}}_{(n)} = \frac{1}{n} \left(\mathbf{U}^n - \mathbf{I} \right) = \sum_{i=1,3} \frac{1}{n} \left(\lambda_i^n - 1 \right) \underline{\mathbf{e}}_i \otimes \underline{\mathbf{e}}_i,$$

$$\mathbf{E}^{\mathrm{E}}_{(n)} = \frac{1}{n} \left(\mathbf{I} - \mathbf{V}^{-n} \right) = \sum_{i=1,3} \frac{1}{n} \left(1 - \lambda_i^{-n} \right) \mathbf{e}_i \otimes \mathbf{e}_i. \tag{2.5.8}$$

[2.2] A more general spectral analysis is provided in Section 2.2.

The logarithmic strain is associated with the limit case $n = 0$ (deduced from the above by L'Hospital rule),

$$
\begin{aligned}
\mathbf{E}_{(0)}^{\mathrm{L}} &= \mathrm{Ln}\,\mathbf{U} = \sum_{i=1,3} \mathrm{Ln}\,\lambda_i\, \underline{\mathbf{e}}_i \otimes \underline{\mathbf{e}}_i\,, \\
\mathbf{E}_{(0)}^{\mathrm{E}} &= \mathrm{Ln}\,\mathbf{V} = \sum_{i=1,3} \mathrm{Ln}\,\lambda_i\, \mathbf{e}_i \otimes \mathbf{e}_i\,.
\end{aligned}
\tag{2.5.9}
$$

Note the relations, resulting from the spectral decomposition,

$$
\mathbf{V} = \mathbf{Q}\cdot\mathbf{U}\cdot\mathbf{Q}^{\mathrm{T}}, \quad \mathbf{V}^2 = \mathbf{Q}\cdot\mathbf{U}^2\cdot\mathbf{Q}^{\mathrm{T}}, \quad \mathrm{Ln}\,\mathbf{V} = \mathbf{Q}\cdot\mathrm{Ln}\,\mathbf{U}\cdot\mathbf{Q}^{\mathrm{T}}\,.
\tag{2.5.10}
$$

To $n = 2$ correspond a Lagrangian measure $\mathbf{E}_{\mathrm{G}} = \frac{1}{2}(\mathbf{C} - \mathbf{I})$ called Green strain and a Eulerian measure $\mathbf{E}_{\mathrm{A}}^{\mathrm{co}} = \frac{1}{2}(\mathbf{I} - \mathbf{B}^{-1})$ called covariant Almansi strain. Similarly to $n = -2$ correspond the contravariant Almansi strain $\mathbf{E}_{\mathrm{A}}^{\mathrm{ct}} = \frac{1}{2}(\mathbf{I} - \mathbf{C}^{-1})$, a Lagrangian measure, and the Finger strain $\frac{1}{2}(\mathbf{B} - \mathbf{I})$, a Eulerian measure. The deformation gradient itself does not in general belong to that family while Biot strain $\mathbf{E}_{\mathrm{B}} = \mathbf{U} - \mathbf{I}$ corresponds to $n = 1$.

More general Lagrangian strain measures suggested and analyzed by Hill [1968],

$$
\mathbf{E} = \sum_{i=1,3} f(\lambda_i)\,\underline{\mathbf{e}}_i \otimes \underline{\mathbf{e}}_i\,,
\tag{2.5.11}
$$

may be defined through the smooth monotone function $f(\lambda)$ satisfying $f(0) = 1$, $(df/d\lambda)(1) = 1$, but otherwise arbitrary.

2.5.4   Rate of deformation and rates of strain measures

The infinitesimal strain and rotation tensors are associated with the (gradient of) displacement. Similarly with the velocity gradient,

$$
\mathbf{L} = \frac{d\mathbf{F}}{dt}\cdot\mathbf{F}^{-1} = \mathbf{v}\overleftarrow{\nabla}, \quad L_{ij} = \frac{\partial v_i}{\partial x_j}\,,
\tag{2.5.12}
$$

are associated the rate of deformation $\mathbf{D}$ and the spin $\mathbf{W}$,

$$
\begin{aligned}
\mathbf{D} &= \tfrac{1}{2}(\mathbf{L} + \mathbf{L}^{\mathrm{T}}) = \frac{1}{2}(\mathbf{v}\overleftarrow{\nabla} + \overrightarrow{\nabla}\mathbf{v})\,, \quad D_{ij} = \tfrac{1}{2}\Big(\frac{\partial v_i}{\partial x_j} + \frac{\partial v_j}{\partial x_i}\Big)\,, \\
\mathbf{W} &= \tfrac{1}{2}(\mathbf{L} - \mathbf{L}^{\mathrm{T}}) = \frac{1}{2}(\mathbf{v}\overleftarrow{\nabla} - \overrightarrow{\nabla}\mathbf{v})\,, \quad W_{ij} = \tfrac{1}{2}\Big(\frac{\partial v_i}{\partial x_j} - \frac{\partial v_j}{\partial x_i}\Big)\,.
\end{aligned}
\tag{2.5.13}
$$

Note the relations,

$$
\frac{d\mathbf{E}_{\mathrm{G}}}{dt} = \mathbf{F}^{\mathrm{T}}\cdot\mathbf{D}\cdot\mathbf{F}, \quad \frac{d\mathbf{B}}{dt} = \mathbf{L}\cdot\mathbf{B} + \mathbf{B}\cdot\mathbf{L}^{\mathrm{T}}, \quad \frac{1}{2}\Big(\frac{d\mathbf{U}}{dt}\cdot\mathbf{U}^{-1} + \mathbf{U}^{-1}\cdot\frac{d\mathbf{U}}{dt}\Big) = \mathbf{Q}^{\mathrm{T}}\cdot\mathbf{D}\cdot\mathbf{Q}\,.
\tag{2.5.14}
$$

The rates of rotation of the right and left eigenvectors, respectively, $\underline{\mathbf{e}}_i$ and $\mathbf{e}_i$, $i \in [1,3]$, also referred to as rates of rotation of the Lagrangian and Eulerian strain ellipsoids[2.3],

$$
\begin{aligned}
\frac{d\underline{\mathbf{e}}_i}{dt} &= \boldsymbol{\Omega}^{\mathrm{L}}\cdot\underline{\mathbf{e}}_i = \sum_{j=1,3} \Omega_{ji}^{\mathrm{L}}\,\underline{\mathbf{e}}_j\,, \\
\frac{d\mathbf{e}_i}{dt} &= \boldsymbol{\Omega}^{\mathrm{E}}\cdot\mathbf{e}_i = \sum_{j=1,3} \Omega_{ji}^{\mathrm{E}}\,\mathbf{e}_j\,,
\end{aligned}
\tag{2.5.15}
$$

are linked by the relation,

$$
\boldsymbol{\Omega}^{\mathrm{E}} = \frac{d\mathbf{Q}}{dt}\cdot\mathbf{Q}^{-1} + \mathbf{Q}\cdot\boldsymbol{\Omega}^{\mathrm{L}}\cdot\mathbf{Q}^{-1}\,.
\tag{2.5.16}
$$

[2.3]Let the three vectors $\mathbf{e}_i$, $i \in [1,3]$, form a cartesian triad. That the second order tensor $\boldsymbol{\Omega}$ such that $d\mathbf{e}_i/dt = \boldsymbol{\Omega}\cdot\mathbf{e}_i$, $i \in [1,3]$, is skew-symmetric is easily proved. Indeed, differentiating $\mathbf{e}_i\cdot\mathbf{e}_i = 1$ yields $\mathbf{e}_i\cdot d\mathbf{e}_i/dt = \Omega_{ii} = 0$ (no sum over i). Also differentiating $\mathbf{e}_i\cdot\mathbf{e}_j = 0$ for $i \neq j$ yields $\Omega_{ij} + \Omega_{ji} = 0$.

As a consequence of the spectral decomposition of the right stretch $\mathbf{U}$, Eqn (2.5.4),

$$\frac{d\mathbf{U}}{dt} = \sum_{i,j} \left(\frac{d\lambda_i}{dt} I_{ij} + (\lambda_i - \lambda_j) \Omega_{ji}^{\mathrm{L}} \right) \underline{\mathbf{e}}_i \otimes \underline{\mathbf{e}}_j . \tag{2.5.17}$$

The rate of the general Lagrangian strain measure (2.5.11) follows by analogy with (2.5.17), namely,

$$\frac{d\mathbf{E}}{dt} = \sum_{i,j} \left(\frac{df(\lambda_i)}{d\lambda_i} \frac{d\lambda_i}{dt} I_{ij} + \left(f(\lambda_i) - f(\lambda_j) \right) \Omega_{ji}^{\mathrm{L}} \right) \underline{\mathbf{e}}_i \otimes \underline{\mathbf{e}}_j . \tag{2.5.18}$$

The components D_{ij} of the rate of deformation which is a Eulerian tensor,

$$\mathbf{D} = \sum_{i,j} D_{ij} \, \mathbf{e}_i \otimes \mathbf{e}_j , \quad \mathbf{Q}^{\mathrm{T}} \cdot \mathbf{D} \cdot \mathbf{Q} = \sum_{i,j} D_{ij} \, \underline{\mathbf{e}}_i \otimes \underline{\mathbf{e}}_j , \tag{2.5.19}$$

may be expressed, in view of $(2.5.14)_2$ and of (2.5.17), in terms of the rates of the stretches and of the Lagrangian rate of rotation[2.4], namely Eqn (1.50) in Hill [1978],

$$D_{ij} = \frac{1}{2} \left(\frac{1}{\lambda_i} + \frac{1}{\lambda_j} \right) \frac{d\lambda_i}{dt} I_{ij} + \frac{1}{2} \left(\frac{\lambda_i}{\lambda_j} - \frac{\lambda_j}{\lambda_i} \right) \Omega_{ji}^{\mathrm{L}} . \tag{2.5.20}$$

Indeed,

$$D_{ii} = \frac{1}{\lambda_i} \frac{d\lambda_i}{dt}; \quad \Omega_{ji}^{\mathrm{L}} = \frac{2 \lambda_i \lambda_j}{\lambda_i^2 - \lambda_j^2} D_{ij}, \quad i \neq j . \tag{2.5.21}$$

The Lagrangian components of the rate of strain may thus be expressed in terms of the Eulerian components of the rate of deformation,

$$\frac{d\mathbf{E}}{dt} = \sum_{i,j} \left(\frac{df(\lambda_i)}{d\lambda_i} \lambda_i D_{ii} I_{ij} + 2 \lambda_i \lambda_j \frac{f(\lambda_i) - f(\lambda_j)}{\lambda_i^2 - \lambda_j^2} D_{ij} \right) \underline{\mathbf{e}}_i \otimes \underline{\mathbf{e}}_j . \tag{2.5.22}$$

The cases of repeated eigenvalues may be considered as limit cases.

2.5.5 Representation theorems for rotations

A second order tensor $\mathbf{R}$ is called a *rotation* or *proper orthogonal tensor* if

$$\mathbf{R} \cdot \mathbf{R}^{\mathrm{T}} = \mathbf{R}^{\mathrm{T}} \cdot \mathbf{R} = \mathbf{I}, \quad \det \mathbf{R} = +1 . \tag{2.5.23}$$

The set of rotations has the structure of a group and it is called the proper orthogonal group and denoted by $\mathcal{SO}(3)$. It is a subgroup of the complete orthogonal group $\mathcal{O}(3)$, Section 5.2.3.1.

As a consequence of (2.5.23), the number of independent scalars necessary to describe a rotation is equal to 3, namely 9 components—6 constraints. Several algebraic structures may be devised to embed these three scalars:

1. the standard representation uses the three Euler angles, e.g., Goldstein [1980], p. 146;

2. alternatively, one may consider a unit vector and an angle;

3. one may also consider a skew-symmetric second order tensor, or equivalently a unit vector and a magnitude. Emphasis is laid on the Cayley and exponential representations which both use a skew-symmetric tensor. The second point of view above is useful to distinguish finite and infinitesimal rotations.

[2.4]The formulas (2.5.20) and (2.5.21) assume no summation over repeated indices.

2.5.5.1 Cayley representation

There exists a one-to-one correspondence between skew-symmetric second order tensors $\mathbf{\Omega}$ and proper orthogonal tensors $\mathbf{R}$, for which -1 is not an eigenvalue, that is, $\det(\mathbf{I} + \mathbf{R}) \neq 0$, e.g., Gantmacher [1959], p. 288:

$$\text{Cayley formula}: \quad \mathbf{R} = (\mathbf{I} - \mathbf{\Omega})^{-1} \cdot (\mathbf{I} + \mathbf{\Omega}), \quad \mathbf{\Omega} = (\mathbf{R} - \mathbf{I}) \cdot (\mathbf{R} + \mathbf{I})^{-1}. \tag{2.5.24}$$

The condition $\det(\mathbf{I} + \mathbf{R}) \neq 0$ excludes a rotation of angle equal to π. Clearly $\mathbf{R}^{-1}(\mathbf{\Omega}) = \mathbf{R}(-\mathbf{\Omega})$.

Consider the component form in cartesian axes of the skew-symmetric tensor $\mathbf{\Omega}$ and of its square $\mathbf{\Omega}^2$,

$$\mathbf{\Omega} = \begin{bmatrix} 0 & \Omega_{12} & \Omega_{13} \\ -\Omega_{12} & 0 & \Omega_{23} \\ -\Omega_{13} & -\Omega_{23} & 0 \end{bmatrix}, \quad \mathbf{\Omega}^2 = \begin{bmatrix} -\Omega_{12}^2 - \Omega_{13}^2 & -\Omega_{13}\Omega_{23} & \Omega_{12}\Omega_{23} \\ -\Omega_{13}\Omega_{23} & -\Omega_{12}^2 - \Omega_{23}^2 & -\Omega_{12}\Omega_{13} \\ \Omega_{12}\Omega_{23} & -\Omega_{12}\Omega_{13} & -\Omega_{13}^2 - \Omega_{23}^2 \end{bmatrix} = (\mathbf{\Omega}^2)^{\mathrm{T}}. \tag{2.5.25}$$

Therefore,

$$\det \mathbf{\Omega} = 0, \quad \operatorname{tr} \mathbf{\Omega} = 0, \quad \operatorname{tr} \mathbf{\Omega}^2 = -|\mathbf{\Omega}|^2, \tag{2.5.26}$$

so that the characteristic polynomial (2.2.6) simplifies to

$$\det(\lambda \mathbf{I} - \mathbf{\Omega}) = \lambda^3 - \operatorname{tr}\mathbf{\Omega}\,\lambda^2 - (\operatorname{tr}\mathbf{\Omega}^2 - (\operatorname{tr}\mathbf{\Omega})^2)\frac{\lambda}{2} - \det\mathbf{\Omega} = \lambda^3 + \frac{|\mathbf{\Omega}|^2}{2}\lambda. \tag{2.5.27}$$

Consequently,

$$\det(\mathbf{I} - \mathbf{\Omega}) = 1 + \tfrac{1}{2}|\mathbf{\Omega}|^2 = \det(\mathbf{I} + \mathbf{\Omega}) > 0, \tag{2.5.28}$$

i.e., $\mathbf{I} \pm \mathbf{\Omega}$ is never singular.

One easily checks the relation $(\mathbf{I} + \mathbf{\Omega}) \cdot (\mathbf{I} - \mathbf{\Omega}) = (\mathbf{I} - \mathbf{\Omega}) \cdot (\mathbf{I} + \mathbf{\Omega})$. Therefore, $\mathbf{R}$ defined as $(\mathbf{I} - \mathbf{\Omega})^{-1} \cdot (\mathbf{I} + \mathbf{\Omega})$ satisfies the property $\mathbf{R} \cdot \mathbf{R}^{\mathrm{T}} = \mathbf{R}^{\mathrm{T}} \cdot \mathbf{R} = \mathbf{I}$ and $\det \mathbf{R} = 1$ since $\det(\mathbf{I} - \mathbf{\Omega}) = \det(\mathbf{I} + \mathbf{\Omega})$. Thus, $\mathbf{R}$ is proper orthogonal.

Using (2.5.26) and (2.5.28), one deduces,

$$(\mathbf{I} - \mathbf{\Omega})^{-1} = \mathbf{I} + \frac{1}{1 + |\mathbf{\Omega}|^2/2}(\mathbf{\Omega} + \mathbf{\Omega}^2). \tag{2.5.29}$$

and the more explicit expression of the Cayley representation,

$$\mathbf{R} = (\mathbf{I} - \mathbf{\Omega})^{-1} \cdot (\mathbf{I} + \mathbf{\Omega}) = \mathbf{I} + \frac{2}{1 + |\mathbf{\Omega}|^2/2}(\mathbf{\Omega} + \mathbf{\Omega}^2). \tag{2.5.30}$$

We knew already, before calculating the characteristic polynomial, that the determinant of a skew-symmetric matrix was zero. Indeed,

- the property results from the representation (2.1.13), namely $\mathbf{\Omega} \cdot \mathbf{x} = \mathbf{x} \wedge \vec{\mathbf{\Omega}}$ for any vector $\mathbf{x}$. Here $\vec{\mathbf{\Omega}}$ with Euclidean norm $|\vec{\mathbf{\Omega}}| = |\mathbf{\Omega}|/\sqrt{2}$ is the pseudo-vector adjoint to $\mathbf{\Omega}$;
- the property also follows from Jacobi theorem for skew-symmetric matrices of any odd order n: $\det \mathbf{\Omega} = \det \mathbf{\Omega}^{\mathrm{T}} = \det(-\mathbf{\Omega}) = (-1)^n \det \mathbf{\Omega} = -\det \mathbf{\Omega}$.

Thus one eigenvalue of $\mathbf{\Omega}$ is zero and the other two are complex conjugate as indicated by (2.5.27),

$$\lambda = 0; \quad \lambda = \pm i\,|\vec{\mathbf{\Omega}}|. \tag{2.5.31}$$

The Cayley-Hamilton theorem (2.3.1) simplifies to

$$\mathbf{\Omega}^3 + |\vec{\mathbf{\Omega}}|^2\,\mathbf{\Omega} = \mathbf{0}, \tag{2.5.32}$$

and therefore,

$$\mathbf{\Omega}^{2k+1} = \left(-|\vec{\mathbf{\Omega}}|^2\right)^k \mathbf{\Omega}, \quad \mathbf{\Omega}^{2k+2} = \left(-|\vec{\mathbf{\Omega}}|^2\right)^k \mathbf{\Omega}^2, \quad k \geq 0. \tag{2.5.33}$$

The half rotation is easily obtained,

$$\mathbf{R} = (\mathbf{I} - \mathbf{\Omega})^{-1} \cdot (\mathbf{I} + \mathbf{\Omega}) \quad \Leftrightarrow \quad \mathbf{R}^{\frac{1}{2}} = (\mathbf{I} - \alpha\,\mathbf{\Omega})^{-1} \cdot (\mathbf{I} + \alpha\,\mathbf{\Omega}), \tag{2.5.34}$$

either by a direct check, yielding $\alpha = (-1 + \sqrt{1 + |\vec{\mathbf{\Omega}}|^2})/|\vec{\mathbf{\Omega}}|^2 > 0$ or using the rotation angle (2.5.43), namely $\psi = 2\arctan|\vec{\mathbf{\Omega}}| = 2 \times 2\arctan|\alpha\,\vec{\mathbf{\Omega}}|$.

2.5.5.2 Exponential representation

For any proper orthogonal tensor $\mathbf{R}$, there exists a skew-symmetric second order tensor $\boldsymbol{\Omega}$ such that, e.g., Gantmacher [1959], p. 287,

$$\mathbf{R} = \exp \boldsymbol{\Omega}\,. \tag{2.5.35}$$

The exponential of the second order tensor $\mathbf{A}$ is defined through a series expansion like for scalars. Indeed

$$\exp \mathbf{A} = \sum_{k=0}^{\infty} \frac{\mathbf{A}^k}{k!}\,, \tag{2.5.36}$$

is absolutely convergent since $|\exp \mathbf{A}| \le \exp |\mathbf{A}|$. As an alternative definition, note the limit

$$\exp \mathbf{A} = \lim_{k \to \infty}\ \Big(\mathbf{I} + \frac{\mathbf{A}}{k}\Big)^k\,. \tag{2.5.37}$$

The estimation of the exponential of a skew-symmetric tensor $\boldsymbol{\Omega}$ is greatly simplified through the expression [Schwerdtfeger 1961, p. 238][2.5],

$$\mathbf{R} = \exp \boldsymbol{\Omega} = \mathbf{I} + \frac{\sin |\vec{\boldsymbol{\Omega}}|}{|\vec{\boldsymbol{\Omega}}|}\, \boldsymbol{\Omega}\ +\ \frac{1 - \cos |\vec{\boldsymbol{\Omega}}|}{|\vec{\boldsymbol{\Omega}}|^2}\, \boldsymbol{\Omega}^2\,. \tag{2.5.38}$$

The proof starts with the exponential expansion,

$$\exp \boldsymbol{\Omega} = \mathbf{I} + \sum_{k=0}^{\infty} \frac{\boldsymbol{\Omega}^{2k+1}}{(2k+1)!} + \sum_{k=0}^{\infty} \frac{\boldsymbol{\Omega}^{2k+2}}{(2k+2)!}\,, \tag{2.5.39}$$

and uses the Cayley-Hamilton expressions (2.5.33),

$$\begin{aligned}
\mathbf{R} = \exp \boldsymbol{\Omega} &= \mathbf{I} + \Big(\sum_{k=0}^{\infty} (-1)^k \frac{|\vec{\boldsymbol{\Omega}}|^{2k+1}}{(2k+1)!}\Big) \frac{\boldsymbol{\Omega}}{|\vec{\boldsymbol{\Omega}}|} + \Big(1 - \sum_{k=-1}^{\infty} (-1)^{k+1} \frac{|\vec{\boldsymbol{\Omega}}|^{2(k+1)}}{(2(k+1))!}\Big) \frac{\boldsymbol{\Omega}^2}{|\vec{\boldsymbol{\Omega}}|^2} \\
&= \mathbf{I} + \Big(\sum_{k=0}^{\infty} (-1)^k \frac{|\vec{\boldsymbol{\Omega}}|^{2k+1}}{(2k+1)!}\Big) \frac{\boldsymbol{\Omega}}{|\vec{\boldsymbol{\Omega}}|} + \Big(1 - \sum_{k'=0}^{\infty} (-1)^{k'} \frac{|\vec{\boldsymbol{\Omega}}|^{2k'}}{(2k')!}\Big) \frac{\boldsymbol{\Omega}^2}{|\vec{\boldsymbol{\Omega}}|^2}\,.
\end{aligned} \tag{2.5.40}$$

The Cayley and exponential representations of a proper orthogonal tensor feature two skew-symmetric tensors which are parallel but have distinct magnitudes as indicated by (2.5.43).

2.5.5.3 Finite versus infinitesimal rotations

The eigenvalues λ of a proper orthogonal tensor $\mathbf{R}$ have modulus one, namely $\lambda \overline{\lambda} = 1$, $\overline{\lambda}$ denoting the complex conjugate of λ. This property results from the definition (2.5.23).

Let $\mathbf{x}$ be an eigenvector associated with the eigenvalue λ. Then $\mathbf{R} \cdot \mathbf{x} = \lambda \mathbf{x}$, and thus $\overline{\mathbf{x}}^{\mathrm{T}} \cdot \mathbf{R}^{\mathrm{T}} \cdot \mathbf{R} \cdot \mathbf{x} = \overline{\lambda}\lambda\, \overline{\mathbf{x}}^{\mathrm{T}} \cdot \mathbf{x}$, whence $\overline{\lambda}\lambda = 1$ since $\mathbf{R}^{\mathrm{T}} \cdot \mathbf{R} = \mathbf{I}$. Actually, since $\mathbf{R}$ is real, there should be one real and two complex conjugate eigenvalues. In fact, the real eigenvalue is equal to one, and its associated eigenvector $\mathbf{n} = \vec{\boldsymbol{\Omega}}/|\vec{\boldsymbol{\Omega}}|$ is the axis of the rotation, while the other two eigenvalues can be expressed as $\exp(\pm i\,\psi)$. The angle ψ is the angle of rotation in a plane orthogonal to the axis $\mathbf{n}$, counted positive clockwise. Indeed, the image $\mathbf{R} \cdot \mathbf{x}$ of any vector $\mathbf{x}$ may be cast in the format [Goldstein 1980, p. 164],

$$\mathbf{x} \quad \longrightarrow \quad \mathbf{R} \cdot \mathbf{x} = (\mathbf{x} \cdot \mathbf{n})\,\mathbf{n} + (\mathbf{x} - (\mathbf{x} \cdot \mathbf{n})\,\mathbf{n}) \cos \psi + \mathbf{x} \wedge \mathbf{n} \sin \psi\,. \tag{2.5.41}$$

A geometrical representation is displayed in Fig. 2.5.1. The angle of rotation can be estimated quite simply from the sum of the eigenvalues of $\mathbf{R}$,

$$\operatorname{tr} \mathbf{R}\ =\ 1 + \exp(i\,\psi) + \exp(-i\,\psi) = 1 + 2 \cos \psi\,, \tag{2.5.42}$$

[2.5]The estimation of the exponential of an arbitrary matrix is another task, see Moler C. and Van Loan C. (2003). Nineteen Dubious Ways to Compute the Exponential of a Matrix, Twenty-Five Years later, *SIAM Review*, 45(1), 3-46.

finite rotation   infinitesimal rotation

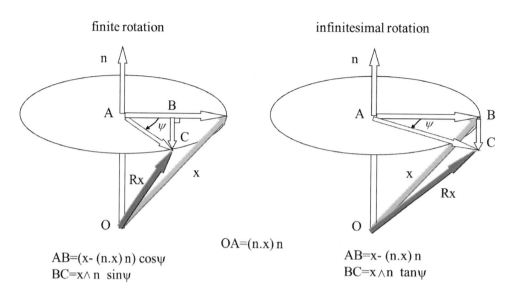

$$OA = (n.x)\,n$$

AB=(x- (n.x) n) cosψ   AB=x- (n.x) n
BC=x∧n sinψ   BC=x∧n tanψ

Fig. 2.5.1: Geometrical representations of a finite rotation and of an infinitesimal rotation.

and therefore using (2.5.30) and (2.5.38),

$$\psi = \arccos \tfrac{1}{2}\left(\operatorname{tr}\mathbf{R} - 1\right) = 2\arctan|\vec{\mathbf{\Omega}}_C| = |\vec{\mathbf{\Omega}}_E|, \tag{2.5.43}$$

the subscripts C and E referring to the Cayley and exponential representations, respectively.

A proper orthogonal tensor conserves the norm, i.e., $|\mathbf{R}\cdot\mathbf{a}| = |\mathbf{a}|$ for any vector $\mathbf{a}$[2.6]. Such is not the case for infinitesimal rotations defined via a vector product. Indeed, let $\mathbf{\Omega}$ be a skew-symmetric tensor, $\vec{\mathbf{\Omega}}$ its adjoint pseudo-vector, and $\mathbf{n} = \vec{\mathbf{\Omega}}/|\vec{\mathbf{\Omega}}|$. The transformation

$$\mathbf{x} \quad\longrightarrow\quad (\mathbf{I}+\mathbf{\Omega})\cdot\mathbf{x} = \mathbf{x} + \mathbf{x}\wedge\mathbf{n}\,|\vec{\mathbf{\Omega}}|, \tag{2.5.45}$$

both rotates $\mathbf{x}$ in the plane orthogonal to the axis of rotation by a clockwise angle $\psi = \arctan|\vec{\mathbf{\Omega}}|$ and changes its length. Fig. 2.5.1 contrasts the geometrical representations of a finite rotation and of an infinitesimal rotation.

2.5.5.4 Rotations versus projections

Anisotropic properties are often characterized by one or several unit vectors. Let $\mathbf{e} = \mathbf{e}(t)$ be a unit vector with value $\underline{\mathbf{e}} = \mathbf{e}(t = 0)$ at time $t = 0$. The deformation gradient, with polar decomposition $\mathbf{F} = \mathbf{R}\cdot\mathbf{U}$, is reckoned from the time $t = 0$. The evolution in time of the unit vector $\mathbf{e}$ is to be traced to its physical interpretation. For example, consider the two evolution rules,

$$\mathbf{e} = \mathbf{R}\cdot\underline{\mathbf{e}}, \quad \mathbf{e} = \frac{\mathbf{F}\cdot\underline{\mathbf{e}}}{|\mathbf{F}\cdot\underline{\mathbf{e}}|}. \tag{2.5.46}$$

Since $\mathbf{e}$ is a unit vector, it remains orthogonal to its rate $d\mathbf{e}/dt$. It may be expressed in the form $\boldsymbol{\mathcal{E}}\cdot\mathbf{e}$ but the second order tensor $\boldsymbol{\mathcal{E}}$ is not unique: dyadic terms $\mathbf{e}\otimes\mathbf{e}'$ with $\mathbf{e}'$ orthogonal to $\mathbf{e}$ are elusive. For the first rule, the entity $\boldsymbol{\mathcal{E}}$,

$$\boldsymbol{\mathcal{E}} = \frac{d\mathbf{R}}{dt}\cdot\mathbf{R}^{\mathrm{T}}, \tag{2.5.47}$$

is a skew-symmetric tensor. Let $\mathbf{L} = d\mathbf{F}/dt\cdot\mathbf{F}^{-1}$ be the velocity gradient and $\mathbf{D}$ and $\mathbf{W}$ respectively its symmetric and skew-symmetric parts. Then, for the second rule, the rate $\boldsymbol{\mathcal{E}}\cdot\mathbf{e}$, with

$$\boldsymbol{\mathcal{E}} = \left(\mathbf{I} - \mathbf{e}\otimes\mathbf{e}\right)\cdot\mathbf{L}, \tag{2.5.48}$$

involves the projection of $\mathbf{L}\cdot\mathbf{e}$ on the plane tangent to the unit sphere at $\mathbf{e} = \mathbf{e}(t)$. Note that the rate $d\mathbf{e}/dt$ may also be expressed in terms of the skew-symmetric tensor,

$$\boldsymbol{\mathcal{E}} = \mathbf{W} + \mathbf{D}\cdot\mathbf{e}\otimes\mathbf{e} - \mathbf{e}\otimes\mathbf{e}\cdot\mathbf{D}. \tag{2.5.49}$$

[2.6]As a slightly more general property, if $\mathbf{R}$ is an orthogonal tensor, as defined in Section 5.2.3.1, then

$$|\mathbf{R}\cdot\mathbf{a}| = |\mathbf{a}|, \quad |\mathbf{R}\cdot\mathbf{A}\cdot\mathbf{R}^{\mathrm{T}}| = |\mathbf{A}|, \tag{2.5.44}$$

for any vector $\mathbf{a}$ and any second order tensor $\mathbf{A}$.

2.5.5.5   A differential equation for the rotation tensor

If $\mathbf{R} = \mathbf{R}(t)$ is a proper orthogonal tensor, then $d\mathbf{R}/dt \cdot \mathbf{R}^{\mathrm{T}}$ is skew-symmetric as indicated by (2.5.2). Then if the skew-symmetric tensors $\mathbf{W}(t)$ and $\int_{t_0}^{t} \mathbf{W}(\tau) \, d\tau$ commute, the solution $\mathbf{R}(t)$ of the differential equation,

$$\frac{d\mathbf{R}}{dt}(t) \cdot \mathbf{R}^{\mathrm{T}}(t) = \mathbf{W}(t), \quad t \in [t_0, \infty], \tag{2.5.50}$$

has the analytical expression,

$$\mathbf{R}(t) = \exp\Big(\int_{t_0}^{t} \mathbf{W}(\tau) d\tau\Big) \cdot \mathbf{R}(t_0), \quad t \in [t_0, \infty]. \tag{2.5.51}$$

Indeed, for small Δt,

$$
\begin{aligned}
\mathbf{R}(t + \Delta t) - \mathbf{R}(t) &= \Big(\exp\Big(\int_{t_0}^{t} \mathbf{W}(\tau) d\tau + \int_{t}^{t+\Delta t} \mathbf{W}(\tau) d\tau\Big) - \exp\Big(\int_{t_0}^{t} \mathbf{W}(\tau) d\tau\Big) \Big) \cdot \mathbf{R}(t_0) \\
&\simeq \Big(\exp\Big(\int_{t_0}^{t} \mathbf{W}(\tau) d\tau + \mathbf{W}(t)\Delta t\Big) - \exp\Big(\int_{t_0}^{t} \mathbf{W}(\tau) d\tau\Big) \Big) \cdot \mathbf{R}(t_0) \\
&= \Big(\exp\big(\mathbf{W}(t)\Delta t\big) \cdot \exp\Big(\int_{t_0}^{t} \mathbf{W}(\tau) d\tau\Big) - \exp\Big(\int_{t_0}^{t} \mathbf{W}(\tau) d\tau\Big) \Big) \cdot \mathbf{R}(t_0) \\
&\simeq \mathbf{W}(t)\Delta t \cdot \exp\Big(\int_{t_0}^{t} \mathbf{W}(\tau) d\tau\Big) \cdot \mathbf{R}(t_0) \\
&= \mathbf{W}(t)\Delta t \cdot \mathbf{R}(t),
\end{aligned}
\tag{2.5.52}
$$

which shows that (2.5.51) is a solution of (2.5.50). The step from line 2 to line 3 has required $\mathbf{W}(t)$ and $\int_{t_0}^{t} \mathbf{W}(\tau) \, d\tau$ to commute: indeed, the relation $\exp(\mathbf{A} + \mathbf{B}) = \exp\mathbf{A} \cdot \exp\mathbf{B}$ holds only if the two second order tensors $\mathbf{A}$ and $\mathbf{B}$ commute. The step from line 3 to line 4 has used the series expansion of the exponential.

2.5.6   Infinitesimal strains

Assume the deformation gradient $\mathbf{F}$ associated with a displacement field $\mathbf{u}$ to be sufficiently small, so that the norm $|\mathbf{H}|$ of $\mathbf{H} = \mathbf{F} - \mathbf{I} = \mathbf{u}\overset{\leftarrow}{\nabla}$ is much smaller than 1. The infinitesimal strain $\boldsymbol{\epsilon}$ and infinitesimal rotation $\boldsymbol{\omega}$ are defined respectively as the symmetric and skew-symmetric parts of $\mathbf{H}$, namely

$$\mathbf{H} = \boldsymbol{\epsilon} + \boldsymbol{\omega}, \quad \boldsymbol{\epsilon} = \frac{1}{2}(\mathbf{H} + \mathbf{H}^{\mathrm{T}}), \quad \boldsymbol{\omega} = \frac{1}{2}(\mathbf{H} - \mathbf{H}^{\mathrm{T}}). \tag{2.5.53}$$

The Green strain expresses exactly in terms of $\boldsymbol{\epsilon}$ and $\boldsymbol{\omega}$,

$$\mathbf{E}_{\mathrm{G}} = \frac{1}{2}(\mathbf{F}^{\mathrm{T}} \cdot \mathbf{F} - \mathbf{I}) = \boldsymbol{\epsilon} + \frac{1}{2}\mathbf{H}^{\mathrm{T}} \cdot \mathbf{H} = \boldsymbol{\epsilon} + \frac{1}{2}(\boldsymbol{\epsilon}^2 + \boldsymbol{\epsilon} \cdot \boldsymbol{\omega} - \boldsymbol{\omega} \cdot \boldsymbol{\epsilon} - \boldsymbol{\omega}^2), \tag{2.5.54}$$

and an exact expansion of $\det(\mathbf{I} + \mathbf{H})$ is derived in (2.2.34).

The polar decomposition $\mathbf{F} = \mathbf{R} \cdot \mathbf{U} = \mathbf{V} \cdot \mathbf{R}$ is approximated to first order by the expressions,

$$\mathbf{U} = \mathbf{I} + \boldsymbol{\epsilon} + O(|\mathbf{H}|^2), \quad \mathbf{V} = \mathbf{I} + \boldsymbol{\epsilon} + O(|\mathbf{H}|^2), \quad \mathbf{R} = \mathbf{I} + \boldsymbol{\omega} + O(|\mathbf{H}|^2), \tag{2.5.55}$$

and $\det \mathbf{F} = 1 + \operatorname{tr}\boldsymbol{\epsilon} + O(|\mathbf{H}|^2)$.

Second order approximations may be derived as well, e.g., Casey [1985],

$$
\begin{aligned}
\mathbf{F}^{-1} &= \mathbf{I} - \mathbf{H} + \mathbf{H}^2 + (O(|\mathbf{H}|^3), \\[4pt]
\mathbf{U} &= \mathbf{I} + \boldsymbol{\epsilon} + \frac{1}{2}(\boldsymbol{\epsilon} \cdot \boldsymbol{\omega} - \boldsymbol{\omega} \cdot \boldsymbol{\epsilon} - \boldsymbol{\omega}^2) + (O(|\mathbf{H}|^3), \\[4pt]
\mathbf{V} &= \mathbf{I} + \boldsymbol{\epsilon} - \frac{1}{2}(\boldsymbol{\epsilon} \cdot \boldsymbol{\omega} - \boldsymbol{\omega} \cdot \boldsymbol{\epsilon} + \boldsymbol{\omega}^2) + O(|\mathbf{H}|^3), \\[4pt]
\mathbf{R} &= \mathbf{I} + \boldsymbol{\omega} - \frac{1}{2}(\boldsymbol{\epsilon} \cdot \boldsymbol{\omega} + \boldsymbol{\omega} \cdot \boldsymbol{\epsilon} - \boldsymbol{\omega}^2) + O(|\mathbf{H}|^3).
\end{aligned}
\tag{2.5.56}
$$

Exercise 2.2 provides an instance where two motions have identical infinitesimal strains but distinct infinitesimal rotations.

2.5.7 Volume change and incompressibility

The relative volume change under a deformation gradient $\mathbf{F}$ is equal to $\det\mathbf{F}$, or $\sqrt{\det\mathbf{C}}$, with $\mathbf{C}=\mathbf{F}^{\mathrm{T}}\cdot\mathbf{F}=\mathbf{I}+2\,\mathbf{E}_{\mathrm{G}}$ the right Cauchy-Green tensor. According to (2.2.34),

$$\det\mathbf{C}=\det\mathbf{F}^2=\det\left(\mathbf{I}+2\,\mathbf{E}_{\mathrm{G}}\right)=1+I_1(2\,\mathbf{E}_{\mathrm{G}})+I_2(2\,\mathbf{E}_{\mathrm{G}})+I_3(2\,\mathbf{E}_{\mathrm{G}})\,,\tag{2.5.57}$$

where the I's are the isotropic invariants of $2\,\mathbf{E}_{\mathrm{G}}$ defined in Section 2.2.2. For infinitesimal strains, $\det\mathbf{F}$ is thus equal to $1+\mathrm{tr}\,\mathbf{E}_{\mathrm{G}}$ to first order wrt to $|\mathbf{E}_{\mathrm{G}}|$, that is, neglecting terms of the order of $|\mathbf{E}_{\mathrm{G}}|^2$ or higher. Therefore, for infinitesimal strains, the relative volume change is estimated by $\mathrm{tr}\,\mathbf{E}_{\mathrm{G}}$.

Let us turn now to finite strains, and consider an incompressible solid, namely one for which $\det\mathbf{F}=1$ at any time. In this case, the above approximation and interpretation of $\mathrm{tr}\,\mathbf{E}_{\mathrm{G}}$ are no longer valid. Indeed, even the inequality $\mathrm{tr}\,\mathbf{E}_{\mathrm{G}}\geq0$ does not imply that the volume change is positive.

But, why does this inequality hold? In fact, it results from the classical theorem of algebra that the arithmetic mean of n real positive numbers a_i is larger than their geometric mean,

$$\frac{1}{n}\left(a_1+a_2\cdots+a_n\right)\geq\sqrt[n]{a_1\,a_2\cdots a_n}\,,\tag{2.5.58}$$

with equality only if all the a's are equal. Applied to the three (positive) eigenvalues of the right Cauchy-Green tensor, this inequality yields,

$$\frac{1}{3}\,\mathrm{tr}\,\mathbf{C}=\frac{1}{3}\left(3+2\,\mathrm{tr}\,\mathbf{E}_{\mathrm{G}}\right)\geq\sqrt[3]{\det\mathbf{C}}=1\quad\Rightarrow\quad\mathrm{tr}\,\mathbf{C}\geq3,\quad\mathrm{tr}\,\mathbf{E}_{\mathrm{G}}\geq0\,.\tag{2.5.59}$$

Actually, we know more: for positive real numbers, the arithmetic mean is larger than the geometric mean, which itself is larger than the harmonic mean,

$$\frac{1}{n}\left(a_1+a_2\cdots+a_n\right)\geq\sqrt[n]{a_1\,a_2\cdots a_n}\geq\frac{n}{\dfrac{1}{a_1}+\dfrac{1}{a_2}\cdots+\dfrac{1}{a_n}}\,.\tag{2.5.60}$$

These inequalities have interesting consequences for the scalar invariants of second order tensors.

2.6 Transports between reference and current configurations

The motion of the points $\mathbf{x}$ of the solid animated with the velocity $\mathbf{v}_s$ is tracked by the deformation gradient $\mathbf{F}$ measured with respect to a reference configuration. Entities and operators reckoned from the reference configuration are underlined. The motion is considered to shift the point $\underline{\mathbf{x}}=(\underline{x}_I)$ in the reference configuration to its current position $\mathbf{x}=(x_i)$. Differential operators are reckoned from either configuration. For example, for a generic entity ψ, the gradient with respect to the current configuration $\nabla\psi=\partial\psi/\partial\mathbf{x}$ becomes with respect to the reference configuration $\underline{\nabla}\psi=\partial\psi/\partial\underline{\mathbf{x}}=\nabla\psi\cdot\mathbf{F}$.

- **Elementary surface and volume:**

The vector product of any two vectors $\mathbf{a}$ and $\mathbf{b}$ represents the oriented area of the underlying parallelepiped. In the cartesian axes $\{\mathbf{e}_i,\,i\in[1,3]\}$, it is defined componentwise via the permutation symbol e_{ijk},

$$\mathbf{a}\wedge\mathbf{b}=e_{ijk}\,a_i\,b_j\,\mathbf{e}_k\,.\tag{2.6.1}$$

Let $(d\underline{\mathbf{x}}_1,d\underline{\mathbf{x}}_2,d\underline{\mathbf{x}}_3)$ be a triad of infinitesimal vectors, and $(d\mathbf{x}_1,d\mathbf{x}_2,d\mathbf{x}_3)$ their image under the transformation of gradient $\mathbf{F}$. Then the elementary volumes defined by these triads transform as follows,

$$d\underline{V}=(d\underline{\mathbf{x}}_1\wedge d\underline{\mathbf{x}}_2)\cdot d\underline{\mathbf{x}}_3\quad\overset{\mathbf{F}}{\longmapsto}\quad dV=(d\mathbf{x}_1\wedge d\mathbf{x}_2)\cdot d\mathbf{x}_3\,.\tag{2.6.2}$$

- **Transformation of a material volume and relations with determinants:**

According to (2.1.15), the determinant of the second order tensor $\mathbf{F}$ with components F_{iI} in cartesian axes satisfies, for any $(I,J,K)\in[1,3]$, (note: $e_{IJK}e_{IJK}=3!$),

$$e_{IJK}\det\mathbf{F}=e_{ijk}\,F_{iI}\,F_{jJ}\,F_{kK}\Rightarrow\det\mathbf{F}=e_{ijk}\,F_{i1}\,F_{j2}\,F_{k3}\text{ and }\det\mathbf{F}=\frac{1}{3!}\,e_{ijk}\,e_{IJK}\,F_{iI}\,F_{jJ}\,F_{kK}\,.\tag{2.6.3}$$

From the definitions of an area and of a determinant results the rule of transformation of volume,

$$dV = \det \mathbf{F} \, d\underline{V}. \tag{2.6.4}$$

Indeed

$$
\begin{aligned}
(\mathbf{a} \wedge \mathbf{b}) \cdot \mathbf{c} &= e_{ijk} \, a_i \, b_j \, c_k \\
&= e_{ijk} \, F_{iI} \, \underline{a}_I \, F_{jJ} \, \underline{b}_J \, F_{kK} \, \underline{c}_K \\
&= e_{IJK} \det \mathbf{F} \, \underline{a}_I \, \underline{b}_J \, \underline{c}_K \\
&= \det \mathbf{F} \, (\underline{a} \wedge \underline{b}) \cdot \underline{c} \quad \square \, .
\end{aligned}
\tag{2.6.5}
$$

Therefore,

$$\int_V \psi \, dV = \int_{\underline{V}} \det \mathbf{F} \, \psi \, d\underline{V}. \tag{2.6.6}$$

Two additional differential relations are recorded, see, e.g., Bowen [1989], p. 45, for a proof:

$$\underline{\operatorname{div}} \left((\det \mathbf{F}) \, \mathbf{F}^{-\mathrm{T}} \right) = \mathbf{0}, \quad \operatorname{div} \left((\det \mathbf{F}^{-1}) \, \mathbf{F}^{\mathrm{T}} \right) = \mathbf{0}. \tag{2.6.7}$$

Here the differential operators are the right divergence, namely componentwise,

$$\frac{\partial}{\partial \underline{x}_I} \left(\det \mathbf{F} \, \frac{\partial \underline{x}_I}{\partial x_i} \right) = 0, \quad \frac{\partial}{\partial x_i} \left(\det \mathbf{F}^{-1} \, \frac{\partial x_i}{\partial \underline{x}_I} \right) = 0. \tag{2.6.8}$$

- **Transformation of a material surface, Nanson's rule:**

Let $d\underline{\mathbf{S}}$ be the elementary material area and $d\mathbf{S}$ the transformed area under the transformation of gradient $\mathbf{F}$. Then

$$d\mathbf{S} = \det \mathbf{F} \, \mathbf{F}^{-\mathrm{T}} \cdot d\underline{\mathbf{S}} = \det \mathbf{F} \, d\underline{\mathbf{S}} \cdot \mathbf{F}^{-1}. \tag{2.6.9}$$

Proof: Let $d\underline{\mathbf{S}} = d\underline{\mathbf{a}} \wedge d\underline{\mathbf{b}}$ and $d\mathbf{S} = d\mathbf{a} \wedge d\mathbf{b}$ with $d\mathbf{a} = \mathbf{F} \cdot d\underline{\mathbf{a}}$ and $d\mathbf{b} = \mathbf{F} \cdot d\underline{\mathbf{b}}$. Then

$$
\begin{aligned}
d\mathbf{S} \cdot \mathbf{F} &= (e_{ijk} \, da_i \, db_j \, \mathbf{e}_k) \cdot (F_{lL} \mathbf{e}_l \otimes \mathbf{e}_L) \\
&= e_{ijk} \, F_{iI} F_{jJ} F_{kK} \, d\underline{a}_I \, d\underline{b}_J \, \mathbf{e}_K \\
&= e_{IJK} \det \mathbf{F} \, d\underline{a}_I \, d\underline{b}_J \, \mathbf{e}_K \\
&= \det \mathbf{F} \, d\underline{\mathbf{a}} \wedge d\underline{\mathbf{b}} \\
&= \det \mathbf{F} \, d\underline{\mathbf{S}} \quad \square \, .
\end{aligned}
\tag{2.6.10}
$$

Alternatively,

$$
\begin{aligned}
dV &= \det \mathbf{F} \, d\underline{V} \\
d\mathbf{S} \cdot d\mathbf{c} &= \det \mathbf{F} \, d\underline{\mathbf{S}} \cdot d\underline{\mathbf{c}} \\
&= \det \mathbf{F} \, d\underline{\mathbf{S}} \cdot \mathbf{F}^{-1} \cdot d\mathbf{c}
\end{aligned}
\tag{2.6.11}
$$

which holds for any $d\mathbf{c}$ $\square$.

- **Conservation of flux:**

Let $\underline{\mathbf{M}}$ and $\mathbf{M}$ be the fluxes of matter per unit reference area and per unit current area, respectively. The conservation of the flux,

$$\underline{\mathbf{M}} \cdot d\underline{\mathbf{S}} = \mathbf{M} \cdot d\mathbf{S}, \tag{2.6.12}$$

together with the transformation rule for areas (2.6.9), implies the relation,

$$\underline{\mathbf{M}} = \det \mathbf{F} \, \mathbf{F}^{-1} \cdot \mathbf{M} = \det \mathbf{F} \, \mathbf{M} \cdot \mathbf{F}^{-\mathrm{T}}, \quad \mathbf{M} = \det \mathbf{F}^{-1} \, \mathbf{F} \cdot \underline{\mathbf{M}}. \tag{2.6.13}$$

Note also the conservation relation for the divergence of a flux,

$$\underline{\operatorname{div}} \, \underline{\mathbf{M}} = \det \mathbf{F} \operatorname{div} \mathbf{M}. \tag{2.6.14}$$

Indeed, using (2.6.13),

$$
\begin{aligned}
\underline{\mathrm{div}}\,\underline{\mathbf{M}} &= \underline{\mathrm{div}}\left(\mathbf{M}\cdot(\det\mathbf{F}\,\mathbf{F}^{\text{-T}})\right)\\[4pt]
&= \frac{\partial}{\partial\underline{x}_I}\left(M_i\det\mathbf{F}\,(\mathbf{F}^{-1})_{iI}\right)\\[4pt]
&= \frac{\partial M_i}{\partial\underline{x}_I}\det\mathbf{F}\,(\mathbf{F}^{-1})_{iI} + M_i\frac{\partial}{\partial\underline{x}_I}\left(\det\mathbf{F}\,(\mathbf{F}^{-1})_{iI}\right)\\[4pt]
&\overset{(2.6.7)}{=} \frac{\partial M_i}{\partial\underline{x}_I}\det\mathbf{F}\,\frac{\partial\underline{x}_I}{\partial x_i}\\[4pt]
&= \frac{\partial M_i}{\partial x_i}\det\mathbf{F}\\[4pt]
&= \det\mathbf{F}\,\mathrm{div}\,\mathbf{M}\quad\square\,.
\end{aligned}
\tag{2.6.15}
$$

- **Conservation of traction**:

The balance of momentum in the reference configuration is phrased in terms of the first Piola-Kirchhoff stress $\underline{\tau}$ which is such that traction is conserved,

$$
\boldsymbol{\sigma}\cdot d\mathbf{S} = \underline{\tau}\cdot d\underline{\mathbf{S}},
\tag{2.6.16}
$$

and therefore with the rule of transformation of areas,

$$
\underline{\tau} = \det\mathbf{F}\,\boldsymbol{\sigma}\cdot\mathbf{F}^{\text{-T}},\quad \boldsymbol{\sigma} = \det\mathbf{F}^{-1}\,\underline{\tau}\cdot\mathbf{F}^{\text{T}}\,.
\tag{2.6.17}
$$

Similarly to (2.6.14), the transformation rule for the divergence operator takes the form,

$$
\underline{\mathrm{div}}\,\underline{\tau} = \det\mathbf{F}\,\mathrm{div}\,\boldsymbol{\sigma}\,,
\tag{2.6.18}
$$

which is a direct consequence of (2.6.7).

- **Conservation of mass and energy**:

Let us envisage a pure change of configuration, at constant mass,

$$
dM = \rho\,dV = d\underline{M} = \underline{\rho}\,d\underline{V}\,,
\tag{2.6.19}
$$

so that the current and reference mass densities vary as indicated by the volume change,

$$
\rho \mapsto \underline{\rho} = \rho\det\mathbf{F}\,.
\tag{2.6.20}
$$

Material properties, like the energies, can be defined per unit current volume V, unit reference (or initial) volume $\underline{V}$, unit current mass M or unit reference mass $\underline{M}$. They are denoted respectively by Ψ, $\underline{\Psi}$ and ψ and $\underline{\psi}$. Conservation of energy under a change of configuration implies

$$
\Psi\,V = \underline{\Psi}\,\underline{V} = \psi\,M = \underline{\psi}\,\underline{M}\,.
\tag{2.6.21}
$$

Under a change of volume only, the entities measured per unit volume change like the mass density while the entities measured per unit mass are not modified, e.g.,

$$
\Psi \mapsto \underline{\Psi} = \Psi\det\mathbf{F},\quad \psi \mapsto \underline{\psi} = \psi\,.
\tag{2.6.22}
$$

In a mixture, both internal mass and energy exchanges between species and mass and energy exchanges with the surroundings may take place. These phenomena are considered in Chapter 4. While certain properties Ψ^k attached to the species k are measured per unit (current or reference) volume of the mixture, the corresponding properties ψ_k are measured per unit (current or reference) mass of the species. Instances are internal energy, free energy, enthalpy, entropy, etc. The notation (2.6.21) is extended so as to involve the total volume of the mixture and not the volume of the species[2.7],

$$
\Psi^k\,V = \underline{\Psi}^k\,\underline{V} = \psi_k\,M_k = \underline{\psi}_k\,\underline{M}_k\,.
\tag{2.6.23}
$$

[2.7]The convention of summation over repeated indices does not apply in Eqns (2.6.23), (2.6.24) and (2.6.25).

• Mass and volume contents:

In order to avoid confusion of notation, for any species k, Biot's approach to mixtures introduces besides the intrinsic and partial densities, ρ_k and ρ^k respectively, and *volume fractions* n^k,

$$\rho_k \equiv \frac{M_k}{V_k}, \quad \rho^k \equiv \frac{M_k}{V}, \quad n^k \equiv \frac{V_k}{V}, \quad \rho^k = n^k \rho_k, \tag{2.6.24}$$

the *mass contents* m^k and *volume contents* v^k which are defined per unit *reference* volume,

$$m^k \equiv \frac{M_k}{\underline{V}} = \rho^k \det \mathbf{F}, \quad v^k \equiv \frac{V_k}{\underline{V}} = n^k \det \mathbf{F}, \quad m^k = v^k \rho_k. \tag{2.6.25}$$

The mass content of the mixture as a whole differs from its mass density,

$$m \equiv \frac{M}{\underline{V}} = \rho \det \mathbf{F} \neq \rho \equiv \frac{M}{V}. \tag{2.6.26}$$

Leaving aside diffusion, the density $\underline{\rho}^k$ of the species k in the reference configuration, say at time $t = 0$, will be different from the mass content m^k, i.e., the current mass per unit reference volume, if internal mass transfer and/or growth take place. Similarly, the density of the mixture in the reference configuration $\underline{\rho}$ is different from the mass content m if growth takes place.

2.7 Time derivatives and Reynolds theorems

Let $\Psi = \Psi(\mathbf{x}, t)$ be a sufficiently smooth field in a volume V so that there is no *discontinuity surface* of the gradient $\partial \Psi / \partial \mathbf{x}$ crossing the volume.

• Material time derivatives:

In a mixture context, besides the partial time derivative $\partial / \partial t$, two types of material time derivatives will be used, namely D / Dt and d^k / dt following respectively the mass center of the mixture with velocity $\mathbf{v}$ and the particles of species k with velocity $\mathbf{v}_k$,

$$\frac{D\Psi}{Dt} = \frac{\partial \Psi}{\partial t} + \boldsymbol{\nabla} \Psi \cdot \mathbf{v}; \quad \frac{d^k \Psi}{dt} = \frac{\partial \Psi}{\partial t} + \boldsymbol{\nabla} \Psi \cdot \mathbf{v}_k, \quad k \in \mathcal{K}. \tag{2.7.1}$$

To simplify the notation, d^s / dt is noted d / dt.

Repeated use will be made of the rate formula,

$$\frac{d}{dt} \det \mathbf{F} = \det \mathbf{F} \operatorname{tr} \frac{d\mathbf{F}}{dt} \cdot \mathbf{F}^{-1} = \det \mathbf{F} \operatorname{div} \mathbf{v}_\mathrm{s}. \tag{2.7.2}$$

To prove this relation, let us consider the spectral decomposition of $\mathbf{F}$. Then $\det \mathbf{F} = F_1 F_2 F_3$. Since $\mathbf{F}$ is assumed invertible, it makes sense to consider the time derivative of $\operatorname{Ln} \det \mathbf{F}$. The result (2.7.2) follows directly. As a direct consequence of (2.7.2),

$$\frac{\partial \det \mathbf{F}}{\partial \mathbf{F}} = \det \mathbf{F} \, \mathbf{F}^{-\mathrm{T}}, \quad \frac{\partial \det \mathbf{F}}{\partial \mathbf{E}_\mathrm{G}} = \det \mathbf{F} \, \mathbf{F}^{-1} \cdot \mathbf{F}^{-\mathrm{T}}, \quad \frac{\partial \det \mathbf{C}}{\partial \mathbf{C}} = \det \mathbf{C} \, \mathbf{C}^{-1}. \tag{2.7.3}$$

As an immediate consequence for any scalar X,

$$\frac{\partial X}{\partial \mathbf{E}_\mathrm{G}} = \frac{1}{\det \mathbf{F}} \frac{\partial (\det \mathbf{F} \, X)}{\partial \mathbf{E}_\mathrm{G}} - X \, \mathbf{F}^{-1} \cdot \mathbf{F}^{-\mathrm{T}}, \tag{2.7.4}$$

which is instrumental when expressing energies in distinct configurations.

• Generalized Reynolds theorem:

Let
- $V = V(t)$ be a mobile volume; the points of its surface $\partial V(t)$ have the velocity $\mathbf{v}_k$;
- $\mathbf{v}_\mathrm{s}$ be the velocity of the material point $\mathbf{x}$.

Then

$$
\begin{aligned}
\frac{d}{dt}\int_V \Psi\, dV &= \frac{d}{dt}\int_{\underline{V}} \Psi \det \mathbf{F}\, d\underline{V}\\[4pt]
&= \int_{\underline{V}} \frac{d}{dt}\left(\Psi \det \mathbf{F}\right) d\underline{V} + \int_{\partial \underline{V}} \Psi \det \mathbf{F}\left(\mathbf{F}^{-1}\cdot(\mathbf{v}_k - \mathbf{v}_\mathrm{s})\right)\cdot d\underline{\mathbf{S}}\\[4pt]
&= \int_V \frac{d\Psi}{dt} + \Psi \operatorname{div}\mathbf{v}_\mathrm{s}\, dV + \int_{\partial V} \Psi\left(\mathbf{v}_k - \mathbf{v}_\mathrm{s}\right)\cdot d\mathbf{S}\\[4pt]
&= \int_V \frac{\partial\Psi}{\partial t} + \operatorname{div}\left(\Psi\,\mathbf{v}_\mathrm{s}\right) dV + \int_{\partial V} \Psi\left(\mathbf{v}_k - \mathbf{v}_\mathrm{s}\right)\cdot d\mathbf{S}\\[4pt]
&= \int_V \frac{\partial\Psi}{\partial t}\, dV + \int_{\partial V} \Psi\,\mathbf{v}_k\cdot d\mathbf{S}\\[4pt]
&= \int_V \frac{\partial\Psi}{\partial t} + \operatorname{div}\left(\Psi\,\mathbf{v}_k\right) dV
\end{aligned}
\tag{2.7.5}
$$

Indeed, the second line and the line before last can be interpreted as follows: the rate of change of the integral is equal to the integral of the rate of change of the integrand plus the variation due to the incoming and outgoing fluxes. Green's theorem has been used repeatedly.

- **Reynolds theorem**:

If V is a *material volume*, then $\mathbf{v}_k = \mathbf{v}_\mathrm{s}$, and the above relation simplifies to Reynolds theorem:

$$
\frac{d}{dt}\int_V \Psi\, dV = \int_V \frac{\partial\Psi}{\partial t}\, dV + \int_{\partial V} \Psi\,\mathbf{v}_\mathrm{s}\cdot d\mathbf{S}\,.
\tag{2.7.6}
$$

- **Time derivatives of a material line/surface**:

Along a closed *material* curve $\mathcal{C}$ and along a *material* surface S where points $\mathbf{x}$ have velocity $\mathbf{v}$ and velocity gradient $\mathbf{L} = \partial\mathbf{v}/\partial\mathbf{x} = \boldsymbol{\nabla}\mathbf{v}$, the material time derivatives of the differential increment $d\mathbf{x} = \mathbf{v}\,dt$ and of the differential area $d\mathbf{S}$ write

$$
\frac{d}{dt}\left(d\mathbf{x}\right) = d\mathbf{v},\quad \frac{d}{dt}\left(d\mathbf{S}\right) = \left(\operatorname{div}\mathbf{v}\,\mathbf{I} - \mathbf{L}^{\mathrm{T}}\right)\cdot d\mathbf{S}\,,
\tag{2.7.7}
$$

due to (2.6.9) and (2.7.2). Whence the material time derivatives of a material curve and material surface,

$$
\begin{aligned}
\frac{d}{dt}\oint_{\mathcal{C}} \Phi\, d\mathbf{x} &= \oint_{\mathcal{C}}\left(\frac{d\Phi}{dt}\mathbf{I} + \Phi\,\mathbf{L}\right)\cdot d\mathbf{x} = \oint_{\mathcal{C}} \frac{d\Phi}{dt}\, d\mathbf{x} + \Phi\, d\mathbf{v},\\[4pt]
\frac{d}{dt}\int_S \Phi\, d\mathbf{S} &= \int_S \left(\left(\frac{d\Phi}{dt} + \Phi\operatorname{div}\mathbf{v}\right)\mathbf{I} - \Phi\,\mathbf{L}^{\mathrm{T}}\right)\cdot d\mathbf{S}\,,
\end{aligned}
\tag{2.7.8}
$$

for Φ sufficiently regular scalar, or arbitrary tensor with appropriate notations.

- **Circulation, theorems of Kelvin and Helmholtz**:

As an application of (2.7.8), we may consider the concept of *circulation*,

$$
C(t) \equiv \oint_{\mathcal{C}} \mathbf{v}\cdot d\mathbf{x}\,,
\tag{2.7.9}
$$

along the closed material curve $\mathcal{C}$. For an inviscid fluid, subjected to conservative body forces as defined by (2.4.19), the *theorem of Kelvin* provides an estimate of this line integral in terms of the pressure p and mass density ρ,

$$
\frac{dC(t)}{dt} = -\oint_{\mathcal{C}} \frac{dp}{\rho}\,,
\tag{2.7.10}
$$

while the *theorem of Helmholtz* indicates that the integral vanishes,

$$
\frac{dC(t)}{dt} = 0\,,
\tag{2.7.11}
$$

if the fluid is inviscid barotropic (its mass density depends only on the pressure $\rho = \rho(p)$). Therefore the circulation is constant in time, $C(t) = C(t_0)$. If this constant is zero, the velocity $\mathbf{v}$ derives from a potential, and therefore the vorticity vector $\boldsymbol{\nabla}\wedge\mathbf{v}$ vanishes due to Stokes theorem $(2.4.13)_2$ and therefore the flow is irrotational. These theorems are addressed in Section 15.7.1.

2.8   Work-conjugate stress-strain pairs

The first Piola-Kirchhoff stress $\underline{\tau} = (\underline{\tau}_{iJ})$ defined by (2.6.17) is not symmetric (the first Piola-Kirchhoff stress is sometimes defined as the transpose of the entity used here), but the symmetry of the Cauchy stress implies relations between its components and those of the deformation gradient. The Kirchhoff stress τ,

$$\tau = \det \mathbf{F}\, \sigma = \underline{\tau} \cdot \mathbf{F}^{\mathrm{T}} = \mathbf{F} \cdot \underline{\tau}^{\mathrm{T}}, \tag{2.8.1}$$

is indeed symmetric.

Various stress measures are obtained by considering the purely mechanical incremental work $\mathcal{W}$ measured per unit *reference volume* which can be expressed in terms of the generic work-conjugate stress-strain pair $(\mathbf{T}, \mathbf{E})$ as

$$\delta\mathcal{W} = \mathbf{T} : \delta\mathbf{E}, \quad \text{or} \quad \frac{d\mathcal{W}}{dt} = \mathbf{T} : \frac{d\mathbf{E}}{dt}\,. \tag{2.8.2}$$

The most common work-conjugate pairs are the following:
   - the 1st Piola-Kirchhoff stress $\underline{\tau}$ and the deformation gradient $\mathbf{F}$;
   - the 2nd Piola-Kirchhoff stress

$$\underline{\underline{\tau}} = \mathbf{F}^{-1} \cdot \underline{\tau} = \mathbf{F}^{-1} \cdot \tau \cdot \mathbf{F}^{-\mathrm{T}}, \tag{2.8.3}$$

   and the Green strain

$$\mathbf{E}_{\mathrm{G}} = \frac{1}{2}\left(\mathbf{F}^{\mathrm{T}} \cdot \mathbf{F} - \mathbf{I}\right), \tag{2.8.4}$$

   - the covariant Piola-Kirchhoff stress

$$\underline{\underline{\tau}}_c = \mathbf{F}^{\mathrm{T}} \cdot \tau \cdot \mathbf{F}, \tag{2.8.5}$$

   and the contravariant Almansi strain

$$\mathbf{E}_{\mathrm{A}}^{\mathrm{ct}} = \frac{1}{2}\left(\mathbf{I} - (\mathbf{F}^{\mathrm{T}} \cdot \mathbf{F})^{-1}\right). \tag{2.8.6}$$

Indeed

$$\frac{d\mathcal{W}}{dt} = \underline{\tau} : \frac{d\mathbf{F}}{dt} = \underline{\underline{\tau}} : \frac{d\mathbf{E}_{\mathrm{G}}}{dt} = \underline{\underline{\tau}}_c : \frac{d\mathbf{E}_{\mathrm{A}}^{\mathrm{ct}}}{dt} = \tau : \mathbf{D}\,. \tag{2.8.7}$$

2.8.1   Stress conjugate to a given Lagrangian strain measure

Hill [1970] has provided a method to define the stress $\mathbf{T}$ conjugate to a given Lagrangian strain $\mathbf{E}$ of the class defined by the function f as in (2.5.11). Indeed, consider the symmetric stress with components $\underline{\underline{T}}_{ij}$ on the Lagrangian strain ellipsoid,

$$\mathbf{T} = \sum_{i,j} \underline{\underline{T}}_{ij}\, \mathbf{e}_i \otimes \mathbf{e}_j\,. \tag{2.8.8}$$

The components of the strain rate $d\mathbf{E}/dt$ being given in explicit form by (2.5.22), work-conjugacy $\mathbf{T} : d\mathbf{E}/dt = \tau : \mathbf{D}$ implies,

$$\sum_{i,j} \underline{\underline{T}}_{ij} \left(\frac{df(\lambda_i)}{d\lambda_i}\, \lambda_i\, D_{ii}\, I_{ij} + 2\, \lambda_i\, \lambda_j\, \frac{f(\lambda_i) - f(\lambda_j)}{\lambda_i^2 - \lambda_j^2}\, D_{ij} \right) = \sum_{i,j} \tau_{ij}\, D_{ij}\,, \tag{2.8.9}$$

from which follow the relations, Eqn (24) in Hill [1970],

$$\underline{\underline{T}}_{ii} = \frac{\tau_{ii}}{\lambda_i\, df(\lambda_i)/d\lambda_i}, \quad i \in [1,3]; \quad \underline{\underline{T}}_{ij} = \begin{cases} \dfrac{\lambda_i^2 - \lambda_j^2}{f(\lambda_i) - f(\lambda_j)} \dfrac{\tau_{ij}}{2\,\lambda_i\,\lambda_j}, & \lambda_i \neq \lambda_j, \\[2mm] \dfrac{\tau_{ij}}{\lambda_i\, df(\lambda_i)/d\lambda}, & \lambda_i = \lambda_j, \end{cases} \quad i \neq j \in [1,3]. \tag{2.8.10}$$

The convention of summation over repeated indices does not apply in (2.8.10). The proof rests on the assumption that the principal axes of the strain ellipsoid are well defined, that is, that the principal stretches are distinct.

The general explicit expression of the stress conjugate to $\mathrm{Ln}\,\mathbf{U}$ has been derived by Hoger [1987]: it is unfortunately hardly tractable.

2.8.2   Issues related to the logarithmic strain

There is no measure of strain whose material derivative $d\mathbf{E}/dt$ is equal to an arbitrary rate of deformation $\mathbf{D}$. Therefore there is no measure of strain directly conjugated to Kirchhoff stress $\boldsymbol{\tau}$. Still, the logarithm of the left stretch of the deformation gradient $\mathrm{Ln}\,\mathbf{V}$ is a serious candidate. The conditions under which it is indeed a strain measure conjugate to Kirchhoff stress have been considered along several lines.

Perhaps the simplest condition is that of coaxiality between the Kirchhoff stress and the logarithmic strain. This property holds in particular for isotropic elasticity where the Kirchhoff stress depends linearly on $\mathbf{I}$, $\mathrm{Ln}\,\mathbf{V}$ and $(\mathrm{Ln}\,\mathbf{V})^2$, namely

$$\boldsymbol{\tau} = a_1\,\mathbf{I} + a_2\,\mathrm{Ln}\,\mathbf{V} + a_3\,(\mathrm{Ln}\,\mathbf{V})^2\,, \tag{2.8.11}$$

with coefficients a_1, a_2, and a_3 that are functions of the three scalar isotropic invariants of $\mathrm{Ln}\,\mathbf{V}$.

The logarithm of the left stretch $\mathrm{Ln}\,\mathbf{V}$ is a member of the Eulerian measures defined in Section 2.5.3. Some useful relations follow from its spectral decomposition over the principal axes of the Eulerian strain ellipsoid, whose rate of rotation is denoted by $\boldsymbol{\Omega}^{\mathrm{E}}$,

$$\mathrm{Ln}\,\mathbf{V} = \sum_{i=1,3} \mathrm{Ln}\,\lambda_i\;\mathbf{e}_i \otimes \mathbf{e}_i\,,$$

$$\frac{d}{dt}\mathrm{Ln}\,\mathbf{V} = \sum_{i=1,3} \frac{1}{\lambda_i}\frac{d\lambda_i}{dt}\;\mathbf{e}_i \otimes \mathbf{e}_i + \mathrm{Ln}\,\lambda_i\;\boldsymbol{\Omega}^{\mathrm{E}}\cdot\mathbf{e}_i \otimes \mathbf{e}_i + \mathrm{Ln}\,\lambda_i\;\mathbf{e}_i \otimes \boldsymbol{\Omega}^{\mathrm{E}}\cdot\mathbf{e}_i\,, \tag{2.8.12}$$

$$\mathbf{e}_i \cdot \frac{d}{dt}\mathrm{Ln}\,\mathbf{V}\cdot\mathbf{e}_i = \frac{1}{\lambda_i}\frac{d\lambda_i}{dt}\,, \quad i \in [1,3]\,.$$

Now assume the Kirchhoff stress $\boldsymbol{\tau} = \sum_i \tau_{ii}\,\mathbf{e}_i \otimes \mathbf{e}_i$ to be coaxial with the Eulerian strain ellipsoid. Consequently $\boldsymbol{\tau}:\mathbf{D} = \sum_i \tau_{ii}\,D_{ii}$ and $\boldsymbol{\tau}:d\mathrm{Ln}\,\mathbf{V}/dt = \sum_i \tau_{ii}\,\mathbf{e}_i\cdot d\mathrm{Ln}\,\mathbf{V}/dt\cdot\mathbf{e}_i$. In view of (2.5.21), $d\mathrm{Ln}\,\lambda_i/dt$ is equal to the Eulerian component D_{ii} of the rate of deformation, and, according to (2.8.12), $\mathbf{e}_i \cdot d\mathrm{Ln}\,\mathbf{V}dt \cdot \mathbf{e}_i = d\mathrm{Ln}\,\lambda_i/dt$. Therefore,

$$\boldsymbol{\tau}:\mathbf{D} = \boldsymbol{\tau}:\frac{d\mathrm{Ln}\,\mathbf{V}}{dt}\,, \tag{2.8.13}$$

that is, the logarithmic strain is work-conjugate to the Kirchhoff stress, Xiao and Chen [2003].

Here is another proof of (2.8.13) that does not use the relation (2.5.21). Gurtin and Spear [1983] have tried to link the rate of deformation to a corotational derivative. In fact, they obtain the following relation,

$$\mathbf{D} = \frac{d\mathrm{Ln}\,\mathbf{V}}{dt} + \mathrm{Ln}\,\mathbf{V}\cdot\boldsymbol{\Omega} - \boldsymbol{\Omega}\cdot\mathrm{Ln}\,\mathbf{V} - \frac{1}{2}\big(\mathbf{F}\cdot\boldsymbol{\Omega}^{\mathrm{L}}\cdot\mathbf{F}^{-1} + (\mathbf{F}\cdot\boldsymbol{\Omega}^{\mathrm{L}}\cdot\mathbf{F}^{-1})^{\mathrm{T}}\big)\,. \tag{2.8.14}$$

Consequently,

$$\begin{aligned}
\boldsymbol{\tau}:\mathbf{D} &= \boldsymbol{\tau}:\frac{d\mathrm{Ln}\,\mathbf{V}}{dt} \\
&+ (\boldsymbol{\tau}\cdot\mathrm{Ln}\,\mathbf{V} - \mathrm{Ln}\,\mathbf{V}\cdot\boldsymbol{\tau}):\boldsymbol{\Omega} + \frac{1}{2}\big(\mathbf{F}^{-1}\cdot\boldsymbol{\tau}\cdot\mathbf{F} - (\mathbf{F}^{-1}\cdot\boldsymbol{\tau}\cdot\mathbf{F})^{\mathrm{T}}\big):\boldsymbol{\Omega}^{\mathrm{L}}\,.
\end{aligned} \tag{2.8.15}$$

As we have seen just above, the second scalar product on the rhs cancels if the coaxiality property holds. The third does as well because $\mathbf{F}^{-1}\cdot\boldsymbol{\tau}\cdot\mathbf{F} = \sum_i \tau_{ii}\,\mathbf{e}_i \otimes \mathbf{e}_i$ is symmetric.

Finally, here is another consequence of coaxiality. Taking the time derivative of (2.5.10), the rotated Kirchhoff stress is seen to be work-conjugate to the Lagrangian logarithmic strain. Indeed,

$$\begin{aligned}
\boldsymbol{\tau}:\frac{d\mathrm{Ln}\,\mathbf{V}}{dt} &= (\mathbf{Q}^{\mathrm{T}}\cdot\boldsymbol{\tau}\cdot\mathbf{Q}):\frac{d\mathrm{Ln}\,\mathbf{U}}{dt} + \boldsymbol{\tau}:(\boldsymbol{\Omega}\cdot\mathrm{Ln}\,\mathbf{V} - \mathrm{Ln}\,\mathbf{V}\cdot\boldsymbol{\Omega}) \\
&= (\mathbf{Q}^{\mathrm{T}}\cdot\boldsymbol{\tau}\cdot\mathbf{Q}):\frac{d\mathrm{Ln}\,\mathbf{U}}{dt} + (\boldsymbol{\tau}\cdot\mathrm{Ln}\,\mathbf{V} - \mathrm{Ln}\,\mathbf{V}\cdot\boldsymbol{\tau}):\boldsymbol{\Omega} \\
&= (\mathbf{Q}^{\mathrm{T}}\cdot\boldsymbol{\tau}\cdot\mathbf{Q}):\frac{d\mathrm{Ln}\,\mathbf{U}}{dt}\,,
\end{aligned} \tag{2.8.16}$$

where $\boldsymbol{\Omega} = d\mathbf{Q}/dt\cdot\mathbf{Q}^{-1}$. The last line is due to the fact, since $\boldsymbol{\tau}$ and $\mathrm{Ln}\,\mathbf{V}$ are coaxial, they commute.

The above discussion was concerned with conjugate stress-strain measures. Interesting properties related to rates of the logarithmic strain may be recorded. Gurtin and Spear [1983] have shown that, when the axes of the Lagrangian strain ellipsoid are fixed, namely $\boldsymbol{\Omega}^{\mathrm{L}} = \mathbf{0}$, there exist at least two corotational derivatives which are equal to the rate of deformation. The corresponding spins are the rate of rotation of the Eulerian strain ellipsoid $\mathbf{\Omega}^{\mathrm{E}}$ and $\mathbf{W} = (\mathbf{L} - \mathbf{L}^{\mathrm{T}})/2$ associated with the Jaumann rate. As a more general result, Xiao et al. [1997] have established that there always exist a single corotational derivative and a single strain measure, namely the logarithmic strain $\mathrm{Ln}\,\mathbf{V}$, such that the corotational derivative of this strain is equal to the rate of deformation $\mathbf{D}$. However the spin defining the derivative is complex, their Eqn (38). By contrast, Bruhns et al. [2004] show that there exist infinitely many non corotational derivatives of all strain measures of the class (2.5.11) which are equal to the rate of deformation $\mathbf{D}$. The Oldroyd rates of the Finger strain and of the covariant Almansi strain are noticeable examples.

2.9   Small upon large

Small perturbations upon a finitely deformed solid may be used to gain information on the dependence of its mechanical properties with respect to its deformation. Indeed, assume a thermoelastic material to be governed by linear isotropic constitutive equations phrased in terms of the 2nd Piola-Kirchhoff stress and Green strain,

$$\underline{\underline{\tau}} = \underline{\underline{\tau}}(\mathbf{E}_{\mathrm{G}}, T) = \underline{\underline{\tau}}_i + \mathbb{E} : (\mathbf{E}_{\mathrm{G}} - c_{\mathrm{T}}\,\Delta T\,\mathbf{I}) = \underline{\underline{\tau}}_i + \lambda\,\mathrm{tr}\,\mathbf{E}_{\mathrm{G}}\,\mathbf{I} + 2\,\mu\,\mathbf{E}_{\mathrm{G}} - 3\,K\,c_{\mathrm{T}}\,\Delta T\,\mathbf{I}, \tag{2.9.1}$$

with $\mathbb{E} = \lambda\,\mathbf{I}\otimes\mathbf{I} + 2\,\mu\,\mathbf{I}\,\overline{\underline{\otimes}}\,\mathbf{I}$ the tensor of isotropic elasticity defined by the Lamé moduli λ and μ, $K = \lambda + 2\,\mu/3$ the bulk modulus, and c_{T} the coefficient of thermal expansion.

The temperature changes from T_0 to $T = T_0 + \delta T$ and the deformation gradient $\mathbf{F}$ undergoes a slight perturbation $\delta\mathbf{F}$ from $\mathbf{F}_0$ to $\mathbf{F}$,

$$\mathbf{F} = (\mathbf{I} + \delta\mathbf{F})\cdot\mathbf{F}_0. \tag{2.9.2}$$

Then the Green strain and the 2nd Piola-Kirchhoff stress become to first order in $\delta\mathbf{F}$,

$$\mathbf{E}_{\mathrm{G}} = \mathbf{E}_{\mathrm{G}0} + \mathbf{F}_0^{\mathrm{T}}\cdot\boldsymbol{\epsilon}\cdot\mathbf{F}_0, \tag{2.9.3}$$

and

$$\underline{\underline{\tau}}(\mathbf{E}_{\mathrm{G}}, T) = \underline{\underline{\tau}}(\mathbf{E}_{\mathrm{G}0}, T_0) + \lambda\,\mathrm{tr}\,(\mathbf{F}_0^{\mathrm{T}}\cdot\boldsymbol{\epsilon}\cdot\mathbf{F}_0)\,\mathbf{I} + 2\,\mu\,\mathbf{F}_0^{\mathrm{T}}\cdot\boldsymbol{\epsilon}\cdot\mathbf{F}_0 - 3\,K\,c_{\mathrm{T}}\,\delta T\,\mathbf{I}, \tag{2.9.4}$$

where $\mathbf{E}_{\mathrm{G}0} = \mathbf{E}_{\mathrm{G}}(\mathbf{F}_0)$ is the Green strain associated with $\mathbf{F}_0$, $\boldsymbol{\epsilon} = \frac{1}{2}\,(\delta\mathbf{F} + \delta\mathbf{F}^{\mathrm{T}})$ is the infinitesimal strain associated with the change in the deformation gradient $\delta\mathbf{F}$, and

$$\underline{\underline{\tau}}(\mathbf{E}_{\mathrm{G}0}, T_0) = \underline{\underline{\tau}}_i + \underline{\underline{\mathbb{E}}} : (\mathbf{E}_{\mathrm{G}0} - c_{\mathrm{T}}\,\Delta T_0\,\mathbf{I}). \tag{2.9.5}$$

The change of the Kirchhoff stress $\boldsymbol{\tau} = \mathbf{F}\cdot\underline{\underline{\tau}}\cdot\mathbf{F}^{\mathrm{T}}$ and Cauchy stress from their values at $(\mathbf{F}_0, T_0)$,

$$\boldsymbol{\tau}(\mathbf{E}_{\mathrm{G}0}, T_0) = \mathbf{F}_0\cdot\underline{\underline{\tau}}(\mathbf{E}_{\mathrm{G}0}, T_0)\cdot\mathbf{F}_0^{\mathrm{T}} = \det\mathbf{F}_0\,\boldsymbol{\sigma}(\mathbf{E}_{\mathrm{G}0}, T_0), \tag{2.9.6}$$

may be estimated to first order in $\delta\mathbf{F}$ and δT as,

$$\begin{aligned} \boldsymbol{\tau} &= \boldsymbol{\tau}(\mathbf{E}_{\mathrm{G}0}, T_0) + \boldsymbol{\tau}(\mathbf{E}_{\mathrm{G}0}, T_0)\cdot\delta\mathbf{F}^{\mathrm{T}} + \delta\mathbf{F}\cdot\boldsymbol{\tau}(\mathbf{E}_{\mathrm{G}0}, T_0) + \mathbb{E}(\mathbf{B}_0) : \boldsymbol{\epsilon} - 3\,K\,c_{\mathrm{T}}\,\delta T\,\mathbf{B}_0, \\ \boldsymbol{\sigma} &= (1 - \mathrm{tr}\,\boldsymbol{\epsilon})\,\boldsymbol{\sigma}(\mathbf{E}_{\mathrm{G}0}) + \boldsymbol{\sigma}(\mathbf{E}_{\mathrm{G}0})\cdot\delta\mathbf{F}^{\mathrm{T}} + \delta\mathbf{F}\cdot\boldsymbol{\sigma}(\mathbf{E}_{\mathrm{G}0}) \\ &\quad + (\det\mathbf{F}_0)^{-1}\left(\mathbb{E}(\mathbf{B}_0) : \boldsymbol{\epsilon} - 3\,K\,c_{\mathrm{T}}\,\delta T\,\mathbf{B}_0\right). \end{aligned} \tag{2.9.7}$$

The tensor of linear elasticity $\mathbb{E}(\mathbf{B}_0)$ is modified by the left Cauchy-Green tensor $\mathbf{B}_0 = \mathbf{F}_0\cdot\mathbf{F}_0^{\mathrm{T}}$,

$$\mathbb{E}(\mathbf{B}_0) = \lambda\,\mathbf{B}_0\otimes\mathbf{B}_0 + 2\,\mu\,\mathbf{B}_0\,\overline{\underline{\otimes}}\,\mathbf{B}_0, \tag{2.9.8}$$

and, with the notation of Section 2.1.2,

$$\mathbb{E}(\mathbf{B}_0) : \boldsymbol{\epsilon} = \lambda\,(\mathbf{B}_0 : \boldsymbol{\epsilon})\,\mathbf{B}_0 + 2\,\mu\,\mathbf{B}_0\cdot\boldsymbol{\epsilon}\cdot\mathbf{B}_0. \tag{2.9.9}$$

Clearly, large pre-deformations $\mathbf{F}_0$ induce an anisotropic thermoelastic response in the current configuration, even if the material response is isotropic in the reference configuration. However, in the special case where

$$\underline{\underline{\tau}}_i = X_i\,\mathbf{I}, \quad \mathbf{F}_0 = F_0\,\mathbf{I}, \tag{2.9.10}$$

the response remains isotropic but the Lamé moduli and the coefficient of thermal expansion are modified,

$$\boldsymbol{\tau} = X\,\mathbf{I} + F_0^4\,\lambda\,\mathrm{tr}\,\boldsymbol{\epsilon}\,\mathbf{I} + 2\,(F_0^4\,\mu + X)\,\boldsymbol{\epsilon} - 3\,K\,c_{\mathrm{T}}\,F_0^2\,\delta T\,\mathbf{I}\,, \tag{2.9.11}$$

where

$$\boldsymbol{\tau}(\mathbf{E}_{\mathrm{G}0}, T_0) = X\,\mathbf{I}\,, \quad X = F_0^2\,X_i + \frac{3}{2}\,K\,(F_0^2 - 1)\,F_0^2 - 3\,K\,c_{\mathrm{T}}\,F_0^2\,\Delta T_0\,. \tag{2.9.12}$$

The linearized response in terms of Cauchy stress $\boldsymbol{\sigma} \sim \boldsymbol{\tau}\,(1 - \mathrm{tr}\,\boldsymbol{\epsilon})/F_0^3$ has the isotropic format,

$$\boldsymbol{\sigma} = \frac{X}{F_0^3}\,\mathbf{I} + \left(F_0\,\lambda - \frac{X}{F_0^3}\right)\,\mathrm{tr}\,\boldsymbol{\epsilon}\,\mathbf{I} + 2\left(F_0\,\mu + \frac{X}{F_0^3}\right)\,\boldsymbol{\epsilon} - 3\,K\,\frac{c_{\mathrm{T}}}{F_0}\,\delta T\,\mathbf{I}\,. \tag{2.9.13}$$

For arbitrary anisotropic thermoelasticity, namely $\underline{\underline{\boldsymbol{\tau}}} = \underline{\underline{\boldsymbol{\tau}}}_i + \mathbb{E} : (\mathbf{E}_{\mathrm{G}} - \mathbf{c}_{\mathrm{T}}\,\Delta T)$, the linearized expression of the Kirchhoff stress becomes,

$$\boldsymbol{\tau} = \boldsymbol{\tau}(\mathbf{E}_{\mathrm{G}0}, T_0) + \boldsymbol{\tau}(\mathbf{E}_{\mathrm{G}0}, T_0)\cdot\delta\mathbf{F}^{\mathrm{T}} + \delta\mathbf{F}\cdot\boldsymbol{\tau}(\mathbf{E}_{\mathrm{G}0}, T_0) + \mathbb{E}(\mathbf{F}_0) : \boldsymbol{\epsilon} - \mathbb{E}_T(\mathbf{F}_0) : \mathbf{c}_{\mathrm{T}}\,\delta T\,, \tag{2.9.14}$$

with the modified thermoelastic moduli,

$$\left(\mathbb{E}(\mathbf{F}_0)\right)_{ijkl} = F_{0im}\,F_{0jn}\,F_{0kp}\,F_{0lq}\,\mathbb{E}_{mnpq}\,, \quad \left(\mathbb{E}_T(\mathbf{F}_0)\right)_{ijpq} = F_{0im}\,F_{0jn}\,\mathbb{E}_{mnpq}\,. \tag{2.9.15}$$

2.10 Kinematical constraints and reaction stresses

Some materials may not undergo arbitrary strains and some equality constraints $f(\mathbf{E}_{\mathrm{G}}) = 0$ or some inequality constraints $f(\mathbf{E}_{\mathrm{G}}) \geq 0$ may have to be to satisfied. These restrictions influence the format of the constitutive equations. Only equality constraints are addressed below.

The kinematical constraints are assumed to hold for any thermomechanical processes. For example we accept that a body which does not change volume under arbitrary mechanical loads does not undergo thermal expansion. And similarly for the inextensibility condition.

Let $\mathbf{L} = (d\mathbf{F}/dt)\cdot\mathbf{F}^{-1}$ be the velocity gradient and $\mathbf{D} = (\mathbf{L} + \mathbf{L}^{\mathrm{T}})/2$ the rate of deformation. Let the free energy be independent of $\mathbf{D}$ as for an elastic material. Typically, the stress is obtained through the dissipation inequality,

$$\boldsymbol{\sigma} : \mathbf{D} - \frac{dE}{dt} + \cdots \geq 0\,, \tag{2.10.1}$$

by collecting the coefficients of the components of rate of deformation $\mathbf{D}$. If the six components of $\mathbf{D}$ are independent, six relations will be obtained for the six components of the symmetric stress tensor, by giving arbitrary values to the components of $\mathbf{D}$ and still requiring the inequality to be satisfied (so-called Coleman-Noll device). Therefore the stress is completely determined by the free energy.

Such is not the case if some components of $\mathbf{D}$ are subjected to one or several constraints. In that case, the standard procedure is to split the stress in two components, the reaction stress $\mathbf{r}$ and the effective stress $\boldsymbol{\sigma}'$ [Spencer 1972],

$$\boldsymbol{\sigma} = \mathbf{r} + \boldsymbol{\sigma}'\,. \tag{2.10.2}$$

The reaction stress does not work in a deformation that satisfies the constraint. As a first example, let us consider incompressibility, namely $\mathbf{I} : \mathbf{D} = 0$. The candidate reaction stress is $-p\,\mathbf{I}$, where p is to be determined, not by constitutive equations, but by satisfaction of field and boundary conditions in a boundary value problem[2.8]. The total stress may be further recast by requiring, without loss of generality[2.9] the effective stress to be orthogonal to the reaction stress [Spencer 1972],

$$\boldsymbol{\sigma} = \mathbf{r} + \boldsymbol{\sigma}'\,, \quad \mathbf{r} = -p\,\mathbf{I}\,, \quad \mathbf{I} : \boldsymbol{\sigma}' = 0\,. \tag{2.10.3}$$

[2.8] Several examples are worked out, e.g., in Sections 7.5, 21.7.1.2 and 24.8.1.

[2.9] In practice, however, this orthogonality condition might not be applied to ease the interpretation of the Lagrange multiplier. For example, for an elastic porous solid saturated by a fluid, a constraint arises when both the solid and fluid constituents are incompressible. The reaction stress is interpreted as the fluid pressure, say $\mathbf{r} = -p\,\mathbf{I}$, while the effective stress derives from the skeleton strain through the drained properties of the solid skeleton which includes both a shear component and an isotropic component, see Section 7.3.6. For an elementary volume at equilibrium in a bath at pressure p_{B}, the total stress $\boldsymbol{\sigma}$ is equal to $-p_{\mathrm{B}}\,\mathbf{I}$, the fluid pressure p inside the porous solid is equal to the external bath pressure since there is no meniscus wrapping the porous solid, and the effective stress vanishes.

Therefore only five components of the effective stress should be provided by constitutive equations.

A material which is inextensible in the current direction $\mathbf{a}$, namely $\mathbf{a} \cdot \mathbf{D} \cdot \mathbf{a} = 0$, is a second example. The associated reaction stress is $\mathbf{r} = T\,\mathbf{a} \otimes \mathbf{a}$, where T is to be defined in a boundary value problem. Again the effective stress and reaction stress may be redefined so as to be orthogonal,

$$\boldsymbol{\sigma} = \mathbf{r} + \boldsymbol{\sigma}', \quad \mathbf{r} = T\,\mathbf{a} \otimes \mathbf{a}, \quad (\mathbf{a} \otimes \mathbf{a}) : \boldsymbol{\sigma}' = 0. \tag{2.10.4}$$

If the material is *both* incompressible and inextensible along the direction $\mathbf{a}$, then,

$$\boldsymbol{\sigma} = \mathbf{r} + \boldsymbol{\sigma}', \quad \mathbf{r} = -p\,\mathbf{I} + T\,\mathbf{a} \otimes \mathbf{a}, \quad \mathbf{I} : \boldsymbol{\sigma}' = 0, \quad (\mathbf{a} \otimes \mathbf{a}) : \boldsymbol{\sigma}' = 0. \tag{2.10.5}$$

As expected four components of the effective stress are independent and are to be defined by constitutive equations.

Let $\mathbf{E}_{\mathrm{G}} = (\mathbf{F}^{\mathrm{T}} \cdot \mathbf{F} - \mathbf{I})/2$ be the Green strain. In view of the relation linking the Kirchhoff stress $\boldsymbol{\tau} = \det \mathbf{F}\,\boldsymbol{\sigma}$ and the 2nd Piola-Kirchhoff stress $\underline{\underline{\tau}} = \mathbf{F}^{-1} \cdot (\det \mathbf{F}\,\boldsymbol{\sigma}) \cdot \mathbf{F}^{-\mathrm{T}}$, one may explore the stress split in the reference configurations implied by the stress split in the current configuration. For example, the incompressibility condition $\det \mathbf{F} = 1$ expresses in rate form in terms of the Green strain via (2.7.3) and in terms of the rate of deformation via (2.5.14),

$$\frac{d}{dt} \det \mathbf{F} = (\det \mathbf{F}\,\mathbf{F}^{-1} \cdot \mathbf{F}^{-\mathrm{T}}) : \frac{d\mathbf{E}_{\mathrm{G}}}{dt} = \det \mathbf{F}\,\mathbf{I} : \mathbf{D}, \tag{2.10.6}$$

and similarly for the inextensibility condition $\mathbf{a}_0 \cdot \mathbf{E}_{\mathrm{G}} \cdot \mathbf{a}_0$ constant, with help of (2.5.14),

$$\frac{d}{dt} (\mathbf{a}_0 \cdot \mathbf{E}_{\mathrm{G}} \cdot \mathbf{a}_0) = \mathbf{a}_0 \cdot (\mathbf{F}^{\mathrm{T}} \cdot \mathbf{D} \cdot \mathbf{F}) \cdot \mathbf{a}_0 = \mathbf{a} \cdot \mathbf{D} \cdot \mathbf{a}. \tag{2.10.7}$$

Note that the constraint has to be objective and it does not include temperature: mechanical incompressibility being understood to imply zero thermal expansion as discussed in the case of fluids in Section 3.3.2.

For the incompressibility condition, the 2nd Piola-Kirchhoff stress split writes,

$$\underline{\underline{\tau}} = -p \det \mathbf{F}\,\mathbf{F}^{-1} \cdot \mathbf{F}^{-\mathrm{T}} + \underline{\underline{\tau}}', \quad (\mathbf{F}^{\mathrm{T}} \cdot \mathbf{F}) : \underline{\underline{\tau}}' = 0. \tag{2.10.8}$$

If $\mathbf{a}_0 \sim \mathbf{F}^{-1} \cdot \mathbf{a}$ is the reference direction of inextensibility, the split of the 2nd Piola-Kirchhoff stress becomes

$$\underline{\underline{\tau}} = T \det \mathbf{F}\,\mathbf{a}_0 \otimes \mathbf{a}_0 + \underline{\underline{\tau}}', \quad (\mathbf{F}^{\mathrm{T}} \cdot \mathbf{F} \cdot \mathbf{a}_0 \otimes \mathbf{F}^{\mathrm{T}} \cdot \mathbf{F} \cdot \mathbf{a}_0) : \underline{\underline{\tau}}' = 0. \tag{2.10.9}$$

Quite generally, with help of (2.5.14), the constraint $f(\mathbf{E}_{\mathrm{G}}) = 0$ may be expressed in terms of Lagrangian and Eulerian forms,

$$\frac{df}{dt} = \frac{\partial f}{\partial \mathbf{E}_{\mathrm{G}}} : \frac{d\mathbf{E}_{\mathrm{G}}}{dt} = \left(\mathbf{F} \cdot \frac{\partial f}{\partial \mathbf{E}_{\mathrm{G}}} \cdot \mathbf{F}^{\mathrm{T}} \right) : \mathbf{D} = 0, \tag{2.10.10}$$

with the stress splits,

$$\begin{aligned} \boldsymbol{\sigma} &= -p\,\mathbf{F} \cdot \frac{\partial f}{\partial \mathbf{E}_{\mathrm{G}}} \cdot \mathbf{F}^{\mathrm{T}} + \boldsymbol{\sigma}', \quad \left(\mathbf{F} \cdot \frac{\partial f}{\partial \mathbf{E}_{\mathrm{G}}} \cdot \mathbf{F}^{\mathrm{T}} \right) : \boldsymbol{\sigma}' = 0, \\ \underline{\underline{\tau}} &= -p \det \mathbf{F}\,\frac{\partial f}{\partial \mathbf{E}_{\mathrm{G}}} + \underline{\underline{\tau}}', \quad \left(\mathbf{F}^{\mathrm{T}} \cdot \mathbf{F} \cdot \frac{\partial f}{\partial \mathbf{E}_{\mathrm{G}}} \cdot \mathbf{F}^{\mathrm{T}} \cdot \mathbf{F} \right) : \underline{\underline{\tau}}' = 0. \end{aligned} \tag{2.10.11}$$

The power relation,

$$\boldsymbol{\tau} : \mathbf{D} = \underline{\underline{\tau}} : \frac{d\mathbf{E}_{\mathrm{G}}}{dt} \left(= \underline{\underline{\tau}} : \frac{d\mathbf{F}}{dt} \right), \tag{2.10.12}$$

shows that neither the Lagrangian nor the Eulerian reaction stress does work if the kinematical constraint is satisfied. Here the orthogonality between the reaction stress and the effective stress has been postulated in the current configuration. Alternatively, the orthogonality condition may be required in the reference configuration: it does not hold simultaneously in the two configurations.

2.11 Invariance, objectivity, isotropy

The implications on the form of the thermomechanical constitutive equations of a change of reference configuration or of current coordinate axes are explored successively. Additional issues arise when the thermomechanical constitutive equations are formulated in terms of rates.

2.11.1 Tensor invariance

Constitutive equations should be invariant under changes of coordinate axes which are independent of time. This requirement is addressed by using tensors and tensor rules. Equivalently, all tensor quantities may be subjected to an orthogonal transformation.

Moreover, Lagrangian constitutive equations should be form invariant under changes of the *reference* configuration. This invariance does not preclude the fact that certain symmetries may be displayed in particular reference configurations. The consequences for elastic-growing material with directors are addressed in Section 23.4.2.

As an example, let us consider the thermoelastic solid $\underline{\tau} = \underline{\underline{\mathcal{F}}}(\mathbf{E}_\mathrm{G}, T)$ for which the 2nd Piola-Kirchhoff stress $\underline{\tau}$ is a one to one function of the Green strain $\mathbf{E}_\mathrm{G}$, at given temperature T. The stress and strain are assumed to refer to the same configuration noted 1. A new configuration 2 is defined by the deformation $\mathbf{F}_2 \cdot \mathbf{R} = \mathbf{F}_1$ with $\mathbf{R}$ a rotation. For any $\mathbf{R}$, the constitutive function should satisfy the relation,

$$\underline{\underline{\tau}}_2 = \mathbf{R} \cdot \underline{\tau}_1 \cdot \mathbf{R}^\mathrm{T} = \underline{\underline{\mathcal{F}}}(\mathbf{R} \cdot \mathbf{E}_{\mathrm{G}1} \cdot \mathbf{R}^\mathrm{T}, T) = \mathbf{R} \cdot \underline{\underline{\mathcal{F}}}(\mathbf{E}_{\mathrm{G}1}, T) \cdot \mathbf{R}^\mathrm{T}. \tag{2.11.1}$$

Thus the constitutive responses in the two configurations deduce as,

$$\underline{\underline{\mathcal{F}}}_2(\mathbf{E}_{\mathrm{G}2}, T) = \mathbf{R} \cdot \underline{\underline{\mathcal{F}}}_1(\underbrace{\mathbf{R}^\mathrm{T} \cdot \mathbf{E}_{\mathrm{G}2} \cdot \mathbf{R}}_{\mathbf{E}_{\mathrm{G}1}}, T) \cdot \mathbf{R}^\mathrm{T}. \tag{2.11.2}$$

The ensuing relations for the Cauchy stress response $\det \mathbf{F}\,\boldsymbol{\sigma} = \mathbf{F} \cdot \underline{\tau} \cdot \mathbf{F}^\mathrm{T}$

$$
\begin{aligned}
\det \mathbf{F}_2\,\boldsymbol{\sigma}_2 \;&=\; \mathbf{F}_2 \cdot \underline{\underline{\mathcal{F}}}_2(\mathbf{E}_{\mathrm{G}2}, T) \cdot \mathbf{F}_2^\mathrm{T} \;\overset{(2.11.2)}{=}\; \mathbf{F}_2 \cdot \mathbf{R} \cdot \underline{\underline{\mathcal{F}}}_1(\mathbf{E}_{\mathrm{G}1}, T) \cdot \mathbf{R}^\mathrm{T} \cdot \mathbf{F}_2^\mathrm{T} \\[2mm]
&=\; \mathbf{F}_1 \cdot \underline{\underline{\mathcal{F}}}_1(\mathbf{E}_{\mathrm{G}1}, T) \cdot \mathbf{F}_1^\mathrm{T} \\[2mm]
&=\; \det \mathbf{F}_2\,\mathcal{F}_2(\mathbf{F}_2, T) \;=\; \det \mathbf{F}_1\,\mathcal{F}_1(\mathbf{F}_1, T).
\end{aligned}
\tag{2.11.3}
$$

imply, with $\det \mathbf{R} = 1$, the invariance of the constitutive response in the *current* configuration,

$$\mathcal{F}_2(\mathbf{F}_2, T) = \mathcal{F}_1(\mathbf{F}_1, T). \tag{2.11.4}$$

Usually the reference configuration is a position that the body can actually possess, and the working axes in the reference and current configurations are both right handed. Then the current configuration deduces from the reference configuration by a deformation gradient $\mathbf{F}$ with $\det \mathbf{F} > 0$ and the transformation is said *proper*. If either the current configuration is deduced from the one mentioned above by a symmetry, or if the working axes in the reference and current configurations are, respectively, right-handed and left-handed, or conversely, then $\det \mathbf{F} < 0$, and the transformation is said *improper*.

2.11.2 Material frame indifference or material objectivity

The reference configuration being held fixed, we now question the effects of a time changing frame in the current configurations.

2.11.2.1 Objective constitutive equations

Consider a body undergoing some loading process. In one case, the body is at rest with respect to the fixed laboratory axes. In another case, it undergoes some time dependent rotation. The underlying idea is that the nature of the *contact forces* does not change. Constitutive equations should reflect this property. In other words, even if they are moving with respect to one another, two observers should record the same material response, e.g., the same stress, in a body. On the other hand, Newton's second law does not satisfy this frame indifference principle without recourse to inertia forces.

While the frame in the initial configuration is assumed to be common to the two observers, the rates of change of the current frames may be different. As a particular case, the current frames may be assumed to coincide at any time t, but not their rates. The question of frame indifference does not arise if the constitutive equations are phrased in terms of Lagrange variables, e.g., for a thermoelastic solid $\underline{\tau} = \underline{\underline{\mathcal{F}}}(\mathbf{E}_\mathrm{G}, T)$ for which the 2nd Piola-Kirchhoff stress $\underline{\tau}$ is a function of the Green strain $\mathbf{E}_\mathrm{G}$ and temperature T. A Lagrangian tensor is identical to the two observers unlike an Eulerian tensor. Consequently, a constitutive equation involving a Lagrangian entity and an Eulerian entity will not be objective.

Consider two observers moving with the frames $\mathbf{x}(t)$ and $\mathbf{x}_*(t)$,

$$\mathbf{x}_*(t) = \mathbf{R}(t) \cdot \mathbf{x}(t) + \mathbf{t}(t) \,, \tag{2.11.5}$$

related by a rotation $\mathbf{R}(t)$ and a translation $\mathbf{t}(t)$. Think of two runners, possibly starting from the same point, and heading along different directions.

The representations of tensor quantities in the two frames express following tensor rules, e.g., $\boldsymbol{\sigma}_* = \mathbf{R} \cdot \boldsymbol{\sigma} \cdot \mathbf{R}^{\mathrm{T}}$ for the stress matrix, $\mathbf{Q}_* = \mathbf{R} \cdot \mathbf{Q}$ for the heat flux vector, see Exercise 2.1 for details. Gradients wrt the two frames require some derivations (the right gradient of a vector $\mathbf{v}\overset{\leftarrow}{\nabla}$ is defined componentwise as $\partial v_i/\partial x_j$, $\overset{\leftarrow}{\nabla}_*$ denotes the gradient wrt the coordinate $\mathbf{x}_*$, and $\underaccent{\leftarrow}{\nabla}$ the gradient wrt the coordinate $\underline{\mathbf{x}}$), namely

$$\mathbf{x}_*(t) = \mathbf{R}(t) \cdot \mathbf{x}(t) + \mathbf{t}(t), \quad \mathbf{v}_*(t) = \mathbf{R}(t) \cdot \mathbf{v}(t) + \frac{d\mathbf{R}}{dt}(t) \cdot \mathbf{x}(t) + \frac{d\mathbf{t}(t)}{dt} \,;$$
$$\mathbf{v}_*\overset{\leftarrow}{\underaccent{\leftarrow}{\nabla}} = \mathbf{R} \cdot (\mathbf{v}\overset{\leftarrow}{\nabla}) + \frac{d\mathbf{R}}{dt}, \quad \mathbf{v}_*\overset{\leftarrow}{\nabla}_* = \mathbf{v}_*\overset{\leftarrow}{\nabla} \cdot \mathbf{R}^{\mathrm{T}} = \mathbf{R} \cdot (\mathbf{v}\overset{\leftarrow}{\nabla}) \cdot \mathbf{R}^{\mathrm{T}} + \frac{d\mathbf{R}}{dt} \cdot \mathbf{R}^{\mathrm{T}} \,. \tag{2.11.6}$$

Consider a fluid for which the stress $\boldsymbol{\sigma} = \mathcal{F}(\mathbf{v}\overset{\leftarrow}{\nabla}, \rho, T)$ is a function of the velocity gradient $\mathbf{v}\overset{\leftarrow}{\nabla}$, mass density ρ and temperature T. Thus the matrix of the stress of the second observer $\boldsymbol{\sigma}_* = \mathbf{R} \cdot \boldsymbol{\sigma} \cdot \mathbf{R}^{\mathrm{T}}$ expresses as,

$$\boldsymbol{\sigma}_* = \mathbf{R} \cdot \mathcal{F}(\mathbf{v}\overset{\leftarrow}{\nabla}, \rho, T) \cdot \mathbf{R}^{\mathrm{T}} = \mathcal{F}(\mathbf{R} \cdot (\mathbf{v}\overset{\leftarrow}{\nabla}) \cdot \mathbf{R}^{\mathrm{T}} + \frac{d\mathbf{R}}{dt} \cdot \mathbf{R}^{\mathrm{T}}, \rho, T) \,. \tag{2.11.7}$$

Since the rate of the rotation is arbitrary, it can be chosen so as to eliminate the skew-symmetric part $\mathbf{W}$ of the velocity gradient $\mathbf{L} \equiv \mathbf{v}\overset{\leftarrow}{\nabla} = \mathbf{D} + \mathbf{W}$, namely such that $\mathbf{R} \cdot \mathbf{W} \cdot \mathbf{R}^{\mathrm{T}} + (d\mathbf{R}/dt) \cdot \mathbf{R}^{\mathrm{T}} = \mathbf{0}$. Therefore, the constitutive equation should have the format $\boldsymbol{\sigma} = \mathcal{F}(\mathbf{D}, \rho, T)$ and the constitutive function $\mathcal{F}$ should satisfy the constraint,

$$\mathbf{R} \cdot \mathcal{F}(\mathbf{D}, \rho, T) \cdot \mathbf{R}^{\mathrm{T}} = \mathcal{F}(\mathbf{R} \cdot \mathbf{D} \cdot \mathbf{R}^{\mathrm{T}}, \rho, T) \,, \tag{2.11.8}$$

for any rotation $\mathbf{R}$. Actually, if the property holds for $\mathbf{R}$, it does for $-\mathbf{R}$. Then the condition is to be satisfied for any orthogonal transformation $\mathbf{R} \in \mathcal{O}(3)$: the function $\mathcal{F}$ is said to be isotropic with respect to its first argument.

Perhaps, assuming the current frames at any time t to be identical for the two observers, i.e., $\mathbf{R}(t) = \mathbf{I}$ while $d\mathbf{R}/dt \neq \mathbf{0}$, provides some further hint why the constitutive relation should be phrased in terms of the rate of deformation and not in terms of the velocity gradient. Indeed then the stress matrices $\boldsymbol{\sigma}$ and $\boldsymbol{\sigma}_*$ are identical for the two observers and, according to (2.11.7), $\mathcal{F}(\mathbf{D} + \mathbf{W}, \rho, T) = \mathcal{F}(\mathbf{D} + \mathbf{W} + d\mathbf{R}/dt, \rho, T)$. The equality is satisfied by symmetrizing the first argument of $\mathcal{F}$.

For a Newtonian fluid, the stress depends linearly on the rate of deformation $\mathbf{D}$. If the constitutive function $\mathcal{F}$ has no other tensor argument than $\mathbf{D}$, this dependence expresses via two viscosity coefficients λ_v and μ_v, Section 2.11.3.2. Actually, the stress is contributed by a static pressure $p = p(\rho, T)$ and the viscous response, namely $\boldsymbol{\sigma} = -p(\rho, T)\mathbf{I} + \lambda_v \operatorname{tr}\mathbf{D}\,\mathbf{I} + 2\mu_v \mathbf{D}$. Dissipation and invertibility require $3\lambda_v + 2\mu_v > 0$, $\mu_v > 0$. Invertibility does not hold for a Stokes fluid for which $3\lambda_v + 2\mu_v = 0$.

As another example of Eulerian constitutive equation, consider Fourier's heat law expressing the heat flux $\mathbf{Q} = -\mathbf{k}_{\mathrm{T}}(\mathbf{F}) \cdot (T\overset{\leftarrow}{\nabla})$ in terms of temperature gradient $T\overset{\leftarrow}{\nabla}$ and tensor of thermal conductivity $\mathbf{k}_{\mathrm{T}}(\mathbf{F})$ assumed to depend on the deformation gradient $\mathbf{F}$. With help of the relations,

$$\mathbf{F}_* = \mathbf{x}_* \underaccent{\leftarrow}{\nabla} = \mathbf{R} \cdot (\mathbf{x}\underaccent{\leftarrow}{\nabla}) = \mathbf{R} \cdot \mathbf{F}, \quad T\overset{\leftarrow}{\nabla}_* = \mathbf{R} \cdot (T\overset{\leftarrow}{\nabla}) \,, \tag{2.11.9}$$

the equation expresses in the second frame in the format $\mathbf{R} \cdot \mathbf{Q} = -\mathbf{k}_{\mathrm{T}}(\mathbf{R} \cdot \mathbf{F}) \cdot (\mathbf{R} \cdot (T\overset{\leftarrow}{\nabla}))$. Therefore, $\mathbf{k}_{\mathrm{T}}(\mathbf{F}) = \mathbf{R}^{\mathrm{T}} \cdot \mathbf{k}_{\mathrm{T}}(\mathbf{R} \cdot \mathbf{F}) \cdot \mathbf{R}$. If further the rotation $\mathbf{R}$ is taken equal to the transpose of the rotation involved in the polar decomposition of the deformation gradient, namely $\mathbf{F} = \mathbf{R}^{\mathrm{T}} \cdot \mathbf{U}$, then the tensor of thermal conductivity should depend on the deformation gradient through the format,

$$\mathbf{k}_{\mathrm{T}}(\mathbf{F}) = \mathbf{R} \cdot \mathbf{k}_{\mathrm{T}}(\mathbf{U}) \cdot \mathbf{R}^{\mathrm{T}} \,. \tag{2.11.10}$$

Hence, even if the conductivity does not depend on the pure deformation, its dependence wrt the deformation gradient should be accounted for because the orientation of the body is observer dependent.

For an elastic solid, the Cauchy stress is a priori a function of the deformation gradient, and temperature, and (2.11.7) becomes,

$$\boldsymbol{\sigma}_* = \mathbf{R} \cdot \boldsymbol{\mathcal{F}}(\mathbf{F}, T) \cdot \mathbf{R}^{\mathrm{T}} = \boldsymbol{\mathcal{F}}(\mathbf{R} \cdot \mathbf{F}, T), \tag{2.11.11}$$

or $\boldsymbol{\sigma} = \boldsymbol{\mathcal{F}}(\mathbf{F}, T) = \mathbf{R}^{\mathrm{T}} \cdot \boldsymbol{\mathcal{F}}(\mathbf{R} \cdot \mathbf{F}, T) \cdot \mathbf{R}$. Again, with the rotation $\mathbf{R}$ taken equal to the transpose of the rotation involved in the polar decomposition of the deformation gradient, namely $\mathbf{F} = \mathbf{R}^{\mathrm{T}} \cdot \mathbf{U}$, the objectivity requirement implies the constitutive relation to adopt the following reduced formats, namely

$$
\begin{aligned}
\boldsymbol{\sigma} &= \mathbf{R}^{\mathrm{T}} \cdot \boldsymbol{\mathcal{F}}(\mathbf{R} \cdot \mathbf{F}, T) \cdot \mathbf{R} \\
&= \mathbf{R}^{\mathrm{T}} \cdot \boldsymbol{\mathcal{F}}(\mathbf{U}, T) \cdot \mathbf{R} \\
&= \mathbf{F} \cdot \left(\mathbf{U}^{-1} \cdot \boldsymbol{\mathcal{F}}(\mathbf{U}, T) \cdot \mathbf{U}^{-1} \right) \cdot \mathbf{F}^{\mathrm{T}} \\
&= \mathbf{F} \cdot \left(\mathbf{U}^{-1} \cdot \boldsymbol{\mathcal{F}}(\mathbf{U}, T) \cdot \mathbf{U} \right) \cdot \mathbf{F}^{-1} \\
&= \mathbf{F}^{-\mathrm{T}} \cdot \left(\mathbf{U} \cdot \boldsymbol{\mathcal{F}}(\mathbf{U}, T) \cdot \mathbf{U} \right) \cdot \mathbf{F}^{-1} \\
&= \mathbf{F}^{-\mathrm{T}} \cdot \left(\mathbf{U} \cdot \boldsymbol{\mathcal{F}}(\mathbf{U}, T) \cdot \mathbf{U}^{-1} \right) \cdot \mathbf{F}^{\mathrm{T}}.
\end{aligned}
\tag{2.11.12}
$$

Alternatively, starting from a constitutive relation in terms of the 2nd Piola-Kirchhoff stress, the latter should be independent of the change of current frame, i.e. $\underline{\boldsymbol{\tau}} = \underline{\boldsymbol{\mathcal{F}}}(\mathbf{F}, T) = \underline{\boldsymbol{\mathcal{F}}}(\mathbf{R} \cdot \mathbf{F}, T)$, and therefore $\underline{\boldsymbol{\tau}} = \underline{\boldsymbol{\mathcal{F}}}(\mathbf{U}, T)$, which is consistent with the expression of $\det \mathbf{F} \, \mathbf{F}^{-1} \cdot \boldsymbol{\sigma} \cdot \mathbf{F}^{-\mathrm{T}}$ associated with the 3rd line of (2.11.12), namely $\det \mathbf{U} \, \mathbf{U}^{-1} \cdot \boldsymbol{\mathcal{F}}(\mathbf{U}, T) \cdot \mathbf{U}^{-1}$ since $\det \mathbf{F} = \det \mathbf{U}$. The associated requirement for the free energy relative to the reference configuration $\underline{E}(\mathbf{F}, T)$ implies the dependence in $\mathbf{F}$ to be realized via the pure deformation $\mathbf{U}$ or by the right Cauchy-Green tensor $\mathbf{C} = \mathbf{U}^2 = \mathbf{F}^{\mathrm{T}} \cdot \mathbf{F}$ or by the Green strain $\mathbf{E}_{\mathrm{G}}$. Then one can prove the following relation between partial derivatives (with an abuse of notation),

$$\frac{\partial \underline{E}}{\partial \mathbf{F}}(\mathbf{C}, T) = 2 \, \mathbf{F} \cdot \frac{\partial \underline{E}}{\partial \mathbf{C}}(\mathbf{C}, T); \quad \frac{\partial \underline{E}}{\partial \mathbf{F}}(\mathbf{E}_{\mathrm{G}}, T) = \mathbf{F} \cdot \frac{\partial \underline{E}}{\partial \mathbf{E}_{\mathrm{G}}}(\mathbf{E}_{\mathrm{G}}, T). \tag{2.11.13}$$

The proof is simple, namely $d\underline{E} = \partial \underline{E}/\partial \mathbf{E}_{\mathrm{G}} : d\mathbf{E}_{\mathrm{G}} = \partial \underline{E}/\partial \mathbf{E}_{\mathrm{G}} : (d\mathbf{F}^{\mathrm{T}} \cdot \mathbf{F}) = \mathbf{F} \cdot \partial \underline{E}/\partial \mathbf{E}_{\mathrm{G}} : d\mathbf{F}$.

2.11.2.2 Objective entities and time derivatives

Consider a body submitted to the constant stress $\boldsymbol{\sigma}$. The stress associated with the moving frame expresses as $\boldsymbol{\sigma}_* = \mathbf{R} \cdot \boldsymbol{\sigma} \cdot \mathbf{R}^{\mathrm{T}}$. The time derivative of $\boldsymbol{\sigma}_*$ does not vanish: it should not be introduced as such in constitutive equations. Therefore we have to define *objective* time derivatives.

Taking a step backward, let us first observe that the matrix representations of the velocity gradient, its symmetric and skew-symmetric parts in the frame 1, namely $\mathbf{L}$, $\mathbf{D}$ and $\mathbf{W}$, and in the frame 2, namely $\mathbf{L}_*$, $\mathbf{D}_*$ and $\mathbf{W}_*$, relate according to (2.11.6),

$$\mathbf{L}_* = \mathbf{R} \cdot \mathbf{L} \cdot \mathbf{R}^{\mathrm{T}} + \boldsymbol{\Omega}, \quad \mathbf{D}_* = \mathbf{R} \cdot \mathbf{D} \cdot \mathbf{R}^{\mathrm{T}}, \quad \mathbf{W}_* = \mathbf{R} \cdot \mathbf{W} \cdot \mathbf{R}^{\mathrm{T}} + \boldsymbol{\Omega}, \tag{2.11.14}$$

where $\boldsymbol{\Omega} = (d\mathbf{R}/dt) \cdot \mathbf{R}^{\mathrm{T}}$.

Scalars, vectors and tensors are said to be *frame indifferent* or *objective* if they follow tensor rules under changes of frames. Thus the velocity $\mathbf{v}$ and the acceleration are not objective according to (2.11.6), neither are the velocity gradient $\mathbf{L}$ and the spin $\mathbf{W}$ according to (2.11.14), but the rate of deformation $\mathbf{D}$ is. Moreover, the relations (2.11.15) below show implicitly that the time derivatives $d\mathbf{a}/dt$ and $d\mathbf{A}/dt$ of a vector $\mathbf{a}$ or a second order tensor $\mathbf{A}$ are not objective either.

There are several ways to build objective derivatives. A first method introduces *corotational derivatives*. Consider a vector $\mathbf{a}$ and a second order tensor $\mathbf{A}$ which express respectively as $\mathbf{a}_* = \mathbf{R} \cdot \mathbf{a}$ and $\mathbf{A}_* = \mathbf{R} \cdot \mathbf{A} \cdot \mathbf{R}^{\mathrm{T}}$ in the moving frame. Then, upon differentiation of these expressions and re-arrangement of the results, one may display so-called corotational derivatives denoted by the symbol D/Dt to make a distinction with the material time derivative d/dt, namely with $\boldsymbol{\Omega} = (d\mathbf{R}/dt) \cdot \mathbf{R}^{\mathrm{T}}$,

$$
\begin{aligned}
\mathbf{a}_* = \mathbf{R} \cdot \mathbf{a}; && \frac{D\mathbf{a}_*}{Dt} &= \frac{d\mathbf{a}_*}{dt} - \boldsymbol{\Omega} \cdot \mathbf{a}_* &&= \mathbf{R} \cdot \frac{d\mathbf{a}}{dt}; \\
\mathbf{A}_* = \mathbf{R} \cdot \mathbf{A} \cdot \mathbf{R}^{\mathrm{T}}; && \frac{D\mathbf{A}_*}{Dt} &= \frac{d\mathbf{A}_*}{dt} - \boldsymbol{\Omega} \cdot \mathbf{A}_* + \mathbf{A}_* \cdot \boldsymbol{\Omega} &&= \mathbf{R} \cdot \frac{d\mathbf{A}}{dt} \cdot \mathbf{R}^{\mathrm{T}}.
\end{aligned}
\tag{2.11.15}
$$

Note the relations based on the property (2.5.44) for $\mathbf{R}$ an orthogonal tensor,

$$
\begin{aligned}
\mathbf{A}_* &= \mathbf{R} \cdot \mathbf{A} \cdot \mathbf{R}^{\mathrm{T}} \quad \Rightarrow \quad |\mathbf{A}_*| = |\mathbf{A}|, \\
\frac{D\mathbf{A}_*}{Dt} &= \mathbf{R} \cdot \frac{d\mathbf{A}}{dt} \cdot \mathbf{R}^{\mathrm{T}} \quad \Rightarrow \quad \left|\frac{D\mathbf{A}_*}{Dt}\right| = \left|\frac{d\mathbf{A}}{dt}\right|.
\end{aligned}
\tag{2.11.16}
$$

When $\mathbf{R}$ is the rotation involved in the polar decomposition of the deformation gradient, the derivative is called Green-Naghdi rate. Now using $(2.11.14)_3$, one may transform $(2.11.15)_2$ to build another corotational derivative, called Jaumann derivative, namely

$$
\begin{aligned}
\mathbf{A}_* &= \mathbf{R} \cdot \mathbf{A} \cdot \mathbf{R}^{\mathrm{T}}; \\
\frac{D\mathbf{A}_*}{Dt} &= \frac{d\mathbf{A}_*}{dt} - \mathbf{W}_* \cdot \mathbf{A}_* + \mathbf{A}_* \cdot \mathbf{W}_* = \mathbf{R} \cdot \left(\frac{d\mathbf{A}}{dt} - \mathbf{W} \cdot \mathbf{A} + \mathbf{A} \cdot \mathbf{W}\right) \cdot \mathbf{R}^{\mathrm{T}} = \mathbf{R} \cdot \frac{D\mathbf{A}}{Dt} \cdot \mathbf{R}^{\mathrm{T}}.
\end{aligned}
\tag{2.11.17}
$$

The notion of convective derivative provides another means to construct objective derivatives. Indeed, with any second order tensor $\mathbf{A}_*(t)$ may be associated a second order tensor $\mathbf{A}(t)$ relative to the reference configuration through various transfer rules based on the deformation gradient $\mathbf{F}$. For example,

$$
\begin{aligned}
\mathbf{A}_* &= \mathbf{F}^{-\mathrm{T}} \cdot \mathbf{A} \cdot \mathbf{F}^{-1}, & \frac{D\mathbf{A}_*}{Dt} &= \frac{d\mathbf{A}_*}{dt} + \mathbf{A}_* \cdot \mathbf{L} + \mathbf{L}^{\mathrm{T}} \cdot \mathbf{A}_* = \mathbf{F}^{-\mathrm{T}} \cdot \frac{d\mathbf{A}}{dt} \cdot \mathbf{F}^{-1}, \\
\mathbf{A}_* &= \mathbf{F} \cdot \mathbf{A} \cdot \mathbf{F}^{\mathrm{T}}, & \frac{D\mathbf{A}_*}{Dt} &= \frac{d\mathbf{A}_*}{dt} - \mathbf{A}_* \cdot \mathbf{L}^{\mathrm{T}} - \mathbf{L} \cdot \mathbf{A}_* = \mathbf{F} \cdot \frac{d\mathbf{A}}{dt} \cdot \mathbf{F}^{\mathrm{T}},
\end{aligned}
\tag{2.11.18}
$$

where $\mathbf{L} = d\mathbf{F}/dt \cdot \mathbf{F}^{-1}$ is the velocity gradient. These derivatives are referred to as Oldroyd rates. The convective derivative is seen to be the image in the current configuration of the time derivative in the reference configuration.

The procedure applies to vectors as well, e.g.,

$$
\mathbf{a}_* = \mathbf{F} \cdot \mathbf{a}, \quad \frac{D\mathbf{a}_*}{Dt} = \frac{d\mathbf{a}_*}{dt} - \mathbf{L} \cdot \mathbf{a}_* = \mathbf{F} \cdot \frac{d\mathbf{a}}{dt}.
\tag{2.11.19}
$$

The right and left Cauchy-Green tensors or Green strain are not objective: this statement follows from (2.5.14) and the fact that the rate of deformation $\mathbf{D}$ is not the derivative of a strain measure. However, the Rivlin-Ericksen tensors are objective tensors related to the derivative of the right Cauchy-Green tensor. They are defined through the scheme, e.g., Malvern [1969],

$$
\begin{aligned}
d\mathbf{x} \cdot \mathbf{A}^{(0)} \cdot d\mathbf{x} &= d\underline{\mathbf{x}} \cdot \mathbf{C} \cdot d\underline{\mathbf{x}} \\
d\mathbf{x} \cdot \mathbf{A}^{(n)} \cdot d\mathbf{x} &= d\underline{\mathbf{x}} \cdot \frac{d^n \mathbf{C}}{dt^n} \cdot d\underline{\mathbf{x}},
\end{aligned}
\tag{2.11.20}
$$

so that $d^n \mathbf{C}/dt^n = \mathbf{F}^{\mathrm{T}} \cdot \mathbf{A}^{(n)} \cdot \mathbf{F}$, $n = 0, 1, 2 \cdots$ with the convention $\mathbf{A}^{(0)} = \mathbf{I}$ and the recursion formula $\mathbf{A}^{(n+1)} = d\mathbf{A}^{(n)}/dt + \mathbf{L}^{\mathrm{T}} \cdot \mathbf{A}^{(n)} + \mathbf{A}^{(n)} \cdot \mathbf{L}$ deduced by derivation of $(2.11.20)_2$.

2.11.3 Material symmetries

Constitutive equations should embody the properties and in particular the symmetry properties of the body. Indeed, some constitutive functions remain invariant under certain changes of axes $\mathbf{R}$ of the reference configuration κ_1. While the matrix of the deformation gradient changes from $\mathbf{F}_1 = \mathbf{F} \cdot \mathbf{R}$ to $\mathbf{F}$, the constitutive response relative to the configuration κ_1, say $\mathcal{F}_1$, remains unchanged, namely

$$
\mathcal{F}_1(\mathbf{F} \cdot \mathbf{R}, T) = \mathcal{F}_1(\mathbf{F}, T).
\tag{2.11.21}
$$

The set of these changes of axes has the structure of a group, an algebraic structure defined in Section 5.2.3.1, and it is termed the symmetry group of the material relative to the reference configuration κ_1. Indeed, first the dot product can be viewed as an internal composition which is associative: if $\mathbf{R}_1$ and $\mathbf{R}_2$ are two such changes of axes, then $\mathbf{R}_1 \cdot \mathbf{R}_2$ belongs to the set,

$$
\mathcal{F}_1(\mathbf{F} \cdot \mathbf{R}_1 \cdot \mathbf{R}_2, T) = \mathcal{F}_1(\mathbf{F} \cdot \mathbf{R}_1, T) = \mathcal{F}_1(\mathbf{F}, T).
\tag{2.11.22}
$$

Second the identity $\mathbf{I}$ belongs to the set and plays the role of the neutral element. Third and finally, each element $\mathbf{R}$ has an inverse that leaves invariant the constitutive response, i.e., it belongs to the set. Indeed,

$$\mathcal{F}_1(\mathbf{F} \cdot \mathbf{R} \cdot \mathbf{R}^{-1}, T) = \mathcal{F}_1(\mathbf{F}, T) = \mathcal{F}_1(\mathbf{F} \cdot \mathbf{R}, T) \stackrel{\mathbf{F}' \equiv \mathbf{F} \cdot \mathbf{R}}{\Rightarrow} \mathcal{F}_1(\mathbf{F}' \cdot \mathbf{R}^{-1}, T) = \mathcal{F}_1(\mathbf{F}', T). \tag{2.11.23}$$

Consider now the change of reference configuration $\kappa_1 \to \kappa_2$, with $\mathbf{F}_1 = \mathbf{F}_2 \cdot \mathbf{R}$. Then if $\mathbf{R}_1$ belongs to the isotropy group of the material relative to the reference configuration κ_1, the isotropy group relative to the configuration κ_2 is of the form $\mathbf{R}_2 = \mathbf{R} \cdot \mathbf{R}_1 \cdot \mathbf{R}^{-1}$. Indeed,

$$\mathcal{F}_2(\mathbf{F}_2, T) \stackrel{(2.11.4)}{=} \mathcal{F}_1(\mathbf{F}_1, T) \stackrel{(2.11.21)}{=} \mathcal{F}_1(\mathbf{F}_1 \cdot \mathbf{R}_1, T) \stackrel{(2.11.4)}{=} \mathcal{F}_2(\mathbf{F}_1 \cdot \mathbf{R}_1 \cdot \mathbf{R}^{-1}, T) \stackrel{\mathbf{F}_1 = \mathbf{F}_2 \cdot \mathbf{R}}{=} \mathcal{F}_2(\mathbf{F}_2 \cdot \mathbf{R}_2, T). \tag{2.11.24}$$

The proper orthogonal group $\mathcal{SO}(3)$ which includes only rotations and the complete orthogonal group $\mathcal{O}(3)$ which includes in addition symmetries play a prominent role in this context and are defined in Section 5.2.3.1. They are both subgroups of the unimodular group whose elements have a determinant equal to ± 1. A function which is left invariant under the complete orthogonal group $\mathcal{O}(3)$ is termed *isotropic*. It is termed *hemitropic* if it is left invariant under the proper orthogonal group $\mathcal{SO}(3)$ only.

Under the change of axes (2.11.5), the deformation gradient changes from $\mathbf{F}$ to $\mathbf{R} \cdot \mathbf{F}$, and the stress from $\boldsymbol{\sigma} = \boldsymbol{\sigma}(\mathbf{F}, T)$ to $\boldsymbol{\sigma}(\mathbf{R} \cdot \mathbf{F}, T) = \mathbf{R} \cdot \boldsymbol{\sigma}(\mathbf{F}, T) \cdot \mathbf{R}^{\mathrm{T}}$ along tensor rules. Moreover, if $\mathbf{R}$ belongs to the group, so does $\mathbf{R}^{\mathrm{T}}$. Then,

$$\mathbf{R} \cdot \mathcal{F}(\mathbf{F}, T) \cdot \mathbf{R}^{\mathrm{T}} = \mathcal{F}(\mathbf{R} \cdot \mathbf{F}, T) \stackrel{(2.11.21)}{=} \mathcal{F}(\mathbf{R} \cdot \mathbf{F} \cdot \mathbf{R}^{\mathrm{T}}, T). \tag{2.11.25}$$

This relation can be extended regarding the nature of both the constitutive function and its arguments. Consider a scalar-valued function f, a vector-valued function $\mathbf{f}$ and a tensor-valued function $\mathcal{F}$ of the set $(\mathbf{A}, \mathbf{a}, a)$ which involves a second order tensor $\mathbf{A}$, a vector $\mathbf{a}$ and a scalar a. These functions are invariant relative to the symmetry group $\mathcal{S}$ if, for any element $\mathbf{R}$ of $\mathcal{S}$,

$$
\begin{aligned}
\mathbf{R} \cdot \mathcal{F}(\mathbf{A}, \mathbf{a}, a) \cdot \mathbf{R}^{\mathrm{T}} &= \mathcal{F}(\mathbf{R} \cdot \mathbf{A} \cdot \mathbf{R}^{\mathrm{T}}, \mathbf{R} \cdot \mathbf{a}, a); \\
\mathbf{R} \cdot \mathbf{f}(\mathbf{A}, \mathbf{a}, a) &= \mathbf{f}(\mathbf{R} \cdot \mathbf{A} \cdot \mathbf{R}^{\mathrm{T}}, \mathbf{R} \cdot \mathbf{a}, a); \\
f(\mathbf{A}, \mathbf{a}, a) &= f(\mathbf{R} \cdot \mathbf{A} \cdot \mathbf{R}^{\mathrm{T}}, \mathbf{R} \cdot \mathbf{a}, a).
\end{aligned}
\tag{2.11.26}
$$

Isotropy requires the above relations to hold for any orthogonal tensor $\mathbf{R}$, that is, for $\mathcal{S} = \mathcal{O}(3)$. The symmetry group associated with material isotropy with a center of symmetry is the full orthogonal group $\mathcal{O}(3)$ while the symmetry group associated with material isotropy with no center of symmetry (hemitropy) is the proper orthogonal group $\mathcal{SO}(3)$.

2.11.3.1 Scalar-valued isotropic functions

The coefficients of the characteristic equation $\det(\lambda \mathbf{I} - \mathbf{A}) = 0$ are termed principal invariants because they are isotropic functions of $\mathbf{A}$. Indeed, for $\mathbf{R}$ an orthogonal transformation, $\det(\lambda \mathbf{I} - \mathbf{R}^{\mathrm{T}} \cdot \mathbf{A} \cdot \mathbf{R}) = \det(\mathbf{R}^{\mathrm{T}} \cdot (\lambda \mathbf{I} - \mathbf{A}) \cdot \mathbf{R}) = \det(\lambda \mathbf{I} - \mathbf{A}) = 0$. Due to isotropy, the scalar invariants are symmetric functions of the eigenvalues, Eqn (2.2.13). Any symmetric function of the eigenvalues of $\mathbf{A}$ is also a function of its scalar invariants.

Let $\mathbf{A}$ be a second order tensor. A scalar-valued function $f(\mathbf{A})$ is isotropic if, for any orthogonal transformation $\mathbf{R} \in \mathcal{O}(3)$ [2.10],

$$f(\mathbf{A}) = f(\underline{\mathbf{A}} = \mathbf{R} \cdot \mathbf{A} \cdot \mathbf{R}^{\mathrm{T}}). \tag{2.11.27}$$

If the function f is smooth, then differentiation for arbitrary $\mathbf{A}$ and $\mathbf{R}$ yields,

$$
\begin{aligned}
\frac{\partial f(\mathbf{A})}{\partial \mathbf{A}} &= \mathbf{R}^{\mathrm{T}} \cdot \frac{\partial f(\underline{\mathbf{A}})}{\partial \underline{\mathbf{A}}} \cdot \mathbf{R}, \\
\mathbf{A} \cdot \frac{\partial f}{\partial \mathbf{A}^{\mathrm{T}}} + \mathbf{A}^{\mathrm{T}} \cdot \frac{\partial f}{\partial \mathbf{A}} &= \left(\mathbf{A} \cdot \frac{\partial f}{\partial \mathbf{A}^{\mathrm{T}}} + \mathbf{A}^{\mathrm{T}} \cdot \frac{\partial f}{\partial \mathbf{A}} \right)^{\mathrm{T}}.
\end{aligned}
\tag{2.11.28}
$$

[2.10]When it comes to constitutive relations, the fact that the function f depends on the point and on the configuration spills on the definition of isotropy. Linear elastic constitutive equations, which are isotropic when phrased in terms of Lagrangian variables, become anisotropic when expressed in the current configuration, as shown in Section 2.9. This dependence wrt the reference configuration is not conflicting with the fact that Lagrangian constitutive equations keep their structure upon a change of reference configuration.

The last result simplifies if the second order tensor $\mathbf{A}$ is symmetric,

$$\mathbf{A} \cdot \frac{\partial f}{\partial \mathbf{A}} = \frac{\partial f}{\partial \mathbf{A}} \cdot \mathbf{A} \,. \tag{2.11.29}$$

Then, the further following implication results using standard algebra,

$$
\begin{aligned}
\frac{df(\mathbf{A})}{dt} &= \frac{\partial f}{\partial \mathbf{A}} : \frac{d\mathbf{A}}{dt} \\
&= \frac{\partial f}{\partial \mathbf{A}} : \left(\frac{D\mathbf{A}}{Dt} + \boldsymbol{\Omega} \cdot \mathbf{A} - \mathbf{A} \cdot \boldsymbol{\Omega} \right) \\
&= \frac{\partial f}{\partial \mathbf{A}} : \frac{D\mathbf{A}}{Dt} + \boldsymbol{\Omega} : \left(\frac{\partial f}{\partial \mathbf{A}} \cdot \mathbf{A}^{\mathrm{T}} - \mathbf{A}^{\mathrm{T}} \cdot \frac{\partial f}{\partial \mathbf{A}} \right) \\
&= \frac{\partial f}{\partial \mathbf{A}} : \frac{D\mathbf{A}}{Dt} \,,
\end{aligned}
\tag{2.11.30}
$$

where $D\mathbf{A}/Dt$ denotes the corotational derivative (2.11.15),

$$\frac{D\mathbf{A}}{Dt} = \frac{d\mathbf{A}}{dt} - \boldsymbol{\Omega} \cdot \mathbf{A} + \mathbf{A} \cdot \boldsymbol{\Omega} \,. \tag{2.11.31}$$

The properties extend to a scalar function with two or more arguments, e.g., for two symmetric second order tensors,

$$f(\mathbf{A}_1, \mathbf{A}_2) = f(\underline{\mathbf{A}}_1 = \mathbf{R} \cdot \mathbf{A}_1 \cdot \mathbf{R}^{\mathrm{T}}, \underline{\mathbf{A}}_2 = \mathbf{R} \cdot \mathbf{A}_2 \cdot \mathbf{R}^{\mathrm{T}}) \,. \tag{2.11.32}$$

Then,

$$
\begin{aligned}
\frac{\partial f}{\partial \mathbf{A}_i} &= \mathbf{R}^{\mathrm{T}} \cdot \frac{\partial f}{\partial \underline{\mathbf{A}}_i} \cdot \mathbf{R}, \quad i = 1, 2, \\
\sum_{i=1,2} \mathbf{A}_i \cdot \frac{\partial f}{\partial \mathbf{A}_i} &= \sum_{i=1,2} \frac{\partial f}{\partial \mathbf{A}_i} \cdot \mathbf{A}_i \,.
\end{aligned}
\tag{2.11.33}
$$

Relation (2.11.30) again with the corotational derivative (2.11.31) extends in the form,

$$
\begin{aligned}
\frac{df}{dt}(\mathbf{A}_1, \mathbf{A}_2) &= \frac{\partial f}{\partial \mathbf{A}_1} : \frac{d\mathbf{A}_1}{dt} + \frac{\partial f}{\partial \mathbf{A}_2} : \frac{d\mathbf{A}_2}{dt} \\
&= \frac{\partial f}{\partial \mathbf{A}_1} : \frac{D\mathbf{A}_1}{Dt} + \frac{\partial f}{\partial \mathbf{A}_2} : \frac{D\mathbf{A}_2}{Dt} \,.
\end{aligned}
\tag{2.11.34}
$$

2.11.3.2   Tensor-valued isotropic functions

Let $\mathbf{A}$ be a second order tensor. A tensor-valued function $\boldsymbol{\mathcal{F}}(\mathbf{A})$ is isotropic if, for any orthogonal transformation $\mathbf{R} \in \mathcal{O}(3)$,

$$\mathbf{R} \cdot \boldsymbol{\mathcal{F}}(\mathbf{A}) \cdot \mathbf{R}^{\mathrm{T}} = \boldsymbol{\mathcal{F}}(\mathbf{R} \cdot \mathbf{A} \cdot \mathbf{R}^{\mathrm{T}}) \,. \tag{2.11.35}$$

As an example, Exercise 2.9 shows that the square root of a positive (semi-)definite second order tensor is isotropic.

The important following result should be recorded, Mandel [1974], Gurtin [1981]: a second order isotropic function $\boldsymbol{\mathcal{F}}(\mathbf{A}, a)$ of a symmetric second order tensor $\mathbf{A}$ and scalar a expresses in the format,

$$\boldsymbol{\mathcal{F}}(\mathbf{A}, a) = \sum_{i=0}^{m-1} b_i \, \mathbf{A}^i \,, \tag{2.11.36}$$

where m is the number of distinct eigenvalues of $\mathbf{A}$ and the coefficients b depend on the scalar invariants of $\mathbf{A}$ and on a.

A proof in $\mathbb{R}^3$ is detailed in Exercise 2.8 but the result holds for any space dimension n. The sum in (2.11.36) can be extended to $i = n$ but the coefficients are not unique if $m < n$.

As a consequence of (2.11.36), an isotropic function of a scalar a is of the form $\boldsymbol{\mathcal{F}}(a) = b_0 \, \mathbf{I}$. An affine function of a symmetric second order tensor reads $\boldsymbol{\mathcal{F}}(\mathbf{A}) = b_0 \, \mathbf{I} + b_1 \, \mathbf{A}$ with $b_0 = b_{00} + b_{01} \operatorname{tr} \mathbf{A}$ depending on the linear invariant $\operatorname{tr} \mathbf{A}$, and b_{00}, b_{01} and b_1 being constants. Finally it is worth recording the explicit expression (2.11.36) in $\mathbb{R}^3$, namely $\boldsymbol{\mathcal{F}}(\mathbf{A}, a) = b_0 \, \mathbf{I} + b_1 \, \mathbf{A} + b_2 \, \mathbf{A}^2$ with the coefficients b depending on the three scalar invariants of $\mathbf{A}$ and on a.

2.11.3.3   From anisotropy to isotropy via a change of arguments

Available tables of representation concern a priori isotropic functions. A theorem of Liu [1982] shows how to transform anisotropic functions into isotropic functions by augmenting the set of variables and therefore allows to capitalize upon available representation tables.

Let f be a scalar-valued, vector-valued or tensor-valued function of a tensor $\mathbf{A}$ and vector $\mathbf{a}$ which is invariant under the group $\mathcal{S}$ as detailed by (2.11.26). Consider a material that is characterized by preferred directions represented by the unit vector $\mathbf{m}$ and the tensor $\mathbf{M}$ and let $s = \{\mathbf{R} \in \mathcal{S}, \ \mathbf{R} \cdot \mathbf{m} = \mathbf{m}; \ \mathbf{R} \cdot \mathbf{M} \cdot \mathbf{R}^{\mathrm{T}} = \mathbf{M}\}$.

Liu's theorem asserts that the function $f(\mathbf{A}, \mathbf{a})$ is invariant relative to s iff it can be represented by the function $F(\mathbf{A}, \mathbf{a}, \mathbf{M}, \mathbf{m})$ invariant relative to $\mathcal{S}$.

The theorem will be applied to two particularly useful symmetries, namely transverse isotropy and orthotropy [2.11].

First, consider the *transversely isotropic groups*

$$s_{\mathrm{trans}'} = \{\mathbf{R} \in \mathcal{O}_3, \ \mathbf{R} \cdot \mathbf{m} = \mathbf{m}\}, \tag{2.11.37}$$

and

$$s_{\mathrm{trans}''} = \{\mathbf{R} \in \mathcal{O}_3, \ \mathbf{R} \cdot \mathbf{m} = \mathbf{m} \text{ or } \mathbf{R} \cdot \mathbf{m} = -\mathbf{m}\}. \tag{2.11.38}$$

Then the function $f(\mathbf{A}, \mathbf{a})$ which is invariant under $s_{\mathrm{trans}'}$ can be represented by the isotropic function $F(\mathbf{A}, \mathbf{a}, \mathbf{m})$, and the function $f(\mathbf{A}, \mathbf{a})$ which is invariant under $s_{\mathrm{trans}''}$ can be represented by the isotropic function $F(\mathbf{A}, \mathbf{a}, \mathbf{M} = \mathbf{m} \otimes \mathbf{m})$.

Next, consider the orthotropic symmetry group

$$s_{\mathrm{ortho}} = \{\mathbf{R} \in \mathcal{O}_3, \ \mathbf{R} \cdot \mathbf{m}_i = \mathbf{m}_i \text{ or } \mathbf{R} \cdot \mathbf{m}_i = -\mathbf{m}_i, \ i \in [1,3]\}, \tag{2.11.39}$$

where the $\mathbf{m}$'s constitute a cartesian triad. Then the function $f(\mathbf{A}, \mathbf{a})$ which is invariant under s_{ortho} can be represented by the isotropic function $F(\mathbf{A}, \mathbf{a}, \mathbf{M}_1 = \mathbf{m}_1 \otimes \mathbf{m}_1, \mathbf{M}_2 = \mathbf{m}_2 \otimes \mathbf{m}_2)$.

It is worth to come back to transverse isotropy which is characterized by a preferred direction $\mathbf{m}$. In fact, transverse isotropy covers five groups of symmetry which can be defined either in extenso [Spencer 1971, p. 251], or by their properties [Liu 1982]. They are listed in Table 2.11.1 from the smallest s_1 to the largest s_5 in terms of the rotation $\mathbf{R}_\theta$ of angle θ about $\mathbf{m}$ and *reflections* $\mathbf{R}_i = \mathbf{I} - 2\,\mathbf{m}_i \otimes \mathbf{m}_i$ about planes of normal $\mathbf{m}_i$, $i \in [1,3]$, which can the expressed componentwise in the cartesian triad $(\mathbf{m}_1, \mathbf{m}_2, \mathbf{m}_3 = \mathbf{m})$,

$$\mathbf{R}_\theta = \begin{bmatrix} \cos\theta & \sin\theta & 0 \\ -\sin\theta & \cos\theta & 0 \\ 0 & 0 & 1 \end{bmatrix}, \quad \mathbf{R}_1 = \begin{bmatrix} -1 & 0 & 0 \\ 0 & 1 & 0 \\ 0 & 0 & 1 \end{bmatrix}, \quad \mathbf{R}_2 = \begin{bmatrix} 1 & 0 & 0 \\ 0 & -1 & 0 \\ 0 & 0 & 1 \end{bmatrix}, \quad \mathbf{R}_3 = \begin{bmatrix} 1 & 0 & 0 \\ 0 & 1 & 0 \\ 0 & 0 & -1 \end{bmatrix}. \tag{2.11.40}$$

Table 2.11.1: The five subgroups of transverse isotropy about the vector $\mathbf{m}$ [Spencer 1971, Liu 1982]. In the definition of the group s_3, the tensor $\mathbf{M}^- = \mathbf{m}_1 \otimes \mathbf{m}_2 - \mathbf{m}_2 \otimes \mathbf{m}_1$ is the skew-symmetric tensor associated with the vector $\mathbf{m}$, Eqn (2.1.11).

Definition of the symmetry group	Elements of the symmetry group
$s_1 = \{\mathbf{R} \in \mathcal{SO}(3), \ \mathbf{R} \cdot \mathbf{m} = \mathbf{m}\}$	$\mathbf{I}, \mathbf{R}_\theta$
$s_2 = \{\mathbf{R} \in \mathcal{O}(3), \ \mathbf{R} \cdot \mathbf{m} = \mathbf{m}\}$	$\mathbf{I}, \mathbf{R}_\theta, \mathbf{R}_1$
$s_3 = \{\mathbf{R} \in \mathcal{O}(3), \ \mathbf{R} \cdot \mathbf{M}^- \cdot \mathbf{R}^{\mathrm{T}} = \mathbf{M}^-\}$	$\mathbf{I}, \mathbf{R}_\theta, \mathbf{R}_3$
$s_4 = \{\mathbf{R} \in \mathcal{SO}(3), \ \mathbf{R} \cdot (\mathbf{m} \otimes \mathbf{m}) \cdot \mathbf{R}^{\mathrm{T}} = \mathbf{m} \otimes \mathbf{m}\}$	$\mathbf{I}, \mathbf{R}_\theta, -\mathbf{R}_2 = \mathbf{R}_1 \cdot \mathbf{R}_3$
$s_5 = \{\mathbf{R} \in \mathcal{O}(3), \ \mathbf{R} \cdot (\mathbf{m} \otimes \mathbf{m}) \cdot \mathbf{R}^{\mathrm{T}} = \mathbf{m} \otimes \mathbf{m}\}$	$\mathbf{I}, \mathbf{R}_\theta, \mathbf{R}_1, \mathbf{R}_3, -\mathbf{R}_2 = \mathbf{R}_1 \cdot \mathbf{R}_3$

Both laminates and materials with woven fabrics are reinforced by fibers. In laminates, the sense of the fibers is elusive while it matters for woven fabrics. In other words, laminates exhibit material symmetry with respect to planes orthogonal to the fiber directions. The property does not hold for certain woven fabrics. The mechanical implications of the presence of arrows on fibers are touched upon for lamellar tissues in Sections 19.19.2.1 and 19.19.2.5.

[2.11] In the following text, we assume $\mathcal{S} = \mathcal{O}_3$ and invoke isotropic functions. For $\mathcal{S} = \mathcal{SO}_3$, we would invoke hemitropic functions.

2.11.3.4 Representation theorems for scalar-valued isotropic functions

Scalar and tensor invariant generators of sets of vectors and second order symmetric and skew-symmetric tensors have been elaborated in the 1960s and 1970s, [Truesdell and Noll 1965; Smith 1970; Wang 1970; Smith 1971; Spencer 1971; Boehler 1978, 1987]. The sets of generators were also required to be irreducible so as to constitute bases. A distinction should be made between *functional* bases and *integrity* bases. Integrity bases assume the constitutive functions to be polynomial invariants of the set of variables. At variance, the lists displayed here are functional bases.

A list of irreducible isotropic scalar invariants of a few vectors and tensors is displayed in Table 2.11.2. To compact the representation, only the elements specific to each line have been listed. For example, the complete list of scalar invariants of the set $(\mathbf{A}, \mathbf{a})$ consists of the invariants pertaining to $\mathbf{A}$, in the invariants pertaining to $\mathbf{a}$ and in the mixed invariants.

Table 2.11.2: Irreducible and complete list of isotropic scalar invariants of symmetric second order tensors $\mathbf{A}$, of a skew-symmetric second order tensor $\mathbf{W}$, vectors $\mathbf{a}$, and of some combinations of these elements [Wang 1970].

Variables	Invariants
$\mathbf{A}$	$\operatorname{tr}\mathbf{A}$, $\operatorname{tr}\mathbf{A}^2$, $\operatorname{tr}\mathbf{A}^3$
$\mathbf{W}$	$\operatorname{tr}\mathbf{W}^2$
$\mathbf{a}$	$\operatorname{tr}\mathbf{a}\otimes\mathbf{a} = \mathbf{a}\cdot\mathbf{a}$
$\mathbf{a}_1, \mathbf{a}_2$	$\mathbf{a}_1\cdot\mathbf{a}_2$
$\mathbf{A}, \mathbf{a}$	$\mathbf{a}\cdot\mathbf{A}\cdot\mathbf{a}$, $\mathbf{a}\cdot\mathbf{A}^2\cdot\mathbf{a}$
$\mathbf{A}_1, \mathbf{A}_2$	$\operatorname{tr}\mathbf{A}_1\cdot\mathbf{A}_2$, $\operatorname{tr}\mathbf{A}_1^2\cdot\mathbf{A}_2$, $\operatorname{tr}\mathbf{A}_1\cdot\mathbf{A}_2^2$, $\operatorname{tr}\mathbf{A}_1^2\cdot\mathbf{A}_2^2$
$\mathbf{A}_1, \mathbf{A}_2, \mathbf{A}_3$	$\operatorname{tr}\mathbf{A}_1\cdot\mathbf{A}_2\cdot\mathbf{A}_3$

Table 2.11.3 displays two equivalent irreducible representations of orthotropic scalar invariants of the symmetric second order tensor $\mathbf{A}$. The preferred directions of the material form a cartesian triad $(\mathbf{m}_1, \mathbf{m}_2, \mathbf{m}_3)$. The symmetry group

$$\mathcal{S}_{\text{ortho}} = \{\pm\mathbf{I}, \mathbf{R}_1, \mathbf{R}_2, \mathbf{R}_3\}, \qquad (2.11.41)$$

consists, besides of the identity $\mathbf{I}$ (neutral element) and its opposite $-\mathbf{I}$ (centrosymmetry), of the reflections $\mathbf{R}_1$, $\mathbf{R}_2$ and $\mathbf{R}_3$ about the respective planes $(\mathbf{m}_2, \mathbf{m}_3)$, $(\mathbf{m}_3, \mathbf{m}_1)$ and $(\mathbf{m}_1, \mathbf{m}_2)$, namely $\mathbf{R}_i = \mathbf{I} - 2\,\mathbf{m}_i\otimes\mathbf{m}_i$ for $i \in [1,3]$. Since the sense of the preferred directions is elusive, the second order tensors $\mathbf{M}_i = \mathbf{m}_i\otimes\mathbf{m}_i$, $i \in [1,3]$, are involved in the representations rather than the vectors $\mathbf{m}_i$ themselves. Ad hoc equivalent representations are trivially obtained by playing with the identity,

$$\mathbf{M}_1 + \mathbf{M}_2 + \mathbf{M}_3 = \mathbf{I}. \qquad (2.11.42)$$

Table 2.11.3: Two equivalent irreducible and complete lists of orthotropic scalar invariants of a symmetric second order tensor $\mathbf{A}$ [Boehler 1987]. $\mathbf{M}_i = \mathbf{m}_i\otimes\mathbf{m}_i$ with $\cup_{i\in[1,3]}\mathbf{m}_i$ cartesian triad of preferred directions.

Variable	Invariants	Variable	Invariants
$\mathbf{A}$	$\operatorname{tr}\mathbf{M}_1\cdot\mathbf{A}$, $\operatorname{tr}\mathbf{M}_1\cdot\mathbf{A}^2$, $\operatorname{tr}\mathbf{A}^3$,	$\mathbf{A}$	$\operatorname{tr}\mathbf{A}$, $\operatorname{tr}\mathbf{A}^2$, $\operatorname{tr}\mathbf{A}^3$,
	$\operatorname{tr}\mathbf{M}_2\cdot\mathbf{A}$, $\operatorname{tr}\mathbf{M}_2\cdot\mathbf{A}^2$		$\operatorname{tr}\mathbf{M}_1\cdot\mathbf{A}$, $\operatorname{tr}\mathbf{M}_1\cdot\mathbf{A}^2$
	$\operatorname{tr}\mathbf{M}_3\cdot\mathbf{A}$, $\operatorname{tr}\mathbf{M}_3\cdot\mathbf{A}^2$		$\operatorname{tr}\mathbf{M}_2\cdot\mathbf{A}$, $\operatorname{tr}\mathbf{M}_2\cdot\mathbf{A}^2$

Table 2.11.4 displays an irreducible representation of transversely isotropic scalar invariants of the symmetric second order tensor $\mathbf{A}$. A cartesian triad $\{\mathbf{e}_1, \mathbf{e}_2, \mathbf{e}_3 = \mathbf{m}\}$ is defined from the preferred direction $\mathbf{m}$. The symmetry group

$$\mathcal{S}_{\text{trans}} = \{\pm\mathbf{I}, \mathbf{R}_1, \mathbf{R}_2, \mathbf{R}_3, \mathbf{R}_\theta\}, \qquad (2.11.43)$$

consists of the same elements as the orthotropy group plus arbitrary rotations $\mathbf{R}_\theta$, $\theta \in [0, 2\pi[$, about the preferred axis. Again, since the sense of the preferred direction is elusive, the second order tensor $\mathbf{M} = \mathbf{m}\otimes\mathbf{m}$ is involved in the representations rather than the vector $\mathbf{m}$ itself.

Table 2.11.4: Irreducible and complete list of transversely isotropic scalar invariants of a symmetric second order tensor $\mathbf{A}$ [Boehler 1987]. $\mathbf{M} = \mathbf{m} \otimes \mathbf{m}$ with $\mathbf{m}$ unit vector along the preferred direction.

Variable	Invariants
$\mathbf{A}$	$\operatorname{tr} \mathbf{A}$, $\operatorname{tr} \mathbf{A}^2$, $\operatorname{tr} \mathbf{A}^3$,
	$\operatorname{tr} \mathbf{M} \cdot \mathbf{A}$, $\operatorname{tr} \mathbf{M} \cdot \mathbf{A}^2$

2.11.3.5 Representation tables for symmetric and skew-symmetric isotropic second order tensors

Given a list of vectors and symmetric and skew-symmetric tensors, tools to build isotropic second order tensors are listed in the tables below. Table 2.11.5 displays generators of symmetric second order tensors built from various vector and tensor variables. Table 2.11.6 displays generators of symmetric and skew-symmetric second order tensors built from two symmetric second order tensors. These tables can be applied to obtain Table 2.11.8 pertaining to transverse isotropy, accounting for simplifications due to $\mathbf{M}^2 = \mathbf{M}$. They need to be extended to include the set of variables $(\mathbf{A}, \mathbf{M}_1, \mathbf{M}_2)$ to obtain, with help of Liu's theorem, Table 2.11.7 pertaining to orthotropic symmetry. Simplifications arise due to the identities $\mathbf{M}_i \cdot \mathbf{M}_j = \mathbf{M}_i \, I_{ij}$ (no sum on i), $i, j \in [1, 3]$, and (2.11.42).

Table 2.11.5: Irreducible list of isotropic symmetric second order generators of a symmetric second order tensor $\mathbf{A}$, a skew-symmetric second order tensor $\mathbf{W}$, vectors $\mathbf{a}$, and some combinations of these elements [Wang 1970].

Variables	Generators
	$\mathbf{I}$
$\mathbf{A}$	$\mathbf{A}$, $\mathbf{A}^2$
$\mathbf{W}$	$\mathbf{W}^2$
$\mathbf{a}$	$\mathbf{a} \otimes \mathbf{a}$
$\mathbf{a}_1, \mathbf{a}_2$	$\mathbf{a}_1 \otimes \mathbf{a}_2 + \mathbf{a}_2 \otimes \mathbf{a}_1$
$\mathbf{A}, \mathbf{a}$	$\mathbf{a} \otimes \mathbf{A} \cdot \mathbf{a} + \mathbf{A} \cdot \mathbf{a} \otimes \mathbf{a}$, $\mathbf{a} \otimes \mathbf{A}^2 \cdot \mathbf{a} + \mathbf{A}^2 \cdot \mathbf{a} \otimes \mathbf{a}$

Table 2.11.6: Irreducible list of isotropic symmetric and skew-symmetric second order generators of two symmetric second order tensors $\mathbf{A}_1$ and $\mathbf{A}_2$ [Wang 1970]. The symmetric generator $\mathbf{A}_1^2 \cdot \mathbf{A}_2^2 + \mathbf{A}_2^2 \cdot \mathbf{A}_1^2$ listed by Wang [1970]-II, p. 215, is redundant [Boehler 1987].

Variables	Symmetric Generators
$\mathbf{A}_1, \mathbf{A}_2$	$\mathbf{A}_1 \cdot \mathbf{A}_2 + \mathbf{A}_2 \cdot \mathbf{A}_1$, $\mathbf{A}_1 \cdot \mathbf{A}_2^2 + \mathbf{A}_2^2 \cdot \mathbf{A}_1$, $\mathbf{A}_1^2 \cdot \mathbf{A}_2 + \mathbf{A}_2 \cdot \mathbf{A}_1^2$
Variables	**Skew-Symmetric Generators**
$\mathbf{A}_1, \mathbf{A}_2$	$\mathbf{A}_1 \cdot \mathbf{A}_2 - \mathbf{A}_2 \cdot \mathbf{A}_1$, $\mathbf{A}_1 \cdot \mathbf{A}_2^2 - \mathbf{A}_2^2 \cdot \mathbf{A}_1$, $\mathbf{A}_1^2 \cdot \mathbf{A}_2 - \mathbf{A}_2 \cdot \mathbf{A}_1^2$,
	$\mathbf{A}_1 \cdot \mathbf{A}_2 \cdot \mathbf{A}_1^2 - \mathbf{A}_1^2 \cdot \mathbf{A}_2 \cdot \mathbf{A}_1$, $\mathbf{A}_2 \cdot \mathbf{A}_1 \cdot \mathbf{A}_2^2 - \mathbf{A}_2^2 \cdot \mathbf{A}_1 \cdot \mathbf{A}_2$

Table 2.11.7: List of orthotropic symmetric and skew-symmetric second order generators of a symmetric second order tensor $\mathbf{A}$. $\mathbf{M}_i = \mathbf{m}_i \otimes \mathbf{m}_i$ with $\cup_{i \in [1,3]} \mathbf{m}_i$ cartesian triad of preferred directions.

Variable	Irreducible Symmetric Generators [Boehler 1987]
	$\mathbf{M}_1, \mathbf{M}_2, \mathbf{M}_3$
$\mathbf{A}$	$\mathbf{M}_1 \cdot \mathbf{A} + \mathbf{A} \cdot \mathbf{M}_1, \mathbf{A}^2$
	$\mathbf{M}_2 \cdot \mathbf{A} + \mathbf{A} \cdot \mathbf{M}_2,$
	$\mathbf{M}_3 \cdot \mathbf{A} + \mathbf{A} \cdot \mathbf{M}_3$
Variable	(nonirreducible) Skew-Symmetric Generators [Wang 1970]
$\mathbf{A}$	$\mathbf{M}_1 \cdot \mathbf{A} - \mathbf{A} \cdot \mathbf{M}_1, \mathbf{M}_2 \cdot \mathbf{A} - \mathbf{A} \cdot \mathbf{M}_2,$
	$\mathbf{M}_1 \cdot \mathbf{A}^2 - \mathbf{A}^2 \cdot \mathbf{M}_1, \mathbf{M}_2 \cdot \mathbf{A}^2 - \mathbf{A}^2 \cdot \mathbf{M}_2,$
	$\mathbf{A} \cdot \mathbf{M}_1 \cdot \mathbf{A}^2 - \mathbf{A}^2 \cdot \mathbf{M}_1 \cdot \mathbf{A}, \mathbf{A} \cdot \mathbf{M}_2 \cdot \mathbf{A}^2 - \mathbf{A}^2 \cdot \mathbf{M}_2 \cdot \mathbf{A},$
	$\mathbf{M}_1 \cdot \mathbf{A} \cdot \mathbf{M}_2 - \mathbf{M}_2 \cdot \mathbf{A} \cdot \mathbf{M}_1$

Table 2.11.8: Irreducible list of transversely isotropic symmetric and skew-symmetric second order generators of a symmetric second order tensor $\mathbf{A}$. $\mathbf{M} = \mathbf{m} \otimes \mathbf{m}$ with $\mathbf{m}$ unit vector along the preferred direction.

Variable	Symmetric Generators [Boehler 1987]
	$\mathbf{I}, \mathbf{M}$
$\mathbf{A}$	$\mathbf{A}, \mathbf{M} \cdot \mathbf{A} + \mathbf{A} \cdot \mathbf{M},$
	$\mathbf{A}^2, \mathbf{M} \cdot \mathbf{A}^2 + \mathbf{A}^2 \cdot \mathbf{M}$
Variable	Skew-Symmetric Generators [Wang 1970]
$\mathbf{A}$	$\mathbf{M} \cdot \mathbf{A} - \mathbf{A} \cdot \mathbf{M},$
	$\mathbf{M} \cdot \mathbf{A}^2 - \mathbf{A}^2 \cdot \mathbf{M},$
	$\mathbf{A} \cdot \mathbf{M} \cdot \mathbf{A}^2 - \mathbf{A}^2 \cdot \mathbf{M} \cdot \mathbf{A}$

Exercises on Chapter 2

Exercise 2.1 Change of coordinate axes

   Obtain the components of the stress tensor $\boldsymbol{\sigma}$ under a rotation of the coordinate axes.

Answer: The right handed set of coordinate vectors $(\mathbf{e}_j, \ j \in [1,3])$ is changed to the right handed set $(\mathbf{e}_i^*, \ i \in [1,3])$,

$$\mathbf{e}_i^* = R_{ij}\,\mathbf{e}_j, \quad \mathbf{e}_i = R_{ji}\,\mathbf{e}_j^*, \tag{2.E.1}$$

where $R_{ij} = \cos(\mathbf{e}_i^*, \mathbf{e}_j)$. Then the stress $\boldsymbol{\sigma} = \sigma_{ij}\,\mathbf{e}_i \otimes \mathbf{e}_j$ may also be expressed as $\boldsymbol{\sigma} = R_{ki}\,\sigma_{ij}\,R_{lj}\,\mathbf{e}_k^* \otimes \mathbf{e}_l^*$, i.e. $\sigma_{kl}^* = R_{ki}\,\sigma_{ij}\,R_{lj}$, or $\boldsymbol{\sigma}^* = \mathbf{R} \cdot \boldsymbol{\sigma} \cdot \mathbf{R}^{\mathrm{T}}$.

   Note that the coordinates x_i change to x_i^*,

$$x_i^* = R_{ij}\,x_j, \quad x_i = R_{ji}\,x_j^*. \tag{2.E.2}$$

The components of a tensor $\mathbf{A}$ of arbitrary order transform similarly,

$$A_{ijk\ldots}^* = R_{ip}\,R_{jq}\,R_{kr}\,A_{pqr\ldots}^*, \quad A_{pqr\ldots} = R_{ip}\,R_{jq}\,R_{kr}\,A_{ijk\ldots}^*. \tag{2.E.3}$$

Note also the tensor forms of the transformations of the coordinate and basis vectors,

$$\mathbf{x}^* = \mathbf{R} \cdot \mathbf{x}, \quad \mathbf{e}^* = \mathbf{R} \cdot \mathbf{e}, \quad \mathbf{x} = \mathbf{R}^{\mathrm{T}} \cdot \mathbf{x}^*, \quad \mathbf{e} = \mathbf{R}^{\mathrm{T}} \cdot \mathbf{e}^*. \tag{2.E.4}$$

Exercise 2.2 Kinematics that differ by a rotation

   Compare the kinematics,

$$u_1^{(1)} = k\,x_2, \quad u_2^{(1)} = u_3^{(1)} = 0, \tag{2.E.5}$$

and

$$u_1^{(2)} = \frac{k}{2}\,x_2, \quad u_2^{(2)} = \frac{k}{2}\,x_1, \quad u_3^{(2)} = 0. \tag{2.E.6}$$

Answer: The infinitesimal strains are identical,

$$\boldsymbol{\epsilon}^{(i)} = \frac{1}{2}\left(\mathbf{u}^{(i)}\overset{\leftarrow}{\nabla} + \overset{\rightarrow}{\nabla}\mathbf{u}^{(i)}\right) = \begin{bmatrix} 0 & \frac{k}{2} & 0 \\ \frac{k}{2} & 0 & 0 \\ 0 & 0 & 0 \end{bmatrix}, \quad i = 1, 2, \tag{2.E.7}$$

but the two kinematics differ by a rotation around the x_3-axis,

$$\mathbf{u}^{(1)}\overset{\leftarrow}{\nabla} = \begin{bmatrix} 0 & k & 0 \\ 0 & 0 & 0 \\ 0 & 0 & 0 \end{bmatrix} = \overbrace{\begin{bmatrix} 0 & \frac{k}{2} & 0 \\ \frac{k}{2} & 0 & 0 \\ 0 & 0 & 0 \end{bmatrix}}^{\boldsymbol{\epsilon}^{(1)}} + \overbrace{\begin{bmatrix} 0 & \frac{k}{2} & 0 \\ -\frac{k}{2} & 0 & 0 \\ 0 & 0 & 0 \end{bmatrix}}^{\boldsymbol{\omega}^{(1)}}; \quad \mathbf{u}^{(2)}\overset{\leftarrow}{\nabla} = \boldsymbol{\epsilon}^{(2)}; \quad \boldsymbol{\omega}^{(2)} = \mathbf{0}. \tag{2.E.8}$$

Exercise 2.3 Polar decomposition of a simple shear

   Obtain the polar decomposition associated with the simple shear kinematics $\mathbf{x} = \underline{\mathbf{x}} + \mathbf{u}$ [Malvern 1969, p. 182] et seq.,

$$u_1 = k\,\underline{x}_2, \quad u_2 = u_3 = 0. \tag{2.E.9}$$

Answer: From the deformation gradient,

$$\mathbf{F} = \begin{bmatrix} 1 & k & 0 \\ 0 & 1 & 0 \\ 0 & 0 & 1 \end{bmatrix}, \tag{2.E.10}$$

follows the right Cauchy-Green tensor

$$\mathbf{C} = \begin{bmatrix} 1 & k & 0 \\ k & 1+k^2 & 0 \\ 0 & 0 & 1 \end{bmatrix}_{\mathbf{e}_1, \mathbf{e}_2, \mathbf{e}_3} = \begin{bmatrix} \lambda_1^2 & 0 & 0 \\ 0 & \lambda_2^2 & 0 \\ 0 & 0 & 1 \end{bmatrix}_{\mathbf{e}_1^*, \mathbf{e}_2^*, \mathbf{e}_3}, \tag{2.E.11}$$

which has the eigenvalues,

$$\lambda_{1,2}^2 = \tfrac{1}{2}\left(2 + k^2 \pm k\sqrt{4 + k^2}\right),$$ (2.E.12)

and the first eigendirection $\mathbf{e}_1^*$ in the plane 1-2 makes an angle ϕ with the 1-axis, with $\tan 2\phi = -2/k$. Further

$$\lambda_1 = \tfrac{1}{2}\left(k + \sqrt{4 + k^2}\right), \quad \lambda_2 = \frac{1}{\lambda_1} = \tfrac{1}{2}\left(-k + \sqrt{4 + k^2}\right),$$ (2.E.13)

and

$$\cos^2\phi = \tfrac{1}{2}\left(1 - \frac{k}{\sqrt{4 + k^2}}\right), \quad \sin^2\phi = \tfrac{1}{2}\left(1 + \frac{k}{\sqrt{4 + k^2}}\right).$$ (2.E.14)

The right stretch deduces,

$$\mathbf{U} = \begin{bmatrix} \lambda_1 & 0 & 0 \\ 0 & \lambda_2 & 0 \\ 0 & 0 & 1 \end{bmatrix}_{\mathbf{e}_1^*,\mathbf{e}_2^*,\mathbf{e}_3} = \frac{1}{\sqrt{4 + k^2}} \begin{bmatrix} 2 & k & 0 \\ k & k^2 + 2 & 0 \\ 0 & 0 & \sqrt{4 + k^2} \end{bmatrix}_{\mathbf{e}_1,\mathbf{e}_2,\mathbf{e}_3},$$ (2.E.15)

and the rotation follows as

$$\mathbf{Q} = \mathbf{F} \cdot \mathbf{U}^{-1} = \frac{1}{\sqrt{4 + k^2}} \begin{bmatrix} 2 & k & 0 \\ -k & 2 & 0 \\ 0 & 0 & \sqrt{4 + k^2} \end{bmatrix}_{\mathbf{e}_1,\mathbf{e}_2,\mathbf{e}_3}.$$ (2.E.16)

Exercise 2.4 A simple example of polar decomposition

Consider the polar decomposition $\mathbf{F} = \mathbf{Q} \cdot \mathbf{U}$ of the deformation gradient $\mathbf{F}$,

$$\mathbf{U} = \begin{bmatrix} 1 & 0 \\ 0 & 2 \end{bmatrix}, \quad \mathbf{Q} = \begin{bmatrix} \cos\alpha & \sin\alpha \\ -\sin\alpha & \cos\alpha \end{bmatrix},$$ (2.E.17)

where α is an arbitrary angle. Obtain the left stretch $\mathbf{V}$, perform its spectral decomposition, and check the expressions (2.5.4).
Answer: The associated deformation gradient and left stretch are obtained as,

$$\mathbf{F} = \begin{bmatrix} \cos\alpha & 2\sin\alpha \\ -\sin\alpha & 2\cos\alpha \end{bmatrix}, \quad \mathbf{V} = \mathbf{Q} \cdot \mathbf{U} \cdot \mathbf{Q}^{\mathrm{T}} = \begin{bmatrix} 1 + \sin^2\alpha & \cos\alpha\sin\alpha \\ \cos\alpha\sin\alpha & 1 + \cos^2\alpha \end{bmatrix}.$$ (2.E.18)

The eigenvalues are 1 and 2 and, with $\underline{\mathbf{e}}_1$ and $\underline{\mathbf{e}}_2$ the eigenvectors of the right stretch $\mathbf{U}$, the eigenvectors of the left stretch $\mathbf{V}$ write $\mathbf{e}_1 = \cos\alpha\,\underline{\mathbf{e}}_1 - \sin\alpha\,\underline{\mathbf{e}}_1$ and $\mathbf{e}_2 = \sin\alpha\,\underline{\mathbf{e}}_1 + \cos\alpha\,\underline{\mathbf{e}}_1$.
The expressions (2.5.4) are then easily checked.

Exercise 2.5 Proof of the generalized Cayley-Hamilton theorem [Rivlin 1955]

- start from the Cayley-Hamilton theorem,

$$\mathbf{CH}(\mathbf{D}) = \mathbf{D}^3 - I_1\,\mathbf{D}^2 + I_2\,\mathbf{D} - I_3\,\mathbf{I} = \mathbf{0},$$ (2.E.19)

with

$$I_1 = \mathrm{tr}\,\mathbf{D}, \quad I_2 = \tfrac{1}{2}\left((\mathrm{tr}\,\mathbf{D})^2 - \mathrm{tr}\,\mathbf{D}^2\right), \quad I_3 = \tfrac{1}{3}\,\mathrm{tr}\,\mathbf{D}^3 - \tfrac{1}{2}\,\mathrm{tr}\,\mathbf{D}\,\mathrm{tr}\,\mathbf{D}^2 + \tfrac{1}{6}\,(\mathrm{tr}\,\mathbf{D})^3\,;$$ (2.E.20)

- apply the Cayley-Hamilton theorem to $\mathbf{D} = \mathbf{A} + \mathbf{B}$ and to $\mathbf{D} = \mathbf{A} - \mathbf{B}$. Substract the expressions to obtain $\mathbf{RCH}(\mathbf{A}, \mathbf{B}) = \mathbf{CH}(\mathbf{A} + \mathbf{B}) - \mathbf{CH}(\mathbf{A} - \mathbf{B})$;
- use the identity $\mathrm{tr}\,\mathbf{A} \cdot \mathbf{B} \cdot \mathbf{A} = \mathrm{tr}\,\mathbf{A}^2 \cdot \mathbf{B}$ and the Cayley-Hamilton theorem applied to $\mathbf{B}$, to eliminate $\mathbf{B}^3$ from $\mathbf{RCH}(\mathbf{A}, \mathbf{B})$ and to retrieve the relation $\mathbf{SCH}(\mathbf{A}, \mathbf{B})$ defined by (2.3.9);
- apply the Cayley-Hamilton theorem to $\mathbf{D} = \mathbf{A} + \mathbf{B} + \mathbf{C}$, and to $\mathbf{A}$, $\mathbf{B}$ and $\mathbf{C}$, and use relations $\mathbf{SCH}(\mathbf{A}, \mathbf{B})$, $\mathbf{SCH}(\mathbf{B}, \mathbf{C})$ and $\mathbf{SCH}(\mathbf{C}, \mathbf{A})$.

Exercise 2.6 Inverse of a rank two modification of the identity

Show by inspection that the inverse of the rank two modification of the identity (2.3.29) is given by (2.3.30).

Hint: A heuristic proof seeks the inverse as the sum $a_0\,\mathbf{I} + \mathbf{a} \otimes \mathbf{b}' + \mathbf{c} \otimes \mathbf{d}'$ with $\mathbf{b}' = a_1\,\mathbf{b} + a_2\,\mathbf{d}$ and $\mathbf{d}' = a_3\,\mathbf{b} + a_4\,\mathbf{d}$ and identify the coefficients a's simply by requiring that the product of the rank two modification by this inverse is indeed equal to the identity.

Exercise 2.7 Derivatives with respect to the stress and the stress deviator

Let $F = F(I_1, J_2, J_3)$ be a scalar function of the isotropic scalar invariants $I_1 = \operatorname{tr} \boldsymbol{\sigma} = \mathbf{I} : \boldsymbol{\sigma}$ of the stress $\boldsymbol{\sigma}$, and $J_2 = \frac{1}{2} \operatorname{tr} \mathbf{s}^2$ and $J_3 = \frac{1}{3} \operatorname{tr} \mathbf{s}^3$ of the stress deviator $\mathbf{s} = \boldsymbol{\sigma} - \frac{1}{3} I_1 \mathbf{I}$. Calculate $\partial F / \partial \boldsymbol{\sigma}$.

Answer: First observe the relations

$$\frac{\partial \sigma_{kl}}{\partial \sigma_{ij}} = I_{ijkl}^{(4)}, \qquad \frac{\partial s_{kl}}{\partial s_{ij}} = I_{ijkl}^{(4)}, \tag{2.E.21}$$

where $\mathbf{I}^{(4)} = \mathbf{I} \,\overline{\otimes}\, \mathbf{I}$ is the fourth order identity over symmetric second order tensors and the operator $\overline{\otimes}$ is defined by (2.1.24). Now

$$I_1 = \boldsymbol{\sigma} : \mathbf{I} \qquad \Rightarrow \qquad \frac{\partial I_1}{\partial \boldsymbol{\sigma}} = \frac{\partial \boldsymbol{\sigma}}{\partial \boldsymbol{\sigma}} : \mathbf{I} = (\mathbf{I} \,\overline{\otimes}\, \mathbf{I}) : \mathbf{I} = \mathbf{I},$$

$$J_2 = \frac{1}{2} s_{kl} s_{kl} \qquad \Rightarrow \qquad \frac{\partial J_2}{\partial s_{ij}} = I_{ijkl}^{(4)} s_{kl} = s_{ij} \qquad \Rightarrow \qquad \frac{\partial J_2}{\partial \mathbf{s}} = \mathbf{s},$$

$$J_3 = \frac{1}{3} s_{kl} s_{lm} s_{mk} \qquad \Rightarrow \qquad \frac{\partial J_3}{\partial s_{ij}} = \frac{1}{3} I_{ijkl}^{(4)} s_{lm} s_{mk} + \frac{1}{3} s_{kl} I_{ijlm}^{(4)} s_{mk} + \frac{1}{3} s_{kl} s_{lm} I_{ijmk}^{(4)} \tag{2.E.22}$$

$$\Rightarrow \qquad \frac{\partial J_3}{\partial \mathbf{s}} = (\mathbf{I} \,\overline{\otimes}\, \mathbf{I}) : (\mathbf{s} \cdot \mathbf{s}) = \mathbf{s} \cdot \mathbf{s}.$$

Therefore

$$\mathbf{s} = \boldsymbol{\sigma} - \frac{1}{3} I_1 \mathbf{I} \qquad \Rightarrow \qquad \frac{\partial \mathbf{s}}{\partial \boldsymbol{\sigma}} = \mathbf{I} \,\overline{\otimes}\, \mathbf{I} - \frac{1}{3} \mathbf{I} \otimes \mathbf{I}. \tag{2.E.23}$$

Consequently

$$\frac{\partial J_2}{\partial \boldsymbol{\sigma}} = \frac{\partial J_2}{\partial \mathbf{s}} : \frac{\partial \mathbf{s}}{\partial \boldsymbol{\sigma}} = \mathbf{s};$$

$$\frac{\partial J_3}{\partial \boldsymbol{\sigma}} = \frac{\partial J_3}{\partial \mathbf{s}} : \frac{\partial \mathbf{s}}{\partial \boldsymbol{\sigma}} = \mathbf{s} \cdot \mathbf{s} - \frac{2}{3} J_2 \mathbf{I}. \tag{2.E.24}$$

Finally,

$$\begin{aligned} \frac{\partial F}{\partial \boldsymbol{\sigma}} &= \frac{\partial F}{\partial I_1} \frac{\partial I_1}{\partial \boldsymbol{\sigma}} + \frac{\partial F}{\partial J_2} \frac{\partial J_2}{\partial \boldsymbol{\sigma}} + \frac{\partial F}{\partial J_3} \frac{\partial J_3}{\partial \boldsymbol{\sigma}} \\[2mm] &= \underbrace{\frac{\partial F}{\partial I_1} \mathbf{I}}_{\substack{\text{volumetric} \\ \text{part}}} + \underbrace{\frac{\partial F}{\partial J_2} \mathbf{s} + \frac{\partial F}{\partial J_3} \left(\mathbf{s} \cdot \mathbf{s} - \frac{2}{3} J_2 \mathbf{I} \right)}_{\substack{\text{deviatoric} \\ \text{part}}}. \end{aligned} \tag{2.E.25}$$

Exercise 2.8 A tensor isotropic function $\mathcal{F}(\mathbf{A}, a)$ of a symmetric second order tensor $\mathbf{A}$ and of a scalar a

The proof in $\mathbb{R}^3$ of (2.11.36) proceeds as follows [Mandel 1974]. First, observe that the eigenvectors of $\mathbf{A}$ are eigenvectors of $\mathcal{F}$. Indeed, let $\mathbf{R}_1$ be the symmetry wrt the axis 1. Then, componentwise in the cartesian basis $(\mathbf{e}_1, \mathbf{e}_2, \mathbf{e}_3)$,

$$\mathbf{R}_1 \cdot \mathbf{A} \cdot \mathbf{R}_1^{\mathrm{T}} = \begin{bmatrix} -1 & 0 & 0 \\ 0 & 1 & 0 \\ 0 & 0 & 1 \end{bmatrix} \begin{bmatrix} A_{11} & A_{12} & A_{13} \\ A_{21} & A_{22} & A_{23} \\ A_{31} & A_{32} & A_{33} \end{bmatrix} \begin{bmatrix} -1 & 0 & 0 \\ 0 & 1 & 0 \\ 0 & 0 & 1 \end{bmatrix} = \begin{bmatrix} A_{11} & -A_{12} & -A_{13} \\ -A_{21} & A_{22} & A_{23} \\ -A_{31} & A_{32} & A_{33} \end{bmatrix}. \tag{2.E.26}$$

If $\mathbf{e}_1$ is an eigenvector of $\mathbf{A}$, then the components $A_{1i} = A_{i1}$, $i \neq 1$, vanish, and the above transformation is of course elusive. The requirement (2.11.35) imposes that $\mathcal{F}$ and $\mathbf{R}_1 \cdot \mathcal{F} \cdot \mathbf{R}_1^{\mathrm{T}}$ are identical, that is, in view of (2.E.26) applied to $\mathcal{F}$, $\mathcal{F}_{1i} = -\mathcal{F}_{1i} = 0$, $i \neq 1$, so that $\mathbf{e}_1$ is an eigenvector of $\mathcal{F}$.

Therefore, in a cartesian basis formed from the eigenvectors of $\mathbf{A}$, $\mathcal{F}$ is principal and its eigenvalues are functions of the eigenvalues of $\mathbf{A}$.

Moreover, if the eigenvalues A_2 and A_3 of $\mathbf{A}$ are equal, so are the eigenvalues $\mathcal{F}_2$ and $\mathcal{F}_3$ of $\mathcal{F}$. Indeed, let $\mathbf{R}_{23}$ be the transformation that rotates the eigenvector $\mathbf{e}_2$ to the eigenvector $\mathbf{e}_3$. Since $\mathbf{R}_{23} \cdot \mathbf{A} \cdot \mathbf{R}_{23}^{\mathrm{T}} = \mathbf{A}$, then according to (2.11.35), a similar relation should hold for $\mathcal{F}$, that is, the axes $\mathbf{e}_2$ and $\mathbf{e}_3$ are undistinguishable, and the eigenvalues $\mathcal{F}_2$ and $\mathcal{F}_3$ are equal.

Assume that $\mathbf{A}$ has m distinct eigenvalues. The associated eigenvectors constitute a basis and the components b_i, $i \in [1, m]$, in (2.11.36) can be obtained by solving the system of m equations, $\mathcal{F}_j = \sum_{i=0}^{m-1} b_i A_j^i$, $j \in [1, m]$, obtained by projecting (2.11.36) on these eigenvectors. The system can be solved uniquely since the eigenvectors constitute a basis of $\mathbb{R}^m$.

The tensor expression (2.11.36) results from the above system expressed on the eigenvectors of $\mathbf{A}$ or of $\mathcal{F}$.

Exercise 2.9 The square root of a positive (semi-)definite symmetric second order tensor
Show that the square root of a positive (semi-)definite symmetric second order tensor $\mathbf{B}$ is isotropic.

Proof: Let $\mathbf{R}$ be an orthogonal transformation. Using the spectral representation, the successive relations below follow,

$$\mathbf{B} = \sum_{i=1}^{3} b_i \, \mathbf{b}_i \otimes \mathbf{b}_i$$

$$\mathbf{B}^{\frac{1}{2}} = \sum_{i=1}^{3} b_i^{\frac{1}{2}} \, \mathbf{b}_i \otimes \mathbf{b}_i \tag{2.E.27}$$

$$\mathbf{R} \cdot \mathbf{B}^{\frac{1}{2}} \cdot \mathbf{R}^{\mathrm{T}} = \sum_{i=1}^{3} b_i^{\frac{1}{2}} \, \mathbf{R} \cdot \mathbf{b}_i \otimes (\mathbf{R} \cdot \mathbf{b}_i)^{\mathrm{T}} = \Big(\sum_{i=1}^{3} b_i \, (\mathbf{R} \cdot \mathbf{b}_i) \otimes (\mathbf{R} \cdot \mathbf{b}_i)^{\mathrm{T}} \Big)^{\frac{1}{2}} \,,$$

because the triad $\{\mathbf{R} \cdot \mathbf{b}_i, \ i \in [1,3]\}$ is a cartesian basis like $\{\mathbf{b}_i, \ i \in [1,3]\}$. Hence

$$\mathbf{R} \cdot \mathbf{B}^{\frac{1}{2}} \cdot \mathbf{R}^{\mathrm{T}} = (\mathbf{R} \cdot \mathbf{B} \cdot \mathbf{R})^{\frac{1}{2}} \,. \tag{2.E.28}$$

Exercise 2.10 The square root of a positive definite symmetric second order tensor, Morman [1986]
Calculate the coefficients α, $i \in [1,3]$, involved in the isotropic expansion of $\mathbf{B}^{\frac{1}{2}}$, Eqn (2.3.31), where $\mathbf{B}$ is a positive definite symmetric second order tensor.

Answer:
The square-root $\mathbf{B}^{\frac{1}{2}}$ is an isotropic function of $\mathbf{B}$ and can be expressed as a linear combination of $\mathbf{I}$, $\mathbf{B}$ and $\mathbf{B}^2$,

$$\left\{ \begin{array}{l} b_1^{\frac{1}{2}} = \alpha_0 + \alpha_1 \, b_1 + \alpha_2 \, b_1^2 \\[4pt] b_2^{\frac{1}{2}} = \alpha_0 + \alpha_1 \, b_2 + \alpha_2 \, b_2^2 \\[4pt] b_3^{\frac{1}{2}} = \alpha_0 + \alpha_1 \, b_3 + \alpha_2 \, b_3^2 \,, \end{array} \right. \tag{2.E.29}$$

whence

$$\alpha_0 = \sum_{k=1}^{3} I_3 \frac{b_k^{-\frac{1}{2}}}{\Delta_k}, \quad \alpha_1 = \sum_{k=1}^{3} \frac{b_k^{\frac{1}{2}} \, (b_k - I_1)}{\Delta_k}, \quad \alpha_2 = \sum_{k=1}^{3} \frac{b_k^{\frac{1}{2}}}{\Delta_k}, \tag{2.E.30}$$

with I_1 and I_3 are scalar invariants of $\mathbf{B}$ and

$$\Delta_k = 2 \, b_k^2 - I_1 \, b_k + I_3 \, b_k^{-1}, \quad k \in [1,3]\,. \tag{2.E.31}$$

Chapter 3

Thermodynamic properties of fluids

3.1 Basic thermodynamic definitions

The basic thermodynamic notions below address essentially pure substances and mixtures of fluids, which are in the realm of chemical engineering, e.g., the comprehensive introduction of Smith et al. [1996]. However, steps toward continuum systems are considered in a transport perspective along Bird et al. [1960] and Haase [1990], and in a thermomechanics perspective along Kestin [1966] [1968].

3.1.1 Open and closed systems. Intensive and extensive properties

An *open* system exchanges mass, momentum, energy and/or entropy with the surroundings, while a *closed* system does not. In practice, an open system can be made closed by including the sources with which the exchanges take place. Conversely, it might be of interest for the analysis to exclude some portions of a closed system, thus turning it open. Fig. 3.1.1 illustrates the concept.

As another example, consider a solid skeleton containing two types of cavities, say (a) and (b), which however host the same type of fluid. Assume the two types of cavities to be connected. There are several points of view to address the mechanical and transport properties of such a system. The analysis could concentrate only on the fluid of the cavities (a) which would be viewed as a system open to mass and momentum exchanges. On the other hand, the standard mixture approach considers the ternary system formed by the solid skeleton and the two fluids: while the system as a whole is closed, none of the three species does so.

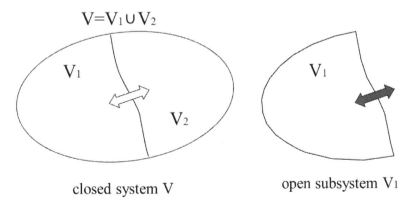

$$V = V_1 \cup V_2$$

V_1 V_2 V_1

closed system V   open subsystem V_1

Fig. 3.1.1: A closed system may be turned open by excluding a part with which exchanges of mass, momentum, etc. take place, and conversely.

A property is said *intensive* if it is independent of the quantity of matter. It is *extensive* if it depends on the quantity of matter. Volume, energy, enthalpy, etc. are extensive properties, while pressure p, temperature T, specific energy, specific enthalpy, etc. (measured per unit mass or unit volume) are intensive properties [3.1].

At this point, it is worth mentioning Euler's theorem for homogeneous functions. Let f be a sufficiently smooth function of two vectors, say $\mathbf{x}$ and $\mathbf{y}$. The function f is said to be *homogeneous of degree* m[3.2] with respect to its argument $\mathbf{x}$ if

$$f(\lambda \mathbf{x}, \mathbf{y}) = \lambda^m f(\mathbf{x}, \mathbf{y}), \quad \forall \lambda. \tag{3.1.1}$$

Differentiating with respect to λ, namely $\delta f / \delta \lambda = \partial f / \partial(\lambda \mathbf{x}) \cdot \mathbf{x} = m \lambda^{m-1} f(\mathbf{x}, \mathbf{y})$, yields

Euler's theorem for homogeneous functions : $\quad \dfrac{\partial f}{\partial \mathbf{x}}(\mathbf{x}, \mathbf{y}) \cdot \mathbf{x} = m f(\mathbf{x}, \mathbf{y})$. $\qquad$ (3.1.2)

This implication, obtained for $\lambda = 1$, is known as Euler's theorem for homogeneous functions. It can be shown that the converse holds as well, namely that a function which satisfies Euler's theorem with a coefficient m is homogeneous of degree m. The homogeneity property may not hold relative to the vector $\mathbf{y}$ or, if it does, it might be of different degree.

Intensive thermodynamic properties are described by homogeneous functions of degree $m = 0$, and extensive properties by homogeneous functions of degree $m = 1$. A typical application will be shown to lead to the Gibbs-Duhem relation[3.3], Section 3.1.4.

3.1.2 Energy, free energy, enthalpy, free enthalpy, Legendre duals

Basic thermodynamic measures of the state of a *closed* fluid system with temperature T, pressure p and volume V are shown in Table 3.1.1. The free enthalpy of a system is also referred to as Gibbs energy. These measures may also be viewed as molar measures, that is, relative to a mole, when the volume V is the molar volume of the system. For a reversible process, $d\mathbb{Q} = T \, d\mathbb{S}$ is the incremental heat exchange.

The expression of the internal energy in Table 3.1.1 is motivated by the balance of energy which, in absence of kinetic energy, states that the increment of internal energy is equal to the sum of the increment of reversible heat and of the work done by the pressure in the volume change.

Table 3.1.1: Thermodynamic measures of the state of a system [unit : J].

Measure	Variables	Arbitrary increment δ
Internal energy	$\mathbb{U} = \mathbb{U}(V, \mathbb{S})$	$\delta \mathbb{U} = -p \, \delta V + T \, \delta \mathbb{S}$
Free energy	$\mathbb{E} = \mathbb{E}(V, T) = \mathbb{U} - T \mathbb{S}$	$\delta \mathbb{E} = -p \, \delta V - \mathbb{S} \, \delta T$
Enthalpy	$\mathbb{H} = \mathbb{H}(p, \mathbb{S}) = \mathbb{U} + p V$	$\delta \mathbb{H} = V \, \delta p + T \, \delta \mathbb{S}$
Free enthalpy	$\mathbb{G} = \mathbb{G}(p, T) = \mathbb{H} - T \mathbb{S}$	$\delta \mathbb{G} = V \, \delta p - \mathbb{S} \, \delta T$

The thermodynamic functions in Table 3.1.1 appear as Legendre transforms[3.4] of one another. Consider a smooth function $\phi = \phi(x)$. The variable x is viewed as a state variable while $y = \partial \phi / \partial x$ is termed the control variable. The Legendre transform (dual) $\phi^*(y)$ of the function $\phi(x)$ is defined through the relation $\phi^*(y) + \phi(x) = x y$, and therefore $x = \partial \phi^* / \partial y$. In fact, the construction of the thermodynamic functions makes use of the negative of this definition. For a function of two or more variables, one may consider a partial Legendre transform, eg $\phi^*(y_1, x_2) + \phi(x_1, x_2) = x_1 y_1$, and then $y_1 = \partial \phi / \partial x_1$, $y_2 = \partial \phi / \partial x_2 = -\partial \phi^* / \partial x_2$, $x_1 = \partial \phi^* / \partial y_1$, or a total Legendre transform, $\phi^*(y_1, y_2) + \phi(x_1, x_2) = x_1 y_1 + x_2 y_2$, and then $y_i = \partial \phi / \partial x_i$, $x_i = \partial \phi^* / \partial y_i$, $i = 1, 2$.

[3.1]The French engineer and thermodynamicist François Massieu (1832-1896) was the first to have introduced the description of thermodynamic functions in terms of conjugate intensive and extensive variables, at least one extensive variable being required to appear for the size of the system to be accounted for. While nowadays seldom quoted for this fundamental contribution [Massieu 1869, 1876], the negatives of thermodynamic functions divided by absolute temperature bear his name in the current thermodynamics literature: they can be viewed as Legendre transforms in which an extensive variable is replaced by its conjugate intensive counterpart, see e.g., Tschoegl [2000], p. 62, for a presentation.

[3.2]Two slightly distinct properties are sometimes introduced. A function $f(\mathbf{x})$ is said *absolutely* homogeneous of degree m if $f(\lambda \mathbf{x}) = |\lambda|^m f(\mathbf{x})$. A function $f(\mathbf{x})$ is said *positively* homogeneous of degree m if $f(\lambda \mathbf{x}) = \lambda^m f(\mathbf{x})$ for $\lambda \geq 0$. The latter property is useful on cones, a cone C being a vector space for which, if $\mathbf{x} \in C$, then $\lambda \mathbf{x} \in C$ for any $\lambda > 0$. Cones may be convex, e.g., a planar fan, or not convex, e.g., two distinct fans emanating from the same origin.

[3.3]Named after the American physicist Josiah Willard Gibbs (1839-1903) and the French physicist and historian of science Pierre Maurice Marie Duhem (1861-1916).

[3.4]Named after the French mathematician Adrien Marie Legendre (1752-1833).

3.1.3 Chemical equilibrium between two multi-species phases

Let us consider a solution with two phases, a and b, separated by some device. For the sake of simplicity the two phases contain the same species which transfer from one phase to the other. The number of moles of the species $k \in [1, n]$ in phase α, is denoted by $N_{\alpha k}$.

Chemical equilibrium for all species k between the two phases is defined by the equality of their chemical potentials,

$$g_{ak} = g_{bk}, \quad k \in [1, n]. \tag{3.1.3}$$

Indeed altogether the two phase system is assumed to be closed so that its change of Gibbs energy is equal to $d\mathbb{G} = V \, dp - \mathbb{S} \, dT$ (a). Individually on the other hand, each phase is an open system and the increment of its Gibbs energy is defined as $d\mathbb{G}_\alpha = V_\alpha \, dp - \mathbb{S}_\alpha \, dT + \sum_k g_{\alpha k} \, dN_{\alpha k}$, $\alpha = a, b$, where it has been assumed that the equilibrium pressure and temperature are uniform over the system. The Gibbs energy is additive, i.e. the Gibbs energy of the system is obtained by summation of the Gibbs energies of the phases, $\mathbb{G} = \sum_\alpha \mathbb{G}_\alpha$. Therefore, since the volume and the entropy are additive, then $\sum_\alpha V_\alpha \, dp - \mathbb{S}_\alpha \, dT = V \, dp - \mathbb{S} \, dT$. For consistency with the relation (a) above, the constraint $\sum_k g_{ak} \, dN_{ak} + g_{bk} \, dN_{bk} = 0$ should be satisfied. The mass of any species k leaving the phase a enters phase b, and mass conservation implies $dN_{ak} + dN_{bk} = 0$. Thus $\sum_k (g_{ak} - g_{bk}) \, dN_{ak} = 0$. Assuming that the dN_{ak}'s are independent and arbitrary finishes the proof □.

Note that the relations (3.1.3) apply only to species that can transfer from one phase to the other. For a species whose transfer is partially or totally hindered, the chemical potentials on each side of the interface are likely to undergo a transient or permanent discontinuity. In some biological systems, e.g., in membrane channels, the transfer is triggered by the build up of the discontinuity once the hindrance has been removed.

The system has been considered closed for simplicity. More generally, it could (1) contain more than two phases, or (2) be open so that the variation of the mole number of a species could be due to phase transfer and exchange with the surroundings, or (3) involve mass exchanges between distinct species. Chemical equilibrium is then defined by appropriate combinations of chemical potentials.

3.1.4 The Gibbs-Duhem relation

Let us consider a fluid phase which is open, namely the number of moles may change. Let $N = \sum_k N_k$ be the total number of moles, and $\mathbb{G}$ be the Gibbs energy for $N = 1$.

The Gibbs energy $N\mathbb{G} = N\mathbb{G}(p, T, \cup_k \{N_k\})$ is an extensive property of a system, while the chemical potentials are intensive properties of the species. Indeed,

$$d(N\mathbb{G}) = \frac{\partial (N\mathbb{G})}{\partial p}\bigg|_{T,N} dp + \frac{\partial (N\mathbb{G})}{\partial T}\bigg|_{p,N} dT + \sum_k g_k \, dN_k, \tag{3.1.4}$$

with the subscript '$|_N$' indicating a derivative at fixed chemical content. The chemical potential per unit mole g_k of the species k,

$$g_k = \frac{\partial (N\mathbb{G})}{\partial N_k}\bigg|_{p,T}, \tag{3.1.5}$$

has unit J, while its counterpart per unit mass μ_k has unit J/kg $\times$ m $=$ m^2/s^2. These two equivalent properties are linked by the molar mass $\widehat{m}_k$ of the species k,

$$g_k = \widehat{m}_k \, \mu_k. \tag{3.1.6}$$

Note that $d(N\mathbb{G}) = dN \, \mathbb{G} + N \, d\mathbb{G}$. If $x_k = N_k / N$ is the molar fraction of species k, then $dN_k = x_k \, dN + dx_k \, N$. Inserting this expression in (3.1.4) and using the arbitrariness of N yields,

$$\mathbb{G} = \sum_k g_k \, x_k \quad \Rightarrow \quad N\mathbb{G} = \sum_k g_k \, N_k, \tag{3.1.7}$$

and

$$d\mathbb{G} = \frac{\partial \mathbb{G}}{\partial p}\bigg|_{T,N} dp + \frac{\partial \mathbb{G}}{\partial T}\bigg|_{p,N} dT + \sum_k g_k \, dx_k, \quad d(N\mathbb{G}) = \frac{\partial (N\mathbb{G})}{\partial p}\bigg|_{T,N} dp + \frac{\partial (N\mathbb{G})}{\partial T}\bigg|_{p,N} dT + \sum_k g_k \, dN_k. \tag{3.1.8}$$

Differentiation of G in the form (3.1.7) yields in turn,

$$dG = \sum_k dg_k \, x_k + g_k \, dx_k, \quad d(NG) = \sum_k dg_k \, N_k + g_k \, dN_k \,. \tag{3.1.9}$$

Comparison of (3.1.8) and (3.1.9) provides the Gibbs-Duhem relation for a mixture in two equivalent forms, e.g., Kestin [1966], p. 332, Smith et al. [1996], p. 320,

Gibbs-Duhem relation:
$$\frac{\partial G}{\partial p}_{|T,x} \, dp \; + \; \frac{\partial G}{\partial T}_{|p,x} \, dT \; - \; \sum_k x_k \, dg_k \; = \; 0 \,;$$
$$\frac{\partial (NG)}{\partial p}_{|T,N} \, dp \; + \; \frac{\partial (NG)}{\partial T}_{|p,N} \, dT \; - \; \sum_k N_k \, dg_k \; = \; 0 \,. \tag{3.1.10}$$

In the first line, the partial derivatives at fixed chemical content have been legitimately replaced by partial derivatives at fixed molar fractions.

This relation can be derived, and generalized, as a direct consequence of Euler's theorem. Indeed, assume the function $f(\mathbf{x}, \mathbf{y})$ to be homogeneous of degree one in $\mathbf{x}$ and of degree 0 in $\mathbf{y}$. Then

Generalized Gibbs-Duhem relation $\quad \dfrac{\partial f}{\partial \mathbf{y}} \cdot d\mathbf{y} - \mathbf{x} \cdot d\left(\dfrac{\partial f}{\partial \mathbf{x}}\right) = 0 \,. \tag{3.1.11}$

Indeed

$$\text{(a)} \quad df \;\; = \;\; \frac{\partial f}{\partial \mathbf{x}} \cdot d\mathbf{x} + \frac{\partial f}{\partial \mathbf{y}} \cdot d\mathbf{y}$$

$$\text{(b)} \quad f \;\; = \;\; \frac{\partial f}{\partial \mathbf{x}} \cdot \mathbf{x} \qquad\qquad : \text{homogeneity of degree 1 in } \mathbf{x}, \; 0 \text{ in } \mathbf{y} \tag{3.1.12}$$

$$\text{(c)} \quad df \;\; = \;\; d\left(\frac{\partial f}{\partial \mathbf{x}}\right) \cdot \mathbf{x} + \frac{\partial f}{\partial \mathbf{x}} \cdot d\mathbf{x} \;\; : \text{differentiation of (b)} \,,$$

and the result follows by the difference (a)-(c) $\square$. The Gibbs-Duhem relation corresponds to $f \to NG$, $x_k \to N_k$, $\mathbf{y} \to (p, T)$.

3.1.5 Chemical affinity associated with a chemical reaction

Let us consider the set of m chemical reactions involving the n chemicals X_j [Kestin 1968, p. 271],

$$\nu_1^1 X_1 + \nu_2^1 X_2 + \cdots + \nu_n^1 X_n \quad \underset{k_1^-}{\overset{k_1^+}{\rightleftharpoons}} \quad \kappa_1^1 X_1 + \kappa_2^1 X_2 + \cdots + \kappa_n^1 X_n$$

$$\nu_1^2 X_1 + \nu_2^2 X_2 + \cdots + \nu_n^2 X_n \quad \underset{k_2^-}{\overset{k_2^+}{\rightleftharpoons}} \quad \kappa_1^2 X_1 + \kappa_2^2 X_2 + \cdots + \kappa_n^2 X_n$$

$$\vdots \qquad\qquad\qquad\qquad\qquad\qquad \vdots \tag{3.1.13}$$

$$\nu_1^m X_1 + \nu_2^m X_2 + \cdots + \nu_n^m X_n \quad \underset{k_m^-}{\overset{k_m^+}{\rightleftharpoons}} \quad \kappa_1^m X_1 + \kappa_2^m X_2 + \cdots + \kappa_n^m X_n$$

The ν's and the κ's are called the stoichiometric coefficients and the k^+'s and k^-'s the forward and backward rate constants.

For the species $j \in [1, n]$, the rate of the concentration $c_j = N_j/V$, defined as the number of moles N_j in the volume of the mixture V, is governed by the kinetic equation,

$$\frac{dc_j}{dt} = \sum_{i=1}^m \left(-\nu_j^i + \kappa_j^i\right) \left(k_i^+ \, c_1^{\nu_1^i} \, c_2^{\nu_2^i} \cdots c_n^{\nu_n^i} - k_i^- \, c_1^{\kappa_1^i} \, c_2^{\kappa_2^i} \cdots c_n^{\kappa_n^i}\right), \quad j \in [1, n] \,. \tag{3.1.14}$$

Mass conservation for the reaction i implies, with $\widehat{m}_j$ the molar mass of species j,

$$\sum_{j=1}^n \left(\nu_j^i - \kappa_j^i\right) \widehat{m}_j = 0, \quad i \in [1, m] \,. \tag{3.1.15}$$

For the reaction i, the *advancement*, or *extent*, of the reaction is defined, in terms of the number of moles of the species, as (the index r denotes a reactant and the index p a product)

$$d\xi_i = \frac{dN_j^{i,r}}{-\nu_j^i} = \frac{dN_j^{i,p}}{\kappa_j^i} = \frac{dN_j^i}{-\nu_j^i + \kappa_j^i}, \quad i \in [1, m].$$ (3.1.16)

The sum of the chemical potentials weighted by the (signed) stoichiometric coefficients,

$$\mathcal{G}_i = \sum_{j=1}^n -g_j^r \nu_j^i + g_j^p \kappa_j^i,$$ (3.1.17)

is termed chemical affinity of the reaction i[3.5]. The chemical work associated with the change of the number of moles dN_j of a species is equal to the product of its chemical potential g_j by this change of mole number, say $g_j \, dN_j$. With help of (3.1.16), the work associated with the chemical reaction i is equal to the work done by the chemical affinity in the variation of the reaction advancement,

$$\sum_{j=1}^n g_j^r \, dN_j^{i,r} + g_j^p \, dN_j^{i,p} = \mathcal{G}_i \, d\xi_i.$$ (3.1.18)

Therefore the change of free enthalpy $G \equiv N\mathbb{G}$, or Gibbs energy, of a phase during the chemical reaction i at constant pressure and temperature which is defined by (3.1.4) can be brought to the format,

$$dG = \sum_{j=1}^n g_j^r \, dN_j^{i,r} + g_j^p \, dN_j^{i,p} = \mathcal{G}_i \, d\xi_i.$$ (3.1.19)

3.1.6 Chemical equilibrium versus steady state

A distinction should made between *chemical* (or *thermodynamic*) equilibrium and steady state (or *kinetic* equilibrium). At steady state, the time derivatives of concentrations vanish. Chemical, or thermodynamic, equilibrium is defined by the equality of the chemical potentials as stated by (3.1.3).

As an example, consider the autocatalytic reaction [Othmer 1981]

$$M_1 + M_2 \quad \underset{k^-}{\overset{k^+}{\rightleftharpoons}} \quad 2\,M_1.$$ (3.1.20)

The rate of change of the concentration c_1 is described by the kinetic equation,

$$\frac{dc_1}{dt} = k^+ c_1 c_2 - k^- c_1^2,$$ (3.1.21)

in such a way that steady state takes place for

$$\frac{c_2}{c_1} = \frac{k^-}{k^+}.$$ (3.1.22)

For a species in a mixture, the chemical potential includes several contributions, in particular a pressure term, a formation term and an electrical term if the species is electrically charged. For an ideal mixture, it will be sufficient to additively decompose the chemical potential g in a *standard* term g^0 and a chemical term $RT \, \mathrm{Ln}\, x$, with R [unit : J/mole/K] the gas constant, T [unit : K] temperature and x [unit : 1] molar fraction[3.6]. Therefore, the chemical affinity (3.1.17),

$$\mathcal{G} = g_1 - g_2 = \mathcal{G}^0 + RT \, \mathrm{Ln}\frac{c_1}{c_2} = RT \, \mathrm{Ln}\Big(\frac{1}{K^{\mathrm{eq}}} \frac{c_1}{c_2}\Big),$$ (3.1.23)

expresses in terms of the concentrations and dimensionless equilibrium constant $K^{\mathrm{eq}} = \exp(-\mathcal{G}^0/RT)$. Chemical equilibrium requires

$$\frac{c_2}{c_1} = \frac{1}{K^{\mathrm{eq}}},$$ (3.1.24)

Thus, unless $K^{\mathrm{eq}} = k^+/k^-$, chemical equilibrium and steady states do not coincide.

[3.5]This quantity is sometimes defined with an opposite sign.

[3.6]In a general context, the standard chemical potential is the chemical potential at zero pressure, concentration equal to one, and vanishing electrical potential.

3.1.7   Flux- and energy-driven chemical reactions

It is of interest to delineate chemical reactions that are *flux-driven* and chemical reactions that are *energy-driven*. In Section 19.3, these notions are elaborated in the context of *active transport* across the corneal endothelium where the transfers of chemicals through membranes are seen as physico-chemical reactions.

3.1.7.1   Flux-driven chemical reactions

For a flux-driven chemical reaction in an *open system*, a species A, which is continuously supplied with a flux J_A, undergoes a reversible physico-chemical reaction to the species B, which is continuously extracted with a flux J_B,

$$\xrightarrow{J_A} A \quad \underset{k^-}{\overset{k^+}{\rightleftharpoons}} \quad B \xrightarrow{J_B} \tag{3.1.25}$$

The rate equations governing this reaction result in a matrix-vector format in terms of the forward and backward rate constants, respectively k^+ and k^- [unit : s^{-1}],

$$\frac{d}{dt}\begin{bmatrix} [A] \\ [B] \end{bmatrix} = \begin{bmatrix} -k^+ & k^- \\ k^+ & -k^- \end{bmatrix}\begin{bmatrix} [A] \\ [B] \end{bmatrix} + \begin{bmatrix} J_A \\ -J_B \end{bmatrix}. \tag{3.1.26}$$

Here the steady state, where time derivatives vanish, is not a state of *chemical equilibrium*, as elaborated now. Indeed, along Qian et al. [2003], the analysis is now restricted to *steady states*. Then the total concentration $c_t = [A] + [B]$ is constant, namely $c_t = c_0$. Obviously, a steady state is possible only if the incoming and outgoing fluxes are equal, $J_A = J_B = J$. It is customary to introduce the forward and backward fluxes, $J_+ = k^+ [A]$, $J_- = k^- [B]$, and thus the concentrations of the reactant and product are linked by the relation $J = J_+ - J_-$.

When the system is *closed*, $J = 0$, the steady state is a state of *chemical equilibrium*, and then $[B]/[A] = K^{eq}$, if the equilibrium constant K^{eq} is equal to k^+/k^-. Even so, an open system allows a steady state to be a *non equilibrium* state as elaborated in Section 3.1.6.

Indeed, first note that $J_+/k^+ + J_-/k^-$ is equal to the initial concentration c_0. Thus,

$$J_- = \frac{c_0 - (k^+)^{-1}J}{(k^+)^{-1} + (k^-)^{-1}}, \quad J_+ = \frac{c_0 + (k^-)^{-1}J}{(k^+)^{-1} + (k^-)^{-1}}. \tag{3.1.27}$$

The reaction tends to be irreversible if $k^- \to 0$, and then a steady state is defined by $J = k^+ [A]$. Thus equilibrium can be reached only by exhaustion of the species A.

The chemical affinity may be expressed in terms of concentrations or fluxes,

$$\mathcal{G} = RT \operatorname{Ln} \frac{1}{K^{eq}} \frac{[B]}{[A]} = RT \operatorname{Ln} \frac{J_-}{J_+}. \tag{3.1.28}$$

The total flux J can thus be obtained in terms of the chemical affinity and of the total concentration, as

$$\frac{J}{c_0} = \frac{e^{-X} - 1}{(k^+)^{-1}e^{-X} + (k^-)^{-1}} \quad \text{with} \quad X \equiv \frac{\mathcal{G}}{RT}. \tag{3.1.29}$$

The dissipation inequality, which can be cast in the form, Eqn (4.8.8),

$$-J\mathcal{G} = c_0 RT \frac{X(1 - e^{-X})}{(k^+)^{-1}e^{-X} + (k^-)^{-1}} \geq 0, \tag{3.1.30}$$

is satisfied whatever the sign of the chemical affinity, and the flux is bounded, Fig. 3.1.2. This format has motivated the exponential mass transfer relations of Section 10.2.3.

For small deviations from chemical equilibrium $X = 0$, the flux (3.1.29) can be linearized to

$$\frac{J}{c_0} = \frac{-X}{(k^+)^{-1} + (k^-)^{-1}}, \tag{3.1.31}$$

which is nothing else than an Onsager type of relation between flux and driving force, $J = -K\mathcal{G}$, where the proportionality coefficient $K > 0$ has been linked to the characteristics of the underlying physico-chemical reaction.

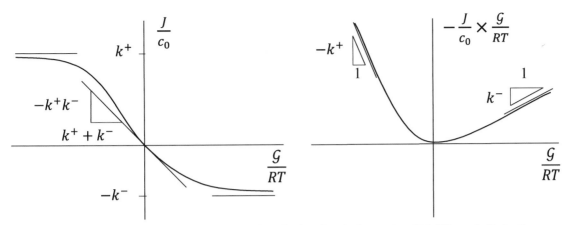

Fig. 3.1.2: Nonlinear and linearized fluxes associated with the chemical reaction (3.1.25), and dissipation associated with the nonlinear transfer law.

3.1.7.2 Energy-driven chemical reactions

An example of an energy-driven chemical reaction may be exhibited via two coupled reactions, say

$$A + X \; \underset{k^-}{\overset{k^+}{\rightleftharpoons}} \; B + Y \,. \tag{3.1.32}$$

The concentrations of the species X and Y are *clamped* (fixed) and the reaction $X \to Y$ provides energy to the reaction $A \to B$. Assuming a first order reaction, the kinetic relations write,

$$\frac{d}{dt}\begin{bmatrix} [A] \\ [B] \end{bmatrix} = \begin{bmatrix} -k^+ [X] & k^- [Y] \\ k^+ [X] & -k^- [Y] \end{bmatrix} \begin{bmatrix} [A] \\ [B] \end{bmatrix}. \tag{3.1.33}$$

In contrast to flux-driven reactions, the steady state is now a state of *chemical equilibrium*, endowed with an equilibrium constant K^{eq},

$$K^{\mathrm{eq}} = \frac{k^+}{k^-}\frac{[X]}{[Y]} = \frac{[B]^{\mathrm{eq}}}{[A]^{\mathrm{eq}}}, \tag{3.1.34}$$

In and out of equilibrium, the flux $J = d[B]/dt$ and the variation of free enthalpy in the reaction $\mathcal{G}$, referred to as chemical affinity and defined by (3.1.17), can be expressed in terms of the equilibrium constant,

$$J = k^- [Y][B]\left(K^{\mathrm{eq}}\frac{[A]}{[B]} - 1\right), \quad \mathcal{G} = RT\,\mathrm{Ln}\,\frac{1}{K^{\mathrm{eq}}}\frac{[B]}{[A]}, \tag{3.1.35}$$

so that

$$J = k^- [Y][B]\left(e^{-\mathcal{G}/RT} - 1\right). \tag{3.1.36}$$

Section 20.2.2 displays a typical instance of energy-driven reaction in a biological context where ATP serves as the energy source.

3.2 Phase change

3.2.1 Phase change and the Clapeyron formulas

Consider a two phase system where a liquid is in equilibrium with its vapor at temperature T and saturating pressure p_{sat}. Assume some infinitesimal amount of liquid to be added to the system, so that continuing equilibrium can be assumed to hold.

Evaporation at constant pressure and temperature leaves the Gibbs energy of the *system* unchanged, $d\mathbb{G} = V\,dp - \mathbb{S}\,dT = 0$, and the chemical potentials of the liquid and vapor remain equal, $\mu_l = \mu_v$ and $d\mu_l = d\mu_v$. Here we shall make use of enthalpies, entropies and chemical potentials per unit mass, displayed in Table 3.2.1.

Table 3.2.1: Thermodynamic measures of the state per unit mass [unit : J/kg = m^2/s^2] for a fluid endowed with temperature T, specific volume v and pressure p.

Measure	Variables	Arbitrary Increment δ
Internal energy	$u = u(v, s)$	$\delta u = -p\,\delta v + T\,\delta s$
Free energy	$e = e(v, T) = u - T\,s$	$\delta e = -p\,\delta v - s\,\delta T$
Enthalpy	$h = h(p, s) = u + p\,v$	$\delta h = v\,\delta p + T\,\delta s$
Free enthalpy	$\mu = \mu(p, T) = h - T\,s$	$\delta\mu = v\,\delta p - s\,\delta T$

However, while the chemical potentials of the liquid and vapor are equal, their molar volumes are not, and neither are the internal energies, enthalpies and other thermodynamic properties. Their jumps help characterize the phase change. For example, the Clapeyron formula[3.7] provides the jump of enthalpy across the phase transition curve. Indeed, with s_α, $\alpha = l, v$, the entropy of species α, and ρ_α its mass density,

$$d\mu_l = \frac{dp}{\rho_l} - s_l\,dT\,,$$

$$d\mu_v = \frac{dp}{\rho_v} - s_v\,dT\,,$$

(3.2.1)

continuing chemical equilibrium $d\mu_l = d\mu_v$ implies the jump of entropy results from the slope of the phase transition curve in the plane (T, p),

1st Clapeyron formula : $\quad s_v - s_l = \left(\frac{1}{\rho_v} - \frac{1}{\rho_l}\right)\frac{dp_{\text{sat}}}{dT}\,.$ (3.2.2)

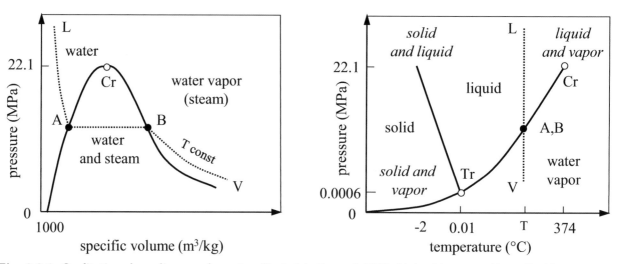

Fig. 3.2.1: Qualitative phase diagrams for water. Typical isotherm LABV obtained by expanding a liquid at constant temperature. The phase transformation A $\rightleftarrows$ B takes place at constant pressure p and temperature T but involves a jump of specific volume v.

Cr critical point; Tr triple point. There exist several analytical approximations of the phase transformation boundaries, e.g., Smith et al. [1996], p. 198. Table 3.2.2 provides accurate values of remarkable points in the phase diagram of the water substance. Under atmospheric pressure, ice melts at $0\,^\circ$C, a temperature lower than the triple point, water being an anomalous substance.

Phase transformation takes place at constant pressure p and temperature T but with a jump in mass density, Fig. 3.2.1, and therefore the difference of enthalpies is equal to the reversible heat associated with the phase transition,

$$dh_\alpha = \frac{dp}{\rho_\alpha} + T\,ds_\alpha\,, \quad \alpha = l, v \quad \Rightarrow \quad h_v - h_l = T\,(s_v - s_l)\,.$$

(3.2.3)

[3.7]Named after the French physicist Benoît Paul Emile Clapeyron (1799-1864).

Table 3.2.2: Remarkable points along the equilibrium curve of pure water and steam. From IAPWS [2002] and IAPWS [2009].

	Temperature		Density (kg/m³)		Pressure
	K	°C	Liquid	Vapor	MPa
0°C	273.15	0	999.79184	0.00485123	611.2104×10^{-6}
Triple point	273.16	0.01	999.79252	0.00485458	611.6548×10^{-6}
Boiling point	373.124	99.974	958.367	0.597657	0.101325
Critical point	647.096	373.946	322.00	322.00	22.064000

This relation can also be arrived at by requiring the continuity across the phase line of the chemical potentials $\mu_\alpha = h_\alpha - T s_\alpha$, $\alpha = l, v$. The jump of enthalpy,

$$\textbf{2nd Clapeyron formula}: \quad \mathcal{L}_{\text{vap}} = h_v - h_l = \left(\frac{1}{\rho_v} - \frac{1}{\rho_l} \right) T \frac{dp_{\text{sat}}}{dT}, \quad (3.2.4)$$

is called *latent heat of evaporation* (or *vaporization*) per unit mass at constant pressure and temperature. The latent heat of condensation is the negative of the latent heat of evaporation.

If the phase change takes place at temperature much lower than the critical temperature, the specific volume of the liquid may be neglected in front of the specific volume of the vapor. The latter being considered as a perfect gas, its specific volume $1/\rho_v$ is equal to $RT/(\widehat{m}_v p)$, with $\widehat{m}_v$ its molar mass. The 2nd Clapeyron formula then provides the saturation pressure of vapor in a differential form,

$$\textbf{Clausius-Clapeyron formula}: \quad \frac{dp}{p} = -\frac{\mathcal{L}_{\text{vap}} \widehat{m}_v}{R} d\left(\frac{1}{T} \right), \quad (3.2.5)$$

that can be integrated if the latent heat of vaporization is constant (or a known function of temperature),

$$\text{Ln} \frac{p_{\text{sat}}(T)}{p_{\text{sat}}(T_0)} = -\frac{\mathcal{L}_{\text{vap}} \widehat{m}_v}{R} \left(\frac{1}{T} - \frac{1}{T_0} \right). \quad (3.2.6)$$

Several analytical approximations of the saturation pressure of water vapor have been proposed in the literature.

Note that the slope dp_{sat}/dT of the saturating pressure is positive, Fig. 3.2.1. Collecting the two Clapeyron formulas,

$$\frac{\mathcal{L}_{\text{vap}}}{T} = \frac{1}{T} \left(h_v - h_l \right) = s_v - s_l = \left(\frac{1}{\rho_v} - \frac{1}{\rho_l} \right) \frac{dp_{\text{sat}}}{dT} > 0, \quad (3.2.7)$$

we therefore observe that the entropy of water vapor is larger than the entropy of the liquid at identical temperature and pressure. This observation agrees with the general concept that the entropy of gases is larger than the entropy of liquids which in turn is larger than the entropy of solids. Indeed, the entropy is linked to the number of energy microstates and degrees of freedom that molecules/particles can assume: particles have fixed positions in a solid lattice while molecules may vibrate much more freely in a gas. So phase changes are accompanied with entropy jumps.

Relations similar to (3.2.7) exist for the latent heats of melting (or fusion) and sublimation. These energies are positive: the body absorbs energy. The converse latent heats of condensation and freezing are negative: the body releases energy. These results are gathered in a qualitative proposition that substantiates the idea that entropy increases as the spatial organization of the matter becomes looser:

Proposition 3.1 on the entropies per unit mass of the three phases of a substance.

The entropy increases as the substance transforms from solid to liquid to gas.

Fig. 3.2.1 also shows that the slope dp_{melt}/dT along the melting boundary is negative. Moreover a (pressure, specific volume, temperature) diagram shows that the specific volume of ice is larger than that of the liquid at the same temperature and pressure. These two unusual properties, while keeping the latent heat of melting positive, have led water to be coined an 'anomalous substance'.

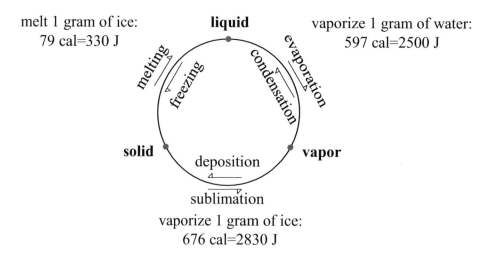

Fig. 3.2.2: Terminology defining phase changes and associated energies required for the phase changes (latent heats) under atmospheric pressure.

Table 3.2.3: Energy required to increase by $1\,^\circ$C the temperature of 1 gram of the water substance under atmospheric pressure.

State	cal	J
Ice just below $0\,^\circ$C	0.47	1.96
Water between $0\,^\circ$C and $100\,^\circ$C	1.008	4.22
Vapor just above $100\,^\circ$C	0.50	2.08

For water at $35\,^\circ$C, the mass densities of vapor and liquid, saturation pressure and latent heat of vaporization take the following values [IAPWS 2009]

$$\rho_v = 0.0396\,\text{kg/m}^3, \quad \rho_l = 994\,\text{kg/m}^3, \quad p_{\text{sat}} = 5.623\,\text{kPa},$$
$$\mathcal{L}_{\text{vap}} = 2418\,\text{kJ/kg} = 43.52\,\text{kJ/mole}. \tag{3.2.8}$$

A specific terminology is used for the phase changes as indicated in Fig. 3.2.2. Numbers indicate that phase changes are path independent. Table 3.2.3 displays some typical heat capacities for the water substance.

Clapeyron formula can also be derived by considering a mixture of saturated vapor and liquid in equilibrium, with mass proportions of λ and $1-\lambda$ respectively. A reversible heat supply is used mainly to produce vapor. However, generally heat supply is also used to increase the temperature of the system,

$$T\,ds = \mathcal{L}_{\text{vap}}\,d\lambda + (c_{pl}\,(1-\lambda) + c_{pv}\,\lambda)\,dT\,, \tag{3.2.9}$$

where the c_p's are the heat capacities at constant pressure. The associated variation of enthalpy per unit mass of the mixture can be cast in the format,

$$dh \;=\; \frac{dp_{\text{sat}}}{\rho_l}\,(1-\lambda) + \frac{dp_{\text{sat}}}{\rho_v}\,\lambda + T\,ds\,. \tag{3.2.10}$$

The phase transition pressure $p_{\text{sat}} = p_{\text{sat}}(T)$ depends only on temperature T. Requiring the two expressions of entropy and enthalpy above to define exact differentials with respect to temperature T and mass ratio λ yields the two relations,

$$c_{pl} - c_{pv} \;=\; T\,\frac{d}{dT}\Big(\frac{\mathcal{L}_{\text{vap}}}{T}\Big)$$
$$c_{pl} - c_{pv} + \Big(\frac{1}{\rho_v} - \frac{1}{\rho_l}\Big)\frac{dp_{\text{sat}}}{dT} \;=\; \frac{d}{dT}\big(\mathcal{L}_{\text{vap}}\big)\,. \tag{3.2.11}$$

Subtraction of these two relations yields again the 2nd Clapeyron formula. The first equation in (3.2.11) is of interest in itself.

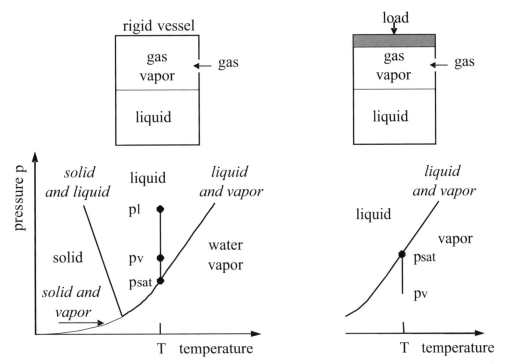

Fig. 3.2.3: Filling a rigid vessel (at constant volume) by an inert gas leads to metastable vapor bubbles (left) or, if the pressure of the gas mixture is constant, to evaporation (right). Since the liquid and vapor are in equilibrium along a flat interface, the liquid pressure is equal to the pressure of the gas mixture. The latter is the sum of the vapor pressure p_v and gas pressure. In absence of inert gas, equilibrium of the liquid and its vapor along a flat interface implies their pressures to be equal to the saturating pressure.

3.2.2 The principle of Spite and the Kelvin-Helmholtz formula

The principle of Spite formulated by Kestin [1968] provides qualitative information on the response of a system to chemical or mechanical loadings. It states that the response to a stimulus is such as to weaken this stimulus.

Let us for example consider a rigid vessel containing a liquid at temperature T pressurized by an inert gas in such a way that the state lies on the phase transformation curve, Fig. 3.2.3. The liquid pressure p_l and vapor pressure p_v are thus equal to the saturating pressure $p_l = p_v = p_{\text{sat}}(T)$ in absence of water drops and air bubbles. What happens if the pressure of the gas is increased at constant temperature, and constant volume?

The principle of Spite indicates that, to contract, the liquid evaporates. This general principle is illustrated by the Kelvin-Helmholtz formula.

(a) *Chemical equilibrium*:

Indeed, the liquid and its vapor are in continuing equilibrium along the phase boundary, so that $d\mu_l = d\mu_v$. Since $d\mu_\alpha = dp_\alpha/\rho_\alpha - s_\alpha\,dT$, $\alpha = l, v$, and since the temperature is kept constant, then $dp_l/\rho_l = dp_v/\rho_v$. The vapor with molar mass $\widehat{m}_v = \widehat{m}_l$ can be considered a perfect gas, $p_v/\rho_v = RT/\widehat{m}_l$. Therefore,

$$\frac{dp_l}{\rho_l} = \frac{dp_v}{\rho_v} = \frac{dp_v}{p_v}\frac{RT}{\widehat{m}_l}\,. \tag{3.2.12}$$

(b) *Mechanical equilibrium*:

Considering the mass density of the liquid to remain constant, (3.2.12) may be integrated from the initial equilibrium to a current state,

Kelvin-Helmholtz formula : $\operatorname{Ln}\dfrac{p_v}{p_{v0}} = \dfrac{\widehat{v}_l}{RT}\left(p_l - p_{l0}\right) \simeq \dfrac{\widehat{v}_v}{RT}\left(p_v - p_{v0}\right) > 0\,,$ (3.2.13)

where the relation $dp_l/\rho_l = dp_v/\rho_v$ or $\widehat{v}_l\,dp_l = \widehat{v}_v\,dp_v$ has been used in finite form.

Now, the liquid is assumed to be in equilibrium with the gas mixture. In the gas mixture, vapor and inert gas occupy the total volume, and, according to Dalton's law, each one contributes by its partial pressure, p_v and p_g

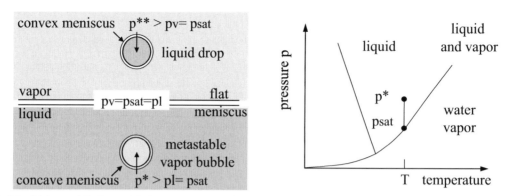

Fig. 3.2.4: Liquid in equilibrium with its vapor over a flat interface: the pressure is continuous across the interface due to Kelvin's law and equal to the saturating pressure, $p_l = p_v = p_{sat}$. Because the meniscus can withstand only tension, pressures in liquid drop and vapor bubble are larger than in the vapor and liquid they bath in. Therefore the vapor bubble is in a metastable state.

respectively, to the pressure $p_v + p_g$ of the mixture. According to Laplace's law, the difference of pressures between two non miscible species is equal to the surface tension T_{lv} of the interface film (meniscus) times the total curvature $2/r$ of the interface. If the pressure in the liquid is p_l, the pressure of the inert gas p_g is thus equal to $p_l - p_v + 2\,T_{lv}/r$. The interface being flat ($r = \infty$), the pressures of the gas p_g is equal to $p_l - p_v$.

How do we interpret this experiment at constant volume? Since $dp_g = dp_l - dp_v = (\rho_l/\rho_v - 1)\,dp_v > 0$, the vapor pressure has increased while the volume available has remained practically identical, since the vessel is rigid. Therefore the mass of vapor has increased. However, since its pressure is larger than the saturating pressure at the prescribed temperature, it is in a so-called 'metastable state'. The concept is illustrated as well in Fig. 3.2.4.

The relation (3.2.13) can be viewed as providing the vapor pressure p_v in terms of the liquid pressure p_l. Now the pressure increase of the vapor is much smaller than the pressure increase of the liquid since $dp_l/dp_v = \rho_l/\rho_v \sim 25 \times 10^3$ at 35 °C according to the data above. Therefore, the vapor pressure remains close to the saturating pressure. Furthermore, if the initial equilibrium is realized over a flat surface, namely $p_{g0} + p_{v0} = p_{l0}$, then $p_{v0} = p_{l0} - p_{g0}$ and $p_v = p_l - p_g$ while $p_v \simeq p_{v0}$.

Kestin [1968], p. 353, also envisages the case where the inert gas is pumped into the vessel such that the pressure of the gas mixture is constant. Since this process reduces the vapor pressure, initially equal to p_{sat}, the liquid will evaporate, Fig. 3.2.3.

3.2.3 The time course of a desiccation process

When the center of curvature of the liquid-air menisci is interior to the liquid (convex menisci), the contact angle between the liquid and the solid is larger than 90°. According to Laplace's law, the liquid pressure is larger than the vapor pressure and the suction $s \equiv p_v - p_l$ is negative.

Drying of a fluid saturated porous medium may be controlled by reducing the vapor pressure. A drying process typically proceeds along the following stages:

- in the initial configuration, the center of curvature of the liquid-air menisci is exterior to the liquid as sketched in Fig. 3.2.5: the contact angle θ between the liquid and the solid is smaller than 90° and, by Laplace's law, the suction $s = 2\,T_{lv}\cos\theta/r$ is positive. Here T_{lv} [unit : N/m] is the vapor-liquid tension. Then the liquid wets the pores and vapor begins penetrating the pores of representative radius r.
- drying proceeds with decreasing vapor pressure. For a vapor pressure much smaller (in algebraic value) than the saturation pressure p_{sat} (at 35 °C, p_{sat}=5.6 kPa), the Kelvin-Helmholtz formula, namely $\widehat{v}_l\,(p_l - p_{l0}) = RT\,\mathrm{Ln}\,(p_v/p_{sat})$ indicates that drying implies increasing liquid 'tension', i.e. negative pressure;
- the period during which the evaporation rate is virtually constant is termed constant rate period (CRP). This period ends up when the porous solid stops shrinking, evaporation decreases, vapor menisci enter the solid and suction increases;
- as the drying process continues by reducing the vapor pressure p_v, the suction increases by the Kelvin-Helmholtz formula. Since both the latter formula and Laplace's law should be satisfied, menisci enter the pore space from

the evaporation surface, the liquid recedes to the body and vapor invades pores of radii $2\,T_{lv}\cos\theta/s$ smaller and smaller.

According to this description, the larger pores drain first. For an actual porous solid, this general trend may be altered. Throats to the (larger) pores may be small, so that these throats constitute a bottleneck for evaporation. The large pores associated with these small throats drain only once the throats have drained.

If the body is kinematically constrained, fracture is likely because suction may create high tensile stresses. This description of desiccation qualitatively applies to gels [Scherer and Smith 1995] and to geomaterials like desiccating cements [Henkensiefken et al. 2009].

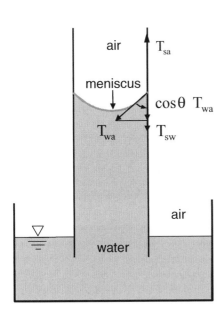

Fig. 3.2.5: The interface between a solid and a fluid develops a tension parallel to the interface. A similar tension parallel to the interface exists between a liquid, e.g., water, and a gas, e.g., air, which are not miscible. When the two non miscible fluids are in contact with a solid, the equilibrium configuration depends on the relative values of tension. For water and air, the actual configuration is indicated on the sketch: equilibrium implies that the projection of the water-air tension T_{wa} on the solid interface is equal to the tension resulting from the solid-water and solid-air interfaces, namely $T_{wa}\cos\theta = T_{sa} - T_{sw}$ where θ is the contact angle between the water-air interface and the solid. The upward vertical force exerted on water inside a capillary (cylindrical tube) of radius r is thus equal to $2\,\pi\,T_{wa}\cos\theta\,r$. This force is equilibrated by the weight of water in the tube, say $\rho_w\,g\,h\,\pi\,r^2$. Thus the height of water is equal to $h = 2\,T_{wa}\cos\theta/(\rho_w\,g\,r)$ and the suction $p_a - p_w = \rho_w\,g\,h$ is equal to $2\,T_{wa}\cos\theta/r$. The water-air tension at ambient temperature is equal to $0.072\,\mathrm{Pa}\times\mathrm{m}$ and the contact angle θ is equal to $0°$. For a capillary of radius $r = 10\,\mathrm{nm}$, the suction is thus equal to 14.4 MPa and water may theoretically (in absence of cavitation) ascent the tube over a height equal to $1440\,\mathrm{m}$ (the pressure due to a column of water of $10\,\mathrm{m}$ is equal to the atmospheric pressure).

3.2.4 Negative pressures, metastability and cavitation

In a porous medium, water is subjected to three types of physical forces: (1.) gravity, (2.) osmosis due to the presence of solutes, especially in fine pores and capillaries, and (3.) tension at the interfaces with air and solids.

Therefore, within tissues where water molecules establish strong bonds with other substances, some physical properties may be different from a pure substance. Water droplet suspended in air shows a convex meniscus (positive radius of curvature), as its pressure is larger than that of air. Conversely, capillary water bound to a solid displays a concave interface with air when its pressure is lower. Intrinsically negative pressures have been measured in water and liquids. The ascent of sap in the xylem conduits of trees is believed to be due to negative pressures, minus ten times the atmospheric pressure or more, generated by surface tension at the water-air interface of leaves. Indeed, the hydrostatic pressure decreases by about one atmosphere per dekameter. Under these circumstances, water is metastable with respect to its vapor (it remains liquid until a vapor bubble catastrophically develops), and the probability of cavitation (explosive liquid to vapor phase change) is large [Tardy and Duplay 1992, Mercury and Tardy 2001]. One way for nature to prevent cavitation might be to make use of a large number of redundant conduits with small diameters. However, too small conduits become hydraulically inefficient, because the permeability of tubes depends on their radius to the power 4, Table 15.7.1. The actual diameter of conduits is thought to be a trade-off between small conduits which are cavitation free (bubbles smaller than a certain critical size can not develop while larger ones do) but hydraulically inefficient, and large conduits which are hydraulically efficient but susceptible of cavitation [Tyree 2003].

The existence of negative pressures and the notion of metastability can be illustrated in the phase diagram [Kestin 1968, p. 238]. Indeed, let us follow an isotherm along which a liquid is expanded at constant temperature. When the experiment is performed in absence of foreign substances (impurities), the phase change actually does not take place at constant pressure, but at decreasing pressure up to some critical value of the specific volume, point A' in Fig. 3.2.6. Conversely, if the vapor is compressed, the phase change takes place at increasing pressure down to some

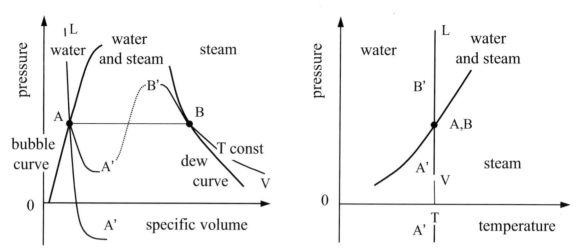

Fig. 3.2.6: In absence of impurities, part of the phase change at constant temperature and increasing specific volume takes place at decreasing pressure AA'. The values of pressure reached by water during volume expansion may be largely negative. Conversely, isothermal phase change at decreasing specific volume takes place in part at increasing pressure BB'. Along AA', the liquid is metastable with respect to its vapor, while, along BB', the vapor is metastable with respect to water.

critical value of the specific volume, point B' in Fig. 3.2.6. The path between A' and B' is *unstable* (not observable) since it would correspond to a negative isothermal bulk modulus. The paths AA' and BB' are termed *metastable*, that is, stable against small disturbances but instable again large ones. These notions of stability could be grasped by a colloquial example: a ball at the apex of a wedge $\wedge$ is in unstable equilibrium, it is in stable equilibrium in the inner edge of a V. Now if zooming the apex of the wedge, we discover a (even tiny) v, the ball will be said in metastable equilibrium at this apex.

Gas bubbles develop in liquids under very high negative pressure, or high suction. Cavitation may be viewed as the fracture of the liquid. An expression of the rate of nucleation of bubbles in terms of the liquid-vapor surface tension T_{lv} [unit : N/m], suction s, temperature T and vapor pressure p_v is proposed by Scherer and Smith [1995]. Inception of cavitation deduces by assuming a small rate of bubble formation, which provides the critical suction s_{cav} and the associated critical bubble radius r_{cav} by Laplace's law[3.8],

$$s_{\text{cav}} = 3.9\sqrt{\frac{T_{lv}^3}{T}}, \quad r_{\text{cav}} = 2\,\frac{T_{lv}}{s_{\text{cav}}} = 0.51\sqrt{\frac{T}{T_{lv}}}. \tag{3.2.14}$$

In these formulas, the suction s is measured in MPa, the surface tension T_{lv} in erg/cm^2 and the radius in nm. For water and a particular gel, the liquid-vapor surface tension T_{lv} at $T = 300$ K is typically equal to 72 erg/cm^2=0.072 N/m. Therefore s_{cav}=137.5 MPa and r_{cav}=1.05 nm. Slightly different values are obtained for solvents other than water, like ethanol, etc.

3.3   Thermodynamic functions of fluids

The first part of this section is a short introduction to the thermomechanical properties for a single fluid species. The reader is referred to Kestin [1968] and Drumheller [1998], p. 321, for theoretical elaboration. The aim of the second part is to formulate constitutive restrictions that allow explicit expressions of the thermodynamic functions to be obtained.

The thermodynamic properties used here, namely internal energy u, free energy e, entropy s, are measured per unit mass. Let $v = 1/\rho$ be the specific volume of the species [unit : m^3/kg]. Since $-p\,dv - s\,dT$ is the differential of the free energy e, then

$$p = -\frac{\partial e}{\partial v}\Big|_T, \quad s = -\frac{\partial e}{\partial T}\Big|_v, \tag{3.3.1}$$

[3.8]The bubbles cannot form if the capillaries have a size smaller than $2\,r_{\text{cav}}$.

and, assuming the free energy to be sufficiently smooth, the second order cross derivatives are equal,

$$\frac{\partial p}{\partial T}_{|v} = \frac{\partial s}{\partial v}_{|T}.$$ (3.3.2)

If one sets

$$T\,ds = \rho\,\mathcal{L}\,dv + c_{\mathrm{v}}\,dT\,,$$ (3.3.3)

with,
- $\mathcal{L}$ *latent heat of deformation* per unit mass [unit : J/kg],
- c_{v} heat capacity per unit mass at constant volume [unit : J/kg/K],

then

$$\rho\,\mathcal{L} = T\,\frac{\partial s}{\partial v}_{|T}\,,\quad c_{\mathrm{v}} = T\,\frac{\partial s}{\partial T}_{|v} = \frac{\partial u}{\partial T}_{|v}\,.$$ (3.3.4)

The last equality in the above relation results from the incremental definition of the internal energy $du = -p\,dv + T\,ds$. The terminology for the 'latent heat of deformation' is in analogy with the 'latent heat of phase change', a phenomenon that takes place at constant temperature. Truesdell [1984], p. 85, illustrates the idea by a simple example. If one compresses the air of a bike pump, its temperature increases and the pump itself becomes warmer. Now, one may submerge the pump in a large bath. Compressing the air no longer increases its temperature because the (latent) heat $dQ = T\,ds = \rho\,\mathcal{L}\,dv$ is absorbed by the bath.

The thermodynamic functions being Legendre transforms of one another, the knowledge of the free energy provides the energy, the enthalpy and the free enthalpy. However, we shall proceed in a different manner to build approximations of these functions for compressible and incompressible liquids.

3.3.1 The perfect or ideal gas

For a *perfect gas*[3.9], the constitutive relation may be expressed in the form $p\,\widehat{v} = RT$ or $p\,v = RT/\widehat{m}$, with $v = 1/\rho$ specific volume, $\widehat{m}$ the molar mass (equal to about 28.97 g/mole for dry air), $\widehat{v}$ the molar volume, and R the gas constant [unit : J/mole/K]. Then

$$\mathcal{L} = p\,v,\quad ds = \frac{R}{\widehat{m}}\frac{dv}{v} + c_{\mathrm{v}}\frac{dT}{T},\quad du = -p\,dv + T\,ds = c_{\mathrm{v}}\,dT\,.$$ (3.3.8)

For u to be a total differential, c_{v} can depend only on T, $c_{\mathrm{v}} = c_{\mathrm{v}}(T)$. In addition, if c_{v} can be considered a constant, then

$$u - u_0 = c_{\mathrm{v}}\,(T - T_0),\quad s - s_0 = \frac{R}{\widehat{m}}\,\mathrm{Ln}\,\frac{v}{v_0} + c_{\mathrm{v}}\,\mathrm{Ln}\,\frac{T}{T_0}\,,$$ (3.3.9)

with the subscript index 0 denoting a reference configuration.

[3.9]The history of the ideal gas equation starts as far as from the 17th century and culminates in the first half of the 19th century with the current analytical expression by Clapeyron. Soon afterwards, measurements of the sound speed at high pressures indicated that the equation was inaccurate.

A significant contribution to a more realistic adjustment to data is due to the Dutch physicist Johannes van der Waals (1837-1923) who, in 1873, introduced two positive material constants a and b with appropriate dimensions that account for molecular attraction and molar volume respectively, so as to modify the state equation to the format,

$$\left(p + \frac{a}{\widehat{v}^2}\right)(\widehat{v} - b) = RT\,.$$ (3.3.5)

Using (3.3.1), the free energy, the internal energy and the enthalpy of the van de Waals gas deduce successively,

$$\begin{aligned}
s &= s_0 + c_{\mathrm{v}}\,\mathrm{Ln}\,\frac{T}{T_0} + \frac{R}{\widehat{m}}\,\mathrm{Ln}\,\frac{\widehat{v} - b}{\widehat{v}_0 - b};\\[4pt]
e &= e_0 - s_0\,(T - T_0) - c_{\mathrm{v}}\left(T\,\mathrm{Ln}\,\frac{T}{T_0} - T + T_0\right) - \frac{RT}{\widehat{m}}\,\mathrm{Ln}\,\frac{\widehat{v} - b}{\widehat{v}_0 - b} - \frac{a}{\widehat{m}}\left(\frac{1}{\widehat{v}} - \frac{1}{\widehat{v}_0}\right);\\[4pt]
u &= u_0 + c_{\mathrm{v}}\,(T - T_0) - \frac{a}{\widehat{m}}\left(\frac{1}{\widehat{v}} - \frac{1}{\widehat{v}_0}\right);\\[4pt]
h &= h_0 + c_{\mathrm{v}}\,(T - T_0) + \frac{RT}{\widehat{m}}\left(\frac{\widehat{v}}{\widehat{v} - b} - \frac{\widehat{v}_0}{\widehat{v}_0 - b}\right) - 2\,\frac{a}{\widehat{m}}\left(\frac{1}{\widehat{v}} - \frac{1}{\widehat{v}_0}\right).
\end{aligned}$$ (3.3.6)

A modern alternative to the van der Waals relation that introduces two constants as well goes by the name of Redlich-Kwong [Smith et al. 1996, p. 82], namely

$$\left(p + \frac{a}{\sqrt{T}\,\widehat{v}\,(\widehat{v} + b)}\right)(\widehat{v} - b) = RT\,.$$ (3.3.7)

The constants a and b express in terms of the pressure and temperature at the critical point [Smith et al. 1996, p. 83].
For *real* gases, the compressibility factor $Z = p\,\widehat{v}/RT$ takes values both below and above 1. It is larger than 1 for hydrogen.

It is instrumental to introduce the heat capacity at constant pressure $c_{\mathrm{p}} \equiv T\,(\partial s/\partial T)_{|p}$ and the adiabatic coefficient γ,

$$c_{\mathrm{p}}(T) \equiv c_{\mathrm{v}}(T) + \frac{R}{\widehat{m}}, \quad \gamma \equiv \frac{c_{\mathrm{p}}(T)}{c_{\mathrm{v}}(T)} = 1 + \frac{R}{\widehat{m}}\frac{1}{c_{\mathrm{v}}(T)} > 1, \quad c_{\mathrm{p}}(T) = \gamma\,c_{\mathrm{v}}(T) = \frac{R}{\widehat{m}}\frac{\gamma}{\gamma - 1}. \tag{3.3.10}$$

The alternative set of variables (p, T),

$$T\,ds = -v\,dp + c_{\mathrm{p}}\,dT, \quad dh = v\,dp + T\,ds = c_{\mathrm{p}}\,dT, \tag{3.3.11}$$

yields, if c_{p} is a constant, explicit expressions of the enthalpy and entropy,

$$h - h_0 = c_{\mathrm{p}}\,(T - T_0), \quad s - s_0 = -\frac{R}{\widehat{m}}\,\mathrm{Ln}\,\frac{p}{p_0} + c_{\mathrm{p}}\,\mathrm{Ln}\,\frac{T}{T_0} = \frac{R}{\widehat{m}}\,\mathrm{Ln}\left(\frac{p_0}{p}\Big(\frac{T}{T_0}\Big)^{\gamma/(\gamma-1)}\right). \tag{3.3.12}$$

Under isentropic conditions, i.e., s constant, the three sets of equivalent relations are easily established,

$$
\begin{aligned}
\frac{T}{T_0} &= \left(\frac{\rho}{\rho_0}\right)^{\gamma - 1} &= \left(\frac{p}{p_0}\right)^{(\gamma-1)/\gamma}; \\[2mm]
\frac{p}{p_0} &= \left(\frac{\rho}{\rho_0}\right)^{\gamma} &= \left(\frac{T}{T_0}\right)^{\gamma/(\gamma-1)}; \\[2mm]
\frac{\rho}{\rho_0} &= \left(\frac{p}{p_0}\right)^{1/\gamma} &= \left(\frac{T}{T_0}\right)^{1/(\gamma-1)}.
\end{aligned}
\tag{3.3.13}
$$

3.3.2   Compressible and incompressible liquids

More generally, let us consider now fluids which are compressible, with a bulk modulus $K > 0$ [unit : Pa], and dilatable, with a cubic coefficient of thermal expansion c_{T} [unit : $1/\mathrm{K}$]), such that

$$\frac{dv}{v} = -\frac{dp}{K} + c_{\mathrm{T}}\,dT. \tag{3.3.14}$$

For a perfect gas,

$$K = p, \quad c_{\mathrm{T}} = 1/T. \tag{3.3.15}$$

Via (3.3.2), (3.3.4) the latent heat of deformation is found to be a function (in fact product) of the thermomechanical coefficients,

$$\rho\,\mathcal{L} = K\,c_{\mathrm{T}}\,T. \tag{3.3.16}$$

The entropy can be expressed in terms of the other variables once it is known in terms of one pair. Indeed, let

$$T\,ds = \rho\,\mathcal{L}\,dv + c_{\mathrm{v}}\,dT = \xi\,dp + c_{\mathrm{p}}\,dT = \zeta\,dv + \eta\,dp. \tag{3.3.17}$$

Then

$$c_{\mathrm{p}} - c_{\mathrm{v}} = K\,v\,c_{\mathrm{T}}^2\,T, \quad \xi = -v\,c_{\mathrm{T}}\,T, \quad \zeta = \frac{c_{\mathrm{p}}}{c_{\mathrm{T}}\,v}, \quad \eta = \frac{c_{\mathrm{v}}}{K\,c_{\mathrm{T}}}. \tag{3.3.18}$$

One may contrast the *isothermal bulk modulus*,

$$K = K^{\mathrm{isoth}} \equiv -v\,\frac{\partial p}{\partial v}_{|T}, \tag{3.3.19}$$

and the *isentropic* (or *adiabatic*) *bulk modulus*,

$$K^{\mathrm{isen}} \equiv -v\,\frac{\partial p}{\partial v}_{|s}. \tag{3.3.20}$$

Indeed,

$$-v\,\frac{dp}{dv} = K \begin{cases} 1 & \text{isothermal process,} \\ \gamma & \text{isentropic process.} \end{cases} \tag{3.3.21}$$

The adiabatic coefficient γ,

$$\gamma = \frac{c_{\mathrm{p}}}{c_{\mathrm{v}}} = \frac{K^{\mathrm{isen}}}{K^{\mathrm{isoth}}} = 1 + \frac{K\,v\,c_{\mathrm{T}}^2\,T}{c_{\mathrm{v}}}\,, \tag{3.3.22}$$

is about 1.3-1.7 for real gases at room temperature, close to 1 for water (1 at $4\,°\mathrm{C}$, 1.024 at room temperature and 1.08 at $80\,°\mathrm{C}$) and metals.

For a fluid which is neither compressible ($K = \infty$) nor dilatable ($c_{\mathrm{T}} = 0$),

$$\xi = 0; \quad T\,ds = du = c\,dT; \quad c = c_{\mathrm{v}} = c_{\mathrm{p}} = c(T)\,. \tag{3.3.23}$$

It seems physically reasonable to restrict the heat capacities and bulk moduli to positive values,

$$c_{\mathrm{p}} \geq c_{\mathrm{v}} > 0; \quad K^{\mathrm{isen}} \geq K = K^{\mathrm{isoth}} > 0\,. \tag{3.3.24}$$

A positive bulk modulus implies the volume to decrease when the pressure increases. While the concept is not elaborated, these conditions are required for the *stability of the thermodynamic equilibrium* [Mandel 1974, p. 111].

It is important to note that the properties derived so far in this section are *tangent* properties, and not *secant* properties, since they are defined through differentials, and not finite increments.

Compressibility and thermal expansion

Incompressibility requires a vanishing coefficient of thermal expansion. Indeed, for the difference between heat capacities at constant pressure and constant volume to tend to 0, the product $K\,c_{\mathrm{T}}^2$ should tend to 0 as well. On the other hand, since then $dp/dT = K\,c_{\mathrm{T}}$, the product $K\,c_{\mathrm{T}}$ is bounded.

Stability of equilibrium provides another argument against the existence of a fluid which would be incompressible but thermally dilatable. Indeed, then (3.3.14) and (3.3.17) would imply, under isentropic conditions, $dp = c_{\mathrm{p}}/(v\,c_{\mathrm{T}})\,dT/T = c_{\mathrm{p}}/(v\,c_{\mathrm{T}})^2\,dv/T$, that is, $dp/dv > 0$, or equivalently a negative isentropic bulk modulus, an inequality excluded by the stability conditions.

One would also expect the coefficient of thermal expansion to be positive, and, in fact, it does so for most substances. However, the fact that it might be negative or zero does not violate any principle. Indeed, water provides a conspicuous, although anomalous, example as illustrated by Figs. 3.2.1 and 3.3.1: the coefficient of thermal expansion changes sign and vanishes while the bulk modulus remains finite. The coefficient of thermal expansion of vitreous silica is negative at temperatures lower than 289 K. So do also the coefficients of thermal expansion of some tetrahedrally bonded materials at low temperatures, below 90 K for diamond.

3.3.3   Explicit thermodynamic functions for liquids

In order to build the thermodynamic functions in some ranges of pressure and temperature, we consider the specific volume and temperature as independent variables and we make successive assumptions, that will be seen to be appropriate for water [3.10]:

Assumption 3.1: constant bulk modulus K and coefficient of thermal expansion $c_{\mathrm{T}} = c_{\mathrm{T}}(T)$.

Note that the fact that the bulk modulus is constant implies the coefficient of thermal expansion to be independent of pressure, since the specific volume is a state variable, indeed $dp/K = -dv/v + c_{\mathrm{T}}\,dT$. The integrability condition for the entropy $ds = K\,c_{\mathrm{T}}\,dv + (c_{\mathrm{v}}/T)\,dT$ implies

$$c_{\mathrm{v}}(v,T) = c_{\mathrm{v}}(v_0,T) + K\,(v - v_0)\,T\,\frac{dc_{\mathrm{T}}}{dT}\,. \tag{3.3.25}$$

Then, with help of the relation (3.3.2) between partial derivatives,

$$\begin{aligned}
v &= v_0 \exp\left(-(p - p_0)/K + I_T(T, T_0)\right); \\[4pt]
s - s_0 &= c_{\mathrm{T}}\,K\,(v - v_0) + \int_{T_0}^{T} \frac{c_{\mathrm{v}}(v_0, x)}{x}\,dx; \\[4pt]
\mu - \mu_0 &= -K\,(v - v_0) - s_0\,(T - T_0) + K\,v_0\,I_T(T, T_0) - \int_{T_0}^{T}\int_{T_0}^{y} \frac{c_{\mathrm{v}}(v_0, x)}{x}\,dx\,dy\,.
\end{aligned} \tag{3.3.26}$$

[3.10] This simplified construction is left to be compared with the experimental measurements of the equation of state of distilled water and sea water, e.g., Fung [1977], p. 213. A comparison with the accurate formulation provided by the International Association for the Properties of Water and Steam (IAPWS) is presented in Section 3.3.4.

where

$$I_T(T, T_0) = \int_{T_0}^{T} c_{\mathrm{T}}(x)\, dx\,.\tag{3.3.27}$$

The index 0 refers to the state $(p = p_0, T = T_0)$. Therefore, a complete knowledge of the entropy and chemical potential requires the heat capacity $c_{\mathrm{v}} = c_{\mathrm{v}}(T)$ to be prescribed and the reference values of specific volume v_0, entropy s_0, and chemical potential μ_0 to be given. The reference value of chemical potential may be seen as containing some history of the species. In practice, it can be used as a degree of freedom to ensure chemical equilibrium of a species at the interface between two phases.

The status of the reference entropy is different: it affects the *variation* of the free energy and chemical potential as temperature varies. Kestin [1968] discusses the issue at length. He deduces from the third law of thermodynamics that the derivatives with respect to temperature of internal energy, free energy, enthalpy and free enthalpy should vanish as the temperature T tends to absolute zero, and so should the entropy, Kestin [1968], p. 364. Indeed for a reversible process, the change of entropy is equal to the heat supply per unit mass over absolute temperature, $ds = dq/T$. Since the heat capacity c is equal to dq/dT, then $ds = (c/T)\, dT$. For the entropy to be integrable from $T{=}0$ K onward, the heat capacity $c = c(T)$ should indeed vanish at $T{=}0$ K[Kestin 1968, p. 268]. Still, these considerations do not seem to be of much help at ambient temperature. In practice, a zero entropy at triple point is often assumed, e.g., Kestin [1978], p. 520. The current convention used by the thermodynamics community is made explicit in Section 3.3.4.

Assumption 3.2: in addition, the heat capacity $c_{\mathrm{v}}(v_0, T)$ is constant, $c_{\mathrm{v}0} = c_{\mathrm{v}}(v_0, T) = c_{\mathrm{v}}(v_0, T_0)$.

Then

$$s - s_0 = c_{\mathrm{T}}\, K\, (v - v_0) + c_{\mathrm{v}0} \operatorname{Ln} \frac{T}{T_0}\,,$$

$$\mu - \mu_0 = -K\, (v - v_0) + (c_{\mathrm{v}0} - s_0)(T - T_0) + K\, v_0\, I_T(T, T_0) - c_{\mathrm{v}0}\, T \operatorname{Ln} \frac{T}{T_0}\,,$$

$$h - h_0 = -K\, (1 - T\, c_{\mathrm{T}})\, (v - v_0) + c_{\mathrm{v}0}\, (T - T_0) + K\, v_0\, I_T(T, T_0)\,,\tag{3.3.28}$$

$$e - e_0 = -(p - p_0)\, v - (p_0 + K)\, (v - v_0) + (c_{\mathrm{v}0} - s_0)\, (T - T_0) + K\, v_0\, I_T(T, T_0) - c_{\mathrm{v}0}\, T \operatorname{Ln} \frac{T}{T_0}\,,$$

$$u - u_0 = -(p - p_0)\, v - (p_0 + K\, (1 - c_{\mathrm{T}}\, T))\, (v - v_0) + c_{\mathrm{v}0}\, (T - T_0) + K\, v_0\, I_T(T, T_0)\,,$$

with references expressed in terms of the quantities (p_0, T_0), v_0 and (s_0, μ_0), and

$$h_0 = \mu_0 + T_0\, s_0, \quad e_0 = \mu_0 - p_0\, v_0, \quad u_0 = \mu_0 + T_0\, s_0 - p_0\, v_0\,.\tag{3.3.29}$$

The chemical potential shows an affine dependence on the reference entropy, the internal energy shows an affine dependence on the reference pressure, and the free energy shows an affine dependence on both the reference entropy and the reference pressure.

Upon linearization to first order about the reference state (p_0, T_0), while maintaining the exact temperature dependence of the coefficient of thermal expansion,

$$s - s_0 = -c_{\mathrm{T}}\, v_0\, (p - p_0) + \frac{c_{\mathrm{p}}'}{T_0}\, (T - T_0),$$

$$\mu - \mu_0 = v_0\, (p - p_0) - s_0\, (T - T_0),$$

$$h - h_0 = (1 - T_0\, c_{\mathrm{T}})\, v_0\, (p - p_0) + c_{\mathrm{p}}'\, (T - T_0)\,,\tag{3.3.30}$$

$$e - e_0 = \frac{p_0 v_0}{K}\, (p - p_0) - s_0\, (T - T_0) - p_0\, v_0\, I_T(T, T_0)\,,$$

$$u - u_0 = (\frac{p_0}{K} - c_{\mathrm{T}}\, T_0)\, v_0\, (p - p_0) + c_{\mathrm{p}}'\, (T - T_0) - p_0\, v_0\, I_T(T, T_0)\,.$$

The *secant* heat capacity c_{p}'

$$c_{\mathrm{p}}' = c_{\mathrm{v}0} + K\, v_0\, c_{\mathrm{T}}\, T_0\, \frac{I_T(T, T_0)}{T - T_0}\,,\tag{3.3.31}$$

reduces to c_p if the coefficient of thermal expansion is a constant.

Under the above assumptions, the isothermal and isentropic bulk moduli, $K^{isoth} = K$ and K^{isen} respectively, relate as follows,

$$K^{isen} = \frac{c'_p}{c_{v0}} K, \tag{3.3.32}$$

and the pressure increment may be expressed in either format,

$$p - p_0 = \begin{cases} -K \dfrac{v - v_0}{v_0} + K \dfrac{I_T(T, T_0)}{T - T_0} (T - T_0); \\[2ex] -K^{isen} \dfrac{v - v_0}{v_0} + \dfrac{K}{c_{v0}} \dfrac{I_T(T, T_0)}{T - T_0} T_0 (s - s_0). \end{cases} \tag{3.3.33}$$

If, in addition to the two first assumptions,

Assumption 3.3: the material is neither compressible ($K = \infty$) nor dilatable ($c_T = 0$), then, the heat capacities are equal and constant, $c_p = c_v = c_{v0} = c$, and

$$s - s_0 = c \operatorname{Ln} \frac{T}{T_0},$$

$$\mu - \mu_0 = v_0 (p - p_0) + (c - s_0)(T - T_0) - c T \operatorname{Ln} \frac{T}{T_0},$$

$$h - h_0 = v_0 (p - p_0) + c (T - T_0), \tag{3.3.34}$$

$$e - e_0 = (c - s_0)(T - T_0) - c T \operatorname{Ln} \frac{T}{T_0},$$

$$u - u_0 = c (T - T_0).$$

In this limit case, the pressure influence survives in the enthalpy and free enthalpy only.

The actual dependence of the thermodynamic functions with respect to pressure and temperature are quantified through contour levels in subsequent plots.

Table 3.3.1: Some physical properties of pure water at 1 atmosphere (101 325 Pa) calculated from (1) IAPWS [2009]; (2) IAPWS [1984].

Temperature T [°C]	Mass Density ρ [kg/m³] (1)	Heat Capacity c_p [kJ/kg/K] (1)	Thermal Expansion c_T [10^{-3}/K] (1)	Bulk Modulus K [GPa] (1)	Dynamic Viscosity [centipoise=10^{-3} Pa×s] (1), (2)
0 (ice)	916.8	1.96	0.158	~10	$10^{16} - 10^{17}$
0 (liquid)	999.843	4.219	-0.068	1.965	1.791
2	999.943	4.213	-0.0326	1.994	1.674
4	999.975	4.207	0.3×10^{-6}	2.021	1.569
20	998.210	4.184	0.207	2.179	1.003
40	992.218	4.179	0.385	2.261	0.653
60	983.197	4.185	0.523	2.247	0.467
80	971.792	4.197	0.641	2.167	0.355
100 (liquid)	958.350	4.216	0.751	2.040	0.282
100 (steam)	0.598	2.080	2.902	10^{-4}	0.012

Table 3.3.1 displays a few physical properties of pure water over the range 0.01 °C to 100 °C. The dynamic viscosity of pure water is equal to 1 centipoise at 20 °C. The calorie (1 cal=4.185 J) is the energy required to increase the temperature of one gram of water from 14.5 °C to 15.5 °C under atmospheric pressure. The heat capacity of water is, after ammonia, the next largest among liquids. On the other hand, water is a poor conductor of heat, or equivalently a good insulator, Table 9.1.2. These properties confer to water and ice a significant resistance to changes of temperature. Heat capacities of biological tissues are of the same order, although a bit smaller [Duck 1990]:
- for blood: heat capacity c_p=3.84 kJ/kg/K; mass density 1.06 kg/m³;

Table 3.3.2: Mass density and heat capacity of an aqueous solution
at $20\,°C$ and under atmospheric pressure with various concentrations of NaCl,
obtained by interpolation from the data of de Marsily [1986], p. 417.

Concentration	Mass/Volume of Solution	Heat Capacity c_p
(mole/liter)	[kg/m^3]	[kJ/kg/K]
0	998.2	4.182
0.15	1004.4	4.133
0.50	1018.7	4.028
1.0	1038.6	3.891

- for human liver: heat capacity c_p=3.60 kJ/kg/K; mass density $1.06\,\text{kg/m}^3$.

Similarly, heat capacity of sea water is slightly smaller than for pure water, Table 3.3.2.

Note the relation between cross-derivatives, $T\,\partial(-v\,c_T)/\partial T = \partial c_p/\partial p$. Therefore, in the range of temperatures of interest here, $\partial c_p/\partial p$ is about $-2.75\times 10^{-6}\,\text{m}^3/\text{kg}/\text{K}$, and the relative variation $\Delta c_p/c_p$ is equal to $-0.65\times 10^{-9}\,\Delta p$, which is quite tiny for a change of a few atmospheres. Consequently, the model yields a heat capacity c_p fairly independent of pressure in the range of interest. However, this property should be manipulated with care since, if adopted from the start, it would have led to $v\,c_T$ independent of temperature, which is definitely inaccurate. Smith et al. [1996], p. 637, provide polynomial approximations of the heat capacities of a number of gases, liquids and solids, e.g., for water (the unit is that of the gas constant R),

$$\frac{c_p(T)}{R} = 8.712 + 1.25\times 10^{-3}\,T - 0.18\times 10^{-6}\,T^2, \quad T \in]273.16\,\text{K}, 373.125\,\text{K}[. \tag{3.3.35}$$

The error induced by Assumption 3.2 can also be estimated: in view of (3.3.25), it amounts to a maximum error of about 10% on the heat capacities. Here, emphasis has been laid to accommodate an arbitrary temperature dependence of the coefficient of thermal expansion. If, on the other hand, accurate heat capacities are needed, alternative assumptions should be devised.

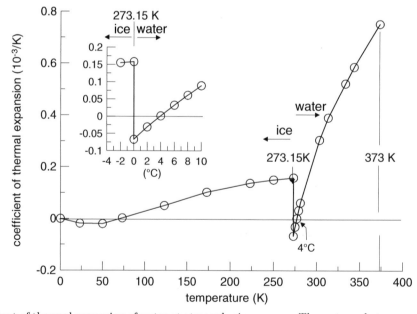

Fig. 3.3.1: Coefficient of thermal expansion of water at atmospheric pressure. The water substance presents an anomalous behavior. As temperature increases, the coefficient of thermal expansion undergoes a jump at $0\,°C$, from a positive value in the solid phase to a negative value in the liquid phase. It keeps this negative value from $0\,°C$ to about $4\,°C$: in this interval, increase of temperature at constant pressure leads to volume decrease. From Kestin [1968], Table XXXIV, p. 541.

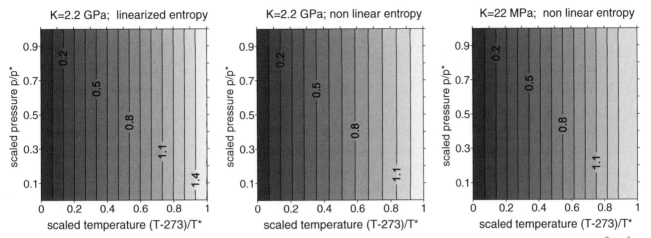

Fig. 3.3.2: Contours of entropy of water [unit : kJ/kg/K] for a coefficient of thermal expansion $c_T = 0.4 \times 10^{-3}\,K^{-1}$, a heat capacity $c_{v0} = 4.02\,kJ/kg/K$, and the reference values for pressure $p_0 = 611\,Pa$, temperature $T_0 = 273.16\,K$, specific volume $v_0 \sim 10^{-3}\,m^3/kg$, entropy $s_0 = 0\,J/kg/K$, and free enthalpy $\mu_0 = 0\,J/kg$. The scaling temperature and pressure are respectively $T* = 100\,K$, and $p* = 10\,MPa$. The left and center contour plots display the linearized and nonlinear entropies respectively, for a bulk modulus $K = 2.2\,GPa$, and the right contour plot displays the nonlinear entropy for $K = 0.022\,GPa$.

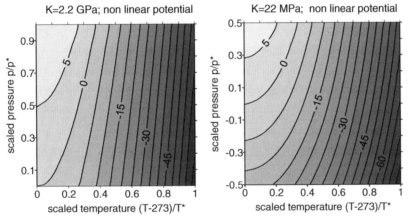

Fig. 3.3.3: Chemical potential of water [unit : kJ/kg] as a function of temperature and pressure.

The coefficient of thermal expansion of the water substance undergoes a negative jump at $0\,°C$, Fig. 3.3.1. It increases monotonically between $0\,°C$ and $100\,°C$, passing through a null value just below $4\,°C$[Kestin 1968, p. 264]. In the liquid state, the affine expression [unit : 10^{-6}/K],

$$c_T(T) = -67 + 8.19\,(T - 273.16), \quad T \in\,]273.16\,K, 373.125\,K[, \tag{3.3.36}$$

is exact at $0\,°C$ and $100\,°C$, vanishes at $8\,°C$, and is below the actual value throughout the temperature interval. The quadratic expression [unit : 10^{-6}/K],

$$c_T(T) = -67 + 15\,(T - 273.16) - 0.068\,(T - 273.16)^2, \quad T \in\,]273.16\,K, 373.125\,K[, \tag{3.3.37}$$

is exact at $0\,°C$ and $100\,°C$, and vanishes at $4.5\,°C$. Overall, it represents a reasonably good approximation between $0\,°C$ and $50\,°C$, but slightly overestimates the actual expansion coefficient above $50\,°C$.

The analytical expressions (3.3.28), and the linearized expressions (3.3.30), are employed to obtain contour plots of the thermodynamic functions in a window of temperature and pressure, Figs. 3.3.2 to 3.3.6. The thermal dependence is much larger than the pressure dependence. This observation applies fully to the entropy but more moderately to

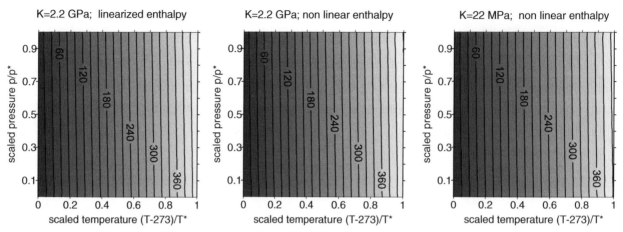

Fig. 3.3.4: Enthalpy of water [unit : kJ/kg] as a function of temperature and pressure.

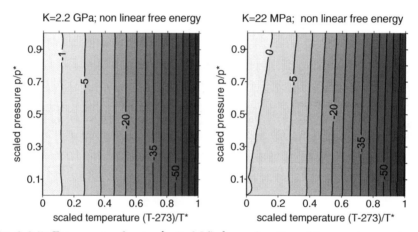

Fig. 3.3.5: Free energy of water [unit : kJ/kg] as a function of temperature and pressure.

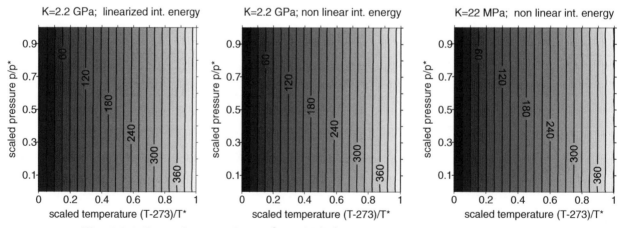

Fig. 3.3.6: Internal energy of water [unit : kJ/kg] as a function of temperature and pressure.

the chemical potential. Plots show that it holds true even if water contains air bubbles that reduce the bulk modulus by two orders of magnitude.

Figs. 3.3.2 to 3.3.6 also allow to assess the accuracy of the linearized expressions (3.3.30) with respect to the nonlinear relations (3.3.28). Entropy, internal energy and enthalpy are reasonably approximated by their linearized expressions. On the other hand, the first order linearized expressions of chemical potential and free energy are quite inaccurate and they are not displayed. These observations justify the developments of thermoelastic models in Chapter 9 where first order linearized expressions of the sole entropy and internal energy are used.

The thermal dependence of the chemical potential is encapsulated in

Proposition 3.2 on the dependence of chemical equilibrium with respect to pressure and temperature.

For temperatures in the range [0.01 °C, 100 °C], and pressures of a few atmospheres, chemical equilibrium is sensitive to disturbances of both pressure and temperature.

In the incompressible and inextensible limit, with reference values $T_0=273.16$ K and $s_0 = 0$, an increase of the temperature from 25 °C to 26 °C and an increase of pressure by 350 kPa have equal and opposite contributions to the chemical potential (3.3.34). The equivalent pressure differential falls to 7 kPa if the temperature changes from 0.01 °C to 1 °C, as can be observed in Fig. 3.3.3.

Indeed, consider a membrane to separate the fluid of a bucket in two parts. Each side of the membrane has its own pressure and temperature. Mass exchange between the two sides is controlled by the difference of chemical potentials (or free enthalpy), Section 3.1.3. A small difference in temperature results in a substantial difference of chemical potentials. If the membrane is permeable to the fluid, a fluid transfer will take place so as to equilibrate the chemical potentials. How exactly ? Well, to answer this question, a constitutive equation of mass transfer should be available. The conditions that control this physical phenomenon and specific linear and nonlinear laws are provided in Chapter 10.

Chemical equilibrium alone, namely equality of the chemical potentials, does not guarantee mechanical and/or thermal equilibria between two compartments. It simply provides a single relation between the two pressures and temperatures. As a very particular example, assume that the temperatures in the two compartments and the pressure in one compartment are controlled. Given reference values (which may well not be the same for the two compartments), chemical equilibrium will then provide the unknown pressure.

Thermal equilibrium may not hold in specific circumstances. Consider a porous medium with two types of porosities, one associated with transport, the other with fluid storage. In the context of soft tissues, vessels provide the first type of porosity[3.11] and pores of the interstitium the second type. Fast transport through the vessels may bring the two fluids at distinct temperatures in sudden local contact. A model of this flavor is developed in Section 10.6.

Thermal equilibrium may be reached if the membrane is a thermal conductor. Similarly, chemical equilibrium can be realized only if the membrane is permeable to the fluid.

At this point, as we are defining equilibria, it is of utmost importance to insist that these equilibria are *local*, that is, they are defined at a geometric point of the porous medium viewed as a continuum.

3.3.4 Liquid-vapor equilibrium curve: IAPWS and simplified formulations

Along the liquid-vapor equilibrium curve, the liquid (l) and vapor (v) phases are in thermal equilibrium (equal temperatures), mechanical equilibrium (equal pressures in absence of vapor bubbles in the liquid phase and liquid drops in the vapor phase) and chemical equilibrium (equal chemical potentials). On the other hand, the specific volumes, or mass densities, are distinct.

Two derivations of the thermodynamic properties of water along the equilibrium curve are considered:
- the first derivation uses the data provided by the International Association for the Properties of Water and Steam (IAPWS), IAPWS [2002] and IAPWS [2009]. The model is accurate over a wide range of temperatures and pressures from the triple point to the critical point, including in particular, but not restricted to, the equilibrium curve. It is unified, in the sense that the liquid and vapor phases use the same constitutive functions which are phrased in terms of mass density (or specific volume) and temperature;
- the second derivation considers vapor as an ideal gas described in Section 3.3.1 and uses the simplified model for the liquid phase exposed in Sections 3.3.2 and 3.3.3. These relations are expected to be reasonably accurate in a restricted domain of temperature and pressure, roughly between 0.01 °C and 100 °C and for a few MPa's;

[3.11]The term porosity is used to denote a certain type of pore geometry as well as the volume fraction associated with all porosity types.

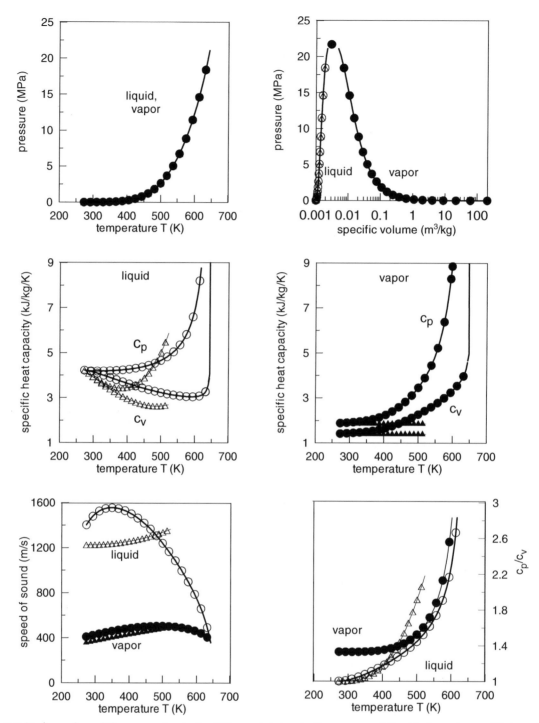

Fig. 3.3.7: A number of thermomechanical entities pertaining to the vapor and liquid along the equilibrium (saturation) curve from 0.01 °C to the critical point.

Open symbols: liquid; solid symbol: vapor; circles: IAPWS unified model; triangles: simplified model. Note that some curves of the IAPWS and simplified models almost superpose while some others are quite distinct.

At the critical point,

- the isobaric and isochoric heat capacities of liquid and vapor both tend to infinity but in such a way that the adiabatic coefficient c_p/c_v, which is always larger than 1, also tends to infinity;
- the isothermal speed of sound, $w^2 = \partial p/\partial \rho_{|T}$, vanishes and the isentropic speed of sound shown on the plot, namely $w_{\text{isen}}^2 = \partial p/\partial \rho_{|s}$, almost vanishes as well ($w_{\text{isen}} = 7.79\,\text{m/s}$).

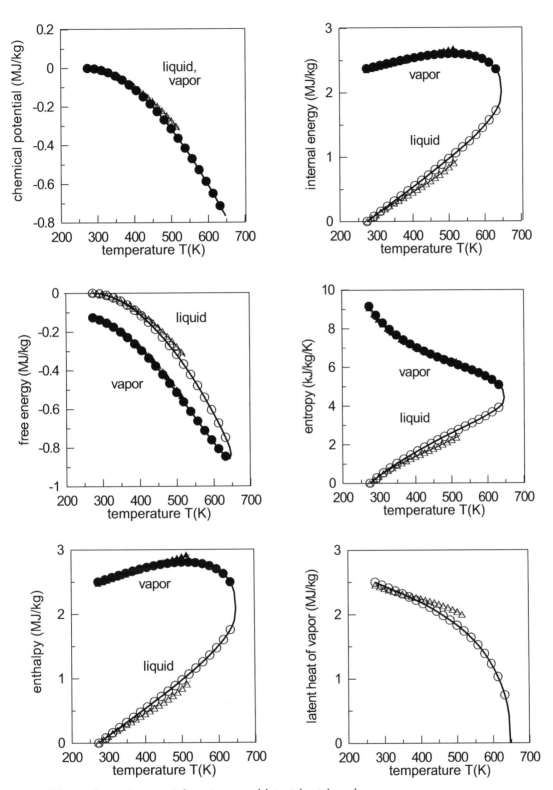

Fig. 3.3.8: Thermodynamic potentials, entropy and latent heat $h_v - h_l$.
- The latent heat decreases moderately with temperature;
- For the liquid, internal energy and entropy vanish at the triple point ($T = 273.16\,\text{K}$). They are slightly negative at $0\,^\circ\text{C}$ ($T = 273.15\,\text{K}$). A sign change between $0\,^\circ\text{C}$ and the triple point also takes place for the enthalpy and chemical potential.

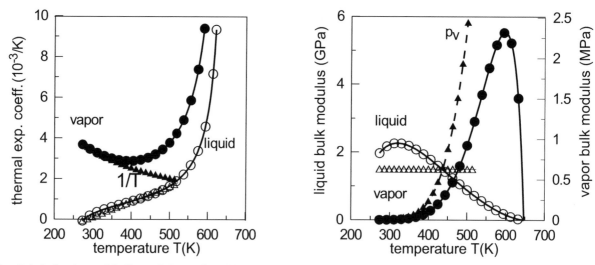

Fig. 3.3.9: Isothermal bulk modulus and coefficient of thermal expansion.
- The coefficient of thermal expansion of the liquid is negative between 0 and about $4\,°C$;
- At the critical point, the coefficient of thermal expansion becomes infinite and the bulk modulus vanishes.
Note: for an ideal gas, the isothermal bulk modulus is equal to the pressure and the coefficient of thermal expansion is equal to the coldness (inverse temperature).

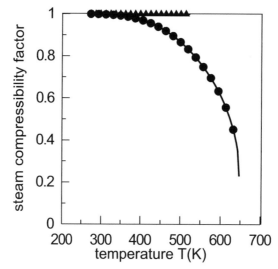

Fig. 3.3.10: The steam compressibility factor $Z \equiv p/(RT\,\widehat{v}_v)$ of the IAPWS model decreases monotonously along the equilibrium curve from the triple point to the critical point. It is equal to 1 for an ideal gas.

Both models provide other entities of interest, namely the isochoric heat capacity $c_v = T\,(ds/dT)_{|v}$ and
- the coefficient of thermal expansion,

$$c_{\mathrm{T}} = -\frac{1}{\rho}\left(\frac{d\rho}{dT}\right)_{|p};$$

- the isothermal and isentropic moduli,

$$K_T = \rho\left(\frac{dp}{d\rho}\right)_{|T}, \quad K_{\mathrm{isen}} = \rho\,(\frac{dp}{d\rho})_{|s} = K_T\left(1 + \frac{K_T\,c_{\mathrm{T}}^2\,T}{\rho\,c_v}\right);$$

- the isobaric heat capacity c_p,

$$c_p = c_v\left(1 + \frac{K_T\,c_{\mathrm{T}}^2\,T}{\rho\,c_v}\right);$$

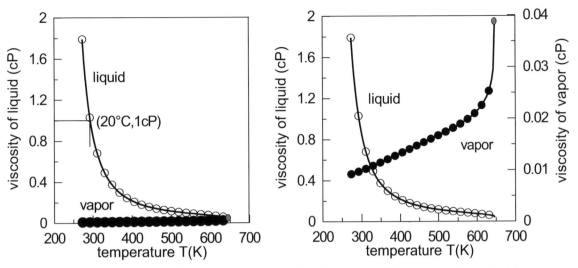

Fig. 3.3.11: The viscosity of the liquid decreases monotonously along the equilibrium curve from the triple point to the critical point. The opposite takes place for the steam. The common value of the viscosity at the critical point is equal to 0.038 990 cP.

- the isothermal and isentropic speeds of sound, defined respectively by their square as

$$w_T^2 = \left(\frac{dp}{d\rho}\right)_{|T} = \frac{K_T}{\rho}, \quad w_{\text{isen}}^2 = \left(\frac{dp}{d\rho}\right)_{|s} = \frac{K_{\text{isen}}}{\rho}.$$

3.3.4.1   The equilibrium curve

The equilibrium curve is sought by enforcing the liquid water and its steam to be simultaneously in thermal, mechanical and chemical equilibria,

$$T_l = T_v, \quad p_l = p_v, \quad \mu_l = \mu_v. \tag{3.3.38}$$

The equilibrium curve is obtained by varying the temperature over a given interval. At each point T of the interval, the constitutive equations for the pressures $p(v_l, T)$ and $p(v_v, T)$ and chemical potentials $\mu(v_l, T)$ and $\mu(v_v, T)$ inserted in the thermal and chemical equilibria provide the specific volumes v_l and v_v, and by the same token the common pressure.

3.3.4.2   Reference thermodynamic potentials along the IAPWS conventions

Along the conventions of the IAPWS, internal energy and entropy of the liquid are set to vanish at the triple point (subscript t),

$$s_{tl} = 0, \quad u_{tl} = 0, \tag{3.3.39}$$

which implies,

$$e_{tl} = 0, \quad h_{tl} = \mu_{tl} = p_t v_{tl}. \tag{3.3.40}$$

The equilibrium of the pressures and chemical potentials at $T_t = 273.16\,°\text{C}$ provides the pressure p_t and specific volumes v_{tl} and v_{tv} at the triple point t reported in Table 3.2.2. The entropy being provided directly by the IAPWS model, the other potentials of vapor at the triple point result in turn as,

$$\mu_{tv} = \mu_{tl}, \quad s_{tv} = 9.156\,\text{kJ/kg/K}, \quad h_{tv} = \mu_{tv} + T_t s_{tv}, \quad u_{tv} = h_{tv} - p_t v_{tv}, \quad e_{tv} = u_{tv} - T_t s_{tv}. \tag{3.3.41}$$

3.3.4.3 The simplified model for water and steam

For the simplified model, the following assumptions are adopted.

Liquid properties

Three properties need to be defined. The bulk modulus K_l is set to 1.5 GPa, and rough approximations are used for the coefficient of thermal expansion $c_{\mathrm{T}l}$ and heat capacity $c_{\mathrm{v}l}(v_0, T)$,

$$c_{\mathrm{T}l} = \big(-67 + 8.19 \times (T - T_t)\big) \times 10^{-6}/\mathrm{K},$$
$$c_{\mathrm{v}l}(v_0, T) = 4217 - 11 \times (T - T_t) + 0.001 \times (T - T_t)^2 \quad \mathrm{J/kg/K}. \tag{3.3.42}$$

The isochoric heat capacity $c_{\mathrm{v}l}(v, T)$ is obtained from $c_{\mathrm{v}l}(v_0, T)$ by (3.3.25) and the isobaric heat capacity $c_{\mathrm{p}l}(v, T)$ derives from the latter by (3.3.18).

Vapor properties

A single entity needs to be fixed: the isochoric heat capacity $c_{\mathrm{v}v}$ is set to the constant value 1.418 kJ/kg/K.

3.3.4.4 Quantitative assessment of the simplified model

Figs. 3.3.7 to 3.3.10 show that some curves of the IAPWS and simplified models nicely superpose. The equilibrium curve provided by the simplified model is acceptable although some error with respect to the IAPWS model is visible. The principal source of error is due to the constant bulk modulus of the liquid. The heat capacities should be approximated more accurately than by a second order polynomial.

The viscosities along the equilibrium curve shown in Fig. 3.3.11 are derived from the expressions suggested in IAPWS [1984].

3.4 Laplace's law of mechanical equilibrium at the interface between two immiscible fluids

Laplace's law links the tension developed in the interface that separates two immiscible fluids to the pressures on each side of this interface. The interface is considered impervious to the fluids at the time scale of application of the law. Laplace's law is advocated in many circumstances. It applies to the meniscus separating two non miscible fluids in a tube, Section 3.2.3, or in a porous medium, Section 3.2.4. Controlling the surface tension of bubbles and droplets in emulsions through surfactants is a key issue in chemical engineering. Formally, it is also used for some solid structures: it provides the orthoradial stress in a solid vessel wall of given radius and thickness, the blood pressure and interstitium pressure being known, Section 15.9.2.

Laplace's law can be proved using work conservation or equilibrium of forces. Work conservation indicates that the work done by the tension T [unit : N/m] in an incremental change of the area of the membrane dS [unit : m^2] is equal to the work done by the pressure differential Δp [unit : N/m^2] across the membrane in the associated incremental change of volume δV [unit : m^3] enclosed by the membrane. Therefore,

$$T\,\delta S - \Delta p\,\delta V = 0 \quad \Rightarrow \quad \Delta p = T\,\frac{\delta S}{\delta V}. \tag{3.4.1}$$

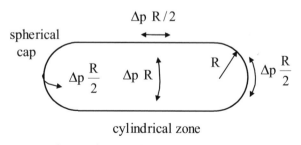

cylindrical zone

Fig. 3.4.1: In an elastic cylinder closed with two spherical caps, the orthoradial tension over the central part is twice the tension over the caps.

The area and volume of a spherical membrane of radius R are respectively equal to $S = 4\pi R^2$ and $V = 4/3\,\pi R^3$. For a cylindrical membrane of unit axial length, these quantities become $S = 2\pi R$ and $V = \pi R^2$. Therefore we have reached explicit expressions of Laplace's law,

$$\text{sphere}: \ \Delta p = 2\,\frac{T}{R}; \quad \text{cylinder}: \ \Delta p = \frac{T}{R}. \tag{3.4.2}$$

As an alternative to the energy method, these formulas can also be arrived at using the force equilibrium of a half sphere or of a half cylinder. Fig. 3.4.1 shows an immediate illustration.

A more general form of Laplace's law results by considering an infinitesimal surface, assuming its radii of curvature R_x and R_y to be known along two orthogonal directions (not necessarily the principal directions). The proof considers the two directions in turn, Fig. 3.4.2. Projecting onto the x-bisector the two forces $T\,dy$ inclined at an angle $dx/(2\,R_x)$ with respect to this bisector yields the component of the tension $T/R_x\,dx\,dy$ along the inward normal to the elementary surface. The orthogonal direction is treated similarly and the two contributions are summed. At equilibrium, the inward force of the membrane $T\,(1/R_x+1/R_y)\,dx\,dy$ counterbalances the outward contribution of the pressure $\Delta p\,dx\,dy$, yielding

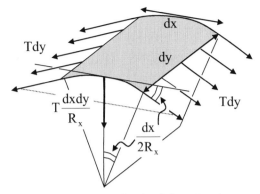

Fig. 3.4.2: Sketch to prove Laplace's law showing an elementary surface cut along two orthogonal directions.

$$\Delta p = T\left(\frac{1}{R_x} + \frac{1}{R_y}\right). \tag{3.4.3}$$

The convention is that the pressure differential Δp is equal to the internal minus the external pressure. The curvature is counted positive if the center of curvature is located in the inner side delineated by the membrane.

Actually, Laplace's law assumes that the tension is identical in all directions of the surface. Therefore, it does not apply to an elastic cylinder: it provides the correct circumferential stress simply because the axial stress is elusive due to an infinite axial radius of curvature. As it has been stated, Laplace's law also assumes a continuous radius of curvature, excluding a priori the interface between a cylinder and a sphere. In fact, energy minimization implies small water droplets to be spherical, Eqn (3.4.1). Gravity and viscous friction tend to modify this shape. The aspiration of vesicles bounded by a fluid lipid membrane through micropipette techniques induces significant changes of curvature in the entrance of the pipette: a correct description of the induced geometry requires a modification of Laplace's law by the introduction of adhesion and curvature energies.

3.5 Henry's law and Raoult's law of chemical equilibrium between a liquid and a gas

Consider first a mixture of gases in a volume V_G. Each gas occupies the *total* volume V_G. Let us consider the gas noted g with N_{gG} moles, and let N_G be the total number of moles of gas in the volume V_G. According to **Dalton's law**, each gas of the mixture contributes additively to the total gas pressure. The contribution p_{gG} of gas g, referred to as partial pressure, to the total pressure p_G is proportional to its molar fraction $x_{gG} = N_{gG}/N_G$,

$$p_{gG} = x_{gG}\,p_G. \tag{3.5.1}$$

For example, the molar fraction of oxygen in air is about 21% so that, under standard conditions, the pressure of oxygen is 0.21 atmosphere$\sim$160 mmHg.

Now let us consider the above gas mixture in equilibrium with a liquid phase L that contains the gas in dissolved form. Let $x_{gL} = N_{gL}/N_L$ be the molar fraction of dissolved gas and $c_{gL} = N_{gL}/V_L$ its concentration.

Henry's law sketched in Fig. 3.5.1 states that, at low gas solute molar fraction (less than 0.03), the molar fraction and concentration of the dissolved gas are proportional to its partial pressure in the gas phase and the gas concentrations in the liquid and gas phase are proportional,

$$x_{gL} = H_{gx}\,p_{gG}, \quad c_{gL} = H_{gc}\,p_{gG} = H_{gcc}\,c_{gG}. \tag{3.5.2}$$

Here H_{gx} [unit : Pa^{-1}], $H_{gc} = H_{gx}/\widehat{v}_L$ [unit : mole/m^3/Pa] and $H_{gcc} = RT\,H_{gc}$ [unit : 1] are Henry's coefficients for this gas and this liquid and $\widehat{v}_L$ is the molar volume of the liquid, i.e., 18 cm^3/mole for water. At 25 °C, Henry's coefficient H_{gc} associated with the dissolution of oxygen in water is equal to 0.013 mmole/l/kPa=1.3×10^{-5} mole/m^3/Pa while it is more than 20 times higher for carbon dioxide, namely 0.34 mmole/l/kPa=34×10^{-5} mole/m^3/Pa. The associated coefficients H_{gcc} are thus equal to 0.032 and 0.83 respectively.

The partial pressure of carbon dioxide in carbonated drinks is high during bottling. When the can is opened, the gas pressure decreases and a great part of the dissolved gas is released. When the ascent of divers is too fast (larger

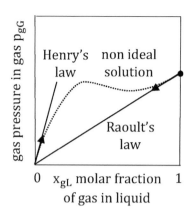
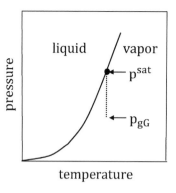

Fig. 3.5.1: At an interface between a gas phase and a liquid phase, the pressure differential is dictated by Laplace's law. At low value, the molar fraction of gas dissolved in the liquid follows Henry's law. At high value, according to Raoult's law, it is equal to the ratio of the pressure of the gas component in the gas phase divided by its saturating value.

than the popular 10 m per minute), bubbles of nitrogen, coming from the air that they breathe, form in their blood and can not vaporize. In fact, with $H_{gc} = 0.60 \times 10^{-5}$ mole/m^3/Pa Henry's coefficient for nitrogen dissolution in water/blood at 25 °C, the concentration of dissolved nitrogen decreases from 5.5 mM at a depth of 100 m to 0.48 mM at sea level, where the partial pressure of nitrogen is equal to 0.80 atmospheric pressure.

Henry's coefficient decreases with temperature: boiled water tastes 'flat' as it has lost its oxygen gas. The coefficient is roughly halved from 0 °C to 40 °C. Around $T_0 = 273.15$ K, the decrease is described by the analytic formula $H_{gx}(T) = H_{gx}(T_0) \exp(\beta (T^{-1} - T_0^{-1}))$ with β equal to 1700 K for oxygen and 2400 K for carbon dioxide. In fact, the exponent β may be seen as linked to the change of enthalpy of the solution, $\beta = -(h_{gL} - h_{gG})/R = d\mathrm{Ln}\, H_{gx}(T)/d(1/T)$. For most gases, this exponent is positive, the enthalpy change during dissolution is negative and the reaction is exothermic.

Since H_{gc} is the nb of moles of gas per unit gas pressure (in the gas phase) per unit volume of liquid, and the molar volume of gas is equal to RT/p_{gG}, the volume of the gas at the pressure p_{gG} that is dissolved in the volume V_L of liquid is equal to

$$V_g^{\mathrm{dissolved}} = H_{gc} \times RT/p_{gG} \times p_{gG} \times V_L = H_{gcc} \times V_L .\tag{3.5.3}$$

At $T = 25$ °C, $RT = 8.3145 \times 298 = 2477$ J/mole and therefore $H_{gcc} = 2477 \times 1.3 \times 10^{-5} = 0.00322$ for oxygen dissolution in water. Therefore, about 16.1 ml of oxygen is dissolved in the 5 l of blood of the human body. This quantity is far from being sufficient to ensure survival as the human consumption of oxygen is about 250 ml per second. In fact, Section 15.11.1 highlights the fact that most of oxygen is transported by blood through devoted carriers, namely the red blood cells.

Raoult's law states that, even at high gas concentration, the partial gas pressure in the gas phase is proportional to the molar fraction of the dissolved gas times the saturating gas pressure,

$$p_{gG} = x_{gL}\, p^{\mathrm{sat}} .\tag{3.5.4}$$

Except for special ideal mixtures, Henry's law applies to low concentrations and Raoult's law to high concentrations so that H_{gx} is in general not the inverse of the saturating pressure $p^{\mathrm{sat}}(T)$. Henry's law provides an approximation of the real behavior of the solute while Raoult's law provides an approximation of the real behavior of the solvent.

For a more thorough thermodynamic treatment of adsorption of gases to a solid, the reader is directed to Smith, van Ness and Abbott [1996], p. 545.

3.6   Composition of fluids in plasma, interstitium, cell

Blood participates in
- the transport of gases, oxygen and carbon dioxide, nutrients, metabolic substances, vitamins, etc.;
- the transport of heat;
- the transport of signals, like hormones;
- maintaining a constant pH by providing a large buffer capacity;
- the defense of the body via plasma proteins.

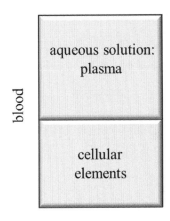

Fig. 3.6.1: Composition of blood, APB [1999], p. 261.
- Plasma proteins 60-80 g/l:
    albumin 45 g/l, α_1-, α_2-, β-, γ-globulins
- Serum = plasma - clotting factors (e.g., fibrinogen)
- Hematocrit = cellular elements/blood   45% vol.
- Cellular elements:
    - Erythrocytes (red blood cells, anucleated in blood)
    - Leukocytes (white blood cells):
        monocytes 6%, lymphocytes 27%, granulocytes 67%
    - Thrombocytes (platelets, anucleated)

Fig. 3.6.1 indicates that the aqueous part of blood, also called *plasma*, occupies about 55% of its volume. Plasma is obtained by centrifugation. Plasma contains water within which are dissolved charged proteins, neutral species (glucose, urea, etc.) and metallic ions. The composition of plasma, interstitium and cell are compared in Fig. 3.6.2.

The total concentration of plasma proteins is about 60 to 80 g/l. Albumin is the plasma protein with the largest concentration, Tables 3.6.1 and 3.6.2. The plasma proteins participate to the defense of the body, define the oncotic pressure, help in the transport of substances which are not soluble in water, and interact with drugs. Note that, for

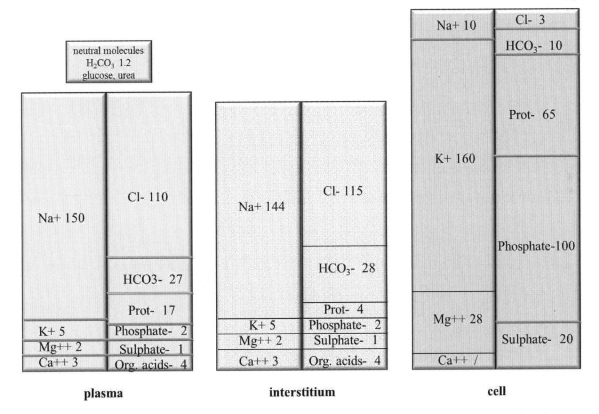

Fig. 3.6.2: Molar concentration [unit : mmole/l] in plasma, interstitium and cells, based on APP [2000], p. 65. Note that the calcium concentration depends on the cell type: even if it plays a major role in excitable cells (muscles), its concentration is tiny in the cytoplasm (10^{-7} mole/l) and it is stored in the sarcoplasmic reticulum (10^{-3} mole/l). Fung [1990], p. 276, gives similar information for the extravascular, extracellular and intracellular fluids of skeletal muscle cell and the normal blood plasma.

charged macromolecules, the entity that matters is not the concentration, but the effective concentration, that is the concentration times the electrical charge.

The concentrations of immunoglobulins which are γ-globulins are very low at birth, except that of IgG, which is easily transmitted by the mother, and after birth, by her milk.

Table 3.6.1: Some plasmic proteins.

Protein	g/l H_2O
Albumin	35-45
α_1-globulins	2.5
α_2-globulins	5
β-globulins	7.5
γ-globulins	10

Table 3.6.2: Immunoglobulins $\in \gamma$-globulins.

Protein	g/l H_2O
IgA	2.25
IgD	0.03
IgE	0.0002
IgG	11.00
IgM	1.150

To alleviate the low concentration of dissolved oxygen and carbon dioxide in blood, nature has created specific transporters dedicated to the transport of these gases. Indeed, erythrocytes are mainly aimed at transporting oxygen and carbon dioxide between lungs and tissues. They link the gases via an enzyme, the carbonic anhydrase, and a buffer, the hemoglobin.

Erythrocytes, usually called red blood cells, are disk-shaped of dimensions about $7.5\,\mu m \times 2\,\mu m$. They can diffuse through small capillaries and reach the interstitium. Their formation in the bone marrow is controlled by a hormone, the erythropoietin, and by the oxygen concentration of blood. While they have a nucleus during their formation, they become anucleated when circulating in blood.

The leukocytes participate to the immune defense of the body by secreting immunoglobulins: monocytes and granulocytes are involved in the non specific defense while lymphocytes address specific defense. Immunoglobulins are called *antibodies* while the foreign particles on the receptors of which they can fix are called *antigens* ($\equiv$ *antibody generators*).

Thrombocytes participate to hemostasis, a phenomenon by which a possible damage of the endothelium of blood vessels can be controlled quickly.

Chapter 4

Multi-species mixtures as thermodynamically open systems

4.1 The thermodynamically open system

Within a mixture, each species may in principle exchange mass, momentum and energy with the other species: these exchanges are referred to as (internal) *transfers*. Here, the mixture is viewed as a thermodynamically *open* system. Indeed, each point in the mixture or at the boundary may also exchange mass, momentum and energy directly with the surroundings: the latter exchanges correspond to *growth* processes. These ideas are sketched in Fig. 4.1.1 which is inspired by the model developed for articular cartilages in Loret and Simões [2004] [2005]a.

The balance equations of mass, momentum, energy, entropy for each species and for the mixture as a whole, as well as the dissipation inequality, need to be reviewed. This chapter proposes a general framework which serves as a repository where the various forms of these equations are exposed in view of their use in the subsequent chapters.

Various forms of energy can be exchanged between a tissue and its surroundings. Irradiation is known to increase the collagen content and also to decrease the permeability [Znati et al. 2003]. Recent fabrication methods of tissues have used external sources of heat or of light. Heat may be used for *in vitro* prefabrication: the temperature of the construct is elevated above its glass value and pressure is applied to assemble parts in a desired pattern. Light energy might be used under physiological temperature and pH in various techniques, e.g., photopolymerization, photocrosslinking, stereolithography. The energy radiated usually by ultraviolet frequencies is absorbed by the chromophore of a photolabile compound. The latter initiates next a series of reactions. *Photopolymerization* concerns discrete chains of polymers while *photocrosslinking* addesses a polymer network. Much effort has been devoted to form hydrogels by photopolymerization, e.g., the hydrophilic polyethylene glycol (PEG), and hard tissues for dental and orthopedic applications. The idea is to inject the compound in a soluble form and to turn it insoluble by photopolymerization. The compound may be a biodegradable scaffold used to help cell proliferation or to heal leaking tissues after surgery. As another use of light, radiation of an encapsulated gel in a damaged cartilage has been shown to excite extracellular matrix production, Elisseeff et al. [1999]. Thus light may be used for triggering both growth and structuration (crosslinking).

4.2 Chemomechanical behavior and growth of soft tissues

The developments presented here are intended to be used in a multi-phase multi-species mixture. Species can be solid particles, fluids, ions or macromolecules. However, at this stage, the separation of *species*, also referred to as *constituents*[4.1], into *phases* is not addressed. All species are considered here on an equal foot, except the solid skeleton which plays a special role. Tacitly, a chemical present in n phases will thus be modeled by n species.

For indications on species segregation into phases, the interested reader may consult Loret et al. [2004]a where the examples of chemically sensitive heteroionic clays and of articular cartilages are compared. Within phases, species

---

[4.1]The terms "species" and "constituent" are used almost interchangeably. Still, while the term "species" is generic, we may occasionally refer to the "solid constituent" and to the "fluid constituent" to contrast these species from species like salts, ions or macromolecules that diffuse in water.

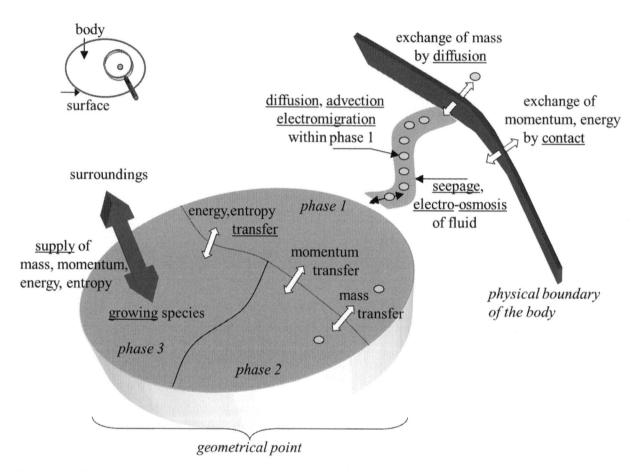

Fig. 4.1.1: Sketch of the exchanges that are accounted for in the multi-species multi-phase open system. For illustration, we have assumed that, at any geometrical point, the species can be segregated in three phases. (1) *Exchanges at the boundary.* Exchanges of various natures can take place at the boundary of the mixture. These exchanges affect only the species of phase 1, which is viewed as a fluid phase, that communicates with the exterior of the mixture. (2) *Transports.* Species of phase 1, are transported to, and from, the boundary by diffusion, advection, and electromigration, while the fluid itself flows through the solid skeleton by seepage and electroosmosis. (3) *Transfers.* Phases can exchange mass, momentum and energy. These exchanges are termed transfers. The characteristics of each exchange depend on both the concerned species (its molar mass, steric characteristics, electrical charge) and of the properties of the real or artificial membranes that separate the phases. Thus, typically, macromolecules are associated with a large transfer time. An infinite transfer time implies impermeability. A certain hierarchy of exchanges may be implied by considering that species of phase 2 that leave their phase have to transfer to phase 1: thus phase 2 can not be controlled directly from the boundary. (4) *Growth in an open system.* If, in addition, the system is thermodynamically open, supply and/or removal, directly from the surroundings, of mass, momentum and energy can take place at each geometrical point: these exchanges are referred to as growth processes. Interest is on the growth of the extracellular matrix, that is, essentially of collagen and proteoglycans.

diffuse. Across phases, they undergo a physico-chemical reaction, also termed mass transfer. Mass transfer can be viewed as a way for the tissue to adapt to mechanical and chemical loadings by modifying the internal repartition of its species.

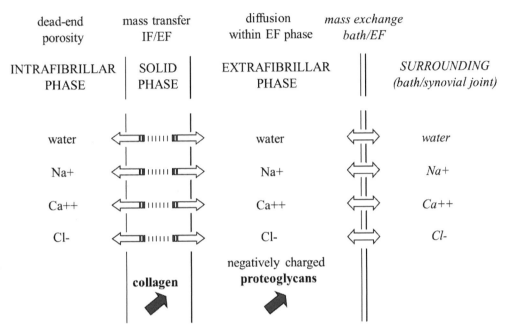

Fig. 4.2.1: An example of multi-phase multi-species biological material [Loret and Simões 2004, 2005]a. Any geometrical point, or representative elementary volume, of articular cartilage is partitioned in three phases: one solid phase and two fluid phases. Each fluid phase contains several species. Some of these species are mobile, at least partially: water and ions can enter and leave the intrafibrillar space defined by collagen fibrils. Proteoglycans which are macromolecules are too large to be admitted into that space, at least in absence of osteoarthritis. Water and ions can also be exchanged between the extrafibrillar phase and the exterior. Within their phase, extrafibrillar species *diffuse*. Across phases, species *transfer*. Mass transfer amounts to a physico-chemical reaction. While the mass transfer mechanism can be seen as a macroscopic mechanism, a zoom over the interface would show that it results from the upscaling of diffusion micromechanisms that take place at lower spatial scales. The modeling of growth (solid arrows) considers that the growing species, namely collagen and proteoglycans, are contributed directly by the surroundings. More generally, mass transfer within the mixture can contribute to growth as well.

In order to fix the ideas, the definition of phases in articular cartilages adopted by Loret and Simões [2004] [2005]a is shown in Fig. 4.2.1. The set of species in the solid phase is denoted by $\mathcal{S}$. The set of species that grow, e.g., collagen and proteoglycans in an articular cartilage, is denoted by $\mathcal{S}^*$. Some species which move with the collagen are not part of the solid phase, e.g., species of the intrafibrillar space: the set of all species which move with the velocity of the collagen is denoted by $\mathcal{S}^{**}$, so that $\mathcal{S} \subset \mathcal{S}^* \subset \mathcal{S}^{**}$.

Moreover, the set of all species is denoted by $\mathcal{K}$. The complements in $\mathcal{K}$ of the sets $\mathcal{S}^*$ and $\mathcal{S}^{**}$ are denoted by $\mathcal{K}^*$ and $\mathcal{K}^{**}$, and thus

$$\underbrace{\mathcal{K}}_{\substack{\text{complete} \\ \text{mixture}}} = \mathcal{K}^* + \underbrace{\mathcal{S}^*}_{\substack{\text{species} \\ \text{that grow}}} = \underbrace{\mathcal{K}^{**}}_{\substack{\text{species} \\ \text{that flow through} \\ \text{or diffuse wrt fluid phase}}} + \underbrace{\mathcal{S}^{**}}_{\substack{\text{species} \\ \text{that move} \\ \text{with the solid}}} . \qquad (4.2.1)$$

The motivations for the developments below can be better appreciated if placed in a more general project:

(1) We have in mind to provide the thermodynamic arguments that have led to the modeling of multi-phase multi-species articular cartilages which accounts for deformation (chemomechanical couplings), mass transfer, momentum transfer, that generates generalized diffusion, as well as cation interaction. Constitutive equations are developed and

tested along equilibrium paths and in transient situations in the form of initial and boundary value problems solved either through analytical techniques or through the finite element method.

(2) An additional physical phenomenon is introduced, namely *growth*. This enlargement has two main consequences with respect to the previous model developments in Loret and Simões [2004] [2005]a:

(2.1) first, the system considered is *open*: the surroundings may contribute to the balance equations of mass, momentum, energy and entropy;

(2.2) as a second fundamental modification, the strain is decomposed into a growth component and an elastic (accommodating) component and the mechanical constitutive equations are based on the elastic part of the strain only, much like in elastoplasticity.

(3) Balance equations in the multi-species mixture context account for internal transfers across phases, and external contributions in terms of mass, momentum, energy and entropy. All four types of interactions have been considered within closed systems. As for open systems, most applications consider that they are open with respect to mass only. The external energy contributions are more seldom accounted for, with exceptions like Klisch and Hoger [2003], Lubarda and Hoger [2002]. In their pioneering continuum model of bone remodeling, Cowin and Hegedus [1976] describe a two species mixture that is open with respect to the momentum only. On the other hand, they view the bone matrix as another smaller system, open to the four types of interactions above, and they require balance and constitutive equations for this species only. By contrast, we will not view species as open systems with respect to other species. Species interact with the rest of the mixture via *transfer*. The whole mixture, and consequently each species, is open with respect to the surroundings and we generically refer to these latter interactions as *growth*. The constitutive equations that control the two phenomena are quite distinct.

(4) The constitutive equations of growth are restricted by the thermodynamic framework and indeed satisfaction of the dissipation inequality by growth motivates the structure of these equations.

(5) Four mechanisms of distinct physical natures are delineated to contribute to the dissipation inequality and therefore enforced to be positive individually. This segregation of the sources of dissipation simplifies the analysis and leads to specific relations, namely

(5.1) chemomechanical hyperelastic relations;

(5.2) coupled generalized diffusion including Darcy's law of seepage through the porous medium, Fick's law of diffusion of ions in water, Ohm's law of electrical flow, Fourier's law of thermal diffusion, electroosmosis, etc.;

(5.3) mass transfer of ions and water across fluid phases;

(5.4) growth in a multi-phase multi-species context where each species is able to evolve according to its own growth rate.

In a sense, the thermodynamic analysis is just an extension of the classical papers on mixtures by Truesdell and Toupin [1960], de Groot and Mazur [1962], Eringen and Ingram [1965], Atkin and Craine [1975], Bowen [1976], Truesdell [1984], Haase [1990]. They do not consider growth, while we precisely need to concentrate on the contributions by the surroundings in all the balance equations.

(6) The analysis is performed along Biot's [1973] point of view where the deformable solid plays a special role.

As an example, consider structuration of a tissue during the growth process. The phenomenon is governed by rate equations that involve mechanics and growth, that is, where mechanics and growth are the stimuli. In addition, chemical variables, indicating the concentrations of nutrients, cytokines and growth factors, may moderate or enhance the intensity of the process. At a later stage of the lifespan of the tissue, glycation induced stiffening of the cornea and annulus fibrosus provides a further example of chemomechanical coupling. The concentrations, or rather the mass contents, should be part of the state variables that define the chemomechanical state. Chemical fluxes have to be defined by generalized diffusion equations, to ensure that, in a boundary value problem, the local chemical content can be obtained by solving the field equations of mass balance.

Couplings versus interdependences

The segregation of the sources of dissipation is part of the framework that is developed. Attention should be paid to the fact this *segregation*, which we will avoid to term *decoupling*, does by no means imply that the four physical mechanisms *do not interact*, in the sense that they could for example be solved in independent initial and boundary value problems. For example, mechanics influences the actual diffusion processes via deformation since it is known that the hydraulic conductivity depends on the void space through which fluid can flow. These *interactions*, or *interdependences*, between the four physical mechanisms and the direct *couplings* that take place within each of these mechanisms as indicated in point (5.2) are distinct from a hierarchical point of view. They also differ from a formal point of view: indeed the couplings connect a set of rate equations associated with a set of driving forces.

4.3 General form of balance equations

In the standard approach to mixtures, the volume considered is the total volume of the mixture, the motion is described by the barycentric velocity $\mathbf{v}$ and by the associated deformation gradient. In Biot's approach, the solid skeleton is paid a special attention, and species are viewed as flowing through the solid skeleton. The volume considered is the volume V of the solid skeleton, the motion of a point $\mathbf{x}$ of the solid is described by the velocity $\mathbf{v}_\mathrm{s}$ and by the associated deformation gradient $\mathbf{F}$ with velocity gradient $\mathbf{L} = (d\mathbf{F}/dt)\cdot\mathbf{F}^{-1}$, $d(\cdot)/dt$ denoting the time derivative following the solid particles.

Entities and operators reckoned from the reference configuration are underlined. The motion is considered to bring the point $\underline{\mathbf{x}} = (\underline{x}_i)$ in the reference configuration to its current position $\mathbf{x} = (x_i) = \mathbf{x}(\underline{\mathbf{x}}, t)$ at time t. Differential operators are reckoned from either configuration. For example, for a generic entity Ψ, the gradient with respect to the current configuration $\boldsymbol{\nabla}\Psi = \partial\Psi/\partial\mathbf{x}$ becomes $\underline{\boldsymbol{\nabla}}\Psi = \partial\Psi/\partial\underline{\mathbf{x}} = \boldsymbol{\nabla}\Psi\cdot\mathbf{F}$ with respect to the reference configuration.

Notation: in this chapter, unless stated otherwise, the convention of summation over the repeated index k which represents species of the mixture does *not* apply.

4.3.1 Partial, apparent and intrinsic properties

A generic species is denoted routinely by the index k. Some properties of the species refer either to the volume of the species or to the volume of the mixture. They are termed
- *intrinsic* when they refer to the volume or mass of the species;
- *partial* or *apparent*, when they refer to the volume or mass of the mixture.

The first situation is denoted by a **subscript** and the second one by a **superscript**. Key instances include
- the apparent mass density $\rho^k = n^k\,\rho_k$ versus the intrinsic mass density ρ_k;
- the partial stress $\boldsymbol{\sigma}^k = n^k\,\boldsymbol{\sigma}_k$ versus the intrinsic stress $\boldsymbol{\sigma}_k$;
- the thermodynamic functions, e.g., the internal energy per unit mass u_k versus the internal energy per unit volume of the mixture $U^k = \rho^k\,u_k$.

Apparent and intrinsic entities are linked by the volume fraction n^k defined below.

4.3.2 Mass, geometry and kinematical descriptors

At any point $\mathbf{x}$ of the porous medium, the present macroscopic approach endows the species k of mass M_k and volume V_k with properties like its absolute velocity $\mathbf{v}_k$, its volume fraction n^k, its *intrinsic* density ρ_k and its *partial* (or *apparent*) density ρ^k:

$$n^k = \frac{V_k}{V}\,; \quad \rho_k = \frac{M_k}{V_k}\,; \quad \rho^k = \frac{M_k}{V} = n^k\,\rho_k\,. \tag{4.3.1}$$

The lineal, areal and volume fractions are usually assumed to be equal, see e.g., Biot [1955], his Eqn (2.2), while the strict equality holds only when the species are distributed randomly. This assumption may be questionable for anisotropic materials in which the directional distribution of voids is not expected to be random. Oda et al. [1985] have attempted to characterize this directional distribution through a fabric tensor.

Therefore the mass flux per unit current area through the solid can be defined via the apparent mass density,

$$\mathbf{M}_k = \rho^k\,(\mathbf{v}_k - \mathbf{v}_\mathrm{s}), \quad k \in \mathcal{K}\,. \tag{4.3.2}$$

By pure formalism, we also define $\mathbf{M}_\mathrm{s} = \mathbf{0}$ so that summations that include the fluxes can be taken over all species $\mathcal{K}$ or only over species flowing through the solid $k = \mathrm{s}$.

For any species $k \in \mathcal{K}$, we should distinguish the *filtration velocity* with respect to the solid which is involved in Darcy's law of seepage from the *diffusion velocity* with respect to the fluid phase that comes into the picture in Fick's law of diffusion, and from the *barycentric filtration velocity* with respect to the mass center,

$$\underbrace{\mathbf{v}_k - \mathbf{v}_\mathrm{s}}_{\text{filtration velocity}} \quad ; \quad \underbrace{\mathbf{v}_k - \mathbf{v}_\mathrm{w}}_{\text{diffusion velocity}} \quad ; \quad \overset{(b)}{\mathbf{v}}{}_k = \underbrace{\mathbf{v}_k - \mathbf{v}}_{\substack{\text{barycentric} \\ \text{filtration velocity}}} \quad . \tag{4.3.3}$$

To the mixture as a whole are attached entities like its density ρ, its barycentric velocity $\mathbf{v}$ and its barycentric body force $\mathbf{b}$, namely,

$$\rho = \sum_{k \in \mathcal{K}} \rho^k, \quad \rho\,\mathbf{v} = \sum_{k \in \mathcal{K}} \rho^k\,\mathbf{v}_k, \quad \rho\,\mathbf{b} = \sum_{k \in \mathcal{K}} \rho^k\,\mathbf{b}_k, \tag{4.3.4}$$

and therefore the barycentric filtration velocities satisfy the closure relation,

$$\sum_{k \in \mathcal{K}} \rho^k\, \overset{(b)}{\mathbf{v}}_k = \mathbf{0}. \tag{4.3.5}$$

Given a property of a species, there are different ways to define the property associated with the mixture as a whole. One way is by simple summation. A second way is by mass averaging, e.g., the barycentric velocity above and the body force per unit mass, Eqn (4.3.4). A third way is by volume averaging, e.g., Eqn (4.5.8) below for the stress. The property obtained by (weighted or unweighted) summation over species is sometimes called the *inner part* of the property in question, e.g., Eqn (4.5.8). In general, an *overall*, or global, property of the mixture as a whole may have to be defined through more involved procedures than simple averaging rules, e.g., the total stress of the mixture defined by Eqn (4.5.7). In fact the definition of overall, or global, properties is addressed in different ways in different theories of mixtures.

Finally, we have to reiterate the following convention stated in Section 2.6. Certain properties Ψ of the mixture as a whole are measured per unit (current or reference) volume, while the corresponding properties ψ_k attached to the species $k \in \mathcal{K}$ are measured per unit mass. Instances are internal energy, enthalpy, entropy, etc. For these entities, the property at the level of the mixture is defined *not* by $\rho\,\psi = \sum_{k \in \mathcal{K}} \rho^k\,\psi_k$ *but by* $\Psi = \sum_{k \in \mathcal{K}} \rho^k\,\psi_k$.

4.3.3   Integral and local balance equations

The generalized Reynolds theorem of Section 2.7.1 states that the rate of change of an integral is equal to the rate of change as if the volume was moving with the solid velocity to which should be added the contributions in terms of fluxes of the individual species through the solid surface. The solid being taken as control, Reynolds theorem, applied to the generic field $\Psi_k = \Psi_k(\mathbf{x}, t)$, $k \in \mathcal{K}$, can be cast in the format,

$$\begin{aligned}
\frac{d}{dt} \int_V \Psi_k\, dV &= \int_V \frac{\partial \Psi_k}{\partial t}\, dV + \int_{\partial V} \Psi_k\,\mathbf{v}_k \cdot d\mathbf{S} \\
&= \int_V \frac{d\Psi_k}{dt} + \Psi_k \operatorname{div} \mathbf{v}_s + \operatorname{div}\left(\frac{\Psi_k}{\rho^k}\,\mathbf{M}_k\right) dV.
\end{aligned} \tag{4.3.6}$$

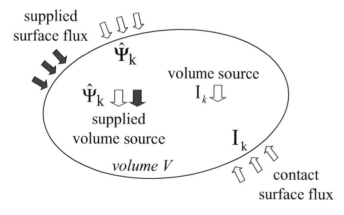

Fig. 4.3.1: The rate of change of an arbitrary function is generically set equal to the sum of two volume and two surface integrals. One set of integrals is standard while the other set represents supplies, which will be further split into internal transfers and external contributions.

This rate of change is generically set to the sum of two volume and two surface integrals, Fig. 4.3.1,

$$\frac{d}{dt} \int_V \Psi_k\, dV = \int_V I_k\, dV + \int_{\partial V} \mathbf{I}_k \cdot d\mathbf{S} + \int_V \hat{\Psi}_k\, dV + \int_{\partial V} \hat{\boldsymbol{\Psi}}_k \cdot d\mathbf{S}, \tag{4.3.7}$$

with

$$\Psi_k = \Psi_k(\mathbf{x}, t) \quad \text{generic field;}$$

$$I_k = I_k(\mathbf{x}, t) \quad \text{a volume source per unit current volume;}$$

$$\mathbf{I}_k = \mathbf{I}_k(\mathbf{x}, t) \quad \text{a surface source (flux) per unit current area;}$$

$$\hat{\Psi}_k = \hat{\Psi}_k(\mathbf{x}, t) \quad \text{a volume source due to a mass transferred between phases within the mixture,}$$
$$\text{and/or to a mass which is deposited in a growth process;}$$

$$\hat{\boldsymbol{\Psi}}_k = \hat{\boldsymbol{\Psi}}_k(\mathbf{x}, t) \quad \text{a surface source (flux) per unit current area contributed}$$
$$\text{by the rest of the mixture and/or the surroundings.}$$

Thus the overall contribution per unit current volume by the surroundings to the growth process sums to,

$$\sum_{k \in \mathcal{K}} \hat{\Psi}_k + \operatorname{div} \hat{\boldsymbol{\Psi}}_k = \hat{\Psi} \,. \tag{4.3.8}$$

In other words, $\hat{\Psi}$ is an entity that characterizes growth: it vanishes in absence of growth, that is when only transfers between phases take place.

In general, the velocities and the generic fields considered may undergo discontinuities within the volume V. Eringen and Ingram [1965] present a general treatment of discontinuities, used for example by Jabbour and Bhattacharya [2003] for the modeling of thin film deposition. Here however no such *discontinuity surfaces* crossing the volume V are assumed to exist. Therefore the local form of the balance equation for the species k may be cast in the following format,

$$\frac{d\Psi_k}{dt} + \Psi_k \operatorname{div} \mathbf{v}_{\mathrm{s}} + \operatorname{div}\left(\frac{\Psi_k}{\rho^k}\mathbf{M}_k\right) - I_k - \operatorname{div} \mathbf{I}_k = \hat{\Psi}_k + \operatorname{div} \hat{\boldsymbol{\Psi}}_k, \quad k \in \mathcal{K}\,. \tag{4.3.9}$$

This equation can be expressed in a format that involves the derivative following the species k,

$$\frac{d^k \Psi_k}{dt} + \Psi_k \operatorname{div} \mathbf{v}_k - I_k - \operatorname{div} \mathbf{I}_k = \hat{\Psi}_k + \operatorname{div} \hat{\boldsymbol{\Psi}}_k, \quad k \in \mathcal{K}\,, \tag{4.3.10}$$

associated with the global balance statement akin to the species $k \in \mathcal{K}$,

$$\frac{d^k}{dt} \int_V \Psi_k \, dV = \int_V I_k \, dV + \int_{\partial V} \mathbf{I}_k \cdot d\mathbf{S} + \int_V \hat{\Psi}_k \, dV + \int_{\partial V} \hat{\boldsymbol{\Psi}}_k \cdot d\mathbf{S}\,, \tag{4.3.11}$$

with

$$\frac{d^k}{dt} \int_V \Psi_k \, dV = \int_V \frac{d^k \Psi_k}{dt} + \Psi_k \operatorname{div} \mathbf{v}_k \, dV\,. \tag{4.3.12}$$

Passman and McTigue [1989] and Manoussaki [2003] have used a flux supply in the momentum balance. Kuhl and Steinmann [2003] introduce flux supplies of mass and entropy, Section 21.6.3.3. Elements of comparison between volume growth and surface growth in the context of bone growth are presented in Rouhi et al. [2004], as indicated in Note 21.1. In the following however, these flux supplies $\hat{\boldsymbol{\Psi}}_k$, $k \in \mathcal{K}$, will be neglected to the exception of Section 4.5.4.

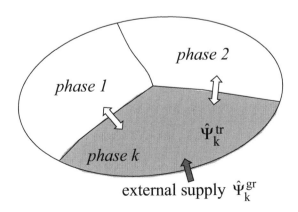

Fig. 4.3.2: Supplies to a phase k are split into internal transfers and external contributions.

As sketched in Fig. 4.3.2, the total volume supply to an individual species k,

$$\hat{\Psi}_k = \hat{\Psi}_k^{\mathrm{tr}} + \hat{\Psi}_k^{\mathrm{gr}}, \tag{4.3.13}$$

is contributed by (internal) transfers which are balanced per se, while external contributions contribute to growth (or resorption),

$$\sum_{k \in \mathcal{K}} \hat{\Psi}_k^{\mathrm{tr}} = 0, \qquad \sum_{k \in \mathcal{K}} \hat{\Psi}_k = \sum_{k \in \mathcal{K}} \hat{\Psi}_k^{\mathrm{gr}} \begin{cases} \hat{\Psi} \neq 0 & \text{growth or resorption,} \\ = 0 & \text{closed system}. \end{cases} \tag{4.3.14}$$

The transport relations established in Section 2.6 assume a pure change of configuration at constant mass. For growth processes in a mixture context, the evolution of mass is a central issue and it is important to delineate the following situations:

- if no mass transfer nor growth takes place, the mass of each species and the mass of the mixture are conserved;
- if there are mass transfers within the mixture, then the mass of each species is *not* conserved while the mass of the mixture does;
- if growth, or degradation, takes place, then neither the masses of the species nor the mass of the mixture are conserved.

The above considerations may be summarized in the invariance relations, for a species $k \in \mathcal{K}$, or for the mixture as whole,

$$M_k = \rho^k V = m^k \underline{V} \begin{cases} = \underline{M}_k = \underline{\rho}^k \underline{V} & \text{no transfer nor growth;} \\ \neq \underline{M}_k & \text{transfer or growth.} \end{cases} \tag{4.3.15}$$

The balance equation (4.3.9) can now be expressed in the reference configuration. Multiplying (4.3.9) by $\det \mathbf{F}$, using the transport relations (2.6.20) for the scalars I_k, Ψ_k, and $\hat{\Psi}_k$,

$$(\cdot) \quad \rightarrow \quad \underline{(\cdot)} = \det \mathbf{F} \, (\cdot),$$

and (2.6.13) for the vectors $\mathbf{I}_k$, $\mathbf{M}_k$ and $\hat{\boldsymbol{\Psi}}_k$,

$$(\cdot) \quad \rightarrow \quad \underline{(\cdot)} = \det \mathbf{F} \, \mathbf{F}^{-1}(\cdot),$$

and the rate relations (2.6.14) and (2.7.2), the local balance equation becomes in the reference configuration,

$$\frac{d\underline{\Psi}_k}{dt} + \underline{\mathrm{div}} \left(\frac{\underline{\Psi}_k}{m^k} \mathbf{M}_k \right) - \underline{I}_k - \underline{\mathrm{div}} \, \mathbf{I}_k = \hat{\underline{\Psi}}_k + \underline{\mathrm{div}} \, \hat{\boldsymbol{\Psi}}_k, \quad k \in \mathcal{K}. \tag{4.3.16}$$

The generic relations listed in this section are applied in turn to the balances of mass, momentum, energy and entropy.

4.4 Balance of mass

The equations of balance of mass are stated in turn for the species and for the mixture. The phenomena of diffusion, mass transfer and growth are segregated subsequently. The kinematics of growth for the species in $\mathcal{S}^*$ susceptible of growth is deferred to the Section 21.3.4.

4.4.1 Balance of mass for species

The balance of mass for the species k is obtained by setting,

$$\underbrace{\Psi_k = \rho^k}_{\substack{\text{partial} \\ \text{mass density}}} , \quad I_k = 0, \quad \mathbf{I}_k = \mathbf{0}, \quad \underbrace{\hat{\Psi}_k = \hat{\rho}^k}_{\substack{\text{rate of} \\ \text{volume supply}}} , \quad \underbrace{\hat{\boldsymbol{\Psi}}_k = \mathbf{0}}_{\substack{\text{rate of} \\ \text{surface supply}}} . \tag{4.4.1}$$

The rate of mass supply $\hat{\rho}^k$ to the species $k \in \mathcal{K}$,

$$\underbrace{\hat{\rho}^k}_{\substack{\text{total rate} \\ \text{of mass supply}}} = \underbrace{\hat{\rho}_{\mathrm{tr}}^k}_{\substack{\text{internal supply} \\ \text{due to transfer}}} + \underbrace{\hat{\rho}_{\mathrm{gr}}^k}_{\substack{\text{external supply} \\ \text{due to growth}}} , \tag{4.4.2}$$

includes two contributions, namely

- the supply rate $\hat{\rho}_{\mathrm{tr}}^k$ by the rest of the mixture due to a *mass transfer*, which can be viewed as a physico-chemical reaction, and
- the supply rate $\hat{\rho}_{\mathrm{gr}}^k$ by the surroundings that participate to the growth process.

The contributions due to mass transfer sum to zero, while the total rate of mass supply due to growth does not vanish a priori,

$$\sum_{k\in\mathcal{K}} \hat{\rho}_{\mathrm{tr}}^k = 0, \quad \hat{\rho} \equiv \sum_{k\in\mathcal{K}} \hat{\rho}^k = \sum_{k\in\mathcal{K}} \hat{\rho}_{\mathrm{gr}}^k. \tag{4.4.3}$$

The rate of mass supply $\hat{\rho}^k$ is sometimes noted $\rho^k\,\hat{\rho}^k$, or $\rho\,\hat{\rho}^k$ by other authors.

The global balance of mass deduces from (4.3.11),

$$\frac{d^k}{dt} \int_V \rho^k \, dV = \int_V \hat{\rho}^k \, dV, \tag{4.4.4}$$

while Eqn (4.3.10) yields the local balance of mass in the current configuration,

$$\frac{d^k \rho^k}{dt} + \rho^k \operatorname{div} \mathbf{v}_k = \hat{\rho}^k, \quad k\in\mathcal{K}. \tag{4.4.5}$$

This local balance of mass can be used to obtain another form of the balance relation for the generic entity Ψ_k as,

$$\rho^k \frac{d^k}{dt}\left(\frac{\Psi_k}{\rho^k}\right) - I_k - \operatorname{div} \mathbf{I}_k = \hat{\Psi}_k + \operatorname{div} \hat{\mathbf{\Psi}}_k - \frac{\Psi_k}{\rho^k} \hat{\rho}^k, \quad k\in\mathcal{K}. \tag{4.4.6}$$

4.4.2 Balance of mass for the mixture

Let us sum (4.4.5) over species, using the definitions of overall density ρ and barycentric velocity $\mathbf{v}$, Eqn (4.3.4), the relation between the time derivatives (2.7.1) and the closure relations (4.4.3). Then the balance of mass of the whole mixture is seen to have the same format as that of a single body of density ρ, animated with the barycentric velocity $\mathbf{v}$, and endowed with the rate of mass supply $\hat{\rho}$,

$$\frac{D\rho}{Dt} + \rho \operatorname{div} \mathbf{v} = \hat{\rho}, \tag{4.4.7}$$

the barycentric time derivative $D(\cdot)/Dt$ being defined by (2.7.1).

4.4.3 Diffusion, mass transfer and growth

The following notations are introduced for the rates of mass supplies of the species and of the mixture

$$\hat{m}^k \equiv \hat{\rho}^k \det \mathbf{F}, \quad k\in\mathcal{K}; \quad \hat{m} \equiv \hat{\rho} \det \mathbf{F}. \tag{4.4.8}$$

With the mass flux defined by (4.3.2), insertion of (4.4.1) in (4.3.16) highlights the fact that the mass change is due to two physical phenomena namely flow, or filtration, of the species through the solid skeleton, and mass transfer or growth,

$$\frac{dm^k}{dt} = -\underline{\operatorname{div}} \, \mathbf{M}_k + \hat{m}^k = \det \mathbf{F} \left(-\operatorname{div} \mathbf{M}_k + \hat{\rho}^k\right), \quad k\in\mathcal{K}. \tag{4.4.9}$$

The last relation is obtained via (2.6.14) and (4.4.8)$_1$.

Upon summation over *all* species, the rate of change of the mass content m of the mixture, Eqn (2.6.26), is seen to be due to fluxes on the boundary and to a growth contribution,

$$\frac{dm}{dt} = -\underline{\operatorname{div}} \left(\sum_{k\in\mathcal{K}^{**}} \mathbf{M}_k \right) + \hat{m} = -\det \mathbf{F} \operatorname{div} \left(\sum_{k\in\mathcal{K}^{**}} \mathbf{M}_k \right) + \hat{\rho} \det \mathbf{F}. \tag{4.4.10}$$

The sum in Eqn (4.4.10) is restricted to the set $\mathcal{K}^{**}$ of species whose flux does not vanish a priori. Indeed, the analysis adopts the following assumption.

Assumption 4.1 : The species that grow move with the solid.

Then, the mass balances in terms of mass-contents, or masses, involve only the mass supplies. An explicit quantification of the rate of mass supply $\hat{\rho}^k$ is obtained in terms of the mass of the species $M_k = \rho^k V$, namely using the mass balance (4.4.9) and the rule of transformation of volumes,

$$\frac{\hat{\rho}^k}{\rho^k} = \frac{d}{dt} \operatorname{Ln} M_k \quad \Leftrightarrow \quad \frac{m^k(t)}{m^k(0)} = \frac{M_k(t)}{M_k(0)} = \exp\left(\int_0^t \frac{\hat{\rho}^k}{\rho^k}\, d\tau\right), \quad k \in \mathcal{S}^*. \tag{4.4.11}$$

A similar relation holds for the mass $M = \sum_{k \in \mathcal{K}} M_k = \rho V$ of the mixture: with ρ the overall density, Eqn (4.3.4), and $\hat{\rho}$ the rate of mass supply, Eqn (4.4.3),

$$\frac{\hat{\rho}}{\rho} = \frac{d}{dt} \operatorname{Ln} M \quad \Leftrightarrow \quad \frac{m(t)}{m(0)} = \frac{M(t)}{M(0)} = \exp\left(\int_0^t \frac{\hat{\rho}}{\rho}\, d\tau\right). \tag{4.4.12}$$

Restrictions on the actual expression of the rate of mass supply may result from thermodynamic considerations. On another side, Drozdov and Khanina [1997] have suggested practical restrictions required to prevent the growth process to yield unexpected results. Indeed, they observed that a rate of mass supply depending linearly of the von Mises equivalent stress may lead to unbounded growth. A more general discussion on the dynamic behavior of the growth equation is presented in Hegedus and Cowin [1976]. In Chapters 22 and 23, bounded growth is achieved by allowing the homeostatic state to adapt.

4.5   Balance of momentum

The balance of momentum is considered, first for individual species and next for the mixture as a whole. The forms of the momentum supplies, of the interaction momenta and of the interaction stresses are commented.

4.5.1   Balance of momentum for species

The balance of momentum for the species $k \in \mathcal{K}$ is obtained by setting, Eringen and Ingram [1965],

$$
\begin{aligned}
\Psi_k &= \underbrace{\rho^k \mathbf{v}_k}_{\text{momentum}} \\[2mm]
I_k &= \underbrace{\rho^k \mathbf{b}_k}_{\text{body force}} \\[2mm]
\mathbf{I}_k &= \underbrace{\boldsymbol{\sigma}^k}_{\text{partial stress}} \\[2mm]
\hat{\Psi}_k &= \underbrace{\hat{\rho}^k \tilde{\mathbf{v}}_k}_{\substack{\text{supply due to} \\ \text{mass supplied}}} + \underbrace{\hat{\boldsymbol{\pi}}^k}_{\substack{\text{supply} \\ \text{by the rest} \\ \text{of the mixture}}} \\[2mm]
\hat{\boldsymbol{\Psi}}_k &= \underbrace{\mathbf{0}}_{\substack{\text{supply rate of} \\ \text{surface momentum}}}
\end{aligned}
\tag{4.5.1}
$$

Just before being transferred and deposited, the masses are endowed with their own velocities $\tilde{\mathbf{v}}_k$, $k \in \mathcal{K}$. These velocities have to be specified by constitutive equations.

The net momentum supply,

$$\hat{\mathbf{p}} \equiv \sum_{k \in \mathcal{K}} \hat{\mathbf{p}}^k, \tag{4.5.2}$$

where

$$\hat{\mathbf{p}}^k \equiv \hat{\rho}^k \tilde{\mathbf{v}}_k + \hat{\boldsymbol{\pi}}^k, \tag{4.5.3}$$

is contributed by the surroundings in the growth process. The term $\hat{\boldsymbol{\pi}}^k$ represents the momentum supply to the species k by the rest of the mixture, aside from the momentum $\hat{\rho}^k \tilde{\mathbf{v}}_k$ due to mass transfer and growth (once again, notation varies with authors, and some authors embody the two contributions in a single term, e.g., Truesdell [1984], p. 219). Note 4.2 presents two ways of specifying the momentum supplies $\hat{\boldsymbol{\pi}}^k$, $k \in \mathcal{K}$.

The global balance of momentum of the species $k \in \mathcal{K}$ deduces from (4.3.11),

$$\frac{d^k}{dt} \int_V \rho^k \mathbf{v}_k \, dV = \int_V \rho^k \mathbf{b}_k \, dV + \int_{\partial V} \boldsymbol{\sigma}^k \cdot d\mathbf{S} + \int_V \hat{\rho}^k \tilde{\mathbf{v}}_k + \hat{\boldsymbol{\pi}}^k \, dV \,, \qquad (4.5.4)$$

while Eqn (4.4.6) yields, with a change of sign, the local balance of momentum in the current configuration, in terms of the partial stress $\boldsymbol{\sigma}^k$ and body force per unit mass $\mathbf{b}_k$,

$$\operatorname{div} \boldsymbol{\sigma}^k + \rho^k \left(\mathbf{b}_k - \frac{d^k \mathbf{v}_k}{dt}\right) = \hat{\rho}^k(\mathbf{v}_k - \tilde{\mathbf{v}}_k) - \hat{\boldsymbol{\pi}}^k \,. \qquad (4.5.5)$$

On the right-hand side of (4.5.5), the *interaction* momentum due to mass transfer and growth, that contributes to the momentum balance, is an objective vector that is *not* the whole momentum supply. In fact, if the mass supplied is endowed with the velocity of the targeted species k, that is $\tilde{\mathbf{v}}_k = \mathbf{v}_k$, the first term of the interaction momentum vanishes, while the momentum of the transferred or deposited mass itself does not.

Note 4.1 on strongly interacting media.

In a theory of strongly interacting media applied to articular cartilages, a single balance of momentum is enforced for a phase where all species have the same velocity, but the balance of mass is imposed for each species, Loret and Simões [2004] [2005]a. Some papers, e.g., Klisch and Hoger [2003], Klisch et al. [2003]a [2003]b, impose the balance of momentum as such to all species: the requirement might be too constraining.

Note 4.2 on momentum supplies and generalized diffusion.

Momentum supplies are dealt with in two standard ways:
- when solving boundary value problems, the balances of momentum of individual species may be used directly if constitutive equations have been provided for the momentum supplies, e.g., Section 7.3.2;
- alternatively, the generalized diffusion equations are postulated first, motivated by and satisfying the Clausius-Duhem inequality, e.g., Eqn (4.7.32) below: in fact, the diffusion equation of the species k is nothing else than the balance of momentum of this species. The momentum supplies are then elusive, but, if necessary, they can be retrieved from the momentum balances.

4.5.2 Balance of momentum for the mixture

The balance of momentum for *the mixture as a whole* is obtained by summing (4.5.5) over all species,

$$\operatorname{div} \boldsymbol{\sigma} + \rho \left(\mathbf{b} - \frac{D\mathbf{v}}{Dt}\right) = \hat{\rho} \mathbf{v} - \hat{\mathbf{p}} \,, \qquad (4.5.6)$$

where $\mathbf{v}$ and $\mathbf{b}$ are respectively the barycentric velocity and the overall body force defined by (4.3.4), and $\boldsymbol{\sigma}$ is the Cauchy stress of the mixture,

$$\boldsymbol{\sigma} = \sum_{k \in \mathcal{K}} \boldsymbol{\sigma}^k - \rho^k \, \overset{(b)}{\mathbf{v}}_k \otimes \overset{(b)}{\mathbf{v}}_k \,. \qquad (4.5.7)$$

A detailed proof is deferred to the Appendix. The diffusion contribution to the stress is often neglected in applications so that the (total) stress $\boldsymbol{\sigma}$ is approximated by the 'inner' part obtained by summation of partial stresses

$$\boldsymbol{\sigma}^I = \sum_{k \in \mathcal{K}} \boldsymbol{\sigma}^k \,. \qquad (4.5.8)$$

Another equivalent form is deduced by summing the individual balances of momentum (4.5.5) over all species and using the closure relation (4.5.2),

$$\operatorname{div} \boldsymbol{\sigma}^I + \rho \, \mathbf{b} - \sum_{k \in \mathcal{K}} \rho^k \frac{d^k \mathbf{v}_k}{dt} + \hat{\rho}^k \mathbf{v}_k = -\hat{\mathbf{p}} \,. \qquad (4.5.9)$$

With help of the relation (2.6.18), the balance of momentum for the mixture as a whole in the reference configuration can be cast in the form,

$$\begin{aligned}
\underline{\operatorname{div}} \, \underline{\boldsymbol{\tau}} + m \left(\mathbf{b} - \frac{D\mathbf{v}}{Dt}\right) &= \hat{m} \mathbf{v} - \underline{\hat{\mathbf{p}}} \,, \\[4pt]
\frac{\partial \underline{\tau}_{iJ}}{\partial \underline{x}_J} + m \left(b_i - \frac{Dv_i}{Dt}\right) &= \hat{m} \, v_i - \underline{\hat{p}}_i \,, \quad i \in [1,3] \,.
\end{aligned} \qquad (4.5.10)$$

Note 4.3 on the symmetry of the stresses.

The current framework assumes that the partial stresses $\boldsymbol{\sigma}^k$, $k \in \mathcal{K}$, are symmetric. Unsymmetric stresses are introduced in some theories of granular or polycrystalline materials, where couple stresses are intended to represent the mechanical interactions at the particle level, Cosserat and Cosserat [1909].

In the context of mixture theories, several authors, e.g., Truesdell and Toupin [1960], Section 215, Eringen and Ingram [1965], Bowen [1976], suggest that species may also interact via moment of momentum supplies.

The stress of the mixture $\boldsymbol{\sigma}$ or $\boldsymbol{\sigma}^I$ is symmetric in a closed system if the sum over all species of the moments of momentum vanishes. For an open system, the surroundings may contribute a moment of momentum supply. In this text, all moment of momentum supplies are assumed to be null, so that the Cauchy stress tensors representing the partial stresses of individual species and the stress of the mixture are symmetric.

4.5.3 Interaction momenta

One might perhaps assume that matter supplied to growing species either internally, by mass transfer, or externally, is endowed with the velocity of the solid skeleton, i.e. $\mathbf{v}_k = \mathbf{v}_s$, $k \in \mathcal{S}^*$.

In general however, the interaction momentum $\hat{\rho}^k (\mathbf{v}_k - \tilde{\mathbf{v}}_k)$ in the momentum balance (4.5.5) due to mass transfer or growth does not vanish. For example, our model of articular cartilage defines three phases, one solid phase and two fluid phases, Loret and Simões [2004]. A species of the intrafibrillar phase is assumed to move with the same velocity as the solid, $\tilde{\mathbf{v}}_k = \mathbf{v}_{kI} = \mathbf{v}_s$, while a species k in the extrafibrillar phase is endowed with its own velocity, $\tilde{\mathbf{v}}_k = \mathbf{v}_{kE}$ so as to be able to diffuse in water through the mixture. Thus upon transfer from the extrafibrillar phase to the intrafibrillar phase, the velocity of a species k undergoes the discontinuity $\mathbf{v}_s - \mathbf{v}_{kE}$.

4.5.4 Interaction stresses

When loosely spread over a scaffold, cells do not interact mechanically. On the other hand, as the cell density increases, a mechanical interaction develops that ultimately, as cells get into contact, reaches a large value that limits the macroscopic cellular density. To model this phenomenon, a cell-cell interaction stress can be introduced that, in absence of external loading, would imply cells to get apart and the tissue to expand. To be specific, let us consider an elastic material in a one-dimensional setting whose stress σ is equal to the product of the Young's modulus E by the strain ϵ. Let us introduce the compressive interaction stress $\hat{\sigma} < 0$ and modify the constitutive equation to $\sigma - \hat{\sigma} = E\,\epsilon$. At zero applied stress σ, the strain due to interaction is positive and the material expands. Conversely, at zero strain, the applied stress is compressive. Under the notation $-\hat{\sigma} = E\,\hat{\epsilon}$, the interaction stress can be transformed into a growth strain $\hat{\epsilon} > 0$, and the constitutive equation then becomes $\sigma = E\,(\epsilon - \hat{\epsilon})$.

Breward et al. [2002] introduce directly such an interaction term between tumor cells viewed as linear viscoelastic materials. In their study, this type of interaction concerns a single species, the tumor cells in the example, and has purely mechanical effects.

In this mixture framework, we consider now the existence of an interaction stress due to momentum supplies. For that purpose, we shall adapt an idea of Passman and McTigue [1989] used to model non-homogeneous mixtures where the supply terms do not balance each other. Specifically let us first rewrite the closure of the momentum supplies in a modified form, namely

$$\sum_{k \in \mathcal{K}} \hat{\rho}^k \tilde{\mathbf{v}}_k + \hat{\boldsymbol{\pi}}^k + \operatorname{div} \hat{\boldsymbol{\sigma}} = \hat{\mathbf{p}}, \tag{4.5.11}$$

which involves the interaction stress $\hat{\boldsymbol{\sigma}}$. To discover the effects of this stress, let us consider two situations (a) and (b) defined by the stresses $\hat{\boldsymbol{\sigma}}_{(a)} = \mathbf{0}$ and $\hat{\boldsymbol{\sigma}}_{(b)} = \hat{\boldsymbol{\sigma}}$, respectively, the other data being identical. Then the momentum supplies $\hat{\boldsymbol{\pi}}^k$ and partial stresses $\boldsymbol{\sigma}^k$, $k \in \mathcal{K}$, in the two systems can be obtained by the correspondence,

$$\hat{\boldsymbol{\pi}}^k_{(a)} \to \hat{\boldsymbol{\pi}}^k_{(b)} = \hat{\boldsymbol{\pi}}^k_{(a)} - \operatorname{div}(\lambda_k\,\hat{\boldsymbol{\sigma}}), \quad \boldsymbol{\sigma}^k_{(a)} \to \boldsymbol{\sigma}^k_{(b)} = \boldsymbol{\sigma}^k_{(a)} + \lambda_k\,\hat{\boldsymbol{\sigma}}, \quad k \in \mathcal{K}, \tag{4.5.12}$$

where the λ_k's are arbitrary quantities that satisfy $\sum_{k \in \mathcal{K}} \lambda_k = 1$. Thus, for the mixture as a whole,

$$\hat{\boldsymbol{\pi}}_{(a)} \to \hat{\boldsymbol{\pi}}_{(b)} = \hat{\boldsymbol{\pi}}_{(a)} - \operatorname{div} \hat{\boldsymbol{\sigma}}, \quad \boldsymbol{\sigma}_{(a)} \to \boldsymbol{\sigma}_{(b)} = \boldsymbol{\sigma}_{(a)} + \hat{\boldsymbol{\sigma}}. \tag{4.5.13}$$

All data being fixed, let us consider variations of the λ_k's that satisfy $\sum_{k \in \mathcal{K}} \Delta\,\lambda_k = 0$. Then we see, by (4.5.12), that variations of the momentum supplies imply variations of the partial stresses even if the mixture properties (4.5.13) are kept constant.

Incidentally, note that the momentum supply for the species k in situation (b) associated with (4.5.11) can be cast, in integral form, as the sum of a volume integral and a surface integral,

$$\int_V \hat{\rho}^k \tilde{\mathbf{v}}_k + \hat{\boldsymbol{\pi}}^k \, dV + \int_{\partial V} \lambda_k \hat{\boldsymbol{\sigma}} \cdot \, d\mathbf{S} \,, \tag{4.5.14}$$

and therefore, with the notation set in Section 4.3.3,

$$\hat{\boldsymbol{\Psi}}_k = \hat{\rho}^k \tilde{\mathbf{v}}_k + \hat{\boldsymbol{\pi}}^k \,, \quad \hat{\boldsymbol{\Psi}}_k = \lambda_k \hat{\boldsymbol{\sigma}} \,. \tag{4.5.15}$$

4.6 Balance of energy

The balance of energy is considered first for each species, and next for the mixture as a whole. Electrolytes are addressed as well.

4.6.1 Notation of thermodynamic functions in a mixture context

In agreement with the notations set by (2.6.23), the internal energy of an elementary volume may be expressed through several equivalent representations,

$$U^k V = \underline{U}^k \underline{V} = u_k M_k \,, \quad U^k = \rho^k u_k \,, \quad \underline{U}^k = m^k u_k \,. \tag{4.6.1}$$

In words, for any species k,
- u_k is the internal energy per unit current mass;
- U^k the internal energy per current volume of the mixture, and
- $\underline{U}^k$ the internal energy per reference volume of the mixture.

The internal energy of the body $\mathbb{U}$ [unit : J] is defined by integration over the volume of the internal energy U measured per current volume of the mixture, namely $\mathbb{U} = \int_V U \, dV$. The internal energy U is the sum over species k of the apparent internal energies $U^k = \rho^k u_k$. In summary, in the current and reference configurations,

$$\underbrace{u_k}_{\substack{\text{internal energy} \\ \text{of the species } k \\ \text{per unit mass} \\ \text{of that species}}} \mapsto \underbrace{\begin{cases} U^k = \rho^k u_k \\[2mm] \underline{U}^k = m^k u_k \end{cases}}_{\substack{\text{internal energy} \\ \text{of the species } k \text{ per} \\ \text{unit current/reference} \\ \text{volume of the mixture}}} \mapsto \underbrace{\begin{cases} U = \sum_{k \in \mathcal{K}} U^k \\[2mm] \underline{U} = \sum_{k \in \mathcal{K}} \underline{U}^k \end{cases}}_{\substack{\text{internal energy} \\ \text{of the mixture per} \\ \text{unit current/reference} \\ \text{volume of the mixture}}} \mapsto \mathbb{U} = \underbrace{\begin{cases} \int_V U \, dV \\[2mm] \int_{\underline{V}} \underline{U} \, d\underline{V} \end{cases}}_{\substack{\text{internal energy} \\ \text{of the body}}} \,. \tag{4.6.2}$$

An identical notation applies to the free energy, enthalpy, free enthalpy and entropy. Let us record the dimensions of internal energy, free energy, enthalpy and free enthalpy,

$$u \, [\text{unit} : \text{J/kg} = \text{m}^2/\text{s}^2], \quad U \, [\text{unit} : \text{Pa} = \text{J/m}^3 = \text{kg/m/s}^2] \,, \tag{4.6.3}$$

and of entropy,

$$s \, [\text{unit} : \text{J/kg/K} = \text{m}^2/\text{s}^2/\text{K}], \quad S \, [\text{unit} : \text{Pa/K} = \text{J/m}^3/\text{K} = \text{kg/m/s}^2/\text{K}] \,. \tag{4.6.4}$$

Additional notation is required to distinguish e.g., the free enthalpy per unit mass and per unit mole. Incidentally, the free enthalpy is also referred to as chemical potential.

4.6.2   First principle of thermodynamics

The concept of state variables is at the root of the thermomechanical analysis that is pursued in this work. Rather than attempting a general definition, we shall be content to proceed via examples. It suffices for now to indicate that they provide in an approximative manner the history of the material in terms of stress, of some representative strain, of temperature, etc. Memory effects may reduce to the current instant, or may be of short range or may require a time integral representation. In a *local state* or *local action* approach, only entities that take values at the geometrical point of interest are involved. Spatial interactions between a point and its neighborhood are introduced in a so called *non local action* approach through space derivatives, gradients of various orders or space integrals.

The first principle of thermodynamics relies on the existence of a function of the state of the body, called internal energy. It can be enounced as follows: for any incremental process, the sum of this internal energy and of the kinetics energy is balanced by the sum of the mechanical energy supplied by the surroundings and of the heat supply,

$$\underbrace{d\,\mathbb{U}}_{\substack{\text{internal}\\\text{energy}}} + \underbrace{d\mathbb{C}}_{\substack{\text{kinetic}\\\text{energy}}} = \underbrace{\delta\mathbb{W}_{\text{ext}}}_{\substack{\text{energy supplied}\\\text{by the surroundings}}} + \underbrace{\delta\mathbb{Q}}_{\substack{\text{heat}\\\text{supply}}}. \tag{4.6.5}$$

The differential in front of the internal energy and kinetic energy is meant to stress the fact that these entities are functions of the state of the material, while the entities on the right-hand side are not. Indeed, while, for a cycle $C = C(\mathcal{V})$ in the state space $\mathcal{V}$,

$$0 = \oint_C d\,\mathbb{U} + \oint_C d\mathbb{C} = \oint_C \delta\mathbb{W}_{\text{ext}} + \oint_C \delta\mathbb{Q}, \tag{4.6.6}$$

each of the two integrals of the left-hand side vanishes, the two integrals on the right-hand side do not, but they cancel each other.

4.6.3   Balance of energy for species

The balance of energy for the species $k \in \mathcal{K}$ is obtained by setting, Eringen and Ingram [1965] (the sign of the heat fluxes used here is opposite to that of these authors),

$$
\begin{aligned}
\Psi_k &= \underbrace{\rho^k\,u_k}_{\text{internal energy}} &&+&& \underbrace{\tfrac{1}{2}\,\rho^k\,\mathbf{v}_k^2}_{\text{kinetic energy}} \\[4pt]
I_k &= \underbrace{\Theta_k}_{\text{heat source}} &&+&& \underbrace{\rho^k\,\mathbf{b}_k\cdot\mathbf{v}_k}_{\text{mechanical source}} \\[4pt]
\mathbf{I}_k &= \underbrace{-\mathbf{Q}_k}_{\text{heat flux}} &&+&& \underbrace{\boldsymbol{\sigma}^k\cdot\mathbf{v}_k}_{\text{mechanical flux}} \\[4pt]
\hat{\Psi}_k &= \underbrace{\hat{\rho}^k\,(\tilde{u}_k + \tfrac{1}{2}\,\tilde{\mathbf{v}}_k^2)}_{\substack{\text{supply rate due to}\\\text{mass supply}}} &&+&& \underbrace{\hat{\boldsymbol{\pi}}^k\cdot\mathbf{v}_k}_{\substack{\text{supply rate due to}\\\text{momentum supply}}} && + && \underbrace{\hat{\mathcal{U}}^k}_{\substack{\text{supply rate}\\\text{in other forms}}} \\[4pt]
\hat{\boldsymbol{\Psi}}_k &= \underbrace{\mathbf{0}}_{\substack{\text{supply rate}\\\text{of surface energy}}}
\end{aligned}
\tag{4.6.7}
$$

The net power supply, namely

$$\hat{U} \equiv \sum_{k\in\mathcal{K}} \hat{U}^k, \tag{4.6.8}$$

with

$$\hat{U}^k \equiv \hat{\rho}^k\,(\tilde{u}_k + \tfrac{1}{2}\,\tilde{\mathbf{v}}_k^2) + \hat{\boldsymbol{\pi}}^k\cdot\mathbf{v}_k + \hat{\mathcal{U}}^k, \tag{4.6.9}$$

characterizes the growth process, i.e. $\hat{U}$ is zero if only (internal) energy transfers take place.

The contribution due to mass transfer and growth is endowed with an independent velocity $\tilde{\mathbf{v}}_k$ and an independent internal energy $\tilde{u}_k$. Cowin [2004], p. 94, presents physical arguments intended to justify that the deposited bone, when interpreted in the proper time scale, can be endowed with the same internal energy as the existing bone at the deposition site. Energy supply to the species k by the rest of the mixture and by the surroundings may take place also via kinetic energy and momentum supply, as well as in other arbitrary forms gathered in the term $\hat{\mathcal{U}}^k$.

The global balance of energy for the species $k \in \mathcal{K}$ deduces from (4.3.11),

$$
\frac{d^k}{dt} \int_V \rho^k \, u_k + \tfrac{1}{2} \rho^k \, \mathbf{v}_k^2 \; dV
$$
$$
= \int_V \Theta_k + \rho^k \, \mathbf{b}_k \cdot \mathbf{v}_k \; dV + \int_{\partial V} \left(-\mathbf{Q}_k + \boldsymbol{\sigma}^k \cdot \mathbf{v}_k \right) \cdot d\mathbf{S} + \int_V \hat{\rho}^k \left(\tilde{u}_k + \tfrac{1}{2} \tilde{\mathbf{v}}_k^2 \right) + \hat{\boldsymbol{\pi}}^k \cdot \mathbf{v}_k + \hat{\mathcal{U}}^k \; dV \,, \tag{4.6.10}
$$

while its local expression can be cast in the format,

$$
\rho^k \frac{d^k u_k}{dt} - \boldsymbol{\sigma}^k : \boldsymbol{\nabla} \mathbf{v}_k + \operatorname{div} \mathbf{Q}_k - \Theta_k = \hat{\rho}^k \left(\tilde{u}_k - u_k + \tfrac{1}{2} \left(\tilde{\mathbf{v}}_k - \mathbf{v}_k \right)^2 \right) + \hat{\mathcal{U}}^k \,. \tag{4.6.11}
$$

Eqn (4.6.11) is obtained by insertion of the terms appearing in (4.6.7) in the generic balance equation (4.4.6) and use of the balance of momentum of species (4.5.5). As observed for the latter relation, the energy (rate) interaction is not equal to the total supply of energy rate.

An equivalent expression, obtained by using the balances of mass and momentum, which highlights the internal energy per unit volume,

$$
\frac{d}{dt} (\rho^k \, u_k) + \rho^k \, u_k \operatorname{div} \mathbf{v}_{\mathrm{s}} + \operatorname{div} \left(-\boldsymbol{\sigma}^k \cdot \mathbf{v}_{\mathrm{s}} + \mathbf{Q}_k + \mathbf{h}_k \cdot \mathbf{M}_k \right) - \Theta_k - \rho^k \left(\mathbf{b}_k - \frac{d^k \mathbf{v}_k}{dt} \right) \cdot \mathbf{v}_k + \frac{1}{2} \hat{\rho}^k \, \mathbf{v}_k^2 - \hat{\mathcal{U}}^k = 0 \,, \tag{4.6.12}
$$

makes use of the enthalpy tensor $\mathbf{h}_k$ introduced by Bowen [1976],

$$
\mathbf{h}_k \equiv u_k \, \mathbf{I} - \frac{\boldsymbol{\sigma}^k}{\rho^k} \,. \tag{4.6.13}
$$

If the partial stress $\boldsymbol{\sigma}^k = -n^k \, p_k \, \mathbf{I}$ is spherical and proportional to the intrinsic pressure p_k, then $\mathbf{h}_k = h_k \, \mathbf{I}$ with h_k the enthalpy per unit mass,

$$
h_k \equiv u_k + \frac{p_k}{\rho_k} \,. \tag{4.6.14}
$$

4.6.4   Balance of energy for the mixture: direct derivation

As already mentioned, in the standard approach of balance equations for the whole mixture, attention is paid to the motion of the mass center of the mixture which is endowed with the barycentric velocity. Along Eringen and Ingram [1965], the balance of energy for the mixture can be obtained by summation of the individual equations (4.6.11), a detailed derivation being provided in the Appendix. Here instead, the motion of the solid plays a central role. The alternative derivation shown below is in fact slightly more general as it includes internal body forces. The entities that appear in (4.6.5) are defined and considered in turn. This procedure motivates the definitions (4.6.7), as in fact it could be used to derive the balance of energy of individual species.

(1) Internal energy:

The time derivative, following the solid particles, of the internal energy of the body $\mathbb{U}$ is obtained via (4.3.6). If, in addition, internal body forces per unit mass are included, namely $\mathbf{b}_{k,\mathrm{int}}$ for species k, then

$$
\frac{d\mathbb{U}}{dt} = \int_V \frac{dU}{dt} + U \operatorname{div} \mathbf{v}_{\mathrm{s}} + \sum_{k \in \mathcal{K}} \operatorname{div} \left(u_k \, \mathbf{M}_k \right) - \rho^k \, \mathbf{b}_{k,\mathrm{int}} \cdot \mathbf{v}_k \; dV \,. \tag{4.6.15}
$$

(2) Kinetic energy:

The kinetic energy of the body is obtained according to the scheme displayed by (4.6.2),

$$
\underbrace{c_k = \tfrac{1}{2} \rho^k \mathbf{v}_k^2}_{\substack{\text{kinetic energy} \\ \text{of the species k}}} \quad \mapsto \quad \underbrace{C = \sum_{k \in \mathcal{K}} c_k}_{\substack{\text{kinetic energy} \\ \text{of the mixture} \\ \text{per current volume}}} \quad \mapsto \quad \underbrace{\mathbb{C} = \int_V C \, dV}_{\substack{\text{kinetic energy} \\ \text{of the body}}} \,. \tag{4.6.16}
$$

The time derivative following the solid particles can be transformed to, see the Appendix for a details,

$$
\begin{aligned}
\frac{d\mathbb{C}}{dt} &= \int_V -\boldsymbol{\sigma}^I : \boldsymbol{\nabla}\mathbf{v}_s + \mathbf{v}_s \cdot (\rho\,\mathbf{b} + \hat{\mathbf{p}} - \hat{\rho}\,\frac{\mathbf{v}_s}{2})\, dV \\
&\quad + \sum_{k\in\mathcal{K}} \int_V \mathbf{M}_k \cdot \frac{d^k\mathbf{v}_k}{dt} + \frac{\hat{\rho}^k}{2}\left(\frac{\mathbf{M}_k}{\rho^k}\right)^2 + \operatorname{div}(\mathbf{v}_s \cdot \boldsymbol{\sigma}^k)\, dV\,.
\end{aligned}
\tag{4.6.17}
$$

(3) Energy supplied by the surroundings:

The power supplied by the surroundings includes, next to the standard contributions of external body forces, namely $\mathbf{b}_{k,\mathrm{ext}}$ for species k, and of the traction over the boundary, the power supply $\hat{U}$,

$$
\frac{\delta\mathbb{W}_{\mathrm{ext}}}{\delta t} = \int_V \hat{U}\, dV + \sum_{k\in\mathcal{K}} \int_V \rho^k\,\mathbf{b}_{k,\mathrm{ext}} \cdot \mathbf{v}_k\, dV + \sum_{k\in\mathcal{K}} \int_{\partial V} \mathbf{v}_k \cdot \boldsymbol{\sigma}^k \cdot d\mathbf{S}\,.
\tag{4.6.18}
$$

(4) Heat supply:

The heat power is contributed by
- a heat source per unit current volume Θ,
- the heat transmitted by conduction.

By application of the lemma of the tetrahedron, the latter is obtained in the format $-\mathbf{Q} \cdot \mathbf{n}$ where $\mathbf{Q}$ is the heat flux and $\mathbf{n}$ is the local outward unit normal to ∂V. Hence

$$
\frac{\delta\mathbb{Q}}{\delta t} = \int_V \Theta\, dV - \int_{\partial V} \mathbf{Q} \cdot d\mathbf{S}\,.
\tag{4.6.19}
$$

For this direct derivation to be consistent with the sum of the individual balances of energy (4.6.11), the heat source and heat flux of the mixture have to be the inner parts, namely

$$
\mathbf{Q} = \mathbf{Q}^I = \sum_{k\in\mathcal{K}} \mathbf{Q}_k, \quad \Theta = \Theta^I = \sum_{k\in\mathcal{K}} \Theta_k\,.
\tag{4.6.20}
$$

The energy equation will involve two additional entities, namely the total body force per unit mass.

$$
\mathbf{b} = \mathbf{b}_{\mathrm{int}} + \mathbf{b}_{\mathrm{ext}}\,,
\tag{4.6.21}
$$

and the enthalpy tensor $\mathbf{h}_k$ defined by (4.6.13).

Collecting Eqns (4.6.15)–(4.6.21), the local forms of the first principle (4.6.5) in the current and reference configurations can be cast in the respective formats,

$$
\begin{aligned}
\frac{dU}{dt} &+ U \operatorname{div}\mathbf{v}_s - \boldsymbol{\sigma}^I : \boldsymbol{\nabla}\mathbf{v}_s - \Theta + \operatorname{div}\mathbf{Q} + \sum_{k\in\mathcal{K}} \operatorname{div}(\mathbf{M}_k \cdot \mathbf{h}_k) + \mathbf{M}_k \cdot \left(\frac{d^k\mathbf{v}_k}{dt} - \mathbf{b}_k\right) \\
&+ \sum_{k\in\mathcal{K}} \frac{1}{2}\hat{\rho}^k\left(\mathbf{v}_k - \mathbf{v}_s\right)^2 + \hat{\mathbf{p}} \cdot \mathbf{v}_s - \frac{1}{2}\hat{\rho}\mathbf{v}_s^2 = \hat{U}\,,
\end{aligned}
\tag{4.6.22}
$$

and

$$
\begin{aligned}
\frac{d\underline{U}}{dt} &- \underline{\underline{\boldsymbol{\tau}}}^I : \frac{d\mathbf{E}_{\mathrm{G}}}{dt} - \underline{\Theta} + \underline{\operatorname{div}}\,\mathbf{Q} + \sum_{k\in\mathcal{K}} \underline{\operatorname{div}}(\underline{\mathbf{M}}_k \cdot \mathbf{h}_k) + \mathbf{F} \cdot \underline{\mathbf{M}}_k \cdot \left(\frac{d^k\mathbf{v}_k}{dt} - \mathbf{b}_k\right) \\
&+ \sum_{k\in\mathcal{K}} \frac{1}{2}\hat{m}^k\left(\mathbf{v}_k - \mathbf{v}_s\right)^2 + \underline{\hat{\mathbf{p}}} \cdot \mathbf{v}_s - \frac{1}{2}\hat{m}\mathbf{v}_s^2 = \underline{\hat{U}}\,.
\end{aligned}
\tag{4.6.23}
$$

A proof of this expression in the current configuration is detailed in the Appendix. The result in the reference configuration follows by multiplying the balance of energy in the current configuration (4.6.22) by $\det\mathbf{F}$, using the transformation rule (2.6.20) for U and Θ, the transformation rule (2.6.13) for the fluxes $\mathbf{M}_k$'s and $\mathbf{Q}$, the definitions of mass supplies (4.4.8), the work invariance relation (2.8.7), and the differential relations (2.7.2).

Note 4.4 on alternative definitions for a mixture.

Here we have defined the internal energy, free energy, entropy, heat flux and volume heat source of the mixture per unit current or initial volume. These entities are defined by simple summations. Moreover, the rate involved in the balances of the mixture is the time derivative following the solid particles.

There are other ways of defining global entities at the mixture level, each designed to cast the balance equations of energy and entropy in the familiar forms of a solid. In relation with the derivative associated with the barycentric velocity, Eringen and Ingram [1965] define the following entities,

$$\rho\, u = \sum_{k \in \mathcal{K}} \rho^k \left(u_k + \tfrac{1}{2} \overset{(b)}{\mathbf{v}}_k \cdot \overset{(b)}{\mathbf{v}}_k \right) = \sum_{k \in \mathcal{K}} \rho^k \left(u_k + \tfrac{1}{2} \mathbf{v}_k^2 \right) - \tfrac{1}{2} \rho \mathbf{v}^2 \,,$$

$$\mathbf{Q} = \sum_{k \in \mathcal{K}} \mathbf{Q}_k - \boldsymbol{\sigma}^k \cdot \overset{(b)}{\mathbf{v}}_k + \rho^k \left(u_k + \tfrac{1}{2} \overset{(b)}{\mathbf{v}}_k \cdot \overset{(b)}{\mathbf{v}}_k \right) \overset{(b)}{\mathbf{v}}_k \,, \qquad (4.6.24)$$

$$\Theta = \sum_{k \in \mathcal{K}} \Theta_k + \rho^k \, \mathbf{b}_k \cdot \overset{(b)}{\mathbf{v}}_k \,,$$

and provide similar redefinitions for the entities entering the entropy balance[4.2].

4.6.5 The case of an electrolyte

Some species, like ions and charged macromolecules, are not electrically neutral. Let ϕ be the electrical potential, ζ_k the valence of the species k and N_k the number of moles of this species in the volume V. An electrical contribution has to be added to the external work (4.6.18), namely

$$- \int_{\partial V} \phi \, \mathrm{F} \, \frac{N_k}{V} \, \zeta_k \mathbf{v}_k \cdot d\mathbf{S} \,. \qquad (4.6.25)$$

Here F is Faraday's equivalent charge. We shall assume the mixture as a whole to be electrically neutral,

$$\sum_{k \in \mathcal{K}} N_k \, \zeta_k = 0 \,. \qquad (4.6.26)$$

Therefore the electrical work can be cast in the format

$$- \int_{\partial V} \phi \, \mathrm{F} \, \frac{\zeta_k}{\widehat{m}_k} \, \mathbf{M}_k \cdot d\mathbf{S} \,, \qquad (4.6.27)$$

where $\widehat{m}_k$ is the molar mass of the species k. Consequently, a charged species is endowed with the additional specific energy $\phi \, \mathrm{F} \, \zeta_k / \widehat{m}_k$ (specific $\equiv$ per unit mass).

More generally, as described in Loret et al. [2004]a for both geological materials and biological tissues, the existence of natural or artificial membranes that separate phases may imply as many independent electroneutrality conditions as there are electrical potentials.

4.7 Balance of entropy and Clausius-Duhem inequality

The balance of entropy is stated first for each species. However, the temperature $T > 0$ is considered to be uniform over all species and a single condition on the entropy production is required for the whole mixture. The terms entering the inequality are next split into three contributions of distinct physical natures, each of which is required to be positive. These three contributions concern the mechanical and growth aspects, generalized diffusion and mass transfer. The unravelling of the mechanics and growth aspects are considered in Section 21.4. The framework built in that way motivates and structures constitutive equations associated with the four physical phenomena of interest. Within each of them, couplings are allowed, while some degrees of interdependence between them are possible.

The assumption of *local thermal equilibrium* (identical temperature for all species of the mixture) is relaxed in Chapter 10. A constitutive thermomechanical model where the solid skeleton and two immiscible fluids are endowed each with its own temperature is developed as an extension of so-called doubled porosity model.

[4.2]While writing his balance equations in terms of redefined entities (4.6.24) and barycentric time derivative, Bowen [1976] finally uses the free energy per unit current volume as a thermodynamic potential. Bowen [1982], Eqns 3.14 et seq., also notes that the use of the free energy per unit initial volume simplifies the expressions.

4.7.1   Balance of entropy for species

The balance of entropy for the species $k \in \mathcal{K}$ is obtained by setting, Eringen and Ingram [1965],

$$
\begin{aligned}
\Psi_k &= \underbrace{\rho^k\, s_k}_{\text{entropy}} \\[2mm]
I_k &= \underbrace{\Theta_k/T}_{\text{heat source}} + \underbrace{\hat{S}^k_{(\mathrm{i})}}_{\substack{\text{internal}\\\text{entropy production}}} \\[2mm]
\mathbf{I}_k &= \underbrace{-\mathbf{Q}_k/T}_{\text{heat flux}} \\[2mm]
\hat{\Psi}_k &= \underbrace{\hat{\rho}^k\, \tilde{s}_k}_{\substack{\text{supply rate}\\\text{due to mass supply}}} + \underbrace{\hat{\mathcal{S}}^k}_{\substack{\text{supply rate}\\\text{in other forms}}} \\[2mm]
\hat{\mathbf{\Psi}}_k &= \underbrace{\mathbf{0}}_{\substack{\text{supply rate}\\\text{of surface entropy}}}
\end{aligned}
\tag{4.7.1}
$$

The contribution to species k due to mass transfer and growth is endowed with its own entropy per unit mass $\tilde{s}_k$. Entropy supply to the species k by the rest of the mixture and by the surroundings arises also in other arbitrary forms gathered in the term $\hat{\mathcal{S}}^k$. The net rate of entropy supply

$$
\hat{S} \equiv \sum_{k \in \mathcal{K}} \hat{S}^k \,,
\tag{4.7.2}
$$

with

$$
\hat{S}^k \equiv \hat{\rho}^k\, \tilde{s}_k + \hat{\mathcal{S}}^k \,,
\tag{4.7.3}
$$

is contributed by the surroundings in the growth process: $\hat{S}$ is zero if only (internal) mass transfers occur.

Inserting the above quantities (4.7.1) in the generic balance equation (4.4.6), the balance of entropy for the species $k \in \mathcal{K}$ can be cast in the formats,

$$
\begin{aligned}
\hat{S}^k_{(\mathrm{i})} &= \rho^k \frac{d^k s_k}{dt} + \operatorname{div} \frac{\mathbf{Q}_k}{T} - \frac{\Theta_k}{T} + \hat{\rho}^k \left(s_k - \tilde{s}_k \right) - \hat{\mathcal{S}}^k \\[2mm]
&= \frac{d^k}{dt}(\rho^k s_k) + \rho^k s_k \operatorname{div} \mathbf{v}_k + \operatorname{div} \frac{\mathbf{Q}_k}{T} - \frac{\Theta_k}{T} - \hat{\rho}^k \tilde{s}_k - \hat{\mathcal{S}}^k \,,
\end{aligned}
\tag{4.7.4}
$$

the latter expression being readily deduced from the global statement,

$$
\frac{d^k}{dt} \int_V \rho^k s_k \, dV = \int_V \frac{\Theta_k}{T} + \hat{S}^k_{(\mathrm{i})} \, dV - \int_{\partial V} \frac{\mathbf{Q}_k}{T} \cdot d\mathbf{S} + \int_V \hat{\rho}^k \tilde{s}_k + \hat{\mathcal{S}}^k \, dV \,.
\tag{4.7.5}
$$

As observed for the balances of momentum and of energy, the entropy interaction (rate) is not equal to the total entropy supply rate. With the balance of mass (4.4.5) and the relation between time derivatives (2.7.1), the local balance of entropy (4.7.4) for the species $k \in \mathcal{K}$ may also be cast in the format,

$$
\hat{S}^k_{(\mathrm{i})} = \frac{d}{dt}(\rho^k s_k) + \rho^k s_k \operatorname{div} \mathbf{v}_{\mathrm{s}} + \operatorname{div} \left(\frac{\mathbf{Q}_k}{T} + s_k \mathbf{M}_k \right) - \frac{\Theta_k}{T} - \hat{\rho}^k \tilde{s}_k - \hat{\mathcal{S}}^k \,.
\tag{4.7.6}
$$

4.7.2   Internal and external entropy productions for a species

In view of comparison with the case of a single solid, it is useful to record the alternative forms of entropy production of a species k using either a free energy per unit (current) mass or a free energy per unit (current) volume (at this point, we refrain to remove the index k because of the time derivative). Substituting $\operatorname{div} \mathbf{Q}_k - \Theta_k$ obtained from the balance of energy (4.6.11) in the balance of entropy (4.7.4) yields with help of the balance of mass (4.4.5) two possible expressions of the entropy production,

$$
\begin{aligned}
T\hat{S}^k_{(\mathrm{i})} &= -\rho^k \frac{d^k e_k}{dt} + \boldsymbol{\sigma}^k : \boldsymbol{\nabla}\mathbf{v}_k - \rho^k s_k \frac{d^k T}{dt} - \frac{\mathbf{Q}_k}{T}\cdot\boldsymbol{\nabla}T \\
&\quad + \hat{\rho}^k\left(\tilde{e}_k - \underline{\underline{e_k}} + \tfrac{1}{2}(\tilde{\mathbf{v}}_k - \mathbf{v}_k)^2\right) + \hat{\mathcal{U}}^k - T\hat{\mathcal{S}}^k\,; \\
T\hat{S}^k_{(\mathrm{i})} &= -\frac{d^k}{dt}(\rho^k e_k) - \underline{\underline{\rho^k e_k \operatorname{div}\mathbf{v}_k}} + \boldsymbol{\sigma}^k : \boldsymbol{\nabla}\mathbf{v}_k - \rho^k s_k \frac{d^k T}{dt} - \frac{\mathbf{Q}_k}{T}\cdot\boldsymbol{\nabla}T \\
&\quad + \hat{\rho}^k\left(\tilde{e}_k + \tfrac{1}{2}(\tilde{\mathbf{v}}_k - \mathbf{v}_k)^2\right) + \hat{\mathcal{U}}^k - T\hat{\mathcal{S}}^k\,.
\end{aligned}
\tag{4.7.7}
$$

For the mixture as a whole, where a common derivative with respect to the solid is used, the rate of the specific free energy introduces a convective term,

$$
\begin{aligned}
\rho^k \frac{d^k e_k}{dt} &= \rho^k \frac{d e_k}{dt} + \mathbf{M}_k\cdot\boldsymbol{\nabla}e_k\,, \\
\frac{d^k}{dt}(\rho^k e_k) + \rho^k e_k \operatorname{div}\mathbf{v}_k &= \frac{d}{dt}(\rho^k e_k) + \operatorname{div}(\mathbf{M}_k\, e_k) + \rho^k e_k \operatorname{div}\mathbf{v}_{\mathrm{s}}\,.
\end{aligned}
\tag{4.7.8}
$$

The last line of the above relation (4.7.7) introduces the external entropy production $\hat{S}^k_{(\mathrm{e})}$, namely

$$
-T\hat{S}^k_{(\mathrm{e})} = \hat{\rho}^k\left(\tilde{e}_k + \tfrac{1}{2}(\tilde{\mathbf{v}}_k - \mathbf{v}_k)^2\right) + \hat{\mathcal{U}}^k - T\hat{\mathcal{S}}^k\,.
\tag{4.7.9}
$$

Now let us assume that the net supplies of momentum, of power and of rate of entropy to the species k, defined by (4.5.2), (4.6.8) and (4.7.2), respectively, are formally known. Then, in turn,

$$
\begin{aligned}
\hat{\mathbf{p}}^k &\equiv \hat{\rho}^k\tilde{\mathbf{v}}_k + \hat{\boldsymbol{\pi}}^k && \Rightarrow && \hat{\boldsymbol{\pi}}^k = \hat{\mathbf{p}}^k - \hat{\rho}^k\tilde{\mathbf{v}}_k\,, \\
\hat{U}^k &\equiv \hat{\rho}^k\left(\tilde{u}_k + \tfrac{1}{2}\tilde{\mathbf{v}}_k^2\right) + \hat{\boldsymbol{\pi}}^k\cdot\mathbf{v}_k + \hat{\mathcal{U}}^k && \Rightarrow && \hat{\mathcal{U}}^k = \hat{U}^k - (\hat{\mathbf{p}}^k - \hat{\rho}^k\tilde{\mathbf{v}}_k)\cdot\mathbf{v}_k - \hat{\rho}^k\tilde{u}_k - \tfrac{1}{2}\hat{\rho}^k\tilde{\mathbf{v}}_k^2\,, \\
\hat{S}^k &\equiv \hat{\rho}^k\tilde{s}_k + \hat{\mathcal{S}}^k && \Rightarrow && \hat{\mathcal{S}}^k = \hat{S}^k - \hat{\rho}^k\tilde{s}_k\,.
\end{aligned}
\tag{4.7.10}
$$

Consequently the external entropy production (4.7.9) may be rephrased in terms of the total supply rates,

$$
-T\hat{S}^k_{(\mathrm{e})} = \tfrac{1}{2}\hat{\rho}^k\mathbf{v}_k^2 - \hat{\mathbf{p}}^k\cdot\mathbf{v}_k + \hat{U}^k - T\hat{S}^k\,.
\tag{4.7.11}
$$

The expressions (4.7.9) and (4.7.11) are strictly equivalent. The former expression displays the rates of mass supply and the rates of independent supply of internal energy and entropy not associated with mass supply nor momentum supply. While included in the formulation, the independent supply of momentum $\hat{\boldsymbol{\pi}}^k$ is elusive in (4.7.9). Expression (4.7.11) highlights the four net supplies associated with the balance equations. It should be emphasized that these supplies are contributed by mass, momentum, energy and entropy.

As an example, a species might be open to mass only, in the sense that it may gain/loose mass, while its *independent* supplies in momentum $\hat{\boldsymbol{\pi}}^k$, internal energy $\hat{\mathcal{U}}^k$ and entropy $\hat{\mathcal{S}}^k$ are zero. By the same token, the species gains/looses momentum, energy and entropy,

$$
\begin{aligned}
\hat{\mathbf{p}}^k &= \hat{\rho}^k\tilde{\mathbf{v}}_k\,, \\
\hat{U}^k &= \hat{\rho}^k\left(\tilde{u}_k + \tfrac{1}{2}\tilde{\mathbf{v}}_k^2\right)\,, \\
\hat{S}^k &= \hat{\rho}^k\tilde{s}_k\,,
\end{aligned}
\tag{4.7.12}
$$

which together contribute to the total external entropy production,

$$
-T\hat{S}^k_{(\mathrm{e})} = \hat{\rho}^k\tilde{e}_k + \tfrac{1}{2}\hat{\rho}^k(\tilde{\mathbf{v}}_k - \mathbf{v}_k)^2\,,
\tag{4.7.13}
$$

a result which was blatant from (4.7.9).

4.7.3   Second principle of thermodynamics

The second principle is stated as follows: There exist
- a scalar $T > 0$ [unit : K] called temperature and
- a function of the state of the material $\mathbb{S} = \int_V S\,dV$, called entropy of the body and obtained by integration of the local entropy $S = S(\mathbf{x}, t)$ [unit : Pa/K = J/m³/K = kg/m/s²/K],

such that, for any process consistent with the balance equations,

$$\int_V \hat{S}_{(\mathrm{i})}\,dV = \frac{d}{dt}\int_V S\,dV - \int_V \frac{\Theta}{T}\,dV + \int_{\partial V} \frac{\mathbf{Q}}{T}\cdot d\mathbf{S} - \int_V \hat{S}\,dV \geq 0\,. \tag{4.7.14}$$

Summing the derivatives (4.3.6), the ensuing local inequality may be phrased in terms of the local entropy production $\hat{S}_{(\mathrm{i})} = \hat{S}_{(\mathrm{i})}(\mathbf{x}, t)$, namely in the current and reference configurations, respectively,

$$\hat{S}_{(\mathrm{i})} = \frac{dS}{dt} + S\,\mathrm{div}\,\mathbf{v}_\mathrm{s} + \mathrm{div}\left(\frac{\mathbf{Q}}{T} + \sum_{k\in\mathcal{K}} s_k\,\mathbf{M}_k\right) - \frac{\Theta}{T} - \hat{S} \geq 0\,,$$

$$\underline{\hat{S}}_{(\mathrm{i})} = \frac{d\underline{S}}{dt} + \underline{\mathrm{div}}\left(\frac{\underline{\mathbf{Q}}}{T} + \sum_{k\in\mathcal{K}} s_k\,\underline{\mathbf{M}}_k\right) - \frac{\underline{\Theta}}{T} - \underline{\hat{S}} \geq 0\,. \tag{4.7.15}$$

The global statement (4.7.14) of the second principle stands on its own. The local statement (4.7.15) is also obtained by summing the species entropy productions (4.7.6), provided that

1. The entropy is obtained along the same scheme as the internal energy, namely in the current and reference configurations,

$$\underbrace{s_k}_{\substack{\text{entropy} \\ \text{of the species } k \\ \text{per unit mass}}} \mapsto \underbrace{\begin{cases} S^k = \rho^k\,s_k \\ \underline{S}^k = m^k\,s_k \end{cases}}_{\substack{\text{entropy} \\ \text{of the species } k \\ \text{per unit current/reference} \\ \text{volume of the mixture}}} \mapsto \underbrace{\begin{cases} S = \displaystyle\sum_{k\in\mathcal{K}} \rho^k\,s_k \\ \underline{S} = \displaystyle\sum_{k\in\mathcal{K}} m^k\,s_k \end{cases}}_{\substack{\text{entropy of the mixture} \\ \text{per unit current/reference} \\ \text{volume of the mixture}}} \mapsto \mathbb{S} = \underbrace{\begin{cases} \displaystyle\int_V S\,dV \\ \displaystyle\int_{\underline{V}} S\,d\underline{V} \end{cases}}_{\substack{\text{entropy} \\ \text{of the body}}}. \tag{4.7.16}$$

2. The total internal entropy production is built additively from the internal entropy production of the species,

$$\hat{S}_{(\mathrm{i})} = \sum_{k\in\mathcal{K}} \hat{S}_{(\mathrm{i})}^k, \quad \underline{\hat{S}}_{(\mathrm{i})} = \sum_{k\in\mathcal{K}} \underline{\hat{S}}_{(\mathrm{i})}^k\,, \tag{4.7.17}$$

with the convention $\underline{\hat{S}}_{(\mathrm{i})} = \det\mathbf{F}\,\hat{S}_{(\mathrm{i})}$ and $\underline{\hat{S}}_{(\mathrm{i})}^k = \det\mathbf{F}\,\hat{S}_{(\mathrm{i})}^k$, $k\in\mathcal{K}$.

3. The heat flux $\mathbf{Q}$ and heat source Θ in (4.7.15) are equal to the inner parts of their species properties as described by (4.6.20).

Note 4.5 on the number of dissipation inequalities for a mixture.

We require a single inequality for the whole mixture. This requirement seems consistent with the local thermal equilibrium assumption (a uniform temperature over all species). Theories assuming a local thermal nonequilibrium assumption may or may not be more demanding. Eringen and Ingram [1965] require an entropy inequality per temperature/species $\hat{S}_{(\mathrm{i})}^k \geq 0$, their Eqn (8.5), which of course implies $\hat{S}_{(\mathrm{i})} \geq 0$ at the mixture level. On the other hand, Atkin and Craine [1975] require a single inequality for the whole mixture, their Eqn (2.40), namely $\hat{S}_{(\mathrm{i})} \geq 0$.

Note 4.6 on local thermal non equilibrium.

For a system in local thermal equilibrium where the temperatures of all species are equal, $T_k = T$, $\forall k\in\mathcal{K}$, the second principle $\hat{S}_{(\mathrm{i})}$ is often expressed in the equivalent format $T\hat{S}_{(\mathrm{i})} \geq 0$. For mixtures where local thermal equilibrium does not hold, where each species is endowed with its own temperature, the entropy inequality at the mixture level is contributed by the species in the format $\sum_k \hat{S}_{(\mathrm{i})}^k \geq 0$, and not $\sum_k T_k\hat{S}_{(\mathrm{i})}^k \geq 0$.

4.7.4 The Clausius-Duhem inequality

4.7.4.1 Free energy of the mixture, electrochemical potentials of species

The Clausius-Duhem inequality (CD) is obtained by insertion of the balance of energy into the second principle. It expresses in terms of
- the free energy of the mixture per unit current volume

$$E = U - TS, \tag{4.7.18}$$

- and of the free enthalpies, also called electrochemical potentials, of the species per unit mass,

$$\mu_k^{ec} = h_k - T s_k = e_k + \frac{p_k}{\rho_k}, \quad k \in \mathcal{K}, \tag{4.7.19}$$

where

$$e_k = u_k - T s_k, \quad k \in \mathcal{K}, \tag{4.7.20}$$

is the free energy per unit mass of the species k.

For further reference, it is instrumental to define the electrochemical potential per unit mole g_k,

$$g_k^{ec} = \widehat{m}_k \, \mu_k^{ec} = \rho_k \, \widehat{v}_k \, \mu_k^{ec}, \quad k \in \mathcal{K}, \tag{4.7.21}$$

where $\widehat{m}_k$ and $\widehat{v}_k$ are respectively the molar mass and molar volume of the species k.

If the partial stress is not isotropic, then, instead of the scalar electrochemical potential (4.7.19), the electrochemical potential tensor $\boldsymbol{\mu}_k^{ec}$ appears,

$$\boldsymbol{\mu}_k^{ec} = \mathbf{h}_k - T s_k \, \mathbf{I} = e_k \, \mathbf{I} - \frac{\boldsymbol{\sigma}^k}{\rho^k}, \quad k \in \mathcal{K}. \tag{4.7.22}$$

The tensor $-\rho^k \, \boldsymbol{\mu}_k^{ec} = \boldsymbol{\sigma}^k - \rho^k \, e_k \, \mathbf{I}$ is sometimes referred to as the Eshelby stress of the species.

The electrochemical potentials have still to be defined by constitutive equations. This issue is pursued in Chapters 13 and 14. The electrochemical potentials contain, besides a mechanical and an electrical component, a chemical component as well as a term that represents an energy of formation. The latter is capital for the transfer equations, as it generates equilibrium constants. Instances for clays and articular cartilages can be found in our previous works, Gajo et al. [2002], Loret and Simões [2007].

The chemical potentials are denoted by the symbol μ and the *electro*chemical potentials by the symbol μ^{ec}. Of course, the chemical and electrochemical potentials are one and the same if the species is electrically neutral.

4.7.4.2 Internal entropy production, equilibrium, stationary state

The (CD) inequality states that the internal entropy production should be positive,

$$\hat{S}_{(i)} \geq 0. \tag{4.7.23}$$

The external entropy production $\hat{S}_{(e)}$, that is the entropy production contributed by the surroundings to the body, and the total entropy production $\hat{S} = \hat{S}_{(i)} + \hat{S}_{(e)}$ may have any sign.

The internal entropy production is a sum of work-conjugated fluxes and driving forces. It is convenient to delineate the fluxes for which no constraint is placed on the associated driving forces: their asymptotic values vanish. On the other hand, the fluxes whose driving forces are subjected to a constraint tend asymptotically to a constant (non zero) value. The asymptotic state may be called an *equilibrium* in the first case, and a *stationary state* in the second case. In both cases, since the state is constant, so do functions of the state, and in particular the (total) entropy,

$$\hat{S} = 0 \quad \text{for a stationary state and for an equilibrium}. \tag{4.7.24}$$

However, both internal and external entropy productions vanish in an equilibrium,

$$\hat{S}_{(i)} = 0, \quad \hat{S}_{(e)} = 0 \quad \text{for an equilibrium}, \tag{4.7.25}$$

while

$$\hat{S}_{(i)} \geq 0, \quad \hat{S}_{(e)} = -\hat{S}_{(i)} \leq 0 \quad \text{for a non equilibrium stationary state}, \tag{4.7.26}$$

that is, entropy is *supplied* to the surroundings. Consequently, non equilibrium stationary states associated with, e.g., active transports require the system to be thermodynamically open, Katchalsky and Kedem [1962], p. 59.

Thus Clausius-Duhem implies an increasing disorder, Fig. 4.7.1. On the other hand, biological systems tend to increase order, e.g., by active transports, and thus to decrease entropy. These two points of view seem opposite. According to Glansdorff and Prigogine [1971], they correspond to different settings: close to equilibrium $\hat{S}_{(e)} = 0$, structures disappear according to (CD), namely $\hat{S}_{(i)} \geq 0$. On the other hand, biological structures are created through nonlinear kinetics far from equilibrium, where $\hat{S}_{(e)} \leq 0$ and $\hat{S} \leq 0$.

negative fixed charge

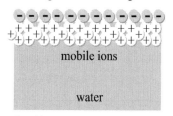

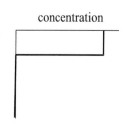

concentration

(a) maximal heterogeneous spatial distribution
minimal energy

negative fixed charge

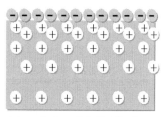

concentration

Fig. 4.7.1: Negative fixed electrical charges attract mobile cations. The three thought spatial distributions of mobile ions illustrate the notions of energy, entropy and the intermediate actual solution. Modified from Appelo and Postma [1993], p. 184.

(b) uniform spatial distribution
maximal entropy

negative fixed charge

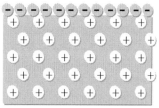

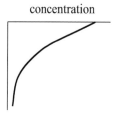

concentration

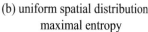

(c) graded spatial distribution
actual intermediate state

4.7.4.3   Additive split of the (CD) inequality

In deriving the (CD) inequality, the thermal contribution $\Theta - \operatorname{div} \mathbf{Q}$ is first obtained from the first principle (4.6.22), and next substituted in the local statement of the second principle (4.7.15).

The (CD) inequality can be split into three terms of distinct physical natures,

$$\hat{S}_{(i)} = \hat{S}_{(i,1,4)} + \hat{S}_{(i,2)} + \hat{S}_{(i,3)} \geq 0. \tag{4.7.27}$$

It is sufficient (but by no means necessary a priori) to require each of them to be positive or zero. As indicated in Section 4.2, this point of view certainly restricts the interactions between the physical phenomena but it does not preclude some interdependence:

1. $\hat{S}_{(i,1,4)}$ represents the contributions to $\hat{S}_{(i)}$ due to **mechanics** and **growth**:

$$
\begin{aligned}
T\hat{S}_{(i,1,4)} &= -\frac{dE}{dt} - E \operatorname{div} \mathbf{v}_s + \boldsymbol{\sigma}^I : \boldsymbol{\nabla}\mathbf{v}_s - S\frac{dT}{dt} + \frac{1}{\det \mathbf{F}} \sum_{k \in \mathcal{K}^*} \mu_k^{ec} \frac{dm^k}{dt} - T\,\hat{S}_{(e)}\,, \\
T\underline{\hat{S}}_{(i,1,4)} &= -\frac{dE}{dt} + \underline{\underline{\boldsymbol{\tau}}}^I : \frac{d\mathbf{E}_{\mathrm{G}}}{dt} - \underline{S}\frac{dT}{dt} + \sum_{k \in \mathcal{K}^*} \mu_k^{ec} \frac{dm^k}{dt} - T\,\underline{\hat{S}}_{(e)}\,,
\end{aligned}
\tag{4.7.28}
$$

with the external entropy production per unit current volume $\hat{S}_{(e)}$ and per unit reference volume $\underline{\hat{S}}_{(e)}$,

$$
\begin{aligned}
-T\,\hat{S}_{(e)} &= \tfrac{1}{2}\,\hat{\rho}\,\mathbf{v}_s^2 - \hat{\mathbf{p}} \cdot \mathbf{v}_s + \hat{U} - T\hat{S}\,, \\
-T\,\underline{\hat{S}}_{(e)} &= \tfrac{1}{2}\,\hat{m}\,\mathbf{v}_s^2 - \underline{\hat{\mathbf{p}}} \cdot \mathbf{v}_s + \underline{\hat{U}} - T\underline{\hat{S}}\,.
\end{aligned}
\tag{4.7.29}
$$

The fact that the system is open is blatant in $\hat{S}_{(e)}$, where the contributions $\hat{\rho}$, $\hat{\mathbf{p}}$, and $\hat{U}$, $\hat{S}$ by the surroundings to the mass, momentum, internal energy and entropy respectively, are recognized.

The unravelling between mechanics and growth is realized in Section 21.4. A vanishing mechanical dissipation $\hat{S}_{(i,1)} = 0$ gives rise to coupled chemo-hyperelasticity in terms of a generalized stress and a generalized strain that are motivated by the equality. Examples are provided in the modeling of articular cartilages in Chapter 14 and of cornea in Chapter 19.

The surface contributions to the growth process by the surroundings have not been considered, except in Section 4.5.4. If they are accounted for, the overall supplies of mass, momentum, energy and entropy should be understood as accounting for the volume and surface contributions, namely along (4.3.8),

$$
\begin{aligned}
\hat{\rho}^k &\rightarrow \hat{\rho}^k + \operatorname{div}\hat{\mathbf{R}}^k, \\
\hat{\boldsymbol{\pi}}^k &\rightarrow \hat{\boldsymbol{\pi}}^k + \operatorname{div}\hat{\boldsymbol{\Pi}}^k, \\
\hat{\mathcal{U}}^k &\rightarrow \hat{\mathcal{U}}^k + \operatorname{div}\hat{\mathbf{U}}^k, \\
\hat{\mathcal{S}}^k &\rightarrow \hat{\mathcal{S}}^k + \operatorname{div}\hat{\mathbf{S}}^k.
\end{aligned}
\tag{4.7.30}
$$

2. $\hat{S}_{(i,2)}$ is the contribution due to (internal) **mass transfers**:

$$
\begin{aligned}
T\hat{S}_{(i,2)} &= -\sum_{k\in\mathcal{K}^*}\left(\mu_k^{\mathrm{ec}} + \tfrac{1}{2}\left(\mathbf{v}_k - \mathbf{v}_\mathrm{s}\right)^2\right)\hat{\rho}^k, \\
T\hat{\underline{S}}_{(i,2)} &= -\sum_{k\in\mathcal{K}^*}\left(\mu_k^{\mathrm{ec}} + \tfrac{1}{2}\left(\mathbf{v}_k - \mathbf{v}_\mathrm{s}\right)^2\right)\hat{m}^k.
\end{aligned}
\tag{4.7.31}
$$

Constitutive equations of mass transfer are guided by satisfaction of the inequality $\hat{S}_{(i,2)} \geq 0$. Thus mass transfers are driven by differences in electrochemical potentials. To realize this point, consider a closed system with two species/compartments: then $\hat{\rho}^1 + \hat{\rho}^2 = 0$, and therefore $T\hat{S}_{(i,2)} = -(\mu_1 - \mu_2)\,\hat{\rho}^1 \geq 0$. In words, mass transfers from the compartment of higher electrochemical potential to the compartment of lower electrochemical potential. Linear and nonlinear constitutive equations of mass transfer are proposed in Chapter 10.

3. $\hat{S}_{(i,3)}$ gives rise to **coupled generalized diffusion**:

$$
\begin{aligned}
T\hat{S}_{(i,3)} &= -\left(\frac{\mathbf{Q}}{T} + \sum_{k\in\mathcal{K}^{**}} s_k\,\mathbf{M}_k\right)\cdot\boldsymbol{\nabla}T - \sum_{k\in\mathcal{K}^{**}}\mathbf{M}_k\cdot\left(\boldsymbol{\nabla}\mu_k^{\mathrm{ec}} + \frac{d^k\mathbf{v}_k}{dt} - \mathbf{b}_k\right), \\
T\hat{\underline{S}}_{(i,3)} &= -\left(\frac{\mathbf{Q}}{T} + \sum_{k\in\mathcal{K}^{**}} s_k\,\underline{\mathbf{M}}_k\right)\cdot\boldsymbol{\nabla}T - \sum_{k\in\mathcal{K}^{**}}\underline{\mathbf{M}}_k\cdot\boldsymbol{\nabla}\mu_k^{\mathrm{ec}} + \mathbf{F}\cdot\underline{\mathbf{M}}_k\cdot\left(\frac{d^k\mathbf{v}_k}{dt} - \mathbf{b}_k\right).
\end{aligned}
\tag{4.7.32}
$$

The species which move with the solid velocity are definitely not involved in (4.7.32), so that summation extends over the set $\mathcal{K}^{**}$. The term

$$
\frac{\mathbf{Q}}{T} + \sum_{k\in\mathcal{K}^{**}} s_k\,\mathbf{M}_k,
\tag{4.7.33}
$$

is referred to as total entropy flux. Uniform body forces $\mathbf{b}_k$ can be viewed as introducing a sedimentation contribution into the electro-chemomechanical potentials.

The species that have the same velocity as the solid have to be eliminated, or kept, in (4.7.28) and (4.7.31) according to the same rule. A priori, we would like to involve the maximum number of species, even those that move with the solid velocity. One possibility is to consider only the species that transfer, and then the set for the summation in (4.7.28) is $\mathcal{K}^*$. But in that way, we tacitly imply that the species that are not involved in (4.7.31) do not transfer, because we will not have constitutive equations for their transfer. Conversely, if the latter were subjected to transfer, they would have to be endowed with a chemical potential, either in scalar form, or in tensorial form (Eshelby stress). In the framework we have in mind, Eshelby stresses would appear for species that grow. The latter are assumed to move with the same velocity as the solid, so that their fluxes relative to the solid vanish. Thus Eshelby stresses do not appear here, because they are conjugated to vanishing fluxes, Section 6.3 of Loret and Simões [2005]b. We will summarize the state of affairs in Section 21.4.1 later.

The inequality $\hat{\underline{S}}_{(i,3)} \geq 0$ is used to formulate the constitutive equations of generalized diffusion, with possible couplings between Darcy's, Fick's, Fourier's and Ohm's laws. Examples are provided in Section 9.3.3 which addresses thermoosmosis and in Chapter 16 which addresses coupled fluid flow, ionic diffusion and electromigration.

Note 4.7 on a more general formulation with electrochemical potentials in tensor form.

In the above decomposition of the dissipation in three terms, use has been made of the fact that the non growing species are endowed with an isotropic partial stress, leading to an isotropic chemical potential tensor. If this assumption is not used, then we are to be content with a decomposition in two parts only,

$$
\begin{aligned}
T\hat{S}_{(i,1,2,4)} &= -\frac{dE}{dt} - E\,\mathrm{div}\,\mathbf{v}_s + \boldsymbol{\sigma}^I : \boldsymbol{\nabla}\mathbf{v}_s - S\frac{dT}{dt} - \sum_{k\in\mathcal{K}^{**}}\left(\boldsymbol{\nabla}\mathbf{M}_k : \boldsymbol{\mu}_k^{ec} + \tfrac{1}{2}\,\hat{\rho}^k\left(\mathbf{v}_k - \mathbf{v}_s\right)^2\right) - T\,\hat{S}_{(e)}\,, \\
T\hat{S}_{(i,3)} &= -\left(\frac{\mathbf{Q}}{T} + \sum_{k\in\mathcal{K}^{**}} s_k\,\mathbf{M}_k\right)\cdot\boldsymbol{\nabla}T - \sum_{k\in\mathcal{K}^{**}}\mathbf{M}_k\cdot\left(\mathrm{div}\,\boldsymbol{\mu}_k^{ec} + \frac{d^k\mathbf{v}_k}{dt} - \mathbf{b}_k\right).
\end{aligned}
\tag{4.7.34}
$$

For the species whose partial stress is isotropic, the electrochemical potential tensor $\boldsymbol{\mu}_k^{ec}$ becomes equal to $\mu_k^{ec}\,\mathbf{I}$. Then the term $\mathbf{M}_k\cdot\mathrm{div}\,\boldsymbol{\mu}_k^{ec}$ becomes $\mathbf{M}_k\cdot\boldsymbol{\nabla}\mu_k^{ec}$ while $\boldsymbol{\nabla}\mathbf{M}_k : \boldsymbol{\mu}_k^{ec}$ simplifies to $\mu_k^{ec}\,\mathrm{div}\,\mathbf{M}_k$. The divergence term $\mathrm{div}\,\mathbf{M}_k$ can be eliminated in favor of dm^k/dt and $\hat{\rho}^k$ using the balance of mass (4.4.9), and this manipulation gives rise to the decomposition of the dissipation in three parts.

Note 4.8 on a more general formulation with active transports.

We have split the (single) entropy inequality into three or four parts, each one being in turn required to be positive or zero. This formalism is not as drastic as it could be thought a priori. Indeed, a redefinition of the various entities can be called for, if necessary. For example, we have tacitly assumed that the contribution by the surroundings $\hat{S}_{(e)}$ is completely diverted to mechanics and growth.

Let us consider active transfer or active diffusion, with energy providers (cells delivering ATP) being exterior to the system. Then the individual contributions to mechanics, transfer, diffusion and growth emanating from the total $\hat{S}_{(e)}$ should be delineated. This split embodies a competition for energy among the physical phenomena at hand.

4.8   Dissipation mechanisms

Constitutive assumptions are needed to split the thermomechanical part from the growth contribution in the dissipation (4.7.28). Constitutive assumptions for the growth rate can then be deduced.

In presence of chemomechanical couplings in geomechanics and biomechanics, the entropy production is contributed by mechanics via viscous effects, and plastic effects in the case of geomaterials, by mass transfer and by generalized diffusion. Following the standard thermodynamic approach, e.g., Kestin [1968], each of these phenomena contributes additively to the entropy production in the form of the product of a thermodynamic force f and of a flux $d\xi/dt$, of appropriate tensorial nature,

$$
\hat{S}_{(i)} = \cdots - f\,\frac{d\xi}{dt}\cdots.
\tag{4.8.1}
$$

The entropy productions $\hat{S}_{(i,2)}$ due to mass transfer and $\hat{S}_{(i,3)}$ due to generalized diffusion are almost in this standard format, namely a sum of such products. Constitutive equations motivated by these expressions have been developed in Gajo and Loret [2003] [2004], Loret et al. [2004]b for geomaterials and in Loret and Simões [2004] [2005]a for articular cartilages.

4.8.1   Dissipation due to generalized diffusion and transfer

Generalized diffusion includes thermal effects (Fourier's law), seepage of water through the solid skeleton (Darcy's law), diffusion of species in water (Fick's law) and electrical flow (Ohm's law). The couplings between the different diffusion phenomena that occur in an electrolyte can not be neglected. As an algebraic consequence, the associated diffusion matrix is not diagonal. In fact, parameter identification requires a rewriting of the dissipation due to diffusion in terms of entities more accessible to measurements than electrochemical potentials, as stressed in the above papers and in Chapter 16. The coupled constitutive equations for generalized diffusion express the fluxes in terms of the driving forces via a symmetric matrix which has to be positive (semi-)definite, but that may well depend on the material state.

These equations apply in continua but they may be used as well to describe passive transport across a cellular membrane. If only a single ionic species is allowed to cross the membrane, equilibrium implies the electrochemical potentials on both sides of the membrane to be equal and the rest potential is the Nernst potential. On the other hand, for an electrogenic membrane crossed by several ionic species, the rest potential is provided either by the Huxley-Hodgkin-Katz (HHK) formula phrased in terms of permeabilities and concentrations, or by the chord conductance equation, phrased in terms of conductances and Nernst potentials, Sperelakis [2001]. In a passive transport, the motion takes place *against* a gradient. On the other hand, active transport, where the motion using available energy (in the form of ATP) may take place *along* a gradient, is usually described empirically by kinematics associated with reversible or irreversible enzymatic reactions, Section 15.12.2.

The dissipation associated with mass transfer simply indicates that the rate of mass transfer is a function of the out-of-balance between electrochemical potentials. In fact, due to electroneutrality, the out-of-balance driving force can be expressed in terms of chemical potentials of electroneutral salts, Loret and Simões [2007]. In our previous works, linear and exponential constitutive equations of mass transfer have been proposed, and the coupling between species has been ignored, so that the transfer matrix was diagonal.

4.8.2 Spontaneous reactions and stable configurations

It is appropriate to highlight here the connections between these continuum thermodynamic developments and chemical thermodynamics and in particular the interpretation of mass transfer as a physico-chemical reaction, possibly involving several species endowed with distinct free enthalpies of formation, resulting in an equilibrium constant nonequal to unity.

In agreement with the above constitutive equation of mass transfer, chemical equilibrium between two thermodynamic states is defined by a vanishing variation of *free enthalpy* ΔG. If the free enthalpy of the final state is lower than initially, $\Delta G < 0$, the reaction is said exergonic, and *spontaneous* at fixed pressure and temperature. Otherwise, i.e., if $\Delta G > 0$, some form of energy should be provided for the reaction to occur and the reaction is said endergonic[4.3]. In continuum thermodynamics, the free enthalpy of species is called (electro)chemical potential because it is the chemical work that a substance can provide.

The variation of free enthalpy ΔG, at constant temperature T, is equal to $\Delta H - T\,\Delta S$, where ΔH is the variation of *enthalpy* and ΔS the variation of *entropy*. Enthalpy is a measure of heat content. Depending whether the reaction takes place with $\Delta H < 0$ or $\Delta H > 0$, it is said exothermic or endothermic. For example, the reaction $H_2 + \frac{1}{2}O_2 \rightarrow H_2O$ is exothermic and $\Delta H = -287\,kJ/mole$. When NaCl is dropped in a reservoir filled with water, the reservoir cools because it has given heat for the dissolution to take place, $\Delta H = +3.8\,kJ/mole$.

Entropy is usually viewed as a measure of *molecular disorder*. It is larger for systems whose order is low, as shown in Section 3.2.1 in relation to phase change. If, during a reaction, the order increases, then the associated $\Delta S < 0$ is negative. For example, the gas mixture of H_2 and O_2 has a larger entropy than water H_2O and the reaction $H_2 + \frac{1}{2}O_2 \rightarrow H_2O$ has $-T\,\Delta S = 49\,kJ/mole$ at $T=20\,°C$. Therefore the reaction is spontaneous since $\Delta G = -238\,kJ/mole$ is negative.

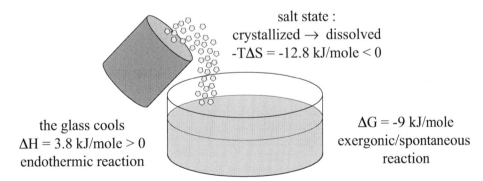

salt state :
crystallized $\rightarrow$ dissolved
$-T\Delta S = -12.8\ kJ/mole < 0$

the glass cools
$\Delta H = 3.8\ kJ/mole > 0$
endothermic reaction

$\Delta G = -9\ kJ/mole$
exergonic/spontaneous
reaction

Fig. 4.8.1: An endothermic chemical reaction may be spontaneous if it is sufficiently entropy-driven.

[4.3] A heat engine receives heat from a heat source (like a boiler, or a device burning fossil or nuclear fuels), produces mechanical work, and transfers some remaining heat to a heat sink. Conversely, heat pumps receive work and transfer heat from a source at lower temperature to a source at higher temperature: there is no contradiction with the second principle of thermodynamics because the thermal flow is no longer spontaneous, but mechanically activated. Refrigerators and geothermal heat pumps are instances of heat pumps. Refrigerators aim primarily at extracting heat from the source at lower temperature, while geothermal heat pumps are designed to transfer heat to the source at higher temperature.

Let us consider again the dissolution of NaCl in water, Fig. 4.8.1. NaCl initially crystallized has a high order of organization while, when dissolved in water, the ions Na^+ and Cl^- move randomly. Therefore dissolution of NaCl in water is associated with $\Delta S > 0$, in fact, at ambient temperature, $-T\Delta S = -12.8\,kJ/mole$.

This example of dissolution of salt in water is a bit peculiar because it is exergonic, i.e. spontaneous, $\Delta G = -9\,kJ/mole < 0$, while endothermic, $\Delta H = +3.8\,kJ/mole > 0$. Therefore, even if quite often $\Delta G \sim \Delta H$, *an endothermic reaction may occur spontaneously if it is (sufficiently) entropy-driven.*

Table 4.8.1: Variations of enthalpy ΔH and free enthalpy ΔG [unit : kJ/mole] associated with the formation of two sequences of amino-acids, Pálfi and Perczel [2008].

	ΔH		ΔG	
Sequence	Triple Helix	β-Pleated Sheet	Triple Helix	β-Pleated Sheet
-Gly-Gly-Gly	-129.3	-286.7	113.8	-68.6
-Gly-Hyp-Pro	-169.9	-65.7	-5.4	54.8

The existence and stability of certain biological structures in preference to others are often dictated by energetic considerations. These structures may differ by their composition or by their spatial conformation. Collagen chains are formed by the repetition of basic units of three amino-acids, -Gly-X-Y, where Gly is the amino-acid glycine, and X and Y may be one of the twenty amino-acids. For example, we consider the sequences -Gly-Gly-Gly, and -Gly-Hyp-Pro where Hyp is hydroxyproline and Pro proline. These two sequences aggregate into chains that form triple helices or triple stranded β-pleated sheets. The changes of enthalpy and free enthalpy associated with the formation of these structures starting from the individual chains are listed in Table 4.8.1.

The entropy variation corresponding to a higher organization is negative. The sequence -Gly-Gly-Gly is more likely to exist under the form of β-pleated sheets, while the sequence -Gly-Hyp-Pro is more likely to exist as a triple helix. β-pleated sheets are known to be associated with diseases that affect the central nervous system.

At a higher scale, the assembly of collagen molecules into fibrils is again both endothermic and exergonic (spontaneous), possibly due to entropy increase associated with loss of water from bound monomers or to hydrophilic interactions. At still higher scales, collagen fibrils organize into various spatial structures that serve best the physiological role of the tissues, e.g., parallel arrays in tendons and ligaments. The cornea and annulus fibrosus are made of a number of lamellae in which the collagen fibers adopt a strictly identical direction. This direction changes abruptly from lamella to lamella. This organization is thought to optimize transparency in the cornea and to resist clockwise and anticlockwise torsion in the annulus fibrosus. In the central zone of articular cartilages the directional distribution of collagen fibers is roughly random, while fibers are aligned normal to the subchondral bone in the lower zone and parallel to the synovial joint in the upper zone.

In a thermodynamic approach to mixture, the key role of the free enthalpy, or chemical potential, of a chemical species is made clear when we introduce the second principle and the Clausius-Duhem inequality. Chemical potentials are basic entities to the development of the chemomechanical constitutive equations and of the constitutive equations of mass transfer, Chapter 10, and for generalized diffusion, Chapter 16.

4.8.3   Coupled chemical reactions

While it may not be spontaneous by itself, a chemical reaction may turn so when coupled with one or several other reactions. In the context of chemistry, coupling between two reactions implies that the reactions share, perhaps temporarily, some chemicals. Indeed, the existence of a source of energy is not enough. A key issue is to find out a device to make use of this energy (perhaps only a part of it, as its efficiency will probably, and even certainly, not be complete).

The concept of *coupling* may be also explained via a mechanical analogy. Consider two identical masses 1 and 2, which are subjected to gravity, Fig. 4.8.2. The issue is to lift mass 1: this experiment requires energy to be provided to the mass. The question is: how to use the energy provided by dropping mass 2? Dropping mass 2 alone won't help. Some device should be invented to transfer directly this energy (or perhaps part of it only due to dissipation) to mass 1. Connecting the two masses through a rope attached to a pulley will solve the problem ... The energy released by mass 2 is used by mass 1.

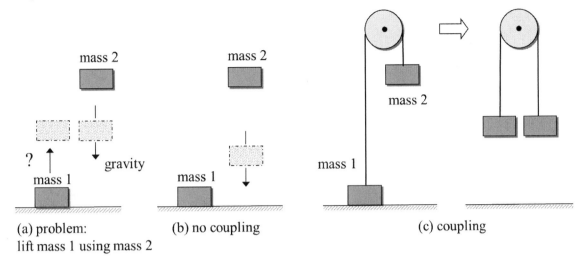

(a) problem:
lift mass 1 using mass 2

(b) no coupling

(c) coupling

Fig. 4.8.2: Illustrative strategy of mechanical coupling, from APB [1999], p. 15.

As an example, consider two independent reactions,

$$A \; \rightleftharpoons \; B, \quad \Delta g_1^0 > 0,$$
$$C \; \rightleftharpoons \; D, \quad \Delta g_2^0 < 0,$$
(4.8.2)

the first one being endergonic (and therefore not spontaneous), while the second one is exergonic (and therefore spontaneous)[4.4]. This information is delivered by the sign of the change of the standard chemical potential Δg_k^0, $k = 1, 2$. Indeed, the equilibria of these two reactions expressed in terms of concentrations point to the left and to the right respectively,

$$\frac{c_B}{c_A} = K_1^{\mathrm{eq}} = \exp\left(-\frac{\Delta g_1^0}{RT}\right) < 1, \quad \frac{c_D}{c_C} = K_2^{\mathrm{eq}} = \exp\left(-\frac{\Delta g_2^0}{RT}\right) > 1.$$
(4.8.3)

Assume that we can couple the two reactions (ideally, that is without loss of energy),

$$A + C \; \rightleftharpoons B + D, \quad \Delta g_{1+2}^0 = \Delta g_1^0 + \Delta g_2^0.$$
(4.8.4)

The *new* equilibrium is defined by

$$\frac{c_B}{c_A}\frac{c_D}{c_C} = K_{1+2}^{\mathrm{eq}} = \exp\left(-\frac{\Delta g_{1+2}^0}{RT}\right) = K_1^{\mathrm{eq}} K_2^{\mathrm{eq}}.$$
(4.8.5)

This equilibrium is shifted to the right with respect to the first independent reaction,

$$\frac{c_B}{c_A} = K_{1+2}^{\mathrm{eq}} \frac{c_C}{c_D} > K_1^{\mathrm{eq}},$$
(4.8.6)

if the concentrations of the species involved in the second reaction which are thought to be clamped satisfy the condition,

$$\frac{c_C}{c_D} > \frac{1}{K_2^{\mathrm{eq}}}.$$
(4.8.7)

Typically, in biological systems, the second reaction involves ATP. The inequality means that the concentration of ATP should be sufficiently large so that ATPase can take place, as argued in Section 20.2.2. Another example with an electrogenic ionic pump is exposed in Section 19.3.3.3.

[4.4]The distinction between endergonic and endothermic reactions, and exergonic and exothermic reactions, is highlighted in Section 4.8.2.

4.8.4 Dissipation associated with a chemical reaction

The contribution due to mass transfers to the dissipation inequality (4.7.31) can be used to estimate the dissipation associated with chemical reactions. Indeed, if the kinetic terms are neglected and if the (electro)chemical potentials per unit mole are used rather than per unit mass, refer to (3.1.6) or (4.7.21) for definitions, the dissipation (4.7.31) may be brought to the format (3.1.19),

$$-\sum_{j=1}^{n} g_j^{\text{ec}} \, dN_j = -\mathcal{G} \, d\xi \geq 0 \,, \tag{4.8.8}$$

where n is the number of chemicals involved in the reaction, and $\mathcal{G}$ and ξ are respectively the chemical affinity and advancement of the reaction defined in Section 3.1.5.

The dissipation inequality requires this quantity to be positive. According to (3.1.4) and (3.1.19), this requirement is equivalent to a negative change of the Gibbs energy during the reaction at constant pressure and temperature.

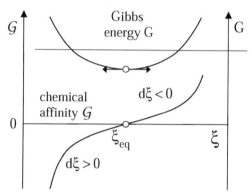

Fig. 4.8.3: Assumed thermodynamic properties of the Gibbs energy G and chemical affinity $\mathcal{G}$ in the vicinity of an equilibrium.

Moreover from (3.1.19),

$$\mathcal{G} = \frac{\partial G}{\partial \xi}\bigg|_{p,T} \,, \qquad \frac{\partial \mathcal{G}}{\partial \xi}\bigg|_{p,T} = \frac{\partial^2 G}{\partial \xi^2}\bigg|_{p,T} \,. \tag{4.8.9}$$

Requiring a stable equilibrium $\xi = \xi_{\text{eq}}$ amounts to a positive second derivative $\partial^2 G / \partial \xi^2 > 0$, which, together with the inequality $-\mathcal{G} \, d\xi \geq 0$ leads, in the vicinity of an equilibrium, to the configuration sketched in Fig. 4.8.3. A particular case is considered in Section 3.1.7.1 and illustrated by Fig. 3.1.2 where $J = d\xi/dt$. For a quantified example, see Kestin [1968], p. 295, where the Gibbs energy and chemical affinity are plotted during the dissociation of water vapor $2\,\mathrm{H_2O} \leftrightharpoons 2\,\mathrm{H_2} + \mathrm{O_2}$ at constant pressure and temperature (notice that his chemical affinity is the negative of the one used here).

4.8.5 Are growth and structuration dissipative processes?

A useful hint for the thermodynamic analysis of a mixture is the two-step application of the second principle, Krishnaswamy and Batra [1997], Klisch and Hoger [2003]. A set of state variables $\mathbf{A}$ is defined as well as a set of all variables $\mathbf{B}$ including diffusion fluxes, thermal and velocity gradients, etc. Equilibrium paths for which $\mathbf{B} \sim \mathbf{A}$ are assumed to be path-independent, or reversible: this assumption turns the second principle into an equality for equilibrium paths and allows to define the entropy, a constructed entity rather than a primitive function, as a function of $\mathbf{A}$. This property is adopted for arbitrary processes, e.g., when diffusion occurs due to distinct velocities of the species. However arbitrary processes are a priori not reversible and they are constrained by the entropy inequality. This viewpoint is in line with Ericksen [1991] who showed that (thermal or hydraulic) diffusion can not be considered as a reversible process.

In relation with the above point, Krishnaswamy and Batra [1997] and Klisch and Hoger [2003] define materials that can not experience reversible growth and materials that can. As an example worked out by Klisch and Hoger [2003], their section 3.3, reversible growth is possible under equilibrium paths if the growth law $\mathbf{L}^{\mathrm{g}} = \mathcal{L}(\mathbf{L}^{\mathrm{e}}, \ldots)$ which provides the rate of growth in terms of the elastic velocity 'gradient' $\mathbf{L}^{\mathrm{e}}$ verifies the property,

$$\mathcal{L}(-\mathbf{L}^{\mathrm{e}}, \cdots) = -\mathcal{L}(\mathbf{L}^{\mathrm{e}}, \cdots). \tag{4.8.10}$$

Now, it remains to assess what is the practical significance of the notion of (ir)reversible growth. Are the natural and engineered growths of cartilage reversible or irreversible in the above sense?

In fact, even if metabolic reactions are nowadays commonly given the possibility to be two-way, in practice some reactions are mainly in one direction. Indeed, anabolism and catabolism, and growth and degradation, usually do not follow the same paths and require or provide energy. They are to be definitely irreversible. For example, the energetic profile of glycolysis displayed in Fig. 20.2.3 shows three steps with significant decrease of free enthalpy, namely

- hexokinase/glucokinase that controls the rate of entrance of glucose into the pathway;
- phospho-fructokinase 1 which gives the limiting rate to the glycolysis;
- pyruvate kinase that controls the production of the glycolysis.

These steps are irreversible and they are bypassed during neo-glycogenesis. The other steps of glycolysis are practically reversible and indeed they are followed in reverse direction during neo-glycogenesis. The formation of highly energetic triglycerides from fatty acids in adipocytes, and their hydrolysis, are similarly considered irreversible.

Still, there might exist growth mechanisms that can be considered to be reversible, perhaps depending on the definition of reversibility.

Another issue to be addressed arises in presence of mechanisms that contribute negatively to the dissipation. Self-healing phenomena of geological and engineering materials belong to this category, Section 23.8. In the framework of growth of tissues, mechanisms that endow the tissues with a structure, a concept that takes several forms, are potential candidates to enter the issue. The consequences on the growth rate are considered in the framework of elastic-growing mixtures in Chapter 23.

4.8.6 Onsager relations and rate potential

The entropy production of the mixture associated with growth, remodeling, structuration, deformation and dissipation $\hat{S}_{(\mathrm{i},1,4)}$ is postulated in the format of Eqn (4.8.1),

$$T\hat{S}_{(\mathrm{i},1,4)} = -\sum_{k \in \mathcal{S}^*} \mathbf{f}_k : \mathbf{L}_k^{\mathrm{g}} - \sum_{k \in \mathcal{S}^*} R_k \frac{dr^k}{dt} - \sum_{n \in \mathcal{N}} X_n \frac{d\xi_n}{dt} \geq 0. \tag{4.8.11}$$

This expression implies that growth and remodeling, corresponding to the first and second terms in (4.8.11), contribute independently to dissipation. The dissipation mechanism due to mechanical viscosity is accounted for by $n = 1$, and the mechanisms $n \in \mathcal{N}$, $n \neq 1$, account for other sources of structuration and dissipation. They might be of scalar, vector or tensor natures.

In order to ensure positiveness of the entropy production associated with mechanics and growth, one may use, close to equilibrium, the Onsager formalism. An alternative nonlinear form is proposed later for scalar variables.

Assuming for simplicity that couplings occur only within each of the three groups displayed in (4.8.11), the driving forces may be considered linear with respect to the rates,

$$-\mathbf{f}_k = \sum_{l \in \mathcal{S}^*} \mathbf{L}_l^{\mathrm{g}} : \mathcal{F}_{lk}, \quad k \in \mathcal{S}^*; \quad -R_k = \sum_{l \in \mathcal{S}^*} \frac{dr^l}{dt} \mathcal{R}_{lk}, \quad k \in \mathcal{S}^*; \quad -X_n = \sum_{l \in \mathcal{N}} \frac{d\xi_l}{dt} \mathcal{X}_{ln}, \quad n \in \mathcal{N}, \tag{4.8.12}$$

where $\mathcal{F}$, $\mathcal{R}$ and $\mathcal{X}$ are *symmetric* and *(semi-)positive definite* 'matrices' of appropriate size and nature. The symmetry is a choice, motivated by the fact that the skew-symmetric parts of these matrices are elusive in the dissipation rate. Uncoupling of the physical phenomenon k from the others is obtained by deleting all terms except the diagonal one, eg. $\mathcal{R}_{lk} = 0$, for all $l \neq k$, and, due to symmetry, $\mathcal{R}_{kl} = 0$, for all $l \neq k$.

The above matrices may be functions of the state variables, but they are a priori independent of the rates they multiply. Then the rate potential per unit reference volume $\Omega^{(\mathrm{i})}$,

$$\Omega^{(\mathrm{i})} = \sum_{k,l \in \mathcal{S}^*} \frac{1}{2} \mathbf{L}_l^{\mathrm{g}} : \mathcal{F}_{lk} : \mathbf{L}_k^{\mathrm{g}} + \sum_{k,l \in \mathcal{S}^*} \frac{1}{2} \frac{dr^l}{dt} \mathcal{R}_{lk} \frac{dr^k}{dt} + \sum_{l,n \in \mathcal{N}} \frac{1}{2} \frac{d\xi_l}{dt} \mathcal{X}_{ln} \frac{d\xi_n}{dt} \geq 0, \tag{4.8.13}$$

is homogeneous of degree 2 with respect to the dissipation rates, so that

$$T\hat{S}_{(i,1,4)} = 2\,\Omega^{(i)} = \sum_{k \in \mathcal{S}^*} \frac{\partial \Omega^{(i)}}{\partial \mathbf{L}_k^g} : \mathbf{L}_k^g + \cdots \geq 0\,. \tag{4.8.14}$$

Under these assumptions, equilibrium, defined by the vanishing of the rates $\mathbf{L}_k^g \cdots$, clearly corresponds to the minimum of the entropy production. That is, out-of-equilibrium excursions necessarily increase the entropy production, in agreement with the arguments of Section 4.7.4.2.

Note 4.9 on nonlinear internal entropy production.

The nonlinear dissipation law proposed in Chapter 10 aims at bounding the rates (fluxes) when the thermodynamic forces (gradients) become large.

Another type of nonlinearity may be introduced through the existence of a surface in the space of thermodynamic forces. Vanishing fluxes may imply the thermodynamic forces to vanish or simply to remain in some prescribed domain.

4.9   General constitutive principles

In an axiomatic or rational (in a Noll and Truesdell perspective) analysis, there are five basic principles that have to be satisfied by constitutive equations, whether for a single body or for mixtures, e.g., Mandel [1974], p. 79, Truesdell [1984], p. 230. They are conceived as 'rules to guide us when we come to set up constitutive relations in the first place', Wang and Truesdell [1973], p. 135. They are briefly stated below:

1. *Determinism* asserts that natural phenomena can be described and predicted. There exists a set of variables such that the knowledge of their past and present values determines the present response of the mixture.

2. *Equipresence* states that any 'quantity present as an independent variable in one constitutive relation must be assumed so present in all', Wang and Truesdell [1973], p. 141. Passman et al., p. 301 in Truesdell [1984], find the principle applicable to single bodies and homogeneous mixtures, but they replace it by the *Principle of phase separation* for inhomogeneous mixtures. The latter states that the principle of equipresence applies to individual phases in terms of their *own* set of independent variables, while interaction and exchange terms follow the general principle of equipresence in terms of *all* independent variables. An example is shown in Section 4.5.4.

3. *Local action* implies that the constitutive response of the species at a point of space depends only of the thermokinetic processes in a close neighborhood of that point. This principle is removed for non local constitutive equations of the integral type where the material response at a point depends on the thermokinetic processes in a domain of non zero measure around that point. Whether constitutive equations involving higher order gradients estimated at the point of interest follow the principle or not is a bit more debatable.

   Non local effects have been advocated to smooth out discontinuities and failure processes in engineering materials, as they regularize the mathematical partial differential equations and restore the well-posedness of boundary value problems. Typically, the thickness of intense straining in cohesionless granular materials is observed to extend over a few grains. Long range interactions are also advocated in biological tissues modeling. Murray [2003] points out the role of long sensors (the filopodia) that are attached to the cells and that extend far, further than next neighbors, in the extracellular matrix. Their role seems significant in providing information on the matrix mass density. They are also advocated to signal the presence of a neighboring cell: upon reception of this information, the cell stops, deflects, or even reverses, its motion.

4. *Material frame-indifference* requires that the constitutive equations of the species are independent of the observer.

5. *Dissipation* requires some form of the second principle to be satisfied for any thermokinetic processes.

To these five statements, Wang and Truesdell [1973], p. 135, add a sixth physical principle that allows for specialization, namely

6. *Material symmetry.* There exists a symmetry group, or group of invariance, describing the properties of the body, which leaves unchanged the constitutive equations. The simplest case is that of a mixture of one solid and several unstructured fluids: the mixture inherits the material symmetries of the solid.

When exploiting the dissipation inequality, and developing diffusion constitutive equations, it is a standard practice to use the *Onsager reciprocity relations* to express the 'fluxes' in terms of the 'forces', e.g., de Groot and Mazur [1962]. Having chosen what we call fluxes and what we call forces, the fluxes are expressed in terms of the forces via a *symmetric* (semi-)definite diffusion matrix. Once a choice has been made, one can embark in consistent modifications of the pairs that leave invariant the dissipation, Exercise 16.1.

Truesdell [1984], p. 365, criticizes at length the Onsagerist approach, that claims that the definition of fluxes and forces is dictated by the fact that the fluxes are the time derivatives of 'something' (a void statement: indeed let $f(t) = \int_0^t g(\tau)\, d\tau$, then $g(t) = df(t)/dt$). Truesdell [1984], p. 387 et seq., also smashes the so-called *Curie principle* according to which coupling between quantities of different tensorial character would be forbidden. In fact, representation theorems of isotropic functions circumvent, or solve, the issue.

The dissipation inequality is a powerful tool to develop constitutive equations and to restrict the number of arguments of the constitutive functions. The idea loosely referred to as *Coleman-Noll device*, Coleman [1964], is of pure algebraic nature. Let a and b be two independent dimensionless scalar variables and $A = A(a,b)$ and $B = B(a, b, db/dt)$ two constitutive functions that should satisfy the inequality

$$A(a,b)\, da + B(a, b, db/dt)\, db \geq 0, \quad \forall\, da,\, db. \tag{4.9.1}$$

Since da and db are arbitrary, let us choose first $db = 0$, $da = 1$, and next $db = 0$, $da = -1$, so that $A(a,b) \geq 0$ and $-A(a,b) \geq 0$. Then

$$A(a,b) = 0, \quad B(a, b, db/dt)\, db \geq 0, \quad \forall\, db. \tag{4.9.2}$$

The above equality on the left is a very strong result. The remaining inequality on the right is less clear-cut at first glance, it leaves more degrees of freedom to introduce viscous effects, but it usually provides tractable restrictions on the material coefficients, once an explicit form for $B(a, b, db/dt)$ has been chosen. Typically, $B(a, b, db/dt) = c\, db/dt$, where the material function c is independent of db/dt. Then $c \geq 0$. Examples of viscous tissues undergoing growth are considered in Section 21.6.4.

More advanced use of the inequality can be done to address a number of constitutive issues in the mechanics of materials, like irreversibilities or damage. Indeed, the idea is to postulate the existence of surfaces whose convexity is exploited. As an example, growth rate laws that comply with the dissipation inequality are worked out in Section 22.5, postulating the existence of convex homeostatic surfaces in stress space.

Appendix: Algebraic derivations for Chapter 4

- Proof of Eqn (4.5.6):

Take the barycentric derivative of $(4.3.4)_2$, namely $\rho\, \mathbf{v} = \sum_k \rho^k\, \mathbf{v}_k$, to obtain

$$\frac{D\rho}{Dt}\, \mathbf{v} + \rho\, \frac{D\mathbf{v}}{Dt} = \sum_{k \in \mathcal{K}} \rho^k\, \frac{d^k \mathbf{v}_k}{dt} + \frac{d^k \rho^k}{dt}\, \mathbf{v}_k - \boldsymbol{\nabla}(\rho^k \mathbf{v}_k) \cdot \overset{(b)}{\mathbf{v}}_k \quad \text{by (2.7.1), (4.3.5),} \tag{4.A.1}$$

and therefore, by (4.4.5), (4.4.7),

$$
\begin{aligned}
\rho\, \frac{D\mathbf{v}}{Dt} &= \sum_{k \in \mathcal{K}} \rho^k\, \frac{d^k \mathbf{v}_k}{dt} + \hat{\rho}^k\, \mathbf{v}_k - \rho^k \mathbf{v}_k\, \mathrm{div}\, \mathbf{v}_k - \boldsymbol{\nabla}(\rho^k \mathbf{v}_k) \cdot \overset{(b)}{\mathbf{v}}_k - (\hat{\rho}\, \mathbf{v} - \rho\, \mathbf{v}\, \mathrm{div}\, \mathbf{v}) \\
&= \sum_{k \in \mathcal{K}} \rho^k\, \frac{d^k \mathbf{v}_k}{dt} + \hat{\rho}^k\, \overset{(b)}{\mathbf{v}}_k - \rho^k \mathbf{v}_k\, \mathrm{div}\, \overset{(b)}{\mathbf{v}}_k - \boldsymbol{\nabla}(\rho^k \mathbf{v}_k) \cdot \overset{(b)}{\mathbf{v}}_k \quad \text{by } (4.3.4)_2, (4.3.5) \\
&= \sum_{k \in \mathcal{K}} \rho^k\, \frac{d^k \mathbf{v}_k}{dt} + \hat{\rho}^k\, \overset{(b)}{\mathbf{v}}_k - \sum_{k \in \mathcal{K}} \mathrm{div}\, (\rho^k\, \mathbf{v}_k \otimes \overset{(b)}{\mathbf{v}}_k) \\
&= \sum_{k \in \mathcal{K}} \rho^k\, \frac{d^k \mathbf{v}_k}{dt} + \hat{\rho}^k\, \overset{(b)}{\mathbf{v}}_k - \sum_{k \in \mathcal{K}} \mathrm{div}\, (\rho^k\, \overset{(b)}{\mathbf{v}}_k \otimes \overset{(b)}{\mathbf{v}}_k) \quad \text{by (4.3.5).}
\end{aligned}
\tag{4.A.2}
$$

Insertion of this relation in the sum over all species $k \in \mathcal{K}$ of the individual balances of momentum (4.5.5) yields

$$\mathrm{div}\, \boldsymbol{\sigma} + \rho\, (\mathbf{b} - \frac{D\mathbf{v}}{Dt}) + \sum_{k \in \mathcal{K}} \hat{\rho}^k\, (\tilde{\mathbf{v}}_k - \mathbf{v}) + \hat{\boldsymbol{\pi}}^k = \mathbf{0}. \tag{4.A.3}$$

Use of the closure relation (4.5.2) provides finally (4.5.6) □ .

- Proof of Eqn (4.6.17):

$$
\begin{aligned}
\frac{d\mathbb{C}}{dt} &= \sum_{k\in\mathcal{K}} \int_V \frac{\partial c_k}{\partial t} + \operatorname{div}(c_k\, \mathbf{v}_k)\, dV \quad \text{by } (2.7.5)_6 \\
&= \sum_{k\in\mathcal{K}} \int_V \frac{d^k c_k}{dt} + c_k \operatorname{div} \mathbf{v}_k\, dV \quad \text{by } (2.7.1) \\
&= \sum_{k\in\mathcal{K}} \int_V \rho^k \mathbf{v}_k \cdot \frac{d^k \mathbf{v}_k}{dt} + \frac{1}{2}\hat{\rho}^k \mathbf{v}_k^2\, dV \quad \text{by } (4.4.5) \\
&= \sum_{k\in\mathcal{K}} \int_V \mathbf{v}_{\mathrm{s}}\cdot\Big(\rho^k \frac{d^k \mathbf{v}_k}{dt} + \hat{\rho}_k\, \mathbf{v}_k - \hat{\rho}_k \frac{\mathbf{v}_{\mathrm{s}}}{2}\Big) + \mathbf{M}_k\cdot\frac{d^k \mathbf{v}_k}{dt} + \frac{\hat{\rho}^k}{2}\Big(\frac{\mathbf{M}_k}{\rho^k}\Big)^2 dV \quad \text{by } (4.3.2),(4.4.3) \\
&= \int_V \mathbf{v}_{\mathrm{s}}\cdot \operatorname{div}\boldsymbol{\sigma}^I + \mathbf{v}_{\mathrm{s}}\cdot(\rho\,\mathbf{b}+\hat{\mathbf{p}}-\hat{\rho}\frac{\mathbf{v}_{\mathrm{s}}}{2})\, dV \\
&\quad + \sum_{k\in\mathcal{K}} \int_V \mathbf{M}_k\cdot\frac{d^k \mathbf{v}_k}{dt} + \frac{\hat{\rho}^k}{2}\Big(\frac{\mathbf{M}_k}{\rho^k}\Big)^2 dV \quad \text{by } (4.5.9) \\
&= \int_V -\boldsymbol{\sigma}^I : \boldsymbol{\nabla}\mathbf{v}_{\mathrm{s}} + \mathbf{v}_{\mathrm{s}}\cdot(\rho\,\mathbf{b}+\hat{\mathbf{p}}-\hat{\rho}\frac{\mathbf{v}_{\mathrm{s}}}{2})\, dV + \sum_{k\in\mathcal{K}} \int_V \mathbf{M}_k\cdot\frac{d^k \mathbf{v}_k}{dt} + \frac{\hat{\rho}^k}{2}\Big(\frac{\mathbf{M}_k}{\rho^k}\Big)^2 dV \\
&\quad + \sum_{k\in\mathcal{K}} \int_{\partial V} \Big(\mathbf{v}_k\cdot\boldsymbol{\sigma}^k - \mathbf{M}_k\cdot\frac{\boldsymbol{\sigma}^k}{\rho^k}\Big)\cdot d\mathbf{S} \ \square.
\end{aligned}
$$

(4.A.4)

- Proof of the balance of energy (4.6.22) for the mixture by summation:
  We start by deriving the local energy for the mixture $U = \sum_{k\in K}\rho^k u_k$.

$$
\begin{aligned}
\frac{dU}{dt} &= \sum_{k\in K} \frac{d^k \rho^k}{dt} u_k + \rho^k \frac{d^k u_k}{dt} - \boldsymbol{\nabla}(\rho^k u_k)\cdot(\mathbf{v}_k-\mathbf{v}_{\mathrm{s}}) \quad \text{by } (2.7.1) \\
&= \sum_{k\in K} -\rho^k u_k \operatorname{div}\mathbf{v}_k + \boldsymbol{\sigma}^k : \boldsymbol{\nabla}\mathbf{v}_k - \operatorname{div}\mathbf{Q}_k + \Theta_k - \boldsymbol{\nabla}(\rho^k u_k)\cdot(\mathbf{v}_k-\mathbf{v}_{\mathrm{s}}) \\
&\quad + \hat{\rho}^k\big(\tilde{u}_k + \tfrac{1}{2}(\tilde{\mathbf{v}}_k-\mathbf{v}_k)^2\big) + \hat{\mathcal{U}}^k \quad \text{by } (4.4.5),(4.6.11) \\
&= -U \operatorname{div}\mathbf{v}_{\mathrm{s}} + \boldsymbol{\sigma}^I : \boldsymbol{\nabla}\mathbf{v}_{\mathrm{s}} - \operatorname{div}\mathbf{Q}^I + \Theta^I + \sum_{k\in K} -\operatorname{div}(u_k\mathbf{M}_k) + \boldsymbol{\sigma}^k : \boldsymbol{\nabla}(\mathbf{v}_k-\mathbf{v}_{\mathrm{s}}) \\
&\quad + \sum_{k\in K} \hat{\rho}^k\big(\tilde{u}_k + \tfrac{1}{2}(\tilde{\mathbf{v}}_k-\mathbf{v}_k)^2\big) + \hat{\mathcal{U}}^k \\
&= -U \operatorname{div}\mathbf{v}_{\mathrm{s}} + \boldsymbol{\sigma}^I : \boldsymbol{\nabla}\mathbf{v}_{\mathrm{s}} - \operatorname{div}\mathbf{Q}^I + \Theta^I + \sum_{k\in K} -\operatorname{div}(h_k\mathbf{M}_k) - \operatorname{div}\boldsymbol{\sigma}^k\cdot(\mathbf{v}_k-\mathbf{v}_{\mathrm{s}}) \\
&\quad + \sum_{k\in K} \hat{\rho}^k\big(\tfrac{1}{2}\mathbf{v}_k^2 - \tilde{\mathbf{v}}_k\cdot\mathbf{v}_k\big) - \hat{\boldsymbol{\pi}}^k\cdot\mathbf{v}_k + \hat{U} \quad \text{by } (4.6.8),(4.6.9),(4.6.14) \\
&= -U \operatorname{div}\mathbf{v}_{\mathrm{s}} + \boldsymbol{\sigma}^I : \boldsymbol{\nabla}\mathbf{v}_{\mathrm{s}} - \operatorname{div}\mathbf{Q}^I + \Theta^I + \sum_{k\in K} -\operatorname{div}(h_k\mathbf{M}_k) - \mathbf{M}_k\cdot\big(\frac{d^k \mathbf{v}_k}{dt}-\mathbf{b}_k\big) \\
&\quad + \sum_{k\in K} \hat{\rho}^k\big(\tfrac{1}{2}\mathbf{v}_k^2 - \tilde{\mathbf{v}}_k\cdot\mathbf{v}_k - (\mathbf{v}_k-\tilde{\mathbf{v}}_k)\cdot(\mathbf{v}_k-\mathbf{v}_{\mathrm{s}})\big) - \hat{\boldsymbol{\pi}}^k\cdot\mathbf{v}_{\mathrm{s}} + \hat{U} \quad \text{by } (4.5.5) \\
&= -U \operatorname{div}\mathbf{v}_{\mathrm{s}} + \boldsymbol{\sigma}^I : \boldsymbol{\nabla}\mathbf{v}_{\mathrm{s}} - \operatorname{div}\mathbf{Q}^I + \Theta^I + \sum_{k\in K} -\operatorname{div}(h_k\mathbf{M}_k) - \mathbf{M}_k\cdot\big(\frac{d^k \mathbf{v}_k}{dt}-\mathbf{b}_k\big) \\
&\quad + \sum_{k\in K} \hat{\rho}^k\big(-\tfrac{1}{2}\mathbf{v}_k^2 + \mathbf{v}_{\mathrm{s}}\cdot\mathbf{v}_k\big) - (\hat{\rho}^k \tilde{\mathbf{v}}_k + \hat{\boldsymbol{\pi}}^k)\cdot\mathbf{v}_{\mathrm{s}} + \hat{U} \\
&= -U \operatorname{div}\mathbf{v}_{\mathrm{s}} + \boldsymbol{\sigma}^I : \boldsymbol{\nabla}\mathbf{v}_{\mathrm{s}} - \operatorname{div}\mathbf{Q}^I + \Theta^I + \sum_{k\in K} -\operatorname{div}(h_k\mathbf{M}_k) - \mathbf{M}_k\cdot\big(\frac{d^k \mathbf{v}_k}{dt}-\mathbf{b}_k\big) \\
&\quad + \sum_{k\in K} -\frac{\hat{\rho}^k}{2}\Big(\frac{\mathbf{M}_k}{\rho^k}\Big)^2 + \hat{\rho}\frac{\mathbf{v}_{\mathrm{s}}^2}{2} - \hat{\mathbf{p}}\cdot\mathbf{v}_{\mathrm{s}} + \hat{U} \quad \text{by } (4.5.2) \ \square.
\end{aligned}
$$

(4.A.5)

Chapter 5

Anisotropic and conewise elasticity

Given that many biological tissues are reinforced by a collagen network, it is appropriate to emphasize their mechanical anisotropy. Since the collagen fiber response differs much according to the compressive or tensile nature of the load, this aspect, coined conewise mechanical response, is worth to be scrutinized. The first section serves to illustrate the different types of elasticity that are used to reproduce aspects of the mechanical response of solids.

5.1 Hypoelasticity, elasticity and hyperelasticity

Based on experimental data or theoretical conceptions, a number of elastic models with stress or strain dependent moduli have been proposed. Attention is focused here on the consequences of this dependency on the recoverability of strain and work done in a closed stress path. Necessary and sufficient conditions for an elastic material to be hyperelastic are given. The purpose is illustrated by examples and requires only the simplest setting, namely isotropic materials and infinitesimal strains. The exposition follows closely Loret [1985].

It is instrumental to decompose the strain ϵ and stress σ into their isotropic and deviatoric parts, namely

$$\epsilon = \epsilon_v \frac{\mathbf{I}}{3} + \mathbf{e}, \quad \text{with} \quad \epsilon_v = \operatorname{tr} \epsilon, \tag{5.1.1}$$

and

$$\sigma = p\,\mathbf{I} + \mathbf{s}, \quad \text{with} \quad p = \frac{1}{3} \operatorname{tr} \sigma. \tag{5.1.2}$$

Besides the first order scalar invariants, namely the volume change ϵ_v and the mean stress p, the second order invariants, referred to as shear stress q and shear strain ϵ_q, namely

$$q = (\tfrac{3}{2}\,\mathbf{s} : \mathbf{s})^{1/2}, \quad \epsilon_q = (\tfrac{2}{3}\,\mathbf{e} : \mathbf{e})^{1/2}, \tag{5.1.3}$$

will be needed. The increments of strain energy $\mathcal{W}$ and complementary energy $\mathcal{W}_c$ can then be decomposed as follows,

$$\delta\mathcal{W} = \sigma : \delta\epsilon = p\,\delta\epsilon_v + \mathbf{s} : \delta\mathbf{e},$$

$$\delta\mathcal{W}_c = \epsilon : \delta\sigma = \epsilon_v\,\delta p + \mathbf{e} : \delta\mathbf{s}. \tag{5.1.4}$$

Consider now the incremental stress-strain relationship,

$$d\epsilon = \frac{dp}{3\,K_t}\,\mathbf{I} + \frac{d\mathbf{s}}{2\,G_t}, \tag{5.1.5}$$

where K_t and G_t are referred to as *tangent* moduli. This relationship embodies three types of constitutive response defined by Table 5.1.1, Fung [1965], Mandel [1974].

The whole issue centers around the fact that the tangent elastic moduli may depend on the stress and strain invariants. To simplify the exposition, the elastic moduli will be assumed to depend on the stress invariants only (and not on the strain invariants).

Now let us consider a particular class of isotropic elastic materials that satisfy

Table 5.1.1: Definitions of hypoelasticity, elasticity and hyperelasticity.

Hypoelasticity	The relationship (5.1.5) is not integrable
Elasticity	The relationship (5.1.5) is integrable and there exists a one to one relation $\boldsymbol{\epsilon} = \boldsymbol{\epsilon}(\boldsymbol{\sigma})$
Hyperelasticity	The relationship (5.1.5) is integrable and there exists a potential $V_c(\boldsymbol{\sigma})$ such that $\boldsymbol{\epsilon} = \partial V_c / \partial \boldsymbol{\sigma}$

Assumption 5.1: the stress-strain relationship of these materials is defined by the *secant* moduli $K_s(p,q)$ and $G_s(p,q)$ and the reference stress $p_a \mathbf{I}$ corresponds to an undeformed hydrostatic state:

$$\boldsymbol{\epsilon} = \frac{p - p_a}{3\,K_s}\,\mathbf{I} + \frac{\mathbf{s}}{2\,G_s}. \tag{5.1.6}$$

For this class, the incremental energy and complementary energy can be expressed in terms of the pairs of scalar invariants (ϵ_v, p) and (ϵ_q, q), which appear as work-conjugate pairs,

$$\delta\mathcal{W} = p\,\delta\epsilon_v + q\,\delta\epsilon_q,$$
$$\delta\mathcal{W}_c = \epsilon_v\,\delta p + \epsilon_q\,\delta q. \tag{5.1.7}$$

Denoting by $\delta\mathcal{W}_{\text{ext}}$ the incremental external work, dU the increment of internal energy, dC the increment of kinetic energy, dS the increment of entropy, all measured per unit volume, and by T the temperature, the first and second laws of thermodynamics may be expressed in the form,

$$\delta\mathcal{W}_{\text{ext}} + \delta Q = dU + dC,$$
$$\delta Q = T\,dS - \delta f, \tag{5.1.8}$$

where δf is the dissipated work. Note that a distinction is made between the differential operators d and δ. For an arbitrary function $A(\mathbf{B})$ depending on a vector or tensor field $\mathbf{B}$, this notation aims at stressing the fact $dA(\mathbf{B})$ depends only on the end points $\mathbf{B}$ and $\mathbf{B} + d\mathbf{B}$, while $\delta A(\mathbf{B})$ may depend on the actual path followed to join the end points $\mathbf{B}$ and $\mathbf{B} + d\mathbf{B}$.

Using the balance of momentum, the incremental work of deformation (strain energy) $\delta\mathcal{W}$ can be shown to be equal to the incremental external work $\delta\mathcal{W}_{\text{ext}}$ minus the incremental kinetic energy dC, namely $\delta\mathcal{W} = \delta\mathcal{W}_{\text{ext}} - dC$. For the purpose of this section, the inertia contribution can be neglected. Moreover the examples examined below satisfy one of the following assumptions:

Assumption 5.2: the thermomechanical process is adiabatic:

$$\delta\mathcal{W} = \delta\mathcal{W}_{\text{ext}} = dU. \tag{5.1.9}$$

Assumption 5.3: the thermomechanical process is isothermal and no dissipation takes place:

$$\delta\mathcal{W} = \delta\mathcal{W}_{\text{ext}} = d(U - T\,S), \quad \delta f = 0. \tag{5.1.10}$$

In order to characterize the mechanical responses, the residual strain and residual energy are calculated after a stress cycle $\mathcal{C}$, i.e., a closed stress path, in the plane (p,q). Three typical responses result as defined by Tables 5.1.1 and 5.1.2, and illustrated below. The cycles considered in these tables are assumed to have some 'thickness', i.e., to encircle a non vanishing area.

If the stress cycle is followed *exactly* in inverse direction, then, for all three responses, the initial strain is retrieved. Similarly, the energy input is entirely recovered, as long as no additional thermomechanical mechanism gives rise to dissipation. This observation justifies the terminology of stored and recoverable energy associated with the stress cycle.

5.1.1   A hypoelastic material

Consider the mechanical response defined by the following scalar incremental relations where both the tangent moduli depend on the mean stress, Coon and Evans [1971], Hueckel and Drescher [1975],

$$d\epsilon_v = \frac{dp}{K_t(p)}, \quad d\epsilon_q = \frac{dq}{3\,G_t(p)}. \tag{5.1.11}$$

For a closed path C_σ in the stress plane (p,q), e.g., a rectangular path with sides parallel to the axes,
- the strain path C_ϵ in the plane (ϵ_v, ϵ_q) is in general not closed, that is a residual strain appears;
- the integral $\int_{C_\sigma} \delta\mathcal{W}$ does not vanish. Therefore, as indicated by Eqn (5.1.8)$_1$, energy may have been extracted from or stored into the material, depending on the clockwise or anticlockwise direction of the path.

Table 5.1.2: Residual strain $\Delta\epsilon$ and residual energy $\Delta\mathcal{W}$ after a stress cycle.

Type of Response	Residual Strain $\Delta\epsilon$	Residual Energy $\Delta\mathcal{W}$
Hypoelasticity	$\neq 0$	$\neq 0$
Elasticity	$= 0$	$\neq 0$
Hyperelasticity	$= 0$	$= 0$

5.1.2 An elastic material

The mechanical response defined by the following scalar incremental relations has relevance to granular materials, Nova and Wood [1978],

$$d\epsilon_v = \kappa\, d\mathrm{Ln}\left|\frac{p}{p_a}\right|, \quad d\epsilon_q = \Gamma\, d\left(\frac{q}{p}\right), \tag{5.1.12}$$

with κ and Γ two material constants. These relations define an elastic material according to Table 5.1.1. The stress-strain relation can be integrated, and the secant moduli result as

$$K_s = \frac{p - p_a}{\kappa\, \mathrm{Ln}\left|\dfrac{p}{p_a}\right|}, \quad G_s = \frac{p}{3\,\Gamma}. \tag{5.1.13}$$

On the other hand, like for hypoelastic materials, a residual energy appears after a closed stress cycle.

5.1.3 A particular hyperelastic material

The elastic potential $V_c(\boldsymbol{\sigma})$ can be identified with the complementary energy $\mathcal{W}_c$. To a closed stress path C_σ corresponds a closed strain path, like for elastic materials. But in addition, the residual complementary energy and the residual energy vanish,

$$\int_{C_\sigma} d\mathcal{W}_c = 0, \quad \int_{C_\sigma} d\mathcal{W} = \int_{C_\sigma} d\left(\boldsymbol{\sigma} : \boldsymbol{\epsilon} - \mathcal{W}_c\right) = 0. \tag{5.1.14}$$

Since the materials are isotropic, the complementary energy depends on the stress $\boldsymbol{\sigma}$ only through the three scalar invariants p, q, and J_3. To simplify the exposition, the complementary energy is first assumed to depend only on p and q. Then the expressions (5.1.7) and (5.1.14) imply,

$$\epsilon_v = \frac{\partial \mathcal{W}_c}{\partial p}, \quad \epsilon_q = \frac{\partial \mathcal{W}_c}{\partial q}. \tag{5.1.15}$$

A necessary and sufficient condition for the elastic potential $\mathcal{W}_c(p, q)$ to exist is that the cross-derivatives are equal,

$$\frac{\partial \epsilon_v}{\partial q} = \frac{\partial \epsilon_q}{\partial p}, \quad \frac{\partial \epsilon_v}{\partial J_3} = 0, \quad \frac{\partial \epsilon_q}{\partial J_3} = 0. \tag{5.1.16}$$

Thus the elastic materials defined by (5.1.6) are hyperelastic iff

$$\frac{\partial}{\partial q}\left(\frac{p}{K_s}\right) = \frac{\partial}{\partial p}\left(\frac{q}{3\,G_s}\right), \quad \frac{\partial K_s}{\partial J_3} = 0, \quad \frac{\partial G_s}{\partial J_3} = 0. \tag{5.1.17}$$

Exercise 5.3 extends the analysis to nonlinear isotropic elastic materials and gives the conditions for these materials to be hyperelastic.

An interesting particular case consists in assuming the Poisson's ratio ν to be constant, or equivalently the ratio $K_s/3G_s$ to be constant, say equal to $K_s^0/3G_s^0$. Then

$$\begin{bmatrix} -\dfrac{K_s^0}{3G_s^0}\, q & p \\[2mm] dp & dq \end{bmatrix} \begin{bmatrix} \partial K_s/\partial p \\[2mm] \partial K_s/\partial q \end{bmatrix} = \begin{bmatrix} 0 \\[2mm] dK_s \end{bmatrix}. \tag{5.1.18}$$

Along the method of characteristics, the determinant of the above matrix,

$$\frac{K_s^0}{3G_s^0}\, q\, dq + p\, dp\,,$$

(5.1.19)

is zeroed. Hence the secant moduli K_s and G_s are functions of

$$\left(\frac{p}{p_a}\right)^2 + \frac{K_s^0}{3G_s^0}\left(\frac{q}{p_a}\right)^2.$$

(5.1.20)

A power expression defined by an exponent $n < 1$ is often adopted, Vermeer [1978],

$$K_s = K_s^0\, p_a \left[\left(\frac{p}{p_a}\right)^2 + \frac{K_s^0}{3G_s^0}\left(\frac{q}{p_a}\right)^2\right]^{(1-n)/2}.$$

(5.1.21)

Exercise 5.19 develops a nonlinear anisotropic hyperelastic model, extending the idea built into (5.1.18).

An alternative particular case assumes the secant shear modulus to depend on p only, namely $G_s = G_s(p)$. Then (5.1.17) implies that K_s is defined up to an arbitrary function $A(p)$,

$$\frac{1}{K_s} = \frac{q^2}{2\, p}\frac{\partial}{\partial p}\frac{1}{3\, G_s} + A(p)\,.$$

(5.1.22)

As a third particular case, the following equivalence holds

$$K_s = K_s(p) \quad\Leftrightarrow\quad G_s = G_s(q)\,.$$

(5.1.23)

Incidentally, this equivalence confirms that the material defined by the secant moduli (5.1.13) is merely elastic, and not hyperelastic.

5.2   Anisotropic linear elasticity

5.2.1   Matrix representation

The linear elastic stress-strain relation is phrased in terms of the elastic *stiffness* $\mathbb{E}$, also referred to as elastic *tensor moduli*. It may be expressed in arbitrary cartesian axes,

$$\boldsymbol{\sigma} = \mathbb{E} : \boldsymbol{\epsilon}, \quad \sigma_{ij} = E_{ijkl}\,\epsilon_{kl}\,.$$

(5.2.1)

It may also be cast in matrix form $[\boldsymbol{\sigma}] = [\mathbb{E}]\,[\boldsymbol{\epsilon}]$, via the elastic stiffness matrix $[\mathbb{E}]$,

$$\begin{bmatrix} \sigma_{11} \\ \sigma_{22} \\ \sigma_{33} \\ \sqrt{2}\,\sigma_{23} \\ \sqrt{2}\,\sigma_{13} \\ \sqrt{2}\,\sigma_{12} \end{bmatrix} = \begin{bmatrix} E_{1111} & E_{1122} & E_{1133} & \sqrt{2}E_{1123} & \sqrt{2}E_{1113} & \sqrt{2}E_{1112} \\ E_{2211} & E_{2222} & E_{2233} & \sqrt{2}E_{2223} & \sqrt{2}E_{2213} & \sqrt{2}E_{2212} \\ E_{3311} & E_{3322} & E_{3333} & \sqrt{2}E_{3323} & \sqrt{2}E_{3313} & \sqrt{2}E_{3312} \\ \sqrt{2}E_{2311} & \sqrt{2}E_{2322} & \sqrt{2}E_{2333} & 2E_{2323} & 2E_{2313} & 2E_{2312} \\ \sqrt{2}E_{1311} & \sqrt{2}E_{1322} & \sqrt{2}E_{1333} & 2E_{1323} & 2E_{1313} & 2E_{1312} \\ \sqrt{2}E_{1211} & \sqrt{2}E_{1222} & \sqrt{2}E_{1233} & 2E_{1223} & 2E_{1213} & 2E_{1212} \end{bmatrix} \begin{bmatrix} \epsilon_{11} \\ \epsilon_{22} \\ \epsilon_{33} \\ \sqrt{2}\,\epsilon_{23} \\ \sqrt{2}\,\epsilon_{13} \\ \sqrt{2}\,\epsilon_{12} \end{bmatrix}.$$

(5.2.2)

This matrix form is referred to as isometric representation. It follows tensor rules under changes of the coordinate axes, Walpole [1984], Mehrabadi and Cowin [1990]. It conserves energy, namely $\boldsymbol{\sigma} : \boldsymbol{\epsilon} = [\boldsymbol{\sigma}]^{\mathrm{T}}[\boldsymbol{\epsilon}]$ as well as euclidean norms, e.g., for the stress tensor $\boldsymbol{\sigma} : \boldsymbol{\sigma} = [\boldsymbol{\sigma}]^{\mathrm{T}}[\boldsymbol{\sigma}]$.

If the elastic stiffness $\mathbb{E}$ is symmetric, it is invertible iff it is positive definite (PD). The conditions for positive definiteness will be explored for different kinds of anisotropy in the sequel. Whether symmetric or not, if the elastic stiffness $\mathbb{E}$ is invertible, the inverse relation involves the elastic compliance $\mathbb{S}$,

$$\boldsymbol{\epsilon} = \mathbb{S} : \boldsymbol{\sigma}, \quad \epsilon_{ij} = S_{ijkl}\,\sigma_{kl}\,.$$

(5.2.3)

The matrix form $[\epsilon] = [\mathbb{S}] : [\sigma]$ introduces the compliance matrix $[\mathbb{S}]$,

$$
\begin{bmatrix}
\epsilon_{11} \\
\epsilon_{22} \\
\epsilon_{33} \\
\sqrt{2}\,\epsilon_{23} \\
\sqrt{2}\,\epsilon_{13} \\
\sqrt{2}\,\epsilon_{12}
\end{bmatrix}
=
\begin{bmatrix}
S_{1111} & S_{1122} & S_{1133} & \sqrt{2}S_{1123} & \sqrt{2}S_{1113} & \sqrt{2}S_{1112} \\
S_{2211} & S_{2222} & S_{2233} & \sqrt{2}S_{2223} & \sqrt{2}S_{2213} & \sqrt{2}S_{2212} \\
S_{3311} & S_{3322} & S_{3333} & \sqrt{2}S_{3323} & \sqrt{2}S_{3313} & \sqrt{2}S_{3312} \\
\sqrt{2}S_{2311} & \sqrt{2}S_{2322} & \sqrt{2}S_{2333} & 2S_{2323} & 2S_{2313} & 2S_{2312} \\
\sqrt{2}S_{1311} & \sqrt{2}S_{1322} & \sqrt{2}S_{1333} & 2S_{1323} & 2S_{1313} & 2S_{1312} \\
\sqrt{2}S_{1211} & \sqrt{2}S_{1222} & \sqrt{2}S_{1233} & 2S_{1223} & 2S_{1213} & 2S_{1212}
\end{bmatrix}
\begin{bmatrix}
\sigma_{11} \\
\sigma_{22} \\
\sigma_{33} \\
\sqrt{2}\,\sigma_{23} \\
\sqrt{2}\,\sigma_{13} \\
\sqrt{2}\,\sigma_{12}
\end{bmatrix}.
\tag{5.2.4}
$$

The compliance tensor and matrix enjoy the same symmetry properties as their respective inverses. The isometric representation is at variance with the so-called Voigt representation, where the out-of diagonal terms are multiplied by 2 in the strain pseudo-vector $[\epsilon]$, and by 1 in the stress pseudo-vector $[\sigma]$. The Voigt representation does not follow tensor rules under changes of the coordinate axes but it conserves energy. The two representations are contrasted in Exercise 5.4.

Warning: The matrix form (5.2.2) assumes that the components of the vectors and matrix gather components of entities which are written on a single cartesian basis. Later, it will appear advantageous to use the axes linked to the material symmetry, referred to as orthotropy axes. Let $\mathbf{e}_i$, $\mathbf{e}_i^\epsilon$ and $\mathbf{e}_i^\sigma$, $i \in [1,3]$, be the cartesian axes associated with the orthotropy axes, with the principal axes of strain and with the principal axes of stress respectively,

$$
\begin{aligned}
\boldsymbol{\epsilon} &= \epsilon_{ii}^* \, \mathbf{e}_i^\epsilon \otimes \mathbf{e}_i^\epsilon = \epsilon_{ij} \, \mathbf{e}_i \otimes \mathbf{e}_j, \\
\boldsymbol{\sigma} &= \sigma_{ii}^* \, \mathbf{e}_i^\sigma \otimes \mathbf{e}_i^\sigma = \sigma_{ij} \, \mathbf{e}_i \otimes \mathbf{e}_j, \\
\mathbb{E} &= E_{ijkl} \, \mathbf{e}_i \otimes \mathbf{e}_j \otimes \mathbf{e}_k \otimes \mathbf{e}_l.
\end{aligned}
\tag{5.2.5}
$$

For general anisotropy, the principal axes of strain and stress do not coincide, see Exercise 5.15.

5.2.2 Minor and major symmetries

The elastic moduli enjoy *minor* symmetries,

$$
\text{minor symmetries} \begin{cases} \text{1. by necessity}: & E_{ijkl} = E_{jikl}, \quad i,j,k,l \in [1,3]; \\ \text{2. by choice}: E_{ijkl} = E_{ijlk}, & i,j,k,l \in [1,3]. \end{cases}
\tag{5.2.6}
$$

Under the two first indices, the symmetry is a *necessity*, due to the symmetry of the Cauchy stress. Under the two last indices, it is a *choice* dictated by simplicity. That it is a choice and not a requirement is commented in Exercise 5.2.

The major symmetry of the elastic tensor moduli, or equivalently compliance, amounts to the symmetry of the elastic matrix, or equivalently compliance matrix,

$$
\text{major symmetries} \quad E_{ijkl} = E_{klij}, \quad i,j,k,l \in [1,3].
\tag{5.2.7}
$$

The major symmetry property amounts to the existence of an elastic potential, also called strain energy function when it depends on the strain.

5.2.3 Microstructure and symmetry groups

5.2.3.1 The group of symmetry or invariance

The mathematical structure which is appropriate to define the anisotropic mechanical and transport properties of materials is the *group*. A group $\mathcal{G}$ is a set of elements endowed with an internal composition rule denoted here by the symbol $\cdot$, namely if $\mathbf{A}$ and $\mathbf{B}$ belong to $\mathcal{G}$, so does $\mathbf{A} \cdot \mathbf{B}$. Elements of the group obey the following three axioms:

1. *associativity*: for any $\mathbf{A}$, $\mathbf{B}$, $\mathbf{C} \in \mathcal{G}$, then
$$(\mathbf{A} \cdot \mathbf{B}) \cdot \mathbf{C} = \mathbf{A} \cdot (\mathbf{B} \cdot \mathbf{C});$$
2. *neutral element*: there exists an element $\mathbf{E} \in \mathcal{G}$ such that,
$$\text{for any } \mathbf{A} \in \mathcal{G}, \ \mathbf{A} \cdot \mathbf{E} = \mathbf{E} \cdot \mathbf{A} = \mathbf{A};$$
3. *inverse element*: for any $\mathbf{A} \in \mathcal{G}$, there exists an element noted $\mathbf{A}^{-1}$ in $\mathcal{G}$ such that
$$\mathbf{A} \cdot \mathbf{A}^{-1} = \mathbf{A}^{-1} \cdot \mathbf{A} = \mathbf{E}.$$
The group is said *abelian*[5.1] or commutative if in addition
$$\text{for any } \mathbf{A}, \mathbf{B} \in \mathcal{G}, \text{ then } \mathbf{A} \cdot \mathbf{B} = \mathbf{B} \cdot \mathbf{A}.$$

A set $\mathcal{G}'$ is a *subgroup* of $\mathcal{G}$, and one notes $\mathcal{G}' \subset \mathcal{G}$, if, for any $\mathbf{A}$, $\mathbf{B} \in \mathcal{G}'$, then $\mathbf{A} \cdot \mathbf{B}^{-1}$ belongs to $\mathcal{G}'$. This condition implies in particular that the neutral element belongs to $\mathcal{G}'$.

For the issue at hand, the elements of the groups are second order tensors and the internal composition rule is the inner product. The neutral element is the identity tensor and the groups are abelian.

The *symmetry group*, or *invariance group*, is the group of changes of axes that leaves invariant the stress-strain relation, that is
$$\underbrace{\boldsymbol{\sigma} = \mathcal{L}(\boldsymbol{\epsilon})}_{\text{reference axes}} \iff \underbrace{\boldsymbol{\sigma}* = \mathcal{L}(\boldsymbol{\epsilon}*)}_{\text{axes } *}, \tag{5.2.8}$$
or equivalently, $\mathcal{L} = \mathcal{L}*$. Consequently, the invariance properties for a linear elastic material are those of the elastic stiffness, or compliance.

The invariance groups contain only *isometries*: for any element $\mathbf{R}$, its norm subordinate to a vector norm, namely $\| \mathbf{R} \| = \sup_{\mathbf{x}} \| \mathbf{R} \cdot \mathbf{x} \| / \| \mathbf{x} \|$, is equal to one. In fact, all elements of the *complete orthogonal group* $\mathcal{O}(3)$ satisfy the orthogonality property,
$$\mathbf{R} \cdot \mathbf{R}^{\mathrm{T}} = \mathbf{R}^{\mathrm{T}} \cdot \mathbf{R} = \mathbf{I}, \tag{5.2.9}$$
and they are
1. either *rotations* (or *proper orthogonal tensors*) such that $\det \mathbf{R} = +1$;
2. or *symmetries* which enjoy the property $\mathbf{R} = \mathbf{R}^{\mathrm{T}}$, so that $\mathbf{R}^2 = \mathbf{I}$. One distinguishes symmetries (or reflections) with respect to a plane of unit normal $\mathbf{n}$, namely $\mathbf{I} - 2\,\mathbf{n} \otimes \mathbf{n}$, such that $\det \mathbf{R} = -1$, and symmetries with respect to a point, such that $\det \mathbf{R} = -1$ in $\mathbb{R}^3$.

The *proper orthogonal group* $\mathcal{SO}(3)$ involves only rotations as defined in Section 2.5.5. The larger the symmetry group, the more the material displays symmetries, and the smaller is the number of independent coefficients required to define the elastic stiffness. If $\mathcal{G}$ is the complete orthogonal group $\mathcal{O}(3)$, then the material response is said to be *isotropic*. If $\mathcal{G}$ is the *proper orthogonal group* $\mathcal{SO}(3)$, then the material response is said to be *hemitropic*.

If the elastic stiffness enjoys the major symmetry property, that is, if the material is in fact hyperelastic, then the most general anisotropy requires $6 \times (6+1)/2 = 21$ elastic coefficients, as can be checked on the matrix representation (5.2.2). This number boils down to 2 for isotropy.

Remark: For a finite strain analysis, the symmetry group depends on the reference configuration, Section 2.11.3. This point should be highlighted for example for a material prestressed by an (an)isotropic initial stress.

5.2.3.2 Structural and fabric anisotropies

A mechanical response is said to be *isotropic* if the response to a load does not depend on the direction of this load. Otherwise, the mechanical response is said to be *anisotropic*.

Elastic anisotropy may be due to different phenomena:
- it may be *inherent* or induced by a loading process, which is accompanied by large strains, damage associated with cracks, nucleation, etc.;
- in biological tissues like in many engineered materials, the anisotropy corresponds to a sought directional property. For example, collagen fibers are expected to exist in a significant density and to be aligned with the traction direction in a tissue whose physiological role is to undergo tension, e.g., ligaments, muscles, arteries, etc. Similar directional compressive or tensile reinforcement is sought in honeycombs and fiber reinforced composites;
- sometimes anisotropy arises simply as a result of the formation process. Such is the case of sedimentary rocks which are deposited under gravity. Clays become progressively anisotropic as the vertical or horizontal tectonic loads increase;

[5.1]Abelian groups are named after the Norwegian mathematician N.H. Abel (1802-1829).

- a material, whether biological, geological or engineered, may be looked at several scales. Let us consider a material formed by a set of elements. One should distinguish the anisotropy due to the *morphology* and the *anisotropy of organization*. In reference to granular materials, the former property concerns the fact that the elements may be spherical, elliptical or else polyhedral. The latter property describes the direction and intensity of contact between elements. A random directional distribution of anisotropic elements results in an isotropic material. Conversely, a directionally oriented distribution of isotropic elements results in an anisotropic material.

In Section 7.4.3, we consider a method of determination of the coefficients of orthotropic and transversely isotropic poroelastic materials: the issue there is to define experiments that are deemed *independent*. A basic tenet of the analysis is that the symmetry group and the axes of material symmetry are known. In fact, the determination of the macroscopic symmetry axes for heterogeneous tissues, possibly reinforced by fibers or other curvilinear objects with a random orientation, is a difficult issue, especially when no spatial scale in the microstructure is prevalent. Moreover, measurements based on stereological methods, Kanatani [1985], may show dispersion or may leave uncertainties. A potential model to tackle these aspects might consider, instead of a tissue reinforced by one or two families of fibers, a smoothed orientation distribution function, Section 12.3.

An approach to define the symmetry group and the axes of symmetry of an elastic material consists

(a) in obtaining in arbitrary cartesian axes the stiffness tensor $\mathbb{E}^{\mathrm{exp}}$ via an experimental program aimed at delivering the 21 elastic constants;

(b) in detecting to which of the nine elastic symmetry groups/classes [5.2] the group of the measured elastic stiffness is closest;

(c) in defining the elastic stiffness $\mathbb{E}^*$ belonging to the chosen symmetry group the measured stiffness is closest.

Point (a) uses mechanical or acoustical tests. Since the symmetry directions are unknown, mechanical tests may easily involve off-axis loadings which are prone to induce a heterogeneous stress and strain state. Ultrasonic measurements along the direction give rise to a quasi-longitudinal wave and two quasi-transversal waves, yielding three wave-speeds. Determining the 21 elastic constants thus requires a priori to probe along at least 7 directions. However, probing along different directions generally provides redundant rather than independent information so that, even for an orthotropic material, it is necessary to probe along more than three directions. Redundant data might be used to estimate errors of measurements.

Points (b) and (c) require the definitions of distances over the set of fourth order elasticity tensors via norms, e.g., the Euclidean norm. Directional averaging provides an element $\mathbb{E}^*$ of the symmetry group $\mathcal{S}_*$ which can be seen as a Voigt bound (associated with uniform strain). Proceeding similarly with the compliance $\mathbb{S}^{\mathrm{exp}}$ yields a Reuss bound $\mathbb{S}^*$ (associated with uniform stress). In fact, the Euclidean norm is not invariant under inversion: it will not deliver the same outcome to the minimization problem depending whether the stiffness or compliance are targeted. Other norms like the log-Euclidean norm will yield a unique result, see, e.g., Norris [2006] for references.

For point (b), François et al. [1998] have proposed to plot the pole figure of the experimental stiffness $\mathbb{E}^{\mathrm{exp}}$. In this crystallographic construction, the normals to the material symmetry planes reveal as dark spots or bands (no spot for triclinic, 1 spot for monoclinic, 3 spots at $120°$ for trigonal, etc.). The closest symmetry group $\mathcal{S}_*$ is decided by comparing *de visu* the pole figure of the experimental stiffness with the pole figures of the nine symmetry groups. In practice, the quality of the image depends on the accuracy of the measurements.

Fedorov [1968] has addressed point (c) with the particular view that the chosen symmetry group is the isotropic group. Minimizing the distances between the slowness surfaces of the measured and isotropic materials, he found the bulk and shear moduli as $K* = E^{\mathrm{exp}}_{iijj}/9$ and $\mu* = E^{\mathrm{exp}}_{ijij}/10 - E^{\mathrm{exp}}_{iijj}/30$. This result is also obtained by minimizing the distance between the stiffness tensors. In fact, point (c) may be seen as projecting the measured stiffness onto the set of stiffness tensors belonging to the chosen symmetry group. Norris [2006] generalized the approach of Fedorov by targeting symmetry groups other than isotropy: he found that the slowness surface approach and the projection of the stiffness using the Euclidean norm yield the same result.

As a further issue, tissue engineering is concerned with finding the optimal orientation of an anisotropic elastic material, by minimizing the strain energy over rotation of the material axes under a fixed load, e.g., Rovati and Taliercio [2003].

[5.2]The terminology in crystallography is different from the one used in Section 5.2.4.1: the nine groups of elastic symmetry are named triclinic (21), monoclinic (13), trigonal (6 and 7), tetragonal (6 and 7), hexagonal (5), cubic (3) and isotropic (2), the numbers in parentheses indicating the number of independent elastic coefficients, Lekhnitskii [1963], p. 26. In fact, there are thirty two classes of *geometric* symmetry in crystals, but only nine groups of *elastic* symmetry.

5.2.4   From anisotropic to isotropic linear hyperelasticity

Under a rotation of the coordinate axes $\mathbf{e}_i$ to $\mathbf{e}_i^* = R_{ij}\,\mathbf{e}_j$, $i \in [1,3]$, the components of the stress $\boldsymbol{\sigma}$ change to $\boldsymbol{\sigma}^* = \mathbf{R}\cdot\boldsymbol{\sigma}\cdot\mathbf{R}^{\mathrm{T}}$, Exercise 2.1. The relation (5.2.8) becomes

$$\mathbf{R}\cdot(\mathbb{E}:\boldsymbol{\epsilon})\cdot\mathbf{R}^{\mathrm{T}} = \mathbb{E}:(\mathbf{R}\cdot\boldsymbol{\epsilon}\cdot\mathbf{R}^{\mathrm{T}}), \ \forall\boldsymbol{\epsilon},\ \forall\mathbf{R}\in\mathcal{G} \quad \Leftrightarrow \quad E_{ijkl} = R_{mi}\,R_{nj}\,R_{pk}\,R_{ql}\,E_{mnpq}\,. \tag{5.2.10}$$

5.2.4.1   Fourth order tensors with minor and major symmetries

A number of coordinate transformations are now concatenated:

1. Symmetry with respect to the coordinate plane $x_3 = 0$, referred to as monoclinic symmetry in crystallography. Let us consider a material whose elastic properties mirror with respect to the plane $x_3 = 0$. Thus

$$\mathbf{R}_3 = \begin{bmatrix} 1 & 0 & 0 \\ 0 & 1 & 0 \\ 0 & 0 & -1 \end{bmatrix} \in \mathcal{G}, \tag{5.2.11}$$

belongs to its symmetry group $\mathcal{G}$. Satisfaction of the relation (5.2.10) implies the components E_{ijkl} of the elastic tensor that contains an odd number of indices 3 to vanish, Exercise 5.5. The 13 remaining coefficients are arbitrary, Eqn (5.2.12),

$$\begin{bmatrix} \bullet & \bullet & \bullet & 0 & 0 & \bullet \\ \bullet & \bullet & \bullet & 0 & 0 & \bullet \\ \bullet & \bullet & \bullet & 0 & 0 & \bullet \\ 0 & 0 & 0 & \bullet & \bullet & 0 \\ 0 & 0 & 0 & \bullet & \bullet & 0 \\ \bullet & \bullet & \bullet & 0 & 0 & \bullet \end{bmatrix} \quad ; \quad \begin{bmatrix} \bullet & \bullet & \bullet & 0 & 0 & 0 \\ \bullet & \bullet & \bullet & 0 & 0 & 0 \\ \bullet & \bullet & \bullet & 0 & 0 & 0 \\ 0 & 0 & 0 & \bullet & 0 & 0 \\ 0 & 0 & 0 & 0 & \bullet & 0 \\ 0 & 0 & 0 & 0 & 0 & \bullet \end{bmatrix} \ . \tag{5.2.12}$$

$$\underbrace{}_{\substack{\text{symmetry} // x_3 = 0 \\ \text{13 independent coefficients}}} \qquad \underbrace{}_{\substack{\text{orthotropy} \\ \text{9 independent coefficients}}}$$

2. *Orthotropy*: symmetry with respect to the coordinate planes $x_1 = 0$ and $x_3 = 0$.

   The transformation

$$\mathbf{R}_1 = \begin{bmatrix} -1 & 0 & 0 \\ 0 & 1 & 0 \\ 0 & 0 & 1 \end{bmatrix}, \tag{5.2.13}$$

and $\mathbf{R}_3$ belong to the group $\mathcal{G}$, then $\mathbf{R}_1 \cdot \mathbf{R}_3 = -\mathbf{R}_2$ belongs to $\mathcal{G}$. The changes in components of $\boldsymbol{\sigma}$ and $\boldsymbol{\epsilon}$ under $-\mathbf{R}_2$ and $\mathbf{R}_2$ are identical, so that $\mathbf{R}_2$ belongs to $\mathcal{G}$. In words, symmetry with respect to two coordinate planes implies symmetry with respect to the three coordinate planes, a property called *orthotropy*, or orthorhombic symmetry in crystallography. Satisfaction of the relation (5.2.10) implies that only nine components remain arbitrary as illustrated on Eqn (5.2.12)$_2$.

3. *Quadratic symmetry*: orthotropy + equivalence of two planes of coordinate axes $x_1 = 0$ and $x_2 = 0$.

   Then the transformation

$$\mathbf{R}_{12} = \begin{bmatrix} 0 & 1 & 0 \\ 1 & 0 & 0 \\ 0 & 0 & 1 \end{bmatrix}, \tag{5.2.14}$$

belongs to the invariance group. Satisfaction of the relation (5.2.10) yields the three restrictions,

$$E_{1111} = E_{2222}, \quad E_{1133} = E_{2233}, \quad E_{2323} = E_{3131}\,, \tag{5.2.15}$$

so that the number of independent coefficients of the elastic stiffness reduces to $9 - 3 = 6$,

$$\begin{bmatrix} \square & \circ & \odot & 0 & 0 & 0 \\ \circ & \square & \odot & 0 & 0 & 0 \\ \odot & \odot & \diamond & 0 & 0 & 0 \\ 0 & 0 & 0 & \triangle & 0 & 0 \\ 0 & 0 & 0 & 0 & \triangle & 0 \\ 0 & 0 & 0 & 0 & 0 & \sqcup \end{bmatrix}. \tag{5.2.16}$$

$$\underbrace{\qquad\qquad\qquad\qquad\qquad}_{\substack{\text{quadratic symmetry} \\ \text{6 independent coefficients}}}$$

The quadratic symmetry property is referred to as tetragonal symmetry in crystallography. The homogenization process of a brick devoid of a central cylinder as a unit cell yields such a quadratic symmetry. The mechanical properties mirror the geometrical lack of radial/cylindrical symmetry.

4. *Cubic symmetry*: orthotropy + equivalence of the three coordinate axes.

   Equivalence of the axes 1 and 3 implies the transformation

$$\mathbf{R}_{13} = \begin{bmatrix} 0 & 0 & 1 \\ 0 & 1 & 0 \\ 1 & 0 & 0 \end{bmatrix}, \tag{5.2.17}$$

to belong to the symmetry group. The resulting further three constraints,

$$E_{1111} = E_{2222} = E_{3333}, \quad E_{1122} = E_{1133} = E_{2233}, \quad E_{2323} = E_{3131} = E_{1212}, \tag{5.2.18}$$

imply the number of independent elastic coefficients to be $6 - 3 = 3$,

$$\begin{bmatrix} \square & \odot & \odot & 0 & 0 & 0 \\ \odot & \square & \odot & 0 & 0 & 0 \\ \odot & \odot & \square & 0 & 0 & 0 \\ 0 & 0 & 0 & \triangle & 0 & 0 \\ 0 & 0 & 0 & 0 & \triangle & 0 \\ 0 & 0 & 0 & 0 & 0 & \triangle \end{bmatrix}. \tag{5.2.19}$$

$$\underbrace{\qquad\qquad\qquad\qquad\qquad}_{\substack{\text{cubic symmetry} \\ \text{3 independent coefficients}}}$$

5. *Transverse isotropy*: rotational invariance about a coordinate axis, e.g., $x_3 = 0$.

   Then

$$\mathbf{R}_\alpha = \begin{bmatrix} \cos\alpha & \sin\alpha & 0 \\ -\sin\alpha & \cos\alpha & 0 \\ 0 & 0 & 1 \end{bmatrix} \in \mathcal{G}, \quad \forall \alpha \in [0, 2\pi[. \tag{5.2.20}$$

Quadratic symmetry in the plane 1-2 can be shown to be verified. Moreover, Eqn (5.2.10) implies in particular,

$$\begin{aligned} E_{1212} &= R_{1p}\, R_{2q}\, [-\sin\alpha\,\cos\alpha\, E_{11pq} + (\cos^2\alpha - \sin^2\alpha)\, E_{12pq} + \sin\alpha\,\cos\alpha\, E_{22pq}] \\ &= (\cos^2\alpha\,\sin^2\alpha)\,(E_{1111} + E_{2222} - 2\, E_{1122}) + (\cos^2\alpha - \sin^2\alpha)^2\, E_{1212}. \end{aligned} \tag{5.2.21}$$

Therefore

$$E_{1212} = \frac{1}{2}\left(E_{1111} - E_{1122}\right). \tag{5.2.22}$$

The number of independent coefficients is thus $6 - 1 = 5$,

$$\begin{bmatrix} \square & \circ & \odot & 0 & 0 & 0 \\ \circ & \square & \odot & 0 & 0 & 0 \\ \odot & \odot & \diamond & 0 & 0 & 0 \\ 0 & 0 & 0 & \triangle & 0 & 0 \\ 0 & 0 & 0 & 0 & \triangle & 0 \\ 0 & 0 & 0 & 0 & 0 & \square - \circ \end{bmatrix}. \tag{5.2.23}$$

$$\underbrace{}$$
transverse isotropy
5 independent coefficients

Transverse isotropy is referred to as hexagonal symmetry in crystallography.

6. *Isotropy*: $\mathcal{G} = \mathcal{O}(3)$ complete orthogonal group

Concatenating the restrictions associated with cubic symmetry and transverse isotropy, only two independent coefficients remain arbitrary, say λ and μ, and $E_{1111} = \lambda + 2\mu$, $E_{1122} = \lambda$ and $E_{2323} = \mu$,

$$\begin{bmatrix} \square & \circ & \circ & 0 & 0 & 0 \\ \circ & \square & \circ & 0 & 0 & 0 \\ \circ & \circ & \square & 0 & 0 & 0 \\ 0 & 0 & 0 & \square - \circ & 0 & 0 \\ 0 & 0 & 0 & 0 & \square - \circ & 0 \\ 0 & 0 & 0 & 0 & 0 & \square - \circ \end{bmatrix}. \tag{5.2.24}$$

$$\underbrace{}$$
isotropy
2 independent coefficients

5.2.4.2　Isotropic tensors of various orders

Isotropic tensors are invariant under the complete orthogonal group $\mathcal{O}(3)$. An isotropic tensor with an odd number of indices necessarily vanishes: apply the symmetry with respect to a plane, as done for fourth order tensors in Exercise 5.5. Moreover, the components are invariant under cyclic permutation of their indices, which corresponds to changes of the coordinate axes. Consequently, isotropic second order tensors are parallel to the identity tensor $\mathbf{I}$.

A careful proof also shows that isotropic fourth order tensors are defined by three coefficients,

$$E_{ijkl} = \lambda\, I_{ij}\, I_{kl} + \mu_1\, I_{ik}\, I_{jl} + \mu_2\, I_{il}\, I_{jk}. \tag{5.2.25}$$

The rewriting shown in Table 5.2.1 highlights the symmetry, or lack of symmetry, under the two first and last indices. Indeed, $\sigma_{ij} = E_{ijkl}\,\epsilon_{kl}$ is equal to $\lambda\,(I_{kl}\epsilon_{kl})\,I_{ij} + \mu\,(\epsilon_{ij} + \epsilon_{ji}) + \mu'\,(\epsilon_{ij} - \epsilon_{ji})$, with $\mu = \frac{1}{2}\,(\mu_1 + \mu_2)$ and $\mu' = \frac{1}{2}\,(\mu_1 - \mu_2)$. If the strain ϵ is symmetric, the coefficient μ' is elusive.

A compact proof is shown in Jeffreys and Jeffreys [1980], p. 88. A slightly different method is used by Fung [1977], p. 192. Here is a brief summary. To prove that the components of a tensor $\mathbf{A}$ in the coordinate system $\mathbf{x}$ and in the rotated coordinates $\mathbf{x}^* = \mathbf{R} \cdot \mathbf{x}$, $x_i^* = R_{ij}\,x_j$, $i \in [1,3]$, are equal, say $A_{ijk\ldots} = A_{ijk\ldots}^*$, one may consider the infinitesimal rotation of angle θ (counted positive in the trigonometric sense) about the axis $\mathbf{n}$ that rotates the coordinates $\mathbf{x}$ to $\mathbf{x}^* = \mathbf{x} + \theta\,\mathbf{n} \wedge \mathbf{x}$, namely componentwise, $x_i^* = R_{ij}\,x_j$, with $R_{ij} = I_{ij} + \theta\,n_z\,e_{zji}$, from which $A_{ijk\ldots}^* = R_{ip}\,R_{jq}\,R_{kr}\cdots A_{pqr\ldots}$. The procedure yields, to first order,

$$A_{ij}^* = A_{ij} + \theta\,n_z\,(e_{zqj}\,A_{iq} + e_{zpi}\,A_{pj});$$

$$A_{ijk}^* = A_{ijk} + \theta\,n_z\,(e_{zqj}\,A_{iqk} + e_{zpi}\,A_{pjk} + e_{zrk}\,A_{ijr}); \tag{5.2.26}$$

$$A_{ijkl}^* = A_{ijkl} + \theta\,n_z\,(e_{zqj}\,A_{iqks} + e_{zpi}\,A_{pjks} + e_{zrk}\,A_{ijrl} + e_{zsl}\,A_{ijks}).$$

Table 5.2.1: Non trivial isotropic tensors from order 0 to four, Fung [1977].

Order		Isotropic Tensors
0	Scalars	All
1	Vectors	None
2	Tensors	$\alpha I_{ij}, \ \forall \alpha \in \mathbb{R}$
3	Tensors	Permutation symbol under $\mathcal{SO}(3)$
4	Tensors	$\lambda I_{ij} I_{kl} + \mu (I_{ik} I_{jl} + I_{il} I_{jk}) + \mu' (I_{ik} I_{jl} - I_{il} I_{jk}), \ \forall \lambda, \mu, \mu' \in \mathbb{R}$

That the elements listed in Table 5.2.1 are invariant under these infinitesimal rotations can be checked via the formulas (2.1.10). The proof that only the elements in this table satisfy the invariance under infinitesimal rotations is more tedious. To finish the proof, it remains to check that the elements of Table 5.2.1 are invariant under reflections and arbitrary rotations.

The permutation symbol is isotropic with respect to rotations only, not under reflections: therefore, it is not isotropic under the complete orthogonal group, and it should therefore be referred to as hemitropic. As an extension, the tensor $\mathbf{e} \otimes \mathbf{I} = (e_{ijk} I_{lm}$ is a fifth order hemitropic tensor.

5.2.5 Linear isotropic elastic materials

For linear isotropic elastic materials, the strain energy $\mathcal{W}(\boldsymbol{\epsilon})$, stress and stiffness $\mathbb{E}$,

$$
\left\{
\begin{aligned}
\mathcal{W}(\boldsymbol{\epsilon}) &= \frac{\lambda}{2} (\mathbf{I} : \boldsymbol{\epsilon})^2 + \mu \, \mathbf{I} : \boldsymbol{\epsilon}^2 \,, \\
\boldsymbol{\sigma} &= \lambda (\mathbf{I} : \boldsymbol{\epsilon}) \mathbf{I} + 2 \, \mu \, \boldsymbol{\epsilon} \,, \\
\mathbb{E} &= \lambda \mathbf{I} \otimes \mathbf{I} + 2 \, \mu \, \mathbf{I} \, \overline{\underline{\otimes}} \, \mathbf{I} \,,
\end{aligned}
\right.
\tag{5.2.27}
$$

are defined by the Lamé moduli λ and μ. The inverse relations for the complementary energy, strain and compliance tensor are best expressed in terms of the Young's modulus E and Poisson's ratio ν,

$$
\left\{
\begin{aligned}
\mathcal{W}_c(\boldsymbol{\sigma}) &= -\frac{\nu}{2\,E} (\mathbf{I} : \boldsymbol{\sigma})^2 + \frac{1+\nu}{2\,E} \mathbf{I} : \boldsymbol{\sigma}^2 \,, \\
\boldsymbol{\epsilon} &= -\frac{\nu}{E} (\mathbf{I} : \boldsymbol{\sigma}) \mathbf{I} + \frac{1+\nu}{E} \boldsymbol{\sigma} \,, \\
\mathbb{E}^{-1} &= -\frac{\nu}{E} \mathbf{I} \otimes \mathbf{I} + \frac{1+\nu}{E} \mathbf{I} \, \overline{\underline{\otimes}} \, \mathbf{I} \,.
\end{aligned}
\right.
\tag{5.2.28}
$$

For the record, let us note the following generic relations (Malvern [1969], p. 293, provides a number of additional useful relations of this type),

$$
\begin{aligned}
\frac{1}{E} &= \frac{\lambda + \mu}{\mu (3\,\lambda + 2\,\mu)} \,, \quad \frac{\nu}{E} = \frac{\lambda}{2\,\mu (3\,\lambda + 2\,\mu)} \,, \quad K = \frac{E}{3(1 - 2\,\nu)} = \lambda + \frac{2}{3}\,\mu \,, \\
E &= \frac{9\,K\,\mu}{3\,K + \mu} \,, \quad \nu = \frac{3\,K - 2\,\mu}{2\,(3\,K + \mu)} \,, \quad \lambda = K - \frac{2}{3}\,\mu \,, \\
\lambda &= \frac{E\,\nu}{(1+\nu)\,(1-2\,\nu)} \,, \quad \mu = \frac{E}{2\,(1+\nu)} \,, \\
\lambda + 2\,\mu &= \frac{E\,(1-\nu)}{(1+\nu)\,(1-2\,\nu)} = 2\,\mu\,\frac{1-\nu}{1-2\,\nu} = 3\,K\,\frac{1-\nu}{1+\nu} \,.
\end{aligned}
\tag{5.2.29}
$$

The conditions of positive definiteness (or of invertibility of the stiffness), and of strong ellipticity (or of existence of two real strictly positive body wave-speeds), read[5.3],

$\mathbb{E}$ Positive Definite (PD) $\Leftrightarrow 3\,\lambda + 2\,\mu > 0$ and $\mu > 0 \Leftrightarrow E > 0$ and $\nu \in \,]-1, \tfrac{1}{2}[\,$.

$\mathbb{E}$ Strongly Elliptic (SE) $\Leftrightarrow \lambda + 2\,\mu > 0$ and $\mu > 0 \Leftrightarrow E > 0$ and $\nu \in \,]-1, \tfrac{1}{2}[\cup]1, \infty[\,$.

$$\tag{5.2.30}$$

[5.3]Note that $E > 0$ and $\nu \in]1, \infty[$ implies $\lambda + 2\,\mu > 0 > \lambda + (2/3)\,\mu$, while $E > 0$ and $\nu \in]-1, 1/2[$ implies $\lambda + 2\,\mu > \lambda + (2/3)\,\mu > 0$.

In cartesian axes, the tensorial product $\overline{\otimes}$ is defined componentwise as in (2.1.24),

$$(\mathbf{A}\,\overline{\otimes}\,\mathbf{B})_{ijkl} = \frac{1}{2}\left(A_{ik}\,B_{jl} + A_{il}\,B_{jk}\right), \quad (\mathbf{A}\,\overline{\otimes}\,\mathbf{B})[\mathbf{X}] = \frac{1}{2}\left(\mathbf{A}\cdot\mathbf{X}\cdot\mathbf{B}^{\mathrm{T}} + \mathbf{A}\cdot\mathbf{X}^{\mathrm{T}}\cdot\mathbf{B}^{\mathrm{T}}\right), \tag{5.2.31}$$

which, for $\mathbf{A}$, $\mathbf{B}$ and $\mathbf{X}$ symmetric second order tensors, simplifies, to

$$(\mathbf{A}\,\overline{\otimes}\,\mathbf{B})[X] = \mathbf{A}\cdot\mathbf{X}\cdot\mathbf{B}. \tag{5.2.32}$$

It is useful to record the component forms in arbitrary cartesian axes of the basic generators for isotropic materials,

$$[\mathbf{I}\otimes\mathbf{I}] = \begin{bmatrix} 1 & 1 & 1 & 0 & 0 & 0 \\ 1 & 1 & 1 & 0 & 0 & 0 \\ 1 & 1 & 1 & 0 & 0 & 0 \\ 0 & 0 & 0 & 0 & 0 & 0 \\ 0 & 0 & 0 & 0 & 0 & 0 \\ 0 & 0 & 0 & 0 & 0 & 0 \end{bmatrix}, \quad [\mathbf{I}\,\overline{\otimes}\,\mathbf{I}] = \begin{bmatrix} 1 & 0 & 0 & 0 & 0 & 0 \\ 0 & 1 & 0 & 0 & 0 & 0 \\ 0 & 0 & 1 & 0 & 0 & 0 \\ 0 & 0 & 0 & 1 & 0 & 0 \\ 0 & 0 & 0 & 0 & 1 & 0 \\ 0 & 0 & 0 & 0 & 0 & 1 \end{bmatrix}. \tag{5.2.33}$$

Note that $\mathbf{I}\,\overline{\otimes}\,\mathbf{I}$ is the identity tensor of order four over symmetric second order tensors,

$$\mathbf{I}\,\overline{\otimes}\,\mathbf{I} = \mathbf{I}_4. \tag{5.2.34}$$

The matrix form of the isotropic elastic stress-strain relation can thus be cast, in arbitrary cartesian axes, in the form,

$$\begin{bmatrix} \sigma_{11} \\ \sigma_{22} \\ \sigma_{33} \\ \sqrt{2}\,\sigma_{23} \\ \sqrt{2}\,\sigma_{13} \\ \sqrt{2}\,\sigma_{12} \end{bmatrix} = \begin{bmatrix} \lambda+2\mu & \lambda & \lambda & 0 & 0 & 0 \\ \lambda & \lambda+2\mu & \lambda & 0 & 0 & 0 \\ \lambda & \lambda & \lambda+2\mu & 0 & 0 & 0 \\ 0 & 0 & 0 & 2\mu & 0 & 0 \\ 0 & 0 & 0 & 0 & 2\mu & 0 \\ 0 & 0 & 0 & 0 & 0 & 2\mu \end{bmatrix} \begin{bmatrix} \epsilon_{11} \\ \epsilon_{22} \\ \epsilon_{33} \\ \sqrt{2}\,\epsilon_{23} \\ \sqrt{2}\,\epsilon_{13} \\ \sqrt{2}\,\epsilon_{12} \end{bmatrix}. \tag{5.2.35}$$

5.2.6   Linear orthotropic elastic materials

Let $\mathbf{e}_a$, $a \in [1,3]$, be three orthonormal vectors indicating the directions of orthotropy of the elastic material. For infinitesimal strains, no distinction is made between the reference and current values of these entities, respectively $\underline{\mathbf{e}}_a$ and $\mathbf{e}_a$, $a \in [1,3]$. The second order tensors $\mathbf{A}_a = \mathbf{e}_a \otimes \mathbf{e}_a$, $a \in [1,3]$, are called *structure tensors*. They satisfy the identity,

$$\mathbf{A}_1 + \mathbf{A}_2 + \mathbf{A}_3 = \mathbf{I}, \tag{5.2.36}$$

with $\mathbf{I}$ the 2nd order identity tensor.

Convention: In this section, the convention of summation over the mute indices a and b applies. A subscript bar prevents this summation.

When the strain is the independent variable, the elastic potential $\mathcal{W}$, stress $\boldsymbol{\sigma} = \partial\mathcal{W}/\partial\boldsymbol{\epsilon} = \mathbb{E}{:}\boldsymbol{\epsilon}$ and elastic tensor moduli $\mathbb{E} = \partial^2\mathcal{W}/\partial\boldsymbol{\epsilon}\partial\boldsymbol{\epsilon}$ can be cast in the following format in terms of the generalized Lamé moduli $\lambda_{\mathrm{ab}} = \lambda_{\mathrm{ba}}$, μ_{a}, $a,b \in [1,3]$, Curnier et al. [1994],

$$\begin{cases} \mathcal{W}(\boldsymbol{\epsilon}) &= \dfrac{\lambda_{\mathrm{ab}}}{2}\,(\mathbf{A}_a{:}\boldsymbol{\epsilon})\,(\mathbf{A}_b{:}\boldsymbol{\epsilon}) + \mu_{\mathrm{a}}\,\mathbf{A}_a{:}\boldsymbol{\epsilon}^2\,, \\[2mm] \boldsymbol{\sigma} &= \dfrac{\lambda_{\mathrm{ab}}}{2}\left((\mathbf{A}_a{:}\boldsymbol{\epsilon})\,\mathbf{A}_b + (\mathbf{A}_b{:}\boldsymbol{\epsilon})\,\mathbf{A}_a\right) + \mu_{\mathrm{a}}\left(\mathbf{A}_a\cdot\boldsymbol{\epsilon} + \boldsymbol{\epsilon}\cdot\mathbf{A}_a\right), \\[2mm] \mathbb{E} &= \dfrac{\lambda_{\mathrm{ab}}}{2}\left(\mathbf{A}_a\otimes\mathbf{A}_b + \mathbf{A}_b\otimes\mathbf{A}_a\right) + \mu_{\mathrm{a}}\left(\mathbf{A}_a\,\overline{\otimes}\,\mathbf{I} + \mathbf{I}\,\overline{\otimes}\,\mathbf{A}_a\right). \end{cases} \tag{5.2.37}$$

Elements of proof are deferred to Exercises 5.8, 5.11, and 6.5.

The number of independent coefficients depends on the material symmetries:

orthotropy $\quad 3\,\lambda_{\underline{aa}}, \qquad 3\,\lambda_{ab}(=\lambda_{ba}),\ a<b,\quad 3\,\mu_a, \qquad$ 9 independent coefficients
transverse $\quad 2\,\lambda_{\underline{aa}}, \qquad 1\,\lambda_{ab}(=\lambda_{ba}),\ a<b,\quad 2\,\mu_a, \qquad$ 5 independent coefficients
  isotropy
isotropy $\qquad 1\,\lambda_{\underline{aa}}=\lambda, \qquad\qquad\qquad\qquad\quad 1\,\mu_a=\mu,\quad$ 2 independent coefficients $\hfill (5.2.38)$

The components of the tensor moduli can be made explicit,

$$
\begin{aligned}
\mathbb{E} \;&=\; \sum_{a,b}\frac{\lambda_{ab}}{2}\left(\mathbf{A}_a\otimes\mathbf{A}_b+\mathbf{A}_b\otimes\mathbf{A}_a\right)+\sum_a \mu_a\left(\mathbf{A}_a\,\overline{\otimes}\,\mathbf{I}+\mathbf{I}\,\underline{\otimes}\,\mathbf{A}_a\right),\\[4pt]
E_{ijkl} \;&=\; \sum_{a,b}\frac{\lambda_{ab}}{2}\left(A_{aij}A_{bkl}+A_{bij}A_{akl}\right)+\sum_a \frac{\mu_a}{2}\left(A_{aik}I_{jl}+A_{ail}I_{jk}+A_{ajl}I_{ik}+A_{ajk}I_{il}\right),\\[4pt]
E_{\underline{iiii}} \;&=\; \sum_{a,b}\lambda_{ab}\,A_{a\underline{ii}}A_{b\underline{ii}}+\sum_a 2\,\mu_a\,A_{a\underline{ii}} \quad\text{for i}=\text{j}=\text{k}=\text{l},\\[4pt]
E_{\underline{iikk}} \;&=\; \sum_{a,b}\frac{\lambda_{ab}}{2}\left(A_{a\underline{ii}}A_{b\underline{kk}}+A_{a\underline{kk}}A_{b\underline{ii}}\right)\quad\text{for i}=\text{j}\neq\text{k}=\text{l},\\[4pt]
E_{\underline{ijij}} \;&=\; \sum_a \frac{\mu_a}{2}\left(A_{a\underline{ii}}+A_{a\underline{jj}}\right)\qquad\qquad\text{for i}=\text{k}\neq\text{j}=\text{l}.
\end{aligned}
\tag{5.2.39}
$$

In the axes of orthotropy, the elastic relation $(5.2.37)_2$ takes the following form,

$$
\begin{bmatrix}
\sigma_{11}\\[4pt] \sigma_{22}\\[4pt] \sigma_{33}\\[4pt] \sqrt{2}\,\sigma_{23}\\[4pt] \sqrt{2}\,\sigma_{13}\\[4pt] \sqrt{2}\,\sigma_{12}
\end{bmatrix}
=
\begin{bmatrix}
\lambda_{11}+2\,\mu_1 & \lambda_{12} & \lambda_{13} & 0 & 0 & 0\\[4pt]
\lambda_{12} & \lambda_{22}+2\,\mu_2 & \lambda_{23} & 0 & 0 & 0\\[4pt]
\lambda_{13} & \lambda_{23} & \lambda_{33}+2\,\mu_3 & 0 & 0 & 0\\[4pt]
0 & 0 & 0 & \mu_2+\mu_3 & 0 & 0\\[4pt]
0 & 0 & 0 & 0 & \mu_1+\mu_3 & 0\\[4pt]
0 & 0 & 0 & 0 & 0 & \mu_1+\mu_2
\end{bmatrix}
\begin{bmatrix}
\epsilon_{11}\\[4pt] \epsilon_{22}\\[4pt] \epsilon_{33}\\[4pt] \sqrt{2}\,\epsilon_{23}\\[4pt] \sqrt{2}\,\epsilon_{13}\\[4pt] \sqrt{2}\,\epsilon_{12}
\end{bmatrix}.
\tag{5.2.40}
$$

For the sake of reference, the coefficients of the stiffness matrix in the orthotropy axes will be denoted through the components $c_{ij}=c_{ji}$, $i,j\in[1,3]$, and $c_{\underline{ii}}$, $i=4,5,6$,

$$
[\mathbb{E}]=
\begin{bmatrix}
c_{11} & c_{12} & c_{13} & 0 & 0 & 0\\[4pt]
c_{12} & c_{22} & c_{23} & 0 & 0 & 0\\[4pt]
c_{13} & c_{23} & c_{33} & 0 & 0 & 0\\[4pt]
0 & 0 & 0 & 2\,c_{44} & 0 & 0\\[4pt]
0 & 0 & 0 & 0 & 2\,c_{55} & 0\\[4pt]
0 & 0 & 0 & 0 & 0 & 2\,c_{66}
\end{bmatrix}.
\tag{5.2.41}
$$

Alternatively, and equivalently, when the stress is the independent variable, the complementary energy $\mathcal{W}_c$, the strain $\epsilon=\partial\mathcal{W}_c/\partial\boldsymbol{\sigma}=\mathbb{E}^{-1}:\boldsymbol{\sigma}$ and the elastic compliance tensor $\mathbb{E}^{-1}$ can be expressed in terms of the Young's moduli E's and generalized Poisson's ratios ν's, linked by the symmetry relations[5.4]

$$
\frac{\nu_{ba}}{E_b}=\frac{\nu_{ab}}{E_a},\qquad a\neq b\in[1,3]. \tag{5.2.42}
$$

[5.4] As a consequence of these three constraints, we have the relation $\nu_{12}\,\nu_{23}\,\nu_{31}=\nu_{21}\,\nu_{13}\,\nu_{32}$.

Indeed, Curnier et al. [1994],

$$
\begin{cases}
\mathcal{W}_c(\boldsymbol{\sigma}) &= -\dfrac{\nu_{ab}}{2\,E_a}\,(\mathbf{A}_a\!:\!\boldsymbol{\sigma})\,(\mathbf{A}_b\!:\!\boldsymbol{\sigma}) + \dfrac{1+\nu_{\underline{aa}}}{2\,E_a}\,\mathbf{A}_a\!:\!\boldsymbol{\sigma}^2\,, \\[2mm]
\boldsymbol{\epsilon} &= -\dfrac{\nu_{ab}}{2\,E_a}\,\big((\mathbf{A}_a\!:\!\boldsymbol{\sigma})\,\mathbf{A}_b + (\mathbf{A}_b\!:\!\boldsymbol{\sigma})\,\mathbf{A}_a\big) + \dfrac{1+\nu_{\underline{aa}}}{2\,E_a}\,(\mathbf{A}_a\cdot\boldsymbol{\sigma} + \boldsymbol{\sigma}\cdot\mathbf{A}_a)\,, \\[2mm]
\mathbb{E}^{-1} &= -\dfrac{\nu_{ab}}{2\,E_a}\,(\mathbf{A}_a\otimes\mathbf{A}_b + \mathbf{A}_b\otimes\mathbf{A}_a) + \dfrac{1+\nu_{\underline{aa}}}{2\,E_a}\,(\mathbf{A}_a\,\overline{\underline{\otimes}}\,\mathbf{I} + \mathbf{I}\,\underline{\overline{\otimes}}\,\mathbf{A}_a)\,.
\end{cases}
\tag{5.2.43}
$$

Recall that the convention of summation over repeated indices does not apply to underlined indices.

As in (5.2.38), the number of independent coefficients depends on the material symmetries:

$$
\begin{array}{llll}
\text{orthotropy} & 3\,\nu_{ab},\ a<b,\ \ 3\,\mu_{ab},\ a<b,\ \ 3\,E_a, & 9\ \text{independent coefficients} \\
\text{transverse isotropy} & 2\,\nu_{ab},\ a<b,\ \ 1\,\mu_{ab},\ a<b,\ \ 2\,E_a, & 5\ \text{independent coefficients} \\
\text{isotropy} & 1\,\nu_{ab}=\nu, & 1\,E_a=E,\ \ 2\ \text{independent coefficients}
\end{array}
\tag{5.2.44}
$$

Like for any material symmetry, the Young's modulus E_a associated with the direction $\mathbf{e}_a$ is defined from the compliance tensor,

$$
\frac{1}{E_a} = (\mathbf{e}_a\otimes\mathbf{e}_a):\mathbb{E}^{-1}:(\mathbf{e}_a\otimes\mathbf{e}_a)\,.
\tag{5.2.45}
$$

The elastic relation $(5.2.43)_2$ can be cast in the following matrix form in the orthotropy axes,

$$
\begin{bmatrix}
\epsilon_{11} \\[2mm]
\epsilon_{22} \\[2mm]
\epsilon_{33} \\[2mm]
\sqrt{2}\,\epsilon_{23} \\[2mm]
\sqrt{2}\,\epsilon_{13} \\[2mm]
\sqrt{2}\,\epsilon_{12}
\end{bmatrix}
=
\begin{bmatrix}
\dfrac{1}{E_1} & -\dfrac{\nu_{21}}{E_2} & -\dfrac{\nu_{31}}{E_3} & 0 & 0 & 0 \\[2mm]
-\dfrac{\nu_{12}}{E_1} & \dfrac{1}{E_2} & -\dfrac{\nu_{32}}{E_3} & 0 & 0 & 0 \\[2mm]
-\dfrac{\nu_{13}}{E_1} & -\dfrac{\nu_{23}}{E_2} & \dfrac{1}{E_3} & 0 & 0 & 0 \\[2mm]
0 & 0 & 0 & \dfrac{1}{2\,\mu_{23}} & 0 & 0 \\[2mm]
0 & 0 & 0 & 0 & \dfrac{1}{2\,\mu_{13}} & 0 \\[2mm]
0 & 0 & 0 & 0 & 0 & \dfrac{1}{2\,\mu_{12}}
\end{bmatrix}
\begin{bmatrix}
\sigma_{11} \\[2mm]
\sigma_{22} \\[2mm]
\sigma_{33} \\[2mm]
\sqrt{2}\,\sigma_{23} \\[2mm]
\sqrt{2}\,\sigma_{13} \\[2mm]
\sqrt{2}\,\sigma_{12}
\end{bmatrix},
\tag{5.2.46}
$$

where

$$
\frac{1}{\mu_{ab}} = \frac{1}{\mu_{ba}} = \frac{1+\nu_{\underline{aa}}}{E_a} + \frac{1+\nu_{\underline{bb}}}{E_b}, \quad a\neq b\in[1,3]\,.
\tag{5.2.47}
$$

The above forms (5.2.43), (5.2.46) and (5.2.47) introduce the (additional) generalized Poisson's ratios $\nu_{ab}, a\geq b$, $a,b\in[1,3]$. Let us insist that the latter are not independent coefficients. They are obtained from the three generalized Young's moduli E_a, $a\in[1,3]$ and three Poisson's ratios $\nu_{ab}, a<b$, $a,b\in[1,3]$, via the symmetry relations (5.2.42).

The Young's moduli and Poisson's ratios can be calculated in terms of the generalized Lamé moduli, and conversely. The necessary algebras are left as an exercise, Exercise 5.9.

Convexity of the elastic potential of a linear elastic material can be shown to be equivalent to the positive definiteness of the elastic tensor moduli. The explicit conditions are explored in Exercise 5.10.

5.2.7   Prestress/prestrain in linear hyperelastic materials

The elastic potentials in the above section are homogeneous of degree two with respect to the strain or stress, so that stress and strain vanish simultaneously. The existence of an initial stress $\boldsymbol{\sigma}_0$ or initial strain $\boldsymbol{\epsilon}_0 = -\mathbb{E}^{-1}:\boldsymbol{\sigma}_0$

emanates from a linear term, namely

$$
\begin{cases}
\mathcal{W}(\epsilon) & = \dfrac{1}{2}\,\epsilon:\mathbb{E}:\epsilon+\sigma_0:\epsilon \Leftrightarrow \sigma=\mathbb{E}:\epsilon+\sigma_0, \\[2ex]
\mathcal{W}_c(\sigma) & = \dfrac{1}{2}\,\sigma:\mathbb{E}^{-1}:\sigma+\epsilon_0:\sigma \Leftrightarrow \epsilon=\mathbb{E}^{-1}:\sigma+\epsilon_0\,.
\end{cases}
\tag{5.2.48}
$$

The next version has another flavor, towards nonlinear elasticity and/or conewise response, but it is strictly equivalent to (5.2.48) for linear elasticity,

$$
\begin{cases}
\mathcal{W}(\epsilon) & = \dfrac{1}{2}\,(\epsilon-\epsilon_0):\mathbb{E}:(\epsilon-\epsilon_0) \Leftrightarrow \sigma=\mathbb{E}:(\epsilon-\epsilon_0), \\[2ex]
\mathcal{W}_c(\sigma) & = \dfrac{1}{2}\,(\sigma-\sigma_0):\mathbb{E}^{-1}:(\sigma-\sigma_0) \Leftrightarrow \epsilon=\mathbb{E}^{-1}:(\sigma-\sigma_0)
\end{cases}
\tag{5.2.49}
$$

5.2.8 Linear transversely isotropic elastic materials

For transversely isotropic materials, with the axis 3 as symmetry axis, the elastic characteristics specialize as indicated below. Details are provided in Exercise 5.11.

The stiffness expresses in terms of five independent generalized Lamé moduli λ_T, λ_L, λ_{LT}, μ_T and μ_L, namely in the orthotropy axes,

$$
[\mathbb{E}] =
\begin{bmatrix}
\lambda_T+2\,\mu_T & \lambda_T & \lambda_{LT} & 0 & 0 & 0 \\[1.5ex]
\lambda_T & \lambda_T+2\,\mu_T & \lambda_{LT} & 0 & 0 & 0 \\[1.5ex]
\lambda_{LT} & \lambda_{LT} & \lambda_L+2\,\mu_L & 0 & 0 & 0 \\[1.5ex]
0 & 0 & 0 & 2\,\mu_L & 0 & 0 \\[1.5ex]
0 & 0 & 0 & 0 & 2\,\mu_L & 0 \\[1.5ex]
0 & 0 & 0 & 0 & 0 & 2\,\mu_T
\end{bmatrix}.
\tag{5.2.50}
$$

Similarly to the orthotropic case, the stiffness matrix in the orthotropy axes is recorded through the components of the associated 6×6 pseudo-matrix,

$$
[\mathbb{E}] =
\begin{bmatrix}
c_{11} & c_{12} & c_{13} & 0 & 0 & 0 \\[1.5ex]
c_{12} & c_{11} & c_{13} & 0 & 0 & 0 \\[1.5ex]
c_{13} & c_{13} & c_{33} & 0 & 0 & 0 \\[1.5ex]
0 & 0 & 0 & 2\,c_{44} & 0 & 0 \\[1.5ex]
0 & 0 & 0 & 0 & 2\,c_{44} & 0 \\[1.5ex]
0 & 0 & 0 & 0 & 0 & c_{11}-c_{12}
\end{bmatrix}.
\tag{5.2.51}
$$

Conversely, the compliance matrix expresses in terms of two generalized Young's moduli, a shear modulus and two Poisson's ratios,

- E_L longitudinal, or axial, elastic modulus;
- E_T transverse, or cross-axial, elastic modulus;
- μ_L longitudinal, or zonal, shear modulus;
- ν_T transverse Poisson's ratio;
- ν_L longitudinal Poisson's ratio,

namely,

$$
[\mathbb{E}]^{-1} =
\begin{bmatrix}
\dfrac{1}{E_{\mathrm{T}}} & -\dfrac{\nu_{\mathrm{T}}}{E_{\mathrm{T}}} & -\dfrac{\nu_{\mathrm{L}}}{E_{\mathrm{L}}} & 0 & 0 & 0 \\[2mm]
-\dfrac{\nu_{\mathrm{T}}}{E_{\mathrm{T}}} & \dfrac{1}{E_{\mathrm{T}}} & -\dfrac{\nu_{\mathrm{L}}}{E_{\mathrm{L}}} & 0 & 0 & 0 \\[2mm]
-\dfrac{\nu_{\mathrm{L}}}{E_{\mathrm{L}}} & -\dfrac{\nu_{\mathrm{L}}}{E_{\mathrm{L}}} & \dfrac{1}{E_{\mathrm{L}}} & 0 & 0 & 0 \\[2mm]
0 & 0 & 0 & \dfrac{1}{2\,\mu_{\mathrm{L}}} & 0 & 0 \\[2mm]
0 & 0 & 0 & 0 & \dfrac{1}{2\,\mu_{\mathrm{L}}} & 0 \\[2mm]
0 & 0 & 0 & 0 & 0 & \dfrac{1}{2\,\mu_{\mathrm{T}}}
\end{bmatrix},
\tag{5.2.52}
$$

where the transverse shear modulus expresses in terms of E_{T} and ν_{T},

$$
\frac{1}{\mu_{\mathrm{T}}} = \frac{2\,(1+\nu_{\mathrm{T}})}{E_{\mathrm{T}}}.
\tag{5.2.53}
$$

The following identifications have been used (the indices 1 and 2 may be exchanged):

$$
E_1 = E_{\mathrm{T}}, \quad E_3 = E_{\mathrm{L}}, \quad \nu_{12} = \nu_{\mathrm{T}}, \quad \nu_{31} = \nu_{\mathrm{L}}, \quad \frac{\nu_{13}}{E_{\mathrm{T}}} = \frac{\nu_{\mathrm{L}}}{E_{\mathrm{L}}}, \quad \mu_{L} = \mu_{13}.
\tag{5.2.54}
$$

The generalized Lamé moduli of the elastic stiffness may be calculated in terms of the coefficients of the elastic compliance matrix, and conversely, Exercise 5.12.

Positive definiteness of the stiffness matrix expresses as, Exercise 5.13,

$$
\mu_{\mathrm{T}} > \max(-\lambda_{\mathrm{T}}/2, 0), \quad (\lambda_{\mathrm{T}} + \mu_{\mathrm{T}})(\lambda_{\mathrm{L}} + 2\,\mu_{\mathrm{L}}) > \lambda_{\mathrm{LT}}^2, \quad \mu_{\mathrm{L}} > 0,
\tag{5.2.55}
$$

while positive definiteness of the compliance matrix can be cast in the format,

$$
E_{\mathrm{L}} > 0, \quad E_{\mathrm{T}} > 0, \quad -1 < \nu_{\mathrm{T}} < 1, \quad \frac{1-\nu_{\mathrm{T}}}{2}\frac{E_{\mathrm{L}}}{E_{\mathrm{T}}} - (\nu_{\mathrm{L}})^2 > 0, \quad \mu_{\mathrm{L}} > 0.
\tag{5.2.56}
$$

These two sets of conditions can be checked to be equivalent using the relations between the coefficients of the stiffness and compliance matrices established in Exercise 5.12.

The strong ellipticity condition is defined by Eqn (11.E.7) with the notation of (5.E.30). It bounds the generalized Lamé moduli,

$$
\mu_{\mathrm{T}} > \max(-\lambda_{\mathrm{T}}/2, 0), \quad \mu_{\mathrm{L}} > \max(-\lambda_{\mathrm{L}}/2, 0), \quad \mu_{\mathrm{L}} + \sqrt{(\lambda_{\mathrm{L}} + 2\,\mu_{\mathrm{L}})(\lambda_{\mathrm{T}} + 2\,\mu_{\mathrm{T}})} > |\lambda_{\mathrm{T}} + \mu_{\mathrm{T}}|.
\tag{5.2.57}
$$

Once again, using the relations between the coefficients of the stiffness and compliance matrices established in Exercise 5.12, the strong ellipticity conditions can be expressed in terms of the generalized Young's moduli and Poisson's ratios.

A particular case will serve to illustrate the restrictions implied by positive definiteness and strong ellipticity, and their differences, Loret and Rizzi [1997]. Starting from the general expression (5.E.27), transverse isotropy is defined by a single parameter deviation from the underlying isotropic backbone, namely $a_1 = \lambda$, $a_2 = 2\,\mu$, $a_3 = 0$, $a_4 \neq 0$, $a_5 = 0$, so that $c_{11} = \lambda + 2\,\mu$, $c_{12} = c_{13} = \lambda$, $c_{44} = \mu$, but $c_{33} = \lambda + 2\,\mu + a_4$.

Assuming the stiffness of the underlying isotropic backbone to be positive definite, the positive definiteness of the transversely isotropic material yields, in addition to $(5.2.30)_1$,

$$
\lambda + \frac{2}{3}\mu > 0, \quad \mu > 0, \quad a_4 > a_4^{\mathrm{LPD}} \equiv -\mu\,\frac{3\lambda + 2\mu}{\lambda + \mu} = -E.
\tag{5.2.58}
$$

On the other hand, the strong ellipticity of the transversely isotropic material implies, in addition to $(5.2.30)_2$,

$$
\lambda + 2\mu > 0, \quad \mu > 0, \quad a_4 > a_4^{\mathrm{LSE}} \equiv
\begin{cases}
-4\mu\,\dfrac{\lambda + \mu}{\lambda + 2\mu} & \text{if } \lambda \geq 0, \\[3mm]
-(\lambda + 2\mu) & \text{if } \lambda \leq 0.
\end{cases}
\tag{5.2.59}
$$

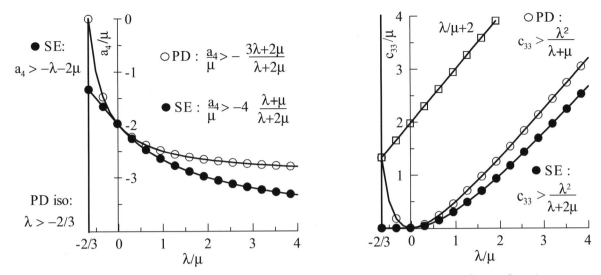

Fig. 5.2.1: Limit values of the parameter a_4 measuring a transversely isotropic perturbation from isotropy, according to (5.2.58), (5.2.59), (5.2.60). Reprinted with permission from Loret and Rizzi [1997].

Interestingly, these conditions relax the lower bound of the coefficient c_{33},

$$c_{33} > c_{33}^{\text{LPD}} \equiv \frac{\lambda^2}{\lambda + \mu}, \quad c_{33} > c_{33}^{\text{LSE}} \equiv \begin{cases} \dfrac{\lambda^2}{\lambda + 2\,\mu} & \text{if } \lambda \geq 0, \\[2mm] 0 & \text{if } \lambda \leq 0. \end{cases} \tag{5.2.60}$$

The limit curves are plotted in Fig. 5.2.1. Note that the condition for positive definiteness is always strictly stronger than the condition for strong ellipticity, except at $\lambda = 0$, where the losses of positive definiteness and of strong ellipticity coincide.

5.2.9 Linear cubic elasticity

A cubic material is defined by three independent coefficients λ, λ' and μ,

$$\begin{cases} \mathcal{W}(\epsilon) & = & \dfrac{\lambda'}{2}\,(\mathbf{I} : \epsilon)^2 + \mu\,\mathbf{I} : \epsilon^2 + \dfrac{1}{2}\,(\lambda - \lambda') \sum_a (\mathbf{A}_a : \epsilon)^2, \\[3mm] \boldsymbol{\sigma} & = & \lambda'\,(\mathbf{I} : \epsilon)\,\mathbf{I} + 2\,\mu\,\epsilon + (\lambda - \lambda') \sum_a (\mathbf{A}_a : \epsilon)\,\mathbf{A}_a, \\[3mm] \mathbb{E} & = & \lambda'\,\mathbf{I} \otimes \mathbf{I} + 2\,\mu\,\mathbf{I}\,\overline{\underline{\otimes}}\,\mathbf{I} + (\lambda - \lambda') \sum_a \mathbf{A}_a \otimes \mathbf{A}_a. \end{cases} \tag{5.2.61}$$

The stiffness may also be expressed componentwise in the symmetry axes as

$$E_{ijkl} = \lambda'\,I_{ij}\,I_{kl} + \mu\,(I_{ik}\,I_{jl} + I_{il}\,I_{jk}) + (\lambda - \lambda')\,I'_{ijkl}, \tag{5.2.62}$$

where all the components if $\mathbf{I}'$ vanish but $I'_{1111} = I'_{2222} = I'_{3333} = 1$. In the axes of material symmetry, the matrix form of the stress-strain relation reads

$$\begin{bmatrix} \sigma_{11} \\ \sigma_{22} \\ \sigma_{33} \\ \sqrt{2}\,\sigma_{23} \\ \sqrt{2}\,\sigma_{13} \\ \sqrt{2}\,\sigma_{12} \end{bmatrix} = \begin{bmatrix} \lambda + 2\,\mu & \lambda' & \lambda' & 0 & 0 & 0 \\ \lambda' & \lambda + 2\,\mu & \lambda' & 0 & 0 & 0 \\ \lambda' & \lambda' & \lambda + 2\,\mu & 0 & 0 & 0 \\ 0 & 0 & 0 & 2\,\mu & 0 & 0 \\ 0 & 0 & 0 & 0 & 2\,\mu & 0 \\ 0 & 0 & 0 & 0 & 0 & 2\,\mu \end{bmatrix} \begin{bmatrix} \epsilon_{11} \\ \epsilon_{22} \\ \epsilon_{33} \\ \sqrt{2}\,\epsilon_{23} \\ \sqrt{2}\,\epsilon_{13} \\ \sqrt{2}\,\epsilon_{12} \end{bmatrix}. \tag{5.2.63}$$

The compliance matrix expresses in terms of the three coefficients E, ν' and μ,

$$
\begin{bmatrix}
\epsilon_{11} \\
\epsilon_{22} \\
\epsilon_{33} \\
\sqrt{2}\,\epsilon_{23} \\
\sqrt{2}\,\epsilon_{13} \\
\sqrt{2}\,\epsilon_{12}
\end{bmatrix}
=
\begin{bmatrix}
\dfrac{1}{E} & -\dfrac{\nu'}{E} & -\dfrac{\nu'}{E} & 0 & 0 & 0 \\
-\dfrac{\nu'}{E} & \dfrac{1}{E} & -\dfrac{\nu'}{E} & 0 & 0 & 0 \\
-\dfrac{\nu'}{E} & -\dfrac{\nu'}{E} & \dfrac{1}{E} & 0 & 0 & 0 \\
0 & 0 & 0 & \dfrac{1}{2\,\mu} & 0 & 0 \\
0 & 0 & 0 & 0 & \dfrac{1}{2\,\mu} & 0 \\
0 & 0 & 0 & 0 & 0 & \dfrac{1}{2\,\mu}
\end{bmatrix}
\begin{bmatrix}
\sigma_{11} \\
\sigma_{22} \\
\sigma_{33} \\
\sqrt{2}\,\sigma_{23} \\
\sqrt{2}\,\sigma_{13} \\
\sqrt{2}\,\sigma_{12}
\end{bmatrix}
,
\tag{5.2.64}
$$

with

$$
E = \lambda + 2\,\mu - \frac{2\,(\lambda')^2}{\lambda + 2\,\mu + \lambda'}\,, \quad \nu' = \frac{\lambda'}{\lambda + 2\,\mu + \lambda'}\,.
\tag{5.2.65}
$$

The conditions of positive definiteness of the elastic stiffness or compliance matrices boil down to

$$
\left\{\lambda + 2\,\mu > \lambda' > -\frac{1}{2}(\lambda + 2\,\mu)\,, \quad \mu > 0\right\} \Leftrightarrow \left\{E > 0, \quad -1 < \nu' < \frac{1}{2}, \quad \mu > 0\right\}.
\tag{5.2.66}
$$

See Exercise 5.14 for a direct proof.

5.2.10 Saint-Venant's relations for linear elasticity

In absence of data on the elastic coefficients, algebraic assumptions are sometimes requested. Saint-Venant's relations[5.5] should be understood as such, Saint-Venant [1863], Lekhnitskii [1963]. They provide non trivial information for orthotropic and transversely isotropic materials, and should boil down to the standard isotropic relations as the material tends to an isotropic material.

For orthotropic materials, the Saint-Venant's relations read,

$$
f_{\mathrm{SV}}(\mathbb{E}) = \frac{1}{\mu_{\mathrm{ab}}} - \frac{1 + \nu_{\mathrm{ab}}}{E_{\mathrm{a}}} - \frac{1 + \nu_{\mathrm{ba}}}{E_{\mathrm{b}}} = 0, \quad a \neq b \in [1,3]\,.
\tag{5.2.67}
$$

Other equivalent forms can be derived via the symmetry relations (5.2.42).

For transversely isotropic materials, with help of (5.2.54), the Saint-Venant's relation in the longitudinal direction ($a = 1, 2$; $b = 3$) specializes to,

$$
f_{\mathrm{SV}}(\mathbb{E}) = \frac{1}{\mu_{\mathrm{L}}} - \frac{1}{E_{\mathrm{T}}} - \frac{1}{E_{\mathrm{L}}} - 2\,\frac{\nu_{\mathrm{L}}}{E_{\mathrm{L}}} = 0,
\tag{5.2.68}
$$

while it is an identity in the plane ($a = 2$; $b = 1$), namely $1/\mu_{\mathrm{T}} = 2\,(1 + \nu_{\mathrm{T}})/E_{\mathrm{T}}$.

5.2.11 A particular elastic material with fabric anisotropy

There are several ways to introduce fabric into the elastic properties. One way is, along Cowin [1985], to assume that the elastic potential $\mathcal{W} = \mathcal{W}(\epsilon, \mathbf{G})$ depends on both the infinitesimal strain tensor ϵ and of the traceless second order fabric tensor $\mathbf{G}$ introduced in Section 12.5. The second step consists in applying representation theorems for scalar-valued isotropic functions, Section 2.11.3. For linear elasticity, the elastic potential involves 9 coefficients c_i, $i \in [1, 9]$, which depend on the isotropic scalar invariants of $\mathbf{G}$, namely:

$$
\begin{aligned}
\mathcal{W} =\ & \frac{a_1}{2}(\mathrm{tr}\,\epsilon)^2 + \frac{a_2}{2}\mathrm{tr}\,\epsilon^2 + \frac{a_3}{2}(\mathrm{tr}\,\epsilon \cdot \mathbf{G})^2 + a_4\,\mathrm{tr}\,\epsilon^2 \cdot \mathbf{G} + \frac{a_5}{2}(\mathrm{tr}\,\epsilon \cdot \mathbf{G}^2)^2 + \frac{a_6}{2}\mathrm{tr}\,(\epsilon \cdot \mathbf{G})^2 + \\
& +\ a_7\,\mathrm{tr}\,\epsilon\,\mathrm{tr}\,\epsilon \cdot \mathbf{G} + a_8\,\mathrm{tr}\,\epsilon \cdot \mathbf{G}\,\mathrm{tr}\,\epsilon \cdot \mathbf{G}^2 + a_9\,\mathrm{tr}\,\epsilon\,\mathrm{tr}\,\epsilon \cdot \mathbf{G}^2\,.
\end{aligned}
\tag{5.2.69}
$$

[5.5]Named after the French engineer, physicist and mathematician Adhémar Jean Claude Barré de Saint-Venant (1797-1886).

The resulting fourth order elastic tensor $\mathbb{E} = \partial^2 \mathcal{W}/\partial \epsilon \, \partial \epsilon$, endowed with minor and major symmetries, takes the form:

$$
\begin{aligned}
\mathbb{E} &= a_1 \, \mathbf{I} \otimes \mathbf{I} + a_2 \, \mathbf{I} \,\overline{\underline{\otimes}}\, \mathbf{I} + a_3 \, \mathbf{G} \otimes \mathbf{G} + a_4 \, (\mathbf{G} \,\overline{\underline{\otimes}}\, \mathbf{I} + \mathbf{I} \,\overline{\underline{\otimes}}\, \mathbf{G}) + a_5 \, \mathbf{G}^2 \otimes \mathbf{G}^2 + a_6 \, \mathbf{G} \,\overline{\underline{\otimes}}\, \mathbf{G} + \\
&+ \; a_7 \, (\mathbf{I} \otimes \mathbf{G} + \mathbf{G} \otimes \mathbf{I}) + a_8 \, (\mathbf{G} \otimes \mathbf{G}^2 + \mathbf{G}^2 \otimes \mathbf{G}) + a_9 \, (\mathbf{I} \otimes \mathbf{G}^2 + \mathbf{G}^2 \otimes \mathbf{I}) \,.
\end{aligned}
\tag{5.2.70}
$$

The fact that $\mathbf{G}$ is traceless affects the formal representation (5.2.69) only through the functional dependence of the scalar functions c_i upon the invariants of $\mathbf{G}$.

An alternative, particular and more compact, form of anisotropic elasticity based on a second order fabric tensor has been proposed by Valanis [1990] and Zysset and Curnier [1995] for the analysis of damage in metals and biomaterials. Its particularly appealing format makes possible the existence of a correspondence principle in the analysis of instability in geomaterials, Bigoni and Loret [1999], Loret et al. [2001].

The derivation is essentially heuristic, as follows. Let us first record the elastic isotropic tensor $\mathbb{E}_{\mathrm{iso}}$ which will be viewed as a reference:

$$
\mathbb{E}_{\mathrm{iso}} = \lambda \, \mathbf{I} \otimes \mathbf{I} + 2 \, \mu \, \mathbf{I} \,\overline{\underline{\otimes}}\, \mathbf{I} \,,
\tag{5.2.71}
$$

where λ and μ are the standard Lamé moduli. Given a symmetric second order tensor $\mathbf{B}$, the homogeneous *de visu* modification of $\mathbb{E}_{\mathrm{iso}}$[5.6],

$$
\mathbb{E} = \lambda \, \mathbf{B} \otimes \mathbf{B} + 2 \, \mu \, \mathbf{B} \,\overline{\underline{\otimes}}\, \mathbf{B} \,,
\tag{5.2.72}
$$

associates with a strain ϵ the effective stress $\boldsymbol{\sigma}$ (recall the notation $(\mathbf{B} \,\overline{\underline{\otimes}}\, \mathbf{B})_{ijkl} = (B_{ik} \, B_{jl} + B_{il} \, B_{jk})/2$),

$$
\boldsymbol{\sigma} = \mathbb{E} : \boldsymbol{\epsilon} \equiv \lambda \, (\mathbf{B} : \boldsymbol{\epsilon}) \, \mathbf{B} + 2 \, \mu \, \mathbf{B} \cdot \boldsymbol{\epsilon} \cdot \mathbf{B} \,.
\tag{5.2.73}
$$

Clearly, for this transformation to be meaningful, $\mathbf{B}$ should satisfy some restrictions. For the elasticity tensor $\mathbb{E}$ to be positive definite, $\mathbf{B}$ should be definite and, without loss of generality, we shall assume $\mathbf{B}$ to be positive definite. Consequently, $\mathbf{B}$ can not be traceless. There is no loss of generality either in scaling $\mathbf{B}$ to facilitate comparison with respect to the isotropic reference $\mathbb{E}_{\mathrm{iso}}$. Summarizing, $\mathbf{B}$ is built from a traceless fabric tensor $\mathbf{G}$ and satisfies the following conditions:

$$
\mathbf{B} = g \, \mathbf{I} + \mathbf{G} \quad \text{Positive Definite}\,, \qquad \mathrm{tr} \, \mathbf{B}^2 = 3 \,.
\tag{5.2.74}
$$

Note that the eigendirections of $\mathbf{B}$ are the axes of orthotropy of the material elasticity. An alternative format could use an angle α and a unit traceless second order tensor $\hat{\mathbf{G}}$,

$$
\mathbf{B} = \cos \alpha \, \mathbf{I} + \sqrt{3} \, \sin \alpha \, \hat{\mathbf{G}} \,.
\tag{5.2.75}
$$

As a second consequence of the positive definiteness and scaling of $\mathbf{B}$, g belongs to the interval $]0, 1]$. As a third consequence, the necessary and sufficient conditions for positive definiteness and strong ellipticity of $\mathbb{E}$ turn out to be the same as for the isotropic reference $\mathbb{E}_{\mathrm{iso}}$:

$$
\mathbb{E} \text{ Positive Definite} \; \Leftrightarrow \; 3 \, \lambda + 2 \, \mu > 0 \text{ and } \mu > 0 \,,
\tag{5.2.76}
$$

$$
\mathbb{E} \text{ Strongly Elliptic} \; \Leftrightarrow \; \lambda + 2 \, \mu > 0 \text{ and } \mu > 0 \,.
\tag{5.2.77}
$$

Positive definiteness results from the relation,

$$
\mathbf{X} : \mathbb{E} : \mathbf{X} = \tilde{\mathbf{X}} : \mathbb{E}_{\mathrm{iso}} : \tilde{\mathbf{X}} \,,
$$
$$
\text{for } \mathbf{X} = \mathbf{X}^{\mathrm{T}}, \quad \tilde{\mathbf{X}} = \mathbf{B}^{1/2} \cdot \mathbf{X} \cdot \mathbf{B}^{1/2} \,,
\tag{5.2.78}
$$

that holds for any symmetric second order tensor $\mathbf{X}$. Similarly, the strong ellipticity property results from Eqn (11.2.30) that will be shown later.

Moreover, the elastic compliance tensor has also the same format as its isotropic counterpart,

$$
\mathbb{E}^{-1} = -\frac{\nu}{E} \, \mathbf{B}^{-1} \otimes \mathbf{B}^{-1} + \frac{1 + \nu}{E} \, \mathbf{B}^{-1} \,\overline{\underline{\otimes}}\, \mathbf{B}^{-1} \,,
\tag{5.2.79}
$$

[5.6]Note also that a material endowed with linear isotropic elasticity in a reference configuration turns anisotropic in the current configuration and its elastic tensor moduli undergo a transformation associated with the left Cauchy-Green tensor of the very form (5.2.72). See Eqn (2.9.8) and Section 2.9 for a detailed derivation.

so that

$$\boldsymbol{\epsilon} = \mathbb{E}^{-1} : \boldsymbol{\sigma} = -\frac{\nu}{E}\left(\mathbf{B}^{-1} : \boldsymbol{\sigma}\right)\mathbf{B}^{-1} + \frac{1+\nu}{E}\mathbf{B}^{-1}\cdot\boldsymbol{\sigma}\cdot\mathbf{B}^{-1}, \tag{5.2.80}$$

where the Young's modulus E and Poisson's ratio ν are linked to the Lamé moduli in the usual way, Eqn (5.2.29).

Let $\mathbf{B} = \sum_a b_a\,\mathbf{B}_a$ be the spectral decomposition of the fabric tensor $\mathbf{B}$. If the latter is considered independent of strain, then the material is in fact hyperelastic and

$$
\begin{cases}
\mathcal{W}(\boldsymbol{\epsilon}) &= \dfrac{\lambda}{2}\left(\mathbf{B}:\boldsymbol{\epsilon}\right)^2 + \mu\,(\mathbf{B}\cdot\boldsymbol{\epsilon}):(\boldsymbol{\epsilon}\cdot\mathbf{B})\,, \\[2mm]
\boldsymbol{\sigma} &= \dfrac{\lambda}{2}\,b_a\,b_b\left((\mathbf{B}_a:\boldsymbol{\epsilon})\,\mathbf{B}_b + (\mathbf{B}_b:\boldsymbol{\epsilon})\,\mathbf{B}_a\right) + \mu\,b_a\,b_b\left(\mathbf{B}_a\cdot\boldsymbol{\epsilon}\cdot\mathbf{B}_b + \mathbf{B}_b\cdot\boldsymbol{\epsilon}\cdot\mathbf{B}_a\right), \\[2mm]
\mathbb{E} &= \dfrac{\lambda}{2}\,b_a\,b_b\,(\mathbf{B}_a\otimes\mathbf{B}_b + \mathbf{B}_b\otimes\mathbf{B}_a) + \mu\,b_a\,b_b\,(\mathbf{B}_a\,\overline{\underline{\otimes}}\,\mathbf{B}_b + \mathbf{B}_b\,\overline{\underline{\otimes}}\,\mathbf{B}_a)\,.
\end{cases}
\tag{5.2.81}
$$

In the orthotropy axes, the stiffness and compliance matrices can be cast in the respective forms,

$$
[\mathbb{E}] =
\begin{bmatrix}
(\lambda+2\,\mu)\,b_1^2 & \lambda\,b_1\,b_2 & \lambda\,b_1\,b_3 & 0 & 0 & 0 \\[2mm]
\lambda\,b_1\,b_2 & (\lambda+2\,\mu)\,b_2^2 & \lambda\,b_2\,b_3 & 0 & 0 & 0 \\[2mm]
\lambda\,b_1\,b_3 & \lambda\,b_2\,b_3 & (\lambda+2\,\mu)\,b_3^2 & 0 & 0 & 0 \\[2mm]
0 & 0 & 0 & 2\,\mu\,b_2\,b_3 & 0 & 0 \\[2mm]
0 & 0 & 0 & 0 & 2\,\mu\,b_1\,b_3 & 0 \\[2mm]
0 & 0 & 0 & 0 & 0 & 2\,\mu\,b_1\,b_2
\end{bmatrix},
\tag{5.2.82}
$$

and

$$
[\mathbb{E}]^{-1} =
\begin{bmatrix}
\dfrac{1}{E}\dfrac{1}{b_1^2} & -\dfrac{\nu}{E}\dfrac{1}{b_1 b_2} & -\dfrac{\nu}{E}\dfrac{1}{b_1 b_3} & 0 & 0 & 0 \\[3mm]
-\dfrac{\nu}{E}\dfrac{1}{b_1 b_2} & \dfrac{1}{E}\dfrac{1}{b_2^2} & -\dfrac{\nu}{E}\dfrac{1}{b_2 b_3} & 0 & 0 & 0 \\[3mm]
-\dfrac{\nu}{E}\dfrac{1}{b_1 b_3} & -\dfrac{\nu}{E}\dfrac{1}{b_2 b_3} & \dfrac{1}{E}\dfrac{1}{b_3^2} & 0 & 0 & 0 \\[3mm]
0 & 0 & 0 & \dfrac{1}{2\,\mu}\dfrac{1}{b_2 b_3} & 0 & 0 \\[3mm]
0 & 0 & 0 & 0 & \dfrac{1}{2\,\mu}\dfrac{1}{b_1 b_3} & 0 \\[3mm]
0 & 0 & 0 & 0 & 0 & \dfrac{1}{2\,\mu}\dfrac{1}{b_1 b_2}
\end{bmatrix}.
\tag{5.2.83}
$$

When the eigenvalues $b_i > 0$, $i \in [1,3]$, of $\mathbf{B}$ are distinct, the above elastic anisotropy involves four scalar parameters, namely λ, μ, and two of the three eigenvalues of $\mathbf{B}$, in addition to the eigenvectors $\mathbf{b}_i$, $i \in [1,3]$, of $\mathbf{B}$ which play the role of orthotropy directions. Now, two questions may arise:

1. Is this anisotropy a particular case of the nine parameter family defined earlier, Eqn (5.2.69)?
2. Is this anisotropy a particular case of structural anisotropy?

Zysset and Curnier [1995] have given an affirmative answer to the first of the above questions and identified the coefficients a_i, $i \in [1,9]$, in relation (5.2.69):

$$a_1 = \lambda\,g^2, \quad a_2 = 2\,\mu\,g^2, \quad a_3 = \lambda, \quad a_4 = 2\,\mu\,g, \quad a_6 = 2\,\mu, \quad a_7 = \lambda\,g, \quad a_5 = a_8 = a_9 = 0\,. \tag{5.2.84}$$

The answer to the second question is affirmative as well: a proof is sketched in Exercise 5.17.

When two of the eigenvalues of **B** are equal, the material response is defined by three independent scalar parameters, and it is transversely isotropic. The proof is deferred to Exercise 5.18.

Otherwise, due to the scaling Eqn (5.2.74), **B** = **I** and the material response is isotropic as expected.

Remark: the fabric elasticity above does not satisfy the Saint-Venant's relations .

On comparing (5.2.46) and (5.2.83), the Young's moduli and Poisson's ratios are identifed to

$$E_{\mathrm{a}} = E\, b_a^2, \quad \nu_{\mathrm{ab}} = \nu\, \frac{b_a}{b_b}, \quad a \neq b \in [1,3]. \tag{5.2.85}$$

The Saint-Venant's relation can be recast in the format,

$$\frac{1}{\mu} = \frac{b_a^2 + b_b^2}{b_a\, b_b} \frac{1}{E} + 2\frac{\nu}{E}, \quad a \neq b \in [1,3]. \tag{5.2.86}$$

Since $1/\mu = 2/E + 2\nu/E$, it is not satisfied, unless $b_a = b_b$, which indeed, as already observed, takes place for a transversely isotropic material in the symmetry plane.

5.3 Spectral analysis of the elastic matrix

The spectral analysis of the elasticity tensor has been addressed, among the firsts, by Mehrabadi and Cowin [1990] and Luehr and Rubin [1990].

The point is to search for the eigenvalues and eigenmodes of the elasticity stiffness or compliance tensors. The isometric representation makes it possible to perform this spectral analysis on the matrix representation itself, namely

$$\boldsymbol{\sigma} = \mathbb{E} : \boldsymbol{\epsilon} = \Lambda\, \boldsymbol{\epsilon} \quad \Leftrightarrow \quad [\boldsymbol{\sigma}] = [\mathbb{E}][\boldsymbol{\epsilon}] = \Lambda\,[\boldsymbol{\epsilon}]. \tag{5.3.1}$$

The maximum number of eigenvalues and eigenmodes is thus six. Since the stiffness matrix is symmetric, the eigenvalues and eigenmodes are real. Let $\Lambda^{(k)}$ and $\boldsymbol{\epsilon}^{(k)}$ ($[\boldsymbol{\epsilon}^{(k)}]$ in vector form) be the k-th eigenvalue and unit eigenmode, respectively, $\boldsymbol{\epsilon}^{(k)} : \boldsymbol{\epsilon}^{(k)} = 1$. Then, if the eigenvalues are distinct, the stiffness can be decomposed as the sum

$$\mathbb{E} = \sum_{k=1}^{6} \Lambda^{(k)}\, \mathbb{P}^{(k)}, \tag{5.3.2}$$

where the fourth order *projection operators* on the associated subspaces,

$$\mathbb{P}^{(k)} = \boldsymbol{\epsilon}^{(k)} \otimes \boldsymbol{\epsilon}^{(k)}, \quad k \in [1,6], \tag{5.3.3}$$

are orthonormal,

$$\mathbb{P}^{(k)} : \mathbb{P}^{(l)} = \mathbb{P}^{(k)}\, I_{\underline{k}l}, \quad k, l \in [1,6], \tag{5.3.4}$$

and sum to the fourth order identity tensor,

$$\sum_{k=1}^{6} \mathbb{P}^{(k)} = \mathbf{I}_4. \tag{5.3.5}$$

5.3.1 Spectral analysis: linear elastic isotropic materials

In the particular case of an isotropic material, $2\,\mu$ is the shear eigenvalue of multiplicity 5 with eigenmodes defined by the expresssions,

$$[\boldsymbol{\epsilon}^{(1)}] = \frac{1}{\sqrt{2}} \begin{bmatrix} 0 & 0 & 0 \\ 0 & 0 & 1 \\ 0 & 1 & 0 \end{bmatrix}, \quad [\boldsymbol{\epsilon}^{(2)}] = \frac{1}{\sqrt{2}} \begin{bmatrix} 0 & 0 & 1 \\ 0 & 0 & 0 \\ 1 & 0 & 0 \end{bmatrix}, \quad [\boldsymbol{\epsilon}^{(3)}] = \frac{1}{\sqrt{2}} \begin{bmatrix} 0 & 1 & 0 \\ 1 & 0 & 0 \\ 0 & 0 & 0 \end{bmatrix}; \tag{5.3.6}$$

$$[\boldsymbol{\epsilon}^{(4)}] = \frac{1}{\sqrt{2}} \begin{bmatrix} 1 & 0 & 0 \\ 0 & -1 & 0 \\ 0 & 0 & 0 \end{bmatrix}, \quad [\boldsymbol{\epsilon}^{(5)}] = \frac{1}{\sqrt{2}} \begin{bmatrix} 1 & 0 & 0 \\ 0 & 0 & 0 \\ 0 & 0 & -1 \end{bmatrix}, \tag{5.3.7}$$

while the bulk eigenvalue $K = \lambda + (2/3)\,\mu$ is associated with the eigenmode $\boldsymbol{\epsilon}^{(6)} = \mathbf{I}/\sqrt{3}$. We thus have a spectral decomposition of the elastic tensor,

$$\mathbb{E} = 2\,\mu \overbrace{\sum_{k \in [1,5]} \mathbb{P}^{(k)}}^{\text{isochoric part}} + \overbrace{K\, \mathbb{P}^{(6)}}^{\substack{\text{dilatational} \\ \text{part}}}. \tag{5.3.8}$$

Table 5.3.1: Spectral decomposition of the isotropic elastic tensor.

Eigenvalue	Multiplicity	Eigenmode	Property
$\Lambda^{(1 \to 5)} = 2\,\mu$	5	$\epsilon^{(1 \to 5)}$, Eqns (5.3.6)-(5.3.7)	Isochoric
$\Lambda^{(6)} = K$	1	$\epsilon^{(6)} = \mathbf{I}/\sqrt{3}$	Dilatational

5.3.2 Spectral analysis: linear elastic transversely isotropic materials

5.3.2.1 Transverse isotropy: PD and SE domains

Let us start from the transversely isotropic elastic matrix (5.2.51). When four eigenvalues of $\mathbb{E}$ are distinct, two of them are double, $\Lambda^{(1,2)}$ and $\Lambda^{(3,4)}$, and the other two are single, $\Lambda^{(5)}$ and $\Lambda^{(6)}$,

$$\Lambda^{(5,6)} = \frac{1}{2}\left(c_{11} + c_{12} + c_{33} \pm \sqrt{(c_{11} + c_{12} - c_{33})^2 + 8\,c_{13}^2}\right),\tag{5.3.9}$$

associated with the eigenmodes

$$\epsilon^{(5,6)} = \mathbf{diag}\left[c_{12}(\Lambda^{(5,6)} - c_{33}) + c_{13}^2,\ c_{12}(\Lambda^{(5,6)} - c_{33}) + c_{13}^2,\ c_{13}(\Lambda^{(5,6)} - c_{11} + c_{12})\right].\tag{5.3.10}$$

The spectral decomposition of the elastic tensor takes the form,

$$\mathbb{E} = \overbrace{2\,c_{44}\sum_{k=1,2}\mathbb{P}^{(k)} + (c_{11} - c_{12})\sum_{k=3,4}\mathbb{P}^{(k)}}^{\text{isochoric part}} + \overbrace{\Lambda^5\,\mathbb{P}^5 + \Lambda^6\,\mathbb{P}^6}^{\text{dilatational part}}.\tag{5.3.11}$$

The eigenvalues and associated eigenmodes are listed in Table 5.3.2.

Table 5.3.2: Spectral decomposition of the transversely isotropic elastic tensor.

Eigenvalue	Multiplicity	Eigenmode	Property
$\Lambda^{(1,2)} = 2\,c_{44}$	2	$\epsilon^{(1,2)}$, Eqn (5.3.6)	Isochoric
$\Lambda^{(3,4)} = c_{11} - c_{12}$	2	$\epsilon^{(3,4)}$, Eqn (5.3.7)	Isochoric
$\Lambda^{(5,6)}$, Eqn (5.3.9)	1	$\epsilon^{(5,6)}$, Eqn (5.3.10)	Dilatational

5.3.2.2 Transverse isotropy: limit of the PD domain

A special case of transverse isotropy defined by a one-parameter deviation with respect to isotropy is considered in section Section 5.2.8. The spectral analysis of the elastic stiffness at loss of positive definiteness for $a_4 = a_4^{\mathrm{LPD}}$ specializes as follows when $\lambda/\mu \neq 0$:

– eigenvalues $\Lambda^{(1,2,3,4)} = 2\,\mu$ of multiplicity 4;

$$\Lambda^{(5)} = \frac{3\,\lambda^2 + 4\,\lambda\mu + 2\,\mu^2}{\lambda + \mu} > 0; \quad \Lambda^{(6)} = 0 \text{ of multiplicity } 1;$$

– eigenmodes $\epsilon^{(1,2,3,4)}$ given by Eqns (5.3.6) $-$ (5.3.7) ;

$$[\epsilon^{(5)}] = \sqrt{\tilde{\epsilon}}\begin{bmatrix} 1 & 0 & 0 \\ 0 & 1 & 0 \\ 0 & 0 & \tilde{\lambda} \end{bmatrix}; \quad [\epsilon^{(6)}] = \frac{\sqrt{\tilde{\epsilon}}}{\sqrt{2}}\begin{bmatrix} -\tilde{\lambda} & 0 & 0 \\ 0 & -\tilde{\lambda} & 0 \\ 0 & 0 & 2 \end{bmatrix},\tag{5.3.12}$$

with $\tilde{\epsilon} = (\lambda + \mu)/\Lambda^{(5)}$ and $\tilde{\lambda} = \lambda/(\lambda + \mu)$. Therefore, the spectral decomposition of the elastic stiffness remains formally as indicated by (5.3.11).

5.3.2.3 Transverse isotropy: limit of the PD and SE domains

The case where $\lambda = 0$ is peculiar because it is the only value for which losses of strong ellipticity and positive definiteness take place simultaneously, Section 5.2.8.

According to Section 2.2.4, the null space of $\mathbb{E}$, namely the eigenmodes of $\mathbb{E}$ associated with its zero eigenvalue, contains a symmetrized dyadic form $\epsilon^{\otimes}$ only for this value. Indeed, then the spectral analysis of ϵ is modified as follows:

- eigenvalues $\Lambda^{(1,2,3,4,5)} = 2\,\mu$ of algebraic multiplicity 5; $\Lambda^{(6)} = 0$ of multiplicity 1;

- eigenmodes $\epsilon^{(1)}, \epsilon^{(2)}, \epsilon^{(3)}$ given by Eqns (5.3.6) – (5.3.7) and

$$[\epsilon^{(4)}] = \begin{bmatrix} 1 & 0 & 0 \\ 0 & 0 & 0 \\ 0 & 0 & 0 \end{bmatrix}, \quad [\epsilon^{(5)}] = \begin{bmatrix} 0 & 0 & 0 \\ 0 & 1 & 0 \\ 0 & 0 & 0 \end{bmatrix}; \quad [\epsilon^{(6)}] = \begin{bmatrix} 0 & 0 & 0 \\ 0 & 0 & 0 \\ 0 & 0 & 1 \end{bmatrix}. \tag{5.3.13}$$

As alluded for earlier, the eigenmode $\epsilon^{(6)}$ has indeed a dyadic form, namely $\epsilon^{(6)} = \mathbf{e}_3 \otimes \mathbf{e}_3$, while its counterpart in the case $\lambda \neq 0$ does not have such a form (even at loss of positive definiteness). Thus, for $\lambda = 0$, the spectral decomposition of $\mathbb{E}$ reads:

$$\mathbb{E} = \overbrace{2\,\mu \sum_{k \in [1,3]} \mathbb{P}^{(k)}}^{\substack{\text{isochoric} \\ \text{contribution}}} + \overbrace{2\,\mu \sum_{k=4,5} \mathbb{P}^{(k)}}^{\substack{\text{dilatational} \\ \text{contribution}}} + \overbrace{[\Lambda^6\,\mathbb{P}^{(6)}]}^{\substack{\text{dilatational} \\ \text{contribution}}}. \tag{5.3.14}$$

5.3.3 Spectral analysis: linear elastic orthotropic materials

The spectral analysis is performed on the matrix form (5.2.41). The three real positive eigenvalues $\Lambda = 2\,c_{ii}$, $i = 4, 5$ and 6, are associated with the deviatoric eigenmodes $\epsilon^{(1)}$, $\epsilon^{(2)}$ and $\epsilon^{(3)}$ of Eqn (5.3.6) in that respective order. The three eigenvalues Λ stemming from the 3×3 principal minor $[\mathbf{c}]$, which are solutions of the secular equation,

$$-\Lambda^3 + I_1([\mathbf{c}])\,\Lambda^2 - I_2([\mathbf{c}])\,\Lambda + I_3([\mathbf{c}]) = 0, \tag{5.3.15}$$

are real positive as well since the symmetric stiffness is assumed to be positive definite.

A particular eigenvalue Λ is associated with an isochoric eigenmode if it satisfies simultaneously the relation,

$$\Lambda^2 - (c_{11} + c_{22} - c_{13} - c_{23})\,\Lambda + (c_{11} - c_{13})(c_{22} - c_{23}) - (c_{12} - c_{13})(c_{12} - c_{23}) = 0. \tag{5.3.16}$$

This equation results from the following manipulations. The eigenvalue problem is expressed in the format $([\mathbf{c}] - \Lambda\,[\mathbf{I}])\,[\epsilon] = \mathbf{0}$, with $[\epsilon] = [\epsilon_{11}\ \epsilon_{22}\ \epsilon_{33}]^\mathsf{T}$. Insertion of $\epsilon_{33} = -\epsilon_{11} - \epsilon_{22}$ in the two first of these equations yields a system for ϵ_{11} and ϵ_{22} with zero rhs. For the solution to be non trivial, the determinant of this system should vanish, resulting in (5.3.16). Two equivalent relations can be obtained by permutation of the indices. Summing the three relations delivers a quadratic relation for Λ with symmetric coefficients.

5.4 Fiber-reinforced materials: extension-contraction response

Fibers are sustaining tensile loads. They buckle under compressive loads. One may also be tempted to consider that the fibers are active in extension, not in contraction. Actually, these two points of view, the stress and the strain points of view, are not equivalent in a general three-dimensional analysis, nor even in a one-dimensional context.

The sets of strains which correspond to extension and contraction will be defined through the notion of *cone*, which is defined in the footnote on p. 70. This section first summarizes the basic relations that any conewise mechanical response has to satisfy. Several forms of anisotropy are examined. A second part reviews the experimental data available to date in the literature for a particular material, which is reinforced by fibers, namely articular cartilage.

Alternative approaches to deal with fiber uncrimping and fiber reinforcement are exposed in Chapters 12 and 19.

5.4.1   Basic remarks on the mechanical response of fiber-reinforced tissues

Articular cartilage is a porous medium saturated by an electrolyte. Compressive loads are sustained by negatively charged macromolecules, the proteoglycans. An interconnected network of collagen fibers allows cartilage to resist tensile and shear loads. Therefore, the apparent mechanical properties of cartilage are bound to be anisotropic *and* distinct in extension and contraction.

The mechanical response of articular cartilage is the result of electro-chemomechanical couplings. In fact, the definition of a purely mechanical response is not unambiguous. Actually, the most reliable sources of information are those who present rough data, free of interpretation, showing applied or measured stress and strain components: such data will reveal the *apparent*, or coupled, or macroscopic, material response of cartilage.

The picture is actually still more complex because chemomechanical couplings are accompanied with diffusion, a phenomenon that per se introduces spatial heterogeneity. It is possible to free the data of this heterogeneity by collecting only measurements at steady states, at which diffusion has come to an end. However, there are several characteristic times in the electro-chemomechanical response of cartilage due to its compartmental structure. Indeed, deformation may become steady before diffusion stops, so that, in practice, data that are reported to correspond to steady states may be biased by a slowly ongoing diffusion. The latter may marginally affect deformation but might, in the long range or when the loading conditions change, lead to difficulties of interpretation.

Another source of difficulty is due to the fact that articular cartilage has a heterogeneous composition in depth. Thus, measurements will depend on the location, and they may also be biased by the size (thickness) of the specimens tested.

5.4.2   Conewise constitutive response: generalities

5.4.2.1   Extension-contraction and tension-compression dependent responses

A distinction should be made between two types of conewise responses, Fig. 5.4.1:
- the mechanical response of the tissue may be *strain driven*. In particular, the moduli along the orthotropy directions may assume different values depending on the sign of the strains along these directions, so that the tensor moduli are said to be **extension-contraction** dependent.
- alternatively, the response may be *stress driven*, so that the moduli depend on the sign of some stress component along the orthotropy axes, and the tensor moduli are said to be **tension-compression** dependent.

This second possibility is not pursued below, and emphasis is placed on the extension-contraction conewise response. Still, a strain based criterion has to face the crucial choice of the reference configuration (where the strain vanishes). For articular cartilages, the hypertonic state is the natural candidate.

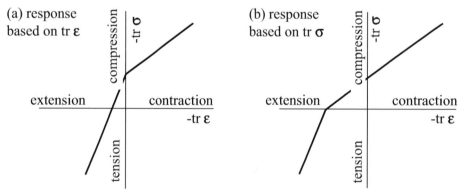

Fig. 5.4.1: For an isotropic elastic material, the sole bulk modulus may take distinct values whether (a) in an extension-contraction strain based conewise response, or (b) in a tension-compression stress based conewise response.

With this terminology in place, the sign conventions for strains and stresses can be defined. An extension gives rise to a positive strain and tension is associated with a positive stress.

5.4.2.2 Analysis of an extension-contraction dependent response

Let us note $g(\epsilon) = 0$ the interface in strain-space on both sides of which the tensor moduli differ. For definiteness, the side $g(\epsilon) > 0$ will be denoted by the superscript '+' and the side $g(\epsilon) < 0$ by the superscript '−'. Two basic properties are required:
- the stress $\boldsymbol{\sigma}$ is continuous across the interface $g(\epsilon) = 0$;
- the stress is differentiable on each side of the interface, in fact $d\boldsymbol{\sigma}/d\epsilon$ is equal to the tensor moduli $\mathbb{E}$ on each side of the interface.

Then by an extension of the Hadamard's compatibility conditions exposed in Section 11.1.3, the difference $\mathbb{E}^{+} - \mathbb{E}^{-}$ should be dyadic as encapsulated by

Proposition 5.1 The tensor moduli on each side of the interface $g(\epsilon) = 0$ differ by a symmetric dyadic product,

$$\mathbb{E}^{+} = \mathbb{E}^{-} + s \frac{dg}{d\epsilon} \otimes \frac{dg}{d\epsilon}, \tag{5.4.1}$$

where $s = s(g(\epsilon))$ denotes the amplitude of the jump across the interface, equivalently

$$\begin{cases} \boldsymbol{\sigma} = \mathbb{E}^{-} : \epsilon & \text{on } g(\epsilon) \le 0, \\[2mm] \boldsymbol{\sigma} = \mathbb{E}^{+} : \epsilon = (\mathbb{E}^{-} + s \dfrac{dg}{d\epsilon} \otimes \dfrac{dg}{d\epsilon}) : \epsilon & \text{on } g(\epsilon) \ge 0. \end{cases} \tag{5.4.2}$$

Note 5.1 on rank one modifications.
There are other instances in continuum mechanics where the traces of two fourth order tensors are required to coincide along some interface, e.g., the elastic and elastoplastic tensor moduli. As another example, one may cite the drained and undrained elastic tensors in fluid-saturated porous media, e.g., Loret et al. [2001] and Section 7.4.

Note 5.2 on piecewise and global hyperelasticity.
A constitutive law which is piecewise hyperelastic and satisfies the above condition (5.4.1) is (globally) hyperelastic, Curnier et al. [1994].

Note 5.3 on convex domains.
One may wish to have both sides of the interface convex in order to apply the classical results of functional analysis. Then the interface should be a hyperplane, i.e., $dg/d\epsilon$ constant.

Note 5.4 on conewise symmetry groups.
The material symmetry groups on each side of the interface might be distinct, e.g., isotropy on one side and orthotropy on the other side.

5.4.3 Application to materials with one family of fibers

Let $\mathbf{m} = \mathbf{e}_1$ be the unit vector that indicates the direction of the fibers in a reference configuration, and

$$\mathbf{A}_1 = \mathbf{M} = \mathbf{m} \otimes \mathbf{m}. \tag{5.4.3}$$

The fibers are assumed to be mechanically active only if they undergo extension, namely

$$g(\epsilon) \equiv \epsilon : \mathbf{M} \ge 0. \tag{5.4.4}$$

Thus the moduli corresponding to extended fibers and contracted fibers differ only by the dyadic product,

$$\mathbb{E} = \mathbb{E}^{-} + s\,\mathcal{H}(\mathbf{M} : \epsilon)\,\mathbf{M} \otimes \mathbf{M}, \tag{5.4.5}$$

where $s = s(\epsilon : \mathbf{M})$ is the magnitude of the jump, and $\mathcal{H}$ is the Heaviside step function.
The matrix form of this relation in the cartesian axes ($\mathbf{e}_1 = \mathbf{m}, \mathbf{e}_2, \mathbf{e}_3$) shows that the two elastic stiffnesses differ by a single diagonal term,

$$[\mathbb{E}] = [\mathbb{E}^{-}] + s\,\mathcal{H}(\mathbf{M} : \epsilon) \begin{bmatrix} 1 & 0 & 0 & 0 & 0 & 0 \\ 0 & 0 & 0 & 0 & 0 & 0 \\ 0 & 0 & 0 & 0 & 0 & 0 \\ 0 & 0 & 0 & 0 & 0 & 0 \\ 0 & 0 & 0 & 0 & 0 & 0 \\ 0 & 0 & 0 & 0 & 0 & 0 \end{bmatrix}. \tag{5.4.6}$$

5.4.4   Application to materials with two families of fibers

Let $\mathbf{m}_1$ and $\mathbf{m}_2$ be the unit vectors (non necessarily orthogonal) that indicate the directions of the fibers in a reference configuration. The fibers are assumed to be mechanically active only if they undergo extension, namely

$$g_i(\boldsymbol{\epsilon}) \equiv \boldsymbol{\epsilon} : \mathbf{M}_i \geq 0, \quad i = 1, 2.\tag{5.4.7}$$

Thus the moduli corresponding to extended fibers and contracted fibers differ only by the dyadic products,

$$\mathbb{E} = \mathbb{E}^- + s_1\,\mathcal{H}(\mathbf{M}_1 : \boldsymbol{\epsilon})\,\mathbf{M}_1 \otimes \mathbf{M}_1 + s_2\,\mathcal{H}(\mathbf{M}_2 : \boldsymbol{\epsilon})\,\mathbf{M}_2 \otimes \mathbf{M}_2,\tag{5.4.8}$$

where $s = s_i(\boldsymbol{\epsilon} : \mathbf{M}_i)$, $i = 1, 2$, are the magnitudes of the jumps.

The two families of fibers are said *mechanically equivalent*, if their mechanical properties are identical and if their densities are identical. Then the mechanical response becomes *locally* orthotropic: the orthogonal axes on which the fiber axes mirror are the orthotropy axes.

5.4.5   Application to linear orthotropic materials: octantwise response

5.4.5.1   The number of independent coefficients

The above relations are now applied to an orthotropic material whose response is extension-contraction dependent. There are three interfaces $g(\boldsymbol{\epsilon}) = \mathbf{A}_a : \boldsymbol{\epsilon} = 0$, $a \in [1, 3]$, defined by the structure tensors $\mathbf{A}$, so that there are a priori $2^3 = 8$ possible stiffnesses, as indicated by Table 5.4.1 and illustrated by Fig. 5.4.5.1.

Table 5.4.1: Algebraic definition of the octants.

Octant nb.	$\mathbf{A}_1 : \boldsymbol{\epsilon}$	$\mathbf{A}_2 : \boldsymbol{\epsilon}$	$\mathbf{A}_3 : \boldsymbol{\epsilon}$
1	+	+	+
2	-	+	+
3	+	+	-
4	-	+	-
5	+	-	+
6	-	-	+
7	+	-	-
8	-	-	-

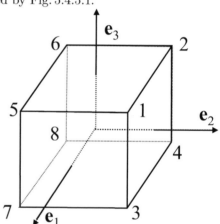

Fig. 5.4.2: Visual definition of the octants.

Let us show that there are only three possible discontinuities. This point is formalized in
Proposition 5.2 The eight tensor moduli defining the response in the eight octants differ at most by **three** scalars.

There are several ways to show this proposition. It is clear, from Proposition 5.1, that the sole discontinuities are represented by symmetric dyadic terms. Looking at the expression of the tensor moduli (5.2.37), we thus see that there are only three such terms, namely $\lambda_{\underline{aa}}\,\mathbf{A}_a \otimes \mathbf{A}_a$, $a \in [1, 3]$. This observation is formalized by
Proposition 5.3 on the discontinuity pattern, Chaboche et al. [1995], Curnier et al. [1994].

The sole possible discontinuities between two arbitrary octants are defined by the jumps $[\![\lambda_{\underline{aa}}]\!]$, $a \in [1, 3]$, of the three coefficients $\lambda_{\underline{aa}}$, $a \in [1, 3]$,

$$[\![\mathbb{E}^+]\!] = [\![\mathbb{E}^-]\!] + [\![\lambda_{11}]\!]\,\mathbf{A}_1 \otimes \mathbf{A}_1 + [\![\lambda_{22}]\!]\,\mathbf{A}_2 \otimes \mathbf{A}_2 + [\![\lambda_{33}]\!]\,\mathbf{A}_3 \otimes \mathbf{A}_3,\tag{5.4.9}$$

and, in the orthotropy axes,

$$[\mathbb{E}^+] = [\mathbb{E}^-] + \begin{bmatrix} [\![\lambda_{11}]\!] & 0 & 0 & 0 & 0 & 0 \\ 0 & [\![\lambda_{22}]\!] & 0 & 0 & 0 & 0 \\ 0 & 0 & [\![\lambda_{33}]\!] & 0 & 0 & 0 \\ 0 & 0 & 0 & 0 & 0 & 0 \\ 0 & 0 & 0 & 0 & 0 & 0 \\ 0 & 0 & 0 & 0 & 0 & 0 \end{bmatrix}.\tag{5.4.10}$$

The six other coefficients λ_{ab}, $a < b \in [1, 3]$, and μ_a, $a \in [1, 3]$, cannot undergo jumps. Therefore the complete octantwise extension-contraction dependent response is defined by $9+3=12$ independent coefficients.

5.4.5.2 Consequences for the Young's moduli and Poisson's ratios

At this point, a <u>warning</u> should be issued.

According to Proposition 5.3, had we considered a tension-compression dependent response, we would have observed, from the elastic compliance (5.2.46), that the sole three coefficients that undergo jumps are the three Young's moduli E_a, $a \in [1,3]$, while the six other coefficients ν_{ab}, $a < b \in [1,3]$, and ν_{aa}, $a \in [1,3]$, cannot undergo jumps.

Thus, the tension-compression dependent response and the extension-contraction dependent response are not correlated. As indicated by (5.E.17)–(5.E.24), the modifications of the generalized Young's moduli and Poisson's ratios *deduced* from the modifications of the generalized Lamé moduli will *not* satisfy in general the tension-compression relations above.

Now, let us return to the extension-contraction response and let us consider the determination of the coefficients.

If the generalized Lamé moduli are measured, then there is no problem of methodology and the subsequent discussion is of no concern. On the other hand, if the components of the compliance matrix are measured in some octants, a problem of methodology arises.

Indeed, according to (5.E.18)–(5.E.19), the three Young's moduli E_a, $a \in [1,3]$, and the three Poisson's ratios ν_{ab}, $a < b \in [1,3]$, undergo jumps, since they depend on the λ_{aa}, $a \in [1,3]$. On the other hand, according to (5.E.21), the coefficients $(1 + \nu_{aa})/E_a$, $a \in [1,3]$, are continuous, so that the three Poisson's ratios ν_{aa}, $a \in [1,3]$, undergo jumps as well. The situation is summarized in

Proposition 5.4 number of independent coefficients.

For an extension-contraction response, *all nine coefficients*, namely the Young's moduli E_a, $a \in [1,3]$, and the six Poisson's ratios ν_{ab}, $a \le b \in [1,3]$, *undergo jumps across octants*, since they depend on the λ_{aa}'s, $a \in [1,3]$. The compliance matrices of the eight octants are in general different. Clearly these jumps are not arbitrary, they are dictated by the three jumps λ_{aa}, $a \in [1,3]$. Between the $8 \times 9 = 72$ coefficients defining the eight octants, there exist relations so that the total number of independent coefficients defining the complete octantwise response is equal to twelve. As an example of such relations, Eqn (5.E.21) indicates, since the μ_a, $a \in [1,3]$, do not undergo jumps, that the coefficients

$$\frac{1 + \nu_{aa}}{E_a}, \quad a \in [1,3], \tag{5.4.11}$$

are the same in all eight octants. Other relations are obtained by imposing that the three coefficients λ_{ab}, $a < b \in [1,3]$, deduced from the ν_{ab}, $a \le b \in [1,3]$, and E_a, $a \in [1,3]$, should be identical in all octants. Thus, from (5.E.24), each of the three terms

$$\begin{cases} \Delta_\mathrm{U} \left(\dfrac{\nu_{13}}{E_1} \dfrac{\nu_{23}}{E_2} + \dfrac{\nu_{12}}{E_1} \dfrac{1}{E_3} \right), \\[2ex] \Delta_\mathrm{U} \left(\dfrac{\nu_{12}}{E_1} \dfrac{\nu_{23}}{E_2} + \dfrac{\nu_{13}}{E_1} \dfrac{1}{E_2} \right), \\[2ex] \Delta_\mathrm{U} \left(\dfrac{\nu_{12}}{E_1} \dfrac{\nu_{13}}{E_1} + \dfrac{\nu_{23}}{E_2} \dfrac{1}{E_1} \right), \end{cases} \tag{5.4.12}$$

is the same over all eight octants, $\Delta_\mathrm{U} = \det \mathbb{E}^\mathrm{U}_{3\times3}$ being the determinant of the upper 3×3 principal minor $\mathbb{E}^\mathrm{U}_{3\times3}$ of the tensor moduli (5.2.40). On the other hand, we have from (5.E.23), that each of the three terms

$$\begin{cases} \Delta_\mathrm{U} \left(\dfrac{1}{E_2} \dfrac{1}{E_3} - \left(\dfrac{\nu_{23}}{E_2} \right)^2 \right), \\[2ex] \Delta_\mathrm{U} \left(\dfrac{1}{E_1} \dfrac{1}{E_3} - \left(\dfrac{\nu_{13}}{E_1} \right)^2 \right), \\[2ex] \Delta_\mathrm{U} \left(\dfrac{1}{E_1} \dfrac{1}{E_2} - \left(\dfrac{\nu_{12}}{E_1} \right)^2 \right), \end{cases} \tag{5.4.13}$$

can take only two values over the eight octants.

Let us reiterate that only twelve independent measurements [5.7] of the Poisson's and Young's moduli need be performed to define the complete octantwise response. For example, one might measure the nine coefficients in a certain octant and any three other coefficients in the other octants.

[5.7]Independent in the sense that they are not equivalent to (5.4.11) and (5.4.12).

5.4.6 Application to transversely isotropic materials

For transversely isotropic materials, defined in terms of 5 elastic coefficients, as indicated in Exercise 5.11,

$$
\begin{aligned}
\boldsymbol{\sigma} &= a_1\,(\mathrm{tr}\,\boldsymbol{\epsilon})\,\mathbf{I} + a_2\,\boldsymbol{\epsilon} + a_3\left((\mathrm{tr}\,\mathbf{M}\cdot\boldsymbol{\epsilon})\,\mathbf{I} + (\mathrm{tr}\,\boldsymbol{\epsilon})\,\mathbf{M}\right) + \\
&\quad + a_4\,(\mathrm{tr}\,\mathbf{M}\cdot\boldsymbol{\epsilon})\,\mathbf{M} + a_5\,(\mathbf{M}\cdot\boldsymbol{\epsilon} + \boldsymbol{\epsilon}\cdot\mathbf{M})\,; \\
\mathbb{E} &= a_1\,\mathbf{I}\otimes\mathbf{I} + a_2\,\mathbf{I}\,\overline{\underline{\otimes}}\,\mathbf{I} + a_3\,(\mathbf{M}\otimes\mathbf{I} + \mathbf{I}\otimes\mathbf{M}) + \\
&\quad + a_4\,\mathbf{M}\otimes\mathbf{M} + a_5\,(\mathbf{M}\,\overline{\underline{\otimes}}\,\mathbf{I} + \mathbf{I}\,\overline{\underline{\otimes}}\,\mathbf{M})\,,
\end{aligned}
\tag{5.4.14}
$$

there exist two interfaces, delineating the sign of the volume change, namely $\mathbf{I}:\boldsymbol{\epsilon}=0$, across which only a_1 can undergo a jump, and the sign of the strain along the symmetry axis $\mathbf{M}:\boldsymbol{\epsilon}=0$, across which only a_4 can undergo a jump. The elastic coefficients may therefore differ by the two dyadic products,

$$
\mathbb{E}^+ - \mathbb{E}^- = [\![a_1]\!]\,\mathbf{I}\otimes\mathbf{I} + [\![a_4]\!]\,\mathbf{M}\otimes\mathbf{M}\,.
\tag{5.4.15}
$$

5.4.7 Application to isotropic materials

Since for isotropic materials the stiffness has the form,

$$
\mathbb{E} = \lambda\,\mathbf{I}\otimes\mathbf{I} + 2\,\mu\,\mathbf{I}\,\overline{\underline{\otimes}}\,\mathbf{I}\,,
\tag{5.4.16}
$$

there exists a single interface, delineating the sign of the volume change, namely $\mathbf{I}:\boldsymbol{\epsilon}=0$, across which only λ can undergo a jump, while μ is continuous. Thus both E and ν undergo jumps but in such a way that $\mu = E/2\,(1+\nu)$ is continuous across the interface.

Note that, for linear isotropic materials, the *extension-contraction* and *tension-compression* dependent responses are identical because the interfaces $\mathbf{I}:\boldsymbol{\epsilon}=0$ and $\mathbf{I}:\boldsymbol{\sigma}=0$, are clearly linked, indeed $\mathbf{I}:\boldsymbol{\epsilon}=\mathbf{I}:\boldsymbol{\sigma}/K$, with $K = E/3\,(1-2\,\nu) = \lambda + \frac{2}{3}\,\mu > 0$ the bulk modulus.

The situation is different in presence of an initial stress, say $\boldsymbol{\sigma}-\boldsymbol{\sigma}_0 = \mathbb{E}:\boldsymbol{\epsilon}$, or of an initial strain, say $\boldsymbol{\sigma} = \mathbb{E}:(\boldsymbol{\epsilon}-\boldsymbol{\epsilon}_0)$, as displayed by Figs. 5.4.1-(a) and (b), respectively, and of both initial stress and initial strain, $\boldsymbol{\sigma}-\boldsymbol{\sigma}_0 = \mathbb{E}:(\boldsymbol{\epsilon}-\boldsymbol{\epsilon}_0)$.

5.4.8 Cartilage as a tissue with cubic symmetry

In order to interpret an experimental study of articular cartilage, Soltz and Ateshian [2000] view the tissue as a cubic material as described in Section 5.2.9. The confined compression modulus along direction 1 (for which $\epsilon_{22} = \epsilon_{33} = 0$) is equal to $\lambda + 2\,\mu$ while the unconfined compression modulus (for which $\sigma_{22} = \sigma_{33} = 0$) is equal to $E = \sigma_{11}/\epsilon_{11}$. The confined compression modulus $\lambda + 2\,\mu$ is discontinuous along the interface $\mathrm{tr}\,\boldsymbol{\epsilon} = 0$ while the Lamé modulus λ' and the Poisson's ratio ν' should be continuous.

Extension-contraction response, Soltz and Ateshian [2000]

Measurements provide the Lamé modulus $\lambda' = 480\,\mathrm{kPa}$ and the shear modulus $\mu = 170\,\mathrm{kPa}$. From the data extracted from confined and unconfined compression tests shown in Table 5.4.2, the apparent Poisson's ratio ν'^- appears to be large. Still, the measured values satisfy the conditions (5.2.66) of positive definiteness of the elastic stiffness (and compliance).

Table 5.4.2: Confined and unconfined compression and torsion tests on humeral head articular cartilage plugs of 6-month-old calves, Soltz and Ateshian [2000].

Unconfined compression	$\lambda^+ + 2\,\mu = 13200\,\mathrm{kPa}$	$E^+ = 13166\,\mathrm{kPa}$	$\nu'^+ = 0.035$
Confined compression	$\lambda^- + 2\,\mu = 640\,\mathrm{kPa}$	$E^- = 229\,\mathrm{kPa}$	$\nu'^- = 0.429$

5.4.9 Cartilage as an orthotropic tissue

Orthotropy axes in articular cartilage are defined by the directions of the split-lines obtained by puncturing the joint with a needle, Fig. 5.4.3.

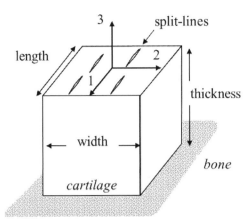

3 split-lines

length 2

thickness

width

bone

cartilage

Fig. 5.4.3: In the articular cartilage layer, the orthotropy axes are defined via the split-lines appearing on the upper zone.

5.4.9.1 Unconfined compression tests along the orthotropy axes

C.C.B. Wang et al. [2003] perform unconfined compression tests over parallelepiped specimens of size $0.91\,\text{mm} \times 0.89\,\text{mm} \times 0.82\,\text{mm}$ (length × width × thickness). Although the point is not mentioned explicitly, the tests have probably been performed at physiological salinities.

For unconfined compression along the axis x, $\sigma_{xx} < 0$, $\epsilon_{xx} < 0$, while $\sigma_{yy} = \sigma_{zz} = 0$, $\epsilon_{yy} > 0$, $\epsilon_{zz} > 0$. The elastic constitutive equations (5.2.40) can be cast in the format,

$$
\begin{bmatrix}
\sigma_{xx} < 0 \\[4pt]
\sigma_{yy} = 0 \\[4pt]
\sigma_{zz} = 0 \\[4pt]
\sqrt{2}\,\sigma_{yz} = 0 \\[4pt]
\sqrt{2}\,\sigma_{xz} = 0 \\[4pt]
\sqrt{2}\,\sigma_{xy} = 0
\end{bmatrix}
=
\begin{bmatrix}
H_x^- & \lambda_{xy} & \lambda_{xz} & 0 & 0 & 0 \\[4pt]
\lambda_{xy} & H_y^+ & \lambda_{yz} & 0 & 0 & 0 \\[4pt]
\lambda_{xz} & \lambda_{yz} & H_z^+ & 0 & 0 & 0 \\[4pt]
0 & 0 & 0 & \mu_y + \mu_z & 0 & 0 \\[4pt]
0 & 0 & 0 & 0 & \mu_x + \mu_z & 0 \\[4pt]
0 & 0 & 0 & 0 & 0 & \mu_x + \mu_y
\end{bmatrix}
\begin{bmatrix}
\epsilon_{xx} < 0 \\[4pt]
\epsilon_{yy} > 0 \\[4pt]
\epsilon_{zz} > 0 \\[4pt]
\sqrt{2}\,\epsilon_{yz}(\Rightarrow = 0) \\[4pt]
\sqrt{2}\,\epsilon_{xz}(\Rightarrow = 0) \\[4pt]
\sqrt{2}\,\epsilon_{xy}(\Rightarrow = 0)
\end{bmatrix},
\tag{5.4.17}
$$

with the notation of the Columbia group $H \equiv \lambda + 2\,\mu$. The non trivial inverse relations provide the strain components along the orthotropy axes,

$$
\begin{bmatrix}
\epsilon_{xx} \\[4pt]
\epsilon_{yy} \\[4pt]
\epsilon_{zz}
\end{bmatrix}
= \frac{\sigma_{xx}}{E_x^-}
\begin{bmatrix}
1 \\[4pt]
-\nu_{xy}^+ \\[4pt]
-\nu_{xz}^+
\end{bmatrix},
\tag{5.4.18}
$$

where, according to (5.E.18)–(5.E.19),

$$
E_x^- = \frac{H_x^- H_y^+ H_z^+ - H_x^- \lambda_{yz}^2 - H_y^+ \lambda_{xz}^2 - H_z^+ \lambda_{xy}^2 + 2\,\lambda_{yz}\,\lambda_{xz}\,\lambda_{xy}}{H_y^+ H_z^+ - \lambda_{yz}^2},
$$

$$
\nu_{xy}^+ = \frac{\lambda_{xy} H_z^+ - \lambda_{xz}\,\lambda_{yz}}{H_y^+ H_z^+ - \lambda_{yz}^2}, \qquad \nu_{xz}^+ = \frac{\lambda_{xz} H_y^+ - \lambda_{xy}\,\lambda_{yz}}{H_y^+ H_z^+ - \lambda_{yz}^2}.
\tag{5.4.19}
$$

These relations are applied for $x = 1$, $y = 2$, $z = 3$, for $x = 2$, $y = 3$, $z = 1$, and for $x = 3$, $y = 1$, $z = 2$.

Data for shoulder joints of young calves, C.C.B. Wang et al. [2003]

C.C.B. Wang et al. [2003] provide data for shoulder joints of young calves. For compression normal to the cartilage surface, strains in the depth are not homogeneous, so that the results are split in two sets, one for the top layer and one for the bottom layer of each specimen, Table 5.4.3. The vertical modulus is larger at the bottom due to the collagen network and other microstructural features close to the bone.

Table 5.4.3: *Apparent* elastic properties measured during unconfined compression tests along the three directions of material symmetry at 5-10% axial contraction, C.C.B. Wang et al. [2003].

Top Layer			Bottom Layer	
$\nu_{21}^+=0.045$	$\nu_{13}^+=0.293$	$\nu_{23}^+=0.307$		
$\nu_{12}^+=0.053$	$\nu_{31}^+=0.048$	$\nu_{32}^+=0.043$	$\nu_{31}^+=0.361$	$\nu_{32}^+=0.363$
$E_1^-=385\,\text{kPa}$	$E_2^-=377\,\text{kPa}$	$E_3^-=229\,\text{kPa}$	$E_3^-=1509$	

One, quite deceptive, key ouput of the analysis is that the compliance matrix (5.2.46) is not symmetric. The lack of symmetry is in contradiction with the hyperelastic response, and many relations derived above. Actually, exploiting the measurements of the three unconfined compression tests is not an easy task. Indeed, each test provides one Young's modulus and two Poisson's ratios corresponding to a specific octant. Moreover, the octantwise orthotropic response requires twelve independent coefficients while only nine values are measured. To proceed, C.C.B. Wang et al. [2003] make a number of simplifications in such a way that they are left with a free modulus. A parameter analysis is performed on this modulus, by varying its values between the bounds that ensure positive definiteness of the stiffness matrix[5.8].

5.4.9.2   Data from uniaxial traction

Let us consider uniaxial traction, along the axis 1, so $\sigma_{11} > 0$, $\epsilon_{11} > 0$, while $\sigma_{22} = \sigma_{33}=0$, $\epsilon_{22} < 0$, $\epsilon_{33} < 0$. The elastic relation (5.2.40) specializes to, with $H \equiv \lambda + 2\,\mu$,

$$
\begin{bmatrix}
\sigma_{11} > 0 \\
\sigma_{22} = 0 \\
\sigma_{33} = 0 \\
\sqrt{2}\,\sigma_{23} = 0 \\
\sqrt{2}\,\sigma_{13} = 0 \\
\sqrt{2}\,\sigma_{12} = 0
\end{bmatrix}
=
\begin{bmatrix}
H_1^+ & \lambda_{12} & \lambda_{13} & 0 & 0 & 0 \\
\lambda_{12} & H_2^- & \lambda_{23} & 0 & 0 & 0 \\
\lambda_{13} & \lambda_{23} & H_3^- & 0 & 0 & 0 \\
0 & 0 & 0 & \mu_2+\mu_3 & 0 & 0 \\
0 & 0 & 0 & 0 & \mu_1+\mu_3 & 0 \\
0 & 0 & 0 & 0 & 0 & \mu_1+\mu_2
\end{bmatrix}
\begin{bmatrix}
\epsilon_{11} > 0 \\
\epsilon_{22} < 0 \\
\epsilon_{33} < 0 \\
\sqrt{2}\,\epsilon_{23}(\Rightarrow=0) \\
\sqrt{2}\,\epsilon_{13}(\Rightarrow=0) \\
\sqrt{2}\,\epsilon_{12}(\Rightarrow=0)
\end{bmatrix}.
\tag{5.4.20}
$$

The non trivial inverse relations write

$$
\begin{bmatrix}
\epsilon_{11} \\
\epsilon_{22} \\
\epsilon_{33}
\end{bmatrix}
= \frac{\sigma_{11}}{E_1^+}
\begin{bmatrix}
1 \\
-\nu_{12}^- \\
-\nu_{13}^-
\end{bmatrix},
\tag{5.4.21}
$$

where, according to (5.E.18)–(5.E.19),

$$
\begin{cases}
E_1^+ = \dfrac{H_1^+ H_2^- H_3^- - H_1^+ \lambda_{23}^2 - H_2^- \lambda_{13}^2 - H_3^- \lambda_{12}^2 + 2\,\lambda_{23}\,\lambda_{13}\,\lambda_{12}}{H_2^- H_3^- - \lambda_{23}^2} \\[2mm]
\nu_{12}^- = \dfrac{\lambda_{12}\,H_3^- - \lambda_{13}\,\lambda_{23}}{H_2^- H_3^- - \lambda_{23}^2}, \quad \nu_{13}^- = \dfrac{\lambda_{13}\,H_2^- - \lambda_{12}\,\lambda_{23}}{H_2^- H_3^- - \lambda_{23}^2}.
\end{cases}
\tag{5.4.22}
$$

Data from Chang et al. [1999] on aged human patellar cartilage, Table 5.4.4

Cartilage strips parallel to the strip-line directions have been subjected to tensile loading, and the strains along the split-line, perpendicular and depth directions have been recorded. The measurements display a significant volumetric contraction. The data are probably affected by the fibrillated nature of the aged cartilages. The tests are expected to have been performed at physiological salinities.

[5.8]An analysis of the same flavor that aims at estimating the elastic properties of a transversely isotropic fluid-saturated porous solid in presence of an incomplete data set is reported in Section 7.4.5.1.

Table 5.4.4: Tensile tests on aged human patellar cartilage using surface strips, Chang et al. [1999]. The split-line, perpendicular and depth directions are denoted by the indices 1, 2 and 3, respectively.

Surface to mid zone	$E_1^+ = 2800\,\text{kPa}$	$\nu_{12}^- = 0.9$	$\nu_{13}^- = 1.8$
Bottom	$E_1^+ = 2900\,\text{kPa}$	$\nu_{12}^- = 0.5$	$\nu_{13}^- = 0.7$

Data from Huang et al. [1999] on human glenohumeral cartilage, Table 5.4.5

The cartilage was harvested from shoulders of cadavers (average age 59 years). The cartilage layers were cut into thin slices parallel to the surface (0.25 mm thick for tension slices). Slices in the surface and in the middle zone were subjected to tensile loading parallel and perpendicular to the split-line direction.

Table 5.4.5: Tensile tests on the humerus surface zone, Huang et al. [1999]. The split-line and perpendicular directions are denoted by the indices 1 and 2, respectively.

$E_1^+ = 7840\,\text{kPa}$ (toe region)	$E_1^+ = 42800\,\text{kPa}$ (higher strain)	$\nu_{12}^- = 1.31$
$E_2^+ = 5930\,\text{kPa}$ (toe region)	$E_2^+ = 26300\,\text{kPa}$ (higher strain)	$\nu_{21}^- = 1.33$

Data from Elliott et al. [2002] on aged human patellas, Table 5.4.6

Along a similar testing protocol as in the above references, cartilage strips harvested from the surface and mid zones were loaded along the split-line direction. The effects of age were scrutinized and the surface moduli were found to decrease strongly with age.

Table 5.4.6: Tensile tests parallel to the split-line direction for subjects younger than 70 years old, Elliott et al. [2002].

Surface	$E_1^+ = 5810\,\text{kPa}$ (toe region)	$E_1^+ = 23920\,\text{kPa}$ (higher strain)	$\nu_{12}^- = 2.38$
Mid zone	$E_1^+ = 1920\,\text{kPa}$ (toe region)	$E_1^+ = 3180\,\text{kPa}$ (higher strain)	$\nu_{12}^- = 0.70$

Since the stress that is involved in the constitutive equations has been assumed to be the total stress, the material properties that have been deduced are *apparent*. The measured Poisson's ratios are found very large. An anisotropic conewise response might not be incompatible with these values. However, the positive definiteness of the stiffnesses and/or compliances has not been checked in any octant. Still, actual constitutive equations should be phrased in terms of an *effective* stress and of an *effective* strain. Accounting for these crucial elements allows to simulate the tissue response to mechanical loadings with an elastic stiffness endowed with moduli and Poisson's ratios that remain within the bounds of positive definiteness, e.g., Loret and Simões [2005]a.

Exercises on Chapter 5

Exercise 5.1 Decomposition of the nonlinear isotropic elastic relationship

The general isotropic stress-strain relationship has the form

$$\boldsymbol{\epsilon} = a_1 \mathbf{I} + a_2 \boldsymbol{\sigma} + a_3 \boldsymbol{\sigma}^2, \tag{5.E.1}$$

where the scalar coefficients a_i, $i \in [1,3]$, depend on the three scalar invariants $p = \operatorname{tr} \boldsymbol{\sigma}/3$, $q = ((3/2)\,\mathbf{s} : \mathbf{s})^{1/2}$ and $J_3 = \operatorname{tr} \boldsymbol{\sigma}^3$, where $\mathbf{s} = \boldsymbol{\sigma} - p\,\mathbf{I}$ is the stress deviator. Define an additive decomposition of the strain in a purely volumetric contribution and a purely deviatoric contribution.

Answer:

$$\boldsymbol{\epsilon} = \underbrace{b_1 \mathbf{I}}_{\substack{\text{volumetric} \\ \text{contribution}}} + \underbrace{b_2 \overbrace{(\boldsymbol{\sigma} - p\,\mathbf{I})}^{\mathbf{s}} + b_3 \left(\boldsymbol{\sigma}^2 - \operatorname{tr} \boldsymbol{\sigma}^2 \frac{\mathbf{I}}{3}\right)}_{\text{deviatoric contribution}}, \tag{5.E.2}$$

with $b_1 = a_1 + a_2\,p + a_3 \operatorname{tr} \boldsymbol{\sigma}^2/3$, $b_2 = a_2$, $b_3 = a_3$. Note the relation

$$\boldsymbol{\sigma}^2 - \operatorname{tr} \boldsymbol{\sigma}^2 \frac{\mathbf{I}}{3} = \mathbf{s}^2 - \operatorname{tr} \mathbf{s}^2 \frac{\mathbf{I}}{3} + 2\,p\,\mathbf{s}. \tag{5.E.3}$$

Exercise 5.2 Minor symmetry of the linear elastic tensor

The minor symmetry of the fourth order elasticity tensor with respect to its two last indices is a choice. Indeed, it stems from the fact that the scalar product of an arbitrary second tensor $\mathbf{A}$ by a symmetric second order tensor $\mathbf{B}$ is independent of the skew-symmetric part of the former, namely

$$\mathbf{A} : \mathbf{B} = \frac{1}{2}(\mathbf{A} + \mathbf{A}^{\mathrm{T}}) : \mathbf{B}. \tag{5.E.4}$$

In fact, it can be shown that the most general fourth order isotropic elastic tensor $\mathbb{E}'$ depends on three arbitrary coefficients λ, μ and μ', Malvern [1969], p. 277,

$$E_{ijkl} = \lambda\,I_{ij}\,I_{kl} + \mu\,(I_{ik}\,I_{jl} + I_{il}\,I_{jk}), \quad E'_{ijkl} = E_{ijkl} + \mu'\,(I_{ik}\,I_{jl} - I_{il}\,I_{jk}). \tag{5.E.5}$$

The infinitesimal strain $\boldsymbol{\epsilon}$ being symmetric, then $\mathbb{E} : \boldsymbol{\epsilon} = \mathbb{E}' : \boldsymbol{\epsilon}$, and the skew-symmetric part of $\mathbb{E}'$ is elusive in evaluating the stress.

Exercise 5.3 Conditions of hyperelasticity

The general isotropic stress-strain relationship has the form (5.E.1). Find the conditions that ensure this elastic material to be hyperelastic.

Answer:

The complementary energy then can be cast in the format,

$$d\mathcal{W}_c = \boldsymbol{\epsilon} : d\boldsymbol{\sigma} = 3\,(a_1 + a_2\,p)\,dp + \frac{2}{3}\,a_2\,dq + \frac{a_3}{3}\,dJ_3. \tag{5.E.6}$$

Necessary and sufficient conditions for the elastic material (5.E.1) to be hyperelastic are

$$\begin{aligned}
\frac{\partial}{\partial q}\Big(3\,(a_1 + a_2\,p)\Big) &= \frac{\partial}{\partial p}\Big(\frac{2}{3}\,a_2\Big); \\
\frac{\partial}{\partial J_3}\Big(3\,(a_1 + a_2\,p)\Big) &= \frac{\partial}{\partial p}\Big(\frac{1}{3}\,a_3\Big); \\
\frac{\partial}{\partial J_3}\Big(\frac{2}{3}\,a_2\Big) &= \frac{\partial}{\partial q}\Big(\frac{1}{3}\,a_3\Big).
\end{aligned} \tag{5.E.7}$$

Exercise 5.4 The Voigt representation

In the Voigt representation of the elasticity relations, the out-of diagonal terms are multiplied by 2 in the strain vector $[\epsilon]$, and by 1 in the stress vector $[\sigma]$. In fact, then neither $[\sigma]$ nor $[\epsilon]$ are vectors, and $[\mathbb{E}]$ is not a matrix. To stress this point, they are referred to as *pseudo*-vectors and *pseudo*-matrix. Indeed, under a change of cartesian axes, these entities do not change as if they were vectors and matrix in an euclidian vector space of dimension 6. To obtain the transformed representation, the components of the stress σ, strain ϵ and elastic moduli $\mathbb{E}$ have to be calculated, viewing these entities as second and fourth order tensors respectively. The resulting components can then be inserted in the pseudo-vectors and pseudo-matrix.

Obtain the matrix form of the elasticity relation $[\sigma] = [\mathbb{E}][\epsilon]$ and its inverse $[\epsilon] = [\mathbb{S}][\sigma]$ with the Voigt representation. Contrast these two matrix forms with their counterparts in the isometric representation. Show that both the Voigt and isometric representations conserve energy, namely $\sigma : \epsilon = [\sigma]^{\mathrm{T}}[\epsilon]$.

Answer:

$$
\begin{bmatrix} \sigma_{11} \\ \sigma_{22} \\ \sigma_{33} \\ \sigma_{23} \\ \sigma_{13} \\ \sigma_{12} \end{bmatrix} = \begin{bmatrix} E_{1111} & E_{1122} & E_{1133} & E_{1123} & E_{1113} & E_{1112} \\ E_{2211} & E_{2222} & E_{2233} & E_{2223} & E_{2213} & E_{2212} \\ E_{3311} & E_{3322} & E_{3333} & E_{3323} & E_{3313} & E_{3312} \\ E_{2311} & E_{2322} & E_{2333} & E_{2323} & E_{2313} & E_{2312} \\ E_{1311} & E_{1322} & E_{1333} & E_{1323} & E_{1313} & E_{1312} \\ E_{1211} & E_{1222} & E_{1233} & E_{1223} & E_{1213} & E_{1212} \end{bmatrix} \begin{bmatrix} \epsilon_{11} \\ \epsilon_{22} \\ \epsilon_{33} \\ 2\,\epsilon_{23} \\ 2\,\epsilon_{13} \\ 2\,\epsilon_{12} \end{bmatrix} .
\tag{5.E.8}
$$

$$
\begin{bmatrix} \epsilon_{11} \\ \epsilon_{22} \\ \epsilon_{33} \\ 2\,\epsilon_{23} \\ 2\,\epsilon_{13} \\ 2\,\epsilon_{12} \end{bmatrix} = \begin{bmatrix} S_{1111} & S_{1122} & S_{1133} & 2\,S_{1123} & 2\,S_{1113} & 2\,S_{1112} \\ S_{2211} & S_{2222} & S_{2233} & 2\,S_{2223} & 2\,S_{2213} & 2\,S_{2212} \\ S_{3311} & S_{3322} & S_{3333} & 2\,S_{3323} & 2\,S_{3313} & 2\,S_{3312} \\ 2\,S_{2311} & 2\,S_{2322} & 2\,S_{2333} & 4\,S_{2323} & 4\,S_{2313} & 4\,S_{2312} \\ 2\,S_{1311} & 2\,S_{1322} & 2\,S_{1333} & 4\,S_{1323} & 4\,S_{1313} & 4\,S_{1312} \\ 2\,S_{1211} & 2\,S_{1222} & 2\,S_{1233} & 4\,S_{1223} & 4\,S_{1213} & 4\,S_{1212} \end{bmatrix} \begin{bmatrix} \sigma_{11} \\ \sigma_{22} \\ \sigma_{33} \\ \sigma_{23} \\ \sigma_{13} \\ \sigma_{12} \end{bmatrix} .
\tag{5.E.9}
$$

Exercise 5.5 Symmetry with respect to the plane $x_3 = 0$

Consider an anisotropic linear elastic material. Assume its orthotropy directions to be known. The plane $x_3 = 0$ is a symmetry plane for the elastic properties. Analyze the consequences on some of the elastic coefficients.

Answer: The reflection matrix $\mathbf{R}_3$

$$
\mathbf{R}_3 = \begin{bmatrix} 1 & 0 & 0 \\ 0 & 1 & 0 \\ 0 & 0 & -1 \end{bmatrix} ,
\tag{5.E.10}
$$

belongs to the symmetry group of the material. Let $\sigma*$ be the transformed components of the stress σ in the mirror axes:

$$
\sigma* = \mathbf{R} \cdot \sigma \cdot \mathbf{R}^{\mathrm{T}} = \begin{bmatrix} 1 & 0 & 0 \\ 0 & 1 & 0 \\ 0 & 0 & -1 \end{bmatrix} \begin{bmatrix} \sigma_{11} & \sigma_{12} & \sigma_{13} \\ \sigma_{12} & \sigma_{22} & \sigma_{23} \\ \sigma_{13} & \sigma_{23} & \sigma_{33} \end{bmatrix} \begin{bmatrix} 1 & 0 & 0 \\ 0 & 1 & 0 \\ 0 & 0 & -1 \end{bmatrix} = \begin{bmatrix} \sigma_{11} & \sigma_{12} & -\sigma_{13} \\ \sigma_{12} & \sigma_{22} & -\sigma_{23} \\ -\sigma_{13} & -\sigma_{23} & \sigma_{33} \end{bmatrix} .
\tag{5.E.11}
$$

The stress-strain relations linking the pairs $([\epsilon], [\sigma])$ and $([\epsilon*], [\sigma*])$ should be identical. Consequently, the elastic relation writes also $[\boldsymbol{\sigma}*] = [\mathbb{E}][\boldsymbol{\epsilon}*]$, namely

$$
\begin{bmatrix}
\sigma_{11} \\[4pt]
\sigma_{22} \\[4pt]
\sigma_{33} \\[4pt]
-\sqrt{2}\,\sigma_{23} \\[4pt]
-\sqrt{2}\,\sigma_{13} \\[4pt]
\sqrt{2}\,\sigma_{12}
\end{bmatrix}
=
\begin{bmatrix}
E_{1111} & E_{1122} & E_{1133} & \sqrt{2}E_{1123} & \sqrt{2}E_{1113} & \sqrt{2}E_{1112} \\[4pt]
E_{2211} & E_{2222} & E_{2233} & \sqrt{2}E_{2223} & \sqrt{2}E_{2213} & \sqrt{2}E_{2212} \\[4pt]
E_{3311} & E_{3322} & E_{3333} & \sqrt{2}E_{3323} & \sqrt{2}E_{3313} & \sqrt{2}E_{3312} \\[4pt]
\sqrt{2}E_{2311} & \sqrt{2}E_{2322} & \sqrt{2}E_{2333} & 2E_{2323} & 2E_{2313} & 2E_{2312} \\[4pt]
\sqrt{2}E_{1311} & \sqrt{2}E_{1322} & \sqrt{2}E_{1333} & 2E_{1323} & 2E_{1313} & 2E_{1312} \\[4pt]
\sqrt{2}E_{1211} & \sqrt{2}E_{1222} & \sqrt{2}E_{1233} & 2E_{1223} & 2E_{1213} & 2E_{1212}
\end{bmatrix}
\begin{bmatrix}
\epsilon_{11} \\[4pt]
\epsilon_{22} \\[4pt]
\epsilon_{33} \\[4pt]
-\sqrt{2}\,\epsilon_{23} \\[4pt]
-\sqrt{2}\,\epsilon_{13} \\[4pt]
\sqrt{2}\,\epsilon_{12}
\end{bmatrix},
\qquad (5.\mathrm{E}.12)
$$

which can be rearranged to

$$
\begin{bmatrix}
\sigma_{11} \\[4pt]
\sigma_{22} \\[4pt]
\sigma_{33} \\[4pt]
\sqrt{2}\,\sigma_{23} \\[4pt]
\sqrt{2}\,\sigma_{13} \\[4pt]
\sqrt{2}\,\sigma_{12}
\end{bmatrix}
=
\begin{bmatrix}
E_{1111} & E_{1122} & E_{1133} & -\sqrt{2}E_{1123} & -\sqrt{2}E_{1113} & \sqrt{2}E_{1112} \\[4pt]
E_{2211} & E_{2222} & E_{2233} & -\sqrt{2}E_{2223} & -\sqrt{2}E_{2213} & \sqrt{2}E_{2212} \\[4pt]
E_{3311} & E_{3322} & E_{3333} & -\sqrt{2}E_{3323} & -\sqrt{2}E_{3313} & \sqrt{2}E_{3312} \\[4pt]
-\sqrt{2}E_{2311} & -\sqrt{2}E_{2322} & -\sqrt{2}E_{2333} & 2E_{2323} & 2E_{2313} & -2E_{2312} \\[4pt]
-\sqrt{2}E_{1311} & -\sqrt{2}E_{1322} & -\sqrt{2}E_{1333} & 2E_{1323} & 2E_{1313} & -2E_{1312} \\[4pt]
\sqrt{2}E_{1211} & \sqrt{2}E_{1222} & \sqrt{2}E_{1233} & -2E_{1223} & -2E_{1213} & 2E_{1212}
\end{bmatrix}
\begin{bmatrix}
\epsilon_{11} \\[4pt]
\epsilon_{22} \\[4pt]
\epsilon_{33} \\[4pt]
\sqrt{2}\,\epsilon_{23} \\[4pt]
\sqrt{2}\,\epsilon_{13} \\[4pt]
\sqrt{2}\,\epsilon_{12}
\end{bmatrix}.
\qquad (5.\mathrm{E}.13)
$$

Clearly, for the above elasticity matrix to be identical to the elasticity matrix $[\mathbb{E}]$, the coefficients in which the index 3 appears an odd number of times should vanish, as made explicit in (5.2.12).

Exercise 5.6 An invariance property

Show that the two sums of elastic coefficients E_{iikk} and E_{ikik} are invariants under changes of coordinate axes.

Proof:

Starting from the invariance relation (5.2.10) namely $E_{ijkl} = R_{im}\,R_{jn}\,R_{kp}\,R_{lq}\,E_{mnpq}$, and the property (5.2.9), namely $R_{im}\,R_{in} = I_{mn}$, the following relations results,

$$
\begin{aligned}
&1.\ E_{iikk} = R_{im}\,R_{in}\,R_{kp}\,R_{kq}\,E_{mnpq} = E_{mnpq}, \\[4pt]
&2.\ E_{ikik} = R_{im}\,R_{kn}\,R_{ip}\,R_{kq}\,E_{mnpq} = E_{mnmn}\,.
\end{aligned}
\qquad (5.\mathrm{E}.14)
$$

Exercise 5.7 Interpretation of the Young's moduli and Poisson's ratios

Hint: Consider an orthotropic material (5.2.46), and assume its orthotropy axes to be known. Apply a uniaxial traction σ_{11}. Then $E_1 = \sigma_{11}/\epsilon_{11}$, while $\nu_{1i} = -\epsilon_{ii}/\epsilon_{11}$ for $i = 2, 3$.

Exercise 5.8 Linear orthotropic elasticity using representation theorems

Using the theorem of representation for an orthotropic scalar function of the strain $\boldsymbol{\epsilon}$, Section 2.11.3, obtain the expressions of the elastic potential shown in Section 5.2.6.

Proof:

Let $\{\mathbf{e}_1, \mathbf{e}_2, \mathbf{e}_3\}$ be an orthonormal triad defining the directions of orthotropy. Then the elastic potential is an orthotropic function of the strain $\boldsymbol{\epsilon}$, or equivalently an isotropic function of the augmented set $\{\boldsymbol{\epsilon}, \mathbf{A}_1 \equiv \mathbf{e}_1 \otimes \mathbf{e}_1, \mathbf{A}_2 \equiv \mathbf{e}_2 \otimes \mathbf{e}_2\,\}$, Section 2.11.3. Retaining only invariants of first and second degrees, and building a homogeneous quadratic expression, the elastic potential for linear orthotropic elasticity is seen to involve, as expected, nine constant coefficients a_i, $i \in [1,9]$,

$$
\begin{cases}
\mathcal{W} = \dfrac{a_1}{2}(\operatorname{tr}\mathbf{A}_1 \cdot \boldsymbol{\epsilon})^2 + \dfrac{a_2}{2}(\operatorname{tr}\mathbf{A}_2 \cdot \boldsymbol{\epsilon})^2 + \dfrac{a_3}{2}(\operatorname{tr}\mathbf{A}_3 \cdot \boldsymbol{\epsilon})^2 + \\
\qquad + \; a_4 \operatorname{tr}\mathbf{A}_1 \cdot \boldsymbol{\epsilon} \operatorname{tr}\mathbf{A}_2 \cdot \boldsymbol{\epsilon} + a_5 \operatorname{tr}\mathbf{A}_1 \cdot \boldsymbol{\epsilon} \operatorname{tr}\mathbf{A}_3 \cdot \boldsymbol{\epsilon} + a_6 \operatorname{tr}\mathbf{A}_2 \cdot \boldsymbol{\epsilon} \operatorname{tr}\mathbf{A}_3 \cdot \boldsymbol{\epsilon} + \\
\qquad + \; a_7 \operatorname{tr}\mathbf{A}_1 \cdot \boldsymbol{\epsilon}^2 + a_8 \operatorname{tr}\mathbf{A}_2 \cdot \boldsymbol{\epsilon}^2 + a_9 \operatorname{tr}\mathbf{A}_3 \cdot \boldsymbol{\epsilon}^2 \, ; \\[4pt]
\boldsymbol{\sigma} = (\operatorname{tr}\mathbf{A}_1 \cdot \boldsymbol{\epsilon})\,(a_1\,\mathbf{A}_1 + a_4\,\mathbf{A}_2 + a_5\,\mathbf{A}_3) + \\
\qquad + \; (\operatorname{tr}\mathbf{A}_2 \cdot \boldsymbol{\epsilon})\,(a_4\,\mathbf{A}_1 + a_2\,\mathbf{A}_2 + a_6\,\mathbf{A}_3) + \\
\qquad + \; (\operatorname{tr}\mathbf{A}_3 \cdot \boldsymbol{\epsilon})\,(a_5\,\mathbf{A}_1 + a_6\,\mathbf{A}_2 + a_3\,\mathbf{A}_3) + \\
\qquad + \; a_7\,(\mathbf{A}_1 \cdot \boldsymbol{\epsilon} + \boldsymbol{\epsilon} \cdot \mathbf{A}_1) + a_8\,(\mathbf{A}_2 \cdot \boldsymbol{\epsilon} + \boldsymbol{\epsilon} \cdot \mathbf{A}_2) + a_9\,(\mathbf{A}_3 \cdot \boldsymbol{\epsilon} + \boldsymbol{\epsilon} \cdot \mathbf{A}_3) \, ; \\[4pt]
\mathbb{E} = \mathbf{A}_1 \otimes (a_1\,\mathbf{A}_1 + a_4\,\mathbf{A}_2 + a_5\,\mathbf{A}_3) + \\
\qquad + \; \mathbf{A}_2 \otimes (a_4\,\mathbf{A}_1 + a_2\,\mathbf{A}_2 + a_6\,\mathbf{A}_3) + \\
\qquad + \; \mathbf{A}_3 \otimes (a_5\,\mathbf{A}_1 + a_6\,\mathbf{A}_2 + a_3\,\mathbf{A}_3) + \\
\qquad + \; a_7\,(\mathbf{A}_1 \,\overline{\otimes}\, \mathbf{I} + \mathbf{I} \,\overline{\otimes}\, \mathbf{A}_1) + a_8\,(\mathbf{A}_2 \,\overline{\otimes}\, \mathbf{I} + \mathbf{I} \,\overline{\otimes}\, \mathbf{A}_2) + a_9\,(\mathbf{A}_3 \,\overline{\otimes}\, \mathbf{I} + \mathbf{I} \,\overline{\otimes}\, \mathbf{A}_3) \, .
\end{cases}
\tag{5.E.15}
$$

The compact expression of the elastic potential (5.2.37) corresponds to the following notation:

$$
\mu_a = a_{6+a}, \; a \in [1,3], \quad \lambda_{aa} = a_a, \; a \in [1,3], \quad \lambda_{12} = a_4, \; \lambda_{13} = a_5, \; \lambda_{23} = a_6 \,.
\tag{5.E.16}
$$

Exercise 5.9 Inversion of the linear orthotropic elastic relations

The linear elastic relations for an orthotropic solid $\boldsymbol{\sigma} = \mathbb{E} : \boldsymbol{\epsilon}$, Eqn (5.2.40), can be inverted to the relations $\boldsymbol{\epsilon} = \mathbb{E}^{-1} : \boldsymbol{\sigma}$, Eqn (5.2.46), at the price of some algebras, as indicated below.

1. Calculate the Young's moduli and Poisson's ratios in terms of the generalized Lamé moduli

Answer: Let us introduce a simplified notation for the components of the compliance tensor,

$$
\begin{aligned}
\mathbb{E}^{-1} &= \Big[\sum_{a \leq b} \frac{\lambda_{ab}}{2}\,(\mathbf{A}_a \otimes \mathbf{A}_b + \mathbf{A}_b \otimes \mathbf{A}_a) + \sum_a \mu_a\,(\mathbf{A}_a \,\overline{\otimes}\, \mathbf{I} + \mathbf{I} \,\overline{\otimes}\, \mathbf{A}_a)\Big]^{-1} \\
&= \sum_{a \leq b} \frac{x_{ab}}{2}\,(\mathbf{A}_a \otimes \mathbf{A}_b + \mathbf{A}_b \otimes \mathbf{A}_a) + \sum_a y_a\,(\mathbf{A}_a \,\overline{\otimes}\, \mathbf{I} + \mathbf{I} \,\overline{\otimes}\, \mathbf{A}_a)
\end{aligned}
\tag{5.E.17}
$$

with $x_{ab} = x_{ba} = -\nu_{\underline{ab}}/E_{\underline{a}}$, $y_a = (1 + \nu_{\underline{aa}})/2\,E_{\underline{a}}$, $a, b \in [1,3]$, through (5.2.43)$_3$. Inversion of the upper 3×3 principal minor $\mathbb{E}^{\mathrm{U}}_{3 \times 3}$ of the tensor moduli (5.2.40) with determinant $\Delta_{\mathrm{U}} = \det \mathbb{E}^{\mathrm{U}}_{3 \times 3}$ results in,

$$
\begin{cases}
\dfrac{1}{E_1} = \dfrac{1}{\Delta_{\mathrm{U}}}\left((\lambda_{22} + 2\,\mu_2)(\lambda_{33} + 2\,\mu_3) - \lambda_{23}^2\right) \\[6pt]
\dfrac{1}{E_2} = \dfrac{1}{\Delta_{\mathrm{U}}}\left((\lambda_{11} + 2\,\mu_1)(\lambda_{33} + 2\,\mu_3) - \lambda_{13}^2\right) \\[6pt]
\dfrac{1}{E_3} = \dfrac{1}{\Delta_{\mathrm{U}}}\left((\lambda_{11} + 2\,\mu_1)(\lambda_{22} + 2\,\mu_2) - \lambda_{12}^2\right)
\end{cases}
\tag{5.E.18}
$$

and

$$
\begin{cases}
-\dfrac{\nu_{12}}{E_1} = -\dfrac{\nu_{21}}{E_2} = \dfrac{1}{\Delta_{\mathrm{U}}}\left(\lambda_{13}\,\lambda_{23} - \lambda_{12}\,(\lambda_{33} + 2\,\mu_3)\right) \\[6pt]
-\dfrac{\nu_{13}}{E_1} = -\dfrac{\nu_{31}}{E_3} = \dfrac{1}{\Delta_{\mathrm{U}}}\left(\lambda_{12}\,\lambda_{23} - \lambda_{13}\,(\lambda_{22} + 2\,\mu_2)\right) \\[6pt]
-\dfrac{\nu_{23}}{E_2} = -\dfrac{\nu_{32}}{E_3} = \dfrac{1}{\Delta_{\mathrm{U}}}\left(\lambda_{12}\,\lambda_{13} - \lambda_{23}\,(\lambda_{11} + 2\,\mu_1)\right)
\end{cases}
\tag{5.E.19}
$$

Inversion of the lower 3×3 principal minor of the tensor moduli (5.2.40) yields

$$y_1 + y_2 = \frac{1}{\mu_1 + \mu_2} \qquad \Rightarrow y_1 + y_2 = \frac{1}{\mu_1 + \mu_2}$$

$$\left.\begin{array}{l} y_2 + y_3 = \dfrac{1}{\mu_2 + \mu_3} \\[2mm] y_3 + y_1 = \dfrac{1}{\mu_3 + \mu_1} \end{array}\right\} \quad \Rightarrow y_1 - y_2 = \frac{1}{\mu_3 + \mu_1} - \frac{1}{\mu_2 + \mu_3} \tag{5.E.20}$$

and therefore

$$2\,y_a = \frac{1 + \nu_{aa}}{E_a} = \frac{1}{\mu_a + \mu_b} + \frac{1}{\mu_a + \mu_c} - \frac{1}{\mu_b + \mu_c}, \quad a \neq b \neq c. \tag{5.E.21}$$

Therefore, the nine moduli λ_{ab}, μ_a, $a \leq b \in [1,3]$, being known, the generalized Young's moduli and Poisson's ratios are obtained in the following sequence:
  - calculate first the three E_a's from (5.E.18);
  - calculate the three ν_{ab}'s, $a < b$, from (5.E.19);
  - calculate the three ν_{aa}'s from (5.E.21).

2. Calculate the generalized Lamé moduli in terms of the Young's moduli and Poisson's ratios

Answer: Inversion of the lower principal minor and upper principal minor yields,

$$\mu_a = \mu_{ab} + \mu_{ac} - \mu_{bc}, \quad a \neq b \neq c \in [1,3], \tag{5.E.22}$$

$$\begin{cases} \lambda_{11} + 2\,\mu_1 = \Delta_U \left(\dfrac{1}{E_2} \dfrac{1}{E_3} - \left(\dfrac{\nu_{23}}{E_2}\right)^2 \right) \\[3mm] \lambda_{22} + 2\,\mu_2 = \Delta_U \left(\dfrac{1}{E_1} \dfrac{1}{E_3} - \left(\dfrac{\nu_{13}}{E_1}\right)^2 \right) \\[3mm] \lambda_{33} + 2\,\mu_3 = \Delta_U \left(\dfrac{1}{E_1} \dfrac{1}{E_2} - \left(\dfrac{\nu_{12}}{E_1}\right)^2 \right) \end{cases} \tag{5.E.23}$$

and

$$\begin{cases} \lambda_{12} = \lambda_{21} = \Delta_U \left(\dfrac{\nu_{13}}{E_1} \dfrac{\nu_{23}}{E_2} + \dfrac{\nu_{12}}{E_1} \dfrac{1}{E_3} \right) \\[3mm] \lambda_{13} = \lambda_{31} = \Delta_U \left(\dfrac{\nu_{12}}{E_1} \dfrac{\nu_{23}}{E_2} + \dfrac{\nu_{13}}{E_1} \dfrac{1}{E_2} \right) \\[3mm] \lambda_{23} = \lambda_{32} = \Delta_U \left(\dfrac{\nu_{12}}{E_1} \dfrac{\nu_{13}}{E_1} + \dfrac{\nu_{23}}{E_2} \dfrac{1}{E_1} \right) \end{cases} \tag{5.E.24}$$

Here Δ_U is the determinant of upper principal minor of the compliance matrix (5.2.46).

Exercise 5.10 Positive definiteness of the elastic *orthotropic* stiffness and compliance matrices

Convexity of the elastic potential of an orthotropic linear elastic material is ensured by the positive definiteness of the stiffness $\mathbb{E}$ (or compliance tensor $\mathbb{E}^{-1}$), that is by the fact that the six eigenvalues of $\mathbb{E}$ are strictly positive.

The elastic stiffness is defined by (5.2.41). The following six Routh-Hurwitz conditions ensure the positive definiteness of the three eigenvalues of the upper principal minor and of the three eigenvalues of the lower diagonal minor,

$$\Delta_1 > 0, \ \Delta_1 \Delta_2 - \Delta_3 > 0, \ \Delta_3 > 0, \ c_{44} > 0, \ c_{55} > 0, \ c_{66} > 0, \tag{5.E.25}$$

where

$$\begin{aligned} \Delta_1 &= c_{11} + c_{22} + c_{33}, \\ \Delta_2 &= c_{11}\,c_{22} + c_{22}\,c_{33} + c_{33}\,c_{11} - c_{23}^2 - c_{13}^2 - c_{12}^2, \\ \Delta_3 &= c_{11}\,c_{22}\,c_{33} - c_{11}\,c_{23}^2 - c_{22}\,c_{13}^2 - c_{33}\,c_{12}^2 + 2\,c_{23}\,c_{13}\,c_{12}. \end{aligned} \tag{5.E.26}$$

Apply the Routh-Hurwitz criterion, Gantmacher [1959], vol. II, p. 194, knowing that the eigenvalues are real since the matrix is symmetric.

Exercise 5.11 Linear transversely isotropic elasticity using representation theorems

Using the theorem of representation for a transversely isotropic scalar function of the strain ϵ, Section 2.11.3, obtain the expression of the elastic potential shown in Section 5.2.6.

Proof: Let $\mathbf{m}$ be the axis of material symmetry. The elastic potential is assumed to be a transversely isotropic function of ϵ, or equivalently an isotropic function of $\{\epsilon, \mathbf{M} = \mathbf{m} \otimes \mathbf{m}\}$. As expected, the representation theorems show that the elastic potential of linear elasticity involves 5 constant coefficients a_i, $i \in [1, 5]$,

$$
\begin{cases}
\mathcal{W} &= \dfrac{a_1}{2}(\operatorname{tr}\boldsymbol{\epsilon})^2 + \dfrac{a_2}{2}\operatorname{tr}\boldsymbol{\epsilon}^2 + a_3\,(\operatorname{tr}\boldsymbol{\epsilon})\,(\operatorname{tr}\mathbf{M}\cdot\boldsymbol{\epsilon}) + \dfrac{a_4}{2}(\operatorname{tr}\mathbf{M}\cdot\boldsymbol{\epsilon})^2 + a_5\operatorname{tr}\mathbf{M}\cdot\boldsymbol{\epsilon}^2\,; \\[2mm]
\boldsymbol{\sigma} &= a_1\,(\operatorname{tr}\boldsymbol{\epsilon})\,\mathbf{I} + a_2\,\boldsymbol{\epsilon} + a_3\left((\operatorname{tr}\mathbf{M}\cdot\boldsymbol{\epsilon})\,\mathbf{I} + (\operatorname{tr}\boldsymbol{\epsilon})\,\mathbf{M}\right) + \\[2mm]
&\quad + a_4\,(\operatorname{tr}\mathbf{M}\cdot\boldsymbol{\epsilon})\,\mathbf{M} + a_5\,(\mathbf{M}\cdot\boldsymbol{\epsilon} + \boldsymbol{\epsilon}\cdot\mathbf{M})\,; \\[2mm]
\mathbb{E} &= a_1\,\mathbf{I}\otimes\mathbf{I} + a_2\,\mathbf{I}\,\overline{\underline{\otimes}}\,\mathbf{I} + a_3\left(\mathbf{M}\otimes\mathbf{I} + \mathbf{I}\otimes\mathbf{M}\right) + \\[2mm]
&\quad + a_4\,\mathbf{M}\otimes\mathbf{M} + a_5\left(\mathbf{M}\,\overline{\underline{\otimes}}\,\mathbf{I} + \mathbf{I}\,\overline{\underline{\otimes}}\,\mathbf{M}\right).
\end{cases}
\tag{5.E.27}
$$

The coefficients c's of the stiffness matrix (5.2.51) are identified in terms of the a's defined by (5.E.27),

$$
\begin{cases}
c_{11} = \lambda_{\mathrm{T}} + 2\,\mu_{\mathrm{T}} = a_1 + a_2, \\[2mm]
c_{12} = \lambda_{\mathrm{T}} = a_1, \\[2mm]
c_{13} = \lambda_{\mathrm{LT}} = a_1 + a_3, \\[2mm]
c_{33} = \lambda_{\mathrm{L}} + 2\,\mu_{\mathrm{L}} = a_1 + a_2 + 2a_3 + a_4 + 2a_5, \\[2mm]
c_{44} = \mu_{\mathrm{L}} = \tfrac{1}{2}\left(a_2 + a_5\right).
\end{cases}
\tag{5.E.28}
$$

The a's identify in terms of the generalized Lamé moduli, Eqn (5.2.50), and conversely,

$$
\begin{cases}
\mu_{\mathrm{T}} = a_2/2, \\[2mm]
\mu_{\mathrm{L}} = (a_2 + a_5)/2, \\[2mm]
\lambda_{\mathrm{T}} = a_1, \\[2mm]
\lambda_{\mathrm{L}} = a_1 + 2a_3 + a_4 + a_5, \\[2mm]
\lambda_{\mathrm{LT}} = a_1 + a_3,
\end{cases}
\quad\Leftrightarrow\quad
\begin{cases}
a_1 = \lambda_{\mathrm{T}}, \\[2mm]
a_2 = 2\,\mu_{\mathrm{T}}, \\[2mm]
a_3 = \lambda_{\mathrm{LT}} - \lambda_{\mathrm{T}}, \\[2mm]
a_4 = \lambda_{\mathrm{T}} + \lambda_{\mathrm{L}} - 2\lambda_{\mathrm{LT}} + 2\,\mu_{\mathrm{T}} - 2\,\mu_{\mathrm{L}}, \\[2mm]
a_5 = 2\,\mu_{\mathrm{L}} - 2\,\mu_{\mathrm{T}}.
\end{cases}
\tag{5.E.29}
$$

Exercise 5.12 Inversion of the linear transversely elastic relations

Calculate the inverse of the stiffness matrix (5.2.51), and link its coefficients with those of the compliance matrix (5.2.52), and conversely.

Answer:

Direct inversion of the elastic stiffness and of the elastic compliance leads to the equivalences:

$$
\begin{cases}
c_{11} = (2\mu_{\mathrm{T}})^2\,\dfrac{E_{\mathrm{L}} - \nu_{\mathrm{L}}^2 E_{\mathrm{T}}}{4\mu_{\mathrm{T}}(E_{\mathrm{L}} - \nu_{\mathrm{L}}^2 E_{\mathrm{T}}) - E_{\mathrm{L}} E_{\mathrm{T}}} \\[4mm]
c_{12} = -2\mu_{\mathrm{T}}\,\dfrac{2\mu_{\mathrm{T}}(E_{\mathrm{L}} - \nu_{\mathrm{L}}^2 E_{\mathrm{T}}) - E_{\mathrm{L}} E_{\mathrm{T}}}{4\mu_{\mathrm{T}}(E_{\mathrm{L}} - \nu_{\mathrm{L}}^2 E_{\mathrm{T}}) - E_{\mathrm{L}} E_{\mathrm{T}}} \\[4mm]
c_{13} = 2\mu_{\mathrm{T}}\,\dfrac{\nu_{\mathrm{L}} E_{\mathrm{L}} E_{\mathrm{T}}}{4\mu_{\mathrm{T}}(E_{\mathrm{L}} - \nu_{\mathrm{L}}^2 E_{\mathrm{T}}) - E_{\mathrm{L}} E_{\mathrm{T}}} \\[4mm]
c_{33} = E_{\mathrm{L}}^2\,\dfrac{4\mu_{\mathrm{T}} - E_{\mathrm{T}}}{4\mu_{\mathrm{T}}(E_{\mathrm{L}} - \nu_{\mathrm{L}}^2 E_{\mathrm{T}}) - E_{\mathrm{L}} E_{\mathrm{T}}} \\[4mm]
c_{44} = \mu_{\mathrm{L}}
\end{cases}
\quad\Leftrightarrow\quad
\begin{cases}
E_{\mathrm{L}} = \dfrac{(c_{11} + c_{12})\,c_{33} - 2\,c_{13}^2}{c_{11} + c_{12}} \\[4mm]
E_{\mathrm{T}} = \dfrac{c_{11} - c_{12}}{c_{11}c_{33} - c_{13}^2}\left((c_{11} + c_{12})\,c_{33} - 2\,c_{13}^2\right) \\[4mm]
\mu_{\mathrm{L}} = c_{44} \\[4mm]
\mu_{\mathrm{T}} = \dfrac{1}{2}\left(c_{11} - c_{12}\right) \\[4mm]
\nu_{\mathrm{L}} = \dfrac{c_{13}}{c_{11} + c_{12}}
\end{cases}
\tag{5.E.30}
$$

Exercise 5.13 Positive definiteness of the linear elastic *transverse isotropic* matrix

The transversely isotropic elastic matrix has the form (5.2.51). Using Exercise 5.10, or through a direct analysis, show that the conditions that ensure the linear elastic transverse isotropic stiffness, or compliance, matrix to be positive definite, are, Chadwick [1989],

$$
c_{11} > |c_{12}|, \quad (c_{11} + c_{12})\,c_{33} - 2\,c_{13}^2 > 0, \quad c_{44} > 0.
\tag{5.E.31}
$$

Still another proof is provided by Exercise 5.20.

Exercise 5.14 Positive definiteness of the linear elastic *cubic* stiffness and compliance matrices

Prove directly that the inequalities (5.2.66) are necessary and sufficient conditions to ensure the linear elastic cubic compliance matrix (5.2.64) to be positive definite.

Answer: A necessary and sufficient condition for a (symmetric) matrix to be positive definite (PD) is that all its principal sub-matrices be PD as well, that is, all their eigenvalues should be positive. This condition implies in particular that the diagonal terms should be positive. Thus, from (5.2.41), $c_{44} = c_{55} = c_{66} > 0$. Moreover

$$c_{11} > 0 \, ;$$

$$\begin{bmatrix} c_{11} & c_{12} \\ c_{12} & c_{11} \end{bmatrix} \text{(PD)} \quad \Rightarrow \quad c_{11} > 0, \; c_{11} - \frac{c_{12}^2}{c_{11}} > 0 \, ;$$

$$\begin{bmatrix} c_{11} & c_{12} & c_{12} \\ c_{12} & c_{11} & c_{12} \\ c_{12} & c_{12} & c_{11} \end{bmatrix} \text{(PD)} \quad \Rightarrow \quad c_{11} - c_{12} > 0 \text{ double eigenvalue}, \; c_{11} + 2 c_{12} > 0 \, .$$

$$(5.E.32)$$

Exercise 5.15 Symmetry property sufficient for coaxiality

Two second order tensors are said to be *coaxial* if they have the same eigenspace. If the eigenvalues are distinct, the principal directions, or eigenvectors, are uniquely defined, and they coincide, say $\mathbf{e}_i^\epsilon = \mathbf{e}_i^\sigma$, $i \in [1,3]$, in (5.2.5).

Find the material symmetries that ensure the stress and strain to be coaxial in linear elasticity.

Exercise 5.16 A property of cubic symmetry

Show that for cubic symmetry and higher, the stress and strain are isotropic simultaneously.

Proof: Use (5.2.62) or (5.2.63).

Exercise 5.17 Structural and fabric orthotropies

Show that the elastic anisotropy defined in Section 5.2.11 by the fabric tensor $\mathbf{B}$ is a particular case of orthotropic elasticity, when the eigenvalues of $\mathbf{B}$ are distinct, Bigoni and Loret [1999].

Answer:

The nine coefficients defining the orthotropic response are easily obtained by comparing the matrix forms (5.2.40) and (5.2.82) of the elastic relations expressed in the orthotropy axes $\mathbf{e} = \mathbf{b}$,

$$\mu_i/\mu = b_i \, (b_j + b_k) - b_j \, b_k \, , \quad i \neq j \neq k \in [1,3] \, ;$$

$$\lambda_{\underline{ii}} = (\lambda + 2\,\mu) \, b_i^2 - 2\,\mu_i \, , \quad i \in [1,3]; \quad \lambda_{ij} = \lambda \, b_i \, b_j \, , \quad i \neq j \in [1,3] \, .$$

$$(5.E.33)$$

Exercise 5.18 Structural and fabric transverse isotropies

Show that the elastic anisotropy defined in Section 5.2.11 by the fabric tensor $\mathbf{B}$ is a particular case of transverse isotropic elasticity, when the two eigenvalues b_1 and b_2 of $\mathbf{B}$ are equal, Bigoni and Loret [1999].

Answer:

With help of the relation $\mathbf{B}_1 + \mathbf{B}_2 + \mathbf{B}_3 = \mathbf{I}$, the fabric tensor $\mathbf{B} = b_1 \, \mathbf{B}_1 + b_2 \, \mathbf{B}_2 + b_3 \, \mathbf{B}_3$ can then be cast in the form $\mathbf{B} = b_T \, \mathbf{I} + (b_L - b_T) \, \mathbf{M}$, where the index T denotes a transverse direction and the index L the longitudinal direction, and with $\mathbf{M} = \mathbf{B}_3$.

The strain energy (5.2.81) can then be recast in the transversely isotropic format (5.E.27), and with the following coefficients a_i, $i \in [1,5]$:

$$a_1 = \lambda \, b_T^2 \, , \quad a_2 = 2\,\mu \, b_T^2 \, , \quad a_3 = \lambda \, b_T \, (b_L - b_T) \, , \quad a_4 = (\lambda + 2\mu)(b_L - b_T)^2 \, , \quad a_5 = 2\mu \, b_T (b_L - b_T) \, .$$

$$(5.E.34)$$

Exercise 5.19 A versatile form of anisotropic nonlinear hyperelasticity, Loret and Simões [2005]a

The elastic anisotropic model exposed in Section 5.2.11 is defined by a fabric tensor $\mathbf{B}$. It is easily tractable. Use this model as a backbone to build a nonlinear hyperelastic model, such that the secant anisotropic elastic operator retains the linear format, up to a multiplicative scalar.

Answer:

A way to reach this property is to assume a constant Poisson's ratio ν, but a stress-dependent Young's modulus E. Thus the nonlinearity affects the axial stress-axial strain curve, and the strains in the lateral directions plot essentially linearly as functions of the axial strain. This approach extends to anisotropic materials the earlier analysis of Loret [1985] devoted to isotropic elasticity at infinitesimal strains, Section 5.1.

Algebraic manipulations simplify if the bulk modulus $K = E/3 (1 - 2\nu)$ and the shear modulus $\mu = E/2(1 + \nu)$ are used. So $K/\mu = K_r/\mu_r$, the subscript 'r' denoting the reference state with zero stress. The nonlinear compliance tensor $\mathbb{E}^{-1}$ is built so that

$$\mathbb{E}^{-1} = \frac{E_r}{E} \, \mathbb{E}_r^{-1} \, . \tag{5.E.35}$$

Let $\mathcal{W}_r$ be the elastic energy associated with the moduli in the reference state:

$$\mathcal{W}_r(\boldsymbol{\sigma}) = \frac{1}{18 \, K_r} \, (\mathrm{tr} \, \mathbf{X})^2 + \frac{1}{4 \, \mu_r} \, \mathrm{tr} \, (\mathrm{dev} \, \mathbf{X} \cdot \mathrm{dev} \, \mathbf{X}) \, , \tag{5.E.36}$$

where $\mathbf{X} \equiv \mathbf{B}^{-1/2} \cdot \boldsymbol{\sigma} \cdot \mathbf{B}^{-1/2}$. The nonlinear potential $\mathcal{W}_{\mathrm{nl}}(\boldsymbol{\sigma})$ is assumed to be a function of the linear potential $\mathcal{W}_r(\boldsymbol{\sigma})$,

$$\mathcal{W}_{\mathrm{nl}} = \mathcal{W}_{\mathrm{nl}}(\mathcal{W}_r) \, . \tag{5.E.37}$$

The secant operator deriving from this potential has indeed the requested form, and,

$$\frac{\partial \mathcal{W}_{\mathrm{nl}}}{\partial \boldsymbol{\sigma}} = \mathbb{E}^{-1} : \boldsymbol{\sigma} = -\frac{\nu}{E} \, (\mathbf{B}^{-1} : \boldsymbol{\sigma}) \, \mathbf{B}^{-1} + \frac{1+\nu}{E} \, \mathbf{B}^{-1} \cdot \boldsymbol{\sigma} \cdot \mathbf{B}^{-1} \, . \tag{5.E.38}$$

The relations $\partial \, \mathrm{tr} \, \mathbf{X}/\partial \boldsymbol{\sigma} = \mathbf{B}^{-1}$ and $\partial(\mathrm{dev} \, \mathbf{X})_{ij}/\partial \sigma_{kl} = \frac{1}{2}((\mathbf{B}^{-\frac{1}{2}})_{ik} (\mathbf{B}^{-\frac{1}{2}})_{jl} + (\mathbf{B}^{-\frac{1}{2}})_{il} (\mathbf{B}^{-\frac{1}{2}})_{jk}) - \frac{1}{3} I_{ij} (\mathbf{B}^{-1})_{kl}$ have been used to derive (5.E.38). The Young's modulus is such that E_r/E is equal to the derivative of $\mathcal{W}_{\mathrm{nl}}$ with respect to $\mathcal{W}_r$. If we set

$$E = E_r \, \phi(\mathcal{W}_r) \, , \tag{5.E.39}$$

then $\mathcal{W}_{\mathrm{nl}}$ is equal to $\int_0^{\mathcal{W}_r} dy/\phi(y)$.

To close the nonlinear model, a specific function $\phi(\mathcal{W}_r)$ has to be postulated based on experimental data. The simplest assumption is to consider a power form, e.g.,

$$\phi(\mathcal{W}_r) = 1 + \frac{E_{rr}}{E_r} \left(\frac{\mathcal{W}_r}{p_a} \right)^{\frac{n}{2}} \, , \tag{5.E.40}$$

where p_a is an arbitrary reference pressure, and $E_{rr} \geq 0$ [unit : Pa] and $n \geq 0$ are parameters to be calibrated from experimental data. As it is, this formula indicates that the moduli K, μ and E take their minimum values, K_r, μ_r and E_r respectively, at zero stress. This point needs to be assessed, based on measurements.

Exercise 5.20 Spectral decomposition of the transversely isotropic elastic tensor, Walpole [1984]

Walpole [1984] has presented a decomposition of the transversely isotropic elastic tensor that can be easily manipulated. Let $\mathbf{e}_3$ be the material symmetry axis. He defines two projection operators, one on this axis and another one on the basal plane,

$$\mathbf{M} = \mathbf{e}_3 \otimes \mathbf{e}_3, \quad \mathbf{N} = \mathbf{I} - \mathbf{e}_3 \otimes \mathbf{e}_3 \, , \tag{5.E.41}$$

which are further used to build the generators,

$$\begin{bmatrix} \mathbf{E}_1 = \mathbf{M} \otimes \mathbf{M} & \mathbf{E}_3 = \frac{1}{\sqrt{2}} \, \mathbf{M} \otimes \mathbf{N} \\[2mm] \mathbf{E}_4 = \frac{1}{\sqrt{2}} \, \mathbf{N} \otimes \mathbf{M} & \mathbf{E}_2 = \frac{1}{2} \, \mathbf{N} \otimes \mathbf{N} \end{bmatrix}, \quad \mathbf{F} = \mathbf{N} \, \overline{\underline{\otimes}} \, \mathbf{N} - \frac{1}{2} \, \mathbf{N} \otimes \mathbf{N}, \quad \mathbf{G} = \mathbf{M} \, \overline{\underline{\otimes}} \, \mathbf{N} + \mathbf{N} \, \overline{\underline{\otimes}} \, \mathbf{M} \, . \tag{5.E.42}$$

1. Prove the basic properties

$$\mathbf{M} \cdot \mathbf{M} = \mathbf{M}, \quad \mathbf{N} \cdot \mathbf{N} = \mathbf{N}, \quad \mathbf{M} \cdot \mathbf{N} = \mathbf{N} \cdot \mathbf{M} = \mathbf{0}, \quad \mathbf{M} : \mathbf{M} = 1, \quad \mathbf{N} : \mathbf{N} = 1, \quad \mathbf{M} : \mathbf{N} = 0 \, . \tag{5.E.43}$$

2. Show that $\mathbf{E}_1$, $\mathbf{E}_2$, $\mathbf{E}_3$ and $\mathbf{E}_4$ are linearly independent and, with help of (5.2.34), that they constitute a partition of the unit tensor of order four $\mathbf{I}_4$,

$$\mathbf{E}_1 + \mathbf{E}_2 + \mathbf{E}_3 + \mathbf{E}_4 = \mathbf{I}_4 \, . \tag{5.E.44}$$

3. Prove the multiplication table

	$\mathbf{E}_1$	$\mathbf{E}_2$	$\mathbf{E}_3$	$\mathbf{E}_4$	$\mathbf{F}$	$\mathbf{G}$
$\mathbf{E}_1$	$\mathbf{E}_1$	$\mathbf{0}$	$\mathbf{E}_3$	$\mathbf{0}$	$\mathbf{0}$	$\mathbf{0}$
$\mathbf{E}_2$	$\mathbf{0}$	$\mathbf{E}_2$	$\mathbf{0}$	$\mathbf{E}_4$	$\mathbf{0}$	$\mathbf{0}$
$\mathbf{E}_3$	$\mathbf{0}$	$\mathbf{E}_3$	$\mathbf{0}$	$\mathbf{E}_1$	$\mathbf{0}$	$\mathbf{0}$
$\mathbf{E}_4$	$\mathbf{E}_4$	$\mathbf{0}$	$\mathbf{E}_2$	$\mathbf{0}$	$\mathbf{0}$	$\mathbf{0}$
$\mathbf{F}$	$\mathbf{0}$	$\mathbf{0}$	$\mathbf{0}$	$\mathbf{0}$	$\mathbf{F}$	$\mathbf{0}$
$\mathbf{G}$	$\mathbf{0}$	$\mathbf{0}$	$\mathbf{0}$	$\mathbf{0}$	$\mathbf{0}$	$\mathbf{G}$

$$\tag{5.E.45}$$

4. Let the linear combination,

$$\mathbf{A} = w_1\,\mathbf{E}_1 + w_2\,\mathbf{E}_2 + w_3\,\mathbf{E}_3 + w_4\,\mathbf{E}_4 + f\,\mathbf{F} + g\,\mathbf{G}\,. \qquad (5.\text{E}.46)$$

Show that

$$w_1 = \mathbf{M}:\mathbf{A}:\mathbf{M}, \quad w_2 = \tfrac{1}{2}\,\mathbf{N}:\mathbf{A}:\mathbf{N}, \quad w_3 = \tfrac{1}{\sqrt{2}}\,\mathbf{M}:\mathbf{A}:\mathbf{N}, \quad w_4 = \tfrac{1}{\sqrt{2}}\,\mathbf{N}:\mathbf{A}:\mathbf{M},$$

$$f = \tfrac{1}{2}\,A_{ijkl}\,F_{ijkl}\,, \quad g = \tfrac{1}{2}\,A_{ijkl}\,G_{ijkl}\,. \qquad (5.\text{E}.47)$$

5. The above set of components will be denoted through the format,

$$\mathbf{A} = \left\{ \begin{bmatrix} w_1 & w_3 \\ w_4 & w_2 \end{bmatrix},\, f,\, g \right\}. \qquad (5.\text{E}.48)$$

Let $\mathbf{A}'$ be another arbitrary transversely isotropic fourth order tensor. Prove the following relations:

$$\mathbf{A}:\mathbf{A}' = \left\{ \begin{bmatrix} w_1\,w_1' + w_3\,w_4' & w_1\,w_3' + w_3\,w_2' \\ w_2\,w_4' + w_4\,w_1' & w_2\,w_2' + w_4\,w_3' \end{bmatrix},\, f\,f',\, g\,g' \right\}; \qquad (5.\text{E}.49)$$

$$\mathbf{A}:\mathbf{A}^{-1} = \mathbf{I}_4 = \left\{ \begin{bmatrix} 1 & 0 \\ 0 & 1 \end{bmatrix},\, 1,\, 1 \right\}; \quad \mathbf{A}^{-1} = \left\{ \frac{1}{w_1\,w_2 - w_3\,w_4} \begin{bmatrix} w_2 & -w_3 \\ -w_4 & w_1 \end{bmatrix},\, \frac{1}{f},\, \frac{1}{g} \right\}; \qquad (5.\text{E}.50)$$

$$\det \mathbf{A} = (w_1\,w_2 - w_3\,w_4)\,f\,g\,. \qquad (5.\text{E}.51)$$

6. Deduce the characteristic polynomial,

$$\det\big(\mathbf{A} - \lambda\,\mathbf{I}_4\big) = \big((w_1 - \lambda)\,(w_2 - \lambda) - w_3\,w_4\big)\,(f - \lambda)\,(g - \lambda)\,, \qquad (5.\text{E}.52)$$

the generalized Cayley-Hamilton relation, namely that a tensor satisfies its characteristic polynomial,

$$\big(\mathbf{A}^2 - (w_1 + w_2)\,\mathbf{A} + (w_1\,w_2 - w_3\,w_4)\,\mathbf{I}\big)\,(f\,\mathbf{I} - \mathbf{A})\,(g\,\mathbf{I} - \mathbf{A}) = \mathbf{0}\,; \qquad (5.\text{E}.53)$$

and the conditions of positive definiteness,

$$w_1 + w_2 > 0, \quad w_1\,w_2 - w_3\,w_4 > 0, \quad f > 0, \quad g > 0\,. \qquad (5.\text{E}.54)$$

7. Introducing the elastic stiffness $\mathbf{A} = \mathbb{E}$ expressed by (5.E.27) in (5.E.47), calculate the coefficients w's, f, and g first in terms of the coefficients a's, and next in terms of the coefficients c's using (5.E.28).

For that purpose, it is instrumental to write the tensors $\mathbf{F}$ and $\mathbf{G}$ in matrix form in the orthotropy axes:

$$[\mathbf{F}] = \begin{bmatrix} \tfrac{1}{2} & -\tfrac{1}{2} & 0 & 0 & 0 & 0 \\ -\tfrac{1}{2} & \tfrac{1}{2} & 0 & 0 & 0 & 0 \\ 0 & 0 & 0 & 0 & 0 & 0 \\ 0 & 0 & 0 & 0 & 0 & 0 \\ 0 & 0 & 0 & 0 & 0 & 0 \\ 0 & 0 & 0 & 0 & 0 & 1 \end{bmatrix}, \quad [\mathbf{G}] = \begin{bmatrix} 0 & 0 & 0 & 0 & 0 & 0 \\ 0 & 0 & 0 & 0 & 0 & 0 \\ 0 & 0 & 0 & 0 & 0 & 0 \\ 0 & 0 & 0 & 1 & 0 & 0 \\ 0 & 0 & 0 & 0 & 1 & 0 \\ 0 & 0 & 0 & 0 & 0 & 0 \end{bmatrix}. \qquad (5.\text{E}.55)$$

Then, since $\mathbf{F}:\mathbf{F} = \mathbf{G}:\mathbf{G} = \mathbf{N}:\mathbf{N} = 2$,

$$w_1 = c_{33}, \quad w_2 = c_{11} + c_{12}, \quad w_3 = w_4 = \sqrt{2}\,c_{13}, \quad f = c_{11} - c_{12}, \quad g = c_{44}\,. \qquad (5.\text{E}.56)$$

The conditions (5.E.54) can then be shown to be equivalent to the conditions (5.E.31).

Chapter 6

Hyperelasticity, a purely mechanical point of view

Most studies on the constitutive behavior of biological tissues, whether isotropic or anisotropic, hyperelastic or visco-hyperelastic, use a strain energy per reference volume, or per reference mass, where various mechanical and biomechanical effects are built in. For example, for the modeling of the passive behavior of arterial walls, Holzapfel et al. [2002] decompose additively the strain energy in three parts representing respectively the volumetric, isochoric and viscous responses. To model the mechanical behavior of the human anterior cruciate ligament, Limbert and Middleton [2004] devise a two-branch parallel model: both the elastic and the viscous stresses derive from strain energies per reference volume. Gathering the ideas of a parallel assembly of Maxwell models, of visco-hyperelasticity and of quasi-linear viscoelasticity (QLV) of Fung, a more general model consists in adding a long range memory that obeys the fading memory principle. Such a model is built by Pioletti and Rakotomanana [2000] to describe the mechanical response of isotropic tendons. For a two-component mixture, Kannan and Rajagopal [2004] make use of a free energy per unit mass, that is built from the individual free energies weighted by the mass fraction.

Before entering the mathematical details of the analysis, we should take a step backward and issue a warning. Soft tissues are porous media saturated by an electrolyte. Porous media intrinsically give rise to time dependent processes and overall nonlinearities. These phenomena are due to interactions between phases and can be produced even if the phases themselves display an individual linear and time independent constitutive response. As a simplification, some studies disregard the multiphase aspects and engage in an analysis as if the tissue was a single phase solid. Then to recover the above experimental effects, the solid should be endowed with intrinsic time effects (viscoelasticity) and nonlinear effects. Thus an apparent simplification may lead not only to undue complexities, but also to errors in interpretation. The time change of the volume fractions is a key point that plagues the single phase analysis. These remarks should nevertheless not deter us to address nonlinear elasticity and intrinsically time dependent mechanical responses, that are indeed observed at certain material scales.

6.1 Restrictions to the strain energy function

While the number of degrees of freedom in devising nonlinear elastic constitutive equations is large, certain requirements of mathematical nature should always be satisfied. Further requirements of mechanical nature in line with the stress-strain response that is sought may be placed on the elements (strain energy function, constitutive tensors, etc.) that are used to develop the equations.

6.1.1 Mathematical requirements

The strain energy function $\underline{\mathcal{W}}(\mathbf{F})$ may be set to vanish in the reference configuration, i.e., where the deformation gradient $\mathbf{F}$ is equal to the identity $\mathbf{I}$ tensor,

$$(1) \quad \underline{\mathcal{W}}(\mathbf{F} = \mathbf{I}) = 0 \, . \tag{6.1.1}$$

Like the strain energy of a spring, the strain energy of three-dimensional elastic solids is considered to increase with the deformation. Consequently, it should be positive or zero,

$$(2) \quad \underline{\mathcal{W}}(\mathbf{F}) \geq 0 \,. \tag{6.1.2}$$

The strain energy should also satisfy growth conditions: it should be infinite when $\det \mathbf{F}$ tends to either $+\infty$ or to 0^+,

$$(3) \quad \lim_{\det \mathbf{F} \to +\infty} \underline{\mathcal{W}}(\mathbf{F}) = +\infty;$$

$$(4) \quad \lim_{\det \mathbf{F} \to 0+} \underline{\mathcal{W}}(\mathbf{F}) = +\infty \,. \tag{6.1.3}$$

These conditions tend to imply that the energy necessary to expand a finite volume infinitely or to shrink it to a tiny volume should be infinite.

The strain energy is required to be objective, i.e., independent of rigid body motions in the current configuration,

$$(5) \ \text{objectivity} \quad \underline{\mathcal{W}}(\mathbf{Q} \cdot \mathbf{F}) = \underline{\mathcal{W}}(\mathbf{F}), \quad \forall \ \mathbf{Q} \ \text{rotation} \,. \tag{6.1.4}$$

As a consequence of (5), $\underline{\mathcal{W}}$ depends only on the stretching part of the deformation gradient $\mathbf{F}$, i.e., it depends only on the symmetric right stretch tensor $\mathbf{U}$ or, equivalently, on the right Cauchy-Green tensor $\mathbf{C}$ or Green strain $\mathbf{E}_\mathrm{G}$. For isotropic materials, the strain energy $\underline{\mathcal{W}}$ should also be independent of rigid body motions of the reference configuration,

$$(6) \ \text{isotropic material} \quad \underline{\mathcal{W}}(\mathbf{F} \cdot \mathbf{Q}) = \underline{\mathcal{W}}(\mathbf{F}), \quad \forall \ \mathbf{Q} \ \text{rotation} \,. \tag{6.1.5}$$

Combination of the relations (5) and (6) yields,

$$(7) \ \text{objectivity } + \text{ isotropic material} \quad \underline{\mathcal{W}}(\mathbf{Q} \cdot \mathbf{U} \cdot \mathbf{Q}^\mathrm{T}) = \underline{\mathcal{W}}(\mathbf{U}), \quad \forall \ \mathbf{Q} \ \text{rotation} \,. \tag{6.1.6}$$

Motivations for these restrictions and further developments may be found in textbooks like Bowen [1989] or Holzapfel [2000].

6.1.2 Constitutive requirements of mechanical nature

A number of optional restrictions with an immediate mechanical interpretation have been devised to restrict the constitutive properties of specific materials. Among them, we may mention for isotropic solids,

- tension-extension inequalities: two principal stretches being fixed $\lambda_b = \lambda_b^*$, an increase of the third principal stretch λ_a to λ_a^* leads to an increase of the principal stress along that direction from σ_a to σ_a^*,

$$\left(\sigma_a^* - \sigma_a\right)\left(\lambda_a^* - \lambda_a\right) > 0, \quad \lambda_a^* \neq \lambda_a, \quad \lambda_b^* = \lambda_b, \quad a \neq b \,. \tag{6.1.7}$$

- the Baker-Ericksen inequalities which state that the principal stress is larger in the direction of the larger stretch, i.e., here

$$\left(\sigma_a - \sigma_b\right)\left(\lambda_a - \lambda_b\right) > 0, \quad \lambda_a \neq \lambda_b \,. \tag{6.1.8}$$

The inequality in terms of the 1st Piola-Kirchhoff stress has been questioned by Ball [1977]. Strong ellipticity implies the Baker-Ericksen inequalities, Wang and Truesdell [1973], p. 247.

- the generalized Coleman-Noll condition (CGN), Truesdell and Toupin [1960], Section 52,

$$\left(\underline{\boldsymbol{\tau}}(\mathbf{F}^*) - \underline{\boldsymbol{\tau}}(\mathbf{F})\right) : \left(\mathbf{F}^* - \mathbf{F}\right) > 0 \,, \tag{6.1.9}$$

with $\underline{\boldsymbol{\tau}}$ the 1st Piola-Kirchhoff stress and the deformation gradients $\mathbf{F}$ and $\mathbf{F}^* = \mathbf{S} \cdot \mathbf{F}$ differ by a symmetric positive definite tensor $\mathbf{S} \neq \mathbf{I}$. It indicates that the transformation from the deformation gradient to the 1st Piola-Kirchhoff stress should be monotone for pairs of deformations that differ by a pure stretch. For a hyperelastic solid, the CGN property and its relation to a convex energy function are considered in Wang and Truesdell [1973], p. 218.

Further properties of the constitutive relations and associated partial differential equations of equilibrium, of an a priori mathematical nature but with mechanical consequences, have been scrutinized, among which are

- positive definiteness
- ellipticity and strong ellipticity
- convexity
- polyconvexity

These properties are briefly introduced in the sections below. The links to the invertibility of the stress-strain relation, algebraic properties of the acoustic tensor and of existence and nature of wave-speeds are pointed out. Local existence and uniqueness theorems in nonlinear elasticity rely on strong ellipticity while global results are based on convexity. Section 5 of Marsden and Hughes [1983] displays the inter-relations between a number of these properties.

6.2 Algebraic properties of constitutive functions and partial differential equations

The definitions are first given in the context of infinitesimal strains. They will be restated in the context of finite deformations in subsequent sections.

Consider a general constitutive law in rate, or rather incremental, form linking an increment of strain $\delta\epsilon$ to an increment of stress $\delta\sigma$,

$$\delta\boldsymbol{\sigma} = \mathbb{E}^t : \delta\boldsymbol{\epsilon}\,. \tag{6.2.1}$$

The increments of stress and strain are considered symmetric second order tensors, so that the tangent stiffness $\mathbb{E}^t$ may, without loss of generality, be considered to be endowed with the minor symmetries under the two first and two last indices, see Section 5.2.2. The constitutive tensor is assumed to be independent of the velocity gradient $\boldsymbol{\nabla}\mathbf{v}$, but it might depend on a set of state variables $\mathcal{V}$.

The next sections address solids. Positive definiteness of the stiffness matrix and strong ellipticity of the momentum balance in an infinite body for saturated porous media are considered in Chapters 7 and 11 respectively. Moreover emphasis is laid here on the quasi-static interpretations of these properties, while Chapter 11 focuses on their dynamic aspects. In fact, during a loading process in a standard boundary value problem, one typically can assign to a certain material quantity the role of a time-like parameter, under the proviso that this parameter varies monotonically. Interest is then in defining the critical values of this time-like parameter at which the properties of positive definiteness, strong ellipticity or ellipticity cease to hold. The mechanical consequences and interpretations of the loss of these properties then depend on the context. For example, loss of strong ellipticity corresponds to the accumulation of deformations along bands of characteristic directions: the phenomenon is termed *strain localization*. In a dynamic context, loss of strong ellipticity corresponds to the onset of a vanishing wave-speed, and it is referred to as a *stationary discontinuity*.

6.2.1 Positive definiteness

Positive definiteness of the stiffness $\mathbb{E}^t$ amounts to

$$(\text{PD}) \quad \boldsymbol{\epsilon} : \mathbb{E}^t : \boldsymbol{\epsilon} = \epsilon_{ij}\, E^t_{ijkl}\, \epsilon_{kl} > 0, \quad \forall \boldsymbol{\epsilon} \neq \mathbf{0}\,. \tag{6.2.2}$$

The major symmetry of the stiffness is defined as the symmetry under the two first and two last indices, namely $E^t_{ijkl} = E^t_{klij}$. The mathematical and mechanical implications of this property are explored in Section 5.1. Let $(\mathbb{E}^t)^{\mathrm{T}}$ be the transpose of the stiffness $\mathbb{E}^t$, namely componentwise $(E^t)^{\mathrm{T}}_{ijkl} = E^t_{klij}$. Positive definiteness depends only on the symmetric part of $\mathbb{E}^t$, namely $\frac{1}{2}\,(\mathbb{E}^t + (\mathbb{E}^t)^{\mathrm{T}})$, and the skew-symmetric part $\frac{1}{2}\,(\mathbb{E}^t - (\mathbb{E}^t)^{\mathrm{T}})$ is elusive. Although the focus here is on stiffness matrices endowed with major symmetry, it is of interest to have a look at Exercise 9.1 that reminds some basic results of algebra on non symmetric matrices. In particular, the existence of real positive eigenvalues is equivalent to positive definiteness only for symmetric matrices. Quite generally, invertibility requires non zero eigenvalues.

For linear isotropic elastic materials, the stiffness (5.2.27) may be decomposed in an isotropic (volume) part and a deviatoric (shear) part,

$$\begin{aligned}
\mathbb{E} &= (3\,K)\,\frac{\mathbf{I}\otimes\mathbf{I}}{3} + 2\,\mu\,\Big(\mathbf{I}\,\overline{\underline{\otimes}}\,\mathbf{I} - \frac{\mathbf{I}\otimes\mathbf{I}}{3}\Big), \\
\mathbb{E}^{-1} &= \frac{1}{3\,K}\,\frac{\mathbf{I}\otimes\mathbf{I}}{3} + \frac{1}{2\,\mu}\,\Big(\mathbf{I}\,\overline{\underline{\otimes}}\,\mathbf{I} - \frac{\mathbf{I}\otimes\mathbf{I}}{3}\Big),
\end{aligned} \tag{6.2.3}$$

and

$$\boldsymbol{\epsilon} : \mathbb{E} : \boldsymbol{\epsilon} = K\,(\mathbf{I} : \boldsymbol{\epsilon})^2 + 2\,\mu\,\operatorname{dev}\boldsymbol{\epsilon} : \operatorname{dev}\boldsymbol{\epsilon}\,. \tag{6.2.4}$$

The conditions of positive definiteness (and hence of invertibility of the stiffness) restrict the bulk, Lamé and Young's moduli and Poisson's ratio,

$$\mathbb{E}\ \text{Positive Definite (PD)} \quad K = \lambda + \frac{2}{3}\mu > 0 \ \text{and} \ \mu > 0 \Leftrightarrow E > 0 \ \text{and} \ \nu \in\,]-1, \tfrac{1}{2}[\,. \tag{6.2.5}$$

The spectral analysis of the isotropic elastic stiffness has been addressed in Section 5.3. The elastic stiffness has components $(3\,K,\,2\,\mu)$ on the basis $(\frac{1}{3}\,\mathbf{I}\otimes\mathbf{I},\mathbf{I}\,\underline{\overline{\otimes}}\,\mathbf{I}-\frac{1}{3}\,\mathbf{I}\otimes\mathbf{I})$ while a tensor $x\,\mathbf{I}\,\underline{\overline{\otimes}}\,\mathbf{I}$ proportional to the identity tensor $\mathbf{I}\,\underline{\overline{\otimes}}\,\mathbf{I}=\frac{1}{3}\,\mathbf{I}\otimes\mathbf{I}+\mathbf{I}\,\underline{\overline{\otimes}}\,\mathbf{I}-\frac{1}{3}\,\mathbf{I}\otimes\mathbf{I}$, has components $(x,\,x)$ on this basis. Then,

$$(\mathbb{E}-x\,\mathbf{I}\,\underline{\overline{\otimes}}\,\mathbf{I})^{-1}=\frac{1}{3\,K-x}\,\frac{\mathbf{I}\otimes\mathbf{I}}{3}+\frac{1}{2\,\mu-x}\,\Big(\mathbf{I}\,\underline{\overline{\otimes}}\,\mathbf{I}-\frac{1}{3}\,\mathbf{I}\otimes\mathbf{I}\Big)\,. \tag{6.2.6}$$

The actual constitutive stiffness of a number of materials, e.g., elastic-plastic solids, is a symmetric rank one modification of their elastic stiffness, namely

$$\begin{aligned}
\mathbb{E}^{t} &= \mathbb{E}-\frac{1}{H}\,\mathbb{E}:\mathbf{P}\otimes\mathbb{E}:\mathbf{P}, \quad H=h+\mathbf{P}:\mathbb{E}:\mathbf{P}>0\,, \\
(\mathbb{E}^{t})^{-1} &= \mathbb{E}^{-1}+\frac{1}{h}\,\mathbf{P}\otimes\mathbf{P}\,,
\end{aligned} \tag{6.2.7}$$

with $\mathbf{P}$ a symmetric second order tensor and the modulus $H>0$ is assumed to be strictly positive. Given the positive definiteness of the symmetric elastic stiffness $\mathbb{E}$, positive definiteness of the stiffness $\mathbb{E}^{t}$,

$$H\,\boldsymbol{\epsilon}:\mathbb{E}^{t}:\boldsymbol{\epsilon} = h\,\underbrace{\boldsymbol{\epsilon}:\mathbb{E}:\boldsymbol{\epsilon}}_{\geq 0}+\underbrace{(\boldsymbol{\epsilon}:\mathbb{E}:\boldsymbol{\epsilon})\,(\mathbf{P}:\mathbb{E}:\mathbf{P})-(\boldsymbol{\epsilon}:\mathbb{E}:\mathbf{P})^{2}}_{\geq 0}\geq 0\,, \tag{6.2.8}$$

requires the modulus $h\geq 0$ to be positive or 0. Indeed, the positive definite elastic stiffness $\mathbb{E}$ may be used to define a norm over symmetric second order tensors, e.g., $|\boldsymbol{\epsilon}|=\sqrt{\boldsymbol{\epsilon}:\mathbb{E}:\boldsymbol{\epsilon}}$. The last term in (6.2.8) is therefore positive as a consequence of the Cauchy-Schwarz inequality,

$$(\boldsymbol{\epsilon}_{1}:\mathbb{E}:\boldsymbol{\epsilon}_{1})\,(\boldsymbol{\epsilon}_{2}:\mathbb{E}:\boldsymbol{\epsilon}_{2})-(\boldsymbol{\epsilon}_{1}:\mathbb{E}:\boldsymbol{\epsilon}_{2})^{2}\geq 0,\quad \forall\boldsymbol{\epsilon}_{1},\boldsymbol{\epsilon}_{2}\,. \tag{6.2.9}$$

The eigenmode $\boldsymbol{\epsilon}$ associated with loss of positive definiteness implies the Cauchy-Schwarz inequality to be an equality, which implies $\boldsymbol{\epsilon}\sim\mathbf{P}$. Invertibility of the stiffness requires strict positive definiteness, namely $h>0$, which is indeed mandatory as shown in the inverse displayed in (6.2.7). This condition holds for any type of elastic anisotropy.

The spectral analysis of the tensor $\mathbb{E}^{t}$ amounts to define its eigenvalues x and (left or right) eigenvectors $\boldsymbol{\epsilon}$ such that $\mathbb{E}^{t}:\boldsymbol{\epsilon}=x\,\boldsymbol{\epsilon}$, or

$$\boldsymbol{\epsilon}=\frac{\mathbf{P}:\mathbb{E}:\boldsymbol{\epsilon}}{H}\,(\mathbb{E}-x\,\mathbf{I}\,\underline{\overline{\otimes}}\,\mathbf{I})^{-1}:(\mathbb{E}:\mathbf{P})\,. \tag{6.2.10}$$

For isotropic elasticity, this expression derives easily using (6.2.6), and, multiplied by $\mathbb{E}:\mathbf{P}$, it yields the quadratic equation for the eigenvalues x,

$$H\,x^{2}-\big(h\,(3\,K+2\,\mu)+6\,K\,\mu\,\mathbf{P}:\mathbf{P}\big)\,x+6\,K\,\mu\,h=0\,, \tag{6.2.11}$$

and the associated eigenmode,

$$\boldsymbol{\epsilon}\sim 6\,K\,\mu\,\mathbf{P}-x\,\mathbb{E}:\mathbf{P}\,. \tag{6.2.12}$$

Since the eigenproblem involves a symmetric matrix, the two eigenvalues are real. They are both positive if $h>0$, one is positive and one negative for $h<0$, and one is zero for $h=0$. In fact, the product of the eigenvalues can be obtained directly with help of (2.3.26), namely

$$\det\mathbb{E}^{t}=\det\mathbb{E}\,\det\Big(\mathbf{I}\,\underline{\overline{\otimes}}\,\mathbf{I}-\frac{1}{H}\,\mathbf{P}\otimes\mathbb{E}:\mathbf{P}\Big)=\det\mathbb{E}\,\Big(1-\frac{\mathbf{P}:\mathbb{E}:\mathbf{P}}{H}\Big)=(3\,K)\,(2\,\mu)\,\frac{h}{H}\,, \tag{6.2.13}$$

in agreement with (6.2.11).

6.2.2   Ellipticity and Strong Ellipticity

The quasi-static balance of momentum, or its rate, can typically be expressed as a quasi-linear partial differential equation, wrt the space variable $\mathbf{x}$, for the displacement $\mathbf{u}$, or velocity $\mathbf{v}$,

$$\frac{\partial}{\partial x_{i}}\Big(E^{t}_{ijkl}\,\frac{\partial v_{l}}{\partial x_{k}}\Big)=0\,, \tag{6.2.14}$$

that is satisfied pointwise inside the body (V). Focus is on the solutions $\mathbf{v}=\mathbf{v}(\mathbf{x})$ which are continuous and whose partial derivatives are piecewise continuous.

6.2.2.1 Ellipticity

The partial differential equation (6.2.14) is termed elliptic[6.1] at state $\mathcal{V}$ if

$$\text{(E)} \quad \det(n_i\, E^t_{ijkl}\, n_k) \neq 0, \quad \forall\, \mathbf{n},\ |\mathbf{n}| = 1\,. \tag{6.2.15}$$

If there exists a vector $\mathbf{n}$ such that $\det(n_i\, E^t_{ijkl}\, n_k) = 0$, then $\mathbf{n}$ is normal to a curve, called characteristic curve, across which $\mathbf{v}$ shows a discontinuity in its normal derivative, $[\![\boldsymbol{\nabla}\mathbf{v}]\!] \cdot \mathbf{n} \neq \mathbf{0}$, in other words, there exists a vector $\boldsymbol{\eta} \neq \mathbf{0}$ such that $[\![\boldsymbol{\nabla}\mathbf{v}]\!] = \boldsymbol{\eta} \otimes \mathbf{n}$.

6.2.2.2 Strong Ellipticity

The partial differential equation (6.2.14) is termed strongly elliptic at $\mathcal{V}$ if[6.2]

$$\text{(SE)} \quad \boldsymbol{\epsilon}^{\otimes} : \mathbb{E}^t : \boldsymbol{\epsilon}^{\otimes} = m_j\, n_i\, E^t_{ijkl}\, n_k\, m_l > 0\,, \quad \forall\, \boldsymbol{\epsilon}^{\otimes} = \frac{1}{2}\,(\mathbf{m} \otimes \mathbf{n} + \mathbf{n} \otimes \mathbf{m}) \neq \mathbf{0}, \quad |\mathbf{m}| = |\mathbf{n}| = 1\,. \tag{6.2.16}$$

Since strong ellipticity is defined as positive definiteness over a restricted set of second order tensors, namely tensors of dyadic form, strong ellipticity is a weaker requirement than positive definiteness.

Moreover, if the stiffness $\mathbb{E}^t$ displays the major symmetry, the following implications hold,

$$\text{(PD)} \quad \Rightarrow \quad \text{(SE)} \quad \Rightarrow \quad \text{(E)}\,. \tag{6.2.17}$$

With ρ the mass density, the second order tensor $\rho\, A_{jl} = n_i\, E^t_{ijkl}\, n_k$ that emerges from these definitions is termed *acoustic tensor*. Strong ellipticity in presence of a symmetric acoustic tensor implies the existence of three real positive eigenvalues, interpreted as wave-speeds in Chapter 11, and associated with three orthogonal eigenvectors.

Strong ellipticity of the isotropic elastic tensor amounts to

$$\mathbb{E} \text{ Strongly Elliptic (SE)} \quad \lambda + 2\mu > 0 \text{ and } \mu > 0 \ \Leftrightarrow\ E > 0 \text{ and } \nu \in\,]-1, \tfrac{1}{2}[\,\cup\,]1, \infty[\,, \tag{6.2.18}$$

while ellipticity is defined by the conditions,

$$\mathbb{E} \text{ Elliptic (E)} \quad \lambda + 2\mu \neq 0 \text{ and } \mu \neq 0 \ \Leftrightarrow\ E \neq 0,\ \nu \neq 1\,. \tag{6.2.19}$$

The acoustic tensor $\mathbf{A}(\mathbf{n})$ associated with the isotropic elastic stiffness and its inverse express as,

$$\begin{aligned}
\mathbf{A}(\mathbf{n}) &= \frac{\lambda + 2\mu}{\rho}\, \mathbf{n} \otimes \mathbf{n} + \frac{\mu}{\rho}\, (\mathbf{I} - \mathbf{n} \otimes \mathbf{n})\,, \\
\mathbf{A}^{-1}(\mathbf{n}) &= \frac{\rho}{\lambda + 2\mu}\, \mathbf{n} \otimes \mathbf{n} + \frac{\rho}{\mu}\, (\mathbf{I} - \mathbf{n} \otimes \mathbf{n})\,,
\end{aligned} \tag{6.2.20}$$

while the acoustic tensor $\mathbf{A}^t(\mathbf{n})$ associated with the rank one modification (6.2.7) is also a rank one modification of its elastic counterpart,

$$\mathbf{A}^t(\mathbf{n}) = \mathbf{A}(\mathbf{n}) - \frac{1}{H}\, \mathbf{a} \otimes \mathbf{a}, \quad \mathbf{a} = \frac{1}{\sqrt{\rho}}\, (\mathbb{E} : \mathbf{P}) \cdot \mathbf{n}\,. \tag{6.2.21}$$

[6.1]The underlying constitutive operator itself is routinely said to be *elliptic*.

[6.2] Alternatively, strong ellipticity is also defined through the existence of a strictly positive scalar $\kappa > 0$ such that

$$\text{(SE}')\quad m_j\, n_i\, E^t_{ijkl}\, n_k\, m_l \geq \kappa\, |\mathbf{m}|\, |\mathbf{n}|,$$

for any vectors $\mathbf{m}$ and $\mathbf{n}$. The property is called *semi-ellipticity* if $\kappa = 0$, namely

$$\text{(sE)}\quad m_j\, n_i\, E^t_{ijkl}\, n_k\, m_l \geq 0,$$

for any vectors $\mathbf{m}$ and $\mathbf{n}$. This property is also referred to as *Legendre-Hadamard condition* (LH). Although the subject of stability is not entered here, it is worth mentioning *Hadamard's theorem*, according to which semi-ellipticity (sE) is a necessary condition of local stability of an equilibrium point.

Furthermore, if $\mathbb{E}^t$ is constant in a domain Ω at the boundary of which the displacement field $\mathbf{u}$ vanishes, then, we have the implication referred as *van Hove theorem*,

$$\text{(SE)}\quad \Rightarrow\quad \int_{\Omega} \boldsymbol{\nabla}\mathbf{u} : \mathbb{E}^t : \boldsymbol{\nabla}\mathbf{u}\, d\Omega > 0,$$

which, in absence of body forces, is a sufficient stability condition, see, e.g., Mandel [1974], p. 114, for an analysis.

Loss of ellipticity amounts to the existence of at least one direction $\mathbf{n}$ and one eigenvector $\boldsymbol{\eta} \neq \mathbf{0}$ such that,

$$\mathbf{A}^t \cdot \boldsymbol{\eta} = \mathbf{A} \cdot \left(\mathbf{I} - \frac{1}{H}\mathbf{A}^{-1} \cdot \mathbf{a} \otimes \mathbf{a}\right) \cdot \boldsymbol{\eta} = \mathbf{0}. \tag{6.2.22}$$

The solution is obtained by inspection,

$$\boldsymbol{\eta} \sim \mathbf{A}^{-1} \cdot \mathbf{a}, \quad H = \mathbf{a} \cdot \mathbf{A}^{-1} \cdot \mathbf{a}, \tag{6.2.23}$$

or by calculating the determinant of $\mathbf{A}^t$,

$$\det \mathbf{A}^t = \left(1 - \frac{1}{H}\mathbf{a} \cdot \mathbf{A}^{-1} \cdot \mathbf{a}\right)\det \mathbf{A}, \tag{6.2.24}$$

leading to

$$h = -\mathbf{P} : \mathbb{E} : \mathbf{P} + \mathbf{P} : \left((\mathbb{E} \cdot \mathbf{n}) \cdot (\mathbf{n} \cdot \mathbb{E} \cdot \mathbf{n})^{-1} \cdot (\mathbf{n} \cdot \mathbb{E})\right) : \mathbf{P}. \tag{6.2.25}$$

This modulus, which may be made more explicit using the inverse elastic acoustic tensor $(6.2.20)_2$, can not be positive as a consequence of the implications (6.2.17), since positive definiteness of $\mathbb{E}^t$ holds for $h > 0$. In an actual loading process, assuming that h decreases monotonously, the issue consists in maximizing $h = h(\mathbf{n})$ to discover the most critical normal $\mathbf{n}$ that leads to loss of ellipticity.

6.2.3 Convexity

A few basic definitions and properties of convex sets and convex functions in $\mathbb{R}^n$, $n \geq 1$, are collected below. For details and proofs, the reader may refer to, e.g., Luenberger [1984], Fletcher [1990].

A generic element $\mathbf{x}$ of the sets considered in the definitions below may be viewed as a geometrical point. When it comes to mechanical constitutive equations, this generic element will be viewed as representing a measure of stress or of strain.

Convex set:

A set $\mathbf{C}$ is convex if, given any two points $\mathbf{x}$ and $\mathbf{y}$ in or on $\mathbf{C}$, any point $\lambda_x \mathbf{x} + \lambda_y \mathbf{y}$ on the line connecting $\mathbf{x}$ and $\mathbf{y}$, with $\lambda_x + \lambda_y = 1$, $0 \leq \lambda_x, \lambda_y \leq 1$, belongs to $\mathbf{C}$.

Convex function:

A function $\phi(\mathbf{x})$ defined on $\mathbb{R}^n$ is said convex iff

$$\forall (\mathbf{x}, \mathbf{y}) \in \mathbb{R}^n \times \mathbb{R}^n, \quad \forall (\lambda_x \geq 0, \lambda_y \geq 0) \in \mathbb{R} \times \mathbb{R}, \text{ such that } \lambda_x + \lambda_y = 1,$$

$$\text{then } \lambda_x \phi(\mathbf{x}) + \lambda_y \phi(\mathbf{y}) \geq \phi(\lambda_x \mathbf{x} + \lambda_y \mathbf{y}). \tag{6.2.26}$$

When the function ϕ is defined over second order tensors, a weaker definition, termed *rank one convexity*, requires that the inequality (6.2.26) holds, not for arbitrary $\mathbf{x}$ and $\mathbf{y}$, but only for $\mathbf{x} - \mathbf{y}$ being of rank one.

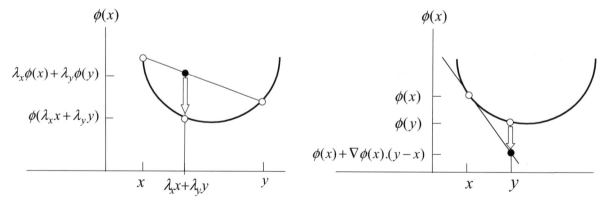

Fig. 6.2.1: Convexity of a differentiable function of the real variable: two equivalent conditions. The curve is below the chord (left), and the tangent defines a supporting hyperplane (right).

The definition is illustrated for a scalar function ($n = 1$) by Fig. 6.2.1. Qualitatively, one may think of a convex function if its graph is bowl-shaped.

A sketch of the multidimensional case ($n = 2$) is displayed in Fig. 6.2.2.

Let us consider the special case of an ellipse $\phi(x, y) = x^2/a^2 + y^2/b^2 - 1 = 0$. For this quadratic function, the inequality (6.2.26) may be transformed to $\phi(\mathbf{x}^{(1)}) - \phi(\mathbf{x}^{(2)}) - \nabla\phi(\mathbf{x}^{(2)}) \cdot (\mathbf{x}^{(1)} - \mathbf{x}^{(2)}) \geq 0$. Actually this inequality holds for any convex function as stated below. If the two points $\mathbf{x}^{(1)}$ and $\mathbf{x}^{(2)}$ are on the ellipse, then $\nabla\phi(\mathbf{x}^{(2)}) \cdot (\mathbf{x}^{(1)} - \mathbf{x}^{(2)}) \leq 0$. As a geometrical implication, all the points $\mathbf{x}^{(1)}$ of the ellipse are located on the inner side of the hyperplane orthogonal to the outer normal $\nabla\phi(\mathbf{x}^{(2)})$ at $\mathbf{x}^{(2)}$.

Convexity property 1:

If ϕ is two times continuously differentiable, it is convex at a point $\mathbf{x}$ iff the quadratic form associated with the Hessian

$$\frac{\partial^2 \phi}{\partial \mathbf{x} \, \partial \mathbf{x}}(\mathbf{x}) \tag{6.2.27}$$

estimated at this point is positive (semi-)definite.

Consequently, a quadratic form $\frac{1}{2}\mathbf{x}^\mathrm{T}\mathbf{A}\mathbf{x} + \mathbf{x}^\mathrm{T}\mathbf{a} + a$ is convex iff the symmetric part of the square matrix $\mathbf{A}$ is positive (semi-)definite, irrespective of the affine contribution $\mathbf{x}^\mathrm{T}\mathbf{a} + a$.

Convexity property 2:

If ϕ is continuously differentiable, it is convex iff $\forall\, \mathbf{x} \neq \mathbf{y}$,

$$\phi(\mathbf{y}) - \phi(\mathbf{x}) \geq \nabla\phi(\mathbf{x}) \cdot (\mathbf{y} - \mathbf{x}). \tag{6.2.28}$$

The property is illustrated by Fig. 6.2.1 for a one-dimensional space $n = 1$, and by Fig. 6.2.2 for a two-dimensional space $n = 2$. Observe in particular that $\phi(\mathbf{y}) - \phi(\mathbf{x})$ is positive while $\nabla\phi(\mathbf{x}) \cdot (\mathbf{y} - \mathbf{x})$ is negative in the fan just 'below' the tangent hyperplane.

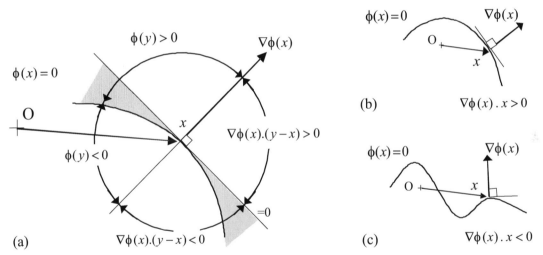

Fig. 6.2.2: Convexity of a differentiable function in $\mathbb{R}^2$. In a certain fan below the hyperplane defined by the tangent at a point $\mathbf{x}$, the function $\phi(\mathbf{y})$ is positive while $\nabla\phi(\mathbf{x}) \cdot (\mathbf{y} - \mathbf{x})$ is negative, (a). The set defined by the function $\phi(\mathbf{x}) \leq 0$ may a priori be convex in the configuration (b) but definitely not in the configuration (c) if it includes the origin O.

The proof of the implication (6.2.26) $\Rightarrow$ (6.2.28) goes as follows. Rewrite the inequality (6.2.26) as

$$\lambda_y \big(\phi(\mathbf{y}) - \phi(\mathbf{x})\big) \geq \phi\big(\mathbf{x} + \lambda_y\,(\mathbf{y} - \mathbf{x})\big) - \phi(\mathbf{x}), \tag{6.2.29}$$

divide both sides by λ_y and let λ_y go to 0^+ $\square$. To prove the converse implication (6.2.28) $\Rightarrow$ (6.2.26), write the inequality (6.2.28) with the pairs $(\mathbf{y}_1, \mathbf{x})$ and $(\mathbf{y}_2, \mathbf{x})$, sum the first inequality multiplied by λ_x and the second inequality multiplied by λ_y and finally set $\mathbf{x} = \lambda_x\,\mathbf{y}_1 + \lambda_y\,\mathbf{y}_2$ and $\lambda_x + \lambda_y = 1$ $\square$.

A proof of the implication (6.2.26) $\Rightarrow$ (6.2.27) starting from (6.2.28) can now be provided. For two points $\mathbf{x}$ and $\mathbf{y}$ sufficiently close, $\nabla\phi(\mathbf{x}) \simeq \nabla\phi(\mathbf{y}) + (\mathbf{x} - \mathbf{y}) \cdot \nabla^2\phi(\mathbf{y})$. Substituting in the inequality (6.2.28), one now has $(\mathbf{x} - \mathbf{y}) \cdot \nabla^2\phi(\mathbf{y}) \cdot (\mathbf{x} - \mathbf{y}) \geq \phi(\mathbf{x}) - \phi(\mathbf{y}) - \nabla\phi(\mathbf{y}) \cdot (\mathbf{x} - \mathbf{y}) \geq 0$. $\square$

Exchanging the roles of $\mathbf{x}$ and $\mathbf{y}$ in the convexity property 2 for the function ϕ and summing the two resulting inequalities implies the monotonicity of its gradient $\nabla\phi$. The latter property implies in turn the convexity property 1. In summary,

Convexity property 3:

If ϕ is continuously differentiable, it is convex iff $\forall\,\mathbf{x}\neq\mathbf{y}$,

$$\big(\nabla\phi(\mathbf{y})-\nabla\phi(\mathbf{x})\big)\cdot(\mathbf{y}-\mathbf{x})\geq 0\,. \tag{6.2.30}$$

Strict convexity:

The above properties hold on convex sets. They apply in the *weak sense*, namely, the inequalities are not strict. *Strict convexity* assumes a strict inequality in (6.2.26) for $\mathbf{x}\neq\mathbf{y}$ and $\lambda\in]0,1[$. Then property 2 with a strict inequality is equivalent to strict convexity. As for convexity property 1, $\partial^2\phi/\partial\mathbf{x}\,\partial\mathbf{x}$ positive definite implies strict convexity, but the converse does not hold: a function may be strictly convex while its Hessian is semi-definite positive. As an example, the real function $\phi(x)=x^4$ is strictly convex, but its second derivative vanishes at $x=0$, Exercise 6.1.

6.2.4   Polyconvexity

The existence of solutions to a particular boundary value problem in nonlinear elasticity depends on whether or not there exist deformations which minimize a given functional (the total potential energy) subjected to specific boundary conditions. Several conditions on the strain energy functions $\underline{W}$ guaranteeing existence of solutions to boundary value problems have been proposed. The simplest condition is convexity. However, convexity is often oversufficient and even overrestrictive: on one hand, it conflicts with the requirement of objectivity, Truesdell and Noll [1965], p. 163; on the other hand, convexity rules out the nonuniqueness essential for the description of some phenomena like, for example, buckling.

A less restrictive condition on the strain energy function, that is also a sufficient condition for existence of solutions in nonlinear elasticity, was introduced by Morrey [1952] and termed *quasi-convexity*. If $\underline{W}$ is quasi-convex, then it satisfies the strong ellipticity condition, that is, the acoustic tensor is positive definite. However the quasi-convexity condition is not a pointwise condition and it is therefore difficult to verify in particular cases.

Another important concept for practical use is the notion of *polyconvexity* introduced by Ball [1977]. Ball established the existence of minimizers of the total elastic energy when $\underline{W}$ has the form,

$$\underline{W}(\underline{\mathbf{x}},\mathbf{F})=g(\underline{\mathbf{x}},\mathbf{F},\mathrm{adj}\,\mathbf{F},\det\mathbf{F})\,, \tag{6.2.31}$$

with g convex function of $(\mathbf{F},\mathrm{adj}\,\mathbf{F},\det\mathbf{F})$ at each geometrical point $\underline{\mathbf{x}}$. He termed such functions $\underline{W}$ as polyconvex. In (6.2.31), $\mathrm{adj}\,\mathbf{F}=\det\mathbf{F}\,\mathbf{F}^{-1}$ denotes the adjoint of the deformation gradient $\mathbf{F}$. Some continuity and growth conditions (coerciveness)[6.3] are also required. Strict polyconvexity implies strong ellipticity. Polyconvexity is equivalent to a sufficient condition for quasi-convexity, that is, it is implied by quasi-convexity, but it is a local condition that it is easier to handle. Polyconvexity of $\underline{W}(\mathbf{F})$ does not imply convexity of the function $\underline{W}=\underline{W}(\mathbf{F})$: as a counterexample, consider the function $\underline{W}(\mathbf{F})=g(\det\mathbf{F})$ with g convex in $\det\mathbf{F}$.

While the physical implications of polyconvexity are not established, a wide variety of realistic models of nonlinear elastic materials satisfy the hypotheses of polyconvexity: the neo-Hookean model, the Mooney-Rivlin model and Ogden's model, Ball [1977]. On the other hand, some models like the Saint Venant-Kirchhoff model or models involving the logarithmic strain tensor do not satisfy the polyconvexity condition, e.g., Raoult [1986].

Some algebraic tools ease the construction of a variety of polyconvex strain energy functions. For example, polyconvexity is preserved under summation and multiplication by positive real numbers: therefore, a large variety of polyconvex strain energy functions may be assembled from a limited set of functions of the invariants of the right Cauchy-Green tensor and of the structural tensors describing the material anisotropy which are initially known/proved to be polyconvex. Notice that, in addition to polyconvexity, these functions must also be coercive and satisfy the condition of stress free reference configuration.

For isotropic materials, I_1^k, I_2^k, I_3^{-1} and $-\mathrm{Ln}\,I_3$, for any real $k\geq 1$, involving the invariants $I_1=\mathrm{tr}\,\mathbf{C}$, $I_2=\mathrm{tr}\,(\mathrm{adj}\,\mathbf{C})$ and $I_3=\det\mathbf{C}$ are examples of polyconvex functions, Kambouchev et al. [2006]. Schröder and

---

[6.3]A function is said to be *coercive* if it grows rapidly at the boundary of the domain over which it is defined. For a vector-valued function of a vector, the condition takes the form $\lim_{|\mathbf{x}|\to\infty}\mathbf{f}(\mathbf{x})\cdot\mathbf{x}/|\mathbf{x}|\to\infty$. Coerciveness is sometimes (in particular for bilinear forms) viewed as synonymous of the strong ellipticity condition defined in the footnote on page 177.

Neff [2003] constructed polyconvex anisotropic strain energy functions, particularly for transverse isotropic materials with structural tensor $\mathbf{M} = \mathbf{m} \otimes \mathbf{m}$, $\mathbf{m}$ being a unit vector in the preferred direction. They give a large list of polyconvex terms involving the mixed invariants $I_4 = \mathbf{C} : \mathbf{M}$ and $I_5 = \mathbf{C}^2 : \mathbf{M}$ such as I_4^k, $(I_5 - I_1 I_4 + I_2)^k$, $(I_1 - I_4)^k$ and $(I_1 I_4 - I_5)^k$, for any real $k \geq 1$. On the other hand, the invariants I_5 and $I_1 I_4$ often used to build transversely isotropic strain energy functions are not polyconvex terms. In Hartmann and Neff [2003], a strain energy function for nearly incompressible solids based on polyconvexity and coerciveness is proposed. A set of orthotropic and transversely isotropic strain energy functions satisfying polyconvexity is also proposed by Itskov and Aksel [2004] including the case of incompressible materials.

6.3 Total potential energy

The total potential energy is equal to the sum of the elastic energy plus the energy contributed by the conservative external forces. It is an ubiquitous entity in continuum mechanics, for instance to obtain the energy release rate in fracture mechanics. It may be expressed in terms of the strain energy function in at least two forms,

$$
\begin{aligned}
&\int_{\underline{V}} \mathcal{W}(\mathbf{F})\, d\underline{V} - \int_{\underline{V}} \mathbf{x} \cdot \underline{\rho}\, \mathbf{b}\, d\underline{V} - \int_{\partial \underline{V}_T} \mathbf{x} \cdot \underline{\boldsymbol{\tau}} \cdot \mathbf{n}\, d\underline{S} \\
&= \int_{\underline{V}} \mathcal{W}(\mathbf{F})\, d\underline{V} - \int_{\underline{V}} \mathbf{F} : \underline{\boldsymbol{\tau}}\, d\underline{V} + \int_{\partial \underline{V}_x} \mathbf{x} \cdot \underline{\boldsymbol{\tau}} \cdot \mathbf{n}\, d\underline{S} .
\end{aligned}
\tag{6.3.1}
$$

Here $\underline{\rho} = \rho \det \mathbf{F}$, $\mathbf{b}$ is the body force, $\underline{\boldsymbol{\tau}}$ the 1st Piola-Kirchhoff stress, and $\partial \underline{V}_T$ and $\partial \underline{V}_x$ denote the parts of the boundary which are subjected to traction and displacement respectively.

Indeed the incremental work done by the boundary traction on the boundary expresses as,

$$
\int_{\partial \underline{V}} \mathbf{x} \cdot \underline{\boldsymbol{\tau}} \cdot d\underline{\mathbf{A}} = \int_{\partial \underline{V}} \mathbf{x} \cdot \underline{\boldsymbol{\tau}} \cdot \mathbf{n}\, d\underline{S} = \int_{\partial V} \mathbf{x} \cdot \boldsymbol{\sigma} \cdot d\mathbf{A} .
\tag{6.3.2}
$$

By the divergence theorem

$$
\int_{\underline{V}} \frac{\partial f}{\partial \underline{x}_i}\, d\underline{V} = \int_{\partial \underline{V}} f\, \underline{n}_i\, d\underline{S} ,
\tag{6.3.3}
$$

the work of body forces and boundary traction under quasi-static conditions transforms as follows:

$$
\begin{aligned}
\int_{\underline{V}} \mathbf{x} \cdot \underline{\rho}\, \mathbf{b}\, d\underline{V} + \int_{\partial \underline{V}} \mathbf{x} \cdot \underline{\boldsymbol{\tau}} \cdot \mathbf{n}\, d\underline{S} &= \int_{\underline{V}} x_j \underline{\rho}\, b_j + \frac{\partial}{\partial \underline{x}_i}(x_j \underline{\tau}_{ji})\, d\underline{V} \\
&= \int_{\underline{V}} \frac{\partial x_j}{\partial \underline{x}_i} \underline{\tau}_{ji} + x_j \Big(\underbrace{\frac{\partial \underline{\tau}_{ji}}{\partial \underline{x}_i} + \underline{\rho}\, b_j}_{=0} \Big)\, d\underline{V} \\
&= \int_{\underline{V}} \frac{\partial x_j}{\partial \underline{x}_i}\, \underline{\tau}_{ji}\, d\underline{V} \\
&= \int_{\underline{V}} \mathbf{F} : \underline{\boldsymbol{\tau}}\, d\underline{V} .
\end{aligned}
\tag{6.3.4}
$$

6.4 Isotropic hyperelasticity

While the solid part of biological tissues remains to be defined, preliminary efforts have been made that concentrate on single phase elastic materials. Typically, the 2nd Piola-Kirchhoff stress $\underline{\underline{\boldsymbol{\tau}}}$ is expressed in terms of the strain energy $\mathcal{W}(\mathbf{C}) = \hat{\mathcal{W}}(\mathbf{E}_{\mathrm{G}})$ which depends on the right Cauchy-Green tensor $\mathbf{C} = \mathbf{F}^{\mathrm{T}} \cdot \mathbf{F}$, or on the Green strain $\mathbf{E}_{\mathrm{G}} = \frac{1}{2}(\mathbf{C} - \mathbf{I})$,

$$
\underline{\underline{\boldsymbol{\tau}}} = 2 \frac{\partial \mathcal{W}}{\partial \mathbf{C}} = \frac{\partial \hat{\mathcal{W}}}{\partial \mathbf{E}_{\mathrm{G}}} .
\tag{6.4.1}
$$

Let

$$
I_1 = \operatorname{tr} \mathbf{C}, \quad I_2 = \tfrac{1}{2}(\operatorname{tr} \mathbf{C})^2 - \tfrac{1}{2}\operatorname{tr} \mathbf{C}^2, \quad I_3 = \det \mathbf{C},
\tag{6.4.2}
$$

be the isotropic scalar invariants of the right and left Cauchy-Green tensors $\mathbf{C} = \mathbf{F}^{\mathrm{T}} \cdot \mathbf{F}$ and $\mathbf{B} = \mathbf{F} \cdot \mathbf{F}^{\mathrm{T}}$, with derivatives,

$$\frac{\partial I_1}{\partial \mathbf{C}} = \mathbf{I}, \quad \frac{\partial I_2}{\partial \mathbf{C}} = I_1 \mathbf{I} - \mathbf{C}, \quad \frac{\partial I_3}{\partial \mathbf{C}} = I_3 \mathbf{C}^{-1}, \tag{6.4.3}$$

the third relation being established by (2.7.3).

For a compressible material defined by the general isotropic strain energy $\underline{\mathcal{W}} = \underline{\mathcal{W}}(I_1, I_2, I_3)$, the 2nd Piola-Kirchhoff stress is expressed in the form,

$$\begin{aligned}
\underline{\underline{\tau}} &= 2 \frac{\partial \underline{\mathcal{W}}}{\partial \mathbf{C}} \\
&= 2 \underline{\mathcal{W}}_1 \mathbf{I} + 2 \underline{\mathcal{W}}_2 (I_1 \mathbf{I} - \mathbf{C}) + 2 \underline{\mathcal{W}}_3 I_3 \mathbf{C}^{-1} \\
&= 2 (\underline{\mathcal{W}}_1 + \underline{\mathcal{W}}_2 I_1) \mathbf{I} - 2 \underline{\mathcal{W}}_2 \mathbf{C} + 2 \underline{\mathcal{W}}_3 I_3 \mathbf{C}^{-1},
\end{aligned} \tag{6.4.4}$$

where $\underline{\mathcal{W}}_i = \partial \underline{\mathcal{W}} / \partial I_i$, $i \in [1,3]$, while the Kirchhoff stress $\boldsymbol{\tau} = \det \mathbf{F}\, \boldsymbol{\sigma}$ becomes,

$$\begin{aligned}
\boldsymbol{\tau} &= 2\, \mathbf{F} \cdot \frac{\partial \underline{\mathcal{W}}}{\partial \mathbf{C}} \cdot \mathbf{F}^{\mathrm{T}} \\
&= 2 \underline{\mathcal{W}}_1 \mathbf{B} + 2 \underline{\mathcal{W}}_2 (I_1 \mathbf{B} - \mathbf{B}^2) + 2 \underline{\mathcal{W}}_3 I_3 \mathbf{I} \\
&= 2 (\underline{\mathcal{W}}_2 I_2 + \underline{\mathcal{W}}_3 I_3) \mathbf{I} + 2 \underline{\mathcal{W}}_1 \mathbf{B} - 2 \underline{\mathcal{W}}_2 I_3 \mathbf{B}^{-1}.
\end{aligned} \tag{6.4.5}$$

Again use of the Cayley-Hamilton theorem exposed in Section 2.3.1, namely $\mathbf{C}^3 - I_1 \mathbf{C}^2 + I_2 \mathbf{C} - I_3 \mathbf{I} = \mathbf{0}$, has been made to reach the third line.

Incompressibility, namely $I_3 = 1$, gives rise to a reaction pressure p, Section 2.10. The so-called neo-Hookean solids and Mooney-Rivlin solids are popular special cases,

neo-Hookean solid	Mooney-Rivlin solid	

$$\underline{\mathcal{W}} = b_1 (I_1 - 3) \qquad\qquad \underline{\mathcal{W}} = b_1 (I_1 - 3) + b_2 (I_2 - 3)$$

$$\underline{\underline{\tau}} = -p\, \mathbf{C}^{-1} + 2\, b_1 \mathbf{I} \qquad\qquad \underline{\underline{\tau}} = -p\, \mathbf{C}^{-1} + 2\, b_1 \mathbf{I} + 2\, b_2 (I_1 \mathbf{I} - \mathbf{C}) \tag{6.4.6}$$

$$\boldsymbol{\sigma} = -p\, \mathbf{I} + 2\, b_1 \mathbf{B} \qquad\qquad \boldsymbol{\sigma} = -p\, \mathbf{I} + 2\, b_1 \mathbf{B} + 2\, b_2 (\underbrace{I_1 \mathbf{B} - \mathbf{B}^2}_{=I_2 \mathbf{I} - I_3 \mathbf{B}^{-1}})$$

While the neo-Hookean solid is a rather acceptable model of the mechanical response of rubberlike materials, the Mooney-Rivlin solid is the most general model that relates linearly the shear strain and the shear stress. The compressible version of the neo-Hookean solid is usually expressed in the format,

$$\underline{\mathcal{W}} = \frac{\mu}{2} \left(I_1 - 3 + \frac{1}{\gamma} \left((I_3)^{-\gamma} - 1 \right) \right), \quad \underline{\underline{\tau}} = -\mu\, (I_3)^{-\gamma} \mathbf{C}^{-1} + \mu\, \mathbf{I}, \tag{6.4.7}$$

with the material constants $\mu > 0$ and $\gamma = \nu/(1 - 2\nu)$, where ν can be seen as a Poisson's ratio in the infinitesimal strain limit. Since $\nu = \gamma/(1 + 2\gamma)$, slight compressibility is introduced by a Poisson's ratio ν close to 0.5 or a large exponent γ.

The compressible Blatz-Ko solid includes the three isotropic invariants,

$$\underline{\mathcal{W}} = \frac{\mu\, f}{2} \left(I_1 - 3 + \frac{1}{\gamma} \left((I_3)^{-\gamma} - 1 \right) \right) + \frac{\mu\, (1-f)}{2} \left(\frac{I_2}{I_3} - 3 + \frac{1}{\gamma} \left((I_3)^{\gamma} - 1 \right) \right), \tag{6.4.8}$$

and involves the three material constants $\mu > 0$, $f \in\,]0,1]$ and $\gamma = \nu/(1 - 2\nu)$. For $\nu = 1/2$, the solid becomes incompressible $I_3 = 1$ and it reduces to the Mooney-Rivlin solid. The stresses are defined by (6.4.4) and (6.4.5) with

$$\frac{\partial \underline{\mathcal{W}}}{\partial I_1} = \frac{\mu\, f}{2}, \quad \frac{\partial \underline{\mathcal{W}}}{\partial I_2} = \frac{\mu\, (1-f)}{2} \frac{1}{I_3}, \quad \frac{\partial \underline{\mathcal{W}}}{\partial I_3} = -\frac{\mu\, f}{2} (I_3)^{-\gamma-1} + \frac{\mu\, (1-f)}{2} (I_3)^{\gamma-1}. \tag{6.4.9}$$

Some hyperelastic models satisfy the so-called Valanis-Landel separation hypothesis, according to which the strain energy function can be expressed as an additive symmetric function (due to isotropy) of the principal stretches λ_1, λ_2 and λ_3, actually the positive square-roots of the right or left Cauchy-Green tensors,

$$\underline{\mathcal{W}}(\lambda_1, \lambda_2, \lambda_3) = \underline{w}(\lambda_1) + \underline{w}(\lambda_2) + \underline{w}(\lambda_3). \tag{6.4.10}$$

The experimental validity of this hypothesis has been verified in restricted ranges of stretch on vulcanized rubbers, see Ogden [1982] for an account. Ogden's strain energy function for an incompressible material, Ogden [1972]a, adopts this additive decomposition,

$$\underline{\mathcal{W}}(\lambda_1, \lambda_2, \lambda_3) = \sum_n \frac{\mu_n}{\alpha_n} \left(\lambda_1^{2\alpha_n} + \lambda_2^{2\alpha_n} + \lambda_3^{2\alpha_n} - 3 \right), \tag{6.4.11}$$

where the μ_n's and the exponents α_n may be positive or negative real material constants. Ogden's strain energy contains the neo-Hookean energy for $n = 1$, $\alpha_1 = 1$, and the Mooney-Rivlin energy for $n = 2$, $\alpha_1 = 1$ and $\alpha_2 = -1$. Indeed, recall the relations between the invariants I_1 and I_2 of the Cauchy-Green tensors and the principal stretches: $I_1 = \lambda_1^2 + \lambda_2^2 + \lambda_3^2$, $I_2 = (\lambda_1 \lambda_2 \lambda_3)^2 (1/\lambda_1^2 + 1/\lambda_2^2 + 1/\lambda_3^2)$. The format was extended to compressible materials in Ogden [1972]b through addition of a function of the volume change,

$$\underline{\mathcal{W}}(\lambda_1, \lambda_2, \lambda_3) = \sum_n \frac{\mu_n}{\alpha_n} \left(\lambda_1^{2\alpha_n} + \lambda_2^{2\alpha_n} + \lambda_3^{2\alpha_n} - 3 \right) + \underline{\mathcal{W}}_v(\lambda_1 \lambda_2 \lambda_3), \tag{6.4.12}$$

with $\underline{\mathcal{W}}_v(x)$ a convex function having a minimum at $x = 1$ and satisfying the asymptotic conditions (6.1.3). For $n = 1$ and $\alpha_1 = 1$, this model is also referred to as Hadamard's solid.

Prange and Margulies [2002] introduce the time dependence of the white and gray brain matter in the μ_n via two characteristic times much like in Section 8.3.6.

Another standard example of isotropic nonlinear hyperelastic equations is due to Murnaghan [1951]. The strain energy,

$$\hat{\underline{\mathcal{W}}}(\mathbf{E}_{\mathrm{G}}) = \frac{1}{2} \left(\lambda + 2\mu \right) I_1^2 - 2\mu I_2 + \frac{1}{3} \left(l + 2m \right) I_1^3 - 2m I_1 I_2 + n I_3, \tag{6.4.13}$$

expresses in terms of the invariants of the Green strain,

$$I_1 = \mathrm{tr}\,\mathbf{E}_{\mathrm{G}}, \quad I_2 = \tfrac{1}{2} \left(\mathrm{tr}\,\mathbf{E}_{\mathrm{G}} \right)^2 - \tfrac{1}{2}\,\mathrm{tr}\,\mathbf{E}_{\mathrm{G}}^2, \quad I_3 = \det \mathbf{E}_{\mathrm{G}} = \frac{1}{3}\,\mathrm{tr}\,\mathbf{E}_{\mathrm{G}}^3 - \frac{1}{2}\,\mathrm{tr}\,\mathbf{E}_{\mathrm{G}}\,\mathrm{tr}\,\mathbf{E}_{\mathrm{G}}^2 + \frac{1}{6} \left(\mathrm{tr}\,\mathbf{E}_{\mathrm{G}} \right)^3, \tag{6.4.14}$$

and of the material constants λ, μ, l, m and n. The last relation in (6.4.14) uses (2.2.33)$_6$. The derivatives,

$$\frac{\partial I_1}{\partial \mathbf{E}_{\mathrm{G}}} = \mathbf{I}, \quad \frac{\partial I_2}{\partial \mathbf{E}_{\mathrm{G}}} = I_1 \mathbf{I} - \mathbf{E}_{\mathrm{G}}, \quad \frac{\partial I_3}{\partial \mathbf{E}_{\mathrm{G}}} = I_2 \mathbf{I} - I_1 \mathbf{E}_{\mathrm{G}} + \mathbf{E}_{\mathrm{G}}^2, \tag{6.4.15}$$

provide the 2nd Piola-Kirchhoff stress,

$$\underline{\underline{\tau}} = \frac{\partial \hat{\underline{\mathcal{W}}}}{\partial \mathbf{E}_{\mathrm{G}}} = \left(\lambda I_1 + (n - 2m) I_2 + l I_1^2 \right) \mathbf{I} + \left(2\mu - (n - 2m) I_1 \right) \mathbf{E}_{\mathrm{G}} + n \mathbf{E}_{\mathrm{G}}^2. \tag{6.4.16}$$

In fact, for $n - 2m = 0$, the factor of n in the strain energy simplifies to $\frac{1}{3}\,\mathrm{tr}\,\mathbf{E}_{\mathrm{G}}^3$.

6.5 Anisotropic hyperelasticity and Fung's strain energy

Debes and Fung [1995] report a linear elastic behavior of the pulmonary artery in the physiological range from zero to the homeostatic stress. At variance, Zhou and Fung [1997] quantify nonlinear effects in blood vessel elasticity and they observe a significant degree of nonlinearity in the canine aorta. In both the linear and nonlinear ranges, the mechanical response displayed anisotropy.

6.5.1   Exponential strain energy

Fung's strain energy adopts the form,

$$\underline{\mathcal{W}}(\mathbf{C}) = \hat{\mathcal{W}}(\mathbf{E}_{\mathrm{G}}) = b\, \mathrm{e}^{Q(\mathbf{E}_{\mathrm{G}})}, \tag{6.5.1}$$

with $Q(\mathbf{E}_{\mathrm{G}})$ a quadratic function of the components of the strain tensor in the orthotropy axes,

$$Q(\mathbf{E}_{\mathrm{G}}) = b_1\, E_{11}^2 + b_2\, E_{22}^2 + b_3\, E_{33}^2 + 2\, b_4\, E_{11}E_{22} + 2\, b_5\, E_{11}E_{33} + 2\, b_6\, E_{22}E_{33}\,, \tag{6.5.2}$$

the b's being material constants. Humphrey [1995] has introduced a more complete type of anisotropy, with three additional material constants, than in (6.5.2),

$$Q(\mathbf{E}_{\mathrm{G}}) \to Q(\mathbf{E}_{\mathrm{G}})_{\mathrm{Fung}} + 4\, b_7\, E_{23}^2 + 4\, b_8\, E_{13}^2 + 4\, b_9\, E_{12}^2\,. \tag{6.5.3}$$

In view of generalization, Fung's strain energy (6.5.3) is cast in the form $\hat{\gamma}(Q) = b \exp(Q)$, $b > 0$. The matrix form in the orthotropic axes of (6.5.3) reads in isometric notation,

$$
Q(\mathbf{E}_{\mathrm{G}}) =
\begin{bmatrix}
E_{11} \\
E_{22} \\
E_{33} \\
\sqrt{2}\,E_{23} \\
\sqrt{2}\,E_{13} \\
\sqrt{2}\,E_{12}
\end{bmatrix}^{\mathrm{T}}
\begin{bmatrix}
b_1 & b_4 & b_5 & 0 & 0 & 0 \\
b_4 & b_2 & b_6 & 0 & 0 & 0 \\
b_5 & b_6 & b_3 & 0 & 0 & 0 \\
0 & 0 & 0 & 2\,b_7 & 0 & 0 \\
0 & 0 & 0 & 0 & 2\,b_8 & 0 \\
0 & 0 & 0 & 0 & 0 & 2\,b_9
\end{bmatrix}
\begin{bmatrix}
E_{11} \\
E_{22} \\
E_{33} \\
\sqrt{2}\,E_{23} \\
\sqrt{2}\,E_{13} \\
\sqrt{2}\,E_{12}
\end{bmatrix}. \tag{6.5.4}
$$

An extension of the quadratic form of Fung,

$$\hat{\mathcal{W}}(\mathbf{E}_{\mathrm{G}}) = \hat{\gamma}(Q(\mathbf{E}_{\mathrm{G}})), \quad Q(\mathbf{E}_{\mathrm{G}}) = \mathbf{E}_{\mathrm{G}} : \mathcal{A} : \mathbf{E}_{\mathrm{G}}\,, \tag{6.5.5}$$

involves the constant fourth order tensor $\mathcal{A}$ endowed with minor and major symmetries (21 independent material constants), and assumes

$$\hat{\gamma}' \equiv \frac{\partial \hat{\gamma}}{\partial Q} > 0, \quad \hat{\gamma}'' \equiv \frac{\partial^2 \hat{\gamma}}{\partial Q \partial Q} > 0\,. \tag{6.5.6}$$

Actually, while the modification below might appear as a formal issue, it turns highly advantageous to formulate the constitutive equation in the form,

$$\underline{\mathcal{W}}(\mathbf{C}) = \gamma(Q(\mathbf{C})), \quad Q(\mathbf{C}) = \mathbf{C} : \mathcal{A} : \mathbf{C}\,. \tag{6.5.7}$$

The motivation stems from the analysis of the strong ellipticity condition by Wilber and Walton [2002] which is exposed in Section 6.5.3. Note that, when developed in terms of the right Cauchy-Green tensor, Q as defined by (6.5.5) is not quadratic as it contains a linear term.

6.5.2   Locking versus softening materials: exponentials and energy limiters

Assuming

$$\lim_{\mathbf{E}_{\mathrm{G}} \to \infty} Q(\mathbf{E}_{\mathrm{G}}) \to \infty, \tag{6.5.8}$$

the idea in the exponential form,

$$\hat{\mathcal{W}}(\mathbf{E}_{\mathrm{G}}) \simeq \exp Q(\mathbf{E}_{\mathrm{G}}) - 1, \tag{6.5.9}$$

is to represent the stiffening of the material as deformation proceeds, that is *locking*. Stress and stiffness have the form,

$$\frac{\partial \hat{\mathcal{W}}}{\partial \mathbf{E}_{\mathrm{G}}} = \exp\left(Q(\mathbf{E}_{\mathrm{G}})\right) \frac{\partial Q}{\partial \mathbf{E}_{\mathrm{G}}}, \quad \frac{\partial^2 \hat{\mathcal{W}}}{\partial \mathbf{E}_{\mathrm{G}}^2} = \exp\left(Q(\mathbf{E}_{\mathrm{G}})\right) \left(\frac{\partial^2 Q}{\partial \mathbf{E}_{\mathrm{G}}^2} + \frac{\partial Q}{\partial \mathbf{E}_{\mathrm{G}}} \otimes \frac{\partial Q}{\partial \mathbf{E}_{\mathrm{G}}} \right). \tag{6.5.10}$$

Conversely, an *energy limiter* can be introduced via the form, Volokh [2007] [2008],

$$\hat{\mathcal{W}}(\mathbf{E}_{\mathrm{G}}) \simeq 1 - \exp\left(-Q(\mathbf{E}_{\mathrm{G}})\right). \tag{6.5.11}$$

Stress and stiffness become now,

$$\frac{\partial \hat{\mathcal{W}}}{\partial \mathbf{E}_{\mathrm{G}}} = \exp\left(-Q(\mathbf{E}_{\mathrm{G}})\right) \frac{\partial Q}{\partial \mathbf{E}_{\mathrm{G}}}, \quad \frac{\partial^2 \hat{\mathcal{W}}}{\partial \mathbf{E}_{\mathrm{G}}^2} = \exp\left(-Q(\mathbf{E}_{\mathrm{G}})\right) \left(\frac{\partial^2 Q}{\partial \mathbf{E}_{\mathrm{G}}^2} - \frac{\partial Q}{\partial \mathbf{E}_{\mathrm{G}}} \otimes \frac{\partial Q}{\partial \mathbf{E}_{\mathrm{G}}} \right). \tag{6.5.12}$$

Fig. 6.5.1 features the differences between the expressions (6.5.9) and (6.5.11) in a one-dimensional context.

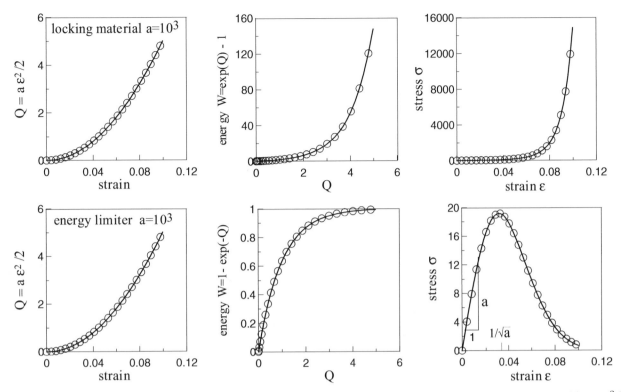

Fig. 6.5.1: Locking and softening materials (in a small strain context). Consider a uniaxial strain, and take $Q(\epsilon) = a\,\epsilon^2/2$ with $a > 0$. The stress reaches a maximum at $\epsilon = 1/\sqrt{a}$ for the softening material associated with a limited energy.

6.5.3 Strong Ellipticity of Fung's like strain energies (SE)

Let us consider a compressible material for which the constitutive equation provides equivalently the 1st Piola-Kirchhoff stress $\underline{\tau} = \mathbf{F} \cdot \underline{\underline{\tau}}$ or the 2nd Piola-Kirchhoff stress $\underline{\underline{\tau}}$ in terms of the deformation gradient $\mathbf{F}$ via a strain energy $\mathcal{W}(\mathbf{C})$,

$$\underline{\tau} = 2\,\mathbf{F} \cdot \frac{\partial \mathcal{W}}{\partial \mathbf{C}}, \quad \underline{\underline{\tau}} = 2\,\frac{\partial \mathcal{W}}{\partial \mathbf{C}}. \tag{6.5.13}$$

A small dyadic change of the deformation gradient $\mathbf{F}_0 \to \mathbf{F} = \mathbf{F}_0 + \delta\lambda\,\mathbf{H}$, with $\mathbf{H} = \mathbf{m} \otimes \mathbf{n}$, $\mathbf{m}$ and $\mathbf{n}$ being two unit vectors, induces a change of the 1st Piola-Kirchhoff stress which is equal to,

$$\delta\underline{\tau} = \delta\lambda\,\frac{\partial \underline{\tau}}{\partial \mathbf{F}}(\mathbf{F}_0) : \mathbf{H}. \tag{6.5.14}$$

Strong ellipticity of the balance of momentum at the deformation gradient $\mathbf{F}_0$,

$$\mathbf{H} : \frac{\partial \underline{\tau}}{\partial \mathbf{F}}(\mathbf{F}_0) : \mathbf{H} > 0, \quad \forall\,\mathbf{H} = \mathbf{m} \otimes \mathbf{n}, \quad |\mathbf{m}| = 1, \;|\mathbf{n}| = 1, \tag{6.5.15}$$

consists in restricting the component of the change of the stress along the direction of change of the deformation gradient to a strictly positive value.

For an incompressible material $\det \mathbf{F} = 1$, the constitutive equation decomposes into a part that involves the pressure p and that does no work if the kinematical constraint is satisfied, namely $\delta\det \mathbf{F} = \det \mathbf{F}\,\mathrm{tr}\,\delta\mathbf{F} \cdot \mathbf{F}^{-1} = \det \mathbf{F}\,\mathbf{F}^{-T} : \delta\mathbf{F}$, and in a reaction part, often referred to as effective stress,

$$\underline{\tau}' = \underline{\tau} + p\,\mathbf{F}^{-T} = 2\,\mathbf{F} \cdot \frac{\partial \mathcal{W}}{\partial \mathbf{C}}, \quad \underline{\underline{\tau}}' = \underline{\underline{\tau}} + p\,\mathbf{C}^{-1} = 2\,\frac{\partial \mathcal{W}}{\partial \mathbf{C}}. \tag{6.5.16}$$

Then, depending whether the inequality is strict or not, the following terminology along Antman [1995] is adopted,

$$\mathbf{H} : \frac{\partial \underline{\tau}}{\partial \mathbf{F}}(\mathbf{F}_0) : \mathbf{H} = \mathbf{H} : \frac{\partial \underline{\tau}'}{\partial \mathbf{F}}(\mathbf{F}_0) : \mathbf{H} \begin{cases} > 0, & \text{strong ellipticity (SE)} \\ \geq 0, & \text{Legendre} - \text{Hadamard condition (LH)} \end{cases} \tag{6.5.17}$$

for any $\mathbf{H} = \mathbf{m} \otimes \mathbf{n} \neq \mathbf{0}$ such that $\mathbf{F}_0^{-T} : \mathbf{H} = 0$.

The last restriction in (6.5.17) ensures that the change of the deformation gradient respects (in the small) the incompressibility condition. It also makes it possible for the strong ellipticity condition to apply for both the effective stress $\underline{\tau}' = \underline{\tau} + p\,\mathbf{F}^{-\mathrm{T}}$ and the total stress $\underline{\tau}$. Indeed,

$$
\begin{aligned}
&\mathbf{H} : \underline{\tau}(\mathbf{F}) \\
={}& \mathbf{H} : \left(\underline{\tau}'(\mathbf{F}) - p\,\mathbf{F}^{-\mathrm{T}}\right) \\
={}& \mathbf{H} : \left(\underline{\tau}'(\mathbf{F}_0) + \delta\lambda\,\frac{\partial\underline{\tau}'}{\partial\mathbf{F}}(\mathbf{F}_0) : \mathbf{H} - p\left(\mathbf{F}_0^{-\mathrm{T}} - \delta\lambda\,(\mathbf{F}_0^{-\mathrm{T}}\cdot\mathbf{n})\otimes(\mathbf{F}_0^{-1}\cdot\mathbf{m})\right)\right) + O((\delta\lambda)^2) \\
={}& \mathbf{H} : \left(\underline{\tau}'(\mathbf{F}_0) - p\,\mathbf{F}_0^{-\mathrm{T}}\right) + \delta\lambda\,\mathbf{H} : \frac{\partial\underline{\tau}'}{\partial\mathbf{F}}(\mathbf{F}_0) : \mathbf{H} + \delta\lambda\,p\,\underbrace{(\mathbf{F}_0^{-\mathrm{T}}\cdot\mathbf{H})^2}_{=0} + O((\delta\lambda)^2) \\
={}& \mathbf{H} : \underline{\tau}(\mathbf{F}_0) + \delta\lambda\,\mathbf{H} : \frac{\partial\underline{\tau}'}{\partial\mathbf{F}}(\mathbf{F}_0) : \mathbf{H} + O((\delta\lambda)^2)\,.
\end{aligned}
\tag{6.5.18}
$$

Note that

$$
\mathbf{H} : \frac{\partial\underline{\tau}'}{\partial\mathbf{F}} : \mathbf{H} = \mathbf{H} : \frac{\partial}{\partial\mathbf{F}}\left(2\,\mathbf{F}\cdot\frac{\partial\mathcal{W}}{\partial\mathbf{C}}\right) : \mathbf{H} = 2\,(\mathbf{H}^{\mathrm{T}}\cdot\mathbf{H}) : \frac{\partial\mathcal{W}}{\partial\mathbf{C}} + 4\,(\mathbf{F}^{\mathrm{T}}\cdot\mathbf{H}) : \frac{\partial^2\mathcal{W}}{\partial\mathbf{C}^2} : (\mathbf{F}^{\mathrm{T}}\cdot\mathbf{H})\,.
\tag{6.5.19}
$$

Indeed, using the fact that $\partial/\partial\mathbf{F} = 2\,\mathbf{F}\cdot\partial/\partial\mathbf{C}$,

$$
\begin{aligned}
\frac{\partial}{\partial F_{ij}}\left(2\,F_{kl}\,\frac{\partial\mathcal{W}}{\partial C_{lm}}\right) &= 2\,\frac{\partial F_{kl}}{\partial F_{ij}}\,\frac{\partial\mathcal{W}}{\partial C_{lm}} + 2\,F_{kl}\,\frac{\partial}{\partial F_{ij}}\,\frac{\partial\mathcal{W}}{\partial C_{lm}} \\
&= 2\,I_{ik}\,\frac{\partial\mathcal{W}}{\partial C_{jm}} + 4\,F_{ip}\,\frac{\partial}{\partial C_{pj}}\left(\frac{\partial\mathcal{W}}{\partial C_{lm}}\right)F_{kl}\,.
\end{aligned}
\tag{6.5.20}
$$

For the Fung's like model (6.5.5),

$$
2\,\frac{\partial\mathcal{W}}{\partial\mathbf{C}} = \frac{\partial\mathcal{W}}{\partial\mathbf{E}_{\mathrm{G}}} = 2\,\hat{\gamma}'\,\mathcal{A} : \mathbf{E}_{\mathrm{G}}, \quad 4\,\frac{\partial^2\mathcal{W}}{\partial\mathbf{C}^2} = 2\,\hat{\gamma}'\,\mathcal{A} + 4\,\hat{\gamma}''\,(\mathcal{A} : \mathbf{E}_{\mathrm{G}})\otimes(\mathcal{A} : \mathbf{E}_{\mathrm{G}})\,,
\tag{6.5.21}
$$

and

$$
\begin{aligned}
\mathbf{H} : \frac{\partial\underline{\tau}'}{\partial\mathbf{F}} : \mathbf{H} ={}& 2\,\hat{\gamma}'\,(\mathbf{H}^{\mathrm{T}}\cdot\mathbf{H}) : (\mathcal{A} : \mathbf{E}_{\mathrm{G}}) \\
&+ 4\,\hat{\gamma}''\left((\mathcal{A} : \mathbf{E}_{\mathrm{G}}) : (\mathbf{F}^{\mathrm{T}}\cdot\mathbf{H})\right)^2 + 2\,\hat{\gamma}'\,(\mathbf{F}^{\mathrm{T}}\cdot\mathbf{H}) : \mathcal{A} : (\mathbf{F}^{\mathrm{T}}\cdot\mathbf{H})\,.
\end{aligned}
\tag{6.5.22}
$$

Wilber and Walton [2002] obtain the following surprising results:
1. Fung's strain energy (6.5.1) satisfies the Legendre-Hadamard condition only if all the coefficients of the upper principal matrix of $\mathcal{A}$ are equal and positive, i.e.,

$$
\mathcal{A} = b\,\mathbf{I}\otimes\mathbf{I} \quad\Leftrightarrow\quad b_1 = b_2 = b_3 = b_4 = b_5 = b_6 = b \geq 0\,.
\tag{6.5.23}
$$

2. Fung's strain energy (6.5.1) satisfies the strong ellipticity condition at $\mathbf{F}\neq\mathbf{I}$ only if all the coefficients are equal and positive. Even then, the condition fails at $\mathbf{F}=\mathbf{I}$, i.e., $\mathbf{E}_{\mathrm{G}}=\mathbf{0}$, as can be checked on (6.5.22). Failure of (SE) at $\mathbf{F}=\mathbf{I}$ is interpreted as follows. Let $\mathbf{H} = \mathbf{e}_1\otimes\mathbf{e}_2$, with $\mathbf{e}_1\cdot\mathbf{e}_2 = 0$, so that the last condition in (6.5.17) associated with incompressibility is satisfied. The condition $\mathbf{H} : \partial\underline{\tau}/\partial\mathbf{F} : \mathbf{H} < 0$ implies that shear along $\mathbf{H}$ continues at decreasing Kirchhoff stress $\underline{\tau}$;
3. they give necessary, but not sufficient, conditions for the model (6.5.3) to satisfy the Legendre-Hadamard and strong ellipticity conditions, their Eqns (5.8).
4. the restrictions on the coefficients of the strain energy of the form (6.5.7) are far less drastic than for the form (6.5.5). Instead of (6.5.23), the necessary and sufficient conditions for the Legendre-Hadamard condition to hold for the Fung's model (6.5.2) are

$$
b_i \geq 0,\; i\in[1,6]; \quad b_1 \geq \tfrac{1}{2}\,(b_4 + b_5); \quad b_2 \geq \tfrac{1}{2}\,(b_4 + b_6); \quad b_3 \geq \tfrac{1}{2}\,(b_5 + b_6)\,.
\tag{6.5.24}
$$

If, in addition, the diagonal terms b_i, $i\in[1,3]$, are strictly positive, the model is strongly elliptic.

6.5.4 Convexity of Fung's like strain energies

Convexity of a scalar or vector function over $\mathbb{R}^n$ has been defined in Section 6.2.3. The definition and properties are simply restated for functions of a second order tensor. In fact, convexity of the strain energy $\underline{W} = \underline{W}(\mathbf{C})$ will be considered over the six-dimensional space of independent components of the right Cauchy-Green tensor. It takes the form

$$\lambda\,\underline{W}(\mathbf{C}_1) + (1-\lambda)\,\underline{W}(\mathbf{C}_2) \geq \underline{W}(\lambda\,\mathbf{C}_1 + (1-\lambda)\,\mathbf{C}_2), \quad \forall\,\mathbf{C}_1,\,\mathbf{C}_2, \quad \lambda \in [0,1]\,. \tag{6.5.25}$$

If the function $\underline{W}$ admits a continuous derivative, then it is convex iff the following inequality is satisfied

$$\underline{W} \quad \text{convex} \quad \Leftrightarrow \quad \underline{W}(\mathbf{C}_1) - \underline{W}(\mathbf{C}_2) \geq \frac{\partial \underline{W}}{\partial \mathbf{C}}(\mathbf{C}_2) : (\mathbf{C}_1 - \mathbf{C}_2), \quad \forall\,\mathbf{C}_1,\,\mathbf{C}_2\,. \tag{6.5.26}$$

If the function $\underline{W}$ admits a continuous second derivative, then it is convex iff its Hessian matrix is positive (semi-)definite,

$$\underline{W} \quad \text{convex} \quad \Leftrightarrow \quad \frac{\partial^2 \underline{W}}{\partial \mathbf{C}^2} \quad \text{positive (semi-)definite}\,. \tag{6.5.27}$$

The analysis of the convexity of the elastic strain energies of the Fung's type, or extended Fung's type, takes advantage of the following property:

Convexity property 3:
If $f : \mathbb{R}^n \to \mathbb{R}$ is convex, and $g : \mathbb{R} \to \mathbb{R}$ is convex and nondecreasing, then $g \circ f : \mathbb{R}^n \to \mathbb{R}$ is convex.

Indeed,

$$\begin{aligned}
&(g \circ f)(\mathbf{C}_1) - (g \circ f)(\mathbf{C}_2) \\
=\ & g\big(f(\mathbf{C}_1)\big) - g\big(f(\mathbf{C}_2)\big) \\
\geq\ & \underbrace{\frac{\partial g}{\partial f}\big(f(\mathbf{C}_2)\big)}_{\geq 0} \big(f(\mathbf{C}_1) - f(\mathbf{C}_2)\big) \geq \underbrace{\frac{\partial g}{\partial f}\big(f(\mathbf{C}_2)\big)\,\frac{\partial f}{\partial \mathbf{C}}(\mathbf{C}_2)}_{\big(\partial(g \circ f)/\partial \mathbf{C}\big)(\mathbf{C}_2)} : (\mathbf{C}_1 - \mathbf{C}_2) \quad \Box\,.
\end{aligned} \tag{6.5.28}$$

Here $f(\mathbf{C}) = \mathbf{C} : \mathcal{A} : \mathbf{C}$, and $g(f) = b\,\exp f$, $b > 0$. Therefore, a sufficient condition of convexity of the Fung's like strain energies is that the tensor $\mathcal{A}$ is definite positive. That $\mathcal{A}$ positive definite implies convexity of Fung's like strain energies can be also realized by considering directly the Hessian (6.5.10).

In contrast to the strong ellipticity condition, the convexity is not affected by the form $\underline{W}(\mathbf{C}) = \gamma(\mathbf{C} : \mathcal{A} : \mathbf{C})$, or $\hat{\underline{W}}(\mathbf{E}_{\mathrm{G}}) = \hat{\gamma}(\mathbf{E}_{\mathrm{G}} : \mathcal{A} : \mathbf{E}_{\mathrm{G}})$.

Pioletti and Rakotomanana [2000] show a practical list of isotropic elastic strain energies and the conditions (convexity, etc.) they satisfy.

6.6 Elastic potentials with the elastic-growth multiplicative decomposition

When the deformation gradient is decomposed into an elastic part and a growth part, namely $\mathbf{F} = \mathbf{F}^{\mathrm{e}} \cdot \mathbf{F}^{\mathrm{g}}$, the state variables are relative to the intermediate configuration (κ) implicitly introduced by the multiplicative decomposition. The deformation enters the set of state variables through its elastic part. Incompressibility is now phrased in terms of $\det \mathbf{F}^{\mathrm{e}}$, rather than $\det \mathbf{F}$. For a compressible solid, the stress-strain relation provides the Kirchhoff stress in terms of the elastic Green strain, while, for an incompressible solid, it provides the effective Kirchhoff stress. The stress-strain relation is cast in terms of the elastic potential (or elastic strain energy function) per unit reference volume $\underline{E}_{\kappa}$ in the form, Section 22.5,

$$\boldsymbol{\tau}' = \det \mathbf{F} \left\{ \begin{array}{ll} \text{elastic compressibility} & \boldsymbol{\sigma} \\ \text{elastic incompressibility} & \boldsymbol{\sigma}' \equiv \boldsymbol{\sigma} + p\,\mathbf{I} \end{array} \right\} = \mathbf{F}^{\mathrm{e}} \cdot \frac{\partial \underline{E}_{\kappa}}{\partial \mathbf{E}_{\mathrm{G}}^{\mathrm{e}}} \cdot (\mathbf{F}^{\mathrm{e}})^{\mathrm{T}}\,, \tag{6.6.1}$$

with $\mathbf{E}_{\mathrm{G}}^{\mathrm{e}} = \frac{1}{2}((\mathbf{F}^{\mathrm{e}})^{\mathrm{T}} \cdot \mathbf{F}^{\mathrm{e}} - \mathbf{I})$ the elastic Green strain.

When growth takes place with the deposited material having the same properties as the already existing material at the deposition site, or generally for nongrowing materials, then

$$\underline{E}_\kappa = E_\kappa \det \mathbf{F}^\mathrm{g} \,, \tag{6.6.2}$$

Thus, in this case, the Cauchy stress (6.6.1) recovers a more familiar form, e.g., for elastic compressibility,

$$\boldsymbol{\sigma} = \frac{1}{\det \mathbf{F}^\mathrm{e}} \mathbf{F}^\mathrm{e} \cdot \frac{\partial E_\kappa}{\partial \mathbf{E}^\mathrm{e}_\mathrm{G}} \cdot (\mathbf{F}^\mathrm{e})^\mathrm{T} \,. \tag{6.6.3}$$

6.6.1 Isotropic Fung's like potentials

For materials that are isotropic in their intermediate configuration, the elastic potential $\underline{E}_\kappa = \underline{E}_\kappa(I^\mathrm{e}_1, I^\mathrm{e}_2, I^\mathrm{e}_3)$ depends on the strain via the three scalar invariants of the Cauchy-Green tensor with eigenvalues c_i, $i \in [1,3]$,

$$\mathbf{C}^\mathrm{e} = (\mathbf{F}^\mathrm{e})^\mathrm{T} \cdot \mathbf{F}^\mathrm{e} = 2 \, \mathbf{E}^\mathrm{e}_\mathrm{G} + \mathbf{I} \,, \tag{6.6.4}$$

namely,

$$I^\mathrm{e}_1 = \operatorname{tr} \mathbf{C}^\mathrm{e} = c_1 + c_2 + c_3$$

$$I^\mathrm{e}_2 = \tfrac{1}{2} \left((\operatorname{tr} \mathbf{C}^\mathrm{e})^2 - \operatorname{tr}(\mathbf{C}^\mathrm{e})^2 \right) = c_1 \, c_2 + c_2 \, c_3 + c_3 \, c_1 \tag{6.6.5}$$

$$I^\mathrm{e}_3 = \det \mathbf{C}^\mathrm{e} = c_1 \, c_2 \, c_3$$

Note that

$$\frac{\partial}{\partial \mathbf{E}^\mathrm{e}_\mathrm{G}} = 2 \, \frac{\partial}{\partial \mathbf{C}^\mathrm{e}} \,. \tag{6.6.6}$$

Moreover, since[6.4]

$$\frac{\partial I^\mathrm{e}_1}{\partial \mathbf{C}^\mathrm{e}} = \mathbf{I}, \quad \frac{\partial I^\mathrm{e}_2}{\partial \mathbf{C}^\mathrm{e}} = I^\mathrm{e}_1 \, \mathbf{I} - \mathbf{C}^\mathrm{e}, \quad \frac{\partial I^\mathrm{e}_3}{\partial \mathbf{C}^\mathrm{e}} = I^\mathrm{e}_3 \, (\mathbf{C}^\mathrm{e})^{-1} \,, \tag{6.6.8}$$

then

$$\begin{aligned}
\frac{\partial E_\kappa}{\partial \mathbf{C}^\mathrm{e}} &= \frac{\partial E_\kappa}{\partial I^\mathrm{e}_1} \mathbf{I} + \frac{\partial E_\kappa}{\partial I^\mathrm{e}_2} \left(I^\mathrm{e}_1 \, \mathbf{I} - \mathbf{C}^\mathrm{e} \right) + \frac{\partial E_\kappa}{\partial I^\mathrm{e}_3} I^\mathrm{e}_3 \, (\mathbf{C}^\mathrm{e})^{-1} \\
&= \left(\frac{\partial E_\kappa}{\partial I^\mathrm{e}_1} + I^\mathrm{e}_1 \frac{\partial E_\kappa}{\partial I^\mathrm{e}_2} \right) \mathbf{I} - \frac{\partial E_\kappa}{\partial I^\mathrm{e}_2} \mathbf{C}^\mathrm{e} + \frac{\partial E_\kappa}{\partial I^\mathrm{e}_3} I^\mathrm{e}_3 \, (\mathbf{C}^\mathrm{e})^{-1} \,.
\end{aligned} \tag{6.6.9}$$

Note that, for a vanishing elastic deformation, i.e., $\mathbf{F}^\mathrm{e} = \mathbf{I}$,

$$\frac{\partial E_\kappa}{\partial \mathbf{C}^\mathrm{e}} = \left(\frac{\partial E_\kappa}{\partial I^\mathrm{e}_1} + 2 \frac{\partial E_\kappa}{\partial I^\mathrm{e}_2} + \frac{\partial E_\kappa}{\partial I^\mathrm{e}_3} \right) \mathbf{I} \,. \tag{6.6.10}$$

Thus the scalar coefficient in this expression should vanish if the stress is to be zero in the unstrained state. For a counterexample, see (6.6.17) below.

Fung's potential originally addresses elastic materials. Applied to elastic-growing incompressible vascular tissues, it takes the form,

$$\underline{E}_\kappa = \frac{a_0}{2} \left(\exp(Q) - Q - 1 \right) + \frac{q}{2} \,, \tag{6.6.11}$$

where, the a's and b's being material constants,

$$Q = a_1 \, (I^\mathrm{e}_1 - 3) + a_2 \, (I^\mathrm{e}_2 - 3) + a_3 \, (I^\mathrm{e}_1 - 3)^2,$$

$$q = b_1 \, (I^\mathrm{e}_1 - 3) + b_2 \, (I^\mathrm{e}_2 - 3) + b_3 \, (I^\mathrm{e}_1 - 3)^2 \,. \tag{6.6.12}$$

[6.4]The last relation uses (2.7.3),

$$\frac{\partial \det \mathbf{F}^\mathrm{e}}{\partial \mathbf{F}^\mathrm{e}} = \det \mathbf{F}^\mathrm{e} \, (\mathbf{F}^\mathrm{e})^{-\mathrm{T}} \,, \quad \frac{\partial \det \mathbf{F}^\mathrm{e}}{\partial \mathbf{E}^\mathrm{e}_\mathrm{G}} = \det \mathbf{F}^\mathrm{e} \, (\mathbf{F}^\mathrm{e})^{-1} \cdot (\mathbf{F}^\mathrm{e})^{-\mathrm{T}} \,. \tag{6.6.7}$$

The second part of the potential addresses the low stiffness at small strain, while the exponential is intended to represent the stiffening at larger strains. Typically, elastin fibers in arteries contribute to sustain loads at small strains while the stiffer collagen fibers are recruited progressively at increasing tensile strains. The derivatives are recorded,

$$\frac{\partial \underline{E}_\kappa}{\partial I_1^e} = \frac{a_0}{2} \left(\exp(Q) - 1 \right) \left(a_1 + 2\, a_3 \left(I_1^e - 3 \right) \right) + \frac{b_1}{2} + b_3 \left(I_1^e - 3 \right)$$
$$\frac{\partial \underline{E}_\kappa}{\partial I_2^e} = \frac{a_0}{2} \left(\exp(Q) - 1 \right) a_2 + \frac{b_2}{2} \, . \tag{6.6.13}$$

Klisch et al. [2001]b consider two strain energy functions, initially developed for purely elastic materials, which exhibit a large difference of stiffnesses in contraction and extension, and which are the isotropic versions of the elastic potentials used respectively by Almeida and Spilker [1998] for articular cartilages, namely

$$\underline{E}_\kappa = \frac{a_0}{2} \left(I_3^e \right)^{-\beta} \exp Q, \quad Q = a_1 \left(I_1^e - 3 \right) + a_2 \left(I_2^e - 3 \right) + a_3 \left(I_1^e - 3 \right)^2, \tag{6.6.14}$$

and by Klisch and Lotz [1999] for annulus fibrosus,

$$\underline{E}_\kappa = \frac{a_0}{2} Q, \quad Q = \exp(a_1 \left(I_1^e - 3 \right)) + \exp(a_2 \left(I_2^e - 3 \right)) + \exp(a_3 \left(I_3^e - 1 \right)). \tag{6.6.15}$$

The exponent β and the a's are material constants. Note that (6.6.14) is a particular case of Fung's potential (the exponential of a sum of invariants), while (6.6.15) is a sum of exponentials.

Rodriguez et al. [1994] use two types of potential, their Eqns 20, 21 and 30,

$$\underline{E}_\kappa = \frac{a_0}{2} \left(\exp(Q) - 1 \right), \quad Q = a \operatorname{tr} \mathbf{E}_G^e, \quad \text{or} \quad \underline{E}_\kappa = a \operatorname{tr} \left(\mathbf{E}_G^e \right)^2. \tag{6.6.16}$$

Omens et al. [2003] use a linear term in the elastic potential of a purely elastic aorta, so as to obtain a residual stress in the reference (unstrained) configuration. If the material was isotropic, the elastic potential for an elastic-growing solid would be of the form, with a and b material constants,

$$\underline{E}_\kappa = a \operatorname{tr} \mathbf{E}_G^e + b \operatorname{tr} \left(\mathbf{E}_G^e \right)^2. \tag{6.6.17}$$

6.6.2 The multiplicative volumetric-isochoric decomposition of the deformation gradient

An extension of the *additive* decomposition of the infinitesimal strain into a spherical part and a deviatoric part is obtained via the *multiplicative* decomposition of the deformation gradient, Flory [1961],

$$\mathbf{F}^e = J_e^{\frac{1}{3}} \mathbf{F}_{iso}^e, \quad \text{with} \quad J_e = \det \mathbf{F}^e, \quad \det \mathbf{F}_{iso}^e = 1. \tag{6.6.18}$$

The subscript 'iso' reflects the fact that the deformation gradient $\mathbf{F}_{iso}^e$ is isochoric. Hence

$$\mathbf{C}_{iso}^e = (\mathbf{F}^e)_{iso}^T \cdot \mathbf{F}_{iso}^e = (J_e)^{-\frac{2}{3}} \mathbf{C}^e, \tag{6.6.19}$$

and

$$\frac{d\mathbf{C}_{iso}^e}{d\mathbf{C}^e} = (J_e)^{-\frac{2}{3}} \left(\mathbf{I} \underline{\overline{\otimes}} \mathbf{I} - \tfrac{1}{3} \mathbf{C}^e \otimes (\mathbf{C}^e)^{-1} \right) = (J_e)^{-\frac{2}{3}} \left(\mathbf{I} \underline{\overline{\otimes}} \mathbf{I} - \tfrac{1}{3} \mathbf{C}_{iso}^e \otimes (\mathbf{C}_{iso}^e)^{-1} \right). \tag{6.6.20}$$

Indeed,

$$\mathbf{C}_{iso}^e(\mathbf{C}^e + d\mathbf{C}^e) - \mathbf{C}_{iso}^e(\mathbf{C}^e) = \left(\det(\mathbf{C}^e + d\mathbf{C}^e) \right)^{-\frac{1}{3}} (\mathbf{C}^e + d\mathbf{C}^e) - \mathbf{C}_{iso}^e(\mathbf{C}^e)$$

$$= (J_e)^{-\frac{2}{3}} \left(\det \left(\mathbf{I} + (\mathbf{C}^e)^{-1} \cdot d\mathbf{C}^e \right) \right)^{-\frac{1}{3}} (\mathbf{C}^e + d\mathbf{C}^e) - \mathbf{C}_{iso}^e(\mathbf{C}^e)$$

$$= (J_e)^{-\frac{2}{3}} \left(1 - \tfrac{1}{3} (\mathbf{C}^e)^{-1} : d\mathbf{C}^e \right) (\mathbf{C}^e + d\mathbf{C}^e) - \mathbf{C}_{iso}^e(\mathbf{C}^e) + o(d\mathbf{C}^e)$$

$$= (J_e)^{-\frac{2}{3}} \left(d\mathbf{C}^e - \tfrac{1}{3} (\mathbf{C}^e)^{-1} : d\mathbf{C}^e\, \mathbf{C}^e \right) + o(d\mathbf{C}^e) \tag{6.6.21}$$

$$= (J_e)^{-\frac{2}{3}} \left(\mathbf{I} \underline{\overline{\otimes}} \mathbf{I} - \tfrac{1}{3} \mathbf{C}^e \otimes (\mathbf{C}^e)^{-1} \right) : d\mathbf{C}^e + o(d\mathbf{C}^e)$$

$$= (J_e)^{-\frac{2}{3}} \left(\mathbf{I} \underline{\overline{\otimes}} \mathbf{I} - \tfrac{1}{3} \mathbf{C}_{iso}^e \otimes (\mathbf{C}_{iso}^e)^{-1} \right) : d\mathbf{C}^e + o(d\mathbf{C}^e) \, \square.$$

The third line has used (2.2.34).

6.6.3   Additive decomposition of anisotropic Fung's like potentials

A number of modifications of Fung's potential has been proposed, based on either experimental measurements or theoretical derivations. A few of these extensions are listed below.

A review of experimental details of biaxial testing of soft tissues including questions related to strain measurement by optical techniques, to homogeneity, to anisotropy, to boundary tethering or clamping and to orthotropy directions can be found in Sacks [2000].

An experimental technique to determine the axes of orthotropy has been proposed by Choi and Vito [1990]. Instead of the exponential of a sum, they use the sum of exponentials to build the elastic potential of canine pericardium, like in (6.6.15). The differences between these two forms remain to be explored.

Holzapfel and Weizsäcker [1998] introduce an anisotropic extension of Fung's potential to compressible solids. The isotropic and anisotropic terms contribute additively and are *decoupled*, namely

$$\underline{E}_\kappa = \underbrace{\frac{a_0}{2}\left(\exp Q(\mathbf{E}_G^e) - 1\right)}_{\text{anisotropic response}} + \underbrace{b_0 \operatorname{tr} \mathbf{E}_G^e}_{\text{isotropic response}}, \tag{6.6.22}$$

where,

$$Q(\mathbf{E}_G^e) = a_1\left(E_{G1}^e\right)^2 + a_2\left(E_{G2}^e\right)^2 + a_3\left(E_{G3}^e\right)^2 + 2\,a_4\,E_{G1}^e\,E_{G2}^e + 2\,a_5\,E_{G1}^e\,E_{G3}^e + 2\,a_6\,E_{G2}^e\,E_{G3}^e\,, \tag{6.6.23}$$

with b_0 and the a's material constants. Here the E_{Gi}^e's, for i=1 to 3, are the components of the elastic Green strain along the fixed orthotropy (symmetry) axes of the material ($\mathbf{e}_1$, $\mathbf{e}_2$, $\mathbf{e}_3$).

van Dyke and Hoger [2002] develop a method to determine the opening angle induced by a residual stress field in an elastic-growing arterial wall. For illustration, they use the above elastic potential per unit reference volume initially developed for purely elastic tissues by Holzapfel and Weizsäcker [1998]. For aortic growth, Taber and Eggers [1996] use (6.6.22) and (6.6.23) for an elastically incompressible elastic-growing material, with $b_0 = 0^{6.5}$.

In fact, for an orthotropic material with symmetry axes ($\mathbf{e}_1$, $\mathbf{e}_2$, $\mathbf{e}_3$), there are seven independent scalar invariants of the Green strain listed in Table 2.11.3. They may be expressed in component form in the orthotropy axes and recombined,

$$E_{G11}^e,\ E_{G22}^e,\ E_{G33}^e,\ \left(E_{G12}^e\right)^2,\ \left(E_{G23}^e\right)^2,\ \left(E_{G13}^e\right)^2, E_{G12}^e\,E_{G23}^e\,E_{G13}^e\,. \tag{6.6.24}$$

Indeed, Humphrey [1995] accounts for the additional quadratic shear terms in Q, namely $a_7\left(E_{G23}^e\right)^2 + a_8\left(E_{G13}^e\right)^2 + a_9\left(E_{G12}^e\right)^2$, that vanish if the orthotropy axes and Green strain are coaxial.

The two external layers of the arterial walls, namely the media and the adventitia, are reinforced by collagen fibers, mainly aligned circumferentially. These fibers can be considered to form two families of fixed directions $\mathbf{m}_1$ and $\mathbf{m}_2$. Moreover, only the isochoric part of the deformation gradient $\mathbf{F}_{\text{iso}}^e$ is considered to act on the fibers. Along these ideas, Holzapfel et al. [2000] change (6.6.22) to

$$\underline{E}_\kappa = \frac{a_1}{2\,a_2}\sum_{i=1,2}\underbrace{\exp\left(a_2\left(I_{mi}^{\text{iso}} - 1\right)^2\right)}_{\substack{\text{anisotropic}\\\text{response}}} + \frac{b_0}{2}\underbrace{\left(\operatorname{tr}\mathbf{C}_{\text{iso}}^e - 3\right)}_{\substack{\text{isotropic}\\\text{response}}}, \tag{6.6.25}$$

with $a_1 > 0$ and $a_2 > 0$ material constants, and

$$I_{mi}^{\text{iso}} = \left(\mathbf{F}_{\text{iso}}^e \cdot \mathbf{m}_i\right)^2 = \mathbf{m}_i \cdot \mathbf{C}_{\text{iso}}^e \cdot \mathbf{m}_i\,, \tag{6.6.26}$$

is the square of the material stretches along the direction $\mathbf{m}_i$ of the fiber i. Holzapfel et al. [2002] modify the argument of the exponential in order to account for the fact that the collagen fibers do not sustain contractile strains,

$$\underline{E}_\kappa = \frac{a_1}{2\,a_2}\sum_{i=1,2}\underbrace{\exp\left(a_2\langle I_{mi}^{\text{iso}} - 1\rangle^2\right)}_{\substack{\text{traceless anisotropic}\\\text{response}}} + \frac{b_0}{2}\underbrace{\left(\operatorname{tr}\mathbf{C}_{\text{iso}}^e - 3\right)}_{\substack{\text{traceless}\\\text{isotropic}\\\text{response}}} + \underbrace{U(\det\mathbf{F}^e)}_{\substack{\text{isotropic}\\\text{volumetric}\\\text{response}}}, \tag{6.6.27}$$

[6.5]Concerning volume change under infinitesimal and finite strains, see however Section 2.5.7.

where the symbol $\langle \cdot \rangle$ denotes the positive part of its argument. Note that the exponent 2 inside the exponentials allows for the stress to be continuous at the contraction-extension interface $I_{mi}^{\mathrm{iso}} - 1 = 0$. In fact, any exponent larger than 1 is sufficient for that purpose.

Since

$$df = \frac{\partial f}{\partial \mathbf{C}_{\mathrm{iso}}^{\mathrm{e}}} : d\mathbf{C}_{\mathrm{iso}}^{\mathrm{e}} = \frac{\partial f}{\partial \mathbf{C}_{\mathrm{iso}}^{\mathrm{e}}} : \frac{d\mathbf{C}_{\mathrm{iso}}^{\mathrm{e}}}{d\mathbf{C}^{\mathrm{e}}} : d\mathbf{C}^{\mathrm{e}} = \frac{\partial f}{\partial (C_{\mathrm{iso}}^{\mathrm{e}})_{ij}} \frac{d(C_{\mathrm{iso}}^{\mathrm{e}})_{ij}}{dC_{kl}^{\mathrm{e}}} \, dC_{kl}^{\mathrm{e}}, \tag{6.6.28}$$

then

$$\mathbf{I} : d\mathbf{C}_{\mathrm{iso}}^{\mathrm{e}} = \mathbf{I} : \frac{d\mathbf{C}_{\mathrm{iso}}^{\mathrm{e}}}{d\mathbf{C}^{\mathrm{e}}} : d\mathbf{C}^{\mathrm{e}} = (J_{\mathrm{e}})^{-\frac{2}{3}} \left(\mathbf{I} - \tfrac{1}{3}(\mathbf{I} : \mathbf{C}^{\mathrm{e}}) \, (\mathbf{C}^{\mathrm{e}})^{-1} \right) : d\mathbf{C}^{\mathrm{e}}, \tag{6.6.29}$$

and

$$\begin{aligned} d\left(\underline{\mathbf{m}} \cdot \mathbf{C}_{\mathrm{iso}}^{\mathrm{e}} \cdot \underline{\mathbf{m}} \right) &= (J_{\mathrm{e}})^{-\frac{2}{3}} \, \underline{\mathbf{m}} \cdot \left(\left(\mathbf{I} \,\overline{\otimes}\, \mathbf{I} - \tfrac{1}{3} \mathbf{C}^{\mathrm{e}} \otimes (\mathbf{C}^{\mathrm{e}})^{-1} \right) : d\mathbf{C}^{\mathrm{e}} \right) \cdot \underline{\mathbf{m}} \\ &= (J_{\mathrm{e}})^{-\frac{2}{3}} \left(\underline{\mathbf{m}} \otimes \underline{\mathbf{m}} - \tfrac{1}{3}(\underline{\mathbf{m}} \cdot \mathbf{C}^{\mathrm{e}} \cdot \underline{\mathbf{m}}) \, (\mathbf{C}^{\mathrm{e}})^{-1} \right) : d\mathbf{C}^{\mathrm{e}}. \end{aligned} \tag{6.6.30}$$

The 2nd Piola-Kirchhoff stress becomes,

$$\begin{aligned} (\mathbf{F}^{\mathrm{e}})^{-1} \cdot \boldsymbol{\tau} \cdot (\mathbf{F}^{\mathrm{e}})^{-\mathrm{T}} &= 2 \frac{\partial E_{\kappa}}{\partial \mathbf{C}^{\mathrm{e}}} \\ &= \left(J_{\mathrm{e}} \frac{dU}{dJ_{\mathrm{e}}} \right) (\mathbf{C}^{\mathrm{e}})^{-1} + b_0 \, (J_{\mathrm{e}})^{-\frac{2}{3}} \left(\mathbf{I} - \tfrac{1}{3}(\mathbf{I} : \mathbf{C}^{\mathrm{e}}) \, (\mathbf{C}^{\mathrm{e}})^{-1} \right) \\ &\quad + 2 \, a_1 \sum_{i=1,2} \langle I_{mi}^{\mathrm{iso}} - 1 \rangle \exp \left(a_2 \langle I_{mi}^{\mathrm{iso}} - 1 \rangle^2 \right) (J_{\mathrm{e}})^{-\frac{2}{3}} \left(\underline{\mathbf{m}}_i \otimes \underline{\mathbf{m}}_i - \tfrac{1}{3}(\underline{\mathbf{m}}_i \cdot \mathbf{C}^{\mathrm{e}} \cdot \underline{\mathbf{m}}_i) \, (\mathbf{C}^{\mathrm{e}})^{-1} \right), \end{aligned} \tag{6.6.31}$$

whence the Kirchhoff stress $\boldsymbol{\tau} = \det \mathbf{F} \, \boldsymbol{\sigma}$,

$$\begin{aligned} \boldsymbol{\tau} &= \left(J_{\mathrm{e}} \frac{dU}{dJ_{\mathrm{e}}} \right) \mathbf{I} + b_0 \left(\mathbf{C}_{\mathrm{iso}}^{\mathrm{e}} - \tfrac{1}{3}(\mathbf{I} : \mathbf{C}_{\mathrm{iso}}^{\mathrm{e}}) \, \mathbf{I} \right) + 2 \, a_1 \sum_{i=1,2} \langle I_{mi}^{\mathrm{iso}} - 1 \rangle \exp \left(a_2 \langle I_{mi}^{\mathrm{iso}} - 1 \rangle^2 \right) \\ &\quad \times \left((\mathbf{F}_{\mathrm{iso}}^{\mathrm{e}} \cdot \underline{\mathbf{m}}_i) \otimes (\mathbf{F}_{\mathrm{iso}}^{\mathrm{e}} \cdot \underline{\mathbf{m}}_i) - \tfrac{1}{3}(\underline{\mathbf{m}}_i \cdot \mathbf{C}_{\mathrm{iso}}^{\mathrm{e}} \cdot \underline{\mathbf{m}}_i) \, \mathbf{I} \right). \end{aligned} \tag{6.6.32}$$

A particular isotropic case of (6.6.27) reads,

$$E_{\kappa} = \frac{a_0}{4} \underbrace{\left(\det \mathbf{C}^{\mathrm{e}} - 1 - \mathrm{Ln} \det \mathbf{C}^{\mathrm{e}} \right)}_{\substack{\text{isotropic} \\ \text{volumetric response}}} + \frac{b_0}{2} \underbrace{\left(\mathrm{tr}\, \mathbf{C}_{\mathrm{iso}}^{\mathrm{e}} - 3 \right)}_{\substack{\text{traceless} \\ \text{isotropic response}}}, \tag{6.6.33}$$

so that the Kirchhoff stress has the form,

$$\boldsymbol{\tau} = \frac{a_0}{2} \left(\det \mathbf{C}^{\mathrm{e}} - 1 \right) \mathbf{I} + b_0 \left(\mathbf{C}_{\mathrm{iso}}^{\mathrm{e}} - \tfrac{1}{3}(\mathbf{I} : \mathbf{C}_{\mathrm{iso}}^{\mathrm{e}}) \, \mathbf{I} \right). \tag{6.6.34}$$

For transversely isotropic fiber-reinforced ligaments, Limbert and Middleton [2004] define a piecewise potential according to the value of the fiber extension indicator $I_m - 1 = \mathbf{m} \cdot \mathbf{C}^{\mathrm{e}} \cdot \mathbf{m} - 1$,

$$E_{\kappa} = \frac{b_0}{2} \left(\mathrm{tr}\, \mathbf{C}^{\mathrm{e}} - 3 \right) + \begin{cases} 0, & I_m \leq 1, \\[2mm] \dfrac{a_1}{2\,a_2} \left(\exp(a_3 (I_m - 1)^2) - 1 \right), & I_m \leq I_m^*, \\[2mm] a_4 \sqrt{I_m} + a_5 \, \mathrm{Ln}\, I_m, & I_m \geq I_m^*. \end{cases} \tag{6.6.35}$$

The motivation is the following. The collagen fibers do not sustain load in contraction. Under progressive extension, they are recruited progressively until the value $I_m = I_m^* > 1$ is reached. The material constants a are chosen so as to ensure continuity of the stress at the critical value $I_m = I_m^*$.

Taber and Chabert [2002] consider the more complex picture of cardiac tissues viewed as elastic-growing materials with muscle activation. If we neglect the circumferential and axial reinforcements due to collagen fibers, the passive behavior is modeled through the elastic potential,

$$E_{\kappa} = \frac{a_0}{2} \left(\exp Q - 1 \right), \quad Q = \frac{\mu}{2} \left(\mathrm{tr}\, \mathbf{C}^{\mathrm{e}} - 3 \right) + \frac{\mu}{2\,\gamma} \left((\det \mathbf{C}^{\mathrm{e}})^{-\gamma} - 1 \right), \tag{6.6.36}$$

that involves the neo-Hookean function Q used for compressible foam rubbers, as defined by (6.4.7), with the material constants a_0, μ and $\gamma = \nu/(1-2\nu)$, where ν can be seen as a Poisson's ratio in the infinitesimal strain limit. Note that the linear term $\operatorname{tr} \mathbf{C}^e$ in Q implies the volumetric and shear response to be coupled, in contrast to (6.6.27).

For a transversely isotropic fiber-reinforced myocardium, the five independent isotropic scalar invariants of the Green strain are listed in Table 2.11.4. They may be expressed in component form in the orthotropy axes (1 direction of the fibers) and recombined,

$$E^e_{\mathrm{G}11}, \; E^e_{\mathrm{G}22} + E^e_{\mathrm{G}33}, \; (E^e_{\mathrm{G}12})^2 + (E^e_{\mathrm{G}13})^2, (E^e_{\mathrm{G}22})^2 + (E^e_{\mathrm{G}33})^2 + 2\,(E^e_{\mathrm{G}23})^2,$$

$$E^e_{\mathrm{G}22}\,(E^e_{\mathrm{G}12})^2 + E^e_{\mathrm{G}33}\,(E^e_{\mathrm{G}13})^2 + 2\,E^e_{\mathrm{G}12}\,E^e_{\mathrm{G}13}\,E^e_{\mathrm{G}23}\,. \tag{6.6.37}$$

An application to the ventricular myocardium determined from a cylindrical model is shown in Guccione et al. [1991], namely

$$\underline{E}_\kappa = \frac{a_0}{2}\left(\exp(Q) - 1\right), \quad Q(\mathbf{E}^e_{\mathrm{G}}) = a_1 \operatorname{tr} \mathbf{E}^e_{\mathrm{G}} + a_2\,(E^e_{\mathrm{G}11})^2 + a_3\left((E^e_{\mathrm{G}22})^2 + (E^e_{\mathrm{G}33})^2 + 2\,(E^e_{\mathrm{G}23})^2\right), \tag{6.6.38}$$

where the a's are material constants. Note that the constitutive stress is not traceless.

6.7 Return to the conewise response in presence of fibers

We have listed three methods that address the conewise constitutive response of fiber-reinforced materials:

1. linear conewise elasticity: Section 5.4.2 defines the concept of admissible constitutive cones. The idea is applied to materials endowed with structural and fabric anisotropic properties in the subsequent sections;

2. additive contribution of a ground substance and fibers, and direct incorporation of the influence of the sign of the strain along fibers in the strain energy of the latter. In (6.6.27), the transition is sharp while it is smoothed over a short step in (6.6.35);

3. a smooth variation of the stiffness at the transition between contraction and extension, associated with largely distinct moduli on each side of the transition, Eqns (6.6.14) and (6.6.15).

In all cases, the continuity of the stress at the transition between constitutive cones should be taken care of.

6.8 Elasticity and workless constitutive stress

Two instances where the constitutive equations are not hyperelastic are reported below.

For a spherical shell, made of an incompressible elastic-growing material, Chen and Hoger [2000] do not assume the existence of an elastic potential. Since the motion is a pure deformation, the polar decomposition of the elastic transformation simplifies to $\mathbf{F}^e = \mathbf{U}^e = \mathbf{V}^e$. The effective stress can be expressed in terms of e.g., the left elastic stretch $\mathbf{V}^e$,

$$\boldsymbol{\sigma} + p\,\mathbf{I} = f_1(I^e_1, I^e_2)\,\mathbf{V}^e + f_2(I^e_1, I^e_2)\,(\mathbf{V}^e)^2\,, \tag{6.8.1}$$

where f_1 and f_2 are scalar functions of the invariants I^e's of $\mathbf{V}^e$, not of $\mathbf{C}^e = (\mathbf{V}^e)^2$,

$$I^e_1 = \operatorname{tr} \mathbf{V}^e, \quad I^e_2 = \tfrac{1}{2}\left((\operatorname{tr} \mathbf{V}^e)^2 - \operatorname{tr}(\mathbf{V}^e)^2\right). \tag{6.8.2}$$

The components of the effective Cauchy stress in the radial and tangential axes may be cast in the form,

$$\boldsymbol{\sigma} + p\,\mathbf{I} = \operatorname{\mathbf{diag}}\left[f_1\,v_r + f_2\,v_r^2,\; f_1\,v_\theta + f_2\,v_\theta^2,\; f_1\,v_\theta + f_2\,v_\theta^2\right]. \tag{6.8.3}$$

The constitutive equation is enforced to satisfy the Baker-Ericksen inequality, which states that the stress is largest in the direction of the largest stretch, i.e., here

$$\frac{\sigma_r - \sigma_\theta}{v_r - v_\theta} = f_1 + (v_r + v_\theta)\,f_2 > 0\,. \tag{6.8.4}$$

Criscione et al. [2003] consider that the stress is contributed by a component that derives from an elastic potential and by a symmetric workless component[6.6]. Section 2.10 introduces a component of the stress that does no work if

---

[6.6]The authors name this stress as 'hypoelastic'. As indicated in Chapter 5, hyperelasticity corresponds to a one-to-one stress-strain relation and the stress derives from a potential, that is, the work done on a closed strain cycle is zero. Elasticity corresponds to one-to-one stress-strain relation, but the stress does not derive from a potential, that is, the work done on a closed strain cycle may not be zero. Hypoelasticity implies a stress-strain relation which is not one-to-one, and the work done in a closed strain cycle does not vanish. Still, on the reverse strain cycle, the stress path is recovered as well as the work done, i.e., the initial stress is recovered, there is no irreversibility in stress and there is no dissipation.

the kinematical constraint is satisfied. This reaction stress is defined by boundary conditions. In contrast here, the component of the stress that does no work is constitutive, namely,

$$\boldsymbol{\tau} = (\det \mathbf{F})\,\boldsymbol{\sigma} = \mathbf{F} \cdot \frac{\partial E}{\partial \mathbf{E}_{\mathrm{G}}} \cdot \mathbf{F}^{\mathrm{T}} + \boldsymbol{\tau}_R, \tag{6.8.5}$$

with $\boldsymbol{\tau}_R : \mathbf{L} = 0$, $\mathbf{L}$ being the velocity gradient. The analysis is tailored to biomembranes. Let $(\underline{\mathbf{m}}, \underline{\mathbf{s}}, \underline{\mathbf{n}})$ be a reference cartesian triad, with the two first vectors in the plane of the membrane. Under loading, this reference triad undergoes the rotation $\mathbf{Q}$ and transforms to $(\mathbf{m}, \mathbf{s}, \mathbf{n})$. The deformation gradient $\mathbf{F}$, local deformation gradient $\mathbf{f} = \mathbf{Q}^{\mathrm{T}} \cdot \mathbf{F}$ and velocity gradient $\mathbf{L} = (d\mathbf{F}/dt) \cdot \mathbf{F}^{-1} = \mathbf{Q} \cdot (d\mathbf{f}/dt) \cdot \mathbf{f}^{-1} \cdot \mathbf{Q}^{\mathrm{T}} + (d\mathbf{Q}/dt) \cdot \mathbf{Q}^{\mathrm{T}}$ express in these axes as,

$$
\begin{aligned}
\mathbf{F} &= \lambda_M \, \mathbf{m} \otimes \underline{\mathbf{m}} + \lambda_M \, \phi_{MS} \, \mathbf{m} \otimes \underline{\mathbf{s}} + \lambda_S \, \mathbf{s} \otimes \underline{\mathbf{s}} \\
\mathbf{f} &= \lambda_M \, \underline{\mathbf{m}} \otimes \underline{\mathbf{m}} + \lambda_M \, \phi_{MS} \, \underline{\mathbf{m}} \otimes \underline{\mathbf{s}} + \lambda_S \, \underline{\mathbf{s}} \otimes \underline{\mathbf{s}} \\
\mathbf{L} &= \frac{d\mathrm{Ln}\lambda_M}{dt} \, \mathbf{m} \otimes \mathbf{m} + \frac{\lambda_M}{\lambda_S} \frac{d\phi_{MS}}{dt} \, \mathbf{m} \otimes \mathbf{s} + \frac{d\mathrm{Ln}\lambda_S}{dt} \, \mathbf{s} \otimes \mathbf{s} + \frac{d\mathbf{Q}}{dt} \cdot \mathbf{Q}^{\mathrm{T}}.
\end{aligned} \tag{6.8.6}
$$

For biaxial stretch under a radial path, namely for $\mathrm{Ln}\lambda_M / \mathrm{Ln}\lambda_S$ constant and $\phi_{MS} = 0$, Criscione et al. [2003] deduce that the workless stress should be of the form,

$$\boldsymbol{\tau}_R = a(\lambda_M, \lambda_S) \left(\mathrm{Ln}\lambda_S \, \mathbf{m} \otimes \mathbf{m} - \mathrm{Ln}\lambda_M \, \mathbf{s} \otimes \mathbf{s} \right), \tag{6.8.7}$$

with a multiplicative factor a which depends on the stretches λ_M and λ_S.

6.9 Use of right and left Cauchy-Green tensors

For an elastic material, the mechanical dissipation expresses as

$$-\frac{dE}{dt} - E\,\mathbf{I} : \mathbf{D} + \boldsymbol{\sigma} : \mathbf{D} \geq 0, \tag{6.9.1}$$

in terms of the free energy (elastic potential) per current volume E, Cauchy stress $\boldsymbol{\sigma}$ and strain rate $\mathbf{D}$.

If the free energy depends on the right Cauchy-Green tensor $\mathbf{C} = \mathbf{F}^{\mathrm{T}} \cdot \mathbf{F}$, or Green strain $\mathbf{E}_{\mathrm{G}} = \frac{1}{2}(\mathbf{C} - \mathbf{I})$, the inequality becomes with help of $(2.5.14)_1$,

$$\left(-\mathbf{F} \cdot \frac{dE}{d\mathbf{E}_{\mathrm{G}}} \cdot \mathbf{F}^{\mathrm{T}} - E\,\mathbf{I} + \boldsymbol{\sigma} \right) : \mathbf{D} \geq 0, \tag{6.9.2}$$

and, therefore, in absence of constraints on the deformation gradient, the constitutive Eshelby stress expresses as

$$
\begin{aligned}
\boldsymbol{\sigma} - E\,\mathbf{I} &= \mathbf{F} \cdot \frac{dE}{d\mathbf{E}_{\mathrm{G}}} \cdot \mathbf{F}^{\mathrm{T}} \\
&= 2\,\mathbf{F} \cdot \frac{dE}{d\mathbf{C}} \cdot \mathbf{F}^{\mathrm{T}}.
\end{aligned} \tag{6.9.3}
$$

As an example, if the free energy may depend only on the first and second scalar isotropic invariants of the strain $\mathbf{E}_{\mathrm{G}}$ via the material constants a and $\mu > 0$,

$$E = a\,\mathrm{tr}\,\mathbf{E}_{\mathrm{G}} + \mu\,\mathrm{tr}\,\mathbf{E}_{\mathrm{G}}^2, \tag{6.9.4}$$

the stress turns to,

$$\boldsymbol{\sigma} - E\,\mathbf{I} = a\,\mathbf{F} \cdot \mathbf{F}^{\mathrm{T}} + 2\,\mu\,\mathbf{F} \cdot \mathbf{E}_{\mathrm{G}} \cdot \mathbf{F}^{\mathrm{T}}. \tag{6.9.5}$$

Now if the free energy depends on the left Cauchy-Green tensor $\mathbf{B} = \mathbf{F} \cdot \mathbf{F}^{\mathrm{T}}$, or on the covariant Almansi strain $\mathbf{E}_{\mathrm{A}}^{\mathrm{co}} = \frac{1}{2}(\mathbf{I} - \mathbf{B}^{-1})$, the inequality becomes with help of $(2.5.14)_2$,

$$\left(-\frac{dE}{d\mathbf{B}} \cdot \mathbf{B} - \mathbf{B} \cdot \frac{dE}{d\mathbf{B}} - E\,\mathbf{I} + \boldsymbol{\sigma} \right) : \mathbf{D} \geq 0, \tag{6.9.6}$$

and, therefore, in absence of constraints on the deformation gradient, the constitutive Eshelby stress expresses as

$$\boldsymbol{\sigma} - E\,\mathbf{I} \;=\; \frac{dE}{d\mathbf{B}} \cdot \mathbf{B} + \mathbf{B} \cdot \frac{dE}{d\mathbf{B}} \,. \tag{6.9.7}$$

As an example, the free energy (elastic potential) may depend only on the first and second scalar isotropic invariants of the strain $\mathbf{E}_A^{co}$,

$$E = a\,\mathrm{tr}\,\mathbf{E}_A^{co} + \mu\,\mathrm{tr}\,(\mathbf{E}_A^{co})^2 \,. \tag{6.9.8}$$

Since to first order in $\mathbf{H}$,

$$\mathbf{E}_A^{co}(\mathbf{B}+\mathbf{H}) - \mathbf{E}_A^{co}(\mathbf{B}) = -\frac{1}{2}\mathbf{B}^{-1}\cdot\mathbf{H}\cdot\mathbf{B}^{-1},$$

$$\left(\mathbf{E}_A^{co}(\mathbf{B}+\mathbf{H})\right)^2 - \left(\mathbf{E}_A^{co}(\mathbf{B})\right)^2 = \frac{1}{2}\left(\mathbf{E}_A^{co}(\mathbf{B})\cdot\mathbf{B}^{-1}\cdot\mathbf{H}\cdot\mathbf{B}^{-1} + \mathbf{B}^{-1}\cdot\mathbf{H}\cdot\mathbf{B}^{-1}\cdot\mathbf{E}_A^{co}(\mathbf{B})\right), \tag{6.9.9}$$

then

$$\frac{d}{d\mathbf{B}}\mathrm{tr}\,\mathbf{E}_A^{co} = \frac{1}{2}\mathbf{B}^{-2}, \quad \frac{d}{d\mathbf{B}}\mathrm{tr}\,(\mathbf{E}_A^{co})^2 = \mathbf{B}^{-1}\cdot\mathbf{E}_A^{co}\cdot\mathbf{B}^{-1}, \tag{6.9.10}$$

and the Elshelby stress becomes now,

$$\begin{aligned}
\boldsymbol{\sigma} - E\,\mathbf{I} &= a\,\mathbf{B}^{-1} + \mu\left(\mathbf{B}^{-1}\cdot\mathbf{E}_A^{co} + \mathbf{E}_A^{co}\cdot\mathbf{B}^{-1}\right) \\
&= a\,\mathbf{F}^{-T}\cdot\mathbf{F}^{-1} + 2\,\mu\,\mathbf{V}^{-1}\cdot\mathbf{E}_A^{co}\cdot\mathbf{V}^{-1},
\end{aligned} \tag{6.9.11}$$

with $\mathbf{V}$ the left stretch in the polar decomposition of $\mathbf{F}$.

Exercises on Chapter 6

Exercise 6.1 Strict convexity and curvature

A positive definite Hessian implies the function it derives from to be strictly convex. On the other hand, a function may be strictly convex while its Hessian is not positive definite, but only positive semi-definite. As an example, show that the real function $\phi(x) = x^4$ is strictly convex while its second derivative vanishes at $x = 0$, Fletcher [1990].

Answer:

$$\lambda\,\phi(x) + (1 - \lambda)\,\phi(y) - \phi(\lambda\,x + (1 - \lambda)\,y)$$

$$= \lambda\,x^4 + (1 - \lambda)\,y^4 - \left(\lambda\,x + (1 - \lambda)\,y\right)^4 \tag{6.E.1}$$

$$= \lambda\,(1 - \lambda)\,\left(\lambda^2\,(x - y)^4 + \lambda\,(x - y)^3(x + 3\,y) + (x - y)^2\,(x^2 + 2\,x\,y + 3\,y^2)\right).$$

The discriminant of the quadratic polynomial in λ,

$$- 3\,(x - y)^6\,(x^2 + \frac{2}{3}\,x\,y + y^2), \tag{6.E.2}$$

is strictly negative for $x \neq y$. The expression (6.E.1) does not change sign and, for $x \neq 0$, $y = 0$, $\lambda \in\,]0, 1[$, it is equal to $\lambda\,(1 - \lambda^3)\,x^4$ and therefore strictly positive, while the 2nd derivative $12\,x^2$ clearly vanishes at $x = 0$.

Exercise 6.2 Derivatives of $\mathbf{A}^n$, $n > 0$

Let $\mathbf{A}$ be a symmetric second order tensor. Calculate the derivatives of $\mathbf{A}^2$ and $\mathbf{A}^3$ with respect to $\mathbf{A}$. Deduce the derivative of $\mathbf{A}^n$, for $n > 0$ integer.

Answer:

1. Derivative of $\mathbf{A}^2$:

$$(\mathbf{A} + \mathbf{H})^2 = \mathbf{A}^2 + \mathbf{A} \cdot \mathbf{H} + \mathbf{H} \cdot \mathbf{A} + \mathbf{H}^2$$

$$(\mathbf{A} + \mathbf{H})^2 - \mathbf{A}^2 = \mathbf{A} \cdot \mathbf{H} + \mathbf{H} \cdot \mathbf{A} + O(\mathbf{H}^2) = \frac{d\mathbf{A}^2}{d\mathbf{A}} : \mathbf{H} + O(\mathbf{H}^2)$$

$$\frac{d\mathbf{A}^2}{d\mathbf{A}} = \mathbf{A}\,\overline{\underline{\otimes}}\,\mathbf{I} + \mathbf{I}\,\underline{\otimes}\,\mathbf{A} \tag{6.E.3}$$

$$\left(\frac{d\mathbf{A}^2}{d\mathbf{A}}\right)_{ijkl} = A_{ik}\,I_{jl} + I_{ik}\,A_{lj}\,.$$

2. Derivative of $\mathbf{A}^3$:

$$(\mathbf{A} + \mathbf{H})^3 = \mathbf{A}^3 + \mathbf{A}^2 \cdot \mathbf{H} + \mathbf{A} \cdot \mathbf{H} \cdot \mathbf{A} + \mathbf{H} \cdot \mathbf{A}^2 + \mathbf{A} \cdot \mathbf{H}^2 + \mathbf{H} \cdot \mathbf{A} \cdot \mathbf{H} + \mathbf{H}^2 \cdot \mathbf{A} + \mathbf{H}^3$$

$$(\mathbf{A} + \mathbf{H})^3 - \mathbf{A}^3 = \mathbf{A}^2 \cdot \mathbf{H} + \mathbf{A} \cdot \mathbf{H} \cdot \mathbf{A} + \mathbf{H} \cdot \mathbf{A}^2 + O(\mathbf{H}^2) = \frac{d\mathbf{A}^3}{d\mathbf{A}} : \mathbf{H} + O(\mathbf{H}^2)$$

$$\frac{d\mathbf{A}^3}{d\mathbf{A}} = \mathbf{A}^2\,\overline{\underline{\otimes}}\,\mathbf{I} + \mathbf{A}\,\overline{\underline{\otimes}}\,\mathbf{A} + \mathbf{I}\,\underline{\otimes}\,\mathbf{A}^2 \tag{6.E.4}$$

$$\left(\frac{d\mathbf{A}^3}{d\mathbf{A}}\right)_{ijkl} = (\mathbf{A}^2)_{ik}\,I_{jl} + A_{ik}\,A_{jl} + I_{ik}\,(\mathbf{A}^2)_{jl}\,.$$

The component form of the last line above has used the property that, if $\mathbf{A}$ is a symmetric second order tensor, so is $\mathbf{A}^2$.

3. Derivative of $\mathbf{A}^n$, $n \geq 0$ positive integer:

$$\frac{d\mathbf{A}^n}{d\mathbf{A}} = \sum_{p=0}^{n-1} \mathbf{A}^{n-1-p}\,\overline{\underline{\otimes}}\,\mathbf{A}^p$$

$$\left(\frac{d\mathbf{A}^n}{d\mathbf{A}}\right)_{ijkl} = \sum_{p=0}^{n-1} (\mathbf{A}^{n-1-p})_{ik}\,(\mathbf{A}^p)_{jl}\,. \tag{6.E.5}$$

Exercise 6.3 Derivatives of $\mathbf{A}^n$, $n < 0$

Let $\mathbf{A}$ be a symmetric second order tensor. Calculate the derivatives of $\mathbf{A}^{-1}$ and $\mathbf{A}^{-2}$ with respect to $\mathbf{A}$. Deduce the derivative of $\mathbf{A}^n$, for $n < 0$ integer.

Answer:
1. Calculation of $(\mathbf{A} + \mathbf{H})^{-1}$:

$$
\begin{aligned}
\mathbf{A} + \mathbf{H} &= \mathbf{A} \cdot (\mathbf{I} + \mathbf{A}^{-1}\mathbf{H}) \\
(\mathbf{A} + \mathbf{H})^{-1} &= (\mathbf{I} + \mathbf{A}^{-1}\mathbf{H})^{-1} \cdot \mathbf{A}^{-1} \\
(\mathbf{I} + \mathbf{A}^{-1}\mathbf{H})^{-1} &= \mathbf{I} - \mathbf{A}^{-1} \cdot \mathbf{H} + (\mathbf{A}^{-1} \cdot \mathbf{H})^2 - (\mathbf{A}^{-1} \cdot \mathbf{H})^3 + \cdots \\
(\mathbf{A} + \mathbf{H})^{-1} &= \mathbf{A}^{-1} - \mathbf{A}^{-1} \cdot \mathbf{H} \cdot \mathbf{A}^{-1} + \mathbf{A}^{-1} \cdot \mathbf{H} \cdot \mathbf{A}^{-1} \cdot \mathbf{H} \cdot \mathbf{A}^{-1} + \cdots .
\end{aligned}
\tag{6.E.6}
$$

2. Derivative of $\mathbf{A}^{-1}$:

$$
\begin{aligned}
(\mathbf{A} + \mathbf{H})^{-1} - \mathbf{A}^{-1} &= -\mathbf{A}^{-1} \cdot \mathbf{H} \cdot \mathbf{A}^{-1} + O(\mathbf{H}^2) \\
\frac{d\mathbf{A}^{-1}}{d\mathbf{A}} &= -\mathbf{A}^{-1} \,\overline{\otimes}\, \mathbf{A}^{-1} \\
\Big(\frac{d\mathbf{A}^{-1}}{d\mathbf{A}}\Big)_{ijkl} &= -(\mathbf{A}^{-1})_{ik}\,(\mathbf{A}^{-1})_{jl} .
\end{aligned}
\tag{6.E.7}
$$

3. Derivative of $\mathbf{A}^{-2}$:
   Using $(6.\text{E}.6)_4$,

$$
\begin{aligned}
(\mathbf{A} + \mathbf{H})^{-2} - \mathbf{A}^{-2} &= (\mathbf{A} + \mathbf{H})^{-1} \cdot (\mathbf{A} + \mathbf{H})^{-1} - \mathbf{A}^{-2} \\
&= -\mathbf{A}^{-2} \cdot \mathbf{H} \cdot \mathbf{A}^{-1} - \mathbf{A}^{-1} \cdot \mathbf{H} \cdot \mathbf{A}^{-2} + O(\mathbf{H}^2) \\
\frac{d\mathbf{A}^{-2}}{d\mathbf{A}} &= -\mathbf{A}^{-2} \,\overline{\otimes}\, \mathbf{A}^{-1} - \mathbf{A}^{-1} \,\overline{\otimes}\, \mathbf{A}^{-2} \\
\Big(\frac{d\mathbf{A}^{-2}}{d\mathbf{A}}\Big)_{ijkl} &= -(\mathbf{A}^{-2})_{ik}\,(\mathbf{A}^{-1})_{jl} - (\mathbf{A}^{-1})_{ik}\,(\mathbf{A}^{-2})_{jl} .
\end{aligned}
\tag{6.E.8}
$$

Exercise 6.4 Derivative $\partial\mathbf{E}_{\mathrm{G}}/\partial\mathbf{F}$ of Green strain wrt the deformation gradient
Calculate the derivative of Green strain $\mathbf{E}_{\mathrm{G}} = \frac{1}{2}(\mathbf{F}^{\mathrm{T}} \cdot \mathbf{F} - \mathbf{I})$ wrt the deformation gradient $\mathbf{F}$.

Answer:

$$
\begin{aligned}
\mathbf{E}_{\mathrm{G}}(\mathbf{F} + \mathbf{H}) - \mathbf{E}_{\mathrm{G}}(\mathbf{F}) &= \frac{1}{2}(\mathbf{F}^{\mathrm{T}} \cdot \mathbf{H} + \mathbf{H}^{\mathrm{T}} \cdot \mathbf{F} + \mathbf{H}^{\mathrm{T}} \cdot \mathbf{H}), \\
&= \frac{1}{2}(\mathbf{F}^{\mathrm{T}} \cdot \mathbf{H} + \mathbf{H}^{\mathrm{T}} \cdot \mathbf{F}) + O(\mathbf{H}^2), \\
\Big(\frac{\partial\mathbf{E}_{\mathrm{G}}}{\partial\mathbf{F}}\Big)_{ijkl} &= \frac{1}{2}\big(F_{ki}\,I_{jl} + F_{kj}\,I_{il}\big) .
\end{aligned}
\tag{6.E.9}
$$

Exercise 6.5 An elementary differential relation
If $\mathbf{A}$ is a symmetric second order tensor, and $\boldsymbol{\epsilon}$ the infinitesimal strain, show that

$$
\frac{d}{d\boldsymbol{\epsilon}}(\mathbf{A} : \boldsymbol{\epsilon}^2) = \mathbf{A} \cdot \boldsymbol{\epsilon} + \boldsymbol{\epsilon} \cdot \mathbf{A} .
\tag{6.E.10}
$$

Proof: Let $\mathbf{H}$ be an arbitrary symmetric second order tensor. Then

$$
\frac{d(\mathbf{A} : \boldsymbol{\epsilon}^2)}{d\boldsymbol{\epsilon}} : \mathbf{H} = \lim_{\mathbf{H} \to 0} \mathbf{A} : (\boldsymbol{\epsilon} + \mathbf{H})^2 - \mathbf{A} : \boldsymbol{\epsilon}^2 .
\tag{6.E.11}
$$

To first order in $\mathbf{H}$, the latter is equal to $\mathbf{A} : (\mathbf{H} \cdot \boldsymbol{\epsilon} + \boldsymbol{\epsilon} \cdot \mathbf{H})$, or componentwise $A_{ij}(H_{ik}\epsilon_{kj} + \epsilon_{ik}H_{kj})$ i.e., to $(\mathbf{A} \cdot \boldsymbol{\epsilon} + \boldsymbol{\epsilon} \cdot \mathbf{A}) : \mathbf{H}$, componentwise $(\epsilon_{ij}A_{jk} + A_{ij}\epsilon_{jk})H_{ik}$.

Chapter 7

Poroelasticity with a single porosity

This chapter addresses deformation and seepage in isothermal poroelastic tissues with a single porosity and devoid of an electric fixed charge. Thermal effects in fluid-saturated porous media are introduced in Chapter 9. Tissues in which two distinct porosities may be highlighted are considered in Sections 10.6 and 24.8.2. In fact, Chapter 9 assumes local thermal equilibrium, namely a temperature which is uniform over constituents, while Chapter 10 enlarges the landscape and allows each constituent to be endowed with its own temperature.

Emphasis is laid here on the anisotropic properties of the fluid-saturated porous media and on their identification through a minimal number of *independent* laboratory experiments. As an example, a simple test, namely the unconfined compression of cylindrical sample, is worked out and solved through analytical tools. Along a guiding principle in this textbook, the test is formatted as an initial and boundary value problem that provides a framework in which the roles of the various developments of this chapter and of the previous chapters can be appreciated.

7.1 Geometrical, kinematical and mechanical descriptors

Let a porous medium be constituted by a solid skeleton and a set of miscible or non miscible fluid species. In this chapter, the porous medium is viewed as a *closed* thermodynamic system: there may well be internal exchanges between species, or phases, of mass, momentum, energy and entropy. However the sum of the contributions over species vanishes.

The set of species of the porous medium is denoted by $\mathcal{K}$. Thus $\mathcal{K} = \{s\} \cup \mathcal{K}^{**}$ where s denotes the solid skeleton and $\mathcal{K}^{**}$ the set of non miscible fluid species.

Phases represent the species when viewed as part of the mixture. The solid phase is also referred to as the solid skeleton. Following the classical approach of mixture theory, e.g., Bowen [1982], the phases are viewed as independent overlapping continua as sketched in Fig. 7.1.1. In fact, defining the phases from the set of species is part of the modeling. A phase may contain several species: the case of a fluid phase containing solutes (dissolved species) is addressed in relation with articular cartilages and cornea in Chapters 14 et seq. Conversely, a single species may be present in two or several phases, like in the double porosity model where the two fluid phases serve different transport purposes.

The simplest setting is standard poroelasticity where a solid is circulated by a non-reacting ideal fluid: here, species and phases are one and the same.

The volume occupied by the fluid phases is also called porosity. Volume fractions sum up to one,

$$\sum_{k \in \mathcal{K}} n^k = 1 \, . \tag{7.1.1}$$

Within the species, strains and stresses are called *intrinsic* while strains and stresses attached to the phases of the porous medium are said *apparent* or *partial*. Intrinsic and apparent quantities act at different levels or, say, scales. A distinction is made between species and mixture variables.

The intrinsic strain of the species k is denoted by ϵ_k, and its intrinsic stress by σ_k. Their apparent counterparts are denoted by superscripts, ϵ^k and $\sigma^k = n^k \sigma_k$ respectively. The intrinsic and apparent fluid stresses are respectively $\sigma_k = -p_k \mathbf{I}$ and $\sigma^k = -p^k \mathbf{I}$, p_k and $p^k = n^k p_k$ denoting the intrinsic and apparent fluid pressures. As a standard

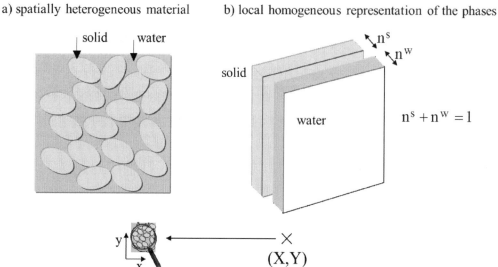

Fig. 7.1.1: While, in an actual material, a single species exists at any geometrical point, the mixture theory considers that all species are present at any point but in proportion with their volume fractions.

approximation to (4.5.7), the diffusive part of the total stress is neglected, and the total stress $\boldsymbol{\sigma}$ is then equal to the spatial (or areal) average of the intrinsic stresses, namely

$$
\boldsymbol{\sigma} = \begin{cases} n^{\mathrm{s}} \boldsymbol{\sigma}_{\mathrm{s}} + \displaystyle\sum_{k \in \mathcal{K}^{**}} n^{k} \boldsymbol{\sigma}_{k} &= n^{\mathrm{s}} \boldsymbol{\sigma}_{\mathrm{s}} - \displaystyle\sum_{k \in \mathcal{K}^{**}} n^{k} p_{k} \mathbf{I}, \\[2ex] \boldsymbol{\sigma}^{\mathrm{s}} + \displaystyle\sum_{k \in \mathcal{K}^{**}} \boldsymbol{\sigma}^{k} &= \boldsymbol{\sigma}^{\mathrm{s}} - \displaystyle\sum_{k \in \mathcal{K}^{**}} p^{k} \mathbf{I}. \end{cases}
\tag{7.1.2}
$$

The infinitesimal apparent strains are defined as the symmetrized gradients of the displacement fields $\mathbf{u}_k$,

$$
\boldsymbol{\epsilon} = \boldsymbol{\epsilon}^{\mathrm{s}} = \frac{1}{2} (\mathbf{u}_{\mathrm{s}} \overleftarrow{\boldsymbol{\nabla}} + \overrightarrow{\boldsymbol{\nabla}} \mathbf{u}_{\mathrm{s}}) ; \quad \boldsymbol{\epsilon}^{k} = \frac{1}{2} (\mathbf{u}_{k} \overleftarrow{\boldsymbol{\nabla}} + \overrightarrow{\boldsymbol{\nabla}} \mathbf{u}_{k}), \quad k \in \mathcal{K}^{**} .
\tag{7.1.3}
$$

The apparent strain of the solid $\boldsymbol{\epsilon}^{\mathrm{s}}$ is also the strain of the solid skeleton $\boldsymbol{\epsilon}$, or mixture as a whole. Apparent strains and intrinsic strains of the species serve different purposes:

- the apparent strain $\boldsymbol{\epsilon} = \boldsymbol{\epsilon}^{\mathrm{s}}$ can be measured by observing, at the macroscopic scale, the deformation of a grid painted on the *solid skeleton*, with nodes moving at velocity $\mathbf{v}_{\mathrm{s}}$;
- intrinsic strains are attached to the *species*; no displacement field is defined to follow these microscopic motions. The intrinsic strains of the solid constituent and of the fluid species are not treated on the same footing. For example, in absence of mass transfer and in an isothermal context, the incompressibility of the solid constituent implies a zero volumetric strain $\mathrm{tr}\,\boldsymbol{\epsilon}_{\mathrm{s}}$. The same cannot be said for fluid species: the volume changes because of the fluid compressibility, and because the fluid diffuses through the solid skeleton. Under these circumstances, the mass change vanishes only if the fluid remains attached to the solid skeleton, so-called *undrained condition*, which implies zero volume change only if the fluid is incompressible and thermally not dilatable.

Upscaling of the intrinsic strains results in

Proposition 7.1 The strain averaging relation.

The strain of the mixture, or solid skeleton, is obtained by spatial integration of the intrinsic strains of the constituents over a representative volume,

$$
n^{\mathrm{s}} \,\mathrm{tr}\,\boldsymbol{\epsilon}_{\mathrm{s}} + \sum_{k \in \mathcal{K}^{**}} n^{k} \,\mathrm{tr}\,\boldsymbol{\epsilon}_{k} = \mathrm{tr}\,\boldsymbol{\epsilon}, \quad n^{\mathrm{s}} \,\mathrm{dev}\,\boldsymbol{\epsilon}_{\mathrm{s}} + \sum_{k \in \mathcal{K}^{**}} n^{k} \,\mathrm{dev}\,\boldsymbol{\epsilon}_{k} = \mathrm{dev}\,\boldsymbol{\epsilon} .
\tag{7.1.4}
$$

Consider a process where the solid skeleton undergoes a deformation of gradient $\mathbf{F}$. Then the intrinsic volumetric strain and variation of volume content (2.6.25) of each species are linked by the relation,

$$\det \mathbf{F}^{-1} \, \Delta v^k = n^k \operatorname{tr} \epsilon_k, \quad k \in \mathcal{K}. \tag{7.1.5}$$

7.2 Elements of mixture theory: the level/scale of the species

The spatial and directional organizations at various length scales of the species constituting the mixture are of paramount significance for the overall response of the mixture. Still the physical properties of the species themselves are an important ingredient to this response. Table 7.2.1 serves to appreciate the typical orders of magnitude of the elastic moduli of the species and of the mixture of a few materials.

Table 7.2.1: Typical mechanical properties of some materials at room temperature and atmospheric pressure. t tensile; c: compressive; r: radial; or: orthoradial

Material	Volume Fraction of Water [-]	Modulus of Main Constituent [GPa]	Overall Bulk Modulus [MPa]	Shear Modulus [MPa]
Water	-	2.2	-	0
Articular cartilages	0.7 - 0.9	t: 0.1	t: 10; c: 0.2	0.1
Cornea	0.8 - 0.9	t: 0.1	or: 1; r: 10^{-3}	or: 10^{-2} ; r: 10^{-3}
Interstitium	0.2	-	0.1 - 0.5	10^{-2}
Sand	0 - 0.5	40	100	100
Marine Clay [a]	0.71	50	20	0.072
Halite (salt) [a]	0.001	23.5	20.7×10^3	12.4×10^3

[a] McTigue [1986]

7.2.1 Intrinsic properties of species

Let Δ denote the variation of any time dependent quantity $a = a(t)$ from its value $a_0 = a(0)$ in the initial state denoted by a subscript or superscript 0: $\Delta a \equiv a(t) - a_0$. In a linearized analysis about an unstressed and unstrained initial state, then simply $\Delta \boldsymbol{\sigma} = \boldsymbol{\sigma}$, $\Delta p_{\mathrm{w}} = p_{\mathrm{w}}$, $\Delta \boldsymbol{\epsilon} = \boldsymbol{\epsilon}$.

The mass transfers may be viewed as introducing a formal modification of the strain of the species,

$$\tilde{\epsilon}^k = \left(\operatorname{tr} \epsilon^k - \int_0^t \frac{\hat{\rho}^k}{\rho^k} \, d\tau \right) \frac{\mathbf{I}}{3} + \operatorname{dev} \epsilon^k \,, \quad k \in \mathcal{K}. \tag{7.2.1}$$

Fluid mass transfers are governed by (electro)chemical potentials, as indicated by (4.7.31). To develop a constitutive equation of mass transfer for the solid, the notion of chemical potential tensor alluded for in Chapter 4 should be worked out. An alternative interpretation of the modified strains arises in the context of growth, addressed in Chapter 21: the rates of mass supply of the solid constituent are governed by growth laws.

The change of partial mass density can then be linearized to

$$\frac{\Delta \rho^k}{\rho_0^k} = -\operatorname{tr} \tilde{\epsilon}^k \,, \quad k \in \mathcal{K}. \tag{7.2.2}$$

The incremental change of mass content results, namely

$$\zeta^k \equiv \frac{\Delta m^k}{\rho_k^0} = n^k \left(\operatorname{tr} \epsilon - \operatorname{tr} \tilde{\epsilon}^k \right), \quad k \in \mathcal{K}. \tag{7.2.3}$$

For the record, let us note several equivalent forms of the equation, emanating from the balance of mass of species (4.4.5), that governs the time evolution of the volume fraction n^k of a species $k \in \mathcal{K}$,

$$
\begin{aligned}
&\frac{d^k n^k}{dt} + \frac{n^k}{\rho_k} \frac{d^k \rho_k}{dt} + n^k \Big(\operatorname{div} \mathbf{v}_k - \frac{\hat{\rho}^k}{\rho^k}\Big) = 0, \\
&\frac{d^k n^k}{dt} + \frac{n^k}{\rho_k} \frac{d^k \rho_k}{dt} + n^k \operatorname{tr} \frac{d\tilde{\boldsymbol{\epsilon}}^k}{dt} = 0, \\
&\frac{dn^k}{dt} + n^k \operatorname{div} \mathbf{v}_{\mathrm{s}} + \operatorname{div}\big(n^k (\mathbf{v}_k - \mathbf{v}_{\mathrm{s}})\big) + \frac{n^k}{\rho_k} \frac{d^k \rho_k}{dt} - \frac{\hat{\rho}^k}{\rho_k} = 0, \\
&\frac{dn^k}{dt} + n^k \operatorname{div} \mathbf{v}_{\mathrm{s}} + \frac{n^k}{\rho_k} \frac{d\rho_k}{dt} - \frac{n^k}{m^k} \frac{dm^k}{dt} = 0 \,.
\end{aligned}
\tag{7.2.4}
$$

If a constitutive equation governing the evolution of a volume fraction n need to be postulated, it should ensure that the volume fraction remains in the interval $[0, 1]$. A simple way to satisfy this constraint is to set the rate dn/dt proportional to (some positive power of) $n(1 - n)$.

7.2.1.1   Intrinsic properties of the fluid constituents

Fluid species are assumed to be a priori compressible and thermally dilatable. These properties are embedded in the relative change of the intrinsic mass density

$$
\frac{\Delta \rho_k}{\rho_k^0} = \frac{\Delta p_k}{K_k} - c_{\mathrm{T}k} \Delta T_k, \quad k \in \mathcal{K}^{**},
\tag{7.2.5}
$$

where
- K_k [unit : Pa] is the isothermal bulk modulus of the fluid k,

$$
K_k = \rho_k^0 \frac{\Delta p_k}{\Delta \rho_k}\Big|_{T_k}, \quad k \in \mathcal{K}^{**} ;
\tag{7.2.6}
$$

- $c_{\mathrm{T}k}$ [unit : K^{-1}] its cubic coefficient of thermal expansion.

Fluid mass contents and volume-contents introduced in (2.6.25) are used as dependent or independent variables,

$$
m^k = \rho^k \det \mathbf{F} = n^k \rho_k \det \mathbf{F} = \rho_k v^k, \quad v^k = n^k \det \mathbf{F}, \quad k \in \mathcal{K}^{**},
\tag{7.2.7}
$$

from which follow the relations between rates,

$$
\begin{aligned}
&\frac{1}{m^k} \frac{dm^k}{dt} = \frac{1}{\rho^k} \frac{d\rho^k}{dt} + \operatorname{div} \mathbf{v}_{\mathrm{s}} = \frac{1}{n^k} \frac{dn^k}{dt} + \frac{1}{\rho_k} \frac{d\rho_k}{dt} + \operatorname{div} \mathbf{v}_{\mathrm{s}} = \frac{1}{v^k} \frac{dv^k}{dt} + \frac{1}{\rho_k} \frac{d\rho_k}{dt}, \\
&\frac{1}{v^k} \frac{dv^k}{dt} = \frac{1}{n^k} \frac{dn^k}{dt} + \operatorname{div} \mathbf{v}_{\mathrm{s}}, \quad k \in \mathcal{K}^{**} \,.
\end{aligned}
\tag{7.2.8}
$$

Therefore, since $dm^k/\rho_k = dv^k + v^k \, d\rho_k/\rho_k$, the (dimensionless) incremental change of the mass content of species k may be expressed in terms of the incremental changes of volume content, pressure and temperature of that species,

$$
\begin{aligned}
\zeta^k \equiv \frac{\Delta m^k}{\rho_k^0} &= \Delta v^k + v^k \Big(\frac{\Delta p_k}{K_k} - c_{\mathrm{T}k} \Delta T_k\Big), \\
&\simeq \Delta v^k + n^k \Big(\frac{\Delta p_k}{K_k} - c_{\mathrm{T}k} \Delta T_k\Big), \quad k \in \mathcal{K}^{**} \,.
\end{aligned}
\tag{7.2.9}
$$

The basic thermodynamic functions, their inter-relations and differentials are listed in Table 7.2.2.
The expressions of the entropy, chemical potential, or free enthalpy, and internal energy of the compressible and thermally expansible fluid k are defined by (3.3.28), under mild restrictions on the thermomechanical coefficients. Upon linearization around a reference state (p_k^0, T_k^0), the incremental change of entropy is obtained from (3.3.30) and (7.2.5),

$$
\rho_k^0 (s_k - s_k^0) = -c_{\mathrm{T}k} (p_k - p_k^0) + \rho_k^0 c_{\mathrm{p}k} \frac{T_k - T_k^0}{T_k^0}, \quad k \in \mathcal{K}^{**} \,.
\tag{7.2.10}
$$

Table 7.2.2: Thermodynamic measures of the state per unit mass [unit : J/kg = m^2/s^2] for the fluid constituent k endowed with temperature T_k, specific volume $1/\rho_k$ and pressure p_k.

Measure	Variables	Arbitrary Increment δ
Internal energy	$u_k = u_k(1/\rho_k, s_k)$	$\delta u_k = p_k\,\delta\rho_k/\rho_k^2 + T_k\,\delta s_k$
Free energy	$e_k = e_k(1/\rho_k, T_k) = u_k - T_k\,s_k$	$\delta e_k = p_k\,\delta\rho_k/\rho_k^2 - s_k\,\delta T_k$
Enthalpy	$h_k = h_k(p_k, s_k) = u_k + p_k/\rho_k$	$\delta h_k = \delta p_k/\rho_k + T_k\,\delta s_k$
Free enthalpy	$\mu_k = \mu_k(p_k, T_k) = h_k - T_k\,s_k$	$\delta \mu_k = \delta p_k/\rho_k - s_k\,\delta T_k$

In (7.2.10), $c_{\mathrm{p}k}$ [unit : J $\times$ kg^{-1} $\times$ K^{-1}] is the heat capacity at constant pressure per unit mass, linked to the heat capacity at constant volume per unit mass $c_{\mathrm{v}k}$ by $(3.3.18)_1$. This expression will be further approximated,

$$
\begin{aligned}
v_0^k\,\rho_k^0\,(s_k - s_k^0) &= -v_0^k\,c_{\mathrm{T}k}\,(p_k - p_k^0) \;+\; v_0^k\,\rho_k^0\,c_{\mathrm{p}k}\,\frac{T_k - T_k^0}{T_k^0}, \\[4pt]
S^k - S_0^k &= -n^k\,c_{\mathrm{T}k}\,(p_k - p_k^0) \;+\; C_{\mathrm{p}}^k\,\frac{T_k - T_k^0}{T_k^0},
\end{aligned}
\tag{7.2.11}
$$

where

$$
S^k = \rho^k s_k, \quad C_{\mathrm{p}}^k = \rho^k c_{\mathrm{p}k},
\tag{7.2.12}
$$

are respectively the apparent entropy and apparent heat capacity of the fluid $k \in \mathcal{K}^{**}$ at constant pressure per unit (current) volume of mixture [unit : J $\times$ m^{-3} $\times$ K^{-1} = kg $\times$ m^{-1} $\times$ s^{-2} $\times$ K^{-1}].

Note also the rate relations between the internal energy and entropy, per unit mass and per unit volume, for any $k \in \mathcal{K}^{**}$,

$$
\begin{aligned}
\frac{du_k}{dt} &= T_k\,\frac{ds_k}{dt} - p_k\,\frac{d}{dt}\!\left(\frac{1}{\rho_k}\right), \\[4pt]
\frac{dU^k}{dt} &= T_k\,\frac{dS^k}{dt} - E^k\,\mathrm{div}\,\mathbf{v}_\mathrm{s} + \frac{\mu_k}{\det \mathbf{F}}\,\frac{dm^k}{dt} - \frac{p_k}{\det \mathbf{F}}\,\frac{dv^k}{dt}.
\end{aligned}
\tag{7.2.13}
$$

Here E^k is the free energy per unit current volume of mixture, and μ_k the chemical potential, or free enthalpy, per unit current mass of the species k.

7.2.1.2 Intrinsic properties of the solid constituent

The solid constituent is endowed with the elastic tensor moduli $\mathbb{E}_\mathrm{s}$ and with the tensor of thermal expansion $\mathbf{c}_{\mathrm{Ts}}$. The constitutive equation,

$$
\boldsymbol{\sigma}_\mathrm{s} = \mathbb{E}_\mathrm{s} : \boldsymbol{\epsilon}_\mathrm{s}^\mathrm{e},
\tag{7.2.14}
$$

links the intrinsic stress $\boldsymbol{\sigma}_\mathrm{s}$ to the 'effective' strain $\boldsymbol{\epsilon}_\mathrm{s}^\mathrm{e}$,

$$
\boldsymbol{\epsilon}_\mathrm{s}^\mathrm{e} = \boldsymbol{\epsilon}_\mathrm{s} - \mathbf{c}_{\mathrm{Ts}}\,\Delta T_\mathrm{s} - \left(\int_0^t \frac{\hat{\rho}^\mathrm{s}}{\rho^\mathrm{s}}\right) d\tau\,\frac{\mathbf{I}}{3}.
\tag{7.2.15}
$$

Since the intrinsic stress $\boldsymbol{\sigma}_\mathrm{s}$ is related to the partial stress $\boldsymbol{\sigma}^\mathrm{s}$ by $\boldsymbol{\sigma}^\mathrm{s} = n^\mathrm{s}\,\boldsymbol{\sigma}_\mathrm{s}$, the change of volume content of the solid can be expressed in terms of total stress and pressures, namely using (7.1.2),

$$
\begin{aligned}
\det \mathbf{F}^{-1}\,\Delta v^\mathrm{s} = n^\mathrm{s}\,\mathrm{tr}\,\boldsymbol{\epsilon}_\mathrm{s} &= n^\mathrm{s}\,\mathbf{I} : \mathbb{E}_\mathrm{s}^{-1} : \boldsymbol{\sigma}_\mathrm{s} + n^\mathrm{s}\,\mathrm{tr}\,\mathbf{c}_{\mathrm{Ts}}\,\Delta T_\mathrm{s} + n^\mathrm{s}\int_0^t \frac{\hat{\rho}^\mathrm{s}}{\rho^\mathrm{s}}\,d\tau \\[4pt]
&= \mathbf{I} : \mathbb{E}_\mathrm{s}^{-1} : \boldsymbol{\sigma} + \frac{1}{K_\mathrm{s}}\sum_{k \in \mathcal{K}^{**}} n^k\,p_k + n^\mathrm{s}\,\mathrm{tr}\,\mathbf{c}_{\mathrm{Ts}}\,\Delta T_\mathrm{s} + n^\mathrm{s}\int_0^t \frac{\hat{\rho}^\mathrm{s}}{\rho^\mathrm{s}}\,d\tau.
\end{aligned}
\tag{7.2.16}
$$

Here K_s is the bulk modulus of the solid constituent,

$$
\frac{1}{K_\mathrm{s}} = \mathbf{I} : \mathbb{E}_\mathrm{s}^{-1} : \mathbf{I}.
\tag{7.2.17}
$$

Note that the 'effective', or 'elastic', strain involved in the constitutive equations (7.2.14) is not the total strain of the solid constituent. The total strain is contributed additively by

- a mechanical strain, the isotropic part of which vanishes if the solid is elastically incompressible, namely $K_s \to \infty$;
- a thermal strain, and
- a strain due to mass change.

The rate of change of the intrinsic mass density $\rho_s = M_s/V_s$ of the solid constituent may be expressed in terms of the rate of change of the volume and of the rate of mass supply,

$$
\begin{aligned}
-\frac{1}{\rho_s}\frac{d\rho_s}{dt} &= \frac{1}{V_s}\frac{dV_s}{dt} - \frac{1}{M_s}\frac{dM_s}{dt} \\
&= \frac{d}{dt}\mathrm{tr}\,\boldsymbol{\epsilon}_s - \frac{\hat{\rho}^s}{\rho^s},
\end{aligned} \tag{7.2.18}
$$

where use has been made in differential form of the relation between the mass of the solid and the rate of mass supply,

$$
\frac{M_s(t)}{M_s(0)} = \int_0^t \frac{\hat{\rho}^s}{\rho^s}\,d\tau . \tag{7.2.19}
$$

Moreover

$$
\begin{aligned}
\frac{V_s(t)}{V(0)} &= \frac{V_s(0)}{V(0)}\det\mathbf{F}_s(t) = n^s(0)\det\mathbf{F}_s(t) \\
&= \frac{V_s(t)}{V(t)}\frac{V(t)}{V(0)} = n^s(t)\det\mathbf{F}(t),
\end{aligned} \tag{7.2.20}
$$

where $\mathbf{F}_s = \mathbf{F}_s(t)$ is the deformation gradient of the solid *species* giving rise to the infinitesimal strain $\boldsymbol{\epsilon}_s$, as opposed to the deformation gradient $\mathbf{F} = \mathbf{F}(t)$ of the solid *skeleton* associated with the infinitesimal strain $\boldsymbol{\epsilon}$. Consequently, the volume fraction of the solid constituent varies according to the difference between the (macroscopic) strain of the solid skeleton and the (microscopic) strain of the solid constituent,

$$
\frac{n^s(t)}{n^s(0)} = \frac{\det\mathbf{F}_s}{\det\mathbf{F}} \simeq 1 + \mathrm{tr}\,\boldsymbol{\epsilon}_s - \mathrm{tr}\,\boldsymbol{\epsilon}. \tag{7.2.21}
$$

As a related result, an expression linking the change of the volume fraction of the fluid of a single porosity mixture to Terzaghi's effective stress is provided by Proposition 9.6, p. 295. Actual reports of measurements at the scale of the solid constituent are scarce. Moreover, the stiffness of the solid constituent is usually orders of magnitude larger than that of the porous medium. Therefore, the simplification below is sometimes adopted:

Proposition 7.2 The micro-isotropy assumption.

The solid particles are assumed to be randomly oriented and positioned in space so that the tensor of elastic moduli of the solid constituent is isotropic, with bulk modulus K_s and shear modulus μ_s, and so does the tensor of thermal expansion $\mathbf{c}_{TS} = \frac{1}{3}c_{TS}\mathbf{I}$,

$$
\frac{1}{3}\mathrm{tr}\,\boldsymbol{\sigma}_s = K_s\,\mathrm{tr}\,\boldsymbol{\epsilon}_s^e, \quad \mathrm{dev}\,\boldsymbol{\sigma}_s = 2\,\mu_s\,\mathrm{dev}\,\boldsymbol{\epsilon}_s. \tag{7.2.22}
$$

Warning: linear and cubic coefficients of thermal expansion

The cubic or volumetric coefficient of thermal expansion is denoted by c_{TS} for the solid constituent and by c_{Tk} for the fluid k.

7.2.1.3 Incompressible and thermally not dilatable species

Mechanical incompressibility and absence of thermal expansion of one species[7.1], namely $\rho_k = \rho_k^0$, can be viewed as a special case. However, if all species are incompressible and thermally nondilatable, the velocity fields satisfy

---

[7.1] Actually, for a given fluid or elastic solid, the two properties are expected to be simultaneous. The case of fluids is considered in Section 3.3.2. Incompressibility $K = \infty$ is seen from (3.3.18) to imply a vanishing thermal dilatation coefficient c_T, in as far as the difference between heat capacities at constant pressure and constant volume is expected to remain finite. Still, the coefficient of thermal expansion of water changes sign at around $4\,°C$ where the bulk modulus is finite, while the former becomes large and the latter small at the critical point as can be observed from Fig. 3.3.9.

Proposition 7.3 Constraint for incompressible and thermally nondilatable species.

$$\sum_{k \in \mathcal{K}} \text{div}\left(n^k \, \mathbf{v}_k\right) = \text{div}\,\mathbf{v}_\text{s} + \sum_{k \in \mathcal{K}} \text{div}\left(\underbrace{n^k\left(\mathbf{v}_k - \mathbf{v}_\text{s}\right)}_{\mathbf{J}_k}\right) = \sum_{k \in \mathcal{K}} \frac{\hat{\rho}^k}{\rho_k} \,. \tag{7.2.23}$$

This relation results from the summation of $(7.2.4)_3$ over species and use of the identity $(7.1.1)$. It represents a kinematical constraint, which gives rise to a reaction stress (a pressure). The reaction stress is not defined by constitutive equations, but it is obtained in a boundary value problem, see Section 2.10.

An alternative linearized form, considering all time derivatives to be identical,

$$\sum_{k \in \mathcal{K}} n^k \, \text{tr}\, \tilde{\boldsymbol{\epsilon}}^k = 0 \,, \tag{7.2.24}$$

results from the summation of $(7.2.4)_2$ over species and use of the identity $(7.1.1)$.

Quite generally, the volume change of the mixture may stem from
- the volume change of species themselves, due to mechanical or thermal loading;
- mass fluxes $\mathbf{J}_k$ crossing the boundary of the mixture;
- internal mass transfers, the change of volume being due to the fact that the densities of species are not identical;
- mass supplies from/towards the surroundings if the mixture is open.

For a solid saturated by a single fluid, the incremental change of fluid mass content is equal to,

$$\zeta^\text{w} = \int_0^t \frac{\hat{\rho}^\text{w}}{\rho_\text{w}} - \text{div}\,\mathbf{J}_\text{w}\, d\tau = \text{tr}\,\boldsymbol{\epsilon} - \int_0^t \frac{\hat{\rho}^\text{s}}{\rho_\text{s}}\, d\tau \,. \tag{7.2.25}$$

This relation is derived by introducing the constraint $(7.2.23)$ in the balance of mass $(4.4.9)$.

When mass transfers take place between species at identical intrinsic mass density, no overall volume change results. Then the right hand side of $(7.2.23)$ vanishes, and $(7.2.24)$ holds for the actual strains,

$$\sum_{k \in \mathcal{K}} n^k \, \text{tr}\, \boldsymbol{\epsilon}^k = 0 \,. \tag{7.2.26}$$

Consider again a solid saturated by a single fluid. Assume the solid skeleton to be endowed with the homogeneous velocity field $\mathbf{v}_\text{s} = \mathbf{L} \cdot \mathbf{x}$, where the velocity gradient $\mathbf{L}$ is constant in space. The fluid flux $\mathbf{J}_\text{w}$ is, to within a constant vector and a curl, equal to $-\mathbf{L} \cdot \mathbf{x}$.

7.2.2 Entropy production for a species

In order to structure the constitutive equations, we intend to derive first the contribution of a generic species $k \in \mathcal{K}$ to the entropy production. Although equivalent, the procedure is at variance with the derivation in Chapter 4 which concerned the mixture itself, that is, the contributions of all species were summed up directly.

The derivation uses the balance of mass in the forms $(4.4.5)$ and $(4.4.9)$, the balance of momentum $(4.5.5)$, and of energy $(4.6.11)$. Eliminating the heat term $\text{div}\,\mathbf{Q}_k - \Theta_k$ with help of the energy equation, the entropy production $(4.7.4)$ can be cast as the sum of three contributions,

$$\hat{S}^k_{(\text{i})} = \hat{S}^k_{(\text{i},1)} + \hat{S}^k_{(\text{i},2)} + \hat{S}^k_{(\text{i},3)} \,, \tag{7.2.27}$$

where

$$\hat{S}^k_{(\text{i},1)} = \frac{d}{dt}(\rho^k s_k) - \frac{1}{T_k}\frac{d}{dt}(\rho^k u_k) + \frac{1}{T_k}(\boldsymbol{\sigma}^k - \rho^k e_k\,\mathbf{I}) : \boldsymbol{\nabla}\mathbf{v}_\text{s} + \frac{1}{\det \mathbf{F}}\frac{\mu_k}{T_k}\frac{dm^k}{dt}\,;$$

$$\hat{S}^k_{(\text{i},2)} = -\frac{\hat{\rho}^k}{T_k}\left(\mu_k + \frac{1}{2}(\mathbf{v}_k - \mathbf{v}_\text{s})^2 - \frac{1}{2}\mathbf{v}_\text{s}^2\right) - \frac{\hat{\mathbf{p}}^k}{T_k} \cdot \mathbf{v}_\text{s} + \frac{\hat{U}^k}{T_k} - \hat{S}^k\,; \tag{7.2.28}$$

$$\hat{S}^k_{(\text{i},3)} = -\left(\frac{\mathbf{Q}_k}{T_k} + s_k\,\mathbf{M}_k\right) \cdot \frac{\boldsymbol{\nabla}T_k}{T_k} - \frac{\mathbf{M}_k}{T_k} \cdot \left(\boldsymbol{\nabla}\mu_k + \frac{d^k \mathbf{v}_k}{dt} - \mathbf{b}_k\right).$$

Here $\hat{\rho}^k$, $\hat{\mathbf{p}}^k$, $\hat{U}^k$ and $\hat{S}^k$ are the total volume supply rates of mass, momentum, energy and entropy to the species k as defined by Eqns (4.4.2), (4.5.3), (4.6.9), and (4.7.3), respectively, namely

$$\hat{\mathbf{p}}^k = \hat{\rho}^k \tilde{\mathbf{v}}_k + \hat{\boldsymbol{\pi}}^k,$$

$$\hat{U}^k = \hat{\rho}^k \left(\tilde{u}_k + \tfrac{1}{2}\,\tilde{\mathbf{v}}_k^2 \right) + \hat{\boldsymbol{\pi}}^k \cdot \mathbf{v}_k + \hat{\mathcal{U}}^k, \qquad (7.2.29)$$

$$\hat{S}^k = \hat{\rho}^k \tilde{s}_k + \hat{\mathcal{S}}^k.$$

For a closed system, and in absence of surface supply rates, the volume supply rates of mass, momentum, energy and entropy sum to zero, Eqn (4.3.8),

$$\sum_{k \in \mathcal{K}} \hat{\rho}^k = 0, \quad \sum_{k \in \mathcal{K}} \hat{\mathbf{p}}^k = 0, \quad \sum_{k \in \mathcal{K}} \hat{U}^k = 0, \quad \sum_{k \in \mathcal{K}} \hat{S}^k = 0. \qquad (7.2.30)$$

In (7.2.28), the terms involving the chemical potentials apply to the fluid species $k \in \mathcal{K}^{**}$ only, which actually have been assumed to be ideal fluids.

The superimposed tilde characterizes a property of the mass just before or at the time of transfer, while the superimposed hat denotes a supply rate in the form of mass, momentum or in other forms.

Remark on the rate of dissipation due to generalized diffusion:

Using the increment of chemical potential of fluid species, Table 7.2.2, namely

$$\delta\mu_k = \frac{\delta p_k}{\rho_k} - s_k\,\delta T_k, \qquad (7.2.31)$$

and the volume flux $\mathbf{J}_k = \mathbf{M}_k/\rho_k$ instead of the mass flux $\mathbf{M}_k$, the contribution associated with diffusion of fluid species can be recast as,

$$\hat{S}^k_{(i,3)} = -\frac{\mathbf{Q}_k}{T_k} \cdot \frac{\boldsymbol{\nabla} T_k}{T_k} - \frac{\mathbf{J}_k}{T_k} \cdot \left(\boldsymbol{\nabla} p_k + \rho_k \left(\frac{d^k \mathbf{v}_k}{dt} - \mathbf{b}_k \right) \right). \qquad (7.2.32)$$

Remark on the mechanical entropy production:

The mechanical entropy production is dealt with in two manners:
- either in the current configuration, as indicated on (7.2.28)$_1$. It then features Eshelby stress $\boldsymbol{\sigma}^{\mathrm{E}k} = \boldsymbol{\sigma}^k - \rho^k e_k\,\mathbf{I}$. However, most often, the distinction between Eshelby stress and Cauchy stress is neglected. Still, the issue is considered in Section 9.1.5;
- or in a reference configuration.

Upon multiplication of (7.2.28)$_1$ by $\det\mathbf{F}$, the entities that come into picture are
- mass contents $m^k = \det\mathbf{F}\,\rho^k$, $k \in \mathcal{K}^{**}$, rather than mass densities ρ^k, $k \in \mathcal{K}^{**}$;
- volume contents $v^k = \det\mathbf{F}\,n^k$, $k \in \mathcal{K}^{**}$ rather than volume fractions n^k, $k \in \mathcal{K}^{**}$;
- apparent pressures $v^k p_k = \det\mathbf{F}\,p^k$, $k \in \mathcal{K}^{**}$, rather than apparent pressures $n^k p_k = p^k$, $k \in \mathcal{K}^{**}$;
- Kirchhoff stresses $\boldsymbol{\tau} = \det\mathbf{F}\,\boldsymbol{\sigma}$ and $\boldsymbol{\tau}^k = \det\mathbf{F}\,\boldsymbol{\sigma}^k$, $k \in \mathcal{K}$, rather than Cauchy stresses $\boldsymbol{\sigma}$ and $\boldsymbol{\sigma}^k$, $k \in \mathcal{K}$;
- entropy per unit reference volume of mixture $\underline{S}^k = m^k s_k$, $k \in \mathcal{K}$, rather than entropy per unit current volume of mixture $S^k = \rho^k s_k$, $k \in \mathcal{K}$;
- entropy production per unit reference volume of mixture $\underline{\hat{S}}^k_{(i)}$, $k \in \mathcal{K}$, rather than entropy production per unit current volume of mixture $\hat{S}^k_{(i)}$, $k \in \mathcal{K}$;
- the free energy per reference volume $\underline{E}$ instead of the free energy per current volume E.

In particular, the relations (7.1.2) turn to,

$$\boldsymbol{\tau} = \boldsymbol{\tau}^{\mathrm{s}} + \sum_{k \in \mathcal{K}^{**}} \boldsymbol{\tau}^k = \boldsymbol{\tau}^{\mathrm{s}} - \sum_{k \in \mathcal{K}^{**}} v^k p_k\,\mathbf{I}. \qquad (7.2.33)$$

In the theoretical developments, we adopt

Definition (ULC) *Updated Lagrangian configuration:*

When the current configuration is taken as reference, that is $\det\mathbf{F}$ is set equal to 1, these two families of entities are equal. However, their increments and rates are *not* identical.

Isothermal and nonisothermal mixtures in local thermal equilibrium or nonequilibrium, including a solid matrix and one or two fluids, are considered in Chapter 10 as particular cases of the framework described here.

Remark on the mechanical entropy production for a perfect fluid:

For a perfect fluid, the mechanical entropy production $\hat{S}_{(i,1)}$, Eqn (7.2.28)$_1$, simplifies to $p_k/T_k\,dn^k/dt$.

7.3   Anisotropic poroelasticity: a composite material format

7.3.1   Mechanical, viscous and inertial couplings

The mixture theory of fluid-saturated porous media involves three types of coupling between the fluids and the solid:
- mechanical coupling described in terms of stress-strain constitutive equations where the apparent stress of each phase (species) depends on the apparent strains of all phases (species);
- viscous coupling, described in terms of relative velocities of a fluid phase with respect to the solid, and defined through Darcy's law and other possible effects, and
- inertial coupling, described in terms of accelerations, Biot [1956].

Viscous coupling is not involved in the analysis of acceleration waves which yields information on the nature of the regime of the field equations. On the other hand, inertial coupling comes into the picture, changing not that much the wave-speeds themselves, rather the kinematics. The "raison d'être" of inertial coupling is the following: to accelerate a body in a fluid, one must exert a force as if the fluid would not provide resistance, plus the additional force to move part of the fluid. For a two phase porous medium, the balances of linear momentum can be cast in the following format, Eqn (4.5.5):

$$
\begin{array}{ccccccc}
\overbrace{\text{div } \boldsymbol{\sigma}^{\mathrm{s}}}^{\text{mechanical coupling}} & + & \overbrace{\hat{\rho}^{\mathrm{s}}(\tilde{\mathbf{v}}_{\mathrm{s}} - \mathbf{v}_{\mathrm{s}}) + \hat{\boldsymbol{\pi}}^{\mathrm{s}}}^{\text{viscous coupling}} & + & \rho^{\mathrm{s}}\,\mathbf{b}_{\mathrm{s}} & = & \rho^{\mathrm{ss}}\,\mathbf{a}_{\mathrm{s}} & + & \overbrace{\rho^{\mathrm{sw}}\,\mathbf{a}_{\mathrm{w}}}^{\text{inertial coupling}} \\[2mm]
\text{div } \boldsymbol{\sigma}^{\mathrm{w}} & + & \hat{\rho}^{\mathrm{w}}(\tilde{\mathbf{v}}_{\mathrm{w}} - \mathbf{v}_{\mathrm{w}}) + \hat{\boldsymbol{\pi}}^{\mathrm{w}} & + & \underbrace{\rho^{\mathrm{w}}\,\mathbf{b}_{\mathrm{w}}}_{\text{body force}} & = & \underbrace{\rho^{\mathrm{ww}}\,\mathbf{a}_{\mathrm{w}}}_{\text{acceleration}} & + & \rho^{\mathrm{ws}}\,\mathbf{a}_{\mathrm{s}}
\end{array}
\tag{7.3.1}
$$

Here the **a**'s are the accelerations. For each phase, the viscous term represents the momentum supply to that phase by the rest of the mixture. It is provided through constitutive equations, Section 7.3.2. For a closed mixture, momentum supplies satisfy the constraint (4.5.2), namely,

$$
\sum_{k=\mathrm{s,w}} \hat{\rho}^{k}\,\tilde{\mathbf{v}}_{k} + \hat{\boldsymbol{\pi}}^{k} = \mathbf{0} \, .
\tag{7.3.2}
$$

Inertial coupling is introduced through the out-of-diagonal coefficients ρ^{sw} and ρ^{ws}. Biot [1956] presents physical arguments that restrict the range of these coefficients, which are equal because the kinetic energy is assumed to be a symmetric quadratic function of the velocities. Consistency with a process with equal accelerations of the two phases requires,

$$
\rho^{\mathrm{ss}} = \rho^{\mathrm{s}} - \rho^{\mathrm{sw}}, \quad \rho^{\mathrm{ww}} = \rho^{\mathrm{w}} - \rho^{\mathrm{ws}} \, .
\tag{7.3.3}
$$

Moreover, if the solid was accelerated while the fluid was kept motionless, the force to be exerted on the fluid would be in opposite direction to the acceleration of the solid: hence

$$
\rho^{\mathrm{sw}} = \rho^{\mathrm{ws}} \le 0 \, .
\tag{7.3.4}
$$

As a consequence,

$$
\rho^{\mathrm{ss}} > 0 \, , \quad \rho^{\mathrm{ww}} > 0 \, , \quad \rho^{\mathrm{ss}}\rho^{\mathrm{ww}} - (\rho^{\mathrm{sw}})^{2} > 0 \, ,
\tag{7.3.5}
$$

and the mass matrix $\mathbf{M}$, or equivalently the kinetic energy, is positive definite (P.D.):

$$
\mathbf{M} = \begin{bmatrix} \rho^{\mathrm{ss}}\,\mathbf{I} & \rho^{\mathrm{sw}}\,\mathbf{I} \\ \rho^{\mathrm{ws}}\mathbf{I} & \rho^{\mathrm{ww}}\mathbf{I} \end{bmatrix} = \mathbf{M}^{\mathrm{T}} \quad \text{P.D.}
\tag{7.3.6}
$$

Notice that the above inequalities do not furnish a bound for the coupling coefficient $\rho^{\mathrm{sw}} = \rho^{\mathrm{ws}}$: they are actually implied by the negative sign of this coefficient.

There is no definitive method to estimate the coupling coefficient. It is usually related to a *structural factor* or *tortuosity factor* τ in the form,

$$
\rho^{\mathrm{sw}} = -(\frac{1}{\tau} - 1)\,\rho^{\mathrm{w}} \, .
\tag{7.3.7}
$$

When the porosity n^{w} is not too small, the tortuosity factor might be correlated to n^{w} through a power law, Sen et al. [1981], Johnson et al. [1982], Gajo [1996],

$$
\tau = \sqrt{n^{\mathrm{w}}} \, .
\tag{7.3.8}
$$

7.3.2   Momentum supplies and Darcy's law

The approach followed in this work consists in trying to unveil the ingredients that allow to satisfy the entropy production. Generalized diffusion may be introduced through a diffusion matrix which should be positive (semi-)definite, but is otherwise arbitrary: the key equation is the entropy production associated with diffusion, Eqn (4.7.32). Examples, in presence of generalized diffusion, are worked out in Chapter 16. The momentum supplies are elusive in this approach, but they can be recovered once the constitutive equations governing mechanics, generalized diffusion, transfers and growth have been defined.

An alternative, pursued by Bowen in his papers, e.g., Bowen [1976], consists in
1. defining the primary variables,
2. expanding the dependent constitutive functions in terms of the latter;
3. restricting the ranges of the parameters by ensuring a positive entropy production.

To illustrate these two approaches, the ways Darcy's law is postulated by Bowen and in this work are contrasted.

In presence of body forces $\mathbf{b}_{\mathrm{w}}$, e.g., gravity $\mathbf{b}_{\mathrm{w}} = \mathbf{g}$, the momentum balance for a fluid phase with intrinsic pressure p_{w} and volume fraction n^{w}, is expressed in the format, Eqn (4.5.5),

$$-\boldsymbol{\nabla}(n^{\mathrm{w}} p_{\mathrm{w}}) + \hat{\rho}^{\mathrm{w}}(\tilde{\mathbf{v}}_{\mathrm{w}} - \mathbf{v}_{\mathrm{w}}) + \hat{\boldsymbol{\pi}}^{\mathrm{w}} + \rho^{\mathrm{w}}\left(\mathbf{b}_{\mathrm{w}} - \frac{d^{\mathrm{w}}\mathbf{v}_{\mathrm{w}}}{dt}\right) = \mathbf{0}, \tag{7.3.9}$$

where the extended momentum supply $\hat{\rho}^{\mathrm{w}}(\tilde{\mathbf{v}}_{\mathrm{w}} - \mathbf{v}_{\mathrm{w}}) + \hat{\boldsymbol{\pi}}^{\mathrm{w}}$ is taken as the sum of a buoyancy term and of a Stokes' drag proportional to the velocity of the solid relative to the fluid $\mathbf{v}_{\mathrm{s}} - \mathbf{v}_{\mathrm{w}}$,

$$\hat{\rho}^{\mathrm{w}}(\tilde{\mathbf{v}}_{\mathrm{w}} - \mathbf{v}_{\mathrm{w}}) + \hat{\boldsymbol{\pi}}^{\mathrm{w}} = p_{\mathrm{w}}\boldsymbol{\nabla} n^{\mathrm{w}} + k_{\mathrm{Sd}}(\mathbf{v}_{\mathrm{s}} - \mathbf{v}_{\mathrm{w}}). \tag{7.3.10}$$

The momentum balance of the fluid becomes

$$-n^{\mathrm{w}}\boldsymbol{\nabla} p_{\mathrm{w}} + k_{\mathrm{Sd}}(\mathbf{v}_{\mathrm{s}} - \mathbf{v}_{\mathrm{w}}) + \rho^{\mathrm{w}}\left(\mathbf{b}_{\mathrm{w}} - \frac{d^{\mathrm{w}}\mathbf{v}_{\mathrm{w}}}{dt}\right) = \mathbf{0}, \tag{7.3.11}$$

which leads to the standard form of Darcy's law for the fluid flux,

$$\mathbf{J}_{\mathrm{w}} \equiv n^{\mathrm{w}}(\mathbf{v}_{\mathrm{w}} - \mathbf{v}_{\mathrm{s}}) = -k_{\mathrm{H}}\left(\boldsymbol{\nabla} p_{\mathrm{w}} + \rho_{\mathrm{w}}\left(\frac{d^{\mathrm{w}}\mathbf{v}_{\mathrm{w}}}{dt} - \mathbf{b}_{\mathrm{w}}\right)\right). \tag{7.3.12}$$

In fact, Darcy's law may be expressed in various forms that are linked through the relations,

$$k_{\mathrm{H}} = \frac{K_{\mathrm{H}}}{\rho_{\mathrm{w}} g} = \frac{k_{\mathrm{in}}}{\eta_{\mathrm{w}}} = \frac{(n^{\mathrm{w}})^2}{k_{\mathrm{Sd}}}, \tag{7.3.13}$$

which involve
- the *intrinsic hydraulic permeability* k_{in} [unit : m^2], a property of the porous medium architecture, also expressed in darcy, or millidarcy, 1 darcy=0.987×10^{-12} m^2;
- the *hydraulic conductivity* K_{H} [unit : m $\times$ s^{-1}];
- the *hydraulic permeability* k_{H} [unit : m^2 $\times$ Pa^{-1} $\times$ s^{-1} = kg^{-1} $\times$ m^3 $\times$ s];
- the *coefficient of Stokes' drag* k_{Sd} [unit : kg $\times$ m^{-3} $\times$ s^{-1}],

some properties of the fluid,
- the intrinsic density of water ρ_{w} [unit : kg/m^3];
- the *dynamic viscosity* of water η_{w} [unit : Pa$\times$s=kg/m/s=10^3 centipoise cP]. It is equal to 1 cP at ambient temperature, and decreases considerably as the temperature increases as shown in Table 3.3.1. The dynamic viscosity should be distinguished from the *kinematic viscosity* [unit : m^2/s],

$$\nu_{\mathrm{w}} = \frac{\eta_{\mathrm{w}}}{\rho_{\mathrm{w}}}, \tag{7.3.14}$$

as well as the acceleration of gravity $g = 9.81$ m/s^2.

The time scale of seepage is quantified via the hydraulic diffusivity [unit : m^2/s] which is defined as the product of the hydraulic permeability by an elastic modulus: a specific example is worked out in Section 9.5. Typical coefficients pertaining to seepage in porous media are displayed in Table 7.3.1.

Table 7.3.1: Indicative properties linked to seepage through some porous materials.

Material	Hydraulic Permeability [m^2/Pa/s]	Elastic Modulus [Pa]	Hydraulic Diffusivity [m^2/s]	Reference
Articular cartilages	10^{-15}	10^6	10^{-9}	Armstrong et al. [1984]
Tumor tissues	3×10^{-14}	10^5	3×10^{-9}	Netti et al. [1995]
Clay	2×10^{-13}	2×10^7	4×10^{-6}	McTigue [1986]
Clay liner	$10^{-18} - 10^{-17}$		$10^{-12} - 10^{-11}$	
Halite (salt)	10^{-18}	37×10^9	0.16×10^{-6}	McTigue [1986]

The dependence of the intrinsic permeability on the void geometry is worked out in Section 15.7. and values for soft tissues are provided in Chapters 14 and 19.

Remark on the gravity contribution:

Under static conditions, the flow is driven by the stimulus $\nabla p_w - \rho_w \, \mathbf{g}$. Over small vertical distances, one may often neglect the gravity component with respect to the pressure gradient. Therefore, one may conclude that water flows from locations of large pressures to locations of low pressures. Now, consider a glass filled of water at rest. Even if the pressure increases downwards, water does not move up. In fact the reason is that the hydraulic head $p_w/(\rho_w \, g) + x_3$, with x_3 vertical coordinate counted positive upwards, vanishes. Therefore, one should actually be more accurate and refer to the hydraulic head, rather than to the pressure, as the driving gradient. See Section 15.7.3 for elaboration in the case of a compressible fluid.

7.3.3 Temperature dependence of the viscosity coefficient of liquids and gases

The dynamic viscosity of liquids decreases significantly as the temperature increases:
- as indicated in Table 3.3.1, the dynamic viscosity of *water* decreases from 1.791 cP at 0 °C to 0.282 cP at 100 °C. It can be approximated from the triple point to a neighborhood of the critical point through the expression (η_w in cP, T in K), Burger et al. [1985],

$$\eta_w = 10^{247.8/(T-140)-1.62} \, ; \tag{7.3.15}$$

- the dynamic viscosity of *blood* increases in an affine way with the hematocrit content, Stammers et al. [2003]. At 37 °C, it is about 3 to 4 cP. A decrease of temperature of 1 °C results in an increase of about 2%;
- the dynamic viscosity of *brine* departs from that of water in proportion of its salt content. It is a bit larger than that of pure water but it follows the same trend;
- the decrease of *bitumen* viscosity with temperature is considerable, from 10^7 cP at 10 °C to 10^1 cP at 200 °C.

As for *gases*, their viscosity increases as temperature increases. Mandel [1966], p. 297, refers to the approximate formula $\eta = A \sqrt{T}/(1 + B/T)$. For dry air: $A = 1.42 \times 10^{-3}$ cP, $B = 102.5$ K so that, at 100 °C, η=0.021 cP. These opposite trends for liquids and gases may be observed in Fig. 3.3.11 which displays the viscosity of liquid water and steam along the equilibrium curve.

The effect of pressure on the dynamic viscosity of liquids is significant at very high pressures only: for instance, the dynamic viscosity of oils is multiplied by 2 for a pressure change from 0 to 30 MPa and by 10 for a pressure change from 0 to 100 MPa.

7.3.4 Momentum supplies, the viscous fluid and Brinkman's law

The derivation of Darcy's law was concerned with a macroscopically inviscid fluid. Still, fluid viscosity is accounted for at a microscale to satisfy the boundary conditions of the flow. The more general case of a macroscopically viscous fluid is considered now.

The apparent fluid stress,

$$\boldsymbol{\sigma}^w = -n^w \, p_w \, \mathbf{I} + 2 \, n^w \, \eta_w \, \frac{d^w \boldsymbol{\epsilon}^w}{dt} \, , \tag{7.3.16}$$

expresses in terms of the intrinsic fluid pressure p_w, coefficient of viscosity η_w and strain rate,

$$\frac{d^w \boldsymbol{\epsilon}^w}{dt} = \frac{1}{2} \left(\nabla \mathbf{v}_w + (\nabla \mathbf{v}_w)^T \right) . \tag{7.3.17}$$

The balance of momentum,

$$\mathrm{div}\,\boldsymbol{\sigma}^{\mathrm{w}} + \hat{\rho}^{\mathrm{w}}(\tilde{\mathbf{v}}_{\mathrm{w}} - \mathbf{v}_{\mathrm{w}}) + \hat{\boldsymbol{\pi}}^{\mathrm{w}} + \rho^{\mathrm{w}}\left(\mathbf{b}_{\mathrm{w}} - \frac{d^{\mathrm{w}}\mathbf{v}_{\mathrm{w}}}{dt}\right) = \mathbf{0}\,, \tag{7.3.18}$$

becomes, neglecting the change of volume fraction in the viscous term,

$$-p_{\mathrm{w}}\boldsymbol{\nabla}n^{\mathrm{w}} - n^{\mathrm{w}}\boldsymbol{\nabla}p_{\mathrm{w}} + n^{\mathrm{w}}\eta_{\mathrm{w}}\left(\boldsymbol{\nabla}^2\mathbf{v}_{\mathrm{w}} + \boldsymbol{\nabla}\left(\mathrm{div}\,\mathbf{v}_{\mathrm{w}}\right)\right) + \hat{\rho}^{\mathrm{w}}(\tilde{\mathbf{v}}_{\mathrm{w}} - \mathbf{v}_{\mathrm{w}}) + \hat{\boldsymbol{\pi}}^{\mathrm{w}} + \rho^{\mathrm{w}}\left(\mathbf{b}_{\mathrm{w}} - \frac{d^{\mathrm{w}}\mathbf{v}_{\mathrm{w}}}{dt}\right) = \mathbf{0}\,. \tag{7.3.19}$$

The momentum supply $\hat{\boldsymbol{\pi}}^{\mathrm{w}}$ is modified with respect to (7.3.10) due to the additional viscous term via the tortuosity factor $\tau < 1$,

$$\hat{\rho}^{\mathrm{w}}(\tilde{\mathbf{v}}_{\mathrm{w}} - \mathbf{v}_{\mathrm{w}}) + \hat{\boldsymbol{\pi}}^{\mathrm{w}} = p_{\mathrm{w}}\boldsymbol{\nabla}n^{\mathrm{w}} + k_{\mathrm{Sd}}(\mathbf{v}_{\mathrm{s}} - \mathbf{v}_{\mathrm{w}}) + \left(\frac{1}{\tau} - 1\right)n^{\mathrm{w}}\eta_{\mathrm{w}}\left(\boldsymbol{\nabla}^2\mathbf{v}_{\mathrm{w}} + \boldsymbol{\nabla}\left(\mathrm{div}\,\mathbf{v}_{\mathrm{w}}\right)\right). \tag{7.3.20}$$

The balance of momentum of the fluid turns to,

$$-n^{\mathrm{w}}\boldsymbol{\nabla}p_{\mathrm{w}} + \frac{n^{\mathrm{w}}}{\tau}\eta_{\mathrm{w}}\left(\boldsymbol{\nabla}^2\mathbf{v}_{\mathrm{w}} + \boldsymbol{\nabla}\left(\mathrm{div}\,\mathbf{v}_{\mathrm{w}}\right)\right) + k_{\mathrm{Sd}}(\mathbf{v}_{\mathrm{s}} - \mathbf{v}_{\mathrm{w}}) + \rho^{\mathrm{w}}\left(\mathbf{b}_{\mathrm{w}} - \frac{d^{\mathrm{w}}\mathbf{v}_{\mathrm{w}}}{dt}\right) = \mathbf{0}\,, \tag{7.3.21}$$

or, with the definitions (7.3.13),

$$n^{\mathrm{w}}\left(\mathbf{v}_{\mathrm{w}} - \mathbf{v}_{\mathrm{s}}\right) = -k_{\mathrm{H}}\left(\boldsymbol{\nabla}p_{\mathrm{w}} + \rho_{\mathrm{w}}\left(\frac{d^{\mathrm{w}}\mathbf{v}_{\mathrm{w}}}{dt} - \mathbf{b}_{\mathrm{w}}\right)\right) + \frac{k_{\mathrm{in}}}{\tau}\left(\boldsymbol{\nabla}^2\mathbf{v}_{\mathrm{w}} + \boldsymbol{\nabla}\left(\mathrm{div}\,\mathbf{v}_{\mathrm{w}}\right)\right). \tag{7.3.22}$$

Analyses accounting for this macroscopic viscosity usually consider a rigid solid skeleton $\mathrm{div}\,\mathbf{v}_{\mathrm{s}} = 0$ and an incompressible fluid $\mathrm{div}\,\mathbf{v}_{\mathrm{w}} = 0$, which simplifies somehow the modification with respect to Darcy's law. The resulting expression of the fluid flux is referred to as Brinkman's law or Brinkman's modification of Darcy's law. This terminology is in fact misleading because the formula merely introduces a viscous fluid. Besides, the paper of Brinkman [1947] actually targeted the dependence of the intrinsic permeability with respect to the fluid volume fraction as indicated by (15.7.40).

7.3.5   Derivation of the momentum equations via Hamilton's principle

An alternative derivation of the equations of momentum balance (7.3.1) may be obtained via the Hamilton's principle, or directly via the resulting Lagrange equations. The energies which come into picture should first be listed, namely,

- the kinetic energy of the body V,

$$\mathbb{C} = \sum_{k,l}^{\mathrm{s,w}}\int_V \frac{1}{2}\rho^{kl}\mathbf{v}_l^2\,dV\,; \tag{7.3.23}$$

- the total potential energy, equal to the elastic energy minus the potential of the external conservative forces, namely body force $\mathbf{b}_k$ and traction $\overline{\mathbf{t}}^k$ over the surface $\partial V_{\overline{t}}$,

$$\begin{aligned}
\mathbb{W} &= \sum_{k}^{\mathrm{s,w}}\int_V \boldsymbol{\sigma}^k : \boldsymbol{\epsilon}^k - \rho^k\,\mathbf{b}_k\cdot\mathbf{u}_k\,dV - \int_{\partial V_{\overline{t}}}\overline{\mathbf{t}}^k\cdot\mathbf{u}_k\,dS \\
&= \sum_{k}^{\mathrm{s,w}}\int_V -\mathrm{div}\,\boldsymbol{\sigma}^k\cdot\mathbf{u}_k - \rho^k\,\mathbf{b}_k\cdot\mathbf{u}_k\,dV + \int_{\partial V}\mathbf{t}^k\cdot\mathbf{u}_k\,dS - \int_{\partial V_{\overline{t}}}\overline{\mathbf{t}}^k\cdot\mathbf{u}_k\,dS\,;
\end{aligned} \tag{7.3.24}$$

- since $\boldsymbol{\nabla}\mu_{\mathrm{w}} = \boldsymbol{\nabla}p_{\mathrm{w}}/\rho_{\mathrm{w}}$ according to Table 7.2.2, the pointwise dissipation $T\hat{S}_{(\mathrm{i},3)}$ due to seepage (4.7.32) simplifies to

$$T\hat{S}_{(\mathrm{i},3)} = -\mathbf{M}_k\cdot\left(\frac{\boldsymbol{\nabla}p_{\mathrm{w}}}{\rho_{\mathrm{w}}} + \frac{d^{\mathrm{w}}\mathbf{v}_{\mathrm{w}}}{dt} - \mathbf{b}_{\mathrm{w}}\right), \tag{7.3.25}$$

and the dissipation potential $\mathbb{D}$ becomes in view of (7.3.11).

$$\mathbb{D} = \int_V \frac{1}{2}k_{\mathrm{Sd}}\left(\mathbf{v}_{\mathrm{s}} - \mathbf{v}_{\mathrm{w}}\right)^2\,dV\,. \tag{7.3.26}$$

With $\mathbf{q} = [\mathbf{u}_{\mathrm{s}}\,\mathbf{u}_{\mathrm{w}}]^{\mathrm{T}}$ the generalized coordinate vector, and $\dot{\mathbf{q}} = [\mathbf{v}_{\mathrm{s}}\,\mathbf{v}_{\mathrm{w}}]^{\mathrm{T}}$ its derivative following the solid, the Lagrange equations,

$$\frac{d}{dt}\frac{\partial\mathbb{C}}{\partial\dot{\mathbf{q}}} - \frac{\partial\mathbb{C}}{\partial\mathbf{q}} + \frac{\partial\mathbb{W}}{\partial\mathbf{q}} = -\frac{\partial\mathbb{D}}{\partial\dot{\mathbf{q}}}\,, \tag{7.3.27}$$

yield, besides traction boundary conditions, a simplified form of the pointwise momentum equations (7.3.1), where the convective contributions to the acceleration terms and the spatial heterogeneity of the volume fraction have been neglected, Biot [1962].

7.3.6 Anisotropic poroelasticity in terms of apparent stresses and strains

There are several ways to introduce linear poroelasticity. One method, following Bowen [1976], consists in choosing apparent stresses (σ^s, σ^w) and strains (ϵ^s, ϵ^w) as work-conjugate variables. Let us recall that apparent (or partial) stresses σ^k are linked to the intrinsic stresses σ_k through the volume fractions n^k, namely $\sigma^k = n^k \sigma_k$, and similarly for the apparent fluid pressure $p^k = n^k p_k$ and the intrinsic fluid pressure p_k. Such an approach is appropriate when the velocities and balances of momentum of all species are required in explicit form. For example, when the inertial terms of both solid and fluid matter, e.g., for the analysis of the propagation of second order waves, Loret and Prévost [1991], or of first order waves, Radi and Loret [2007] [2008].

Another method, usually referred to as Biot's approach, uses the total stress $\sigma = \sigma^s + \sigma^w$ and intrinsic fluid pressure p_w which are work-conjugate to the strain of the solid skeleton ϵ and fluid volume content v^w, respectively. The field equations which are usually solved in this context are the balance of momentum of the mixture and the balance of mass of the fluid.

The differences between the two approaches are more formal than fundamental, at least in the linear static case. Biot's approach is presented in Section 7.4. The first approach is superior when inertial terms can not be neglected. The analysis below follows Bowen's approach, it is isothermal and assumes the mass supplies to vanish.

7.3.6.1 Work-conjugate apparent stresses and strains

The constitutive relations for the elastic mixture with an anisotropic solid phase, also referred to as solid skeleton, are postulated in a form which enjoys the major symmetry property so that they may be derived from a quadratic strain energy function $\mathcal{W}(\epsilon^s, \epsilon^w)$:

$$
\begin{bmatrix} \sigma^s \\ \sigma^w \end{bmatrix} = \begin{bmatrix} \mathbb{E}^{ss} & \mathbb{E}^{sw} \\ \mathbb{E}^{ws} & \mathbb{E}^{ww} \end{bmatrix} \begin{bmatrix} \epsilon^s \\ \epsilon^w \end{bmatrix} = \begin{bmatrix} \mathbb{E}^{ss} & \lambda^{sw} \boldsymbol{\lambda}_w \otimes \mathbf{I} \\ \mathbf{I} \otimes \lambda^{sw} \boldsymbol{\lambda}_w & \lambda^{ww} \mathbf{I} \otimes \mathbf{I} \end{bmatrix} \begin{bmatrix} \epsilon^s \\ \epsilon^w \end{bmatrix} . \tag{7.3.28}
$$

Note that the apparent strain of the solid ϵ^s is also the strain of the solid skeleton ϵ. However, the notations in this section are kept distinct so as to highlight the fact that the work-conjugate pairs are apparent quantities, that refer to the volume of the mixture.

The symmetry of the overall tensor moduli is ensured by the major symmetry of the diagonal moduli $\mathbb{E}^{ss}$ and $\mathbb{E}^{ww} = \lambda^{ww}\mathbf{I} \otimes \mathbf{I}$ and by the fact that the coupling moduli are the transpose of one another: $\mathbb{E}^{sw} = \lambda^{sw}\boldsymbol{\lambda}_w \otimes \mathbf{I} = (\mathbb{E}^{sw})^T$. The diagonal moduli $\mathbb{E}^{ss}$ display the minor symmetry on the two first indices by necessity and on the two last indices by choice. The second order tensor $\boldsymbol{\lambda}_w$ is symmetric by necessity as well.

The apparent constitutive parameters introduced in Eqns (7.3.28), namely the symmetric fourth order tensor $\mathbb{E}^{ss}$, the (dimensionless) symmetric second order tensor $\boldsymbol{\lambda}_w$, normalized to $\| \boldsymbol{\lambda}_w \| = \sqrt{3}$, and the scalars λ^{sw}, λ^{ww} are unspecified at the present stage but they may be identified from intrinsic material parameters measured in appropriate experimental tests as shown in Section 7.3.6.4.

7.3.6.2 The underlying drained solid and the strain energy function

The *underlying drained elastic solid* corresponds to the mixture where the fluid stress σ^w, or fluid pressure p_w, vanishes. In an incremental process, it can also be defined by a constant intrinsic fluid pressure p_w. Elimination of the fluid strain $\mathbf{I} : \epsilon^w$ from the second elastic constitutive equation (7.3.28) yields the drained moduli $\mathbb{E}$,

$$
\sigma = \sigma^s = \mathbb{E} : \epsilon^s \quad \text{where} \quad \mathbb{E} = \mathbb{E}^{ss} - \frac{\lambda^{sw} \boldsymbol{\lambda}_w \otimes \lambda^{sw} \boldsymbol{\lambda}_w}{\lambda^{ww}} . \tag{7.3.29}
$$

For

$$
\mathbb{E} \text{ positive definite} \quad \text{and} \quad \lambda^{ww} > 0 , \tag{7.3.30}
$$

the strain energy function $\mathcal{W}(\epsilon^s, \epsilon^w)$

$$
\begin{aligned}
\mathcal{W}(\epsilon^s, \epsilon^w) &= \frac{1}{2} \sigma^s : \epsilon^s + \frac{1}{2} \sigma^w : \epsilon^w \\
&= \frac{1}{2} \epsilon^s : \mathbb{E} : \epsilon^s + \frac{1}{2\lambda^{ww}} (\lambda^{sw} \boldsymbol{\lambda}_w : \epsilon^s + \lambda^{ww} \mathbf{I} : \epsilon^w)^2 ,
\end{aligned} \tag{7.3.31}
$$

is positive definite so that the stress-strain relation is invertible. Incidentally note another work-conjugate pair,

$$
\mathcal{W}(\epsilon^s, \epsilon^w) = \frac{1}{2} \sigma : \epsilon^s + \frac{1}{2} p_w n^w (\mathbf{I} : \epsilon^s - \mathbf{I} : \epsilon^w) . \tag{7.3.32}
$$

7.3.6.3   The elastic effective stress of Biot or Nur-Byerlee

With any (dimensionless) symmetric second order tensor $\boldsymbol{\alpha}$, we associate a generalized elastic effective stress through the following formula (in the framework of a continuum mechanics approach, tensile stresses are positive):

$$\boldsymbol{\sigma}' \equiv \boldsymbol{\sigma} + p_{\mathrm{w}}\, \boldsymbol{\alpha}\, . \tag{7.3.33}$$

For a specific choice of tensor $\boldsymbol{\alpha}$, which allows *decoupling* of the elastic response, $\boldsymbol{\sigma}'$ is referred to as the effective stress of Biot or Nur-Byerlee. It is said to be effective in the sense that it characterizes completely the elastic response of the solid skeleton. The physical concepts associated with the notions of partial, total and effective stresses are illustrated in Fig. 7.3.1.

The effective stress expresses in terms of the apparent strains via the constitutive equations (7.3.28),

$$
\begin{aligned}
\boldsymbol{\sigma}' &= \boldsymbol{\sigma}^{\mathrm{s}} - n^{\mathrm{w}}\, p_{\mathrm{w}}\left(\mathbf{I} - \frac{\boldsymbol{\alpha}}{n^{\mathrm{w}}}\right) \\
&= \left(\mathbb{E}^{\mathrm{ss}} + \left(\mathbf{I} - \frac{\boldsymbol{\alpha}}{n^{\mathrm{w}}}\right) \otimes \lambda^{\mathrm{sw}}\,\boldsymbol{\lambda}_{\mathrm{w}}\right) : \boldsymbol{\epsilon}^{\mathrm{s}} + \left(\lambda^{\mathrm{sw}}\,\boldsymbol{\lambda}_{\mathrm{w}} + \lambda^{\mathrm{ww}}\left(\mathbf{I} - \frac{\boldsymbol{\alpha}}{n^{\mathrm{w}}}\right)\right)(\mathbf{I} : \boldsymbol{\epsilon}^{\mathrm{w}})\, .
\end{aligned}
\tag{7.3.34}
$$

Decoupling takes place for the specific choice,

$$\mathbf{I} - \frac{\boldsymbol{\alpha}}{n^{\mathrm{w}}} = -\frac{\lambda^{\mathrm{sw}}}{\lambda^{\mathrm{ww}}}\,\boldsymbol{\lambda}_{\mathrm{w}}\, , \tag{7.3.35}$$

that is for

$$\boldsymbol{\alpha} = \boldsymbol{\alpha}_{\mathrm{w}} \equiv n^{\mathrm{w}}\left(\mathbf{I} + \frac{\lambda^{\mathrm{sw}}}{\lambda^{\mathrm{ww}}}\,\boldsymbol{\lambda}_{\mathrm{w}}\right)\, , \tag{7.3.36}$$

and then the effective tensor moduli turn out to be the drained moduli,

$$\boldsymbol{\sigma}' = \mathbb{E} : \boldsymbol{\epsilon}^{\mathrm{s}}\, . \tag{7.3.37}$$

7.3.6.4   Identification of the apparent coefficients

The material properties introduced in the constitutive equations may be deduced from
   - the drained moduli $\mathbb{E}$;
   - the bulk moduli of the fluid K_{w}, Eqn (7.2.5), and of the solid constituent $K_{\mathrm{s}} = 1/\mathbf{I} : \mathbb{E}_{\mathrm{s}}^{-1} : \mathbf{I}$, Eqn (7.2.17);
   - a volume fraction, n^{s} or n^{w}.

Various procedures may be followed, e.g., Biot and Willis [1957] and Nur and Byerlee [1971] for isotropic mixtures and Carroll [1979] and Thompson and Willis [1991] for anisotropic solid skeletons. Here, the volumetric strain of the solid constituent is obtained in terms of the strain $\boldsymbol{\epsilon}^{\mathrm{s}}$ and fluid pressure, using the relation $n^{\mathrm{s}}\,\boldsymbol{\sigma}_{\mathrm{s}} = \boldsymbol{\sigma}^{\mathrm{s}}$, and Eqns (7.2.14), (7.3.34)$_1$ and (7.3.37). The volume change of the fluid is obtained in terms of the apparent strains and fluid pressure through (7.2.3) and (7.2.9).

Spatial averaging of the intrinsic strains yields the apparent strain of the solid, Eqn (7.1.4). The fluid pressure is provided in terms of the apparent strains by the constitutive equation (7.3.28)$_2$. Therefore, we have two equations to be satisfied by three quantities, namely the apparent strains and the fluid pressure,

$$
\begin{aligned}
(\mathbf{I} : \mathbb{E}_{\mathrm{s}}^{-1} : \mathbb{E} - n^{\mathrm{s}}\,\mathbf{I}) : \boldsymbol{\epsilon}^{\mathrm{s}} - n^{\mathrm{w}}\,\mathbf{I} : \boldsymbol{\epsilon}^{\mathrm{w}} + \left(\mathbf{I} : \mathbb{E}_{\mathrm{s}}^{-1} : \left(\mathbf{I} - \frac{\boldsymbol{\alpha}}{n^{\mathrm{w}}}\right) - \frac{1}{K_{\mathrm{w}}}\right) n^{\mathrm{w}}\, p_{\mathrm{w}} &= 0, \\
-\lambda^{\mathrm{ww}}\left(\mathbf{I} - \frac{\boldsymbol{\alpha}}{n^{\mathrm{w}}}\right) : \boldsymbol{\epsilon}^{\mathrm{s}} + \lambda^{\mathrm{ww}}\,\mathbf{I} : \boldsymbol{\epsilon}^{\mathrm{w}} &= -n^{\mathrm{w}}\, p_{\mathrm{w}}\, ,
\end{aligned}
\tag{7.3.38}
$$

or, equivalently, one equation to be satisfied by two arbitrary quantities. Two relations result,

$$
\begin{aligned}
\boldsymbol{\alpha}_{\mathrm{w}} &= \mathbf{I} - \mathbb{E} : \mathbb{E}_{\mathrm{s}}^{-1} : \mathbf{I}, \\
\frac{(n^{\mathrm{w}})^2}{\lambda^{\mathrm{ww}}} &= \frac{n^{\mathrm{s}}}{K_{\mathrm{s}}} + \frac{n^{\mathrm{w}}}{K_{\mathrm{w}}} - (\mathbf{I} : \mathbb{E}_{\mathrm{s}}^{-1}) : \mathbb{E} : (\mathbb{E}_{\mathrm{s}}^{-1} : \mathbf{I})\, .
\end{aligned}
\tag{7.3.39}
$$

a) partial or apparent stresses

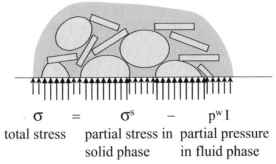

$$\sigma \quad = \quad \sigma^s \quad - \quad p^w\, I$$

total stress · partial stress in · partial pressure
solid phase · in fluid phase

b) effective stress

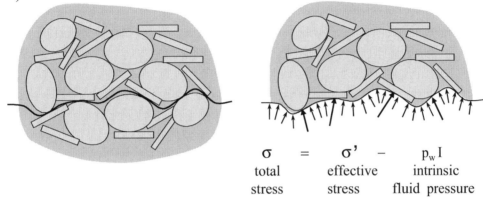

$$\sigma \quad = \quad \sigma' \quad - \quad p_w\, I$$

total · effective · intrinsic
stress · stress · fluid pressure

Fig. 7.3.1: (a) The sum of the forces exerted by a species over a plane surface divided by the total area of this surface provides the apparent stress of that species; (b) consider a granular material with pointwise contacts between rigid grains. A path in the water phase gives access to the intrinsic water pressure. The intergranular stress obtained by summing the forces at the contacts between grains is understood to govern the deformation of the solid skeleton. Reprinted with permission from Loret et al. [2004]a.

In turn,

$$\lambda^{sw}\,\boldsymbol{\lambda}_w \quad = \quad \frac{\lambda^{ww}}{n^w}\left(n^s\,\mathbf{I} - \mathbf{I} : \mathbb{E}_s^{-1} : \mathbb{E}\right),$$

$$\mathbb{E}^{ss} \quad = \quad \mathbb{E} + \frac{\lambda^{sw}\,\boldsymbol{\lambda}_w \otimes \lambda^{sw}\,\boldsymbol{\lambda}_w}{\lambda^{ww}}\,. \tag{7.3.40}$$

Under undrained conditions, namely constant mass content, that is $\operatorname{tr}\boldsymbol{\epsilon}^s - \operatorname{tr}\boldsymbol{\epsilon}^w = 0$, the fluid pressure and total stress are linked by the Skempton tensor $\mathbf{S}$,

$$p_w = -\mathbf{S} : \boldsymbol{\sigma}\,, \tag{7.3.41}$$

whose expression is deduced from the constitutive equations,

$$\mathbf{S} = \frac{\lambda^{ww}}{(n^w)^2}\,\boldsymbol{\alpha}_w : \left(\mathbb{E} + \frac{\lambda^{ww}}{(n^w)^2}\,\boldsymbol{\alpha}_w \otimes \boldsymbol{\alpha}_w\right)^{-1}. \tag{7.3.42}$$

7.3.6.5 Isotropic mixture

In the present framework, the anisotropic properties of the solid constituent and of the solid skeleton are formally independent of one another.

For a porous medium defined by the isotropic solid constituent tensor moduli $\mathbb{E}_s$, with bulk modulus $1/K_s = \mathbf{I} : \mathbb{E}_s^{-1} : \mathbf{I}$, and the isotropic drained tensor moduli $\mathbb{E}$, with bulk modulus $1/K = \mathbf{I} : (\mathbb{E})^{-1} : \mathbf{I}$ and shear modulus μ, all tensor properties are isotropic, in particular,

$$\boldsymbol{\alpha}_w = \overbrace{\left(1 - \frac{K}{K_s}\right)}^{\alpha_w =}\mathbf{I}, \quad \frac{1}{S} = \frac{1}{\operatorname{tr}\mathbf{S}} = 1 + \frac{n^w}{\alpha_w}\left(\frac{K}{K_w} - \frac{K}{K_s}\right). \tag{7.3.43}$$

Let $(\lambda = K - \frac{2}{3}\mu, \mu)$ be the Lamé moduli associated with the drained tensor moduli $\mathbb{E}$, and (λ^{ss}, μ^{ss}) be the Lamé moduli associated with the tensor moduli $\mathbb{E}^{ss}$. Then the coefficients in the format (7.3.28) can be identified in the order below,

$$\lambda^{ww} = \frac{(n^w)^2}{\alpha_w} \frac{K\,S}{1 - \alpha_w\,S} = \frac{(n^w)^2\,K_s^2}{\left(\dfrac{1 - n^w}{K_s} + \dfrac{n^w}{K_w}\right) K_s^2 - K}, \quad \boldsymbol{\lambda}_w = \mathbf{I},$$

$$\frac{\lambda^{sw}}{\lambda^{ww}} = \frac{\alpha_w}{n^w} - 1,$$

$$\lambda^{ss} = \lambda + \frac{(\lambda^{sw})^2}{\lambda^{ww}}, \quad \mu^{ss} = \mu,$$

$$S = \frac{\alpha_w\,\lambda^{ww}}{(n^w)^2 K + \alpha_w^2\,\lambda^{ww}} = \frac{1}{n^w} \frac{\lambda^{ww} + \lambda^{sw}}{\lambda^{ss} + \lambda^{ww} + 2\lambda^{sw} + \dfrac{2}{3}\mu}.$$

(7.3.44)

Relations between the coefficients λ's and undrained moduli are considered in Section 11.5.2.3.

The positive definiteness of the strain energy function, Eqn (7.3.30), takes the usual explicit form in terms of the drained properties K and μ and of the representative water bulk modulus λ^{ww},

$$K = \lambda + \frac{2}{3}\mu = \lambda^{ss} - \frac{(\lambda^{sw})^2}{\lambda^{ww}} + \frac{2}{3}\mu > 0, \quad \mu > 0, \quad \lambda^{ww} > 0.$$

(7.3.45)

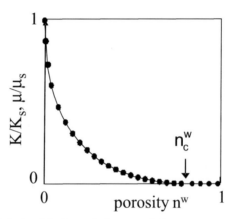

Fig. 7.3.2: Expected dependence of drained bulk and shear moduli in terms of porosity. The moduli may vanish at porosity smaller than 1, as if it was possible to define a path through the material entirely within the air constituent. For cancellous bones, the critical porosity n_c^w can reach 1.

7.3.6.6   Recovering the solid limit as $n^w \to 0$

Although the analysis addresses primarily fluid-saturated porous media, one might think to establish contact with elastic solids, for which $n^w = 0$. For that purpose the bulk modulus and shear modulus of the drained material are assumed to depend on the porosity n^w as sketched by Fig. 7.3.2. A wide range of data displaying the dependence in porosity of both dry moduli and dry wave-speeds of several cemented materials are reported by Knackstedt et al. [2005]. These data are approximated according to the simple, but realistic relation,

$$\frac{K}{K_s} = \frac{\mu}{\mu_s} = \left\langle 1 + \frac{1}{1-q}\left(q\,\frac{n^w}{n_c^w} - (\frac{n^w}{n_c^w})^q\right)\right\rangle,$$

(7.3.46)

where $\langle\cdot\rangle$ denotes the positive part of its argument. Thus the moduli K_s and μ_s of the solid constituent are recovered as the porosity vanishes, and Poisson's ratio ν of the drained material remains equal to Poisson's ratio ν_s of the solid constituent, irrespective of the porosity, namely

$$\nu = \frac{3\,K - 2\,\mu}{6\,K + 2\,\mu} = \nu_s = \frac{3\,K_s - 2\,\mu_s}{6\,K_s + 2\,\mu_s}.$$

(7.3.47)

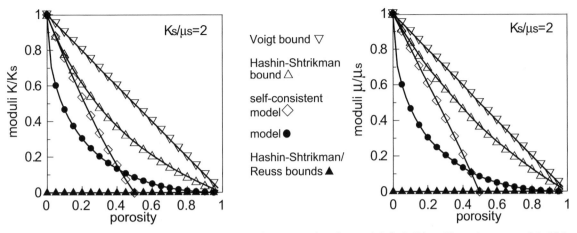

Fig. 7.3.3: Drained bulk and shear moduli in terms of porosity for the model (7.3.48), self-consistent model, Voigt and Reuss bounds, and Hashin-Shtrikman bounds. In presence of air constituent, the Reuss and lower Hashin-Shtrikman bounds vanish, Voigt bound is linear in terms of porosity, the self consistent model fails for porosities larger than 0.5.

Moreover, the drained moduli vanish smoothly at the upper limit $n^{\mathrm{w}} = n_c^{\mathrm{w}} \leq 1$. The slopes of the function that relate the moduli to the porosity are enforced to vanish at $n^{\mathrm{w}} = n_c^{\mathrm{w}}$, whatever the value of the exponent $q \in]0, 1[$. This empirical relation for the shear modulus remains below the upper bound of Voigt $\mu/\mu_{\mathrm{s}} = 1 - n^{\mathrm{w}}$ defined by Proposition 7.4.

The experimental value of the upper limit porosity n_c^{w} may be as low as 0.5. The consequences of this dependence in porosity for the body wave-speeds are analyzed in Chapter 11.

Table 7.3.2 displays the moduli of typical constituents of geological materials. These values may be contrasted to the bulk modulus of de-aired water, namely 2.2 GPa.

Table 7.3.2: Approximative bulk and shear moduli of quartz, dolomite, calcite and clay.

	K_{s} (GPa)	μ_{s} (GPa)	$K_{\mathrm{s}}/\mu_{\mathrm{s}}$
Quartz	36.0	40.0	0.9
Dolomite	76.4	49.7	1.5
Calcite	74.8	36.0	2.1
Clay	20.8	6.9	3.0

As another consequence of this relation, the coefficient of Biot α_{w} defined by (7.3.43) tends continuously to zero with the porosity. As an example illustrated by Fig. 7.3.3, consider the parameter $q = 1/2$ and a limit porosity $n_c^{\mathrm{w}} = 1$ so that the bulk and shear moduli, and the coefficient of Biot vary with porosity along the simple form,

$$\frac{K}{K_{\mathrm{s}}} = \frac{\mu}{\mu_{\mathrm{s}}} = (1 - \sqrt{n^{\mathrm{w}}})^2, \quad \alpha_{\mathrm{w}} = 1 - (1 - \sqrt{n^{\mathrm{w}}})^2. \tag{7.3.48}$$

Moreover, by inserting the above relation for α_{w} in (7.3.44), this dependence implies the modulus of elastic coupling to be positive, or zero for $n^{\mathrm{w}} = 0$,

$$\frac{\lambda^{\mathrm{sw}}}{\lambda^{\mathrm{ww}}} = \frac{2}{\sqrt{n^{\mathrm{w}}}} \left(1 - \sqrt{n^{\mathrm{w}}}\right) \geq 0. \tag{7.3.49}$$

7.3.6.7 Incompressible constituents

If only the solid constituent is incompressible, namely

$$K_{\mathrm{s}} \to \infty, \tag{7.3.50}$$

then (7.3.40) implies the asymptotic relation,

$$\frac{\lambda^{\mathrm{sw}}}{\lambda^{\mathrm{ww}}} \, \boldsymbol{\lambda}_{\mathrm{w}} \to \frac{n^{\mathrm{s}}}{n^{\mathrm{w}}} \, \mathbf{I}. \tag{7.3.51}$$

If both solid and fluid constituents become incompressible, namely

$$K_{\mathrm{s}} \to \infty, \quad K_{\mathrm{w}} \to \infty, \tag{7.3.52}$$

the apparent strains are subjected to the constraint (7.2.26), namely,

$$n^{\mathrm{s}} \operatorname{tr} \boldsymbol{\epsilon}^{\mathrm{s}} + n^{\mathrm{w}} \operatorname{tr} \boldsymbol{\epsilon}^{\mathrm{w}} = 0. \tag{7.3.53}$$

The following asymptotic conditions should be satisfied,

$$\lambda^{\mathrm{ww}} \to \infty, \quad \lambda^{\mathrm{sw}} \to \infty, \quad \frac{\lambda^{\mathrm{sw}}}{\lambda^{\mathrm{ww}}} \boldsymbol{\lambda}_{\mathrm{w}} \to \frac{n^{\mathrm{s}}}{n^{\mathrm{w}}} \mathbf{I}. \tag{7.3.54}$$

The fluid pressure becomes indeterminate and the constitutive equations provide only the effective stress, Eqn (7.3.37). Since then $\boldsymbol{\alpha}_{\mathrm{w}} = \mathbf{I}$, the effective stress of Nur-Byerlee boils down to Terzaghi's effective stress, Terzaghi [1936].

$$\boldsymbol{\sigma}' = \boldsymbol{\sigma} + p_{\mathrm{w}} \left(\mathbf{I} - \mathbb{E} : \mathbb{E}_{\mathrm{s}}^{-1} : \mathbf{I} \right) \quad \to \quad \boldsymbol{\sigma}' = \boldsymbol{\sigma} + p_{\mathrm{w}} \mathbf{I}. \tag{7.3.55}$$

Skempton tensor tends to $\mathbf{I}/3$, and Skempton coefficient to one, and therefore, according to (7.3.41), $p_{\mathrm{w}} = -\mathbf{I} : \boldsymbol{\sigma}/3$. This relation is checked in laboratory experiments to verify that specimens tested are saturated by water. Any significant departure of the Skempton coefficient from 1 would signal the presence of a gas.

7.4 Anisotropic poroelasticity: drained and undrained properties

A general isothermal theory for anisotropic elastic porous solids saturated by a fluid has been exposed by Biot [1955]. Although many fluid infiltrated biological and geological porous solids display anisotropic mechanical properties, applications are scant. An analytical solution of Mandel's consolidation problem for a material with transverse isotropy under plane strain condition has been presented by Abousleiman et al. [1996]. Satisfaction of particular boundary conditions may require the development of an out-of-plane strain: this situation, referred to as generalized plane strain, has been analyzed analytically by Cheng [1998] on specific examples.

One reason that has delayed the application of the anisotropic poroelastic theory might come from the difficulties in measuring the material coefficients which characterize the mechanical behavior. Macroscopic homogeneity conditions that allow for space averaging have been examined by Carroll [1979] and Thompson and Willis [1991]. The identification of the material parameters uses generally, following Biot and Willis [1957], information from different scales, namely coefficients identifying the mechanical properties of the species (microscale) and coefficients defining the mechanical properties of the solid skeleton (macroscale). However, accurate information at the microscale may not be easily accessible. Therefore another route has been proposed, e.g., by Rice and Cleary [1976], which consists in using only information gathered at the macroscale along particular loadings, for instance drained paths corresponding to a constant pore pressure and undrained paths corresponding to a constant fluid mass content.

The number of material coefficients defining the most general form of anisotropy is 28, Biot [1955], Biot and Willis [1957], Cheng [1997], corresponding to a fourth order tensor with minor and major symmetries (21 coefficients), a symmetric second order tensor (6 coefficients) and a scalar (1 coefficient). The number of parameters follows also from a brute force count: the linear elastic relations involve $n = 6 + 1 = 7$ stress and strain components and so 49 material parameters, that, due to major symmetry, reduce to $\frac{1}{2} n (n + 1) = 28$. Clearly, for the identification of these 28 parameters, the information provided together by the drained and undrained tensor moduli (42 coefficients) is overabundant. The above triplet of coefficients is (9,3,1) for orthotropy, amounting to 13 coefficients, (5,2,1) for transverse isotropy, amounting to 8 coefficients, and (2,1,1) for isotropy, that is 4 coefficients. Assuming the drained tensor moduli and the scalar referred to, in the isotropic case, as the Skempton coefficient to be known, it is important to assess, for specific classes of anisotropy, *which undrained parameters need to be measured and which need not*. Conversely, due to the symmetric roles played by the drained and undrained moduli in the following analysis, a simple switch of the results is feasible if the undrained moduli are known instead of the drained moduli.

Section 7.4 is based on Loret et al. [2001]. The micro-isotropy assumption is not used in this section. Still its implications are scrutinized in Section 7.4.5.2.

7.4.1 Constitutive equations in terms of drained properties

A basic issue in establishing constitutive equations is to choose the primary (or independent) variables. While the issue may be thought to be pointless for linear constitutive equations, the identification of coefficients might be easier in terms of a specific choice of variables. As for nonlinear constitutive equations, the decision has consequences, because the inversion of the equations, namely obtaining the independent variables in terms of the dependent variables, may be algebraically a burden if not impossible.

7.4.1.1 Generalized strains as primary variables

Constitutive equations are established using the free energy per unit reference volume $\underline{E}$, which can be expressed in terms of the work conjugate pairs $(\boldsymbol{\tau}, \cup_{k \in \mathcal{K}^{**}} \mu_k)$ and $(\boldsymbol{\epsilon}, \cup_{k \in \mathcal{K}^{**}} m^k)$. Here $\boldsymbol{\epsilon}$ is the infinitesimal strain of the solid skeleton, $\boldsymbol{\tau}$ is the Kirchhoff stress of the mixture, m^k is the fluid mass content per unit reference volume of fluid k, and $\mu_k = \int_0^{p_k} dp'_k/\rho_k(p'_k)$ its chemical potential which, in an isothermal context and in absence of solutes, reduces to the 'pressure function', Biot [1973] and Table 7.2.2.

We proceed as follows. Upon multiplication by $T \det \mathbf{F}$, the mechanical entropy production $(7.2.28)_1$ of the fluid k is modified to

$$T \hat{\underline{S}}^k_{(i,1)} = -\frac{d}{dt}(m^k e_k) + \boldsymbol{\tau}^k : \boldsymbol{\nabla}\mathbf{v}_{\mathrm{s}} + \mu_k \frac{dm^k}{dt}, \quad k \in \mathcal{K}^{**}. \tag{7.4.1}$$

The mechanical entropy production of the solid constituent has the form (7.4.1) but it does not involve the chemical potential term. Therefore, the mechanical entropy production of the wole mixture,

$$T \hat{\underline{S}}_{(i,1)} = T \sum_{k \in \mathcal{K}} \hat{\underline{S}}^k_{(i,1)} = -\frac{d\underline{E}}{dt} + \boldsymbol{\tau} : \boldsymbol{\nabla}\mathbf{v}_{\mathrm{s}} + \sum_{k \in \mathcal{K}^{**}} \mu_k \frac{dm^k}{dt} = 0, \tag{7.4.2}$$

expressed in terms of the free energy per reference volume of the mixture,

$$\underline{E} = \underline{E}(\boldsymbol{\epsilon}, \cup_{k \in \mathcal{K}^{**}} m^k) = \sum_{k \in \mathcal{K}} m^k e_k, \tag{7.4.3}$$

is set to zero. Constitutive equations result for the following entities,

$$\boldsymbol{\tau} = \boldsymbol{\tau}^{\mathrm{s}} + \sum_{k \in \mathcal{K}^{**}} \boldsymbol{\tau}^k = \frac{\partial \underline{E}}{\partial \boldsymbol{\epsilon}}; \quad \mu_k = \frac{\partial \underline{E}}{\partial m^k}, \quad k \in \mathcal{K}^{**}. \tag{7.4.4}$$

7.4.1.2 Generalized stresses as primary variables

As an alternative, we can recover the stress and fluid pressures as primary variables, with an appropriate Legendre transform, namely,

$$\underline{E} \longrightarrow \underline{\mathcal{W}}(\boldsymbol{\tau}, \cup_{k \in \mathcal{K}^{**}} p_k) = -\underline{E} + \sum_{k \in \mathcal{K}} \boldsymbol{\tau}^k : \boldsymbol{\epsilon} + \sum_{k \in \mathcal{K}^{**}} \mu_k m^k. \tag{7.4.5}$$

Eqn (7.4.5) transforms to

$$T \hat{\underline{S}}_{(i,1)} = \frac{d\underline{\mathcal{W}}}{dt} - \frac{d\boldsymbol{\tau}}{dt} : \boldsymbol{\epsilon} - \sum_{k \in \mathcal{K}^{**}} v^k \frac{dp_k}{dt} = 0, \tag{7.4.6}$$

and provides constitutive equations in the format,

$$\boldsymbol{\epsilon} = \frac{\partial \underline{\mathcal{W}}}{\partial \boldsymbol{\tau}}; \quad v^k = \frac{\partial \underline{\mathcal{W}}}{\partial p_k}, \quad k \in \mathcal{K}^{**}. \tag{7.4.7}$$

7.4.1.3   Linear poroelasticity with a single fluid constituent

The analysis is now restricted to linear poroelasticity, and the current configuration is taken as reference. The generalized strain $(\epsilon,\ \cup_{k\in\mathcal{K}^{**}} v^k)$ is obtained in terms of the generalized stress $(\sigma,\ \cup_{k\in\mathcal{K}^{**}} p_k)$ in a form which enjoys the major symmetry property. Moreover, the linearized relation $m^k = \rho_k\, v^k$, Eqn (7.2.9), simplifies in the isothermal case to,

$$\zeta^{\mathrm{k}} \equiv \frac{\Delta m^k}{\rho_k^0} = \Delta v^k + n^k\,\frac{p_k}{K_k}\,. \tag{7.4.8}$$

Consequently, the major symmetry property is conserved when the constitutive equations express the generalized strain $(\epsilon,\ \cup_{k\in\mathcal{K}^{**}} \zeta^k)$ in terms of the generalized stress $(\tau,\ \cup_{k\in\mathcal{K}^{**}} p_k)$, *even if these pairs are not work-conjugate.* Actually, we would like to write the constitutive equations in the matrix form,

$$[\mathbf{E} - \mathbf{E}_0] = [\mathbb{C}]\,([\boldsymbol{\Sigma} - \boldsymbol{\Sigma}_0])\,, \tag{7.4.9}$$

where $[\mathbf{E}]$ and $[\boldsymbol{\Sigma}]$ are the generalized strain and stress, respectively, $[\mathbb{C}]$ is the poroelastic compliance matrix, and the subscript 0 indicates a reference value. Therefore the stress potential $\mathcal{W}$ should include both linear and quadratic terms,

$$\mathcal{W}\,([\boldsymbol{\Sigma} - \boldsymbol{\Sigma}_0]) = ([\boldsymbol{\Sigma} - \boldsymbol{\Sigma}_0])^{\mathrm{T}}[\mathbf{E}_0] + \frac{1}{2}([\boldsymbol{\Sigma} - \boldsymbol{\Sigma}_0])^{\mathrm{T}}\,[\mathbb{C}]\,[\boldsymbol{\Sigma} - \boldsymbol{\Sigma}_0])\,. \tag{7.4.10}$$

For a mixture with a single fluid constituent, the matrix form of the constitutive equations adopts the format,

$$\begin{bmatrix} \epsilon \\[4pt] \zeta^{\mathrm{w}} \end{bmatrix} = \begin{bmatrix} \mathbb{E}^{-1} & \mathbb{E}^{-1}\!:\boldsymbol{\alpha}_{\mathrm{w}} \\[6pt] \boldsymbol{\alpha}_{\mathrm{w}}:\mathbb{E}^{-1} & \mathbf{I}:\dfrac{\mathbb{E}^{-1}}{S}:\boldsymbol{\alpha}_{\mathrm{w}} \end{bmatrix} \begin{bmatrix} \sigma \\[4pt] p_{\mathrm{w}} \end{bmatrix}\,. \tag{7.4.11}$$

The constitutive equations (7.4.11) involve a fourth order tensor moduli $\mathbb{E}$ endowed with minor and major symmetries, a symmetric second order tensor $\boldsymbol{\alpha}_{\mathrm{w}}$ and the dimensionless Skempton coefficient S, for a total of $21 + 6 + 1 = 28$ parameters. The format and notations used to cast these equations are now justified.

When full drainage is allowed, that is $p_{\mathrm{w}} = 0$, then $\sigma^{\mathrm{s}} = \mathbb{E}\!:\!\epsilon$, which shows that the tensor moduli $\mathbb{E}$ are the elastic moduli of the *Drained Solid*. These tensor moduli, which inherit the major symmetry property from the existence of the potential $\mathcal{W}$, are accessible to macroscopic measurements at the scale of the porous medium. Eqn $(7.4.11)_1$ introduces implicitly a Nur-Byerlee or Biot *effective stress* σ',

$$\sigma' = \sigma + p_{\mathrm{w}}\,\boldsymbol{\alpha}_{\mathrm{w}}\,. \tag{7.4.12}$$

The effective stress σ' and the strain of the solid skeleton ϵ are linked through the drained stiffness,

$$\sigma' = \mathbb{E}:\epsilon,\quad \epsilon = \mathbb{E}^{-1}:\sigma'\,. \tag{7.4.13}$$

The tensor $\boldsymbol{\alpha}_{\mathrm{w}}$ can be defined in terms of the macroscopic moduli $\mathbb{E}$ and of the moduli of the solid constituent $\mathbb{E}_{\mathrm{s}}$; indeed

$$\boldsymbol{\alpha}_{\mathrm{w}} = \mathbf{I} - \mathbb{E}:\mathbb{E}_{\mathrm{s}}^{-1}:\mathbf{I}\,. \tag{7.4.14}$$

The material identification procedure is sketched in Fig. 7.4.1. In fact, let us assume the intrinsic stresses in the solid constituent σ_{s} and in the fluid constituent σ_{w} to be equal, say $\sigma_{\mathrm{s}} = \sigma_{\mathrm{w}} = -p\,\mathbf{I}$; then if we assume the porous medium to behave homogeneously at the scale of the solid constituent, its strain will be $\epsilon = \mathbb{E}_{\mathrm{s}}^{-1}:(-p\,\mathbf{I})$ as if there was no porosity; equating this expression with Eqn $(7.4.11)_1$, and accounting for the fact that the total stress σ is now equal to $-p\,\mathbf{I}$ as well, yields the above identification, Eqn (7.4.14).

For the above loading, the assumption of a homogeneous response is equivalent to proportional volume changes Δv and $\Delta v^{\mathrm{w}} = n^{\mathrm{w}}\Delta v$, where $\Delta v \equiv \Delta V/V_0 = \mathrm{tr}\,\epsilon$, now equal to $\mathbf{I}:\mathbb{E}_{\mathrm{s}}^{-1}:(-p\,\mathbf{I})$, is the relative volume change of the solid skeleton. Then Eqns (7.4.8) and $(7.4.11)_2$ together with Eqn (7.4.14) yield the Skempton coefficient S, e.g., Bishop and Hight [1977], Eqn (23),

$$\frac{1}{S} = 1 + n^{\mathrm{w}}\left(\frac{1}{K_{\mathrm{w}}} - \frac{1}{K_{\mathrm{s}}}\right)\left(\frac{1}{K} - \frac{1}{K_{\mathrm{s}}}\right)^{-1}, \tag{7.4.15}$$

in terms of the porosity n^{w}, of the bulk moduli of the fluid and solid constituents, respectively K_{w} and K_{s}, and of the macroscopic bulk modulus K,

$$\frac{1}{K_{\mathrm{s}}} \equiv \mathbf{I}:\mathbb{E}_{\mathrm{s}}^{-1}:\mathbf{I},\quad \frac{1}{K} \equiv \mathbf{I}:\mathbb{E}^{-1}:\mathbf{I}\,. \tag{7.4.16}$$

The formalism (7.4.11) is intended for visual purposes. A more operational expression of the coefficients can now be deduced,

$$\mathbf{A} \equiv \mathbb{E}^{-1}:\boldsymbol{\alpha}_{\mathrm{w}} = \mathbb{E}^{-1}:\mathbf{I} - \mathbb{E}_{\mathrm{s}}^{-1}:\mathbf{I},\quad \mathbf{I}:\frac{\mathbb{E}^{-1}}{S}:\boldsymbol{\alpha}_{\mathrm{w}} = \frac{1}{K} - \frac{1+n^{\mathrm{w}}}{K_{\mathrm{s}}} + \frac{n^{\mathrm{w}}}{K_{\mathrm{w}}}\,. \tag{7.4.17}$$

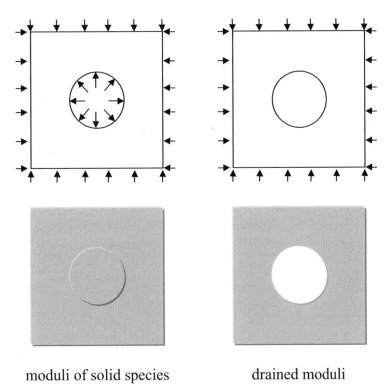

moduli of solid species   drained moduli

Fig. 7.4.1: Elastic constants are accessible to measurements by appropriate loadings of the pores and of the solid.

7.4.2 Constitutive equations in terms of undrained properties

A potential difficulty inherent in the above expressions of $\boldsymbol{\alpha}_{\mathrm{w}}$ and S comes from the fact that they involve material properties of the species which may not be accurately accessible, due for example to an imperfect homogeneity at the corresponding scale. To alleviate this problem, it might be more appropriate to use moduli all measured at the same scale. Following Rice and Cleary [1976], a good candidate is the *Undrained tensor moduli* $\mathbb{E}^{\mathrm{U}}$ corresponding to a constant fluid mass content, namely $\zeta^{\mathrm{w}}=0$. Indeed, inverting the constitutive relations (7.4.11), we get the generalized stresses $(\boldsymbol{\sigma}, p_{\mathrm{w}})$ in terms of the generalized strains $(\boldsymbol{\epsilon}, \zeta^{\mathrm{w}})$,

$$\begin{bmatrix} \boldsymbol{\sigma} \\ p_{\mathrm{w}} \end{bmatrix} = \begin{bmatrix} \mathbb{E}^{\mathrm{U}} & -M\,\boldsymbol{\alpha}_{\mathrm{w}} \\ -M\,\boldsymbol{\alpha}_{\mathrm{w}} & M \end{bmatrix} \begin{bmatrix} \boldsymbol{\epsilon} \\ \zeta^{\mathrm{w}} \end{bmatrix}, \tag{7.4.18}$$

where

$$\begin{aligned} \frac{1}{M} &= \frac{1}{S}\,\mathbf{I} : \mathbb{E}^{-1} : \boldsymbol{\alpha}_{\mathrm{w}} - \boldsymbol{\alpha}_{\mathrm{w}} : \mathbb{E}^{-1} : \boldsymbol{\alpha}_{\mathrm{w}} \\ &= \frac{1-n^{\mathrm{w}}}{K_{\mathrm{s}}} + \frac{n^{\mathrm{w}}}{K_{\mathrm{w}}} - (\mathbf{I} : \mathbb{E}_{\mathrm{s}}^{-1}) : \mathbb{E} : (\mathbb{E}_{\mathrm{s}}^{-1} : \mathbf{I}). \end{aligned} \tag{7.4.19}$$

Then, the undrained stress-strain relationships can be recast in the format

$$\boldsymbol{\epsilon} = (\mathbb{E}^{\mathrm{U}})^{-1} : \boldsymbol{\sigma}, \quad p_{\mathrm{w}} = -\mathbf{S} : \boldsymbol{\sigma}. \tag{7.4.20}$$

Here $\mathbf{S} = M\,(\mathbb{E}^{\mathrm{U}})^{-1} : \boldsymbol{\alpha}_{\mathrm{w}}$ is the *Skempton tensor* (see Thompson and Willis [1991], Eqn (16), and, up to a factor $1/3$, Cheng [1997], Eqn (17)),

$$\frac{\mathbf{S}}{\operatorname{tr}\mathbf{S}} = \frac{\mathbf{A}}{\operatorname{tr}\mathbf{A}}, \quad S = \operatorname{tr}\mathbf{S}, \tag{7.4.21}$$

which generalizes to the anisotropic case the Skempton pore pressure parameter $B = S$, Skempton [1954]. For isotropy, $\mathbf{S}$ reduces to $S\,\mathbf{I}/3$ and (7.4.20) becomes $p_{\mathrm{w}} = -S\operatorname{tr}\boldsymbol{\sigma}/3$. Moreover

Property 1: The drained and undrained compliance tensors and the drained and undrained tensor moduli differ by a rank one term:

$$(\mathbb{E}^{\mathrm{U}})^{-1} \equiv \mathbb{E}^{-1} - S\,\frac{\mathbf{A} \otimes \mathbf{A}}{\operatorname{tr}\mathbf{A}} \quad \Leftrightarrow \quad \mathbb{E}^{\mathrm{U}} = \mathbb{E} + M\,\boldsymbol{\alpha}_{\mathrm{w}} \otimes \boldsymbol{\alpha}_{\mathrm{w}}. \tag{7.4.22}$$

The strain energy function $\underline{\mathcal{W}}(\boldsymbol{\epsilon}, \boldsymbol{\epsilon}^{\mathrm{w}})$ obtained from the free energy $\underline{E}(\boldsymbol{\epsilon}, \zeta^{\mathrm{w}})$ through the change of strain variables (7.4.8),

$$\underline{E}\left(\boldsymbol{\epsilon}, \zeta^{\mathrm{w}}\right) = \frac{1}{2}\,\boldsymbol{\sigma}:\boldsymbol{\epsilon} + \frac{1}{2}\,p_{\mathrm{w}}\,\zeta^{\mathrm{w}} = \underline{\mathcal{W}}\left(\boldsymbol{\epsilon}, \boldsymbol{\epsilon}^{\mathrm{w}}\right), \tag{7.4.23}$$

can be recast in the following alternative formats:

$$\underline{\mathcal{W}}\left(\boldsymbol{\epsilon}, \boldsymbol{\epsilon}^{\mathrm{w}}\right) = \begin{cases} \dfrac{1}{2}\,\boldsymbol{\epsilon} : \mathbb{E} : \boldsymbol{\epsilon} + \dfrac{1}{2}\,M\left((\boldsymbol{\alpha}_{\mathrm{w}} - n^{\mathrm{w}}\,\mathbf{I}) : \boldsymbol{\epsilon} + n^{\mathrm{w}}\,\mathbf{I} : \boldsymbol{\epsilon}^{\mathrm{w}}\right)^2, \\[2mm] \dfrac{1}{2}\,(\boldsymbol{\epsilon} - \zeta^{\mathrm{w}}\,\mathbf{S}) : \mathbb{E}^{\mathrm{U}} : (\boldsymbol{\epsilon} - \zeta^{\mathrm{w}}\,\mathbf{S}) + \dfrac{1}{2}\,(M - \mathbf{S} : \mathbb{E}^{\mathrm{U}} : \mathbf{S})\,(\zeta^{\mathrm{w}})^2. \end{cases} \tag{7.4.24}$$

Therefore, we have reached

Property 2: Positive definiteness (PD) of the strain energy function takes one of the following equivalent forms:

$$\underline{\mathcal{W}}(\boldsymbol{\epsilon}, \boldsymbol{\epsilon}^{\mathrm{w}}) \quad \text{PD} \quad \Leftrightarrow \quad \mathbb{E}\ \text{PD and}\ M > 0 \quad \Leftrightarrow \quad \mathbb{E}^{\mathrm{U}}\ \text{PD and}\ \frac{\operatorname{tr}\mathbf{A}}{S} = \frac{1}{M - \mathbf{S} : \mathbb{E}^{\mathrm{U}} : \mathbf{S}} > 0. \tag{7.4.25}$$

Notice that Eqns (7.4.14), (7.4.19) and (7.4.25) are consistent with the inequalities,

$$\operatorname{tr}\mathbf{A} > 0 \quad \text{and} \quad \operatorname{tr}\mathbf{S} = S > 0, \tag{7.4.26}$$

the latter result being expected on physical grounds as it will be seen below in relation with Eqn (7.4.20)$_2$. In view of Eqns (7.4.22)$_{1,2}$, since $M > 0$ and $\operatorname{tr}\mathbf{A}/S > 0$, the undrained response appears to be *stiffer* than the drained response.

Later, it will be convenient to rewrite relation (7.4.22) in terms of the second order tensor $\mathbf{E}$ of unit norm ($\|\mathbf{E}\| = \sqrt{\mathbf{E}:\mathbf{E}} = 1$) proportional to $\mathbf{A}$ and $\mathbf{S}$ and defined implicitly through the following formulas,

$$\mathbf{A} = \frac{\Lambda}{S}\,(\operatorname{tr}\mathbf{E})\,\mathbf{E}, \quad \Lambda > 0, \quad \frac{\mathbf{S}}{\operatorname{tr}\mathbf{S}} = \frac{\mathbf{A}}{\operatorname{tr}\mathbf{A}} = \frac{\mathbf{E}}{\operatorname{tr}\mathbf{E}}. \tag{7.4.27}$$

The sign of Λ is implied by $\operatorname{tr}\mathbf{A}/S > 0$. Then *Property 1* can be viewed in a slightly different perspective:

Property 3: The difference $\mathbb{C}$ of compliance tensors, Eqn (7.4.22)$_1$,

$$\mathbb{C} \equiv \mathbb{E}^{-1} - (\mathbb{E}^{\mathrm{U}})^{-1} = \Lambda\,\mathbf{E} \otimes \mathbf{E} \tag{7.4.28}$$

has a single nonzero eigentensor $\mathbf{E}$ which is parallel to $\mathbf{A}$ and $\mathbf{S}$. Since $\operatorname{tr}\mathbf{A}$ is strictly positive, this eigentensor is non-isochoric.

Another slightly different form of Eqn (7.4.22)$_1$ is also recorded for later use:

Property 4: Assuming the triplet $(\mathbb{E}^{\mathrm{U}}, \mathbf{S}, M)$ corresponding to the undrained moduli $\mathbb{E}^{\mathrm{U}}$, the Skempton tensor $\mathbf{S}$ and the scalar M to be known, the drained compliance tensor is obtained explicitly as

$$\mathbb{E}^{-1} = (\mathbb{E}^{\mathrm{U}})^{-1} + \frac{1}{M - \mathbf{S} : \mathbb{E}^{\mathrm{U}} : \mathbf{S}}\,\mathbf{S} \otimes \mathbf{S}. \tag{7.4.29}$$

To prove this relation, observe first that $\boldsymbol{\alpha}_{\mathrm{w}}$ can be expressed in terms of the above triplet, using Eqns (7.4.18) and (7.4.20),

$$\boldsymbol{\alpha}_{\mathrm{w}} = \frac{1}{M}\,\mathbf{S} : \mathbb{E}^{\mathrm{U}}. \tag{7.4.30}$$

Further, Eqns (7.4.17), (7.4.19), (7.4.21) and (7.4.30) provide

$$\mathbf{A} = \frac{1}{M - \mathbf{S} : \mathbb{E}^{\mathrm{U}} : \mathbf{S}}\,\mathbf{S}, \tag{7.4.31}$$

which, introduced in Eqn (7.4.22)$_1$, yields the announced result, Eqn (7.4.29). Combining (7.4.19) and (7.4.17) with (7.4.31) yields in turn,

$$M = \frac{\mathbf{S} : \mathbb{E}^{\mathrm{U}} : \mathbf{S}}{\boldsymbol{\alpha}_{\mathrm{w}} : \mathbf{S}}. \tag{7.4.32}$$

The constitutive relations (7.4.18) can then be recast in a format,

$$
\begin{bmatrix} \boldsymbol{\sigma} \\ p_{\mathrm{w}} \end{bmatrix} = \begin{bmatrix} \mathbb{E}^{\mathrm{U}} & -\mathbb{E}^{\mathrm{U}} : \mathbf{S} \\ -\mathbf{S} : \mathbb{E}^{\mathrm{U}} & M \end{bmatrix} \begin{bmatrix} \boldsymbol{\epsilon} \\ \zeta^{\mathrm{w}} \end{bmatrix}. \tag{7.4.33}
$$

Therefore, at zero total stress,

$$
\boldsymbol{\epsilon} = \zeta^{\mathrm{w}} \, \mathbf{S}, \quad p_{\mathrm{w}} = \zeta^{\mathrm{w}} \, (\frac{1}{\alpha_{\mathrm{w}} : \mathbf{S}} - 1) \, \mathbf{S} : \mathbb{E}^{\mathrm{U}} : \mathbf{S}. \tag{7.4.34}
$$

Given the positive definiteness of the undrained moduli $\mathbb{E}^{\mathrm{U}}$, the positive definiteness of the strain energy function Eqn (7.4.25) is ensured by the inequality,

$$
M > M_{\mathrm{PD}} = \mathbf{S} : \mathbb{E}^{\mathrm{U}} : \mathbf{S}. \tag{7.4.35}
$$

So, since the scalar coefficient in the denominator of Eqn (7.4.29) is strictly positive, this relation is another manifestation that the undrained response is *stiffer* than the drained response.

Let K^{U} be the undrained bulk modulus, namely

$$
\frac{1}{K^{\mathrm{U}}} = \mathbf{I} : (\mathbb{E}^{\mathrm{U}})^{-1} : \mathbf{I}. \tag{7.4.36}
$$

In view of the definition of $\mathbf{A}$, Eqn (7.4.14), and using the expression of K in terms of the triplet $(\mathbb{E}^{\mathrm{U}}, \mathbf{S}, M)$ provided by (7.4.29), namely

$$
\frac{1}{K} = \frac{1}{K^{\mathrm{U}}} + \frac{(\mathrm{tr}\, \mathbf{S})^2}{M - \mathbf{S} : \mathbb{E}^{\mathrm{U}} : \mathbf{S}}, \tag{7.4.37}
$$

the bulk modulus of the solid constituent K_{s} follows in terms of this triplet,

$$
\frac{1}{K_{\mathrm{s}}} = \frac{1}{K^{\mathrm{U}}} + \frac{(\mathrm{tr}\, \mathbf{S})^2 - \mathrm{tr}\, \mathbf{S}}{M - \mathbf{S} : \mathbb{E}^{\mathrm{U}} : \mathbf{S}}. \tag{7.4.38}
$$

Since, as noted below, it is expected that $0 < S = \mathrm{tr}\, \mathbf{S} < 1$, we have a more stringent requirement than Eqn (7.4.35); indeed

Property 5: The positiveness of K_{s} requires a higher lower bound for M, namely,

$$
M > M_{\mathrm{inf}} = \mathbf{S} : \mathbb{E}^{\mathrm{U}} : \mathbf{S} + (\mathrm{tr}\, \mathbf{S} - (\mathrm{tr}\, \mathbf{S})^2) \, K^{\mathrm{U}}, \tag{7.4.39}
$$

than that required by the mere positive definiteness of the strain energy function (7.4.35).

7.4.3   Independent measurements

Henceforth we assume the drained moduli $\mathbb{E}$ (or the undrained moduli $\mathbb{E}^{\mathrm{U}}$) and the Skempton coefficient S to be known and measurable through macroscopic experiments. To measure S, one may perform an experiment with an isotropic total stress $\boldsymbol{\sigma} = -p\, \mathbf{I}$: relations $(7.4.20)_2$ and $(7.4.21)_2$ yield the Skempton coefficient $S = p_{\mathrm{w}}/p$, which is expected to be strictly positive. Moreover, according to the identification (7.4.15), since $K_{\mathrm{s}} > K$, the inequality $S < 1$ is equivalent to $K_{\mathrm{s}} > K_{\mathrm{w}}$: the latter inequality holds for example for all the geological materials listed by Cheng and Detournay [1993]. In that context, discovering the *compatibility relations* between drained and undrained material parameters in view of *Properties 1 to 5* is an issue. The knowledge of these compatibility relations is crucial to define, not the number of measurements that we need to perform (this we already know) but which measurements are *independent*, in a sense that will become clear through examples. Of course, we have already some knowledge of $\mathbb{E}^{\mathrm{U}}$ (or of $\mathbb{E}$): since water can not sustain shear stresses, the undrained shear moduli should be equal to the drained shear moduli. This information can also be read directly from Eqns (7.4.22) and (7.4.28),(7.4.29). The existence of other relations is less intuitive, especially in the presence of anisotropy.

A general and systematic approach has been adopted in Loret et al. [2001], where materials endowed with isotropy, general transverse isotropy and general orthotropy are examined in turn. These authors also consider the situation where the Skempton tensor $\mathbf{S}$ is known and then define which information should be extracted from the undrained (or drained) properties.

To discover the compatibility relations that link the drained and undrained stiffness and compliance tensors, the point of view delivered by *Property 1* is adopted for orthotropic materials and the perspective contained in *Property 3* is envisaged for isotropic and transversely isotropic materials. In all cases, the drained and undrained tensor stiffness and compliance tensors are assumed from the start to belong to the same class of anisotropy and to be endowed with the same directions of orthotropy.

7.4.4 Compatibility conditions for isotropic porous media

The drained and undrained isotropic compliance tensors are defined by their respective pairs of Young's moduli and Poisson's ratios (E, ν) and $(E^{\mathrm{U}}, \nu^{\mathrm{U}})$:

$$
\begin{aligned}
\mathbb{E}^{-1} &= \frac{1+\nu}{E}\,\mathbf{I}\,\overline{\otimes}\,\mathbf{I} - \frac{\nu}{E}\,\mathbf{I}\otimes\mathbf{I}, \\
(\mathbb{E}^{\mathrm{U}})^{-1} &= \frac{1+\nu^{\mathrm{U}}}{E^{\mathrm{U}}}\,\mathbf{I}\,\overline{\otimes}\,\mathbf{I} - \frac{\nu^{\mathrm{U}}}{E^{\mathrm{U}}}\,\mathbf{I}\otimes\mathbf{I}.
\end{aligned}
\tag{7.4.40}
$$

The meaning of the operator $\overline{\otimes}$ is read from (2.1.24). The tensor $\mathbf{A}$ defined by (7.4.17) can be obtained by a simple direct calculation. However, the derivation is less straightforward for anisotropic materials and recourse to a spectral decomposition of the elasticity tensors is an alternative. This method can also be used in the present context. Indeed, for isotropy as indicated in Section 5.3.1, the difference $\mathbb{C}$, Eqn (7.4.28), has a single non-isochoric eigenmode $\mathbf{E}=\mathbf{I}/\sqrt{3}$ associated with the eigenvalue $1/3K - 1/3K^{\mathrm{U}}$, while the five other eigenmodes are isochoric and associated with the eigenvalue $1/2\,\mu - 1/2\,\mu^{\mathrm{U}}$: in terms of Young's moduli and Poisson's ratios, drained and undrained bulk and shear moduli are equal to $K=E/3\,(1-2\nu)$, $K^{\mathrm{U}}=E^{\mathrm{U}}/3\,(1-2\nu^{\mathrm{U}})$, and $\mu = E/2(1+\nu)$, $\mu^{\mathrm{U}} = E^{\mathrm{U}}/2(1+\nu^{\mathrm{U}})$, respectively. Then Eqn (7.4.27)$_1$ implies, as expected, that $\mathbf{A}$ is isotropic. As a further consequence of Eqn (7.4.22), the drained and undrained shear moduli are equal as expected as well, $\mu = \mu^{\mathrm{U}}$. Therefore, together with the drained Young's modulus and Poissons' ratio, a *single undrained parameter needs to be measured*, say Poisson's ratio ν^{U}. Then,

$$
E^{\mathrm{U}} = E\,\frac{1+\nu^{\mathrm{U}}}{1+\nu} = 2\,\mu\,(1+\nu^{\mathrm{U}}), \quad K^{\mathrm{U}} = 2\,\mu\,\frac{1+\nu^{\mathrm{U}}}{3\,(1-2\,\nu^{\mathrm{U}})},
\tag{7.4.41}
$$

and Eqn (7.4.27)$_1$ yields, e.g., Rice and Cleary [1976],

$$
\boldsymbol{\alpha}_{\mathrm{w}} = \frac{3}{S}\,\frac{\nu^{\mathrm{U}}-\nu}{(1+\nu^{\mathrm{U}})(1-2\nu)}\,\mathbf{I}, \quad \mathbf{A} = \mathbb{E}^{-1}{:}\,\boldsymbol{\alpha}_{\mathrm{w}} = \frac{3}{S\,E}\,\frac{\nu^{\mathrm{U}}-\nu}{1+\nu^{\mathrm{U}}}\,\mathbf{I}.
\tag{7.4.42}
$$

The positive definiteness of the strain energy $\underline{\mathcal{W}}$, Eqn (7.4.25), restricts the ranges of E^{U} and ν^{U}, indeed

$$
-1 < \nu < \nu^{\mathrm{U}} < \frac{1}{2} \quad \text{and} \quad 0 < E < E^{\mathrm{U}} < \frac{3}{2}\,\frac{E}{1+\nu},
\tag{7.4.43}
$$

and therefore $K^{\mathrm{U}} > 0$. Indeed, the undrained Poisson's ratio can also be cast in the following forms:

$$
\nu^{\mathrm{U}} = \nu + \frac{\alpha_{\mathrm{w}}\,S\,(1-2\,\nu)(1+\nu)}{3 - \alpha_{\mathrm{w}}\,S\,(1-2\,\nu)} = \frac{1}{2} - \frac{3}{2}\,\frac{(1-\alpha_{\mathrm{w}}\,S)\,(1-2\,\nu)}{3 - \alpha_{\mathrm{w}}\,S\,(1-2\,\nu)}.
\tag{7.4.44}
$$

It is worth recording the particular form of some constitutive relations for an isotropic solid skeleton. Indeed then $\boldsymbol{\alpha}_{\mathrm{w}}$ simplifies to the classical expressions, Nur and Byerlee [1971], Rice and Cleary [1976],

$$
\boldsymbol{\alpha}_{\mathrm{w}} = \alpha_{\mathrm{w}}\,\mathbf{I}, \quad \alpha_{\mathrm{w}} = 1 - \frac{K}{K_{\mathrm{s}}}.
\tag{7.4.45}
$$

The constitutive equations (7.4.11) take the form,

$$
\begin{bmatrix} \operatorname{tr}\boldsymbol{\epsilon} \\ \zeta^{\mathrm{w}} \end{bmatrix} = \frac{1}{K}\begin{bmatrix} 1 & \alpha_{\mathrm{w}} \\ \alpha_{\mathrm{w}} & \alpha_{\mathrm{w}}/S \end{bmatrix}\begin{bmatrix} \tfrac{1}{3}\operatorname{tr}\boldsymbol{\sigma} \\ p_{\mathrm{w}} \end{bmatrix}, \quad \operatorname{dev}\boldsymbol{\epsilon} = \frac{1}{2\,\mu}\operatorname{dev}\boldsymbol{\sigma},
\tag{7.4.46}
$$

where μ is the (drained) shear modulus of the solid skeleton. The inverse form (7.4.18),

$$
\begin{bmatrix} \tfrac{1}{3}\operatorname{tr}\boldsymbol{\sigma} \\ p_{\mathrm{w}} \end{bmatrix} = \begin{bmatrix} K^{\mathrm{U}} & -M\,\alpha_{\mathrm{w}} \\ -M\,\alpha_{\mathrm{w}} & M \end{bmatrix}\begin{bmatrix} \operatorname{tr}\boldsymbol{\epsilon} \\ \zeta^{\mathrm{w}} \end{bmatrix}, \quad \operatorname{dev}\boldsymbol{\sigma} = 2\,\mu\operatorname{dev}\boldsymbol{\epsilon},
\tag{7.4.47}
$$

features the undrained bulk modulus K^{U} that can be expressed in terms of the drained bulk modulus K, and of the coefficients α_{w} and S,

$$
K^{\mathrm{U}} = \frac{K}{1 - \alpha_{\mathrm{w}}\,S} = \frac{M\,\alpha_{\mathrm{w}}}{S}.
\tag{7.4.48}
$$

Another form of the elastic relations may be recorded,

$$\begin{bmatrix} \frac{1}{3}\,\mathrm{tr}\,\boldsymbol{\sigma} \\[4pt] \zeta^{\mathrm{w}} \end{bmatrix} = \begin{bmatrix} K & -\alpha_{\mathrm{w}} \\[4pt] \alpha_{\mathrm{w}} & 1/M \end{bmatrix} \begin{bmatrix} \mathrm{tr}\,\boldsymbol{\epsilon} \\[4pt] p_{\mathrm{w}} \end{bmatrix}, \quad \mathrm{dev}\,\boldsymbol{\sigma} = 2\,\mu\,\mathrm{dev}\,\boldsymbol{\epsilon}\,, \tag{7.4.49}$$

which provides a further mixed expression of the energy,

$$\frac{1}{2}\,\boldsymbol{\sigma}:\boldsymbol{\epsilon} + p_{\mathrm{w}}\,\zeta^{\mathrm{w}} = \frac{1}{2}\,\boldsymbol{\sigma}':\boldsymbol{\epsilon} + \frac{1}{2}\,\frac{p_{\mathrm{w}}^2}{M} = \frac{1}{2}\,K\,(\mathrm{tr}\,\boldsymbol{\epsilon})^2 + \mu\,(\mathrm{dev}\,\boldsymbol{\epsilon})^2 + \frac{1}{2}\,\frac{p_{\mathrm{w}}^2}{M}\,. \tag{7.4.50}$$

in terms of strain and pore pressure, since $\boldsymbol{\sigma}' = \boldsymbol{\sigma} + \alpha_{\mathrm{w}}\,\mathbf{I} = K\,\mathrm{tr}\,\boldsymbol{\epsilon}\,\mathbf{I} + 2\,\mu\,\mathrm{dev}\,\boldsymbol{\epsilon}$.

Finally, since water does not resist shear stresses, note

Proposition 7.4 Voigt bound for the shear modulus.

For an isotropic solid skeleton, the estimation of the shear modulus of the solid skeleton in terms of the shear modulus of the solid constituent and its volume fraction,

$$\mu = n^{\mathrm{s}}\,\mu_{\mathrm{s}}\,, \tag{7.4.51}$$

represents an upper bound known as Voigt bound, Fig. 7.3.3.

Indeed, the deviatoric parts of the constitutive laws at the scale of the solid constituent and porous medium write respectively, $\mathrm{dev}\,\boldsymbol{\sigma}_{\mathrm{s}} = 2\,\mu_{\mathrm{s}}\,\mathrm{dev}\,\boldsymbol{\epsilon}_{\mathrm{s}}$, Eqn (7.2.22), and $\mathrm{dev}\,\boldsymbol{\sigma} = 2\,\mu\,\mathrm{dev}\,\boldsymbol{\epsilon}$, Eqn (7.4.47). Moreover $\mathrm{dev}\,\boldsymbol{\sigma} = n^{\mathrm{s}}\mathrm{dev}\,\boldsymbol{\sigma}_{\mathrm{s}}$, Eqn (7.1.2), since the fluid does not resist shear. The relation follows if the deviatoric strain is assumed to be uniform over constituents, namely $\mathrm{dev}\,\boldsymbol{\epsilon} = \mathrm{dev}\,\boldsymbol{\epsilon}_{\mathrm{s}}$.

7.4.5 Compatibility conditions for transversely isotropic porous media

We consider now a transversely isotropic elastic porous medium with rotational symmetry about the unit vector $\mathbf{b}$, and we make use of a cartesian material frame $(\mathbf{b}^{(1)}, \mathbf{b}^{(2)}, \mathbf{b}^{(3)}{=}\mathbf{b})$. The drained elastic compliance tensor $\mathbb{E}^{-1}$ are expressed in terms of five independent parameters, namely longitudinal Young's modulus E_{L} and Poisson's ratio ν_{L} (contraction in transverse direction due to traction in longitudinal direction), transverse Young's modulus E_{T} and Poisson's ratio ν_{T}, and longitudinal shear modulus μ_{L}, the transverse shear modulus μ_{T} being equal to $E_{\mathrm{T}}/2(1+\nu_{\mathrm{T}})$.

In order to perform the spectral analysis of the drained elastic compliance tensor $\mathbb{E}^{-1}$, it is convenient to use the matrix representation in the format (5.2.52). Positive definiteness of $\mathbb{E}^{-1}$ is ensured by the inequalities (5.2.56). A similar representation in terms of undrained moduli and Poisson's ratios is available for the undrained compliance tensor $(\mathbb{E}^{\mathrm{U}})^{-1}$ which are also assumed to be positive definite.

The spectral analysis of the elastic stiffness $\mathbb{E}$ of a transversely isotropic material is reported in Section 5.3.2.1. This information can be used for the difference $\mathbb{C}$ of elastic compliance tensors, Eqn (7.4.28). $\mathbb{C}$ has in general four distinct eigenvalues, namely $\Lambda^{(1,2)} = 1/2\,\mu_{\mathrm{L}} - 1/2\,\mu_{\mathrm{L}}^{\mathrm{U}}$ and $\Lambda^{(3,4)} = 1/2\,\mu_{\mathrm{T}} - 1/2\,\mu_{\mathrm{T}}^{\mathrm{U}}$ each of multiplicity two, and two eigenvalues $\Lambda^{(5)}$ and $\Lambda^{(6)}$ of single multiplicity. Since from *Property 3*, the difference $\mathbb{C}$ should have a single non zero eigenvalue of multiplicity one associated with a dilatational eigenmode, the following four cases are available:

$$\begin{array}{llllll} 1) \text{ either} & \mu_{\mathrm{L}} - \mu_{\mathrm{L}}^{\mathrm{U}} \neq 0 & \text{while} & \mu_{\mathrm{T}} - \mu_{\mathrm{T}}^{\mathrm{U}} = 0, & \Lambda^{(5,6)} = 0, & \\ 2) \text{ or} & \mu_{\mathrm{T}} - \mu_{\mathrm{T}}^{\mathrm{U}} \neq 0 & \text{while} & \mu_{\mathrm{L}} - \mu_{\mathrm{L}}^{\mathrm{U}} = 0, & \Lambda^{(5,6)} = 0, & \\ 3) \text{ or} & \Lambda^{(5)} \neq 0 & \text{while} & \mu_{\mathrm{L}} - \mu_{\mathrm{L}}^{\mathrm{U}} = 0, & \mu_{\mathrm{T}} - \mu_{\mathrm{T}}^{\mathrm{U}} = 0, & \Lambda^{(6)} = 0, \\ 4) \text{ or} & \Lambda^{(6)} \neq 0 & \text{while} & \mu_{\mathrm{L}} - \mu_{\mathrm{L}}^{\mathrm{U}} = 0, & \mu_{\mathrm{T}} - \mu_{\mathrm{T}}^{\mathrm{U}} = 0, & \Lambda^{(5)} = 0. \end{array} \tag{7.4.52}$$

However, $\Lambda^{(1,2)}$ and $\Lambda^{(3,4)}$ are of double multiplicity and, moreover, their associated eigenmodes are isochoric: therefore, cases 1) and 2) above are excluded. Using Eqn (5.3.9), one may show that cases 3) and 4) yield formally similar constraints. Consequently, there are three compatibility relations to be satisfied by the drained and undrained coefficients for the difference $\mathbb{E}^{-1} - (\mathbb{E}^{\mathrm{U}})^{-1}$ to be dyadic, namely,

$$\mu_{\mathrm{L}}^{\mathrm{U}} = \mu_{\mathrm{L}}\,, \quad \mu_{\mathrm{T}}^{\mathrm{U}} = \frac{E_{\mathrm{T}}^{\mathrm{U}}}{2\,(1+\nu_{\mathrm{T}}^{\mathrm{U}})} = \mu_{\mathrm{T}} = \frac{E_{\mathrm{T}}}{2\,(1+\nu_{\mathrm{T}})}\,, \quad C_{11}C_{33} - C_{13}^2 = 0\,, \tag{7.4.53}$$

with

$$C_{11} = \frac{1}{E_{\mathrm{T}}} - \frac{1}{E_{\mathrm{T}}^{\mathrm{U}}}, \quad C_{33} = \frac{1}{E_{\mathrm{L}}} - \frac{1}{E_{\mathrm{L}}^{\mathrm{U}}}, \quad C_{13} = -\frac{\nu_{\mathrm{L}}}{E_{\mathrm{L}}} + \frac{\nu_{\mathrm{L}}^{\mathrm{U}}}{E_{\mathrm{L}}^{\mathrm{U}}}. \tag{7.4.54}$$

Cases 3) and 4) differ only by the sign of C_{11} and C_{33}: for case 3), both these scalars are negative while they are positive for case 4). In both cases, the eigenvalue Λ and unit eigentensor $\mathbf{E}$ of $\mathbb{C}$ are equal to,

$$\Lambda = 2\,C_{11} + C_{33}, \quad \mathbf{E} = \frac{\mathbf{diag}\,[C_{11}, C_{11}, C_{13}]}{\sqrt{2\,C_{11}^2 + C_{13}^2}}. \tag{7.4.55}$$

In view of Eqns $(7.4.53)_3$ and $(7.4.55)_1$, C_{11} and C_{33} should be of the same sign as Λ, that is, according to $(7.4.27)_1$, positive,

$$C_{11} > 0, \quad C_{33} > 0. \tag{7.4.56}$$

Also, the constraints (7.4.53) and (7.4.54) and the positive definiteness of the strain energy function (7.4.25) imply the inequality

$$1 + \frac{(\nu_{\mathrm{L}}^{\mathrm{U}} - \nu_{\mathrm{L}})^2}{\dfrac{1 - \nu_{\mathrm{T}}}{2}\dfrac{E_{\mathrm{L}}}{E_{\mathrm{T}}} - (\nu_{\mathrm{L}})^2} < \frac{E_{\mathrm{L}}^{\mathrm{U}}}{E_{\mathrm{L}}}, \tag{7.4.57}$$

and, as a further consequence of (7.4.53) and (7.4.57), the following inequalities hold as well:

$$-1 < \nu_{\mathrm{T}} < \nu_{\mathrm{T}}^{\mathrm{U}} < 1 \quad \text{and} \quad 0 < E_{\mathrm{T}} < E_{\mathrm{T}}^{\mathrm{U}} < \frac{2\,E_{\mathrm{T}}}{1 + \nu_{\mathrm{T}}}. \tag{7.4.58}$$

Since there are three constraints, Eqns (7.4.53), on the five undrained parameters $E_{\mathrm{L}}^{\mathrm{U}}$, $\nu_{\mathrm{L}}^{\mathrm{U}}$, $E_{\mathrm{T}}^{\mathrm{U}}$, $\nu_{\mathrm{T}}^{\mathrm{U}}$, $\mu_{\mathrm{L}}^{\mathrm{U}}$, two *independent undrained measurements* are, as expected, still required to define completely the eight constitutive parameters.

The question of which measurements are the most satisfactory is discussed further in Loret et al. [2001]. The notion of *independent* measurements can now be illustrated: for example, due to $(7.4.53)_2$, the measurements of the undrained Poisson's ratio $\nu_{\mathrm{T}}^{\mathrm{U}}$ and undrained Young's modulus $E_{\mathrm{T}}^{\mathrm{U}}$ are not independent.

7.4.5.1   Further applications to incomplete data sets

The above compatibility analysis is useful to estimate lacking data. An incomplete data set of material constants is provided by Aoki et al. [1993] for a transversely isotropic shale. They give values for only seven of the eight material coefficients, namely the undrained moduli $\mathbb{E}^{\mathrm{U}}$ and the Skempton tensor $\mathbf{S}$, Table 7.4.1.

Table 7.4.1: Data set of Aoki et al. [1993] for a transversely isotropic shale.

$E_{\mathrm{L}}^{\mathrm{U}}$	$E_{\mathrm{T}}^{\mathrm{U}}$	G_{L}^{U}	$\nu_{\mathrm{L}}^{\mathrm{U}}$	$\nu_{\mathrm{T}}^{\mathrm{U}}$	S_{L}	S_{T}
18.8 GPa	22.0 GPa	7.23 GPa	0.34	0.27	0.21	0.17

With one lacking measurement, bounds to the poroelastic constants are obtained by choosing the coefficient M, Eqn (7.4.19), as the missing parameter. This coefficient is left to take values at least larger than the lower bound ensuring positive definiteness of the tensor moduli, namely $M > M_{\mathrm{PD}} = 6.41\,\mathrm{GPa}$, Eqn (7.4.35). At the upper bound of the interval, namely for $M \to \infty$, the drained and undrained elastic compliance tensors are identical, Eqn (7.4.29). Actually, since $\mathrm{tr}\,\mathbf{S} = 2\,S_{\mathrm{T}} + S_{\mathrm{L}} = 0.55 < 1$, the positiveness of the tensor moduli $\mathbb{E}_{\mathrm{s}}$ of the solid constituent requires the higher lower bound $M > M_{\mathrm{inf}} = 11.65\,\mathrm{GPa}$, which is obtained from (7.4.39) by substituting the undrained bulk modulus,

$$K^{\mathrm{U}} = \frac{E_{\mathrm{L}}^{\mathrm{U}}\,E_{\mathrm{T}}^{\mathrm{U}}}{2\,(1 - \nu_{\mathrm{T}}^{\mathrm{U}})\,E_{\mathrm{L}}^{\mathrm{U}} + (1 - 4\,\nu_{\mathrm{L}}^{\mathrm{U}})\,E_{\mathrm{T}}^{\mathrm{U}}} = 21.18\,\mathrm{GPa}. \tag{7.4.59}$$

For any value of $M > M_{\mathrm{inf}}$, the drained compliance tensor is obtained explicitly from Eqn (7.4.29), the second order tensor $\mathbf{A}$ from Eqn (7.4.31) and the bulk moduli K and K_{s} from Eqns (7.4.37) and (7.4.38), respectively. Notice that, as it appears from Eqns (7.4.29) and (7.4.38), for $M \to \infty$, $(\mathbb{E})^{-1} \to (\mathbb{E}^{\mathrm{U}})^{-1}$ and K, $K_{\mathrm{s}} \to K^{\mathrm{U}}$, while, from Eqn (7.4.31), $\mathbf{A} \to \mathbf{0}$.

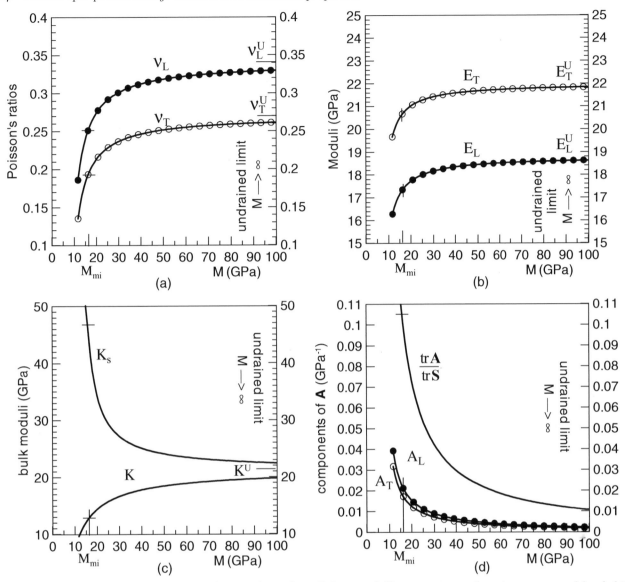

Fig. 7.4.2: Transverse isotropic shale whose undrained coefficients and Skempton tensor have been measured by Aoki et al. [1993] as indicated by Table 7.4.1. (a) Drained longitudinal and transverse Poisson's ratios ν_L and ν_T as a function of the unknown modulus M varying in the admissible range $]M_{inf}, +\infty[$. The lower bound $M_{inf}=11.65$ GPa ensures positiveness of the bulk modulus $K_s > 0$ of the solid constituent, and is oversufficient to ensure the positive definiteness of the tensor of drained moduli. The value $M = M_{mi}$ reported by Cheng [1997] for an isotropic solid constituent is equal to 15.76 GPa. (b) Drained longitudinal and transverse Young's moduli E_L and E_T. (c) Drained bulk modulus K and bulk modulus K_s of the solid constituent. (d) Longitudinal and transverse components A_L, A_T of the second order tensor **A** and coefficient $\operatorname{tr}\mathbf{A}/S$. Redrawn with permission from Loret et al. [2001].

The poroelastic constants involved in the constitutive Eqns (7.4.11), namely the drained compliance tensor $\mathbb{E}^{-1}$, given in terms of ν_L, ν_T, E_L, E_T, and μ_L, the second order tensor **A** and the scalar $\operatorname{tr}\mathbf{A}/S$, together with the bulk moduli K and K_s, are plotted in Fig. 7.4.2. As shown in Fig. 7.4.2(b), the ranges of variation of the drained moduli E_L and E_T are quite narrow and, as implied by inequalities (7.4.57) and (7.4.58), E_L and E_T are smaller than their undrained counterparts. On the other hand, the ranges of the other parameters are wider. Fig. 7.4.2(a) shows that the Poisson's ratios vary more significantly than the Young's moduli and, consistently with Eqn (7.4.58), $\nu_T < \nu_T^U$. The range of the bulk modulus K_s shown in Fig. 7.4.2(c) is very wide, starting from $+\infty$ for $M \to M_{inf}$, to K^U for $M \to \infty$, while the variation of K, which also converges to K^U for $M \to \infty$, is less pronounced, since $K \to (1-S)K^U$ for $M \to M_{inf}$. Finally, Fig. 7.4.2(d) shows how the Skempton coefficient S amplifies the interval of variation of the scalar $\operatorname{tr}\mathbf{A}/S$ with respect to the ranges of the two components A_L, A_T, which are rather close to each other; as expected from Eqn (7.4.31), the components of $\mathbf{A}$ tend to zero for $M \to \infty$, while they increase rapidly when $M \to M_{\text{inf}}$ even if this value is still larger than M_{PD}, Eqns (7.4.31) and (7.4.35).

A different solution procedure is also proposed in Loret et al. [2001]: all material coefficients may be solved explicitly in terms of one of the drained parameters E_{L}, E_{T}, ν_{L} and ν_{T}, ranging between the bounds ensuring positiveness of the strain energy and of K_{s}.

While this approach only bounds the elastic parameters, the two alternatives below allow to recover some definite value of the missing datum.

7.4.5.2 Micro-isotropy

Cheng [1997] assumes the microscale elastic response to be isotropic: macroscale anisotropy emanates from non-random directional and spatial distributions of structural members.

This micro-isotropy assumption results in a reduced number of independent coefficients. For example, for transverse isotropy, we know that the constitutive equations are defined by 8 independent constants, e.g., from Eqns (7.4.18) and (7.4.30), by $\mathbb{E}^{\text{U}}$ (5 constants), $\boldsymbol{\alpha}_{\text{w}}$ or $\mathbf{S}$ (2 constants) and M. However, for micro-isotropy, $\mathbb{E}_{\text{s}}^{-1} : \mathbf{I} = \mathbf{I}/3K_{\text{s}}^{\text{mi}}$, and K_{s}^{mi} and M_{mi} can be expressed in terms of $\mathbb{E}^{\text{U}}$ and $\mathbf{S}$. Therefore only 7 independent measurements are required. In fact using Eqns (7.4.14), (7.4.22)$_2$, and (7.4.30) yields the (Biot) tensor,

$$\boldsymbol{\alpha}_{\text{w}}^{\text{mi}} = \frac{1}{1 - \dfrac{1}{3K_{\text{s}}^{\text{mi}}} \mathbf{S} : \mathbb{E}^{\text{U}} : \mathbf{I}} \left(\mathbf{I} - \frac{1}{3K_{\text{s}}^{\text{mi}}} \mathbb{E}^{\text{U}} : \mathbf{I} \right), \tag{7.4.60}$$

which, again with (7.4.30), provides K_{s}^{mi} and M_{mi} in terms of $\mathbb{E}^{\text{U}}$ and $\mathbf{S}$:

$$\frac{1}{3K_{\text{s}}^{\text{mi}}} = \frac{3}{X} \operatorname{dev} \mathbf{S} : \mathbb{E}^{\text{U}} : \mathbf{S}, \qquad \frac{1}{M_{\text{mi}} - \mathbf{S} : \mathbb{E}^{\text{U}} : \mathbf{S}} = \frac{3}{X} \frac{\operatorname{dev} \mathbf{S} : \mathbb{E}^{\text{U}} : \mathbf{I}}{\operatorname{tr} \mathbf{S} - 1}, \tag{7.4.61}$$

where $\operatorname{dev} \mathbf{S}$ is the deviatoric part of $\mathbf{S}$, namely $\operatorname{dev} \mathbf{S} = \mathbf{S} - (\operatorname{tr} \mathbf{S}/3) \, \mathbf{I}$, which is expected to be non zero for true anisotropy, and then

$$X = (\mathbf{S} : \mathbb{E}^{\text{U}} : \mathbf{S}) (\mathbf{I} : \mathbb{E}^{\text{U}} : \mathbf{I}) - (\mathbf{S} : \mathbb{E}^{\text{U}} : \mathbf{I})^2, \tag{7.4.62}$$

is positive as a consequence of the Cauchy-Schwarz inequality.

7.4.5.3 On the Saint-Venant's relations for transverse isotropy

In place of the micro-isotropy assumption, one might have thought to use the so-called Saint-Venant's relations, Section 5.2.10, to obtain the missing coefficient M. For arbitrary anisotropy, Eqn (7.4.29) yields M in the following format,

$$\frac{1}{M - \mathbf{S} : \mathbb{E}^{\text{U}} : \mathbf{S}} = \frac{f_{\text{SV}}(\mathbb{E}^{\text{U}}) - f_{\text{SV}}(\mathbb{E})}{(S_{\text{T}} - S_{\text{L}})^2}, \tag{7.4.63}$$

where $f_{\text{SV}}(\mathbb{E})$, Eqn (5.2.68), is the Saint-Venant's function associated with the drained moduli,

$$f_{\text{SV}}(\mathbb{E}) \equiv \frac{1}{\mu_{\text{L}}} - \frac{1}{E_L} - \frac{1}{E_T} - 2 \frac{\nu_{\text{L}}}{E_L}, \tag{7.4.64}$$

and similarly for $f_{\text{SV}}(\mathbb{E}^{\text{U}})$. Consequently, the Saint-Venant's relation $f_{\text{SV}} = 0$ can not hold *strictly* for *both* the drained and undrained coefficients, if these two sets of coefficients are to be different and to correspond to a positive definite strain energy function.

7.4.6 Compatibility conditions for orthotropic porous media

Let us consider an orthotropic porous medium with axes of elastic orthotropy denoted by $(\mathbf{b}^{(1)}, \mathbf{b}^{(2)}, \mathbf{b}^{(3)})$. The matrix representation in these axes of the drained elastic compliance tensor $\mathbb{E}^{-1}$ is displayed on Eqn (5.2.46). It expresses in terms of three drained Young's moduli E_1, E_2, E_3, three shear moduli μ_{12}, μ_{13}, μ_{23} and three Poisson's ratios ν_{12}, ν_{13}, ν_{23}. The conditions of positive definiteness are provided by (5.E.25).

By the principle of equipresence, the symmetric second order tensor $\mathbf{E}$, Eqns $(7.4.27)_1$ and $(7.4.28)$, expresses along the orthotropy axes,

$$\mathbf{E} = \sum_{i=1}^{3} \gamma_i \, \mathbf{b}^{(i)} \otimes \mathbf{b}^{(i)} \quad \text{with} \quad \sum_{i=1}^{3} \gamma_i^2 = 1 \, . \tag{7.4.65}$$

Consequently the lower principal $3{\times}3$ minor of the matrix representation of the fourth order tensor $\mathbb{C}$ vanishes: the drained and undrained shear coefficients are equal as expected,

$$\mu_{12} = \mu_{12}^{\text{U}}, \quad \mu_{23} = \mu_{23}^{\text{U}}, \quad \mu_{13} = \mu_{13}^{\text{U}} \, . \tag{7.4.66}$$

Let us now have a closer look at the upper principal minor of $\mathbb{C}$,

$$\mathbf{C} = \begin{bmatrix} C_{11} & C_{12} & C_{13} \\ C_{12} & C_{22} & C_{23} \\ C_{13} & C_{23} & C_{33} \end{bmatrix} = \Lambda \begin{bmatrix} \gamma_1^2 & \gamma_1\gamma_2 & \gamma_1\gamma_3 \\ \gamma_1\gamma_2 & \gamma_2^2 & \gamma_2\gamma_3 \\ \gamma_1\gamma_3 & \gamma_2\gamma_3 & \gamma_3^2 \end{bmatrix}, \tag{7.4.67}$$

where

$$\begin{aligned} C_{11} &= \frac{1}{E_1} - \frac{1}{E_1^{\text{U}}}, \quad C_{22} = \frac{1}{E_2} - \frac{1}{E_2^{\text{U}}}, \quad C_{33} = \frac{1}{E_3} - \frac{1}{E_3^{\text{U}}}, \\ C_{12} &= -\frac{\nu_{12}}{E_1} + \frac{\nu_{12}^{\text{U}}}{E_1^{\text{U}}}, \quad C_{13} = -\frac{\nu_{13}}{E_1} + \frac{\nu_{13}^{\text{U}}}{E_1^{\text{U}}}, \quad C_{23} = -\frac{\nu_{23}}{E_2} + \frac{\nu_{23}^{\text{U}}}{E_2^{\text{U}}}. \end{aligned} \tag{7.4.68}$$

Notice that if $\{\gamma_1, \gamma_2, \gamma_3\}$ is a solution set of $(7.4.67)$, so is $\{-\gamma_1, -\gamma_2, -\gamma_3\}$; therefore, only one of these sets needs to be considered. Now, for solutions to $(7.4.67)$ to exist, the coefficient matrix $\mathbf{C}$ should satisfy the following compatibility requirements, for $i \neq j \neq k$ (no sum over repeated indices):

$$\text{sgn}\, C_{ii} = \text{sgn}\, \Lambda = 1; \quad C_{ii}\, C_{jj} - C_{ij}^2 = 0; \quad \text{if } C_{ij}\, C_{jk}\, C_{ki} \neq 0, \quad \text{then } \text{sgn}\, C_{ij}\, C_{jk}\, C_{ki} = 1 \, . \tag{7.4.69}$$

Then, since $\gamma_i^2 = C_{ii}/\Lambda$, $i \in [1,3]$, (no sum over i), the signs of the γ's are defined according to the following rules:

$$\begin{aligned} &- \text{ if } \gamma_1 \neq 0, \text{ then } \text{sgn}\, \gamma_1 = 1 \text{ and, for } j \neq 1: \begin{cases} \text{if } C_{1j} \neq 0, & \text{then } \text{sgn}\, \gamma_j = \text{sgn}\, (C_{1j}/\Lambda), \\ \text{if } C_{1j} = 0, & \text{then } \gamma_j = 0; \end{cases} \\ &- \text{ if } \gamma_1 = 0, \text{ then } \begin{cases} \text{if } C_{23} \neq 0, & \text{then } \text{sgn}\, \gamma_2 \, \text{sgn}\, \gamma_3 = \text{sgn}\, (C_{23}/\Lambda), \\ \text{if } C_{23} = 0, & \text{then } \gamma_2 = 0 \text{ or } \gamma_3 = 0 \, . \end{cases} \end{aligned} \tag{7.4.70}$$

In other words, the eigenmode $\mathbf{E}$ of the compliance difference $\mathbb{C}$, Eqn $(7.4.28)$, may be defined by one of the following three equivalent forms:

$$\mathbf{E} = \mathbf{E}_{(1)} = \mathbf{E}_{(2)} = \mathbf{E}_{(3)} \quad \text{with} \quad \mathbf{E}_{(i)} = \frac{\text{diag}\,[\, C_{1i}, C_{2i}, C_{3i}\,]}{\sqrt{C_{1i}^2 + C_{2i}^2 + C_{3i}^2}}, \quad i \in [1,3]. \tag{7.4.71}$$

Summarizing, notwithstanding the sign restrictions, there are six constraints, Eqns $(7.4.66)$ and $(7.4.69)_2$, on the nine undrained coefficients E_1^{U}, E_2^{U}, E_3^{U}, μ_{12}^{U}, μ_{13}^{U}, μ_{23}^{U}, ν_{12}^{U}, ν_{13}^{U}, ν_{23}^{U}: *three independent undrained measurements required to define the 13 constitutive parameters can be proposed*, Loret et al. [2001]:

- three undrained coefficients may be measured directly;
- alternatively, the Skempton tensor $\mathbf{S}$ may be measured together with one undrained coefficient.

7.4.7 Porous media with fabric anisotropy

The particular form of anisotropic elasticity based on a symmetric second order fabric tensor $\mathbf{B}$ has been described in Section 5.2.11. The relations $(7.4.13)$ linking the strain and the effective stress express through the drained stiffness $(5.2.72)$ or its inverse $(5.2.79)$. Positive definiteness of the drained stiffness expresses, through Eqn $(5.2.30)$, like for isotropic elasticity, either in terms of the drained Lamé moduli or in terms of Young's modulus and Poisson's ratio.

When the eigenvalues $b_i > 0$, $i \in [1,3]$, of $\mathbf{B}$ are distinct, the material is orthotropic. When only two of the eigenvalues of $\mathbf{B}$ are equal, the material is endowed with transverse isotropy. Otherwise, the material is isotropic, since then $\mathbf{B} = \mathbf{I}$.

7.4.7.1   Compatibility between drained and undrained fabrics

A representation similar to that of the drained compliance (and stiffness) holds for its undrained counterpart: it expresses in terms of the undrained Young's modulus E^{U} and Poisson's ratio ν^{U}, and of the undrained fabric tensor $\mathbf{B}^{\mathrm{U}}$:

$$
\begin{aligned}
(\mathbb{E})^{-1} &= \frac{1+\nu}{E}\,(\mathbf{B})^{-1}\,\overline{\underline{\otimes}}\,(\mathbf{B})^{-1} \;-\; \frac{\nu}{E}\,(\mathbf{B})^{-1}\otimes(\mathbf{B})^{-1}, \\
(\mathbb{E}^{\mathrm{U}})^{-1} &= \frac{1+\nu^{\mathrm{U}}}{E^{\mathrm{U}}}\,(\mathbf{B}^{\mathrm{U}})^{-1}\,\overline{\underline{\otimes}}\,(\mathbf{B}^{\mathrm{U}})^{-1} \;-\; \frac{\nu^{\mathrm{U}}}{E^{\mathrm{U}}}\,(\mathbf{B}^{\mathrm{U}})^{-1}\otimes(\mathbf{B}^{\mathrm{U}})^{-1}\,.
\end{aligned}
\tag{7.4.72}
$$

The undrained fabric tensor $\mathbf{B}^{\mathrm{U}}$ should satisfy conditions analogous to (5.2.74). Also, since we have assumed the drained and undrained moduli to share the same directions of anisotropy, the undrained fabric tensor needs to be coaxial with $\mathbf{B}$. Indeed, more than that, the compatibility relations (7.4.22) and (7.4.28) require the two fabric tensors to be equal,

$$
\mathbf{B}^{\mathrm{U}} = \mathbf{B}\,.
\tag{7.4.73}
$$

In fact, since the diagonal shear terms of the compliance difference $\mathbb{C}$ corresponding to the $\overline{\underline{\otimes}}$ products must vanish, from Eqn (7.4.72) we get:

$$
\frac{1+\nu^{\mathrm{U}}}{E^{\mathrm{U}}}(\mathbf{B}^{\mathrm{U}})^{-1}\,\overline{\underline{\otimes}}\,(\mathbf{B}^{\mathrm{U}})^{-1} = \frac{1+\nu}{E}\mathbf{B}^{-1}\,\overline{\underline{\otimes}}\,\mathbf{B}^{-1}\,.
\tag{7.4.74}
$$

The inverse of $\mathbf{B}^{-1}\overline{\underline{\otimes}}\mathbf{B}^{-1}$ is $\mathbf{B}\,\overline{\underline{\otimes}}\,\mathbf{B}$, which has matrix components along the diagonal only, namely $[\,b_1^2, b_2^2, b_3^2, b_2\,b_3,\, b_1\,b_3,\, b_1\,b_2\,]$, and similarly for the undrained fabric. Then the normalization $(5.2.74)_2$ implies the same relation (7.4.41) which holds for isotropy, namely $E^{\mathrm{U}}/(1+\nu^{\mathrm{U}})=E/(1+\nu)$.

Finally, to satisfy the dyadic forms (7.4.22) and (7.4.28), the tensor $\mathbf{A}$, Eqn $(7.4.27)_1$, must be proportional to $\mathbf{B}^{-1}$ and can be cast in a form that generalizes its isotropic counterpart, Eqn (7.4.42):

$$
\mathbf{A} = \mathbb{E}^{-1}\colon \boldsymbol{\alpha}_{\mathrm{w}} = \frac{1}{S\,E}\frac{\nu^{\mathrm{U}}-\nu}{1+\nu^{\mathrm{U}}}\,(\operatorname{tr}\mathbf{B}^{-1})\,\mathbf{B}^{-1}\,.
\tag{7.4.75}
$$

Consequently, together with the drained Young's modulus and Poisson's ratio, *a single undrained parameter needs to be measured*, e.g., ν^{U}. The condition $M>0$, Eqns (7.4.19) and (7.4.25), implies the same restrictions on the measured coefficient ν^{U} and on E and E^{U} as in the isotropic case, Eqn (7.4.43). The general properties outlined here are confirmed below by separate derivations for the particular forms of transverse isotropy and orthotropy.

7.4.7.2   Compatibility between drained and undrained fabrics: transverse isotropy

Let us assume the axis of rotational symmetry to be the axis $\mathbf{b}=\mathbf{b}^{(3)}$, i.e., $b_1=b_2$, and hence $2\,b_1^2 + b_3^2 = 3$. The drained compliance being expressed in the axes of anisotropy, the identification of the five drained coefficients in terms of, e.g., E, ν and b_1 (and b_3) yields, the indices T and L indicating the transverse and longitudinal directions, respectively,

$$
E_T = E\,b_1^2,\quad \nu_{\mathrm{T}}=\nu,\quad E_{\mathrm{L}}=E\,b_3^2,\quad \nu_{\mathrm{L}}=\nu\,\frac{b_3}{b_1},\quad \mu_{\mathrm{L}}=\frac{E}{2(1+\nu)}\,b_1\,b_3\,.
\tag{7.4.76}
$$

Similar relations hold for the undrained coefficients:

$$
E_{\mathrm{T}}^{\mathrm{U}} = E^{\mathrm{U}}\,(b_1^U)^2,\quad \nu_{\mathrm{T}}^{\mathrm{U}}=\nu^{\mathrm{U}},\quad E_{\mathrm{L}}^{\mathrm{U}}=E^{\mathrm{U}}\,(b_3^U)^2,\quad \nu_{\mathrm{L}}^{\mathrm{U}}=\nu^{\mathrm{U}}\,\frac{b_3^U}{b_1^U},\quad \mu_{\mathrm{L}}^{\mathrm{U}}=\frac{E^{\mathrm{U}}}{2\,(1+\nu^{\mathrm{U}})}\,b_1^U b_3^U\,.
\tag{7.4.77}
$$

Out of the three constraints in Eqn (7.4.53), only two are independent. As expected, they imply the tensors $\mathbf{B}$ and $\mathbf{B}^{\mathrm{U}}$ to be equal; also the undrained coefficients can be expressed in terms of E, ν, b_1 (and b_3) and of a single undrained parameter to be measured, say ν^{U}:

$$
E_{\mathrm{T}}^{\mathrm{U}} = E\,\frac{1+\nu^{\mathrm{U}}}{1+\nu}\,b_1^2,\quad \nu_{\mathrm{T}}^{\mathrm{U}}=\nu^{\mathrm{U}},\quad E_{\mathrm{L}}^{\mathrm{U}}=E\,\frac{1+\nu^{\mathrm{U}}}{1+\nu}\,b_3^2,\quad \nu_{\mathrm{L}}^{\mathrm{U}}=\nu^{\mathrm{U}}\,\frac{b_3}{b_1},\quad \mu_{\mathrm{L}}^{\mathrm{U}}=\mu_{\mathrm{L}}=\frac{E}{2(1+\nu)}\,b_1 b_3.
\tag{7.4.78}
$$

Notice that positive definiteness of $\mathbb{E}^{\mathrm{U}}$ requires ν^{U} to belong to the interval $]-1,1/2[$. Then, using (7.4.55), it is an easy matter to calculate the eigenvalue Λ and eigenmode $\mathbf{E}$, from which the tensor $\mathbf{A}$, Eqn $(7.4.27)_1$, may be cast in the form (7.4.75).

In summary, elastic fluid-saturated porous media embedding this particular form of transverse isotropy based on a fabric tensor are characterized by five parameters, e.g., E, ν, b_1, ν^{U}, S, as opposed to eight for general transversely isotropic porous media.

7.4.7.3   Compatibility between drained and undrained fabrics: orthotropy

Proceeding as above, one may first identify the orthotropic drained parameters in terms of E, ν and the eigenvalues of $\mathbf{B}$:

$$E_i = E\,b_i^2 \quad \text{for } i = 1, 2, 3\,;$$

$$\nu_{ij} = \nu\,\frac{b_i}{b_j}, \quad \mu_{ij} = \mu\,b_i\,b_j \quad \text{for } (i, j) = (1, 2),\,(1, 3),\,(2, 3)\,. \tag{7.4.79}$$

Similar relations hold for the undrained coefficients:

$$E_i^{\mathrm{U}} = E^{\mathrm{U}}\,(b_i^{\mathrm{U}})^2 \quad \text{for } i = 1, 2, 3\,;$$

$$\nu_{ij}^{\mathrm{U}} = \nu^{\mathrm{U}}\,\frac{b_i^{\mathrm{U}}}{b_j^{\mathrm{U}}}\,; \quad \mu_{ij}^{\mathrm{U}} = \mu^{\mathrm{U}}\,b_i^{\mathrm{U}}\,b_j^{\mathrm{U}} \quad \text{for} \quad (i, j) = (1, 2),\,(1, 3),\,(2, 3)\,. \tag{7.4.80}$$

Positive definiteness of $\mathbb{E}^{\mathrm{U}}$ requires ν^{U} to belong to the interval $\,] - 1, 1/2[$. As expected, the six constraints, Eqns (7.4.66) and (7.4.69)$_2$, imply $\mathbf{B} = \mathbf{B}^{\mathrm{U}}$; the remaining relations reduce to a single independent constraint which makes it possible to express the nine undrained parameters in terms of E, ν, b_1, b_2 (and b_3) and of a single undrained parameter to be measured, say ν^{U}:

$$E_i^{\mathrm{U}} = E\,\frac{1 + \nu^{\mathrm{U}}}{1 + \nu}\,b_i^2 \quad \text{for } i = 1, 2, 3\,;$$

$$\nu_{ij}^{\mathrm{U}} = \nu^{\mathrm{U}}\,\frac{b_i}{b_j}, \quad \mu_{ij}^{\mathrm{U}} = \mu_{ij} = \frac{E}{2(1 + \nu)}\,b_i\,b_j \quad \text{for } (i, j) = (1, 2),\,(1, 3),\,(2, 3)\,. \tag{7.4.81}$$

The ensuing expression of $\mathbf{A}$ is still given by Eqn (7.4.75).

In summary, elastic fluid-saturated porous media embedding this particular form of orthotropy based on the fabric tensor are characterized by 6 parameters, e.g., E, ν, b_1, b_2, ν^{U}, S, as opposed to 13 for general orthotropic porous media.

7.5   Unconfined compression of a poroelastic cylinder

The unconfined compression of a poroelastic cylinder is analysed as an initial and boundary value problem. The constitutive equations, the field equations and boundary conditions, and the initial conditions that govern the problem in space and time are first highlighted. An analytical solution is next derived.

7.5.1   Poroelastic field and constitutive equations

The equations of standard poroelasticity with incompressible species are cast in the format,

$$\text{Balance of total mass} \qquad \operatorname{div}\,(n^{\mathrm{s}}\,\mathbf{v}_{\mathrm{s}} + n^{\mathrm{w}}\,\mathbf{v}_{\mathrm{w}}) = 0,$$

$$\text{Balance of momentum of solid phase} \quad \operatorname{div}\,\boldsymbol{\sigma}^{\mathrm{s}} + k_{\mathrm{Sd}}(\mathbf{v}_{\mathrm{w}} - \mathbf{v}_{\mathrm{s}}) - p_{\mathrm{w}}\,\boldsymbol{\nabla} n^{\mathrm{w}} = \mathbf{0}\,,$$

$$\text{Balance of momentum of fluid phase} \quad \operatorname{div}\,\boldsymbol{\sigma}^{\mathrm{w}} - k_{\mathrm{Sd}}(\mathbf{v}_{\mathrm{w}} - \mathbf{v}_{\mathrm{s}}) + p_{\mathrm{w}}\,\boldsymbol{\nabla} n^{\mathrm{w}} = \mathbf{0}\,, \tag{7.5.1}$$

$$\text{Elastic constitutive equation} \qquad \boldsymbol{\sigma}' = \boldsymbol{\sigma} + p_{\mathrm{w}}\,\mathbf{I} = \lambda\,e\,\mathbf{I} + 2\,\mu\,\boldsymbol{\epsilon}, \quad e \equiv \operatorname{tr}\boldsymbol{\epsilon}\,.$$

Several stresses come into picture: the partial stress in the fluid $\boldsymbol{\sigma}^{\mathrm{w}} = -n^{\mathrm{w}}\,p_{\mathrm{w}}\,\mathbf{I}$, p_{w} being the intrinsic fluid pressure; the total stress $\boldsymbol{\sigma} = \boldsymbol{\sigma}^{\mathrm{s}} + \boldsymbol{\sigma}^{\mathrm{w}}$ which is the sum of the partial stresses in the solid and fluid phases; and the effective stress $\boldsymbol{\sigma}' = \boldsymbol{\sigma} + p_{\mathrm{w}}\,\mathbf{I}$. The velocity of phase α is denoted by $\mathbf{v}_\alpha$, $\alpha = \mathrm{s}, \mathrm{w}$. The volume fractions n^α, $\alpha = \mathrm{s}, \mathrm{w}$, are tacitly assumed to remain fixed. The elastic constitutive equations feature the Lamé moduli λ and μ of the drained solid skeleton, or matrix, which can be expressed in terms of the drained Young's modulus and Poisson's ratio, E and ν respectively. The two balances of momentum are equivalent to the two equations,

$$\text{Balance of momentum of porous medium} \quad \operatorname{div}\,\boldsymbol{\sigma} = \mathbf{0}\,,$$

$$\text{Darcy's law} \qquad\qquad n^{\mathrm{w}}\,(\mathbf{v}_{\mathrm{w}} - \mathbf{v}_{\mathrm{s}}) = -k_{\mathrm{H}}\,\boldsymbol{\nabla} p_{\mathrm{w}}, \tag{7.5.2}$$

where the hydraulic permeability $k_{\mathrm{H}} = (n^{\mathrm{w}})^2/k_{\mathrm{Sd}}$ [unit : $\mathrm{m^2/Pa/s}$] is equal to the intrinsic permeability [unit : $\mathrm{m^2}$] of the porous medium divided by the fluid viscosity [unit : $\mathrm{Pa \times s}$].

The field and constitutive equations should be complemented by appropriate initial and boundary conditions in order to obtain a well-posed problem.

7.5.2   Axisymmetric deformations independent of the axial coordinate

The analysis features a cylindrical specimen of radius a and the boundary conditions result in axisymmetric strains that depend on the radial position r but are independent of the axial z-coordinate as sketched in Fig. 7.5.1.

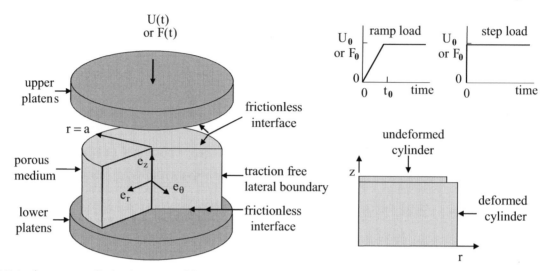

Fig. 7.5.1: A porous cylinder is squeezed between two frictionless platens, so that it keeps its cylindrical shape under application of either an axial force or an axial displacement, while the lateral boundary remains traction free, at atmospheric pressure.

The tests mimic unconfined compression, where the interface between the upper and lower platens and the specimen are frictionless. Therefore the radial, circumferential and axial components of the solid displacement in the cylindrical coordinates (r, θ, z) are of the form,

$$u_{sr} = u_{sr}(r, t), \quad u_{s\theta} = 0, \quad u_{sz} = z\,\epsilon(t), \tag{7.5.3}$$

resulting in the nonzero strain components,

$$\epsilon_{rr} = \frac{\partial u_{sr}}{\partial r}, \quad \epsilon_{\theta\theta} = \frac{u_{sr}}{r}, \quad \epsilon_{zz} = \epsilon(t), \tag{7.5.4}$$

and dilatation of the porous medium,

$$e = \frac{\partial u_{sr}}{\partial r} + \frac{u_{sr}}{r} + \epsilon(t). \tag{7.5.5}$$

The displacement is reckoned from the configuration at time $t = 0$.

In view of the isotropic material behavior, both shear stresses and shear strains vanish in the whole specimen. The principal stresses depend on r and t due to the poroelastic equations.

There is no relative fluid flow along the axial direction, so that the components of the fluid velocity have the format,

$$v_{wr} = v_{wr}(r, t), \quad v_{w\theta} = 0, \quad v_{wz} = z\,\dot{\epsilon}. \tag{7.5.6}$$

The z-component of the pressure gradient $\nabla p_w = (\partial p_w / \partial r, \partial p_w / r \partial \theta, \partial p_w / \partial z)$ vanishes in view of Darcy's law, and the fluid pressure depends only on r and t.

The lateral surface $r = a$ is traction free and the pore pressure vanishes (the fluid pressure being reckoned from the atmospheric pressure):

$$\sigma_{rr}(a, t) = 0, \quad p_w(a, t) = 0, \quad t \geq 0. \tag{7.5.7}$$

Of course, the radial solid displacement and fluid velocity vanish on the symmetry axis,

$$u_{sr}(r = 0, t) = 0, \quad v_{wr}(r = 0, t) = 0, \quad t \geq 0. \tag{7.5.8}$$

The field equations that are not satisfied automatically are the balance of mass $(7.5.1)_1$,

$$\frac{1}{r}\frac{\partial}{\partial r}\Big(r\,(n^w\,v_{wr} + n^s\,v_{sr})\Big) + \frac{\partial}{\partial z}(n^w\,v_{wz} + n^s\,v_{sz}) = 0, \tag{7.5.9}$$

and the balance of momentum $(7.5.2)_1$ in the r direction

$$\frac{\partial \sigma_{rr}}{\partial r} + \frac{1}{r}(\sigma_{rr} - \sigma_{\theta\theta}) = 0. \tag{7.5.10}$$

Since the axial solid and fluid velocities are equal, the equation of mass balance can be integrated in space with help of the condition (7.5.8) on the symmetry axis,

$$n^{\mathrm{w}} v_{\mathrm{w}r} = -n^{\mathrm{s}} v_{sr} - \frac{r}{2}\dot{\epsilon}. \tag{7.5.11}$$

Thus Darcy's law $(7.5.2)_2$ yields the fluid pressure in terms of the radial solid velocity,

$$k_{\mathrm{H}}\frac{\partial p_{\mathrm{w}}}{\partial r} = v_{sr} + \frac{r}{2}\dot{\epsilon}. \tag{7.5.12}$$

Introducing the elastic equation $(7.5.1)_4$ in the field equation (7.5.10), and making use of (7.5.12), yields the field equation satisfied by the radial displacement,

$$\frac{\partial^2 u_{sr}}{\partial r^2} + \frac{1}{r}\frac{\partial u_{sr}}{\partial r} - \frac{u_{sr}}{r^2} = \frac{1}{H\,k_{\mathrm{H}}}\left(v_{sr} + \frac{r}{2}\dot{\epsilon}\right), \tag{7.5.13}$$

where

$$H = \lambda + 2\,\mu = 2\,\mu\,\frac{1-\nu}{1-2\,\nu}, \tag{7.5.14}$$

is the drained confined compression modulus. Armstrong et al. [1984] provide an analytical solution for this problem that is detailed below.

7.5.3   Solution in the Laplace domain

The configuration at time $t = 0$ being taken as reference, the Laplace transform [7.2] $\overline{u}_{sr}(r, s)$ of $u_{sr}(r, t)$ satisfies the non homogeneous modified Bessel equation,

$$\frac{\partial^2 \overline{u}_{sr}}{\partial R^2} + \frac{1}{R}\frac{\partial \overline{u}_{sr}}{\partial R} - \left(1 + \frac{1}{R^2}\right)\overline{u}_{sr} = \frac{1}{2}\,b\,\overline{\epsilon}\,R, \tag{7.5.15}$$

with $R = r/b$, and

$$b = \sqrt{H\,k_{\mathrm{H}}/s}. \tag{7.5.16}$$

It is useful to introduce the characteristic seepage time t_{H} (note: the Darcy's seepage time is often defined as this quantity divided by π^2),

$$t_{\mathrm{H}} = \frac{a^2}{H\,k_{\mathrm{H}}}, \tag{7.5.17}$$

and then

$$\frac{a}{b} = \sqrt{t_{\mathrm{H}}\,s}. \tag{7.5.18}$$

The solution of the inhomogeneous differential equation is sought in the form,

$$\overline{u}_{sr}(r, s) = A(R, s)\,I_1(R) + B(R, s)\,K_1(R), \tag{7.5.19}$$

where the functions $A(R, s)$ and $B(R, s)$ are required to satisfy the condition

$$\frac{dA(R, s)}{dR}\,I_1(R) + \frac{dB(R, s)}{dR}\,K_1(R) = 0. \tag{7.5.20}$$

Here I_m and K_m are the modified Bessel functions of order m of the first and second kind respectively. For $m = 1$, they satisfy the homogeneous equation (7.5.15) defined by a zero rhs.

[7.2] The solution process based on the Laplace transform technique implicitly assumes that the transforms of the functions $f(t)$ that will be operated on are bounded, which, in practice, amounts to assume these functions to 1. vanish for a negative argument, $f(t) = 0$, $t < 0$; 2. have an finite number of points of discontinuity; 3. to be of exponential order, that is, to decay sufficiently fast towards infinity.

Backsubstitution of (7.5.19), of its derivative with respect to R and of (7.5.20) in (7.5.15), and use of the relation $(7.5.75)_2$ with $m = 1$ results in the system

$$\begin{bmatrix} dI_1/dR & dK_1/dR \\ I_1 & K_1 \end{bmatrix} \begin{bmatrix} dA/dR \\ dB/dR \end{bmatrix} = \begin{bmatrix} b\,\bar{\epsilon}\,R/2 \\ 0 \end{bmatrix} \implies \begin{bmatrix} dA/dR \\ dB/dR \end{bmatrix} = b\,\bar{\epsilon}/2 \begin{bmatrix} R^2\,K_1 \\ -R^2\,I_1 \end{bmatrix}. \tag{7.5.21}$$

Using in turn (7.5.71) with $m = 2$ yields,

$$A(R, s) = A_0(s) - b\,\frac{\bar{\epsilon}}{2}\,R^2\,K_2(R), \quad B(R, s) = B_0(s) - b\,\frac{\bar{\epsilon}}{2}\,R^2\,I_2(R). \tag{7.5.22}$$

Reporting these expressions in (7.5.19) results in,

$$\bar{u}_{sr}(r, s) = A_0(s)\,I_1(R) + B_0(s)\,K_1(R) - b\,\frac{\bar{\epsilon}}{2}\,R^2\,(I_1(R)\,K_2(R) + I_2(R)\,K_1(R)), \tag{7.5.23}$$

which, via (7.5.76), simplifies to,

$$\bar{u}_{sr}(r, s) = A_0(s)\,I_1(R) + B_0(s)\,K_1(R) - b\,\frac{\bar{\epsilon}}{2}\,R. \tag{7.5.24}$$

Since $K_1(R)$ is unbounded at $R = 0$, the function $B_0(s)$ should vanish for the solution to remain bounded at $R = 0$. The function $A_0(s)$ is obtained via satisfaction of the boundary conditions $p_w = 0$ and $\sigma_{rr} = -p_w + \lambda e + 2\,\mu\,\epsilon_{rr} = 0$ at $r = a$, and use of $(7.5.73)_1$,

$$A_0(s) = \frac{\bar{\epsilon}(s)}{2}\,\frac{b}{D(s)}, \tag{7.5.25}$$

where

$$D(s) = \frac{H}{2\,\mu}\,I_0(\frac{a}{b}) - \frac{b}{a}\,I_1(\frac{a}{b}). \tag{7.5.26}$$

Backsubstitution in (7.5.24) provides the Laplace transform of the radial solid displacement,

$$\bar{u}_{sr}(r, s) = \frac{\bar{\epsilon}(s)}{2}\,r\,\frac{\frac{b}{r}\,I_1(\frac{r}{b})}{D(s)} - \frac{\bar{\epsilon}(s)}{2}\,r. \tag{7.5.27}$$

The Laplace transform of the pressure results, upon integration in space of (7.5.12), accounting for the boundary condition $p_w(a, t) = 0$, and using (7.5.72) with $m = 0$,

$$\frac{\overline{p_w}}{\mu}(r, s) = \frac{H}{2\,\mu}\,\bar{\epsilon}(s)\,\frac{I_0(\frac{r}{b}) - I_0(\frac{a}{b})}{D(s)}. \tag{7.5.28}$$

Other Laplace transforms follow readily, namely the volume change and the radial flux,

$$\bar{e}(r, s) = \frac{\bar{\epsilon}(s)}{2}\,\frac{I_0(\frac{r}{b})}{D(s)}, \quad n^w\,(\bar{v}_{wr} - \bar{v}_{sr})(r, s) = -\frac{\bar{\epsilon}(s)}{2}\,\frac{s\,\frac{b}{r}\,I_1(\frac{r}{b})}{D(s)}, \tag{7.5.29}$$

and the total radial and axial stress components,

$$\begin{aligned}
\bar{\sigma}_{rr}(r, s) &= \frac{\mu\,\bar{\epsilon}(s)}{D(s)}\,\left(\frac{H}{2\,\mu}\,I_0(\frac{a}{b}) - \frac{b}{r}\,I_1(\frac{r}{b})\right) - \mu\,\bar{\epsilon}(s), \\
\bar{\sigma}_{zz}(r, s) &= \frac{\mu\,\bar{\epsilon}(s)}{D(s)}\,\left(-I_0(\frac{r}{b}) + 3\,\frac{H}{2\,\mu}\,I_0(\frac{a}{b}) - 2\,\frac{b}{a}\,I_1(\frac{a}{b})\right).
\end{aligned} \tag{7.5.30}$$

Therefore the Laplace transform of the axial load $F(t) = 2\,\pi\,\int_0^a \sigma_{zz}(r, t)\,r\,dr$ takes the form,

$$\frac{\overline{F}(s)}{\mu\,\pi\,a^2} = \frac{\bar{\epsilon}(s)}{D(s)}\,\left(3\,\frac{H}{2\,\mu}\,I_0(\frac{a}{b}) - 4\,\frac{b}{a}\,I_1(\frac{a}{b})\right). \tag{7.5.31}$$

7.5.4 Relaxation tests: applied axial strain

The axial strain $\epsilon(t)$ is imposed either as a step load or as a ramp load,

$$
\epsilon(t) = \begin{cases}
-\epsilon_0\, \mathcal{H}(t) & \text{step load,} \\[2ex]
-\dot{\epsilon}_0 \left(t\, \mathcal{H}(t) - (t - t_0)\, \mathcal{H}(t - t_0) \right) & \text{ramp load,}
\end{cases} \tag{7.5.32}
$$

where $\mathcal{H}$ is the Heaviside step function, ϵ_0 is a positive reference strain, and $\dot{\epsilon}_0$ a positive strain rate. In practice, the primary entity which is controlled is not the strain, but the axial displacement.

7.5.4.1 Relaxation tests: step strain

Instantaneous response:

The instantaneous response is obtained by the initial value theorem: for any generic function g, $\lim_{t \to 0^+} g(t) = \lim_{s \to \infty} s\, \bar{g}(s)$. For that purpose, we use the asymptotic property (7.5.77) of Bessel functions, defining $\sqrt{s}$ such that Re $\sqrt{s} > 0$.

The instantaneous strains result spatially uniform. Their values can be retrieved based on the basic assumption, consistent with application of the initial value theorem to (7.5.27), that the radial solid displacement is linear in r. Indeed

- seepage flow is not instantaneous, and it has not yet started at $t = 0^+$, thus $n^{\mathrm{w}}\, (v_{\mathrm{wr}} - v_{\mathrm{sr}})(r, 0^+) = 0$. Moreover, since both the solid and fluid constituents are incompressible, the only way for the porous medium to change volume its by seepage. Thus $e(r, 0^+) = 0$;
- admitting the radial displacement to be linear in r, the strains in the radial and circumferential directions are uniform and equal;
- gathering the two previous arguments implies $\epsilon_{rr}(r, 0^+) = \epsilon_{\theta\theta}(r, 0^+) = \epsilon_0/2$;
- equilibrium in the radial and circumferential directions yields $\sigma_{rr}(r, 0^+) = \sigma_{\theta\theta}(r, 0^+) = 0$. Thus the elastic equations in these directions, and the stress decomposition, yield $p_{\mathrm{w}}(r, 0^+) = \mu\, \epsilon_0$;
- the strains being known, the elastic equations yield $\sigma'_{zz}(r, 0^+) = -2\,\mu\,\epsilon_0$, $\sigma'_{rr}(r, 0^+) = \sigma'_{\theta\theta}(r, 0^+) = \mu\,\epsilon_0$. The total axial stress results from the effective stress and pressure as $\sigma_{zz}(r, 0^+) = -3\,\mu\,\epsilon_0$. Consequently, the axial force is equal to $F(0^+)/(\mu\,\pi\,a^2) = -3\,\epsilon_0$.

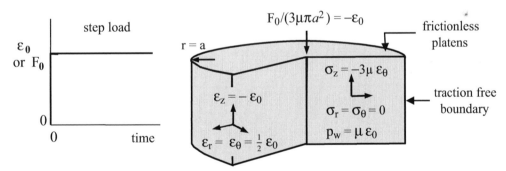

Fig. 7.5.2: Instantaneous response of the cylinder subjected to an axial step loading.

These results are sketched in Fig. 7.5.2 and summarized in Table 7.5.1 which displays the instantaneous response after an axial step load in terms of strain, $\epsilon(t) = -\epsilon_0\, \mathcal{H}(t)$, and the asymptotic (drained) response at large times.

A point is puzzling, still: the fact that the pressure at $t = 0^+$ is everywhere equal to a non zero value is in contradiction with the boundary condition at the lateral surface. In fact, the material will quickly manage to restore the boundary conditions, namely $\sigma_{rr}(r, 0^{++}) = 0$, $p_{\mathrm{w}}(r, 0^{++}) = 0$, so that $\sigma'_{rr}(r, 0^{++}) = 0$. If, following Berry et al. [2006], we admit that, close to the surface, $\epsilon_{\theta\theta}(r, 0^{++})$ remains equal to $\epsilon_0/2$, the elastic relation results in $\epsilon_{rr}(r, 0^{++})/\epsilon_0 = \nu/(2(1 - \nu)) < \nu$. Thus immediately after loading, the strain close to the boundary drops to a value lower than the asymptotic value $\nu\,\epsilon_0$ corresponding to the drained response.

For plane strain, the same reasoning would provide $\epsilon_{xx}(x, 0^+)/\epsilon_0 = 1$, and $\epsilon_{xx}(x, 0^{++})/\epsilon_0 = \nu/(1 - \nu)$, which again is the drained response.

Table 7.5.1: Porous cylinder loaded by frictionless platens with traction free lateral boundary subjected to an instantaneous axial strain $\epsilon(t) = -\epsilon_0\,\mathcal{H}(t)$. For a step force $F(t) = -F_0\,\mathcal{H}(t)$, these instantaneous and asymptotic responses hold when ϵ_0 is replaced respectively by $\epsilon_c^{0+} = F_0/(3\mu\,\pi\,a^2)$, and by $\epsilon_c = F_0/(E\,\pi\,a^2)$.

Response at Time $t = 0^+$			Response at Time $t = +\infty$		
$\epsilon_{zz} = -\epsilon_0;$	$\epsilon_{rr} = \epsilon_{\theta\theta} = \epsilon_0/2$		$\epsilon_{zz} = -\epsilon_0;$	$\epsilon_{rr} = \epsilon_{\theta\theta} = \nu\,\epsilon_0$	
$\sigma_{zz} = -3\,\mu\,\epsilon_0;$	$\sigma_{rr} = \sigma_{\theta\theta} = 0$		$\sigma_{zz} = -E\,\epsilon_0;$	$\sigma_{rr} = \sigma_{\theta\theta} = 0$	
$\sigma'_{zz} = -2\,\mu\,\epsilon_0;$	$\sigma'_{rr} = \sigma'_{\theta\theta} = \mu\,\epsilon_0$		$\sigma'_{zz} = -E\,\epsilon_0;$	$\sigma'_{rr} = \sigma'_{\theta\theta} = 0$	
$p_{\mathrm{w}} = \mu\,\epsilon_0$			$p_{\mathrm{w}} = 0$		

Transient response in the time domain

To calculate the inverse Laplace transforms through integration in the complex domain, the roots of the denominator $D(s)$ in (7.5.27) are needed. Note first that the functions in the complex domain of the Laplace variable s are uniform (in the sense of a function of the complex variable!), due to the parity of the modified Bessel functions I_0 and I_1. Using the relation between the Bessel and modified Bessel functions indicated in (7.5.69)$_2$, the roots of the equation below where $X = i\,a/b$,

$$\frac{1}{X}\frac{J_1(X)}{J_0(X)} = \frac{H}{2\,\mu}, \tag{7.5.33}$$

are requested. The roots with large magnitude can be sought by using the asymptotic expansion (7.5.77). Then

$$X_j = \frac{\pi}{4} + \tan^{-1}\left(\frac{H}{2\,\mu}X_j\right) + j\,\pi, \quad j \in]-\infty, +\infty[. \tag{7.5.34}$$

For X_j large, this formula simplifies further to

$$X_j = \frac{3}{4}\pi + j\,\pi, \quad j \in [0, +\infty[. \tag{7.5.35}$$

This approximation is quite correct, except for the first two or three roots. Indeed, for $\nu = 0.25$, the actual roots scaled by π are 0.658710, 1.71752, 2.72976, $\cdots$, 9.74436, $\cdots$ For $\nu = 0.0$, the above sequence becomes 0.586067, 1.69705, 2.71719, $\cdots$, 9.74089, $\cdots$ The accuracy increases as ν gets closer to 0.5.

Since Eqn (7.5.33) is even in X, it is sufficient to collect the positive roots. For each root,

$$\frac{a}{b} = \sqrt{t_{\mathrm{H}}\,s_j} = -i\,X_j, \tag{7.5.36}$$

so that

$$s_j = -\frac{X_j^2}{t_{\mathrm{H}}}, \quad j \in [0, +\infty[. \tag{7.5.37}$$

As just remarked, the roots $j = 0, 1, 2$ above are not accurate, but they can be used as a starting point to solve the nonlinear equation (7.5.33). The roots $j \geq 3$ can be verified to be quite accurate (the accuracy improves with $H/2\,\mu$, i.e., with Poisson's ratio ν), in the sense that the asymptotic expansion provides a quite acceptable estimation of the actual roots of (7.5.33). Moreover, it can also be verified numerically that the actual number of solutions is the same as the number of approximate roots.

The different functions whose inverse Laplace transforms are to be estimated can be expressed formally as $\overline{g}(s) = \overline{\epsilon}(s)\,N(s)/D(s)$, omitting the dependence in r for simplicity. For the step load, $\overline{\epsilon}(s) = -\epsilon_0/s$ and $s = 0$ is a pole of order 1, while it is a pole of order two for the ramp load. The other poles are simple. Thus, the Laplace inverse can be cast in the generic format,

$$\frac{g(t)}{-\epsilon_0} = \frac{N(0)}{D(0)} + \sum_j \frac{1}{s_j}\frac{N(s_j)}{D'(s_j)}\,e^{t\,s_j}. \tag{7.5.38}$$

The appropriate functions required to obtain the radial displacement have the following explicit expression,

$$D(s) = \frac{H}{2\,\mu}I_0(\sqrt{t_{\mathrm{H}}\,s}) - \frac{I_1(\sqrt{t_{\mathrm{H}}\,s})}{\sqrt{t_{\mathrm{H}}\,s}}, \quad N(s) = \frac{I_1((r/a)\sqrt{t_{\mathrm{H}}\,s})}{(r/a)\sqrt{t_{\mathrm{H}}\,s}}. \tag{7.5.39}$$

Using $(7.5.69)_2$, $(7.5.72)$, $(7.5.73)$, and the iterated fact that the s_j's are the roots of the denominator $D(s)$, Eqn $(7.5.33)$, Eqn $(7.5.39)$ yields for each root,

$$\frac{dD}{ds}(s_j) = \frac{t_{\mathrm{H}}}{2} \frac{(1-\nu)^2 X_j^2 - (1-2\nu)}{(1-\nu)(1-2\nu)} \frac{J_1(X_j)}{X_j^3}, \quad N(s_j) = \frac{J_1((r/a)X_j)}{(r/a)X_j}. \tag{7.5.40}$$

Note, for small R, the approximations $I_0(R) \sim 1$ and $I_1(R)/R \sim 1/2$. Thus the radial displacement becomes

$$\frac{u_{sr}(r,t)}{a\,\epsilon_0} = \frac{r}{a}\nu + \sum_j \frac{(1-\nu)(1-2\nu)}{(1-\nu)^2 X_j^2 - (1-2\nu)} \frac{J_1((r/a)X_j)}{J_1(X_j)} e^{-X_j^2 t/t_{\mathrm{H}}}. \tag{7.5.41}$$

Thus t_{H} appears as *a time characterizing the seepage phenomenon* governed by Darcy's law. It is referred to as consolidation time in soil mechanics. Observe that it is proportional to the *square* of the seepage length and inversely proportional to the stiffness and permeability of the porous medium. Numerical results show that a large part of the seepage has taken place at $t = t_{\mathrm{H}}$. In fact, the smallest value of the X_j^2's, namely X_0^2, is about 4.5 for $\nu = 0.3$ and does not vary much with ν.

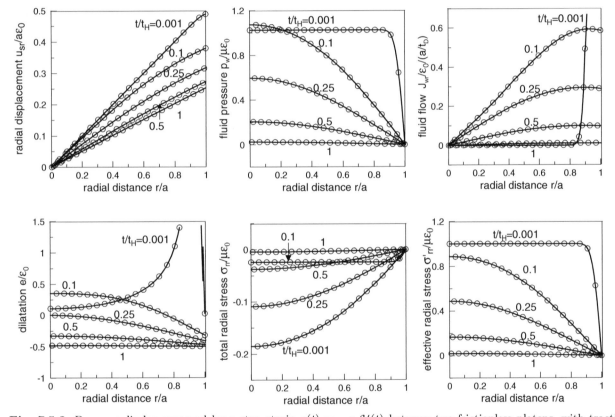

Fig. 7.5.3: Porous cylinder squeezed by a step strain $\epsilon(t) = -\epsilon_0 \, \mathcal{H}(t)$ between two frictionless platens, with traction free lateral boundary: Poisson's ratio $\nu = 0.25$, radius a=5 mm. For a shear modulus $\mu = 0.33$ MPa and a hydraulic permeability equal to 10^{-15} m^2/Pa/s, Darcy's characteristic time t_{H} is equal to 25×10^3 s. The value of the fluid volume fraction n^{w} is elusive for all quantities plotted here. On the other hand, it comes into picture to obtain the particle fluid velocity, see Eqn $(7.5.2)$. The temporary increase of fluid pressure in the center of the specimen is referred to as Mandel-Cryer effect.

Other quantities are derived similarly, e.g., the load,

$$\frac{F(t)}{\pi\mu\,a^2\,\epsilon_0} = -2(1+\nu) - \sum_j \frac{2(1-\nu)(1-2\nu)}{(1-\nu)^2 X_j^2 - (1-2\nu)} e^{-X_j^2 t/t_{\mathrm{H}}}, \tag{7.5.42}$$

the pressure,

$$\frac{p_{\mathrm{w}}(r,t)}{\mu\,\epsilon_0} = \sum_j \frac{2\,(1-\nu)\,(1-2\,\nu)}{(1-\nu)^2\,X_j^2 - (1-2\,\nu)} \left(\frac{J_0(\frac{r}{a}X_j)}{J_0(X_j)} - 1\right) e^{-X_j^2\,t/t_{\mathrm{H}}}, \tag{7.5.43}$$

the radial flux $J_{\mathrm{w}r} = n^{\mathrm{w}}\,(v_{\mathrm{w}r} - v_{sr})$,

$$\frac{J_{\mathrm{w}r}(r,t)}{(a/t_{\mathrm{H}})\,\epsilon_0} = \sum_j \frac{(1-2\,\nu)^2\,X_j}{(1-\nu)^2\,X_j^2 - (1-2\,\nu)} \frac{J_1(\frac{r}{a}X_j)}{J_0(X_j)} e^{-X_j^2\,t/t_{\mathrm{H}}}, \tag{7.5.44}$$

the dilatation,

$$\frac{e(r,t)}{\epsilon_0} = -(1-2\,\nu) + \sum_j \frac{(1-2\,\nu)^2}{(1-\nu)^2\,X_j^2 - (1-2\,\nu)} \frac{J_0((r/a)\,X_j)}{J_0(X_j)} e^{-X_j^2\,t/t_{\mathrm{H}}}, \tag{7.5.45}$$

and the total radial stress,

$$\frac{\sigma_{rr}(r,t)}{\mu\,\epsilon_0} = \sum_j \frac{2\,(1-\nu)\,(1-2\,\nu)}{(1-\nu)^2\,X_j^2 - (1-2\,\nu)} \left(1 - \frac{\frac{a}{r}J_1(\frac{r}{a}X_j)}{J_1(X_j)}\right) e^{-X_j^2\,t/t_{\mathrm{H}}}. \tag{7.5.46}$$

Steady state:

At steady state, i.e., when the flow has practically vanished because the pressure homogenizes, the load is supported completely by the solid skeleton, i.e., the behavior is elastic and dictated by the shear modulus μ and drained Poisson's ratio ν. Thus $\epsilon_{zz} = -\epsilon_0$, $\epsilon_{rr} = \nu\,\epsilon_0$, $\sigma_{rr} = 0$, $\sigma_{zz} = -E\,\epsilon_0$, with $E = 2\,(1+\nu)\,\mu$ drained Young's modulus.

The plots shown in Fig. 7.5.3 display the transient response that covers the time interval between time $t = 0^+$ where the material responds as if it was incompressible and the steady state where the material response is contributed entirely by the underlying drained solid skeleton. Steady state is nearly reached at time $t = t_{\mathrm{H}}$. The total radial stress is compressive while the effective radial stress is tensile due to the initial radial extension, which, later on, decreases as water is expelled from the sample. The flow begins close to the lateral boundary, it penetrates progressively the specimen, and finally dies out over time. Expulsion of water leads to a simultaneous decrease in radial strain and volume.

The calculations have been performed with double precision using the expressions of Bessel functions provided in Numerical Recipes, Press et al. [1992]. The number of poles used in the estimation of the inverse Laplace transforms should be larger than 20 to avoid oscillations at small times.

7.5.4.2 Relaxation tests: ramp strain

Since $s = 0$ is a pole of order two, the inverse Laplace transforms can be expressed in terms of the generic function $g_1(s) = N(s)/D(s)$ in the format,

$$\begin{aligned}
\frac{g_1(t)}{-\dot{\epsilon}_0} &= \frac{1}{1!}\frac{d}{ds}\left(\frac{N(s)}{D(s)}e^{t\,s}\right)_{|s=0} + \sum_j \frac{1}{s_j^2}\frac{N(s_j)}{D'(s_j)}e^{t\,s_j}, \\
&= \frac{d}{ds}\left(\frac{N(s)}{D(s)}\right)_{|s=0} + t\,\frac{N(0)}{D(0)} + \sum_j \frac{1}{s_j^2}\frac{N(s_j)}{D'(s_j)}e^{t\,s_j}.
\end{aligned} \tag{7.5.47}$$

The generic inverse of $\overline{g}(s) = \overline{\epsilon}(s)\,g_1(s)/(-\dot{\epsilon}_0)$ is equal to $g(t) = g_1(t)\mathcal{H}(t) - g_1(t-t_0)\mathcal{H}(t-t_0)$. For instance, the function $F_1(t)$ associated with the axial load expresses as,

$$\frac{t_0}{t_{\mathrm{H}}}\frac{F_1(t)}{\mu\,\pi\,a^2\,\epsilon_0} = -\frac{1}{4}\,(1-\nu)\,(1-2\nu) - 2\,(1+\nu)\,\frac{t}{t_{\mathrm{H}}} + \sum_j \frac{2\,(1-\nu)\,(1-2\,\nu)}{(1-\nu)^2\,X_j^2 - (1-2\,\nu)}\frac{1}{X_j^2} e^{-X_j^2\,t/t_{\mathrm{H}}}. \tag{7.5.48}$$

To estimate the influence of the strain rate, simulations corresponding to different loading periods t_0, but to the same strain $\epsilon_0 = \dot{\epsilon}_0\,t_0$, are displayed in Fig. 7.5.4. The maximum force increases with strain rate: the maximum scaled force $F/(\pi\,a^2)$ varies from $-2\,(1+\nu)\,\mu\,\epsilon_0$ for very low loading rate, or steady state, to $-3\,\mu\,\epsilon_0$ for step loading.

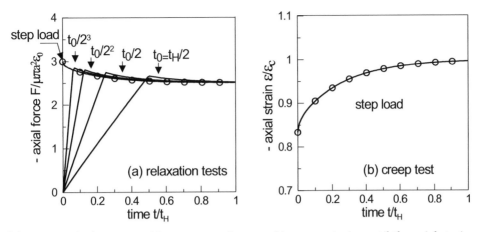

Fig. 7.5.4: (a) Porous cylinder squeezed by a step strain ϵ_0, and by ramp strains until the axial strain ϵ_0 is reached. The highest the loading rate, the largest is the maximum axial force supported by the loading device. (b) Creep test under a step load F_0.

7.5.4.3 Instantaneous and ramp loading: use of the convolution property

The convolution property of linear transforms may be used to deduce, from the response to a simple load function, the response to a more complex load history. Indeed, let $\mathcal{T}(x,s)$ be a linear partial differential operator in the Laplace domain. Omitting the space dependence for simplicity of notation, let $\bar{u}_i(s)$ be the response to the datum $\bar{a}_i(s)$, in the sense that $\mathcal{T}\bar{u}_i(s) = \bar{a}_i(s)$, $i = 1, 2$. Then $\bar{u}_2(s) = \bar{u}_1(s)\,\bar{a}_2(s)/\bar{a}_1(s)$, and therefore $u_2(t) = \int_0^t u_1(t - \tau)\,\mathcal{L}^{-1}\{\bar{a}_2(s)/\bar{a}_1(s)\}(\tau)\,d\tau$. Special cases for a_1 may be exploited, for example,

$$
\begin{aligned}
&a_1(t) = \delta(t) \;\Rightarrow\; u_2(t) = \int_0^t u_1(t - \tau)\,a_2(\tau)\,d\tau; \\
&a_1(t) = \mathcal{H}(t) \;\Rightarrow\; u_2(t) = \int_0^t u_1(t - \tau)\,\frac{da_2(\tau)}{d\tau}\,d\tau \;=\; u_1(t)\,a_2(0)\,\mathcal{H}(t) + \int_0^t u_1(t - \tau)\,\frac{d^+ a_2(\tau)}{d\tau}\,d\tau\,.
\end{aligned}
\tag{7.5.49}
$$

In the second line, the operator d/dt denotes the derivative in the sense of distributions, $da_2/dt = d^+ a_2/dt + a_2(0^+)\,\delta(t)$, with $d^+ a_2/dt$ the right (standard) derivative and $\delta(t)$ the Dirac delta function. Note that the Laplace transform of da_2/dt is equal to $s\,\bar{a}_2(s)$, while the Laplace transform of $d^+ a_2/dt$ is equal to $s\,\bar{a}_2(s) - a_2(0^+)$.

7.5.5 Creep tests: applied axial load

The sole case considered for creep tests is that of an axial step load $F(t)$,

$$
F(t) = -F_0\,\mathcal{H}(t)\,,
\tag{7.5.50}
$$

where $F_0 > 0$ corresponds to a compression.

Instantaneous response:

The instantaneous response is the same as for the relaxation tests, with ϵ_0 replaced by $-\epsilon(0^+) = F_0/(3\mu\,\pi\,a^2)$.

Transient response in the Laplace domain

The roots involved in calculating the inverse Laplace transforms are those of the numerator $N(s)$ in (7.5.31), namely

$$
N(s) = 3\,\frac{H}{2\,\mu}\,I_0\Big(\frac{a}{b}\Big) - 4\,\frac{b}{a}\,I_1\Big(\frac{a}{b}\Big)\,.
\tag{7.5.51}
$$

Therefore, instead of (7.5.33), the roots are now the solutions of the equation,

$$
\frac{1}{X}\,\frac{J_1(X)}{J_0(X)} = \frac{3}{4}\,\frac{H}{2\,\mu}\,.
\tag{7.5.52}
$$

Substitution of $\overline{F}(s)$ for $\overline{\epsilon}(s)$ yields the Laplace transforms of the axial strain, radial displacement, pressure and dilatation,

$$\overline{\epsilon}(s) = \frac{\overline{F}(s)}{\mu \pi a^2} \frac{D(s)}{N(s)} ; \tag{7.5.53}$$

$$\overline{u}_{sr}(r,s) = r \frac{\overline{F}(s)}{2\mu \pi a^2} \frac{\frac{b}{r} I_1(\frac{r}{b}) - \frac{H}{2\mu} I_0(\frac{a}{b}) + \frac{b}{a} I_1(\frac{a}{b})}{N(s)} ; \tag{7.5.54}$$

$$\frac{\overline{p}}{\mu}(r,s) = \frac{H}{2\mu} \frac{\overline{F}(s)}{\mu \pi a^2} \frac{I_0(\frac{r}{b}) - I_0(\frac{a}{b})}{N(s)} ; \tag{7.5.55}$$

$$\overline{e} = \frac{\overline{F}(s)}{2\mu \pi a^2} \frac{I_0(\frac{r}{b})}{N(s)} . \tag{7.5.56}$$

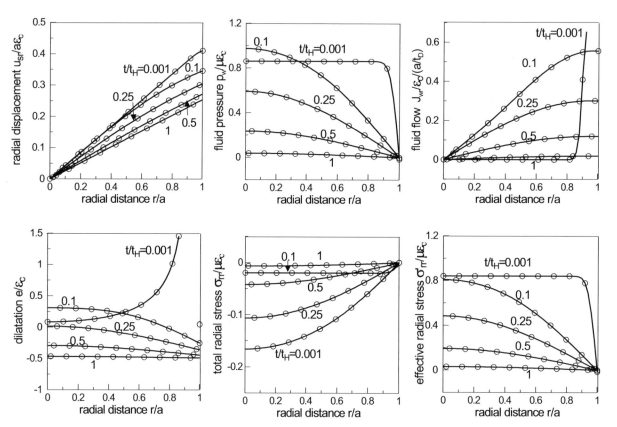

Fig. 7.5.5: Porous cylinder squeezed by a step force $F(t) = -F_0 \mathcal{H}(t)$ between two frictionless platens, with traction free lateral boundary.

Transient response in the time domain

Instead of (7.5.40), the derivative of the function $N(s)$ with respect to the Laplace variable s can be expressed, for each root $s = s_j$, in the form,

$$N'(s_j) = \frac{t_H}{2} \frac{9(1-\nu)^2 X_j^2 - 8(1+\nu)(1-2\nu)}{3(1-\nu)(1-2\nu)} \frac{J_1(X_j)}{X_j^3} . \tag{7.5.57}$$

Let us note for the record $D(0) = 1/(2(1-2\nu))$, $N(0) = (1+\nu)/(1-2\nu)$. The radial displacement and axial strain become, respectively,

$$\frac{u_{sr}(r,t)}{a\,\epsilon_c} = \nu \frac{r}{a} + \sum_j \frac{2(1-\nu^2)(1-2\nu)}{9(1-\nu)^2 X_j^2 - 8(1+\nu)(1-2\nu)} (3\frac{J_1(\frac{r}{a}X_j)}{J_1(X_j)} - \frac{r}{a}) e^{-X_j^2 t/t_H} , \tag{7.5.58}$$

and

$$\frac{\epsilon(t)}{\epsilon_c} = -1 + \sum_j \frac{4\,(1-\nu^2)\,(1-2\,\nu)}{9\,(1-\nu)^2\,X_j^2 - 8\,(1+\nu)\,(1-2\,\nu)}\, e^{-X_j^2\,t/t_{\mathrm{H}}}\,, \tag{7.5.59}$$

where the asymptotic axial strain $\epsilon_{zz}(\infty) = -\epsilon_c$ is defined by

$$\epsilon_c = \frac{F_0}{E\,\pi\,a^2}\,. \tag{7.5.60}$$

The pressure,

$$\frac{p_{\mathrm{w}}(r,t)}{\mu\,\epsilon_c} = \sum_j \frac{16\,(1-\nu^2)\,(1-2\,\nu)}{9\,(1-\nu)^2\,X_j^2 - 8\,(1+\nu)\,(1-2\,\nu)}\left(\frac{J_0(\frac{r}{a}X_j)}{J_0(X_j)} - 1\right) e^{-X_j^2\,t/t_{\mathrm{H}}}\,, \tag{7.5.61}$$

provides the radial flux $J_{\mathrm{w}r} = n^{\mathrm{w}}\,(v_{\mathrm{w}r} - v_{sr})$, with help of (7.5.74),

$$\frac{J_{\mathrm{w}r}(r,t)}{(a/t_{\mathrm{H}})\,\epsilon_c} = \sum_j \frac{8\,(1+\nu)\,(1-2\,\nu)^2\,X_j}{9\,(1-\nu)^2\,X_j^2 - 8\,(1+\nu)\,(1-2\,\nu)}\,\frac{J_1(\frac{r}{a}X_j)}{J_0(X_j)}\, e^{-X_j^2\,t/t_{\mathrm{H}}}\,. \tag{7.5.62}$$

Finally the dilatation takes the form

$$\frac{e(r,t)}{\epsilon_c} = -(1-2\,\nu) + \sum_j \frac{8\,(1+\nu)\,(1-2\,\nu)^2}{9\,(1-\nu)^2\,X_j^2 - 8\,(1+\nu)\,(1-2\,\nu)}\,\frac{J_0(\frac{r}{a}X_j)}{J_0(X_j)}\, e^{-X_j^2\,t/t_{\mathrm{H}}}\,, \tag{7.5.63}$$

while the total radial stress becomes,

$$\frac{\sigma_{rr}(r,t)}{\mu\,\epsilon_0} = \sum_j \frac{2\,(1-\nu)\,(1-2\,\nu)}{(1-\nu)^2\,X_j^2 - (1-2\,\nu)}\left(1 - \frac{\frac{a}{r}J_1(\frac{r}{a}X_j)}{J_1(X_j)}\right) e^{-X_j^2\,t/t_{\mathrm{H}}}\,. \tag{7.5.64}$$

The spatial profiles are plotted in Fig. 7.5.5 at a few instants. The scaled entities corresponding to the step relaxation and step creep tests have very similar values. However, the axial strain $-\epsilon(t)$ increases during creep from its instantaneous value $-\epsilon(0^+) = \epsilon_c(0^+)$ to its steady state value $-\epsilon(\infty) = \epsilon_c$, Fig. 7.5.4,

$$\epsilon_c(0^+) = \frac{F_0}{3\mu\,\pi\,a^2} > \epsilon_c = \frac{F_0}{E\,\pi\,a^2} = \frac{3}{2\,(1+\nu)}\,\epsilon_c(0^+)\,. \tag{7.5.65}$$

Since the scaling refers to steady state, scaled entities differ somehow at short term from their counterparts in the relaxation tests.

7.5.6   Compendium of Bessel functions

Some properties of Bessel functions required in the text above are collected from the books of Carslaw and Jaeger [1959], p. 488, and Sneddon [1980]. The Bessel functions of order $m = 0, 1, 2 \cdots$ of the first kind J_m and the modified Bessel functions of the first kind I_m are defined by series expansions,

$$J_m(R) = \sum_{j=0}^{\infty} \frac{(-1)^j}{j!(m+j)!}\left(\frac{R}{2}\right)^{m+2j}, \quad I_m(R) = \sum_{j=0}^{\infty} \frac{1}{j!(m+j)!}\left(\frac{R}{2}\right)^{m+2j}\,. \tag{7.5.66}$$

Note also the integral representation,

$$J_m(R) = \frac{1}{2\,\pi}\int_0^{2\pi} \exp\left(i(R\sin\theta - m\,\theta)\right)d\theta = \frac{1}{\pi}\int_0^{\pi} \cos(R\sin\theta - m\,\theta)d\theta\,. \tag{7.5.67}$$

The Bessel function of the first kind J_m satisfies the Bessel equation $R^2\,d^2y/dR^2 + R\,dy/dR + (R^2 - m^2)\,y = 0$. If m was not an integer, J_m and J_{-m} would be two independent solutions of the above equation. For m an integer, the second independent solution is the Bessel function of the second kind Y_m.

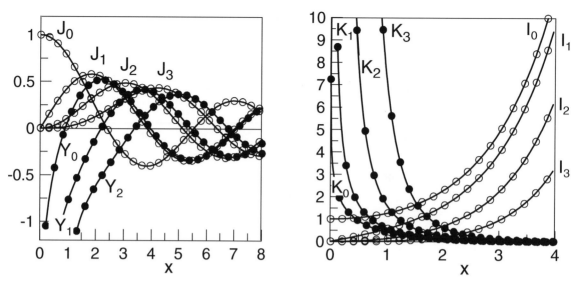

Fig. 7.5.6: Bessel functions of the first kind and second kind, J and Y respectively, and modified Bessel functions of the first kind and second kind, I and K respectively.

The modified Bessel function of the first kind I_m satisfies the Bessel equation $R^2 \, d^2y/dR^2 + R \, dy/dR - (R^2 + m^2) \, y = 0$. If m was not an integer, I_m and I_{-m} would be two independent solutions of the above equation. For m an integer, the second independent solution is the modified Bessel function of the second kind K_m, namely

$$K_m(R) = \frac{\pi}{2} \, i^{m+1} \left(J_m(iR) + i \, Y_m(iR) \right).$$

(7.5.68)

Fig. 7.5.6 shows the variations of the Bessel and modified Bessel functions of the first kind and second kind over interval of interest. The Bessel function and modified Bessel function of the second kind, $Y_m(R)$ and $K_m(R)$ respectively, are unbounded at $R = 0$.

The relations needed in the derivations are listed below. Let us record first the basic properties,

$$(-1)^m \, J_{-m}(R) = J_m(R), \quad I_{-m}(R) = I_m(R), \quad I_m(R) = i^{-m} \, J_m(i \, R) \quad (i = \sqrt{-1}),$$

(7.5.69)

and the recurrence relations,

$$J_{m+1}(R) = -J_{m-1}(R) + \frac{2m}{R} \, J_m(R),$$

$$I_{m+1}(R) = I_{m-1}(R) - \frac{2m}{R} \, I_m(R),$$

$$K_{m+1}(R) = K_{m-1}(R) + \frac{2m}{R} \, K_m(R).$$

(7.5.70)

Note also the relations for the derivatives,

$$\frac{d}{dR}(R^m \, I_m(R)) = R^m \, I_{m-1}(R), \quad \frac{d}{dR}(R^m \, K_m(R)) = -R^m \, K_{m-1}(R),$$

(7.5.71)

$$\frac{d}{dR}(R^{-m} \, I_m(R)) = R^{-m} \, I_{m+1}(R).$$

(7.5.72)

The relations below are equivalent to (7.5.71),

$$\frac{d}{dR} I_m(R) = I_{m-1}(R) - \frac{m}{R} \, I_m(R), \quad \frac{d}{dR} K_m(R) = -K_{m-1}(R) - \frac{m}{R} \, K_m(R).$$

(7.5.73)

On the other hand,

$$\frac{d}{dR}(R^m \, J_m(R)) = R^m \, J_{m-1}(R), \quad \frac{d}{dR} J_m(R) = \frac{m}{R} \, J_m(R) - J_{m+1}(R) = J_{m-1}(R) - \frac{m}{R} \, J_m(R).$$

(7.5.74)

The cross-relations between the Bessel functions of the first kind and second kind on one side, and between the modified Bessel functions of the first kind and second kind on the other side are quite remarkable,

$$J_m(R)\,\frac{dY_m}{dR}(R) - \frac{dJ_m}{dR}(R)\,Y_m(R) = \frac{2}{\pi}\frac{1}{R},$$

$$I_m(R)\,\frac{dK_m}{dR}(R) - \frac{dI_m}{dR}(R)\,K_m(R) = -\frac{1}{R}.$$

(7.5.75)

Substitution of (7.5.73) in (7.5.75)$_2$ yields,

$$I_m(R)\,K_{m-1}(R) + I_{m-1}(R)\,K_m(R) = \frac{1}{R}.$$

(7.5.76)

The asymptotic relations to first order for large R are instrumental to apply the initial value theorem,

$$J_m(R) \sim \sqrt{\frac{2}{\pi R}}\,\cos\left(R - (2\,m+1)\,\frac{\pi}{4}\right); \quad Y_m(R) \sim \sqrt{\frac{2}{\pi R}}\,\sin\left(R - (2\,m+1)\,\frac{\pi}{4}\right);$$

$$I_m(R) \sim \frac{e^R}{\sqrt{2\,\pi R}}; \qquad\qquad K_m(R) \sim e^{-R}\sqrt{\frac{\pi}{2\,R}}.$$

(7.5.77)

For small R, the following approximations hold,

$$J_m(R) \sim \frac{1}{m!}\frac{R^m}{2^m}; \quad Y_0(R) \sim \frac{2}{\pi}\left(\operatorname{Ln}\frac{R}{2} + \gamma\right), \quad Y_m(R) \sim -\frac{(m-1)!}{\pi}\frac{2^m}{R^m}, \quad m \neq 0;$$

$$I_m(R) \sim \frac{1}{m!}\frac{R^m}{2^m}; \quad K_0(R) \sim -\left(\operatorname{Ln}\frac{R}{2} + \gamma\right), \quad K_m(R) \sim \frac{(m-1)!}{2}\frac{2^m}{R^m}, \quad m \neq 0,$$

(7.5.78)

where $\gamma \sim 0.5772$ is the Euler constant.

Chapter 8

Viscoelasticity and poro-viscoelasticity

8.1 Viscoelasticity, poroelasticity and poro-viscoelasticity

Overall incompressibility of tissues can be a good approximation only for short term responses. For moderate and long term responses, fluid circulates in and out the tissue, and this diffusion endows the tissues with an overall compressibility. In fact, a clear distinction should be made between overall, or *apparent*, properties of the whole tissue, and *intrinsic* properties of the solid skeleton. Experiments sometimes impose a pre-compression in order to minimize the amount of blood in the vessels before testing. Preconditioning of natural and artificial materials ensures a valid comparison of experimental results from different specimens.

On the other hand, some testing devices are designed to actually measure apparent properties: Parsons and Coger [2002] specifically designed a shear rheometer that keeps the *in situ* tissue hydration, by minimizing the actuator contact area. Common shear plate covers the whole biological layer and therefore modifies the *in situ* hydration conditions, rendering the contact area practically dry.

Both fluid diffusion and intrinsic rate dependence of the mechanical constitutive relations contribute to energy dissipation, even if they serve distinct physiological purposes. Viscous sliding between fibrils of collagen allows their alignment with the load axis, contributing to a progressive increase of stiffness and limited deformation of an initially compliant tissue.

In words, time effects due to fluid flows (*poroelasticity*) and due to the interactions between the (dry) components of the tissues (*intrinsic viscoelasticity*) should be distinguished. The issue arises in acute form in the Poisson's ratio of articular cartilages. For example, *apparent* Poisson's ratios are sometimes measured to vary with time and to be much larger than 1/2 while an appropriate analysis, recognizing the multiphase nature of the tissue, would identify an intrinsic elastic Poisson's ratio in an admissible range and constant in time. The issue is more complex if the solid skeleton displays a time dependent response, as viscoelastic Poisson's ratio may vary over a wider range than its elastic counterpart.

This chapter represents only an introduction to time effects in soft tissues. It adopts the *rule of time translation invariance* and addresses essentially linear viscoelasticity of non aging tissues while real tissues display both intrinsic time effects and nonlinearity at large strains. Still nonlinear effects in a visco-hyperelastic framework are briefly touched upon at the end of the chapter.

8.2 Intrinsic time dependent behavior of collagen fibrils

Among the first, Mak [1986] developed a continuum model of tissue behavior where rate effects are due to both poroelasticity and viscoelasticity. Intrinsic rate dependence of tissues emanates mainly from the collagen fibrils. Actually, collagen viscoelasticity is thought to account for most of the stress relaxation in tension while poroelastic effects associated with fluid diffusion dominates the relaxation effects in compression, Li et al. [2005].

While the analysis of Wang et al. [2002] considers compression, Charlebois et al. [2004] and Lei and Szeri [2007] address rate effects under tension. Wilson et al. [2004] develop a finite element model of osteoarthritis using a poro-viscoelastic fibril-reinforced model, accounting for the directional non random distribution of fibrils and their spatial inhomogeneities across the depth. In fact, unconfined compression tests exhibit contraction and a poroelastic behavior dominated by the interactions of water and proteoglycans along the compression direction while the lateral direction undergoes an extension resisted by the activation of collagen fibrils. Di Silvestro and Suh [2001] endow the collagen fibrils with a viscous behavior to better describe the fibril extension along the lateral direction. As an improvement, Huang et al. [2001] [2003] attempt to identify the parameters defining the quasi-linear viscosity (QLV) theory of Fung in conjunction with a linear or conewise linear behavior of articular cartilages. The identification uses results from confined and unconfined compression tests including stress relaxation.

Boyce et al. [2007] address the tensile response of collagen fibrils of the bovine cornea with a QLV theory based on creep tests and tension tests over a range of loading rates. The verification is hampered by the complex directional distribution of collagen fibrils across the stroma and material nonlinearities.

Lei and Szeri [2007] introduced a rate dependent response of collagen fibrils through the standard nonlinear solid, namely a Maxwell model in parallel with a spring with a strain dependent stiffness. In the simulations of confined and unconfined tests on articular cartilages, the relaxation time of fibrils ranges from 100 to 200 s.

The viscous properties addressed here assume that the tissues do not undergo aging, that is, the tissues do not modify their mechanical properties over time. While the validity of this assumption may be rescued by considering restricted time windows, its actual range of applicability remains to be assessed. Charlebois et al. [2004] found that the equilibrium and peak tensile modulus of bovine cartilages from 6 months to 6 years increase significantly with maturation and age, from 0 to 15 MPa at equilibrium and from 10 to 28 MPa at peak. This aging might be linked with the age-dependent content of enzymatic and nonenzymatic collagen crosslinks. The time translation invariance adopted in most of this chapter does not fit with osteoarthritis (OA) which is associated with a damaged collagen network resulting in swelling.

8.3   Complex modulus

Rate effects are associated with energy dissipation which is traditionally detected through hysteresis loops. Quantification of these rate effects is complicated by the nonlinearity of the stress-strain relation that develop at moderate deformations. Moreover, repeatable results require preconditioning of the tissues. The details of the preconditioning (intensity of the stretch, number of cycles, duration of the rest period, etc.) are significant for biological tissues and even more for bio-artificial tissues which display a more pronounced hysteresis curve than native tissues, Wagenseil et al. [2003].

The dissipation in biological materials is relatively constant over a wide range of frequencies: this observation has led to the introduction of a continuous spectrum of relaxation times, Fung [1993], p. 281 et seq. As a discrete alternative, Holzapfel et al. [2002] used a generalized linear solid including 5 parallel Maxwell branches with relaxation times spanning from 10^{-3} s to 10 s. In application to arterial walls, the resulting hysteresis loops and loss modulus are relatively insensitive to strain rates in a frequency range from 0.01 to 100 Hz.

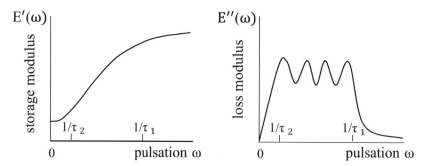

Fig. 8.3.1: Qualitative variation of the storage and loss moduli as a function of pulsation $\omega = 2\pi f$ or frequency f for biological materials. The storage modulus increases monotonically to an asymptotic value at large frequencies. The loss modulus remains pretty constant over a pulsation window and it vanishes at zero and infinite frequencies.

The traditional treatment of viscoelasticity introduces a complex modulus whose real and imaginary parts display over frequencies the key features highlighted in Fig. 8.3.1. For an elastic solid, the complex modulus actually reduces to a frequency independent real number. For a viscoelastic solid, the real part called storage modulus is typically increasing with frequency while the imaginary part called loss modulus vanishes at very low and very high frequencies and it shows a sort of plateau in a certain frequency range.

8.3.1 Storage and loss moduli

Let us start with a 1D linear context. The configuration at time $t = -\infty$ is taken as reference so that the strain $q(-\infty)$ vanishes. Let $E(t) = E(\infty) + E(t) - E(\infty)$ be the viscoelastic modulus, or relaxation modulus, in which the long range component $E(\infty)$ is isolated. The stress response is obtained as a convolution integral where the viscoelastic modulus plays the role of the kernel,

$$Q(t) = \int_{-\infty}^{t} E(t - u)\, dq(u) = E(\infty)\, q(t) + \int_{-\infty}^{t} \big(E(t - u) - E(\infty)\big)\, dq(u)\,. \tag{8.3.1}$$

For a harmonic strain with pulsation ω,

$$q(t) = \hat{q}(t) = q_0\, e^{i\omega t}\,, \tag{8.3.2}$$

the stress

$$\hat{Q}(t) = E^*(i\,\omega)\, \hat{q}(t)\,, \tag{8.3.3}$$

expresses in terms of the complex modulus which decomposes additively into a storage modulus E' and a loss modulus E'',

$$E^*(i\,\omega) = E'(\omega) + i\, E''(\omega)\,. \tag{8.3.4}$$

With the change of variable $t - u = v$, the storage and loss moduli become,

$$E'(\omega) = E(\infty) + \omega \int_0^\infty \sin(\omega v)\, \big(E(v) - E(\infty)\big)\, dv\,, \quad E''(\omega) = \omega \int_0^\infty \cos(\omega v)\, \big(E(v) - E(\infty)\big)\, dv\,, \tag{8.3.5}$$

which, upon integration by parts, can also be cast in the format,

$$E'(\omega) = E(0) + \int_0^\infty \cos(\omega v)\, \frac{dE(v)}{dv}\, dv\,, \quad E''(\omega) = -\int_0^\infty \sin(\omega v)\, \frac{dE(v)}{dv}\, dv\,. \tag{8.3.6}$$

It will be shown that the complex modulus can be interpreted as the Laplace-Carson transform $t \to p = i\,\omega$ of the viscoelastic modulus. Then by the initial and final value theorems,

$$E(0) = E^*(i\infty) = E'(\infty) + i\, E''(\infty)\,,$$
$$E(\infty) = E^*(0) = E'(0) + i\, E''(0)\,. \tag{8.3.7}$$

Therefore the loss modulus typically vanishes at very low and very high pulsations,

$$E''(\omega = 0) = 0\,, \quad E''(\omega = \infty) = 0\,. \tag{8.3.8}$$

In other words, the material then behaves like an elastic material. In addition,

$$E(0) = E'(\infty)\,, \quad E(\infty) = E'(0)\,. \tag{8.3.9}$$

Introducing the modulus and argument of the complex modulus,

$$E^*(i\,\omega) = E'(\omega) + i\, E''(\omega) = |E^*(i\,\omega)|\, e^{i\phi}\,, \tag{8.3.10}$$

we see that the stress response is harmonic with a shift of phase,

$$\hat{Q}(t) = E^*(i\,\omega)\, \hat{q}(t) = \overbrace{|E^*(i\,\omega)|\, q_0}^{Q_0}\, e^{i(\omega\, t + \phi)}\,. \tag{8.3.11}$$

The Maxwell model and Kelvin model are defined by a spring of stiffness E [unit : Pa] and a dashpot of viscosity η [unit : Pa×s], respectively in series and in parallel, as indicated in Fig. 8.3.2. The differential equations linking the strain, stress, and their derivatives result by adopting the two basic rules:
    (a) the stresses of two parallel branches sum up while the strains are identical;

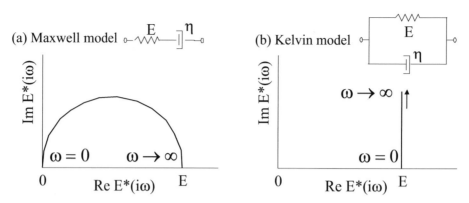

Fig. 8.3.2: Locus in the complex plane of the complex modulus of the models of Maxwell and Kelvin.

(b) the stress is uniform along a branch while the strains of the elements sum up.
From the differential equations governing these models,

$$\text{Maxwell model} : \ \frac{dq}{dt} = \frac{1}{E}\frac{dQ}{dt} + \frac{Q}{\eta}; \quad \text{Kelvin model} : \ Q = E\,q + \eta\,\frac{dq}{dt}, \tag{8.3.12}$$

follow their complex moduli and the identifications (8.3.9),

$$\text{Maxwell model} : \ E^{*}(i\,\omega) = \frac{E\,i\,\omega}{i\,\omega + E/\eta} = \frac{E}{2} + \frac{E}{2}\frac{i\,\omega - E/\eta}{i\,\omega + E/\eta}, \quad E(0) = 0, \ E(\infty) = E; \tag{8.3.13}$$

$$\text{Kelvin model} : \ E^{*}(i\,\omega) = E + i\,\omega\,\eta, \quad E(0) = E, \ E(\infty) = E + i\,\infty.$$

When the pulsation ω varies from to 0 to ∞, the locus in the upper complex plane described by the complex modulus is a half circle for the Maxwell model while it is a half segment for the Kelvin model. The phase shift ϕ is equal to $\arctan\big((E/\eta)/\omega\big)$ for the Maxwell model and to $\arctan\big(\omega/(E/\eta)\big)$ for the Kelvin model.

Given the storage modulus, or the loss modulus, the viscoelastic modulus itself can be obtained by inverse sine or cosine transform[8.1],

$$E(t) - E(\infty) = \frac{2}{\pi}\int_{0}^{\infty}\frac{E'(\omega) - E(\infty)}{\omega}\sin(\omega t)\,d\omega, \tag{8.3.19}$$

or, equivalently,

$$E(t) - E(\infty) = \frac{2}{\pi}\int_{0}^{\infty}\frac{E''(\omega)}{\omega}\cos(\omega t)\,d\omega\,. \tag{8.3.20}$$

Here $E'(\omega)$ is assumed to be an even function of its argument, while $E''(\omega)$ is odd.

[8.1]Some basic properties of the sine and cosine transforms are required. First the definitions of the sine transform $F_s(\omega) = \mathcal{F}_s\{f(t)\}(\omega)$ of the odd function $f(t)$ and of its inverse,

$$F_s(\omega) = \int_{0}^{\infty}\sin(\omega t)\,f(t)\,dt, \quad f(t) = \frac{2}{\pi}\int_{0}^{\infty}F_s(\omega)\sin(\omega t)\,d\omega, \tag{8.3.14}$$

and of the cosine transform $F_c(\omega) = \mathcal{F}_c\{f(t)\}(\omega)$ of the even function $f(t)$ and inverse,

$$F_c(\omega) = \int_{0}^{\infty}\cos(\omega t)\,f(t)\,dt, \quad f(t) = \frac{2}{\pi}\int_{0}^{\infty}F_c(\omega)\cos(\omega t)\,d\omega\,. \tag{8.3.15}$$

The transforms of the exponential function e^{-at}, $a > 0$, are worth recording,

$$\mathcal{F}_s\{e^{-at}\}(\omega) = \frac{\omega}{\omega^2 + a^2}, \quad \mathcal{F}_c\{e^{-at}\}(\omega) = \frac{a}{\omega^2 + a^2}\,. \tag{8.3.16}$$

Also useful are the integral representations of the Heaviside step function $\mathcal{H}$ and Dirac delta function δ,

$$\mathcal{H}(t) = \frac{2}{\pi}\int_{0}^{\infty}\frac{\sin(\omega t)}{\omega}\,d\omega, \quad \delta(t) = \frac{2}{\pi}\int_{0}^{\infty}\cos(\omega t)\,d\omega\,. \tag{8.3.17}$$

Thus

$$\frac{1}{\omega} = \int_{0}^{\infty}\sin(\omega t)\,\mathcal{H}(t)\,dt, \quad 1 = \int_{0}^{\infty}\cos(\omega t)\,\delta(t)\,d\omega\,. \tag{8.3.18}$$

As an example, let us start from the complex modulus of the Maxwell model (8.3.13). The inverse transform (8.3.19) of the real part gives the viscoelastic modulus via $(8.3.14)_2$ and $(8.3.16)_1$,

$$\frac{E'(\omega)}{\omega} = E\frac{\omega}{\omega^2 + (E/\eta)^2} \quad \Leftrightarrow \quad E(t) = E\exp\left(-t\frac{E}{\eta}\right). \tag{8.3.21}$$

The inverse transform (8.3.20) of the imaginary part gives the (same!) viscoelastic modulus via $(8.3.15)_2$ and $(8.3.16)_2$,

$$\frac{E''(\omega)}{\omega} = E\frac{E/\eta}{\omega^2 + (E/\eta)^2} \quad \Leftrightarrow \quad E(t) = E\exp\left(-t\frac{E}{\eta}\right). \tag{8.3.22}$$

Starting from the viscoelastic modulus $E(t) = E + \eta\,\delta(t)$ of the Kelvin model, the storage and loss moduli derive from (8.3.5) with help of (8.3.18),

$$E'(\omega) - E = \omega \int_0^\infty \sin(\omega v)\,(E(v) - E)\,dv = 0, \quad E''(\omega) = \omega \int_0^\infty \cos(\omega v)\,(E(v) - E)\,dv = \omega\,\eta. \tag{8.3.23}$$

Conversely, $E(t) - E$ is retrieved equal to $\eta\,\delta(t)$ by (8.3.20) with help of $(8.3.17)_2$.

Clearly, since the two inverse transforms (8.3.19) and (8.3.20) give the same output, the storage and loss moduli should be related. In fact, it is possible to calculate one in terms of the other, Gross [1953], e.g.,

$$\begin{aligned}
E'(\omega) - E(\infty) &= \frac{\omega}{\pi}\,\mathrm{PV}\int_0^\infty \frac{E''(u)}{u}\left(\frac{1}{u+\omega} - \frac{1}{u-\omega}\right)du; \\
E''(\omega) &= \frac{\omega}{\pi}\,\mathrm{PV}\int_0^\infty \frac{E'(u) - E(\infty)}{u}\left(\frac{1}{u+\omega} + \frac{1}{u-\omega}\right)du.
\end{aligned} \tag{8.3.24}$$

The first relation is obtained by introducing (8.3.20) in the first relation of (8.3.5) and by using the trigonometric relation $2\sin a\,\cos b = \sin(a+b) + \sin(a-b)$ and $(8.3.18)_1$. The second line of (8.3.24) is obtained by introducing (8.3.19) in the second relation of (8.3.5).

These two relations form a sort of Hilbert transform pair known as Kramers-Kronig relations: with $E'(\omega)$ even and $E''(\omega)$ odd, they result from the fact that the complex modulus $\left(E^*(i\,\omega) - E(\infty)\right)/\omega$ is analytic in the lower half-plane $\mathrm{Im}\,\omega < 0$. The integration is performed in the direct sense along a semi-circle in the lower half-plane closed along the real axis, and accounts for the fact that ω lies on the integration axis, resulting in a Cauchy principal value (PV):

$$\begin{aligned}
\frac{1}{\omega}\left(E'(\omega) - E(\infty) + i\,E''(\omega)\right) &= -\frac{1}{i\,\pi}\,\mathrm{PV}\int_{-\infty}^\infty \frac{E'(x) - E(\infty) + i\,E''(x)}{x\,(x-\omega)}\,dx \\
&= \frac{i}{\pi}\,\mathrm{PV}\int_0^\infty \frac{1}{x}\left(\frac{1}{x+\omega} + \frac{1}{x-\omega}\right)\left(E'(x) - E(\infty)\right)dx \\
&\quad - \frac{1}{\pi}\,\mathrm{PV}\int_0^\infty \frac{1}{x}\left(-\frac{1}{x+\omega} + \frac{1}{x-\omega}\right)E''(x)\,dx,
\end{aligned} \tag{8.3.25}$$

which implies the relations (8.3.24).

As an application, the storage modulus of the Maxwell model (8.3.21) may be retrieved from its loss modulus (8.3.22), and conversely, by additive decomposition of the integrands into simple elements and integration in the complex plane.

8.3.2 Time translation invariance

The above derivation has considered the real and imaginary parts of the complex modulus. If the analysis is performed wholly in the complex plane, then the complex modulus appears as the Laplace-Carson transform $t \to p$ of the viscoelastic modulus evaluated at the point $p = i\,\omega$. Indeed, the stress response (8.3.1) to the stimulus (8.3.2) can be cast in the format

$$\hat{Q}(t) = \int_{-\infty}^t E(t-u)\frac{d\big(q_0\,e^{i\omega u}\big)}{du}\,du \overset{u \to t-v}{=} \left(i\,\omega \int_0^\infty e^{-i\omega v}\,E(v)\,dv\right)\hat{q}(t). \tag{8.3.26}$$

The Laplace-Carson transform $E(t) \to E^*(p)$ at a complex point p is defined as the product of p by the Laplace transform of $E(t)$,

$$E^*(p) = p \int_0^\infty e^{-t\,p}\, E(t)\, dt\,. \tag{8.3.27}$$

In fact, integration by parts of (8.3.1) shows that the response $Q(t)$ in terms of the viscoelastic modulus $E(t)$ and its dual $q(t)$ in terms of viscoelastic, or creep, compliance $J(t)$ can be brought in the form of a convolution that we shall refer to as a Stieltjes convolution integral,

$$Q(t) = \frac{d}{dt} \int_0^t E(t-u)\, q(u)\, du; \quad q(t) = \frac{d}{dt} \int_0^t J(t-u)\, Q(u)\, du\,. \tag{8.3.28}$$

The proof is elementary and proceeds as follows. At time u, let us apply a stress $Q(u)$ and let us note $q(t)$ the strain at time $t \ge u$,

$$Q(t) = Q(u)\, \mathcal{H}(t-u) \quad \to \quad q(t) = J(t,u)\, Q(u)\, \mathcal{H}(t-u)\,. \tag{8.3.29}$$

Here $\mathcal{H} = \mathcal{H}(t)$ denotes the Heaviside step function. The creep function $J(t,u)$ depends (temporarily) on two arguments, the 'source' time u and the 'receptor' time t: in fact, $J(t,u)$ is the strain at time t due to a unit stress applied at time u. The load may actually be changed through infinitesimal increments,

$$dQ(t) = dQ(u)\, \mathcal{H}(t-u) \quad \to \quad dq(t) = J(t,u)\, dQ(u)\, \mathcal{H}(t-u)\,. \tag{8.3.30}$$

Upon integration, the strain at time t may be given various forms, e.g.,

$$\begin{aligned}
q(t) &= J(t,0)\, Q(0)\, \mathcal{H}(t-0) &&+ \int_{0+}^t J(t,u)\, \frac{dQ(u)}{du}\, \mathcal{H}(t-u)\, du \\
&= J(t,t)\, Q(t) &&- \int_{0+}^t \frac{\partial J(t,u)}{\partial u}\, Q(u)\, du\,.
\end{aligned} \tag{8.3.31}$$

The assumption of time translation invariance is henceforth adopted, which states that the response to a stimulus depends not on the receptor and source times individually but only on their difference

$$\text{time translation invariance} \quad J(t,u) = J(t-u,0) \tag{8.3.32}$$

With an abuse of notation, we shall note $J(t-u)$ instead of $J(t-u,0)$. The assumption of time translation invariance is very strong. Although it does not hold for most materials from fabrication to degradation, it is quite acceptable in time windows where the microstructure evolves little. Materials which obey this invariance are also termed 'non-aging'. Early-age concrete, say before 200 days, does not enter this family while mature concrete does at least over a few years. Even if the structural modifications of articular cartilages over time are not so significant as in concrete, aging is mentioned to increase its stiffness, Charlebois et al. [2004].

The strain may now be cast in the format of a Stieltjes convolution integral, and the dual relation that provides the stress as a function of the strain time history may be derived in a similar manner,

$$\begin{aligned}
q(t) &= J(0)\, Q(t) + \int_{0+}^t \frac{dJ(t-u)}{d(t-u)}\, Q(u)\, du &&= \frac{d}{dt} \int_0^t J(t-u)\, Q(u)\, du; \\
Q(t) &= E(0)\, q(t) + \int_{0+}^t \frac{dE(t-u)}{d(t-u)}\, q(u)\, du &&= \frac{d}{dt} \int_0^t E(t-u)\, q(u)\, du\,.
\end{aligned} \tag{8.3.33}$$

Denoting the Stieltjes convolution integral by the symbol $\otimes$, we have reached the relations,

$$q(t) = \big(J \otimes Q\big)(t), \quad Q(t) = \big(E \otimes q\big)(t)\,. \tag{8.3.34}$$

The fact that the Laplace-Carson transform of the Stieltjes convolution integral is equal to the product of the Laplace-Carson transforms,

$$q^*(p) = J^*(p)\, Q^*(p); \quad Q^*(p) = E^*(p)\, q^*(p)\,, \tag{8.3.35}$$

implies the relation $E^*(p)\, J^*(p) = 1$.

This essential property can be recovered by proceeding somehow formally. Indeed, we shall admit the following properties of the Stieltjes convolution integral for arbitrary time dependent functions f, g and h (for which the convolution exists):

$$\text{Commutativity}: \quad f \otimes g = g \otimes f,$$

$$\text{Associativity}: \quad f \otimes (g \otimes h) = (f \otimes g) \otimes h, \tag{8.3.36}$$

$$\mathcal{H} \text{ single neutral element}: \quad f \otimes \mathcal{H} = \mathcal{H} \otimes f = f.$$

The time argument has been dropped for simplicity. Combining the two results (8.3.34) and the three above properties,

$$q = J \otimes Q = J \otimes (E \otimes q) = (J \otimes E) \otimes q$$
$$Q = E \otimes q = E \otimes (J \otimes Q) = (E \otimes J) \otimes Q, \tag{8.3.37}$$

implies, since $\mathcal{H}$ is the single neutral element, the nice result that the modulus and compliance are the inverse of one another in the transformed space, namely

$$E \otimes J = E \otimes J = \mathcal{H} \quad \Rightarrow \quad E^*(p)\, J^*(p) = 1. \tag{8.3.38}$$

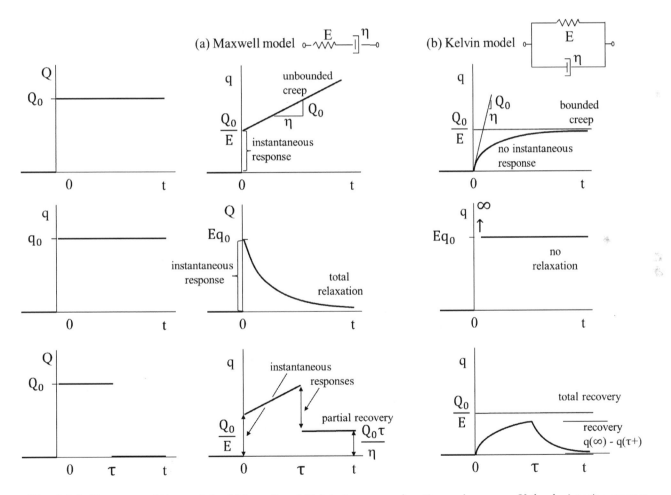

Fig. 8.3.3: Responses of the models of Maxwell and Kelvin to creep, relaxation and recovery. If the deviatoric response of a viscoelastic material is governed by a Maxwell model, its deviatoric stress in a relaxation test vanishes at large time: the Maxwell model is said 'fluid-like' while the Kelvin model is said 'solid-like'.

As a simple application, from the complex modulus of the Maxwell model obtained from (8.3.13) follow in turn its viscoelastic modulus by inverse Laplace-Carson transform, its complex compliance by (8.3.38) and its real compliance by inverse Laplace-Carson transform,

$$E^*(p) = E \, \frac{p}{E/\eta + p}; \quad E(t) = E \exp\left(-t\,\frac{E}{\eta}\right); \quad J^*(p) = \frac{1}{E} + \frac{1}{\eta}\frac{1}{p}; \quad J(t) = \frac{1}{E} + \frac{t}{\eta}, \tag{8.3.39}$$

and, in the same mood, but in a different order for the Kelvin model,

$$J(t) = \frac{1}{E}\left(1 - \exp\left(-t\,\frac{E}{\eta}\right)\right); \quad J^*(p) = \frac{1}{\eta}\,\frac{1}{E/\eta + p}; \quad E^*(p) = E + \eta\,p; \quad E(t) = E + \eta\,\delta(t). \tag{8.3.40}$$

One may proceed similarly with another common viscoelastic compliance, or creep function, namely the power law $J(t) = J_0\,t^\alpha$ with $\alpha \in \,]0,1[$:

$$J(t) = J_0\,t^\alpha; \quad J^*(p) = J_0\,\frac{\Gamma(1+\alpha)}{p^\alpha}; \quad E^*(p) = \frac{1}{J_0}\,\frac{p^\alpha}{\Gamma(1+\alpha)}; \quad E(t) = \frac{1}{J_0}\,\frac{\sin(\pi\alpha)}{\pi\alpha}\,t^{-\alpha}. \tag{8.3.41}$$

Finally, here is another way due to Mandel [1966] to recover the expression (8.3.3) for a harmonic strain via the Laplace-Carson formalism. With help of (8.3.45) and of the Laplace-Carson transform (8.3.35),

$$q(t) = q_0\,e^{i\omega t} \quad \rightarrow \quad q^*(p) = q_0\,\frac{p}{p - i\,\omega} \quad \rightarrow \quad Q^*(p) = E^*(p)\,q^*(p) = q_0\,E^*(p)\,\frac{p}{p - i\,\omega}. \tag{8.3.42}$$

The stress response $Q(t)$ may be obtained by inversion of the Laplace-Carson transform via a Bromwich contour $\mathcal{C}$ in the complex plane, by adapting the expression known for the inverse Laplace transform, namely

$$Q(t) = \frac{1}{2\,i\,\pi}\int_{\mathcal{C}} \frac{Q^*(p)}{p}\,e^{t\,p}\,dp = \frac{1}{2\,i\,\pi}\int_{\mathcal{C}} q_0\,\frac{E^*(p)}{p - i\,\omega}\,e^{t\,p}\,dp. \tag{8.3.43}$$

We shall assume the poles of $E^*(p)$ to have strictly negative real parts so that their contributions die out over time. Then the sole contribution at large time will be due to the imaginary pole $p = i\,\omega$ and, by the Cauchy formula, $Q(t) = E^*(i\,\omega)\,(q_0\,e^{i\omega\,t})$.

8.3.3   The fading memory principle

Whether they obey the property of time translation invariance or not, the relaxation function $E(t)$ and the creep function $J(t)$ agree with the *fading memory principle*. In essence, the fading memory principle states that the past history of the strain has less effects on the current stress than its more recent history, Coleman and Noll [1960], Coleman [1964]. Thermodynamic considerations imply that the time derivative of the relaxation function is a non-negative $E(t) \geq 0$, continuously decreasing function of time $dE(t)/dt \leq 0$, e.g., Christensen [1982], p. 87. As can be appreciated from the second line of (8.3.33), the fading memory principle implies in addition that the absolute value of the time derivative of the relaxation function should be a decreasing function of time and therefore the 2nd derivative of the relaxation function should be nonnegative $d^2E(t)/dt^2 \geq 0$. In summary,

$$E(t) \geq 0, \quad \frac{dE(t)}{dt} \leq 0, \quad \frac{d^2E(t)}{dt^2} \geq 0. \tag{8.3.44}$$

8.3.4   Basics of Laplace-Carson transform

The Laplace-Carson transform $f(t) \rightarrow f^*(p)$ enjoys properties similar, but not exactly identical, to the one-sided Laplace transform. With the convention that all real functions are tacitly multiplied by the Heaviside step function $\mathcal{H}(t)$, some basic properties are worth recording, namely

$$t^n\,\mathcal{H}(t) \rightarrow \frac{n!}{p^n}, \quad \frac{d^n f}{dt^n} \rightarrow p^n\,f^*(p), \quad n \geq 0 \text{ integer}; \quad \exp(-a\,t) \rightarrow \frac{p}{a + p}, \text{ Re } a \geq 0; \quad \delta(t) \rightarrow p. \tag{8.3.45}$$

Also, for $\alpha \in \,]0,1[$,

$$t^\alpha\,\mathcal{H}(t) \rightarrow \frac{\Gamma(1+\alpha)}{p^\alpha}, \tag{8.3.46}$$

Γ being the Gamma function, some information on which is provided in Section 15.3.2.3.

The transform of the Stieltjes convolution integral is most noticeable,

$$\frac{d}{dt} \int_0^t f(t-u)\, g(u)\, du \rightarrow f^*(p)\, g^*(p)\,. \tag{8.3.47}$$

In (8.3.46) and (8.3.47), the operator $d\,/dt$ denotes the derivative in the sense of distributions, $df/dt = d^+ f/dt + f(0^+)\,\delta(t)$, with $d^+ f/dt$ the right derivative and $\delta(t)$ the Dirac delta function, Section 7.5.4.3.

The Laplace-Carson transform is a useful tool to apply *the elastic viscoelastic correspondence principle*. Roughly speaking, this principle states that a viscoelastic boundary value problem expressed in the Laplace-Carson domain has the same form as the (associated) elastic boundary value problem expressed in the time domain (the time scale here being possibly given by the loading process). It assumes that the viscoelastic response obeys time translation invariance. Provided that the partition of the boundary defining the zones where either the displacement or the traction are prescribed is fixed, it is a quite handy and powerful tool for applications in strength of materials and continuum mechanics.

8.3.5 Loss coefficient and hysteresis loops

The long-term response to a harmonic stress stimulus may be expressed in real form,

$$Q(t) = Q_0 \sin(\omega\, t), \quad q(t) = q_0 \sin(\omega\, t - \phi)\,. \tag{8.3.48}$$

The trajectory in the $(q/q_0, Q/Q_0)$ plane is sketched in Fig. 8.3.4.

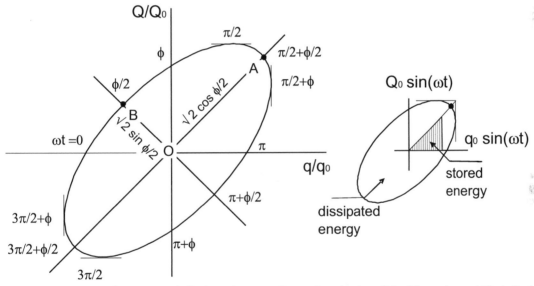

Fig. 8.3.4: Trajectory, stored energy and dissipated energy for a viscoelastic solid with a phase shift ϕ. In biological tissues, the hysteresis loops are believed to depend weakly on the loading rate, Fung [1993], p. 283 et seq.

It is an ellipse with semi-axes equal to $\sqrt{2}\,\cos(\phi/2)$ and $\sqrt{2}\,\sin(\phi/2)$ oriented at $\pi/4$ and $3\,\pi/4$ respectively. Indeed,

$$\frac{1}{(\sqrt{2}\,\cos(\phi/2))^2}\left(\frac{1}{\sqrt{2}}\frac{q}{q_0} + \frac{1}{\sqrt{2}}\frac{Q}{Q_0}\right)^2 + \frac{1}{(\sqrt{2}\,\sin(\phi/2))^2}\left(-\frac{1}{\sqrt{2}}\frac{q}{q_0} + \frac{1}{\sqrt{2}}\frac{Q}{Q_0}\right)^2 = 1\,. \tag{8.3.49}$$

The energy spent

$$W(t) = \int_0^t Q\, dq = W_{\mathrm{e}}(t) + W_{\mathrm{loss}}(t)\,, \tag{8.3.50}$$

may be split in an elastic stored contribution and a dissipated contribution. The in phase stored contribution

$$W_{\mathrm{e}}(t) = \int_0^{\omega t} Q_0 \sin(\omega\, u)\, d\big(q_0 \sin(\omega\, u)\big) \cos\phi = \frac{Q_0\, q_0}{2}\cos\phi \sin^2(\omega\, t)\,, \tag{8.3.51}$$

corresponds to the triangular area below the curve $(q_0 \sin(\omega t), Q_0 \cos(\omega t))$. Its maximum $W_e^{\max} = \frac{1}{2} Q_0 q_0 \cos\phi$ takes place for $\omega t = \frac{1}{2}\pi$. It has an average value of

$$\overline{W}_e = \frac{1}{T} \int_0^T W_e(t)\,dt = \frac{Q_0 q_0}{4} \cos\phi \,, \tag{8.3.52}$$

over a cycle $T = 2\pi/\omega$. The dissipated energy stems from the out of phase response,

$$W_{\text{loss}}(t) = \int_0^{\omega t} Q_0 \sin(\omega u)\,d\big(q_0 \cos(\omega u)\big)\,(-\sin\phi) = \frac{Q_0 q_0}{2} \sin\phi \left(\omega t - \frac{1}{2} \sin(2\omega t)\right). \tag{8.3.53}$$

The dissipated energy over a cycle

$$\Delta W_{\text{loss}} = W_{\text{loss}}(T) - W_{\text{loss}}(0) = Q_0 q_0 \pi \sin\phi\,. \tag{8.3.54}$$

is, as expected, equal to the area of the hysteresis loop (the area of an ellipse is equal to π times the product of the semi-axes). The loss coefficient L is defined in terms of the out of phase angle,

$$\frac{1}{L} = \frac{1}{4\pi} \frac{\Delta W_{\text{loss}}}{\overline{W}_e} = \tan\phi = \frac{E''(\omega)}{E'(\omega)}\,, \tag{8.3.55}$$

the last relation resulting from (8.3.10).

8.3.6   Complex modulus of the standard linear solid model

The standard linear solid is defined by a spring E_P in parallel with a Maxwell model as shown in Fig. 8.3.5. The differential equation of the model,

$$\frac{1}{E_P} \left[Q + t_q \frac{dQ}{dt} \right] = q + t_Q \frac{dq}{dt}, \tag{8.3.56}$$

is phrased in terms of two characteristic times,
  - $t_q = \eta/E_S$ the characteristic time at constant strain,
  - $t_Q = (1 + E_S/E_P)\,t_q$ the characteristic time at constant stress.
In this context, $t_q = \eta/E_S$, $t_Q = \infty$ for a Maxwell model ($E_P = 0$), while $t_q = 0$, $t_Q = \eta/E_P$ for a Kelvin model ($E_S = \infty$).

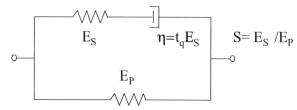

Fig. 8.3.5: The standard linear solid model defined by a Maxwell model in parallel with a spring.

The differential equation (8.3.56) can be transformed in integral form through the relaxation modulus $E(t)$ or through the compliance $J(t)$ as

$$Q(t) = \int_{-\infty}^t E(t-u)\,dq(u)\,, \quad q(t) = \int_{-\infty}^t J(t-u)\,dQ(u)\,, \tag{8.3.57}$$

with

$$\frac{E(t)}{E_P} = 1 + S\,e^{-t/t_q}\,, \quad S = \frac{t_Q}{t_q} - 1 = \frac{E_S}{E_P}\,, \tag{8.3.58}$$

and

$$E_P\,J(t) = 1 - \frac{S}{1+S}\,e^{-t/t_Q}\,. \tag{8.3.59}$$

Alternatively, still assuming $q(-\infty) = 0$, and with the change of variable $t - u = v$,

$$\frac{Q(t)}{E_P} = q(t) - S \int_0^\infty e^{-v/t_q} \frac{dq(t-v)}{dv} \, dv \, . \tag{8.3.60}$$

Note that a purely elastic response $Q(t) = (E_P + E_S) \, q(t)$ is characterized by $\eta \to \infty$ and $t_q \to \infty$ and not by $S = 0$.

The response $\hat{Q}(t) = E^*(i\,\omega) \, \hat{q}(t)$ to a harmonic strain $\hat{q}(t) = q_0 \, e^{i\omega t}$ introduces the complex modulus,

$$\frac{E(t)}{E_P} = 1 + S \, e^{-t/t_q} \longmapsto \frac{E^*(i\,\omega)}{E_P} = 1 + S \frac{i\,\omega\,t_q}{1 + i\,\omega\,t_q} = 1 + S \frac{\omega\,t_q + i}{(\omega\,t_q)^{-1} + \omega\,t_q} \, , \tag{8.3.61}$$

whence the storage and loss moduli,

$$\frac{E'(\omega)}{E_P} = 1 + S \frac{\omega\,t_q}{(\omega\,t_q)^{-1} + \omega\,t_q}, \qquad \frac{E''(\omega)}{E_P} = S \frac{1}{(\omega\,t_q)^{-1} + \omega\,t_q} \, . \tag{8.3.62}$$

The loss angle,

$$\tan\phi = \frac{E''(\omega)}{E'(\omega)} = \frac{S\,\omega\,t_q}{1 + (1+S)\,(\omega\,t_q)^2} \, , \tag{8.3.63}$$

vanishes at zero and infinite frequencies and has a maximum equal to $S/(2\sqrt{1+S})$ at $\omega\,t_q = 1/\sqrt{1+S}$. It tends to zero at high frequencies for a Maxwell model, namely $E''(\omega)/E'(\omega) = E_S/(\eta\omega)$, while it tends to infinity for a Kelvin model, namely $E''(\omega)/E'(\omega) = \eta\,\omega/E_P$.

8.4 Relaxation spectrum

8.4.1 A discrete spectrum

The complex modulus of a model constituted by n parallel Maxwell models plus a spring branch is obtained by simple summation,

$$\frac{E(t)}{E_P} = 1 + \sum_{k=1}^n S_k \, e^{-t/t_{qk}}, \qquad \frac{E^*(i\,\omega)}{E_P} = 1 + \sum_{k=1}^n S_k \frac{i\,\omega\,t_{qk}}{1 + i\,\omega\,t_{qk}} \, , \tag{8.4.1}$$

where $S_k = E_{Sk}/E_P$ and $t_{qk} = \eta_k/E_{Sk}$ is the relaxation time at constant strain associated with the branch $k \in [1, n]$. The proof is left as an exercise, Exercise 8.1. The real part of the complex modulus,

$$\frac{E'(i\,\omega)}{E_P} = 1 + \sum_{k=1}^n S_k \frac{(\omega\,t_{qk})^2}{1 + (\omega\,t_{qk})^2} = 1 + S - \sum_{k=1}^n S_k \frac{1}{1 + (\omega\,t_{qk})^2} \, , \tag{8.4.2}$$

with $S = \sum_{k=1}^n S_k$ is clearly increasing continuously with frequency,

$$\frac{1}{E_P} \frac{dE'(i\,\omega)}{d\omega} = \sum_{k=1}^n 2\,S_k\,t_{qk} \frac{\omega\,t_{qk}}{(1 + (\omega\,t_{qk})^2)^2}, \qquad \frac{1}{E_P} \frac{d^2 E'(i\,\omega)}{d\omega^2} = \sum_{k=1}^n 2\,S_k\,(t_{qk})^2 \frac{1 - 3\,(\omega\,t_{qk})^2}{(1 + (\omega\,t_{qk})^2)^3} \, , \tag{8.4.3}$$

as sketched in Figs. 8.3.1 and 8.4.1. On the other hand, the imaginary part,

$$\frac{E''(i\,\omega)}{E_P} = \sum_{k=1}^n S_k \frac{\omega\,t_{qk}}{1 + (\omega\,t_{qk})^2}, \qquad \frac{1}{E_P} \frac{dE''(i\,\omega)}{d\omega} = \sum_{k=1}^n S_k\,t_{qk} \frac{1 - (\omega\,t_{qk})^2}{(1 + (\omega\,t_{qk})^2)^2} \, , \tag{8.4.4}$$

has maxima at the relaxation times, namely for $\omega \sim (t_{qk})^{-1}$, $k \in [1, n]$, and minima on each side of the relaxation times, as sketched in Fig. 8.4.1. To prove this pattern, let us assume the relaxation times numbered in increasing values, namely $t_{q1} < t_{q2} < \cdots$ to be well separated, say $t_{qj} \ll t_{qk} \ll t_{ql}$ with $j = k-1$ and $l = k+1$. For frequencies around $(t_{qk})^{-1}$,

$$\omega\,t_{qj} \ll 1; \quad \omega\,t_{qk} \sim 1; \quad 1 \ll \omega\,t_{ql}, \tag{8.4.5}$$

the subsequent implications follow,

$$\frac{(\omega\,t_{qj})^2}{1 + (\omega\,t_{qj})^2} \ll 1; \quad \frac{(\omega\,t_{qk})^2}{1 + (\omega\,t_{qk})^2} \sim \frac{1}{2}; \quad \frac{(\omega\,t_{ql})^2}{1 + (\omega\,t_{ql})^2} \sim 1;$$

$$\frac{\omega\,t_{qj}}{1 + (\omega\,t_{qj})^2} \ll 1; \quad \frac{\omega\,t_{qk}}{1 + (\omega\,t_{qk})^2} \sim \frac{1}{2}; \quad \frac{\omega\,t_{ql}}{1 + (\omega\,t_{ql})^2} \ll 1 \, . \tag{8.4.6}$$

Consequently the value of the storage modulus $E'(i\,\omega)/E_P$ at $\omega = (t_{qk})^{-1}$ is about $1 + S_k/2 + \sum_{l>k} S_l$. Conversely, a single term dominates the loss modulus around this frequency. As indicated by (8.4.3), the derivative is increasing for $\omega < (t_{qk})^{-1}$, it vanishes at $\omega = (t_{qk})^{-1}$ and it is negative for $\omega > (t_{qk})^{-1}$. This pattern repeats around each frequency $\omega \sim (t_{qk})^{-1}$, $k \in [1, n]$. The value of the loss modulus $E''(i\,\omega)/E_P$ at $\omega = (t_{qk})^{-1}$ is about $S_k/2$.

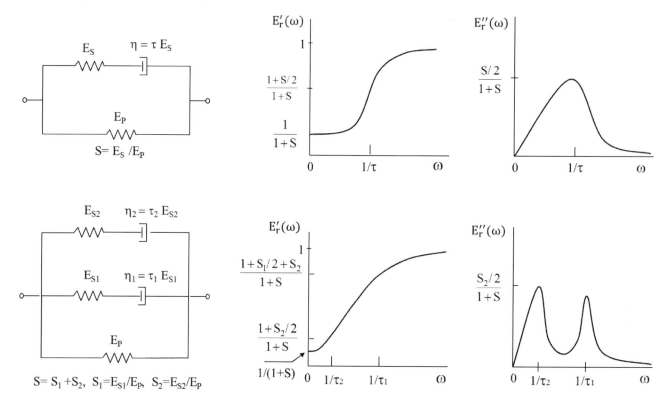

Fig. 8.4.1: Reduced storage and loss moduli of two viscoelastic materials. Parallel branches endowed with their own distinct relaxation times at constant strain allow for dissipation for a large range of frequencies. The local maxima of the loss modulus take place at pulsations linked to the inverses of the relaxation times.

8.4.2   A continuous spectrum

The discrete generalization of the previous section motivates the continuous (integral) generalization,

$$
\begin{aligned}
\frac{E(t)}{E(\infty)} &= 1 + \int_0^\infty S(\tau)\, e^{-t/\tau} d\tau, \\
\frac{E^*(i\,\omega)}{E(\infty)} &= 1 + \int_0^\infty \frac{i\,\omega\,\tau\, S(\tau)}{1 + i\,\omega\,\tau}\, d\tau, \\
&= 1 + \int_0^\infty \frac{\omega\tau\, S(\tau)}{(\omega\tau)^{-1} + \omega\tau}\, d\tau + i \int_0^\infty \frac{S(\tau)}{(\omega\tau)^{-1} + \omega\tau}\, d\tau\,.
\end{aligned}
\tag{8.4.7}
$$

The discrete spectra for the standard linear solid and the generalization above are recovered for $E(\infty) = E_P$ and, respectively,

$$
S(\tau) = S\,\delta(\tau - t_q), \quad S(\tau) = \sum_{k=1}^n S_k\,\delta(\tau - t_{qk})\,.
\tag{8.4.8}
$$

For biological tissues, the loss moduli and energy dissipation are widely believed to be significant and virtually constant over a wide range of frequency, Fung [1993], p. 283 et seq. This observation motivates the *relaxation spectrum*,

$$
S(\tau) = a_v
\begin{cases}
\dfrac{1}{\tau}, & \tau \in [\tau_1, \tau_2] \\[2mm]
0, & \text{elsewhere,}
\end{cases}
\tag{8.4.9}
$$

with τ_1 and τ_2 the short and long relaxation times and $a_v > 0$ a dimensionless material constant. With $E_1(u) = \int_u^\infty e^{-v}/v\, dv^{8.2}$, the relaxation modulus and the complex modulus can be given an explicit form, namely

$$\frac{E(t)}{E(\infty)} = 1 + a_v \left(E_1(\frac{t}{\tau_2}) - E_1(\frac{t}{\tau_1}) \right), \tag{8.4.11}$$

and

$$\frac{E^*(i\,\omega)}{E(\infty)} = 1 + \frac{a_v}{2} \mathrm{Ln} \frac{1+(\omega\tau_2)^2}{1+(\omega\tau_1)^2} + i\,a_v \left(\tan^{-1}(\omega\tau_2) - \tan^{-1}(\omega\tau_1) \right). \tag{8.4.12}$$

Here we have defined the viscoelastic modulus and complex modulus from the relaxation spectrum. The converse point of view may also be considered. Setting $\lambda = 1/\tau$ in (8.4.7), then $E(t)/E(\infty) - 1$ appears as the Laplace transform of $S(1/\lambda)/\lambda^2$. The converse relation is obtained by inversion of the Laplace transform via an appropriate Bromwich contour $\mathcal{C}$ in the complex plane,

$$\frac{1}{\lambda^2} S(\frac{1}{\lambda}) = \frac{1}{2\,i\pi} \int_{\mathcal{C}} e^{\lambda\,p} \left(\frac{E(p)}{E(\infty)} - 1 \right) dp. \tag{8.4.13}$$

The spectrum may also be derived from the complex modulus, Gross [1953], Christensen [1982].

8.4.3   A specific relaxation spectrum

Note that, with a logarithmic time scale,

$$\frac{E(t)}{E(\infty)} = 1 + \int_0^\infty \tilde{S}(\tau)\, e^{-t/\tau} d\,\mathrm{Ln}\,\tau, \tag{8.4.14}$$

the relaxation spectrum $\tilde{S}(\tau) \equiv \tau\, S(\tau) = a_v \left(\mathcal{H}(\tau - \tau_1) - \mathcal{H}(\tau - \tau_2) \right)$ corresponding to (8.4.9) is piecewise constant. This observation leads to define the *relaxation spectrum S* via $\tilde{S}(\xi) = \xi\, S(\xi)$ which is now centered at τ and has half-width $\Delta\tau$, Fung [1993],

$$\tilde{S}(\xi) = \begin{cases} \dfrac{\tau}{\Delta\tau}, & \xi \in [\tau - \Delta\tau, \tau + \Delta\tau], \\ 0, & \text{otherwise.} \end{cases} \tag{8.4.15}$$

The *memory function $m(t)$*,

$$m(t) = \int_0^{+\infty} \tilde{S}(\xi)\, e^{-t/\xi} d\,\mathrm{Ln}\,\xi, \tag{8.4.16}$$

has Laplace transform $t \to p^{8.3}$,

$$\overline{m}(p) = \int_0^{+\infty} e^{-tp}\, m(t)\, dt = \int_0^{+\infty} \frac{\tilde{S}(\xi)}{\xi\, p + 1}\, d\xi = \frac{\tau}{\Delta\tau}\, \frac{1}{p}\, \mathrm{Ln}\, \frac{p\,(\tau + \Delta\tau) + 1}{p\,(\tau - \Delta\tau) + 1}. \tag{8.4.17}$$

Note the limits for infinitely short and infinitely large relaxation times τ,

$$\lim_{\tau \to 0} p\,\overline{m}(p) = 0; \quad \lim_{\tau \to \infty} p\,\overline{m}(p) = 2. \tag{8.4.18}$$

These limits can be checked on the relaxation modulus in time space (8.4.11) using the properties of the function E_1 provided in footnote 8.2: at small and large τ_1 and τ_2, the viscous contribution to the relaxation modulus $E(t)/E(\infty)$ for $t > 0$ is equal, respectively, to 0 and to $2\,a_v\,\Delta\tau/\tau = 2$ for $a_v = \tau/\Delta\tau$.

[8.2] The function $E_1(t)$ is defined for any real $t \neq 0$ as,

$$E_1(t) = \int_t^\infty \frac{e^{-v}}{v}\, dv = -\gamma - \mathrm{Ln}\,|t| - \sum_{k=1,\infty} \frac{(-t)^k}{k \times k!}, \tag{8.4.10}$$

with $\gamma = 0.5772\cdots$ the Euler's constant. As t varies from $-\infty$ to 0, it increases monotonically from $-\infty$ to $+\infty$ with a single zero at $t \sim -0.37250$. As t varies from 0 to $+\infty$, it decreases monotonically from $+\infty$ to 0. It behaves like e^{-t}/t at large positive or negative t.

[8.3] If p was real, the argument of the logarithm should be in absolute value. However, the Laplace transform is a priori complex and, in fact, to be accurate, a definition of the logarithm in the complex plane should be given.

8.5   The quasi-linear viscoelasticity QLV model of Fung

8.5.1   Elastic response and reduced relaxation function

The above relations for the standard linear solid are rewritten along the scheme proposed by Fung [1993]. First the elastic (instantaneous) response is introduced, e.g., for the example above,

$$Q^{\mathrm{e}}(t) = E_P \, (1 + S) \, q(t) = (E_P + E_S) \, q(t) \,. \tag{8.5.1}$$

The viscoelastic relation can be recast in integral form through the *reduced* dimensionless relaxation modulus $E_r(t)$

$$Q(t) = \int_{-\infty}^{t} E_r(t - u) \, dQ^{\mathrm{e}}(u) \,, \tag{8.5.2}$$

with

$$E_r(t) = \frac{1 + S \, e^{-t/t_q}}{1 + S}, \quad S = \frac{t_Q}{t_q} - 1 = \frac{E_S}{E_P} \,. \tag{8.5.3}$$

Note the particular values,

$$E_r(0) = 1, \quad E_r(\infty) = \frac{1}{1 + S} < 1 \,. \tag{8.5.4}$$

As for the reduced complex modulus,

$$E_r(t) = \frac{1}{1 + S} \Big(1 + S \, e^{-t/t_q} \Big) \longmapsto E_r^*(i\,\omega) = \frac{1}{1 + S} \Big(1 + S \, \frac{i\,\omega\,t_q}{1 + i\,\omega\,t_q} \Big), \tag{8.5.5}$$

the variations with the pulsation ω of its real and imaginary parts,

$$E_r'(\omega) = \frac{1}{1 + S} \Big(1 + S \, \frac{\omega\,t_q}{(\omega\,t_q)^{-1} + \omega\,t_q} \Big), \quad E_r''(\omega) = \frac{S}{1 + S} \Big(\frac{1}{(\omega\,t_q)^{-1} + \omega\,t_q} \Big), \tag{8.5.6}$$

are sketched in Fig. 8.4.1 for the standard linear solid (with two parallel branches) and for a generalized model (with three parallel branches).

A *de visu* extension of the reduced relaxation modulus can be suggested,

$$E_r(t) = \frac{1 + \displaystyle\int_0^{\infty} S(\tau) \, e^{-t/\tau} d\tau}{1 + \displaystyle\int_0^{\infty} S(\tau) \, d\tau}, \tag{8.5.7}$$

from which follows the associated complex modulus,

$$E_r^*(i\,\omega) = \frac{1 + \displaystyle\int_0^{\infty} \frac{i\,\omega\,\tau\,S(\tau)}{1 + i\,\omega\,\tau} d\tau}{1 + \displaystyle\int_0^{\infty} S(\tau) \, d\tau} = \frac{1 + \displaystyle\int_0^{\infty} \frac{\omega\,\tau\,S(\tau)}{(\omega\tau)^{-1} + \omega\tau} d\tau}{1 + \displaystyle\int_0^{\infty} S(\tau) \, d\tau} + i \, \frac{\displaystyle\int_0^{\infty} \frac{S(\tau)}{(\omega\tau)^{-1} + \omega\tau} d\tau}{1 + \displaystyle\int_0^{\infty} S(\tau) \, d\tau} \,. \tag{8.5.8}$$

For the continuous spectrum (8.4.9),

$$\int_0^{\infty} S(\tau) \, d\tau = a_v \, \mathrm{Ln}\frac{\tau_2}{\tau_1} \,, \tag{8.5.9}$$

the reduced relaxation modulus and the reduced complex modulus specialize to,

$$E_r(t) = \frac{1 + a_v \left(E_1(t/\tau_2) - E_1(t/\tau_1) \right)}{1 + a_v \, \mathrm{Ln}(\tau_2/\tau_1)} \,, \tag{8.5.10}$$

and

$$E_r^*(i\,\omega) = \frac{1 + \dfrac{a_v}{2} \mathrm{Ln}(\omega^2 \tau_2^2 + 1)/(\omega^2 \tau_1^2 + 1)}{1 + a_v \, \mathrm{Ln}(\tau_2/\tau_1)} + i \, a_v \, \frac{\tan^{-1}(\omega\tau_2) - \tan^{-1}(\omega\tau_1)}{1 + a_v \, \mathrm{Ln}(\tau_2/\tau_1)} \,. \tag{8.5.11}$$

An identification procedure for the viscoelastic parameters is proposed by Sarver et al. [2003]. The unconfined compression tests of articular cartilages in Huang et al. [2003] use different parameters depending whether the tissue is viewed as a biphasic-conewise linear elastic or as a biphasic poro-viscoelastic. These parameters are reported in Table 8.5.1.

Tissue	Model	Reference	a_v (-)	τ_1 (s)	τ_2 (s)
Articular cartilage	Biphasic-conewise elastic	Huang et al. [2003]	0.51 ± 0.23	0.77 ± 0.54	162 ± 90
Articular cartilage	Biphasic poro-viscoelastic	Huang et al. [2003]	5.98 ± 2.31	0.38 ± 0.39	448 ± 327
Cornea	Viscoelastic	Boyce et al. [2007]	0.2	1.0	$10\,000$

8.5.2 The three-dimensional reduced relaxation tensor

The three-dimensional format of the one-dimensional model defined by (8.5.2),

$$\boldsymbol{\sigma}(t) = \int_{-\infty}^{t} \mathbb{E}_r(t-u) : d\boldsymbol{\sigma}^{\mathrm{e}}(u) , \qquad (8.5.12)$$

is expressed via the fourth order *reduced* relaxation tensor $\mathbb{E}_r(t)$ and the 'elastic' stress $\boldsymbol{\sigma}^{\mathrm{e}}$. The elastic stress is obtained via a potential $W^{\mathrm{e}} = W^{\mathrm{e}}(\boldsymbol{\epsilon})$ as $\boldsymbol{\sigma}^{\mathrm{e}} = \partial W^{\mathrm{e}}/\partial \boldsymbol{\epsilon}$. It is viewed as embodying strain nonlinearity, and typically defined in terms of the strain invariants. Consistency implies that the *reduced* relaxation tensor satisfies the relation,

$$\mathbb{E}_r(0) = \text{identity tensor of order four} . \qquad (8.5.13)$$

In a relaxation experiment at a fixed strain $\boldsymbol{\epsilon}_0$, the equilibrium (asymptotic) stress is seen as the product of the reduced relaxation tensor times the elastic stress,

$$\boldsymbol{\sigma}(\infty) = \mathbb{E}_r(\infty) : \boldsymbol{\sigma}^{\mathrm{e}}(\boldsymbol{\epsilon}_0) . \qquad (8.5.14)$$

If the material enjoys isotropic viscoelastic properties, the reduced relaxation tensor is defined by two scalar dimensionless relaxation functions $\lambda_r(t)$ and $\mu_r(t)$,

$$
\begin{aligned}
\mathbb{E}_r(t) &= \lambda_r(t)\,\mathbf{I} \otimes \mathbf{I} + 2\,\mu_r(t)\,\mathbf{I}\,\overline{\otimes}\,\mathbf{I}, \\
\boldsymbol{\sigma}(t) &= \int_{-\infty}^{t} \lambda_r(t-u)\,\mathrm{tr}\,d\boldsymbol{\sigma}^{\mathrm{e}}(u)\,\mathbf{I} + 2\,\mu_r(t-u)\,d\boldsymbol{\sigma}^{\mathrm{e}}(u) .
\end{aligned}
\qquad (8.5.15)
$$

If the material is incompressible, the isotropic constitutive equations specify the effective stress

$$\boldsymbol{\sigma}(t) + p_i(t)\,\mathbf{I} = \int_{-\infty}^{t} 2\,\mu_r(t-u)\,d\boldsymbol{\sigma}^{\mathrm{e}}(u) , \qquad (8.5.16)$$

the incompressibility pressure p_i being defined by boundary conditions.

8.6 Overall viscoelastic properties: ECM and a contractile cell

In the parallel model of Fig. 8.6.1, the upper branch represents the extracellular matrix (M) and the lower branch the cells (C). For each branch individually, the strain is denoted by q and the stress by Q.

1. The matrix branch:

The governing differential equation of the upper branch,

$$\frac{dQ_{\mathrm{M}}}{dt} + \frac{Q_{\mathrm{M}}}{t_{\mathrm{M}}} = (E_{\mathrm{PM}} + E_{\mathrm{SM}})\frac{dq_{\mathrm{M}}}{dt} + E_{\mathrm{PM}}\frac{q_{\mathrm{M}}}{t_{\mathrm{M}}}, \qquad (8.6.1)$$

with $t_{\mathrm{M}} = \eta_{\mathrm{M}}/E_{\mathrm{SM}}$ the relaxation time at constant strain, can be recast in the integral form,

$$Q_{\mathrm{M}}(t) = \int_{-\infty}^{t} \left((E_{\mathrm{PM}} + E_{\mathrm{SM}}) - E_{\mathrm{SM}}\left(1 - e^{-(t-u)/t_{\mathrm{M}}}\right) \right) \frac{dq_{\mathrm{M}}(u)}{du}\,du . \qquad (8.6.2)$$

Upon comparison with the general integral representation (8.3.1) of the relaxation modulus E_{M}, we see that

$$E_{\mathrm{PM}} = E_{\mathrm{M}}(\infty), \quad E_{\mathrm{SM}} = E_{\mathrm{M}}(0) - E_{\mathrm{M}}(\infty), \quad \eta_{\mathrm{M}} = t_{\mathrm{M}}\left(E_{\mathrm{M}}(0) - E_{\mathrm{M}}(\infty)\right). \qquad (8.6.3)$$

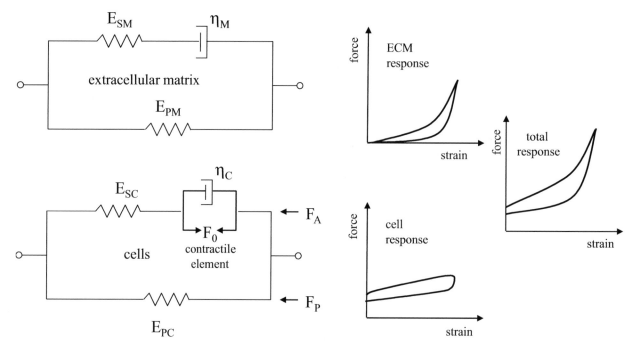

Fig. 8.6.1: The extracellular matrix and the cells are assumed to work in parallel in the model of Zahalak et al. [2000]. The cells contain a contractile element. Physiologically, the force due to cells can be activated by serum and abolished by drugs, e.g., cytochalasin D. Similarly papaverine, a calcium antagonist, abolishes the contractive activity of the gastro-intestinal tractus. These properties are used to measure the tissue (ECM+cells) and ECM contributions independently. Similarly, low concentration of cytochalasin D inhibits the *spreading* of cells, Wakatsuki et al. [2003].

The active branch displays a force that increases approximately linearly with the strain while the passive branch displays an exponential increase, Wakatsuki et al. [2000]. The relative contributions of the two parallel branches depend on the number of cells per unit volume and volume fraction of ECM. In a growing tissue, hyperplasia and ECM growth are unlikely to keep constant proportional contributions.

2. The cell branch:

The differential equation of the lower branch writes,

$$\frac{dQ_C}{dt} + \frac{Q_C}{t_C} = \frac{F_0}{t_C} + (E_{PC} + E_{SC})\frac{dq_C}{dt} + E_{PC}\frac{q_C}{t_C}, \tag{8.6.4}$$

again with $t_C = \eta_C/E_{SC}$ the relaxation time. The single difference with respect to the matrix branch comes from *the contractile isometric force* F_0 which depends on time through various physiological sources.

Now compare the integral representation,

$$Q_C(t) = \int_{-\infty}^{t} e^{-(t-u)/t_C} \frac{F_0(u)}{t_C} \, du + \int_{-\infty}^{t} \left[(E_{PC} + E_{SC}) - E_{SC}\left(1 - e^{-(t-u)/t_C}\right) \right] \frac{dq_C(u)}{du} \, du, \tag{8.6.5}$$

with a general integral representation of the relaxation modulus E_C. Then, similarly to (8.6.3),

$$E_{PC} = E_C(\infty), \ E_{SC} = E_C(0) - E_C(\infty), \ \eta_C = t_C\left(E_C(0) - E_C(\infty)\right). \tag{8.6.6}$$

8.7   An electrical analogy applied to the corneal endothelium

Consider the viscoelastic model displayed in Fig. 8.7.1(a). Let Q and q be respectively the stress and strain at the bounds of the model. The governing differential equation,

$$\left(\frac{t_a}{E_b} + \frac{t_b}{E_a}\right)\frac{dQ}{dt} + \left(\frac{1}{E_b} + \frac{1}{E_a}\right)Q =$$
$$= t_a\, t_b \frac{d^2 q}{dt^2} + \left(t_a + t_b + \left(\frac{t_a}{E_b} + \frac{t_b}{E_a}\right)E_S\right)\frac{dq}{dt} + \left(1 + \left(\frac{1}{E_a} + \frac{1}{E_b}\right)E_S\right)q, \tag{8.7.1}$$

(a) viscoelastic model

(b) electrical properties of corneal endothelium

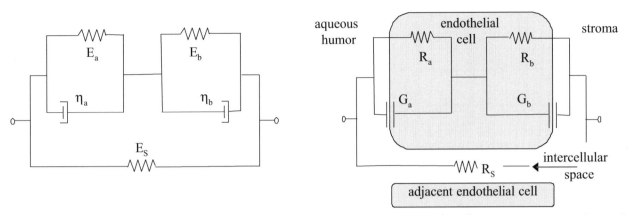

Fig. 8.7.1: (a) A viscoelastic model; (b) the analog used by Lim and Fischbarg [1981] to characterize the electrical properties of the corneal endothelium.

is established in Exercise 8.3. Here $t_k = \eta_k/E_k$, $k = a, b$, is a characteristic time.

The complex modulus results by assuming the strain q of the form (8.3.2),

$$E^*(i\omega) = E_S + E_a E_b \frac{(1 + i\omega t_a)(1 + i\omega t_b)}{E_a(1 + i\omega t_a) + E_b(1 + i\omega t_b)}. \tag{8.7.2}$$

The viscoelastic model can be associated with an electrical analog through the following correspondences[8.4],

$$
\begin{array}{rcl}
\text{strain } q & \rightleftharpoons & \text{electrical potential } \phi \\
\text{stress } Q & \rightleftharpoons & \text{electrical current } I \\
\text{viscosity } \eta & \rightleftharpoons & \text{areal capacitance } G \\
\text{elastic modulus } E & \rightleftharpoons & \text{electrical conductance } \sigma
\end{array} \tag{8.7.3}
$$

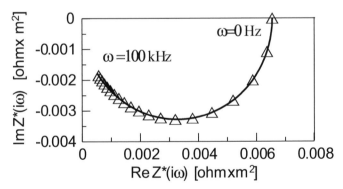

Fig. 8.7.2: Locus in the complex plane of the impedance of the electrical circuit shown in Fig. 8.7.1.-(b), with the following physical constants, Lim and Fischbarg [1981]: resistances $R_a = 0.158\,\text{Ohm}\times\text{m}^2$, $R_b = 0.153\,\text{Ohm}\times\text{m}^2$, $R_S = 0.0067\,\text{Ohm}\times\text{m}^2$, capacitances $G_a = G_b = 0.0103\,\text{F/m}^2$, and associated relaxation times $t_a \sim t_b = 1.6\,\text{ms}$.

As an alternative to the last correspondence, one may use

$$\text{elastic compliance } 1/E \quad \rightleftharpoons \quad \text{electrical resistance } R = 1/\sigma \tag{8.7.4}$$

[8.4]Note that another analog is also used based on the correspondences: strain rate $\dot{q} \rightleftharpoons$ electric current I, stress $Q \rightleftharpoons$ electric potential ϕ, viscosity $\eta \rightleftharpoons$ electric resistance R, compliance $1/E \rightleftharpoons$ capacitance G.

In this context, the impedance $Z^* = Z^*(i\,\omega) \equiv \hat{\phi}(t)/\hat{I}(t)$ is defined as the ratio of the electrical potential and current assumed in harmonic form, namely $\hat{I}(t) = I_0 \, \exp(i\,\omega t)$,

$$\frac{1}{Z^*(i\,\omega)} = \frac{1}{R_S} + \frac{1}{Z_1(i\,\omega)}, \quad Z_1(i\,\omega) \equiv \frac{R_a}{1 + i\,\omega\,t_a} + \frac{R_b}{1 + i\,\omega\,t_b}. \tag{8.7.5}$$

The characteristic times express now as $t_k = G_k\,R_k = G_k/\sigma_k$, $k = a, b$. The electrical resistances R_a and R_b and capacitances G_a and G_b are shown on the electrical analog of Fig. 8.7.1. The locus of this impedance in the complex plane with material data associated with the corneal endothelium is displayed in Fig. 8.7.2.

8.8   Fluid infusion in a viscoelastic polymer gel

A localized fluid infusion in a viscoelastic polymer gel is considered in Netti et al. [2003]. The model aims at describing the drug delivery process in a dynamically changing mechanical environment as the drug transport affects, and is affected by, the mechanical state. In fact, while polymer gels are engineered to deliver drugs according to a specific time course, the mechanical state *in vivo* has to be accounted for. An application could be intratumoral drug injection, which, unlike systemic drug delivery, has not to overcome the high interstitial fluid pressure barrier. Practical issues for such a goal are that a fast infusion may lead to tissue fracture (by tensile hoop stress) around the injection point, which may compromise the homogeneous delivery of the drug in the tumor. On the other hand, cyclic mechanical compression of a polymer gel with bound macromolecules (VEGF, etc.) results in larger release than static compression, K.Y. Lee et al. [2000].

A viscoelastic analysis addresses the relaxation properties of the polymer chains. The memory function is chosen according to Fung's formalism in which the imaginary part of the relaxation modulus shows a weak dependence in frequency, Fung [1993], p. 284. The two characteristic times are the polymer relaxation time t_v and the percolation time t_H defined by (8.8.40). While Péclet number Eqn (15.5.3) characterizes the relative effects of convection and diffusion, the Deborah number is defined as the ratio of a characteristic relaxation time over the seepage (fluid diffusion) time,

$$D_{\mathrm{eb}} = \frac{t_v}{t_\mathrm{H}}. \tag{8.8.1}$$

The modulus involved in the percolation time is affected by the viscoelastic properties of the tissue. With the spectrum of relaxation centered at τ and of half-width $\Delta\tau$ and the memory function m defined in Section 8.4.3, the Laplace-Carson transform of the viscoelastic modulus (with variable p),

$$\lambda + 2\,\mu + (\lambda_{ve} + 2\,\mu_{ve})\,p\,\overline{m}(p) = \lambda + 2\,\mu + (\lambda_{ve} + 2\,\mu_{ve})\,\frac{\tau}{\Delta\tau}\,\mathrm{Ln}\,\frac{p(\tau + \Delta\tau) + 1}{p(\tau - \Delta\tau) + 1}, \tag{8.8.2}$$

simplifies to $\lambda + 2\,\mu$ for poroelasticity ($\tau = 0$), and to $\lambda + 2\,\mu + 2\,(\lambda_{ve} + 2\,\mu_{ve})$ for very large τ.

The characteristic time t_v is expected to be a multiple of the relaxation time τ. From the numerical results obtained in the following boundary value problem, rate effects are observed to influence the constitutive response up to $t_v \simeq 50\,\tau$.

However, for some loadings that the poroelastic gel may have difficulties to accommodate, these effects remain significant as shown in Section 8.9 until $t = t_\mathrm{H}$, partly due to large values of the viscoelastic moduli λ_{ve} and μ_{ve} relative to the elastic Lamé moduli.

The viscoelastic constitutive equations, field equations, and boundary conditions defining the infusion problem are described in the next sections. The solutions are calculated analytically in the Laplace domain, and next transformed back to the time domain either analytically or using numerical inversion, e.g., Durbin [1974], Hoog et al. [1982], Honig and Hirdes [1984], Hollenbeck [1998]. The influence of the tissue stiffness on fluid flow and pressure profiles is seen only in the dynamic analysis, the steady state being identical to a rigid tissue. Of course, the stresses differ in the transient period and at steady state.

8.8.1   Poro-viscoelastic field and constitutive equations

Netti et al. [2000] [2003] considered the infusion of a polymer gel by a point source. The gel is viewed as a viscoelastic porous medium. The *effective stress* $\boldsymbol{\sigma} + p_i\,\mathbf{I}$, where p_i is the interstitial, or tissue, pressure, obeys the quasi-linear viscoelastic Fung's format. The following field and constitutive equations govern the process,

Balance of mass $\qquad$ div $(n^{\mathrm{s}}\,\mathbf{v}_{\mathrm{s}} + n^{\mathrm{w}}\,\mathbf{v}_{\mathrm{w}}) = 0,$

Balance of momentum $\quad$ div $\boldsymbol{\sigma} = \mathbf{0}\,,$

Darcy's law $\qquad n^{\mathrm{w}}\,(\mathbf{v}_{\mathrm{w}} - \mathbf{v}_{\mathrm{s}}) = -k_{\mathrm{H}}\,\boldsymbol{\nabla} p_i,$ $\qquad$ (8.8.3)

Viscoelastic equation $\quad \boldsymbol{\sigma} + p_i\,\mathbf{I} = \lambda\,e\,\mathbf{I} + 2\,\mu\,\boldsymbol{\epsilon} + \displaystyle\int_{-\infty}^{t}\,m(t-u)\,\frac{\partial}{\partial u}(\lambda_{ve}\,e\,\mathbf{I} + 2\,\mu_{ve}\,\boldsymbol{\epsilon})(u)\,du\,,$

where $e \equiv \operatorname{tr}\boldsymbol{\epsilon}$ is the dilatation. The $\mathbf{v}_{\alpha}$'s are the velocities of the solid ($\alpha = \mathrm{s}$) and fluid ($\alpha = \mathrm{w}$) phases of the mixture. The volume fractions n^{s} and n^{w} sum up to 1, $n^{\mathrm{s}} + n^{\mathrm{w}} = 1$, The species are all incompressible.

The poro-viscoelastic model is isotropic in both its transport and mechanical properties. The poroelastic response is defined by the tissue hydraulic permeability k_{H} [unit : $\mathrm{m}^2/\mathrm{s}/\mathrm{Pa}$] and the Lamé moduli of the drained tissue λ and μ [unit : Pa]. The viscoelastic response is defined by the constants λ_{ve} and μ_{ve} and by the memory function $m(t)$, e.g., that given by Eqn (8.4.16), that does not need to be specified for the analytical derivations.

The solution of the problem in terms of displacement and pressure can be obtained by complementing the field and constitutive equations (8.8.3) by initial and boundary conditions as described below.

8.8.2 The initial and boundary value problem

A polymer gel is injected by a point source at constant pressure or constant flow, Fig. 8.8.1(a). The problem displays spherical symmetry with radial motions and flows about a fixed center. In practice, the source has "radius" a and the gel has radius $R \gg a$.

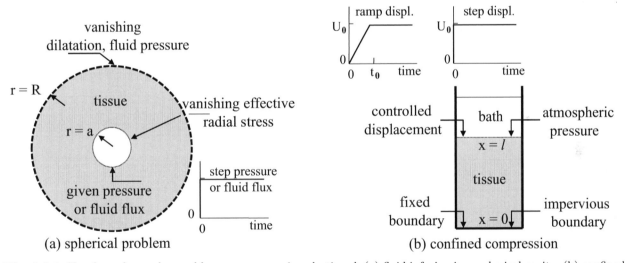

Fig. 8.8.1: Two boundary value problems on a poro-viscoelastic gel: (a) fluid infusion in a spherical cavity; (b) confined compression with controlled displacement.

As the ratio a/R is small, the pressure p_i and dilation e at the outer boundary remain equal to 0,

$$p_i(r = R, t) = 0; \quad e(r = R, t) = 0, \quad t \geq 0\,. \qquad (8.8.4)$$

The displacements and strains are reckoned from time $t = 0$, namely, for $t \leq 0$,

$$\mathbf{u}_{\alpha}(r, t) = \mathbf{0}, \quad \alpha = \mathrm{s}, \mathrm{w}; \quad e(r, t) = 0, \quad \forall r \in [a, R]\,. \qquad (8.8.5)$$

As for the infusion source, either a constant pressure

$$p_i(r = a, t) = PI\,\mathcal{H}(t)\,, \qquad (8.8.6)$$

where $\mathcal{H}(t)$ is Heaviside step function, or a constant flow rate Q are imposed. In the latter case, the radial fluid velocity at $r = a$ is constant in time,

$$v_{\mathrm{wr}}(r = a, t) = \frac{Q}{4\pi a^2}\,\mathcal{H}(t)\,. \tag{8.8.7}$$

The infusion surface is not loaded mechanically, so that the effective radial stress vanishes,

$$\sigma_r(r = a, t) + p_i(r = a, t) = 0\,. \tag{8.8.8}$$

8.8.3 The differential equation for the dilatation, or pressure

For irrotational motions, the field equations (11.2.42) simplify. Combining Eqns $(8.8.3)_{2,4}$ yields

$$\boldsymbol{\nabla}p_i = (\lambda + 2\,\mu)\,\boldsymbol{\nabla}e + (\lambda_{ve} + 2\,\mu_{ve})\int_{-\infty}^{t} m(t - u)\,d\,\boldsymbol{\nabla}e(u)\,. \tag{8.8.9}$$

Insertion of Eqn $(8.8.3)_1$ into Eqn $(8.8.3)_3$ provides Darcy's law in terms of the dilation rate, namely

$$\mathrm{div}\,\mathbf{v}_{\mathrm{s}} = \frac{de}{dt} = k_{\mathrm{H}}\,\mathrm{div}\,\boldsymbol{\nabla}p_i\,. \tag{8.8.10}$$

Elimination of the pressure in (8.8.9) yields a partial differential equation for the dilation e,

$$\frac{de}{dt}(r, t) = k_{\mathrm{H}}\,(\lambda + 2\,\mu)\,\mathrm{div}\,\boldsymbol{\nabla}e(r, t) + k_{\mathrm{H}}\,(\lambda_{ve} + 2\,\mu_{ve})\int_{-\infty}^{t} m(t - u)\,\frac{\partial}{\partial u}(\mathrm{div}\,\boldsymbol{\nabla}e(r, u))\,du\,. \tag{8.8.11}$$

The Laplace transform $e(r, t) \to \overline{e}(r, p)$ of this equation,

$$p\,\overline{e}(r, p) - e(r, 0) = k_{\mathrm{H}}\,(\lambda + 2\,\mu)\,\mathrm{div}\,\boldsymbol{\nabla}\overline{e}(r, p) + k_{\mathrm{H}}\,(\lambda_{ve} + 2\,\mu_{ve})\,\overline{m}(p)\,\mathrm{div}\,\boldsymbol{\nabla}\big(p\,\overline{e}(r, p) - \overline{e}(r, 0)\big)\,, \tag{8.8.12}$$

simplifies with help of (8.8.5) to,

$$\beta^2\,\overline{e}(r, p) = R^2\,\mathrm{div}\,\boldsymbol{\nabla}\overline{e}(r, p)\,, \tag{8.8.13}$$

with,

$$\beta^2 = \frac{p}{\gamma^2}\,, \quad \gamma^2 = \frac{k_{\mathrm{H}}}{R^2}\,\overline{H}(p)\ [\text{unit} : \mathrm{s}^{-1}]\,, \quad \overline{H}(p) = \lambda + 2\,\mu + (\lambda_{ve} + 2\,\mu_{ve})\,p\,\overline{m}(p)\,. \tag{8.8.14}$$

The relation (8.8.20) below shows that the pressure satisfies the differential equation (8.8.13) as well.

Table 8.8.1: Parameters used in the simulations, extracted from Netti et al. [2003].

Entity	Value or Range	Unit
Seepage properties		
Fluid volume fraction in gel n^{w}	0.2	-
Hydraulic permeability k_{H}	$7.6 \times 10^{-13} - 7.6 \times 10^{-15}$	$\mathrm{m}^2/\mathrm{Pa/s}$
Viscoelastic properties		
Drained elastic moduli $\lambda = 2\,\mu$	13×10^3	Pa
Drained viscoelastic moduli $\lambda_{ve} = 2\,\mu_{ve}$	$0 - 52 \times 10^3$	Pa
Relaxation time τ	300	s
Half-width $\Delta\tau$	$\tau/10$	s
Geometry		
Inner diameter a	0.35×10^{-3}	m
Outer diameter R	1×10^{-2}	m
Loading		
Infusion pressure PI	$0.65 - 2.6 \times 10^3$	Pa
Infusion flux Q	$0.1 - 10 \times 10^{-9}/60$	m^3/s

8.8.4 Radial displacement in a spherical setting

A scaled spatial coordinate is introduced as $\hat{r} = r/R$. The relations simplify by introduction of the entity $E(\hat{r}, t) = \hat{r}\, e(\hat{r}, t)$. Eqn (8.8.13), which takes the form,

$$\beta^2 \overline{E} - \frac{d^2}{d\hat{r}^2}\overline{E} = 0\,, \tag{8.8.15}$$

has solution

$$\overline{E}(\hat{r}, p) \equiv \hat{r}\,\overline{e}(\hat{r}, p) = A(p)\,\sinh \beta\hat{r} + B(p)\,\cosh \beta\hat{r}\,, \tag{8.8.16}$$

where $A(p)$ and $B(p)$ are fixed by the boundary conditions, e.g., $(8.8.4)_2$ for the outer boundary,

$$\overline{E}(1, p) = \overline{e}(1, p) = A(p)\,\sinh \beta + B(p)\,\cosh \beta = 0\,, \tag{8.8.17}$$

and therefore,

$$\overline{E}(\hat{r}, p) = \hat{r}\,\overline{e}(\hat{r}, p) = \frac{A(p)}{\cosh \beta}\,\sinh\left(\beta\,(\hat{r} - 1)\right)\,. \tag{8.8.18}$$

With the outer boundary conditions (8.8.4), the differential equation (8.8.9) can be integrated in space,

$$p_i = (\lambda + 2\,\mu)\,e + (\lambda_{ve} + 2\,\mu_{ve}) \int_{-\infty}^{t} m(t - u)\,\frac{\partial}{\partial u}e(u)\,du\,, \tag{8.8.19}$$

and transformed to Laplace domain,

$$\overline{p}_i(\hat{r}, p) = \overline{H}(p)\,\overline{e}(\hat{r}, p)\,. \tag{8.8.20}$$

Over the infusion surface $r = a$, the effective radial stress $\sigma_r + p_i$ is vanishing, thus providing a link between the dilatation e and the radial and circumferential strains, $\epsilon_r = du_{sr}/dr$ and $\epsilon_\theta = \epsilon_\phi = u_{sr}/r$, respectively,

$$\frac{d\overline{u}_{sr}}{dr}(\hat{a}, p) = -\frac{\overline{\lambda}(p)}{2\,\overline{\mu}(p)}\,\overline{e}(\hat{a}, p), \qquad \frac{\overline{u}_{sr}}{a}(\hat{a}, p) = \frac{1}{2}\,\frac{\overline{H}(p)}{2\,\overline{\mu}(p)}\,\overline{e}(\hat{a}, p)\,, \tag{8.8.21}$$

with $\overline{\lambda}(p)$ and $2\,\overline{\mu}(p)$ the viscoelastic moduli in the Laplace domain,

$$\overline{\lambda}(p) = \lambda + \lambda_{ve}\,p\,\overline{m}(p), \qquad 2\,\overline{\mu}(p) = 2\,\mu + 2\,\mu_{ve}\,p\,\overline{m}(p)\,. \tag{8.8.22}$$

Two types of conditions may be considered on the inner boundary.

8.8.4.1 Given infusion pressure

With help of (8.8.20), the boundary condition (8.8.6) takes the following form in the Laplace domain,

$$\frac{PI}{p} = \overline{H}(p)\,\overline{e}(\hat{a}, p)\,. \tag{8.8.23}$$

Thus using (8.8.17), Eqn (8.8.16) becomes,

$$\overline{E}(\hat{a}, p) = \hat{a}\,\overline{e}(\hat{a}, p) = \frac{PI}{\overline{H}(p)}\,\frac{\hat{a}}{p} = \frac{A(p)}{\cosh \beta}\,\sinh\left(\beta\,(\hat{a} - 1)\right)\,, \tag{8.8.24}$$

which allows to fix $A(p)$, and finally,

$$\overline{e}(\hat{r}, p) = \frac{X(p)}{\hat{r}}\,\sinh\left(\beta\,(1 - \hat{r})\right)\,, \tag{8.8.25}$$

with

$$X(p) = \frac{PI}{\overline{H}(p)}\,\frac{\hat{a}}{p}\,\frac{1}{\sinh\left(\beta(1 - \hat{a})\right)}\,. \tag{8.8.26}$$

Given $\overline{e}(\hat{r}, p)$, the Laplace transforms at arbitrary $(\hat{r}, p)$ of the following entities can be obtained in turn.

First, the fluid pressure is defined by (8.8.20). Next, the Laplace transform of the radial solid displacement u_{sr} and of its radial derivative are known at $r = a$ by (8.8.21). The radial solid displacement $\overline{u}_{sr}$ then follows by integration of $\overline{e} = r^{-2}d(r^2\overline{u}_{sr})/dr$, with the condition $(8.8.21)_2$ at $\hat{r} = \hat{a}$,

$$\frac{\overline{u}_{sr}}{R}(\hat{r},p) = \frac{\hat{a}^2}{\hat{r}^2}\frac{\overline{u}_{sr}}{R}(\hat{a},p) - \frac{1}{\hat{r}}\frac{X(p)}{\beta}\left(\eta(\hat{r},p) - \frac{\hat{a}}{\hat{r}}\eta(\hat{a},p)\right). \tag{8.8.27}$$

The fluid velocity results from Darcy's law and $(8.8.5)_1$,

$$\frac{\overline{v}_{wr}}{R}(\hat{r},p) = \frac{\hat{a}^2}{\hat{r}^2}p\frac{\overline{u}_{sr}}{R}(\hat{a},p) + \frac{p}{\hat{r}}\frac{X(p)}{\beta}\left(\frac{n^s}{n^w}\eta(\hat{r},p) + \frac{\hat{a}}{\hat{r}}\eta(\hat{a},p)\right). \tag{8.8.28}$$

In the above relations, the function $\eta(\hat{\rho},p)$ is defined as

$$\eta(\hat{\rho},p) \equiv -\frac{\hat{\rho}}{\beta}\frac{d}{d\hat{\rho}}\left(\frac{1}{\hat{\rho}}\left(\sinh\left(\beta\left(1-\hat{\rho}\right)\right)\right)\right) = \cosh\left(\beta\left(1-\hat{\rho}\right)\right) + \frac{1}{\beta\hat{\rho}}\sinh\left(\beta\left(1-\hat{\rho}\right)\right). \tag{8.8.29}$$

Note that[8.5]

$$\eta(\hat{\rho},p \sim 0) = \frac{1}{\hat{\rho}} + \frac{\beta^2}{6\hat{\rho}}\left(1-\hat{\rho}\right)^2\left(2\hat{\rho}+1\right), \tag{8.8.30}$$

and it can be verified that $\beta^2 = 0$ does not imply a branch point.

Laplace transforms of the strain and stress components follow readily, e.g.,
- the radial strain $\overline{\epsilon}_r(\hat{r},p) = d\overline{u}_{sr}(\hat{r},p)/(R\,d\hat{r})$,
- the circumferential strain $\overline{\epsilon}_\theta(\hat{r},p) = \overline{u}_{sr}(\hat{r},p)/(R\,\hat{r})$,
- the total radial stress $\overline{\sigma}_r(\hat{r},p) = -2\,\overline{\mu}(p) \times 2\,\overline{\epsilon}_\theta(\hat{r},p)$, and
- the total circumferential stress $\overline{\sigma}_\theta(\hat{r},p) = -2\,\overline{\mu}(p)\left(\overline{\epsilon}_r(\hat{r},p) + \overline{\epsilon}_\theta(\hat{r},p)\right)$.

8.8.4.2   Given infusion flux

Inserting the boundary condition (8.8.7) and $(8.8.21)_2$ in Darcy's law, accounting for (8.8.20) and $u_{sr}(\hat{r},0) = 0$, yields an equation for $\overline{e}(\hat{a},p)$,

$$\frac{d\overline{e}}{d\hat{r}}(\hat{a},p) - n^w\,\hat{a}\,\frac{p}{2\,\gamma_2^2}\,\overline{e}(\hat{a},p) + n^w\,\frac{Q}{4\,\pi\,a^2}\,\frac{R}{p}\,\frac{1}{k_H\,\overline{H}(p)} = 0, \tag{8.8.31}$$

that in fact provides $X(p)$. The Laplace transform of the dilatation still assumes the form (8.8.25), with now

$$X(p) = n^w\,\frac{Q}{4\,\pi\,a}\,\frac{1}{\beta\,p}\,\frac{1}{k_H\,\overline{H}(p)}\,\frac{1}{\cosh\left(\beta(1-\hat{a})\right) + \dfrac{x}{\beta\,\hat{a}}\sinh\left(\beta(1-\hat{a})\right)}, \tag{8.8.32}$$

with $x = 1 + \hat{a}^2\,n^w\,p/(2\,\gamma_2^2)$ and $\gamma_2^2 = 2\,\overline{\mu}(p)\,k_H/R^2$. The Laplace transforms of the entities of interest derive as for the case of given infusion pressure.

8.8.5   The solution in time domain

If the relaxation time τ is very small or very large and if the memory function satisfies the associated relation in (8.4.18), then the characteristic time that governs the transient process can be extracted from (8.8.15) as if the mixture was elastic, namely $l^2/\overline{H}(0)\,k_H$, with $\overline{H}(0)$ equal to $\lambda + 2\,\mu$ and $l = R - a$ the drainage length. For finite non vanishing relaxation time τ, the details of the relaxation function affects the transient process.

For given infusion flux, the complete solution in time is best obtained through numerical inversion of the Laplace transforms. For given infusion pressure, the analytical inversion is manageable.

[8.5]$\cosh x = 1 + x^2/2! + x^4/4! + \cdots$, $\sinh x = x + x^3/3! + \cdots$

8.8.5.1 Steady state

The steady state at large time is directly obtained by the final value theorem, e.g., for the dilatation e,

$$\lim_{t \to \infty} e(\hat{r}, t) = \lim_{p \to 0} p\,\bar{e}(\hat{r}, p)\,, \tag{8.8.33}$$

provided all poles of $p\,\bar{e}(\hat{r}, p)$ are in the left half-plane. The results below are independent of the relaxation function provided the latter satisfies the limit condition $\lim_{p \to 0} p\,\overline{m}(p) = 0$. The relaxation function of Section 8.4.3 satisfies this condition, namely $p\,\overline{m}(p) \sim 2\,\tau\,p$ for small p.

The results hold for both the pressure controlled infusion and the flux controlled infusion and indicate that the spatial profiles are identical in the two cases,

$$e(\hat{r}, \infty) = \frac{p_i(\hat{r}, \infty)}{\lambda + 2\,\mu} = \frac{Y}{\lambda + 2\,\mu} \frac{1 - \hat{r}}{\hat{r}} \geq 0\,; \tag{8.8.34}$$

$$\frac{u_{sr}}{R}(\hat{r}, \infty) = \frac{Y}{4\,\mu}\,(1 - \hat{a})\,\frac{\hat{a}^2}{\hat{r}^2} - \frac{Y}{\lambda + 2\,\mu}\,\frac{1}{6\,\hat{r}^2}\left[2\,(\hat{r}^3 - \hat{a}^3) - 3\,(\hat{r}^2 - \hat{a}^2)\right]\,; \tag{8.8.35}$$

$$v_{wr}(\hat{r}, \infty) = \frac{k_{\mathrm{H}}}{n^{\mathrm{w}}}\,\frac{Y}{R\,\hat{r}^2} \geq 0\,, \tag{8.8.36}$$

with

$$Y = \begin{cases} \dfrac{\hat{a}}{1 - \hat{a}}\,PI & \text{pressure controlled infusion,} \\[3mm] \dfrac{n^{\mathrm{w}}}{k_{\mathrm{H}}}\,\dfrac{Q}{4\,\pi\,R} & \text{flux controlled infusion}\,. \end{cases} \tag{8.8.37}$$

Note that the (visco)elastic behavior affects the steady state of the dilatation and solid displacement, but not the pressure nor fluid flow.

From the ensuing steady state values of the strains,

$$\begin{aligned} \epsilon_r(\hat{r}, \infty) &= -\frac{\hat{a}^2}{\hat{r}^3}\,(1 - \hat{a})\,\frac{Y}{2\,\mu} + \frac{Y}{\lambda + 2\,\mu}\,\frac{1}{3\,\hat{r}^3}\left[2\,(\hat{r}^3 - \hat{a}^3) - 3\,(\hat{r}^2 - \hat{a}^2)\right] + \frac{Y}{\lambda + 2\,\mu}\,\frac{1 - \hat{r}}{\hat{r}}\,; \\[2mm] \epsilon_\theta(\hat{r}, \infty) &= \epsilon_\phi(\hat{r}, \infty) = \frac{\hat{a}^2}{\hat{r}^3}\,(1 - \hat{a})\,\frac{Y}{4\,\mu} - \frac{Y}{\lambda + 2\,\mu}\,\frac{1}{6\,\hat{r}^3}\left[2\,(\hat{r}^3 - \hat{a}^3) - 3\,(\hat{r}^2 - \hat{a}^2)\right]\,, \end{aligned} \tag{8.8.38}$$

follow in turn the effective and total stresses, e.g., for the latter,

$$\sigma_r(\hat{r}, \infty) = -4\,\mu\,\epsilon_\theta(\hat{r}, \infty)\,; \quad \sigma_\theta(\hat{r}, \infty) = -2\,\mu\,\big(\epsilon_r(\hat{r}, \infty) + \epsilon_\theta(\hat{r}, \infty)\big)\,. \tag{8.8.39}$$

8.8.5.2 The transient solution for given infusion pressure

The solution is obtained by inversion of the Laplace transform in the complex plane, where the following entities are involved:

- Darcy's seepage time t_{H} associated with the purely elastic porous mixture of modulus $\overline{H}(0) = \lambda + 2\,\mu$,

$$t_{\mathrm{H}} = \frac{l^2}{\overline{H}(0)\,k_{\mathrm{H}}}\,, \tag{8.8.40}$$

with the drainage length l equal to $R - a = (1 - \hat{a})\,R$;
- the poles p_j, $j \geq 1$, that contribute to the inverse transform which are the real negative solutions of the equation [8.6],

$$p_j = -\frac{X_j^2}{t_{\mathrm{H}}}\,, \quad X_j = j\,\pi\,\left(\frac{\overline{H}(p_j)}{\overline{H}(0)}\right)^{\frac{1}{2}}\,; \tag{8.8.41}$$

[8.6] For the record: 1. for β real, $\sinh(i\,\beta) = i\,\sin\beta$ and $\cosh(i\,\beta) = \cos\beta$; 2. $\sinh\beta = 0$, for $\beta = i\,j\,\pi$ and then $\cosh\beta = \cos(j\,\pi) = (-1)^j$; 3. $\cosh\beta = 0$, for $\beta = i\,\pi/2 + i\,j\,\pi$ and then $\sinh\beta = i\,\cos(j\,\pi) = i\,(-1)^j$.

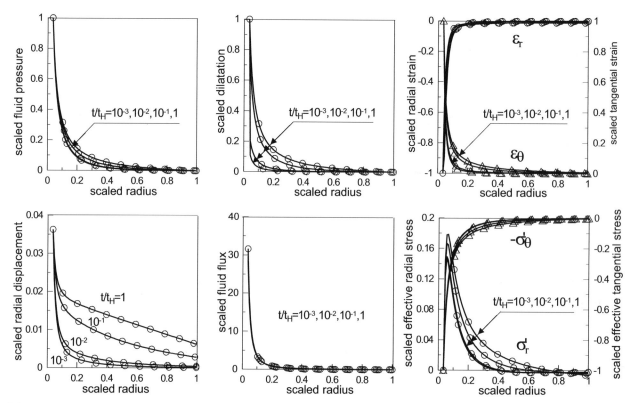

Fig. 8.8.2: Poro-viscoelastic hollow sphere subjected to a pressure jump on its inner radius as sketched in Fig. 8.8.1. Scaled spatial profiles at four instants. The scaling rules for displacement, velocity, deformation, pressure and stress are indicated in (8.8.43).

- terms peculiar to the viscous behavior, associated with the derivative $d\beta/dp = \Delta\beta/(2p)$, estimated at the poles p_j, namely when the relaxation function of Section 8.4.3 is used,

$$\Delta(p) = 1 - \frac{p}{\overline{H}(p)}\frac{d\overline{H}}{dp}(p), \quad \frac{d\overline{H}}{dp}(p) = \frac{2\tau(\lambda_{ve} + 2\mu_{ve})}{(p(\tau + \Delta\tau) + 1)(p(\tau - \Delta\tau) + 1)}. \tag{8.8.42}$$

The fluid pressure p_i, dilatation e, solid displacement u_{sr}, solid velocity v_{sr} and fluid velocity v_{wr} are contributed by a stationary value and by a sum of transient terms,

$$\frac{\overline{H}(0)}{PI}\begin{bmatrix} \dfrac{p_i(\hat{r},t)}{\overline{H}(0)} \\[2ex] e(\hat{r},t) \\[2ex] \dfrac{u_{sr}(\hat{r},t)}{l} \\[2ex] \dfrac{v_{sr}(\hat{r},t)}{l/t_{\mathrm{H}}} \\[2ex] \dfrac{v_{wr}(\hat{r},t)}{l/t_{\mathrm{H}}} \end{bmatrix} = \begin{bmatrix} \dfrac{\hat{a}}{\hat{r}}\dfrac{1-\hat{r}}{1-\hat{a}} \\[2ex] \dfrac{\hat{a}}{\hat{r}}\dfrac{1-\hat{r}}{1-\hat{a}} \\[2ex] \dfrac{u_{sr}(\hat{r},\infty)}{l\,PI/\overline{H}(0)} \\[2ex] 0 \\[2ex] \dfrac{1}{n^{\mathrm{w}}}\dfrac{\hat{a}}{\hat{r}^2} \end{bmatrix} + 2\frac{\hat{a}}{\hat{r}}\sum_{j=1}^{\infty}\frac{(-1)^j}{\Delta(p_j)}\begin{bmatrix} \sin\left(j\pi\dfrac{1-\hat{r}}{1-\hat{a}}\right)\dfrac{1}{j\pi} \\[2ex] \sin\left(j\pi\dfrac{1-\hat{r}}{1-\hat{a}}\right)\dfrac{1}{j\pi}\dfrac{\overline{H}(0)}{\overline{H}(p_j)} \\[2ex] \left(\xi_j(\hat{r}) - \dfrac{\hat{a}}{\hat{r}}(-1)^j\right)\dfrac{1}{(j\pi)^2}\dfrac{\overline{H}(0)}{\overline{H}(p_j)} \\[2ex] -\xi_j(\hat{r}) + \dfrac{\hat{a}}{\hat{r}}(-1)^j \\[2ex] \dfrac{n^{\mathrm{s}}}{n^{\mathrm{w}}}\xi_j(\hat{r}) + \dfrac{\hat{a}}{\hat{r}}(-1)^j \end{bmatrix}e^{-X_j^2\,t/t_{\mathrm{H}}}, \tag{8.8.43}$$

with

$$\xi_j(\hat{r}) = \cos\left(j\pi\frac{1-\hat{r}}{1-\hat{a}}\right) + \frac{1-\hat{a}}{j\pi\hat{r}}\sin\left(j\pi\frac{1-\hat{r}}{1-\hat{a}}\right). \tag{8.8.44}$$

Fig. 8.8.2 displays spatial profiles of the solution while Fig. 8.8.3 compares the time courses of the poroelastic and poro-viscoelastic gels.

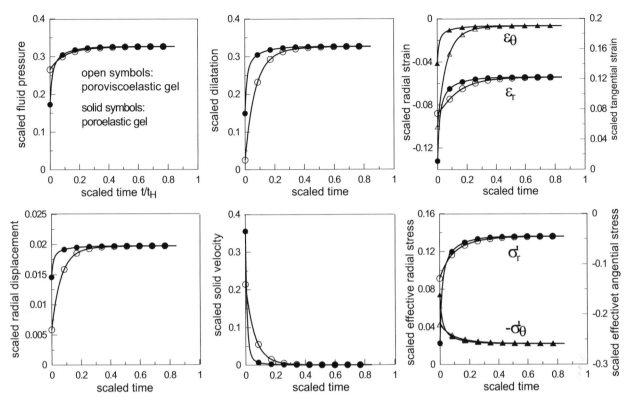

Fig. 8.8.3: Hollow sphere subjected to a pressure jump on its inner radius as sketched in Fig. 8.8.1. Scaled time profiles at radius $r = 2\,a$ for the poro-viscoelastic gel (open symbols) with parameters as indicated in Table 8.8.1 and for the poroelastic gel with $\lambda_{ve} = \mu_{ve} = 0$. Seepage time $t_{\mathrm{H}} = 50 \times 10^3$ s, relaxation time of the poro-viscoelastic gel $\tau = 300$ s so that $\tau/t_{\mathrm{H}} = 0.006$. The viscous effect shows up at early times but it affects the constitutive response up to time $t/t_{\mathrm{H}} = 0.3$, i.e. $50\,\tau$.

For the poroelastic gel, the fluid pressure p_i is equal to $\lambda + 2\,\mu$ times the dilatation e at any point of space and at any time.

The results are insensitive to the width of the relaxation interval defined by the time $\Delta\tau \in [0, \tau]$.

For $k_{\mathrm{H}} = 7.6 \times 10^{-14}\,\mathrm{m}^2/\mathrm{Pa/s}$ and a drainage length $l = 9.65 \times 10^{-3}$ m, the seepage times $t_{\mathrm{H}} = l^2/(H\,k_{\mathrm{H}})$ range as follows depending on the relaxation time τ:

- for a very small relaxation time τ corresponding to a poroelastic response, the confined compression modulus $H = \lambda + 2\,\mu$ is equal to 26 kPa, and t_{H} to about 50×10^3 s ~ 14 hours;
- for a very large relaxation time τ, the confined compression modulus $H = \lambda + 2\,\mu + 2\,(\lambda_{ve} + 2\,\mu_{ve})$ is equal to 130 kPa, and t_{H} to about 10×10^3 s ~ 2.8 hour.

Following a sudden loading, viscous effects are expected mainly at early times, up to some multiples of τ, actually up to $50\,\tau$ as shown in Fig. 8.8.3. Therefore, the Deborah number (8.8.1) should be defined as $50\,\tau/t_{\mathrm{H}}$.

8.9 Confined compression of a viscoelastic polymer gel

Netti et al. [2000] address the confined compression of the poro-viscoelastic medium defined in Section 8.8. The material is confined laterally as sketched in Fig. 8.8.1(b) so that the motion is one-dimensional. The solid displacement is prescribed at the boundaries $\hat{x} \equiv x/l = 0$ and $\hat{x} = 1$, namely

$$u_{\mathrm{s}}(\hat{x} = 0, t) = 0, \quad u_{\mathrm{s}}(\hat{x} = 1, t) = U(t), \quad t \geq 0. \tag{8.9.1}$$

The bottom of the column is impervious while the top is in chemical equilibrium with a bath at atmospheric pressure,

$$u_{\mathrm{w}}(\hat{x} = 0, t) = 0, \quad p_i(\hat{x} = 1, t) = 0, \quad t \geq 0. \tag{8.9.2}$$

The analysis takes advantage of the developments of Section 8.8.

8.9.1 The solution in the Laplace domain

The dilatation $e = du_{\mathrm{s}}/dx$ (not the displacement) satisfies the differential equation (8.8.13),

$$\beta^2 \, \overline{e}(\hat{x}, p) - \frac{d^2}{d\hat{x}^2} \overline{e}(\hat{x}, p) = 0 \,, \tag{8.9.3}$$

with $\beta^2 = p/\gamma^2$, $\gamma^2 = k_{\mathrm{H}} \, \overline{H}(p)/l^2$, so that its solution expresses as

$$\overline{e}(\hat{x}, p) = A(p) \sinh \beta \hat{x} + B(p) \cosh \beta \hat{x} \,, \tag{8.9.4}$$

where $A(p)$ and $B(p)$ are fixed by the boundary conditions. The Laplace transform of Eqn (8.8.9) yields a relation between pressure and dilatation similar to (8.8.20), but, this time, for the gradients,

$$\frac{d}{d\hat{x}} \overline{p}_i(\hat{x}, p) = \overline{H}(p) \, \frac{d}{d\hat{x}} \overline{e}(\hat{x}, p) \,. \tag{8.9.5}$$

Since the bottom $\hat{x} = 0$ is impervious, the gradient of pressure should vanish at this point according to Darcy's law. So does the dilatation gradient by (8.9.5), and therefore $A(p) = 0$. Since the operators of spatial differentiation and Laplace transform commute, one thus has $d\overline{u}_{\mathrm{s}}/d\hat{x} = l \, B(p) \cosh \beta \hat{x}$. Accounting for the conditions at the bottom and top boundaries, the Laplace transform of the solid displacement takes the form

$$\overline{u}_{\mathrm{s}}(\hat{x}, p) = \overline{U}(p) \, \frac{\sinh \beta \, \hat{x}}{\sinh \beta} \,. \tag{8.9.6}$$

All the boundary conditions of this coupled problem are now satisfied. The dilatation deduces from the displacement by simple differentiation, and the pressure by integration in space of (8.9.5),

$$\overline{e}(\hat{x}, p) = \beta \, \frac{\overline{U}(p)}{l} \, \frac{\cosh \beta \, \hat{x}}{\sinh \beta} \,, \quad \overline{p}_i(\hat{x}, p) = \overline{H}(p) \, \beta \, \frac{\overline{U}(p)}{l} \, \frac{\cosh \beta \, \hat{x} - \cosh \beta}{\sinh \beta} \,. \tag{8.9.7}$$

Also since the strain is uniaxial,

$$\overline{\sigma}'_x(\hat{x}, p) = \overline{\sigma}_x(\hat{x}, p) + \overline{p}_i(\hat{x}, p) = \overline{H}(p) \, \overline{e}(\hat{x}, p) \,. \tag{8.9.8}$$

8.9.2 The steady state solution

For ramp loading, the applied displacement and its Laplace transform read,

$$U(t) = \frac{U_0}{t_0} \left(t \, \mathcal{H}(t) - (t - t_0) \, \mathcal{H}(t - t_0) \right) \quad \Rightarrow \quad \overline{U}(p) = \frac{U_0}{t_0} \, \frac{1 - e^{-t_0 p}}{p^2} \,. \tag{8.9.9}$$

In the limit case of a step loading, namely $t_0 \to 0$, then $U(t) = U_0 \, \mathcal{H}(t)$ and $\overline{U}(p) = U_0/p$.

The steady state solution follows trivially from the final value theorem, given that $\lim_{p \to 0} p \, \overline{U}(p) = U_0$ for both loading cases:

$$\frac{\sigma'_x(\hat{x}, \infty)}{\overline{H}(0)} = \frac{\sigma_x(\hat{x}, \infty)}{\overline{H}(0)} = e(\hat{x}, \infty) = \frac{U_0}{l} \,; \quad p_i(\hat{x}, \infty) = 0 \,; \quad u_{\mathrm{s}}(\hat{x}, \infty) = U_0 \, \hat{x} \,; \quad v_{\mathrm{w}}(\hat{x}, \infty) = 0 \,. \tag{8.9.10}$$

8.9.3 The transient solution

For the two loading functions defined in Section 8.9.2, the scaled displacement, dilatation, fluid pressure, effective stress, stress and fluid velocity express as follows:

$$
\begin{bmatrix}
\dfrac{u_{\mathrm{s}}(\hat{x}, t)}{U_0} \\[2ex]
\dfrac{e(\hat{x}, t)}{U_0/l} \\[2ex]
\dfrac{p_i(\hat{x}, t)}{\overline{H}(0) \, U_0/l} \\[2ex]
\dfrac{\sigma'_x(\hat{x}, t)}{\overline{H}(0) \, U_0/l} \\[2ex]
\dfrac{\sigma_x(\hat{x}, t)}{\overline{H}(0) \, U_0/l} \\[2ex]
\dfrac{v_{\mathrm{w}}(\hat{x}, t)}{U_0/t_{\mathrm{H}}}
\end{bmatrix}
=
\begin{bmatrix}
\hat{x} \\[2ex]
1 \\[2ex]
0 \\[2ex]
1 \\[2ex]
1 \\[2ex]
0
\end{bmatrix}
+ 2 \sum_{j=1}^{\infty} \frac{(-1)^j}{\Delta(p_j)} \, A(p_j)
\begin{bmatrix}
\dfrac{\sin(j \, \pi \, \hat{x})}{j \, \pi} \\[2ex]
\cos(j \, \pi \, \hat{x}) \\[2ex]
\dfrac{\overline{H}(p_j)}{\overline{H}(0)} \left(\cos(j \, \pi \, \hat{x}) - (-1)^j \right) \\[2ex]
\dfrac{\overline{H}(p_j)}{\overline{H}(0)} \cos(j \, \pi \, \hat{x}) \\[2ex]
\dfrac{\overline{H}(p_j)}{\overline{H}(0)} (-1)^j \\[2ex]
\dfrac{n^{\mathrm{s}}}{n^{\mathrm{w}}} \dfrac{\overline{H}(p_j)}{\overline{H}(0)} \, j \, \pi \, \sin(j \, \pi \, \hat{x})
\end{bmatrix}
e^{-X_j^2 \, t/t_{\mathrm{H}}} \,, \tag{8.9.11}
$$

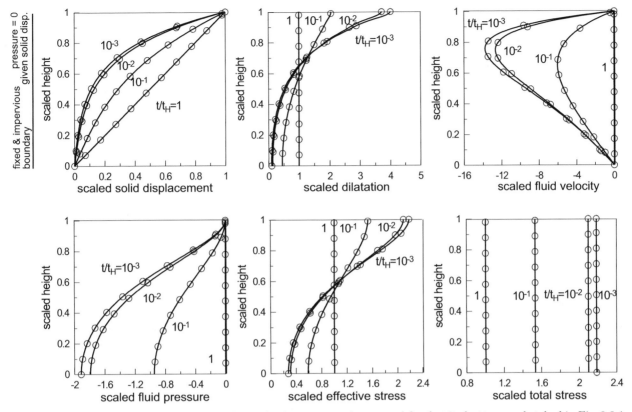

Fig. 8.9.1: Poro-viscoelastic layer subjected to a displacement at the top and fixed at its bottom as sketched in Fig. 8.8.1. Scaled quantities as indicated in (8.9.11) at different instants. The seepage time t_H appears as actually indicating the time to equilibrium. The very early time response does not differ much from the response at $t/t_H = 10^{-3}$.

where the definitions (8.8.40) to (8.8.42) still hold and

$$A(p) = p\,\frac{\overline{U}(p)}{U_0} = \begin{cases} 1, & \text{step load,} \\[2mm] \dfrac{1 - e^{-t_0 p}}{t_0\,p}, & \text{ramp load.} \end{cases} \qquad (8.9.12)$$

The plots shown in Fig. 8.9.1 aim at highlighting
  - the influence of the loading rate U_0/t_0 on a purely poroelastic medium;
  - the additional influence of the loading rate due to the intrinsic viscous response of the skeleton;
  - the influence of the amplitude of the viscous parameter $(\lambda_{ve} + 2\,\mu_{ve})/(\lambda + 2\,\mu)$;
  - the relative influence of the relaxation and Darcy's seepage times, namely τ/t_H.

8.10   Compression of a poroelastic layer: displacement control

The solution of the displacement controlled confined compression developed in Section 8.9, sketched in Fig. 8.8.1(b), applies as a limit case to a poroelastic layer where the modulus $H = \overline{H}(0) = \lambda + 2\,\mu$ is constant.

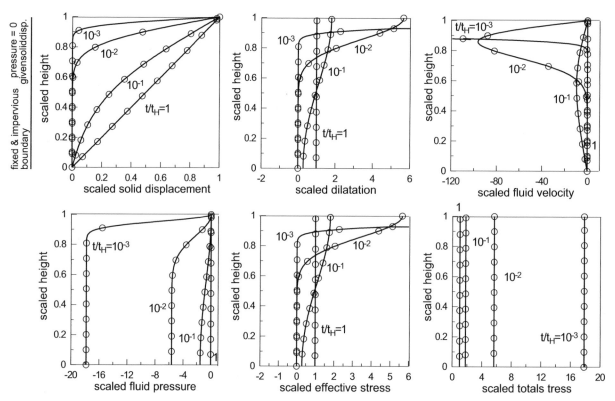

Fig. 8.10.1: Same as Fig. 8.9.1 but for a poroelastic layer. The mechanical response at early times to the instantaneous loading is significantly higher than for the poro-viscoelastic gel. At very early times, the fluid velocity in a skin close to the top surface and the fluid pressure on the complementary depth are large. Since the strain is one-dimensional, the scaled dilatation and the scaled effective stress have identical spatial profiles: this feature does not apply to the poro-viscoelastic gel.

8.10.1   The analytical solution

For both the ramp loading and step loading defined in Section 8.9.2, the scaled solution can be cast in the format,

$$
\begin{bmatrix}
\dfrac{u_{\mathrm{s}}(\hat{x},t)}{U_0} \\[2mm]
\dfrac{e(\hat{x},t)}{U_0/l} \\[2mm]
\dfrac{p_i(\hat{x},t)}{H\,U_0/l} \\[2mm]
\dfrac{\sigma'_x(\hat{x},t)}{\overline{\overline{H}}(0)\,U_0/l} \\[2mm]
\dfrac{\sigma_x(\hat{x},t)}{\overline{\overline{H}}(0)\,U_0/l} \\[2mm]
\dfrac{v_{\mathrm{w}}(\hat{x},t)}{U_0/t_{\mathrm{H}}}
\end{bmatrix}
=
\begin{bmatrix}
\hat{x} \\[2mm]
1 \\[2mm]
0 \\[2mm]
1 \\[2mm]
1 \\[2mm]
0
\end{bmatrix}
+ 2 \sum_{j=1}^{\infty} (-1)^j\, A(p_j)
\begin{bmatrix}
\dfrac{\sin(j\,\pi\,\hat{x})}{j\,\pi} \\[2mm]
\cos(j\,\pi\,\hat{x}) \\[2mm]
\cos(j\,\pi\,\hat{x}) - (-1)^j \\[2mm]
\cos(j\,\pi\,\hat{x}) \\[2mm]
(-1)^j \\[2mm]
\dfrac{n^{\mathrm{s}}}{n^{\mathrm{w}}}\, j\,\pi\,\sin(j\,\pi\,\hat{x})
\end{bmatrix}
e^{-X_j^2\, t/t_{\mathrm{H}}},
\tag{8.10.1}
$$

where

$$
t_{\mathrm{H}} = \frac{l^2}{H\,k_{\mathrm{H}}}, \quad p_j = -\frac{X_j^2}{t_{\mathrm{H}}}, \quad X_j = j\,\pi,
\tag{8.10.2}
$$

and $A(p)$ is defined by (8.9.12).

8.10.2 The elastic gel as a limit of the viscous gel with a zero relaxation time

Formally, the poroelastic response is obtained as the limit of the poro-viscoelastic response while the relaxation time τ tends to 0 based on (8.4.18). The spatial profiles are shown in Fig. 8.10.1. The actual rate of convergence towards the poroelastic response is quite slow and a good approximation of the poroelastic gel via the poro-viscoelastic gel requires tiny relaxation times. Fig. 8.10.2 shows the position of the first thousand poles involved in the inversion of the Laplace transform. The x- and y-axes are scaled so that the curve is a straight segment for the poroelastic gel. A visual inspection requires the actual relaxation time to be scaled down by a factor 10^{-7} for the inviscid limit to be reached.

8.10.3 Inversion of the Laplace transform: the minimum number of poles

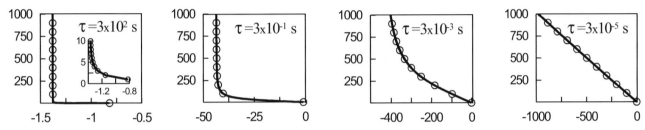

Fig. 8.10.2: The poles p_j, $j \geq 1$, involved in the inversion of the Laplace transform in the complex plane are real negative:

$$\mathrm{x-axis} \equiv \frac{\sqrt{-p_j\, t_{\mathrm{H}}}}{\pi} = j\,\sqrt{\frac{\overline{H}(p_j)}{\overline{H}(0)}}, \quad \mathrm{y-axis} = \text{pole nb } 1, 2 \cdots, 1000$$

Their actual positions depend on the poro-viscoelastic properties. For a non zero relaxation time τ, the poles accumulate at the lower bound. For a zero relaxation time, the lower bound does not exist. Since the range covered by the poles increases as the relaxation time decreases, the number of poles required for a smooth solution increases as well. The fact that the lower bound of the range of the poles increases towards negative values is not an issue since the contributions of these terms are damped by the exponential involved in the inverse Laplace transform.

The poles involved in the inversion of the Laplace transform p_j's are the negative real roots of the equation,

$$p_j = -\frac{(j\,\pi)^2}{t_{\mathrm{H}}}\,\frac{\overline{H}(p_j)}{\overline{H}(0)}, \quad \frac{\overline{H}(p_j)}{\overline{H}(0)} = 1 + \frac{\lambda_{ve} + 2\,\mu_{ve}}{\overline{H}(0)}\,\frac{\tau}{\Delta\tau}\,\mathrm{Ln}\,\frac{p_j(\tau + \Delta\tau) + 1}{p_j(\tau - \Delta\tau) + 1}. \tag{8.10.3}$$

The minimum number of poles required to obtain a smooth response should be considered. Fig. 8.10.2 shows that the poles of the poro-viscoelastic gel range over an interval whose size increases inversely with the relaxation time. The poles accumulate at the lower bound of this interval. As the relaxation time is decreased, the neighbor to neighbor distance homogenizes gradually while the interval becomes larger and actually turns unbounded at the poroelastic limit. The minimum number of poles necessary for a smooth response depends on the time at which the response is sought. For the results shown in Figs. 8.9.1 and 8.10.1, at least 20 poles should be used for the poro-viscoelastic gel and 50 poles for the poroelastic gel. The mechanical response at earlier time $t/t_{\mathrm{H}} = 10^{-4}$ requires more than 100 poles for the poroelastic gel: an insufficient number of poles is witnessed by spurious wavy curves for some quantity, typically the fluid velocity.

8.11 Visco-hyperelasticity and other memory effects

In this section, a derivative with respect to time t is denoted either by the symbol $d(\cdot)/dt$ or by a superimposed dot $(\dot{\cdot})$, when the former symbol is too cumbersome.

8.11.1 Linear viscous fluids

A thermodynamic presentation of viscous models embedding a growth strain is proposed in Sections 21.6.4.1 and 21.6.4.2. The short introduction below addresses a barotropic fluid, namely an isotropic compressible viscous fluid with a pressure $p = p(\rho, T)$ that may depend on density ρ and temperature T,

$$\boldsymbol{\sigma} + p\,\mathbf{I} = \boldsymbol{\sigma}_v = \mathbb{E}_v : \dot{\boldsymbol{\epsilon}} = \lambda_v \operatorname{tr} \dot{\boldsymbol{\epsilon}}\, \mathbf{I} + 2\,\mu_v\,\dot{\boldsymbol{\epsilon}}, \tag{8.11.1}$$

where $\dot{\boldsymbol{\epsilon}}$ is the strain rate and λ_v and μ_v [unit : Pa$\times$s] are the viscosity coefficients.

The free energy $e = e(\rho, T)$ is considered a function of mass density ρ and temperature T. Ensuring a positive mechanical dissipation $\hat{S}_{(\mathrm{i},1)}$,

$$\begin{aligned}
T\,\hat{S}_{(\mathrm{i},1)} &= -\rho\,\dot{e} + \boldsymbol{\sigma} : \dot{\boldsymbol{\epsilon}} - \rho\,s\,\dot{T} \\
&= \left(-p + \rho^2\,\frac{\partial e}{\partial \rho} \right) \operatorname{tr} \dot{\boldsymbol{\epsilon}} - \rho \left(\frac{\partial e}{\partial T} + s \right) \dot{T} + \boldsymbol{\sigma}_v : \dot{\boldsymbol{\epsilon}},
\end{aligned} \tag{8.11.2}$$

poses restrictions on the constitutive equation. Note that use has been made here of the balance of mass $\dot{\rho} = -\rho \operatorname{tr} \dot{\boldsymbol{\epsilon}}$. Since the free energy is independent of the strain rate and of the temperature rate, a standard argument yields the constitutive relations for the pressure p and entropy s,

$$p = \rho^2\,\frac{\partial e}{\partial \rho}, \quad s = -\frac{\partial e}{\partial T}, \tag{8.11.3}$$

and we are left with the reduced inequality to be satisfied,

$$\begin{aligned}
T\,\hat{S}_{(\mathrm{i},1)} &= \boldsymbol{\sigma}_v : \dot{\boldsymbol{\epsilon}} \\
&= \left(\lambda_v + \frac{2}{3}\,\mu_v \right) (\operatorname{tr} \dot{\boldsymbol{\epsilon}})^2 + 2\,\mu_v \operatorname{dev} \dot{\boldsymbol{\epsilon}} : \operatorname{dev} \dot{\boldsymbol{\epsilon}} \ge 0.
\end{aligned} \tag{8.11.4}$$

The constraint requires the viscosity coefficients to satisfy the inequalities,

$$\lambda_v + \frac{2}{3}\,\mu_v \ge 0, \quad \mu_v \ge 0. \tag{8.11.5}$$

These coefficients may depend on density and temperature. Actually, they are mostly independent of density although they may increase at very high pressures. As indicated in Section 7.3.3, the shear viscosity coefficient μ_v of liquids decreases as the temperature increases and the converse holds for gases.

8.11.2 Maxwell and Kelvin models in a dynamic context

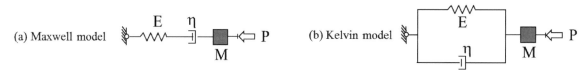

Fig. 8.11.1: Rheological representation of Maxwell and Kelvin models.

Consider a mass M attached to a Kelvin model with a fixed end and subjected to a load P at the other end as sketched in Fig. 8.11.1. Let q be the generalized coordinate describing the motion of the mass. Then the spring sustains a stress $Q_e = E\,q$ and the dashpot the stress $Q_v = \eta\,\dot{q}$, yielding the total stress

$$Q = \eta\,\dot{q} + E\,q. \tag{8.11.6}$$

Let us record the kinetic energy T, the elastic energy V and the dissipation potential $\mathcal{D} \ge 0$, namely

$$T(\dot{q}) = \frac{1}{2}\,M\,\dot{q}^2, \quad V(q) = \frac{1}{2}\,E\,q^2, \quad \mathcal{D}(\dot{q}) = \frac{1}{2}\,\eta\,\dot{q}^2 \ge 0, \tag{8.11.7}$$

so that the Lagrange equation, which then reduces to

$$\frac{d}{dt}\left(\frac{\partial T}{\partial \dot{q}} \right) - \frac{\partial T}{\partial q} + \frac{\partial V}{\partial q} = P + Q_v, \quad Q_v \equiv -\frac{\partial \mathcal{D}}{\partial \dot{q}}, \tag{8.11.8}$$

takes the form,

$$M \ddot{q} + \eta \dot{q} + E q = P \quad \Leftrightarrow \quad M \ddot{q} = P - Q . \tag{8.11.9}$$

Upon multiplying $M \ddot{q} = P - Q$ by $\dot{q}$ and replacing Q in the product $\dot{q} Q$ by (8.11.6) follows the general result that the rate of the sum of the kinetic and elastic energies is equal to the external power supply minus the power dissipated by the dashpot,

$$\frac{d}{dt}(T + V) = -2 \mathcal{D} + P \dot{q} . \tag{8.11.10}$$

For a Maxwell model, the strain on the spring is equal to $q_e = Q/E$ while the strain rate on the dashpot is equal to $\dot{q}_v = Q/\eta$, yielding the total strain rate,

$$\dot{q} = \frac{\dot{Q}}{E} + \frac{Q}{\eta} . \tag{8.11.11}$$

However, we have a problem here because the energy of the spring $E q_e^2/2$ depends on the elastic part of the generalized variable and we do not know how to take its partial differential. In fact, the same problem arises in all models where the constitutive equations involve an 'effective' coordinate, e.g., thermoelasticity, growth, etc.

The equation of motion is simply $M \ddot{q} = P - Q$. It can be integrated in time as well as the constitutive equation (8.11.11), yielding

$$\frac{\ddot{u}}{E} + \frac{\dot{u}}{\eta} + \frac{u}{M} = \dot{q}(0) + \frac{1}{M} \int_0^t P(\tau) \, d\tau \quad \text{with} \quad u(t) = \int_0^t Q(\tau) \, d\tau . \tag{8.11.12}$$

Following Sivaselvan and Reinhorn [2006], by analogy with (8.11.9), we redefine the energies in terms of the pair $(u, \dot{u})$ in place of the pair $(q, \dot{q})$ for the Kelvin model,

$$T(\dot{u}) = \frac{\dot{u}^2}{2E}, \quad V(u) = \frac{u^2}{2M}, \quad \mathcal{D}(\dot{u}) = \frac{\dot{u}^2}{2\eta} \geq 0 , \tag{8.11.13}$$

and use the Lagrange equation,

$$\frac{d}{dt}\left(\frac{\partial T}{\partial \dot{u}}\right) - \frac{\partial T}{\partial u} + \frac{\partial V}{\partial u} = \dot{q}(0) + \frac{1}{M} \int_0^t P(\tau) \, d\tau + U_v, \quad U_v \equiv -\frac{\partial \mathcal{D}}{\partial \dot{u}} . \tag{8.11.14}$$

Upon multiplying $M \ddot{q} = P - Q$ by $\dot{q}$ and replacing $\dot{q}$ in the product $\dot{q} Q$ by (8.11.11), the balance of energy takes now the form,

$$\frac{d}{dt}\left(\frac{1}{2} M \dot{q}^2 + \frac{Q^2}{2E}\right) = -\frac{Q^2}{\eta} + P \dot{q} = -2 \mathcal{D} + P \dot{q} . \tag{8.11.15}$$

8.11.3 Rate effects as non-equilibrium thermodynamically consistent processes

The thermodynamic potentials per unit volume shown in Table 9.1.1, namely the internal energy U, the free energy E, the enthalpy H and the free enthalpy (or Gibbs potential) G, deduce from one another through a partial Legendre transform. The dissipation inequality may be expressed in terms of either one,

$$
\begin{aligned}
T \hat{S}_{(i)} &= -\dot{U} + \boldsymbol{\sigma} : \dot{\boldsymbol{\epsilon}} + T \dot{S} + \mathbf{A} \cdot \dot{\mathbf{a}} \\
&= -\dot{E} + \boldsymbol{\sigma} : \dot{\boldsymbol{\epsilon}} - S \dot{T} + \mathbf{A} \cdot \dot{\mathbf{a}} \\
&= -\dot{H} - \boldsymbol{\epsilon} : \dot{\boldsymbol{\sigma}} + T \dot{S} + \mathbf{A} \cdot \dot{\mathbf{a}} \\
&= -\dot{G} - \boldsymbol{\epsilon} : \dot{\boldsymbol{\sigma}} - S \dot{T} + \mathbf{A} \cdot \dot{\mathbf{a}} \\
&= -\dot{F} - \boldsymbol{\epsilon} : \dot{\boldsymbol{\sigma}} - S \dot{T} - \mathbf{a} \cdot \dot{\mathbf{A}} .
\end{aligned}
\tag{8.11.16}
$$

Here $(\mathbf{a}, \mathbf{A})$ is an arbitrary formal conjugate pair, possibly associated with damage/healing, plasticity or some other dissipative processes.

With the generic potential $\phi(\mathbf{x})$ can be associated the generalized potential $\phi^*(\mathbf{x}, \mathbf{y})$,

$$\phi^*(\mathbf{x}, \mathbf{y}) = \phi(\mathbf{x}) - \mathbf{y} \cdot \mathbf{x} . \tag{8.11.17}$$

Here $\mathbf{x}$ denotes the set of state variables and $\mathbf{y}$ is the set of control variables. For example, the internal energy U and the potential F feature the following conjugate sets of variables,

$$\text{state variables}: \ \mathbf{x} = (\epsilon, S, \mathbf{a}); \quad \text{control variables}: \ \mathbf{y} = (\boldsymbol{\sigma}, T, \mathbf{A}),\tag{8.11.18}$$

and

$$F = U^* = U - \boldsymbol{\sigma} : \epsilon - T\,S - \mathbf{A} \cdot \mathbf{a}.\tag{8.11.19}$$

From (8.11.17) derives the *affinity* $\mathbf{X}$,

$$\mathbf{X} = \frac{\partial \phi^*(\mathbf{x}, \mathbf{y})}{\partial \mathbf{x}} = \frac{\partial \phi(\mathbf{x})}{\partial \mathbf{x}} - \mathbf{y},\tag{8.11.20}$$

which vanishes at equilibrium, namely $\mathbf{X} = \mathbf{0}$. Thus at equilibrium, the elements of sets $\mathbf{x}$ and $\mathbf{y}$ can be termed respectively *independent* variables and *dependent* variables.

If $\mathbf{H} = \partial \mathbf{X}/\partial \mathbf{x} = \partial^2 \phi/\partial \mathbf{x}\partial \mathbf{x}$ is invertible, the relation (8.11.20) can be used to express the set of state variables $\mathbf{x}$ in terms of the affinity $\mathbf{X}$ and control variables $\mathbf{y}$, namely $\delta \mathbf{X} = \mathbf{H} \cdot \delta \mathbf{x} - \delta \mathbf{y}$. Then the generalized thermodynamic function $\phi^*(\mathbf{x}, \mathbf{y})$ can be cast in the format $\Phi^*(\mathbf{X}, \mathbf{y})$ and

$$\mathbf{X} = \frac{\partial \Phi^*(\mathbf{X}, \mathbf{y})}{\partial \mathbf{X}} \cdot \frac{\partial \mathbf{X}}{\partial \mathbf{x}} \quad \Rightarrow \quad \frac{\partial \Phi^*(\mathbf{X}, \mathbf{y})}{\partial \mathbf{X}} = \mathbf{X} \cdot \mathbf{H}^{-1}.\tag{8.11.21}$$

Haslach [2005] developed a procedure to generate out-of-equilibrium rate equations for the state variables that respect the dissipation inequality. Haslach postulates the rate of the affinity as a gradient of a non equilibrium process,

$$\dot{\mathbf{X}} = -k_v\, \frac{\partial \Phi^*(\mathbf{X}, \mathbf{y})}{\partial \mathbf{X}},\tag{8.11.22}$$

with $k_v \geq 0$ a positive constitutive parameter. With help of (8.11.21), this rate may be expressed as

$$\dot{\mathbf{X}} = -k_v\, \mathbf{X} \cdot \mathbf{H}^{-1}.\tag{8.11.23}$$

Moreover $\dot{\mathbf{X}} = (\partial \mathbf{X}/\partial \mathbf{x}) \cdot \dot{\mathbf{x}}$ at $\mathbf{y}$ fixed. Hence

$$\dot{\mathbf{x}} = -k_v\, \mathbf{H}^{-1} \cdot (\mathbf{X} \cdot \mathbf{H}^{-1}).\tag{8.11.24}$$

This relation is given the following interpretation. If, starting from equilibrium, the control variables $\mathbf{y}$ are changed suddenly to a value which is maintained fixed, the material relaxes to the new equilibrium $\dot{\mathbf{x}} = \mathbf{0}$ which corresponds to $\mathbf{X} = \mathbf{0}$. The dissipation inequality may now be shown to be non negative,

$$\begin{aligned}
T\,\hat{S}_{(\mathrm{i})} &= -\dot{\phi}^*(\mathbf{x}, \mathbf{y}) - \mathbf{x} \cdot \dot{\mathbf{y}} \\
&= -\dot{\phi}(\mathbf{x}) + \mathbf{y} \cdot \dot{\mathbf{x}} \qquad \text{by (8.11.17)} \\
&= -\mathbf{X} \cdot \dot{\mathbf{x}} \qquad\qquad \text{by (8.11.20)} \\
&= k_v\,(\mathbf{X} \cdot \mathbf{H}^{-1}) \cdot (\mathbf{X} \cdot \mathbf{H}^{-1}) \geq 0 \quad \text{by (8.11.24)}.
\end{aligned}\tag{8.11.25}$$

As a basic purely mechanical one-dimensional example, x is viewed as the strain ϵ and y as the stress σ. Denoting the stiffness by k_σ [unit : Pa], the internal energy ϕ and associated energy ϕ^* adopt the format $\phi = \frac{1}{2}\,k_\sigma\,\epsilon^2$ and $\phi^* = \frac{1}{2}\,k_\sigma\,\epsilon^2 - \sigma\,\epsilon$, whence the affinity $X = \partial \phi^*/\partial \epsilon = k_\sigma\,\epsilon - \sigma$ and $H = k_\sigma$. At equilibrium $X = 0$, the formalism retrieves the elastic relation $\sigma = k_\sigma\,\epsilon$. The potential $\phi^*(\epsilon, \sigma)$ can now be expressed as $\Phi^*(X, \sigma)$, namely $\Phi^* = (X + \sigma)^2/(2\,k_\sigma) - \sigma\,(X + \sigma)/k_\sigma$. Now at constant σ, (8.11.23) becomes $k_\sigma\,\dot{\epsilon} = -k_v/k_\sigma\,(k_\sigma\,\epsilon - \sigma)$ which may rephrased in the format $\sigma = k_\sigma\,\epsilon + k_\sigma^2/k_v\,\dot{\epsilon}$. The constitutive coefficient k_v has dimension Pa/s. The viscoelastic Kelvin model is recovered by interpreting the coefficient k_σ^2/k_v as a viscosity coefficient with dimension Pa$\times$s. On the other hand, the standard linear solid model does not fit in this construction. Haslach [2005] used this approach to introduce rate effects consistent with the dissipation inequality in several elastic constitutive relations of biological tissues. The extension of Fung's QLV model accounts for the equilibrium relations (8.5.14), namely $\mathbf{X} \equiv \mathbb{E}_r(\infty) : \boldsymbol{\sigma}^{\mathrm{e}}(\epsilon) - \boldsymbol{\sigma}$.

8.11.4 Visco-hyperelasticity

Visco-hyperelasticity expresses the 2nd Piola-Kirchhoff stress in terms of the Green strain $\mathbf{E}_G$ and its rate via an elastic potential and a viscoelastic potential per unit initial volume, $\underline{\mathcal{W}}^e(\mathbf{E}_G)$ and $\underline{\mathcal{W}}^v(\dot{\mathbf{E}}_G)$ respectively,

$$\underline{\underline{\tau}} = \underline{\underline{\tau}}^e + \underline{\underline{\tau}}^v = \frac{\partial \underline{\mathcal{W}}^e(\mathbf{E}_G)}{\partial \mathbf{E}_G} + \frac{\partial \underline{\mathcal{W}}^v(\dot{\mathbf{E}}_G)}{\partial \dot{\mathbf{E}}_G} \,. \tag{8.11.26}$$

The viscoelastic potential describes short term dissipative effects. The dissipation inequality

$$T\,\underline{\hat{S}}^1_{(i)} = \underline{\underline{\tau}} : \dot{\mathbf{E}}_G - \underline{\dot{\mathcal{W}}}^e \geq 0, \tag{8.11.27}$$

which simplifies to

$$T\,\underline{\hat{S}}^1_{(i)} = \frac{\partial \underline{\mathcal{W}}^v(\dot{\mathbf{E}}_G)}{\partial \dot{\mathbf{E}}_G} : \dot{\mathbf{E}}_G \geq 0, \tag{8.11.28}$$

can be ensured by appropriate convexity properties of $\underline{\mathcal{W}}^v(\dot{\mathbf{E}}_G)$ as detailed in Section 6.2.3.

The tissue may be endowed with anisotropic properties. Their variation gives rise to dissipation. Explicit expressions of the potentials may be obtained using representation theorems of isotropic scalar invariants that depend on tensorial arguments, see, e.g., Limbert and Middleton [2004] for transversely isotropic biological materials.

8.11.5 Long range memory

Collecting the ideas of the parallel assembly of Maxwell models, of the QLV concept of Fung and of visco-hyperelasticity, a more general model consists in considering the additive contributions of an elastic response $\underline{\underline{\tau}}^e$, of a short range viscous memory $\underline{\underline{\tau}}^{vs}$ and of a long range memory $\underline{\underline{\tau}}^{vl}$ that obeys the fading memory principle,

$$\underline{\underline{\tau}} = \underline{\underline{\tau}}^e + \underline{\underline{\tau}}^{vs} + \underline{\underline{\tau}}^{vl} = \frac{\partial \underline{\mathcal{W}}^e(\mathbf{E}_G)}{\partial \mathbf{E}_G} + \frac{\partial \underline{\mathcal{W}}^v(\dot{\mathbf{E}}_G)}{\partial \dot{\mathbf{E}}_G} + \int_{0+}^{t} \underline{\underline{\tau}}^e\big(\mathbf{E}_G(t-u)\big)\,\frac{d\phi(u)}{du}\,du\,. \tag{8.11.29}$$

Pioletti and Rakotomanana [2000] apply such a relation to isotropic tendons, including in addition a coupling in the short range memory $\underline{\mathcal{W}}^v(\dot{\mathbf{E}}_G, \mathbf{E}_G)$. They define the reduced relaxation function $\phi(t)$ by an exponential Prony series with n terms, the term k having a modulus E_k and a relaxation time t_k,

$$\Big(\sum_{k=1,n} E_k\Big)\,\phi(t) = \sum_{k=1,n} E_k\,e^{-t/t_k} = \sum_{k=1,n} E_k - \sum_{k=1,n} E_k\,\big(1 - e^{-t/t_k}\big)\,. \tag{8.11.30}$$

8.11.6 Fading memories in a mixture context

Humphrey and Rajagopal [2002] [2003] apply the QLV idea but, instead of a convolution integral for the viscous term, they use an ordinary mass weighted time integration.

They address the stress in a mixture of incompressible constituents, specifically a fluid/proteoglycan-dominated gel matrix, an elastin-dominated gel matrix (referred by the index gel), a smooth muscle (m) and the collagen (c). The two last constituents grow at much faster pace than the two first ones.

Time $t = 0$ separates development ($t < 0$) and regular growth and maintenance ($t > 0$). For $t < 0$, the constitutive relations use the individual gradient decomposition (21.3.32) with a common homogenized intermediate configuration. The stress is taken in the form

$$
\begin{aligned}
\boldsymbol{\sigma}(t) \;=\; & -p(t)\,\mathbf{I} + \phi_{gel}(t)\,\boldsymbol{\sigma}^{e,gel}(t) && \text{homogenized solid} \\[2mm]
& + \sum_{k=m,c} \phi_k(t)\,\boldsymbol{\sigma}^{e,k}(t) && \text{removed early smooth muscle, collagen} \\[2mm]
& + \sum_{k=m,c} \int_0^t \frac{\det \mathbf{F}^e_k(\tau)}{\det \mathbf{F}^e_k(t)}\,\boldsymbol{\sigma}^{e,k}\big(\mathbf{F}^e_k(\tau)\big)\,S^k(\tau)\,\frac{\hat{\rho}^k(\tau)}{\rho(t)}\,d\tau && \text{growing smooth muscle, collagen}\,.
\end{aligned}
\tag{8.11.31}
$$

The elastic stress $\boldsymbol{\sigma}^{e,k}(\mathbf{F}_k^e)$ is obtained from an elastic potential for collagen and elastin. It is contributed by the passive and active responses for the muscle. The mass fraction of constituent k is defined from the relative mass densities, namely $\phi_k(t) = \rho^k(t)/\rho(t)$. The mass fractions of smooth muscle and collagen are decaying exponentials that indicate the pace at which these materials, which are produced during development, are removed. The 'survival functions' S, which pertain to each constituent, indicate the time period during which a constituent exists, typically $S(t) = \mathcal{H}(t - t_\alpha) - \mathcal{H}(t - t_\omega)$, where t_α and t_ω are the times at which the constituent is respectively produced and removed, with $\mathcal{H}$ the Heaviside step function. The rates of mass supply $\hat{\rho}^k$ are functions of the total stress and the time constants reflect levels of nutrients, growth factors and regulators like nitric oxide NO, a vasodilator, and endothelin-1, a vasoconstrictor. The overall density is assumed to remain constant. If this property applies to a growing constituent, the dilatational part of its growth velocity 'gradient' is known from the rate of mass supply. The deviatoric part of response remains to be provided.

8.11.7 Nonlinear viscoelastic models for collagen gels

In Fung's quasi-linear model (8.5.12), the nonlinearities are included in the 'elastic' stress $\boldsymbol{\sigma}^e = \boldsymbol{\sigma}^e(\boldsymbol{\epsilon})$. These nonlinearities correspond to a stiffened behavior at larger strains. Moreover, the model decouples memory effects (through the reduced relaxation function) and the nonlinear elastic response.

1. Nonlinear collagen gels:

Pryse et al. [2003] consider a model with n parallel units (spring $\mathbb{E}_i(\boldsymbol{\epsilon})$, dashpot with relaxation time $\tau_i(\boldsymbol{\epsilon})$, $i \in [1, n]$) where the stiffnesses $\mathbb{E}_i$ and relaxation times τ_i depend on strain,

$$\mathbb{E}(t, \boldsymbol{\epsilon}) = \sum_{i=1}^n \mathbb{E}_i(\boldsymbol{\epsilon})\, e^{-t/\tau_i(\boldsymbol{\epsilon})} = \sum_{i=1}^n \mathbb{E}_i(\boldsymbol{\epsilon}) - \sum_{i=1}^n \mathbb{E}_i(\boldsymbol{\epsilon})\left(1 - e^{-t/\tau_i(\boldsymbol{\epsilon})}\right). \tag{8.11.32}$$

They test the model on self-assembled solutions of pure collagen gels and on solutions of collagen and fibroblasts. Note that this model,

$$\boldsymbol{\sigma}(t) = \int_{-\infty}^t \mathbb{E}(t - u, \boldsymbol{\epsilon}) : \frac{d\boldsymbol{\epsilon}(u)}{du}\, du \tag{8.11.33}$$

generalizes Fung's model

$$\boldsymbol{\sigma}(t) = \int_{-\infty}^t \mathbb{E}_r(t - u) : \left(\frac{d\boldsymbol{\sigma}^e(\boldsymbol{\epsilon})}{d\boldsymbol{\epsilon}} : \frac{d\boldsymbol{\epsilon}(u)}{du}\right) du\,, \tag{8.11.34}$$

in which the relaxation times are constant while the reduced stiffnesses depend on strain but their ratios do not.

2. Wiener-Volterra kernels:

Along an idea attributed to Volterra, a mathematical scheme to represent nonlinear viscoelasticity consists in developing the response in series using a number of kernels, e.g., below j_0, $j_1(\tau)$ and $j_2(\tau_1, \tau_2)$,

$$\boldsymbol{\epsilon}(t) = j_0 + \int_0^\infty j_1(\tau)\, \boldsymbol{\sigma}(t - \tau)\, d\tau + \int_0^\infty j_2(\tau_1, \tau_2)\, \boldsymbol{\sigma}(t - \tau_1) \cdot \boldsymbol{\sigma}(t - \tau_2)\, d\tau_1 d\tau_2\,. \tag{8.11.35}$$

The zeroth order term provides the mean strain response, the first order term represents the viscoelastic response through a convolution integral and the second order term the nonlinear response. As such, the series has a weakness: the terms are not orthogonal, so that a refinement would modify all terms. The orthogonalization of the Volterra expansion proposed by Wiener [1958] is based on a Gram-Schmidt orthogonalization procedure requiring the covariance between any two terms to vanish for a Gaussian white noise stimulus. The Laguerre functions are used to build the basis. The Wiener expansion involves at time increment n (time increments have a uniform size), besides the kernels j_0, j_1 and j_2, the memory length R,

$$\boldsymbol{\epsilon}(n) = j_0 + \sum_{k=0}^{R-1} j_1(k)\, \boldsymbol{\sigma}(n - k) + \sum_{k_1=0}^{R-1} \sum_{k_2=0}^{R-1} j_2(k_1, k_2)\, \boldsymbol{\sigma}(n - k_1) \cdot \boldsymbol{\sigma}(n - k_2)\,. \tag{8.11.36}$$

Identification procedures are tested to the viscoelastic behavior of diabetic rat skin and ligaments in Hoffman and Grigg [2002].

Exercises on Chapter 8

Exercise 8.1 Complex modulus of series and parallel groups of Maxwell models

The set of Maxwell models (E_k, η_k), $k \in [1, n]$, assembled in series shown at the top of Fig. 8.E.1 is still a Maxwell model with modulus E and viscosity η, such that $E^{-1} = \sum_{k=1}^{n} E_k^{-1}$, and $\eta^{-1} = \sum_{k=1}^{n} \eta_k^{-1}$. The complex modulus is given by (8.3.39).

Below we address the parallel assembly of viscoelastic elements shown at the bottom of Fig. 8.E.1.

1. Consider a viscoelastic model composed by $n + 1$ parallel branches, namely n models of Maxwell defined by the elastic moduli E_{Sk} and viscosities η_k, $k \in [1, n]$, and a spring E_P. Write the differential equation governing the response of this model.

Answer: Let q be the strain and $Q = E_P q + \sum_{k=1}^{n} Q_k$ the stress at the boundaries of the model. Let

$$M_k = \frac{1}{E_{Sk}} \frac{d}{dt} + \frac{1}{\eta_k}, \quad k \in [1, n], \tag{8.E.1}$$

be the differential operator associated with the Maxwell model k. Then the strain q and stress Q are linked by the relations,

$$\frac{dq}{dt} = M_k Q_k, \quad k \in [1, n]. \tag{8.E.2}$$

Then,

$$\prod_{j=1, j \neq k}^{n} M_j \frac{dq}{dt} = (\prod_{j=1}^{n} M_j) Q_k, \quad k \in [1, n],$$
$$\sum_{k=1}^{n} \prod_{j=1, j \neq k}^{n} M_j \frac{dq}{dt} = \prod_{j=1}^{n} M_j \sum_{k=1}^{n} Q_k, \tag{8.E.3}$$

so that the governing equation is a linear differential equation of order n,

$$\sum_{k=1}^{n} \prod_{j=1, j \neq k}^{n} M_j \frac{dq}{dt} = \prod_{j=1}^{n} M_j \; (Q - E_P q), \tag{8.E.4}$$

or

$$\Big(\sum_{k=1}^{n} \prod_{j=1, j \neq k}^{n} M_j\Big) \frac{dq}{dt} + E_P \Big(\prod_{j=1}^{n} M_j\Big) q = \Big(\prod_{j=1}^{n} M_j\Big) Q. \tag{8.E.5}$$

To derive the above relations, the fact that the operators M commute has been accounted for, so that the products can be written in arbitrary orders.

For $n = 1$, the relation (8.E.4) should be understood as

$$\frac{dq}{dt} = M_1 \; (Q - E_P q). \tag{8.E.6}$$

For $n = 2$, the differential equation is of second order,

$$\Big(\big(\frac{1}{E_{S1}} + \frac{1}{E_{S2}}\big) \frac{d}{dt} + \frac{1}{\eta_1} + \frac{1}{\eta_2}\Big) \frac{dq}{dt} = \Big(\frac{d^2}{dt^2} + \big(\frac{E_{S1}}{\eta_1} + \frac{E_{S2}}{\eta_2}\big) \frac{d}{dt} + \frac{E_{S1}}{\eta_1} \frac{E_{S2}}{\eta_2}\Big) \frac{Q - E_P q}{E_{S1} E_{S2}}. \tag{8.E.7}$$

2. Obtain the complex modulus $E^*(i\omega)$ of this model and the relaxation modulus $E(t)$.

Answer: For a harmonic strain with pulsation ω, the complex modulus is obtained from (8.E.5) as defined by (8.3.3),

$$\frac{E^*(i\omega)}{E_P} = 1 + \sum_{k=1}^{n} S_k \frac{i\omega}{i\omega + \dfrac{1}{t_{qk}}}, \tag{8.E.8}$$

where $S_k = E_{Sk}/E_P$ and $t_{qk} = \eta_k/E_{Sk}$ is the relaxation time at constant strain associated with the branch k, $k \in [1, n]$.

If the above expression is interpreted as a Laplace-Carson transform at $p = i\omega$, then its inverse is

$$\frac{E(t)}{E_P} = 1 + \sum_{k=1}^{n} S_k e^{-t/t_{qk}}. \tag{8.E.9}$$

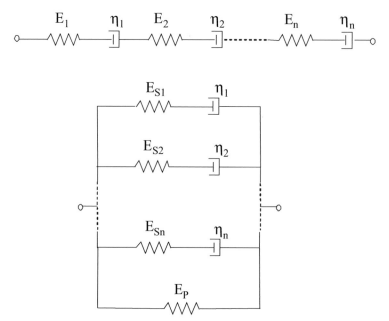

Fig. 8.E.1: Maxwell models in series and parallel.

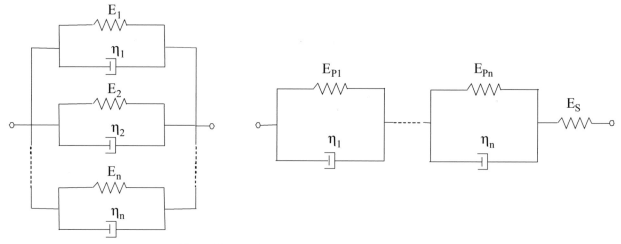

Fig. 8.E.2: Kelvin models in parallel and in series.

Exercise 8.2 Complex modulus of series and parallel groups of Kelvin models

The set of Kelvin models (E_k, η_k), $k \in [1, n]$, assembled in parallel shown on the left side of Fig. 8.E.2 is still a Kelvin model with modulus E and viscosity η such that $E = \sum_{k=1}^{n} E_k$ and $\eta = \sum_{k=1}^{n} \eta_k$. The complex modulus is given by (8.3.40).

Below we address the series assembly of viscoelastic elements shown on the right side of Fig. 8.E.2.

1. Consider a viscoelastic model composed by n models of Kelvin defined by the elastic moduli E_{Pk} and viscosities η_k, $k \in [1, n]$, in series with a spring E_S. Write the differential equation governing the response of this model.

Answer: Let Q be the stress and $q = Q/E_S + \sum_{k=1}^{n} q_k$ be the strain at the boundaries of the model. Let

$$K_k = \eta_k \frac{d}{dt} + E_{Pk}, \quad k \in [1, n], \tag{8.E.10}$$

be the differential operator associated with the Kelvin model k. Then,

$$Q = K_k\, q_k, \quad k \in [1, n],$$

$$\prod_{j=1,j\neq k}^{n} K_j\, Q = (\prod_{j=1}^{n} K_j)\, q_k, \quad k \in [1, n], \tag{8.E.11}$$

$$\sum_{k=1}^{n} \prod_{j=1,j\neq k}^{n} K_j\, Q = (\prod_{j=1}^{n} K_j)\, \Big(\sum_{k=1}^{n} q_k\Big),$$

and the differential equation of order n that governs the model response can be cast in the form,

$$\sum_{k=1}^{n} \prod_{j=1,j\neq k}^{n} K_j\, Q = (\prod_{j=1}^{n} K_j)\, \Big(q - \frac{Q}{E_S}\Big), \tag{8.E.12}$$

or

$$\Big(\sum_{k=1}^{n} \prod_{j=1,j\neq k}^{n} K_j\Big)\, Q + \Big(\prod_{j=1}^{n} K_j\Big)\, \frac{Q}{E_S} = \Big(\prod_{j=1}^{n} K_j\Big)\, q. \tag{8.E.13}$$

For $n = 1$, the relation (8.E.12) is interpreted as

$$Q = K_1\, \Big(q - \frac{Q}{E_S}\Big). \tag{8.E.14}$$

2. Obtain the complex compliance $J^*(i\,\omega)$ of this model and the creep function $J(t)$.

Answer: For a harmonic strain with pulsation ω, the complex modulus is obtained from (8.E.13) as defined by (8.3.3),

$$E_S\, J^*(i\,\omega) = 1 + \sum_{k=1}^{n} \frac{1}{P_k\, t_{qk}}\, \frac{1}{i\,\omega + \dfrac{1}{t_{qk}}}, \tag{8.E.15}$$

where $P_k = E_{Pk}/E_S$ and $t_{qk} = \eta_k/E_{Pk}$ is the relaxation time associated with the model k, $k \in [1, n]$.

If the above expression is interpreted as a Laplace-Carson transform at $p = i\,\omega$, then its inverse is

$$E_S\, J(t) = 1 + \sum_{k=1}^{n} \frac{1}{P_k}\, (1 - e^{-t/t_{qk}}). \tag{8.E.16}$$

Exercise 8.3 Governing equation of a model with a spring in parallel with two Kelvin models in series

Deduce the governing equation of the viscoelastic model of Fig. 8.7.1.

Answer:

Let Q_S ad Q_P be the stresses in the spring and Kelvin branches, respectively. The strain in the parallel branch is the sum of the strains in the Kelvin models, say q_a and q_b. Then

$$Q = Q_S + Q_P,$$

$$Q_S = E_S\, q,$$

$$Q_P = E_a\, q_a + \eta_a\, \frac{dq_a}{dt} = E_b\, q_b + \eta_b\, \frac{dq_b}{dt}, \tag{8.E.17}$$

$$q = q_a + q_b.$$

The strain q_b may be eliminated from $(8.E.17)_{3,4}$,

$$(E_a + E_b)\, q_a + (\eta_a + \eta_b)\, \frac{dq_a}{dt} = E_b\, q + \eta_b\, \frac{dq}{dt}. \tag{8.E.18}$$

Next dq_a/dt and dq_b/dt can be calculated from $(8.E.17)_3$,

$$\frac{dq_a}{dt} = \frac{Q_P}{\eta_a} - \frac{E_a}{\eta_a}\, q_a, \quad \frac{dq_b}{dt} = \frac{Q_P}{\eta_b} - \frac{E_b}{\eta_b}\, q_b. \tag{8.E.19}$$

Summing up and using $(8.E.17)_4$ yields

$$\Big(\frac{E_a}{\eta_a} - \frac{E_b}{\eta_b}\Big)\, q_a = \Big(\frac{1}{\eta_a} + \frac{1}{\eta_b}\Big)\, Q_P - \frac{E_b}{\eta_b}\, q - \frac{dq}{dt}. \tag{8.E.20}$$

Back-substituting in (8.E.18) with $Q_P = Q - E_S\, q$ yields the governing equation (8.7.1).

Chapter 9

Thermoelasticity and thermo-poroelasticity

This chapter addresses first thermoelastic constitutive equations of a solid. The analysis is next extended to include thermo-poroelasticity which involves three independent fields, namely the displacement of the solid, the pressure of the fluid and the temperature. In formulations which account for dynamic effects, the fluid pressure is substituted by the fluid velocity. The solid and porous medium under consideration here are in local thermal equilibrium, namely the temperature is identical over all species. Local thermal nonequilibrium is considered in Chapter 10.

Here are a few motivations for a thermal analysis of biological tissues:

- detection of unusual physical activities. Vessels and tissue exchange both mass (blood) and heat. Maintenance of an acceptable temperature is a main function of the central nervous system. In that sense, the consequences of deviations wrt the standard conditions are of interest. In particular, the difference in energy consumption between normal and tumor tissues may be used to detect tumors;
- cure of tumors through cryotherapy (use of a cold source) or conversely through a localized heat source (hyperthermia). For thermal models of hyperthermia, see e.g., Atkinson [1979]; for a review on cryosurgery, cryotherapy and cryopreservation, see Stánczyk and Telega [2002,2003];
- thermal ablation techniques like short pulse laser ablation. Ablation takes place at much shorter times than the thermal relaxation time (which is about 10 to 20 s for flesh, see below) so that thermal expansion has no time to develop and ablation takes place at constant volume. The photomechanical effect plays a major role: the photoenergy is used to create tension and to damage the tissue;
- analysis of thermal injury processes such as burns and frostbites, e.g., Moritz and Henriques [1947];
- temporary cooling of brain tissue following brain ischemia due to traumatic injury, stroke or cardiac arrest. Fast moderate cooling and slow re-heating requires careful protocols to be favorable, Diller and Zhu [2009];
- applications including food preservation, fertility programs and medical device implants, e.g., Polge et al [1949], Whittingham et al. [1972];
- application of cryo-protective agents in tissue preservation. Preservation of organs between extraction and transplant requires the knowledge of the physiological response of the tissues under hypothermal conditions;
- change of thermal conductivity of tissues with respect to temperature, zones of reversibility, Bhattacharya and Mahajan [2003].

9.1   Thermomechanical properties of elastic solids

Since $\boldsymbol{\sigma} : d\boldsymbol{\epsilon} - S\,dT$ is the differential of the free energy E, Table 9.1.1, then

$$\boldsymbol{\sigma} = \frac{\partial E}{\partial \boldsymbol{\epsilon}}\Big|_{T}, \quad S = -\frac{\partial E}{\partial T}\Big|_{\boldsymbol{\epsilon}}, \tag{9.1.1}$$

and

$$\frac{\partial \boldsymbol{\sigma}}{\partial T}\Big|_{\boldsymbol{\epsilon}} = -\frac{\partial S}{\partial \boldsymbol{\epsilon}}\Big|_{T}. \tag{9.1.2}$$

In view of (9.1.1), the knowledge of the free energy provides the other thermodynamic functions.

Table 9.1.1: Thermodynamical measures per unit (reference or current) volume [unit : Pa = J/m^3] of the state of a solid endowed with stress $\boldsymbol{\sigma}$, strain $\boldsymbol{\epsilon}$ and temperature T.

Measure	Variables	Arbitrary Increment
Internal energy	$U = U(\boldsymbol{\epsilon}, S)$	$dU = \boldsymbol{\sigma} : d\boldsymbol{\epsilon} + T\,dS$
Free energy	$E = E(\boldsymbol{\epsilon}, T) = U - T\,S$	$dE = \boldsymbol{\sigma} : d\boldsymbol{\epsilon} - S\,dT$
Enthalpy	$H = H(\boldsymbol{\sigma}, S) = U - \boldsymbol{\sigma} : \boldsymbol{\epsilon}$	$dH = -\boldsymbol{\epsilon} : d\boldsymbol{\sigma} + T\,dS$
Free enthalpy	$G = G(\boldsymbol{\sigma}, T) = H - T\,S$	$dG = -\boldsymbol{\epsilon} : d\boldsymbol{\sigma} - S\,dT$

9.1.1   Isothermal and isentropic moduli

An expression of the thermodynamic functions is obtained through an expansion up to order two in terms of the primary variables, e.g., Mandel [1974], p. 115. If the primary variables are strain $\boldsymbol{\epsilon}$ and temperature change $T - T_0$, then the free energy,

$$E = E_0 + \boldsymbol{\sigma}_0 : \boldsymbol{\epsilon} - (S_0 + \boldsymbol{\epsilon} : \mathbb{E} : \mathbf{c}_{\mathrm{T}})(T - T_0) + \frac{1}{2}\,\boldsymbol{\epsilon} : \mathbb{E} : \boldsymbol{\epsilon} - \frac{1}{2}\,\frac{C_{0\epsilon}}{T_0}(T - T_0)^2, \qquad (9.1.3)$$

expresses in terms of
  - the symmetric fourth order isothermal elastic stiffness $\mathbb{E}$ [unit : Pa];
  - the symmetric second order tensor of thermal expansion $\mathbf{c}_{\mathrm{T}}$ [unit : 1/K];
  - the heat capacity per unit volume at constant strain $C_{0\epsilon}$ [unit : kg/m/s^2/K].

The subscript 0 indicates a reference value corresponding to vanishing primary variables $\boldsymbol{\epsilon} = \mathbf{0}$, $T - T_0 = 0$. The thermodynamic functions follow from (9.1.1) and the definitions shown in Table 9.1.1,

$$
\begin{aligned}
S &= S_0 + \mathbf{c}_{\mathrm{T}} : \mathbb{E} : \boldsymbol{\epsilon} + \frac{C_{0\epsilon}}{T_0}(T - T_0); \\[4pt]
U &= E_0 + T_0\,S_0 + (\boldsymbol{\sigma}_0 + \mathbb{E} : \mathbf{c}_{\mathrm{T}}\,T_0) : \boldsymbol{\epsilon} + \frac{1}{2}\,\boldsymbol{\epsilon} : \mathbb{E} : \boldsymbol{\epsilon} + \frac{1}{2}\,\frac{C_{0\epsilon}}{T_0}(T^2 - T_0^2); \\[4pt]
H &= E_0 + T_0\,S_0 + \boldsymbol{\epsilon} : \mathbb{E} : \mathbf{c}_{\mathrm{T}}\,T - \frac{1}{2}\,\boldsymbol{\epsilon} : \mathbb{E} : \boldsymbol{\epsilon} + \frac{1}{2}\,\frac{C_{0\epsilon}}{T_0}(T^2 - T_0^2); \\[4pt]
G &= E_0 - \frac{1}{2}\,\boldsymbol{\epsilon} : \mathbb{E} : \boldsymbol{\epsilon} - (T - T_0)\,S_0 - \frac{1}{2}\,\frac{C_{0\epsilon}}{T_0}(T - T_0)^2.
\end{aligned}
\qquad (9.1.4)
$$

The thermoelastic constitutive equations (9.1.1) can be cast in the matrix form,

$$
\begin{bmatrix} \boldsymbol{\sigma} - \boldsymbol{\sigma}_0 \\[6pt] -(S - S_0) \end{bmatrix} =
\begin{bmatrix} \mathbb{E} & -\mathbb{E} : \mathbf{c}_{\mathrm{T}} \\[6pt] -\mathbf{c}_{\mathrm{T}} : \mathbb{E} & -C_{0\epsilon}/T_0 \end{bmatrix}
\begin{bmatrix} \boldsymbol{\epsilon} \\[6pt] T - T_0 \end{bmatrix}. \qquad (9.1.5)
$$

A partial inversion of these relations,

$$
\begin{bmatrix} \boldsymbol{\sigma} - \boldsymbol{\sigma}_0 \\[6pt] T - T_0 \end{bmatrix} =
\begin{bmatrix} \mathbb{E}^{\mathrm{isEn}} & -\mathbb{E} : \mathbf{c}_{\mathrm{T}}\,(T_0/C_{0\epsilon}) \\[6pt] -(T_0/C_{0\epsilon})\mathbf{c}_{\mathrm{T}} : \mathbb{E} & T_0/C_{0\epsilon} \end{bmatrix}
\begin{bmatrix} \boldsymbol{\epsilon} \\[6pt] S - S_0 \end{bmatrix}, \qquad (9.1.6)
$$

displays the *isentropic* (or *adiabatic*) elastic tensor moduli $\mathbb{E}^{\mathrm{isEn}} = (\partial^2 U/\partial\boldsymbol{\epsilon}\,\partial\boldsymbol{\epsilon})_{|S}$,

$$\mathbb{E}^{\mathrm{isEn}} = \mathbb{E} + \mathbf{c}_{\mathrm{T}} : \mathbb{E} \otimes \mathbb{E} : \mathbf{c}_{\mathrm{T}}\,\frac{T_0}{C_{0\epsilon}}, \qquad (9.1.7)$$

rather than the *isothermal* elastic tensor moduli $\mathbb{E} = (\partial^2 E/\partial\boldsymbol{\epsilon}\partial\boldsymbol{\epsilon})_{|T}$

The alternative partial inversion,

$$
\begin{bmatrix} \boldsymbol{\epsilon} \\[6pt] S - S_0 \end{bmatrix} =
\begin{bmatrix} \mathbb{E}^{-1} & \mathbf{c}_{\mathrm{T}} \\[6pt] \mathbf{c}_{\mathrm{T}} & C_{0\sigma}/T_0 \end{bmatrix}
\begin{bmatrix} \boldsymbol{\sigma} - \boldsymbol{\sigma}_0 \\[6pt] T - T_0 \end{bmatrix}, \qquad (9.1.8)
$$

displays the heat capacity per unit volume at *constant stress*,

$$C_{0\sigma} = C_{0\epsilon} + \mathbf{c}_{\mathrm{T}} : \mathbb{E} : \mathbf{c}_{\mathrm{T}} \, T_0 \,, \tag{9.1.9}$$

rather than the heat capacity per unit volume at *constant strain* $C_{0\epsilon}$.

For an isotropic material, both the isothermal and isentropic tensor moduli are endowed with the standard structure,

$$\mathbb{E} = \lambda \, \mathbf{I} \otimes \mathbf{I} + 2\,\mu\, \mathbf{I} \,\overline{\otimes}\, \mathbf{I}, \quad \mathbb{E}^{\mathrm{isEn}} = \lambda^{\mathrm{isEn}} \, \mathbf{I} \otimes \mathbf{I} + 2\,\mu^{\mathrm{isEn}} \, \mathbf{I} \,\overline{\otimes}\, \mathbf{I}. \tag{9.1.10}$$

With the isothermal Lamé moduli λ and μ, and bulk modulus $K = \lambda + 2\,\mu/3$, are associated the isentropic counterparts,

$$\lambda^{\mathrm{isEn}} = \lambda + K^2 \,(\mathrm{tr}\,\mathbf{c}_{\mathrm{T}})^2 \,\frac{T_0}{C_{0\epsilon}}, \quad \mu^{\mathrm{isEn}} = \mu, \quad K^{\mathrm{isEn}} = K + K^2 \,(\mathrm{tr}\,\mathbf{c}_{\mathrm{T}})^2 \,\frac{T_0}{C_{0\epsilon}} \,, \tag{9.1.11}$$

and the heat capacities at constant strain and constant stress obey the relation,

$$C_{0\sigma} = C_{0\epsilon} + K \,(\mathrm{tr}\,\mathbf{c}_{\mathrm{T}})^2 \, T_0 \,. \tag{9.1.12}$$

Stability of equilibrium requires the heat capacity at constant strain $C_{0\epsilon}$ to be positive and the isothermal elastic moduli to be positive definite, Mandel [1974], p. 115, and these conditions in turn imply the heat capacity at constant stress to be positive and the isentropic elastic moduli to be positive definite,

$$C_{0\sigma} \geq C_{0\epsilon} > 0, \quad \mathbb{E}^{\mathrm{isEn}} \text{ P.D.}, \quad \mathbb{E}^{\mathrm{isEn}} - \mathbb{E} \text{ P.sD.} \tag{9.1.13}$$

Actual values of the isothermal and isentropic moduli differ little. Typically, for rocks or cortical bones, the bulk modulus has magnitude $10^9\,\mathrm{Pa}$, the coefficient of thermal expansion $10^{-4}\,/\mathrm{K}$, the heat capacity per unit volume $10^6\,\mathrm{Pa/K}$, so that the relative difference of isothermal and isentropic moduli is not larger than 10^{-3}.

An analogy between linear poroelasticity and thermoelasticity can be observed through the correspondences between entropy and fluid mass content on one side and between temperature and pressure on the other side. The point is elaborated and used to cross-check macroscopic properties for spatially heterogeneous materials in Norris [1992].

9.1.2 Temperature dependence of heat capacity

Mimicking the formalism introduced for fluids by (3.3.3), let

$$T \, dS = \rho \mathcal{L} : d\epsilon + C_\epsilon \, dT \,, \tag{9.1.14}$$

with
   - $\mathcal{L}$ second order symmetric tensor of latent heat of deformation per unit mass [unit : J/kg];
   - C_ϵ heat capacity per unit volume at constant strain [unit : J/m³/K].
Both entities appear linear in absolute temperature,

$$\rho \mathcal{L} = \mathbb{E} : \mathbf{c}_{\mathrm{T}} \, T, \quad C_\epsilon = \frac{T}{T_0} \, C_{0\epsilon} \,. \tag{9.1.15}$$

Relations similar to (3.3.17) can be obtained for a thermoelastic solid,

$$T \, dS = \rho \mathcal{L} : d\epsilon + C_\epsilon \, dT = \boldsymbol{\xi} : d\boldsymbol{\sigma} + C_\sigma \, dT = \boldsymbol{\zeta} : d\epsilon + \boldsymbol{\eta} : d\boldsymbol{\sigma} \,. \tag{9.1.16}$$

In that vein, the four thermodynamic state functions U, E, H and G, may be expressed in terms of the four pairs of variables, (ϵ, T), $(\boldsymbol{\sigma}, T)$, (ϵ, S) and $(\boldsymbol{\sigma}, S)$. A method proposed in Lubarda [2004]b uses the equality of cross derivatives and alleviates lengthy algebraic derivations.

If one insists that the heat capacity should be a constant, $C_\epsilon = C_{0\epsilon}$, then the free energy should be modified, e.g., Chadwick [1960], p. 275,

$$E = E_0 + \boldsymbol{\sigma}_0 : \epsilon - (S_0 + \epsilon : \mathbb{E} : \mathbf{c}_{\mathrm{T}}) \,(T - T_0) + \frac{1}{2}\,\epsilon : \mathbb{E} : \epsilon + C_{0\epsilon}\,(T - T_0) - C_{0\epsilon}\,T \,\mathrm{Ln}\,\frac{T}{T_0} \,, \tag{9.1.17}$$

and then,

$$S = S_0 + \mathbf{c}_{\mathrm{T}} : \mathbb{E} : \boldsymbol{\epsilon} + C_{0\epsilon}\,\mathrm{Ln}\,\frac{T}{T_0};$$

$$U = E_0 + S_0\,T_0 + (\boldsymbol{\sigma}_0 + \mathbb{E} : \mathbf{c}_{\mathrm{T}}\,T_0) : \boldsymbol{\epsilon} + \frac{1}{2}\,\boldsymbol{\epsilon} : \mathbb{E} : \boldsymbol{\epsilon} + C_{0\epsilon}\,(T - T_0);$$

$$H = E_0 + S_0\,T_0 + \boldsymbol{\epsilon} : \mathbb{E} : \mathbf{c}_{\mathrm{T}}\,T - \frac{1}{2}\,\boldsymbol{\epsilon} : \mathbb{E} : \boldsymbol{\epsilon} + C_{0\epsilon}\,(T - T_0);$$

$$G = E_0 - \frac{1}{2}\,\boldsymbol{\epsilon} : \mathbb{E} : \boldsymbol{\epsilon} - S_0\,(T - T_0) + C_{0\epsilon}\,(T - T_0) - C_{0\epsilon}\,T\,\mathrm{Ln}\,\frac{T}{T_0}\,.$$

$$(9.1.18)$$

It is useful to record the expansion to second order,

$$T - T_0 - T\,\mathrm{Ln}\,\frac{T}{T_0} = -\frac{1}{2\,T_0}\,(T - T_0)^2\,. \tag{9.1.19}$$

9.1.3 The energy equation of thermoelasticity

The constitutive equations (9.1.1) have been obtained assuming the linearized mechanical entropy production (4.7.28),

$$T\hat{S}_{(\mathrm{i},1)} = -\frac{dE}{dt} + \boldsymbol{\sigma} : \frac{d\boldsymbol{\epsilon}}{dt} - S\,\frac{dT}{dt}\,, \tag{9.1.20}$$

to vanish,

$$T\hat{S}_{(\mathrm{i},1)} = 0\,. \tag{9.1.21}$$

Eliminating the rate of work $\boldsymbol{\sigma} : d\boldsymbol{\epsilon}/dt$ from the entropy production (9.1.20) and from the linearized balance of energy (4.6.22),

$$\frac{dU}{dt} - \boldsymbol{\sigma} : \frac{d\boldsymbol{\epsilon}}{dt} + \mathrm{div}\,\mathbf{Q} - \Theta = 0\,, \tag{9.1.22}$$

where $\mathbf{Q}$ is the heat flux [unit : W/m^2=kg/m/s^3] and Θ is a heat source per unit volume [unit : W/m^3 = kg/m^2/s^3], yields in view of (9.1.21),

$$T\hat{S}_{(\mathrm{i},1)} = T\,\frac{dS}{dt} + \mathrm{div}\,\mathbf{Q} - \Theta = 0\,. \tag{9.1.23}$$

Fourier's law, expressing the heat flux $\mathbf{Q}$ in terms of the temperature gradient $\boldsymbol{\nabla}T$ via the symmetric positive (semi-)definite second order tensor of thermal conductivity $\mathbf{k}_{\mathrm{T}}$ [unit : W/m/K],

$$\mathbf{Q} = -\mathbf{k}_{\mathrm{T}} \cdot \boldsymbol{\nabla}T\,, \tag{9.1.24}$$

implies the thermal contribution to the internal entropy production $T\hat{S}_{(\mathrm{i},3)} = -\mathbf{Q} \cdot \boldsymbol{\nabla}T/T$ to be nonnegative.

The energy equation (9.1.23) can then be phrased in terms of primary variables, say strain and temperature, in the format,

$$\underbrace{T\,\mathbf{c}_{\mathrm{T}} : \mathbb{E} : \frac{d\boldsymbol{\epsilon}}{dt}}_{\substack{\text{thermoelastic}\\\text{heat}}} + \underbrace{C_\epsilon\,\frac{dT}{dt}}_{\substack{\text{purely thermal}\\\text{contribution}}} - \underbrace{\mathrm{div}\,(\mathbf{k}_{\mathrm{T}} \cdot \boldsymbol{\nabla}T)}_{\substack{\text{conductive}\\\text{heat flux}}} - \underbrace{\Theta}_{\substack{\text{heat}\\\text{source}}} = 0\,, \tag{9.1.25}$$

which holds whether the heat capacity is variable, Eqn (9.1.4), or constant, Eqn (9.1.18).

The thermoelastic coupling term in the energy equation can be neglected under the conditions indicated in the following proposition:

Proposition 9.1 on the thermoelastic decoupling of the energy equation.

As indicated by Fig. 1.8.1 and Table 9.1.2, for many solids, the entities involved have the following orders of magnitude,

$$T = O(10^2)\,\mathrm{K}, \quad \mathbf{c}_{\mathrm{T}} = O(10^{-4})\,\mathrm{K}^{-1}, \quad \mathbb{E} \le O(10^9)\,\mathrm{Pa}, \quad C_\epsilon = O(10^6)\,\mathrm{J\,m^{-3}\,K^{-1}}\,. \tag{9.1.26}$$

Moreover, the total strain rate $d\boldsymbol{\epsilon}/dt$ and the thermal contribution $\mathbf{c}_{\mathrm{T}}\,dT/dt$ to the strain rate may be assumed to be of the same order of magnitude. Therefore the ratio of the thermoelastic and purely thermal contributions is quite small,

$$\frac{T}{C_\epsilon}\,\mathbf{c}_{\mathrm{T}} : \mathbb{E} : \mathbf{c}_{\mathrm{T}} \le O(10^{-3})\,. \tag{9.1.27}$$

Further in absence of heat source, and if the thermal conductivity is spatially homogeneous, the energy equation reduces to the canonical form of a parabolic equation,

$$\left(\frac{\partial}{\partial t} - D_{\mathrm{T}} \frac{\partial^2}{\partial x^2} \right) T = 0 \,, \tag{9.1.28}$$

where the coefficient $D_{\mathrm{T}} = k_{\mathrm{T}}/C_{\epsilon}$ [unit : m^2/s] is called *thermal diffusivity*.

The same parabolic operator appears in many physical problems. In particular, for fluid flow across a porous medium governed by Darcy's law, it stems from the balance of mass: the coefficient $D = D_{\mathrm{H}}$ is then called *hydraulic diffusivity* and the unknown is a pressure or volumetric strain.

While thermal diffusivity involves thermal conductivity and heat capacity, hydraulic diffusivity of deformable tissues involves hydraulic permeability and elastic moduli. The time scale for equilibrium under seepage is equal to L^2/D_{H}, L seepage distance, $D_{\mathrm{H}} = k_{\mathrm{H}} (\lambda + 2\,\mu)$ diffusivity, k_{H} hydraulic permeability, λ and μ Lamé moduli. Modifications of hydraulic diffusivity are indicators of modifications of the latter properties. In tumors, the ECM volume fraction is larger than in sound tissues, so that the bulk modulus is expected to be smaller and the shear modulus larger than normal. Detection of tumors by palpation is indeed based on spatially heterogeneous shear moduli. Secomb and El-Kareh [2001] suggest the following values: 1st Lamé modulus $\lambda = 4 \times 10^3 \,\mathrm{Pa}$, shear modulus $\mu = 0.17 \times 10^3 \,\mathrm{Pa}$, $k_{\mathrm{H}} = 3.1 \times 10^{-14}\,\mathrm{m}^2/\mathrm{Pa}/\mathrm{s}$ leading to a diffusivity $D_{\mathrm{H}} = 1.4 \times 10^{-10}\,\mathrm{m}^2/\mathrm{s}$. For $L = 100\,\mu\mathrm{m}$ (typical spacing of micro-vessels), fluid seepage should be accounted for after about $71\,\mathrm{s}$. For brain white matter, the hydraulic conductivity $k_{\mathrm{H}} = 750 \times 10^{-14}\,\mathrm{m}^2/\mathrm{Pa}/\mathrm{s}$ is larger by two orders of magnitude, yielding a diffusivity $D_{\mathrm{H}} = 330 \times 10^{-10}\,\mathrm{m}^2/\mathrm{s}$ so that the characteristic time of diffusion over $100\,\mu\mathrm{m}$ falls down to a third of a second.

A parabolic operator governs also the diffusion of a chemical dissolved in a fluid which may itself be at rest or may move through a porous medium. Section 15.3 presents a comparative introduction to heat diffusion in a solid and to diffusion of a chemical convected by a fluid through a porous medium. A detailed example of chemical diffusion through a poroelastic medium is worked out in Section 13.13.

It might be worth to enlarge slightly the perspective and mention that diffusion may be seen as a transmission mode where the information moves at infinite speed. While the point of view is quite relevant to describe the physical aspects of the examples listed above, a more accurate description is desirable in particular circumstances, and especially when the dynamic response is of interest. Models along that vein are briefly introduced in Section 9.1.4.

Typical values of hydraulic and thermal diffusivities are displayed in Tables 7.3.1 and 9.1.2. The values listed in these tables are indicative only. Indeed, some properties may vary with temperature and display anisotropy. For example, thermal conductivity is larger across muscle fibers than along fibers. For a measurement technique based on pulse-decay method, the reader may consult M.M. Chen et al. [1981].

The thermal conductivities of a number of biological tissues have been collected by K.R. Holmes and posted on the website http://users.ece.utexas.edu/~valvano/research/Thermal.pdf (last visit: May 13, 2015). The mass density of soft tissues is often approximated to $1050\,\mathrm{kg/m}^3$ and their specific heat capacities c can be considered to increase linearly with their water content $w \in [0, 1]$, namely

$$c = 4.18 \times (0.40 + 0.60\,w) \quad [\text{unit} : \mathrm{kJ/kg/K}] \,, \tag{9.1.29}$$

at physiological temperature.

The heat capacity per unit volume of a porous medium which is in local thermal equilibrium is equal to the spatial average of its counterparts over the constituents, say for a solid and a fluid,

$$C = n^{\mathrm{s}} C_{\mathrm{s}} + n^{\mathrm{w}} C_{\mathrm{w}} \,. \tag{9.1.30}$$

The thermal conductivity of the porous medium k_{T},

$$k_{\mathrm{T}} = \sum_{k=\mathrm{s,w}} n^k \, k_{\mathrm{T}k} \,, \tag{9.1.31}$$

is usually taken as the volume average of the thermal conductivities $k_{\mathrm{T}k}$ of the constituents. This arithmetic average assumes that heat flows in parallel in the constituents and the heat fluxes sum up. Still, the effective thermal conductivity is bounded by the arithmetic mean and the harmonic mean over constituents,

$$\sum_{k=\mathrm{s,w}} n^k \, k_{\mathrm{T}k} \geq \prod_{k=\mathrm{s,w}} (k_{\mathrm{T}k})^{n^k} \geq \frac{1}{\displaystyle\sum_{k=\mathrm{s,w}} \frac{n^k}{k_{\mathrm{T}k}}} \,. \tag{9.1.32}$$

Table 9.1.2: Indicative thermal properties at ambient (20 °C) or physiological (37.5 °C) temperature of some biological tissues, biomaterials, geomaterials and metals.

• Liquids have a higher coefficient of thermal expansion than solids, $\sim 10^{-3}$/K versus $\sim 10^{-5}$/K;

• Water has the next largest heat capacity among liquids, after ammonia. The energy necessary to increase by 1 °C the temperature of 1 kg of water is equivalent to the energy delivered by gravity when this mass falls by 416 m. Water is a poor conductor of heat or equivalently, it is a good insulator. Average heat capacity of human body is about[4] $c = 3.470 \times 10^3$ J/kg/K.

• The negative coefficient of thermal expansion of cornea is attributed to the fact that chloride binding decreases with temperature, which implies a decreases of swelling pressure, see Section 19.2.2.

• The water content of human soft tissues listed here ranges from 70 to 85%, except for fat, where it does not exceed 15 to 20%.

Material	Density [kg/m³]	Heat Capacity C[10^6 J/m³/K] c[10^3 J/kg/K]	Volumetric Thermal Expansion Coeff. [10^{-3}/K]	Thermal Conductivity [W/m/K]	Thermal Diffusivity [10^{-6} m²/s]
Air (1 atm)	1.2	c=1.005	3.4	0.026[3]	21.5
Ice (0°C)	916.8 [1]	C=1.919 [1]	0.158 [1]	2.25 [3]	1.17
Water	997.1 [1]	c=4.18	0.21[a]; 0.36[b]	0.60 [3]	0.14
Human blood	1050-1600	C=3.6-3.9	0.27[a]; 0.38[b]	0.49-0.55	∼0.13
Alcohols	800	C= 2	0.75-0.95	0.15-0.25	0.075-0.125
Subcutaneous fat	920-940	c=2.2-3.0	-	0.20-0.25	∼0.10
Human skeletal muscle	1100-1300	c=3.1-3.8	0.01-0.1	0.45-0.55	∼0.16
Human liver	1060	c=3.6		0.50-0.55	∼0.14
Articular cartilage	1260	C=4		0.6	0.15
Porcine cornea	1060	C=4	-1.8[5]		
Cortical bone	1700	C=2.7	0.02	2.3	0.85
Bone tissues	1-3000	C=1-2.5		0.2-0.6	0.08-0.6
PMMA	1200	C=1.8	0.067	0.17	0.09
Silicon	2330	C=2.74	0.0026	1.4	0.5
Granite	2670	c=0.79[4]	0.01-0.02	2.8	1.3
Saturated soil	2100	C=1.3	0.01-0.05	0.25-2.5 [3]	0.19-1.9
Clay	2000-2700	C=3.92 [2]	0.034 [2]	1.02 [2]	0.26 [2]
Halite (salt)	2170	C=1.89 [2]	0.120 [2]	6.60 [2]	3.5 [2]
Tantalum	16 650	C=2.3	0.0065	54	23.5
Copper	8930	c=0.39	0.051	389 [3]	112
Steel	7850	c=0.49	0.033-0.039	46	12

[1] Kestin [1968], p. 541; [2] McTigue [1986]; [3] Mitchell [1993], p. 247; [4] Engineering Toolbox; [5] Kwok and Klyce [1990]; [a] 20 °C; [b] 37.5 °C.

In the harmonic mean, the constituents are thermally in series, while the geometric mean correspond to an intermediate constituent distribution.

The thermal conductivity k_{TS} of most tissues is not far from the thermal conductivity k_{Tw} of water. The thermal conductivity of most solid constituents is one to five times that of water. On the other hand, the heat capacity of water is the next largest among liquids after ammonia. The averaging relations (9.1.30) and (9.1.31) are in strong contrast with Proposition 9.4 that implies the thermal expansion of the porous *skeleton* to be equal to its counterpart in the solid constituent, irrespective of the volume fractions.

Characteristic diffusion times (9.5.14) are inversely proportional to the diffusivity. Therefore typical diffusion processes in water and soft tissues take place in identical time windows, while they are about an order of magnitude faster in bones and geological materials, and about two to three orders of magnitude faster in metals.

9.1.4 Cattaneo's heat conductor

The energy equation for a heat conductor obeying Fourier's law is of a parabolic type, that is, temperature propagates at infinite speed. The thermoelastic models of Cattaneo [1948,1949,1958] and Green and Laws [1972] aim at describing temperature propagation at finite speed. Fundamental differences between partial differential equations of parabolic and hyperbolic types are exposed in Chapter 15.

Green and Laws [1972] formulate the entropy inequality with help of a dependent variable that is required to be equal to the temperature only at equilibrium. For a conductor, Fourier's heat law remains formally untouched but the thermal conductivity depends on both the temperature and its rate. The idea has been used by Green and Lindsay [1972] to formulate a thermoelastic model, whose dynamic response in a finite element framework has been examined by Prévost and Tao [1983].

In Cattaneo's model on the other hand, Fourier's law is modified so as to include a relaxation time. The short presentation below is a simple extension to deformable anisotropic solids of the formulation presented in Bowen [1989], p. 195, which was restricted to rigid conductors. According to the original idea of Müller [1967], to the set of state variables, namely strain ϵ and temperature T, are added the temperature gradient $\mathbf{G} = \boldsymbol{\nabla} T$ and the heat flux $\mathbf{Q}$. The total entropy production,

$$
\begin{aligned}
T \hat{S}_{(i)} &= -\frac{dE}{dt} + \boldsymbol{\sigma} : \frac{d\epsilon}{dt} - S \frac{dT}{dt} - \mathbf{Q} \cdot \frac{\mathbf{G}}{T} \\
&= \left(\boldsymbol{\sigma} - \frac{\partial E}{\partial \epsilon}\right) : \frac{d\epsilon}{dt} - \left(S + \frac{\partial E}{\partial T}\right) \frac{dT}{dt} - \frac{\partial E}{\partial \mathbf{G}} \cdot \frac{d\mathbf{G}}{dt} - \frac{\partial E}{\partial \mathbf{Q}} \cdot \frac{d\mathbf{Q}}{dt} - \mathbf{Q} \cdot \frac{\mathbf{G}}{T} \geq 0,
\end{aligned}
\tag{9.1.33}
$$

should be positive. The constitutive equations (9.1.1) for the stress and entropy dispose of the first two terms on the left hand side of the above inequality. The free energy is assumed to be independent of the temperature gradient, but it depends on strain and temperature as in (9.1.3) or (9.1.17), and it also depends quadratically on the heat flux via a symmetric second order tensor $\mathbf{A}_{\mathrm{Cat}}$,

$$
E(\epsilon, T, \mathbf{Q}) = E_1(\epsilon, T) + \frac{1}{2} \mathbf{Q} \cdot \mathbf{A}_{\mathrm{Cat}} \cdot \mathbf{Q} .
\tag{9.1.34}
$$

Moreover in this approach, the rate of thermal flux is a dependent variable, for which a constitutive equation is required, say

$$
t_{\mathrm{Cat}} \frac{d\mathbf{Q}}{dt} + \mathbf{Q} = -\mathbf{k}_{\mathrm{T}} \cdot \mathbf{G} ,
\tag{9.1.35}
$$

with t_{Cat} [unit : s] a characteristic relaxation time and $\mathbf{k}_{\mathrm{T}}$ [unit : W/m/K] the positive definite thermal conductivity tensor. Clearly Fourier's heat law appears as a limit case of (9.1.35), namely for $t_{\mathrm{Cat}} = 0$. Then the reduced entropy production,

$$
T \hat{S}_{(i)} = \frac{1}{t_{\mathrm{Cat}}} \mathbf{Q} \cdot \mathbf{A}_{\mathrm{Cat}} \cdot \mathbf{Q} + \mathbf{Q} \cdot \left(\mathbf{A}_{\mathrm{Cat}} - \frac{t_{\mathrm{Cat}}}{T} \mathbf{k}_{\mathrm{T}}^{-1}\right) \cdot \frac{\mathbf{k}_{\mathrm{T}}}{t_{\mathrm{Cat}}} \cdot \mathbf{G} \geq 0,
\tag{9.1.36}
$$

is satisfied by setting $\mathbf{A}_{\mathrm{Cat}} = t_{\mathrm{Cat}} (T \mathbf{k}_{\mathrm{T}})^{-1}$ which is indeed positive definite.

The thermoelastic constitutive equation (9.1.5) for the stress is not modified while an additional nonlinear term contributes to the entropy,

$$
\begin{bmatrix} \boldsymbol{\sigma} - \boldsymbol{\sigma}_0 \\ -(S - S_0) \end{bmatrix} = \begin{bmatrix} \mathbb{E} & -\mathbb{E} : \mathbf{c}_{\mathrm{T}} \\ -\mathbf{c}_{\mathrm{T}} : \mathbb{E} & -C_{0\epsilon}/T_0 \end{bmatrix} \begin{bmatrix} \epsilon \\ T - T_0 \end{bmatrix} - \frac{t_{\mathrm{Cat}}}{2 T^2} \mathbf{Q} \cdot \mathbf{k}_{\mathrm{T}}^{-1} \cdot \mathbf{Q} \begin{bmatrix} 0 \\ 1 \end{bmatrix} .
\tag{9.1.37}
$$

A linearized energy equation is obtained by adding the production of mechanical entropy (9.1.23) and its time derivative multiplied by the relaxation time t_{Cat}, yielding with help of (9.1.35),

$$
T \mathbf{c}_{\mathrm{T}} : \mathbb{E} : \left(t_{\mathrm{Cat}} \frac{\partial^2 \epsilon}{\partial t^2} + \frac{\partial \epsilon}{\partial t}\right) + C_\epsilon \left(t_{\mathrm{Cat}} \frac{\partial^2 T}{\partial t^2} + \frac{\partial T}{\partial t}\right) - \operatorname{div}\left(\mathbf{k}_{\mathrm{T}} \cdot \boldsymbol{\nabla} T\right) - t_{\mathrm{Cat}} \frac{\partial \Theta}{\partial t} - \Theta = 0 ,
\tag{9.1.38}
$$

which, for $t_{\mathrm{Cat}} > 0$ and a positive definite thermal conductivity $\mathbf{k}_{\mathrm{T}}$, is recognized as a hyperbolic equation for the thermal field. The issue is considered in Section 11.3: the relaxation time t_{Cat} is shown to be linked to the speed of heat propagation and thermal diffusivity, Eqn (11.3.10).

9.1.5 Cauchy stress versus Eshelby stress

Thermo-(hyper)elastic constitutive equations are built starting from a vanishing mechanical dissipation (4.7.28), which for a solid takes the form

$$T\hat{S}_{(i,1)} = -\frac{dE}{dt} + \boldsymbol{\sigma}^{\mathrm{E}} : \frac{d\boldsymbol{\epsilon}}{dt} - S\frac{dT}{dt} = 0\,, \tag{9.1.39}$$

where $\boldsymbol{\sigma}^{\mathrm{E}} = \boldsymbol{\sigma} - E\,\mathbf{I}$ is Eshelby stress and the free energy per unit volume E is a function of strain $\boldsymbol{\epsilon}$ and temperature T. The point here is to explore the consequences of not neglecting the difference between Cauchy and Eshelby stresses.

Consider the quadratic free energy,

$$E = E_0 + \boldsymbol{\sigma}_0^{\mathrm{E}} : \boldsymbol{\epsilon} - (S_0 + \boldsymbol{\epsilon} : \mathbb{E} : \mathbf{c}_{\mathrm{T}})\,(T - T_0) + \frac{1}{2}\,\boldsymbol{\epsilon} : \mathbb{E} : \boldsymbol{\epsilon} - \frac{1}{2}\frac{C_{0\epsilon}}{T_0}\,(T - T_0)^2\,. \tag{9.1.40}$$

The constitutive equations for Eshelby stress and entropy in terms of the primary variables can be cast in the matrix form,

$$\begin{bmatrix} \boldsymbol{\sigma}^{\mathrm{E}} - \boldsymbol{\sigma}_0^{\mathrm{E}} \\ -(S - S_0) \end{bmatrix} = \begin{bmatrix} \mathbb{E} & -\mathbb{E} : \mathbf{c}_{\mathrm{T}} \\ -\mathbf{c}_{\mathrm{T}} : \mathbb{E} & -C_{0\epsilon}/T_0 \end{bmatrix} \begin{bmatrix} \boldsymbol{\epsilon} \\ T - T_0 \end{bmatrix}\,. \tag{9.1.41}$$

Since the free energy is quadratic with respect to the primary variables, Eshelby stress and entropy are affine functions of the latter. On the other hand, unlike Eshelby stress, Cauchy stress $\boldsymbol{\sigma} = \boldsymbol{\sigma}^{\mathrm{E}} + E\,\mathbf{I}$ is quadratic with respect to the primary variables.

The energy equation differs from the linearized analysis of Section 9.1.3. Indeed, the balance of energy,

$$\frac{dU}{dt} + U\,\mathbf{I} : \frac{d\boldsymbol{\epsilon}}{dt} - \boldsymbol{\sigma} : \frac{d\boldsymbol{\epsilon}}{dt} + \operatorname{div}\mathbf{Q} - \Theta = 0\,, \tag{9.1.42}$$

and the dissipation,

$$\begin{aligned} T\hat{S}_{(i,1)} &= -\frac{dE}{dt} - E\,\mathbf{I} : \frac{d\boldsymbol{\epsilon}}{dt} + \boldsymbol{\sigma} : \frac{d\boldsymbol{\epsilon}}{dt} - S\frac{dT}{dt} \\ &= T\frac{dS}{dt} + T\,S\,\mathbf{I} : \frac{d\boldsymbol{\epsilon}}{dt} + \operatorname{div}\mathbf{Q} - \Theta\,, \end{aligned} \tag{9.1.43}$$

include an additional term so that the energy equation becomes

$$\underbrace{T\,\mathbf{c}_{\mathrm{T}} : \mathbb{E} : \frac{d\boldsymbol{\epsilon}}{dt}}_{\substack{\text{thermoelastic} \\ \text{heat}}} + \underbrace{C_\epsilon\frac{dT}{dt}}_{\substack{\text{purely thermal} \\ \text{contribution}}} + \underbrace{T\,S\,\mathbf{I} : \frac{d\boldsymbol{\epsilon}}{dt}}_{\substack{\text{nonlinear} \\ \text{term}}} - \underbrace{\operatorname{div}(\mathbf{k}_{\mathrm{T}} \cdot \boldsymbol{\nabla} T)}_{\substack{\text{conductive} \\ \text{heat flux}}} - \underbrace{\Theta}_{\substack{\text{heat} \\ \text{source}}} = 0\,. \tag{9.1.44}$$

9.1.6 Free energy versus Massieu function

Let us recall again the expression of mechanical dissipation (4.7.28) for a solid, namely

$$T\hat{S}_{(i,1)} = -\frac{dE}{dt} + (\boldsymbol{\sigma} - E\,\mathbf{I}) : \frac{d\boldsymbol{\epsilon}}{dt} - S\frac{dT}{dt} = 0\,, \tag{9.1.45}$$

where the free energy per unit volume E is a function of strain $\boldsymbol{\epsilon}$ and temperature T. Constitutive equations result from the expression (9.1.45), namely

$$\boldsymbol{\sigma} - E\,\mathbf{I} = \frac{\partial E}{\partial \boldsymbol{\epsilon}}, \quad S = -\frac{\partial E}{\partial T}, \quad U = E - T\frac{\partial E}{\partial T}\,. \tag{9.1.46}$$

Partial or total Legendre transforms allow to change the set of primary (independent) variables. Examples are given in Section 14.8.

The issue discussed in this section circles on the use of either $T\,\hat{S}_{(i,1)} \geq 0$ or $\hat{S}_{(i,1)} \geq 0$ to ensure the mechanical entropy production to be non negative. The difference seems a priori harmless for solids, or for mixtures which are in local thermal equilibrium. On the other hand, the question will be more acute for mixtures which are *not* in local thermal equilibrium. Indeed, the entropy production of the mixture is the sum of the individual contributions and not the sum of coldness weighted contributions.

If the entropy production is not multiplied by T, the potential that naturally appears is the Massieu function associated with free energy, Bowen and Garcia [1970],

$$\Xi = -\frac{E}{T} = S - \frac{U}{T}, \tag{9.1.47}$$

which depends on strain and coldness $\nu = 1/T$. Indeed,

$$\hat{S}_{(i,1)} = \frac{d\Xi}{dt} + (\nu\,\boldsymbol{\sigma} + \Xi\,\mathbf{I}) : \frac{d\boldsymbol{\epsilon}}{dt} + U\frac{d\nu}{dt} = 0, \tag{9.1.48}$$

which yields the stress, internal energy U and entropy S, namely

$$\boldsymbol{\sigma} = -\frac{\Xi}{\nu}\mathbf{I} - \frac{1}{\nu}\frac{\partial\Xi}{\partial\boldsymbol{\epsilon}}, \quad U = -\frac{\partial\Xi}{\partial\nu}, \quad S = \Xi - \nu\frac{\partial\Xi}{\partial\nu}. \tag{9.1.49}$$

For example, an expansion to second order in terms of strain $\boldsymbol{\epsilon}$ and coldness $\Delta\nu = \nu - \nu_0$ starting from the reference $\boldsymbol{\epsilon} = \mathbf{0}$ and $\nu = \nu_0$,

$$\Xi = \Xi_0 - \alpha_s\,\Delta\nu - \sigma_s\,\nu_0\,\mathrm{tr}\,\boldsymbol{\epsilon} + \frac{1}{2}\,v_s\,\frac{(\Delta\nu)^2}{\nu_0} + \tau_s\,\Delta\nu\,\mathrm{tr}\,\boldsymbol{\epsilon} - \frac{\nu_0}{2}\,\boldsymbol{\epsilon} : \mathbb{E} : \boldsymbol{\epsilon}, \tag{9.1.50}$$

with α_s, v_s, σ_s and τ_s [unit : Pa] material constants, yields

$$\begin{aligned} U &= \alpha_s - v_s\,\frac{\Delta\nu}{\nu_0} - \tau_s\,\mathrm{tr}\,\boldsymbol{\epsilon}, \\ S &= \Xi_0 + \alpha_s\,\nu_0 - \nu_0\,(\tau_s + \sigma_s)\,\mathrm{tr}\,\boldsymbol{\epsilon} - \frac{1}{2}\frac{v_s}{\nu_0}\,(\nu^2 - \nu_0^2) - \frac{\nu_0}{2}\,\boldsymbol{\epsilon} : \mathbb{E} : \boldsymbol{\epsilon}. \end{aligned} \tag{9.1.51}$$

The heat capacity at constant strain and latent heat of deformation may be deduced from the resulting expression,

$$T\,dS = \frac{v_s}{\nu_0}\frac{dT}{T^2} - \frac{T}{T_0}\left(\mathbb{E} : \boldsymbol{\epsilon} + (\tau_s + \sigma_s)\,\mathbf{I}\right) : d\boldsymbol{\epsilon}. \tag{9.1.52}$$

A 2nd order expansion of the stress results as,

$$\begin{aligned} \boldsymbol{\sigma} &= \frac{\nu_0}{\nu}\Bigg(\Big(\underbrace{\sigma_s - \frac{\Xi_0}{\nu_0}}_{\sigma_0}\Big)\mathbf{I} + \mathbb{E} : \boldsymbol{\epsilon} + \sigma_s\,\mathrm{tr}\,\boldsymbol{\epsilon}\,\mathbf{I} + \frac{\Delta\nu}{\nu_0}\,(\alpha_s - \tau_s)\,\mathbf{I} \\ &\qquad + \Big(\tfrac{1}{2}\,\boldsymbol{\epsilon} : \mathbb{E} : \boldsymbol{\epsilon} - \tfrac{1}{2}\,v_s\,\frac{(\Delta\nu)^2}{\nu_0^2} - \tau_s\,\frac{\Delta\nu}{\nu_0}\,\mathrm{tr}\,\boldsymbol{\epsilon}\Big)\,\mathbf{I}\Bigg) \\ &= \sigma_0\,\mathbf{I} + \mathbb{E} : \boldsymbol{\epsilon} + \sigma_s\,\mathrm{tr}\,\boldsymbol{\epsilon}\,\mathbf{I} + \frac{\Delta\nu}{\nu_0}\,(\alpha_s - \tau_s - \sigma_0)\,\mathbf{I} \\ &\qquad + \Big(\tfrac{1}{2}\,\boldsymbol{\epsilon} : \mathbb{E} : \boldsymbol{\epsilon} - (\alpha_s - \tau_s + \tfrac{1}{2}\,v_s)\,\frac{(\Delta\nu)^2}{\nu_0^2}\Big)\,\mathbf{I} - \frac{\Delta\nu}{\nu_0}\,\big(\mathbb{E} : \boldsymbol{\epsilon} + (\tau_s + \sigma_s)\,\mathrm{tr}\,\boldsymbol{\epsilon}\,\mathbf{I}\big). \end{aligned} \tag{9.1.53}$$

Clearly, the structure of the thermodynamic functions and stress have quite different features than when the starting potential is the free energy.

9.2 Thermoelastic heat, stored energy and dissipated energy

The notions of external, thermoelastic and inelastic heats, plastic work and recoverable and irrecoverable stored energies are illustrated through a one-dimensional thermomechanical model.

9.2.1   A thermomechanical model including an internal variable

The model below developed by Rosakis et al. [2000] assumes small strains and it is one-dimensional. Let E and S be the free energy and entropy per unit current or reference volume. The state variables are the elastic strain ϵ^{e}, an internal variable ξ that describes some physical property of the material and the temperature T,

$$\mathcal{V} = \{\epsilon^{\mathrm{e}}, \, \xi, \, T\} \, . \tag{9.2.1}$$

The exact interpretation of the elastic strain, as opposed to the total strain, need not to be specified here. Suffices to say for now that it is the entity that is involved in the stress-strain relation. Then, as a consequence of Clausius-Duhem inequality (4.7.28),

$$\sigma = \frac{\partial E}{\partial \epsilon^{\mathrm{e}}}, \quad S = -\frac{\partial E}{\partial T} \, . \tag{9.2.2}$$

The specific heat capacity at constant strain and internal variable is therefore equal to,

$$C_v = C_v(\mathcal{V}) = -T \, \frac{\partial^2 E}{\partial T^2} \, . \tag{9.2.3}$$

Three assumptions are adopted in order to obtain a minimal framework:

Assumption 9.1: the heat capacity C_v is independent of the internal variable ξ.

Then

$$\frac{\partial C_v}{\partial \xi} = 0 \Rightarrow \frac{\partial E}{\partial \xi} = a(\epsilon^{\mathrm{e}}, \xi) \, T + b(\epsilon^{\mathrm{e}}, \xi) \, , \tag{9.2.4}$$

where $a(\epsilon^{\mathrm{e}}, \xi)$ and $b(\epsilon^{\mathrm{e}}, \xi)$ are arbitrary functions of their arguments.

Assumption 9.2: the elastic moduli are independent of the internal variable ξ, i.e. $\partial \sigma / \partial \xi = 0$.

Since $\partial \sigma / \partial \xi = \partial (\partial E / \partial \epsilon^{\mathrm{e}}) / \partial \xi = \partial (\partial E / \partial \xi) / \partial \epsilon^{\mathrm{e}}$, then, with help of (9.2.2), $\partial E / \partial \xi$ is independent of ϵ^{e}. This condition restricts (9.2.4) further:

$$\frac{\partial \sigma}{\partial \xi} = 0 \Rightarrow \frac{\partial E}{\partial \xi} = a(\xi) \, T + b(\xi) \, . \tag{9.2.5}$$

Thus

$$E(\mathcal{V}) = E_1(\epsilon^{\mathrm{e}}, T) + \overbrace{U_0(\xi) - T \, S_0(\xi)}^{\substack{\text{stored free energy} \\ \text{of cold work}}}, \quad S(\mathcal{V}) = -\frac{\partial E_1}{\partial T}(\epsilon^{\mathrm{e}}, T) + S_0(\xi) \, . \tag{9.2.6}$$

Rosakis et al. [2000] provide experimental determination of the energy of cold work for two metals in terms of an internal variable, taken as the plastic strain.

Assumption 9.3: as a further restriction, the heat capacity C_v is constant and the stress depends linearly on the strain.

Then

$$E(\mathcal{V}) = U_0(\xi) - T_0 \, S_0(\xi) - (S_0(\xi) + \mu \, \epsilon^{\mathrm{e}} c_{\mathrm{T}}) \, (T - T_0) + \frac{\mu}{2} \, (\epsilon^{\mathrm{e}})^2 + C_v \, (T - T_0) - C_v \, T \, \mathrm{Ln} \, \frac{T}{T_0} \, , \tag{9.2.7}$$

from which follows

$$\sigma = \mu \left(\epsilon^{\mathrm{e}} - c_{\mathrm{T}} \, (T - T_0) \right) , \tag{9.2.8}$$

and

$$S(\mathcal{V}) = S_0(\xi) + \mu \, c_{\mathrm{T}} \epsilon^{\mathrm{e}} + C_v \, \mathrm{Ln} \, \frac{T}{T_0}, \quad U(\mathcal{V}) = U_0(\xi) + \mu \, c_{\mathrm{T}} \, T_0 \, \epsilon^{\mathrm{e}} + \frac{\mu}{2} \, (\epsilon^{\mathrm{e}})^2 + C_v \, (T - T_0) \, , \tag{9.2.9}$$

with μ the constant elastic modulus and c_{T} the constant coefficient of thermal expansion.

9.2.2 Plastic work and inelastic heat

The strain is decomposed into an elastic and a plastic part, $\epsilon = \epsilon^{\mathrm{e}} + \epsilon^{\mathrm{p}}$. From the mechanical part of the linearized internal entropy production (the contributions due to diffusion and mass transfer are excluded here),

$$T\,\hat{S}_{(\mathrm{i},1)} = -\frac{dE}{dt} + \boldsymbol{\sigma} : \frac{d\epsilon}{dt} - S\,\frac{dT}{dt} \geq 0\,, \tag{9.2.10}$$

follow the constitutive equations,

$$\boldsymbol{\sigma} = \frac{\partial E}{\partial \epsilon^{\mathrm{e}}}, \quad S = -\frac{\partial E}{\partial T}\,, \tag{9.2.11}$$

and the reduced inequality,

$$T\,\hat{S}_{(\mathrm{i},1)} = \boldsymbol{\sigma} : \frac{d\epsilon^{\mathrm{p}}}{dt} - \frac{dE}{d\xi}\frac{d\xi}{dt}\,. \tag{9.2.12}$$

The linearized balance of energy may be expressed in terms of the internal energy per unit volume U, heat flux $\mathbf{Q}$ and heat source Θ,

$$\frac{dU}{dt} - \boldsymbol{\sigma} : \frac{d\epsilon}{dt} + \operatorname{div}\mathbf{Q} - \Theta = 0\,. \tag{9.2.13}$$

The energy equation may also be phrased in terms of the free energy $E = U - T\,S$, and the resulting expression of $-dE/dt + \boldsymbol{\sigma} : d\epsilon/dt$ inserted in (9.2.10) yields an alternative form of the dissipation,

$$T\,\hat{S}_{(\mathrm{i},1)} = T\,\frac{dS}{dt} + \operatorname{div}\mathbf{Q} - \Theta\,. \tag{9.2.14}$$

Comparison with (9.2.12) results in the relation,

$$-\operatorname{div}\mathbf{Q} + \Theta + \boldsymbol{\sigma} : \frac{d\epsilon^{\mathrm{p}}}{dt} - \frac{dE}{d\xi}\frac{d\xi}{dt} = T\,\frac{dS}{dt}\,, \tag{9.2.15}$$

that, using (9.2.11), can be split in physically meaningful parts,

$$C_v\,\frac{dT}{dt} = \overbrace{-\operatorname{div}\mathbf{Q} + \Theta}^{\text{external heat}} + \overbrace{\underbrace{\boldsymbol{\sigma} : \frac{d\epsilon^{\mathrm{p}}}{dt}}_{\substack{\text{plastic}\\\text{power } dW^{\mathrm{pl}}/dt}} + \Big(T\,\frac{\partial^2 E}{\partial T\,\partial \xi} - \frac{\partial E}{\partial \xi}\Big)\frac{d\xi}{dt}}^{\text{inelastic heat } dQ^{\mathrm{pl}}/dt} + \overbrace{T\,\frac{\partial \boldsymbol{\sigma}}{\partial T} : \frac{d\epsilon^{\mathrm{e}}}{dt}}^{\text{thermoelastic heat}}\,. \tag{9.2.16}$$

For the free energy (9.2.7), this expression becomes, Rosakis et al. [2000],

$$C_v\,\frac{dT}{dt} = \overbrace{-\operatorname{div}\mathbf{Q} + \Theta}^{\text{external heat}} + \overbrace{\underbrace{\boldsymbol{\sigma} : \frac{d\epsilon^{\mathrm{p}}}{dt}}_{\substack{\text{plastic}\\\text{power } dW^{\mathrm{pl}}/dt}} - \frac{\partial U_0}{\partial \xi}\frac{d\xi}{dt}}^{\text{inelastic heat } dQ^{\mathrm{pl}}/dt} + \overbrace{T\,\frac{\partial \boldsymbol{\sigma}}{\partial T} : \frac{d\epsilon^{\mathrm{e}}}{dt}}^{\text{thermoelastic heat}}\,. \tag{9.2.17}$$

For many metals, the thermoelastic heat is negligible with respect to the inelastic heat. With $\xi = \epsilon^{\mathrm{p}}$, Rosakis et al. [2000] provide measurements of the ratio $\beta = dQ^{\mathrm{pl}}/dW^{\mathrm{pl}}$ of inelastic versus plastic heats: they find β around 0.5 for alumina and 0.8-0.9 for α-titanium. For more recent measurements using infrared thermography, see Badulescu et al. [2011].

Another viewpoint:

Another perspective consists in viewing the rate of internal entropy as dissipated power, e.g., by introducing a potential of dissipation $\Omega \geq 0$. The work done by the plastic strain rate is then the sum of the dissipated heat and a contribution due to the stored energy,

$$\boldsymbol{\sigma} : \frac{d\epsilon^{\mathrm{p}}}{dt} = \overbrace{\Omega}^{\geq 0} + \overbrace{\frac{dE}{d\xi}\frac{d\xi}{dt}}^{\geq 0 \text{ or } \leq 0}\,. \tag{9.2.18}$$

This point of view is ambiguous when one assumes for example $\xi = \epsilon^{\mathrm{p}}$, because the plastic strain rate appears on both the lhs and rhs. However, the models below precise the recoverable nature of the stored energy and give a physical illustration of the internal variables ξ.

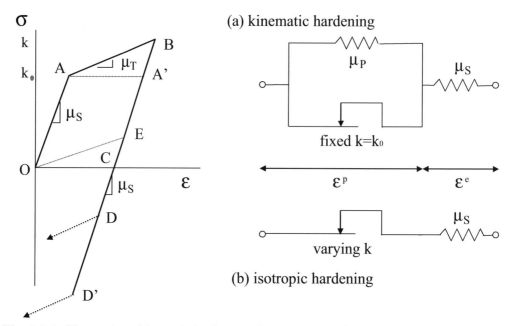

Fig. 9.2.1: Kinematic and isotropic hardenings. Stored energy AA'B=OCE, plastic work OEBA.

9.2.3 Plastic work, recoverable and irrecoverable stored energies

The issues below disregard temperature effects on the mechanical behavior and are one-dimensional. They consider elastic-plastic models, where $\epsilon = \epsilon^{\mathrm{e}} + \epsilon^{\mathrm{p}}$.

9.2.3.1 Kinematic hardening

Here $\xi = \epsilon^{\mathrm{p}}$. Let

$$E = \overbrace{\tfrac{1}{2}\,\mu_S\,(\epsilon^{\mathrm{e}})^2}^{\text{elastic energy}} + \overbrace{\tfrac{1}{2}\,\mu_P\,(\epsilon^{\mathrm{p}})^2}^{\text{recoverable stored energy}}, \qquad (9.2.19)$$

where $\mu_S > 0$ and $\mu_P \geq 0$ are the constant elastic modulus and plastic hardening modulus. Thus,

$$\sigma = \frac{\partial E}{\partial \epsilon^{\mathrm{e}}} = \mu_S\,\epsilon^{\mathrm{e}}, \quad \hat{S}_{(\mathrm{i})} = \sigma \frac{d\epsilon^{\mathrm{p}}}{dt} - \frac{\partial E}{\partial \xi} \frac{d\xi}{dt} = (\sigma - \mu_P\,\epsilon^{\mathrm{p}}) \frac{d\epsilon^{\mathrm{p}}}{dt} \geq 0. \qquad (9.2.20)$$

In the framework of the generalized normality rule, Chapter 22, the inequality above is ensured by the existence of a convex function $f = f(\sigma - \mu_P\,\epsilon^{\mathrm{p}})$ defined on a domain that contains the origin $\sigma - \mu_P\,\epsilon^{\mathrm{p}} = 0$ if

$$\frac{d\epsilon^{\mathrm{p}}}{dt} = \frac{d\lambda}{dt} \frac{\partial f}{\partial(\sigma - \mu_P\,\epsilon^{\mathrm{p}})} = \frac{d\lambda}{dt} \frac{\partial f}{\partial \sigma}, \quad \frac{d\lambda}{dt} \geq 0, \qquad (9.2.21)$$

the derivative being estimated at the current point $\sigma - \mu_P\,\epsilon^{\mathrm{p}}$. Let

$$f(\sigma - \mu_P\,\epsilon^{\mathrm{p}}) = |\sigma - \mu_P\,\epsilon^{\mathrm{p}}| - k_0 \leq 0, \qquad (9.2.22)$$

be the convex of plasticity, with k_0 the constant plastic threshold.

In order to interpret the energies, let us consider the scheme shown in Fig. 9.2.1. The material is subjected to a load cycle controlled in stress. For simplicity of explanation, the unloading point B is assumed to be such that unloading from B to C occurs elastically. The plastic strain at point B, or C, is denoted by ϵ^{p}.

The following relations are easily obtained:
- stress-strain relation:

$$\epsilon = \epsilon^{\mathrm{e}} + \epsilon^{\mathrm{p}}; \quad \frac{d\epsilon^{\mathrm{e}}}{d\sigma} = \frac{1}{\mu_S}; \quad \frac{d\epsilon^{\mathrm{p}}}{d\sigma} = \frac{1}{\mu_P}; \quad \frac{d\epsilon}{d\sigma} = \frac{1}{\mu_T} = \frac{1}{\mu_S} + \frac{1}{\mu_P} \quad \text{on AB}; \qquad (9.2.23)$$

$$de = \frac{d\epsilon}{d\sigma}\frac{d\sigma}{d\epsilon^{\mathrm{P}}} \quad \Rightarrow \quad \epsilon = \epsilon_A + (1 + \frac{\mu_P}{\mu_S})\,\epsilon^{\mathrm{P}} \quad \text{on AB}; \tag{9.2.24}$$

- work done by the spring μ_P : area AA$'$B = OCE :

$$W_{\mathrm{r}} = \int_0^{\epsilon^{\mathrm{P}}} (\sigma - k_0)\,d\epsilon^{\mathrm{P}} = \int_0^{\epsilon^{\mathrm{P}}} \mu_T\,(\epsilon - \epsilon_A)\,d\epsilon^{\mathrm{P}} = \int_0^{\epsilon^{\mathrm{P}}} \mu_P\,\epsilon^{\mathrm{P}}\,d\epsilon^{\mathrm{P}} = \frac{1}{2}\,\mu_P\,(\epsilon^{\mathrm{P}})^2\,; \tag{9.2.25}$$

- work done by the slider : area OEBA = OCA$'$A :

$$W_{\mathrm{nr}} = \int_0^{\epsilon^{\mathrm{P}}} k_0\,d\epsilon^{\mathrm{P}} = k_0\,\epsilon^{\mathrm{P}}\,. \tag{9.2.26}$$

Thus the stored energy is the work done by the spring μ_P. It begins to be recovered upon unloading when the path hits the plastic limit at E, with $|\sigma_E - \sigma_B| = 2\,k_0$.

9.2.3.2 Isotropic hardening

Let

$$E = \overbrace{\tfrac{1}{2}\,\mu_S\,(\epsilon^{\mathrm{e}})^2}^{\text{elastic energy}} + \overbrace{E_h(\xi)}^{\substack{\text{irrecoverable}\\\text{stored energy}}}\,, \tag{9.2.27}$$

where $\mu_S > 0$ is the constant elastic modulus. Thus,

$$\sigma = \frac{\partial E}{\partial \epsilon^{\mathrm{e}}} = \mu_S\,\epsilon^{\mathrm{e}}, \quad \hat{S}_{(\mathrm{i})} = \sigma\,\frac{d\epsilon^{\mathrm{P}}}{dt} - \frac{\partial E}{\partial \xi}\frac{d\xi}{dt} = \sigma\,\frac{d\epsilon^{\mathrm{P}}}{dt} + A\,\frac{d\xi}{dt} \geq 0, \quad A = -\frac{\partial E_h}{\partial \xi}\,. \tag{9.2.28}$$

Again, the inequality above is ensured by the existence of a convex function $f = f(\sigma, \xi)$ defined on a domain that contains the origin $\sigma = 0$,

$$f(\sigma, \xi) = |\sigma| - k \leq 0, \quad k = k_0 - A\,, \tag{9.2.29}$$

with k_0 the initial plastic threshold and k the current plastic threshold, if

$$\frac{d\epsilon^{\mathrm{P}}}{dt} = \frac{d\lambda}{dt}\frac{\partial f}{\partial \sigma}, \quad \frac{d\xi}{dt} = \frac{d\lambda}{dt}\frac{\partial f}{\partial A} = \frac{d\lambda}{dt}, \quad \frac{d\lambda}{dt} \geq 0\,, \tag{9.2.30}$$

the derivative being estimated at the current point (σ, ξ). Thus

$$\frac{d\epsilon^{\mathrm{P}}}{dt} = \frac{d\lambda}{dt}\frac{\partial f}{\partial \sigma} = \frac{d\lambda}{dt}\frac{\sigma}{|\sigma|}, \quad \frac{d\xi}{dt} = \frac{d\lambda}{dt} \Rightarrow \xi = \int_0^{\epsilon^{\mathrm{P}}} d|\epsilon^{\mathrm{P}}|\,. \tag{9.2.31}$$

The interpretations of the stored energy and plastic work are identical to the kinematic model, as long as the points where the unloading path hits the plastic yield limit correspond both to stress of the same sign (negative if the loading involves traction, and conversely positive if the loading involves compression). The point D' at which plastic loading resumes upon unloading is still such that $|\sigma'_D - \sigma_B| = 2\,k$. But now k has increased, assuming the plastic modulus $\mu_P = d^2 E_h/d\xi^2$ positive (plastic hardening) or zero (perfect plasticity) rather than negative (plastic softening). If μ_P is constant, then $E_h = \frac{1}{2}\mu_P\,\xi^2$.

However, there are differences with respect to kinematic hardening. There the stored energy was due to the strain in the spring μ_P, and, upon unloading, the stored energy could be recovered. Such is not the case for the isotropic model above, due to the fact that the threshold of the slider increases, for $\mu_P \geq 0$ constant. In other words, the stored energy is connected with modifications of the microstructure of the model. It can be recovered if the initial microstructure can be recovered as well.

9.3 Anisotropic thermo-poroelasticity: mechanics, transport, energy

This section considers a mixture composed of an anisotropic solid skeleton $k = \mathrm{s}$ circulated by one or several fluid species $k \in \mathcal{K}^{**}$. These species obey
Proposition 9.2 on local thermal equilibrium (LTE).

All species have identical temperature,

$$T_k = T, \quad k \in \mathcal{K}\,. \tag{9.3.1}$$

The property is referred to as *local* thermal equilibrium in order to emphasize the fact that it holds at the level of a representative elementary volume.

9.3.1 Thermodynamic derivation

For convenience, the mechanical entropy productions are now expressed in the reference configuration. The expression for the solid constituent $k = s$ has the form $(7.2.28)_1$ but it does not involve the chemical potential term. Upon multiplication by $T \det \mathbf{F}$, the mechanical entropy productions $(7.2.28)_1$ are modified to

$$
\begin{aligned}
T \hat{\underline{S}}^{\mathrm{s}}_{(\mathrm{i},1)} &= -\frac{d}{dt}(m^{\mathrm{s}} e_{\mathrm{s}}) + \boldsymbol{\tau}^{\mathrm{s}} : \boldsymbol{\nabla} \mathbf{v}_{\mathrm{s}} - m^{\mathrm{s}} s_{\mathrm{s}} \frac{dT}{dt}, \\
T \hat{\underline{S}}^{k}_{(\mathrm{i},1)} &= -\frac{d}{dt}(m^{k} e_{k}) + \boldsymbol{\tau}^{k} : \boldsymbol{\nabla} \mathbf{v}_{\mathrm{s}} + \mu_k \frac{dm^k}{dt} - m^k s_k \frac{dT}{dt}.
\end{aligned}
\tag{9.3.2}
$$

The mechanical entropy production of the mixture,

$$
T \hat{\underline{S}}_{(\mathrm{i},1)} = -\frac{d\underline{E}}{dt} + \boldsymbol{\tau} : \boldsymbol{\nabla} \mathbf{v}_{\mathrm{s}} + \sum_{k \in \mathcal{K}^{**}} \mu_k \frac{dm^k}{dt} - \underline{S} \frac{dT}{dt},
\tag{9.3.3}
$$

expressed in terms of the free energy per reference volume of the mixture,

$$
\underline{E} = \underline{E}(\boldsymbol{\epsilon}, \cup_{k \in \mathcal{K}^{**}} m^k, T) = \sum_{k \in \mathcal{K}} m^k e_k,
\tag{9.3.4}
$$

is set to zero. Constitutive equations result for the Kirchhoff stress $\boldsymbol{\tau}$, the chemical potentials μ_k of the fluid species, and the entropy of the mixture $\underline{S}$,

$$
\boldsymbol{\tau} = \boldsymbol{\tau}^{\mathrm{s}} + \sum_{k \in \mathcal{K}^{**}} \boldsymbol{\tau}^k = \frac{\partial \underline{E}}{\partial \boldsymbol{\epsilon}}; \quad \mu_k = \frac{\partial \underline{E}}{\partial m^k}, \quad k \in \mathcal{K}^{**}; \quad \underline{S} = \sum_{k \in \mathcal{K}} m^k s_k = -\frac{\partial \underline{E}}{\partial T}.
\tag{9.3.5}
$$

If one wishes to change the primary variables, one can
  - either perform a partial or total Legendre transform at the mixture level by changing the strain energy function,
  - or perform the corresponding manipulations directly on the mechanical entropy production of each species.
For example, in order to recover the stress and fluid pressures as primary variables, the appropriate Legendre transform reads,

$$
\underline{E} \longrightarrow \underline{W} = -\underline{E} + \sum_{k \in \mathcal{K}} \boldsymbol{\tau}^k : \boldsymbol{\epsilon} + \sum_{k \in \mathcal{K}^{**}} \mu_k m^k.
\tag{9.3.6}
$$

The mechanical entropy productions (9.3.2) are modified to

$$
\begin{aligned}
T \hat{\underline{S}}^{\mathrm{s}}_{(\mathrm{i},1)} &= \frac{d}{dt}(-m^{\mathrm{s}} e_{\mathrm{s}} + \boldsymbol{\tau}^{\mathrm{s}} : \boldsymbol{\epsilon}) && -\frac{d\boldsymbol{\tau}^{\mathrm{s}}}{dt} : \boldsymbol{\epsilon} - m^{\mathrm{s}} s_{\mathrm{s}} \frac{dT}{dt}, \\
T \hat{\underline{S}}^{k}_{(\mathrm{i},1)} &= \frac{d}{dt}(-m^k e_k + \boldsymbol{\tau}^k : \boldsymbol{\epsilon} + \mu_k m^k) && -\frac{d\boldsymbol{\tau}^k}{dt} : \boldsymbol{\epsilon} - v^k \frac{dp_k}{dt}, \quad k \in \mathcal{K}^{**}.
\end{aligned}
\tag{9.3.7}
$$

Summing up these contributions and setting the resulting mechanical entropy production of the mixture to zero provides constitutive equations in the format,

$$
\boldsymbol{\epsilon} = \frac{\partial \underline{W}}{\partial \boldsymbol{\tau}}; \quad v^k = \frac{\partial \underline{W}}{\partial p_k}, \quad k \in \mathcal{K}^{**}; \quad \underline{S}^{\mathrm{s}} = m^{\mathrm{s}} s_{\mathrm{s}} = \frac{\partial \underline{W}}{\partial T}.
\tag{9.3.8}
$$

9.3.2 Linear thermo-poroelasticity

The analysis is now restricted to linear thermo-poroelastic materials, with a single fluid constituent. The current configuration is taken as reference. The poroelastic part of the constitutive equations is given by (7.4.11), with however a slight modification indicated by (7.4.8) due to the fact that the strain measure is now the volume content, and not the (dimensionless) mass content. The thermo-poroelastic constitutive equations can be cast in the matrix form,

$$
\begin{bmatrix}
\boldsymbol{\epsilon} \\
v^{\mathrm{w}} - v^{\mathrm{w}}_0 \\
S^{\mathrm{s}} - S^{\mathrm{s}}_0
\end{bmatrix}
=
\begin{bmatrix}
\mathbb{E}^{-1} & \mathbb{E}^{-1} : \boldsymbol{\alpha}_{\mathrm{w}} & \mathbf{c}_{\mathrm{T}} \\
\boldsymbol{\alpha}_{\mathrm{w}} : \mathbb{E}^{-1} & \mathbf{I} : \mathbb{E}^{-1} : \boldsymbol{\alpha}_{\mathrm{w}}/S - n^{\mathrm{w}}/K_{\mathrm{w}} & n^{\mathrm{w}} \mathbf{I} : \mathbf{c}_{\mathrm{T}} \\
\mathbf{c}_{\mathrm{T}} & n^{\mathrm{w}} \mathbf{I} : \mathbf{c}_{\mathrm{T}} & C^{\mathrm{s}}_{\sigma p}/T_0
\end{bmatrix}
\begin{bmatrix}
\boldsymbol{\sigma} \\
p_{\mathrm{w}} \\
T - T_0
\end{bmatrix}.
\tag{9.3.9}
$$

Here $\mathbf{c}_T$ [unit : K^{-1}] denotes the thermal expansion tensor of the drained porous medium, and $C_{\sigma p}^s$ [unit : $J/m^3/K$] is the *apparent heat capacity* of the solid, at constant stress and fluid pressure, per unit volume of porous medium, i.e., heat capacity times volume fraction of the solid.

The first line of the matrix form is motivated by the effective stress introduced in (9.3.10), the strain coming into picture now being the elastic strain, that is, the total strain minus the thermal expansion,

$$\boldsymbol{\sigma}' = \boldsymbol{\sigma} + p_w\,\boldsymbol{\alpha}_w = \mathbb{E} : \left(\boldsymbol{\epsilon} - \mathbf{c}_T\,(T - T_0)\right), \tag{9.3.10}$$

with $\boldsymbol{\alpha}_w$ defined by (7.4.14).

The format and notations used to cast the thermo-poroelastic constitutive coefficients appearing in the third line and column of the matrix are now justified. The identification procedure uses the following property:

Proposition 9.3 on homogeneous deformation under purely thermal loading.

For a purely thermal loading, i.e. $\Delta T \neq 0$ while the total stress and pressures vanish, $\boldsymbol{\sigma} = \mathbf{0}$, $p_k = 0$, $k \in \mathcal{K}^{**}$, the deformation should be homogeneous throughout the phases: the contribution of each phase is proportional to its volume fraction, $\Delta v^k = n^k\,\mathrm{tr}\,\boldsymbol{\epsilon}$, $k \in \mathcal{K}$.

The above proposition does not violate the strain averaging proposition 7.1. In fact, since the deformation is homogeneous, the strain in all species is equal to the strain of the solid skeleton, i.e. $\mathrm{tr}\,\boldsymbol{\epsilon}_k = \mathrm{tr}\,\boldsymbol{\epsilon}$, and the strain averaging relation trivially follows from the constraint (7.1.1). In addition, the identification of the thermal expansion tensor of the solid skeleton uses the property epitomized by Proposition 9.4 below. Indeed, under the above purely thermal loading, i.e. all intrinsic and apparent stresses and pressures vanish, the effective strain of the solid constituent given by (7.2.14) and (7.2.15) vanishes, namely $\boldsymbol{\epsilon}_s = \mathbf{c}_{Ts}\,\Delta T$, while the strain of the porous medium follows from (9.3.9) as $\boldsymbol{\epsilon} = \mathbf{c}_T\,\Delta T$. The proposition follows from the fact that, for purely thermal loading, the strains are homogeneous:

Proposition 9.4 on tensors of thermal expansion of the solid and of the solid skeleton.

The tensors of thermal expansion of the solid skeleton $\mathbf{c}_T$ and of the solid constituent $\mathbf{c}_{Ts}$ are identical:

$$\mathbf{c}_T = \mathbf{c}_{Ts}. \tag{9.3.11}$$

The variation of fluid mass expresses in terms of the variations of fluid volume content, fluid pressure and temperature via (7.2.9). Then it appears that the major symmetry property is *not* conserved when the constitutive equations express the 'strain' triplet $(\boldsymbol{\epsilon}, \zeta^w, S^s)$ in terms of the 'stress' triplet $(\boldsymbol{\sigma}, p_w, T)$. On the other hand, the major symmetry property is conserved for the triplet $(\boldsymbol{\epsilon}, \zeta^w, S)$, S being the entropy of the mixture per unit volume,

$$
\begin{bmatrix} \boldsymbol{\epsilon} \\[1mm] \zeta^w \\[1mm] S - S_0 \end{bmatrix}
=
\begin{bmatrix}
\mathbb{E}^{-1} & \mathbb{E}^{-1} : \boldsymbol{\alpha}_w & \mathbf{c}_T \\[1mm]
\boldsymbol{\alpha}_w : \mathbb{E}^{-1} & \mathbf{I} : \mathbb{E}^{-1} : \boldsymbol{\alpha}_w / S & n^w\,(\mathbf{I} : \mathbf{c}_T - c_{Tw}) \\[1mm]
\mathbf{c}_T & n^w\,(\mathbf{I} : \mathbf{c}_T - c_{Tw}) & C_{\sigma p} / T_0
\end{bmatrix}
\begin{bmatrix} \boldsymbol{\sigma} \\[1mm] p_w \\[1mm] T - T_0 \end{bmatrix}, \tag{9.3.12}
$$

where $C_{\sigma p}$ [unit : $J/m^3/K$] is the *apparent heat capacity* of the porous medium at constant stress and pressure per unit volume of porous medium. In fact,

$$
\begin{aligned}
C_{\sigma p} &= & C_{\sigma p}^s & + & C_{\sigma p}^w \\
&= & n^s\,C_{\sigma p,s} & + & n^w\,C_{\sigma p,w} \\
&= & n^s\,\rho_s\,c_{\sigma p,s} & + & n^w\,\rho_w\,c_{\sigma p,w},
\end{aligned} \tag{9.3.13}
$$

with $C_{\sigma p}^k$ [unit : $J/m^3/K$] being the *apparent heat capacity* of the species k, at constant stress and fluid pressure, per unit volume of porous medium, and $c_{\sigma p,k}$ [unit : $J/kg/K$] the *heat capacity* of the species k, at constant stress and fluid pressure as well but per unit mass of species k.

Thermal loading $T - T_0 > 0$ implies a volume expansion of the solid skeleton, and a proportional expansion of the pore space. The change of fluid mass under purely thermal loading can now be interpreted:

Proposition 9.5 on influx and efflux of fluid mass due to thermal loading.

The change of fluid mass in a porous medium in local thermal equilibrium due to a change of temperature, at constant stress and fluid pressure, is governed by the difference of the coefficients of thermal expansion of the solid and fluid constituents. Table 9.1.2 displays values for a few solids and liquids.

If the strain, instead of the total stress, is considered as a primary variable, then (9.3.12) becomes,

$$
\begin{bmatrix} -\boldsymbol{\sigma} \\ \zeta^{\mathrm{w}} \\ S - S_0 \end{bmatrix} = \begin{bmatrix} -\mathbb{E} & \boldsymbol{\alpha}_{\mathrm{w}} & \mathbb{E} : \mathbf{c}_{\mathrm{T}} \\ \boldsymbol{\alpha}_{\mathrm{w}} & 1/M & n^{\mathrm{w}} \left(\mathbf{I} : \mathbf{c}_{\mathrm{T}} - c_{\mathrm{Tw}} \right) - \boldsymbol{\alpha}_{\mathrm{w}} : \mathbf{c}_{\mathrm{T}} \\ \mathbb{E} : \mathbf{c}_{\mathrm{T}} & n^{\mathrm{w}} \left(\mathbf{I} : \mathbf{c}_{\mathrm{T}} - c_{\mathrm{Tw}} \right) - \boldsymbol{\alpha}_{\mathrm{w}} : \mathbf{c}_{\mathrm{T}} & C_{\epsilon p}/T_0 \end{bmatrix} \begin{bmatrix} \boldsymbol{\epsilon} \\ p_{\mathrm{w}} \\ T - T_0 \end{bmatrix}, \tag{9.3.14}
$$

where the scalar modulus M has been defined by (7.4.19), and $C_{\epsilon p}$ [unit : J/m^3/K] is the apparent heat capacity of the porous medium at constant strain and fluid pressure,

$$
C_{\sigma p} = C_{\epsilon p} + T_0 \, \mathbf{c}_{\mathrm{T}} : \mathbb{E} : \mathbf{c}_{\mathrm{T}} \,. \tag{9.3.15}
$$

One may further use the fluid mass content as primary variable instead of the fluid pressure. The constitutive relations then involve the undrained tensor of thermal expansion $\mathbf{c}_{\mathrm{T}}^{\mathrm{U}}$,

$$
\mathbf{c}_{\mathrm{T}}^{\mathrm{U}} = \mathbf{c}_{\mathrm{T}} + n^{\mathrm{w}} \left(c_{\mathrm{Tw}} - \mathbf{I} : \mathbf{c}_{\mathrm{T}} \right) \mathbf{S} \,, \tag{9.3.16}
$$

where the Skempton tensor $\mathbf{S}$ defined by (7.4.20) satisfies the relation (7.4.30). Moreover, with help of (7.4.22) and (7.4.30), the following relation may be obtained,

$$
\left(n^{\mathrm{w}} \left(\mathbf{I} : \mathbf{c}_{\mathrm{T}} - c_{\mathrm{Tw}} \right) - \boldsymbol{\alpha}_{\mathrm{w}} : \mathbf{c}_{\mathrm{T}} \right) M \, \boldsymbol{\alpha}_{\mathrm{w}} = \mathbf{c}_{\mathrm{T}} : \mathbb{E} - \mathbf{c}_{\mathrm{T}}^{\mathrm{U}} : \mathbb{E}^{\mathrm{U}} \,. \tag{9.3.17}
$$

With repeated use of (7.4.30), the constitutive equations can be cast in a sort of *undrained* format,

$$
\begin{bmatrix} \boldsymbol{\sigma} \\ p_{\mathrm{w}} \\ -(S - S_0) \end{bmatrix} = \begin{bmatrix} \mathbb{E}^{\mathrm{U}} & -\mathbb{E}^{\mathrm{U}} : \mathbf{S} & -\mathbb{E}^{\mathrm{U}} : \mathbf{c}_{\mathrm{T}}^{\mathrm{U}} \\ -\mathbf{S} : \mathbb{E}^{\mathrm{U}} & M & -(\mathbf{c}_{\mathrm{T}} : \mathbb{E} - \mathbf{c}_{\mathrm{T}}^{\mathrm{U}} : \mathbb{E}^{\mathrm{U}}) : \boldsymbol{\alpha}_{\mathrm{w}}^{-1} \\ -\mathbf{c}_{\mathrm{T}}^{\mathrm{U}} : \mathbb{E}^{\mathrm{U}} & -(\mathbf{c}_{\mathrm{T}} : \mathbb{E} - \mathbf{c}_{\mathrm{T}}^{\mathrm{U}} : \mathbb{E}^{\mathrm{U}}) : \boldsymbol{\alpha}_{\mathrm{w}}^{-1} & -C_{\epsilon\zeta}^{\mathrm{U}}/T_0 \end{bmatrix} \begin{bmatrix} \boldsymbol{\epsilon} \\ \zeta^{\mathrm{w}} \\ T - T_0 \end{bmatrix}, \tag{9.3.18}
$$

which involves the heat capacity of the porous medium at constant strain and fluid mass,

$$
C_{\epsilon\zeta}^{\mathrm{U}} = C_{\epsilon p} - \left((\mathbf{c}_{\mathrm{T}} : \mathbb{E} - \mathbf{c}_{\mathrm{T}}^{\mathrm{U}} : \mathbb{E}^{\mathrm{U}}) : \boldsymbol{\alpha}_{\mathrm{w}}^{-1} \right)^2 \frac{T_0}{M} \,, \tag{9.3.19}
$$

with M given by one of the expressions proposed in Section 7.4.2, for example (7.4.32).

These constitutive equations are a generalization to anisotropic materials of the equations derived by Mc-Tigue [1986]. Yet, McTigue [1986] contains two additional qualitative features, which are nevertheless not quantified:
- he does not adopt neither Proposition 9.3 nor Proposition 9.4. According to (7.2.21), the change of the volume fraction of the solid constituent Δn^{s} is equal to $\mathrm{tr}\,\boldsymbol{\epsilon}_{\mathrm{s}} - \mathrm{tr}\,\boldsymbol{\epsilon}$, which, for a purely thermal loading, becomes $\Delta n^{\mathrm{s}} = n^{\mathrm{s}} \mathbf{I} : (\mathbf{c}_{\mathrm{Ts}} - \mathbf{c}_{\mathrm{T}}) \Delta T$. Propositions 9.3 and 9.4 amount to set this change of volume fraction to zero;
- he assumes that under uniform pressure loading, the volume change Δv is controlled by a second tensor moduli $\mathbb{E}_{\mathrm{s}}^{\prime\prime}$ different from the tensor of the solid constituent $\mathbb{E}_{\mathrm{s}}$. Then $\boldsymbol{\alpha}_{\mathrm{w}}$, Eqn (7.4.14), remains unchanged but the Skempton coefficient (7.4.15) becomes,

$$
\frac{1}{S} = 1 + n^{\mathrm{w}} \left(\frac{1}{K_{\mathrm{w}}} - \frac{1}{K_{\mathrm{s}}^{\prime\prime}} \right) \left(\frac{1}{K} - \frac{1}{K_{\mathrm{s}}} \right)^{-1}, \tag{9.3.20}
$$

where $1/K_{\mathrm{s}}^{\prime\prime} \equiv \mathbf{I} : (\mathbb{E}_{\mathrm{s}}^{\prime\prime})^{-1} : \mathbf{I}$.

These two features are highlighted in the change of volume fraction $\Delta n^{\mathrm{w}} = \Delta v^{\mathrm{w}} - n^{\mathrm{w}} \mathrm{tr}\,\boldsymbol{\epsilon}$. Indeed, with insertion of (7.4.14) and (9.3.20) in (9.3.9), the change of volume fraction becomes,

$$
n^{\mathrm{w}} - n_0^{\mathrm{w}} = (\boldsymbol{\alpha}_{\mathrm{w}} - n^{\mathrm{w}} \mathbf{I}) : \mathbb{E}^{-1} : (\boldsymbol{\sigma} + p_{\mathrm{w}} \mathbf{I}) + (\frac{n^{\mathrm{w}}}{K_{\mathrm{s}}} - \frac{n^{\mathrm{w}}}{K_{\mathrm{s}}^{\prime\prime}}) p_{\mathrm{w}} + n^{\mathrm{w}} \mathbf{I} : (\mathbf{c}_{\mathrm{Ts}} - \mathbf{c}_{\mathrm{T}}) \Delta T \,. \tag{9.3.21}
$$

By contrast, two effective stresses epitomize the model developed in this section:

Proposition 9.6 on effective stresses.

While the thermoelastic strain is driven by the Nur-Byerlee effective stress,

$$\boldsymbol{\epsilon} - \mathbf{c}_{\mathrm{T}}\,(T - T_0) \;=\; \mathbb{E}^{-1} : \overbrace{(\boldsymbol{\sigma} + p_{\mathrm{w}}\,\boldsymbol{\alpha}_{\mathrm{w}})}^{\text{Nur}-\text{Byerlee stress}}\;, \tag{9.3.22}$$

the volume fractions are governed by Terzaghi's effective stress, e.g., for the fluid volume fraction,

$$n^{\mathrm{w}} - n_0^{\mathrm{w}} \;=\; (\boldsymbol{\alpha}_{\mathrm{w}} - n^{\mathrm{w}}\,\mathbf{I}) : \mathbb{E}^{-1} : \overbrace{(\boldsymbol{\sigma} + p_{\mathrm{w}}\,\mathbf{I})}^{\text{Terzaghi's stress}}\;. \tag{9.3.23}$$

If the solid constituent is incompressible[9.1], then $\boldsymbol{\alpha}_{\mathrm{w}} = \mathbf{I}$. Using (9.3.22), the latter relation (9.3.23) becomes $n^{\mathrm{w}} - n_0^{\mathrm{w}} = n^{\mathrm{s}}\,\mathrm{tr}\,(\boldsymbol{\epsilon} - \mathbf{c}_{\mathrm{T}}\,\Delta T)$ as it should. Indeed then $n^{\mathrm{s}} = V_{\mathrm{s}}/V$, and $n^{\mathrm{s}} - n_0^{\mathrm{s}} = -n^{\mathrm{s}}\,\mathrm{tr}\,(\boldsymbol{\epsilon} - \mathbf{c}_{\mathrm{Ts}}\,\Delta T)$. The changes of the two volume fractions are opposite, as they should, due to Proposition 9.4.

9.3.3 Coupled flows: thermoosmosis

Let us consider a porous medium saturated by water to be in contact with two reservoirs in which the temperature and the pressure are controlled. The interface between the porous medium and the baths can be permeable or impermeable to water flow and/or heat flow.

The thermoosmotic properties are assumed to be isotropic, so as to simplify the exposition. Neglecting acceleration and body force in (7.2.32), the internal dissipation associated with the diffusion of water and heat expresses in terms of the heat flux $\mathbf{Q}$ [unit : W/m^2=kg/s^3], volume hydraulic flux $\mathbf{J}_{\mathrm{w}}$ [unit : m/s], and gradients of the temperature T [unit : K], and of the pressure p_{w} [unit : Pa],

$$T\hat{S}_{(\mathrm{i},3)} = -\mathbf{Q} \cdot \frac{\boldsymbol{\nabla} T}{T} - \mathbf{J}_{\mathrm{w}} \cdot \boldsymbol{\nabla} p_{\mathrm{w}} \geq 0\,. \tag{9.3.24}$$

By choosing the fluxes $\mathcal{J}$ to be linearly related to the gradients $\mathcal{F}$ through a positive (semi-)definite thermoosmotic matrix $\mathcal{K}$,

$$\mathcal{J} = -\mathcal{K}\,\mathcal{F}, \qquad \begin{bmatrix} \mathbf{J}_{\mathrm{w}} \\ \mathbf{Q} \end{bmatrix} = - \begin{bmatrix} k_{\mathrm{H}} & k_{\mathrm{TO}} \\ k_{\mathrm{MC}} & T\,k_{\mathrm{T}} \end{bmatrix} \begin{bmatrix} \boldsymbol{\nabla} p_{\mathrm{w}} \\ \boldsymbol{\nabla} T/T \end{bmatrix}, \tag{9.3.25}$$

the dissipation inequality can be satisfied. The actual dissipation depends only on the symmetric part of the matrix, Exercise 9.1. This observation is used as an argument to assume the matrix to be symmetric. The symmetry property is conserved by consistent scalings of the fluxes and driving gradients, Exercise 16.1. On the other hand, there is no argument why the fluxes themselves should display a symmetric coupling: while, in many cases, symmetry is used mainly as a starting convenient assumption, it should *in fine* be verified, or disproved experimentally.

Thus if the thermoosmotic matrix is symmetric, $k_{\mathrm{MC}} = k_{\mathrm{TO}}$, the conditions of positive definiteness become,

$$k_{\mathrm{H}} > 0, \quad k_{\mathrm{T}} > 0, \quad \sqrt{T\,k_{\mathrm{T}}\,k_{\mathrm{H}}} \geq k_{\mathrm{TO}} \geq -\sqrt{T\,k_{\mathrm{T}}\,k_{\mathrm{H}}}\,. \tag{9.3.26}$$

Indeed,

$$\begin{aligned} \mathcal{F}^{\mathrm{T}}\mathcal{K}\,\mathcal{F} &= k_{\mathrm{H}}\,\mathcal{F}_{\mathrm{w}}^2 + 2\,k_{\mathrm{TO}}\,\mathcal{F}_{\mathrm{w}} \cdot \mathcal{F}_{\mathrm{T}} + T\,k_{\mathrm{T}}\,\mathcal{F}_{\mathrm{T}}^2, \\ &= k_{\mathrm{H}}\,(\mathcal{F}_{\mathrm{w}} + \frac{k_{\mathrm{TO}}}{k_{\mathrm{H}}}\,\mathcal{F}_{\mathrm{T}})^2 + (T\,k_{\mathrm{T}} - \frac{k_{\mathrm{TO}}^2}{k_{\mathrm{H}}})\,\mathcal{F}_{\mathrm{T}}^2\,. \end{aligned} \tag{9.3.27}$$

If the matrix is not symmetric, just replace k_{TO} by $(k_{\mathrm{TO}} + k_{\mathrm{MC}})/2$ in (9.3.26) and (9.3.27).

Let us record the units of the coefficients of the thermoosmotic matrix:

$$\text{units:} \quad k_{\mathrm{H}}\text{: m}^3\times\text{s/kg}; \quad k_{\mathrm{TO}}, k_{\mathrm{MC}}\text{: m}^2\text{/s}; \quad k_{\mathrm{T}}\text{: kg}\times\text{m/s}^3\text{/K}.$$

[9.1] Actually, we have seen in Section 3.3.2 that incompressibility is accompanied by thermal inextensibility.

9.3.3.1   Experimental identification of the thermoosmotic coefficients

Since the flows are coupled, the experiments to be used to identify the diffusion coefficients need to be described accurately. A similar issue is encountered for electroosmosis in Section 16.2.5. Indeed,

(1) isothermal flow $\nabla T = \mathbf{0}$:

Then the transfer of matter is accompanied with a transfer of heat. If vaporization takes place, the proportionality coefficient is the latent heat of vaporization $\mathcal{L}_{\mathrm{vap}}$ [unit : $\mathrm{m}^2/\mathrm{s}^2$],

$$\mathbf{J}_{\mathrm{w}} = -k_{\mathrm{H}} \cdot \nabla p_{\mathrm{w}}, \quad \mathbf{Q} = -k_{\mathrm{MC}} \cdot \nabla p_{\mathrm{w}} \Rightarrow \mathbf{Q} = \rho_{\mathrm{w}} \mathcal{L}_{\mathrm{vap}} \mathbf{J}_{\mathrm{w}}, \quad \rho_{\mathrm{w}} \mathcal{L}_{\mathrm{vap}} = \frac{k_{\mathrm{MC}}}{k_{\mathrm{H}}}. \qquad (9.3.28)$$

According to Spanner [1964], p. 246, at 35 °C, $\mathcal{L}_{\mathrm{vap}} = 2.417\,\mathrm{MJ/kg}$, $\rho_{\mathrm{w}} = 994\,\mathrm{kg/m}^3$, so that $k_{\mathrm{MC}}/k_{\mathrm{H}} = 2403 \times 10^6\,\mathrm{kg/m/s}^2$.

(2) thermally insulated flow $\mathbf{Q} = \mathbf{0}$:

Measurement of hydraulic permeability in absence of heat flux provides the 'thermally insulated' hydraulic permeability

$$\nabla T = -\frac{k_{\mathrm{MC}}}{k_{\mathrm{T}}} \nabla p_{\mathrm{w}}, \quad \mathbf{J}_{\mathrm{w}} = -\underbrace{(k_{\mathrm{H}} - \frac{k_{\mathrm{TO}}\,k_{\mathrm{MC}}}{T\,k_{\mathrm{T}}})}_{\substack{\text{'thermally insulated'}\\ \text{hydraulic permeability} \geq 0}} \nabla p_{\mathrm{w}}. \qquad (9.3.29)$$

(3) thermoosmotic pressure in absence of water flow $\mathbf{J}_{\mathrm{w}} = \mathbf{0}$:

Measurement of thermal conductivity in absence of water flux provides the 'hydraulically open circuit' thermal conductivity

$$\nabla p_{\mathrm{w}} = -\frac{k_{\mathrm{TO}}}{k_{\mathrm{H}}} \frac{\nabla T}{T}, \quad \mathbf{Q} = -\underbrace{(k_{\mathrm{T}} - \frac{k_{\mathrm{MC}}\,k_{\mathrm{TO}}}{T\,k_{\mathrm{H}}})}_{\substack{\text{'hydraulically open circuit'}\\ \text{thermal conductivity} \geq 0}} \nabla T. \qquad (9.3.30)$$

Thus at 35°C, assuming $k_{\mathrm{MC}} = k_{\mathrm{TO}}$, a gradient of temperature of 1°C gives rise to a gradient of pressure equal to -7.8 MPa: note that the signs of the gradients are opposite.

9.3.3.2   The sign of the thermoosmotic coefficient

Whether in membranes, biomaterials or geomaterials, thermoosmosis is strongly related to the presence of a fixed electrical charge. For positively charged (hydrophobic) membranes, the thermoosmotic coefficient is negative so that the thermoosmotic flow is from the cold side to the hot side. For negatively charged membranes, like Nafion®, endowed with hydrophilic sulfonic groups, and for negatively charged geomaterials, experimental measurements seem to depend on the properties of the electrolyte and, in particular, of the counterions. Both positive and negative thermoosmotic coefficients are reported.

Constitutive equations of generalized diffusion with thermal effects in an electrolyte are developed in Section 16.5.

9.3.4   Energy equation of a porous medium in local thermal equilibrium

Since the two species have identical temperature, a single energy equation associated with the whole mixture is needed. The following simplifying assumptions are adopted:
- no exchange of mass between the fluid and solid constituents takes place;
- no supply of mass, momentum and internal energy emanates from the surroundings.

In a small strain context, the energy equation (4.6.22) reduces to

$$\frac{dU}{dt} + U \operatorname{div} \mathbf{v}_{\mathrm{s}} + \operatorname{div}(h_{\mathrm{w}} \mathbf{M}_{\mathrm{w}}) + \mathbf{M}_{\mathrm{w}} \cdot (\frac{d^{\mathrm{w}} \mathbf{v}_{\mathrm{w}}}{dt} - \mathbf{b}_{\mathrm{w}}) - \boldsymbol{\sigma} : \frac{d\boldsymbol{\epsilon}}{dt} + \operatorname{div} \mathbf{Q} - \Theta = 0, \qquad (9.3.31)$$

where U [unit : Pa] is the internal energy of the porous medium per unit volume, and Θ [unit : W/m^3] a heat source per unit volume. The mechanical entropy production (4.7.28) simplifies to,

$$T\hat{S}_{(i,1)} = -\frac{dE}{dt} - E\,\mathrm{div}\,\mathbf{v}_s + \boldsymbol{\sigma} : \frac{d\boldsymbol{\epsilon}}{dt} - S\frac{dT}{dt} + \mu_w\frac{dm^w}{dt}\,. \tag{9.3.32}$$

It vanishes by construction of the thermoelastic constitutive equations. Upon elimination of the rate of work $\boldsymbol{\sigma} : d\boldsymbol{\epsilon}/dt$, the energy equation can be phrased in terms of the entropy of the porous medium,

$$T\hat{S}_{(i,1)} = T\frac{dS}{dt} + TS\,\mathrm{div}\,\mathbf{v}_s + \mu_w\frac{dm^w}{dt} + \mathrm{div}\,(\mathbf{Q} + h_w\,\mathbf{M}_w) + \mathbf{M}_w \cdot (\frac{d^w\mathbf{v}_w}{dt} - \mathbf{b}_w) - \Theta = 0\,. \tag{9.3.33}$$

With the entropy now given by the constitutive equations (9.3.18), the energy equation becomes,

$$\underbrace{T\,\mathbf{c}_T^U : \mathbb{E}^U : \frac{d\boldsymbol{\epsilon}}{dt} + T\,(\mathbf{c}_T : \mathbb{E} - \mathbf{c}_T^U : \mathbb{E}^U) : \boldsymbol{\alpha}_w^{-1}\frac{d\zeta^w}{dt}}_{\text{thermoelastic heat}} + \underbrace{\frac{T}{T_0}\,C_{\epsilon\zeta}^U\frac{dT}{dt}}_{\substack{\text{purely thermal} \\ \text{contribution}}}$$

$$+ \underbrace{TS\,\mathbf{I} : \frac{d\boldsymbol{\epsilon}}{dt} + \rho_w\mu_w\frac{d\zeta^w}{dt}}_{\text{nonlinear terms}} + \underbrace{\mathrm{div}\,(\mathbf{Q} + h_w\,\mathbf{M}_w)}_{\substack{\text{conductive+convective} \\ \text{heat fluxes}}} + \mathbf{M}_w \cdot (\frac{d^w\mathbf{v}_w}{dt} - \mathbf{b}_w) - \underbrace{\Theta}_{\substack{\text{heat} \\ \text{source}}} = 0\,. \tag{9.3.34}$$

The specific entropy and specific enthalpy of water, s_w and h_w respectively, are given in explicit form in terms of pressure and temperature in Section 3.3.3.

The orders of magnitude of the various terms may be estimated as indicated by Proposition 9.1. Fluid mechanics and diffusion models introduce dimensionless numbers, e.g., Péclet, Reynolds and Prandtl numbers, that indicate the relative weights of various physical phenomena: examples are shown in Sections 15.3.4 and 15.6.

It thus appears that the heat transported by the porous medium is composed of
- the *conductive heat* $\mathbf{Q}$ and of
- the heat $h_w\,\mathbf{M}_w$ convected by the fluid.

With help of the relation (4.4.9), an alternative format would display the convective contribution in the form of a product of a flux times a gradient of a scalar entity (pressure, thermodynamic function $\cdots$), for example,

$$\underbrace{T\,\mathbf{c}_T^U : \mathbb{E}^U : \frac{d\boldsymbol{\epsilon}}{dt} + T\,(\mathbf{c}_T : \mathbb{E} - \mathbf{c}_T^U : \mathbb{E}^U) : \boldsymbol{\alpha}_w^{-1}\frac{d\zeta^w}{dt}}_{\text{thermoelastic heat}} + \underbrace{\frac{T}{T_0}\,C_{\epsilon\zeta}^U\frac{dT}{dt}}_{\substack{\text{purely thermal} \\ \text{contribution}}} + \underbrace{\mathbf{M}_w \cdot \boldsymbol{\nabla}h_w}_{\substack{\text{convective} \\ \text{term}}}$$

$$+ \underbrace{TS\,\mathbf{I} : \frac{d\boldsymbol{\epsilon}}{dt} - T\rho_w s_w\frac{d\zeta^w}{dt}}_{\substack{\text{nonlinear} \\ \text{terms}}} + \underbrace{\mathrm{div}\,\mathbf{Q}}_{\substack{\text{conductive} \\ \text{heat flux}}} + \mathbf{M}_w \cdot (\frac{d^w\mathbf{v}_w}{dt} - \mathbf{b}_w) - \underbrace{\Theta}_{\substack{\text{heat} \\ \text{source}}} = 0\,. \tag{9.3.35}$$

This form is more appropriate to the case where the flux is either a datum or strongly constrained, like for forced convection. The current exposition is more in line with free convection: the hydraulic and thermal fluxes are not considered as primary variables, but rather as dependent variables given by the constitutive equations of generalized diffusion. In a finite element perspective, prescribing the hydraulic flux at arbitrary points within the body requires that this flux, or the velocities it is built on, is a primary variable. As an example, the finite element scheme used in a dynamic context by Loret and Prévost [1991] makes use of solid and fluid displacements as primary variables. The associated constitutive framework is exposed in Section 7.3.6. By contrast, the approach adopted in the other sections of this chapter is in line with a *mixed* finite element method with the solid displacement, fluid pressure and temperature as primary variables.

9.4 Generalized compatibility conditions for the stress

The compatibility conditions for infinitesimal strains (21.8.12),

$$\mathbf{A} \equiv \overset{\rightarrow}{\boldsymbol{\nabla}}^2 \mathrm{tr}\,\boldsymbol{\epsilon} - \overset{\rightarrow}{\boldsymbol{\nabla}}(\overset{\rightarrow}{\boldsymbol{\nabla}}\cdot\boldsymbol{\epsilon}) - \left(\overset{\rightarrow}{\boldsymbol{\nabla}}(\overset{\rightarrow}{\boldsymbol{\nabla}}\cdot\boldsymbol{\epsilon})\right)^{\mathrm{T}} + \Delta\boldsymbol{\epsilon} = \mathbf{0}\,,$$

$$B \equiv \Delta\mathrm{tr}\,\boldsymbol{\epsilon} - \overset{\rightarrow}{\boldsymbol{\nabla}}\cdot(\overset{\rightarrow}{\boldsymbol{\nabla}}\cdot\boldsymbol{\epsilon}) = 0\,, \tag{9.4.1}$$

may be phrased in terms of the stress. The resulting restrictions are useful if a solution to a boundary value problem is sought in terms of stress, rather than strain.

9.4.1 Compatibility conditions for the stress for linear isotropic elasticity

For isotropic elasticity, introduction of the strain $E\,\boldsymbol{\epsilon} = -\nu\,\mathrm{tr}\,\boldsymbol{\sigma}\,\mathbf{I} + (1+\nu)\,\boldsymbol{\sigma}$ in the above relations yields the two conditions,

$$\mathbf{A} \equiv \frac{1}{E}\,\vec{\nabla}^2\,\mathrm{tr}\,\boldsymbol{\sigma} - \frac{\nu}{E}\,\Delta\mathrm{tr}\,\boldsymbol{\sigma}\,\mathbf{I} - \frac{1+\nu}{E}\left(\vec{\nabla}(\vec{\nabla}\cdot\boldsymbol{\sigma}) + ((\vec{\nabla}(\vec{\nabla}\cdot\boldsymbol{\sigma}))^{\mathsf{T}}\right) + \frac{1+\nu}{E}\,\Delta\boldsymbol{\sigma} = \mathbf{0}\,,$$

$$B \equiv \frac{1-\nu}{E}\,\Delta\mathrm{tr}\,\boldsymbol{\sigma} - \frac{1+\nu}{E}\,\vec{\nabla}\cdot(\vec{\nabla}\cdot\boldsymbol{\sigma}) = 0\,, \tag{9.4.2}$$

which are combined to $\mathbf{A} + \nu/(1-\nu)\,B\,\mathbf{I} = \mathbf{0}$. Together with the three equations of balance of momentum $\vec{\nabla}\cdot\boldsymbol{\sigma} + \mathbf{f} = \mathbf{0}$, $\sigma_{ij,i} + f_j = 0$, with body force $\mathbf{f}$, the six compatibility conditions,

$$\frac{1}{E}\,\vec{\nabla}^2\,\mathrm{tr}\,\boldsymbol{\sigma} + \frac{1+\nu}{E}\,\Delta\boldsymbol{\sigma} + \frac{\nu}{E}\frac{1+\nu}{1-\nu}\,\vec{\nabla}\cdot\mathbf{f}\,\mathbf{I} + \frac{1+\nu}{E}\left(\vec{\nabla}\mathbf{f} + \mathbf{f}\overleftarrow{\nabla}\right) = \mathbf{0}\,,$$

$$\frac{1}{E}\,\sigma_{kk,ij} + \frac{1+\nu}{E}\,\sigma_{ij,kk} + \frac{\nu}{E}\frac{1+\nu}{1-\nu}\,f_{k,k}\,I_{ij} + \frac{1+\nu}{E}\left(f_{j,i} + f_{i,j}\right) = 0\,, \tag{9.4.3}$$

require nine equations to be satisfied by the six stress components, and therefore represent three functionally independent constraints. They are referred to as Beltrami-Mitchell compatibility equations.

Special case: body force deriving from a harmonic potential
    As a special case, consider a body force that derives from a harmonic potential,

$$\mathbf{f} = -\vec{\nabla}F\,, \quad \Delta F = 0\,. \tag{9.4.4}$$

Then, the compatibility equations simplify to,

$$\frac{1}{E}\,\vec{\nabla}^2\,\mathrm{tr}\,\boldsymbol{\sigma} + \frac{1+\nu}{E}\,\Delta\boldsymbol{\sigma} - 2\frac{1+\nu}{E}\,\vec{\nabla}^2 F = \mathbf{0}\,,$$

$$\frac{1}{E}\,\sigma_{kk,ij} + \frac{1+\nu}{E}\,\sigma_{ij,kk} - 2\frac{1+\nu}{E}\,F_{,ij} = 0\,. \tag{9.4.5}$$

The trace of these relations implies (since $2 + \nu \neq 0$) that the first invariant of the stress $\mathrm{tr}\,\boldsymbol{\sigma}$ is harmonic,

$$\Delta\mathrm{tr}\,\boldsymbol{\sigma} = 0\,. \tag{9.4.6}$$

Therefore, if it vanishes on the boundary of a simply connected body, it vanishes within the body. Moreover, applying the Laplacian operator to the compatibility relations, and accounting for the fact that both $\mathrm{tr}\,\boldsymbol{\sigma}$ and the potential F are harmonic implies that the stress is biharmonic,

$$\Delta\Delta\boldsymbol{\sigma} = 0\,, \quad \sigma_{ij,kkll} = 0\,. \tag{9.4.7}$$

9.4.2 Compatibility conditions for the stress in presence of an inelastic strain

Assume now that the strain contains an inelastic component $\boldsymbol{\epsilon}^*$,

$$\boldsymbol{\epsilon} = -\frac{\nu}{E}\,\mathrm{tr}\,\boldsymbol{\sigma}\,\mathbf{I} + \frac{1+\nu}{E}\,\boldsymbol{\sigma} + \boldsymbol{\epsilon}^*\,. \tag{9.4.8}$$

The inelastic strain $\boldsymbol{\epsilon}^*$ may have various origins, e.g., it can represent a thermal strain or a growth strain. The inelastic strain $\boldsymbol{\epsilon}^*$ corresponding to an anisotropic linear thermo-poroelastic material with a single porosity can be read from (9.3.9). The special case where both the elastic and the thermal properties are isotropic is given by (9.5.2).
    Then the following terms should be added to the lhs of the compatibility relations (9.4.3),

$$\vec{\nabla}^2\,\mathrm{tr}\,\boldsymbol{\epsilon}^* - \vec{\nabla}(\vec{\nabla}\cdot\boldsymbol{\epsilon}^*) - \left(\vec{\nabla}(\vec{\nabla}\cdot\boldsymbol{\epsilon}^*)\right)^{\mathsf{T}} + \Delta\boldsymbol{\epsilon}^* + \frac{\nu}{1-\nu}(\Delta\mathrm{tr}\,\boldsymbol{\epsilon}^* - \vec{\nabla}\cdot(\vec{\nabla}\cdot\boldsymbol{\epsilon}^*))\,\mathbf{I}\,,$$

$$\epsilon^*_{kk,ij} - \epsilon^*_{ik,jk} - \epsilon^*_{jk,ik} + \epsilon^*_{ij,kk} + \frac{\nu}{1-\nu}\left(\epsilon^*_{mm,kk} - \epsilon^*_{km,km}\right)I_{ij}\,. \tag{9.4.9}$$

If the inelastic strain is isotropic, the above terms simplify to

$$
\frac{1}{3} \vec{\nabla}^2 \operatorname{tr} \boldsymbol{\epsilon}^* + \frac{1}{3}\frac{1+\nu}{1-\nu} \Delta \operatorname{tr} \boldsymbol{\epsilon}^* \, \mathbf{I},
$$
$$
\frac{1}{3} \epsilon^*_{kk,ij} + \frac{1}{3}\frac{1+\nu}{1-\nu} \epsilon^*_{kk,mm} \, I_{ij}.
$$

(9.4.10)

Taking the trace of the compatibility relations yields, up to a multiplicative factor $(2+\nu)/E \neq 0$,

$$
\Delta \operatorname{tr} \boldsymbol{\sigma} + \frac{1+\nu}{1-\nu} \vec{\nabla} \cdot \mathbf{f} + 2 \frac{E}{3\,(1-\nu)} \Delta \operatorname{tr} \boldsymbol{\epsilon}^* = 0.
$$

(9.4.11)

As indicated above, the inelastic strain may include a pressure contribution associated with the fluid pressure p_{w}, a thermal contribution associated with the isotropic thermal expansion tensor $\frac{1}{3}\operatorname{tr} \mathbf{c}_{\mathrm{T}}\,\mathbf{I}$ and temperature change $T - T_0$, an isotropic growth contribution $\frac{1}{3}\operatorname{tr} \boldsymbol{\epsilon}^{\mathrm{g}}\,\mathbf{I}$,

$$
2 \frac{E}{3\,(1-\nu)} \operatorname{tr} \boldsymbol{\epsilon}^* = 2 \frac{1-2\nu}{1-\nu} \left(\left(1 - \frac{K}{K_{\mathrm{s}}}\right) p_{\mathrm{w}} + K \operatorname{tr} \mathbf{c}_{\mathrm{T}} \,(T - T_0) + K \operatorname{tr} \boldsymbol{\epsilon}^{\mathrm{g}} \right),
$$

(9.4.12)

and possibly other effects. The pressure coefficient in (9.4.12) has been rewritten using (7.4.42) and (7.4.45).

The results (9.4.6) and (9.4.7) hold if the body force derives from a harmonic potential and if the trace of the inelastic strain is harmonic.

9.4.3 Compatibility conditions for spherical problems with radial symmetry

The particular situation where the displacement of the solid constituent of a linear isotropic poroelastic mixture is radial and subjected to a radial body force $\mathbf{f} = f_r\,\hat{\mathbf{e}}_r$ is detailed using the spherical coordinates (r, θ, ϕ) and the associated unit axes $(\hat{\mathbf{e}}_r, \hat{\mathbf{e}}_\theta, \hat{\mathbf{e}}_\phi)$. The required definitions are exposed in Section 2.4.3.1.

The stress and the strain tensors are principal in these axes, and $\sigma_{\theta\theta} = \sigma_{\phi\phi}$, $\epsilon_{\theta\theta} = \epsilon_{\phi\phi}$, so that

$$
\boldsymbol{\sigma} = \sigma_{rr}\,\hat{\mathbf{e}}_r \otimes \hat{\mathbf{e}}_r + \sigma_{\theta\theta}\,(\mathbf{I} - \hat{\mathbf{e}}_r \otimes \hat{\mathbf{e}}_r), \quad \boldsymbol{\epsilon} = \epsilon_{rr}\,\hat{\mathbf{e}}_r \otimes \hat{\mathbf{e}}_r + \epsilon_{\theta\theta}\,(\mathbf{I} - \hat{\mathbf{e}}_r \otimes \hat{\mathbf{e}}_r).
$$

(9.4.13)

A number of preliminary differential relations are needed, namely

$$
\vec{\nabla}\operatorname{tr}\boldsymbol{\sigma} = \frac{\hat{\mathbf{e}}_k}{h_k}\frac{\vec{\partial}\operatorname{tr}\boldsymbol{\sigma}}{\partial x_k} = \frac{d\operatorname{tr}\boldsymbol{\sigma}}{dr}\,\hat{\mathbf{e}}_r,
$$

$$
\vec{\nabla}(\vec{\nabla}\operatorname{tr}\boldsymbol{\sigma}) = \frac{\hat{\mathbf{e}}_k}{h_k}\frac{\vec{\partial}}{\partial x_k}\left((\vec{\nabla}\operatorname{tr}\boldsymbol{\sigma})\right) = \frac{d^2\operatorname{tr}\boldsymbol{\sigma}}{dr^2}\,\hat{\mathbf{e}}_r \otimes \hat{\mathbf{e}}_r + \frac{1}{r}\frac{d\operatorname{tr}\boldsymbol{\sigma}}{dr}\,(\mathbf{I} - \hat{\mathbf{e}}_r \otimes \hat{\mathbf{e}}_r),
$$

$$
\vec{\nabla}\boldsymbol{\sigma} = \frac{\hat{\mathbf{e}}_k}{h_k}\frac{\vec{\partial}\boldsymbol{\sigma}}{\partial x_k}
$$
$$
= \left(\frac{d}{dr}(\sigma_{rr} - \sigma_{\theta\theta}) - \frac{\sigma_{rr} - \sigma_{\theta\theta}}{r}\right)\hat{\mathbf{e}}_r \otimes \hat{\mathbf{e}}_r \otimes \hat{\mathbf{e}}_r
$$
$$
+ \frac{d\sigma_{\theta\theta}}{dr}\hat{\mathbf{e}}_r \otimes \mathbf{I} + \frac{\sigma_{rr} - \sigma_{\theta\theta}}{r}\mathbf{I} \otimes \hat{\mathbf{e}}_r + \frac{\sigma_{rr} - \sigma_{\theta\theta}}{r}(\hat{\mathbf{e}}_\theta \otimes \hat{\mathbf{e}}_r \otimes \hat{\mathbf{e}}_\theta + \hat{\mathbf{e}}_\phi \otimes \hat{\mathbf{e}}_r \otimes \hat{\mathbf{e}}_\phi),
$$

$$
\Delta\boldsymbol{\sigma} = \vec{\nabla}\cdot(\vec{\nabla}\boldsymbol{\sigma}) = \frac{\hat{\mathbf{e}}_k}{h_k}\frac{\vec{\partial}}{\partial x_k}\cdot(\vec{\nabla}\boldsymbol{\sigma})
$$
$$
= \left(\frac{d^2}{dr^2}(\sigma_{rr} - \sigma_{\theta\theta}) + \frac{2}{r}\frac{d}{dr}(\sigma_{rr} - \sigma_{\theta\theta}) - \frac{6}{r^2}(\sigma_{rr} - \sigma_{\theta\theta})\right)\hat{\mathbf{e}}_r \otimes \hat{\mathbf{e}}_r
$$
$$
+ \left(\frac{d^2\sigma_{\theta\theta}}{dr^2} + \frac{2}{r}\frac{d\sigma_{\theta\theta}}{dr} + \frac{2}{r^2}(\sigma_{rr} - \sigma_{\theta\theta})\right)\mathbf{I}.
$$

(9.4.14)

Collecting the above relations, the compatibility relations can be cast in the format,

$$
\hat{\mathbf{e}}_r \otimes \hat{\mathbf{e}}_r \Big[\frac{1}{E} \Big(\frac{d^2 \mathrm{tr}\, \boldsymbol{\sigma}}{dr^2} - \frac{1}{r} \frac{d \mathrm{tr}\, \boldsymbol{\sigma}}{dr} \Big)
$$

$$
+ \frac{1+\nu}{E} \Big(\frac{d^2}{dr^2} (\sigma_{rr} - \sigma_{\theta\theta}) + \frac{2}{r} \frac{d}{dr} (\sigma_{rr} - \sigma_{\theta\theta}) - \frac{6}{r^2} (\sigma_{rr} - \sigma_{\theta\theta}) \Big)
$$

$$
+ 2 \frac{1+\nu}{E} \Big(\frac{df_r}{dr} - \frac{f_r}{r} \Big) + \frac{1}{3} \frac{d^2 \mathrm{tr}\, \boldsymbol{\epsilon}^*}{dr^2} - \frac{1}{3} \frac{1}{r} \frac{d \mathrm{tr}\, \boldsymbol{\epsilon}^*}{dr} \Big] \tag{9.4.15}
$$

$$
+ \mathbf{I} \Big[\frac{1}{E} \Big(\frac{1}{r} \frac{d \mathrm{tr}\, \boldsymbol{\sigma}}{dr} \Big) + \frac{1+\nu}{E} \Big(\frac{d^2 \sigma_{\theta\theta}}{dr^2} + \frac{2}{r} \frac{d\sigma_{\theta\theta}}{dr} + \frac{2}{r^2} (\sigma_{rr} - \sigma_{\theta\theta}) \Big) + \frac{\nu}{E} \frac{1+\nu}{1-\nu} \Big(\frac{df_r}{dr} + 2 \frac{f_r}{r} \Big)
$$

$$
+ 2 \frac{1+\nu}{E} \frac{f_r}{r} + \frac{1}{3} \frac{1}{r} \frac{d \mathrm{tr}\, \boldsymbol{\epsilon}^*}{dr} + \frac{1}{3} \frac{1+\nu}{1-\nu} \Big(\frac{d^2 \mathrm{tr}\, \boldsymbol{\epsilon}^*}{dr^2} + \frac{2}{r} \frac{d \mathrm{tr}\, \boldsymbol{\epsilon}^*}{dr} \Big) \Big] = \mathbf{0} \,.
$$

Of course, the trace of the left hand-side is a particular case of (9.4.11).

9.5 Heating a thermo-poroelastic medium

A porous half-space is heated uniformly over its surface and the only non zero component of the strain tensor is the strain along the depth of the half-space of axis x. The medium is saturated by a fluid, and both the solid constituent and the fluid are compressible. The initial and boundary value problem exposed below in Sections 9.5.1 to 9.5.8 draws from McTigue [1986]. A number of simplifications are adopted to obtain a tractable analytical solution. The simulations highlight the influence of the ratio of the hydraulic and thermal diffusivities on the thermo-hydromechanical response. Stronger thermohydraulic interactions are expected when this ratio is close to one.

9.5.1 Thermo-poroelastic field and constitutive equations

The linearized field equations governing the thermo-poroelastic problem addressed here consist of

$$
\text{the balance of momentum of the porous medium} \quad \mathrm{div}\, \boldsymbol{\sigma} = \mathbf{0} \,,
$$

$$
\text{the balance of mass of fluid} \quad \frac{dm^{\mathrm{w}}}{dt} + \mathrm{div}\, \mathbf{M}_{\mathrm{w}} = 0 \,, \tag{9.5.1}
$$

$$
\text{the energy equation of the porous medium} \quad T \frac{dS_{\mathrm{pm}}}{dt} + \mathrm{div}\, \mathbf{Q} = 0 \,.
$$

These equations justify as follows:
  - the quasi-static balance of momentum phrased in terms of the total stress $\boldsymbol{\sigma}$ assumes a zero body force;
  - unlike in some other models, like the double porosity model of Section 10.6 or the models of Section 24.8, there are no internal mass transfers. Therefore the mass balance of fluid indicates that the rate of change of the fluid mass content m^{w} is compensated by the exchange of mass with the surroundings represented by the mass flux $\mathbf{M}_{\mathrm{w}}$;
  - the energy equation assumes a zero heat source and neglects the convective and nonlinear terms: the sole contributions due to the rate of change of entropy S_{pm} of the porous medium and heat flux $\mathbf{Q}$ survive from (9.3.33).

The porous medium under consideration is isotropic in its elastic, flow and thermal properties. Thus the tensor of elastic moduli is a fourth order isotropic tensor defined by two material parameters, while the tensors providing the heat and fluid fluxes in terms of the gradients of temperature and fluid pressure respectively, and the tensor of thermal expansion are second order isotropic tensors, each of them being defined by a single material parameter.

The thermo-poroelastic constitutive equations (9.3.12) for the strain and dimensionless fluid mass may be recast, with help of (7.4.42), in the format,

$$
\boldsymbol{\epsilon} = \frac{1}{2\mu} \Big(\boldsymbol{\sigma} - \frac{\nu}{1+\nu} \mathrm{tr}\, \boldsymbol{\sigma}\, \mathbf{I} \Big) + \frac{3}{2\mu} \frac{1}{S} \frac{\nu^{\mathrm{U}} - \nu}{(1+\nu)(1+\nu^{\mathrm{U}})} p_{\mathrm{w}} \mathbf{I} + \frac{1}{3} \mathrm{tr}\, \mathbf{c}_{\mathrm{T}}\, \theta \mathbf{I} \,,
$$

$$
\zeta^{\mathrm{w}} = \frac{3}{2\mu} \frac{1}{S} \frac{\nu^{\mathrm{U}} - \nu}{(1+\nu)(1+\nu^{\mathrm{U}})} \Big(\mathrm{tr}\, \boldsymbol{\sigma} + \frac{3}{S} p_{\mathrm{w}} \Big) + n^{\mathrm{w}} (\mathrm{tr}\, \mathbf{c}_{\mathrm{T}} - c_{\mathrm{Tw}}) \theta \,. \tag{9.5.2}
$$

The constitutive expression (9.3.18) for the entropy has been introduced in the energy equation (9.3.33) to yield the differential equation (9.3.34). For this energy equation to become a partial differential equation for the sole temperature field $\theta(x,t) = T(x,t) - T_0$, where T_0 is a reference uniform temperature, the thermoelastic heat should be neglected. Thermoosmosis, considered in Section 9.3.3, should also be neglected, in such a way that the volume fluid flux $\mathbf{J}_{\mathrm{w}}$ [unit : m/s] and heat flux $\mathbf{Q}$ [unit : W/m^2] are uncoupled,

$$\mathbf{J}_{\mathrm{w}} = -k_{\mathrm{H}} \, \boldsymbol{\nabla} \, p_{\mathrm{w}}, \quad \mathbf{Q} = -k_{\mathrm{T}} \, \boldsymbol{\nabla} \, \theta, \tag{9.5.3}$$

with $k_{\mathrm{H}} > 0$ hydraulic permeability [unit : m^2/Pa/s], and $k_{\mathrm{T}} > 0$ thermal conductivity [unit : W/m/K]. The energy equation (9.3.34) then simplifies to

$$\frac{\partial \theta}{\partial t} - D_{\mathrm{T}} \, \Delta \theta = 0, \tag{9.5.4}$$

where the thermal diffusivity D_{T} [unit : m^2/s],

$$D_{\mathrm{T}} = \frac{k_{\mathrm{T}}}{C^{\mathrm{U}}_{\epsilon \zeta}}, \tag{9.5.5}$$

is built from the thermal conductivity k_{T} and the heat capacity of the porous medium at constant strain and fluid mass defined by (9.3.19). In fact, consistent with the linearization of the energy equation, all three heat capacities appearing in the constitutive expressions (9.3.12), (9.3.14) and (9.3.18) should be close. As indicated by (9.1.30), the thermal conductivity of a porous medium which is in local thermal equilibrium is equal to the spatial average of the thermal conductivities of its constituents.

Since the stress components will be calculated first, the compatibility conditions in terms of stress derived in Section 9.4 with the thermo-poroelastic strain $\boldsymbol{\epsilon}^*$ given by (9.5.2),

$$\overset{\rightarrow}{\boldsymbol{\nabla}}{}^2 \operatorname{tr} \boldsymbol{\sigma} + (1+\nu)\, \Delta \boldsymbol{\sigma} + \frac{3}{S} \frac{\nu^{\mathrm{U}} - \nu}{1 + \nu^{\mathrm{U}}} \left(\overset{\rightarrow}{\boldsymbol{\nabla}}{}^2 p_{\mathrm{w}} + \frac{1+\nu}{1-\nu} \Delta p_{\mathrm{w}} \, \mathbf{I} \right) + \frac{2}{3} \mu \,(1+\nu) \operatorname{tr} \mathbf{c}_{\mathrm{T}} \left(\overset{\rightarrow}{\boldsymbol{\nabla}}{}^2 \theta + \frac{1+\nu}{1-\nu} \Delta \theta \, \mathbf{I} \right) = \mathbf{0}, \tag{9.5.6}$$

should be satisfied. Use will be made of the trace of this tensorial relation, which, factoring out the coefficient $(2+\nu)(1+\nu)/(1-\nu)$, can be cast in the format,

$$\frac{1-\nu}{1+\nu} \Delta \left(\operatorname{tr} \boldsymbol{\sigma} + \frac{3}{S} p_{\mathrm{w}} \right) - \frac{3}{S} \frac{1-\nu^{\mathrm{U}}}{1+\nu^{\mathrm{U}}} \Delta p_{\mathrm{w}} + \frac{4}{3} \mu \operatorname{tr} \mathbf{c}_{\mathrm{T}} \, \Delta \theta = 0. \tag{9.5.7}$$

Upon linearization, the balance of fluid mass becomes $\partial \zeta^{\mathrm{w}} / \partial t + \operatorname{div} \mathbf{J}_{\mathrm{w}} = 0$. Insertion of the elastic constitutive equation and flow equation and elimination of the Laplacians of pressure and temperature via (9.5.4) and (9.5.7), respectively, yield an inhomogeneous equation of diffusion type phrased in terms of a hydraulic (Darcy's) diffusivity D_{H} [unit : m^2/s],

$$D_{\mathrm{H}} = \frac{2}{9} S^2 \frac{(1+\nu^{\mathrm{U}})^2 (1-\nu)}{(\nu^{\mathrm{U}} - \nu)(1 - \nu^{\mathrm{U}})} \mu \, k_{\mathrm{H}}, \tag{9.5.8}$$

for $\operatorname{tr} \boldsymbol{\sigma} + (3/S) \, p_{\mathrm{w}}$, namely

$$\left(\frac{\partial}{\partial t} - D_{\mathrm{H}} \frac{\partial^2}{\partial x^2} \right) \left(\operatorname{tr} \boldsymbol{\sigma} + \frac{3}{S} p_{\mathrm{w}} \right) = \frac{4}{3} \mu \frac{1+\nu}{1-\nu} \left(\left(\frac{D_{\mathrm{H}}}{D_{\mathrm{T}}} - 1 \right) \operatorname{tr} \mathbf{c}_{\mathrm{T}} + \frac{9}{4S} \frac{b_1}{\mu} \frac{1-\nu^{\mathrm{U}}}{1+\nu^{\mathrm{U}}} \right) \frac{\partial \theta}{\partial t}, \tag{9.5.9}$$

or, via (9.5.2), for the dimensionless mass content ζ^{w},

$$\left(\frac{\partial}{\partial t} - D_{\mathrm{H}} \frac{\partial^2}{\partial x^2} \right) \zeta^{\mathrm{w}} = \frac{D_{\mathrm{H}}}{D_{\mathrm{T}}} b_2 \frac{\partial \theta}{\partial t}. \tag{9.5.10}$$

The two positive coefficients b_1 [unit : Pa/K] and b_2 [unit : 1/K],

$$b_1 \, k_{\mathrm{H}} = b_2 \, D_{\mathrm{H}},$$

$$\frac{b_1}{b_0} = \frac{4}{9} \mu \frac{1+\nu^{\mathrm{U}}}{1-\nu^{\mathrm{U}}} S, \quad \frac{b_2}{b_0} = \frac{2}{S} \frac{\nu^{\mathrm{U}} - \nu}{(1+\nu^{\mathrm{U}})(1-\nu)}, \tag{9.5.11}$$

$$b_0 = \operatorname{tr} \mathbf{c}_{\mathrm{T}} + \frac{S}{2} \frac{(1+\nu^{\mathrm{U}})(1-\nu)}{\nu^{\mathrm{U}} - \nu} n^{\mathrm{w}} \left(c_{\mathrm{Tw}} - \operatorname{tr} \mathbf{c}_{\mathrm{T}} \right),$$

have been introduced so as to comment the results in terms of thermomechanical properties, and thermal and hydraulic diffusivities. It is also instrumental to introduce the ratio of the diffusivities,

$$\chi \equiv \frac{\sqrt{D_{\mathrm{H}}}}{\sqrt{D_{\mathrm{T}}}}, \qquad (9.5.12)$$

and the dimensionless coordinates,

$$X_{\mathrm{T}} \equiv \frac{x}{2\sqrt{D_{\mathrm{T}}\,t}}, \qquad X_{\mathrm{H}} \equiv \frac{x}{2\sqrt{D_{\mathrm{H}}\,t}}. \qquad (9.5.13)$$

Even if the medium is semi-infinite, one may focus attention on the diffusion processes over a finite distance L. Then, t_{T} and t_{H}, defined by

$$t_{\mathrm{T}} \equiv \frac{L^2}{D_{\mathrm{T}}}, \qquad t_{\mathrm{H}} \equiv \frac{L^2}{D_{\mathrm{H}}}, \qquad (9.5.14)$$

appear as times characterizing respectively thermal diffusion and hydraulic diffusion.

9.5.2 Thermo-poroelastic field equations in a 1D context

In the one-dimensional context envisaged here, the only nonzero displacement component u_x is along the axis x, and all fields are independent of the lateral coordinates y and z. Therefore, the only nonzero component of the strain tensor $\boldsymbol{\epsilon}$ is the axial strain $\epsilon_{xx} = \partial u_x/\partial x$. The constraint $\epsilon_{yy} = \epsilon_{zz} = 0$ provides the lateral stresses $\sigma_{yy} = \sigma_{zz}$ in terms of σ_{xx}, p_{w} and θ. With help of the energy equation (9.5.4), the field equation (9.5.9) transforms to

$$\left(\frac{\partial}{\partial t} - D_{\mathrm{H}}\frac{\partial^2}{\partial x^2}\right)\left(\frac{S}{3}\frac{1+\nu^{\mathrm{U}}}{1-\nu^{\mathrm{U}}}\,\sigma_{xx} + p_{\mathrm{w}}\right) = b_1\frac{\partial\theta}{\partial t}. \qquad (9.5.15)$$

In a one-dimensional context, the compatibility condition for the strain is automatically satisfied since there is a single nonzero strain component. In fact, the constitutive relations yield directly,

$$\frac{1-\nu}{1+\nu}\left(\mathrm{tr}\,\boldsymbol{\sigma} + \frac{3}{S}\,p_{\mathrm{w}}\right) - \sigma_{xx} - \frac{3}{S}\frac{1-\nu^{\mathrm{U}}}{1+\nu^{\mathrm{U}}}\,p_{\mathrm{w}} + \frac{4}{3}\mu\,\mathrm{tr}\,\mathbf{c}_{\mathrm{T}}\,\theta = 0. \qquad (9.5.16)$$

To simplify the analysis, the mechanical loading is specified in terms of traction,

$$
\begin{aligned}
&(\mathrm{FE})_{\mathrm{M}}\ \text{field equation} & \frac{\partial\sigma_{xx}}{\partial x} &= 0, \quad t > 0,\ x > 0;\\[2mm]
&(\mathrm{IC})_{\mathrm{M}}\ \text{initial condition} & \sigma_{xx}(x,0) &= 0;\\[2mm]
&(\mathrm{BC})_{\mathrm{M}}\ \text{boundary condition} & \sigma_{xx}(0,t) &= \sigma_{xx}^0\,\mathcal{H}(t);\\[2mm]
&(\mathrm{RC})_{\mathrm{M}}\ \text{radiation condition} & u_x(x\to\infty,t) &= 0, \quad t > 0.
\end{aligned}
\qquad (9.5.17)
$$

Consequently the axial stress is uniform in both space and time, $\sigma_{xx}(x,t) = \sigma_{xx}^0\,\mathcal{H}(t)$. The field equation (9.5.15) becomes a decoupled equation for the fluid pressure. It would not be so if the displacement, rather than the traction, was prescribed at the boundary $x = 0$.

The scheme to obtain a complete analytical solution simplifies, due to the partial couplings:

    1. the thermal problem is solved first;

    2. the flow problem, driven by the thermal loading, can then be solved;

    3. finally, the axial displacement and all components of stress follow.

    The analytical treatment via the Laplace transform technique capitalizes upon the elementary solutions of diffusion problems worked out in Section 15.3.2.

    The four thermo-hydromechanical boundary conditions at the left boundary $x = 0$ envisaged below are sketched in Fig. 9.5.1.

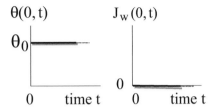

(a) prescribed temperature and pressure

(b) prescribed temperature and fluid flux

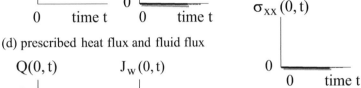

(a) to (d) prescribed traction

$\sigma_{xx}(0,t)$

(c) prescribed heat flux and pressure

(d) prescribed heat flux and fluid flux

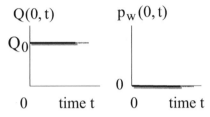

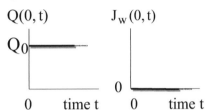

Fig. 9.5.1: Boundary conditions for the coupled heat diffusion problem, fluid diffusion problem and mechanical problem.

9.5.3 Prescribed temperature and drained boundary

The solution to the one-dimensional Initial and Boundary Value Problem (IBVP) for the temperature,

$$
\begin{aligned}
&\text{(FE)}_\text{T} \text{ field equation} && \frac{\partial \theta}{\partial t} - D_\text{T} \frac{\partial^2 \theta}{\partial x^2} = 0, \quad t > 0,\ x > 0; \\
&\text{(IC)}_\text{T} \text{ initial condition} && \theta(x,0) = 0\,; \\
&\text{(BC)}_\text{T} \text{ boundary condition} && \theta(0,t) = \theta_0\,\mathcal{H}(t); \\
&\text{(RC)}_\text{T} \text{ radiation condition} && -\infty < \theta(x \to \infty, t) < \infty, \quad t > 0\,,
\end{aligned}
\tag{9.5.18}
$$

is detailed in Section 15.3.2.1, and reads

$$
\theta(x,t) = \theta_0\,\mathrm{erfc}(X_\text{T}), \quad Q(x,t) = -k_\text{T}\frac{\partial \theta}{\partial x}(x,t) = \frac{k_\text{T}\,\theta_0}{\sqrt{\pi D_\text{T}\,t}}\exp(-X_\text{T}^2)\,. \tag{9.5.19}
$$

For any coefficient α,

$$
\frac{\partial \theta}{\partial t} = D_\text{T}\,\Delta\theta = \alpha\,\frac{\partial \theta}{\partial t} + (1-\alpha)\,D_\text{T}\,\Delta\theta\,, \tag{9.5.20}
$$

the field equation (9.5.15) may be transformed to,

$$
\frac{\partial}{\partial t}\left(p_\text{w} - \alpha\,b_1\,\theta\right) - D_\text{H}\frac{\partial^2}{\partial x^2}\left(p_\text{w} + (1-\alpha)\,b_1\,\frac{D_\text{T}}{D_\text{H}}\,\theta\right) = 0\,. \tag{9.5.21}
$$

Then, with

$$
\alpha = \frac{1}{1 - \chi^2}\,, \tag{9.5.22}
$$

the IBVP for the fluid pressure may be conveniently cast in a format which is identical to the thermal IBVP,

$$
\begin{aligned}
&\text{(FE)}_\text{H} \text{ field equation} && \left(\frac{\partial}{\partial t} - D_\text{H}\frac{\partial^2}{\partial x^2}\right)(p_\text{w} - \alpha\,b_1\,\theta) = 0, \quad t > 0,\ x > 0; \\
&\text{(IC)}_\text{H} \text{ initial condition} && (p_\text{w} - \alpha\,b_1\,\theta)(x,0) = 0\,; \\
&\text{(BC)}_\text{H} \text{ boundary condition} && (p_\text{w} - \alpha\,b_1\,\theta)(0,t) = -\alpha\,b_1\,\theta_0\,\mathcal{H}(t); \\
&\text{(RC)}_\text{H} \text{ radiation condition} && -\infty < (p_\text{w} - \alpha\,b_1\,\theta)(x \to \infty, t) < \infty, \quad t > 0\,.
\end{aligned}
\tag{9.5.23}
$$

The solution for the pressure results readily,

$$p_{\mathrm{w}}(x,t) = \alpha\, b_1\, \theta_0 \left(\mathrm{erfc}(X_{\mathrm{T}}) - \mathrm{erfc}(X_{\mathrm{H}})\right). \tag{9.5.24}$$

Whether the thermal diffusivity is smaller or larger than the hydraulic diffusivity, namely $\chi > 1$ or $\chi < 1$, the pressure has a maximum for

$$X_{\mathrm{H}} = \frac{x}{2\sqrt{D_{\mathrm{H}}\,t}} = \sqrt{\frac{\mathrm{Ln}\,\chi}{\chi^2 - 1}}. \tag{9.5.25}$$

This maximum is a constant and its position moves inward. It is interesting to note that, even if it is triggered by a thermal boundary condition, the fluid flow may precede the thermal flow at some depth within the body, if the hydraulic diffusivity is larger than the thermal diffusivity.

The fluid flux,

$$J_{\mathrm{w}}(x,t) = -b_2\,\theta_0\, \frac{\sqrt{D_{\mathrm{H}}}\sqrt{D_{\mathrm{T}}}}{\sqrt{D_{\mathrm{H}}} + \sqrt{D_{\mathrm{T}}}}\, \frac{1}{\sqrt{\pi t}}\, \frac{\exp(-X_{\mathrm{H}}^2) - \chi\exp(-X_{\mathrm{T}}^2)}{1 - \chi}, \tag{9.5.26}$$

has distinct sign on each side of the maximum of the pressure, and vanishes at large times. Its value at the boundary $x = 0$,

$$J_{\mathrm{w}}(0,t) = -b_2\,\theta_0\, \frac{\sqrt{D_{\mathrm{H}}}\sqrt{D_{\mathrm{T}}}}{\sqrt{D_{\mathrm{H}}} + \sqrt{D_{\mathrm{T}}}}\, \frac{1}{\sqrt{\pi t}}, \tag{9.5.27}$$

is controlled by the lowest of the thermal and hydraulic diffusivities. It has a sign opposite to the temperature change: if the temperature is increased, the flux at the boundary is directed outwards.

9.5.4 Prescribed temperature and impervious boundary

For this boundary condition, the only change in the IBVP defined by (9.5.23) reads,

$$(\mathrm{BC})_{\mathrm{H}} \text{ boundary condition } \quad \frac{\partial}{\partial x}(p_{\mathrm{w}} - \alpha\,b_1\,\theta)(0,t) = -\alpha\,b_1\, \frac{\partial\theta}{\partial x}(0,t). \tag{9.5.28}$$

The solution is obtained by the Laplace transform technique with respect to time. The solution in the Laplace domain derives from the field equation, radiation condition and boundary condition,

$$(\mathrm{FE}) \ \mathcal{L}\{(p_{\mathrm{w}} - \alpha\,b_1\,\theta)(x,t)\}(p) = A(p)\,\exp\left(-\sqrt{\frac{p}{D_{\mathrm{H}}}}\,x\right) + \underbrace{B(p)\,\exp\left(\sqrt{\frac{p}{D_{\mathrm{H}}}}\,x\right)}_{B(p)=0,\ (\mathrm{RC})_{\mathrm{H}}} \tag{9.5.29}$$

$$(\mathrm{BC})_{\mathrm{H}} \ A(p) = -\alpha\,b_1\, \frac{\sqrt{D_{\mathrm{H}}}}{\sqrt{p}}\, \mathcal{L}\left\{\frac{\partial\theta}{\partial x}(0,t)\right\}(p) = -\chi\,\alpha\, \frac{b_1\,\theta_0}{p}.$$

The inverse transform is obtained by using Table 15.3.1, yielding finally,

$$p_{\mathrm{w}}(x,t) = b_1\,\theta_0\, \frac{\sqrt{D_{\mathrm{T}}}}{\sqrt{D_{\mathrm{H}}} + \sqrt{D_{\mathrm{T}}}}\, \frac{\mathrm{erfc}(X_{\mathrm{T}}) - \chi\,\mathrm{erfc}(X_{\mathrm{H}})}{1 - \chi}. \tag{9.5.30}$$

Its value at the boundary $x = 0$,

$$p_{\mathrm{w}}(0,t) = b_1\,\theta_0\, \frac{\sqrt{D_{\mathrm{T}}}}{\sqrt{D_{\mathrm{H}}} + \sqrt{D_{\mathrm{T}}}}, \tag{9.5.31}$$

is constant in time. The fluid flux,

$$J_{\mathrm{w}}(x,t) = -\alpha\,b_2\,\theta_0\, \frac{D_{\mathrm{H}}}{\sqrt{D_{\mathrm{T}}}}\, \frac{1}{\sqrt{\pi t}}\left(\exp(-X_{\mathrm{H}}^2) - \exp(-X_{\mathrm{T}}^2)\right), \tag{9.5.32}$$

has the sign of the temperature change. It duly vanishes at the boundary $x = 0$. At any time $t > 0$, the flux is maximum at the position

$$X_{\mathrm{H}} = \frac{x}{2\sqrt{D_{\mathrm{H}}\,t}} = \sqrt{\frac{\mathrm{Ln}\,\chi^2}{\chi^2 - 1}}, \tag{9.5.33}$$

and the value of this maximum,

$$J_{\mathrm{w}}(x,t) = b_2\,D_{\mathrm{T}}\,\theta_0\, \frac{\exp(-X_{\mathrm{H}}^2)}{\sqrt{\pi D_{\mathrm{T}}\,t}} = b_2\,D_{\mathrm{H}}\,\theta_0\, \frac{\exp(-X_{\mathrm{T}}^2)}{\sqrt{\pi D_{\mathrm{T}}\,t}}, \tag{9.5.34}$$

is infinitely large at short times but decreases as time goes by. In fact, at large times, the pressure homogenizes and the flux vanishes.

9.5.5 Prescribed thermal flux and drained boundary

The solution to the IBVP corresponding to a sudden heat flux Q_0 [unit: W/m^2],

$$(FE)_T \text{ field equation} \qquad \frac{\partial \theta}{\partial t} - D_T \frac{\partial^2 \theta}{\partial x^2} = 0, \quad t > 0, \ x > 0;$$

$$(IC)_T \text{ initial condition} \qquad \theta(x, 0) = 0;$$

$$(BC)_T \text{ boundary condition} \quad Q(0, t) = -k_T \frac{\partial \theta}{\partial x}(0, t) = Q_0 \, \mathcal{H}(t);$$

$$(RC)_T \text{ radiation condition} \quad \theta(x \to \infty, t) < \infty, \quad t > 0,$$

(9.5.35)

is detailed in Section 15.3.2.2, and reads

$$\theta(x, t) = 2 \frac{Q_0}{k_T} \sqrt{D_T \, t} \, F_1(X_T), \quad Q(x, t) = Q_0 \operatorname{erfc}(X_T), \qquad (9.5.36)$$

where $F_1(X) \equiv \exp(-X^2)/\sqrt{\pi} - X \operatorname{erfc}(X)$, and $F_1(0) = 1/\sqrt{\pi}$. The heat flux tends to homogenize progressively in the body. On the other hand, the temperature increases to large values,

$$\theta(x = 0, t) = \theta(x, t \to \infty) = 2 \frac{Q_0}{k_T} \sqrt{\frac{D_T}{\pi} t}. \qquad (9.5.37)$$

The IBVP for the fluid pressure is still given by (9.5.23) but with the temperature (9.5.36). It is solved once again through the Laplace transform,

$$(FE) \ \mathcal{L}\{(p_w - \alpha \, b_1 \, \theta)(x, t)\}(p) = A(p) \exp\left(-\sqrt{\frac{p}{D_H}} x\right) + \overbrace{B(p) \exp\left(\sqrt{\frac{p}{D_H}} x\right)}^{B(p)=0, \ (RC)_H}$$

$$(BC)_H \ A(p) = -\alpha \, b_1 \, \mathcal{L}\{\theta(0, t)\}(p) = -\alpha \, b_1 \frac{Q_0}{k_T} \frac{\sqrt{D_T}}{p^{3/2}},$$

(9.5.38)

yielding,

$$p_w(x, t) = 2 \, \alpha \, b_1 \frac{Q_0}{k_T} \sqrt{D_T \, t} \left(F_1(X_T) - F_1(X_H) \right). \qquad (9.5.39)$$

Therefore, the pressure vanishes at any point at times $t = 0$ and $t = \infty$. It has a maximum for X_H solution of the equation $\chi \operatorname{erfc}(X_T) = \operatorname{erfc}(X_H)$, and this maximum, which moves inward, increases in time like $t^{1/2}$. The fluid flux,

$$J_w(x, t) = -\frac{Q_0}{k_T} b_2 \frac{\sqrt{D_H} \, D_T}{\sqrt{D_H} + \sqrt{D_T}} \frac{\operatorname{erfc}(X_H) - \chi \operatorname{erfc}(X_T)}{1 - \chi}, \qquad (9.5.40)$$

is finite and constant in time at the boundary, and tends to homogenize at large times,

$$J_w(x = 0, t) = J_w(x, t \to \infty) = -\frac{Q_0}{k_T} b_2 \frac{\sqrt{D_H} \, D_T}{\sqrt{D_H} + \sqrt{D_T}}. \qquad (9.5.41)$$

9.5.6 Prescribed thermal flux and impervious boundary

The IBVP is again defined by (9.5.23) and (9.5.28). The solution is obtained by the Laplace transform technique with respect to time. The solution in the Laplace domain derives from the field equation, radiation condition and boundary condition,

$$(FE) \ \mathcal{L}\{(p_w - \alpha \, b_1 \, \theta)(x, t)\}(p) = A(p) \exp\left(-\sqrt{\frac{p}{D_H}} x\right) + \overbrace{B(p) \exp\left(\sqrt{\frac{p}{D_H}} x\right)}^{B(p)=0, \ (RC)_H},$$

$$(BC)_H \ \mathcal{L}\left\{\frac{\partial}{\partial x}(p_w - \alpha \, b_1 \, \theta)(0, t)\right\}(p) = -\alpha \, b_1 \, \mathcal{L}\left\{\frac{\partial \theta}{\partial x}(0, t)\right\}(p) = -A(p) \sqrt{\frac{p}{D_H}},$$

(9.5.42)

$$\Rightarrow \quad A(p) = -\alpha \, b_1 \frac{Q_0}{k_T} \frac{\sqrt{D_H}}{p^{3/2}}.$$

The inverse transform is obtained by using Table 15.3.1, yielding finally,

$$p_{\mathrm{w}}(x,t) = 2\,b_1 \frac{Q_0}{k_{\mathrm{T}}} \frac{D_{\mathrm{T}}\sqrt{t}}{\sqrt{D_{\mathrm{H}}} + \sqrt{D_{\mathrm{T}}}} \frac{\mathrm{F}_1(X_{\mathrm{T}}) - \chi \mathrm{F}_1(X_{\mathrm{H}})}{1 - \chi}\,, \tag{9.5.43}$$

and

$$p_{\mathrm{w}}(x=0,t) = p_{\mathrm{w}}(x, t \to \infty) = 2\,b_1 \frac{Q_0}{k_{\mathrm{T}}} \frac{D_{\mathrm{T}}\sqrt{t}}{\sqrt{D_{\mathrm{H}}} + \sqrt{D_{\mathrm{T}}}} \frac{1}{\sqrt{\pi}}\,. \tag{9.5.44}$$

The fluid flux,

$$J_{\mathrm{w}}(x,t) = -\frac{Q_0}{k_{\mathrm{T}}}\,\alpha\, b_2\, D_{\mathrm{H}} \left(\mathrm{erfc}(X_{\mathrm{H}}) - \mathrm{erfc}(X_{\mathrm{T}}) \right), \tag{9.5.45}$$

has the sign of the prescribed heat flux. It tends to vanish at large times all over the body.

9.5.7 Mechanical output, strain, stress and displacement

Given the pressure and temperature fields, the lateral stresses and axial strain deduce from the vanishing lateral strain through the elastic constitutive equations (9.5.2),

$$\sigma_{yy} = \sigma_{zz} = \frac{\nu}{1-\nu}\sigma_{xx} - \frac{3}{S}\frac{\nu^{\mathrm{U}} - \nu}{(1+\nu^{\mathrm{U}})(1-\nu)}p_{\mathrm{w}} - \frac{2}{3}\mu \frac{1+\nu}{1-\nu}\,\mathrm{tr}\,\mathbf{c}_{\mathrm{T}}\,\theta,$$

$$\epsilon_{xx} = \underbrace{\frac{1-2\nu}{1-\nu}\frac{\sigma_{xx}}{2\mu}}_{a_0} + \underbrace{\frac{3}{S}\frac{\nu^{\mathrm{U}} - \nu}{(1+\nu^{\mathrm{U}})(1-\nu)}\frac{p_{\mathrm{w}}}{2\mu}}_{a_1>0} + \underbrace{\frac{1}{3}\frac{1+\nu}{1-\nu}\,\mathrm{tr}\,\mathbf{c}_{\mathrm{T}}\,\,\theta}_{a_2>0}\,. \tag{9.5.46}$$

For the displacement to vanish at $x = \infty$, the prescribed traction σ_{xx}^0 should itself vanish. The fact the total stress should be zero for the displacement at infinity to vanish is a one-dimensional feature. For two- and three-dimensional problems, the applied traction can spread over a surface or a volume and dies out quickly.

When the temperature is prescribed at the boundary $x = 0$ which is drained, the displacement integrates to

$$u_x(x,t) = 2\sqrt{D_{\mathrm{T}}\,t}\,\theta_0 \left(a_1 \frac{b_1}{2\mu} \frac{\chi\,\mathrm{ierfc}(X_{\mathrm{H}}) - \mathrm{ierfc}(X_{\mathrm{T}})}{1 - \chi^2} - a_2\,\mathrm{ierfc}(X_{\mathrm{T}}) \right), \tag{9.5.47}$$

where the integral ierfc defined by (15.3.36) satisfies $\mathrm{ierfc}(0) = 1/\sqrt{\pi}$. The displacement at the boundary,

$$u_x(0,t) = -2\frac{\sqrt{D_{\mathrm{T}}\,t}}{\sqrt{\pi}}\,\theta_0 \left(a_1 \frac{b_1}{2\mu} \frac{1}{1 + \chi} + a_2 \right), \tag{9.5.48}$$

varies in time like $t^{1/2}$, and has the sign opposite to the temperature change.

When the heat flux is prescribed at the boundary $x = 0$ which is drained, the displacement integrates to

$$u_x(x,t) = 4\frac{Q_0}{k_{\mathrm{T}}}\,D_{\mathrm{T}}\,t \left(a_1 \frac{b_1}{2\mu} \frac{\chi\,\mathrm{i}^2\mathrm{erfc}(X_{\mathrm{H}}) - \mathrm{i}^2\mathrm{erfc}(X_{\mathrm{T}})}{1 - \chi^2} - a_2\,\mathrm{i}^2\mathrm{erfc}(X_{\mathrm{T}}) \right). \tag{9.5.49}$$

The function $\mathrm{i}^2\mathrm{erfc}$ defined by (15.3.36) is the integral of ierfc, that satisfies $\mathrm{i}^2\mathrm{erfc}(0) = 1/4$, and, with help of the expansion (15.3.32), $\mathrm{i}^2\mathrm{erfc}(\infty) = 0$. The following relations have been used, $\int_x^\infty \mathrm{ierfc}(X_A(y))dy = 2\sqrt{D_A\,t}\,\mathrm{i}^2\mathrm{erfc}(X_A(x))$, for $A = H, T$. The displacement $u_x(0,t)$ at the boundary,

$$u_x(0,t) = -\frac{Q_0}{k_{\mathrm{T}}}\,D_{\mathrm{T}}\,t \left(a_1 \frac{b_1}{2\mu} \frac{1}{1 + \chi} + a_2 \right), \tag{9.5.50}$$

varies linearly in time, and has the sign opposite to the heat flux.

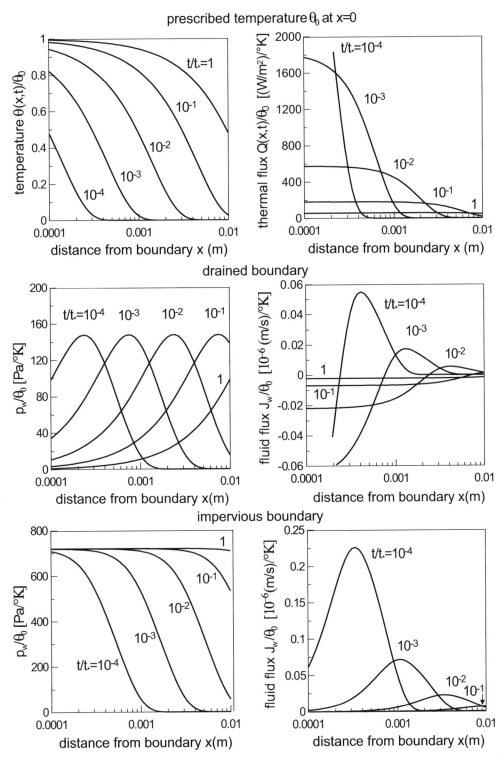

Fig. 9.5.2: One-dimensional poroelastic medium subjected to a prescribed temperature θ_0 at the left boundary $x = 0$. Temperature and thermal flux profiles in the neighborhood of this boundary over a depth $L = 10^{-2}$ m at selected times t/t_*. Thermal diffusion time $t_{\mathrm{T}} = L^2/D_{\mathrm{T}} = 384.3$ s, hydraulic diffusion time $t_{\mathrm{H}} = L^2/D_{\mathrm{H}} = 27.8$ s, and $t_* = \max(t_{\mathrm{T}}, t_{\mathrm{H}})$.

Fluid pressure and fluid flux for a drained boundary, and for an impervious boundary.

For a drained boundary, and for $\theta_0 > 0$, the pressure is positive, its maximum moves to the interior of the medium, and therefore water flows *out* close to the boundary (which is wet) but, deeper in the medium, it tends to flow *in*.

When the boundary is impervious, the pressure is again positive, and the water tends to move to the interior of the medium. As time goes by, the pressure tends to homogenize in space, and the water flux consequently dies out.

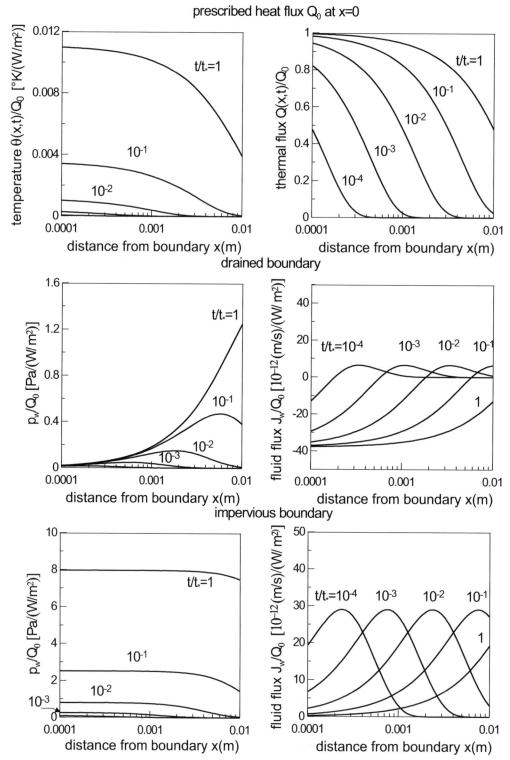

Fig. 9.5.3: Same as Fig. 9.5.2, but with heat flux Q_0 prescribed at the boundary $x = 0$. When the boundary is drained, it is wet for $Q_0 > 0$, and the flux tends to homogenize throughout the medium to its boundary value. When the boundary is impervious, water tends initially to move to the interior of the medium for $Q_0 > 0$, before coming to rest: the neighborhood of the boundary tends to dry out.

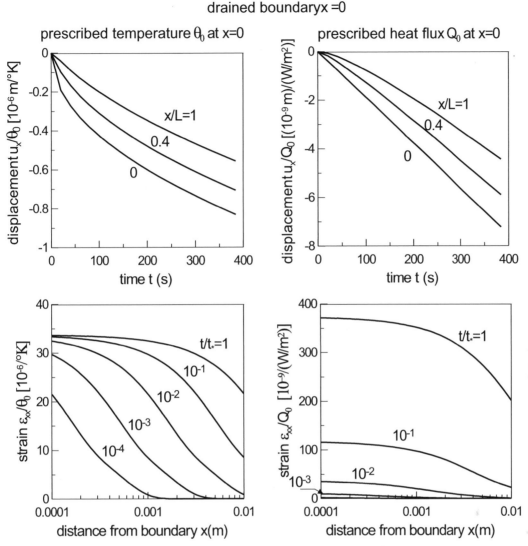

Fig. 9.5.4: Same as Fig. 9.5.2. Time profiles of displacement at selected depths x/L, and spatial profiles of the strain at selected times t/t_* for a drained boundary.

9.5.8 Simulations, drained and undrained limits

9.5.8.1 Simulations with thermal and hydraulic diffusivities of like orders

The parameters used pertain to a clay, McTigue [1986],
- thermal conductivity of porous medium $k_{\rm T} = 1.02\,{\rm W/m/K}$;
- heat capacity of porous medium $C = 3.92 \times 10^6\,{\rm J/m^3/K}$;
- cubic thermal expansion of the solid skeleton ${\rm tr}\,{\bf c}_{\rm T} = 0.034 \times 10^{-3}\,{\rm K^{-1}}$;
- cubic thermal expansion of the fluid $c_{\rm Tw} = 0.3 \times 10^{-3}\,{\rm K^{-1}}$;
- hydraulic permeability $k_{\rm H} = 2.0 \times 10^{-13}\,{\rm m^2/Pa/s}$;
- shear modulus of solid skeleton $\mu = 72 \times 10^3\,{\rm Pa}$;
- drained Poisson's ratio $\nu = 0.498$;
- bulk modulus of solid constituent $K_{\rm s} = 50\,{\rm GPa}$;
- bulk modulus of fluid $K_{\rm w} = 2\,{\rm GPa}$;
- volume fraction of fluid $n^{\rm w} = 0.71$.

Mechanical properties of interest defined in Section 7.3.6.5 follow, e.g., the drained bulk modulus $K = 18\,{\rm MPa}$, Skempton coefficient $S = 0.994$, undrained Poisson's ratio $\nu^{\rm U} = 0.49999$.

The ensuing thermal diffusivity $D_{\mathrm{T}} = 0.26 \times 10^{-6}\,\mathrm{m^2/s}$ is an order of magnitude smaller than the hydraulic diffusivity $D_{\mathrm{H}} = 3.59 \times 10^{-6}\,\mathrm{m^2/s}$.

Figs. 9.5.2 and 9.5.3 display the spatial profiles of temperature, pressure and heat and fluid fluxes over a depth $L = 10^{-2}\,\mathrm{m}$, at different times. The thermal diffusion time $t_{\mathrm{T}} = L^2/D_{\mathrm{T}}$=384.3 s, which is larger than the hydraulic diffusion time $t_{\mathrm{H}} = L^2/D_{\mathrm{H}}$=27.8 s, controls the time evolution of the processes.

Fig. 9.5.4 displays mechanical output for a drained boundary. The displacement at the boundary increases like the square root of time when the temperature is prescribed, and linearly when the heat flux is prescribed. This effect is a typical one-dimensional feature.

With the above material properties, the hydraulic diffusivity is about 13.8 times larger than the thermal diffusivity, corresponding to $\chi \simeq 3.7$. The thermal information, while diffusing into the body, triggers hydraulic diffusion through thermomechanical coupling. In fact, hydraulic diffusion being faster than thermal diffusion, fluid pressure penetrates the medium ahead of temperature. Both fields contribute to the deformation: at a point inside the medium, the contribution due to pressure takes place first. In the examples considered, the two contributions have the same sign, corresponding to a dilatation if $\theta_0 > 0$ and $Q_0 > 0$. For example, at point $x = L/10$, the pressure reaches its maximum at time about $t_{\mathrm{T}}/10^3$ when the boundary is drained, while the temperature has virtually not increased yet, Fig. 9.5.2. At that time, the deformation has reached about 25% of its limit value, Fig. 9.5.4.

9.5.8.2 Thermal and hydraulic diffusivities of different orders

Let us consider first the case where the thermal diffusivity is much smaller than the hydraulic diffusivity, corresponding to $\chi \gg 1$, and $\alpha \ll 1$. Then, the IBVP (9.5.23) indicates that the thermal loading does not trigger significantly the fluid pressure, which is expected to remain small. The problem is therefore close to a drained problem.

Conversely, the problem is approximately undrained when the thermal diffusivity is much larger than the hydraulic diffusivity. Indeed, let $\tilde{t} = t/t_{\mathrm{T}}$ and $\tilde{x} = x/L$. Then the field equation (9.5.10) for the dimensionless mass content ζ^{w} may be rephrased in terms of dimensionless variables associated with the thermal process, namely

$$\left(\frac{\partial}{\partial \tilde{t}} - \frac{D_{\mathrm{H}}}{D_{\mathrm{T}}} \frac{\partial^2}{\partial \tilde{x}^2} \right) \zeta^{\mathrm{w}} = \frac{D_{\mathrm{H}}}{D_{\mathrm{T}}} b_2 \frac{\partial \theta}{\partial \tilde{t}} \,. \tag{9.5.51}$$

Therefore, for $D_{\mathrm{H}}/D_{\mathrm{T}} \ll 1$, the fluid mass content is constant at the thermal time scale. Similarly, $p_{\mathrm{w}} = b_1 \theta$ up to a constant according to (9.5.15). The axial strain (9.5.46) can then be cast in terms of the undrained cubic coefficient of thermal expansion $\mathrm{tr}\,\mathbf{c}_{\mathrm{T}}^{\mathrm{U}} = \mathrm{tr}\,\mathbf{c}_{\mathrm{T}} + S\,n^{\mathrm{w}}\,(c_{\mathrm{Tw}} - \mathrm{tr}\,\mathbf{c}_{\mathrm{T}})$ defined by (9.3.16),

$$\epsilon_{xx}(x,t) = \frac{1+\nu^{\mathrm{U}}}{1-\nu^{\mathrm{U}}} \frac{1}{3} \mathrm{tr}\,\mathbf{c}_{\mathrm{T}}^{\mathrm{U}}\, \theta(x,t)\,. \tag{9.5.52}$$

9.5.9 Extension of the above model including thermoosmosis

Zhou et al. [1998] consider a one-dimensional saturated porous medium where one boundary is drained, subjected to a thermal shock and to a constant traction, while the other boundary is drained but and has a fixed temperature and zero displacement. They include diffusion couplings, namely the thermoosmotic effect and the symmetric mechanocaloric effect, also referred to as thermal filtration.

They derived an analytical solution through Laplace transform technique. For low hydraulic permeability, the thermal diffusivity is larger than the hydraulic diffusivity. In absence of diffusion couplings, spatial thermal equilibrium is established first. Even for a coupling coefficient k_{TO} lower than one percent of the limit $\sqrt{T\,k_{\mathrm{T}}\,k_{\mathrm{H}}}$ indicated by (9.3.26), they observe that the magnitudes and time profiles of the pore pressure and mechanical outputs are significantly modified with respect to the uncoupled case $k_{\mathrm{TO}} = 0$. On the other hand, this coupling is observed to affect little the thermal field.

In other words, diffusion coupling through k_{TO} affects significantly the hydraulic flow by thermoosmosis, but coupling through k_{MC} affects little the thermal flow. In order to obtain a tractable analytical solution, one might therefore be tempted to abandon Onsager symmetry, so as to be able to decouple the thermal field from the hydraulic flow. This position is adopted by Ghassemi and Diek [2002] in a thermo-poro-chemoelastic analysis of swelling materials. When introduced in the balance of mass, the thermal field contributes due to mechanical coupling and thermoosmosis.

Zhou et al. [1998] and Ghassemi and Diek [2002] address materials that are negatively charged at neutral pH. The former authors use a positive thermoosmotic coefficient and the latter a negative one.

9.5.10 Chemo-poroelasticity versus thermo-poroelasticity

With respect to the chemo-poroelastic problem considered in Section 13.13, the thermal field plays the role of the chemical concentration. The compressibility of the two constituents implies that a specific constitutive equation should be provided for the fluid constituent. In the case of incompressible constituents, this constitutive equation is replaced by a condition of incompressibility, which can be interpreted as a balance of mass within the porous medium.

The mass balance of the chemical and the energy equation play the role of field equation. Both are parabolic diffusion equations. In absence of thermoosmosis, they are decoupled from the flow and deformation and can be solved first. While the method to obtain the concentration profile uses Fourier series, the temperature is calculated via the Laplace transform technique. The methods could be interchanged. As another difference, the chemical diffuses over a finite thickness while the thermal problem considers a semi-infinite body.

Once known, the chemical concentration, or temperature, is used to trigger the flow and deformation.

Viscous coupling through thermoosmosis

In presence of thermoosmosis, as described in Section 9.3.3, the energy equation involves both temperature and fluid pressure. Therefore, an analytical/numerical solution should address both fields simultaneously, Section 9.5.9.

Exercises on Chapter 9

Exercise 9.1 Symmetric versus non symmetric matrices
1. Consider the 2×2 real matrix $\mathbf{A}$,

$$\mathbf{A} = \begin{bmatrix} a & b \\ c & d \end{bmatrix}. \tag{9.E.1}$$

Calculate the eigenvalues and show that positive-definiteness involves the symmetric part of the matrix. Show that the existence of real positive eigenvalues does not imply positive definiteness,

$$\text{all eigenvalues} \quad \lambda(\mathbf{A}) \quad \text{real positive} \quad \not\Rightarrow \quad \mathbf{A} \quad \text{positive definite} \quad \text{if} \quad \mathbf{A} \neq \mathbf{A}^{\mathrm{T}}, \tag{9.E.2}$$

and that the converse proposition does not hold either,

$$\mathbf{A} \quad \text{positive definite} \quad \not\Rightarrow \quad \lambda(\mathbf{A}) \quad \text{real positive} \quad \text{if} \quad \mathbf{A} \neq \mathbf{A}^{\mathrm{T}}. \tag{9.E.3}$$

Answer:
    The eigenvalues of $\mathbf{A}$, namely $\lambda_{\pm} = (a+d)/2 \pm \sqrt{(a-d)^2/4 + bc}$, are the roots of the polynomial $\lambda^2 - (a+d)\lambda + ad - bc$. They may be real or complex. Moreover,

$$[x \; y] \, \mathbf{A} \begin{bmatrix} x \\ y \end{bmatrix} = a x^2 + (b+c)\,x y + d y^2 = a\Big(x + \frac{b+c}{2a}\,y\Big)^2 + \Big(d - \frac{(b+c)^2}{4a}\Big) y^2. \tag{9.E.4}$$

The conditions of positive definiteness are therefore

$$a > 0; \quad d - \frac{(b+c)^2}{4a} > 0. \tag{9.E.5}$$

Now consider the special case $a > 0$, $d = 0$, $a^2/4 + bc > 0$, $bc < 0$: the eigenvalues are real positive but positive definiteness does not hold. Conversely, consider the case $a = d > 0$, $b = -c$. Positive definiteness holds, but the eigenvalues are not real.

2. By contrast, for a symmetric matrix $\mathbf{A}$, namely $b = c$, verify that the conditions for positive definiteness and positive eigenvalues are equivalent.

Answer:
    The eigenvalues are real, $\lambda_{\pm} = (a+d)/2 \pm \sqrt{(a-d)^2/4 + b^2}$. They are positive and positive definiteness holds if

$$a > 0; \quad d - \frac{b^2}{a} > 0. \tag{9.E.6}$$

3. A positive determinant does not imply positive definiteness, even for symmetric matrices.
Hint: consider the matrix $\mathbf{diag}[-1 \;\; -1]$.

4. Let $\mathbf{A}$ and $\mathbf{D}$ be two square symmetric matrices, of respective sizes $m \times m$ and $n \times n$, and $\mathbf{B}$ a matrix of size $m \times n$. Show that the block matrix

$$\begin{bmatrix} \mathbf{A} & \mathbf{B} \\ \mathbf{B}^{\mathrm{T}} & \mathbf{D} \end{bmatrix}, \tag{9.E.7}$$

is positive definite (PD) iff

$$\mathbf{A} \; (\text{PD}) \quad \text{and} \quad \mathbf{D} - \mathbf{B}^{\mathrm{T}}\mathbf{A}^{-1}\mathbf{B} \; (\text{PD}). \tag{9.E.8}$$

Hint: $\mathbf{A}$ should clearly be positive definite. Now

$$\begin{aligned} [\mathbf{x} \; \mathbf{y}] \begin{bmatrix} \mathbf{A} & \mathbf{B} \\ \mathbf{B}^{\mathrm{T}} & \mathbf{D} \end{bmatrix} \begin{bmatrix} \mathbf{x} \\ \mathbf{y} \end{bmatrix} &= \mathbf{x}^{\mathrm{T}}\mathbf{A}\mathbf{x} + \mathbf{x}^{\mathrm{T}}\mathbf{B}\mathbf{y} + \mathbf{y}^{\mathrm{T}}\mathbf{B}^{\mathrm{T}}\mathbf{x} + \mathbf{y}^{\mathrm{T}}\mathbf{D}\mathbf{y} \\ &= \big(\mathbf{x} + \mathbf{A}^{-1}\mathbf{B}\mathbf{y}\big)^{\mathrm{T}} \mathbf{A} \big(\mathbf{x} + \mathbf{A}^{-1}\mathbf{B}\mathbf{y}\big) + \mathbf{y}^{\mathrm{T}}\big(\mathbf{D} - \mathbf{B}^{\mathrm{T}}\mathbf{A}^{-1}\mathbf{B}\big)\mathbf{y} \quad \Box. \end{aligned} \tag{9.E.9}$$

5. Consider now the case where $\mathbf{A}$ and $\mathbf{D}$ may not be symmetric.
Hint: The condition of positive definiteness (9.E.8) still holds but with $\mathbf{A}$ and $\mathbf{D}$ replaced by their symmetric parts.

Chapter 10

Transfers of mass, momentum, and energy

Diffusion is associated with spatial heterogeneity. In a macroscopic analysis, the geometrical points may store information that pertains to lower geometrical scales. Zooming on a point, we may uncover several phases with a definite spatial organization at these lower scales. As a typical example, the lower scale may display unit cells composed of a solid containing pores and capillaries. The role of the pores is to store the liquid while the capillaries are devoted to its circulation. At this scale, the exchange of liquid between the two cavities takes place by diffusion through the solid. At the macroscopic scale, the exchange is realized through *local transfer*: while most of the geometrical details of the lower scales are wiped out, the transfer coefficients attempt to keep some tracks of the physical properties, eg the diffusion coefficients, and of the characteristic diffusion times or lengths.

Both transfer of mass and transfer of heat are targeted in this chapter.

10.1 Summary of balance equations

The supplies contributing to the equations of balance considered in Chapter 4 may be due to exchanges between two or several constituents of the mixture or with the surroundings. It is instrumental to recall these equations and the constraints that the supplies satisfy. With $\mathcal{K}$ the set of all species in the mixture, the four balance equations hold for any species $k \in \mathcal{K}$,

- Balance of mass (4.4.5) : $\dfrac{d^k \rho^k}{dt} + \rho^k \operatorname{div} \mathbf{v}_k = \hat{\rho}^k$;

- Balance of momentum (4.5.5) : $\operatorname{div} \boldsymbol{\sigma}^k + \rho^k \left(\mathbf{b}_k - \dfrac{d^k \mathbf{v}_k}{dt} \right) = \begin{cases} \hat{\rho}^k (\mathbf{v}_k - \tilde{\mathbf{v}}_k) - \hat{\boldsymbol{\pi}}^k; \\ \hat{\rho}^k \, \mathbf{v}_k - \hat{\mathbf{p}}^k; \end{cases}$

  Net momentum supply (4.5.3) : $\hat{\mathbf{p}}^k \equiv \hat{\rho}^k \tilde{\mathbf{v}}_k + \hat{\boldsymbol{\pi}}^k$;

- Balance of energy (4.6.11) :

$$\dfrac{d^k}{dt}(\rho^k u_k) - (\boldsymbol{\sigma}^k - \rho^k u_k \, \mathbf{I}) : \boldsymbol{\nabla} \mathbf{v}_k + \operatorname{div} \mathbf{Q}_k - \Theta_k = \begin{cases} \hat{\rho}^k \left(\tilde{u}_k + \frac{1}{2}\,(\tilde{\mathbf{v}}_k - \mathbf{v}_k)^2 \right) + \hat{\mathcal{U}}^k; \\ \hat{U}^k - \hat{\mathbf{p}}^k \cdot \mathbf{v}_k + \frac{1}{2}\,\hat{\rho}^k\,\mathbf{v}_k^2; \end{cases} \qquad (10.1.1)$$

  Net energy supply (4.6.9) : $\hat{U}^k \equiv \hat{\rho}^k \left(\tilde{u}_k + \frac{1}{2}\,\tilde{\mathbf{v}}_k^2 \right) + \hat{\boldsymbol{\pi}}^k \cdot \mathbf{v}_k + \hat{\mathcal{U}}^k$;

- Balance of entropy (4.7.4) : $\hat{S}^k_{(\mathrm{i})} = \dfrac{d^k}{dt}(\rho^k s_k) + \rho^k s_k \operatorname{div} \mathbf{v}_k + \operatorname{div}\left(\dfrac{\mathbf{Q}_k}{T_k} \right) - \dfrac{\Theta_k}{T_k} - \hat{S}^k$;

  Net entropy supply (4.7.3) : $\hat{S}^k \equiv \hat{\rho}^k \, \tilde{s}_k + \hat{\mathcal{S}}^k$.

If the mixture is closed to mass, momentum, energy and entropy supplies, the net supplies of mass $\hat{\rho}$, of momentum

$\hat{\mathbf{p}}$, of energy $\hat{U}$, and of entropy $\hat{S}$ vanish,

$$\hat{\rho} \equiv \sum_{k \in \mathcal{K}} \hat{\rho}^k = 0, \quad \hat{\mathbf{p}} \equiv \sum_{k \in \mathcal{K}} \hat{\mathbf{p}}^k = \mathbf{0}, \quad \hat{U} \equiv \sum_{k \in \mathcal{K}} \hat{U}^k = 0 \quad \hat{S} \equiv \sum_{k \in \mathcal{K}} \hat{S}^k = 0. \tag{10.1.2}$$

The above equations use the current configuration as reference. Their counterparts using a different reference configuration may be found in Chapter 4. These equations apply over a volume. Their specialization across a first order discontinuity may be found in Section 11.1.4 for a solid, and in Section 11.10 for a mixture. The corresponding issues across acceleration waves in elastic and thermoelastic solids and mixtures are dealt with in Sections 11.2 to 11.6. While the above supplies do not contribute across a first or second order discontinuity, there may exist additional surface supplies for a first order discontinuity, Section 11.1.5

The properties of the supplied mass, namely its velocity $\tilde{\mathbf{v}}$, internal energy $\tilde{u}$ and entropy $\tilde{s}$ are denoted through a superimposed tilde. These entities need to be specified. A convenient assumption consists in setting their values equal to their counterparts, namely velocity $\mathbf{v}$, internal energy u and entropy s, at the location where they are supplied. The assumption is challenged in the context of the growth of soft tissues in Section 22.6.

10.2 Intercompartment mass transfers

10.2.1 Linear mass transfer

Let us consider mass transfers between the n species of a mixture which is closed to external mass supplies. Then the rates of all n mass transfers phrased in terms of the mass contents m sum to zero,

$$\sum_{k=1,n} \frac{dm_k}{dt} = 0, \tag{10.2.1}$$

and the associated dissipation per unit reference volume $(4.7.31)_2$ may be written as,

$$T \underline{\hat{S}}_{(\mathrm{i},2)} = \sum_{k=1,n} -\mu_k \frac{dm_k}{dt}. \tag{10.2.2}$$

Let $\boldsymbol{\kappa} = (\kappa_{jk})$ be a symmetric matrix of size $n \times n$, the diagonal components being in fact not involved for the present purpose. The linear rate of mass transfer is adopted in the format,

$$\frac{d}{dt} \begin{bmatrix} m_1 \\ m_2 \\ m_3 \\ \vdots \\ m_n \end{bmatrix} = - \begin{bmatrix} \sum_{j \neq 1} \kappa_{1j} & -\kappa_{12} & -\kappa_{13} & \cdots & -\kappa_{1n} \\ -\kappa_{12} & \sum_{j \neq 2} \kappa_{2j} & -\kappa_{23} & \cdots & -\kappa_{2n} \\ -\kappa_{13} & -\kappa_{23} & \sum_{j \neq 3} \kappa_{3j} & \cdots & -\kappa_{3n} \\ \vdots & \vdots & \vdots & \ddots & \cdots \\ -\kappa_{1n} & -\kappa_{2n} & -\kappa_{3n} & \cdots & \sum_{j \neq n} \kappa_{nj} \end{bmatrix} \begin{bmatrix} \mu_1 \\ \mu_2 \\ \mu_3 \\ \vdots \\ \mu_n \end{bmatrix}. \tag{10.2.3}$$

The dissipation may be recast in a format in which its sign can be easily assessed,

$$\begin{aligned} T \underline{\hat{S}}_{(\mathrm{i},2)} &= \sum_{k=1,n} -\mu_k \frac{dm_k}{dt} \\ &= \sum_{k=1,n} \mu_k \underbrace{\sum_{j \neq k} \kappa_{jk} (\mu_k - \mu_j)}_{A} \\ &= \sum_{j \neq k=1,n} \kappa_{jk} (\mu_k - \mu_j) \mu_k \\ &= \sum_{j \neq k=1,n} \kappa_{jk} \left(\underbrace{(\mu_k - \mu_j) (\mu_k - \mu_j)}_{B} + \underbrace{(\mu_k - \mu_j) \mu_j}_{A} \right) \\ &= 2 \underbrace{\sum_{j < k=1,n} \kappa_{jk} (\mu_k - \mu_j) (\mu_k - \mu_j)}_{B} - \underbrace{\sum_{j \neq k=1,n} \kappa_{jk} (\mu_k - \mu_j) \mu_k}_{A} \\ &= \sum_{j < k=1,n} \kappa_{jk} (\mu_k - \mu_j) (\mu_k - \mu_j). \end{aligned} \tag{10.2.4}$$

A positive contribution to the entropy production results if the $(n-1)\,n/2$ coefficients κ_{kj}, $k \neq j$, are positive.

As a special case, the mixture is now considered to contain the two phases I and E. The two species 1 and 2 are present in the two phases. The mass exchange is driven by the out of balances of individual chemical potentials,

$$\frac{d}{dt}\begin{bmatrix} m_{1\text{I}} \\ m_{1\text{E}} \\ m_{2\text{I}} \\ m_{2\text{E}} \end{bmatrix} = -\begin{bmatrix} \kappa_1 & -\kappa_1 & \kappa' & -\kappa' \\ -\kappa_1 & \kappa_1 & -\kappa' & \kappa' \\ \kappa' & -\kappa' & \kappa'' & -\kappa'' \\ -\kappa' & \kappa' & -\kappa'' & \kappa'' \end{bmatrix}\begin{bmatrix} \mu_{1\text{I}} \\ \mu_{1\text{E}} \\ \mu_{2\text{I}} \\ \mu_{2\text{E}} \end{bmatrix}. \tag{10.2.5}$$

This constitutive equation of transfer has introduced a coupling between an out-of-balance of the chemical potentials of species 2 and a mass transfer of species 1, and conversely. The coupling may be justified if the interface between the phases I and E may be viewed as a semi-permeable membrane that gives rise to the phenomenon of osmosis described in Section 13.7. If $\kappa' \neq 0$, the above rate of transfer can actually be split in two independent parts.

10.2.2 Mass transfer of electrically charged species

If the species are not electrically neutral, the independence of the mass transfers is lost. With ζ_k the valence of species k and $\widehat{m}_k$ its molar mass, electroneutrality constraints the changes of mass contents in each phase K,

$$\zeta_1 \frac{dm_{1\text{K}}}{\widehat{m}_1} + \zeta_2 \frac{dm_{2\text{K}}}{\widehat{m}_2} = 0. \tag{10.2.6}$$

Since the mass changes satisfy the additional constraint,

$$dm_{k\text{I}} + dm_{k\text{E}} = 0, \quad k = 1, 2, \tag{10.2.7}$$

the dissipation,

$$\begin{aligned} &-(\mu_{1\text{I}}^{\text{ec}} - \mu_{1\text{E}}^{\text{ec}})\frac{dm_{1\text{I}}}{dt} - (\mu_{2\text{I}}^{\text{ec}} - \mu_{2\text{E}}^{\text{ec}})\frac{dm_{2\text{I}}}{dt} \\ &= -\left(\widehat{m}_1\,(\mu_{1\text{I}}^{\text{ec}} - \mu_{1\text{E}}^{\text{ec}}) - \frac{\zeta_1}{\zeta_2}\,\widehat{m}_2\,(\mu_{2\text{I}}^{\text{ec}} - \mu_{2\text{E}}^{\text{ec}})\right)\frac{1}{\widehat{m}_1}\frac{dm_{1\text{I}}}{dt} \\ &= -(\mu_{\text{sI}} - \mu_{\text{sE}})\frac{dm_{1\text{I}}}{dt}, \end{aligned} \tag{10.2.8}$$

now expresses in terms of a single independent variable and of the jump of the chemical potentials of the neutral salt associated with the ionic species 1 and 2,

$$\widehat{m}_1\,\mu_{\text{sK}} = \widehat{m}_1\,\mu_{1\text{K}}^{\text{ec}} - \frac{\zeta_1}{\zeta_2}\,\widehat{m}_2\,\mu_{2\text{K}}^{\text{ec}}. \tag{10.2.9}$$

Note that the fact that this definition of the chemical potential of the salt breaks somehow the symmetry among the two ions is not an issue. Particular cases pertaining to articular cartilages are worked out in more explicit form in Section 16.1.2.4.

These relations may be extended to more general settings, e.g., a neutral species n and three ions 1, 2 and 3. While the exchange (10.2.7) still holds for the four species, the electroneutrality condition changes to,

$$\zeta_1 \frac{dm_{1\text{K}}}{\widehat{m}_1} + \zeta_2 \frac{dm_{2\text{K}}}{\widehat{m}_2} + \zeta_3 \frac{dm_{3\text{K}}}{\widehat{m}_3} = 0. \tag{10.2.10}$$

and the dissipation which may be brought in the format,

$$\begin{aligned} &-(\mu_{n\text{I}} - \mu_{n\text{E}})\frac{dm_{n\text{I}}}{dt} - (\mu_{1\text{I}}^{\text{ec}} - \mu_{1\text{E}}^{\text{ec}})\frac{dm_{1\text{I}}}{dt} - (\mu_{2\text{I}}^{\text{ec}} - \mu_{2\text{E}}^{\text{ec}})\frac{dm_{2\text{I}}}{dt} - (\mu_{3\text{I}}^{\text{ec}} - \mu_{3\text{E}}^{\text{ec}})\frac{dm_{3\text{I}}}{dt} \\ &= -(\mu_{n\text{I}} - \mu_{n\text{E}})\frac{dm_{n\text{I}}}{dt} - (\mu_{\text{s}_1\text{I}} - \mu_{\text{s}_1\text{E}})\frac{dm_{1\text{I}}}{dt} - (\mu_{\text{s}_2\text{I}} - \mu_{\text{s}_2\text{E}})\frac{dm_{2\text{I}}}{dt}, \end{aligned} \tag{10.2.11}$$

now expresses in terms of three independent variables and of the jump of the chemical potential of the neutral species and of the chemical potentials of the two neutral salts associated with the ionic species (1,3) and (2,3),

$$\widehat{m}_k \, \mu_{\mathrm{s}_k \mathrm{K}} = \widehat{m}_k \, \mu_{k\mathrm{K}}^{\mathrm{ec}} - \frac{\zeta_k}{\zeta_3} \, \widehat{m}_3 \, \mu_{3\mathrm{K}}^{\mathrm{ec}}, \quad k = 1, 2 \, . \tag{10.2.12}$$

The linear constitutive equation of transfer may be provided in the expanded format (10.2.5), or in the equivalent compacted format involving a symmetric positive semi-definite matrix,

$$\frac{d}{dt} \begin{bmatrix} m_{n\mathrm{I}} \\ m_{1\mathrm{I}} \\ m_{2\mathrm{I}} \end{bmatrix} = - \begin{bmatrix} \kappa_{nn} & \kappa_{n1} & \kappa_{n2} \\ \kappa_{n1} & \kappa_{11} & \kappa_{12} \\ \kappa_{n2} & \kappa_{12} & \kappa_{22} \end{bmatrix} \begin{bmatrix} \mu_{n\mathrm{I}} - \mu_{n\mathrm{E}} \\ \mu_{\mathrm{s}_1\mathrm{I}} - \mu_{\mathrm{s}_1\mathrm{E}} \\ \mu_{\mathrm{s}_2\mathrm{I}} - \mu_{\mathrm{s}_2\mathrm{E}} \end{bmatrix} = - \frac{d}{dt} \begin{bmatrix} m_{n\mathrm{E}} \\ m_{1\mathrm{E}} \\ m_{2\mathrm{E}} \end{bmatrix} . \tag{10.2.13}$$

If the mass transfers are uncoupled by setting the out-of-diagonal terms to 0, the remaining diagonal terms may be scaled via a representative mass density ρ_k and a characteristic transfer time $t_{\mathrm{tr},k}$,

$$\kappa_{kk} = \frac{\rho_k}{t_{\mathrm{tr},k}} \, A_k \, , \ \ k = n, \mathrm{s}_1, \mathrm{s}_2 \, , \tag{10.2.14}$$

where $A_k \geq 0$ [unit : $\mathrm{s}^2/\mathrm{m}^2$] is a material parameter.

The approach presented here has used electroneutrality to formulate the transfer equations in terms of neutral species. The electrical potentials have been temporary eliminated from the problem. Still, the difference of electrical potentials between the two compartments can be recovered at equilibrium by requiring the equality of the electro-chemical potentials of any species. Alternatively, if the electrical potentials are of key interest, the explicit use of the electroneutrality might be abandoned, and the rate transfer equations of *all* species are better phrased in terms of the electro-chemical potentials. This method involves one additional unknown (the electrical potential) and one additional equation due to the fact that the rate transfer equations are required for the charged and uncharged species, rather than for the associated neutral species.

10.2.3   Nonlinear constitutive equation of mass transfer

More general constitutive equations of mass transfer that satisfy the dissipation inequality can be postulated if the above uncoupling holds. Indeed, consider the transfer of the species k from phase I to phase E. The dissipation associated with the transfer should be positive, namely

$$- [\mu] \, \frac{dm}{dt} \geq 0 \, , \tag{10.2.15}$$

where $m = m_{k\mathrm{I}}$ is the mass content of species k in the phase I and $[\mu]$ is understood as the difference $\mu_{k\mathrm{I}} - \mu_{k\mathrm{E}}$.

The analysis below simplifies if the mass content m and time t are thought as dimensionless. The linear rate equation $dm/dt = -A\,[\mu]$ satisfies the inequality if A is a positive scalar. So does the rate equation $dm/dt = (\exp(-\epsilon A[\mu]) - 1)\,\epsilon$, with $\epsilon = \pm 1$. While the latter rate equation reduces to the linear equation at small $[\mu]$, its behavior at large positive and negative $[\mu]$ depends strongly on ϵ, Fig. 10.2.1.

Consider first the transfer of a species with $\epsilon = 1$. If its chemical potential in the phase I is large with respect to the phase E, the above $[\mu]$ is positive, the species tends to leave the phase I but the trend to exhaustion of the species in phase I is limited. On the other hand, if the chemical potential in the phase I is much smaller than in phase E, the rate of replenishment becomes exponential. The value $\epsilon = -1$ has converse effects on the transfer rates.

Now, observe that the rate relation

$$\frac{dm}{dt} = \big(\exp(-A|[\mu]|) - 1 \big) \, s_\mu, \tag{10.2.16}$$

with $s_\mu = \mathrm{sign}[\mu]$, is an example of constitutive equation of transfer that
  - is motivated by and satisfies the inequality dissipation;
  - ensures a finite rate of transfer at any concentration for A a positive scalar, and
  - reduces to a linear constitutive equation of transfer close to equilibrium.

With ρ_k the mass density [unit : kg/m^3] of species k and $t_{\mathrm{tr},k}$ a characteristic transfer time [unit : s], the above formula may be written in the more explicit form,

$$\frac{dm_{k\mathrm{I}}}{dt} = \frac{\rho_k}{t_{\mathrm{tr},k}} \big(\exp(-A_k|\mu_{k\mathrm{I}} - \mu_{k\mathrm{E}}|) - 1 \big) \, s_\mu, \tag{10.2.17}$$

with $s_\mu = \mathrm{sign}(\mu_{k\mathrm{I}} - \mu_{k\mathrm{E}})$ and A_k [unit : $\mathrm{s}^2/\mathrm{m}^2$] a positive scalar.

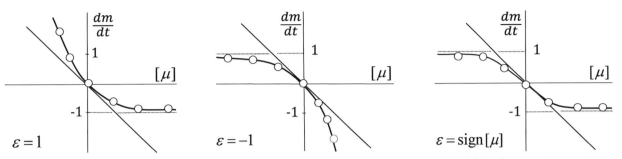

Fig. 10.2.1: Exponential constitutive equation of transfer of the form $dm/dt = (\exp(-\epsilon A[\mu]) - 1)\,\epsilon$, with $A > 0$, behave differently depending on the sign of ϵ but, close to equilibrium, they are tangent to the linear constitutive equation $dm/dt = -A\,[\mu]$. For $\epsilon = \text{sign}\,[\mu]$, the mass rate is bounded whatever the driving jump in chemical potential $[\mu]$.

10.3 Momentum supply in a poroelastic context

In Section 7.3.2, the momentum supply to the fluid phase associated with diffusion through the porous skeleton (seepage) has been identified so as to comply with the standard expression of Darcy's law.

The net momentum supply of water $\hat{\mathbf{p}}^{\mathrm{w}}$ as defined by (4.5.3) has been shown to be contributed by three terms, namely a Stokes' drag proportional to the velocity of the solid relative to the fluid $\mathbf{v}_{\mathrm{s}} - \mathbf{v}_{\mathrm{w}}$, a buoyancy term due to the spatial heterogeneity of the porosity n^{w} and a term associated with the momentum of the supplied mass,

$$\hat{\mathbf{p}}^{\mathrm{w}} = \hat{\rho}^{\mathrm{w}}\,\tilde{\mathbf{v}}_{\mathrm{w}} + \hat{\boldsymbol{\pi}}^{\mathrm{w}} = -k_{\mathrm{Sd}}\,(\mathbf{v}_{\mathrm{w}} - \mathbf{v}_{\mathrm{s}}) + p_{\mathrm{w}}\boldsymbol{\nabla}n^{\mathrm{w}} + \hat{\rho}^{\mathrm{w}}\,\mathbf{v}_{\mathrm{w}}\,, \tag{10.3.1}$$

where the *coefficient of Stokes' drag* $k_{\mathrm{Sd}} \geq 0$ [unit : $\mathrm{kg} \times \mathrm{m}^{-3} \times \mathrm{s}^{-1}$] is proportional to the inverse of the permeability coefficient.

In absence of external supply, the net momentum supply to the solid $\hat{\mathbf{p}}^{\mathrm{s}}$ is equal to $-\hat{\mathbf{p}}^{\mathrm{w}}$.

10.4 The bioheat equation

The bioheat equation is intended to account for the heat transferred in the interstitium by the blood flow. The thermoelastic model developed in Section 10.6 offers a larger constitutive framework which embodies the bioheat equation as a particular case but also accounts for thermomechanical couplings. Still even if the tissue is considered rigid, the bioheat equation is worth consideration *per se*.

10.4.1 The bioheat energy equation

The heat convected by the blood is equal to the specific heat capacity of blood c_b [unit : J/kg/K] times the blood perfusion rate w_b [unit : kg/m³/s] times the blood (arterial) temperature T_b. In the model of Pennes [1948], the transfer of heat is due to the difference of temperature $T_b - T$ between blood and tissue. This heat is instantaneously and totally available to the tissue. The details of the interface between the microvessels and the tissue, like the specific surface area or the interfacial heat transfer coefficient, are elusive here unlike in Section 10.8.

The total heat source Θ [unit : W/m³] may involve in addition the metabolic heat rate Θ_m as well as other local heat sources Θ_l,

$$\Theta = c_b\,w_b\,(T_b - T) + \Theta_m + \Theta_l\,. \tag{10.4.1}$$

The typical blood perfusion rate w_b is estimated to $0.5\,\mathrm{kg/m^3/s}$ and the metabolic rate Θ_m to $33.8\,\mathrm{kW/m^3}$. The energy equation of the tissue (9.1.23) becomes

$$T\hat{S}_{(\mathrm{i},1)} = T\frac{dS}{dt} + \mathrm{div}\,\mathbf{Q} - c_b\,w_b\,(T_b - T) - \Theta_m - \Theta_l = 0\,. \tag{10.4.2}$$

The heat flux $\mathbf{Q}$ [unit : W/m²] is governed by Fourier's law (15.3.10), namely $\mathbf{Q} = -\mathbf{k}_{\mathrm{T}}\cdot\boldsymbol{\nabla}T$, where $\mathbf{k}_{\mathrm{T}}$ [unit : W/m/K] is the thermal conductivity. For a rigid tissue, the increment of entropy per unit volume $T\,dS$ is equal to the heat capacity of the tissue $C_{\mathrm{T}} > 0$ [unit : J/m³/K] times the increment of temperature dT. Therefore the bioheat equation,

$$C_{\mathrm{T}}\frac{\partial T}{\partial t} - \mathrm{div}\,(\mathbf{k}_{\mathrm{T}}\cdot\boldsymbol{\nabla}T) - c_b\,w_b\,(T_b - T) - \Theta_m - \Theta_l = 0\,, \tag{10.4.3}$$

appears as a reactive parabolic equation. It is inhomogeneous, that is, it involves source terms and one of these terms includes the unknown temperature.

10.4.2   The bioheat energy equation with a relaxation time

Following the scheme developed in Section 9.1.4, the parabolic equation may be transformed to the hyperbolic equation (9.1.38),

$$
C_{\mathrm{T}} \left(t_{\mathrm{Cat}} \frac{\partial^2 T}{\partial t^2} + \frac{\partial T}{\partial t} \right) - \mathrm{div}\left(\mathbf{k}_{\mathrm{T}} \cdot \boldsymbol{\nabla} T \right) - t_{\mathrm{Cat}} \frac{\partial \Theta}{\partial t} - \Theta = 0 \,,
\tag{10.4.4}
$$

with the source Θ given by (10.4.1) including now a reactive term. The relaxation time t_{Cat} is linked to the wave-speed and thermal conductivity, Eqn (11.3.10). For metals, the relaxation time $t_{\mathrm{Cat}} \in [10^{-14}\,\mathrm{s}, 10^{-8}\,\mathrm{s}]$ is quite small with respect to typical durations of heating processes, Kaminski [1990]. It ranges between 20 to 30 s for biological tissues, Liu [2008].

10.4.3   Thermal boundary conditions

The bioheat equation may be used to simulate boundary value problems aimed at modeling heat treatments of tissues, especially of tumor tissues. Heat may be applied internally, pointwise or distributed, or it may be applied at the surface of the body either through a given normal flux, namely $\mathbf{Q} \cdot \mathbf{n} = q_n$ given with $\mathbf{n}$ outward unit normal to the body, or through a convective boundary condition,

$$
\mathbf{Q} \cdot \mathbf{n} = h_{\mathrm{s}\beta} \left(T - T_{\mathrm{ext}} \right),
\tag{10.4.5}
$$

with T_{ext} the given external temperature and $h_{\mathrm{s}\beta}$ [unit : W/m^2/K] the interfacial heat transfer coefficient between the solid and the fluid β. The heat may be supplied as well by radiation,

$$
\mathbf{Q} \cdot \mathbf{n} = \epsilon \, \sigma \left(T^4 - T_{\mathrm{rad}}^4 \right),
\tag{10.4.6}
$$

with T_{rad} the radiant temperature, $\sigma = 5.6704 \times 10^{-8}\,\mathrm{W/m^2/K^4}$ the Stefan-Boltzmann constant and ϵ a coefficient lumping the (dimensionless) emissivity and a view (shape) factor. The wave-lengths of thermal radiation range from 10^{-1} to $10^2\,\mathrm{\mu m}$, Bejan [1993], p. 505[10.1]. For modeling purposes, the radiative boundary condition (10.4.6) may be recast in the convective format (10.4.5) via a radiative coefficient h_{rad} [unit : W/m^2/K],

$$
\mathbf{Q} \cdot \mathbf{n} = h_{\mathrm{rad}} \left(T - T_{\mathrm{rad}} \right).
\tag{10.4.7}
$$

Whole-body measurements in dry air conditions indicate a natural convection coefficient h_{sa} of about 3.3 W/m^2/K and a radiative coefficient h_{rad} of about 4.5 W/m^2/K, de Dear et al. [1997].

   Unsolved key issues in hyperthermia include the intensity and duration of the thermal loading.

10.4.4   Significance of the Biot number

Let us consider a porous medium with a single porosity, namely a solid s bathed in a fluid β which can be water ($\beta = \mathrm{w}$) or air ($\beta = \mathrm{a}$). The thermal gradients are assumed to be negligible and only heat conduction and heat transfer are accounted for. With help of (10.8.1), the energy equation of the solid constituent (10.8.2) can be cast in the form,

$$
C^{\mathrm{s}} \frac{dT_{\mathrm{s}}}{dt} + h_{\mathrm{s}\beta} \, a_{\mathrm{s}\beta} \left(T_{\mathrm{s}} - T_{\beta} \right) = 0 \,,
\tag{10.4.8}
$$

where C^{s} [unit : J/m^3/K] is the apparent heat capacity of the solid, $h_{\mathrm{s}\beta}$ [unit : W/m^2/K] is the interfacial heat transfer coefficient and $a_{\mathrm{s}\beta}$ [unit : m^2/m^3] is the specific interfacial area. If the above material properties are constant in time as well as the fluid temperature T_{β}, the solid temperature,

$$
\frac{T_{\mathrm{s}} - T_{\beta}}{T_{\mathrm{s}}(t_*) - T_{\beta}} = \exp\left(-\frac{t - t_*}{t_{\mathrm{T}}} \right),
\tag{10.4.9}
$$

is found to decrease exponentially with a characteristic time t_{T} equal to $C^{\mathrm{s}}/h_{\mathrm{s}\beta}\, a_{\mathrm{s}\beta}$. Here t_* is a reference time beyond which the assumption of negligible gradient holds.

[10.1]The complete electromagnetic spectrum is displayed in Fig. 24.3.7.

The validity of the assumption of a negligible thermal gradient inside the body can be checked as suggested by Bejan [1993], p. 146. At a regular boundary point, the normal heat flux is given in two ways, by Fourier's law applied over a characteristic length L and by the boundary condition (10.4.5),

$$\mathbf{Q} \cdot \mathbf{n} \simeq -k_{\mathrm{TS}} \frac{T_b - T_s}{L} = h_{s\beta} \left(T_b - T_\beta\right), \tag{10.4.10}$$

with T_s and T_b respectively temperatures inside the solid at distance L from the boundary and at the boundary. Therefore, the gradient of temperature in a skin of width L inside the solid,

$$\frac{T_s - T_b}{L} = \frac{\mathrm{Bi}}{1 + \mathrm{Bi}} \frac{T_s - T_\beta}{L}, \tag{10.4.11}$$

will be negligible only if the Biot number $\mathrm{Bi} = L\,h_{s\beta}/k_{\mathrm{TS}}$ defined by (10.8.8) is sufficiently small.

10.5 Mass and energy transfers between a liquid and a solid

In Section 10.2, it has been shown how mass exchange between liquid phases can be dealt with in terms of their out-of-balance chemical potentials. Solids are endowed with a chemical potential which is a symmetric second order tensor defined by (4.7.22). However the split of the dissipation into a transfer contribution and a diffusion contribution is easily tractable only for liquid phases, as stressed in Note 4.7. The Massieu function,

$$\Xi_k = -\frac{\rho^k e_k}{T_k} = \rho^k \left(s_k - \frac{u_k}{T_k}\right), \tag{10.5.1}$$

provides an alternative to the chemical potential. The analysis below considers a mixture of a viscous liquid $k = \mathrm{w}$ and of an elastic solid s which can exchange mass, momentum and energy, the latter exchange being a consequence of the fact that they are in local thermal non equilibrium (LTNE), Fig. 10.5.1.

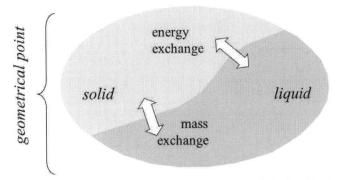

Fig. 10.5.1: Exchange of mass and of energy between the solid and the liquid of a porous medium.

10.5.1 Constitutive equations for the solid and liquid

The entropy production (10.1.1) for the species $k \in \mathcal{K} \equiv \{\mathrm{s}, \mathrm{w}\}$ may be expressed in terms of the Massieu function and of the net supplies defined in Section 10.1. Manipulating the balance of energy and the entropy production (10.1.1) to eliminate the term $\mathrm{div}\,\mathbf{Q}_k - \Theta_k$ yields a relation involving the strain rate $\mathbf{D}_k$,

$$\hat{S}_{(\mathrm{i})}^k = \frac{d^k \Xi_k}{dt} + \left(\frac{\boldsymbol{\sigma}^k}{T_k} + \Xi_k\,\mathbf{I}\right) : \mathbf{D}_k + \rho^k\,u_k\,\frac{d^k}{dt}\left(\frac{1}{T_k}\right) + \mathbf{Q}_k \cdot \boldsymbol{\nabla}\left(\frac{1}{T_k}\right) + \frac{\hat{U}^k}{T_k} - \hat{S}^k - \frac{\hat{\mathbf{p}}^k}{T_k} \cdot \mathbf{v}_k + \frac{\hat{\rho}^k}{T_k} \frac{\mathbf{v}_k^2}{2}. \tag{10.5.2}$$

Constitutive assumptions are needed to exploit the dissipation inequality. The Massieu function of the elastic solid will be assumed to depend on the right Cauchy-Green tensor $\mathbf{C}_s$, of its coldness T_s^{-1} and of its apparent mass density ρ^s,

$$\Xi_s = \Xi_s(\mathbf{C}_s, T_s^{-1}, \rho^s). \tag{10.5.3}$$

With help of the differential relation (2.5.14), the entropy production becomes,

$$
\hat{S}^{s}_{(i)} = \left(\frac{\boldsymbol{\sigma}^{s}}{T_{s}} + (\Xi_{s} - \rho^{s} \frac{\partial \Xi_{s}}{\partial \rho^{s}}) \mathbf{I} + 2\,\mathbf{F}^{T}_{s} \cdot \frac{\partial \Xi_{s}}{\partial \mathbf{C}_{s}} \cdot \mathbf{F}_{s} \right) : \mathbf{D}_{s} + \left(\rho^{s}\, u_{s} + \frac{\partial \Xi_{s}}{\partial T^{-1}_{s}} \right) \frac{d^{s}}{dt} \left(\frac{1}{T_{s}} \right) + \mathbf{Q}_{s} \cdot \boldsymbol{\nabla} \left(\frac{1}{T_{s}} \right)
$$
$$
+\; \hat{\rho}^{s} \left(\frac{\partial \Xi_{s}}{\partial \rho^{s}} + \frac{\mathbf{v}^{2}_{s}}{2T_{s}} \right) + \frac{\hat{U}^{s}}{T_{s}} - \hat{S}^{s} - \frac{\hat{\mathbf{p}}^{s}}{T_{s}} \cdot \mathbf{v}_{s} \,.
$$

(10.5.4)

The Massieu function of the viscous liquid will be assumed to depend on its coldness T^{-1}_{w} and of its apparent mass density ρ^{w},

$$
\Xi_{w} = \Xi_{w}(T^{-1}_{w}, \rho^{w})\,,
$$

(10.5.5)

in such a way that the entropy production becomes,

$$
\hat{S}^{w}_{(i)} = \left(\frac{\boldsymbol{\sigma}^{w}}{T_{w}} + (\Xi_{w} - \rho^{w} \frac{\partial \Xi_{w}}{\partial \rho^{w}}) \mathbf{I} \right) : \mathbf{D}_{w} + \left(\rho^{w}\, u_{w} + \frac{\partial \Xi_{w}}{\partial T^{-1}_{w}} \right) \frac{d^{w}}{dt} \left(\frac{1}{T_{w}} \right) + \mathbf{Q}_{w} \cdot \boldsymbol{\nabla} \left(\frac{1}{T_{w}} \right)
$$
$$
+\; \hat{\rho}^{w} \left(\frac{\partial \Xi_{w}}{\partial \rho^{w}} + \frac{\mathbf{v}^{2}_{w}}{2T_{w}} \right) + \frac{\hat{U}^{w}}{T_{w}} - \hat{S}^{w} - \frac{\hat{\mathbf{p}}^{w}}{T_{w}} \cdot \mathbf{v}_{w} \,.
$$

(10.5.6)

In line with Note 4.5, the entropy production of the mixture as a whole is requested to be positive. This procedure does not preclude interesting constraints for the individual species. Indeed, arguing of the independence of velocity gradient of the solid and of the time rate and spatial gradient of its temperature, the following constitutive equations result,

$$
\frac{\boldsymbol{\sigma}^{s}}{T_{s}} = -(\Xi_{s} - \rho^{s} \frac{\partial \Xi_{s}}{\partial \rho^{s}}) \mathbf{I} - 2\,\mathbf{F}^{T}_{s} \cdot \frac{\partial \Xi_{s}}{\partial \mathbf{C}_{s}} \cdot \mathbf{F}_{s},
$$
$$
\rho^{s}\, s_{s} = \Xi_{s} - T^{-1}_{s} \frac{\partial \Xi_{s}}{\partial T^{-1}_{s}} = \Xi_{s} + \frac{\rho^{s}\, u_{s}}{T_{s}}
$$
$$
\frac{\mathbf{Q}_{s}}{T^{2}_{s}} = \mathbf{k}_{T_{s}} \cdot \boldsymbol{\nabla} \left(\frac{1}{T_{s}} \right),
$$

(10.5.7)

where $\mathbf{k}_{T_{s}}$ is the positive definite second order tensor of thermal conductivity in the solid [unit : W/m/K]. The liquid being viscous, its stress depends on the strain rate $\mathbf{D}_{w}$, in such a way that the independence of $\mathbf{D}_{w}$ can not be exploited directly as pointed out in Section 4.9. The way out of this difficulty consists in postulating the dissipation associated with this strain rate in a specific format. For a Newtonian liquid, the dissipation would be of the form $\mathbf{D}_{w} : \mathbb{E}_{v} : \mathbf{D}_{w}$ where $\mathbb{E}_{v}$ [unit : Pa×s] is a (semi-)positive definite fourth order tensor. For a Newtonian isotropic liquid, this tensor is defined in terms of the Lamé viscosities λ_{v} and μ_{v} obeying the inequalities $\lambda_{v} + (2/3)\,\mu_{v} \geq 0$ and $\mu_{v} \geq 0$. The constitutive equations are then compatible with a positive entropy production,

$$
\frac{\boldsymbol{\sigma}^{w}}{T_{w}} = -(\Xi_{w} - \rho^{w} \frac{\partial \Xi_{w}}{\partial \rho^{w}}) \mathbf{I} + \mathbb{E}_{v} : \mathbf{D}_{w},
$$
$$
\rho^{w}\, s_{w} = \Xi_{w} - T^{-1}_{w} \frac{\partial \Xi_{w}}{\partial T^{-1}_{w}} = \Xi_{w} + \frac{\rho^{w}\, u_{w}}{T_{w}}
$$
$$
\frac{\mathbf{Q}_{w}}{T^{2}_{w}} = \mathbf{k}_{T_{w}} \boldsymbol{\nabla} \left(\frac{1}{T_{w}} \right),
$$

(10.5.8)

where $\mathbf{k}_{T_{w}}$ is the positive definite second order tensor of thermal conductivity in the liquid.

10.5.2 Constitutive equations for mass and heat transfers

Given the closure relations (10.1.2) in absence of external supplies, the sum of the reduced dissipations results in the format,

$$
\hat{S}_{(i)} = \hat{\rho}^{s} \left(\frac{\partial \Xi_{s}}{\partial \rho^{s}} - \frac{\partial \Xi_{w}}{\partial \rho^{w}} + \frac{1}{2} \left(\frac{\mathbf{v}^{2}_{s}}{T_{s}} - \frac{\mathbf{v}^{2}_{w}}{T_{w}} \right) \right) + \hat{U}^{s} \left(\frac{1}{T_{s}} - \frac{1}{T_{w}} \right) - \hat{\mathbf{p}}^{s} \cdot \left(\frac{\mathbf{v}_{s}}{T_{s}} - \frac{\mathbf{v}_{w}}{T_{w}} \right).
$$

(10.5.9)

Neglecting the buoyancy term, the identification of the net momentum supply of the liquid (10.3.1) may be used to modify this expression,

$$
\hat{S}_{(i)} = \hat{\rho}^{s} \left(\frac{\partial \Xi_{s}}{\partial \rho^{s}} - \frac{\partial \Xi_{w}}{\partial \rho^{w}} + \frac{1}{2} \frac{(\mathbf{v}_{w} - \mathbf{v}_{s})^{2}}{T_{w}} \right) + \left(\hat{U}^{s} - \hat{\mathbf{p}}^{s} \cdot \mathbf{v}_{s} + \hat{\rho}^{s} \frac{\mathbf{v}^{2}_{s}}{2} \right) \left(\frac{1}{T_{s}} - \frac{1}{T_{w}} \right)
$$
$$
+\; k_{Sd} \frac{(\mathbf{v}_{w} - \mathbf{v}_{s})^{2}}{T_{w}} \,.
$$

(10.5.10)

The coefficient of Stokes' drag k_{Sd} being positive, the entropy production is ensured to be positive for the following linear constitutive equations of mass transfer and energy transfer,

$$
\begin{aligned}
\hat{\rho}^{\mathrm{s}} &= \kappa_\rho \left(\frac{\partial \Xi_{\mathrm{s}}}{\partial \rho^{\mathrm{s}}} - \frac{\partial \Xi_{\mathrm{w}}}{\partial \rho^{\mathrm{w}}} + \tfrac{1}{2} \frac{(\mathbf{v}_{\mathrm{w}} - \mathbf{v}_{\mathrm{s}})^2}{T_{\mathrm{w}}} \right), \\
\hat{U}^{\mathrm{s}} - \hat{\mathbf{p}}^{\mathrm{s}} \cdot \mathbf{v}_{\mathrm{s}} + \hat{\rho}^{\mathrm{s}} \frac{\mathbf{v}_{\mathrm{s}}^2}{2} &= \kappa_T \left(\frac{1}{T_{\mathrm{s}}} - \frac{1}{T_{\mathrm{w}}} \right),
\end{aligned}
\tag{10.5.11}
$$

if the transfer coefficients satisfy the inequalities,

$$
\kappa_\rho \geq 0, \quad \kappa_T \geq 0. \tag{10.5.12}
$$

The energy transfer equation features the rhs of the energy equation of the solid (10.1.1) and indicates that the energy of the solid increases if its temperature is lower than that of the liquid.

In absence of external mass supply, the rate of mass transfer to the fluid is the negative of its solid counterpart, $\hat{\rho}^{\mathrm{w}} = -\hat{\rho}^{\mathrm{s}}$. However, attention should be paid to the fact that the solid and liquid do not play symmetric roles in the energy equation due to the identification of the momentum transfer of the liquid. The energy exchange for the liquid writes,

$$
\hat{U}^{\mathrm{w}} - \hat{\mathbf{p}}^{\mathrm{w}} \cdot \mathbf{v}_{\mathrm{s}} + \hat{\rho}^{\mathrm{w}} \frac{\mathbf{v}_{\mathrm{s}}^2}{2} = \kappa_T \left(\frac{1}{T_{\mathrm{w}}} - \frac{1}{T_{\mathrm{s}}} \right), \tag{10.5.13}
$$

and the entity involved in this equation is *not* the rhs of the energy equation (10.1.1) of the liquid.

An energy exchange between a solid and two liquids is derived in Section 10.6.3: the constitutive relation features the same lhs as in (10.5.11) and (10.5.13) to within the fact that the term $\hat{\rho}^{\mathrm{s}}$ does not appear as the solid does not exchange mass in that section. An estimation of the heat transfer coefficient in terms of physical and geometrical properties of the solid and liquid is proposed in Section 10.8.

The interpretation of the constitutive equation of mass transfer is less immediate. In the analysis of isothermal solidification/fluidification, Henson [2013] works out the Massieu function for a neo-Hookean elastic solid with the ratio of $\rho^{\mathrm{s}} (\det \mathbf{C}_{\mathrm{s}})^{1/2}$ over the reference mass density $\underline{\rho}^{\mathrm{s}}$ (at $\mathbf{C}_{\mathrm{s}} = \mathbf{I}$) as an additional argument, as allowed by the formulation (10.5.3). Note that this ratio would vanish in absence of mass exchange. Similarly, the Massieu function of the liquid contains the ratio of the current and reference mass densities, the current mass density being obtained from the mass balance. The sense of the mass exchange is driven both by the difference of mass densities between the solid and the liquid and by the strain. Note that the elastic constitutive equations should be such that increase of mass leads to a compressive stress in agreement with (7.2.14) and (7.2.15). The above derivations closely follows Henson [2013], except in the identification of the momentum supply of the liquid, which leads to somehow different transfer equations.

10.6 A porous medium with two porosities and three temperatures

Specialized cavities:

Porous media with double porosity contain two types of cavities with distinct roles, Fig. 10.6.1. Pores provide most of the space to store fluid while vessels provide most of the permeability. Barenblatt et al. [1960] appear to be the first to have introduced the double porosity concept in the study of non-deformable fissured rock aquifers. For many applications, the restriction to rigid skeletons is not admissible: it has been alleviated by Duguid and Lee [1977] who however considered an incompressible solid constituent, and by Aifantis [1977] in a more general approach. Wilson and Aifantis [1984] presented analytical solutions for the borehole problem with various boundary conditions. A mixed finite element formulation with the solid displacement and the fluid pressures as unknowns is presented by Khaled et al. [1984] and applied to display the quantitative effects of double porosity in one- and two-dimensional consolidation analyses.

The deformation coupling used by Aifantis and coworkers was not complete and more general coupling equations are presented in Khalili and Valliappan [1996] and Tuncay and Corapcioglu [1996]a [1996]b.

The influence of a single porosity on the number, amplitude and attenuation of harmonic waves has received much more attention than the effect of double porosity. Biot [1956]I-II has evidenced the existence of two, one fast and the other slow, longitudinal waves due to the coupling between the solid skeleton and a fluid phase. The slow wave is attenuated so that its experimental measurement has been made by Plona [1980] more than twenty years after

Biot's papers. For double porosity mixtures, Wilson and Aifantis [1984] report the existence of three longitudinal waves, with much attenuation for the second and third waves, and of one rotational wave. The analysis of Tuncay and Corapcioglu [1996]a [1996]b is more general as it concerns actually a mixture with four species: the pores are saturated by air and water, capillary effects are included and the pore pressure is approximated through volume averaging. They report the effects of frequency, saturation and volume fractions on the speed and attenuation of the four longitudinal and single rotational waves.

<table>
<tr><td>mass exchange
between the two cavities</td><td>energy exchange
between the solid and the two cavities</td></tr>
</table>

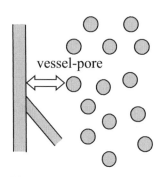
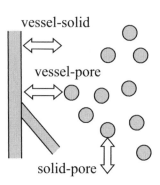

Fig. 10.6.1: The double porosity concept highlights two cavities with distinct roles.

Spatial scale separation and time scale separation:

Pore spacing is assumed to be much smaller than vessel spacing and therefore the representative elementary volume should be at least an order of magnitude larger than the vessel spacing. The method suggested to identify the mechanical constants uses this spatial scale separation.

Another method to identify material constants has been suggested by Wilson and Aifantis [1984]: it exploits the time scale separation implied by the contrast in hydraulic conductivities of the pores and vessels. Berryman and Wang [1995] propose a mixed method using the existence of two scales in both time and space. Both works concern isotropic solid skeletons under isothermal conditions.

Thermomechanical analysis, mass transfer and energy transfer:

The fluids of the two cavities flow through the solid skeleton essentially according to Darcy's law for the pore fluid. Moreover, pores and vessels exchange fluid (phenomenon referred to as *mass transfer*). The mass of each fluid may vary due to mass transfer with the sister cavity and due to diffusion from/to the surroundings. On the other hand, the solid constituent does not undergo a mass change, which could be due to a mass transfer or a growth/resorption. The porous medium itself is closed in the thermodynamic sense defined in Chapter 4.

Here, the thermomechanical analysis considers local thermal non equilibrium (LTNE) and the three species are endowed with their own temperatures. Convection of heat by the fluid of the vessels is an important aspect that generates/makes possible LTNE. Therefore, the solid and fluid constituents exchange energy (phenomenon referred to as *energy transfer*).

The approach capitalizes upon attempts in the framework of unsaturated porous media, Loret and Khalili [2000], Khalili and Loret [2001], and in clays, Khalili [2003], Khalili and Selvadurai [2003]. The model has been used to solve initial and boundary value problems through a finite element approach, Gelet [2011].

Anisotropy in mechanical, diffusion and thermal properties:

All the above works concern isothermal elastic isotropic solid skeletons. The formulation of Loret and Rizzi [1999] includes mechanical anisotropy in the solid constituent and in the solid skeleton. Here, diffusion properties and the thermal expansion properties may a priori be anisotropic as well.

10.6.1 Thermomechanical equations

10.6.1.1 Thermodynamic derivation

The expression $(7.2.28)_1$ of the mechanical entropy production and the definitions of the thermodynamic functions in Table 7.2.2 stand as our starting point. Repeated use is also made of the rate relation (2.7.2).

For the solid, upon multiplication by $T_s \det \mathbf{F}$, Eqn $(7.2.28)_1$ is modified so as to display entropy and free energy,

$$T_s \hat{\underline{S}}_{(\mathrm{i},1)}^s = -\frac{d}{dt}(m^s e_s) + \boldsymbol{\tau}^s : \boldsymbol{\nabla}\mathbf{v}_s - m^s s_s \frac{dT_s}{dt}. \tag{10.6.1}$$

Recall that terms in the mechanical entropy production involving a chemical potential do not pertain to solid constituent.

For a fluid constituent $k \in \mathcal{K}^{**}$, upon multiplication by $T_s \det \mathbf{F}$, use of the identity $\mu_k \, dm^k = d(\mu_k \, m^k) - d\mu_k \, m^k$, and of (7.2.31), Eqn $(7.2.28)_1$ takes the form,

$$T_s \hat{\underline{S}}_{(\mathrm{i},1)}^k = \frac{d}{dt}\left(v^k \, \tilde{p}_k\right) + \tilde{\boldsymbol{\tau}}^k : \boldsymbol{\nabla}\mathbf{v}_s - v^k \frac{d\tilde{p}_k}{dt}, \tag{10.6.2}$$

where $\tilde{p}_k$ is a modified intrinsic pressure and $\tilde{\boldsymbol{\tau}}^k$ a modified partial Kirchhoff stress,

$$\tilde{\boldsymbol{\tau}}^k = -v^k \, \tilde{p}_k \, \mathbf{I}, \quad \tilde{p}_k = p_k \frac{T_s}{T_k}, \quad k \in \mathcal{K}^{**}. \tag{10.6.3}$$

Upon summation over the solid constituent and all fluid constituents, the mechanical part of entropy production for the mixture multiplied by T_s,

$$T_s \hat{\underline{S}}_{(\mathrm{i},1)} = -\frac{d\mathcal{W}}{dt} + \tilde{\boldsymbol{\tau}} : \boldsymbol{\nabla}\mathbf{v}_s - m^s s_s \frac{dT_s}{dt} - \sum_{k \in \mathcal{K}^{**}} v^k \frac{d\tilde{p}_k}{dt}, \tag{10.6.4}$$

expresses in terms of the thermo-poroelastic energy of the mixture $\mathcal{W}$,

$$\underline{\mathcal{W}} = \underline{\mathcal{W}}(\boldsymbol{\epsilon}, \cup_{k \in \mathcal{K}^{**}} \tilde{p}_k, T_s) = m^s e_s - \sum_{k \in \mathcal{K}^{**}} v^k \, \tilde{p}_k. \tag{10.6.5}$$

The second order tensor $\tilde{\boldsymbol{\tau}}$ is a modified total Kirchhoff stress,

$$\tilde{\boldsymbol{\tau}} = \boldsymbol{\tau}^s + \sum_{k \in \mathcal{K}^{**}} \tilde{\boldsymbol{\tau}}^k. \tag{10.6.6}$$

The total Kirchhoff stress deduces as,

$$\boldsymbol{\tau} = \boldsymbol{\tau}^s + \sum_{k \in \mathcal{K}^{**}} \boldsymbol{\tau}^k = \tilde{\boldsymbol{\tau}} - \sum_{k \in \mathcal{K}^{**}} v^k \, (p_k - \tilde{p}_k) \, \mathbf{I}. \tag{10.6.7}$$

Setting the entropy production to zero, constitutive equations result for the dependent variables,

$$\tilde{\boldsymbol{\tau}} = \frac{\partial \mathcal{W}}{\partial \boldsymbol{\epsilon}}; \quad -v^k = \frac{\partial \underline{\mathcal{W}}}{\partial \tilde{p}_k}, \quad k \in \mathcal{K}^{**}; \quad -\underline{S}^s = -m^s s_s = \frac{\partial \mathcal{W}}{\partial T_s}. \tag{10.6.8}$$

The above derivation is quite distinct from that of Bowen and Garcia [1970] who, for each species, introduced a Massieu function Ξ_k as defined by (10.5.1). They obtain constitutive equations for the internal energies and partial stresses. Each Massieu function is developed up to second order in terms of the main variables, which are the strains of the species and their temperatures. An application to a porous medium including a solid matrix and a fluid is presented in Bowen and Chen [1975].

By contrast, the present derivation aims at highlighting some specific symmetries and decouplings of thermomechanical properties as can be observed already in (10.6.8).

The analysis is now restricted:

    1. the porous medium is constituted by a solid skeleton saturated by two non miscible fluids, referred to by the indices 1 and 2;

    2. the temperatures of the species are distinct, but their ratios are close enough to 1 that the differences between the pressures p_k and $\tilde{p}_k$ defined by (10.6.3) are neglected;

    3. the thermomechanical equations to be developed are linear and the current configuration is taken as reference.

The thermomechanical constitutive relations simplify to,

$$\boldsymbol{\sigma} = \frac{\partial \mathcal{W}}{\partial \boldsymbol{\epsilon}}; \quad -v^k = \frac{\partial \mathcal{W}}{\partial p_k}; \quad k \in \mathcal{K}^{**} = \{1, 2\}, \quad -S^s = -\rho^s s_s = \frac{\partial \mathcal{W}}{\partial T_s}. \tag{10.6.9}$$

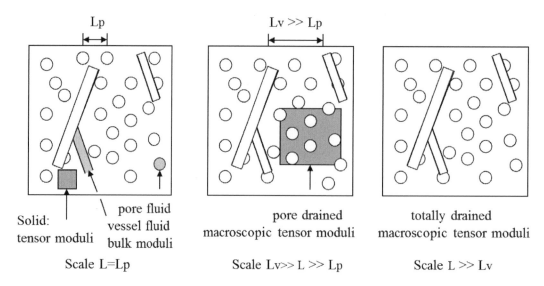

Fig. 10.6.2: The measurement of the elastic constants uses the spatial scale separation and considers three scales: the smallest scales devoid of pores, an intermediate scale devoid of vessels and the largest scale with both pores and vessels.

10.6.1.2   The generalized effective stress

The effective stress introduced in (9.3.10) is extended so as to include two pressures, the strain coming into picture being the elastic strain, that is, the total strain minus the thermal expansion,

$$\boldsymbol{\sigma}' = \boldsymbol{\sigma} + \boldsymbol{\alpha}_1\, p_1 + \boldsymbol{\alpha}_2\, p_2 = \mathbb{E} : \left(\boldsymbol{\epsilon} - \mathbf{c}_\mathrm{T}\, (T_\mathrm{s} - T_\mathrm{s}^0) \right). \tag{10.6.10}$$

This effective stress involves three types of constitutive tensors:

- when full drainage is allowed, $p_1 = p_2 = 0$ and, under isothermal conditions, the effective stress is then equal to the partial stress of the solid skeleton, $\boldsymbol{\sigma}' = \boldsymbol{\sigma} = \boldsymbol{\sigma}^\mathrm{s} = \mathbb{E} : \boldsymbol{\epsilon}$, which shows that the tensor moduli $\mathbb{E}$ are the elastic moduli of the *underlying drained solid*;
- the symmetric second order tensor $\mathbf{c}_\mathrm{T}$ [unit : K^{-1}] is clearly the tensor of thermal expansion of the drained solid skeleton, equal to the tensor of thermal expansion of the solid constituent, as indicated by Proposition 9.4;
- the two symmetric second order tensors $\boldsymbol{\alpha}_1$ and $\boldsymbol{\alpha}_2$ will be identified below.

The existence of a thermo-poroelastic energy,

$$
\begin{aligned}
\mathcal{W}(\boldsymbol{\epsilon},\, \cup_{k=1,2}\, p_k,\, T_\mathrm{s}) \;=\;\; & \frac{1}{2}\, \boldsymbol{\epsilon} : \mathbb{E} : \boldsymbol{\epsilon} - (T_\mathrm{s} - T_\mathrm{s}^0)\, \mathbf{c}_\mathrm{T} : \mathbb{E} : \boldsymbol{\epsilon} - (p_1\, \boldsymbol{\alpha}_1 + p_2\, \boldsymbol{\alpha}_2) : \boldsymbol{\epsilon} \\[4pt]
& - \frac{1}{2}\, (a_{11}\, p_1^2 + 2\, a_{12}\, p_1\, p_2 + a_{22}\, p_2^2) - v_0^1\, p_1 - v_0^2\, p_2 \\[4pt]
& - (a_{1T}\, p_1 + a_{2T}\, p_2 + S_0^\mathrm{s})\, (T_\mathrm{s} - T_\mathrm{s}^0) - \frac{1}{2}\, \frac{C_{\epsilon p}^\mathrm{s}}{T_\mathrm{s}^0}\, (T_\mathrm{s} - T_\mathrm{s}^0)^2,
\end{aligned}
\tag{10.6.11}
$$

implies the matrix form of the thermomechanical equations to be endowed with the major symmetry,

$$
\begin{bmatrix} -\boldsymbol{\sigma} \\[6pt] v^1 - v_0^1 \\[6pt] v^2 - v_0^2 \\[6pt] S^\mathrm{s} - S_0^\mathrm{s} \end{bmatrix}
=
\begin{bmatrix}
-\mathbb{E} & \boldsymbol{\alpha}_1 & \boldsymbol{\alpha}_2 & \mathbb{E} : \mathbf{c}_\mathrm{T} \\[6pt]
\boldsymbol{\alpha}_1 & a_{11} & a_{12} & a_{1T} \\[6pt]
\boldsymbol{\alpha}_2 & a_{12} & a_{22} & a_{2T} \\[6pt]
\mathbf{c}_\mathrm{T} : \mathbb{E} & a_{1T} & a_{2T} & C_{\epsilon p}^\mathrm{s}/T_\mathrm{s}^0
\end{bmatrix}
\begin{bmatrix} \boldsymbol{\epsilon} \\[6pt] p_1 \\[6pt] p_2 \\[6pt] T_\mathrm{s} - T_\mathrm{s}^0 \end{bmatrix}.
\tag{10.6.12}
$$

Here $C_{\epsilon p}^\mathrm{s}$ [unit : J/m^3/K] is the *apparent heat capacity* of the solid, at constant strain and fluid pressures, per unit volume of porous medium, i.e. heat capacity times volume fraction.

The coefficients defining the thermal response, and the entropy of the solid, result from Proposition 9.3 (where $T = T_\mathrm{s}$) namely

$$a_{kT} = (n^k\, \mathbf{I} - \boldsymbol{\alpha}_k) : \mathbf{c}_\mathrm{T}, \quad k = 1, 2. \tag{10.6.13}$$

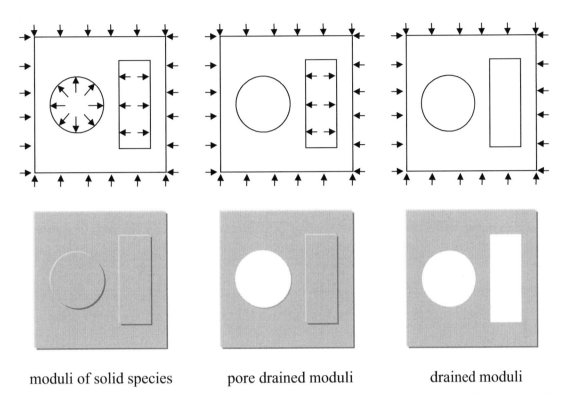

moduli of solid species   pore drained moduli   drained moduli

Fig. 10.6.3: The elastic constants are accessible to measurements by well chosen loadings of the pores, of the vessels and of the solid.

The remaining mechanical coefficients may be calculated in terms of three types of constitutive moduli that are measurable through experiments performed at different scales, as sketched in Fig. 10.6.2, namely:

- the tensor moduli $\mathbb{E}_s$ of the solid constituent, that link its intrinsic strain $\boldsymbol{\epsilon}_s$ and its intrinsic stress $\boldsymbol{\sigma}_s$, Eqn (7.2.14). The involved scale corresponds to that provided by the solid constituent devoid of the two porosities;
- the tensor moduli $\mathbb{E}^{\mathrm{PD}}$ corresponding to a zero pressure in the pores of type 1 ($p_1{=}0$), the acronym 'PD' standing for *Pore Drained mixture*. The experiments should be performed on specimens an order of magnitude larger than the spacing of the smaller pores but devoid of the larger pores;
- the tensor moduli $\mathbb{E}$ of the underlying Drained Solid, or solid skeleton, corresponding to zero pressures in the two types of pores, ($p_1{=}p_2{=}0$). The experiments should be performed at the largest scale, say an order of magnitude larger than the spacing of the vessels.

The identification makes use of

Proposition 10.1 on homogeneous deformation under homogeneous isothermal mechanical loading.

For an isothermal loading, i.e. $\Delta T_s = 0$, and a homogeneous mechanical loading, $\boldsymbol{\sigma} = -p\,\mathbf{I}$, $p_k = p$, $k \in \mathcal{K}^{**}$, the material responds as if it was devoid of porosity, and the volume change of each species is proportional to its volume fraction, $\Delta v^k = n^k \operatorname{tr} \boldsymbol{\epsilon}$, $k \in \mathcal{K}$.

The proposition is illustrated by Fig. 10.6.3. Then the strain is governed by both the effective stress law Eqn (10.6.10) and the mechanical response of the solid constituent, Eqn (7.2.14) with $\boldsymbol{\sigma}_s = -p\,\mathbf{I}$, namely,

$$\boldsymbol{\epsilon} = \mathbb{E}^{-1} : (-p\,\mathbf{I} + \boldsymbol{\alpha}_1\,p + \boldsymbol{\alpha}_2\,p) = \mathbb{E}_s^{-1} : (-p\,\mathbf{I})\,, \tag{10.6.14}$$

while the volume changes of the pores are in proportion with their volume fractions,

$$\Delta v = \mathbf{I} : \boldsymbol{\epsilon}\,, \quad \Delta v^1 = n^1 \Delta v\,, \quad \Delta v^2 = n^2 \Delta v\,. \tag{10.6.15}$$

On the other hand, for the loading defined by $(\boldsymbol{\sigma},\,p_1,\,p_2){=}(-p\,\mathbf{I},\,0,\,p)$, the strain is governed by both the effective stress law Eqn (10.6.10) and the moduli $\mathbb{E}^{\mathrm{PD}}$,

$$\boldsymbol{\epsilon} = \mathbb{E}^{-1} : (-p\,\mathbf{I} + \boldsymbol{\alpha}_2\,p) = (\mathbb{E}^{\mathrm{PD}})^{-1} : (-p\,\mathbf{I}) \tag{10.6.16}$$

and

$$\Delta v = \mathbf{I} : \boldsymbol{\epsilon} \,, \quad \Delta v^2 = n^2 \Delta v \,. \tag{10.6.17}$$

The tensors $\boldsymbol{\alpha}_1$ and $\boldsymbol{\alpha}_2$ result from Eqns (10.6.15) and (10.6.17), namely

$$\boldsymbol{\alpha}_1 = \mathbb{E} : (\mathbb{E}^{\mathrm{PD}})^{-1} : \mathbf{I} - \mathbb{E} : \mathbb{E}_{\mathrm{s}}^{-1} : \mathbf{I}, \quad \boldsymbol{\alpha}_2 = \mathbf{I} - \mathbb{E} : (\mathbb{E}^{\mathrm{PD}})^{-1} : \mathbf{I} \,. \tag{10.6.18}$$

It is convenient to define the bulk moduli associated with the tensor moduli $\mathbb{E}_{\mathrm{s}}$, $\mathbb{E}^{\mathrm{PD}}$ and $\mathbb{E}$ as

$$\frac{1}{K_{\mathrm{s}}} = \mathbf{I} : \mathbb{E}_{\mathrm{s}}^{-1} : \mathbf{I}, \quad \frac{1}{K^{\mathrm{PD}}} = \mathbf{I} : (\mathbb{E}^{\mathrm{PD}})^{-1} : \mathbf{I}, \quad \frac{1}{K} = \mathbf{I} : \mathbb{E}^{-1} : \mathbf{I} \,. \tag{10.6.19}$$

Then, the coefficients a_{11}, a_{12} and a_{11} result by comparing the fluid volume contents given by the constitutive equations (10.6.12) and by equations (10.6.15) and (10.6.17), for the two above particular loadings, namely

$$
\begin{aligned}
a_{11} &= \frac{1 - n^2}{K^{\mathrm{PD}}} - \frac{1 + n^1 - n^2}{K_{\mathrm{s}}} - \boldsymbol{\alpha}_1 : \mathbb{E}^{-1} : \boldsymbol{\alpha}_1 \,; \\[2mm]
a_{12} &= \frac{n^2}{K^{\mathrm{PD}}} - \frac{n^2}{K_{\mathrm{s}}} - \boldsymbol{\alpha}_1 : \mathbb{E}^{-1} : \boldsymbol{\alpha}_2 \,; \\[2mm]
a_{22} &= \frac{1}{K} - \frac{1 + n^2}{K^{\mathrm{PD}}} - \boldsymbol{\alpha}_2 : \mathbb{E}^{-1} : \boldsymbol{\alpha}_2 \,.
\end{aligned}
\tag{10.6.20}
$$

In Wilson and Aifantis [1984], see their Eqn (10), the coupling coefficient a_{12}, which links the volume changes of the fluids, vanishes. For an isotropic solid skeleton, a_{12} can be cast in the format $\left((1/K^{\mathrm{PD}} - 1/K_{\mathrm{s}}) \left(n^2 - (1 - K/K^{\mathrm{PD}}) \right) \right)$. Its vanishing yields the bulk modulus $K^{\mathrm{PD}} = K/(1 - n^2)$. This uncoupling is somehow incompatible with the incompressibility of the fluids: together they give rise to discontinuous pressure response under specific loadings, Khalili [2003].

The mixed matrix form (10.6.12) can be partially inverted to obtain a generalized strain vector in terms of a generalized stress vector,

$$
\begin{bmatrix}
\boldsymbol{\epsilon} \\
v^1 - v_0^1 \\
v^2 - v_0^2 \\
S^{\mathrm{s}} - S_0^{\mathrm{s}}
\end{bmatrix}
=
\begin{bmatrix}
\mathbb{E}^{-1} & \mathbb{E}^{-1} : \boldsymbol{\alpha}_1 & \mathbb{E}^{-1} : \boldsymbol{\alpha}_2 & \mathbf{c}_{\mathrm{T}} \\
\boldsymbol{\alpha}_1 : \mathbb{E}^{-1} & b_{11} & b_{12} & n^1 \mathbf{I} : \mathbf{c}_{\mathrm{T}} \\
\boldsymbol{\alpha}_2 : \mathbb{E}^{-1} & b_{12} & b_{22} & n^2 \mathbf{I} : \mathbf{c}_{\mathrm{T}} \\
\mathbf{c}_{\mathrm{T}} & n^1 \mathbf{I} : \mathbf{c}_{\mathrm{T}} & n^2 \mathbf{I} : \mathbf{c}_{\mathrm{T}} & C_{\sigma p}^{\mathrm{s}}/T_{\mathrm{s}}^0
\end{bmatrix}
\begin{bmatrix}
\boldsymbol{\sigma} \\
p_1 \\
p_2 \\
T_{\mathrm{s}} - T_{\mathrm{s}}^0
\end{bmatrix}
, \tag{10.6.21}
$$

where

$$b_{11} = a_{11} + \boldsymbol{\alpha}_1 : \mathbb{E}^{-1} : \boldsymbol{\alpha}_1 \,, \quad b_{12} = a_{12} + \boldsymbol{\alpha}_1 : \mathbb{E}^{-1} : \boldsymbol{\alpha}_2 \,, \quad b_{22} = a_{22} + \boldsymbol{\alpha}_2 : \mathbb{E}^{-1} : \boldsymbol{\alpha}_2 \,, \tag{10.6.22}$$

and $C_{\sigma p}^{\mathrm{s}}$ is the heat capacity of the solid at constant stress and pressures, per unit volume of porous medium,

$$C_{\sigma p}^{\mathrm{s}} - C_{\epsilon p}^{\mathrm{s}} = T_{\mathrm{s}}^0 \, \mathbf{c}_{\mathrm{T}} : \mathbb{E} : \mathbf{c}_{\mathrm{T}} \,. \tag{10.6.23}$$

Similar differences between the heat capacities at constant stress and constant strain have been obtained for a fluid in Eqn (3.3.18), for a solid in Eqn (9.1.9), and for a porous medium in local thermal equilibrium in Eqn (9.3.15).

Again, the choice of primary and dependent variables is a key decision, both for the major symmetry of the constitutive matrix and parameter identification. For example, the use of the volume contents, as opposed to the mass contents, does not involve the bulk moduli of the fluids.

10.6.1.3 The complete thermomechanical constitutive equations

The changes of fluid mass contents in a dimensionless form $\zeta^k = \Delta m^k / \rho_k^0$, $k = 1, 2$, derive from (7.2.9). They involve additional material parameters, namely the bulk moduli K_1 and K_2, the coefficients of thermal expansion $c_{\mathrm{T}1}$ and $c_{\mathrm{T}2}$, which characterize the relative changes of the intrinsic mass densities of the fluids, as defined by Eqn (7.2.5). The apparent entropies of the fluids S^1 and S^2 given by (7.2.11) involve in addition the apparent heat capacities at constant pressure of each fluid C_p^1 and C_p^2. The comprehensive thermomechanical constitutive equations in mixed form can be cast in the form,

$$
\begin{bmatrix} -\boldsymbol{\sigma} \\ \zeta^1 \\ \zeta^2 \\ \Delta S^{\mathrm{s}} \\ \Delta S^1 \\ \Delta S^2 \end{bmatrix} = \begin{bmatrix} -\mathbb{E} & \boldsymbol{\alpha}_1 & \boldsymbol{\alpha}_2 & \mathbb{E}:\mathbf{c}_{\mathrm{T}} & 0 & 0 \\ \boldsymbol{\alpha}_1 & a_{11}+n^1/K_1 & a_{12} & a_{1\mathrm{T}} & -n^1 c_{\mathrm{T}1} & 0 \\ \boldsymbol{\alpha}_2 & a_{12} & a_{22}+n^2/K_2 & a_{2\mathrm{T}} & 0 & -n^2 c_{\mathrm{T}2} \\ \mathbf{c}_{\mathrm{T}}:\mathbb{E} & a_{1\mathrm{T}} & a_{2\mathrm{T}} & C^{\mathrm{s}}_{\epsilon p}/T^0_{\mathrm{s}} & 0 & 0 \\ 0 & -n^1 c_{\mathrm{T}1} & 0 & 0 & C^1_p/T^0_1 & 0 \\ 0 & 0 & -n^2 c_{\mathrm{T}2} & 0 & 0 & C^2_p/T^0_2 \end{bmatrix} \begin{bmatrix} \boldsymbol{\epsilon} \\ p_1 \\ p_2 \\ \Delta T_{\mathrm{s}} \\ \Delta T_1 \\ \Delta T_2 \end{bmatrix} . \tag{10.6.24}
$$

This format is aimed at finite element simulations where the primary unknowns are the solid displacement, the fluid pressures and the temperatures. The three first lines are already in appropriate form. The field equations to be used are

- the balance of momentum of the mixture, Eqn (4.5.9);
- the balances of mass of the fluids, Eqn (4.4.9);
- the energy equations for the three species, Eqn (4.6.11), expressed either in terms of the entropies, or in terms of the internal energies. If the latter option is chosen, the internal energies should be recovered first.

The mixed representation is endowed with two nice features: first, the matrix form is symmetric and, second, it displays decouplings which are not easily realized when the equations are phrased in terms of other primary variables.

Indeed, as another equivalent form of the thermomechanical equations, consider the primary variables to be the apparent strains defined by (7.2.1), and the temperatures, while the dependent variables are the apparent stresses and entropies. Then

$$
\begin{bmatrix} \boldsymbol{\sigma}^{\mathrm{s}} \\ -p^1 \\ -p^2 \\ -\Delta S^{\mathrm{s}} \\ -\Delta S^1 \\ -\Delta S^2 \end{bmatrix} = \begin{bmatrix} \mathbb{E}^{\mathrm{s}} & \boldsymbol{\lambda}_1 & \boldsymbol{\lambda}_2 & -\mathbb{E}^{\mathrm{s}}:\mathbf{c}_{\mathrm{T}} & -\boldsymbol{\lambda}_1 c_{\mathrm{T}1} & -\boldsymbol{\lambda}_2 c_{\mathrm{T}2} \\ \boldsymbol{\lambda}_1 & \lambda_{11} & \lambda_{12} & -\boldsymbol{\lambda}_1:\mathbf{c}_{\mathrm{T}} & -\lambda_{11} c_{\mathrm{T}1} & -\lambda_{12} c_{\mathrm{T}2} \\ \boldsymbol{\lambda}_2 & \lambda_{12} & \lambda_{22} & -\boldsymbol{\lambda}_2:\mathbf{c}_{\mathrm{T}} & -\lambda_{12} c_{\mathrm{T}1} & -\lambda_{22} c_{\mathrm{T}2} \\ -\mathbf{c}_{\mathrm{T}}:\mathbb{E}^{\mathrm{s}} & -\boldsymbol{\lambda}_1:\mathbf{c}_{\mathrm{T}} & -\boldsymbol{\lambda}_2:\mathbf{c}_{\mathrm{T}} & -C^{\mathrm{s}}_m/T^0_{\mathrm{s}} & (\boldsymbol{\lambda}_1:\mathbf{c}_{\mathrm{T}}) c_{\mathrm{T}1} & (\boldsymbol{\lambda}_2:\mathbf{c}_{\mathrm{T}}) c_{\mathrm{T}2} \\ -c_{\mathrm{T}1}\boldsymbol{\lambda}_1 & -\lambda_{11} c_{\mathrm{T}1} & -\lambda_{12} c_{\mathrm{T}1} & (\boldsymbol{\lambda}_1:\mathbf{c}_{\mathrm{T}}) c_{\mathrm{T}1} & -C^1_v/T^0_1 & \lambda_{12} c_{\mathrm{T}1} c_{\mathrm{T}2} \\ -c_{\mathrm{T}2}\boldsymbol{\lambda}_2 & -\lambda_{12} c_{\mathrm{T}2} & -\lambda_{22} c_{\mathrm{T}2} & (\boldsymbol{\lambda}_2:\mathbf{c}_{\mathrm{T}}) c_{\mathrm{T}2} & \lambda_{12} c_{\mathrm{T}2} c_{\mathrm{T}1} & -C^2_v/T^0_2 \end{bmatrix} \begin{bmatrix} \boldsymbol{\epsilon} \\ \mathrm{tr}\,\tilde{\boldsymbol{\epsilon}}^1 \\ \mathrm{tr}\,\tilde{\boldsymbol{\epsilon}}^2 \\ \Delta T_{\mathrm{s}} \\ \Delta T_1 \\ \Delta T_2 \end{bmatrix}, \tag{10.6.25}
$$

with associated moduli,

$$
\frac{1}{\Gamma} = \left(a_{11}+\frac{n^1}{K_1}\right)\left(a_{22}+\frac{n^2}{K_2}\right)-(a_{12})^2, \quad \Lambda=(n^1 n^2)^2\,\Gamma;
$$

$$
\lambda_{11}=(n^1)^2\left(a_{22}+\frac{n^2}{K_2}\right)\Gamma, \quad \lambda_{12}=-n^1 n^2\, a_{12}\,\Gamma, \quad \lambda_{22}=(n^2)^2\left(a_{11}+\frac{n^1}{K_1}\right)\Gamma;
$$

$$
\boldsymbol{\lambda}_1=-\lambda_{11}\left(\mathbf{I}-\frac{\boldsymbol{\alpha}_1}{n^1}\right)-\lambda_{12}\left(\mathbf{I}-\frac{\boldsymbol{\alpha}_2}{n^2}\right), \quad \boldsymbol{\lambda}_2=-\lambda_{12}\left(\mathbf{I}-\frac{\boldsymbol{\alpha}_1}{n^1}\right)-\lambda_{22}\left(\mathbf{I}-\frac{\boldsymbol{\alpha}_2}{n^2}\right); \tag{10.6.26}
$$

$$
\mathbb{E}^{\mathrm{s}}=\mathbb{E}+\frac{1}{\Lambda}\left(\lambda_{22}\,\boldsymbol{\lambda}_1\otimes\boldsymbol{\lambda}_1+\lambda_{11}\,\boldsymbol{\lambda}_2\otimes\boldsymbol{\lambda}_2-\lambda_{12}\left(\boldsymbol{\lambda}_1\otimes\boldsymbol{\lambda}_2+\boldsymbol{\lambda}_2\otimes\boldsymbol{\lambda}_1\right)\right),
$$

and heat capacities,

$$
C^{\mathrm{s}}_m=C^{\mathrm{s}}_{\epsilon p}+\mathbf{c}_{\mathrm{T}}:(\mathbb{E}-\mathbb{E}^{\mathrm{s}}):\mathbf{c}_{\mathrm{T}}\,T^0_{\mathrm{s}}=C^{\mathrm{s}}_{\sigma p}-\mathbf{c}_{\mathrm{T}}:\mathbb{E}^{\mathrm{s}}:\mathbf{c}_{\mathrm{T}}\,T^0_{\mathrm{s}};
$$

$$
C^1_v=C^1_p-\lambda_{11}\,(c_{\mathrm{T}1})^2\,T^0_1, \quad C^2_v=C^2_p-\lambda_{22}\,(c_{\mathrm{T}2})^2\,T^0_2. \tag{10.6.27}
$$

The explicit expressions of the above coefficients are provided by Loret and Rizzi [1999] for an isotropic solid skeleton. In fact, in their isotropic form, the above constitutive equations are identical to those given in Tuncay and Corapcioglu [1996]a [1996]b.

The couplings, uncouplings and symmetry properties of the constitutive matrices involved in this analysis are encapsulated in the two propositions below:

Proposition 10.2 *on thermomechanical constitutive couplings and uncouplings.*

- the thermomechanical constitutive equations (10.6.12) or (10.6.21) yielding the changes of volume contents do not involve the temperatures of the fluids. Indeed, the volume changes of the fluids are here entirely controlled by the solid skeleton, as already alluded for in Section 9.3.2;
- these uncouplings may be concealed by arbitrary choices of dependent and independent variables, e.g., Eqn (10.6.25);
- in contrast to the volume contents, the changes of mass contents of the fluids depend on their own temperatures, Eqn (10.6.24).

Proposition 10.3 *on thermomechanical constitutive symmetries.*

- the thermomechanical constitutive equations (10.6.12) or (10.6.21) yielding the changes of volume contents express in terms of a 4×4 symmetric matrix;
- the comprehensive thermomechanical constitutive equations (10.6.24) yielding the changes of mass contents and of entropies of the fluids express in terms of a 6×6 symmetric matrix;
- the comprehensive thermomechanical constitutive equations (10.6.25) yielding the apparent stresses and entropies express also in terms of a 6×6 symmetric matrix;
- the comprehensive thermomechanical constitutive equations of the form (10.6.24) but involving the changes of volume contents, instead of mass contents, would express in terms of a 6×6 *non* symmetric matrix.

The conditions of positive definiteness (PD) of the purely mechanical part of the equations, associated with the 3×3 principal minor, namely

$$
\begin{bmatrix} \boldsymbol{\sigma}^{\mathrm{s}} \\ -p^1 \\ -p^2 \end{bmatrix} = \begin{bmatrix} \mathbb{E}^{\mathrm{s}} & \boldsymbol{\lambda}_1 & \boldsymbol{\lambda}_2 \\ \boldsymbol{\lambda}_1 & \lambda_{11} & \lambda_{12} \\ \boldsymbol{\lambda}_2 & \lambda_{12} & \lambda_{22} \end{bmatrix} \begin{bmatrix} \boldsymbol{\epsilon} \\ \operatorname{tr} \tilde{\boldsymbol{\epsilon}}^1 \\ \operatorname{tr} \tilde{\boldsymbol{\epsilon}}^2 \end{bmatrix} .
\tag{10.6.28}
$$

may be expressed in the form, Loret and Rizzi [1999],

$$
\mathbb{E} \ \mathrm{PD}; \quad \left(a_{11} + \frac{n^1}{K_1} \right) \left(a_{22} + \frac{n^2}{K_2} \right) - (a_{12})^2 > 0; \quad a_{11} + \frac{n^1}{K_1} > 0 .
\tag{10.6.29}
$$

Indeed, the mechanical strain energy can be cast in the following format,

$$
\begin{aligned}
& \frac{1}{2} \boldsymbol{\sigma}^{\mathrm{s}} : \boldsymbol{\epsilon} - \frac{1}{2} p^1 \operatorname{tr} \tilde{\boldsymbol{\epsilon}}^1 - \frac{1}{2} p^2 \operatorname{tr} \tilde{\boldsymbol{\epsilon}}^2 \\
& = \frac{1}{2} \boldsymbol{\epsilon} : \mathbb{E} : \boldsymbol{\epsilon} \\
& + \frac{1}{2\Lambda} \left(\boldsymbol{\lambda}_1 : \boldsymbol{\epsilon} \sqrt{\lambda_{22}} \cos \omega + \boldsymbol{\lambda}_2 : \boldsymbol{\epsilon} \sqrt{\lambda_{11}} \cos \theta + \operatorname{tr} \tilde{\boldsymbol{\epsilon}}^1 \sqrt{\lambda_{11} \Lambda} \sin \theta - \operatorname{tr} \tilde{\boldsymbol{\epsilon}}^2 \sqrt{\lambda_{22} \Lambda} \sin \omega \right)^2 \\
& + \frac{1}{2\Lambda} \left(\boldsymbol{\lambda}_1 : \boldsymbol{\epsilon} \sqrt{\lambda_{22}} \sin \omega + \boldsymbol{\lambda}_2 : \boldsymbol{\epsilon} \sqrt{\lambda_{11}} \sin \theta - \operatorname{tr} \tilde{\boldsymbol{\epsilon}}^1 \sqrt{\lambda_{11} \Lambda} \cos \theta + \operatorname{tr} \tilde{\boldsymbol{\epsilon}}^2 \sqrt{\lambda_{22} \Lambda} \cos \omega \right)^2 ,
\end{aligned}
\tag{10.6.30}
$$

where the two real angles θ and ω are defined by the relations,

$$
\cos(\omega - \theta) = -\frac{\lambda_{12}}{\sqrt{\lambda_{11} \lambda_{22}}}, \quad \sin(\omega - \theta) = -\frac{\sqrt{\Lambda}}{\sqrt{\lambda_{11} \lambda_{22}}} .
\tag{10.6.31}
$$

10.6.1.4   The volume fractions

The relations $(7.2.4)_4$ and $(7.2.5)$ provide the rate of volume fraction dn^k/dt of the fluid k,

$$
\frac{dn^k}{dt} = -n^k \mathbf{I} : \frac{d\boldsymbol{\epsilon}}{dt} - n^k \left(\frac{1}{K_k} \frac{dp_k}{dt} - c_{\mathrm{T}k} \frac{dT_k}{dt} \right) + \frac{n^k}{m^k} \frac{dm^k}{dt}, \quad k = 1, 2 .
\tag{10.6.32}
$$

Upon insertion of the rate of mass content (10.6.24) in (10.6.32) and elimination of the thermoelastic strain via (10.6.10), the fluid volume fractions express in terms of the total stress and fluid pressures only,

$$
\begin{aligned}
\frac{dn^1}{dt} &= (\boldsymbol{\alpha}_1 - n^1 \mathbf{I}) : \mathbb{E}^{-1} : \frac{d}{dt}(\boldsymbol{\sigma} + p_1 \mathbf{I}) + (n^1 (\boldsymbol{\alpha}_2 - n^2 \mathbf{I}) - n^2 (\boldsymbol{\alpha}_1 - n^1 \mathbf{I})) : \mathbb{E}^{-1} : \mathbf{I} \frac{d}{dt}(p_1 - p_2), \\
\frac{dn^2}{dt} &= (\boldsymbol{\alpha}_2 - n^2 \mathbf{I}) : \mathbb{E}^{-1} : \frac{d}{dt}(\boldsymbol{\sigma} + p_2 \mathbf{I}) .
\end{aligned}
\tag{10.6.33}
$$

The single porosity limit (9.3.23) is retrieved along the schemes indicated in Section 10.6.8.1.

The rate of change of the volume fraction $n^s = v^s / \det \mathbf{F}$ of the solid constituent follows from (7.2.16),

$$\frac{dn^s}{dt} = (-n^s \mathbf{I} : \mathbb{E}^{-1} + \mathbf{I} : \mathbb{E}_s^{-1}) : \frac{d\boldsymbol{\sigma}}{dt} + \Big(\frac{n^1}{K_s} - n^s \mathbf{I} : \mathbb{E}^{-1} : \boldsymbol{\alpha}_1\Big)\frac{dp_1}{dt} + \Big(\frac{n^2}{K_s} - n^s \mathbf{I} : \mathbb{E}^{-1} : \boldsymbol{\alpha}_2\Big)\frac{dp_2}{dt}\,, \tag{10.6.34}$$

or alternatively from the compatibility condition $n^s + n^1 + n^2 = 1$.

10.6.2 Generalized diffusion

The internal entropy production due to diffusion through the whole porous medium is obtained by summation of the contributions (7.2.32) associated with individual species,

$$\hat{S}_{(i,3)} = - \sum_{k=1,2,s} \mathbf{Q}_k \cdot \frac{\boldsymbol{\nabla} T_k}{T_k^2} - \sum_{k=1,2} \mathbf{J}_k \cdot \frac{\boldsymbol{\nabla} P_k}{T_k} \geq 0, \quad \boldsymbol{\nabla} P_k \equiv \boldsymbol{\nabla} p_k + \rho_k \Big(\frac{d^k \mathbf{v}_k}{dt} - \mathbf{b}_k\Big). \tag{10.6.35}$$

The pores occupy distinct positions in the matrix. Therefore, diffusion coupling between the two fluids is neglected. On the other hand, thermoosmosis may take place within each pore fluid, as described in Section 9.3.3. The fluid fluxes and heat fluxes $\mathcal{J} = -\mathcal{K}\mathcal{F}$ are related linearly to the gradients of pressure and temperature $\mathcal{F}$ through the matrix $\mathcal{K}$,

$$\begin{bmatrix} \mathbf{J}_1 \\ \mathbf{J}_2 \\ \mathbf{Q}_s \\ \mathbf{Q}_1 \\ \mathbf{Q}_2 \end{bmatrix} = - \begin{bmatrix} T_1\,\mathbf{k}_{H1} & 0 & 0 & T_1\,n^1\,\mathbf{k}_{TO1} & 0 \\ 0 & T_2\,\mathbf{k}_{H2} & 0 & 0 & T_2\,n^2\,\mathbf{k}_{TO2} \\ 0 & 0 & (T_s)^2\,n^s\,\mathbf{k}_{Ts} & 0 & 0 \\ T_1\,n^1\,\mathbf{k}_{MC1} & 0 & 0 & (T_1)^2\,n^1\,\mathbf{k}_{T1} & 0 \\ 0 & T_2\,n^2\,\mathbf{k}_{MC2} & 0 & 0 & (T_2)^2\,n^2\,\mathbf{k}_{T2} \end{bmatrix} \begin{bmatrix} \boldsymbol{\nabla} P_1/T_1 \\ \boldsymbol{\nabla} P_2/T_2 \\ \boldsymbol{\nabla} T_s/(T_s)^2 \\ \boldsymbol{\nabla} T_1/(T_1)^2 \\ \boldsymbol{\nabla} T_2/(T_2)^2 \end{bmatrix}. \tag{10.6.36}$$

As stressed in Section 4.9, there is a certain arbitrariness in defining the 'fluxes' and 'forces'. The diffusion properties are assumed to inherit the anisotropic properties of the solid skeleton. The $\mathbf{k}$'s are symmetric second order tensors, with units as indicated in Section 9.3.3. They satisfy the conditions (9.E.8),

$$\mathbf{k}_{Ts}\ (\mathrm{PD}); \quad \mathbf{k}_{Hk}\ (\mathrm{PD}); \quad \text{and} \quad T_k\,n^k\,\mathbf{k}_{Tk} - \frac{n^k}{2}\,(\mathbf{k}_{TOk} + \mathbf{k}_{MCk}) \cdot \mathbf{k}_{Hk}^{-1} \cdot \frac{n^k}{2}\,(\mathbf{k}_{TOk} + \mathbf{k}_{MCk})\ (\mathrm{PD}), \quad k = 1, 2, \tag{10.6.37}$$

so as to comply with the entropy inequality $\hat{S}_{(i,3)} \geq 0$.

If the temperatures are spatially uniform, and the fluid pressures are equal, say $P_1 = P_2 = P$, the fluid flux $\mathbf{J}_1 + \mathbf{J}_2$ is equal to $-\mathbf{k}_H \boldsymbol{\nabla} P$, with an equivalent hydraulic permeability $\mathbf{k}_H$ equal to $\mathbf{k}_{H1} + \mathbf{k}_{H2}$,

$$(\mathbf{k}_{H1}, \mathbf{k}_{H2}) \overset{P_1 = P_2 = P}{\Longrightarrow} \mathbf{k}_H = \mathbf{k}_{H1} + \mathbf{k}_{H2}\,. \tag{10.6.38}$$

If the pressures are spatially uniform, and the temperatures are equal, say $T_s = T_1 = T_2 = T$, the heat flux through the porous medium $\mathbf{Q}_s + \mathbf{Q}_1 + \mathbf{Q}_2$ is equal to $-\mathbf{k}_T \cdot \boldsymbol{\nabla} T$, with an equivalent thermal conductivity $\mathbf{k}_T$ equal to the volume (surface) average over the species of the mixture, namely $\mathbf{k}_T = n^s\,\mathbf{k}_{Ts} + n^1\,\mathbf{k}_{T1} + n^2\,\mathbf{k}_{T2}$,

$$(\mathbf{k}_{Ts}, \mathbf{k}_{T1}, \mathbf{k}_{T2}) \overset{T_s = T_1 = T_2 = T}{\Longrightarrow} \mathbf{k}_T = n^s\,\mathbf{k}_{Ts} + n^1\,\mathbf{k}_{T1} + n^2\,\mathbf{k}_{T2}\,. \tag{10.6.39}$$

10.6.3 Mass and energy transfers

The particular status of mass and energy exchanges in the model of interest here is sketched in Fig. 10.6.1. Summing the individual contributions (7.2.28)$_2$ over species, and accounting for the definitions (7.2.29) and conditions (7.2.30), the entropy production associated with transfers reduces to,

$$\hat{S}_{(i,2)} = \sum_{k=1,2,s} -x_k^{(M)}\,y_k^{(M)} - x_k^{(E)}\,y_k^{(E)} \geq 0\,, \tag{10.6.40}$$

where some temporary notation has been introduced, namely for mass transfer,

$$x_k^{(M)} \equiv \hat{\rho}^k; \quad y_k^{(M)} \equiv \big(\mu_k + \tfrac{1}{2}\,(\mathbf{v}_k - \mathbf{v}_s)^2\big)\frac{1}{T_k}\,, \tag{10.6.41}$$

and for energy transfer,

$$x_k^{(E)} \equiv \hat{U}^k - \hat{\mathbf{p}}^k \cdot \mathbf{v}_s + \tfrac{1}{2}\,\hat{\rho}^k\,\mathbf{v}_s^2; \quad y_k^{(E)} \equiv -\frac{1}{T_k}\,. \tag{10.6.42}$$

10.6.4   Linear constitutive equations of mass transfer

No mass exchange between the solid and the fluids is considered to take place. Consequently, the rates of mass exchanges between the two fluids sum up to zero, since the system is closed. The resulting entropy production can be cast in the form,

$$-\sum_{k=1,2,\mathrm{s}} x_k^{(\mathrm{M})}\, y_k^{(\mathrm{M})} = -x_1^{(\mathrm{M})}\,(y_1^{(\mathrm{M})} - y_2^{(\mathrm{M})})\,. \qquad \qquad (10.6.43)$$

The linear rate of mass transfer,

$$\begin{bmatrix} x_1^{(\mathrm{M})} \\ x_2^{(\mathrm{M})} \end{bmatrix} = -\kappa \begin{bmatrix} 1 & -1 \\ -1 & 1 \end{bmatrix} \begin{bmatrix} y_1^{(\mathrm{M})} \\ y_2^{(\mathrm{M})} \end{bmatrix}, \quad \kappa \ge 0\,, \qquad (10.6.44)$$

yields a positive contribution to the entropy production,

$$-\sum_{k=1,2,\mathrm{s}} x_k^{(\mathrm{M})}\, y_k^{(\mathrm{M})} = \kappa\,(y_1^{(\mathrm{M})} - y_2^{(\mathrm{M})})^2 \ge 0\,, \qquad (10.6.45)$$

provided the coefficient of mass transfer κ [unit : kg $\times$ m^{-5} $\times$ s $\times$ K] is positive.

To get an insight into the sense the mass will transfer (from the tissue pores to vessels or conversely), it is useful to analyze the limit case of an incompressible and not dilatable fluid. Then the chemical potential of the generic fluid reduces to, Eqn (3.3.34),

$$\mu = \mu_0 + v_0\,(p - p_0) + (c - s_0)\,(T - T_0) - c\,T\,\mathrm{Ln}\,\frac{T}{T_0}\,, \qquad (10.6.46)$$

where v [unit : m^3/kg] is the specific volume and c [unit : kJ/kg/K] the heat capacity. While the reference value of the chemical potential μ_0 [unit : J/kg] does not affect its variations in a (p,T) plane, the reference value of entropy s_0 [unit : J/kg/K] does.

Now, the rate of mass transfer is governed by (10.6.44) which, neglecting the kinetic energy terms, simplifies to [10.2],

$$\hat{\rho}^1 = -\kappa\left(\frac{\mu_1}{T_1} - \frac{\mu_2}{T_2}\right) = -\hat{\rho}^2\,, \quad \kappa \ge 0\,. \qquad (10.6.48)$$

The chemical potentials increase linearly with pressure. For isothermal analyses, or when the two fluids are in thermal equilibrium, *mass will thus transfer from the high pressure cavities to the low pressure cavities*.

The situation is more complex for thermal non equilibrium. However, the partial derivatives,

$$\frac{\partial \mu}{\partial T} = -s_0 - c\,\mathrm{Ln}\,\frac{T}{T_0}, \quad \frac{\partial}{\partial T}\left(\frac{\mu}{T}\right) = -\frac{1}{T^2}\left(\mu_0 + s_0\,T_0 + v_0\,(p - p_0) + c\,(T - T_0)\right). \qquad (10.6.49)$$

are both expected to be negative for pressure and temperature higher than the reference values. So is indeed the case as observed in Fig. 10.6.4 at least in the ranges of pressure and temperature displayed on the figure and for the chosen reference value of entropy. Therefore, if the pressures in the two cavities are identical, *the mass will transfer from the cold cavities to the hot cavities*. Conversely, if both pressure and temperature are smaller than the reference values, the direction of mass transfer is opposite[10.3].

[10.2] Phrased in terms of the Massieu chemical potential $\psi = -\mu/T$, the constitutive equation of mass transfer,

$$\hat{\rho}^1 = -\kappa\,(\psi_2 - \psi_1) = -\hat{\rho}^2\,, \quad \kappa \ge 0\,, \qquad (10.6.47)$$

indicates that mass transfer takes place at increasing Massieu chemical potential.

[10.3] This outcome may be compared with the flows that are induced by thermoosmosis, a phenomenon of coupled transport through a membrane. At identical pressures, fluid flow may be directed either to the hot side or to the cold side, depending on the properties of the fluid and of the membrane, Sections 9.3.3 and 9.5.9.

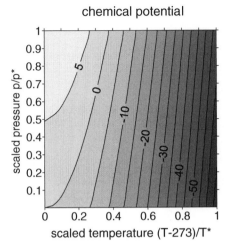
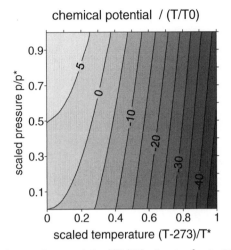

Fig. 10.6.4: Contours of the chemical potential μ and of the chemical potential $\mu/(T/T_0)$ of water [unit : MJ/kg] for a heat capacity $c = 4.18$ kJ/kg/K, and the reference values of entropy $s_0 = 0$ J/kg/K, and chemical potential $\mu_0 = p_0\,v_0$ J/kg at triple point of water, namely pressure $p_0 = 611$ Pa, temperature $T_0 = 273.16$ K, specific volume $v_0 \sim 10^{-3}$ m^3/kg. The scaling temperature and pressure are respectively $T* = 100$ K, and $p* = 10$ MPa.

10.6.5   Linear constitutive equations of energy transfer

The rates of energy transfer sum to zero,

$$\sum_{k=1,2,\mathrm{s}} x_k^{(\mathrm{E})} = 0\,, \tag{10.6.50}$$

so that the associated dissipation rate may be brought to the form,

$$-x_1^{(\mathrm{E})}\left(y_1^{(\mathrm{E})} - y_\mathrm{s}^{(\mathrm{E})}\right) - x_2^{(\mathrm{E})}\left(y_2^{(\mathrm{E})} - y_\mathrm{s}^{(\mathrm{E})}\right)\,. \tag{10.6.51}$$

The linear rate of energy transfer,

$$
\begin{bmatrix} x_\mathrm{s}^{(\mathrm{E})} \\ x_1^{(\mathrm{E})} \\ x_2^{(\mathrm{E})} \end{bmatrix}
= -
\begin{bmatrix}
\kappa_{\mathrm{s}1} + \kappa_{\mathrm{s}2} & -\kappa_{\mathrm{s}1} & -\kappa_{\mathrm{s}2} \\
-\kappa_{\mathrm{s}1} & \kappa_{\mathrm{s}1} + \kappa_{12} & -\kappa_{12} \\
-\kappa_{\mathrm{s}2} & -\kappa_{12} & \kappa_{\mathrm{s}2} + \kappa_{12}
\end{bmatrix}
\begin{bmatrix} y_\mathrm{s}^{(\mathrm{E})} \\ y_1^{(\mathrm{E})} \\ y_2^{(\mathrm{E})} \end{bmatrix}\,, \tag{10.6.52}
$$

with positive coefficients of energy transfer [unit : W $\times$ m^{-3} $\times$ K = kg $\times$ m^{-1} $\times$ s^{-3} $\times$ K],

$$\kappa_{\mathrm{s}1} \geq 0\,, \quad \kappa_{\mathrm{s}2} \geq 0\,, \quad \kappa_{12} \geq 0\,, \tag{10.6.53}$$

yields a positive contribution to the entropy production,

$$\sum_{k=1,2,\mathrm{s}} -x_k^{(\mathrm{E})}\,y_k^{(\mathrm{E})} = \kappa_{\mathrm{s}1}\,(y_1^{(\mathrm{E})} - y_\mathrm{s}^{(\mathrm{E})})^2 + \kappa_{\mathrm{s}2}\,(y_2^{(\mathrm{E})} - y_\mathrm{s}^{(\mathrm{E})})^2 + \kappa_{12}\,(y_1^{(\mathrm{E})} - y_2^{(\mathrm{E})})^2 \geq 0\,. \tag{10.6.54}$$

In essence, the rate equation indicates, as expected, that *energy transfers from the hot species to the cold species.*

10.6.6   The energy equations

The balance of energy (4.6.11) of any species $k = \mathrm{s}, 1, 2$, phrased in terms of specific internal energies is first transformed and phrased in terms of the internal energies per unit volume,

$$\rho^k \frac{d^k u_k}{dt} + \hat{\rho}^k\,u_k - \boldsymbol{\sigma}^k : \boldsymbol{\nabla}\mathbf{v}_k - \hat{\rho}^k\left(\tilde{u}_k + \tfrac{1}{2}\,(\tilde{\mathbf{v}}_k - \mathbf{v}_k)^2\right) - \hat{\mathcal{U}}^k + \operatorname{div}\mathbf{Q}_k - \Theta_k = 0\,,$$

$$\frac{dU^k}{dt} + \operatorname{div}(u_k\,\mathbf{M}_k) + U^k\operatorname{div}\mathbf{v}_\mathrm{s} - \boldsymbol{\sigma}^k : \boldsymbol{\nabla}\mathbf{v}_k - \hat{\rho}^k\left(\tilde{u}_k + \tfrac{1}{2}\,(\tilde{\mathbf{v}}_k - \mathbf{v}_k)^2\right) - \hat{\mathcal{U}}^k + \operatorname{div}\mathbf{Q}_k - \Theta_k = 0\,. \tag{10.6.55}$$

Given the thermo-poroelastic energy $\underline{W}$, the internal energy of the solid results from (10.6.5),

$$\underline{U}^s = \underline{W} + T_s\,\underline{S}^s + \sum_{k=1,2} v^k\,p_k\,, \tag{10.6.56}$$

and, with help of (10.6.8), its rate form simplifies to,

$$\begin{aligned}
\frac{dU^s}{dt} &= \boldsymbol{\tau} : \frac{d\boldsymbol{\epsilon}}{dt} + \sum_{k=1,2} p_k\,\frac{dv^k}{dt} + T_s\,\frac{dS^s}{dt}\\[4pt]
\frac{dU^s}{dt} &= (\boldsymbol{\sigma}^s - E^s\,\mathbf{I}) : \frac{d\boldsymbol{\epsilon}}{dt} + \sum_{k=1,2} p_k\,\frac{dn^k}{dt} + T_s\,\frac{dS^s}{dt}\,.
\end{aligned} \tag{10.6.57}$$

Therefore, the balance of energy of the solid is obtained by inserting (10.6.57) into (10.6.55), namely,

$$\frac{dS^s}{dt} + S^s\,\mathrm{div}\,\mathbf{v}_s + \frac{1}{T_s}\sum_{k=1,2} p_k\,\frac{dn^k}{dt} + \frac{1}{T_s}\,\mathrm{div}\,\mathbf{Q}_s - \frac{\Theta_s}{T_s} + \underbrace{x_s^{(E)}\,y_s^{(E)}}_{\text{energy transfer}} = 0\,. \tag{10.6.58}$$

As for the fluids, insertion of (7.2.13) into (10.6.55) with help of the balance momentum (4.5.5) yields an expression that displays the total heat flux,

$$\frac{dS^k}{dt} + S^k\,\mathrm{div}\,\mathbf{v}_s + \frac{\mu_k}{T_k}\,\frac{1}{\det\mathbf{F}}\,\frac{dm^k}{dt} - \frac{p_k}{T_k}\,\frac{dn^k}{dt} + \frac{1}{T_k}\,\mathrm{div}\,(\;\overbrace{\mathbf{Q}_k + h_k\,\mathbf{M}_k}^{\substack{\text{conductive + convective}\\\text{heat fluxes}}}\;) - \frac{\Theta_k}{T_k}$$
$$+ \frac{1}{2}\,(\mathbf{v}_k - \mathbf{v}_s)^2\,\frac{\hat{\rho}^k}{T_k} + \underbrace{x_k^{(E)}\,y_k^{(E)}}_{\text{energy transfer}} + \frac{\mathbf{M}_k}{T_k}\cdot\Big(\frac{d^k\mathbf{v}_k}{dt} - \mathbf{b}_k\Big) = 0\,. \tag{10.6.59}$$

Using the balance of mass (4.4.9) and the differential expression (7.2.31), the alternative expression, more akin to the expression (10.6.58) for the solid, highlights mass and energy transfers,

$$\frac{dS^k}{dt} + S^k\,\mathrm{div}\,\mathbf{v}_s + s_k\,\mathrm{div}\,\mathbf{M}_k - \frac{p_k}{T_k}\,\frac{dn^k}{dt} + \frac{1}{T_k}\,\mathrm{div}\,\mathbf{Q}_k - \frac{\Theta_k}{T_k}$$
$$+ \underbrace{x_k^{(M)}\,y_k^{(M)}}_{\text{mass transfer}} + \underbrace{x_k^{(E)}\,y_k^{(E)}}_{\text{energy transfer}} + \frac{\mathbf{M}_k}{T_k}\cdot\Big(\nabla h_k + \frac{d^k\mathbf{v}_k}{dt} - \mathbf{b}_k\Big) = 0\,. \tag{10.6.60}$$

The expression below in terms of intrinsic entropy per unit mass,

$$\rho^k\,\frac{ds_k}{dt} - \frac{p_k}{T_k}\,\frac{dn^k}{dt} + \frac{1}{T_k}\,\mathrm{div}\,\mathbf{Q}_k - \frac{\Theta_k}{T_k}$$
$$+ \underbrace{\Big(h_k + \frac{1}{2}\,(\mathbf{v}_k - \mathbf{v}_s)^2\Big)\frac{\hat{\rho}^k}{T_k}}_{\text{mass transfer}} + \underbrace{x_k^{(E)}\,y_k^{(E)}}_{\text{energy transfer}} + \frac{\mathbf{M}_k}{T_k}\cdot\Big(\nabla h_k + \frac{d^k\mathbf{v}_k}{dt} - \mathbf{b}_k\Big) = 0\,, \tag{10.6.61}$$

is more appropriate to phase change as it highlights the role of the enthalpy. Summing the contributions of the two phases, e.g., liquid water l and vapor v, displays the latent heat of evaporation $h_v - h_l$.

The next equivalent expression highlights mass transfer, energy transfer and seepage in the very format in which the constitutive equations of these phenomena are formulated,

$$\frac{dS^k}{dt} + S^k\,\mathrm{div}\,\mathbf{v}_s + \mathrm{div}\,(s_k\,\mathbf{M}_k) - \frac{p_k}{T_k}\,\frac{dn^k}{dt} + \frac{1}{T_k}\,\mathrm{div}\,\mathbf{Q}_k - \frac{\Theta_k}{T_k}$$
$$+ \underbrace{x_k^{(M)}\,y_k^{(M)}}_{\text{mass transfer}} + \underbrace{x_k^{(E)}\,y_k^{(E)}}_{\text{energy transfer}} + \underbrace{\frac{\mathbf{M}_k}{T_k}\cdot\Big(\frac{\nabla p_k}{\rho_k} + \frac{d^k\mathbf{v}_k}{dt} - \mathbf{b}_k\Big)}_{\text{seepage}} = 0\,. \tag{10.6.62}$$

The energy equations may also be expressed in a format that involves the entropy productions associated with generalized diffusion and transfers, e.g., for the fluid k,

$$\frac{dS^k}{dt} + S^k \operatorname{div} \mathbf{v}_s - \frac{p_k}{T_k} \frac{dn^k}{dt} + \operatorname{div} \left(\frac{\mathbf{Q}_k}{T_k} + s_k \mathbf{M}_k \right) - \frac{\Theta_k}{T_k} - \left(\hat{S}^k_{(i,2)} + \hat{S}^k_{(i,3)} + \hat{S}^k \right) = 0 \,. \tag{10.6.63}$$

If the species are in local thermal equilibrium, summing up the energy equations over species, accounting for the closure relations (7.2.30), the energy equation for the mixture boils down to,

$$\begin{aligned}
\frac{dS}{dt} + S \operatorname{div} \mathbf{v}_s &+ \frac{1}{\det \mathbf{F}} \sum_{k=1,2} \frac{\mu_k}{T} \frac{dm^k}{dt} + \frac{1}{T} \sum_{k=s,1,2} \operatorname{div} (\mathbf{Q}_k + h_k \mathbf{M}_k) - \frac{\Theta_k}{T} \\
&+ \sum_{k=s,1,2} \frac{1}{2} (\mathbf{v}_k - \mathbf{v}_s)^2 \frac{\hat{\rho}^k}{T} + \frac{\mathbf{M}_k}{T} \cdot (\frac{d^k \mathbf{v}_k}{dt} - \mathbf{b}_k) = 0 \,,
\end{aligned} \tag{10.6.64}$$

which further simplifies to (9.3.33) under the conditions stated in Section 9.3.4. An equivalent form deduced from (10.6.62) writes,

$$\begin{aligned}
\frac{dS}{dt} + S \operatorname{div} \mathbf{v}_s &+ \sum_{k=1,2} \operatorname{div} (s_k \mathbf{M}_k) + \frac{1}{T} \sum_{k=s,1,2} \operatorname{div} \mathbf{Q}_k - \frac{\Theta_k}{T} \\
&+ \sum_{k=1,2} x_k^{(M)} y_k^{(M)} + \sum_{k=1,2} \frac{\mathbf{M}_k}{T} \cdot (\frac{\nabla p_k}{\rho_k} + \frac{d^k \mathbf{v}_k}{dt} - \mathbf{b}_k) = 0 \,.
\end{aligned} \tag{10.6.65}$$

For the record, let us note that the dimension of the energy equations above are actually that of the dissipation rate, namely $\mathrm{W/m^3/K}$.

10.6.7 The complete set of field equations

The six field equations associated with the six primary unknowns, namely $(\epsilon, p_1, p_2, T_s, T_1, T_2)$, include
  - the balance of momentum of the mixture as a whole, Eqn (4.5.6),

$$\operatorname{div} \boldsymbol{\sigma} + \sum_{k=s,1,2} \rho^k \left(\mathbf{b}_k - \frac{d\mathbf{v}_k}{dt} - \boldsymbol{\nabla} \mathbf{v}_k \cdot (\mathbf{v}_k - \mathbf{v}_s) \right) + \hat{\rho}^k \mathbf{v}_k = \mathbf{0} \,; \tag{10.6.66}$$

  - the two balances of mass of the fluids, Eqn (4.4.9),

$$\frac{1}{\det \mathbf{F}} \frac{dm^k}{dt} = -\operatorname{div} \mathbf{M}_k + \hat{\rho}^k, \quad k = 1, 2 \,; \tag{10.6.67}$$

  - the three energy equations, namely Eqn (10.6.58) for the solid and Eqns (10.6.59) for the two fluids.
The dependent entities are expressed in terms of the six primary variables $(\epsilon, p_1, p_2, T_s, T_1, T_2)$ through
 - the thermomechanical constitutive equations (10.6.24) for the dependent variables $(\boldsymbol{\sigma}, \zeta^1, \zeta^2, S^s, S^1, S^2)$;
 - the constitutive generalized diffusion equations (10.6.36) for the fluid fluxes $\mathbf{J}_1$, $\mathbf{J}_2$ and heat fluxes $\mathbf{Q}_s$, $\mathbf{Q}_1$, $\mathbf{Q}_2$;
 - the constitutive equations of mass transfer and energy transfer, (10.6.44) and (10.6.52), respectively;
 - the relations (10.6.32) that provide the rates of volume fraction dn^k/dt of the fluids;
 - the nonlinear definitions (3.3.28), or linearized approximations (3.3.30), of the specific thermodynamic functions s_k, u_k, h_k, and nonlinear definitions of the specific free energy e_k and chemical potential μ_k of the fluid k. Recall that, by construction, the entropies of the fluids (7.2.11) given by the constitutive equations (10.6.24) are equivalent to the linearized relations (3.3.30). The accuracy of the linearized approximations of the specific thermodynamic functions is commented at the end of Section 3.3.3. The fluids of the cavities have been assumed to be identical. If the fluids are not miscible, they are separated by a meniscus, whose temperature and thermal properties have to be defined. Accordingly, the thermodynamic functions of the fluids are altered.

10.6.8   Increasing and decreasing the degree of sophistication

The two pressure, three temperature model embodies simpler models analyzed in this chapter, namely by decreasing order of complexity,

- the porous medium with a single type of cavity but still in local thermal non equilibrium is retrieved in Section 10.7;
- the latter is specialized to the porous medium with a single porosity and in local thermal equilibrium (LTE) exposed in Section 9.3;
- the thermo-poroelastic model reduces to the purely poroelastic model of Section 7.3 for isothermal processes.

10.6.8.1   From double to single porosity

The single porosity, single temperature model developed in Section 9.3 can be recovered in two ways:

- by letting the volume fraction n^2 of the porosity 2 tend to zero. Then the pore drained stiffness $\mathbb{E}^{\mathrm{PD}}$ tends to the drained stiffness $\mathbb{E}$, the matrix form (10.6.21) can be checked to collapse to the form (9.3.9), and the mass balance of the fluid of porosity 2 becomes elusive;
- alternatively, the contributions of the two cavities may be summed, under the assumption that their pressures are identical. The summation procedure applies to the lines and columns of the constitutive matrices and to the two balances of fluid mass.

Conversely, the two cavities may be isolated from one another by zeroing their mass exchange, which amounts to set the coefficient κ to 0 in the constitutive equation of mass transfer exposed in Section 10.6.4. Then, setting to zero the hydraulic conductivity of one of the cavities retrieves an occluded porosity.

10.6.8.2   From local thermal non equilibrium to local thermal equilibrium

A model of local thermal equilibrium is retrieved from the model of local thermal non equilibrium in two ways:

- by a penalty method, where the coefficients of energy transfer in the constitutive equation exposed in Section 10.6.5 are set to a large value, the complete structure of the local thermal non equilibrium model being preserved;
- by setting the temperatures to a common value, summing the corresponding lines and columns in the constitutive matrices as well as the energy equations.

10.7   A porous medium with a single porosity and two temperatures

It is worth recording the linearized thermomechanical relations in the simpler case of a porous medium with a single porosity but where the solid and fluid constituents are endowed with their own temperatures. Starting from (10.6.21) where $T_1 = T_2 = T_{\mathrm{w}}$, the thermomechanical relations may be cast in the format,

$$
\begin{bmatrix}
\boldsymbol{\sigma}^{\mathrm{s}} \\
-p^{\mathrm{w}} \\
-\Delta S^{\mathrm{s}} \\
-\Delta S^{\mathrm{w}}
\end{bmatrix}
=
\begin{bmatrix}
\mathbb{E}^{\mathrm{s}} & \boldsymbol{\lambda}_{\mathrm{w}} & -\mathbb{E}^{\mathrm{s}}:\mathbf{c}_{\mathrm{T}} & -\boldsymbol{\lambda}_{\mathrm{w}}\,c_{\mathrm{Tw}} \\
\boldsymbol{\lambda}_{\mathrm{w}} & M\,(n^{\mathrm{w}})^2 & -\boldsymbol{\lambda}_{\mathrm{w}}:\mathbf{c}_{\mathrm{T}} & -M\,(n^{\mathrm{w}})^2\,c_{\mathrm{Tw}} \\
-\mathbf{c}_{\mathrm{T}}:\mathbb{E}^{\mathrm{s}} & -\boldsymbol{\lambda}_{\mathrm{w}}:\mathbf{c}_{\mathrm{T}} & -C_m^{\mathrm{s}}/T_{\mathrm{s}}^0 & (\boldsymbol{\lambda}_{\mathrm{w}}:\mathbf{c}_{\mathrm{T}})\,c_{\mathrm{Tw}} \\
-c_{\mathrm{Tw}}\,\boldsymbol{\lambda}_{\mathrm{w}} & -M\,(n^{\mathrm{w}})^2\,c_{\mathrm{Tw}} & (\boldsymbol{\lambda}_{\mathrm{w}}:\mathbf{c}_{\mathrm{T}})\,c_{\mathrm{Tw}} & -C_v^{\mathrm{w}}/T_{\mathrm{w}}^0
\end{bmatrix}
\begin{bmatrix}
\boldsymbol{\epsilon} \\
\operatorname{tr}\boldsymbol{\epsilon}^{\mathrm{w}} \\
\Delta T_{\mathrm{s}} \\
\Delta T_{\mathrm{w}}
\end{bmatrix},
\tag{10.7.1}
$$

where use as been made of (7.2.3), (7.2.9), $\boldsymbol{\alpha}_{\mathrm{w}}$ is defined by (7.4.14), M by (7.4.17) and (7.4.19), and [10.4]

$$
\frac{1}{M} = \frac{1}{S}\,\mathbf{I}:\mathbb{E}^{-1}:\boldsymbol{\alpha}_{\mathrm{w}} - \boldsymbol{\alpha}_{\mathrm{w}}:\mathbb{E}^{-1}:\boldsymbol{\alpha}_{\mathrm{w}}, \quad \boldsymbol{\lambda}_{\mathrm{w}} = -M\,(n^{\mathrm{w}})^2\left(\mathbf{I} - \frac{\boldsymbol{\alpha}_{\mathrm{w}}}{n^{\mathrm{w}}}\right);
$$

$$
\mathbb{E}^{\mathrm{s}} = \mathbb{E} + M\,(n^{\mathrm{w}})^2\left(\mathbf{I} - \frac{\boldsymbol{\alpha}_{\mathrm{w}}}{n^{\mathrm{w}}}\right) \otimes \left(\mathbf{I} - \frac{\boldsymbol{\alpha}_{\mathrm{w}}}{n^{\mathrm{w}}}\right),
\tag{10.7.2}
$$

and

$$
C_m^{\mathrm{s}} = C_{\epsilon p}^{\mathrm{s}} + \mathbf{c}_{\mathrm{T}}:(\mathbb{E} - \mathbb{E}^{\mathrm{s}}):\mathbf{c}_{\mathrm{T}}\,T_{\mathrm{s}}^0 = C_{\sigma p}^{\mathrm{s}} - \mathbf{c}_{\mathrm{T}}:\mathbb{E}^{\mathrm{s}}:\mathbf{c}_{\mathrm{T}}\,T_{\mathrm{s}}^0,
$$

$$
C_v^{\mathrm{w}} = C_p^{\mathrm{w}} - M\,(n^{\mathrm{w}})^2\,(c_{\mathrm{Tw}})^2\,T_{\mathrm{w}}^0.
\tag{10.7.3}
$$

[10.4] $\boldsymbol{\lambda}_{\mathrm{w}}$ was denoted by $\lambda^{\mathrm{sw}}\,\boldsymbol{\lambda}_{\mathrm{w}}$ in Section 7.3.6.

For an isotropic solid skeleton,

$$\frac{1}{M} = \frac{1-n^{\mathrm{w}}}{K_{\mathrm{s}}} + \frac{n^{\mathrm{w}}}{K_{\mathrm{w}}} - \frac{K}{K_{\mathrm{s}}^2}, \quad \boldsymbol{\lambda}_{\mathrm{w}} = -M\,n^{\mathrm{w}}\left(-n^{\mathrm{s}} + \frac{K}{K_{\mathrm{s}}}\right)\mathbf{I}. \tag{10.7.4}$$

The thermomechanical equations (10.6.24) reduce to

$$\begin{bmatrix} -\boldsymbol{\sigma} \\ \zeta^{\mathrm{w}} \\ \Delta S^{\mathrm{s}} \\ \Delta S^{\mathrm{w}} \end{bmatrix} = \begin{bmatrix} -\mathbb{E} & \boldsymbol{\alpha}_{\mathrm{w}} & \mathbb{E}:\mathbf{c}_{\mathrm{T}} & 0 \\ \boldsymbol{\alpha}_{\mathrm{w}} & 1/M & a_{\mathrm{wT}} & -n^{\mathrm{w}}c_{\mathrm{Tw}} \\ \mathbf{c}_{\mathrm{T}}:\mathbb{E} & a_{\mathrm{wT}} & C_{\epsilon p}^{\mathrm{s}}/T_{\mathrm{s}}^0 & 0 \\ 0 & -n^{\mathrm{w}}c_{\mathrm{Tw}} & 0 & C_p^{\mathrm{w}}/T_{\mathrm{w}}^0 \end{bmatrix} \begin{bmatrix} \boldsymbol{\epsilon} \\ p_{\mathrm{w}} \\ \Delta T_{\mathrm{s}} \\ \Delta T_{\mathrm{w}} \end{bmatrix}, \tag{10.7.5}$$

where $a_{\mathrm{wT}} = (n^{\mathrm{w}}\mathbf{I} - \boldsymbol{\alpha}_{\mathrm{w}}):\mathbf{c}_{\mathrm{T}}$ in agreement with (10.6.13).

These constitutive relations further boil down to (9.3.14) when the solid and fluid are in local thermal equilibrium.

10.8   Interfacial transfer coefficients and specific surface areas

In a porous medium with two porosities, heat exchange takes place across the interface between the solid and the fluids, and across the menisci that separates the two fluids. Therefore the coefficients of heat exchange $\kappa_{\alpha\beta}$ between species α and β introduced in (10.6.52) are assumed to take the multiplicative form

$$\frac{\kappa_{\alpha\beta}}{T_0^2} = h_{\alpha\beta}\,a_{\alpha\beta}, \tag{10.8.1}$$

where T_0 is a reference temperature and
- $h_{\alpha\beta}$ [unit : $\mathrm{W/m^2/K}$] is the interfacial heat transfer coefficient, a property of the interface, i.e. of the two species;
- $a_{\alpha\beta}$ [unit : $\mathrm{m^2/m^3}$] is the specific surface area, i.e. the area $A_{\alpha\beta}$ of the interface between the two species in the volume V of porous medium.

A boundary condition of the convective type, say between a solid and air, involves the interfacial heat transfer coefficient h_{sa}, Section 10.4.

10.8.1   Interfacial heat transfer coefficients

Experimental determination of the solid-fluid interfacial heat transfer coefficient h_{sw}

Consider a porous medium with a single porosity, and assume that the thermal gradients can be neglected. If only heat conduction and heat transfer are accounted for, the energy equation of the solid constituent boils down to,

$$\frac{C^{\mathrm{s}}}{T_{\mathrm{s}}}\frac{dT_{\mathrm{s}}}{dt} + \frac{\kappa_{\mathrm{sw}}}{T_{\mathrm{s}}}\left(\frac{1}{T_{\mathrm{w}}} - \frac{1}{T_{\mathrm{s}}}\right) = 0, \tag{10.8.2}$$

and therefore, for $T_{\mathrm{s}} \sim T_{\mathrm{w}} \sim T_0$,

$$\frac{\kappa_{\mathrm{sw}}}{T_0^2} = -C^{\mathrm{s}}\frac{dT_{\mathrm{s}}/dt}{T_{\mathrm{s}} - T_{\mathrm{w}}}. \tag{10.8.3}$$

A basic experimental setup to measure the solid-fluid heat transfer coefficient consists in circulating a fluid through a relatively thin porous wall, so that the inlet and outlet temperatures are quite close. The heat transfer coefficient is then defined by (10.8.3) where the temperatures are approximated by their averages over the wall boundaries.

In the context of porous media with double porosity, the above experiment would correspond to the tissue-vessel (porosity 2) heat transfer coefficient κ_{s2}. The interfacial tissue-water heat transfer coefficient $h_{\mathrm{sw}} = \kappa_{\mathrm{s2}}/T_0^2/a_{\mathrm{s2}}$ may be deduced, assuming the specific surface area of the vessel network a_{s2} to be known. The tissue-pore heat transfer coefficient $\kappa_{\mathrm{s1}} = T_0^2\,h_{\mathrm{sw}}\,a_{\mathrm{s1}}$ deduces in turn, assuming the specific surface area a_{s1} of the pore network (porosity 1) to be known as well. Alternatively, if at all possible, one may repeat the experiment by isolating a piece of the matrix with a single type of porosity.

Theoretical determination of the solid-fluid interfacial heat transfer coefficient h_{sw}

The interfacial solid-fluid heat transfer coefficient has been calculated for certain geometries, in particular for a cylindrical capillary, in terms of the thermal conductivities of the solid and fluid and of geometrical characteristics, Kaviany [1995], p. 398, or for in-line periodic arrays of three-dimensional cubes, Hsu [1999]. Typically, if L is a characteristic length of the porous network, the heat transfer coefficient may be estimated to be of the order of the thermal conductivity of the fluid divided by this length,

$$h_{sw} \sim \frac{k_{Tw}}{L} . \tag{10.8.4}$$

The medium we consider is a fluid-infiltrated solid. For the somehow dual configuration, namely for spherical particles bathed in a Newtonian fluid, a general expression of h_{sw} is obtained via the Nusselt number, which has been related to the Reynolds and Prandtl numbers, Kaviany [1995], p. 403, namely

$$N_u = 2 + 1.1\, R_e^{0.6}\, P_r^{0.33} . \tag{10.8.5}$$

The Nusselt N_u number is a measure of the relative weight of convective and conductive heats, the Reynolds number R_e a measure of the relative weight of the inertial and viscous contributions to the momentum balance, and the Prandtl number P_r quantifies the ratio of momentum diffusivity (kinematic viscosity) over thermal diffusivity,

$$R_e = \frac{v_w\, L}{\nu_w} , \quad P_r = \frac{C^w \nu_w}{k_{Tw}} = \frac{\nu_w}{D_{Tw}} , \quad N_u = \frac{h_{sw}\, L}{k_{Tw}} , \tag{10.8.6}$$

with

- L [unit : m] typical dimension of the problem, e.g., that of the pores;
- v_w [unit : m/s] fluid velocity (Darcy's velocity in a porous medium);
- ν_w [unit : m^2/s] kinematic viscosity defined by (7.3.14);
- k_{Tw} [unit : W/m/K] thermal conductivity of the fluid;
- C^w [unit : J/m^3/K] thermal capacity of the fluid;
- $D_{Tw} = k_{Tw}/C^w$ [unit : m^2/s] thermal diffusivity of the fluid;
- h_{sw} [unit : W/m^2/K=kg/s^3/K] interfacial solid-fluid heat transfer coefficient.

In hydrodynamics, the transition from laminar to turbulent flows is considered to take place typically at a Reynolds number of 2000. For fluid-saturated porous media, using a microstructure based definition, Kaviany [1995], p. 48, suggests to reduce this transition value to 300. Actually, a definition of a Reynolds number for porous media is suggested in Section 15.7.9.2. The Prandtl number is low for mercury, about 7.07 for water at 20 °C, and much higher for engine oil, Bejan [1993]. The Péclet number defined by (15.5.3) is a relative measure of the convective and diffusive characters of the equation that controls the diffusion of temperature or of particles in a fluid. It is in fact the product of the Reynolds and Prandtl numbers,

$$P_e = R_e \times P_r . \tag{10.8.7}$$

Perhaps, it is worth to delineate the Nusselt and Biot numbers. These two numbers are formally identical but they have quite distinct meanings. For a solid in contact with a fluid β,

$$N_u = \frac{h_{s\beta}\, L}{k_{T\beta}} , \quad Bi = \frac{h_{s\beta}\, L}{k_{Ts}} . \tag{10.8.8}$$

The Nusselt number involves the thermal conductivity of the fluid. The Biot number, which involves the thermal conductivity of the solid, gives an indication of the relative influence of the thermal resistance at the surface and inside a body. See Section 10.4.4 for more details.

As Reynolds number tends to 0, Nusselt number tends to 2 according to (10.8.5), and therefore the interfacial heat transfer coefficient tends to $2\, k_{Tw}/L$.

Fluid-fluid heat transfer coefficient κ_{12}

Heat transfer through the meniscus separating two immiscible fluids may be modeled via heat conduction. Indeed, if the two fluids have not too dissimilar properties, the heat conduction of the meniscus k_T may be approximated as $\frac{1}{2}(k_{T1} + k_{T2})$. Then the transfer of power through the meniscus of width L is, via Fourier's law, equal to,

$$- k_T \frac{T_2 - T_1}{L} \times \frac{1}{L} . \tag{10.8.9}$$

This transfer may also be expressed via the constitutive relations for heat transfer (10.6.42) and (10.6.52). Indeed, the power transferred to fluid 2 from fluid 1 expresses as,

$$\hat{U}^2 = -\hat{U}^1 = -\kappa_{12}\left(\frac{1}{T_1} - \frac{1}{T_2}\right) \simeq -\frac{\kappa_{12}}{T_0^2}\left(T_2 - T_1\right), \tag{10.8.10}$$

assuming the two temperatures T_1 and T_2 to be close to the reference value T_0. Moreover in this thought experiment, the specific surface area is equal to $1/L$. Then

$$\frac{\kappa_{12}}{T_0^2} = \frac{k_{T1} + k_{T2}}{2\,L^2}, \quad h_{12} = \frac{k_{T1} + k_{T2}}{2\,L}\,. \tag{10.8.11}$$

The width over which the heat transfer between the two fluids takes place may be estimated as a percentage of the diameters of the pores and vessels, whether the fluids are miscible or immiscible.

The above considerations address heat transfer as a limit case of heat conduction without phase change. Heat transfer due for example to film condensation over vertical tubes or walls involves the latent heat of vaporization, see, e.g., Bird et al. [1960].

10.8.2 Specific surface area

Since the heat exchanges are directly proportional to the specific surface area on the interfaces, it is of interest to explore the influence of basic geometrical characteristics[10.5].

1. Decreasing the pore size:

As a particular setting, consider a porous medium containing spherical pores of uniform diameter d, the entire surface of which is available for solid-pore heat transfer. Therefore, each pore is completely embedded in the matrix. The maximum volume fraction is equal to the ratio of the volume of a sphere over the volume of the cube it is contained in, namely $n \le \pi/6 \sim 0.52$. The volume fraction n of the pores is equal to the number N of pores over the volume V of porous medium times the volume $\pi d^3/6$ of a pore. Hence $N/V = 6\,n/(\pi\,d^3)$. The specific surface area a_{s1}, namely the area of contact between the matrix and the pores in a volume V, is equal to N/V times $\pi\,d^2$, that is $a_{s1} = 6\,n/d$[10.6]. The specific surface area of a cube of size d is larger, namely $6/d$, but smaller than the specific surface area of a parallelepiped of sides $d_1 \times d_2 \times d_3$, namely $2\,(d_1\,d_2 + d_2\,d_3 + d_3\,d_1)/(d_1\,d_2\,d_3)$, which, for small d_3, behaves like $2/d_3$. Thus particulate materials which are foliated are expected to have a larger specific surface area than granular materials: in fact, typical values for sand, fine sandstone and montmorillonite increase from $1.5 \times 10^4\,\mathrm{m^2/m^3}$ to $1.5 \times 10^5\,\mathrm{m^2/m^3}$ and $1.5 \times 10^9\,\mathrm{m^2/m^3}$, de Marsily [1986], p. 22.

For cylindrical tubes of uniform diameter, the solid-vessel specific surface area becomes $a_{s2} = 4\,n/d$, and the maximum volume fraction is equal to $\pi/4 \sim 0.78$.

The specific surface area a_{12} of the two fluids may be assumed, as a first approximation, to be equal to n^1 times the tissue-vessel specific surface area a_{s2}.

2. Optimization under a physical constraint:

One may reconsider the above construction in a two-dimensional context, by concentrating on cross-sections. Consider a circular cylinder of radius R and unit height. Its cross-section S and circumference L are respectively equal to $\pi\,R^2$ and $2\,\pi\,R$. Now, consider N identical cylinders of radius r. When increasing the number of cylinders, one may require the volume fraction n to remain constant. Thus the radius r is equal to $R/N^{1/2}$, and the sum of all circumferences $2\,\pi\,R\,N^{1/2}$. The specific surface area is then equal to $2\,n/r$, that is $2\,n/R$ times $N^{1/2}$.

Alternatively, the axial flow may be required to remain constant. The flow rate of a newtonian fluid in a cylinder is known to be proportional to the fourth power of the radius, Eqn (15.6.13). Thus the radius r is equal to $R/N^{1/4}$, the sum of all sections is $\pi\,R^2\,N^{1/2}$, and the sum of all circumferences to $2\,\pi\,R\,N^{3/4}$. Thus the volume fraction varies like $N^{1/2}$ and the specific surface area like $N^{3/4}$. Therefore, increasing the number of cylinders appears favorable to exchanges that depend (linearly) on the length (or area) of the interface. Now, there are physiological limitations to a too large N. In particular, the volume fraction has certainly to be bounded for structural reasons. Moreover, a minimum size of the cylinders is required so that small particles can be convected by the flow: red blood cells are disk-like with a diameter about $7.5\,\mu\mathrm{m}$, while the smallest micro-vessels have a diameter about $9\,\mu\mathrm{m}$.

[10.5]Typical values of the specific surface areas associated with vessels in normal tissues and tumor tissues are shown in Table 24.8.1. In some geological materials, namely montmorillonites, the specific surface area reaches very high values, up to $700\,\mathrm{m^2}$ per gram, which correlate with their propensity to attract water.

[10.6] The formula holds by reversing the roles of solid and fluid, i.e., the specific surface area of a porous medium containing solid spheres of diameter d bathed in a fluid is equal to $6\,n/d$ where n is now the volume fraction of the solid.

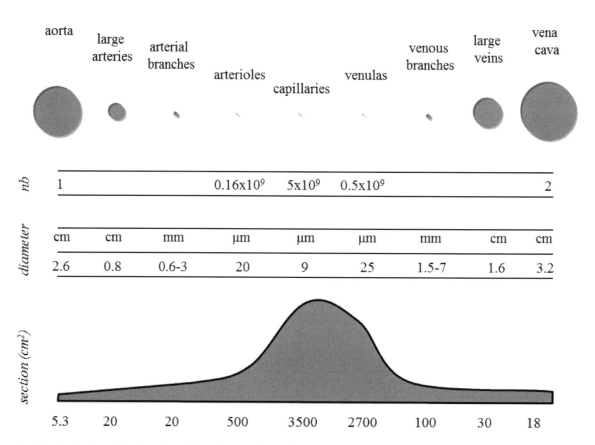

Fig. 10.8.1: Indicative distribution of blood vessel number, diameter and section in human adults, adapted from APP [2000], p. 157.

The actual distribution of the size of the vessels in humans does not follow exactly these relations, although the trends are correct, with a total cross-section that increases considerably as the diameter of the vessels decreases, Fig. 10.8.1. In fact, the above considerations have assumed that the driving gradient was constant along the cylinders and independent of their size. Such is not the case along the blood circuit: the resistance to flow due to the Hagen-Poiseuille relation varies like $1/r^4$, and therefore becomes quite large in the capillaries, so that the gradient of pressure drops, both axially and transmurally. Moreover, the average blood velocity varies like the square of the vessel diameter, Eqn (15.6.13): the low velocity in the capillaries, about $0.3\,\mathrm{mm/s}$ against $0.2\,\mathrm{m/s}$ in the aorta, allows sufficient contact time for exchanges of mass and heat across the capillary wall to actually take place. The surface offered by the capillaries for these exchanges is about $300\,\mathrm{m}^2$, APP [2000], p. 156.

3. Optimization via a fractal construction:

The above process has consisted in creating a number of objects, similar in shape to the initiator (the cylinder of radius R), but of smaller size. Alternatively, one may keep considering the initiator, but insist in working its boundary. Indeed, it is worth recalling that there are two-dimensional mathematical objects with finite area but infinite circumference. The fractal called island of von Koch constitutes a famous instance. The construction goes as follows:

- start from an equilateral triangle of side ℓ;
- remove the central third of each of the three sides, and replace it by an inverted V, each side having length $L/3$;
- repeat the above generating process.

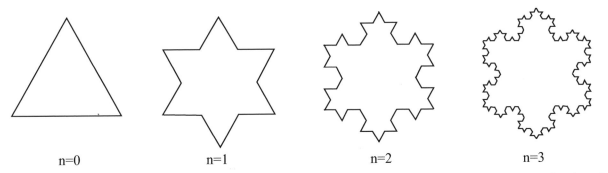

Fig. 10.8.2: The fractal construction of so-called von Koch island ends up with an object with a finite surface but the boundary has infinite length. The sketch shows four pre-fractal objects associated with the first steps of the construction.

Fig. 10.8.2 displays the generator and three pre-fractals $n = 1, 2$ and 3. At step n, the length L_n of the boundary and the area S_n of the pre-fractal domain have simple explicit expressions,

$$L_n = 3\ell \left(\frac{4}{3}\right)^n, \quad S_n = \ell^2 \frac{\sqrt{3}}{4} \left(1 + \frac{1}{3} \frac{1 - \left(\frac{4}{9}\right)^n}{1 - \frac{4}{9}} \right), \tag{10.8.12}$$

and therefore the length L_∞ of the boundary of the fractal domain is infinite while the area S_∞ remains finite[10.7],

$$L_\infty = \infty, \quad S_\infty = \ell^2 \frac{\sqrt{3}}{4} \left(1 + \frac{3}{5} \right). \tag{10.8.13}$$

Perhaps, one may wish to alter the fractal generator of von Koch island by introducing some additional constraint, like maintaining a constant area, etc.

Actually, maintaining a single main channel has pros and cons. The large interfacial area offered by pre-fractals participates to increased transmural exchanges. On the other hand, the advantages associated with small circular cylinders mentioned above (reduced velocity, etc.) should be re-examined for the pre-fractal boundary. Moreover, even if exchanges take actually place, heat or mass which has crossed the vessel wall should diffuse through the tissue. The dense micro-vessel network was an ideal tool for that purpose.

10.9 The coefficients of transfer: two scale derivation

In Section 10.8, the coefficients of transfer have been factorized based on qualitative arguments. As an alternative, a two scale derivation is sketched here. The underlying motivation is to account for the diffusion properties at a scale lower than the macroscopic scale, Fig. 10.9.1.

10.9.1 Small scale boundary value problems

10.9.1.1 Sphere bathed in a fluid at fixed given temperature

A solid sphere Ω of radius R is suddenly bathed in a fluid. The fluid temperature is constant in time and uniform in space, it is not influenced by the presence of the solid sphere. Let $\theta_s(r, t)$ and θ_w be the temperatures of the solid and fluid relative to some common reference.

The heat diffusion equation in the sphere,

$$C_s \frac{\partial \theta_s}{\partial t} - \mathrm{div}(k_{Ts} \boldsymbol{\nabla} \theta_s) = 0, \tag{10.9.1}$$

[10.7]On a similar mood, consider a staircase of stringer length L inclined at $45°$, with steps of equal run and rise. A path that follows the steps has length $L\sqrt{2}$ independently of the size of the steps.

three typical unit cells :

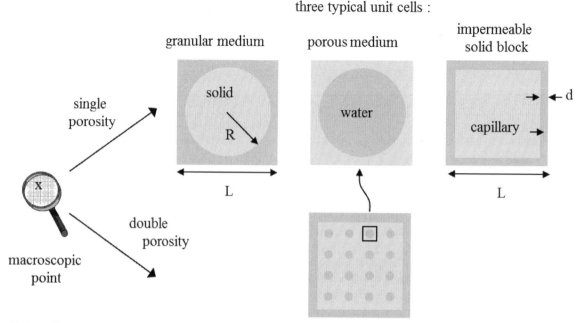

Fig. 10.9.1: Zooming on a geometrical point unveils the unit cells that deliver information on the geometrical arrangements of the phases at a lower scale either in a single porosity model or in a double porosity model.

simplifies to $\partial\theta_s/\partial t - D_{Ts}\,\mathrm{div}\boldsymbol{\nabla}\theta_s = 0$ if the thermal conductivity k_{Ts} [unit : W/m/K] is spatially uniform. Here C_s [unit : J/m^3/K] is the heat capacity of the solid (at constant pressure), and $D_{Ts} = k_{Ts}/C_s$ [unit : m^2/s] the thermal diffusivity. Integration of (10.9.1) and use of the Green's theorem implies,

$$\int_\Omega C_s \frac{\partial\theta_s}{\partial t}\,d\Omega = \int_\Omega \mathrm{div}(k_{Ts}\boldsymbol{\nabla}\theta_s)\,d\Omega = -q_s(t)\,, \qquad (10.9.2)$$

where $q_s(t)$ [unit : W] is the heat flux that exits the sphere Ω,

$$q_s(t) = -\int_{\partial\Omega} k_{Ts}\boldsymbol{\nabla}\theta_s \cdot \mathbf{e}_r\,dS \left(= -\int_{\partial\Omega} k_{Ts}\frac{\partial\theta_s}{\partial r}\,dS = -4\,\pi\,R^2\,k_{Ts}\frac{\partial\theta_s}{\partial r}(R,t)\right). \qquad (10.9.3)$$

Since the thermal capacity is spatially uniform, (10.9.2) writes also

$$\Omega\,C_s\,\frac{d\theta_s^{\mathrm{av}}}{dt}(t) = -q_s(t)\,, \qquad (10.9.4)$$

where $\Omega = (4/3)\,\pi\,R^3$ is the volume of the sphere and $\theta_s^{\mathrm{av}}(t) = \Omega^{-1}\int_\Omega \theta_s(r,t)\,d\Omega$ is the spatial average of the temperature in the sphere.

The heat transfer coefficient is defined as

$$\beta(t) = \frac{-q_s(t)}{\theta_w - \theta_s^{\mathrm{av}}(t)} = \frac{\Omega\,C_s}{\theta_w - \theta_s^{\mathrm{av}}(t)}\frac{d\theta_s^{\mathrm{av}}}{dt}(t)\,. \qquad (10.9.5)$$

The solution to the boundary value problem (10.9.1) with $\theta_s(R,t) = \theta_w$ is given by Carslaw and Jaeger [1959], p. 233. They also provide the spatial average, which is the only information that is needed here,

$$\frac{\theta_s^{\mathrm{av}}(t)}{\theta_w} = 1 - \frac{6}{\pi^2}\sum_{n=1}^{\infty}\frac{1}{n^2}\,e^{-n^2\,\tilde{t}}\,, \qquad (10.9.6)$$

where $\tilde{t} \equiv t/t_D$ is the dimensionless time and t_D the characteristic time,

$$t_D = \frac{1}{\pi^2}\frac{R^2}{D_{Ts}}\,. \qquad (10.9.7)$$

10.9.1.2   Sphere bathed in a fluid at prescribed temperature evolving in time

Now, let us consider a small change of the fluid temperature at time τ in the form

$$\delta\theta_{\rm w}(t) = \mathcal{H}(t-\tau)\,\delta\theta_{\rm w}(\tau)\,, \tag{10.9.8}$$

with $\mathcal{H}$ the Heaviside step function. The resulting thermal change in the solid sphere is given in the form,

$$\delta\theta_{\rm s}^{\rm av}(t) = \mathcal{T}(t-\tau)\,\mathcal{H}(t-\tau)\,\delta\theta_{\rm w}(\tau)\,, \tag{10.9.9}$$

where the function $\mathcal{T}$ is given by (10.9.6), namely,

$$\mathcal{T}(u) = 1 - \frac{6}{\pi^2}\sum_{n=1}^{\infty}\frac{1}{n^2}e^{-n^2\,\tilde{u}}\,. \tag{10.9.10}$$

Then upon integration, the average temperature in the sphere,

$$
\begin{aligned}
\theta_{\rm s}^{\rm av}(t) - \theta_{\rm s}^{\rm av}(0) &= \mathcal{T}(t-0)\,\mathcal{H}(t-0)\,\theta_{\rm w}(0) + \int_0^t \mathcal{T}(t-\tau)\,\mathcal{H}(t-\tau)\,\frac{\delta\theta_{\rm w}(\tau)}{\delta\tau}\,\delta\tau \\
&= \mathcal{T}(t-0)\,\theta_{\rm w}(0) + \int_0^t \mathcal{T}(t-\tau)\,\frac{\delta\theta_{\rm w}(\tau)}{\delta\tau}\,\delta\tau \\
&= \mathcal{T}(t-t)\,\theta_{\rm w}(t) - \int_0^t \frac{\partial\mathcal{T}(t-\tau)}{\partial\tau}\,\theta_{\rm w}(\tau)\,\delta\tau \\
&= \frac{d}{dt}\int_0^t \mathcal{T}(t-\tau)\,\theta_{\rm w}(\tau)\,\delta\tau\,,
\end{aligned}
\tag{10.9.11}
$$

adopts the format of a Stieltjes convolution integral. Clearly, if $\theta_{\rm w}$ is constant, the result (10.9.6) is recovered. Therefore by (10.9.4),

$$-q_{\rm s}(t) = \Omega\,C_{\rm s}\,\frac{d\theta_{\rm s}^{\rm av}}{dt}(t) = \Omega\,C_{\rm s}\,\frac{d^2}{dt^2}\int_0^t \mathcal{T}(t-\tau)\,\theta_{\rm w}(\tau)\,\delta\tau\,. \tag{10.9.12}$$

10.9.2   Heat transfer with a characteristic time

A modification of the constitutive equation of heat transfer may be formulated in a macroscopic context, i.e., referring to a single spatial scale, using the idea presented in Section 9.1.4 that led from Fourier's heat conduction law to Cattaneo's heat conduction law in a thermodynamic framework.

For both the solid ($k = {\rm s}$) and the liquid ($k = {\rm w}$), the starting point is to enlarge the set of state variables $\mathcal{V}_k$ by including the rate of energy transfer $\hat{U}^k$ [unit : W/m^3]. The Massieu function Ξ_k is then postulated in the format,

$$\Xi_k(\mathcal{V}_k,\hat{U}^k) = \Xi_{k1}(\mathcal{V}_k) - \frac{1}{2}\,A_{Tk}(\hat{U}^k)^2\,, \tag{10.9.13}$$

where A_{Tk} is a constant. Manipulating the balance of energy and the entropy production (10.1.1) to eliminate the term div $\mathbf{Q}_k - \Theta_k$, the energy exchange process is seen to contribute to the entropy production of the porous medium by the terms,

$$
\begin{aligned}
\hat{S}_{\rm (i)} &= \frac{\partial\Xi_{\rm s}}{\partial\hat{U}^{\rm s}}\frac{d\hat{U}^{\rm s}}{dt} + \frac{\hat{U}^{\rm s}}{T_{\rm s}} \quad + \quad \frac{\partial\Xi_{\rm w}}{\partial\hat{U}^{\rm w}}\frac{d\hat{U}^{\rm w}}{dt} + \frac{\hat{U}^{\rm w}}{T_{\rm w}} \\
&= -A_{T{\rm s}}\,\hat{U}^{\rm s}\frac{d\hat{U}^{\rm s}}{dt} + \frac{\hat{U}^{\rm s}}{T_{\rm s}} \quad - \quad A_{T{\rm w}}\,\hat{U}^{\rm w}\frac{d\hat{U}^{\rm w}}{dt} + \frac{\hat{U}^{\rm w}}{T_{\rm w}} \\
&= -(A_{T{\rm s}} + A_{T{\rm w}})\,\hat{U}^{\rm s}\frac{d\hat{U}^{\rm s}}{dt} \quad + \quad \hat{U}^{\rm s}\left(\frac{1}{T_{\rm s}} - \frac{1}{T_{\rm w}}\right)\,,
\end{aligned}
\tag{10.9.14}
$$

the last line accounting for the fact that the energy exchange is assumed to be closed, namely $\hat{U}^{\rm s} + \hat{U}^{\rm w} = 0$.

Moreover, in this approach the rate $d\hat{U}^{\mathrm{s}}/dt = -d\hat{U}^{\mathrm{w}}/dt$ is viewed as a dependent variable that satisfies the relation,

$$t_{\hat{U}}\,\frac{d\hat{U}^{\mathrm{s}}}{dt} + \hat{U}^{\mathrm{s}} = \kappa_{\mathrm{sw}}\left(\frac{1}{T_{\mathrm{s}}} - \frac{1}{T_{\mathrm{w}}}\right). \tag{10.9.15}$$

Here $\kappa_{\mathrm{sw}} \geq 0$ [unit : W/m³× K] is the coefficient of energy transfer while $t_{\hat{U}}$ [unit : s] is a characteristic time of energy transfer. For $t_{\hat{U}} = 0$, the process is instantaneous.

Insertion of the rate $d\hat{U}^{k}/dt$ obtained from (10.9.15) in the entropy production (10.9.14) yields,

$$\hat{S}_{(\mathrm{i})} = \left(A_{T\mathrm{s}} + A_{T\mathrm{w}}\right)\frac{(\hat{U}^{\mathrm{s}})^2}{t_{\hat{U}}} + \hat{U}^{\mathrm{s}}\left(\frac{1}{T_{\mathrm{s}}} - \frac{1}{T_{\mathrm{w}}}\right)\left(1 - \kappa_{\mathrm{sw}}\,\frac{A_{T\mathrm{s}} + A_{T\mathrm{w}}}{t_{\hat{U}}}\right). \tag{10.9.16}$$

Setting

$$\frac{A_{T\mathrm{s}} + A_{T\mathrm{w}}}{t_{\hat{U}}} = \frac{1}{\kappa_{\mathrm{sw}}} \geq 0\,, \tag{10.9.17}$$

ensures the entropy production to be positive or zero,

$$\hat{S}_{(\mathrm{i})} = \frac{(\hat{U}^{\mathrm{s}})^2}{\kappa_{\mathrm{sw}}} \geq 0\,. \tag{10.9.18}$$

The rate of energy transfer emanating from (10.9.15) is a convolution integral with an exponential kernel,

$$\hat{U}^{\mathrm{s}}(t) = \exp\left(-\frac{t}{t_{\hat{U}}}\right)\hat{U}^{\mathrm{s}}(0) + \int_0^t \kappa\,\exp\left(-\frac{t-\tau}{t_{\hat{U}}}\right)\left(T_{\mathrm{w}}(\tau) - T_{\mathrm{s}}(\tau)\right)\frac{d\tau}{t_{\hat{U}}}\,, \tag{10.9.19}$$

where the following notation was introduced,

$$\kappa = \frac{\kappa_{\mathrm{sw}}}{T_{\mathrm{s}}\,T_{\mathrm{w}}}\,. \tag{10.9.20}$$

The correspondence between the two scale analysis of Section 10.9.1.1 which involves a sphere of volume Ω and this macroscopic analysis is realized by the relations

- $\hat{U}^{\mathrm{s}} = -q_{\mathrm{s}}/\Omega$ rate of energy transfer per unit volume of solid [unit : W/m³];
- $\kappa = \kappa_{\mathrm{sw}}/(T_{\mathrm{s}}\,T_{\mathrm{w}}) = \beta/\Omega$ energy transfer coefficient [unit : W/m³/K] .

The constitutive equation of transfer endowed with a characteristic time sketched above can certainly match more accurately time course data than an instantaneous formulation. Still it is also more heavy on a computational point of view. While the constitutive equations of transfer adopted in this chapter are easily amenable to computational implementation and consistent with a thermodynamic analysis, leading to positive dissipation, they are also known to be inaccurate at early times. The nonlinear Vermeulen scheme has been adopted by Zimmerman et al. [1993] in the analysis of fractured geothermal reservoirs where, at each point of the fracture continuum, a porous block of spherical shape is attached: the fluid diffuses in the block and the net flow through its boundary is viewed as a source/sink term for the fracture continuum. Lu and Connell [2007] have devised a one-dimensional semi-analytical scheme that provides the time course of the transferred mass in a gas reservoir. At early times, the rate of mass transfer tends to vanish for the latter model, it approaches a constant for the linear transfer scheme and infinity for the Vermeulen scheme. Correspondingly, the mass transferred depends linearly on time in the linear transfer scheme, but on the square root of time in the schemes of Vermeulen and Lu and Connell, albeit with distinct scaling factors. A simple, accurate while computationally efficient, model of transfer that avoids delving with a convolution integral, is yet to come.

10.9.3 Identification of the transfer coefficients

Warren and Root [1963] identify the coefficient of transfer using a two scale analysis in the mood of Section 10.9.1.1. They address a fractured network where the solid blocks are parallelepipeds. In the case where the latter have identical dimension L in all directions, they define the heat transfer coefficient κ [unit : W/m³/K] in the form,

$$\kappa = \overline{\alpha}\,k_{T\mathrm{s}}\,, \quad \overline{\alpha} = 4\,\frac{n\,(n+2)}{L^2}\,, \tag{10.9.21}$$

where $n = 1, 2$ or 3 is the number of orthogonal fracture planes, which can be associated with the dimension of the space. Actually, several interpretations have been proposed for the coefficient $\overline{\alpha}$.

For a *cubic solid block*,
- Warren and Root [1963] take $\overline{\alpha} = 60/L^2$;
- Zimmerman et al. [1993] suggest to take $\overline{\alpha} = \pi^2/\tilde{L}^2$, with $\tilde{L} = 3/a$, where a is the specific surface area. For a cube, $a = 6/L$ so that $\tilde{L} = L/2$ and $\overline{\alpha} = 4\,\pi^2/L^2$;
- the first term in the analytical solution of the diffusion problem is equal to $\overline{\alpha} = 3\,\pi^2/L^2$, Carslaw and Jaeger [1959], p. 184.

For a *spherical solid block* of radius R, the specific surface area is equal to $3/R$, so that the Zimmerman suggestion yields $\overline{\alpha} = \pi^2/R^2$. Now, if we retain only the first term in the solution (10.9.6),

$$\frac{T_{\mathrm{s}} - T_0}{T_{\mathrm{w}} - T_0} = 1 - \frac{6}{\pi^2}\,e^{-\tilde{t}}\,, \tag{10.9.22}$$

we obtain from (10.9.5) and (10.9.7),

$$\overline{\alpha} = \frac{\pi^2}{R^2}\,, \tag{10.9.23}$$

which agrees with the suggestion by Zimmerman et al. [1993] related with the specific surface area.

A characteristic time of energy transfer $t_{\mathscr{U}}$ might be obtained by considering thermal diffusion in the solid as in Section 10.9.2, and by setting $t_{\mathscr{U}}$ equal to the diffusion time t_{D} defined by (10.9.7). The radius R is to be viewed as half of the characteristic length of the transfer process.

This section has considered energy transfer. Still, the results concerning the transfer coefficients apply in essence to mass transfer. For example, consider the transfer of mass between the compartments 1 and 2 driven by the pressure differential $p^1 - p^2$ and ruled by the constitutive relation $\hat{\rho}^1/\rho^1 = -\kappa\,(p^1 - p^2)$. The mass transfer coefficient κ [unit : 1/Pa/s] adopts the Warren and Root format (10.9.21),

$$\kappa = \overline{\alpha}\,k_{\mathrm{H}}, \quad \overline{\alpha} = 4\,\frac{n\,(n+2)}{L^2}\,, \tag{10.9.24}$$

with k_{H} [unit : $\mathrm{m}^2 \times \mathrm{Pa}^{-1} \times \mathrm{s}^{-1}$] the *hydraulic permeability* of the porous medium.

Chapter 11

Waves in thermoelastic solids and saturated porous media

Dynamic methods are used to probe engineering materials, and biological tissues in particular. Efforts are deployed to efficiently tune ultrasonic techniques and magnetic-resonance elastography to living tissues. Their deformability and spatial heterogeneity and the presence of water pose considerable challenges. Nevertheless, dynamic methods are expected to help the diagnosis as they are seen as quick tools able to contrast the mechanical properties of normal and pathological tissues.

On another side, extracorporeal shock wave lithotripsy and intracorporeal ultrasound lithotripsy using lower frequencies but higher intensities than for medical diagnosis are routinely used to trigger the comminution of kidney stones. Further potential applications of shock waves, such as *in vitro* and *in vivo* damaging of tumor cells and healing of muskulo-skeletal disorders, have been suggested. The high amplitude waves generate positive and negative pressures that are thought to induce cavitation. The mechanisms through which cavitation acts to succeed in the healing process are not yet understood. Neo-vascularization induced by micro-injuries and enhancement of cellular reactions are potential candidates.

As yet another venue, damage induced by shock waves in specific organs, e.g., the inner ear, the optical nerve, brain vessels, is being investigated with a view to improve the protection against blasts.

11.1 Basic definitions

11.1.1 Phase wave and discontinuity wave

A wave is a mode of propagation in which some property is maintained on a surface that moves through the medium. If this property is a phase, the phenomenon is referred to as a *phase wave*. As an example, consider the motion along the axis x defined by the displacement $\phi(x,t) = a \sin(2\pi (x - ct)/L)$, where a characterizes the amplitude of the motion, L is the wave-length and c the wave-speed. When we analyze the motion, we pay attention to the motion of the maxima $2\pi (x - ct)/L = \pi/2$, Fig. 11.1.1. As a generalization to three dimensions, a *plane wave* is defined by a motion in a direction $\mathbf{n}$, all points of the planes orthogonal to $\mathbf{n}$ being disturbed equally, say $\phi(\mathbf{x},t) = a \sin(2\pi (\mathbf{n} \cdot \mathbf{x} - ct)/L)$.

When the property that moves through the medium is a discontinuity, we speak of a *discontinuity wave*. Let $\mathcal{S}(t)$ be a smooth surface separating two domains $V_-(t)$ and $V_+(t)$ across which some function ϕ is discontinuous. $\mathcal{S}$ is referred to as a *singular surface*. The discontinuity, or jump, of the function $\phi(\mathbf{x},t)$ at $\mathbf{x} \in \mathcal{S}$ is denoted via the symbol,

$$[\phi](\mathbf{x},t) = \lim_{\mathbf{x}_+ \in V_+ \to \mathbf{x}} \phi(\mathbf{x}_+,t) - \lim_{\mathbf{x}_- \in V_- \to \mathbf{x}} \phi(\mathbf{x}_-,t). \tag{11.1.1}$$

By convention, V_- and V_+ are respectively the domains behind and ahead of the wave. Please notice that, as defined by (11.1.1), the jump of ϕ is the opposite of its time change. The jump of a sum is equal to the sum of the jumps and the jump of a product is given by the expression,

$$[\![\phi\,\psi]\!](\mathbf{x},t) = [\![\phi]\!](\mathbf{x},t) \times \tfrac{1}{2}(\psi_- + \psi_+) + \tfrac{1}{2}(\phi_- + \phi_+) \times [\![\psi]\!](\mathbf{x},t), \tag{11.1.2}$$

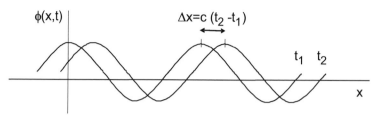

Fig. 11.1.1: A phase wave propagating at speed c captured at two instants t_1 and $t_2 > t_1$: the top of the wave has moved the distance $c(t_2 - t_1)$.

where $\phi_\pm(\mathbf{x}, t) = \lim_{\mathbf{x}_\pm \in V_\pm \to \mathbf{x}} \phi(\mathbf{x}_\pm, t)$, and similarly for $\psi_\pm$. The formula applies to scalars and tensors in the indicated order.

If the singular surface is composed of the same material points at any time, the phenomenon is referred to as a *stationary discontinuity*, and the domains V_- and V_+ might be assigned arbitrarily.

If across the singular surface, all derivatives of the function $\phi(x, t)$ are continuous up to order $m - 1$, while some derivatives are discontinuous at order m, we refer

- to a *strong discontinuity* or *first order wave* if $m = 1$;
- to an *ordinary wave* if $m \geq 2$, and in particular to a second order wave or *acceleration wave* if $m = 2$.

There does not exist a wave of order $m = 0$ which would correspond to a discontinuous displacement, and therefore to an infinite discontinuity of velocity. The term *shock wave* is usually restricted to an accumulation of acceleration waves giving rise to a higher order discontinuity.

11.1.2 Lagrangian and Eulerian analyses

Let the singular surfaces at times t and $t + dt$ be denoted respectively by $\mathcal{S}(t)$ and $\mathcal{S}(t + dt)$. The sets of material points, which form the surfaces $\mathcal{S}(t)$ and $\mathcal{S}(t + dt)$, are located on the surfaces $\underline{\mathcal{S}}(t)$ and $\underline{\mathcal{S}}(t + dt)$ in the reference configuration, say at time $t = 0$.

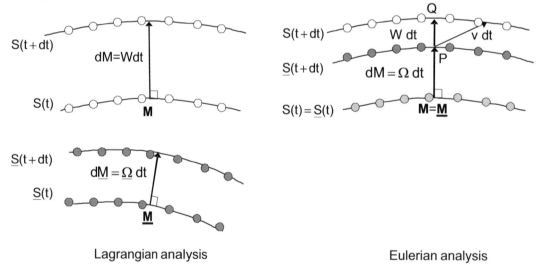

Fig. 11.1.2: Lagrangian and Eulerian analyses of wave propagation. Sketch illustrating the difference between Ω, the normal speed of propagation of the wave with respect to the material, and W, the normal speed of displacement of the wave.

Referring to Fig. 11.1.2, consider a particular point $\mathbf{M}$ on $\mathcal{S}(t)$ and its position $\underline{\mathbf{M}}(t) \in \underline{\mathcal{S}}(t)$ in the reference configuration. This point sits at time $t + dt$ on the surface $\mathcal{S}(t + dt)$ and its position in the reference configuration is $\underline{\mathbf{M}}(t + dt) \in \underline{\mathcal{S}}(t + dt)$. A distinction should be made between the *normal speed of propagation of the wave with respect to the material in its reference configuration,*

$$\underline{\Omega} = \frac{d\mathbf{M}}{dt} \cdot \underline{\mathbf{n}},$$ (11.1.3)

and the *normal speed of displacement of the wave,*

$$W = \frac{d\mathbf{M}}{dt} \cdot \mathbf{n}.$$ (11.1.4)

Here $\underline{\mathbf{n}} = \underline{\mathbf{n}}(t)$ and $\mathbf{n} = \mathbf{n}(t)$ are the unit vectors that indicate the local direction of propagation of the surface in the reference configuration and current configuration, respectively. Both $\underline{\Omega}$ and W may vary in time and space, and $\underline{\Omega}$ depends clearly on the choice of the reference configuration.

In an updated Lagrangian analysis, the current configuration is taken as reference. Then, with $\mathbf{v}$ the particle velocity,

$$W = \Omega + \mathbf{v} \cdot \mathbf{n},$$ (11.1.5)

where $\Omega = d\mathbf{M}/dt \cdot \mathbf{n}$ is the *normal speed of propagation of the wave with respect to the material in its current configuration.* Since the normal speed W is an absolute entity, its discontinuity vanishes,

$$[W](\mathbf{x}, t) = [\Omega + \mathbf{v} \cdot \mathbf{n}](\mathbf{x}, t) = 0.$$ (11.1.6)

From now on, to simplify the notation, the space and time dependences of jumps will be indicated only if necessary to the understanding.

11.1.3 Hadamard's compatibility relations in the current configuration

Hadamard's compatibility relations[11.1] are derived in an updated Lagrangian formalism, where the current configuration is taken as reference.

Let $\phi(\mathbf{x}, t)$ be a scalar function which is *continuous across the wave front* that propagates at speed Ω in the direction $\mathbf{n}(\mathbf{x}, t)$, and differentiable on both sides of the front. However, some of its derivatives suffer a discontinuity across the wave front.

Hadamard's compatibility relations indicate that there exists a scalar η such that

$$\left[\!\!\left[\frac{\partial \phi}{\partial \mathbf{x}} \right]\!\!\right] = \eta\, \mathbf{n}, \quad \left[\!\!\left[\frac{d\phi}{dt} \right]\!\!\right] = -\Omega\, \eta,$$ (11.1.7)

and therefore the discontinuities of the space and time derivatives of a continuous function obey the relation,

$$\left[\!\!\left[\frac{\partial \phi}{\partial \mathbf{x}} \right]\!\!\right] = -\frac{1}{\Omega} \left[\!\!\left[\frac{d\phi}{dt} \right]\!\!\right] \mathbf{n}.$$ (11.1.8)

These relations extend to a *continuous* vector or tensor $\phi(\mathbf{x}, t)$,

$$\left[\!\!\left[\frac{\partial \boldsymbol{\phi}}{\partial \mathbf{x}} \right]\!\!\right] = \boldsymbol{\eta} \otimes \mathbf{n}, \quad \left[\!\!\left[\frac{d\boldsymbol{\phi}}{dt} \right]\!\!\right] = -\Omega\, \boldsymbol{\eta}, \quad \left[\!\!\left[\frac{\partial \boldsymbol{\phi}}{\partial \mathbf{x}} \right]\!\!\right] = -\frac{1}{\Omega} \left[\!\!\left[\frac{d\boldsymbol{\phi}}{dt} \right]\!\!\right] \otimes \mathbf{n}.$$ (11.1.9)

Indeed, let us consider the differential increments of $\phi(\mathbf{x}, t)$ on both sides of the singular surface,

$$d\phi_{\pm} = \frac{\partial \phi}{\partial \mathbf{x}_{\pm}} \cdot d\mathbf{x} + \frac{d\phi}{dt}_{\pm} dt.$$ (11.1.10)

For points along the singular surface $\mathbf{n} \cdot d\mathbf{x} - \Omega\, dt = 0$, the difference between the two differentials in front of and behind the singular surface yields, since ϕ is continuous,

$$\begin{aligned}
[\![d\phi]\!] = 0 &= \left[\!\!\left[\frac{\partial \phi}{\partial \mathbf{x}} \right]\!\!\right] \cdot d\mathbf{x} + \left[\!\!\left[\frac{d\phi}{dt} \right]\!\!\right] dt \\
&= \left(\left[\!\!\left[\frac{\partial \phi}{\partial \mathbf{x}} \right]\!\!\right] + \frac{1}{\Omega} \left[\!\!\left[\frac{d\phi}{dt} \right]\!\!\right] \mathbf{n} \right) \cdot d\mathbf{x},
\end{aligned}$$ (11.1.11)

which holds for any $d\mathbf{x}$, whence the relations (11.1.8) □.

[11.1] Named after the French mathematician Jacques Salomon Hadamard (1865-1963).

The Eulerian version of (11.1.7) applied to the continuous scalar function $\psi(\mathbf{x}, t)$,

$$\left[\!\!\left[\frac{\partial \psi}{\partial \mathbf{x}}\right]\!\!\right] = \eta\, \mathbf{n}\,, \quad \left[\!\!\left[\frac{\partial \psi}{\partial t}\right]\!\!\right] = -W\, \eta\,, \tag{11.1.12}$$

deduces similarly by considering points on the current position of the singular surface such that $\mathbf{n} \cdot d\mathbf{x} - W\, dt = 0$. In fact, for $\phi(\underline{\mathbf{x}}, t) = \psi(\mathbf{x}, t)$, then $\partial \phi(\underline{\mathbf{x}}, t)/\partial t = d\psi(\mathbf{x}, t)/dt$ with $d\psi(\mathbf{x}, t)/dt$ the material derivative, namely $d\psi(\mathbf{x}, t)/dt = \partial \psi(\mathbf{x}, t)/\partial t + \partial \psi(\mathbf{x}, t)/\partial \mathbf{x} \cdot \mathbf{v}$. Therefore in an updated Lagrangian formalism, (11.1.12) implies (11.1.7), namely $[\![d\psi(\mathbf{x}, t)/dt]\!] = -\Omega\, \eta$.

Note that relations similar to Hadamard's compatibility relations hold also for phase waves. Indeed, let $\psi(\mathbf{x}, t) = \psi(\mathbf{x} \cdot \mathbf{n} - W\, t)$. Then $\partial \psi/\partial \mathbf{x} = \psi' \mathbf{n}$ and $\partial \psi/\partial t = -W\, \psi'$, with ψ' denoting the derivative of ψ with respect to its argument.

The existence of 3 or more kinematical constraints on the displacement vector may prevent the existence of propagating discontinuities: indeed, these constraints spill over the 3-component vector $\boldsymbol{\eta}$.

11.1.4   Conservation laws in a solid domain with a crossing discontinuity

Consider a singular surface $\mathcal{S}$ moving through a solid of volume V with absolute normal speed W and with a speed Ω with respect to the material in its current configuration. Material points move with velocity $\mathbf{v}$, in such way that $W = \Omega + \mathbf{v} \cdot \mathbf{n}$. As indicated in Fig. 11.1.3, V_- denotes the part of the volume behind the singular surface, V_+ denotes the part of the volume ahead of the singular surface.

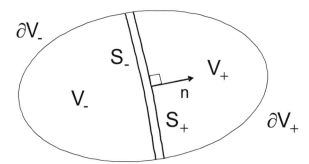

Fig. 11.1.3: A singular surface moving across a body V in the direction $\mathbf{n}$.

The generalized Reynolds theorem (2.7.5) can be applied separately to the volumes V_- and V_+ to any field $\phi = \phi(\mathbf{x}, t)$ which is differentiable on each side of the surface. Let $dS = \mathbf{n} \cdot d\mathbf{S}$ be the increment of area over the singular surface. For the volume behind the singular surface,

$$\begin{aligned}
\frac{d}{dt} \int_{V_-} \phi\, dV &= \int_{V_-} \frac{\partial \phi}{\partial t}\, dV + \int_{\partial V_- \cup S_-} \phi\, \mathbf{v} \cdot d\mathbf{S} + \int_{\mathcal{S}} \phi\, \Omega\, dS \\
&= \int_{V_-} \frac{\partial \phi}{\partial t} + \operatorname{div}(\phi\, \mathbf{v})\, dV + \int_{\mathcal{S}} \phi\, \Omega\, dS\,,
\end{aligned} \tag{11.1.13}$$

while changing the algebraic incremental area dS to $-dS$ for the volume in front of the singular surface,

$$\frac{d}{dt} \int_{V_+} \phi\, dV = \int_{V_+} \frac{\partial \phi}{\partial t} + \operatorname{div}(\phi\, \mathbf{v})\, dV - \int_{\mathcal{S}} \phi\, \Omega\, dS\,. \tag{11.1.14}$$

The integral over the whole volume $V = V_- \cup V_+$, obtained by summation of the two above partial integrals,

$$\frac{d}{dt} \int_{V} \phi\, dV = \int_{V} \frac{\partial \phi}{\partial t} + \operatorname{div}(\phi\, \mathbf{v})\, dV - \int_{\mathcal{S}} [\![\phi\, \Omega]\!]\, dS\,, \tag{11.1.15}$$

highlights the contribution of the singular surface.

We now explore in a thermodynamically closed solid the consequences of this result for the balances of mass, momentum, energy and entropy.

11.1.4.1 Conservation of mass across a strong discontinuity

Let ϕ be equal to the mass density ρ. The volume being assumed to be closed, global conservation of mass implies the left-hand side of (11.1.15) to vanish. Local conservation of mass $\partial \rho / \partial t + \operatorname{div}(\rho \mathbf{v}) = 0$ implies the volume integral in the right-hand side of (11.1.15) to vanish as well. Therefore, across a strong discontinuity, $\rho \Omega$ is continuous,

$$[\![\rho \, \Omega]\!] = 0. \tag{11.1.16}$$

Note 11.1 on the contribution of mass supplies.

If the solid was open, a field (volume) mass supply $\hat{\rho}$ could be present, but it would not contribute to the mass balance across a first order surface, even if it was discontinuous, as long as the discontinuity is finite. Indeed, in a one-dimensional context, let v be the velocity. The contribution of a generic term ϕ to the mass balance $\partial \rho / \partial t + d(\rho \, v)/dx = \hat{\rho}$ is obtained by integration across the surface, namely $\lim_{dx \to 0} \phi \, dx$: (a) the term $\phi = d(\rho \, v)/dx$ yields an integral equal to $(\rho v)_+ - (\rho v)_-$; (b) the singular surface has moved an absolute distance $dx = W \, dt$ in a time interval dt. The term $\phi = \partial \rho / \partial t$ integrates to $\phi \, dx = \phi \, W \, dt = (\rho_- - \rho_+) \, W$; (c) finally $\lim_{dx \to 0} \hat{\rho} \, dx = 0$, whence the relation $[\![\rho \, \Omega]\!] = 0$.

11.1.4.2 Conservation of momentum across a strong discontinuity

Let ϕ be equal to the momentum $\rho \mathbf{v}$. Let us first observe that local conservation of mass simplifies (11.1.15) to,

$$\frac{d}{dt} \int_V \rho \, \mathbf{v} \, dV = \int_V \rho \, \frac{d\mathbf{v}}{dt} \, dV - \int_{\mathcal{S}} [\![(\rho \, \Omega) \, \mathbf{v}]\!] \, dS. \tag{11.1.17}$$

Next, the local conservation of momentum $\operatorname{div} \boldsymbol{\sigma} + \rho \, \mathbf{b} = \rho \, d\mathbf{v}/dt$, with $\mathbf{b}$ body force per unit mass, is integrated by parts separately over the volumes V_- and V_+. The sum of the two integrals,

$$\int_V \rho \, \frac{d\mathbf{v}}{dt} \, dV = \int_V \rho \, \mathbf{b} \, dV + \int_{\partial V} \boldsymbol{\sigma} \cdot d\mathbf{S} - \int_{\mathcal{S}} [\![\boldsymbol{\sigma}]\!] \cdot \mathbf{n} \, dS, \tag{11.1.18}$$

may be contrasted with the global conservation of momentum over the volume V as a whole, Eqn (4.5.4)

$$\frac{d}{dt} \int_V \rho \, \mathbf{v} \, dV = \int_V \rho \, \mathbf{b} \, dV + \int_{\partial V} \boldsymbol{\sigma} \cdot d\mathbf{S}. \tag{11.1.19}$$

The combination of these three equations provides the jump of the traction across a first order wave as a function of the velocity jump,

$$[\![\boldsymbol{\sigma}]\!] \cdot \mathbf{n} + (\rho \, \Omega) \, [\![\mathbf{v}]\!] = \mathbf{0}. \tag{11.1.20}$$

The continuity of $\rho \, \Omega$ across the wave front has been used to rewrite the second term on the left hand side of the above relation.

Note that the jump of $\partial \mathbf{u} / \partial \mathbf{x}$ is equal to the jump of the deformation gradient $\mathbf{F} = \mathbf{I} + \partial \mathbf{u} / \partial \mathbf{x}$. Then, as a consequence of (11.1.9) and (11.1.20), the jump of the traction and the jump of the deformation gradient obey the Rankine-Hugoniot relations[11.2],

$$[\![\boldsymbol{\sigma}]\!] \cdot \mathbf{n} = (\rho \, \Omega^2) \, [\![\mathbf{F}]\!] \cdot \mathbf{n}, \quad [\![\mathbf{F}]\!] = \frac{1}{\rho \, \Omega^2} \, [\![\boldsymbol{\sigma}]\!] \cdot \mathbf{n} \otimes \mathbf{n}. \tag{11.1.21}$$

Note 11.2 on a direct proof of the momentum balance across a first order surface.

The result (11.1.20) can also be obtained by the following qualitative argument. Consider a piece of material of unit area. During the time interval dt, the wave front has moved a distance $\Omega \, dt$ with respect to the matter, and it has changed the velocity $\mathbf{v}$ of a mass $\rho \, \Omega \, dt$, which was ahead at time t and is behind at time $t + dt$, to $\mathbf{v} - [\![\mathbf{v}]\!]$. The resulting acceleration of this mass is thus $-[\![\mathbf{v}]\!]/dt$. The force per unit area required to induce this acceleration is equal to the mass $\rho \, \Omega \, dt$ times the acceleration $-[\![\mathbf{v}]\!]/dt$, namely $-(\rho \, \Omega) \, [\![\mathbf{v}]\!]$.

Note 11.3 on the contribution of body forces and momentum supplies.

A body force does not contribute to the relation (11.1.20) even if it is discontinuous, as long as the discontinuity is finite. The rule would apply to field (volume) momentum supplies which could be present if the solid was open. Indeed, in a one-dimensional context, let σ be the stress and v the velocity. The contribution of a generic term ϕ to the momentum balance $d\sigma/dx + \rho \, b - \rho \, dv/dt = 0$ is obtained by integration across the surface, namely $\lim_{dx \to 0} \phi \, dx$. (a) the term $\phi = d\sigma/dx$ yields an integral equal to $\sigma_+ - \sigma_-$; (b) the singular surface has moved a relative distance $dx = \Omega \, dt$ with respect to the matter in a time interval dt. Let v_- be the velocity of a point in front of the wave, and v_+ the velocity of this same point after it has been crossed by the wave. Then the term $\phi = -\rho \, dv/dt$ integrates to $\phi \, dx = \phi \, \Omega \, dt = -\big((\rho \, \Omega \, v)_- - (\rho \, \Omega \, v)_+ \big)$, whence, using (11.1.16), the one-dimensional relation $[\![\sigma]\!] + (\rho \, \Omega) \, [\![v]\!] = 0$.

[11.2] Named after the Scot engineer and physicist William John Macquorn Rankine (1820-1872) and the French engineer Pierre Henri Hugoniot (1851-1887).

11.1.4.3   Conservation of energy across a strong discontinuity

Let ϕ be equal to the sum of the internal and kinetic energies, $\phi = \rho\,(u + \frac{1}{2}\mathbf{v}^2)$. Let us first observe that local conservation of mass simplifies (11.1.15) to,

$$\frac{d}{dt}\int_V \rho\,(u + \tfrac{1}{2}\mathbf{v}^2)\,dV \;=\; \int_V \rho\,\frac{d}{dt}\,(u + \tfrac{1}{2}\mathbf{v}^2)\,dV - \int_S [\![(\rho\,\Omega)\,(u + \tfrac{1}{2}\mathbf{v}^2)]\!]\,dS. \tag{11.1.22}$$

Next, the local conservation of energy $\rho\,du/dt - \boldsymbol{\sigma} : \boldsymbol{\nabla}\mathbf{v} - \Theta + \mathrm{div}\,\mathbf{Q} = 0$ is integrated by parts separately over the volumes V_- and V_+. The sum of the two integrals,

$$\int_V \rho\,\frac{du}{dt}\,dV \;=\; \int_V -\mathrm{div}\,\boldsymbol{\sigma}\cdot\mathbf{v} + \Theta\,dV + \int_{\partial V} (\mathbf{v}\cdot\boldsymbol{\sigma} - \mathbf{Q})\cdot d\mathbf{S} - \int_S [\![\mathbf{v}\cdot\boldsymbol{\sigma} - \mathbf{Q}]\!]\cdot\mathbf{n}\,dS, \tag{11.1.23}$$

may be transformed, with help of the local balance of momentum, to

$$\int_V \rho\,\frac{du}{dt}\,dV \;=\; \int_V \rho\,(\mathbf{b} - \frac{d\mathbf{v}}{dt})\cdot\mathbf{v} + \Theta\,dV + \int_{\partial V} (\mathbf{v}\cdot\boldsymbol{\sigma} - \mathbf{Q})\cdot d\mathbf{S} - \int_S [\![\mathbf{v}\cdot\boldsymbol{\sigma} - \mathbf{Q}]\!]\cdot\mathbf{n}\,dS. \tag{11.1.24}$$

Finally, global conservation of energy on the volume V as a whole (4.6.10) simplifies for a closed solid to,

$$\frac{d}{dt}\int_V \rho\,(u + \tfrac{1}{2}\mathbf{v}^2)\,dV \;=\; \int_V \rho\,\mathbf{b}\cdot\mathbf{v} + \Theta\,dV + \int_{\partial V} (\mathbf{v}\cdot\boldsymbol{\sigma} - \mathbf{Q})\cdot d\mathbf{S}. \tag{11.1.25}$$

Combining the three equations (11.1.22), (11.1.24) and (11.1.25) provides the jump of energy across a first order wave as a function of the jumps of mechanical and thermal fluxes,

$$(\rho\,\Omega)\,[\![u + \tfrac{1}{2}\mathbf{v}^2]\!] + [\![\mathbf{v}\cdot\boldsymbol{\sigma} - \mathbf{Q}]\!]\cdot\mathbf{n} = 0, \tag{11.1.26}$$

accounting for the continuity of $\rho\,\Omega$ across the wave front.

The displacement is continuous across a strong discontinuity. Thus the gradient of displacement might be discontinuous but it remains finite. As a consequence, we have

Proposition 11.1 on the continuity of temperature across a singular surface in a Fourier's heat conductor.

If the internal energy depends only on the gradient of displacement and temperature, and if the material is a conductor (at least in the direction of the propagation), then the temperature itself can not undergo a jump across the singular surface.

Therefore, for the temperature to undergo a jump across the singular surface, the thermal conductivity across the surface should vanish.

The proof of the proposition rests on the fact that all terms in the energy conservation (11.1.26) are bounded except possibly the heat exchange which is the ratio of the thermal conductivity times the jump of temperature over the width of the singular surface, which actually vanishes. Therefore either the temperature jump or the thermal conductivity along the normal to the singular surface should vanish.

11.1.4.4   Entropy inequality across a strong discontinuity

Let ϕ be equal to the entropy per unit volume, $\phi = \rho\,s$. The local conservation of mass simplifies (11.1.15) to,

$$\frac{d}{dt}\int_V \rho\,s\,dV \;=\; \int_V \rho\,\frac{ds}{dt}\,dV - \int_S [\![(\rho\,\Omega)\,s]\!]\,dS, \tag{11.1.27}$$

Next, the global entropy inequality on the volume V as a whole (4.7.14) is integrated by parts over the two subvolumes,

$$\begin{aligned}
\frac{d}{dt}\int_V \rho\,s\,dV \;&\geq\; \int_V \frac{\Theta}{T}\,dV - \int_{\partial V} \frac{\mathbf{Q}}{T}\cdot d\mathbf{S} \\
&\geq\; \int_V \frac{\Theta}{T} - \mathrm{div}\,\frac{\mathbf{Q}}{T}\,dV - \int_S [\![\frac{\mathbf{Q}}{T}]\!]\cdot\mathbf{n}\,dS.
\end{aligned} \tag{11.1.28}$$

Combining the two above relations yields the inequality,

$$\int_V \rho\,\frac{ds}{dt} - \frac{\Theta}{T} + \mathrm{div}\,\frac{\mathbf{Q}}{T}\,dV + \int_S [\![\frac{\mathbf{Q}}{T}]\!]\cdot\mathbf{n} - [\![(\rho\,\Omega)\,s]\!]\,dS \;\geq 0, \tag{11.1.29}$$

whence the inequality at any regular point in the volume $\hat{S}_{(i)} = \rho\,ds/dt - \Theta/T + \mathrm{div}\,(\mathbf{Q}/T) \geq 0$, and the inequality,

$$\hat{S}_{(i)\|} = [\![\frac{\mathbf{Q}}{T}]\!]\cdot\mathbf{n} - (\rho\,\Omega)\,[\![s]\!] \geq 0, \tag{11.1.30}$$

at a point on the singular surface. The thermal source Θ does not contribute as long as its jump is finite. The very definition of the discontinuity operator (11.1.1) matters to interpret this inequality. Indeed, $[\![a]\!]$ denotes the difference between the values of a ahead and behind the wave front. Therefore, under adiabatic conditions, the entropy increases through the wave front.

11.1.4.5 Incremental work on a singular surface

Let $\delta\mathbf{u}$ be a virtual displacement field which is continuous and differentiable on both sides of a surface $\mathcal{S}$ that separates the volume V in the two subdomains V_- and V_+ as displayed in Fig. 11.1.3. The actual velocity field $\mathbf{v}$ is in dynamic equilibrium with the stress field $\boldsymbol{\sigma}$ and the body force $\mathbf{b}$, and therefore,

$$\int_V \left(\operatorname{div} \boldsymbol{\sigma} + \rho \left(\mathbf{b} - \frac{d\mathbf{v}}{dt} \right) \right) \cdot \delta\mathbf{u}\, dV = 0\,. \tag{11.1.31}$$

Integration by parts over each subdomain highlights the contribution of the singular surface,

$$\int_V -\boldsymbol{\sigma} : \boldsymbol{\nabla}(\delta\mathbf{u}) + \rho\left(\mathbf{b} - \frac{d\mathbf{v}}{dt} \right) \cdot \delta\mathbf{u}\, dV + \int_{\partial V} \delta\mathbf{u} \cdot \boldsymbol{\sigma} \cdot d\mathbf{S} - \int_{\mathcal{S}} [\![\delta\mathbf{u} \cdot \boldsymbol{\sigma}]\!] \cdot \mathbf{n}\, dS = 0\,. \tag{11.1.32}$$

11.1.4.6 Lagrangian version of the discontinuity relations

The (totally) Lagrangian version of the Hadamard's compatibility relations for a scalar (11.1.7) and (11.1.8) can be cast in the format,

$$\left[\!\left[\frac{\partial \phi}{\partial \underline{\mathbf{x}}} \right]\!\right] = \underline{\eta}\, \underline{\mathbf{n}}\,, \quad \left[\!\left[\frac{d\phi}{dt} \right]\!\right] = -\underline{\Omega}\,\underline{\eta} \quad \Rightarrow \quad \left[\!\left[\frac{\partial \phi}{\partial \underline{\mathbf{x}}} \right]\!\right] = -\frac{1}{\underline{\Omega}} \left[\!\left[\frac{d\phi}{dt} \right]\!\right] \underline{\mathbf{n}}\,, \tag{11.1.33}$$

with $\underline{\Omega}$ defined as follows,

$$\frac{\Omega}{\underline{\Omega}} = |\underline{\mathbf{n}} \cdot \mathbf{F}^{-1}| = \frac{1}{|\mathbf{n} \cdot \mathbf{F}|}\,. \tag{11.1.34}$$

The relation (11.1.33) is obtained by observing that points along the singular surface obey the relation $\mathbf{n} \cdot d\mathbf{x} - \Omega\, dt = 0$ which, using Nanson's rule of change of the normal to a surface $\mathbf{n} = \underline{\mathbf{n}} \cdot \mathbf{F}^{-1}/|\underline{\mathbf{n}} \cdot \mathbf{F}^{-1}|$, transforms to $\underline{\mathbf{n}} \cdot d\underline{\mathbf{x}}/|\underline{\mathbf{n}} \cdot \mathbf{F}^{-1}| - \Omega\, dt = 0$.

Moreover, Nanson's rule of change of the surface $d\mathbf{S}$, Eqn (2.6.9), yielding $d\mathbf{S}_\pm = \det \mathbf{F}_\pm\, d\underline{\mathbf{S}} \cdot \mathbf{F}_\pm^{-1}$ implies that the area $d\mathbf{S}$ maps identically on both sides of the discontinuity even if the deformation gradient may be discontinuous, namely, Bowen [1989], p. 91,

$$[\![(\det \mathbf{F})\, \underline{\mathbf{n}} \cdot \mathbf{F}^{-1}]\!] = \mathbf{0}, \quad [\![d\mathbf{S}]\!] = \mathbf{0}, \quad [\![\mathbf{n}]\!] = \mathbf{0}\,. \tag{11.1.35}$$

The proof relies on the fact that the discontinuity is dyadic, as indicated by (11.1.33) applied to a vector, say $\mathbf{F}_+ = \mathbf{F}_- + \underline{\boldsymbol{\eta}} \otimes \underline{\mathbf{n}}$, and on Sherman-Morrison formula (2.3.20) and (2.3.21). As a consequence of (11.1.34), $\rho\,\Omega = \rho\,\underline{\Omega}\,(\det \mathbf{F})\,|\underline{\mathbf{n}} \cdot \mathbf{F}^{-1}|$ and therefore, as a consequence of $(11.1.35)_1$, $[\![\rho\,\underline{\Omega}]\!] = 0$ $\Box$.

The extension of the relation (11.1.33) to a vector or a tensor is immediate, for example, for the deformation gradient,

$$[\![\mathbf{F}]\!] = \underline{\boldsymbol{\eta}} \otimes \underline{\mathbf{n}}, \quad [\![\mathbf{v}]\!] = -\underline{\Omega}\,\underline{\boldsymbol{\eta}}, \quad [\![\mathbf{F}]\!] = -\frac{1}{\underline{\Omega}}\, [\![\mathbf{v}]\!] \otimes \underline{\mathbf{n}}\,. \tag{11.1.36}$$

The three other discontinuity relations deduce similarly from their updated Lagrangian versions,

$$\begin{aligned} \text{balance of momentum} \quad & [\![\underline{\boldsymbol{\tau}}]\!] \cdot \underline{\mathbf{n}} + (\rho\,\underline{\Omega})\, [\![\mathbf{v}]\!] = \mathbf{0}\,, \\[4pt] \text{balance of energy} \quad & (\rho\,\underline{\Omega})\, [\![u + \tfrac{1}{2}\,\mathbf{v}^2]\!] + [\![\mathbf{v} \cdot \underline{\boldsymbol{\tau}} - \underline{\mathbf{Q}}]\!] \cdot \underline{\mathbf{n}} = 0\,, \\[4pt] \text{entropy inequality} \quad & [\![\underline{\mathbf{Q}}/T]\!] \cdot \underline{\mathbf{n}} - (\rho\,\underline{\Omega})\, [\![s]\!] \geq 0\,. \end{aligned} \tag{11.1.37}$$

Here $\underline{\boldsymbol{\tau}} = \det \mathbf{F}\, \boldsymbol{\sigma} \cdot \mathbf{F}^{-\mathrm{T}}$ is the 1st Piola-Kirchhoff stress, and the heat flux $\underline{\mathbf{Q}} = \det \mathbf{F}\, \mathbf{F}^{-1} \cdot \mathbf{Q}$ is transported from the current configuration as indicated by (2.6.13). The updated Lagrange and (totally) Lagrange versions of the Rankine-Hugoniot relation (11.1.21) compare as,

$$\rho\,\Omega^2 = \frac{\mathbf{n} \cdot [\![\boldsymbol{\sigma}]\!] \cdot \mathbf{n}}{\mathbf{n} \cdot [\![\mathbf{F}]\!] \cdot \mathbf{n}}, \quad \rho\,\underline{\Omega}^2 = \frac{\underline{\mathbf{n}} \cdot [\![\underline{\boldsymbol{\tau}}]\!] \cdot \underline{\mathbf{n}}}{\underline{\mathbf{n}} \cdot [\![\mathbf{F}]\!] \cdot \underline{\mathbf{n}}}\,. \tag{11.1.38}$$

11.1.4.7   Discontinuity relations of order 1 for a solid: summary

We start by listing the kinematic relation linking the wave-speeds (11.1.5), the three equations of balance across the strong discontinuity (11.1.16), (11.1.20), (11.1.26), and the entropy constraint (11.1.30), namely

$$
\begin{aligned}
&\text{kinematic relation} && W = \Omega + \mathbf{v} \cdot \mathbf{n}, \quad [\![W]\!] = 0, \\
&\text{balance of mass} && [\![\rho\,\Omega]\!] = 0, \\
&\text{balance of momentum} && [\![\boldsymbol{\sigma}]\!] \cdot \mathbf{n} + (\rho\,\Omega)\,[\![\mathbf{v}]\!] = \mathbf{0}, \\
&\text{balance of energy} && (\rho\,\Omega)\,[\![u + \tfrac{1}{2}\,\mathbf{v}^2]\!] + [\![\mathbf{v} \cdot \boldsymbol{\sigma} - \mathbf{Q}]\!] \cdot \mathbf{n} = 0, \\
&\text{entropy inequality} && [\![\mathbf{Q}/T]\!] \cdot \mathbf{n} - (\rho\,\Omega)\,[\![s]\!] \geq 0.
\end{aligned}
\tag{11.1.39}
$$

Here, $\mathbf{n}$ indicates the direction of the wave displacement, W is the normal speed of displacement of the wave, Ω is the normal speed of propagation of the wave relative to the material in its current configuration, ρ the mass density, T the temperature, $\mathbf{v}$ the particle velocity, u and s the internal energy and the entropy per unit mass respectively. Under adiabatic conditions, the heat flux $\mathbf{Q}$ vanishes.

For definiteness, the domains ahead and behind the wave are denoted by 1 and 2, respectively. With help of (11.1.2) and (11.1.39)$_3$, the jump of internal internal energy (11.1.39)$_4$ may be expressed in the format,

$$
(\rho\,\Omega)\,[\![u]\!] + \tfrac{1}{2}\,[\![\mathbf{v}]\!] \cdot (\boldsymbol{\sigma}_1 + \boldsymbol{\sigma}_2) \cdot \mathbf{n} - [\![\mathbf{Q}]\!] \cdot \mathbf{n} = 0,
\tag{11.1.40}
$$

an expression that can be sharpened for an ideal fluid. In fact, when the velocity is decomposed into a contribution along the wave direction $\mathbf{n}$ and a contribution along the orthogonal direction $\mathbf{t}$, then,

$$
(\rho\,\Omega)\,[\![u]\!] + \tfrac{1}{2}\Big((\rho\,\Omega)\,(w_2 - w_1)\,\mathbf{n} + ([\![\mathbf{v}]\!] \cdot \mathbf{t})\,\mathbf{t}\Big) \cdot (\boldsymbol{\sigma}_1 + \boldsymbol{\sigma}_2) \cdot \mathbf{n} - [\![\mathbf{Q}]\!] \cdot \mathbf{n} = 0,
\tag{11.1.41}
$$

where $w = 1/\rho$ is the specific volume. Since $\underline{\boldsymbol{\tau}} : [\![\mathbf{F}]\!] = -\underline{\Omega}^{-1}\,[\![\mathbf{v}]\!] \cdot \underline{\boldsymbol{\tau}} \cdot \mathbf{n}$, the Lagrangian version of the relation (11.1.40) has the form,

$$
[\![u]\!] - \frac{1}{2\,\rho}\,(\underline{\boldsymbol{\tau}}_1 + \underline{\boldsymbol{\tau}}_2) : [\![\mathbf{F}]\!] - \frac{[\![\mathbf{Q}]\!] \cdot \mathbf{n}}{\rho\,\underline{\Omega}} = 0.
\tag{11.1.42}
$$

Given a state 1 ahead of the wave, the states 2 that satisfy the relations (11.1.40) or (11.1.42) are said to lie on 'the Hugoniot'.

11.1.4.8   Adiabatic Hugoniot curve for an ideal compressible fluid

For an ideal barotropic fluid, the stress $\boldsymbol{\sigma} = -p\,\mathbf{I}$ is isotropic, with $p = p(\rho, T)$ the fluid pressure. Consequently, the particle velocity jump is aligned with the propagation direction, namely $[\![\mathbf{v}]\!] = -[\![\Omega]\!]\,\mathbf{n}$. If the working axes are taken to be attached to the wave, namely $W = 0$, then $\mathbf{v} \cdot \mathbf{n} = -\Omega$, and in fact $\mathbf{v} = -\Omega\,\mathbf{n}$. The four entities, namely pressure p, mass density ρ, normal speed Ω and particle velocity $\mathbf{v}$, are thus subjected to the jump relations,

$$
\begin{aligned}
&\text{balance of mass} && [\![\rho\,\Omega]\!] = 0, \\
&\text{balance of momentum} && [\![\mathbf{v} + \Omega\,\mathbf{n}]\!] = \mathbf{0}, \quad [\![p + \rho\,\Omega^2]\!] = 0, \\
&\text{balance of energy} && [\![h + \tfrac{1}{2}\,\Omega^2]\!] = 0, \\
&\text{entropy inequality} && [\![s]\!] \leq 0.
\end{aligned}
\tag{11.1.43}
$$

where $h = u + p/\rho$ is the enthalpy per unit mass.

The slope of the 'Rayleigh line' that links the points 1 and 2 is obtained by some manipulations as,

$$
-(\rho\,\Omega)^2 = \frac{p_2 - p_1}{w_2 - w_1} = 2\,\frac{h_2 - h_1}{w_2^2 - w_1^2}.
\tag{11.1.44}
$$

The first equality in (11.1.44) involves only mass and momentum balances, and is in fact a particular expression of the Rankine-Hugoniot relation (11.1.38) for a fluid. The changes of enthalpy and internal energy across the wave deduce from (11.1.44),

$$h_2 - h_1 = \frac{1}{2}\left(p_2 - p_1\right)\left(w_1 + w_2\right), \quad u_2 - u_1 = \frac{1}{2}\left(p_2 + p_1\right)\left(w_1 - w_2\right). \tag{11.1.45}$$

Since $dh = w\,dp + T\,ds$, then as a further consequence of (11.1.44),

$$\int_1^2 T\,ds = \frac{1}{2}\left(p_2 - p_1\right)\left(w_2 + w_1\right) - \int_1^2 w\,dp = \frac{1}{12}\,w_1''\,\left(p_2 - p_1\right)^3 + O\big((p_2 - p_1)^4\big). \tag{11.1.46}$$

The estimation in the right-hand side is obtained through a Taylor expansion of the specific volume $w = w(p) = w_1 + w_1'\,(p - p_1) + w_1''\,(p - p_1)^2/2 + \cdots$ around the pressure p_1. Now, the constitutive relation $w = w(p)$ is assumed to be convex with respect to the origin and it is characterized by the inequalities,

$$0 > \frac{dw}{dp} = -\frac{1}{\rho^2}\frac{d\rho}{dp}, \quad 0 < \frac{d^2w}{dp^2} = \frac{2}{\rho^3}\Big(\frac{d\rho}{dp}\Big)^2 - \frac{1}{\rho^2}\frac{d^2\rho}{dp^2}\,. \tag{11.1.47}$$

The entropy inequality (11.1.43) requires the jump $[\![s]\!] = s_1 - s_2$ to be negative, that is, the entropy *increases* across the wave, $s_2 > s_1$, and in fact the relation applies between any slice of the shock, so that $ds \geq 0$. Therefore, the integral (11.1.46) should be positive, which implies that

only compression waves ($p_2 > p_1$, $\rho_2 > \rho_1$) propagate in a stable manner in a barotropic fluid.

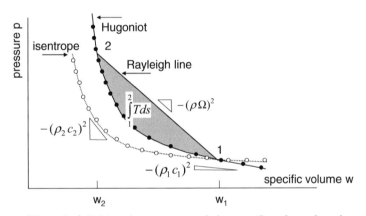

Fig. 11.1.4: Typical relation (Hugoniot) linking the pressure and the specific volume for a barotropic fluid, and Rayleigh line. c speed of sound, Ω wave-speed relative to the material. The grey area delimited by the Rayleigh line and the Hugoniot is equal to $\int_1^2 T\,ds$. Under a compression path where the specific volume decreases from w_1, the Hugoniot is above the isentrope. Under a release where the specific volume increases from w_2, the Hugoniot is below the isentrope.

Let c be the isentropic speed of sound in the fluid, namely $c^2 = dp/d\rho$. The slopes of the curve $p = p(w)$ at points 1 and 2 and of the chord between the two points (the Rayleigh line) in Fig. 11.1.4 may be shown to range in the following order,

$$\frac{dp}{dw}(p_2) = -(\rho\,c)_2^2 < \frac{p_2 - p_1}{w_2 - w_1} = -(\rho\,\Omega)^2 < \frac{dp}{dw}(p_1) = -(\rho\,c)_1^2. \tag{11.1.48}$$

Hence, since $\rho\,\Omega$ is continuous, the speed of propagation of the wave is supersonic ahead of the wave, i.e. faster than the sound speed, $\Omega_1 > c_1$, and subsonic behind the wave, i.e., slower than the sound speed, $\Omega_2 < c_2$.

For a perfect gas, the change of enthalpy h across the wave can be expressed as,

$$h_2 - h_1 = c_{\mathrm{p}}\,(T_2 - T_1) = \frac{\gamma}{\gamma - 1}\,(p_2\,w_2 - p_1\,w_1), \tag{11.1.49}$$

with c_{p} the heat capacity at constant pressure, and $\gamma > 1$ the adiabatic coefficient, see Section 3.3.2. Insertion of this relation in (11.1.44) yields,

$$\frac{w_2}{w_1} = \frac{(\gamma - 1)\,p_2 + (\gamma + 1)\,p_1}{(\gamma + 1)\,p_2 + (\gamma - 1)\,p_1}, \quad \frac{p_2}{p_1} = \frac{(\gamma + 1)\,w_1 - (\gamma - 1)\,w_2}{(\gamma + 1)\,w_2 - (\gamma - 1)\,w_1}\,. \tag{11.1.50}$$

A Taylor expansion shows that the Hugoniot $p = p_{\mathrm{H}}(w)$ and the isentrope $p = p_{\mathrm{I}}(w)$ differ at 3rd order in the neighborhood $w_2 = w_1 + dw$ of the reference common point (w_1, p_1),

$$p_{\mathrm{H}} - p_{\mathrm{I}} = \frac{1}{12} \gamma (\gamma^2 - 1) \left(\frac{dw}{w_1} \right)^3. \tag{11.1.51}$$

The Hugoniot is above the isentrope for specific volumes lower (pressures higher) than the reference.

11.1.4.9   Nonadiabatic shock wave

If the propagation is not adiabatic, the energy conservation across the wave front,

$$\text{balance of energy} \quad [\![h + \tfrac{1}{2} \Omega^2]\!] + q_n = 0, \tag{11.1.52}$$

involves the normal heat flux released by the shock per unit mass $q_n = -[\![\mathbf{Q}]\!] \cdot \mathbf{n}/(\rho \, \Omega)$ [unit : J/kg].

The energy output of explosives and fuel oil/air burn ranges from 4 to 5 MJ/kg. Detonation through an explosive involves the propagation of a shock which transforms the unreacted explosive (in a solid form) in front of the wave to gas products (behind the wave). What distinguishes explosives from other materials (say otherwise, explosions from other exothermic reactions) is the rate at which this energy is released, namely the quantity of interest is $[\![\mathbf{Q}]\!] \cdot \mathbf{n}$ [unit : W/m^2] rather than q_n. In fact, for typical explosives, the speed of the shock reaches 6000 to 8000 m/s and the released power is as high as 10^{13} W/m^2 while it is only 10^5 W/m^2 for fuel oil/air burn.

11.1.5   Conservation laws in a mixture with a crossing discontinuity

Across a first order wave front, the equations of balance of mass, momentum, energy and entropy for a saturated porous medium are deduced from the relations for a solid exposed in Sections 11.1.4.1 to 11.1.4.4. As indicated in Section 11.1.4, field (volume) supplies do not contribute to the discontinuity relations as long as their discontinuity (if any) is finite. Still, for each constituent k, additional surface supplies *along* the discontinuity may exist, namely

- a supply of mass $\hat{\rho}_{\parallel}^k$;

- a supply of momentum $\hat{\boldsymbol{\pi}}_{\parallel}^k$;

- a supply of internal energy $\hat{\mathcal{U}}_{\parallel}^k$;

- a supply of entropy $\hat{\mathcal{S}}_{\parallel}^k$.

If the mixture is a closed system, these supplies represent actually internal exchanges among constituents, and they are subjected to closure relations. The supplied mass of the constituent k comes with a velocity $\tilde{\mathbf{v}}_k$, an internal energy $\tilde{u}_k$, and an entropy $\tilde{s}_k$ which are not necessarily equal to their existing counterparts $\mathbf{v}_k$, u_k, and s_k at the point of the discontinuity line where they are supplied. These quantities should be specified.

Hadamard's compatibility relations (11.1.9) applied to a *continuous* vector or tensor $\boldsymbol{\phi}_k(\mathbf{x}, t)$ transform to,

$$\left[\!\!\left[\frac{\partial \boldsymbol{\phi}_k}{\partial \mathbf{x}} \right]\!\!\right] = \boldsymbol{\eta}_k \otimes \mathbf{n}, \quad \left[\!\!\left[\frac{d^k \boldsymbol{\phi}_k}{dt} \right]\!\!\right] = -\Omega_k \, \boldsymbol{\eta}_k, \quad \left[\!\!\left[\frac{\partial \boldsymbol{\phi}_k}{\partial \mathbf{x}} \right]\!\!\right] = -\frac{1}{\Omega_k} \left[\!\!\left[\frac{d^k \boldsymbol{\phi}_k}{dt} \right]\!\!\right] \otimes \mathbf{n}. \tag{11.1.53}$$

In the above relations and in the balance equations listed below for each constituent, the speed of propagation of the discontinuity relative to constituent k is denoted by $\Omega_k = W - \mathbf{v}_k \cdot \mathbf{n}$, W being the absolute speed of displacement of the discontinuity.

The conservation relations are listed below for each constituent:

○ Balance of mass for constituent k

$$[\![\rho^k \, \Omega_k]\!] + \hat{\rho}_{\parallel}^k = 0, \tag{11.1.54}$$

subjected to the closure relation

$$\sum_k \hat{\rho}_{\parallel}^k = 0. \tag{11.1.55}$$

○ Balance of momentum for constituent k

$$[\![\boldsymbol{\sigma}^k]\!] \cdot \mathbf{n} + (\rho^k \, \Omega_k) \, [\![\mathbf{v}_k]\!] + \hat{\rho}_{\parallel}^k \, \tilde{\mathbf{v}}_k + \hat{\boldsymbol{\pi}}_{\parallel}^k = \mathbf{0}, \tag{11.1.56}$$

subjected to the closure relation

$$\sum_k \hat{\rho}_\parallel^k \, \tilde{\mathbf{v}}_k + \hat{\boldsymbol{\pi}}_\parallel^k = \mathbf{0} \,. \tag{11.1.57}$$

○ Balance of energy for constituent k

$$[\![(\rho^k \, \Omega_k) \, (u_k + \tfrac{1}{2} \, \mathbf{v}_k^2)]\!] + [\![\mathbf{v}_k \cdot \boldsymbol{\sigma}^k - \mathbf{Q}]\!] \cdot \mathbf{n} + \hat{\rho}_\parallel^k \, (\tilde{u}_k + \tfrac{1}{2} \, \tilde{\mathbf{v}}_k^2) + \hat{\boldsymbol{\pi}}_\parallel^k \cdot \mathbf{v}_k + \hat{\mathcal{U}}_\parallel^k = 0 \,, \tag{11.1.58}$$

subjected to the closure relation,

$$\sum_k \hat{\rho}_\parallel^k \, (\tilde{u}_k + \tfrac{1}{2} \, \tilde{\mathbf{v}}_k^2) + \hat{\boldsymbol{\pi}}_\parallel^k \cdot \mathbf{v}_k + \hat{\mathcal{U}}_\parallel^k = 0 \,. \tag{11.1.59}$$

○ Balance of entropy for constituent k

$$\hat{S}_{(i)\parallel}^k = [\![\frac{\mathbf{Q}}{T}]\!] \cdot \mathbf{n} - [\![(\rho^k \, \Omega_k) \, s_k]\!] - \hat{\rho}_\parallel^k \, \tilde{s}_k - \hat{\mathcal{S}}_\parallel^k \geq 0 \,, \tag{11.1.60}$$

subjected to the closure relation

$$\sum_k \hat{\rho}_\parallel^k \, \tilde{s}_k + \hat{\mathcal{S}}_\parallel^k = 0 \,. \tag{11.1.61}$$

Eringen and Ingram [1965] provide in addition the conservation equations at the mixture level.

If the mixture is open, then the right hand sides of the closure relations do not vanish and they are contributed by the surroundings.

11.1.6 Propagation of discontinuities in a solid domain

A singular surface moves through the medium, and carries a certain discontinuity that evolves pointwise in time and space. The topic is examined with care in, e.g., Truesdell and Toupin [1960] and Bowen [1976], from where the brief summary below is extracted.

11.1.6.1 Normal speed of displacement

A moving surface is defined by a sufficiently smooth function $f(\mathbf{x}, t)$ and a constant α, say $f(\mathbf{x}, t) = \alpha$. If $\boldsymbol{\nabla} f \neq \mathbf{0}$, the unit normal $\mathbf{n} = \mathbf{n}(\mathbf{x}, t)$ and the normal speed of displacement $W = W(\mathbf{x}, t)$ are defined via the space gradient and time derivative of the function f, namely

$$\mathbf{n}(\mathbf{x}, t) = \frac{\boldsymbol{\nabla} f(\mathbf{x}, t)}{|\boldsymbol{\nabla} f(\mathbf{x}, t)|}, \quad W(\mathbf{x}, t) = -\frac{1}{|\boldsymbol{\nabla} f(\mathbf{x}, t)|} \frac{\partial f}{\partial t}(\mathbf{x}, t) \,. \tag{11.1.62}$$

If both $\partial f / \partial t \neq 0$ and $\boldsymbol{\nabla} f \neq \mathbf{0}$, the normal speed W is non zero and bounded, and it can be assumed to be positive.

11.1.6.2 Displacement derivative

If $\partial f / \partial t \neq 0$ at $(\mathbf{x}, t)$, then the relation $f(\mathbf{x}, t) = \alpha$ may be inverted to $t = t^*(\mathbf{x}, \alpha)$. Now let $\phi(\mathbf{x}, t)$ be an arbitrary function of space and time, and let

$$\phi^*(\mathbf{x}, \alpha) = \phi(\mathbf{x}, t^*) \,. \tag{11.1.63}$$

Then, since $f(\mathbf{x}, t) = \alpha$,

$$\begin{aligned} \frac{\partial \phi}{\partial t}(\mathbf{x}, t) &= \frac{\partial \phi^*}{\partial \alpha}(\mathbf{x}, \alpha) \, \frac{\partial f}{\partial t}(\mathbf{x}, t) \,, \\ \boldsymbol{\nabla} \phi(\mathbf{x}, t) &= \boldsymbol{\nabla} \phi^*(\mathbf{x}, \alpha) + \frac{\partial \phi^*}{\partial \alpha}(\mathbf{x}, \alpha) \, \boldsymbol{\nabla} f(\mathbf{x}, t) \,. \end{aligned} \tag{11.1.64}$$

Combining the two relations together with the definitions (11.1.62) yields the gradient of $\phi^*(\mathbf{x}, \alpha)$,

$$\begin{aligned} \boldsymbol{\nabla} \phi^*(\mathbf{x}, \alpha) &= \boldsymbol{\nabla} \phi(\mathbf{x}, t) + \frac{\partial \phi}{\partial t}(\mathbf{x}, t) \, \frac{\mathbf{n}}{W} \,, \\ \boldsymbol{\nabla} \phi^*(\mathbf{x}, \alpha) \cdot (W \, \mathbf{n}) &= \boldsymbol{\nabla} \phi(\mathbf{x}, t) \cdot (W \, \mathbf{n}) + \frac{\partial \phi}{\partial t}(\mathbf{x}, t) \,. \end{aligned} \tag{11.1.65}$$

The *displacement derivative* is defined as the time derivative following a surface α, namely Truesdell and Toupin [1960], Eqn (179.8),

$$\begin{aligned} \frac{\delta \phi}{\delta t}(\mathbf{x}, t) &= \boldsymbol{\nabla} \phi^*(\mathbf{x}, \alpha) \cdot (W \, \mathbf{n}) \\ &= \frac{\partial \phi}{\partial t}(\mathbf{x}, t) + \boldsymbol{\nabla} \phi(\mathbf{x}, t) \cdot (W \, \mathbf{n}) \,. \end{aligned} \tag{11.1.66}$$

11.1.6.3   Gradient of the jump and jump of the gradient

For any field $\phi(\mathbf{x}, t)$ which is differentiable on each side of the discontinuity, the gradient of the jump is equal to the jump of the gradient,

$$(\boldsymbol{\nabla}[\![\phi]\!])(\mathbf{x}, t) = [\![\boldsymbol{\nabla}\phi]\!](\mathbf{x}, t). \tag{11.1.67}$$

Indeed, let $\mathbf{x}_1$ and $\mathbf{x}_2$ be two points on the discontinuity, and let the domain behind and ahead of the front be denoted by $-$ and $+$, respectively. Then

$$
\begin{aligned}
& (\phi(\mathbf{x}_2^+, t) - \phi(\mathbf{x}_2^-, t)) - (\phi(\mathbf{x}_1^+, t) - \phi(\mathbf{x}_1^-, t)) \\
& = \quad (\phi(\mathbf{x}_2^+, t) - \phi(\mathbf{x}_1^+, t)) - (\phi(\mathbf{x}_2^-, t) - \phi(\mathbf{x}_1^-, t)) \; \square.
\end{aligned}
\tag{11.1.68}
$$

11.1.6.4   Kinematical condition of compatibility

Application of the property (11.1.67) to $\phi^*(\mathbf{x}, \alpha)$, with the gradient $\boldsymbol{\nabla}\phi^*(\mathbf{x}, \alpha)$ defined by (11.1.65) yields,

$$
\begin{aligned}
\boldsymbol{\nabla}[\![\phi^*]\!] &= [\![\boldsymbol{\nabla}\phi^*]\!], \\
&= [\![\boldsymbol{\nabla}\phi]\!] + \left[\!\!\left[\frac{\partial\phi}{\partial t}\right]\!\!\right]\frac{\mathbf{n}}{W}.
\end{aligned}
\tag{11.1.69}
$$

From (11.1.66) and (11.1.67),

$$\frac{\delta[\![\phi]\!]}{\delta t} = \boldsymbol{\nabla}[\![\phi^*]\!] \cdot (W\,\mathbf{n}) = [\![\boldsymbol{\nabla}\phi^*]\!] \cdot (W\,\mathbf{n}) = \left[\!\!\left[\frac{\delta\phi}{\delta t}\right]\!\!\right], \tag{11.1.70}$$

results the *kinematical condition of compatibility*, Truesdell and Toupin [1960], Eqn (180.2),

$$\frac{\delta[\![\phi]\!]}{\delta t} = \left[\!\!\left[\frac{\partial\phi}{\partial t}\right]\!\!\right] + [\![\boldsymbol{\nabla}\phi]\!] \cdot (W\,\mathbf{n}). \tag{11.1.71}$$

11.1.6.5   Iterated kinematical compatibility equation for a continuous field

If ϕ is continuous across the wave front, Hadamard's compatibility relations are recovered from (11.1.69) and (11.1.71), namely

$$[\![\boldsymbol{\nabla}\phi]\!] + \left[\!\!\left[\frac{\partial\phi}{\partial t}\right]\!\!\right]\frac{\mathbf{n}}{W} = \mathbf{0}, \qquad \left[\!\!\left[\frac{\partial\phi}{\partial t}\right]\!\!\right] + [\![\boldsymbol{\nabla}\phi]\!] \cdot (W\,\mathbf{n}) = 0. \tag{11.1.72}$$

If the wave front travels at constant speed W in a fixed direction $\mathbf{n}$, the *iterated kinematical compatibility equation* applied to a continuous field $\phi(\mathbf{x}, t)$ reads, Truesdell and Toupin [1960], Eqn (181.8),

$$2\frac{\delta}{\delta t}\left[\!\!\left[\frac{\partial\phi}{\partial t}\right]\!\!\right] = \left[\!\!\left[\frac{\partial^2\phi}{\partial t^2}\right]\!\!\right] - W^2\,\mathbf{n} \cdot [\![\boldsymbol{\nabla}^2\phi]\!] \cdot \mathbf{n}. \tag{11.1.73}$$

The proof starts from (11.1.71),

$$
\begin{aligned}
\frac{\delta}{\delta t}\left[\!\!\left[\frac{\partial\phi}{\partial t}\right]\!\!\right] &= \left[\!\!\left[\frac{\partial^2\phi}{\partial t^2}\right]\!\!\right] + \left[\!\!\left[\boldsymbol{\nabla}\!\left(\frac{\partial\phi}{\partial t}\right)\right]\!\!\right] \cdot (W\,\mathbf{n}) \\
\frac{\delta[\![\boldsymbol{\nabla}\phi]\!]}{\delta t} &= \left[\!\!\left[\frac{\partial}{\partial t}(\boldsymbol{\nabla}\phi)\right]\!\!\right] + [\![\boldsymbol{\nabla}^2\phi]\!] \cdot (W\,\mathbf{n}) \\
\frac{\delta}{\delta t}\left[\!\!\left[\frac{\partial\phi}{\partial t}\right]\!\!\right] - \frac{\delta[\![\boldsymbol{\nabla}\phi]\!]}{\delta t} \cdot (W\,\mathbf{n}) &= \left[\!\!\left[\frac{\partial^2\phi}{\partial t^2}\right]\!\!\right] - W^2\,\mathbf{n} \cdot [\![\boldsymbol{\nabla}^2\phi]\!] \cdot \mathbf{n}.
\end{aligned}
\tag{11.1.74}
$$

The last step of the proof consists in using Hadamard's compatibility relation $(11.1.72)_2$ which applies since ϕ is continuous and to take its displacement derivative at constant W and $\mathbf{n}$.

11.1.6.6 Induced discontinuity for a continuous field

The displacement $\mathbf{u}$, its time derivative $\mathbf{v} = \partial\mathbf{u}/\partial t$ and space derivative $\boldsymbol{\nabla}\mathbf{u}$ remain continuous across the front of an acceleration wave. Repetitive applications of Hadamard's compatibility relations to the velocity and displacement gradient, namely

$$[\![\mathbf{v}]\!] = \mathbf{0}; \qquad [\![\boldsymbol{\nabla}\mathbf{u}]\!] = \mathbf{0};$$

$$[\![\boldsymbol{\nabla}\mathbf{v}]\!] = -\left[\!\!\left[\frac{\partial\mathbf{v}}{\partial t}\right]\!\!\right] \otimes \frac{\mathbf{n}}{W},$$

$$[\![\boldsymbol{\nabla}(\boldsymbol{\nabla}\mathbf{u})]\!] = -\left[\!\!\left[\frac{\partial(\boldsymbol{\nabla}\mathbf{u})}{\partial t}\right]\!\!\right] \otimes \frac{\mathbf{n}}{W} = -\left[\!\!\left[\boldsymbol{\nabla}\left(\frac{\partial\mathbf{u}}{\partial t}\right)\right]\!\!\right] \otimes \frac{\mathbf{n}}{W} = \left[\!\!\left[\frac{\partial^2\mathbf{u}}{\partial t^2}\right]\!\!\right] \otimes \frac{\mathbf{n}}{W} \otimes \frac{\mathbf{n}}{W},$$

(11.1.75)

lead to the definition of the induced discontinuity,

$$[\![\boldsymbol{\nabla}^2\mathbf{u}]\!] = \left[\!\!\left[\frac{\partial^2\mathbf{u}}{\partial t^2}\right]\!\!\right] \otimes \frac{\mathbf{n}}{W} \otimes \frac{\mathbf{n}}{W}, \qquad \left[\!\!\left[\frac{\partial^2 u_k}{\partial x_i \partial x_j}\right]\!\!\right] = \left[\!\!\left[\frac{\partial^2 u_k}{\partial t^2}\right]\!\!\right] \frac{n_i}{W} \frac{n_j}{W},$$

(11.1.76)

and to the relations,

$$\left[\!\!\left[\frac{\partial^2\mathbf{u}}{\partial t^2}\right]\!\!\right] = W^2\, \mathbf{n} \cdot [\![\boldsymbol{\nabla}^2\mathbf{u}]\!] \cdot \mathbf{n},$$

$$[\![\boldsymbol{\nabla}\mathrm{div}\,\mathbf{u}]\!] = \left(\left[\!\!\left[\frac{\partial^2\mathbf{u}}{\partial t^2}\right]\!\!\right] \cdot \frac{\mathbf{n}}{W}\right) \frac{\mathbf{n}}{W},$$

(11.1.77)

$$[\![\mathrm{div}\,\boldsymbol{\nabla}\mathbf{u}]\!] = \left[\!\!\left[\frac{\partial^2\mathbf{u}}{\partial t^2}\right]\!\!\right] \frac{1}{W^2}.$$

11.2 Waves in elastic solids

In a linearized analysis, no difference is made between material and partial derivatives, $d(\ )/dt$ and $\partial(\ )/\partial t$ respectively, and they will be denoted collectively by a superimposed dot $(\dot{\ })$. As a consequence, the normal speed of displacement W, the speed of propagation relative to the material Ω and the wave-speed c are not distinguished either, and they will be denoted collectively by c.

11.2.1 Jump relations for acceleration waves

Consider a second order, or acceleration, wave front with normal $\mathbf{n}(\mathbf{x}, t)$ which moves with speed c relative to the material. At any point $\mathbf{x}$ and at any time t, the velocity $\mathbf{v}(\mathbf{x}, t)$, the stress $\boldsymbol{\sigma}(\mathbf{x}, t)$, and the mass density $\rho(\mathbf{x}, t)$ remain continuous across the wave front. Therefore, Hadamard's compatibility relations apply to these entities.

11.2.1.1 Hadamard's compatibility relations for second order discontinuities

Hadamard's compatibility relations imply that the jump $[\![\partial\mathbf{v}/\partial\mathbf{x}]\!]$ of the velocity gradient and the jump $[\![\dot{\mathbf{v}}]\!]$ of the acceleration across the wave front are not independent: there exists a discontinuity vector $\boldsymbol{\eta} = \boldsymbol{\eta}(\mathbf{x}, t)$ which can be viewed as the jump of the velocity gradient along the normal $\mathbf{n}$, i.e., $\boldsymbol{\eta} = [\![\partial\mathbf{v}/\partial(\mathbf{x} \cdot \mathbf{n})]\!]$, such that

$$\left[\!\!\left[\frac{\partial\mathbf{v}}{\partial\mathbf{x}}\right]\!\!\right] = \boldsymbol{\eta} \otimes \mathbf{n}, \quad [\![\dot{\mathbf{v}}]\!] = -c\,\boldsymbol{\eta}.$$

(11.2.1)

Similarly with $\boldsymbol{\zeta} = [\![\partial\boldsymbol{\sigma}/\partial(\mathbf{x} \cdot \mathbf{n})]\!]$, the normal jump of the stress gradient satisfies the relations,

$$\left[\!\!\left[\frac{\partial\boldsymbol{\sigma}}{\partial\mathbf{x}}\right]\!\!\right] = \boldsymbol{\zeta} \otimes \mathbf{n}, \quad [\![\dot{\boldsymbol{\sigma}}]\!] = -c\,\boldsymbol{\zeta}.$$

(11.2.2)

Consequently, if the body force per unit mass $\mathbf{b}$ is continuous, taking the jump of the balance of linear momentum,

$$\mathrm{div}\,\boldsymbol{\sigma} + \rho\,(\mathbf{b} - \dot{\mathbf{v}}) = \mathbf{0},$$

(11.2.3)

yields $\mathbf{n} \cdot \boldsymbol{\zeta} = -\rho\,c\,\boldsymbol{\eta}$, that is, the jump of the traction rate is aligned with the normal jump of the velocity gradient,

$$\mathbf{n} \cdot [\![\dot{\boldsymbol{\sigma}}]\!] = \rho\,c^2\,\boldsymbol{\eta}.$$

(11.2.4)

11.2.1.2   Acoustic tensor, wave-speeds and eigenmodes

Let us consider a material whose constitutive equation in rate form writes

$$\dot{\boldsymbol{\sigma}} = \mathbb{E} : \dot{\boldsymbol{\epsilon}} \,. \tag{11.2.5}$$

Insertion of the rate constitutive equations (11.2.5) in the jump relation (11.2.4) yields an eigenvalue problem for the jump $\boldsymbol{\eta}$ that features the wave-speed c:

$$[\mathbf{A}(\mathbf{n}) - c^2 \, \mathbf{I}] \cdot \boldsymbol{\eta} = \mathbf{0} \,. \tag{11.2.6}$$

Here $\mathbf{A} = \mathbf{A}(\mathbf{n})$ denotes the acoustic tensor[11.3],

$$\mathbf{A}(\mathbf{n}) = \mathbf{n} \cdot \frac{\mathbb{E}}{\rho} \cdot \mathbf{n} \,, \quad A_{ij} = \frac{1}{\rho} \, n_k \, \mathbb{E}_{kilj} \, n_l \,. \tag{11.2.7}$$

11.2.2   Acceleration waves in linear isotropic materials

If the rate tensor moduli $\mathbb{E}$ in (11.2.5) is constant, the material response is linear. The acoustic tensor $\mathbf{A}(\mathbf{n})$ associated with linear isotropic elasticity takes the form,

$$\mathbf{A}(\mathbf{n}) = (c_{\mathrm{L}})^2 \, \mathbf{n} \otimes \mathbf{n} + (c_{\mathrm{S}})^2 \, (\mathbf{I} - \mathbf{n} \otimes \mathbf{n}) \,, \tag{11.2.8}$$

which shows explicitly that there exists a single longitudinal wave with eigenmode $\boldsymbol{\eta} \sim \mathbf{n}$ and with speed c_{L}, and two shear waves with $\boldsymbol{\eta} \sim \mathbf{n}^\perp$ with common speed c_{S}, where

$$c_{\mathrm{L}}^2 = \frac{\lambda + 2\,\mu}{\rho} > c_{\mathrm{S}}^2 = \frac{\mu}{\rho} > 0 \,. \tag{11.2.9}$$

The above inequalities are consequences of the positive definiteness of the elastic tensor $\mathbb{E}$, Eqn (5.2.30). As far as terminology is concerned, the longitudinal wave is also termed dilatational, and the shear wave is also called rotational or distortional.

Setting $\boldsymbol{\epsilon} = \frac{1}{2}(\mathbf{m} \otimes \mathbf{n} + \mathbf{n} \otimes \mathbf{m})$ in $\boldsymbol{\epsilon} : \mathbb{E} : \boldsymbol{\epsilon}$ shows that positive definiteness of the elastic tensor implies positive definiteness of the elastic acoustic tensor, a property referred to as *strong ellipticity* (SE), and the wave-speeds are real positive. This property holds independently of the material symmetries. For isotropic elasticity,

$$(\mathrm{SE}) \ \text{of} \ \mathbb{E} \ \Leftrightarrow \ \mathbf{m} \cdot \mathbf{A}(\mathbf{n}) \cdot \mathbf{m} = c_{\mathrm{S}}^2 + \left(c_{\mathrm{L}}^2 - c_{\mathrm{S}}^2\right) (\mathbf{m} \cdot \mathbf{n})^2 > 0 \,, \ \forall \mathbf{m} \text{ and } \mathbf{n} \text{ unit vectors} \,. \tag{11.2.10}$$

Since $\mathbb{E}$ is endowed with the major symmetry, its strong ellipticity is equivalent to its *ellipticity* (E), or, in other words, the positive definiteness of the elastic acoustic tensor implies its invertibility:

$$(\mathrm{E}) \ \text{of} \ \mathbb{E} \ \Leftrightarrow \ \det \mathbf{A}(\mathbf{n}) = c_{\mathrm{L}}^2 \, (c_{\mathrm{S}}^2)^2 \neq 0 \,, \ \forall \mathbf{n} \text{ unit vector} \,. \tag{11.2.11}$$

Note that the elastic acoustic tensor (11.2.8) is, to within a multiplicative term, a rank one modification of the identity, as defined in Section 2.3.5. Therefore it remains form-invariant when raised to any power $a \in \mathbb{R}$,

$$\left(\mathbf{A}(\mathbf{n})\right)^a = (c_{\mathrm{L}}^2)^a \, \mathbf{n} \otimes \mathbf{n} + (c_{\mathrm{S}}^2)^a \, (\mathbf{I} - \mathbf{n} \otimes \mathbf{n}) \,. \tag{11.2.12}$$

11.2.3   Acoustic tensor of a linear elastic orthotropic material

With the stiffness matrix in component form (5.2.39), the acoustic tensor (11.2.7) expresses componentwise in the orthotropy axes $\mathbf{e}_a$,

$$\rho \, A_{jl} = \frac{\lambda_{ab}}{2} \, n_a \, n_b \, (e_{aj} \, e_{bl} + e_{bj} \, e_{al}) + \frac{\mu_a}{2} \, (n_a^2 \, I_{jl} + n_a \, (n_j \, e_{al} + n_l \, e_{aj}) + e_{aj} \, e_{al}) \,, \tag{11.2.13}$$

or in matrix form, via the notation (5.2.41),

$$\rho \, \mathbf{A} = \begin{bmatrix} c_{11} \, n_1^2 + c_{66} \, n_2^2 + c_{55} \, n_3^2 & (c_{12} + c_{66}) \, n_1 \, n_2 & (c_{13} + c_{55}) \, n_1 \, n_3 \\ (c_{12} + c_{66}) \, n_1 \, n_2 & c_{22} \, n_2^2 + c_{66} \, n_1^2 + c_{44} \, n_3^2 & (c_{23} + c_{44}) \, n_2 \, n_3 \\ (c_{13} + c_{55}) \, n_1 \, n_3 & (c_{23} + c_{44}) \, n_2 \, n_3 & c_{55} \, n_1^2 + c_{44} \, n_2^2 + c_{33} \, n_3^2 \end{bmatrix} \,. \tag{11.2.14}$$

[11.3] At first glance, the acoustic tensor does not look like the product of a fourth order tensor by a second order tensor. Now consider the tensor $\rho \, \tilde{A}_{ij} = \tilde{E}_{ijkl} \, n_k \, n_l$ with $\tilde{E}_{ijkl} = \frac{1}{2} \, (E_{ikjl} + E_{iljk})$. Then $\rho \, \tilde{\mathbf{A}} = \tilde{\mathbb{E}} : (\mathbf{n} \otimes \mathbf{n})$ and $\mathbf{A} = \tilde{\mathbf{A}}$, Norris [2006].

11.2.4   Acoustic tensor of a linear transversely isotropic material

The elastic material displays transverse isotropy around the axis $\mathbf{e}_3$, and the stiffness $\mathbb{E}$ is assumed to be positive definite. With the notation (5.2.51), the acoustic tensor (11.2.14) specializes to

$$
\rho\,\mathbf{A} = \begin{bmatrix} c_{11}\,n_1^2 + \tfrac{1}{2}(c_{11}-c_{12})\,n_2^2 + c_{44}\,n_3^2 & \tfrac{1}{2}(c_{11}+c_{12})\,n_1\,n_2 & (c_{13}+c_{44})\,n_1\,n_3 \\[2mm] \tfrac{1}{2}(c_{11}+c_{12})\,n_1\,n_2 & c_{11}\,n_2^2 + \tfrac{1}{2}(c_{11}-c_{12})\,n_1^2 + c_{44}\,n_3^2 & (c_{13}+c_{44})\,n_2\,n_3 \\[2mm] (c_{13}+c_{44})\,n_1\,n_3 & (c_{13}+c_{44})\,n_2\,n_3 & c_{44}\,(n_1^2+n_2^2) + c_{33}\,n_3^2 \end{bmatrix} . \tag{11.2.15}
$$

Alternatively, one may start from the expression of $\mathbb{E}$ given in (5.E.27) in terms of the coefficients a's. The positive definiteness conditions (5.E.31) where the coefficients c's are defined by (5.E.28) restrict the coefficients a's. The acoustic tensor can be cast into the tensor form,

$$
\rho\,\mathbf{A} = \alpha_1\,\mathbf{I} + \alpha_2\,\mathbf{n}\otimes\mathbf{n} + \alpha_3\,(\mathbf{e}_3\otimes\mathbf{n} + \mathbf{n}\otimes\mathbf{e}_3) + \alpha_4\,\mathbf{e}_3\otimes\mathbf{e}_3 \tag{11.2.16}
$$

which involves the four scalars α_i, $i\in[1,4]$:

$$
\alpha_1 = \frac{a_2}{2} + \frac{a_5}{2}(\mathbf{e}_3\cdot\mathbf{n})^2 > 0, \quad \alpha_2 = a_1 + \frac{a_2}{2}, \quad \alpha_3 = \left(a_3 + \frac{a_5}{2}\right)(\mathbf{e}_3\cdot\mathbf{n}), \quad \alpha_4 = \frac{a_5}{2} + a_4(\mathbf{e}_3\cdot\mathbf{n})^2 . \tag{11.2.17}
$$

The sign of α_1 is implied by the positive definiteness of the elastic stiffness. The determinant of the acoustic tensor can be factorized as $\det(\rho\,\mathbf{A}) = \alpha_1\,\Delta$, and thus

$$
\Delta = (\alpha_1 + \alpha_4 + \alpha_3\,(\mathbf{e}_3\cdot\mathbf{n}))\,(\alpha_1 + \alpha_2 + \alpha_3\,(\mathbf{e}_3\cdot\mathbf{n})) - (\alpha_3 + \alpha_4\,(\mathbf{e}_3\cdot\mathbf{n}))\,(\alpha_3 + \alpha_2\,(\mathbf{e}_3\cdot\mathbf{n})) > 0 , \tag{11.2.18}
$$

is strictly positive as well for all directions $\mathbf{n}$.

The inverse of the elastic acoustic tensor takes a form similar to that of the acoustic tensor itself:

$$
\det(\rho\,\mathbf{A})\,(\rho\,\mathbf{A})^{-1} = \beta_1\,\mathbf{I} + \beta_2\,\mathbf{n}\otimes\mathbf{n} + \beta_3\,(\mathbf{e}_3\otimes\mathbf{n} + \mathbf{n}\otimes\mathbf{e}_3) + \beta_4\,\mathbf{e}_3\otimes\mathbf{e}_3 \tag{11.2.19}
$$

with

$$
\beta_1 = \Delta, \quad \beta_2 = -\alpha_2\,(\alpha_1 + \alpha_4) + \alpha_3^2,
$$
$$
\beta_3 = -\alpha_1\,\alpha_3 + (\alpha_2\,\alpha_4 - \alpha_3^2)\,(\mathbf{e}\cdot\mathbf{n}), \quad \beta_4 = -\alpha_4\,(\alpha_1 + \alpha_2) + \alpha_3^2. \tag{11.2.20}
$$

11.2.5   Acceleration waves in materials with fabric anisotropy

11.2.5.1   Materials with fabric anisotropy: the eigenvalue problem

Materials with fabric anisotropy defined by a symmetric, positive definite, second order tensor $\mathbf{B}$ are described in Section 5.2.11. The acoustic tensor $\mathbf{A}(\mathbf{n})$ associated with the elastic tensor $\mathbb{E}$, Eqn (5.2.72), takes the form,

$$
\mathbf{A}(\mathbf{n}) = \frac{\lambda+\mu}{\rho}\,\mathbf{B}\cdot\mathbf{n}\otimes\mathbf{B}\cdot\mathbf{n} + \frac{\mu}{\rho}\,(\mathbf{n}\cdot\mathbf{B}\cdot\mathbf{n})\,\mathbf{B} . \tag{11.2.21}
$$

For $\mathbf{B} = \mathbf{I}$, $\mathbf{A}(\mathbf{n})$ boils down to the isotropic acoustic tensor $\mathbf{A}^{\mathrm{iso}}(\mathbf{n})$, associated with the isotropic tensor moduli $\mathbb{E}^{\mathrm{iso}}$. It is given by (11.2.8), and can be recast in the form,

$$
\mathbf{A}^{\mathrm{iso}}(\mathbf{n}) = \frac{\lambda+\mu}{\rho}\,\mathbf{n}\otimes\mathbf{n} + \frac{\mu}{\rho}\,\mathbf{I} . \tag{11.2.22}
$$

Since the elastic tensors $\mathbb{E}^{\mathrm{iso}}$ and $\mathbb{E}$ are assumed to be positive definite, they are strongly elliptic and therefore the associated acoustic tensors are positive definite. Thus, they are invertible, indeed

$$
\det\mathbf{A}(\mathbf{n}) = \det\mathbf{B}\,(\mathbf{n}\cdot\mathbf{B}\cdot\mathbf{n})^3\,(\frac{\mu}{\rho})^2\,\frac{\lambda+2\mu}{\rho} \tag{11.2.23}
$$

and

$$
\mathbf{A}^{-1}(\mathbf{n}) = -\frac{\rho}{\mu}\frac{\lambda+\mu}{\lambda+2\mu}\frac{\mathbf{n}\otimes\mathbf{n}}{(\mathbf{n}\cdot\mathbf{B}\cdot\mathbf{n})^2} + \frac{\rho}{\mu}\frac{\mathbf{B}^{-1}}{\mathbf{n}\cdot\mathbf{B}\cdot\mathbf{n}} . \tag{11.2.24}
$$

Let us consider an acceleration wave that propagates through the elastic material in a direction $\mathbf{n}$. The propagation is governed by the eigenvalue problem, Eqn (11.2.6),

$$[\mathbf{A}(\mathbf{n}) - c^2\,\mathbf{I}] \cdot \boldsymbol{\eta} = \mathbf{0}, \tag{11.2.25}$$

where c is the normal wave-speed and $\boldsymbol{\eta}$ the mode of propagation.

It is instrumental to introduce the modified entities,

$$\tilde{\mathbf{n}} = \frac{\mathbf{B}^{\frac{1}{2}} \cdot \mathbf{n}}{\parallel \mathbf{B}^{\frac{1}{2}}\,\mathbf{n} \parallel}, \quad \tilde{\boldsymbol{\eta}} = \mathbf{B}^{\frac{1}{2}} \cdot \boldsymbol{\eta}. \tag{11.2.26}$$

Let

$$\mathbf{B} = \sum_{i=1}^{3} b_i\,\mathbf{B}_i, \quad \text{with } \mathbf{B}_i \equiv \mathbf{b}_i \otimes \mathbf{b}_i, \tag{11.2.27}$$

be the spectral representation of the fabric tensor $\mathbf{B}$. Its square root $\mathbf{B}^{\frac{1}{2}}$ can be given an explicit expression,

$$\mathbf{B} = \mathbf{B}^{\frac{1}{2}} \cdot \mathbf{B}^{\frac{1}{2}}, \quad \mathbf{B}^{\frac{1}{2}} = \sum_{i=1}^{3} \sqrt{b_i}\,\mathbf{B}_i, \tag{11.2.28}$$

but an alternative expression has been proposed in Section 2.3.6.

The elastic acoustic tensor can then be recast in terms of its underlying isotropic counterpart,

$$\mathbf{A}(\mathbf{n}) = (\mathbf{n} \cdot \mathbf{B} \cdot \mathbf{n})\,\mathbf{B}^{\frac{1}{2}} \cdot \mathbf{A}^{\mathrm{iso}}(\tilde{\mathbf{n}}) \cdot \mathbf{B}^{\frac{1}{2}}. \tag{11.2.29}$$

The strong ellipticity property of the underlying isotropic tensor is inherited by the anisotropic tensor. Indeed, let $\mathbf{x} \neq \mathbf{0}$ be an arbitrary vector, and $\tilde{\mathbf{x}} = \mathbf{B}^{\frac{1}{2}} \cdot \mathbf{x}$. Then, from (11.2.29),

$$\mathbf{x} \cdot \mathbf{A}(\mathbf{n}) \cdot \mathbf{x} = (\mathbf{n} \cdot \mathbf{B} \cdot \mathbf{n})\,\tilde{\mathbf{x}} \cdot \mathbf{A}^{\mathrm{iso}}(\tilde{\mathbf{n}}) \cdot \tilde{\mathbf{x}} > 0. \tag{11.2.30}$$

With help of (11.2.29), the eigenvalue problem for the wave-speeds (11.2.25) can be transformed to a generalized eigenvalue problem relying on the *isotropic* elastic acoustic tensor, Bigoni and Loret [1999],

$$\left[\mathbf{A}^{\mathrm{iso}}(\tilde{\mathbf{n}}) - c^2 \frac{\mathbf{B}^{-1}}{\mathbf{n} \cdot \mathbf{B} \cdot \mathbf{n}} \right] \cdot \tilde{\boldsymbol{\eta}} = \mathbf{0}. \tag{11.2.31}$$

11.2.5.2 Materials with fabric anisotropy: wave-speeds

In the orthotropy axes, i.e. eigendirections of $\mathbf{B}$, the acoustic tensor (11.2.14) becomes,

$$\rho\,\mathbf{A} = \begin{bmatrix} (\lambda+\mu)\,b_1^2\,n_1^2 + (\mathbf{n}\cdot\mathbf{B}\cdot\mathbf{n})\,\mu\,b_1 & (\lambda+\mu)\,b_1\,b_2\,n_1\,n_2 & (\lambda+\mu)\,b_1\,b_3\,n_1\,n_3 \\ (\lambda+\mu)\,b_1\,b_2\,n_1\,n_2 & (\lambda+\mu)\,b_2^2\,n_2^2 + (\mathbf{n}\cdot\mathbf{B}\cdot\mathbf{n})\,\mu\,b_2 & (\lambda+\mu)\,b_2\,b_3\,n_2\,n_3 \\ (\lambda+\mu)\,b_1\,b_3\,n_1\,n_3 & (\lambda+\mu)\,b_2\,b_3\,n_2\,n_3 & (\lambda+\mu)\,b_3^2\,n_3^2 + (\mathbf{n}\cdot\mathbf{B}\cdot\mathbf{n})\,\mu\,b_3 \end{bmatrix}. \tag{11.2.32}$$

The squares of the wave-speeds are the roots of the cubic equation,

$$(c^2)^3 - a_1\,(c^2)^2 + a_2\,c^2 - a_3 = 0, \tag{11.2.33}$$

where the coefficients a are defined by the expressions,

$$\rho\,a_1 = (\lambda+\mu)\,(\mathbf{n}\cdot\mathbf{B}^2\cdot\mathbf{n}) + \mu\,(\mathbf{n}\cdot\mathbf{B}\cdot\mathbf{n})\,\mathrm{tr}\,\mathbf{B},$$

$$\rho^2\,a_2 = \mu\,(\mathbf{n}\cdot\mathbf{B}\cdot\mathbf{n})\,\big((\lambda+2\,\mu)\,(\mathbf{n}\cdot\mathbf{B}\cdot\mathbf{n})\,I_2(\mathbf{B}) - (\lambda+\mu)\,I_3(\mathbf{B})\big), \tag{11.2.34}$$

$$\rho^3\,a_3 = \mu^2\,(\mathbf{n}\cdot\mathbf{B}\cdot\mathbf{n})^3\,(\lambda+2\,\mu)\,I_3(\mathbf{B}).$$

The Cayley-Hamilton relation has been used to establish these expressions, namely

$$\mathbf{B}^3 - I_1(\mathbf{B})\,\mathbf{B}^2 + I_2(\mathbf{B})\,\mathbf{B} - I_3(\mathbf{B})\,\mathbf{I} = \mathbf{0}\,, \tag{11.2.35}$$

where $I_1(\mathbf{B}) = \operatorname{tr}\mathbf{B}$, $I_2(\mathbf{B}) = ((\operatorname{tr}\mathbf{B})^2 - \operatorname{tr}\mathbf{B}^2)/2$, and $I_3(\mathbf{B}) = \det\mathbf{B}$ are the three isotropic scalar invariants of $\mathbf{B}$, Section 2.2.

The eigenvectors $\boldsymbol{\eta}$ satisfy the relations,

$$
\begin{aligned}
\rho\,\mathbf{A}\cdot\boldsymbol{\eta} &= (\lambda+\mu)\,(\boldsymbol{\eta}\cdot\mathbf{B}\cdot\mathbf{n})\,\mathbf{B}\cdot\mathbf{n} + \mu\,(\mathbf{n}\cdot\mathbf{B}\cdot\mathbf{n})\,\mathbf{B}\cdot\boldsymbol{\eta} = \rho\,c^2\,\boldsymbol{\eta}\,,\\
\boldsymbol{\eta} &= (\lambda+\mu)\,(\boldsymbol{\eta}\cdot\mathbf{B}\cdot\mathbf{n})\,\Big(\rho\,c^2\,\mathbf{I} - \mu\,(\mathbf{n}\cdot\mathbf{B}\cdot\mathbf{n})\,\mathbf{B}\Big)^{-1}\cdot\mathbf{B}\cdot\mathbf{n}\,.
\end{aligned}
\tag{11.2.36}
$$

The tensor to be inverted can be expressed, once again via the Cayley-Hamilton relation, in terms of $\mathbf{I}$, $\mathbf{B}$, $\mathbf{B}^2$. Thus the eigenmode $\boldsymbol{\eta}$ can be expressed as $(\alpha\,\mathbf{I} + \beta\,\mathbf{B} + \gamma\,\mathbf{B}^2)\cdot\mathbf{n}$. The coefficients α, β, γ are obtained by re-insertion of the latter expression in $(11.2.36)_1$, yielding finally

$$\boldsymbol{\eta} \simeq \Big(I_3(\mathbf{B})\,(\mathbf{n}\cdot\mathbf{B}\cdot\mathbf{n})\,\frac{\mu}{\rho\,c^2}\,\mathbf{I} + (\frac{\rho\,c^2}{\mu}\,\frac{1}{(\mathbf{n}\cdot\mathbf{B}\cdot\mathbf{n})} - I_1(\mathbf{B}))\,\mathbf{B} + \mathbf{B}^2\Big)\cdot\mathbf{n}\,. \tag{11.2.37}$$

Observe, from $(11.2.36)_1$, that, unless $\mathbf{B} = \mathbf{I}$, there is no longitudinal wave in directions distinct from the orthotropy axes, since $\rho\,\mathbf{A}\cdot\mathbf{n} = (\lambda + 2\,\mu)\,(\mathbf{n}\cdot\mathbf{B}\cdot\mathbf{n})\,\mathbf{B}\cdot\mathbf{n}$.

In the orthotropy directions $\mathbf{b}_i$, $i \in [1,3]$, the analysis is elementary,

$$\mathbf{n} = \mathbf{b}_i,\ i \in [1,3]\ :\ \begin{cases} \boldsymbol{\eta} = \mathbf{b}_i, & \rho\,c^2 = (\lambda + 2\,\mu)\,b_i^2\ : & \text{longitudinal wave-speed,}\\[4pt] \boldsymbol{\eta} = \mathbf{b}_j,\ j \neq i, & \rho\,c^2 = \mu\,b_i\,b_j\ : & \text{shear wave-speed.} \end{cases} \tag{11.2.38}$$

11.2.6 Wave motions and Helmholtz potentials in isotropic elastic solids

This section addresses infinitesimal deformations of an elastic solid with Lamé moduli λ and μ and makes no distinction between the partial and total derivatives with respect to time which are both denoted by a superimposed dot. In presence of a body force per unit mass $\mathbf{b}$, the equations of balance of momentum express in terms of the displacement vector $\mathbf{u}$ in the format,

$$(\lambda+\mu)\,\boldsymbol{\nabla}\operatorname{div}\mathbf{u} + \mu\,\Delta\mathbf{u} + \rho\,\mathbf{b} = \rho\,\ddot{\mathbf{u}}\,, \tag{11.2.39}$$

where $\Delta = \boldsymbol{\nabla}\operatorname{div} - \boldsymbol{\nabla}\wedge(\boldsymbol{\nabla}\wedge)$ is the Laplacian operator.

As indicated in Section 2.4.1.1, the displacement vector can be defined in terms of two potentials, referred to as Helmholtz potentials, namely a scalar potential ϕ contributing an irrotational vector field, and a vector potential $\boldsymbol{\psi}$ contributing the solenoidal part,

$$\mathbf{u} = \boldsymbol{\nabla}\phi + \boldsymbol{\nabla}\wedge\boldsymbol{\psi},\quad u_k = \frac{\partial\phi}{\partial x_k} + e_{krs}\,\frac{\partial\psi_s}{\partial x_r}\,. \tag{11.2.40}$$

Since the displacement vector has three independent components, a restriction is imposed on the two potential functions, namely $\operatorname{div}\boldsymbol{\psi} = 0$.

With help of (2.4.4) and (2.4.5),

$$
\begin{aligned}
& e = \operatorname{div}\mathbf{u} = \Delta\phi,\quad \boldsymbol{\omega} = \frac{1}{2}\,\boldsymbol{\nabla}\wedge\mathbf{u} = -\frac{1}{2}\,\Delta\boldsymbol{\psi}\\[4pt]
& \Delta\mathbf{u} = (\boldsymbol{\nabla}\operatorname{div} - \boldsymbol{\nabla}\wedge\boldsymbol{\nabla}\wedge)\,\mathbf{u} = \boldsymbol{\nabla}\Delta\phi + \boldsymbol{\nabla}\wedge(\Delta\boldsymbol{\psi})\,.
\end{aligned}
\tag{11.2.41}
$$

In an infinitesimal strain context, e is interpreted as the volume change. We may also adopt the Helmholtz decomposition for the body force $\mathbf{b} = \boldsymbol{\nabla}b_{\mathrm{L}} + \boldsymbol{\nabla}\wedge\mathbf{b}_{\mathrm{s}}$ with $\operatorname{div}\mathbf{b}_{\mathrm{s}} = 0$, see Achenbach [1975] for justifications. Then the fact that the balance of momentum may be recast in terms of the potentials,

$$(\lambda+2\,\mu)\,\boldsymbol{\nabla}\Delta\phi + \mu\,\boldsymbol{\nabla}\wedge\Delta\boldsymbol{\psi} + \rho\,(\boldsymbol{\nabla}b_{\mathrm{L}} + \boldsymbol{\nabla}\wedge\mathbf{b}_{\mathrm{s}}) = \rho\,(\boldsymbol{\nabla}\ddot{\phi} + \boldsymbol{\nabla}\wedge\ddot{\boldsymbol{\psi}})\,, \tag{11.2.42}$$

motivates a split in a longitudinal (irrotational) part and in a rotational (solenoidal) part, namely

$$\ddot{\phi} - c_{\mathrm{L}}^2\,\Delta\phi = b_{\mathrm{L}}\,,\quad \ddot{\boldsymbol{\psi}} - c_{\mathrm{S}}^2\,\Delta\boldsymbol{\psi} = \mathbf{b}_{\mathrm{s}}\,. \tag{11.2.43}$$

Note that the above equations hold as well for e and $\boldsymbol{\omega}$,

$$\ddot{e} - c_{\mathrm{L}}^2\,\Delta\,e = \Delta\,b_{\mathrm{L}}\,,\quad \ddot{\boldsymbol{\omega}} - c_{\mathrm{S}}^2\,\Delta\boldsymbol{\omega} = -\frac{1}{2}\,\Delta\,\mathbf{b}_{\mathrm{s}}\,. \tag{11.2.44}$$

These relations indicate that the dilatational wave associated with the scalar potential ϕ propagates at speed c_{L} while the shear wave propagates at speed c_{S}. The split implies that the two motions are independent. In general however, the two potentials are coupled via boundary conditions.

11.3 Acceleration waves in thermoelastic solids

The state variables defining the thermoelastic solids considered are the strain and the temperature. Heat conductors and non conductors enjoy distinct, although parallel, dynamic properties which may be compared systematically.

11.3.1 Acceleration waves in solid conductors obeying Fourier's heat law

The thermal conduction obeys Fourier's heat law, namely $\mathbf{Q} = -\mathbf{k}_\mathrm{T} \cdot \boldsymbol{\nabla} T$, where the thermal conductivity $\mathbf{k}_\mathrm{T}$ is a strictly positive definite symmetric second order tensor.

According to Proposition 11.1, the temperature is continuous across a first order wave front propagating at speed $c \neq 0$ in a Fourier's heat conductor. The property holds as well across a second order wave front, by which we are concerned here. Consequently, the state variables are continuous, and so are the internal energy, entropy, stress and velocity. The jump of energy (11.1.26) reduces to,

$$[\![\mathbf{Q}]\!] \cdot \mathbf{n} = 0 \,. \tag{11.3.1}$$

Since the temperature is continuous, Hadamard's compatibility relations (11.1.7) apply and imply that the jump of its gradient is aligned with the normal to the wave front, namely $[\![\boldsymbol{\nabla} T]\!] = \eta\, \mathbf{n}$, and consequently $[\![\mathbf{Q}]\!] = -\eta\, \mathbf{k}_\mathrm{T} \cdot \mathbf{n}$. Therefore if the normal conductivity $\mathbf{k}_\mathrm{T} \cdot \mathbf{n}$ does not vanish, then $\eta = 0$, and, again due to Hadamard's compatibility relations, $[\![\dot{T}]\!] = -c\,\eta = 0$.

We have thus reached *Duhem's first theorem* for a solid conductor obeying Fourier's heat law: the temperature, its spatial and time derivatives remain continuous across a second order wave front,

$$[\![T]\!] = 0, \quad [\![\boldsymbol{\nabla} T]\!] = \mathbf{0}, \quad [\![\dot{T}]\!] = 0 \,. \tag{11.3.2}$$

Hadamard's compatibility relations apply to the gradient of temperature which is continuous, and there exists a vector $\boldsymbol{\eta}_T = \mathbf{n} \cdot [\![\boldsymbol{\nabla}(\boldsymbol{\nabla} T)]\!]$ which can be interpreted as the normal jump of the second gradient of the temperature such that

$$[\![\boldsymbol{\nabla}(\boldsymbol{\nabla} T)]\!] = \boldsymbol{\eta}_T \otimes \mathbf{n}, \quad [\![\overline{\dot{\boldsymbol{\nabla} T}}]\!] = -c\,\boldsymbol{\eta}_T \,. \tag{11.3.3}$$

Similarly, since $\dot{T}$ is continuous, there exists a scalar η_T such that,

$$[\![\boldsymbol{\nabla} \dot{T}]\!] = \eta_T\, \mathbf{n}, \quad [\![\ddot{T}]\!] = -c\,\eta_T \,. \tag{11.3.4}$$

The partial derivatives with respect to time and space commute, $\overline{\boldsymbol{\nabla} T} = \boldsymbol{\nabla} \dot{T}$, and therefore $-c\,\boldsymbol{\eta}_T = \eta_T\, \mathbf{n}$. Consequently,

$$[\![\boldsymbol{\nabla}(\boldsymbol{\nabla} T)]\!] = [\![\ddot{T}]\!]\, \frac{\mathbf{n}}{c} \otimes \frac{\mathbf{n}}{c} \,. \tag{11.3.5}$$

Note incidentally that a result of the same flavor is derived in Section 11.1.6.6.

As an example, let us consider the thermoelastic material defined by (9.1.5). The jumps of time and spatial derivatives of the strain and temperature gradient should satisfy the jump equations deriving from the balance of momentum and energy conservation. Let $\boldsymbol{\eta} = \mathbf{n} \cdot [\![\boldsymbol{\nabla} \mathbf{v}]\!]$ be the normal jump of the velocity gradient, so that $[\![\boldsymbol{\nabla} \mathbf{v}]\!] = \boldsymbol{\eta} \otimes \mathbf{n}$ and $[\![\dot{\mathbf{v}}]\!] = -c\,\boldsymbol{\eta}$. Since the jump of the time rate of temperature vanishes, the jump of the balance of momentum (11.2.4) keeps its isothermal format, and reduces to an eigenvalue problem for the normal jump $\boldsymbol{\eta}$, namely

$$(\mathbf{A} - c^2\, \mathbf{I}) \cdot \boldsymbol{\eta} = \mathbf{0} \,, \tag{11.3.6}$$

and the acoustic tensor $\mathbf{A} = \mathbf{A}^{\mathrm{isoth}} = \mathbf{n} \cdot (\rho^{-1}\, \mathbb{E}) \cdot \mathbf{n}$ is built from the isothermal tensor moduli $\mathbb{E}$.

The jump of the energy equation (9.1.25),

$$-T\,(\mathbf{c}_\mathrm{T} : \mathbb{E} \cdot \mathbf{n}) \cdot \frac{[\![\dot{\mathbf{v}}]\!]}{c} - \frac{[\![\ddot{T}]\!]}{c^2}\, \mathbf{n} \cdot \mathbf{k}_\mathrm{T} \cdot \mathbf{n} = 0 \,, \tag{11.3.7}$$

couples the mechanical and thermal jumps. The wave-speeds being known from the isothermal eigenvalue problem as well as the directions of the jump of acceleration, the above equation should be seen as providing the associated jumps of the second time derivative of temperature, given that $\mathbf{n} \cdot \mathbf{k}_\mathrm{T} \cdot \mathbf{n}$ is strictly positive. This jump next enters the spatial jump in (11.3.5).

The picture is different for a rigid conductor, or when the mechanical contribution can be neglected in the energy equation. Then only the second term survives in the energy equation,

$$-\frac{[\![\ddot{T}]\!]}{c^2}\, \mathbf{n} \cdot \mathbf{k}_\mathrm{T} \cdot \mathbf{n} = 0 \,. \tag{11.3.8}$$

For the jump not to vanish, the wave-speed is unboundedly large, namely $c \to \infty$, which is typical of a partial differential equation of parabolic type. This eigenmode is purely thermal, in the sense that the associated mechanical jump vanishes, that is $\boldsymbol{\eta} = \mathbf{0}$.

11.3.2 Acceleration waves in solid conductors obeying Cattaneo's heat law

The models briefly exposed in Section 9.1.4 aim at transforming the parabolic equation of heat energy to a hyperbolic equation, in which thermal and mechanical signals propagate at finite speeds.

Indeed, for a rigid conductor or in the uncoupled case, the jump of the energy equation associated with Cattaneo's heat conductor (9.1.38),

$$C_\epsilon \, t_{\text{Cat}} \, [\![\ddot{T}]\!] \; - \; \frac{[\![\ddot{T}]\!]}{c^2} \, \mathbf{n} \cdot \mathbf{k}_{\text{T}} \cdot \mathbf{n} = 0 \,, \tag{11.3.9}$$

delivers a finite wave-speed with square,

$$c^2 = \frac{1}{t_{\text{Cat}}} \, \mathbf{n} \cdot \frac{\mathbf{k}_{\text{T}}}{C_\epsilon} \cdot \mathbf{n} > 0 \,, \tag{11.3.10}$$

proportional to the thermal diffusivity $\mathbf{k}_{\text{T}}/C_\epsilon$ and inversely proportional to the relaxation time t_{Cat}.

11.3.3 Acceleration waves in nonconductor thermoelastic solids

For a nonconductor (actually, if the conductivity in the direction of propagation vanishes), the normal heat flux vanishes, namely $\mathbf{Q} \cdot \mathbf{n} = 0$. For thermoelastic solids, the mechanical entropy production $\hat{S}_{(\text{i},1)} = \rho \dot{s} - \Theta/T + (\text{div } \mathbf{Q})/T = 0$ vanishes. In absence of heat source, namely $\Theta = 0$, the vanishing of the jump of mechanical entropy production $[\![\hat{S}_{(\text{i},1)}]\!] = 0$ implies the rate of entropy to vanish,

$$[\![\dot{s}]\!] = 0 \,, \tag{11.3.11}$$

since the mass density is continuous across a second order wave front. This result is sometimes referred to as *Duhem's second theorem*.

As indicated by Proposition 11.1, the temperature might not be continuous in a non conductor. If it is, so will be the entropy of a thermoelastic solid. Then, according to Duhem's second theorem and Hadamard's compatibility relation, both the time rate and spatial gradient of the entropy vanish across a non stationary acceleration wave, namely $[\![\dot{s}]\!] = 0$, $[\![\boldsymbol{\nabla} s]\!] = \mathbf{0}$.

For the thermoelastic material defined by (9.1.5), in view of (11.3.11), the jump of the time rate of temperature expresses in terms of the jump of strain rate. Consequently the jump of the stress rate expresses in terms of the jump of the strain rate via the isentropic tensor moduli $\mathbb{E}^{\text{isen}}$ given by (9.1.7). The eigenvalue problem yielding the wave-speeds and mechanical eigenmode has the format (11.3.6) but the acoustic tensor $\mathbf{A} = \mathbf{A}^{\text{isen}} = \mathbf{n} \cdot (\rho^{-1} \, \mathbb{E}^{\text{isen}}) \cdot \mathbf{n}$ is built from the isentropic tensor moduli $\mathbb{E}^{\text{isen}}$. Accurate developments would distinguish a process at constant entropy per unit volume from a process at constant entropy per unit mass. The distinction is not addressed in this linearized analysis.

Isothermal and isentropic wave-speeds are shown to interlace in Section 11.5.4.2.

11.4 Acceleration waves and acoustic waves in fluids

11.4.1 Acceleration waves in a frictionless ideal fluid

For a frictionless ideal fluid, also referred to as barotropic fluid, the stress $\boldsymbol{\sigma}$,

$$\boldsymbol{\sigma} = -p \, \mathbf{I} \,, \tag{11.4.1}$$

expresses in terms of the thermodynamic pressure $p = p(\rho, T)$ which is a function of the mass density ρ if the fluid is compressible, and of temperature T if the fluid is thermally expansible. Therefore the normal jump of the stress rate,

$$\dot{\boldsymbol{\sigma}} = -\Big(\frac{\partial p}{\partial \rho} \dot{\rho} + \frac{\partial p}{\partial T} \dot{T} \Big) \mathbf{I} \,, \tag{11.4.2}$$

across a second order wave, reduces to,

$$\mathbf{n} \cdot [\![\dot{\boldsymbol{\sigma}}]\!] = K [\![\text{tr} \, \dot{\boldsymbol{\epsilon}}]\!] \, \mathbf{n} \,, \tag{11.4.3}$$

where

$$K = \rho \Big(\frac{\partial p}{\partial \rho} \Big)_{|T} \,, \tag{11.4.4}$$

is the isothermal bulk modulus of the fluid. Indeed $\dot{\rho}/\rho = -\operatorname{tr}\dot{\epsilon}$ due to the mass balance, and, for a conductor fluid, $[\![\dot{T}]\!] = 0$ according to (11.3.2). Therefore the acoustic tensor $\mathbf{A}(\mathbf{n}) = K\,\mathbf{n}\otimes\mathbf{n}$ that governs the eigenvalue problem $(\mathbf{A} - \rho\,c^2\,\mathbf{I})\cdot\boldsymbol{\eta} = \mathbf{0}$ for the jump of the velocity gradient is singular. Dilatational waves with $\boldsymbol{\eta} \sim \mathbf{n}$ propagate at speed c with square,

$$c^2 = \frac{K}{\rho}\,, \tag{11.4.5}$$

while shear waves, with $\boldsymbol{\eta} \perp \mathbf{n}$, $\boldsymbol{\eta} \neq \mathbf{0}$ but $[\![\dot{\mathbf{v}}]\!] = \mathbf{0}$, are stationary, i.e. $c = 0$.

The same conclusion would be reached for a non conductor fluid with constitutive equation of the form $p = p(\rho, s)$, in view of (11.3.11).

This phenomenon is at variance with the situation in elastic solids. Indeed, it is shown in Section 11.5.3 that stationary discontinuities are excluded for thermoelastic solids and mixtures as long as their tensor moduli, and acoustic tensors, are symmetric definite positive. On the other hand, stationary discontinuities are a well-known mode of failure in a number of engineering materials, also referred to as strain-localization: as the mechanical information can no longer propagate in some direction, it is deflected orthogonally, and its pile up gives rise to shear bands of intense straining. The failure of the positive definiteness of the acoustic tensors in these materials is to be traced to a strong constitutive or geometrical nonlinearity.

Returning to the wave-speed c, which for a barotropic fluid is usually referred to as *speed of sound*, some further information may be gained for a perfect gas under isentropic condition, namely from (3.3.13),

$$c^2 = \gamma\,\frac{p_0}{\rho_0}\left(\frac{\rho}{\rho_0}\right)^{\gamma-1} = \gamma\,\frac{p}{\rho}\,. \tag{11.4.6}$$

For example, for dry air at temperature $T_0 = 20\,°\mathrm{C}$ and under atmospheric pressure $p_0 = 0.101\,\mathrm{MPa}$, the mass density ρ_0 is equal to $1.20\,\mathrm{kg/m^3}$ and the adiabatic coefficient to $\gamma = 1.4$. The speed of sound c_0 is therefore about $343.3\,\mathrm{m/s}$, while, under a pressure $p = 1.01\,\mathrm{MPa}$, c is equal to $477\,\mathrm{m/s}$.

The speed of sound in water depends on pressure, temperature and salinity. In distilled water, it increases from $1407\,\mathrm{m/s}$ at $0\,°\mathrm{C}$ up to about $1560\,\mathrm{m/s}$ at $70\,°\mathrm{C}$, but it is anomalous, e.g., Fung [1977], p. 215. At $20\,°\mathrm{C}$, it is equal to $1484\,\mathrm{m/s}$: with a mass density $\rho \sim 10^3\,\mathrm{kg/m^3}$, the associated isothermal bulk modulus $K = \rho\,c^2$ is found equal to $2.2\,\mathrm{GPa}$, which is indeed the standard static value. Interestingly, both the bulk modulus and mass density of water are orders of magnitude larger than in air but the speeds of sound result to be comparable.

11.4.2 Acoustic waves in a frictionless ideal fluid

The wave-speed c defined by (11.4.5) governs also the speed of propagation of small acoustic waves in barotropic fluids where $p = p(\rho)$. Indeed, let us consider small perturbations around the mass density ρ, and linearize the balance of momentum $-\boldsymbol{\nabla}p - \rho\,\dot{\mathbf{v}} = \mathbf{0}$, and the balance of mass $\dot{\rho} + \rho\operatorname{div}\mathbf{v} = 0$. Since around the reference state $dp = c^2\,d\rho$, then both $\psi = p$ and $\psi = \rho$ are found to satisfy the wave equation,

$$c^2\,\Delta\psi - \ddot{\psi} = 0\,, \tag{11.4.7}$$

while the velocity satisfies the equation $c^2\,\boldsymbol{\nabla}\operatorname{div}\mathbf{v} - \ddot{\mathbf{v}} = 0$. However for irrotational motions, $\mathbf{v}(\mathbf{x}, t) = \boldsymbol{\nabla}\phi(\mathbf{x}, t)$, this equation for the velocity transforms to $\boldsymbol{\nabla}(c^2\,\Delta\phi - \ddot{\phi}) = \mathbf{0}$. Hence $c^2\,\Delta\phi - \ddot{\phi}$ is a function of time only, say, equal to $\ddot{\phi}_1(t)$, but $\phi(\mathbf{x}, t)$ and $\phi(\mathbf{x}, t) + \phi_1(t)$ yield the same velocity irrespective of the time dependent function ϕ_1. Therefore the potential ϕ is also a solution of (11.4.7).

The solutions of (11.4.7) are contributed by the sum of waves traveling at speeds c and $-c$ along the axis x, namely $\phi = \phi_+(x - c\,t) + \phi_-(x + c\,t)$, with $\phi_\pm$ arbitrary twice differentiable functions.

11.4.3 Regimes in the steady irrotational flow in a barotropic fluid

Consider steady state flow where the partial derivatives with respect to time vanish. Then the steady state flow of a barotropic fluid is governed by

$$\text{its balance of mass}\qquad \operatorname{div}\rho\,\mathbf{v} = \mathbf{0};$$

$$\text{its balance of momentum}\qquad -\frac{\boldsymbol{\nabla}p}{\rho} + \mathbf{b} = \mathbf{v}\cdot\boldsymbol{\nabla}\mathbf{v}; \tag{11.4.8}$$

$$\text{its mechanical behavior}\qquad p = p(\rho)\,.$$

To simplify the presentation, the flow is assumed to take place in the plane (x, y) and to be irrotational, and the body force $\mathbf{b}$ vanishes. For an irrotational flow, the velocity $\mathbf{v}$ may be taken equal to the gradient $\boldsymbol{\nabla}\phi$ of a scalar potential ϕ. With c the wave-speed defined by (11.4.4) and (11.4.5), namely $c^2 = dp/d\rho$, the balance of mass may be recast in the format,

$$c^2 \operatorname{div} \mathbf{v} - \mathbf{v} \cdot (\mathbf{v} \cdot \boldsymbol{\nabla}\mathbf{v}) = 0;$$

$$\left(c^2 - (\frac{\partial \phi}{\partial x})^2\right) \frac{\partial^2 \phi}{\partial x^2} - 2 \frac{\partial \phi}{\partial x} \frac{\partial \phi}{\partial y} \frac{\partial^2 \phi}{\partial x \partial y} + \left(c^2 - (\frac{\partial \phi}{\partial y})^2\right) \frac{\partial^2 \phi}{\partial y^2} = 0 \,. \tag{11.4.9}$$

Along standard mathematical textbooks, e.g., Courant and Hilbert [1953,1962], the last equation is recognized as a quasi-linear second order partial differential equation, of the form

$$A \frac{\partial^2 \phi}{\partial x^2} + 2 B \frac{\partial^2 \phi}{\partial x \partial y} + C \frac{\partial^2 \phi}{\partial y^2} = 0 \,, \tag{11.4.10}$$

where the coefficients A, B and C may depend on the coordinates, on ϕ and on its first derivatives. The mathematical type of this equation relies on the sign of $B^2 - AC$, which here is equal to $c^2 (|\mathbf{v}|^2 - c^2)$. The partial differential equation is said to be hyperbolic, elliptic or parabolic according to the conventions shown in Table 11.4.1.

Table 11.4.1: Classification of second order partial differential equations.

Type of the Equation	$B^2 - AC$	Mach nb. $|\mathbf{v}|/c$	Regime of the Flow
Hyperbolic	> 0	> 1	Supersonic
Elliptic	< 0	< 1	Subsonic
Parabolic	$= 0$	$= 1$	Transonic

The implications of this classification are fundamental for an understanding of the physical meaning of partial differential equations. Basic examples of parabolic and hyperbolic equations describing diffusion of heat and mass are worked out in Chapter 15. The issue is also of key concern for the energy equation, Section 11.3.2: when the material obeys Fourier's heat law of conduction, the energy equation is of parabolic type, while it is of hyperbolic type when the material obeys Cattaneo's heat law, which endows the diffusion phenomenon with a finite relaxation time.

In a gas, there is a single speed of sound c, namely $c^2 = \gamma p/\rho$ according to Eqn (11.4.6). Interestingly, subsonic flows modify little the air pressure. Indeed, according to Section (15.7.2), the relative change in pressure p due to flow $\frac{1}{2}\rho\mathbf{v}^2 = \frac{1}{2}\gamma p (v/c)^2$ is less than 1% if the Mach number is less than $0.1\sqrt{2/\gamma} \sim 0.1$.

Elastic solids and elastic saturated porous media display respectively two and three body wave-speeds. Therefore the classification of the regimes is somehow more complex. For saturated porous media, the classification is reported in Table 11.5.1 and the dependence on porosity and the relative order of the three body wave-speeds are displayed in Fig. 11.5.2. The flow regime is also addressed in Section 11.10, where the entity that plays the role of the particle velocity is the speed of a crack tip that moves through the material.

11.4.4 A Newtonian viscous fluid

For a Newtonian viscous fluid, the effective stress $\boldsymbol{\sigma} + p\,\mathbf{I}$ expresses linearly in terms of the strain rate $\dot{\boldsymbol{\epsilon}}$,

$$\boldsymbol{\sigma} + p\,\mathbf{I} = K_{\mathrm{v}} \operatorname{tr}\dot{\boldsymbol{\epsilon}} + 2\,\mu_{\mathrm{v}} \operatorname{dev}\dot{\boldsymbol{\epsilon}} \,, \tag{11.4.11}$$

via the bulk viscosity $K_{\mathrm{v}} \geq 0$ and the shear viscosity $\mu_{\mathrm{v}} \geq 0$. The thermodynamic pressure $p = p(\rho, T)$ is in general a function of the mass density ρ and temperature T. A thermodynamic exposition of the constitutive equation is found in Bowen [1989], p. 96.

Thus we are lead to distinguish the following subcases:

(a) bulk viscosity $K_{\mathrm{v}} \neq 0$ and shear viscosity $\mu_{\mathrm{v}} \neq 0$.

In order to address the existence of acceleration waves, it is instrumental to attempt to rewrite the constitutive equation in the format $\dot{\boldsymbol{\epsilon}} = \dot{\boldsymbol{\epsilon}}(\boldsymbol{\sigma}, \rho, T)$. The constitutive equation is indeed invertible,

$$\dot{\boldsymbol{\epsilon}} = \frac{1}{K_{\mathrm{v}}} \left(\frac{\operatorname{tr}\boldsymbol{\sigma}}{3} + p\right) \frac{\mathbf{I}}{3} + \frac{1}{2\,\mu_{\mathrm{v}}} \operatorname{dev}\boldsymbol{\sigma} \,. \tag{11.4.12}$$

Across an acceleration wave front, the right hand side of (11.4.12) is continuous, so that the jump of the strain rate vanishes.

(b) bulk viscosity $K_v = 0$ but shear viscosity $\mu_v \neq 0$.

Then the constitutive equation is invertible in the form,

$$\operatorname{dev} \dot{\boldsymbol{\epsilon}} = \frac{1}{2\,\mu_v} \operatorname{dev} \boldsymbol{\sigma}, \quad p = -\frac{\operatorname{tr} \boldsymbol{\sigma}}{3}\,. \tag{11.4.13}$$

The jump of the deviatoric part of the strain rate vanishes according to the first relation (11.4.13), which actually implies the jump of the velocity gradient to vanish[11.4]. As a consequence, a Newtonian viscous fluid with non vanishing shear viscosity cannot propagate acceleration waves.

11.5 Acceleration waves in saturated porous media

The constitutive equations are derived within the general framework of mixture theory, e.g., Truesdell and Toupin [1960], Green and Naghdi [1969], Bowen [1976]. The body of the theory has been exposed in Chapters 4 and 7, but the basic elements used in the dynamic analysis are summarized below. Emphasis is placed on second order (acceleration) waves which are discontinuity waves. First order waves are considered in Section 11.10. The speeds of propagation and eigenmodes of acceleration waves are controlled by the mechanical constitutive equations. On the other hand, the diffusion phenomenon comes into picture in the propagation of these waves, Section 11.6, as well as in harmonic waves, Section 11.9. Diffusion introduces a viscous effect in the field equations, and by the same token, a length scale.

11.5.1 The characteristic equation for the wave-speeds

We consider here porous media which are made up of two constituents, a solid and a fluid, namely water, which are viewed as two independent overlapping continua. Phases (solid (s) and fluid (w)) represent the constituents when viewed as part of the mixture. The solid phase is usually referred to as *solid skeleton*. Each constituent has a mass M_α and a volume V_α, $\alpha =$ s, w, which make up the total mass $M = M_s + M_w$ and the total volume $V = V_s + V_w$ of the mixture.

At each point of each phase, we define *intrinsic* or *true* quantities, labeled by subscripts, and *apparent* or *partial* quantities, labeled by superscripts. For example, the intrinsic mass density is defined as $\rho_\alpha = M_\alpha/V_\alpha$, whereas the apparent mass density is defined by $\rho^\alpha = M_\alpha/V$; hence $\rho^\alpha = n^\alpha \rho_\alpha$ where $n^\alpha = V_\alpha/V$ is the volume fraction of phase α. Since we consider fluid-saturated porous media, the volume fractions satisfy the constraint,

$$n^s + n^w = 1\,. \tag{11.5.1}$$

The equations of balance of momentum of the phases involve two (symmetric) *partial stress tensors* $\boldsymbol{\sigma}^s$ and $\boldsymbol{\sigma}^w$, which make up the total stress tensor of the mixture,

$$\boldsymbol{\sigma} = \boldsymbol{\sigma}^s + \boldsymbol{\sigma}^w\,, \tag{11.5.2}$$

when the contribution due to diffusion is neglected. With each partial stress tensor $\boldsymbol{\sigma}^\alpha$ is associated an *intrinsic stress tensor* $\boldsymbol{\sigma}_\alpha$ defined as $\boldsymbol{\sigma}_\alpha = \boldsymbol{\sigma}^\alpha/n^\alpha$. The partial stress tensor of the fluid phase, associated with a linear compressible fluid, is defined in terms of its partial p^w or intrinsic p_w pressure as,

$$\boldsymbol{\sigma}^w = -p^w\,\mathbf{I} = -n^w p_w\,\mathbf{I}\,. \tag{11.5.3}$$

Each phase α is endowed with its own apparent (infinitesimal) strain tensor $\boldsymbol{\epsilon}^\alpha$ defined as the symmetrized gradient of its macroscopic displacement field. As for constituents, they are generally compressible. Incompressibility of one or both constituents can be viewed as a limit case. If both solid and fluid constituents are incompressible, the equations of mass balance imply that the rates of volume change must satisfy the constraint (7.2.23), which for a binary mixture simplifies to,

$$n^s \operatorname{div} \mathbf{v}_s + n^w \operatorname{div} \mathbf{v}_w = \frac{\hat{\rho}^s}{\rho_s} + \frac{\hat{\rho}^w}{\rho_w}\,. \tag{11.5.4}$$

[11.4]Indeed let $[\![\operatorname{dev} \dot{\boldsymbol{\epsilon}}]\!] = \frac{1}{2}(\boldsymbol{\eta} \otimes \mathbf{n} + \mathbf{n} \otimes \boldsymbol{\eta}) - \frac{1}{3}\,\boldsymbol{\eta} \cdot \mathbf{n}\,\mathbf{I} = \mathbf{0}$. Scalar product by $\mathbf{n} \otimes \mathbf{n}$ implies $\boldsymbol{\eta} \cdot \mathbf{n} = 0$. Hence $\boldsymbol{\eta} \otimes \mathbf{n} + \mathbf{n} \otimes \boldsymbol{\eta} = \mathbf{0}$. Since $\boldsymbol{\eta}$ is orthogonal to $\mathbf{n}$, scalar product by $\mathbf{n}$ implies in turn $\boldsymbol{\eta} = \mathbf{0}$.

For a closed mixture, the mass supplies $\hat{\rho}^\alpha$ satisfy the condition (4.4.3), namely here $\hat{\rho}^s + \hat{\rho}^w = 0$.

For each phase of the porous medium, the balance of momentum (4.5.5),

$$\text{div } \boldsymbol{\sigma}^\alpha + \rho^\alpha \left(\mathbf{b}_\alpha - \frac{d^\alpha \mathbf{v}_\alpha}{dt} \right) = \hat{\rho}^\alpha \left(\mathbf{v}_\alpha - \tilde{\mathbf{v}}_\alpha \right) - \hat{\boldsymbol{\pi}}^\alpha, \quad \alpha = \mathrm{s}, \mathrm{w}, \tag{11.5.5}$$

involves, in addition to the usual terms present in single phase solids, namely the divergence of the stress tensor, the body force per unit mass $\mathbf{b}_\alpha$ and the acceleration $d^\alpha \mathbf{v}_\alpha / dt$, the apparent mass density ρ^α and the momentum supply $\hat{\boldsymbol{\pi}}^\alpha$ to the phase α by the rest of the mixture. Momentum supplies, defined by (7.3.10), introduce the diffusion of the fluid through the solid skeleton via the coefficient of Stokes' drag k_{Sd},

$$\hat{\rho}^w (\tilde{\mathbf{v}}_w - \mathbf{v}_w) + \hat{\boldsymbol{\pi}}^w = p_w \boldsymbol{\nabla} n^w + k_{\mathrm{Sd}} (\mathbf{v}_s - \mathbf{v}_w). \tag{11.5.6}$$

For a closed mixture, momentum supplies satisfy the constraint (4.5.2), namely,

$$\sum_{\alpha=\mathrm{s},\mathrm{w}} \hat{\rho}^\alpha \tilde{\mathbf{v}}_\alpha + \hat{\boldsymbol{\pi}}^\alpha = \mathbf{0}, \tag{11.5.7}$$

where $\tilde{\mathbf{v}}_\alpha$ is the velocity of the mass transferred to phase α.

The balance of momentum holds pointwise, except along a singular surface of order 1 where it is replaced by the jump relation (11.1.56). Singular surfaces of order 1 display a velocity discontinuity, while, across singular surfaces of order 2, the velocity is continuous but the acceleration and the velocity gradient are not.

The analysis below involves a number of simplifications. First, it assumes that no mass transfer takes place between constituents, namely $\hat{\rho}^\alpha = 0$, $\alpha = \mathrm{s}, \mathrm{w}$. In addition, since we envisage a linearized analysis around an equilibrium state, the volume fractions and mass densities are fixed to their reference values. Moreover, we shall make no difference between partial and material time derivatives, and denote them by a superimposed dot. Therefore we shall not distinguish the speed of displacement W along its direction of propagation $\mathbf{n}$, and the two speeds of propagation $c_\alpha = W - \mathbf{v}_\alpha \cdot \mathbf{n}$ of an acceleration wave with respect to the particles of each of the two phases $\alpha = \mathrm{s}, \mathrm{w}$.

Note also that, in a linearized analysis, the buoyancy term associated with the gradient of volume fraction disappears from the momentum supplies. Therefore the remaining terms in the momentum supplies involve only a velocity difference, which is continuous across singular surfaces of order 2. Finally, the rates of the body forces $\mathbf{b}_\alpha$, $\alpha = \mathrm{s}, \mathrm{w}$, are also assumed to be continuous across the wave front.

Consider an acceleration wave that propagates through the porous medium at speed c. Hadamard's compatibility relations (11.1.9) apply to the two phases independently. e.g., for $\phi = \boldsymbol{\sigma}^\alpha$ the partial stress of phase $\alpha = \mathrm{s}, \mathrm{w}$,

$$\left[\!\!\left[\frac{\partial \boldsymbol{\sigma}^\alpha}{\partial \mathbf{x}} \right]\!\!\right] = -\frac{1}{c} [\dot{\boldsymbol{\sigma}}^\alpha] \otimes \mathbf{n}. \tag{11.5.8}$$

and, for $\phi = \mathbf{v}_\alpha$ the velocity of phase α,

$$\left[\!\!\left[\frac{\partial \mathbf{v}_\alpha}{\partial \mathbf{x}} \right]\!\!\right] = \boldsymbol{\eta}_\alpha \otimes \mathbf{n}, \quad [\![\dot{\mathbf{v}}_\alpha]\!] = -c \boldsymbol{\eta}_\alpha. \tag{11.5.9}$$

As a consequence of the above assumptions on the entities involved in momentum balances, the jump across the front of an acceleration wave of the rate of traction is proportional to the normal jump of the velocity gradient,

$$\mathbf{n} \cdot [\![\dot{\boldsymbol{\sigma}}^\alpha]\!] = \rho^\alpha c^2 \boldsymbol{\eta}_\alpha, \quad \alpha = \mathrm{s}, \mathrm{w}. \tag{11.5.10}$$

As an immediate consequence, we have

Proposition 11.2 on the kinematics of the acceleration wave in the fluid phase.

The normal jump of the velocity gradient in the fluid phase $\boldsymbol{\eta}_w$, Eqn (11.2.1), is aligned with the wave normal $\mathbf{n}$,

$$\boldsymbol{\eta}_w = \eta_w \, \mathbf{n}. \tag{11.5.11}$$

The proof simply uses Hadamard's compatibility relation for the fluid phase, (11.5.10), accounting for the fact that the fluid stress is isotropic, that is

$$\mathbf{n} \cdot [\![-\dot{p}^w \mathbf{I}]\!] = \rho^w c^2 \boldsymbol{\eta}_w. \tag{11.5.12}$$

Still it might be worth mentioning that Proposition 11.2 does not hold in presence of inertial coupling: the issue is elaborated in Section 11.5.5.

The form of the poroelastic constitutive equations that is convenient in the analysis of acceleration waves is detailed in Section 7.3.6 together with the central concept of underlying drained solid. Insertion of the constitutive equations (7.3.28) in the discontinuity relations (11.5.10) provides a 4×4 generalized eigenvalue problem for the wave-speeds c^2 in terms of a constitutive matrix $[\mathbf{A}^{\#}]$, of a diagonal mass matrix $[\mathbf{R}^{\#}] = \mathrm{diag}\,[\rho^s,\ \rho^s,\ \rho^s,\ \rho^w]$ and of an eigenvector $[\boldsymbol{\eta}^{\#}] = [\boldsymbol{\eta}_s,\ \eta_w]^T$,

$$\left[\mathbf{A}^{\#} - c^2 \mathbf{R}^{\#}\right]\left[\boldsymbol{\eta}^{\#}\right] = \mathbf{0}\,,$$

$$\begin{bmatrix} \mathbf{n} \cdot \mathbb{E}^{ss} \cdot \mathbf{n} - c^2\,\rho^s \mathbf{I} & \mathbf{n} \cdot \lambda^{sw}\,\boldsymbol{\lambda}_w \\ \lambda^{sw}\,(\boldsymbol{\lambda}_w \cdot \mathbf{n})^T & \lambda^{ww} - c^2\rho^w \end{bmatrix} \begin{bmatrix} \boldsymbol{\eta}_s \\ \eta_w \end{bmatrix} = \begin{bmatrix} \mathbf{0} \\ 0 \end{bmatrix}. \qquad (11.5.13)$$

Through simple manipulations, the generalized eigenvalue problem can be transformed into a standard eigenvalue problem,

$$\left[\mathbf{A} - c^2 \mathbf{I}_4\right]\left[\boldsymbol{\eta}\right] = [\mathbf{0}]\,, \qquad (11.5.14)$$

where $[\mathbf{I}_4]$ is the identity matrix over the set of 4×4 matrices. The block structure of the problem emanates from the mechanical constitutive equations of the constituents,

$$\begin{bmatrix} \mathbf{A}^{ss} - c^2\,\mathbf{I} & \mathbf{a}^{sw} \\ (\mathbf{a}^{sw})^T & c_w^2 - c^2 \end{bmatrix} \begin{bmatrix} \sqrt{\rho^s}\,\boldsymbol{\eta}_s \\ \sqrt{\rho^w}\,\eta_w \end{bmatrix} = \begin{bmatrix} \mathbf{0} \\ 0 \end{bmatrix}, \qquad (11.5.15)$$

and, in explicit form,

$$\begin{bmatrix} \mathbf{n} \cdot \dfrac{\mathbb{E}^{ss}}{\rho^s} \cdot \mathbf{n} - c^2\,\mathbf{I} & \mathbf{n} \cdot \boldsymbol{\lambda}_w\,\dfrac{\lambda^{sw}}{\sqrt{\rho^s\,\rho^w}} \\ \dfrac{\lambda^{sw}}{\sqrt{\rho^s\,\rho^w}}\,(\boldsymbol{\lambda}_w \cdot \mathbf{n})^T & \dfrac{\lambda^{ww}}{\rho^w} - c^2 \end{bmatrix} \begin{bmatrix} \sqrt{\rho^s}\,\boldsymbol{\eta}_s \\ \sqrt{\rho^w}\,\eta_w \end{bmatrix} = \begin{bmatrix} \mathbf{0} \\ 0 \end{bmatrix}. \qquad (11.5.16)$$

Consequently, there are at most four possible modes of propagation, each one associated with a solution c^2 of (11.5.13) or of (11.5.16). In addition, (11.5.16) shows clearly that the major symmetry of the constitutive equations is inherited by the acoustic matrix $[\mathbf{A}]$. Thus, the eigenvalue problem is symmetric and, therefore, the squares of the wave-speeds and the eigenvectors are real.

Mechanical coupling between the solid and fluid phases assumes $\lambda^{sw} \neq 0$. Then the eigenmodes affect in general both phases. An eigenmode that does not affect the fluid phase, namely $\eta_w = 0$, satisfies the relation $\boldsymbol{\eta}_s \cdot \boldsymbol{\lambda}_w \cdot \mathbf{n} = 0$.

Given that the fluid jump is necessarily parallel to the normal $\mathbf{n}$, a wave is said to be *longitudinal* if the solid jump $\boldsymbol{\eta}_s$ is also parallel to the normal $\mathbf{n}$, and *transversal* if the solid jump is orthogonal to $\mathbf{n}$.

For isotropic mixtures with compressible constituents, the analysis below ranks the elastic wave-speeds, according to the material parameters. The wave-speeds correspond to a shear wave-speed of double multiplicity which affects only the solid phase, and to two longitudinal wave-speeds, so-called Biot's first and second waves, which affect both the solid and fluid phases. When both solid and fluid constituents are incompressible, the largest longitudinal wave-speed becomes unbounded, so that there are only two finite wave-speeds.

11.5.2   Acceleration wave-speeds in isotropic porous media

11.5.2.1   Relative order of acceleration wave-speeds

For an isotropic elastic mixture, the constitutive equations of Section 7.3.6 linking the partial stresses to the partial strains simplify to,

$$\begin{aligned} \sigma^s &= \lambda^{ss}\ \mathrm{tr}\ \epsilon^s \mathbf{I}\ +\ 2\mu\,\epsilon^s\ +\ \lambda^{sw}\ \mathrm{tr}\ \epsilon^w \mathbf{I}\,, \\ -n^w\,p_w &= \lambda^{sw}\ \mathrm{tr}\ \epsilon^s\ \ \ \ \ \ \ \ \ \ \ \ \ \ \ +\ \lambda^{ww}\ \ \mathrm{tr}\ \epsilon^w\,. \end{aligned} \qquad (11.5.17)$$

The four constitutive moduli have been identified in the above section in terms of measurable quantities. The eigenvalue problem,

$$\begin{bmatrix} c_s^2\,\mathbf{n} \otimes \mathbf{n} + c_3^2\,(\mathbf{I} - \mathbf{n} \otimes \mathbf{n}) - c^2\,\mathbf{I} & c_{sw}^2\,\mathbf{n} \\ c_{sw}^2\,\mathbf{n}^T & c_w^2 - c^2 \end{bmatrix} \begin{bmatrix} \sqrt{\rho^s}\,\boldsymbol{\eta}_s \\ \sqrt{\rho^w}\,\eta_w \end{bmatrix} = \begin{bmatrix} \mathbf{0} \\ 0 \end{bmatrix}, \qquad (11.5.18)$$

is advantageously expressed in the cartesian axes $(\mathbf{n}, \mathbf{t}_1, \mathbf{t}_2)$,

$$
\begin{bmatrix}
c_{\mathrm{s}}^2 - c^2 & 0 & 0 & c_{\mathrm{sw}}^2 \\
0 & c_3^2 - c^2 & 0 & 0 \\
0 & 0 & c_3^2 - c^2 & 0 \\
c_{\mathrm{sw}}^2 & 0 & 0 & c_{\mathrm{w}}^2 - c^2
\end{bmatrix}
\begin{bmatrix}
\sqrt{\rho^{\mathrm{s}}}\,\eta_{\mathrm{s}1} \\
\sqrt{\rho^{\mathrm{s}}}\,\eta_{\mathrm{s}2} \\
\sqrt{\rho^{\mathrm{s}}}\,\eta_{\mathrm{s}3} \\
\sqrt{\rho^{\mathrm{w}}}\,\eta_{\mathrm{w}}
\end{bmatrix}
=
\begin{bmatrix}
0 \\ 0 \\ 0 \\ 0
\end{bmatrix}.
\tag{11.5.19}
$$

Some notation has been introduced, namely

$$
c_{\mathrm{s}}^2 = \frac{\lambda^{\mathrm{ss}} + 2\mu}{\rho^{\mathrm{s}}} > 0, \quad
c_{\mathrm{w}}^2 = \frac{\lambda^{\mathrm{ww}}}{\rho^{\mathrm{w}}} > 0, \quad
c_{\mathrm{ms}}^2 = \frac{\lambda^{\mathrm{sw}}}{\rho^{\mathrm{s}}}, \quad
c_{\mathrm{mw}}^2 = \frac{\lambda^{\mathrm{sw}}}{\rho^{\mathrm{w}}}, \quad
c_{\mathrm{sw}}^2 = \frac{\lambda^{\mathrm{sw}}}{\sqrt{\rho^{\mathrm{s}}\rho^{\mathrm{w}}}}.
\tag{11.5.20}
$$

The signs of c_{s}^2 and c_{w}^2 are implied by the positive definiteness of the poroelastic energy, Eqn (7.3.45). Note that only the squares c_{ms}^2, c_{mw}^2 and c_{sw}^2 are used, not the square-roots themselves, and the sign of the coefficient of mechanical coupling λ^{sw}, although significant from a mechanical point of view, is arbitrary. Actually, for most materials, the coefficient λ^{sw} is positive. In particular, the property holds when the solid constituent is incompressible, Eqn (7.3.51).

As a general property, positive definiteness of the poroelastic energy (7.3.45) implies strong ellipticity, and the existence of a real shear wave-speed c_3 of double multiplicity, with a non zero jump in the solid phase only,

$$
c_3^2 = \frac{\mu}{\rho^{\mathrm{s}}}, \quad \boldsymbol{\eta}_{\mathrm{s}} \perp \mathbf{n}, \quad \boldsymbol{\eta}_{\mathrm{w}} = \mathbf{0},
\tag{11.5.21}
$$

and of two real longitudinal wave-speeds, $c_1 \geq c_2$, of single multiplicity,

$$
\left.\begin{array}{c} c_1^2 \\ c_2^2 \end{array}\right\}
= \tfrac{1}{2}\left(c_{\mathrm{s}}^2 + c_{\mathrm{w}}^2\right) \pm \tfrac{1}{2}\left(\left(c_{\mathrm{s}}^2 - c_{\mathrm{w}}^2\right)^2 + 4\,c_{\mathrm{sw}}^4\right)^{1/2},
\tag{11.5.22}
$$

with eigenmodes,

$$
\boldsymbol{\eta}_{\mathrm{s}} = (\boldsymbol{\eta}_{\mathrm{s}} \cdot \mathbf{n})\,\mathbf{n}, \quad
\boldsymbol{\eta}_{\mathrm{w}} = \frac{\sqrt{\rho^{\mathrm{s}}}}{\sqrt{\rho^{\mathrm{w}}}}\,\frac{c^2 - c_{\mathrm{s}}^2}{c_{\mathrm{sw}}^2}\,\boldsymbol{\eta}_{\mathrm{s}} = \frac{\sqrt{\rho^{\mathrm{s}}}}{\sqrt{\rho^{\mathrm{w}}}}\,\frac{c_{\mathrm{sw}}^2}{c^2 - c_{\mathrm{w}}^2}\,\boldsymbol{\eta}_{\mathrm{s}}.
\tag{11.5.23}
$$

The squares of the longitudinal wave-speeds are solutions of the polynomial

$$
F(c^2) = (c^2)^2 - (c_{\mathrm{s}}^2 + c_{\mathrm{w}}^2)\,c^2 + c_{\mathrm{s}}^2\,c_{\mathrm{w}}^2 - c_{\mathrm{sw}}^4 = 0,
\tag{11.5.24}
$$

and they obey the relations,

$$
c_1^2 + c_2^2 = c_{\mathrm{s}}^2 + c_{\mathrm{w}}^2,
\tag{11.5.25}
$$

and

$$
c_{\mathrm{sw}}^4 = c_{\mathrm{s}}^2\,c_{\mathrm{w}}^2 - c_1^2\,c_2^2 = \left(c_1^2 - c_{\mathrm{s}}^2\right)\left(c_{\mathrm{s}}^2 - c_2^2\right) = \left(c_1^2 - c_{\mathrm{w}}^2\right)\left(c_{\mathrm{w}}^2 - c_2^2\right).
\tag{11.5.26}
$$

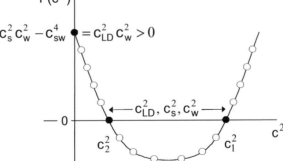

Fig. 11.5.1: Relative positions of entities involved in the analysis of the longitudinal wave-speeds in an isotropic poroelastic body.

Irrespective of the material parameters, the wave-speeds satisfy the inequalities that can be read from Fig. 11.5.1,

$$
\max\{c_2, c_3\} \leq c_1, \quad c_2 \leq \min\{c_{\mathrm{s}}, c_{\mathrm{w}}\} \leq \max\{c_{\mathrm{s}}, c_{\mathrm{w}}\} \leq c_1.
\tag{11.5.27}
$$

Indeed,

$$F(c_{\mathrm{w}}^2) = F(c_{\mathrm{s}}^2) = -c_{\mathrm{sw}}^4 \leq 0\,, \tag{11.5.28}$$

and

$$F(c_3^2) = \frac{1}{\rho^{\mathrm{s}}\rho^{\mathrm{w}}} \left((\lambda_0^{\mathrm{sw}})^2 - (\lambda^{\mathrm{sw}})^2 \right), \tag{11.5.29}$$

with $(\lambda_0^{\mathrm{sw}})^2$ defined by the relation,

$$\frac{(\lambda_0^{\mathrm{sw}})^2}{\rho^{\mathrm{s}}\rho^{\mathrm{w}}} = (c_{\mathrm{s}}^2 - c_3^2)\,(c_{\mathrm{w}}^2 - c_3^2)\,. \tag{11.5.30}$$

Then, the more detailed inequalities result,

$$
\begin{aligned}
c_3 < c_2 = c_1 \quad &\text{if} \quad c_{\mathrm{s}} = c_{\mathrm{w}} \quad \text{and} \quad \lambda^{\mathrm{sw}} = 0\,, \\[4pt]
c_3 < c_2 < c_1 \quad &\text{if} \quad (\lambda_0^{\mathrm{sw}})^2 > (\lambda^{\mathrm{sw}})^2\,, \\[4pt]
c_3 = c_2 < c_1 \quad &\text{if} \quad (\lambda_0^{\mathrm{sw}})^2 = (\lambda^{\mathrm{sw}})^2\,, \\[4pt]
c_2 < c_3 < c_1 \quad &\text{if} \quad (\lambda_0^{\mathrm{sw}})^2 < (\lambda^{\mathrm{sw}})^2\,.
\end{aligned}
\tag{11.5.31}
$$

The relative order of the three poroelastic wave-speeds delineates, in the plane of porosity n^{w} versus speed c, three intersonic regions denoted by (i), (ii) and (iii), defined in explicit form in Table 11.5.1.

Table 11.5.1: The subsonic, intersonic and supersonic regions in a saturated porous medium.

Region	Range of Speed c
Subsonic	$c < \min(c_2, c_3)$
Intersonic (i)	$\max(c_2, c_3) < c < c_1$
Intersonic (ii)	$c_3 < c < c_2 < c_1$
Intersonic (iii)	$c_2 < c < c_3 < c_1$
Supersonic	$c_1 < c$

11.5.2.2   Dry materials and drained materials

For dry materials, the moduli involved are those of the solid skeleton, say bulk modulus $K = \lambda + 2\,\mu$ and shear modulus μ. Then the longitudinal and shear wave-speeds boil down to their counterparts in the single phase solid defined by the above moduli and the *apparent* mass density of the skeleton,

$$c_{\mathrm{LD}}^2 = \frac{\lambda + 2\,\mu}{\rho^{\mathrm{s}}} = \frac{K + 4\,\mu/3}{\rho^{\mathrm{s}}}, \quad c_{\mathrm{SD}}^2 = \frac{\mu}{\rho^{\mathrm{s}}}\,. \tag{11.5.32}$$

These wave-speeds are also the wave-speeds of the underlying drained solid. Indeed setting the pressure to zero in (11.5.17), and eliminating the fluid strain yields a relation linking the partial solid stress and apparent solid strain,

$$\boldsymbol{\sigma}^{\mathrm{s}} = \lambda \operatorname{tr} \boldsymbol{\epsilon}^{\mathrm{s}}\,\mathbf{I} + 2\,\mu\,\boldsymbol{\epsilon}^{\mathrm{s}}\,, \tag{11.5.33}$$

with help of the relation $\lambda = \lambda^{\mathrm{ss}} - (\lambda^{\mathrm{sw}})^2/\lambda^{\mathrm{ww}}$, Eqn (7.3.44). Insertion of this constitutive equation in the jump of the momentum balance of the solid phase (11.5.10) yields an eigenvalue problem for the drained acoustic tensor. The elastic acoustic tensor of the drained solid is defined as for a single phase solid,

$$\mathbf{A}^{\mathrm{D}} = \mathbf{n} \cdot \frac{\mathbb{E}}{\rho^{\mathrm{s}}} \cdot \mathbf{n} = (c_{\mathrm{LD}})^2\, \mathbf{n} \otimes \mathbf{n} + (c_{\mathrm{SD}})^2\,(\mathbf{I} - \mathbf{n} \otimes \mathbf{n})\,, \tag{11.5.34}$$

which delivers the longitudinal wave-speed c_{LD} with single multiplicity and the shear wave-speed c_{SD} with double multiplicity.

The longitudinal wave-speed c_{LD} of the drained solid is smaller than the first longitudinal wave-speed c_1 of the porous medium, and larger than the second longitudinal wave-speed c_2. Indeed, $c_{\mathrm{s}}^2\, c_{\mathrm{w}}^2 - c_{\mathrm{sw}}^4$ can be shown to be equal to $c_{\mathrm{LD}}^2\, c_{\mathrm{w}}^2$, and the function $F(c_{\mathrm{LD}}^2)$ defined by (11.5.24) is negative, indeed equal to $-c_{\mathrm{sw}}^4\, c_{\mathrm{LD}}^2/c_{\mathrm{w}}^2$. This separation property is generalized to any type of mechanical anisotropy in Section 11.5.4.3.

11.5.2.3 Undrained wave-speeds

Two additional wave-speeds are relevant, namely

$$c_{\mathrm{LU}}^2 = \frac{\lambda^{\mathrm{U}} + 2\mu}{\rho} = \frac{K^{\mathrm{U}} + 4\mu/3}{\rho}, \quad c_{\mathrm{SU}}^2 = \frac{\mu}{\rho}, \tag{11.5.35}$$

where $\rho = \rho^{\mathrm{s}} + \rho^{\mathrm{w}}$ is the mass density of the porous medium, and

$$\lambda^{\mathrm{U}} = \lambda^{\mathrm{ss}} + \lambda^{\mathrm{ww}} + 2\lambda^{\mathrm{sw}}, \quad K^{\mathrm{U}} = \lambda^{\mathrm{U}} + 2\mu/3, \tag{11.5.36}$$

are undrained moduli. Using the definitions (11.5.20), c_{LU}^2 may also be cast in the format,

$$c_{\mathrm{LU}}^2 = \frac{\rho^{\mathrm{s}}}{\rho}(c_{\mathrm{s}}^2 + c_{\mathrm{ms}}^2) + \frac{\rho^{\mathrm{w}}}{\rho}(c_{\mathrm{w}}^2 + c_{\mathrm{mw}}^2). \tag{11.5.37}$$

Indeed, the speeds defined by (11.5.35) are the undrained wave-speeds. They can be viewed also as the low frequency limits of harmonic waves-speeds, i.e. the viscosity of the fluid implying the solid and fluid to move together, their displacements $\mathbf{u} = \mathbf{u}_{\mathrm{s}} = \mathbf{u}_{\mathrm{w}}$, and therefore their strains $\boldsymbol{\epsilon} = \boldsymbol{\epsilon}(\mathbf{u}) = \boldsymbol{\epsilon}(\mathbf{u}_{\mathrm{s}}) = \boldsymbol{\epsilon}(\mathbf{u}_{\mathrm{w}})$ being equal. The total stress $\boldsymbol{\sigma}$ obtained by summation of the two partial stresses in (11.5.17) expresses in terms of the strain $\boldsymbol{\epsilon}$,

$$\boldsymbol{\sigma} = \lambda^{\mathrm{U}} \operatorname{tr} \boldsymbol{\epsilon} \, \mathbf{I} + 2\mu \boldsymbol{\epsilon}, \tag{11.5.38}$$

and the balance of momentum reduces to a single equation,

$$\operatorname{div} \boldsymbol{\sigma} - \rho \, \ddot{\mathbf{u}} = \mathbf{0}. \tag{11.5.39}$$

Thus the medium behaves like a solid, and admits one longitudinal wave-speed c_{LU} and one shear wave-speed c_{SU} with double multiplicity, and its acoustic tensor reads,

$$\mathbf{A}^{\mathrm{U}} = \mathbf{n} \cdot \frac{\mathbb{E}^{\mathrm{U}}}{\rho} \cdot \mathbf{n} = c_{\mathrm{LU}}^2 \, \mathbf{n} \otimes \mathbf{n} + c_{\mathrm{SU}}^2 \, (\mathbf{I} - \mathbf{n} \otimes \mathbf{n}). \tag{11.5.40}$$

Note that the total mass density ρ is involved since the fluid and solid are endowed with identical kinematics. The above undrained relations are sometimes referred to as Gassmann relations.

Using $(7.3.44)_4$ for λ^{ss}, $(7.3.44)_3$ for λ^{sw}, and $(7.3.44)_1$ for λ^{ww}, the undrained bulk modulus may be phrased in terms of these moduli,

$$K^{\mathrm{U}} = K + (\frac{\lambda^{\mathrm{sw}}}{\lambda^{\mathrm{ww}}} + 1)^2 \, \lambda^{\mathrm{ww}} = K + \frac{(K_{\mathrm{s}} - K)^2}{(\dfrac{1 - n^{\mathrm{w}}}{K_{\mathrm{s}}} + \dfrac{n^{\mathrm{w}}}{K_{\mathrm{w}}}) K_{\mathrm{s}}^2 - K}. \tag{11.5.41}$$

Note that

$$\lim_{n^{\mathrm{w}} \to 0} K^{\mathrm{U}} = K_{\mathrm{s}}, \quad \lim_{n^{\mathrm{w}} \to 1} K^{\mathrm{U}} = K_{\mathrm{w}}, \tag{11.5.42}$$

the second limit assuming that $\lim_{n^{\mathrm{w}} \to 1} K = 0$.

As an alternative format, the undrained bulk modulus K^{U} and undrained Poisson's ratio ν^{U} might be considered as measurable entities, and then

$$\lambda^{\mathrm{U}} = 3\frac{\nu^{\mathrm{U}}}{1 + \nu^{\mathrm{U}}} K^{\mathrm{U}}, \quad \mu = \frac{3}{2}\frac{1 - 2\nu^{\mathrm{U}}}{1 + \nu^{\mathrm{U}}} K^{\mathrm{U}}, \quad \lambda^{\mathrm{U}} + 2\mu = 3\frac{1 - \nu^{\mathrm{U}}}{1 + \nu^{\mathrm{U}}} K^{\mathrm{U}}. \tag{11.5.43}$$

The inequalities (7.4.43) ensure the undrained wave-speeds (11.5.35) to be real and the undrained longitudinal wave-speed to be larger than the undrained shear wave-speed, namely $c_{\mathrm{LU}} > c_{\mathrm{SU}}$.

11.5.2.4   The frozen mixture

Let c_0 be the wave-speed defined by the relation,

$$\min(c_s^2 + c_{ms}^2, \, c_w^2 + c_{mw}^2) \le c_0^2 = \tfrac{1}{2}\left(c_s^2 + c_{ms}^2 + c_w^2 + c_{mw}^2\right) \le \max(c_s^2 + c_{ms}^2, \, c_w^2 + c_{mw}^2)\,. \tag{11.5.44}$$

When the relation of *dynamic compatibility* is satisfied, Biot [1956],

$$c_s^2 + c_{ms}^2 = c_w^2 + c_{mw}^2\,, \tag{11.5.45}$$

then, the common value c_0^2 is easily checked to be a root of (11.5.24), and therefore c_0 is a longitudinal wave-speed, in fact the largest one. Under this condition, this wave-speed is also easily checked to be equal to the undrained wave-speed c_{LU},

$$c_1^2 = c_0^2 = c_s^2 + c_{ms}^2 = c_w^2 + c_{mw}^2 = c_{LU}^2\,, \quad c_2^2 = c_s^2 - c_{mw}^2 = c_w^2 - c_{ms}^2\,. \tag{11.5.46}$$

According to (11.5.23), the solid and fluid jumps on the longitudinal wave enjoy the relations,

$$\begin{aligned} c = c_1 : \quad & \boldsymbol{\eta}_w = \boldsymbol{\eta}_s, \\[6pt] c = c_2 : \quad & \boldsymbol{\eta}_w = -\frac{\rho^s}{\rho^w}\,\boldsymbol{\eta}_s\,. \end{aligned} \tag{11.5.47}$$

Then the fastest longitudinal wave turns out to be non dissipative, the solid and fluid jumps being equal. This argument led Biot to call the wave-speed c_0 the wave-speed of the *frozen mixture*. The second longitudinal wave is, on the other hand, strongly dissipative, i.e., it dies out at short distances.

11.5.2.5   Incompressible constituents

When both constituents are incompressible, the second order discontinuities $\boldsymbol{\eta}_\alpha$ are constrained by the rate form of the incompressibility condition (11.5.4), namely:

$$n^s\left(\boldsymbol{\eta}_s \cdot \mathbf{n}\right) + n^w\left(\boldsymbol{\eta}_w \cdot \mathbf{n}\right) = 0\,. \tag{11.5.48}$$

Therefore the number of independent modes of propagation is now at most three. The associated eigenvalue problem can be obtained using the limit procedure (7.3.54), or directly following Loret and Harireche [1991]: first the jump of the stress $\boldsymbol{\sigma}' = \boldsymbol{\sigma}^s + n^s\, p_w\, \mathbf{I}$ is formed by combining the two relations (11.5.10), and, second, the jump of the fluid velocity gradient $\boldsymbol{\eta}_w$ is eliminated with help of the constraint (11.5.48). The rate constitutive equations $\dot{\boldsymbol{\sigma}}' = \mathbb{E} : \dot{\boldsymbol{\epsilon}}^s$ then yield a generalized eigenvalue problem for the solid jump,

$$\left(\mathbf{A}^D - c^2\, \mathbf{I}(r)\right) \cdot \boldsymbol{\eta}_s = \mathbf{0}\,, \tag{11.5.49}$$

which involves the elastic acoustic tensor of the drained solid. However, the second order tensor $\mathbf{I}(r)$ differs from the identity $\mathbf{I} = \mathbf{I}(1)$ due to the scalar r greater than 1 (Section 2.3.5 shows a few properties of this rank one modification of the identity):

$$\mathbf{I}(r) = \mathbf{I} + (r-1)\,\mathbf{n} \otimes \mathbf{n}\,, \quad r = 1 + \left(\frac{n^s}{n^w}\right)^2 \frac{\rho^w}{\rho^s} > 1\,. \tag{11.5.50}$$

The characteristic equation $\det\left(\mathbf{A}^D - c^2\,\mathbf{I}(r)\right) = 0$ differs from its counterpart for single phase solids. Indeed, for isotropic elasticity,

$$\det\left(\mathbf{A}^D - c^2\,\mathbf{I}(r)\right) = r\,(c^2 - c_3^2)^2\,(c^2 - c_{2I}^2)\,, \tag{11.5.51}$$

in which we have anticipated the fact that there exist one shear wave-speed c_3 of double multiplicity and a single longitudinal wave with speed c_{2I},

$$c_{2I}^2 = \frac{1}{r}\frac{\lambda + 2\,\mu}{\rho^s}\,, \quad c_3^2 = \frac{\mu}{\rho^s}\,. \tag{11.5.52}$$

In contrast to the case of single phase solids, Eqn (11.2.9), and despite the inequalities (7.3.45), the presence of the coefficient $r > 1$ makes it possible for the elastic shear wave-speed to be larger than the elastic longitudinal wave-speed. This phenomenon is not at odds with the classification established in Section 11.5.2.1 for compressible porous media. In fact, one may consider incompressible porous media as a limit case of compressible media, with the bulk moduli of the solid and of the fluid tending to infinity along the scheme defined by (7.3.54). The largest longitudinal wave-speed c_1 is then observed to become unbounded while the second longitudinal wave-speed c_2 tends to the above value of c_{2I}. And, indeed, the relative order provided in Section 11.5.2.1 indicates that the second longitudinal wave-speed c_2 may be larger or smaller than the shear wave-speed c_3, depending on the material parameters.

    As a final remark, let us mention that the dynamic condition of compatibility which results in (11.5.47) and the incompressibility condition (11.5.48) are clearly at odds.

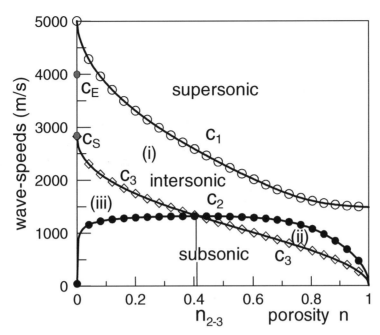

Fig. 11.5.2: Variations with porosity n of the three acceleration body wave-speeds, c_1 fast longitudinal wave-speed, c_2 slow longitudinal wave-speed and c_3 shear wave-speed, and corresponding subsonic region, supersonic region and intersonic regions (i), (ii), and (iii).

The parameters used for the solid constituent are: shear modulus $\mu_s = 20\,\mathrm{GPa}$, bulk modulus $K_s = 36\,\mathrm{GPa}$ (Poisson's ratio $\nu_s = 0.265$), mass density $\rho_s = 2500\,\mathrm{kg/m^3}$, porosity dependence $n_c = 1$, $q = 0.5$; and for the fluid constituent: bulk modulus $K_w = 2.2\,\mathrm{GPa}$, mass density $\rho_w = 1000\,\mathrm{kg/m^3}$.

For these parameters, the wave-speed c_2 is equal to the shear wave-speed c_3 for a porosity $n = n_{2-3} \sim 0.415$.

For the solid ($n = 0$), the longitudinal wave-speed c_L is equal to $5006.7\,\mathrm{m/s}$, Eshelby wave-speed $c_E = \sqrt{2}\,c_S$ to $4000\,\mathrm{m/s}$ and the shear wave-speed c_S to $2828.4\,\mathrm{m/s}$. These theoretical variations of the wave-speeds agree qualitatively with the available experimental data shown in Fig. 11.5.3.

11.5.2.6 Wave-speeds with porosity dependent moduli

As for the dependence of the moduli in porosity, a parameter $q = 1/2$ and a limit porosity $n_c^w = 1$ are assumed so that the relation (7.3.46) simplifies to $K/K_s = \mu/\mu_s = (1 - \sqrt{n^w})^2$. This dependence implies the modulus of elastic coupling $\lambda^{sw} \geq 0$ to be positive as indicated by (7.3.49), which justifies the two definitions of c_{ms} and c_{mw} in (11.5.20). Fig. 11.5.2 illustrates the typical variations of the wave-speeds with porosity for stiff materials, like bones. Fig. 11.5.3 shows experimental data for an artificial material and for a cancellous bone.

Irrespective of the values of $q \in\,]0, 1[$ and n_c^w, the wave-speeds c_1 and c_3 tend, as the porosity vanishes, to their counterparts in the solid constituent,

$$c_1^2 \to c_L^2 = \frac{K_s}{\rho_s} + \frac{4}{3}\frac{\mu_s}{\rho_s}, \quad c_2^2 \to 0, \quad c_3^2 \to c_S^2 = \frac{\mu_s}{\rho_s}. \tag{11.5.53}$$

At small porosity, the slow longitudinal wave-speed is proportional to $(n^w)^{(1-q)/2}$, so that the initial slope dc_2/dn^w is quite large. On the other hand, in his description of ultrasonic waves, as opposed to acceleration waves of interest here, Berryman [1995] obtains a much smoother relation $c_2(n^w)$, through the high-frequency device of Biot [1956].

For linear elastic solids, the so-called Eshelby speed $c_E = \sqrt{2}\,c_S$ is below the longitudinal wave-speed c_L if Poisson's ratio ν_s is positive: indeed $c_L/c_E = (1 + \nu_s/(1 - 2\,\nu_s))^{1/2}$.

At the upper limit $n^w = n_c^w$, the situation becomes similar to that of a compressible fluid with a single longitudinal wave-speed,

$$c_1^2 \to \frac{\dfrac{1 - n_c^w}{\rho_s} + \dfrac{n_c^w}{\rho_w}}{\dfrac{1 - n_c^w}{K_s} + \dfrac{n_c^w}{K_w}}, \quad c_2^2 \to 0, \quad c_3^2 \to 0. \tag{11.5.54}$$

These expressions are established using the relations of Section 7.3.6.5.

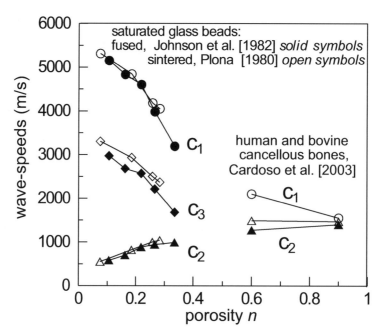

Fig. 11.5.3: Wave-speeds in porous materials in dry conditions or saturated by water: c_1 and c_2 fast and second (slow) longitudinal wave-speeds, c_3 shear wave-speed, Plona [1980], Johnson et al. [1982], Lakes et al. [1983], Cardoso et al. [2003].

11.5.3   Stationary discontinuity

Now that we have established the equations for the acceleration wave-speeds, we have the tools to study the existence of a stationary discontinuity where $c = 0$ is solution of the generalized eigenvalue problem (11.5.13). The dynamical interpretation of the phenomenon, also referred to as strain localization, is as follows: since it can not propagate any longer in the directions along which the wave-speed vanishes, the mechanical information accumulates in zones orthogonal to these directions and gives rise to a high strain gradient.

According to (11.5.10), the jump of the traction rate of each phase vanishes across a stationary discontinuity, namely $\mathbf{n} \cdot [\![\dot{\boldsymbol{\sigma}}^\alpha]\!] = \mathbf{0}$, $\alpha = \mathrm{s, w}$. While the tangent components of the rate of the solid partial stress $\dot{\boldsymbol{\sigma}}^\mathrm{s}$ may be discontinuous, the fact that the fluid stress is isotropic implies trivially,

Proposition 11.3 on the jump of the rate of fluid pressure.

The jump of the rate of fluid pressure across a stationary discontinuity vanishes:

$$[\![-\dot{p}^\mathrm{w}]\!] = 0 . \tag{11.5.55}$$

This simple observation indicates that, across a stationary discontinuity, the material behaves as if it was drained. In fact, setting $c = 0$ in (11.5.16), the condition for non zero jumps $\boldsymbol{\eta}_\alpha$, $\alpha = \mathrm{s, w}$, is that the determinant of the 4×4 acoustic matrix $\mathbf{A}$ vanishes,

$$\det \mathbf{A} = \det \begin{bmatrix} \mathbf{A}^\mathrm{ss} & \mathbf{a}^\mathrm{sw} \\ (\mathbf{a}^\mathrm{sw})^\mathrm{T} & c_\mathrm{w}^2 \end{bmatrix} = 0 . \tag{11.5.56}$$

Now, since the diagonal scalar c_w^2 is strictly positive, Exercise 11.1 shows that the singularity of $\mathbf{A}$ reduces to the singularity of its Schur complement in $\mathbf{A}$, which, due to (7.3.29), is equal to the acoustic tensor $\mathbf{A}^\mathrm{D}$ of the drained solid defined by (11.5.34),

$$\det \mathbf{A} = c_\mathrm{w}^2 \det \mathbf{A}^\mathrm{D} . \tag{11.5.57}$$

Therefore we have proved that stationary discontinuities are simultaneous in the porous medium and in its underlying drained solid. However, positive definiteness of the drained stiffness implies the drained wave-speeds to be strictly positive, and therefore prevents stationary discontinuities in the underlying drained solid, and consequently in the porous medium. This result was expected since the material parameters have been constrained so as to ensure the poroelastic energy to be positive definite, e.g., Section 7.4.2.

11.5.4 Interlacing and separation properties

With help of the two powerful theorems of linear algebra enounced in Section 2.2.1, the relative order of the wave-speeds of a saturated poroelastic medium, of its underlying drained solid and of its underlying undrained medium, are sorted out, Loret [1990]. The results hold for any type of mechanical anisotropy. By the same token, the isothermal and isentropic wave-speeds of a thermoelastic material are shown to interlace.

11.5.4.1 Interlacing property: drained solid and undrained medium

Let $\mathbf{A}$ be a symmetric positive definite matrix of size $m \times m$, $\mathbf{a}$ a vector of size m, and $a \geq 0$ a scalar, and let $\mathbf{A}(a)$ be a rank one modification of $\mathbf{A} = \mathbf{A}(0)$,

$$\mathbf{A}(a) = \mathbf{A} + a\,\mathbf{a} \otimes \mathbf{a}. \tag{11.5.58}$$

Then the eigenvalues $e_i^2(0)$ and $e_i^2(a)$, $i \in [1, m]$, of $\mathbf{A}(0)$ and $\mathbf{A}(a)$ respectively interlace, namely for $a \geq 0$,

$$e_m^2(0) \leq e_m^2(a) \leq e_{m-1}^2(0) \leq e_{m-1}^2(a) \cdots \leq e_1^2(0) \leq e_1^2(a). \tag{11.5.59}$$

The proof goes as follows. Let $\mathbf{I}_m$ be the identity matrix of size $m \times m$. Consider the eigenvalue problem,

$$\mathbf{Z}(a) \cdot \mathbf{e} = \mathbf{0}, \quad \text{with} \quad \mathbf{Z}(a) = \mathbf{A}(a) - e^2\,\mathbf{I}_m, \tag{11.5.60}$$

and assume the eigenvalues to be ordered in decreasing order, namely $e_m^2(a) \leq e_{m-1}^2(a) \cdots \leq e_1^2(a)$. Now, using the property (2.3.20), the determinant of $\mathbf{Z}(a)$ may be cast in the useful format,

$$
\begin{aligned}
\mathbf{Z}(a) &= \mathbf{Z}(0) \cdot (\mathbf{I}_m + a\,\mathbf{Z}(0)^{-1} \cdot \mathbf{a} \otimes \mathbf{a}) \\
\det \mathbf{Z}(a) &= \det \mathbf{Z}(0) + a\,\mathbf{a} \cdot \mathrm{adj}\,\mathbf{Z}(0) \cdot \mathbf{a},
\end{aligned}
\tag{11.5.61}
$$

where $\mathrm{adj}\,\mathbf{Z}(0)$ is the matrix adjoint to $\mathbf{Z}(0)$, namely $\det \mathbf{Z}(0) \times \mathbf{Z}(0)^{-1}$. When expressed in the eigenspace of $\mathbf{Z}(0)$, the determinant $\det \mathbf{Z}(a)$ viewed as a function of e^2 may be further expanded as, Mandel [1962],

$$F(e^2) = \prod_{i=1}^{m} (e_i^2(0) - e^2) + a \sum_{i=1}^{m} a_i^2 \prod_{j=1, j \neq i}^{m} (e_j^2(0) - e^2). \tag{11.5.62}$$

Consequently,

$$F(e_m^2(0)) \geq 0, \quad \cdots \quad (-1)^k\,F(e_{m-k}^2(0)) \geq 0, \quad \cdots \quad (-1)^{m-1}\,F(e_1^2(0)) \geq 0, \tag{11.5.63}$$

and this completes the proof in view of the fact that F is a polynomial of degree m in e^2 $\square$.

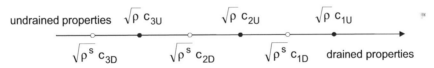

Fig. 11.5.4: Interlacing of drained and undrained characteristics.

This theorem can be applied to interlace the spectra of the drained and undrained acoustic tensors times an appropriate mass density, namely $\rho^s\,\mathbf{A}^D = \mathbf{n} \cdot \mathbb{E} \cdot \mathbf{n}$ and $\rho\,\mathbf{A}^U = \mathbf{n} \cdot \mathbb{E}^U \cdot \mathbf{n}$, respectively. Indeed, we have seen in Section 11.5.2.2 that the three drained wave-speeds are obtained by solving the eigenvalue problem,

$$(\mathbf{n} \cdot \mathbb{E} \cdot \mathbf{n} - \rho^s c^2 \mathbf{I}) \cdot \boldsymbol{\eta}_s^D = \mathbf{0}, \tag{11.5.64}$$

and in Section 11.5.2.3 that the three undrained wave-speeds are associated with the eigenvalue problem,

$$(\mathbf{n} \cdot \mathbb{E}^U \cdot \mathbf{n} - \rho c^2 \mathbf{I}) \cdot \boldsymbol{\eta}_s^U = \mathbf{0}. \tag{11.5.65}$$

Now, according to Section 7.4.2, the difference $\mathbf{n} \cdot \mathbb{E}^U \cdot \mathbf{n} - \mathbf{n} \cdot \mathbb{E} \cdot \mathbf{n}$ is dyadic with a positive coefficient $a = M$. Therefore,

$$\rho^s c_{3D}^2 \leq \rho\,c_{3U}^2 \leq \rho^s c_{2D}^2 \leq \rho\,c_{2U}^2 \leq \rho^s c_{1D}^2 \leq \rho\,c_{1U}^2. \tag{11.5.66}$$

The interlacing property is illustrated in Fig. 11.5.4. It holds for the square-roots of the entities involved. On the other hand, the inequalities do not hold a priori for the wave-speeds by themselves.

For isotropic poroelastic properties, the shear wave-speed has multiplicity two, and the inequalities simplify to,

$$\rho^s c_{SD}^2 = \rho\,c_{SU}^2 \leq \rho^s c_{LD}^2 \leq \rho\,c_{LU}^2, \tag{11.5.67}$$

as can be checked in Sections 7.4.2 and 11.5.2.3.

11.5.4.2   Interlacing property: isothermal and isentropic wave-speeds

In Section 11.3, the thermoelastic wave-speeds associated with a Fourier's heat conductor have been shown to be governed by the acoustic tensor of the underlying isothermal tensor moduli. By contrast, the thermoelastic wave-speeds associated with a non conductor have been shown to be governed by the acoustic tensor of the underlying isentropic tensor moduli.

The isothermal and isentropic tensor moduli are related by (9.1.7) so that their acoustic tensors differ only by a symmetric dyadic product. The interlacing property established in the previous section applies and actually can be sharpened since the mass densities associated with the two solids are identical. Indeed,

$$c_{3\text{isoth}} \le c_{3\text{isEn}} \le c_{2\text{isoth}} \le c_{2\text{isEn}} \le c_{1\text{isoth}} \le c_{1\text{isEn}}\,. \qquad (11.5.68)$$

For isotropic thermoelastic properties, the elastic shear wave-speed c_S has double multiplicity for both solids, and the inequalities simplify to,

$$c_\text{S} \le c_{1\text{isoth}} \le c_{1\text{isEn}}\,. \qquad (11.5.69)$$

11.5.4.3   Separation property: drained solid and porous medium

The acoustic matrix $\mathbf{A}$ defining the eigenvalue problem that yields the (real) wave-speeds of the porous medium (11.5.16) has the format considered in Exercise 11.1, namely

$$\mathbf{A} = \begin{bmatrix} \mathbf{A}^{\text{ss}} - c^2\,\mathbf{I} & \mathbf{a}^{\text{sw}} \\ (\mathbf{a}^{\text{sw}})^{\text{T}} & c_{\text{w}}^2 - c^2 \end{bmatrix}, \qquad (11.5.70)$$

and therefore

$$F(c^2) \equiv \det\left[\mathbf{A} - c^2\mathbf{I}_4\right] = (c_{\text{w}}^2 - c^2)\,\det(\mathbf{A}^{\text{ss}} - c^2\,\mathbf{I}) - \mathbf{a}^{\text{sw}} \cdot \text{adj}\,(\mathbf{A}^{\text{ss}} - c^2\,\mathbf{I}) \cdot \mathbf{a}^{\text{sw}}\,. \qquad (11.5.71)$$

Due to (7.3.29), the drained acoustic tensor can also be rephrased in terms of $\mathbf{A}^{\text{ss}}$, $\mathbf{a}^{\text{sw}}$ and c_{w}^2, namely

$$\mathbf{A}^{\text{D}} = \mathbf{A}^{\text{ss}} - \frac{\mathbf{a}^{\text{sw}} \otimes \mathbf{a}^{\text{sw}}}{c_{\text{w}}^2}\,. \qquad (11.5.72)$$

Therefore, using again the property (2.3.20),

$$F_{\text{D}}(c_{\text{D}}^2) \equiv \det\left[\mathbf{A}^{\text{D}} - c_{\text{D}}^2\mathbf{I}_3\right] = \frac{1}{c_{\text{w}}^2}\left(c_{\text{w}}^2\,\det(\mathbf{A}^{\text{ss}} - c_{\text{D}}^2\,\mathbf{I}) - \mathbf{a}^{\text{sw}} \cdot \text{adj}\,(\mathbf{A}^{\text{ss}} - c_{\text{D}}^2\,\mathbf{I}) \cdot \mathbf{a}^{\text{sw}}\right)\,. \qquad (11.5.73)$$

The wave-speeds c_i, $i \in [1,4]$, of the porous medium satisfy the secular equation $F(c_i^2) = 0$. Combining (11.5.71) and (11.5.73) yields,

$$F_{\text{D}}(c_i^2) = \frac{c_i^2}{c_{\text{w}}^2}\,\det(\mathbf{A}^{\text{ss}} - c_i^2\,\mathbf{I})\,. \qquad (11.5.74)$$

The sign of $\det(\mathbf{A}^{\text{ss}} - c_i^2\,\mathbf{I})$ is deduced from the theorem of algebra referred to as separation theorem enounced in Section 2.2.1:

$$F_{\text{D}}(c_4^2) \ge 0, \quad F_{\text{D}}(c_3^2) \le 0, \quad F_{\text{D}}(c_2^2) \ge 0, \quad F_{\text{D}}(c_1^2) \le 0\,. \qquad (11.5.75)$$

Consequently, the (squares of the) three wave-speeds of the drained solid separate the (squares of the) four wave-speeds of the porous medium as illustrated by Fig. 11.5.5,

$$c_4 \le c_{3\text{D}} \le c_3 \le c_{2\text{D}} \le c_2 \le c_{1\text{D}} \le c_1\,. \qquad (11.5.76)$$

For isotropic poroelastic properties, the drained shear wave-speed is smaller than the longitudinal wave-speed and it has multiplicity two, namely $c_{2\text{D}} = c_{3\text{D}} = c_{\text{SD}} < c_{1\text{D}}$. Consequently, the wave-speed c_3 of the porous medium is necessarily equal to c_{SD},

$$c_4 \le c_3 = c_{\text{SD}} \le c_2 \le c_{\text{LD}} \le c_1\,. \qquad (11.5.77)$$

In fact, the shear wave-speed has also multiplicity two for the porous medium, and either it is the smallest wave-speed,

$$c_4 = c_3 = c_{\text{SD}} \le c_2 \le c_{\text{LD}} \le c_1\,, \qquad (11.5.78)$$

or the second longitudinal wave-speed is the smallest one,

$$c_4 \le c_3 = c_{\text{SD}} = c_2 \le c_{\text{LD}} \le c_1\,. \qquad (11.5.79)$$

The notation is different from that of Section 11.5.2.1 where the second longitudinal wave-speed was denoted by c_2 and the shear wave-speed by c_3, while the numbers here correspond to the rank.

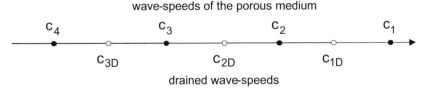

wave-speeds of the porous medium

Fig. 11.5.5: The drained wave-speeds separate the wave-speeds of the saturated porous medium.

11.5.5 Inertial coupling: influence on acceleration waves

Inertial coupling has been introduced by Biot [1956] to induce an added mass effect. The idea is motivated in Section 7.3.1. Inertial coupling is often neglected because it is difficult to quantify. However, it might be necessary to take it into account, for instance for an accurate interpretation of wave-speeds in the resonant column. Actually, due to viscous coupling, the response of porous media to a general impulse is dispersive. Gajo [1996] has analyzed the relative influence of inertial and viscous couplings according to porosity and permeability.

Implications of inertial coupling on the speeds and modes of propagation of acceleration waves are reported in Loret and Rizzi [1998]. In essence, inertial coupling does not alter the structures neither of the secular polynomial yielding the wave-speeds, nor of the eigenvalue problem for the solid eigenmode. On the other hand, the structure of the fluid eigenmode is modified by inertial coupling as transverse motions in the solid phase carry over to the fluid phase. Consequently, Proposition 11.2 does not hold in presence of inertial coupling.

As a consequence of inertial coupling, the discontinuity relations across the acceleration front (11.5.10) modify to,

$$\mathbf{n} \cdot [\![\dot{\boldsymbol{\sigma}}^{\alpha}]\!] = \sum_{\beta = \mathrm{s}, \mathrm{w}} \rho^{\alpha\beta} \, c^2 \, \boldsymbol{\eta}_{\beta} \,, \quad \alpha = \mathrm{s}, \mathrm{w} \,. \tag{11.5.80}$$

We consider in turn compressible and incompressible constituents. The analysis is restricted to isotropic mechanical properties.

11.5.5.1 Inertial coupling with compressible constituents

The generalized eigenvalue problem (11.5.13) modifies to,

$$\begin{bmatrix} (\lambda^{\mathrm{ss}} + 2\mu) \, \mathbf{n} \otimes \mathbf{n} + \mu \, (\mathbf{I} - \mathbf{n} \otimes \mathbf{n}) & \lambda^{\mathrm{sw}} \, \mathbf{n} \otimes \mathbf{n} \\ \lambda^{\mathrm{sw}} \, \mathbf{n} \otimes \mathbf{n} & \lambda^{\mathrm{ww}} \, \mathbf{n} \otimes \mathbf{n} \end{bmatrix} \begin{bmatrix} \boldsymbol{\eta}_{\mathrm{s}} \\ \boldsymbol{\eta}_{\mathrm{w}} \end{bmatrix} = c^2 \begin{bmatrix} \rho^{\mathrm{ss}} \, \mathbf{I} & \rho^{\mathrm{sw}} \, \mathbf{I} \\ \rho^{\mathrm{sw}} \, \mathbf{I} & \rho^{\mathrm{ww}} \, \mathbf{I} \end{bmatrix} \begin{bmatrix} \boldsymbol{\eta}_{\mathrm{s}} \\ \boldsymbol{\eta}_{\mathrm{w}} \end{bmatrix} \,. \tag{11.5.81}$$

Positive definiteness of the elastic energy of the porous medium implies positive definiteness of the symmetric constitutive array (7.3.28), and consequently of the constitutive array appearing at the left hand side of (11.5.81). Moreover, Section 7.3.1 has restricted the coefficients of inertial coupling so that the kinetic energy and the mass array appearing at the right-hand side of (11.5.81) are positive definite as well. The generalized eigenvalue problem can then be transformed to a standard eigenvalue problem, featuring a symmetric positive definite matrix. Thus positive definiteness of the strain and kinetic energies ensures the elastic wave-speeds to be real and strictly positive.

The jump in the fluid phase $\boldsymbol{\eta}_{\mathrm{w}}$ can be eliminated in favor of the solid jump $\boldsymbol{\eta}_{\mathrm{s}}$,

$$\boldsymbol{\eta}_{\mathrm{w}} = -\Big(\frac{\rho^{\mathrm{sw}} c^2 - \lambda^{\mathrm{sw}}}{\rho^{\mathrm{ww}} c^2 - \lambda^{\mathrm{ww}}} \, \mathbf{n} \otimes \mathbf{n} + \frac{\rho^{\mathrm{sw}}}{\rho^{\mathrm{ww}}} (\mathbf{I} - \mathbf{n} \otimes \mathbf{n}) \Big) \cdot \boldsymbol{\eta}_{\mathrm{s}} \,, \tag{11.5.82}$$

so as to obtain an equation in terms of the jump in the solid phase $\boldsymbol{\eta}_{\mathrm{s}}$ only,

$$\Big(a \, \mathbf{n} \otimes \mathbf{n} + b \, (\mathbf{I} - \mathbf{n} \otimes \mathbf{n}) \Big) \cdot \boldsymbol{\eta}_{\mathrm{s}} = \mathbf{0} \,, \tag{11.5.83}$$

where

$$a = \lambda^{\mathrm{ss}} + 2\,\mu - \rho^{\mathrm{ss}} \, c^2 + \frac{(\rho^{\mathrm{sw}} c^2 - \lambda^{\mathrm{sw}})^2}{\rho^{\mathrm{ww}} c^2 - \lambda^{\mathrm{ww}}} \,, \quad b = \mu - \tilde{\rho}^{\mathrm{ss}} \, c^2 \,, \tag{11.5.84}$$

with

$$\tilde{\rho}^{\mathrm{ss}} \equiv \rho^{\mathrm{ss}} \, \Delta \,, \quad \Delta \equiv 1 - \frac{(\rho^{\mathrm{sw}})^2}{\rho^{\mathrm{ss}} \rho^{\mathrm{ww}}} > 0 \,. \tag{11.5.85}$$

Now

$$\det \left(a \, \mathbf{n} \otimes \mathbf{n} \, + \, b \, (\mathbf{I} - \mathbf{n} \otimes \mathbf{n}) \right) = a \, b^2 . \tag{11.5.86}$$

Thus there exist in general one shear wave-speed c_3 of double multiplicity corresponding to $b^2 = 0$, and two longitudinal wave-speeds $c_2 \le c_1$ corresponding to $a = 0$, Biot [1956]. As a characteristic of inertial coupling, the shear wave gives rise to in-phase motions in both the solid and fluid phases (remember that ρ^{sw} is negative),

$$c_3 = \left(\frac{\mu}{\bar{\rho}^{\mathrm{ss}}} \right)^{1/2} , \quad \boldsymbol{\eta}_{\mathrm{s}} \perp \mathbf{n} , \quad \boldsymbol{\eta}_{\mathrm{w}} = -\frac{\rho^{\mathrm{sw}}}{\rho^{\mathrm{ww}}} \, \boldsymbol{\eta}_{\mathrm{s}} . \tag{11.5.87}$$

The longitudinal wave-speeds are obtained from the sum and product of their squares:

$$\begin{aligned} c_1^2 + c_2^2 &= \frac{1}{\Delta} \left(\frac{\lambda^{\mathrm{ss}} + 2\mu}{\rho^{\mathrm{ss}}} + \frac{\lambda^{\mathrm{ww}}}{\rho^{\mathrm{ww}}} - 2 \frac{\rho^{\mathrm{sw}} \lambda^{\mathrm{sw}}}{\rho^{\mathrm{ss}} \rho^{\mathrm{ww}}} \right) , \\ c_1^2 \, c_2^2 &= \frac{1}{\Delta} \frac{\lambda + 2\mu}{\rho^{\mathrm{ss}}} \frac{\lambda^{\mathrm{ww}}}{\rho^{\mathrm{ww}}} , \end{aligned} \tag{11.5.88}$$

with the usual notation $\lambda = \lambda^{\mathrm{ss}} - (\lambda^{\mathrm{sw}})^2 / \lambda^{\mathrm{ww}}$ and μ for the Lamé moduli of the underlying drained solid. Indeed, with a defined by (11.5.84), the longitudinal wave-speeds are found to be the roots of a polynomial of degree two in c^2:

$$F(c^2) = \Delta \, (c^2)^2 - \left(\frac{\lambda^{\mathrm{ss}} + 2\mu}{\rho^{\mathrm{ss}}} + \frac{\lambda^{\mathrm{ww}}}{\rho^{\mathrm{ww}}} - 2 \frac{\rho^{\mathrm{sw}} \lambda^{\mathrm{sw}}}{\rho^{\mathrm{ss}} \rho^{\mathrm{ww}}} \right) c^2 + \frac{\lambda + 2\mu}{\rho^{\mathrm{ss}}} \frac{\lambda^{\mathrm{ww}}}{\rho^{\mathrm{ww}}} . \tag{11.5.89}$$

The associated longitudinal jumps are linked by Eqn (11.5.82),

$$\boldsymbol{\eta}_{\mathrm{s}} = (\boldsymbol{\eta}_{\mathrm{s}} \cdot \mathbf{n}) \, \mathbf{n} , \quad \boldsymbol{\eta}_{\mathrm{w}} = -\frac{\rho^{\mathrm{sw}} c^2 - \lambda^{\mathrm{sw}}}{\rho^{\mathrm{ww}} c^2 - \lambda^{\mathrm{ww}}} \, \boldsymbol{\eta}_{\mathrm{s}} . \tag{11.5.90}$$

Information about the relative location of the scalar x with respect to the wave-speeds may be obtained from the sign of the function $F(x^2)$:

$$F\left(\frac{\lambda^{\mathrm{sw}}}{\rho^{\mathrm{sw}}} \right) = \left(\frac{\lambda^{\mathrm{sw}}}{\rho^{\mathrm{sw}}} - \frac{\lambda^{\mathrm{ss}} + 2\mu}{\rho^{\mathrm{ss}}} \right) \left(\frac{\lambda^{\mathrm{sw}}}{\rho^{\mathrm{sw}}} - \frac{\lambda^{\mathrm{ww}}}{\rho^{\mathrm{ww}}} \right) , \tag{11.5.91}$$

$$F\left(\frac{\lambda^{\mathrm{ww}}}{\rho^{\mathrm{ww}}} \right) = -\frac{(\rho^{\mathrm{sw}})^2}{\rho^{\mathrm{ss}} \rho^{\mathrm{ww}}} \left(\frac{\lambda^{\mathrm{ww}}}{\rho^{\mathrm{ww}}} - \frac{\lambda^{\mathrm{sw}}}{\rho^{\mathrm{sw}}} \right)^2 \le 0 , \tag{11.5.92}$$

$$F\left(\frac{\lambda^{\mathrm{ss}} + 2\mu}{\rho^{\mathrm{ss}}} \right) = -\frac{(\rho^{\mathrm{sw}})^2}{\rho^{\mathrm{ss}} \rho^{\mathrm{ww}}} \left(\frac{\lambda^{\mathrm{ss}} + 2\mu}{\rho^{\mathrm{ss}}} - \frac{\lambda^{\mathrm{sw}}}{\rho^{\mathrm{sw}}} \right)^2 \le 0 , \tag{11.5.93}$$

$$F(c_0^2) = -\frac{(\rho^{\mathrm{s}} \rho^{\mathrm{w}})^2}{(\rho^{\mathrm{s}} + \rho^{\mathrm{w}})^2 \rho^{\mathrm{ss}} \rho^{\mathrm{ww}}} \left(\frac{\lambda^{\mathrm{ss}} + \lambda^{\mathrm{sw}} + 2\mu}{\rho^{\mathrm{s}}} - \frac{\lambda^{\mathrm{ww}} + \lambda^{\mathrm{sw}}}{\rho^{\mathrm{w}}} \right)^2 \le 0 , \tag{11.5.94}$$

where c_0,

$$c_0^2 = \frac{\lambda^{\mathrm{ss}} + 2\mu + \lambda^{\mathrm{ww}} + 2\lambda^{\mathrm{sw}}}{\rho^{\mathrm{s}} + \rho^{\mathrm{w}}} , \tag{11.5.95}$$

is the *frozen mixture* wave-speed, Section 11.5.2.4. Therefore,

$$c_2^2 \le \min \left\{ \frac{\lambda^{\mathrm{ss}} + 2\mu}{\rho^{\mathrm{ss}}}, \frac{\lambda^{\mathrm{ww}}}{\rho^{\mathrm{ww}}}, c_0^2 \right\} \le \max \left\{ \frac{\lambda^{\mathrm{ss}} + 2\mu}{\rho^{\mathrm{ss}}}, \frac{\lambda^{\mathrm{ww}}}{\rho^{\mathrm{ww}}}, c_0^2 \right\} \le c_1^2 . \tag{11.5.96}$$

The mass densities involved in the above relations are linked by the restrictions detailed in Section 7.3.1.
    When the dynamic condition of compatibility,

$$\frac{\lambda^{\mathrm{ss}} + \lambda^{\mathrm{sw}} + 2\mu}{\rho^{\mathrm{s}}} = \frac{\lambda^{\mathrm{ww}} + \lambda^{\mathrm{sw}}}{\rho^{\mathrm{w}}} , \tag{11.5.97}$$

is satisfied, then c_0 is a wave-speed as indicated by (11.5.95), actually the larger longitudinal wave-speed c_1,

$$c_0^2 = c_1^2 = \frac{\lambda^{\mathrm{ss}} + \lambda^{\mathrm{sw}} + 2\mu}{\rho^{\mathrm{s}}} = \frac{\lambda^{\mathrm{ww}} + \lambda^{\mathrm{sw}}}{\rho^{\mathrm{w}}} = \frac{\lambda^{\mathrm{ss}} + 2\mu + \lambda^{\mathrm{ww}} + 2\lambda^{\mathrm{sw}}}{\rho^{\mathrm{s}} + \rho^{\mathrm{w}}} . \tag{11.5.98}$$

The faster longitudinal wave then results to be non dissipative, since (11.5.82) yields $\boldsymbol{\eta}_{\mathrm{w}} = \boldsymbol{\eta}_{\mathrm{s}}$.

11.5.5.2 Inertial coupling with incompressible constituents

When both the solid and the fluid constituents are incompressible, the fluid pressure is indeterminate, the normal components of the solid and fluid jumps satisfy the relation (11.5.48), and the constitutive equations provide the effective stress as a function of the solid strain. Upon elimination of the jump of the rate of fluid pressure with help of (11.5.80) follows a relation involving the fluid and solid jumps,

$$\left((\lambda + 2\,\mu)\,\mathbf{n} \otimes \mathbf{n} + \mu\,(\mathbf{I} - \mathbf{n} \otimes \mathbf{n}) \right) \cdot \boldsymbol{\eta}_s = c^2 \left(\left(\rho^{ss} - \frac{n^s}{n^w} \rho^{sw} \right) \boldsymbol{\eta}_s + \left(\rho^{sw} - \frac{n^s}{n^w} \rho^{ww} \right) \boldsymbol{\eta}_w \right) . \tag{11.5.99}$$

From the discontinuity equation (11.5.48),

$$\boldsymbol{\eta}_w = -\frac{n^s}{n^w} (\boldsymbol{\eta}_s \cdot \mathbf{n})\,\mathbf{n} + \boldsymbol{\eta}_w^\perp , \quad \text{with} \quad \boldsymbol{\eta}_w^\perp \cdot \mathbf{n} = 0 . \tag{11.5.100}$$

If $c \neq 0$, Eqn (11.5.80) for the fluid phase shows the sum $\rho^{sw} \boldsymbol{\eta}_s + \rho^{ww} \boldsymbol{\eta}_w$ to be parallel to $\mathbf{n}$. Therefore, the fluid jump can be expressed in terms of the solid jump through the relation,

$$\boldsymbol{\eta}_w = -\left(\frac{n^s}{n^w} \mathbf{n} \otimes \mathbf{n} + \frac{\rho^{sw}}{\rho^{ww}} (\mathbf{I} - \mathbf{n} \otimes \mathbf{n}) \right) \cdot \boldsymbol{\eta}_s , \tag{11.5.101}$$

which could have been derived directly from Eqn (11.5.82) using the limit process (7.3.54). Elimination of the fluid jump from (11.5.99) gives a generalized eigenvalue problem for the solid jump, which holds for any c,

$$\left(\frac{\lambda + 2\,\mu}{\tilde{\rho}^{ss}}\,\mathbf{n} \otimes \mathbf{n} + \frac{\mu}{\tilde{\rho}^{ss}}\,(\mathbf{I} - \mathbf{n} \otimes \mathbf{n}) \right) \cdot \boldsymbol{\eta}_s = c^2\,\mathbf{I}(\tilde{r}) \cdot \boldsymbol{\eta}_s , \tag{11.5.102}$$

where $\mathbf{I}(\tilde{r}) = \mathbf{I} + (\tilde{r} - 1)\,\mathbf{n} \otimes \mathbf{n}$ and

$$\tilde{r} \equiv 1 + \frac{\rho^{ww}}{\tilde{\rho}^{ss}} \left(\frac{n^s}{n^w} - \frac{\rho^{sw}}{\rho^{ww}} \right)^2 > 1 . \tag{11.5.103}$$

This eigenvalue problem has exactly the same form as its counterpart without inertial coupling, as described in Section 11.5.2.5. The uncoupled wave-speeds are retrieved from the above relations by the substitutions $\tilde{\rho}^{ss} \to \rho^s$ and $\tilde{r} \to r \equiv 1 + (\rho^w/\rho^s)(n^s/n^w)^2$. Like for compressible constituents, the shear wave-speed has double multiplicity, its value and the eigenmodes are still given by (11.5.87). However there is a single finite longitudinal wave with speed c_2,

$$c_2 = \left(\frac{1}{\tilde{r}}\,\frac{\lambda + 2\,\mu}{\tilde{\rho}^{ss}} \right)^{1/2} , \tag{11.5.104}$$

and the associated fluid and solid jumps are linked by the relation $\boldsymbol{\eta}_w = -\boldsymbol{\eta}_s\,n^s/n^w$.

11.5.6   Waves in fluid saturated porous media: pushing the window ajar

The analysis of the existence of isothermal acceleration waves so far may be extended to include fluid saturated porous media with double porosity introduced in Section 10.6.

Table 11.5.2: Acceleration wave-speeds for isotropic solids and fluid saturated porous media. The longitudinal wave-speeds have in general single multiplicity and the shear wave-speeds double multiplicity.

Material	nb. Longitudinal Wave-Speed	nb. Shear Wave-Speed
Solid	1	1
Porous solid with single porosity	2	1
Porous solid with double porosity	3	1

Table 11.5.2 applies to fluid saturated porous media in the following context: (1) their mechanical constitutive equations are linear; (2) their mechanical constitutive equations are isotropic; (3) the fluids filling the pores are non viscous.

Harmonic waves provide a more general tool to probe mechanical and transport properties than acceleration waves. In fact, acceleration waves can be viewed as the limit of harmonic waves where the frequency tends to infinity.

11.6   Propagation of 2nd order waves in saturated porous media

In Section 11.5, we have formally defined the modes under which plane acceleration waves propagate. We shall now study how their amplitude varies in time as they propagate. We begin by establishing the differential equation governing the evolution in time of the jumps of the velocity gradients. Next, we will analyze the coefficients of growth or decay depending on the details of the constitutive equations.

To simplify the presentation, we consider only incompressible constituents, and direct the reader to Loret et al. [1997] for the compressible case. Moreover, the body forces are assumed to vanish, and the momentum supplies reduce to Darcy's law.

11.6.1   Equations of propagation

Elements of the differential equations that govern the propagation of discontinuities on a moving surface have been summarized in Section 11.1.6. They will be used in the present context of second order waves.

Upon derivation with respect to time, the linearized equations of balance of momentum (11.5.5) yield the discontinuity equations,

$$\rho^\alpha [\![\ddot{\mathbf{v}}_\alpha]\!] = [\![\operatorname{div}\dot{\boldsymbol{\sigma}}^\alpha]\!] - \epsilon_\alpha\, k_{\mathrm{Sd}}\left([\![\dot{\mathbf{v}}_{\mathrm{s}}]\!] - [\![\dot{\mathbf{v}}_{\mathrm{w}}]\!]\right), \tag{11.6.1}$$

with $\epsilon_{\mathrm{s}} = 1$ and $\epsilon_{\mathrm{w}} = -1$. The evolution of the amplitude of the acceleration wave front is expressed through a differential equation in terms of the *displacement derivative*, Eqn (11.1.66),

$$\frac{\delta[\![\dot{\mathbf{v}}_\alpha]\!]}{\delta t} = \frac{\partial[\![\dot{\mathbf{v}}_\alpha]\!]}{\partial t} + \boldsymbol{\nabla}[\![\dot{\mathbf{v}}_\alpha]\!]\cdot(c\,\mathbf{n}). \tag{11.6.2}$$

Now, since the velocity is continuous across the front that travels at constant speed c in a given direction $\mathbf{n}$, the *iterated kinematical compatibility equation* (11.1.73) simplifies to

$$2\,\frac{\delta[\![\dot{\mathbf{v}}_\alpha]\!]}{\delta t} = [\![\ddot{\mathbf{v}}_\alpha]\!] - c^2\,\mathbf{n}\cdot\left[\![\boldsymbol{\nabla}^2\mathbf{v}_\alpha]\!\right]\cdot\mathbf{n}. \tag{11.6.3}$$

Eliminating the discontinuity $[\![\ddot{\mathbf{v}}_\alpha]\!]$, $\alpha = \mathrm{s},\mathrm{w}$, from (11.6.1) and (11.6.3) and using the rate constitutive equations $\dot{\boldsymbol{\sigma}}' = \mathbb{E}:\dot{\boldsymbol{\varepsilon}}^{\mathrm{s}}$ in terms of the effective stress $\boldsymbol{\sigma}' = \boldsymbol{\sigma}^{\mathrm{s}} - n^{\mathrm{s}}/n^{\mathrm{w}}\,\boldsymbol{\sigma}_{\mathrm{w}}$, the evolution equations for the discontinuities of acceleration $[\![\dot{\mathbf{v}}_\alpha]\!]$, $\alpha = \mathrm{s},\mathrm{w}$, follow,

$$\begin{aligned}
2\,\rho^{\mathrm{s}}\,\frac{\delta[\![\dot{\mathbf{v}}_{\mathrm{s}}]\!]}{\delta t} &= \rho^{\mathrm{s}}\,(\mathbf{A}^{\mathrm{D}} - c^2\,\mathbf{I})\cdot\boldsymbol{\Delta}_{\mathrm{s}} &+&\ \frac{n^{\mathrm{s}}}{n^{\mathrm{w}}}[\![\operatorname{div}\dot{\boldsymbol{\sigma}}^{\mathrm{w}}]\!] &-&\ k_{\mathrm{Sd}}\left([\![\dot{\mathbf{v}}_{\mathrm{s}}]\!] - [\![\dot{\mathbf{v}}_{\mathrm{w}}]\!]\right), \\
2\,\rho^{\mathrm{w}}\,\frac{\delta[\![\dot{\mathbf{v}}_{\mathrm{w}}]\!]}{\delta t} &= -\rho^{\mathrm{w}}c^2\,\boldsymbol{\Delta}_{\mathrm{w}} &+&\ [\![\operatorname{div}\dot{\boldsymbol{\sigma}}^{\mathrm{w}}]\!] &+&\ k_{\mathrm{Sd}}\left([\![\dot{\mathbf{v}}_{\mathrm{s}}]\!] - [\![\dot{\mathbf{v}}_{\mathrm{w}}]\!]\right),
\end{aligned} \tag{11.6.4}$$

where $\mathbf{A}^{\mathrm{D}}$ is the drained acoustic tensor and $\boldsymbol{\Delta}_\alpha = \mathbf{n}\cdot\left[\![\boldsymbol{\nabla}^2\mathbf{v}_\alpha]\!\right]\cdot\mathbf{n}$, $\alpha = \mathrm{s},\mathrm{w}$, are the induced discontinuities. Due to Hadamard's compatibility relation (11.2.1), the relation (11.5.11) and the incompressibility condition (11.5.48) imply the discontinuity of acceleration in the fluid phase to be aligned with the wave normal $\mathbf{n}$, and

$$[\![\dot{\mathbf{v}}_{\mathrm{w}}]\!] = -\frac{n^{\mathrm{s}}}{n^{\mathrm{w}}}\left([\![\dot{\mathbf{v}}_{\mathrm{s}}]\!]\cdot\mathbf{n}\right)\mathbf{n}. \tag{11.6.5}$$

As another consequence of Hadamard's compatibility relation, the jump $[\![\operatorname{div}\dot{\boldsymbol{\sigma}}^{\mathrm{w}}]\!]$ is aligned with $\mathbf{n}$. Upon space differentiation of the incompressibility condition (11.5.4), the induced discontinuities are seen to be balanced by the condition,

$$n^{\mathrm{s}}(\boldsymbol{\Delta}_{\mathrm{s}}\cdot\mathbf{n}) + n^{\mathrm{w}}(\boldsymbol{\Delta}_{\mathrm{w}}\cdot\mathbf{n}) = 0. \tag{11.6.6}$$

Therefore, the projection of (11.6.4)$_2$ along $\mathbf{n}$ can be expressed as,

$$-2\,\rho^{\mathrm{w}}\,\frac{\delta}{\delta t}\left(\frac{n^{\mathrm{s}}}{n^{\mathrm{w}}}([\![\dot{\mathbf{v}}_{\mathrm{s}}]\!]\cdot\mathbf{n})\,\mathbf{n}\right) = [\![\operatorname{div}\dot{\boldsymbol{\sigma}}^{\mathrm{w}}]\!] + \rho^{\mathrm{w}}c^2\,\frac{n^{\mathrm{s}}}{n^{\mathrm{w}}}(\boldsymbol{\Delta}_{\mathrm{s}}\cdot\mathbf{n})\,\mathbf{n} + \frac{k_{\mathrm{Sd}}}{n^{\mathrm{w}}}([\![\dot{\mathbf{v}}_{\mathrm{s}}]\!]\cdot\mathbf{n})\,\mathbf{n}. \tag{11.6.7}$$

Subtraction of Eqn (11.6.7) multiplied by $n^{\mathrm{s}}/n^{\mathrm{w}}$ from Eqn (11.6.4) yields the evolution equation for the 3-component vector $\mathbf{V} = [\![\dot{\mathbf{v}}_{\mathrm{s}}]\!]$ in the solid phase,

$$2\,\mathbf{I}(r)\cdot\frac{\delta\mathbf{V}}{\delta t} = \left(\mathbf{A}^{\mathrm{D}} - c^2\,\mathbf{I}(r)\right)\cdot\boldsymbol{\Delta}_{\mathrm{s}} - k_{\mathrm{Sd}}\,\frac{\mathbf{I}(e)}{\rho^{\mathrm{s}}}\cdot\mathbf{V}, \tag{11.6.8}$$

where, for any real x, the matrix $\mathbf{I}(x) = \mathbf{I} + (x-1)\,\mathbf{n}\otimes\mathbf{n}$, r is defined by (11.5.50), and $e = 1/(n^{\mathrm{w}})^2 > 1$.

Once the solid jump is known, the wave front in the fluid-phase is then given by (11.6.5).

11.6.2   The decaying wave-front

Since we assume the mechanical state on the wave front to be fixed, the unit left and right eigenvectors of the generalized eigenvalue problem (11.5.49), namely $\mathbf{e}^L$ and $\mathbf{e}^R$, are constant. Therefore, multiplication of the evolution equation (11.6.8) by the unit left eigenvector $\mathbf{e}^L$ yields a differential equation for the amplitude $|\mathbf{V}|$ that is integrated to,

$$\frac{|\mathbf{V}|(t)}{|\mathbf{V}|(0)} = \exp\left(-\frac{k_{\mathrm{Sd}}}{2\rho^{\mathrm{s}}}\,\zeta\,t\right), \quad \zeta = \frac{\mathbf{e}^L \cdot \mathbf{I}(e) \cdot \mathbf{e}^R}{\mathbf{e}^L \cdot \mathbf{I}(r) \cdot \mathbf{e}^R}. \tag{11.6.9}$$

Notice that the products by $\mathbf{e}^L$ are not inner products in the usual sense, that is, they remain unchanged even if $\mathbf{e}^L$ is complex, Wilkinson [1965], Eqn (3.10), p. 4.

Since the coefficient k_{Sd} is a real positive number, growth or decay of the wave front is governed by the sign of the dimensionless *decay* coefficient ζ. For a poroelastic skeleton, the generalized eigenvalue problem (11.5.49) is symmetric, the wave-speeds are real and the left and right eigenvectors are one and the same. Since the matrices $\mathbf{I}(r)$ and $\mathbf{I}(e)$ are symmetric positive definite, the coefficient ζ is then positive and propagation is accompanied by a wave front whose amplitude strictly decays: the fact that the two constituents are incompressible excludes the possibility that the fastest wave can propagate without dissipation, a situation referred to as *dynamic compatibility* by Biot [1956] and considered in Section 11.5.2.4. In fact, this remark can be sharpened.

11.6.2.1   The decay coefficient as a Rayleigh quotient

Indeed, let us transform the generalized eigenvalue problem (11.5.49) into a standard eigenvalue problem,

$$\big((\mathbf{A}^{\mathrm{D}})^{\#} - c^2\,\mathbf{I}\big) \cdot \boldsymbol{\eta}_{\mathrm{s}}^{\#} = \mathbf{0}\,, \tag{11.6.10}$$

with

$$(\mathbf{A}^{\mathrm{D}})^{\#} = \mathbf{I}(r)^{-\frac{1}{2}} \cdot \mathbf{A}^{\mathrm{D}} \cdot \mathbf{I}(r)^{-\frac{1}{2}}, \quad \boldsymbol{\eta}_{\mathrm{s}}^{\#} = \mathbf{I}(r)^{\frac{1}{2}} \cdot \boldsymbol{\eta}_{\mathrm{s}}\,. \tag{11.6.11}$$

If one denotes by $\mathbf{e}^{L\#} = \mathbf{I}(r)^{\frac{1}{2}} \cdot \mathbf{e}^L$ and $\mathbf{e}^{R\#} = \mathbf{I}(r)^{\frac{1}{2}} \cdot \mathbf{e}^R$ the new eigenvectors, then one obtains **Proposition 11.4** on the decay coefficient as a Rayleigh quotient, Loret et al. [1997].

The decay coefficient ζ appears in the form of a Rayleigh quotient,

$$\zeta = \frac{\mathbf{e}^{R\#} \cdot \mathbf{I}^{\#}(e) \cdot \mathbf{e}^{R\#}}{\mathbf{e}^{R\#} \cdot \mathbf{e}^{R\#}}\,, \tag{11.6.12}$$

associated with the symmetric positive definite matrix $\mathbf{I}^{\#}(e) = \mathbf{I}(r)^{-\frac{1}{2}} \cdot \mathbf{I}(e) \cdot \mathbf{I}(r)^{-\frac{1}{2}}$, which, according to (2.3.23) and (2.3.24), is equal to $\mathbf{I}(e/r)$. Consequently, it is bounded below and above by the smallest and largest eigenvalues of this matrix,

$$1 \leq \zeta \leq \frac{1}{r(n^{\mathrm{w}})^2}\,. \tag{11.6.13}$$

As indicated by (2.3.25), the largest eigenvalue is associated with a longitudinal eigenmode, while the smallest one is associated with transverse eigenmodes.

The evolution equation (11.6.8) yields not only the decay coefficient but also the part of the induced discontinuity which is orthogonal to the eigenmode $\mathbf{e}^R$: in fact, when the wave-speed has single multiplicity, this part can be shown to be uniquely defined.

11.6.2.2   An alternative direct evaluation of the decay coefficient ζ

To obtain the decay coefficient for a given wave-speed c, the natural approach is to calculate first the associated left and right eigenvectors $\mathbf{e}^R$ and $\mathbf{e}^L$. However, if c is a wave-speed of single multiplicity, considerable algebraic simplifications result from the alternative procedure below. Only the eigenvalues, not the eigenmodes, of an extended problem are required.

Let $c^2(0) = c^2$ be an eigenvalue with single multiplicity of the generalized eigenproblem (11.5.49) and let $\mathbf{e}^R(0) = \mathbf{e}^R$ and $\mathbf{e}^L(0) = \mathbf{e}^L$ be the associated right and left eigenvectors. Consider now the *extended* generalized eigenvalue problem:

$$\Big(\mathbf{A}^{\mathrm{D}} + \varphi\,\mathbf{I}(e) - c^2(\varphi)\,\mathbf{I}(r)\Big) \cdot \mathbf{e}^R(\varphi) = 0\,. \tag{11.6.14}$$

For φ a sufficiently small real number, the eigenvalue $c^2(\varphi)$ is an analytic function of φ, Stoer and Bulirsch [1980], p. 389. Therefore scalar multiplication of (11.6.14) by the left eigenvector $\mathbf{e}^L(0) = \mathbf{e}^L$, differentiation with respect to φ of the resulting product and estimation of the derivative at $\varphi = 0$ lead to

Proposition 11.5 on a direct evaluation of the decay coefficient ζ, Stoer and Bulirsch [1980].

Let $c(0)$ be a solution of single multiplicity of the eigenvalue problem (11.5.49) with $\mathbf{e}^R$ and $\mathbf{e}^L$ the respective right and left eigenvectors. Then the decay coefficient associated with $c(0)$ and defined by (11.6.9) is given by the derivative of the square of the generalized wave-speed,

$$\zeta = \frac{dc^2}{d\varphi}(\varphi = 0)\,. \tag{11.6.15}$$

The procedure breaks down at an eigenvalue of multiplicity greater than one, where the nature of the eigenspace may be more complex. Let us simply consider the case of a double eigenvalue, for $\mathbf{A}^D$ a non-symmetric matrix, a possibility that takes place for some irreversible mechanical behaviors. Then, the eigenvectors may not be unique or, if the eigenspace is defective, the left and right eigenvectors may be well-defined and then they are orthogonal with respect to the matrix $\mathbf{I}(r)$, that is, Wilkinson [1965], Eqn (7.5), p.10:

$$\mathbf{e}^L \cdot \mathbf{I}(r) \cdot \mathbf{e}^R = 0\,, \tag{11.6.16}$$

which is in strong contrast with the case of single multiplicity, where this equality occurs only for $\mathbf{e}^L$ and $\mathbf{e}^R$ associated with distinct eigenvalues. This result suggests that, in the vicinity of a wave-speed of multiplicity greater than 1, the decay coefficient for materials with a non-symmetric acoustic tensor may become unbounded, Loret et al. [1997].

11.7   Plane motions and Helmholtz potentials in poroelasticity

11.7.1   Displacements and strains in terms of Helmholtz potentials

The apparent strains $\boldsymbol{\epsilon}^\alpha$ of each phase derive from the corresponding displacements $\mathbf{u}_\alpha$,

$$\boldsymbol{\epsilon}^\alpha = \frac{1}{2}(\boldsymbol{\nabla}\mathbf{u}_\alpha + (\boldsymbol{\nabla}\mathbf{u}_\alpha)^{\mathrm{T}})\,, \quad \alpha = \mathrm{s}, \mathrm{w}\,, \tag{11.7.1}$$

so that

$$\operatorname{tr}\boldsymbol{\epsilon}^\alpha = \operatorname{div}\mathbf{u}_\alpha\,, \quad \alpha = \mathrm{s}, \mathrm{w}\,. \tag{11.7.2}$$

Henceforth, the analysis considers displacements in the plane (x_1, x_2) endowed with cartesian axes. Let $\mathbf{e}_3$ be the unit vector orthogonal to this plane. Then, as a generalization of the approach exposed for single phase solids in Section 11.2.6, the displacements are defined in terms of four potentials,

$$\mathbf{u}_\alpha = \boldsymbol{\nabla}\phi_\alpha + \boldsymbol{\nabla}\wedge(\psi_\alpha\mathbf{e}_3) \quad \text{with} \quad \operatorname{div}(\psi_\alpha\mathbf{e}_3) = 0\,, \quad \alpha = \mathrm{s}, \mathrm{w}\,, \tag{11.7.3}$$

and, in component form, for $\alpha = \mathrm{s}, \mathrm{w}$,

$$\begin{aligned} u_{\alpha 1} &= \frac{\partial\phi_\alpha}{\partial x_1} + \frac{\partial\psi_\alpha}{\partial x_2} \\ u_{\alpha 2} &= \frac{\partial\phi_\alpha}{\partial x_2} - \frac{\partial\psi_\alpha}{\partial x_1} \end{aligned} \quad \text{and} \quad \frac{\partial\psi_\alpha}{\partial x_1} + \frac{\partial\psi_\alpha}{\partial x_2} = 0\,. \tag{11.7.4}$$

The resulting strains take the form,

$$\begin{aligned} \epsilon_{11}^\alpha &= \frac{\partial^2\phi_\alpha}{\partial x_1^2} + \frac{\partial^2\psi_\alpha}{\partial x_1\partial x_2}\,, \\ \epsilon_{22}^\alpha &= \frac{\partial^2\phi_\alpha}{\partial x_2^2} - \frac{\partial^2\psi_\alpha}{\partial x_1\partial x_2}\,, \\ \epsilon_{12}^\alpha &= \frac{\partial^2\phi_\alpha}{\partial x_1\partial x_2} + \frac{1}{2}\frac{\partial^2\psi_\alpha}{\partial x_2^2} - \frac{1}{2}\frac{\partial^2\psi_\alpha}{\partial x_1^2}\,. \end{aligned} \tag{11.7.5}$$

11.7.2 Poroelastic stresses in terms of Helmholtz potentials

For linear poroelasticity, the partial stresses deduce easily in terms of the Helmholtz potentials,

$$
\begin{aligned}
\sigma_{11}^{\mathrm{s}} &= (\lambda^{\mathrm{ss}} + 2\,\mu)\,\frac{\partial^2 \phi_{\mathrm{s}}}{\partial x_1^2} + \lambda^{\mathrm{ss}}\,\frac{\partial^2 \phi_{\mathrm{s}}}{\partial x_2^2} + \lambda^{\mathrm{sw}}\,\Delta \phi_{\mathrm{w}} + 2\,\mu\,\frac{\partial^2 \psi_{\mathrm{s}}}{\partial x_1 \partial x_2}\,, \\
\sigma_{22}^{\mathrm{s}} &= \lambda^{\mathrm{ss}}\,\frac{\partial^2 \phi_{\mathrm{s}}}{\partial x_1^2} + (\lambda^{\mathrm{ss}} + 2\,\mu)\,\frac{\partial^2 \phi_{\mathrm{s}}}{\partial x_2^2} + \lambda^{\mathrm{sw}}\,\Delta \phi_{\mathrm{w}} - 2\,\mu\,\frac{\partial^2 \psi_{\mathrm{s}}}{\partial x_1 \partial x_2}\,, \\
\sigma_{12}^{\mathrm{s}} &= \mu\left(2\,\frac{\partial^2 \phi_{\mathrm{s}}}{\partial x_1 \partial x_2} + \frac{\partial^2 \psi_{\mathrm{s}}}{\partial x_2^2} - \frac{\partial^2 \psi_{\mathrm{s}}}{\partial x_1^2}\right), \\
-n^{\mathrm{w}}\,p_{\mathrm{w}} &= \lambda^{\mathrm{sw}}\,\Delta\,\phi_{\mathrm{s}} + \lambda^{\mathrm{ww}}\,\Delta\,\phi_{\mathrm{w}}\,,
\end{aligned}
\tag{11.7.6}
$$

as well as the total stresses,

$$
\begin{aligned}
\sigma_{11} &= (\lambda^{\mathrm{ss}} + \lambda^{\mathrm{sw}} + 2\,\mu)\,\frac{\partial^2 \phi_{\mathrm{s}}}{\partial x_1^2} + (\lambda^{\mathrm{ss}} + \lambda^{\mathrm{sw}})\,\frac{\partial^2 \phi_{\mathrm{s}}}{\partial x_2^2} + (\lambda^{\mathrm{ww}} + \lambda^{\mathrm{sw}})\,\Delta \phi_{\mathrm{w}} + 2\,\mu\,\frac{\partial^2 \psi_{\mathrm{s}}}{\partial x_1 \partial x_2}\,, \\
\sigma_{22} &= (\lambda^{\mathrm{ss}} + \lambda^{\mathrm{sw}})\,\frac{\partial^2 \phi_{\mathrm{s}}}{\partial x_1^2} + (\lambda^{\mathrm{ss}} + \lambda^{\mathrm{sw}} + 2\,\mu)\,\frac{\partial^2 \phi_{\mathrm{s}}}{\partial x_2^2} + (\lambda^{\mathrm{ww}} + \lambda^{\mathrm{sw}})\,\Delta \phi_{\mathrm{w}} - 2\,\mu\,\frac{\partial^2 \psi_{\mathrm{s}}}{\partial x_1 \partial x_2}\,, \\
\sigma_{12} &= \mu\left(2\,\frac{\partial^2 \phi_{\mathrm{s}}}{\partial x_1 \partial x_2} + \frac{\partial^2 \psi_{\mathrm{s}}}{\partial x_2^2} - \frac{\partial^2 \psi_{\mathrm{s}}}{\partial x_1^2}\right).
\end{aligned}
\tag{11.7.7}
$$

The relation,

$$
\sigma_{11} + \sigma_{22} = 2\,(\lambda^{\mathrm{ss}} + \lambda^{\mathrm{sw}} + \mu)\,\Delta\,\phi_{\mathrm{s}} + 2\,(\lambda^{\mathrm{sw}} + \lambda^{\mathrm{ww}})\,\Delta\,\phi_{\mathrm{w}}\,,
\tag{11.7.8}
$$

is worth recording as well as the divergence and Laplacian of the displacement vectors defined by (2.4.5), namely for $\alpha = \mathrm{s}, \mathrm{w}$,

$$
\begin{aligned}
\operatorname{div}\mathbf{u}_\alpha &= \Delta\phi_\alpha\,, \\
\Delta\mathbf{u}_\alpha &= \boldsymbol{\nabla}\operatorname{div}\mathbf{u}_\alpha - \boldsymbol{\nabla}\wedge(\boldsymbol{\nabla}\wedge\mathbf{u}_\alpha) = \boldsymbol{\nabla}\Delta\phi_\alpha + \boldsymbol{\nabla}\wedge(\Delta\psi_\alpha\mathbf{e}_3)\,.
\end{aligned}
\tag{11.7.9}
$$

11.7.3 Field equations in terms of Helmholtz potentials

In absence of body forces, the equations of balance of momentum of the solid and fluid phases expressed in terms of displacements may be linearized to

$$
\begin{aligned}
(\lambda^{\mathrm{ss}} + \mu)\,\boldsymbol{\nabla}\operatorname{div}\mathbf{u}_{\mathrm{s}} + \mu\,\Delta\mathbf{u}_{\mathrm{s}} + \lambda^{\mathrm{sw}}\,\boldsymbol{\nabla}\operatorname{div}\mathbf{u}_{\mathrm{w}} &- k_{\mathrm{Sd}}\,(\dot{\mathbf{u}}_{\mathrm{s}} - \dot{\mathbf{u}}_{\mathrm{w}}) = \rho^{\mathrm{s}}\,\ddot{\mathbf{u}}_{\mathrm{s}}\,, \\
\lambda^{\mathrm{sw}}\,\boldsymbol{\nabla}\operatorname{div}\mathbf{u}_{\mathrm{s}} \qquad\qquad + \lambda^{\mathrm{ww}}\,\boldsymbol{\nabla}\operatorname{div}\mathbf{u}_{\mathrm{w}} &+ k_{\mathrm{Sd}}\,(\dot{\mathbf{u}}_{\mathrm{s}} - \dot{\mathbf{u}}_{\mathrm{w}}) = \rho^{\mathrm{w}}\,\ddot{\mathbf{u}}_{\mathrm{w}}\,,
\end{aligned}
\tag{11.7.10}
$$

the buoyancy term $p_{\mathrm{w}}\,\boldsymbol{\nabla}n^{\mathrm{w}}$ disappearing from the momentum supplies, see Sections 4.5.1, 7.3.1 and 7.3.2. Moreover, no distinction is made between the material time derivatives and the partial derivative with respect to time, which are all denoted by a superimposed dot.

Via (11.7.9), the two vector equations of balance of momentum (11.7.10) can be split into their longitudinal part

$$
\begin{aligned}
(\lambda^{\mathrm{ss}} + 2\,\mu)\,\Delta\,\phi_{\mathrm{s}} + \lambda^{\mathrm{sw}}\,\Delta\,\phi_{\mathrm{w}} - k_{\mathrm{Sd}}\,(\dot{\phi}_{\mathrm{s}} - \dot{\phi}_{\mathrm{w}}) &= \rho^{\mathrm{s}}\,\ddot{\phi}_{\mathrm{s}}\,, \\
\lambda^{\mathrm{sw}}\,\Delta\,\phi_{\mathrm{s}} + \lambda^{\mathrm{ww}}\,\Delta\,\phi_{\mathrm{w}} + k_{\mathrm{Sd}}\,(\dot{\phi}_{\mathrm{s}} - \dot{\phi}_{\mathrm{w}}) &= \rho^{\mathrm{w}}\,\ddot{\phi}_{\mathrm{w}}\,,
\end{aligned}
\tag{11.7.11}
$$

and rotational part,

$$
\begin{aligned}
\mu\,\Delta\,\psi_{\mathrm{s}} - k_{\mathrm{Sd}}\,(\dot{\psi}_{\mathrm{s}} - \dot{\psi}_{\mathrm{w}}) &= \rho^{\mathrm{s}}\,\ddot{\psi}_{\mathrm{s}}\,, \\
k_{\mathrm{Sd}}\,(\dot{\psi}_{\mathrm{s}} - \dot{\psi}_{\mathrm{w}}) &= \rho^{\mathrm{w}}\,\ddot{\psi}_{\mathrm{w}}\,.
\end{aligned}
\tag{11.7.12}
$$

Note that the four field equations are still pairwise coupled.

11.7.4   Boundary conditions in a dynamic non dissipative context

As an example, let us consider the lower half-plane whose surface is free of traction and permeable to the fluid. An alternative boundary condition pertaining to the fluid phase consists in assuming that the fluid flux in the direction normal to the surface is prescribed, say, typically equal to zero.

The analysis is greatly simplified if the dissipative terms are omitted from the field equations, either because the permeability is very high, or because the fluid and solid move together for some reason.

An incident wave-motion impinges the surface, giving rise to reflected waves of all possible types in general, namely, two longitudinal waves and one shear wave, Fig. 11.7.1. It is convenient here to associate the Helmholtz potentials, not with the phases as done above, but with the waves. For example the incident motion may be due to the first longitudinal wave propagating in the direction $\mathbf{n}^I = \sin\beta_I\,\mathbf{e}_1 - \cos\beta_I\,\mathbf{e}_2$ with pulsation ω [unit: 1/s] and with associated potential,

$$\phi_I = \exp\left(i\,\omega\,(t - \frac{\mathbf{x}\cdot\mathbf{n}^I}{c_1})\right). \tag{11.7.13}$$

Each of the three reflected waves has its own direction of propagation $\mathbf{n}^{Rj} = \sin\beta_{Rj}\,\mathbf{e}_1 + \cos\beta_{Rj}\,\mathbf{e}_2$ and its own speed c_j, $j = 1, 2, 3$, but the pulsation may be shown to be inherited from the incident wave. The associated potentials assume the formats,

$$\phi_{Rj} = \exp\left(i\,\omega\,(t - \frac{\mathbf{x}\cdot\mathbf{n}^{Rj}}{c_j})\right),\ j = 1, 2; \quad \psi_{R3} = \exp\left(i\,\omega\,(t - \frac{\mathbf{x}\cdot\mathbf{n}^{R3}}{c_3})\right). \tag{11.7.14}$$

The displacements are contributed by the incident field and by the two solenoidal and single rotational reflected fields,

$$\mathbf{u}_\alpha = a_\alpha\,\boldsymbol{\nabla}\phi_I + \sum_{j=1,2} b_{\alpha j}\,\boldsymbol{\nabla}\phi_{Rj} + b_{\alpha 3}\boldsymbol{\nabla}\wedge(\psi_{R3}\mathbf{e}_3), \quad \alpha = \text{s}, \text{w}. \tag{11.7.15}$$

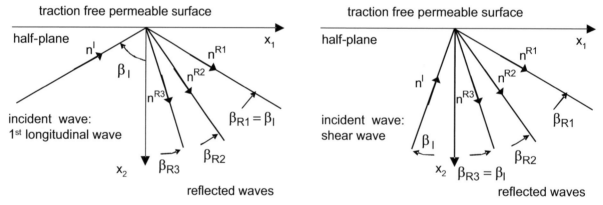

Fig. 11.7.1: To meet the boundary conditions along the free surface of the poroelastic lower half-plane, an incident plane wave triggers in general several reflected waves. The sketches assume the shear wave-speed to be smaller than the second longitudinal wave-speed so that $\beta_{R3} < \beta_{R2}$. Otherwise, the inequality is reversed.

If the incident field is associated with the second longitudinal wave, just change the index 1 to 2 in (11.7.13). If the incident wave is the shear wave, then change the incident potential (11.7.13) to

$$\psi_I = \exp\left(i\,\omega\,(t - \frac{\mathbf{x}\cdot\mathbf{n}^I}{c_3})\right), \tag{11.7.16}$$

as well as the displacements to

$$\mathbf{u}_\alpha = a_\alpha\,\boldsymbol{\nabla}\wedge(\psi_I\mathbf{e}_3) + \sum_{j=1,2} b_{\alpha j}\,\boldsymbol{\nabla}\phi_{Rj} + b_{\alpha 3}\boldsymbol{\nabla}\wedge(\psi_{R3}\mathbf{e}_3), \quad \alpha = \text{s}, \text{w}. \tag{11.7.17}$$

From the displacements deduce the strains as in (11.7.1), and next the stresses through the poroelastic constitutive equations.

In the nondissipative case, the field equations (11.7.11) and (11.7.12) yield a relation between the amplitudes of the solid and fluid displacements of each of the four types. If the incident field is the longitudinal wave with speed c_k, $k = 1, 2$, then,

$$\frac{b_{wj}}{b_{sj}} = \frac{c_j^2 - c_s^2}{c_{ms}^2} = \frac{c_{mw}^2}{c_j^2 - c_w^2}, \quad j = 1, 2; \quad \frac{a_w}{a_s} = \frac{b_{wk}}{b_{sk}}; \quad b_{w3} = 0. \tag{11.7.18}$$

If the incident field is the shear wave, then,

$$\frac{b_{wj}}{b_{sj}} = \frac{c_j^2 - c_s^2}{c_{ms}^2} = \frac{c_{mw}^2}{c_j^2 - c_w^2}, \quad j = 1, 2; \quad a_w = b_{w3} = 0. \tag{11.7.19}$$

In words, the displacements in the fluid can be considered known as soon as the solid displacements are defined. The needed information is provided by the boundary conditions, as follows.

If for example the surface is traction free and permeable, the boundary conditions at the surface $x_2 = 0$,

$$\sigma_{22}^s(x_1, x_2 = 0, t) = 0, \quad \sigma_{12}^s(x_1, x_2 = 0, t) = 0, \quad -n^w p_w(x_1, x_2 = 0, t) = 0, \tag{11.7.20}$$

should be satisfied for any $x_1 \in]-\infty, \infty[$. Whence the directions of the reflected waves, namely the angles β_{Rj}, $j = 1, 2, 3$. Indeed if the incident wave is oriented at the angle β_I with respect to the axis $\mathbf{e}_2$, and has speed c_j, then

$$\beta_{Rj} = \beta_I, \quad \frac{\sin \beta_{R1}}{c_1} = \frac{\sin \beta_{R2}}{c_3} = \frac{\sin \beta_{R3}}{c_3}. \tag{11.7.21}$$

This simple relation indicates that the angles of reflection are in proportion with the wave-speeds. For some critical incident angle(s), it might happen that one, or two, reflected angles reach 90°. Larger incident angles imply the associated wave to be reflected, no longer as a plane wave, but as a surface wave with an amplitude decreasing with depth. The phenomenon does not take place if the incident wave is the first longitudinal wave. Surface waves are addressed in Section 11.9.

Finally, the three boundary conditions constitute a linear system which provides the amplitude of the reflected waves relative to the incident amplitude, namely b_{sj}/a_s, $j = 1, 2, 3$.

The only case where an incident wave reflects only itself is for a normal incidence. Conversely, there exist conditions for which an incident wave is completely converted into its two sister waves. Some details may be found in Deresiewicz [1960].

More advanced configurations might be considered. If the porous medium includes two porosities, a wave impinging on a free surface will in general trigger four reflected waves, namely three longitudinal waves and one transversal wave. Reflections also take place at the interface between two layers endowed with distinct material properties, or with distinct saturations. The properties of the interface come into picture: actual situations are expected to be intermediate between a perfect interface (ensuring perfect bonding) and a frictionless interface.

11.8   Silent poroelastic boundaries

Section 11.7.4 indicates that, in a dynamic context, a signal emanating from an inner part of the body triggers one or several reflected signals in order to satisfy boundary conditions. In that line, the boundary conditions (11.7.20) may be rewritten in the form of an incident traction and of reflected traction(s),

$$\mathbf{t}^I(x_1, x_2 = 0, t) + \mathbf{t}^R(x_1, x_2 = 0, t) = \mathbf{0}. \tag{11.8.1}$$

Often, in analytical or computer models, only a piece of a large body is of direct interest. The boundary generated by this procedure is artificial, and it may generate spurious reflections. A number of methods aim at alleviating this undesirable feature.

Among them are methods that add supplementary tractions $\mathbf{t}^{supp}$ along the boundary, in view of minimizing the reflected signals $\mathbf{t}^R$,

$$\mathbf{t}^I(x_1, x_2 = 0, t) + \mathbf{t}^R(x_1, x_2 = 0, t) + \mathbf{t}^{supp}(x_1, x_2 = 0, t) = \mathbf{0}. \tag{11.8.2}$$

The formulation below intends to obtain a silent boundary whose viscous tractions are frequency independent. This feature makes it possible to solve the equations of motion in the time domain and to take into account a possible nonlinear constitutive response at the interior of the body.

The viscous boundary method developed below follows Zerfa and Loret [2004] where a short review of the pertinent literature is presented and the theoretical performances of the method have been tested. The method has been used in a finite element context in Zerfa and Loret [2003]. Computations show that, at no additional cost, this *drained* method is more accurate for all permeabilities than the *undrained* method, which misses the existence of the second dilatational wave.

This viscous boundary method has been challenged by the so-called paraxial methods. These methods are motivated by the following observation. The second order hyperbolic partial differential equations, that govern the field equations of solids and saturated porous media, allow waves to travel forward and backward. The idea is to replace these field equations, in regions adjacent to the boundary, by partial differential equations that generate only waves that exit the body. While the idea seems exciting, numerical issues and practical difficulties in applications in a multidimensional finite element context limit the performances of these methods.

11.8.1   Impedance matrix, general case $\lambda^{\mathrm{sw}} \neq 0$

The starting point of the viscous boundary method is the observation that Hadamard's compatibility relations for acceleration waves hold also for phase waves. Indeed, let $\phi(\mathbf{x}, t) = \phi(\mathbf{x} \cdot \mathbf{n} - c\,t)$ be a scalar function. Then $\partial \phi / \partial \mathbf{x} = \eta\,\mathbf{n}$ and $\partial \phi / \partial t = -c\,\eta$, with $\eta = \phi'$ denoting the derivative of ϕ with respect to its argument. Therefore harmonic waves in a medium with *infinite* permeability propagate at the speeds of acceleration waves and along the same eigenmodes. The method can thus capitalize upon the spectral analysis of Section 11.5.2.

For the displacements fields $\mathbf{u}_\alpha(\mathbf{x} \cdot \mathbf{n} - c\,t)$, $\alpha = \mathrm{s}, \mathrm{w}$, the gradients and velocities obey the following relations,

$$\boldsymbol{\nabla} \mathbf{u}_\alpha = -\frac{1}{c}\,\mathbf{v}_\alpha \otimes \mathbf{n}, \quad \mathbf{v}_\alpha = -c\,\boldsymbol{\nabla} \mathbf{u}_\alpha \cdot \mathbf{n}, \tag{11.8.3}$$

and the partial stresses (11.5.17) express in terms of the velocities,

$$
\begin{aligned}
\boldsymbol{\sigma}^{\mathrm{s}} &= -\frac{1}{c}\left(\lambda^{\mathrm{ss}}\,\mathbf{v}_{\mathrm{s}} \cdot \mathbf{n}\,\mathbf{I} + \mu\,(\mathbf{v}_{\mathrm{s}} \otimes \mathbf{n} + \mathbf{n} \otimes \mathbf{v}_{\mathrm{s}}) + \lambda^{\mathrm{sw}}\,\mathbf{v}_{\mathrm{w}} \cdot \mathbf{n}\,\mathbf{I}\right), \\
-n^{\mathrm{w}}\,p_{\mathrm{w}} &= -\frac{1}{c}\left(\lambda^{\mathrm{sw}}\,\mathbf{v}_{\mathrm{s}} \cdot \mathbf{n} \qquad\qquad\qquad + \lambda^{\mathrm{ww}}\,\mathbf{v}_{\mathrm{w}} \cdot \mathbf{n}\right).
\end{aligned}
\tag{11.8.4}
$$

The eigenmodes associated with the two longitudinal wave-speeds c_1 and c_2, and single shear wave-speed with double multiplicity given in explicit form in Section 11.5.2 express in terms of the unit direction of propagation $\mathbf{n}$, and of the two orthogonal unit vectors $\mathbf{t}^{(1)}$ and $\mathbf{t}^{(2)}$:

$$
\begin{aligned}
\text{longitudinal mode} \quad c = c_1, c_2 \;&:\; \mathbf{v}_{\mathrm{s}} \sim c_{\mathrm{ms}}^2\,\mathbf{n}, \qquad \mathbf{v}_{\mathrm{w}} \sim (c^2 - c_{\mathrm{s}}^2)\,\mathbf{n}, \\
\text{shear mode} \quad c = c_3 \;&:\; \mathbf{v}_{\mathrm{s}} \sim \mathbf{t}^{(1)}, \mathbf{t}^{(2)}, \qquad \mathbf{v}_{\mathrm{w}} = \mathbf{0}.
\end{aligned}
\tag{11.8.5}
$$

Conversely, the velocities can be calculated in terms of the two longitudinal and two shear eigenmodes,

$$
\begin{aligned}
\mathbf{v}_{\mathrm{s}} &= a_1\,c_{\mathrm{ms}}^2\,\mathbf{n} + a_2\,c_{\mathrm{ms}}^2\,\mathbf{n} + (\mathbf{v}_{\mathrm{s}} \cdot \mathbf{t}^{(1)})\,\mathbf{t}^{(1)} + (\mathbf{v}_{\mathrm{s}} \cdot \mathbf{t}^{(2)})\,\mathbf{t}^{(2)}, \\
\mathbf{v}_{\mathrm{w}} &= a_1\,(c_1^2 - c_{\mathrm{s}}^2)\,\mathbf{n} + a_2\,(c_2^2 - c_{\mathrm{s}}^2)\,\mathbf{n},
\end{aligned}
\tag{11.8.6}
$$

which may be rearranged to

$$
\begin{aligned}
\mathbf{v}_{\mathrm{s}} &= (V_1 + V_2)\,\mathbf{n} + (\mathbf{v}_{\mathrm{s}} \cdot \mathbf{t}^{(1)})\,\mathbf{t}^{(1)} + (\mathbf{v}_{\mathrm{s}} \cdot \mathbf{t}^{(2)})\,\mathbf{t}^{(2)} \\
\mathbf{v}_{\mathrm{w}} &= (W_1 + W_2)\,\mathbf{n},
\end{aligned}
\tag{11.8.7}
$$

where

$$
\begin{aligned}
V_1 &= \frac{1}{c_1^2 - c_2^2}\left((c_{\mathrm{s}}^2 - c_2^2)\,\mathbf{v}_{\mathrm{s}} \cdot \mathbf{n} + c_{\mathrm{ms}}^2\,\mathbf{v}_{\mathrm{w}} \cdot \mathbf{n}\right), \quad W_1 = \frac{c_1^2 - c_{\mathrm{s}}^2}{c_{\mathrm{ms}}^2}\,V_1, \\
V_2 &= \frac{1}{c_2^2 - c_1^2}\left((c_{\mathrm{s}}^2 - c_1^2)\,\mathbf{v}_{\mathrm{s}} \cdot \mathbf{n} + c_{\mathrm{ms}}^2\,\mathbf{v}_{\mathrm{w}} \cdot \mathbf{n}\right), \quad W_2 = \frac{c_2^2 - c_{\mathrm{s}}^2}{c_{\mathrm{ms}}^2}\,V_2.
\end{aligned}
\tag{11.8.8}
$$

Substituting these expressions in the partial stresses (11.8.4) yields the supplementary tractions of viscous nature in (11.8.2),

$$
\begin{bmatrix} \mathbf{h}^{\mathrm{s}} \\ \mathbf{h}^{\mathrm{w}} \end{bmatrix} = \begin{bmatrix} \boldsymbol{\sigma}^{\mathrm{s}} \cdot \mathbf{n} \\ \boldsymbol{\sigma}^{\mathrm{w}} \cdot \mathbf{n} \end{bmatrix} = - \begin{bmatrix} \mathbf{A}^{\mathrm{ss}} & \mathbf{A}^{\mathrm{sw}} \\ (\mathbf{A}^{\mathrm{sw}})^{\mathrm{T}} & \mathbf{A}^{\mathrm{ww}} \end{bmatrix} \begin{bmatrix} \mathbf{v}_{\mathrm{s}} \\ \mathbf{v}_{\mathrm{w}} \end{bmatrix},
\tag{11.8.9}
$$

in terms of the velocities via the impedance matrix $\mathbf{A}$ with block components,

$$
\begin{aligned}
\mathbf{A}^{\mathrm{ss}} &= \frac{\lambda^{\mathrm{ss}} + 2\,\mu + \rho^{\mathrm{s}}\,c_1\,c_2}{c_1 + c_2}\ \mathbf{n} \otimes \mathbf{n} \ + \ \frac{\mu}{c_3}\,(\overbrace{\mathbf{t}^{(1)} \otimes \mathbf{t}^{(1)} + \mathbf{t}^{(2)} \otimes \mathbf{t}^{(2)}}^{\mathbf{I}-\mathbf{n}\otimes\mathbf{n}}), \\
\mathbf{A}^{\mathrm{ww}} &= \frac{\lambda^{\mathrm{ww}} + \rho^{\mathrm{w}}\,c_1\,c_2}{c_1 + c_2}\ \mathbf{n} \otimes \mathbf{n}, \\
\mathbf{A}^{\mathrm{sw}} &= \frac{\lambda^{\mathrm{sw}}}{c_1 + c_2}\ \mathbf{n} \otimes \mathbf{n}.
\end{aligned}
\tag{11.8.10}
$$

11.8.2 Impedance matrix: no poromechanical coupling, $\lambda^{\mathrm{sw}} = 0$

When the poroelastic coupling parameter λ^{sw} vanishes, the two longitudinal modes decouple. The eigenmode associated with the wave-speed c_{s} of square $(\lambda^{\mathrm{ss}} + 2\,\mu)/\rho^{\mathrm{s}}$ affects only the solid, while the eigenmode with wave-speed c_{w} of square $\lambda^{\mathrm{ww}}/\rho^{\mathrm{w}}$ affects only the fluid.

The impedance matrix can be recalculated following the same steps as in the coupled case, or deduced as a limit case, yielding

$$
\mathbf{A}^{\mathrm{ss}} = \rho^{\mathrm{s}}\,c_{\mathrm{s}}\,\mathbf{n} \otimes \mathbf{n} + \rho^{\mathrm{s}}\,c_3\,(\mathbf{I} - \mathbf{n} \otimes \mathbf{n}), \quad \mathbf{A}^{\mathrm{ww}} = \rho^{\mathrm{w}}\,c_{\mathrm{w}}\,\mathbf{n} \otimes \mathbf{n}, \quad \mathbf{A}^{\mathrm{sw}} = \mathbf{0}.
\tag{11.8.11}
$$

As expected, the coefficients involved in the impedance are equal to the product of an apparent mass density times a wave-speed, or to a ratio of a modulus over a wave-speed, namely $\rho^{\mathrm{s}}\,c_{\mathrm{s}} = (\lambda^{\mathrm{ss}} + 2\,\mu)/c_{\mathrm{s}}$, $\rho^{\mathrm{w}}\,c_{\mathrm{w}} = \lambda^{\mathrm{ww}}/c_{\mathrm{w}}$, and $\rho^{\mathrm{s}}\,c_3 = \mu/c_3$.

11.8.3 Impedance matrix: dry/drained medium

When the pore pressure vanishes over a finite time interval, the velocities are linked by the condition $\lambda^{\mathrm{sw}}\,\mathbf{v}_{\mathrm{s}} \cdot \mathbf{n} + \lambda^{\mathrm{ww}}\,\mathbf{v}_{\mathrm{w}} \cdot \mathbf{n} = 0$. Substituting the normal fluid velocity in the partial solid stress yields,

$$
\boldsymbol{\sigma}^{\mathrm{s}} = -\frac{1}{c}\left(\lambda\ \mathbf{v}_{\mathrm{s}} \cdot \mathbf{n}\,\mathbf{I} \ + \ \mu\,(\mathbf{v}_{\mathrm{s}} \otimes \mathbf{n} + \mathbf{n} \otimes \mathbf{v}_{\mathrm{s}})\right).
\tag{11.8.12}
$$

with $\lambda = \lambda^{\mathrm{ss}} - (\lambda^{\mathrm{sw}})^2/\lambda^{\mathrm{ww}}$ the drained Lamé modulus. According to Section 11.5.2.2, the drained medium is endowed with a longitudinal wave-speed c_{LD} and a shear wave-speed $c_{\mathrm{SD}} = c_3$ of double multiplicity. The viscous traction can then be cast in the format $\mathbf{h}^{\mathrm{s}} = \boldsymbol{\sigma}^{\mathrm{s}} \cdot \mathbf{n} = -\mathbf{A}^{\mathrm{ss}} \cdot \mathbf{v}_{\mathrm{s}}$ with

$$
\mathbf{A}^{\mathrm{ss}} = \rho^{\mathrm{s}}\,c_{\mathrm{LD}}\,\mathbf{n} \otimes \mathbf{n} + \rho^{\mathrm{s}}\,c_{\mathrm{SD}}\,(\mathbf{I} - \mathbf{n} \otimes \mathbf{n}),
\tag{11.8.13}
$$

where $\rho^{\mathrm{s}}\,c_{\mathrm{LD}} = (\lambda + 2\,\mu)/c_{\mathrm{LD}}$ and $\rho^{\mathrm{s}}\,c_{\mathrm{SD}} = \mu/c_{\mathrm{SD}}$.

11.8.4 Impedance matrix: undrained medium

When the permeability vanishes, the motions of the solid and fluid phases are identical, and the flow turns out undrained. The converse is not true, since the undrained property concerns only the volumes of the phases. In both cases, the velocities are linked by the condition $\mathbf{v}_{\mathrm{s}} \cdot \mathbf{n} = \mathbf{v}_{\mathrm{w}} \cdot \mathbf{n}$. Substituting the normal fluid velocity in the total stress yields,

$$
\boldsymbol{\sigma} = -\frac{1}{c}\left(\lambda^{\mathrm{U}}\ \mathbf{v}_{\mathrm{s}} \cdot \mathbf{n}\,\mathbf{I} \ + \ \mu\,(\mathbf{v}_{\mathrm{s}} \otimes \mathbf{n} + \mathbf{n} \otimes \mathbf{v}_{\mathrm{s}})\right)
\tag{11.8.14}
$$

with λ^{U} the undrained Lamé modulus. According to Section 11.5.2.3, the undrained medium is endowed with a longitudinal wave-speed c_{LU} and a shear wave-speed c_{SU} of double multiplicity. The total viscous traction can then be cast in the format,

$$
\mathbf{h} = \boldsymbol{\sigma} \cdot \mathbf{n} = -\mathbf{A} \cdot \mathbf{v}_{\mathrm{s}}, \quad \mathbf{A} = \rho\,c_{\mathrm{LU}}\,\mathbf{n} \otimes \mathbf{n} + \rho\,c_{\mathrm{SU}}\,(\mathbf{I} - \mathbf{n} \otimes \mathbf{n}).
\tag{11.8.15}
$$

11.9   Surface waves in a saturated porous half-plane

The analyses of the preceding sections, except Section 11.8, have considered the propagation of waves in infinite media. The corresponding waves are termed *body waves*. When the waves encounter a boundary, the mechanical conditions at the boundary should be satisfied, Fig. 11.9.1. If the body is finite, e.g., a finite bar, the waves may travel back and fourth and the motion will be periodic if there is no energy dissipation, as in elastic solids. When the body is semi-infinite, e.g., a half-plane or half-space, the fact that the total energy that generates the motion is finite implies that the amplitude of the wave should decay away from the source. As an example, reflection of a plane shear wave impinging a free surface at large incident angles may imply the wave to be reflected, no longer as a plane wave, but as a wave with an amplitude decreasing with depth. Such waves are referred to as *surface waves*.

Surface waves in solids are covered in several treatises, and in particular in Achenbach [1975] or Graff [1991]. The analysis below addresses surface waves in a saturated porous half-plane endowed with isotropic properties in both its elastic and permeability properties. Biot [1956], Jones [1961], Deresiewicz [1962] and Staroszczyk [1992] inter alii have contributed to the subject matter. Beskos et al. [1989] consider a refined analysis where the saturated porous half-space involves two porosities.

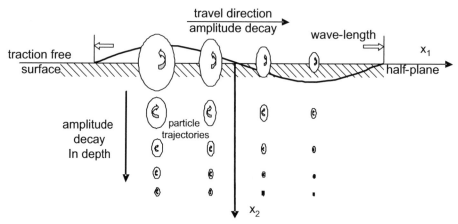

Fig. 11.9.1: Axes used for the analysis of surface waves in the poroelastic lower half-plane. The amplitude of the motion decays both in depth and in the travel direction. Actually, the latter decay is present for a poroelastic half-plane but absent for an elastic half-plane. Conversely, the particle elliptical trajectories as shown are typical of an elastic half-plane, being retrograde at distances smaller than 0.2 wave-length and clockwise deeper.

11.9.1   Time harmonic solutions decreasing with depth

The displacements are sought in the form of harmonic waves in the plane (x_1, x_2) traveling in the x_1-direction and decreasing in depth, i.e., for a generic function f that depends on space and time,

$$f(x_1, x_2, t) = \hat{f}(x_2)\, e^{i(\omega\, t - k\, x_1)} = \hat{f}(x_2)\, e^{i(\omega\, t - \mathrm{Re}\, k\, x_1) + \mathrm{Im}\, k\, x_1}\, . \tag{11.9.1}$$

The pulsation $\omega > 0$ [unit : Hz] is a real number, the wave-number k [unit : 1/m] may be complex. The direction of propagation of the Rayleigh wave, namely towards increasing or decreasing x_1, depends on the sign of the real part of k. Actually, one may choose the positive sign, since the secular equation is phrased in terms of the square of the wave-number. On the other hand, the ratio $-\mathrm{Im}\, k / \mathrm{Re}\, k$ should be positive in order the amplitude to remain bounded in the direction of propagation. In particular, the Helmholtz potentials ϕ_α and ψ_α in the two phases $\alpha = \mathrm{s}, \mathrm{w}$, adopt the form (11.9.1),

$$\begin{bmatrix} \phi_\alpha(x_1, x_2, t) \\ \psi_\alpha(x_1, x_2, t) \end{bmatrix} = \begin{bmatrix} \hat{\phi}_\alpha(x_2) \\ \hat{\psi}_\alpha(x_2) \end{bmatrix} e^{i(\omega\, t - k\, x_1)}\, . \tag{11.9.2}$$

The Laplacian and time derivative operators transform to

$$\Delta = \frac{\partial^2}{\partial x_1^2} + \frac{\partial^2}{\partial x_2^2} \to -k^2 + \frac{\partial^2}{\partial x_2^2}\, ; \qquad \frac{\partial}{\partial t} \to i\,\omega\, ; \qquad \frac{\partial^2}{\partial t^2} \to -\,\omega^2\, . \tag{11.9.3}$$

The field equations (11.7.11) and (11.7.12) become

$$
\left(\begin{bmatrix} \lambda^{ss} + 2\mu & \lambda^{sw} \\ \lambda^{sw} & \lambda^{ww} \end{bmatrix} \frac{\partial^2}{\partial x_2^2} + \begin{bmatrix} -k^2(\lambda^{ss}+2\mu)+\rho^s\omega^2 - iw\,k_{Sd} & -k^2\lambda^{sw}+iw\,k_{Sd} \\ -k^2\lambda^{sw}+iw\,k_{Sd} & -k^2\lambda^{ww}+\rho^w\omega^2-iw\,k_{Sd} \end{bmatrix} \right) \begin{bmatrix} \hat{\phi}_s \\ \hat{\phi}_w \end{bmatrix} = \begin{bmatrix} 0 \\ 0 \end{bmatrix}, \quad (11.9.4)
$$

and

$$
\left(\begin{bmatrix} \mu & 0 \\ 0 & 0 \end{bmatrix} \frac{\partial^2}{\partial x_2^2} + \begin{bmatrix} -k^2\mu+\rho^s\omega^2-iw\,k_{Sd} & iw\,k_{Sd} \\ i\,\omega\,k_{Sd} & \rho^w\omega^2-iw\,k_{Sd} \end{bmatrix} \right) \begin{bmatrix} \hat{\psi}_s \\ \hat{\psi}_w \end{bmatrix} = \begin{bmatrix} 0 \\ 0 \end{bmatrix}. \quad (11.9.5)
$$

11.9.2   The complete formal solution

The depth-decreasing dilatational potentials can be shown to be of the form,

$$
\begin{bmatrix} \hat{\phi}_s \\ \hat{\phi}_w \end{bmatrix} = \begin{bmatrix} a_{s1} \\ a_{w1} \end{bmatrix} e^{-A_1 x_2} + \begin{bmatrix} a_{s2} \\ a_{w2} \end{bmatrix} e^{-A_2 x_2}, \quad \mathrm{Re}\,A_j > 0\,, \ j = 1, 2\,, \quad (11.9.6)
$$

and, similarly for the shear potentials,

$$
\begin{bmatrix} \hat{\psi}_s \\ \hat{\psi}_w \end{bmatrix} = \begin{bmatrix} a_{s3} \\ a_{w3} \end{bmatrix} e^{-A_3 x_2}, \quad \mathrm{Re}\,A_3 > 0\,. \quad (11.9.7)
$$

Thus the displacements $\mathbf{u}_\alpha$, $\alpha = $ s, w, (11.7.4) become with help of (11.9.2), (11.9.6), and (11.9.7),

$$
u_{\alpha 1}(x_1, x_2, t) = [-i\,k\,(a_{\alpha 1}\,e^{-A_1 x_2} + a_{\alpha 2}\,e^{-A_2 x_2}) - A_3\,a_{\alpha 3}\,e^{-A_3 x_2}]\,e^{i(\omega t - k\,x_1)},
$$
$$
u_{\alpha 2}(x_1, x_2, t) = [\,-A_1\,a_{\alpha 1}\,e^{-A_1 x_2} - A_2\,a_{\alpha 2}\,e^{-A_2 x_2} + i\,k\,a_{\alpha 3}\,e^{-A_3 x_2}]\,e^{i(\omega t - k\,x_1)}\,. \quad (11.9.8)
$$

The coefficients of depth dependence A_j, $j \in [1,3]$, may be complex, but their real parts are required to be strictly positive, so as to ensure a vanishing displacement as $x_2 \to \infty$. They are defined in Section 11.9.4.

The depth dependent part of the stresses can be calculated either via (11.7.6), (11.9.1), (11.9.6), and (11.9.7), or via (11.5.17), (11.7.5), and (11.9.8), as

$$
\begin{bmatrix} \hat{\sigma}_{22}^s(x_2) \\ \hat{\sigma}_{12}^s(x_2) \\ -n^w\hat{p}^w(x_2) \end{bmatrix} = \mathcal{R} \begin{bmatrix} a_{s1}\,e^{-A_1 x_2} \\ a_{s2}\,e^{-A_2 x_2} \\ a_{s3}\,e^{-A_3 x_2} \end{bmatrix}, \quad (11.9.9)
$$

with the Rayleigh matrix,

$$
\mathcal{R} = \begin{bmatrix} 2\,\mu\,k^2 + (\lambda^{ss}+2\mu+\lambda^{sw}\frac{a_{w1}}{a_{s1}})(A_1^2-k^2) & 2\,\mu\,k^2+(\lambda^{ss}+2\mu+\lambda^{sw}\frac{a_{w2}}{a_{s2}})(A_2^2-k^2) & -2\,\mu\,i\,k\,A_3 \\ 2\,\mu\,i\,k\,A_1 & 2\,\mu\,i\,k\,A_2 & \mu(A_3^2+k^2) \\ (\lambda^{sw}+\lambda^{ww}\frac{a_{w1}}{a_{s1}})(A_1^2-k^2) & (\lambda^{sw}+\lambda^{ww}\frac{a_{w2}}{a_{s2}})(A_2^2-k^2) & 0 \end{bmatrix}. \quad (11.9.10)
$$

11.9.3   Frequency dependent dissipation

Along the definitions of Section 7.3.2, the viscous coefficient k_{Sd} [unit : kg/m^3/s] is inversely proportional to the hydraulic conductivity K_H [unit : m/s],

$$
k_{Sd} = \frac{n^w \rho^w g}{K_H}, \quad (11.9.11)
$$

with g the acceleration of gravity. Biot [1956] has introduced a frequency dependent dissipation, through the function $F = F(\kappa)$, submitted to $F(0) = 1$ and increasing linearly for a large argument. The actual function $F(\kappa)$ used in the calculations is detailed in Section 11.9.7. Here κ is a dimensionless frequency,

$$\kappa = \delta \sqrt{\frac{\omega \, K_{\mathrm{H}}(\omega=0)}{n^{\mathrm{w}} \, g}}, \tag{11.9.12}$$

phrased in terms of actual properties, of the pulsation ω and of an arbitrary dimensionless frequency coefficient δ. Therefore

$$\frac{k_{\mathrm{Sd}}}{\rho^{\mathrm{w}} \, \omega} = G(\kappa) \quad \text{with} \quad G(\kappa) = \left(\frac{\delta}{\kappa}\right)^2 F(\kappa), \tag{11.9.13}$$

and the hydraulic conductivity vanishes at large frequencies,

$$\frac{K_{\mathrm{H}}(\omega)}{K_{\mathrm{H}}(\omega=0)} = \frac{1}{F(\kappa)}. \tag{11.9.14}$$

The linear increase of the function $F(\kappa)$ at high dimensionless frequencies enlarges the set of frequencies for which Rayleigh waves exist. Indeed, for the material properties used, there exists, in a certain domain of porosity, a frequency cutoff above which Rayleigh waves are not available. The virtue of the function F is to increase this frequency cutoff. Thus the following equivalent interpretations hold,

$$
\begin{array}{rcllclclll}
\text{large } k_{\mathrm{Sd}} & \rightleftharpoons & \text{small permeability} & \rightleftharpoons & \text{large frequency } \omega & \rightleftharpoons & \text{large } \kappa, \\
\text{small } k_{\mathrm{Sd}} & \rightleftharpoons & \text{large permeability} & \rightleftharpoons & \text{small frequency } \omega & \rightleftharpoons & \text{small } \kappa.
\end{array} \tag{11.9.15}
$$

Actually, the situation is somehow complex: in addition to frequency, the hydraulic conductivity depends on porosity, as detailed in Section 16.4.1,

$$\frac{K_{\mathrm{H}}(e,\omega)}{K_{\mathrm{H}}(e_0,0)} = \left(\frac{e}{e_0}\right)^3 \frac{1+e_0}{1+e} \frac{1}{F(\kappa)}. \tag{11.9.16}$$

where $e = n^{\mathrm{w}}/(1 - n^{\mathrm{w}})$ is the void ratio and e_0 a reference value. Moreover, like the conductivity, the mechanical moduli depend on porosity, Section 7.3.6.6.

11.9.4   Determination of the coefficients of depth dependence A_j, $j \in [1,3]$

The unit of the depth dependence coefficients and of the wave-number k is m^{-1}. Next to a dimensionless wave-number,

$$\tilde{k} = \frac{k}{k_r} \quad \text{with} \quad k_r = \frac{\omega}{c_{\mathrm{SU}}}, \tag{11.9.17}$$

it is also instrumental to introduce dimensionless depth dependence coefficients $\tilde{A}$ and eigenvalues $\tilde{X}$,

$$\tilde{A}_j = \frac{A_j}{k_r}, \; j = 1, 3; \quad \tilde{X}_j^2 = \frac{H^{\mathrm{U}}}{\mu} (\tilde{A}_j^2 - \tilde{k}^2), \quad j = 1, 2; \quad \tilde{X}_3^2 = \tilde{A}_3^2 - \tilde{k}^2, \tag{11.9.18}$$

in terms of the longitudinal undrained modulus $H^{\mathrm{U}} = \lambda^{\mathrm{U}} + 2\,\mu$ defined in Section 11.5.2.3.

The definitions and relative order of the wave-speeds and wave-speed like quantities discussed in Section 11.5.2.1 are used repeatedly.

11.9.4.1   Dilatation potentials

Insertion of the solution (11.9.6) in the two equations (11.9.4) yields an eigenvalue problem,

$$\begin{bmatrix} (\lambda^{\mathrm{ss}} + 2\,\mu)\,(A^2 - k^2) + \rho^{\mathrm{s}}\omega^2 - i\,\omega\,k_{\mathrm{Sd}} & \lambda^{\mathrm{sw}}\,(A^2 - k^2) & +\,i\,\omega\,k_{\mathrm{Sd}} \\ \lambda^{\mathrm{sw}}\,(A^2 - k^2) & +\,i\,\omega\,k_{\mathrm{Sd}} & \lambda^{\mathrm{ww}}\,(A^2 - k^2) + \rho^{\mathrm{w}}\omega^2 - i\,\omega\,k_{\mathrm{Sd}} \end{bmatrix} \begin{bmatrix} a_{\mathrm{s}} \\ a_{\mathrm{w}} \end{bmatrix} = \begin{bmatrix} 0 \\ 0 \end{bmatrix}. \tag{11.9.19}$$

The characteristic equation associated with this eigenvalue problem is viewed as the equation that defines the coefficients A_1 and A_2 of depth dependence. Equivalently, with the normalization introduced by (11.9.17) and (11.9.18), this equation can be viewed as defining the two solutions $\tilde{X}_j = \tilde{X}_j(\kappa)$, $j \in [1,2]$,

$$(\tilde{X}^2)^2 \frac{c_1^2 c_2^2}{c_{\mathrm{LU}}^4} + \tilde{X}^2 \left(\frac{c_1^2 + c_2^2}{c_{\mathrm{LU}}^2} - i\, G(\kappa) \frac{\rho}{\rho^{\mathrm{s}}} \right) + 1 - i\, G(\kappa) \frac{\rho}{\rho^{\mathrm{s}}} = 0 \,. \tag{11.9.20}$$

The eigenvector associated with each of the two solutions of the above equation is defined as,

$$\frac{a_{\mathrm{w}j}}{a_{\mathrm{s}j}} = -\frac{\dfrac{c_{\mathrm{mw}}^2}{c_{\mathrm{LU}}^2} \tilde{X}_j^2 + i\, G(\kappa)}{\dfrac{c_{\mathrm{w}}^2}{c_{\mathrm{LU}}^2} \tilde{X}_j^2 + 1 - i\, G(\kappa)} = -\frac{\dfrac{c_{\mathrm{s}}^2}{c_{\mathrm{LU}}^2} \tilde{X}_j^2 + 1 - i\, \dfrac{\rho^{\mathrm{w}}}{\rho^{\mathrm{s}}} G(\kappa)}{\dfrac{c_{\mathrm{ms}}^2}{c_{\mathrm{LU}}^2} \tilde{X}_j^2 + i\, \dfrac{\rho^{\mathrm{w}}}{\rho^{\mathrm{s}}} G(\kappa)} \,, \quad j = 1,2 \,. \tag{11.9.21}$$

The wave-speeds c_1 and c_2 are defined by (11.5.22), the undrained longitudinal wave-speed c_{LU} by (11.5.35), and the wave-speed like quantities c_{s}, c_{w}, c_{ms} and c_{mw} by (11.5.20). Repeated use is made of the relations (11.5.25) and (11.5.26).

11.9.4.2 Shear potential

Insertion of the solution (11.9.7) in the two equations (11.9.5) yields an eigenvalue problem,

$$\begin{bmatrix} \mu\,(A^2 - k^2) + \rho^{\mathrm{s}}\omega^2 - i\omega\, k_{\mathrm{Sd}} & i\omega\, k_{\mathrm{Sd}} \\ i\,\omega\, k_{\mathrm{Sd}} & \rho^{\mathrm{w}}\omega^2 - i\,\omega\, k_{\mathrm{Sd}} \end{bmatrix} \begin{bmatrix} a_{\mathrm{s}} \\ a_{\mathrm{w}} \end{bmatrix} = \begin{bmatrix} 0 \\ 0 \end{bmatrix} \,. \tag{11.9.22}$$

The characteristic equation associated with this eigenvalue problem is viewed as the equation that defines the coefficient A_3 of depth dependence. Equivalently, with the normalization introduced by (11.9.17) and (11.9.18), this equation can be viewed as defining the solution $\tilde{X}_3 = \tilde{X}_3(\kappa)$,

$$\tilde{X}_3^2 = -\frac{\rho^{\mathrm{s}}/\rho - i\, G(\kappa)}{1 - i\, G(\kappa)} \,, \tag{11.9.23}$$

with associated eigenmode,

$$\frac{a_{\mathrm{w}3}}{a_{\mathrm{s}3}} = -\frac{i\, G(\kappa)}{1 - i\, G(\kappa)} \,. \tag{11.9.24}$$

11.9.5 Permeable and traction free surface

Since the pore pressure vanishes on the boundary, the boundary conditions along the surface $x_2 = 0$ write equivalently in terms of the total stress $\boldsymbol{\sigma} = \boldsymbol{\sigma}^{\mathrm{s}} - n^{\mathrm{w}} p_{\mathrm{w}}\, \mathbf{I}$,

$$\sigma_{22}(x_1, x_2 = 0, t) = 0, \quad \sigma_{12}(x_1, x_2 = 0, t) = 0, \quad -n^{\mathrm{w}} p_{\mathrm{w}}(x_1, x_2 = 0, t) = 0, \quad \forall x_1 \in]-\infty, \infty[\,, \tag{11.9.25}$$

or of the partial stresses $\boldsymbol{\sigma}^{\mathrm{s}}$,

$$\sigma_{22}^{\mathrm{s}}(x_1, x_2 = 0, t) = 0, \quad \sigma_{12}^{\mathrm{s}}(x_1, x_2 = 0, t) = 0, \quad -n^{\mathrm{w}} p_{\mathrm{w}}(x_1, x_2 = 0, t) = 0, \quad \forall x_1 \in]-\infty, \infty[\,. \tag{11.9.26}$$

11.9.5.1 The dispersion relation

With the normalization (11.9.17) and (11.9.18), and the tractions (11.9.9) and (11.9.10), the secular equations can be cast in the matrix form,

$$\begin{bmatrix} \hat{\sigma}_{22}^{\mathrm{s}}(x_2) \\ \hat{\sigma}_{12}^{\mathrm{s}}(x_2) \\ -n^{\mathrm{w}} \hat{p}^{\mathrm{w}}(x_2) \end{bmatrix} = \mu\, k_r^2\, \mathcal{R}_{\mathrm{per}} \begin{bmatrix} a_{\mathrm{s}1}\, e^{-A_1 x_2} \\ a_{\mathrm{s}2}\, e^{-A_2 x_2} \\ a_{\mathrm{s}3}\, e^{-A_3 x_2} \end{bmatrix} = \begin{bmatrix} 0 \\ 0 \\ 0 \end{bmatrix} \tag{11.9.27}$$

with the permeable Rayleigh matrix,

$$
\mathcal{R}_{\mathrm{per}} =
\begin{bmatrix}
2\,\tilde{k}^2 + (\lambda^{\mathrm{ss}} + 2\,\mu + \lambda^{\mathrm{sw}}\,\dfrac{a_{\mathrm{w}1}}{a_{\mathrm{s}1}})\,\dfrac{\tilde{X}_1^2}{H^{\mathrm{U}}} & 2\,\tilde{k}^2 + (\lambda^{\mathrm{ss}} + 2\,\mu + \lambda^{\mathrm{sw}}\,\dfrac{a_{\mathrm{w}2}}{a_{\mathrm{s}2}})\,\dfrac{\tilde{X}_2^2}{H^{\mathrm{U}}} & -2\,i\,\tilde{k}\,\tilde{A}_3 \\[2ex]
2\,i\,\tilde{k}\,\tilde{A}_1 & 2\,i\,\tilde{k}\,\tilde{A}_2 & 2\,\tilde{k}^2 + \tilde{X}_3^2 \\[2ex]
(\lambda^{\mathrm{sw}} + \lambda^{\mathrm{ww}}\,\dfrac{a_{\mathrm{w}1}}{a_{\mathrm{s}1}})\,\dfrac{\tilde{X}_1^2}{H^{\mathrm{U}}} & (\lambda^{\mathrm{sw}} + \lambda^{\mathrm{ww}}\,\dfrac{a_{\mathrm{w}2}}{a_{\mathrm{s}2}})\,\dfrac{\tilde{X}_2^2}{H^{\mathrm{U}}} & 0
\end{bmatrix}.
\tag{11.9.28}
$$

In view of comparison with an impermeable free surface, note that the determinant of this matrix is the same as the determinant of the following matrix obtained by summing the first and third lines,

$$
\begin{bmatrix}
2\,\tilde{k}^2 + \beta_1\,\tilde{X}_1^2 & 2\,\tilde{k}^2 + \beta_2\,\tilde{X}_2^2 & -2\,i\,\tilde{k}\,\tilde{A}_3 \\[1.5ex]
2\,i\,\tilde{k}\,\tilde{A}_1 & 2\,i\,\tilde{k}\,\tilde{A}_2 & 2\,\tilde{k}^2 + \tilde{X}_3^2 \\[1.5ex]
\gamma_1\,\tilde{X}_1^2 & \gamma_2\,\tilde{X}_2^2 & 0
\end{bmatrix},
\tag{11.9.29}
$$

where we have introduced some notations to compact the components,

$$
\tau_j \equiv \frac{a_{\mathrm{w}j}}{a_{\mathrm{s}j}} - 1, \quad j \in [1,3]; \quad \beta_j \equiv 1 + \frac{\lambda^{\mathrm{sw}} + \lambda^{\mathrm{ww}}}{H^{\mathrm{U}}}\,\tau_j, \quad \gamma_j \equiv \frac{\lambda^{\mathrm{sw}} + \lambda^{\mathrm{ww}}}{H^{\mathrm{U}}} + \frac{\lambda^{\mathrm{ww}}}{H^{\mathrm{U}}}\,\tau_j, \quad j \in [1,2].
\tag{11.9.30}
$$

All coefficients of the above matrix are functions of the dimensionless frequency κ, Eqn (11.9.12). The vanishing of the determinant of this matrix defines the **dispersion relation** that links the wave-number k to the pulsation ω,

$$
\tilde{k} = \tilde{k}(\kappa) \quad \Leftrightarrow \quad k = k(\omega),
\tag{11.9.31}
$$

and the **Rayleigh wave-speed**

$$
c_R = c_R(\omega) = \frac{c_{\mathrm{SU}}}{\mathrm{Re}\,\tilde{k}} = \frac{\omega}{\mathrm{Re}\,k}.
\tag{11.9.32}
$$

If this ratio is independent of frequency, the wave is said to be non dispersive. Body waves and Rayleigh waves in elastic solids and body waves in saturated poroelastic media are non dispersive. At variance, Rayleigh waves in saturated poroelastic media will be observed to be dispersive.

11.9.5.2   Results in graphical form

The material data shown in Table 11.9.1 are representative of a compact bone (or of a rock). The hydraulic conductivity depends on porosity n^{w} and frequency ω as indicated by (11.9.16). The poroelastic moduli depend on porosity as defined in Section 7.3.6.6.

Table 11.9.1: Material parameters used in the graphical outputs.

Shear modulus of solid constituent	$\mu_{\mathrm{s}} = 20\,\mathrm{GPa}$
Bulk modulus of solid constituent	$K_{\mathrm{s}} = 36\,\mathrm{GPa}$
Mass density of solid constituent	$\rho_{\mathrm{s}} = 2500\,\mathrm{kg/m}^3$
Reference void ratio (porosity)	$e_0 = 0.25\;(n_0^{\mathrm{w}} = 0.2)$
Hydraulic conductivity	$K_{\mathrm{H}}(e_0,0) = 10^{-10}\,\mathrm{m/s}$
Bulk modulus of water	$K_{\mathrm{w}} = 2.2\,\mathrm{GPa}$
Mass density of water	$\rho_{\mathrm{w}} = 1000\,\mathrm{kg/m}^3$
Frequency coefficient	$\delta = 3.0$
Acceleration of gravity	$g = 9.81\,\mathrm{m/s}^2$

Given the intrinsic bulk moduli of water K_{w} and of the solid constituent K_{s}, and the shear modulus of the solid constituent μ_{s}, the drained bulk modulus $K = \lambda + \frac{2}{3}\mu$, the shear modulus μ and the coefficient of Biot α_{w} deduce as indicated by (7.3.48),

$$
\frac{K}{K_{\mathrm{s}}} = \frac{\mu}{\mu_{\mathrm{s}}} = (1 - \sqrt{n^{\mathrm{w}}})^2, \quad \alpha_{\mathrm{w}} = 1 - (1 - \sqrt{n^{\mathrm{w}}})^2,
\tag{11.9.33}
$$

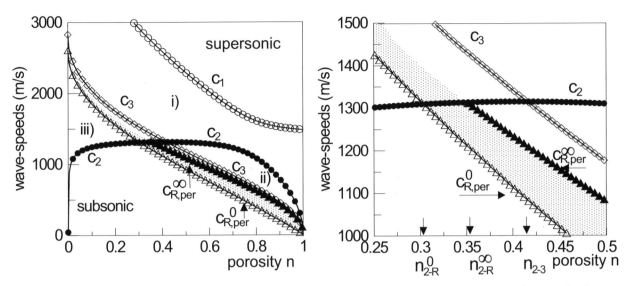

Fig. 11.9.2: Permeable free surface. Longitudinal body wave-speeds c_1, c_2, shear body wave-speed c_3, Rayleigh wave-speeds at vanishingly small and high frequencies, $c_{R\text{per}}^0$ and $c_{R\text{per}}^\infty$, respectively, and a close-up around the critical porosity range.

The relative order of the three body wave-speeds depends on the porosity as sketched in Fig. 11.5.2. For $n^w < n_{2-3}$ (respectively, $n^w > n_{2-3} \simeq 0.415$), the shear wave-speed c_3 is larger (respectively smaller) than the second longitudinal wave-speed c_2. Furthermore, as explained in Sections (11.9.5.4) and (11.9.5.3),

- there exists a porosity $n_{2-R}^0 \simeq 0.30$ at which the second longitudinal wave-speed c_2 and the Rayleigh wave-speed for low frequency $c_{R\text{per}}^0$ are equal;
- there exists also a porosity $n_{2-R}^\infty \simeq 0.35$, at which the second longitudinal wave-speed c_2 and the Rayleigh wave-speed for infinitely large frequencies $c_{R\text{per}}^\infty$ are equal;
- above this porosity, the Rayleigh waves exist for all frequencies, and they are subsonic;
- below this porosity, there exists a frequency cutoff above which the Rayleigh waves do not exist. This cutoff depends on porosity, Fig. 11.9.3;
- below the porosity n_{2-R}^0, the existing Rayleigh waves are intersonic;
- the range of variation of the Rayleigh wave-speeds with the frequency, at given porosity, is shown dotted: it is modest. For higher porosity $n^w > n_{2-R}^\infty$, this range is bounded below by the zero frequency speed $c_{R\text{per}}^0$, and above by the infinite frequency speed $c_{R\text{per}}^\infty$. For lower porosity $n^w < n_{2-R}^\infty$, it is bounded below by the zero frequency speed $c_{R\text{per}}^0$, and above by the wave-speed associated with the frequency cutoff. Although the allowable frequency range becomes larger as the porosity tends to vanish, Fig. 11.9.3, the zero frequency and the cutoff frequency wave-speeds become closer. The latter collapse to the Rayleigh wave-speed for a solid which is indeed known to be frequency independent and slightly smaller than the shear wave-speed.

whence the mixture coefficients by (7.3.44),

$$\lambda^{ww} = \frac{(n^w)^2 K_s^2}{\left(\dfrac{1-n^w}{K_s} + \dfrac{n^w}{K_w}\right) K_s^2 - K} > 0, \quad \frac{\lambda^{sw}}{\lambda^{ww}} = \frac{\alpha_w}{n^w} - 1 \geq 0, \quad \lambda^{ss} = K - \frac{2}{3}\mu + \frac{(\lambda^{sw})^2}{\lambda^{ww}}. \tag{11.9.34}$$

The calculations are performed according to the following scheme:
1. a porosity n^w is selected;
2. a range of frequencies of interest either in terms of ω or in terms of κ is defined;
3. given a frequency κ, the eigenvalues $\tilde{X}^2$'s are obtained first by solving (11.9.20) and (11.9.23);
4. the eigenmodes are then known by (11.9.21) and (11.9.24), and parts of the coefficients of the secular matrix by (11.9.28), except those that involve the $\tilde{A}$'s themselves, not their squares;
5. the determinant of the secular matrix is expanded in polynomial form, so as to eliminate the $\tilde{A}$'s by squaring, in order to make appear the $\tilde{A}^2$'s, and be able to use (11.9.18). The squarings induce spurious solutions that will be eliminated later. The dispersion relation appears as a polynomial of order 14 in $\tilde{k}$, actually of order 7 in $\tilde{k}^2$,

$$\sum_{j=1,7} s_j^{\text{per}} \tilde{k}^{2j} = 0. \tag{11.9.35}$$

The coefficients of the polynomial are in general complex. The equation is solved using an algorithm that finds all solutions of such a polynomial. Out of the 14 complex solutions, only the ones with positive real part and negative imaginary part are retained, at maximum 7 $\tilde{k}$'s;

6. given the $\tilde{X}^2$'s and the set of $\tilde{k}$'s, the $\tilde{A}_j^2$, $j \in [1,3]$, are deduced by (11.9.18): the latter should not be real negative so as to satisfy the decay in depth. The real parts of the admissible $\tilde{A}$'s are chosen strictly positive;

7. an additional screening is required over the set of $\tilde{k}$'s to eliminate the spurious solutions induced by the algebraic manipulations (squaring) in step 5 above, that is, to eliminate the wave-numbers $\tilde{k}$'s that do not zero the secular determinant $\det \mathcal{R}_{\mathrm{per}}$.

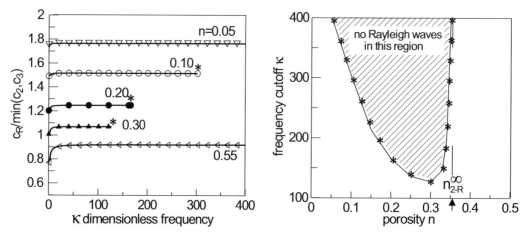

Fig. 11.9.3: Permeable free surface. (Left) Rayleigh wave-speeds as a function of frequency, for several porosities. (Right) Cutoff frequency, above which the Rayleigh waves do not exist, as a function of porosity, in the range $n^{\mathrm{w}} < n_{2-\mathrm{R}}^{\infty}$. In this range, the Rayleigh waves are intersonic. For larger porosities, the Rayleigh waves exist for all frequencies and are subsonic.

The results highlight the following features, which can be observed in Figs. 11.9.2 to 11.9.5:

1. the low frequency limit Rayleigh wave exists at all porosities. There exists a porosity $n_{2-\mathrm{R}}^0$ at which the low frequency Rayleigh wave-speed $c_R(0)$ is equal to the slow longitudinal wave-speed c_2. For a porosity smaller (respectively larger) than $n_{2-\mathrm{R}}^0$, the low frequency wave-speed is intersonic (respectively subsonic);

2. there exists a porosity for $n_{2-\mathrm{R}}^{\infty}$ at which the Rayleigh wave-speed $c_R(\infty)$ at large frequencies is equal to the second longitudinal wave-speed c_2;

3. for a porosity $n^{\mathrm{w}} \geq n_{2-\mathrm{R}}^{\infty}$, the Rayleigh waves exist at all frequencies, and they are subsonic;

4. for a porosity $n^{\mathrm{w}} < n_{2-\mathrm{R}}^{\infty}$, there exists a frequency cutoff above which the Rayleigh wave does not exist, Fig. 11.9.3. Below the frequency cutoff, the Rayleigh waves are intersonic for $n^{\mathrm{w}} < n_{2-\mathrm{R}}^0$, and, in the range $n_{2-\mathrm{R}}^0 < n^{\mathrm{w}} < n_{2-\mathrm{R}}^{\infty}$, they are subsonic at the lowest frequencies and intersonic at the largest available frequencies;

5. the frequency cutoff is correlative with the vanishing of the real part of the coefficient of depth dependence A_1, Figs. 11.9.4 and 11.9.5.

For any porosity and any frequency, there exists at maximum a single Rayleigh wave associated with the permeable surface. The existence is guaranteed at any frequency for a porosity larger than $n_{2-\mathrm{R}}^{\infty}$.

11.9.5.3 Low frequency limit, $\kappa \to 0$

Since $F(0) = 1$, it follows from (11.9.13) that, at low frequency, namely for small κ,

$$\frac{k_{\mathrm{Sd}}}{\rho^{\mathrm{w}} \omega} = G(\kappa) = \left(\frac{\delta}{\kappa}\right)^2 = O(\kappa^{-2}) \to \infty \,. \tag{11.9.36}$$

Solving (11.9.20) and (11.9.23) leads to

$$\tilde{X}_1^2 = -1 + O(\kappa^2) \,, \quad \tilde{X}_2^2 = i\, G(\kappa) \frac{c_3^2}{c_{\mathrm{SU}}^2} \frac{c_{\mathrm{LU}}^4}{c_1^2\, c_2^2} + 1 - c_{\mathrm{LU}}^2 \frac{c_1^2 + c_2^2}{c_1^2\, c_2^2} = O(\kappa^{-2}) \,, \quad \tilde{X}_3^2 = -1 + O(\kappa^2) \,, \tag{11.9.37}$$

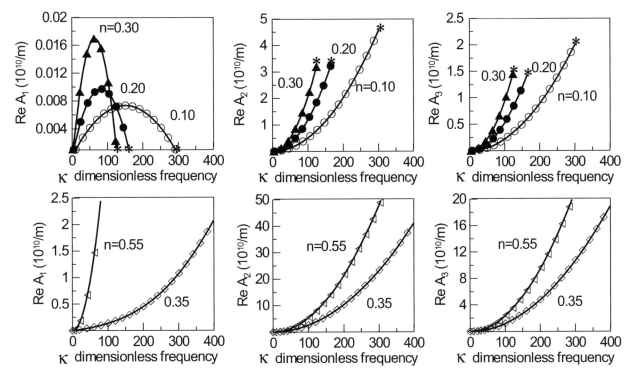

Fig. 11.9.4: Permeable free surface. Real parts of the coefficients of depth dependence as a function of frequency, for several porosities. According to (11.9.12) and (11.9.18), the depth dependence coefficients scale inversely with the hydraulic conductivity. These coefficients tend to vanish at low frequencies, see Section 11.9.5.3. Conversely, they are large at high frequencies, especially in the zone $n^{\mathrm{w}} > n_{2-\mathrm{R}}^{\infty}$. Rayleigh waves cease to exist at high frequencies in the range $n^{\mathrm{w}} < n_{2-\mathrm{R}}^{\infty}$ when one depth dependence coefficient vanishes.

$$\tilde{A}_1^2 = \tilde{k}^2 - \frac{\mu}{H^{\mathrm{U}}} + O(\kappa^2)\,, \quad \tilde{A}_2^2 = \tilde{k}^2 + i\,G(\kappa)\,\frac{c_{\mathrm{LU}}^2\,c_3^2}{c_1^2\,c_2^2} + (1 - c_{\mathrm{LU}}^2\,\frac{c_1^2 + c_2^2}{c_1^2\,c_2^2})\,\frac{c_{\mathrm{SU}}^2}{c_{\mathrm{LU}}^2}\,, \quad \tilde{A}_3^2 = \tilde{k}^2 - 1 + O(\kappa^2)\,, \qquad (11.9.38)$$

and

$$\tau_j \equiv \frac{a_{\mathrm{w}j}}{a_{\mathrm{s}j}} - 1 = O(\kappa^2)\,, \quad j \in [1,3]\,. \qquad (11.9.39)$$

The determinant of the secular matrix (11.9.29) can be shown to be proportional to main order to

$$\Phi_0(\tilde{k}^2) = (2\,\tilde{k}^2 - 1)^2 - 4\,\tilde{k}^2\,\tilde{A}_1\,\tilde{A}_3\,, \qquad (11.9.40)$$

which can be transformed by squaring to a cubic polynomial in $\tilde{k}^2$,

$$P_0(\tilde{k}^2) = 16\,(1 - \frac{\mu}{H^{\mathrm{U}}})\,(\tilde{k}^2)^3 - 8\,(3 - 2\,\frac{\mu}{H^{\mathrm{U}}})\,(\tilde{k}^2)^2 + 8\,\tilde{k}^2 - 1 = 0\,. \qquad (11.9.41)$$

The secular matrix and the cubic polynomial are identical to their counterparts for an elastic solid, to within the fact the undrained modulus H^{U}, associated with the undrained Poisson's ratio ν^{U}, takes the place of the longitudinal modulus H, associated with the drained Poisson's ratio ν. Since the undrained Poisson's ratio is bounded by -1 and $1/2$, Eqn (7.4.43), the ratio $\mu/H^{\mathrm{U}} = (1 - 2\,\nu^{\mathrm{U}})/(2\,(1 - \nu^{\mathrm{U}}))$ varies between the same bounds as for a solid. Therefore the analysis pertaining to the elastic solid of Section 11.9.8 applies to a saturated porous medium in the limit case of low frequency. In essence, the single admissible dimensionless wave-number $\tilde{k}$ defined by (11.9.31) and (11.9.32) is larger than 1, so that the real wave-speed is smaller than $c_{\mathrm{SU}} = \sqrt{\mu/\rho}$, and hence smaller than the shear wave-speed $c_3 = \sqrt{\mu/\rho^{\mathrm{s}}}$, but it can be larger than c_2, Fig. 11.9.2. Since $\tilde{k}$ is finite, it follows from (11.9.38) that the coefficients of depth decay tend to vanish as the frequency goes to 0, and, in fact, $A_1 = O(\kappa^2)$, $A_2 = O(\kappa)$, $A_3 = O(\kappa^2)$.

11.9.5.4 High frequency limit, $\kappa \to \infty$

The comments below assume the wave-number k to be real at large frequency,

$$\operatorname{Im} k(\kappa \to \infty) = 0\,, \qquad (11.9.42)$$

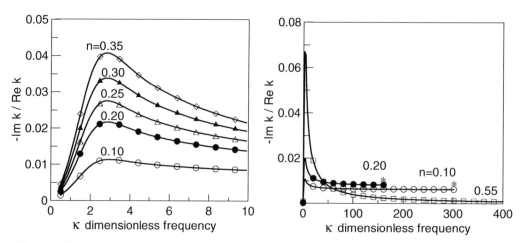

Fig. 11.9.5: Permeable free surface. Wave-number as a function of frequency, for several porosities. The ratio $-\operatorname{Im} k/\operatorname{Re} k$ should remain negative for the amplitude of the wave to remain bounded in the direction of propagation. The wave-number is real at the low and high frequency limits.

so that Eqn (11.9.32) yields $c_R(\infty) = \omega/k$.

From (11.9.13) and the fact that $F(\kappa \gg 1) = O(\kappa)$,

$$\frac{k_{\mathrm{Sd}}}{\rho^{\mathrm{w}} \omega} = G(\kappa) = O(\kappa^{-1}) \,. \tag{11.9.43}$$

Solving (11.9.20) and (11.9.23) provides (recall that $c_{\mathrm{LU}} = \sqrt{H^{\mathrm{U}}/\rho}$),

$$\tilde{X}_1^2 = -\frac{c_{\mathrm{LU}}^2}{c_1^2} \,, \quad \tilde{X}_2^2 = -\frac{c_{\mathrm{LU}}^2}{c_2^2} \,, \quad \tilde{X}_3^2 = -\frac{\rho^{\mathrm{s}}}{\rho} \,, \tag{11.9.44}$$

and the eigenmodes (11.9.21) and (11.9.24) simplify somehow,

$$\frac{a_{\mathrm{w}j}}{a_{\mathrm{s}j}} = \frac{c_j^2 - c_{\mathrm{s}}^2}{c_{\mathrm{ms}}^2} \,, \ j = 1, 2; \quad \frac{a_{\mathrm{w}3}}{a_{\mathrm{s}3}} = 0 \,. \tag{11.9.45}$$

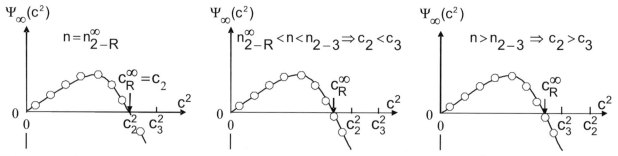

Fig. 11.9.6: Permeable free surface. Expected behavior of the secular determinant $\Psi_\infty(c^2)$ at high frequency as a function of the square of the wave-speed. The critical porosity n_{2-R}^∞ is associated with the Rayleigh wave-speed c_R^∞ being equal to the second longitudinal wave speed c_2. Intersonic wave-speeds, potentially associated with porosities smaller than the critical porosity, are ruled out because they do not satisfy the decay conditions.

These results can be checked to hold as well in the case of a vanishingly small dissipation coefficient k_{Sd}, i.e., of a large permeability. The coefficients of depth dependence A's result from (11.9.17) and (11.9.18), and, as another consequence of (11.9.42), they are either real positive, or imaginary. A decay of the wave in depth excludes the latter possibility and, in fact, *constraints the Rayleigh wave to be subsonic*,

$$c_R(\infty) < c_j \,, \quad j \in [1, 3] \,. \tag{11.9.46}$$

Then the three coefficients α's, defined by their squares,

$$\alpha_j^2(c^2) \equiv \frac{A_j^2}{k^2} = 1 - \frac{c^2}{c_j^2}, \quad j \in [1,3], \tag{11.9.47}$$

with $c = c_R(\infty)$, are real. The determinant of the secular matrix (11.9.28) becomes,

$$\Psi_\infty(c^2) \simeq \det \begin{bmatrix} 1 + \alpha_3^2 & 1 + \alpha_3^2 & -2\,\alpha_3\,i \\ 2\,\alpha_1\,i & 2\,\alpha_2\,i & 1 + \alpha_3^2 \\ c_2^2 - c_{\rm w}^2 & c_1^2 - c_{\rm w}^2 & 0 \end{bmatrix}, \tag{11.9.48}$$

or

$$\Psi_\infty(c^2) \simeq 4\,\alpha_1\,\alpha_3\,(c_1^2 - c_{\rm w}^2) + 4\,\alpha_2\,\alpha_3\,(c_{\rm w}^2 - c_2^2) - (1 + \alpha_3^2)^2\,(c_1^2 - c_2^2), \tag{11.9.49}$$

and

$$\Psi_\infty(0) = 0, \quad \frac{d\,\Psi_\infty}{dc^2}(0) > 0, \quad \Psi_\infty(c_3^2) < 0. \tag{11.9.50}$$

Consequently, as sketched in Fig. 11.9.6,

1. if $c_3 = \min\{c_2, c_3\}$, that is for $n^{\rm w} > n_{2-3}$, then there exists at least one real (subsonic) Rayleigh speed $c_R(\infty) < \min\{c_2, c_3\}$;

2. if $c_2 = \min\{c_2, c_3\}$, then, for $n_{2-R}^\infty < n^{\rm w} < n_{2-3}$, there exists at least one real (subsonic) Rayleigh speed $c_R(\infty) < c_2$ if $\Psi(c_2^2) < 0$. Otherwise, that is for $n^{\rm w} < n_{2-R}^\infty$, a Rayleigh wave associated with strictly decaying amplitude down the half-space might not exist, and indeed, this hypothesis is confirmed by the numerical results (but we have not proved it).

The equation that defines the porosity n_{2-R}^∞ at which $c_R(\infty) = c_2$ is thus

$$4\,\alpha_1\,\alpha_3\,(c_1^2 - c_{\rm w}^2) - (1 + \alpha_3^2)^2\,(c_1^2 - c_2^2) = 0, \tag{11.9.51}$$

with α_1 and α_3 defined by (11.9.47) with $c = c_R(\infty) = c_2$.

11.9.6 Impermeable free surface

The boundary conditions along the surface $x_2 = 0$ now involve the total traction and the normal velocity, or displacement, of the fluid relative to the solid,

$$\sigma_{22}(x_1, x_2 = 0) = 0, \quad \sigma_{12}(x_1, x_2 = 0) = 0, \quad (u_{\rm s2} - u_{\rm w2})(x_1, x_2 = 0) = 0, \quad \forall x_1 \in\,] -\infty, \infty[. \tag{11.9.52}$$

11.9.6.1 The dispersion relation

With the normalization (11.9.17), (11.9.18), since $\sigma_{22} = \sigma_{22}^{\rm s} - n^{\rm w} p_{\rm w}$ and $\sigma_{12} = \sigma_{12}^{\rm s}$, the stresses (11.9.10) and displacements (11.9.8) combine to yield the secular equation in matrix form as,

$$\begin{bmatrix} \hat{\sigma}_{22}(x_2) \\ \hat{\sigma}_{12}(x_2) \\ \mu\,k_r\,(u_{\rm s2} - u_{\rm w2})(x_2) \end{bmatrix} = \mu\,k_r^2\,\mathcal{R}_{\rm imp} \begin{bmatrix} a_{\rm s1}\,e^{-A_1 x_2} \\ a_{\rm s2}\,e^{-A_2 x_2} \\ a_{\rm s3}\,e^{-A_3 x_2} \end{bmatrix} = \begin{bmatrix} 0 \\ 0 \\ 0 \end{bmatrix}, \tag{11.9.53}$$

with the secular Rayleigh matrix,

$$\mathcal{R}_{\rm imp} = \begin{bmatrix} 2\,\tilde{k}^2 + \beta_1\,\tilde{X}_1^2 & 2\,\tilde{k}^2 + \beta_2\,\tilde{X}_2^2 & -2\,i\,\tilde{k}\,\tilde{A}_3 \\ 2\,i\,\tilde{k}\,\tilde{A}_1 & 2\,i\,\tilde{k}\,\tilde{A}_2 & 2\,\tilde{k}^2 + \tilde{X}_3^2 \\ \tau_1\,\tilde{A}_1 & \tau_2\,\tilde{A}_2 & -i\,\tilde{k}\,\tau_3 \end{bmatrix}, \tag{11.9.54}$$

the β's and τ's being defined by (11.9.30). The procedure to solve the secular equation in terms of the dimensionless frequency κ is identical to the permeable case. The dispersion relation appears as a polynomial of order 18 in $\tilde{k}$, actually of order 9 in $\tilde{k}^2$,

$$\sum_{j=1,9} s_j^{\rm imp}\,\tilde{k}^{2j} = 0. \tag{11.9.55}$$

The use of a symbolic software is required to obtain the complex coefficients $s^{\rm imp}$ which are *very* lengthy.

The results are shown in Figs. 11.9.7 to 11.9.9. In essence

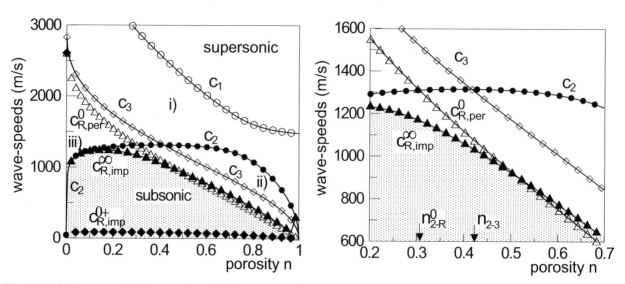

Fig. 11.9.7: Impermeable free surface. Body wave-speeds c_1, c_2 and c_3 and Rayleigh wave-speeds at zero, small and infinite frequencies, and a close-up around the critical porosity.

There exists a single Rayleigh wave at any frequency and any porosity. The Rayleigh wave-speed at zero frequency is equal to its counterpart for the permeable free surface, $c^0_{Rimp} = c^0_{Rper}$, while the Rayleigh wave-speed at small but non zero frequency is vanishingly small. As the frequency increases, the Rayleigh wave-speed increases but it remains subsonic, even if, at small porosity, it tends to the second longitudinal body wave-speed c_2.

The dependence on frequency, at given porosity, is displayed in Fig. 11.9.8. The range of variation, shown dotted here, is large, but the actual variations take place mainly at small frequencies.

1. there always exists a single Rayleigh wave which, except at zero frequency, is subsonic;

2. at zero frequency, the Rayleigh wave-speed is equal to its permeable counterpart;

3. at small, but non zero, frequency, the wave-speed is vanishingly small and it increases monotonically with frequency.

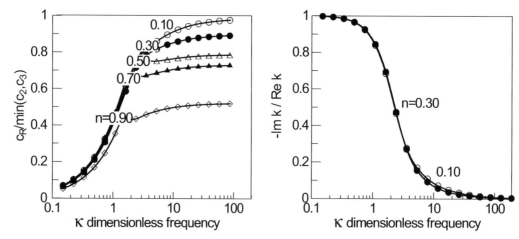

Fig. 11.9.8: Impermeable free surface. (Left) Rayleigh wave-speeds as a function of frequency, for several porosities. The wave-speeds increase from zero towards a subsonic asymptotic speed, which is close to the 2nd longitudinal wave-speed c_2 for small porosities.

(Right) Ratio of the imaginary and real parts of the wave-number as a function of frequency, for several porosities. This ratio should be negative for the amplitude of the wave to remain bounded. The wave-number is real at high frequencies, like for a permeable free surface. On the other hand, at low frequencies, as indicated by (11.9.56) and (11.9.57), the ratio $-\operatorname{Im} k / \operatorname{Re} k \simeq 1 + Y (\kappa/\delta)^2$ is equal to 1 at $\kappa = 0^+$ and it decreases as the frequency increases, the coefficient Y being negative.

Some qualitative explanations of the existence, number, and dependence of the solution in terms of frequency and porosity may be provided at low and high frequencies.

11.9.6.2 Low frequency limit, $\kappa \to 0$

The relations (11.9.36)-(11.9.39) pertaining to the eigenvalues $\tilde{X}$'s and eigenmodes τ's hold irrespective of the boundary conditions. The three τ's almost vanish according to (11.9.39), but $\tau_2 = O(\kappa^2)$ while $\tilde{A}_2 = O(\kappa^{-1})$ is large. Thus the secular determinant is the product of two factors. One factor is exactly as for a permeable free surface, namely Eqn (11.9.40): the vanishing of this factor implies that the zero frequency solution of the permeable case is also a solution of the impermeable case. The second factor is equal to $\tau_2 \tilde{A}_2$. It may be of interest to study the solution of $\tilde{A}_2 = 0$ for small frequencies, even if it does not furnish an appropriate decay. Eqns (11.9.37) and (11.9.38) provide the eigenvalue,

$$\tilde{X}_2^2 = i\left(\frac{\delta}{\kappa}\right)^2 \frac{\rho}{\rho^s} \frac{c_{\text{LU}}^4}{c_1^2 c_2^2}\left(1 - i\,Y\,(\frac{\kappa}{\delta})^2\right), \quad Y \equiv \left(1 - c_{\text{LU}}^2 \frac{c_1^2 + c_2^2}{c_1^2 c_2^2}\right)\frac{c_1^2 c_2^2}{c_{\text{LU}}^4}\frac{\rho^s}{\rho} < 0\,. \tag{11.9.56}$$

The resulting admissible dimensionless wave-number with $(\text{Re}\,\tilde{k} > 0,\ \text{Im}\,\tilde{k} < 0)$,

$$\tilde{k} = \frac{1}{\sqrt{2}}\frac{\delta}{\kappa}\frac{c_{\text{LU}}\,c_3}{c_1\,c_2}\left(1 - \frac{Y}{2}\,(\frac{\kappa}{\delta})^2 - i\,(1 + \frac{Y}{2}\,(\frac{\kappa}{\delta})^2)\right), \tag{11.9.57}$$

is large, actually of the order of δ/κ, while the Rayleigh wave-speed,

$$c_R(0) = \frac{c_{\text{SU}}}{\text{Re}\,\tilde{k}} \simeq \sqrt{2}\,c_{\text{SU}}\frac{c_1\,c_2}{c_{\text{LU}}\,c_3}\frac{\kappa}{\delta}\left(1 + \frac{Y}{2}\,(\frac{\kappa}{\delta})^2\right), \tag{11.9.58}$$

is small, of the order κ/δ. The depth dependence coefficients (11.9.38),

$$A_1 = k_r \tilde{A}_1 = \frac{\kappa}{\delta}\frac{c_{\text{LU}}\,c_3}{c_1\,c_2}\frac{n^{\text{w}}\,g}{c_{\text{SU}} K_{\text{H}}(\omega = 0)}\,e^{-i\pi/4}\left(1 + O(\kappa^2)\right) \simeq A_3\,, \tag{11.9.59}$$

are seen to vanish with the frequency, like for the permeable free surface, Fig. 11.9.9.

Thus at zero frequency, the Rayleigh wave-speed $c_{R\text{imp}}(0)$ is identical to the wave-speed associated with the permeable free surface $c_{R\text{per}}(0)$. At small but not zero frequency, the Rayleigh wave-speed is close to 0 and it increases with frequency, as confirmed by the numerical results displayed in Fig. 11.9.8.

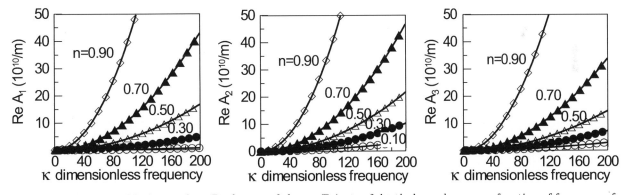

Fig. 11.9.9: Impermeable free surface. Real parts of the coefficients of depth dependence as a function of frequency, for several porosities. These coefficients tend to vanish at low frequencies, and increase with frequency.

11.9.6.3 High frequency limit, $\kappa \to \infty$

The relations (11.9.42)-(11.9.46) pertaining to the eigensolutions still hold. The secular determinant can be shown to be proportional to

$$\Psi_\infty(c^2) = \det\begin{bmatrix} 2 + \chi_1\,(-1 + \alpha_3^2) & 2 + \chi_2\,(-1 + \alpha_3^2) & -2\,\alpha_3\,i \\ 2\,\alpha_1\,i & 2\,\alpha_2\,i & 1 + \alpha_3^2 \\ -\tau_1\,\alpha_1 & -\tau_2\,\alpha_2 & -i \end{bmatrix}. \tag{11.9.60}$$

The α's involved in this subsection are in fact real positive, since it will be shown that the Rayleigh wave is subsonic. The χ's and τ's are defined as follows,

$$\chi_j = \frac{c_j^2 - c_s^2}{c_{mw}^2} + 1, \quad \tau_j = \frac{c_j^2 - c_s^2}{c_{ms}^2} - 1, \quad j = 1, 2. \tag{11.9.61}$$

Therefore

$$\begin{aligned}
\Psi_\infty(c^2) &= 4\,\alpha_1\,\alpha_2\,\alpha_3\,(\tau_1 - \tau_2) \\
&\quad - \alpha_1\left(2 + \chi_2\left(-1 + \alpha_3^2\right)\right)\left(2 + \tau_1\left(1 + \alpha_3^2\right)\right) \\
&\quad + \alpha_2\left(2 + \chi_1\left(-1 + \alpha_3^2\right)\right)\left(2 + \tau_2\left(1 + \alpha_3^2\right)\right).
\end{aligned} \tag{11.9.62}$$

The following relations,

$$\Psi_\infty(0) = 0; \quad \frac{d\,\Psi_\infty}{dc^2}(0) > 0; \quad \Psi_\infty(c_3^2) < 0 \text{ for } c_3 \le c_2; \quad \Psi_\infty(c_2^2) < 0 \text{ for } c_2 \le c_3, \tag{11.9.63}$$

may be established with the proviso that the mechanical coupling coefficient λ^{sw} is positive, Eqn (11.9.34). The proof of the relation $\Phi(c_3^2) < 0$ for $c_3 \le c_2$ goes as follows. First $\Psi_\infty(c_3^2)$ simplifies to,

$$\Psi_\infty(c_3^2) = -\alpha_1\left(2 - \chi_2\right)\left(2 + \tau_1\right) + \alpha_2\left(2 - \chi_1\right)\left(2 + \tau_2\right), \tag{11.9.64}$$

and $2 - \chi_2$ and $2 + \tau_1$ are easily shown to be positive. Furthermore, use of the relation $(c_j^2 - c_w^2)(c_j^2 - c_s^2) = c_{ms}^2 c_{mw}^2$, for $j = 1, 2$, yields $2 - \chi_1 = 1 - 1/X$ and $2 + \tau_2 = 1 - X$, with $X = (c_1^2 - c_w^2)/c_{ms}^2 > 0$, so that $(2 - \chi_1)(2 + \tau_2) < 0$.

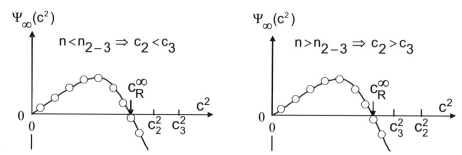

Fig. 11.9.10: Impermeable free surface. Expected shape of the secular determinant Ψ_∞ as a function of the wave-speed, leading to the existence of a single subsonic Rayleigh wave-speed.

As a consequence of (11.9.63), the variation of the secular determinant (11.9.60) as a function of the wave-speed is expected as sketched in Fig. 11.9.10, indicating that there exists at least one Rayleigh wave which is subsonic. Note that we have not proved uniqueness.

11.9.7 The Kelvin functions ber and bei

Biot [1956] makes use of the frequency function $F(\kappa)$,

$$F(\kappa) = \frac{\kappa^2}{4}\,\frac{d\mathrm{be}(\kappa)/d\kappa}{\kappa\,\mathrm{be}(\kappa) + 2\,i\,d\mathrm{be}(\kappa)/d\kappa}, \tag{11.9.65}$$

with

$$\mathrm{be}(\kappa) = \mathrm{ber}(\kappa) + i\,\mathrm{bei}(\kappa) = I_0(i^{\frac{1}{2}}\kappa) = J_0(i^{\frac{3}{2}}\kappa), \tag{11.9.66}$$

where I_0 is the modified Bessel function of zeroth order, and ber and bei are referred to as Kelvin functions. The actual series definitions follow from (7.5.66),

$$\mathrm{ber}(\kappa) = \sum_{n=0}^{\infty} \frac{(-1)^n}{((2n)!)^2}\left(\frac{\kappa^2}{4}\right)^{2n}, \quad \mathrm{bei}(\kappa) = \sum_{n=0}^{\infty} \frac{(-1)^n}{((2n+1)!)^2}\left(\frac{\kappa^2}{4}\right)^{2n+1}. \tag{11.9.67}$$

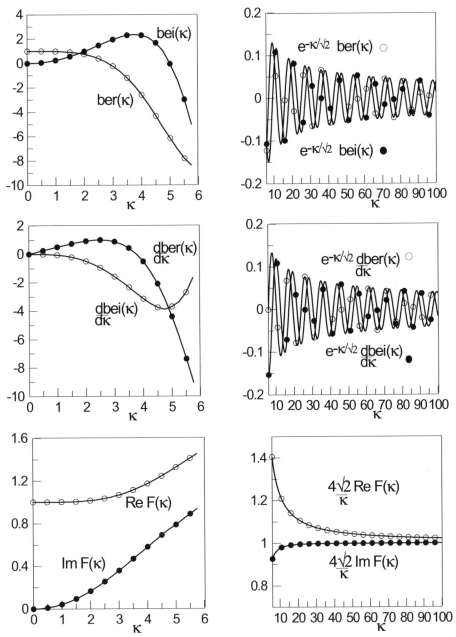

Fig. 11.9.11: The Kelvin functions $\mathrm{ber}(\kappa)$ and $\mathrm{bei}(\kappa)$, their derivatives, and the function $F(\kappa)$ used to introduce a frequency dependent hydraulic conductivity.

These functions are plotted in Fig. 11.9.11. In actual computations, the series are defined recursively, with about 50 terms. For $\kappa > 5$, based on (7.5.77), the following asymptotic expression may be used:

$$\mathrm{be}(\kappa) = \frac{1}{\sqrt{2\,\pi\kappa}}\, \exp\left(\frac{1+i}{\sqrt{2}}\,\kappa - i\,\frac{\pi}{8}\right). \tag{11.9.68}$$

More generally, the Kelvin functions of order $m \geq 0$,

$$\mathrm{be}_m(\kappa) = \mathrm{ber}_m(\kappa) + i\,\mathrm{bei}_m(\kappa) = I_m(i^{\frac{1}{2}}\kappa) = i^{-m}\,J_m(i^{\frac{3}{2}}\kappa), \tag{11.9.69}$$

are solutions bounded at $\kappa = 0$ of the homogeneous differential equation $\kappa^2\,d^2y/d\kappa^2 + \kappa\,dy/d\kappa - (i\kappa^2 + m^2)\,y = 0$.

With help of the relation $(7.5.70)_2$ with $m = 1$, and $(7.5.72)$ with $m = 0$, the function $F(\kappa)$ may be rephrased directly in terms of the modified Bessel functions of the first kind I_1 and I_2,

$$F(\kappa) = \frac{i^{\frac{1}{2}}\kappa}{4}\, \frac{I_1(i^{\frac{1}{2}}\kappa)}{I_2(i^{\frac{1}{2}}\kappa)}\,. \tag{11.9.70}$$

Therefore, with further use of the asymptotic expressions $(7.5.77)$ and $(7.5.78)$ for large and small arguments, respectively,

$$\lim_{\kappa \to 0} F(\kappa) = 1, \quad \lim_{\kappa \to \infty} F(\kappa) = \frac{1}{4}\, i^{\frac{1}{2}}\kappa\,. \tag{11.9.71}$$

11.9.8   Surface waves in an elastic half-plane

Instead of a complete derivation, the analysis for an elastic solid may be obtained from the analysis for a poroelastic medium performed in the previous sections, simply by setting the fluid volume fraction to 0 and disregarding Darcy's law, and the associated balance of momentum and boundary condition. A key difference emerges, namely that the Rayleigh waves are *not dispersive* in an elastic solid unlike in the poroelastic medium, that is, the wave-speed $c_R = \omega/\mathrm{Re}\, k$ which is the ratio between the pulsation ω and the wave-number k is independent of ω. The Rayleigh wave-speed which turns out to be smaller than the shear wave-speed and a standard approximation are plotted in Fig. 11.9.12.

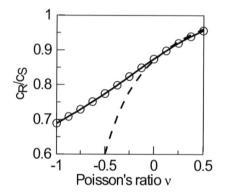

ν	c_R/c_S
-1	0.689
-0.5	0.774
0	0.874
0.25	0.919
0.333	0.932
0.5	0.955

Fig. 11.9.12: Rayleigh wave-speed in an elastic solid as a function of Poisson's ratio. The dotted standard approximation $c_R/c_S = (0.874 + 1.117\,\nu)/(1 + \nu)$ holds for $\nu \geq 0$ only.

The relations for the solid deduce from the analysis of the poroelastic medium by changing the indices 1 and 3 pertaining to the largest longitudinal wave and shear wave, respectively, to L and S. The eigenvalue problems $(11.9.19)$ and $(11.9.22)$ yield

$$X_L^2 = \frac{H}{\mu}\,(A_L^2 - k^2) = -\frac{\omega^2}{c_S^2}, \quad X_S^2 = A_S^2 - k^2 = -\frac{\omega^2}{c_S^2}\,, \tag{11.9.72}$$

with $H = \lambda + 2\,\mu$ the longitudinal modulus. The Rayleigh matrix $(11.9.10)$ simplifies to,

$$\mathcal{R} = \begin{bmatrix} \mu\,(2\,k^2 + X_L^2) & -2\,\mu\,i\,k\,A_S \\[2mm] 2\,\mu\,i\,k\,A_L & \mu\,(2\,k^2 + X_S^2) \end{bmatrix}. \tag{11.9.73}$$

Dimensionless entities denoted by a superimposed tilde result via scaling by the wave-number $k_r = \omega/c_S$ yielding the secular Rayleigh matrix,

$$\mathcal{R} = \mu\,k_r^2 \begin{bmatrix} 2\,\tilde{k}^2 - 1 & -2\,i\,\tilde{k}\,\tilde{A}_S \\[2mm] 2\,i\,\tilde{k}\,\tilde{A}_L & 2\,\tilde{k}^2 - 1 \end{bmatrix}. \tag{11.9.74}$$

The determinant of the secular matrix is viewed as an equation for the dimensionless wave-number $\tilde{k} = c_S/c_R$,

$$\Phi(\tilde{k}^2) = (2\,\tilde{k}^2 - 1)^2 - 4\tilde{k}^2\,\tilde{A}_L\,\tilde{A}_S\,, \qquad (11.9.75)$$

which can be transformed, by squaring, to a cubic polynomial in $\tilde{k}^2$,

$$P(\tilde{k}^2) = 16\,(1 - \frac{\mu}{H})\,(\tilde{k}^2)^3 - 8\,(3 - 2\,\frac{\mu}{H})\,(\tilde{k}^2)^2 + 8\,\tilde{k}^2 - 1 = 0\,. \qquad (11.9.76)$$

The ratio μ/H can be expressed in terms of the Poisson's ratio ν, namely $\mu/H = (1 - 2\nu)/(2\,(1 - \nu))$. Since Poisson's ratio is bounded by -1 and $1/2$, the ratio μ/H decreases from $3/4$ to 0. Therefore, by Descartes's rule of signs, Section 11.9.10, the polynomial has one or three positive roots and no negative root.

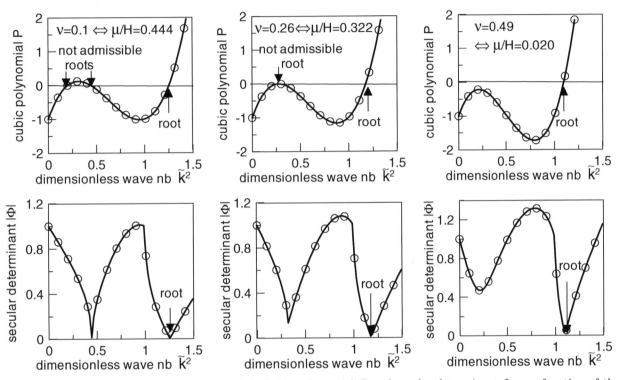

Fig. 11.9.13: Rayleigh waves in an elastic solid. Cubic polynomial P and secular determinant Φ as a function of the square of the dimensionless wave-number $\tilde{k} = c_S/c_R$.

Since $P(0) = P(1) = -1$ and $P(\infty) = \infty$, there exists a real root to $P = 0$ that is larger than 1. This root satisfies the decay conditions and is also a root of the original secular determinant Φ. Therefore, it represents an admissible solution, actually the sole solution, and the associated Rayleigh wave-speed c_R is smaller than the shear wave-speed c_S.

Two other real roots of P exist for $\mu/H \geq 0.322$ but they are smaller than 1 and do not satisfy the decay conditions. Note that the largest of these roots $\tilde{k}'$ is not a root of the secular determinant Φ even if $\Phi(\tilde{k}')$ is quite small for $\nu > 0$. The sole exception is for $\nu = 0$, and then $\tilde{k}' = 1/2$. The two complex conjugate roots of $P = 0$ for $\mu/H < 0.322$ are not roots of the secular determinant either.

While it is easier to obtain the roots of the polynomial equation, the squaring process introduces solutions of the polynomial that are not solutions of the original secular determinant. To help the understanding, the polynomial $P(\tilde{k}^2)$ and the original secular determinant $\Phi(\tilde{k}^2)$ have been plotted in Fig. 11.9.13 for three values of Poisson's ratio.

There are special cases for which the roots can be obtained easily. First, for $\mu/H = 1/2$, i.e. $\nu = 0$, the roots $\tilde{k}^2$ are $1/2$ and $(3 \pm \sqrt{5})/4$, yielding the sole admissible solution $c_R/c_S = \sqrt{3 - \sqrt{5}} \simeq 0.874$. For $\nu = 1/4$, the roots $\tilde{k}^2$ are $1/4$ and $3/4\,(1 \pm 1/\sqrt{3})$, yielding the sole solution $c_R/c_S = \sqrt{2 - 2/\sqrt{3}} \simeq 0.919$.

Since $P(0) = P(1) = -1$ and $P(\infty) \to +\infty$, there always exists a real and positive root $\tilde{k}^2$ of the cubic polynomial (11.9.41), which is larger than 1. This root satisfies the decay conditions in depth (11.9.38), and it is also a root of the original secular equation (11.9.40). Since $\tilde{k} = c_S/c_R$, it follows that the associated Rayleigh wave-speed c_R is smaller than the shear wave-speed c_S.

The discriminant of the cubic polynomial can be shown to be positive if $\mu/H < 0.322$, equivalent to $\nu > 0.263$. Then the above root is the single real root[11.5].

If $\mu/H \geq 0.322$, the two other real roots of the cubic polynomial are both either smaller or larger than 1 because $P(1) = -1$ and $P(\infty) \to +\infty$. The product of the roots is equal to $1/16 \, (1 - \mu/H)$, which is bounded above by $1/4$. Therefore, the two additional roots fall in the interval $]0, 1[$, that is, the decay condition $\tilde{k}^2 - 1 > 0$ is not satisfied.

It remains to consider the two complex roots for $\mu/H < 0.322$. In fact, they are not roots of the secular determinant as indicated in Section 11.9.9 below.

The motion of particles near the surface and the strains and stresses are described in textbooks. The horizontal and vertical displacements are 90° out of phase. Close to the surface, the motion is retrograde, counterclockwise for a wave traveling to the right, and the vertical displacement is about 1.5 larger than the horizontal displacement. The horizontal displacement changes sign at a depth about 0.2 wave-length c_R/ω. The motion penetrates slightly more than one wave-length below the surface.

11.9.9   The principle of the argument

The principle of the argument gives the number of roots minus the number of poles (multiplicity included) of a function of the complex variable over a domain defined by a closed contour and inside which it is single-valued. It has been used to determine the number of roots of the Rayleigh secular determinant by, e.g., Achenbach [1975] and Freund [1990], accounting of the fact that the latter has no pole.

The wave-number $\tilde{k} = c_S/c$ may also be viewed as a scaled slowness s/s_S, the slowness s being the reciprocal of wave-speed c. The secular determinant (11.9.75) may be recast in the explicit format,

$$R(\tilde{k}) = (2\,\tilde{k}^2 - 1)^2 - 4\,\tilde{k}^2 \sqrt{\tilde{k}^2 - 1}\sqrt{\tilde{k}^2 - \tilde{k}_L^2} \, , \tag{11.9.77}$$

or equivalently

$$R(\tilde{k}) = (2\,\tilde{k}^2 - 1)^2 + 4\,\tilde{k}^2 \sqrt{1 - \tilde{k}^2} \, \sqrt{\tilde{k}_L^2 - \tilde{k}^2} \, , \tag{11.9.78}$$

with $\tilde{k}_L = c_S/c_L = s_L/s_S = \sqrt{1 - 2\nu}/\sqrt{2 - 2\nu}$. The square-roots involve four branch points $\pm\tilde{k}_L$ and $\pm\tilde{k}_S$. The branch cuts may in a first step be chosen to run on the real axis from the branch points to $\pm\infty$, say $]\tilde{k}_S, +\infty[$ and $] - \infty, -\tilde{k}_S[$ on one hand and $]\tilde{k}_L, +\infty[$ and $] - \infty, -\tilde{k}_L[$ on the other hand. However, since these two branch cuts overlap, the branch cuts of the secular determinant may be restricted to contours around the pairs $(\tilde{k}_L, 1)$ and $(-1, -\tilde{k}_L)$.

The square-root,

$$\sqrt{1 - \tilde{k}^2} = \sqrt{|\tilde{k} - (-1)|} \, \sqrt{|\tilde{k} - 1|} \, \exp\left(\frac{i}{2}\,(\theta_a + \theta_b) - i\,\frac{\pi}{2}\right), \tag{11.9.79}$$

is defined in terms of the angles $\theta_a = \text{angle}(\text{real axis}, \tilde{k} - 1)$ and $\theta_b = \text{angle}(\text{real axis}, \tilde{k} - (-1))$. Therefore the square-root has positive real parts for points on the real segment from -1 to $+1$ where the angles θ_a and θ_b are equal to π and 0 respectively. A similar determination is used for the second square-root involved in the Rayleigh determinant (11.9.78). Therefore there are no roots of (11.9.78) between $\tilde{k}_L$ and 1 nor between -1 and $-\tilde{k}_L$. The number of roots in the remaining of the complex plane is obtained by using the principle of the argument on a closed contour C that respects the cuts and that includes an outer circle whose radius may be made arbitrary large, Fig. 11.9.14. Fig. 11.9.15 displays the evolution of the integral $(2\,i\,\pi)^{-1} \int_C dR(\tilde{k})/R$ along this contour. The end value is 2, indicating two roots. The latter are real and belong to the interval $]1, \infty$ and $] - \infty, -1[$ since $R(1) > 0$ and $R(\pm\infty) < 0$: indeed, at large $|\tilde{k}|$,

$$R(\tilde{k}) \simeq -2\,\tilde{k}^2 \,(1 - \tilde{k}_L^2)\,. \tag{11.9.80}$$

11.9.10   Descartes's rule of signs

Descartes's rule of signs is a useful tool that provides information on the number of real roots of a real polynomial which is arranged in increasing, or decreasing, power. It states that the number p of positive roots is less than or equal to the number v of sign changes in the coefficients, ignoring the powers that do not appear. Moreover, the

---

[11.5]The discriminant of the cubic equation $x^3 + a\,x^2 + b\,x + c = 0$ with real coefficients is equal to $R^2 - Q^3$ with $Q = (a^2 - 3\,b)/9$ and $R = (2\,a^3 - 9\,a\,b + 27\,c)/54$. The equation has three real roots if the discriminant is negative.

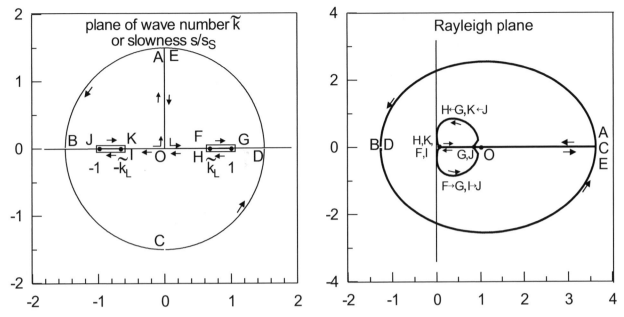

Fig. 11.9.14: The number of roots of the Rayleigh secular determinant in the complex plane is obtained through the principle of the argument by counting the number of tours (here equal to 2) about the origin of the Rayleigh plane associated with a closed path in the slowness plane that respects the branch cuts. Isotropic elastic solid with Poisson's ratio $\nu = 0.1$. A small ν implies a smaller outer contour in the R-plane, and therefore lends itself more easily to scrutinize the details of the path.

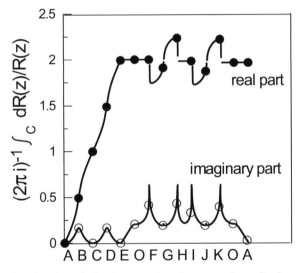

Fig. 11.9.15: Evolution along the closed path in the complex slowness plane displayed in Fig. 11.9.14 of the integral associated with the principle of the argument that yields the number of roots of the Rayleigh determinant, equal to 2 here.

difference $v - p$ is a non negative even integer, namely $v - p = 0, 2, 4 \cdots$. The multiplicity of the root is included: $f(x) = x^2 - 2x + 1 = 0$ has $v = 2$, hence p equal 0 or 2. The rule can be applied to the polynomial $P(x)$ to get information on the positive roots of $P(x)$, and to the polynomial $P(-x)$ to get information on the negative roots of $P(x)$.

11.10   First order waves in saturated porous media

Key features of first order waves in saturated porous media are illustrated via the propagation, at constant speed $c > 0$, of a plane crack along a rectilinear path in an infinite medium. Besides a fixed cartesian system $(0, X_1, X_2, X_3)$, a second cartesian coordinate system $(0, x_1, x_2, x_3)$ centered at the crack tip and moving with it in the X_1 direction is used, with the out-of-plane x_3-axis along the straight crack front. Under steady state crack propagation, an arbitrary scalar field $\varphi = \varphi(X_1, X_2, t)$ depends on time t through the x-coordinates, and, although an abuse of notation, it will be denoted by $\varphi = \varphi(x_1 = X_1 - ct, x_2 = X_2)$. Its first and second material time derivatives are defined by the relations $d\varphi/dt = -c\,\partial\varphi/\partial x_1$, and $d^2\varphi/dt^2 = c^2\,\partial^2\varphi/\partial^2 x_1$.

The regime of the flow equations is governed by the relative position of the crack tip speed c with respect to the three body wave-speeds of the porous medium c_1, c_2 and c_3 based on the classification shown in Table 11.5.1. Given that the motion considered here is a steady state motion, the issue deserves to be considered in this context.

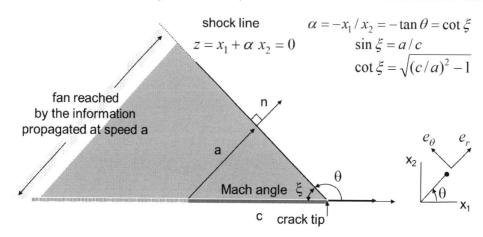

Fig. 11.10.1: A crack tip propagates in the saturated porous medium at speed c to the right along the axis x_1, while the mechanical information propagates at speed a. The problem being symmetric with respect to the crack line, only the upper half-plane is displayed. A first order discontinuity exists if $c > a$. It is referred to as shock line (curve) or Mach line. The direction of the discontinuity is defined by the Mach angle ξ such that $\sin \xi = a/c$. The first order discontinuity is characterized by the fact that only a fan in the wake of the crack tip, between the crack line and the Mach line, is reached by the shear, or longitudinal, information propagated by the moving crack tip.

In the intersonic range, the material wave-speed a can be the second longitudinal wave-speed c_2 or the shear wave-speed c_3. Thus there may be two discontinuity lines, as indeed displayed in the region i) in Fig. 11.10.4.

Emphasis is on intersonic propagation of the crack tip, and on the type and on the properties of the first order discontinuities that may propagate through the medium across the characteristic lines, as sketched in Fig. 11.10.1. The derivations of the elements reported below are detailed in Radi and Loret [2007], [2008].

11.10.1   Regime of the field equations

Along a standard terminology, a partial differential equation, involving second order derivatives in both time and space, e.g., $\partial^2\phi/\partial x_1^2 + \partial^2\phi/\partial x_2^2 - a^{-2}\,\partial^2\phi/\partial t^2 = 0$, is said to be hyperbolic if there exist real characteristics which allow for the mechanical information to propagate at finite speed, e.g., for non zero real a. Otherwise, for a purely imaginary, it is said elliptic. Here, the partial time derivative is given by $\partial\phi/\partial t = -c\,\partial\phi/\partial x_1$, so that the above equation becomes $(1 - c^2/a^2)\,\partial^2\phi/\partial x_1^2 + \partial^2\phi/\partial x_2^2 = 0$.

Thus according to the above terminology, the subsonic propagation of the crack tip $c < a$ corresponds to an elliptic equation. In the intersonic regions, the material wave-speed a can be either the second longitudinal wave-speed c_2 or the shear wave-speed c_3. A first order discontinuity exists if $c > a$. Its trace is referred to as shock line (curve) or Mach line. Its direction is defined by the Mach angle ξ or by the polar angle θ,

$$\sin \xi_j = \frac{c_j}{c}, \quad \cot \xi_j = \sqrt{\left(\frac{c}{c_j}\right)^2 - 1}, \quad \theta_j = \pi - \xi_i, \quad j \in [1,3]. \tag{11.10.1}$$

The displacement field derives from the potentials ϕ which obey second order partial differential equations: a real characteristic for ϕ in the hyperbolic regime bears a first order discontinuity, i.e. a discontinuity of the displacement gradient, and an accompanying velocity discontinuity. According to Hadamard's compatibility relations, there exists a vector $\boldsymbol{\eta}$ such that $[\![\nabla \mathbf{u}]\!] = \boldsymbol{\eta} \otimes \mathbf{n}$, $[\![\mathbf{v}]\!] = -a\,\boldsymbol{\eta}$, with $\mathbf{n}$ the unit normal to the wave front. The displacement gradient being discontinuous, the poroelastic strain and stress result discontinuous as well.

For poroelastic media under plane deformation, there are four equations of balance of momentum, namely two for each phase, as described in Section 11.7. Equivalently, the field equations involve four equations for four potentials. There exist four body wave-speeds, accounting for the multiplicity of degree two of the shear wave. The solution will be expressed in terms of the three complex variables $z_j = x_1 + \Omega_j x_2$, $j \in [1, 3]$. The Ω's, with moduli α, can be real or imaginary numbers. They are (not unambiguously) defined by their squares,

$$\Omega_j^2 = \left(\frac{c}{c_j}\right)^2 - 1, \quad \alpha_j = |\Omega_j|, \quad j \in [1, 3]. \tag{11.10.2}$$

The regime of the field equations thus depends on the region of the plane (porosity n^{w}, crack tip speed c), as shown in Fig. 11.5.2.
- region (i): the equation associated with the fastest longitudinal wave is elliptic, while the three others are hyperbolic, and $z_1 = x_1 + i\,\alpha_1 x_2$, $z_2 = x_1 \pm \alpha_2 x_2$, $z_3 = x_1 \pm \alpha_3 x_2$;
- region (ii): the two equations associated with the longitudinal waves are elliptic while those two associated with the shear waves are hyperbolic and $z_1 = x_1 + i\,\alpha_1 x_2$, $z_2 = x_1 + i\,\alpha_2 x_2$, $z_3 = x_1 \pm \alpha_3 x_2$;
- region (iii): the equation associated with the second longitudinal wave is hyperbolic while the three others are elliptic and $z_1 = x_1 \pm i\,\alpha_1 x_2$, $z_2 = x_1 \pm \alpha_2 x_2$, $z_3 = x_1 + i\,\alpha_3 x_2$;
- subsonic region: the four equations are elliptic, and $z_j = x_1 + i\,\alpha_j x_2$, $j \in [1, 3]$.

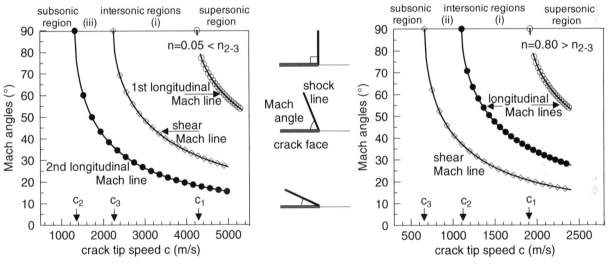

Fig. 11.10.2: Variations of Mach angles with the crack tip speed for two porosities. In region (i), the Mach angle defining the first order shear discontinuity is smaller or larger than the Mach angle of the first order longitudinal discontinuity depending on the position of the porosity with respect to $n_{2-3} \simeq 0.415$, indicated in Fig. 11.5.2, where the second longitudinal wave-speed and the shear wave-speed are equal. Reprinted with permission from Radi and Loret [2007].

In the present context, a first order discontinuity corresponds to the existence of a ray emanating from the crack tip that limits the zone of extension of the mechanical information propagated by the wave, as sketched in Fig. 11.10.1. Fig. 11.10.2 displays the variations of Mach angles with the crack tip speed for two porosities. The closer the crack tip speed to the critical wave-speed, i.e., c_3 for the shear Mach line, c_2 for the longitudinal Mach line, the closer the Mach angle is to $90°$. Conversely, the farther the crack tip speed from the critical wave-speed, the more oblique is the Mach line.

11.10.2 Form of the solution

The analysis follows the initial steps adopted in Section 11.9.1 for surface waves. The displacements in the solid and fluid phases are sought in terms of potentials which are obtained as functions of the three complex variables

$z_j = x_1 + \Omega_j x_2$, $j \in [1,3]$. Satisfaction of the field equations gives rise to eigenvalue problems similar to those described in Section 11.9.4 that delivers the Ω's shown in (11.10.2). The counterpart of the complete formal solution of Section 11.9.2 may be expressed in terms of three complex functions $F_j(z_j)$ of the complex variables z_j,

$$\phi_{\mathrm{s}}(x_1, x_2) = \operatorname{Re} F_1(z_1) + \operatorname{Re} F_2(z_2),$$

$$\phi_{\mathrm{w}}(x_1, x_2) = c_{\mathrm{ms}}^{-2} \left((c_1^2 - c_{\mathrm{s}}^2) \operatorname{Re} F_1(z_1) - (c_{\mathrm{s}}^2 - c_2^2) \operatorname{Re} F_2(z_2) \right), \qquad (11.10.3)$$

$$\psi_{\mathrm{s}}(x_1, x_2) = \operatorname{Im} F_3(z_3), \quad \psi_{\mathrm{w}}(x_1, x_2) = 0.$$

For example, in the intersonic region (i), z_1 is complex while z_2 and z_3 are real, so that the functions $\operatorname{Re} F_2(z_2)$ and $\operatorname{Im} F_3(z_3)$ may be substituted by the respective real functions $f(z_2)$ and $g(z_3)$. The components of the displacements in the solid and fluid phases result from (11.7.4),

$$
\begin{aligned}
u_{\mathrm{s}1} &= \operatorname{Re} F_1'(z_1) + f'(z_2) + \alpha_3\, g'(z_3), \\
u_{\mathrm{s}2} &= -\alpha_1 \operatorname{Im} F_1'(z_1) + \alpha_2\, f'(z_2) - g'(z_3),
\end{aligned}
\qquad (11.10.4)
$$

and

$$
\begin{aligned}
u_{\mathrm{w}1} &= c_{\mathrm{ms}}^{-2} \left((c_1^2 - c_{\mathrm{s}}^2) \operatorname{Re} F_1'(z_1) - (c_{\mathrm{s}}^2 - c_2^2)\, f'(z_2) \right), \\
u_{\mathrm{w}2} &= -c_{\mathrm{ms}}^{-2} \left(\alpha_1\,(c_1^2 - c_{\mathrm{s}}^2) \operatorname{Im} F_1'(z_1) + \alpha_2\,(c_{\mathrm{s}}^2 - c_2^2) f'(z_2) \right),
\end{aligned}
\qquad (11.10.5)
$$

and similarly for the apparent solid stresses and pore pressure using (11.7.6),

$$
\begin{aligned}
\mu^{-1} \sigma_{11}^{\mathrm{s}} &= (1 + \alpha_3^2 + 2\,\alpha_1^2) \operatorname{Re} F_1''(z_1) + (1 + \alpha_3^2 - 2\,\alpha_2^2)\, f''(z_2) + 2\,\alpha_3\, g''(z_3), \\
\mu^{-1} \sigma_{22}^{\mathrm{s}} &= (\alpha_3^2 - 1)\,(\operatorname{Re} F_1''(z_1) + f''(z_2)) - 2\,\alpha_3\, g''(z_3), \\
\mu^{-1} \sigma_{12}^{\mathrm{s}} &= -2\,\alpha_1 \operatorname{Im} F_1''(z_1) + 2\,\alpha_2\, f''(z_2) + (\alpha_3^2 - 1)\, g''(z_3), \\
-\mu^{-1} n^{\mathrm{w}} p_{\mathrm{w}} &= (1 + \alpha_3^2)\, c_{\mathrm{mw}}^{-2} \left((c_1^2 - c_{\mathrm{s}}^2) \operatorname{Re} F_1''(z_1) - (c_{\mathrm{s}}^2 - c_2^2)\, f''(z_2) \right).
\end{aligned}
\qquad (11.10.6)
$$

The calculations of the strains and velocities are facilitated by the formulas below that express the spatial derivatives of the real and imaginary parts of a complex function $F(z)$ of the complex variable $z = x_1 + i\,\alpha\,x_2$ in terms of its derivative $F' \equiv dF/dz$, namely

$$
\begin{aligned}
\frac{\partial \operatorname{Re} F}{\partial x_1} &= \operatorname{Re} F'; \quad & \frac{\partial \operatorname{Re} F}{\partial x_2} &= -\alpha \operatorname{Im} F'; \\[4pt]
\frac{\partial^2 \operatorname{Re} F}{\partial x_1^2} &= \operatorname{Re} F''; \quad & \frac{\partial^2 \operatorname{Re} F}{\partial x_1 \partial x_2} &= -\alpha \operatorname{Im} F''; \quad & \frac{\partial^2 \operatorname{Re} F}{\partial x_2^2} &= -\alpha^2 \operatorname{Re} F'',
\end{aligned}
\qquad (11.10.7)
$$

and

$$
\begin{aligned}
\frac{\partial \operatorname{Im} F}{\partial x_1} &= \operatorname{Im} F'; \quad & \frac{\partial \operatorname{Im} F}{\partial x_2} &= \alpha \operatorname{Re} F'; \\[4pt]
\frac{\partial^2 \operatorname{Im} F}{\partial x_1^2} &= \operatorname{Im} F''; \quad & \frac{\partial^2 \operatorname{Im} F}{\partial x_1 \partial x_2} &= \alpha \operatorname{Re} F''; \quad & \frac{\partial^2 \operatorname{Im} F}{\partial x_2^2} &= -\alpha^2 \operatorname{Im} F''.
\end{aligned}
\qquad (11.10.8)
$$

The actual expressions of the three primary unknown functions deduce from the boundary conditions on the crack line. The latter naturally delineate points ahead of the crack and points behind the crack tip. These conditions involve the gradient along the propagation direction of some displacement components, the total normal stress, the shear stress and either the pore pressure for a permeable crack surface or the fluid flux for an impervious crack surface. In addition, Mode I and Mode II crack propagations are distinguished by distinct symmetry conditions with respect to the crack line. In the context of the functions of a complex variable, these mixed boundary conditions give rise to inhomogeneous Riemann-Hilbert problems, whose solutions express as integrals over the cohesive zone of singular functions with a characteristic singularity exponent γ. The existence of a cohesive zone in the vicinity of the crack tip allows the energy release to be positive, an energetic requirement.

While the intersonic propagation of a non cohesive crack is energetically admissible only for some special speeds of the crack tip, much like for linear elastic solids, it is still worth interest, as it is likely to display limit features of a cohesive crack as the cohesive length tends to zero, Radi and Loret [2007] [2008]. Then, in the vicinity of the crack tip, the dissipative terms associated with seepage provide higher order contributions with respect to the stress divergence and acceleration. The basic primary functions are found to be singular, with a singularity exponent $\gamma \in [0, 1/2]$ that depends on the body wave-speeds and crack tip speed, namely

$$F_1''(z_1) \simeq \frac{1}{i\,z_1^\gamma}, \quad f''(z_2) \simeq \frac{\mathcal{H}(-z_2)}{|z_2|^\gamma}, \quad g''(z_3) \simeq \frac{\mathcal{H}(-z_3)}{|z_3|^\gamma}, \tag{11.10.9}$$

where $\mathcal{H}$ denotes the unit step function. Due to symmetry, it will be sufficient to consider only the upper half-plane $x_2 > 0$. Given that the Ω's are defined by their squares, there are a priori two variables $z_j = x_1 \pm \Omega_j x_2$ for each of the three Ω_j's. However, in the upper complex plane $x_2 > 0$, the complex variable is definitely $z_1 = x_1 + i\,\alpha_1 x_2$. On the other hand, the segments $z_2 = x_1 + \alpha_1 x_2 = 0$ and $z_2 = x_1 - \alpha_1 x_2 = 0$ both involve points of the upper half-plane. The lines $z_2 = x_1 + \alpha_2 x_2 = 0$ and $z_3 = x_1 + \alpha_3 x_2 = 0$ define two waves that propagate to the left of the crack tip in the upper half-plane, as sketched in Fig. 11.10.1. The lines $z_2 = x_1 - \alpha_2 x_2$ and $z_3 = x_1 - \alpha_3 x_2$, which correspond to two waves travelling to the right of the crack tip in the upper half-plane, are excluded because the material particles ahead of the crack tip have not yet felt the disturbance caused by the crack propagation.

11.10.3 Discontinuities across the shear Mach line in regions (i) and (ii)

A priori both a shear and a longitudinal, or dilatational, Mach lines may exist in the region (i), while the sole shear Mach line may take place in region (ii), and the sole longitudinal Mach line in region (iii).

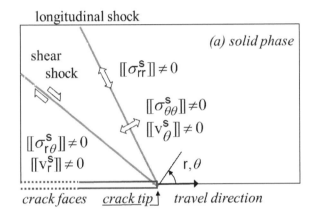
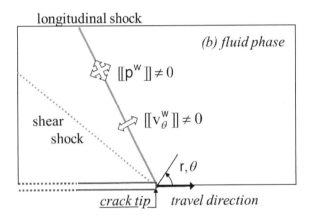

Fig. 11.10.3: Components of the stress and velocity which undergo a discontinuity across the Mach lines for a non cohesive crack. The Mach rays emanating from the crack tip are directed towards the rear, with respect to the direction of propagation of the crack. The entities which undergo a non zero first order discontinuity across the Mach lines depend on the nature, shear or longitudinal, of the discontinuity. A first order shear discontinuity affects only the solid phase, while a first order longitudinal discontinuity affects both the solid and fluid phases. The sketch indicates a longitudinal Mach angle larger than the shear Mach angle, and thus corresponds to a porosity larger than n_{2-3} in region (i), as shown in Fig. 11.10.2(b). Reprinted with permission from Radi and Loret [2007].

In these regions, the issue concerns the continuity of the real function $g''(z_3)$ across the ray $z_3 = x_1 + \alpha_3 x_2 = 0$, with $z_3 = |z_3|$ ahead of the Mach line and $z_3 = |z_3|\,e^{i\pi}$ behind the Mach line. It is instrumental to introduce polar coordinates (r, θ) centered at the crack tip, and to express the stress and velocity in these coordinates. The ray $z_3 = 0$ is defined by the polar angle $\theta_3 = \pi/2 + \arctan\alpha_3$, Figs. 11.10.1 and 11.10.2. Let $[\![g'']\!]$ be the discontinuity of g'' across the Mach line at a current point (r, θ_3) of the Mach line. The contributions of $[\![g'']\!]$ to the stress components, and to the velocity components, e.g., Eqns (11.10.4), (11.10.5) and (11.10.6) in region i), are collected. As expected, the sole shear stress is observed to undergo a discontinuity,

$$[\![\sigma_{rr}^{\text{s}}]\!] = 0, \quad [\![\sigma_{\theta\theta}^{\text{s}}]\!] = 0, \quad [\![\sigma_{r\theta}^{\text{s}}]\!] = \mu\,\frac{c^2}{c_3^2}\,[\![g'']\!], \quad [\![n^{\text{w}} p_{\text{w}}]\!] = 0, \tag{11.10.10}$$

and, similarly the sole discontinuous velocity component is the solid radial velocity,

$$[\![v_r^{\text{s}}]\!] = \frac{c^2}{c_3}\,[\![g'']\!], \quad [\![v_\theta^{\text{s}}]\!] = 0, \quad [\![v_r^{\text{w}}]\!] = 0, \quad [\![v_\theta^{\text{w}}]\!] = 0. \tag{11.10.11}$$

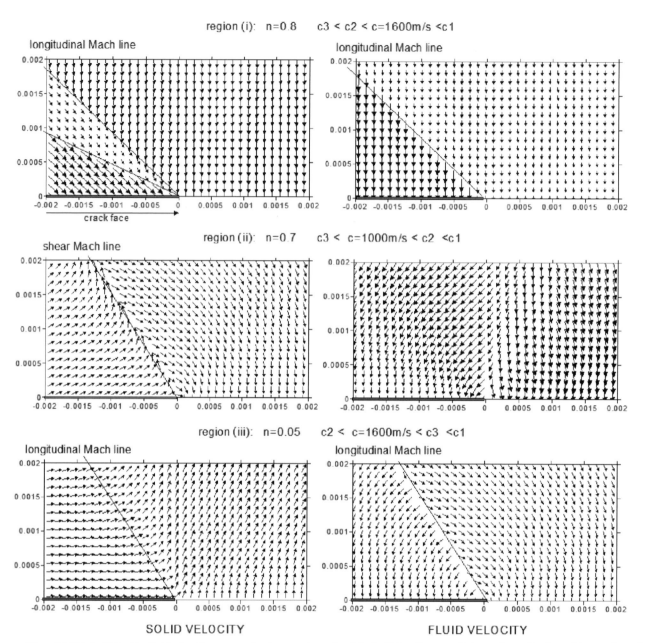

Fig. 11.10.4: Velocity fields for the non cohesive crack under Mode II loading conditions, from Radi and Loret [2007]. The velocity fields have been obtained by analytical integration of singular integrals in the complex plane. For graphical purpose, the velocity is scaled by r^γ, with r distance to the crack tip and γ singularity exponent. The size of the arrows pertains to each plot. A first order shear discontinuity gives rise to velocity discontinuities in the solid phase only, while a first order longitudinal discontinuity affects both phases, as indicated in Fig. 11.10.3. In region (i), the crack tip speed is larger than both the shear wave-speed and the 2nd longitudinal wave-speed, so that there are two shock lines. In region (ii), there is only a shear shock and, in region (iii), a longitudinal shock. Reprinted with permission from Radi and Loret [2007].

Since $z_3 = 0$ on the Mach line, the function g'' undergoes an infinite jump for the noncohesive crack according to (11.10.9). Consequently, the stress and velocity fields undergo, across the shear Mach line, an infinite jump that is *parallel* to, and extends all along the Mach line, Fig. 11.10.3. The presence of cohesion in the vicinity of the crack removes the jump, which is in fact smoothed out. The actual smoothing is obviously expected to depend on the properties of the cohesion. Still, it has been observed in Radi and Loret [2007], [2008] to deteriorate considerably for large singularity exponents.

11.10.4 Discontinuities across the longitudinal Mach line in regions (i) and (iii)

In these regions, the concern is the continuity of the real function $f''(z_2)$ across the ray $z_2 = x_1 + \alpha_2 x_2 = 0$, defined by the polar angle $\theta_2 = \pi/2 + \arctan \alpha_2$. Collecting the contributions of the discontinuity of f'' in the expressions of stress and pore pressure, e.g., Eqn (11.10.6) for region (i), results in

$$[\![\sigma_{rr}^{\rm s}]\!] = (-2\frac{c^2}{c_2^2} + \frac{c^2}{c_3^2})\,\mu\,[\![f'']\!], \quad [\![\sigma_{\theta\theta}^{\rm s}]\!] = \mu\,\frac{c^2}{c_3^2}\,[\![f'']\!], \quad [\![\sigma_{r\theta}^{\rm s}]\!] = 0, \quad [\![n^{\rm w}\,p_{\rm w}]\!] = \frac{c_{\rm s}^2 - c_2^2}{c_{\rm mw}^2}\frac{c^2}{c_3^2}\,\mu\,[\![f'']\!]. \tag{11.10.12}$$

Similarly, the discontinuity of f'', in, e.g., Eqns (11.10.4), (11.10.5) for region (i), yields discontinuities of the solid and fluid velocities, namely

$$[\![v_r^{\rm s}]\!] = 0, \quad [\![v_\theta^{\rm s}]\!] = \frac{c^2}{c_2}\,[\![f'']\!], \quad [\![v_r^{\rm w}]\!] = 0, \quad [\![v_\theta^{\rm w}]\!] = -\frac{c_{\rm s}^2 - c_2^2}{c_{\rm ms}^2}\frac{c^2}{c_2}\,[\![f'']\!]. \tag{11.10.13}$$

The function f'' undergoes an infinite jump across the longitudinal Mach line according to (11.10.9), and so do the stress and velocity. The velocity jump is *normal* to the Mach line.

11.10.5 Jumps of strains across first order waves

Consider a first order wave that propagates at speed a relative to the phase $\alpha = {\rm s, w}$. The displacement being continuous, the strain jump across the wave front is obtained from the compatibility relation (11.1.9), namely $[\![\nabla \mathbf{u}_\alpha]\!] = -a^{-1}\,[\![\mathbf{v}_\alpha]\!] \otimes \mathbf{n}$. Therefore, in the local polar axes where $\mathbf{n} = -\mathbf{e}_\theta$, $\mathbf{e}_\theta(\theta)$ being the orthoradial vector $(-\sin\theta, \cos\theta)$, as indicated in Fig. 11.10.1, the sole non zero discontinuous components of the strain in phase α write,

$$[\![\epsilon_{r\theta}^\alpha]\!] = \frac{[\![v_r^\alpha]\!]}{2\,a}, \quad [\![\epsilon_{\theta\theta}^\alpha]\!] = \frac{[\![v_\theta^\alpha]\!]}{a}, \quad \alpha = {\rm s, w}. \tag{11.10.14}$$

The first order shear discontinuity is thus clearly isochoric. Across the first order longitudinal discontinuity on the other hand, the solid and fluid phases undergo volume changes that are of opposite signs, but do not compensate.

In fact, if the solid and fluid constituents were incompressible, the volume changes in the body would satisfy the constraint $n^{\rm s}\,{\rm tr}\,\epsilon^{\rm s} + n^{\rm w}\,{\rm tr}\,\epsilon^{\rm w} = 0$, and thus the jumps of the normal velocities across the discontinuity would be such that $n^{\rm s}\,[\![v_\theta^{\rm s}]\!] + n^{\rm w}\,[\![v_\theta^{\rm w}]\!] = 0$. That this relation is indeed satisfied by the jumps (11.10.13) may easily be checked. Indeed, as detailed in Sections (7.3.6.5) and (7.3.6.7), incompressibility of the two constituents implies the two moduli $\lambda^{\rm sw}$ and $\lambda^{\rm ww}$ to become infinite but their ratio $\lambda^{\rm sw}/\lambda^{\rm ww}$ tends to $n^{\rm s}/n^{\rm w}$ and the drained Lamé modulus $\lambda = \lambda^{\rm ss} - (\lambda^{\rm sw})^2/\lambda^{\rm ww}$ remains finite. Thus, the ratio $c_{\rm s}^2/c_{\rm ms}^2$ tends to $n^{\rm s}/n^{\rm w}$, as expected.

11.10.6 Balances of momentum across first order waves

For the first order shear discontinuity that propagates at speed c_3, the stress and velocity discontinuities (11.10.10) and (11.10.11) can be checked to satisfy the generic balance of momentum (11.1.56) that applies across a first order wave, namely $[\![\boldsymbol{\sigma}^\alpha]\!] \cdot \mathbf{n} + (\rho^\alpha\,a)\,[\![\mathbf{v}_\alpha]\!] = \mathbf{0}$, for $\alpha = {\rm s, w}$, and with $a = c_3$ and $\mathbf{n}$ equal to $-\mathbf{e}_\theta(\theta_3)$.

For the first order longitudinal discontinuity that propagates at speed c_2, the relation (11.1.56) is satisfied as well by the stress and velocity discontinuities (11.10.12) and (11.10.13), with $a = c_2$ and $\mathbf{n} = -\mathbf{e}_\theta(\theta_2)$.

Section 7.3.2 indicates the general form of the momentum supplies. The momentum supply $\hat{\boldsymbol{\pi}}^\alpha$ to phase α does not contribute to the jump of the balance of momentum as long as it is finite. Still, such is not the case here, but the field analysis reported by Radi and Loret [2007], [2008] applies close to the crack tip and the dissipative terms can be neglected from the outset, Loret and Radi [2001].

Velocity fields, associated with a crack steadily propagating at speed c, are displayed in Figs. 11.10.4, at a certain point $(n^{\rm w}, c)$ of the plane (porosity, propagation speed) of each of the three intersonic regions.

Exercises on Chapter 11

Exercise 11.1 LU factorization

1. Consider the symmetric block matrix,

$$\begin{bmatrix} \mathbf{A} & \mathbf{a} \\ \mathbf{a}^\mathrm{T} & a \end{bmatrix}, \tag{11.E.1}$$

where the individual components have the following respective sizes, namely, matrix $\mathbf{A}$: $n \times n$, vector $\mathbf{a}$: n and $a \neq 0$ a scalar. Let

$$\mathbf{A}^* = \mathbf{A} - \frac{1}{a}\mathbf{a}\mathbf{a}^\mathrm{T}. \tag{11.E.2}$$

be the Schur complement of a. Check that the block matrix admits the LU factorization,

$$\begin{bmatrix} \mathbf{A}_{[n,n]} & \mathbf{a}_{[n,1]} \\ \mathbf{a}_{[1,n]}^\mathrm{T} & a \end{bmatrix} = \begin{bmatrix} \mathbf{I}_{[n,n]} & \dfrac{\mathbf{a}}{a} \\ \mathbf{0} & 1 \end{bmatrix}\begin{bmatrix} \mathbf{A}_{[n,n]}^* & \mathbf{0}_{[n,1]} \\ \mathbf{a}^\mathrm{T} & a \end{bmatrix}. \tag{11.E.3}$$

The sizes of some elements are displayed to ease the reading.

2. Consider the symmetric block matrix,

$$\begin{bmatrix} \mathbf{A} & \mathbf{a} & \mathbf{b} \\ \mathbf{a}^\mathrm{T} & a & b \\ \mathbf{b}^\mathrm{T} & b & c \end{bmatrix}, \tag{11.E.4}$$

where the individual components have the following respective sizes, namely, matrix $\mathbf{A}$: $n \times n$, vector $\mathbf{a}$: n, vector $\mathbf{b}$: n, and a, b and c are scalars. Let

$$\Delta = a\,c - b^2, \quad \mathbf{A}^* = \mathbf{A} - \frac{1}{\Delta}\left(c\,\mathbf{a}\mathbf{a}^\mathrm{T} + a\,\mathbf{b}\mathbf{b}^\mathrm{T} - b\,(\mathbf{a}\mathbf{b}^\mathrm{T} + \mathbf{b}\mathbf{a}^\mathrm{T})\right), \tag{11.E.5}$$

If $c \neq 0$ and $\Delta \neq 0$, then check that the block matrix admits the LU factorization,

$$\begin{bmatrix} \mathbf{A}_{[n,n]} & \mathbf{a}_{[n,1]} & \mathbf{b}_{[n,1]} \\ \mathbf{a}_{[1,n]}^\mathrm{T} & a & b \\ \mathbf{b}_{[1,n]}^\mathrm{T} & b & c \end{bmatrix} = \begin{bmatrix} \mathbf{I}_{[n,n]} & \dfrac{c}{\Delta}\mathbf{a} - \dfrac{b}{\Delta}\mathbf{b} & \dfrac{\mathbf{b}}{c} \\ \mathbf{0} & 1 & \dfrac{b}{c} \\ \mathbf{0} & 0 & 1 \end{bmatrix}\begin{bmatrix} \mathbf{A}_{[n,n]}^* & \mathbf{0}_{[n,1]} & \mathbf{0}_{[n,1]} \\ \mathbf{a}^\mathrm{T} - \dfrac{b}{c}\mathbf{b}^\mathrm{T} & \dfrac{\Delta}{c} & 0 \\ \mathbf{b}^\mathrm{T} & b & c \end{bmatrix}. \tag{11.E.6}$$

Exercise 11.2 Strong ellipticity of the linear elastic *transverse isotropic* matrix

Using the elastic matrix (5.2.51), show that the condition for strong ellipticity (6.2.16) is equivalent to, Payton [1983],

$$c_{11} > 0, \quad c_{11} > c_{12}, \quad c_{33} > 0, \quad c_{44} > 0, \quad \sqrt{c_{11}\,c_{33}} + c_{44} > |c_{13} + c_{44}|. \tag{11.E.7}$$

Exercise 11.3 Spectral analysis of the material defined by a fabric anisotropy in Section 5.2.11

Calculate the wave-speeds c and associated propagation modes $\boldsymbol{\eta}$ of the material defined by a fabric anisotropy.

Hint: use Eqns (11.2.25) or (11.2.31).

Exercise 11.4 Extended generalized eigenvalue problem

Consider the generalized eigenvalue problem $(\mathbf{A} - \lambda\,\mathbf{B}) \cdot \mathbf{e}^R = \mathbf{0}$ where $\mathbf{A}$ and $\mathbf{B}$ are 3×3 matrices, and $\mathbf{e}^R$ is the single right eigenvector associated with the eigenvalue λ. The associated left eigenvector, assumed to be unique, is denoted by $\mathbf{e}^L$. Now let $\mathbf{C}$ be another arbitrary 3×3 matrix. Find a way to estimate the ratio $(\mathbf{e}^L \cdot \mathbf{C} \cdot \mathbf{e}^R)/(\mathbf{e}^L \cdot \mathbf{B} \cdot \mathbf{e}^R)$.

Answer: the following procedure has been introduced by Stoer and Bulirsch [1980]. Consider the *extended* generalized eigenvalue problem:

$$\left(\mathbf{A} + \varphi\,\mathbf{C} - \lambda(\varphi)\,\mathbf{B}\right) \cdot \mathbf{e}^R(\varphi) = 0. \tag{11.E.8}$$

For φ a sufficiently small real number, the eigenvalue $\lambda(\varphi)$ is an analytic function of φ, Stoer and Bulirsch [1980], p. 389. Scalar multiplication of (11.E.8) by the left eigenvector $\mathbf{e}^L(0) = \mathbf{e}^L$, differentiation with respect to φ of the resulting product and estimation of the derivative at $\varphi = 0$ lead to

$$\frac{d\lambda}{d\varphi}(\varphi = 0) = \frac{\mathbf{e}^L \cdot \mathbf{C} \cdot \mathbf{e}^R}{\mathbf{e}^L \cdot \mathbf{B} \cdot \mathbf{e}^R}. \tag{11.E.9}$$

The procedure breaks down if the right and left eigenvectors associated with the same eigenvalue λ are orthogonal with respect to the matrix $\mathbf{B}$. See Section 11.6.2.2 for more details.

Part II

Electro-chemomechanical couplings in tissues with a fixed electrical charge

Chapter 12

Fiber-reinforced tissues: directional averaging

This chapter is essentially concerned with the presence in tissues of collagen fibers which endow their mechanical and transport properties with a directional dependence. Indeed, while the collagen fibers in the middle zone of articular cartilages may be considered to develop in a three-dimensional isotropic network, they tend to be aligned tangentially to the synovial joint in the upper zone and orthogonally to the subchondral bone in the lower zone. They constitute planar lamellae with alternate directions in the corneal stroma and annulus fibrosus and form one-dimensional structures in ligaments and tendons.

Here focus is on directional averaging over fibers. Still, the mechanical properties of individual fibers present issues of their own. Under an extension the fibers progressively uncrimp. They can be considered fully activated if their extension is larger than some characteristic value. Fibers aligned with a tensile load are certainly in extension, slightly inclined fibers may uncrimp later as the load intensity increases while fibers along orthogonal directions are likely to undergo contraction and to be mechanically inactive.

At physiological pH the compressive stiffness of articular cartilages is contributed mostly by the ground substance via the electrical repulsion between charged GAGs. The collagen network contributes as well at non physiological pHs where collagen molecules become charged. The tensile stiffness is provided mostly by collagen fibers even if the ground substance may not be negligible.

12.1 Directional analysis

12.1.1 Spatial and directional averaging

When defining the properties of material, reference is made to a representative elementary volume, which becomes a material point at the scale at which the constitutive equations are built. This scale is referred to as macroscopic level.

In general, it is not possible to keep track at this macroscopic level of all details of the representative element, which involve both geometry and physical properties. The characteristic features which survive in the upscaling procedure are smoothed images of the fine scale structure.

A purely macroscopic model has tools on its own that do not necessarily emanate from an upscaling procedure. For example, consider a network of collagen fibers. One issue is to describe the properties of individual fibers. Another issue is to consider the assembly of the fibers. Each fiber is endowed with a direction. A set of fibers may be assembled together in such a way that the network of fibers may also be endowed with an *orientation*. There are indications that the lamellar structure of the corneal stroma may be endowed with interlacing features that are displayed by woven fabrics, Sections 19.1.2.4 and 19.19.2.5. Of course, a macroscopic model may not be able to keep track of the details of the weave fabric. Still, the fact that the fibers are oriented may be made effective by devising a theory that respects the basic principles of material modeling listed in Section 4.9 and that allows the unit vector $\mathbf{m}$ indicating the direction and *sense* of the fiber to enter the constitutive equations. By contrast, the sense is elusive in the dyadic product $\mathbf{m} \otimes \mathbf{m}$.

Typically, a macroscopic property is obtained by averaging over the points $\mathbf{X}$ of the REV of volume V, and at each point, over the directions $\mathbf{m}$ of the unit sphere $\mathbb{S}^2$,

$$\int_V \int_{\mathbb{S}^2} f(\mathcal{V}, \mathbf{m}, \mathbf{X}) \frac{dS}{4\pi} \frac{dV(\mathbf{X})}{V}, \tag{12.1.1}$$

where the property f depends, next to the fiber directions $\mathbf{m}$ at $\mathbf{X}$, on the set of variables $\mathcal{V} = \mathcal{V}(\mathbf{X})$. In a mixture theory, the property to be integrated would express as the product of this property $f_k(\mathcal{V}, \mathbf{m}, \mathbf{X})$ of the species k by the characteristic function $\chi_k(\mathbf{X})$ associated with this species,

$$f(\mathcal{V}, \mathbf{m}, \mathbf{X}) = f_k(\mathcal{V}, \mathbf{m}, \mathbf{X}) \, \chi_k(\mathbf{X}). \tag{12.1.2}$$

The latter is equal to one at a point $\mathbf{X}$ belonging to the species k, and zero at points that belong to other species. In the simplest case where the property $f_k(\mathcal{V}, \mathbf{m})$ does not depend on the point $\mathbf{X}$, the spatial integration displays the volume fraction $n^k = V_k/V$. Then, we are left with a directional integration,

$$\int_V \int_{\mathbb{S}^2} f_k(\mathcal{V}, \mathbf{m}, \mathbf{X}) \, \chi_k(\mathbf{X}) \frac{dS}{4\pi} \frac{dV(\mathbf{X})}{V} = n^k \int_{\mathbb{S}^2} f_k(\mathcal{V}, \mathbf{m}) \frac{dS}{4\pi}. \tag{12.1.3}$$

As a particular case, consider $f_k(\mathcal{V}, \mathbf{m}) = \mathbf{m} \otimes \mathbf{m}$. Then the directional integration may be performed analytically to yield,

$$\int_V \int_{\mathbb{S}^2} \mathbf{m} \otimes \mathbf{m} \, \chi_k(\mathbf{X}) \frac{dS}{4\pi} \frac{dV(\mathbf{X})}{V} = n^k \int_{\mathbb{S}^2} \mathbf{m} \otimes \mathbf{m} \frac{dS}{4\pi} = n^k \frac{\mathbf{I}}{3}. \tag{12.1.4}$$

Scaling coefficients have been introduced in the averaging operator so that

$$\int_V \int_{\mathbb{S}^2} \frac{dS}{4\pi} \frac{dV(\mathbf{X})}{V} = 1. \tag{12.1.5}$$

12.1.2   Directional integration

The issue is the integration of a function $f = f(\mathbf{m})$ over the directions $\mathbf{m}$ of two- and three-dimensional domains, typically the unit disc $\mathbb{S}^1$ and the unit sphere $\mathbb{S}^2$, or parts of these domains.

Integration over the unit disc:

The unit vector $\mathbf{m}$ is defined by the angle θ, namely componentwise in cartesian axes,

$$\underbrace{\mathbf{m} = \begin{bmatrix} \cos\theta \\ \sin\theta \end{bmatrix}}_{\text{unit direction}}, \quad \underbrace{dl = dS = |\frac{\partial \mathbf{m}}{\partial \theta}| \, d\theta = d\theta}_{\text{differential length}}, \tag{12.1.6}$$

where dl is the incremental integration length, Fig. 12.1.1. The integration over all directions of the plane is then cast in the format[12.1],

$$\int_{\mathbb{S}^1} f(\mathbf{m}) \, dS = \int_0^{2\pi} f(\theta) \, d\theta. \tag{12.1.7}$$

If the vector $\mathbf{m}$ is defined by the angle $\theta \in [0, 2\pi[$, the vector $-\mathbf{m}$ is defined by the angle $\theta + \pi$ modulo 2π. The actual range of integration might take advantage of the possible symmetries of the integrand, e.g., if $f(\pi - \theta) = f(\theta)$, and the integration may be performed over a half disc only.

If $f(\mathbf{m})$ is a distribution density, say $\mathcal{D}$, then it satisfies the constraint

$$\int_{\mathbb{S}^1} \mathcal{D}(\mathbf{m}) \, dS = 1, \quad \int_{\mathbb{S}^1} dS = 2\pi. \tag{12.1.8}$$

[12.1] As an abuse of notation $f(\theta)$ stands for $f(\mathbf{m}(\theta))$.

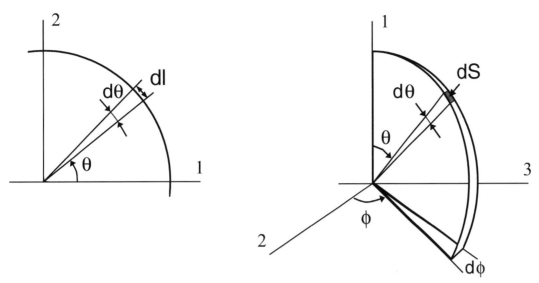

Fig. 12.1.1: Angular parametrization of the unit disc $\mathbb{S}^1 = \mathbb{S}^1(\theta \in [0, 2\pi[)$ and of the unit sphere $\mathbb{S}^2 = \mathbb{S}^2(\theta \in [0, \pi[, \phi \in [0, 2\pi[)$.

Integration over the unit sphere:

The unit vector $\mathbf{m}$ expresses componentwise in cartesian axes in terms of the angles (θ, ϕ), Fig. 12.1.1,

$$\mathbf{m} = \begin{bmatrix} \cos\theta \\ \sin\theta \ \cos\phi \\ \sin\theta \ \sin\phi \end{bmatrix}. \tag{12.1.9}$$

The differential vector area and differential area result as,

$$d\vec{\mathbf{S}} = \frac{\partial \mathbf{m}}{\partial \theta} \wedge \frac{\partial \mathbf{m}}{\partial \phi} \, d\theta \, d\phi = \mathbf{m} \ \sin\theta \, d\theta \, d\phi, \quad dS \equiv |d\vec{\mathbf{S}}| = \sin\theta \, d\theta \, d\phi, \tag{12.1.10}$$

and the integration over the unit sphere is cast in the form,

$$\int_{\mathbb{S}^2} f(\mathbf{m}) \, dS = \int_0^{2\pi} d\phi \int_0^\pi f(\theta, \phi) \, \sin\theta \, d\theta. \tag{12.1.11}$$

If $f(\mathbf{m})$ is a distribution density, say $\mathcal{D}$, then it satisfies the constraint

$$\int_{\mathbb{S}^2} \mathcal{D}(\mathbf{m}) \, dS = 1, \quad \int_{\mathbb{S}^2} dS = 4\pi. \tag{12.1.12}$$

While the vector $\mathbf{m}$ is defined by the angles $\theta \in [0, \pi[$ and $\phi \in [0, 2\pi[$, the vector $-\mathbf{m}$ is defined by the angle $\pi - \theta$ and $\pi + \phi$ modulo 2π.

Cubature over the unit sphere:

The previous integration methods were scanning over directions $\mathbf{m}$. Alternatively, and equivalently, the methods below scan on points $\mathbf{M}$ on the unit sphere. The cubature, i.e., approximative integration over the unit sphere, of $\int_{\mathbb{S}^2} f(\mathbf{M}) \, dS$ may be defined through various ways:

- *The latitude-longitude rule.* Instead of integration with respect to the angles $\theta \in [0, \pi]$ and $\phi \in [0, 2\pi]$, use the latitude $x_1 = \cos\theta \in [-1, 1]$ and longitude $\phi \in [0, 2\pi]$. Integrate the latitude with a Gauss-Legendre rule with m points x_1^i and weights w_i which is exact for polynomials up to degree $2m - 1$. Integrate the longitude with the generalized mid-point rule with $L + 1$ points $\phi_j = 2\pi j/(L+1)$ which is exact for trigonometric polynomials up to degree L, namely $1, \cos\phi, \sin\phi \, cdots, \cos(L\phi), \sin(L\phi)$. The cubature over the unit sphere is finally obtained by tensor product of these two one-dimensional rules,

$$\int_{\mathbb{S}^2} f(\mathbf{M}) \, dS = \frac{2\pi}{L+1} \sum_{j=0}^L \sum_{i=1}^m w_i \, f(\mathbf{M}(x_1^i, \phi_j)). \tag{12.1.13}$$

- *Spherical designs.* Spherical t-designs, which consist in carefully selected P sampling points on the unit sphere and use of a uniform weight $4\pi/P$,

$$\int_{\mathbb{S}^2} f(\mathbf{M})\,dS = \frac{4\pi}{P} \sum_{i=1}^{P} f(\mathbf{M}^{(i)}), \tag{12.1.14}$$

achieve exact integration of spherical polynomials of degree t or less. The number P of points should be larger than $(t+2)^2/4$ if t is even, and $(t+1)(t+3)/4$ if t is odd. The points in spherical designs are more evenly distributed than with the Gauss-Legendre rule which involves a number of points near the poles with small weights. Tabulated designs with limited P may be found in Hardin and Sloane [1996].
- *Voronoi tessellation.* Hesse et al. [2010] provide a review of methods of cubature over the unit sphere, including integration of scattered data, namely when the integrand $f(\mathbf{M})$ is known only at certain points $\mathbf{M}^{(i)}$, $i \in [1, P]$. Then a Voronoi tessellation partitions the unit sphere over sets T_i of area $|T_i|$, and

$$\int_{\mathbb{S}^2} f(\mathbf{M})\,dS = \sum_{i=1}^{P} f(\mathbf{M}^{(i)})\,|T_i|. \tag{12.1.15}$$

- *Delaunay triangulation.* An alternative method is to use the Delaunay triangulation $\mathcal{T}$, associated with the scattered data points. The generic spherical triangles T_{ijk} with apices $(\mathbf{m}_i, \mathbf{m}_j, \mathbf{m}_k)$ are formed by three intersecting circular arcs. They have area $|T_{ijk}| = \alpha_i + \alpha_j + \alpha_k - \pi$, where the α's are the angles at the apices (for a planar triangle, the above sum vanishes). Then

$$\int_{\mathbb{S}^2} f(\mathbf{M})\,dS = \sum_{T_{ijk} \in \mathcal{T}} \frac{1}{3}\left(f(\mathbf{M}^{(i)}) + f(\mathbf{M}^{(j)}) + f(\mathbf{M}^{(k)})\right)|T_{ijk}|. \tag{12.1.16}$$

Differential operators over the unit sphere:

The definitions expressed in the spherical axes $\hat{\mathbf{e}}_r$, $\hat{\mathbf{e}}_\theta$ and $\hat{\mathbf{e}}_\phi$ of a few differential operators over the unit sphere are recorded as particular cases of the general definitions provided in Section 2.4.3.1.

$\boldsymbol{\nabla}_{\mathbb{S}}\,u$ denotes the left or right gradient of a scalar u, namely Eqn (2.4.57),

$$\boldsymbol{\nabla}_{\mathbb{S}}\,u = \overrightarrow{\boldsymbol{\nabla}}_{\mathbb{S}}\,u = u\,\overleftarrow{\boldsymbol{\nabla}}_{\mathbb{S}} = \frac{\partial u}{\partial \theta}\,\hat{\mathbf{e}}_\theta + \frac{1}{\sin\theta}\frac{\partial u}{\partial \phi}\,\hat{\mathbf{e}}_\phi, \tag{12.1.17}$$

while the Laplacian of a scalar follows from (2.4.66),

$$\Delta_{\mathbb{S}}u = \frac{1}{\sin\theta}\frac{\partial}{\partial\theta}\left(\sin\theta\,\frac{\partial u}{\partial\theta}\right) + \frac{1}{\sin^2\theta}\frac{\partial^2 u}{\partial\phi^2}. \tag{12.1.18}$$

The (left) gradient $\boldsymbol{\nabla}_{\mathbb{S}}\mathbf{u} = \overrightarrow{\boldsymbol{\nabla}}_{\mathbb{S}}\,\mathbf{u}$ and divergence $\mathrm{div}_{\mathbb{S}}\mathbf{u} = \mathrm{tr}\,(\boldsymbol{\nabla}_{\mathbb{S}}\mathbf{u})$ of the vector $\mathbf{u}$ simplify to,

$$\boldsymbol{\nabla}_{\mathbb{S}}\mathbf{u} = \begin{bmatrix} 0 & 0 & 0 \\[2mm] \dfrac{\partial u_r}{\partial\theta} - u_\theta & \dfrac{\partial u_\theta}{\partial\theta} + u_r & \dfrac{\partial u_\phi}{\partial\theta} \\[3mm] \dfrac{1}{\sin\theta}\dfrac{\partial u_r}{\partial\phi} - u_\phi & \dfrac{1}{\sin\theta}\dfrac{\partial u_\theta}{\partial\phi} - u_\phi\cot\theta & \dfrac{1}{\sin\theta}\dfrac{\partial u_\phi}{\partial\phi} + u_r + u_\theta\cot\theta \end{bmatrix}, \tag{12.1.19}$$

and

$$\mathrm{div}_{\mathbb{S}}\mathbf{u} = 2\,u_r + \frac{1}{\sin\theta}\frac{\partial}{\partial\theta}(u_\theta\,\sin\theta) + \frac{1}{\sin\theta}\frac{\partial u_\phi}{\partial\phi}. \tag{12.1.20}$$

For the record, let us note also

$$\boldsymbol{\nabla}_{\mathbb{S}}\hat{\mathbf{e}}_r = \mathbf{I} - \hat{\mathbf{e}}_r \otimes \hat{\mathbf{e}}_r, \tag{12.1.21}$$

and, using the rates (2.4.48) and gradients (12.1.19),

$$\frac{d\hat{\mathbf{e}}_r}{dt} \cdot \boldsymbol{\nabla}_{\mathbb{S}}\hat{\mathbf{e}}_r = \frac{d\hat{\mathbf{e}}_r}{dt}, \quad \frac{d\hat{\mathbf{e}}_r}{dt} \cdot \boldsymbol{\nabla}_{\mathbb{S}}\hat{\mathbf{e}}_\theta = \frac{d\hat{\mathbf{e}}_\theta}{dt}, \quad \frac{d\hat{\mathbf{e}}_r}{dt} \cdot \boldsymbol{\nabla}_{\mathbb{S}}\hat{\mathbf{e}}_\phi = \frac{d\hat{\mathbf{e}}_\phi}{dt}. \tag{12.1.22}$$

The component form of the unit vector $\mathbf{m}$ in the principal axes of the Cauchy stress tensor $\boldsymbol{\sigma}$ being given by (12.1.9), the normal stress $\mathbf{m} \cdot \boldsymbol{\sigma} \cdot \mathbf{m}$ is equal to $\sigma_1 \cos^2 \theta + \sin^2 \theta \, (\sigma_2 \cos^2 \phi + \sigma_3 \sin^2 \phi)$, and its Laplacian,

$$\Delta_S \, (\mathbf{m} \cdot \boldsymbol{\sigma} \cdot \mathbf{m}) = 2 \, (1 - 3 \cos^2 \theta) \, (\sigma_1 - \frac{\sigma_2 + \sigma_3}{2}) + 3 \cos(2 \, \phi) \sin^2 \theta \, (\sigma_3 - \sigma_2), \qquad (12.1.23)$$

vanishes for all directions $\mathbf{m}$ only if the stress is isotropic.

The eigenfunctions $Y_{l,m}$ with eigenvalues $\lambda_{l,m}$ of the Laplacian, that satisfy the relation $\Delta_S Y = \lambda Y$, namely

$$\lambda_{l,m} = -l \, (l+1), \quad Y_{l,m}(\theta,\phi) = \sqrt{\frac{2\,l+1}{4\,\pi} \frac{(l-m)!}{(l+m)!}} \, P_{l,m}(\cos\theta) \, e^{i \, m \, \phi}, \quad -l \le m \le l, \quad l = 0, 1, \cdots, \qquad (12.1.24)$$

are termed *spherical harmonics* and they constitute an orthonormal basis of functions on the unit sphere,

$$\int_{\mathbb{S}^2} Y_{l,m}(\theta,\phi) \, \overline{Y}_{l',m'}(\theta,\phi) \, dS = I_{ll'} \, I_{mm'}, \qquad (12.1.25)$$

$\overline{Y}$ denoting the complex conjugate of Y. The associated Legendre polynomials $P_{l,m}(x)$, $x \in [0,1]$,

$$P_{l,m}(x) = (-1)^m \, (1-x^2)^{m/2} \, \frac{d^m}{dx^m} P_l(x), \quad m \ge 0, \quad P_{l,m}(x) = (-1)^m \, \frac{(l+m)!}{(l-m)!} \, P_{l,-m}(x), \quad m < 0, \qquad (12.1.26)$$

derive from the ordinary Legendre polynomials $P_l(x) = 1/(2^l \, l!) \, d^l/dx^l (x^2 - 1)^l$. Note the addition theorem for $\mathbf{m} \equiv (\theta,\phi)$ and $\mathbf{n} \equiv (\theta',\phi')$ two unit vectors,

$$P_l(\mathbf{m} \cdot \mathbf{n}) = \frac{4\,\pi}{2\,l+1} \sum_{m=-l}^{l} Y_{l,m}(\mathbf{m}) \, \overline{Y}_{l,m}(\mathbf{n}). \qquad (12.1.27)$$

For efficient and accurate numerical evaluation of the associated Legendre polynomials, refer to Press et al. [1992].

12.2 A simple fiber model for the collagen network

Emphasis is laid below on the interactions between the directional properties of the fiber network and chemomechanical couplings. Indeed, chemomechanical couplings, namely swelling and shrinking, modify the sets of directions in which the fibers are mechanically active.

To simplify the analysis, full activation of individual fibers is considered to take place as soon as they undergo extension. Progressive uncrimping is introduced through a nonlinear stress-strain relation. However the model, and some interesting properties, are first introduced in a conewise linear context.

12.2.1 A single family of collagen fibers

Let $\underline{\mathbf{m}}_c$ be a unit vector indicating the direction of a collagen fiber in a reference configuration where the deformation gradient is equal to the identity, $\mathbf{F} = \mathbf{I}$. Let $\mathbf{C} = \mathbf{F}^T \cdot \mathbf{F}$ be the right stretch associated with the deformation gradient $\mathbf{F}$. It is instrumental to introduce the operator $\langle \cdot \rangle$ which denotes the positive part of its argument. The collagen fibers are mechanically active only in extension, namely when $\langle \mathbf{m}_c \cdot \mathbf{C} \cdot \mathbf{m}_c - 1 \rangle$ is non zero. Then the strain energy of a collagen fiber per unit volume of collagen expresses in terms of the right stretch $\mathbf{C} = 2\,\mathbf{E}_G + \mathbf{I}$ or Green strain $\mathbf{E}_G$ in the format,

$$w_c = w_c(L^2), \qquad (12.2.1)$$

where

$$L = \tfrac{1}{2} \langle \mathbf{C} : \underline{\mathbf{M}}_c - 1 \rangle = \underbrace{\langle \mathbf{E}_G : \underline{\mathbf{M}}_c \rangle}_{x}, \qquad (12.2.2)$$

and $\underline{\mathbf{M}}_c = \mathbf{m}_c \otimes \mathbf{m}_c$. In (12.2.1), the square, actually any exponent strictly greater than 1, is required to ensure continuity of the stress at the extension-contraction transition $L = 0$ where, in contrast, the stiffness may or may not be discontinuous.

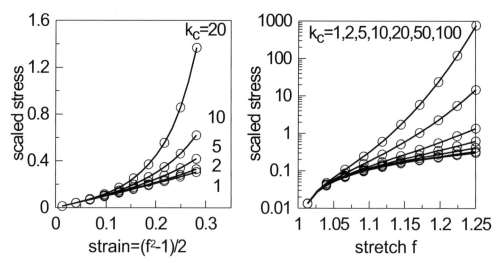

Fig. 12.2.1: The exponential law with a power 2 ensures continuity of the stress at the balance point where the strain becomes contractive. Effect of the exponent k_c on the scaled stress $\exp(k_c\,\epsilon^2)\,\epsilon$ as a function of the strain $\epsilon = (f^2 - 1)/2$ or stretch f.

Note that use of the deviatoric deformation in the above energy would lead isotropic extension, like swelling, to leave collagen fibers unactivated. We adopt the exponential form of energy, defined by the modulus K_c [unit : MPa] of the collagen fibers and the dimensionless exponent k_c,

$$w_c = \frac{K_c}{2\,k_c}\left(\exp(k_c\,L^2) - 1\right),$$

$$\frac{\partial w_c}{\partial \mathbf{E}_G} = K_c\,\exp(k_c\,L^2)\,L\,\underline{\mathbf{M}}_c,$$

$$\frac{\partial^2 w_c}{\partial \mathbf{E}_G \partial \mathbf{E}_G} = K_c\,\exp(k_c\,L^2)\left(2\,k_c\,\underbrace{L^2}_{\text{continuous}} + \underbrace{\mathcal{H}(x)}_{\text{discontinuous}}\right)\underline{\mathbf{M}}_c \otimes \underline{\mathbf{M}}_c.$$

(12.2.3)

Fig. 12.2.1 displays the quantitative influence of the exponent k_c.

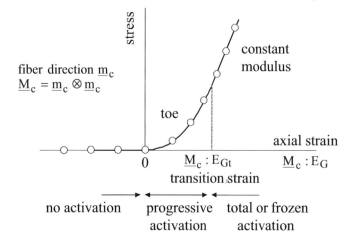

Fig. 12.2.2: One-dimensional sketch showing a progressive increase of the elastic modulus (toe region) followed by a region of constant modulus that avoids locking at large strains.

The above expression gives rise to infinite stiffness at large extension. In order to keep a constant stiffness for a strain larger than $\mathbf{E}_{Gt} : \underline{\mathbf{M}}_c$ as sketched in Fig. 12.2.2, consider the following strain energy density,

$$w_c = \left(1 - \mathcal{H}(x_2)\right) w_{cf1} + \mathcal{H}(x_2)\,w_{cf2},$$

$$\underline{\underline{\tau}} = \left(1 - \mathcal{H}(x_2)\right) \underline{\underline{\tau}}_1 + \mathcal{H}(x_2)\,\underline{\underline{\tau}}_2,$$

(12.2.4)

$$\mathbb{E} = \left(1 - \mathcal{H}(x_2)\right) \mathbb{E}_1 + \mathcal{H}(x_2)\,\mathbb{E}_2,$$

$$w_{\text{cf1}} = \frac{K_{\text{c}}}{2\,k_{\text{c}}} \left(\exp(k_{\text{c}}\,L_1^2) - 1 \right),$$

$$\underline{\underline{\tau}}_1 = \frac{\partial w_{\text{cf1}}}{\partial \mathbf{E}_{\text{G}}} = K_{\text{c}}\,\exp(k_{\text{c}}\,L_1^2)\,L_1\,\underline{\mathbf{M}}_{\text{c}}\,, \qquad (12.2.5)$$

$$\mathbb{E}_1 = \frac{\partial^2 w_{\text{cf1}}}{\partial \mathbf{E}_{\text{G}}\partial \mathbf{E}_{\text{G}}} = K_{\text{c}}\,\exp(k_{\text{c}}\,L_1^2)\left(2\,k_{\text{c}}\,L_1^2 + \mathcal{H}(x_1)\right)\underline{\mathbf{M}}_{\text{c}} \otimes \underline{\mathbf{M}}_{\text{c}}\,,$$

$$w_{\text{cf2}} = \underline{\underline{\tau}}_t : \mathbf{E}_{\text{G}} + \frac{1}{2}\left(\mathbf{E}_{\text{G}} - \mathbf{E}_{\text{G}t}\right) : \mathbb{E}_t : \left(\mathbf{E}_{\text{G}} - \mathbf{E}_{\text{G}t}\right),$$

$$\underline{\underline{\tau}}_2 = \frac{\partial w_{\text{cf2}}}{\partial \mathbf{E}_{\text{G}}} = \underline{\underline{\tau}}_t + \mathbb{E}_t : \left(\mathbf{E}_{\text{G}} - \mathbf{E}_{\text{G}t}\right), \qquad (12.2.6)$$

$$\mathbb{E}_2 = \frac{\partial^2 w_{\text{cf2}}}{\partial \mathbf{E}_{\text{G}}\partial \mathbf{E}_{\text{G}}} = \mathbb{E}_t\,,$$

where

$$L_1 = \langle \underbrace{\mathbf{E}_{\text{G}} : \underline{\mathbf{M}}_{\text{c}}}_{x_1} \rangle, \quad L_2 = \langle \underbrace{(\mathbf{E}_{\text{G}} - \mathbf{E}_{\text{G}t}) : \underline{\mathbf{M}}_{\text{c}}}_{x_2} \rangle, \qquad (12.2.7)$$

and $\underline{\underline{\tau}}_t = \underline{\underline{\tau}}_1(\mathbf{E}_{\text{G}t})$ and $\mathbb{E}_t = \mathbb{E}_1(\mathbf{E}_{\text{G}t})$.

An even more drastic departure from locking is exposed in Section 6.5.2.

12.2.2 Random directional distribution of collagen fibers

Henceforth we have in mind a tissue that can be assumed to be isotropic. For example, we may think of a specimen taken in the middle zone of articular cartilages. Otherwise, the directional distribution of fibers should be specified. The fibers are assumed to be independent: there are no crosslinks between fibers nor direct interactions with the matrix.

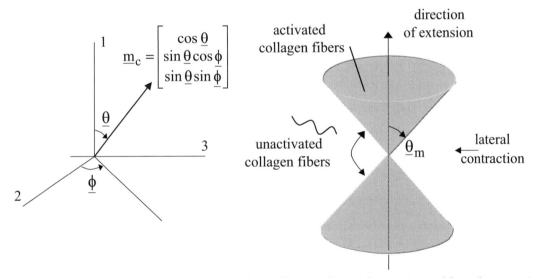

Fig. 12.2.3: Cones of activation of collagen fibers under axial extension and lateral contraction.

The unit vector $\underline{\mathbf{m}}_{\text{c}}$ indicating the direction of a collagen fiber is expressed in terms of the angles $\underline{\phi} \in [0,\,2\,\pi[$ and $\underline{\theta} \in [0,\,\pi[$, Fig. 12.2.3,

$$\underline{\mathbf{m}}_{\text{c}} = \begin{bmatrix} \cos\underline{\theta} \\ \sin\underline{\theta}\,\cos\underline{\phi} \\ \sin\underline{\theta}\,\sin\underline{\phi} \end{bmatrix}. \qquad (12.2.8)$$

Let us record the directional integrations over space directions of the second order tensor $\underline{\mathbf{M}}_c$,

$$\int_{\mathbb{S}^2} \underline{\mathbf{m}}_c \otimes \underline{\mathbf{m}}_c \, \frac{dS}{4\pi} = \frac{1}{4\pi} \int_0^{2\pi} d\underline{\phi} \int_0^{\pi} \underline{\mathbf{m}}_c \otimes \underline{\mathbf{m}}_c \, \sin\underline{\theta} \, d\underline{\theta} = \frac{\mathbf{I}}{3}, \tag{12.2.9}$$

and of the fourth order tensor $\underline{\mathbf{M}}_c \otimes \underline{\mathbf{M}}_c$,

$$\int_{\mathbb{S}^2} \underline{\mathbf{M}}_c \otimes \underline{\mathbf{M}}_c \, \frac{dS}{4\pi} = \frac{1}{15} \left(\mathbf{I} \otimes \mathbf{I} + 2\,\mathbf{I} \,\overline{\underline{\otimes}}\, \mathbf{I} \right). \tag{12.2.10}$$

The tensor product $\overline{\underline{\otimes}}$ is defined componentwise as $(\mathbf{A}\,\overline{\underline{\otimes}}\,\mathbf{B})_{ijkl} = (A_{ik}\,B_{jl} + A_{il}\,B_{jk})/2$. The expression (12.2.10) results from the component relations,

$$\frac{1}{4\pi} \int_0^{2\pi} d\underline{\phi} \int_0^{\pi} \underline{m}_{ci}\, \underline{m}_{cj}\, \underline{m}_{ck}\, \underline{m}_{cl}\, \sin\underline{\theta} \, d\underline{\theta} = \begin{cases} 1/5, & i=j=k=l, \\ 1/15, & i=j\neq k=l, \\ 0, & \text{otherwise}. \end{cases} \tag{12.2.11}$$

Note the useful closed form integral for $m \geq 0$ and $n \geq 0$ integers,

$$\frac{1}{2\pi} \int_0^{2\pi} (\cos\underline{\phi})^{2m} \, (\sin\underline{\phi})^{2n} \, d\underline{\phi} = \frac{(2\,m-1)!!\,(2\,n-1)!!}{2^{m+n}\,(m+n)!}, \tag{12.2.12}$$

where $n! = (n)\,(n-1)\cdots(2)\,(1)$ denotes the factorial of n and $(2\,n-1)!!$ is the double factorial,

$$(2\,n-1)!! = (2\,n-1)\,(2\,n-3)\,\cdots\,(3)\,(1). \tag{12.2.13}$$

Note also the generalization for even n, Kanatani [1984],

$$\frac{1}{4\pi} \int_0^{2\pi} d\underline{\phi} \int_0^{\pi} \underline{m}_{ci_1}\, \underline{m}_{ci_2}\, \cdots \underline{m}_{ci_n}\, \sin\underline{\theta} \, d\underline{\theta} = \frac{1}{n+1}\, I_{(i_1 i_2}\, I_{i_3 i_4} \cdots I_{i_{n-1} i_n)}, \tag{12.2.14}$$

where $\mathbf{I}$ is the second order identity tensor and the parentheses imply symmetrization of the indices, involving $(n-1)!!$ independent terms, namely 1 for $n=2$, 3 for $n=4$, 15 for $n=6$, 105 for $n=8\cdots$, e.g.,

$$\frac{1}{4\pi} \int_0^{2\pi} d\underline{\phi} \int_0^{\pi} \underline{m}_{ci_1}\, \underline{m}_{ci_2}\, \cdots \underline{m}_{ci_n}\, \sin\underline{\theta} \, d\underline{\theta} =$$

$$= \begin{cases} \dfrac{1}{3}\, I_{i_1 i_2}, & n=2; \\[2mm] \dfrac{1}{15}\, (I_{i_1 i_2}\, I_{i_3 i_4} + I_{i_1 i_3}\, I_{i_2 i_4} + I_{i_1 i_4}\, I_{i_2 i_3}), & n=4; \\[2mm] \dfrac{1}{105}\, (\, I_{i_1 i_2}\, I_{i_3 i_4}\, I_{i_5 i_6} + I_{i_1 i_3}\, I_{i_2 i_4}\, I_{i_5 i_6} + I_{i_1 i_4}\, I_{i_2 i_3}\, I_{i_5 i_6} & n=6 \\[1mm] \quad + I_{i_1 i_2}\, I_{i_3 i_5}\, I_{i_4 i_6} + I_{i_1 i_2}\, I_{i_3 i_6}\, I_{i_4 i_5} + I_{i_1 i_5}\, I_{i_3 i_4}\, I_{i_2 i_6} \\[1mm] \quad + I_{i_1 i_6}\, I_{i_3 i_4}\, I_{i_2 i_5} + I_{i_1 i_5}\, I_{i_2 i_3}\, I_{i_4 i_6} + I_{i_1 i_6}\, I_{i_2 i_3}\, I_{i_4 i_5} \\[1mm] \quad + I_{i_1 i_4}\, I_{i_3 i_5}\, I_{i_2 i_6} + I_{i_1 i_4}\, I_{i_3 i_6}\, I_{i_2 i_5} + I_{i_1 i_5}\, I_{i_2 i_4}\, I_{i_3 i_6} \\[1mm] \quad + I_{i_1 i_3}\, I_{i_4 i_6}\, I_{i_2 i_5} + I_{i_1 i_6}\, I_{i_2 i_4}\, I_{i_3 i_5} + I_{i_1 i_3}\, I_{i_4 i_5}\, I_{i_2 i_6}). \end{cases} \tag{12.2.15}$$

12.2.3   Conewise linear stress-strain response of collagen fibers

In the limit of a small strain $\mathbf{C} = \mathbf{I} + 2\,\boldsymbol{\epsilon} + \cdots$, the effective elongation L writes $\langle \boldsymbol{\epsilon} : \underline{\mathbf{M}}_c \rangle$, and the energy of a single collagen fiber and its first and second derivatives may be linearized to

$$\begin{aligned} w_c &= \frac{K_c}{2}\, \langle \boldsymbol{\epsilon} : \underline{\mathbf{M}}_c \rangle^2, \\[2mm] \frac{\partial w_c}{\partial \boldsymbol{\epsilon}} &= K_c\, \langle \boldsymbol{\epsilon} : \underline{\mathbf{M}}_c \rangle\, \underline{\mathbf{M}}_c, \\[2mm] \frac{\partial^2 w_c}{\partial \boldsymbol{\epsilon}\,\partial \boldsymbol{\epsilon}} &= K_c\, \underbrace{\mathcal{H}(\boldsymbol{\epsilon} : \underline{\mathbf{M}}_c)}_{\text{discontinuous}}\, \underline{\mathbf{M}}_c \otimes \underline{\mathbf{M}}_c. \end{aligned} \tag{12.2.16}$$

12.2.3.1 All fibers in extension

If *all* fibers are in extension, then integrations over space and directions of the apparent strain energy and of its first and second derivatives,

$$
\left\{
\begin{array}{l}
n^{\mathrm{c}} \displaystyle\int_{\mathbb{S}^2} w_{\mathrm{c}} \, \dfrac{dS}{4\pi} \\[2ex]
n^{\mathrm{c}} \displaystyle\int_{\mathbb{S}^2} \dfrac{\partial w_{\mathrm{c}}}{\partial \boldsymbol{\epsilon}} \, \dfrac{dS}{4\pi} \\[2ex]
n^{\mathrm{c}} \displaystyle\int_{\mathbb{S}^2} \dfrac{\partial^2 w_{\mathrm{c}}}{\partial \boldsymbol{\epsilon}\,\partial \boldsymbol{\epsilon}} \, \dfrac{dS}{4\pi}
\end{array}
\right\}
= \dfrac{K^{\mathrm{c}}}{15}
\left\{
\begin{array}{l}
\dfrac{1}{2}\,(\mathbf{I}:\boldsymbol{\epsilon})^2 + \boldsymbol{\epsilon}:\boldsymbol{\epsilon}, \\[2ex]
(\boldsymbol{\epsilon}:\mathbf{I})\,\mathbf{I} + 2\,\boldsymbol{\epsilon}, \\[2ex]
\mathbf{I}\otimes\mathbf{I} + 2\,\mathbf{I}\,\overline{\otimes}\,\mathbf{I},
\end{array}
\right.
\tag{12.2.17}
$$

yield an isotropic apparent stiffness with equal apparent Lamé moduli,

$$
\lambda^{\mathrm{c}} = \mu^{\mathrm{c}} = \dfrac{K^{\mathrm{c}}}{15} \quad \text{with} \quad K^{\mathrm{c}} = n^{\mathrm{c}}\,K_{\mathrm{c}},
\tag{12.2.18}
$$

that is, with a Poisson's ratio equal to 0.25.

12.2.3.2 Rotational symmetry about the traction direction

However, in general, the directions for which extension takes place are not known beforehand and the integration can not be performed analytically, except in particular cases.

For example, let us consider extension along the direction 1 with symmetry about this axis. Of course if the transverse directions undergo extension, we recover the isotropic case. However the fibers in the transverse directions may undergo contraction. Then the fibers which undergo extension,

$$
\boldsymbol{\epsilon}:\underline{\mathbf{M}}_{\mathrm{c}} \geq 0 \quad \Rightarrow \quad \cot^2 \underline{\theta} = \dfrac{m_{c1}^2}{m_{c2}^2 + m_{c3}^2} \geq \cot^2 \underline{\theta}_m \equiv \dfrac{-\epsilon_{22}}{\epsilon_{11}},
\tag{12.2.19}
$$

form two symmetric cones of half angle $\underline{\theta}_m$ about the extension direction,

$$
\underline{\theta} \in [0, \underline{\theta}_m] \cup [\pi - \underline{\theta}_m, \pi].
\tag{12.2.20}
$$

In the principal axes of strain, we just need to calculate four components of the stiffness in view of the symmetry about the axis 1. Therefore the contribution of the collagen network to the total stress expresses in terms of a symmetric matrix,

$$
\boldsymbol{\sigma}^{\mathrm{c}} = \mathbb{E}^{\mathrm{c}}(\boldsymbol{\epsilon}):\boldsymbol{\epsilon}, \qquad
\begin{bmatrix} \sigma_{11} \\ \sigma_{22} \\ \sigma_{33} \end{bmatrix}
=
\begin{bmatrix}
E_{11}^{\mathrm{c}} & E_{12}^{\mathrm{c}} & E_{12}^{\mathrm{c}} \\
E_{21}^{\mathrm{c}} & E_{22}^{\mathrm{c}} & E_{23}^{\mathrm{c}} \\
E_{21}^{\mathrm{c}} & E_{32}^{\mathrm{c}} & E_{22}^{\mathrm{c}}
\end{bmatrix}
\begin{bmatrix} \epsilon_{11} \\ \epsilon_{22} \\ \epsilon_{33} \end{bmatrix}
\tag{12.2.21}
$$

with coefficients,

$$
E_{11}^{\mathrm{c}} = \dfrac{K^{\mathrm{c}}}{80}\left(16 - 10\cos\underline{\theta}_m - 5\cos(3\underline{\theta}_m) - \cos(5\underline{\theta}_m)\right) = \dfrac{K^{\mathrm{c}}}{5}\left(1 - (\cos\underline{\theta}_m)^5\right)
$$

$$
E_{22}^{\mathrm{c}} = E_{33}^{\mathrm{c}} = 3\,E_{23}^{\mathrm{c}} = \dfrac{K^{\mathrm{c}}}{10}\left(19 + 18\cos\underline{\theta}_m + 3\cos(2\underline{\theta}_m)\right)\left(\sin\dfrac{\underline{\theta}_m}{2}\right)^6
\tag{12.2.22}
$$

$$
E_{12}^{\mathrm{c}} = E_{13}^{\mathrm{c}} = \dfrac{K^{\mathrm{c}}}{480}\left(32 - 30\cos\underline{\theta}_m - 5\cos(3\underline{\theta}_m) + 3\cos(5\underline{\theta}_m)\right).
$$

The efficiency of a reinforcement

Let us assume a planar reinforcement defined by a distribution density function. Assume the material to be subjected to uniaxial traction. The efficiency of the reinforcement may be defined by the ratio of the Young's modulus of the anisotropic material over the Young's modulus of the associated isotropic reinforcement.

12.2.3.3 Interplay between the stiffness of the fiber network and chemical effects

The model may display *a swelling induced stiffening* by fiber recruitment and, conversely, *a shrinking induced softening* by fiber deactivation.

Indeed, consider the material to undergo uniaxial traction while swelling emanates from some chemomechanical process. Swelling will imply the lateral strains to become less contractive. Therefore, more collagen fibers will be activated and the angle θ_m of the cone of active fibers will be larger. Consequently, the axial stiffness is certainly going to increase. Simulations in Chapter 17 show that this self-contained stiffening/softening effect is significant: a piece of articular cartilage in contact with a bath of controlled chemical composition is subjected to a given axial strain. When the ionic strength of the bath increases, the lateral strain and the volume of the cartilage decrease, and the cone of mechanically active fibers reduces, Figs. 17.13.2–17.13.4.

12.2.3.4 The apparent Poisson's ratio of the dry tissue

Stress in soft tissues is considered to be contributed additively by the matrix (ground substance) and the collagen network. Therefore the stiffness of the cartilage is the sum of the apparent stiffnesses of the ground substance and of the collagen network, $\mathbb{E}^{\mathrm{car}} = \mathbb{E}^{\mathrm{gs}} + \mathbb{E}^{\mathrm{c}}$. The apparent isotropic tensor moduli $\mathbb{E}^{\mathrm{gs}}$ of the ground substance are defined by the Lamé moduli,

$$\lambda = n^{\mathrm{gs}}\,\Lambda_{\mathrm{gs}}, \quad \mu = n^{\mathrm{gs}}\,M_{\mathrm{gs}}\,. \tag{12.2.23}$$

Here Λ_{gs} and M_{gs} are the intrinsic Lamé moduli of the matrix, and n^{gs} its volume fraction. Similarly, the stiffness of the collagen network $\mathbb{E}^{\mathrm{c}}$ is given by (12.2.22), where K^{c},

$$K^{\mathrm{c}} = n^{\mathrm{c}}\,K_{\mathrm{c}}\,, \tag{12.2.24}$$

is the apparent modulus of collagen: K_{c} is the intrinsic modulus of a collagen fiber and n^{c} the volume fraction of collagen.

Leaving aside for now chemomechanical couplings, the constitutive equations in the principal strain axes may be cast in the format,

$$\boldsymbol{\sigma} = \mathbb{E}^{\mathrm{car}} : \boldsymbol{\epsilon}, \quad
\begin{bmatrix} \sigma_{11} \\ \sigma_{22} \\ \sigma_{33} \end{bmatrix} =
\begin{bmatrix}
\lambda + 2\mu + E_{11}^{\mathrm{c}} & \lambda + E_{12}^{\mathrm{c}} & \lambda + E_{12}^{\mathrm{c}} \\
\lambda + E_{21}^{\mathrm{c}} & \lambda + 2\mu + E_{22}^{\mathrm{c}} & \lambda + E_{23}^{\mathrm{c}} \\
\lambda + E_{21}^{\mathrm{c}} & \lambda + E_{32}^{\mathrm{c}} & \lambda + 2\mu + E_{22}^{\mathrm{c}}
\end{bmatrix}
\begin{bmatrix} \epsilon_{11} \\ \epsilon_{22} \\ \epsilon_{33} \end{bmatrix}. \tag{12.2.25}$$

Under symmetric loading about the axis 1, the strain energy associated with the stiffness $\mathbb{E}^{\mathrm{car}}$ defined by (12.2.25),

$$\tfrac{1}{2}\,E_{11}^{\mathrm{car}}\,\epsilon_{11}^2 + 2\,E_{12}^{\mathrm{car}}\,\epsilon_{11}\,\epsilon_{22} + (E_{22}^{\mathrm{car}} + E_{23}^{\mathrm{car}})\,\epsilon_{22}^2\,, \tag{12.2.26}$$

is positive if

$$E_{11}^{\mathrm{car}} > 0, \quad 2\,(E_{12}^{\mathrm{car}})^2 - E_{11}^{\mathrm{car}}\,(E_{22}^{\mathrm{car}} + E_{23}^{\mathrm{car}}) < 0\,. \tag{12.2.27}$$

The latter inequality can be rephrased as,

$$\Delta = 2\,(\nu^{\mathrm{car}})^2 - \frac{E_{11}^{\mathrm{car}}}{E_{22}^{\mathrm{car}} + E_{23}^{\mathrm{car}}} < 0\,, \tag{12.2.28}$$

in terms of an aggregate Poisson's ratio,

$$\nu^{\mathrm{car}} = \frac{E_{12}^{\mathrm{car}}}{E_{22}^{\mathrm{car}} + E_{23}^{\mathrm{car}}}\,. \tag{12.2.29}$$

For K^{c}/μ below 43.21, $1 - 2\,\nu^{\mathrm{car}}$ remains positive, Fig. 12.2.4. Conversely for K^{c}/μ larger than 43.21, $1 - 2\,\nu^{\mathrm{car}}$ becomes negative for some range of angles θ_m, that is, for some range of lateral contraction defined by the ratio $-\epsilon_{22}/\epsilon_{11} = (\cot\theta_m)^2$. Said otherwise, the ratio $-\epsilon_{22}/\epsilon_{11}$ should be larger than some threshold, about 0.12, for $1 - 2\,\nu^{\mathrm{car}}$ to have a chance to be negative.

In fact, the smallest value of K^{c}/μ for which $1 - 2\,\nu^{\mathrm{car}}$ vanishes corresponds to uniaxial traction, and then $-\epsilon_{22}/\epsilon_{11} = (\cot\theta_m)^2 = 1/2$, that is $\theta_m = 54.74°$.

In all circumstances, the quantity Δ defined by (12.2.28) is negative, so that the strain energy function, and by the same token the aggregate stiffness $\mathbb{E}^{\mathrm{car}}$ remain positive definite.

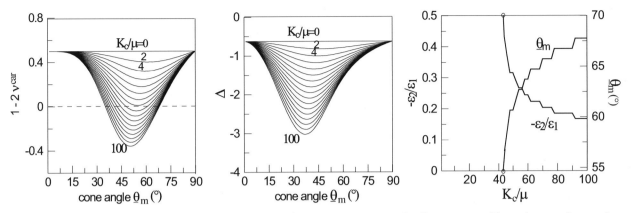

Fig. 12.2.4: While the Poisson's ratio of the cartilage (aggregate matrix and collagen network) can become larger than 1/2, for a sufficiently large relative stiffness of collagen with respect to matrix, the strain energy function remains positive. Poisson's ratio of matrix $\nu = 0.25$.

12.2.4 Conewise nonlinear stress-strain response of collagen fibers

We now return to the energy (12.2.3) in exponential form. For uniaxial traction (symmetry about the axis 1), the activation criterion relies on the elongation L,

$$L = \langle \boldsymbol{\epsilon} : \mathbf{M}_c \rangle = \langle \epsilon_{11} (\cos \underline{\theta})^2 + \epsilon_{22} (\sin \underline{\theta})^2 \rangle . \tag{12.2.30}$$

The strain energy per unit volume of tissue,

$$\mathcal{W}^c(\boldsymbol{\epsilon}) = n^c \int_{\mathbb{S}^2} w_c(L) \, \frac{dS}{4\pi}, \tag{12.2.31}$$

provides the stress of the collagen network,

$$\begin{aligned} \boldsymbol{\sigma}^c &= \frac{\partial \mathcal{W}^c(\boldsymbol{\epsilon})}{\partial \boldsymbol{\epsilon}} \\ &= \int_{\mathbb{S}^2} K^c \exp(k_c L^2) \, L \, \mathbf{M}_c \, \frac{dS}{4\pi} \\ &= \frac{K^c}{4\pi} \int_0^\pi d\underline{\theta} \, \sin \underline{\theta} \, \exp(k_c L^2) \, L \int_0^{2\pi} d\underline{\phi} \, \mathbf{M}_c, \end{aligned} \tag{12.2.32}$$

namely in component form,

$$\sigma_{ij}^c = K^c \int_0^{\pi/2} d\underline{\theta} \, \sin \underline{\theta} \, \exp(k_c L^2) \, L \begin{cases} (\cos \underline{\theta})^2, & i = j = 1; \\ (\sin \underline{\theta})^2/2, & i = j = 2,3; \\ 0, & i \neq j . \end{cases} \tag{12.2.33}$$

In order to perform a Gauss-Legendre integration, the stress is cast in the format,

$$\sigma_{ij}^c = K^c \int_0^1 dx \, \exp(k_c L^2) \, L \begin{cases} x^2, & i = j = 1; \\ (1 - x^2)/2, & i = j = 2,3; \\ 0, & i \neq j , \end{cases} \tag{12.2.34}$$

where

$$L = \langle \boldsymbol{\epsilon} : \mathbf{M}_c \rangle = \langle \epsilon_{11} \, x^2 + \epsilon_{22} \, (1 - x^2) \rangle . \tag{12.2.35}$$

12.2.5 Random planar directional distribution of collagen fibers

The collagen fibers are now assumed to be randomly oriented in the plane 2-3. The unit vector $\underline{\mathbf{m}}_c$ indicating the direction of a collagen fiber is expressed in terms of the angle $\underline{\phi} \in [0, 2\pi[$,

$$\underline{\mathbf{m}}_c = \begin{bmatrix} 0 \\ \cos \underline{\phi} \\ \sin \underline{\phi} \end{bmatrix} . \tag{12.2.36}$$

Let us record the directional integrations over space directions of the second order tensor $\underline{\mathbf{M}}_c = \underline{\mathbf{m}}_c \otimes \underline{\mathbf{m}}_c$,

$$\int_{\mathbb{S}^1} \underline{\mathbf{m}}_c \otimes \underline{\mathbf{m}}_c \frac{d\underline{\phi}}{2\pi} = \int_0^{2\pi} \underline{\mathbf{m}}_c \otimes \underline{\mathbf{m}}_c \frac{d\underline{\phi}}{2\pi} = \frac{\mathbf{I}^{\parallel}}{2} , \tag{12.2.37}$$

and of the fourth order tensor $\underline{\mathbf{M}}_c \otimes \underline{\mathbf{M}}_c$,

$$\int_{\mathbb{S}^1} \underline{\mathbf{M}}_c \otimes \underline{\mathbf{M}}_c \frac{d\underline{\phi}}{2\pi} = \frac{1}{8} \left(\mathbf{I}^{\parallel} \otimes \mathbf{I}^{\parallel} + 2\, \mathbf{I}^{\parallel} \,\overline{\underline{\otimes}}\, \mathbf{I}^{\parallel} \right) . \tag{12.2.38}$$

Here $\mathbf{I}^{\parallel}$ is the planar second order identity tensor. The expression (12.2.38) results from the component relations,

$$\frac{1}{2\pi} \int_0^{2\pi} \underline{m}_{ci}\, \underline{m}_{cj}\, \underline{m}_{ck}\, \underline{m}_{cl}\, d\underline{\phi} = \begin{cases} 3/8, & i = j = k = l, \\ 1/8, & i = j \neq k = l, \\ 0, & \text{otherwise}. \end{cases} \tag{12.2.39}$$

12.2.5.1 Conewise linear stress-strain response of planar collagen fibers

If *all* fibers are in extension, then integration over space directions of the strain energy and of its first and second derivatives,

$$\begin{cases} n^c \displaystyle\int_{\mathbb{S}^1} w_c \, \frac{d\underline{\phi}}{2\pi} \\[2mm] n^c \displaystyle\int_{\mathbb{S}^1} \frac{\partial w_c}{\partial \boldsymbol{\epsilon}} \, \frac{d\underline{\phi}}{2\pi} \\[2mm] n^c \displaystyle\int_{\mathbb{S}^1} \frac{\partial^2 w_c}{\partial \boldsymbol{\epsilon}\, \partial \boldsymbol{\epsilon}} \, \frac{d\underline{\phi}}{2\pi} \end{cases} = \frac{K^c}{8} \begin{cases} \frac{1}{2} \left(\mathbf{I}^{\parallel} : \boldsymbol{\epsilon} \right)^2 + \boldsymbol{\epsilon} : \boldsymbol{\epsilon}, \\[2mm] \left(\boldsymbol{\epsilon} : \mathbf{I}^{\parallel} \right) \mathbf{I}^{\parallel} + 2\, \boldsymbol{\epsilon}, \\[2mm] \mathbf{I}^{\parallel} \otimes \mathbf{I}^{\parallel} + 2\, \mathbf{I}^{\parallel} \,\overline{\underline{\otimes}}\, \mathbf{I}^{\parallel}, \end{cases} \tag{12.2.40}$$

yields an isotropic stiffness with equal Lamé moduli,

$$\lambda^c = \mu^c = \frac{K^c}{8} , \tag{12.2.41}$$

that is, with a Poisson's ratio equal to 0.25.

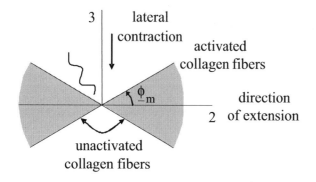

Fig. 12.2.5: Fans of activation of a planar network of collagen fibers in axial extension and lateral contraction.

Let us consider now extension along the direction 2, with symmetry about this axis. Of course, if the transverse directions undergo extension, we recover the isotropic stiffness. If the transverse directions undergo contraction then, the fibers which undergo extension,

$$\epsilon : \underline{\mathbf{M}}_{\mathrm{c}} \geq 0 \quad \Rightarrow \quad \cot^2 \underline{\phi} = \frac{m_{c2}^2}{m_{c3}^2} \geq \cot^2 \underline{\phi}_m = \frac{-\epsilon_{33}}{\epsilon_{22}}, \tag{12.2.42}$$

form two symmetric cones of half angle $\underline{\phi}_m$ about the extension direction, as shown in Fig. 12.2.5,

$$\underbrace{\underline{\phi} \in [-\underline{\phi}_m, \underline{\phi}_m] \cup [\pi - \underline{\phi}_m, \pi + \underline{\phi}_m]}_{\text{fans of extension}}. \tag{12.2.43}$$

In the principal axes of strain, we just need to calculate three components of the stiffness in view of the symmetry about the axis 2. Therefore, the contribution of the collagen network to the total stress expresses in terms of a symmetric matrix,

$$\boldsymbol{\sigma}^{\mathrm{c}} = \mathbb{E}^{\mathrm{c}}(\boldsymbol{\epsilon}) : \boldsymbol{\epsilon}, \quad \begin{bmatrix} \sigma_{11} \\ \sigma_{22} \\ \sigma_{33} \end{bmatrix} = \begin{bmatrix} 0 & 0 & 0 \\ 0 & E_{22}^{\mathrm{c}} & E_{23}^{\mathrm{c}} \\ 0 & E_{32}^{\mathrm{c}} & E_{33}^{\mathrm{c}} \end{bmatrix} \begin{bmatrix} \epsilon_{11} \\ \epsilon_{22} \\ \epsilon_{33} \end{bmatrix}, \tag{12.2.44}$$

with coefficients,

$$E_{22}^{\mathrm{c}} = \frac{K^{\mathrm{c}}}{16\pi} \left(12\,\underline{\phi}_m + 8\,\sin(2\underline{\phi}_m) + \sin(4\underline{\phi}_m) \right)$$

$$E_{33}^{\mathrm{c}} = \frac{K^{\mathrm{c}}}{16\pi} \left(12\,\underline{\phi}_m - 8\,\sin(2\underline{\phi}_m) + \sin(4\underline{\phi}_m) \right) \tag{12.2.45}$$

$$E_{23}^{\mathrm{c}} = E_{32}^{\mathrm{c}} = \frac{K^{\mathrm{c}}}{16\pi} \left(4\underline{\phi}_m - \sin(4\underline{\phi}_m) \right).$$

12.2.5.2 Conewise nonlinear stress-strain response of planar collagen fibers

We now return to the energy (12.2.3) in exponential form. For uniaxial traction (symmetry about the axis 2), the activation criterion relies on the elongation L,

$$L = \langle \boldsymbol{\epsilon} : \underline{\mathbf{M}}_{\mathrm{c}} \rangle = \langle \epsilon_{22} (\cos \underline{\phi})^2 + \epsilon_{33} (\sin \underline{\phi})^2 \rangle. \tag{12.2.46}$$

The strain energy per unit volume of tissue,

$$\mathcal{W}^{\mathrm{c}}(\boldsymbol{\epsilon}) = n^{\mathrm{c}} \int_{\mathbb{S}^1} w_{\mathrm{c}}(L) \frac{d\underline{\phi}}{2\pi}, \tag{12.2.47}$$

provides the stress of the collagen network,

$$\begin{aligned} \boldsymbol{\sigma}^{\mathrm{c}} &= \frac{\partial \mathcal{W}^{\mathrm{c}}(\boldsymbol{\epsilon})}{\partial \boldsymbol{\epsilon}}, \\ &= \frac{K^{\mathrm{c}}}{2\pi} \int_0^{2\pi} d\underline{\phi} \, \exp(k_{\mathrm{c}} L^2) \, L \, \underline{\mathbf{M}}_{\mathrm{c}}, \end{aligned} \tag{12.2.48}$$

$$\sigma_{ij}^{\mathrm{c}} = 2\frac{K^{\mathrm{c}}}{\pi} \int_0^{\pi/2} d\underline{\phi} \, \exp(k_{\mathrm{c}} L^2) \, L \begin{cases} (\cos \underline{\phi})^2, & i = j = 2; \\ (\sin \underline{\phi})^2, & i = j = 3; \\ 0, & i = 2, j = 3. \end{cases}$$

In order to perform a Gauss-Legendre integration, the stress is expressed in the format,

$$\sigma_{ij}^{\mathrm{c}} = 2\frac{K^{\mathrm{c}}}{\pi} \int_0^1 \frac{\sqrt{2}\,dx}{\sqrt{1 - \frac{x^2}{2}}} \exp(k_{\mathrm{c}} L^2) \, L \begin{cases} (1 - x^2)^2, & i = j = 2; \\ x^2 (2 - x^2), & i = j = 3; \\ 0, & i = 2, j = 3, \end{cases} \tag{12.2.49}$$

where

$$L = \langle \boldsymbol{\epsilon} : \underline{\mathbf{M}}_{\mathrm{c}} \rangle = \langle (1 - x^2)^2 \, \epsilon_{22} + x^2 (2 - x^2) \, \epsilon_{33} \rangle. \tag{12.2.50}$$

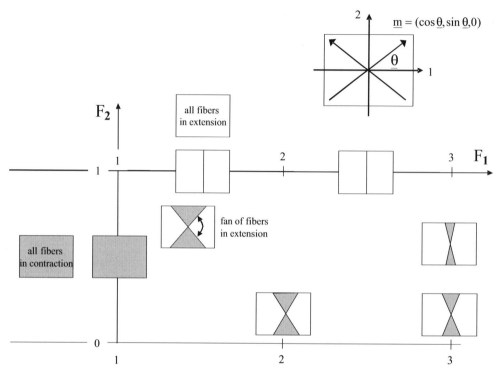

Fig. 12.2.6: Biaxial loading in the plane of the fibers with $F_1 = 1 + \epsilon_1 \geq 1$ and $0 < F_2 = 1 + \epsilon_2 \leq 1$ components of the deformation gradient along the axes (1,2). Close to the reference state, the size of the fans of fibers in extension depends discontinuously on the exact departure from this reference state, namely $\sin \underline{\theta} = 1/\sqrt{1 - \epsilon_2/\epsilon_1}$.

12.2.5.3   Fans in extension: nonlinearity due to progressive fiber recruitment

Assume the material to undergo biaxial extension along the symmetry axes as sketched in Fig. 12.2.6, with $F_1 \geq 1$ and $F_2 \leq 1$ the components of the deformation gradient along these axes. The fibers oriented towards the axis 1 are likely to undergo extension while those along the axis 2 are likely to undergo contraction. The transformed length of a unit segment $\underline{m} = (\cos \theta, \sin \theta, 0)$ has square $\ell^2 = \underline{m} \cdot \mathbf{C} \cdot \underline{m} = F_1^2 + \sin^2 \theta \, (F_2^2 - F_1^2)$. The deformed segments within the fan $(-\underline{\theta}, \underline{\theta})$ with $\sin^2 \underline{\theta} = (F_1^2 - 1)/(F_1^2 - F_2^2)$ have a length greater than 1.

Therefore, at the beginning of a biaxial experiment, the fibers are not all active. The fibers get recruited progressively as the deformation F_1 increases. There is thus a purely geometrical nonlinear response, explaining in part the existence of a toe.

12.3   Directional models of tissues

12.3.1   Directional distribution of collagen fibers in cornea

In some tissues, the directional distribution of collagen fibers is strongly heterogeneous. For example, the corneal stroma and the annulus fibrosus are built by the superposition of lamellae. In each lamella, the fibers are strictly parallel. In the center of the tissues, the fiber directions alternate from one lamella to the other. At the boundaries, the arrangement may not be so regular. More details are provided in Chapter 19.

Perhaps it is appropriate here to highlight the two concepts of spatial heterogeneity and anisotropy. The first step is the choice of a representative elementary volume to which we refer, in a continuum context, as a point. Anisotropy is a pointwise, or local, property. If the size of the RVE is equal to, or smaller than, the width of a lamella, then the material considered can be seen as a fiber reinforced material. It is described as transversely isotropic, the symmetry axis being the fiber direction. If the size of the RVE encompasses two lamellae, the material we consider now is reinforced by two families of fibers which are not parallel: such a material is referred to as clinotropic. Incidentally, note that these two families of fibers may or may not be mechanically equivalent.

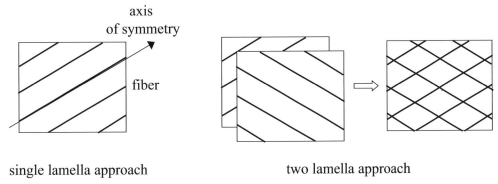

Fig. 12.3.1: The corneal stroma is formed by the superposition of lamellae. The direction of the fiber reinforcement alternates over two adjacent lamellae. In the single lamella approach, the material is transversely isotropic at the lamellar scale, while, in the two lamella approach, it is clinotropic at the associated scale.

As a further example, outside the realm of the corneal structure, one may consider a stack of layers, or deck of cards, which are parallel and individually isotropic. If the RVE encompasses several layers, the material at hand is again transversely isotropic but the symmetry axis is now the normal to the layers. While a fiber reinforced material deforms by shear parallel to the fibers (which define the axis of material symmetry), the deck of cards deforms by shear parallel to the plane of the cards (which is orthogonal to the axis of material symmetry).

When the size of the RVE is equal to, or smaller than, the width of a lamella, a strong spatial heterogeneity is encountered along a path going from one lamella to its neighbor. On the other hand, the spatial heterogeneity along a path connecting two successive RVEs is much smaller when the latter encompasses two lamellae. These two points of view sketched in Fig. 12.3.1 are often referred to respectively as the *single lamella approach* and the *two lamella approach*. As an approximation, it is often assumed that the fibers in the next neighbor lamellae are parallel and mechanically equivalent. Then, in the two lamella approach, no heterogeneity across the depth is encountered.

12.3.2 Continuous planar directional distributions of collagen fibers

While Section 12.2 addresses isotropic directional distributions of collagen fibers, here are instances of non isotropic distributions specifically tailored to lamellar tissues. Emphasis is laid on three points:
- the directional distribution describing the alternate directions of adjacent lamellae, including some dispersion;
- the change of the directional distribution in the plane of the cornea from the corneal center to the periphery;
- the depth dependence along the corneal axis from the anterior side to the posterior side.

As a typical instance along this line, Pinsky et al. [2005] start from the lamellar energy and deduce the stromal energy by a directional planar distribution of fibers. For that purpose, they define a projection plane $\underline{x}-\underline{y}$ orthogonal to the corneal axis $\underline{z}$. At a point $\mathbf{x}$, the probability to have a fiber in the direction $\underline{\theta}$ is denoted by $\mathcal{D}(\mathbf{x}, \theta)$. The stromal energy is contributed by the matrix energy which depends on the point through the local mechanical properties and the local strain and by the fibril energy which depends in addition on the fibril direction,

$$\underline{\mathcal{W}}_{\text{stroma}} = \underline{\mathcal{W}}_{\text{matrix}}(\underline{\mathbf{x}}) + \int_0^\pi \underline{\mathcal{W}}_{\text{fibril}}(\underline{\mathbf{x}}, \theta)\, \mathcal{D}(\underline{x}, \underline{y}, \underline{\theta})\, d\underline{\theta}\,. \tag{12.3.1}$$

In a polar coordinate system $(\underline{x}, \underline{y}) \to (R, \phi)$, the probability density function $\mathcal{D}$ adopts the format of a sort of normal distribution,

$$\mathcal{D}(R, \underline{\phi}, \underline{\theta}) = \frac{1}{\pi} \begin{cases} \left(\cos\theta\right)^{2n} + \left(\sin\theta\right)^{2n} + c_1, & 0 \leq R \leq 4, \\ \left(\sin(\underline{\theta} - \underline{\phi})\right)^{2n} + c_2, & 5.5 \leq R \leq 6.5, \end{cases} \tag{12.3.2}$$

where c_1 and c_2 are such that the cumulative distribution function $D(R, \underline{\phi}, \theta) = \int_0^\theta \mathcal{D}(R, \underline{\phi}, \theta)\, d\theta$ satisfies the consistency condition $D(R, \underline{\phi}, \pi) = 1$. The complete relation for the radius R [unit : mm] over $]4, 5.5]$ and $]6.5, 8]$ ensures a smooth transition from an orthogonal to a tangential fibril distribution. A planar isotropic density is retrieved for the exponent n equal to 0. Higher values of n increase the non uniformity of the angular fibril distribution. A value of $n = 4$ is found to fit best the data by Aghamohammadzadeh et al. [2004].

A general comparison of discrete and continuous directional distributions of fibers is exposed in Bischoff [2006]. He delineates the single lamellar approach and the multilamellar approach that are characterized by unimodal distributions and bimodal distributions, respectively. The considered distribution densities are a priori non planar and expressed as $\mathcal{D} = \mathcal{D}(\theta, \phi)$ in spherical coordinates. However the examples that are worked out are independent of ϕ and symmetric in $\underline{\theta}$ over the range $-\pi/2 \leq \underline{\theta} \leq \pi/2$. Typically unimodal normal distribution densities centered around $\underline{\theta}_0$ and with standard deviation σ adopt the format,

$$\frac{\mathcal{D}(\underline{\theta})}{\mathcal{D}_N} = \frac{1}{\sigma\sqrt{2\pi}} \exp\left(-\frac{(\underline{\theta} - \underline{\theta}_0)^2}{(\sqrt{2}\,\sigma)^2}\right), \quad \int_{-\pi/2}^{\pi/2} \mathcal{D}(\underline{\theta})\, d\underline{\theta} = 1\,, \tag{12.3.3}$$

where

$$\int_{\underline{\theta}_0 - \infty}^{\underline{\theta}_0 + \infty} \frac{\mathcal{D}(\underline{\theta})}{\mathcal{D}_N}\, d\underline{\theta} = 1, \quad \int_{\underline{\theta}_0 - \sigma}^{\underline{\theta}_0 + \sigma} \frac{\mathcal{D}(\underline{\theta})}{\mathcal{D}_N}\, d\underline{\theta} = 0.683, \quad \frac{1}{\mathcal{D}_N} = \frac{1}{2}\left(\mathrm{erf}\left(\frac{\pi/2 - \underline{\theta}_0}{\sqrt{2}\,\sigma}\right) + \mathrm{erf}\left(\frac{\pi/2 + \underline{\theta}_0}{\sqrt{2}\,\sigma}\right)\right). \tag{12.3.4}$$

For bimodal normal distribution densities centered around the angles $\underline{\theta}_1$ and $\underline{\theta}_2 > \underline{\theta}_1$,

$$\frac{\mathcal{D}(\underline{\theta})}{\mathcal{D}_N} = \frac{1}{\sigma\sqrt{2\pi}} \exp\left(-\frac{(\underline{\theta} - \underline{\theta}_1)^2}{(\sqrt{2}\,\sigma)^2}\right) + \frac{1}{\sigma\sqrt{2\pi}} \exp\left(-\frac{(\underline{\theta} - \underline{\theta}_2)^2}{(\sqrt{2}\,\sigma)^2}\right), \quad \int_{-\pi/2}^{\pi/2} \mathcal{D}(\underline{\theta})\, d\underline{\theta} = 1\,, \tag{12.3.5}$$

where

$$\int_{\underline{\theta}_1 - \infty}^{\underline{\theta}_2 + \infty} \frac{\mathcal{D}(\underline{\theta})}{\mathcal{D}_N}\, d\underline{\theta} = 2, \quad \frac{1}{\mathcal{D}_N} = \frac{1}{2}\left(\mathrm{erf}\left(\frac{\pi/2 - \underline{\theta}_1}{\sqrt{2}\,\sigma}\right) + \mathrm{erf}\left(\frac{\pi/2 + \underline{\theta}_1}{\sqrt{2}\,\sigma}\right) + \mathrm{erf}\left(\frac{\pi/2 - \underline{\theta}_2}{\sqrt{2}\,\sigma}\right) + \mathrm{erf}\left(\frac{\pi/2 + \underline{\theta}_2}{\sqrt{2}\,\sigma}\right)\right). \tag{12.3.6}$$

Li and Tighe [2006]b use an angular distribution density around the angles $0°$ and $90°$ defined by three parameters, namely N_t the total number of lamellae, N_c a normalizing constant and C a constant affecting the pattern of the distribution. The number of lamellae around the angle $\underline{\theta}$ is defined in an exponential format,

$$N(\underline{\theta}) = \begin{cases} \mathrm{int}\left\{\dfrac{N_t}{2N_c} \exp\left(-\dfrac{\theta^2}{C}\right)\right\}, & |\underline{\theta}| \leq \pi/4, \\[3mm] \mathrm{int}\left\{\dfrac{N_t}{2N_c} \exp\left(-\dfrac{(\underline{\theta} - \pi/2)^2}{C}\right)\right\}, & |\underline{\theta} - \pi/2| \leq \pi/4, \end{cases} \tag{12.3.7}$$

where the operator 'int' means integer part of its argument.

Note that the distribution densities above actually refer to a reference configuration and the directional integrations are performed with respect to this configuration.

12.3.3 A directional model of active rod-like cells

In a series of papers, Zahalak and co-workers present experimental results and propose models of tissues reconstituted from cells and ECM that aim to simulate natural tissues, in particular their mechanical changes during development and wound healing. Zahalak et al. [2000] address the mechanical behavior of a tissue composed of a matrix interspersed with fibroblasts. Cells with generic direction $\mathbf{m}$ are oriented spatially according to the directional distribution density $\mathcal{D}(\mathbf{m})$. The overall viscoelastic properties of the tissue matrix-contractile cells composite derive from the basic assumptions below:
  - the total stress $\boldsymbol{\sigma}$ is the sum of the matrix and cell stresses, $\boldsymbol{\sigma}^M$ and $\boldsymbol{\sigma}^C$, respectively;
  - the strain is homogeneous in the matrix and cells, and equal to the macroscopic strain $\boldsymbol{\epsilon}$. The fact that, at high cell density, cells tend to form a continuum suggests that the assumption is acceptable;
  - the cells and the ECM are viscoelastic. The ECM stress is defined by Fung's quasi-linear viscoelastic model, which is detailed in Section 8.6;
  - the cells (fibroblasts) are rods, developing active contractile forces, characterized mechanically by a Hill muscle model.

The cells behave as rods of initial length l_0 and section S_0. The force F each of them can sustain is aligned with their direction $\mathbf{m}$ and its intensity depends on length change $l/l_0 - 1$ which may be expressed via the strain $\boldsymbol{\epsilon}$ as $\boldsymbol{\epsilon} : (\mathbf{m} \otimes \mathbf{m})$. Let N_C be the number of cells in the reference volume V_0. Then the stress they carry is obtained by integration over the unit sphere,

$$\boldsymbol{\sigma}^C = \frac{N_C}{V_0} l_0 \int_{\mathbb{S}^2} \mathbf{m} \otimes \mathbf{m}\, F(\mathbf{m} \cdot \boldsymbol{\epsilon} \cdot \mathbf{m})\, \mathcal{D}(\mathbf{m})\, dS\,. \tag{12.3.8}$$

In fact, based on their background, Zahalak et al. [2000] consider that the elongated cells behave mechanically as muscle cells, which are described by a passive elasticity (PE) in parallel with a branch containing in series an elastic element (SE) and a contractile viscous element (CE), as sketched in Fig. 8.6.1 where the notation is displayed. The intensity of the contractile force is actually related to the concentration of a diffusible stimulant/activator like bovine serum. Note that the matrix phase is not present in muscles.

Let F_P and F_A be the forces acting respectively on the passive branch and active branch of the parallel model, and F_0 be the isometric contractile force. In linearized form,

$$F_P = E_P \, (l - l_0), \quad F_A = E_S \, (l_{SE} - l_{SE0}), \quad \frac{d}{dt} l_{CE} = \frac{1}{\eta_C} \, (F_A - F_0), \tag{12.3.9}$$

with η_C cell's viscosity, that may also be cast as the product of a relaxation time τ_C and of the Young's modulus E_S, and

$$F = F_P + F_A, \quad l = l_{SE} + l_{CE} \, . \tag{12.3.10}$$

Combining these relations yields a first order differential equation linking the force F acting over the cell to its length l,

$$\frac{dF}{dt} + \frac{F}{\tau_C} = \frac{F_0}{\tau_C} + l_0 (E_S + E_P) \frac{d}{dt} \frac{l}{l_0} + \frac{l_0 E_P}{\tau_C} \Big(\frac{l}{l_0} - 1 \Big) \, . \tag{12.3.11}$$

The tissue fabric is assumed not to change, or equivalently, no re-structuration takes place. Since $l/l_0 - 1 = \boldsymbol{\epsilon} :$ $(\mathbf{m} \otimes \mathbf{m})$, integration over cells of the above differential equation provides now a first order differential equation linking the cell stress $\boldsymbol{\sigma}^C$ and the strain $\boldsymbol{\epsilon}$,

$$\frac{d\boldsymbol{\sigma}^C}{dt} + \frac{\boldsymbol{\sigma}^C}{\tau_C} = \frac{3\sigma_0}{\tau_C} \mathbf{A} + \Big(\kappa \frac{d}{dt} + \frac{\omega}{\tau_C} \Big) \mathbf{B} : \boldsymbol{\epsilon}, \tag{12.3.12}$$

with $\sigma_0 = (N_C/V_0) \, l_0 \, F_0/3$, $\kappa = (N_C/V_0) \, l_0^2 (E_S + E_P)$, $\omega = (N_C/V_0) \, l_0^2 \, E_P$, and $\mathbf{A}$ and $\mathbf{B}$ respectively second order and fourth order fabric tensors,

$$\mathbf{A} = \int_{\mathbb{S}^2} \mathbf{m} \otimes \mathbf{m} \, \mathcal{H}(\mathbf{m} \cdot \boldsymbol{\epsilon} \cdot \mathbf{m}) \, \mathcal{D}(\mathbf{m}) \, dS, \quad \mathbf{B} = \int_{\mathbb{S}^2} \mathbf{m} \otimes \mathbf{m} \otimes \mathbf{m} \otimes \mathbf{m} \, \mathcal{H}(\mathbf{m} \cdot \boldsymbol{\epsilon} \cdot \mathbf{m}) \, \mathcal{D}(\mathbf{m}) \, dS \, , \tag{12.3.13}$$

where $\mathcal{H}$ denotes the Heaviside step function. When the cell directional distribution is isotropic, the fabric tensors simplify as indicated in Section 12.2.2. The stress can be expressed in integral form,

$$\boldsymbol{\sigma}^C(t) - \boldsymbol{\sigma}^C(-\infty) = \int_{-\infty}^{t} e^{-(t-u)/\tau_C} \Big(\frac{3\sigma_0}{\tau_C} \mathbf{A}(u) + \Big(\kappa \frac{d}{du} + \frac{\omega}{\tau_C} \Big) \mathbf{B}(u) : \boldsymbol{\epsilon}(u) \Big) \, du \, , \tag{12.3.14}$$

its reference value $\boldsymbol{\sigma}^C(t = -\infty)$ presumably vanishing.

In a similar approach that does not refer to hyperelasticity, Sacks [2003] calculates the stress along the scheme (12.3.8) by integrating over space the uniaxial contribution of each fiber of planar collagenous tissues. Still, his analysis does not involve rate effects.

12.4 The mechanical response of individual fibers

12.4.1 The toe region, nonlinearity and fiber uncrimping

Fiber activation is considered by Lanir [1979], [1983] via the *uncrimping strain*[12.2][12.3]. His analysis is used in particular by Sacks [2003] and Bischoff [2006].

[12.2]Miyazaki and Hayashi [1999] isolate individual fibers (1 μm in diameter) from fascicles (300 μm in diameter). While the individual fibers show almost linear behavior, the fascicles and bulk tendons where the fibrils and fibers are crimped display a toe region. A mechanical isolation method of fibers was observed not to introduce damage, and preferred to enzymatic digestion.
[12.3]From the tensile experiments of Kempson et al. [1973] on articular cartilages, Silver and Bradica [2002] deduce *linear* stress-strain curves for strains larger than a few percent both in the direction of the fibrils and in the orthogonal direction, with a modulus linking the axial strain and stress of about 7 GPa and 2 GPa, respectively. They indicate that the former value is similar to that of tendons of rat and turkey and twice that of skin.

Let L_0, L_s and L be respectively the initial (crimped), straightened (taut) and actual fiber lengths. The fiber becomes stretched when its length L reaches the value L_s. With the stretch ratios,

$$\frac{L}{L_0} = \frac{L}{L_s} \times \frac{L_s}{L_0} \quad \Leftrightarrow \quad \lambda = \lambda_f \, \lambda_s \,, \tag{12.4.1}$$

are associated the strains $E = (\lambda^2 - 1)/2$, $E_f = (\lambda_f^2 - 1)/2$ and $E_s = (\lambda_s^2 - 1)/2$ which are linked together,

$$2\,(E - E_s) = \lambda^2 - \lambda_s^2 = \lambda_s^2\,(\lambda_f^2 - 1) = 2\,E_f\,(1 + 2\,E_s)\,. \tag{12.4.2}$$

The nonlinear response under tension is wholly attributed to the progressive fiber recruitment, that is, uncrimping. The load necessary to reach uncrimping is considered negligible. In other words, the response is linear once all the fibers are uncrimped. Thus the fiber stress is taken as a linear function of the strain $\langle E_f \rangle$,

$$\sigma_f = K \, \langle E_f \rangle \,, \tag{12.4.3}$$

where K is the fiber modulus.

The stretch at which the fiber uncrimps is described by a probability density function $\mathcal{P}(E_s)$. Sacks [2003] and Bischoff [2006] use the Gamma function,

$$\mathcal{P}(E_s) = \frac{1}{\beta}\,\frac{1}{\Gamma(\alpha)}\,\left(\frac{E_s}{\beta}\right)^{\alpha-1}\,\exp(-\frac{E_s}{\beta})\,, \tag{12.4.4}$$

with α and β positive parameters, and $\Gamma(\alpha) = \int_0^\infty \exp(-x)\,x^{\alpha-1}\,dx$ the Gamma function. The density function $\mathcal{P}$ has mean value $\int_0^\infty E_s\,\mathcal{P}(E_s)\,dE_s = \alpha\,\beta$ and standard deviation $\int_0^\infty (E_s - \alpha\,\beta)^2\,\mathcal{P}(E_s)\,dE_s = \alpha\,\beta^2$. The density function $\mathcal{P}(E_s)$ vanishes at $E_s = 0$ and at $E_s = \infty$. It has a maximum $(\alpha - 1)^{\alpha-1}\,\exp(1 - \alpha)/\beta/\Gamma(\alpha)$ inversely proportional to β at a strain $E_s = \beta\,(\alpha - 1)$ proportional to β.

Sverdlik and Lanir [2002] use a beta probability density function, and Hurschler et al. [2003] a Weibull probability density function,

$$\mathcal{P}(\lambda_s) = \mathcal{H}(\frac{\lambda_s - \gamma}{\delta})\,\frac{\beta}{\delta}\,\left(\frac{\lambda_s - \gamma}{\delta}\right)^{\beta-1}\,\exp\left(-(\frac{\lambda_s - \gamma}{\delta})^\beta\right)\,, \tag{12.4.5}$$

defined by three parameters: $\beta > 0$ and $\delta > 0$ are shape and scale parameters related to the toe, $\gamma > 0$ is the *location parameter*, that is the stretch at which the first fiber begins to bear load. The cumulative distribution function associated with the Weibull distribution density,

$$P(\lambda) = \int_\gamma^\lambda \mathcal{P}(\lambda_s)\,d\lambda_s = 1 - \exp\left(-(\frac{\lambda - \gamma}{\delta})^\beta\right) \in [0, 1[\,, \tag{12.4.6}$$

represents the percentile of fibers that carry load. Note that this relation can be inverted, yielding the stretch for which a given percentile of fibers is under load.

The stress contributed by a collagen fiber is

$$\sigma(E) = \int_0^E \sigma_f(\langle E_f \rangle)\,\mathcal{P}(E_s)\,dE_s\,, \tag{12.4.7}$$

or, for Hurschler et al. [2003],

$$\sigma(\lambda) = \mathcal{H}(\lambda - \gamma)\,\int_\gamma^\lambda \sigma_f(\lambda_f)\,\mathcal{P}(\lambda_s)\,d\lambda_s\,. \tag{12.4.8}$$

The extraction of the parameters and reference length for 'slack' (crimped) and preloaded fibers is discussed in Hurschler et al. [2003]. Given that the parameters characterize the toe region, a too large pre-load decreases their accuracy and significance. Additional information on small-deformation modulus of calf skin may be found in Kronick [1988].

The analysis is extended to include viscous effects by assuming a time-dependent straightened length L_s in e.g., Sverdlik and Lanir [2002] and de Vita and Slaughter [2006].

Coupled with a fiber directional distribution $\mathcal{D}(\underline{\mathbf{m}}_c)$, the collagen stress $\underline{\underline{\tau}}^c(\mathbf{E}_G)$ at the overall strain $\mathbf{E}_G$ has the tensor form,

$$\underline{\underline{\tau}}^c(\mathbf{E}_G) = n^c \int_{\mathbb{S}^2} \underline{\mathbf{m}}_c \otimes \underline{\mathbf{m}}_c\,\sigma\big(E(\mathbf{E}_G, \underline{\mathbf{m}}_c)\big)\,\mathcal{D}(\underline{\mathbf{m}}_c)\,dS\,, \tag{12.4.9}$$

where n^c is the collagen volume fraction. More generally, this formula can be extended to account for a strain energy $\mathcal{W}_c = \mathcal{W}_c(\mathbf{E}_G, \underline{\mathbf{m}}_c)$ expressed in terms of anisotropic scalar invariants,

$$\underline{\underline{\tau}}^c(\mathbf{E}_G) = n^c \int_{\mathbb{S}^2} \frac{\partial \mathcal{W}_c}{\partial \mathbf{E}_G}(\mathbf{E}_G, \underline{\mathbf{m}}_c)\,\mathcal{D}(\underline{\mathbf{m}}_c)\,dS\,. \tag{12.4.10}$$

12.4.2 Intrinsic viscous behavior of collagen fibrils

Time dependence of the mechanical response of soft tissues emanates from several sources. A first source is due to the fact that soft tissues are fluid saturated. Another source is the intrinsic viscous response of collagen fibers. The latter has been analyzed for various tissues where it is expected to play a key role, e.g., muscles and tendons. It has been also considered for cornea and articular cartilages.

Boyce et al. [2007] detect a rate-dependent response in bovine cornea submitted to various stress rates. For the same axial stress of 400 kPa, the strain is equal to 0.05 for a stress rate of 3.5 kPa/s and to 0.04 for a stress rate of 350 kPa/s. A quasi-linear viscous stress-strain constitutive model (QLV) is used in an attempt to display the observed rate-dependency. Specifically, the stress response is assumed in the form (8.3.33),

$$\underline{\underline{\tau}}(t) = \underline{\underline{\tau}}^{e}(t) + \int_{0}^{t} \frac{dE_r}{dt}(t-u)\,\underline{\underline{\tau}}^{e}(u)\,du\,, \tag{12.4.11}$$

in terms of the reduced relaxation function $E_r(t)$ defined by (8.5.10), namely

$$E_r(t) = \frac{1 + a_v\,(E_1(t/\tau_2) - E_1(t/\tau_1))}{1 + a_v\,\mathrm{Ln}\,(\tau_2/\tau_1)}\,, \tag{12.4.12}$$

with $E_1(t) = \int_{t}^{\infty} \exp(-u)/u\,du$, a_v a dimensionless positive constant and τ_1 and τ_2 two relaxation times. As an alternative choice that also ensures the consistency condition $E_r(0) = 1$, one may consider the expression,

$$E_r(t) = \frac{a_{v0} + a_{v1}\,\exp(-t/\tau_1) + a_{v2}\,\exp(-t/\tau_2)}{a_{v0} + a_{v1} + a_{v2}}\,, \tag{12.4.13}$$

with a_{v0}, a_{v1} and a_{v2} dimensionless positive constants.

The elastic response $\underline{\underline{\tau}}^{e} = 2\,\partial\underline{\mathcal{W}}/\partial\mathbf{C}$ follows from the energy decomposition (12.3.1). The matrix energy is isotropic neo-Hookean as indicated by (6.4.7),

$$\underline{\mathcal{W}}_{\mathrm{matrix}} = \frac{\mu}{2}\,(\mathrm{tr}\,\mathbf{C} - 3) + \frac{\mu}{2\,\gamma}\,\big((\det\mathbf{C})^{-\gamma} - 1\big)\,, \tag{12.4.14}$$

with μ a constant stiffness and $\gamma > 0$ a positive exponent. The strain energy of individual fibrils is integrated over space directions as indicated by (12.3.1) with the fibril energy $\underline{\mathcal{W}}_{\mathrm{fibril}}$ in a format slightly modified with respect to (19.19.72) through the introduction of the material parameter β,

$$\underline{\mathcal{W}}_{\mathrm{fibril}}(I_4 = \mathbf{C} : \underline{\mathbf{M}}_{\mathrm{c}}) = \alpha\,\big(\exp(\beta(I_4 - 1)) - \beta\,I_4\big)\,. \tag{12.4.15}$$

For cornea strips, Boyce et al. [2007] obtain the following elastic parameters, namely $\mu = 10\,\mathrm{kPa}$, $\alpha = 2.32\,\mathrm{kPa}$, $\beta = 30.4$, $\gamma = 0.49$, and the viscoelastic parameters are listed in Table 8.5.1.

12.5 Fabric tensors

A number of physical properties involve not only scalars, but also vectors and tensors of various orders. Indeed, at any point of a material, or biological tissue, the thermal, hydraulic and elastic properties may depend on the space direction. A proper model traces these directional properties back to the microstructure at a certain length scale. Fabric tensors aim at representing the heterogeneous directional features that give rise to anisotropy at a larger scale.

12.5.1 Elementary examples of fabric tensors

Instead of the familiar *structural* anisotropy defined in terms of planes and axes of symmetry, the anisotropic character of the mechanical and transport properties may be defined by a traceless symmetric second order tensor $\mathbf{G}$. This tensor is a fabric tensor, namely a macroscopic description of microstructural features,

$$\text{microstructural properties} \xrightarrow{\ \mathbf{G}\ } \text{macroscopic level}$$

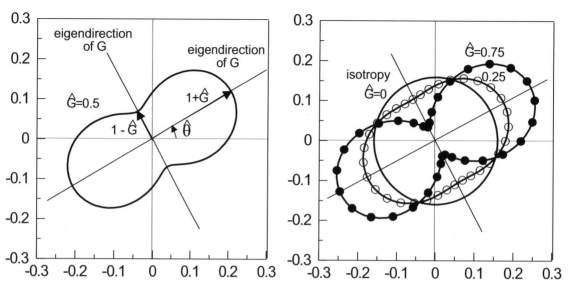

Fig. 12.5.1: 2D representation of the directional distribution density $D(\mathbf{m})$, Eqn (12.5.2), for $\hat{\theta} = \pi/6$, $\hat{G} = 0$, 0.25, 0.5, 0.75. The isotropic reference corresponds to $\hat{G} = 0$.

such as the directional distribution of contact forces in granular materials, Oda et al. [1982], or of damage in the form of continuously distributed voids or micro-cracks, Valanis [1990], or the average distance between two interfaces of a constituent in a composite material such as bone, Harrigan and Mann [1984].

In Section 5.2.11, an anisotropic elastic model based on such a fabric tensor is shown to be a special case of orthotropic elasticity. The eigendirections of the fabric tensor fix the orthotropy directions of the elastic properties and its eigenvalues are related to the directional heterogeneity of the elastic moduli.

In a stereological analysis of a scalar quantity $\mathcal{D}(\mathbf{m})$, directionally dependent but symmetric with respect to the origin, $\mathcal{D}(-\mathbf{m}) = \mathcal{D}(\mathbf{m})$, the fabric tensor defines the spherical harmonics of order 2,

$$\mathcal{D}(\mathbf{m}) = \mathcal{D}_0 + \mathbf{m} \cdot \mathbf{G} \cdot \mathbf{m}, \tag{12.5.1}$$

where $\mathcal{D}_0$ is the unweighted average of $\mathcal{D}$. The fabric tensor is required to be a deviator (or traceless tensor) so as to ensure the independence, or orthogonality in a certain sense, of the terms in the series.

In a plane, the directional distribution (12.5.1) may be recast in the format,

$$D(\mathbf{m}) = \frac{1}{2\pi}[1 + \hat{G}\cos(2(\theta - \hat{\theta}))] \tag{12.5.2}$$

If the distribution is isotropic, then $\hat{G} = 0$. Since the second order tensor $\mathbf{G}$ is traceless and symmetric, its eigenvalues are defined by a single scalar, say $\hat{G}$, with $-1 \leq \hat{G} \leq 1$, and its eigenvectors by a single angle, say $\hat{\theta}$, Fig. 12.5.1. The extrema of $2\pi\mathcal{D}(\mathbf{m})$, namely $1 \pm \hat{G} \in [0, 2]$, are reached along the orthogonal eigendirections of $\mathbf{G}$.

How does this type of anisotropy affect the physical properties of materials? As an illustration, Section 5.2.11 explores the implications for elastic properties. Section 12.5.2 below introduces finer descriptions that involve higher harmonics of even order, and accompanying traceless tensors of like orders.

Data of various physical origins help capture the notion of fabric tensors:

- in order to characterize the evolving anisotropy in granular materials, Bathurst and Rothenburg [1990] sample the directional distribution of contact normals at three stages, namely in the initial isotropic assembly of elliptical particles, after biaxial compression and after shearing;
- Nakayama et al. [1996] consider plates of thickness e which are perforated along a regular rectangular grid by cylindrical voids of diameter d. The plate can be covered by the repetition along the axes of the unit cell shown in Fig. 12.5.2. Let $\mathbf{e}_i$, $i \in [1, 2]$, be the axes of the perforation, $\mathbf{e}_3$ be the normal to the plate, and Ω_i be the surface ratio along a plane of normal $\mathbf{e}_i$, $i \in [1, 3]$,

$$\Omega_1 = \frac{de}{2L_2 e} = \frac{1}{2}\frac{d}{L_2}, \quad \Omega_2 = \frac{de}{2L_1 e} = \frac{1}{2}\frac{d}{L_1}, \quad \Omega_3 = \frac{\pi(d/2)^2 e}{(2L_1)(2L_2)e} = \frac{\pi}{16}\frac{d^2}{L_1 L_2}. \tag{12.5.3}$$

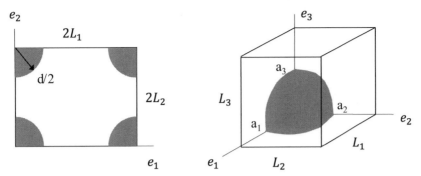

Fig. 12.5.2: Illustration of the notion of fabric tensors on a representative elementary volume ("unit cell") of perforated plates and porous solids.

A fabric tensor $\mathbf{B}$ may be defined from the 'damage' tensor $\boldsymbol{\Omega} = \sum_{i=1}^{3} \Omega_i\,\mathbf{e}_i \otimes \mathbf{e}_i$ as follows:

$$\mathbf{B} = g\,\mathbf{I} + \operatorname{dev}\boldsymbol{\Omega}, \quad g^2 = 1 - \frac{1}{3}\operatorname{dev}\boldsymbol{\Omega} : \operatorname{dev}\boldsymbol{\Omega}, \tag{12.5.4}$$

with the normalization $\operatorname{tr}\mathbf{B}^2 = 3$.

- For a three-dimensional solid with elliptical voids along three orthogonal directions, the eigenvalues of the damage tensor read, Nagaki et al. [1993],

$$\Omega_1 = \frac{\pi}{4}\frac{a_2\,a_3}{L_2\,L_3}, \quad \Omega_2 = \frac{\pi}{4}\frac{a_1\,a_3}{L_1\,L_3}, \quad \Omega_3 = \frac{\pi}{4}\frac{a_1\,a_2}{L_1\,L_2}. \tag{12.5.5}$$

12.5.2 Higher order fabric tensors

Consider $\pm\mathbf{m}^{(1)}$, $\pm\mathbf{m}^{(2)}$, $\cdots$ $\pm\mathbf{m}^{(N)}$ to be N observed data in the form of unit vectors. In order to characterize these raw data, various moments and directional distribution functions may be defined along the terminology of the seminal work of Kanatani [1984] of which the following section provides a brief summary. These representations are invariant with respect to coordinate transformations and, due to the assumed symmetry with respect to the origin, they involve only even tensors. Although the directional dependence of a typical distribution function f is denoted by $f(\mathbf{m})$, it should be understood as $f(\mathbf{m} \otimes \mathbf{m})$ in view of the symmetry with respect to the origin.

Three kinds of fabric tensors of arbitrary ranks are considered. While the actual data $\mathcal{M}$ collected in the tensors of first kind are irregular, the fabric tensors of second kind and third kind propose smoother directional distributions. The former tensors are based on a single rank. At variance, the tensors of the third kind involve several ranks and they are especially tailored to further refinements of higher ranks. Much like in Fourier series, estimating the coefficients involved in these approximations in terms of the actual data $\mathcal{M}$ requires an inversion of the formulas.

12.5.2.1 Fabric tensors of the first kind

The moment tensor, or fabric tensor, of the first kind and even rank n,

$$\mathcal{M}_{i_1 i_2 \cdots i_n} = \frac{1}{N}\sum_{\alpha=1}^{N} m_{i_1}^{(\alpha)}\, m_{i_2}^{(\alpha)} \cdots m_{i_n}^{(\alpha)} = \int_{\mathbb{S}^2} m_{i_1}\, m_{i_2} \cdots m_{i_n}\,\mathsf{d}(\mathbf{m})\,dS, \tag{12.5.6}$$

is symmetric under exchange of any two indices. The trace of $\mathcal{M}_{i_1 i_2 \cdots i_n}$ obtained by contracting indices pairwise is equal to 1 so that the actual distribution function $\mathsf{d}(\mathbf{m})$ is normalized[12.4],

$$\int_{\mathbb{S}^2} \mathsf{d}(\mathbf{m})\,dS = 1. \tag{12.5.7}$$

[12.4]The sum (12.5.6) extends over the N data α of the sample. The task of performing the integration over the unit sphere of the right-hand side of (12.5.6) is considered in Section 12.1.2.

This sample mean is of course distinct from the *unweighted* average obtained by integration over the unit sphere,

$$\frac{1}{4\pi} \int_{\mathbb{S}^2} m_{i_1} m_{i_2} \cdots m_{i_n} \, dS \,. \tag{12.5.8}$$

In place of the algebraic sum (12.5.6), one may think to express the directional information via a distribution function, the prototype being the Dirac delta function,

$$\mathsf{d}(\mathbf{m}) = \frac{1}{N} \sum_{\alpha=1}^{N} \delta(\mathbf{m} - \mathbf{m}^{(\alpha)}), \quad \int_{\mathbb{S}^2} \mathsf{d}(\mathbf{m}) \, dS = 1, \quad \int_{\mathbb{S}^2} m_{i_1} m_{i_2} \cdots m_{i_n} \, \mathsf{d}(\mathbf{m}) \, dS = \mathcal{M}_{i_1 i_2 \cdots i_n} \,. \tag{12.5.9}$$

Note the expression of the Dirac distribution in spherical coordinates, $\delta(\mathbf{m} - \mathbf{m}^{(\alpha)}) = \delta(\theta - \theta^{(\alpha)}) \, \delta(\phi - \phi^{(\alpha)}) / \sin \theta^{(\alpha)}$.

12.5.2.2   Fabric tensors of the second kind

Instead of the actual distribution $\mathsf{d}(\mathbf{m})$ which is very irregular, alternative descriptions of the observed data derive by smoothing. As an example, mimicking a truncated Fourier expansion, let

$$\mathcal{D}(\mathbf{m}) = \frac{1}{4\pi} \left(E_0 + E_{i_1 i_2} m_{i_1} m_{i_2} + \cdots + E_{i_1 i_2 \cdots i_n} m_{i_1} m_{i_2} \cdots m_{i_n} \right). \tag{12.5.10}$$

The approximation $\mathcal{D}(\mathbf{m})$ to $\mathsf{d}(\mathbf{m})$ is defined in the sense of the method of least squares, that is, the coefficients of $\mathcal{D}(\mathbf{m})$ are chosen so as to minimize the integral,

$$\int_{\mathbb{S}^2} \left(\mathcal{D}(\mathbf{m}) - \mathsf{d}(\mathbf{m}) \right)^2 dS \,. \tag{12.5.11}$$

Clearly, the coefficients $E_{\cdots}$ are not unique since $m_{i_1} m_{i_2} I_{i_1 i_2} = 1$, etc. In fact, 1, $m_{i_1} m_{i_2}$, $\cdots$ and $m_{i_1} m_{i_2} \cdots m_{i_n}$ are not independent. To obtain an approximation of the nth order, it is sufficient to consider the highest order term only,

$$\mathcal{D}(\mathbf{m}) = \frac{1}{4\pi} F_{i_1 i_2 \cdots i_n} m_{i_1} m_{i_2} \cdots m_{i_n} \,. \tag{12.5.12}$$

The coefficient $F_{i_1 i_2 \cdots i_n}$ of the fabric tensor of the second kind of rank n is defined via the minimization (12.5.11) which yields, with help of (12.5.9),

$$F_{j_1 j_2 \cdots j_n} \int_{\mathbb{S}^2} m_{j_1} m_{j_2} \cdots m_{j_n} m_{i_1} m_{i_2} \cdots m_{i_n} \frac{dS}{4\pi} = \mathcal{M}_{i_1 i_2 \cdots i_n} \,, \tag{12.5.13}$$

with $\mathcal{M}_{i_1 i_2 \cdots i_n}$ defined by (12.5.6). With help of (12.2.14), the above relation can be inverted to yield $F_{j_1 j_2 \cdots j_n}$, e.g., in the simplest case $n = 2$, $F_{i_1 i_2} = (15/2) \left(\mathcal{M}_{i_1 i_2} - I_{i_1 i_2}/5 \right)$. A general inversion formula has been derived by Kanatani [1984],

$$F_{i_1 i_2 \cdots i_n} = \frac{2n+1}{2^n} \binom{2n}{n} \left(\mathcal{M}_{i_1 i_2 \cdots i_n} + a_{n-2}^n I_{(i_1 i_2} \mathcal{M}_{i_3 \cdots i_n)} + \cdots + a_0^n I_{(i_1 i_2} I_{i_3 i_4} \cdots I_{i_{n-1} i_n)} \right), \tag{12.5.14}$$

the parentheses meaning 'totally symmetric part', see, e.g., (12.2.15), and $\binom{p}{q} = p!/(q!(p-q)!)$ denoting the binomial coefficient, and the coefficients $a_{\cdot}^{\cdot}$ being defined by (12.5.27) below.

Tensors of the second kind are compact, but they are strictly tied to the rank n, and they provide no clue to the higher approximation of rank $n+2$. Ideally, a formulation should be both compact and lend itself to easy higher order approximations. For that purpose, the expansion (12.5.10) is reconsidered by constraining the successive terms to be independent. Then each term can be defined independently.

12.5.2.3 Fabric tensors of the third kind

If the terms of the polynomial expansion (12.5.10) are independent, it is easy to define successive approximations of increasing order. By independent, we mean orthogonal in the sense of the unweighted integral (12.5.8). For example, independence of the terms of order $n-2$ and order n holds if

$$G_{j_1 j_2 \cdots j_n} \int_{\mathbb{S}^2} m_{j_1} m_{j_2} \cdots m_{j_n} m_{i_1} m_{i_2} \cdots m_{i_{n-2}} \frac{dS}{4\pi} = 0. \tag{12.5.15}$$

The condition implies that the symmetric tensor coefficients $G_\cdots$ are *deviators*, *i.e., any contraction (and summation over mute indices) over two indices makes the coefficients to vanish.* Indeed, setting $i_1 = i_2, \cdots, i_{n-3} = i_{n-2}$ and summation over the repeated indices yields $G_{j_1 j_2 \cdots j_n} \int_{\mathbb{S}^2} m_{j_1} m_{j_2} \cdots m_{j_n} \, dS = 0$. With help of (12.2.14) which provides an explicit expression of the latter integral, and in view of the symmetry of $G_{j_1 j_2 \cdots j_n}$, the constraint becomes $G_{j_1 j_1 \cdots j_{n-1} j_{n-1}} = 0$. Contraction of $i_3 = i_4, \cdots, i_{n-3} = i_{n-2}$ in (12.5.15) yields $G_{j_1 j_2 \cdots j_n} \int_{\mathbb{S}^2} m_{j_1} m_{j_2} \cdots m_{j_n} m_{i_1} m_{i_2} \, dS = 0$, that, with help of (12.2.14), reduces to $G_{i_1 i_2 i_3 i_3 \cdots j_{n-1} j_{n-1}} = 0$. Again contraction of $i_5 = i_6, \cdots, i_{n-3} = i_{n-2}$ in (12.5.15) yields $G_{j_1 j_2 \cdots j_n} \int_{\mathbb{S}^2} m_{j_1} m_{j_2} \cdots m_{j_n} m_{i_1} m_{i_2} m_{i_3} m_{i_4} \, dS = 0$, that, with help of (12.2.14), reduces to $G_{i_1 i_2 i_3 i_4 i_5 i_5 \cdots j_{n-1} j_{n-1}} = 0$, and so on until

$$G_{i_1 i_2 i_3 i_4 \cdots j_{n-4} j_{n-3} j_{n-1} j_{n-1}} = 0, \tag{12.5.16}$$

the position of the indices being arbitrary in view of the symmetry of the tensor coefficients $G_\cdots$. $\square$. The deviator tensor coefficient $G_{j_1 j_2 \cdots j_n}$ has $2n+1$ independent components. The component form of the deviator part $\mathbf{G} = \operatorname{dev} \mathbf{E}$ of a symmetric tensor $\mathbf{E}$ is denoted by the index symbol $\{ \}$, Kanatani [1984],

$$G_{\{i_1 i_2 \cdots i_n\}} = c_0^n E_{i_1 i_2 \cdots i_n} + c_2^n I_{(i_1 i_2} E_{i_3 \cdots i_n) j_1 j_1} + \cdots + c_n^n I_{(i_1 i_2} I_{i_3 i_4} \cdots I_{i_{n-1} i_n)} E_{j_1 j_1 \cdots j_{n-1} j_{n-1}}, \tag{12.5.17}$$

the coefficients $c_0^n = 1$ and c_m^n, $m > 0$, being defined so that the above expression (12.5.17) actually yields a deviator, namely

$$\begin{pmatrix} 2n-1 \\ m \end{pmatrix} c_m^n = (-1)^{m/2} \begin{pmatrix} n \\ m \end{pmatrix} \begin{pmatrix} n-1 \\ m/2 \end{pmatrix}. \tag{12.5.18}$$

For example, for $n = 2$ and $n = 4$,

$$\operatorname{dev} \boldsymbol{\mathcal{M}} = \boldsymbol{\mathcal{M}} - \frac{1}{3} \mathbf{I},$$
$$\operatorname{dev} \boldsymbol{\mathcal{M}} \otimes \boldsymbol{\mathcal{M}} = \boldsymbol{\mathcal{M}} \otimes \boldsymbol{\mathcal{M}} - \frac{1}{7}(\mathbf{I} \otimes \boldsymbol{\mathcal{M}} + \boldsymbol{\mathcal{M}} \otimes \mathbf{I} + 2\mathbf{I} \overline{\otimes} \boldsymbol{\mathcal{M}} + 2\boldsymbol{\mathcal{M}} \overline{\otimes} \mathbf{I}) + \frac{1}{35}(\mathbf{I} \otimes \mathbf{I} + 2\mathbf{I} \overline{\otimes} \mathbf{I}), \tag{12.5.19}$$

accounting for the definition (12.5.6), which implies in particular $\mathcal{M}_{i_3 i_4 j_1 j_1} = \mathcal{M}_{i_3 i_4}$, $\mathcal{M}_{j_1 j_1 j_3 j_3} = 1$. The tensor product $\overline{\otimes}$ is defined componentwise as $(\mathbf{A} \overline{\otimes} \mathbf{B})_{i_1 i_2 i_3 i_4} = (A_{i_1 i_3} B_{i_2 i_4} + A_{i_1 i_4} B_{i_2 i_3})/2$.

When the $G_\cdots$ are deviators, the two approximations below are equivalent,

$$\mathcal{D}(\mathbf{m}) = \frac{G^{(0)}}{4\pi} \begin{cases} 1 + G_{i_1 i_2} m_{i_1} m_{i_2} + \cdots + G_{i_1 i_2 \cdots i_n} m_{i_1} m_{i_2} \cdots m_{i_n}, \\ 1 + G_{i_1 i_2} m_{\{i_1} m_{i_2\}} + \cdots + G_{i_1 i_2 \cdots i_n} m_{\{i_1} m_{i_2} \cdots m_{i_n\}}. \end{cases} \tag{12.5.20}$$

In view of the orthogonality of the terms in (12.5.20)$_1$, the coefficients are defined by (12.5.13). Inversion of this relation with help of (12.2.14) yields now, instead of (12.5.14), Kanatani [1984],

$$G_{i_1 i_2 \cdots i_n} = \frac{2n+1}{2^n} \begin{pmatrix} 2n \\ n \end{pmatrix} \mathcal{M}_{\{i_1 i_2 \cdots i_n\}}, \qquad \mathcal{M}_{i_1 i_2 \cdots i_n} = \frac{1}{G^{(0)}} \int_{\mathbb{S}^2} m_{i_1} m_{i_2} \cdots m_{i_n} \mathcal{D}(\mathbf{m}) \, dS. \tag{12.5.21}$$

Then, using (12.5.17) and (12.5.21), the three first deviators $G_\cdots$ associated with the tensor coefficients $\mathcal{M}_\cdots$ may be listed, namely for $n = 0$,

$$G^{(0)} = \int_{\mathbb{S}^2} \mathcal{D}(\mathbf{m}) \, dS = 1, \tag{12.5.22}$$

for $n = 2$,

$$G^{(2)}_{i_1 i_2} = \frac{15}{2}\left(\mathcal{M}_{i_1 i_2} - \frac{1}{3} I_{i_1 i_2}\right),$$
$$\mathbf{G}^{(2)} = \frac{15}{2}\left(\boldsymbol{\mathcal{M}} - \frac{1}{3}\mathbf{I}\right), \tag{12.5.23}$$
$$\mathbf{G}^{(2)} = \frac{15}{2}\int_{\mathbb{S}^2} \operatorname{dev}(\mathbf{m} \otimes \mathbf{m}) \, \mathcal{D}(\mathbf{m}) \, dS,$$

and, for $n = 4$,

$$
\begin{aligned}
G^{(4)}_{i_1 i_2 i_3 i_4} &= \frac{315}{8}\Big(\mathcal{M}_{i_1 i_2 i_3 i_4} - \frac{6}{7}\, I_{(i_1 i_2}\,\mathcal{M}_{i_3 i_4)} + \frac{3}{35}\, I_{(i_1 i_2}\, I_{i_3 i_4)}\Big) \\[2mm]
&= \frac{315}{8}\Big(\mathcal{M}_{i_1 i_2 i_3 i_4} \\[1mm]
&\quad -\frac{1}{7}\,\big(I_{i_1 i_2}\,\mathcal{M}_{i_3 i_4} + I_{i_1 i_3}\,\mathcal{M}_{i_2 i_4} + I_{i_1 i_4}\,\mathcal{M}_{i_2 i_4} + \mathcal{M}_{i_1 i_2}\, I_{i_3 i_4} + \mathcal{M}_{i_1 i_3}\, I_{i_2 i_4} + \mathcal{M}_{i_1 i_4}\, I_{i_2 i_4}\big) \\[1mm]
&\quad +\frac{1}{35}\,\big(I_{i_1 i_2}\, I_{i_3 i_4} + I_{i_1 i_3}\, I_{i_2 i_4} + I_{i_1 i_4}\, I_{i_2 i_3}\big)\Big),
\end{aligned}
\tag{12.5.24}
$$

$$
\mathbf{G}^{(4)} = \frac{315}{8}\Big(\boldsymbol{\mathcal{M}} \otimes \boldsymbol{\mathcal{M}} - \frac{1}{7}\big(\mathbf{I}\otimes\boldsymbol{\mathcal{M}} + \boldsymbol{\mathcal{M}}\otimes\mathbf{I} + 2\,\mathbf{I}\,\overline{\otimes}\,\boldsymbol{\mathcal{M}} + 2\,\boldsymbol{\mathcal{M}}\,\overline{\otimes}\,\mathbf{I}\big) + \frac{1}{35}\big(\mathbf{I}\otimes\mathbf{I} + 2\,\mathbf{I}\,\overline{\otimes}\,\mathbf{I}\big)\Big),
$$

$$
\mathbf{G}^{(4)} = \frac{315}{8}\int_{\mathbb{S}^2} \mathrm{dev}\,(\mathbf{m}\otimes\mathbf{m}\otimes\mathbf{m}\otimes\mathbf{m})\,\mathcal{D}(\mathbf{m})\,dS.
$$

The distribution function with two spherical harmonics may then be cast in tensor format,

$$
\mathcal{D}(\mathbf{m}) = \frac{G^{(0)}}{4\,\pi}\Big(1 + \mathbf{G}^{(2)} : (\mathbf{m}\otimes\mathbf{m}) + (\mathbf{m}\otimes\mathbf{m}) : \mathbf{G}^{(4)} : (\mathbf{m}\otimes\mathbf{m})\Big).
\tag{12.5.25}
$$

Once the fabric tensor $G_{...}$ of the third kind of rank n is known, the fabric tensor $F_{...}$ of the second kind of identical rank follows,

$$
F_{i_1 i_2 \cdots i_n} = G_{i_1 i_2 \cdots i_n} + I_{(i_1 i_2}\, G_{i_3 \cdots i_n)} + \cdots + I_{(i_1 i_2}\, I_{i_3 i_4} \cdots I_{i_{n-1} i_n)}.
\tag{12.5.26}
$$

Comparison of (12.5.14) and (12.5.21) provides the coefficients a,

$$
\frac{2\,n+1}{2^n}\binom{2\,n}{n}\, a^n_m = \sum_{k=m\ \mathrm{even}}^{n} \frac{2\,k+1}{2^k}\binom{2\,k}{k}\, c^k_{k-m}.
\tag{12.5.27}
$$

12.6   Spatial homogenization over cells and ECM

In order to derive mechanical properties, the solid skeleton which occupies only a volume fraction of the overall tissue is often considered itself as a two-phase material, Weng [1984], Jain et al. [1988]. One phase is the matrix phase (extracellular matrix ECM consisting mainly of collagen fibers and other macromolecules), the other phase is the inclusion phase (biological cell). Typically, mathematical cells composed of the inclusion and surrounding ECM are given various geometrical forms, from cubes to 14-sided polyhedron, that repeat by periodicity over the three-dimensional space. El-Kareh et al. [1993] use cubic cells to estimate the diffusivity of solutes in tissues while Secomb and El-Kareh [2001] use 14-sided polyhedrons to estimate elastic properties. While the cells are formed by isotropic inclusions and matrix, the resulting material displays cubic symmetry. When the above cells are randomly oriented, it is a simple matter to obtain the Voigt and Reuss bounds of the isotropic elastic moduli and transport coefficients, and an additional effort provides the Hashin-Shtrikman bounds.

Secomb and El-Kareh [2001] consider cells of very soft tissues to be incompressible because the hydraulic permeability of their membrane is low, say about $1.35 \times 10^{-13}\,\mathrm{m/Pa/s}$. On the other hand, their shear modulus is about a tenth of the shear modulus of the extracellular matrix, typically 100 Pa for the cells of glands and brain matter. Thus the overall bulk modulus of these tissues is controlled by the cell and cell volume fraction and the shear modulus by the ECM and ECM volume fraction. Assuming a bulk modulus of the cell about $4.4 \times 10^3\,\mathrm{Pa}$, a specific surface area of the cell membrane about $10^{-6}\,\mathrm{m}$ (assuming a cell size of $10\,\mu\mathrm{m}$), the characteristic time of diffusion across the cell membrane is as high as 28 min.

Chapter 13

Electro-chemomechanical couplings

13.1 Chemomechanical couplings in engineering and biology

Chemomechanical couplings occur in many engineering and biological fields. A typical instance is swelling that has relevance in active clays and polymers inter alia. *Gels* are also highly susceptible of swelling. Gels are aqueous environments that contain soluble components, e.g., metallic ions, and insoluble structural molecules. Articular cartilages and corneal stroma are instances of *extracellular* biological gels in which the structural molecules are located outside the cells, forming the extracellular matrix. By contrast, the muscle is an *intracellular* gel in which the myosin filament has a diameter comparable to that of collagen in articular cartilage.

Swelling in non-ionic polymers is traditionally traced to the volume exclusion concept according to which, in a very dilute solution in a good solvent, each molecule tends to exclude all others from the volume it occupies, Flory [1953]. In ionic polymers, electrical repulsion between fixed charges provides another source of swelling. The presence of mobile ions, required by electroneutrality, shields these repulsive forces but is not sufficient to abolish their effects, unless at very high concentrations, a configuration coined *hypertonic state*.

Electro-chemomechanical couplings are also exploited by engineers. For example, electromigration is used to remediate soils polluted by heavy metals and electroosmosis to consolidate soils.

In biological tissues, couplings arise as processes to maintain a certain equilibrium: swelling of articular cartilages opposes the water depletion that would occur due to mechanical compression. In some instances, they are driving processes: the cardiac function is a typical process of electro-chemomechanical coupling. A common feature to most, if not all, these examples is the presence of *fixed electrical charges* that play a key role in promoting the couplings.

Reverse couplings, e.g., mechanoelectric feedback, are less studied. They appear nevertheless to be at work in many instances of the development of biological organs. Indeed, while *in vitro* cultures of engineered soft tissues are able to induce cell proliferation, the mechanical properties of the tissues are not satisfactory. *Mechanobiology* aims at defining mechanical loading programs to be applied to the cultures so as to improve these properties.

13.2 Molar volumes, electrostriction

13.2.1 Molar masses and molar volumes

The molar masses of a few chemicals are recorded in Table 13.2.1. They can be contrasted with the large molar masses of macromolecules of biological interest recorded in Table 13.2.2.

Table 13.2.1: Molar masses (gm) and apparent molar volumes (cm^3) of some ions at low ionic strength.

Species	H_2O	Na^+	Ca^{2+}	K^+	H^+	OH^-	Cl^-
Molar mass	18	23	40.1	39.1	1	17	35.5
Molar volume	18	2.37	1.7	12.8	4.1	-7.8	15.4

Table 13.2.2: Molar masses (kg) of some macromolecules of biological interest.

Species	PGs	Collagen	Albumin	Dextran	IGF-1	Insulin
Molar mass	2000	285	66	10-70	8	5

The apparent molar volumes of dissolved species are not constant, even under isothermal conditions. The apparent molar volumes of electrically neutral species, like NaCl and KCl, depend on their own molalities, Lobo [1990]. In a solution, due to *electrostriction*, a charged ion is attracted so strongly by counterions that, at low concentration, the ionic bond implies a reduction in volume with respect to the two independent species. The apparent molar volume of some ions can result to be negative. As the concentration increases, the phenomenon is counteracted by the repulsion between ions whose electrical charges are of the same sign.

As a rough approximation, the molar volumes in Table 13.2.1 are equal to the values corresponding to quasi-distilled water, which can be considered meaningful at low ionic concentrations (up to 0.1 M).

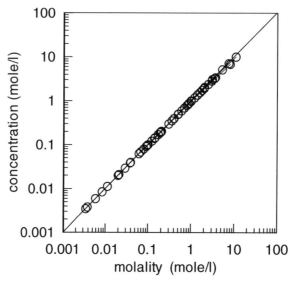

Fig. 13.2.1: Molality and concentration as defined by Eqn (13.2.1) differ only at large concentrations of sodium chloride. The molar volumes are shown in Fig. 13.2.2.

13.2.2   The molar volume of dissolved salts

An electrolyte contains a volume of water $V_{\rm w}$=1 liter=1000 cm^3 in which N moles of salt are dissolved in dissociated form. This amount of salt is measured through its molality m [unit : mole/liter], or concentration c [unit : mole/liter],

$$\text{molality } m = \frac{\text{nb. of moles } N}{\text{volume of water } V_{\rm w}}; \quad \text{concentration } c = \frac{\text{nb. of moles } N}{\text{volume of solution } V}. \tag{13.2.1}$$

Fig. 13.2.1 shows that molality and concentration differ only at large concentrations.

Let $\hat{v}$ be the molar volume of the solute. Since

$$V = V_{\rm w} + N\,\hat{v}, \tag{13.2.2}$$

then molality and concentration can be expressed in terms of each other,

$$c = \frac{m}{1 + m\,\hat{v}}, \quad m = \frac{c}{1 - c\,\hat{v}}. \tag{13.2.3}$$

Based on data in Lobo [1990], the analytical expressions below provide approximations to the molar volumes [unit : cm^3/mole], of three compounds, namely sodium chloride NaCl, hydrochloric acid HCl and sodium hydroxide NaOH, as a function of their molality [unit : mole/liter]:

$$\begin{aligned}
\hat{v}_{\rm NaCl} &= 17.04 + 1.58\,m_{\rm NaCl} - 0.192\,m_{\rm NaCl}^2; \\
\hat{v}_{\rm HCl} &= 18.18 + 0.55\,m_{\rm HCl} - 0.035\,m_{\rm HCl}^2; \\
\hat{v}_{\rm NaOH} &= -4.61 + 1.98\,m_{\rm NaOH} - 0.085\,m_{\rm NaOH}^2.
\end{aligned} \tag{13.2.4}$$

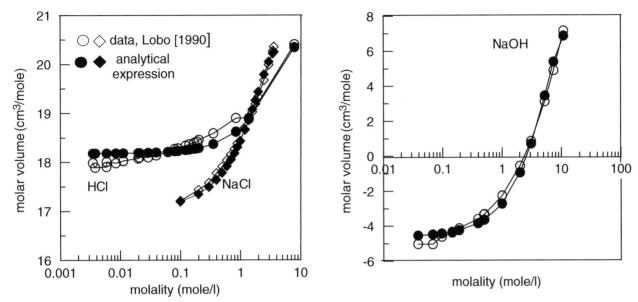

Fig. 13.2.2: Molar volumes of salts as a function of molality. Data at 25°C (open symbols) by Lobo [1990], p. 532 for HCl, p. 1640 for NaCl and p. 1807 for NaOH, and approximations (solid symbols) by Eqn (13.2.4).

The above expressions displayed in Fig. 13.2.2 hold for molalities below the solubility limit, that is, they assume the salt to be completely dissolved.

13.2.3 The molar volume of dissolved ions

The molar volume of the chloride anion $\hat{v}_{Cl}$ may be considered to be fairly constant, namely $\hat{v}_{Cl} \simeq 15.4 \, cm^3$. As a first approximation, it can be assumed to be known. Then

$$
\begin{aligned}
\hat{v}_{Na} &= \hat{v}_{NaCl}(m_{NaCl}) - \hat{v}_{Cl}\,; \\
\hat{v}_{H} &= \hat{v}_{HCl}(m_{HCl}) - \hat{v}_{Cl}\,; \\
\hat{v}_{OH} &= \hat{v}_{NaOH}(m_{NaOH}) - \hat{v}_{NaCl}(m_{NaCl}) + \hat{v}_{Cl}\,.
\end{aligned}
\tag{13.2.5}
$$

The ionic molar volumes, based on the approximations (13.2.4), can be observed to have significative variations with ionic strength, Fig. 13.2.3. The range of the molar volume of the hydroxyl OH^- ion is especially large. The increase of the molar volume with the concentration of sodium chloride seems to indicate that attraction between hydroxyl anion and sodium cation is more important than the repulsion between hydroxyl anion and chloride anion.

13.3 Electrokinetic processes

Electrokinetics has been used in various directions that take advantage, enhance or decrease some properties of media with fixed charges. The modeling of electrokinetic remediation processes has so far mainly assumed a rigid solid skeleton. Advection of contaminants by water flow is accounted for, but models differ in the way they include ionic mobility of the various species present in the pore fluid and in the number of electrochemical reactions included. A terminology has emerged that highlights the key physical properties that are triggered. *Electroosmosis* denotes the process of transport of water into low permeability media in order to remove soluble electrically neutral species, Fig. 13.3.1. *Electromigration* on the other hand addresses the removal of ionic species. Note that higher concentrations enroll more electromigration than electroosmosis.

In order to improve the efficiency of remediation processes, enhancement techniques are tested, like
- addition of acid at cathode to desorb cations,

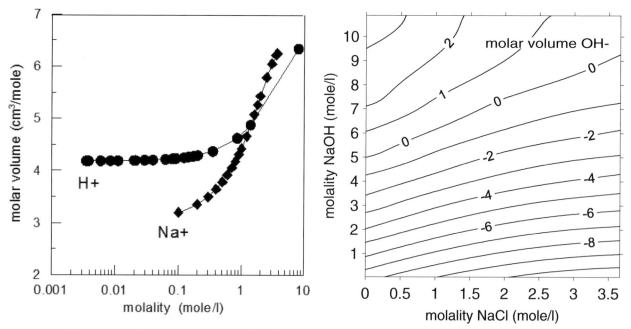

Fig. 13.2.3: Molar volumes of ions (cm^3) as a function of molality as described by Eqn (13.2.5).

- addition of EDTA[13.1] that enhances complexation of desorbed cations, Yeung et al. [1996], Alshawabkeh et al. [1999], and

- use of semi-permeable membranes close to cathode and optimization of the position of electrodes, Alshawabkeh et al. [1999].

Indeed, there are two keys to successful electrokinetic remediation of metal-contaminated media: first to extract the cations from solid phase, second to keep the desorbed cations in fluid phase so as to move them to electrodes either by electromigration to cathode or by electroosmosis when complexated. Of course, addition of acid is of theoretical help in both steps, but the quantity required increases prohibitively for media with high buffer capacity[13.2]. Moreover acids are hazardous substances. That is one of the reasons of the use of a strong anionic complexing agent by Yeung et al. [1996].

Alternative substances, like polyphosphates, are used to increase the negative charge for specific purpose, e.g., enhance the adsorption of heavy metals.

13.3.1   Laboratory setups

Electrodes are endowed with reservoirs that allow to circulate an electrolyte of controlled chemical composition. This separation between electrodes and medium avoids heating and desiccation, desaturation and cracking. Reservoirs are separated from the medium by a filter permeable to water and ions. In fact, electroosmosis depletes water from the anode region and moves it towards the cathode, Fig. 13.3.1: control of the electrolyte circulating in the reservoir allows to provide water at the anode and dispose of water at the cathode. Additional passive electrodes positioned along the sample allow measurements.

A typical electrokinetic experiment proceeds as follows. The initial state is usually homogeneous and defined by a given concentration of cations, anions and pH. A difference of electrical potential is applied between the electrodes. The pore pressures at the electrodes are maintained at a reference value, say zero, assuming a one-dimensional

---

[13.1]EDTA = ethylene-diamine-tetra-acetate.

[13.2]The pKa of an acid is the value of pH at which it is half dissociated, the concentrations of the acid and its conjugate base being equal: according to the Henderson-Hasselblach equation, pH=pKa+log([acid]/[base]). The titration curve displays an inflection at the pKa. The *buffer capacity* of a system is defined as the amount of strong acid that causes this system to have a pH increase of one unit. A substance is an efficient buffer in a range of pH if it has its pKa in this zone. Indeed, then the amount of substance necessary to change the pH is large. For example, the plasma has a pH between 7.35 and 7.45, corresponding to a concentration of hydrogen ions of 40 nM, and it is buffered efficiently by phosphate and carbonate ions which have their pKa's in this range, namely pKa(H_2CO_3)=6.35, pKa($H_2PO_4^-$)=6.82. Incidentally, note that the cytoplasm is somehow more acidic, with a pH between 7 and 7.30, the stomach is very acidic, with a pH about 2, while the small intestine is basic, with a pH about 8.

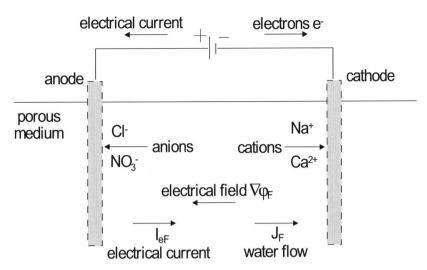

Fig. 13.3.1: Sketch of electrokinetic remediation. By convention, the electrical current in the circuit flows in the direction opposite to electrons. By *electromigration*, cations move towards the cathode and anions move towards the anode with velocities depending on individual *ionic mobilities*. *Electroosmosis* implies water flow towards the cathode, when the acidity is not too high, i.e., overall electrical charge above isoelectric point. For higher acidity, a *reverse* osmotic flow occurs.

analysis. The boundary conditions for the ionic species at the electrodes are of two types: either the concentrations are prescribed initially and, during the experiments, the ionic fluxes are free, or the concentrations are continuously monitored in the reservoirs.

If the electrodes are metallic, oxidation-reduction reactions take place (Me$\equiv$ metal):

$$\text{Me} \rightarrow \text{Me}^{z^+} + z\ e^- \quad \text{oxidation at anode,}$$
$$\text{Me} \leftarrow \text{Me}^{z^+} + z\ e^- \quad \text{reduction at cathode.}$$

Laboratory experimental setups use graphite passive electrodes to avoid spreading of parasite ions into the medium.

13.3.2 Development of an acid front

A significant step forward in the understanding of electrokinetic processes is due to the contribution of Acar et al. [1994]. They show that the most prominent phenomenon is the development of an acid front starting close to the anode and moving towards the cathode, Fig. 13.3.2. An electrolytic decomposition of water in hydrogen and hydroxyl ions takes place at the electrodes[13.3],

$$
\begin{aligned}
2\,\text{H}_2\text{O} - 4\,e^- &\rightarrow \text{O}_2 \uparrow + \ 4\,\text{H}^+ \quad \text{oxidation at anode} \\
4\,\text{H}_2\text{O} + 4\,e^- &\rightarrow 2\,\text{H}_2 \uparrow + 4\,\text{OH}^- \quad \text{reduction at cathode}
\end{aligned}
\tag{13.3.1}
$$

The mobility of these ions is high with respect to that of other ions and, once generated, they move in the medium by saltation from one water molecule to the other, advection of the pore fluid, fickian diffusion and migration due to electrical field. In fact, the hydrogen ion H$^+$ is not stable in aqueous solutions and it would be more appropriate to consider the ion hydronium H$_3$O$^+$, although even this form does not represent the complexity of aqueous solutions, Alberty [2003].

The acid front created by the motion of hydrogen ions H$^+$ towards the cathode and the basic front created by the motion of hydroxyl ions OH$^-$ towards the anode are *not symmetric*:

- first, because the mobility of the former is much larger, and,
- second, because the motion of the H$^+$'s (respectively OH$^-$) is assisted (respectively opposed) by electroosmosis, Fig. 13.3.2.

[13.3]According to Faraday's law of equivalence of mass and charge, if the efficiency of the reaction is one, the flux per unit area of ion i [unit: mole/s/m^2] created at the electrodes is equal to $\mathbf{J}_i = \mathbf{I}_{ew}/(z_i\,F)$, with $i = \text{H}^+$, $z_i = 1$ at anode, $i = \text{OH}^-$, $z_i = -1$ at cathode, where $\mathbf{I}_{ew}$ is the electrical current density [unit: A/m^2] and F Faraday's equivalent charge [unit: C/mole].

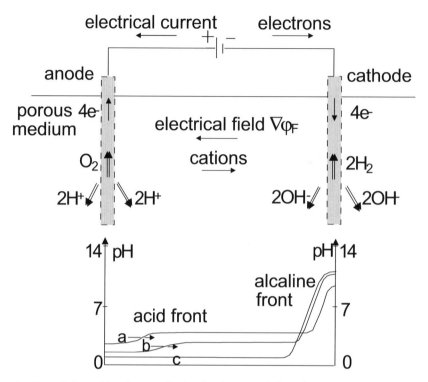

Fig. 13.3.2: Schematization of electrokinetic remediation by Acar et al. [1994]. An acid front due to freed hydrogen ions H^+ is created at the anode and it propagates towards the cathode. A basic front due to freed hydroxyl ions OH^- is also created at the cathode and it propagates towards the anode. However, the motions are not symmetric and the acid front is faster than the basic front: first, because the mobility of hydrogen ions is 1.76 higher than that of hydroxyl ions; second because electroosmosis implies, below the isoelectric point, a water flow towards the cathode, therefore amplifying the motion of cations and opposing that of anions. However, high pH prevalent close to the cathode might jeopardize the efficiency of the process and precipitate metal hydroxides.

Table 13.3.1[13.4] shows the diffusion coefficients and ionic mobilities of common ions in a blank (dilute) solution. The actual diffusion is hindered in porous media in general, and in the extracellular matrix in particular. This hindrance is quantified by a tortuosity factor. The *effective* diffusion coefficients and ionic mobilities are obtained by multiplication of their counterparts in a blank solution by the tortuosity factor.

Table 13.3.1: Ionic diffusion coefficients D [unit: $10^{-10}\,\mathrm{m^2/s}$] and mobilities $u = D\,|\zeta|\mathrm{F}/RT$ [unit: $10^{-9}\,\mathrm{m^2/s/V}$] in water at $25\,^\circ\mathrm{C}$ ($\mathrm{F}/RT = 0.02568\,1/\mathrm{volt}$).

Ion	Na^+	Ca^{2+}	Cl^-	H^+	OH^-	NO_3^-	SO_4^-
D	13.3	7.92	20.3	93.09	52.77	6.33	5.30
u	51.75	61.75	79.0	362.5	205.5	74	82.5

13.3.3　Electroinjection

Electroinjection through electrical fields may be attempted to improve mechanical properties. Various types of *electrokinetic stabilization* may be devised:

　　i) electroosmotic consolidation where an electrical gradient is applied, without mechanical load, in order to extract pore fluid;

　　ii) use of electroosmotic flow to draw stabilizing chemicals into the medium;

[13.4]Voet and Voet [1998], p. 34, give slightly different values.

iii) triggering ionic migration by adding an acid at the cathode, extracting a minimum of fluid in order to minimize advection that opposes motion of anions towards the anode.

Modifications of transport and mechanical properties by pH may, or may not, be simultaneous. Acidification may change the sign of the electroosmotic coefficient and give rise to reverse osmosis.

13.4 Nanoscopic aspects

13.4.1 The fixed electrical charge

Many tissues, like articular cartilage and cornea, and geological materials, like clays, are not electrically neutral due to uncompensated termination sites. Since these sites are usually negatively charged at standard pH, cations desorb from the water phase. A list of common exchangeable cations includes

$$Ca^{2+}, Mg^{2+}, H^+, K^+, NH_4^+, Na^+, \tag{13.4.1}$$

while

$$SO^{2-}, Cl^-, PO_4^{3-}, NH_4^-, \tag{13.4.2}$$

are common exchangeable anions. The ability of a material to exchange cations (respectively anions) is characterized by its *cation exchange capacity* (respectively, *anion retention capacity*).

The terminology adopted to quantify the electrical charges depends on the volume accounted for, Fang [1997], p. 101. In any event, all structural charges, within and on the solid, as well as adsorbed H^+ are included. When the volume is limited by the Stern layer[13.5], the transition pH between a negative and a positive overall charge is termed *point of zero electrical charge*. When, as in the definition of the ζ-potential, the reference volume extends further to the plane of shear between solid and fluid, the charge changes sign together with the ζ-potential at the pH called *isoelectric point*.

Fixed charge, mobile ions, permanent charge, variable charge

To fix the terminology, let us note that the charge of the solid is called *fixed charge*, as opposed to the *mobile charge* of ions. The part of the charge of the solid which is insensitive to pH is referred to as *permanent fixed charge*, while the part which changes with pH is called *variable fixed charge*.

The negatively charged proteoglycans

The proteoglycans are responsible for the swelling behavior of biological tissues as they repel themselves and originate a 'chemical' pressure. The repulsion force is however shielded by the presence of ions: in the hypertonic state (high ionic concentration), the repulsive force is minimum and the structure tends to collapse. The proteoglycans are macromolecules of molar mass 2×10^6 gm, with *effective* concentration (=concentration×valence) of 0.1 to 0.2 M. Their mass is contributed for 85% by glycosaminoglycans (GAGs) and for 15% by proteins. GAGs contain 80% of chondroitin sulfates of charge -2 and 20% of keratan sulfate of charge -1, so that the valence of PGs varies from -6000 to -8000.

13.4.2 Mechanisms due to pH

As they move away from the anode, hydrogen ions H^+ in the electrolyte solution extract cations absorbed in or on the fixed charge. Desorbed into the electrolyte, these cations will be attracted to the cathode where they can be disposed of. The basic property in force is that, at pH 7 or higher, the materials are negatively charged on their surface. So in a superficial view, cations H^+ are much more likely to absorb to the materials than anions OH^-, which are a priori electrically repelled. Said otherwise, acid solutions are able to dissolve metallic cations. pH decrease, i.e. acidification of the tissue, is required to desorb metallic ions and solubilize them in pore fluid, while precipitation of metallic cations occurs as the pH increases.

A more precise description of the effects of hydrogen ions on negatively charged structures follows. At pH 7 or higher, hydroxyl OH^- termination sites are present on their surfaces. As pH increases, the dissociation/deprotonation $Me - OH \rightarrow Me - O^- + H^+$, or $Me - OH^- + OH^- \rightarrow Me - O^{2-} + H_2O$, that turns the charge more negative, is favored (Me is a metal, e.g., Si).

[13.5]The Stern layer consists of tightly bound cations surrounding the GAGs, see Fig. 13.4.1.

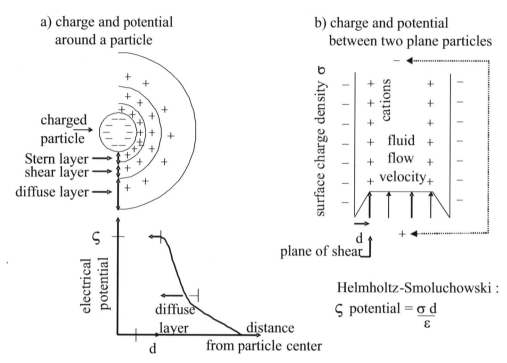

Fig. 13.4.1: The measure of the electrical charge depends on the distance d from the particle surface, σ surface charge density, ϵ permittivity. Some notions of electrostatics are provided in Section 13.6.

On the other hand, as pH decreases, fixation of H^+/protonation, makes the overall charge less negative, and even positive. In fact, fixation of H^+ on these sites, $Me - OH^- + H^+ \rightarrow Me - HOH$, further weakens internal bonds and leads to the release in the solution (dissolution) of multivalent ions, which become exchangeable, and possibly might further precipitate. Small radius multivalent ions, like Al^{3+}, play a prominent role here in reducing the absorbed water content for several reasons. First, they require less volume and, second, the repulsive electrical forces that would be due to three monovalent cations vanish.

The desorption into the solution of a metallic ion with valence n decreases the acidity of the solution, while the absorption increases the acidity:

$$\begin{aligned}\text{desorption}: \text{pH} \nearrow \quad X - \text{tissue} + n\,H^+ &\longrightarrow H_n - \text{tissue} + X^{n+} \\ \text{absorption}: \text{pH} \searrow \quad X - \text{tissue} + n\,H^+ &\longleftarrow H_n - \text{tissue} + X^{n+}\end{aligned} \qquad (13.4.3)$$

Alternatively, protons might only replace a univalent cation which is released into solution, e.g.,

$$K - \text{tissue} + H^+OH^- \longrightarrow H - \text{tissue} + K^+OH^- . \qquad (13.4.4)$$

While complexation of Na^+ and K^+ is known to be small, it dominates the aqueous behavior of Al^{3+}, see Section 13.5.2. Therefore, only a part of the released cations is uncomplexed. Only this part can become exchangeable, that is, can enter the interlayer space of the charged structure.

13.5   Hydration, hydrolysis, complexation and solubility

In water, a cation of charge z exists
  - in hydrated form, a bond with the water molecule being established;
  - in free (uncomplexed) form M^{z+} and
  - in complexed forms with n hydroxyl anions OH^-, namely $(M(OH)_n)^{z+-n^-}$, $n = 1, 2, \cdots, z, \cdots$
Being solid, the complexed form which is neutral plays a special role, e.g., gibbsite $Al(OH)_3$ or lead hydroxide $Pb(OH)_2$. Each complexed form is in equilibrium with its dissociated form.

(Nonhydrated) ionic radii of cations are smaller than their parent atoms, and the converse holds for anions. Moreover, upon hydration the size of some metallic ions may increase significantly, as indicated by Table 13.5.1 and this large size may lead to steric exclusion from certain domains which can be reached through dedicated channels of definite size.

Table 13.5.1: Ranges of non hydrated and hydrated ionic radii [unit : Å=0.1 nm], from several sources including Mitchell [1993], p. 40. As a reference, the size of the water molecule is about 3.6 Å.

Ion	Na^+	K^+	Ca^{2+}	Mg^{2+}	O^{2-}	Cl^-	OH^-	H^+	H_3O^+
Nonhydrated radius	1	1.4-1.6	1-1.3	0.6-0.9	1.4	1.81	1.33	0.8	1.15
Hydrated radius	3.6-7.9	3- 5.33	4.1-9.6	4.3-11.8	–	3	3	9	2.8

13.5.1   Solvation and hydration

The water molecule is highly polar. Two of the six electrons of the outer shell of the oxygen atom participate to create covalent bonds with the two hydrogen atoms: in a covalent bond the atoms share pairs of electrons. Due to the remaining six electrons on the inner and outer shells, the covalent bond O-H has a ionic character as if the end O was negatively charged with a charge $-2\delta = -0.6$ and the ends H positively charged with a charge $\delta = 0.3$ as sketched in Fig. 13.5.1. Altogether, the spatial distribution of electrical charges turns the water molecule into a dipole of dipolar moment equal to 6.2×10^{-30} C×m.

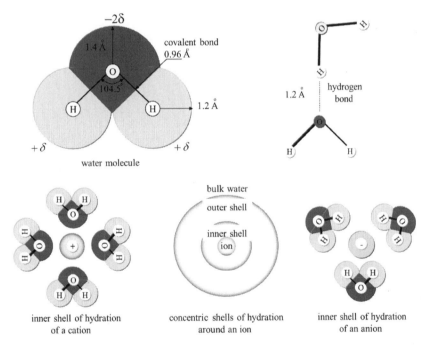

Fig. 13.5.1: The water molecule has polar properties. The hydrogen and oxygen atoms retain a part of the charge which is at the root of the formation of hydrogen bonds and hydration of cations and anions. Here only the inner hydration shells are shown while an outer less ordered hydration shell serves as a transition with the random arrangement of bulk water.

In pure water, the hydrogen ions establish a *hydrogen bond* with oxygen atoms of other water molecules. Hydrogen bonds are much weaker than covalent bonds and, although their lifetime is short (order of a few picoseconds), their numbers may strengthen considerably the stability of water. Chemists have tried to elucidate, through molecular dynamics simulations, the existence of structures, that is, of specific arrangements, or *clusters*, of water molecules. The idea revolves of course on the time scale and volume scale under consideration. The end result is that *pure* water does not seem to show signs of the existence of stable clusters in the sense that hydrogen bonds are continually broken and rebuilt in a few picoseconds.

The dipolar property and the hydrogen bonds are thought to be responsible for the large latent heat of vaporization of water equal to $43.5\,\mathrm{kJ/mole}$, as indicated in Fig. 3.2.2. In absence of hydrogen bonds, the boiling temperature is estimated to about be $-90\,^{\circ}\mathrm{C}$.

Polar substances and ions *hydrate* when immersed in water. Loosely speaking, water is sometimes referred to as the universal solvent. This solvation property is due to the high dielectric constant of water: the electric force between charged species is screened by water and so is their interaction energy. Section 13.6.1 provides some details on this issue. Since the cohesion force is low in water, ionic species attract much less than in other solvents and may remain separated.

In addition, solvated ions interact with a polar solvent like water. In presence of ions, the poles of the solvent are attracted by the opposite charge: ions are thus surrounded by water molecules, a feature termed hydration.

The standard hydration energy is measured as the enthalpy ΔH yielded (in the form of heat) when the ion in gaseous form is immersed in water. The total energy released (when $\Delta H < 0$) however should involve also the change of internal structure of the chemical. The quantity that gives a measure of the order of the structure is the entropy S. Entropy increases as disorder increases. Therefore the variation of entropy from the gaseous phase to the hydrated state where motions are restrained by the ion-dipole bonds is certainly negative, $\Delta S < 0$. The total measure of energy that is released (absorbed) during the process is the variation of free enthalpy ΔG (also called chemical potential) equal, at constant temperature, to $\Delta H - T\,\Delta S$. A process is said to be spontaneous, and exergonic, if $\Delta G < 0$. Despite the fact that $\Delta S < 0$, hydration is exergonic. For a cation M of valence z,

$$\begin{array}{c}\textbf{hydration}\\ \text{spontaneous if } \Delta G < 0\end{array}\qquad \mathrm{M}^{z^{+}} + n\,\mathrm{H_2O} \;\rightleftharpoons\; \mathrm{M(H_2O)}_n^{z^{+}}, \qquad (13.5.1)$$

and ΔG [unit : kJ/mole] is estimated as $-686\,z^2/r' + 5.44 < 0$, r' [unit : Å] being the modified ionic radius.

13.5.2   Hydrolysis and complexation

When the energy provided by immersion is larger, it can be sufficient to delete $n = 1,\,2\cdots z$ bonds O-H and, instead of hydration, *hydrolysis* occurs[13.6]:

$$\textbf{hydrolysis}:\quad \mathrm{M}^{z^{+}} + n\,\mathrm{H_2O} \;\rightleftharpoons\; \mathrm{X}_n + n\,\mathrm{H}^{+}, \qquad (13.5.2)$$

with $\mathrm{X}_n \equiv (\mathrm{M(OH)}_n)^{z^{+}-n^{-}}$. The concentration of water, $[\mathrm{H_2O}]= 55$ mole/liter, is inserted into the constant of equilibrium to yield

$$K_n = [\mathrm{X}_n][\mathrm{H}^{+}]^n/[\mathrm{M}^{z^{+}}]. \qquad (13.5.3)$$

The above reaction is itself equivalent to the sum of the two reactions

$$n\,\mathrm{H_2O} \;\rightleftharpoons\; n\,\mathrm{OH}^{-} + n\,\mathrm{H}^{+}, \quad (K_{\mathrm{dis}}^{\mathrm{c}})^n = ([\mathrm{H}^{+}]\,[\mathrm{OH}^{-}])^n = 10^{-14\,n}$$

$$\mathrm{M}^{z^{+}} + n\,\mathrm{OH}^{-} \rightleftharpoons \mathrm{X}_n, \qquad K_n' = [\mathrm{X}_n]/([\mathrm{M}^{z^{+}}]\,[\mathrm{OH}^{-}]^n) \qquad\qquad (13.5.4)$$

so that $K_n = (K_{\mathrm{dis}}^{\mathrm{c}})^n\,K_n'$. If we consider that $\mathrm{M}^{z^{+}}$ comes from the dissolution of $\mathrm{X}_z = \mathrm{M(OH)}_z$, then

$$(\mathrm{X}_z)_{\mathrm{solid}} \;\rightleftharpoons\; \mathrm{M}^{z^{+}} + z\,\mathrm{OH}^{-}, \qquad K_z = [\mathrm{M}^{z^{+}}]\,[\mathrm{OH}^{-}]^z, \qquad (13.5.5)$$

where the concentration X_z has been set to 1 because it is solid. The concentration of free $[\mathrm{M}^{z^{+}}]$ can be calculated as a function of the pH of the solution using $Y \equiv 10^{-\mathrm{pH}} = [\mathrm{H}^{+}]$, and then

$$[\mathrm{M}^{z^{+}}]_{\mathrm{free}} = K_z\,(K_{\mathrm{dis}}^{\mathrm{c}})^{-z}\,Y^z. \qquad (13.5.6)$$

[13.6] A chemical reaction where $\mathrm{H_2O}$ is formed is called a *condensation* reaction, e.g., formation of a glycosidic bond by the junction of two monosaccharide units. Hydrolysis corresponds to the reverse reaction where bonds are broken by addition of water.

The total concentration of ions M^{z+} in solution is the sum of the free (uncomplexed) M^{z+} and of the hydroxy complexes X_n,

$$[M^{z+}]_{\text{solution}} = [\,Y^z + \sum_{i=1,n}^{i \neq z} K_i\, Y^{z-i}\,]\, K_z\, (K^c_{\text{dis}})^{-z}\,. \tag{13.5.7}$$

Electroneutrality provides the equation for Y,

$$K_z\, (K^c_{\text{dis}})^{-z}\, [\,z\,Y^z + \sum_{i=1,n}^{i \neq z} (z-i)\, K_i\, Y^{z-i}\,] + Y - K^c_{\text{dis}}\, Y^{-1} = 0\,. \tag{13.5.8}$$

The example $M^{z+} = Al^{3+}$ is considered in Section 13.5.2.1. The free Al^{3+} in solution is minimal and very low in the range of pH from 4 to 9, Appelo and Postma [1993], p. 208, Mitchell [1993], p. 44. Solubility is high at low or high pH, Fig. 13.5.2(a).

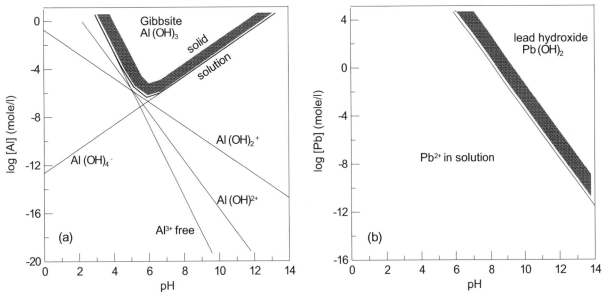

Fig. 13.5.2: (a) Gibbsite is in equilibrium with Al^{3+} in solution. The latter forms hydroxy complexes. The amount of free uncomplexed aluminum and hydroxy complexes is strongly dependent on pH. Equilibrium constants K_n defined by Eqn (13.5.3): $\log K_1 = -5$, $\log K_2 = -10$, $\log K_G = -32.6$, $\log K_4 = -22$.

(b) Lead hydroxide is in equilibrium with Pb^{2+} in solution with an equilibrium constant $K_z = 2.8 \times 10^{-16}$ $(\text{mole/l})^3$. The amount of free Pb^{2+} in solution is strongly pH-dependent, high in acid solutions and low in alkaline solutions.

13.5.2.1 Hydroxy complexes of aluminum

Complexation of dissolved Na^+ and K^+ is negligible but complexation dominates the aqueous solutions of Al^{3+}. Hydroxyalumina are the principal interlayer materials in acid geological materials, and hydroxymagnesium $Mg(OH)_2$ in alkaline geological materials.

Dissolution of gibbsite provides hydroxyl ions OH^- and so turns the solution more alkaline. Complexations have an opposite effect. Indeed, as an example, let us consider the dissolution of gibbsite $Al(OH)_3$,

$$Al(OH)_3 \;\rightleftharpoons\; Al^{3+} + 3\,OH^-\,, \quad K_G = [Al^{3+}]\,[OH^-]^3 = 10^{-32.6}\,. \tag{13.5.9}$$

In solution, Al^{3+} has a tendency to form hydroxy complexes $X_n = Al^{3+}(OH^-)_n$, $n = 1, 2, 4$, according to the reaction,

$$Al^{3+} + n\,H_2O \;\rightleftharpoons\; X_n + n\,H^+\,, \quad K_n = [X_n]\,[H^+]^n/[Al^{3+}]\,. \tag{13.5.10}$$

Each of these reactions is itself equivalent to the sum of the two reactions,

$$n\, H_2O \; \rightleftharpoons n\, OH^- + n\, H^+, \quad (K^c_{dis})^n = ([H^+]\,[OH^-])^n = 10^{-14n}$$

$$Al^{3+} + n\, OH^- \; \rightleftharpoons X_n, \qquad K'_n = [X_n]/([Al^{3+}]\,[OH^-]^n) \tag{13.5.11}$$

so that $K_n = (K^c_{dis})^n\, K'_n$.

Case A: the equilibrium constants K_n are known, $\log K_1 = -5$, $\log K_2 = -10$, $\log K_G = -32.6$, $\log K_4 = -22$. The equilibrium constants are measured in mole per liter, e.g., K_n [unit : $(\text{mole}/l)^n$] and K_G [unit : $(\text{mole}/l)^3$].

Let $Y \equiv 10^{-pH} = [H^+]$. For the complex n,

$$[X_n] = K_n\, K_G/(K^c_{dis})^3\, Y^{3-n} \quad \Leftrightarrow \quad \log[X_n] = \log K_n\, K_G/(K^c_{dis})^3 + (n-3)\, Y. \tag{13.5.12}$$

The concentration of free $[Al^{3+}]$ can be calculated from the pH using $K_G = [Al^{3+}]\,[OH^-]^3$, and then

$$[Al^{3+}]_{\text{free}} = K_G/(K^c_{dis})^3\, Y^3. \tag{13.5.13}$$

The total concentration of aluminum in solution is the sum of dissolved (free, uncomplexed) Al^{3+} and hydroxy complexes, Appelo and Postma [1993], p. 208,

$$[Al^{3+}]_{\text{solution}} = K_G/(K^c_{dis})^3[Y^3 + K_1\, Y^2 + K_2\, Y + K_4\, Y^{-1}]. \tag{13.5.14}$$

Electroneutrality of the solution

$$3\,[Al^{3+}] + 2\,[X_1] + [X_2] - [X_4] + [H^+] - [OH^-] = 0, \tag{13.5.15}$$

provides the equation for Y,

$$K_G/(K^c_{dis})^3[3\,Y^3 + 2\,K_1\, Y^2 + K_2\, Y - K_4\, Y^{-1}] + Y - K^c_{dis}\, Y^{-1} = 0. \tag{13.5.16}$$

Case B: the equilibrium constants K'_n are known.

Let $Y = 10^{14-pH} = [OH^-]^{-1}$. For the complex n,

$$[X_n] = K_G\, K'_n\, Y^{3-n} \quad \Leftrightarrow \quad \log[X_n] = \log K_G\, K'_n + (3-n)\, Y. \tag{13.5.17}$$

The concentration of free $[Al^{3+}]$ can be calculated from the pH using $K_G = [Al^{3+}]\,[OH^-]^3$, and then

$$[Al^{3+}]_{\text{free}} = K_G\, Y^3. \tag{13.5.18}$$

The total concentration of aluminum in solution is the sum of dissolved (free, uncomplexed) Al^{3+} and hydroxy complexes, Appelo and Postma [1993], p. 208,

$$[Al^{3+}]_{\text{solution}} = K_G[Y^3 + K'_1\, Y^2 + K'_2\, Y + K'_4\, Y^{-1}]. \tag{13.5.19}$$

Electroneutrality of the solution,

$$3\,[Al^{3+}] + 2\,[X_1] + [X_2] - [X_4] + [H^+] - [OH^-] = 0, \tag{13.5.20}$$

provides the equation for $Y = 10^{14-pH} = [OH^-]^{-1}$,

$$K_G[3\,Y^3 + 2\,K'_1\, Y^2 + K'_2\, Y - K'_4\, Y^{-1}] + 10^{-14}\, Y - Y^{-1} = 0. \tag{13.5.21}$$

13.5.2.2 Hydrolysis with single complex

Some low valence cations do not form complexes. If once again M^z is seen as coming from the dissolution of $X_z = M(OH)_z$, then

$$(X_z)_{\text{solid}} \; \rightleftharpoons M^{z+} + z\, OH^-, \quad K_z = [M^{z+}]\,[OH^-]^z, \tag{13.5.22}$$

where the concentration of X_z has been set to 1 because it is solid. The concentration of free $[M^{z+}]$ can be calculated from the pH using $Y \equiv 10^{-pH} = [H^+]$, and then

$$[M^{z+}]_{\text{free}} = K_z\, (K^c_{dis})^{-z}\, Y^z, \quad \log[M^{z+}]_{\text{free}} = \log K_z\, (K^c_{dis})^{-z} - z\, Y. \tag{13.5.23}$$

For example for lead $M^{z+} = Pb^{2+}$, $z = 2$ and $K_z = 2.8 \times 10^{-16}$ $(\text{mole}/l)^3$. The solubility thus decreases when pH increases, i.e. the solubility is high at low pH only, Fig. 13.5.2(b).

13.5.3 Individual and simultaneous solubilities of NaCl and KCl

Under standard conditions of ambient temperature and pressure (temperature $=25°C=298.15$ K, pressure $=10^5$ Pa), the maximum mass of NaCl dissolved *in one liter of distilled water* is 360 gm, that is 6.15 moles, corresponding to the molar fraction $x_{Na}^{sat} = 0.091$, i.e. $x_w^{Na-sat} = 0.819$, since the molar masses of the sodium ion $\widehat{m}_{Na}$ and of the chloride ion $\widehat{m}_{Cl}$ are equal to 23 gm and 35.5 gm respectively. The molar volume of dissolved salt varies from about 17 cm^3 at low molarity to about 21 cm^3 at high molarity, Fig. 13.2.3. Here, we admit a fixed value of $\hat{v}_{NaCl}=20$ cm^3. The molar volume of the fluid phase varies from 18 cm^3 for distilled water to $0.91 \times 18 + 0.09 \times \hat{v}_{NaCl}=18.2$ cm^3, which gives a saturated salt concentration equal to 5 moles, or also 115 gm of Na$^+$ *per liter of solution* ($\equiv$ fluid phase, or electrolyte).

The maximum mass of KCl dissolved in one liter of distilled water is 342 gm, that is 4.58 moles, since $\widehat{m}_K = 39.1$ gm. Admitting also a molar volume of dissolved salt of $\hat{v}_{KCl}=20$ cm^3, we obtain now: molar fraction of potassium $x_K^{sat} = 0.071$, i.e. $x_w^{K-sat} = 0.858$, molar volume of the fluid phase $\hat{v}_F=18.15$ cm^3, concentration of potassium equal to 3.91 moles, that is 153 gm of K$^+$ per liter of solution.

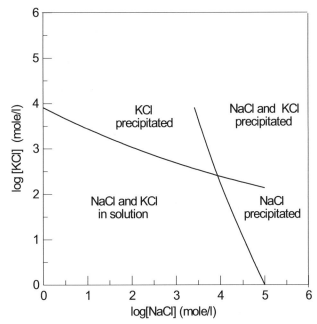

Fig. 13.5.3: Solubility limits for an electrolyte with the two salts NaCl and KCl under standard conditions of temperature and pressure.

When they are dissolved together, the saturated values vary with the concentrations as displayed by Fig. 13.5.3. Indeed, the above remarks can be rephrased in terms of *solubility constants* K_{NaCl} and K_{KCl} of individual salts. For example, NaCl dissolves spontaneously or saturates in distilled water, that is, an addition of salt in water precipitates, according to the following rule,

$$
\begin{aligned}
\text{NaCl} &\rightarrow \text{Na}^+ + \text{Cl}^- &&\text{if } [\text{Na}^+][\text{Cl}^-] < K_{NaCl}, \\
\text{NaCl} &\rightleftharpoons \text{Na}^+ + \text{Cl}^- &&\text{if } [\text{Na}^+][\text{Cl}^-] = K_{NaCl}.
\end{aligned}
\tag{13.5.24}
$$

Accounting for electroneutrality, the solubility constant K_{NaCl} for NaCl is equal to $5.0^2 = 25.0$ (mole/liter of solution)^2, and the solubility constant K_{KCl} for KCl is equal to $3.91^2 = 15.29$ (mole/liter of solution)^2.

Now assume water to be NaCl-saturated, that is $[\text{Na}^+][\text{Cl}^-] = K_{NaCl}$, with $[\text{Cl}^-] = [\text{Na}^+] + [\text{K}^+]$, but not KCl-saturated. The concentration of dissolved Na$^+$ is

$$
[\text{Na}^+] = -\frac{[\text{K}^+]}{2} + \left(\frac{[\text{K}^+]^2}{4} + K_{NaCl}\right)^{1/2}.
\tag{13.5.25}
$$

Addition of KCl results in precipitation of Na$^+$ and in dissolution of K$^+$ because the concentration of chloride increases. The concentrations of dissolved ions vary as $([\text{K}^+] + 2[\text{Na}^+])\,d[\text{Na}^+] + [\text{Na}^+]\,d[\text{K}^+]=0$.

Conversely, if water is first KCl-saturated, any addition of NaCl will result in precipitation of K^+ and in dissolution of Na^+. The concentration of dissolved K^+ is

$$[K^+] = -\frac{[Na^+]}{2} + \left(\frac{[Na^+]^2}{4} + K_{KCl}\right)^{1/2} . \tag{13.5.26}$$

The concentrations of dissolved ions vary as $([Na^+] + 2\,[K^+])\,d[K^+] + [K^+]\,d[Na^+] = 0$. In both cases, water will end up to be KCl- and NaCl-saturated when $[Na^+]\,[Cl^-] = K_{NaCl}$ and $[K^+]\,[Cl^-] = K_{KCl}$, that is for

$$[K^+] = \frac{K_{KCl}}{(K_{KCl} + K_{NaCl})^{1/2}}, \quad [Na^+] = \frac{K_{NaCl}}{(K_{KCl} + K_{NaCl})^{1/2}} , \tag{13.5.27}$$

and then the precipitates are in equilibrium with the dissolved salts,

$$KCl + Na^+ \; \rightleftharpoons \; NaCl + K^+ . \tag{13.5.28}$$

The above remarks have implicitly assumed that salt precipitation/dissolution is instantaneous. Alternatively, if the precipitation/dissolution is not instantaneous, the above relations can be viewed as steady states or equilibria. At a sufficiently small time scale, precipitation/dissolution is governed by a kinetic equation of identical order n_{NaCl} in Na^+ and Cl^+ and defined by the precipitation constant k_{NaCl}, namely

$$\frac{d}{dt}[NaCl] = -\frac{d}{dt}[Na^+] = -\frac{d}{dt}[Cl^-] = k_{NaCl}\left(([Na^+]\,[Cl^-])^{n_{NaCl}} - (K_{NaCl})^{n_{NaCl}}\right). \tag{13.5.29}$$

13.6   Some basic notions of electrostatics

13.6.1   Permittivity and relative permittivity

Charged molecules exert a mutual force on each other. The permittivity ϵ [unit : F/m or C^2/J/m] measures the influence on this force by the medium in which they are bathed. Indeed, according to Coulomb's law, the electrostatic force $F = Q\,Q'/(4\,\pi\,\epsilon\,r^2)$ between two charges Q and Q' is proportional to the square of the inverse of their distance r. The interaction energy $Q\,Q'/(4\,\pi\,\epsilon\,r)$ is proportional to the inverse of their distance. This interaction energy is equal to the work done by the charge Q' in presence of the electrical potential ϕ (equal to $\phi(r)$ at the position of Q') generated by the charge Q, or by the charge Q in presence of the electrical potential ϕ' (equal to $\phi'(r)$ at the position of Q) generated by the charge Q'.

The permittivity of the vacuum ϵ_0 is equal to 8.8542×10^{-12} C^2/J/m. The permittivity of solvents is introduced through the relative permittivity (or dielectric constant) ϵ_r as $\epsilon = \epsilon_0\,\epsilon_r$. Table 13.6.1 indicates that the relative permittivity is affected by temperature. The high value of the relative permittivity of water with respect to a few organic solvents can be appreciated from Table 13.6.2. The consequences on its high solvation properties are commented in Section 13.5.1.

Table 13.6.1: Dependence of the relative permittivity of water ϵ_r with respect to temperature.

Temperature (°C)	0	20	25	60
ϵ_r	88	80.1	78.5	67

Table 13.6.2: Relative permittivity of some organic compounds at 20 °C.

Water H_2O	Ethylene Glycol $C_2H_6O_2$	Ethanol C_2H_6O	Acetic acid $C_2H_4O_2$	Benzene C_6H_6	Heptane C_7H_{16}
80.1	37	24.3	6.1	2.3	1.9

13.6.2 Gauss's law and Poisson's equation

As indicated above, the electrical potential originated by a pointwise electrical charge Q located at $\mathbf{x}_Q$ is equal, at a distance r from the charge, to

$$\phi(r) = \frac{Q}{4\,\pi\,\epsilon\,r}\,.\tag{13.6.1}$$

The electrical field $\mathbf{E}$ associated with the electrical potential (13.6.1) is defined as $-\boldsymbol{\nabla}\phi$,

$$\mathbf{E}(r) = \frac{Q}{4\,\pi\,\epsilon\,r^2}\,\hat{\mathbf{r}}\,,\tag{13.6.2}$$

with $\hat{\mathbf{r}} = (\mathbf{x} - \mathbf{x}_Q)/|\mathbf{x} - \mathbf{x}_Q|$ the unit vector indicating the direction of the charge. **Gauss's law** states that the total electric flux through a closed surface ∂V is equal to the total charge inside the volume V,

$$\int_{\partial V} \epsilon\,\mathbf{E}\cdot d\mathbf{A} = \int_V I_e\,dV\,,\tag{13.6.3}$$

with I_e the electrical density [unit : C/m^3]. This equation is easily proved for a pointwise charge. Indeed, on a sphere of radius r surrounding the charge, and parameterized by the angles $\alpha \in [0, 2\pi[$ and $\beta \in [0, \pi[$ as indicated in Section 12.1.2, the element of area is equal to $d\mathbf{A} = r^2 \sin\beta\,d\alpha\,d\beta\,\hat{\mathbf{r}}$, and therefore $\mathbf{E}\cdot d\mathbf{A} = (Q/4\pi)\sin\beta\,d\alpha\,d\beta$.

With help of the Green's theorem, the surface integral transforms into a volume integral yielding finally, under sufficiently smooth conditions, **Poisson's equation** as a pointwise version of Gauss's law,

$$\operatorname{div}(\epsilon\,\mathbf{E}) = -\operatorname{div}(\epsilon\,\boldsymbol{\nabla}\phi) = I_e\,.\tag{13.6.4}$$

As another consequence of Gauss's law, the potential due to a sphere of radius r_0 with a charge Q on its surface,

$$\phi(r) = \begin{cases} \dfrac{Q}{4\,\pi\,\epsilon\,r} & r \geq r_0, \\[2mm] \text{constant} & r \leq r_0, \end{cases}\tag{13.6.5}$$

is observed to be identical, outside the sphere, to that of a point charge. Indeed, according to Gauss's law (13.6.3), the electrical field depends only on the charge, not on the details of the spatial repartition of the charge inside the volume defined by the closed surface ∂V.

13.6.3 Characteristic time of electric relaxation

The length scale of interest in this textbook assumes that electroneutrality holds pointwise. As the analysis of the diffuse double layer in Exercise 13.1 shows, the property does not apply at the nanometer scale but we have no access to this scale in a macroscopic analysis.

Still, even at a macroscopic spatial scale, displacements of charges perturb local electroneutrality over a certain characteristic time t_{elec}. This electric relaxation time is much smaller than the other characteristic times of interest in this work, in such a way that electroneutrality can be considered to hold pointwise and instantaneously.

Indeed, let I_e [unit : coulomb/m^3] be the electrical density and $\mathbf{I}_e$ [unit : coulomb/m^2/s] the electrical current density. The electrical density is linked to the electrical potential ϕ [unit : volt] through the capacitance G_e [unit : F=coulomb/volt], namely $V\,I_e = G_e\,\phi$, where $V = L^3$ is the volume of interest. The capacitance depends on the permittivity ϵ of the solvent, and on the characteristic length of the capacitor, Eqn (13.E.15), say $G_e \sim \epsilon\,L$. Moreover, the electrical current density depends linearly on the electrical field through the electrical conductivity σ_e [unit : ampère A/volt V/m], namely $\mathbf{I}_e = -\sigma_e\,\boldsymbol{\nabla}\phi$.

The conservation of the charge,

$$\frac{\partial I_e}{\partial t} + \operatorname{div}\mathbf{I}_e = 0\,,\tag{13.6.6}$$

becomes an equation of diffusion for the electrical potential,

$$\frac{G_e}{V}\,\frac{\partial \phi}{\partial t} - \sigma_e\,\operatorname{div}\boldsymbol{\nabla}\phi = 0\,,\tag{13.6.7}$$

which can be phrased in terms of a dimensionless space variable x/L and of a dimensionless time t/t_{elec},

$$t_{\mathrm{elec}} = \frac{G_{\mathrm{e}}\, L^2}{V\, \sigma_{\mathrm{e}}} \sim \frac{\epsilon}{\sigma_{\mathrm{e}}}\,. \tag{13.6.8}$$

For water at ambient temperature, $\epsilon \sim 7 \times 10^{-10}\,\mathrm{C^2/J/m}$ according to Table 13.6.1. Moreover, according to Table 16.4.3, the electrical conductivity of articular cartilages takes values in the range $10^{-2} - 1\,\mathrm{A/V/m}$. Therefore, the electrical relaxation time ranges in the interval

$$t_{\mathrm{elec}} \in [10^{-9},\ 10^{-7}]\,\mathrm{s}\,. \tag{13.6.9}$$

This time scale is much smaller than the characteristic times of the other phenomena we are concerned with when studying the electro-chemomechanical couplings of soft tissues in a continuum perspective. Therefore, in this macroscopic analysis, electroneutrality can be considered safely to hold pointwise and instantaneously.

13.7   Semi-permeable membrane and osmotic effect

The notion of chemical potential has been shown in Chapter 10 to control the exchanges of mass between two compartments. It serves also as a vehicle to explain the phenomenon of osmosis.

Consider a bucket separated in two parts by a membrane impermeable to solute as sketched in Fig. 13.7.1[13.7]. Initially, the two sides of the membrane, numbered 1 and 2, have identical chemical content, water w (the solvent) and the neutral solute s, and are submitted to the same external pressure p. At any time, the molar fractions of water and salt satisfy the identities $x_1^{\mathrm{w}} + x_1^{\mathrm{s}} = 1$ and $x_2^{\mathrm{w}} + x_2^{\mathrm{s}} = 1$.

The *chemical potential* of water at pressure p and molar fraction x^{w} expresses as[13.8]

$$g^{\mathrm{w}} = \widehat{v}_{\mathrm{w}}\, p + R\, T \, \mathrm{Ln}\, x^{\mathrm{w}}, \tag{13.7.1}$$

with $\widehat{v}_{\mathrm{w}}$ molar volume of water, R is the universal gas constant and T the absolute temperature. Since only water can transfer through the membrane, chemomechanical *equilibrium* across the membrane is phrased in terms of the chemical potential of water only. The initial concentration of solute is very small, so that:

$$\text{Initial equilibrium}: \quad p_1 = p_2,\ x_1^{\mathrm{w}} = x_2^{\mathrm{w}} \sim 1,\ x_1^{\mathrm{s}} = x_2^{\mathrm{s}} \sim 0. \tag{13.7.2}$$

Solute is then added on side 1:

$$\text{Add solute}: \quad \Delta x_1^{\mathrm{s}} > 0 \ \Rightarrow \Delta x_1^{\mathrm{w}} = -\Delta x_1^{\mathrm{s}} < 0. \tag{13.7.3}$$

The short term rate of mass transfer of water through the membrane is assumed to be governed by the following law defined by a constant $k_{\mathrm{H}} > 0$ representative of the permeability of the membrane to water (motivation for and generalization of this law will be addressed later),

$$\text{Law of transfer}: \quad \frac{dm_1^{\mathrm{w}}}{dt} \;=\; k_{\mathrm{H}}\,(\Delta g_2^{\mathrm{w}} - \Delta g_1^{\mathrm{w}})$$
$$\;=\; k_{\mathrm{H}}\,(\,0\, - (-R\,T\,\Delta x_1^{\mathrm{s}})\,) > 0 \quad \text{instantaneously}. \tag{13.7.4}$$

Consequently observation leads to
Rule 1: *flow is directed towards the region of higher solute concentration.*

In the long range, a final equilibrium corresponding to a vanishing flow through the membrane gets established when the chemical potentials on each side of the membrane of water tend to a common value,

$$g_1^{\mathrm{w}} \sim \widehat{v}_{\mathrm{w}}\, p_1 - R\,T\, x_1^{\mathrm{s}} \;=\; g_2^{\mathrm{w}} \sim \widehat{v}_{\mathrm{w}}\, p_2 - R\,T\, x_2^{\mathrm{s}} \tag{13.7.5}$$

This result is encapsulated as

[13.7]Unlike the usual presentation which uses a thin U-shaped tube, gravity is neglected here and the mechanical loads are applied by external agents. In the usual presentation, the external load on both sides of the tube is the atmospheric pressure and the difference of pressures gives rise to different elevations of water on the two sides of the tube.
[13.8]In this section, we use *mole based* chemical potentials following the usual convention of textbooks of physical chemistry.

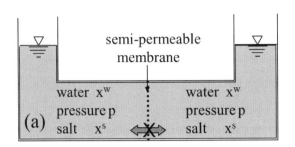

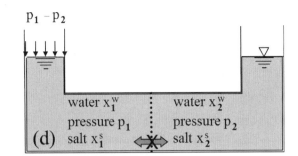

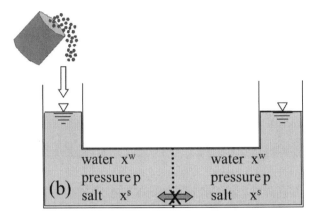

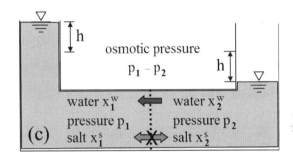

Fig. 13.7.1: Osmotic phenomenon via a semi-permeable membrane. (a) A membrane permeable to water only separates a bucket in two parts. Initially, the chemical contents and pressures are identical on both sides. (b) Addition of solute in side 1 triggers a transfer of water from side 2 to side 1. (c) The height of water, and its pressure, increases in side 1. The converse takes place in side 2. With g the acceleration of gravity and $2h$ the differential of water height, equilibrium (zero flow) requires the mechanical pressure difference $p_1 - p_2 = (p + \rho_w\, g\, h) - (p - \rho_w\, g\, h)$ to be equal to the chemical pressure difference $\pi \equiv RT\,(x_1^s - x_2^s)/\widehat{v}_w$. The osmotic effect is a typical instance of *chemomechanical coupling*. (d) An alternative way of reaching equilibrium at fixed height of water is to apply directly the osmotic pressure π to side 1. Then $p_1 = p + \pi$ and $p_2 = p$. If the applied pressure is larger than π, the *reverse osmosis* phenomenon briefly touched upon in Section 13.8 takes place: this phenomenon represents a basic example of *mechanochemical coupling*.

<u>Rule 2</u> or van't Hoff law: *at equilibrium, pressure is higher where solute concentration is higher, namely*

$$p_1 - p_2 = RT\,(x_1^s - x_2^s)/\widehat{v}_w > 0. \tag{13.7.6}$$

Hence, osmotic effect has generated a counterflow to usual Darcy's flow.

The chemical potential may be phrased in terms of concentration $c^s = x^s/\widehat{v}_w$ instead of molar fraction. The quantity,

$$\pi = RT\,c^s. \tag{13.7.7}$$

is referred to as the osmotic pressure associated with the concentration c^s.

So far the membrane was assumed to be completely impermeable to solute. More generally the permeability of the membrane for the solute can be introduced via the *reflection coefficient* ω also called *osmotic coefficient* or *osmotic efficiency*. This coefficient is a property of both the membrane and the solute. It varies between 0 and 1: it is equal to 0 for a completely permeant solute and to 1 for an non-permeant solute (as assumed above). Then the law of transfer of water has to be modified. For that purpose we need the chemical potential of the solute. We adopt an approximative expression for dilute solution,

$$g^s \simeq R\,T\,\mathrm{Ln}\,x^s. \tag{13.7.8}$$

The general coupled mass transfer law is postulated as a linear combination of the differentials of the chemical potentials,

$$
\begin{aligned}
\frac{dm_1^w}{dt} &= k_H(\Delta g_2^w - \Delta g_1^w) + k_{ws}(\Delta g_2^s - \Delta g_1^s) \\
&= k_H \, \widehat{v}_w \big(\Delta p_2 - \Delta p_1 - \omega \, (\Delta \pi_2 - \Delta \pi_1) \big) ,
\end{aligned}
\tag{13.7.9}
$$

with the notation $k_{ws} = k_H \, x^s \, (1 - \omega)$, assuming the solute concentrations on the two sides to be close. Therefore for any noncompletely permeant solute, i.e., $\omega \neq 0$, a pressure differential proportional to the osmotic coefficient should be exerted to reach a steady state, that is, to stop water flow:

$$
\Delta p_2 - \Delta p_1 = \omega(\Delta \pi_2 - \Delta \pi_1) .
\tag{13.7.10}
$$

The law of transfer for the solute involves its diffusion properties through the membrane,

$$
\begin{aligned}
\frac{dm_1^s}{dt} &= k_{sw}(\Delta g_2^w - \Delta g_1^w) + k_s(\Delta g_2^s - \Delta g_1^s) \\
&= k_{sw} \, \widehat{v}_w \big(\Delta p_2 - \Delta p_1 - Y \, (\Delta \pi_2 - \Delta \pi_1) \big) ,
\end{aligned}
\tag{13.7.11}
$$

with $Y \equiv 1 - k_s/(x^s \, k_{sw})$. If $k_s/k_{sw} = k_{ws}/k_H$, then $Y = \omega$ and the solute is convected by the liquid through the membrane, with the proviso that k_{sw} is proportional to the solute concentration, as exposed in Section 15.8.3 which addresses the extravasation of blood and macromolecules through the vascular wall.

Complete equilibrium is defined by the equality of the chemical potentials of both water and salt and more generally of all (even partially) permeant species.

We have seen that osmosis requires a membrane to develop. Crucially, membranes are selective with respect to species. A species is endowed with a transfer time. An infinite transfer time practically prevents a species to cross the membrane. Transfer times depend inter alia on the size, polarity, molar mass of the species. Membranes may be well defined objects, natural like the ones that limit cells or engineered clay barriers to prevent leakage of pollutants. In cartilage, steric considerations prevent macromolecules to enter the space defined by collagen fibers. Electrical repulsion creates fictitious membranes which generate significant ionic disequilibria.

The osmosis phenomenon may be illustrated by the process of intravenous feeding. Hypertonic solutions lead to shrinking of the cells from which water is desorbed. Therefore administration of a hypertonic solution may be a cure against oedema (e.g., swelling of ankles or feet). Conversely, a hypotonic solution leads the cells to absorb water. A number of drinks like tea, apple juice, etc. are hypotonic.

13.8   Reverse or Inverse osmosis

Filtration, or inverse osmosis, actually known as reverse osmosis in the anglo-saxon world, is a process where water and solutes move through a membrane due to a pressure differential.

According to (13.7.9), the flows of liquid and solute, if the latter is permeant, are directed from the side of large $p - \omega \, \pi$ to the side of small $p - \omega \, \pi$.

This principle rules the mass exchanges in kidneys. Water and small solutes move from the capillaries to the tubules because blood pressure in the former is larger than fluid pressure in the latter. This process is not highly selective, and only the largest blood proteins or blood cells are not admitted in the tubules. In fact, only a part of the substances that transfer through this process are eliminated, e.g., urea. Indeed a percentage of water is next re-absorbed passively through osmosis, while glucose and amino acids are re-absorbed through dedicated active transporters.

Inverse osmosis is also an industrial process to desalinize water. It consists in applying a pressure of several atmospheres to the saline water (side 1 of the membrane) while the other side is at atmospheric pressure. As of spring 2008, the French company Degrémont was running over 250 desalinizing plants all over the world, using this principle with a pressure differential of about 10 atmospheres. Membranes used in that process should of course be highly selective with respect to sodium ions and chloride anions, and be inert with respect to acid-base reactions and chlorine.

Inverse osmosis is also at work in trees to drive water from salty soils. The solute concentration of xylem water is low, much lower than that of the soil, the roots coming into picture as a sort of semi-permeable membrane. Under equal fluid pressure, osmosis would dictate that water leaves the tree. To reverse this trend, the above industrial process would apply a positive pressure on the soil water. The alternative choice by nature in botany is to apply a negative pressure in the tree. Equilibrium of the sap between the roots and the leaves, namely $p_r - \pi_r = \rho\,g\,h + p_l - \pi_l$ combined with equilibrium between the soil and the roots, namely $p_r - \pi_r = p_s - \pi_s$, together with the fact that the leaves and soil are at atmospheric pressure, namely $p_l = p_s(= p_{\text{atm}})$, implies $\rho\,g\,h = \pi_l - \pi_s > 0$, which is positive because the solute concentration of the cells of the leaves is high, much higher than the solute concentration of the water at the roots, Scholander et al. [1964].

13.9 Electrical repulsion and electrical shielding

Proteoglycans are negatively charged. Therefore adjacent branches repel one another. Electroneutrality is ensured by the presence of *counterions*, that is, ions of electrical charge opposite to that of the fixed charge. This repulsion tends to increase the volume of the tissue which is penetrated by water, as sketched in Fig. 13.9.1.

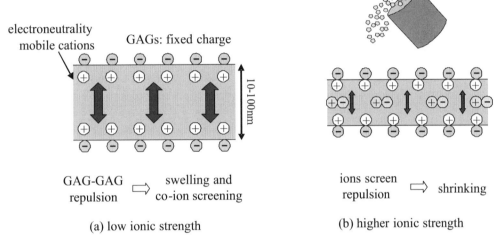

Fig. 13.9.1: Electrical repulsion maintains the fixed charges away from one another. Electroneutrality is ensured at some length scale by mobile ions that diffuse in the fluid. While cations sodium do not bind to GAGs, cations calcium do to some extent and hydrogen ions may even change the sign of the fixed charge. The presence of counterions decreases the electrical repulsion between fixed charges of identical sign and leads to macroscopic contraction.

Electrical repulsion is due to fixed charges, while electrical shielding is due to mobile ions. Observations show that
- the magnitude of the repulsion, and the distance between branches, decrease as the ionic concentration of the electrolyte, also called *ionic strength*, increases. In other words, counterions screen or shield the repulsion force;
- the magnitude of the shielding is ion-dependent, valence and hydrated size of the ions playing a role;
- however at sufficiently high ionic strength in the so-called hypertonic state, the ultimate shielding and distance between branches are essentially ion-independent.

The repulsion between two fixed electrical charges of identical signs depends on the dielectric constant of the electrolyte they bath in. Pure fluids have their own dielectric constants as indicated in Tables 13.6.1 and 13.6.2. Replacement of permeants in fixed charge materials leads to volume changes consistent with their respective dielectric constant. Data tend to show that the presence, and actual concentration, of ions in the permeant fluid modifies the dielectric constant of the electrolyte, Di Maio et al. [2004].

13.10 The pore composition in materials with fixed charge

Mechanical loading consists in monitoring some components of the load vector and the complementary components of the displacement vector. During quasi-static laboratory experiments, the mechanical load is applied either at once or incrementally. It would be transmitted instantaneously to any point of the specimen if the material was behaving like an elastic solid. However time dependence delays somehow the mechanical information applied at the boundary. This time dependence is due to either an intrinsic viscous behavior or to the fact that the specimen is a multiphase material.

Chemical loading is applied by putting the specimen in contact with a bath or reservoir of controlled chemical composition. The contact is not direct and a porous layer is often used as an interface. The interface being stiffer than the tissue is aimed at homogenizing the mechanical load and at smoothing the time profile of the changes of chemical composition in the bath.

The chemical information never transfers instantaneously to the specimen. The delay depends on the properties of the porous layer and on the size and geometry of the specimen. Indeed, chemicals need to move throughout the porous layer and next through the specimen before a steady state can be reached. The motion of the chemicals takes place within the fluid phase, under several modes, like diffusion, advection, electromigration, etc. Modeling of the transport issues is considered in Chapter 16.

The examples described below intend to
- quantify the effect of the overall ionic concentration;
- highlight the effects of the ionic concentration of individual species;
- contrast the influence of chemical loadings at *given mechanical load* and at *given strain*. The distinction between these two types of experimental conditions is obtained in situations as close as possible to a one-dimensional context. However, the actual mechanical behavior is always three-dimensional: in the oedometer, that provides the conditions for confined compression, lateral displacements are prevented so that the lateral strains vanish and therefore stresses develop as part of the material response.

These examples feature a geological material, namely *a chemically active clay*, and articular cartilages. Interestingly enough, the structural elements that control the electro-chemomechanical coupling have, for both materials, a nanometric length scale. The driving engine is the fixed electrical charge which is negative at neutral pH. Its sensitivity to the pH of the electrolyte depends on the details of the electrical charge. A lesson from these examples can be encapsulated in the following

Property: chemomechanical coupling versus pH sensitivity

Some materials with fixed charge can be highly sensitive to electro-chemomechanical coupling while displaying a low pH dependence, and conversely.

The first property is linked to the effective concentration of the charge, a volumetric property, while the second property is tied to the nature of the fixed charge, a surface property in geological materials.

13.10.1   Chemical consolidation and swelling in clays

Smectites are clays formed by structural units, between which the bonds are weak, so that layers of absorbed water and cations may enter the interlayer space and produce swelling. Typically, when two ions have non-hydrated sizes that do not differ by more than 15% and valences by more than unity they can exchange in the clay sheets: this phenomenon called *isomorphous substitution* leaves the clay with a fixed negative charge at neutral pH.

13.10.1.1   Nanostructure of clays

Fundamental to the understanding and modeling of chemically driven swelling of expansive clays are their microstructure and the organization of their pore space and water. For smectites, this microstructure is dominated by clusters of parallel platelets of clay mineral separated by 10-20 Å interlamellar pores filled with few one-molecule layers of water, called interlayer water or internal absorbed water. This water has properties such as density or viscosity which are slightly different from those of free water. Interlayer water does not flow even when subjected to high hydraulic gradients and it deforms together with the solid part of the clusters. Clusters are enveloped by external adsorbed water, up to seven molecular diameters in thickness. Montmorillonite clusters of width 1 to 10 μm may contain from 5 to 15 three-layer units and up to 1000 units in compacted clays. Clusters in kaolinites have width from 0.1 to 4 μm and thickness usually from 0.05 to 0.1 μm equivalent to 60 to 120 two-layer units.

Usually clusters are separated by pores with the characteristic size of the order of 1 micrometer and more. Water that resides in such pores is called free water. Indeed, this water can be displaced by ordinary hydraulic gradients.

Swelling of compacted clays develops in two distinct regimes. During swelling at the lowest water contents, called *crystalline swelling*, interlamellar space adsorbs water until it is filled with three or four molecular water layers. After the interlamellar space has been saturated, water starts to adsorb to the external surfaces of clay particles, producing *osmotic swelling*. However in the comments below we shall gather crystalline and osmotic swelling under the term *chemical swelling* and make no distinction between *adsorbed* and *absorbed* water, both being parts of the solid phase.

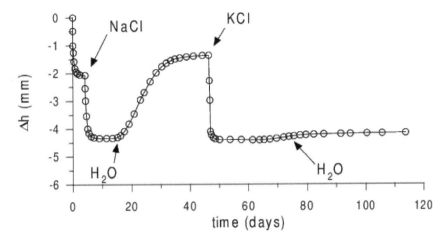

(a) Experimental data on Bisaccia clay, redrawn from Di Maio and Fenelli (1997).

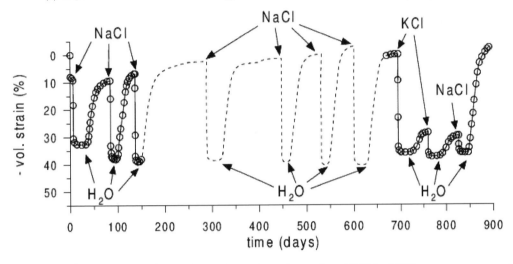

(b) Experimental data on Ponza bentonite, redrawn from Di Maio (1998).

Fig. 13.10.1: Successive replacements of a pore solution by NaCl and KCl saturated solutions during oedometric tests at constant vertical stress of 40 kPa. Evolution of (a) the settlement and (b) the volumetric strain. Increase of salt concentration implies chemical consolidation. KCl solutions lead to larger contraction than NaCl. The volume change associated with cation exchange seems mostly reversible. Still, replacement of cations initially present in Ponza bentonite by cations Na^+ is slow, progressive, and several cycles are needed for the settlement to get steady. Reprinted with permission from Loret et al. [2004]a.

13.10.1.2 Modification of the ionic strength in homoionic clays

Montmorillonites are particularly sensitive to the composition of their pore water. Fig. 13.10.1(a) shows the time evolution of the vertical settlement in an oedometric test. The sample of 2 cm height is in contact with a reservoir of controlled chemical composition. Initially the reservoir is filled with distilled water (dw). The sample is first mechanically compressed by the application of an instantaneous load which results in a *mechanically induced consolidation* which tends to stabilize in few days. Next, while the load is kept unchanged, the water of the reservoir is NaCl-saturated. A significant *chemically induced consolidation* develops. The consolidation time is of about the same duration as for mechanical consolidation. When the reservoir electrolyte is refreshed, the sample undergoes a volume expansion termed *chemical swelling* that requires a few weeks to stabilize.

The volume change upon the chemical cycle turns out here to be positive (expansion). In general, it is the result of mechanisms which might compete or cooperate:
   - cations are attracted by the *negatively charged clay clusters* and they are hydrated by absorbed water;
   - the cations initially present around clusters may be 'displaced' by other ions which have a higher affinity for

the clay charge. Later in the context of the theory of mixtures, we shall use the terms 'transfer' and 'exchange' rather than 'displacement'. Ion exchange may result in an expansion or contraction;

- as a result of the osmotic effect, absorbed water between clay platelets is desorbed when the salt concentration in pore water increases. Desorption is accompanied by volume decrease and an increase of the overall stiffness of the clay. The converse phenomena take place when the ionic strength of pore water is decreased;
- irreversible contraction may develop, especially in presence of large mechanical stresses.

13.10.1.3   Ionic replacements in clays

Actually, the *reversible* or *irreversible* character of the material response to a chemical loading is difficult to be assessed because it involves a number of physical aspects, as well as microstructural features of the clays. Reversibility can be assumed as a first rough approximation. Let us illustrate this statement by an example.

Fig. 13.10.1(a) shows that saturation of the fresh reservoir by KCl results in about the same consolidation as that induced by NaCl yielding a unique hypertonic state. However later refreshment of the reservoir implies a very small swelling. The phenomenon is due to replacement of ions Na$^+$ by ions K$^+$, and this replacement contracts considerably the material behavior.

To recover approximately the initial volume, a complete *symmetric* chemical cycle should be performed, that is dw NaCl - dw KCl dw - NaCl dw, as witnessed by Fig. 13.10.1(b).

The latter figure shows that a number of cycles dw NaCl dw are necessary to reach steady state. This might be traced to the fact that the transfer times of ions from the interlayer water to the free water, and conversely, is long with respect to the time scale of the chemical cycles.

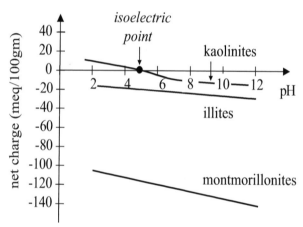

Fig. 13.10.2: Qualitative evolution of the electrical charge of clays as a function of pH, redrawn with permission from Loret et al. [2004]a, gathering information from several sources:
- the electrical charge is quantified through various measures, e.g., as a volumetric charge Q_e in milliequivalent (meq) per 100 gm of dry material=centimole/kg, or as surface charge σ_e in C/m^2;
- these two measures are related through the specific surface of dry material a [unit : m^2/kg] and Faraday's constant F = 96485 C/mole by $\sigma_e = Q_e \, F/a$;
- the valence ζ is equal to $10^{-5} \, Q_e \, \hat{m}$ where $\hat{m}$ is the molar mass of the dry material in gm;
- the specific surface a is about 10×10^3 to 20×10^3 m^2/kg for kaolinites, 65×10^3 to 100×10^3 m^2/kg for illites, 700×10^3 to 840×10^3 m^2/kg for montmorillonites.
For a montmorillonite of molar mass $\hat{m} = 382$ gm, an electrical charge Q_e equal to -100 meq/100 gm=1 mole/kg is equivalent to $\sigma_e = -0.125$ C/m^2 and to a valence $\zeta = -0.382$.

13.10.1.4   pH, electrical charge and isoelectric point

Kaolinites are more sensitive to pH than montmorillonites because the former have OH$^-$ termination sites on the octahedral faces and edges while OH$^-$ termination sites exist only on the edges for the latter. The faces are negatively charged at any pH. The edges contain silica and alumina and their charge is positive at low pH and negative at high pH. Montmorillonites and illites whose negative charge is due to isomorphous substitutions are actually sometimes called *fixed charge clays* while kaolinites whose negative charge is due to broken bonds are known as *variable charge clays*. At low pH the charge of the latter becomes positive. A direct consequence is that contaminant cations like Pb^{2+} are more easily extracted/desorbed from kaolinites than from illites at low pH. Cation exchange for illites is practically independent of pH in contrast to kaolinites. Therefore acidification is not an efficient technique for the remediation of illites.

Montmorillonites are very active clays in the sense that a change of the pore liquid composition implies large volume changes at constant load. Kaolinites are much less affected by pore liquid composition.

Fig. 13.10.2 shows the change of the electrical charge of clays as a function of pH. The material properties that depend directly on the clay charge, like the ζ-potential and the electroosmotic coefficient, are pH-dependent: the dependence is strong for kaolinites where the charge may change sign but quite moderate for illites and montmorillonites where the charge is always negative.

These observations provide some elements of comparison with biological tissues endowed with a fixed charge. The charge of proteoglycans changes according to pH but it remains negative. At variance, collagen are positively charged at low pH, neutral at physiological pH, and negatively charged at high pH. The total charge of articular cartilage and corneal stroma changes sign much like in kaolinites.

13.10.2 Modification of the electrolyte circulating an articular cartilage

Much like for active clays, emphasis is laid on the effects of the overall ionic strength of the electrolyte circulating the cartilage and on the individual effects of cations. While the issue is addressed later in Chapter 14, it is important to make clear that the chemical composition of this electrolyte is not directly controllable. The cartilage can be thought of as being wrapped around by a semi-permeable membrane. In fact, only the chemical composition of the bath the cartilage sample is in contact with can be controlled. The composition of the electrolyte circulating the cartilage is the result of electro-chemomechanical couplings.

13.10.2.1 Homoionic cartilage

Articular cartilage is a porous medium that provides normal compliance so as to minimize friction at joints. The negatively charged proteoglycans (PGs) are a key constituent. The skeleton of cartilage is formed by collagen fibers which resist both the external loads and the chemical pressure exerted by PGs.

The latter depends much on the repulsive forces between PGs, which are themselves function of the ionic content of the electrolyte that circulates the cartilage. This electrolyte communicates with the synovial fluid. The macroscopic outcome of these nanoscale interactions is a compressive stress that decreases when the electrolyte changes from distilled water to a salt-saturated solution if the axial strain is kept constant.

Eisenberg and Grodzinsky [1985] performed confined compression tests on cylindrical samples, 9 mm in diameter and about 500 μm in thickness, in contact with a bath of controlled concentration in NaCl, Fig. 13.10.3. Mechanical loading is performed stepwise and, after each load increment, sufficient time is left to reach equilibrium: equilibrium requires about 15 mn for bovine articular cartilage. A similar incremental procedure is followed for changes in composition of the bath.

The axial stress-axial strain measurements of Eisenberg and Grodzinsky [1985] show that mechanical tests occur at approximately constant apparent moduli. Free swelling strains under confined compression are obtained by prolongating the linear stress-strain relation to zero axial stress. Of course, the amplitude of the swelling strain decreases when the salt concentration in the bath increases. Since collagen fibers are activated essentially under extension, the apparent elastic moduli even for small extensive strains are much larger than for contractive strains.

13.10.2.2 Tensile experiments

Akizuki et al. [1986], their Fig. 4, show chemical experiments where the strain is fixed in the direction of traction. They observe that, at relatively small strain, increasing the ionic strength of the bath implies an increase of the tension, while the opposite holds at larger strain. Owens et al. [1991] perform purely chemical tests at constant strain. Increase of the salt concentration in the bath reduces the stress: this feature is attributed to an increased *shielding* of the charged proteoglycans by the dissolved salt. For the point of view adopted below to be consistent with the experiments of both Akizuki et al. [1986] and Owens et al. [1991], the axial strain at which the latter perform their experiments should be relatively large.

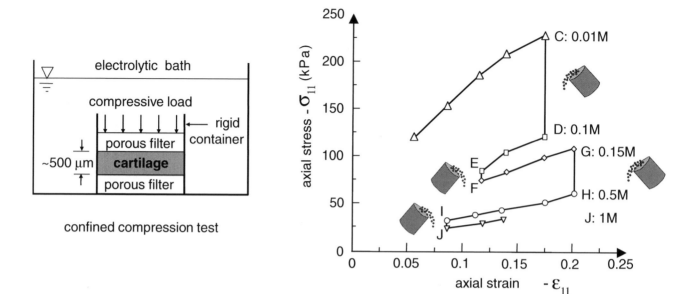

Fig. 13.10.3: Confined compression test on bovine articular cartilage by Eisenberg and Grodzinsky [1985], redrawn with permission from Loret et al. [2004]a. The sample is in contact with a bath of controlled ionic strength. Sufficient time is left after each modification of the ionic strength of the bath to reach steady state. The complex loading path involves successive confined compression at fixed ionic strength of the bath and increase in ionic strength of the bath at fixed strain. *Shielding* of negative charges of proteoglycans by the dissolved salt reduces the repulsive forces and the overall compressive stress. The apparent moduli are thus largest when the salt concentration of the electrolyte is minimal and lowest when the salt concentration of the electrolyte is maximal.

13.10.2.3   Ionic replacements in articular cartilages

Let us reconsider the experiments of Owens et al. [1991] under a different perspective. Owens et al. [1991] have performed uniaxial traction tests on specimens of bovine articular cartilages, harvested from the superficial zone of the medial patello-femoral groove, parallel to the split-line orientation. Subsequent to stress free swelling strain in quasi-distilled water, pre-equilibration is performed at a constant axial tensile strain of 5%. The ionic strength of the bath is then modified as indicated by Fig. 13.10.4. Data show that shielding by calcium ions is more important than by sodium ions, in the sense, that the induced mechanical effects are larger.

On the other hand, a number of experiments tend to indicate that the hypertonic state is universal: the nature of ions matter in the shielding process at low to moderate ionic strengths only. Consequently, except perhaps close to the hypertonic state, in articular cartilages like in clays, what matters is not only the overall strength of the electrolyte including all ions, but also the very concentration of each ionic species. Fig. 13.10.4 shows that the feature is particularly conspicuous at low ionic strength, point E in presence of calcium chloride versus point G in presence of sodium chloride. The tests of Owens et al. [1991] are used to infer the dependence of the effective pressure $p_{\rm eff}$ in terms of cations shown in Figs. 13.10.5 and 13.10.6.

13.10.2.4   A picture highlighting chemomechanical couplings

These observations prompt us to sketch the tensile response shown in Figs. 13.10.7 and 13.10.8. The existence of an intersection point, noted Ω in these sketches, between all mechanical tests at fixed ionic strength of the bath explains nicely the mixed chemomechanical tests of Akizuki et al. [1986], [1987].

In confined compression, the axial stress-axial strain relation at given ionic strength of the bath displays linearity up to 15% in the experiments of Eisenberg and Grodzinsky [1985]. On the other hand, uniaxial traction shows a nonlinear relation at small strain. This aspect is not displayed in the sketch of Fig. 13.10.8.

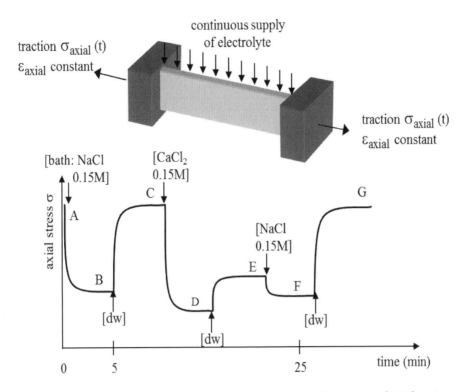

Fig. 13.10.4: Uniaxial traction experiment on bovine articular cartilage by Owens et al. [1991], redrawn with permission from Loret et al. [2004]a. At a given axial strain of 5% counted from the isotropic free swelling configuration, the chemical composition of the bath is modified instantaneously at times A, B, C, D, E and F. Sufficient time is left however for equilibrium to be reached before the next modification. Note that the stress of point A is recovered only after calcium chloride is substituted by sodium chloride.

13.10.2.5 The effects of pH in articular cartilage and corneal stroma

As pH decreases, the hydrogen ions H^+ fix onto the proteoglycans and decrease their electrical charge. At given ionic concentration, the electrical repulsion between PGs decreases and therefore so does the macroscopic chemical pressure, Fig. 13.10.9(a). However the picture is complicated by the fact that, at low and high pH, the collagen molecules become charged, while at physiological pH they display nearly equal numbers of amino and carboxyl groups, NH_3^+ and COO^-, respectively, and hence their net charge is virtually nil.

The corneal stroma consists mainly of a collagen fiber network and intermingled proteoglycans like articular cartilages. The presence of a fixed charge implies qualitatively similar electro-chemomechanical couplings. Still, the intensity of the charge is much less and the structural arrangements of these two components are quite apart. In the stroma, the collagen fibers run in alternate directions in adjacent lamellae presumably to maximize the transparency. The stroma has an innate tendency to swell without limit because the crosslinks between lamellae are not strong enough to oppose the chemical pressure unlike in cartilage. However a too large hydration would tend to generate an inhomogeneous repartition of the absorbed water and to form 'lakes' that would decrease the transparency. The latter is ensured by endothelial pumps that drive bicarbonate ions HCO_3^+ out of the cornea, thus maintaining ionic strength and hydration to acceptable levels.

The variation of the swelling properties of corneal stroma with pH provides another illustration of the electromechanical coupling. At neutral pH, collagen is almost neutral. At high pH, amino-groups loose their positive charge, and the collagen molecules become negatively charged. Conversely, at low pH, the negatively charged carboxyl groups become neutral, and the collagen molecules become positively charged. As a consequence of the electrical repulsion between proteoglycans and collagen fibrils, macroscopic swelling is observed as the pH of the tissue electrolyte departs from the isoelectric point. This strongly pH-dependent swelling is illustrated by the change of hydration sketched in Fig. 13.10.9(b).

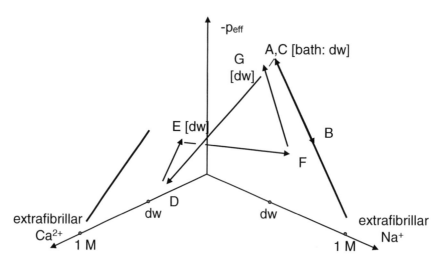

Fig. 13.10.5: Interpretation via the effective pressure p_{eff} of the test by Owens et al. [1991] at fixed axial strain of 5% shown in Fig. 13.10.4, reproduced with permission from Loret and Simões [2005]a. The basic assumption is the notion of effective stress, to be developed in Chapter 14: the stress-strain relation links the strain to this effective stress $\sigma + p_{eff} \mathbf{I}$. Indeed, the presence of proteoglycans in the cartilage gives rise to a chemical pressure p_{eff}. For the strain of interest, the chemical pressure p_{eff} is maximum (in absolute value) for a fresh bath and decreases as ionic strength increases. The collagen fibers resist the external load σ and the pressure p_{eff} by stretching. Since the strain is prescribed (in fact in the axial direction only), decrease of p_{eff} implies the external load to decrease as well. The reverse phenomenon occurs when the water of the bath is refreshed. The difference in response to the two salts can be traced to the fact that the chemical pressure induced by calcium ions is smaller than by sodium ions.

13.11   The heart muscle and cell electrophysiology

The human heart beats about 60 to 80 times per minute at rest. Functionally, each beat involves a passive stage (intake of blood) and an active stage (expulsion of blood). For the organ itself, the process involves the following steps:

- step 1: *electrical conduction*: the sino-atrial node, which has a larger eigenfrequency than the other parts of the heart, is the cardiac pacemaker. It sends a signal that travels through a dedicated net: the signal reaches first the atria and next, through the His bundle and the Purkinje net, it circulates the ventricles, from the endocardium to the epicardium: the epicardium has received the electrical impulsion after about 200 ms;

- step 2: *action potential and calcium release*: the electrical signal induces a depolarizing current across the membrane of the cardiac cell (the myocyte) which gives rise to the action potential: the resting potential difference of -90 mV increases temporarily, a small amount of plasmic ions Na^+ enters the cell through dedicated channels, and this intake in turn produces a considerable modification of the internal repartition of ions Ca^{2+}, the latter being temporarily released from the sarcoplasmic reticulum;

- step 3: *establishment of cross-bridges and force development*: calcium is used immediately. Binding to the low affinity site of Troponin C, calcium induces a change of conformation of the tropomyosin that makes possible myosin heads to establish a *bond* with actin sites: the relative sliding of the myosin and actin filaments is the vehicle that develops a force.

Cell electrophysiology models aim at describing the many currents and cytoplasmic fluxes that develop, Fig. 13.11.1. Some of these currents and fluxes are driven by ionic pumps that require energy. The biological energy unit ATP is produced by the metabolism. In fact, the proportion of energy used by ionic pumps is small with respect to that necessary to ensure the bond of the myosin head on the actin sites for force development. Many models of muscle behavior use the *cross-bridge theory* of Huxley as a basic ingredient.

It is important to highlight that this description corresponds to

- an electrochemical coupling, i.e., the electrical current induces a change of ionic concentrations in the cardiomyocytes, followed by

- a chemomechanical coupling, i.e., the temporary increase of calcium concentration allows a relative motion of the myosin and actin filaments that implies force development.

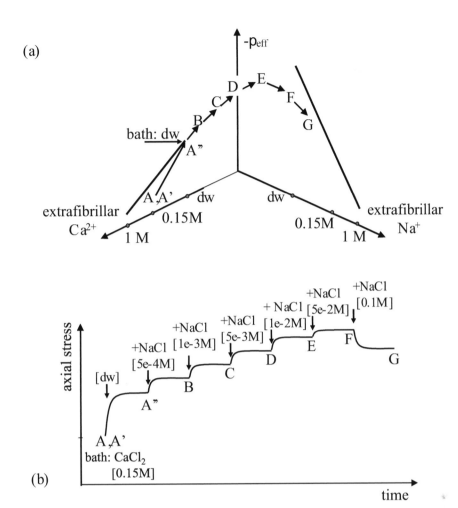

Fig. 13.10.6: Interpretation of a second test by Owens et al. [1991] via the effective pressure, reproduced with permission from Loret and Simões [2005]a. At a given axial strain of 10% counted from the isotropic free swelling configuration, the bath composition is changed periodically as indicated by the sketch (b). Sufficient time is left however for equilibrium to be reached before the subsequent modification. Initially, the bath contains only calcium cations, but the extrafibrillar water may also contain sodium cations: that is why point A has been located at short distance from the Ca-plane. The experiment starts by refreshing the bath to quasi-distilled water.

The concentration of sodium cations in the bath is next increased and, when equilibrium is reached, the water of the bath is refreshed. This sequence is repeated several times with increasing sodium concentrations.

As explained in the caption of Fig. 13.10.5, for this strain, the pressure $-p_{\text{eff}}$ is zero at maximal ionic strength and increases as ionic strength decreases. This dependence is indicated in the sketch (a) by the thick lines on the vertical planes. The quantitative variation of p_{eff} depends on the cations, and it is assumed to be larger for sodium cations than for calcium cations. The overall picture is due to the competition of two phenomena. Indeed, as the concentration of sodium ions increases, they overweigh the influence of calcium ions ($-p_{\text{eff}}$ increases). However, by the same token, electrical shielding begins to be significant, and the pressure $-p_{\text{eff}}$ finally decreases.

It is worth to recall that the effective pressure p_{eff} does not depend directly on the bath composition but on the extrafibrillar electrolyte. Still, the change of chemical composition of the bath indicated in sketch (b) is guessed to induce the qualitative path shown in sketch (a) and the associated change in tension shown in sketch (b).

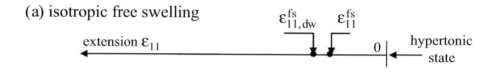

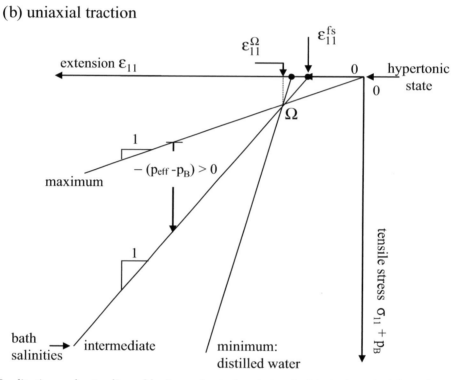

Fig. 13.10.7: Qualitative understanding of basic mechanical and chemical tests in uniaxial tension, reproduced with permission from Loret and Simões [2005]a. Specimens are in equilibrium with a bath of controlled ionic strength and pressure p_B. The presence of sodium or calcium cations in the extrafibrillar water shields the repulsive forces due to negatively charged proteoglycans, and, at given strain, modifies the total stress.

(a) Isotropic free swelling consists in decreasing the ionic strength of the bath starting from the hypertonic state with a zero external traction. The fresher the bath, the larger the free swelling strain: it vanishes in the hypertonic state, and it is maximal for distilled water (denoted by dw).

(b) At small strains, uniaxial traction shows that the stress varies nonlinearly with the strain, e.g., Kempson et al. [1968], [1973]. This nonlinearity is not shown here to simplify the understanding of the chemomechanical coupling phenomenon. The apparent moduli decrease with the ionic strength. Isotropic free-swelling strains are then obtained by extrapolation to zero axial stress (a). The sketch is in qualitative agreement with the experimental data shown in Fig. 4 of Akizuki et al. [1986]: in particular, there exists a balance strain, indicated here with the superimposed index Ω, which separates opposite effects associated with chemomechanical coupling. Above this balance strain, decreasing the ionic strength of the bath makes $p_{eff} - p_B$ more negative and thus induces a larger stress. The opposite phenomenon occurs below the balance strain.

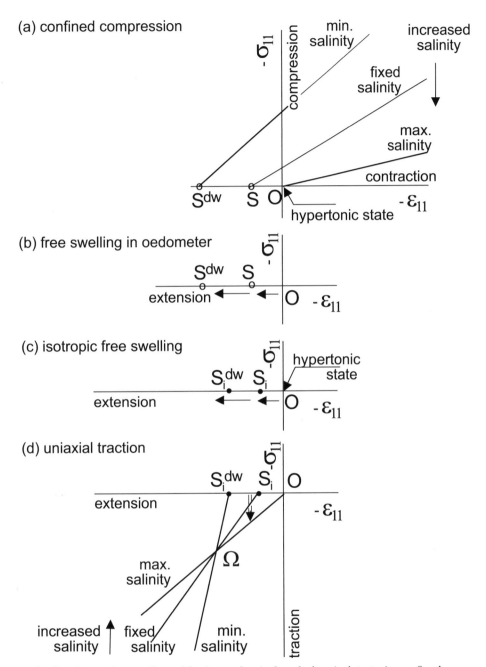

Fig. 13.10.8: Qualitative understanding of basic mechanical and chemical tests in confined compression, oedometric free swelling, isotropic free swelling and uniaxial traction, reproduced with permission from Loret and Simões [2004].

Specimens are in equilibrium with a bath of fixed ionic strength and pressure p_B taken as reference. Confined compression tests show apparent linear elasticity with modulus decreasing with the ionic strength (a). Oedometric free-swelling strains are then obtained by extrapolation to zero axial stress (b). Isotropic free-swelling with all stress components vanishing induces smaller strain components, which are equal only if the material behavior is isotropic (c). If starting from the hypertonic state, isotropic free-swelling is followed by uniaxial traction, the in-axis apparent moduli are observed to be larger than for confined compression at the same ionic strength of the bath, and to decrease with ionic strength of the bath due to the shielding effect.

The presence of sodium cations in extrafibrillar water shields the repulsive forces due to negatively charged proteoglycans and reduces the absolute value of the axial stress.

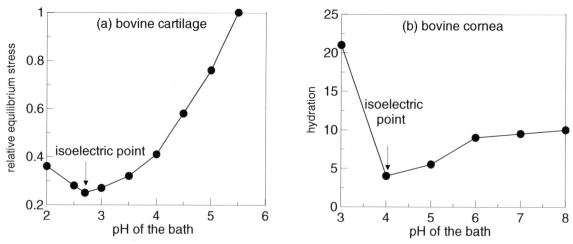

Fig. 13.10.9: pH affects the electrical charge, and therefore the electrostatic repulsion. At the isoelectric point, where the electrical charge vanishes, the repulsion is at its minimum. (a): axial stress relative to its value in distilled water in a uniaxial traction test at given axial strain, data on bovine articular cartilage redrawn from Grodzinsky et al. [1981]; (b): relative hydration (≡ wet weight/dry weight - 1) at low ionic strength of sodium chloride (30 mM) and under zero stress, data on bovine cornea from Huang and Meek [1999]. Reproduced with permission from Loret et al. [2004]a.

But none of these two steps involves a priori a feedback, that is

- neither a mechanochemical effect, i.e., the force developed by the muscle does not influence the chemical intra- or extracellular concentrations,

- nor a chemoelectrical feedback, that is the electrical current created by the ions moving in and out of the cells does not modify the electric signal originated from the sino-atrial node.

However, due to some physiological observations, e.g., commotio cordis, there has been recently a renewed interest in so-called stretch-activated ionic channels through which mechanics impacts directly on ionic currents. Along another avenue, in view of mechanical tests on muscle fibers, we will come to implicate rate-effects in the mechanochemical feedback.

13.12   Reverse couplings

The above examples address essentially the effects of the chemical composition on the mechanical properties. Conversely, the mechanical environment may influence the time course of the chemical composition. However, significant reverse mechanochemical effects are much more seldom considered than the direct chemomechanical effects. Besides stretch-dependent ionic currents, one may cite,

- the influence of confinement on the breakthrough time of a diffusion process;

- the influence of the intraocular pressure on the fixed charge density in the stroma, Section 19.11.4;

- the fact that the kinetics of some enzymes, like the matrix metalloproteinases that affect collagen degradation, are sometimes suggested to be strain-dependent.

13.12.1   Mechanoelectric feedback in the heart muscle

13.12.1.1   Commotio cordis and stretch-gated ionic channels

Stretch-activated ionic channels located on the membrane of cardiac cells (SACs) are still subject to debate because they have not been cloned in adult mammalian myocytes but observed in cultured cells and in young cells. Interest in these channels has been renewed due to their link to *commotio cordis* which is observed in young adults, because they may originate arrhythmia and fibrillation. The Oxford group has contributed to suggest models and to stress the physiological importance of the phenomemon, Kohl and Sachs [2001]. Commotio cordis is associated with the impact of a projectile on the chest leading to sudden cardiac death, but without creating material damage. In fact,

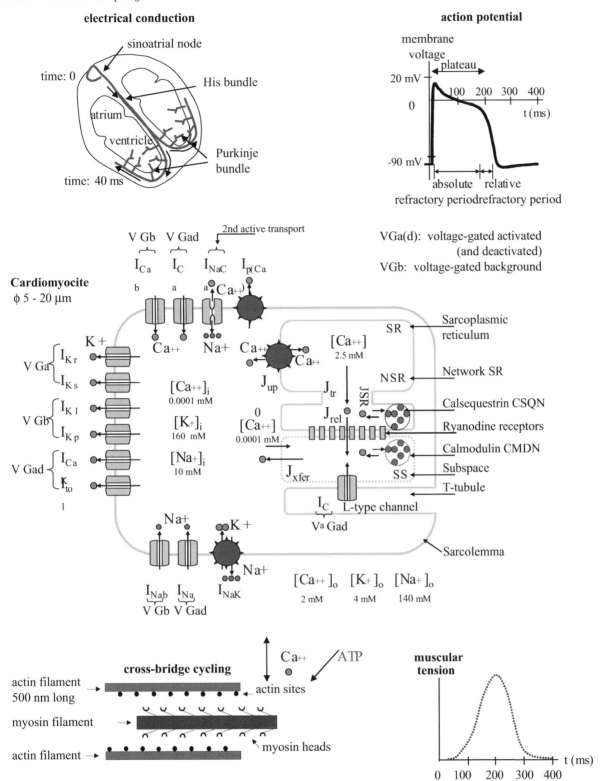

electrical conduction

action potential

VGa(d): voltage-gated activated
(and deactivated)

VGb: voltage-gated background

Fig. 13.11.1: Sketch of the successive steps of a cardiac period. The signal starts from the sinoatrial node and reaches quickly the endocardium and next the pericardium. It gives rise to a potential difference across the cell membrane, which in turn triggers complex currents between the cytoplasm and the extracellular milieu as well as intracellular fluxes. The main effect is a temporary release of calcium which is necessary to ensure cross-bridge cycling and force development. The central sketch illustrates the membrane currents and intracellular fluxes in the cardiomyocyte electrophysiology model of Winslow et al. [1999]. Redrawn with permission from Loret et al. [2004]a.

the potential danger of such impacts depends much on the precise time during the heart beat at which they occur: they are much more dangerous if applied during diastole.

Quite generally, SACs are thought to contribute to depolarize cell membrane, essentially during diastole, alter/prolongate action potential, raise the rest potential, increase the active force and induce premature ventricular excitations and arrhythmias.

In practice, SACs just create another additional current I to be added to the list of membrane currents. They are controlled by some representative of the stretch of the muscle fibers. Stretch λ, or increase of sarcomere length SL, is believed to increase the electric conductance. Several models, ion-*insensitive* or ion-sensitive, have been proposed. A general expression has the form $I = g(\lambda)(V - E_r)$ where E_r is the reverse potential. For $V < E_r$, this model indicates an inward/depolarizing current, while the opposite applies for $V > E_r$. That is, the maximum potential is lowered while the rest potential is increased and the descending branch of the voltage is modified/raised if the reverse potential is sufficiently high.

Stretch-activated channels participate to the mechanoelectric feedback. However, there has been doubt that stretch itself is the sole agent responsible for phenomena that can be traced to belong to this kind of effect. Indeed, strain rate or tension are other candidates to enter the game.

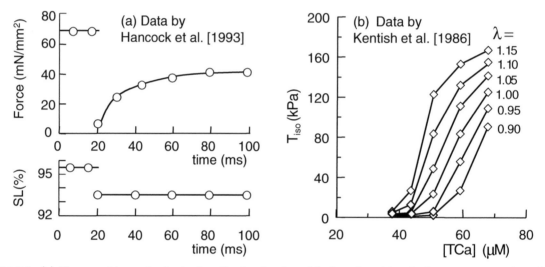

Fig. 13.12.1: (a) Upon rapid muscle shortening, the tension is suddenly reduced but it does not recover its value after transients are wiped off, even if the elastic correction is applied. (b) The increase of the isometric tension with the concentration of calcium bound to Troponin C is 'length-modulated'. Redrawn with permission from Loret et al. [2004]a.

13.12.1.2   Rate-dependent cell electrophysiology

Upon rapid shortening the isometric tension T_{iso} of muscle fibers decreases and, after the transient effects have been wiped off, it returns to a lower value, Fig. 13.12.1(a). This *force deficit* phenomenon is attributed to an increased (rate of) Ca-detachment from Troponin C. This force deficit is too large to be due only to an elastic parallel contribution. That elastic component is on the other hand sufficient to define the force at its steady state for instantaneous (re-)stretching. Therefore a *force recovery* phenomenon does not exist during (re-)stretching.

One way to describe the force deficit phenomenon is to consider a cell-electrophysiology model that depends on the rate of stretch or on the tension transients represented by the ratio T/T_{iso} of the tension over its isometric value. Hunter et al. [1998] modify the kinetics of the Ca^{2+} binded to Troponin C,

$$\frac{d[TCa]}{dt} = k_{+1}[Ca^{2+}]_i\Big([TCa]_{max} - [TCa]\Big) - k_{-1}[TCa], \qquad (13.12.1)$$

with $[Ca^{2+}]_i$ the cytoplasmic calcium concentration, k_{+1} and k_{-1} the rates of attachment and detachment of calcium on Troponin C. The rate of detachment $k_{-1} = k_{-1}^0\big(1 - (T/T_{iso})/\gamma\big)$, with $\gamma > 0$, increases as the tension T decreases, so that a lower concentration $[TCa]$ of Ca^{2+} binded to Troponin C induces a lower isometric tension, Fig. 13.12.1(b).

13.12.2 Mechanobiology and engineered cartilages

Immature articular cartilage contains vessels that circulate nutrients. On the other hand, adult articular cartilage is aneural, avascular and alymphatic. Nutrients from the synovial joint are transported by the extracellular fluid through diffusion and convection. Therefore the self-healing capacities of adult articular cartilage are practically nil. An introduction to tissue engineering is presented in Chapter 20.

Although experiments tend to quantify the variations in the rates of stimulation and inhibition of cartilage synthesis, the numbers should be taken with care. Indeed, the depth-distribution of strain in a cartilage layer submitted to compression at its top surface is highly inhomogeneous. Strains are much larger near the surface and thus the percentage of tissue undergoing mechanical stimuli strongly depends on the layer thickness. These heterogeneities on the other hand may imply that there are optimal geometries for the loaded cultures. The biological environment should also be controlled, e.g., hypoxic conditions are also known to stimulate regeneration of cartilage.

Special bioreactors have been developed to apply unconfined compression on cartilage cultures. The following points have been highlighted, Buschmann et al. [1992] [1995], Mauck et al. [2000]:
- static (prolongated) compression has a detrimental influence on the synthesis of the ECM;
- dynamic (time variable) compression on the other hand contributes to matrix synthesis. The actual efficiency of dynamic compression depends on the frequency and on the details of the loading program (amplitude, mean value, etc.).

Unconfined compression, which is accompanied by lateral extension, is thought to represent correctly the *in vivo* state of stress. On the other hand, the loading programs are highly arbitrary and in particular the pros and cons of strain- over stress-controlled loadings are not elucidated. Therefore, there are plenty of degrees of freedom for optimization.

Mauck et al. [2000] loaded dynamically disks about 1 mm thick with a strain amplitude of 10%, at a frequency of 1 Hz, 3 times 1 hour on, 1 hour off, 5 days a week during 4 weeks. At the end of this period, the elastic modulus is increased sixfold over unloaded disks. They also observed an increase in glycosaminoglycans (GAGs) and hydroxyproline (measurements use [^{35}S]sulfate and [^{3}H]proline radiolabel incorporation). However, the GAG content showed a peak at week 3, and further studies showed that it is later substantially reduced. On the other hand, Buschmann et al. [1992] observed, on agarose-seeded calf chondrocytes, an increase of GAGs until day 70. Another conclusion of the studies of Mauck et al. [2000] is that agarose gels are more prone than alginate gels to improve the mechanical properties and matrix proliferation.

Whether mechanics and metabolism cooperate or compete in the growth and resorption of cartilage is an open question which probably does not have a global answer. The intensity and mode of application of the mechanical load and the respective concentrations of macromolecules are key factors. One possible way in which mechanics stimulates biological synthesis is by generation of *advective flow* that particularly enhances the transport of large soluble molecules (e.g., growth factors) that are thus more accessible to the chondrocytes and their immediate neighborhood, Garcia et al. [1996]. A second mode in which mechanics acts is by *altering the metabolism*. Deforming the chondrocytes and their organelles modifies intracellular signaling and communication between cells.

The effects of shear-stress have also been tested. M.S. Lee et al. [2002] observe that a fluid-induced shear-stress (1.64 Pa) on human chondrocytes during periods of 2 to 24 hours increases more than twofold the production of nitric oxide (NO), an intra- and extracellular messenger present at increased levels in osteoarthritis, and inhibits aggrecan and type II collagen mRNA levels by about a third. These observations confirm the alternative point of view that cartilage degeneration and ossification are stimulated by shear stress but inhibited by dynamic compression, Carter and Wong [1990].

13.13 A partially coupled chemo-poroelastic model

Interactions between chemical content and mechanics are introduced in the chemo-poroelastic model below due to Myers et al. [1984]: the basic motivation is that increase of the concentration of the solute at constant stress implies contraction of the porous medium by expulsion of water. The underlying physical phenomena are investigated in detail in relation with articular cartilages in Chapter 14.

13.13.1   The chemo-poroelastic field and constitutive equations

The constituents being incompressible, the following field equations should be satisfied pointwise,

Balance of mass of the porous medium $\text{div } (n^{\text{s}}\,\mathbf{v}_{\text{s}} + n^{\text{w}}\,\mathbf{v}_{\text{w}}) = 0,$

Balance of mass of the chemical $\dfrac{\partial C}{\partial t} + \text{div } \mathbf{J} = 0,$

Balance of momentum of solid phase $\text{div } \boldsymbol{\sigma}^{\text{s}} + k_{\text{Sd}}(\mathbf{v}_{\text{w}} - \mathbf{v}_{\text{s}}) - p_{\text{w}}\boldsymbol{\nabla} n^{\text{w}} = \mathbf{0}\,,$

Balance of momentum of fluid phase $\text{div } \boldsymbol{\sigma}^{\text{w}} - k_{\text{Sd}}(\mathbf{v}_{\text{w}} - \mathbf{v}_{\text{s}}) + p_{\text{w}}\boldsymbol{\nabla} n^{\text{w}} = \mathbf{0}\,,$

$(13.13.1)$

where $\mathbf{v}_k$, $k = \text{s}, \text{w}$, denotes the velocity of phase k. The partial stresses in the solid and fluid phases sum up to the total stress, $\boldsymbol{\sigma} = \boldsymbol{\sigma}^{\text{s}} + \boldsymbol{\sigma}^{\text{w}}$. The partial stress in the fluid is equal to $\boldsymbol{\sigma}^{\text{w}} = -n^{\text{w}}\,p_{\text{w}}\,\mathbf{I}$, p_{w} being the intrinsic fluid pressure. The volume fractions n^k, $k = \text{s}, \text{w}$, are implicitly assumed to remain fixed. The concentration of the chemical $C = C(x,t)$ [unit : mole/m^3] obeys a diffusion equation, that is controlled by the coefficient of molecular diffusion (or chemical diffusivity) D [unit : m^2/s]. Indeed, the diffusion phenomenon is governed by Fick's law $\mathbf{J} = -D\,\boldsymbol{\nabla} C$ that relates the flux $\mathbf{J} = C\,(\mathbf{v}_c - \mathbf{v}_{\text{w}})$, where $\mathbf{v}_c$ is the absolute velocity of the chemical, to the gradient of concentration $\boldsymbol{\nabla} C$, leading to the decoupled equation of mass balance of the chemical,

$$\frac{\partial C}{\partial t} - D\,\frac{\partial^2 C}{\partial x^2} = 0\,. \tag{13.13.2}$$

The chemomechanical constitutive equation,

Chemo $-$ poroelastic equation $\boldsymbol{\sigma}' = \lambda\,\text{tr } \boldsymbol{\epsilon}'\,\mathbf{I} + 2\,\mu\,\boldsymbol{\epsilon}'\,,$ $(13.13.3)$

involves the poroelastic *effective* stress $\boldsymbol{\sigma}'$, that accounts for poroelastic effects, and the chemoelastic *effective* strain $\boldsymbol{\epsilon}'$, that accounts for chemical effects.

$$\boldsymbol{\sigma}' = \boldsymbol{\sigma} + p_{\text{w}}\,\mathbf{I}\,, \quad \boldsymbol{\epsilon}' = \boldsymbol{\epsilon} + \alpha_c\,C\,\mathbf{I}\,. \tag{13.13.4}$$

The poroelastic equation is phrased in terms of the Lamé moduli of the drained solid skeleton, or matrix, namely λ and μ, or equivalently in terms of its drained Young's modulus E and Poisson's ratio ν.

By convention, $C = 0$ for a fresh solution (quasi-distilled water). At zero effective stress, the effective strain vanishes as well and the total strain $\boldsymbol{\epsilon}$ is equal to $-\alpha_c\,C\,\mathbf{I}$: increase of the chemical concentration C from a fresh solution leads to contraction. The coefficient of chemical contraction $\alpha_c \geq 0$ [unit : m^3/mole] plays a role similar to the coefficient $c_{\text{T}} > 0$ of thermal expansion: the thermoelastic strain associated with a change of temperature ΔT would be $\boldsymbol{\epsilon}' = \boldsymbol{\epsilon} - c_{\text{T}}\,\Delta T\,\mathbf{I}$.

The two balances of momentum are equivalent to

Balance of momentum of porous medium $\text{div } \boldsymbol{\sigma} = \mathbf{0}\,,$

Darcy's law $n^{\text{w}}\,(\mathbf{v}_{\text{w}} - \mathbf{v}_{\text{s}}) = -k_{\text{H}}\,\boldsymbol{\nabla} p_{\text{w}},$

$(13.13.5)$

where the hydraulic permeability $k_{\text{H}} = (n^{\text{w}})^2/k_{\text{Sd}}$ [unit : m^2/Pa/s] is equal to the intrinsic permeability [unit : m^2] of the porous medium divided by the viscosity [unit : Pa×s].

13.13.2   Uniaxial deformation under chemical loading

Myers et al. [1984] consider a strip of cartilage of sizes $L_1 \times L_2 \times L_3$ with $L_1 \ll L_2 \ll L_3$. At time $t = 0$, the strip is bathed in a solution of given concentration C_0. It is maintained between two grips that prevent deformation along the main direction of the strip. Close to the mid-height in the intermediate direction, the strain is mainly unidirectional along the thin direction of thickness $2\,l = L_1$.

The assumption of a unidirectional solid displacement $u_{\text{s}} = u_{\text{s}}(x)$, for $x \in [-l, l]$, is adopted subject to the initial conditions,

$$(\text{IC})_u \quad u_{\text{s}}(x, t = 0) = 0\,, \tag{13.13.6}$$

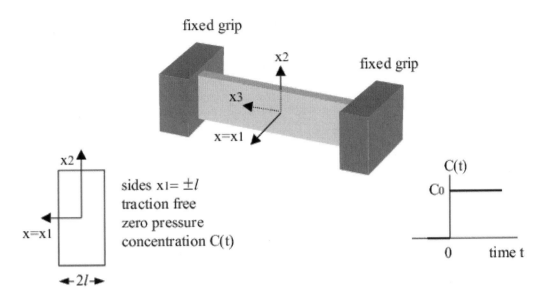

Fig. 13.13.1: A cartilage strip is maintained by two grips. In the central part, only the lateral deformation along x_1 can be considered to be nonzero. At time $t = 0$, the tissue is put in contact with a solution of controlled chemical composition.

and boundary conditions for the displacement and total (and effective) traction,

$$\text{(BC)}_u : \quad u_{\mathrm{s}}(x = 0, t) = 0; \quad \text{(BC)}_{\sigma_x} : \quad \sigma'_{xx}(x = \pm l, t) = \left(H \frac{\partial u_{\mathrm{s}}}{\partial x} + 3 K \alpha_c C\right)(x = \pm l, t) = 0, \qquad (13.13.7)$$

where H and K are respectively the confined compression modulus and the drained bulk modulus,

$$H = \lambda + 2\mu = 2\mu \frac{1 - \nu}{1 - 2\nu}, \quad K = \lambda + \frac{2}{3}\mu = \frac{2}{3}\mu \frac{1 + \nu}{1 - 2\nu}. \qquad (13.13.8)$$

The second condition (13.13.7) results from the fact that both the normal stress and pore pressure vanish on the sides of the strip,

$$\text{(BC)}_{p_{\mathrm{w}}} : \quad p_{\mathrm{w}}(x = \pm l, t) = 0, \qquad (13.13.9)$$

and accounts for symmetry with respect to the axis $x = 0$. The balance of mass $(13.13.1)_1$ can be integrated in space with help of the symmetry conditions,

$$n^{\mathrm{w}} v_{\mathrm{w}} = -n^{\mathrm{s}} \frac{\partial u_{\mathrm{s}}}{\partial t}. \qquad (13.13.10)$$

Therefore Darcy's law $(13.13.5)_2$ yields the gradient of fluid pressure in terms of the solid velocity,

$$k_{\mathrm{H}} \frac{\partial p_{\mathrm{w}}}{\partial x} = \frac{\partial u_{\mathrm{s}}}{\partial t}. \qquad (13.13.11)$$

Introduction of the elastic equation (13.13.3) in the field equation $(13.13.5)_1$ and use of (13.13.11) yields the field equation satisfied by the displacement,

$$\text{(FE)}_u \quad \frac{\partial^2 u_{\mathrm{s}}}{\partial x^2} - \frac{1}{H k_{\mathrm{H}}} \frac{\partial u_{\mathrm{s}}}{\partial t} + 3 \alpha_c \frac{K}{H} \frac{\partial C}{\partial x} = 0. \qquad (13.13.12)$$

Elimination of the solid velocity between (13.13.11) and (13.13.12) and integration in space, accounting for the vanishing of pore pressure on the boundaries $\pm l$, provides the pore pressure,

$$p_{\mathrm{w}}(x, t) = H \left(\frac{\partial u_{\mathrm{s}}}{\partial x}(x, t) - \frac{\partial u_{\mathrm{s}}}{\partial x}(l, t)\right) + 3 \alpha_c K \left(C(x, t) - C(l, t)\right). \qquad (13.13.13)$$

The diffusion of the chemical uncouples from deformation,

$$\text{(FE)}_C \quad \frac{\partial C}{\partial t} - D \frac{\partial^2 C}{\partial x^2} = 0,$$

$$\text{(IC)}_C \quad C(x, t = 0) = 0, \qquad (13.13.14)$$

$$\text{(BC)}_C : \quad C(\pm l, t) = C_0 \, \mathcal{H}(t).$$

The chemical loading is the sole loading source. Diffusing through the medium, the chemical will be shown to give rise to shrinking, due to chemomechanical coupling, and to an accompanying axial tensile force.

13.13.3 Diffusion of the chemical

In order to solve the diffusion equation for the concentration C, both a change of spatial variable $x \to y = x + l$, and a change of unknown $C(x,t) \to \tilde{C}(y,t) = C(y,t) - C_0$ are defined.

Elementary solutions are sought in separable form $\tilde{C}(y,t) = Y(y) T(t)$, so that $Y''/Y = T'/(D T)$ has to be a negative constant, say $-a^2$, so as to ensure that the solution does not grow unboundedly in time. Thus $T(t) \sim \exp(-a^2 D t)$, and $Y(y) = b \cos(ay) + c \sin(ay)$. Requiring the boundary conditions to be satisfied yields $b = 0$, $a = n\pi/(2l)$, with n an arbitrary integer. Thus the elementary solutions,

$$\tilde{C}_n(y,t) = c_n \sin(n\pi \frac{y}{2l}) \exp(-n^2 \frac{\pi^2}{(2l)^2} D t), \tag{13.13.15}$$

satisfy the field and boundary conditions for any c_n. The latter coefficients are obtained by imposing the initial condition $\tilde{C}(y,0) = -C_0$. For that purpose, the odd function, equal to C_0 over $[-2l,0[$, and to $-C_0$ over $]0,2l]$, is defined. Its Fourier series involves only terms of the form (13.13.15). The coefficients c_n are obtained as $c_n = (1/l) \int_0^{2l} (-C_0) \sin(n\pi y/(2l)) \, dy$, namely $c_{2n} = 0$, $c_{2n+1} = -2 C_0/X_n$ with $X_n = (n+1/2)\pi$ for $n \geq 0$. Consequently the concentration has the following analytical expression,

$$\frac{C(x,t)}{C_0} = 1 - 2 R(\frac{x}{l}, \frac{t}{t_C}), \tag{13.13.16}$$

where, for $t_X = t_C$, t_H,

$$R(\frac{x}{l}, \frac{t}{t_X}) = \sum_{n=0}^{\infty} \frac{(-1)^n}{X_n} \cos\left(X_n \frac{x}{l}\right) e^{-X_n^2 t/t_X}. \tag{13.13.17}$$

The characteristic time of chemical diffusion,

$$t_C = \frac{l^2}{D}, \tag{13.13.18}$$

is to be compared with the characteristic time of seepage,

$$t_H = \frac{l^2}{H k_H}. \tag{13.13.19}$$

This initial and boundary value problem for the chemical can also be solved through the Laplace transform technique, e.g., Mei [1995], p. 267.

13.13.4 The solution for the displacement

13.13.4.1 The solution in the Laplace domain

If the displacement is measured from time $t = 0$, namely $u_s(x,0) = 0$, the Laplace transform $\bar{u}_s(x,s)$ of $u_s(x,t)$ satisfies the non homogeneous differential equation,

$$\frac{\partial^2 \bar{u}_s}{\partial x^2} - \frac{\bar{u}_s}{b^2} = \overline{F}(x,s), \tag{13.13.20}$$

with $b = \sqrt{H k_H/s}$. The rhs $\overline{F}$ can be read from (13.13.12). The solution is sought in the form,

$$\bar{u}_s(x,s) = A(x,s) \cosh \frac{x}{b} + B(x,s) \sinh \frac{x}{b}, \tag{13.13.21}$$

where the functions $A(x,s)$ and $B(x,s)$ are required to satisfy the constraint,

$$\frac{dA(x,s)}{dx} \cosh \frac{x}{b} + \frac{dB(x,s)}{dx} \sinh \frac{x}{b} = 0. \tag{13.13.22}$$

Backsubstitution of (13.13.21) and (13.13.22), and its derivative wrt x in (13.13.20), results in the system

$$\begin{bmatrix} \cosh(x/b) & \sinh(x/b) \\ \sinh(x/b) & \cosh(x/b) \end{bmatrix} \begin{bmatrix} dA/dx \\ dB/dx \end{bmatrix} = \begin{bmatrix} 0 \\ b\,\overline{F} \end{bmatrix} \implies \begin{bmatrix} dA/dx \\ dB/dx \end{bmatrix} = b\,\overline{F} \begin{bmatrix} -\sinh(x/b) \\ \cosh(x/b) \end{bmatrix}. \tag{13.13.23}$$

Since $\overline{F} = -3\,\alpha_c\,K/H\,\partial\overline{C}/\partial x$,

$$A(x,s) = A_0(s) + 3\,b\,\alpha_c\,\frac{K}{H}\int_0^x \frac{\partial\overline{C}}{\partial x}\,\sinh\frac{x}{b}\,dx, \quad B(x,s) = B_0(s) - 3\,b\,\alpha_c\,\frac{K}{H}\int_0^x \frac{\partial\overline{C}}{\partial x}\,\cosh\frac{x}{b}\,dx. \tag{13.13.24}$$

The standard algebraic relations below will be used,

$$\int_0^x \sin(\alpha\,y)\,\sinh(\beta\,y)\,dy = \frac{\beta\,\sin(\alpha\,x)\,\cosh(\beta\,x) - \alpha\,\cos(\alpha\,x)\,\sinh(\beta\,x)}{\alpha^2 + \beta^2},$$
$$\int_0^x \sin(\alpha\,y)\,\cosh(\beta\,y)\,dy = \frac{-\alpha\,\cos(\alpha\,x)\,\cosh(\beta\,x) + \beta\,\sin(\alpha\,x)\,\sinh(\beta\,x)}{\alpha^2 + \beta^2}\,. \tag{13.13.25}$$

From (13.13.16),

$$\frac{\partial\overline{C}}{\partial x}(x,s) = \frac{2\,C_0}{l}\sum_{n=0}^{\infty}(-1)^n\,\frac{\sin(X_n\,x/l)}{s + X_n^2/t_{\mathrm{C}}}\,. \tag{13.13.26}$$

Thus with $\beta = 1/b$,

$$A(x,s) = A_0(s) + 6\,C_0\,\alpha_c\,b\,k_{\mathrm{H}}\,\frac{K}{l}\sum_{n=0}^{\infty}(-1)^n\,\frac{\beta\,\sin(X_n\,x/l)\,\cosh(\beta\,x) - X_n/l\,\cos(X_n\,x/l)\,\sinh(\beta\,x)}{(s + X_n^2/t_{\mathrm{C}})\,(s + X_n^2/t_{\mathrm{H}})},$$
$$B(x,s) = B_0(s) - 6\,C_0\,\alpha_c\,b\,k_{\mathrm{H}}\,\frac{K}{l}\sum_{n=0}^{\infty}(-1)^n\,\frac{-X_n/l\,\cos(X_n\,x/l)\,\cosh(\beta\,x) + \beta\,\sin(X_n\,x/l)\,\sinh(\beta\,x)}{(s + X_n^2/t_{\mathrm{C}})\,(s + X_n^2/t_{\mathrm{H}})}\,. \tag{13.13.27}$$

Insertion of these expressions in (13.13.21) yields

$$\overline{u}_{\mathrm{s}}(x,s) = A_0(s)\,\cosh(\beta\,x) + B_0(s)\,\sinh(\beta\,x) + 6\,C_0\,\alpha_c\,k_{\mathrm{H}}\,\frac{K}{l}\sum_{n=0}^{\infty}\frac{(-1)^n\,\sin(X_n\,x/l)}{(s + X_n^2/t_{\mathrm{C}})\,(s + X_n^2/t_{\mathrm{H}})}\,. \tag{13.13.28}$$

The constants $A_0(s)$ and $B_0(s)$ are obtained via the boundary conditions (13.13.7). The Laplace transform of the displacement finally simplifies to,

$$\overline{u}_{\mathrm{s}}(x,s) = -3\,C_0\,\alpha_c\,\frac{K}{H}\,\frac{b}{s}\,\frac{\sinh(\beta\,x)}{\cosh(\beta\,l)} + 6\,C_0\,\alpha_c\,k_{\mathrm{H}}\,\frac{K}{l}\sum_{n=0}^{\infty}\frac{(-1)^n\,\sin(X_n\,x/l)}{(s + X_n^2/t_{\mathrm{C}})\,(s + X_n^2/t_{\mathrm{H}})}\,. \tag{13.13.29}$$

13.13.4.2   The solution in the time domain

Using the retardation formula, and decomposing the denominator in the sum of two terms, the inverse transform of the second part of the solution is simply

$$6\,C_0\,\alpha_c\,\frac{K}{H}\,\frac{l}{t_{\mathrm{H}}/t_{\mathrm{C}} - 1}\left(S(\frac{x}{l}, \frac{t}{t_{\mathrm{H}}}) - S(\frac{x}{l}, \frac{t}{t_{\mathrm{C}}})\right), \tag{13.13.30}$$

where generically, for $t_{\mathrm{x}} = t_{\mathrm{C}}, t_{\mathrm{H}}$,

$$S(\frac{x}{l}, \frac{t}{t_{\mathrm{x}}}) = \sum_{n=0}^{\infty}\frac{(-1)^n}{X_n^2}\,\sin(X_n\,\frac{x}{l})\,e^{-X_n^2\,t/t_{\mathrm{x}}}\,. \tag{13.13.31}$$

The first part of displacement can be inverted through integration in the complex domain. Note first that the functions in the complex domain of the Laplace variable s are uniform, due to the parity of the function sinh and the definition $b = 1/\beta = \sqrt{H\,k_{\mathrm{H}}/s}$, so that

$$\frac{l}{b} = \pi\,\sqrt{t_{\mathrm{H}}\,s}\,. \tag{13.13.32}$$

The roots of the denominator are all simple, namely $s = 0$ and $\sqrt{t_{\mathrm{H}}\,s_n} = i\,(n + 1/2)$, that is $s_n = -(n + 1/2)^2/t_{\mathrm{H}}$, $n \geq 0$. The contribution of each root is calculated as indicated in (7.5.38), yielding

$$-3\,C_0\,\alpha_c\,\frac{K}{H}\,x + 6\,C_0\,\alpha_c\,\frac{K}{H}\,l\,S(\frac{x}{l}, \frac{t}{t_{\mathrm{H}}})\,. \tag{13.13.33}$$

From the displacement,

$$\frac{1}{C_0\,\alpha_c}\,\frac{u_s(x,t)}{l} = -3\,\frac{K}{H}\,\frac{x}{l} + 6\,\frac{K}{H}\,\frac{1}{t_H - t_C}\left(t_H\,S(\frac{x}{l},\frac{t}{t_H}) - t_C\,S(\frac{x}{l},\frac{t}{t_C})\right),\tag{13.13.34}$$

follows the strain,

$$\frac{1}{C_0\,\alpha_c}\,\frac{\partial u_s}{\partial x}(x,t) = -3\,\frac{K}{H} + 6\,\frac{K}{H}\,\frac{1}{t_H - t_C}\left(t_H\,R(\frac{x}{l},\frac{t}{t_H}) - t_C\,R(\frac{x}{l},\frac{t}{t_C})\right),\tag{13.13.35}$$

with R defined by (13.13.17). Thus the strain at the boundaries $\pm l$ takes instantaneously the final value, to which all points of the strip will ultimately converge, namely

$$\frac{\partial u_s}{\partial x}(\pm l,t) = -3\,C_0\,\alpha_c\,\frac{K}{H}\,.\tag{13.13.36}$$

This value agrees with a vanishing normal stress and a vanishing pore pressure, as indicated by (13.13.7).

13.13.5   The solution for the pore pressure

The pressure (13.13.13) is now deduced from (13.13.16) and (13.13.35), namely

$$\frac{1}{C_0\,\alpha_c}\,\frac{p_w(x,t)}{H} = 6\,\frac{K}{H}\,\frac{t_H}{t_H - t_C}\left(R(\frac{x}{l},\frac{t}{t_H}) - R(\frac{x}{l},\frac{t}{t_C})\right).\tag{13.13.37}$$

The fluid flux $J_w = n^w\,(v_w - v_s)$ results as

$$\frac{1}{C_0\,\alpha_c}\,\frac{J_w}{l/t_H} = 6\,\frac{K}{H}\,\frac{t_H}{t_H - t_C}\sum_{n=0}^{\infty}(-1)^n\,\sin(X_n\,\frac{x}{l})\left(e^{-X_n^2\,t/t_H} - e^{-X_n^2\,t/t_C}\right).\tag{13.13.38}$$

Other entities follow from the above expressions, like the effective axial strain $\epsilon'_{xx} = \epsilon_{xx} + C\,\alpha_c$, the effective axial stress $\sigma'_{xx} = H\,\epsilon_{xx} + 3\,K\,C\,\alpha_c$ or $\sigma'_{xx} = H\,\epsilon'_{xx} + (3\,K - H)\,C\,\alpha_c$. In fact, the total axial stress $\sigma_{xx} = \sigma'_{xx} - p_w$ vanishes everywhere for the 1D problem.

13.13.6   The axial force

The fact that the strains are prevented along the axis 3 gives rise to a reaction over the section of area $L_1 \times L_2$,

$$F(t) = L_2 \int_{-l}^{l}\sigma_{33}^s(x,t)\,dx\,.\tag{13.13.39}$$

The force is transmitted only by the solid phase, on the ground that the material is actually in tension along the axial direction. The stress-strain relation implies,

$$\sigma_{33}^s = \lambda\,\frac{\partial u_s}{\partial x} + 3\,K\,\alpha_c\,C - n^s\,p_w\,.\tag{13.13.40}$$

The following intermediate quantities are needed:

$$u_s(l,t) - u_s(-l,t) = 6\,C_0\,\alpha_c\,\frac{K}{H}\,\frac{l}{t_H - t_C}\left(-t_H\,Q(\frac{t}{t_H}) + t_C\,Q(\frac{t}{t_C})\right),$$

$$\int_{-l}^{l}C(x,t)\,dx = 2\,C_0\,l\,Q(\frac{t}{t_C}),\tag{13.13.41}$$

$$\int_{-l}^{l}p_w(x,t)\,dx = 6\,C_0\,\alpha_c\,K\,l\left(-Q(\frac{t}{t_H}) + Q(\frac{t}{t_C})\right)\frac{t_H}{t_H - t_C}\,,$$

where

$$Q(\frac{t}{t_x}) = 1 - 2\,S(1,\frac{t}{t_x}) = 1 - 2\sum_{n=0}^{\infty}\frac{e^{-X_n^2\,t/t_x}}{X_n^2}\,.\tag{13.13.42}$$

Note that $Q(0) = 0$ because $\sum_{n=0}^{\infty}1/(n+1/2)^2 = \pi^2/2$, and $Q(\infty) = 1$. The force can then be expressed in the format,

$$\frac{F(t)}{F_\infty} = (1 - \frac{H}{2\,\mu}\,n^w)\,\frac{t_H}{t_H - t_C}\,Q(\frac{t}{t_H}) - (\frac{t_C}{t_H} - \frac{H}{2\,\mu}\,n^w)\,\frac{t_H}{t_H - t_C}\,Q(\frac{t}{t_C}),\tag{13.13.43}$$

with F_∞ the value at large times,

$$F_\infty = 6\,\mu\,L_1\,L_2\,C_0\,\alpha_c\,\frac{K}{H}\,.\tag{13.13.44}$$

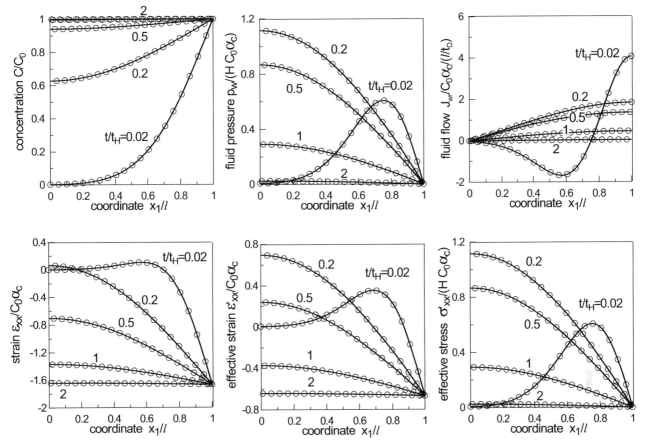

Fig. 13.13.2: Strip of cartilage submitted to step chemical loading along the boundary $x_1/l = 1$. Spatial profiles at several times.

13.13.7   Simulations of chemical loadings

Simulations using the values shown in Table 13.13.1, which are typical of articular cartilages, are displayed in Figs. 13.13.2 and 13.13.3. The associated seepage time is $t_H = 10\,\mathrm{s}$ and the minimum time for chemical diffusion is $t_C = 4\,\mathrm{s}$.

The profiles of various entities displayed in Fig. 13.13.2 correspond to the ratio $t_H/t_C = 2.5$. Consequently, the trend towards equilibrium of the concentration which is governed solely by Fick's time is faster than for the other entities whose transient behavior depends on both seepage and chemical diffusion.

In other words, chemical diffusion takes place first and the chemical penetrates the tissue progressively. Chemo-mechanical coupling implies contraction and increase of the pore fluid pressure. Since the latter is maintained to zero at the boundary, the transient pressure profile displays a maximum which enters the tissue progressively. This non monotonous profile leads to a temporary divergent flow process associated with Darcy's law.

Since the total stress is permanently zero, and the pore pressure ultimately vanishes, so does the effective stress. On the other hand, the steady states of the effective strain and total strain are contractile, in agreement with the expressions shown at the end of Section 13.13.5 since $3K - H > 0$.

The time profile of the force shown in Fig. 13.13.3 depends much on the ratio of the characteristic times. If t_H/t_C is small, i.e., seepage is fast, the force is monotonous, corresponding to the progressive penetration of the chemical into the strip. The situation is more complex when the seepage is slow, and the force displays a maximum if t_H/t_C is larger than about 0.5. In fact when this ratio is large, only the second term of (13.13.43) contributes to the force at times much larger than t_C but much smaller than t_H. Since $Q(0) = 0$ and $Q(\infty) = 1$, the maximum is bounded by $n^w H/2\mu$. Myers et al. [1984] observe that a ratio $t_H/t_C = 2.5$ and a Poisson's ratio $\nu = 0.44$ describes best their experimental time profiles.

Table 13.13.1: Constitutive parameters used in the simulations, Myers et al. [1984].

Entity	Symbol	Value	Unit
Fluid volume fraction	n^{w}	0.80	-
Drained Poisson's ratio	ν	0.25	-
Elastic shear modulus	μ	333	kPa
Hydraulic permeability	k_{H}	10^{-15}	$\mathrm{m}^2/\mathrm{Pa}/\mathrm{s}$
Coefficient of chemical diffusion	D	$0.5\text{-}25 \times 10^{-10}$	m^2/s
Coefficient of chemical contraction	α_c	0.093	liter/mole
Initial chemical concentration	C_0	0.15	mole/liter
Dimensions of specimen	$L_1 = 2\,l$	2×10^{-4}	m
	L_2	17×10^{-4}	m

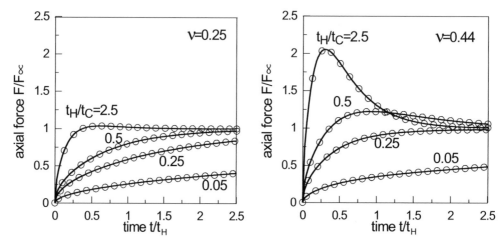

Fig. 13.13.3: Axial force along the direction 3, as indicated in Fig. 13.13.1, due to the increase of concentration of the bath, the cartilage is put in contact with at time $t = 0$. The relative values of the characteristic times for seepage and chemical diffusion and Poisson's ratio are key parameters affecting the transient profile of the force.

Exercises on Chapter 13

Exercise 13.1 Single diffuse double layer

The spatial repartition of a fixed electrical charge bathed in an electrolyte has been for long idealized, at the nanoscale, by a linear fixed attractor, charged negatively at neutral pH, adjacent to a diffuse layer of mobile ions: due to the presence of the two parallel charge distributions, the model has been called single diffuse double layer theory.

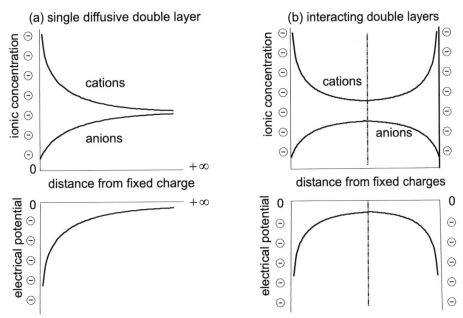

Fig. 13.E.1: Spatial profiles of the ionic concentrations and electrical potentials across (a) a single diffuse double layer and (b) interacting double layers in presence of a negative fixed charge.

A more elaborated configuration considers two parallel and identical attractors separated by a symmetric diffuse layer of electrolyte, so called interacting double layers, Fig. 13.E.1. Mitchell [1993], p. 111 et seq., contains a compact summary of developments in the area, as well as references to the subject.

Here we consider only a single diffuse double layer theory.

Assumptions:

A1. Electroneutrality is satisfied globally, not locally. Indeed, it is not satisfied pointwise in the diffuse layer, but the sum of all charges in the diffuse layer and in the attractor vanishes. On the other hand, pointwise electroneutrality is satisfied far from the attractor at $z = \infty$.

A2. Electrochemical equilibrium holds pointwise;

A3. The ionic strength $I(\infty)$ and electrical potential $\phi(\infty)$ are known far from the attractor.

The electrical potential $\phi(0)$ at the surface of the attractor $z = 0$ is obtained as a result of the analysis, if the charge of the attractor is given, and conversely: the issue is addressed in Exercise 13.2.

1. Prove the Boltzmann relation linking the molar fraction x, or concentration c, to the electrical potential ϕ,

$$\frac{x(z)}{x_\infty} = \frac{c(z)}{c_\infty} = \exp\left(-\frac{F\zeta}{RT}[\phi](z)\right), \quad [\phi](z) \equiv \phi(z) - \phi(\infty). \tag{13.E.1}$$

Hint: Electrochemical equilibrium of a species with valence ζ implies its electrochemical potential $g^{ec} = RT \operatorname{Ln} x + \zeta F\phi$ to be spatially uniform, in particular equal to its value at $z = +\infty$. Whence the Boltzmann relation.

2. Assume two species of valences ζ_+ and ζ_- to be present so that their respective electrical densities are $I_e^+ = F(\zeta c)_+$ and $I_e^- = F(\zeta c)_-$. Show that the electric density expresses in terms of the electrical potential as

$$
\begin{aligned}
I_e &= I_e^+ + I_e^- \\
&= I_e^+(\infty)\left(\exp\left(-\frac{F\zeta_+}{RT}[\phi]\right) - \exp\left(-\frac{F\zeta_-}{RT}[\phi]\right)\right).
\end{aligned}
\tag{13.E.2}
$$

Hint: Use of the relations (13.E.1) applied to both species, namely

$$\frac{c_\pm}{c_\pm(\infty)} = \exp\left(-\frac{F\zeta_\pm}{RT}[\phi]\right),$$

(13.E.3)

and of the electroneutrality condition far from the attractor,

$$I_e(\infty) = I_e^+(\infty) + I_e^-(\infty) = 0,$$

(13.E.4)

provides (13.E.2).

3. Applying Poisson's equation (13.6.4),

$$\frac{d^2\phi}{d^2z} = -\frac{I_e}{\epsilon},$$

(13.E.5)

obtain the Poisson-Boltzmann differential equation for the sole electrical potential,

$$\frac{d^2\phi}{d^2z} = -\frac{I_e^+(\infty)}{\epsilon}\left(\exp\left(-\frac{F\zeta_+}{RT}[\phi]\right) - \exp\left(-\frac{F\zeta_-}{RT}[\phi]\right)\right).$$

(13.E.6)

Here ϵ [unit : F/m] is the permittivity of the medium.
Hint: Use relation (13.E.2) in the Poisson's equation.

4. Assuming a small variation of the electrical potential, so that the exponentials can be approximated by a first order Taylor expansion, show that the electrical potential decreases exponentially from the attractor to infinity with a length constant $1/\kappa$ called the Debye length,

$$\frac{\phi(z) - \phi(\infty)}{\phi(0) - \phi(\infty)} = \exp(-\kappa z), \quad \frac{1}{\kappa} = \sqrt{\frac{RT\,\epsilon}{2\,I(\infty)}}\,\frac{1}{F},$$

(13.E.7)

where I is the *ionic strength* of the solution [unit : mole/m^3] (replace I by 1000 I if I is expressed in mole/liter),

$$I = \frac{1}{2}(\zeta_+^2\,c_+ + \zeta_-^2\,c_-).$$

(13.E.8)

Note that the Debye length does not depend on the charge σ_e of the attractor.
Hint: Linearization of the Poisson-Boltzmann differential equation yields

$$\frac{d^2\phi}{d(\kappa z)^2} = [\phi].$$

(13.E.9)

The integration constants are readily obtained from the condition at the surface of the attractor $z = 0$ and at infinity. The ionic strength comes into picture by use of the electroneutrality condition (13.E.4).

The Debye length represents the range of influence of the attractor. It decreases as the ionic strength at infinity increases: the mobile ions in the electrolyte are said to shield the influence of the attractor.

The influence of the solvent is present through its dielectric constant $\epsilon_r = \epsilon/\epsilon_0$. Clearly the larger ϵ_r, the longer the Debye length.

A multivalent cation is also seen to shield more efficiently the attractor. For sodium chloride and calcium chloride, the ionic strength at infinity is $2\,I = 6\,c_{Ca}$ and $2\,I = 2\,c_{Na}$ respectively. Hence, the Debye length for calcium chloride is $1/\sqrt{3}$ smaller than the Debye length for sodium chloride at equal cationic concentration $c_{Ca} = c_{Na}$ at infinity.

At a macroscopic level, increase of ionic strength will imply a reduced Debye length, corresponding to closer PGs, and therefore to overall contraction.

Once the potential ϕ is known, the spatial variations of the concentrations of the two species are read from (13.E.3). The density of counterions, endowed with a charge opposite to the attractor, is expected to decrease from the attractor surface $z = 0$ to infinity where it is given. By contrast, the density of co-ions, of charge of the same sign as the attractor, is expected to be small close to the attractor and to increase to the prescribed value at infinity.

5. Calculate the Debye length as a function of the concentration (or ionic strength) $c^+(\infty) = I_e^+(\infty)/F$ for a binary symmetric electrolyte $\zeta_+ = -\zeta_- = \zeta$.
Answer: With the constants given in Chapter 25, and with $c^+(\infty)$ in mole/liter,

$$\frac{1}{\kappa} = \frac{3.044}{\zeta\sqrt{c^+(\infty)}}\quad[\text{unit}:\text{Å}].$$

(13.E.10)

6. Restricting now the analysis to symmetric binary electrolytes, $\zeta_+ = -\zeta_- = \zeta$, reconsider determining the electrical potential without linearization. Show that the Debye length (13.E.7) remains unchanged and that the solution $[\tilde\phi](z) = (\zeta\,F/RT)\,(\phi(z) - \phi(\infty))$ is given in the format,

$$[\tilde\phi](z) = 2\,\mathrm{Ln}\,\frac{1 + \exp(-\kappa z)\,\tanh([\tilde\phi(0)]/4)}{1 - \exp(-\kappa z)\,\tanh([\tilde\phi(0)]/4)}, \quad \frac{1}{\kappa} = \sqrt{\frac{RT\,\epsilon}{2\,c^+(\infty)}}\,\frac{1}{\zeta\,F}.$$

(13.E.11)

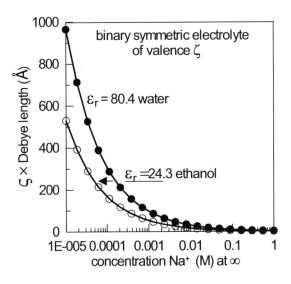

Fig. 13.E.2: Debye length times the valence of a binary symmetric electrolyte as a function of the concentration far from the attractor. The diffuse layer is larger when the solvent is water, rather than ethanol.

Fig. 13.E.2 displays typical ranges for the Debye length as a function of the concentration.

Proof:

Eqn (13.E.6) becomes

$$\frac{d^2[\tilde{\phi}]}{d(\kappa z)^2} = \sinh[\tilde{\phi}].$$ (13.E.12)

Accounting for the flattening of the electrical potential at infinity, the expression below can be checked to satisfy the second order differential equation above,

$$\frac{d(\frac{1}{2}\tilde{\phi})}{d(\kappa z)} = -\sinh\left(\frac{1}{2}[\tilde{\phi}]\right).$$ (13.E.13)

The remaining integration

$$\frac{d(\frac{1}{2}\tilde{\phi})}{\sinh\left(\frac{1}{2}[\tilde{\phi}]\right)} = -d(\kappa z),$$ (13.E.14)

is performed through the change of variable $[\tilde{\phi}]/2 \to \exp([\tilde{\phi}]/2)$, and the boundary condition at $z = 0$.

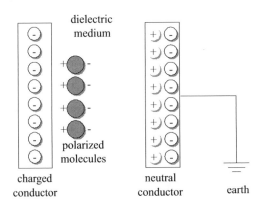

Fig. 13.E.3: Sketch of a typical capacitor device.

Exercise 13.2 The constant capacitance model

1. Basic definitions of capacitors

A capacitor can be represented by two conductor plates separated by an insulating dielectric medium in which the molecules can be polarized to form electric dipoles. One plate has a fixed charge, while the second charge is often neutralized through an earth connection, Fig. 13.E.3. The capacitance G_e [unit : F=C/V] is equal to the charge Q required to increase the potential ϕ by one volt, $G_e = Q/\phi$. It depends only on the geometry of the capacitor and on the dielectric constant of the medium.

The energy of a capacitor is equal to the work done in charging it, namely $\int_0^Q \phi \, dQ = \frac{1}{2} G_e \phi^2$, or $\frac{1}{2} Q^2/G_e$.

Connected in parallel (same potential but individual charges), the capacitance of two capacitors is equal to the sum of the individual capacitances. Conversely, connected in series (same charge but individual potentials), the inverse capacitance of set is equal to the sum of the inverses of the individual capacitances.

Consider a capacitor consisting of two parallel plates of area S separated by a distance d and with respective charges $(0, Q)$ and $(-Q, +Q)$. According to Gauss's law (13.6.3), the difference of potentials between the two plates ϕ and the charge Q are linked by the relation $\epsilon \, (\phi/d) \, S = Q$: hence the capacitance of the two parallel plates is equal to $G_e = \epsilon \, S/d$, where ϵ [unit : F/m] is the permittivity.

An isolated sphere of radius r_0 can be seen as the inner part of a capacitor whose external neutralized part is a sphere whose radius is large with respect to r_0. With help of (13.6.5), its capacitance has the expression,

$$G_e = \frac{Q}{\phi} = 4 \, \pi \, \epsilon \, r_0 \, . \tag{13.E.15}$$

The diffuse double layer can be seen as a (half) capacitor. Indeed, consider an attractor of surface charge σ_e [unit : C/m^2] and the adjacent diffuse layer extending from the attractor $z = 0$ to infinity. As indicated in the assumptions of Exercise 13.1, the sum of all charges vanishes,

$$\sigma_e + \int_0^\infty I_e \, dz = 0 \, . \tag{13.E.16}$$

The electrical charge may be eliminated through the Poisson's relation (13.E.5),

$$\sigma_e = \int_0^\infty \epsilon \frac{d^2\phi}{dz^2} \, dz = -\epsilon \frac{d\phi}{dz}(0) \, , \tag{13.E.17}$$

where the fact that the electrical potential flattens at infinity has been used, Eqn (13.E.7). The second relation above has also assumed the permittivity ϵ to be spatially uniform. This spatial property holds if the permittivity is seen as a property of the solvent only. It is more questionable if ϵ is interpreted as a property of the electrolyte, since the spatial repartition of the ions is not uniform.

2. If the linearization of exponentials, as in Exercise 13.1, is accurate enough, show that the diffuse double layer can be seen as a capacitor of areal capacitance $\kappa \epsilon$ [unit : F/m^2], namely

$$\phi(0) - \phi(\infty) = \frac{\sigma_e}{\kappa \, \epsilon} \, . \tag{13.E.18}$$

Hence, if the attractor is negatively (respectively positively) charged, the potential is lower than at infinity (respectively higher): it attracts cations (respectively anions) and plays the role of a cathode (respectively anode).

Proof: The derivative of the potential can be estimated by Eqn (13.E.7), namely $d\phi/dz(z = 0) = -\kappa \, (\phi(0) - \phi(\infty))$, whence the relation (13.E.18).

3. Reconsider the above problem without linearization and show that

$$[\tilde{\phi}](0) \equiv \frac{\zeta \, F}{RT} \left(\phi(0) - \phi(\infty) \right) = 2 \, \mathrm{Ln} \left(X + \sqrt{X^2 + 1} \right), \quad X \equiv \frac{\sigma_e}{\sqrt{8 \, RT \, \epsilon \, c^+(\infty)}} \, . \tag{13.E.19}$$

Proof: Insertion of (13.E.13) into (13.E.17) yields

$$\sigma_e = \sqrt{8 \, RT \, \epsilon \, c^+(\infty)} \, \sinh \left(\frac{1}{2} \, [\tilde{\phi}](0) \right) . \tag{13.E.20}$$

Solving (13.E.20) for the potential $\phi(0)$ yields the announced result.

The spatial profiles of the electrical potential and ionic concentrations away from the fixed charge are displayed in Fig. 13.E.4 for the fixed charge $\sigma_e = -0.125 \, \mathrm{C/m}^2$.

Exercise 13.3 Triple layer theory

Various degrees of refinement can be developed for the spatial variation of the electrical potential in the neighborhood of an attractor A endowed with fixed charge σ_0, Sposito [1984], Stumm [1992], Petrangeli Papini and Majone [2002]. Three models are sketched in Fig. 13.E.5.

In the layers where the charges are located along discrete planes, the electrical potential varies linearly according to Poisson's equation. The constant capacitance model involves a single constant to be obtained by optimization of some nanometric or macroscopic entities. The triple layer model requires a second capacitance. For the diffuse layer model and the triple layer model, where mobile charges are distributed in the outer layer, the spatial variation of the electrical potential in this layer is the one defined in Exercise 13.1.

The fixed charge σ_0 of the attractor is due to broken bonds, e.g., pH-sensitive terminations A-OH, A-OH-H$^+$, A-O$^-$. For the triple layer model, the plane β gathers the charge σ_β of adsorbed ions, e.g., Cl$^-$ and Na$^+$.

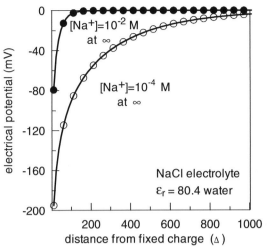
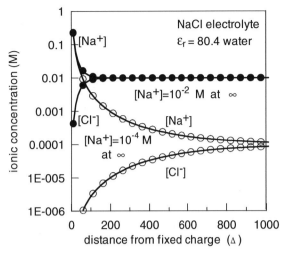

Fig. 13.E.4: Diffuse double layer. Spatial variation of the electrical potential and ionic concentration for sodium chloride and a negative fixed electrical charge $\sigma_e = -0.125\,\text{C/m}^2$.

What are the potentials on the planes of interest?

Answer: The difference of electrical potentials between the layers is linked to the fixed or adsorbed charge through the capacitances, namely for the constant capacitance model,

$$\phi_0 - \phi_\infty = \frac{\sigma_0}{G_{e0}}\,, \tag{13.E.21}$$

for the triple layer model,

$$\phi_0 - \phi_\beta = \frac{\sigma_0}{G_{e0}}\,, \quad \phi_\beta - \phi_s = \frac{\sigma_0 + \sigma_\beta}{G_{es}}\,, \tag{13.E.22}$$

and for the outer layers where ions are solvated,

$$\phi_s - \phi_\infty \text{ given by (13.E.18) or (13.E.19) with } \sigma_e = -\sigma_s\,. \tag{13.E.23}$$

Exercise 13.4 Osmotic pressure in the diffuse double layer model

Show that the osmotic pressure, at low ionic strength $I = I(\infty)$, is equal to

$$\pi = p(0) - p(\infty) = \frac{(\text{F}\,[\phi])^2}{RT}\,I(\infty)\,. \tag{13.E.24}$$

Proof:

The local osmotic pressure is equal to $p = RT\,(c_+ + c_-)$. Therefore the relation (13.E.3) yields

$$\frac{1}{RT\,c_+(\infty)}\left(p(0) - p(\infty)\right) = \exp\left(-\frac{\text{F}\zeta_+}{RT}\,[\phi]\right) - 1 - \frac{\zeta_+}{\zeta_-}\left(\exp\left(-\frac{\text{F}\zeta_-}{RT}\,[\phi]\right) - 1\right). \tag{13.E.25}$$

The result is obtained by linearization to second order of the exponentials, and use of electroneutrality at infinity (13.E.4), with the ionic strength defined by (13.E.8).

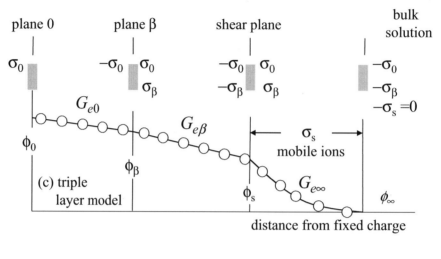

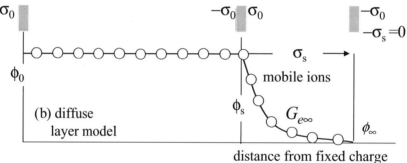

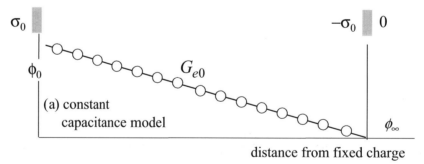

Fig. 13.E.5: Three models of spatial variation of the electrical potential in the neighborhood of the fixed charge, Sposito [1984], Stumm [1992], Petrangeli Papini and Majone [2002].

Chapter 14

Chemomechanical couplings in articular cartilages

14.1 Overview

Articular cartilage is a porous medium, structured by collagen fibers and saturated by an electrolyte, with water as solvent and metallic ions as solutes. Charged macromolecules, the proteoglycans, intermingled with collagen fibers, give rise to electro-chemomechanical couplings that allow moderate deformation to take place, and ensure an optimal adaption of the tissue to physiological loads.

The mechanical significance of the partition of tissue water in intrafibrillar (IF) and extrafibrillar compartments (EF) has been advocated by Maroudas and co-workers. The intrafibrillar compartment is defined as the volume between the collagen fibrils. The presence of ions in the intrafibrillar compartment is constrained by steric considerations: the size of a collagen molecule is $15\,\text{Å}$, and the distance between molecules $5\,\text{Å}$. Thus water molecules of size $3.6\,\text{Å}$, chloride ions Cl^-, sodium cations Na^+, and calcium cations Ca^{2+} with respective non hydrated diameters equal to $3.6\,\text{Å}$, $2\,\text{Å}$ and $2.2\,\text{Å}$, can penetrate the intrafibrillar compartment while larger molecules, like PGs, cannot, Torzilli [1985]. Proteoglycans (PGs) belong to the extrafibrillar compartment. Experimentally, the partition between intrafibrillar and extrafibrillar water is measured by injecting serum albumin whose size (hydrodynamic radius $35\,\text{Å}$) is large with respect to the intrafibrillar characteristic dimension.

Hydrated PGs induce collagen fibers in tension: a mechanical model would consist of two parallel systems, the pressure induced by PGs being resisted by the applied mechanical load *and* collagen in tension, Maroudas et al. [1991], Lai et al. [1991], Basser et al. [1998]. For unloaded cartilage under physiological salinity, intrafibrillar water represents up to 25% of total water, the extrafibrillar water furnishing the complement. The latter can be moved by mechanical loading and osmosis with water external to the cartilage (synovial fluid). The intrafibrillar water is in contact with the extrafibrillar compartment only and it is moved essentially by changes of the chemical composition of the latter. Still, a mechanical loading modifies the relative chemical composition of water, e.g., water being expelled, the concentration of proteoglycans increases and, therefore, indirectly induces a transfer of water between the intrafibrillar and extrafibrillar compartments.

The two-compartment idea is adopted in a hierarchical multi-phase multi-species context. In line with the idea of Maroudas, collagen fibrils behave as a semipermeable membrane, impermeable to macromolecules of molar mass larger than about $4000\,\text{gm}$ and permeable to dissolved metallic ions and water, Li and Katz [1976]. They are viewed as separating the two fluid phases.

Experiments of Eisenberg and Grodzinsky [1985] consider a cartilage specimen in equilibrium with a bath of controlled chemical composition. The stresses and strains induced by changes of the bath composition and of the applied mechanical loads are simulated in Loret and Simões [2004]. The latter reference assumes that only dissolved sodium chloride NaCl is present in the cartilage. In Loret and Simões [2005]a, attention is paid to the mechanical interactions between ions, specifically sodium Na^+ and calcium Ca^{2+}, and simulations of ionic replacements intertwined with mechanical loadings are performed.

The electro-chemomechanical model exposed in this chapter is restricted to mechanical and chemical equilibria. In other words, the chemical and mechanical loadings are performed with characteristic times much larger than the times characterizing the material response. Moreover the electro-chemomechanical constitutive equations are assumed here as time-independent. Therefore time is elusive. The extension to transient loadings where time comes explicitly into picture due to generalized diffusion and mass transfer is exposed in Chapter 16.

The constitutive equations developed here use a thermodynamic framework, that in fact may embody not only purely mechanical aspects but also transfers of masses between the two fluid phases, flow of water through the solid skeleton and diffusion of ions through the extrafibrillar phase, Section 14.7. Still, as just pointed out, diffusion phenomena are not addressed in this chapter. The complete framework is used in Chapter 18 to solve, via the finite element method, initial and boundary value problems in order to simulate mechanical and transport phenomena in laboratory specimens submitted to transient mechanical and chemical loading processes.

14.2 Histological aspects of articular cartilages

14.2.1 The three major types of cartilage in the human body

Cartilage exists in three major types in the body, Fig. 14.2.1:
- elastic cartilage in the epiglottis and Eustachian tube;
- hyaline cartilage in the foetal skeleton and in the growing bones of young animals and humans as well as in diarthroidal joints covering long bones;
- fibrocartilage in the intervertebral disks of the spine and in the meniscus of the knee.

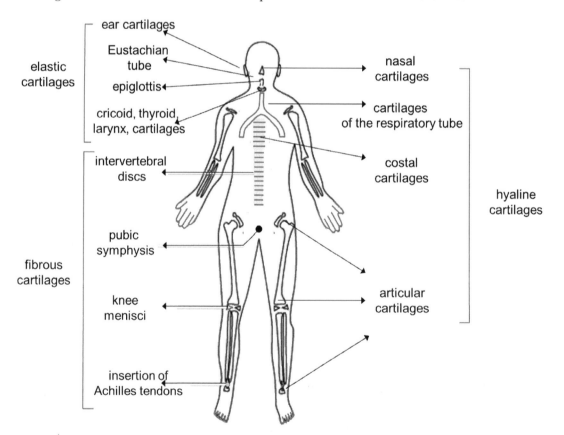

Fig. 14.2.1: Three main types of cartilage exist in the human body: hyaline cartilages, elastic cartilages and fibrous cartilages. The sketch of the human body has been modified from the site https://www.unifr.ch/anatomy/assets/ files/elearning/fr/stuetzgewebe/knorpel/faser/f-faser.php (last visit 2015/11/20)

Hyaline, elastic and fibrous cartilages differ by the relative proportions of their constituents and by the types of collagen they contain, Table 14.2.1.

Table 14.2.1: Composition of various cartilages (percentage in weight).

Tissue	Collagen (% dry weight)	Proteoglycans (% dry weight)	Water (% wet weight)
Articular cartilage	50-73	15-30	58-78
Meniscus	75-80	2-6	70
Intervertebral disc-Nucleus Pulposus	15-25	50	70-90
Intervertebral disc-Annulus Fibrosus	50-70	10-20	60-70

Articular cartilage is a soft hyaline tissue that minimizes friction between diarthroidal joints covering long bones. The thickness of a diarthroidal joint varies typically from 2 to 4 mm, Fig. 14.2.2. Cartilage from adult animals and humans is avascular, aneural and alymphatic. Nutrients are transported from the synovial fluid towards the cells by convection, water diffusing through the articular surface and extracellular matrix. The cells in foetus and young subjects, called chondroblasts, are very active and they produce the extracellular matrix (ECM) composed essentially of fibrillar proteins (elastin, fibronectin and mainly collagen) and highly negatively charged proteoglycans. In adults, the chondroblasts become almost inactive; they are then called chondrocytes and their volume decreases, Stevens and Lowe [1993], p. 52. Mature chondrocytes occupy only 2% of the volume, Fig. 14.2.3. In that respect, cartilage is the opposite extreme of a syncitium, like the heart muscle or onion tissue, where the intercellular space and extracellular matrix are absent, cells occupy most of the volume and are linked together by gap junctions.

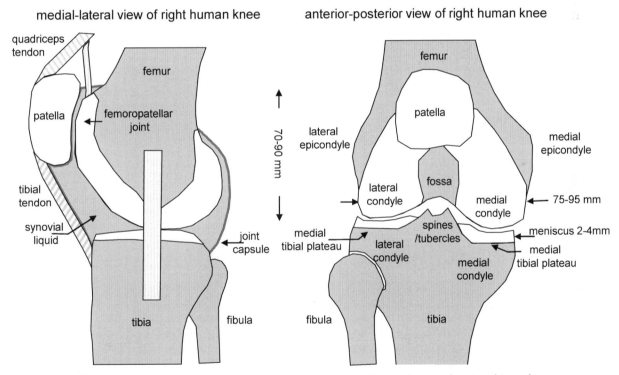

Fig. 14.2.2: Schematic of the human knee anatomy with cartilage surface in white color.

The negatively charged proteoglycans are hydrated. For normal cartilage, the negative fixed charge density ranges from 0.1 to 0.2 mole/liter of water[14.1]. The matrix is circulated by an electrolyte, namely water containing dissolved ionic species, essentially NaCl. These ions ensure electroneutrality, they move in water closeby to the fixed charge and neither Na^+ nor Cl^- bind to the charged groups.

[14.1] C.C.B. Wang et al. [2002] have developed a non-invasive technique based on video microscopy to determine the stiffness and charge density of articular cartilages.

Absorbed water within the intrafibrillar space of collagen can represent as much as 20 to 30% of the mass of the cartilage. Absorbed (intrafibrillar) water and free (extrafibrillar) water represent together 70 to 80% of the mass in articular cartilage and a little bit less in meniscus. Absorption of water results in *swelling* of the cartilage. The amount of swelling depends on the concentration of proteoglycans and on the tensile stiffness of collagen.

At physiological pH, in contrast with PGs, collagen has almost no net charge.

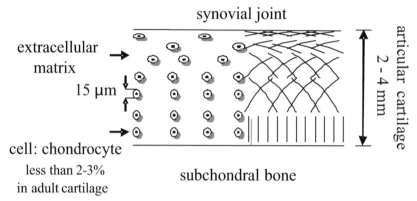

Fig. 14.2.3: Articular cartilage is formed by the interweaving of collagen fibers and proteoglycans (PGs). The spatial orientation of the collagen network aims at optimizing its mechanical role: normal to the subchondral bone to anchor the cartilage into the bone and parallel to the synovial joint to resist tangential tensile stress induced by friction. Reproduced with permission from Loret et al. [2004]a.

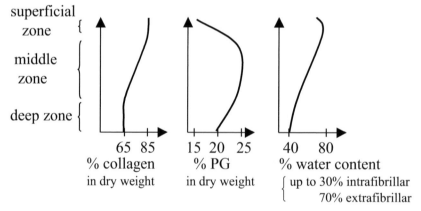

Fig. 14.2.4: The cartilage layer is formed by three main sub-layers (zones) and the content of the various components varies along the depth, presumably so as to optimize the physiological function of the organ. The mass distribution along the depth of human knee cartilage is documented in Muir et al. [1970] and Brocklehurst et al. [1984]. The sketch above redrawn from Mow and Guo [2002] summarizes data for bovine knee cartilage. The depth variation of the intervertebral disc is different from articular cartilage: the collagen content increases with depth while the content of PGs decreases, Muir [1978]. Reproduced with permission from Loret et al. [2004]a.

14.2.2 A layered structure

In the cornea and in the intervertebral disc, collagen fibers are arranged in layers in which they run parallel and along a specific pattern: two adjacent layers are rotated one with respect to the other, Chapter 19. In these two organs, the seemingly similar spatial organizations in rotated layers serve clearly distinct purposes, which are induced by additional features like the presence or absence of crosslinks. The directional distribution of collagen fibers in articular cartilages varies with depth, Fig. 14.2.3. Data are reported in Hiltner et al. [1985], their Fig. 12.

The density of PGs varies depending on the specific joint: it is larger in the femoral head than in the femoral condyle, Maroudas [1975], Fig. 2. Within a joint, it usually is minimal at the surface, maximal on the subsurface and decreases again in the depth, Maroudas [1976], Maroudas et al. [1980], as indicated in Fig. 14.2.4. The depth density of collagen varies as well. This heterogeneity in the concentrations of collagen and PGs carries over to their ratio which seems to correlate with some mechanical properties: the tensile modulus is shown to decrease with depth, see Fig. 2 in Kempson et al. [1973], and Fig. 7 in Akizuki et al. [1986]. On the other hand, strong increase of compressive modulus with depth is reported by A.C. Chen et al. [2001] and S.S. Chen et al. [2001].

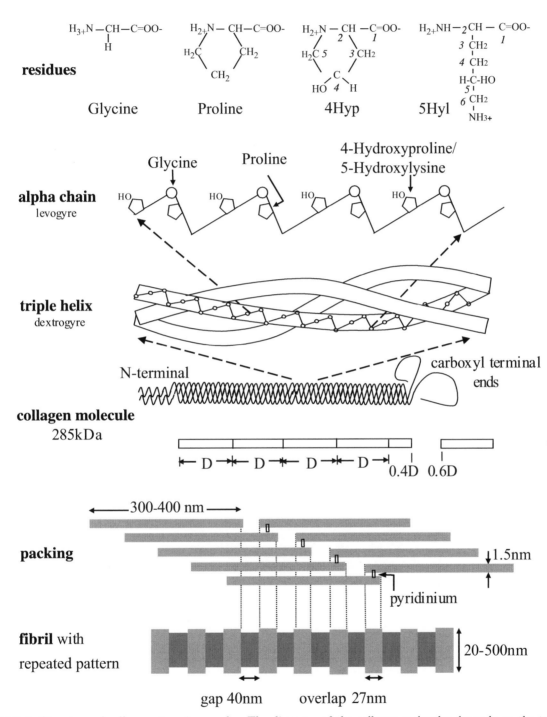

Fig. 14.2.5: Structure of collagen at various scales. The diameter of the collagen molecules depends on the type, e.g., 290 nm for collagen type I. Cartilages and cornea are endowed with fibrils of small diameters, e.g., 30-40 nm for corneal stroma. Collagen fibrils are formed by staggered collagen molecules, shifted by 67 nm and interconnected by pyridinium crosslinks (pyridinoline): the crosslink density increases with maturation.

14.2.3   Composition of collagen

Collagen is the most abundant protein, representing around 25% of proteins in the human body. There are more than twenty types of collagen in vertebrates which are classified into three super families. Some types form cylindrical fibrils (among them types I and II), others form networks, fibril surfaces and transmembrane proteins.

The tissues where the three main types of collagen are found are listed in Table 14.2.2. Adult articular cartilage, human or bovine, and the vitreous humor of the eye[14.2] contain predominantly hyaline collagen of type II (approximately 96% in mass). Fibrocartilage of the meniscus involves collagen of types I and II. Skin, bones, tendons and cornea contain primarily collagen of type I. The inner part of intervertebral discs, called nucleus pulposus, contains collagen of type II, while the outer part, called annulus fibrosus, involves collagen of type I. Blood vessels and skin contain collagen of types I and III. Actually, the inner layer of arteries (intima) changes with age from collagen of type IV to collagen of type III while the outer layer contains collagen of type I. In fact, while collagen types I and II are the most widespread, the presence of other types even in minor amount is important to ensure the fibrillar structure and packing. Collagen of articular cartilages and tendons establishes strong links with proteoglycans.

Table 14.2.2: The main collagen types in some tissues.

Type I	Skin, blood vessel, annulus fibrosus, bone, tendon, cornea
Type II	Nucleus pulposus, articular cartilage, vitreous humor
Type III	Blood vessel

14.2.3.1   The structure at various scales

The collagen molecule is elongated, about 1.5 nm in diameter and 300 to 400 nm in length, Fig. 14.2.5. It is formed by three chains of polypeptides[14.3] which are arranged in a dextrogyre triple helix.

The collagen molecule of type I consists of two $\alpha_1(I)$ chains and one $\alpha_2(I)$ chain, which contains more hydrophobic residues. Collagen types II and III consist of three identical $\alpha_1(II)$ chains and $\alpha_1(III)$ chains respectively.

Each chain bears a polypeptide sequence whose sense of rotation is levogyre. The inverse sense of rotation of the three chains and of the polypeptide sequences in each chain is presumed to oppose unpacking of the molecule through lateral contraction under axial traction, Voet and Voet [1998], p. 159. This remarkable structural arrangement may be compared with other fibrillar organizations. Hollow cylinders can be reinforced by winding consecutively in alternative directions, Rogers [1984]: there, the alternative winding takes place at a single level of the structure. As another example, ropes are built by successive layers of alternate handedness. At variance, the alternate handedness in the collagen molecule takes place at distinct structural levels (polypeptide sequence/triple helix).

The polypeptide of each chain is formed by the repeat of the sequence -Gly-X-Y, where Gly is the amino acid glycine, X is often the amino acid proline (Pro) or alanine, arginine$^+$, glutamine, while Y is the 4- or 3-hydroxyproline[14.4] (4Hyp, 3Hyp), the 5-hydroxylysine (5Hyl), arginine$^+$, or glutamine. Having the hydrogen H as a side chain, glycine is the smallest amino acid. Glycine constitutes 33% of the amino acids, proline and hydroxyproline counting for 10%. The polypeptide sequence is shifted by one unit, from one chain to the other.

For each chain, the sequence of the amino acid residues depends on the collagen type and on the tissue. Mutations may lead to severe diseases. In osteogenesis imperfecta, known as brittle bone disease, a glycine of one of the type I collagen chains is substituted by another amino acid. Ehlers-Danlos syndrome, which affects blood vessels, is due to a mutation in type III collagen.

[14.2]The vitreous humor is a gel formed by collagen (1% in weight) and hyaluronic acid (1% in weight) with a high hydration (98% of water).

[14.3] A *polypeptide* is a macromolecule composed of amino acids linked by covalent peptide bonds. A peptide bond, or amide bond, is a bond between the carboxyl group and the amino group of two amino acids. This bond is obtained by a reaction of condensation, as described in Section 20.2.1, that releases a water molecule, the oxygen coming from the carboxyl group and the two hydrogen molecules coming from the amino group,

$$\underset{R1}{\overset{H}{C=OO^- -CH-NH_3^+}} + \underset{R2}{\overset{H}{C=OO^- -CH-NH_3^+}} \rightarrow \underset{R1}{\overset{H}{C=OO^- -CH-}}NH-\underset{R2}{\overset{H}{C=O -CH-NH_3^+}} + H_2O. \qquad (14.2.1)$$

A *protein* is a macromolecule of molar mass larger than 5000 g containing one or several polypeptide chain(s).

[14.4] Hydroxyproline is typically used to estimate the collagen content, based on a mass ratio of 7.25-7.30, Herbage et al. [1977].

Each chain is split in four blocks, of equal size D, and one fifth block of size 0.4 D. Each D-block contains 234 amino acids, so that, together with the non structured telopeptide ends, the total number of amino acids per collagen chain is about 1050 and, therefore, 3150 per collagen molecule.

The amino acid sequence is clustered in polar and hydrophobic domains. Besides the alternate handedness of the polypeptide sequences and of the triple helix, the thermal and mechanical stability of the collagen packing is ensured by several chemical bonds. In the highly polar domains, the positively and negatively charged amino acids alternate. Hydroxylation (introduction of the hydroxyl group $-OH^-$) of the proline and lysine residues provides crosslinks between the chains: these crosslinks are observed in adult tissues only and they serve as a marker of maturation in cartilages and bones. Conversely, their release in blood is viewed as a testimony of metastatic bone tumor. Since these crosslinks stiffen the tissue, it is not unexpected that they are not observed in skin.

Each chain is formed by the assembly of amino acids through covalent bonds (peptide bonds). The three chains are held together by weaker bonds, as just mentioned. Heat denaturation breaks these bonds, but leaves intact peptide bonds.

14.2.3.2 Amino acids and the electrical charge of collagen

Amino acids are composed of the elements carbon, hydrogen, oxygen, nitrogen and sulfur (C H O N S). They contain at least two ionizable groups (or termini). Their structure is organized around a central carbon and consists of a basic amino terminus NH_3^+, a carboxylic acid, or acidic carboxyl terminus COO^-, a hydrogen atom, and a side chain R. The central carbon and the amino and carboxyl termini are usually referred to as α carbon, α amino and α carboxyl termini. There exist 20 side chains R corresponding to the 20 amino acids, the simpler where the chain R is reduced to hydrogen H being glycine.

Such molecules, with a positive charge and a negative charge, are called zwitterions. Their charge depends on pH:

$$
\begin{array}{ccc}
O=C\text{-}OH & O=C\text{-}O^- & O=C\text{-}O^- \\
| & | & | \\
H_3N^+ - C - H & H_3N^+ - C - H & H_2N - C - H \\
| & | & | \\
R & R & R
\end{array}
$$

$$
\text{pH:} \qquad pK_{cg} \simeq 3.2 \qquad\qquad pK_{ag} \simeq 9.2
$$

$$
\text{charge:} \quad \longleftarrow \quad +1\,|\,0 \quad \text{———} \quad 0\,|\,{-1} \quad \longrightarrow
$$

$$(14.2.2)$$

The carboxyl group COO^- becomes protonated at pH lower than $pK_{cg} \simeq 3.2$. The amino group NH_3^+ becomes neutral at pH higher than $pK_{ag} \simeq 9.2$. At physiological pH, the carboxyl group bears a negative charge, while the amino group bears a positive charge. Therefore, the charge due to carboxyl and amino groups is positive at low pH, zero in a wide interval of pH between 4 to 8, and negative at higher pH.

The above description sketched in (14.2.2) holds if the side chains R remain neutral. However, some side chains are ionizable. Their position in the peptide sequence may be visualized by staining by heavy metal salts. In fact, the D-blocks are divided in 12 bands, whose striation is observed upon staining, due to the reaction of anions on the positively charged amino acids. For collagen of type I, bands e1, e2, a1, a2 are in the gap region, band a3 is at gap/overlap boundary, and bands a4, b1, b2, c1, c2, c3 are in the overlap region, Meek and Quantock [2001]. Negative staining corresponds to zones of stain penetration or exclusion. Positive staining reveals the sites of charged amino acid residues.

Besides, some amino acids can be considered hydrophobic: turned inside the protein, they participate to intramolecular crosslinks. The hydrophilic amino acids are exposed at the surface, Voet and Voet [1998], p. 166 et seq. The percentage of charged amino acids at the surface (termed *exposed* amino acids) is expected to be larger than within the molecule, since water is a polar substance. Still, not all exposed amino acids are charged and not all charged amino acids are at the surface.

Some of these amino acid residues, namely aspartic acid and glutamic acid, are acidic, with a pK about 4:

$$
\begin{array}{cc}
\text{O=C-O}^- & \text{O=C-O}^- \\
| & | \\
\text{H}_3\text{N}^+ - \text{C} - \text{H} & \text{H}_3\text{N}^+ - \text{C} - \text{H} \\
| & | \\
\text{R'-CO-OH} & \text{R'-CO-O}^-
\end{array}
\qquad (14.2.3)
$$

pH : pK_{cg} $\text{pK}_{Ra} \simeq 4$ pK_{ag}

R charge : $\longleftarrow$ ———————— $0\,|\,\text{-}1$ ———————— $\longrightarrow$

Others, namely lysine, hydroxylysine and hydroxyproline, are basic, with a pK about 10.5 :

$$
\begin{array}{cc}
\text{O=C-O}^- & \text{O=C-O}^- \\
| & | \\
\text{H}_2\text{N} - \text{C} - \text{H} & \text{H}_2\text{N} - \text{C} - \text{H} \\
| & | \\
\text{R'-N-H}_3^+ & \text{R'-NH}_2
\end{array}
\qquad (14.2.4)
$$

pH : pK_{ag} $\text{pK}_{Rc} \simeq 10.5$

R charge : $\longleftarrow$ ——— $1\,|\,0$ ———

and a pK of about 12.5 for arginine. At physiological pH, the side chains of aspartic and glutamic acids are negatively charged, the side chains of lysine, hydroxylysine and arginine are positively charged and the side chain of hydroxyproline is neutral.

The carboxyl and amino termini are engaged in covalent peptide bonds and their charge is therefore not available for ionic interactions. On the other hand, available data at physiological pH indicate that about 17% of the carboxyl groups of acidic residues are negatively charged and about the same percentage of amino groups of basic residues are protonated. In fact, collagen has often nearly equal numbers of acidic and basic residues and hence has almost no net charge at physiological pH. In articular cartilages which contain mainly collagen of type II, the number of positively charged residues per collagen molecule is 273 and the number of negatively charged groups is 264, Wachtel and Maroudas [1998]. These numbers fit with the estimate of 17% of charged residues: 273+264=537 versus 17%×3150 = 535.5. Only a dozen of lysine groups per molecule are not exposed to the surface of the molecule. Consequently, at physiological pH, the numbers of functionally active positive and negative charged residues are 261=273-12 and 264, respectively. On the other hand, departure from electroneutrality is more definite in other tissues due to an excess of negative charges, Table 14.2.3.

Table 14.2.3: Number of positively (+) and negatively (-) charged amino acid residues at physiological pH per (tropo-)collagen molecule (3150 residues).

Tissue	Collagen Type (% in mass)	+	-	Ref.
Articular cartilages	Type II (96%)	273	264	Wachtel and Maroudas [1998]
Rat tail tendon	Type I	264	350	Grodzinsky and Melcher [1976]
Human skin	Type I (80%) type III(15%)	268	362	Seehra and Silver [2006]
Bovine corneal stroma	Type I (75%) type V(15%)	281	394	Panjwani and Harding [1978]

Collagen molecules form fibrils, Fig. 14.2.5, bathed in (extrafibrillar) water and containing as well a significant amount of (intrafibrillar) water. The charges, namely carboxyl groups, amino groups and exposed charged residues are partitioned into a part in direct, or close, contact with the extrafibrillar water, and a part in contact with the intrafibrillar water. The charges in contact with the extrafibrillar water are considered to participate directly to the swelling and transport properties of the tissues: they interact with other mobile and fixed charges located in the extrafibrillar water. The intrafibrillar compartment can be viewed as introducing a second porosity associated with a delayed response to mechanical stimuli and it may also have a significant role in chemomechanical interactions.

14.2.3.3   Collagen development, fibrillogenesis

Cells secrete soluble substances that are fibrillated outside the cell. Single or small groups of fibrils transfer from the perinuclear region and Golgi apparatus to the extracellular space via tubules. More complex structures at higher length scales are formed outside the cell by crosslinking that is responsible for their resistance to dissolution, Jackson [1958]. After secretion, collagen molecules and intermediates first extend in length, by linear end-to-end aggregation, and next laterally. End-to-end aggregation is possible once the N- and C-termini have been removed by proteinases. The heterogeneity of structures, marked by e.g., fibril diameter, increases during development. Extracellular linear aggregation is thought to contribute to the tensile strength of the collagen assembly. Extension of a collagen tissue is believed to be due to the viscous sliding between units in the fetal stage and, later, to the elastic deformation of the collagen molecules. The ultimate energy $\int_0^l \tau \, x \, dx = \tau \, l^2/2$ required by the relative parallel sliding of two units of length l undergoing a uniform shear stress τ is therefore proportional to the square of the length of the units. The actual length is limited by the axial strength of the units themselves. Typical tensile strengths vary from a few MPa's in the chick embryo to 50-60 MPa at a month. Measurements indicate that fibril length, rather than fibril diameter, is the key parameter in the elastic properties of collagen.

Many tissues involve heterotypic collagen, e.g., type I and type III in tendons. In a number of tissues, e.g., cartilages and tendons, strong links are established with proteoglycans during the early steps of development.

These structures are expected to depend on the mechanical conditions of the ECM. The fibers are secreted by the cells (fibroblasts) consistently with the existing state of the ECM. Girton et al. [2002] report that muscle cell morphology and fiber orientation evolve according to the stress and/or strain eigendirections. In fact, the fibroblasts are roughly aligned with the fibers and they secrete collagen molecules along their own morphology. Fibers provide guidance to adherent cells whose activity is regulated in particular by the mechanical stimuli of the fibers and ECM.

14.2.3.4 Suprafibrillar architecture of collagen units is function-dependent

The spatial organization of collagen fibrils presumably adapts to the physiological functions of the organs, Fratzl [2003]. In tendons, the fibrils are grouped in successive units of increasing diameters, which are aligned along the traction direction,

$$\underbrace{\text{molecule}}_{\phi\ 1.5\,\text{nm}} \longrightarrow \underbrace{\text{fibril}}_{\phi\ 50-500\,\text{nm}} \longrightarrow \underbrace{\text{fascicle}}_{\phi\ 50-300\,\mu\text{m}} \longrightarrow \underbrace{\text{fiber/tendon}}_{\phi\ 100-500\,\mu\text{m}} \qquad (14.2.5)$$

Another terminology is used as well, namely molecule, microfibril, subfibril, fibril, fiber and tendon,

$$\underbrace{\text{molecule}}_{\phi\ 1.5\,\text{nm}} \longrightarrow \underbrace{\text{microfibril}}_{\phi\ 3.5\,\text{nm}} \longrightarrow \underbrace{\text{subfibril}}_{\phi\ 10-20\,\text{nm}} \longrightarrow \underbrace{\text{fibril}}_{\phi\ 50-500\,\text{nm}} \longrightarrow \underbrace{\text{fiber}}_{\phi\ 50-300\,\mu\text{m}} \longrightarrow \underbrace{\text{tendon}}_{\phi\ 100-500\,\mu\text{m}} \qquad (14.2.6)$$

This hierarchical structure is generally thought to improve the mechanical properties, e.g., stiffnesses at low or high strains and tensile strength. Miyazaki and Hayashi [1999] report considerable increase of the stiffness from fibrils to fascicles and to tendons[14.5]. This situation is in line with that of composite materials where a matrix is reinforced by fibers. In the hierarchical sequence (14.2.5), the geometrical arrangement of the collagen units induces strong interactions and mechanical properties correlate with size.

Still, much remains to be explored, given that the strains are not identical at the different scales. The properties of the units themselves matter: Christiansen et al. [2000] suggest that lateral fusion of small subunits onto the fibrils increases the elastic modulus at low strain while longitudinal fusion increases the modulus at high strain.

At rest, tendons are observed to display waviness or crimp, with a wave-length of 10 to 100 μm, perhaps due to contraction imposed by the ground substance. They uncrimp progressively on extension.

To explain the more compliant mechanical behavior of skin with respect to tendon, Silver et al. [2001] point out to some differences in composition and geometrical arrangement:

- collagen of type III is present in skin in combination with collagen of type I while tendon contains only collagen of type I;
- the length of the collagen fibril is an order of magnitude smaller in skin;
- the D period of 67 nm in tendon is associated with a tilt angle of 7° while the D period of 65 nm in skin requires a larger tilt angle of 16°: this larger default of orientation with respect to the traction axis is expected to decrease significantly the stiffness.

Quite similar spatial organizations sometimes serve different purposes. For example, the annulus fibrosus, which forms the external part of vertebral discs and the cornea are both constituted by thin lamellae: in each of them, collagen fibrils are all aligned but the alignment directions between two successive layers is alternate. This organization in the annulus fibrosus limits the radial extension of the nucleus pulposus, the central part of the disc. In the cornea, the target to be optimized is thought to be transparency.

[14.5] Typical stiffness of fibrils: 0.4 GPa for Sasaki and Odajima [1996], 0.5-1 GPa for Fung [1993].

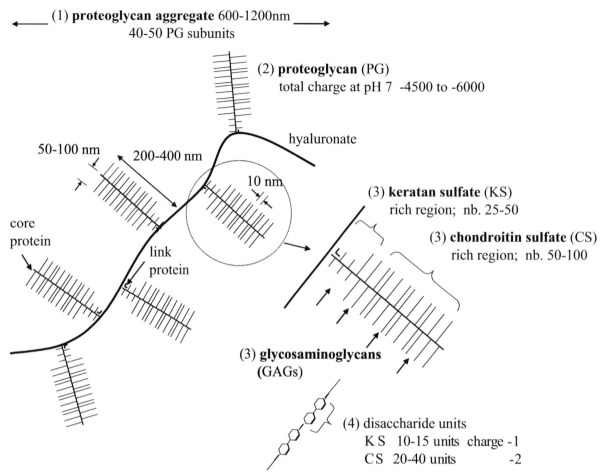

Fig. 14.2.6: Hierarchical organization of the multimolecular units composing proteoglycans. (1) an aggrecan is formed by the gathering of proteoglycans along a hyaluronate; (2) proteoglycan monomer (PG) of molar mass about 2000×10^3 gm; (3) glycosaminoglycans (GAGs), essentially chondroitin sulfates (CS), and keratan sulfates (KS); (4) disaccharide units. The negative charge of proteoglycans gives rise to electrical repulsion at various structural levels, which in turn provides compressive stiffness at the macroscopic level.

14.2.4   Composition and structure of proteoglycans

Proteoglycan aggregates are long molecular structures developing along an axis formed by a hyaluronic acid, a polysaccharide, as sketched in Fig. 14.2.6. On that axis are attached, like leaves, proteoglycan monomers structured around their own axis. Along this axis are attached in turn groups of amino acids called glycosaminoglycans (GAGs). The basic structure of GAGs is composed by disaccharide units containing a uronic acid and an aminoglycan. The uronic acid displays a negatively charged carboxyl group COO^- and the aminoglycan displays at least one sulfonic group SO_3^-. The two main types of GAGs in articular cartilages are chondroitin sulfates (CS) and keratan sulfates (KS), Table 14.2.4. Keratan sulfates are shorter than chondroitin sulfates, and contain about half of the number of disaccharide units, Muir [1978]. They possess no uronic acid (carboxyl group), so that their charge is -1, while chondroitin sulfates have a charge equal to -2. The ratio of chondroitin to keratan sulfates decreases significantly with age. On the other hand, it is much larger in osteoarthritic cartilage than in normal cartilage. The disaccharide units of chondroitin sulfates consist of glucuronic acid and N-acetylgalactosamine, while the disaccharide units of keratan sulfates feature a galactose and an N-acetylglucosamine, Fig. 14.2.7. The ratio of galactosamine over glucosamine is used to obtain the relative content of chondroitin over keratan sulfates.

Proteoglycans, which are negatively charged, are hydrated: the water molecule being an electric dipole, the hydrogen branches are attracted by the negative charges, the oxygen pointing outwards the crowned negative charges.

Wether or not proteoglycans can fix/bind other ions which would act as inhibitors of the hydration process is an issue. Na^+ and Cl^- are known not to bind to charged groups of the extracellular matrix of articular cartilages: this

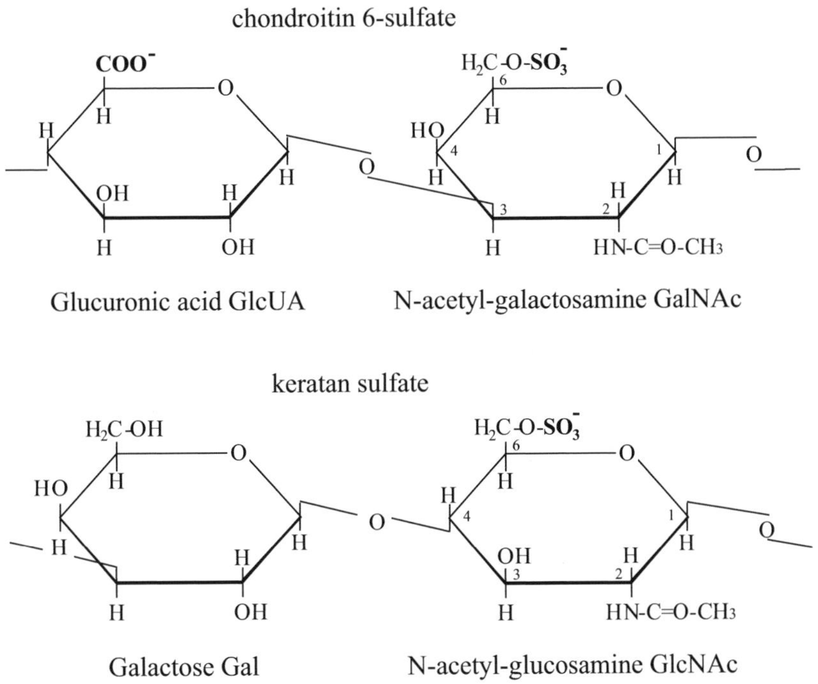

chondroitin 6-sulfate

COO⁻

Glucuronic acid GlcUA   N-acetyl-galactosamine GalNAc

keratan sulfate

Galactose Gal   N-acetyl-glucosamine GlcNAc

Fig. 14.2.7: The two main disaccharide units in articular cartilages are chondroitin sulfates and keratan sulfates. The fact that they are electrically charged is a key to the mechanical behavior of cartilages. Chondroitin 4-sulfate and 6-sulfate are sulfated at carbons 4 and 6, respectively.

Table 14.2.4: Average composition of proteoglycans of articular cartilage (% weight).

	GAGs		Proteins
85-92%	Chondroitin sulfate (charge -2)	80%	8-15%
	Keratan sulfate (charge -1)	20%	

property has been verified through macroscopic experiments by e.g., Frank and Grodzinsky [1987], p. 621. It does not hold in every tissue, Stevens and Lowe [1993], p. 42. Calcium cations bind reversibly onto the carboxyl groups and irreversibly onto the sulfonic groups, Chapter 17. In the corneal stroma, chloride anions bind reversibly to ligands associated with collagen molecules, Chapter 19.

In summary, sodium and chloride ions do not bind to the proteoglycans of articular cartilages: they may diffuse in the extrafibrillar fluid phase, even if a certain number of cations surround the negative charges so as to satisfy the electroneutrality condition.

14.2.5 Physical properties

The molar mass $\widehat{m}_{\rm PG}$ of a proteoglycan monomer is approximately equal to 2×10^6 gm. The main charged glycosaminoglycans of proteoglycans are keratan sulfates (KS) and chondroitin sulfates (CS). The disaccharide units of keratan sulfates have molar mass $\widehat{m}_{\rm KS}^{\rm du} = 464$ gm and valence $\zeta_{\rm KS} = -1$ resulting from the negatively charged sulfonic group. The disaccharide units of chondroitin sulfates have molar mass $\widehat{m}_{\rm CS}^{\rm du} = 513$ gm and valence $\zeta_{\rm CS} = -2$, resulting from the negatively charged carboxyl and sulfonic groups, Urban et al. [1979]. Typical cartilage consists of 80% of CS and 20% of KS, Table 14.2.4. Proteoglycans are composed of GAGs and proteins, and the relative mass of all GAGs wrt to PGs is around 85% according to Sweet et al. [1977], from 85 to 90% according to Muir [1978] and around 92% according to Basser et al. [1998].

The valence $\zeta_{\rm PG}$ of PGs can be obtained from
- the relative number of moles of disaccharide units in KS and CS namely $N_{\rm du}^{\rm KS} N_{\rm KS}$ over $N_{\rm du}^{\rm CS} N_{\rm CS}$,
- the relative mass $M_{\rm GAG}/M_{\rm PG}$ of GAGs with respect to PGs.

Indeed, the fixed charge in a reference volume of cartilage has the format,

$$\zeta_{KS} N_{KS} + \zeta_{CS} N_{CS} = \zeta_{PG} N_{PG},$$ (14.2.7)

while the mass of GAGs is equal to,

$$M_{GAG} = \widehat{m}_{KS}^{du} N_{du}^{KS} N_{KS} + \widehat{m}_{CS}^{du} N_{du}^{CS} N_{CS}.$$ (14.2.8)

Then

$$\zeta_{PG} = \widehat{m}_{PG} \frac{\zeta_{KS} N_{du}^{KS} N_{KS} + \zeta_{CS} N_{du}^{CS} N_{CS}}{\widehat{m}_{KS}^{du} N_{du}^{KS} N_{KS} + \widehat{m}_{CS}^{du} N_{du}^{CS} N_{CS}} \frac{M_{GAG}}{M_{PG}}.$$ (14.2.9)

The lower algebraic value is reached for a CS-proteoglycan cartilage and the upper algebraic value for a KS-proteoglycan cartilage.

With the above data, the valence results in $\zeta_{PG} \sim -6080$ for the mass ratio of Sweet et al. [1977], and in $\zeta_{PG} \sim -6580$ for the mass ratio of Basser et al. [1998]. Since proteoglycans are macromolecules their concentration is necessarily much smaller than those of ions, their valence is much larger and it depends strongly on pH. Electroneutrality will introduce the notion of *effective* concentration.

Table 14.2.5: Representative physical properties of constituents of articular cartilage.

Species	Molar Mass (gm)	Molar Volume cm^3	Density (gm/cm^3)	Valence * at pH=7
CS	18×10^3			-50*
KS	9×10^3			-13*
PGs	2000×10^3	1.1×10^6	1.8	- 6000 *
Collagen	285×10^3	2×10^5	1.4	~ 0 *
H$_2$O	18	18	1	0
Na$^+$	23	2.4-6.2		+1
Ca^{2+}	40.1			+2
Cl$^-$	35.5	14.9		-1

Table 14.2.5 shows the representative values of the molar mass, molar volume, density and valence of the constituents of articular cartilage. Calculating the volume of a mole of cation Na$^+$ and anion Cl$^-$ using their non-hydrated radius posted in Table 13.5.1, yields $\widehat{v}_{Na} = 2.37 \, \text{cm}^3$, $\widehat{v}_{Cl} = 14.9 \, \text{cm}^3$, resulting in a density $\rho_s = 3.6 \, \text{gm/cm}^3$ and molar volume $\widehat{v}_s = 17.3 \, \text{cm}^3$ of dissolved salt NaCl. Molar volumes vary with concentration as displayed in Fig. 13.2.2: the above values correspond to physiological concentration.

The density of collagen ρ_c is equal to $1/0.7 \sim 1.4 \, \text{gm/cm}^3$, and hence its molar volume $\widehat{v}_c$ to $2 \times 10^5 \, \text{cm}^3$, Basser et al. [1998], Table I. The specific volume of GAGs and proteins that compose PGs are equal to 0.54 and 0.74 cm^3/gm respectively, Basser et al. [1998], Table I, and therefore the specific volume of PGs is equal to $0.92 \times 0.54 + 0.08 \times 0.74 = 0.556 \, \text{cm}^3/\text{gm}$, i.e. $\rho_{PG} = 1.8 \, \text{gm/cm}^3$, and therefore $\widehat{v}_{PG} = 1.1 \times 10^6 \, \text{cm}^3$.

The difference of density between constituents is used for separation through density gradient centrifugation, larger and denser particles pelleting at smaller centrifugal forces, Muir [1978].

14.2.6   Electroneutrality in the two compartment context

Most macroscopic studies consider a single condition of electroneutrality for the cartilage as a whole. At variance, Li and Katz [1976] assume electroneutrality of the intrafibrillar compartment, that is, of collagen fibrils plus the species in that compartment. Then a second electroneutrality condition applies to the extrafibrillar compartment. This way of thinking is adopted in the constitutive model of articular cartilage developed in this work. As a preliminary requirement, the electrical charges of collagen under non physiological pHs are partitioned between the intra- and extrafibrillar fluid phases. These charges are considered to be electrically and mechanically active in the phases they are in contact with.

14.2.7   Methods of measurement of the fixed charge

The tracer cation method, Maroudas and Thomas [1973], is the standard method used for the determination of the fixed charge density in articular cartilages and intervertebral discs. When a cartilage sample is equilibrated in a dilute electrolyte solution, typically a 0.015 M NaCl solution, the sole ions that remain in the cartilage are the cations that balance the negative fixed charge of PGs. Their charge is the negative of the total fixed charge of the cartilage, that is, of the PGs at neutral pH. The concentration of these neutralizing cations is obtained by transferring the sample into a radio-labelled 0.015 M NaCl solution. The latter solution is stirred for a long period under low temperature, typically 4°C, to avoid autolysis. The sample is finally dried and the radioactive isotopes Na^{22} are counted by a γ-radiation scintillation detector.

In fact, the spread of the cation tracer method is hampered by the presence of radio-active ions which is subject to regulatory approval. Therefore, other destructive and non destructive methods have been developed.

In the conductivity technique developed by Jackson et al. [2009], the FCD of the nucleus pulposus is accessed by measuring the electrical conductivity. In general, the swelling pressure generated by the fixed charge is sustained by the collagen network in tension. However, the collagen tension is small in the nucleus pulposus. Then the outward flux due to an increase of pressure applied to the sample can be related to the FCD. This property has been used by Sivan et al. [2013] in developing a needle micro-osmometer technique to measure the FCD of excised human nucleus pulposus.

While the cation tracer and the dimethyl methylene blue techniques are destructive, i.e. they require that the tissue should be sacrificed, the micro-osmometer technique might possibly be applied *in vivo*. Magnetic resonance imaging (MRI) has also been used to assess the fixed charge of intervertebral discs and articular cartilages but with moderate success. Key issues with *in vivo* techniques such as MRI, Nuclear Magnetic Resonance imaging, or micro-computed tomography, are the availability and cost of the equipment.

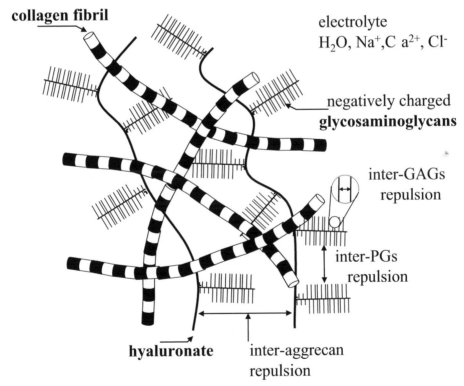

Fig. 14.2.8: Main components of articular cartilages, adapted from Loret and Simões [2004] :
- collagen fibers resist external tensile loads and chemical pressure;
- negatively charged proteoglycans (PGs) provide compressive stiffness;
- these proteins are bathed in an electrolytic solution, consisting of sodium chloride NaCl and calcium chloride $CaCl_2$ dissolved in water.

14.2.8   Mechanical properties

Articular cartilage undergoes repetitive high stresses ranging from 3 to 18 MPa as many as 1 million times a year.

As indicated by Figs. 14.2.3 and 14.2.4, the collagen content increases and the fibers become more and more aligned with the articular surface from the cancellous bone to the articular surface. Close to the bone, the fibers are almost orthogonal to the separating surface. The water content decreases from the surface to be minimal at the subchondral bone. Therefore, as a consequence of this heterogeneous composition, the mechanical properties of articular cartilage vary from point to point and, mainly in the deep and superficial zones, display anisotropy.

Collagen fibers endow cartilage with high *tensile* properties in the fiber directions, Fig. 14.2.8. Tensile stiffness may be 5-10 times higher than compressive stiffness. This difference of mechanical response in tension and compression endows the tissue with a conewise constitutive elasticity. This concept and the concept of anisotropy are, at least a priori, quite distinct.

On the other hand, the *compressive* properties are provided by the proteoglycans who resist compression because GAGs repel each other due to their negative charges. The presence of cations Na^+ *shields* the negative charges of the PGs and the mutual repulsive forces decrease with increasing sodium concentration, Section 13.9. Shielding implies decreasing macroscopic compressive and tensile moduli when the salt content increases, Eisenberg and Grodzinsky [1985], Akizuki et al. [1986][1987]. Correlatively, at higher salt content, a smaller confinement is necessary to expel water, as shown by Fig. 10 in Maroudas [1975] sketched in Fig. 18.5.1.

14.2.9   Typical laboratory experiments

Typical chemomechanical experiments are performed on specimens taken from post mortem or injured bodies[14.6]. Maroudas [1975] uses full-depth plugs of the hip, of diameter $2 L$ about 6 mm. The hydraulic permeability k_H decreases from 5 to about $1 \times 10^{-15} \, m^2/Pa/s$ as the fixed charge density increases from 0.1 to 0.25 M, her Fig. 7. Adopting the drained confined compression modulus $H = 1$ MPa of Armstrong et al. [1984] results in a diffusivity $D_H = H \, k_H$ from 1 to $5 \times 10^{-9} \, m^2/s$. The consolidation time L^2/D_H ranges from 30 mn to 2 1/2 hours, in agreement with the experimental data shown in her Fig. 10.

To minimize the consolidation time, current practice is to slice thin strips parallel to the cartilage layer, of the order of 500 μm. Another motivation to test thin strips is to avoid the heterogeneous deformations (curling) due to the depth-dependent composition of the cartilage layer.

14.2.10   Cell density and spatial organization

Cell density is measured in cell number per wet weight or in cell number per tissue volume. It is estimated via the mass of DNA with a conversion factor of 7.7 pg per cell, Kim et al. [1988]. The order of magnitude of DNA per wet weight is 1 mg/g, Williamson et al. [2003], so that the order of magnitude of the cell density is 100×10^6 cells per gram of wet weight, or 120×10^6 cells per ml of tissue.

For immature articular cartilage, the chondrocyte diameter is typically equal to 10 μm and the cell density 200×10^6 cells per ml of tissue, Wong et al. [1997], so that the volume fraction of the cells is significant, namely 0.12, and it can influence, actually weaken, the mechanical properties of the tissue since the typical modulus of cells is three orders of magnitude smaller than the modulus of the extracellular matrix. The size of the chondrocytes correlates inversely with the cell density, increasing with age. The range of variation is moderate, the diameter being multiplied at maximum by 1.5, Stockwell [1979].

The size of chondrocytes is generally reported to be larger in the middle zone of the cartilage layer, although this observation might be an artefact due to the fact that the chondrocytes loose their spherical shape to become flattened in the lower zone and spindle-shaped near the upper zone.

The earlier the tissue is analyzed the higher the cell density, e.g., 450×10^6 cells/ml in the 2nd trimester of gestation against 290×10^6 cells/ml in the 3rd trimester for the upper zone of bovine distal articular cartilage, Jadin et al. [2007]. Collected data by Stockwell [1979] indicate that the cell density in human femoral condyle decreases from 110×10^6 cells/ml at birth to 25×10^6 cells/ml at year 10 to 15×10^6 cells/ml at year 25.

The cell density decreases with depth in fetal and calf cartilage from 180×10^6 cells/ml in surface to 80×10^6 cells/ml below a depth of 0.5 mm, Klein et al. [2007]. The decrease with depth is stronger according to Stock-

---

[14.6]Usually specimens after excision are kept frozen at -20°C and, before handling, they are soaked in physiological saline. This freezing and thawing protocol has been shown to preserve the mechanical properties of tissues.

well [1979], the cell density varying practically inversely with the distance from the synovial joint. In any case, these trends, namely decrease of the cell density from fetal to adult and from the surface to the depth, are confirmed by a number of studies. The decrease of cell density with depth is to be traced to the facts that the source of nutrients is the synovial fluid and that the mode of transport of these nutrients is passive diffusion or convection by the fluid for large molecules. Still, the subchondral bone marrow has also been suggested to act as a nutrient source during maturation. Maroudas [1968] [1975] provides the diffusion coefficients of some ions and macromolecules of interest to articular cartilages. The diffusion of large molecules is known to be slow, as stressed in Section 15.5.

The spatial disposition of cells varies across the depth of the cartilage layer. Cells of mature cartilage align parallel to the synovial joint in the upper zone, and form vertical columns in the deep zone, while the organization in fetal and calf cartilages is more isotropic, Jadin et al. [2007]. The decrease in cell density during growth is thought to deduce mainly from the production of extracellular matrix, but cell death is also suggested. Still, the latter authors report that the distance between next-neighbors does not change with age, which indicates that the spatial reorganization is three-dimensional.

14.2.11   Depth heterogeneities, split-lines and anisotropy

When a whole cartilage layer is bathed in an electrolyte of ionic strength lower than physiological, it deforms non uniformly and undergoes curling, Setton et al. [1998]. The upper zone resists swelling more efficiently than the middle and lower zones. Indeed, close to the surface, collagen fibrils are oriented along the split-lines as indicated by histological analyses. The directional distribution of collagen fibrils is approximately random in the central zone, while the fibrils anchor perpendicularly to the subchondral bone in the lower zone.

A collagen fiber endows tissues with tensile resistance along its direction. When a small circular hole is punched in a collagenous tissue, its shape changes to elliptical, with the principal axis of the ellipse along the direction of larger tension: Fig. 7 of Silver and Bradica [2002] displays the split-lines in a femoral articular cartilage. These lines may be interpreted as orthotropy directions of the elastic properties: the issue is addressed in Section 5.4.

14.2.12   Natural repair and transport of nutrients

Due to the avascular nature of cartilage, renewal of the PGs in adult animals is very slow, 2-5 years for humans. It is even slower for collagen, typically a lifespan for dogs, Maroudas [1975], DiMicco and Sah [2003].

Repair can take place only for deep lesions that affect the subchondral bone, Jackson et al. [2001]. Then, there is the possibility for the bone marrow cells to migrate to the place occupied by the damaged cartilage and proliferate. Such a repair situation may be natural or resulting from operative treatments. However, the produced collagen is principally fibrous of type I, its mechanical and wear characteristics are inferior to collagen of type II and it deteriorates over time.

Transport of nutrients inside the cartilage takes place by diffusion for small molecules. For large molecules whose diffusivity is low, the "pumping" action due to Darcy's law is thought to be the main engine responsible for their transport.

14.2.13   *In vitro* engineered cartilage

Tissue-engineered cartilage has been able to mimic the biochemical appearance of hyaline cartilage but the mechanical properties of the substitutes are inferior. The idea of triggering mechanically the genesis of vessels (angiogenesis) and activity of the cells (chondrocytes) has been around for some time. It is a fact that young animals and humans produce cartilage while fully active. By contrast, it has been observed that prolonged rest was detrimental to the genesis of cartilage and other tissues. Mechanical loading of chondrocyte-seeded agarose gels (scaffolds) is shown to enhance the content of sulfated GAGs and hydroxyproline as well as the compressive modulus, Mauck et al. [2000]. To speed up the period required to produce mechanically improved *in vitro* cartilage, *bioreactors* have been designed in correlation with cyclic loading. The increase of compressive aggregate modulus reported with respect to free swelling agarose and alginate cultured cartilage is up to fourfold. Mauck et al. [2000] however observed that the positive effect of mechanical loading stops for chondrocyte-seeded cartilage cultured over 28 days. Progress is still needed as the aggregate modulus at day 28 has decreased to about one-fourth of the value of natural cartilage.

14.2.14   Interspecies, age and spatial heterogeneity

The magnitude of the representative moduli does not vary much between species as shown in Athanasiou et al. [1991] where a comparison between bovine, human, dog, monkey and rabbit cartilages is presented[14.7]. On the other hand, there are strong differences depending on
- the age of the subject;
- the function and on the location of the cartilage: Chen and Sah [2001] report moduli of aged human femoral head an order of magnitude larger than the ones of aged femoral condyle;
- the location of the sample tissue and on its depth: Akizuki et al. [1986] report tensile moduli from 1 to 15 MPa: moduli from the upper layer and from the low weight bearing zones are higher.

14.3   Pathologies, osteoarthritis, rheumatoid arthritis

Osteoarthritis (OA) is the main pathology that affects articular cartilages, Fig. 14.3.1. In the United States, OA affects 5% of the population and 70% of the population aged over 65, and costs \$8 billion annually.

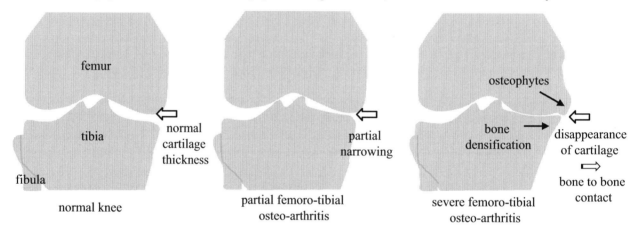

Fig. 14.3.1: Sketches of a normal knee, of a partial internal femoro-tibial osteoarthritis (gonarthrosis) and of a severe pathology. Degeneration of the cartilage layer gives rise to partial contact between femur and tibia, yielding edema and aching. Complete contact leads to a reaction of the bones, in the form of osteophytes resulting in severe pain.

Osteoarthritic cartilage has decreased tensile and compressive stiffness, increased tissue hydration and permeability. The elastic Young's modulus for normal cartilage varies from around 4 MPa in the interior to 8 MPa at the surface of the cartilage layer. These values fall to less than 1 MPa and 1.5 MPa for cartilage undergoing osteoarthritis, Table 14.3.6. Swelling due to proteoglycans is no longer limited by the confinement of collagen which, damaged, loses stiffness. OA is not linked to degradation of the PGs themselves. Since the water content is larger, the density of PGs is smaller and therefore the overall stiffness of the cartilage decreases.

Table 14.3.6: Orders of magnitude of tensile elastic moduli (MPa).

Zone	Normal Cartilage	Osteoarthritic Cartilage
Surface	8	1.5-2
Depth	4	1

One possible cause of OA is traced to the disruption of the crosslinks between collagen fibers. Osteoarthritis is correlated with age because the chondrocytes lose their ability to replace the extracellular matrix (ECM). Martin and Buckwalter [2002] find that senescence of chondrocytes is marked by an increase of β-galactosidase and a decrease in mitotic activity. Therefore, transplantation of chondrocytes in elderly people, to be successful, should also stimulate in place chondrocytes.

[14.7]Isolated moduli should be considered with care. First, it should be made clear whether they are tangent or secant moduli. Besides, the type of experiment (confined compression, traction, etc.) and the strain at which they hold should be known.

Williams et al. [2003] consider that early OA is marked by an increased activity of degradative enzymes and enhanced synthesis of ECM. Clinical treatment of OA includes the use of nonsteroidal anti-inflammatory drugs (NSAIDs): they have unfortunately secondary effects. One of these NSAIDs is the high molar mass hyaluronan, which is present in the synovial fluid, see Section 20.3.3 for more details.

Fernandes et al. [2002] seem to imply that osteoarthritis is not due to a self-degradation of cartilage components, but to proteolytic enzymes[14.8]. These enzymes are generated by the chondrocytes themselves or diffuse into cartilage from an inflamed synovial liquid. Cytokines, interleukin 1 (IL-1) and tumor necrosis factor α (TNF-α) stimulate metalloproteinases gene expression and inhibit the metabolic pathways that generate the receptor antagonists of IL-1 and TNF-α. They suggest that the excess of receptors of IL-1 and TNF-α and lack of receptors of their antagonists may originate overproduction of nitric oxide (NO) in OA tissues. According to Jouzeau et al. [2002], NO mediates most effects of IL-1. Imbalance between pro-inflammatory cytokines such as IL-1 and regulatory or inhibitory systems such as receptors of IL-1 antagonist leads to lesions in cartilage, namely both rheumatoid arthritis (inflammation) and osteoarthritis (disruption of collagen, inhibition of the production of certain components of ECM). Lahiji et al. [2002] mention that the β_1-integrin is associated with the severity of osteoarthritis. A natural target of therapy is thus to increase the levels of antagonists to proteolytic enzymes and of their receptors either by gene therapy or injection.

Electrophysiology of chondrocytes is seldom mentioned but nevertheless interesting. Mechanical stimulation of normal cartilage at 0.33 Hz is reported to induce a membrane hyperpolarization, with secondary activation of Ca^{2+}-activated K-channels, while it produces depolarization in OA cartilage, Salter et al. [2002].

Supra-physiological mechanical loads are also advocated to trigger osteoarthritis. Indeed, clinical and animal studies show that meniscectomy is often followed by osteoarthritic degeneration. W. Wilson et al. [2003] performed an axisymmetric finite element analysis of the knee joint, including part of the subchondral bone, using a biphasic behavior, either isotropic or transversely isotropic. The hydraulic conductivity is assumed to be strain dependent as in (16.4.2). They find that menisectomy increases the stress levels and shear stresses in the articular cartilage. It is in keeping with the role of meniscus which restraints the radial expansion of the cartilage, being itself a transversely isotropic material with a higher stiffness in the circumferential direction. The increase in stress level may lead to damage of either the cartilage or the subchondral bone and finally trigger osteoarthritis. This simulation should however be seen more as a suggestion than as a proof.

Gene therapy is advocated as a promising tool to treat osteoarthritis. Gene therapy is the process by which a protein is produced from a cell whose DNA has been modified. Classical clinical therapy tries to ensure a sufficiently high level of insulin-like growth factor 1 (IGF-I) which is known to improve the metabolism. However the longevity of IGF-I is limited. Gene expression aims at alleviating this problem, Nixon et al. [2000], Grande and Nixon [2000]. One issue concerns the cells to be modified: should they be autologous chondrocytes or chondroprogenitor stem cells? Another issue concerns the genes to be inserted. The idea is to deliver a therapeutic gene to the cells that would inhibit the production, or the action, of inflammatory cells. For example, stimulation of the level of IL-1 receptor antagonists (IL-1Ra) would result in more IL-1 to be bound to these receptors and prevent them from acting on chondrocytes. To transfer the genes to the cells, the vectors used include recombinant retro-viruses and adenoviruses: Frenkel and di Cesare [1999] provide a list of references.

14.4 A brief review of modeling aspects

14.4.1 Intrafibrillar (IF) and extrafibrillar (EF) compartments

In a series of papers, Maroudas and co-workers [1975]-[1991] have argued that the partition of water in intrafibrillar (IF) and extrafibrillar compartments (EF) has mechanical significance. The intrafibrillar compartment is defined as the volume inside the collagen fibrils or, more exactly, between collagen molecules. Proteoglycans belong to the extrafibrillar compartment. At physiological pH, the slightly positive charge of collagen attracts anions while cations are repelled somehow but they are still present in the intrafibrillar space, possibly to ensure its electroneutrality, Li and Katz [1976]. The presence of chemicals within the intrafibrillar compartment is constrained by steric considerations: the distance between collagen molecules being about 5 Å, Fig. 14.4.1, water molecules and chloride anions

---

[14.8]For a biochemical description of proteolysis, see APB [1999], pp. 170, 278. Some enzymes degrade proteins into independent amino acids. They are proteases and peptidases. They cut bonds between molecules. Among proteases are serine-proteases 3.4.21.n, metallo-proteases 3.4.24.n, etc. Among serine-proteases are trypsin, chymotrypsin and elastase used to digest components of cartilage.

Cl^- of size 3.6 Å, sodium cations Na^+ and calcium cations Ca^{2+} of size around 2 Å can penetrate the intrafibrillar compartment while large molecules like sucrose or PGs of size comparable to that of the collagen molecule can not. Experimentally, the partition between intrafibrillar and extrafibrillar waters is performed by injecting serum albumin whose size (hydrodynamic radius 35 Å) prevents them to enter the intrafibrillar compartment.

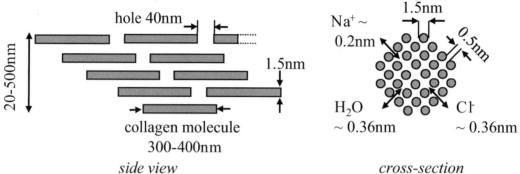

Fig. 14.4.1: Side view and cross-section of a collagen fibril. Collagen fibrils play the role of a semipermeable membrane that allows the transfer, between the intra- and extrafibrillar spaces, of small molecules like water and metallic ions. Macromolecules like proteoglycans are not admitted in the intrafibrillar space. Adapted from Torzilli [1985]. Redrawn with permission from Loret and Simões [2004].

For unloaded cartilage under physiological composition, intrafibrillar water represents up to 25% of total water, the extrafibrillar water furnishing the complement. The latter can be moved by mechanical loading and osmosis with water external to the cartilage (synovial fluid). Since the intrafibrillar water is in contact with the extrafibrillar compartment only, its status with respect to mechanical and chemical loadings is not the same as that of extrafibrillar water. The IF water can be moved by change of chemical composition of the EF compartment. Now a mechanical loading changes the relative chemical composition of the tissue (e.g., water being expelled, the concentration of proteoglycans increases) and therefore indirectly induces a transfer of water from the intrafibrillar compartment. Data are provided by Maroudas et al. [1991] for PG-depleted cartilage. She shows that increase of the concentration in the extrafibrillar compartment depletes the water from the intrafibrillar compartment and so decreases the tension in the collagen. Ultimately, with a large applied compressive load, the fibrils may undergo compression and may buckle, Katz et al. [1986].

Intrafibrillar water and/or hydrated PGs induce collagen fibers in tension, Fig. 14.4.2 : a mechanical model would consist of two parallel systems, collagen and PGs, that resist the applied mechanical load. The tension in the collagen branch is the tensile load carried by the collagen fibers, while the branch of the PGs is compressive, Maroudas et al. [1991], Lai et al. [1991]. The sum of these two components opposes the applied mechanical load, Basser et al. [1998].

The osmotic stress $\sigma_{osm} = -\pi_{osm}\,\mathbf{I}$ due to PGs may be seen as isotropic. It is measured in a solution of proteoglycans at a given fixed charge and therefore at a given volume fraction of water[14.9],

$$\mathrm{FCD} = \frac{|\zeta_{PG}|\,N_{PG}}{M_w}\,,\qquad\qquad (14.4.1)$$

where ζ_{PG}, varying from -6000 to -4000, is the valence of PGs. From the curve displayed in Fig. 3 of Basser et al. [1998] that provides the osmotic pressure $\pi_{osm} = \pi_{osm}(\mathrm{FCD})$ [unit : MPa] as a function of the fixed charge density FCD [unit : mole/kg water],

$$\pi_{osm} = \pi_{osm}(\mathrm{FCD}) = 0.351\,\mathrm{FCD} + 1.93\,\mathrm{FCD}^2\,,\qquad\qquad (14.4.2)$$

follows the tangent modulus,

$$H_{osm} = -\frac{d\pi_{osm}}{d\,\mathrm{tr}\,\epsilon} = -\frac{d\pi_{osm}}{d(\Delta V/V_0)} = -\frac{d\pi_{osm}}{d(\Delta V_w/V_0)} = \mathrm{FCD}\,\frac{V_0}{V_w}\,\frac{d\pi_{osm}}{d\mathrm{FCD}} > 0\,.\qquad (14.4.3)$$

The sign of the modulus H_{osm} ensures the stabilization of the osmotic effect. Indeed in Section 14.9.1, the osmotic pressure will be shown to induce tension in the collagen fibers and to lead to an increase of the volume of the tissue. This expansion tends in turn to decrease the osmotic pressure as a consequence of the positive sign of H_{osm} and therefore counteract the effect.

[14.9]The fixed charge density FCD is measured either in mole of PGs per mass of water or in mole of PGs per volume of water. These two definitions differ only by their units. A distinct definition to be introduced later in this chapter refers to the sole volume of the extrafibrillar fluid phase, and not to the total fluid volume.

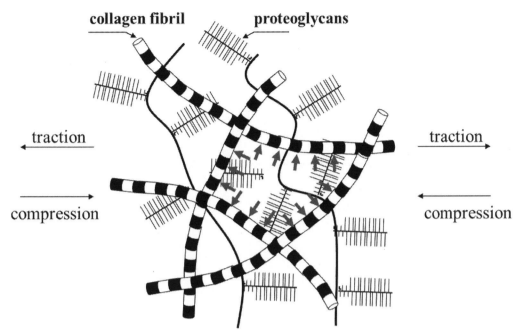

Fig. 14.4.2: The high negative charges of PGs induce repulsive forces at various structural levels. PGs provide compressive stiffness to the cartilage while collagen is put in tension by the water pressure induced by the charged PGs and external loads.

14.4.2 Tensile experiments

Myers et al. [1984] perform tensile tests on bovine articular cartilage on parallelepiped samples of thickness 200 μm bathed in an electrolyte of controlled composition. Under constant axial strain, increase of salt concentration induces a transient tension which is relaxed due to diffusion inside the specimen and perhaps to viscous properties of collagen. The final axial stress is smaller than its initial value. An interpretation of tensile experiments is proposed in the qualitative sketch 13.10.7.

In absence of axial force (isotropic free swelling), increase of salt concentration in the bath implies increased contraction, with in fact a linear dependence of the strain with respect to the salt concentration *in the bath*[14.10]. The latter effects under isotropic free swelling are in agreement with the experiments of Eisenberg and Grodzinsky [1985] and they are due to osmosis which pumps water out of the sample.

Tensile properties of cartilage are mainly due to collagen and in fact to the collagen network. Bader et al. [1981] use an enzyme, the elastase of neutrophil leucocytes, which is known to degrade the non-helical terminal peptides of cartilage collagen molecules and therefore to disrupt the crosslinks between collagen fibrils. In fact this enzyme is present in large quantity in joints affected by rheumatoid arthritis. They show that both tensile stiffness and tensile strength are considerably reduced by leucocyte elastase.

14.4.3 The triphasic model of the Columbia University group

The triphasic model of articular cartilage of Lai et al. [1991] introduces three *phases*, one for the solid that includes collagen and proteoglycans, one for the pore fluid and one for ions, resulting from the dissolution of a single neutral salt, specifically NaCl[14.11]. The chemical potential for the ionic phase is obtained by summation of the individual chemical potentials, i.e.,

$$g_{\mathrm{s}} = g_{\mathrm{Na}} + g_{\mathrm{Cl}} = \widehat{m}_{\mathrm{s}}\,\mu_{\mathrm{s}} = \widehat{m}_{\mathrm{Na}}\,\mu_{\mathrm{Na}}^{\mathrm{ec}} + \widehat{m}_{\mathrm{Cl}}\,\mu_{\mathrm{Cl}}^{\mathrm{ec}} = \widehat{m}_{\mathrm{Na}}\,\mu_{\mathrm{Na}} + \widehat{m}_{\mathrm{Cl}}\,\mu_{\mathrm{Cl}}, \qquad (14.4.4)$$

[14.10] The interpretation of this observation requires some care, since the concentrations in the tissue and in the bath are different due to the membrane effect, as explained in Section 14.6.2.

[14.11] The terminology varies with authors. The notion of *phase* used by Lai et al. [1991] and followers and the one defined later in this chapter and used throughout this work are definitely distinct.

where the g's and the μ's are the *mole based* and *mass based* electrochemical potentials respectively, and the $\widehat{m}_k$ the molar masses of the cation for $k=\mathrm{Na}^+$, of the anion for $k=\mathrm{Cl}^-$ and of the salt NaCl for $k=\mathrm{s}$. The electrochemical potential of the ion involves three contributions, of mechanical, chemical and electrical natures,

$$g_k^{\mathrm{ec}} = \widehat{m}_{\mathrm{Na}}\,\mu_k = \widehat{v}_{\mathrm{Na}}\,p_{\mathrm{F}} + RT\,\mathrm{Ln}\,x_k + \zeta_k\,\mathrm{F}\,\phi\,, \tag{14.4.5}$$

where F is Faraday's equivalent charge, ϕ is the electric potential and p_{F} is the pressure of the pore fluid. Hence the chemical potential of the salt,

$$g_{\mathrm{s}} = \widehat{m}_{\mathrm{s}}\,\mu_{\mathrm{s}} = \widehat{v}_{\mathrm{s}}\,p_{\mathrm{F}} + RT\,\mathrm{Ln}\,x_{\mathrm{Na}}\,x_{\mathrm{Cl}}\,, \tag{14.4.6}$$

with $\rho_{\mathrm{s}} = \widehat{m}_{\mathrm{s}}/\widehat{v}_{\mathrm{s}}$ the mass density of NaCl, $\widehat{v}_{\mathrm{s}} = \widehat{v}_{\mathrm{Na}} + \widehat{v}_{\mathrm{Cl}}$ its molar volume and $\widehat{m}_{\mathrm{s}} = \widehat{m}_{\mathrm{Na}} + \widehat{m}_{\mathrm{Cl}}$ its molar mass.

The chemical potential of water has also the standard form of chemical physics, namely

$$g_{\mathrm{w}} = \widehat{m}_{\mathrm{w}}\,\mu_{\mathrm{w}} = \widehat{v}_{\mathrm{w}}\,p_{\mathrm{F}} + RT\,\mathrm{Ln}\,x_{\mathrm{w}}\,. \tag{14.4.7}$$

The molar fractions involved in the definition of the chemical potentials above are defined with respect to the electrolyte, that involves the fluid and ionic phases,

$$x_k = \frac{N_k}{N_{\mathrm{w}} + N_{\mathrm{Na}} + N_{\mathrm{Cl}}}, \quad k = \mathrm{w, Na, Cl}, \tag{14.4.8}$$

where the N's are the mole numbers.

The concentration [unit: mole/l] is the number of moles per unit volume of electrolyte (or per unit volume of fluid only since both are close at low ionic concentration, $V_{\mathrm{F}} = N_{\mathrm{w}}\,\widehat{v}_{\mathrm{w}} + N_{\mathrm{s}}\,\widehat{v}_{\mathrm{s}} \simeq N_{\mathrm{w}}\,\widehat{v}_{\mathrm{w}}$, i.e.

$$c_k = \frac{N_k}{V_{\mathrm{F}}}, \quad k = \mathrm{w, Na, Cl}\,. \tag{14.4.9}$$

The *effective* concentration e_{PG}, of proteoglycans

$$e_{\mathrm{PG}} = \frac{\text{algebraic nb. of moles of charges on proteoglycans}}{\text{volume of electrolyte}}\,, \tag{14.4.10}$$

plays a key role in the chemomechanical coupling. The fixed charge density FCD is equal to the absolute value of the effective concentration measured in mole of PGs per volume of water. The fluid phase, or electrolyte, is considered to include both intrafibrillar and extrafibrillar water, because this partition is not recognized by the Columbia group. If it was accounted for, only the extrafibrillar water should enter here since the proteoglycans belong to this compartment and are absent from the intrafibrillar compartment. In other words, the molar fractions, or the concentrations, are counted with respect to the tissue water and not to extrafibrillar water.

A *single* electroneutrality condition is enforced for the whole porous medium,

$$c_{\mathrm{Cl}} = c_{\mathrm{Na}} + e_{\mathrm{PG}}\,. \tag{14.4.11}$$

For low ionic concentrations,

$$x_k \simeq \frac{c_k}{c_{\mathrm{w}}}, \quad k = \mathrm{Na, Cl}; \quad x_{\mathrm{w}} \simeq 1 - \frac{c_{\mathrm{Cl}}}{c_{\mathrm{w}}} - \frac{c_{\mathrm{Na}}}{c_{\mathrm{w}}}\,, \tag{14.4.12}$$

and, expanded to first order, the chemical potential of water is affine with respect to the ionic concentrations,

$$g_{\mathrm{w}} = \widehat{m}_{\mathrm{w}}\,\mu_{\mathrm{w}} \simeq \widehat{v}_{\mathrm{w}}\,p_{\mathrm{F}} - RT\left(\frac{c_{\mathrm{Na}}}{c_{\mathrm{w}}} + \frac{c_{\mathrm{Cl}}}{c_{\mathrm{w}}}\right), \tag{14.4.13}$$

while the chemical potential of salt involves the product of the ionic concentrations,

$$g_{\mathrm{s}} = \widehat{m}_{\mathrm{s}}\,\mu_{\mathrm{s}} \simeq \widehat{v}_{\mathrm{s}}\,p_{\mathrm{F}} + RT\,\mathrm{Ln}\,\frac{c_{\mathrm{Na}}}{c_{\mathrm{w}}}\frac{c_{\mathrm{Cl}}}{c_{\mathrm{w}}}\,. \tag{14.4.14}$$

Assuming an incompressible solid matrix, the fixed charged density is modified by the volume change of the porous medium, as detailed in Exercise 14.3,

$$\frac{V_{\mathrm{F}}^0}{V_{\mathrm{F}}} = \frac{n_0^{\mathrm{F}}}{n_0^{\mathrm{F}} + \det\mathbf{F} - 1} \simeq 1 - \frac{\mathrm{tr}\,\boldsymbol{\epsilon}}{n_0^{\mathrm{F}}}\,, \tag{14.4.15}$$

with n^{F} the volume fraction of the fluid, $\mathbf{F}$ the gradient of deformation and $\boldsymbol{\epsilon}$ the infinitesimal strain of the solid skeleton (or porous medium), and the index 0 denoting the reference configuration. As expected, the concentrations of all species and fixed charge decrease if the porous medium expands by pumping water in, and increase if the porous medium contracts by expelling water,

$$c_k = \frac{N_k}{V_{\mathrm{F}0}}\frac{V_{\mathrm{F}}^0}{V_{\mathrm{F}}}, \quad \frac{c_{\mathrm{PG}}}{c_{\mathrm{PG}}^0} = \frac{e_{\mathrm{PG}}}{e_{\mathrm{PG}}^0} = \frac{V_{\mathrm{F}}^0}{V_{\mathrm{F}}}\,. \tag{14.4.16}$$

In fact, the model exposed in Lai et al. [1991] differs slightly with respect to the above exposition: 1. The fluid pressure is neglected in the electrochemical potentials of ions; 2. the chemical contribution to the latter is introduced via concentrations, rather than molar fractions; 3. these concentrations come into picture multiplied by chemical activities. The latter are not quantified and actually the ratios of their values in the bath and in the cartilage are set equal to one in applications.

14.4.4 The quadriphasic model of the Eindhoven University group

Here, cations and all anions are gathered in two separate phases, Frijns et al. [1997]. This approach seems more appropriate from an electrokinetic point of view than the triphasic model. Nevertheless, it treats all ionic species of each of the two phases on the same foothold. The partition between intra- and extrafibrillar waters is not envisaged either. A single electroneutrality condition is applied to the whole porous medium.

Once again, the notion of phase is not mechanically nor kinematically motivated in these models but rather serves to gather species based on their nature.

14.4.5 The experiments of the MIT group

Compression tests: shielding effects and chemical stress, Eisenberg and Grodzinsky [1985]

Eisenberg and Grodzinsky [1985] perform experiments on bovine articular cartilage and corneal stroma of 1- to 2-week-old calves. Cartilage samples are cylindrical, 9 mm in diameter and about 600 μm in thickness. Samples of corneal stroma have similar size, 6.4 mm in diameter and 750 to 850 μm in thickness. The epithelium and endothelium have been removed by scraping the corneal surfaces with a scalpel. The specimens are put in contact with a bath of controlled chemical composition. Mechanical loading is performed stepwise and, after each load increment, time to reach equilibrium is ensured: equilibrium requires about 15 mn for the tested specimens of articular cartilage and 60 mn for corneal stroma.

Their results highlight two effects:
- when the concentration of NaCl in the bath is increased at constant axial strain, the compressive axial stress decreases, their Fig. 5. The drop of the compressive stress is large for low NaCl concentrations, it diminishes as NaCl increases and it tends to a very small value at higher NaCl concentration: the stress decrease between 0.5 M and 1 M of NaCl is almost negligible. This effect is compatible with the chemical consolidation observed in clays, Loret et al. [2002];
- at given concentration of NaCl in the bath, the stress-strain relation is affine, their Fig. 9. The intercept with the unstrained state is viewed as a chemical stress. The stiffness (slope of the stress-strain curves) decreases with salt concentration, stabilization being reached between 0.5 M and 1 M of NaCl.

This increase of compressibility is attributed to an increased "electrical shielding" of PGs by cations. Eisenberg and Grodzinsky [1985] also note that this phenomenon has already been observed by others, e.g., Maroudas [1975], her Fig. 10. They model the stiffness and chemical stress as exponentially decaying with the concentration of NaCl in the bath.

The idea is also retained by Lai et al. [1991] who coin the additional stress *chemical expansion stress*. The latter is expressed as the product of the effective concentration of PGs times a function that decays with the bath concentration, their Eqn (82): this expression clearly displays the shielding effect due to NaCl.

At given bath concentration, the stress-strain relation is affine and approximately isotropic. Therefore, NaCl-dependent Lamé moduli are defined by Eisenberg and Grodzinsky [1985]. They also argue that this linear elasticity over a large range of compressive strains can not be explained by the action of PGs alone: the resulting strains would be isotropic, which is not verified experimentally. They explain their results by (a) either the fact that collagen tension is not completely released in confined compression, (b) or that the PGs form a gel-network that has solid-like properties and thus do not behave like in a solution.

Khalsa and Eisenberg [1997] find that the shear modulus μ of articular cartilage does not depend on bath composition. By contrast, Jin and Grodzinsky [2001] observe a dependence of the shear modulus with respect to ionic strength, their Figs. 6 and 9. These observations are not really in contradiction with the results of Eisenberg and Grodzinsky [1985] where only the apparent aggregate modulus $\lambda + 2\,\mu$ was measured.

Mechanoelectrical couplings, Frank and Grodzinsky [1987]

Frank and Grodzinsky [1987] have performed tests on adult bovine cartilage where the tissue is in chemical equilibrium with a reservoir of controlled composition. Their experiments are intended to reveal and quantify mechano-electrical and electro-mechanical couplings:
- "open circuit test": under a sinusoidal uniaxial compression, they measure the induced streaming potential, the circuit being electrically open;
- "short circuit test": the electrodes being shorted together (uniform electrical field), they measure the electrical current generated by a sinusoidal mechanical compression;
- a harmonic electric current density being applied to the specimen, they measure the generated stress.

Tension and compression, rate effects and anisotropy

As mentioned above, PGs are believed to contribute mainly to the compressive properties. Collagen has an effective role in compression as well where it is put in tension by PGs. However, tensile properties of collagen are mainly highlighted under tensile loadings. Accordingly, rate effects are expected to be revealed by tensile loadings rather than by compressive loadings, Myers et al. [1984], Huang et al. [2001], Lei and Szeri [2007].

Diffusion equations, Eisenberg and Grodzinsky [1987]

The above results are used as a basis to simulate transient loadings including pore pressure build up and dissipation. The diffusive flux of an ionic species involves the gradients of its own concentration and of the electrical potential but the effect of the pore pressure gradient is neglected, their Eqn (12). Several couplings are also neglected in the flux of water which reduces to the standard Darcy's law, their Eqn (4).

Diffusion properties: hydraulic conductivity and diffusivity

Hydraulic conductivity of articular cartilage is low, Maroudas [1975]. Effective diffusivity accounts for the tortuosity of the cartilage, that is, the longer path that ionic species have to follow with respect to diffusion in pure water. Eisenberg and Grodzinsky [1987], their Eqn (17), estimate the tortuosity factor via the Mackie-Meares expression (16.4.5). Section 16.4 provides ranges of values for the diffusion parameters.

14.4.6 Molecular and microstructural models

Nanometric models using Poisson-Boltzmann equation (13.E.6) aim at explaining the compressive behavior due to proteoglycans, Buschmann and Grodzinsky [1995]. Some features of the geometrical configuration of PGs at the nanometer scale are accounted for but adjustable parameters have to be introduced as well. Collagen plays no role in these models.

On the other hand, collagen is viewed to play an essential role in some microstructural models. Soulhat et al. [1999] consider cartilage as a composite material consisting of a matrix reinforced by fibrils. Bursac et al. [2000] idealize cartilage through a network of cables under pre-tension, representing collagen submitted to osmotic pressure. Under small external compression, the cables remain in tension, resulting in a high stiffness. As compression increases, the forces in the cables become compressive, the cables buckle and the overall stiffness drops significantly.

14.5   Interpretation of laboratory experiments

14.5.1   Difficulties of methodology

Most if not all mechanical models of articular cartilage present simultaneously elastic constitutive equations and diffusion equations. While there is nothing wrong in doing so, some confusion might arise as properties that hold at equilibrium only are not clearly delineated.

Another formal difficulty comes from the fact that derivation of constitutive equations is sometimes made in parallel with the presentation of typical experiments where a specimen of cartilage is brought in contact with a bath of known chemical composition: then the concentration of salt in the bath appears in the constitutive equations, while local constitutive equations should involve quantities defined at the point under consideration only. The issue can be resolved formally by introducing a fictitious compartment/bath, whose chemical composition is a function of that of the cartilage, Section 14.5.2. Of course, due to the presence of proteoglycans, salt concentration in cartilage is not expected to be equal to salt concentration in the bath even under equilibrium condition, Section 14.6.2. This discontinuity may easily lead to misinterpretation of experimental results and induce confusion in the identification of parameters, Section 16.3.3.2.

14.5.2   The fictitious bath: a tool for heterogeneous and transient processes

Standard formulations of chemomechanical constitutive equations consider a homogeneous piece of tissue surrounded by a bath of controlled chemical composition, pressure and electrical potential. Equilibrium is assumed between the bath and the cartilage. Some caution is required, especially concerning the use and interpretation of the associated osmotic pressure.

In fact, let us consider now a specimen of cartilage which is nonspatially uniform. One may also consider a point inside the cartilage during the transient period where the bath concentration is changed. In both cases, points inside the specimen are not in equilibrium with the bath, either due to a transient loading or due to spatial non homogeneity. In addition, the bath might not be homogeneous, e.g., the case of a layer between two nonhomogeneous baths. In any case, the composition and properties of the bath are not constitutive: they should *not* be involved in the constitutive equations of the tissue.

Still one may use the above tissue-bath device, but with a different perspective. For that purpose, a *fictitious* bath is introduced which may vary in time and space.

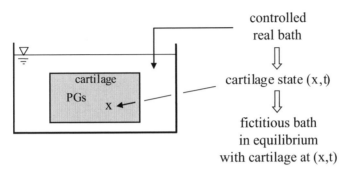

Fig. 14.5.1: The notion of a fictitious bath. Consider a cartilage specimen in contact with a bath of controlled and time varying chemical composition. Points inside the cartilage are not in instantaneous equilibrium with the bath. With each point is associated a time-varying fictitious bath. The properties of the fictitious bath are *deduced* from the properties of the extrafibrillar phase. If the specimen has spatially uniform physical properties and if equilibrium is homogeneous throughout the specimen, the local fictitious baths are identical to the real bath in contact with the specimen.

In fact, the properties of the fictitious bath (chemical composition, pressure and electrical potential) are *deduced* from the actual tissue at the point and time of interest, Fig. 14.5.1. Crucially, the fictitious bath contains the same species as the fluid phase of the cartilage, except PGs, it is electroneutral and in chemical equilibrium with the cartilage at the point and time of interest. Consider for example a single salt to be present, namely sodium chloride. Electroneutrality (one equation) and equilibrium of the electrochemical potentials (two equations) of the two ions yield the molar fractions and the electrical potential (three unknowns) in the fictitious bath in terms of the properties of the tissue. The consistency relation between molar fractions (they sum to one) yields the molar fraction of water, in terms of now known quantities. Equality of the chemical potentials of water of the cartilage and fictitious bath provides an osmotic pressure π_{osm}, in terms of quantities which are functions of the thermodynamic state of the cartilage. This osmotic pressure and the pressure p_{E} of the tissue fluid are used to define the pressure $\tilde{p}_{\text{E}}$ of the fictitious bath,

$$\tilde{p}_{\text{E}} = p_{\text{E}} - \pi_{\text{osm}} \,. \tag{14.5.1}$$

To illustrate further the concept of a fictitious bath, let us assume that the constitutive equation is given through an effective stress format, namely

$$\boldsymbol{\sigma} + \overbrace{(p_{\text{E}} + p_{\text{ch}})}^{p_{\text{eff}}} \mathbf{I} = \sum_k \boldsymbol{\sigma}^k_{\text{ch-mech}} \,, \tag{14.5.2}$$

where p_{ch} is a term of chemical origin, and the rhs represents the apparent chemomechanical contributions of the structural components (collagen, PGs $\cdots$).

Use of this concept depends on the application:

- for example, to develop chemomechanical constitutive equations, we will consider a homogeneous specimen surrounded by a bath of known and controlled properties. The cartilage is in continuing equilibrium with the fictitious bath and with the real bath. Therefore fictitious and real baths are just one and the same in this circumstance. The above procedure still holds but is not used as such. Rather, the chemical composition and properties of the cartilage are *obtained* from those of the bath, which are controlled, e.g., now $\tilde{p}_{\text{E}}$ is equal to the pressure p_{B} of the real bath, and

$$p_{\text{E}} = \tilde{p}_{\text{E}} + \pi_{\text{osm}} \,. \tag{14.5.3}$$

The constitutive equations (14.5.2) may be rewritten in the form,

$$\boldsymbol{\sigma} + \tilde{p}_{\mathrm{E}} \mathbf{I} + (\pi_{\mathrm{osm}} + p_{\mathrm{ch}}) \mathbf{I} = \sum_k \boldsymbol{\sigma}^k_{\mathrm{ch-mech}}. \qquad (14.5.4)$$

- the notion of a fictitious bath is fully exploited during transient states and in the context of boundary value problems through, e.g., the finite element method. The chemical composition of the fictitious bath is deduced from that of cartilage as indicated above. The pressure p_{E} should be specified through some means, e.g., boundary conditions as for an incompressibility constraint. In other words, it is thought as a primary unknown, or at least, easily deducible from primary unknowns. Then, the pressure $\tilde{p}_{\mathrm{E}}$ is obtained as in (14.5.1). Similar comments hold for the electrical potentials. The real bath comes into picture but like in a boundary value problem and not pointwise.

The constitutive equations (14.5.2) may be recast as,

$$\boldsymbol{\sigma} + p_{\mathrm{E}} \mathbf{I} + p_{\mathrm{ch}} \mathbf{I} = \sum_k \boldsymbol{\sigma}^k_{\mathrm{ch-mech}}, \quad p_{\mathrm{ch}} = p_{\mathrm{eff}} - \pi_{\mathrm{osm}} - \tilde{p}_{\mathrm{E}}. \qquad (14.5.5)$$

The standard formulation of porous media without chemical effects, namely $p_{\mathrm{ch}} = 0$, is recovered by interpreting p_{E} as the fluid pressure.

14.6   Partition of the tissue into phases

This section and the subsequent ones present a constitutive model with emphasis on chemomechanical couplings, Loret and Simões [2004] [2005]a. Transient aspects, including time dependent mass transfer and generalized diffusion, are addressed in Chapters 16 to 18.

In fact, several continuum theories have been proposed to model deformation and transport and to describe macroscopic couplings in articular cartilages, e.g., Lai et al. [1991], Gu et al. [1998], Huyghe and Janssen [1999]. However, their structure is different, as far as the intrafibrillar phase is usually not recognized. The intra/extrafibrillar split is considered in Huyghe [1999] and Huyghe et al. [2003], but only the mechanical aspects of cartilages bathed in a binary electrolyte are addressed.

14.6.1   Definition of the phases: a measure of spatial organization

The definition of the phases is mechanically motivated. A kinematical criterion on the other hand would sort species according to their velocities. Here, cartilage is viewed as a three-phase, multi-species, porous medium, Fig. 14.6.1.

The *solid phase* S contains the collagen fibers denoted by the symbol c. The *intrafibrillar fluid phase* I contains intrafibrillar water w, sodium ions Na^+, calcium ions Ca^{2+} and chloride ions Cl^-. The *extrafibrillar fluid phase* E contains proteoglycans, extrafibrillar water, sodium and calcium cations, and chloride anions. The sets of species of the solid, intrafibrillar and extrafibrillar phases can be defined in explicit form,

$$\mathrm{S} = \{\mathrm{c}\}, \quad \mathrm{I} = \{\mathrm{w, Na, Ca, Cl}\}, \quad \mathrm{E} = \{\mathrm{w, PG, Na, Ca, Cl}\}. \qquad (14.6.1)$$

Exchanges of water and ions occur between the fluid phases but only the extrafibrillar phase communicates with the surroundings.

Due to the presence of proteoglycans a minimum number of cations is required to ensure electroneutrality of the extrafibrillar phase. Therefore, mobile cations are endowed with a velocity independent of that of their non mobile counterparts which, like the proteoglycans, move with the velocity of the solid phase. Thus cations in the extrafibrillar space are partitioned into a mobile mo part and a non mobile nm part. The set E_{mo} of extrafibrillar mobile species contains all extrafibrillar species but proteoglycans. Non mobile ions sodium do not bind to PGs. If this distinction between mobile and non mobile ions is not adopted, all ions are able to diffuse through the extrafibrillar fluid phase. The two options are left open in the model of generalized diffusion developed in Chapter 16. Unlike sodium ions, calcium ions may bind to proteoglycans. The issue is addressed in relation with the effects of pH in Chapter 17.

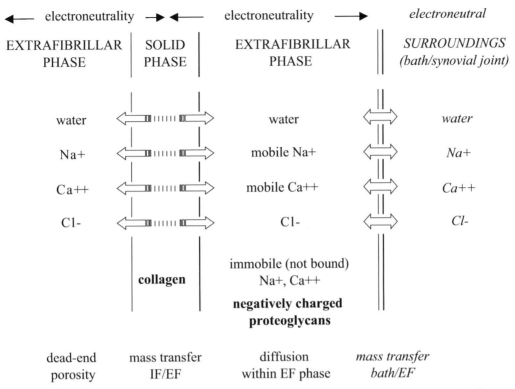

Fig. 14.6.1: Articular cartilage is partitioned in three phases, a solid phase and two fluid phases, Loret and Simões [2005]a. Each fluid phase contains several species. Some of these species are mobile, at least partially: water and ions can enter and leave the intrafibrillar space crossing the collagen fibrils. Proteoglycans which are macromolecules are too large to be admitted into that space. Water and ions of the extrafibrillar phase diffuse within their phase. For points at the surface of the tissue, only the extrafibrillar phase is in contact with the surroundings: species of the intrafibrillar phase should transit through the extrafibrillar fluid to access the synovial joint.

14.6.2   The fictitious membrane surrounding a cartilage specimen

When defining the partition of the species into phases, the fixed charge is of primary concern: should it reside in the solid phase with the collagen fibers, or in the fluid phases, where their effects are of primary importance for the electro-chemomechanical couplings?

   According to a kinematic criterion, it would be part of the solid phase. On the other hand, its mechanical effects are related to its concentration with respect to the volume of extrafibrillar water. Some key differences between the two options are listed below:

1. Fixed charge in the solid phase.
   This option has been adopted in a model of expansive geomaterials in Loret et al. [2004]a [2004]b. Let us consider the chemomechanical equilibrium of a tissue with an external reservoir: pore pressure, concentration and electrical potential are continuous at the boundary between the water phase and the reservoir. Moreover, the electrical conductivity vanishes when the external reservoir is quasi-distilled water. In addition, the electroosmotic coefficient k_e and the osmotic coefficient ω are not constrained by compatibility conditions. Thus they have to be provided by additional constitutive equations.

2. Fixed charge in the fluid phase.
   The presence of the fixed charge in the extrafibrillar phase has two main consequences.
   First, pore pressure, concentrations and electrical potential undergo a jump across the interface that separates the cartilage and a bath, even at equilibrium, Fig. 14.6.2. This observation is of primary importance in the interpretation of laboratory experiments and in the identification procedure of material constants. In the finite element context addressed in Chapter 18, it strongly guides the choice of the primary variables. In fact, the

equilibrium condition is phrased in terms of the electrochemical potentials and not in terms of concentrations, pressures or electrical potentials.

Second, the electroosmotic coefficient k_e and the osmotic coefficient ω are constrained by compatibility conditions: in fact, they are shown in Chapter 16 *to result from the formulation*, in explicit or implicit form.

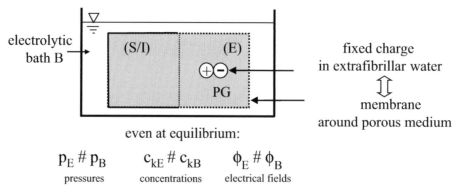

$$p_E \# p_B \qquad c_{kE} \# c_{kB} \qquad \phi_E \# \phi_B$$
$$\text{pressures} \qquad \text{concentrations} \qquad \text{electrical fields}$$

Fig. 14.6.2: The presence of the electrically charged proteoglycans in the extrafibrillar water implies a number of physical entities to be discontinuous across the specimen boundary, as if the specimen was surrounded by a semipermeable membrane. The sketch displays a delocalized partition of cartilage, isolating the extrafibrillar compartment from the intrafibrillar compartment and solid phase.

14.6.3   A strongly interacting model

The main assumptions which underly the three-phase multi-species model follow a *strongly interacting* model. They have been listed in Loret and Simões [2004] in a purely mechanical perspective. Since deformation, mass transfer and generalized diffusion are accounted for, they need to be re-stated in this more general context, namely,

(H1) Mass balance is required for each species in each phase.

(H2) Momentum balance is required for the mixture as a whole. Water and mobile ions in the extrafibrillar phase are endowed with their own velocities so as to allow water to flow through the solid skeleton and ions to diffuse in their phase and satisfy their own balance of momentum.

(H3) The velocity of any species in the intrafibrillar phase is that of the solid phase, i.e., of collagen, $\mathbf{v}_{kI} = \mathbf{v}_S$, $\forall k \in I$. The velocities of the proteoglycans, and of non mobile cations of the extrafibrillar phase, which do not diffuse through the cartilage are also equal to $\mathbf{v}_S$. Thus, the balance of momentum of the above species is not required explicitly, but accounted for by the balance of momentum of the mixture as a whole. Exchange of species between the two fluid phases is viewed as a *mass transfer* and not as a diffusion process.

(H4) Incompressibility of all the constituents gives rise to a Lagrange multiplier interpreted as the intrafibrillar pressure p_I. However, in both the intrafibrillar and extrafibrillar phases, each constituent is endowed, by constitutive equations, with its own "generalized pressure" that contains specific contributions in addition to p_I.

(H5) Electroneutrality is required for the extrafibrillar fluid phase and for intrafibrillar fluid phase independently. Collagen fibers are charged at non physiological pH. Because collagen is in contact with both fluid phases, its charge is partitioned in the two phases: the charge partition is addressed in Chapter 17.

Although the point is seldom mentioned, the definition of phases in articular cartilages is not unambiguous because the mechanical, chemical and electrical roles of proteoglycans dictate contradictory choices. The consequences of two options are commented further in Section 14.6.2. In particular, proteoglycans move with the solid but their electrical charge plays a key role in the extrafibrillar phase. One way to go around this issue is to consider a single (fluid) phase and a single electroneutrality condition, as in Lai et al. [1991]. However, topological considerations dictate electroneutrality separately for the two fluid phases. Indeed, when electroneutrality is disturbed, typical times to recover equilibrium are expected to be much smaller than transfer times, so that electroneutrality gets established independently in the two fluid compartments, Section 13.6.3. As a direct consequence of the two electroneutrality conditions, the electrical potentials in the two fluid phases are distinct.

14.6.4 Measures of the chemical composition

Various macroscopic measures of mass and volume are used to formulate the constitutive equations. They are defined below. Since a species can be present in more than a single phase, it is endowed with two indices, one referring to the species itself, the other to the phase, e.g., V_{kK} denotes the volume of species k in phase K. The only exceptions to this convention are proteoglycans and collagen which unambiguously belong to a single phase. Entities at the phase level are referred by the index of the phase, e.g., V_K is the volume of phase K.

The current number of moles, volume and mass of the species k in phase K are denoted by N_{kK}, V_{kK} and M_{kK}. Let the initial volume of the porous medium be V_0 and let $V = V(t) = V_0 \det \mathbf{F}$ be its current volume, $\mathbf{F}$ denoting the deformation gradient, and M its current mass.

Various entities are attached to species:
- some are intrinsic, like the intrinsic density ρ_k, the molar volume $\widehat{v}_k$ and the molar mass $\widehat{m}_k$ linked by $\widehat{m}_k = \rho_k \widehat{v}_k$[14.12]. The molar mass is a constant, while the molar volume of compressible species depends on strain and the molar volume of dissolved ions is varying with the ionic strength of the electrolyte;
- some refer to the current volume, like the *volume fraction* n^{kK} of a species, and n^K of a phase,

$$n^{kK} = \frac{V_{kK}}{V}, \quad n^K = \sum_{k \in K} n^{kK}, \quad \sum_{K=S,I,E} n^K = 1, \tag{14.6.2}$$

and the *apparent* density ρ^{kK},

$$\rho^{kK} = \frac{M_{kK}}{V} = n^{kK} \rho_k \text{ (no sum on k)}, \quad \rho^K = \frac{M_K}{V} = \sum_{k \in K} \rho^{kK}, \quad \rho = \frac{M}{V} = \sum_{K=S,I,E} \rho^K; \tag{14.6.3}$$

- some refer to the initial volume like the *volume content* v^{kK},

$$v^{kK} = \frac{V_{kK}}{V_0}, \quad v^K = \frac{V_K}{V_0} = \sum_{k \in K} v^{kK}, \quad \sum_{K=S,I,E} v^K = \frac{V}{V_0} = \det \mathbf{F}, \tag{14.6.4}$$

the *mass content* m^{kK},

$$m^{kK} = \frac{M_{kK}}{V_0} = \frac{N_{kK} \widehat{m}_k}{V_0} = \rho_k v^{kK}, \quad m^K = \frac{M_K}{V_0} = \sum_{k \in K} m^{kK}, \quad m = \frac{M}{V_0} = \sum_{K=S,I,E} m^K, \tag{14.6.5}$$

and the *mole content*,

$$\mathcal{N}_{kK} = \frac{N_{kK}}{V_0}. \tag{14.6.6}$$

From $(14.6.4)_3$ follow the incremental and rate relations,

$$\sum_{K=S,I,E} dv^K = \frac{dV}{V_0} = d \det \mathbf{F}, \quad \sum_{K=S,I,E} \frac{dv^K}{dt} = \frac{d(\det \mathbf{F})}{dt} = \det \mathbf{F} \operatorname{div} \mathbf{v}_S. \tag{14.6.7}$$

The above entities are *intensive*: as indicated, their phase counterparts are defined by algebraic summation of individual contributions. In turn, the phase entities themselves contribute additively at the level of the porous medium.

Other entities live in their phase, e.g., the molar fractions and the concentrations. The *molar fraction* x_{kK} of the species k in phase K is defined by the ratio of the mole number N_{kK} of that species over the total number of moles N_K within the phase,

$$x_{kK} = \frac{N_{kK}}{N_K}, \quad \sum_{k \in K} x_{kK} = 1, \quad K = S, I, E. \tag{14.6.8}$$

[14.12] As just stated, the same species is usually endowed with distinct properties in the intra- and extrafibrillar phases. Therefore, any property should be referred to by the indices of both the species and the phase. However, because we have in mind incompressible species, the intrinsic mass density might dispense of the phase index. The notation is ambiguous for compressible species. Now, even for incompressible species, *apparent* properties have to refer to both the species and the phase because the volume fractions are certainly distinct in the two phases.

Mole content, mole number and mass content are equivalent variables,

$$\mathcal{N}_{kK} = \frac{N_{kK}}{V_0} = \frac{m^{kK}}{\widehat{m}_k}, \quad k \in K, \quad K = S, I, E,$$ (14.6.9)

and therefore the molar fractions can also be expressed in terms of mass contents,

$$x_{kK} = \frac{m^{kK}/\widehat{m}_k}{\sum_{l \in K} m^{lK}/\widehat{m}_l}, \quad k \in K, \quad K = S, I, E.$$ (14.6.10)

The *concentration* of an extrafibrillar species is equal to its number of moles referred to the volume of extrafibrillar phase,

$$c_{kE} = \frac{N_{kE}}{V_E} = \frac{N_{kE}}{N_E \widehat{v}_E} = \frac{1}{\widehat{v}_k} \frac{V_{kE}}{V_E} = \frac{1}{\widehat{v}_k} \frac{n^{kE}}{n^E} = \frac{1}{\widehat{v}_k} \frac{v^{kE}}{n^E} \frac{V_0}{V} = \frac{x_{kE}}{\widehat{v}_E}, \quad k \in E,$$ (14.6.11)

with $\widehat{v}_E = \sum_{k \in E} x_{kE} \widehat{v}_k \sim \widehat{v}_w$ the molar volume of the extrafibrillar fluid phase.

Collagen and proteoglycans are macromolecules with a large molar mass, 0.285×10^6 gm for collagen and 2×10^6 gm for PGs. The molar fraction $x_{PG} = N_{PG}/N_E$ and concentration $c_{PG} = N_{PG}/V_E$ of proteoglycans are thus quite small with respect to the other species of the extrafibrillar phase. On the other hand, the valence ζ_{PG} of proteoglycans is large at neutral pH. The *effective* molar fraction y_{PG} and *effective* concentration e_{PG} are commensurable with molar fractions and concentrations of other species, and they are key parameters of the biochemical and biomechanical responses of PGs. To stress this issue, the intrinsic and effective measures used for PGs are listed below,

$$\underbrace{x_{PG} = N_{PG}/N_E}_{\substack{\text{molar} \\ \text{fraction}}}, \quad \underbrace{c_{PG} = N_{PG}/V_E}_{\text{concentration}}, \quad \underbrace{y_{PG} = \zeta_{PG}\, x_{PG}}_{\substack{\text{effective} \\ \text{molar fraction}}}, \quad \underbrace{e_{PG} = \zeta_{PG}\, c_{PG}}_{\substack{\text{effective} \\ \text{concentration}}}.$$ (14.6.12)

The definitions relative to PGs purposely refer to the extrafibrillar phase: indeed, according to Maroudas, the effects of PGs is primarily felt in that compartment. However, unlike in the formula (14.6.12), the effective concentration reported in the literature usually refers to the total volume or mass of the fluid and, with an opposite sign, it is referred to as fixed charge density FCD, Eqn (14.4.1). Maroudas [1975] reports values of FCD between 0.05 and 0.20 mole per liter of *total* water. According to Gu et al. [1997], the typical value of FCD is about 0.15 M for sound articular cartilage and 0.05 M for osteoarthritic cartilage.

For different zones of the human hip cartilage, Maroudas et al. [1991] indicate values ranging from 0.109 to 0.210 and 0.228 mole per liter of *total* water, equivalent to 0.14, 0.28 and 0.32 mole per liter of EF water. Basser et al. [1998] report a mean FCD of 0.37 mole per liter of EF water for normal hip cartilage. This value drops to 0.167 for osteoarthritic cartilage.

14.6.5   Work and electrochemical potentials

14.6.5.1   Incremental work

The incremental work done per unit initial volume by the total stress $\mathbf{T}$ in the incremental strain $d\mathbf{E}$ of the porous medium and by the *electrochemical potentials* μ_{kK}^{ec} during the addition of mass dm^{kK} of the species k to the phase K is obtained by summation as,

$$\begin{aligned} d\underline{E} &= \mathbf{T} : d\mathbf{E} + \sum_{k,K} \mu_{kK}^{ec}\, dm^{kK} \\ &= \mathbf{T} : d\mathbf{E} + \sum_{k,K} g_{kK}^{ec}\, d\mathcal{N}_{kK}. \end{aligned}$$ (14.6.13)

Here the electrochemical potentials μ_{kK}^{ec} [unit: m^2/s^2] are mass based and the fluid mass contents per unit initial volume of the porous medium m^{kK} are measured in kg/m^3. On the other hand, the electrochemical potentials $g_{kK}^{ec} = \widehat{m}_k\, \mu_{kK}^{ec}$ [unit: kg$\times$m^2/s^2/mole] are mole based and the mole contents per unit initial volume of the porous medium $\mathcal{N}_{kK}$'s are measured in mole/m^3.

The sum in (14.6.13) involves all species but in practice only the species that change mass contribute and need to be endowed with an electrochemical potential. Collagen and proteoglycans are assumed to maintain their mass and therefore do not contribute to work.

The chemical potential $\mu_{k\mathrm{K}}$ of a species k in phase K identifies a generalized pressure $p_{k\mathrm{K}}$ and a chemical contribution which accounts for the molar fraction $x_{k\mathrm{K}}$. For a charged species in presence of the electrical potential ϕ_{K} [unit volt V=kg$\times$m^2/s^3/A], the electrochemical potential involves in addition an electrical contribution.

For compressible species, the algebraic expression involves an integration which is performed from a reference state to the current state,

$$g_{k\mathrm{K}}^{\mathrm{ec}} = \widehat{m}_k \, \mu_{k\mathrm{K}}^{\mathrm{ec}} = \int \widehat{v}_k \, dp_{k\mathrm{K}} + RT \operatorname{Ln} x_{k\mathrm{K}} + \zeta_k \operatorname{F} \phi_{\mathrm{K}}, \quad k \in \mathrm{K}. \tag{14.6.14}$$

For incompressible species, the expression simplifies to,

$$g_{k\mathrm{K}}^{\mathrm{ec}} = \widehat{m}_k \, \mu_{k\mathrm{K}}^{\mathrm{ec}} = \widehat{v}_k \, p_{k\mathrm{K}} + RT \operatorname{Ln} x_{k\mathrm{K}} + \zeta_k \operatorname{F} \phi_{\mathrm{K}}, \quad k \in \mathrm{K}. \tag{14.6.15}$$

In this formula, $R = 8.31451 \, \mathrm{J/mole/K}$ is the universal gas constant, T [unit : K] the absolute temperature, and $\mathrm{F} = 96\,485 \, \mathrm{coulomb/mole}$ is Faraday's equivalent charge. The ζ's are the valences.

The extrafibrillar generalized pressures $p_{k\mathrm{E}}$ require constitutive equations: they include in particular
- a purely mechanical contribution p_{I};
- a term aimed at satisfying the chemomechanical equilibrium between the EF and IF compartments in the hypertonic state, and
- a term representing the enthalpy of formation, or dissolution into water, of ionic species and their affinity to PGs.

The intrafibrillar species are subjected to the pressure p_{I}. In addition, an adhesion mechanism tends to oppose the osmotic flow of intrafibrillar water towards the extrafibrillar compartment where the PGs reside.

Clearly, since the valences of electrically neutral species vanish, the chemical and electrochemical potentials of these species are one and the same. The electrochemical potentials are detailed in the subsequent sections and in Section 18.2.2.

14.6.5.2 Chemical affinity, enthalpy of formation, and equilibrium constant

Chemical and physico-chemical reactions involve a linear combination of chemical potentials termed chemical affinity defined by (3.1.17). Its vanishing means equilibrium of the reaction. Chemical affinity includes, besides mechanical, chemical and electrical terms, a constant referred to as equilibrium constant. For example, for the reaction

$$A \rightleftharpoons B, \tag{14.6.16}$$

the variations of the numbers of moles of species involved are linked,

$$- dN_{\mathrm{A}} = dN_{\mathrm{B}}. \tag{14.6.17}$$

Consequently, if the reaction does not give rise to deformation, the incremental work done during this reaction is defined by the chemical affinity $\mathcal{G}$,

$$g_{\mathrm{A}} \, dN_{\mathrm{A}} + g_{\mathrm{B}} \, dN_{\mathrm{B}} = \left(\mathcal{G}^0 + RT \operatorname{Ln} \frac{x_{\mathrm{B}}}{x_{\mathrm{A}}} + \mathrm{F} \left(\zeta_{\mathrm{B}} \phi_{\mathrm{B}} - \zeta_{\mathrm{A}} \phi_{\mathrm{A}} \right) \right) dN_{\mathrm{B}} = \mathcal{G} \, dN_{\mathrm{B}}. \tag{14.6.18}$$

The equilibrium of the reaction is defined via the enthalpy $\mathcal{G}^0 = -RT \operatorname{Ln} K^{\mathrm{eq}}$, which is expressed through a dimensionless equilibrium constant K^{eq}, or conversely $K^{\mathrm{eq}} = \exp(-\mathcal{G}^0/RT)$. At 25°C, a change of K^{eq} of an order of magnitude corresponds to a change of affinity of 5.7 kJ/mole. For acid-base reactions, the equilibrium constant is replaced by a pK, easily interpreted in a pH-scale,

$$\mathcal{G}^0 = -RT \operatorname{Ln} K^{\mathrm{eq}}, \quad K^{\mathrm{eq}} = \frac{10^{-\mathrm{pK}}}{\widehat{v}_{\mathrm{w}}}. \tag{14.6.19}$$

In biochemistry, the standard thermodynamic state corresponds to atmospheric pressure and a temperature of 25°C. The standard enthalpy of formation of substances is the energy required to obtain the substance from its elements, under standard conditions. Therefore the affinity $\mathcal{G}^0$ associated with the reaction (14.6.16) can be obtained from the enthalpies of formation of its reactant and product.

The concentration of water, about $55.55\,\mathrm{mole/liter}$, or its molar volume, $\widehat{v}_\mathrm{w} = 18\,\mathrm{cm}^3/\mathrm{mole}$, are often included in the equilibrium constants. The dimension of these constants helps to decipher whether it is the case or not.

In summary, vanishing of the chemical affinity,

$$\mathcal{G} = RT\,\mathrm{Ln}\left(\frac{1}{K^\mathrm{eq}}\,\frac{x_\mathrm{B}}{x_\mathrm{A}}\right) + \mathrm{F}\left(\zeta_\mathrm{B}\phi_\mathrm{B} - \zeta_\mathrm{A}\phi_\mathrm{A}\right), \tag{14.6.20}$$

defines chemical equilibrium. If the electrical potentials ϕ are known, the ratio of molar fractions results. Conversely, if the concentrations are known, the difference of electrical potentials required to imply equilibrium follows. When $\zeta_\mathrm{A} = \zeta_\mathrm{B} = \zeta$, this difference is referred to as Nernst potential,

$$\text{Nernst potential}\quad \phi_\mathrm{B} - \phi_\mathrm{A} = -\frac{RT}{\zeta\,\mathrm{F}}\,\mathrm{Ln}\left(\frac{1}{K^\mathrm{eq}}\,\frac{x_\mathrm{B}}{x_\mathrm{A}}\right). \tag{14.6.21}$$

If the reactant and product are electrically neutral, equilibrium simplifies to $x_\mathrm{B}/x_\mathrm{A} = K^\mathrm{eq}$. Formally, the reaction has been assumed to be reversible. Still, if K^eq is small, then it is actually essentially from right to left, and conversely from left to right if K^eq is large.

14.6.5.3 Phasewise electroneutrality

In phase K, the *electrical density* I_eK [unit : coulomb/m^3] is defined as the number of moles in the volume V weighted by their charges,

$$I_\mathrm{eK} = \frac{\mathrm{F}}{V}\sum_{k\in\mathrm{K}} \zeta_k\,N_{k\mathrm{K}}\ \left[= \mathrm{F}\,n^\mathrm{E}\sum_{k\in\mathrm{E}}\zeta_k\,c_{k\mathrm{E}}\ \text{for K} = \mathrm{E}\right]. \tag{14.6.22}$$

According to Assumption (H5), both the intrafibrillar and extrafibrillar fluid phases are electrically neutral,

$$I_\mathrm{eI} = 0;\quad I_\mathrm{eE} = 0. \tag{14.6.23}$$

For pH close to 7, the collagen can be considered neutral, Li and Katz [1976]. Given that the PGs are negatively charged at any pH, electroneutrality requires a minimal number of extrafibrillar cations, that we refer to as *non mobile* cations : the charge of these cations balances the large negative charge of the proteoglycans. Let us reiterate that the non mobile sodium cations *do not bind* to the PGs. The situation is more complex for calcium cations, as indicated in Section 16.1.2.6.

Pointwise electroneutrality in each phase has far-reaching consequences which are cast in two propositions:

Proposition 14.1 on dependent and independent chemical variables.

As a further consequence of electroneutrality, the number of moles of chloride anions is no longer an independent variable and it can be eliminated in favor of the numbers of moles of the cations,

$$dN_\mathrm{ClK} = \sum_{i\in\mathrm{K_{in}}} \zeta_i\,dN_{i\mathrm{K}},\quad dm^\mathrm{ClK} = \sum_{i\in\mathrm{K_{in}}} \zeta_i\,\frac{\widehat{m}_\mathrm{Cl}}{\widehat{m}_i}\,dm^{i\mathrm{K}},\quad \mathrm{K} = \mathrm{I,\,E}, \tag{14.6.24}$$

with $\mathrm{K_{in}}$ the set of independent variables.

The relation holds only in incremental form in the extrafibrillar phase due to the presence of PGs. However it does in total form in the intrafibrillar phase,

$$N_\mathrm{ClI} = \sum_{i\in\mathrm{I_{in}}} \zeta_i\,N_{i\mathrm{I}},\quad m^\mathrm{ClI} = \sum_{i\in\mathrm{I_{in}}} \zeta_i\,\frac{\widehat{m}_\mathrm{Cl}}{\widehat{m}_i}\,m^{i\mathrm{I}}. \tag{14.6.25}$$

The incremental relations between the molar fractions and the independent variables can now be made explicit, Exercise 14.5. First, let us note the differential form of (14.6.10),

$$\frac{dx_{k\mathrm{K}}}{x_{k\mathrm{K}}} = \sum_{l\in\mathrm{K}} \frac{dm^{l\mathrm{K}}}{m^{l\mathrm{K}}}\left(I_{kl} - x_{l\mathrm{K}}\right),\quad k\in\mathrm{K}. \tag{14.6.26}$$

As another consequence of electroneutrality, the partial derivatives of the molar fractions with respect to the independent mass contents take the form,

$$\frac{m^{i\mathrm{K}}}{x_{k\mathrm{K}}}\,\frac{\partial x_{k\mathrm{K}}}{\partial m^{i\mathrm{K}}} = I_{ki} - x_{i\mathrm{K}}\left(1 + \zeta_i\right) + \zeta_i\,\frac{x_{i\mathrm{K}}}{x_\mathrm{ClK}}\,I_{k\mathrm{Cl}},\quad k\in\mathrm{K},\ i\in\mathrm{K_{in}}. \tag{14.6.27}$$

In explicit form, with K = I or E, $K_{in} = \{w, Na, Ca\}$,

$$
\begin{aligned}
dx_{wK} &= (1 - x_{wK})\frac{x_{wK}}{m^{wK}}\,dm^{wK} & -2\,x_{wK}\frac{x_{NaK}}{m^{NaK}}\,dm^{NaK} & -3\,x_{wK}\frac{x_{CaK}}{m^{CaK}}\,dm^{CaK}, \\[4pt]
dx_{NaK} &= -x_{NaK}\frac{x_{wK}}{m^{wK}}\,dm^{wK} & +(1 - 2\,x_{NaK})\frac{x_{NaK}}{m^{NaK}}\,dm^{NaK} & -3\,x_{NaK}\frac{x_{CaK}}{m^{CaK}}\,dm^{CaK}, \\[4pt]
dx_{CaK} &= -x_{CaK}\frac{x_{wK}}{m^{wK}}\,dm^{wK} & -2\,x_{CaK}\frac{x_{NaK}}{m^{NaK}}\,dm^{NaK} & +(1 - 3\,x_{CaK})\frac{x_{CaK}}{m^{CaK}}\,dm^{CaK}, \\[4pt]
dx_{ClK} &= -x_{ClK}\frac{x_{wK}}{m^{wK}}\,dm^{wK} & +(1 - 2\,x_{ClK})\frac{x_{NaK}}{m^{NaK}}\,dm^{NaK} & +(2 - 3\,x_{ClK})\frac{x_{CaK}}{m^{CaK}}\,dm^{CaK}.
\end{aligned}
\tag{14.6.28}
$$

The second proposition follows:

Proposition 14.2 The incremental work expresses in terms of the independent variables.

The electrical potential does not enter the elastic constitutive equations, that can be phrased in terms of chemical potentials rather than *electro*chemical potentials: indeed the electrical potential does no work due to electroneutrality. In fact, the incremental energy (14.6.13) can be recast in terms of the chemical potentials of water w and salts $s_1 = NaCl$ and $s_2 = CaCl_2$ conjugated to the mass contents of water and (mobile) sodium and calcium cations, Exercise 14.6,

$$
\begin{aligned}
d\underline{E} &= \mathbf{T}:d\mathbf{E} + \sum_{k,K} \mu_{kK}^{ec}\,dm^{kK} \\[4pt]
&= \mathbf{T}:d\mathbf{E} + \sum_{k,K} \mu_{kK}\,dm^{kK} \\[4pt]
&= \mathbf{T}:d\mathbf{E} + \sum_{K=I,E}\ \sum_{(i,n)\in(K_{in},K_{ne})} \mu_{nK}\,dm^{iK},
\end{aligned}
\tag{14.6.29}
$$

with K_{in} the set of *independent* variables, and K_{ne} the set of neutral species in phase K = I, E,

$$
\underbrace{K_{in} = \{w, Na, Ca\}}_{\substack{\text{set of} \\ \text{independent variables}}}, \quad \underbrace{K_{ne} = \{w, s_1 = NaCl, s_2 = CaCl_2\}}_{\text{set of neutral species}}.
\tag{14.6.30}
$$

The entities μ_{nK}, $n = s_1, s_2$, can be viewed as the chemical potentials of the dissociated salts in phase K, namely in accordance with (14.6.15),

$$
\widehat{m}_i\,\mu_{nK} = \begin{cases} \widehat{m}_i\,\mu_{iK}^{ec} + \zeta_i\,\widehat{m}_{Cl}\,\mu_{ClK}^{ec}, \\[6pt] \widehat{v}_n\,p_{nK} + RT\,\mathrm{Ln}\,x_{iK}\,(x_{ClK})^{\zeta_i}. \end{cases}
\tag{14.6.31}
$$

The molar volumes $\widehat{v}_n$ and densities ρ_n are defined by the relations,

$$
\widehat{v}_n = \rho_n^{-1}\,\widehat{m}_i = \widehat{v}_i + \zeta_i\,\widehat{v}_{Cl}, \quad (i,n) \in (K_{in}, K_{ne}).
\tag{14.6.32}
$$

It should be stressed that the electrochemical potentials of ions involve the molar fractions of all ions. The molar fraction of chloride ion deduces from the electroneutrality condition, which indicates that it might not be equal to the molar fraction of sodium cations,

- either due to the presence of the fixed charge in the extrafibrillar phase, for example for the sodium electrolyte, $x_{ClE} = x_{NaE} + y_{PG}$;
- or in the presence of a ternary electrolyte, e.g., sodium chloride and calcium chloride: then in the intrafibrillar phase and bath, $x_{ClK} = x_{NaK} + 2\,x_{CaK}$, K = I, B;
- or if water is considered to dissociate into hydrogen and hydroxyl ions.

14.6.5.4 Absolute fluxes, diffusive fluxes and electrical current density

Balance and constitutive equations are phrased in terms of several fluxes, namely mass and volume fluxes, absolute and diffusive fluxes. Due to the incompressibility of the species, mass fluxes and volume fluxes can be viewed as entities that differ only by their units.

The *absolute mass flux* through the solid skeleton $\mathbf{M}_{kK}$ [unit : kg/m^2/s] and the associated *volume flux* $\mathbf{J}_{kK}$ [unit : m/s] of the species k in phase K involve a velocity relative to the solid skeleton,

$$
\rho_k^{-1}\,\mathbf{M}_{kK} = \mathbf{J}_{kK} = n^{kK}\,(\mathbf{v}_{kK} - \mathbf{v}_S), \quad k \in K.
\tag{14.6.33}
$$

Therefore the fluxes of the species in the solid phase and intrafibrillar phase vanish, since these species move with the solid velocity as a consequence of Assumption (H3),

$$\rho_k^{-1} \mathbf{M}_{kK} = \mathbf{J}_{kK} = \mathbf{0}, \quad k \in K = S, I. \tag{14.6.34}$$

The sum of the fluxes $\mathbf{J}_{kK}$, $k \in K$, defines the volume averaged flux $\mathbf{J}_K$ of phase K through the solid skeleton,

$$\mathbf{J}_S = \mathbf{0}, \quad \mathbf{J}_I = \mathbf{0}, \quad \mathbf{J}_E = \sum_{k \in E} n^{kE} \left(\mathbf{v}_{kE} - \mathbf{v}_S \right). \tag{14.6.35}$$

The *diffusive flux* with respect to extrafibrillar water is denoted by $\mathbf{J}_{kE}^d$,

$$\mathbf{J}_{kE}^d = n^{kE} \left(\mathbf{v}_{kE} - \mathbf{v}_{wE} \right), \quad k \in E. \tag{14.6.36}$$

The absolute flux of PGs is zero as a consequence of Assumption (H3), but their diffusive flux does *not* vanish,

$$\mathbf{J}_{PG} = \mathbf{0}, \quad \mathbf{J}_{PG}^d = n^{PG} \left(\mathbf{v}_S - \mathbf{v}_{wE} \right) = -\frac{n^{PG}}{n^{wE}} \mathbf{J}_{wE}. \tag{14.6.37}$$

Conversely the diffusive flux of water vanishes while its absolute flux $\mathbf{J}_{wE}$ does not. Relations between the different fluxes are detailed in Exercise 14.1.

The *electrical current density* $\mathbf{I}_{eK}$ in phase K [unit : A/m^2] is defined as the sum of the constituent velocities weighted by their valences and molar densities,

$$\mathbf{I}_{eK} = F \sum_{k \in K} \zeta_k \frac{N_{kK}}{V} \mathbf{v}_{kK}. \tag{14.6.38}$$

A uniform velocity for all species of a phase satisfying electroneutrality is seen to be a sufficient condition for the electrical current density to vanish in that phase. Therefore, the electrical current density vanishes in the intrafibrillar phase,

$$\mathbf{I}_{eI} = \mathbf{0}. \tag{14.6.39}$$

Due to electroneutrality, the electrical current density in the extrafibrillar phase $\mathbf{I}_{eE}$ may be viewed as a sum of either interphase or diffusive fluxes, namely

$$\mathbf{I}_{eE} = F \sum_{k \in E} \zeta_k \frac{\mathbf{J}_{kE}}{\widehat{v}_k} = F \sum_{k \in E} \zeta_k \frac{\mathbf{J}_{kE}^d}{\widehat{v}_k}. \tag{14.6.40}$$

14.6.6 Balance equations

In line with a strongly interacting multi-species continuum, rules governing the balance equations may be stated :
- the balances of mass are required for all mobile species in the fluid phases but extrafibrillar water, and for the extrafibrillar fluid phase as a whole;
- the balances of mass for the intrafibrillar species are accounted for by the transfer relations;
- the balances of momentum of mobile species of the extrafibrillar phase are accounted for indirectly through the generalized diffusion relations, as explained in a poroelastic context in Section 7.3.2;
- the balance of momentum for the mixture as a whole is required in a standard format.

The balances of momentum of species whose velocities are equal to the solid velocity, namely intrafibrillar species, collagen and PGs, are not required individually. Instead they are included in the balance of momentum of the whole mixture.

14.6.6.1 Balance of mass

A change of mass of the species k of phase K is due a priori to both *transfer*, i.e., a physico-chemical reaction, and *diffusion* within its phase, see Exercise 14.4,

$$\frac{1}{\det \mathbf{F}} \frac{dm^{k\mathrm{K}}}{dt} = \underbrace{\hat{\rho}^{k\mathrm{K}}}_{\substack{\text{mass}\\\text{transfer}}} - \underbrace{\operatorname{div} \mathbf{M}_{k\mathrm{K}}}_{\text{diffusion}}, \ k \in \mathrm{K}. \tag{14.6.41}$$

The symbol div denotes the divergence operator, and d/dt represents the derivative following the solid phase whose velocity is $\mathbf{v}_\mathrm{S}$.

The mass changes in the species of the intrafibrillar fluid phase are purely reactive and they are due to transfer, between the fluid phases, of water and ionic species, Fig. 14.6.1,

$$\frac{1}{\det \mathbf{F}} \frac{dm^{k\mathrm{I}}}{dt} = \underbrace{\hat{\rho}^{k\mathrm{I}}}_{\text{mass transfer}}, \ k \in \mathrm{I}. \tag{14.6.42}$$

These transfers are governed by the constitutive equations which are developed in Section 10.2.

On the other hand, the extrafibrillar compartment is in contact with both the intrafibrillar phase and the surroundings,

$$\frac{1}{\det \mathbf{F}} \frac{dm^{k\mathrm{E}}}{dt} = \underbrace{\hat{\rho}^{k\mathrm{E}}}_{\text{mass transfer}} - \underbrace{\operatorname{div} \mathbf{M}_{k\mathrm{E}}}_{\text{diffusion}}, \ k \in \mathrm{E}. \tag{14.6.43}$$

In the present context, a transfer concerns the same species in both compartments,

$$\hat{\rho}^{k\mathrm{I}} + \hat{\rho}^{k\mathrm{E}} = 0, \ k \in \mathrm{I}. \tag{14.6.44}$$

For a species that does not transfer we simply set formally its rate of mass transfer to zero. Therefore,

$$\frac{1}{\det \mathbf{F}} \frac{dm^{k\mathrm{E}}}{dt} = - \underbrace{\frac{1}{\det \mathbf{F}} \frac{dm^{k\mathrm{I}}}{dt}}_{\text{mass transfer}} - \underbrace{\operatorname{div} \mathbf{M}_{k\mathrm{E}}}_{\text{diffusion}}, \ k \in \mathrm{E}. \tag{14.6.45}$$

A detailed derivation is provided in Exercise 14.4.

As a consequence of (14.6.45) and of the electroneutrality in the two fluid phases, the electrical current density $\mathbf{I}_{\mathrm{eE}}$ defined by (14.6.38) or (14.6.40), may be shown to be divergence free, Exercise 14.2,

$$\operatorname{div} \mathbf{I}_{\mathrm{eE}} = 0. \tag{14.6.46}$$

14.6.6.2 Equilibrium, continuing equilibrium and time scale granularity

A number of time scales are involved in laboratory tests on articular cartilages, namely
  - external time scales which are representative of loadings of mechanical, chemical and electrical natures;
  - internal time scales characterizing the material response.
The material response embeds
  - the time dependent mechanical behavior;
  - the time dependence of mass transfer;
  - the time dependence of diffusion processes, including seepage of water through the extrafibrillar phase and ionic diffusion;
  - departure from electroneutrality.
The notion of equilibrium is linked to some time scale. To explore the issue, let us consider chemical equilibrium of a species k at the interface between the two fluid compartments. Chemical equilibrium corresponds to equal chemical potentials. The term "chemical" should be understood as "electrochemical" for a charged species.

The mass transfers are not instantaneous, with respect to the other representative physical phenomena of interest listed above, and the constitutive equations of mass transfer involve specific transfer times. The mass transfers will be shown to be driven by the out-of-balance $\mu_{k\mathrm{I}} - \mu_{k\mathrm{E}}$ in electrochemical potentials across the IF-EF interface.

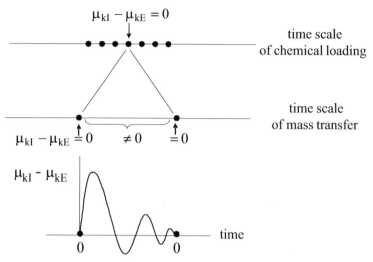

Fig. 14.6.3: Time scale granularity. Time scales may differ by orders of magnitude, depending on the associated physical phenomena. If the time scale of loading is much larger than the transfer time, continuing equilibrium applies at the time scale of the experimentalist: chemical potentials are equal but mass transfer is nevertheless possible.

Continuing equilibrium should be contrasted with equilibrium. Let us consider a chemical laboratory experiment in which the specimen is in contact with a bath whose chemical composition is changed according to a controlled process while the mechanical loading is kept fixed. In the context of continuing equilibrium, mass can transfer even if the chemical potentials μ_{kI} and μ_{kE} in the two fluid phases adopt a common time profile. This point can be realized by considering that the transfer times are infinitesimally small with respect to the time scale, of granularity Δt, along which the chemical potentials are considered to be equal, Fig. 14.6.3. Indeed, let us assume the latter to be equal at time t_1 and $t_2 = t_1 + \Delta t$. Experimentally they are certainly not equal *within* this interval: this departure allows the transfer to take place, as far as the transfer times are assumed to be smaller than Δt.

As an immediate consequence, transfer equations will not be needed in this chapter where only continuing equilibria are considered: the exchanged masses are then obtained by constraining the material to adapt to the mechanical and chemical loading path. Alternatively, transfer equations should be used if the granularity of the loading path is of the same order of magnitude as, or smaller than, the transfer times. The issue is considered in Section 10.2.

14.6.6.3 Balance of momentum

In absence of external supplies the balance of momentum of the porous medium as a whole reduces to the standard format,

$$\text{div } \boldsymbol{\sigma} + \rho\,\mathbf{b} = \mathbf{0}, \tag{14.6.47}$$

where $\rho\,\mathbf{b} = \sum_{k,K} \rho^{kK}\,\mathbf{b}_{kK}$ is the sum of the body forces and acceleration terms. Under quasi-static loading and with the sole gravity $\mathbf{g}$ of intensity g as body force, then $\rho = \sum_{k,K} \rho^{kK} = \sum_K \rho^K$ is identified as the mass density of the porous medium as a whole.

For most laboratory tests, inertia has negligible effects on stress and strain with respect to the mechanical and chemical loadings. Since specimens are of small size, the effects of gravity can be neglected in mechanical and chemical tests.

14.6.7   Incompressibility of species

The incompressibility of all individual species is commonly accepted, even if the loads to which soft tissues are subjected may be high. In algebraic form,

$$d\rho_k = 0, \quad \forall k \in \{I, E\}. \tag{14.6.48}$$

If all species are incompressible, the change of volume of the solid skeleton, which is the same as that of porous medium, is equal to the change of volume of the extrafibrillar fluid phase due to diffusion,

$$\text{div } \mathbf{v}_S + \text{div } \mathbf{J}_E = 0 \,. \tag{14.6.49}$$

The above relation is self-explanatory (see, however, Exercise 14.4 for a detailed proof). Incompressibility leads also to

Proposition 14.3 the volume change of the mixture expresses in terms of the independent chemical variables.

The volume change of the whole porous medium dV/V_0 is equal to the sum of volume changes due to addition/subtraction of individual species, Eqn (14.6.7). It can be expressed in terms of independent variables, Loret and Simões [2005]a,

$$
\begin{aligned}
\det \mathbf{F} \, \text{tr} \left(d\mathbf{F} \cdot \mathbf{F}^{-1} \right) &= \sum_{K=I,E} \sum_{k \in K} dv^{kK} \\
&= \sum_{K=I,E} \sum_{k \in K} \rho_k^{-1} \, dm^{kK} \\
&= \sum_{K=I,E} \sum_{(i,n) \in (K_{in}, K_{ne})} \rho_n^{-1} \, dm^{iK} \,,
\end{aligned}
\tag{14.6.50}
$$

where ρ_n is the mass density of the neutral species $n \in K_{ne}$ defined by (14.6.32). The second relation assumes incompressible species, and the last one is a consequence of phasewise electroneutrality, Exercise 14.6.

14.7 The constitutive structure: deformation, mass transfer, diffusion

The constitutive equations are developed in a thermodynamic framework *à la Biot* where the solid skeleton is taken as reference. Such an approach has been derived in Chapter 4 in a more general context including in addition the growth phenomenon. It is specialized in the current context where the tissue has been partitioned into phases.

A single inequality for the internal entropy production is required for the porous medium as a whole,

$$T\hat{\underline{S}}_{(i)} = -\frac{dE}{dt} + \mathbf{T} : \frac{d\mathbf{E}}{dt} - \sum_{k,K} \underline{\text{div}} \left(\mu_{kK}^{ec} \, \mathbf{M}_{kK} \right) - \mathbf{F} \cdot \mathbf{M}_{kK} \cdot \mathbf{b}_{kE} \geq 0 \,. \tag{14.7.1}$$

With help of the balance of mass (14.6.41), the dissipation $T\hat{\underline{S}}_{(i)}$ may advantageously be recast in a form that highlights the contributions due to deformation, mass transfer and diffusion, namely,

$$T\hat{\underline{S}}_{(i)} = -\frac{dE}{dt} + \mathbf{T} : \frac{d\mathbf{E}}{dt} + \sum_{k,K} \mu_{kK}^{ec} \left(\frac{dm^{kK}}{dt} - \det \mathbf{F} \, \hat{\rho}^{kK} \right) - \sum_{k,K} \boldsymbol{\nabla} \mu_{kK}^{ec} \cdot \mathbf{M}_{kK} - \mathbf{F} \cdot \mathbf{M}_{kK} \cdot \mathbf{b}_{kE} \geq 0 \,. \tag{14.7.2}$$

The inequality is viewed/partitioned as the sum of three terms of distinct natures, and which consequently are required to be positive separately,

$$
\begin{aligned}
\text{mechanics} \qquad & T\hat{\underline{S}}_{(i,1)} = -\frac{dE}{dt} + \mathbf{T} : \frac{d\mathbf{E}}{dt} + \sum_{k,K} \mu_{kK}^{ec} \frac{dm^{kK}}{dt} \geq 0 \,, \\
\text{mass transfer} \qquad & T\hat{\underline{S}}_{(i,2)} = -\sum_{k \in I} (\mu_{kI}^{ec} - \mu_{kE}^{ec}) \frac{dm^{kI}}{dt} \geq 0 \,, \\
\text{generalized diffusion} \qquad & T\hat{\underline{S}}_{(i,3)} = -\sum_{k \in E} \boldsymbol{\nabla} \mu_{kK}^{ec} \cdot \underline{\mathbf{M}}_{kK} - \mathbf{F} \cdot \underline{\mathbf{M}}_{kK} \cdot \mathbf{b}_{kE} \,.
\end{aligned}
\tag{14.7.3}
$$

Due to electroneutrality in each of the fluid phases, the electrical potentials do not work, and the inequalities associated with mechanics and mass transfer may be rephrased in terms of the chemical potentials rather than in terms of the *electro*chemical potentials,

$$
\begin{aligned}
\text{mechanics} \qquad & T\hat{\underline{S}}_{(i,1)} = -\frac{dE}{dt} + \mathbf{T} : \frac{d\mathbf{E}}{dt} + \sum_{k,K} \mu_{kK} \frac{dm^{kK}}{dt} \geq 0 \,, \\
\text{mass transfer} \qquad & T\hat{\underline{S}}_{(i,2)} = -\sum_{k \in I} (\mu_{kI} - \mu_{kE}) \frac{dm^{kI}}{dt} \geq 0 \,, \\
\text{generalized diffusion} \qquad & T\hat{\underline{S}}_{(i,3)} = -\sum_{k \in E} \boldsymbol{\nabla} \mu_{kK}^{ec} \cdot \underline{\mathbf{M}}_{kK} - \mathbf{F} \cdot \underline{\mathbf{M}}_{kK} \cdot \mathbf{b}_{kE} \,.
\end{aligned}
\tag{14.7.4}
$$

The chemo-hyperelastic behavior is constructed in order for the mechanical entropy production $\hat{S}_{(i,1)}$ to exactly vanish, Loret and Simões [2004] [2005]a. Since the inequalities associated with mechanics and mass transfer involve chemical potentials, the mechanical relations and the transfer relations do not depend directly on the electrical potentials as already mentioned in a different format in Section 14.6.5.3.

Satisfaction of the second and third inequalities motivates generalized transfer equations and generalized diffusion equations respectively: they are developed in Chapters 10 and 16, respectively. For quasi-static analyses, the body force densities $\mathbf{b}_{kE}$ are equal to the acceleration of gravity $\mathbf{g}$, and individual accelerations are neglected.

14.8   Chemoelastic energy of the tissue

In order to build the mechanical constitutive equations, two main steps are delineated:
- the chemoelastic energy associated with the solid skeleton and extrafibrillar fluid phase are first obtained, eliminating the chemical energy of the intrafibrillar phase using a modified version of the Gibbs-Duhem relation;
- the resulting chemoelastic energy is next expressed in terms of independent variables using the global constraint of incompressibility and the phasewise constraint of electroneutrality.

14.8.1   The Gibbs-Duhem relation for the intrafibrillar fluid phase

For a fluid phase, the Gibbs-Duhem relation (3.1.10) provides the change of fluid pressure in terms of the changes of the electrochemical potentials or chemical potentials of the species, e.g., Haase [1990]. If the intrafibrillar water molecules does not adhere to the collagen fibers, then

$$v^{\mathrm{I}}\, dp_{\mathrm{I}} = \sum_{k \in \mathrm{I}} m^{k\mathrm{I}}\, d\mu_{k\mathrm{I}}^{\mathrm{ec}}\,. \tag{14.8.1}$$

In the present continuum mechanics framework, a formula of the same flavor can be derived for the intrafibrillar fluid phase which is an electrolyte, in contrast to the extrafibrillar fluid phase which contains an insoluble species. However, for that purpose, two kinds of modifications are needed:
- modified electrochemical potentials $\tilde{\mu}_{kK}^{\mathrm{ec}}$ are introduced whose generalized pressures are equal to the fluid pressure p_K. Indeed, anticipating subsequent developments, an adhesion contribution μ_{adh} is removed from the chemical potential of IF water, while the electrochemical potentials of IF ions need not be modified,

$$\widehat{m}_k\, \tilde{\mu}_{k\mathrm{I}}^{\mathrm{ec}} = \widehat{m}_k\, \mu_{k\mathrm{I}}^{\mathrm{ec}} - \widehat{m}_k\, \mu_{\mathrm{adh}}\, I_{k\mathrm{w}} = \widehat{v}_k\, p_{\mathrm{I}} + RT\, \mathrm{Ln}\, x_{k\mathrm{I}} + \mathrm{F}\, \zeta_k\, \phi_{\mathrm{I}}, \quad k \in \mathrm{I}\,; \tag{14.8.2}$$

- the second modification is that only species for which an electrochemical potential has been defined are considered.

A relation, similar in form to Gibbs-Duhem relation for a fluid phase, can then be derived in terms of the modified electrochemical potentials:

Proposition 14.4 on a Gibbs-Duhem relation for the intrafibrillar fluid phase.

The intrafibrillar fluid pressure expresses in terms of the modified (electro)chemical potentials of the intrafibrillar species,

$$v^{\mathrm{I}}\, dp_{\mathrm{I}} = \sum_{k \in \mathrm{I}} m^{k\mathrm{I}}\, d\tilde{\mu}_{k\mathrm{I}}^{\mathrm{ec}}\,. \tag{14.8.3}$$

The proof, detailed in Exercise 14.7, rests on the electroneutrality of both the intrafibrillar and extrafibrillar fluid phases and on the fact that the molar fractions are entities that live in a single phase.

Due to the electroneutrality of the intrafibrillar fluid phase, the electrical contribution is elusive in the right-hand side of (14.8.3) which in fact can be expressed in terms of independent variables and the chemical potentials of neutral salts, namely via (14.6.29),

$$v^{\mathrm{I}}\, dp_{\mathrm{I}} = \sum_{k \in \mathrm{I}} m^{k\mathrm{I}}\, d\tilde{\mu}_{k\mathrm{I}} = \sum_{(i,n) \in (\mathrm{I}_{\mathrm{in}}, \mathrm{I}_{\mathrm{ne}})} m^{i\mathrm{I}}\, d\tilde{\mu}_{n\mathrm{I}}\,. \tag{14.8.4}$$

Exercise 14.6 proposes a proof.

14.8.2   Electrochemical energy of the intrafibrillar phase

In order to define the mechanical constitutive equations of the tissue, we need to isolate the chemical effects in the fluid. For that purpose, the electrochemical energy per reference unit volume of the IF phase is required. It is contributed by chemical terms due to the work of the (electro)chemical potentials in mass exchanges and by a mechanical term due to the work of the pressure in the variation of the volume,

$$d\underline{E}_{\mathrm{I}} = \sum_{k\in\mathrm{I}} \tilde{\mu}_{k\mathrm{I}}^{\mathrm{ec}}\, dm^{k\mathrm{I}} - p_{\mathrm{I}}\, dv^{\mathrm{I}} = \sum_{k\in\mathrm{I}} \tilde{\mu}_{k\mathrm{I}}\, dm^{k\mathrm{I}} - p_{\mathrm{I}}\, dv^{\mathrm{I}} = \sum_{(i,n)\in(\mathrm{I}_{\mathrm{in}},\mathrm{I}_{\mathrm{ne}})} \tilde{\mu}_{n\mathrm{I}}\, dm^{i\mathrm{I}} - p_{\mathrm{I}}\, dv^{\mathrm{I}}, \qquad (14.8.5)$$

which, via the Gibbs-Duhem relation (14.8.4), can be integrated, up to a constant, to

$$\underline{E}_{\mathrm{I}} = \sum_{k\in\mathrm{I}} \tilde{\mu}_{k\mathrm{I}}^{\mathrm{ec}}\, m^{k\mathrm{I}} - p_{\mathrm{I}}\, v^{\mathrm{I}} = \sum_{k\in\mathrm{I}} \tilde{\mu}_{k\mathrm{I}}\, m^{k\mathrm{I}} - p_{\mathrm{I}}\, v^{\mathrm{I}} = \sum_{(i,n)\in(\mathrm{I}_{\mathrm{in}},\mathrm{I}_{\mathrm{ne}})} \tilde{\mu}_{n\mathrm{I}}\, m^{i\mathrm{I}} - p_{\mathrm{I}}\, v^{\mathrm{I}}. \qquad (14.8.6)$$

14.8.3   Chemoelastic energy in terms of generalized strains

14.8.3.1   The structure of the constitutive relations

The chemical energy of the intrafibrillar phase $\underline{E}_{\mathrm{I}}$ does not affect *directly* the mechanical behavior of the cartilage, in the sense that the chemoelastic energy $\underline{\mathcal{W}}$ that dictates the elastic behavior is defined as $\underline{E} - \underline{E}_{\mathrm{I}}$, Heidug and Wong [1996]. Gathering the total free energy (14.7.3) and the electrochemical energy (14.8.5), the incremental chemoelastic energy adopts the format,

$$d\underline{\mathcal{W}} = d\underline{E} - d\underline{E}_{\mathrm{I}} = \mathbf{T} : d\mathbf{E} + \sum_{(i,n)\in(\mathrm{E}_{\mathrm{in}},\mathrm{E}_{\mathrm{ne}})} \mu_{n\mathrm{E}}\, dm^{i\mathrm{E}} + p_{\mathrm{I}}\, dv^{\mathrm{I}} + \mu_{\mathrm{adh}}\, dm^{w\mathrm{I}}. \qquad (14.8.7)$$

The last term represents the energy necessary to extract water from the intrafibrillar space. Elimination of the intrafibrillar volume content through the incompressibility condition (14.6.50) introduces the intrafibrillar pressure in the generalized shifted stresses,

$$\overline{\mathbf{T}} = \mathbf{T} + p_{\mathrm{I}} \det \mathbf{F}\, \mathbf{F}^{-1} \cdot \mathbf{F}^{-\mathrm{T}}; \quad \overline{\mu}_{n\mathrm{E}} = \mu_{n\mathrm{E}} - \frac{p_{\mathrm{I}}}{\rho_n}, \ n \in \mathrm{E}_{\mathrm{ne}}. \qquad (14.8.8)$$

This *shifted* stress $\overline{\mathbf{T}}$ differs from the *effective* stress, due to chemomechanical coupling, to be obtained in Eqn (14.8.22). Consequently, the chemoelastic energy $\underline{\mathcal{W}}$ of the cartilage depends on strain, on the independent mass contents of the extrafibrillar fluid phase, and on the mass content of intrafibrillar water,

$$d\underline{\mathcal{W}} = \overline{\mathbf{T}} : d\mathbf{E} + \sum_{(i,n)\in(\mathrm{E}_{\mathrm{in}},\mathrm{E}_{\mathrm{ne}})} \overline{\mu}_{n\mathrm{E}}\, dm^{i\mathrm{E}} + \mu_{\mathrm{adh}}\, dm^{w\mathrm{I}}. \qquad (14.8.9)$$

The elastic relations link generalized strains to generalized shifted stresses,

$$\boldsymbol{\mathcal{S}} = \begin{bmatrix} \overline{\mathbf{T}} \\ \overline{\mu}_{w\mathrm{E}} \\ \overline{\mu}_{\mathrm{NaE}} \\ \overline{\mu}_{\mathrm{CaE}} \\ \mu_{\mathrm{adh}} \end{bmatrix} \qquad \begin{matrix} \leftarrow \text{mechanics} \rightarrow \\ \leftarrow \text{hydromechanical coupling} \rightarrow \\ \leftarrow \text{chemomechanical coupling} \rightarrow \\ \leftarrow \text{chemomechanical coupling} \rightarrow \\ \leftarrow \text{adhesion coupling} \rightarrow \end{matrix} \qquad \boldsymbol{\mathcal{E}} = \begin{bmatrix} \mathbf{E} \\ m^{w\mathrm{E}} \\ m^{\mathrm{NaE}} \\ m^{\mathrm{CaE}} \\ m^{w\mathrm{I}} \end{bmatrix} \qquad (14.8.10)$$

with the headings "generalized shifted stresses" over $\boldsymbol{\mathcal{S}}$ and "generalized strains" over $\boldsymbol{\mathcal{E}}$,

in the format,

$$\boldsymbol{\mathcal{S}} = \frac{\partial \underline{\mathcal{W}}}{\partial \boldsymbol{\mathcal{E}}}, \qquad (14.8.11)$$

or, more explicitly,

$$\overline{\mathbf{T}} = \frac{\partial \underline{\mathcal{W}}}{\partial \mathbf{E}}; \quad \overline{\mu}_{n\mathrm{E}} = \frac{\partial \underline{\mathcal{W}}}{\partial m^{i\mathrm{E}}}, \ (i,n)\in(\mathrm{E}_{\mathrm{in}},\mathrm{E}_{\mathrm{ne}}); \quad \mu_{\mathrm{adh}} = \frac{\partial \underline{\mathcal{W}}}{\partial m^{w\mathrm{I}}}. \qquad (14.8.12)$$

The chemoelastic energy is additively contributed by a mixed chemomechanical part $\underline{\mathcal{W}}_{\mathrm{ch-mech}}$, a purely chemical part $\underline{\mathcal{W}}_{\mathrm{ch}}$, and a part $\underline{\mathcal{W}}_{\mathrm{ef}}$ associated with the enthalpies of formation to be defined in turn,

$$\underline{\mathcal{W}} = \underline{\mathcal{W}}_{\mathrm{ch-mech}} + \underline{\mathcal{W}}_{\mathrm{ch}} + \underline{\mathcal{W}}_{\mathrm{ef}}. \qquad (14.8.13)$$

14.8.3.2   Chemoelastic energy: the purely chemical contribution

The purely chemical part is chosen so that the chemical potentials (14.8.12) retrieve a chemical part in logarithmic form as in (14.6.15) and (14.6.31), namely

$$\underline{W}_{\mathrm{ch}}(N_{kE}, k \in E - \{\mathrm{PG}\}) = \frac{RT}{V_0} \sum_{k \in E} N_{kE} \, \mathrm{Ln} \, N_{kE} - \frac{RT}{V_0} \left(\sum_{k \in E} N_{kE} \right) \mathrm{Ln} \left(\sum_{l \in E} N_{lE} \right). \tag{14.8.14}$$

Partial derivation with respect to mass contents has to account for the electroneutrality constraint (14.6.24), as detailed in Exercise 14.8.

14.8.3.3   Chemoelastic energy: the chemomechanical contribution

The mixed part of the chemoelastic energy contains a purely mechanical contribution and a chemomechanical term,

$$\underline{W}_{\mathrm{ch-mech}}(\boldsymbol{\mathcal{E}}) = -p_{\mathrm{ch}}(\boldsymbol{\mathcal{E}}) \, \epsilon_{\mathrm{ch}} + \underline{W}_{\mathrm{mech}}(\mathbf{E}) + \int_0^{m^{\mathrm{wI}}} P_{\mathrm{adh}}(m) \, \frac{dm}{\rho_{\mathrm{w}}}. \tag{14.8.15}$$

The pressure p_{ch} depends on the strain through the isotropic scalar,

$$\epsilon_{\mathrm{ch}} = \det \mathbf{F} - 1, \tag{14.8.16}$$

and the coupled term gives rise to the chemical stress $\mathbf{T}^{\mathrm{ch}}$,

$$\mathbf{T}^{\mathrm{ch}} = -\left(p_{\mathrm{ch}} + \epsilon_{\mathrm{ch}} \frac{\partial p_{\mathrm{ch}}}{\partial \epsilon_{\mathrm{ch}}} \right) \det \mathbf{F} \, \mathbf{F}^{-1} \cdot \mathbf{F}^{-\mathrm{T}}. \tag{14.8.17}$$

More generally, the chemical pressure p_{ch} can depend on strain through any symmetric second order structural tensor that characterizes the material symmetries of the anisotropic cartilage. This set contains the identity tensor for any material symmetry. For an isotropic material, it in fact reduces to the identity tensor.

The mechanical stress,

$$\mathbf{T}^{\mathrm{mech}} = \mathbf{T}^{\mathrm{gs}} + \mathbf{T}^{\mathrm{c}} = \frac{\partial \underline{W}_{\mathrm{mech}}(\mathbf{E})}{\partial \mathbf{E}}, \tag{14.8.18}$$

is contributed by the ground substance, which provides much of the compressive stiffness, and by collagen which endows the cartilage with most of its tensile stiffness. These stresses are independent of the chemical state of the tissue and, therefore, they can be seen as representative of the hypertonic state.

Henceforth the work-conjugate pair $(\mathbf{T}, \mathbf{E})$ is identified with the 2nd Piola-Kirchhoff stress and Green strain. The mechanical constitutive equations (14.8.12),

$$\mathbf{T} + (p_{\mathrm{I}} + p_{\mathrm{ch}} + \epsilon_{\mathrm{ch}} \frac{\partial p_{\mathrm{ch}}}{\partial \epsilon_{\mathrm{ch}}}) \det \mathbf{F} \, \mathbf{F}^{-1} \cdot \mathbf{F}^{-\mathrm{T}} = \mathbf{T}^{\mathrm{gs}} + \mathbf{T}^{\mathrm{c}}, \tag{14.8.19}$$

may be rephrased in terms of the Cauchy stresses $\boldsymbol{\sigma}^{\mathrm{gs}}$ and $\boldsymbol{\sigma}^{\mathrm{c}}$ associated with the second Piola-Kirchhoff stresses $\mathbf{T}^{\mathrm{gs}}$ and $\mathbf{T}^{\mathrm{c}}$ so as to display the effective stress,

$$\boldsymbol{\sigma} + \overbrace{\left(p_{\mathrm{I}} + p_{\mathrm{ch}} + \epsilon_{\mathrm{ch}} \frac{\partial p_{\mathrm{ch}}}{\partial \epsilon_{\mathrm{ch}}} \right)}^{p_{\mathrm{eff}}} \mathbf{I} = \boldsymbol{\sigma}^{\mathrm{gs}}(\mathbf{E}) + \boldsymbol{\sigma}^{\mathrm{c}}(\mathbf{E}). \tag{14.8.20}$$

The constitutive function P_{adh} in (14.8.15) represents the contribution to the adhesion pressure in the intrafibrillar compartment and remains to be postulated.

14.8.3.4   Chemoelastic energy: the contribution of the enthalpies of formation

The energy $\underline{W}_{\mathrm{ef}}$ accounts for the history of formation of the cartilage and of loadings up to the time $t = 0$ where the experiment starts. It is assumed to be linear with respect to the independent extrafibrillar mass contents,

$$\underline{W}_{\mathrm{ef}}(\{m^{i\mathrm{E}}, i \in \mathrm{E}_{\mathrm{in}}\}) = \sum_{(i,n) \in (\mathrm{E}_{\mathrm{in}}, \mathrm{E}_{\mathrm{ne}})} \frac{p_{fn}}{\rho_n} \, m^{i\mathrm{E}} = \frac{p_{fw}}{\rho_{\mathrm{w}}} \, m^{\mathrm{wE}} + \frac{p_{fs_1}}{\rho_{s_1}} \, m^{\mathrm{NaE}} + \frac{p_{fs_2}}{\rho_{s_2}} \, m^{\mathrm{CaE}}. \tag{14.8.21}$$

The pressures $p_{f\mathrm{w}}$, p_{fs_1} and p_{fs_2} are referred to as pressures of formation.

14.8.3.5   The formal constitutive relations

With these assumptions, the elastic equations (14.8.12) can now be recast in the more familiar format,

$$
\begin{cases}
\boldsymbol{\sigma} + p_{\text{eff}}\,\mathbf{I} &= \boldsymbol{\sigma}^{\text{gs}}(\mathbf{E}) &+ \boldsymbol{\sigma}^{\text{c}}(\mathbf{E}), \\[4pt]
\widehat{m}_{\text{w}}\,\mu_{\text{wE}} &= \widehat{v}_{\text{w}}\,p_{\text{wE}} &+ RT\,\text{Ln}\,x_{\text{wE}}, \\[4pt]
\widehat{m}_{\text{Na}}\,\mu_{\text{s}_1\text{E}} &= \widehat{v}_{\text{s}_1}\,p_{\text{s}_1\text{E}} &+ RT\,\text{Ln}\,x_{\text{NaE}}\,x_{\text{ClE}}, \\[4pt]
\widehat{m}_{\text{Ca}}\,\mu_{\text{s}_2\text{E}} &= \widehat{v}_{\text{s}_2}\,p_{\text{s}_2\text{E}} &+ RT\,\text{Ln}\,x_{\text{CaE}}\,(x_{\text{ClE}})^2, \\[4pt]
\widehat{m}_{\text{w}}\,\mu_{\text{adh}} &= \widehat{v}_{\text{w}}\,p_{\text{adh}}\,.
\end{cases}
\tag{14.8.22}
$$

The pressure terms in the chemical potentials gather contributions due to incompressibility, tissue formation and of some form of chemomechanical coupling. In explicit form,

$$
\begin{cases}
p_{\text{eff}} &= p_{\text{I}} + p_{\text{ch}} &+ \dfrac{\partial p_{\text{ch}}}{\partial\,\epsilon_{\text{ch}}}\,\epsilon_{\text{ch}}\,, \\[10pt]
p_{\text{wE}} &= p_{\text{I}} + p_{fw} &- \rho_{\text{w}}\,\dfrac{\partial p_{\text{ch}}}{\partial m^{\text{wE}}}\,\epsilon_{\text{ch}}\,, \\[10pt]
p_{\text{s}_1\text{E}} &= p_{\text{I}} + p_{fs_1} &- \rho_{\text{s}_1}\,\dfrac{\partial p_{\text{ch}}}{\partial m^{\text{NaE}}}\,\epsilon_{\text{ch}}\,, \\[10pt]
p_{\text{s}_2\text{E}} &= p_{\text{I}} + p_{fs_2} &- \rho_{\text{s}_2}\,\dfrac{\partial p_{\text{ch}}}{\partial m^{\text{CaE}}}\,\epsilon_{\text{ch}}\,, \\[10pt]
p_{\text{adh}} &= P_{\text{adh}} &- \rho_{\text{w}}\,\dfrac{\partial p_{\text{ch}}}{\partial m^{\text{wI}}}\,\epsilon_{\text{ch}}\,.
\end{cases}
\tag{14.8.23}
$$

The adhesion pressure p_{adh} is not involved in the mechanical constitutive equations $(14.8.22)_1$, while it contributes to the chemical potential of *intra*fibrillar water, Eqn (14.8.2). Consequently it affects crucially the intrafibrillar pressure p_{I}.

14.8.4   Chemoelastic energy in terms of mixed generalized variables

As an alternative to the strain formulation of the preceding sections, stress may be used as a primary variable,

$$
\boldsymbol{\mathcal{E}} = \{\overline{\mathbf{T}},\,\{m^{i\text{E}}, i \in \text{E}_{\text{in}}\},\,m^{\text{wI}}\},\quad \boldsymbol{\mathcal{S}} = \{\mathbf{E},\,\{\overline{\mu}_{n\text{E}}, i \in \text{E}_{\text{in}}\},\,\mu_{\text{adh}}\}.
\tag{14.8.24}
$$

For that purpose, a mixed energy is defined by a partial Legendre transform of the energy $\underline{\mathcal{W}} \to \underline{\mathcal{W}} - \overline{\mathbf{T}} : \mathbf{E}$, so that

$$
d\underline{\mathcal{W}} = -\mathbf{E} : d\overline{\mathbf{T}} + \sum_{(i,n)\in(\text{E}_{\text{in}},\text{E}_{\text{ne}})} \overline{\mu}_{n\text{E}}\,dm^{i\text{E}} + \mu_{\text{adh}}\,dm^{\text{wI}}\,.
\tag{14.8.25}
$$

The mixed elastic relations result in the format,

$$
\mathbf{E} = -\frac{\partial \underline{\mathcal{W}}}{\partial\overline{\mathbf{T}}},\qquad \overline{\mu}_{n\text{E}} = \frac{\partial \underline{\mathcal{W}}}{\partial m^{i\text{E}}},\ (i,n) \in (\text{E}_{\text{in}},\text{E}_{\text{ne}});\quad \mu_{\text{adh}} = \frac{\partial \underline{\mathcal{W}}}{\partial m^{\text{wI}}}\,.
\tag{14.8.26}
$$

The mixed part $\underline{\mathcal{W}}_{\text{ch}-\text{mech}}$ of the chemoelastic energy mimics its counterpart in the strain formulation,

$$
\underline{\mathcal{W}}_{\text{ch}-\text{mech}}(\boldsymbol{\mathcal{S}}) = -p_{\text{ch}}\,\epsilon_{\text{ch}} - \frac{1}{2}\,\overline{\mathbf{T}} : \mathbb{E}^{-1} : \overline{\mathbf{T}} + \int_0^{m^{\text{wI}}} P_{\text{adh}}(m)\,\frac{dm}{\rho_{\text{w}}}\,,
\tag{14.8.27}
$$

where

$$
\epsilon_{\text{ch}} = \mathbf{I} : \mathbb{E}^{-1} : \overline{\mathbf{T}}\,,
\tag{14.8.28}
$$

and the pressure p_{ch} depends on stress via ϵ_{ch}. The strain is now given in the form,

$$
\mathbf{E} = \mathbb{E}^{-1} : \left(\mathbf{T} + p_{\text{I}}\,\det \mathbf{F}\,\mathbf{F}^{-1}\cdot\mathbf{F}^{-\text{T}} + (p_{\text{ch}} + \frac{\partial p_{\text{ch}}}{\partial\,\epsilon_{\text{ch}}}\,\epsilon_{\text{ch}})\,\mathbf{I}\right),
\tag{14.8.29}
$$

in such a way that the Cauchy effective stress,

$$
\boldsymbol{\sigma} + p_{\text{I}}\,\mathbf{I} + (p_{\text{ch}} + \frac{\partial p_{\text{ch}}}{\partial\,\epsilon_{\text{ch}}}\,\epsilon_{\text{ch}})\,\det \mathbf{F}^{-1}\,\mathbf{F}\cdot\mathbf{F}^{\text{T}}\,,
\tag{14.8.30}
$$

involves a nonisotropic chemical contribution. Still, in the small strain limit, the elastic equations (14.8.26) can be recast in the format (14.8.22).

14.9   Features of the constitutive framework

14.9.1   The effective stress format

The role of the electro-chemomechanical couplings is easier commented in a small strain context where the elastic
stress depends linearly on the strain,

$$\boldsymbol{\sigma} + p_{\text{eff}}\,\mathbf{I} = \mathbb{E} : \boldsymbol{\epsilon}\,. \tag{14.9.1}$$

The elastic moduli $\mathbb{E} = \mathbb{E}^{\text{PG}} + \mathbb{E}^{\text{c}}$ include the contributions of the ground substance and collagen, which undergo the
same strain. They differ in extension and in contraction since the collagen network behaves differently along these
two constitutive cones. However, in each of these cones, the moduli are independent of chemical composition of the
tissue. Indeed, they are representative of the *hypertonic state*,

$$\mathbb{E} \equiv \mathbb{E}^{\text{sat}}\,. \tag{14.9.2}$$

Then the effective stress format (14.9.1) is in agreement with the point of view of Maroudas and co-workers sketched
in Section 14.4.1, that hydrated PGs induce collagen fibers in extension even under compressive loads. The basic
property is that the effective pressure p_{eff} is positive under compression as sketched in Fig. 13.10.8 and negative
under large tensile load, Fig. 13.10.7. As a consequence, the effective pressure is tuned to minimize the strain, both
under compression and tension. Indeed,

- under zero external stress, since p_{eff} is positive, the collagen is induced in extension, and then $\mathbb{E}$ should be the
  tensile moduli $\mathbb{E}^{+}$:

$$\underbrace{\boldsymbol{\sigma}}_{0} + \underbrace{p_{\text{eff}}}_{>0}\,\mathbf{I} = \underbrace{\mathbb{E}^{+} : \boldsymbol{\epsilon}}_{\Rightarrow\ \substack{\text{extension}\\\text{of collagen}}}; \tag{14.9.3}$$

- for a moderate compressive load, the positive pressure p_{eff} induced by PGs is resisted by the applied mechanical
  stress $\boldsymbol{\sigma}$ and the extension of collagen:

$$\underbrace{\underbrace{\boldsymbol{\sigma}}_{\substack{\text{small}\\\text{compression}}} + \underbrace{p_{\text{eff}}}_{>0}\,\mathbf{I}}_{\text{tension}} = \underbrace{\mathbb{E}^{+} : \boldsymbol{\epsilon}}_{\Rightarrow\ \substack{\text{extension}\\\text{of collagen}}}; \tag{14.9.4}$$

- under larger contraction, the collagen fibers buckle, the PGs play a key role to resist large compression and
  the moduli assume the much smaller value $\mathbb{E}^{-}$:

$$\underbrace{\underbrace{\boldsymbol{\sigma}}_{\substack{\text{large}\\\text{compression}}} + \underbrace{p_{\text{eff}}}_{>0}\,\mathbf{I}}_{\text{compression}} = \underbrace{\mathbb{E}^{-} : \boldsymbol{\epsilon}}_{\Rightarrow\ \substack{\text{contraction}\\\text{of collagen}}}; \tag{14.9.5}$$

- on the other hand, under large tensile stresses, the negative pressure p_{eff} decreases the effective stress and
  therefore minimizes the elongation:

$$\underbrace{\boldsymbol{\sigma}}_{\substack{\text{large}\\\text{tension}}} + \underbrace{p_{\text{eff}}}_{<0}\,\mathbf{I} = \underbrace{\mathbb{E}^{+} : \boldsymbol{\epsilon}}_{\Rightarrow\ \substack{\text{moderated extension}\\\text{of collagen}}}, \tag{14.9.6}$$

14.9.2   Equilibrium of the extrafibrillar phase with a sodium chloride bath

In laboratory tests, a specimen of tissue is brought in contact with an electrolytic bath. In the simplest setting, the
bath is a binary sodium chloride electrolyte. Electroneutrality and consistency of molar fractions in the bath provide
the latter in terms of the molar fraction of the sodium cation, namely $x_{\text{ClB}} = x_{\text{NaB}}$, $x_{\text{wB}} = 1 - 2\,x_{\text{NaB}}$. The chemical
composition, namely molar fraction x_{NaB}, pressure p_{B}, and electrical potential ϕ_{B} of the bath are controlled.

The counterparts of these quantities in the extrafibrillar fluid E are partially defined by requesting

$$\text{electroneutrality in phase E} \quad x_{\text{NaE}} - x_{\text{ClE}} + y_{\text{PG}} = 0;$$

$$\text{consistency in phase E} \quad x_{\text{wE}} + x_{\text{NaE}} + x_{\text{ClE}} + x_{\text{PG}} = 1; \quad (14.9.7)$$

$$\text{chemical equilibrium} \quad \mu_{n\text{E}} = \mu_{n\text{B}}, \quad n = \text{w}, \text{NaCl}.$$

These four equations yield the four unknowns,

$$\text{molar fraction of sodium ion} \quad \frac{x_{\text{NaE}}}{x_{\text{NaB}}} = -\frac{y_{\text{PG}}}{2\,x_{\text{NaB}}} + \sqrt{1 + \left(\frac{y_{\text{PG}}}{2\,x_{\text{NaB}}}\right)^2};$$

$$\text{molar fraction of chloride ion} \quad \frac{x_{\text{ClE}}}{x_{\text{NaB}}} = \frac{y_{\text{PG}}}{2\,x_{\text{NaB}}} + \sqrt{1 + \left(\frac{y_{\text{PG}}}{2\,x_{\text{NaB}}}\right)^2};$$

$$\quad (14.9.8)$$

$$\text{molar fraction of water} \quad x_{\text{wE}} = 1 - 2\,x_{\text{NaB}}\sqrt{1 + \left(\frac{y_{\text{PG}}}{2\,x_{\text{NaB}}}\right)^2};$$

$$\text{osmotic pressure} \quad \frac{1}{RT}(p_{\text{wE}} - p_{\text{B}}) \simeq 2\,\frac{x_{\text{NaB}}}{\widehat{v}_{\text{w}}}\left(\sqrt{1 + \left(\frac{y_{\text{PG}}}{2\,x_{\text{NaB}}}\right)^2} - 1\right) > 0.$$

The expressions (14.9.8) have neglected the pressure term with respect to the chemical term in the chemical potentials of ions. The calculation of the osmotic pressure has used first order expansion,

$$\text{Ln}\, x_{\text{wK}} \simeq -\left(x_{\text{NaK}} + x_{\text{ClK}} + x_{\text{PG}}\, I_{\text{KE}}\right). \quad (14.9.9)$$

These two approximations are acceptable as far as the ionic molar fractions and effective molar fraction y_{PG} are small. Moreover, the molar fraction x_{PG} of PGs, but *not* their effective molar fraction y_{PG}, can be neglected with respect to the other molar fractions.

The jump in electrical potential deduces from the equilibrium of the electrochemical potentials of either the sodium cation or the chloride anion,

$$\text{electrical field} \quad \frac{\text{F}}{RT}(\phi_{\text{E}} - \phi_{\text{B}}) = -\text{Ln}\left(-\frac{y_{\text{PG}}}{2\,x_{\text{NaB}}} + \sqrt{1 + \left(\frac{y_{\text{PG}}}{2\,x_{\text{NaB}}}\right)^2}\right) = \text{argsinh}\,\frac{y_{\text{PG}}}{2\,x_{\text{NaB}}} < 0. \quad (14.9.10)$$

The above formulas seem explicit but they are not since the volume of water in the cartilage is still unknown. This volume certainly depends on the applied pressure and load. A mechanical constitutive relation, to be used to solve the mechanical equilibrium, is needed. This setting is a typical instance of chemomechanical coupling.

Chemical equilibrium between the bath and the cartilage implies a number of useful relations, equivalent to the above expressions, e.g.,

$$x_{\text{NaB}} = x_{\text{ClB}} = \sqrt{x_{\text{NaE}}\,x_{\text{ClE}}}, \quad x_{\text{ClE}} = x_{\text{NaE}} + y_{\text{PG}}, \quad x_{\text{NaE}}^2 - x_{\text{NaB}}^2 = -y_{\text{PG}}\,x_{\text{NaE}} > 0. \quad (14.9.11)$$

As direct consequences of the negative sign of the fixed charge,
- the molar fraction of the counterion, namely sodium cation, in the cartilage is as expected to be always larger than the molar fraction of the co-ion, namely $x_{\text{NaE}} > x_{\text{ClE}}$;
- the molar fraction of the counterion in the cartilage is always larger than its counterpart in the bath, namely $x_{\text{NaE}} > x_{\text{NaB}}$.

14.9.3 Equilibrium between the two fluid phases

As long as the intrafibrillar fluid does not involve a fixed charge, the equations governing the chemical equilibrium between the two fluid phases adopt the same format as for equilibrium between the extrafibrillar phase and the bath, the latter being replaced formally by the intrafibrillar phase. Still, the analogy is mostly formal because the volume of the intrafibrillar fluid matters and it interacts with its extrafibrillar counterpart.

14.9.4   The reference hypertonic state

In order to develop the nonlinear constitutive equations, it is crucial to define a configuration, or state, that can serve as reference.

Let us consider a cartilage specimen to be in equilibrium with a bath, at pressure p_B, saturated with sodium chloride NaCl. The reference configuration assumes by definition a zero strain, while the single mechanical loading is due to the bath pressure, and mechanical equilibrium simplifies to $\boldsymbol{\sigma} + p_B \mathbf{I} = \mathbf{0}$.

Since the bath is saturated, the specimen is in a *hypertonic state*. Actually, bath ionic strengths from $1\,M(\equiv \text{mole/liter})$ to NaCl saturation, that is $6.15\,M$ at $20\,^\circ C$, are observed to have almost identical mechanical effects so that, in actual laboratory experiments, it is sufficient to consider a $1\,M$ bath to define the hypertonic state. The high salt content overweighs the presence of PGs, as shielding of their fixed charge by mobile ions is at its maximum. Therefore,

Proposition 14.5 The electrochemical effects are at their minimum at high salt content.

Then the chemical pressure p_{ch}, and the adhesion pressure p_{adh} vanish. Moreover, the chemical composition in the IF compartment is quite close to that of the bath, and therefore $p_{eff} = p_I = p_B$.

On the other hand, even if quite tiny, the concentration of PGs is still felt in the EF compartment. The corresponding osmotic pressure p_{fw}, defined as the difference between extrafibrillar water pressure p_{wE} and bath pressure p_B, is expected to be at its minimum, although still positive, that is $p_{fw} = p_{wE} - p_B \geq 0$. Decreasing the ionic strength of the bath necessarily increases the osmotic pressure.

Observations tend to imply that the hypertonic state is ion-independent, that is, at sufficiently high concentrations, all ions have the same mechanical effects in shielding the electrical repulsion due to the fixed charge. Then, the following proposition is adopted:

Proposition 14.6 The hypertonic state is ion-independent.

On the other hand, the property does not hold at moderate to low concentrations:

Proposition 14.7 The shielding effect is ion-dependent.

This observation gives rise to an immediate issue for constitutive equations. Indeed, the osmotic pressure considers all ions on the same footing. Thus, it has pros and cons: it does not require the identification of parameters for each ion but, by the same token, it does not discriminate among the respective effects of the various ions.

For large ionic strength in the bath, the ratio $x \equiv -y_{PG}/(2\,x_{NaB})$ may be considered to be small so that the expressions (14.9.8) may be expanded in terms of x,

$$
\begin{array}{lll}
\text{molar fraction sodium ion} & \dfrac{x_{NaE}}{x_{NaB}} = 1 + x + \dfrac{1}{2}\,x^2 - \dfrac{1}{8}\,x^4 \cdots \; ; & \\[3mm]
\text{molar fraction chloride ion} & \dfrac{x_{ClE}}{x_{NaB}} = 1 - x + \dfrac{1}{2}\,x^2 - \dfrac{1}{8}\,x^4 \cdots \; ; & \\[3mm]
\text{molar fraction water} & x_{wE} = 1 - x_{NaB}\left(2 + x^2 - \dfrac{1}{4}\,x^4 \cdots\right); & (14.9.12) \\[3mm]
\text{osmotic pressure} & p_{wE} - p_B = RT\,\dfrac{x_{NaB}}{\widehat{v}_w}\left(x^2 - \dfrac{1}{4}\,x^4 + \cdots\right) & \\[3mm]
\text{electrical potential} & \phi_E - \phi_B = \dfrac{RT}{F}\left(-x + \dfrac{1}{6}\,x^3 + \cdots\right). &
\end{array}
$$

The expansion of the osmotic pressure is demonstrative and it is not expected to be accurate due to the approximation (14.9.9).

14.9.5   The limit case of a fresh bath

As a particular case consider a cartilage specimen in equilibrium with a sodium chloride electrolyte. Equilibrium of the electrochemical potentials of the sodium and chloride ions and of water yields the chemical composition of the extrafibrillar fluid phase of the cartilage in terms of molar fractions or concentrations. We are particularly interested in the limit of fresh bath where the molar fractions of ions $x \equiv x_{NaB} = x_{ClB}$ in the bath tend to zero. The following estimations of the molar fractions of ions, jump of fluid pressure and jump of electrical potential between the cartilage and the bath are obtained,

molar fraction sodium ion $\quad x_{\mathrm{NaE}} = -y_{\mathrm{PG}} - \dfrac{x^2}{y_{\mathrm{PG}}} + \cdots;$

molar fraction chloride ion $\quad x_{\mathrm{ClE}} = -\dfrac{x^2}{y_{\mathrm{PG}}} + \cdots;$

$$(14.9.13)$$

osmotic pressure $\quad p_{\mathrm{wE}} - p_{\mathrm{B}} = -\dfrac{RT}{\widehat{v}_{\mathrm{w}}} \left(\mathrm{Ln}\,(1 + y_{\mathrm{PG}}) + 2\,x + \cdots \right);$

electrical potential $\quad \phi_{\mathrm{E}} - \phi_{\mathrm{B}} = -\dfrac{RT}{\mathrm{F}} \,\mathrm{Ln} \left(-\dfrac{y_{\mathrm{PG}}}{x} - \dfrac{x}{y_{\mathrm{PG}}} \right).$

The expansion of the osmotic pressure has not used the approximation (14.9.9).

14.9.6 Exhaustion and recharge of the intrafibrillar phase

It might happen that the intrafibrillar water and ions tend to be completely pumped out into the extrafibrillar phase. This event may possibly take place during loading processes that imply a large volume decrease like uniaxial traction on a specimen in equilibrium with a fresh bath. Alternatively, one might be interested in analyzing either the potential effects of very large transfer times between the intra- and extrafibrillar compartments, or simply the inhibition of mass transfers.

The corresponding incremental constitutive equations are obtained as a limit of the general constitutive equations.

The analysis parallels that of Section 14.8.3, to within the fact that now the (change of) intrafibrillar mass-content m^{wI} vanishes. The constitutive equations (14.8.20), (14.8.22)$_{1-4}$ and (14.8.23)$_{1-4}$ hold unchanged. The memory of the existence of the intrafibrillar phase is apparent through the pressures $p_{f\mathrm{w}}$, $p_{f s_1}$ and $p_{f s_2}$.

In actual computations, e.g., finite element computations, the exhaustion of the intrafibrillar phase does not require a change of structure of the code: the residuals associated with the intrafibrillar degrees of freedom are zeroed and the appropriate diagonal components of the tangent matrix are given large values. This permanent structure makes it easier to address the reverse phenomenon of re-charge of an empty intrafibrillar phase.

14.10 Remarks on constitutive frameworks and constraints

In this chapter the tissue is modeled by introducing two fluid phases, namely the intrafibrillar phase and the extrafibrillar fluid phase. The extrafibrillar phase bathes the solid skeleton. To build the mechanical constitutive equations, we attempt to eliminate the chemical energy of the intrafibrillar fluid phase using the Gibbs-Duhem relation. The attempt is not complete due to the adhesion of intrafibrillar water on the collagen fibrils. On the other hand, intrafibrillar ionic species are completely eliminated from the mechanical equations and they are involved only in the transfer mechanisms between the two fluid phases. The diffusion processes through the extrafibrillar phase thus appear as reaction-diffusion processes.

Two constraints are to be accounted for:
- the two fluid phases are electrically neutral. The constraint is satisfied by reducing the number of independent variables associated with ionic species and eliminating the mass content of chloride ion in favor of the other ionic species;
- all species are incompressible, so that the volume change of the mixture is equal to the volume of species flowing through its boundary. This condition provides the change of the volume of the intrafibrillar compartment in terms of the volume changes of the mixture and of the extrafibrillar compartment.

Reducing the number of independent variables requires to single out a species: the procedure may be somehow unfortunate, breaking symmetries and introducing awkward algebras. As an alternative, the incompressibility and the electroneutrality constraints may be satisfied by introducing Lagrange multipliers which can be interpreted as pressure and electrical potential respectively. Instances are worked out in Sections 17.11.1, 19.16.2 and 23.1.1.3.

Exercises on Chapter 14

Exercise 14.1 Relations between the various measures of fluxes

Check the following relations between the absolute, diffusive, volume and mass fluxes, starting from the definitions given in Section 14.6.5.4.

1. Check the relations between the absolute and diffusive fluxes,

$$\mathbf{J}_{k\mathrm{E}} = \mathbf{J}_{k\mathrm{E}}^d + n^{k\mathrm{E}}\left(\mathbf{v}_{\mathrm{wE}} - \mathbf{v}_{\mathrm{S}}\right) = \mathbf{J}_{k\mathrm{E}}^d + \frac{n^{k\mathrm{E}}}{n^{\mathrm{wE}}}\mathbf{J}_{\mathrm{wE}}\,,\quad k \in \mathrm{E}\,. \tag{14.E.1}$$

2. Check the relation

$$\mathbf{J}_{\mathrm{E}} = \sum_{k\in\mathrm{E}}\mathbf{J}_{k\mathrm{E}}^d + \frac{n^{\mathrm{E}}}{n^{\mathrm{wE}}}\mathbf{J}_{\mathrm{wE}}\,. \tag{14.E.2}$$

Exercise 14.2 The electrical current density $\mathbf{I}_{\mathrm{eE}}$ in the extrafibrillar fluid phase is divergence free

Prove Eqn (14.6.46).

Proof: The proof consists in 1. using the decomposition (14.6.45) of the change of mass of a species in the fluid phase into a reactive part and a diffusive part, 2. dividing by the molar mass, 3. multiplying by the valence, and 4. summing over extrafibrillar species,

$$\sum_{k\in\mathrm{E}}\frac{\zeta_k}{\det\mathbf{F}}\frac{d\mathcal{N}_{k\mathrm{E}}}{dt} = -\sum_{k\in\mathrm{I}}\frac{\zeta_k}{\det\mathbf{F}}\frac{d\mathcal{N}_{k\mathrm{I}}}{dt} - \frac{1}{\mathrm{F}}\operatorname{div}\mathbf{I}_{\mathrm{eE}}\,. \tag{14.E.3}$$

We have accounted for the facts that
  - the mass exchange concerns the same species in both IF and EF compartments;
  - if a species does not transfer, whether it is in the IF or EF compartments, then we simply set $d\mathcal{N}_{k\mathrm{I}} = 0$.

As a consequence of electroneutrality in both fluid phases, the sums on the left-hand-side, and on the right-hand-side of (14.E.3) vanish.

Exercise 14.3 Variation of the fixed charge density with the volume of the porous medium

Prove Eqn (14.4.16).

Proof: An infinitesimal volume V_0 changes to $V = V_0 \det\mathbf{F}$ when the deformation gradient varies from $\mathbf{I}$ to $\mathbf{F}$. If the variation is sufficiently small, then $\mathbf{F} \simeq \mathbf{I} + \boldsymbol{\epsilon} + \boldsymbol{\omega}$, where $\boldsymbol{\epsilon}$ and $\boldsymbol{\omega}$ are respectively symmetric and skew-symmetric second order tensors. Then $\det\mathbf{F} \simeq 1 + \operatorname{tr}\boldsymbol{\epsilon}$.

Assuming an incompressible solid matrix, the change in volume of the electrolyte is $V_{\mathrm{F}} - V_{\mathrm{F}}^0 = V - V_0 = V_0 \operatorname{tr}\boldsymbol{\epsilon}$. Therefore $V_{\mathrm{F}}/V_{\mathrm{F}}^0 = 1 + \operatorname{tr}\boldsymbol{\epsilon}/n^{\mathrm{F}}$, where $n^{\mathrm{F}} = V_{\mathrm{F}}/V$ is the volume fraction of the fluid phase, and $n_0^{\mathrm{F}} = V_{\mathrm{F}}^0/V_0$ its initial value. Whence the relation $V_{\mathrm{F}}^0/V_{\mathrm{F}} \simeq 1 - \operatorname{tr}\boldsymbol{\epsilon}/n_0^{\mathrm{F}}$.

Exercise 14.4 Mass balance and volume change for incompressible species

1. For the species k of the phase K, the balance (4.4.5) of mass reads,

$$\frac{d^{k\mathrm{K}}}{dt}\rho^{k\mathrm{K}} + \rho^{k\mathrm{K}}\operatorname{div}\mathbf{v}_{k\mathrm{K}} = \hat{\rho}^{k\mathrm{K}}\,, \tag{14.E.4}$$

where $\hat{\rho}^{k\mathrm{K}}$ is the rate of mass supply and the symbol $d^{k\mathrm{K}}/dt$ denotes the time derivative following the particles of species k in phase K. This equation can be expressed in terms of the time derivative following the solid skeleton,

$$\frac{d^{k\mathrm{K}}}{dt}(\cdot) = \frac{d^{\mathrm{s}}}{dt}(\cdot) + \boldsymbol{\nabla}(\cdot)\cdot(\mathbf{v}_{k\mathrm{K}} - \mathbf{v}_{\mathrm{S}}),\quad k\in\mathrm{K}\,. \tag{14.E.5}$$

Using the definition of the apparent mass density $\rho^{k\mathrm{K}} = n^{k\mathrm{K}}\rho_k$ and the incompressibility condition, ρ_k=constant, the balance of mass may be recast in terms of the volume fraction $n^{k\mathrm{K}}$,

$$\frac{d^{k\mathrm{K}}}{dt}n^{k\mathrm{K}} + n^{k\mathrm{K}}\operatorname{div}\mathbf{v}_{k\mathrm{K}} = \frac{\hat{\rho}^{k\mathrm{K}}}{\rho_k}\,,\quad k\in\mathrm{K}\,. \tag{14.E.6}$$

2. Alternatively, the balance of mass may be recast in terms of the mass content $m^{k\mathrm{K}} = \rho^{k\mathrm{K}}\det\mathbf{F}$,

$$\begin{aligned}
\frac{1}{\det\mathbf{F}}\frac{d^{\mathrm{s}}m^{k\mathrm{I}}}{dt} &= \hat{\rho}^{k\mathrm{I}} &, k\in\mathrm{I}\\[2mm]
\frac{1}{\det\mathbf{F}}\frac{d^{\mathrm{s}}m^{k\mathrm{E}}}{dt} &= \underbrace{\hat{\rho}^{k\mathrm{E}}}_{\text{reactive part}} - \underbrace{\operatorname{div}\mathbf{M}_{k\mathrm{E}}}_{\text{diffusive part}} &, k\in\mathrm{E}
\end{aligned} \tag{14.E.7}$$

with $\mathbf{M}_{k\mathrm{E}}=\rho^{k\mathrm{E}}\,(\mathbf{v}_{k\mathrm{E}}-\mathbf{v}_{\mathrm{S}})$ the mass flux defined by (14.6.33). This format highlights our assumptions that the intrafibrillar species are only reactive while the extrafibrillar species are both diffusive and reactive, that is, extrafibrillar mass changes are due both to transfer into/from the intrafibrillar phase and to exchange with the surroundings of the representative elementary volume.

3. Summing up the relations (14.E.6) over the species of phase K, and using the volume flux $\mathbf{J}_{\mathrm{K}}=\sum_{k\in\mathrm{K}} n^{k\mathrm{K}}\,(\mathbf{v}_{k\mathrm{E}}-\mathbf{v}_{\mathrm{S}})$ of phase K through the porous skeleton, the mass balance of phase K may be phrased in terms of the volume content (14.6.4) of the phase,

$$\frac{1}{\det\mathbf{F}}\frac{d^{s}v^{\mathrm{K}}}{dt}+\operatorname{div}\mathbf{J}_{\mathrm{K}}=\sum_{k\in\mathrm{K}}\frac{\hat{\rho}^{k\mathrm{K}}}{\rho_{k}}\,. \tag{14.E.8}$$

4. Summing the relations (14.E.8) over the phases yields

$$\frac{1}{\det\mathbf{F}}\overbrace{\sum_{\mathrm{K=S,I,E}}\frac{d^{s}v^{\mathrm{K}}}{dt}}^{\det\mathbf{F}\,\operatorname{div}\mathbf{v}_{\mathrm{S}},\ \mathrm{Eqn}\ (14.6.7)}+\operatorname{div}\overbrace{\mathbf{J}_{\mathrm{S}}}^{=0}+\operatorname{div}\overbrace{\mathbf{J}_{\mathrm{I}}}^{=0}+\operatorname{div}\mathbf{J}_{\mathrm{E}}=\sum_{k,\mathrm{K}}\frac{\hat{\rho}^{k\mathrm{K}}}{\rho_{k}}=\sum_{k\in\mathrm{I}}\frac{1}{\rho_{k}}\overbrace{(\hat{\rho}^{k\mathrm{I}}+\hat{\rho}^{k\mathrm{E}})}^{=0}\,. \tag{14.E.9}$$

The absolute fluxes $\mathbf{J}_{\mathrm{S}}$ and $\mathbf{J}_{\mathrm{I}}$ vanish since all the species in the solid and intrafibrillar phases move with the velocity of the solid. Moreover the sum of the mass rates $\hat{\rho}^{k\mathrm{I}}+\hat{\rho}^{k\mathrm{E}}$ vanishes as well because the exchanges concern individual species independently of one another. Hence

$$\operatorname{div}\mathbf{v}_{\mathrm{S}}+\operatorname{div}\mathbf{J}_{\mathrm{E}}=0\,. \tag{14.E.10}$$

In the text, the derivative following the solid phase d^{s}/dt is denoted by d/dt.

Exercise 14.5 Partial derivatives of the molar fractions

1. Prove Eqn (14.6.26).

Take the differential of the logarithm of (14.6.8), and use (14.6.9),

$$\begin{aligned}\frac{dx_{k\mathrm{K}}}{x_{k\mathrm{K}}}&=\frac{dm^{k\mathrm{K}}}{m^{k\mathrm{K}}}-\frac{V_{0}}{N_{\mathrm{K}}}\sum_{l\in\mathrm{K}}\frac{dm^{l\mathrm{K}}}{\widehat{m}_{l}}\\&=\frac{dm^{k\mathrm{K}}}{m^{k\mathrm{K}}}-\sum_{l\in\mathrm{K}}x_{l\mathrm{K}}\frac{dm^{l\mathrm{K}}}{m^{l\mathrm{K}}}\\&=\sum_{l\in\mathrm{K}}\frac{dm^{l\mathrm{K}}}{m^{l\mathrm{K}}}\left(I_{kl}-x_{l\mathrm{K}}\right),\quad k\in\mathrm{K}\,.\end{aligned} \tag{14.E.11}$$

2. Prove Eqn (14.6.27).

Insert Eqn (14.6.24) into (14.E.11):

$$\begin{aligned}\frac{dx_{k\mathrm{K}}}{x_{k\mathrm{K}}}&=\sum_{i\in\mathrm{K}_{\mathrm{in}}}\frac{dm^{i\mathrm{K}}}{m^{i\mathrm{K}}}\left(I_{ki}-x_{i\mathrm{K}}\right)+\sum_{i\in\mathrm{K}_{\mathrm{in}}}\zeta_{i}\frac{dm^{i\mathrm{K}}}{m^{\mathrm{ClK}}}\frac{\widehat{m}_{\mathrm{Cl}}}{\widehat{m}_{i}}\left(I_{k\mathrm{Cl}}-x_{\mathrm{ClK}}\right)\\&=\sum_{i\in\mathrm{K}_{\mathrm{in}}}\frac{dm^{i\mathrm{K}}}{m^{i\mathrm{K}}}\left(I_{ki}-x_{i\mathrm{K}}\left(1+\zeta_{i}\right)+\zeta_{i}\frac{x_{i\mathrm{K}}}{x_{\mathrm{ClK}}}I_{k\mathrm{Cl}}\right),\quad k\in\mathrm{K}\,.\end{aligned} \tag{14.E.12}$$

Exercise 14.6 Relations in terms of independent variables

1. Prove Eqn (14.6.50).

$$\begin{aligned}\sum_{k\in\mathrm{K}}\frac{dm^{k\mathrm{K}}}{\rho_{k}}&\overset{(14.6.24)}{=}\sum_{i\in\mathrm{K}_{\mathrm{in}}}\frac{dm^{i\mathrm{K}}}{\rho_{i}}+\sum_{\mathrm{K}}\sum_{i\in\mathrm{K}_{\mathrm{in}}}\zeta_{i}\frac{dm^{i\mathrm{K}}}{\rho_{\mathrm{Cl}}}\frac{\widehat{m}_{\mathrm{Cl}}}{\widehat{m}_{i}}\\&=\sum_{i\in\mathrm{K}_{\mathrm{in}}}\underbrace{\left(\frac{1}{\rho_{i}}+\frac{\zeta_{i}}{\rho_{\mathrm{Cl}}}\frac{\widehat{m}_{\mathrm{Cl}}}{\widehat{m}_{i}}\right)}_{\rho_{n\mathrm{K}}^{-1},\ \mathrm{Eqn}\ (14.6.32)}dm^{i\mathrm{K}}=\sum_{i\in\mathrm{K}_{\mathrm{in}}}\left(1+\zeta_{i}\frac{\widehat{v}_{\mathrm{Cl}}}{\widehat{v}_{i}}\right)dv^{i\mathrm{K}}\,.\end{aligned} \tag{14.E.13}$$

or, equivalently in terms of volume contents,

$$\sum_{k\in\mathrm{K}}dv^{k\mathrm{K}}=\sum_{i\in\mathrm{K}_{\mathrm{in}}}\left(1+\zeta_{i}\frac{\widehat{v}_{\mathrm{Cl}}}{\widehat{v}_{i}}\right)dv^{i\mathrm{K}}\,. \tag{14.E.14}$$

2. Prove Eqn (14.6.29).

$$
\begin{aligned}
\sum_{k \in \mathrm{K}} \mu_{k\mathrm{K}}^{\mathrm{ec}}\, dm^{k\mathrm{K}}
&= \sum_{k \in \mathrm{K}} \left(\mu_{k\mathrm{K}} + \mathrm{F}\,\frac{\zeta_k}{\widehat{m}_k}\,\phi_\mathrm{K} \right) dm^{k\mathrm{K}} \\[2mm]
&= \sum_{k \in \mathrm{K}} \mu_{k\mathrm{K}}\, dm^{k\mathrm{K}} + \mathrm{F}\,\phi_\mathrm{K} \overbrace{\sum_{k \in \mathrm{K}} \zeta_k\, dN_{k\mathrm{K}}/V_0}^{=0,\ \mathrm{Eqn}\ (14.6.22)} \\[2mm]
&\overset{(14.6.24)}{=} \sum_{i \in \mathrm{K_{in}}} \mu_{i\mathrm{K}}\, dm^{i\mathrm{K}} + \sum_{i \in \mathrm{K_{in}}} \mu_{\mathrm{ClK}}\,\zeta_i\,\frac{\widehat{m}_{\mathrm{Cl}}}{\widehat{m}_i}\, dm^{i\mathrm{K}} \\[2mm]
&= \sum_{i \in \mathrm{K_{in}}} \left(\widehat{m}_i\,\mu_{i\mathrm{K}} + \zeta_i\,\widehat{m}_{\mathrm{Cl}}\,\mu_{\mathrm{ClK}} \right) \frac{dm^{i\mathrm{K}}}{\widehat{m}_i} \\[2mm]
&\overset{(14.6.31)}{=} \sum_{i \in \mathrm{K_{in}}} \mu_{n\mathrm{K}}\, dm^{i\mathrm{K}} .
\end{aligned}
\tag{14.E.15}
$$

3. Prove Eqn (14.8.4).

$$
\begin{aligned}
\sum_{k \in \mathrm{I}} m^{k\mathrm{I}}\, d\mu_{k\mathrm{I}}^{\mathrm{ec}}
&= \sum_{k \in \mathrm{I}} m^{k\mathrm{I}}\, d\mu_{k\mathrm{I}} + \mathrm{F}\,\phi_\mathrm{I} \overbrace{\sum_{k \in \mathrm{I}} \zeta_k\, dN_{k\mathrm{I}}/V_0}^{=0,\ (14.6.22)} \\[2mm]
&\overset{(14.6.25)}{=} \sum_{i \in \mathrm{I_{in}}} m^{i\mathrm{I}}\, d\mu_{i\mathrm{I}} + \sum_{i \in \mathrm{I_{in}}} \zeta_i\,\frac{\widehat{m}_{\mathrm{Cl}}}{\widehat{m}_i}\, m^{i\mathrm{I}}\, d\mu_{\mathrm{ClI}} \\[2mm]
&= \sum_{i \in \mathrm{I_{in}}} \frac{m^{i\mathrm{I}}}{\widehat{m}_i}\left(\widehat{m}_i\, d\mu_{i\mathrm{I}} + \zeta_i\,\widehat{m}_{\mathrm{Cl}}\, d\mu_{\mathrm{ClI}} \right) \\[2mm]
&\overset{(14.6.31)}{=} \sum_{i \in \mathrm{I_{in}}} m^{i\mathrm{I}}\, d\mu_{n\mathrm{I}} .
\end{aligned}
\tag{14.E.16}
$$

The relation holds for the chemical potentials, whether modified along (14.8.2) or not.

Exercise 14.7   The Gibbs-Duhem relation for the fluid phases

Prove the Gibbs-Duhem relation (14.8.3) for the intrafibrillar fluid phase.

Proof: The idea is to form a linear combination of the electrochemical potentials (14.6.15) that eliminates the chemical contribution via the compatibility condition $(14.6.8)_2$ satisfied by the molar fractions, and the electrical contribution via the electroneutrality condition (14.6.23). These requirements motivate the coefficients α_k, $k \in \mathrm{K_{mo}} = \mathrm{K} - \{\mathrm{PG}\}$, of the linear combination to be equal to $N_{k\mathrm{K}}/V_0$, to within an arbitrary multiplicative factor. Then

$$
\begin{aligned}
\sum_{k \in \mathrm{K_{mo}}} \alpha_k\, \widehat{m}_k\, d\tilde{\mu}_{k\mathrm{K}}^{\mathrm{ec}}
&= \sum_{k \in \mathrm{K_{mo}}} \alpha_k\, \widehat{v}_k\, dp_\mathrm{K} + RT \sum_{k \in \mathrm{K_{mo}}} \alpha_k\,\frac{dx_{k\mathrm{K}}}{x_{k\mathrm{K}}} + \mathrm{F}\, d\phi_\mathrm{K} \sum_{k \in \mathrm{K_{mo}}} \alpha_k\, \zeta_k \\[2mm]
&= \sum_{k \in \mathrm{K_{mo}}} v^{k\mathrm{K}}\, dp_\mathrm{K} + RT\, v^\mathrm{K} \sum_{k \in \mathrm{K_{mo}}} dx_{k\mathrm{K}} + \mathrm{F}\, \mathcal{N}_\mathrm{K}\, d\phi_\mathrm{K} \sum_{k \in \mathrm{K_{mo}}} \zeta_k\, c_{k\mathrm{K}} .
\end{aligned}
\tag{14.E.17}
$$

Hence for $\mathrm{K} = \mathrm{I}$,

$$
\sum_{k \in \mathrm{K_{mo}}} m^{k\mathrm{K}}\, d\tilde{\mu}_{k\mathrm{K}}^{\mathrm{ec}} = v^\mathrm{I}\, dp_\mathrm{I} ,
\tag{14.E.18}
$$

and for $\mathrm{K} = \mathrm{E}$,

$$
\sum_{k \in \mathrm{K_{mo}}} m^{k\mathrm{K}}\, d\tilde{\mu}_{k\mathrm{K}}^{\mathrm{ec}} = (v^\mathrm{E} - v_{\mathrm{PG}})\, dp_\mathrm{E} - RT\, \mathcal{N}_\mathrm{E}\, dx_{\mathrm{PG}} - \mathrm{F}\, v^\mathrm{E}\, d\phi_\mathrm{E}\, e_{\mathrm{PG}} .
\tag{14.E.19}
$$

Exercise 14.8   Constrained partial derivative

Calculate the partial derivatives of (14.8.14) taking account of the constraint (14.6.24).

Answer: For $i \in \mathrm{E_{in}}$,

$$
\begin{aligned}
\frac{V_0}{RT}\,\frac{\partial \mathcal{W}_{\mathrm{ch}}}{\partial N_{i\mathrm{E}}}
&= \mathrm{Ln}\, N_{i\mathrm{E}} + 1 - (\mathrm{Ln}\, N_\mathrm{E} + 1) + \zeta_i \left(\mathrm{Ln}\, N_{\mathrm{ClE}} + 1 - (\mathrm{Ln}\, N_\mathrm{E} + 1) \right) \\[2mm]
&= \mathrm{Ln}\, x_{i\mathrm{E}}\, (x_{\mathrm{ClE}})^{\zeta_i} .
\end{aligned}
\tag{14.E.20}
$$

Chapter 15

Passive transport in the interstitium and circulation: basics

Passive transports are dissipative processes where the flux is directed *against* a certain gradient. By contrast, biology displays a number of examples where the flux takes place *along* the gradient. For example, cells move in directions where the density of adhesive sites is larger since they get a stronger grip to the matrix in these regions: the phenomenon is referred to as haptotaxis. Much in the same flavor, chemotaxis consists of the motion of cells towards a zone of higher nutrient content. Active transport across the corneal endothelium avoids swelling of the cornea and contributes to maintain its transparency. Active transports require energy. The topic is touched in Section 19.3.

This chapter is devoted solely to passive transports. Emphasis is laid on the fact that a number of passive transport processes in engineering are governed by parabolic equations which might possibly be coupled. The couplings are important only when the characteristic times of the underlying physical processes are of like order. Coupled transports in tissues endowed with a fixed electrical charge are addressed in Chapter 16: Darcy's law of seepage, Fick's law of solute diffusion, Ohm's law of electrical conduction and Fourier's law of heat conduction are seen to interact.

Parabolic equations are associated with an infinite speed of propagation. This simplification is usually acceptable but it might be overcome by modifying the constitutive equations that govern the underlying physical processes so as to turn the partial differential equations into hyperbolic equations which propagate information at finite speed.

15.1 Diffusion as a mode of passive transport

Seepage, that is flow of water through a solid porous skeleton, is governed by Darcy's law: water flows against the gradient of water pressure. According to Fick's law, a solute diffuses against the gradient of its concentration in a fluid so as to homogenize its spatial distribution. The electrical flow takes place against the gradient of the electrical potential along Ohm's law. Fourier's law of heat conduction is of the same vein and indicates that heat flux is directed against the temperature gradient.

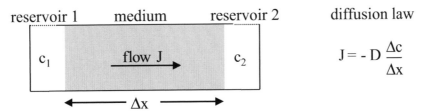

Fig. 15.1.1: One-dimensional sketch illustrating passive transport by diffusion. The spatial disequilibrium of the quantity c plays the role of a driving gradient. Flow takes place in the direction opposite to the gradient, $c_2 > c_1$.

In these four physical phenomena, a flux $\mathbf{J}$ arises as a result of a non homogeneous spatial distribution of some quantity c, Fig. 15.1.1. The associated internal entropy production is of the form, see (4.7.32),

$$T\hat{S}_{(i,3)} = -\mathbf{J} \cdot \boldsymbol{\nabla} c \geq 0 \,. \tag{15.1.1}$$

531

One way to satisfy the dissipation inequality is ensured by an Onsager type of constitutive assumption discussed in Section 4.8,

$$\mathbf{J} = -D\,\boldsymbol{\nabla}c, \tag{15.1.2}$$

where the scalar $D > 0$ is positive. Since the physical entities are innate to each diffusion process, it is worth to list the conjugate variables separately:

• Darcy's law of seepage of a fluid through a porous skeleton:

$$\mathbf{J}_\mathrm{H} = -K_\mathrm{H}\,\boldsymbol{\nabla}h \quad \begin{cases} \mathbf{J}_\mathrm{H} : \text{flux of fluid relative to solid skeleton [unit : m/s]} \\ h : \text{hydraulic head [unit : m]} \\ K_\mathrm{H} : \text{hydraulic conductivity [unit : m/s]} \end{cases} \tag{15.1.3}$$

• Fick's law of diffusion of a solute in a fluid phase:

$$\mathbf{J}_\mathrm{s}^d = -D_\mathrm{s}\,\boldsymbol{\nabla}c \quad \begin{cases} \mathbf{J}_\mathrm{s}^d : \text{molar flux of solute relative to fluid [unit : mole/m}^2\text{/s]} \\ c : \text{concentration of solute [unit : mole/m}^3\text{]} \\ D_\mathrm{s} : \text{diffusion coefficient or chemical diffusivity [unit : m}^2\text{/s]} \end{cases} \tag{15.1.4}$$

• Ohm's law of electrical flow through a conductor of current:

$$\mathbf{I}_\mathrm{e} = -\sigma_\mathrm{e}\,\boldsymbol{\nabla}\phi \quad \begin{cases} \mathbf{I}_\mathrm{e} : \text{electrical current density [unit : C/m}^2\text{/s]} \\ \phi : \text{electrical potential [unit : volt]} \\ \sigma_\mathrm{e} : \text{electrical conductivity [unit : siemens/m]} \end{cases} \tag{15.1.5}$$

• Fourier's law of thermal diffusion through a heat conductor:

$$\mathbf{Q} = -k_\mathrm{T}\,\boldsymbol{\nabla}T \quad \begin{cases} \mathbf{Q} : \text{thermal flux [unit : W/m}^2 = \text{J/m}^2\text{/s} = \text{kg/s}^3\text{]} \\ T : \text{temperature [unit : K]} \\ k_\mathrm{T} : \text{thermal conductivity [unit : W/m/K} = \text{kg} \times \text{m/s}^3\text{/K]} \end{cases} \tag{15.1.6}$$

In the simplest context, the above physical phenomena are *uncoupled*. They are also expressed through linear constitutive equations. This linearity is sometimes questioned, e.g., note 4.7. On the other hand, couplings can not be neglected in a number of situations, especially in presence of fixed electrical charges as detailed in Chapter 16.

15.2   Coupled flows

A number of direct and coupled diffusion phenomena are shown in Table 15.2.1. Thermoosmosis is briefly discussed in Section 9.3.3. Couplings related to the presence of ions and to a fixed electrical charge are considered in Chapters 16 and 17. Thermal couplings are introduced in Section 16.5.

Table 15.2.1: Direct diffusion phenomena and reciprocal coupled flows.

Gradient ⇒ Flux ⇓	Hydraulic Head	Temperature	Solute concentration	Electrical potential
Solvent flux	Hydraulic conduction **Darcy's law**	Thermoosmosis	Chemical osmosis	Electroosmosis
Heat flux	Mechanocaloric effect	Thermal conduction **Fourier's law**	Dufour effect	Peltier effect
Diffusive solute flux	Pressure diffusion ultrafiltration	Thermal diffusion Soret effect	Ordinary diffusion **Fick's law**	Electrophoresis
Electrical current	Streaming conduction	Thermoelectricity Seebeck effect	Membrane potential	Electrical conduction **Ohm's law**

The Seebeck and Peltier effects are analyzed in the context of the thermocouple in Callen [1960], p. 316, Kestin [1968], p. 603 and Haase [1990], p. 346. The Seebeck effect describes the generation of an electrical current density by a thermal gradient while, in the Peltier effect, an electrical field gives rise to a heat flux,

$$
\begin{bmatrix} \mathbf{Q} \\ \mathbf{I}_e \end{bmatrix} = - \begin{bmatrix} T\,k_{\mathrm{T}} & P \\ \sigma_e\,T\,S & \sigma_e \end{bmatrix} \begin{bmatrix} \boldsymbol{\nabla} T/T \\ \boldsymbol{\nabla}\phi \end{bmatrix} .
\tag{15.2.1}
$$

The Seebeck coefficient S [unit : V/K] may take positive or negative values. It measures the intensity of the electrical field due to a thermal gradient in an electrically open circuit, namely $\mathbf{I}_e = \mathbf{0}$, i.e., $\boldsymbol{\nabla}\phi = -S\,\boldsymbol{\nabla} T$. Eliminating the electrical field from the first line of (15.2.1), the heat flux can be recast in the format,

$$
\mathbf{Q} = - \overbrace{k_{\mathrm{T}}\left(1 - \frac{P\,S}{k_{\mathrm{T}}}\right)}^{k_{\mathrm{TI}}} \boldsymbol{\nabla} T + \frac{P}{\sigma_e}\,\mathbf{I}_e ,
\tag{15.2.2}
$$

which displays the thermal conductivity at vanishing electrical current density, namely k_{TI}, as opposed to k_{T}, the thermal conductivity at vanishing electrical field. The coefficient P has dimension [unit : A/m]. The coefficient $\Pi = T\,S$ [unit : V] is referred to as Peltier coefficient.

When $P = \sigma_e\,T\,S$, the constitutive matrix is symmetric. Then the coefficient $P\,S/k_{\mathrm{T}} = T\,\sigma_e\,S^2/k_{\mathrm{T}}$ is called 'figure of merit': a large electrical conductivity minimizes Joule heating and a low thermal conductivity retains heat and maintains a large thermal gradient. The divergence of the heat flux can then be decomposed so as to highlight typical effects,

$$
\begin{aligned}
\operatorname{div}\mathbf{Q} &= \operatorname{div}\left(-k_{\mathrm{TI}}\boldsymbol{\nabla} T\right) + T\,\boldsymbol{\nabla} S\cdot\mathbf{I}_e + S\,\boldsymbol{\nabla} T\cdot\mathbf{I}_e + T\,S\operatorname{div}\mathbf{I}_e \\[2mm]
&= \operatorname{div}\left(-k_{\mathrm{TI}}\boldsymbol{\nabla} T\right) + \underbrace{\left(\boldsymbol{\nabla}(T\,S) - S\boldsymbol{\nabla} T\right)}_{\substack{\text{Peltier-Thomson}\\ \text{effect}}}\cdot\mathbf{I}_e - \underbrace{\frac{\mathbf{I}_e^2}{\sigma_e}}_{\substack{\text{Joule}\\ \text{heating}}} - \mathbf{I}_e\cdot\boldsymbol{\nabla}\phi + T\,S\operatorname{div}\mathbf{I}_e .
\end{aligned}
\tag{15.2.3}
$$

The first term on the right-hand side represents the heat produced due to non-isothermal heat transfer, the second and third term embeds the Peltier-Thomson effect, the fourth term denotes the dissipated electrical work (Joule heating) and the fifth term the produced electrical work. In addition, in the applications to tissues endowed with a fixed electrical charge developed in this textbook, the electrical current density will be shown to be divergence free. W. Thomson (Lord Kelvin) has showed connections between the earlier works of Seebeck and Peltier and has given his name to a coefficient, namely $T\,dS/dT$, emanating from the second term in the 2nd line of (15.2.3).

While the Joule heating effect is irreversible ($\sigma_e \geq 0$), no such restriction of sign applies to the Seebeck and Peltier coefficients, S and $\Pi = T\,S$ respectively, even if their magnitude is restricted by the fact that k_{TI} should be positive for the dissipation inequality to be satisfied. In that sense, the Seebeck and Peltier effects are reversible.

Thermal diffusion, or thermophoresis, has been coined Soret effect in honor of the Swiss physicist Charles Soret (1854-1904) who analyzed the phenomenon in liquid mixtures, although it had been already observed both in liquid and gas mixtures. Soret and Dufour effects can be displayed in a form essentially similar to that used for thermoosmosis in Section 9.3.3,

$$
\begin{bmatrix} \mathbf{J}_{\mathrm{s}}^d \\ \mathbf{Q} \end{bmatrix} = - \begin{bmatrix} D_{\mathrm{s}} & k_{\mathrm{ST}} \\ k_{\mathrm{TS}} & T\,k_{\mathrm{T}} \end{bmatrix} \begin{bmatrix} \boldsymbol{\nabla} c \\ \boldsymbol{\nabla} T/T \end{bmatrix} .
\tag{15.2.4}
$$

The coefficients involved in the above matrix are subjected to inequalities similar to (9.3.26) so as to ensure a positive dissipation.

Consider an aqueous solution in a tube with ends at different temperatures, Fig. 15.2.1. Soret effect indicates that a diffusion of salt takes place and thus that the salt concentration is not spatially uniform. For hydrocarbon mixtures, the Soret coefficient $k_{\mathrm{ST}}/D_{\mathrm{s}}$ is reported to be positive, Platten [2006]. Thus, at steady state, the solute concentration is larger at the cold end. Soret effect might be thought of as a transport vehicle of solutes across biological membranes separating tissues at different temperatures.

According to the *pressure diffusion effect*, a pressure gradient generates a solute flux. It is generally small although it is enhanced in centrifuge separation through large pressures gradients, e.g., Bird et al. [1960], p. 575.

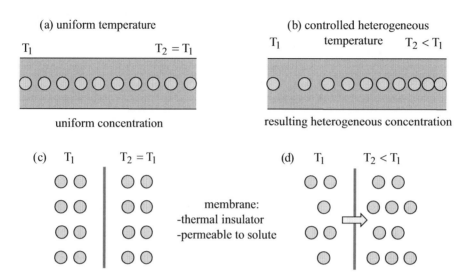

Fig. 15.2.1: Two qualitative illustrations of Soret effect, in a tube containing an aqueous solution, and across a membrane. (a),(b): at steady state, a thermal gradient in a continuum is accompanied by a chemical gradient; (c),(d): consider a membrane which is a thermal insulator but permeable to solute. To a thermal jump across the membrane corresponds a chemical jump.

Quite generally, the diffusion properties certainly inherit some of the attributes of the microstructure of the materials and, in particular, anisotropic properties. The anisotropic form of (15.1.2) involves a symmetric positive (semi-)definite (sPD) tensor of diffusion, say

$$\mathbf{J} = -\mathbf{D} \cdot \boldsymbol{\nabla} c, \quad \mathbf{D} = \mathbf{D}^{\mathrm{T}} \text{ sPD}. \tag{15.2.5}$$

Associated with a balance of mass, these transports give rise to partial differential equations (PDEs) in time and space of the parabolic type. In this context, it is instructive to compare the different types of PDEs that arise in mathematical physics. The prototype operator of elliptic equations is the Laplacian. Elliptic equations arise, for example, in elastostatics. Heat conduction serves as a prototype of parabolic equations, and the wave equation as a prototype of a hyperbolic equation. While only the higher derivatives decide of the type of a PDE, a PDE may sometimes display, to some degrees, mixed features: such is the case of convective diffusion where propagation and diffusion take place simultaneously: which phenomenon takes over the other depends on a characteristic dimensionless number.

15.3 Propagation, diffusion, convection

The one-dimensional examples below intend to display basic features and pinpoint differences between parabolic and hyperbolic partial differential equations. Examples of convective diffusion and reactive diffusion which are of interest in tissue biomechanics are also considered. While focus is here on the mathematical nature of the partial differential field equations, it is important to introduce (even briefly) the underlying constitutive assumptions, so as to conform to the general concept of initial and boundary value problem (IBVP) where the constitutive material response is of prime concern.

15.3.1 An example of hyperbolic PDE: propagation of a shock wave

A semi-infinite elastic plane, or half-space, is subjected to a load normal to its boundary. The motion is one-dimensional and could also be viewed as due to an axial load on the end of an elastic bar with vanishing Poisson's ratio. The bar is initially at rest and the loading takes the form of an arbitrary velocity discontinuity $v = v_0(t)$.

The issue is to derive the axial displacement $u(x)$ and axial velocity $v(x)$ of the points of the bar while the mechanical information propagates along the bar with a wave-speed $c = \sqrt{E/\rho}$, with E and ρ respectively the Young's modulus and mass density of the material. The governing equations of dynamic linear elasticity include

(FE) a field equation $\quad \dfrac{\partial^2 u}{\partial x^2} - \dfrac{1}{c^2}\dfrac{\partial^2 u}{\partial t^2} = 0, \quad t > 0,\ x > 0;$

(IC) initial conditions $\quad u(x,0) = 0; \quad \dfrac{\partial u}{\partial t}(x,0) = 0\,;$

(BC) boundary conditions $\quad \dfrac{\partial u}{\partial t}(0,t) = v_0(t)\,;$

(RC) a radiation condition.

$$(15.3.1)$$

The field equation is obtained by combining the balance of momentum and the constitutive equation,

momentum balance $\quad \dfrac{\partial \sigma}{\partial x} - \rho\,\dfrac{\partial^2 u}{\partial t^2} = 0,$

elasticity $\quad \sigma = E\,\dfrac{\partial u}{\partial x}\,,$

$$(15.3.2)$$

where σ is the axial stress.

The first initial condition (IC) means that the displacement is measured from time $t = 0$, or said otherwise, that the configuration (geometry) at time $t = 0$ is used as reference. The second initial condition simply states that the bar is initially at rest.

The radiation condition is intended to imply that the mechanical information propagates in a single direction, and that the bar is either of infinite length or, at least, that the signal has not the time to reach its right boundary in the time window of interest. Indeed, any function of the form $f_1(x - c\,t) + f_2(x + c\,t)$ satisfies the field equation. A function of $x - c\,t$ represents a signal that propagates towards increasing x. To grasp this point, let us keep the eyes on some given value of f_1, corresponding to $x - c\,t$ equal to some constant. Then clearly the point we follow moves in time towards increasing x.

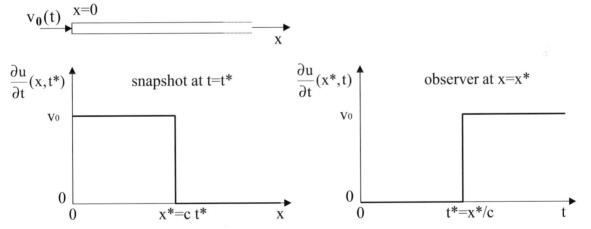

Fig. 15.3.1: A semi-infinite elastic bar is subjected to a velocity discontinuity. Spatial and time profiles of the particle velocity. The discontinuity propagates along the bar at the wave-speed $c = \sqrt{E/\rho}$, where E is Young's modulus and ρ the mass density.

The solution is obtained through the Laplace transform in time,

$$u(x,t) \to U(x,p) = \mathcal{L}\{u(x,t)\}(p), \qquad (15.3.3)$$

and we admit that the operators Laplace transform and partial derivative wrt space commute,

$$\frac{\partial}{\partial x}\mathcal{L}\{u(x,t)\}(p) = \mathcal{L}\left\{\frac{\partial}{\partial x}u(x,t)\right\}(p)\,. \qquad (15.3.4)$$

Therefore,

$$(FE) \quad \frac{\partial^2 U}{\partial x^2}(x,p) - \frac{1}{c^2}\left(p^2\,U(x,p) - p\,\overbrace{u(x,0)}^{=0,\ (IC)} - \overbrace{\frac{\partial u}{\partial t}(x,0)}^{=0,\ (IC)}\right) = 0$$

$$\Rightarrow \quad U(x,p) = a(p)\,\exp\left(-\frac{p}{c}x\right) + \underbrace{b(p)\,\exp\left(+\frac{p}{c}x\right)}_{b(p)=0,\ (RC)}.$$

(15.3.5)

The term $\exp(-p\,x/c)$ gives rise to a wave which propagates towards increasing value of x, and the term $\exp(p\,x/c)$ gives rise to a wave which propagates towards decreasing value of x: this interpretation can be checked on the result to be obtained. Thus the radiation condition implies to set $b(p)$ equal to 0. In turn,

$$(BC) \quad p\,U(0,p) - \overbrace{u(0,0)}^{=0,\ (IC)} = \mathcal{L}\{v_0(t)\}(p)$$

$$\Rightarrow \quad U(x,p) = \frac{1}{p}\,\exp\left(-\frac{p}{c}x\right)\mathcal{L}\{v_0(t)\}(p).$$

(15.3.6)

The inverse Laplace transform is a convolution integral,

$$u(x,t) = \int_0^t \mathcal{H}\left(t - \frac{x}{c} - \tau\right)v_0(\tau)\,d\tau\,,$$

(15.3.7)

which, for a shock $v_0(t) = V_0\,\mathcal{H}(t)$, simplifies to

$$u(x,t) = V_0\left(t - \frac{x}{c}\right)\mathcal{H}\left(t - \frac{x}{c}\right),\qquad \frac{\partial u}{\partial t}(x,t) = V_0\,\mathcal{H}\left(t - \frac{x}{c}\right).$$

(15.3.8)

The analysis of the velocity of the particles reveals two main characteristics of a partial differential equation of the hyperbolic type in nondissipative materials, Fig 15.3.1:

(H1) the mechanical information propagates at finite speed, namely c, which is therefore termed elastic wave-speed;

(H2) the wave front carrying the mechanical information propagates undistorted and with a constant amplitude.
Besides, if that is needed, this example contrasts the respective meanings of the elastic wave-speed c and of the velocity of the particles $\partial u/\partial t$.

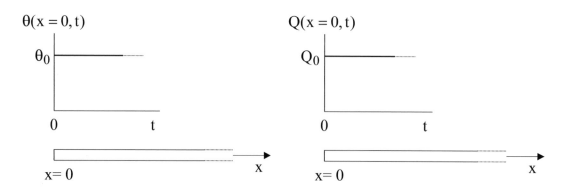

(a) prescribed temperature   (b) prescribed heat flux

Fig. 15.3.2: Heat diffusion in a semi-infinite (rigid) body submitted to (a) a heat shock, (b) a heat flux at the boundary $x = 0$.

15.3.2 A parabolic PDE: diffusion of heat

A semi-infinite elastic bar, or plane or half-space, is subjected to a given temperature $T = T_0(t)$ or to a heat flux $\mathbf{Q}_0(t)$ at its left boundary $x = 0$, Fig. 15.3.2. The thermal diffusion is one-dimensional. The initial temperature along the bar is uniform, $T(x,t) = T_\infty$. It is a bit more convenient to work with the field $\theta(x,t) = T(x,t) - T_\infty$ than with the temperature itself.

For a rigid material, in absence of heat source, the energy equation links the spatial variation of the heat flux $\mathbf{Q}$ [unit : $\mathrm{W/m^2 = kg/s^3}$] and the time rate of change of the sensible heat, namely

$$\text{energy equation} \quad \operatorname{div} \mathbf{Q} + C_\mathrm{T} \frac{\partial \theta}{\partial t} = 0, \tag{15.3.9}$$

where $C_\mathrm{T} > 0$ [unit : $\mathrm{J/m^3/K = kg/m/s^2/K}$] is the heat capacity per unit volume. The heat flux is related to the temperature gradient by Fourier's law,

$$\text{Fourier's law} \quad \mathbf{Q} = -k_\mathrm{T} \boldsymbol{\nabla} \theta, \tag{15.3.10}$$

where $k_\mathrm{T} > 0$ [unit : $\mathrm{W/m/K}$] is the thermal conductivity. Eqn (15.3.9) simply states that the energy leaving the volume V through its surface ∂V is compensated by a change of enthalpy inside the volume, namely $\int_{\partial V} \mathbf{Q} \cdot \mathbf{n} \, dS + \int_V C_\mathrm{T} \partial\theta/\partial t \, dV = 0$. See Section 9.1.3 for elaboration.

Inserting Fourier's law in the energy equation shows that the field equation can be phrased in terms of a single material coefficient D_T [unit : $\mathrm{m^2/s}$],

$$\text{thermal diffusivity} \quad D_\mathrm{T} = \frac{k_\mathrm{T}}{C_\mathrm{T}}. \tag{15.3.11}$$

15.3.2.1 Heat shock at the boundary of a semi-infinite body

The initial and boundary value problem (IBVP) describing a heat shock is thus governed by the following set of equations[15.1]:

$$\text{(FE) field equation} \quad D_\mathrm{T} \frac{\partial^2 \theta}{\partial x^2} - \frac{\partial \theta}{\partial t} = 0, \quad t > 0, \ x > 0;$$

$$\text{(IC) initial conditions} \quad \theta(x,0) = 0;$$

$$\text{(BC) boundary conditions} \quad \theta(0,t) = \theta_0(t);$$

$$\text{(RC) a radiation condition}. \tag{15.3.12}$$

The radiation condition is intended to imply that the thermal information diffuses in a single direction, namely towards increasing x, and that the bar is either of infinite length, or at least, that the signal does not get reflected back at its right boundary in the time window of interest.

The solution is obtained through the Laplace transform which, with the time variable t, associates the complex variable p,

$$\theta(x,t) \to \Theta(x,p) = \mathcal{L}\{\theta(x,t)\}(p). \tag{15.3.13}$$

Therefore,

$$\text{(FE)} \quad D_\mathrm{T} \frac{\partial^2 \Theta}{\partial x^2}(x,p) - \left(p\,\Theta(x,p) - \overbrace{\theta(x,0)}^{=0,\ (\text{IC})} \right) = 0,$$

$$\Rightarrow \quad \Theta(x,p) = a(p) \exp\left(-\sqrt{\frac{p}{D_\mathrm{T}}}\, x \right) + \underbrace{b(p) \exp\left(\sqrt{\frac{p}{D_\mathrm{T}}}\, x \right)}_{b(p)=0,\ (\text{RC})} \tag{15.3.14}$$

$$\text{(BC)} \quad \Theta(0,p) = \mathcal{L}\{\theta_0(t)\}(p)$$

$$\Rightarrow \quad \Theta(x,p) = \exp\left(-\sqrt{\frac{p}{D_\mathrm{T}}}\, x \right) \mathcal{L}\{\theta_0(t)\}(p).$$

[15.1] To derive (15.3.12), the thermal conductivity k_T has been assumed independent of space. If $k_\mathrm{T} = k_\mathrm{T}(\theta)$, it is instrumental to introduce the intermediate function $A(\theta)$ with differential $dA/d\theta = k_\mathrm{T}(\theta)/k_\mathrm{T}(0)$. Then $\boldsymbol{\nabla}(k_\mathrm{T}\boldsymbol{\nabla}\theta)$ becomes $k_\mathrm{T}(0)\,\boldsymbol{\nabla}(\boldsymbol{\nabla}A)$.

The multiform complex function $\sqrt{p}$ has been made uniform by introducing a branch cut along the negative axis Re $p \leq 0$, and by defining $p = |p| \exp(i\theta)$ with $\theta \in]-\pi, \pi]$, and $\sqrt{p} = \sqrt{|p|} \exp(i\theta/2)$ so that Re $\sqrt{p} \geq 0$. Then the term $\exp(\sqrt{p/D_T}\,x)$ gives rise to heat propagation towards decreasing x, in contradiction with the radiation condition, and this justifies why $b(p)$ has been set to 0 in $(15.3.14)_2$. Conversely, the term $\exp(-\sqrt{p/D_T}\,x)$ gives rise to heat propagation towards increasing x: this interpretation can be checked once the solution has been obtained.

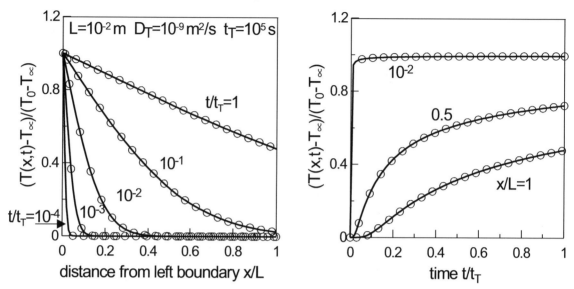

Fig. 15.3.3: Heat diffusion on a semi-infinite bar subjected to a heat shock on its left boundary $x = 0$: spatial profiles of the temperature at a few times (left); time profiles at a few positions (right). The characteristic time t_T that can be used to describe the information that reaches the position $x = L$ is equal to L^2/D_T.

The inverse transform is a convolution integral,

$$\theta(x,t) = \int_0^t \theta_0(t-u)\, \mathcal{L}^{-1}\Big\{ \exp\Big(-x\sqrt{\frac{p}{D_T}}\Big) \Big\}(u)\, du\,. \qquad (15.3.15)$$

Using the relation, Carslaw and Jaeger [1959], p. 62,

$$\mathcal{L}^{-1}\Big\{ \exp\Big(-x\sqrt{\frac{p}{D_T}}\Big) \Big\}(u) = \frac{x}{2\sqrt{\pi D_T u^3}}\, \exp\Big(-\frac{x^2}{4 D_T u}\Big)\,, \qquad (15.3.16)$$

then

$$\theta(x,t) = \frac{x}{2\sqrt{D_T \pi}} \int_0^t \theta_0(t-u)\, \frac{1}{u^{3/2}}\, \exp\Big(-\frac{x^2}{4 D_T u}\Big)\, du\,. \qquad (15.3.17)$$

With the further change of variable

$$u \to v\,, \quad \text{with} \quad v^2 = \frac{x^2}{4 D_T u}\,, \qquad (15.3.18)$$

the inverse Laplace transform takes the form,

$$\theta(x,t) = \frac{2}{\sqrt{\pi}} \int_{x/2\sqrt{D_T t}}^{\infty} \theta_0\Big(t - \frac{x^2}{4 D_T v^2}\Big)\, \exp(-v^2)\, dv\,, \qquad (15.3.19)$$

which, for a shock $\theta_0(t) = (T_0 - T_\infty)\, \mathcal{H}(t)$, simplifies to

$$T(x,t) - T_\infty = (T_0 - T_\infty)\, \text{erfc}\Big(\frac{x}{2\sqrt{D_T t}}\Big)\,, \qquad (15.3.20)$$

where erfc denotes the complementary error function. A few properties of the error and complementary error functions are listed in Section 15.3.2.3.

The plot of the spatial profiles of the temperature reveals two main characteristics of a partial differential equation of the parabolic type, Fig. 15.3.3:

(P1) the thermal information diffuses at infinite speed: indeed, the fact that the boundary $x = 0$ has been heated is known instantaneously at any point of the bar;

(P2) however, the amplitude of the thermal shock applied at the boundary requires time to fully develop in the bar. In fact, the temperature needs an infinite time to become spatially homogeneous.

The heat flux may be expressed in two forms to ease the interpretation at $x = 0$ and at other points $x > 0$,

$$Q(x,t) = -k_{\mathrm{T}} \frac{\partial T}{\partial x}(x,t) = \frac{k_{\mathrm{T}}\, \theta_0}{\sqrt{\pi\, D_{\mathrm{T}}\, t}}\, \exp(-X^2) = 2\, \frac{k_{\mathrm{T}}\, \theta_0}{x\, \sqrt{\pi}}\, X \exp(-X^2)\,, \tag{15.3.21}$$

where $X \equiv x/(2\sqrt{D_{\mathrm{T}}\, t})$ is a dimensionless variable. Thus the heat flux varies like the inverse square root of time at the boundary $x = 0$. At any other point, it increases to a maximum with magnitude inversely proportional to x corresponding to $X = 1/\sqrt{2}$ or $t = x^2/(2\, D_{\mathrm{T}})$ for which $X \exp(-X^2) \sim 0.429$ and it dies out at large times as indicated in Fig. 15.3.5.

15.3.2.2 Prescribed heat flux at the boundary of a semi-infinite body

The initial and boundary value problem (IBVP) describing the effect of a prescribed heat flux at the boundary $x = 0$ is still governed by the set of equations (15.3.12), except the boundary condition

$$\text{(BC) boundary condition} \quad Q(0,t) = Q_0\, \mathcal{H}(t)\,. \tag{15.3.22}$$

Due to Fourier's law, the gradient of temperature is thus prescribed at the boundary, $\partial\theta/\partial x(x = 0, t) = -Q_0\, \mathcal{H}(t)/k_{\mathrm{T}}$. The solution in the Laplace domain is still of the form (15.3.14) with $b(p) = 0$ but now the Laplace transform of the boundary condition yields,

$$\text{(BC)} \quad \mathcal{L}\{Q(0,t)\}(p) = k_{\mathrm{T}}\, a(p)\, \sqrt{\frac{p}{D_{\mathrm{T}}}} = \frac{Q_0}{p}$$

$$\Rightarrow \quad \Theta(x,p) = \frac{Q_0}{k_{\mathrm{T}}}\, \sqrt{D_{\mathrm{T}}}\, \frac{1}{p^{3/2}}\, \exp\!\left(-\sqrt{\frac{p}{D_{\mathrm{T}}}}\, x\right)\,. \tag{15.3.23}$$

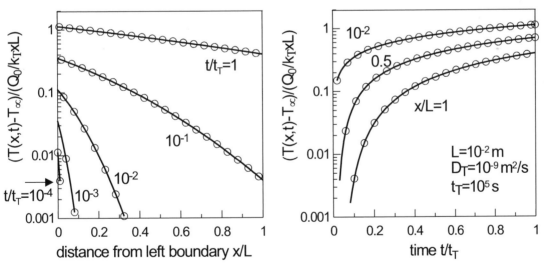

Fig. 15.3.4: Heat diffusion on a semi-infinite bar subjected to a given flux Q_0 on its left boundary $x = 0$. Spatial profiles of temperature at a few times (left) and time profiles at a few locations (right). At the boundary $x = 0$, the temperature increases like the square root of time.

From the inverse Laplace transform,

$$\mathcal{L}^{-1}\!\left\{\frac{1}{\sqrt{p}}\, \exp\!\left(-x\sqrt{\frac{p}{D_{\mathrm{T}}}}\right)\right\}(u) = \frac{1}{\sqrt{\pi\, u}}\, \exp\!\left(-\frac{x^2}{4 D_{\mathrm{T}}\, u}\right)\,, \tag{15.3.24}$$

provided in Table 15.3.1, the further inverse below follows by the convolution theorem,

$$
\mathcal{L}^{-1}\left\{\frac{1}{p}\frac{1}{\sqrt{p}}\exp\left(-x\sqrt{\frac{p}{D_{\mathrm{T}}}}\right)\right\}(t) \quad = \quad \int_0^t \mathcal{H}(t-u)\,\frac{1}{\sqrt{\pi\,u}}\exp\left(-\frac{x^2}{4D_{\mathrm{T}}\,u}\right)du
$$

$$
\overset{\text{Eqn }(15.3.18)}{=}\quad \frac{x}{\sqrt{\pi\,D_{\mathrm{T}}}}\int_{x/2\sqrt{D_{\mathrm{T}}t}}^{\infty}\frac{e^{-v^2}}{v^2}\,dv \tag{15.3.25}
$$

$$
\overset{\text{Eqn }(15.3.34)}{=}\quad 2\sqrt{t}\;\mathrm{ierfc}\left(\frac{x}{2\sqrt{D_{\mathrm{T}}t}}\right).
$$

The temperature field,

$$
\theta(x,t) = 2\,\frac{Q_0}{k_{\mathrm{T}}}\sqrt{D_{\mathrm{T}}\,t}\;\mathrm{ierfc}\left(\frac{x}{2\sqrt{D_{\mathrm{T}}t}}\right), \tag{15.3.26}
$$

is displayed in Fig. 15.3.4. It increases smoothly at the boundary like the square root of time,

$$
\theta(0,t) = 2\,\frac{Q_0}{k_{\mathrm{T}}}\frac{\sqrt{D_{\mathrm{T}}t}}{\sqrt{\pi}}. \tag{15.3.27}
$$

The heat flux profile due to the prescribed heat flux at the boundary $Q(0,t) = Q_0\,\mathcal{H}(t)$,

$$
Q(x,t) = -k_{\mathrm{T}}\frac{\partial\theta}{\partial x}(x,t) = Q_0\,\mathrm{erfc}\left(\frac{x}{2\sqrt{D_{\mathrm{T}}t}}\right), \tag{15.3.28}
$$

is identical to the temperature profile resulting from a heat shock $\theta(0,t) = \theta_0\,\mathcal{H}(t)$. Indeed it obeys exactly the same field equation deduced by derivation with respect to space of the equation of heat diffusion and it is prescribed at the boundary.

15.3.2.3 Useful functions and relations used in diffusion problems

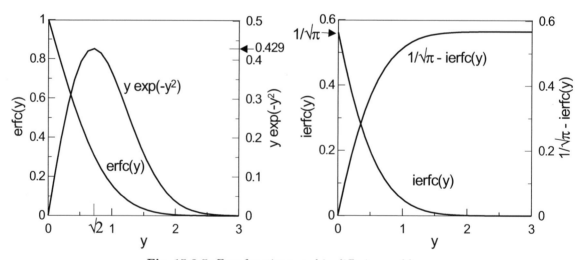

Fig. 15.3.5: Four functions used in diffusion problems.

Definition and basic properties of the error function

The error function erf and the complementary error function erfc,

$$
\mathrm{erf}(y) = \frac{2}{\sqrt{\pi}}\int_0^y e^{-v^2}\,dv\,, \quad \mathrm{erfc}(y) = \frac{2}{\sqrt{\pi}}\int_y^{\infty} e^{-v^2}\,dv\,, \tag{15.3.29}
$$

satisfy the relations,

$$
\mathrm{erf}(y) + \mathrm{erfc}(y) = 1\,, \quad \mathrm{erf}(-y) = -\mathrm{erf}(y)\,, \quad \mathrm{erfc}(-y) = 2 - \mathrm{erfc}(y)\,. \tag{15.3.30}
$$

Integrating term by term the uniformly convergent series expansion of the exponential yields a series expansion of the error function itself,

$$\text{erf}(y) = \frac{2}{\sqrt{\pi}} \sum_{k=0}^{\infty} \frac{(-1)^k}{k!} \frac{y^{2k+1}}{2k+1} = \frac{2}{\sqrt{\pi}} \left(y - \frac{y^3}{3} + \cdots \right). \tag{15.3.31}$$

For large y on the other hand, the complementary error function is approximated by the series expansion,

$$\text{erfc}(y) = \frac{e^{-y^2}}{\sqrt{\pi}} \left(\frac{1}{y} - \frac{1}{2} \frac{1}{y^3} + \cdots + (-1)^{n-1} \frac{1 \times 3 \cdots \times (2n-3)}{2^{n-1} y^{2n-1}} + \cdots \right). \tag{15.3.32}$$

Note also the derivatives,

$$\frac{d}{dy} \text{erfc}(y) = -\frac{2}{\sqrt{\pi}} e^{-y^2}, \quad \frac{d^2}{dy^2} \text{erfc}(y) = \frac{4}{\sqrt{\pi}} y e^{-y^2}, \tag{15.3.33}$$

and the relations,

$$\frac{e^{-v^2}}{v^2} = -\frac{d}{dv} \frac{e^{-v^2}}{v} - 2 e^{-v^2}, \quad \int_y^\infty \frac{e^{-v^2}}{v^2} dv = \frac{e^{-y^2}}{y} - \sqrt{\pi} \, \text{erfc}(y) = \frac{\sqrt{\pi}}{y} \text{ierfc}(y), \tag{15.3.34}$$

and

$$\text{erfc}(v) = \frac{2}{\sqrt{\pi}} v e^{-v^2} + \frac{d}{dv} \left(v \, \text{erfc}(v) \right), \tag{15.3.35}$$

established by integration by parts. Integrals of the complementary error functions are useful to be recorded,

$$\text{ierfc}(y) \equiv \int_y^\infty \text{erfc}(v) \, dv = \frac{e^{-y^2}}{\sqrt{\pi}} - y \, \text{erfc}(y),$$

$$\text{i}^2\text{erfc}(y) = \int_y^\infty \text{ierfc}(v) \, dv = \left(\frac{1}{4} + \frac{y^2}{2} \right) \text{erfc}(y) - \frac{y}{2\sqrt{\pi}} e^{-y^2}. \tag{15.3.36}$$

Note also the estimation for $\alpha > 0$,

$$\int_{-\infty}^\infty y^{2n} \exp(-\alpha y^2) \, dy = (-1)^n \frac{d^n}{d\alpha^n} \left(\int_{-\infty}^\infty \exp(-\alpha u^2) \, du \right) = (-1)^n \frac{d^n}{d\alpha^n} \left(\sqrt{\frac{\pi}{\alpha}} \right). \tag{15.3.37}$$

Definition and basic properties of the Gamma function

The Gamma function,

$$\Gamma(y) = \int_0^\infty u^{y-1} e^{-u} \, du, \quad y > 0, \tag{15.3.38}$$

satisfies the relation, proved by integration by parts,

$$\Gamma(y+1) = y \, \Gamma(y), \quad y > 0. \tag{15.3.39}$$

Therefore, for n integer,

$$\Gamma(n+1) = n!, \quad n \geq 0. \tag{15.3.40}$$

Moreover,

$$\Gamma(y) \, \Gamma(1-y) = \frac{\pi}{\sin(\pi y)}, \quad 0 < y < 1, \tag{15.3.41}$$

and therefore,

$$\Gamma(\frac{1}{2}) = \sqrt{\pi}, \quad \Gamma(\frac{3}{2}) = \frac{\sqrt{\pi}}{2} \quad \cdots. \tag{15.3.42}$$

Carslaw and Jaeger [1959], p. 482 et seq, provide additional properties of the error and complementary error functions as well as a list of Laplace transforms.

Table 15.3.1: A few inverse Laplace transforms involved in diffusion problems, $a > 0$.

Laplace transform $\mathcal{L}\{f(u)\}(p)$	Inverse $f(u)\,\mathcal{H}(u)$
$\dfrac{1}{p^a}$	$\dfrac{u^{a-1}}{\Gamma(a)}$
$\exp\left(-a\sqrt{p}\right)$	$\dfrac{a}{2\sqrt{\pi\,u^3}}\exp\left(-\dfrac{a^2}{4\,u}\right)$
$\dfrac{1}{\sqrt{p}}\exp\left(-a\sqrt{p}\right)$	$\dfrac{1}{\sqrt{\pi\,u}}\exp\left(-\dfrac{a^2}{4u}\right)$
$\dfrac{1}{p}\exp\left(-a\sqrt{p}\right)$	$\operatorname{erfc}\left(\dfrac{a}{2\sqrt{u}}\right)$
$\dfrac{1}{p^{3/2}}\exp\left(-a\sqrt{p}\right)$	$2\,\dfrac{\sqrt{u}}{\sqrt{\pi}}\exp\left(-\dfrac{a^2}{4u}\right)-a\operatorname{erfc}\left(\dfrac{a}{2\sqrt{u}}\right)$
$\dfrac{1}{p-b}\exp\left(-a\sqrt{p}\right)$	$\frac{1}{2}\exp(b\,u)\ \left(\exp\left(a\sqrt{b}\right)\operatorname{erfc}\left(\dfrac{a}{2\sqrt{u}}+\sqrt{b\,u}\right)\right.$ $\left.+\exp\left(-a\sqrt{b}\right)\operatorname{erfc}\left(\dfrac{a}{2\sqrt{u}}-\sqrt{b\,u}\right)\right)$

15.3.3 Diffusion of a solute in a medium of finite length

The diffusion of a chemical species dissolved in a fluid *at rest* obeys the same field equation as for thermal diffusion. Diffusion takes place so as to homogenize the concentration $c = c(x,t)$ of the solute in space. The diffusion phenomenon is governed

- by Fick's law (15.1.4) that relates the *diffusive* flux $\mathbf{J}^d = c\,(\mathbf{v_s} - v)$ to the gradient of concentration via the coefficient of molecular diffusion (or chemical diffusivity) D_C [unit : m^2/s],

$$\text{Fick's law}\quad \mathbf{J}^d = -D_C\,\boldsymbol{\nabla} c\,, \tag{15.3.43}$$

- by the mass balance (4.4.5) which may be phrased in terms of the *absolute* flux $\mathbf{J}$,

$$\text{balance of mass}\quad \frac{\partial c}{\partial t} + \operatorname{div}\mathbf{J} = 0\,. \tag{15.3.44}$$

Since the fluid is at rest $(v = 0)$, the *absolute* flux $\mathbf{J}$ is equal to the diffusive flux $\mathbf{J}^d$.

At $t > 0$, the concentration at the extremities of the medium $x = -L$ to $x = L$ is set to the constant value c_L. Because of the symmetry with respect to $x = 0$, we may restrict the analysis to the domain $x \in [0, L]$. The one-dimensional equations of diffusion analyzed in Section 15.3.2 modify to

$$\text{(FE) field equation}\quad D_C\frac{\partial^2 c}{\partial x^2} - \frac{\partial c}{\partial t} = 0\,,\quad t > 0,\ L > x > 0\,;$$

$$\text{(IC) initial conditions}\quad c(x,0) = c_i\,; \tag{15.3.45}$$

$$\text{(BC) boundary conditions}\quad c(L,t) = c_L\,\mathcal{H}(t);\quad \nabla c(0,t) = 0\,.$$

The solution $c(x,t)$ is obtained through the Laplace transform $c(x,t) \to C(x,p)$ of the field equation,

$$\text{(FE)}\quad \left(D_C\frac{d^2}{dx^2} - p\right)\left(C(x,p) - \frac{c_i}{p}\right) = 0\,. \tag{15.3.46}$$

The solution,

$$C(x,p) - \frac{c_i}{p} = a(p)\,\exp(q\,x) + b(p)\,\exp(-q\,x)\,, \tag{15.3.47}$$

involves two unknown functions $a(p)$ and $b(p)$ and $q = \sqrt{p/D_C}$. The boundary condition at $x = 0$ transformed in the Laplace domain yields $a(p) = b(p)$. The unknown function $a(p)$ results from the boundary condition at $x = L$:

$$C(x,p) = \mathcal{L}\{c(L,t)\}(p) = \frac{c_L}{p} = \frac{c_i}{p} + a(p)\left(\exp(q\,L) + \exp(-q\,L)\right)\,. \tag{15.3.48}$$

The resulting solution in the Laplace domain is finally,

$$C(x,p) = \frac{c_i}{p} + \frac{c_L - c_i}{p} \frac{\cosh(q\,x)}{\cosh(q\,L)}.$$ (15.3.49)

The inverse transform is furnished by Carslaw and Jaeger [1959], pp. 100 and 309, in two formats,

$$
\begin{aligned}
c(x,t) &= c_L - \frac{4\,(c_L - c_i)}{\pi} \sum_{n=0}^{\infty} \frac{(-1)^n}{2\,n+1} \cos\left((2\,n+1)\frac{\pi}{2}\frac{x}{L}\right) \exp\left(-D_{\mathrm{c}}\,(2\,n+1)^2\,\pi^2\,t/(4\,L^2)\right) \\
&= c_i + (c_L - c_i) \sum_{n=0}^{\infty} (-1)^n \operatorname{erfc}\left(\frac{(2\,n+1)\,L - x}{2\sqrt{D_{\mathrm{c}}\,t}}\right) + (-1)^n \operatorname{erfc}\left(\frac{(2\,n+1)\,L + x}{2\sqrt{D_{\mathrm{c}}\,t}}\right).
\end{aligned}
$$ (15.3.50)

The first series converges rapidly at large times (actually large $\sqrt{D_{\mathrm{c}}\,t}/L$) and the second expression converges rapidly at small times.

The average concentration over the medium which is obtained from the second series, Carslaw and Jaeger [1959], p. 310,

$$\frac{1}{L} \int_0^L c(x,t)\,dx = c_i + 2\,(c_L - c_i) \frac{\sqrt{D_{\mathrm{c}}\,t}}{L} \left(\frac{1}{\sqrt{\pi}} + 2 \sum_{n=1}^{\infty} (-1)^n \operatorname{ierfc} \frac{n\,L}{\sqrt{D_{\mathrm{c}}\,t}}\right),$$ (15.3.51)

behaves like $\sqrt{t}$ at short times.

15.3.4 A parabolic PDE displaying diffusion and convection

15.3.4.1 A linear equation including diffusion and convection of a chemical

The previous example has considered that the solute was diffusing in a fluid at rest. Usually however, the fluid itself moves due to different physical phenomena: for example, seepage of the fluid through a porous medium is triggered by a gradient of fluid pressure and governed by Darcy's law. For the time being, let us restrict our considerations to uncoupled flows and let us assume the velocity $\mathbf{v}$ of the fluid, referred to as *convective velocity*, to be a given constant. Most of the time, a solute is convected (or advected) by the flow, Fig. 15.3.6, and it also diffuses so as to homogenize the concentration, as a turtle colony would possibly enjoy to do, either by separating from one another or by getting closer.

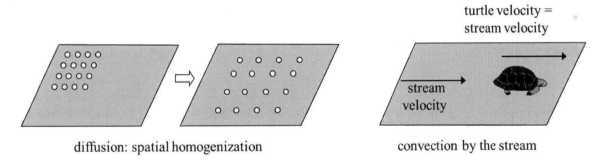

Fig. 15.3.6: Diffusion leads ultimately to a spatially uniform repartition of particles. In a stream, a lazy turtle is convected by the flow: its velocity is that of the flow.

Thus, we assume the fluid to move with velocity $\mathbf{v}$ and the solute with velocity $\mathbf{v}_{\mathrm{s}}$. In order to highlight that diffusion is relative to the fluid, two fluxes are introduced, a diffusive flux $\mathbf{J}^d$ and an absolute flux $\mathbf{J}$,

$$\underbrace{\mathbf{J} = c\,\mathbf{v}_{\mathrm{s}}}_{\text{absolute flux}}, \quad \underbrace{\mathbf{J}^d = c\,(\mathbf{v}_{\mathrm{s}} - \mathbf{v})}_{\text{diffusive flux}},$$ (15.3.52)

whence,

$$\mathbf{J} = \mathbf{J}^d + c\,\mathbf{v}.$$ (15.3.53)

The diffusion phenomenon is governed

- by Fick's law (15.1.4) that relates the *diffusive* flux $\mathbf{J}^d = c\,(\mathbf{v}_s - \mathbf{v})$ to the gradient of concentration via the coefficient of molecular diffusion (or chemical diffusivity) D_c [unit : m^2/s],

$$\text{Fick's law}\quad \mathbf{J}^d = -D_c\,\boldsymbol{\nabla}c\,, \tag{15.3.54}$$

- by the mass balance (4.4.5) which, in terms of concentration and *absolute* flux $\mathbf{J} = c\,\mathbf{v}_s$, writes,

$$\text{balance of mass}\quad \frac{\partial c}{\partial t} + \operatorname{div}\mathbf{J} = 0\,. \tag{15.3.55}$$

The one-dimensional equations of diffusion analyzed in Section 15.3.2 modify to

$$
\begin{array}{lll}
\text{(FE) field equation} & \overbrace{D_c\,\dfrac{\partial^2 c}{\partial x^2} - \dfrac{\partial c}{\partial t}}^{\substack{\text{diffusive}\\\text{terms}}} = \overbrace{v\,\dfrac{\partial c}{\partial x}}^{\substack{\text{convective}\\\text{term}}} & ,\quad t>0,\ x>0; \\[2ex]
\text{(IC) initial conditions} & c(x,0) = c_i\,; & \\[1ex]
\text{(BC) boundary conditions} & c(0,t) = c_0(t); & \\[1ex]
\text{(RC) a radiation condition}\,. &
\end{array}
\tag{15.3.56}
$$

A concentration discontinuity is imposed at $x=0$, namely $c(0,t) = c_0\,\mathcal{H}(t)$.

The solution $c(x,t)$ is obtained through the Laplace transform $c(x,t) \to C(x,p)$. First, the field equation becomes an ordinary differential equation wrt space where the Laplace variable p is viewed as a parameter,

$$\text{(FE)}\quad D_c\,\frac{d^2\tilde{C}}{dx^2} - v\,\frac{d\tilde{C}}{dx} - p\,\tilde{C} = 0,\quad \tilde{C}(x,p) \equiv C(x,p) - \frac{c_i}{p}\,. \tag{15.3.57}$$

The solution,

$$\tilde{C}(x,p) = a(p)\,\exp\left(\frac{v\,x}{2D_c} - \sqrt{\left(\frac{v}{2D_c}\right)^2 + \frac{p}{D_c}}\;x\right) + b(p)\,\exp\left(\frac{v\,x}{2D_c} + \sqrt{\left(\frac{v}{2D_c}\right)^2 + \frac{p}{D_c}}\;x\right), \tag{15.3.58}$$

involves two unknown functions $a(p)$ and $b(p)$. The function $b(p)$ is set to 0, because it multiplies a function that would give rise to diffusion from right to left: the radiation condition (RC) intends to prevent this phenomenon. The second unknown $a(p)$ results from the (BC):

$$\text{(RC)}\quad b(p) = 0, \quad \text{(BC)}\quad a(p) = \frac{c_0 - c_i}{p}\,. \tag{15.3.59}$$

The resulting complete solution in the Laplace domain is finally,

$$C(x,p) = \frac{c_i}{p} + (c_0 - c_i)\,\exp\left(\frac{v\,x}{2D_c}\right) F(p)\,, \tag{15.3.60}$$

where

$$F(p) = \frac{1}{p}\,\exp\left(-x\,\sqrt{\left(\frac{v}{2D_c}\right)^2 + \frac{p}{D_c}}\right) = \frac{\exp(-x\,\sqrt{P/D_c})}{P - \alpha}\,, \tag{15.3.61}$$

with $P \equiv p+\alpha$ and $\alpha = v^2/(4\,D_c)$. The inverse transform of the second term displayed in Table 15.3.1 is established in Exercise 15.2,

$$F(p) = \frac{\exp(-x\,\sqrt{P/D_c})}{P - \alpha} = \mathcal{L}\big\{\exp(\alpha\,t)\,f(t)\big\}(P) = \mathcal{L}\big\{f(t)\big\}(P - \alpha = p)\,, \tag{15.3.62}$$

whence

$$\mathcal{L}^{-1}\big\{F(p)\big\}(t) = f(t)\,, \tag{15.3.63}$$

where

$$f(t) = \frac{1}{2} \left(\exp\left(-x\sqrt{\frac{\alpha}{D_\mathrm{C}}} \right) \mathrm{erfc}\left(\frac{x}{2\sqrt{D_\mathrm{C}t}} - \sqrt{\alpha\,t} \right) + \exp\left(x\sqrt{\frac{\alpha}{D_\mathrm{C}}} \right) \mathrm{erfc}\left(\frac{x}{2\sqrt{D_\mathrm{C}t}} + \sqrt{\alpha\,t} \right) \right). \tag{15.3.64}$$

The solution can be cast in the format, which holds for positive or negative velocity v,

$$c(x,t) = c_i + \frac{1}{2}(c_0 - c_i) \left(\mathrm{erfc}\left(\frac{x - v\,t}{2\sqrt{D_\mathrm{C}\,t}} \right) + \exp\left(\frac{v\,x}{D_\mathrm{C}} \right) \mathrm{erfc}\left(\frac{x + v\,t}{2\sqrt{D_\mathrm{C}\,t}} \right) \right). \tag{15.3.65}$$

Once the concentration c is known, the solute velocity follows by (15.3.54),

$$\mathbf{v}_\mathrm{s} = \mathbf{v} - D_\mathrm{C}\,\frac{\boldsymbol{\nabla}c}{c}\,. \tag{15.3.66}$$

Note that any arbitrary function $f = f(x - v\,t)$ leaves unchanged the first order part of the field equation (15.3.56). The solution can thus be seen as displaying a front propagating towards $x = \infty$, which is smoothed out by the diffusion phenomenon. The dimensionless Péclet number Pe, named after the French physicist Jean Claude Eugène Péclet (1793-1857), quantifies the relative weight of convection and diffusion, e.g. Bear [1972],

$$\text{Péclet number} \quad \mathrm{P_e} = \frac{L\,v}{D_\mathrm{C}} \begin{cases} \ll 1 & \text{diffusion dominated flow} \\ \gg 1 & \text{convection dominated flow} \end{cases} \tag{15.3.67}$$

Its actual influence can be appreciated from the plots of the solution displayed in Fig. 15.3.7. The length L is a characteristic length of the problem, e.g., mean grain size in granular media, or length of the column for breakthrough tests in a column of finite length.

For transport of species, the Péclet number is related to the ratio of the mass of particles transported by convection and of the mass of particles transported by diffusion. For heat transport, the Péclet number gives an indication of the ratio of the heat transported by convection and of the heat transported by conduction.

Note 15.1 on numerical issues associated with convection.

Convection causes much trouble in actual numerical solutions obtained with finite difference or finite element methods. The streamline upwind Petrov-Galerkin method (SUPG), distinguishing shape functions and weight functions, gives rise to non symmetric elementary matrices, e.g., Johan and Hughes [1991]. This unpleasant lack of symmetry seems to be the price to pay to suppress wiggles that would otherwise appear when the wave front hits a boundary where the concentration of the chemical or its flux are prescribed, see, e.g., Gajo and Loret [2003] for simulations of the injection of a neutral solute in a deformable fluid-saturated solid.

15.3.4.2 A linear equation with heat convection

The previous example considered diffusion and convection of a chemical. A similar approach applies in the analysis of the diffusion and convection of heat. The diffusive heat flux is now recast in the form,

$$\mathbf{Q} - C_\mathrm{T}\,\theta\,\mathbf{v} = -k_\mathrm{T}\,\boldsymbol{\nabla}\theta\,. \tag{15.3.68}$$

Upon introduction of the heat flux in the energy equation (15.3.9), the equation governing diffusion and convection of heat deduces in the format,

$$\frac{\partial\theta}{\partial t} + \mathbf{v}\cdot\boldsymbol{\nabla}\theta = D_\mathrm{T}\,\mathrm{div}\,\boldsymbol{\nabla}\theta\,, \tag{15.3.69}$$

with $D_\mathrm{T} = k_\mathrm{T}/C_\mathrm{T}$ [unit : m^2/s] thermal diffusivity.

15.3.4.3 A nonlinear convection-diffusion equation: Burgers equation

The Burgers equation has been studied in several fields of mathematical physics, starting in fluid turbulence. It is usually expressed in the following format for the unknown $u = u(x,t)$,

$$D\,\frac{\partial^2 u}{\partial x^2} = \frac{\partial u}{\partial t} + u\,\frac{\partial u}{\partial x}\,. \tag{15.3.70}$$

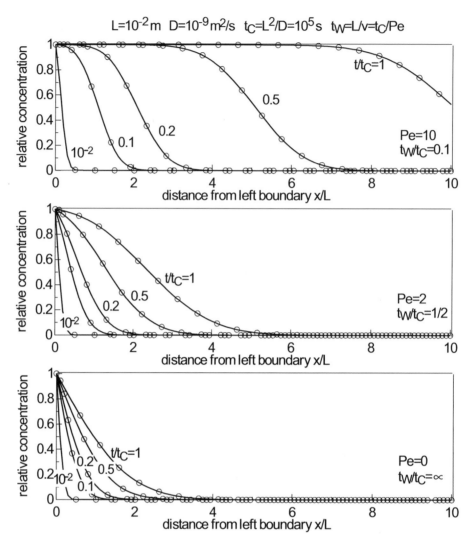

Fig. 15.3.7: Convection-diffusion along a semi-infinite axis of a species whose concentration is subjected at time $t = 0$ to a sudden increase at the left boundary $x = 0$. The fluid is animated with a velocity v such that the Péclet number is equal to 10, 2 and 0 (pure diffusion), respectively. Spatial profiles of the relative concentration $(c(x, t) - c_i)/(c_0 - c_i)$ at different times. Focus is on the events that take place at point $x = L$. The time t_w which characterizes the propagation phenomenon is equal to L/v while the characteristic time of chemical diffusion t_C is equal to L^2/D_C with D_C the diffusion coefficient. Therefore $t_w = t_C/P_e$.

When $D = 0$, the equation can be interpreted as a.conservation equation and its solution displays a shock. Therefore, the higher order term that implies the equation to be a parabolic PDE is seen as smoothing the shock.

Consider the nonlinear equation,

$$D \frac{\partial^2 \phi}{\partial x^2} + \left(\frac{\partial \phi}{\partial x}\right)^2 = \frac{\partial \phi}{\partial t} . \tag{15.3.71}$$

With $\phi = D \ln u$, it transforms to a heat equation $D \, \partial^2 u/\partial x^2 = \partial u/\partial t$ if the diffusion coefficient D is constant. As for the equation,

$$D \frac{\partial^2 \phi}{\partial x^2} - \left(\frac{\partial \phi}{\partial x}\right)^2 = \frac{\partial \phi}{\partial t} , \tag{15.3.72}$$

it transforms, upon derivation with respect to the space variable, to a Burgers equation $D' \, \partial^2 u/\partial x^2 = \partial u/\partial t' + u \, \partial u/\partial x$ for the function $u = \partial \phi/\partial x$, noting $D' = D/2$ and $t' = 2 \, t$.

15.3.5 Hyperbolic heat equation

In hyperbolic PDEs of the wave propagation type, time and space are involved in linear expressions of the form $x \pm c\,t$, where c is the wave-speed. By contrast, in parabolic PDEs of the heat diffusion type, the space variable is associated with the square root of the time variable and the key space and time dependence takes place via the term $x/\sqrt{D\,t}$ where D is a diffusivity. Therefore diffusion over a distance $2L$ requires a time interval four times larger than diffusion over a length L.

A modification of Fourier's heat law tailored to deliver propagation of heat at finite speed has been exposed in Section 9.1.4 and the hyperbolic character of the modified energy equation is checked in Section 11.3.2. The scheme is applied to the bioheat equation in Section 10.4.

15.3.6 Molecular diffusion

The derivation of the theory of molecular diffusion, actually diffusion of any 'small' particle, might be worth recording because it provides a means to obtain a value of the diffusion coefficient if the motions of the particles are known over a finite time window. Let $c(x, t)$ be the concentration of particles at (x, t). The motion of the particles obeys a random walk: the probability for a particle located at position $x - y$ at time t to be at x at time $t + \Delta t$ is $\phi(y)\,dy$, the probability function ϕ being even, $\phi(-y) = \phi(y)$ and normalized such that $\int_{-\infty}^{+\infty} \phi(y)\,dy = 1$. Therefore summing over all particles provides $c(x, t + \Delta t)$,

$$c(x, t + \Delta t) = \int_{-\infty}^{+\infty} c(x - y, t)\,\phi(y)\,dy\,. \tag{15.3.73}$$

Linearization of the left-hand side to first order and of the right-hand side to second order, Einstein [1956],

$$c(x, t) + \Delta t\,\frac{\partial c}{\partial t}(x, t) = \int_{-\infty}^{+\infty} \left(c(x, t) - y\,\frac{\partial c}{\partial x}(x, t) + \frac{y^2}{2}\,\frac{\partial^2 c}{\partial x^2}(x, t) \right) \phi(y)\,dy\,, \tag{15.3.74}$$

yields the diffusion equation $\partial c / \partial t = D\,\partial^2 c / \partial x^2$ and the diffusion coefficient may be calculated by estimating the integral,

$$D = \frac{1}{2\,\Delta t} \int_{-\infty}^{+\infty} y^2\,\phi(y)\,dy\,, \tag{15.3.75}$$

of the squares of the fluctuations y^2 between t and $t + \Delta t$ of the particles which are located at x at time t. The expression also applies to the space of dimension $d = 3$ by dividing the right-hand side of (15.3.75) by d.

As an alternative derivation, the solution $\psi(x, t) = 1/(2\sqrt{\pi D\,t})\,\exp(-(x - x_0)^2/(4D\,t))$ that satisfies the diffusion equation $\partial \psi / \partial t = D\,\partial^2 \psi / \partial x^2$ and the initial condition $\psi(x, 0) = \delta(x - x_0)$ is obtained via a Fourier transform. Then, the moment of order 2 defined as $\int_{-\infty}^{\infty} y^2\,\psi(y, t)\,dy$ is found equal to $2D\,t + x_0^2$ using the relation (15.3.37).

15.4 Physico-chemical processes involving diffusion, convection and reaction

A species may diffuse through a porous medium, or be convected by a fluid, and, at the same time, be captured or released along its path by external agents. As an example worked out in Section 15.4.2, oxygen diffuses through the tissue and is partly consumed by cells. More generally, a balance equation is said to be reactive when the source term depends on the variable, e.g., the oxygen concentration in the above case. The bioheat equation which indicates that the heat transferred to the tissue by the blood flow depends on the relative temperature of the tissue and of the blood is another conspicuous example. Chapter 10 provides instances of mass and heat transfers in fluid-saturated porous media that give rise to equations that include diffusion, convection and reaction.

The reactions that are involved may be of physico-chemical nature, e.g., exchange of fluid, but they may also be chemical reactions catalyzed by enzymes, as elaborated in Section 15.12.

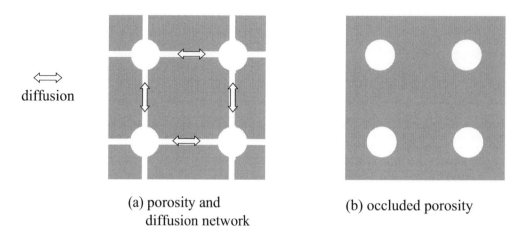

Fig. 15.4.1: A porous network may be endowed with a diffusion network to which corresponds a non zero permeability, or the porosity may be totally occluded and therefore its associated permeability vanishes.

15.4.1   Reaction-diffusion in a porous medium context

In the simplest context of a deformable fluid-saturated porous medium, two phenomena take place, namely deformation, which falls in the realm of mechanics, and seepage, a transport process which allows water to move from one pore to the other, Fig. 15.4.1. Additional phenomena may become important for multi-component mixtures. For example, in chapter 14 devoted to articular cartilages, arguments sketched in Fig. 14.6.1 are presented to justify a physical partition of water in two compartments or phases. The collagen fibers bath in the extrafibrillar water and, themselves, they contain intrafibrillar water. In the model by Loret and Simões [2004], extrafibrillar water diffuses through the solid skeleton made by collagen fibers and other proteins and communicates with the surroundings. Its seepage properties are characterized by a hydraulic conductivity. On the other hand, intrafibrillar water acts as an occluded (dead-end) porosity: it does not flow through the porous medium and its only way in and out is by transfer from and to the extrafibrillar compartment. This situation is illustrated by Fig. 15.4.2-(b).

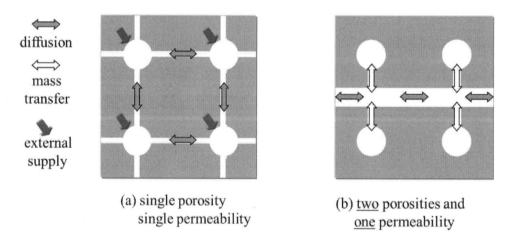

Fig. 15.4.2: Two models that display reactive diffusion (a) by external supply and (b) by internal mass transfer with a secondary porosity, which in this case is occluded. The double porosity model sketched in Fig. 15.4.3 differs from model (b) because the two porosities are connected, and thus each of them has its own permeability.

More general mixture theories that go by the name of double porosity theories assume the two fluid compartments to be both endowed with their own seepage networks and permeabilities, Fig. 15.4.3. One of the two porosities (vessels in biological tissues, fractures in rocks) is endowed with a small volume fraction and it serves mainly to transport the fluid. On the other hand, the other porosity (the pores) with a larger volume fraction serves mainly to store the fluid although it has its own, smaller, seepage properties.

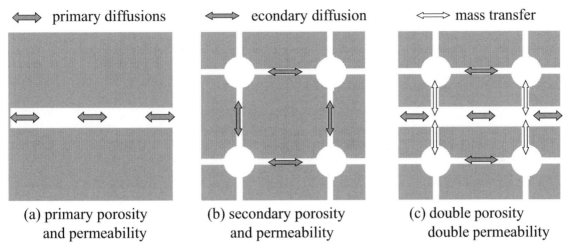

Fig. 15.4.3: In the double porosity model, the two porosities are both endowed with their own seepage network. In addition, fluid exchanges between the two porosities.

The balance of mass of the fluid involves then both a seepage contribution and a transfer, or reactive, contribution as indicated by (4.4.9). Seepage is governed by Darcy's law with a (scaled) hydraulic permeability $\hat{k}$. Mass transfer is driven by the pressure difference $p_i - p_c$ between the tissue and the capillaries with a (scaled) proportionality coefficient $\hat{\chi}$. Typical times characterizing the kinetics of the diffusion and transfer processes can be introduced, namely $t_\mathrm{H} = L^2/\hat{k}$ for seepage, with L a relevant length, and $t_\mathrm{tr} = 1/\hat{\chi}$ for mass transfer. Given that the two phenomena take place in parallel, the actual time response is essentially the inverse of the sum $1/t_\mathrm{H} + 1/t_\mathrm{tr}$ and it is therefore controlled by the fastest phenomenon.

The porous specimen of half-width L analyzed in Section 24.8.2 is drained at its both ends $x = \pm L$. The capillary pressure is considered constant and it is used as a reference ($p_c = 0$). The mass balance of water can be phrased in terms of the tissue pressure p_i in the format,

$$\frac{L^2}{t_\mathrm{H}} \frac{\partial^2 p_i}{\partial x^2} - \frac{\partial p_i}{\partial t} - \frac{p_i}{t_\mathrm{tr}} = 0, \quad x \in]0, L[, \ t > 0. \tag{15.4.1}$$

Upon introduction of the intermediate function $q(x,t) = \exp(t/t_\mathrm{tr}) \, p_i(x,t)$, the standard format of a diffusion equation is recovered,

$$\frac{L^2}{t_\mathrm{H}} \frac{\partial^2 q}{\partial x^2} - \frac{\partial q}{\partial t} = 0, \ x \in]0, L[, \ t > 0. \tag{15.4.2}$$

15.4.2 Diffusion of oxygen in the tissue

Soluble and nonsoluble species

Oxygen is transported in the blood in dissolved and mainly bound forms. Oxygen is lipid-soluble so that it can be considered to penetrate freely the capillary wall and to enter the tissue in dissolved form. The oxygen flux at the blood/tissue interface is thus continuous.

Oxygen extravasation through the capillary wall

Oxygen extravasation and diffusion through the tissue may be considered in a section orthogonal to a set of parallel vessels. Accounting for spatial periodicity, it is appropriate to center a hexagonal domain around a vessel of radius a although a circular domain that results in radial diffusion may be adopted as an approximation. Note that hexagonal domains fill the section. Three boundary conditions for the concentration are needed, typically with c_b and c_t the oxygen concentrations in blood and tissue, respectively,

1. zero gradient $\partial c_b/\partial r(r = 0) = 0$ at the center of the capillary;
2. zero normal gradient: $\boldsymbol{\nabla} c_t \cdot \mathbf{n} = 0$ at the outer boundary of the circular (or hexagonal) domain of unit outer normal $\mathbf{n}$;

3. continuous flux of oxygen, $D_b \, \partial c_b/\partial r(r = a) = D_t \, \partial c_t/\partial r(r = a)$ at the interface tissue-blood, i.e., across the capillary wall that oxygen crosses freely.

The continuous flux at the interface is dictated by the fact that oxygen is lipid-soluble. On the other hand, some substances like water and albumin are lipid-insoluble. As explained in Sections 15.8.3 and 15.9.3, Starling's law should be used for these substances.

Reaction-diffusion of oxygen in the tissue

Let $\mathbf{J}^d = c \, (\mathbf{v} - \mathbf{v}_a)$ be the *diffusive* molar flux of oxygen at concentration c, moving with the relative velocity $\mathbf{v} - \mathbf{v}_a$ with respect to a solvent endowed itself with the velocity $\mathbf{v}_a$. The concentration $c = N/V_a$ is defined as the number of moles of oxygen over the volume of the solvent. The absolute flux $\mathbf{J} \equiv c \, \mathbf{v} = \mathbf{J}^d + c \, \mathbf{v}_a$ is thus contributed by a diffusive part and a convective part. The diffusive molar flux $\mathbf{J}^d = c \, (\mathbf{v} - \mathbf{v}_a)$ obeys Fick's law, Eqn (15.1.4), namely $\mathbf{J}^d = -\mathbf{D} \cdot \boldsymbol{\nabla} c$, where the diffusion tensor $\mathbf{D}$ may be assumed to be isotropic, $\mathbf{D} = D \, \mathbf{I}$. The balance of mass of oxygen which may be produced/deleted by a chemical reaction at a rate $\hat{c}$ may be brought in the format,

$$\frac{\partial(n^a \, c)}{\partial t} + \text{div}\,(n^a \, c \, \mathbf{v}) = \hat{c}. \tag{15.4.3}$$

Here $n^a = V_a/V$ is the volume fraction of the solvent with respect to the tissue and $\hat{c} = \hat{N}/V$ measures the number of moles produced/deleted per unit volume of tissue as well. This equation derives from the balance of mass of a species (4.4.5) with apparent mass density equal to $n^a \, c \, \widehat{m}$, $\widehat{m}$ being its molar mass.

Upon insertion of Fick's law $\mathbf{J}^d = -D \, \boldsymbol{\nabla} c$ in (15.4.3), the concentration of oxygen is found to obey an equation involving diffusion, convection and reaction,

$$\frac{\partial(n^a \, c)}{\partial t} + \overbrace{\text{div}\,(n^a \, c \, \mathbf{v}_a)}^{\text{convection}} = \overbrace{\text{div}(n^a \, D \nabla c)}^{\text{diffusion}} + \overbrace{\hat{c}}^{\text{reaction}}. \tag{15.4.4}$$

Let $n_{\text{cell}} = V_{\text{cell}}/V$ be the volume fraction occupied by the cells, $c_{\text{cell}} = N_{\text{cell}}/V$ the cell density [unit : mole/m^3] and $\widehat{v}_{\text{cell}}$ the molar volume of cells. The oxygen release rate in the tissue is actually the oxygen consumption rate by cells. It is often assumed to adopt the Michaelis-Menten format,

$$\hat{c} = -n_{\text{cell}} \, \frac{\lambda \, c}{c_h + c} = -c_{\text{cell}} \, \frac{\lambda' \, c}{c_h + c}, \quad \text{with } \lambda' = \widehat{v}_{\text{cell}} \, \lambda, \tag{15.4.5}$$

with λ' maximal consumption rate and c_h concentration of oxygen at half maximal consumption rate.

Oxygen consumption in normal mammalian cells is about 0.5×10^{-12} mole O$_2$/cell/hr or 0.14×10^{-15} mole/cell/s, Cartwright [1994]. Oxygen consumption per unit volume is thus equal to this value times the cell density.

For cartilage cultures, where the typical oxygen concentration is 0.1 mmole/l=0.1 mole/m^3, Obradovic et al. [2000] consider $c_h = 6 \times 10^{-3}$ mole/m^3 and the maximum oxygen consumption rate λ' equal to 1.86×10^{-18} mole O$_2$/cell/s which slightly exceeds the range $0.6 - 1.6 \times 10^{-18}$ mole O$_2$/cell/s reported for chondrocytes of bovine articular cartilages. The chondrocyte density in articular cartilages centers around 50×10^6 cells per ml of medium.

Table 15.4.1: Parameters of the Michaelis-Menten relation governing the nutrient consumption rate defined by (15.4.5) with c concentration of nutrient(=oxygen) in mole per volume of fluid.

Tissue	Reference	c (mole/l)	n_{cell} -	λ' (mole/cell/s)	λ (mole/l/s)	c_h (mole/l)
Cartilage	Obradovic et al. [2000]	10^{-4}	-	1.86×10^{-18}	-	6×10^{-6}
Tumor tissue	Roose et al. [2003]	2.8×10^{-4}	0.6	-	10^{-5}	4.64×10^{-6}

For tumor tissues, based on data by Casciari et al. [1992], Roose et al. [2003] consider a steady state concentration of oxygen in tissue equal to 0.28 mole/m^3=0.28 mmole/l, $\lambda = 10^{-2}$ mole/m^3/s the maximum consumption rate, $c_h = 4.64 \times 10^{-3}$ mole/m^3 the concentration corresponding to half of the maximum consumption rate. These data are collected in Table 15.4.1 while Table 15.5.1 shows the diffusion coefficients of some macromolecules and gases.

15.4.3 The reaction-diffusion equations of Turing [1952]

The reaction-diffusion equations of Turing [1952] aim at accounting for the fact that growth factors or other chemicals involved in wound healing, growth of tumors and engineered tissues have to diffuse across the existing tissue before acting through a chemical reaction. For a vector of growth factors $\mathbf{X}$, the equations write

$$\frac{\partial \mathbf{X}}{\partial t} = \mathbf{R}(\mathbf{X}) + \mathbf{D}\,\Delta\mathbf{X}\,, \tag{15.4.6}$$

where $\mathbf{R}$ is the reaction term and $\mathbf{D}$ a diffusion matrix. At equilibrium, $\mathbf{R}(\mathbf{X}^{\mathrm{eq}}) = \mathbf{0}$ and, close to equilibrium, $\mathbf{R}(\mathbf{X}) = \mathbf{M}\,(\mathbf{X} - \mathbf{X}^{\mathrm{eq}})$ where $\mathbf{M}$ is a reaction matrix.

More generally, the reaction terms are not linear in $\mathbf{X}$. One example is the Schnakenberg reaction-diffusion equations used to model the growth of chicken limb in the Compucell model (last visit May 29, 2015). Other models adopt the Michaelis-Menten kinetics.

The dynamics of the set of nonlinear equations coined "Belousov-Zhabotinski equations" has been made famous by Prigogine and co-workers in the 1960s–1970s. They showed that diffusion can trigger instability: this effect is at variance with the smoothing effect of diffusion in the Burgers equation. Whether or not the Belousov-Zhabotinski equations can be adopted to described the growth of tissues is a debatable issue, Cowin [2000].

In modern organogenesis, numerical simulations using various techniques (finite differences, cellular automata) proceed as follows. Cells are distributed at certain nodes of a grid (lattice). The unknown vector usually contains an activator (e.g. TGF-α), an inhibitor, and the key protein (PGs, collagen, fibronectin, etc.)[15.2][15.3]. Of course the reaction takes place at the nodes of the grid(lattice) where cells are located. Cells produce the activator itself (possibly the inhibitor) and the key protein. Depending on models, the protein diffuses slightly from the cell and is deposited in the ECM, or does not diffuse at all. The exact chemical reactions are not described. The mass conservation is formally expressed (essentially) in the form of a reaction equation. It is assumed that the ingredients necessary for the production are available locally. In fact they come from the medium, which is not included in the domain to be analyzed. When diffusion is accounted for in the equation of the key protein, it aims at moving the product in a neighborhood of the producing cell. Diffusion of the ingredients towards the producing cell is not considered.

At equilibrium, the sole term to survive in (15.4.6) is the term involving the spatial derivative. Solutions of the reaction-diffusion equation are obtained by addressing sequentially the space dependence and the time dependence. Spatial patterns are built from the solutions $(k, \mathbf{E}_k(\mathbf{x}))$ of the eigenvalue problem,

$$k^2\,\mathbf{E}_k + \Delta\mathbf{E}_k = \mathbf{0}\,, \tag{15.4.7}$$

that meet the boundary conditions. Here the k's are wave-numbers [unit : m^{-1}].

In fact, the idea is the following. First, linearize the equations of reaction-diffusion about an equilibrium state. Second, view the departure from the equilibrium state $\mathbf{Y}(\mathbf{x},t) = \mathbf{X}(\mathbf{x},t) - \mathbf{X}^{\mathrm{eq}}$ as the solution of the linearized equations, to be sought as a combination of these eigenmodes, say

$$\mathbf{Y}(\mathbf{x},t) = \sum_k \exp(\lambda\,t)\,c_k\,\mathbf{E}_k(\mathbf{x})\,, \tag{15.4.8}$$

with the coefficients of the expansion c_k defined at $t = 0$. Since the eigenmodes are linearly independent, the coefficient λ of time dependence is obtained by requiring the eigenvalue problem,

$$\left(\lambda\,\mathbf{I} - \mathbf{M} + k^2\mathbf{D}\right)\mathbf{E}_k(\mathbf{x}) = \mathbf{0}\,, \tag{15.4.9}$$

to be satisfied, i.e., $\det\left(\lambda\,\mathbf{I} + \mathbf{M} - k^2\mathbf{D}\right) = 0$. The resulting *dispersion relation* $\lambda = \lambda(k)$ between the time dependence coefficient λ and the wave-number k yields information on the equilibrium state: the latter is said to be stable, in absence of spatial disturbance, if $\mathrm{Re}\,\lambda(k=0) < 0$ and unstable, with respect to spatial disturbances, if there exists at least one k such that $\mathrm{Re}\,\lambda(k) > 0$. If there are more than one such wave-numbers, the mode that will grow fastest is the one corresponding to the largest λ. These unstable modes are interpreted as giving rise to those spatially non uniform patterns which are actually observed in animal coat patterns, Murray [2003], p. 141.

[15.2]Some models also consider the dynamics of different types of cells, apoptosis, cell differentiation, haptotaxis i.e., directed motion of cells along a gradient of adhesion, etc.

[15.3]In a model of cartilage growth, the unknowns are an activator (oxygen or glucose) and the product (PGs).

15.5 Diffusion versus convection

Small molecules of diameter a few Å, whether apolar or polar uncharged like H_2O, O_2, N_2, urea, etc. *diffuse* in the interstitium according to *Fick's law*. Diffusion distances are proportional to the product of the square root of the diffusion time by the diffusion coefficient, say $\delta x \simeq \sqrt{D \, \delta t}$. A molecule with $D = 10^{-5} \, cm^2/s$ would travel the small distance $\delta x = 1 \, \mu m$ very fast in $\delta t = 1 \, ms$ but, for the longer distance $\delta x = 1 \, cm$, it needs an exceedingly long time $\delta t = 28 \, hr$. In practice, a species has a spatial **diffusion limit**, e.g., 100 to 200 μm for oxygen as shown in Fig. 24.3.6. The issue of the diffusion length in a poroelastic tissue is analyzed in Section 24.8.3.

An important notion for the diffusion through porous media of both charged and uncharged species is that of **tortuosity**. The path that a particle has to travel in a porous medium is longer than in an infinitely dilute solution due to the complex geometry of the pores and of the pore network. The **effective** diffusion coefficient $D* = \tau D$ is scaled down with respect to its value D in a blank solution by a tortuosity factor $\tau < 1$. In the chemistry jargon, an infinitely dilute solution is often referred to as a *blank* solution.

In practice, several phenomena, which are difficult to quantify, are included in a multiplicative way in the tortuosity factor. They include electrical repulsion, binding, conformation or viscous hindrance. Electrical repulsion of negatively charged globular albumin molecules by negatively charged tissue proteins is advocated to explain that its effective diffusion coefficient is smaller than that of the linear dextran molecules of similar molar mass. In a demonstrative approach, its tortuosity factor τ is defined as the product $\tau_1 \tau_2$ where τ_1 accounts for the tissue stereology and τ_2 for the electrical repulsion between tissue and albumin. This repulsion is said to exclude albumin from a part of the tissue volume surrounding the fixed charges (*volume exclusion* phenomenon).

Cells represent obstacles to a number of diffusing species, thus increasing their effective path. The interstitium is considered to oppose diffusion through frictional forces, e.g., Pluen et al. [2001]. The differences of the diffusion coefficients exhibited by dextran and albumin in a blank solution, in normal tissues and in tumor tissues, Nugent and Jain [1984] and Jain [1990], his Fig. 4, serve to illustrate the influence of the medium.

Next to its electrical properties, the diffusion coefficient of a species depends on its concentration. Mostly, it decreases with its molar mass $\widehat{m}$, being inversely proportional to some power of $\widehat{m}$, Jain [1987]a [1987]b,

$$D = D_1 \left(\frac{\widehat{m}_1}{\widehat{m}}\right)^{\alpha}, \quad \alpha = 0.5 \text{ in free solution, } 0.6 - 3 \text{ in tissues}, \tag{15.5.1}$$

where $\widehat{m}_1 = 1 \, g/mole$ is a scaling constant and D_1 has dimension [unit : m^2/s]. Swabb et al. [1974] suggest $D_1 = 1.8 \times 10^{-8} \, m^2/s$ and $\alpha = 3/4$ over a large range of molar masses, namely $32 \le \widehat{m}/\widehat{m}_1 \le 69,000$.

Another relation referred to as Stokes-Einstein relation relates the diffusion coefficient D in a blank solution to the radius a of the solute and to the dynamic viscosity η [unit : Pa×s] of the fluid,

$$D = \frac{k_B \, T}{6 \, \pi \, \eta \, a}, \tag{15.5.2}$$

with T absolute temperature, $k_B = R/N_A$ Boltzmann constant [unit : J/K], R perfect gas constant and N_A Avogadro's number [unit : molecules/mole]. This relation applies to spherical particles in absence of interactions of the diffusing species with ground species[15.4][15.5]. The formula has motivated a number of models that link the diffusion coefficient of polymers to their hydrodynamic radii. The hydrodynamic radius, or Stokes radius, or Stokes-Einstein radius, of a molecule is defined as the radius of a sphere whose diffusion coefficient given by the Stokes-Einstein relation (15.5.2) is equal to the *effective* diffusion coefficient of the molecule in the medium. Still, sometimes the hydrodynamic radius

---

[15.4] This formula (15.5.2) is established by equating two expressions of the force F exerted on particles of radius a moving in a viscous fluid of dynamic viscosity η and behaving like a perfect gas. Indeed, the balance of momentum of the fluid indicates that the gradient of the pressure p is equal to the force exerted over the particles, $dp/dx = c \, N_A \, F$, with c the number of moles of particles per unit volume of fluid and N_A Avogadro's number. In absence of applied external pressure, the fluid pressure is equal to $p = RT \, c$ due to van't Hoff law, Section 13.7. Therefore $c \, N_A \, F = RT \, dc/dx$.

A steady state is reached when this pressure driven flow is balanced by the viscous force exerted on particles by the fluid which according to (15.E.9) is equal to $-6 \, \pi \, a \, \eta \, U$ with U the particle velocity. Now, Fick's law indicates that the particle flux per unit volume $c \, U$ is equal to $-D \, dc/dx$. Therefore, $c \, N_A \, F = 6 \, \pi \, a \, \eta \, N_A \, D \, dc/dx$.

Equating the two above expressions of the force provides the diffusion coefficient (15.5.2). The French physicist and Nobel Prize winner Jean Perrin calculated Avogadro's number in 1908 using this formula by observing the motion of small particles in suspension in a viscous fluid.

[15.5] Historical remarks on the origins of the formulas (15.5.1) and (15.5.2) and some variations may be found in Stiles W. (1921). Permeability - Chapter 4 - Diffusion. *The New Phytologist*, XX(4), 137.

is chosen to match the diffusion coefficient in a blank solution.. Some data for macromolecules, BSA, Dextran, etc. diffusing in agarose gels are reported in Pluen at al. [1999].

The formula (15.5.1) introduces the size of the diffusing species through its molar mass and the formula (15.5.2) view them as spheres. This approximation is questionable for diffusing species whose shape departs significantly from the sphere. The reptation theory considers flexible polymer chains like DNA whose diffusion is constrained by entanglement points (fixed obstacles) representing the gel chains. During diffusion of the polymer by brownian motion, the obstacles appear as forming a sort of tube. Lateral motion of the diffusant can take place only at the extremities of the tube while its central part moves essentially along the tube. As for the diffusion coefficient $D \sim 1/\widehat{m}^{\alpha}$, with $\widehat{m}$ the molar mass of the polymer, two regimes are distinguished according to the ratio of the length of the polymer chains versus the tube size. For a small ratio (short chains), the behavior is said Rouse-like and $\alpha \in [0.5, 1]$, while, for a large ratio > 10 (long chains), the reptation model corresponds to $\alpha \in [2, 3]$.

According to the measurements of Comper and Williams [1987], the diffusion coefficients of rat chondrosarcoma GAG units, PGs and aggregates of PGs, of molar mass 20×10^3, 2.6×10^6 and 211×10^6 g/mole respectively, are quite close for concentrations above 20 mg/ml (the physiological GAG concentration in articular cartilage ranges from about 20 to 80 mg/ml). They attribute this coincidence to the fact that GAGs dominate the interactions with water. Above this bound, the diffusion coefficients increase linearly with concentration, with a slope of 4×10^{-12} m^2/s per mg/ml. At lower concentrations, the diffusion coefficient of GAGs is larger than that of PGs and PG aggregates.

Table 15.5.1: Diffusion coefficients D of chemicals of biological interest (at 37 °C).
(a) human adenocarcinomas from excised tissue; (b) grown in mice.

Species	Molar mass (g/mole)	Reference	Medium	D (m^2/s)
Oxygen	16		Water	32×10^{-10}
		Casciari et al. [1992]	Interstitium	15×10^{-10}
		Obradovic et al. [2000]	Cartilage constructs	15×10^{-10}
Glucose	180.2	Jain [1987]a	Plasma	8.75×10^{-10}
			Tissues	$1\text{-}2 \times 10^{-10}$
ATP	507.2	-	Blood plasma	5×10^{-10}
		-	Cytoplasm	1×10^{-10}
BSA	66×10^3	Brown et al. [2004]	Water	12.5×10^{-11}
			Tumor tissues (a)	9×10^{-11}
			Tumor tissues (b)	1.2×10^{-11}
IgG	150×10^3	Netti et al. [2000]	Human glioblastomas	1×10^{-11}
			Human/murine carcinomas	2×10^{-11}
IgM	970×10^3	Brown et al. [2004]	Water	5.6×10^{-11}
			Tumor tissues (a)	4.2×10^{-11}
			Tumor tissues (b)	0.8×10^{-11}
IGF-I	7.6×10^3	Garcia et al. [2003]	Cartilage	$2\text{-}4 \times 10^{-11}$
GAG (CS)	20×10^3	Comper and Williams [1987]	Rat cartilage (dilute solution)	4×10^{-11}
GAG	$10 - 20 \times 10^3$	Obradovic et al. [2000]	Cartilage constructs	7×10^{-15}
GAG (CS)	30×10^3	Jen [1995]	Rat cartilage	$0.5\text{-}1 \times 10^{-12}$
PG monomer	2.6×10^6		Bovine adult cartilage	$5\text{-}10 \times 10^{-12}$
Chondrocyte	-	Galban and Locke [1999]ab	Cartilage constructs	3×10^{-16}

The diffusion coefficients of a few biological substances at 37 °C are shown in Table 15.5.1. The values at arbitrary temperature can be retrieved via the Stokes-Einstein formula (15.5.2), the temperature dependence on the viscosity of water being provided by Table 3.3.1. For the species like oxygen and carbon dioxide that diffuse freely through cell membranes so that their intra- and extracellular concentrations are equal, the diffusion coefficients are averaged over the tissues. Other species like glucose or lactate are extracellular.

As an important consequence of (15.5.1), **free diffusion becomes inefficient for larger molecules, which thus are transported by convection.** Convection, or advection, is the transport phenomenon by which a molecule moves due to the motion of the fluid in which it is bathed or in which it is dissolved. The relative importance of convection with respect to diffusion is measured by the Péclet number as already indicated by (15.3.67),

$$\text{P}_e \sim \frac{\text{fluid velocity} \times \text{length}}{\text{diffusivity}}. \tag{15.5.3}$$

Buschmann et al. [1995] observe a fluid velocity of $1\,\mu m/s$ in cultures of chondrocytes over agarose gels subjected to an oscillatory displacement of $30\,\mu m$, resulting in a Péclet number smaller than 1 for diffusivities larger than $30 \times 10^{-12}\,m^2/s$. Therefore the flow is mostly diffusive for molecules of molar mass smaller than $5000\,gm$ based on the expression (15.5.1) with the coefficients suggested by Swabb et al. [1974].

An information similar to the Péclet number is provided for many substances in terms of their molar mass and of the tissue's GAG content in Fig. 5 of Jain [1987]a.

15.6   Newtonian viscous fluids and Reynolds number

Newtonian viscous fluids, where the stress is an affine function of the strain rate, are traditionally split in two subsets depending on their compressibility properties.

15.6.1   A compressible viscous fluid

For a compressible viscous fluid, the fluid stress $\boldsymbol{\sigma}$,

$$\boldsymbol{\sigma} = -p\,\mathbf{I} + \zeta \operatorname{tr} \frac{d\boldsymbol{\epsilon}}{dt}\,\mathbf{I} + 2\,\eta\,\frac{d\boldsymbol{\epsilon}}{dt}\,, \tag{15.6.1}$$

expresses in terms of the fluid pressure p and strain rate $d\boldsymbol{\epsilon}/dt$,

$$\frac{d\boldsymbol{\epsilon}}{dt} = \frac{1}{2}\left(\boldsymbol{\nabla}\mathbf{v} + (\boldsymbol{\nabla}\mathbf{v})^{\mathrm{T}}\right), \tag{15.6.2}$$

via the coefficients of dynamic viscosity ζ and η [unit: Pa×s]. The balance of momentum,

$$\operatorname{div}\boldsymbol{\sigma} + \rho\left(\mathbf{b} - \frac{d\mathbf{v}}{dt}\right) = \mathbf{0}\,, \tag{15.6.3}$$

becomes the famous Navier-Stokes equation,

$$
\begin{aligned}
-\boldsymbol{\nabla}p + \boldsymbol{\nabla}\left(\zeta\operatorname{div}\mathbf{v}\right) + \operatorname{div}\left(\eta\left(\boldsymbol{\nabla}\mathbf{v} + (\boldsymbol{\nabla}\mathbf{v})^{\mathrm{T}}\right)\right) + \rho\,\mathbf{b} - \rho\,\frac{d\mathbf{v}}{dt} &= \mathbf{0}, \\
-\frac{\partial p}{\partial x_i} + \frac{\partial}{\partial x_i}\left(\zeta\,\frac{\partial v_j}{\partial x_j}\right) + \frac{\partial}{\partial x_j}\left(\eta\left(\frac{\partial v_i}{\partial x_j} + \frac{\partial v_j}{\partial x_i}\right)\right) + \rho\,b_i - \rho\,\frac{dv_i}{dt} &= 0, \quad i \in [1,3].
\end{aligned}
\tag{15.6.4}
$$

Note that this equation is nonlinear due to the convective terms arising in the acceleration.

15.6.2   An incompressible viscous fluid

The special case of an incompressible viscous fluid, for which $\operatorname{div}\mathbf{v} = 0$, is considered now. The fluid stress,

$$\boldsymbol{\sigma} = -p\,\mathbf{I} + 2\,\eta\,\frac{d\boldsymbol{\epsilon}}{dt}\,, \tag{15.6.5}$$

expresses in terms of a single coefficient of dynamic viscosity η. If the latter is spatially uniform, the balance of momentum can be phrased in terms of the kinematic viscosity $\nu = \eta/\rho$ [unit: m^2/s], namely

$$-\frac{\boldsymbol{\nabla}p}{\rho} + \underbrace{\nu\,\boldsymbol{\nabla}^2\mathbf{v}}_{\simeq\,\nu\,\frac{U}{L^2}} + \mathbf{b} - \underbrace{\frac{d\mathbf{v}}{dt}}_{\simeq\,\frac{U}{L/|\mathbf{v}|} = \frac{U^2}{L}} = \mathbf{0}\,. \tag{15.6.6}$$

If L is a characteristic length of the flow, e.g., pore radius, then, for any quantity $a = a(x,t)$, $\boldsymbol{\nabla}a$ has the order of magnitude of a/L. Moreover if U is a characteristic velocity, the time derivative da/dt amounts to $a\,U/L$. Whence the orders of magnitude indicated in (15.6.6). The dimensionless Reynolds number quantifies the relative weight of the inertial and viscous contributions to the momentum balance,

$$
\text{Reynolds number} \quad \mathrm{R_e} = \frac{\text{inertia terms}}{\text{viscous terms}} = \frac{U\,L}{\nu} = \frac{\rho\,U\,L}{\eta} \quad
\begin{cases}
\ll 1 & \begin{array}{l} \text{viscosity dominated flow :} \\ \quad \text{laminar flow} \end{array} \\[2ex]
\gg 1 & \begin{array}{l} \text{acceleration dominated flow :} \\ \quad \text{turbulent flow} \end{array}
\end{cases}
\tag{15.6.7}
$$

It is standard to cast the Navier-Stokes equations in a dimensionless format by dividing by U^2/L,

$$
-\tilde{\boldsymbol{\nabla}}\tilde{p}\,\frac{p_0}{\rho\,U^2} + \tilde{\boldsymbol{\nabla}}^2\,\tilde{\mathbf{v}}\,\frac{\nu}{L\,U} + \tilde{\mathbf{b}}\,\frac{g\,L}{U^2} - \frac{d\tilde{\mathbf{v}}}{d\tilde{t}}\,\frac{L}{U\,T} = \mathbf{0}\,,
\tag{15.6.8}
$$

where the dimensionless gradient, time, velocity, pressure and body force,

$$
\tilde{\boldsymbol{\nabla}} = L\,\boldsymbol{\nabla}\,, \quad \tilde{t} = \frac{t}{T}\,, \quad \tilde{\mathbf{v}} = \frac{\mathbf{v}}{U}\,, \quad \tilde{p} = \frac{p}{p_0}\,, \quad \tilde{\mathbf{b}} = \frac{\mathbf{b}}{g}\,,
\tag{15.6.9}
$$

have been defined through scaling factors, namely L length, T time, U velocity, p_0 pressure and g gravity.

Besides the Reynolds number, the dimensionless Navier-Stokes equations display further dimensionless numbers, namely,

- Euler number $\dfrac{p_0}{\rho\,U^2} = \dfrac{\text{pressure}}{\text{steady flow inertia}}$;

- Froude number $\dfrac{U^2}{g\,L} = \dfrac{\text{steady flow inertia}}{\text{gravity}}$;

- Strouhal number $\dfrac{L}{U\,T} = \dfrac{\text{transient inertia}}{\text{steady flow inertia}}$.

The delineation between laminar and turbulent flows is a key concept of fluid mechanics, and the reader is invited to consider examples described in textbooks on the subject area. Actually most, if not all, derivations in this work assume a laminar regime.

15.6.3 Hagen-Poiseuille flows along rigid and elastic walls

The quasi-static flow[15.6] of an incompressible viscous fluid in simple boundary value problems serves to provide an estimation of the permeability of porous media.

15.6.3.1 Viscous flow in a horizontal channel

Consider the steady flow of an incompressible Newtonian fluid in a rigid horizontal channel of height $2a$, infinite along the horizontal directions (x, z) sketched in Fig. 15.6.1-a. This flow is termed Couette's flow, after the French physicist Maurice Marie Alfred Couette (1858-1943). The flow is one-dimensional with velocity components ($u_x \neq 0, u_y = 0, u_z = 0$), the axial velocity depends only on the vertical coordinate $u_x = u(y)$ and the non zero components of the stress are $\sigma_{xx} = \sigma_{yy} = \sigma_{zz} = -p$, p fluid pressure, and $\sigma_{xy} = \eta\,du/dy$. The no-slip boundary conditions require $u(\pm a) = 0$.

The balance of momentum implies pointwise inside the channel: $-\partial p/\partial x + \eta\,d^2 u/dy^2 = 0$, $-\partial p/\partial y = 0$, $-\partial p/\partial z = 0$. Thus $p = p(x)$ and, from the first of the three momentum balance equations, dp/dx is a constant, that is, the pressure p varies linearly along the axis x. Integrating this equation yields $u = A + By + y^2/(2\eta) \times dp/dx$. The constants A and B result from the boundary conditions. From the velocity profile which is parabolic along the vertical direction,

$$
u(y) = -\frac{a^2 - y^2}{2\eta}\,\frac{dp}{dx}\,,
\tag{15.6.10}
$$

follow the flow rate Q [unit : m^2/s] and the average velocity U across the section,

$$
Q = \int_{-a}^{a} u\,dy = -\frac{2}{3}\,\frac{a^3}{\eta}\,\frac{dp}{dx}\,, \quad U = \frac{1}{2\,a}\int_{-a}^{a} u\,dy = -\frac{1}{3}\,\frac{a^2}{\eta}\,\frac{dp}{dx}\,.
\tag{15.6.11}
$$

[15.6]Incidentally, it might be worthwhile to precise the terminology. The standard balance of momentum writes, with standard notations, $\operatorname{div}\boldsymbol{\sigma} + \rho\,(\mathbf{b} - d^2\mathbf{u}/dt^2) = \mathbf{0}$. The quasi-static solution is obtained by neglecting the acceleration term in the balance of momentum, $\operatorname{div}\boldsymbol{\sigma}^{\mathrm{qs}} + \rho\,\mathbf{b} = \mathbf{0}$. Still, the acceleration $d^2\mathbf{u}^{\mathrm{qs}}/dt^2$ associated with the quasi-static solution may *not* be zero.

(a) horizontal channel

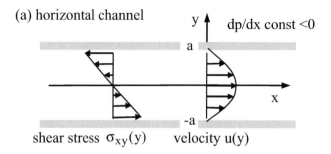

shear stress $\sigma_{xy}(y)$ velocity $u(y)$

(b) cylindrical tube

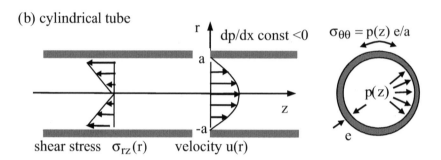

shear stress $\sigma_{rz}(r)$ velocity $u(r)$

(c) establishment of parabolic flow

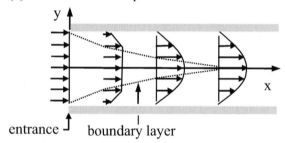

entrance   boundary layer

Fig. 15.6.1: Flows of an incompressible Newtonian fluid in (a) a channel and (b) a cylindrical tube; (c) establishment of the parabolic flow at the entrance of a channel.

15.6.3.2   Viscous flow in a cylindrical horizontal tube

The flow takes place in a cylindrical tube of radius a and axis z. It is one-dimensional with velocity components $(u_r = 0, u_\theta = 0, u_z \neq 0)$, the axial velocity depends only on the radius $u_z = u(r)$ and the non zero components of the stress are $\sigma_{rr} = \sigma_{\theta\theta} = \sigma_{zz} = -p$, p fluid pressure, and $\sigma_{rz} = \eta \, du/dr$. The no-slip boundary condition requires $u(a) = 0$.

The balance of momentum implies pointwise inside the channel: $-\partial p/\partial r = 0$, $-\partial p/\partial \theta = 0$, $-\partial p/\partial z + d(r\sigma_{rz})/(r dr) = 0$. Thus $p = p(z)$ and, from the last of the three momentum balance equations, dp/dz is a constant, that is, the pressure p varies linearly along the axis z. Integrating this equation yields a parabolic profile as well,

$$u(r) = -\frac{a^2 - r^2}{4\eta} \frac{dp}{dz}. \tag{15.6.12}$$

The flow rate Q [unit : m^3/s] and the average velocity U across the surface S of the tube obey the relations,

$$Q = \int_S u \, dS = 2\pi \int_0^a u \, r \, dr = -\frac{\pi}{8} \frac{a^4}{\eta} \frac{dp}{dz}, \quad U = \frac{1}{S} \int_S u \, dS = -\frac{1}{8} \frac{a^2}{\eta} \frac{dp}{dz}, \tag{15.6.13}$$

which have been observed in experiments by Hagen and Poiseuille[15.7].

[15.7] The French physicist and medical doctor Jean Léonard Marie Poiseuille (1797-1869) pioneered hemodynamics. The experiments of the German engineer Gotthilf Heinrich Ludwig Hagen (1797-1884) highlighted the entrance effects and the transition to turbulence.

One may now contemplate the problem as a flow through a thin vessel of thickness $e \ll a$. The circumferential stress on the inner surface of the vessel depends on the pressure $\sigma_{\theta\theta}(r = a, z) = p(z) \, a/e$. The shear stress and shear strain depend on the axial gradient of the pressure, or, equivalently, on the axial flow rate Q,

$$\frac{\sigma_{rz}}{\eta}(r) = \frac{du}{dr}(r) = \frac{r}{2\,\eta} \frac{dp}{dz} = -\frac{4}{\pi} \frac{Q\,r}{a^4} \,. \tag{15.6.14}$$

The power required to overcome the viscous drag over a length L reads,

$$L \int_0^a \sigma_{rz} \frac{du}{dr} \, 2\,\pi\,r\,dr = \frac{8}{\pi} \frac{\eta}{a^4} \, Q^2 \, L \,. \tag{15.6.15}$$

Observations indicate that, due to hypertension, the wall e of the vessel thickens, so as to maintain an approximatively constant circumferential stress. On the other hand, increased flow rate Q at constant pressure gradient leads to increased lumen a.

15.6.3.3 Establishment of the parabolic flow and development of a boundary layer

The above derivations have assumed a channel and a tube of infinite length. Phenomena that take place at the entrance and exit have not been accounted for.

Indeed, assume that fluid particles enter the channel all animated with the same velocity parallel its axis. This uniform flow does not subsist due to the no-slip condition along the walls, Fig. 15.6.1-(c): the no-slip condition gives rise to a boundary layer of increasing thickness $\delta = \delta(x)$. The parabolic regime is fully established throughout the section only at a distance from the entrance. In the case of the channel, an order of magnitude L of this distance can be estimated as follows, Malvern [1969], p. 476:
- first, define scaled velocity components, say via the average velocity of the Poiseuille flow, $\tilde{u}_i = u_i/U$, $i = x, y$;
- next, over the transition zone where the axial gradient is significant, define a length scale L such that $\partial \tilde{u}_x / \partial \tilde{x} = O(1)$, $\tilde{x} = x/L$, $\tilde{y} = y/L$ being the scaled coordinates;
- by the condition of incompressibility, $\partial \tilde{u}_y / \partial \tilde{y} = O(1)$: hence the velocity $\tilde{u}_y$ has maximum magnitude δ/L, given that $\tilde{u}_y = 0$ at $\tilde{y} = a$;
- the scaling pressure p_0 is chosen equal to $\rho \, U^2$ and the scaling time is set to L/U. The dimensionless Navier-Stokes equation (15.6.8) simplifies to $-\tilde{\boldsymbol{\nabla}}\tilde{p} + (\partial^2/\partial \tilde{x}^2 + \partial^2/\partial \tilde{y}^2)\tilde{\mathbf{v}}/\mathrm{R_e} - d\tilde{\mathbf{v}}/d\tilde{t} = \mathbf{0}$;
- in the x-component of this equation, the orders of magnitude of the inertial terms and of the viscous terms can be checked to be equal to 1 and $(L/\delta)^2$, respectively; along the y-component, these numbers are δ/L and L/δ respectively.

For the inertial and viscous contributions to be of identical order, the Reynolds number should be equal to the above ratio of inertial and viscous terms, namely $(L/\delta)^2$. The boundary layer is fully developed when its thickness δ is equal to the half-width a of the tube. Therefore, since $\mathrm{R_e} = U\,L/\nu$, the length L can be estimated to be $O(a^2\,U/\nu)$.

15.6.3.4 Viscous flow in an elastic cylindrical horizontal tube

The previous analyses assumed in fact the channel and tube to be rigid. When the tube is elastic, its section will change according to the pressure, so that its low pressure end will have a smaller section than its high pressure entrance, Fig. 15.6.2.

An instructive example is exposed by Fung [1990], p. 174. The radius of pulmonary arteries and veins may be assumed to depend on the pressure p in an affine way,

$$a(p) = a(0) + \alpha\,p \,, \tag{15.6.16}$$

with $\alpha > 0$ a constant. If there is no loss across the capillary, the flow rate Q is constant. Integration of (15.6.13) over a tube extending from $z = 0$ to $z = L$ yields

$$a^5(0) - a^5(L) = \frac{40}{\pi} \eta\,\alpha\,L\,Q \,, \tag{15.6.17}$$

and the velocity,

$$u(r, z) = 2 \frac{Q}{\pi\,a^4(z)} \left(a^2(z) - r^2\right) , \tag{15.6.18}$$

increases along the axis tube essentially like $a^{-2}(z)$. Thus if the ratio $a(L)/a(0) < 1/2$, then the contribution to the flux of the end term is negligible. This observation introduces the *waterfall phenomenon* where the end conditions do not influence the flux, see Section 15.10.2.

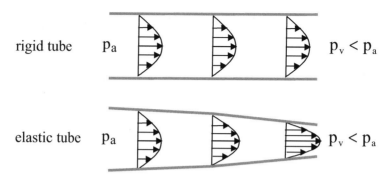

Fig. 15.6.2: Flow in a capillary with the pressure at the arteriole end larger than at the venous end. If the walls of the capillary are elastic, the pressure profile is no longer linear and the velocity field varies along the tube to accommodate a constant flow rate and a decreasing section.

15.6.3.5 Driving engines, pressure versus difference of potential

Water can be moved parallel to a plane or along a cylindrical pipe in different ways. One way is by applying a pressure gradient as described by Poiseuille flow for a viscous fluid. If the plane or pipe bears a fixed electrical charge, another way is by applying a difference of potential, a phenomenon referred to as electroosmosis.

McLaughlin and Mathias [1985] propose a nice comparison of the two phenomena in the context of re-absorption of fluid in mammalian renal proximal tubules. Some information related to the presence of a fixed charge is reported in Chapter 13, and especially in Sections 13.4, 13.6 and in Exercise 13.1. The electroosmotic phenomenon is included in the coupled constitutive equations of transport developed in Chapter 16.

15.7   Hydraulic conductivity

The rationale for the theoretical formulation of Darcy's law in a thermodynamic framework has been exposed in Section 7.3.2. If the fluid is inviscid and barotropic, Darcy's law can be interpreted as a potential flow. The introduction of the potential goes along standard theorems of fluid mechanics that are briefly described first. The actual expression of the intrinsic hydraulic permeability is next derived for simple geometries of the pores circulated by a Newtonian fluid.

The fact that the fluid circulating the porosity is considered either inviscid or viscous seems paradoxical and deserves to be explained. It should be realized that the analysis takes place at two spatial scales:

- the analytical expressions of the hydraulic permeability are worked out at the scale of the porosity where the microstructure is available and the fluid is considered viscous. Indeed, a realistic analysis can not neglect the *no-slip condition* along the walls;
- the hydraulic permeability is a macroscopic property, representative of an elementary volume. At this scale, which is the one we are mostly concerned, the sole features of the microstructure that have survived upscaling are embodied in the *intrinsic permeability* which may be a scalar or a tensor. The properties of the fluid come into play in the *hydraulic permeability* and *conductivity* via the mass density and dynamic viscosity.

Most often, the fluid is considered to be inviscid at the macroscale. When it is viscous, Darcy's law is replaced by Brinkman's law which involves a viscous term, Section 7.3.4.

There are two ways to look at the permeability of a porous medium:

- the porous medium may be contemplated as a modification of a solid, from which matter has been extracted. Then the permeability will stem from the channels which are pathways for the fluid: the larger their volume fraction, the larger the permeability;
- alternatively, focus may be on the fluid whose flow is hindered by obstacles: the larger the volume fraction of the latter, the smaller the permeability.

The permeability associated with the microvasculature may be considered along the first point of view. Conversely, the collagen network may be viewed as a hindrance to the flow.

15.7.1   Circulation, theorems of Kelvin and Helmholtz

Consider water as an inviscid fluid with pressure p and density ρ. Its stress $\boldsymbol{\sigma}$ is equal to $-p\,\mathbf{I}$, and, if it is subjected to gravity $\mathbf{g} = g\,\mathbf{e}_g$ with magnitude g, its balance of momentum may be expressed in the format,

$$\mathbf{F} \equiv -\frac{\boldsymbol{\nabla}p}{\rho} + \mathbf{g} - \frac{d\mathbf{v}}{dt} = \mathbf{0} \,. \tag{15.7.1}$$

Let us consider a fluid curve $\mathcal{C}$, that is, a curve that follows the velocity field and therefore is formed by the same particles at any time. The vector joining neighboring points is such that $d\mathbf{x} = \mathbf{v}\,dt$. The theorems we are concerned with address the time derivative of the circulation,

$$C(t) \equiv \oint_{\mathcal{C}} \mathbf{v} \cdot d\mathbf{x} \,, \tag{15.7.2}$$

along a closed fluid curve (run in the direct sense). Let us first observe that the time derivative of this integral is equal to the integral of the time derivative of the integrand since the integration curve lies on the same fluid particles,

$$\frac{dC(t)}{dt} = \oint_{\mathcal{C}} \frac{d\mathbf{v}}{dt} \cdot d\mathbf{x} + \mathbf{v} \cdot d\mathbf{v} \,. \tag{15.7.3}$$

The second term in the integrand arises because the operators of time derivation and integration commute,

$$\frac{d}{dt}d\mathbf{x} = \frac{d}{dt}(\mathbf{v}\,dt) = d\mathbf{v} \,. \tag{15.7.4}$$

Elimination of the acceleration between the equation of momentum balance (15.7.1) and Eqn (15.7.3) yields the *theorem of Kelvin*,

$$\begin{aligned}
\frac{dC(t)}{dt} &= \oint_{\mathcal{C}} \left(-\frac{\boldsymbol{\nabla}p}{\rho} + \mathbf{g}\right) \cdot d\mathbf{x} + \tfrac{1}{2}\,d\mathbf{v}^2 \,, \\
&= -\oint_{\mathcal{C}} \frac{\boldsymbol{\nabla}p}{\rho} \cdot d\mathbf{x} \,, \\
&= -\oint_{\mathcal{C}} \frac{dp}{\rho} \,,
\end{aligned} \tag{15.7.5}$$

the integrals of the two last terms in the first line vanishing since the curve is closed. The derivative of the circulation vanishes if the integrand of the second line is a gradient, that is, if the integrand is irrotational. According to (2.4.4) and (2.4.6), this condition requires that $\boldsymbol{\nabla}p$ and $\boldsymbol{\nabla}\rho$ should be parallel. It holds for barotropic fluids for which $\rho = \rho(p)$,

$$\text{barotropic fluids} \quad \rho = \rho(p) \,. \tag{15.7.6}$$

The integral which then depends only on endpoints of the integration path, is usually termed the *pressure function*,

$$P(p_0, p) = \int_{p_0}^{p} \frac{dp'}{\rho(p')} \,. \tag{15.7.7}$$

For an isentropic (or adiabatic) flow, the pressure function can be identified with the fluid enthalpy per unit mass, since then $dh = dp/\rho$ according to Table 3.2.1.

Note that the integrals (15.7.5) and (15.7.7) are *not* path independent if the density depends, in addition to pressure, on other variables like temperature, chemical concentration or position. The issue is discussed by Bear [1972], p. 160.

For a barotropic fluid, the result

$$\frac{dC(t)}{dt} = 0 \,, \tag{15.7.8}$$

is known as *theorem of Helmholtz*.

15.7.2   Bernoulli equation for steady irrotational flow

Let us assume
- the body force to be conservative, say $\mathbf{b} = -\boldsymbol{\nabla}\Omega$, e.g., for gravity $\Omega = g\,z$, with z counted positively upwards;
- the fluid to be inviscid and barotropic.

It is convenient to adopt the format of the acceleration derived in (2.4.92). The balance of momentum (15.7.1) may then be recast in the format,

$$-\boldsymbol{\nabla}\Big(\Omega + \int_{p_0}^{p} \frac{dp'}{\rho(p')} + \frac{\mathbf{v}^2}{2}\Big) = \frac{\partial \mathbf{v}}{\partial t} + (\boldsymbol{\nabla} \wedge \mathbf{v}) \wedge \mathbf{v}\,. \tag{15.7.9}$$

If the flow is steady, namely $\partial \mathbf{v}/\partial t = \mathbf{0}$, then

$$-\mathbf{v} \cdot \boldsymbol{\nabla}\Big(\Omega + \int_{p_0}^{p} \frac{dp'}{\rho(p')} + \frac{\mathbf{v}^2}{2}\Big) = 0\,, \tag{15.7.10}$$

that is,

$$\Omega + \int_{p_0}^{p} \frac{dp'}{\rho(p')} + \frac{\mathbf{v}^2}{2} \tag{15.7.11}$$

is constant along streamlines $(-d\mathbf{x} \cdot \boldsymbol{\nabla}(\cdot) = -d(\cdot))$.

If the flow is steady, namely $\partial \mathbf{v}/\partial t = \mathbf{0}$, and irrotational, namely $\boldsymbol{\nabla} \wedge \mathbf{v} = \mathbf{0}$, then the expression (15.7.11) which is contributed by the body force potential, the enthalpy of the fluid and its kinetic energy is constant everywhere in the flow and not only along streamlines. This result is known as Bernoulli equation.

In the case of a perfect gas for which the pressure depends on mass density as indicated by (3.3.13), the pressure function is equal to $(c^2 - c_0^2)/(\gamma - 1)$ where the speed of sound c is defined by (11.4.6). Therefore Bernoulli equation takes the form,

$$\Omega + \frac{c^2 - c_0^2}{\gamma - 1} + \frac{\mathbf{v}^2}{2} \quad \text{constant}\,. \tag{15.7.12}$$

15.7.3   Hubbert potential

Darcy's law as expressed by (7.3.12) displays the stimulus $\mathbf{F}$ that drives the apparent velocity of the fluid relative to the solid $\mathbf{J} = n\,(\mathbf{v} - \mathbf{v}_\mathrm{s})$. This stimulus vanishes when the fluid stands by itself, Eqn (15.7.1). On the other hand, as a species of a porous medium, it does not: indeed, the fluid is then subjected to an interaction force by the rest of the mixture.

Let us consider the work done by each of the three terms of (15.7.1) during an incremental motion $d\mathbf{x} = \mathbf{v}\,dt$ along a fluid curve:
- work done by gravity:

$$\mathbf{g} \cdot d\mathbf{x} = -g\,dz\,, \tag{15.7.13}$$

assuming the coordinate z parallel to $\mathbf{g}$ to be counted positively upwards (along $-\mathbf{g}$);
- work done by the acceleration:

$$-\frac{d\mathbf{v}}{dt} \cdot \mathbf{v}\,dt = -d\Big(\frac{\mathbf{v}^2}{2}\Big)\,; \tag{15.7.14}$$

- work done by the pressure:

$$-\frac{\boldsymbol{\nabla}p}{\rho} \cdot d\mathbf{x} = -\frac{dp}{\rho}\,. \tag{15.7.15}$$

The incremental work done by the stimulus $\mathbf{F}$ scaled by the gravity $-g$,

$$dH = -\frac{\mathbf{F}}{g} \cdot d\mathbf{x} = dz + \frac{dp}{\rho\,g} + \frac{d\mathbf{v}^2}{2\,g}\,, \tag{15.7.16}$$

depends on $(z, p, \rho, \mathbf{v})$. Given a reference state denoted by the subscript 0, the integral

$$H(z, p, \rho, \mathbf{v}) - H(z_0, p_0, \rho_0, \mathbf{v}_0) = -\int_{\mathbf{x}_0}^{\mathbf{x}} \frac{\mathbf{F}}{g} \cdot d\mathbf{x} = z - z_0 + \int_{p_0}^{p} \frac{dp'}{\rho(p')\,g} + \frac{\mathbf{v}^2}{2\,g} - \frac{\mathbf{v}_0^2}{2\,g}\,, \tag{15.7.17}$$

can be seen as a potential only for barotropic fluids, as discussed in Section 15.7.1. For a barotropic fluid, Darcy's law as expressed by (7.3.12) can then be interpreted as a potential flow,

$$\mathbf{J} = -K_{\mathrm{H}} \boldsymbol{\nabla} H \,, \tag{15.7.18}$$

where K_{H} [unit : m/s] is the hydraulic conductivity. In absence of inertial terms, the hydraulic head H is termed Hubbert potential after the author of the above derivation, Hubbert [1940].

The terminology associated with Darcy's law is described in Section 7.3.2. The intrinsic permeability can be obtained in explicit form for simple porous geometries as elaborated in Section 15.7.4.

15.7.4 Porosity-permeability relation

The permeability is dictated by the microstructure/geometry of the porosity (=fluid volume fraction). Theoretical models of permeability using microstructure information are reviewed in Levick [1987], Edwards and Prausnitz [1998] and Berryman [2006].

Although questionable, there is a strong tendency to relate directly permeability and porosity. A porosity dependent permeability is legitimate as long as the microstructure is completely characterized by a single scalar. However, in general, higher order tensors are needed for a more representative description of the microstructure.

Some models describe the permeability via power functions of the porosity. For example, Fontainebleau sandstone has a power 3 at high porosity and 7 or more at low porosity, Mok et al. [2002]. In geomaterials, typical mechanical tests suggest that permeability changes result from the competition of dilatant microcracking (increase) and dissolution-precipitation (decrease).

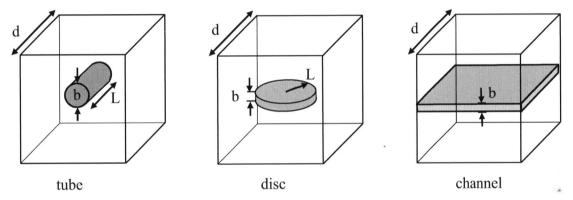

Fig. 15.7.1: Three typical porosities embedded in a representative cubic elementary volume.

Table 15.7.1: Microstructural models with one-dimensional flow illustrating the relations between intrinsic permeability and porosity, Bear [1972], p. 164, Guéguen and Palciauskas [1992], p. 126.

	Porosity n	Permeability k_{in}
Tube	$\dfrac{\pi}{4}\dfrac{b^2 L}{d^3}$	$n\,\dfrac{b^2}{32} = \dfrac{\pi}{128}\dfrac{b^4 L}{d^3}$
Disc	$\pi\dfrac{bL^2}{d^3}$	$n\,\dfrac{b^2}{12} = \dfrac{\pi}{12}\dfrac{b^3 L^2}{d^3}$
Channel	$\dfrac{b}{d}$	$n\,\dfrac{b^2}{12} = \dfrac{1}{12}\dfrac{b^3}{d}$

Simple microstructural models, e.g., tubes and cylindrical fissures sketched in Fig. 15.7.1, yield the relations shown in Table 15.7.1. These relations derive easily from the expressions defining a Poiseuille flow in a channel and in a tube. Darcy's flux is defined as the average fluid velocity, e.g. flow rate Q over the surface of the tube S, times the fluid volume fraction n. For a rectangular channel of aperture b located in a representative element of size d, the relation (15.6.11) yields,

$$J = n\,U = -\frac{k_{\mathrm{in}}}{\eta}\frac{dp}{dx}, \quad k_{\mathrm{in}} = n\,\frac{b^2}{12} = \frac{1}{12}\frac{b^3}{d}, \quad n = \frac{b\,d^2}{d^3}\,. \tag{15.7.19}$$

For a disc-shape channel of aperture b and radius L located in a representative element of size d (that is, the number of discs per unit volume is equal to $1/d^3$), the relation (15.6.11) yields,

$$J = nU = -\frac{k_{\text{in}}}{\eta}\frac{dp}{dx}, \quad k_{\text{in}} = n\,\frac{b^2}{12} = \frac{\pi}{12}\frac{b^3\,L^2}{d^3}, \quad n = \pi\,\frac{b\,L^2}{d^3}\,. \tag{15.7.20}$$

For the tube of diameter b with a length L parallel to flow and located in a representative element of size d (that is, the number of tubes per unit volume is equal to $1/d^3$), Eqn (15.6.13) yields in turn,

$$J = nU = -\frac{k_{\text{in}}}{\eta}\frac{dp}{dz}, \quad k_{\text{in}} = n\,\frac{b^2}{32} = \frac{\pi}{128}\frac{b^4\,L}{d^3}, \quad n = \frac{\pi}{4}\frac{b^2\,L}{d^3}\,. \tag{15.7.21}$$

As a first extension of the channel-like porosity, one may consider that the width and the thickness of the channel differ along the two in-plane orthogonal directions $\mathbf{e}_1$ and $\mathbf{e}_2$. For example, assume a channel of width l_2 along $\mathbf{e}_2$ and aperture l_3 along $\mathbf{e}_3$ to be embedded in a representative elementary volume of section $L_2\,L_3$ orthogonal to $\mathbf{e}_1$. The permeability associated with a gradient along $\mathbf{e}_1$ is then

$$\mathbf{k} = n_1\,\frac{l_3^2}{12}\,\mathbf{e}_1 \otimes \mathbf{e}_1, \quad n_1 = \frac{l_2\,l_3}{L_2\,L_3}\,. \tag{15.7.22}$$

The flow may take place in the two in-plane orthogonal directions $\mathbf{e}_1$ and $\mathbf{e}_2$,

$$\mathbf{k} = n_1\,\frac{l_3^2}{12}\,\mathbf{e}_1 \otimes \mathbf{e}_1 + n_2\,\frac{l_3^2}{12}\,\mathbf{e}_2 \otimes \mathbf{e}_2, \quad n_1 = \frac{l_2\,l_3}{L_2\,L_3}, \quad n_2 = \frac{l_1\,l_3}{L_1\,L_3}\,. \tag{15.7.23}$$

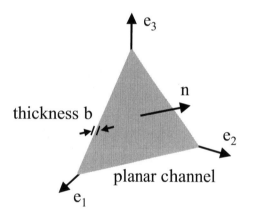

Fig. 15.7.2: A planar channel of uniform aperture b arbitrarily oriented with respect to the fixed cartesian axes $\mathbf{e}_1$, $\mathbf{e}_2$ and $\mathbf{e}_3$.

As a second extension sketched in Fig. 15.7.2, let us assume the channel to be aligned with the cartesian vectors $\hat{\mathbf{e}}_1$ and $\hat{\mathbf{e}}_2$, and the areal fractions to be identical, $n_1 = n_2 = n$. Then the two-dimensional flow expresses as,

$$\mathbf{J} = -\frac{k_{\text{in}}}{\eta}\,(\hat{\mathbf{e}}_1 \otimes \hat{\mathbf{e}}_1 + \hat{\mathbf{e}}_2 \otimes \hat{\mathbf{e}}_2) \cdot \frac{\partial p}{\partial \mathbf{x}} = -\frac{k_{\text{in}}}{\eta}\,(\mathbf{I} - \mathbf{n} \otimes \mathbf{n}) \cdot \boldsymbol{\nabla}p\,, \tag{15.7.24}$$

with $k_{\text{in}} = n\,b^2/12$ and b the aperture of the channel along normal $\mathbf{n} = \hat{\mathbf{e}}_3$.

This relation may be established in an alternative way. Indeed, $\mathbf{J} = -(k_{\text{in}}/\eta)\,(\hat{\mathbf{e}}_1\,\partial p/\partial \hat{x}_1 + \hat{\mathbf{e}}_2\,\partial p/\partial \hat{x}_2)$. The permeability may be expressed in the arbitrary cartesian axes $(\mathbf{e}_1, \mathbf{e}_2, \mathbf{e}_3)$, instead of the local axes $(\hat{\mathbf{e}}_1, \hat{\mathbf{e}}_2, \hat{\mathbf{e}}_3)$. Let $\mathbf{R}$ be the rotation that transforms the $\mathbf{e}$-axes to the local $\hat{\mathbf{e}}$-axes, namely from (2.4.49),

$$\begin{bmatrix} \hat{\mathbf{e}}_1 \\[2mm] \hat{\mathbf{e}}_2 \\[2mm] \hat{\mathbf{e}}_3 \end{bmatrix} = \begin{bmatrix} \cos\theta\,\cos\phi & \cos\theta\,\sin\phi & -\sin\theta \\[2mm] -\sin\phi & \cos\phi & 0 \\[2mm] \sin\theta\,\cos\phi & \sin\theta\,\sin\phi & \cos\theta \end{bmatrix} \begin{bmatrix} \mathbf{e}_1 \\[2mm] \mathbf{e}_2 \\[2mm] \mathbf{e}_3 \end{bmatrix}\,. \tag{15.7.25}$$

The coordinates transform accordingly, namely $\hat{\mathbf{x}} = \mathbf{R} \cdot \mathbf{x}$, and therefore $\partial / \partial \hat{\mathbf{x}} = \mathbf{R} \cdot \partial / \partial \mathbf{x}$. The components of the flow in the $\mathbf{e}$-axes derive as follows,

$$
\begin{aligned}
\mathbf{J} &= -\frac{k_{\mathrm{in}}}{\eta} \sum_{k,l=1,3} \mathbf{e}_l \Big(\sum_{i=1,2} R_{ik} R_{il} \Big) \frac{\partial p}{\partial x_k} \\
&= -\frac{k_{\mathrm{in}}}{\eta} \sum_{k,l=1,3} \mathbf{e}_l \Big(\sum_{i=1,3} R_{ik} R_{il} - R_{3k} R_{3l} \Big) \frac{\partial p}{\partial x_k} \\
&= -\frac{k_{\mathrm{in}}}{\eta} \sum_{k,l=1,3} \mathbf{e}_l \left(I_{kl} - R_{3k} R_{3l} \right) \frac{\partial p}{\partial x_k} \\
&= -\frac{k_{\mathrm{in}}}{\eta} \sum_{k,l=1,3} \mathbf{e}_l \left(I_{kl} - n_k n_l \right) \frac{\partial p}{\partial x_k} \\
&= -\frac{k_{\mathrm{in}}}{\eta} \begin{bmatrix} 1-n_1^2 & -n_1 n_2 & -n_1 n_3 \\ -n_1 n_2 & 1-n_2^2 & -n_2 n_3 \\ -n_1 n_3 & -n_2 n_3 & 1-n_3^2 \end{bmatrix} \begin{bmatrix} \partial p/\partial x_1 \\ \partial p/\partial x_2 \\ \partial p/\partial x_3 \end{bmatrix} .
\end{aligned} \tag{15.7.26}
$$

The components of the unit normal to the planar channel $\mathbf{n} = \hat{\mathbf{e}}_3$ are read from the last line of (15.7.25), namely $n_k = R_{3k}$, $k \in [1,3]$. The consequences on the permeability of the choice of the sense $\mathbf{n}$ or $-\mathbf{n}$ are elusive as long as direct cartesian axes are used throughout.

15.7.5 Planar network

Consider a porous medium endowed with two types of permeability, say a background isotropic permeability k_0 associated with the matrix and a permeability $k_l = k_l(\mathbf{l})$ due to a directional network aligned with the unit vector $\mathbf{l}$ of volume fraction n. The resulting permeability along a direction $\mathbf{m}$ is equal to

$$
k = (1-n) k_0 + k_l(\mathbf{l}) (\mathbf{m} \cdot \mathbf{l})^2 . \tag{15.7.27}
$$

Indeed the fluid flux $\mathbf{J}_l$ in the network associated with the pressure field p is equal to

$$
\begin{aligned}
\mathbf{J}_l &= -\frac{k_l(\mathbf{l})}{\eta} \frac{\partial p}{\partial (\mathbf{x} \cdot \mathbf{l})} \mathbf{l} \\
&= -\frac{k_l(\mathbf{l})}{\eta} \Big(\frac{\partial p}{\partial \mathbf{x}} \cdot \mathbf{l} \Big) \mathbf{l} \\
&= -\Big(\frac{k_l(\mathbf{l})}{\eta} \mathbf{l} \otimes \mathbf{l} \Big) \cdot \frac{\partial p}{\partial \mathbf{x}} .
\end{aligned} \tag{15.7.28}
$$

Therefore, if the pressure gradient is uniform over all channels, the total permeability admits the tensor format,

$$
\mathbf{k} = (1-n) k_0 \mathbf{I} + k_l(\mathbf{l}) \mathbf{l} \otimes \mathbf{l} . \tag{15.7.29}
$$

In presence of a continuous directional distribution of permeability in a plane ($d = 2$) or over all space directions ($d = 3$) defined by the density $k_l = k_l(\mathbf{l})$, the permeability tensor generalizes to,

$$
\mathbf{k} = (1-n) k_0 \mathbf{I} + \int_{\Omega^d} k_l(\mathbf{l}) \mathbf{l} \otimes \mathbf{l} \, d\Omega . \tag{15.7.30}
$$

If the directional permeability is isotropic, namely $k_l(\mathbf{l}) = k_l/(2^{d-1}\pi)$ independent of $\mathbf{l}$, the permeability tensor can be cast in the explicit form, with help of (12.2.9) and (12.2.37),

$$
\mathbf{k} = (1-n) k_0 \mathbf{I} + k_l \frac{\mathbf{I}^{(d)}}{d} , \tag{15.7.31}
$$

where $\mathbf{I}^{(d)}$ is the identity tensor of order $d = 2, 3$.

If the individual flow channel is considered as a linear crack of length $2a$ subjected to a uniform internal pressure p, and embedded in an isotropic elastic medium of Young's modulus E and Poisson's ratio ν, its width $w(x)$ at distance x from the cavity center is equal to $\sqrt{a^2 - x^2} \times 4(1-\nu^2)/E \times p$, Sneddon [1946]. The volume of fluid per unit length of the cavity is then equal to $2ab$, where $b = a \times \pi(1-\nu^2)/E \times p$ is an equivalent aperture.

An important assumption of the upscaling procedure where the representative elements are gathered according to some scheme is that the elements are perfectly interconnected, an extreme situation. The percolation theory is a tool that provides some response to this issue, Guéguen and Palciauskas [1992], p. 128.

15.7.6   Three-dimensional network

Let us consider a penny shaped cavity of unit normal $\mathbf{n}$, radius $r(\mathbf{n})$ and width $b(\mathbf{n})$. The volume associated with this single cavity is equal to $\pi\,r^2(\mathbf{n})\,b(\mathbf{n})$. From (15.7.20) and (15.7.24), the flux per volume V of porous medium associated with the cavity equals,

$$\mathbf{J}(\mathbf{n}) = -n\,\frac{b^2(\mathbf{n})}{12\,\eta}\,(\mathbf{I} - \mathbf{n}\otimes\mathbf{n})\cdot\boldsymbol{\nabla}p, \quad n = \frac{\pi\,r^2(\mathbf{n})\,b(\mathbf{n})}{V}. \tag{15.7.32}$$

Consider now that the porous medium has a background isotropic permeability k_0 and that it contains penny shaped cavities, of overall porosity n, directionally distributed over the unit sphere $\mathbb{S}^2$ but subjected to the same pressure gradient. Then the resulting volume flux per unit volume of porous medium,

$$\mathbf{J} = -\mathbf{k}\cdot\boldsymbol{\nabla}p, \quad \mathbf{k} = (1-n)\,k_0\,\mathbf{I} + \frac{1}{48\,\eta}\,\frac{N}{V}\int_{\mathbb{S}^2} r^2(\mathbf{n})\,b^3(\mathbf{n})(\mathbf{I} - \mathbf{n}\otimes\mathbf{n})\,dS(\mathbf{n}), \tag{15.7.33}$$

expresses in terms of the permeability $\mathbf{k}$ which itself involves the crack density N/V and a directional integral featuring the radii $r(\mathbf{n})$, widths $b(\mathbf{n})$ and unit normals to the cracks $\mathbf{n}$.

   If a penny shaped cavity of radius a is considered as a crack subjected to a uniform internal pressure p and embedded in an isotropic elastic medium of Young's modulus E and Poisson's ratio ν, its width $w(r)$ at radius r is equal to $\sqrt{a^2 - r^2}\times 8\,(1-\nu^2)/(\pi\,E)\,p$, Sneddon [1946]. The volume of fluid contained in the cavity is then equal to $\pi\,a^2\,b$ where $b = a\times(16/3)\,(1-\nu^2)/(\pi\,E)\,p$ is an equivalent aperture.

15.7.7   Dependence of permeability with respect to a characteristic length

The dependence of permeability with respect to the square of a characteristic length may be introduced by the theory of hydraulic radii. An equivalent pore radius is defined as the ratio of the pore volume $n = V_v/V$ over the wetted pore surface area $S_v = \Sigma_v/V$ in a volume V of porous medium, namely $r_v = V_v/\Sigma_v = n/S_v$ (the hydraulic radius is sometimes defined as twice this quantity). For example, the pore volume and pore surface of the tube of diameter b in Fig. 15.7.1 are respectively $V_v = \pi\,(b^2/4)\,L$ and $\Sigma_v = \pi\,b\,L + \pi\,b^2/2$, and therefore the hydraulic radius associated with a long tube ($b \ll L$) is equal to half of its radius, indeed $r_v = b/4$. The permeability (15.7.21) can the be recast in the format,

$$k_{\rm in} = \frac{1}{2}\,n\,r_v^2 = \frac{1}{2}\,\frac{n^3}{S_v^2}. \tag{15.7.34}$$

For a sphere of diameter b, volume $V_v = \pi\,b^3/6$ and wetted surface $\Sigma_v = \pi\,b^2$, the hydraulic radius is equal to $b/6$.

   Let $S_0 = S_v/(1-n)$ be the surface of solid particles in a solid volume $V_{\rm s} = (1-n)\,V$. When phrased in terms of S_0, the intrinsic permeability (15.7.34) is known as Kozeny-Carman permeability,

$$k_{\rm in} = \frac{1}{2}\,\frac{n^3}{S_v^2} = \frac{1}{2}\,\frac{1}{S_0^2}\,\frac{n^3}{(1-n)^2}, \tag{15.7.35}$$

The coefficient $1/2$ in front of the expression (15.7.35) is questionable and an estimation based on experimental data is more reliable.

   The fact that the actual path followed by water is not a straight path but involves a number of zigzags may be introduced: the permeability is decreased by a factor $\tau \le 1$ termed geometric tortuosity which is equal to the square of the ratio of the ideal distance run by the flow over the actual distance,

$$k_{\rm in} = \frac{1}{2}\,\frac{\tau}{S_0^2}\,\frac{n^3}{(1-n)^2}. \tag{15.7.36}$$

For a dense swarm of spherical particles of uniform radius a, the surface area S_0 is equal to $4\pi\,a^2$ over $(4/3)\,\pi\,a^3$, namely $3/a$, and the permeability is seen to depend on the square of particle size,

$$k_{\rm in} = \tau\,\frac{a^2}{18}\,\frac{n^3}{(1-n)^2}. \tag{15.7.37}$$

   As a further refinement, Jain [1987]a lets the hydraulic conductivity in the interstitium depend on the water content in a power law format, his Eqn (J). An improvement of the Kozeny-Carman formula accounting for texture goes by the name of Fair-Hatch relation, Bear [1972], p. 134, Liu et al. [1997]. The presence of the porosity in the actual expression of the permeability introduces a mechanodiffusive coupling. In this vein, Lai and Mow [1980] have considered a strain-dependent hydraulic conductivity,

$$K_{\mathrm{H}} = K_{\mathrm{H0}} \exp(M \operatorname{tr} \boldsymbol{\epsilon}), \tag{15.7.38}$$

where $M > 0$ is a material coefficient: the conductivity increases *strongly* with dilatation. For example, for an injectable collagen matrix, Laude et al. [2000] use $K_{\mathrm{H0}} = 13 \times 10^{-9}\,\mathrm{m/s}$ and $M \simeq 2$.

15.7.8 The drag coefficient k_{Sd} for spherical and cylindrical particles

The relation between intrinsic permeability and the coefficient of Stokes' drag has been introduced in the framework of saturated porous media in Section 7.3.2: indeed, $k_{\mathrm{in}} = \eta\, n^2/k_{\mathrm{Sd}}$. The drag coefficient may be calculated in explicit form in simple contexts. Let us consider an incompressible viscous fluid with viscosity η whose uniform velocity U is perturbed by the presence of a spherical particle of radius a. Both gravity and inertial terms are neglected.

The resistance F [unit : N] exerted by the sphere thus depends on the radius a [unit : m], velocity U [unit : m/s] and dynamic viscosity η [unit : N×s/m²]. According to the Buckingham pi theorem, e.g., Malvern [1969], p. 470, a relation between m' quantities with m independent dimensions can be cast in an equivalent relation between $m' - m$ dimensionless entities. Here we have $m' = 4$ and $m = 3$ and thus F should be proportional to $a\,\eta\,U$.

Analytical expressions of the coefficient of Stokes' drag and intrinsic permeability for a sphere and a cylindrical particle of radius a can be obtained in exact form,

$$k_{\mathrm{Sd}} = n\, \frac{\eta}{a^2} \begin{cases} 9/2 & \text{sphere} \\ 8/\pi & \text{cylinder} \end{cases} \quad ; \quad k_{\mathrm{in}} = n\, a^2 \begin{cases} 2/9 & \text{sphere} \\ \pi/8 & \text{cylinder} \end{cases} \tag{15.7.39}$$

The calculation of the drag of a spherical particle is detailed in Exercise 15.1. The drag is defined as the force exerted on the fluid phase per unit volume, i.e., F/V_{particle}. The resistance coefficient, defined as the ratio of the force F over the inertial contribution $\frac{1}{2}\,\rho_{\mathrm{w}}\,\pi\,a^2\,U^2$, is equal to $12/\mathrm{R_e}$. Experimental data show that the formulas hold only for low Reynolds number $\mathrm{R_e} = \rho_{\mathrm{w}}\,U\,a/\eta$: indeed the inertial terms become more and more significant close to the sphere surface as Reynolds number increases.

The intrinsic permeability obtained by Brinkman [1947] for a number of spherical particles of radius a that occupy the volume fraction V_{s}/V,

$$k_{\mathrm{in}} = a^2 \left(\frac{4\, V/V_{\mathrm{s}} - 6}{9 + 3\, \sqrt{8\, V/V_{\mathrm{s}} - 3}} \right)^2 = \frac{a^2}{1-n} \left(\frac{2\,(n - 1/3)}{3\,\sqrt{1-n} + \sqrt{5 + 3\,n}} \right)^2, \tag{15.7.40}$$

breaks down for fluid porosities $n = 1 - V_{\mathrm{s}}/V$ smaller than $1/3$. It behaves like $(2/9)\,a^2/(1-n)$ at large porosities $n \to 1$. This expression applies to a swarm of particles across which the flow is more constrained than in the analysis of Exercise 15.1 which targets a single sphere.

Based on a review of literature, Jackson and James [1986] proposed an expression of the intrinsic permeability k_{in} of gels composed of cylindrical rods of radius a under random three-dimensional arrangements,

$$k_{\mathrm{in}} = -\frac{3}{20}\, \frac{a^2}{1-n} \left(\operatorname{Ln}(1-n) + 0.931 \right). \tag{15.7.41}$$

As the volume fraction $1-n$ of the rods increases from 0.019 to 0.072, this formula shows a decrease of the permeability that fits qualitatively the experimental results by Johnson and Deen [1996] on agarose gels which are polysaccharide chains.

15.7.9 Modifications of Darcy's law

Although it accounts for the friction of the fluid over the solid walls, Darcy's law (15.7.18) addresses a macroscopic ideal fluid. Brinkman's law defined in Section 7.3.4 includes macroscopic viscous effects.

Deviation with respect to a linear relation between flux and pressure gradients has been observed at high Reynolds number. Reynolds number may become large for gases whose viscosity is small. In fact, non Darcian laws in the format $U \sim -i^{\alpha}$ are used at the two ends of the spectrum of hydraulic gradients i, with an exponent $\alpha > 1$ at low i and < 1 at large i. Here U [unit : m/s] is the fluid particle velocity relative to the solid skeleton and the dimensionless hydraulic gradient i is equal to the pressure gradient scaled by the specific weight of water. The incipience of turbulence is usually considered also to mark the threshold of non Darcian flows at large i.

15.7.9.1   Non Darcian flows

The relation of Forchheimer-Dupuit is an attempt to account for nonlinear effects,

$$-\frac{\Delta p}{\Delta x} = \frac{\eta}{k_{\text{in}}} \frac{Q}{A} + \beta\, \rho \left(\frac{Q}{A}\right)^2, \tag{15.7.42}$$

where η is the dynamic viscosity [unit : Pa×s], k_{in} the intrinsic permeability [unit : m^2], Q the volumetric flow rate [unit : m^3/s], A the section [unit : m^2], $Q/A = J$ Darcy's velocity, i.e. relative fluid velocity times porosity, ρ the mass density of the fluid [unit : kg/m^3] and β an inertia coefficient [unit : 1/m]. In this relation which applies to an incompressible permeating fluid, the standard Darcy's term introduces the viscous damping due to friction of the fluid over the vessels. The nonlinear term aims at introducing the inertial effects due to the converging and diverging flows inside the pores and channels.

Asperities of the vessels are seen as triggering turbulence. This point of view has led to delineate zones in terms of a Reynolds number $R_e = \rho\, U\, L/\eta$ and of the relative roughness $R_r = \epsilon/L$, with L the diameter of the conduit and ϵ the height of the irregularities. Therefore, $\epsilon = 0$ for smooth walls while $\epsilon = L/2$ for an aperture tending to 0. In this context, the incipience of turbulence takes place at about $R_e = 2000$ for smooth vessel walls but decreases to a few hundreds for strongly irregular vessels. Typical flows adopt the format $U \sim -i$ in a laminar regime and $U \sim -i^{1/2}$ in a very turbulent regime. Note that these two expressions are particular cases of (15.7.42) where either the quadratic term or the linear term is neglected.

Non-Darcy's laws have been suggested at the other side of the hydraulic gradient spectrum, namely for low hydraulic gradients: flow seems to require a minimum hydraulic gradient to initiate, or perhaps more accurately, the flow rule is nonlinear for small hydraulic gradients and turns linear at larger hydraulic gradients,

$$U = -\left\{ \begin{array}{ll} k_1\, i^\alpha, & 0 < i < i_1; \\[2mm] k_2\,(i - i_0), & i_1 < i\,. \end{array} \right. \tag{15.7.43}$$

The exponent $\alpha > 1$, the coefficients k_1 and k_2 and the gradients $i_0 < i_1$ should satisfy the continuity of the flow velocity at $i = i_1$. The increase of permeability might be due to a modification of the vessel roughness.

15.7.9.2   Conditions for turbulent flows

The condition for which the non Darcian terms are not negligible at large hydraulic gradients is sometimes stated in terms of a *Reynolds number in porous media*,

$$R_e = \frac{\rho\, n\, U\, k_{\text{in}}^{1/2}}{\eta}, \tag{15.7.44}$$

where $J = n\, U$ is Darcy's velocity (a macroscopic quantity), while the Reynolds number of Section 15.7.9.1 involved the fluid particle velocity U. For porous media, a laminar flow where Darcy's law holds requires the Reynolds number to be lower than about 10, de Marsily [1986], p. 74. At values larger than 100, the flow is considered turbulent and Darcy's law is definitely inaccurate. This critical value is much smaller than the value 2000 admitted in hydrodynamics.

15.7.10   Measurement of pressure and tissue hydraulic conductivity via a manometer

Khosravani et al. [2004] propose a method to measure the hydraulic conductivity of a tissue[15.8]. A horizontal needle is inserted in the tissue and linked to a vertical manometer. Readings on the manometer yield information that can be exploited as follows.

The piece of tissue involved is considered to be located between two microvessels. Moreover the tissue elasticity is neglected and the extravasation process is not accounted for so that the pressure measured is assumed to be both the microvascular pressure and the interstitial fluid pressure[15.9]. The setting is sketched in Fig. 15.7.3. Subscripts m, n and t refer to manometer, needle and tissue, respectively.

[15.8] Fung [1990], p. 323 et seq., reviews methods for measuring the interstitial pressure in tissues. The actual values, even on their sign, are still controversial although the general opinion is that pressures in tissues are mostly positive, about 1-3 mmHg.
[15.9] Terminology: MVP: microvascular pressure; IFP: interstitial fluid pressure.

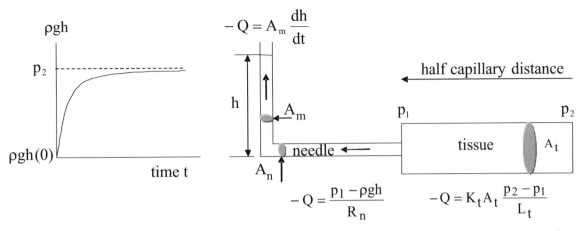

Fig. 15.7.3: Sketch of the setting described in Khosravani et al. [2004] to measure the hydraulic conductivity of a tissue. The tissue specimen is bounded by two capillaries. The pressure p_2 at its center is assumed not to be perturbed by the experiment.

Let us consider first the flux in the needle. According to the Hagen-Poiseuille relation (15.6.13), or equivalently to (15.9.1) and (15.9.2), the blood flow rate Q [unit : m^3/s] is equal to the pressure drop along the needle divided by the resistance R_n along the needle,

$$-Q = \frac{p_1 - \rho\, g\, h}{R_n}, \quad R_n = \frac{8}{\pi}\frac{\eta}{r_n^4} L_n, \tag{15.7.45}$$

where r_n is the radius of the needle, L_n its length, ρ the blood density and η the blood viscosity. Now the flow rate Q can be observed on the manometer as

$$-Q = A_m \frac{dh}{dt}, \quad A_m = \pi\, r_m^2, \tag{15.7.46}$$

where r_m is the manometer's radius. Combining these two equations yields a differential equation for the blood head h, assuming p_1 to be known,

$$\frac{dh}{dt} + \frac{h}{t_1} = \frac{1}{t_1}\frac{p_1}{\rho\, g}, \quad t_1 = \frac{A_m R_n}{\rho\, g}. \tag{15.7.47}$$

Here t_1 is a characteristic time response of the manometer-needle system.

However the tissue pressure p_1 is not known directly. The flow rate in the tissue is obtained by Darcy's law as

$$-Q = A_t\, K_t \frac{p_2 - p_1}{\rho\, g\, L_t}, \quad A_t = \pi\, r_t^2, \tag{15.7.48}$$

where K_t is the hydraulic conductivity of the tissue and the tissue length L_t is assumed to be half of the capillary distance. Eliminating p_1 in favor of the head h and pressure p_2 by combining (15.7.45) and (15.7.48) yields a differential equation for h,

$$\frac{dh}{dt} + \frac{h}{t_2} = \frac{1}{t_2}\frac{p_2}{\rho\, g}, \quad t_2 = t_1 + \frac{A_m}{A_t}\frac{L_t}{K_t}, \tag{15.7.49}$$

where t_2 is the characteristic time response of the system composed of the manometer, needle and tissue. Assuming the pressure p_2 to be constant, the time course of the solution h is governed by the relaxation time t_2 and h is observed to increase up to the asymptotic value $p_2/(\rho\, g)$,

$$h(t) = \frac{p_2}{\rho\, g} + \Big(h(0) - \frac{p_2}{\rho\, g}\Big)\exp(-\frac{t}{t_2}). \tag{15.7.50}$$

These relations are exploited as follows by Khosravani et al. [2004]. The density and viscosity of blood are close to their counterparts in water, namely density $\rho = 10^3\,\mathrm{kg/m^3}$; viscosity $\eta = 10^{-3}\,\mathrm{Pa \times s}$. The radii of the needle and manometer are equal, $r_m = r_n = 0.02\,\mathrm{cm}$, and the length of the needle is $L_n = 6.5\,\mathrm{cm}$. Consequently $t_1 = 1.3\,\mathrm{s}$. Furthermore, readings of the manometer provide $h(0)$ and $h(\infty) = p_2/(\rho\, g)$ as well as the time course of variation of h from which the relaxation time of the system t_2 is estimated equal to $14\,\mathrm{s}$ for a series of cervical tumors.

The intercapillary distance L_t is estimated equal to $0.0025\,\mathrm{cm} = 25\,\mathrm{\mu m}$ and the radius of drained tissue r_t is taken equal to half of the needle side port, namely $0.15\,\mathrm{cm}$. The hydraulic conductivity K_t follows from (15.7.49),

$$K_t = \frac{A_m}{A_t}\frac{L_t}{t_2 - t_1} = 3.5 \times 10^{-8}\,\mathrm{m/s}. \tag{15.7.51}$$

15.8   The steps of drug delivery, extravasation, transport in the interstitium

15.8.1   Delivery of molecules and particles

Therapies of cancer have to overcome the barrier to drug delivery that is opposed by tumors. There are many obstacles for the use of traditional therapeutic agents, e.g., Jain [1990],

- the high pressure that holds all over the tumors;
- heterogeneous blood supply;
- large transport distance in the interstitium;
- only a small fraction of the extravascular space in tumors is accessible to macromolecules, Krol et al. [1999];
- reduced mobile antibodies available due to binding to antigens.

Moreover, while avascular tumors are at atmospheric pressure, tumor interstitial hypertension is associated with angiogenesis, Boucher et al. [1996]. The spatial and temporal heterogeneities of tumor blood flow are not incompatible with some simple features, e.g., almost uniform interstitial fluid pressure (IFP), as shown in Fig. 2 of Eikenes et al. [2004]. The center of the tumor is mainly hypoxic and anoxic and drug delivery is especially difficult, Fig. 15.8.1.

Novel therapies aim at first circumventing the delivery issue, the actual efficiency of the drug being another issue. A review of drug delivery to solid tumors by systemic approaches is presented in Jang et al. [2003]. A number of barriers have to be overcome by drugs injected systemically, in particular the high IFP pressure and the limited size of vessels (100-800 nm). Size is becoming an issue as drugs are now more and more macromolecules. Another solution is to proceed by intratumoral infusion (topical), Netti et al. [2003].

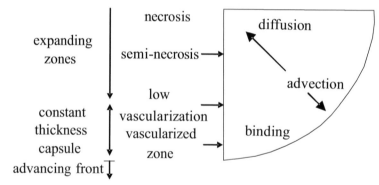

Fig. 15.8.1: Diffusion and convection dominate the transport of macromolecules in distinct zones of the tumor, Jain [1990]. The central zones are diffusion dominated while convection in the outer vascularized zone diverts the flow outside the tumor.

The transport of fluid and solutes in a porous medium depends on the physiological, geometrical and electrical properties of the porous medium as well as of the solutes. Seepage of water and diffusion of ions in saturated porous medium including coupled Darcy's law of seepage, Fick's diffusion and Ohm's law of electrical flow are addressed in Chapter 16. Modeling the various steps of blood-borne drug delivery to tumors sketched in Fig. 24.6.1 presents further challenges.

15.8.2   The successive steps in drug delivery

Let us consider a blood-borne species. It is first **convected** by the blood and, in some targeted or accidental zone (thin vessel, accidental movement of the needle, etc.), it leaves the vessel and enters the tissue by **extravasation**. Finally it diffuses and/or is convected in the extracellular and cellular spaces. In summary, the following steps are involved in blood-borne drug delivery:

  1. administration;
  2. blood convection in vessels;

3. adhesion to vessel wall;

4. extravasation through the wall by convection and diffusion;

5. transport in interstitium;

6. transport across the cellular membrane.

The following sections address essentially extravasation across the vascular wall and transport within the interstitium. Concerning step 6, we will just note that molecules may enter the cells, next to diffusion and convection, by endocytosis, a mode that has to be characterized quantitatively. Pharmacokinetic analyses of step 1 usually assume that the delivery of the drug in the blood results in a concentration whose variation in time can be taken as a sum of exponentials, e.g., Drummond et al. [1999], their scheme 1, and Krol et al. [1999], their Eqn (E).

15.8.3 Extravasation across the vascular wall

Extravasation of a solute from the blood flow b to the interstitium i across the vascular wall involves diffusion and convection:

- in a given volume of tissue, the diffusive solute rate Q_s^d [unit : mole/s] is proportional to the vascular surface S_v and to the difference in concentration of this species between the blood and the interstitium $c_b - c_i$ [unit : mole/m^3] with a proportionality coefficient $L_{v,s}$ [unit : m/s] interpreted as the solute vascular permeability

$$Q_s^d = L_{v,s} S_v (c_b - c_i); \tag{15.8.1}$$

- convection is proportional to the rate of leakage of the fluid from the vessel Q_w [unit : m^3/s] which is itself proportional to the vascular surface S_v and to the difference in pressure between the blood and the interstitium via the hydraulic vascular permeability L_v [unit : m/Pa/s],

$$Q_w = L_v S_v (p_b - p_i - \omega(\pi_b - \pi_i)). \tag{15.8.2}$$

Actually, if osmosis is accounted for, to the pressure difference $p_b - p_i$ (p_b=microvascular pressure MVP, p_i=interstitial fluid pressure IFP) should be subtracted the difference in oncotic pressure $\pi_b - \pi_i$ times the reflection coefficient ω. The reflection coefficient determines the efficiency of the oncotic pressure gradient across the vascular wall. The vascular wall is largely impermeable to large molecules for which $\omega \simeq 1$, while $\omega \simeq 0$ for small molecules. For albumin, reflection coefficient ω is almost 0 for the highly permeable liver, 1 for the quasi-impermeable brain, and about 0.5 for the lung. Because the tumor vasculature is highly permeable, the reflection coefficient of tumors is thought to be close to 0, Tong et al. [2004].

The absolute solute rate J_s is equal to $Q_s^d + c_b Q_w$. Jain [1987]b and Netti et al. [1999] propose modifications that weight the two basic rates by the Péclet number P_e and by a drag-reflection coefficient ω_w,

$$J_s = Q_s^d \frac{P_e}{e^{P_e} - 1} + c_b Q_w (1 - \omega_w). \tag{15.8.3}$$

Extravasation is characterized by four material properties (hydraulic and solute vascular permeabilities, solute and fluid reflection coefficients), the specific vascular surface, the relative pressures and the concentrations.

While the fraction of the extravascular space available to macromolecules in gels is decreasing monotonically and smoothly with their size, it undergoes jumps in tumor tissues, it is much smaller and its spatial variation is large, Krol et al. [1999]. Therefore, the apparent concentration as seen from the plasma during extravasation has to be amplified, *op. cit.*, their Eqn (D).

Jain and co-workers have provided a number of experimental observations on the pressures within the tumors:

- it turns out that VEGF (vascular endothelial growth factor) implies the **hydraulic and solute vascular permeabilities** of tumors, even if highly heterogeneous, **to be high**, Jain [1999]. Anti-VEGF/VPF's significantly reduce the tumor solute vascular permeability, typically from 2 to 1×10^{-7} cm/s, Tables 2 and 3 in Yuan et al. [1996];

- therefore one would expect drug delivery to be facilitated. However, there are two factors that counter drug delivery. First **the interstitial fluid pressure is large in tumors**: the latter is typically about 0 mmHg in most normal tissues while it varies from 15 to 30 mmHg in carcinomas, table 1 in Jain [1999]. This high pressure thus limits fluid leakage and so convection. But precisely, we have seen that large molecules are transported mainly by convection. In fact, it may well be that convection is directed out of the tumor, rather than towards the center, a phenomenon that may further contribute to the spreading out of tumor cells and formation of metastases. In addition, the fact that the exchange area per unit mass of tissue is smaller in large tumors implies that drug delivery will be still more difficult as the tumor grows;

- consequently, we are left with diffusion as the major mode of drug transport. As novel therapies concentrate on the use of large macromolecules (liposomes $\sim$90 nm, gene vectors $\sim$20-300 nm)[15.10] diffusion will have to be enhanced;
- Netti et al. [1999] measure interstitial and microvascular pressures and observe that they are very close, which they attribute to the fact that there is no lymphatic fluid drainage in tumors;
- Netti et al. [1995] [1999], Fig. 1 of the latter, have shown that the pressure transmission across the tumor vasculature takes about 10 s (a value close to the characteristic transcapillary time). During this time interval, the convective contribution is quite small, and most probably, the transmission time of molecules is quite larger, so that extravasation is low;
- in order to enhance convection and extravasation, Netti et al. [1999] use a vasoconstrictor agent: by increasing the systemic blood pressure[15.11], this agent increases the microvascular pressure and thus enhances convection[15.12]. However, they observe that non specific antibodies extravasate but leave the tissue within a blood pressure cycle. They conclude that to improve delivery, they must enhance the fluid filtration towards the interstitium **and** use specific antibodies that bind with cancer cells;
- Stohrer et al. [2000] develop a method to measure the oncotic pressure in the interstitial space of tumors. They observe that it is high and, due to the leaky nature of tumor vessels, sensibly equal to its value in the vessels and that the latter is only slightly affected by the existence of tumors. In fact, while the concentration of larger molecules is lower in the interstitial space, the opposite holds for the smaller molecules, yielding finally equal osmotic pressures. Therefore the oncotic pressure differential is low which hinders extravasation;
- the presence of high collagen levels in the interstitial space decreases the diffusion of large molecules (monoclonal antibodies, viral vectors), Boucher and Jain [1992], Netti et al. [2000]. Znati et al. [2003] measure the evolution of collagen type I and hyaluronan contents during irradiation: they observe that irradiation implies an increase of collagen content and a decrease of tissue hydraulic conductivity by an order of magnitude, from 20 to 2×10^{-8} cm^2/mmHg/s, with the unfortunate outcome of hindering delivery of drugs;
- intravenous or systemic injection of collagenase results in transient and non-synchronous modifications of the IFP and MVP, Figs. 3 and 4 in Eikenes et al. [2004]. The MVP drop is attributed to the disruption of collagen lining the vessels concomitant with an increased extravasation. The degradation of collagen in the interstitium is thought to generate the drop of the IFP. The effects of collagenase are dose-dependent and temporary (less than 1 day for non-lethal doses). Brekken et al. [2000] observe a quite similar IFP drop after intratumor injection of hyaluronidase.

15.8.4   Transport within the interstitium

Once it has been extravasated, a species is transported within the tissue once again by diffusion and convection which are characterized by the material parameters of the tissue. Data reported by Gullino et al. [1962] [1965] and Jain [1987]a [1987]b [1990] indicate that:

- the interstitial space in tumors is large;
- the collagen concentration in tumors is high, unlike GAG content;
- the diffusion coefficient is increased by the tumor, especially for large molecules, chart 2 in Nugent and Jain [1984], and, for several molecules, the tortuosity, that they call "diffusive hindrance", is equal to around one third;
- the hydraulic conductivity depends on water content in a power law format, Eqn (J) in Jain [1987]a;
- the hydraulic conductivity is increased in the tumoral tissue. The macromolecular transport in tumors is hindered less than in a normal tissue, when the interstitial space is larger, Nugent and Jain [1984]. Netti et al. [1995] indicate that the blood flow reaches the center of the tumor in about 1000 s;
- Ramanujan et al. [2002] study the effects of collagen type I on the diffusion and convection. At variance with the formula of Mackie and Meares [1955] used for polymers, they calculate the tortuosity factor (namely the ratio of the actual diffusion coefficient over the diffusion coefficient in a blank solution) through the Brinkman's effective medium model as a function of the ratio of the pore hydrodynamic radius to the square over the permeability. The latter is in turn calculated through a Kozeny-Carman formula where the collagen gel is considered as

---

[15.10]Drugs are encapsulated in macromolecular carriers such as liposomes. These sizes are to be compared with the size of capillaries shown in Table 15.9.1.

[15.11]The *systemic pressure* is defined as the difference (at systole) between the pressures at aorta and vena cava.

[15.12]Tumor vessels are devoid of smooth muscle vessels.

an array of cylinders, see Levick [1987], Phillips [2000]. The hydrodynamic radius is obtained by the Stokes-Einstein relation (15.5.2) which defines the diffusion coefficient in a blank solution. The measurements of Netti et al. [2000] and Ramanujan et al. [2002] show that collagen significantly hinders molecular diffusion, but that the latter is not affected by GAG content. Pluen et al. [2001] consider that the tortuosity factor is contributed by two multiplicative factors: first, geometry (steric hindrance) and second viscous forces (frictional hindrance) due to the fact that molecules undergo friction in the tissue capillaries and this effect increases with the molar mass. More subtly, Netti et al. [2000] suggest that what matters is not solely the collagen content but also the organization of the collagen network: they provide arguments that imply that the degree of organization of the collagen network correlates with the resistance to macromolecular transport. Thus, the relative disorganization of the collagen network of the tumor tissues favors transport of drugs while the higher collagen content has opposite effects.

Alexandrakis et al. [2004] criticize the lack of spatial resolution of previous measurements of diffusion coefficients in tumors: indeed, techniques must be able to reduce sampling volumes so as to detect the different zones of the tumors, namely center and capsule. They claim that previous data concern actually the capsule. Their technique highlights two components of diffusion whose coefficients differ by two orders of magnitude. Collagenase increases the fast diffusing component. The effect of hyaluronidase is opposite. In any case, since the tumor growth is accompanied by collagen synthesis[15.13], their findings indicate that drugs, that target collagen synthesis, not only would decompress the medium and decrease the pressures, thus enhancing convection, but they would also increase the diffusion.

- Brown et al. [2004] provide measurements of the diffusion coefficients of BSA (Bovine Serum Albumin) and IgM (Immunoglobulin M) in the ECM of human adenocarcinomas before and after excision and cooling. They highlight the fact that previous measurements concerned animal tumors: since direct measurements on patients are not currently feasible, they perform the measurements on excised tumors, and correct the data, to obtain an estimation of in situ values. For that purpose, they compare actual measurements in human xenografts on mice and after cooling and excision. They observe that the diffusion coefficients are higher than in xenografts, by a factor 2 to 3, and that the diffusion coefficients follow the temperature dependence of the Stokes-Einstein relationship (15.5.2), with a ratio of 2.55 for a change of temperature from 4 to 37 °C (the temperature change of viscosity is shown in Table 3.3.1).

Let us re-iterate the qualitative argument of Section 15.5 that the time δt for a species with an effective diffusion coefficient $D*$ to diffuse the distance δx is equal to $\delta x^2 / D*$. Thus immunoglobulin G, with a hydrodynamic radius of 35 Å and an effective diffusion coefficient $D* \approx 10^{-11}$ m^2/s, travels 100 μm in 17 mn, and needs almost 28 h to travel 1 mm, and 4 months to travel 1 cm in human glioblastomas or sarcomas. For a typical tumor of size 1 to 10 mm, Fig. 24.3.1 for typical dimensions, this diffusion time is quite long given that the vasculature may have already collapsed before the drug has been delivered. Additionally, because the interstitial pressure is large in the tumor center, convection may in fact not help but oppose the delivery of large molecules.

For the final cellular intake of blood-borne drugs,
- the rate-limiting process is generally the low blood flow;
- for tumors which are highly perfused, the bottle-neck is either the transport across the cellular membrane, or extravasation, Nugent and Jain [1984].

A general introduction to mass transport in capillaries and interstitial space and lymph flow can be found in Fung [1990], p. 309.

15.8.5 Transport of cells and adhesion

Cells constitute another family of therapeutic agents, besides molecules and particles. Two aspects have to be controlled: their transport by the blood vessels and their adhesion to tumor vessels. The strength of the adhesion depends on the adhesion molecules, on the surface of contact and on the deformability of the cells. Studies center on defining which molecules regulate adhesion.

[15.13]Still some data report that the collagen content per unit tumor weight remains constant, see Jain [1987a].

15.9 Blood circulation

15.9.1 The overall blood circulation network

The circulatory system involves two subparts, *the high pressure system* circulated by blood leaving the heart and *the low pressure system* circulated by blood reaching the heart. About 70% of the blood volume is contained in the low pressure part. Another distinction is made between the *large circulatory system* and the *small circulatory system* which includes the vessels that link the heart and the lungs. A simplified sketch of the blood circulation in the human body is displayed in Fig. 15.9.1.

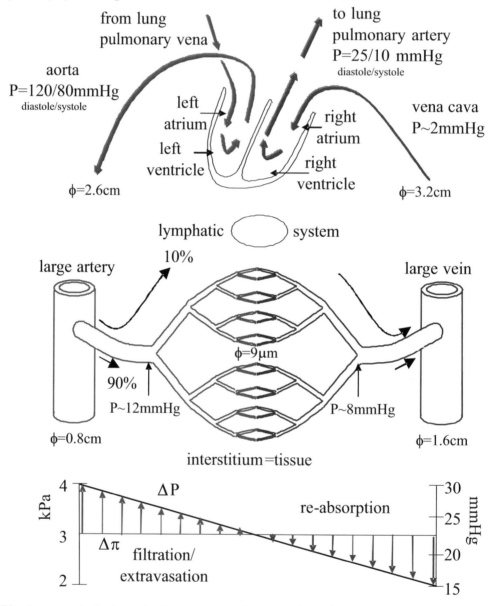

Fig. 15.9.1: Blood pressure in the large circulatory system, diameters of vessels, and transmural hydrostatic and osmotic blood pressures in the region of the capillaries of normal human tissues, $\Delta() \equiv ()_b - ()_i$, the indices b and i denoting respectively the blood and the interstitium, based on APP [2000], p. 155-159. The sign of $\Delta P - \Delta \pi$ is positive in the high pressure system, leading to *fluid* extravasation (filtration), and negative in the low pressure system, leading to re-absorption. The extravasation of *macromolecules* is dictated by Eqn (15.8.3). Pressures at the entrance and exit of capillaries should be corrected to account for their vertical distance to the heart. In tumors, the lymphatic network is not functional and the transmural pressures are practically zero.

In normal conditions, the human heart beats about 70 times per minute. During the systole, blood is ejected from the left ventricle to the aorta where the pressure reaches 16 kPa (120 mmHg). Its lowest value in the aorta is 10.7 kPa (80 mmHg) during the diastole when the blood is re-injected from the vena cava to the right atrium. The pressures are much lower in the low pressure system, from 0 kPa (0 mmHg) to 0.5 kPa (4 mmHg) in the vena cava. Values and variations of the pressure during a heart beat depend on the location of the organ in the body, but are within the range defined above.

The elasticity of aorta and of the arteries moderates the discontinuous blood ejection to a continuous flow, so called *Windkessel effect*. The largest vessels store energy at the systole and release it at the diastole.

While circulating in the high pressure system, the blood enters vessels whose number increases while their radii decrease as quantified in Fig. 10.8.1.

The *Reynolds number* R_e defined by (15.6.7) distinguishes laminar flows from turbulent flows. Large Reynolds numbers indicate a significant inertial effect. In the human aorta, R_e can reach more than 2000 so that turbulence might be possible, while it is quite small in capillaries, $R_e \simeq 10^{-3} - 10^{-2}$.

The loss of pressure ΔP between the aorta and the vena cava may be characterized by the resistance R [unit : Pa $\times$ s/m^3] to the blood flow rate Q [unit : m^3/s], much like for an electrical flow,

$$\text{systemic pressure} = \text{resistance} \times \text{flow rate} : \quad \Delta P = -R\,Q. \tag{15.9.1}$$

The total resistance of the circulatory network is equal to 2.4 kPa$\times$min/l. The resistances of the different parts of the circulatory network are shown in APP [2000], p. 156.

15.9.2   The vascular network

The resistance of a straight cylindrical vessel of radius a and length L circulated by a fluid of viscosity η is given by the Hagen-Poiseuille relation (15.6.13),

$$\text{Hagen} - \text{Poiseuille relation} : \quad R = \frac{8}{\pi}\frac{\eta}{a^4}\, L \quad [\text{unit} : \text{Pa} \times \text{s/m}^3]. \tag{15.9.2}$$

The actual viscosity η depends much on the hematocrit and protein contents. Note that the radius of the vessels appears in the denominator with a power 4: 19% increase of the radius is enough to halve the resistance.

The orthoradial tension T_θ in vessels is equal to the *transmural pressure*, or difference between the inner (blood) pressure p_b and outer (interstitial) pressure p_i, times the radius of the vessel a. The orthoradial stress σ_θ is simply equal to the tension T_θ divided by the thickness e:

$$\text{Laplace}'\text{s law} : \quad T_\theta = (p_b - p_i)\, a, \quad \sigma_\theta = (p_b - p_i)\,\frac{a}{e}. \tag{15.9.3}$$

Indeed, consider a cross-section. The force per unit axial length exerted by the radial pressure $\int_0^\pi (p_b - p_i)\sin\theta\, a\, d\theta = 2\,(p_b - p_i)\, a$ on one half vessel is compensated by the sum of the tensions $2\,T_\theta$ on the two diametrally opposite cuts. A more general proof of Laplace's law is exposed in Section 3.4.

Table 15.9.1: Typical diameters and collective sections of vessels, from APP [2000], p. 157.

	Aorta	Large artery	Arteriole	Capillary	Venula	Large vein	Vena cava
Diameter(cm)	2.6	0.8	0.002	0.0009	0.0025	1.6	3.2
Section(cm^2)	5.3	20	500	3500	2700	30	16

Table 15.9.1 displays typical diameters of arteries and veins. Note that the diameter of the capillaries, about 0.0009 cm (9 μm), is about 3000 times smaller than that of the aorta, about 2.6 cm. The capillary wall can be very thin although a lower bound is requested to limit the orthoradial stress as indicated by Eqn (15.9.3).

As it will be explained later, the decrease of the pressure along the capillary network allows the exchanges of liquid and solutes between blood and the interstitium by filtration from the high pressure system and re-absorption in the low pressure system.

The irrigation of the tissues should be performed through small vessels in order that 1. a great part of the transport of nutrients is ensured by convection and 2. the blood velocity is reduced. Indeed, at any point along the blood path, the flow rate is equal to the product of blood velocity times the total section of the vessels at this point. Actually, it turns out that, even if the individual section of small vessels is low, their number is so large that they offer the largest section to the flow along the blood path as indicated in Table 15.9.1. Still, the largest blood volume is shifted towards the venous side. Considering a constant flow rate, this section repartition along the blood path has the effect of decreasing the blood velocity in small vessels (0.2 m/s in the aorta, 0.3 mm/s in capillaries), a requirement for *exchange of nutrients and oxygen* to take place between blood and tissues.

15.9.3   Capillaries: filtration and re-absorption

Capillary walls have pores with a diameter of about 8 nm that allows small solutes, but not hematocrit or macromolecules, to leave the capillary flow and to enter the interstitium. The presence of pores in the capillary wall does not pose a structural threat since the orthoradial stress is low due to an appropriate ratio radius over thickness. Taken on another view point, the fact that capillaries are small allows them to have pores.

About 1/200 of the flow ejected by the heart crosses the capillaries to the tissue, that is about 20 liters per day. Out of these 20 liters, 18 liters are re-absorbed by the capillaries while 2 liters enter first the lymphatic circulation and latter re-enters the capillaries.

The *fluid* flux per unit area of capillary J_w [unit : m/s] across its wall is governed by the so-called Starling's law,

$$\text{Starling's law for the liquid}: \quad J_w = L_c \left(p_b - p_i - \omega \left(\pi_b - \pi_i \right) \right), \tag{15.9.4}$$

where L_c [unit : m/Pa/s] is the hydraulic permeability of the capillary wall and π is the oncotic pressure. The transcapillar osmotic pressure $\Delta \pi = \pi_b - \pi_i$ is approximately constant from the high pressure system to the low pressure system, Fig. 15.9.1. On the other hand, the effective transcapillar pressure $\Delta p - \omega \Delta \pi$ is positive on the high pressure system, leading to extravasation and blood filtration through the interstitium, Eqn (15.9.9). It is negative on the low pressure system, leading to *re-absorption*. In this formulation, the reflection coefficient $\omega \in [0, 1]$ considers a single type of solutes in the blood. A value $\omega = 1$ means that no solute leaves the capillary while $\omega = 0$ means that the capillary is not seen as a barrier at all by the solutes. Intermediate values imply that the capillary wall allows a part of the solutes to extravasate. Starling's law is motivated in Section 13.7.

Physiological (moderate) variations of the reflection coefficient are accompanied by a *self-regulation of leakage*. Indeed, a decrease of ω will a priori imply an increased extravasation of fluid since $\Delta \pi > 0$. As a result, the concentration of the solutes in the capillary will increase, and conversely in the interstitium, leading to an increase of $\Delta \pi$ and thus to a decrease of the fluid flux. A deregulation of the above balanced loop may lead to oedema when the filtration exceeds re-absorption, e.g., when the reflection coefficient in the high pressure system is decreased by a drug.

The arterial pressures shown in Fig. 15.9.1 correspond to locations above the heart. At lower locations, the blood pressure is increased due to the height of the blood column, typically +12 kPa at the feet. Standing positions are compensated by an increased filtration in the kidney region.

Among the suggested modifications with respect to the above presentation, let us mention that Fu et al. [2003] introduce electrical coupling in the above model. Hu et al. [2000] challenge the current interpretation of Starling's law. They contend that the entities related to the interstitium in the Starling's law, namely p_i and π_i, should be taken just behind the glycocalyx (the inner vessel membrane) and not in the tissue itself. In fact, the glycocalyx is covered externally by a cleft zone that is leaky but across which the concentrations vary.

The above remarks have addressed the filtration of the blood. Let us now consider the blood proteins. If neither the cellular elements (hematocrit) nor the plasma proteins cross the capillary wall, the numbers of proteins crossing the section $z = 0$ and the current section z during the time increment Δt are equal,

$$c_a \left(\underbrace{Q_a}_{\substack{\text{total} \\ \text{volume} \\ \text{at z} = 0}} - \underbrace{h_H \, Q_a}_{\substack{\text{cellular} \\ \text{elements}}} \right) \Delta t = c(z) \left(\underbrace{Q(z)}_{\substack{\text{total} \\ \text{volume} \\ \text{at z}}} - \underbrace{h_H \, Q_a}_{\substack{\text{cellular} \\ \text{elements}}} \right) \Delta t, \tag{15.9.5}$$

where h_H is the hematocrit percentage, the index a denoting the arteriole end $z = 0$. Thus the conservation of proteins yields the concentration $c(z)$ at section z in the format,

$$c(z) = c_a \, \frac{(1 - h_H) \, Q_a}{Q(z) - h_H \, Q_a} \, . \tag{15.9.6}$$

The oncotic pressure is usually expanded as a polynomial of the concentration of the plasma protein c (c=nb. proteins/volume of plasma),

$$\pi(c) = \beta_1 c + \beta_2 c^2 + \beta_3 c^3, \tag{15.9.7}$$

where the β's are constants. For example, Hu et al. [2000] provide values of these β's for human albumin.

While the Starling's law (15.9.4) concerns the liquid flow, the flow of a solute (a priori electrically neutral) is either deduced through a formula like (15.8.3) or postulated in the form,

$$\text{Starling's law for a solute:}\quad J_\text{s} = \alpha_1\,(p_b - p_i) + \alpha_2\,(\pi_b - \pi_i), \tag{15.9.8}$$

where α_1 and α_2 are material coefficients, possibly depending on molar masses, mechanical information, etc.An alternative formulation of the solute extravasation is reported in Section 15.8.3.

The amount of fluid that is filtered per unit time is called filtration rate. The filtration rate along a cylindrical capillary of radius $a(z)$ extending from $z = 0$ (arteriole end) to $z = L$ (venous end) is compensated by the change of flow rate Q [unit : m^3/s] along this path,

$$2\,\pi \int_0^L a(z)\,J_\text{w}(z)\,dz + Q(L) - Q(0) = 0, \tag{15.9.9}$$

the flow rate being proportional to the gradient of the blood pressure according to $(15.6.13)^{15.14}$,

$$a^4 \frac{dp_b}{dz} = -\frac{8\,\eta}{\pi}\,Q. \tag{15.9.10}$$

Determination of all the time and space variables involved in the filtration problem requires some quantities to be fixed while others may vary in time and along the blood flow path. Several particular cases are listed by Fung [1990], p. 312.

15.10   Mechanics of vessels

15.10.1   Constraints in the formation of vessels

Before addressing a few mechanical issues arising in the circulation of blood, it is of interest to record the three main types of constraints that control the formation of vessels during embryogenesis, Gilbert [1996], p. 346 et seq.:

- *physiological constraints*: the growing organs should work and develop while the future machinery is not yet operating: the heart is not yet sending blood and its nutrients, the lungs do not yet provide oxygen, etc. so that temporary systems should substitute these future functions;
- *constraints due to evolution*: a number of organs that are common to fishes, birds and mammals form temporarily and later either take precedence or disappear. For example the six aortic arcs that develop in the early stage give rise to branchies in fishes but disappear to form a single arc that evolve in a carotid artery in mammals;
- *physical constraints*: the Poiseuille relation $Q = -\Delta P/R$ with the resistance to flow R inversely proportional to the radius a of the vessels to the power four implies that *transport is a priori much faster in large vessels*.

15.10.2   Collapse of the veins

Collapse of veins

About two thirds of the blood is contained in the veins. Veins have a larger compliance than arteries because their thickness to diameter ratio is smaller. Moreover the vein blood pressure is smaller than in arteries so that the transmural pressure may transiently be negative as sketched in Fig. 15.9.1.

Lateral buckling is therefore a potential collapse mode. Let us therefore consider a long cylinder of thickness h, radius to midwall R, section up to midwall $A_0 = \pi R^2$, section of the lumen A, made of an elastic material of Young's modulus E and Poisson's ratio ν. The cylinder undergoes an internal pressure p and an external pressure p_e. The conditions for collapse express in terms of the three dimensionless quantities below, Fung [1990], p.185 et seq.,

$$\tilde{p} = \frac{p - p_e}{K_p}, \quad \alpha = \frac{A}{A_0}, \quad K_p = \frac{E\,h^3}{12\,(1 - \nu^2)\,R^3}. \tag{15.10.1}$$

15.14 The capillaries considered here are *porous* to plasma (not to hematocrit) but *rigid*, i.e., its radius is constant, in contrast to Section 15.6.3.4 where the capillaries are not porous but elastic.

As the transmural pressure decreases (in algebraic value), (a) collapse occurs for $\tilde{p} = -3$ and the cross-section leaves the circular shape; (b) at $\tilde{p} = -5.247$ and $\alpha = 0.27$, the opposite walls come into pointwise contact; (c) the contact area increases and the open part of the cross-section decreases. The experimental collapse curve is found to be approximated by the theoretical derivations $-\tilde{p} = \alpha^{-3/2}$ or $-\tilde{p} = \alpha^{-3/2} - 1$.

The actual setting is more complex because the veins are attached to the tissue which is itself elastic: its influence is more complex than a mere constant and uniform compression or tension.

The waterfall phenomenon in large veins (infinite Reynolds), Fung [1990], p.190 et seq.

Let us consider a gas flow in a tunnel, an unsteady frictional one-dimensional flow in a cylinder of section A and unit height $(h = 1)$ and a liquid flow in a horizontal uniform channel of height h and unit section $(A = 1)$. The equation of motion, neglecting gravity, reads in the three cases,

$$\frac{1}{\rho}\frac{\partial p}{\partial x} + \frac{\partial u}{\partial t} + u\frac{\partial u}{\partial x} = 0, \qquad (15.10.2)$$

where p is the pressure, ρ the density and u the velocity. The equations of mass conservation and speeds of sound read, respectively,

$$\text{gas flow} \qquad \frac{\partial \rho}{\partial t} + \frac{\partial(\rho u)}{\partial x} = 0, \quad c = \sqrt{\frac{dp}{d\rho}},$$

$$\text{cylinder} \qquad \frac{\partial A}{\partial t} + \frac{\partial(Au)}{\partial x} = 0, \quad c = \sqrt{\frac{A}{\rho}\frac{dp}{dA}}, \qquad (15.10.3)$$

$$\text{channel flow} \qquad \frac{\partial h}{\partial t} + \frac{\partial(hu)}{\partial x} = 0, \quad c = \sqrt{\frac{h}{\rho}\frac{dp}{dh}}.$$

The equations of mass conservation for the fluid derive by considering a volume $A \times h$ of mass $M = \rho A h$ and a constant mass density ρ, namely $\partial M/\partial t + \operatorname{div}(M u) = 0$. For the channel flow, the system of the two partial differential equations of mass and momentum can be cast in the format $\mathbf{A}\,\partial\mathbf{U}/\partial t + \mathbf{B}\,\partial\mathbf{U}/\partial x = \mathbf{0}$ with

$$\mathbf{A} = \begin{pmatrix} 1 & 0 \\ 0 & 1 \end{pmatrix}, \quad \mathbf{B} = \begin{pmatrix} u & h \\ \rho^{-1}dp/dh & u \end{pmatrix}, \quad \mathbf{U} = \begin{pmatrix} h \\ u \end{pmatrix}. \qquad (15.10.4)$$

The absolute wave-speeds $C = u \pm c$ are result from the secular equation $\det(\mathbf{A}\,C - \mathbf{B}) = 0$. The wave-speeds in the two other cases derive similarly.

Bernoulli equation (15.7.11) indicates that the quantity $p + \frac{1}{2}\rho u^2$ is constant. Therefore, for the flow in the cylinder, $du = -dp/(\rho u)$ and, using the speed of sound (15.10.3), $du/dA = -c^2/(A u)$. The derivative of the flow rate in the cylinder $Q = A u$ may be now expressed in the format,

$$\frac{dQ}{dp} = \frac{A}{\rho u}(S^2 - 1), \quad \text{with } S = \frac{u}{c} \quad \text{Shapiro number}. \qquad (15.10.5)$$

Thus if $u < c$, then $dp < 0$ increases the flux, and conversely for $u > c$. Thus the maximum flux obtainable by decreasing the pressure p is $Q = A c$ which depends neither on the upstream nor on the downstream pressures (waterfall phenomenon). These pressures matter only to have the phenomenon established.

The waterfall phenomenon in capillaries, Fung [1990], p. 215 et seq.

The waterfall phenomenon is observed as well in capillaries where the arteriole influx pressure is larger than the alveola external pressure while the venule outflux pressure is smaller so that capillaries collapse at the venous end. Then the flow is independent of the downstream condition.

15.11 Transport of oxygen and carbon dioxide in blood

15.11.1 Dissolution and binding of oxygen

Oxygen is present in blood in two forms in proportion indicated in Fig. 15.11.1, namely
- a small *dissolved* part;

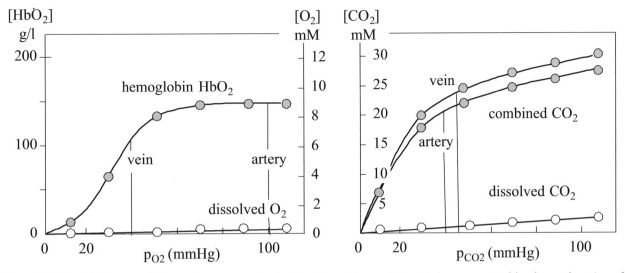

Fig. 15.11.1: (a) Curves of dissolved oxygen and of saturation of hemoglobin Hb by oxygen in blood as a function of oxygen pressure: the actual saturation depends somehow on the Hb content of the blood and of various factors, e.g. it increases as the concentration of carbon dioxide CO_2 decreases and decreases as the pH decreases; (b) curves of dissolved and of combined (total) content of CO_2 in blood. The oxygen and carbon dioxide pressures in veins and arteries are indicated in Fig. 15.11.3. Adapted from APP [2000], pp. 99-101.

- the most part *bound* to hemoglobin as oxyhemoglobin HbO_2.

According to Henry's law phrased in terms of concentration, Section 3.5, the concentration of dissolved oxygen in blood is proportional to its partial pressure,

$$[O_2]_b = \alpha_{O_2}\, p_{O_2}, \tag{15.11.1}$$

with $\alpha_{O_2} = 0.01\,\text{mmole}/1/\text{kPa}$ the coefficient of solubility. Actually, the solubility coefficient for blood (plasma) is typically expressed as $0.003\,\text{ml}\ O_2/100\,\text{ml blood}/\text{mmmHg}\ O_2$. The O_2 pressure in arteries is equal to about 12.7 kPa or 95 mmHg. Consequently the **solubility of O_2 is low**: according to (15.11.1), about $0.13\,\text{mmole}=0.13\times22.4=2.9\,\text{ml}$ of O_2 is dissolved per liter under standard conditions where one mole of O_2 occupies 22.4 l[15.15].

To alleviate the low solubility of O_2, there is another mode by which O_2 is transported by blood, namely via the erythrocytes (red blood cells). The latter fix O_2 through a protein, the **hemoglobin Hb**, which allows to increase the O_2 content 70 times to 210 ml/l of blood[15.16].

The hemoglobin acts as an allosteric enzyme: its saturation curve in O_2 is sigmoidal, Fig. 15.11.1. The human hemoglobin content is about 150 g/l. For a given Hb content, there is a maximum concentration of O_2 in blood, called oxygen binding capacity.

Hemoglobin Hb is a macromolecule with molar mass 64,800 g. One mole of Hb can fix at maximum 4 moles of O_2, equivalently 1 g of Hb fixes 0.062 mmole of O_2, and 150 g of Hb can fix 9.3 mmole or $0.0093\times22.4=0.207\,l$ of O_2.

The oxygen intake and carbon dioxide release by blood at the level of the lungs are allowed by the high partial pressure of oxygen $pO_2 \sim 100$ mmHg, Fig. 15.11.2. Conversely, the oxygen release by blood and carbon dioxide intake at the level of the capillaries are allowed by the low partial pressure of oxygen $pO_2 \sim 40$ mmHg. The sets of reactions in the alveoli and capillaries are shown in Fig. 15.11.2. The reaction that binds O_2 to Hb is affected by pH. The formation of bicarbonate H_2CO_3 is catalyzed by an enzyme, the carbonic anhydrase.

15.11.2   Dissolution and combination of carbon dioxide

Carbon dioxide CO_2 is present in blood in three forms:
- a small *dissolved* part;
- a small part bound to the N-end of Hb (not shown in the sketches);

[15.15] Actually, several conventions are in use. Here, we consider a temperature of 0°C, or 273.15 K, and a pressure of 1 atm=101 325 Pa.

[15.16] Nature has many ways to go around issues of this sort. In methane hydrates, water forms M-cages and S-cages of respectively, 6 and 5 hydrogen-bonded molecules that trap a gas molecule, the typical ratio between the number of M- versus S-cages being 3, yielding 4/23=0.17 molecule of gas per water molecule. By contrast, water can dissolve only about 0.002 molecule of gas per water molecule.

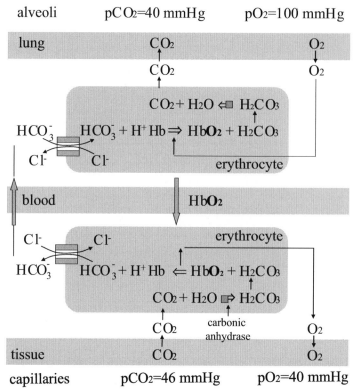

Fig. 15.11.2: Sketch of the reactions that allow the intake of O_2 from the lungs and the release of CO_2 and the converse reactions in the capillaries, as well as the consequent modes of transport of these gases in blood. The capillary pressures actually hold for a systemic vein. Fig. 15.11.3 shows more information on the gas pressures in lungs and tissue. Based on APB [1999], p. 269.

- the most part as *dissolved* bicarbonate (hydrogen carbonate) HCO_3^-.

The coefficient of solubility of carbon dioxide α_{CO_2} is equal to $0.22\,\text{mmole/l/kPa}$, about 20 times higher than that of O_2. Therefore the diffusion rate of CO_2 across the membrane of pulmonary alveolae is about 20 times higher than that of O_2. Even so, the solubility is low and, to be transported by blood, most of CO_2 is transformed in the more soluble form of bicarbonate HCO_3^- as displayed in Fig. 15.11.1. The dissolved form and the HCO_3^- form are transported by erythrocytes for about $1/3$ and by the plasma for $2/3$. The exchange of CO_2 between the erythrocytes and plasma is performed via an antiport that involves the bicarbonate HCO_3^- and the chloride ion (this exchange is termed the *chloride shift*). Venous blood chloride concentration is higher than arterial blood chloride concentration in proportion to the concentration of bicarbonate HCO_3^-.

15.11.3   Respiration and circulation of oxygen

The heart beats about 70 times in a minute and the volume of ejected blood per beat through the aorta is about $0.07\,\text{l}$. Therefore, the **cardiac flux** Q, i.e., the volume of ejected blood per unit time, is about $70 \times 0.07 \sim 5\,\text{l/min}$.

The flux q_{pa} of O_2 coming to the lungs through the pulmonary artery and the flux q_{pv} of O_2 leaving the lungs by the pulmonary vein are proportional to the product of the cardiac flux by their blood concentration,

$$q_{pa} = Q\,[O_2]_{pa}\,, \quad q_{pv} = Q\,[O_2]_{pv}\,. \tag{15.11.2}$$

The difference is compensated by the uptake q_{O_2} of O_2 by tissues,

$$q_{pa} + q_{O_2} = q_{pv} \quad \Rightarrow \quad Q = \frac{q_{O_2}}{[O_2]_{pv} - [O_2]_{pa}}\,. \tag{15.11.3}$$

Physiological measurements, APP [2000], pp. 92 and 154-155, provide the following typical values: $[O_2]_{pa} = 0.15\,\text{l/l}$ of blood, $[O_2]_{pv} = 0.20\,\text{l/l}$ of blood, $q_{O_2} = 250$ ml/min, so that the cardiac flux Q is equal to $5\,\text{l/min}$, as indicated above. Since one mole of O_2 occupies $22.4\,\text{l}$ under standard conditions, these concentrations can be expressed also as $[O_2]_{pa} = 6.7\,\text{mmole/l}$ of blood, $[O_2]_{pv} = 8.9\,\text{mmole/l}$ of blood, which, according to Fig. 15.11.1, should correspond to 40 mmHg and 100 mmHg, respectively.

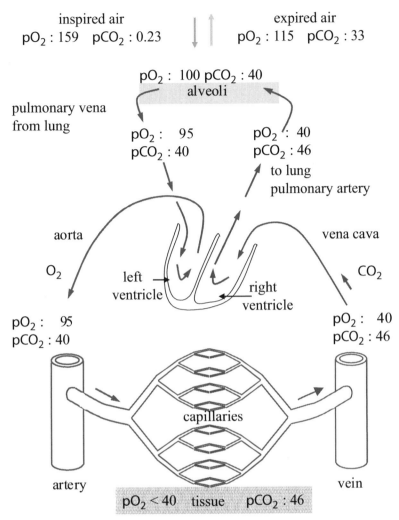

inspired air
$pO_2 : 159$ $pCO_2 : 0.23$

expired air
$pO_2 : 115$ $pCO_2 : 33$

$pO_2 :$ 100 $pCO_2 : 40$
alveoli

pulmonary vena
from lung

$pO_2 :$ 95
$pCO_2 : 40$

$pO_2 :$ 40
$pCO_2 : 46$

to lung
pulmonary artery

aorta

O_2

left
ventricle

right
ventricle

vena cava

CO_2

$pO_2 :$ 95
$pCO_2 : 40$

$pO_2 :$ 40
$pCO_2 : 46$

capillaries

artery

vein

$pO_2 < 40$ tissue $pCO_2 : 46$

Fig. 15.11.3: Pressures of O_2 and CO_2 [unit : mmHg] in lungs and tissues (100 mmHg=13.33 kPa). Note that the pressure of blood depends on the vertical position: pressure increases by about 7.4 mmHg=980 Pa per 10 cm.

15.12   Basics of enzymatic kinetics

Chemical reactions and enzymatic reactions may be brought in contact with a thermodynamic framework where chemical potentials and chemical affinities play a key role. Energy is required to activate a reaction that would lead to an increase of the free enthalpy (non spontaneous reactions). Catalysis concerns both spontaneous and non spontaneous reactions and it aims primarily at increasing the rates of the reactions: the target is to lower the energy path required to realize the reaction and the endpoints of the original reaction are untouched.

The concern here is on individual reactions. The existence of stable steady states in entire portions of the cell metabolism, namely cytoplasmic glycolysis, mitochondrial β-oxidation that degrades fatty acids, Krebs cycle (also called TCA cycle) that degrades products of glycolysis and β-oxidation, and oxidative phosphorylation that consists of the electron transport chain and ATP-ase, is the subject of bioenergetics. A general analysis of *supply and demand* is presented by Hofmeyr and Cornish-Bowden [2000]. They note, with faint surprise, that biochemistry books and models consider rate limiting steps within isolated parts of metabolism. A meaningful analysis should integrate the different parts of metabolism, so as to highlight the relative role (physiological function) of each of them.

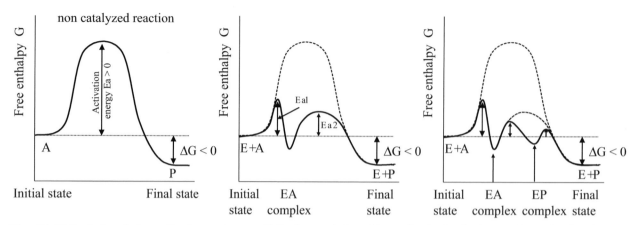

Fig. 15.12.1: A chemical reaction is spontaneous if the free enthalpy G is smaller in the final state than initially. The rate of advancement depends on the activation energy E_a to be overcome. Catalysis aims at finding different pathways, towards the *same final state*, that decrease the activation energy. Troughs (local minima) allow for the temporary stabilization of the complexes enzyme-substrate EA and enzyme-product EP. Adapted from APB [1999], p. 23.

15.12.1   Catalysis and enzymatic catalysis

The rate of advancement of a chemical reaction, whether exergonic or endergonic, depends on the successive thermodynamic states that the system assumes in the course of the reaction. Most reactions of organic chemistry are slow even if they are spontaneous because they need to overcome an energy barrier, the *activation energy*, an entity independent of the total variation of free enthalpy, Fig. 15.12.1. The higher the activation energy, the slower the rate of advancement. Catalysis aims at changing the pathway to the final state by minimizing the activation energy, and so maximizing the rate of advancement, with a multiplicative factor which might be as high as 10^{12}. Note that catalysis changes only the path between the initial state and final state. The latter is not affected by catalysis.

Biological reactants are covered by a film of water (hydration crown) and they move in all directions. Therefore, chemical reactions that require collisions to occur in special directions, have a low probability to be realized. Still when collisions occur in the right directions, they must bear enough energy to overcome the activation energy. This energy is used to remove the hydration crown and move electrical charges to allow contact between the reactants.

Catalysts help in accelerating the reactions by providing an appropriate receptacle to the reactants where they acquire a spatial conformation adequate for binding. Enzymes are powerful catalysts in biochemical reactions. Most enzymes are proteins, to the exception of the nucleic acids involved in ligases. As reactants enter the enzymes, the hydration water is lost, stage of enzyme-substrate complex EA in Fig. 15.12.2. While then all conditions for the actual binding of reactants to occur are completed, still it remains to be performed. Here is another key role played by the enzymes that interact with the reactants through amino acid residues and exchanges of protons, stage of enzyme-product complex EP in Fig. 15.12.2.

Rate constants increase with temperature, according to Arrhenius law,

$$k = p\,Z \exp\left(\frac{-E_a}{RT}\right) = \frac{k_\text{B}\,T}{h} \exp\left(\frac{-E_a}{RT}\right),\tag{15.12.1}$$

where E_a is the activation energy, R is the gas constant and T is the absolute temperature. The product of the steric constant p, indicating the proportion of collisions between reactants that have the correct orientation to allow reaction, and of the collision frequency Z is estimated by $k_\text{B}\,T/h$ with k_B Boltzmann constant and h Planck constant. The activation energy of many chemical reactions is in the range of $100\,\text{kJ/mole}$. For biological processes, rates of advancement approximately double when temperature increases by $10\,°\text{C}$, a phenomenon coined "Q10 rule". The beneficial effect of an increased temperature is however counterbalanced by a *denaturation* of the enzymes that, over 50-60 °C, decreases their catalyzing properties.

The 2000 or so various enzymes identified today are categorized into six principal classes, Table 15.12.1.

15.12.2   Reversible and irreversible enzymatic kinetics

The standard irreversible Michaelis-Menten kinetics is often advocated in a biological context. However, this model has two well-known pitfalls. First, most biochemical reactions are more or less reversible (there are still conspicuous counterexamples in the metabolic pathways, Section 20.2.3). In fact, even if a reaction was irreversible, it could be modeled as a reversible reaction with a very low backward rate constant.

Second, the irreversible Michaelis-Menten kinetics can not be reconciled with thermodynamics. That is why we consider in a second step two reversible formulations of the Michaelis-Menten kinetics. The nomenclature follows Cleland [1963] where a wealth of enzymatic reactions is analyzed.

15.12.3 The irreversible Michaelis-Menten model

Three states are described in the simplest enzymatic reaction. Enzymes E form a bound complex EA with the substrates ($\equiv$ reactants) A:

$$
\left\{
\begin{array}{ccccc}
\text{enzyme + substrate} & \rightleftharpoons & \begin{array}{c}\text{enzyme–substrate} \\ \text{complex}\end{array} & \rightarrow & \text{enzyme + product} \\[2mm]
E + A & \underset{k_1^-}{\overset{k_1^+}{\rightleftharpoons}} & EA & \overset{k_2^+}{\underset{}{\rightarrow}} & E + P
\end{array}
\right.
\tag{15.12.2}
$$

The rate constants adopt values in the following typical ranges: $k_1^+ = 10^5 - 10^8\,\mathrm{M^{-1}\,s^{-1}}$, $k_1^- = 1 - 10^4\,\mathrm{s^{-1}}$, $k_2^+ = 1 - 10^5\,\mathrm{s^{-1}}$, Voet and Voet [1998], p. 353. The rate of advancement, or *flux*, is the rate at which the product P is formed,

$$
\frac{d[P]}{dt} = k_2^+\,[EA],
\tag{15.12.3}
$$

where the symbol [ ] denotes the molar concentration measured in mole/liter of solution($\equiv$ M). The rate of change of the complex EA includes a positive and a negative contribution,

$$
\frac{d[EA]}{dt} = k_1^+\,[E][A] - (k_1^- + k_2^+)[EA]\,.
\tag{15.12.4}
$$

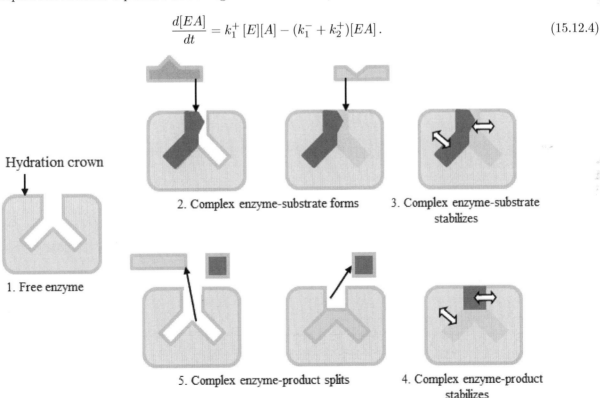

Fig. 15.12.2: Enzymes are powerful catalysts because they provide a receptacle to reactants/substrates where the latter adopt a favorable geometric positioning for binding (2). As they enter this configuration, they also dispose of the hydration crown. Stabilization of the enzyme-substrate complex can then occur (3). Finally, enzymes interact with the stabilized substrate to form the product (4),(5). Simplified from APB [1999], p. 89.

Table 15.12.1: Classes of enzymatic activities, modified from APB [1999], p. 87.

Class (subclasses)	Action	Schematic
oxido-reductases (dehydrogenases, oxidases)	transfer of reducing equivalents between twor edox systems	reducing agent + oxidant $\Leftrightarrow$ + reducer
transferases	transfer of groups between twom olecules	A B + C $\Leftrightarrow$ A + B C
hydrolases (peptidases)	transfer of groups between a molecule and water	A B + H$\,$O$\,$H $\Leftrightarrow$ H A + H$\,$O B
lyases or synthases (C-O-lyases)	catalysis of a reaction wherea double link is present	A + B $\Leftrightarrow$ AB
isomerases	catalysis of a group inside a molecule withoutc hanging the global formula	A $\Leftrightarrow$ isoA
ligases or synthetases (C-O-ligases)	catalysis of an endergonic reaction	X =A,C,G,U X-PPP + A B $\rightarrow$ AB + X-PP

Conservation of the mass of the enzyme should hold during the reaction,
$$[E]_0 = [E] + [EA],$$
(15.12.5)

with $[E]_0 = [E](t = 0)$. Upon substitution of (15.12.5) in (15.12.4), we obtain,
$$\frac{1}{k_1^+} \frac{1}{[A] + K_a} \frac{d[EA]}{dt} + [EA] = \frac{[E]_0 [A]}{[A] + K_a},$$
(15.12.6)

where the Michaelis constant $K_a = (k_1^- + k_2^+)/k_1^+$ ranges from 10^{-5} to 10^{-1} M. At steady state, or more generally, if the first term on the lhs of (15.12.6) is negligible in front of the second term, the reaction (15.12.6) simplifies to the *irreversible* Michaelis-Menten kinetics,
$$\frac{d[P]}{dt} = k_2^+ [EA] = \frac{k_2^+ [E]_0 [A]}{[A] + K_a} \quad \text{or} \quad J = \frac{J_{\max} [A]}{[A] + K_a} \quad \text{or} \quad \frac{[EA]}{[E]_0} = \frac{[EA]}{[E] + [EA]} = \frac{[A]}{[A] + K_a}.$$
(15.12.7)

The maximum rate of advancement $J_{\max} \equiv k_2^+ [E]_0$ [unit : M/s] occurs when the enzyme is completely saturated with substrate, i.e. $[A] \gg K_a$, and it is a characteristic of the *efficiency* of the catalysis. For $[A] = K_a$, $J = J_{\max}/2$: the Michaelis constant K_a indicates the *affinity* of the enzyme for the substrate. It is also the value of $[A]$ for which half of the enzyme molecules binds to the substrate. The rate constant k_2^+ is sometimes noted k_{cat} and termed *turnover number* as it represents the maximum number of substrate molecules that the enzyme can turn over to product per unit time. The rate of advancement at low concentration of substrate is governed by the *specificity constant* k_{cat}/K_a. In practice, K_a and $J_{\max}$ can be calculated by plotting $J = J_{\max} - K_a J/[A]$ as a function of $J/[A]$, so-called Eadie-Hofstee diagram.

15.12.4 Two reversible Michaelis-Menten models

Two reversible Michaelis-Menten models that differ by the number of central complexes are addressed separately first. Common properties are gathered next.

15.12.4.1 Uni uni reaction with a single central complex

The simpler scheme displays a single a central complex:

$$\begin{cases} \text{enzyme} + \text{substrate} & \rightleftharpoons & \begin{matrix}\text{enzyme}\\\text{complexes}\end{matrix} & \rightleftharpoons & \text{enzyme} + \text{product} \\[2mm] E + A & \underset{k_1^-}{\overset{k_1^+}{\rightleftharpoons}} & X = EA = EP & \underset{k_2^-}{\overset{k_2^+}{\rightleftharpoons}} & E + P \end{cases} \tag{15.12.8}$$

The rate equations may be cast in a matrix form,

$$\frac{d}{dt}\begin{bmatrix} [E] \\ [X] \end{bmatrix} = \begin{bmatrix} -k_1^+[A] - k_2^-[P] & k_1^- + k_2^+ \\ k_1^+[A] + k_2^-[P] & -k_1^- - k_2^+ \end{bmatrix}\begin{bmatrix} [E] \\ [X] \end{bmatrix}. \tag{15.12.9}$$

Conservation of mass of the enzyme should be ascertained,

$$[E] + [X] = [E]_0. \tag{15.12.10}$$

At steady state, the flux $J = d[P]/dt$ is constant at any point along the reaction path and it can be calculated in terms of the concentrations of products and reactants and of the total enzyme concentration $[E]_0$[15.17]. Here it may be brought to the form,

$$\frac{J}{[E]_0} = \frac{1}{D}\left(k_1^+ k_2^+[A] - k_1^- k_2^-[P]\right) \tag{15.12.11}$$

with

$$D = k_1^+[A] + k_2^-[P] + k_1^- + k_2^+. \tag{15.12.12}$$

An alternative form of *flux* will be displayed below in terms of *forward* and *backward maximum fluxes*,

$$\frac{J_1}{[E]_0} = k_2^+, \quad \frac{J_2}{[E]_0} = k_1^-, \tag{15.12.13}$$

kinetic constants,

$$K_a = \frac{k_1^- + k_2^+}{k_1^+}, \quad K_p = \frac{k_1^- + k_2^+}{k_2^-}, \tag{15.12.14}$$

and *equilibrium constant,*

$$K_{eq} = \frac{k_1^+ k_2^+}{k_1^- k_2^-}. \tag{15.12.15}$$

The irreversible scheme is retrieved in the limit of $k_2^- = 0$.

15.12.4.2 Uni uni reaction with two central complexes

In the second scheme, two central complexes, namely enzyme-substrate EA and enzyme-product EP, are recognized,

$$\begin{cases} \text{enzyme} + \text{substrate} & \rightleftharpoons & \begin{matrix}\text{enzyme/}\\\text{substrate}\\\text{complex}\end{matrix} & \rightleftharpoons & \begin{matrix}\text{enzyme/}\\\text{product}\\\text{complex}\end{matrix} & \rightleftharpoons & \text{enzyme} + \text{product} \\[2mm] E + A & \underset{k_1^-}{\overset{k_1^+}{\rightleftharpoons}} & EA & \underset{k_2^-}{\overset{k_2^+}{\rightleftharpoons}} & EP & \underset{k_3^-}{\overset{k_3^+}{\rightleftharpoons}} & E + P \end{cases} \tag{15.12.16}$$

[15.17]On the other hand, the knowledge of the kinetic constants alone does not allow in general to obtain the concentrations of the enzyme complexes at steady state. Moreover, it seems that the enzymatic concentrations are not directly involved in the kinetic constants, but they are in particular, the forward and backward fluxes are proportional to the total (=initial) enzyme concentration.

The governing equations include the rate equations,

$$
\frac{d}{dt}
\begin{bmatrix}
[E] \\
[EA] \\
[EP]
\end{bmatrix}
=
\begin{bmatrix}
-k_1^+ [A] - k_3^- [P] & k_1^- & k_3^+ \\
k_1^+ [A] & -k_1^- - k_2^+ & k_2^- \\
k_3^- [P] & k_2^+ & -k_2^- - k_3^+
\end{bmatrix}
\begin{bmatrix}
[E] \\
[EA] \\
[EP]
\end{bmatrix},
\qquad (15.12.17)
$$

and the conservation of mass of the enzyme,

$$
[E] + [EA] + [EP] = [E]_0 . \qquad (15.12.18)
$$

The resulting flux $J = d[P]/dt$ at steady state now writes,

$$
\frac{J}{[E]_0} = \frac{1}{D} \left(k_1^+ k_2^+ k_3^+ [A] - k_1^- k_2^- k_3^- [P] \right) \qquad (15.12.19)
$$

with

$$
D = k_1^- (k_2^- + k_3^+) + k_2^+ k_3^+ + k_1^+ (k_2^+ + k_2^- + k_3^+) [A] + (k_1^- + k_2^+ + k_2^-) k_3^- [P] . \qquad (15.12.20)
$$

The alternative form of *flux* expresses in terms of *forward* and *backward* maximum fluxes,

$$
\frac{J_1}{[E]_0} = \frac{k_2^+ k_3^+}{k_2^+ + k_2^- + k_3^+} , \qquad
\frac{J_2}{[E]_0} = \frac{k_1^- k_2^-}{k_1^- + k_2^+ + k_2^-} , \qquad (15.12.21)
$$

kinetic constants,

$$
K_a = \frac{k_1^- (k_2^- + k_3^+) + k_2^+ k_3^+}{k_1^+ (k_2^+ + k_2^- + k_3^+)} , \qquad
K_p = \frac{k_1^- (k_2^- + k_3^+) + k_2^+ k_3^+}{(k_1^- + k_2^+ + k_2^-) k_3^-} , \qquad (15.12.22)
$$

and *equilibrium constant,*

$$
K_{eq} = \frac{k_1^+ k_2^+ k_3^+}{k_1^- k_2^- k_3^-} . \qquad (15.12.23)
$$

The irreversible scheme is retrieved in the limit of $k_1^- = 0$ or $k_2^- = 0$. The scheme with a single central complex is a particular case of the scheme with two central complexes with $k_2^+ = k_2^- = \infty$.

15.12.4.3 Steady state unified flux for the two reversible reactions

The steady state flux of the two reactions may be expressed in the common format,

$$
J = \frac{J_1 J_2 \left([A] - \dfrac{[P]}{K_{eq}} \right)}{K_a J_2 + J_2 [A] + \dfrac{1}{K_{eq}} J_1 [P]} = \frac{J_1 \left([A] - \dfrac{[P]}{K_{eq}} \right)}{K_a + [A] + \dfrac{K_a}{K_p} [P]} , \qquad (15.12.24)
$$

where however the fluxes J_1 and J_2, kinetic and equilibrium constants are innate to each reaction. Similarly, the equilibrium constant and kinetic coefficients of each reaction satisfy the *Haldane compatibility relationship*:

$$
K_{eq} = \frac{J_1}{J_2} \frac{K_p}{K_a} . \qquad (15.12.25)
$$

With $\mathcal{G}$ the chemical affinity associated with the two reversible reactions and defined by (3.1.17), namely

$$
\mathcal{G} = RT \, \mathrm{Ln} \, \frac{1}{K^{\mathrm{eq}}} \frac{[P]}{[A]} , \qquad (15.12.26)
$$

the flux may be recast in the format,

$$
J = \frac{J_2 [P] \left(e^{-\mathcal{G}/RT} - 1 \right)}{K_p + [P] + \dfrac{K_p}{K_a} [A]} . \qquad (15.12.27)
$$

The models are compatible with a thermodynamic approach: the signs of the flux and chemical affinity are opposite so that the dissipation inequality (4.8.8) is satisfied.

15.12.5 Catalysis inhibition

Many substances can reduce the catalyzing activity of enzymes. Some are non-specific denaturants. Others that act in a specific way are called *inhibitors*. Some inhibitors act reversibly, i.e. the usual enzymatic activity is recovered if the inhibitors are removed. Other inhibitors act irreversibly on the enzymes whose usual activity can not be recovered. Inhibition can be competitive or non competitive. Competitive inhibitors may act as analogs of the substrate A or as analogs of the enzyme-substrate complex EA, e.g., for the irreversible Michaelis-Menten kinetics,

$$
\begin{cases}
E + A \quad \underset{k_1^-}{\overset{k_1^+}{\rightleftharpoons}} \quad EA \quad \overset{k_2^+}{\rightarrow} \quad E + P \\[2mm]
+I \;\uparrow\downarrow K_i \qquad\qquad +I \;\uparrow\downarrow K_i' \\[2mm]
EI \qquad\qquad\qquad EAI
\end{cases}
\tag{15.12.28}
$$

with K_i and K_i' equilibrium constants.

Suicide substrates, e.g., penicillin, are a special case of non competitive inhibitors: they first take the place of substrate and next bind irreversibly to the enzymes. Allosteric inhibitors fix to a site of the enzymes different from the sites used by substrates and modify the conformation of the enzymes whose catalyzing activity is altered. The kinetics does no longer obey the Michaelis-Menten model.

A detailed account of a number of reversible, irreversible, competitive, non competitive inhibitors which act as analogs of the substrates or analogs of the substrate-enzyme complexes is exposed in Cornish-Bowden [1979].

15.13 Acid-base equilibrium

The influence of pH on the transport and mechanical properties of two tissues endowed with a fixed electrical charge, namely articular cartilage and cornea, is addressed in Chapters 17 and 19. Some basic information on the systemic values of pH is given below.

15.13.1 pHs in plasma, interstitium and cytoplasm

The physiological activities are optimized at the standard pH. The molecular form and charge of most proteins depends on pH. pH also affects the transport properties of some ionic channels and of some enzymatic reactions.

In fact, the physiological pH is not uniform and it takes distinct values in blood and cell. While the pHs in plasma and interstitium are very similar, ranging from 7.35 to 7.45, the cytoplasm is slightly more acid, with a pH from 7.1 to 7.3. For the sake of reference, pHs of 7 and 7.40 correspond to concentrations of hydrogen ions $[H^+]$ equal to 100 and 40 nM, respectively.

The main sources of hydrogen ions (or protons) in the human body are
- food which produces acids and proteins whose degradation yields sulfuric acid;
- cetonic bodies. Indeed, diabetes prevents the completion of the tri-carboxylic acid (TCA) cycle. The remaining acetyl-CoA is transformed into cetonic bodies, which may give rise to a severe acidosis leading to diabetic coma;
- metabolism: glucose is transformed into lactate and lactic acid.

15.13.2 Buffers in plasma

Protonation/deprotonation, i.e., absorption/desorption of a proton H^+, is used by plasma to buffer protons and avoid pH to take damaging extreme values. In fact, the systemic pH is strongly regulated, a variation of pH of 0.05 with respect to the physiological value, associated with a change of concentration of hydrogen ions of about 5 nM being considered a deregulation.

Along a standard definition, the pH is equal to minus the decimal logarithm of the concentration of protons measured in mole/liter of blood,

$$
\text{pH} = -\log[H^+] \Leftrightarrow [H^+] = 10^{-\text{pH}}.
$$

The most important buffer in blood is carbon dioxide CO_2/bicarbonate HCO_3^-: the weak carbonic acid H_2CO_3 exists in general under its anhydrous form CO_2 and, for high pH ($pH > pK_1 = 6.1$), it dissociates in an anion and a proton as shown by Fig. 15.11.2,

$$\text{bicarbonate buffer} \begin{cases} CO_2 + H_2O \;\rightleftharpoons\; H_2CO_3 \;\rightleftharpoons\; H^+ + HCO_3^- \\[2mm] \dfrac{d[H^+]}{dt} \;=\; k_1 \left([H_2CO_3] - K_{1,eq}[H^+][HCO_3^-] \right) \end{cases} \tag{15.13.1}$$

The equilibrium constant $K_{1,eq} = 10^{pK_1}$ is defined by the particular value of pH, noted $pK_1 = 6.1$, for which the acid H_2CO_3 and the bicarbonate ion HCO_3^- have identical concentrations. Clearly, for $pH > pK_1$, the 2nd reaction above is from left to right and the carbonic acid dissociates and releases protons.

The next significant buffer is the hemoglobin buffer. The reaction shown in Fig. 15.11.2,

$$\text{hemoglobin buffer} \begin{cases} HbH^+ \;\rightleftharpoons\; H^+ + Hb \\[2mm] \dfrac{d[H^+]}{dt} \;=\; k_2 \left([HbH^+] - K_{2,eq}[H^+][Hb] \right) \end{cases} \tag{15.13.2}$$

has a $pK = pK_2$ equal to 8.25.

The phosphate buffer is less important but it is of particular interest for ATP-synthase. Dephosphorylation of ATP^{4-} yields an inorganic phosphate noted P_i which exists in essentially two ionic forms in equilibrium,

$$\text{phosphate buffer} \begin{cases} H_2PO_4^- \;\rightleftharpoons\; H^+ + HPO_4^{2-} \\[2mm] \dfrac{d[H^+]}{dt} \;=\; k_3 \left([H_2PO_4^-] - K_{3,eq}[H^+][HPO_4^{2-}] \right) \end{cases} \tag{15.13.3}$$

The $pH = pK_3$ is equal to $6.8^{15.18}$.

At equilibrium of the three reactions, the concentration of hydrogen ions and of anions satisfy the relations,

$$[H^+] = \frac{1}{K_{1,eq}} \frac{[H_2CO_3]}{[HCO_3^-]} = \frac{1}{K_{2,eq}} \frac{[HbH^+]}{[Hb]} = \frac{1}{K_{3,eq}} \frac{[H_2PO_4^-]}{[HPO_4^{2-}]}. \tag{15.13.4}$$

At pH=7.40, the ratio $[HCO_3^-]/[H_2CO_3]$ is equal to $20/1$, so that the bicarbonate buffer is very efficient. For a concentration of (dissolved !) CO_2 equal to 1.2 mM, Fig. 15.11.1, the concentration of $[HCO_3^-]$ is equal to 24 mM. At the same pH, the ratio $[HPO_4^{2-}]/[H_2PO_4^-]$ is only equal to 4.

[15.18] At equilibrium, the partition of the two ionic forms is obtained by the relations,

$$[P_i] = [H_2PO_4^-] + [HPO_4^{2-}], \quad [H_2PO_4^-] = [H^+] \times [HPO_4^{2-}] \times K_{3,eq},$$

so that

$$\frac{[H_2PO_4^-]}{[P_i]} = \frac{1}{1 + 10^{pH - pK_3}}, \quad \frac{[HPO_4^{2-}]}{[P_i]} = \frac{10^{pH - pK_3}}{1 + 10^{pH - pK_3}}.$$

An identical representation is feasible with the bicarbonate buffer as well.

Exercises on Chapter 15

Exercise 15.1 Calculation of drag coefficient for spherical particles

Let us consider an incompressible viscous fluid whose uniform motion is perturbed by the presence of a spherical particle of radius a and with center at the origin of the coordinates. Both gravity and inertial terms are neglected. Far from the particle, the velocity of the fluid $\mathbf{w}$ is parallel to the x_1-axis and has components $(U, 0, 0)$. On the other hand, it vanishes on the surface of the particle.

The stress of the fluid is equal to $\boldsymbol{\sigma} = 2\eta\,\dot{\boldsymbol{\epsilon}} - P\mathbf{I}$, with $\dot{\boldsymbol{\epsilon}} = \frac{1}{2}\left(\boldsymbol{\nabla}\mathbf{w} + (\boldsymbol{\nabla}\mathbf{w})^{\mathrm{T}}\right)$ the strain rate, P the fluid pressure and η the coefficient of viscosity. The equations that govern the motion are the balance of mass (incompressibility) and the balance of momentum,

$$\operatorname{div}\mathbf{w} = 0, \quad \operatorname{div}\boldsymbol{\sigma} = \mathbf{0}, \tag{15.E.1}$$

that is, componentwise

$$\sum_j \frac{\partial \sigma_{ij}}{\partial x_j} = \sum_j \eta\,\frac{\partial^2 w_i}{\partial x_j^2} + \eta\,\frac{\partial}{\partial x_i}\left(\frac{\partial w_j}{\partial x_j}\right) - \frac{\partial P}{\partial x_i} = \eta\,\Delta w_i - \frac{\partial P}{\partial x_i} = 0, \quad i \in [1,3]. \tag{15.E.2}$$

To build the solution, the important property that $1/r$ is harmonic is used. A function is harmonic in a certain domain if its Laplacian vanishes in this domain. The velocity is sought in the form

$$\mathbf{w} = \boldsymbol{\nabla}\phi + \mathbf{w}_1, \quad \mathbf{w}_1 \equiv (-c/r, 0, 0), \quad c \text{ constant}, \tag{15.E.3}$$

where r is the distance to the center of the particle. Thus (15.E.2) becomes

$$\eta\,\Delta w_i = \eta\,\frac{\partial\Delta\phi}{\partial x_i} = \frac{\partial P}{\partial x_i}, \quad i \in [1,3], \tag{15.E.4}$$

and therefore

$$\eta\,\Delta\phi = P + d, \quad d \text{ constant}, \tag{15.E.5}$$

and (15.E.1)$_1$ becomes

$$\Delta\phi + \operatorname{div}\mathbf{w}_1 = \Delta\phi + c\,x_1/r^2 = 0. \tag{15.E.6}$$

A solution of this equation can be checked to be

$$\phi = \frac{2}{3}\frac{c\,x_1}{a}\left(1 + \frac{3}{4}\frac{a}{r} - \frac{1}{4}\frac{a^3}{r^3}\right). \tag{15.E.7}$$

Of course, the problem displays symmetry around the x_1-axis. Let (r, θ) be polar coordinates in an arbitrary meridian plane containing the symmetry axis. Then $x_1 = r\cos\theta$, and the velocity is meridian,

$$\begin{aligned}
w_r &= -c\frac{\cos\theta}{r} + \frac{\partial\phi}{\partial r} = \frac{2}{3}\frac{c}{a}\cos\theta\left(1 - \frac{3}{2}\frac{a}{r} + \frac{1}{2}\frac{a^3}{r^3}\right), \\
w_\theta &= c\frac{\sin\theta}{r} + \frac{1}{r}\frac{\partial\phi}{\partial\theta} = \frac{2}{3}\frac{c}{a}\sin\theta\left(-1 + \frac{3}{4}\frac{a}{r} + \frac{1}{4}\frac{a^3}{r^3}\right),
\end{aligned} \tag{15.E.8}$$

and can be checked to vanish on the surface of the sphere $r = a$. Far from the sphere, it is duly parallel to the x_1-axis and equal to $2c/3a$ which should be equal to U. Thus $c = 3/2\,aU$.

The traction $\mathbf{t}$ over the surface of the sphere is equal to $\sigma_{rr}\mathbf{e}_r + \sigma_{r\theta}\mathbf{e}_\theta$. Now $\sigma_{rr} = \eta\,(\partial w_r/\partial r + \partial w_\theta/r\partial\theta) - P$ and $\sigma_{r\theta} = \eta\,(\partial w_\theta/\partial r + \partial w_r/r\partial\theta - w_\theta/r)$. From (15.E.5) and (15.E.6), on the sphere, $P = -3\eta U/(2a)\cos\theta + \text{constant}$. Thus $\sigma_{rr} = -P = 3\eta U/(2a)\cos\theta$ to within a constant and $\sigma_{r\theta} = -3\eta U/(2a)\sin\theta$. Consequently the traction $\mathbf{t}$ is equal to $3\eta U/(2a)\,(\cos\theta\,\mathbf{e}_r - \sin\theta\,\mathbf{e}_\theta)$. Thus, on the sphere, the traction $\mathbf{t}$ is parallel to $\mathbf{e}_1$, uniform and equal to $3\,\eta U/(2a)\,\mathbf{e}_1$. The force exerted by the moving viscous fluid on the sphere results as[15.19]

$$\mathbf{F} = 4\pi\,a^2 \times \frac{3\,\eta U}{2a}\,\mathbf{e}_1 = 6\,\pi\,\eta\,a\,U\,\mathbf{e}_1. \tag{15.E.9}$$

Incidentally, this result due to Stokes provides the steady state velocity of a small sphere (density ρ_{s}) falling in a viscous liquid (mass density ρ_{w}). Indeed, equaling the downward driving force due to gravity g and the upward viscous resistance gives

$$\frac{4}{3}\pi\,a^3\,(\rho_{\mathrm{s}} - \rho_{\mathrm{w}})\,g = 6\pi\eta a U \quad \Rightarrow \quad U = \frac{2}{9}\frac{a^2}{\eta}(\rho_{\mathrm{s}} - \rho_{\mathrm{w}})\,g. \tag{15.E.10}$$

[15.19] For a sphere in a tube of radius b, the force exerted by the fluid is multiplied by a factor $1 + 1.05\,a/b$ with respect to (15.E.9). The forces on some non spherical objects are of interest: for example, for a slender cylinder of length L, the force orthogonal to the cylinder axis is equal to $4\pi\eta L U$.

This expression can be interpreted as providing the body force exerted by the moving sphere on the fluid,

$$(\rho_{\rm s} - \rho_{\rm w})\,g = \frac{9}{2}\frac{\eta}{a^2}\,U\,. \tag{15.E.11}$$

This expression is the *Stokes' drag* that contributes to the momentum supply of the fluid phase. Then U is viewed as the relative velocity of the solid and fluid, $\mathbf{v} - \mathbf{w}$, and the coefficient of U in (15.E.11) is called *drag coefficient*, Section 7.3.2. This derivation can be found in textbooks, e.g., Mandel [1966], p. 423, Malvern [1969], p. 470, Fung [1990], p. 132.

For cylindrical fibrils of radius a and volume fraction $n^{\rm w}$, the Stokes' drag has the expression, Kim and Karrila [1991],

$$(\rho_{\rm s} - \rho_{\rm w})\,g = \frac{8}{\pi}\frac{\eta}{a^2}\frac{U}{{\rm Ln}\,(\pi/n^{\rm w}) - 1} \simeq \frac{8}{\pi}\frac{\eta}{a^2}\,U. \tag{15.E.12}$$

Exercise 15.2 A multipurpose contour integration

The purpose is the inversion of the one-sided Laplace transform $\bar{q}(x,p)$ of the function $q(x,t)$,

$$\bar{q}(x,p) = \frac{\exp(-x\,\sqrt{p/a})}{p - b}\,, \tag{15.E.13}$$

where $a > 0$, $b \geq 0$ are constants, $x \geq 0$ is the space coordinate and p the Laplace variable associated with time t. The integral is provided in textbooks, e.g., Carslaw and Jaeger [1959], p. 495, but we want to obtain the result through elementary analytical properties.

The function is made uniform by introducing a branch cut along the negative axis Re $p \leq 0$, and the definitions,

$$p = |p|\,\exp(i\,\theta), \quad \theta \in]-\pi, \pi], \quad \sqrt{p} = \sqrt{|p|}\,\exp(i\,\theta/2) \quad (\Rightarrow {\rm Re}\,\sqrt{p} \geq 0)\,. \tag{15.E.14}$$

With the (inverse) transforms, Table 15.3.1,

$$\mathcal{L}^{-1}\left\{\exp(-x\,\sqrt{\frac{p}{a}})\right\}(u) = \frac{x}{2\,\sqrt{\pi\,a\,u^3}}\,\exp(-\frac{x^2}{4\,a\,u}), \quad \mathcal{L}^{-1}\left\{\frac{1}{p - b}\right\}(u) = \exp(b\,u)\,\mathcal{H}(u)\,, \tag{15.E.15}$$

the function $q(x,t)$ can be expressed in the form,

$$q(x,t) = \exp(b\,t)\,\frac{2}{\sqrt{\pi}}\int_{x/2\sqrt{a\,t}}^{\infty}\exp\left(-v^2 - \frac{b\,x^2}{4\,a\,v^2}\right)dv\,, \tag{15.E.16}$$

since the Laplace transform of a convolution integral is equal to the product of the Laplace transforms.

This integral seems out of reach, except if $b = 0$, where

$$q(x,t) = \exp(b\,t)\,{\rm erfc}\left(\frac{x}{2\sqrt{a\,t}}\right)\,. \tag{15.E.17}$$

Still it is provided in explicit form in Abramowitz and Stegun [1964], p. 304, for any b.

Since we want to obtain the result on our own, even for $b \neq 0$, we take a step backward and consider the inversion in the complex plane through an appropriate closed contour $C = C(R, \epsilon)$, that respects the branch cut and includes the pole $p = b$. A preliminary finite contour is shown in Fig. 15.E.1. The residue theorem yields,

$$\frac{1}{2\,i\,\pi}\left(\int_{c-i\,R}^{c+i\,R} + \int_{C_R} + \int_{C_\epsilon} + \int_{C+} + \int_{C-}\right)\exp(t\,p)\,\bar{q}(x,p)\,dp$$
$$= \frac{1}{2\,i\,\pi}\oint_C\exp(t\,p)\,\bar{q}(x,p)\,dp = \exp(b\,t - x\,\sqrt{b/a})\,, \tag{15.E.18}$$

where c is an arbitrary real, which is required to be greater than b for a correct definition of the inverse Laplace transform. For vanishingly small radius ϵ, the contour C_ϵ does not contribute as long as $b \neq 0$. If $b = 0$, then the inverse Laplace transform is read directly from (15.E.16). Moreover, since $\bar{q}(x,p)$ tends to 0 for large p in view of (15.E.14), the contour C_R does not contribute either for large R, in view of Jordan's lemma. Consequently,

$$q(x,t) = \lim_{R\to\infty}\frac{1}{2\,i\,\pi}\int_{c-i\,R}^{c+i\,R}\exp(t\,p)\,\bar{q}(x,p)\,dp$$
$$= \exp(b\,t - x\,\sqrt{b/a}) - \lim_{R\to\infty,\,\epsilon\to0}\frac{1}{2\,i\,\pi}\left(\int_{C+} + \int_{C-}\right)\exp(t\,p)\,\bar{q}(x,p)\,dp\,. \tag{15.E.19}$$

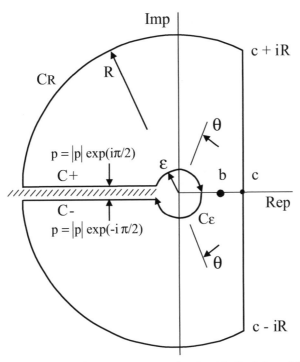

Fig. 15.E.1: Contour of integration associated with the integral (15.E.18).

On the branch cut, $p = -r < 0$, but $\sqrt{p}$ is equal to $i\sqrt{r}$ on the upper part and to $-i\sqrt{r}$ on the lower part. Therefore, the integrals along the branch cut become,

$$
\begin{aligned}
I &= \frac{1}{2i\pi} \int_\infty^0 \exp(-t\,r) \frac{\exp(-i\,x\,\sqrt{r/a})}{-r-b}(-dr) + \frac{1}{2i\pi} \int_0^\infty \exp(-t\,r) \frac{\exp(i\,x\,\sqrt{r/a})}{-r-b}(-dr) \\
&= \frac{1}{2i\pi} \int_0^\infty \frac{\exp(-t\,r)}{r+b}\left(\exp(i\,x\,\sqrt{r/a}) - \exp(-i\,x\,\sqrt{r/a})\right) dr \\
&= \frac{1}{i\pi} \int_0^\infty \frac{\exp(-t\,\rho^2)}{\rho^2+b}\left(\exp(i\,x\,\rho/\sqrt{a}) - \exp(-i\,x\,\rho/\sqrt{a})\right)\rho\,d\rho \\
&= \frac{1}{i\pi} \int_{-\infty}^\infty \exp\left(-t\,\rho^2 + i\,x\,\rho/\sqrt{a}\right) \frac{\rho}{\rho^2+b}\,d\rho .
\end{aligned}
\tag{15.E.20}
$$

In an attempt to see the error function emerging, we make a change of variable that transforms, to within a constant, the argument of the exponential into a square, namely

$$
\rho \longrightarrow v = \sqrt{t}\rho - i\,v_0, \quad v_0 \equiv \frac{x}{2\sqrt{at}} .
\tag{15.E.21}
$$

Then

$$
I = \exp\left(-\left(\frac{x}{2\sqrt{at}}\right)^2\right) \times \frac{1}{2i\pi} \int_{-\infty-iv_0}^{\infty-iv_0} \frac{\exp(-v^2)}{v+iv^+} + \frac{\exp(-v^2)}{v+iv^-}\,dv, \quad v^\pm \equiv v_0 \pm \sqrt{b\,t}.
\tag{15.E.22}
$$

Mei [1995] proposes a contour integration in the complex plane. A slightly different method is followed here.

First observe that, by an appropriate choice of integration path, Fig. 15.E.2, and application of the residue theorem,

$$
\frac{1}{2i\pi} \int_{-\infty-iv_0}^{\infty-iv_0} \frac{\exp(-v^2)}{v+iv^\pm}\,dv = \frac{1}{2i\pi} \int_{-\infty}^\infty \frac{\exp(-v^2)}{v+iv^\pm}\,dv + \begin{cases} 0 & \text{for } v^\pm = v^+ \text{ or } v^- < 0, \\ \exp((v^-)^2) & \text{for } v^\pm = v^- > 0, \end{cases}
\tag{15.E.23}
$$

we are left with integrals on the real line. The basic idea is to insert the following identity,

$$
\int_0^\infty \exp(-u\,X^2)\,du = \frac{1}{X^2}, \quad X \neq 0,
\tag{15.E.24}
$$

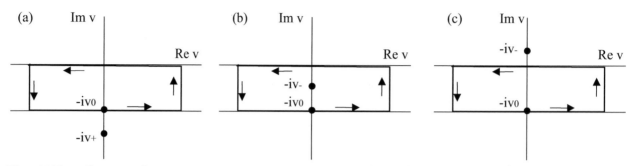

Fig. 15.E.2: Contours of integration associated with the integrals (15.E.23). Note that (a) v^+ is always larger than $v_0 > 0$, while v^- is (b) either between 0 and v^0 or (c) negative.

in the integrals to be estimated and to exchange the order of integration. With this idea in mind, the following straightforward transformations are performed,

$$\frac{1}{2\,i\,\pi}\int_{-\infty}^{\infty}\frac{\exp(-v^2)}{v+i\,v^\pm}\,dv$$

$$= -\frac{v^\pm}{\pi}\int_0^{\infty}\frac{\exp(-v^2)}{v^2+(v^\pm)^2}\,dv \quad \text{(use (15.E.24))}$$

$$= -\frac{v^\pm}{\pi}\int_0^{\infty}\exp\big(-v^2-u\,(v^2+(v^\pm)^2)\big)\,du\,dv, \quad (du\,dv \to dv\,du, \quad v \to v\,\sqrt{1+u}) \tag{15.E.25}$$

$$= -\frac{v^\pm}{2\sqrt{\pi}}\int_0^{\infty}\frac{\exp(-u\,(v^\pm)^2)}{\sqrt{1+u}}\,du \quad (u \to (\sinh v)^2, \quad |v^\pm|\cosh v \to w)$$

$$= -\frac{\operatorname{sgn}(v^\pm)}{2}\exp\big((v^\pm)^2\big)\frac{2}{\sqrt{\pi}}\int_{|v^\pm|}^{\infty}\exp(-w^2)\,dw\,,$$

and therefore

$$\frac{1}{2\,i\,\pi}\int_{-\infty}^{\infty}\frac{\exp(-v^2)}{v+i\,v^\pm}\,dv = -\frac{\operatorname{sgn}(v^\pm)}{2}\exp\big((v^\pm)^2\big)\operatorname{erfc}(|v^\pm|)\,. \tag{15.E.26}$$

Collecting the results (15.E.23) and (15.E.26), and using the property (15.3.30)$_3$,

$$\frac{1}{2\,i\,\pi}\int_{-\infty-i\,v_0}^{\infty-i\,v_0}\frac{\exp(-v^2)}{v+i\,v^\pm}\,dv = \begin{cases} -\frac{1}{2}\exp\big((v^+)^2\big)\operatorname{erfc}(v^+) & \text{for } v^\pm = v^+, \\[2mm] -\frac{1}{2}\exp\big((v^-)^2\big)\operatorname{erfc}(v^-) + \exp\big((v^-)^2\big) & \text{for } v^\pm = v^-. \end{cases} \tag{15.E.27}$$

Finally, the inversion formula deduces from (15.E.19) and (15.E.27),

$$q(x,t) = \frac{1}{2}\exp\Big(x\sqrt{\frac{b}{a}}+b\,t\Big)\operatorname{erfc}\Big(\frac{x}{2\sqrt{at}}+\sqrt{b\,t}\Big) + \frac{1}{2}\exp\Big(-x\sqrt{\frac{b}{a}}+b\,t\Big)\operatorname{erfc}\Big(\frac{x}{2\sqrt{at}}-\sqrt{b\,t}\Big)\,. \tag{15.E.28}$$

Chapter 16

Coupled transports in tissues endowed with a fixed electrical charge

16.1 Macroscopic transports in tissues with a fixed charge: overview and constitutive framework

When chemical and mechanical loadings are performed at so slow rates that they do not elicit the times characterizing the material response, the processes can be considered to involve only chemomechanical equilibria. At variance, the present chapter is devoted to transient and spatially heterogeneous processes.

16.1.1 Overview of transport issues

Species diffuse within their phase due to macroscopic gradients. They may also transfer from one fluid phase to the other one due to *local* out-of-equilibria.

The constitutive equations developed here, mostly with an eye towards articular cartilages, use a framework that in fact embodies not only diffusion of matter through the extrafibrillar phase but also transfers of masses between the intrafibrillar and extrafibrillar fluid phases and purely mechanical aspects. The various sources of dissipation are built in a thermodynamic framework, segregated and decoupled via the Clausius-Duhem inequality. The purpose is, first, to present how the transport phenomena are aggregated to the three-phase context and, second, to detail the constitutive equations of each of these transports.

Several continuum theories have been proposed to model the mechanical and transport responses and to describe macroscopic couplings in articular cartilages, Lai et al. [1991], Gu et al. [1998], Huyghe and Janssen [1999]. However, their structure is different, as far as the intrafibrillar phase is not recognized. The intra/extrafibrillar split is considered in Huyghe [1999] and Huyghe et al. [2003], but only the mechanical aspects of cartilages bathed in a binary electrolyte are addressed.

While the mechanical behavior is time-independent, unless collagen fibers are considered as viscoelastic, the phenomena of mass transfer and diffusion involve typical characteristic times, so that the overall response is time-dependent. Mass transfer relations are directly motivated by the dissipation inequality. and, along an Onsager approach, they are proposed in linear and nonlinear formats, Section 10.2. Electroneutrality reduces by one the number of independent concentrations but it generates/is associated with/is accompanied by a reaction variable, namely the electrical potential.

The generalized diffusion in the extrafibrillar compartment accounts for Darcy's law of seepage through the porous solid skeleton, Fick's law of ionic diffusion, Ohm's law of electrical flow, electroosmosis and the existence of streaming potentials. Strong couplings exist between these phenomena. The internal entropy inequality associated with diffusion is exposed in two formats: one form involves fluxes relative to the solid while the second form involves diffusive fluxes relative to the extrafibrillar fluid. The equivalence between these two formats is used to structure the diffusion matrix, Section 16.2. The general form of the diffusion equations is provided for a ternary electrolyte, with sodium and calcium cations and chloride anions as mobile ions.

Two simple additional assumptions allow to recover arrowhead diffusion matrices, Section 16.2.6. The present approach has the advantage of displaying in explicit form the complete algebraic structure of the diffusion matrices, and therefore to pave the way to embody further phenomena. Besides, the two above additional assumptions concern entities that are standard in the modeling of transport in mixtures. Another approach by Gu et al. [1998] consists in working directly on the momentum equations and in postulating constitutive equations for the momentum transfers between phases: the two methods are mainly equivalent but the latter alternative is more abstract.

The tight algebraic structure of the constitutive equations of diffusion implies that the electroosmotic coefficient results in explicit form in terms of the fixed charge of the proteoglycans. At variance, an osmotic, or reflection, coefficient ω is shown to be hidden in the formulation. Therefore, unlike in some other models of charged porous media where the fixed charge is not part of the fluid phase, e.g., Mitchell [1993], Loret et al. [2004]b, the two above entities, namely the electroosmotic coefficient and the reflection coefficient, emerge naturally from the formulation and need not be provided by additional constitutive equations. Two interpretations of the osmotic coefficient are proposed and negative osmosis is analyzed in detail and shown to take place for both sodium chloride and calcium chloride electrolytes, Section 16.3.2.

Data from an experimental setup aimed at measuring transport properties show that the experimental diffusion coefficient increases, from a tiny value, to a plateau as the ionic strength of the bath the tissue is in contact with is increased, Section 16.3.3. This feature is reproduced by the model and the key effect on the phenomenon of the intensity of the fixed charge is quantified. The data of this setup have been sometimes improperly interpreted as providing a basic constitutive relation where the diffusion coefficients are made functions of the transmission coefficient $1 - \omega$.

The influences of ionic strength and fixed charge density on various key transport properties are discussed, Section 16.4. The interactions with mechanical deformation, through the fluid volume fractions, are also underlined.

In this chapter, the pH of the extrafibrillar fluid phase is assumed neutral, and the sole fixed charge, which is negative, is due to proteoglycans. The effects of a varying pH on the diffusion properties are considered in Chapter 17.

While the analysis targets transport phenomena, which involve their own internal *couplings*, it should be remembered that there exist *interactions* with mechanics as stressed in Chapter 4: indeed a number of transport properties depend on volume fractions. A comprehensive analysis of a boundary value problem involves simultaneously mechanics, mass transfer and diffusion. The complete framework is used in Chapter 18, to solve, via the finite element method, initial and boundary value problems in order to simulate mechanical and transport phenomena in laboratory specimens submitted to transient mechanical and chemical loading processes.

Notation: In this chapter, unless stated otherwise, the convention of summation over repeated indices does <u>not</u> apply.

16.1.2 The constitutive framework

The model is "macroscopic" even if it "sees" a number of details of the tissues at several length scales. The phenomena that are thought to be of significance are collected and quantified through various scalar measures of mass and volume and they are used to formulate the constitutive equations. On the other hand, the spatial organization is missed, and no characteristic length scale appears in the equations. This feature does not mean that interactions between structural elements at the nanoscale are not scrutinized but, if they are accounted for, they come into picture at a single spatial level. Moreover, the pointwise directional organization of the individual elements, e.g., collagen fibers, may be accounted for naturally at this macroscopic level via anisotropic elastic or diffusion properties.

As a typical instance of this line of thinking, let us mention the description of the intrafibrillar compartment. The model recognizes that collagen fibers delineate a volume that only low molar mass species can penetrate: macromolecules, like proteoglycans, are excluded. The model misses the sizes of ions and PGs but it accounts for some physical consequences through appropriate concentrations of species in the intrafibrillar compartment.

The constitutive framework for articular cartilages is described in Chapter 14. A few general aspects are recalled here to make the chapter self-contained. On the other hand, refinements pertaining to coupled transports are introduced and a number of definitions and relations are reconsidered in Sections 16.1.2.1 to 16.1.2.6.

16.1.2.1 Phase partition

Cartilage is viewed as a three-phase, multi-species, porous medium, Fig. 14.6.1. The *solid phase* S is formed by the collagen network denoted by the symbol c. The sets of species of the solid, intrafibrillar (IF) and extrafibrillar (EF) phases are, respectively,

$$S = \{c\}, \quad I = \{w, Na, Ca, Cl\}, \quad E = \{w, PG, Na, Ca, Cl\}. \tag{16.1.1}$$

The set E_{mo} of extrafibrillar mobile species contains all species in E but proteoglycans,

$$E_{mo} = E - \{PG\}. \tag{16.1.2}$$

A minimum number of cations is required to ensure electroneutrality of the extrafibrillar phase where electrically charged proteoglycans are located. Therefore, mobile cations are endowed with a velocity independent of that of their non mobile counterparts, which, like the proteoglycans, move with the velocity of the solid phase. Thus cations in the extrafibrillar space are partitioned into a mobile mo part and a non mobile nm part. Still, the analytic developments of this chapter leaves the partition of ions into a mobile part and a non mobile part as an option. Standard analyses do not consider this partition.

Exchanges of water and ions occur between the fluid phases but only the extrafibrillar phase communicates with the surroundings, Fig. 14.6.1. In that sense the model is hierarchical: intrafibrillar water must flow through the extrafibrillar compartment to reach the surroundings. The exchange of species between the IF and EF compartments takes place via mass transfer, a kind of physico-chemical reaction, and not via diffusion. In some theories of porous media, the intrafibrillar fluid phase would be viewed as an unconnected porosity or as a dead-end porosity.

The main assumptions, which underly the three-phase multi-species model, follow the *strongly interacting* model described in Section 14.6.3. This framework is adopted to within the fact that this chapter offers the possibility to partition the EF cations. A key feature of the model is that the IF and EF compartments obey independent electroneutrality conditions and are therefore endowed with distinct electrical potentials.

16.1.2.2 Description of geometry and mass

Since a species can be present in more than a single phase, it is endowed with two indices, one referring to the species itself, the other to the phase. The only exceptions to this convention are proteoglycans which unambiguously belong to the extrafibrillar phase and collagen. The collagen network constitutes the solid phase (skeleton). Still, the status of collagen is a bit particular. In fact, the water it contains is viewed as part of another phase. Moreover, a partition of its charge will be motivated under non physiological pHs.

The current volume (respectively mass) of the species k in phase K is denoted by V_{kK} (respectively M_{kK}). Let the initial volume of the porous medium be V_0 and let $V = V(t)$ be its current volume. Various entities are attached to species:
   - some are intrinsic like the intrinsic density ρ_k, the molar volume $\widehat{v}_k$ and the molar mass $\widehat{m}_k$ linked by $\widehat{m}_k = \rho_k \widehat{v}_k$;
   - some refer to the current volume, like the *volume fraction* $n^{kK} = V_{kK}/V$, and the *apparent density* $\rho^{kK} = n^{kK} \rho_k$;
   - some refer to the initial volume like the *volume content* $v^{kK} = V_{kK}/V_0 = n^{kK} V/V_0$ and the *mass content* $m^{kK} = M_{kK}/V_0 = \rho_k v^{kK}$.

The associated entities for the phase K are defined by algebraic summation of individual contributions, e.g., the current volume V_K and mass M_K, the volume fraction $n^K = V_K/V$ (volume fractions satisfy the closure relation $n^S + n^I + n^E = 1$), the *apparent density* $\rho^K = M_K/V$, the volume content $v^K = V_K/V_0 = n^K V/V_0$ and the mass content $m^K = M_K/V_0$.

Other entities live in their phase, e.g. the molar fractions and the concentrations. The *molar fraction* x_{kK} of the species k in phase K is defined by the ratio of the mole number N_{kK} of that species over the total number of moles within the phase N_K, $x_{kK} = N_{kK}/N_K$. In each phase, the molar fractions satisfy the closure relation $\sum_{k \in K} x_{kK} = 1$, K = S, I, E. Since $N_{kK}/V_0 = m^{kK}/\widehat{m}_k$, the molar fractions can also be expressed in terms of mass contents.

The *concentration* of an extrafibrillar species is equal to its number of moles referred to the volume of the extrafibrillar phase,

$$c_{kE}^{(mo)} = \frac{N_{kE}^{(mo)}}{V_E} = \frac{N_{kE}^{(mo)}}{N_E \widehat{v}_E} = \frac{1}{\widehat{v}_k} \frac{V_{kE}^{(mo)}}{V_E} = \frac{1}{\widehat{v}_k} \frac{n_{(mo)}^{kE}}{n^E} = \frac{1}{\widehat{v}_k} \frac{v_{(mo)}^{kE}}{n^E} \frac{V_0}{V} = \frac{x_{kE}^{(mo)}}{\widehat{v}_E}, \quad k \in E, \tag{16.1.3}$$

with $\widehat{v}_E = \sum_{k \in E} x_{kE} \widehat{v}_k \sim \widehat{v}_w$ the molar volume of the extrafibrillar fluid phase. The index (mo) indicates that the definition holds for the entities as a whole, their mobile parts and their non mobile parts.

16.1.2.3 The electrical charge

For proteoglycans, both the molar fraction, the concentration and their *effective* counterparts, which include the valence, should be defined,

$$\underbrace{x_{\mathrm{PG}} = \frac{N_{\mathrm{PG}}}{N_{\mathrm{E}}}}_{\text{molar fraction}} , \quad \underbrace{y_{\mathrm{PG}} = \zeta_{\mathrm{PG}}\, x_{\mathrm{PG}}}_{\substack{\text{effective} \\ \text{molar fraction}}} , \quad \underbrace{c_{\mathrm{PG}} = \frac{N_{\mathrm{PG}}}{V_{\mathrm{E}}}}_{\text{concentration}} , \quad \underbrace{e_{\mathrm{PG}} = \zeta_{\mathrm{PG}}\, c_{\mathrm{PG}}}_{\substack{\text{effective} \\ \text{concentration}}} . \tag{16.1.4}$$

The molar fraction x_{PG} and concentration c_{PG} of proteoglycans, which are macromolecules, are quite small with respect to the other species of the extrafibrillar phase. On the other hand, the valence ζ_{PG} of proteoglycans is large at neutral pH. Thus the *effective* concentration is a key parameter of the biochemical and biomechanical responses of PGs and articular cartilages. For the sake of accuracy in the algebraic derivations, we shall also introduce the notations $\tilde{c}_{\mathrm{PG}} = c_{\mathrm{PG}}\, n^{\mathrm{E}}/n^{\mathrm{wE}}$ and $\tilde{e}_{\mathrm{PG}} = e_{\mathrm{PG}}\, n^{\mathrm{E}}/n^{\mathrm{wE}}$, even if the ratio $n^{\mathrm{E}}/n^{\mathrm{wE}}$ is close to one.

While the fixed charge density FCD [unit : M] refers to the amount of extrafibrillar water, the cation exchange capacity (CEC) [unit : coulomb C/kg] refers to the dry mass, namely the mass M_{dry} of collagen *and* PGs :

$$e_{\mathrm{PG}} = \frac{\zeta_{\mathrm{PG}}\, N_{\mathrm{PG}}}{V_{\mathrm{E}}} \simeq \frac{y_{\mathrm{PG}}}{\widehat{v}_{\mathrm{w}}}, \quad \mathrm{FCD} = \frac{|\zeta_{\mathrm{PG}}|\, N_{\mathrm{PG}}}{V_{\mathrm{E}}}, \quad \mathrm{CEC} = \mathrm{F}\, |\zeta_{\mathrm{PG}}| \frac{N_{\mathrm{PG}}}{M_{\mathrm{dry}}} , \tag{16.1.5}$$

and therefore,

$$\mathrm{FCD} = \frac{\mathrm{CEC}}{\mathrm{F}} \times \frac{n^{\mathrm{dry}}}{n}\, \rho_{\mathrm{dry}} . \tag{16.1.6}$$

Here $\mathrm{F} = 96,485$ coulomb/mole is Faraday's equivalent charge, n^{dry} the volume fraction of the dry mass, ρ_{dry} its intrinsic mass density, and n the volume fraction associated with the extrafibrillar fluid.

For a large part of the literature however, the FCD refers, not to the extrafibrillar water, but to the total water. The expressions of FCD in (16.1.5) and (16.1.6) should be modified accordingly.

For ease of interpretation, Table 16.1.1 displays the values of FCD for articular cartilages corresponding to a representative value of CEC, and to a representative fluid volume fraction n. Note that, during a chemical loading process where the bath composition is varied, the CEC is kept fixed, as far as the proteoglycans are not damaged, while the FCD varies, because the volume fractions vary due to diffusion, swelling or shrinking.

Table 16.1.1: Values of FCD [unit : M] for CEC=42.9 kC/kg, $n^{\mathrm{dry}} = 1 - n$, ρ_{dry}=1800 kg/m³.

n	0.7	0.8	0.9
FCD	0.343	0.20	0.089

The electrochemical potentials $\mu_{k\mathrm{K}}^{\mathrm{ec}}$ [unit : m²/s²] are defined in Section 14.6.5. For incompressible species,

$$\widehat{m}_k\, \mu_{k\mathrm{K}}^{\mathrm{ec}} = \widehat{v}_k\, p_{k\mathrm{K}} + RT\, \mathrm{Ln}\, x_{k\mathrm{K}} + \zeta_k\, \mathrm{F}\, \phi_{\mathrm{K}} , \quad k \in \mathrm{K} . \tag{16.1.7}$$

In this formula, $R = 8.31451$ J/mole/K is the universal gas constant, T [unit : K] the absolute temperature, the valence of the species k is denoted by ζ_k, and its pressure by $p_{k\mathrm{K}}$, and ϕ_{K} [unit : V] is the electrical potential of the phase K.

16.1.2.4 Electroneutrality

For pH close to 7, collagen can be considered to be electrically neutral, Li and Katz [1976]. According to Assumption (H5) of Section 14.6.3, both the intrafibrillar and extrafibrillar fluid phases are electrically neutral. Electroneutrality requires a minimal number of extrafibrillar cations, that we refer to as *non mobile* cations : the charge of these cations balances the large negative charge of the proteoglycans. The condition of electroneutrality (14.6.22) and (14.6.23) yields

$$\sum_{k \in \mathrm{E}} \zeta_k\, N_{k\mathrm{E}}^{\mathrm{nm}} = 0, \quad \sum_{k \in \mathrm{E}} \zeta_k\, N_{k\mathrm{E}}^{\mathrm{mo}} = 0 . \tag{16.1.8}$$

Note however that the non mobile sodium cations *do not bind* to the PGs. The situation is more complex for calcium cations, as indicated in Section 16.1.2.6.

Indeed, as a further consequence of electroneutrality, the number of moles of chloride anions is no longer an independent variable and it can be eliminated in favor of the number of moles of the cations,

$$\delta N_{\text{ClK}} = \sum_{i \in \text{K}_{\text{in}}} \zeta_i \, \delta N_{i\text{K}}, \quad \text{K = I, E}. \tag{16.1.9}$$

Therefore, the electrochemical energy can be recast in terms of the chemical potentials of water and salts $\mu_{n\text{K}}$ conjugated to the mass contents $m^{i\text{K}}$ of water and (mobile) sodium and calcium cations. The conjugate pairs (i, n) belong to the respective sets (acronyms "in": independent, "ne": neutral)

$$\text{K}_{\text{in}} = \{\text{w}, \text{Na}, \text{Ca}\}, \quad \text{K}_{\text{ne}} = \{\text{w}, \text{s}_1 = \text{NaCl}, \text{s}_2 = \text{CaCl}_2\}. \tag{16.1.10}$$

The entities $\mu_{n\text{K}}$, $n = \text{s}_1, \text{s}_2$, can be viewed as the chemical potentials of the dissociated salts in phase K, namely in accordance with (16.1.7),

$$\widehat{m}_i \, \mu_{n\text{K}} = \begin{cases} \widehat{m}_i \, \mu_{i\text{K}}^{\text{ec}} + \zeta_i \, \widehat{m}_{\text{Cl}} \, \mu_{\text{ClK}}^{\text{ec}}, \\ \widehat{v}_n \, p_{n\text{K}} + RT \, \text{Ln} \, x_{i\text{K}} \, (x_{\text{ClK}})^{\zeta_i}. \end{cases} \tag{16.1.11}$$

The molar volumes $\widehat{v}_n$, densities ρ_n and pressures $p_{n\text{K}}$ are defined by the relations,

$$\widehat{v}_n = \rho_n^{-1} \, \widehat{m}_i = \widehat{v}_i + \zeta_i \, \widehat{v}_{\text{Cl}}, \quad \widehat{v}_n \, p_{n\text{K}} = \widehat{v}_i \, p_{i\text{K}} + \zeta_i \, \widehat{v}_{\text{Cl}} \, p_{\text{ClK}}, \quad (i, n) \in \{\text{K}_{\text{in}}, \text{K}_{\text{ne}}\}. \tag{16.1.12}$$

16.1.2.5 Fluxes and electrical current density

Due to the distinction between mobile and non mobile ions, the fluxes shown in Section 14.6.5.4 need to be redefined. The *mass flux* through the solid skeleton $\mathbf{M}_{k\text{K}}$ [unit : kg/m^2/s] and the associated *volume flux* $\mathbf{J}_{k\text{K}}$ [unit : m/s] of the species k of phase K are now defined as,

$$\rho_k^{-1} \, \mathbf{M}_{k\text{K}} = \mathbf{J}_{k\text{K}} = n_{\text{mo}}^{k\text{K}} \, (\mathbf{v}_{k\text{K}} - \mathbf{v}_{\text{S}}). \tag{16.1.13}$$

The sum of the fluxes $\mathbf{J}_{k\text{E}}$, $k \in \text{E}$, defines the volume averaged flux $\mathbf{J}_{\text{E}}$ of the extrafibrillar fluid phase through the solid skeleton,

$$\mathbf{J}_{\text{E}} = \sum_{k \in \text{E}} \mathbf{J}_{k\text{K}} = \sum_{k \in \text{E}} n_{\text{mo}}^{k\text{E}} \, (\mathbf{v}_{k\text{E}} - \mathbf{v}_{\text{S}}), \quad k \in \text{E}. \tag{16.1.14}$$

Let $n_{\text{mo}}^{k\text{E}}$ and $n_{\text{nm}}^{k\text{E}}$ be a partition of the volume fraction $n^{k\text{E}}$ into its mobile and non mobile parts. The *diffusive flux* with respect to extrafibrillar water is denoted by $\mathbf{J}_{k\text{E}}^d$,

$$\mathbf{J}_{k\text{E}}^d = n_{\text{mo}}^{k\text{E}} \, (\mathbf{v}_{k\text{E}} - \mathbf{v}_{w\text{E}}) + n_{\text{nm}}^{k\text{E}} \, (\mathbf{v}_{\text{S}} - \mathbf{v}_{w\text{E}}), \quad k \in \text{E}. \tag{16.1.15}$$

Relations between the fluxes relative to the solid skeleton and to the extrafibrillar water, e.g.,

$$\mathbf{J}_{k\text{E}} = \mathbf{J}_{k\text{E}}^d + n^{k\text{E}} \, (\mathbf{v}_{w\text{E}} - \mathbf{v}_{\text{S}}) = \mathbf{J}_{k\text{E}}^d + \frac{n^{k\text{E}}}{n^{w\text{E}}} \mathbf{J}_{w\text{E}}, \quad k \in \text{E}_{\text{mo}}, \tag{16.1.16}$$

are detailed in Exercise 16.1.

The *electrical current density* $\mathbf{I}_{\text{eK}}$ in phase K [unit : A/m^2] is defined as the sum of constituent velocities weighted by their valences and molar densities,

$$\mathbf{I}_{\text{eK}} = \text{F} \sum_{k \in \text{K}} \zeta_k \, \frac{N_{k\text{K}}}{V} \, \mathbf{v}_{k\text{K}} = \text{F} \sum_{k \in \text{K}} \zeta_k \, \frac{N_{k\text{K}}^{\text{mo}}}{V} \, \mathbf{v}_{k\text{K}}. \tag{16.1.17}$$

The second equality is due to the satisfaction of the electroneutrality by the non mobile ions and the PGs. Due to the electroneutrality of both the mobile and non mobile parts in E, $\mathbf{I}_{\text{eE}}$ may be viewed as a sum of either interphase or diffusive fluxes, namely

$$\mathbf{I}_{\text{eE}} = \text{F} \sum_{k \in \text{E}} \zeta_k \, \frac{\mathbf{J}_{k\text{E}}}{\widehat{v}_k} = \text{F} \sum_{k \in \text{E}} \zeta_k \, \frac{\mathbf{J}_{k\text{E}}^d}{\widehat{v}_k}. \tag{16.1.18}$$

16.1.2.6   Calcium partition, screening and binding

Partition of mobile and non mobile cations

The partition between mobile and non mobile cations in the extrafibrillar phase needs to be specified by constitutive relations. If the mobile and non mobile parts of a species were made distinct species belonging to distinct phases, they would be endowed with their own electrochemical potentials, and the interchange between the two species could be viewed as a transfer or chemical reaction, controlled by the differences of their potentials.

However, here, the mechanical effect of a species is considered to be due to the species as a whole and the chemical potential is defined in terms of the total molar fraction. On the other hand, only the mobile part of the species can diffuse through the extrafibrillar water.

If a single cation k is present in the solution, its partition in mobile and non mobile parts is known, namely its non mobile part is defined by the relation $\zeta_k N_{kE}^{nm} + \zeta_{PG} N_{PG} = 0$. When the two sodium and calcium cations are present, their partition,

$$\underbrace{2\,\mathrm{Na}^+}_{\text{mobile}} + \underbrace{\mathrm{Ca}^{2+}}_{\text{non mobile}} \;\rightleftharpoons\; \underbrace{\mathrm{Ca}^{2+}}_{\text{mobile}} + \underbrace{2\,\mathrm{Na}^+}_{\text{non mobile}} \quad, \tag{16.1.19}$$

needs to be defined. One might tentatively endow the mobile and non mobile parts with chemical potentials and the partition will be defined by the equilibrium constant K_{eq} that differentiates the affinities of sodium and calcium cations for proteoglycans, namely in terms of concentrations

$$\frac{(c_{NaE}^{nm})^2}{(c_{NaE}^{mo})^2} = K_{eq}\,\frac{c_{CaE}^{nm}}{c_{CaE}^{mo}}\,. \tag{16.1.20}$$

Therefore the concentrations can be obtained from the single positive solution of the equation,

$$(c_{NaE}^{nm})^2\,\overbrace{\frac{c_{ClE} - c_{NaE}^{mo}}{K_{eq}\,(c_{NaE}^{mo})^2}}^{\ge 0} + c_{NaE}^{nm} + \overbrace{e_{PG}}^{<0} = 0\,. \tag{16.1.21}$$

The electrical charge of PGs is indeed always negative.

Calcium binding to the fixed charge

The fixed charge of polyelectrolyte gels of biological interest and of active clays are known to be pH-sensitive. Indeed, as the pH of the surroundings decreases (respectively increases) the fixed charge becomes less negative (respectively more negative). In active clays, the change of charge is due to surface complexation mechanisms involving hydrogen ions H^+ and hydroxyl ions OH^-. The effects of a change of pH on the transport and mechanical properties of biological tissues are addressed in Chapter 17.

The change of the fixed charge may also be due to the binding of other ions. Calcium cations are known to bind, at least partially, to proteoglycans, while sodium cations do not. At sufficiently high calcium concentration, the binding might become irreversible and modify the structure of the proteoglycans. The ability of proteoglycans to bind calcium has been advocated to contribute to the calcification process of articular cartilages, Werner and Gründer [1999]. Still, whether proteoglycans are inhibitors or initiators of hydroxyapatite is not elucidated, Hunter [1991]. The phenomenon remains to be quantified. Moreover it is known to be very sensitive to a number of histological details: for example chondroitin sulfate 4 is more prone to bind calcium than chondroitin sulfate 6. Binding of calcium ions in articular cartilages is addressed in Section 17.4, while binding of chloride ions in corneal stroma is described in Section 19.10.1 et seq.

16.1.3   Thermodynamic arguments governing mass transfer and diffusion

The constitutive equations are developed in a thermodynamic framework *à la Biot* where the solid skeleton is taken as reference, Chapter 14. In the context of this chapter, it is sufficient to recall the entropy inequalities associated with mass transfer and diffusion, namely

$$T\hat{S}_{(i,2)} = -\sum_{k\in I}(\mu_{kI} - \mu_{kE})\,\frac{1}{\det\mathbf{F}}\,\frac{dm^{kI}}{dt} \ge 0\,,$$

$$T\hat{S}_{(i,3)} = -\sum_{k\in E_{mo}}(\boldsymbol{\nabla}\mu_{kE}^{ec} - \mathbf{b}_{kE})\cdot\mathbf{M}_{kE} \ge 0\,. \tag{16.1.22}$$

Satisfaction of these inequalities motivates generalized mass transfer equations and generalized diffusion equations, respectively.

The balances of mass of intra- and extrafibrillar species are described in Section 14.6.6. The mass transfers contribute to the dissipation which is used to motivate the associated constitutive equations which are developed in Section 10.2.

Generalized diffusion in the extrafibrillar phase is the subject of the subsequent sections. For quasi-static analyses, the body force densities $\mathbf{b}_{kE}$ are equal to the gravity $\mathbf{g}$, and individual accelerations are neglected. Uniform body forces can be viewed as introducing a sedimentation contribution into the electro-chemomechanical potentials.

16.2 Generalized and coupled diffusion: structure of the constitutive equations

The internal entropy inequality $\hat{S}_{(i,3)} \geq 0$, Eqn $(16.1.22)_2$, associated with diffusion is ensured by a generalization of Darcy's law of seepage through the porous medium, Fick's law of diffusion of ions, and Ohm's law of electrical flow in the fluid phase. At least two ways of using the inequality to build the generalized diffusion equations may be envisaged. Indeed, it can be expressed either
- in terms of fluxes relative to the solid or
- in terms of diffusive fluxes relative to the fluid.

However, in Loret and Simões [2007], the end results of the two separate developments have been shown to be equivalent for isotropic diffusion properties. The whole Section 16.2 is an extension of Loret and Simões [2007] to anisotropic diffusion properties in presence of n_{ion} mobile ionic species, namely three ions for a ternary electrolyte. The compatibility of the two lines of thinking restricts the structure of the generalized coupled diffusion model. The relations between the coefficients of the two generalized diffusion matrices introduced, of respective sizes $[3 \times (1 + n_{ion})]^2$ and $[3 \times (2 + n_{ion})]^2$, are provided. As particular by-products of the analysis emerge explicit expressions of
- the electrical conductivity;
- the electroosmotic coefficient;
- the osmotic, or reflection, coefficient: even if it does not appear in explicit form, this coefficient is shown to be hidden in the structure of the generalized diffusion matrix.

16.2.1 Diffusion in terms of fluxes relative to the solid

An immediate way to satisfy the inequality $\hat{S}_{(i,3)} \geq 0$, Eqn $(16.1.22)_2$, is to postulate the existence of a symmetric positive semi-definite matrix (PsD) $\boldsymbol{\kappa}$, of size $[3 \times (1 + n_{ion})]^2$, that provides the volume fluxes as a function of the electrochemical potentials, namely,

$$\mathbf{j} = -\boldsymbol{\kappa}\, \mathbf{f}, \quad \boldsymbol{\kappa} = \boldsymbol{\kappa}^{\mathrm{T}}\, \mathrm{PsD}, \tag{16.2.1}$$

that is, formally,

$$\mathbf{j} = \begin{bmatrix} \mathbf{J}_{wE} \\ \mathbf{J}_{NaE} \\ \mathbf{J}_{CaE} \\ \mathbf{J}_{ClE} \end{bmatrix}, \quad \mathbf{f} = \begin{bmatrix} \rho_{w}\, \boldsymbol{\nabla}\mu_{wE} \\ \rho_{Na}\, \boldsymbol{\nabla}\mu_{NaE}^{ec} \\ \rho_{Ca}\, \boldsymbol{\nabla}\mu_{CaE}^{ec} \\ \rho_{Cl}\, \boldsymbol{\nabla}\mu_{ClE}^{ec} \end{bmatrix}, \quad \boldsymbol{\kappa} = \begin{bmatrix} \boldsymbol{\kappa}_{ww} & \boldsymbol{\kappa}_{w\,Na} & \boldsymbol{\kappa}_{w\,Ca} & \boldsymbol{\kappa}_{w\,Cl} \\ \boldsymbol{\kappa}_{Naw} & \boldsymbol{\kappa}_{NaNa} & \boldsymbol{\kappa}_{NaCa} & \boldsymbol{\kappa}_{NaCl} \\ \boldsymbol{\kappa}_{Caw} & \boldsymbol{\kappa}_{CaNa} & \boldsymbol{\kappa}_{CaCa} & \boldsymbol{\kappa}_{CaCl} \\ \boldsymbol{\kappa}_{Clw} & \boldsymbol{\kappa}_{Cl\,Na} & \boldsymbol{\kappa}_{Cl\,Ca} & \boldsymbol{\kappa}_{Cl\,Cl} \end{bmatrix}. \tag{16.2.2}$$

Each block $\boldsymbol{\kappa}_{kl}$, $k, l \in E_{mo}$, is a 3×3 symmetric positive definite matrix (PD). It is diagonal in the orthotropy axes of the diffusion properties. It would be proportional to the 2nd order identity tensor if the diffusion properties were isotropic. However, such is not the case for many soft tissues like cornea: the fibrous and/or strongly layered structure may reflect on the transport properties.

Still measurements that would allow to quantify the anisotropic properties of the diffusion coefficients are in need. For stroma, radial and circumferential hydraulic permeabilities differ at hydration lower than physiological only, Fig. 19.4.5 from Hedbys and Mishima [1962]. For adult bovine articular cartilages, Reynaud and Quinn [2006] do not detect directional properties in the electroosmotic coefficient, even if the hydraulic permeability of the tested samples displays strain-induced anisotropy.

16.2.2 Diffusion in terms of diffusive fluxes and current density

A more familiar form, where the diffusive fluxes and electric current density appear, emerges by using the explicit expression of the electrochemical potentials (16.1.7), the closure relation satisfied by the molar fractions, and the definitions of fluxes Eqns (16.1.13) and (16.1.15).

Equivalent to $(16.1.22)_2$, the dissipation inequality can be formally expressed as the sum of products of a flux times a driving force,

$$T\hat{S}_{(i,3)} = -\boldsymbol{\mathcal{F}}^{\mathrm{T}}\,\boldsymbol{\mathcal{J}} = -\mathbf{F}_{\mathrm{E}}\cdot\mathbf{J}_{\mathrm{wE}} - \sum_{k\in\mathrm{E}_{\mathrm{ions}}}\mathbf{F}_{k\mathrm{E}}^{d}\cdot\mathbf{J}_{k\mathrm{E}}^{d} - \mathbf{F}_{\mathrm{eE}}\cdot\mathbf{I}_{\mathrm{eE}} \geq 0. \tag{16.2.3}$$

Details of the derivation are provided in Exercises 16.2 and 16.3. The diffusive flux of proteoglycans does not vanish. However, their absolute flux does, which explains why proteoglycans do not contribute to the dissipation inequality. The vector flux $\boldsymbol{\mathcal{J}}$ and its conjugate vector $\boldsymbol{\mathcal{F}}$ have now $n_{\mathrm{ion}} + 2 = 5$ entries,

$$\boldsymbol{\mathcal{J}} = \begin{bmatrix} \mathbf{J}_{\mathrm{wE}} \\ \mathbf{J}_{\mathrm{NaE}}^{d} \\ \mathbf{J}_{\mathrm{CaE}}^{d} \\ \mathbf{J}_{\mathrm{ClE}}^{d} \\ \mathbf{I}_{\mathrm{eE}} \end{bmatrix}, \quad \boldsymbol{\mathcal{F}} = \begin{bmatrix} \mathbf{F}_{\mathrm{E}} \\ \mathbf{F}_{\mathrm{NaE}}^{d} \\ \mathbf{F}_{\mathrm{CaE}}^{d} \\ \mathbf{F}_{\mathrm{ClE}}^{d} \\ \mathbf{F}_{\mathrm{eE}} \end{bmatrix} = \begin{bmatrix} \boldsymbol{\nabla}P_{\mathrm{wE}} - \rho_{*}^{\mathrm{E}}/n_{*}^{\mathrm{E}}\,\mathbf{g} \\ RT/\widehat{v}_{\mathrm{Na}}\,\boldsymbol{\nabla}\mathrm{Ln}\,x_{\mathrm{NaE}} - (\rho_{\mathrm{Na}} - \rho_{*}^{\mathrm{E}}/n_{*}^{\mathrm{E}})\,\mathbf{g} \\ RT/\widehat{v}_{\mathrm{Ca}}\,\boldsymbol{\nabla}\mathrm{Ln}\,x_{\mathrm{CaE}} - (\rho_{\mathrm{Ca}} - \rho_{*}^{\mathrm{E}}/n_{*}^{\mathrm{E}})\,\mathbf{g} \\ RT/\widehat{v}_{\mathrm{Cl}}\,\boldsymbol{\nabla}\mathrm{Ln}\,x_{\mathrm{ClE}} - (\rho_{\mathrm{Cl}} - \rho_{*}^{\mathrm{E}}/n_{*}^{\mathrm{E}})\,\mathbf{g} \\ \boldsymbol{\nabla}\phi_{\mathrm{E}} \end{bmatrix}, \tag{16.2.4}$$

where $\boldsymbol{\nabla}P_{\mathrm{wE}} = \boldsymbol{\nabla}p_{\mathrm{wE}} - RT\,\boldsymbol{\nabla}\tilde{c}_{\mathrm{PG}}$, and the volume fraction n_{*}^{E} and mass density ρ_{*}^{E} are defined in Exercise 16.2.

The generalized diffusion law can be expressed via a symmetric matrix $\boldsymbol{\mathcal{K}}$ made by 5×5 symmetric blocks of size 3×3,

$$\boldsymbol{\mathcal{J}} = -\boldsymbol{\mathcal{K}}\,\boldsymbol{\mathcal{F}}, \quad \boldsymbol{\mathcal{K}} = \boldsymbol{\mathcal{K}}^{\mathrm{T}}\ \mathrm{PsD}, \tag{16.2.5}$$

with

$$\boldsymbol{\mathcal{K}} = \begin{bmatrix} \mathbf{k}_{\mathrm{EE}} & \mathbf{k}_{\mathrm{ENa}}^{d} & \mathbf{k}_{\mathrm{ECa}}^{d} & \mathbf{k}_{\mathrm{ECl}}^{d} & \mathbf{k}_{\mathrm{e}} \\ \mathbf{k}_{\mathrm{NaE}}^{d} & \mathbf{k}_{\mathrm{NaNa}}^{d} & \mathbf{k}_{\mathrm{NaCa}}^{d} & \mathbf{k}_{\mathrm{NaCl}}^{d} & \mathbf{k}_{\mathrm{Nae}}^{d} \\ \mathbf{k}_{\mathrm{CaE}}^{d} & \mathbf{k}_{\mathrm{CaNa}}^{d} & \mathbf{k}_{\mathrm{CaCa}}^{d} & \mathbf{k}_{\mathrm{CaCl}}^{d} & \mathbf{k}_{\mathrm{Cae}}^{d} \\ \mathbf{k}_{\mathrm{ClE}}^{d} & \mathbf{k}_{\mathrm{Cl\,Na}}^{d} & \mathbf{k}_{\mathrm{Cl\,Ca}}^{d} & \mathbf{k}_{\mathrm{Cl\,Cl}}^{d} & \mathbf{k}_{\mathrm{Cle}}^{d} \\ \mathbf{k}_{\mathrm{e}} & \mathbf{k}_{\mathrm{eNa}}^{d} & \mathbf{k}_{\mathrm{eCa}}^{d} & \mathbf{k}_{\mathrm{eCl}}^{d} & \boldsymbol{\sigma}_{\mathrm{e}} \end{bmatrix}. \tag{16.2.6}$$

The components $\mathbf{F}_{\mathrm{NaE}}^{d}$, $\mathbf{F}_{\mathrm{CaE}}^{d}$ and $\mathbf{F}_{\mathrm{ClE}}^{d}$ of the vector $\mathcal{F}$ are linearly independent, even in absence of gravity, due to the presence of PGs. In fact, electroneutrality implies

$$\sum_{k\in\mathrm{E}_{\mathrm{ions}}} \zeta_{k}\,c_{k\mathrm{E}}\,\widehat{v}_{k}\,\mathbf{F}_{k\mathrm{E}}^{d} = -RT\,\boldsymbol{\nabla}e_{\mathrm{PG}}\,. \tag{16.2.7}$$

On the other hand, the components of the vector flux $\mathcal{J}$ are linearly dependent. Indeed the electrical current density $\mathbf{I}_{\mathrm{eE}}$ is a linear combination of the ionic diffusive fluxes $\mathbf{J}_{k\mathrm{E}}^{d}$, Eqn $(16.1.18)_2$, and of the diffusive flux of PGs, which is proportional to the flux of water,

$$\mathbf{I}_{\mathrm{eE}} - \mathrm{F}\sum_{k\in\mathrm{E}_{\mathrm{ions}}}\zeta_{k}\frac{\mathbf{J}_{k\mathrm{E}}^{d}}{\widehat{v}_{k}} = \mathrm{F}\,\zeta_{\mathrm{PG}}\frac{\mathbf{J}_{\mathrm{PG}}^{d}}{\widehat{v}_{\mathrm{PG}}} = -\mathrm{F}\,\tilde{e}_{\mathrm{PG}}\,\mathbf{J}_{\mathrm{wE}}\,. \tag{16.2.8}$$

As a consequence of the above remark, the lines of the diffusion matrix $\boldsymbol{\mathcal{K}}$ are linearly dependent and $\boldsymbol{\mathcal{K}}$ can be at best positive semi-definite (PsD). That the matrices $\boldsymbol{\kappa}$ and $\boldsymbol{\mathcal{K}}$ are actually PsD will be addressed in Note 16.2.

Note 16.1 on identification procedures.

Any identification of the diffusion coefficients uses data from specific experimental processes. Two main methods may be devised. The generalized diffusion coefficients can be obtained from ionic mobilities measured in experiments
- either at *vanishing pore pressure gradient* $\mathbf{F}_{\mathrm{E}} = \mathbf{0}$, in which case the diffusive fluxes simplify to

$$\mathbf{J}_{k\mathrm{E}}^{d} = -\sum_{l\in\mathrm{E}_{\mathrm{ions}}}\mathbf{k}_{kl}^{d}\cdot\mathbf{F}_{l\mathrm{E}}^{d} - \mathbf{k}_{k\mathrm{e}}^{d}\cdot\mathbf{F}_{\mathrm{eE}}\,, \quad k\in\mathrm{E}_{\mathrm{ions}}\,, \tag{16.2.9}$$

- or at *vanishing water flux* $\mathbf{J}_{\mathrm{wE}} = \mathbf{0}$, and then

$$\mathbf{J}_{k\mathrm{E}}^{d} = -\sum_{l\in\mathrm{E}_{\mathrm{ions}}}(\mathbf{k}_{kl}^{d} - \mathbf{k}_{k\mathrm{E}}^{d}\cdot\mathbf{k}_{\mathrm{EE}}^{-1}\cdot\mathbf{k}_{\mathrm{E}l}^{d})\cdot\mathbf{F}_{l\mathrm{E}}^{d} - (\mathbf{k}_{k\mathrm{e}}^{d} - \mathbf{k}_{k\mathrm{E}}^{d}\cdot\mathbf{k}_{\mathrm{EE}}^{-1}\cdot\mathbf{k}_{\mathrm{e}})\cdot\mathbf{F}_{\mathrm{eE}}\,, \quad k\in\mathrm{E}_{\mathrm{ions}}\,. \tag{16.2.10}$$

16.2.3 Relations between the diffusion matrices κ and $\mathcal{K}$

As noted above, the simplest way to define the diffusion properties would be to identify directly the matrix κ. Its coefficients are a priori independent and they are restricted only by symmetry and positive semi-definiteness. However, more information is available on the matrix $\mathcal{K}$. But this matrix $\mathcal{K}$ is not definite, and therefore there exist relations between its coefficients. So there are two ways of proceeding:

- either to postulate directly the matrix $\mathcal{K}$ and unveil the compatibility between its coefficients due to the relation of linear dependence (16.2.8) and symmetry, as done in Loret et al. [2004]a in their analysis of two-phase clays where the fixed charges (clay platelets) belong to the solid phase;
- or to use directly the compatibility relations inferred by the relations between the coefficients of the matrices κ and $\mathcal{K}$.

The latter approach is the adopted here. The matrix κ results as a by-product of the identification.

The coefficients of the matrix $\mathcal{K}$ can be expressed in terms of those of the matrix κ:

$$
\mathcal{K} = \left[
\begin{array}{c|ccc|c}
\mathbf{k}_{\mathrm{EE}} = \kappa_{\mathrm{ww}} & \times & \mathbf{k}_{\mathrm{E}l}^d = \kappa_{l\mathrm{w}}^d & \times & \mathbf{k}_{\mathrm{e}} = \mathrm{F}\sum_{l\in\mathrm{E_{ions}}}\zeta_l\dfrac{\kappa_{\mathrm{w}l}}{\widehat{v}_l} \\
\hline
\times & \times & \times & \times & \times \\
\mathbf{k}_{k\mathrm{E}}^d = \mathbf{k}_{\mathrm{E}k}^d & \times & \begin{array}{c}\mathbf{k}_{kl}^d = \mathbf{k}_{lk}^d = \\ \kappa_{kl}^d - \dfrac{n^{l\mathrm{E}}}{n^{\mathrm{wE}}}\kappa_{k\mathrm{w}}^d\end{array} & \times & \mathbf{k}_{k\mathrm{e}}^d = \mathrm{F}\sum_{l\in\mathrm{E_{ions}}}\zeta_l\dfrac{\kappa_{kl}^d}{\widehat{v}_l} \\
\times & \times & \times & \times & \times \\
\hline
\mathbf{k}_{\mathrm{e}} & \times & \mathbf{k}_{\mathrm{e}l}^d = \mathbf{k}_{l\mathrm{e}}^d & \times & \sigma_{\mathrm{e}} = \mathrm{F}^2\sum_{k,l\in\mathrm{E_{ions}}}\zeta_k\dfrac{\kappa_{kl}}{\widehat{v}_k\widehat{v}_l}\zeta_l
\end{array}
\right].
\tag{16.2.11}
$$

It has been instrumental to introduce the coefficients,

$$
\kappa_{kl}^d \equiv \kappa_{kl} - \frac{n^{k\mathrm{E}}}{n^{\mathrm{wE}}}\kappa_{\mathrm{w}l}, \quad k \in \mathrm{E_{ions}}, \ l \in \mathrm{E_{mo}}.
\tag{16.2.12}
$$

The symmetry of the matrix κ has been used to establish (16.2.11). The generalized diffusion matrix $\mathcal{K}$ inherits this symmetry property, Exercise 16.5.

The $2 + n_{\mathrm{ion}}$ compatibility relations that the coefficients of the matrix $\mathcal{K}$ have to satisfy due to the relation of linear dependence (16.2.8) can now be cast in the format

$$
\left\{
\begin{aligned}
\mathbf{k}_{\mathrm{e}} &= \mathrm{F}\sum_{l\in\mathrm{E_{ions}}}\zeta_l\frac{\mathbf{k}_{l\mathrm{E}}^d}{\widehat{v}_l} - \mathrm{F}\,\tilde{e}_{\mathrm{PG}}\,\mathbf{k}_{\mathrm{EE}}, \\
\mathbf{k}_{\mathrm{e}k}^d &= \mathrm{F}\sum_{l\in\mathrm{E_{ions}}}\zeta_l\frac{\mathbf{k}_{kl}^d}{\widehat{v}_l} - \mathrm{F}\,\tilde{e}_{\mathrm{PG}}\,\mathbf{k}_{k\mathrm{E}}^d, \quad k \in \mathrm{E_{ions}}, \\
\sigma_{\mathrm{e}} &= \mathrm{F}\sum_{l\in\mathrm{E_{ions}}}\zeta_l\frac{\mathbf{k}_{l\mathrm{e}}^d}{\widehat{v}_l} - \mathrm{F}\,\tilde{e}_{\mathrm{PG}}\,\mathbf{k}_{\mathrm{e}}.
\end{aligned}
\right.
\tag{16.2.13}
$$

A particular consequence of these compatibility relations is worth to be recorded, namely

$$
\mathrm{F}^2\sum_{k,l\in\mathrm{E_{ions}}}\zeta_k\frac{\mathbf{k}_{kl}^d}{\widehat{v}_k\widehat{v}_l}\zeta_l = \sigma_{\mathrm{e}} + 2\,\mathbf{k}_{\mathrm{e}}\,\mathrm{F}\,\tilde{e}_{\mathrm{PG}} + \mathbf{k}_{\mathrm{EE}}\,\mathrm{F}^2\,\tilde{e}_{\mathrm{PG}}^2.
\tag{16.2.14}
$$

16.2.4 Procedures of identification of the constitutive functions

The general form of the diffusion coefficients is now restricted. Indeed, these coefficients are constrained by the equivalence of the two forms of the dissipation inequality, Sections 16.2.1 and 16.2.2, by the compatibility conditions

(16.2.13) and (16.2.14), and further by the condition of positive semi-definiteness of the generalized diffusion matrices. In addition, as pointed out in Note 16.1 above, experimental conditions that are used to measure the parameters should be specified: the analysis in Sections 16.2.5 to 16.2.6 below considers that the ionic mobilities are measured at vanishing pore pressure gradient.

The set of ions necessary to balance the charge of the PGs is termed collectively non mobile. This terminology is however not completely appropriate, because the individual cations do not bind to the proteoglycans and they are individually free to move around. It might be more appropriate to use the term "neutralizing ions". Thus, for a sodium chloride electrolyte, $e_{PG} + N_{NaE}^{nm} = 0$, and $N_{NaE}^{mo} - N_{ClE}^{mo} = 0$.

In view of the algebraic manipulations below, it is instrumental to introduce the following pseudo-vectors $\mathbf{N}$ and $\mathbf{Z}$ of length equal to the number n_{ion} of ions:

$$N_k = \frac{1}{\widehat{v}_k} \frac{n^{kE}}{n^{wE}}, \quad Z_k = \zeta_k, \quad k \in E_{ions}. \tag{16.2.15}$$

Repeated use of electroneutrality will be made in the forms

$$\mathbf{Z}^T \mathbf{Z}^\perp = 0, \quad \mathbf{Z}^T \mathbf{N} = \mathbf{Z}^T \mathbf{N}_{nm} = -\widetilde{e}_{PG}. \tag{16.2.16}$$

Here $\mathbf{Z}^\perp$ is a generic element of the vector space of dimension $n_{ion} - 1$ which is orthogonal to $\mathbf{Z}$. When mobile and non mobile ions are distinguished, $\mathbf{Z}^\perp$ may also assume the value $\mathbf{N}_{mo}$.

16.2.5 A general form of diffusion with mobile and non mobile ions

The format of the generalized diffusion equations is extracted in three key steps.

(D1) The open circuit permeability $\mathbf{k}_H$ is known:

Measurement of hydraulic conductivity at uniform concentrations of ions and PGs and at vanishing electrical current $\mathbf{I}_{eE}$, necessarily gives rise to a "streaming potential",

$$\boldsymbol{\nabla}\phi_E = -(\boldsymbol{\sigma}_e)^{-1} \cdot \mathbf{k}_e \cdot \mathbf{F}_E. \tag{16.2.17}$$

Thus

$$\mathbf{J}_{wE} = -\mathbf{k}_H \cdot \mathbf{F}_E, \quad \underbrace{\mathbf{k}_H = \mathbf{k}_{EE} - \mathbf{k}_e \cdot (\boldsymbol{\sigma}_e)^{-1} \cdot \mathbf{k}_e}_{\text{open circuit permeability}}. \tag{16.2.18}$$

The measured entity $\mathbf{k}_H$ [unit: $m^3 \times s/kg$] is thus the "open circuit" permeability. It is expected to be positive definite so that water flows against pore pressure gradient. On the other hand, the water flow is oriented along the streaming potential,

$$\mathbf{J}_{wE} = -\mathbf{k}_H \cdot \mathbf{F}_E = \mathbf{k}_H \cdot \mathbf{k}_e^{-1} \cdot \boldsymbol{\sigma}_e \cdot \boldsymbol{\nabla}\phi_E \quad \text{total water flow}, \tag{16.2.19}$$

while the sole *electroosmotic flow* is oriented against the streaming potential,

$$- \mathbf{k}_e \cdot \boldsymbol{\nabla}\phi_E \quad \text{electroosmotic water flow}, \tag{16.2.20}$$

if $\mathbf{k}_e$ and $\boldsymbol{\sigma}_e$ are positive definite. In this chapter, the pH is assumed neutral and the sole fixed charge, which is negative, is due to proteoglycans so that $\mathbf{k}_e$ will indeed be shown to be positive definite.

Alternatively, measurement of the flow at uniform concentrations of ions and PGs and uniform electrical potential yields the "short circuit" permeability $\mathbf{k}_{EE}$, which is proportional to the hydraulic conductivity $\mathbf{K}_H$, namely

$$\mathbf{J}_{wE} = -\mathbf{k}_{EE} \cdot \mathbf{F}_E, \quad \underbrace{\mathbf{k}_{EE} = \frac{\mathbf{K}_H}{\rho_w \, g} = \kappa_{ww}}_{\text{short circuit permeability}}. \tag{16.2.21}$$

(D2) The effective ionic mobilities $\mathbf{u}_k^*$, $k \in \mathrm{E_{ions}}$, are known:

The velocity relative to water that the ionic component k can reach in presence of an electrical potential ϕ_E at uniform ionic concentrations and vanishing pore pressure gradient is equal to $-\mathrm{sgn}\,\zeta_k\,\mathbf{u}_k^* \cdot \boldsymbol{\nabla}\phi_\mathrm{E}$[16.1]. The sign of the electrical charge indicates that a cation is moving towards the cathode, while an anion is moving towards the anode, i.e., in the direction of increasing electrical potential. In suspension mechanics and environmental geomechanics, this phenomenon known as *electrophoresis* is used to densify fine particle suspensions around the anode; densified materials are removed periodically to allow for the process of anion elimination to continue. In agreement with the usual convention, the electrical current density has a direction opposite to that of electrons. Since the diffusive flux involves both mobile and non mobile parts, there is an ambiguity on the volume fraction involved in the resulting flux, and, temporarily, the volume fraction and concentration are left undecided as $\tilde{n}^{k\mathrm{E}}$ and $\tilde{c}_{k\mathrm{E}}$ respectively, that is, at uniform ionic concentrations, $\mathbf{J}_{k\mathrm{E}}^d = -\tilde{n}^{k\mathrm{E}}\,\mathrm{sgn}\,\zeta_k\,\mathbf{u}_k^* \cdot \boldsymbol{\nabla}\phi_\mathrm{E}$. The coefficient matrices $\mathbf{k}_{k\mathrm{e}}^d$, $k \in \mathrm{E_{ions}}$, result as indicated in $(16.2.26)_2$ below. Then, using the definition of $\mathbf{k}_{k\mathrm{e}}^d$ in (16.2.11), the coefficient matrices $\boldsymbol{\kappa}_{kl}^d$ are obtained to within a matrix $\mathbf{A}_k$ of size 3×3,

$$\frac{\boldsymbol{\kappa}_{kl}^d}{\widehat{v}_k \widehat{v}_l} = n^\mathrm{E}\,\tilde{c}_{k\mathrm{E}}\,\frac{\mathbf{u}_k^*}{\mathrm{F}\,|\zeta_k|}\,I_{kl} + \mathbf{A}_k\,Z_l^\perp\,, \quad k,l \in \mathrm{E_{ions}}\,. \tag{16.2.22}$$

(D3) The matrix $\boldsymbol{\kappa}$ is enforced to be symmetric:

Upon insertion of the relation (16.2.22) in the definition (16.2.12), symmetry of

$$\frac{\boldsymbol{\kappa}_{kl}}{\widehat{v}_k \widehat{v}_l} = n^\mathrm{E}\,\tilde{c}_{k\mathrm{E}}\,\frac{\mathbf{u}_k^*}{\mathrm{F}\,|\zeta_k|}\,I_{kl} + \mathbf{A}_k\,Z_l^\perp + N_k\,\frac{\boldsymbol{\kappa}_{wl}}{\widehat{v}_l}\,, \quad k,l \in \mathrm{E_{ions}}\,, \tag{16.2.23}$$

implies

$$\mathbf{A}_k\,Z_l^\perp - Z_k^\perp\,\mathbf{A}_l = \frac{\boldsymbol{\kappa}_{wk}}{\widehat{v}_k}\,N_l - N_k\,\frac{\boldsymbol{\kappa}_{wl}}{\widehat{v}_l}\,, \quad k,l \in \mathrm{E_{ions}}\,. \tag{16.2.24}$$

A general solution of (16.2.24) involves three arbitrary matrices $\mathbf{a}_i$, $i \in [1,3]$, of size 3×3, and implies a particular structure of the $\boldsymbol{\kappa}_{wk}/\widehat{v}_k$'s, $k \in \mathrm{E_{ions}}$,

$$\mathbf{A}_k = \mathbf{a}_1\,Z_k^\perp - \mathbf{a}_2\,N_k\,, \quad \frac{\boldsymbol{\kappa}_{wk}}{\widehat{v}_k} = \mathbf{a}_2\,Z_k^\perp + \mathbf{a}_3\,N_k\,, \quad k \in \mathrm{E_{ions}}\,. \tag{16.2.25}$$

Hence, with help of (16.2.11), (16.2.23)–(16.2.25), the electrical conductivity matrix $\boldsymbol{\sigma}_\mathrm{e}$ and electroosmotic matrix $\mathbf{k}_\mathrm{e}$ are defined up to the matrix $\mathbf{a}_3$. Moreover (16.2.25) implies $\mathbf{k}_{\mathrm{E}k}^d/\widehat{v}_k = \boldsymbol{\kappa}_{\mathrm{kw}}^d/\widehat{v}_k = \boldsymbol{\kappa}_{\mathrm{wk}}/\widehat{v}_k - N_k\,\boldsymbol{\kappa}_{\mathrm{ww}}$, $k \in \mathrm{E_{ions}}$, and, then $\mathbf{k}_{kl}^d/\widehat{v}_k \widehat{v}_l = \boldsymbol{\kappa}_{kl}^d/\widehat{v}_k \widehat{v}_l - N_l\,\boldsymbol{\kappa}_{\mathrm{kw}}^d/\widehat{v}_k$, $k,l \in \mathrm{E_{ions}}$, results from (16.2.22).

In summary,

$$\begin{cases} \mathbf{k}_\mathrm{EE} = \dfrac{\mathbf{K_H}}{\rho_\mathrm{w}\,\mathrm{g}} > 0, \quad \mathbf{k_H} = \mathbf{k_{EE}} - \mathbf{k_e} \cdot (\boldsymbol{\sigma}_\mathrm{e})^{-1} \cdot \mathbf{k_e}, \\[2ex] \dfrac{\mathbf{k}_{k\mathrm{e}}^d}{\widehat{v}_k} = n^\mathrm{E}\,\tilde{c}_{k\mathrm{E}}\,\mathbf{u}_k^*\,\mathrm{sgn}\,\zeta_k\,, \quad k \in \mathrm{E_{ions}}, \\[2ex] \dfrac{\mathbf{k}_{\mathrm{E}k}^d}{\widehat{v}_k} = \mathbf{a}_2\,Z_k^\perp + (\mathbf{a}_3 - \mathbf{k_{EE}})\,N_k\,, \quad k \in \mathrm{E_{ions}}, \\[2ex] \dfrac{\mathbf{k}_{kl}^d}{\widehat{v}_k \widehat{v}_l} = \mathbf{a}_1\,Z_k^\perp\,Z_l^\perp - \mathbf{a}_2\,(Z_k^\perp\,N_l + N_k\,Z_l^\perp) + (\mathbf{k_{EE}} - \mathbf{a}_3)\,N_k\,N_l \\[2ex] \qquad\qquad + n^\mathrm{E}\,\tilde{c}_{k\mathrm{E}}\,\dfrac{\mathbf{u}_k^*}{\mathrm{F}\,|\zeta_k|}\,I_{kl}\,, \quad k,l \in \mathrm{E_{ions}}, \\[2ex] \mathbf{k_e} = -\mathbf{a}_3\,\mathrm{F}\,\tilde{e}_\mathrm{PG}\,, \quad \boldsymbol{\sigma}_\mathrm{e} = n^\mathrm{E}\,\mathrm{F}\,\displaystyle\sum_{k \in \mathrm{E_{ions}}} |\zeta_k|\,\tilde{c}_{k\mathrm{E}}\,\mathbf{u}_k^* + \mathbf{a}_3\,\mathrm{F}^2\,\tilde{e}_\mathrm{PG}^2\,. \end{cases} \tag{16.2.26}$$

The generalized diffusion matrix is left with three arbitrary matrices $\mathbf{a}_i$, $i \in [1,3]$. The relations $(16.2.26)_{5-7}$ agree with the compatibility condition (16.2.14). A formal way to define these matrices is proposed in Section 16.2.6.

[16.1] The effective ionic mobility refers to the mobility in the porous structure, while the ionic mobility refers to a blank solution.

The diffusion matrix κ is known as soon as the diffusion matrix $\mathcal{K}$ is, using successively $\kappa_{ww} = \mathbf{k}_{EE}$, (16.2.25) and (16.2.23):

$$\begin{cases} \dfrac{\kappa_{wk}}{\widehat{v}_k} = \mathbf{a}_2\,Z_k^\perp + \mathbf{a}_3\,N_k\,, \quad k \in \mathrm{E}_{ions}\,, \\[2mm] \dfrac{\kappa_{kl}}{\widehat{v}_k\widehat{v}_l} = n^E\,\tilde{c}_{kE}\,\dfrac{\mathbf{u}_k^*}{F\,|\zeta_k|}\,I_{kl} + \mathbf{a}_1\,Z_k^\perp\,Z_l^\perp + \mathbf{a}_3\,N_k\,N_l\,, \quad k,\,l \in \mathrm{E}_{ions}\,. \end{cases} \tag{16.2.27}$$

To summarize the analysis at this point, let us emphasize that the only assumption of symmetry of the diffusion matrices and their algebraic equivalence has provided the general structure of the constitutive equations to within three 3×3 matrices and a pseudo-vector of size n_{ion}.

Note that, in the above analysis, a single pseudo-vector $\mathbf{Z}^\perp$ has been used. This covers completely the case of a binary electrolyte. For ternary electrolytes, a complete solution should involve two independent vectors $\mathbf{Z}^\perp$ in (16.2.22): the complete derivation is elaborated in Exercise 16.9.

16.2.6   The particular case of arrowhead diffusion matrices

A particular form of diffusion is derived below under the two simple following assumptions:

- *Assumption* $(\mathcal{A}1)$: the diffusive flux of ions is not affected by a gradient of fluid pressure: thus $\mathbf{a}_2 = \mathbf{0}$ and $\mathbf{a}_3 = \mathbf{k}_{EE}$. Symmetry of the matrix $\mathcal{K}$ implies that the ionic gradients do not affect the fluid flow;

- *Assumption* $(\mathcal{A}2)$: a gradient of concentration of ion k does not affect the diffusive flux of ion $l \neq k$. Then $\mathbf{a}_1 = \mathbf{0}$.

Therefore the diffusion matrix takes the symmetric arrowhead form,

$$\mathcal{K} = \begin{bmatrix} \mathbf{k}_{EE} & 0 & 0 & 0 & \mathbf{k}_e \\ 0 & \mathbf{k}_{NaNa}^d & 0 & 0 & \mathbf{k}_{Nae}^d \\ 0 & 0 & \mathbf{k}_{CaCa}^d & 0 & \mathbf{k}_{Cae}^d \\ 0 & 0 & 0 & \mathbf{k}_{Cl\,Cl}^d & \mathbf{k}_{Cl\,e}^d \\ \mathbf{k}_e & \mathbf{k}_{eNa}^d & \mathbf{k}_{eCa}^d & \mathbf{k}_{eCl}^d & \boldsymbol{\sigma}_e \end{bmatrix}. \tag{16.2.28}$$

Given the matrix $\mathcal{K}$, the matrix κ can be cast in the form,

$$\kappa = \left[\begin{array}{c|ccc} \mathbf{k}_{EE} & N_{Na}\,\mathbf{k}_{EE} & N_{Ca}\,\mathbf{k}_{EE} & N_{Cl}\,\mathbf{k}_{EE} \\ \hline N_{Na}\,\mathbf{k}_{EE} & N_{Na}^2\,\mathbf{k}_{EE} & N_{Na}\,N_{Ca}\,\mathbf{k}_{EE} & N_{Na}\,N_{Cl}\,\mathbf{k}_{EE} \\ N_{Ca}\,\mathbf{k}_{EE} & N_{Ca}\,N_{Na}\,\mathbf{k}_{EE} & N_{Ca}^2\,\mathbf{k}_{EE} & N_{Ca}\,N_{Cl}\,\mathbf{k}_{EE} \\ N_{Cl}\,\mathbf{k}_{EE} & N_{Cl}\,N_{Na}\,\mathbf{k}_{EE} & N_{Cl}\,N_{Ca}\,\mathbf{k}_{EE} & N_{Cl}^2\,\mathbf{k}_{EE} \end{array} \right] +$$

$$+ \left[\begin{array}{c|ccc} 0 & 0 & 0 & 0 \\ \hline 0 & \mathbf{k}_{NaNa}^d & 0 & 0 \\ 0 & 0 & \mathbf{k}_{CaCa}^d & 0 \\ 0 & 0 & 0 & \mathbf{k}_{ClCl}^d \end{array} \right], \tag{16.2.29}$$

the $N's$ being *now* defined as $N_k = n^{kE}/n^{wE}$, $k \in \mathrm{E}_{ions}$. This expression makes it possible to establish contact with the model developed by Gu et al. [1998], Exercise 16.6.

Note 16.2 on the dissipation inequality.

Note that the diffusion matrix $\mathcal{K}$, Eqn (16.2.28), is positive semi-definite because its principal matrix minors $\mathbf{k}_{EE}$, $\mathbf{k}_{kk}^d$, $k \in \mathrm{E}_{ions}$, and $\boldsymbol{\sigma}_e$ are strictly positive while the matrix itself is singular. A formal proof is provided in Exercise 9.1. Therefore the dissipation $(16.1.22)_2$ due to diffusion is indeed positive or zero.

Note 16.3 on the identification procedures.

Since now the matrices $\mathbf{k}_{Ek}^{d}$, $k \in \mathrm{E_{ions}}$, are zero, then the identification procedures at vanishing gradient of pore pressure and at vanishing flux of water are identical, as observed by comparing Eqns (16.2.9) and (16.2.10).

Note 16.4 on commutation properties.

Among the matrices $\mathbf{k}_{\mathrm{EE}}$, $\mathbf{k}_{\mathrm{H}}$, $\mathbf{k}_{\mathrm{e}}$, $\boldsymbol{\sigma}_{\mathrm{e}}$, and $\boldsymbol{\sigma}_{\mathrm{e}}^{\mathrm{ion}}$ involved in the generalized diffusion matrix and defined in Table 16.2.1, only two are linearly independent, say $\mathbf{k}_{\mathrm{EE}}$ as representative of the seepage properties and $\boldsymbol{\sigma}_{\mathrm{e}}^{\mathrm{ion}}$ as representative of the diffusion properties. The following commutation properties may be proved, Exercise 16.7,

$$\mathbf{k}_{\mathrm{H}} \cdot \mathbf{k}_{\mathrm{EE}}^{-1} = \boldsymbol{\sigma}_{\mathrm{e}}^{\mathrm{ion}} \cdot (\boldsymbol{\sigma}_{\mathrm{e}})^{-1}, \quad \mathbf{k}_{\mathrm{EE}}^{-1} \cdot \mathbf{k}_{\mathrm{H}} = (\boldsymbol{\sigma}_{\mathrm{e}})^{-1} \cdot \boldsymbol{\sigma}_{\mathrm{e}}^{\mathrm{ion}} . \tag{16.2.30}$$

Note also the relation,

$$\boldsymbol{\sigma}_{\mathrm{e}}^{\mathrm{ion}} = \boldsymbol{\sigma}_{\mathrm{e}} - \mathbf{k}_{\mathrm{e}} \cdot \mathbf{k}_{\mathrm{EE}}^{-1} \cdot \mathbf{k}_{\mathrm{e}} . \tag{16.2.31}$$

For isotropic properties, these relations take the form,

$$\frac{k_{\mathrm{H}}}{k_{\mathrm{EE}}} = \frac{\sigma_{\mathrm{e}}^{\mathrm{ion}}}{\sigma_{\mathrm{e}}}, \quad \sigma_{\mathrm{e}}^{\mathrm{ion}} = \sigma_{\mathrm{e}} - \frac{k_{\mathrm{e}}^{2}}{k_{\mathrm{EE}}} . \tag{16.2.32}$$

16.2.7 Summary of constitutive equations of transport

The coefficients of the diffusion matrices are expressed in terms of the hydraulic conductivity matrix $\mathbf{K}_{\mathrm{H}}$, effective ionic mobilities $\mathbf{u}_{k}^{*}$, chemical contents and fixed charge $\tilde{e}_{\mathrm{PG}}$ in Table 16.2.1.

Table 16.2.1: Coefficients of generalized diffusion.

○	Hydraulic properties:	• Short circuit permeability: $\mathbf{k}_{\mathrm{EE}} = \mathbf{K}_{\mathrm{H}}/(\rho_{\mathrm{w}}\,\mathrm{g}) > 0$ • Open circuit permeability: $\mathbf{k}_{\mathrm{H}} = \mathbf{k}_{\mathrm{EE}} - \mathbf{k}_{\mathrm{e}} \cdot (\boldsymbol{\sigma}_{\mathrm{e}})^{-1} \cdot \mathbf{k}_{\mathrm{e}}$
○	Electrodiffusive properties:	• Electrodiffusion matrices: $\dfrac{\mathbf{k}_{k\mathrm{e}}^{d}}{\widehat{v}_{k}} = n^{\mathrm{E}}\,\tilde{c}_{k\mathrm{E}}\,\mathbf{u}_{k}^{*}\,\mathrm{sgn}\,\zeta_{k}\,, \; k \in \mathrm{E_{ions}}$ • Self-diffusion matrices: $\dfrac{\mathbf{k}_{kl}^{d}}{\widehat{v}_{k}\widehat{v}_{l}} = n^{\mathrm{E}}\,\tilde{c}_{k\mathrm{E}}\,\dfrac{\mathbf{u}_{k}^{*}}{\mathrm{F}\,|\zeta_{k}|}\,I_{kl}, \; k,l \in \mathrm{E_{ions}}$
○	Electroosmotic matrix:	$\mathbf{k}_{\mathrm{e}} = -\mathbf{k}_{\mathrm{EE}}\,\mathrm{F}\,\tilde{e}_{\mathrm{PG}}$
○	Electrical conductivity: $\boldsymbol{\sigma}_{\mathrm{e}} = \boldsymbol{\sigma}_{\mathrm{e}}^{\mathrm{ion}} + \boldsymbol{\sigma}_{\mathrm{e}}^{\mathrm{PG}}$	• Ionic conductivity: $\boldsymbol{\sigma}_{\mathrm{e}}^{\mathrm{ion}} = n^{\mathrm{E}}\,\mathrm{F} \displaystyle\sum_{k\in\mathrm{E_{ions}}} |\zeta_{k}|\,\tilde{c}_{k\mathrm{E}}\,\mathbf{u}_{k}^{*}$ • PG conductivity: $\boldsymbol{\sigma}_{\mathrm{e}}^{\mathrm{PG}} = \mathbf{k}_{\mathrm{EE}}\,\mathrm{F}^{2}\,\tilde{e}_{\mathrm{PG}}^{2}$

The short circuit permeability matrix $\mathbf{k}_{\mathrm{EE}}$ is naturally positive definite. So are the matrices of ionic mobilities and the electrical conductivity. The open circuit permeability $\mathbf{k}_{\mathrm{H}}$ is proved to be positive definite as well in Exercise 16.8 due to the commutation relation (16.2.30).

To summarize, the constitutive equations of transport can be cast in terms of

- either fluxes relative to the solid,

$$
\underset{\text{generalized gradients}}{}\qquad\qquad\qquad\underset{\text{absolute fluxes}}{}
$$

$$
\mathbf{f} =
\begin{bmatrix}
\rho_{\mathrm{w}}\,\boldsymbol{\nabla}\mu_{\mathrm{wE}} \\[4pt]
\rho_{\mathrm{Na}}\,\boldsymbol{\nabla}\mu_{\mathrm{NaE}}^{\mathrm{ec}} \\[4pt]
\rho_{\mathrm{Ca}}\,\boldsymbol{\nabla}\mu_{\mathrm{CaE}}^{\mathrm{ec}} \\[4pt]
\rho_{\mathrm{Cl}}\,\boldsymbol{\nabla}\mu_{\mathrm{ClE}}^{\mathrm{ec}}
\end{bmatrix}
\begin{matrix}
\leftarrow\ \text{water}\ \rightarrow \\[4pt]
\leftarrow\ \text{sodium ion}\ \rightarrow \\[4pt]
\leftarrow\ \text{calcium ion}\ \rightarrow \\[4pt]
\leftarrow\ \text{chloride ion}\ \rightarrow
\end{matrix}
\qquad
\mathbf{j} =
\begin{bmatrix}
\mathbf{J}_{\mathrm{wE}} \\[4pt]
\mathbf{J}_{\mathrm{NaE}} \\[4pt]
\mathbf{J}_{\mathrm{CaE}} \\[4pt]
\mathbf{J}_{\mathrm{ClE}}
\end{bmatrix}
\tag{16.2.33}
$$

namely in explicit form, $\mathbf{j} = -\boldsymbol{\kappa}\,\mathbf{f}$, the 4×4 matrix $\boldsymbol{\kappa}$,

$$
\boldsymbol{\kappa} =
\begin{bmatrix}
\mathbf{k}_{\mathrm{EE}} & N_{\mathrm{Na}}\,\mathbf{k}_{\mathrm{EE}} & N_{\mathrm{Ca}}\,\mathbf{k}_{\mathrm{EE}} & N_{\mathrm{Cl}}\,\mathbf{k}_{\mathrm{EE}} \\[6pt]
N_{\mathrm{Na}}\,\mathbf{k}_{\mathrm{EE}} & N_{\mathrm{Na}}^{2}\,\mathbf{k}_{\mathrm{EE}} + \mathbf{k}_{\mathrm{NaNa}}^{d} & N_{\mathrm{Na}}\,N_{\mathrm{Ca}}\,\mathbf{k}_{\mathrm{EE}} & N_{\mathrm{Na}}\,N_{\mathrm{Cl}}\,\mathbf{k}_{\mathrm{EE}} \\[6pt]
N_{\mathrm{Ca}}\,\mathbf{k}_{\mathrm{EE}} & N_{\mathrm{Ca}}\,N_{\mathrm{Na}}\,\mathbf{k}_{\mathrm{EE}} & N_{\mathrm{Ca}}^{2}\,\mathbf{k}_{\mathrm{EE}} + \mathbf{k}_{\mathrm{CaCa}}^{d} & N_{\mathrm{Ca}}\,N_{\mathrm{Cl}}\,\mathbf{k}_{\mathrm{EE}} \\[6pt]
N_{\mathrm{Cl}}\,\mathbf{k}_{\mathrm{EE}} & N_{\mathrm{Cl}}\,N_{\mathrm{Na}}\,\mathbf{k}_{\mathrm{EE}} & N_{\mathrm{Cl}}\,N_{\mathrm{Ca}}\,\mathbf{k}_{\mathrm{EE}} & N_{\mathrm{Cl}}^{2}\,\mathbf{k}_{\mathrm{EE}} + \mathbf{k}_{\mathrm{ClCl}}^{d}
\end{bmatrix},
\tag{16.2.34}
$$

being expressed via $N_k = n^{k\mathrm{E}}/n^{\mathrm{wE}}$, $k \in \mathrm{E_{ions}}$,
- or, equivalently, in terms of diffusive fluxes,

$$
\underset{\text{generalized gradients}}{}\qquad\qquad\qquad\underset{\text{diffusive fluxes}}{}
$$

$$
\mathcal{F} =
\begin{bmatrix}
\boldsymbol{\nabla}P_{\mathrm{wE}} \\[4pt]
RT/\widehat{v}_{\mathrm{Na}}\,\boldsymbol{\nabla}\mathrm{Ln}\,x_{\mathrm{NaE}} \\[4pt]
RT/\widehat{v}_{\mathrm{Ca}}\,\boldsymbol{\nabla}\mathrm{Ln}\,x_{\mathrm{CaE}} \\[4pt]
RT/\widehat{v}_{\mathrm{Cl}}\,\boldsymbol{\nabla}\mathrm{Ln}\,x_{\mathrm{ClE}} \\[4pt]
\boldsymbol{\nabla}\phi_{\mathrm{E}}
\end{bmatrix}
\begin{matrix}
\leftarrow\ \text{water seepage}\ -\ \text{Darcy's law}\ \rightarrow \\[4pt]
\leftarrow\ \text{sodium diffusion}\ -\ \text{Fick's law}\ \rightarrow \\[4pt]
\leftarrow\ \text{calcium diffusion}\ -\ \text{Fick's law}\ \rightarrow \\[4pt]
\leftarrow\ \text{chloride diffusion}\ -\ \text{Fick's law}\ \rightarrow \\[4pt]
\leftarrow\ \text{electrical flow}\ -\ \text{Ohm's law}\ \rightarrow
\end{matrix}
\qquad
\mathcal{J} =
\begin{bmatrix}
\mathbf{J}_{\mathrm{wE}} \\[4pt]
\mathbf{J}_{\mathrm{NaE}}^{d} \\[4pt]
\mathbf{J}_{\mathrm{CaE}}^{d} \\[4pt]
\mathbf{J}_{\mathrm{ClE}}^{d} \\[4pt]
\mathbf{I}_{\mathrm{eE}}
\end{bmatrix}
\tag{16.2.35}
$$

namely, $\mathcal{J} = -\mathcal{K}\,\mathcal{F}$ with the 5×5 matrix $\mathcal{K}$,

$$
\mathcal{K} =
\begin{bmatrix}
\mathbf{k}_{\mathrm{EE}} & \mathbf{0} & \mathbf{0} & \mathbf{0} & \mathbf{k}_{\mathrm{e}} \\[4pt]
\mathbf{0} & \mathbf{k}_{\mathrm{NaNa}}^{d} & \mathbf{0} & \mathbf{0} & \mathbf{k}_{\mathrm{Nae}}^{d} \\[4pt]
\mathbf{0} & \mathbf{0} & \mathbf{k}_{\mathrm{CaCa}}^{d} & \mathbf{0} & \mathbf{k}_{\mathrm{Cae}}^{d} \\[4pt]
\mathbf{0} & \mathbf{0} & \mathbf{0} & \mathbf{k}_{\mathrm{ClCl}}^{d} & \mathbf{k}_{\mathrm{Cle}}^{d} \\[4pt]
\mathbf{k}_{\mathrm{e}} & \mathbf{k}_{\mathrm{eNa}}^{d} & \mathbf{k}_{\mathrm{eCa}}^{d} & \mathbf{k}_{\mathrm{eCl}}^{d} & \boldsymbol{\sigma}_{\mathrm{e}}
\end{bmatrix}.
\tag{16.2.36}
$$

If accounted for, gravity contributes to the generalized gradients as indicated by (16.2.4).

16.2.8   Special cases: electrically and hydraulically open circuits

16.2.8.1   Electrically open circuits

For an open electrical circuit, $\mathbf{I}_{\mathrm{eE}} = \mathbf{0}$, the electrical field $\mathbf{F}_{\mathrm{eE}}$ can be expressed in terms of the gradients of fluid pressure $\mathbf{F}_{\mathrm{E}}$ and ionic concentrations $\mathbf{F}_{l\mathrm{E}}^{d}$, namely

$$
\mathbf{F}_{\mathrm{eE}} = -(\boldsymbol{\sigma}_{\mathrm{e}})^{-1}\cdot\mathbf{k}_{\mathrm{e}}\cdot\mathbf{F}_{\mathrm{E}} - (\boldsymbol{\sigma}_{\mathrm{e}})^{-1}\cdot\sum_{l\in\mathrm{E_{ions}}}\frac{\mathbf{k}_{\mathrm{e}l}^{d}}{\widehat{v}_{l}}\cdot\widehat{v}_{l}\,\mathbf{F}_{l\mathrm{E}}^{d}.
\tag{16.2.37}
$$

Then, using the constitutive equations (16.2.4) and (16.2.28), the flux of water becomes

$$\mathbf{J}_{\mathrm{wE}} = -\mathbf{k}_{\mathrm{H}} \cdot \mathbf{F}_{\mathrm{E}} + \mathbf{k}_{\mathrm{e}} \cdot (\boldsymbol{\sigma}_{\mathrm{e}})^{-1} \cdot \sum_{l \in \mathrm{E_{ions}}} \frac{\mathbf{k}_{\mathrm{el}}^d}{\widehat{v}_l} \cdot \widehat{v}_l \, \mathbf{F}_{l\mathrm{E}}^d \, . \tag{16.2.38}$$

With help of the relation (16.1.16) between absolute and diffusive fluxes, the ionic fluxes relative to the solid $\mathbf{J}_{k\mathrm{E}}$ can be recast in the format,

$$\frac{\mathbf{J}_{k\mathrm{E}}}{\widehat{v}_k} = -\mathbf{k}_{k\mathrm{E}}^0 \cdot \mathbf{F}_{\mathrm{E}} - \sum_{l \in \mathrm{E_{ions}}} \mathbf{k}_{kl\mathrm{E}}^0 \cdot \widehat{v}_l \, \mathbf{F}_{l\mathrm{E}}^d \, , \quad l \in \mathrm{E_{ions}} \, , \tag{16.2.39}$$

with coefficients:

$$\begin{cases} \mathbf{k}_{k\mathrm{E}}^0 = -\dfrac{\mathbf{k}_{ke}^d}{\widehat{v}_k} \cdot (\boldsymbol{\sigma}_{\mathrm{e}})^{-1} \cdot \mathbf{k}_{\mathrm{e}} + \dfrac{1}{\widehat{v}_k} \dfrac{n^{k\mathrm{E}}}{n^{\mathrm{wE}}} \mathbf{k}_{\mathrm{H}} \, , \quad k \in \mathrm{E_{ions}}, \\[3mm] \mathbf{k}_{kl\mathrm{E}}^0 = \dfrac{\mathbf{k}_{kl}^d}{\widehat{v}_k \, \widehat{v}_l} - \dfrac{\mathbf{k}_{ke}^d}{\widehat{v}_k} \cdot (\boldsymbol{\sigma}_{\mathrm{e}})^{-1} \cdot \dfrac{\mathbf{k}_{el}^d}{\widehat{v}_l} - \dfrac{1}{\widehat{v}_k} \dfrac{n^{k\mathrm{E}}}{n^{\mathrm{wE}}} \mathbf{k}_{\mathrm{e}} \cdot (\boldsymbol{\sigma}_{\mathrm{e}})^{-1} \cdot \dfrac{\mathbf{k}_{el}^d}{\widehat{v}_l} \, , \quad k, l \in \mathrm{E_{ions}} \, . \end{cases} \tag{16.2.40}$$

16.2.8.2 Electrically and hydraulically open circuits

Let us consider a diffusion experiment with open electrical circuit, $\mathbf{I}_{\mathrm{eE}} = \mathbf{0}$, and no net fluid flux, $\mathbf{J}_{\mathrm{wE}} = \mathbf{0}$. Then, with help of the commutation relation (16.2.30) and of the relation (16.2.31), the pressure gradient $\mathbf{F}_{\mathrm{E}}$ and electrical field $\mathbf{F}_{\mathrm{eE}}^d$ can be calculated in terms of the concentration gradients $\mathbf{F}_{l\mathrm{E}}^d$ as

$$\mathbf{F}_{\mathrm{eE}} = -\mathbf{k}_{\mathrm{EE}} \cdot \mathbf{k}_{\mathrm{e}}^{-1} \cdot \mathbf{F}_{\mathrm{E}} = -(\boldsymbol{\sigma}_{\mathrm{e}}^{\mathrm{ion}})^{-1} \cdot \sum_{l \in \mathrm{E_{ions}}} \frac{k_{el}^d}{\widehat{v}_l} \, \widehat{v}_l \, \mathbf{F}_{l\mathrm{E}}^d \, . \tag{16.2.41}$$

Backsubstituting in the constitutive equations of the diffusive fluxes yields expressions in terms of concentration gradients,

$$-\frac{\mathbf{J}_{k\mathrm{E}}^d}{\widehat{v}_k} = \sum_{l \in \mathrm{E_{ions}}} \mathbf{k}_{kl\mathrm{E}}^* \, \widehat{v}_l \, \mathbf{F}_{l\mathrm{E}}^d \, , \quad k \in \mathrm{E_{ions}} \, , \tag{16.2.42}$$

where

$$\mathbf{k}_{kl\mathrm{E}}^* = \frac{\mathbf{k}_{kl}^d}{\widehat{v}_k \, \widehat{v}_l} - \frac{\mathbf{k}_{ke}^d}{\widehat{v}_k} \cdot (\boldsymbol{\sigma}_{\mathrm{e}}^{\mathrm{ion}})^{-1} \cdot \frac{\mathbf{k}_{el}^d}{\widehat{v}_l} \, , \quad l \in \mathrm{E_{ions}} \, . \tag{16.2.43}$$

If a single cation k is present, then the relation between diffusive fluxes and electric current density (16.2.8), with $\mathbf{I}_{\mathrm{eE}} = \mathbf{0}$ and $\mathbf{J}_{\mathrm{wE}} = \mathbf{0}$, implies the diffusive fluxes of the anion and cation to be parallel:

$$\begin{aligned} -\zeta_k \, \frac{\mathbf{J}_{k\mathrm{E}}^d}{\widehat{v}_k} &= -\frac{\mathbf{J}_{\mathrm{ClE}}^d}{\widehat{v}_{\mathrm{Cl}}} = \\[2mm] = \zeta_k \sum_{l \in \mathrm{E_{ions}}} \mathbf{k}_{kl\mathrm{E}}^* \cdot \widehat{v}_l \, \mathbf{F}_{l\mathrm{E}}^d &= \sum_{l \in \mathrm{E_{ions}}} \mathbf{k}_{\mathrm{Cl}l\mathrm{E}}^* \cdot \widehat{v}_l \, \mathbf{F}_{l\mathrm{E}}^d \, . \end{aligned} \tag{16.2.44}$$

16.3 Generalized diffusion: the membrane effect

The presence of the fixed charge in tissues requires special attention during the processes of identification of parameters. Indeed, the trick lies in the fact that the primary variables in the extrafibrillar fluid phase are not controlled directly. The identification procedure should first obtain the evolution of these variables in terms of the entities which are controlled.

16.3.1　Equilibrium relations between a gel and a bath

Let us consider a gel to be in chemical equilibrium with a bath. Its chemical state is in general spatially non uniform. Still, each material point of the gel can be viewed to be in equilibrium with a fictitious bath. Of course, if the gel state is spatially uniform, then all local fictitious baths become identical to the real bath. We have this situation in mind when we plot the osmotic coefficient and the diffusion coefficients as a function of a bath of controlled chemical composition, pressure and electrical potential.

Henceforth, the gel we refer to is the extrafibrillar fluid phase of an articular cartilage. The two compartments, namely the extrafibrillar fluid phase and the real or fictitious bath, contain the same mobile components but the fixed charge is present only in the extrafibrillar fluid phase. Both compartments satisfy electroneutrality.

Two particular electrolytes are envisaged, namely a binary symmetric electrolyte containing dissolved sodium chloride NaCl, and a binary nonsymmetric electrolyte containing calcium chloride $CaCl_2$, denoted by the index $k=Na$ and $k=Ca$, respectively.

In view of the electroneutrality of each of the two compartments,

$$x_{ClB} = \zeta_k\, x_{kB}, \quad x_{ClE} = \zeta_k\, x_{kE} + y_{PG}, \quad y_{PG} \equiv \zeta_{PG}\, x_{PG}, \tag{16.3.1}$$

the chemical potentials of the salts NaCl or $CaCl_2$ can be brought in the form (16.1.11). The mechanical contribution to the chemical potential of the salt is now considered small with respect to the chemical contribution. Then chemical equilibrium of the salt between the two compartments simplifies to,

$$x_{kE}\, (x_{ClE})^{\zeta_k} = x_{kB}\, (x_{ClB})^{\zeta_k}, \quad k = Na,\ Ca. \tag{16.3.2}$$

The variations of the molar fractions of the cations and anions are connected by the relation,

$$\frac{dx_{kE}}{dx_{kB}} = \frac{dx_{ClE}}{dx_{ClB}} = \frac{1 + \zeta_k}{x_{kB}} \frac{x_{kE}\, x_{ClE}}{\zeta_k^2\, x_{kE} + x_{ClE}}, \quad k = Na,\ Ca. \tag{16.3.3}$$

Equality of the chemical potentials of water provides the osmotic pressure $p_{wE} - p_B$,

$$\frac{\widehat{v}_w}{RT}(p_{wE} - p_B) \simeq y_{PG} + (1 + \zeta_k)\,(x_{kE} - x_{kB}) = -\frac{y_{PG}}{\zeta_k} + \frac{1 + \zeta_k}{\zeta_k}\,(x_{ClE} - x_{ClB}), \quad k = Na,\ Ca, \tag{16.3.4}$$

neglecting the contribution of the concentration c_{PG} of PGs.

The explicit relations between the molar fractions in the two compartments are now detailed. For the NaCl electrolyte, the extrafibrillar molar fractions express in terms of the molar fractions of the bath in the explicit form,

$$x_{NaE} = -\frac{y_{PG}}{2} + \sqrt{\left(\frac{y_{PG}}{2}\right)^2 + x_{NaB}^2}, \quad x_{ClE} = \frac{y_{PG}}{2} + \sqrt{\left(\frac{y_{PG}}{2}\right)^2 + x_{NaB}^2}. \tag{16.3.5}$$

For the $CaCl_2$ electrolyte, the molar fraction of calcium cations is obtained via the single positive solution $y = \sqrt{x_{CaE}}$ of the cubic equation,

$$y^3 + \frac{y_{PG}}{2}\,y - (\sqrt{x_{CaB}})^3 = 0, \quad x_{CaE} = y^2, \quad x_{ClE} = 2\,y^2 + y_{PG}. \tag{16.3.6}$$

Conversely, for both binary electrolytes, the ionic molar fractions in the bath express in terms of the ionic molar fractions of the extrafibrillar phase in the format,

$$x_{kB} = \frac{x_{ClB}}{\zeta_k} = \left(x_{kE}\left(\frac{x_{ClE}}{\zeta_k}\right)^{\zeta_k}\right)^{(1+\zeta_k)^{-1}}. \tag{16.3.7}$$

The membrane effects can be appreciated in Fig. 16.3.1 where the cation concentration and osmotic pressure are displayed as a function of the bath concentration of chloride ion.

16.3.2　The osmotic, or reflection, coefficient/matrix

Once the chemical composition of the bath is known the composition of the extrafibrillar fluid phase can be deduced as indicated in Section 16.3.1. The osmotic coefficient matrix ω is next obtained by comparing the ionic fluxes in the extrafibrillar fluid phase and in the bath in absence of electric current ("open circuit"). This interpretation is shown to be equivalent to another identification where the fluid fluxes in the two compartments are compared.

The osmotic coefficient is viewed as characterizing the *filtration properties* of the fixed charge medium. Indeed at small ionic concentrations of the bath, and at small ionic concentrations of the mobile ions in the extrafibrillar fluid phase, the range of influence of the fixed charge is large and ω is close to identity. The co-ions and, by electroneutrality, the counter-ions are repelled from the fixed charge and they can hardly diffuse through the medium. As the ionic concentrations increase, this range of influence of the fixed charge decreases, the osmotic matrix tends to vanish allowing ions to diffuse more easily.

The fixed charge content, assumed to be spatially uniform, is defined by (16.1.6).

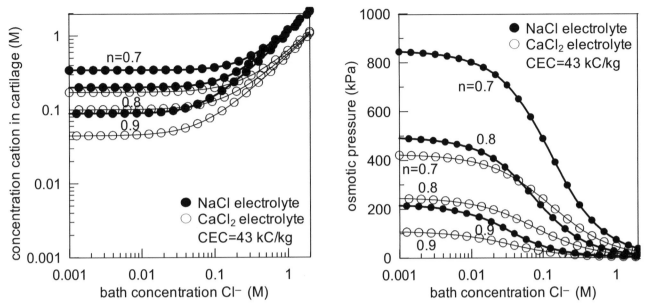

Fig. 16.3.1: Cartilage containing a binary electrolyte, either sodium chloride or calcium chloride, in chemical equilibrium with a bath containing the same electrolyte at varying strength. Concentration of cation and osmotic pressure for three volume fractions of water. The CEC and water volume fraction being fixed, so is the effective concentration y_{PG} of PGs by (16.1.5) and (16.1.6). At low ionic strength of the bath, the osmotic pressure is halved by the valence 2 of calcium in agreement with Eqn (16.3.4).

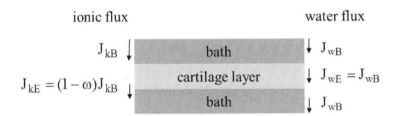

Fig. 16.3.2: The osmotic coefficient can be viewed as a filtration, or reflection, indicator of the ionic flux at uniform water flux. The water flux is uniform over the depth while the ionic flux is lower in the downstream bath.

16.3.2.1 The osmotic coefficient as a filtration indicator

First let us note that the experiment with open circuit sketched in Fig. 16.3.2, involving in addition uniform concentrations of ions and PGs, can be checked to be feasible. In other words, the relations (16.3.8) and (16.3.9) below are indeed consistent. Moreover, the fluid flux $\mathbf{J}_{wE}$ and the pressure gradient $\mathbf{F}_E$ are linked by the relation $\mathbf{J}_{wE} = -\mathbf{k}_H \cdot \mathbf{F}_E$.

Since the electrical current density vanishes in the extrafibrillar fluid phase, Eqn (16.1.18), then the ionic fluxes that are given by (16.2.39) and (16.2.40) satisfy the relation:

$$-\zeta_k \frac{\mathbf{J}_{kE}}{\widehat{v}_k} = -\frac{\mathbf{J}_{ClE}}{\widehat{v}_{Cl}} = -\zeta_k \, \mathbf{k}_{kE}^0 \cdot (\mathbf{k}_H)^{-1} \cdot \mathbf{J}_{wE} = -\mathbf{k}_{ClE}^0 \cdot (\mathbf{k}_H)^{-1} \cdot \mathbf{J}_{wE}, \quad k = \mathrm{Na, \ Ca}. \tag{16.3.8}$$

Similarly in the bath,

$$-\zeta_k \frac{\mathbf{J}_{kB}}{\widehat{v}_k} = -\frac{\mathbf{J}_{ClB}}{\widehat{v}_{Cl}} = -\zeta_k \frac{c_{kB}}{n^{wB}} \, \mathbf{J}_{wB} = -\frac{c_{ClB}}{n^{wB}} \, \mathbf{J}_{wB}. \tag{16.3.9}$$

We are now in position to provide a quantitative definition of the osmotic coefficient via
Definition 1: the osmotic coefficient as a filtration indicator

In a one directional flow context, with open circuit, *the osmotic or reflection coefficient matrix is defined as the relative difference between the cation or anion fluxes in the extrafibrillar fluid and bath compartments*, namely

$$\mathbf{J}_{kE} = (\mathbf{I} - \boldsymbol{\omega}) \cdot \mathbf{J}_{kB}, \quad \mathbf{J}_{ClE} = (\mathbf{I} - \boldsymbol{\omega}) \cdot \mathbf{J}_{ClB}, \quad k = \mathrm{Na, \ Ca}, \tag{16.3.10}$$

at identical fluid fluxes in the bath and extrafibrillar fluid phase, namely $\mathbf{J}_{wB} = \mathbf{J}_{wE}$. Thus, with help of the definitions in Table 16.2.1 and of the coefficients (16.2.40), the above relations (16.3.8) and (16.3.10) yield the osmotic matrix,

$$\boldsymbol{\omega} = \mathbf{I} - \frac{c_{wB}}{c_{wE}} \frac{c_{ClE}}{c_{ClB}} \zeta_k \, c_{kE} \, (\mathbf{u}_k + \mathbf{u}_{Cl}) \cdot (\zeta_k \, c_{kE} \, \mathbf{u}_k + c_{ClE} \, \mathbf{u}_{Cl})^{-1}, \quad k = \text{Na}, \text{Ca}. \qquad (16.3.11)$$

Assumption $(\mathcal{T})$ formulated in Section 16.4.3 has been used to eliminate the tortuosity factor, together with the relation $(\sigma_e)^{-1} \cdot \mathbf{k}_e = -(e_{PG}/n^E) \, (\zeta_k \, c_{kE} \, \mathbf{u}_k + c_{ClE} \, \mathbf{u}_{Cl})^{-1} \cdot \mathbf{k}_H$ deduced from (16.2.30) and from the definition of $\mathbf{k}_e$ in Table 16.2.1.

To interpret this coefficient matrix, we view the bath composition as prescribed, and deduce the chemical content of the extrafibrillar fluid phase as indicated by (16.3.5) or (16.3.6). As the cation, or anion, concentration in the bath increases from zero, the chloride content of the liquid varies much slower than in the bath, as indicated by (16.3.2), (16.3.3) and (16.3.7): thus the osmotic matrix for a fresh bath is the identity matrix (equivalently the osmotic coefficient is equal to one for isotropic diffusion properties).

For infinitesimal fixed charge, the bath and extrafibrillar fluid concentrations are equal and $\zeta_k \, c_{kE}$ is equal to c_{ClE}. Then $\boldsymbol{\omega}$ tends quickly to vanish as soon as the ionic strength of the bath increases. The situation is similar when the ionic content tends to overweigh the fixed charge.

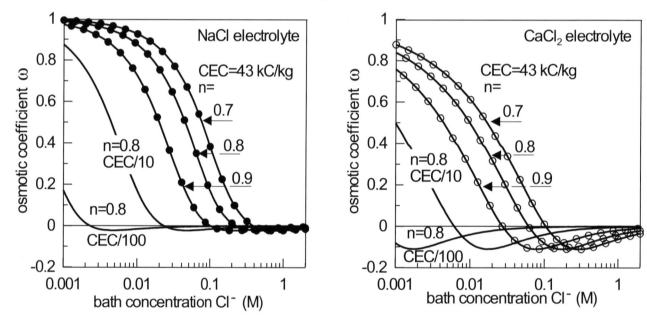

Fig. 16.3.3: Reflection, or osmotic, coefficient ω for articular cartilages in contact with a bath containing either a sodium chloride electrolyte or a calcium chloride electrolyte. The parameters used are: ionic mobilities [unit: $10^{-9} \, \text{m}^2/\text{s/V}$]: $u_{Na} = 51.75$, $u_{Ca} = 61.75$, $u_{Cl} = 79.0$, dry density $\rho_{dry} = 1800 \, \text{kg/m}^3$, fluid volume fraction n and CEC as indicated on the curves.

For the values of ionic mobilities used, there always exists an ionic strength defined by (16.3.16) above which the osmotic coefficient is negative. The negative minimum value of the osmotic coefficient is defined by (16.3.12). Reprinted with permission from Loret and Simões [2007].

Fig. 16.3.3 shows the variations of the osmotic coefficient as a function of the anion strength in the bath, for isotropic diffusive properties. Some effects of the CEC and volume fraction of the extrafibrillar fluid n^E should be highlighted:
- the osmotic coefficient decreases quickly from 1 to 0 as the ionic strength of the bath increases;
- at fixed volume fraction n^E of the extrafibrillar fluid phase and at small ionic strength, the higher the CEC the stronger the filtration effect of the fixed charge and the higher the algebraic value of the osmotic coefficient. The trend is opposite at large ionic strength;
- at fixed CEC, the higher the fluid volume fraction the lower the algebraic value of the osmotic coefficient at small ionic strength; the trend is again opposite at large ionic strength. Indeed, a large fluid volume fraction reduces the concentration of the fixed charge density;

- for sufficiently large values of the bath concentration, the osmotic coefficient becomes negative, a phenomenon referred to as *negative osmosis*. In the above experimental conditions, $\mathbf{J}_{\mathrm{NaE}} = (1 - \boldsymbol{\omega}) \cdot \mathbf{J}_{\mathrm{NaB}}$. Consequently, a negative osmosis implies a larger ionic flux in the extrafibrillar fluid phase than in the bath. Negative osmosis has been observed experimentally in both geological and biological materials with fixed charge.

16.3.2.2 Negative osmosis

The discussion below assumes isotropic diffusion properties.

A negative "anomalous" value of the reflection coefficient is reported in the experimental data of Kemper and Quirk [1972] on kaolinites, bentonites and illites. The phenomenon is analyzed in Olsen et al. [1990] for soil systems and in Gu et al. [1997] for articular cartilages.

With (16.3.3) and (16.3.7), the minimum of the osmotic coefficient ω,

$$\omega^{\mathrm{min}} = 1 - \left(\frac{\beta}{\alpha}\right)^{\beta} \left(\frac{1 - \beta}{1 - \alpha}\right)^{1-\beta}, \tag{16.3.12}$$

takes place for the following concentration of the chloride ions in the bath,

$$c_{\mathrm{ClB}}^{\mathrm{min}} = \frac{e_{\mathrm{PG}}}{\alpha - \beta} \left(\alpha \left(1 - \beta\right)\right)^{\beta} \left((1 - \alpha)\,\beta\right)^{1-\beta}, \tag{16.3.13}$$

where

$$\alpha = \frac{u_k}{u_k + u_{\mathrm{Cl}}}, \quad \beta = \frac{\zeta_k}{\zeta_k + 1}. \tag{16.3.14}$$

Clearly, while the extremum concentration is proportional to the fixed charge density, the minimum osmotic coefficient depends only on α and β, that is, on the relative ionic mobilities and on the valence of the counterion. It is encouraging to observe that the experimental curves obtained for Ca^{2+}-soil systems by Kemper and Quirk [1972] and reported in Bresler [1973] and Mitchell [1993] show precisely the trends produced by the model and displayed in Fig. 16.3.3. Moreover, the experimental negative osmosis is more marked for illites and kaolinites whose fixed charge is much smaller than for montmorillonites. Correspondingly, the experiments and simulations for bentonites in Leroy [2005] do not show negative osmosis. The experimental negative osmosis is more marked as well for Ca^{2+}-soils than for Na^+-soils. These two additional features are displayed by the model. The actual minimum osmotic coefficient is equal to -0.022 for the NaCl electrolyte and to -0.111 for the $CaCl_2$ electrolyte, irrespective of CEC and volume fraction. However, the experimental minima increase, in algebraic value, as the fixed charge increases.

For a negative fixed charge, the condition of existence of this extremum is $\alpha < \beta$, that is

$$u_{\mathrm{Cl}} > \frac{u_k}{\zeta_k}. \tag{16.3.15}$$

This condition is verified for the two electrolytes of interest. The above results on negative osmosis complete and generalize the analysis of Gu et al. [1997] which addresses only a sodium chloride electrolyte.

The condition $\alpha < \beta$ ensures the minimum of ω to be negative and thus ensures the existence of a concentration c_{ClB}^0 at which ω vanishes, e.g., for the sodium chloride electrolyte,

$$c_{\mathrm{ClB}}^0 = e_{\mathrm{PG}} \frac{\alpha \left(1 - \alpha\right)}{2\,\alpha - 1}, \tag{16.3.16}$$

which is blatantly proportional to the fixed charge density and depends on the ratio of ionic mobilities. The algebraic expression for calcium chloride can be obtained as well as a solution of a quadratic equation.

16.3.2.3 The osmotic coefficient via the diffusion-osmosis phenomenon

Another derivation and interpretation of the osmotic matrix $\boldsymbol{\omega}$ can be proposed.

Definition 2: the diffusion-osmosis phenomenon

The electrical circuit being open, the fluid flux is considered to be driven by an osmotic gradient in the underlying baths, namely

$$\mathbf{J}_{\mathrm{wE}} = \mathbf{J}_{\mathrm{wB}} = -\mathbf{k}_{\mathrm{H}} \cdot (\boldsymbol{\nabla} p_{\mathrm{B}} - \boldsymbol{\omega} \cdot \boldsymbol{\nabla} \pi_{\mathrm{B}}), \tag{16.3.17}$$

where π_{K} is the contribution of the compartment K to the osmotic pressure,

$$\pi_{\mathrm{K}} = RT \sum_{l \in \mathrm{K_{ions}}} c_{l\mathrm{K}} \,. \tag{16.3.18}$$

The pressure gradient $\mathbf{F}_{\mathrm{E}}$, which is essentially equal to $\boldsymbol{\nabla} p_{\mathrm{wE}}$, is obtained via (16.3.4), namely

$$\mathbf{F}_{\mathrm{E}} = \boldsymbol{\nabla} p_{\mathrm{wE}} = \boldsymbol{\nabla} p_{\mathrm{B}} + RT\, \boldsymbol{\nabla} e_{\mathrm{PG}} + \boldsymbol{\nabla}(\pi_{\mathrm{E}} - \pi_{\mathrm{B}}) \,. \tag{16.3.19}$$

Assuming a uniform fluid pressure, $\boldsymbol{\nabla} p_{\mathrm{B}} = \mathbf{0}$, and a uniform distribution of PGs, $\boldsymbol{\nabla} e_{\mathrm{PG}} = \mathbf{0}$, we recover the standard presentation of the osmotic coefficient viewed as moderating the effect on the flow of a chemical gradient.

The algebraic expression (16.3.11) of the osmotic matrix is retrieved by equating the two expressions of the fluid flux, provided on one hand by (16.2.38) where $\mathbf{F}_{\mathrm{E}}$ is given by (16.3.19) and on the other hand by (16.3.17), with help of (16.2.30) and (16.3.3).

For isotropic diffusion properties, $\boldsymbol{\omega} = \omega\, \mathbf{I}$: if ω is negative, relation (16.3.17) indicates that water flows against, and not along, the chemical gradient, which has led the phenomenon to be coined also *anomalous osmosis*.

16.3.3 Appropriate interpretation of diffusion tests

Laboratory experiments at vanishing fluid flow and vanishing electrical current provide information on the diffusion coefficients. Plotting the results as a function of a controlled concentration yields quantities that we will refer to as "experimental" diffusion coefficients. They are shown to embody the chemical composition and a number of basic properties of the tested cartilage layer.

16.3.3.1 Diffusion coefficients

The spatial distribution of the fixed charge is assumed to be uniform, in such a way that electroneutrality yields $\zeta_k\, \boldsymbol{\nabla} x_{k\mathrm{E}} = \boldsymbol{\nabla} x_{\mathrm{ClE}}$. Substituting this relation in (16.2.44), and using the definition (16.2.4) of the driving gradients $\mathbf{F}_{l\mathrm{E}}^{d}$, $l = k, \mathrm{Cl}$, the diffusive molar fluxes [unit : mole/m^2/s] of the cation and of the chloride ion can be cast in the format,

$$-\zeta_k\, \frac{\mathbf{J}_{k\mathrm{E}}^{d}}{\widehat{v}_k} = \zeta_k\, \overline{\mathbf{D}}_k \cdot \boldsymbol{\nabla} c_{k\mathrm{E}} = -\frac{\mathbf{J}_{\mathrm{ClE}}^{d}}{\widehat{v}_{\mathrm{Cl}}} = \overline{\mathbf{D}}_{\mathrm{Cl}} \cdot \boldsymbol{\nabla} c_{\mathrm{ClE}} \,. \tag{16.3.20}$$

The (apparent) diffusion coefficients [unit : m^2/s] $\overline{\mathbf{D}}_k$ and $\overline{\mathbf{D}}_{\mathrm{Cl}}$ are equal,

$$\overline{\mathbf{D}}_k = RT \left(\frac{\mathbf{k}_{kk\mathrm{E}}^{*}}{c_{k\mathrm{E}}} + \zeta_k\, \frac{\mathbf{k}_{k\mathrm{ClE}}^{*}}{c_{\mathrm{ClE}}} \right) = \overline{\mathbf{D}}_{\mathrm{Cl}} = RT \left(\frac{1}{\zeta_k}\, \frac{\mathbf{k}_{\mathrm{Cl}\,k\mathrm{E}}^{*}}{c_{k\mathrm{E}}} + \frac{\mathbf{k}_{\mathrm{Cl}\,\mathrm{ClE}}^{*}}{c_{\mathrm{ClE}}} \right), \tag{16.3.21}$$

and they are denoted by $\overline{\mathbf{D}}$,

$$\overline{\mathbf{D}} = \frac{RT}{\mathrm{F}}\, \frac{n^{\mathrm{E}}}{\zeta_k}\, (\zeta_k^{2}\, c_{k\mathrm{E}} + c_{\mathrm{ClE}})\, \mathbf{u}_k^{*} \cdot (\zeta_k\, c_{k\mathrm{E}} \mathbf{u}_k^{*} + c_{\mathrm{ClE}} \mathbf{u}_{\mathrm{Cl}}^{*})^{-1} \cdot \mathbf{u}_{\mathrm{Cl}}^{*}, \quad k = \mathrm{Na,\ Ca} \,. \tag{16.3.22}$$

The following limits are of interest:
- at infinitely small ionic strength of the bath the cartilage is in equilibrium with, the diffusion coefficient $\overline{\mathbf{D}}$ tends to,

$$\lim_{c_{\mathrm{ClB}} = \zeta_k\, c_{k\mathrm{B}} \to 0} \overline{\mathbf{D}} = n^{\mathrm{E}}\, \frac{RT}{\mathrm{F}}\, \mathbf{u}_{\mathrm{Cl}}^{*}, \tag{16.3.23}$$

which, along the Nernst-Einstein relation, is equal to the volume fraction of the fluid n^{E} times the effective diffusion coefficient $\mathbf{D}_{\mathrm{Cl}}^{*}$ of the chloride ion;
- the limit value of the diffusion coefficient $\overline{\mathbf{D}}$ of a cartilage in absence of fixed charge ($y_{\mathrm{PG}} = 0$ or $e_{\mathrm{PG}} = 0$) is independent of the ionic strength of the bath $c_{\mathrm{ClB}} = \zeta_k\, c_{k\mathrm{B}}$, $k = \mathrm{Na}$ or Ca,

$$\lim_{y_{\mathrm{PG}} \to 0} \overline{\mathbf{D}} = n^{\mathrm{E}}\, \frac{RT}{\mathrm{F}}\, \frac{1 + \zeta_k}{\zeta_k}\, \mathbf{u}_k^{*} \cdot (\mathbf{u}_k^{*} + \mathbf{u}_{\mathrm{Cl}}^{*})^{-1} \cdot \mathbf{u}_{\mathrm{Cl}}^{*} \,. \tag{16.3.24}$$

16.3.3.2 Application to a layer of cartilage maintained between two baths

As an application, a layer of articular cartilage of thickness h is placed between two baths of controlled chemical content: a slight concentration difference is imposed between the two baths. The experiment is realized at vanishing fluid flow and electrical current. The diffusion coefficient associated with this experiment, denoted by $\mathbf{D}_B^{exp}$ [unit : m^2/s], is defined through the relation,

$$-\frac{\mathbf{J}_{kE}^d}{\widehat{v}_k} = \overline{\mathbf{D}} \cdot \boldsymbol{\nabla} c_{kE} = \mathbf{D}_B^{exp} \cdot \boldsymbol{\nabla} c_{kB}, \quad k = \text{Na, Ca}. \quad (16.3.25)$$

Here $\boldsymbol{\nabla}$ denotes the difference between the two baths divided by the thickness h. Hence, with help of (16.3.3) and Assumption ($\mathcal{T}$),

$$\mathbf{D}_B^{exp} = \frac{RT}{F} \frac{1+\zeta_k}{\zeta_k} \tau\, n^E \frac{c_{ClE}}{c_{ClB}} \zeta_k\, c_{kE}\, \mathbf{u}_k \cdot (\zeta_k\, c_{kE}\mathbf{u}_k + c_{ClE}\mathbf{u}_{Cl})^{-1} \cdot \mathbf{u}_{Cl}, \quad k = \text{Na, Ca}. \quad (16.3.26)$$

It is worthwhile to observe that the above experimental diffusion coefficient, Eqn (16.3.26), can be expressed in terms of the *transmission* coefficient $\mathbf{I} - \boldsymbol{\omega}$, Eqn (16.3.11), namely

$$\mathbf{D}_B^{exp} = \frac{RT}{F} \frac{1+\zeta_k}{\zeta_k} \tau\, n^E \frac{c_{wE}}{c_{wB}} \mathbf{u}_k \cdot (\mathbf{u}_k + \mathbf{u}_{Cl})^{-1} \cdot (\mathbf{I} - \boldsymbol{\omega}) \cdot \mathbf{u}_{Cl}, \quad k = \text{Na, Ca}. \quad (16.3.27)$$

A relation of this type has been suggested in the literature, either for the effective diffusion coefficient, on the grounds that $\mathbf{D}_B^{exp}$ and $\mathbf{I} - \boldsymbol{\omega}$ both vanish at infinitesimal ionic strength of the baths. The present analysis shows that such a relation does not need to be postulated as a basic constitutive equation: indeed it *results* from the constitutive equations of diffusion.

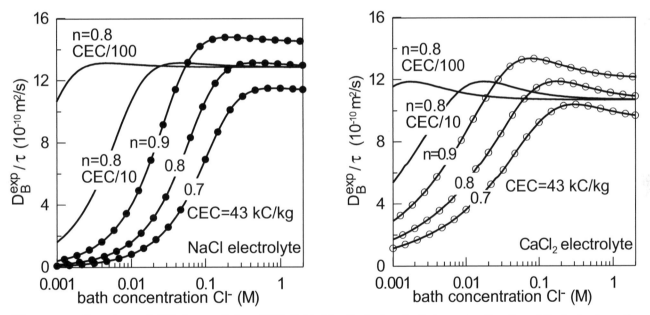

Fig. 16.3.4: Experimental diffusion coefficient D_B^{exp} divided by the tortuosity factor τ as a function of the ionic strength of the bath. The ionic mobilities are indicated in Fig. 16.3.3, the tortuosity factor $\tau = 0.4$ and the hydraulic permeability $k_{EE} = 10^{-15}$ m^2/Pa/s are constant, and the fluid volume fractions n and CEC are indicated on the curves. Reprinted with permission from Loret and Simões [2007].

For isotropic diffusion properties, Fig. 16.3.4 displays the effects of the ionic strength of the bath, of the fluid volume fraction and of the fixed charge density on the experimental diffusion coefficient D_B^{exp} divided by the tortuosity factor τ. The experimental coefficient D_B^{exp} can be observed to increase with the ionic strength of the baths, starting from a null value, and reaching a plateau at larger ionic strength, two features observed in the experiments of Malusis and Shackelford [2002] on a bentonite, a geological material with fixed charge. The latter plays a crucial role here: indeed in absence of fixed charge, the experimental diffusion coefficient D_B^{exp} assumes the constant value indicated by (16.3.24).

Further details may be worth of interest:

- for the values of ionic mobilities used, there always exists an ionic strength at which the experimental diffusion coefficient presents a maximum. The existence and position of this extremum have been investigated in relation with the osmotic coefficient;
- at given fluid volume fraction n^E, the higher the CEC the smaller the experimental diffusion coefficient at small ionic strength; the trend is opposite at large ionic strength;
- at given CEC, the higher the fluid volume fraction the higher the experimental diffusion coefficient;
- the Ca^{2+}-electrolyte is endowed with an experimental diffusion coefficient that is slightly larger, at small ionic strength, than the monovalent electrolyte; the trend is opposite at large ionic strength.

16.4 Generalized diffusion: ranges of the coefficients and refinements

The constitutive equations above introduced several material properties which may or may not be constant. Typical values, gathered from literature, for human articular cartilage, Maroudas [1968], and for bovine cartilages, Mansour and Mow [1976], Frank and Grodzinsky [1987]-I-II, Sachs and Grodzinsky [1989], Gu et al. [1998], Mow and Guo [2002], are listed below. Most of them refer to isotropic material properties.

It is worth reiterating that the model of generalized diffusion exposed above does not introduce explicitly an osmotic coefficient: the latter has been shown to be hidden in the equations. An obvious consequence is that there is no specific material coefficient attached to the osmotic coefficient.

16.4.1 Hydraulic conductivity

Table 16.4.1: Hydraulic conductivity.

Symbol	Unit	Range
K_H	m/s	$[10^{-12}, 70 \times 10^{-12}]$

Typical ranges of values of the hydraulic conductivity are listed in Table 16.4.1. Hydraulic conductivity is sometimes considered as varying with the void ratio according to the Kozeny-Carman formula, namely

$$\frac{K_H}{K_{H0}} = \left(\frac{e}{e_0}\right)^3 \frac{1+e_0}{1+e}, \tag{16.4.1}$$

where the subscript 0 indicates the reference state. This expression is motivated in Section 15.7.7. As for articular cartilages, the data of Mansour and Mow [1976] indicate that the permeability decreases at increasing compressive strain while the bath pressure is fixed. Lai et al. [1981] consider relations of the form

$$\frac{K_H}{K_{H0}} = \left(\frac{1+e}{1+e_0}\right)^M, \quad \text{or} \quad \frac{K_H}{K_{H0}} = \exp(M \operatorname{tr}\epsilon). \tag{16.4.2}$$

In the latter relation, M is a positive parameter that may range up to 20. Holmes and Mow [1990] propose a two-parameter expression, namely,

$$\frac{K_H}{K_{H0}} = \left(\frac{e}{e_0}\right)^{M_1} \exp(M_2 \operatorname{tr}\epsilon), \tag{16.4.3}$$

with $M_1 \sim 0.1$ and $M_2 \sim 4-5$. In a finite deformation extension of this formula, Ateshian et al. [1997] adopt, for bovine articular cartilage as well, the values $M_1 = 2$ and $M_2 = 0.4 - 3.2$. Gu et al. [2004] use $M_2 = 0$ and M_1 around 3 for uncharged agarose gels and around 7 for porcine lumbar annulus fibrosus.

Netti et al. [2000], their Fig. 3, report an exponential decrease of the hydraulic conductivity with the compressive strain. Moreover they provide values of hydraulic conductivity for different tumor tissues, and observe that, at comparable GAG content, the hydraulic conductivity is larger for tumor tissues than for sound tissues.

Indeed, the elastic modulus of cartilage is presumably, and actually, in direct relation with the collagen and GAG content. The hydraulic conductivity is on the other hand in inverse relation with GAG content, but independent of the ionic content, Maroudas [1975] [1979]. Thus, as water is expelled during compression, the relative content of GAG increases and therefore the hydraulic conductivity is expected to decrease. In any case, the wide range of values of the parameters M above indicates that the actual dependence of hydraulic conductivity in terms of strain is not completely resolved by these expressions for at least two reasons. First, an assessment of these relations revolves around the definition of the void ratio! In the present three-phase context, the void ratio involved would be $e = V_E/(V_I + V_S)$, while the void ratio e in biphasic theories is defined as $(V_E + V_I)/V_S$. Second, the actual permeability depends not only on the actual porosity, but also on the microstructure of this porosity. The notion of hydraulic tortuosity is briefly discussed below in relation with ionic mobilities and diffusion coefficients.

Moreover, these relations imply that the mechanical and transport properties should vary across depth of the cartilage layer, since the chemical composition of cartilage does, A.C. Chen et al. [2001] and S.S. Chen et al. [2001].

16.4.2 Diffusion coefficients and ionic mobilities

Table 16.4.2: Ranges of absolute diffusion coefficients D and ionic mobilities u.

Symbol	Unit	Range for metallic ions
D	m^2/s	$[2 \times 10^{-10}, 20 \times 10^{-10}]$
u	$m^2/s/V$	$[8 \times 10^{-9}, 80 \times 10^{-9}]$

The ionic mobility is the distance run per unit time by an ion submitted to a unit difference of electrical potential per unit length. It is linked to the diffusion coefficient by the Nernst-Einstein relation,

$$\text{Nernst} - \text{Einstein relation} \quad u_k = D_k \, |\zeta_k| \frac{F}{RT} \,. \tag{16.4.4}$$

Table 16.4.2 shows the ranges of *absolute* diffusion coefficients and ionic mobilities. The *effective* diffusion coefficient accounts for several aspects that imply diffusion in the porous medium to be slower than diffusion in an infinitely dilute solution, also called blank solution. For any ion, the absolute diffusion coefficients D and of effective diffusion $D^* = \tau D$ are linked by the tortuosity factor $\tau < 1$[16.2]. A similar relation applies to *effective* ionic mobilities, namely $u^* = \tau u$.

16.4.3 Tortuosity factor

One contribution τ_1 to the tortuosity factor stems from geometrical considerations, namely the actual path that an ion has to travel in a porous medium is larger than in a blank solution. The tortuosity factor is defined as the square of the ratio of the ionic path length in a blank solution over the actual path length in the porous medium, Bear [1972], p. 106. However, this estimation is awkward in applications. Approximations in terms of available quantities have been proposed. For a crosslinked polymer membrane, the tortuosity factor for ions that do not interact with the membrane has been estimated, using statistical arguments, by Mackie and Meares [1955], as a function of the solid volume fraction $n^S = V_S/V$, or of the fluid volume fraction $n^E = V_E/V$, or of the relative volumes of fluid and solid $e = V_E/V_S$ namely $(n^E = e/(1+e) \Leftrightarrow e = n^E/(1-n^E))$,

$$\tau_1 = \frac{(1-n^S)^2}{(1+n^S)^2} = \frac{(n^E)^2}{(2-n^E)^2} = \frac{e^2}{(2+e)^2} \,, \tag{16.4.5}$$

which is equal to about 0.44 for a fluid volume fraction $n^E = 0.80$.

This relation would take the format $\tau_1(e)$ with $e = V_E/(V_I + V_S)$ in the present three-phase context. Mechanical compression expels first extrafibrillar water out of the cartilage and therefore, according to this model, it reduces considerably the tortuosity factor and therefore the effective diffusion coefficients. A formula of the same flavor which goes by the name of Archie's law is used for rocks by the geophysical community where the tortuosity factor is postulated as a power relationship

$$\tau_1 = (n^E)^{m-1} \,. \tag{16.4.6}$$

The exponent m varies in the typical range $[1.5, 2.5]$ for cemented, or not cemented, granular materials. It reaches the minimum value 1.5 for a random array of spheres and it is greater than this value for flat particles.

[16.2]This entity is sometimes defined as the inverse of the one used here.

While the diffusion equation features the effective coefficient D^*, some formulations refer to the apparent diffusion coefficient $n^E D^*$ and introduce the formation factor $F = D/(n^E D^*)$, so that $F = 1/(n^E \tau)$, and, according to Archie's law, $F = (n^E)^{-m}$.

As expected, the tortuosity factors defined by the above expressions decrease as the fluid volume fraction decreases, being equal to 1 in a blank solution $n^E = 1$. Again, it may be worth stressing that the phase partition matters in the above expressions. For example, the fluid volume may involve the total fluid present in the tissue or only that part where diffusion takes place, referred to as extrafibrillar fluid phase in the context of articular cartilages, as depicted in Chapter 14.

As such, the above formulas have a main drawback: the tortuosity factor τ_1 accounts solely for geometrical aspects of the pore structure. However, the presence of fixed charges modifies considerably the effective diffusion coefficients and ionic mobilities. Indeed, at small concentrations, the range of action of the fixed charge is large, and the ionic mobilities are reduced. This aspect may be inserted in the tortuosity factor τ which can be viewed as the product of two terms,

$$\tau = \tau_1 \, \tau_2 \,, \tag{16.4.7}$$

the Mackie-Meares's geometrical factor τ_1 and a second factor of electrochemical nature τ_2 that vanishes at infinitely small concentrations of mobile ions. Parker et al. [1988] review some propositions for these two factors. Note however that this standard qualitative description of the tortuosity factor is empirical and may be either conflicting or redundant with some characteristics built in the model, see, e.g., the remark in relation with Eqn (16.3.27).

For metallic ions, it is customary to adopt

Assumption ($\mathcal{T}$): tortuosity affects all ions identically.

An attempt to introduce the size of macromolecules in the tortuosity factor is described by the expression, Yasuda et al. [1968],

$$\tau = \exp\left(-Y \frac{1 - n^E}{n^E}\right), \tag{16.4.8}$$

where the free coefficient $Y > 0$ may be adjusted to the molar mass of the solute. Maroudas [1975], her Table 1, suggests that tortuosity factors for articular cartilages are independent of concentrations, but vary from 0.33 for large molecules to 0.45 for small anions. Such a low tortuosity factor does not seem to apply universally: Table 15.5.1 shows counterexamples for tumor tissues.

16.4.4   Electrical conductivity

Table 16.4.3: Bulk electrical conductivity of the porous medium.

Symbol	Unit	Range
σ_e	S/m=A/V/m	$[10^{-2}, 1]$

The electrical conductivity shown in Table 16.2.1 is not constant, it assumes a minimal non zero value for distilled water and it increases with ionic strength. Its range is displayed in Table 16.4.3. Various constituents of cartilage contribute to the electrical conductivity:

- in the present model the collagen network is considered infinitely resistive. The intrafibrillar ions are not involved either. However, mass transfers and the trend towards chemical equilibrium between the intra- and extrafibrillar phases affect indirectly the extrafibrillar ionic concentrations and volume of water, and therefore the tortuosity factor. So, the possibility that an increase of the extrafibrillar ionic strength be accompanied by a decrease of electrical conductivity cannot be ruled out a priori;
- the contribution of the PGs through the square of its effective concentration is strictly positive: the fixed charge of PGs is strictly negative. This property holds true even under non physiological pHs, an issue addressed in Chapter 17. Then collagen becomes charged, and, at the isoelectric point, the overall fixed charge vanishes. The isoelectric pH is reported to be around 2.5 to 3.0 by Frank and Grodzinsky [1987]-I. Note that, if the PGs were part of the solid phase, they would not enter in the electrical conductivity in Table 16.2.1 nor would the non mobile ions, Loret et al. [2004]a [2004]b;
- ions in the extrafibrillar fluid phase contribute additively, with an intensity proportional to the product of their effective ionic mobilities and of their concentrations. Here, the mobilities are taken constant. However,

according to Kohlrausch's square root law, they decrease with concentration c, say generically as $u = u_0 - (\alpha\, u_0 + \beta)\sqrt{c}$, where u_0 is the ionic mobility at dilute concentration, α and β are positive parameters. Therefore, the contribution of an ion to the overall conductivity, essentially proportional to $c\,u$, would not increase linearly with concentration but would tend to flatten or possibly to decrease at concentrations larger than $\frac{4}{9}\,u_0^2/(\alpha\,u_0 + \beta)^2$;

- non mobile extrafibrillar ions can be made to either contribute or not, depending whether $\tilde{n}^{kE}$ is set to n^{kE} or to n_{mo}^{kE} in Table 16.2.1. Only the former option has been used in Sections 16.3.2 and 16.3.3.

16.4.5 Electroosmotic coefficient

Table 16.4.4: Range of electroosmotic coefficient at neutral pH.

Symbol	Unit	Range
k_e	$\mathrm{m^2/s/V}$	$[0.1 \times 10^{-8}, 5 \times 10^{-8}]$

According to the Helmholtz-Smoluchowski theory, the electroosmotic coefficient should be proportional to the fixed charge, with opposite sign. The expression shown in Table 16.2.1 is in perfect agreement with these requirements.

In this chapter, the sole fixed charge is contributed by proteoglycans: since it is negative, the electroosmotic coefficient is positive.

The range of values shown in Table 16.4.4 assumes a neutral pH. As pH decreases, the charge of PGs increases in algebraic value, even if it remains always negative. At non physiological pH, collagen becomes charged. The overall charge of PGs and collagen vanishes at the isoelectric point and becomes positive at lower pH, Grodzinsky et al. [1981]. Therefore, in absence of pressure and concentration gradients, a difference of electrical potential induces an electroosmotic water flow towards the cathode at neutral and high pH and towards the anode at low pH.

A clear distinction should be made between

- the phenomenon termed *reversal of the electroosmotic flow*, associated with a change of sign of the electroosmotic component of the water flow at the aggregate (overall) isoelectric point;
- *anomalous negative osmosis*, associated with a negative osmotic or reflection coefficient, investigated in Section 16.3.2.2;
- *inverse osmosis* which is a pressure-driven industrial process briefly described in Section 13.8.

16.4.6 Variations and general trends of coefficients of generalized diffusion

The general trends of the electrical conductivity and of some other indicators of hydraulic and electrical diffusion are displayed in Figs. 16.4.1 and 16.4.2. Some features are worthy of notice:

- for the reference CEC=43 kC/kg, the ionic conductivity tends to vary linearly with the ionic strength of the bath c_{ClB} for $c_{\mathrm{ClB}} \geq 0.2$ M. The influence of the fluid volume fraction would be moderated if the tortuosity factor was following Archie's law;
- as CEC decreases to zero, the electrical conductivity tends to $\mathrm{F}\, n^{\mathrm{E}}\, c_{\mathrm{ClB}}\,(u_k^* + u_{\mathrm{Cl}}^*)$;
- the difference in electrical conductivity between the sodium chloride and calcium chloride electrolytes is small;
- as expected, the relative contribution of ions to the electrical conductivity increases when the ionic strength of the bath increases, when the fluid volume fraction increases and when the CEC decreases. The influence of the fluid volume fraction would be amplified if the tortuosity factor was following Archie's law;
- the ratio of open and short circuit hydraulic permeabilities may be significatively smaller than 1 at low ionic strength;
- the ratio k_e/σ_e decreases as the ionic strength of the bath increases, but varies nonmonotonously in terms of fluid volume fraction and CEC. For articular cartilages under physiological conditions, the ratio takes values of the order of 1×10^{-8} V/Pa in agreement with measurements by Frank and Grodzinsky [1987]-II and Huyghe et al. [2002].

The trends may be checked against data available in the literature where both the fixed charge density and fluid volume fractions are known accurately. For example, Huyghe et al. [2002] analyzed a hydrogel[16.3] with fixed charge of -0.15 mole per liter of fluid tissue, with fluid volume fraction of 0.977, in contact with a bath at 0.15 M of NaCl. They report the following measurements: $k_e = 0.88 \times 10^{-8}$ $\mathrm{m^2/s/V}$ and $\sigma_e = 0.76$ S/m, so that the ratio k_e/σ_e is equal to 1.16×10^{-8} V/Pa. These data yield the short circuit permeability $k_{\mathrm{EE}} = -k_e/\mathrm{F}/e_{\mathrm{PG}} = 0.61 \times 10^{-15}$ $\mathrm{m^4/N/s}$

[16.3] *Hydrogels* are crosslinked chains of swellable hydrophilic insoluble polymers.

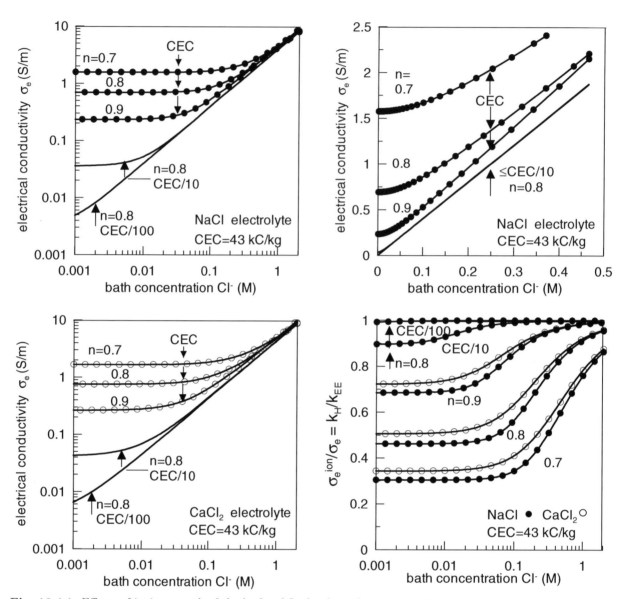

Fig. 16.4.1: Effects of ionic strength of the bath, of fluid volume fraction n and of CEC on the electrical conductivity of articular cartilages. The ionic mobilities are indicated in Fig. 16.3.3, the tortuosity factor $\tau = 0.4$ and the hydraulic permeability $k_{EE} = 10^{-15} \, \text{m}^2/\text{Pa/s}$ are constant, and the fluid volume fractions n and CEC are shown on the curves. Reprinted with permission from Loret and Simões [2007].

and the open circuit permeability $k_H = k_{EE} - k_e^2/\sigma_e = 0.51 \times 10^{-15} \, \text{m}^4/\text{N/s}$, so that $k_H/k_{EE} = 0.83$. Assuming $\rho_{dry} = 1800 \, \text{kg/m}^3$, the corresponding CEC is $347.4 \, \text{kC/kg}$: this value is much larger than in articular cartilages due to the high fluid volume fraction. The model yields values quite in the range of experimental measurements, namely $k_H/k_{EE} = 0.82$, $\sigma_e = 1.15 \, \text{S/m}$, $k_e/\sigma_e = 1.54 \times 10^{-8} \, \text{V/Pa}$.

16.4.7 Perspective

The framework for this analysis of articular cartilages is purely macroscopic, in the sense that the actual details of the geometry and microstructure at the nanoscale are wiped out. On the other hand, the implicit averaging process has kept track of the features that are thought to control both the transport and chemomechanical behaviors, e.g.,

- the presence of the fixed charge on the proteoglycans which influences mechanics through electrical shielding, swelling and shrinking, and transport via the osmotic effect;

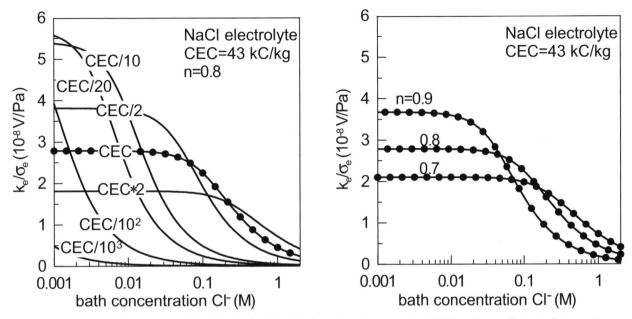

Fig. 16.4.2: Effects of ionic strength of the bath, of fluid volume fraction n and of CEC on the coefficient of streaming potential of articular cartilages. Same parameters as in Fig. 16.4.1. Reprinted with permission from Loret and Simões [2007].

- the existence of two types of water, inside the collagen fibrils and around the proteoglycans, of different chemical compositions. This feature endows both the mechanical and transport properties with instantaneous and delayed responses.

Indeed, even if the mechanical responses of the individual constituents are time independent (non-viscous), the overall behavior of the cartilage is indeed time-dependent and it displays several characteristic times, namely a time associated with seepage, times associated with the diffusion of ions and times associated with mass transfers between the intra- and extrafibrillar compartments.

In Loret and Simões [2004] [2005]a, simulations of a succession of equilibria were performed in order to test the chemomechanical constitutive equations only. Mechanical and chemical equilibria were assumed to hold between the two water phases and between the extrafibrillar phase and the bath. The framework described here endows the material response with intrinsic times. It allows to simulate laboratory experiments with characteristic loading times which may be smaller, of the same order or larger than the various material time scales. The transient response depends strongly on the relative positions of these characteristic times. Transient responses are addressed and illustrated in Chapter 18 via the finite element method.

16.5 Generalized diffusion with thermal effects

The constitutive equations of generalized diffusion established in Section 16.2 are extended to account for thermal effects. Unlike in Section 10.6, the porous medium is assumed to be in local thermal equilibrium, that is, the temperature is identical over all species.

Along Table 3.2.1, the entropy of the species k is defined from the electrochemical potential,

$$\delta\mu_{kE}^{ec} = \delta\mu_{kE|T}^{ec} - s_{kE}\,\delta T, \quad s_{kE} = -\frac{\partial\mu_{kE}^{ec}}{\partial T}, \tag{16.5.1}$$

the derivative assuming all variables but temperature to be fixed. The entropy inequality associated with diffusion (4.7.32), namely

$$T\hat{S}_{(i,3)} = -\sum_{k\in E_{mo}} \mathbf{J}_{kE} \cdot \left(\rho_k\,\boldsymbol{\nabla}\mu_{kE}^{ec} + \rho_k\,\frac{d^k\mathbf{v}_k}{dt} - \rho_k\,\mathbf{b}_k\right) - \mathbf{J}_{TE}\cdot\boldsymbol{\nabla}T \geq 0, \tag{16.5.2}$$

expresses in terms of the total entropy flux $\mathbf{J}_{TE}$ [unit : W/m^2/K],

$$\mathbf{J}_{TE} = \mathbf{J}_T + \sum_{k\in E_{mo}} s_{kE}\,\mathbf{M}_{kE}, \quad \mathbf{J}_T = \frac{\mathbf{Q}}{T}. \tag{16.5.3}$$

To simplify the presentation, accelerations and body forces will be omitted. The constitutive relations $\tilde{\jmath} = -\tilde{\kappa}\,\tilde{\mathbf{f}}$ with $\tilde{\kappa}$ symmetric positive definite become,

$$\tilde{\jmath} = \begin{bmatrix} \mathbf{J}_{wE} \\ \mathbf{J}_{NaE} \\ \mathbf{J}_{CaE} \\ \mathbf{J}_{ClE} \\ \mathbf{J}_{TE} \end{bmatrix}, \quad \tilde{\mathbf{f}} = \begin{bmatrix} \tilde{\mathbf{f}}_{wE} \\ \tilde{\mathbf{f}}_{NaE} \\ \tilde{\mathbf{f}}_{CaE} \\ \tilde{\mathbf{f}}_{ClE} \\ \tilde{\mathbf{f}}_{T} \end{bmatrix} = \begin{bmatrix} \rho_w\,\boldsymbol{\nabla}\mu_{wE} \\ \rho_{Na}\,\boldsymbol{\nabla}\mu_{NaE}^{ec} \\ \rho_{Ca}\,\boldsymbol{\nabla}\mu_{CaE}^{ec} \\ \rho_{Cl}\,\boldsymbol{\nabla}\mu_{ClE}^{ec} \\ \boldsymbol{\nabla}T \end{bmatrix}, \quad \tilde{\kappa} = \begin{bmatrix} \kappa_{ww} & \kappa_{w\,Na} & \kappa_{w\,Ca} & \kappa_{w\,Cl} & \tilde{\kappa}_{w\,T} \\ \kappa_{Naw} & \kappa_{NaNa} & \kappa_{NaCa} & \kappa_{NaCl} & \tilde{\kappa}_{NaT} \\ \kappa_{Caw} & \kappa_{CaNa} & \kappa_{CaCa} & \kappa_{CaCl} & \tilde{\kappa}_{CaT} \\ \kappa_{Clw} & \kappa_{Cl\,Na} & \kappa_{Cl\,Ca} & \kappa_{Cl\,Cl} & \tilde{\kappa}_{ClT} \\ \tilde{\kappa}_{Tw} & \tilde{\kappa}_{T\,Na} & \tilde{\kappa}_{T\,Ca} & \tilde{\kappa}_{T\,Cl} & \tilde{\kappa}_{TT} \end{bmatrix}. \quad (16.5.4)$$

Alternatively, like in Section 9.3.3 devoted to poroelasticity where the convective entropy fluxes are eliminated, the entropy inequality is exploited with the isothermal electrochemical potentials. The entropy inequality (16.5.2) may then be written in the alternative form,

$$T\hat{S}_{(i,3)} = -\sum_{k\in E_{mo}} \mathbf{J}_{kE}\cdot\left(\rho_k\,\boldsymbol{\nabla}\mu_{kE|T}^{ec} + \rho_k\frac{d^k\mathbf{v}_k}{dt} - \rho_k\,\mathbf{b}_k\right) - \mathbf{J}_T\cdot\boldsymbol{\nabla}T \geq 0\,, \quad (16.5.5)$$

and the constitutive equations now take the form $\mathbf{j} = -\kappa\mathbf{f}$ with κ symmetric positive definite,

$$\mathbf{j} = \begin{bmatrix} \mathbf{J}_{wE} \\ \mathbf{J}_{NaE} \\ \mathbf{J}_{CaE} \\ \mathbf{J}_{ClE} \\ \mathbf{J}_{T} \end{bmatrix}, \quad \mathbf{f} = \begin{bmatrix} \mathbf{f}_{wE} \\ \mathbf{f}_{NaE} \\ \mathbf{f}_{CaE} \\ \mathbf{f}_{ClE} \\ \mathbf{f}_{T} \end{bmatrix} = \begin{bmatrix} \rho_w\,\boldsymbol{\nabla}\mu_{wE|T} \\ \rho_{Na}\,\boldsymbol{\nabla}\mu_{NaE|T}^{ec} \\ \rho_{Ca}\,\boldsymbol{\nabla}\mu_{CaE|T}^{ec} \\ \rho_{Cl}\,\boldsymbol{\nabla}\mu_{ClE|T}^{ec} \\ \boldsymbol{\nabla}T \end{bmatrix}, \quad \kappa = \begin{bmatrix} \kappa_{ww} & \kappa_{w\,Na} & \kappa_{w\,Ca} & \kappa_{w\,Cl} & \kappa_{w\,T} \\ \kappa_{Naw} & \kappa_{NaNa} & \kappa_{NaCa} & \kappa_{NaCl} & \kappa_{NaT} \\ \kappa_{Caw} & \kappa_{CaNa} & \kappa_{CaCa} & \kappa_{CaCl} & \kappa_{CaT} \\ \kappa_{Clw} & \kappa_{Cl\,Na} & \kappa_{Cl\,Ca} & \kappa_{Cl\,Cl} & \kappa_{ClT} \\ \kappa_{Tw} & \kappa_{T\,Na} & \kappa_{T\,Ca} & \kappa_{T\,Cl} & \kappa_{TT} \end{bmatrix}. \quad (16.5.6)$$

The two sets of coefficients are related as follows:

$$\kappa_{Tl} = \tilde{\kappa}_{Tl} - \sum_{k\in E_{mo}} \rho_k\,s_{kE}\,\kappa_{lk}, \quad l\in\{w\}\cup E_{ions}; \quad \kappa_{TT} = \tilde{\kappa}_{TT} - \sum_{k,l\in E_{mo}} \rho_k\,s_{kE}\,\kappa_{kl}\,\rho_l\,s_{lE}\,. \quad (16.5.7)$$

The dissipation inequality may also be phrased in terms of diffusive ionic fluxes, electric current density and entropy flux

$$T\hat{S}_{(i,3)} = -\mathcal{F}^{\mathrm{T}}\,\mathcal{J} = -\mathbf{F}_E\cdot\mathbf{J}_{wE} - \sum_{k\in E_{ions}} \mathbf{F}_{kE}^d\cdot\mathbf{J}_{kE}^d - \mathbf{F}_{eE}\cdot\mathbf{I}_{eE} - \mathbf{J}_T\cdot\boldsymbol{\nabla}T \geq 0\,. \quad (16.5.8)$$

The vector flux $\mathcal{J}$ and its conjugate vector $\mathcal{F}$ have now $n_{ion}+3=6$ entries,

$$\mathcal{J} = \begin{bmatrix} \mathbf{J}_{wE} \\ \mathbf{J}_{NaE}^d \\ \mathbf{J}_{CaE}^d \\ \mathbf{J}_{ClE}^d \\ \mathbf{I}_{eE} \\ \mathbf{J}_T \end{bmatrix}, \quad \mathcal{F} = \begin{bmatrix} \mathbf{F}_E \\ \mathbf{F}_{NaE}^d \\ \mathbf{F}_{CaE}^d \\ \mathbf{F}_{ClE}^d \\ \mathbf{F}_{eE} \\ \mathbf{F}_T \end{bmatrix} = \begin{bmatrix} \boldsymbol{\nabla}P_{wE} \\ RT/\widehat{v}_{Na}\,\boldsymbol{\nabla}\mathrm{Ln}\,x_{NaE} \\ RT/\widehat{v}_{Ca}\,\boldsymbol{\nabla}\mathrm{Ln}\,x_{CaE} \\ RT/\widehat{v}_{Cl}\,\boldsymbol{\nabla}\mathrm{Ln}\,x_{ClE} \\ \boldsymbol{\nabla}\phi_E \\ \boldsymbol{\nabla}T \end{bmatrix}, \quad (16.5.9)$$

neglecting once again body forces. The generalized diffusion law can be expressed via a symmetric matrix $\mathcal{K}$, made from 6×6 symmetric blocks of size 3×3,

$$\mathcal{J} = -\mathcal{K}\,\mathcal{F}, \quad \mathcal{K} = \mathcal{K}^{\mathrm{T}}\,\text{PsD}\,. \quad (16.5.10)$$

Therefore the diffusion matrix takes the symmetric arrowhead form

$$\mathcal{K} = \begin{bmatrix} \mathbf{k}_{EE} & \mathbf{0} & \mathbf{0} & \mathbf{0} & \mathbf{k}_e & \mathbf{k}_{ET} \\ \mathbf{0} & \mathbf{k}_{NaNa}^d & \mathbf{0} & \mathbf{0} & \mathbf{k}_{Nae}^d & \mathbf{k}_{NaT} \\ \mathbf{0} & \mathbf{0} & \mathbf{k}_{CaCa}^d & \mathbf{0} & \mathbf{k}_{Cae}^d & \mathbf{k}_{CaT} \\ \mathbf{0} & \mathbf{0} & \mathbf{0} & \mathbf{k}_{Cl\,Cl}^d & \mathbf{k}_{Cl\,e}^d & \mathbf{k}_{ClT} \\ \mathbf{k}_e & \mathbf{k}_{eNa}^d & \mathbf{k}_{eCa}^d & \mathbf{k}_{eCl}^d & \sigma_e & \mathbf{k}_{eT} \\ \mathbf{k}_{TE} & \mathbf{k}_{TNa} & \mathbf{k}_{TCa} & \mathbf{k}_{TCl} & \mathbf{k}_{Te} & \mathbf{k}_{TT} \end{bmatrix}, \quad (16.5.11)$$

while the matrix $\boldsymbol{\kappa}$ becomes

$$
\boldsymbol{\kappa} = \left[\begin{array}{c|cccc}
\mathbf{k}_{\mathrm{EE}} & N_{\mathrm{Na}}\,\mathbf{k}_{\mathrm{EE}} & N_{\mathrm{Ca}}\,\mathbf{k}_{\mathrm{EE}} & N_{\mathrm{Cl}}\,\mathbf{k}_{\mathrm{EE}} & \mathbf{0} \\
\hline
N_{\mathrm{Na}}\,\mathbf{k}_{\mathrm{EE}} & N_{\mathrm{Na}}^2\,\mathbf{k}_{\mathrm{EE}} & N_{\mathrm{Na}}\,N_{\mathrm{Ca}}\,\mathbf{k}_{\mathrm{EE}} & N_{\mathrm{Na}}\,N_{\mathrm{Cl}}\,\mathbf{k}_{\mathrm{EE}} & \mathbf{0} \\
N_{\mathrm{Ca}}\,\mathbf{k}_{\mathrm{EE}} & N_{\mathrm{Ca}}\,N_{\mathrm{Na}}\,\mathbf{k}_{\mathrm{EE}} & N_{\mathrm{Ca}}^2\,\mathbf{k}_{\mathrm{EE}} & N_{\mathrm{Ca}}\,N_{\mathrm{Cl}}\,\mathbf{k}_{\mathrm{EE}} & \mathbf{0} \\
N_{\mathrm{Cl}}\,\mathbf{k}_{\mathrm{EE}} & N_{\mathrm{Cl}}\,N_{\mathrm{Na}}\,\mathbf{k}_{\mathrm{EE}} & N_{\mathrm{Cl}}\,N_{\mathrm{Ca}}\,\mathbf{k}_{\mathrm{EE}} & N_{\mathrm{Cl}}^2\,\mathbf{k}_{\mathrm{EE}} & \mathbf{0} \\
\mathbf{0} & \mathbf{0} & \mathbf{0} & \mathbf{0} & \mathbf{0}
\end{array}\right] +
$$

$$
+ \left[\begin{array}{c|cccc}
\mathbf{0} & \mathbf{0} & \mathbf{0} & \mathbf{0} & \mathbf{0} \\
\hline
\mathbf{0} & \mathbf{k}_{\mathrm{NaNa}}^d & \mathbf{0} & \mathbf{0} & \mathbf{0} \\
\mathbf{0} & \mathbf{0} & \mathbf{k}_{\mathrm{CaCa}}^d & \mathbf{0} & \mathbf{0} \\
\mathbf{0} & \mathbf{0} & \mathbf{0} & \mathbf{k}_{\mathrm{ClCl}}^d & \mathbf{0} \\
\mathbf{0} & \mathbf{0} & \mathbf{0} & \mathbf{0} & \mathbf{0}
\end{array}\right] +
\left[\begin{array}{ccccc}
\mathbf{0} & \mathbf{0} & \mathbf{0} & \mathbf{0} & \boldsymbol{\kappa}_{\mathrm{w\,T}} \\
\mathbf{0} & \mathbf{0} & \mathbf{0} & \mathbf{0} & \boldsymbol{\kappa}_{\mathrm{NaT}} \\
\mathbf{0} & \mathbf{0} & \mathbf{0} & \mathbf{0} & \boldsymbol{\kappa}_{\mathrm{CaT}} \\
\mathbf{0} & \mathbf{0} & \mathbf{0} & \mathbf{0} & \boldsymbol{\kappa}_{\mathrm{ClT}} \\
\boldsymbol{\kappa}_{\mathrm{Tw}} & \boldsymbol{\kappa}_{\mathrm{T\,Na}} & \boldsymbol{\kappa}_{\mathrm{T\,Ca}} & \boldsymbol{\kappa}_{\mathrm{T\,Cl}} & \boldsymbol{\kappa}_{\mathrm{TT}}
\end{array}\right] ,
\tag{16.5.12}
$$

the N's being defined as $N_k = n^{k\mathrm{E}}/n^{\mathrm{wE}}$, $k \in \mathrm{E_{ions}}$.

The coefficients of the matrix $\mathcal{K}$ deduce from the $\boldsymbol{\kappa}$ matrix,

$$
\mathbf{k}_{\mathrm{TE}} = \boldsymbol{\kappa}_{\mathrm{Tw}}, \quad \mathbf{k}_{\mathrm{T}k} = \boldsymbol{\kappa}_{\mathrm{T}k} - \frac{n^{k\mathrm{E}}}{n^{\mathrm{wE}}}\,\boldsymbol{\kappa}_{\mathrm{Tw}}, k \in \mathrm{E_{ions}}, \quad \mathbf{k}_{\mathrm{Te}} = \mathrm{F}\sum_{k \in \mathrm{E_{ions}}} \frac{\zeta_k}{\widetilde{v}_k}\,\boldsymbol{\kappa}_{\mathrm{T}k}, \quad \mathbf{k}_{\mathrm{TT}} = \boldsymbol{\kappa}_{\mathrm{TT}}\,.
\tag{16.5.13}
$$

An assumption similar to *Assumption* $(\mathcal{A}2)$ may be formulated,

Assumption $(\mathcal{A}3)$: a gradient of temperature does not affect the diffusive flux of ions.

Then $\mathbf{k}_{\mathrm{T}k} = \mathbf{0}$ and $\boldsymbol{\kappa}_{\mathrm{T}k} = (n^{k\mathrm{E}}/n^{\mathrm{wE}})\,\boldsymbol{\kappa}_{\mathrm{Tw}}$ for $k \in \mathrm{E_{ions}}$. The thermoelectric coefficient,

$$
\mathbf{k}_{\mathrm{eT}} = \mathbf{k}_{\mathrm{Te}} = \mathrm{F}\sum_{k \in \mathrm{E_{ions}}} \frac{\zeta_k}{\widetilde{v}_k}\,\boldsymbol{\kappa}_{\mathrm{T}k} = -\mathrm{F}\,\tilde{e}_{\mathrm{PG}}\,\mathbf{k}_{\mathrm{TE}}\,,
\tag{16.5.14}
$$

is observed to be proportional to the fixed electrical charge. Assumption $(\mathcal{A}3)$ amounts to neglecting Soret effect and, by symmetry of the diffusion matrix, Dufour effect, as indicated in Table 15.2.1.

A non zero coefficient $\mathbf{k}_{\mathrm{eT}}$ is associated with the Seebeck effect, namely, to an electrical current generated by a temperature gradient in a closed circuit, or equivalently to a potential difference in an open circuit. A non zero coefficient $\mathbf{k}_{\mathrm{Te}}$ is associated with the Peltier effect, namely, to a heat flux generated by an electrical potential difference.

Under *Assumption* $(\mathcal{A}3)$, the diffusion matrix simplifies to

$$
\mathcal{K} = \left[\begin{array}{cccccc}
\mathbf{k}_{\mathrm{EE}} & \mathbf{0} & \mathbf{0} & \mathbf{0} & \mathbf{k}_{\mathrm{e}} & \mathbf{k}_{\mathrm{ET}} \\
\mathbf{0} & \mathbf{k}_{\mathrm{NaNa}}^d & \mathbf{0} & \mathbf{0} & \mathbf{k}_{\mathrm{Nae}}^d & \mathbf{0} \\
\mathbf{0} & \mathbf{0} & \mathbf{k}_{\mathrm{CaCa}}^d & \mathbf{0} & \mathbf{k}_{\mathrm{Cae}}^d & \mathbf{0} \\
\mathbf{0} & \mathbf{0} & \mathbf{0} & \mathbf{k}_{\mathrm{Cl\,Cl}}^d & \mathbf{k}_{\mathrm{Cl\,e}}^d & \mathbf{0} \\
\mathbf{k}_{\mathrm{e}} & \mathbf{k}_{\mathrm{eNa}}^d & \mathbf{k}_{\mathrm{eCa}}^d & \mathbf{k}_{\mathrm{eCl}}^d & \boldsymbol{\sigma}_{\mathrm{e}} & \mathbf{k}_{\mathrm{eT}} \\
\mathbf{k}_{\mathrm{TE}} & \mathbf{0} & \mathbf{0} & \mathbf{0} & \mathbf{k}_{\mathrm{Te}} & \mathbf{k}_{\mathrm{TT}}
\end{array}\right].
\tag{16.5.15}
$$

The absolute fluxes,

$$
\left[\begin{array}{c} \mathbf{J}_{\mathrm{wE}} \\ \mathbf{J}_{\mathrm{NaE}} \\ \mathbf{J}_{\mathrm{CaE}} \\ \mathbf{J}_{\mathrm{ClE}} \end{array}\right] = -
\left[\begin{array}{cccc}
\boldsymbol{\kappa}_{\mathrm{ww}} & \boldsymbol{\kappa}_{\mathrm{w\,Na}} & \boldsymbol{\kappa}_{\mathrm{w\,Ca}} & \boldsymbol{\kappa}_{\mathrm{w\,Cl}} \\
\boldsymbol{\kappa}_{\mathrm{Naw}} & \boldsymbol{\kappa}_{\mathrm{NaNa}} & \boldsymbol{\kappa}_{\mathrm{NaCa}} & \boldsymbol{\kappa}_{\mathrm{NaCl}} \\
\boldsymbol{\kappa}_{\mathrm{Caw}} & \boldsymbol{\kappa}_{\mathrm{CaNa}} & \boldsymbol{\kappa}_{\mathrm{CaCa}} & \boldsymbol{\kappa}_{\mathrm{CaCl}} \\
\boldsymbol{\kappa}_{\mathrm{Clw}} & \boldsymbol{\kappa}_{\mathrm{Cl\,Na}} & \boldsymbol{\kappa}_{\mathrm{Cl\,Ca}} & \boldsymbol{\kappa}_{\mathrm{Cl\,Cl}}
\end{array}\right]
\left\{ \left[\begin{array}{c}
\rho_{\mathrm{w}}\,\boldsymbol{\nabla}\mu_{\mathrm{wE}|T} \\
\rho_{\mathrm{Na}}\,\boldsymbol{\nabla}\mu_{\mathrm{NaE}|T}^{\mathrm{ec}} \\
\rho_{\mathrm{Ca}}\,\boldsymbol{\nabla}\mu_{\mathrm{CaE}|T}^{\mathrm{ec}} \\
\rho_{\mathrm{Cl}}\,\boldsymbol{\nabla}\mu_{\mathrm{ClE}|T}^{\mathrm{ec}}
\end{array}\right] +
\left[\begin{array}{c}
\boldsymbol{\kappa}_{\mathrm{Tw}}^* \\
\boldsymbol{\kappa}_{\mathrm{T\,Na}}^* \\
\boldsymbol{\kappa}_{\mathrm{T\,Ca}}^* \\
\boldsymbol{\kappa}_{\mathrm{T\,Cl}}^*
\end{array}\right] \boldsymbol{\nabla}T \right\},
\tag{16.5.16}
$$

$$\mathbf{J}_\mathrm{T} = - \begin{bmatrix} \boldsymbol{\kappa}_\mathrm{Tw} & \boldsymbol{\kappa}_\mathrm{T\,Na} & \boldsymbol{\kappa}_\mathrm{T\,Ca} & \boldsymbol{\kappa}_\mathrm{T\,Cl} \end{bmatrix} \begin{bmatrix} \rho_\mathrm{w}\,\boldsymbol{\nabla}\mu_\mathrm{wE|T} \\ \rho_\mathrm{Na}\,\boldsymbol{\nabla}\mu^\mathrm{ec}_\mathrm{NaE|T} \\ \rho_\mathrm{Ca}\,\boldsymbol{\nabla}\mu^\mathrm{ec}_\mathrm{CaE|T} \\ \rho_\mathrm{Cl}\,\boldsymbol{\nabla}\mu^\mathrm{ec}_\mathrm{ClE|T} \end{bmatrix} - \boldsymbol{\kappa}_\mathrm{TT}\cdot\boldsymbol{\nabla}T\,, \tag{16.5.17}$$

may be also expressed in terms of the pseudo-vector which is solution of the linear system,

$$\begin{bmatrix} \boldsymbol{\kappa}_\mathrm{ww} & \boldsymbol{\kappa}_\mathrm{w\,Na} & \boldsymbol{\kappa}_\mathrm{w\,Ca} & \boldsymbol{\kappa}_\mathrm{w\,Cl} \\ \boldsymbol{\kappa}_\mathrm{Naw} & \boldsymbol{\kappa}_\mathrm{NaNa} & \boldsymbol{\kappa}_\mathrm{NaCa} & \boldsymbol{\kappa}_\mathrm{NaCl} \\ \boldsymbol{\kappa}_\mathrm{Caw} & \boldsymbol{\kappa}_\mathrm{CaNa} & \boldsymbol{\kappa}_\mathrm{CaCa} & \boldsymbol{\kappa}_\mathrm{CaCl} \\ \boldsymbol{\kappa}_\mathrm{Clw} & \boldsymbol{\kappa}_\mathrm{Cl\,Na} & \boldsymbol{\kappa}_\mathrm{Cl\,Ca} & \boldsymbol{\kappa}_\mathrm{Cl\,Cl} \end{bmatrix} \begin{bmatrix} \boldsymbol{\kappa}^*_\mathrm{Tw} \\ \boldsymbol{\kappa}^*_\mathrm{T\,Na} \\ \boldsymbol{\kappa}^*_\mathrm{T\,Ca} \\ \boldsymbol{\kappa}^*_\mathrm{T\,Cl} \end{bmatrix} = \begin{bmatrix} \boldsymbol{\kappa}_\mathrm{Tw} \\ \boldsymbol{\kappa}_\mathrm{T\,Na} \\ \boldsymbol{\kappa}_\mathrm{T\,Ca} \\ \boldsymbol{\kappa}_\mathrm{T\,Cl} \end{bmatrix}\,. \tag{16.5.18}$$

In component form, the constitutive equations in terms of absolute fluxes,

$$\begin{aligned}
\mathbf{J}_{k\mathrm{E}} &= -\sum_{l\in\mathrm{E_{mo}}} \boldsymbol{\kappa}_{kl}\cdot\mathbf{f}_{l\mathrm{E}} - \boldsymbol{\kappa}_{k\mathrm{T}}\cdot\mathbf{f}_\mathrm{T}, \quad k\in\mathrm{E_{mo}}, \\
\mathbf{J}_\mathrm{T} &= -\sum_{l\in\mathrm{E_{mo}}} \boldsymbol{\kappa}_{\mathrm{T}l}\cdot\mathbf{f}_{l\mathrm{E}} - \boldsymbol{\kappa}_\mathrm{TT}\cdot\mathbf{f}_\mathrm{T},
\end{aligned} \tag{16.5.19}$$

may also be expressed in the form,

$$\begin{aligned}
\mathbf{J}_{k\mathrm{E}} &= -\sum_{l\in\mathrm{E_{mo}}} \boldsymbol{\kappa}_{kl}\cdot\big(\mathbf{f}_{l\mathrm{E}} + \boldsymbol{\kappa}^*_{l\mathrm{T}}\cdot\mathbf{f}_\mathrm{T}\big), \quad k\in\mathrm{E_{mo}}, \\
\mathbf{J}_\mathrm{T} &= \sum_{l\in\mathrm{E_{mo}}} \boldsymbol{\kappa}^*_{\mathrm{T}l}\cdot\mathbf{J}_{l\mathrm{E}} - \Big(\boldsymbol{\kappa}_\mathrm{TT} - \sum_{l\in\mathrm{E_{mo}}} \boldsymbol{\kappa}^*_{\mathrm{T}l}\cdot\boldsymbol{\kappa}_{\mathrm{T}l}\Big)\cdot\mathbf{f}_\mathrm{T},
\end{aligned} \tag{16.5.20}$$

where

$$\sum_{l\in\mathrm{E_{mo}}} \boldsymbol{\kappa}_{kl}\cdot\boldsymbol{\kappa}^*_{\mathrm{T}l} = \boldsymbol{\kappa}_{\mathrm{T}k}, \quad k\in\mathrm{E_{mo}}\,. \tag{16.5.21}$$

Under spatially uniform temperature,

$$\mathbf{Q} = T\,\mathbf{J}_\mathrm{T} = \sum_{l\in\mathrm{E_{mo}}} T\,\boldsymbol{\kappa}^*_{\mathrm{T}l}\cdot\mathbf{J}_{l\mathrm{E}}\,, \tag{16.5.22}$$

and

$$\mathbf{J}_\mathrm{TE} = \mathbf{J}_\mathrm{T} + \sum_{l\in\mathrm{E_{mo}}} \rho_l\,s_{l\mathrm{E}}\,\mathbf{J}_{l\mathrm{E}} = \sum_{l\in\mathrm{E_{mo}}} \big(\boldsymbol{\kappa}^*_{\mathrm{T}l} + \rho_l\,s_{l\mathrm{E}}\,\mathbf{I}\big)\cdot\mathbf{J}_{l\mathrm{E}}\,, \tag{16.5.23}$$

which has led the entities $T\,\boldsymbol{\kappa}^*_{\mathrm{T}l}$ to be called *heats of transport*, and the entities $\boldsymbol{\kappa}^*_{\mathrm{T}l} + \rho_l\,s_{l\mathrm{E}}\,\mathbf{I}$ to be called *transported entropies*, Haase [1990], pp. 186, 329.

If the circuit is electrically open, the fluid flux is equal to the sum of the temperature independent part given by (16.2.38) and of the term

$$-\big(\mathbf{k}_\mathrm{ET} - \mathbf{k}_\mathrm{e}\cdot\boldsymbol{\sigma}^{-1}_\mathrm{e}\cdot\mathbf{k}_\mathrm{eT}\big)\cdot\boldsymbol{\nabla}T\,. \tag{16.5.24}$$

If *Assumption* ($\mathcal{A}$3) holds, then the above term becomes

$$-\mathbf{k}_\mathrm{H}\cdot\mathbf{k}^{-1}_\mathrm{EE}\cdot\mathbf{k}_\mathrm{ET}\cdot\boldsymbol{\nabla}T\,, \tag{16.5.25}$$

with help of (16.2.30) and with the open circuit permeability $\mathbf{k}_\mathrm{H}$ defined by (16.2.26).

Exercises on Chapter 16

Exercise 16.1 Nonuniqueness of the pairs of fluxes and driving gradients

Show that the choice of the conjugate pair of fluxes and forces in $\hat{S}_{(i,3)}$, Eqn (16.1.22), contains two degrees of arbitrariness. *Proof*: Indeed, let α be an arbitrary strictly positive scalar and $\boldsymbol{\Delta}$ be an arbitrary, but invertible, diagonal matrix. Then the pair $(\bar{\mathbf{j}}, \bar{\mathbf{f}})$, defined by

$$\bar{\mathbf{j}} = \alpha \, \boldsymbol{\Delta} \, \mathbf{j}, \quad \bar{\mathbf{f}} = \boldsymbol{\Delta}^{-\mathrm{T}} \, \mathbf{f}, \tag{16.E.1}$$

satisfies

$$T \hat{S}_{(i,3)} = -\mathbf{f}^{\mathrm{T}} \mathbf{j} = -\alpha^{-1} \, \bar{\mathbf{f}}^{\mathrm{T}} \, \bar{\mathbf{j}} \geq 0. \tag{16.E.2}$$

Therefore it is equivalent to postulate the generalized diffusion law on the pairs $(\mathbf{j}, \mathbf{f})$ and $(\bar{\mathbf{j}}, \bar{\mathbf{f}})$. The symmetry and positive definiteness of the diffusion matrix $\boldsymbol{\kappa}$ carry over to the diffusion matrix $\bar{\boldsymbol{\kappa}}$, and conversely,

$$\bar{\mathbf{j}} = -\bar{\boldsymbol{\kappa}} \bar{\mathbf{f}}, \quad \bar{\boldsymbol{\kappa}} = \alpha \, \boldsymbol{\Delta} \, \boldsymbol{\kappa} \, \boldsymbol{\Delta}^{\mathrm{T}}. \tag{16.E.3}$$

Exercise 16.2 Relations between the various measures of fluxes

Check the relations below between volume and mass fluxes, starting from the definitions stated in Section 16.1.2.5.

1. Check the relation (16.1.16) between the absolute and diffusive fluxes.

2. Let

$$n_*^{\mathrm{E}} \equiv \sum_{k \in \mathrm{E_{mo}}} n^{k\mathrm{E}} \neq n_{\mathrm{mo}}^{\mathrm{E}} \equiv \sum_{k \in \mathrm{E_{mo}}} n_{\mathrm{mo}}^{k\mathrm{E}}, \tag{16.E.4}$$

and

$$\rho_*^{\mathrm{E}} \equiv \sum_{k \in \mathrm{E_{mo}}} n^{k\mathrm{E}} \rho_k \neq \rho_{\mathrm{mo}}^{\mathrm{E}} \equiv \sum_{k \in \mathrm{E_{mo}}} n_{\mathrm{mo}}^{k\mathrm{E}} \rho_k. \tag{16.E.5}$$

Check the relations

$$\mathbf{J}_{\mathrm{E}} = \sum_{k \in \mathrm{E_{mo}}} \mathbf{J}_{k\mathrm{E}}^d + \frac{n_*^{\mathrm{E}}}{n^{\mathrm{wE}}} \mathbf{J}_{\mathrm{wE}} = \sum_{k \in \mathrm{E}} \mathbf{J}_{k\mathrm{E}}^d + \frac{n^{\mathrm{E}}}{n^{\mathrm{wE}}} \mathbf{J}_{\mathrm{wE}}, \tag{16.E.6}$$

$$\frac{\mathbf{J}_{\mathrm{wE}}}{n^{\mathrm{wE}}} = \frac{1}{n^{k\mathrm{E}}} (\mathbf{J}_{k\mathrm{E}} - \mathbf{J}_{k\mathrm{E}}^d) = \frac{1}{n_*^{\mathrm{E}}} \sum_{l \in \mathrm{E_{mo}}} \mathbf{J}_{l\mathrm{E}} - \mathbf{J}_{l\mathrm{E}}^d, \quad k \in \mathrm{E}^{(\mathrm{mo})}, \tag{16.E.7}$$

$$\sum_{k \in \mathrm{E_{mo}}} \mathbf{M}_{k\mathrm{E}} = \sum_{k \in \mathrm{E_{mo}}} \rho_k \mathbf{J}_{k\mathrm{E}} = \frac{\rho_*^{\mathrm{E}}}{n_*^{\mathrm{E}}} \mathbf{J}_{\mathrm{E}} + \sum_{k \in \mathrm{E_{mo}}} \left(\rho_k - \frac{\rho_*^{\mathrm{E}}}{n_*^{\mathrm{E}}} \right) \mathbf{J}_{k\mathrm{E}}^d. \tag{16.E.8}$$

Exercise 16.3 A second expression of the dissipation due to diffusion

Using the expression of the electrochemical potentials (16.1.7), the closure relation satisfied by the molar fractions, and Eqns (16.1.13) and (16.1.15), obtain the dissipation due to diffusion in terms of the diffusive fluxes and of the electric current density.

Answer: The four terms arise as follows:

- the pressure term can be simplified. First, the gradients of the generalized pressures $\boldsymbol{\nabla} p_{k\mathrm{E}}$ are all identical in the model developed in Loret and Simões [2004] [2005]a. Second, the total flux $\mathbf{J}_{\mathrm{E}}$ may be approximated by the flux of water $\mathbf{J}_{\mathrm{wE}}$. Then

$$\sum_{k \in \mathrm{E_{mo}}} \boldsymbol{\nabla} p_{k\mathrm{E}} \cdot \mathbf{J}_{k\mathrm{E}} \simeq \boldsymbol{\nabla} p_{\mathrm{wE}} \cdot \mathbf{J}_{\mathrm{wE}}; \tag{16.E.9}$$

- the chemical term can be additively decomposed, using (16.1.16), in the standard term plus a term due to the presence of PGs in the extrafibrillar phase,

$$\sum_{k \in \mathrm{E_{mo}}} RT \, \boldsymbol{\nabla} \mathrm{Ln} \, x_{k\mathrm{E}} \cdot \frac{\mathbf{J}_{k\mathrm{E}}}{\widehat{v}_k} = \sum_{k \in \mathrm{E_{ions}}} RT \, \boldsymbol{\nabla} \mathrm{Ln} \, x_{k\mathrm{E}} \cdot \frac{\mathbf{J}_{k\mathrm{E}}^d}{\widehat{v}_k} - RT \, \boldsymbol{\nabla} \tilde{c}_{\mathrm{PG}} \cdot \mathbf{J}_{\mathrm{wE}}; \tag{16.E.10}$$

- the electrical term is simply

$$\boldsymbol{\nabla} \phi_{\mathrm{E}} \cdot \mathbf{I}_{e\mathrm{E}}; \tag{16.E.11}$$

- in the gravity term, using (16.E.8),

$$\sum_{k \in \mathrm{E_{mo}}} -\rho_k \mathbf{g} \cdot \mathbf{J}_{k\mathrm{E}} = -\mathbf{g} \cdot \left(\frac{\rho_*^{\mathrm{E}}}{n_*^{\mathrm{E}}} \mathbf{J}_{\mathrm{E}} + \sum_{l \in \mathrm{E_{mo}}} \left(\rho_k - \frac{\rho_*^{\mathrm{E}}}{n_*^{\mathrm{E}}} \right) \mathbf{J}_{k\mathrm{E}}^d \right), \tag{16.E.12}$$

the total flux $\mathbf{J}_{\mathrm{E}}$ may be approximated once again by the flux of water $\mathbf{J}_{\mathrm{wE}}$.

To derive the above expressions, repeated use has been made of the relation:

$$\frac{1}{\widetilde{v}_{\mathrm{w}}} \boldsymbol{\nabla} \mathrm{Ln}\, x_{\mathrm{wE}} = -\boldsymbol{\nabla}\widetilde{c}_{\mathrm{PG}} - \sum_{l\in \mathrm{E_{ions}}} \frac{n^{l\mathrm{E}}}{n^{\mathrm{wE}}} \frac{1}{\widetilde{v}_l} \boldsymbol{\nabla} \mathrm{Ln}\, x_{l\mathrm{E}}\,. \tag{16.E.13}$$

Exercise 16.4 On the dependence of the components of $\mathcal{J}$ and $\mathcal{F}$

Prove the relations of linear independence of the driving ionic gradients, Eqn (16.2.7), and the relation of linear dependence of the fluxes, Eqn (16.2.8).

Exercise 16.5 Inheritance of the symmetry property from one diffusion matrix to the other

The symmetry of the matrix $\boldsymbol{\kappa}$ has been used to establish (16.2.11). The diffusion matrix $\boldsymbol{\mathcal{K}}$, Eqn (16.2.11), inherits the symmetry property of the diffusion matrix $\boldsymbol{\kappa}$. Prove this implication.

Proof: Even if a priori $\kappa_{kl}^d \neq \kappa_{lk}^d$, for $k \neq l$, Eqn (16.2.12), the central coefficients of the matrix $\boldsymbol{\mathcal{K}}$ are symmetric. In fact,

$$\mathbf{k}_{kl}^d = \mathbf{k}_{lk}^d = \kappa_{kl} - \frac{n^{k\mathrm{E}}}{n^{\mathrm{wE}}} \kappa_{wl} - \kappa_{kw} \frac{n^{l\mathrm{E}}}{n^{\mathrm{wE}}} + \frac{n^{k\mathrm{E}}}{n^{\mathrm{wE}}} \frac{n^{l\mathrm{E}}}{n^{\mathrm{wE}}} \kappa_{ww}, \quad k,l \in \mathrm{E_{ions}}\,. \tag{16.E.14}$$

Exercise 16.6 Structure of the diffusion matrix

Establish relations that establish contact with the isotropic model of diffusion based on friction coefficients developed by Gu et al. [1998].

1. Obtain the gradients of the electrochemical potentials in terms of the fluxes.

The relations (16.2.1) and (16.2.2) link the fluxes, relative to the solid, of water $\mathbf{J}$ and ions $\mathbf{J}_i$, $i \in [1,3]$, to the gradients of the electrochemical potentials of the associated species $\mathbf{M}$ and ions $\mathbf{M}_i$, $i \in [1,3]$. They may be recast in the simplified format (the matrix below plays the role of the matrix $\boldsymbol{\kappa}$ in (16.2.29)):

$$\begin{bmatrix} \mathbf{J} \\ \mathbf{J}_1 \\ \mathbf{J}_2 \\ \mathbf{J}_3 \end{bmatrix} = \begin{bmatrix} 1 & n^1 & n^2 & n^3 \\ n^1 & c_{11} & c_{12} & c_{13} \\ n^2 & c_{21} & c_{22} & c_{23} \\ n^3 & c_{31} & c_{32} & c_{33} \end{bmatrix} \begin{bmatrix} \mathbf{M} \\ \mathbf{M}_1 \\ \mathbf{M}_2 \\ \mathbf{M}_3 \end{bmatrix}. \tag{16.E.15}$$

Obtain a partial inversion of these relations by expressing the electrochemical potentials in terms of the diffusive fluxes as

$$\begin{bmatrix} \mathbf{M}_1 \\ \mathbf{M}_2 \\ \mathbf{M}_3 \end{bmatrix} = \mathbf{B} \begin{bmatrix} \mathbf{J}_1 - n^1\, \mathbf{J} \\ \mathbf{J}_2 - n^2\, \mathbf{J} \\ \mathbf{J}_3 - n^3\, \mathbf{J} \end{bmatrix}, \quad \mathbf{M} = \mathbf{J} - \sum_{j=1,3} n^j\, \mathbf{M}_j\,. \tag{16.E.16}$$

With $\mathbf{n}$ the vector of components n_i, $i \in [1,3]$, and $\mathbf{C}$ the matrix of components c_{ij}, $i,j \in [1,3]$, the matrix $\mathbf{B}$ above is defined as

$$\mathbf{B} = \left(\mathbf{C} - \mathbf{n} \otimes \mathbf{n}\right)^{-1}. \tag{16.E.17}$$

2. Obtain the complete inverse relation in the format,

$$\begin{bmatrix} \mathbf{M} \\ \mathbf{M}_1 \\ \mathbf{M}_2 \\ \mathbf{M}_3 \end{bmatrix} = \begin{bmatrix} 1 - \sum_{j,k} B_{jk} & -\sum_j B_{j1} & -\sum_j B_{j2} & -\sum_j B_{j3} \\ -\sum_j B_{1j} & B_{11} & B_{12} & B_{13} \\ -\sum_j B_{2j} & B_{21} & B_{22} & B_{23} \\ -\sum_j B_{3j} & B_{31} & B_{32} & B_{33} \end{bmatrix} \begin{bmatrix} \mathbf{J} \\ \mathbf{J}_1 \\ \mathbf{J}_2 \\ \mathbf{J}_3 \end{bmatrix}. \tag{16.E.18}$$

3. Study the article of Gu et al. [1998]. Establish contact between the model presented in this chapter and the model of Gu et al. [1998].

Answer: The above matrix is proportional to the matrix $\mathbf{A}$ in Gu et al. [1998], their Eqn 17. They require the lower major to be diagonal. This restriction implies the inverse of the matrix $\mathbf{C} - \mathbf{n} \otimes \mathbf{n}$, and thus $\mathbf{C} - \mathbf{n} \otimes \mathbf{n}$, to be diagonal. Equivalently the matrix $\mathbf{C}$ has to be the sum of a diagonal part, plus a dyadic part, as indeed can be checked on (16.2.29). Now, the matrix $\mathbf{C}$ represents the submatrix of $\boldsymbol{\kappa}$ defined by (16.2.27)$_2$, and the matrix $\mathbf{C} - \mathbf{n} \otimes \mathbf{n}$ represents the central part of the diffusive matrix noted k_{kl}^d, $k,l \in \mathrm{E_{ions}}$, and defined by (16.2.26)$_5$ for $a_1 = a_2 = 0$, $a_3 = k_{\mathrm{EE}}$.

Exercise 16.7 Commutation properties of the anisotropic diffusion matrices

Establish the commutation relations (16.2.30).

Proof: First, since all matrices involved are symmetric, the two relations in (16.2.30) are simply the transpose of one another. Using the various relations in Table 16.2.1 yields successively,

$$
\begin{aligned}
\mathbf{k_H} \cdot \mathbf{k_{EE}}^{-1} &= (\mathbf{k_{EE}} - \mathbf{k_e} \cdot (\boldsymbol{\sigma}_e)^{-1} \cdot \mathbf{k_e}) \cdot \mathbf{k_{EE}}^{-1} \\
&= \mathbf{I} + (F\,\tilde{e}_{PG})\,\mathbf{k_e} \cdot (\boldsymbol{\sigma}_e)^{-1} \\
&= (\boldsymbol{\sigma}_e + (F\,\tilde{e}_{PG})\,\mathbf{k_e}) \cdot (\boldsymbol{\sigma}_e)^{-1} \\
&= \boldsymbol{\sigma}_e^{\text{ion}} \cdot (\boldsymbol{\sigma}_e)^{-1}.
\end{aligned}
\tag{16.E.19}
$$

Exercise 16.8 Positive definiteness of the open circuit permeability $\mathbf{k_H}$

Establish that the open circuit permeability $\mathbf{k_H}$ is positive definite.

Proof: Using (16.2.30) and the various relations in Table 16.2.1 leads to, successively,

$$
\begin{aligned}
\mathbf{k_H} &= \boldsymbol{\sigma}_e^{\text{ion}} \cdot (\boldsymbol{\sigma}_e)^{-1} \cdot \mathbf{k_{EE}} \\
&= \left(\mathbf{k_{EE}}^{-1} \cdot \boldsymbol{\sigma}_e \cdot (\boldsymbol{\sigma}_e^{\text{ion}})^{-1} \right)^{-1} \\
&= \left(\mathbf{k_{EE}}^{-1} \cdot \left(\boldsymbol{\sigma}_e^{\text{ion}} + (F\,\tilde{e}_{PG})^2\,\mathbf{k_{EE}} \right) \cdot (\boldsymbol{\sigma}_e^{\text{ion}})^{-1} \right)^{-1} \\
&= \left(\mathbf{k_{EE}}^{-1} + (F\,\tilde{e}_{PG})^2\,(\boldsymbol{\sigma}_e^{\text{ion}})^{-1} \right)^{-1}.
\end{aligned}
\tag{16.E.20}
$$

Exercise 16.9 Derivation of the generalized diffusion matrix for ternary electrolytes

Reconsider the derivation of the generalized diffusion matrix for ternary electrolytes presented in Section 16.2.5 for binary electrolytes.

Proof: In the case of a ternary electrolyte, the set of (pseudo-)vectors of length three orthogonal to the vector of valences $\mathbf{Z}$ has dimension two, and it is generated by the orthogonal (pseudo-)vectors $\mathbf{Z}_1^\perp$ and $\mathbf{Z}_2^\perp$.

The solution (16.2.23) involves two (pseudo-)vectors $\mathbf{A}_1$ and $\mathbf{A}_2$ of length three. With simplified notations, the undetermined part of this solution can be cast in the format,

$$
\boldsymbol{\kappa} = \cdots + \mathbf{A}_1 \otimes \mathbf{Z}_1^\perp + \mathbf{A}_2 \otimes \mathbf{Z}_2^\perp + \mathbf{N} \otimes \boldsymbol{\kappa}_w,
\tag{16.E.21}
$$

while

$$
\boldsymbol{\kappa}^d = \cdots + \mathbf{A}_1 \otimes \mathbf{Z}_1^\perp + \mathbf{A}_2 \otimes \mathbf{Z}_2^\perp.
\tag{16.E.22}
$$

Symmetry of $\boldsymbol{\kappa}$ implies

$$
\mathbf{A}_1 \wedge \mathbf{Z}_1^\perp + \mathbf{A}_2 \wedge \mathbf{Z}_2^\perp + \mathbf{N} \wedge \boldsymbol{\kappa}_w = \mathbf{0}.
\tag{16.E.23}
$$

$\mathbf{Z}_1^\perp$, $\mathbf{Z}_2^\perp$ and $\mathbf{N}$ being known vectors of length three, the three vectors $\mathbf{A}_1$, $\mathbf{A}_2$ and $\boldsymbol{\kappa}_w$ are sought subject to the constraint (16.E.23). This solution can thus be expressed in terms of $3 \times 3 - 3 = 6$ arbitrary scalars.

Since electroneutrality implies $\mathbf{Z} \cdot \mathbf{N} = -\tilde{e}_{PG} \neq 0$, the vector $\mathbf{N}$ does not belong to the set which is orthogonal to $\mathbf{Z}$. Therefore $\mathbf{Z}_1^\perp$, $\mathbf{Z}_2^\perp$ and $\mathbf{N}$ constitute a basis for $\mathbf{A}_1$, $\mathbf{A}_2$ and $\boldsymbol{\kappa}_w$, say

$$
\mathbf{A}_1 = a_{11}\,\mathbf{Z}_1^\perp + a_{12}\,\mathbf{Z}_2^\perp + a_{13}\,\mathbf{N},
$$

$$
\mathbf{A}_2 = a_{21}\,\mathbf{Z}_1^\perp + a_{22}\,\mathbf{Z}_2^\perp + a_{23}\,\mathbf{N},
\tag{16.E.24}
$$

$$
\boldsymbol{\kappa}_w = a_{31}\,\mathbf{Z}_1^\perp + a_{32}\,\mathbf{Z}_2^\perp + a_{33}\,\mathbf{N}.
$$

The constraint (16.E.23) becomes

$$
\begin{aligned}
(a_{21} - a_{12})\,\mathbf{Z}_1^\perp \wedge \mathbf{Z}_2^\perp &+ (a_{13} + a_{31})\,\mathbf{N} \wedge \mathbf{Z}_1^\perp &+ (a_{23} + a_{32})\,\mathbf{N} \wedge \mathbf{Z}_2^\perp &= \mathbf{0}, \\
(a_{21} - a_{12})\,\mathbf{Z} &+ (a_{13} + a_{31})\,(\alpha_1\,\mathbf{Z}_1^\perp + \beta_1\,\mathbf{Z}) &+ (a_{23} + a_{32})\,(\alpha_2\,\mathbf{Z}_2^\perp + \beta_2\,\mathbf{Z}) &= \mathbf{0}.
\end{aligned}
\tag{16.E.25}
$$

The coefficients α are different from 0 because $\mathbf{N}$ has a component along $\mathbf{Z}$ due to electroneutrality. Taking the scalar product of the above vector (which vanishes) successively by $\mathbf{Z}_1^\perp$, $\mathbf{Z}_2^\perp$ and $\mathbf{Z}$ yields $a_{13} + a_{31} = 0$, $a_{23} + a_{32} = 0$ and $a_{21} - a_{12} = 0$.

Thus we are left with six unknown scalars, as expected,

$$\mathbf{A}_1 = a_{11}\,\mathbf{Z}_1^{\perp} + a_{12}\,\mathbf{Z}_2^{\perp} + a_{13}\,\mathbf{N},$$

$$\mathbf{A}_2 = a_{12}\,\mathbf{Z}_1^{\perp} + a_{22}\,\mathbf{Z}_2^{\perp} + a_{23}\,\mathbf{N}, \qquad (16.\mathrm{E}.26)$$

$$\boldsymbol{\kappa}_{\mathrm{w}} = -a_{13}\,\mathbf{Z}_1^{\perp} - a_{23}\,\mathbf{Z}_2^{\perp} + a_{33}\,\mathbf{N}.$$

Consequently

$$\mathbf{k}_{\mathrm{E}}^{d} = \boldsymbol{\kappa}_{\mathrm{w}}^{d} = \boldsymbol{\kappa}_{\mathrm{w}} - \kappa_{\mathrm{ww}}\,\mathbf{N} = -a_{13}\,\mathbf{Z}_1^{\perp} - a_{23}\,\mathbf{Z}_2^{\perp} + (a_{33} - \kappa_{\mathrm{ww}})\,\mathbf{N}, \qquad (16.\mathrm{E}.27)$$

while

$$\mathbf{k}^{d} = \cdots \quad + a_{11}\,\mathbf{Z}_1^{\perp} \otimes \mathbf{Z}_1^{\perp} + a_{22}\,\mathbf{Z}_2^{\perp} \otimes \mathbf{Z}_2^{\perp} + (\kappa_{\mathrm{ww}} - a_{33})\,\mathbf{N} \otimes \mathbf{N}$$

$$+ a_{12}\,(\mathbf{Z}_1^{\perp} \otimes \mathbf{Z}_2^{\perp} + \mathbf{Z}_2^{\perp} \otimes \mathbf{Z}_1^{\perp}) + a_{13}\,(\mathbf{Z}_1^{\perp} \otimes \mathbf{N} + \mathbf{N} \otimes \mathbf{Z}_1^{\perp}) + a_{23}\,(\mathbf{Z}_2^{\perp} \otimes \mathbf{N} + \mathbf{N} \otimes \mathbf{Z}_2^{\perp}). \qquad (16.\mathrm{E}.28)$$

Once again, since $\mathbf{Z}_1^{\perp}$, $\mathbf{Z}_2^{\perp}$ and $\mathbf{N}$ are linearly independent, Assumption $(\mathcal{A}1)$ which amounts to $\mathbf{k}_{\mathrm{E}}^{d} = \mathbf{0}$, implies $a_{13} = 0$, $a_{23} = 0$, and $a_{33} - \kappa_{\mathrm{ww}} = 0$. Assumption $(\mathcal{A}2)$ implies $a_{11} = 0$, $a_{22} = 0$ and $a_{12} = 0$, which finally proves the arrowhead format displayed in Section 16.2.6.

Chapter 17

Effects of the pH on the transport and mechanical properties of articular cartilages

Articular cartilages swell and shrink depending on the ionic strength of the electrolyte they are bathed in. This electro-chemomechanical coupling is driven by the fixed electrical charge borne by proteoglycans. The electrolyte contains mainly sodium cations and chloride anions but also, under nonphysiological conditions, hydrogen and hydroxyl ions. In fact, for an electrolyte, it might be more appropriate to refer to the hydronium ion H_3O^+ than to the hydrogen ion H^+. Under non physiological pH, collagen fibers become charged while the charge of PGs varies. Therefore, variation of the pH of the electrolyte has strong implications on the electrical charge of cartilages and, by the same token, on their transport and mechanical properties. In fact, as the pH of the tissue decreases, the overall fixed charge becomes less negative. It vanishes at the so-called isoelectric point and becomes positive as pH decreases further.

Laboratory experiments involve a tissue specimen in contact with a bath of controlled chemical composition. Due to the fixed charge, the ionic concentrations within the tissue are not equal to the concentrations in the bath. This observation holds as well for the pH, except at the isoelectric point. Therefore a clear distinction should be made between bath pH, which is controlled, and tissue pH which entails the response of the tissue to chemical and mechanical loadings.

In line with the point of view developed in Chapter 14, articular cartilage is viewed as a porous medium with a solid skeleton circulated by an electrolyte. Both the solid skeleton and the electrolyte contain a number of species. The constitutive framework is phrased within the theory of thermodynamics of porous media. Acid-base reactions, as well as calcium binding, are embedded in this framework. Although macroscopic in nature, the model accounts for a number of biochemical features associated with collagen and proteoglycans.

Emphasis is laid first on the effects of pH, ionic strength and calcium binding on the transport properties of cartilages. The model is used to simulate laboratory experiments where a cartilage is in contact with a bath of controlled chemical composition. The simulations display the evolutions of the chemical compositions of mobile ions and of the charges of collagen and glycosaminoglycans at constant volume fraction of water. The resulting variations of the transport properties, including the change of sign of the electroosmotic coefficient at the isoelectric point, are highlighted.

The effects of pH on the mechanical properties of cartilages are considered next. The model of generalized diffusion is extended to include mechanical effects in the thermodynamic framework of multi-species deformable porous media. Again, to assess the validity of the model, laboratory specimens are subjected to tests where both the mechanical and chemical boundary conditions are controlled. Chemical loadings, where the ionic composition and pH of the bath the specimen is in contact with are varied, are intermingled with changes of the mechanical load. The variations of the stresses and strains are observed to depend strongly on the ionic strength and ion type present in the bath: sodium chloride leads to a stiffer response than calcium chloride, which in turn is associated with a stiffer response than hydrochloric acid. The change of sign of the fixed charge at the isoelectric point has definite mechanical implications and gives rise to a non monotonous variation of the stress at constant axial strain.

17.1   Overview and laboratory observations

Articular cartilages are porous media, structured by collagen fibers and saturated by an electrolyte, with water as solvent and metallic ions as solutes. Charged macromolecules, the proteoglycans (PGs), intermingled with collagen fibers, give rise to electro-chemomechanical couplings that allow moderate deformation to take place and ensure an optimal adaptation of the tissue to physiological loads. Roughly speaking, under constant load, these tissues swell if the ionic strength of the bath they are in contact with decreases. Conversely they shrink if the ionic strength increases. On the other hand, at given strain, the compressive equilibrium stress increases as the ionic strength decreases, Eisenberg and Grodzinsky [1985].

The electrical charge of proteoglycans is said to be *fixed* in contrast to that of *mobile* ions, typically sodium Na^+, calcium Ca^{2+} and chloride Cl^-. The electrical charges of the two main non soluble components of articular cartilages, namely collagen and proteoglycans, are sensitive to pH. There exists a value of pH, termed isoelectric point (IEP), where the fixed charge of the extrafibrillar fluid of cartilage vanishes. While a number of geological, biological and engineered materials contain a fixed electrical charge, comprehensive sets of data on the physical effects of pH on these materials are not available. Still, sparse information indicates unambiguously that variation of pH affects both the mechanical and transport properties.

For example, the apparent compressibility depends strongly on the magnitude of this fixed charge. Traction tests on articular cartilages by Grodzinsky et al. [1981] show that the axial stress necessary to ensure equilibrium has a minimum at the isoelectric point. A similar observation applies to the relative hydration (defined as the ratio of wet weight over dry weight) of bovine cornea, Huang and Meek [1999]. On crosslinked polymer chains of hydrogels that contain acid and basic sites, De and Aluru [2004] note that variation of pH of the aqueous solution implies reversible swelling or shrinking.

The presence of the hydrogen ions H^+ and hydroxyl ions OH^-, in addition to metallic ions, affects strongly the transport properties as well. Actually, the sign of the electroosmotic coefficient is opposite to that of the fixed electrical charge. Thus, while electroosmosis implies water to move towards the cathode above the isoelectric point, a reversal of the osmotic flow direction takes place at very low pH below the isoelectric point.

Perhaps some caution is required in the definition of the isoelectric point. Here, the charge involved in this definition is the part of the charge of collagen and PGs that is macroscopically active. The associated sites are in contact with the extrafibrillar water. Sites of collagen in contact with the intrafibrillar water are not considered to be involved in the processes to be addressed here.

17.1.1   Indications on the biochemical composition of collagen and PGs

The collagen present in articular cartilages is mainly a hyaline collagen of type II. The three chains of the triple helix of collagen are formed by a sequence -Gly-X-Y. As indicated in Fig. 14.2.5, Gly is the amino acid glycine, while X and Y are hydroxylized amino acids, namely X proline (Pro) while Y is the 4- or 3-hydroxyproline (4Hyp, 3Hyp) or 5-hydroxylysine (5Hyl).

Amino acids contain, around a central carbon, at least two ionizable sites, an amino site NH_3^+ and a carboxy acid site $CO-O^-$ (also referred to as carboxyl site):

$$
\begin{array}{c}
O{=}C\text{-}O^- \\
| \\
H_3N^+ \;\text{---}\; C \;\text{---}\; H \quad . \\
| \\
R
\end{array}
\tag{17.1.1}
$$

Such molecules, with a positive and a negative charge, are called zwitterions. The amino and carboxyl sites, or groups, are often referred to as N-terminus and C-terminus, respectively.

There exist twenty side chains R, corresponding to the twenty amino acids. Some of these chains are ionizable. Among them, aspartic and glutamic acids are acidic with a pK in the range 3 to 4 while others like lysine, hydroxylysine, hydroxyproline and arginine are basic with a pK larger than 9. The pK will be seen to be the pH of the tissue water at which the number of occupied and free sites are equal at equilibrium.

The N- and C-termini of two successive amino acids adopt a typical relative conformation so as to be able to engage in rather stable peptide bonds. Each chain of the collagen triple helix is formed by the polypeptides resulting from the linear sequence of these bonds. Therefore the charge of the collagen molecule is contributed mainly by the side chains, rather than by the N- and C-termini. At physiological pH, the positive and negative charges equilibrate approximately, so that collagen of articular cartilages can be considered electrically neutral. On the other hand, the collagen molecule becomes positively charged at low pH and negatively charged at high pH.

In summary, only a small part of the amino and carboxyl groups are functionally active in the electro-chemomechanical couplings. Indeed, as described in Section 14.2.3.2,
- the N- and C-termini are mostly engaged in strong peptide bonds;
- only parts of the residues are ionizable;
- out of the ionizable residues, a part of them is turned inside the collage molecule and contributes to its internal stability;
- out of the ionizable residues which are turned to the exterior of the molecule, a part is bathed in the intrabrillar water;
- only those residues bathed by the extrafibrillar water are considered functionally active.

Proteoglycans aggregate along a hyaluronate protein to which 40 to 60 subunits are attached through a specific link protein. The subunit, referred to as proteoglycan molecule (PG), is constituted by a number of glycosaminoglycans (GAGs) which are attached along a core protein. GAGs are formed by disaccharide units, with one uronic acid containing a carboxyl group $CO-O^-$, and a variable number of amino sugars containing a sulfonic group SO_3^-.

The two main types of GAGs in articular cartilages are chondroitin sulfates (CS) and keratan sulfates (KS). Keratan sulfates are shorter than chondroitin sulfates, and contain about half of the number of disaccharide units, Muir [1978]. They possess no uronic acid (carboxyl site), so that their charge is -1, while chondroitin sulfates have a charge equal to -2. The ratio of chondroitin to keratan sulfates decreases significantly with age, Sweet et al. [1977], and is also much larger in osteoarthritic than in normal cartilages, Muir [1978].

17.1.2 Laboratory observations

The presence of a fixed electrical charge is a key element of the mechanical and diffusion properties of cartilages. This charge is modified by acid-base reactions, i.e., binding-release of hydrogen/hydronium ions, and reversible or irreversible binding of some metallic ions.

A definite distinction should be made between the pH=pH_B of the bath the cartilage is in contact with, and the pH=pH_E of the electrolyte that circulates the cartilage. Indeed, the fixed electrical charge creates a sort of membrane around the cartilage, which implies discontinuities in a number of mechanical properties (pressure), electrical properties (electrical potential), and chemical properties (concentrations). The point is discussed in Section 14.6.2 and in Chapter 18. Actually, the situation is even more complex at the nanoscale where the fields are inherently inhomogeneous: the spatial heterogeneity of pH is of course associated with the heterogeneity in dissolved and bound hydrogen ions. Despite the membrane effect, the cartilage pH, say extrafibrillar pH, can be modified by a variation of the pH of the bath and, to a minor extent, by a variation of its ionic strength.

A number of experimental observations guide the development of the model:
- measurements of diffusion properties involve the whole (aggregate) piece of cartilage and access to the individual charges of PGs and collagen is not possible through macroscopic experiments. The aggregate isoelectric point corresponds to the negatively charged carboxyl sites $CO-O^-$ and sulfonic sites SO_3^- to be balanced by amino sites NH_3^+. The isoelectric point for bovine articular cartilages has been reported in the range of pH from about 2.5 to 3, Frank and Grodzinsky [1987], p. 626. While the pHs of the bath and cartilage are in general different, they are equal just at the IEP;
- a key manifestation of the modification of the fixed charge, actually its change of sign, is the reversal of the direction of the electroosmotic flow in the neighborhood of the IEP, Fig. 17.1.1. This reversal corresponds to a change of sign of the streaming potential. It has been observed on bones by Pienkowski and Pollack [1983] due to an increase of ionic strength of NaCl. The phenomenon is attributed by Pienkowski and Pollack to a specific adsorption while it is known that neither sodium cation nor chloride anion bind to PGs in articular cartilages;
- for bovine articular cartilages in contact with a bath at concentration of NaCl larger than 1 M, Frank and Grodzinsky [1987], their Fig. 8, observe that the amplitude of the streaming potential approaches zero;
- the model developed here reproduces the above observations. It also indicates that decreasing the ionic strength acidifies the extrafibrillar water, i.e. decreases pH_E;
- another manifestation of the change of the magnitude and sign of the fixed charge is of mechanical nature. In a uniaxial traction test at given axial strain, but varying pH, the equilibrium stress shows a minimum at

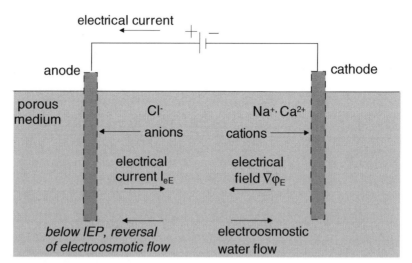

Fig. 17.1.1: By *electromigration*, cations move toward the cathode and anions move toward the anode with velocities depending on individual *ionic mobilities*. *Electroosmosis* implies a water flow toward the cathode, when the acidity is not too high, i.e., pH of tissue water above isoelectric point. For higher acidity, a *reversal* of the electroosmotic flow takes place. The direction of the *total* water flow depends on the boundary conditions, see, e.g., the case of a hydraulically open circuit in Section 16.2.5.

the IEP, Grodzinsky et al. [1981], their Fig. 8. Increase of HCl concentration (in bath) decreases the stress until the extrafibrillar pH reaches a value close to 2.6. Further decrease of pH implies an increase of the stress, corresponding to the fact that the overall cartilage charge becomes positive below the isoelectric point. The compressive axial stress is expected to vary in proportion with the absolute value of the fixed charge, since the latter, whatever its sign, triggers repulsion between cartilage units;

- increasing ionic strength in presence of NaCl and CaCl$_2$ also decreases the equilibrium stress in a monotonic way. In that respect, CaCl$_2$ is more efficient than NaCl, even if the hypertonic values (at large concentrations) are postulated to be identical along Loret and Simões [2005]a;

- a model of reversal electroosmosis subsumes that the electroosmotic coefficient involves the sign and amplitude of the overall electrical charge of the cartilage. Indeed, in the transport model developed here, the electroosmotic coefficient depends linearly on the charge. Other quantitative data on the pH influence on transport properties are scarce. Still, hydraulic conductivity of proteoglycan solutions is reported to be independent of pH over the range 3.2-8.7, Comper and Lyons [1993], their Fig. 3.

17.1.3 Scope

Articular cartilage is considered as a three phase multi-species porous medium. The theoretical framework draws upon Chapters 14 and 16 and will be briefly sketched in Section 17.2. Incorporation of the acid-base reactions into this thermodynamic construction is performed in Section 17.3. Details of the biochemical composition of the collagen, proteoglycans and their glycosaminoglycans are essential ingredients of this model which, although macroscopic in nature, takes account of a number of biochemical properties, Section 17.6. The evolution of the concentrations of the ionizable sites and of the electrical charge can then be displayed as a function of the pH of the extrafibrillar cartilage fluid, Section 17.7.2. The effects of calcium binding on these elements are examined in Section 17.4.

The sites for acid-base reactions and calcium binding which are taken into account in this analysis are displayed in Fig. 17.1.2.

Emphasis is laid first here on the effects of pH on the transport properties. Still, the influence of mechanics is present through the volume fraction of water. The transport model developed in Loret and Simões [2007], and presented in Chapter 16, is extended so as to account for the presence of hydrogen and hydroxyl ions, Section 17.9.

A comprehensive model, including mechanical effects and variation of the volume fraction of water, is the next step. Incorporation of the acid-base reactions into the mechanical part of the constitutive equations is a main task of this chapter. It is performed in Section 17.11. Calcium binding is inserted in this framework as well.

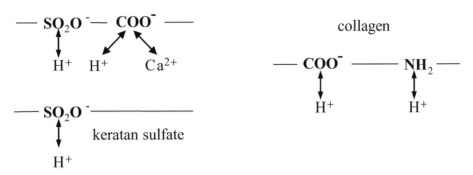

chondroitin sulfate

collagen

keratan sulfate

Fig. 17.1.2: Acid-base reactions at carboxyl and sulfonic sites of GAGs, at carboxyl and amino sites of the side chains of the amino acids of collagen and calcium binding at carboxyl sites of GAGs.

The general framework introduced in Section 17.11 is particularized in Section 17.12, where a prototype of chemo-mechanical model including the influences of ionic strength and pH on the mechanical response (stress, strain) is developed.

In actual laboratory experiments, a cartilage specimen is bathed in a reservoir, whose chemical composition and mechanical conditions are controlled. The ensuing chemical composition of the cartilage in equilibrium with the bath is deduced in Section 17.8. Experiments in which either the ionic strength, or the pH of the bath, or the mechanical conditions (load, stress and strain) are varied can then be mimicked by the model, Sections 17.10 and 17.13.

To simplify the analysis, the species themselves (not the tissue as a whole!) are assumed to be incompressible.

17.2 pH agents embedded in the mixture framework

The definitions of the entities required in this chapter are provided first. Chapter 14 provides a more general exposition, although it does not include information related to pH.

17.2.1 The constituents and phases

Articular cartilage is viewed as a three phase multi-species porous medium. The *solid phase* S contains the collagen fibers denoted by the symbol c. Proteoglycans (PGs) are considered to be bathed in the *extrafibrillar fluid phase* E. The latter contains as well water w, sodium cations Na^+, or/and calcium cations Ca^{2+}, chloride anions Cl^-, hydrogen ions H^+ and hydroxyl ions OH^-,

$$S = \{c\}, \quad E = \{w, PG, Na, Ca, Cl, H, OH\}. \tag{17.2.1}$$

The *intrafibrillar fluid phase* I contains the same species as the *extrafibrillar fluid phase* but proteoglycans. Species of the intrafibrillar fluid phase move with the velocity of the solid phase. They have no access to the surroundings and must first *transfer* to the extrafibrillar phase. Only the extrafibrillar phase communicates with the surroundings. The phase segregation and hierarchical structure adopted are motivated in Chapter 14.

On the other hand, water and ions in the extrafibrillar phase are endowed with their own velocities so as to allow water to flow through the solid skeleton and ions to diffuse in this phase and satisfy their own balance of momentum. Although bathed in the extrafibrillar water, proteoglycans move with the solid phase.

The amount of hydrogen ions is measured via the base-10 logarithm of their concentration expressed in mole/liter, namely pH $= -\log_{10} c_H$, equivalent to $c_H = 10^{-pH}$. For pH close to 7, collagen can be considered electrically neutral, Li and Katz [1976]. However, at nonphysiological pH, collagen becomes charged. Topological considerations seem to indicate that electroneutrality should be ensured separately for the intra- and extrafibrillar compartments. Since collagen is in contact with both compartments, a partition of its charge should be defined.

17.2.2 Macroscopic descriptors of the geometry and mass of the mixture

Various macroscopic measures of mass and volume are used to formulate the constitutive equations. They are defined below.

Let the initial volume of the porous medium be V_0 and let $V = V(t)$ be its current volume. The current mole number, volume and mass of the species l of phase K are denoted by N_{lK}, V_{lK} and M_{lK} respectively. Various additional entities are attached to species:

- some are intrinsic like the intrinsic density ρ_l, the molar volume $\widehat{v}_l$ and molar mass $\widehat{m}_l$ linked by the relation $\widehat{m}_l = \rho_l \widehat{v}_l$;
- some refer to the current volume, like the *volume fraction* $n^{lK} = V_{lK}/V$;
- some refer to the initial volume like the *mole content* $\mathcal{N}_{lK} = N_{lK}/V_0$, the *volume content* $v^{lK} = V_{lK}/V_0 = n^{lK} V/V_0$, and the *mass content* $m^{lK} = M_{lK}/V_0 = \rho_l v^{lK}$.

The corresponding entities associated with the phase K are defined by algebraic summation of individual contributions, e.g. the current volume V_K and mass M_K, and the volume fraction $n^K = V_K/V$. Volume fractions satisfy the compatibility relation $\sum_K n^K = 1$.

Other entities live in their phase, e.g., the molar fractions and the concentrations. The *molar fraction* x_{lK} of the species l in phase K is defined by the ratio of the mole number N_{lK} of that species over the total number of moles N_K within the phase, $x_{lK} = N_{lK}/N_K$. In each phase the molar fractions satisfy the compatibility relation $\sum_{l \in K} x_{lK} = 1$. Since $N_{lK}/V_0 = m^{lK}/\widehat{m}_l$, the molar fractions can also be expressed in terms of mass contents.

The *concentration* of an extrafibrillar species is equal to its number of moles referred to the volume of the extrafibrillar phase,

$$c_{lE} = \frac{N_{lE}}{V_E} = \frac{x_{lE}}{\widehat{v}_E}, \quad l \in E, \tag{17.2.2}$$

with $\widehat{v}_E = \sum_{l \in E} x_{lE} \widehat{v}_l \simeq \widehat{v}_w$ the molar volume of the extrafibrillar fluid phase.

Collagen and proteoglycans are macromolecules with a large molar mass, 0.285×10^6 gm for collagen and 2×10^6 gm for PGs. The molar fractions $x_l = N_l/N_E$ and concentrations $c_l = N_l/V_E$ of proteoglycans ($l = $ PG) and collagen ($l = $ c) are thus quite small with respect to the other species of the extrafibrillar phase. On the other hand, the valence ζ_{PG} of proteoglycans is large at neutral pH, while the extrafibrillar valence of collagen ζ_c is large at nonphysiological pH. Thus the *effective* molar fraction y_l and *effective* concentration $e_l = y_l/\widehat{v}_w$,

$$\underbrace{y_l = \zeta_l x_l}_{\substack{\text{effective} \\ \text{molar fraction}}} \quad , \quad \underbrace{e_l = \zeta_l c_l}_{\substack{\text{effective} \\ \text{concentration}}} \quad , \quad l = \text{PG}, \text{c} , \tag{17.2.3}$$

are key parameters of the biochemical and biomechanical responses of cartilages.

The effective molar fraction y_e and effective concentration $e_e = y_e/\widehat{v}_w$ of the extrafibrillar charge including PGs and collagen are similarly defined in terms of valences and mole numbers,

$$y_e = \zeta_{PG} \frac{N_{PG}}{N_E} + \zeta_c \frac{N_c}{N_E}, \quad e_e = \zeta_{PG} \frac{N_{PG}}{V_E} + \zeta_c \frac{N_c}{V_E} . \tag{17.2.4}$$

Some relations in absence of exchange with the intrafibrillar compartment:

The change of volume of the porous medium modifies the concentrations of species. If the species are incompressible, and if only the extrafibrillar volume varies, then the deformation of the extrafibrillar phase is implied by the deformation gradient $\mathbf{F}$ of the porous medium. The index 0 denoting the reference configuration at which $\mathbf{F} = \mathbf{I}$,

$$\frac{V}{V_0} = \det \mathbf{F}, \quad \frac{V_E}{V_{E0}} = 1 + \frac{1}{n_0^E} (\det \mathbf{F} - 1) , \tag{17.2.5}$$

and

$$n^E = \frac{n_0^E}{\det \mathbf{F}} + 1 - \frac{1}{\det \mathbf{F}}, \quad c_{lE} = \frac{N_{lE}}{V_{E0}} \frac{n_0^E}{n_0^E + \det \mathbf{F} - 1} . \tag{17.2.6}$$

In a small strain context where $\mathbf{F} = \mathbf{I} + \epsilon$ to first order, the volume change can be expressed in terms of the infinitesimal strain ϵ. With help of (2.2.34),

$$\frac{V}{V_0} = 1 + \text{tr}\,\epsilon, \quad \frac{V_E}{V_{E0}} = 1 + \frac{\text{tr}\,\epsilon}{n_0^E} , \tag{17.2.7}$$

and

$$n^E = n_0^E + (1 - n_0^E)\,\text{tr}\,\epsilon, \quad c_{lE} = \frac{N_{lE}}{V_{E0}} \times (1 - \frac{\text{tr}\,\epsilon}{n_0^E}) . \tag{17.2.8}$$

17.3 Acid-base reactions in a thermodynamic framework

The solid skeleton of cartilage is constituted by proteoglycans, collagen and non collagenous proteins circulated by a fluid in which sodium, calcium and chloride ions diffuse. All species of the extrafibrillar phase undergo the same electrical field and an electroneutrality condition is enforced in this phase. An alternative would be to consider the sites on PGs and collagen to be endowed with a distinct electrical potential, and to view the charged solid as behaving like a capacitor, thus providing a link between the difference of the potentials and the electrical charge. This approach has been applied to clays, in a geochemical perspective by Petrangeli Papini and Majone [2002], and in a geomechanical context by Gajo and Loret [2007]. An issue with this approach is that capacitances are not easily accessible to measurements and they have to be obtained by back-analysis. The concept is briefly introduced in Exercise 13.3.

The acid-base reactions are considered to be **reversible**. There are some experimental indications that sustain this assumption. In a uniaxial tension test at given axial strain, subsequent to acidification through hydrochloric acid HCl, Grodzinsky et al. [1981], p. 225, increase the concentration of sodium hydroxyde NaOH : they observe that the stress-pH curve does not show hysteresis.

The chemical affinities associated with the four acid-base reactions are defined in turn.

17.3.1 Electrochemical potentials

The incremental works done per unit initial volume V_0 by the stress $\mathbf{T}$ in the incremental strain $\delta \mathbf{E}$ and by the electrochemical potentials $g^{\text{ec}}_{l\text{K}}$ during the addition/subtraction of the mole contents $\delta \mathcal{N}_{l\text{K}}$ of the species l to/from the phases K sum to

$$\delta \underline{\mathcal{W}} = \mathbf{T} : \delta \mathbf{E} + \sum_{\text{K},\, l \in \text{K}} g^{\text{ec}}_{l\text{K}}\, \delta \mathcal{N}_{l\text{K}}\,. \tag{17.3.1}$$

Here $(\mathbf{T}, \mathbf{E})$ is a work-conjugated stress-strain pair. With $\mathbf{F}$ the deformation gradient, $\mathbf{E}$ is typically the Green strain $\frac{1}{2}(\mathbf{F}^{\text{T}} \cdot \mathbf{F} - \mathbf{I})$, and $\mathbf{T}$ the second Piola-Kirchhoff stress with respect to the reference configuration linked to the Cauchy stress $\boldsymbol{\sigma}$ by the relation $\mathbf{T} = \det \mathbf{F}\, \mathbf{F}^{-1} \cdot \boldsymbol{\sigma} \cdot \mathbf{F}^{-\text{T}}$. The electrochemical potentials $g^{\text{ec}}_{l\text{K}}$ [unit : kg×mole×m^2/s^2/mole] are mole based while the mole contents per unit initial volume of the porous medium $\mathcal{N}_{l\text{K}}$'s are measured in mole/m^3.

For a neutral species, the chemical potential $g_{l\text{K}}$ of a species l identifies a pressure contribution introduced by the intrinsic pressure $p_{l\text{K}}$ and the molar volume $\widehat{v}_l$ and a chemical contribution which accounts for the molar fraction $x_{l\text{K}}$. For a charged species in presence of the electrical potential ϕ_{K} [unit V=kg×m^2/s^3/A], the electrochemical potential $g^{\text{ec}}_{l\text{K}}$ involves in addition an electrical contribution. For incompressible species,

$$g^{\text{ec}}_{l\text{K}} = \widehat{v}_l\, p_{l\text{K}} + RT \operatorname{Ln} x_{l\text{K}} + \zeta_l\, \text{F}\, \phi_{\text{K}}\,, \quad l \in \text{K}\,. \tag{17.3.2}$$

In this formula, $R = 8.31451\,\text{J/mole/K}$ is the universal gas constant, T the absolute temperature, and F $= 96\,485\,\text{coulomb/mole}$ is Faraday's equivalent charge. The ζ's are the valences. In the extrafibrillar phase, chemo-mechanical couplings imply ions and water to be endowed with their own intrinsic pressure $p_{l\text{E}}$. The enthalpy of formation up to the reference configuration, which generates an equilibrium constant, is considered to be contained in $p_{l\text{E}}$, Loret and Simões [2005]a.

17.3.2 Spontaneous dissociation of water

Water dissociates spontaneously into hydrogen and hydroxyl ions,

$$\text{H}_2\text{O} \quad \rightleftharpoons \text{H}^+ + \text{OH}^-\,. \tag{17.3.3}$$

Equilibrium,

$$\frac{c_{\text{HE}} \times c_{\text{OHE}}}{c_{\text{wE}}} = 2 \times 10^{-16}\,\text{mole/liter}\,, \tag{17.3.4}$$

is usually recast in the format, characterized by the dissociation constant $K^{\text{c}}_{\text{dis}}$ at 25 °C,

$$c_{\text{HE}} \times c_{\text{OHE}} = K^{\text{c}}_{\text{dis}} \simeq 1 \times 10^{-14}\,(\text{mole/liter})^2\,, \tag{17.3.5}$$

with $c_{\text{wE}} = 55.55\,\text{mole/liter}$, the concentrations being expressed in mole/liter. With $\widehat{v}_{\text{w}} = 0.018\,\text{liter/mole}$ the molar volume of water, another equivalent form expresses in terms of molar fractions,

$$x_{\text{HE}} \times x_{\text{OHE}} = K^{\text{x}}_{\text{dis}} = \widehat{v}^2_{\text{w}}\, K^{\text{c}}_{\text{dis}} = 4 \times 10^{-18}\,. \tag{17.3.6}$$

17.3.3   The chemical affinities associated with the acid-base reactions

Let us consider first the reactions at the weakly acid carboxyl sites of either collagen or PGs, and at the strongly acid sulfonic sites of PGs,

$$S = CO, SOO: \qquad \underbrace{S-OH}_{\text{occupied site}} \quad \rightleftharpoons \underbrace{S-O^-}_{\text{free site}} + \underbrace{H^+}_{\text{fluid phase}} . \qquad (17.3.7)$$

During this reaction, the variations of the numbers of moles of the sites and species involved are linked,

$$- \delta N_{\text{SOH}} = \delta N_{\text{SO}} = \delta N_{\text{HE}} . \qquad (17.3.8)$$

The equilibrium constant of the reaction is defined by a pK or, equivalently, by an enthalpy of formation $\mathcal{G}^0 = RT \ln (10^{\text{pK}}/\widehat{v}_{\text{w}})$. Consequently, the work done in the volume V_0 during this reaction,

$$g_{\text{SOH}} \, \delta N_{\text{SOH}} + g_{\text{SO}}^{\text{ec}} \, \delta N_{\text{SO}} + g_{\text{HE}}^{\text{ec}} \, \delta N_{\text{HE}}$$

$$= (-g_{\text{SOH}} + g_{\text{SO}}^{\text{ec}} + g_{\text{HE}}^{\text{ec}}) \, \delta N_{\text{HE}}$$

$$= (-RT \ln x_{\text{SOH}} + RT \ln x_{\text{SO}} - F \, \phi_{\text{E}} + RT \ln x_{\text{HE}} + F \, \phi_{\text{E}} + \mathcal{G}^0) \, \delta N_{\text{HE}} \qquad (17.3.9)$$

$$= \mathcal{G} \, \delta N_{\text{HE}} ,$$

can be phrased in terms of the chemical affinity $\mathcal{G}$ associated with the reaction,

$$\mathcal{G} = RT \ln \frac{c_{\text{SO}}}{c_{\text{SOH}}} \frac{c_{\text{HE}}}{10^{-\text{pK}}} . \qquad (17.3.10)$$

Chemical equilibrium is defined by a vanishing affinity. Since $c_{\text{HE}} = 10^{-\text{pH}_{\text{E}}}$, the pK is seen to be the pH_{E} at which the number of occupied and free sites are equal at equilibrium.

Corresponding relations can be deduced for the reactions at the amino sites of collagen,

$$S = NH_2: \qquad \underbrace{S-H^+}_{\text{occupied site}} \quad \rightleftharpoons \underbrace{S}_{\text{free site}} + \underbrace{H^+}_{\text{fluid phase}} . \qquad (17.3.11)$$

During this reaction, the variations of the numbers of moles of the sites and species involved are linked,

$$- \delta N_{\text{SH}} = \delta N_{\text{S}} = \delta N_{\text{HE}} . \qquad (17.3.12)$$

Consequently, the work done in the volume V_0 during this reaction,

$$g_{\text{SH}}^{\text{ec}} \, \delta N_{\text{SH}} + g_{\text{S}} \, \delta N_{\text{S}} + g_{\text{HE}}^{\text{ec}} \, \delta N_{\text{HE}}$$

$$= (-g_{\text{SH}}^{\text{ec}} + g_{\text{S}} + g_{\text{HE}}^{\text{ec}}) \, \delta N_{\text{HE}}$$

$$= (-RT \ln x_{\text{SH}} - F \, \phi_{\text{E}} + RT \ln x_{\text{S}} + RT \ln x_{\text{HE}} + F \, \phi_{\text{E}} + \mathcal{G}^0) \, \delta N_{\text{HE}} \qquad (17.3.13)$$

$$= \mathcal{G} \, \delta N_{\text{HE}} ,$$

is phrased in terms of the chemical affinity $\mathcal{G}$ associated with the reaction,

$$\mathcal{G} = RT \ln \frac{c_{\text{S}}}{c_{\text{SH}}} \frac{c_{\text{HE}}}{10^{-\text{pK}}} . \qquad (17.3.14)$$

For the side chains of the acid amino acids, the pKs of their sulfonic groups are about 2, and the pKs of their carboxyl groups are in the range 3 to 4, as indicated in Fig. 17.3.1. As for the amino groups of the side chains of the basic amino acids, their pKs are larger than 9. Therefore, collagen is expected to become positively charged at pH smaller than 4, and negatively charged at pH larger than 8. Since the pK's of the sulfonic groups are quite low, and

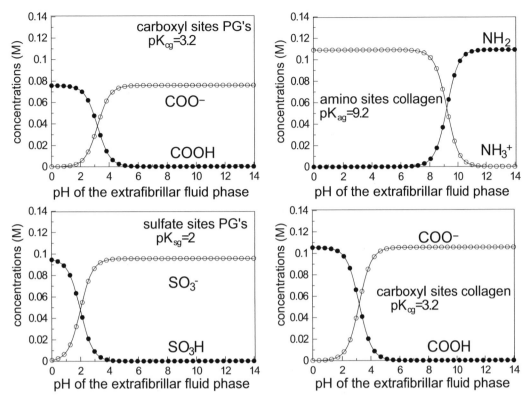

Fig. 17.3.1: Concentrations of carboxyl, sulfonic and amino sites, as a function of the pH of the fluid of the cartilage. The data are provided in the text and the volume fraction of the fluid is $n^E = 0.8$. Collagen involves carboxyl and amino sites. GAGs involve carboxyl and sulfonic sites in equal numbers for chondroitin sulfates but only sulfonic sites for keratan sulfates. Carboxyl sites of hyaluronate are not accounted for.

fall out of the realm of usual laboratory experiments, the sulfonic groups are often not considered to get protonated, e.g., Wachtel and Maroudas [1998].

In order to compact the notation, we number the four acid-base reactions,

$$R_j \;\rightleftharpoons\; P_j + H_{(j)}, \quad j \in [1,4], \tag{17.3.15}$$

with the following conventions for the reactants R, products P, participating hydrogen ions, and pKs:

S_1 carboxyl sites on PGs $\qquad R_1 = CO\text{--}OH \qquad P_1 = CO\text{--}O^- \qquad HE_{(1)} = H_{(1)}^+ \qquad pK_1 = pK_{cg}$

S_2 sulfonic sites on PGs $\qquad R_2 = SOO\text{--}OH \qquad P_2 = SOO\text{--}O^- \qquad HE_{(2)} = H_{(2)}^+ \qquad pK_2 = pK_{sg}$

S_3 amino sites on collagen $\qquad R_3 = NH_2\text{--}H^+ \qquad P_3 = NH_2 \qquad HE_{(3)} = H_{(3)}^+ \qquad pK_3 = pK_{ag}$

S_4 carboxyl sites on collagen $\qquad R_4 = CO\text{--}OH \qquad P_4 = CO\text{--}O^- \qquad HE_{(4)} = H_{(4)}^+ \qquad pK_4 = pK_{cg}$

For each reaction, the variations in the number of moles of reactants and products obey the relations,

$$\delta N_{HE_{(j)}} = \delta N_{P_j} = -\delta N_{R_j}, \quad j \in [1,4]. \tag{17.3.16}$$

The chemical affinities associated with the reactions,

$$\mathcal{G}_j = RT \, \mathrm{Ln} \, \frac{c_{P_j}}{c_{R_j}} \frac{c_{HE}}{10^{-pK_j}}, \quad j \in [1,4], \tag{17.3.17}$$

are defined in terms of an enthalpy of formation $\mathcal{G}_j^0 = RT \, \mathrm{Ln} \left(10^{pK_j}/\widehat{v}_w\right)$ or, equivalently, of a pK=pK$_j$. Again, for each reaction j, since $c_{HE} = 10^{-pH_E}$, the pK$_j$ is seen to be the pH$_E$ at which the numbers of occupied and free sites are equal at equilibrium, corresponding to $\mathcal{G}_j = 0$.

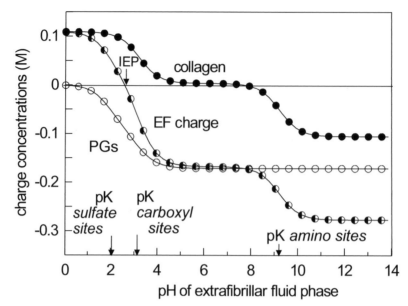

Fig. 17.3.2: Contributions of proteoglycans and collagen to the extrafibrillar charge of cartilage as a function of the pH of the fluid of the cartilage, for a volume fraction $n^E = 0.8$. Since it is due to carboxyl and sulfonic sites, the charge of proteoglycans is always negative, and it tends to vanish at very low pH. On the other hand, the amino sites of collagen counteract its carboxyl sites in a large range of pH, say from 4 to 8, and they take over so as to change the sign of the extrafibrillar charge of the tissue at pHs below the isoelectric point (IEP) which, for most articular cartilages, ranges from 2.5 to 3.5. Reprinted with permission from Loret and Simões [2010]a.

The four chemical reactions (17.3.15), numbered $j \in [1, 4]$, are summarized through the relations,

$$c_{S_j} = c_{R_j} + c_{P_j},$$

$$\frac{c_{P_j} \times c_{HE}}{c_{R_j}} = 10^{-pK_j}, \quad \alpha_j \equiv \frac{c_{R_j}}{c_{P_j}} = 10^{pK_j - pH_E}, \quad (17.3.18)$$

$$\frac{c_{R_j}}{c_{S_j}} = \frac{\alpha_j}{1 + \alpha_j}, \quad \frac{c_{P_j}}{c_{S_j}} = \frac{1}{1 + \alpha_j},$$

where c_P, c_R and c_S are the concentrations of the free sites, occupied sites and total sites respectively.

The contributions of proteoglycans and collagen to the extrafibrillar charge of cartilage as a function of the pH of the fluid of the cartilage are displayed in Fig. 17.3.2 for a given volume fraction.

17.4   Calcium binding in a thermodynamic framework

In the vicinity of fixed electrical charges, the concentrations of co-ions and counterions are quite non uniform. In fact, several models have tried to describe these spatial variations as well as the constraints imposed on the velocities of the ions. In the diffuse layer model and in the triple layer model, Sposito [1984], the mobility of ions in the inner layers is strongly hindered, and a layer ensures the progressive transition between the inner complexes and the bulk solution where ions can diffusive freely.

In articular cartilages, at neutral pH, the fixed charge is negative and the main co-ion is sodium cation. Its presence shields the electrical repulsion induced by charges located on adjacent GAGs. Neither the sodium cations nor the chloride anions *bind* the fixed charge of articular cartilages, Li and Katz [1976], Grodzinsky et al. [1981], p. 230. In other words, they do not change this fixed charge. The same does not apply to other tissues neither to calcium cations.

17.4.1 Binding of metallic ions

Binding of certain metallic ions modifies the fixed charge. Hence, the presence of these ions has two distinct effects: ionic shielding by the free ions and alteration of the fixed charge due to the bound ions.

The effect of ionic strength for cornea is explainable by calling for the binding of Cl^- on some unidentified ligands L which seem to be associated with collagen fibrils, Huang and Meek [1999]. Higher ionic strength increases the negative fixed charge:

$$\underbrace{L-Cl^-}_{\text{occupied site}} \quad \rightleftharpoons \quad \underbrace{L}_{\text{free site}} \quad + \quad \underbrace{Cl^-}_{\substack{\text{extrafibrillar} \\ \text{fluid phase}}} \quad . \tag{17.4.1}$$

On the other hand, cation specific adsorption on certain sites S leads to decrease of the negative fixed charge, e.g., for Ca^{2+},

$$\underbrace{S-OCa^+}_{\text{occupied site}} \quad \rightleftharpoons \quad \underbrace{S-O^-}_{\text{free site}} \quad + \quad \underbrace{Ca^{2+}}_{\substack{\text{extrafibrillar} \\ \text{fluid phase}}} \quad . \tag{17.4.2}$$

In the context of the delivery of ionic drugs, calcium binding has been studied in a theoretical model of pH-dependent membranes, Ramirez et al. [2003]. There, calcium binding inhibits the conductance of nanopores by reducing the fixed charge through a sort of "electrostatic blocking".

To characterize the binding reaction (17.4.2) at steady state, its equilibrium constant is required. It is also necessary to assess its reversible or irreversible nature. Here, we consider reversible binding. On the other hand, calcium binding to the sulfonic sites of the amino sugars of the disaccharide units is highly irreversible, Werner and Gründer [1999]. This mechanism is probably linked to bone formation, although whether proteoglycans are inhibitors or initiators of hydroxyapatite does not seem to be resolved, Hunter [1991]. The reasons why calcium ions can form a bond with the carboxyl or sulfonic sites of PGs while the hydration shell of sodium cations prevents this bond to develop are not elucidated either. A detailed analysis of the conditions of reversible binding to the carboxyl sites and irreversible binding to the sulfonic sites, respectively of the uronic acids and amino sugars of PGs, is a subject of interest for biochemists.

The above comments were concerning binding to PGs. Theis and Jacoby [1943] consider calcium binding onto collagen: their experiments show that calcium chloride denaturates collagen, denaturation being measured by shrinkage and contraction.

17.4.2 Calcium binding to carboxyl sites of PGs

Calcium ions compete with hydrogen ions to bind reversibly onto the carboxyl sites of PGs according to the reaction,

$$\underbrace{CO-OCa^+}_{\text{occupied PGs site}} \quad \rightleftharpoons \quad \underbrace{CO-O^-}_{\text{free PGs site}} \quad + \quad \underbrace{Ca^{2+}}_{\text{fluid phase}} \quad . \tag{17.4.3}$$

The reaction is numbered $1'$, and according to the conventions above

$$S_{1'} \text{ carboxyl sites on PGs} \quad R_{1'} = CO-O-Ca^+ \quad P_{1'} = CO-O^- \quad CaE_{(1')} = Ca^{2+}_{(1')} \quad pK_{1'} = pKCa$$

The variations in the number of moles of reactant and product obey the relation,

$$\delta N_{CaE_{(1')}} = \delta N_{P_{1'}} = -\delta N_{R_{1'}} \ . \tag{17.4.4}$$

The chemical affinity associated with the reaction is (note $S_{1'} = S_1$, $P_{1'} = P_1$),

$$\mathcal{G}_{1'} = RT \operatorname{Ln} \frac{c_{P_{1'}}}{c_{R_{1'}}} \frac{c_{CaE}}{10^{-pK_{1'}}} \ . \tag{17.4.5}$$

The equilibrium constant $pK_{1'} = pKCa$ is not known. Experiments by MacGregor and Bowness [1970] on aggregated and disaggregated proteoglycans at pH=7 show a variation of the pK with ionic strength in the range $1.7 - 2.5$. Absence of calcium binding corresponds to $pKCa = -\infty$. A parameter analysis will be performed in the range $1 - 3$, following Ramirez et al. [2002]. Note that this range falls below the pK of the carboxyl sites, $pK_{cg} = pK_1$. Therefore calcium ions compete with hydrogen ions to bind on the carboxyl sites of PGs. Calcium binding is weak at low pH and develops at pH higher than pK_{cg}, Fig. 17.4.1. Altogether, calcium binding increases the algebraic value of the overall fixed charge. The intensity of the phenomenon increases with the binding constant pKCa.

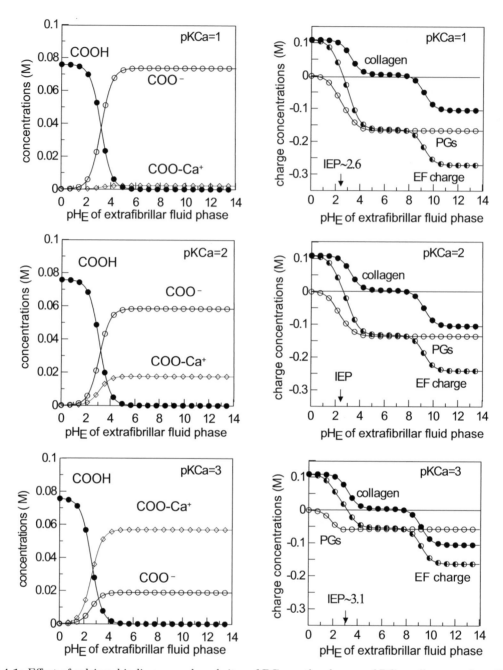

Fig. 17.4.1: Effect of calcium binding on carboxyl sites of PGs on the charges of PGs, collagen and on the extrafibrillar charge of cartilage. Concentrations as a function of the pH of the extrafibrillar fluid for a volume fraction $n^E = 0.8$, several binding constants pKCa and with $c_{CaB} = 3$ mM. Calcium ions compete with hydrogen ions to bind on the carboxyl sites of PGs. Calcium binding on the carboxyl sites of PGs turns the charge of PGs less negative and, consequently, increases significantly the algebraic value of the extrafibrillar charge and the IEP.

17.5 Variation of the fixed electrical charge with pH

The concentrations at the sites for acid-base reactions and calcium binding may be gathered in the following generalized format,

$$\underbrace{c_{S_j}}_{\text{sites}} = \underbrace{c_{R_j}}_{\substack{\text{sites occupied} \\ \text{by hydrogen ions}}} + \underbrace{c_{R_{j'}}}_{\substack{\text{sites occupied} \\ \text{by calcium ions}}} + \underbrace{c_{P_j}}_{\text{free sites}}, \quad j \in [1,4]. \tag{17.5.1}$$

The concentrations of hydrogen ions and calcium ions [unit : mole/liter] are equivalently expressed in terms of a generalized pH,

$$c_{\text{HE}} = 10^{-\text{pH}_{\text{E}}}, \quad \text{pH}_{\text{E}} = -\log_{10} c_{\text{HE}}, \quad c_{\text{CaE}} = 10^{-\text{pCa}_{\text{E}}}, \quad \text{pCa}_{\text{E}} = -\log_{10} c_{\text{CaE}}. \tag{17.5.2}$$

On the other hand, when the concentrations of the generic ion k in phase $\text{K} = \text{E}, \text{B}$ are measured in mole/m^3, these relations should be understood as

$$\frac{c_{k\text{K}}}{10^3} = 10^{-p_{k\text{K}}}, \quad p_{k\text{K}} = -\log_{10} \frac{c_{k\text{K}}}{10^3}. \tag{17.5.3}$$

Let

$$\alpha_j = \begin{cases} 10^{\text{pK}_j - \text{pH}_{\text{E}}}, & j \in [1,4]; \\ 10^{\text{pK}_j - \text{pCa}_{\text{E}}}, & j = 1'; \\ 0, & j \in [2', 4']. \end{cases} \tag{17.5.4}$$

Then the chemical affinities adopt the format,

$$\mathcal{G}_j = RT \, \text{Ln} \, \frac{c_{P_j}}{c_{R_j}} \, \alpha_j, \quad j \in [1,4] \cup \{1'\}. \tag{17.5.5}$$

At equilibrium of the acid-base reactions $\mathcal{G}_j = 0$, $j \in [1,4]$, and of calcium binding $\mathcal{G}_{1'} = 0$, the relative concentrations of the free sites and occupied sites, c_{P_j}/c_{S_j}, c_{R_j}/c_{S_j} and $c_{R_{j'}}/c_{S_j}$ respectively, express in terms of the concentrations of hydrogen and calcium ions in the format, for $j \in [1,4]$,

$$\frac{c_{P_j}}{c_{S_j}} = \frac{1}{1 + \alpha_j + \alpha_{j'}}, \quad \frac{c_{R_j}}{c_{S_j}} = \frac{\alpha_j}{1 + \alpha_j + \alpha_{j'}}, \quad \frac{c_{R_{j'}}}{c_{S_j}} = \frac{\alpha_{j'}}{1 + \alpha_j + \alpha_{j'}}. \tag{17.5.6}$$

The concentration of the electrical charge,

$$\begin{aligned} e_{\text{e}} &= -c_{P_1} - c_{P_2} + c_{R_3} - c_{P_4} + c_{R_{1'}} \\ &= -\sum_{j \in [1,4]} \frac{1 - \alpha_{j'}}{1 + \alpha_j + \alpha_{j'}} \, c_{S_j} + c_{S_3}, \end{aligned} \tag{17.5.7}$$

thus depends on the concentrations of hydrogen and calcium ions via the α's. The electroneutrality of the extrafibrillar fluid phase can be phrased in terms of this concentration,

$$c_{\text{NaE}} + 2 \, c_{\text{CaE}} + e_{\text{e}} + c_{\text{HE}} - c_{\text{OHE}} - c_{\text{ClE}} = 0. \tag{17.5.8}$$

Once the concentration of electrical charge and the volume are known, the mole content of electrical charge $\mathcal{N}_{\text{e}} = V_{\text{E}}/V_0 \times e_{\text{e}}$ deduces readily.

17.6 Biochemical composition of articular cartilages

A number of data defining the species (PGs, collagen, water) that compose the tissue as well as their proportions are required. They are listed and discussed below.

17.6.1   Molar masses and volumes

- molar volumes: $\widehat{v}_{\rm w} = 18\,{\rm cm}^3$, $\widehat{v}_{\rm c} = 2 \times 10^5\,{\rm cm}^3$, $\widehat{v}_{\rm PG} = 1.1 \times 10^6\,{\rm cm}^3$;
- molar masses: $\widehat{m}_{\rm w} = 18\,{\rm gm}$, $\widehat{m}_{\rm c} = 285 \times 10^3\,{\rm gm}$, $\widehat{m}_{\rm PG} = 2000 \times 10^3\,{\rm gm}$;
- molar mass of chondroitin sulfates (CS) and keratan sulfates (KS), Muir [1978]:

$$\widehat{m}_{\rm CS} = 18 \times 10^3{\rm gm} \in [15,\,20] \times 10^3{\rm gm}; \quad \widehat{m}_{\rm KS} = 9 \times 10^3{\rm gm} \in [5,\,10] \times 10^3{\rm gm}. \tag{17.6.1}$$

The above relations indicate a range of values estimated from the literature and the values which are used in the subsequent calculations.

17.6.2   pKs of the acid-base reactions

The exact list and proportion of amino acids are not introduced in the analysis. The pKs of the acidic and basic amino acids of the side chains are not identical but they cluster around two values, Voet and Voet [1998], p. 58. As a further approximation, the pKs of the carboxyl groups of the PGs and collagen have been taken identical. The pKs of the acid-base reactions used in the calculations assume the following values,

$$\underbrace{{\rm pK}_{\rm sg} = 2.0,}_{\text{sulfonic sites}} \quad \underbrace{{\rm pK}_{\rm cg} = 3.2,}_{\text{carboxyl sites}} \quad \underbrace{{\rm pK}_{\rm ag} = 9.2}_{\text{amino sites}}. \tag{17.6.2}$$

17.6.3   Composition of GAGs

The composition of GAGs requires the number of moles $N_{\rm du}^{\rm CS}$ of disaccharide units per mole of chondroitin sulfate and the number of moles of chondroitin sulfates $N_{\rm CS}$, and similarly for keratan sulfates. From these data, the numbers of moles of carboxyl sites $N_{\rm cg}^{\rm PG}$ and of sulfonic sites $N_{\rm sg}^{\rm PG}$ in $N_{\rm PG}$ moles of PGs, and the valence $\zeta_{\rm PG}$ of PGs can be deduced:

$$N_{\rm cg}^{\rm PG} = N_{\rm du}^{\rm CS} \times N_{\rm CS};$$

$$N_{\rm sg}^{\rm PG} = N_{\rm du}^{\rm CS} \times N_{\rm CS} + N_{\rm du}^{\rm KS} \times N_{\rm KS}; \tag{17.6.3}$$

$$\zeta_{\rm PG} = (-2) \times N_{\rm du}^{\rm CS} \times \frac{N_{\rm CS}}{N_{\rm PG}} + (-1) \times N_{\rm du}^{\rm KS} \times \frac{N_{\rm KS}}{N_{\rm PG}}.$$

The data below are extracted from Muir [1978], pp. 68-69, Muir [1983], p. 613,

$$N_{\rm du}^{\rm CS} = 25 \in [25,\,30]; \quad \frac{N_{\rm CS}}{N_{\rm PG}} = 80 \in [50,\,100];$$

$$N_{\rm du}^{\rm KS} = 13 \in [10,\,15]; \quad \frac{N_{\rm KS}}{N_{\rm PG}} = 40 \in [25,\,50], \tag{17.6.4}$$

yielding a valence $\zeta_{\rm PG} = -4520$. For young and adult articular cartilages, Urban et al. [1979] find a fixed charge of PGs of 1.5 to 2 meq/gm of dry weight, which amounts to 3000 to 4000 mole/mole of PGs.

In place of the above mole numbers, an alternative consists in providing
- the relative molar or mass contribution of CS and KS in GAGs:

$$\frac{N_{\rm CS}}{N_{\rm KS}}, \quad {\rm or} \quad \frac{M_{\rm CS}}{M_{\rm KS}} = \frac{N_{\rm CS} \times \widehat{m}_{\rm CS}}{N_{\rm KS} \times \widehat{m}_{\rm KS}}. \tag{17.6.5}$$

The disaccharide units of chondroitin sulfates consist of glucuronic acid and N-acetyl-galactosamine, while the disaccharide units of keratan sulfates consist of galactose and N-acetylglucosamine. Experimentally, the ratio of galactosamine over glucosamine is used to obtain the relative content of chondroitin and keratan sulfates. The molar ratio of chondroitin to keratan sulfates is high in immature articular cartilages but smaller in adult cartilages, Sweet et al. [1977]. This ratio is particularly increased in osteoarthritic cartilages. Urban et al. [1979] deduce a value of about 1 for adult human knee and hip articular cartilages and 10 for young human hip. Sweet et al. [1977] obtain a ratio of 3 for femoral condyles of 10 to 12 weeks old calves. The ratio 2 taken here corresponds to young adult cartilage;

- the relative mass of proteins and GAGs in PGs, Sweet et al. [1977], Muir [1978]:

$$\frac{M_{\mathrm{GAG}}}{M_{\mathrm{PG}}} = 0.90 \in [0.85, 0.92];$$ (17.6.6)

- the number of disaccharide units per chondroitin sulfate and keratan sulfate.

The resulting valence of PGs has the format,

$$\zeta_{\mathrm{PG}} = \widehat{m}_{\mathrm{PG}} \frac{(-2)\, N_{\mathrm{du}}^{\mathrm{CS}} \times N_{\mathrm{CS}} + (-1)\, N_{\mathrm{du}}^{\mathrm{KS}} \times N_{\mathrm{KS}}}{\widehat{m}_{\mathrm{CS}} \times N_{\mathrm{CS}} + \widehat{m}_{\mathrm{KS}} \times N_{\mathrm{KS}}} \frac{M_{\mathrm{GAG}}}{M_{\mathrm{PG}}} .$$ (17.6.7)

17.6.4 Composition and partition of the charge of collagen

The charge of collagen rests on the amino acids of the side chains. For mature collagen of type II, the numbers of moles of charged amino groups $N_{\mathrm{ag}}^{(c)}$ and charged carboxyl groups $N_{\mathrm{cg}}^{(c)}$ in N_{c} moles of collagen are, according to Wachtel and Maroudas [1998], equal to

$$N_{\mathrm{ag}}^{(c)} = 273\, N_{\mathrm{c}} ; \quad N_{\mathrm{cg}}^{(c)} = 264\, N_{\mathrm{c}} .$$ (17.6.8)

At neutral pH, these authors measure a maximum charge of collagen of about 0.9 meq per gram of collagen, or $0.9 \times 10^{-3} \times 285 \times 10^{3} \sim 270$ basic groups per mole of collagen, which is close to the theoretical maximum of 273 amino groups for type II collagen.

The diffusive properties of cartilage are thought to be linked mainly to the extrafibrillar water. Intrafibrillar water is considered to be part of another phase. To account for this water partition, the carboxyl and amino sites of collagen are partitioned as well. The charge partition is designed so as to obtain a tissue isoelectric point (IEP) in the range indicated by experimental data. In other words, the charge that comes into picture in the diffusion properties is not the total charge but the charge acting in the extrafibrillar water.

For the data reported by Frank and Grodzinsky [1987], half of the charge is considered to contribute to the extrafibrillar phase and half to the intrafibrillar phase. Indeed, with this partition, the simulations yield an IEP about 2.6, while the whole collagen charge yielded an IEP about 3. Note that this partition of the charge has consequences on the composition of the intrafibrillar compartment. At low extrafibrillar pH, neutralizing anions should be present in the intrafibrillar water; correspondingly, at high extrafibrillar pH, neutralizing cations should be present. However, the present analysis does not intend to address comprehensively the effects of pH in this two fluid compartment framework and emphasis is laid on the extrafibrillar phase.

17.6.5 Additional relative proportions

Three additional mass and volume proportions are needed to complete the composition:
- relative mass of collagen and PGs, Basser et al. [1998], Mow and Guo [2002]:

$$r_{\mathrm{c}} = \frac{M_{\mathrm{PG}}}{M_{\mathrm{c}}} = \frac{1}{3};$$ (17.6.9)

This ratio is characteristic of the middle zone in the thickness of a cartilage layer;
- mass density $\rho_{\mathrm{nc}} = 1.35 \,\mathrm{gm/cm}^3$ and relative mass of noncollagenous proteins excluding proteins of PGs, Basser et al. [1998]:

$$r_{\mathrm{nc}} = \frac{M_{\mathrm{nc}}}{M_{\mathrm{c}}} = \frac{M_{\mathrm{tot\ prot}}}{M_{\mathrm{c}}} - \frac{M_{\mathrm{PG}} - M_{\mathrm{GAG}}}{M_{\mathrm{PG}}} \frac{M_{\mathrm{PG}}}{M_{\mathrm{c}}} = 0.57 - 0.1 \times 0.33 \sim 0.54;$$ (17.6.10)

- initial extrafibrillar fluid volume fraction, Mow and Guo [2002]:

$$n^{\mathrm{E}} = \frac{V_{\mathrm{E}}}{V} \in [0.7, 0.9] .$$ (17.6.11)

17.6.6 Overall mass and volume composition

With these data, the numbers of moles of collagen and proteoglycans can be calculated as a function of the initial number of moles of water N_{wE},

$$\frac{N_{\mathrm{c}}}{N_{\mathrm{wE}}} = \frac{\widehat{v}_{\mathrm{w}}}{\widehat{v}_{\mathrm{c}}} \frac{1}{\dfrac{n^{\mathrm{E}}}{1-n^{\mathrm{E}}}(1+r_{\mathrm{nc}}\dfrac{\rho_{\mathrm{c}}}{\rho_{\mathrm{nc}}}) - r_{\mathrm{c}}\dfrac{\rho_{\mathrm{c}}}{\rho_{\mathrm{PG}}}}, \qquad \frac{N_{\mathrm{PG}}}{N_{\mathrm{c}}} = r_{\mathrm{c}}\frac{\widehat{m}_{\mathrm{c}}}{\widehat{m}_{\mathrm{PG}}}. \tag{17.6.12}$$

The initial mole number of water is in turn obtained from the initial mass or volume of cartilage,

$$\frac{M}{N_{\mathrm{wE}}} = (1+r_{\mathrm{nc}})\frac{N_{\mathrm{c}}}{N_{\mathrm{wE}}}\widehat{m}_{\mathrm{c}} + \frac{N_{\mathrm{PG}}}{N_{\mathrm{c}}}\frac{N_{\mathrm{c}}}{N_{\mathrm{wE}}}\widehat{m}_{\mathrm{PG}} + \widehat{m}_{\mathrm{w}}, \tag{17.6.13}$$

$$\frac{V}{N_{\mathrm{wE}}} = (1+r_{\mathrm{nc}}\frac{\rho_{\mathrm{c}}}{\rho_{\mathrm{nc}}})\frac{N_{\mathrm{c}}}{N_{\mathrm{wE}}}\widehat{v}_{\mathrm{c}} + \frac{N_{\mathrm{PG}}}{N_{\mathrm{c}}}\frac{N_{\mathrm{c}}}{N_{\mathrm{wE}}}\widehat{v}_{\mathrm{PG}} + \widehat{v}_{\mathrm{w}}. \tag{17.6.14}$$

Given a volume of cartilage $V = V_0 = 1\,\mathrm{cm}^3$, the mass and density resulting from the data in this section are respectively equal to $M = 1.13\,\mathrm{gm}$ and $1.13\,\mathrm{gm/cm}^3$, assuming $n^{\mathrm{E}} = 0.8$. The relative proportions in terms of volume and mass of water, PGs and collagen are displayed in Table 17.6.1.

Table 17.6.1: Relative proportions in terms of volume and mass of water, PGs and collagen used in the simulations, based on a volume fraction of water phase n^{E} equal to 0.80.

	Water	PGs	Collagen	Noncollagenous proteins
Relative volume	76.7	3.3	12.8	7.2
Relative mass	69.3	5.5	16.4	8.8

17.7 Concentrations of ions, sites and charges

17.7.1 Physiological ionic concentrations

The physiological concentration of sodium chloride is about 144 mM in the interstitium, Fig. 3.6.2, and somehow higher in articular cartilages, say 154 mM per liter of water.

The concentration of calcium in the interstitium is about 3 mM. The total concentration, including bound and free ions, is much higher in articular cartilages, 113 mM per kg of dry weight, corresponding to 23 mM per kg of wet weight assuming a volume fraction of water equal to 0.8, and to about 32 mM per liter of water, Hunter [1991]. This concentration can be used to bracket the value of the coefficient pKCa.

17.7.2 Evolution of sites and charges as a function of the extrafibrillar pH

Acid-base reactions:

The concentrations of sites and charges are plotted in Figs. 17.3.1 and 17.3.2. The variations with pH are as expected: in particular, there exists a whole range of pH, say from 4 to 8, in which neither the charge of collagen nor that of PGs changes.

The value of the fixed charge of -0.17 M under physiological conditions, $n^{\mathrm{E}} = 0.80$ and $\mathrm{pH_E} \sim 7$, is in the range reported in the literature, e.g., Maroudas [1975], her Fig. 2, Eisenberg and Grodzinsky [1985], p. 156, and Wachtel and Maroudas [1998], their Fig. 2.

Calcium binding:

Given a concentration of calcium ions, the quantitative effects of the binding constant pKCa on the concentrations of the related sites, on the charge of PGs and extrafibrillar charge are explored in Fig. 17.4.1.

For binding constants pKCa equal to 1, 2 and 3, when the calcium concentration in the bath takes the standard value in the interstitium, namely 3 mM, the concentrations of the carboxyl sites occupied by calcium are respectively about 2 mM, 17.5 mM and 57 mM under neutral pH. Simultaneously the IEP increases from 2.6 to 3.1.

When the calcium concentration in the bath is $10\,\text{mM}$, the latter numbers increase respectively to about $7\,\text{mM}$, $38\,\text{mM}$ and $69\,\text{mM}$, and the IEP changes from 2.6 to 3.5 as the pKCa increases from 1 to 3.

Chondroitin 4 sulfates are known to bind more than Chondroitin 6 sulfates (the denominations 4 and 6 are attached to the positions of the sulfation at carbons 4 or 6 in the six sided saccharide ring), MacGregor and Bowness [1970], Rodriguez-Carvajal et al. [2003]. Accurate proportion of chondroitins 4 and 6 in articular cartilages is not known, although the ratio is varying with age, Muir [1978]. The relative proportion of one type of CS might possibly be kept track of through the pKCa.

It is worth re-iterating that the pHs of the bath and of the (extrafibrillar) water circulating the cartilage are different. In a first step, the evolutions of the charge of the different sites involved in the analysis have been examined as a function of the pH of the extrafibrillar fluid, noted pH_E. The issue of defining pH_E from the chemical composition of the bath surrounding the cartilage is addressed in the next section.

17.8 Chemical equilibrium at the cartilage-bath interface

A specimen of articular cartilage is in electro-chemomechanical equilibrium with a bath of controlled chemical composition, pressure p_{wB} and electrical potential ϕ_B, Fig. 17.8.1. The counterparts of these entities in the extrafibrillar phase are sought. Several cases are considered in turn.

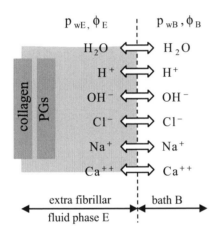

Fig. 17.8.1: Exchanges between the extrafibrillar phase and the bath take place while electrochemical equilibria hold for each species individually.

17.8.1 A binary electrolyte

Let us observe first that equilibrium of hydroxyl ions is implied by equilibrium of hydrogen ions and the dissociation relation (17.3.6). Indeed, the equilibria write,

$$g_{HE}^{ec} = RT\,\text{Ln}\,x_{HE} + F\,\phi_E = g_{HB}^{ec} = RT\,\text{Ln}\,x_{HB} + F\,\phi_B,$$
$$g_{OHE}^{ec} = RT\,\text{Ln}\,x_{OHE} - F\,\phi_E = g_{OHB}^{ec} = RT\,\text{Ln}\,x_{OHB} - F\,\phi_B.$$
(17.8.1)

The variation of the electrical charge is a function of the seven variables, namely a pressure, five molar fractions and an electrical potential,

$$p_{wE},\ x_{wE},\ x_{kE},\ x_{ClE},\ x_{HE},\ x_{OHE},\ \phi_E,$$
(17.8.2)

defined by the seven equations expressing the chemical equilibria of water, ion k, chloride and hydrogen ions, compatibility of the molar fractions, electroneutrality of the extrafibrillar phase and equilibrium of water dissociation, namely

$$g_{wE} = g_{wB},\quad g_{kE}^{ec} = g_{kB}^{ec},\quad g_{ClE}^{ec} = g_{ClB}^{ec},\quad g_{HE}^{ec} = g_{HB}^{ec},$$
$$\sum_{l\in E} x_{lE} = 1,\quad \sum_{l\in E} \zeta_l\,x_{lE} = 0,\quad x_{HE}\,x_{OHE} = K_{dis}^{x}.$$
(17.8.3)

The compatibility relation for the molar fractions is used to define the molar fraction of water,

$$x_{\mathrm{wE}} = 1 - x_{k\mathrm{E}} - x_{\mathrm{ClE}} - x_{\mathrm{HE}} - x_{\mathrm{OHE}} - x_{\mathrm{PG}}\,. \tag{17.8.4}$$

This relation and the equilibrium of water yields the osmotic pressure,

$$p_{\mathrm{wE}} - p_{\mathrm{wB}} = \frac{RT}{\widehat{v}_{\mathrm{w}}}\,\mathrm{Ln}\frac{x_{\mathrm{wB}}}{x_{\mathrm{wE}}} \simeq -\frac{RT}{\widehat{v}_{\mathrm{w}}}\left(x_{\mathrm{wE}} - x_{\mathrm{wB}}\right) = -RT\left(c_{\mathrm{wE}} - c_{\mathrm{wB}}\right). \tag{17.8.5}$$

So we are now left with five unknowns,

$$x_{k\mathrm{E}},\ x_{\mathrm{ClE}},\ x_{\mathrm{HE}},\ x_{\mathrm{OHE}},\ \phi_{\mathrm{E}}\,, \tag{17.8.6}$$

and five equations, namely

$$g_{k\mathrm{E}}^{\mathrm{ec}} = g_{k\mathrm{B}}^{\mathrm{ec}},\quad g_{\mathrm{ClE}}^{\mathrm{ec}} = g_{\mathrm{ClB}}^{\mathrm{ec}},\quad g_{\mathrm{HE}}^{\mathrm{ec}} = g_{\mathrm{HB}}^{\mathrm{ec}},$$

$$\sum_{l\in\mathrm{E}} \zeta_l\, x_{l\mathrm{E}} = 0,\quad x_{\mathrm{HE}}\, x_{\mathrm{OHE}} = K_{\mathrm{dis}}^{\mathrm{x}}\,. \tag{17.8.7}$$

With the (electro)chemical potentials in explicit form, the above set of equations can be cast in the format,

$$\frac{\mathrm{F}}{RT}\left(\phi_{\mathrm{E}} - \phi_{\mathrm{B}}\right) = \mathrm{Ln}\,\frac{x_{\mathrm{HB}}}{x_{\mathrm{HE}}} = -\mathrm{Ln}\,\frac{x_{\mathrm{ClB}}}{x_{\mathrm{ClE}}} = \frac{1}{\zeta_k}\,\mathrm{Ln}\,\frac{x_{k\mathrm{B}}}{x_{k\mathrm{E}}},$$

$$\sum_{l\in\mathrm{E_{ions}}} \zeta_l\, x_{l\mathrm{E}} + y_{\mathrm{e}} = 0,\quad x_{\mathrm{HE}}\, x_{\mathrm{OHE}} = K_{\mathrm{dis}}^{\mathrm{x}}\,. \tag{17.8.8}$$

Given the concentrations of hydrogen ions, the difference of electrical potentials will be seen as defined by the first equality (17.8.8). At given volume, the fixed electrical charge is a function of x_{HE} as long as one admits $\widehat{v}_{\mathrm{E}} \simeq \widehat{v}_{\mathrm{w}}$. Combining the equilibrium relations (17.8.8) yields the equation for the molar fraction x_{HE} of hydrogen ions,

$$f = \zeta_k\,\frac{x_{k\mathrm{B}}}{(x_{\mathrm{HB}})^{\zeta_k}}\left(x_{\mathrm{HE}}\right)^{\zeta_k+1} + \left(x_{\mathrm{HE}}\right)^2 + y_{\mathrm{e}}\, x_{\mathrm{HE}} - x_{\mathrm{HB}}\left(\zeta_k\, x_{k\mathrm{B}} + x_{\mathrm{HB}}\right) = 0\,. \tag{17.8.9}$$

The yet unknown charge $y_{\mathrm{e}} = y_{\mathrm{e}}(c_{\mathrm{HE}}, c_{\mathrm{CaE}})$ is defined by (17.5.7). The latter expression applies also in absence of calcium binding, by setting $\mathrm{pKCa} = -\infty$. In view of the algebraic form of this charge, Eqn (17.8.9) can be transformed into a polynomial in x_{HE}. Alternatively, it can be seen as a non-linear equation to be solved by a Newton scheme. The other unknowns are then successively deduced by $(17.8.8)_{1-3}$, (17.8.4) and (17.8.5).

Trends: for sodium chloride NaCl, Eqn (17.8.9) can be recast in the format,

$$\left(x_{\mathrm{HE}}\right)^2 - \left(x_{\mathrm{HB}}\right)^2 = -y_{\mathrm{e}}\,\frac{x_{\mathrm{HB}}\, x_{\mathrm{HE}}}{x_{\mathrm{NaB}} + x_{\mathrm{HB}}}\,. \tag{17.8.10}$$

With a negative charge y_{e}, clearly the cartilage pH will be smaller than the bath pH. The opposite holds for a positive fixed charge. These trends are representative of states above the isoelectric point and below the isolectric point respectively. They are illustrated in the simulations of the next sections.

17.8.2   A ternary electrolyte

So far, we have considered a single metallic cation to be present in the bath and cartilage. In the simultaneous presence of sodium and calcium cations, the equilibria between all species that can transfer between the bath and the extrafibrillar compartment, and the electroneutrality condition, undergo some changes, namely

$$\frac{\mathrm{F}}{RT}\left(\phi_{\mathrm{E}} - \phi_{\mathrm{B}}\right) = \mathrm{Ln}\,\frac{x_{\mathrm{HB}}}{x_{\mathrm{HE}}} = -\mathrm{Ln}\,\frac{x_{\mathrm{ClB}}}{x_{\mathrm{ClE}}} = \frac{1}{\zeta_k}\,\mathrm{Ln}\,\frac{x_{k\mathrm{B}}}{x_{k\mathrm{E}}},\quad k = \mathrm{Na}, \mathrm{Ca},$$

$$2\, x_{\mathrm{CaE}} + x_{\mathrm{NaE}} - x_{\mathrm{ClE}} + x_{\mathrm{HE}} - x_{\mathrm{OHE}} + y_{\mathrm{e}} = 0,$$

$$2\, x_{\mathrm{CaB}} + x_{\mathrm{NaB}} - x_{\mathrm{ClB}} + x_{\mathrm{HB}} - x_{\mathrm{OHB}} = 0\,, \tag{17.8.11}$$

$$x_{\mathrm{HE}}\, x_{\mathrm{OHE}} = x_{\mathrm{HB}}\, x_{\mathrm{OHB}} = K_{\mathrm{dis}}^{\mathrm{x}}\,.$$

The more explicit relations in terms of molar fractions follow (these relations also hold in terms of concentrations),

$$\frac{x_{\text{NaE}}}{x_{\text{NaB}}} = \frac{\sqrt{x_{\text{CaE}}}}{\sqrt{x_{\text{CaB}}}} = \frac{x_{\text{HE}}}{x_{\text{HB}}} = \frac{x_{\text{ClB}}}{x_{\text{ClE}}}. \tag{17.8.12}$$

The electroneutrality condition still provides an equation for the molar fraction x_{HE} of hydrogen ions,

$$2\,\frac{x_{\text{CaB}}}{(x_{\text{HB}})^2}\,(x_{\text{HE}})^3 + \frac{x_{\text{NaB}}}{x_{\text{HB}}}\,(x_{\text{HE}})^2 + (x_{\text{HE}})^2 + y_{\text{e}}\,x_{\text{HE}} - x_{\text{HB}}\,(2\,x_{\text{CaB}} + x_{\text{NaB}} + x_{\text{HB}}) = 0\,, \tag{17.8.13}$$

where the molar fraction of the charge $y_{\text{e}} = y_{\text{e}}(c_{\text{HE}}, c_{\text{CaE}})$ defined by (17.5.7) can be seen as a function of x_{HE} only since the calcium content derives directly from the hydrogen content in view of (17.8.12).

Let us reiterate that the diffusion properties are thought to be linked to the charge acting in the extrafibrillar compartment, which includes the charge of the PGs and part of the charge of collagen. Measurements on the other hand provide the total charge in the whole cartilage. The total charge yielded by the model matches quantitatively well data from human and bovine articular cartilages, Fig. 17.8.2.

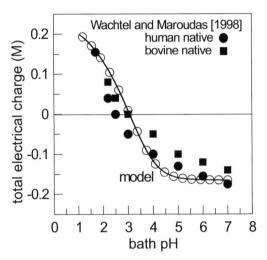 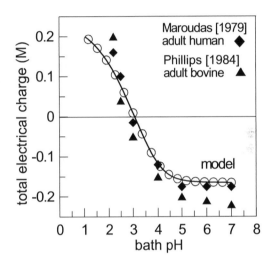

Fig. 17.8.2: Total charge as given by the model assuming physiological concentration of sodium chloride (150 mM), and calcium chloride at 3 mM without binding. Data adapted from Maroudas [1979] and Wachtel and Maroudas [1998] for adult human femoral head cartilage and bovine femoral condyle cartilage of 2-3 year old animals, and from Phillips [1984] for adult bovine femoropatellar groove cartilage (the latter data are extracted from Frank et al. [1990]). Reprinted with permission from Loret and Simões [2010]a.

17.8.3 A ternary electrolyte and enthalpies of formation

The chemical potentials of ionic species in the extrafibrillar cartilage fluid,

$$g_{l\text{E}}^{\text{ec}} = g_{l\text{E}}^{0} + RT\,\text{Ln}\,x_{l\text{E}} + \zeta_l\,\text{F}\,\phi_{\text{E}}\,, \quad l \in \text{E}_{\text{ions}}\,, \tag{17.8.14}$$

are now endowed with an enthalpy of formation $g_{l\text{E}}^{0}$, which is expressed in the format of a dimensionless equilibrium constant κ_l,

$$g_{l\text{E}}^{0} = -RT\,\text{Ln}\,\kappa_l\,, \quad l \in \text{E}_{\text{ions}}\,. \tag{17.8.15}$$

The relations $(17.8.11)_{2-4}$ are unchanged, but Eqn $(17.8.11)_1$ is modified to,

$$\frac{\text{F}}{RT}\,(\phi_{\text{E}} - \phi_{\text{B}}) = \text{Ln}\left(\kappa_{\text{H}}\,\frac{x_{\text{HB}}}{x_{\text{HE}}}\right) = -\text{Ln}\left(\kappa_{\text{Cl}}\,\frac{x_{\text{ClB}}}{x_{\text{ClE}}}\right) = \frac{1}{\zeta_k}\,\text{Ln}\left(\kappa_k\,\frac{x_{k\text{B}}}{x_{k\text{E}}}\right), \quad k = \text{Na}, \text{Ca}, \tag{17.8.16}$$

while the ionic relations (17.8.12) become,

$$\frac{1}{\kappa_{\text{Na}}}\,\frac{x_{\text{NaE}}}{x_{\text{NaB}}} = \frac{1}{\sqrt{\kappa_{\text{Ca}}}}\,\frac{\sqrt{x_{\text{CaE}}}}{\sqrt{x_{\text{CaB}}}} = \frac{1}{\kappa_{\text{H}}}\,\frac{x_{\text{HE}}}{x_{\text{HB}}} = \kappa_{\text{Cl}}\,\frac{x_{\text{ClB}}}{x_{\text{ClE}}}. \tag{17.8.17}$$

In other words, the cartilage "sees" the bath with ionic concentrations scaled by the equilibrium constants.

The equation for the molar fraction x_{HE} of hydrogen ions turns to,

$$2\,\frac{\kappa_{Ca}}{\kappa_H^2}\,\frac{x_{CaB}}{(x_{HB})^2}\,(x_{HE})^3 + \frac{\kappa_{Na}}{\kappa_H}\,\frac{x_{NaB}}{x_{HB}}\,(x_{HE})^2 + (x_{HE})^2 + y_e\,x_{HE} - \kappa_H\,x_{HB}\,(\kappa_{OH}\,x_{OHB} + \kappa_{Cl}\,x_{ClB}) = 0\,, \qquad (17.8.18)$$

where $y_e = y_e(c_{HE}, c_{CaE})$, defined by (17.5.7), is still a function of x_{HE} only, in view of (17.8.17). Observe that, if the relations below are satisfied,

$$\kappa_{Na} = \sqrt{\kappa_{Ca}} = \kappa_H = \frac{1}{\kappa_{OH}} = \frac{1}{\kappa_{Cl}}\,, \qquad (17.8.19)$$

the effects of the equilibrium constants on the chemical composition of the extrafibrillar fluid are elusive.

17.8.4   Sodium chloride electrolyte and enthalpies of formation

For a sodium chloride electrolyte, the above relations simplify somehow. The equation for x_{HE} becomes quadratic with discriminant,

$$\Delta = y_e^2 + 4\,(\kappa_H\,x_{HB} + \kappa_{Na}\,x_{NaB})\,(\kappa_{OH}\,x_{OHB} + \kappa_{Cl}\,x_{ClB})\,, \qquad (17.8.20)$$

so that,

$$x_{HE} = \frac{1}{2}\,\big(-y_e + \sqrt{\Delta}\big)\,\frac{\kappa_H\,x_{HB}}{\kappa_H\,x_{HB} + \kappa_{Na}\,x_{NaB}}\,, \qquad x_{NaE} + x_{HE} = \frac{1}{2}\,\big(-y_e + \sqrt{\Delta}\big)\,, \qquad (17.8.21)$$

and the osmotic pressure may be cast in the format,

$$
\begin{aligned}
p_{wE} - p_{wB} \;\simeq\; & \frac{RT}{\widehat{v}_w}\,(y_e + 2\,x_{NaE} + 2\,x_{HE} - 2\,x_{NaB} - 2\,x_{HB}) \\[4pt]
=\; & \frac{RT}{\widehat{v}_w}\,(\sqrt{\Delta} - 2\,x_{NaB} - 2\,x_{HB})\,,
\end{aligned}
\qquad (17.8.22)
$$

where the fixed charge y_e depends on the cartilage pH via (17.5.7).

At given bath pH ($dx_{HB} = dx_{OHB} = 0$, $dx_{ClB} = dx_{NaB}$), the derivative of the osmotic pressure with respect to the sodium molar fraction expresses as,

$$\frac{\widehat{v}_w}{RT}\,\frac{\partial}{\partial x_{NaB}}\,(p_{wE} - p_{wB}) = \left(\frac{dy_e}{dx_{HE}} + 2 + 2\,\frac{x_{NaE}}{x_{HE}}\right)\frac{\partial x_{HE}}{\partial x_{NaB}} - 2 + 2\,\frac{x_{NaE}}{x_{NaB}}\,, \qquad (17.8.23)$$

where

$$\frac{\partial x_{HE}}{\partial x_{NaB}} = \frac{\kappa_{Cl}\,(\kappa_H\,x_{HB})^2 - \kappa_{Na}\,(x_{HE})^2}{x_{HE}\,\dfrac{dy_e}{dx_{HE}}\,\kappa_H\,x_{HB} + y_e\,\kappa_H\,x_{HB} + 2\,x_{HE}\,(\kappa_H\,x_{HB} + \kappa_{Na}\,x_{NaB})}\,, \qquad (17.8.24)$$

and

$$\frac{dy_e}{dx_{HE}} = \sum_{i=1}^{4}\,\frac{10^{pK_i - 3}}{(1 + \alpha_i)^2}\,c_{Si}\,. \qquad (17.8.25)$$

The α's are defined by (17.5.4) and the convention (17.5.3) has been used.

Alternatively, at given sodium chloride concentration in bath, namely for $dx_{NaB} = 0$, $dx_{OHB} = -(x_{OHB}/x_{HB})\,dx_{HB}$, $dx_{ClB} = (1 + x_{OHB}/x_{HB})\,dx_{HB}$, the derivative of the osmotic pressure with respect to the hydrogen molar fraction expresses as,

$$\frac{\widehat{v}_w}{RT}\,\frac{\partial}{\partial x_{HB}}\,(p_{wE} - p_{wB}) = \left(\frac{dy_e}{dx_{HE}} + 2 + 2\,\frac{x_{NaE}}{x_{HE}}\right)\frac{\partial x_{HE}}{\partial x_{HB}} - 2 - 2\,\frac{x_{NaE}}{x_{HB}}\,, \qquad (17.8.26)$$

where

$$\frac{\partial x_{HE}}{\partial x_{HB}} = \kappa_H\,\frac{\kappa_H\,\kappa_{Cl}\,x_{HB}\,(x_{HB} + x_{OHB} + x_{ClB}) + x_{NaE}\,x_{HE}}{x_{HE}\,\dfrac{dy_e}{dx_{HE}}\,\kappa_H\,x_{HB} + y_e\,\kappa_H\,x_{HB} + 2\,x_{HE}\,(\kappa_H\,x_{HB} + \kappa_{Na}\,x_{NaB})}\,. \qquad (17.8.27)$$

The derivative (17.8.26) of the osmotic pressure vanishes at the isoelectric point if the equilibrium constants satisfy the relations (17.8.19).

17.8.5 Sodium chloride electrolyte and enthalpies of formation in absence of pH effect

In view of the electroneutrality in both bath and extrafibrilar fluid phase, further simplifications arise. Indeed, then

$$\Delta = y_{\mathrm{e}}^2 + 4\,\kappa_{\mathrm{Na}}\,\kappa_{\mathrm{Cl}}\,(x_{\mathrm{NaB}})^2 \,, \tag{17.8.28}$$

so that,

$$x_{\mathrm{NaE}} = \frac{1}{2}\left(-\,y_{\mathrm{e}} + \sqrt{\Delta}\right), \quad \frac{dx_{\mathrm{NaE}}}{dx_{\mathrm{NaB}}} = \frac{2\,\kappa_{\mathrm{Na}}\,\kappa_{\mathrm{Cl}}\,x_{\mathrm{NaB}}}{2\,x_{\mathrm{NaE}} + y_{\mathrm{e}}} \,. \tag{17.8.29}$$

In view of the similarity of (17.8.21) and (17.8.29), the expressions of the osmotic pressure are formally identical to the case where the pH effect on the electrical charge is accounted for. Since now the charge is fixed, the derivative of the osmotic pressure with respect to the sodium molar fraction simplifies to,

$$\begin{aligned}
\frac{\widehat{v}_{\mathrm{w}}}{RT}\,\frac{d}{dx_{\mathrm{NaB}}}\left(p_{\mathrm{wE}} - p_{\mathrm{wB}}\right) &= 2\left(\frac{dx_{\mathrm{NaE}}}{dx_{\mathrm{NaB}}} - 1\right) \\
&= -\frac{2}{\sqrt{\Delta}}\,\frac{y_{\mathrm{e}}^2 + 4\,\kappa_{\mathrm{Na}}\,\kappa_{\mathrm{Cl}}\,(x_{\mathrm{NaB}})^2\,(1 - \kappa_{\mathrm{Na}}\,\kappa_{\mathrm{Cl}})}{2\,\kappa_{\mathrm{Na}}\,\kappa_{\mathrm{Cl}}\,x_{\mathrm{NaB}} + \sqrt{\Delta}} \,.
\end{aligned} \tag{17.8.30}$$

For $\kappa_{\mathrm{Na}}\,\kappa_{\mathrm{Cl}} = 1$, the osmotic pressure is clearly decreasing at increasing ionic strength of the bath.

The derivative $d/d\mathrm{pk_B} = -\mathrm{Log}10\,x_{k\mathrm{B}}\,d/dx_{k\mathrm{B}}$ for a generic species k may be needed if the results are plotted as function of the $\mathrm{pk_B}$ instead of the molar fraction $x_{k\mathrm{B}}$ or concentration $c_{k\mathrm{B}}$.

17.9 Constitutive equations of generalized diffusion

The constitutive equations of diffusion have been derived in a thermodynamic framework *à la Biot* where the solid skeleton is taken as reference, Chapter 16. A single inequality for the internal entropy production is required for the porous medium as a whole. It results in an expression that contains three terms of distinct natures, and which consequently are required to be positive individually. Here only the part of the internal entropy production associated with generalized diffusion is of interest, namely

$$T\hat{S}_{(\mathrm{i},3)} = -\sum_{l \in \mathrm{E}}\left(\frac{1}{\widehat{v}_l}\,\boldsymbol{\nabla}g_{l\mathrm{E}}^{\mathrm{ec}} - \rho_l\,\mathbf{b}_{l\mathrm{E}}\right) \cdot \mathbf{J}_{l\mathrm{E}} \geq 0 \,. \tag{17.9.1}$$

For quasi-static analyses, the body force densities $\mathbf{b}_{l\mathrm{E}}$ are equal to the gravity $\mathbf{g}$ and individual accelerations are neglected.

Inequality $\hat{S}_{(\mathrm{i},3)} \geq 0$, Eqn (17.9.1), is ensured by generalization of Darcy's law of seepage through the porous medium, Fick's law of diffusion of ions in the extrafibrilar fluid phase, and Ohm's law of electrical flow.

Here, we just provide the definitions of the entities, in terms of which the constitutive equations are phrased, and detail the constitutive coefficients that come into picture in this analysis of pH effects. Actually, the analysis of Chapter 16 was concerned with metallic ions only and collagen was electrically neutral. Modifications to include hydrogen and hydroxyl ions and the collagen charge are thus required. To simplify the presentation, a single metallic cation, referred to by the index k=Na, or Ca, is assumed to be present. Still, there is no difficulty to include both cations simultaneously since in a sense they do not interact in the diffusion process. This more general situation is indeed considered below in some simulations.

17.9.1 Fluxes

The *volume flux* $\mathbf{J}_{l\mathrm{E}}$ of the species l [unit : m/s] is defined as,

$$\mathbf{J}_{l\mathrm{E}} = n^{l\mathrm{E}}\,(\mathbf{v}_{l\mathrm{E}} - \mathbf{v}_{\mathrm{S}}), \quad \forall\, l \in \mathrm{E} \,. \tag{17.9.2}$$

The fluxes of proteoglycans and collagen vanish since their velocities are that of the solid. The *diffusive flux* of any species $l \in \mathrm{E}$ with respect to water in the extrafibrilar fluid phase is denoted by $\mathbf{J}_{l\mathrm{E}}^d$,

$$\mathbf{J}_{l\mathrm{E}}^d = n^{l\mathrm{E}}\,(\mathbf{v}_{l\mathrm{E}} - \mathbf{v}_{\mathrm{wE}}) = \mathbf{J}_{l\mathrm{E}} + \frac{n^{l\mathrm{E}}}{n^{\mathrm{wE}}}\,\mathbf{J}_{\mathrm{wE}}, \quad \forall\, l \in \mathrm{E} \,. \tag{17.9.3}$$

The *electrical current density* $\mathbf{I}_{eE}$ [unit: A/m^2] is defined as the sum of constituent velocities weighted by their valences and molar densities,

$$\mathbf{I}_{eE} = F \sum_{l \in E} \zeta_l \frac{N_{lE}}{V} \, \mathbf{v}_{lE} \, . \tag{17.9.4}$$

Due to the electroneutrality in the extrafibrillar fluid phase, $\mathbf{I}_{eE}$ may be viewed as a sum of either interphase or diffusive fluxes, namely

$$\mathbf{I}_{eE} = F \sum_{k \in E} \zeta_l \frac{\mathbf{J}_{lE}}{\widehat{v}_l} = F \sum_{k \in E} \zeta_l \frac{\mathbf{J}_{lE}^d}{\widehat{v}_l} \, . \tag{17.9.5}$$

An alternative form introduces the diffusive fluxes of PGs and collagen, which are proportional to the flux of water,

$$\mathbf{I}_{eE} - F \sum_{l \in E_{ions}} \zeta_l \frac{\mathbf{J}_{lE}^d}{\widehat{v}_l} = F \left(\zeta_{PG} \frac{\mathbf{J}_{PG}^d}{\widehat{v}_{PG}} + \zeta_c \frac{\mathbf{J}_c^d}{\widehat{v}_c} \right) = -F \, \widetilde{e}_e \, \mathbf{J}_{wE} \, . \tag{17.9.6}$$

The right equality, where $\widetilde{e}_e$ defined as $e_e \, n^E/n^{wE}$ is actually quite close to e_e, deduces from (17.2.4). The contribution of collagen follows from the charge partition defined in Section 17.6.

17.9.2   Generalized diffusion in presence of hydrogen and hydroxyl ions

Along the lines of Chapter 16, but neglecting the distinction between mobile and non mobile ions, the dissipation (17.9.1) is expressed in terms of diffusive fluxes and electric current density. The inequality can be cast in the form $-\boldsymbol{\mathcal{J}}^T \boldsymbol{\mathcal{F}} \geq 0$, in terms of a vector flux $\boldsymbol{\mathcal{J}}$ and of its conjugate force $\boldsymbol{\mathcal{F}}$,

$$\boldsymbol{\mathcal{J}} = \begin{bmatrix} \mathbf{J}_{wE} \\ \mathbf{J}_{kE}^d \\ \mathbf{J}_{ClE}^d \\ \mathbf{J}_{HE}^d \\ \mathbf{J}_{OHE}^d \\ \mathbf{I}_{eE} \end{bmatrix}, \quad \boldsymbol{\mathcal{F}} = \begin{bmatrix} \mathbf{F}_E \\ \mathbf{F}_{kE}^d \\ \mathbf{F}_{ClE}^d \\ \mathbf{F}_{HE}^d \\ \mathbf{F}_{OHE}^d \\ \mathbf{F}_{eE} \end{bmatrix} = \begin{bmatrix} \boldsymbol{\nabla} P_{wE} \\ RT/\widehat{v}_k \, \boldsymbol{\nabla} \text{Ln} \, x_{kE} \\ RT/\widehat{v}_{Cl} \, \boldsymbol{\nabla} \text{Ln} \, x_{ClE} \\ RT/\widehat{v}_H \, \boldsymbol{\nabla} \text{Ln} \, x_{HE} \\ RT/\widehat{v}_{OH} \, \boldsymbol{\nabla} \text{Ln} \, x_{OHE} \\ \boldsymbol{\nabla} \phi_E \end{bmatrix} - \begin{bmatrix} \rho^E/n^E \, \mathbf{g} \\ (\rho_k - \rho^E/n^E) \, \mathbf{g} \\ (\rho_{Cl} - \rho^E/n^E) \, \mathbf{g} \\ (\rho_H - \rho^E/n^E) \, \mathbf{g} \\ (\rho_{OH} - \rho^E/n^E) \, \mathbf{g} \\ 0 \end{bmatrix}, \tag{17.9.7}$$

where $\boldsymbol{\nabla} P_{wE} = \boldsymbol{\nabla} p_{wE} - RT \, \boldsymbol{\nabla} c_e \, n^E/n^{wE}$. Gravity effects are shown for the record but they may be neglected in most circumstances. If the material response is isotropic, a generalized law describing coupled seepage of pore water through the solid skeleton, diffusion of ions with respect to the extrafibrillar water and electrical flow is introduced via the symmetric matrix $\boldsymbol{\mathcal{K}}$,

$$\boldsymbol{\mathcal{J}} = -\boldsymbol{\mathcal{K}} \boldsymbol{\mathcal{F}}, \quad \boldsymbol{\mathcal{K}} = \boldsymbol{\mathcal{K}}^T \text{ s.D.P.} \tag{17.9.8}$$

Extending the procedure of Chapter 16 so as to include hydrogen and hydroxyl ions, the diffusion matrix is obtained under the sole two following assumptions:

- *Assumption* $(\mathcal{D}1)$: the diffusive flux of ions is not affected by a gradient of fluid pressure. Symmetry of the diffusion matrix $\boldsymbol{\mathcal{K}}$ implies that the ionic gradients do not affect the fluid flow;

- *Assumption* $(\mathcal{D}2)$: a gradient of concentration of ion n does not affect the diffusive flux of ion $l \neq n$.

The resulting diffusion matrix can be cast in the arrowhead format,

$$\boldsymbol{\mathcal{K}} = \begin{bmatrix} k_{EE} & 0 & 0 & 0 & 0 & k_e \\ 0 & k_{kk}^d & 0 & 0 & 0 & k_{ke}^d \\ 0 & 0 & k_{ClCl}^d & 0 & 0 & k_{Cle}^d \\ 0 & 0 & 0 & k_{HH}^d & 0 & k_{He}^d \\ 0 & 0 & 0 & 0 & k_{OHOH}^d & k_{OHe}^d \\ k_e & k_{ek}^d & k_{eCl}^d & k_{eH}^d & k_{eOH}^d & \sigma_e \end{bmatrix} . \tag{17.9.9}$$

Its coefficients are given in terms of the hydraulic conductivity K_{H} [unit : m/s], effective ionic mobilities u_l^*, $l \in \mathrm{E}_{\mathrm{ions}}$ [unit : m^2/s/V] and effective concentration of fixed charge $\tilde{e}_{\mathrm{e}}$ [unit : mole/m^3] in the format,

$$\begin{cases}
k_{\mathrm{EE}} = \dfrac{K_{\mathrm{H}}}{\rho_{\mathrm{w}}\, \mathrm{g}} > 0, \quad k_{\mathrm{e}} = -k_{\mathrm{EE}}\, \mathrm{F}\, \tilde{e}_{\mathrm{e}}, \quad k_{\mathrm{H}} = k_{\mathrm{EE}} - \dfrac{k_{\mathrm{e}}^2}{\sigma_{\mathrm{e}}} > 0, \\[2mm]
\sigma_{\mathrm{e}} = \sigma_{\mathrm{e}}^{\mathrm{ion}} + \sigma_{\mathrm{e}}^{\mathrm{e}}, \quad \sigma_{\mathrm{e}}^{\mathrm{ion}} = n^{\mathrm{E}}\, \mathrm{F} \displaystyle\sum_{l \in \mathrm{E}_{\mathrm{ions}}} |\zeta_l|\, c_{l\mathrm{E}}\, u_l^*, \quad \sigma_{\mathrm{e}}^{\mathrm{e}} = k_{\mathrm{EE}}\, \mathrm{F}^2\, \tilde{e}_{\mathrm{e}}^2, \\[2mm]
\dfrac{k_{le}^d}{\widehat{v_l}} = n^{\mathrm{E}}\, c_{l\mathrm{E}}\, u_l^*\, \mathrm{sgn}\, \zeta_l, \quad l \in \mathrm{E}_{\mathrm{ions}}, \\[2mm]
\dfrac{k_{ln}^d}{\widehat{v_l}\widehat{v_n}} = n^{\mathrm{E}}\, c_{l\mathrm{E}}\, \dfrac{u_l^*}{\mathrm{F}\, |\zeta_l|}\, I_{ln}, \quad l, n \in \mathrm{E}_{\mathrm{ions}}.
\end{cases} \qquad (17.9.10)$$

Note that the open circuit permeability k_{H} is verified to be positive for actual parameters. The effective ionic mobilities u^* are equal to the ionic mobilities u times a tortuosity factor τ. Crucially, the electroosmotic coefficient k_{e} is proportional to the fixed charge with an opposite sign. The electrical conductivity σ_{e} is contributed by both mobile ions and the fixed charge.

17.10 Simulations of bath-cartilage equilibria

The simulations below consider a piece of cartilage which is in equilibrium with a bath of controlled chemical composition and at reference (zero) pressure. They assume a constant volume fraction n^{E}. Two types of simulations are described: either varying the ionic strength of the bath at constant pH or varying the pH of the bath at constant strength of the metallic cations.

Table 17.10.1: Ionic mobilities [unit : 10^{-9} m^2/s/V] in a blank solution of the ions involved in the simulations.

Ion	Na	Ca	Cl	H	OH
Ionic mobility	51.75	61.75	79.0	362.5	205.5

17.10.1 Material data

The biochemical composition of the cartilage in terms of PGs, collagen and water is described in Section 17.6. Next to the ionic mobilities listed in Table 17.10.1, the following diffusion properties are used:
- constant tortuosity factor $\tau = 0.4$ (a dependence in volume fraction is sometimes accounted for);
- short circuit hydraulic permeability $k_{\mathrm{EE}} = 10^{-15}$ m$^3 \times$s/kg.

At neutral pH, typical values of electrical conductivity range from 10^{-2} to a few siemens/m, while the electroosmotic coefficient k_{e} varies between 0.1 and 5×10^{-8} m^2/s/V.

17.10.2 Variation of ionic strength

The pH of the bath is kept constant while the ionic strength is decreased from a (quasi-)hypertonic state (large ionic concentrations). The actual concentration of sodium cations is decreased and the concentration of chloride ions is adjusted so as to ensure electroneutrality.

17.10.2.1 Variation of ionic strength in presence sodium chloride

The variation of a number of quantities, namely extrafibrillar pH, charges, coefficient of streaming potential, etc. are shown in Figs. 17.10.1 to 17.10.3. As a first general remark, the influence of bath pHs close to physiological is seen to develop only at low ionic strength because high ionic strengths overweigh the influence of the fixed charge.

For physiological concentrations and higher, the cartilage pH is only slightly smaller than in the bath, Fig. 17.10.1. On the other hand, at moderate to low ionic strength, the pH of the cartilage reaches values clearly distinct from the bath. Actually, while it varies with ionic strength, the pH of the cartilage stays on the same side of the IEP as the pH in the bath. Therefore, the extrafibrillar charge varies but does not change sign. The same remark applies to

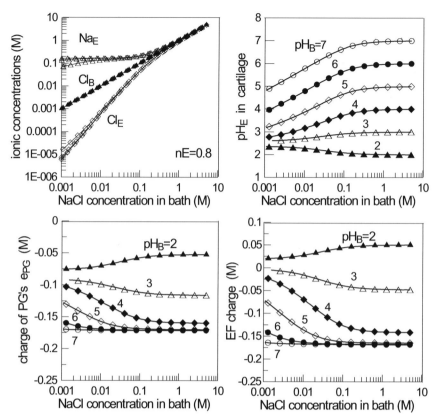

Fig. 17.10.1: Cartilage, of volume fraction $n^E = 0.8$, in equilibrium with a bath containing a sodium chloride electrolyte. The ionic strength is decreased, starting from the hypertonic state at given pH=pH$_B$ of the bath. Ionic concentrations in the cartilage, cartilage pH=pH$_E$, electrical charge of PGs and extrafibrillar charge. Reprinted with permission from Loret and Simões [2010]a.

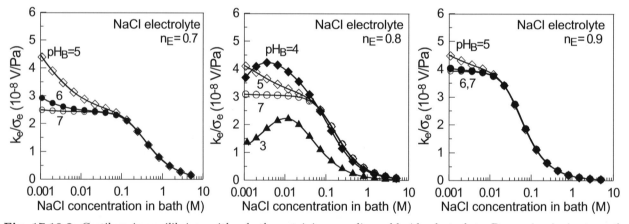

Fig. 17.10.2: Cartilage in equilibrium with a bath containing a sodium chloride electrolyte. Decreasing ionic strength starting from the hypertonic state. Influence of the volume fraction n^E and of the pH=pH$_B$ of the bath on the coefficient of streaming potential. This coefficient depends non monotonously on the bath pH. At given pH of the bath, it does not change sign since the extrafibrillar charge does not either. Reprinted with permission from Loret and Simões [2010]a.

the electroosmotic coefficient k_e and to the coefficient of streaming potential k_e/σ_e, Fig. 17.10.2. At given pH, and decreasing ionic strength, the latter is expected to increase in absolute value.

Note that the shape and values of the coefficient of streaming potential are quite similar to the experimental measurements on a bovine adult articular cartilage maintained at neutral pH by Frank and Grodzinsky [1987], their Fig. 8. However, while in the present model the coefficient of streaming potential becomes very small at large ionic strength, as indicated above, it does not change sign, as alluded for by Frank and Grodzinsky [1987], p. 623. Indeed,

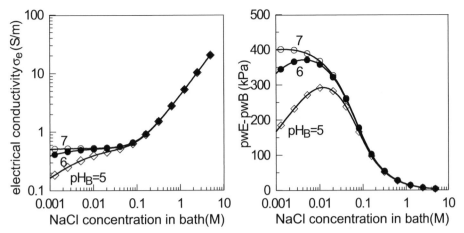

Fig. 17.10.3: Cartilage, of volume fraction $n^E = 0.8$, in equilibrium with a bath containing a sodium chloride electrolyte. Influence of the ionic strength of the bath, for three pH=pH$_B$, on the electrical conductivity and osmotic pressure. Reprinted with permission from Loret and Simões [2010]a.

these authors mention that the phase angle underwent a discontinuity of 180° at 5 M, which would indicate that the IEP has been passed. Specific adsorption, e.g., as in the case of Ca^{2+}, may be necessary for the sole increase of ionic strength to pass the IEP.

Fig. 17.10.2 also indicates that the coefficient of streaming potential increases with the water volume fraction n^E: indeed, at low ionic strength, the ratio k_e/σ_e is essentially inversely proportional to the electrical charge which decreases with n^E.

The standard values of the osmotic pressure under physiological conditions are about 170 kPa according to Lai et al. [1991], p. 245, and in the range 20-250 kPa, Grodzinsky et al. [1981], p. 222. Maroudas [1975] reports osmotic pressures about 350 kPa for human femoral heads. In fact, Fig. 17.10.3 shows that the osmotic pressure varies much with ionic strength and, at moderate to low ionic strength, with pH. Subsequent results will also highlight the strong influence of the water volume fraction n^E.

At high ionic strength, the osmotic pressure is low because the fixed charge is overweighed by mobile ions. As the ionic strength decreases, the presence of the fixed charge is progressively felt and the osmotic pressure increases, Fig. 17.10.3. Still, this increase is limited and the osmotic pressure shows a maximum because the fixed charge decreases at low ionic strength due to hydrogen-binding (this maximum does not appear if water does not dissociate spontaneously, in other words, if the effects of pH are not accounted for). As a further consequence, free swelling is likely to be limited.

17.10.2.2 Variation of ionic strength in presence sodium and calcium chlorides

The concentration of calcium cations in the bath is now fixed to a non zero value, namely 0.01 M. Figs. 17.10.4, 17.10.5 and 17.10.6 intend to illustrate the effects of calcium binding onto the carboxyl sites of proteoglycans.

As expected, calcium binding decreases the amount of free calcium in the cartilage fluid, turns the electrical charge less negative, and therefore modifies the transport properties and osmotic pressure. It is also interesting to consider the variation of pH in the cartilage with ionic strength so as to highlight the influence of the presence of calcium cations. Indeed, comparison of Figs. 17.10.1 and 17.10.5 shows that calcium acidifies the cartilage considerably less than sodium. In fact, at low ionic strength of the bath, the concentration of calcium in the cartilage increases significantly. Electroneutrality does not require so many hydrogen ions to be present as for a sodium chloride electrolyte. This trend is amplified by the valence two of calcium ions.

17.10.3 Variation of pH of the bath

The concentration of the metallic cations in the bath is now fixed and the pH is decreased.

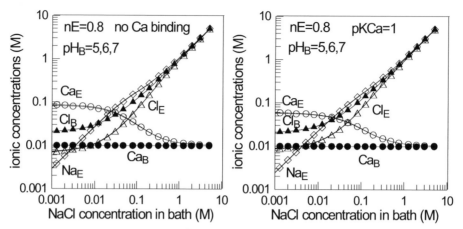

Fig. 17.10.4: Cartilage, of fluid volume fraction $n^E = 0.8$, in equilibrium with a bath containing a ternary electrolyte with sodium and calcium chlorides. Bath: decreasing ionic strength of sodium ion starting from hypertonic state, with constant concentration of calcium ions equal to 10 mM, for three given pHs in bath.

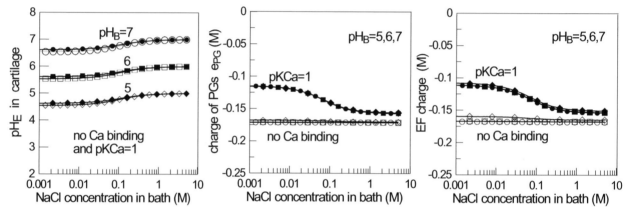

Fig. 17.10.5: Same as Fig. 17.10.4. Calcium binding (solid symbols) affects the electrical charge of PGs and extrafibrillar charge, but little the cartilage pH. Even in absence of binding (open symbols), the presence of calcium reduces the influence of the bath pH on these electrical charges (compare with Fig. 17.10.1). Reprinted with permission from Loret and Simões [2010]a.

17.10.3.1 Variation of pH of a binary electrolyte

The pH in the cartilage and bath are not equal. Indeed, the general trend is that the cartilage is more acid than the bath above the isoelectric point (IEP), and less acid below the IEP, as indicated by Eqn (17.8.10). Note that the pH in the cartilage has values quite close to those indicated by Wachtel and Maroudas [1998]: under a compression of 6 atm, and bath pH of 7.6, they find the pH of the extrafibrillar compartment to be 6.9, definitely more acid. On the other hand, while we do not address here the intrafibrillar compartment, we might still record that, for pH of the bath equal to 2, Wachtel and Maroudas [1998] deduce a pH in this compartment equal to 2.4.

At the isoelectric point of the cartilage, the concentrations of all ions are identical in the bath and cartilage since the concentration of PGs is quite tiny. Therefore the pH in the bath and cartilage are equal, $pH_E = pH_B$. As another consequence, the osmotic pressure vanishes as well as the contribution σ_e^e of the fixed charge to the electrical conductivity. These two entities are thus miminum at the IEP, Fig. 17.10.8.

The concentrations of sodium and chloride ions are also almost equal in the bath above the isoelectric point: in fact, in this range, the concentration of hydrogen ions is small with respect to that of metallic ions, Fig. 17.10.7. The chemical concentrations of anions in the bath and in the cartilage increase below the isoelectric point where the concentration of hydrogen ions becomes high and, as a consequence, the extrafibrillar charge becomes positive.

Note that the higher the volume fraction of water n^E, the smaller is the charge concentration. This observation implies higher osmotic pressures for low volume fractions, Fig. 17.10.8. The volume fraction has identical effects on the contribution σ_e^e of the fixed charge to the electrical conductivity and on the osmotic pressure: both are in direct

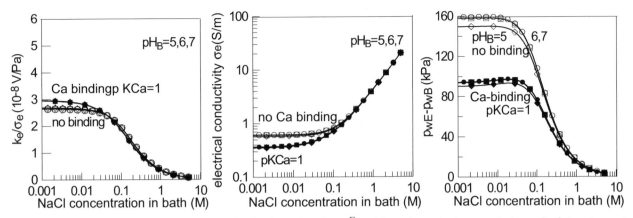

Fig. 17.10.6: Ternary electrolyte, constant fluid volume fraction $n^E = 0.8$, and constant concentration of calcium ions set to 10 mM. Influence of sodium concentration in the bath on the coefficient of streaming potential, electrical conductivity and osmotic pressure, for three given pHs in bath and two calcium binding constants, pKCa$=-\infty$ (no binding), 1. Since, due to its valence 2, calcium reduces significantly the electrical charges, it also reduces the osmotic pressure. Reprinted with permission from Loret and Simões [2010]a.

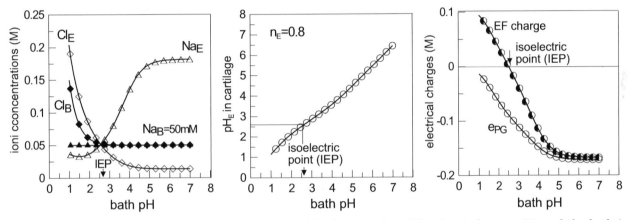

Fig. 17.10.7: Cartilage in equilibrium with a sodium chloride electrolyte. The chemical composition of the bath is controlled: constant concentration of cations Na$^+$ equal to 50 mM and varying pH. In order to satisfy electroneutrality, the amount of cations in the extrafibrillar fluid dominates the amount of anions above the IEP and the converse takes place below the IEP. Reprinted with permission from Loret and Simões [2010]a.

relation with concentrations and therefore in inverse relation with the volume fraction n^E. The effects on the ionic contribution σ_e^{ion} are similar above the isoelectric point and opposite below the IEP, Fig. 17.10.9. These concurrent and adverse influences result in a strong dependence of the electrical conductivity on the bath pH above the IEP only.

The coefficient of streaming potential in Fig. 17.10.10 fits nicely with the experiments of Frank and Grodzinsky [1987], their Fig. 9, which display a plateau above the IEP, a steep decrease in the neighborhood of the IEP and a final increase as the bath pH continues to decrease.

17.10.3.2   Variation of pH of a ternary electrolyte

The effects of calcium binding are analyzed through simulations where the concentrations of sodium and calcium cations in the bath are maintained fixed while the pH is varied.

Calcium binding decreases the electrical charge, leaving less dissolved calcium ions in the fluid compartment, Fig. 17.10.11: the phenomenon is observed mostly above the IEP. The consequences of calcium binding, and therefore of a smaller electrical charge, on the transport properties and osmotic pressure are somehow similar to a larger volume fraction n^E as can be realized by comparing Figs. 17.10.8 and 17.10.12.

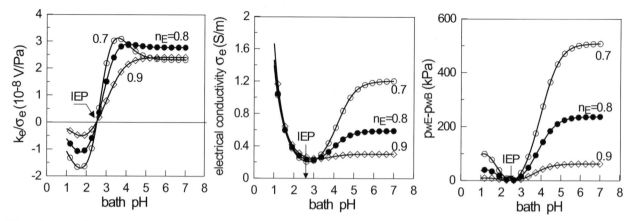

Fig. 17.10.8: Same simulation protocol as for Fig. 17.10.7 with a constant concentration of sodium ions equal to 50 mM. Coefficient of streaming potential, electrical conductivity and osmotic pressure in terms of the pH of the bath for three volume fractions n^{E}. The influence of the volume fraction of water n^{E} on the electrical conductivity and osmotic pressure is explained by the fact that the higher the volume fraction the smaller is the charge concentration. Reprinted with permission from Loret and Simões [2010]a.

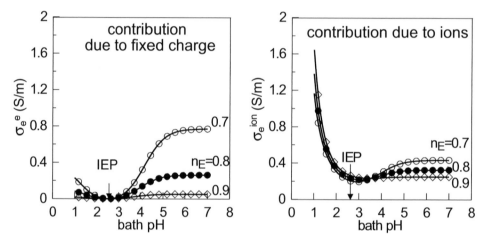

Fig. 17.10.9: Same simulation protocol as for Fig. 17.10.7. Influence of volume fraction on the respective contributions by the fixed charge and ions to the electrical conductivity.

17.10.4   Comments on the transport model

Emphasis has been laid so far on transport issues and the simulations presented have considered constant volume fraction n^{E}. However, in actual experiments, chemomechanical couplings imply the volume fraction of water n^{E} to vary as a function of the chemical composition and pH of the bath. Therefore, actual simulations of laboratory experiments need to take these couplings into account since the transport properties depend not only on chemical composition and pH but also on the volume fraction n^{E}. This topic is addressed in the next sections.

Variations of the volume fraction n^{E} affect the chemical composition: this feature is termed *mechanochemical coupling*. The converse effect, *chemomechanical coupling*, manifests itself by a number of features, among which swelling, whether under given stress or under complementary constraints in stress and strain.

The above considerations were concerned with states of mechanical and electromechanical equilibrium. A complete description of the time evolution of the mechanical and chemical properties of a cartilage specimen during an experiment may be formulated as an initial and boundary value problem to be solved through either analytical or numerical procedures, e.g., the finite element method. It is then possible to highlight the retardation effects due to the acid-base reactions. Notice however that the analysis presented in Chapter 18 already incorporates reaction-diffusion phenomena. Indeed, mass transfers between the intra- and extrafibrillar fluid phases can be interpreted as physico-chemical reactions endowed with a kinetics defined by a characteristic time. Acid-base reactions and calcium binding increase the number of reactions but do not modify the existing theoretical framework.

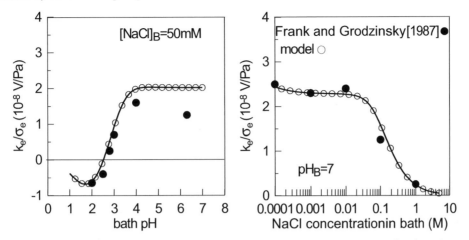

Fig. 17.10.10: Same simulation protocol as for Fig. 17.10.7 with a constant concentration of sodium ions equal to 50 mM. Coefficient of streaming potential versus data. Note that the value at pH=7 of the coefficient should be the same in the two experiments, unless the details of the experimental protocol come into picture. Reprinted with permission from Loret and Simões [2010]a.

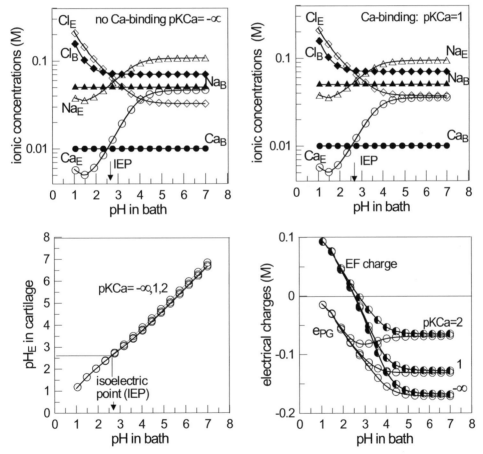

Fig. 17.10.11: Cartilage in equilibrium with a ternary electrolyte with constant concentrations of sodium (50 mM) and calcium (10 mM) and varying pH. Concentrations in the extrafibrillar compartment of the cartilage, cartilage pH, charges of PGs and extrafibrillar charge for three calcium binding constants, $pK_{Ca} = -\infty$ (no binding), 1, 2. The isoelectric point is slightly higher for $pK_{Ca} = 2$ than for $pK_{Ca} = -\infty$, 1. Reprinted with permission from Loret and Simões [2010]a.

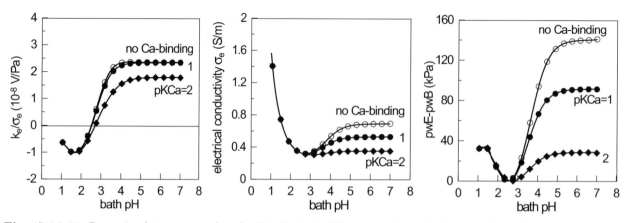

Fig. 17.10.12: Same simulation protocol as for Fig. 17.10.11. Effects of calcium binding on the coefficient of streaming potential, electrical conductivity and osmotic pressure as a function of the pH of the bath. As observed in Fig. 17.10.11, the calcium binding constant has virtually no effect below the IEP. Above the IEP, a higher binding constant decreases the magnitude of the fixed charge and the osmotic pressure. Reprinted with permission from Loret and Simões [2010]a.

17.11 Constitutive mechanical framework including the effects of pH and calcium binding

The acid-base reactions and calcium binding phenomenon described in Sections 17.3 and 17.4 are now embedded in a mechanical framework, where the cartilage is viewed as a porous medium which is saturated by an electrolyte. The acid-base reactions and calcium binding are considered to be reversible. The resulting chemo-hyperelastic equations are expressed in terms of work-conjugate generalized stresses and strains. Chemomechanical couplings imply that the concentrations of metallic ions and the pH of the cartilage affect strongly the mechanical response.

To illustrate the chemomechanical couplings, simulations of laboratory experiments are performed where a piece of cartilage in contact with a bath of controlled chemical composition is subjected simultaneously to a mechanical load path phrased in terms of stress and/or strain. A first step in these simulations consists in obtaining the chemical composition of the cartilage assuming the bath and cartilage to be in chemical and mechanical equilibrium.

17.11.1 Structure of the constitutive equations

All species in the extrafibrillar phase can exchange with the surroundings except PGs. Let $E_{mo} = E - \{PG\}$ denote this set.

The extrafibrillar hydrogen ions participate to five independent reactions, namely exchange with the surroundings (bath), and four internal acid-base reactions, so that the variation of their mole number is the sum of five contributions,

$$\delta \mathcal{N}_{HE} = \delta \mathcal{N}_{HE}^{*} + \sum_{j \in [1,4]} \delta \mathcal{N}_{HE_{(j)}} \,. \tag{17.11.1}$$

Similarly, extrafibrillar calcium ions exchange with the surroundings and react with PGs,

$$\delta \mathcal{N}_{CaE} = \delta \mathcal{N}_{CaE}^{*} + \delta \mathcal{N}_{CaE_{(1')}} \,. \tag{17.11.2}$$

An additional mechanism should be accounted for if exchanges with the intrafibrillar compartment were activated.

The mole content of the extrafibrillar electrical charge can be calculated directly as,

$$\delta \mathcal{N}_{e} = -\delta \mathcal{N}_{P_1} - \delta \mathcal{N}_{P_2} + \delta \mathcal{N}_{R_3} - \delta \mathcal{N}_{P_4} + \delta \mathcal{N}_{R_{1'}} = - \sum_{j \in [1,4] \cup \{1'\}} \delta \mathcal{N}_{XE_{(j)}} \,, \tag{17.11.3}$$

in view of (17.3.16) and (17.4.4). Since the acid-base reactions and calcium binding are electrically neutral, the variation of the electrical density can be expressed in terms of the sole exchanges with the surroundings,

$$\delta I_{eE} = F \sum_{l \in E_{mo}} \zeta_l \, \delta \mathcal{N}_{lE}^{*} \,. \tag{17.11.4}$$

The electroneutrality constraint $\delta I_{eE} = 0$ will be satisfied by introduction of a Lagrange multiplier to be interpreted as the electrical potential ϕ_E.

The volume occupied by ions solvated or bound to the solid sites is assumed to be the same. Therefore, since all constituents are incompressible, the condition of compatibility of volume change of the porous medium expresses in the incremental format,

$$\delta I_{\mathrm{Inc}} = \det \mathbf{F} \, \mathrm{tr}\,(\delta \mathbf{F} \cdot \mathbf{F}^{-1}) - \sum_{l \in \mathrm{E_{mo}}} \widehat{v}_l \, \delta \mathcal{N}_{lE}^* = 0\,. \tag{17.11.5}$$

This constraint is satisfied by introduction of a second Lagrange multiplier to be interpreted as the pressure p_E in the fluid phase.

Along (17.3.1), the work done in a volume V_0 during an incremental process may be cast in a format that highlights the two constraints of incompressibility and electroneutrality,

$$\begin{aligned}
\delta \underline{\mathcal{W}} &= \mathbf{T} : \delta \mathbf{E} + \sum_{l \in \mathrm{E_{mo}}} g_{lE} \, \delta \mathcal{N}_{lE}^* + \sum_{j \in [1,4] \cup \{1'\}} \mathcal{G}_j \, \delta \mathcal{N}_{\mathrm{XE}_{(j)}} + p_E \, \delta I_{\mathrm{inc}} + \phi_E \, \delta I_{eE} \\
&= \overline{\mathbf{T}} : \delta \mathbf{E} + \sum_{l \in \mathrm{E_{mo}}} \overline{g}_{lE}^{\,\mathrm{ec}} \, \delta \mathcal{N}_{lE}^* + \sum_{j \in [1,4] \cup \{1'\}} \mathcal{G}_j \, \delta \mathcal{N}_{\mathrm{XE}_{(j)}}\,,
\end{aligned} \tag{17.11.6}$$

where g_{lE} denotes the chemical potential, as opposed to the electrochemical potential g_{lE}^{ec}, and the overlined quantities are shifted through the fluid pressure,

$$\overline{\mathbf{T}} = \mathbf{T} + p_E \det \mathbf{F} \, \mathbf{F}^{-1} \cdot \mathbf{F}^{-\mathrm{T}}; \quad \overline{g}_{lE}^{\,\mathrm{ec}} = g_{lE} + \mathrm{F} \, \zeta_k \, \phi_E - \widehat{v}_l \, p_E, \quad l \in \mathrm{E_{mo}}\,. \tag{17.11.7}$$

The set of independent variables $\mathbb{V}_E$ consists of
- the strain $\mathbf{E}$ of the solid skeleton (or porous medium);
- the sets of mole contents of chemical species $\mathrm{E_{mo}} \cup \mathrm{E_{rea}}$,

$$\mathrm{E_{mo}} = \mathrm{E} - \{\mathrm{PG}\}, \quad \mathrm{E_{rea}} = \{\mathcal{N}_{\mathrm{XE}_{(j)}}, j \in [1,4] \cup \{1'\}\}\,; \tag{17.11.8}$$

- the fluid pressure p_E and extrafibrillar electrical potential ϕ_E.

Hence the coupled chemo-hyperelastic constitutive equations can be cast in the format,

$$\overline{\mathbf{T}} = \frac{\partial \underline{\mathcal{W}}}{\partial \mathbf{E}}; \quad \overline{g}_{lE}^{\,\mathrm{ec}} = \frac{\partial \underline{\mathcal{W}}}{\partial \mathcal{N}_{lE}^*}, \quad l \in \mathrm{E_{mo}}; \quad \mathcal{G}_j = \frac{\partial \underline{\mathcal{W}}}{\partial \mathcal{N}_{\mathrm{XE}_{(j)}}}, \quad j \in [1,4] \cup \{1'\}\,, \tag{17.11.9}$$

subject to the constraints,

$$I_{\mathrm{inc}} = \frac{\partial \underline{\mathcal{W}}}{\partial p_E} = 0\,; \quad I_{eE} = \frac{\partial \underline{\mathcal{W}}}{\partial \phi_E} = 0\,. \tag{17.11.10}$$

Note that the work-conjugate variables include
- a mechanical pair, involving the work-conjugate stress $\mathbf{T}$ and strain $\mathbf{E}$;
- as many chemical pairs (electrochemical potential, mole content) as species that exchange matter with the surroundings;
- as many chemical pairs (chemical affinity, mole content) as extrafibrillar chemical reactions.

The chemoelastic energy

$$\underline{\mathcal{W}}(\mathbf{E}, \mathrm{E_{mo}}, \mathrm{E_{rea}}) = \underline{\mathcal{W}}_{\mathrm{ch-mech}}(\mathbf{E}, \mathrm{E_{mo}}) + \underline{\mathcal{W}}_{\mathrm{ch}}(\mathrm{E_{mo}}, \mathrm{E_{rea}}) + \underline{\mathcal{W}}_{\mathrm{ef}}(\mathrm{E_{rea}})\,, \tag{17.11.11}$$

is additively decomposed into
- a coupled chemomechanical contribution,

$$\underline{\mathcal{W}}_{\mathrm{ch-mech}}(\mathbf{E}, \mathrm{E_{mo}})\,; \tag{17.11.12}$$

- a chemical contribution,

$$\underline{\mathcal{W}}_{\mathrm{ch}}(\mathrm{E_{mo}}, \mathrm{E_{rea}}) = RT \left(\sum_{l \in \mathrm{E_{mo}}} \mathcal{N}_{lE} \, \mathrm{Ln}\, N_{lE} - \mathcal{N}_E \, \mathrm{Ln}\, N_E \right. \\
\left. + \sum_{j \in [1,4] \cup \{1'\}} \mathcal{N}_{\mathrm{P}_j} \, \mathrm{Ln}\, N_{\mathrm{P}_j} + \mathcal{N}_{\mathrm{R}_j} \, \mathrm{Ln}\, N_{\mathrm{R}_j} \right)\,, \tag{17.11.13}$$

with

$$\mathcal{N}_E = \sum_{l \in \mathrm{E_{mo}}} \mathcal{N}_{lE}\,; \tag{17.11.14}$$

- a term describing the enthalpies of formation,

$$\underline{\mathcal{W}}_{\text{ef}}(\text{E}_{\text{rea}}) = \sum_{j \in [1,4] \cup \{1'\}} \mathcal{G}_j^0 \, \mathcal{N}_{\text{XE}_{(j)}} \,, \tag{17.11.15}$$

with $\mathcal{G}_j^0 = -RT \, \text{Ln} \, (10^{-\text{pK}_j} \, \widehat{v}_{\text{w}})$, $j \in [1,4] \cup \{1'\}$.

The constitutive equations (17.11.9) can be recast in terms of Cauchy stress in the format,

$$\boldsymbol{\sigma} = -p_{\text{E}} \, \mathbf{I} + (\det \mathbf{F})^{-1} \, \mathbf{F} \cdot \frac{\partial \mathcal{W}_{\text{ch−mech}}}{\partial \mathbf{E}} \cdot \mathbf{F}^{\text{T}} \,,$$

$$g_{l\text{E}}^{\text{ec}} = \widehat{v}_l \, p_{\text{E}} + \frac{\partial \mathcal{W}_{\text{ch−mech}}}{\partial \mathcal{N}_{l\text{E}}} + RT \, \text{Ln} \, x_{l\text{E}} + \text{F} \, \zeta_l \, \phi_{\text{E}}, \quad l \in \text{E}_{\text{mo}} - \{\text{H}\} - \{\text{Ca}\},$$

$$g_{\text{HE}}^{\text{ec}} = \widehat{v}_{\text{H}} \, p_{\text{E}} + \frac{\partial \mathcal{W}_{\text{ch−mech}}}{\partial \mathcal{N}_{\text{HE}}^*} + RT \, \text{Ln} \, x_{\text{HE}} + \text{F} \, \phi_{\text{E}}, \tag{17.11.16}$$

$$g_{\text{CaE}}^{\text{ec}} = \widehat{v}_{\text{Ca}} \, p_{\text{E}} + \frac{\partial \mathcal{W}_{\text{ch−mech}}}{\partial \mathcal{N}_{\text{CaE}}^*} + RT \, \text{Ln} \, x_{\text{CaE}} + 2 \, \text{F} \, \phi_{\text{E}},$$

$$\mathcal{G}_j = RT \, \text{Ln} \, \frac{c_{\text{P}_j}}{c_{\text{R}_j}} \frac{c_{\text{XE}}}{10^{-\text{pK}_j}}, \quad j \in [1,4] \cup \{1'\} \,,$$

subject to the conditions (17.11.10) of incompressibility and electroneutrality.

17.11.2   Equilibria

At the time scale of interest here, the acid-base reactions, as well as the exchanges between the cartilage and the surroundings, are considered at equilibrium. Moreover, it is tacitly understood that all species in the set E_{mo} can exchange with the bath and that the latter contains only these species.

Therefore, the equations that control the chemomechanical evolution of the cartilage in contact with a bath of given chemical composition govern

- exchanges with the surroundings:

$$g_{l\text{E}}^{\text{ec}} = g_{l\text{B}}^{\text{ec}}, \quad l \in \text{E}_{\text{mo}}; \tag{17.11.17}$$

- acid-base reactions and calcium binding:

$$\mathcal{G}_j = 0, \quad j \in [1,4] \cup \{1'\}; \tag{17.11.18}$$

- compatibility of molar fractions:

$$\sum_{l \in \text{K}} x_{l\text{K}} = 1, \quad \text{K} = \text{E}, \text{B}; \tag{17.11.19}$$

- electroneutrality in terms of the molar fraction $y_{\text{e}} = \widehat{v}_{\text{w}} \, e_{\text{e}}$:

$$x_{\text{NaK}} + 2 \, x_{\text{CaK}} + y_{\text{e}} \, I_{\text{KE}} + x_{\text{HK}} - x_{\text{OHK}} - x_{\text{ClK}} = 0, \quad \text{K} = \text{E}, \text{B} \,; \tag{17.11.20}$$

- water dissociation in terms of molar fractions:

$$x_{\text{HK}} \, x_{\text{OHK}} = 4 \times 10^{-18}, \quad \text{K} = \text{E}, \text{B} \,. \tag{17.11.21}$$

Mechanical equilibrium should be added to the above conditions to close the problem. The electrochemical potentials of the species in the cartilage are provided by the constitutive equations (17.11.16) while the expressions for the corresponding species in the bath are given by (17.3.2).

17.11.3 A word on the kinetics of acid-base reactions and calcium binding

In this continuum thermodynamics framework, the acid-base reactions and calcium binding may be described through a first order kinetics,

$$\frac{\delta N_{\mathrm{XE}_{(j)}}}{\delta t} = -\frac{1}{t_{\mathrm{T}j}} \frac{\mathcal{G}_j}{RT}, \quad j \in [1,4] \cup \{1'\}, \tag{17.11.22}$$

where the t_{T}'s are the characteristic reaction times. When introduced in the mass balance of the ions, the above mass transfers combine with the constitutive equations of diffusion to yield partial differential equations of reaction-diffusion. The chemical reactions can be shown to result in retardated diffusion times. Grodzinsky et al. [1981], Fig. 9, show in particular that acid-base reactions, and, to a minor extent, calcium binding retard considerably the occurrence of equilibrium.

At this point, it may be worth to mention another source of retardation of equilibrium, namely that due to multiple porosities. Mass transfers between two fluid compartments can be viewed as physico-chemical reactions. The associated retardation effect is examined in Gajo and Loret [2003] in the context of reactive clays, where the two fluid compartments can be interpreted as the counterparts of the intra- and extrafibrillar compartments of articular cartilages.

The format (17.11.22) satisfies automatically the dissipation inequality $-\mathcal{G}_j \, \delta N_{\mathrm{XE}_{(j)}} \geq 0$, which derives from (4.7.31), viewing the chemical reaction as a mass transfer between two phases. The inequality requires that the sum of the chemical works associated with the chemical reactions, e.g., (17.3.9), should be negative.

17.12 Chemical softening/stiffening by fiber de-/activation

At physiological pH, the compressive stiffness of articular cartilages is contributed mostly by the ground substance via the electrical repulsion between charged GAGs. The collagen network contributes as well at non physiological pHs. Indeed, at low and high pHs, the collagen molecules are charged, and their repulsion contributes to a significant compressive stiffness.

The tensile stiffness is provided by the ground substance and mostly by collagen fibers. Under an extension, the fibers progressively uncrimp so as to carry the tensile load. They can be considered fully activated if their extension is larger than some characteristic value. Fibers aligned with a tensile load are certainly in extension, while fibers along orthogonal directions are likely to undergo contraction and therefore to be mechanically inactive.

Emphasis is laid below on the interactions between the directional properties of the fiber network and chemomechanical couplings. The model displays *a swelling induced stiffening* by fiber recruitment and, conversely, *a shrinking induced softening* by fiber deactivation.

Indeed, consider the material to undergo uniaxial traction while swelling emanates from some chemomechanical process. Swelling will imply the lateral strains to become less contractive. Therefore, more collagen fibers will be activated. Consequently, the axial stiffness is certainly bound to increase. Simulations will show that this inherent stiffening/softening effect is significant.

To simplify the analysis, full activation of individual fibers is considered to take place as soon as the fibers undergo an extension. Progressive uncrimping of individual fibers is qualitatively introduced through a macroscopic nonlinear stress-strain relation.

17.12.1 A single family of collagen fibers

The unit vector $\underline{\mathbf{m}}_{\mathrm{c}}$ indicating the direction of a collagen fiber in the reference configuration is expressed in terms of the angles $\underline{\alpha} \in [0, 2\pi[$ and $\underline{\beta} \in [0, \pi[$, Fig. 17.12.1. Let $\underline{\mathbf{M}}_{\mathrm{c}} = \underline{\mathbf{m}}_{\mathrm{c}} \otimes \underline{\mathbf{m}}_{\mathrm{c}}$.

The collagen fibers are mechanically active only in extension, namely when $\langle \underline{\mathbf{m}}_{\mathrm{c}} \cdot \mathbf{C} \cdot \underline{\mathbf{m}}_{\mathrm{c}} - 1 \rangle$ is nonzero, where $\mathbf{C} = \mathbf{F}^{\mathrm{T}} \cdot \mathbf{F}$ is the right stretch associated with the deformation gradient $\mathbf{F}$ and the operator $\langle \cdot \rangle$ denotes the positive part of its argument. Then along the lines of Section 12.2.1, the strain energy of collagen fibers expresses in terms of the right stretch or Green strain $\mathbf{E} = (\mathbf{C} - \mathbf{I})/2$ in the format,

$$\underline{w}_{\mathrm{c}} = \underline{w}_{\mathrm{c}}(L^2), \tag{17.12.1}$$

where

$$L = \tfrac{1}{2} \langle \mathbf{C} : \underline{\mathbf{M}}_{\mathrm{c}} - 1 \rangle = \langle \mathbf{E} : \underline{\mathbf{M}}_{\mathrm{c}} \rangle. \tag{17.12.2}$$

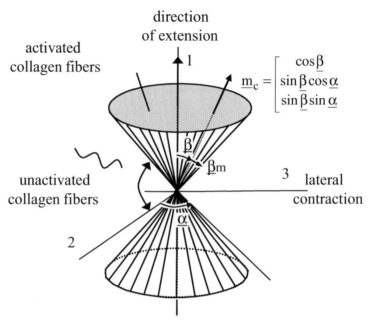

Fig. 17.12.1: Cones of activation of collagen fibers in axial extension and lateral contraction.

In (17.12.1), the square, actually any power strictly greater than 1, is required to ensure continuity of the stress at the extension-contraction transition, while, in contrast, the stiffness may, or may not, be discontinuous. Note that use of the deviatoric deformation in the above energy would lead isotropic extension, like swelling, to leave collagen fibers inactive, which is in contradiction with the chemically induced stiffening/softening effect. The standard exponential form of energy, defined by the modulus K_c [unit : MPa] and the dimensionless exponent k_c, is adopted,

$$
\underline{w}_c = \frac{K_c}{2\,k_c}\left(\exp(k_c\,L^2) - 1\right),
$$

$$
\frac{\partial \underline{w}_c}{\partial \mathbf{E}} = K_c\,\exp(k_c\,L^2)\,L\,\underline{\mathbf{M}}_c,
$$
(17.12.3)

$$
\frac{\partial^2 \underline{w}_c}{\partial \mathbf{E}\partial \mathbf{E}} = K_c\,\exp(k_c\,L^2)\,\big(2\,k_c\,\underbrace{L^2}_{\text{continuous}} + \underbrace{\mathcal{H}(\mathbf{E}:\underline{\mathbf{M}}_c)}_{\text{discontinuous}}\big)\,\underline{\mathbf{M}}_c \otimes \underline{\mathbf{M}}_c.
$$

Along this approach, the fiber orientation matters but its *sense* is elusive, i.e., $\underline{\mathbf{m}}_c$ and $-\underline{\mathbf{m}}_c$ play the same role.

17.12.2 The stress of the collagen network

The fibers are considered to be independent, namely there are no crosslinks between fibers nor interaction with the matrix. The apparent strain energy, namely the energy per unit volume of tissue, obtained by integration over the unit sphere $\mathbb{S}^2$,

$$
\underline{\mathcal{W}}^c(\mathbf{E}) = n^c \int_{\mathbb{S}^2} \underline{w}_c(L^2)\,\Phi(\underline{\alpha},\underline{\beta})\,d\underline{S},
$$
(17.12.4)

provides the stress of the collagen network,

$$
\mathbf{T}^c = \frac{\partial \underline{\mathcal{W}}^c(\mathbf{E})}{\partial \mathbf{E}} = K^c \int_{\mathbb{S}^2} L\,\exp(k_c\,L^2)\,\underline{\mathbf{M}}_c\,\Phi(\underline{\alpha},\beta)\,d\underline{S},
$$
(17.12.5)

with $K^c = n^c\,K_c$. If the directional distribution of collagen fibers shows a rotational symmetry about the axis 1, so that $\Phi = \Phi(\underline{\beta})$, then the stress components can be cast in the format,

$$
T_{ij}^c = K^c \int_0^{\pi/2} d\underline{\beta}\,\sin\beta\,L\,\exp(k_c\,L^2)\,4\,\pi\,\Phi(\underline{\beta})
\begin{cases}
(\cos\beta)^2, & i=j=1; \\
(\sin\beta)^2/2, & i=j=2,3; \\
0, & i \neq j.
\end{cases}
$$
(17.12.6)

For uniaxial traction along the symmetry axis 1, the activation criterion relies on the elongation L,

$$L = \langle \mathbf{E} : \underline{\mathbf{M}}_c \rangle = \langle E_{11} \left(\cos \underline{\beta}\right)^2 + E_{22} \left(\sin \underline{\beta}\right)^2 \rangle. \tag{17.12.7}$$

In general, the directions for which extension takes place are not known beforehand and the integration can not be performed analytically, except in particular cases. Of course, if the transverse directions 2 and 3 undergo extension we recover an isotropic response. If the transverse directions undergo contraction then the fibers which undergo extension $\mathbf{E} : \underline{\mathbf{M}}_c \geq 0$,

$$\cot^2 \underline{\beta} = \frac{m_{c1}^2}{m_{c2}^2 + m_{c3}^2} \geq \cot^2 \underline{\beta}_m \equiv \frac{-E_{22}}{E_{11}}, \tag{17.12.8}$$

form, as sketched in Fig. 17.12.1, two symmetric cones of half angle $\underline{\beta}_m$ about the extension direction,

$$\underline{\beta} \in [0, \underline{\beta}_m] \cup [\pi - \underline{\beta}_m, \pi]. \tag{17.12.9}$$

Henceforth, we have in mind a collagen piece taken in the middle zone of the cartilage layer, as in Grodzinsky et al. [1981], so that the directional distribution of collagen fibers $\Phi(\underline{\alpha}, \underline{\beta})$ can be assumed to be uniform, namely $4\pi \Phi(\underline{\alpha}, \underline{\beta}) = 1$. Nevertheless, the influence of a non isotropic directional distribution of collagen fibers has been tested. The directional distribution $\Phi(\underline{\beta})$ is described by a single harmonic,

$$\Phi(\underline{\beta}) = \frac{\Phi_0}{4\pi} + \frac{3}{4\pi} \left(1 - \Phi_0\right) \left(\cos \underline{\beta}\right)^2, \tag{17.12.10}$$

with a coefficient $\Phi_0 \in]0, 3/2[$. The ratio of the fiber densities along the symmetry axis and in an orthogonal direction is equal to $3/\Phi_0 - 2$. Isotropy corresponds to $\Phi_0 = 1$.

17.12.3 Chemoelastic energy with electro-chemomechanical couplings

The general framework developed in Section 17.11 is now specialized so as to simulate intermingled chemical and mechanical loading paths available in the literature. These data unambiguously show the strong influence of ionic strength and pH on the material response. However, since no sufficient information on the tested articular cartilage is available, only a prototype of mechanical model can be formulated.

The chemomechanical part of the chemoelastic energy

$$\underline{\mathcal{W}}_{\text{ch-mech}}(\mathbf{E}, \text{E}_{\text{mo}}) = \underline{\mathcal{W}}_{\text{ch},1}(\mathbf{E}, \text{E}_{\text{mo}}) + \underline{\mathcal{W}}_{\text{ch},2}(\text{E}_{\text{mo}}) \left(\underline{\mathcal{W}}^{\text{gs}}(\mathbf{E}) + \underline{\mathcal{W}}^{\text{c}}(\mathbf{E})\right), \tag{17.12.11}$$

is contributed by
  - the mechanical energy $\underline{\mathcal{W}}^{\text{gs}}(\mathbf{E})$ of the matrix, PGs and other proteins (ground substance);
  - the mechanical energy $\underline{\mathcal{W}}^{\text{c}}(\mathbf{E})$ of the activated collagen fibers;
  - a coupled chemomechanical term

$$\underline{\mathcal{W}}_{\text{ch},1}(\mathbf{E}, \text{E}_{\text{mo}}) = -p_{\text{ch}}(\text{E}_{\text{mo}}) \left(\det \mathbf{F} - 1\right), \tag{17.12.12}$$

  - a purely chemical term $\underline{\mathcal{W}}_{\text{ch},2}(\text{E}_{\text{mo}})$.
The coupled chemomechanical term gives rise to the chemical stress $\mathbf{T}_{\text{ch}} = -p_{\text{ch}} \det \mathbf{F} \, \mathbf{F}^{-1} \cdot \mathbf{F}^{-\text{T}}$ which results in an isotropic Cauchy stress $\boldsymbol{\sigma}_{\text{ch}} = -p_{\text{ch}} \mathbf{I}$. The constitutive stresses of the ground substance and of the collagen,

$$\mathbf{T}^{\text{gs}} = \frac{\partial \underline{\mathcal{W}}^{\text{gs}}}{\partial \mathbf{E}}, \quad \mathbf{T}^{\text{c}} = \frac{\partial \underline{\mathcal{W}}^{\text{c}}}{\partial \mathbf{E}}, \tag{17.12.13}$$

are to be understood as volume weighted averages. Moreover, both are representative of the stresses of the matrix and collagen in the hypertonic state, that is, in absence of chemical effects. The multiplicative chemical term $\underline{\mathcal{W}}_{\text{ch},2}$ is intended to indicate that a smaller ionic strength amplifies the stiffness, both in confined compression and in uniaxial traction, Loret and Simões [2004], their Fig. 5.

The mechanical constitutive equations (17.11.16),

$$\mathbf{T} + (p_{\text{E}} + p_{\text{ch}}) \det \mathbf{F} \, \mathbf{F}^{-1} \cdot \mathbf{F}^{-\text{T}} = \underline{\mathcal{W}}_{\text{ch},2}(\text{E}_{\text{mo}}) \left(\frac{\partial \underline{\mathcal{W}}^{\text{gs}}}{\partial \mathbf{E}} + \frac{\partial \underline{\mathcal{W}}^{\text{c}}}{\partial \mathbf{E}}\right),$$

$$\mathbf{T} - \mathbf{T}_{\text{inc}} - \mathbf{T}_{\text{ch}} = \underline{\mathcal{W}}_{\text{ch},2}(\text{E}_{\text{mo}}) \left(\mathbf{T}^{\text{gs}} + \mathbf{T}^{\text{c}}\right) \tag{17.12.14}$$

may be rephrased in terms of the Cauchy stresses with $\boldsymbol{\sigma}^{gs}$ and $\boldsymbol{\sigma}^c$ associated with the second Piola-Kirchhoff stresses $\mathbf{T}^{gs}$ and $\mathbf{T}^c$, so as to display an effective stress,

$$\boldsymbol{\sigma} + \overbrace{(p_E + p_{ch})}^{p_{eff}} \mathbf{I} = \underline{\mathcal{W}}_{ch,2}(E_{mo}) \left(\boldsymbol{\sigma}^{gs} + \boldsymbol{\sigma}^c\right). \tag{17.12.15}$$

The cartilage specimen is assumed to be in equilibrium with a fictitious bath at pressure $\tilde{p}_E$. The concept of a fictitious bath is exposed in Section 14.5.2. The constitutive equation (17.12.15) is then recast in the format,

$$\boldsymbol{\sigma} + \tilde{p}_E \mathbf{I} + (\pi_{osm} + p_{ch}) \mathbf{I} = \underline{\mathcal{W}}_{ch,2}(E_{mo}) \left(\boldsymbol{\sigma}^{gs} + \boldsymbol{\sigma}^c\right), \tag{17.12.16}$$

that involves the osmotic pressure $\pi_{osm} = p_E - \tilde{p}_E$. The properties of the fictitious bath are deduced from those of the cartilage. The osmotic pressure expresses in terms of molar fractions, which can themselves be expressed in terms of mole contents. Therefore, the osmotic pressure can be seen as a function of the state variables. It consequently qualifies to enter the constitutive equations.

For a general boundary value problem, the indeterminate pressure p_E is defined by boundary conditions. Here, we consider a homogeneous specimen in equilibrium with a (real) bath. Therefore, the fictitious bath and the real bath can be identified, $\tilde{p}_E$ is understood as the pressure p_B of the real bath, chemical equilibrium provides π_{osm} and then the pressure p_E in the cartilage is equal to $\pi_{osm} + \tilde{p}_E$.

17.12.4   Constitutive parameters

In the hypertonic state, the high ionic strength shields almost completely the electrochemical repulsion forces between fixed charges of identical sign. The hypertonic state is therefore a likely candidate as a reference state.

Hypertonic elastic coefficients of the ground substance

The moduli that come into picture in the macroscopic model are apparent moduli weighted by the volume fractions. The apparent isotropic tensor moduli $\mathbb{E}^{gs}$ of the matrix are defined by the Lamé moduli,

$$\lambda = n^{gs} \Lambda_{gs}, \quad \mu = n^{gs} M_{gs}, \tag{17.12.17}$$

built from the intrinsic moduli of the matrix Λ_{gs} and M_{gs} and scaled by its volume fraction n^{gs}. The order of magnitude of the moduli of the ground substance in the hypertonic state can be estimated from the confined compression tests of Eisenberg and Grodzinsky [1985], namely $\lambda = \mu = 100\,\text{kPa}$.

Hypertonic elastic coefficients of the collagen network

The modulus K_c of collagen fibrils of bovine Achilles tendons is reported to be 430 MPa by Sasaki and Odajima [1996]. Lower values are mentioned for collagen fibrils (of type I) of annulus fibrosus, namely about 50 MPa, Pezowicz et al. [2005]. Federico et al. [2005] report values from Athanasiou et al. [1991] about 10 MPa for articular cartilages. For collagen, the volume fraction n^c expresses as the product of the volume fraction of the fluid phase $n^E = 0.8$, times the concentration of collagen with respect to the fluid phase $c_c = 0.8\,\text{mole/m}^3$, times the molar volume of collagen $\hat{v}_c = 0.2\,\text{m}^3/\text{mole}$. Therefore $n^c = 0.128$. The simulations have been run with a modulus $K^c = n^c K_c = 1\,\text{MPa}$ and an exponent $k_c = 100$. At low ionic strength, these values yield an axial stress of about 1 MPa for an axial strain of 15%.

Next to the uncrimping stage, the apparent moduli of a collagen network increase with deformation up to the stage where all fibers are activated. Since the uncrimping stage is not described here, the moduli mentioned above refer to the fully activated stage, Fig. 12.2.2. The existence of this (nonlinear) uncrimping stage explains in part the wide dispersion of experimental moduli: a common reference configuration and a common strain at which the measurements are reported should be known for a rational comparison of these data.

Coefficients of the chemomechanical couplings

The constitutive chemomechanical couplings are motivated by the following remarks:

- sodium cations Na^+ and chloride anions Cl^- contribute only to the shielding effect in articular cartilages, screening the repulsion between fixed charges of identical sign located on the PGs, and collagen;

- calcium cations Ca^{2+} display the shielding effect as well, which should be more efficient than the one of sodium ions due to the valence 2, and, in addition, they increase the algebraic value of the charge through binding to carboxyl sites of PGs. These two effects cooperate above the isoelectric point (IEP), and compete below the IEP. Note however that binding concerns only part of the sites and calcium binding can not by itself alone change the sign of the fixed charge;

- hydrogen ions both shield the charge and bind to the PGs and collagen sites and are able to change the sign of the fixed charge.

The osmotic pressure is large at low ionic strength and vanishes at large ionic strength and at the IEP. It is therefore a likely candidate to carry the extremum properties that are expected for the constitutive pressure p_{ch} and function $\underline{\mathcal{W}}_{\mathrm{ch},2}$. A tentative format would adopt a linear and an affine dependence with respect to the osmotic pressure, say $p_{\mathrm{ch}} = \alpha_p\,\pi_{\mathrm{osm}}$, and $\underline{\mathcal{W}}_{\mathrm{ch},2} = 1 + \alpha_{\mathrm{w}}\,\pi_{\mathrm{osm}}/\pi_{\mathrm{osm}}^0$ where α_p and α_{w} are dimensionless constants. In the tests to be described below, the only data that are available are the axial stresses relative to their values at a certain stage of the loading history. The coefficients $\alpha_p = 0$ and $\alpha_{\mathrm{w}} = 10$ in (17.12.18) imply this stress ratio to be about $1/4$ in the hypertonic state in agreement with data and the isotropic free swelling strain to be about 4%. The osmotic pressure π_{osm}^0 induced by isotropic free swelling, up to $10^{-4}\,\mathrm{M}$ of NaCl at $\mathrm{pH_B}{=}7$, is estimated to be about $325\,\mathrm{kPa}$.

Assume the material to be strained, subsequently to isotropic free swelling, at fixed bath chemical composition. Then the constitutive equation (17.12.16) expresses the axial stress in the form,

$$\sigma_{11} + \tilde{p}_{\mathrm{E}} = \underline{\mathcal{W}}_{\mathrm{ch},2}\,(\sigma_{11}^{\mathrm{gs}} + \sigma_{11}^{\mathrm{c}}) - (\pi_{\mathrm{osm}} + p_{\mathrm{ch}})\,. \tag{17.12.18}$$

17.13 Mechanical and chemical tests at varying pH

Cartilage strips cut from bovine femoropatellar joints, of 1.5 to 2 year old animals, were tested by Grodzinsky et al. [1981]. Since the strips are taken from the middle zone, the collagen fibers can be assumed to be randomly distributed over all spatial directions.

The cartilage specimens, in contact with a bath of controlled chemical composition, are subjected to intermingled chemical and mechanical loadings. The complete simulations start from a hypertonic state in presence of sodium cations at 1 M and involve three sequences, as sketched in Fig. 17.13.1, namely

- sequence 1: isotropic free swelling OA, i.e. decrease progressively the ionic strength of sodium of the bath from pNaB=0 to pNaB=4;
- sequence 2: uniaxial traction at fixed bath composition AB, i.e. apply an additional axial strain of 10%;
- sequence 3: chemical loading BC, i.e. increase the ionic strength of the bath while maintaining the axial strain constant. The test is actually not isometric since the lateral strains vary.

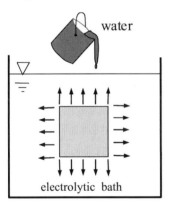
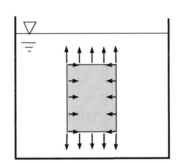
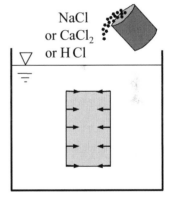

OA: isotropic free swelling

AB: uniaxial traction
<u>imposed</u> axial elongation
<u>resulting</u> lateral contraction

BC: increased ionic strength
at fixed axial length
<u>resulting</u> lateral contraction

Fig. 17.13.1: Cartoon of the three sequences of the chemomechanical tests to which articular cartilages are subjected. The electrolytic bath is kept at atmospheric pressure, but its chemical composition is modified in sequences 1 and 3. OA: isotropic free swelling starting from hypertonic state; AB: uniaxial traction at fixed bath chemical composition; BC: increasing ionic strength of the bath at constant axial strain.

Data are reported as a function of the $\mathrm{p}k_{\mathrm{B}} = -\log_{10} c_{k\mathrm{B}}$ of the bath, the concentration $c_{k\mathrm{B}} = 10^{-\mathrm{p}k_{\mathrm{B}}}$ being measured in mole/l, with k=Na, Ca, H. If $c_{k\mathrm{B}}$ is measured in mole/m^3, these relations write respectively $\mathrm{p}k_{\mathrm{B}} = -\log_{10}(c_{k\mathrm{B}}/10^3)$ and $c_{k\mathrm{B}}/10^3 = 10^{-\mathrm{p}k_{\mathrm{B}}}$.

During sequence 1, the pH of the bath is maintained constant, equal to 7, except for the H-test where the pH is equal to 6.

The third sequence consists in increasing the bath concentrations of sodium chloride (Na-test), calcium chloride (Ca-test) or hydrochloric acid (H-test). For the Na- and Ca-tests, the pH in the bath is maintained equal to 7. For the Na- and H-tests, the amount of calcium is virtually nil. In detail,

- the conditions at B for the Na-test are $pNa_B=4$, $pH_B=7$;
- for the Ca-test, the neutralizing cations at B are Na^+, with $pNa_B=6$, the initial pCa_B is 4 and the initial pH_B is equal to 7;
- for the H-test, the free swelling test stops for a concentration of Na^+ equal to 0.06 M and the pH at B is equal to 6.

For a conewise nonlinear response of collagen fibers, the lateral boundary condition for uniaxial traction along the direction 1,

$$0 = \sigma_{22} + \tilde{p}_E = \underline{\mathcal{W}}_{ch,2}(E_{mo}) \left(\sigma_{22}^{gs} + \sigma_{22}^{c} \right) - \left(\pi_{osm} + p_{ch} \right), \qquad (17.13.1)$$

becomes an implicit equation for the lateral strain, which has to be solved numerically. Moreover, for a randomly distributed fiber network, isotropic free swelling is recovered by constraining the strain to be isotropic.

The simulations below have been performed with both quadratic and exponential strain energy functions for the collagen fibrils. Unlike mechanical variables, the chemical variables (concentrations, fixed charge, etc.) do not show neither qualitative nor quantitatively significant differences for the two models: therefore the mechanochemical coupling is weak. Such is not the case for the corneal stroma, Section 19.11.4.

Only the nonlinear simulations are presented. They are displayed in Figs. 17.13.2 to 17.13.5:

1. during the free swelling sequence, at decreasing concentration of Na^+, the pH in the cartilage fluid decreases much, in agreement with the findings of Loret and Simões [2010]a [2010]b. Therefore the fixed electrical charge decreases (in absolute value). On the other hand, decreasing ionic concentration alone would lead to increase of the osmotic pressure. These two competing effects imply the osmotic pressure to show a maximum. The phenomenon is displayed essentially in the Na-test because the swelling sequence is limited in the H-test. In presence of calcium, the variation of cartilage pH is tiny and the osmotic pressure does not show such a marked maximum;

2. in the Na-test, the concentration of Na^+ in the cartilage returns to its initial value. So do also the other concentrations in the cartilage, as another indication that the *mechanochemical coupling is weak*;

3. on the other hand, strain and stress components do not follow a closed path, since the mechanical loading path is not closed. Further, the mechanical response (stress and strain components) is strongly affected by the chemical loading of sequence 3: this feature is a witness that the *chemomechanical coupling is strong*;

4. during sequence 3, at increasing ionic strength, the stress falls down onto the hypertonic curve at large ionic strength of the bath. This feature is a consequence of the shielding effect, the fixed charge varying little. For the H-test, the strong variation of the fixed charge either collaborates or competes with shielding as explained below;

5. indeed the decrease is not monotonous for the H-test, because the isoelectric state is reached first, at $pH_B=pH_E=2.6$. Let us recall that, at the isoelectric point (IEP), the chemical concentrations in the bath and extrafibrillar fluid are virtually identical (to within the tiny concentration of PGs) and therefore the osmotic pressure vanishes. Once the IEP is passed, subsequent decrease of the bath pH implies an increasing positive fixed charge of the collagen network. This increasing fixed charge and the increasing ionic strength compete to control the magnitude of the stress, which first increases, reaches a local maximum at $pH_B \sim 1$, and finally falls down on to the hypertonic curve at larger ionic strength. In summary, for the H-test, the stress falls down to the hypertonic line twice, first because the fixed charge vanishes at the IEP, and second as a consequence of the strong shielding of the electrical charge by the mobile ions due to the large ionic strength;

6. the volumetric contraction is maximum in the hypertonic state which, as already mentioned, is reached either due to a large ionic strength (Na-, Ca- and H-curves) or at the isoelectric point (H-curve);

7. as indicated by (17.12.16), during isotropic free swelling, the chemical pressure driven by the electrical repulsion between cartilage units is resisted by both the matrix (PGs) and the collagen network. During uniaxial traction, the latter have, in addition, to resist the external load. In fact, during the swelling sequence 1, all collagen fibers are in extension, corresponding to an angle $\underline{\beta}_m=90°$, Fig. 17.12.1. The lateral strain reduces during the traction sequence 2. This decrease continues during the chemical loading sequence 3, associated with shrinking, which, since the axial strain is fixed, can take place only along the lateral direction. During the sequences 2 and 3, the angle $\underline{\beta}_m$ decreases progressively, as soon as the lateral strain becomes negative. The actual range of variation $\underline{\beta}_m$ depends, according to

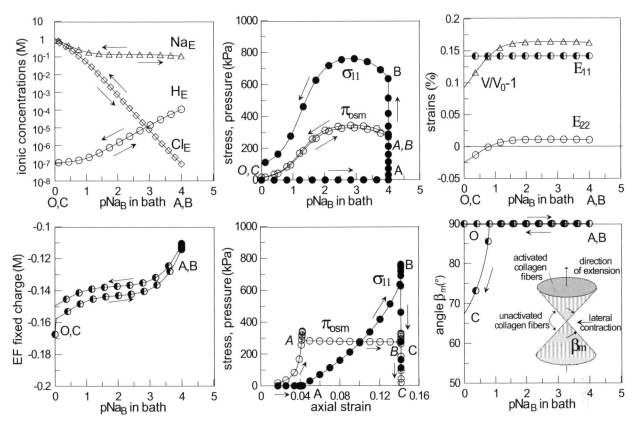

Fig. 17.13.2: Na-test. Simulations of a three sequence loading path OA: isotropic free swelling starting from hypertonic state; AB: uniaxial traction at fixed bath chemical composition; BC: increasing ionic strength of the bath at constant axial strain. Concentrations of ionic species in the cartilage, axial stress, osmotic pressure, strains, extrafibrillar electrical charge and angle $\underline{\beta}_m$ defining the set of fibers in extension, during the three sequences. The ionic strength of the bath is measured through a $pk_B = -\log_{10} c_{kB}$, the concentration $c_{kB} = 10^{-pk_B}$ being measured in mole/l, with k=Na,Ca,H.

The stress varies due to mechanical loading AB and chemical loading BC, because the chemomechanical coupling is strong, while the osmotic pressure varies essentially due to chemical loadings OA and BC, but little due to mechanical loading because the mechanochemical coupling is weak.

Lateral contraction during increase of ionic strength implies the cones of fibers which undergo extension to shrink, namely the angle $\underline{\beta}_m$ decreases, as another witness of the strong chemomechanical coupling.

When the concentration of metallic ions in the bath increases, path BC, the two adverse effects of electrical shielding and increase of fixed charge compete and the former effect takes over, leading to contraction. The situation is somehow different in the H-test because the fixed charge changes sign, leading to a non monotonous evolution of the strains.

In the Na-test, the chemical loading path is closed, while the mechanical loading path is not. Still the ionic concentrations in the cartilage are almost closed as the mechanochemical coupling is weak.

As the concentration in metallic ions of the bath increases, path BC, the cartilage becomes less acidic, so that the fixed charge becomes more negative. The opposite effect takes place when the pH of the bath decreases and the total fixed charge changes sign at the IEP, Fig. 17.13.4. Reprinted with permission from Loret and Simões [2010]b.

the relation (17.14.7) below, on the ratio $(\pi_{osm} + p_{ch})/\mathcal{W}_{ch,2}$ which could be estimated from isotropic free swelling; the larger this ratio, the larger the lateral strain and the larger the angle $\underline{\beta}_m$;

8. calcium binding to the carboxyl sites of PGs lowers the stress and osmotic pressure since it reduces the fixed charge, Fig. 17.13.3. The influence is not that apparent in the stress ratio of the Ca-test in Fig. 17.13.5;

9. the axial stresses relative to their values at the beginning of sequence 3 are the only data provided by Grodzinsky et al. [1981]. The simulations show indeed ratios quite in agreement with these data, Fig. 17.13.5. However, a systematic shift takes place at low ionic strength (large pk_B), as if the model was overestimating small concentrations. Calcium binding improves slightly the match, but large values of the binding constant decrease the stress ratio in the hypertonic state.

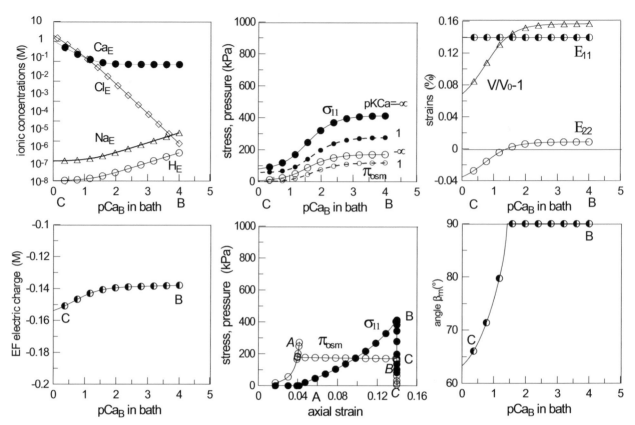

Fig. 17.13.3: Ca-test. Same loading path as in Fig. 17.13.2 but, in the third sequence BC, the concentration of calcium $c_{CaB} = 10^{-pCaB}$ [unit : mole/l], rather than sodium, is increased in the bath. Only this third sequence is displayed, except for the stress-strain curves. Binding of calcium is characterized by the constant pKCa, namely pKCa$=-\infty$ (no binding), pKCa$=1$ (binding at large concentrations only). The osmotic pressure, and resulting stress, are much smaller than for the Na-test, in agreement with the findings of Section 16.3.1 and Fig. 16.3.1. Reprinted with permission from Loret and Simões [2010]b.

17.14   The limit case of a conewise linear collagen response

The simulations above use a conewise nonlinear stress-strain response of the collagen fibers. Still a number of issues related to the couplings are easier exposed in a linear context. In fact, while *chemomechanical* couplings are strong, *mechanochemical* couplings are weak, and the mechanical nonlinearity affects little the chemical concentrations and pH effects.

17.14.1   The apparent Poisson's ratio of the solid skeleton

In the limit of a small strain, the energy of a single fiber may be linearized as indicated by (12.2.16). If *all* fibers are in extension, then the integration of the strain energy over fiber directions yields an isotropic stiffness with equal Lamé moduli, that is, with a Poisson's ratio equal to 0.25,

$$\lambda^c = \mu^c = \frac{K^c}{15}, \tag{17.14.1}$$

In the principal axes of strain, we just need to calculate four components of the stiffness, in view of the symmetry about the axis 1. Therefore, the contribution of the collagen network to the total stress expresses in terms of the symmetric matrix (12.2.21) with coefficients given in explicit form by (12.2.22). The strain dependence of the collagen moduli, namely $\mathbb{E}^c = \mathbb{E}^c(\epsilon)$, implied by its conewise linear response is apparent through the angle β_m.

The strain energy of the solid skeleton, or dry tissue, is associated with the stiffness $\mathbb{E}^{car}(\epsilon) = \mathbb{E}^{gs} + \mathbb{E}^c(\epsilon)$. For loadings along the axis 1 this strain energy can be shown to remain positive, even if the aggregate Poisson's ratio,

$$\nu^{car} = \frac{E_{12}^{car}}{E_{22}^{car} + E_{23}^{car}}, \tag{17.14.2}$$

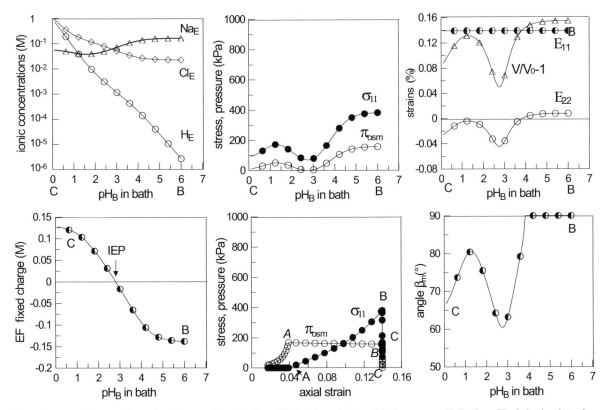

Fig. 17.13.4: H-test. Same loading path as in Fig. 17.13.2 but, in the third sequence BC, the pH of the bath is decreased. The hydrogen concentration in the bath c_{HB} [unit : mole/l] is equal to 10^{-pHB}. The stress reaches the hypertonic value, and the osmotic pressure vanishes, at the IEP$\sim$ 2.6 and at large ionic strength. A number of mechanical entities show an extremum at the isoelectric point. Reprinted with permission from Loret and Simões [2010]b.

may become larger than $1/2$. Actually, for K^c/μ below 43, $1 - 2\,\nu^{car}$ remains positive, Fig. 12.2.4. Conversely for K^c/μ larger than 43, $1 - 2\,\nu^{car}$ becomes negative for some range of angles $\underline{\beta}_m$, that is, for some range of lateral contraction defined by the ratio $-\epsilon_{22}/\epsilon_{11} = \cot^2 \underline{\beta}_m$. Said otherwise, the ratio $-\epsilon_{22}/\epsilon_{11}$ should be larger than some

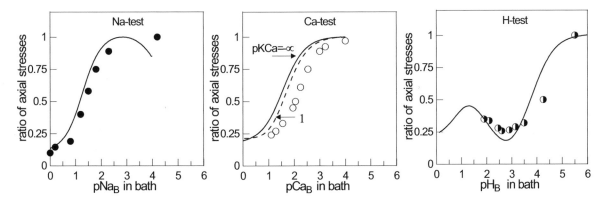

Fig. 17.13.5: Axial stress scaled by its maximum value during the third sequence BC. Comparison with data (symbols) by Grodzinski et al. [1981]. While the stress decrease is monotonic and mostly due to ionic shielding for the Na- and Ca-tests, it reaches twice the minimum hypertonic value for the H-curve, once at the isoelectric point, at pH about 2.6, and later as the ionic strength overweighs the presence of the fixed charge. Reprinted with permission from Loret and Simões [2010]b.

threshold, about 0.12, for $1 - 2\nu^{\text{car}}$ to have a chance to be negative.

17.14.2   Isotropic free swelling

The mechanical constitutive equations (17.12.16) take the form,

$$\boldsymbol{\sigma} + \tilde{p}_{\text{E}}\,\mathbf{I} + (\pi_{\text{osm}} + p_{\text{ch}})\,\mathbf{I} = \underline{\mathcal{W}}_{\text{ch},2}(\mathrm{E}_{\text{mo}})\,\Big(\overbrace{\mathbb{E}^{\text{gs}} + \mathbb{E}^{\text{c}}(\boldsymbol{\epsilon})}^{\mathbb{E}^{\text{car}}(\boldsymbol{\epsilon})}\Big) : \boldsymbol{\epsilon}\,. \tag{17.14.3}$$

Under all around pressure $\tilde{p}_{\text{E}}$, namely $\boldsymbol{\sigma} + \tilde{p}_{\text{E}}\,\mathbf{I} = \mathbf{0}$, the tissue is stressed in all directions identically, all the fibers undergo extension, and since they are distributed randomly, the material response is isotropic. Thus the axial strain along the arbitrary direction 1 during isotropic free swelling,

$$\epsilon_{11}^{\text{ifs}} = \frac{\pi_{\text{osm}} + p_{\text{ch}}}{\underline{\mathcal{W}}_{\text{ch},2}}\,\frac{1}{3\,K_{\text{iso}}^{\text{car}}}\,, \tag{17.14.4}$$

expresses in terms of the bulk modulus $K_{\text{iso}}^{\text{car}}$ associated with activation of all collagen fibers, Eqn (17.14.1),

$$K_{\text{iso}}^{\text{car}} = \lambda^{\text{gs}} + \lambda^{\text{c}} + \frac{2}{3}(\mu^{\text{gs}} + \mu^{\text{c}})\,. \tag{17.14.5}$$

17.14.3   Uniaxial traction along axis 1

The lateral boundary is subjected to the pressure $\tilde{p}_{\text{E}}$,

$$\sigma_{22} + \tilde{p}_{\text{E}} = \sigma_{33} + \tilde{p}_{\text{E}} = 0\,. \tag{17.14.6}$$

Then the axial strain, lateral strain and volume change,

$$\begin{aligned}
\epsilon_{11} &= \frac{\pi_{\text{osm}} + p_{\text{ch}}}{\underline{\mathcal{W}}_{\text{ch},2}}\,\frac{1}{3K^{\text{car}}} + \frac{\sigma_{11} + \tilde{p}_{\text{E}}}{\underline{\mathcal{W}}_{\text{ch},2}}\,\frac{1}{E^{\text{car}}}\,, \\
\epsilon_{22} &= \frac{\pi_{\text{osm}} + p_{\text{ch}}}{\underline{\mathcal{W}}_{\text{ch},2}}\,\frac{1}{M} - \nu^{\text{car}}\,\epsilon_{11}, \\
\operatorname{tr}\boldsymbol{\epsilon} &= \frac{\pi_{\text{osm}} + p_{\text{ch}}}{\underline{\mathcal{W}}_{\text{ch},2}}\,\frac{2}{M} + (1 - 2\,\nu^{\text{car}})\,\epsilon_{11}\,,
\end{aligned} \tag{17.14.7}$$

are clearly contributed by a chemical part and a mechanical part. The aggregate Poisson's ratio ν^{car}, Eqn (17.14.2), and moduli,

$$E^{\text{car}} = E_{11}^{\text{car}} - 2\,E_{12}^{\text{car}}\,\nu^{\text{car}}\,, \quad K^{\text{car}} = \frac{E^{\text{car}}}{3\,(1 - 2\,\nu^{\text{car}})}\,, \quad M = E_{22}^{\text{car}} + E_{23}^{\text{car}}\,, \tag{17.14.8}$$

depend on strain due to the conewise response of the collagen network. Here the E^{car}'s are the components of the aggregate stiffness $\mathbb{E}^{\text{car}}(\boldsymbol{\epsilon})$.

Note the equivalent alternative expression of the axial stress,

$$\sigma_{11} + \tilde{p}_{\text{E}} + (1 - 2\,\nu^{\text{car}})\,(\pi_{\text{osm}} + p_{\text{ch}}) = \underline{\mathcal{W}}_{\text{ch},2}\,E^{\text{car}}\,\epsilon_{11}\,. \tag{17.14.9}$$

The relations for free swelling of Section 17.14.2 can be recovered for a vanishing value of $\sigma_{11} + \tilde{p}_{\text{E}}$ in (17.14.7), noting that the collagen moduli are isotropic in this state of strain.

The hypertonic curve in uniaxial traction is obtained in the limit $\pi_{\text{osm}} + p_{\text{ch}} = 0$ and $\underline{\mathcal{W}}_{\text{ch},2} = 1$,

$$\boldsymbol{\sigma} + \tilde{p}_{\text{E}}\,\mathbf{I} = \mathbb{E}^{\text{car}}(\boldsymbol{\epsilon}) : \boldsymbol{\epsilon}\,, \quad \sigma_{11} + \tilde{p}_{\text{E}} = E^{\text{car}}(\boldsymbol{\epsilon})\,\epsilon_{11}\,, \quad \sigma_{22} + \tilde{p}_{\text{E}} = \sigma_{33} + \tilde{p}_{\text{E}} = 0\,, \tag{17.14.10}$$

and, by (17.14.7),

$$\epsilon_{22} = -\nu^{\text{car}}(\boldsymbol{\epsilon})\,\epsilon_{11}\,, \quad \operatorname{tr}\boldsymbol{\epsilon} = \big(1 - 2\,\nu^{\text{car}}(\boldsymbol{\epsilon})\big)\,\epsilon_{11}\,. \tag{17.14.11}$$

The relative positions of the hypertonic lines, and of the curve of uniaxial traction are sketched in Fig. 17.14.1.

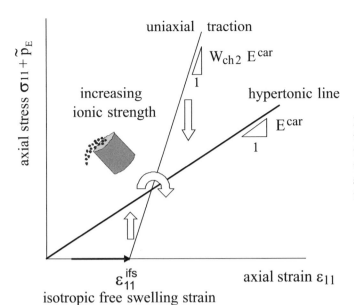

Fig. 17.14.1: Increasing the ionic strength of the bath in contact with the cartilage decreases the slope $\underline{\mathcal{W}}_{ch,2}$ of the line (curve) of uniaxial traction which, in the plane (axial strain, axial stress), ultimately tends to align with the hypertonic line.

17.15   Comments on the mechanical model

Accounting for the effect of pH on deformation has required a number of features to be addressed:

- a constitutive framework has been developed: it is consistent with the thermodynamics of porous media, including electro-chemomechanical couplings in presence of several metallic ions, hydrogen and hydroxyl ions, accounting for water dissociation, calcium binding and resulting in chemo-hyperelastic constitutive equations;

- the fact that the collagen fibers are mechanically active only in extension results in a directional response, even if the directional distribution of the fibers is random. In presence of chemically induced swelling, chemomechanical coupling results in stiffening of the mechanical response due to fiber recruitment. Chemical shrinking has opposite effects;

- in fact two types of nonlinearity are associated with the response of collagen. The overall nonlinearity associated with the collagen network is related to the activation of fibers in extension only (referred to as conewise response). The nonlinear response of individual fibers is aimed at accounting for the progressive uncrimping of the fibers. Indeed, under a progressive extension, the collagen fibers aligned with the extension direction will tend to first uncrimp and next to really carry a tensile stress, once they are straight. The influence of fiber nonlinearity on the chemical variables is small. In fact, a model in which fibers have a linear mechanical response in extension, namely for small k_c in (17.12.3), yields virtually the same results for chemical variables (concentrations, etc.), but of course not for mechanical variables (stress, strain, etc.);

- the chemomechanical response accounts for both electrical shielding at large ionic strength and variation of the fixed electrical charge (the key engine of chemomechanical couplings) due to acid-base reactions and possibly calcium binding;

- the chemomechanical coupling is strong in tissues with a fixed electrical charge like articular cartilages and corneal stroma: changes of the concentrations of ions and fixed charges imply significant changes in strains at given stresses, or in stresses at given strains. The converse mechanochemical coupling is strong only in corneal stroma due to the large compliance of the matrix, Section 19.11.4.

The simulations have assumed a random directional distribution of collagen fibers. Still, the influence of a non isotropic collagen network, displaying rotational symmetry about the traction axis, was tested. While the various mechanical entities depend strongly on the collagen directional distribution, the stress ratios displayed in Fig. 17.13.5 are found to be quite insensitive to even relatively large deviations with respect to an isotropic collagen network, as described by the rotational symmetry (17.12.10).

Here, as a tradeoff, the chemomechanical model has exploited the osmotic pressure. The key advantage is that arbitrary types of ions are accounted for in a straightforward manner, that does not require the determination of parameters. The drawback is certainly its lack of accuracy. Calibration of a more complete chemomechanical model where the weight of individual ions is accounted for requires data along several chemical and mechanical loading paths, Loret and Simões [2005]a. Stresses and strains in the three directions of space would be most welcome, so as to assess the three dimensional validity of the constitutive equations.

Experiments involving changes in pH and the control of the mechanical conditions at the boundaries are necessarily complex. A particular difficulty consists in defining a trustful reference state that can be used to compare the effects of the subsequent mechanical and chemical loading processes. The hypertonic state with large sodium chloride concentration and neutral pH has been used to start the simulations.

Chapter 18

Finite element analysis of electro-chemomechanical couplings

In line with the constitutive developments of the previous chapters, a three-phase multi-species electro-chemomechanical model of articular cartilage that accounts for the effect of two water compartments is pursued. Here a finite element formulation is presented in order to solve initial and boundary value problems and to simulate the time and space heterogeneous fields generated by mechanical and chemical loadings in laboratory samples.

A thin cartilage layer, laterally confined, is loaded through time varying mechanical and chemical conditions applied on the lower and upper boundaries. Two types of chemical loadings are simulated. First the concentration of NaCl in the bath in contact with the cartilage is changed according to several time schemes, including in particular free swelling and cyclic changes. A more complex process consists in replacing NaCl by $CaCl_2$. Chemical loading gives rise to heterogeneous fields during a transient period, whose extent depends on geometry and on a number of characteristic material properties, e.g., Darcy's seepage time, Fick's diffusion time, and typical times for intercompartment mass transfer. Mechanical confined compression and swelling tests are simulated as well. Spatial profiles of mechanical, chemical and electrical entities highlight the influences of the existence of two water compartments.

18.1 Overview of the finite element analysis

18.1.1 The constitutive framework

Articular cartilage is a porous medium, reinforced by collagen fibers, and saturated by an electrolyte, with water as solvent and metallic ions as solutes. Electrically charged macromolecules, the proteoglycans, intermingled with collagen fibers, give rise to electro-chemomechanical couplings.

The formulation capitalizes (1) upon the chemomechanical model in Loret and Simões [2004] [2005]a which was restricted to successive equilibria and from which both time and space were excluded; and (2) upon the model in Loret and Simões [2007], where the equations of mass transfer between the two fluid phases and the generalized diffusion equations in the extrafibrillar phase have been established. These constitutive models are embedded in a three-phase multi-species thermodynamic framework, which includes electro-chemomechanical effects. This chapter is based on Loix et al. [2008].

The mechanical significance of the partition of tissue water in two compartments has been advocated by Maroudas and coworkers. Intrafibrillar water (IF) is stored between collagen fibrils and extrafibrillar water (EF) covers proteoglycans, Fig. 14.2.8. The collagen fibers constitute the solid phase. Intrafibrillar water and dissolved salts NaCl and $CaCl_2$ on one hand and extrafibrillar water, ions Na^+, Ca^{2+} and Cl^- and negatively charged proteoglycans on the other hand form the two fluid phases. The two-compartment structure is hierarchical: the IF compartment communicates only with the EF compartment, while the EF compartment exchanges water and ions with the IF compartment and with the surroundings.

The present developments concentrate on the finite element formulation and on numerical simulations of mechanical and chemical loading processes. For comprehensiveness, the basic equations of the chemomechanical behavior, mass transfer and generalized diffusion are briefly recalled. Mass transfer between the two fluid compartments introduce source terms in the equations of diffusion. This sort of problem is sometimes referred to as a problem of reaction-diffusion, or reactive transport.

The generic initial and boundary value problem (IBVP) to be simulated is as follows. A layer of articular cartilage is placed in a mechanical testing machine and the mechanical loading is such that quasi-static mechanical equilibrium holds continuously in the porous medium. The cartilage piece is surrounded on its upper and lower boundaries by a bath. The chemical composition of the bath may be changed arbitrarily but chemical equilibrium is assumed to hold at any time at the interface between the bath and the cartilage, that is with extrafibrillar phase. The layout, boundary and loading conditions imply that mechanical, hydraulic and chemical equilibria do not get established instantaneously within the porous medium. In fact, the actual duration of the transient period the information transmitted by the boundary takes to reach a pointwise steady state needs to be appreciated. Indeed several characteristic times are involved in the coupled processes of deformation, seepage, diffusion and mass transfer. For spatially homogeneous specimens, the steady state strains, stresses, pressures and the distribution of water and ions in the two compartments obtained by pointwise analyses are recovered.

Yet other time dependent phenomena are not accounted for. For example, since only compression tests are addressed, the possible intrinsic viscous behavior of the collagen fibers does not come into picture. Moreover, electroneutrality is assumed to hold pointwise. Indeed, the typical time to recover electroneutrality is much smaller than the other characteristic times involved here and in particular much smaller than the intercompartment transfer times. Electroneutrality is considered to hold in each of the two water compartments independently.

18.1.2 IBVPs including chemomechanical couplings in articular cartilages

A few analyses have addressed initial and boundary value problems where chemomechanical couplings are included, e.g., Snijders et al. [1995], Simon et al. [1996], Frijns et al. [1997], Levenston et al. [1998]a [1998]b. They all considered a symmetric binary electrolyte with water as the solute and cations Na^+ and anions Cl^- as dissolved ions. Moreover, water is totally contained within a single compartment.

Sun et al. [1999] were the first to develop a finite element formulation which accounts for the fact that many physical entities, like ionic concentrations, fluid pressures, etc. are discontinuous at the boundary of the cartilage specimen. This observation constrains the choice of primary variables. They used the constitutive model of Lai et al. [1991] and presented simulations of confined free swelling and confined compression.

Hon et al. [1999] considered a very restricted situation where, in absence of external electrical field, the velocities of the anion Na^+, cation Cl^- and water can be considered equal, so that ionic diffusion is neglected. Consequently, the applications are restricted to one-dimensional confined compression because change in ionic concentration of the bath is out of reach. Hon et al. [2002] extended the numerical simulations to plane strain and axisymmetric configurations via a domain decomposition technique.

In fact, the prospect of simplifying the electro-chemomechanical coupled behavior of soft tissues with fixed electrical charge was considered previously by Lanir [1987]. Again, if ionic diffusion can be neglected, Lanir [1987] proposed that the coupled behavior could be inserted in the standard saturated porous media context through a pressure term depending on strain. The issue is considered in a finite element context in W. Wilson et al. [2005]b. While Lanir's simplification could apply for small samples, where spatial homogeneity is expected to hold, it is neither appropriate for transient situations, nor for arbitrary geometries and organs, where the diffusion processes are *per se* key phenomena.

Yao and Gu [2004] used essentially the constitutive behavior of Lai et al. [1991] and computational tools of Sun et al. [1999]. However, in addition to Na^+ and Cl^-, the electrolyte contains a neutral macromolecular solute. Their axisymmetric finite element analysis addressed the influence, during cycling mechanical loadings, of the permeability of the platens on the material response of cartilage specimens.

18.1.3 Scope

The model for mass transfer, diffusion and elastic deformation for heteroionic cartilage described in the previous chapters is used to simulate confined compression tests. Complex paths involving interspersed mechanical and chemical loading and unloading can be addressed.

The field equations, as well as the skeleton structure of the constitutive equations for the chemomechanical model, and the constitutive equations of diffusion and mass transfer are first recalled in Section 18.2. Next the finite element formulation is introduced: the choice of the primary variables is guided by the electrochemical boundary conditions. The spatial discretization, the time integration scheme and equation solving method are described in Section 18.3.

The geometrical set up is presented and material data are provided in Section 18.4. The numerical simulations of several initial and boundary value problems are described next in Sections 18.5 and 18.6. Section 18.5 considers a binary symmetric electrolyte where the concentration of the dissolved sodium chloride NaCl may be varied. Section 18.6 considers a ternary non symmetric electrolyte, and addresses the mechanical effects of exchanging sodium chloride NaCl for calcium chloride $CaCl_2$, and conversely.

Since a multi-phase multi-species mixture is used, a species is designated by two indices. The index k corresponds to the species itself, namely w for water, Na for sodium cation, Ca for calcium cation, Cl for chloride anion. The capital index K corresponds to the phase, namely S solid, I intrafibrillar and E extrafibrillar phases, respectively. Since collagen c and proteoglycans PG unambiguously belong to a single phase, the index of that phase is not indicated. For ease of reference, the sets of species used are defined in Table 18.1.1.

Table 18.1.1: Sets of species. EF extrafibrillar fluid; IF intrafibrillar fluid.

Set	Species	Property
E	w, PG, Na, Ca, Cl	EF compartment
E_{ions}	Na, Ca, Cl	EF ions
E_{mo}	E - {PG}	EF mobile species
I_{in}	w, Na, Ca	Independent IF species
I_{ne}	$\{w\} \cup I_{salts}$	Neutral IF species
I_{salts}	$s_1 = NaCl$, $s_2 = CaCl_2$	IF salts

18.2 Field and constitutive equations

The field and constitutive equations are first collected and cast in a format appropriate to finite element simulations. Three types of constitutive equations govern the behavior of the multi-phase multi-species mixture: chemoelastic relations, equations of generalized diffusion in the extrafibrillar phase and equations of mass transfer between the two fluid phases.

18.2.1 Balance equations

Under quasi-static loading and in absence of body forces, thus excluding gravity, the balance of momentum of the porous medium as a whole involves the total stress $\boldsymbol{\sigma}$ only, namely after Eqn (14.6.47),

$$\operatorname{div} \boldsymbol{\sigma} = \mathbf{0}, \tag{18.2.1}$$

where div denotes the divergence operator.

All species are incompressible. The changes in volume due to mass transfer between the intrafibrillar and extrafibrillar phases balance exactly. As a consequence of the hierarchical structure of the model, the change of volume of the porous medium is opposite to the change of volume of the extrafibrillar fluid phase due to diffusion, Eqn (14.6.49),

$$\operatorname{div} \mathbf{v}_S + \operatorname{div} \mathbf{J}_E = 0, \tag{18.2.2}$$

where $\mathbf{v}_S$ is the velocity of the solid skeleton, $\mathbf{J}_E = \sum_{k \in E} \mathbf{J}_{kE}$ the total extrafibrillar flux per unit surface, and $\mathbf{J}_{kE}$ [unit : m/s] the flux of the species k through the porous medium.

Since the species are incompressible, the balance of mass of the extrafibrillar ion k can be phrased in terms of the intra- and extrafibrillar volume contents, v^{kI} and v^{kE} respectively, namely by summation of the two equations (14.E.7),

$$\frac{1}{\det \mathbf{F}} \frac{dv^{kE}}{dt} + \frac{1}{\det \mathbf{F}} \frac{dv^{kI}}{dt} + \operatorname{div} \mathbf{J}_{kE} = 0, \quad k \in E_{ions}. \tag{18.2.3}$$

Here d/dt denotes the time derivative following the solid phase, and $\det \mathbf{F}$ is the determinant of the deformation gradient $\mathbf{F}$.

The electroneutrality condition in the intrafibrillar phase is viewed as providing the amount of chloride ions in terms of intrafibrillar cations. The rates of change of the mass content, or equivalently volume content, of the independent intrafibrillar species are viewed as balance equations. Along (14.6.42), they are postulated in the format,

$$\frac{1}{\det \mathbf{F}} \frac{dm^{iI}}{dt} - O_i = 0, \quad i \in I_{\text{in}},$$

(18.2.4)

where the O's are constitutive functions addressed in Section 10.2

The set of field equations is displayed in Table 18.2.1, together with the primary unknowns that will be used in the finite element formulation. The field equations are coupled through transport and mechanics. In this table, the unknown associated with each field equation is the one that would appear if it was uncoupled.

Table 18.2.1: Set of field equations and list of primary unknowns.
EF extrafibrillar fluid; IF intrafibrillar fluid.

Nature of Equation	Field Equation	Main Unknowns
• Balance of momentum of the whole mixture	$\operatorname{div} \boldsymbol{\sigma} = \mathbf{0}$	Displacement $\mathbf{u}$
• Balance of mass of the whole mixture	$\operatorname{div} \mathbf{v}_S + \operatorname{div} \mathbf{J}_E = 0$	Chemical potential of EF water μ_{wE}
• Balance of mass of EF ions	$\dfrac{1}{\det \mathbf{F}} \left(\dfrac{dv^{kE}}{dt} + \dfrac{dv^{kI}}{dt} \right) + \operatorname{div} \mathbf{J}_{kE} = 0$	Electrochemical potential $\mu_{kE}^{\text{ec}}, \ k \in E_{\text{ions}}$
• Balance of mass in IF-EF exchange	$\dfrac{1}{\det \mathbf{F}} \dfrac{dm^{iI}}{dt} - O_i(\mu_{nI} - \mu_{nE}) = 0$	IF mass content $m^{iI}, \ i \in I_{\text{in}}, \ n \in I_{\text{ne}}$

18.2.2   The electrochemical potentials

The electrochemical potentials of ions μ_{kK}^{ec}, $k \in K_{\text{ions}}$, and chemical potentials of water μ_{wK}, in phase $K = I, E$, enter the three types of constitutive equations above. In the potentials of the ionic species, both the mechanical terms and the formation pressures are retained to prevent loss of accuracy close to the hypertonic state. Indeed the former are often negligible with respect to the term of chemical origin at physiological concentration, but they might become significant at high ionic strength. However, the simulations, with or without these terms, have shown little differences.

In explicit form,
- for the extrafibrillar species:

$$
\begin{aligned}
\widehat{m}_{\text{w}}\, \mu_{\text{wE}} &= \widehat{v}_{\text{w}} \overbrace{(p_{f\text{w}} + p_{\text{I}})}^{p_{\text{wE}}} \ + \ RT \operatorname{Ln} x_{\text{wE}}, \\
\widehat{m}_k\, \mu_{kE}^{\text{ec}} &= \widehat{v}_k\, (p_{fk} + p_{\text{I}}) \ + \ RT \operatorname{Ln} x_{kE} + \mathrm{F}\, \zeta_k\, \phi_{\text{E}}, \quad k \in E_{\text{ions}},
\end{aligned}
$$

(18.2.5)

- for the intrafibrillar species:

$$
\begin{aligned}
\widehat{m}_{\text{w}}\, \mu_{\text{wI}} &= \widehat{v}_{\text{w}} \overbrace{(p_{\text{adh}} + p_{\text{I}})}^{p_{\text{wI}}} \ + \ RT \operatorname{Ln} x_{\text{wI}}, \\
\widehat{m}_k\, \mu_{kI}^{\text{ec}} &= \widehat{v}_k\, p_{\text{I}} \ + \ RT \operatorname{Ln} x_{kI} + \mathrm{F}\, \zeta_k\, \phi_{\text{I}}, \quad k \in I_{\text{ions}}.
\end{aligned}
$$

(18.2.6)

The chemical potentials of the dissolved salts s_1=NaCl and s_2=CaCl$_2$ deduce,
- in the extrafibrillar fluid phase:

$$
\begin{aligned}
\widehat{m}_{\text{Na}}\, \mu_{s_1\text{E}} &= \widehat{v}_{s_1}\, (p_{fs_1} + p_{\text{I}}) + RT \operatorname{Ln} x_{\text{NaE}}\, x_{\text{ClE}}, \\
\widehat{m}_{\text{Ca}}\, \mu_{s_2\text{E}} &= \widehat{v}_{s_2}\, (p_{fs_2} + p_{\text{I}}) + RT \operatorname{Ln} x_{\text{CaE}}\, x_{\text{ClE}}^2,
\end{aligned}
$$

(18.2.7)

with

$$\widehat{v}_{s_1} = \widehat{v}_{\text{Na}} + \widehat{v}_{\text{Cl}}, \quad \widehat{v}_{s_2} = \widehat{v}_{\text{Ca}} + 2\,\widehat{v}_{\text{Cl}},$$

(18.2.8)

and

$$\widehat{v}_{s_1}\, p_{fs_1} = \widehat{v}_{\text{Na}}\, p_{f\text{Na}} + \widehat{v}_{\text{Cl}}\, p_{f\text{Cl}}, \quad \widehat{v}_{s_2}\, p_{fs_2} = \widehat{v}_{\text{Ca}}\, p_{f\text{Ca}} + 2\,\widehat{v}_{\text{Cl}}\, p_{f\text{Cl}};$$

(18.2.9)

- in the intrafibrillar fluid phase:

$$\widehat{m}_{\mathrm{Na}}\,\mu_{\mathrm{s_1I}} = \widehat{v}_{\mathrm{s_1}}\,p_{\mathrm{I}} + R\,T\,\mathrm{Ln}\,x_{\mathrm{NaI}}\,x_{\mathrm{ClI}}\,,$$

$$\widehat{m}_{\mathrm{Ca}}\,\mu_{\mathrm{s_2I}} = \widehat{v}_{\mathrm{s_2}}\,p_{\mathrm{I}} + R\,T\,\mathrm{Ln}\,x_{\mathrm{CaI}}\,(x_{\mathrm{ClI}})^2\,. \tag{18.2.10}$$

Here, the x's are the molar fractions defined in each phase, namely $x_{k\mathrm{K}} = N_{k\mathrm{K}}/N_{\mathrm{K}}$ is the relative number of moles of the species k in the phase K. The x's are constrained by a closure relation and by electroneutrality, Eqn (18.E.7). The $\widehat{v}$'s are molar volumes, the $\widehat{m}$'s are the molar masses, the ζ's are valences, while ϕ_{K} is the electrical potential in phase K. Moreover, $R = 8.31451\,\mathrm{J/mole/K}$ is the universal gas constant, $T = 298\,\mathrm{K}$ the absolute temperature, and $\mathrm{F} = 96485\,\mathrm{coulomb/mole}$ is Faraday's equivalent charge. The chemical potentials are sometimes expressed in terms of the concentrations $c_{k\mathrm{K}}$ which are approximately equal to the molar fractions $x_{k\mathrm{K}}$ divided by the molar volume of water $\widehat{v}_{\mathrm{w}}$.

Three types of pressure come into picture, Chapter 14:
- p_{I} is the pressure due to incompressibility;
- $p_{\mathrm{adh}} = p_{\mathrm{adh}}(m^{\mathrm{wI}})$ is the adhesion pressure that moderates the transfer of intrafibrillar water;
- p_{fw} is a constant that ensures the chemical equilibrium between fluid phases in the initial hypertonic state, Loret and Simões [2004]. The ionic formation pressures p_{fk} are representative of the affinity of the ion k to proteoglycans. They give rise to an equilibrium constant K^{eq} different from 1, namely, Loret and Simões [2005]a,

$$\frac{x_{\mathrm{CaE}}}{(x_{\mathrm{NaE}})^2} = K^{\mathrm{eq}}\,\frac{x_{\mathrm{CaB}}}{(x_{\mathrm{NaB}})^2}\,, \tag{18.2.11}$$

with

$$K^{\mathrm{eq}} = \frac{K_{fs_2}}{(K_{fs_1})^2}\,, \quad K_{fn} \simeq \exp\left(-\frac{\widehat{v}_n\,p_{fn}}{RT}\right), \quad n = \mathrm{s_1,\,s_2}\,. \tag{18.2.12}$$

18.2.3 Chemoelastic relations

18.2.3.1 The effective stress

The stress-strain relation for the present three-phase framework with an isotropic collagen network was developed in Chapter 14 based on Loret and Simões [2004]. It links an effective stress $\boldsymbol{\sigma}'$ to the strain $\boldsymbol{\epsilon}$ of the collagen network, or equivalently solid skeleton or porous medium,

$$\boldsymbol{\sigma}' = \boldsymbol{\sigma} + p_{\mathrm{eff}}\,\mathbf{I} = \mathbb{E}:\boldsymbol{\epsilon}\,, \tag{18.2.13}$$

where the constant moduli $\mathbb{E}$, defined by the Lamé moduli λ and μ, are representative of the hypertonic state. The chemomechanical coupling is thus incorporated in the effective stress.

18.2.3.2 The fictitious bath

The constitutive equations developed in Loret and Simões [2004] considered a sequence of equilibria. In order to generalize these constitutive equations out of equilibrium, a fictitious bath is introduced, Section 14.5.2. This fictitious bath contains the same species as the extrafibrillar compartment but PGs. It is in equilibrium with the extrafibrillar compartment, at any time. The chemical content of the fictitious bath, denoted by the superimposed symbol $\widetilde{}$, is *deduced* from the chemical content of the extrafibrillar phase, by requiring chemical equilibrium of all species. Details are provided in Exercise 18.3.

The difference of the chemical potentials between the water of the cartilage and of the fictitious bath provides an osmotic pressure π_{osm}, that is the difference between two pressures, p_{wE} of the extrafibrillar fluid and $\widetilde{p}_{\mathrm{wE}}$ of the fictitious bath,

$$\pi_{\mathrm{osm}} = p_{\mathrm{wE}} - \widetilde{p}_{\mathrm{wE}}\,. \tag{18.2.14}$$

In this way, π_{osm} depends on the extrafibrillar ionic content and a differential with respect to the primary variables can be defined so as to build the tangent matrix.

The format of the effective pressure $p_{\mathrm{eff}} = p_{\mathrm{I}} - X$ highlights that the strain and chemistry dependent scalar X vanishes in the hypertonic state: standard poroelasticity, associated with the effective stress $\boldsymbol{\sigma}' = \boldsymbol{\sigma} + p_{\mathrm{I}}\,\mathbf{I}$, is then recovered, with p_{I} the incompressibility pressure. Based on the qualitative sketches 13.10.7 and 13.10.8, the scalar X is affine with respect to the strain,

$$X = \Lambda\,(\mathrm{tr}\,\boldsymbol{\epsilon} - \mathrm{tr}\,\boldsymbol{\epsilon}^{\Omega}) + \pi_{\mathrm{osm}} - p_{fw}\,. \tag{18.2.15}$$

The slope Λ depends on the chemical composition of the extrafibrillar compartment and the osmotic pressure is minimal and equal to p_{fw} in the hypertonic state.

18.2.4   Equations of generalized diffusion in the extrafibrillar phase

Within the extrafibrillar phase, species *diffuse* and are advected by water. Their motions are governed by coupled effects including Darcy's law of seepage of water, Fick's law of diffusion of ions and Ohm's law of electrical flow. The vector of volume fluxes $\mathbf{j}$ relative to the solid phase is related to the vector $\mathbf{f}$ of electrochemical gradients weighted by the mass densities,

$$\mathbf{j} = \begin{bmatrix} \mathbf{J}_{\mathrm{wE}} \\ \mathbf{J}_{\mathrm{NaE}} \\ \mathbf{J}_{\mathrm{CaE}} \\ \mathbf{J}_{\mathrm{ClE}} \end{bmatrix}, \quad \mathbf{f} = \begin{bmatrix} \rho_{\mathrm{w}}\, \boldsymbol{\nabla}\mu_{\mathrm{wE}} \\ \rho_{\mathrm{Na}}\, \boldsymbol{\nabla}\mu^{\mathrm{ec}}_{\mathrm{NaE}} \\ \rho_{\mathrm{Ca}}\, \boldsymbol{\nabla}\mu^{\mathrm{ec}}_{\mathrm{CaE}} \\ \rho_{\mathrm{Cl}}\, \boldsymbol{\nabla}\mu^{\mathrm{ec}}_{\mathrm{ClE}} \end{bmatrix}, \tag{18.2.16}$$

by the isotropic generalized diffusion matrix $\boldsymbol{\kappa}$, defined in Loret and Simões [2007], and exposed in Chapter 16,

$$\mathbf{j} = -\boldsymbol{\kappa}\,\mathbf{f}, \quad \boldsymbol{\kappa} = \boldsymbol{\kappa}^{T}\ \mathrm{sDP}. \tag{18.2.17}$$

The generalized diffusion matrix $\boldsymbol{\kappa}$ is required to be positive semi-definite (PsD) in order for the dissipation inequality (16.1.22) to be satisfied.

Table 18.2.2: Physical phenomena and pairs of work-conjugate variables involved in the finite element analysis. EF, IF: extra-, intrafibrillar phase; E^{x}, I^{x} subsets of the set of extra-, intrafibrillar species.

○ **Chemomechanics**	Generalized Strain	Generalized Stress
● Chemo-hyperelasticity	Strain ϵ of porous medium	Total stress $\boldsymbol{\sigma}$
● Coupled osmosis	Mass content of EF water m^{wE}	Chemical potential of water μ_{wE}
● Chemomechanical coupling	Mass content of EF ion, $m^{i\mathrm{E}}$, $i \in \mathrm{E}_{\mathrm{in}}$	Chemical potential of salt $\mu_{n\mathrm{E}}$, $n \in \mathrm{E}^{\mathrm{salts}}$
○ Darcy's, Fick's, Ohm's laws of **coupled diffusion** in extrafibrillar compartment	Flux Relative to Solid	Driving Gradient
● Seepage, electroosmosis	Flux of EF water $\mathbf{J}_{\mathrm{wE}}$	Chemical potential of water $\rho_{\mathrm{w}}\,\boldsymbol{\nabla}\mu_{\mathrm{wE}}$
● Ionic diffusion, electromigration	EF ionic flux $\mathbf{J}_{k\mathrm{E}}$, $k \in \mathrm{E}_{\mathrm{ions}}$	Electrochemical potential of ion $\rho_{k}\,\boldsymbol{\nabla}\mu^{\mathrm{ec}}_{k\mathrm{E}}$
○ Intra-extra fibrillar **mass transfer**	IF mass content	Out-of-balance of chemical potentials
● Transfer of small molecules through the collagen fibrils	IF mass content, $m^{i\mathrm{I}}$, $i \in \mathrm{I}_{\mathrm{in}}$	$\mu_{n\mathrm{I}} - \mu_{n\mathrm{E}}$, $n \in \mathrm{I}_{\mathrm{ne}}$

18.2.5   Equations of mass transfer between the two fluid phases

Species of the intrafibrillar compartment are endowed with the velocity of the solid, so that no diffusion takes place in this compartment. Across the fluid phases, species *transfer* according to a transfer law that provides the rates of mass contents of water and cations in terms of the out-of-balance of chemical potentials. The kinetics of each transfer is governed by a characteristic time.

Let $\mathrm{I}_{\mathrm{salts}} = \{\mathrm{s}_1 = \mathrm{NaCl}, \mathrm{s}_2 = \mathrm{CaCl}_2\}$ be the set of intrafibrillar salts, and $\mathrm{I}_{\mathrm{ne}} = \{\mathrm{w}\} \cup \mathrm{I}_{\mathrm{salts}}$ be the set of neutral intrafibrillar species. The constitutive functions (18.2.4) that define the rate of change of the mass content of each independent intrafibrillar species are postulated in the format (10.2.13),

$$O_i = -\frac{\rho_i}{t_{\mathrm{tr},i}}\, A_i\,(\mu_{n\mathrm{I}} - \mu_{n\mathrm{E}})\,, \tag{18.2.18}$$

with $(i, n(i)) \in (I_{in}, I_{ne})$ one of the three pairs (w, w), (Na, s_1) and (Ca, s_2). The coefficient $A_i > 0$ is defined as $\rho_i / (\lambda + 2\mu)$, with ρ_i the mass density. The entity $t_{tr,i}$ is the characteristic transfer time associated with the transfer of the chemical $n = n(i)$.

Table 18.2.2 presents a schematic sketch of the constitutive equations. Note however that the various *couplings within* each of the three types of constitutive equations, and the *interactions across* the three types, e.g., a strain dependent permeability, do not appear on this sketch.

18.3 Finite element formulation

18.3.1 The primary variables

Due to the presence of the proteoglycans in the extrafibrillar fluid phase, the molar fractions, concentrations, fluid pressure and electrical potentials in general undergo a discontinuity between the articular cartilage and the surroundings, Section 14.6.2. Moreover, like the electrical potential, the electrical current density is in general discontinuous at the boundary of the cartilage specimen. Therefore it is not possible to monitor directly none of these entities.

On the other hand, the boundary conditions on part of the surface of the sample are phrased in terms of electrochemical potentials. Therefore, as in Sun et al. [1999], it is convenient to use the electrochemical potentials as primary variables.

Note that, in finite element analyses of multi-phase multi-species mixtures, the primary variables associated with a given species may be either a scalar (its mass, concentration, chemical potential, etc.), or a vector (its absolute or some relative velocity). In the first case, the field equation to be discretized is the mass balance, and the (relative) velocity is obtained by the generalized diffusion equations. In the second case, the field equation to be discretized is the balance of momentum of the species. The latter method has practical computational advantages in non-reactive porous media when dynamic effects are to be accounted for, because it allows to use time-integration schemes for which stable time-steps are available, e.g., Loret and Prévost [1991]. However, it becomes computationally heavy for multi-phase multi-species mixtures. That is the reason why the former method using scalars as primary variables is preferred for quasi-static analyses.

18.3.2 The semi-discrete equations

In space dimension n_{sd}, the following $n_{sd} + 7$ field equations need to be satisfied, namely
- the balance of momentum of the porous medium as a whole, Eqn (18.2.1);
- the global balance of mass, Eqn (18.2.2);
- the balances of mass of the three extrafibrillar ions, Eqn (18.2.3);
- the three equations of mass transfer, Eqn (18.2.4).

Boundary data can be prescribed in terms of either primary unknowns, as explained in Section 18.3.1, or in terms of the fluxes that appear in the right-hand sides of the weak forms below.

The $n_{sd} + 7$ primary unknown fields, namely
- the solid displacement $\mathbf{u} = \mathbf{N}_u \mathbf{u}^e$;
- the electrochemical potentials of the extrafibrillar species $\mu_{kE}^{ec} = \mathbf{N}_\mu [\boldsymbol{\mu}_{kE}^{ec}]^e$, $k \in E_{mo}$;
- the independent intrafibrillar mass-contents $m^{iI} = \mathbf{N}_m \mathbf{m}_{iI}^e$, $i \in I_{in}$,

are interpolated, within the generic element e, in terms of nodal values through a priori distinct shape functions. Here $E_{mo} = E - \{PG\}$ represents the set of mobile species in the extrafibrillar fluid phase.

Multiplying the field equations by the virtual fields $\delta \mathbf{w}$, $\delta \mu$, δm, and integrating by parts over the body V provides the weak form of the problem as:

- Global balance of momentum
$$\int_V \boldsymbol{\nabla}(\delta \mathbf{w}) : \boldsymbol{\sigma} \, dV = \int_{\partial V} \delta \mathbf{w} \cdot \boldsymbol{\sigma} \cdot \mathbf{n} \, dS,$$

- Global balance of mass
$$\int_V \delta \mu \, \mathrm{div} \, \mathbf{v}_S - \boldsymbol{\nabla}(\delta \mu) \cdot \mathbf{J}_E \, dV = -\int_{\partial V} \delta \mu \, \mathbf{J}_E \cdot \mathbf{n} \, dS,$$

- Balance of mass of EF ion k
$$\int_V \delta \mu \, \beta_k - \boldsymbol{\nabla}(\delta \mu) \cdot \mathbf{J}_{kE} \, dV = -\int_{\partial V} \delta \mu \, \mathbf{J}_{kE} \cdot \mathbf{n} \, dS,$$

- Exchange of IF-EF species $i \in I_{in}$,
$$\int_V \delta m \left(\frac{1}{\det \mathbf{F}} \frac{dm^{iI}}{dt} - O_i \right) dV = 0$$

(18.3.1)

where

$$\beta_k = \frac{1}{\det \mathbf{F}} \left(\frac{dv^{k\mathrm{E}}}{dt} + \frac{dv^{k\mathrm{I}}}{dt} \right), \quad k \in \mathrm{E}_{\mathrm{ions}}, \tag{18.3.2}$$

and $\mathbf{n}$ is the unit outward normal to the boundary ∂V.

A generalized Galerkin procedure is adopted, in as far as the same interpolation functions are used for the primary unknowns and for the variations. In the simulations, the finite element mesh involves quadratic solid displacements (three nodes per element) while the other unknowns vary linearly (two nodes per element). Due to the presence of highly nonlinear terms of chemical origin, the number of integration points is three, for all matrices and all residuals.

The global unknown vector $\mathbb{X}$,

$$\underbrace{\text{unknown vector}}_{\substack{\text{max nodal length} \\ n_{\mathrm{sd}} + 7}} \quad \mathbb{X} = \begin{bmatrix} \mathbf{u} \\ \hline \boldsymbol{\mu}_{\mathrm{wE}}^{\mathrm{ec}} \\ \boldsymbol{\mu}_{\mathrm{NaE}}^{\mathrm{ec}} \\ \boldsymbol{\mu}_{\mathrm{CaE}}^{\mathrm{ec}} \\ \boldsymbol{\mu}_{\mathrm{ClE}}^{\mathrm{ec}} \\ \hline \mathbf{m}^{\mathrm{wI}} \\ \mathbf{m}^{\mathrm{NaI}} \\ \mathbf{m}^{\mathrm{CaI}} \end{bmatrix} \quad \begin{array}{l} \} \text{ displacement vector} \\ \\ \left. \begin{array}{l} \\ \\ \\ \\ \end{array} \right\} \begin{array}{l} \text{extrafibrillar} \\ \text{electrochemical potentials} \\ \boldsymbol{continuous} \text{ at the boundary} \end{array} \\ \\ \left. \begin{array}{l} \\ \\ \end{array} \right\} \text{ intrafibrillar masses} \end{array} \tag{18.3.3}$$

has therefore maximum nodal length $n_{\mathrm{sd}}+7$. The resulting nonlinear first order semi-discrete equations imply the residual $\mathbb{R}$,

$$\mathbb{R} = \mathbb{F}^{\mathrm{surf}}(\mathbb{S}, \mathbb{X}) - \mathbb{F}^{\mathrm{int}}\left(\mathbb{X}, \frac{d\mathbb{X}}{dt} \right) = \mathbb{0}, \tag{18.3.4}$$

to vanish. Here $\mathbb{F}^{\mathrm{surf}}$ is the vector of surface loadings denoted collectively by $\mathbb{S}$ and $\mathbb{F}^{\mathrm{int}}$ is the vector gathering the internal elastic and viscous forces. The contribution of a generic element e to the vector $\mathbb{F}_e^{\mathrm{surf}}$ of boundary data, and the contribution to the vector $\mathbb{F}_e^{\mathrm{int}}$ of internal forces are obtained from the weak form as,

$$\mathbb{F}_e^{\mathrm{surf}} = \begin{bmatrix} \int_{\partial V_e} \mathbf{N}_u^{\mathrm{T}} \, \boldsymbol{\sigma} \cdot \mathbf{n} \, dS \\ -\int_{\partial V_e} \mathbf{N}_\mu^{\mathrm{T}} \mathbf{J}_{\mathrm{E}} \cdot \mathbf{n} \, dS \\ -\int_{\partial V_e} \mathbf{N}_\mu^{\mathrm{T}} \mathbf{J}_{\mathrm{NaE}} \cdot \mathbf{n} \, dS \\ -\int_{\partial V_e} \mathbf{N}_\mu^{\mathrm{T}} \mathbf{J}_{\mathrm{CaE}} \cdot \mathbf{n} \, dS \\ -\int_{\partial V_e} \mathbf{N}_\mu^{\mathrm{T}} \mathbf{J}_{\mathrm{ClE}} \cdot \mathbf{n} \, dS \\ \mathbf{0} \\ \mathbf{0} \\ \mathbf{0} \end{bmatrix}, \quad \mathbb{F}_e^{\mathrm{int}} = \begin{bmatrix} \int_{V_e} \mathbf{B}_u^{\mathrm{T}} \, \boldsymbol{\sigma} \, dV_e \\ \int_{V_e} \mathbf{N}_\mu^{\mathrm{T}} \operatorname{div} \mathbf{v}_S - \boldsymbol{\nabla} \mathbf{N}_\mu^{\mathrm{T}} \, \mathbf{J}_{\mathrm{E}} \quad dV_e \\ \int_{V_e} \mathbf{N}_\mu^{\mathrm{T}} \quad \beta_{\mathrm{Na}} \quad - \boldsymbol{\nabla} \mathbf{N}_\mu^{\mathrm{T}} \, \mathbf{J}_{\mathrm{NaE}} \, dV_e \\ \int_{V_e} \mathbf{N}_\mu^{\mathrm{T}} \quad \beta_{\mathrm{Ca}} \quad - \boldsymbol{\nabla} \mathbf{N}_\mu^{\mathrm{T}} \, \mathbf{J}_{\mathrm{CaE}} \, dV_e \\ \int_{V_e} \mathbf{N}_\mu^{\mathrm{T}} \quad \beta_{\mathrm{Cl}} \quad - \boldsymbol{\nabla} \mathbf{N}_\mu^{\mathrm{T}} \, \mathbf{J}_{\mathrm{ClE}} \, dV_e \\ \int_{V_e} \mathbf{N}_m^{\mathrm{T}} \left(\frac{1}{\det \mathbf{F}} \frac{dm^{\mathrm{wI}}}{dt} - O_w \right) dV_e \\ \int_{V_e} \mathbf{N}_m^{\mathrm{T}} \left(\frac{1}{\det \mathbf{F}} \frac{dm^{\mathrm{NaI}}}{dt} - O_{\mathrm{Na}} \right) dV_e \\ \int_{V_e} \mathbf{N}_m^{\mathrm{T}} \left(\frac{1}{\det \mathbf{F}} \frac{dm^{\mathrm{CaI}}}{dt} - O_{\mathrm{Ca}} \right) dV_e \end{bmatrix}, \tag{18.3.5}$$

where $\mathbf{B}_u$ is the strain-displacement matrix, namely $\delta\boldsymbol{\epsilon}(\mathbf{u}) = \boldsymbol{\epsilon}(\delta\mathbf{u}) = \mathbf{B}_u \, \delta\mathbf{u}$, and the O's are defined by (18.2.18).

18.3.3   Time integration and equation solving

The semi-discrete equations are integrated through a generalized midpoint scheme defined by a scalar $\alpha \in]0, 1]$, Hughes [1987]: at step $n + 1$, the equations are enforced at time $t_{n+\alpha} = t_n + \alpha \Delta t$, where $\Delta t = t_{n+1} - t_n$, namely

$$\mathbb{R}_{n+\alpha} = \mathbb{F}^{\mathrm{surf}}(\mathbb{S}_{n+\alpha}, \mathbb{X}_{n+\alpha}) - \mathbb{F}^{\mathrm{int}}(\mathbb{X}_{n+\alpha}, \mathbb{V}_{n+\alpha}) = 0. \tag{18.3.6}$$

Generically, for $\mathbb{Z} = \mathbb{S}$, $\mathbb{X}$, $\mathbb{V}$, $\mathbb{Z}_{n+\alpha}$ is defined as $(1 - \alpha)\,\mathbb{Z}_n + \alpha\,\mathbb{Z}_{n+1}$, and $\mathbb{X}_{n+1}$ and $\mathbb{V}_{n+1}$ are approximations of $\mathbb{X}(t_{n+1})$ and $d\mathbb{X}/dt(t_{n+1})$ respectively. For any operator $\mathbb{F}$, the subscripts $\mathbb{E}$ and $\mathbb{I}$ denote an additive partitioning into an explicit operator and an implicit operator, $\mathbb{F}_{\mathbb{E}} + \mathbb{F}_{\mathbb{I}} = \mathbb{F}$. Indeed, the system (18.3.6) is solved iteratively by an explicit/implicit partitioning. At iteration $i \geq 1$, the residual is linearized,

$$\mathbb{F}^{\mathrm{surf}}_{\mathbb{E}}(\mathbb{S}^{i-1}_{n+\alpha}, \mathbb{X}^{i-1}_{n+\alpha}) - \mathbb{F}^{\mathrm{int}}_{\mathbb{I}}(\mathbb{X}^{i}_{n+\alpha}, \mathbb{V}^{i}_{n+\alpha}) \simeq \mathbb{R}^{i}_{n+\alpha} - \mathbb{C}^{*}\,(\alpha\,\Delta\mathbb{V}) = 0, \tag{18.3.7}$$

with $\mathbb{R}^{i}_{n+\alpha}$ the residual at step $n + 1$, iteration $i \geq 1$,

$$\mathbb{R}^{i}_{n+\alpha} = \mathbb{F}^{\mathrm{surf}}_{\mathbb{E}}\!\left(\mathbb{S}^{i-1}_{n+\alpha}, \mathbb{X}^{i-1}_{n+\alpha}\right) - \mathbb{F}^{\mathrm{int}}_{\mathbb{I}}\!\left(\tilde{\mathbb{X}}^{i-1}_{n+\alpha}, \mathbb{V}^{i-1}_{n+\alpha}\right), \tag{18.3.8}$$

where,

$$\mathbb{X}^{0}_{n+1} = \mathbb{X}_n + (1 - \alpha)\,\Delta t\,\mathbb{V}_n, \quad \mathbb{V}^{0}_{n+1} = \mathbb{V}_n;$$

$$\tilde{\mathbb{X}}^{i-1}_{n+1} = \mathbb{X}^{0}_{n+1} + \alpha\,\Delta t\,\mathbb{V}^{i-1}_{n+1} \;(= \mathbb{X}^{i-1}_{n+1} \text{ for } i > 1); \tag{18.3.9}$$

$$\mathbb{Z}^{i-1}_{n+\alpha} = (1 - \alpha)\,\mathbb{Z}_n + \alpha\,\mathbb{Z}^{i-1}_{n+1}, \quad \mathbb{Z} = \tilde{\mathbb{X}}, \mathbb{V}, \mathbb{X}, \mathbb{S},$$

with the convention $\tilde{\mathbb{X}}_n = \mathbb{X}_n$. The Newton direction $\Delta\mathbb{V}$ provides the correctors,

$$\mathbb{X}^{i}_{n+1} = \mathbb{X}_n + \Delta t\,\mathbb{V}^{i}_{n+\alpha} = \mathbb{X}^{0}_{n+1} + \alpha\,\Delta t\,\mathbb{V}^{i}_{n+1} = \tilde{\mathbb{X}}^{i-1}_{n+1} + \alpha\,\Delta t\,s\,\Delta\mathbb{V};$$

$$\mathbb{V}^{i}_{n+1} = \mathbb{V}^{i-1}_{n+1} + s\,\Delta\mathbb{V}; \tag{18.3.10}$$

and

$$\mathbb{X}^{i}_{n+\alpha} = \tilde{\mathbb{X}}^{i-1}_{n+\alpha} + \alpha^2\,\Delta t\,s\,\Delta\mathbb{V};$$

$$\mathbb{V}^{i}_{n+\alpha} = \mathbb{V}^{i-1}_{n+\alpha} + \alpha\,s\,\Delta\mathbb{V}. \tag{18.3.11}$$

Here $s \in]0, 1]$ is a search variable. The partitioning shown in (18.3.7) is motivated by the following observations:

- the dependence of the vector of external forces on the solution is weak;
- the vector of internal forces depends linearly on the rate vector $\mathbb{V}$;
- the vector of internal forces depends nonlinearly on the solution in several ways, in particular because the mechanical response is strongly nonlinear with respect to the chemical state. Therefore, the elastic operator will always be treated implicitly.

The Newton direction $\Delta\mathbb{V}$ at iteration $i \geq 1$ is obtained by insertion of the time-integrator (18.3.9) in the residual (18.3.7)$_1$, linearizing and forcing the result to vanish, Eqn (18.3.7)$_2$. Newton's method is known to converge quadratically close to the solution. However, for some nonlinear problems, e.g., stiffening materials, it may not converge if the starting point is too far from the solution. Convergence may be recovered by line search procedures which damp the Newton step far from the solution by the search variable s and which should automatically provide the Newton step $s = 1$ close to the solution.

The effective capacity matrix $\mathbb{C}^{*}$ expresses in terms of the capacity matrix $\mathbb{C}$ and of the diffusion-stiffness matrix $\mathbb{K}$,

$$\mathbb{C}^{*} = \mathbb{C} + \alpha\,\Delta t\,\mathbb{K} \quad \text{with} \quad \mathbb{C} = \frac{\partial \mathbb{F}_{\mathbb{I}}}{\partial \mathbb{V}}\!\left(\tilde{\mathbb{X}}^{i-1}_{n+\alpha}\right), \quad \mathbb{K} = \frac{\partial \mathbb{F}_{\mathbb{I}}}{\partial \mathbb{X}}\!\left(\tilde{\mathbb{X}}^{i-1}_{n+\alpha}\right), \tag{18.3.12}$$

where it has been recognized that the derivatives of $\mathbb{F}_{\mathbb{I}}$ do not depend on $\mathbb{V}$.

The element capacity matrix and diffusion-stiffness matrix are not symmetric, and they do not even have a symmetric profile,

$$
\mathbb{C}^e =
\begin{bmatrix}
\mathbf{0} & \mathbf{0} & \mathbf{0} & \mathbf{0} & \mathbf{0} & \mathbf{0} & \mathbf{0} & \mathbf{0} \\
\mathbf{C}^e_{\mu_w u} & \mathbf{0} & \mathbf{0} & \mathbf{0} & \mathbf{0} & \mathbf{0} & \mathbf{0} & \mathbf{0} \\
\mathbf{C}^e_{\mu_{Na} u} & \mathbf{0} & \mathbf{C}^e_{\mu_{Na}\mu_{Na}} & \mathbf{C}^e_{\mu_{Na}\mu_{Ca}} & \mathbf{C}^e_{\mu_{Na}\mu_{Cl}} & \mathbf{C}^e_{\mu_{Na}m_w} & \mathbf{C}^e_{\mu_{Na}m_{Na}} & \mathbf{C}^e_{\mu_{Na}m_{Ca}} \\
\mathbf{C}^e_{\mu_{Ca} u} & \mathbf{0} & \mathbf{C}^e_{\mu_{Ca}\mu_{Na}} & \mathbf{C}^e_{\mu_{Ca}\mu_{Ca}} & \mathbf{C}^e_{\mu_{Ca}\mu_{Cl}} & \mathbf{C}^e_{\mu_{Ca}m_w} & \mathbf{C}^e_{\mu_{Ca}m_{Na}} & \mathbf{C}^e_{\mu_{Ca}m_{Ca}} \\
\mathbf{C}^e_{\mu_{Cl} u} & \mathbf{0} & \mathbf{C}^e_{\mu_{Cl}\mu_{Na}} & \mathbf{C}^e_{\mu_{Cl}\mu_{Ca}} & \mathbf{C}^e_{\mu_{Cl}\mu_{Cl}} & \mathbf{C}^e_{\mu_{Cl}m_w} & \mathbf{C}^e_{\mu_{Cl}m_{Na}} & \mathbf{C}^e_{\mu_{Cl}m_{Ca}} \\
\mathbf{0} & \mathbf{0} & \mathbf{0} & \mathbf{0} & \mathbf{0} & \mathbf{C}^e_{m_w m_w} & \mathbf{0} & \mathbf{0} \\
\mathbf{0} & \mathbf{0} & \mathbf{0} & \mathbf{0} & \mathbf{0} & \mathbf{0} & \mathbf{C}^e_{m_{Na} m_{Na}} & \mathbf{0} \\
\mathbf{0} & \mathbf{0} & \mathbf{0} & \mathbf{0} & \mathbf{0} & \mathbf{0} & \mathbf{0} & \mathbf{C}^e_{m_{Ca} m_{Ca}}
\end{bmatrix},
\qquad (18.3.13)
$$

$$
\mathbb{K}^e =
\begin{bmatrix}
\mathbf{K}^e_{uu} & \mathbf{K}^e_{u\mu_w} & \mathbf{K}^e_{u\mu_{Na}} & \mathbf{K}^e_{u\mu_{Ca}} & \mathbf{K}^e_{u\mu_{Cl}} & \mathbf{K}^e_{um_w} & \mathbf{K}^e_{um_{Na}} & \mathbf{K}^e_{um_{Ca}} \\
\mathbf{0} & \mathbf{K}^e_{\mu_w\mu_w} & \mathbf{K}^e_{\mu_w\mu_{Na}} & \mathbf{K}^e_{\mu_w\mu_{Ca}} & \mathbf{K}^e_{\mu_w\mu_{Cl}} & \mathbf{0} & \mathbf{0} & \mathbf{0} \\
\mathbf{0} & \mathbf{K}^e_{\mu_{Na}\mu_w} & \mathbf{K}^e_{\mu_{Na}\mu_{Na}} & \mathbf{K}^e_{\mu_{Na}\mu_{Ca}} & \mathbf{K}^e_{\mu_{Na}\mu_{Cl}} & \mathbf{0} & \mathbf{0} & \mathbf{0} \\
\mathbf{0} & \mathbf{K}^e_{\mu_{Ca}\mu_w} & \mathbf{K}^e_{\mu_{Ca}\mu_{Na}} & \mathbf{K}^e_{\mu_{Ca}\mu_{Ca}} & \mathbf{K}^e_{\mu_{Ca}\mu_{Cl}} & \mathbf{0} & \mathbf{0} & \mathbf{0} \\
\mathbf{0} & \mathbf{K}^e_{\mu_{Cl}\mu_w} & \mathbf{K}^e_{\mu_{Cl}\mu_{Na}} & \mathbf{K}^e_{\mu_{Cl}\mu_{Ca}} & \mathbf{K}^e_{\mu_{Cl}\mu_{Cl}} & \mathbf{0} & \mathbf{0} & \mathbf{0} \\
\mathbf{K}^e_{m_w u} & \mathbf{0} & \mathbf{K}^e_{m_w\mu_{Na}} & \mathbf{K}^e_{m_w\mu_{Ca}} & \mathbf{K}^e_{m_w\mu_{Cl}} & \mathbf{K}^e_{m_w m_w} & \mathbf{K}^e_{m_w m_{Na}} & \mathbf{K}^e_{m_w m_{Ca}} \\
\mathbf{0} & \mathbf{0} & \mathbf{K}^e_{m_{Na}\mu_{Na}} & \mathbf{0} & \mathbf{K}^e_{m_{Na}\mu_{Cl}} & \mathbf{K}^e_{m_{Na}m_w} & \mathbf{K}^e_{m_{Na}m_{Na}} & \mathbf{K}^e_{m_{Na}m_{Ca}} \\
\mathbf{0} & \mathbf{0} & \mathbf{0} & \mathbf{K}^e_{m_{Ca}\mu_{Ca}} & \mathbf{K}^e_{m_{Ca}\mu_{Cl}} & \mathbf{K}^e_{m_{Ca}m_w} & \mathbf{K}^e_{m_{Ca}m_{Na}} & \mathbf{K}^e_{m_{Ca}m_{Ca}}
\end{bmatrix}.
\qquad (18.3.14)
$$

The definition of the element tangent matrices involved in the above formulas and details of the linearization process are reported in Exercise 18.1.

The global iteration process uses the Newton-Raphson procedure as described above. While the residuals account fully for all the mechanical and pressure terms in the electrochemical potentials of all species, the linearization process neglects the pressures in the electromechanical potentials of ions. Therefore, the iterative method is not a full Newton method.

The time-step is made variable: its value is decreased in order to keep the number of equilibrium iterations to a reasonable value, or increased to avoid too many steps as consolidation gets established. It takes value within the interval 10^{-5} to $10^{-1} t_H$, where t_H is the characteristic consolidation time, defined by (18.4.2) below.

Iterations are stopped when the criteria below involving both residuals and unknowns are satisfied:

- the residuals are made dimensionless by dividing the lines corresponding to mechanical equilibrium by $\sigma_* V_*/L_*$, to the diffusion equations by V_*/t_*, to the transfer equations by $m_* V_*/t_*$. A dimensionless residual is deemed acceptable if its norm is smaller than 10^{-6};
- the unknowns are considered acceptable if the iterative corrections of the displacements scaled by L_*, of the extrafibrillar chemical potentials divided by μ_*, of the intrafibrillar mass contents divided by m_*, are smaller than 10^{-3}.

Here σ_* is a typical stress, e.g., $100\,\mathrm{kPa}$; L_* a typical length, e.g., the half thickness L of the specimen; V_* a typical volume, e.g., the initial volume of the sample; m_* a typical intrafibrillar mass content, e.g., its initial value for water, and maximum of their collective initial values for ions; t_* a typical time, e.g., Darcy's consolidation time for the problem at hand, and μ_* a typical chemical potential, e.g., the initial value.

18.3.4 Dependent variables

Once the primary variables are known, the variations of the dependent variables, *and* their updated values, can be obtained in the following order:

- Pressures $\qquad$ δp_{wE}, δp_{I} by (18.E.14);

- Volume contents $\qquad$ δv^{I} by (18.E.11) and $\delta v^{\mathrm{E}} = \det \mathbf{F}\, \delta \mathrm{tr}\, \boldsymbol{\epsilon} - \delta v^{\mathrm{I}}$;

- IF chloride ions $\qquad$ $\delta m^{\mathrm{ClI}} = \delta m^{\mathrm{I}} - \sum_{i \in \mathrm{I_{in}}} \delta m^{i\mathrm{I}}$;

- IF mole numbers $\qquad$ $\delta N_{k\mathrm{I}} = V_0\, \delta m^{k\mathrm{I}}/\widehat{m}_k$, $k \in \mathrm{I}$, and $\delta N_{\mathrm{I}} = \sum_{k \in \mathrm{I}} \delta N_{k\mathrm{I}}$;

- IF molar fractions $\qquad$ $x_{k\mathrm{I}} = N_{k\mathrm{I}}/N_{\mathrm{I}}$, $k \in I$;

- EF mass contents $\qquad$ $\delta m^{k\mathrm{E}} = \rho_k\, \delta v^{k\mathrm{E}}$, $k \in \mathrm{E_{ions}}$, by (18.E.16);

- EF water content $\qquad$ $\delta v^{\mathrm{wE}} = \det \mathbf{F}\, \delta \mathrm{tr}\, \boldsymbol{\epsilon} - \delta v^{\mathrm{I}} - \sum_{k \in \mathrm{E_{ions}}} \delta v^{k\mathrm{E}}$;

- EF volume fraction $\qquad$ δn^{E} by (18.E.8);

- EF mole numbers $\qquad$ $\delta N_{k\mathrm{E}} = V_0\, \delta m^{k\mathrm{E}}/\widehat{m}_k$, $k \in \mathrm{E}$, and $\delta N_{\mathrm{E}} = \sum_{k \in \mathrm{E}} \delta N_{k\mathrm{E}}$;

- PG effective concentration $\quad$ δy_{PG} by (18.E.9) and $e_{\mathrm{PG}} = N_{\mathrm{E}}/V_{\mathrm{E}}\, y_{\mathrm{PG}}$;

- EF molar fractions $\qquad$ $x_{k\mathrm{E}} = N_{k\mathrm{E}}/N_{\mathrm{E}}$, $k \in \mathrm{E}$;

- EF electrical potential $\qquad$ $\delta \phi_{\mathrm{E}}$ by (18.E.12).

18.3.5 A slightly different approach

Phasewise electroneutrality has been used to structure and compact the rate transfer equations between the intrafibrillar and extrafibrillar compartments, which have been phrased in terms of neutral species, namely water and salt, and of their chemical potentials.

These relations deduce from the more fundamental relations that describe the transfer of each species between the two compartments in terms of their electro-chemical potentials. This approach would turn the analysis of the IF-EF interface similar to the approach used for the EF-bath interface. Use of the electrochemical potentials of the intrafibrillar species as primary unknowns would allow to access the time histories of the mass contents, or concentrations, of all the intrafibrillar species, together with the intrafibrillar electrical potential.

18.4 Testing setup and material data

18.4.1 Specimen geometry

The cartilage specimen is cylindrical with a thickness $2\,\mathrm{L}{=}600\,\mu\mathrm{m}$ so that the material response to mechanical load reaches a steady state in relatively short times, Fig. 18.4.1. The specimen is supported on both its top and bottom faces by filters which are, first, more rigid than the cartilage so as to homogenize the deformation, and, second, much more permeable so as not to delay significantly the establishment of steady states. The filters and cartilage are placed in an apparatus that does not allow for radial extension. The small thickness of the cartilage minimizes nonuniform strains that may arise on the surface of the oedometer by friction. Therefore, under the assumption that the radial directions undergo neither extension nor contraction, the displacement is one-dimensional along the sole axial direction.

In view of the symmetries of the geometry and boundary conditions, only the upper half sample has been discretized in 30 elements of equal length. A single mesh has been used for all tests. The qualitative accuracy of the results has been checked with a uniform but twice finer mesh.

The load and chemical potentials are prescribed at the top of the specimen in contact with the bath. The axial displacement and fluxes of water and ions vanish on the symmetry line.

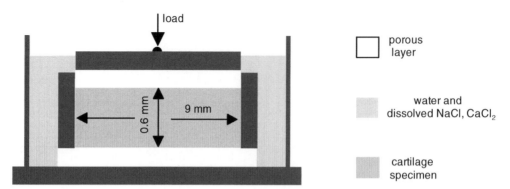

Fig. 18.4.1: Schematic of a confined compression test (not at scale). The specimen is cylindrical, with a small thickness so as to minimize heterogeneity of the deformation at steady states and to avoid an excessively large transient period before equilibrium. Radial strains are prevented. The specimen undergoes instantaneous mechanical loading, which is applied via an inert platen that improves homogeneity of deformation and whose porosity is large so as not to delay diffusion of water and ions. On the other hand, any change of the chemical composition of the bath is accompanied by a transient period during which the state of the material varies in both space and time, before a final homogeneous state is reached.

18.4.2   Initial data

Physical properties of the chemicals involved are shown in Table 18.4.1. The tests are run at ambient temperature T=298 K, and the bath pressure p_B is maintained constant and equal to atmospheric pressure.

Table 18.4.1: Physical constants.

Species	Molar Mass (gm)	Molar Volume (cm^3)	Valence
Water	18	18	0
Cation Na$^+$	23	2.37	+1
Cation Ca^{2+}	40.1	1.75	+2
Anion Cl$^-$	35.5	15.42	-1
Collagen	0.285×10^6	0.2×10^6	0
Proteoglycans	2×10^6	1.11×10^6	-6580

   Initially the specimen is in equilibrium with a hypertonic bath at 1 M of NaCl. The following masses are assumed: 200 mg of intrafibrillar water and 200 mg of extrafibrillar water, 100 mg of collagen, 20 mg of PGs, corresponding to an effective concentration e_{PG} of PGs equal to -0.3 M. Note that this concentration is relative to the extrafibrillar phase only and it is therefore higher than traditional values which refer to the sum of the two fluid phases. The initial repartition of the other species within the phases of the cartilage is calculated in such a way that the initial state realizes a chemomechanical equilibrium, as described in Loret and Simões [2004][2005]a. The resulting distribution of each species k within the three phases K = I, E, S, in terms of mole numbers N_{kK}, molar fractions x_{kK} or effective molar fractions y_{kK}, is shown in Table 18.4.2. The resulting overall initial characteristics are displayed in Table 18.4.3.

   Let us recall that the molar fraction $x_{kK} = N_{kK}/N_K$ of a species k in phase K is the ratio of its mole number N_{kK} over the mole number N_K in that phase. The effective molar fraction $y_{kK} = \zeta_k N_{kK}/N_K$ of a charged species accounts for its valence ζ_k. Proteoglycans being macromolecules, their molar fraction $x_{PG} = N_{PG}/N_E$ and concentration $c_{PG} = N_{PG}/V_E$ are quite small, but their effective molar fraction $y_{PG} = \zeta_{PG} x_{PG}$ and effective concentration $e_{PG} = \zeta_{PG} c_{PG}$ are key entities that control the chemomechanical and diffusion properties. Their presence in the extrafibrillar phase results in a pressure discontinuity $p_{fw} = 66$ kPa across the two fluid phases. The formation pressures of ions have been simply set to zero so that the equilibrium constants for the salts (18.2.12) are equal to 1.

18.4.3   Mechanical and transport parameters

The aggregate hypertonic elastic modulus H is set equal to 500 kPa. A method to identify the chemomechanical parameters is reported in Loret and Simões [2004]. The parameters of the chemomechanical elastic coupling which

Table 18.4.2: Initial repartition of the mole numbers and effective molar fractions.

Species	Solid Phase		IF phase		EF phase	
	mmole nb.	molar frac.	mmole nb.	molar frac.	mmole nb.	molar frac.
Water	-	-	11.111	0.96525	11.111	0.96477
Cation Na$^+$	-	-	0.2	0.01737	0.2357	0.02047
Cation Ca^{2+}	-	-	0.0	0.0	0.0	0.0
Anion Cl$^-$	-	-	0.2	-0.01737	0.17	-0.01476
Proteoglycans	-	-	-	-	0.00001	-0.00571
Collagen	0.000351	1	-	-	-	-

Table 18.4.3: Overall initial characteristics of the cartilage specimen.

Mass M_0	Volume V_0	Density ρ_0
529 mg	480 mm^3	1.10 mg/mm^3

define the interpolation of the modulus in terms of ionic mass contents and the interpolation of the adhesion pressure p_{adh}, Eqn (18.2.6), in terms of the mass content of intrafibrillar water are listed in Loret and Simões [2004][2005]a.

The data related to transport are shown in Table 18.4.4. The hydraulic conductivity K_{H} varies according to the Kozeny-Carman formula $K_{\mathrm{H}}/K_{\mathrm{H0}} = (e/e_0)^3 (1+e)/(1+e_0)$, where the void ratio e is defined as $V_{\mathrm{E}}/(V_{\mathrm{S}}+V_{\mathrm{I}})$ and the subscript 0 indicates the initial state. The values of the diffusion coefficients are extracted from Maroudas [1968] [1975]. The diffusion coefficients D, shown in Table 18.4.4, from which the ionic mobilities $u = D|\zeta|\mathrm{F}/RT$ deduce, apply to a blank solution at infinitesimal dilution. The effective diffusion coefficients D^* and effective ionic mobilities u^* are defined as $D^* = \tau D$ and $u^* = \tau u$, respectively. The tortuosity factor τ is kept constant and equal to 0.40, Maroudas [1975].

Table 18.4.4: Transport coefficients.

Hydraulic conductivity [10^{-12} m/s]	$K_{\mathrm{H0}}{=}5$		
Ionic diffusion coefficients [10^{-10} m^2/s]	$D_{\mathrm{Na}} = 13.3$	$D_{\mathrm{Ca}} = 7.92$	$D_{\mathrm{Cl}} = 20.3$
Ionic mobilities [10^{-9} m^2/s/volt]	$u_{\mathrm{Na}} = 51.8$	$u_{\mathrm{Ca}} = 61.7$	$u_{\mathrm{Cl}} = 79.1$

No specific data are available that would allow a definite determination of transfer times. Therefore, a parameter analysis has been performed to assess their quantitative influence, starting from the reference values shown in Table 18.4.5.

18.4.4   Characteristic times

The processes to be simulated are controlled by several typical times. Indeed, the material response depends on the characteristic times associated with the mechanical and chemical loading processes. Moreover, individually, in absence of couplings, ionic diffusion, seepage and mass transfer introduce each its own characteristic time, namely

- *diffusion of ionic species*: to diffuse over half of the specimen thickness L, the ion k, with effective diffusion coefficient D_k^* [unit : m^2/s], requires the characteristic time

$$t_{\mathrm{C},k} = \frac{L^2}{D_k^*}, \quad k \in \mathrm{E_{ions}}. \tag{18.4.1}$$

Here the slowest characteristic time is that of sodium ions, about 170 s;

- *seepage of water*: let us consider a layer of thickness L, with aggregate elastic modulus H [unit : Pa], and hydraulic permeability $k_{\mathrm{EE}} = K_{\mathrm{H0}}/(\rho_{\mathrm{w}} g)$ [unit : m^2/Pa/s], where g is the acceleration of gravity. If the layer is drained at one side and undrained at the other side, the characteristic time of seepage (consolidation time) is equal to

$$t_{\mathrm{H}} = \frac{L^2}{H k_{\mathrm{EE}}}. \tag{18.4.2}$$

Here the characteristic time is about 400 s;

Table 18.4.5: Standard transfer times.

Species	Water	Salt NaCl	Salt CaCl$_2$
Time (s)	9.9×10^3	3.2×10^8	4.4×10^8

- *mass transfer between the intra- and extrafibrillar fluid phases.* The rate of mass content of species $i \in \mathrm{I}_{\mathrm{in}}$ can be expressed in the format $dm_{i\mathrm{I}}/dt = -T_i\,(\mu_{n\mathrm{I}} - \mu_{n\mathrm{E}})$, Eqns (18.2.4),(18.2.18). Hence the characteristic time of transfer of the species $i \in \mathrm{I}_{\mathrm{in}}$,

$$t_{\mathrm{tr},i} = \frac{\rho_i^2}{H\,T_i}, \quad i \in \mathrm{I}_{\mathrm{in}}\,. \tag{18.4.3}$$

For $T_{\mathrm{w}} = 2 \times 10^{-4}\,\mathrm{kg \times s/m^5}$, the transfer time $t_{\mathrm{tr,w}}$ is equal to $10^4\,\mathrm{s}$.

While these times are indicative, the fact that the field and constitutive equations are nonlinear and coupled implies that the actual material response depends on the internal characteristics times in a more intricate manner than implied by the individual times above.

18.5   Mechanical and chemical loadings with NaCl

The time profiles of the pressure and chemical composition of the bath are controlled. The boundary conditions at the interface between the bath and the cartilage result in the monitoring of the following entities at the top and bottom surfaces of the specimen:
- chemical potential of water μ_{wE} and electrochemical potentials of ions $\mu_{k\mathrm{E}}^{\mathrm{ec}}$, k=Na, Ca, Cl, in the extrafibrillar compartment, and
- either axial displacement or axial applied load.

Due to the symmetry conditions, the displacement and the fluxes of water and ions are required to vanish at the specimen center. The bath is initially hypertonic at $1\,\mathrm{M}$ NaCl and the cartilage specimen is unloaded. Departure from this initial state is performed through either chemical or mechanical loading: mechanical loading rates are piecewise constant, while chemical loading rates vary in a nonlinear way.

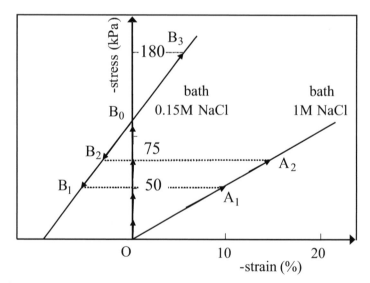

Fig. 18.5.1: Schematic of stress controlled tests, or mechanical consolidation tests, at two given bath chemistries, starting from the hypertonic state ($\sim 1\mathrm{M}$). In the A-tests, the sample is in equilibrium with a bath at $1\,\mathrm{M}$ NaCl. For the B-tests, the bath is first brought from $1\,\mathrm{M}$ to $0.15\,\mathrm{M}$ in $120\,\mathrm{s}$, path OB$_0$, and then left to consolidate. In the mechanical tests OA$_1$, $\cdots$, B$_0$B$_1$, $\cdots$, the cartilage disks are brought, in $120\,\mathrm{s}$, to the target stress, and then left to consolidate. Reprinted with permission from Loix et al. [2008].

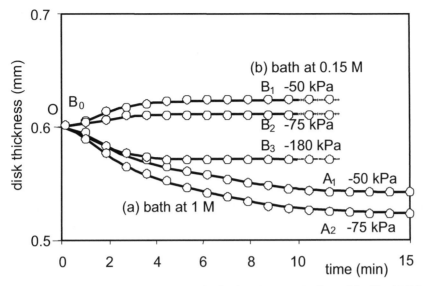

Fig. 18.5.2: Mechanical consolidation test, at given bath chemistry, as indicated in Fig. 18.5.1, (a) bath at 1 M NaCl, (b) bath at 0.15 M NaCl. As expected from the chemomechanical model, for a given applied compressive load, the disks either contract or swell depending on the chemical composition of the bath. Reprinted with permission from Loix et al. [2008].

18.5.1 Mechanical consolidation at different bath chemistries

These tests aim at showing that the mechanical model is able to simulate the experimental data by Maroudas [1975], her Fig. 10. She reports of mechanical consolidation tests at two fixed bath chemistries, 1 M NaCl and 0.15 M NaCl. For each chemistry, she performed tests with different loads. For each load, comparing the effects of chemistry, she observes that the cartilage weight at equilibrium is smaller for the 1 M NaCl solution.

The procedure is as follows, Fig. 18.5.1:
- start from the hypertonic state O, at zero strain;
- stage 1: For the tests with a bath at 0.15 M NaCl, change to targeted chemistry in a chosen time interval (120 s) at constant zero strain, and wait for steady state, path OB_0;
- stage 2: Apply the mechanical load in a chosen short time interval (120 s), and wait for steady state, keeping the chemical composition of the bath fixed, paths OA_1, OA_2, or B_0B_1, B_0B_2, B_0B_3.

It is important not to misinterpret the sketch shown in Fig. 18.5.1. Indeed, the thick curves on this sketch correspond to equilibrium curves, at given bath chemistries: they can be followed for very slow loading rates. They also are reached, pointwise, after a long rest period. The transient paths described in the simulations differ from these equilibrium curves, except at the end point. This remark applies as well to the subsequent sketches.

The evolution of the thickness of the cartilage disk during stage 2 is shown in Fig. 18.5.2. The following points are worth of notice:
- these tests illustrate a striking effect of the chemomechanical coupling that governs the response of articular cartilage. In fact, the disks either contract or swell depending on the chemical composition of the bath and of the targeted compressive stress. At given load, the strain at equilibrium is more contractile for the 1 M NaCl solution in agreement with the data reported by Maroudas [1975] and Eisenberg and Grodzinsky [1985], and with the model developed in Loret and Simões [2004], Fig. 18.5.1;
- the time period for seepage to be completed is longer than the characteristic time (18.4.2) which does not account for the chemomechanical couplings, Section 18.4.4. The Kozeny-Carman formula corresponds to a smaller hydraulic conductivity at larger compressive load. Therefore, at given bath chemical composition, the larger the load, the larger the consolidation time;
- the consolidation times for a bath at 0.15 M are smaller than at 1 M as shown in Fig. 10 of Maroudas [1975].

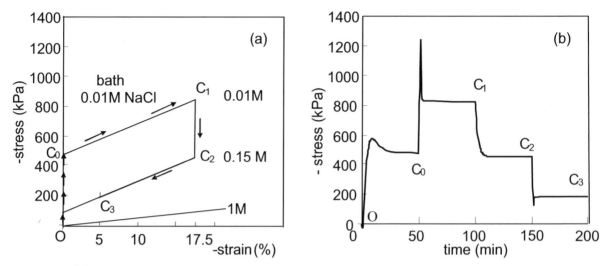

Fig. 18.5.3: (a) Schematic of a test that intermingles mechanical loading at constant chemical composition of the bath, and change of the chemical composition of the bath, at constant strain. The thick curves, shown on the sketch, correspond to equilibrium curves, at given bath chemistries, that can be followed for very slow loading rates. The actual path departs from the equilibrium curve, that is nevertheless reached at the end of each of the four stages. (b) Model simulations: see text for details of loading rates. Reprinted with permission from Loix et al. [2008].

18.5.2 Intermingled mechanical and chemical consolidation

The subsequent program monitors the strain, at variance with the previous mechanical tests which were stress-controlled. The program intermingles chemical and mechanical loadings and involves four stages. It is depicted in Fig. 18.5.3:

- stage 1: At time t=0, the NaCl concentration of the bath is decreased from 1 M to 0.01 M in 120 s, at constant (zero) strain, path OC_0. The setup is left to rest until equilibrium is reached (50 min);
- stage 2: The specimen is then submitted to a contractive strain of -17.5% in 120 s, at fixed bath chemistry, after which the setup is again left to rest until equilibrium, path C_0C_1;
- stage 3: The NaCl concentration of the bath is increased from 0.01 M to 0.15 M, at constant strain, and the setup is again left to rest until equilibrium, path C_1C_2;
- stage 4: The contractive strain is brought from -17.5% to -5% in 120 s, at fixed bath chemistry, path C_2C_3.

This test illustrates the fact that the equilibrium curves have an apparent slope that increases as the NaCl concentration of the bath decreases. The model simulations are displayed in Fig. 18.5.3. A small overshoot is observed when we first change the chemistry of the bath. A larger overshoot takes place when we impose the strain. This phenomenon can be traced to the fact that the fluid pressure initially bears the load change, as indicated in Section 18.5.7.

18.5.3 Reversibility

To check the reversibility of the stress-strain model during transient loadings, two tests are performed.

First, the ionic strength of the bath is decreased to 0.15 M in 120 s at constant zero strain, path OB_0 in Fig. 18.5.4. The chemical composition of the bath is later kept unchanged. Once equilibrium has been reached, a compressive mechanical stress of -240 kPa is applied in 120 s, path B_0B_1. Again once equilibrium has been reached, the strain is reduced to zero in 120 s. The transient paths from B_0 to B_1, and from B_1 to B_0, have no reason to be identical, in particular because a stress is targeted in the first case, and a strain for the return path. On the other hand, reversibility implies that the point B_0 should be recovered at equilibrium. That it is the case can indeed be checked in Fig. 18.5.4.

In the second test, a stress of -75 kPa is applied in 120 s, in the hypertonic state. Keeping the stress constant, the ionic strength of the bath is next decreased to 0.15 M in 120 s, path A_1B_1, and finally returned to 1 M, Fig. 18.5.5. For the three stages of the path, sufficient time is left for equilibrium to be reached, before the bath composition is

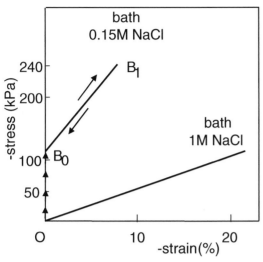
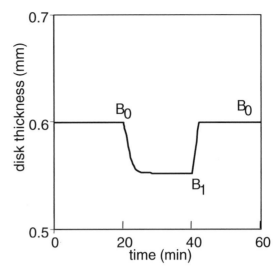

Fig. 18.5.4: (a) Schematic of a cyclic test $B_0B_1B_0$. (b) Evolution in time of the thickness of the specimen subjected to the cyclic test $B_0B_1B_0$ at constant chemical composition of the bath. Reprinted with permission from Loix et al. [2008].

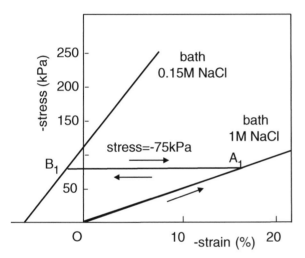
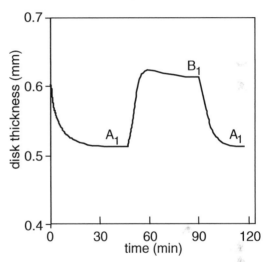

Fig. 18.5.5: (a) Schematic of a cyclic test $OA_1B_1A_1$; (b) Evolution in time of the thickness of the specimen subjected to the cyclic test $OA_1B_1A_1$ at constant axial load of the bath. As expected from the sketch, the disk thickness at B_1 is larger than in the hypertonic state. Reprinted with permission from Loix et al. [2008].

further modified. The issue here is to verify that the equilibrium point A_1 at the end of stage 1 is recovered after the chemical loading-unloading path.

18.5.4   Uniaxial free swelling

Free swelling tests in the oedometer apparatus differ from isotropic free swelling, where the stress tensor vanishes. In the oedometer apparatus, extension is prevented in the lateral direction. Thus free swelling in the oedometer apparatus involves only a vanishing axial stress.

Starting from the hypertonic state, the concentration of the bath is decreased to 0.15 M in 120 s, path OS_1 in Fig. 18.5.6. Once equilibrium has been reached, the concentration of the bath is further decreased to 0.125 M in 10 s, path S_1S_2. Fig. 18.5.7 displays the simulations using the present model with two water compartments as well as using the model without intrafibrillar compartment. The typical features associated with models with a single water compartment and with two water compartments are considered in more details in Section 18.5.5.

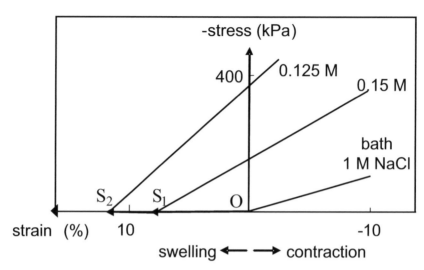

Fig. 18.5.6: Schematic of a free swelling test OS_1S_2 under confined compression. Reprinted with permission from Loix et al. [2008].

18.5.5 Influence of the presence of the intrafibrillar compartment

The presence of the intrafibrillar compartment introduces a double porosity: the two compartments have distinct chemical compositions and distinct physiological roles. However, the model presented here displays a sequential structure: indeed, the intrafibrillar compartments are not interconnected, and only the extrafibrillar species communicate with the surroundings. In that sense, the model differs from so-called double porosity/double permeability models where the two types of pores are endowed with their own permeabilities, e.g., Aifantis [1977].

The intrafibrillar compartment delays the occurrence of equilibrium, and it may either damp or induce oscillations in the transient response of the porous medium. Indeed, mass exchanges between the two fluid compartments are not instantaneous, and their time profiles are regulated by the transfer times. This point is illustrated in Figs. 18.5.7 and 18.5.8.

Fig. 18.5.7 displays the time courses of the thickness for the two compartment model. The confined compression curves (a) and (b), corresponding respectively to two compartments and one compartment, are quite similar as long as the influence of the IF compartment is not felt. Indeed, the key transport phenomenon is seepage through the extrafibrillar compartment. Later the transfers between the compartments take place: the pumping out of the IF water is triggered by the osmotic phenomenon but limited by adhesion forces. The final resulting volume is thus smaller for the two compartment model.

Curves (b) and (c) in Fig. 18.5.7 correspond to the model without IF compartment and to permeabilities differing by an order of magnitude. On comparing the curves (a), (b) and (c), it appears that, with the values of parameters used here, the short term response is controlled by diffusion while the long term response is controlled by transfer. Moreover, it will be seen in Section 18.5.7 that ionic diffusion is slightly faster than water diffusion (seepage).

The influence of the intrafibrillar phase is further analyzed in Figs. 18.5.8 to 18.5.11 during the path OC_0 shown in Fig. 18.5.3. The initial small tensile stress in Fig. 18.5.8 is a bit intriguing. In fact, since the bath is refreshed, the trend to a decreasing ionic strength will take place first at the top of the specimen, where the pressure p_I becomes positive in agreement with the osmotic effect, Fig. 18.5.9. However, the overall total strain is imposed to be zero. Thus, water has to move from the specimen center towards the bath, Fig. 18.5.10. Thus, once again due to the osmotic effect, the pressure p_I tends to turn negative in most of the specimen, leading to an initial overall tensile stress.

The phenomenon is enhanced by the porous layer: there, the sample (specimen plus porous layer) is longer. Thus the pressure will be more negative in the specimen, and it will remain negative for a longer period, Figs. 18.5.8-(A) and 18.5.9-(c). Since it takes place at early times, the phenomenon is not modified by the presence of the IF compartment, Figs. 18.5.8-(A) and 18.5.9-(a),(b).

Additional tests have shown that

- an increase of the hydraulic conductivity by an order of magnitude does not modify neither the maximum value nor the time profile of the stress at early times;

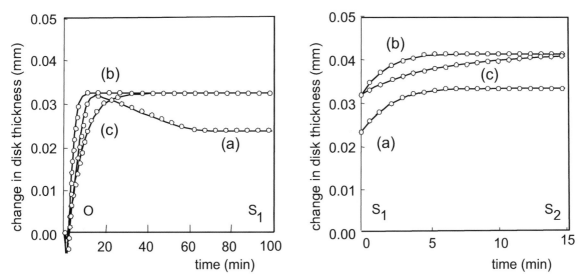

Fig. 18.5.7: Time profile of the thickness of the disk subjected to a two-step free swelling test under confined compression: (a) standard model; (b) model without IF compartment with the standard hydraulic conductivity; (c) model without IF compartment with the standard hydraulic conductivity divided by 10. Reprinted with permission from Loix et al. [2008].

- a faster change of chemical composition of the bath (in 12 s instead of 120 s) leads the maximum tensile stress to take place much earlier but does not change the maximum value.

18.5.6 Influence of the presence of a porous layer at the boundaries

For compression tests, the horizontal boundaries of the cartilage disk are sometimes covered with a porous layer, which is thought to homogenize the strain state in the disk. This layer is much stiffer than the cartilage, with a Young's modulus equal to 25 GPa, and its hydraulic conductivity is much higher, say $10 \times K_{H0}$. The tortuosity factor τ is taken equal to 0.9.

The presence of the porous layer modifies the time profile of chemical loadings, since the chemicals need to diffuse first through the layer. The porous layer is also expected to smooth out the transient effect of chemical loadings. The delay in the material response depends clearly on the diffusion characteristics of the porous layer. On the other hand, the equilibrium states are not affected by its presence. These statements can be checked in Fig. 18.5.8.

18.5.7 Evolution in time of spatial profiles

Figs. 18.5.9 to 18.5.11 display the evolution in time of the spatial profiles of various entities defining the mechanical responses of the extrafibrillar and intrafibrillar compartments of the cartilage. They correspond to path OC_0 in Fig. 18.5.3, where the ionic strength of the bath decreases from 1 M to 0.15 M in 120 s, at fixed disk thickness. The profiles are displayed for the standard model with an intrafibrillar compartment, in presence and absence of the porous layer, and for the model without the intrafibrillar compartment in absence of the porous layer:

- as the bath is refreshed, the ionic strength in the extrafibrillar (EF) compartment at the top of the disk is decreased first. This decrease diffuses progressively down to the center of the disk, Fig. 18.5.10;
- the decrease of the ionic strength in the intrafibrillar (IF) compartment is delayed with respect to that of the EF compartment, since the IF-EF mass transfers are not instantaneous, but characterized by specific transfer times, compare Figs. 18.5.10 and 18.5.11;
- the presence of mobile metallic ions in the extrafibrillar water screens the repulsion between negatively charged PGs. Decrease of ionic strength implies the electrical shielding effect to diminish, so that the distance between PGs increases. In the early times, EF water thus flows towards the top of disk, both from the bath and from the center part of the disk, Fig. 18.5.10;
- in contrast, the spatial profiles of the mass content of IF water are monotonous and smooth, Fig. 18.5.11;

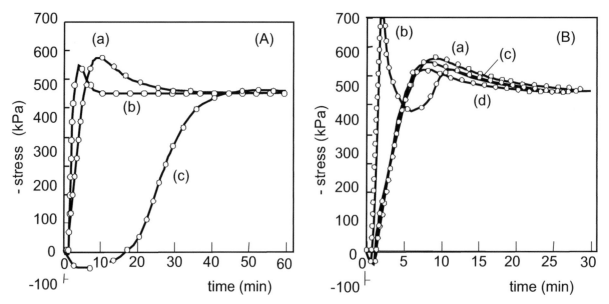

Fig. 18.5.8: Time profile of the axial stress during path OC_0 in sketch 18.5.3, where the ionic strength of the bath is decreased while the disk thickness is maintained fixed. Plot (A) displays (a) the standard model; (b) the model without intrafibrillar compartment; (c) the standard model but in presence of a porous layer. Plot (B) shows the influence of the transfer times: (a) reference parameters; (b) transfer time of water divided by 100; (c) transfer time of sodium ion divided by 100; (d) the two above transfer times divided by 100. Reprinted with permission from Loix et al. [2008].

- in absence of intrafibrillar phase, the trend to equilibrium is slightly stronger for ions than for water, Fig. 18.5.10-(b). In other words, ionic diffusion is slightly faster than seepage. Thus for the mode without IF compartment, the material response would be controlled and delayed by seepage;

- in absence of intrafibrillar phase, as equilibrium progressively settles down, the mass content of extrafibrillar water returns to its initial value as a consequence of the fact that the overall strain is zero and that the species are all incompressible, Fig. 18.5.10-(b);

- in contrast, as shown by comparison of Figs. 18.5.10-(a) and (b), when the intrafibrillar phase is present, it looses part of its water in favor of the extrafibrillar phase, Figs. 18.5.10-(a), 18.5.11-(a). The driving engine here is an osmotic effect due to the presence of PGs in the EF phase. The motion is partially resisted by adhesion forces that "glue" water to the collagen fibers: the phenomenon is introduced in the model through the adhesion pressure p_{adh};

- the evolution of the mass content of EF water is the same whether the IF compartment exists, until about 500 s. The mass increases close to the bath until a time of about 150 s, and later decreases. At about 500 s, IF water is sucked out its compartment, and becomes EF water, Figs. 18.5.10-(a),(b), and 18.5.11-(a).

As expected, the porous layers smooth out the spatial profiles and delay the response to chemical loadings.

18.5.8   The membrane surrounding the cartilage specimen

The presence of the PGs in the extrafibrillar compartment implies that a number of physical quantities are discontinuous across the cartilage surface, namely from the bath to the extrafibrillar compartment. Instances are the fluid pressure $p_{wE} = p_I + p_{fw} \neq p_B$, the concentrations $c_{kE} \neq c_{kB}$ and molar fractions $x_{kE} \neq x_{kB}$, $k \in E_{ions}$, and the electrical potential $\phi_E \neq \phi_B$.

For example, at the equilibrium point C_0 in Fig. 18.5.3, the simulations provide the following values:

- $p_{wE} = p_I + p_{fw} = 530 + 66 = 596 \, \text{kPa} > p_B = 0 \, \text{kPa}$;
- $c_{NaE} = 0.266 \, \text{M} > c_{NaB} = 0.01 \, \text{M}$;
- $c_{ClE} = 3.83 \times 10^{-4} \, \text{M} < c_{ClB} = 0.01 \, \text{M}$;
- $\phi_E = -84 \, \text{mV} < \phi_B = 0 \, \text{mV}$.

These values can be checked to be consistent with the equilibrium of the electrochemical potentials in the EF compartment and in the bath. The time and space evolutions of the electrical potential are shown in Fig. 18.5.9.

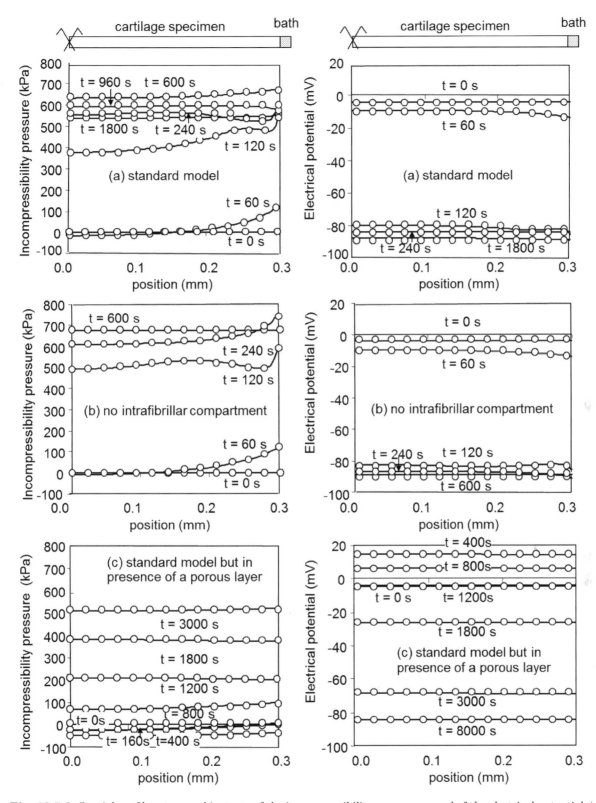

Fig. 18.5.9: Spatial profiles at several instants of the incompressibility pressure p_I and of the electrical potential ϕ_E along path OC$_0$ in Fig. 18.5.3, where the bath is refreshed at constant disk thickness. (a) standard model; (b) no intrafibrillar compartment; (c) standard model but in presence of a porous layer. The center of the specimen is located at the position $z = 0$ and the interface with the bath or the porous layer at $z = L = 0.3$ mm. Reprinted with permission from Loix et al. [2008].

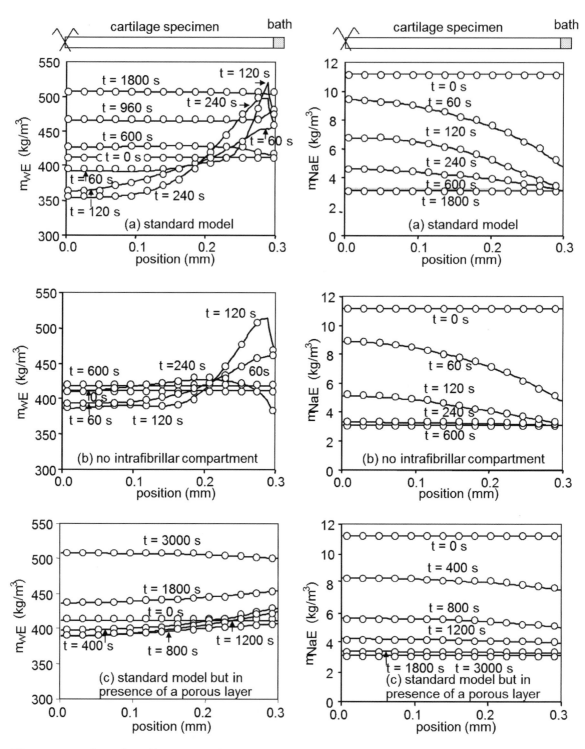

Fig. 18.5.10: Spatial profiles at several instants of the mass contents of extrafibrillar water and extrafibrillar sodium ion during path OC_0 in Fig. 18.5.3. Reprinted with permission from Loix et al. [2008].

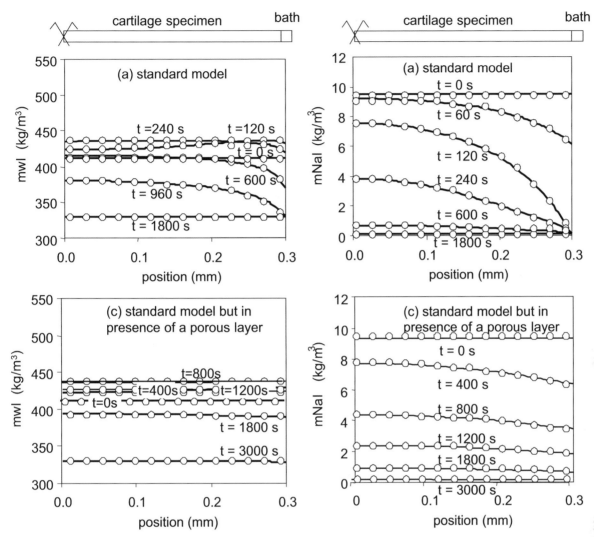

Fig. 18.5.11: Spatial profiles at several instants of the mass contents of intrafibrillar water and intrafibrillar sodium ion during path OC$_0$ in Fig. 18.5.3. Reprinted with permission from Loix et al. [2008].

The spatial profiles are quite smooth and the trend to equilibrium is somehow faster than for the extrafibrillar water content.

Note that the osmotic pressure of about 600 kPa for a bath at 10 mM of NaCl is, as expected, larger than the standard value of osmotic pressure under physiological conditions (150 mM) which takes values around 200 kPa.

18.6 Cyclic substitution of NaCl and CaCl2

Cyclic substitution of NaCl and CaCl$_2$ is simulated through the loading procedure sketched in Fig. 18.6.1:
- start from the hypertonic state O (1 M NaCl, 10^{-11} M CaCl$_2$), at zero strain;
- stage 1: Keeping the chemistry of the bath fixed, increase the compressive stress to -120 kPa in a chosen time interval (120 s), and wait for steady state, path OA;
- the axial strain is henceforth kept fixed;
- stage 2: Refresh the bath to 0.15 M of NaCl in 120 s, and wait for steady state, path AB;
- stage 3: Refresh the bath to 10^{-4} of NaCl in 120 s, and wait for steady state, path BC;

- stage 4: Change the bath composition to 10^{-4} M of NaCl and 0.15 M CaCl$_2$ in 120 s, path CD, and wait for steady state;
- stage 5: Change the bath chemistry to 10^{-4} M of NaCl and 10^{-11} M CaCl$_2$ in 120 s, path DE, and wait for steady state.

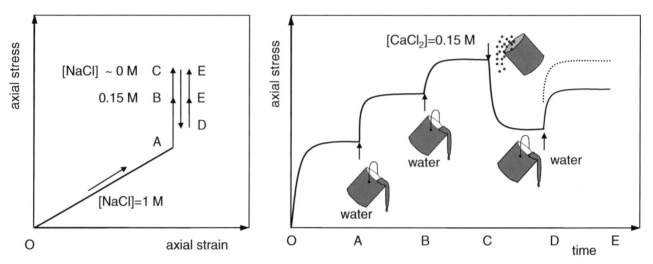

Fig. 18.6.1: Schematic of a very slow test involving two salts at fixed disk thickness. The actual numerical simulations differ qualitatively from this sketch by a nonmonotonous response associated with a finite rate of change of the chemical composition of the bath.

Model simulations have been performed with the standard model with two water compartments and with the model without IF compartment, Figs. 18.6.2 (a) and (b), respectively.

Decreasing the ionic strength of the bath leads to increase of the compressive stress, path ABC, as a consequence of the increased repulsion between the fixed negative charges.

The calcium ions appear clearly more efficient to shield the inter-PGs electrical repulsion, path CD, than sodium ions, path BC. The phenomenon may be attributed to the valence of calcium ions, and to their smaller hydrated diameter. Indeed, while the non hydrated diameters of Na$^+$ and Ca^{2+} are 1.96 Å and 3.30 Å, respectively, the respective hydrated diameters are approximately equal to 14 Å and 8 Å. The phenomenon might be also attributed to the fact that calcium ion binds partially to the PGs. However neither steric considerations nor binding are accounted for directly here: the more efficient shielding effect is simply reproduced by smaller elastic moduli.

Since the chemical composition of the bath is perfectly controlled in this numerical simulation, the path CDE is reversible. The situation is more complex in actual experiments where ionic exchanges between the cartilage and the bath pollutes the latter. Sophisticated setups would be needed to monitor accurately and continuously the chemical composition of the bath.

The temporary plateau just after calcium ions are introduced in the bath, point C, is due to the fact the chemoelastic modulus has been given a finite maximum value based on previous experimental information, Loret and Simões [2005]a. To improve this detail of the constitutive modeling, experimental data are needed.

In contrast with the simulations shown in Fig. 18.5.8, the strains at which the model simulations are performed differ depending whether the IF compartment is included or not. Indeed, this difference results from the initial compression stage OA. Consequently, the stresses at B differ in the two models.

18.7   Improvements of the chemomechanical model

Articular cartilages sustain load through electro-chemomechanical couplings that limit the magnitude of the strains. Because of the presence of the electrically charged proteoglycans, a number of physical entities are discontinuous at the interface between the cartilage and the bath. This phenomenon restricts the choice of primary variables that allow the boundary conditions to be efficiently dealt with.

Other finite element simulations have so far considered a single water compartment and a single dissolved salt, namely NaCl. Here, in contrast, in line with the ideas of Maroudas and coworkers, two compartments are recognized

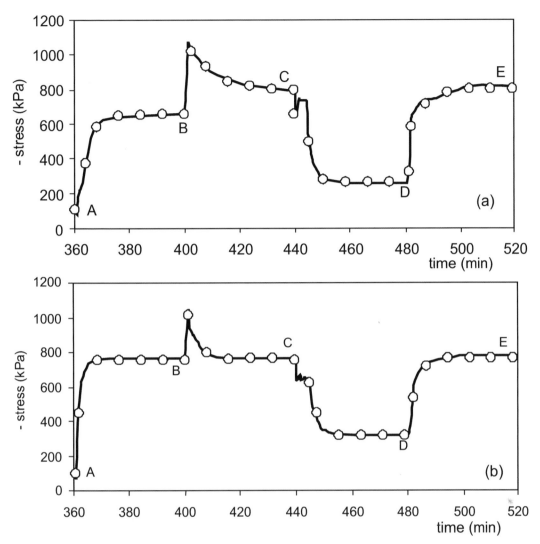

Fig. 18.6.2: (a) Chemical loading, with ionic replacement as indicated in Fig. 18.6.1. With respect to the sketch displayed in Fig. 18.6.1, some stages show a nonmonotonous behavior due to the specific rate at which the change of the chemical composition of the bath is performed. (b) Same as (a) but without IF compartment. Reprinted with permission from Loix et al. [2008].

to have distinct mechanical roles. Water and ions can transfer between the intrafibrillar (IF) and extrafibrillar (EF) compartments. Two salts, namely NaCl and CaCl2, are dissolved in these compartments. While these features are original, they need to be improved to obtain a realistic representation of the tissue. For example, the present analysis considers that the fixed charge is permanent and cation binding is not accounted for. However, calcium cations are known to bind, at least partially, to proteoglycans, while sodium cations do not. At sufficiently high calcium concentration, the binding might become irreversible and modify the structure of the proteoglycans. Moreover, the fixed charge is known to be pH-sensitive. Indeed, as the pH of the surroundings decreases (respectively increases), the fixed charge becomes less negative (respectively more negative).

With the standard data used in the simulations, the transient period towards equilibrium is controlled by the transfer times between the IF and EF compartments, rather than by the Darcy's seepage time or Fick's diffusion time. Very large transfer times result in a model with a single mobile water compartment: the IF compartment can then be seen as an occluded porosity. On the other hand, decreasing the transfer times results in a period towards equilibrium that becomes more and more similar to the period, controlled by seepage and diffusion times, that holds in absence of an IF compartment. However, the smaller the transfer times the larger the oscillations of the masses between the two compartments before equilibrium is established. Damping these oscillations might require to modify the transfer laws which are first order time differential equations, structured so as to satisfy the entropy inequality. Instead of the linear relation (18.2.18), the nonlinear expression suggested in Section 10.2.3,

$$O_i = \frac{\rho_i}{t_{\mathrm{tr},i}} \big(\exp(-A_i \, |\mu_{n\mathrm{I}} - \mu_{n\mathrm{E}}|) - 1 \big) \, s_\mu \,, \tag{18.7.1}$$

with $s_\mu = \mathrm{sign}(\mu_{n\mathrm{I}} - \mu_{n\mathrm{E}})$, is a likely candidate (here i is the exchanged ion and n the associated neutral intrafibrillar salt). It ensures a bounded rate of transfer whatever the magnitude of the driving out-of-equilibrium engines, which are phrased in terms of chemical potentials. Like the linear relation (18.2.18), this nonlinear relation is structured so as to ensure the internal entropy production associated with mass transfers to be positive. Moreover, it boils down to the linear relation for small departures from the equilibria $\mu_{n\mathrm{I}} - \mu_{n\mathrm{E}} = 0$.

The present model and simulations have targeted essentially chemomechanical couplings and addressed only confined compression. Along ideas applied to other soft tissues, like annulus fibrosus, the mechanical behavior should recognize that collagen fibers behave differently in extension and contraction. These features are required so as to envisage simulations of plane strain (unconfined compression) and axisymmetric tests. Attempts along these lines are developed in Section 17.11. Other improvements consist in accounting for an anisotropic directional distribution of the collagen fibers, and heterogeneous densities of proteoglycans and collagen through the depth of the cartilage layer.

Exercises on Chapter 18

Exercise 18.1 The element algorithmic matrices $\mathbb{C}^e$ and $\mathbb{K}^e$

With $\mathbf{B}_u$ the strain-displacement matrix, the nonzero components of the algorithmic element matrices obtained by first order linearization of the residual can be listed and cast in the following format:

- for the element capacity matrix:

$$\mathbf{C}^e_{\mu_w u} = \int_{V_e} \mathbf{N}^T_\mu \, \mathrm{tr}\, \mathbf{B}_u \, dV_e,$$

$$\mathbf{C}^e_{\mu_k u} = \int_{V_e} \mathbf{N}^T_\mu \Big[\frac{n^{kE}}{n^E} z_{kE} \Big] \, \mathrm{tr}\, \mathbf{B}_u \, dV_e, \quad k \in \mathrm{E_{ions}},$$

$$\mathbf{C}^e_{\mu_k \mu_l} = \int_{V_e} \mathbf{N}^T_\mu \Big[\frac{\widehat{m}_l}{RT} z_{klE} \, n^{kE} \Big] \mathbf{N}_\mu \, dV_e, \quad k \in \mathrm{E_{ions}}, \ l \in \mathrm{E_{ions}},$$ $$(18.E.1)$$

$$\mathbf{C}^e_{\mu_k m_j} = \int_{V_e} \mathbf{N}^T_\mu \Big[\frac{I_{kj}}{\rho_j} + \zeta_j \frac{\widehat{v}_{Cl}}{\widehat{v}_j} \frac{I_{kCl}}{\rho_j} - \frac{n^{kE}}{n^E} z_{kE} \frac{z_{iI}}{\rho_j} \Big] \mathbf{N}_m \, dV_e, \quad k \in \mathrm{E_{ions}}, \ j \in \mathrm{I_{in}},$$

$$\mathbf{C}^e_{m_i m_i} = \int_{V_e} \mathbf{N}^T_m \frac{1}{\det \mathbf{F}} \mathbf{N}_m \, dV_e, \quad i \in \mathrm{I_{in}}.$$

- for the element diffusion-stiffness matrix:

$$\mathbf{K}^e_{uu} = \int_{V_e} \mathbf{B}^T_u \Big[\, \mathbb{E} + [\Lambda + \Lambda^*] \, \mathbf{I} \otimes \mathbf{I} \, \Big] \mathbf{B}_u \, dV_e,$$

$$\mathbf{K}^e_{u\mu_l} = \int_{V_e} \mathbf{B}^T_u \Big[\Lambda^*_l \widehat{m}_l \, \mathbf{I} \Big] \mathbf{N}_\mu \, dV_e, \quad l \in \mathrm{E_{mo}},$$

$$\mathbf{K}^e_{u m_j} = \int_{V_e} \mathbf{B}^T_u \Big[- \frac{z_{jI}}{\rho_j} \frac{\Lambda^*}{\det \mathbf{F}} \mathbf{I} \Big] \mathbf{N}_m \, dV_e, \quad j \in \mathrm{I_{in}},$$

$$\mathbf{K}^e_{\mu_w \mu_l} = \int_{V_e} \boldsymbol{\nabla} \mathbf{N}^T_\mu \Big[\rho_l \sum_{k \in \mathrm{E_{mo}}} \kappa_{kl} \Big] \boldsymbol{\nabla} \mathbf{N}_\mu \, dV_e, \quad l \in \mathrm{E_{mo}},$$

$$\mathbf{K}^e_{\mu_k \mu_l} = \int_{V_e} \boldsymbol{\nabla} \mathbf{N}^T_\mu \Big[\rho_l \, \kappa_{kl} \Big] \boldsymbol{\nabla} \mathbf{N}_\mu \, dV_e, \quad k \in \mathrm{E_{ions}}, \ l \in \mathrm{E_{mo}},$$

$$\mathbf{K}^e_{m_w u} = \int_{V_e} \mathbf{N}^T_m \Big[T^*_w \frac{\partial \mu_{wI}}{\partial \mathrm{tr}\, \boldsymbol{\epsilon}} \Big] \, \mathrm{tr}\, \mathbf{B}_u \, dV_e,$$ $$(18.E.2)$$

$$\mathbf{K}^e_{m_w \mu_l} = \int_{V_e} \mathbf{N}^T_m \Big[T^*_w \Big(\frac{\partial \mu_{wI}}{\partial \mu^{ec}_{lE}} - I_{wl} \Big) \Big] \mathbf{N}_\mu \, dV_e, \quad l \in \mathrm{E_{mo}},$$

$$\mathbf{K}^e_{m_w m_j} = \int_{V_e} \mathbf{N}^T_m \Big[T^*_w \frac{\partial \mu_{wI}}{\partial m^{jI}} \Big] \mathbf{N}_m \, dV_e, \quad j \in \mathrm{I_{in}},$$

$$\mathbf{K}^e_{m_i \mu_i} = \int_{V_e} \mathbf{N}^T_m \Big[- T^*_i \Big] \mathbf{N}_\mu \, dV_e, \quad i \in \mathrm{I_{in}} - \{w\},$$

$$\mathbf{K}^e_{m_i \mu_{Cl}} = \int_{V_e} \mathbf{N}^T_m \Big[- T^*_i \zeta_i \frac{\widehat{m}_{Cl}}{\widehat{m}_i} \Big] \mathbf{N}_\mu \, dV_e, \quad i \in \mathrm{I_{in}} - \{w\},$$

$$\mathbf{K}^e_{m_i m_j} = \int_{V_e} \mathbf{N}^T_m \Big[T^*_i \frac{\partial \mu_{iI}}{\partial m_{jI}} \Big] \mathbf{N}_m \, dV_e, \quad i \in \mathrm{I_{in}} - \{w\}, \ j \in \mathrm{I_{in}}.$$

The sets of species are defined by Table 18.1.1. The scalars Λ, Λ^* and Λ^*_l for $l \in \mathrm{E_{mo}}$, are defined in Exercices 18.2. Some additional definitions are introduced to simplify the notation,

$$z_E \equiv \sum_{k \in \mathrm{E_{ions}}} \zeta^2_k \, x_{kE}, \tag{18.E.3}$$

$$z_{kE} = 1 + \zeta_k \frac{y_{PG}}{z_E}, \ k \in \mathrm{E_{mo}}, \quad z_{klE} = I_{kl} - \zeta_k \zeta_l \frac{x_{lE}}{z_E}, \quad (k, l) \in (\mathrm{E_{ions}}, \mathrm{E_{mo}}), \tag{18.E.4}$$

$$z_{iI} = 1 + \zeta_i \frac{\widehat{v}_{Cl}}{\widehat{v}_i}, \ i \in \mathrm{I_{in}}. \tag{18.E.5}$$

The next entities are related to mass transfer, namely for $(i, n(i)) \in (I_{in}, I_{ne})$,

$$T_i = A_i \frac{\rho_i}{t_{tr,i}}, \quad T_i^* = T_i \text{ for linear transfer}; \quad T_i^* = T_i \exp(-A_i|\mu_{nI} - \mu_{nE}|) \text{ for bounded transfer.} \quad (18.E.6)$$

Exercise 18.2 Iterative extraction of intermediate variables from the primary unknowns

The change of the primary variables over two iterations being an output of the finite element analysis, obtain the iterative changes of standard variables and prove the explicit form of the element algorithmic matrices listed in Exercise 18.1.

Proof: Some of the algebraic manipulations necessary to obtain the element effective capacity matrix are provided below. The operator δ can be given arbitrary meaning, but, typically in an iterative process, it will a priori be interpreted as the difference between two iterates.

First let us recall the constraints that are obeyed by the molar fractions,

$$\sum_{k \in E_{mo}} x_{kE} + x_{PG} = 1, \qquad \sum_{k \in I} x_{kI} = 1,$$
$$\sum_{k \in E_{mo}} \zeta_k x_{kE} + y_{PG} = 0, \qquad \sum_{k \in I} \zeta_k x_{kI} = 0. \quad (18.E.7)$$

In the linearization below, the molar fraction x_{PG} of PGs can be neglected, but not their effective molar fraction $y_{PG} \equiv \zeta_{PG} x_{PG} < 0$. The pressure terms are also neglected in the chemical potentials of ions. These approximations should *not* be used in the estimation of the *residual* during the finite element analysis.

Since $\delta v^I + \delta v^E = \det \mathbf{F} \operatorname{div} \delta \mathbf{u}$, then

$$\delta n^E = (1 - n^E) \operatorname{div} \delta \mathbf{u} - \frac{\delta v^I}{\det \mathbf{F}}. \quad (18.E.8)$$

Approximating the molar volume of the extrafibrillar phase by the molar volume of water, the molar fraction x_{kE} of any species can be expressed in terms of the number of moles N_{kE} as $x_{kE} = (\widehat{v}_w/V)(N_{kE}/n^E)$. The variation of the effective molar fraction $y_{PG} = \zeta x_{PG}$ of PGs expresses as

$$\delta y_{PG} = -\frac{y_{PG}}{n^E} \operatorname{div} \delta \mathbf{u} + \frac{y_{PG}}{n^E} \frac{\delta v^I}{\det \mathbf{F}}. \quad (18.E.9)$$

Alternatively, $x_{kE} = (\widehat{v}_w/\widehat{v}_k)(v^{kE}/n^E)(V_0/V)$ can be obtained in terms of the volume content v^{kE} so that

$$\frac{\delta m^{kE}}{m^{kE}} = \frac{\delta v^{kE}}{v^{kE}} = \frac{\delta x_{kE}}{x_{kE}} + \frac{1}{n^E} \operatorname{div} \delta \mathbf{u} - \frac{1}{n^E} \frac{\delta v^I}{\det \mathbf{F}}, \quad k \in E_{mo}. \quad (18.E.10)$$

Electroneutrality of the intrafibrillar phase makes it possible to express the intrafibrillar volume in terms of elements of I_{in} only via the coefficients z_{iI} defined by (18.E.5),

$$v^I = \sum_{i \in I_{in}} z_{iI} v^{iI}. \quad (18.E.11)$$

Using the definitions of the electrochemical potentials (18.2.5) and the constraints on the molar fractions (18.E.7), the variation of the extrafibrillar electrical potential deduces in terms of p_{wE} and of primary variables,

$$y_{PG} F \delta\phi_E = x_{wE} \widehat{v}_w \delta p_{wE} - \sum_{l \in E_{mo}} x_{lE} \widehat{m}_l \delta\mu_{lE}^{ec}. \quad (18.E.12)$$

Backsubstitution in the electrochemical potentials yields, in terms of the same entities, the variation of molar fractions x_{kE}, $k \in E_{ions}$,

$$\delta x_{kE} = -\frac{\zeta_k}{y_{PG}} \frac{x_{kE}}{RT} x_{wE} \widehat{v}_w \delta p_{wE} + \frac{x_{kE}}{RT} \sum_{l \in E_{mo}} \left(I_{kl} + \frac{\zeta_k}{y_{PG}} x_{lE}\right) \widehat{m}_l \delta\mu_{lE}^{ec}. \quad (18.E.13)$$

Substitution of (18.E.9), (18.E.11) and (18.E.13) in (18.E.7)$_2$ provides the variation of extrafibrillar pressure as a function of primary variables,

$$z_E x_{wE} \widehat{v}_w \delta p_{wE} = -RT \frac{y_{PG}^2}{n^E} \operatorname{div} \delta \mathbf{u} + \sum_{l \in E_{mo}} z_E z_{lE} x_{lE} \widehat{m}_l \delta\mu_{lE}^{ec} + RT \frac{y_{PG}^2}{n^E} \sum_{i \in I_{in}} z_{iI} \frac{\delta v^{iI}}{\det \mathbf{F}}. \quad (18.E.14)$$

The 6th line of the element matrices corresponds to the mass transfer of water governed by the rate relations (10.2.13) and (18.2.18). The matrix coefficients of this line are obtained by linearization of these relations and use of $\delta p_I = \delta p_{wE}$, (14.6.28) and (18.E.14).

The 7th and 8th lines correspond to the mass transfer of salts. The chemical potentials of salts are expressed in terms of electrochemical potentials of ions, Eqns (18.2.7)-(18.2.10). Linearization yields the coefficients indicated in (18.E.2).

Substitution of (18.E.14) in (18.E.13) provides in turn, as functions of the primary variables, the variation of the extrafibrillar ionic molar fractions $k \in E_{ions}$,

$$\delta x_{kE} = \frac{y_{PG}}{n^E} \frac{\zeta_k x_{kE}}{z_E} \operatorname{div} \delta \mathbf{u} + \frac{x_{kE}}{RT} \sum_{l \in E_{mo}} z_{klE} \, \widehat{m}_l \, \delta \mu_{lE}^{ec} - \frac{y_{PG}}{n^E} \frac{\zeta_k x_{kE}}{z_E} \sum_{i \in I_{in}} z_{iI} \frac{\delta v^{iI}}{\det \mathbf{F}} . \tag{18.E.15}$$

Use of the relations (18.E.10) and (18.E.15) yields the variation of the mass-content m^{kE}, $k \in E_{ions}$, becomes in terms of the primary variables,

$$\delta m^{kE} = \frac{m^{kE}}{n^E} z_{kE} \operatorname{div} \delta \mathbf{u} + \frac{m^{kE}}{RT} \sum_{l \in E_{mo}} z_{klE} \, \widehat{m}_l \, \delta \mu_{lE}^{ec} - \frac{m^{kE}}{n^E} z_{kE} \sum_{i \in I_{in}} z_{iI} \frac{\delta v^{iI}}{\det \mathbf{F}} . \tag{18.E.16}$$

The coefficients β_k, $k \in E_{ions}$, in Eqn (18.3.2), become in terms of the primary variables,

$$\beta_k = \frac{n^{kE}}{n^E} z_{kE} \operatorname{div} \frac{d\mathbf{u}}{dt} + \frac{n^{kE}}{RT} \sum_{l \in E_{mo}} z_{klE} \, \widehat{m}_l \, \frac{d}{dt} \mu_{lE}^{ec} - \frac{n^{kE}}{n^E} z_{kE} \sum_{i \in I_{in}} \frac{z_{iI}}{\det \mathbf{F}} \frac{dv^{iI}}{dt} + \frac{1}{\det \mathbf{F}} \frac{dv^{kI}}{dt} . \tag{18.E.17}$$

Lines $3, 4$ and 5 of the capacity matrix $\mathbb{C}^e$ are thus read from (18.E.17). The corresponding coefficients of the matrix $\mathbb{K}^e$ follow directly from the generalized diffusion equations for the fluxes $\mathbf{J}_{kE}$, $k \in E_{ions}$.

The mechanical constitutive equations adopt the general format $\boldsymbol{\sigma} + p_{eff} \mathbf{I} = \mathbb{E} : \boldsymbol{\epsilon}$ where $p_{eff} = p_I - X$. The chemomechanical couplings are included in the scalar X, Section 18.2.3.2. The pressure p_I and the extrafibrillar molar fractions and mass contents depend on the primary variables as indicated by Eqns (18.E.14), (18.E.15) and (18.E.16), respectively. Differentiation of the constitutive equations, and use of $\delta p_I = \delta p_{wE}$, yields the stress variation as a function of the primary variables,

$$\delta \boldsymbol{\sigma} = (\overbrace{\mathbb{E} + (\Lambda + \Lambda^*) \mathbf{I} \otimes \mathbf{I}}^{\text{chemomechanics}}) : \delta \boldsymbol{\epsilon} \quad + \quad \overbrace{\sum_{l \in E_{mo}} \Lambda_l \, \widehat{m}_l \, \delta \mu_{lE}^{ec}}^{\substack{\text{surroundings-extrafibrillar} \\ \text{exchanges}}} \mathbf{I} \quad - \quad \overbrace{\sum_{i \in I_{in}} \Lambda^* \frac{z_{iI}}{\det \mathbf{F}} \delta v^{iI}}^{\substack{\text{intra-extrafibrillar} \\ \text{mass exchanges}}} \mathbf{I} , \tag{18.E.18}$$

where, accounting for the constraint $\delta v^E = \det \mathbf{F} \operatorname{tr} \delta \boldsymbol{\epsilon} - \delta v^I$,

$$\Lambda^* = -\det \mathbf{F} \frac{\partial p_{eff}}{\partial v^E} ; \quad \Lambda_l = -\frac{1}{\widehat{m}_l} \frac{\partial p_{eff}}{\partial \mu_{lE}^{ec}} , \quad l \in E_{mo} , \tag{18.E.19}$$

from which derive the coefficients of the first line of the element diffusion-stiffness matrix.

Exercise 18.3 Definition of the fictitious bath and associated Donnan pressure

Given a cartilage with known chemical composition, obtain the chemical composition of the fictitious bath, with which it is in equilibrium.

1. Obtain the chemical composition of the fictitious bath, and calculate the fictitious Donnan pressure p_D between the extrafibrillar compartment and the fictitious bath,

$$p_D = \pi_{osm} - p_{fw} = \frac{RT}{\widehat{v}_w} \operatorname{Ln} \frac{\tilde{x}_{wB}}{x_{wE}} - p_{fw} . \tag{18.E.20}$$

The minimal intercompartment pressure difference p_{fw} takes place in the hypertonic state, so that the fictitious Donnan pressure vanishes in the hypertonic state.

2. Obtain the fictitious Donnan pressure in differential form. Consider the case of two salts NaCl and CaCl$_2$ and the case of a single salt.

1. *Fictitious bath and associated fictitious Donnan pressure:*

Electroneutrality of the extrafibrillar compartment and bath, and equilibrium of the two salts at the interface implies,

$$\tilde{x}_{ClB} = \tilde{x}_{NaB} + 2 \, \tilde{x}_{CaB} , \quad x_{ClE} = x_{NaE} + 2 \, x_{CaE} + y_{PG} , \tag{18.E.21}$$

$$x_{NaE} \, x_{ClE} = K_{fs_1} \tilde{x}_{NaB} \, \tilde{x}_{ClB} , \quad x_{CaE} \, (x_{ClE})^2 = K_{fs_2} \, \tilde{x}_{CaB} \, (\tilde{x}_{ClB})^2 , \tag{18.E.22}$$

where the K_f's and K^{eq} are defined by (18.2.12). Hence

$$\tilde{x}_{CaB} = \frac{x_{CaE}}{K^{eq}} \frac{(\tilde{x}_{NaB})^2}{(x_{NaE})^2} . \tag{18.E.23}$$

Thus $\tilde{x}_{\text{NaB}}$ is the positive root of the equation,

$$2\,\frac{x_{\text{CaE}}}{K^{\text{eq}}}\,(\tilde{x}_{\text{NaB}})^3 + x_{\text{NaE}}^2(\tilde{x}_{\text{NaB}})^2 - \frac{1}{K_{fs_1}}\,x_{\text{NaE}}^3\,x_{\text{ClE}} = 0\,, \tag{18.E.24}$$

and $\tilde{x}_{\text{CaB}}$ is deduced by (18.E.23) and, in turn, the electroneutraality of the bath (18.E.21) yields $\tilde{x}_{\text{ClB}}$. Since

$$\tilde{x}_{\text{wB}} = 1 - (\tilde{x}_{\text{ClB}} + \tilde{x}_{\text{NaB}} + \tilde{x}_{\text{CaB}}),\quad x_{\text{wE}} \simeq 1 - (x_{\text{ClE}} + x_{\text{NaE}} + x_{\text{CaE}}), \tag{18.E.25}$$

the fictitious Donnan pressure can be obtained in either exact or linearized forms,

$$p_D = \frac{RT}{\widehat{v}_{\text{w}}}\,\text{Ln}\,\frac{\tilde{x}_{\text{wB}}}{x_{\text{wE}}} - p_{f\text{w}} \simeq \frac{RT}{\widehat{v}_{\text{w}}}\,\big(y_{\text{PG}} + 2\,(x_{\text{NaE}} - \tilde{x}_{\text{NaB}}) + 3\,(x_{\text{CaE}} - \tilde{x}_{\text{CaB}})\big) - p_{f\text{w}}\,. \tag{18.E.26}$$

2. *Differentiation of Donnan pressure:*
  Differentiation of (18.E.24) yields

$$\delta\tilde{x}_{\text{NaB}} = \sum_{k\in\text{E}_{\text{ions}}} A_{\text{Na}k}\,\delta x_{k\text{E}} \tag{18.E.27}$$

where

$$A\begin{bmatrix} A_{\text{NaNa}} \\ A_{\text{NaCa}} \\ A_{\text{NaCl}} \end{bmatrix} = \frac{K^{\text{eq}}}{K_{fs_1}}\begin{bmatrix} 3\,x_{\text{NaE}}^2\,x_{\text{ClE}} - 2\,K_{fs_1}\,x_{\text{NaE}}\,(\tilde{x}_{\text{NaB}})^2 \\ -2\,(\tilde{x}_{\text{NaB}})^3\,K_{fs_1}/K^{\text{eq}} \\ x_{\text{NaE}}^3 \end{bmatrix}, \tag{18.E.28}$$

and

$$A = 6\,x_{\text{CaE}}\,(\tilde{x}_{\text{NaB}})^2 + 2\,K^{\text{eq}}\,x_{\text{NaE}}^2\,\tilde{x}_{\text{NaB}} \tag{18.E.29}$$

Differentiation of (18.E.23) gives

$$\delta\tilde{x}_{\text{CaB}} = 2\,\frac{\tilde{x}_{\text{CaB}}}{\tilde{x}_{\text{NaB}}}\,\delta\tilde{x}_{\text{NaB}} + \frac{\tilde{x}_{\text{CaB}}}{x_{\text{CaE}}}\,\delta x_{\text{CaE}} - 2\,\frac{\tilde{x}_{\text{CaB}}}{x_{\text{NaE}}}\,\delta x_{\text{NaE}} \tag{18.E.30}$$

and thus

$$\delta\tilde{x}_{\text{CaB}} = \sum_{k\in\text{E}_{\text{ions}}} A_{\text{Ca}k}\,\delta x_{k\text{E}} \tag{18.E.31}$$

with

$$\begin{bmatrix} A_{\text{CaNa}} \\ A_{\text{CaCa}} \\ A_{\text{CaCl}} \end{bmatrix} = \frac{\tilde{x}_{\text{CaB}}}{x_{\text{NaE}}}\begin{bmatrix} -2 \\ x_{\text{NaE}}/x_{\text{CaE}} \\ 0 \end{bmatrix} + 2\,\frac{\tilde{x}_{\text{CaB}}}{\tilde{x}_{\text{NaB}}}\begin{bmatrix} A_{\text{NaNa}} \\ A_{\text{NaCa}} \\ A_{\text{NaCl}} \end{bmatrix}. \tag{18.E.32}$$

Eliminating δy_{PG} from the differential of (18.E.26) using (18.E.21)$_2$ yields

$$\delta p_D \simeq \frac{RT}{\widehat{v}_{\text{w}}}\,\Big(\frac{\delta x_{\text{NaE}} + \delta x_{\text{CaE}} + \delta x_{\text{ClE}}}{x_{\text{wE}}} - \frac{2\,\delta\tilde{x}_{\text{NaB}} + 3\,\delta\tilde{x}_{\text{CaB}}}{\tilde{x}_{\text{wB}}}\Big) = \sum_{\text{E}_{\text{ions}}} P_k\,\delta x_{k\text{E}}\,, \tag{18.E.33}$$

with

$$P_k = \frac{RT}{\widehat{v}_{\text{w}}}\,\Big(\frac{1}{x_{\text{wE}}} - \frac{2\,A_{\text{Na}k}}{\tilde{x}_{\text{wB}}} - \frac{3\,A_{\text{Ca}k}}{\tilde{x}_{\text{wB}}}\Big),\quad k \in \text{E}_{\text{ions}}\,. \tag{18.E.34}$$

Thus, using (18.E.15), and by a simple rewriting of (18.E.14), the differentials of the fictitious Donnan pressure and of the pressure p_{I} can be cast in terms of the primary variables in the format,

$$\begin{bmatrix} \delta p_D \\ \delta p_{\text{I}} \end{bmatrix} \simeq \begin{bmatrix} Q_u \\ R_u \end{bmatrix}\,\text{div}\,\delta\mathbf{u} + \sum_{l\in\text{E}_{\text{mo}}}\begin{bmatrix} Q_l \\ R_l \end{bmatrix}\,\widehat{m}_l\,\delta\mu_{l\text{E}}^{\text{ec}} - \begin{bmatrix} Q_u \\ R_u \end{bmatrix}\,\sum_{i\in\text{I}_{\text{in}}} z_{i\text{I}}\,\frac{\delta v^{i\text{I}}}{\det\mathbf{F}} \tag{18.E.35}$$

with

$$Q_u = \frac{y_{\text{PG}}}{n^{\text{E}}}\,\sum_{k\in\text{E}_{\text{ions}}} P_k\,\frac{\zeta_k\,x_{k\text{E}}}{z_{\text{E}}},\quad Q_l = \sum_{k\in\text{E}_{\text{ions}}} P_k\,\frac{x_{k\text{E}}}{RT}\,z_{kl\text{E}},\quad l \in \text{E}_{\text{mo}}, \tag{18.E.36}$$

and

$$R_u = -\frac{y_{\text{PG}}}{n^{\text{E}}}\,\frac{RT}{\widehat{v}_{\text{w}}}\,\frac{y_{\text{PG}}}{z_{\text{E}}\,x_{\text{wE}}},\quad R_l = \frac{RT}{\widehat{v}_{\text{w}}}\,\frac{x_{l\text{E}}}{RT}\,\frac{z_{l\text{E}}}{x_{\text{wE}}},\quad l \in \text{E}_{\text{mo}}\,. \tag{18.E.37}$$

Help for coding

  Input the equilibrium constants (18.2.12).
  1. solve (18.E.24) for $\tilde{x}_{\text{NaB}}$ (single positive root);

2. obtain $\tilde{x}_{\text{CaB}}$ by (18.E.23), and $\tilde{x}_{\text{wB}}$ and x_{wE} by (18.E.25);
3. obtain the fictitious Donnan pressure (18.E.26);
4. define the A_{Na}'s by (18.E.28) and (18.E.29) and the A_{Ca}'s by (18.E.32);
5. define the P's by (18.E.34) and the Q's by (18.E.36).

3. *Special case: a single salt, k=Na or Ca only:*
In view of the electroneutrality of each of the two compartments,

$$\tilde{x}_{\text{ClB}} = \zeta_k \, \tilde{x}_{k\text{B}}, \quad x_{\text{ClE}} = \zeta_k \, x_{k\text{E}} + y_{\text{PG}} \, . \tag{18.E.38}$$

Equilibrium of the salt between the two compartments implies,

$$x_{k\text{E}} \, (x_{\text{ClE}})^{\zeta_k} = \tilde{x}_{k\text{B}} \, (\tilde{x}_{\text{ClB}})^{\zeta_k} = (\zeta_k)^{\zeta_k} \, (\tilde{x}_{k\text{B}})^{1+\zeta_k} \, . \tag{18.E.39}$$

Hence the fictitious bath content is given in explicit form as,

$$\tilde{x}_{k\text{B}} = \frac{\tilde{x}_{\text{ClB}}}{\zeta_k} = \left(x_{k\text{E}} \, (\frac{x_{\text{ClE}}}{\zeta_k})^{\zeta_k} \right)^{(1+\zeta_k)^{-1}} \tag{18.E.40}$$

Thus

$$A_{kk} = \frac{\tilde{x}_{k\text{B}}}{x_{k\text{E}}} \frac{1}{1+\zeta_k}, \quad A_{k\text{Cl}} = \frac{\tilde{x}_{\text{ClB}}}{x_{\text{ClE}}} \frac{1}{1+\zeta_k}, \tag{18.E.41}$$

the four other A's vanishing. The relations (18.E.34) et seq. apply unchanged.

Help for coding
1. obtain $\tilde{x}_{k\text{B}}$ and $\tilde{x}_{\text{ClB}}$ by (18.E.40);
2. obtain the fictitious Donnan pressure (18.E.26);
3. define the A_k's by (18.E.41) and set the four other A's to 0;
4. define the P's by (18.E.34) and the Q's by (18.E.36).

Chapter 19

Two lamellar tissues: cornea and annulus fibrosus. Active transport

19.1 Function, structure and composition

The eye consists of two spherical parts. The larger sphere, which contains the vitreous humor, is coated by three layers, namely from inside to outside, (a) the retina whose posterior part contains the visual apparatus, (b) the vascularized choroid continued by the iris, and (c) the protective opaque and vascularized sclera which is continued in the anterior part of the eye by the second smaller sphere, namely the cornea which is transparent.

The cornea and the lens, which is attached to the ciliary body via muscles, represent the dioptric apparatus of the eye, whose role is to obtain the proper focal distance for rays entering the eye.

Gels are aqueous environments containing structural insoluble components. These components can be located inside the cell, e.g., the muscle, or outside the cell, e.g., articular cartilage. The cornea is an avascular polyelectrolyte[19.1] *extra*cellular gel. Collagen fibrils, which run parallel to the corneal surface, are its main structural component and they contribute to the overall stability of the eye ball.

The maintenance of epithelial cells, keratocytes, and endothelial cells is ensured by exchanges of oxygen and nutrients through the epithelium and through the endothelium. Cornea is covered by a tear fluid, which serves both as a lubricant and as a vector for physiological molecules and drugs. This fluid is secreted by glands located outside the cornea itself. Repair of injuries may take place only in the epithelial zone where cells undergo mitosis. Corneal wounds are temporarily opaque, because the orientation of the collagen fibers is disorganized. It takes several months to recover a more standard organization and transparency: still the recovery of neither one nor the other is perfect, Muir [1983].

Solely the anterior part of the cornea is innervated, with a nerve sensitivity that decreases from center to periphery.

19.1.1 A layered structure

The cornea can be approximated as a spherical cap, Fig. 19.1.1. In animals, its thickness varies from 0.1 to 1 mm. In humans, the thickness ranges from about 0.52 mm at the center to 0.65 mm on its boundary. The radius of curvature of the external surface is about 7.8 mm while the average diameter is 12 mm.

The organ is structured in five layers along the radial direction. The two outer layers are cellular, the three inner ones being mainly acellular and fibrous. From the anterior to the exterior region,

1. the endothelium, a monolayer of cells, of diameter 20 μm and thickness 4-5 μm, is permeable, and allows the entrance of water, nutrients and the exit of waste, and ions, Fig. 19.1.2. In fact, the cells are separated from one another by spaces of about 20 nm. Transport of water and hydrophilic drugs is thought to take place across these slits. Active transport mechanisms, driven by ionic pumps, maintain transparency of the stroma by controlling its hydration state. Whether active transport is an intracellular or extracellular process or partly intracellular and partly extracellular is still a debatable question.

[19.1]A polyelectrolyte is an ionic *poly*mer which bears a number of interacting ionic groups that, upon dissociation in aqueous environments, attract counterions. Ions and counterions contribute to conduct electricity: this property justifies the terminology "*electrolyte*".

At variance with other mammals, human endothelial cells are contact-inhibited *in vivo* (the cell cycle shown in Fig. 24.3.2 is interrupted at phase G1) and they increase in size to cover damage of neighbor cells. Their density of about $6000 \, \text{cells/mm}^2$ in the first month of life is approximately halved at 60 years of age, Bourne [2003], due to both the enlargement of the cornea and a decrease in their number.

2. Descemet's membrane is fibrous. It is continuously secreted by endothelial cells so that its initial $4 \, \mu\text{m}$ thickness more than doubles during life.

3. the stroma consists of about 250 lamellae, of $1.5 \, \mu\text{m}$-$2.5 \, \mu\text{m}$ thickness, parallel to the corneal surface. The stroma has a *stacked sheet* morphology that is similar to that of an *angle-ply composite*. In each lamella, the collagen fibrils are parallel, but the fibrils of two adjacent lamellae run in alternating directions. This description holds mainly in the central part of the stroma. At the periphery, the collagen bundles are interwoven.

In the central region of the cornea, a large number of lamellae have their fibrils in the vertical and horizontal directions. The human collagen fibrils have a diameter of about $30 \, \text{nm}$. A cross-section of a lamella displays a regular arrangement of collagen fibrils, with a spacing that varies from $20 \, \text{nm}$ in dry condition to $200 \, \text{nm}$ for high hydration, Elliott and Hodson [1998]. This regular arrangement may be perturbed by an abnormally high hydration, leading to a stromal oedema. Understanding the detailed organization of the stroma is of capital

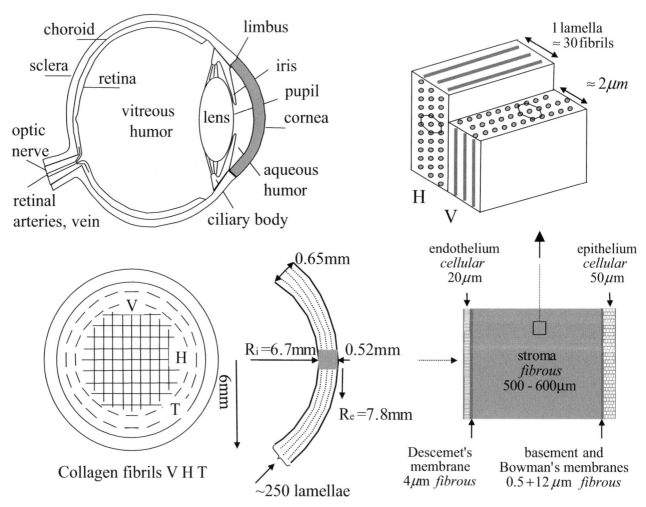

Fig. 19.1.1: Structure of the cornea at several scales. Lengths apply to the human cornea under standard physiological hydration. The cornea is composed of five main layers along the posterior to anterior directions. The central layer, called stroma, is made of lamellae parallel to the corneal membrane. Adjacent lamellae are reinforced by collagen fibrils in alternating directions. The preferred directions are essentially horizontal (H), vertical (V) at the center, and tangential (T) at the periphery, Meek et al. [1987]. In each lamella, the collagen fibrils form a regular pattern, which is thought to contribute to transparency. For a normal cornea, a single fibril is thought to extend from limbus to limbus.

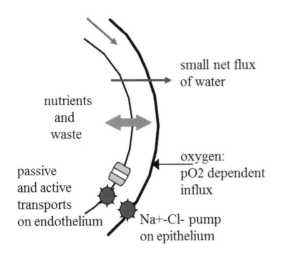

Fig. 19.1.2: Sketch of the **transport role** of the cornea. Exchange of nutrients, cytokines and water takes place across the endothelium. The epithelium is less permeable. According to Friedman [1973], as the eye remains open (between two blinkings), there is a net flux of water from the stroma through the epithelium, and a net flux of sodium ions from the stroma through the endothelium. The first flux would increase the ionic strength, but the second flux occurs to maintain this ionic strength approximately constant, even if the stroma thins slightly. This homeostatic assumption is to be scrutinized, Ruberti and Klyce [2003].

interest to optimize the transparency of engineered stromas, Freund et al. [1986], Orwin et al. [2003].

The fibrils are surrounded by an amorphous non oriented gel (termed ground substance) consisting of other types of macromolecules, among which noncollagenous proteins and proteoglycans. The electrical charge of proteoglycans is negative under physiological pH and it plays a key role by inducing electro-chemomechanical couplings.

The stroma contains a small proportion of quiescent cells, 3-5% in volume, called keratocytes.

4. Bowman's membrane is formed by fibrils of type VII collagen.

5. the epithelium, is composed by three functional strata formed by four to six layers. Cells migrate from the posterior stratum germinatum where they undergo mitosis to the surface stratum where they undergo desquamation. The epithelium has uniform thickness, about 50 μm. It forms a smooth refractive surface on the cornea, which is less permeable than the endothelium: while it allows O_2 to enter, it acts primarily as a tight barrier against external noxious agents.

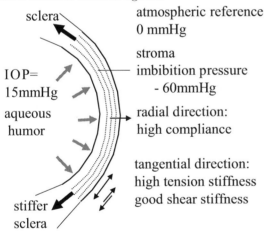

Fig. 19.1.3: Sketch of the **mechanical role** of the cornea. The intraocular pressure (IOP) is transmitted to the sclera through the collagen network. The spatial organization of the collagen leads to stiff tangential directions, and to compliant radial directions, prone to swell. Endowed with a higher stiffness, the sclera is considered as clamping the cornea, and thus as limiting its bending.

Cornea is a soft fibrous tissue that provides protection from external mechanical and chemical agents, and optimal transparency for light to penetrate the ocular globe. Appropriate hydration is mainly due to both the presence of negatively charged proteoglycans in the stroma and to active pumps, located in the endothelium, that expel cations to the aqueous humor. The gel formed by hydrated proteoglycans is reinforced by collagen fibrils. The intraocular pressure of the aqueous humor induces the collagen fibrils in tension, Fig. 19.1.3.

19.1.2   Histology of the cornea

19.1.2.1   The architecture of collagen fibrils

Collagen fibrils form a network that endows cornea with high *tensile* properties along the orthoradial directions, Fig. 19.1.1. Crosslinks between fibrils contribute to the mechanical properties of the network. Experiments by Bader et al. [1981] show that both tensile stiffness and tensile strength are considerably reduced when the crosslinks between collagen fibrils are disrupted.

Cornea, human or bovine, contains predominantly collagen of type I (75%) and type V (10 to 20%). The presence of type V collagen influences the diameter of the fibrils: the higher the percentage of type V collagen, the thinner are the fibrils. In contrast to collagen of type I, collagen of type II of articular cartilage has strong mechanical links with proteoglycans. This remark on the much lower density of crosslinks in cornea than in articular cartilage deserves to be quantified. In fact, crosslinks are considered as an important contributor to the overall stiffness. The relative and possible influences of the components of the cornea (collagen, PGs) and the various possible crosslinks (intrafibrillar, interfibrillar, fibril-matrix) are exposed nicely in Nickerson [2005], Chapter 6. This quasi-absence of links would be highly detrimental in cartilages where the fluid pressure can reach several MPa. The order of magnitude of pressure is much lower in the cornea, Fig. 19.1.3. Indeed, absorption of water from the aqueous humor results in *swelling*. The amount of swelling depends, inter alia, on the concentration of proteoglycans and on the stiffness of the collagen network. An additional mechanism, namely active transport, will be advocated to limit swelling. In absence of this mechanism (located on the endothelium), that is, when the endothelium is removed, the swelling is indeed quite large as detailed in Section 19.2.1.

The mechanical properties of the cornea display macroscopic anisotropy whose intensity and axes vary from the center to the periphery. In addition, the fibrous collagen structure induces much higher elastic moduli in tension than in compression. These composition and structure are presumably in line with the function of the organ which acts as a membrane aimed at sustaining tensile loads in its plane but also bending moments incurred by clamping into the more rigid sclera.

The spatial and directional distribution of collagen fibrils is not homogeneous neither over the surface nor in depth. Data acquisitions use electron microscopy, a technique which can reveal local information, and x-ray diffraction. X-ray beams directed parallel to the optical axis provide the fibrillar arrangement *averaged* over the depth, the scatter intensity being proportional to the number of fibrils. An analysis of the heterogeneity along the depth is still possible with small angle x-ray diffraction by passing beams through the edge of a strip, parallel to the corneal plane and probing at several points along the depth. Using this technique, Quantock et al. [2007] observed that the collagen Bragg spacing is smaller in the anterior region, and varies from about 45 nm to 55 nm, for the human cornea under physiological hydration, so that the collagen density is larger in the anterior region.

In the central part of the human stroma, the fibrils are essentially aligned along two orthogonal directions, the nasal-temporal and superior-anterior meridians. These two families of fibrils become progressively more aligned from the center of the stroma to the limbus where the fibrils run almost parallel in the circumferential direction. However, the number of lamellae increases with the cornea thickness at the periphery and the peripheral collagen fibrils form a diamond-shape network that is thought to ensure anchoring of the cornea in the stiffer sclera, Boote et al. [2006]. The noncircumferential orientation of the latter fibrils implies that the directional distribution of fibers is scattered about the circumferential direction, Fig. 4 in Aghamohammadzadeh et al. [2004], Fig. 6 in Hayes et al. [2007]a. While the cornea thickens at the periphery, the fibril density per unit area of cornea decreases from the center to the periphery by about 25%.

In a synchrotron radiation facility, the x-ray beam produced by electrons when they are deflected has a high intensity, with a typical wave-length about 1 Å and energy 10 keV so that the time of exposure is reduced, say from a tenth to ten seconds. Still, recording information over a grid of typically 30×30 points takes several hours and requires the sample to be inserted in an envelope which is impermeable to water, so as to keep a constant hydration, but permeable to x-rays. X-ray beams can penetrate deep into samples, thus avoiding processing the tissues. In contrast to x-rays, electron microscopy examines ultra-thin sections, about 100 nm, which need to be prepared through a number of steps. Dehydration by immersion in alcohols and other chemicals, infiltration of resin, curing and sectioning are lengthy and inevitably alter the integrity of the tissue.

X-ray diffraction provides quantitative data on fibril diameter, interfibrillar spacing, fibrillar density and orientation. Upon impinging on the tissue, x-rays produce a diffraction pattern on a detector plate located at a tenth of centimeters (wide angle diffraction) or at several meters (small angle diffraction) behind the specimen. Wide angle diffraction reveals information on the molecular structure of collagen. The center of the scattering pattern is the shadow of a lead beam stop placed between the specimen and the detector and aimed at removing the unscattered rays. A uniform directional distribution of fibrils produces a uniform circular pattern, while a single fibril gives rise to two symmetric arcs. The interpretation of the data consists in decomposing the scattering pattern into a uniform contribution and a perturbation which is interpreted as produced by a non random directional distribution of fibrils.

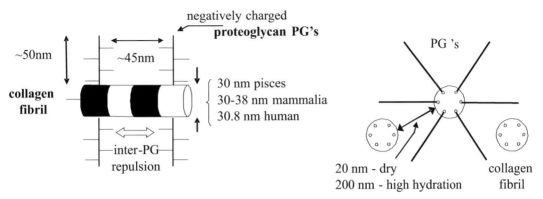

Fig. 19.1.4: Qualitative spatial organization of collagen and proteoglycans in human cornea. The fibril diameters and interfibrillar distances indicated on the sketch are extracted from Elliott and Hodson [1998]. For human cornea, Müller et al. [2004] estimate the fibril diameter to about 23 nm. Proteoglycans attach laterally at specific bands of the D-blocks, Meek and Fullwood [2001]. Keratan sulfates attach to bands a and c in the gap and overlap regions and chondroitin sulfates to bands d and e in the gap region (refer to Fig. 14.2.5).

19.1.2.2 Proteoglycans

Proteoglycan monomers (PGs) are polysacharides, formed by a core protein to which are attached laterally glycosaminoglycan chains (GAGs), Fig. 19.1.4. GAGs are composed of disaccharide units. The two main GAGs that compose corneal proteoglycans are dermatan/chondroitin sulfates (CS) and keratan sulfates (KS). While dermatan sulfates are the major GAGs in sclera, keratan sulfates are the most abundant ones in stroma. The negative charges of PGs induce repulsive forces at various structural levels.

The structure of a proteoglycan monomer of corneal stroma is sketched in Fig. 19.1.5. The disaccharide units of chondroitin sulfates display a uronic acid with a negatively charged carboxyl group $CO-O^-$ and an aminoglycan with none or a single sulfonic group SO_3^-. Keratan sulfates possess no uronic acid. Their sulfation is variable: disaccharide units are either not sulfated, mono-sulfated or di-sulfated. Even if the ratio of CS versus KS varies over species, the variable sulfation is sometimes advocated to provide a fixed charge which is almost identical over species, Scott and Bosworth [1990]. Note also that the proportion of chondroitin and keratan glycosaminoglycans may vary in space across the thickness of the stroma.

With respect to articular cartilages, the proteoglycans of cornea are much smaller with an estimated molar mass of 72×10^3 gm versus 2000×10^3 gm, the mass proportion of GAG-associated proteins is larger up to 75%, Axelsson and Heinegård [1978], versus 10%. Their fixed charge is smaller as well: under physiological concentrations, the effective concentration of PGs is typically around -30 mM, while it varies from -100 to -200 mM in articular cartilages.

Proteoglycan monomers contain only chains of keratan sulfates or only chains of chondroitin sulfates, while proteoglycan monomers in articular cartilages contain both keratan and chondroitin GAGs. The molar mass of chondroitin proteoglycans is larger than that of keratan proteoglycans, say 100 to 150×10^3 gm against 40 to 70×10^3 gm, so that the value 72×10^3 gm should be seen as a weighted average over the two types of proteoglycans.

As for the GAG chains, the molar mass of GAGs of keratan sulfates ranges from 4×10^3 gm to 20×10^3 gm and it is close to this upper bound for GAGs of chondroitin sulfates. The total numbers of disaccharide units (sulfated and unsulfated) in GAG chains of KS are quite similar in articular cartilages and cornea. Chondroitin chains in cornea have more disaccharide units than in cartilage. On the other hand, the number of chains per proteoglycan monomer is much larger in articular cartilages, 25 to 100, against 1 to 10 in cornea.

The charge indicated above holds under physiological pH. An acid environment turns the charge less negative. Since the fixed charge is the key engine to swelling, any change of pH has important consequences on chemomechanical couplings. Indeed, the fixed charge of both articular cartilages and cornea are contributed by PGs and, at non physiological pH, by collagen molecules. In corneal stroma, chloride binding to specific ligands provides an additional negative charge, whose value increases with chloride concentration. By contrast, in articular cartilages, calcium cations compete with hydrogen to bind on the carboxyl sites of chondroitin sulfates.

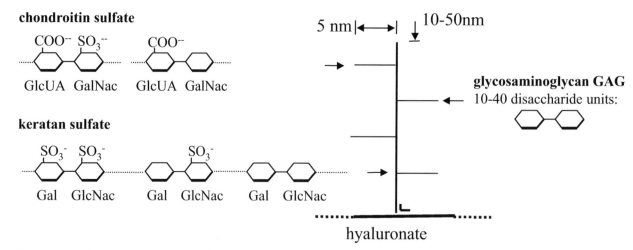

chondroitin sulfate

GlcUA   GalNac   GlcUA   GalNac

keratan sulfate

Gal   GlcNac   Gal   GlcNac   Gal   GlcNac

hyaluronate

5 nm   10-50nm

glycosaminoglycan GAG
10-40 disaccharide units:

Fig. 19.1.5: Structure of a proteoglycan monomer of corneal stroma. Proteoglycans are formed by glycosaminoglycans (GAGs) attached to a central protein. The two main glycosaminoglycans of corneal proteoglycans are dermatan/chondroitin sulfates (DS/CS) and keratan sulfates (KS). CS are formed by disaccharide units of guluronic acid GlcUA, with a negative charge, and N-acetyl-galactosamine GalNAc, with a variable sulfation. The disaccharide units of KS, namely galactose Gal and N-acetyl-glucosamine GlcNAc, have a variable sulfation as well. The number of GAGs in a proteoglycan monomer may vary from one to about ten. Size related numbers are taken from Müller et al. [2004].

19.1.2.3   Intralamellar structure

Within a lamella, the collagen fibrils run parallel. PGs are attached laterally almost at right angle providing some crosslinks between next neighbors. Chondroitin sulfates and keratan sulfates are attached at two specific staining bands of the D-blocks, Meek and Fullwood [2001].

The fibrils are arranged in a hexagonal array, Fig. 19.1.6. Collagen molecules are secreted into the extracellular matrix in a soluble form (procollagen) which displays liquid crystalline order, Hulmes [2002]. When they become insoluble by removal of propeptides, they spontaneously assemble into fibrils, apparently retaining this liquid crystalline order.

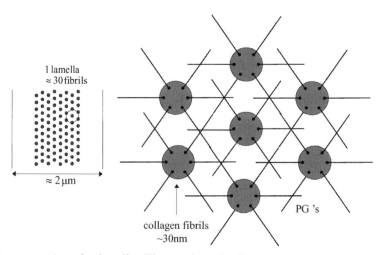

1 lamella
≈ 30 fibrils

≈ 2 μm

collagen fibrils
∼30nm

PG's

Fig. 19.1.6: Idealized cross-section of a lamella. The number of collagen fibrils through the thickness of a lamella is about 30. The collagen fibrils form a hexagonal lattice-like array (center to center distance of fibrils = 1.12 Bragg spacing). According to Müller et al. [2004], the PGs are attached laterally to the fibrils and they extend, not between neighbors, but between next neighbors. The interfibrillar Bragg spacing in human cornea increases slightly in depth, from 45 nm to 55 nm, Quantock et al. [2007]. Actually, this idealized crystalline lattice was thought in the 50's to be linked to the transparency of the cornea. Later, models of scattering of light showed that a liquid like order of the fibrils was compatible with transparency, Hart and Farrell [1969], Benedek [1971].

The uniform diameter of collagen fibrils, high degree of organization and uniformity of the collagen network of the corneal stroma contrast with the heterogeneities of the sclera. This spatial homogeneity is thought to contribute to the transparency of the cornea, e.g., Meek and Fullwood [2001]. The transient disorganization of this regular hexagonal array in scar tissues, accompanied with a loss of transparency, is considered to witness this point of view, Rawe et al. [1994].

19.1.2.4 Interlamellar structure and antero-posterior heterogeneities

Crosslinks between structural units are lacking mainly in the radial direction, thus allowing swelling. Shearing parallel to the corneal surface is resisted by the collagen network. Interlacing and connections between lamellae at the periphery of the stroma are thought to contribute greatly to the mechanical stiffness and strength. Smolek and McCarey [1990] and Smolek [1993] measure the interlamellar adhesive strength which is found larger near the limbus where collagen forms interwoven bundles.

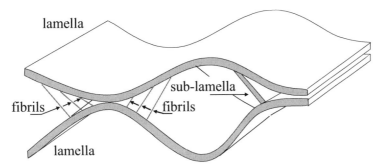

Fig. 19.1.7: The interlamellar space can open by anterior-posterior stretching, creating tunnels and a kind of honeycomb structure, Radner and Mallinger [2002]. But sub-lamellae and fibrils connections limit the relative displacements parallel to the lamellae. Note that swelling takes place both within the lamellae, by increasing the interfibrillar spacing and in the interlamellar space. The latter swelling takes probably place above the physiological hydration, and may lead to the creation of *lakes* that decrease transparency. On the other hand, it seems that the intrafibrillar space is not affected by hydration unlike in articular cartilages.

Experimental data by Radner and Mallinger [2002] support this point of view. They cut pieces 4 mm×4 mm of corneal stroma parallel to the surface and stretch them along the radial (anterior-posterior) direction, so as to open the interlamellar structure which then appears like a honeycomb. Adjacent lamellae are connected by both a number of fibrils and oblique sub-lamellae, Fig. 19.1.7.

The lamellae run almost from limbus to limbus in the posterior part of the stroma. In the anterior part, they form bundles with a marked antero-posterior interweave. This heterogeneity may give rise to curling when a thick part of the stroma is exposed to nonphysiological aqueous solutions. In fact, the anterior zone is observed not to swell in de-ionized water, in contrast to the rest of the stroma where increase of the interfibrillar spacing gives rise to swelling, Huang and Meek [1999], Müller et al. [2001]. Swelling affects the extrafibrillar space only, and fibril diameter appears to be unaffected by ionic strength. Besides mechanical restraint, the absence of swelling of the anterior zone may be due to either a lower density of PGs or to a larger proportion of chondroitin sulfates whose retentive power is larger than that of keratan sulfates. Accordingly, the water content has a positive gradient inwards.

19.1.3 Keratoconus

Keratoconus is the most common dystrophy of the cornea with an incidence of 1 over 2000. It is characterized by a thinning and ectasia of the central cornea, inducing severe astigmatism, Fig. 19.1.8.

The biochemical aspects of keratoconus are reviewed in Rabinovitz [1998]. Hayes et al. [2007]a summarize the current state of knowledge from the ophthalmologic point of view.

Keratoconus is accompanied by a number of structural disturbances in the collagen architecture. Intermolecular collagen spacing is smaller than in normal corneas, typically 1.75 nm versus 1.85 nm, Fullwood et al. [1992]. The preferred collagen directions, which are at 90° and 180° in the normal eye, change to 60° and 120°, Daxer and

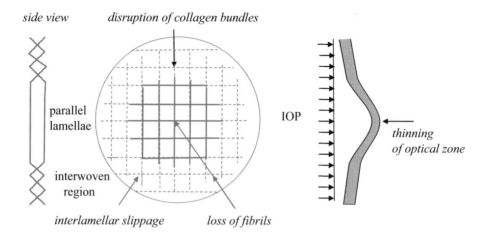

Fig. 19.1.8: In the normal stroma, collagen fibrils run from limbus to limbus. Keratoconus is due to the decrease of strength (partial breakage) and increase relative slip of collagen bundles at the periphery of the anterior part of the stroma, Meek et al. [2005]. The periphery is then less effective in clamping the central part of the cornea to the sclera which results to be more deformable.

Fratzl [1997]. Both the central cornea and the periphery undergo a loss of fibrils, resulting in a more compliant structure, Hayes et al. [2007]a.

The lamellar bundles are discontinuous in the anterior part, Meek et al. [2005]. Weakened interfibrillar glue may leave the bundles to slip relative to one another, resulting in a disorganized orientation of collagen fibrils, Meek et al. [2005]. In that view, keratoconus would be due to *decreased interlamellar shear stiffness* at the circumference of the anterior part of the stroma. Alternatively, rather than interlamellar shear stiffness, the entity to be faulted might be the *interlamellar shear strength*: the lamellae would then break and next slip past one another, Smolek [1993], Smolek and Beekhuis [1997].

Proposed treatments intend to improve collagen crosslinking through chemicals, Spoerl et al. [1998], Wollensack et al. [2003].

19.1.4 Variations over species

Elliott and Hodson [1998], p. 1333, provide a table of interfibrillar spacings, fibril diameters and intermolecular spacings of the stroma for a number of vertebrate species. Even if the proportion of chondroitin and keratan sulfates varies from species to species, Scott and Bosworth [1990] suggest that the sulfation of the keratan sulfates adjusts to achieve an almost uniform fixed charge over species.

In the central part of the human cornea, the directional distribution of collagen fibrils is cross-shaped. This distribution is not linked to the existence of four extraocular muscles in orthogonal planes. Indeed, the fibrillar arrangement is more circumferential in the central part of the stroma of most species, despite the fact that these muscles are present, Hayes et al. [2007]b.

The geometrical characteristics of the cornea of humans and pigs are quite similar. Zeng et al. [2001] perform uniaxial experiments, using strip extensiometry, that indicate that the tensile strength and stress-strain relations of human and porcine corneas are close. However, the porcine cornea relaxes more rapidly than the human cornea. Hoeltzel et al. [1992] use strip extensiometry as well to compare the mechanical responses of bovine, rabbit, and human corneas.

19.2 Biomechanical aspects

The chemomechanical properties of the cornea are complex because the endothelium, stroma and epithelium play their own roles. The stroma has a natural tendency to imbibe water. However, the phenomenon depends much on the relative ionic strengths of the stroma and of the solution it is in contact with. For example, in case of oedema, drops of 5% NaCl are clinically used to drain some water out of the stroma and improve transparency (and vision).

19.2.1   Osmotic swelling

Osmotic swelling is a phenomenon that takes place in the presence of semi-impermeable membranes. Indeed, let us consider a reservoir separated by a membrane permeable to water, but not to (some) ions. Assume that the ionic concentration on one side is suddenly increased. Then, a flow of water takes place towards the region where the ionic concentration is higher, so to say, to attempt to homogenize the ionic concentration. The flow stops when the chemical potentials of water on both sides are equal.

At small ionic concentration c, the chemical potential of water is contributed additively by a mechanical part, the water pressure p, and a chemical part, proportional to the ionic concentration, namely $\mu_w = \mu_{w0} + (p - RT\,c)/\rho_w$, where ρ_w is the density of water, μ_{w0} is some reference value. The difference of water pressure $p_2 - p_1$ between the two sides is called osmotic pressure. This difference is found to have the sign of $c_2 - c_1$. For definiteness, let $c_2 > c_1$. Then the transient osmotic flow has taken place from the region of low pressure to the region of high pressure, in opposition to Darcy's flow.

The phenomenon of osmotic swelling in cornea results from electroneutrality: indeed, the existence of negative fixed charges (on PGs) requires a minimal concentration of positive mobile charges (cations) in their neighborhood. This fixed charge can be seen as endowing the stroma with a fictitious membrane, on each side of which a number of quantities (pressure, concentrations, electrical potential $\cdots$) are distinct.

This description applies fully for solutions. In articular cartilages and many gels, the presence of a solid skeleton restricts the flow of water and leads to additional couplings. The situation seems different in the corneal stroma due to the absence of crosslinks between collagen fibrils. Infinite swelling is avoided by active transport located on the endothelium, Elliott and Hodson [1998].

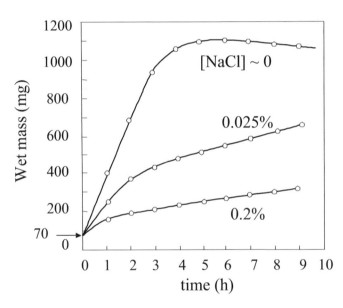

Fig. 19.2.1: A mass of 70 mg of *stroma* is cut from the central part of a bovine cornea, from an animal about 2 years old. It is immersed at 37 °C in a NaCl solution with controlled concentration. A solution with x percent of NaCl has a concentration of $x \times 1000/58.5 = 170 \times x$ mM. The absence of crosslinks along the radial direction leads to *large swelling* of the de-endothelialized corneal stroma. Active transport mechanisms located on the endothelium limit the phenomenon in actual *intact cornea*. Averaged values redrawn from experiments by Doughty [2001]

To display the phenomenon, de-endothelialized square samples of stroma, $8 \times 8\,\text{mm}^2$ are immersed in baths at various NaCl concentrations, Doughty [2001]. The results that illustrate *this quasi-infinite swelling* are sketched in Fig. 19.2.1. Of course, swelling increases at decreasing ionic concentration. Immersed in quasi-distilled water, the sample becomes translucent, as opposed to transparent, and its mass is multiplied by about 1100 mg/70 mg∼16, after 5-6 hours of exposure.

19.2.2   Binding of chloride ions

Under neutral pH, PGs do not bind sodium cations, which remain solvated. On the other hand, chloride ions Cl^- do partly *bind*, that is, they are no longer free, and they contribute to increase negatively the fixed charge associated with the sulfonic and carboxylic groups of PGs.

Chloride ions bind to a ligand of unknown composition, but neutron scattering seems to indicate that the ligand is associated with collagen fibrils, Regini et al. [2004]. According to Hodson et al. [1992], the binding process obeys a first order kinetics with a dissociation constant of 300 mM and a total capacity of 75 meq/l under physiological hydration.

Chloride binding decreases with temperature so that the fixed charge becomes less negative as temperature raises: measurements by Regini and Elliott [2001] show a linear variation with temperature from about -110 mM at 15 °C to -40 mM at 35 °C. As the fixed charge decreases in absolute value at higher temperature, the imbibition tendency decreases, giving rise to an apparent negative (anomalous) coefficient of thermal expansion, an effect referred to as 'temperature inversion phenomenon', Kwok and Klyce [1990], Elliott and Hodson [1998].

19.2.3   Electrical shielding

The electrolyte of the aqueous humor contains dissolved sodium chloride at concentration 150 mM. The existence of PGs in the stroma modifies this concentration. Cations *screen* the negative charges of the PGs, and the mutual repulsive forces decrease with increasing cation concentration, resulting in shrinking. This phenomenon, referred to as *electrical shielding*, competes with osmotic swelling and active transport through the endothelium to establish a steady state volume of electrolyte. Shrinking under increased ionic strength, and conversely swelling under decreasing ionic strength, are typical in articular cartilages and other tissues with a fixed charge.

Chloride binding may also play a role, competing with electrical shielding. Increased ionic strength implies increased shielding but also increased fixed charge, and repulsion. Conversely, decreased ionic strength implies decreased shielding but also decreased fixed charge, and decreased repulsion. Which phenomenon is stronger remains to be fully described.

Regini et al. [2004] indicate that chloride binding takes place on ligands located within the collagen fibrils. They find that the diameter of collagen fibrils is minimal under physiological strength of NaCl. At lower concentration, electrical shielding is low, and repulsion (inducing swelling) between negatively charged PGs takes over, as indicated by the experiments of Doughty [2001], Fig. 19.2.1, and Huang and Meek [1999]. At higher concentration, electrical shielding should take over repulsion between fibrils, but the increase in fixed charge due to chloride binding contributes to the repulsion force. Experiments by Ruberti and Klyce [2003] show that increase of ionic strength may in fact lead to some swelling.

19.2.4   Effects of pH

Carboxyl, sulfonic and amino sites of glycoaminoglycans (GAGs) and collagen bind the hydrogen ion reversibly. The charge of PGs becomes less negative at pH of the stroma lower than physiological. Collagen molecules become positively charged at low pH and negatively charged at pH higher than physiological. Table 2 in Elliott and Hodson [1998] shows the variation of the fixed charge with pH.

The ligand sites for chloride binding are assumed not to be affected by pH in the model developed in Section 19.11 below, but this assumption remains to be assessed.

19.2.5   Dependence of material properties on PG and collagen content

The elastic stiffness is presumably in direct relation with the collagen and PG content.

The hydraulic conductivity is on the other hand in inverse relation with the PG content, but independent of the ionic content, Maroudas [1975] [1979] for articular cartilages. Thus, as water is expelled during compression, the concentration of PGs increases and therefore the hydraulic conductivity is expected to decrease.

The actual dependence of permeability in terms of strain, or void ratio, or hydration, is not completely resolved by the standard analytical expressions, for at least two reasons. First, an assessment of these relations revolves around the definition of the void ratio. Second, the actual permeability depends not only on the actual porosity, but also on the microstructure of this porosity. The notion of hydraulic tortuosity is briefly discussed below in relation with ionic mobilities and diffusion coefficients.

Moreover, the dependence of properties in composition implies that the mechanical and transport properties should vary across the thickness of the cornea, since this composition varies.

19.2.6 Partition of water

The biomechanical and mechanical roles of water are not uniform throughout the cornea. Intracellular water in the endothelium constitutes about 15% of total water in physiological conditions. In the stroma, most water is extracellular and its volume fraction is about 80%. Extracellular water can be partitioned in intrafibrillar and extrafibrillar water. As already mentioned, the fibril diameters, and hence intrafibrillar water, vary little with ionic strength, Huang and Meek [1999]. Extrafibrillar water is the main component which is responsible of volume change of the tissue.

19.3 Active transport in the endothelium: the cell and organ scales

The endothelium contains active transport mechanisms that limit the infinite radial swelling, that might take place due to weak radial crosslinking between collagen fibrils, Mergler and Pleyer [2007]. Swelling is due to the absorption of water by the stroma from the aqueous humor through the leaky endothelium. The epithelium is much more impermeable to chemicals including water and the exchanges with the sclera are quite small.

The dynamic equilibrium that maintains the optimal hydration regarding transparency is evidenced by decreasing the temperature just above freezing so as to silence the metabolic pumps. Then the thickness of the cornea increases but, when the temperature is returned to physiological value, the normal thickness is progressively retrieved after a few hours, Fig. 14 in Elliott and Hodson [1998]. To stress the reversible character of the experiment, the effect has been coined "temperature reversal phenomenon", not to be confused with the "temperature inversion phenomenon" linked to chloride binding as mentioned in Section 19.2.2.

19.3.1 Active transport: the organ scale

These mechanisms continuously expel ions out of the stroma towards the aqueous humor, by reversing the osmotic pressure induced by the presence of the fixed charge. This continuous exchange implies that there is no permanent chemical equilibrium as far as ions are concerned, rather, a *steady state*, sometimes termed *dynamic equilibrium*, takes place. More generally, we may consider flows across a membrane. Flows conjugated to driving forces which are controlled to a non zero value will tend, after a transient period, to a non zero value as well.

The membrane of the endothelial cells contains transport mechanisms on their basolateral (stroma), and on their apical (aqueous humor) sides. Fig. 19.3.1 shows the main transport mechanisms located on the basolateral side according to Bonanno [2003]. In particular, bicarbonate ions are continuously pumped out of the stroma by the Na^+-$2\,HCO_3^-$ symport. In turn, the ionic out-of-equilibrium between stroma and aqueous humor drives an osmotic flow towards the aqueous humor.

If we accept that water transport reaches equilibrium, namely equality of the chemical potentials μ_w, then the pressure p and ionic concentration c are higher in the aqueous humor,

$$\mu_w^{aq} = \mu_w^{stroma} \quad \Rightarrow \quad p_{aq} - p_{stroma} = RT\,(c_{aq} - c_{stroma}) > 0\,. \qquad (19.3.1)$$

Thus at equilibrium of water transport, water flow from the aqueous humor to the stroma due to Darcy's law is counterbalanced by an opposite osmotic flow, Fig. 19.3.2.

The pumping out of ions thus gives rise to a lower pressure in the stroma. There is *no equilibrium in ionic transport*, namely $\mu_{ion,aq}^{ec} \neq \mu_{ion,stroma}^{ec}$. On the other hand, the entering and exiting fluxes compensate *at steady state*. In a first approximation, we may accept that the ionic flux in the stroma due to *passive transport* is proportional to the gradient of the electrochemical potential of this ion, namely $J_+^{(p)} = -k_+ \nabla \mu_+^{ec}$ with $k_+ > 0$ the ionic permeability. Similarly, the ionic flux across the membrane is proportional to the differential electrochemical potential across this membrane, say $J_+^{(p)} = -K_+ (\mu_{+,stroma}^{ec} - \mu_{+,aq}^{ec})$ with $K_+ > 0$ the membrane ionic permeability. Steady state implies the sum of the passive and active ionic flows to counterbalance, namely

$$J_+^{(p)} + J_+^{(a)} = 0\,. \qquad (19.3.2)$$

In fact, while the ion transport does not reach equilibrium, an electrochemical equilibrium exists but it is associated with a reaction that couples the ion transport and a metabolic reaction: an example is provided in Section 19.3.3.3.

Since the active transport is towards the aqueous humor, say $J_+^{(a)} > 0$, the passive transport should be towards the stroma, $J_+^{(p)} < 0$. The electrochemical potential of sodium cations $\widehat{m}_+ \,\mu_+^{ec} = RT \operatorname{Ln} c_+ + F\,\phi$ is defined in terms

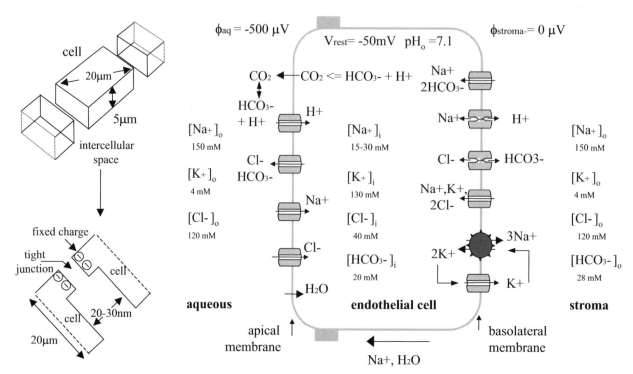

Fig. 19.3.1: The corneal endothelium is a monolayer of flat smooth cells. The resting potential V_{rest} of the endothelial membrane is about -50 mV, determined principally by the high potassium conductance, Rae and Watsky [1996]. The reverse Nernst potential of chloride ions at $[Cl^-]$=40 mM is equal to -28 mV (the reverse Nernst potential is defined in Section 19.14.2). The low transendothelial electrical potential equal to -0.5 mV and low electrical resistance are explained by the leaky tight junctions.

The cell membranes include a number of ionic channels, antiports and symports, on both basolateral and apical sides, some of which are electrogenic, Bonanno [2003]. The endothelial pump which exchanges $2\,K^+$ against $3\,Na^+$ and uses glucose as an energy source is located on the basolateral side. The active transport of sodium is inhibited reversibly or irreversibly by blockers like amiloride or ouabain at a proper concentration.

Among other transporters are ionic channels for K^+, antiports for Na^+-H^+ and Cl^--HCO_3^-, and symports for Na^+-K^+-$2Cl^-$ and the bicarbonate pump Na^+-$2\,HCO_3^-$.

Hydration of carbon dioxide to carbonic acid is accelerated by carbonic anhydrase (CA), while the equilibrium between carbonic acid and bicarbonate ions HCO_3^- is instantaneous, Bonanno [2003],

$$CO_2 + H_2O \overset{CA}{\rightleftharpoons} H_2CO_3 \rightleftharpoons H^+ + HCO_3^-$$

Apical H^+-channels and basolateral Na^+-H^+ antiports contribute to pH regulation.

The thin intercellular space is bounded by a *tight junction*, about 4 nm in width, and thus leaky to small molecules like water, chloride anion (diameter 0.36 nm), sodium cation (0.2 nm): negative fixed charges, located on its surface, give rise to electroosmosis, Rubashkin et al. [2005].

A tentative overall description of the fluxes of sodium chloride and water is as follows:
- the Na-K-ATPase pumps expel sodium cations out of the cells, a part of these ions crossing the endothelial layer through an intercellular path. Chloride anions follow due to electroneutrality;
- lateral intercellular surfaces are negatively charged: water follows sodium chloride in the apical side due to electroosmosis;
- once in the aqueous humor, sodium cations and chloride anions re-enter the endothelial cells via apical channels, due to passive diffusion;
- as a consequence, water also enters the cytoplasm due to osmosis.

A quite similar scheme has been suggested by McLaughlin and Mathias [1985] for renal proximal tubules, another endothelial structure.

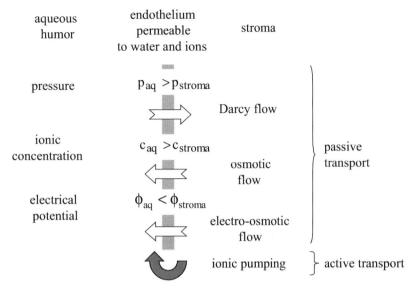

Fig. 19.3.2: Sketch of the *pump leak hypothesis*, Maurice [1951]. The pumps located in the endothelium continuously expel cations out of the stroma so that the (active) concentrations of the latter are higher in the aqueous humor. At equilibrium of water transport, the Darcy's and osmotic flow are opposite. Indeed, water flow is essentially directed against the chemical potential of water, say $J_w = -k_H\,\rho_w\,\Delta\mu_w$, with $k_H > 0$ hydraulic permeability of the endothelium, and $\rho_w\,\Delta\mu_w \simeq \Delta p_w - \Delta\pi_{osm}$, p_w is the water pressure and π_{osm} the osmotic pressure.

Electrochemical couplings should be accounted for. In fact, rather than osmosis, in view of the transendothelial difference of electrical potential, the electroosmotic effect is advocated these days to explain the fluid flow from stroma to aqueous humor.

of the molar mass $\widehat{m}_+$, Faraday's equivalent charge F and the electrical potential ϕ. Thus steady state implies a relation between the concentrations, electrical potentials and active flux, namely

$$
\begin{aligned}
0 = J_+^{(p)} + J_+^{(a)} &= -K_+\,(\underbrace{\mu_{+,\,\text{stroma}}^{\text{ec}} - \mu_{+,\,\text{aq}}^{\text{ec}}}_{>0}) + \underbrace{J_+^{(a)}}_{>0} \\
&= -\frac{K_+}{\widehat{m}_+}\left(RT\,\text{Ln}\,\frac{c_{\text{stroma}}}{c_{\text{aq}}} + \text{F}\,(\phi_{\text{stroma}} - \phi_{\text{aq}})\right) + J_+^{(a)}\,.
\end{aligned}
\tag{19.3.3}
$$

This relation might be used to estimate the transfer coefficient K_+ if the other terms are known, or the active flux if the passive flux is known.

The elastic properties of the stroma are highly anisotropic, the radial direction being much more compliant than the in-plane directions. Along this point of view, the swelling takes place in the radial direction. The crosslinks between lamellae might be aimed at transmitting shear stress, while allowing easy relative normal separation.

This highly anisotropic collagen stiffness introduces a strong difference between articular cartilage and cornea. In cartilage, hydrated PGs induce collagen fibrils in tension: a mechanical model would consist of two parallel systems, the effective pressure p_{eff} induced by PGs being resisted by the applied mechanical stress σ and collagen in tension $\mathbb{E}:\mathbf{E}_G$, Maroudas et al. [1991], Lai et al. [1991], Basser et al. [1998], say $\sigma + p_{\text{eff}}\,\mathbf{I} = \mathbb{E}:\mathbf{E}_G$. Here $\mathbf{E}_G$ is the Green strain and $\mathbb{E}$ the elastic stiffness. On the other hand, the low radial stiffness of cornea requires some additional mechanism to be present so as to limit the radial swelling: this is what active transport is all about.

Sclera, which is the outer white (non transparent) coat of the eye, is more similar to articular cartilage: radial entanglement of collagen fibrils limits the swelling. The price to pay to such stability seems to be the lack of transparency.

19.3.2 Active transport: the cell scale

The pump leak hypothesis sketched in Fig. 19.3.2 is still subject to controversy. In particular, measurements have shown that the concentration of cations is higher in the stroma than in the aqueous humor. The fact that not all cations may be pumped out is advocated to explain this discrepancy, Bryant and McDonnell [1998]. Actually, it seems that a significant part of sodium ions is bound in the stroma while only the free ions contribute to the osmotic effect. Still, in contrast to chloride binding, the binding of sodium has not been investigated yet. According to Stiemke et al. [1991], the total and free concentration of sodium ions in the stroma are about 164 mM and 134 mM respectively while the concentration of this ion in the aqueous humor is about 140 mM.

19.3.2.1   Which species are actively pumped out ?

The apical and basolateral membranes of endothelial cells contain a number of channels, Fig. 19.3.1. Moreover, passive exchanges between the aqueous humor and stroma take place also through intercellular slits. The complete picture of exchanges is complex:
  - ions entering the cells through the basolateral side exit to the aqueous humor, or return to the stroma;
  - a percentage of the latter flows crosses the intercellular slits to enter the aqueous humor and re-enter the cell through the apical side.

The pumping out of water would imply that the associated chemical potentials across the endothelium are not equal. In fact water pumping is only a secondary process. The primary processes are due to transport of ions. This point is demonstrated by Na^+-channels blockers like amiloride. Inhibiting the Na^+-K^+-ATPase pump results in significant hydration increase, Kuang et al. [2004]. The latter indicate that the symports Na^+-$2\,HCO_3^-$ and Na^+-K^+-$2Cl^-$ contribute for 70% and 30% to transendothelial fluid transport. It is not established whether the transport of water through the endothelium is transcellular or paracellular, namely through the slits. Even more, according to Riley et al. [1997], the symport Na^+-K^+-$2Cl^-$ is not exhibited by the corneal endothelium.

To summarize, the existence of a pump attached to a species implies the electrochemical potentials of this species to display a transendothelial jump. On the other hand, whether the species undergoes active transport or not, its *total* flux at steady state or equilibrium vanishes.

19.3.2.2   Net fluxes

The ions which are most likely involved are Na^+ and HCO_3^-. Bicarbonate ions may enter the cell through the electrogenic Na^+-$2\,HCO_3^-$ symport, and exit (at least partially) through an apical channel. Sodium cations enter the cell through basolateral channels and may re-enter the stroma through the Na^+-K^+-ATPase: a part of them exit the endothelium to the aqueous humor through intercellular slits, and re-enter the cells through apical channels.
  - Bryant and McDonnell [1998] consider the pumping out of stroma of the sole sodium cations. They fix the value of this flux to 0.70×10^{-6} mole/m^2/s so as to obtain an imbibition pressure of -60 mmHg;
  - the transmembrane flux J_{Na} is defined as the product of the rate of intracellular concentration dc_{Na}/dt times the ratio volume over surface taken equal, to simplify, to the intercellular distance, 10 to 20 nm, Kuang et al. [2004]. Kuang et al. [2004] quantify the cellular influx of sodium and the net efflux of sodium from stroma to aqueous humor due to the Na^+-pump to 1.5×10^{-6} mole/m^2/s. This efflux is paracellular, while the return flux is transcellular, through apical channels, Fig. 10 of Sanchez et al. [2002];
  - the associated flux of water is calculated by assuming that isotonic ions (namely $c_{Na} = 150$ mM) do not diffuse with respect to water, namely $J_w = J_{Na}/c_{Na} \simeq 10^{-8}$ m/s. This value corresponds to the rate of increase of thickness of the stroma when the sodium channels are blocked;
  - the transendothelial flux of bicarbonate ion, from stroma to aqueous humor, is estimated to be about 1 to 1.5×10^{-6} mole/m^2/s, Fischbarg et al. [2006];
  - Sanchez et al. [2002] consider fluid flow to be purely passive, paracellular and due to electroosmosis, not to local osmosis. They provide data that allow to access the electroosmotic coefficient k_e of the endothelium in absence of coupling, namely 3.3×10^{-5} m/s/V, while the electrical conductivity σ_e is estimated to about 500 siemens/m^2. Fischbarg and Diecke [2005] have developed a biochemical model that accounts for many ionic channels and transendothelial fluid transport through electroosmosis;
  - Hodson [1974] estimates the active salt efflux to 3.3×10^{-6} mole/m^2/s. The total net flux includes in addition a secondary active influx, from aqueous humor to stroma;
  - Swan and Hodson [2004] propose that bicarbonate ions enter the cell, along its chemical gradient (active influx), through the symport Na^+-$2\,HCO_3^-$ on the basolateral side, and exit to the aqueous humor, against its gradient (passive efflux), through an apical channel. They note that the transendothelial current 0.2 A/m^2 is larger than the current of 0.1-0.15 A/m^2 induced by the flux of bicarbonate ions, namely $1 - 1.5 \times 10^{-6}$ mole/m^2/s according to Fischbarg et al. [2006]. They contend that the missing flux is due neither to sodium nor to chloride.

19.3.3 Active transport: energetic coupling

The endothelium is said to pump fluid out of the stroma to the aqueous humor, so as to maintain a constant hydration of the cornea. However, this so-called fluid-pump might be only secondary in the active transport of ions across the basolateral and apical membranes of endothelial cells.

Key to an active transport is a metabolic reaction. The energy produced by this metabolic reaction triggers the flows of some species which are not primary in the reaction. Thus, the basic idea is that of coupling between flows, Katchalsky and Kedem [1962], p. 73.

19.3.3.1 Formalism

The dissipation inequality is the sum of the works done by conjugate pairs of fluxes and driving forces $(\mathbf{J}, \mathbf{F})$ and $(\mathcal{J}, \mathcal{G})$,

$$-\sum_k \mathbf{F}_k \cdot \mathbf{J}_k - \mathcal{G}\,\mathcal{J} \geq 0\,. \tag{19.3.4}$$

Here $\mathcal{G}$ is the affinity of the metabolic reaction (an example of affinity is detailed in Section 19.10.1),

$$\mathcal{G} = \sum_l \nu_l g_l\,, \tag{19.3.5}$$

the ν's being the stoichiometry coefficients, the g's mole based (electro)chemical potentials, while $\mathcal{J}$ is the metabolic rate (also called rate of advancement of the reaction). For an isotropic material, the inequality may be satisfied through the linear relations

$$\begin{aligned}
-\mathbf{F}_k &= \textstyle\sum_l R_{kl}\,\mathbf{J}_l + \mathbf{R}_k\,\mathcal{J}, \quad k=1,\cdots, \\
-\mathcal{G} &= \textstyle\sum_k \mathbf{R}_k^{\mathrm{T}}\,\mathbf{J}_k + R\,\mathcal{J}.
\end{aligned} \tag{19.3.6}$$

Symmetry of the frictional matrix, implicitly introduced by the above relations, is adopted. The dissipation inequality may be satisfied through these linear relations, provided the frictional matrix is positive semi-definite. If it is positive definite, these linear relations may be inverted,

$$\begin{aligned}
-\mathbf{J}_k &= \textstyle\sum_l K_{kl}\,\mathbf{F}_l + \mathbf{K}_k\,\mathcal{G}, \quad k=1,\cdots, \\
-\mathcal{J} &= \textstyle\sum_k \mathbf{K}_k^{\mathrm{T}}\,\mathbf{F}_k + K\,\mathcal{G}.
\end{aligned} \tag{19.3.7}$$

Note that the frictional and diffusion coefficients, R_{kl} in (19.3.6) and K_{kl} in (19.3.7), respectively, are scalars while the line and column that involve the metabolic flux and affinity are vectors.

These equations may be recast in an *energy driven* format,

$$\begin{aligned}
\mathbf{F}_k - \mathbf{R}_k \frac{\mathcal{G}}{R} &= -\sum_l \mathbf{R}_{kl}^{(p)}\,\mathbf{J}_l, \quad k=1,\cdots, \\
-\mathcal{J} &= \frac{\mathcal{G}}{R} + \sum_k \frac{\mathbf{R}_k^{\mathrm{T}}}{R}\,\mathbf{J}_k,
\end{aligned} \tag{19.3.8}$$

which highlights the modifications to the driving forces (electrochemical potentials) induced by the metabolic reaction: in other words, the so-called enthalpies of formation are now exhibited in explicit form,

$$\mathbf{F}_k = \rho_k\,\boldsymbol{\nabla}(\mu_k - \mu_k^0) = \frac{1}{\widehat{v}_k}\,(\widehat{v}_k\,\boldsymbol{\nabla}p_k + RT\,\boldsymbol{\nabla}\mathrm{Ln}\,x_k + \zeta_k\,\mathrm{F}\,\boldsymbol{\nabla}\phi), \quad -\mathbf{R}_k\frac{\mathcal{G}}{R} = \rho_k\,\boldsymbol{\nabla}\mu_k^0\,. \tag{19.3.9}$$

Moreover, in absence of the metabolic energy ($\mathcal{G} = 0$, but a priori $\mathcal{J} \neq 0$ due to couplings), the expressions (19.3.8)$_1$ boil down to the passive frictional equations, so that the coefficients in the rhs are the coefficients of passive diffusion (note that they are a priori second order tensors, not scalars),

$$\mathbf{R}_{kl}^{(p)} = R_{kl}\,\mathbf{I} - \frac{1}{R}\mathbf{R}_k\,\mathbf{R}_l^{\mathrm{T}}\,. \tag{19.3.10}$$

The alternative *flux driven* format,

$$
\mathbf{J}_k = \overbrace{-\sum_l \mathbf{K}_{kl}^{(p)}\,\mathbf{F}_l}^{\mathbf{J}_k^{(p)}} + \overbrace{\mathbf{K}_k \frac{\mathcal{J}}{K}}^{\mathbf{J}_k^{(a)}},\quad k=1,\cdots,
$$

$$
\mathcal{G} = -\frac{\mathcal{J}}{K} - \sum_k \frac{\mathbf{K}_k^{\mathrm{T}}}{K}\,\mathbf{F}_k,
$$

(19.3.11)

highlights the fact that the fluxes are contributed by a passive part and an *active part*. Moreover, in absence of the metabolic flux ($\mathcal{J}=0$, but a priori $\mathcal{G}\neq0$), the expressions $(19.3.11)_1$ boil down to the passive diffusion equations, so that the coefficients in the rhs are the coefficients of passive diffusion (note that they are a priori second order tensors, not scalars),

$$
\mathbf{K}_{kl}^{(p)} = K_{kl}\,\mathbf{I} - \frac{1}{K}\,\mathbf{K}_k\,\mathbf{K}_l^{\mathrm{T}}.
$$

(19.3.12)

As an example, Katchalsky and Kedem [1962] consider an electrolyte of sodium chloride where the active transport concerns only the sodium cation. To simplify the exposition of this example, the flow of water is assumed to vanish, $\mathbf{J}_{\mathrm{w}}=\mathbf{0}$, and the cross-flows are neglected. Then, (19.3.6) simplifies to,

$$
\begin{aligned}
-\mathbf{F}_{\mathrm{Na}} &= R_{\mathrm{NaNa}}\,\mathbf{J}_{\mathrm{Na}} &+&\ \mathbf{R}_{\mathrm{Na}}\,\mathcal{J}\\
-\mathbf{F}_{\mathrm{Cl}} &= R_{\mathrm{ClCl}}\,\mathbf{J}_{\mathrm{Cl}} &&\\
-\mathcal{G} &= \mathbf{R}_{\mathrm{Na}}^{\mathrm{T}}\,\mathbf{J}_{\mathrm{Na}} &+&\ R\,\mathcal{J}
\end{aligned}
$$

(19.3.13)

Let us consider a situation where the ionic fluxes both vanish. Since $\mathbf{F}_k = \rho_k\,\boldsymbol{\nabla}(\mu_k-\mu_k^0) \sim (RT\boldsymbol{\nabla}\mathrm{Ln}\,x_k+\zeta_k\,\mathbf{F}\boldsymbol{\nabla}\phi)/\widehat{v}_k$, so that

$$
-\widehat{v}_{\mathrm{Na}}\,\mathbf{F}_{\mathrm{Na}} - \widehat{v}_{\mathrm{Cl}}\,\mathbf{F}_{\mathrm{Cl}} = \begin{cases} -RT\,\boldsymbol{\nabla}\mathrm{Ln}\ c_{\mathrm{Na}}\,c_{\mathrm{Cl}},\\[2mm] \widehat{v}_{\mathrm{Na}}\,\mathbf{R}_{\mathrm{Na}}\,\mathcal{J}. \end{cases}
$$

(19.3.14)

The above relations can be seen either as involving gradients in a continuum, or discontinuities across an interface. With the latter interpretation, the differences in concentrations on the sides 1 and 2 of an interface are directly related to the metabolic reaction,

$$
RT\,\mathrm{Ln}\,\frac{c_{\mathrm{Na1}}\,c_{\mathrm{Cl1}}}{c_{\mathrm{Na2}}\,c_{\mathrm{Cl2}}} = \widehat{v}_{\mathrm{Na}}\,\mathbf{R}_{\mathrm{Na}}\,\mathcal{J} = -\widehat{v}_{\mathrm{Na}}\,\frac{\mathbf{R}_{\mathrm{Na}}}{R}\,\mathcal{G}.
$$

(19.3.15)

Or, perhaps, the alternative point of view might be more explicative: the metabolic flux, or energy, is the engine that implies the equilibrium constant to be different from one.

19.3.3.2   Equilibrium of a non-electrogenic exchanger: the Na^+-Ca^{2+} exchanger

If $dn_{\mathrm{Ca}} = dn$ moles of Ca^{2+} exit the cell, $dn_{\mathrm{Na}} = -2\,dn$ moles of Na^+ enter from both sides of the membrane to preserve electroneutrality. The works done by moving the ions Ca^{2+} and Na^+ are respectively $(-g_{\mathrm{Ca(i)}} + g_{\mathrm{Ca(o)}})\,dn$, and $(2\,g_{\mathrm{Na(i)}} - 2\,g_{\mathrm{Na(o)}})\,dn$, where the g's are the mole based electrochemical potentials. Absence of dissipation implies the sum of these two works to vanish, or said otherwise, the exchange is workless. If the enthalpies of formation vanish, that is, if the equilibrium constant is equal to one, the inner and outer concentrations satisfy the following equality at equilibrium,

$$
\frac{c_{\mathrm{Ca(o)}}}{c_{\mathrm{Ca(i)}}} = \frac{c_{\mathrm{Na(o)}}^2}{c_{\mathrm{Na(i)}}^2}.
$$

(19.3.16)

19.3.3.3   Electrogenic ionic pumps: the Na^+-K^+-ATPase

When the reaction driven by a pump is not neutral, the resulting equilibrium depends on the difference of electrical potentials between the cytosol (i) and the extracellular electrolyte (o). The Na^+-K^+-ATPase is an example: it pumps into the cell two potassium cations while three sodium cations are expelled to the electrolyte. The ATPase site and resulting products, ADP and P_i, are located inside the cellular membrane. If one admits that the electrical potentials are uniform both within and outside the cell, the reaction,

$$\underbrace{ATP^{4-} + 3\,Na^+}_{\text{intracellular}} + \underbrace{2\,K^+}_{\text{extracellular}} \rightleftharpoons \underbrace{ADP^{3-} + P_i^{2-} + H^+ + 2\,K^+}_{\text{intracellular}} + \underbrace{3\,Na^+}_{\text{extracellular}} , \qquad (19.3.17)$$

has affinity,

$$\mathcal{G} = RT\,\mathrm{Ln}\,\frac{c_{ADP}\,c_{P_i}}{c_{ATP}}\,\frac{c_{Na(o)}^3\,c_{K(i)}^2}{c_{Na(i)}^3\,c_{K(o)}^2}\,\frac{10^{7-pH}}{K_{eq}} + \mathrm{F}\,(\phi_{(o)} - \phi_{(i)})\,, \qquad (19.3.18)$$

where K_{eq} is the equilibrium constant of the ATPase at pH=7, namely about 2.22×10^5 M under physiological concentration of magnesium. This expression is obtained by observing that the numbers of moles of the chemicals involved in this reaction obey the relations,

$$-dN_{ATP} = dN_{ADP} = dN_{Pi} = dN_H = -\frac{1}{3}dN_{Na(i)} = \frac{1}{3}dN_{Na(o)} = \frac{1}{2}dN_{K(i)} = -\frac{1}{2}dN_{K(o)}\,, \qquad (19.3.19)$$

while the affinity $\mathcal{G}$ associated with hydrogen is defined as a linear combination of the electrochemical potentials weighted by the stoichiometric coefficients,

$$\mathcal{G}\,dN_H = \sum_k g_k\,dN_k\,. \qquad (19.3.20)$$

19.4 Physical data: literature review

19.4.1 Measures of the fixed electrical charge

The total fixed charge is contributed by PGs, collagen and ligands that trap chloride ions. The term "fixed" does not mean "constant in time" but points out the fact that PGs, collagen and ligands are moving with the solid, as opposed to mobile ions that diffuse through the solid skeleton. The fixed charge may vary due to changes of pH and ionic strength. The charge of PGs is modified by pH only while the charge of collagen is affected by pH and chloride binding. At neutral pH, the PGs are negatively charged while the collagen molecules are practically neutral. The charge of ligands varies with the ionic strength.

The pH and ionic strength alluded for here refer to the stroma. Consider a specimen of stroma in contact with a bath. The presence of the fixed charge implies that the concentrations of species are different within the stroma and in the bath. Therefore, when referring to a chemical composition, it is important to indicate whether the ionic strength and pH in question refer to the stroma or to the bath.

The *effective concentration* of a charged macromolecule measures its effective charge, i.e., number of moles N times its valence ζ, in a volume of water, or rather solution V_{sol},

$$e_e = \frac{\zeta\,N}{V_{sol}}\,. \qquad (19.4.1)$$

The absolute value of the effective concentration is referred to as *fixed charge density* FCD. While the FCD refers to the volume of solution, the cation exchange capacity (CEC) [unit : coulomb C/kg] refers to the dry mass M_{dry}, namely the mass of the solid constituents :

$$\frac{CEC}{F} = \frac{|\zeta|\,N}{M_{dry}}\,. \qquad (19.4.2)$$

Here F $= 96,485$ coulomb/mole is Faraday's equivalent charge. The relation between the two measures reads,

$$|\zeta|\,N = V_{sol}\,|e_e| = M_{dry}\,\frac{CEC}{F}\,. \qquad (19.4.3)$$

The ratio CEC/F which is measured in mole per kg of dry mass is sometimes loosely referred as cation exchange capacity.

Consider a piece of stroma in equilibrium with a bath of controlled ionic strength and pH. Under physiological ionic concentration, namely 154 mM of NaCl and pH=7, the total charge is about -40 mM, Elliott et al. [1980]. As a matter of comparison, the fixed charge of articular cartilages, which is then contributed only by PGs, is about -150 to -200 mM. Table 2 of Elliott and Hodson [1998] provides values for a range of ionic strengths and pHs of the bath. Under physiological ionic strength, the fixed charge changes from -32 mM to -37 mM when the pH varies from 6 to 8. At pH=7 and ionic strength 20 mM, the effective charge drops to -18 mM: since chloride binding is minimal when the ionic strength of the bath tends to zero, the effective charge is then contributed by PGs only.

Hedbys and Mishima [1966] and Hodson et al. [1992] argue that the product of the total charge e_e and hydration H_w=water weight/dry weight, is constant,

$$|e_e| \times H_w = \text{constant}, \tag{19.4.4}$$

at given ionic concentration and pH. In other words, this product depends only on concentration and pH. Indeed

$$|e_e| \times H_w = \rho_w \frac{\text{CEC}}{\text{F}}. \tag{19.4.5}$$

The relation is used during confined compression tests (constant surface, uniaxial deformation) to relate the hydration and the thickness, Hodson [1974], p. 299. Let x be the thickness of a stromal specimen and x_{dry} be the thickness occupied by the dry part. Then the thickness and total charge are related to their values under a reference state, e.g., the physiological state,

$$|e_e| \times (x - x_{dry}) = |e_e^{physio}| \times (x_{physio} - x_{dry}). \tag{19.4.6}$$

19.4.2   Standard chemical content of stroma

• Ionic content:

The physiological concentration of NaCl in the aqueous humor is equal to 154 mM. The presence of PGs modifies this concentration in the cornea. The actual concentrations of Na^+ and Cl^- depend on the concentration of PGs. In addition to ions Na^+ and Cl^-, other metallic and organic ions are present at smaller concentrations, like the bicarbonate ion which plays a role in active transport. Values in aqueous humor, stroma and endothelial cells are shown in Fig. 19.3.1. Still, these numbers are subject to interpretation: although the topic has not been investigated in detail, it seems that a part of sodium ions are bound in the stroma. They are therefore not available for exchange with the surroundings. According to Stiemke et al. [1991], the total concentration of sodium ions in the stroma is about 164 mM and 30 mM are bound.

• Mass and volume contents:

More than 75% percent of the mass of cornea is water. The solid mass is shared primarily by collagen (71%), free and GAG-associated proteins (24%) and GAGs (5%), Tables 19.4.1 to 19.4.4.

Table 19.4.1: Chemical composition of human corneal *stroma* and sclera, reproduced with permission from Edwards and Prausnitz [1998]. De-epithelization has increased the mass of water in the stroma from 78% to 86%.

Constituent	Stroma (% mass)	Sclera (% mass)	Stroma (% volume)	Sclera (% volume)
Water	86	70	89.7	77.7
Salts	0.64	4.2	0.19	1.3
Collagen	9.6	22.5	7.3	18.4
GAGs	0.63	0.3	0.45	0.23
GAG-associated proteins	a	a	0.68	-
Other proteins	b; a+b=3.2	b; a+b=3	1.70	-

The relative mass content of water is measured through the *hydration* H_w,

$$H_w = \frac{M_w}{M_{dry}}, \tag{19.4.7}$$

Table 19.4.2: Relative mass composition of a specimen of bovine corneal *stroma* at hydration $H_w = 3.2$, Doughty [2001].

Water	Total Protein	PGs
78.1	5.08	2.28

which is equal to 3.2 in physiological conditions, corresponding to 76.2% of mass of water. Hodson et al. [1991] note that mechanical measurements give a higher hydration than osmometry using polyethylene glycol as a non-penetrating solute, typically 3.4 versus 3.2.

- Mass densities:

The molar masses of species involved are needed, namely $\widehat{m}_w = 18\,\mathrm{gm}$ for water, $\widehat{m}_{Na} = 23\,\mathrm{gm}$ for sodium ion, $\widehat{m}_{Cl} = 35.5\,\mathrm{gm}$ for chloride ion, $\widehat{m}_c = 285 \times 10^3\,\mathrm{gm}$ for collagen of type I, and $\widehat{m}_{PG} = 72 \times 10^3\,\mathrm{gm}$ for PGs. Edwards and Prausnitz [1998] assume the density of collagen ρ_c to be $1/0.73 = 1.37\,\mathrm{gm/cm}^3$.

Table 19.4.3: Mass density of bovine corneal *stroma*, Huang and Meek [1999].

Density Collagen	Density Other Proteins	V_c/V_{prot}	Density Dry Part
1.41 gm/ml	1.06 gm/ml	1.25	1.254 gm/ml

Table 19.4.4: Mass densities of stroma components, Edwards and Prausnitz [1998].

Density Collagen	Density Free Proteins	Density PGs
1.37 gm/ml	1.35 gm/ml	1.49 gm/ml

The density of the dry part of the stroma can be estimated, assuming the mass densities of collagen and other proteins and their relative volumes to be known, Tables 19.4.3 and 19.4.4

$$\rho_{\text{dry part}} = \frac{\rho_c V_c + \rho_{\text{prot}} V_{\text{prot}}}{V_c + V_{\text{prot}}}. \tag{19.4.8}$$

The mass density of the stroma,

$$\rho_{\text{stroma}} = \frac{H_w + 1}{H_w + \rho_w/\rho_{\text{dry part}}} \rho_w, \tag{19.4.9}$$

is equal to $1.050\,\mathrm{gm/ml} = 1050\,\mathrm{kg/m}^3$ under physiological hydration $H_w = 3.2$. The corresponding volume fraction of water,

$$n^w = \frac{V_w}{V} = \frac{H_w}{H_w + \rho_w/\rho_{\text{dry part}}}, \tag{19.4.10}$$

is equal to 0.80.

Swelling takes place in the radial direction, and water fills the fibril lattice in the lamellae up to the physiological hydration. At higher hydrations, water forms *lakes* that perturb the spatial organization of the organ and decrease the transparency, Elliott and Hodson [1998].

Sclera swells much less than cornea, about one third, Huang and Meek [1999]. First, the driving engine to swelling, namely the fixed charge concentration, is lower and, second, the fibrils are interconnected so that the mechanical stiffness is higher.

19.4.3 Standard mechanical properties of stroma

The mechanical properties of the cornea result from its two main components that sustain loads, namely collagen and PGs. Collagen and the ground substance (PGs) contribute additively to the stress. The order of magnitude of the stress, whether in tension or in compression, is the hundred of kPa. Collagen fibrils have a tensile modulus of

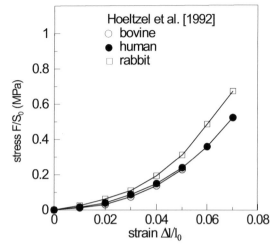

Fig. 19.4.1: Uniaxial traction tests on strips of size 2 mm×10 mm excised in the central part of corneas, redrawn from Hoeltzel et al. [1992]. Three cycles of preconditioning are applied first. At cycle 2, the slack strain (i.e., the strain from which load bearing starts) decreases from rabbit (6%) to bovine (3%) to human (0.9%). Hysteresis was higher for rabbit and bovine corneas and stabilizes at the 3rd cycle. The curves shown correspond to the 3rd cycle only.

The mechanical responses of human, porcine and rabbit corneas are quite similar at low strain rates (elongation rate 0.008 mm/s).

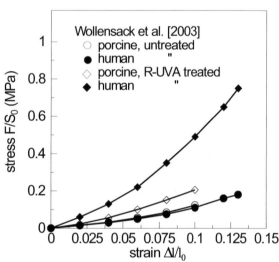

Fig. 19.4.2: Uniaxial traction tests on strips excised in the central part of corneas along the superior-inferior direction, redrawn from Wollensack et al. [2003]. A prestress of 5 kPa is applied to simulate the physiological state. The elongation rate was 0.025 mm/s.

The mechanical responses of normal (untreated) human and porcine corneas are quite similar. On the other hand, the Riboflavin-UVA treatment stiffens considerably more the human corneas than the porcine corneas, by a factor 4.5 against 2. The reason may be the higher density of crosslinking in the human corneas which are also thinner (550μm against 850μm).

Young's modulus for an axial strain of 6%:
  human, untreated: 1.3 MPa; treated: 5.9 MPa
  porcine, untreated: 1.5 MPa; treated: 2.7 MPa

about 0.5-1 GPa, Fung [1993]. They are mechanically active in extension only. PGs can sustain both compressive and tensile stresses but presumably with different moduli.

The *mechanical properties* of the cornea are essentially *planar*, with an out of plane axis along the radial direction. In fact, close to the center, the in-plane stiffness most probably possesses only the *quadratic symmetry*, along the vertical and horizontal directions. A more accurate examination of the spatial distribution of fibers in the plane parallel to the surface indicates that the tensile properties are in fact due to collagen fibrils which are essentially aligned along two directions, so that the tensile properties are those of a clinotropic material.

Uniaxial traction tests on excised corneal specimens are reported in Figs. 19.4.1 to 19.4.4. They clearly indicate a strongly nonlinear response so that the search for a definite value of a tensile modulus is a rather quixotic experience. At least is it necessary to indicate the range of strain or of load to which the value refers. Hoeltzel et al. [1992] obtain tensile in-plane Young's modulus in the range of 0.3 MPa and 4 MPa for equivalent I.O.P.'s of 10 mmHg and 400 mmHg respectively. Still, early finite element simulations used linear elasticity and so do models based on thin or thick shells. In that vein, Vito et al. [1989] and Howland et al. [1992] consider values from 0.5 to 2.1 MPa.

The radial (out-of-plane) stiffness is low due to lack of interlamellar crosslinks. It is not documented so that rough estimations are resorted to. For the sclera, the compressive Young's modulus along the radial direction is postulated 100 times lower than the tensile circumferential modulus by Battaglioli and Kamm [1984]. Bryant and McDonnell [1998] adopt this relation for a cornea endowed with elastic transverse isotropy and assume Poisson's ratios equal to 0.5.

The above remarks concerned the purely mechanical response. Still, electro-chemomechanical couplings are of prime importance in the function of the cornea as for any living tissue. As an example, the apparent elastic properties change with hydration, Jayasuriya et al. [2003]a [2003]b.

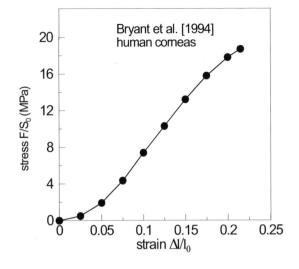

Fig. 19.4.3: Uniaxial traction tests on strips excised in the paracentral part of human corneas and cut along the orthoradial direction, redrawn from Bryant et al. [1994]. The elongation rate was 0.1 mm/s.

The strength appears quite high with respect to the value of 3.8 MPa reported in Fig. 19.4.4.

These tests were used as control to estimate the long term effects of RK on the strength which actually were not found to be significant.

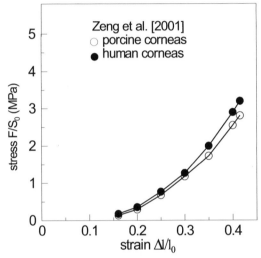

Fig. 19.4.4: Uniaxial traction tests on strips excised in the periphery of the cornea, redrawn from Zeng et al. [2001]. The strips were rectangular with width 2.5 mm over 12 to 15 mm in length. Three cycles of preconditioning are applied first. The experiments are performed at room temperature and the specimens are continuously moisturized.

The mechanical responses of human and porcine corneas are quite similar at low strain rates (elongation rate 0.16 mm/s). On the other hand, relaxation (subsequent to a sudden extension with an elongation rate 4.16 mm/s) is faster in the porcine corneas (not shown). The tensile strength is found equal to 3.8 MPa.

19.4.4 Pressures

• Intraocular pressure IOP:

The pressure exerted by the aqueous humor, referred to as *intraocular pressure* IOP, is under normal physiological conditions equal to 15 to 20 mmHg (2 to 2.66 kPa). The orthoradial stress resulting from Laplace's law is thus about $2/2 \times 7/0.5 = 14$ kPa. For a strain of 2%, this stress level corresponds to a secant modulus of about 1 MPa. Still, these numbers assume a uniform stress distribution: the actual stress varies both in depth and over the surface due to the non spherical geometry of the cornea (radius of curvature and thickness are not uniform) and to the clamping in the more rigid sclera. Therefore, the stress is expected to locally reach 100 kPa or perhaps more, especially at the limbus and in the central region in presence of keratoconus.

• Osmotic pressure:

The *osmotic pressure* is defined by the van't Hoff formula, namely, for sodium chloride and with respect to the stroma-aqueous humor interface,

$$\pi_{\mathrm{osm}} = RT \left(c_{\mathrm{Na,aq}} + c_{\mathrm{Cl,aq}} - c_{\mathrm{Na,stroma}} - c_{\mathrm{Cl,stroma}} \right). \tag{19.4.11}$$

Alternatively, the term is used to denote the contribution of the ions within a *single* compartment (not a difference). It may also be used to address the contribution of single ions.

Moreover, the osmotic pressure is sometimes already scaled down by the osmotic coefficient ω (or reflection coefficient). In addition, bound ions do not participate to the osmotic pressure and only the *free* concentrations should be accounted for.

With data from Stiemke et al. [1991], the osmotic contribution weighted by the osmotic coefficient of the sodium ion across the endothelium is calculated as,

$$\omega \, RT \left(c_{\text{aq}} - c_{\text{stroma,free}} \right) = 0.6 \times 8.31451 \times 310 \times (142.9 - 134.4) = 13\,145\,\text{Pa} = 98.6\,\text{mmHg}. \tag{19.4.12}$$

- Swelling pressure:

Consider two solutions separated by a membrane permeable to water but not to solutes. As sketched in Fig. 13.7.1, the *swelling pressure* is the pressure to be exerted so as to prevent swelling, that is, exchange of water. In fact, the swelling pressure is equal to the osmotic pressure scaled down by the osmotic coefficient, even if reported measurements do not fit accurately.

Measurements of swelling pressure of corneal stromas by Hedbys and Dohlman [1963] are reported in Ethier et al. [2004], their Fig. 3. Defined as,

$$p_{\text{swelling}} = \text{IOP} - p_{\text{stroma}}, \tag{19.4.13}$$

it is a positive quantity. It is about $60\,\text{mmHg} \simeq 8\,\text{kPa}$ for the standard hydration of cornea and $25\,\text{mmHg}$ for the sclera, Dohlman et al. [1962]. More generally it is decreasing with hydration, Fatt and Goldstick [1965],

$$p_{\text{swelling}} = \gamma \, \exp(-H_{\text{w}}), \tag{19.4.14}$$

with $\gamma = 241\,\text{kPa}$. At high hydration, where the ionic concentration is consequently low, the pressure differential dies out.

- Donnan pressure:

Consider the stroma to be in contact with a bath. Donnan pressure associated with sodium chloride,

$$RT \left(\sqrt{e_{\text{e}}^2 + (2\,c_{\text{Na,bath}})^2} - 2\,c_{\text{Na,bath}} \right) \geq 0, \tag{19.4.15}$$

is equal to the difference of fluid pressures $p_{\text{stroma}} - p_{\text{bath}}$ only if the electrochemical potentials of ions do not involve an energy of formation and if the separating membrane does not contain a pump. Therefore the formula does not apply as such to the endothelium. In this relation, e_{e} is the effective concentration of the fixed charge. The formula holds in this circumstance even for chloride binding but then e_{e} depends on the chloride concentration. Electrochemical equilibrium at an interface is considered in detail in Section 19.10.3.

Donnan pressure across the endothelium ranges from 60 to 70 mmHg under physiological conditions.

- Imbibition pressure:

Following Hedbys et al. [1963], the imbibition pressure,

$$p_{\text{imb}} = p_{\text{stroma}} - \text{IOP} - RT \left(\sqrt{e_{\text{e}}^2 + (2\,c_{\text{Na,bath}})^2} - 2\,c_{\text{Na,bath}} \right), \tag{19.4.16}$$

is the result of the nonequilibrium between stroma and aqueous humor due to the active transports. It vanishes in absence of the ionic pumps. Under physiological hydration, Bryant and McDonnell [1998] obtain a fluid pressure in the stroma p_{stroma} equal to 20 to 25 mmHg and a Donnan pressure about 70 mmHg, resulting in an imbibition pressure of about -70 mmHg.

19.4.5 Standard transport properties of endothelium, epithelium and stroma

The passive transport properties of the stroma (a continuum) and of the endothelium and epithelium (two membranes) are quantitatively quite distinct. As we have seen, active transport mechanisms are located only in these membranes, and actually, mainly on the endothelium. Correlatively, actual experiments and simulations delineate whether the specimen of interest is the whole (intact) cornea, or the de-endothelialized and de-epithelialized cornea (stroma), Fig. 19.18.1.

• Coupled diffusion in the endothelium and epithelium

The standard diffusion equations link the volume flux J_M, and molar salt flux J_{sM} to the jumps of pressure Δp_{wM} and salt concentration Δc_{sM} via three parameters, namely the open circuit permeability k_{HM}, the osmotic coefficient ω_M and the salt permeability P_{sM}, e.g., Hodson [1974], Klyce and Russell [1979],

$$J_M = -k_{HM}\left(\Delta p_{wM} - \omega_M\, RT\, \Delta c_{sM}\right),$$
$$J_{sM} = (1-\omega_M)\, c_{sM}\, J_M - P_{sM}\,\Delta c_{sM}. \tag{19.4.17}$$

The structure of these equations will be recovered in the constitutive equations developed below for a non charged membrane circulated by an electrolyte when the circuit is electrically open, Eqn (19.15.57).

No experimental data are available but Klyce and Russell [1979] back-calculated values to fit to some flux measurements on rabbit corneas, Tables 19.4.5 and 19.4.6 (note that the dimensions of the hydraulic permeability and ionic permeability (=diffusion coefficients) of these interfaces are divided by a length with respect to the stroma). Clearly the endothelium is more permeable to water and salt than the epithelium.

Table 19.4.5: Hydraulic and diffusion properties of endothelium and epithelium of rabbit corneas. From (a) Klyce and Russell [1979]; (b) Mishima and Hedbys [1967]; (c) Hodson [1974].

	k_H Open Circuit Hydraulic Permeability $[10^{-13}$ m/Pa/s$]$	P_s salt Permeability $[10^{-8}$m/s$]$	ω Reflection Coefficient -
Endothelium NaCl	42 (a)	80 (a)	0.45(a)-0.60(b)
Endothelium NaHCO$_3$ (c)		54	0.60
Epithelium NaCl (a)	6.1 (a)	0.2(a)	0.80(a)

Bryant and McDonnell [1998] start from simplified diffusion equations in terms of absolute fluxes and gradients of chemical potentials that are integrated across the endothelium. They use a value of the permeability intended for human endothelium which is taken from Maurice [1951], namely $P_{s\,endo} = 20 \times 10^{-8}$ m/s.

Table 19.4.6: Electrical properties of endothelium and epithelium of rabbit corneas. From (a) Hodson [1974]; (b) Klyce [1972].

	Electrical Conductivity σ_e [siemens/m^2]	Electroosmotic Coefficient k_e $[10^{-5}$ m/s/V$]$	Translayer Potential $[10^{-3}$ V$]$	Short Circuit Current $[$A/m$^2]$
Endothelium (a)	400–500	3.3(b)	-0.5	0.2
Epithelium (b)	2-6		+10 – +20	0.02–0.10

The electrical potential decreases from stroma to aqueous humor by -500 μV, Fig. 19.3.1. With a resistance of 20×10^{-4} ohm×m^2, a short circuit current of about 0.250 A/m^2 results. On the other hand, the trans*epi*thelial difference of electrical potentials $\phi_{stroma} - \phi_{tear\,side}$ is found to be positive, about 20 mV, and practically equal to the transcorneal potential difference, $\phi_{aqueous} - \phi_{tear\,side}$, Klyce [1972].

• Active transport in endothelium and epithelium

The high permeability of the endothelium, a very leaky membrane, is compensated by a significant metabolic rate. The active flux of sodium from stroma to aqueous humor is of the order of 1×10^{-6} mole/m^2/s, Section 19.3.2.2.

• Hydraulic permeability of stroma

Li et al. [2004] assume the radial *stroma permeability* [unit : m^2/Pa/s] to increase with the hydration, along with the data of Hedbys and Mishima [1962],

$$k_{Hstroma} = 1.15 \times \left(\frac{H_w}{3.4}\right)^4 \times 10^{-15} \text{ m}^2/\text{Pa/s}. \tag{19.4.18}$$

A formula of the same flavor had been reported by Hodson [1997],

$$k_{Hstroma} = \frac{1.197\,(H_w/3.4)^3}{H_w/3.4 + 0.197} \times 3.74 \times 10^{-15} \text{ m}^2/\text{Pa/s}. \tag{19.4.19}$$

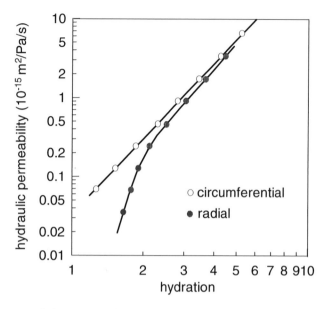

Fig. 19.4.5: Radial and circumferential stromal permeabilities as a function of hydration, redrawn from Hedbys and Mishima [1962]. Experimental data in 0.9% NaCl have been obtained without Descemet membrane, but tests in 0.9% NaCl with Descemet membrane and tests without Descemet membrane in distilled water corresponding to higher hydrations fit with the curves.

This permeability is roughly the same order of magnitude as that of articular cartilage which ranges from 0.1 to $7 \times 10^{-16}\,\mathrm{m^2/Pa/s}$.

Overby et al. [2001] use modified transmission electron micrographs (TEM) to relate the tissue ultrastructure and content to the hydraulic conductivity of the corneal stroma. They find values in the range 0.5 to $10 \times 10^{-10}\,\mathrm{m^2}$, which yields a hydraulic permeability of 5 to $100 \times 10^{-15}\,\mathrm{m^2/Pa/s}$. The modified method, namely quick-freeze/deep etch, preserves better the tissue ultrastructure and PGs than traditional TEM.

Table 19.4.7: Radial permeability [unit : $\mathrm{m^2/Pa/s}$] under physiological conditions, Edwards and Prausnitz [1998].

Stroma	Intact Cornea	Articular Cartilage
4.9×10^{-15}	2.2×10^{-15}	$0.1 - 7.0 \times 10^{-15}$

Edwards and Prausnitz [1998] develop a multi-step scheme to define the stromal permeability:

- they consider first the ground substance composed of two components, $1 \to$ GAGs, $2 \to$ core proteins. Let k_{gs} be the permeability of the ground substance. The fluid velocity $\mathbf{v}$ due to a pressure gradient $\boldsymbol{\nabla} p$ is viewed as contributed by pressure gradients in series through these two components, with permeability k_i and volume fraction n_i, $i = 1, 2$, $(n_1 + n_2 = 1)$, namely $n_i\,\mathbf{v}/k_i = -\boldsymbol{\nabla} p_i$, so that, $n_1\,\mathbf{v}/k_1 + n_2\,\mathbf{v}/k_2 = -\boldsymbol{\nabla} p = \mathbf{v}/k_{gs}$. Thus

$$\frac{1}{k_{gs}} = \frac{n_1}{k_1} + \frac{n_2}{k_2}\,. \tag{19.4.20}$$

The individual permeabilities k_i, $i = 1, 2$, are calculated as if the two components were randomly oriented cylinders, Jackson and James [1986];

- next they consider the hindrance to flow due to the presence of collagen fibrils of volume fraction n^c. The analytical expressions of the hindrance due to an individual sphere or cylinder are worked out in Exercise 15.1. In fact, for any lamella, the fibrils are parallel to the corneal surface so that the radial flow is orthogonal to the fibril axes.

The model predicts permeabilities that increase with hydration, and are in the range of the data from Hedbys and Mishima [1962] reported in Fig. 19.4.5 and from Fatt and Hedbys [1970]a [1970]b.

The relations between the composition and ultrastructure of fibrous tissues and their transport properties are also considered in Levick [1987]. The anisotropic permeability of annulus fibrosus, another lamellar tissue, is considered by Gu et al. [1999] and Sun and Leong [2004].

● **Ionic diffusion and tortuosity factor of stroma**

The ionic mobility u_k of an ion k is linked to the diffusion coefficient D_k by the Nernst-Einstein relation : $u_k = D_k\,|\zeta_k|\mathrm{F}/RT$. The ranges of values in a blank solution are listed in Table 16.4.2 and a few specific values are shown in Table 13.3.1. The coefficients of free diffusion of the sodium cation, chloride anion and undissociated sodium chloride salt are respectively equal to 13.3, 20.3 and $16 \times 10^{-10} \, \text{m}^2/\text{s}$.

The *effective* diffusion coefficient accounts for several aspects that imply diffusion in the porous medium to be slower than diffusion in a free infinitely dilute solution. For any ion, the coefficients of intrinsic diffusion D and of effective diffusion $D^* = \tau D$ are linked by the tortuosity factor $\tau < 1$. A similar relation applies to *effective* ionic mobilities, namely $u^* = \tau u$. Further details are exposed in Sections 16.4.2 and 16.4.3. Maurice [1961] provides an effective diffusion coefficient D^*_{NaCl} of sodium chloride equal to $9 \times 10^{-10} \, \text{m}^2/\text{s}$ for rabbit stromas. This value corresponds to a tortuosity factor of 0.56.

- **Diffusion of salts and drugs**

Edwards and Prausnitz [1998] provide theoretical estimations and experimental measurements of diffusion co-efficients of a number of lipophilic and hydrophilic drug compounds, for the stroma and for the endothelium. The coefficients for the whole cornea are obtained by considering that flows in stroma and endothelium are in series. They indicate that the rate limiting process to drug delivery is the diffusion through the epithelium, not the stroma, unless the latter has been damaged.

For sclera, Ethier et al. [2004] note that, even for macromolecules like albumin, the diffusion flux is at least an order of magnitude larger than the convective flux.

- **Electrical properties of stroma**

According to Hodson [1974], the coefficient of streaming potential, namely k_e/σ_e equal to the ratio of the difference of electrical potential induced by a difference of fluid pressure in an open circuit, is equal to $2.5 \times 10^{-9} \, \text{V/Pa}$.

- **Characteristic times**

The characteristic time for seepage is proportional to the square of the drainage length L and inversely proportional to the representative elastic modulus E and permeability k_H,

$$t_D = \frac{L^2}{E \, k_H} . \tag{19.4.21}$$

For $L = 0.40 \, \text{mm}$, $E = 10 \, \text{kPa}$, $k_{\text{Hstroma}} = 2 \times 10^{-15} \, \text{m}^2/\text{Pa/s}$, the characteristic time is $t_D = 8000 \, \text{s} \sim 2 \, \text{h}$. The modulus here is representative of the radial elastic stiffness under extension: it is not documented but it is certainly small since interlamellar crosslinks are essentially absent.

The characteristic time for ionic diffusion is similarly proportional to the square of the diffusion length L and inversely proportional to the effective diffusion coefficient D^* [unit : m^2/s],

$$t_F = \frac{L^2}{D^*} . \tag{19.4.22}$$

For $L = 0.40 \, \text{mm}$, $D^* = D^*_{\text{NaCl}} = 10.0 \times 10^{-10} \, \text{m}^2/\text{s}$, the characteristic time is $t_F = 160 \, \text{s}$.

Thus seepage is much slower than ionic diffusion: the phenomenon is clearly observed in the experiments reported by Ruberti and Klyce [2003]. The seepage time above should be compared with the time course of the thickness measured by Doughty [2001], in his swelling experiments and reported in Fig. 19.2.1. Still, note that the thickness increases considerably in these experiments and reaches a final value much higher than the initial thickness of the stroma.

19.4.6 Acquisition of mechanical data

Given the heterogeneity of the cornea, testing the whole organ, including part of the sclera, is considered to be representative of its physiological role. A finer analysis requires information on the mechanical and transport roles of the various parts of the cornea. The protocol that controls the biochemical conditions of the tests should be recorded. In particular, tests on the *intact* cornea and on the *de-epithelialized* or *de-endothelialized* cornea are expected to be significantly distinct since absence of epithelium or endothelium allows for unrestricted imbibition to take place.

19.4.6.1 Tests on the whole organ

Testing the whole intact cornea has significance because testing of stromal pieces raises two main issues:

- cutting a cornea into pieces destroys a number of links, at different scales of the structure. In other words, the issue of obtaining a representative sample of the stroma is not trivial, given the heterogeneities;

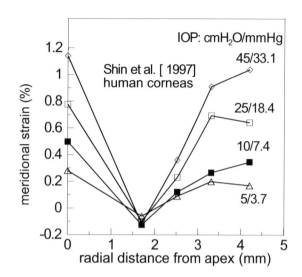

Fig. 19.4.6: Meridional strains during inflation tests on human corneas, redrawn from Shin et al. [1997]. Prior to testing, the corneas are preconditioned by cycling the IOP between 0 and 45 cmH$_2$O slowly (1 cycle per min).

Strains are obtained by tracking adherent particles inserted in the cornea. They display a dip at some distance from the apex.

- laboratory testing should account for the fact that the actual organ is subjected to a *biaxial* loading.

Experimental and finite element simulations of the whole cornea need to assume identical boundary conditions. A portion of the sclera, which is stiffer than the cornea and introduces only an imperfect clamping, is usually included in the mesh. Matching the actual clamping is an issue in itself.

Shin et al. [1997] performed inflation tests, subjecting a whole cornea to a set of IOP's in the physiological range, namely [0-45] cm H$_2$O. The strain range is small, with a maximum of 1.14% at the apex under the largest IOP, Fig. 19.4.6. On the other hand, the strain distribution is highly heterogeneous. It presents a minimum at the third of the distance from apex to limbus. Shell models, whether thin or thick, cannot represent this feature. Simulations of these heterogeneities require a refined finite element mesh that accounts as accurately as possible of the initial geometry and thickness of the cornea as well as of the directional and spatial distributions of the collagen fibers across the depth and over the surface.

High IOP, associated with glaucoma, represents a potential damage to the optic nerve. At the interface with the eye cavity, the optical nerve is covered by a collagen membrane, called lamina cribosa, which is thought to be compliant. Therefore, under high IOP, the lamina cribosa undergoes tension due to the overall expansion of the eye cavity, it thickens, compressing the optical nerve and reducing blood flow, both phenomena leading in the long range to nerve damage and blindness.

Applanation tonometry is considered a good clinical test to estimate the IOP. The test consists in applying a plane disk of 3.06 mm in diameter onto the cornea: the pressure required to obtain contact over the complete area is considered to provide an estimate of the IOP. The test is also used in the laboratory to obtain information on the mechanical properties of the cornea: Orssengo and Pye [1999] develop a shell model while Elsheikh et al. [2006] perform a three-dimensional analysis using the finite element method.

Regional heterogeneities of the mechanical properties of the cornea have been measured by inserting mercury droplet markers, Hjortdal [1996]. Young's modulus is found larger at the center of the cornea, and the highest circumferential modulus is found at the limbus. The meridional directions are reinforced in the para-central region, in agreement with the dip of the meridional strains in Fig. 19.4.6.

19.4.6.2 Strip extensiometry

Uniaxial traction tests reported in Figs. 19.4.1 to 19.4.4 use excised strips of cornea. The position of a strip in the cornea and its direction are information of interest. Preconditioning through cycles eliminates the hysteretic response. Replication of the physiological biaxial stress state is an issue. Sufficiently low elongation rates are required to ensure an elastic response.

Reichel et al. [1989] measure the elastic modulus of the central and peripheral parts of bovine corneas using strip extensiometry. At variance with Hjortdal [1996], they indicate that the elastic modulus is larger at the periphery. However, their analysis probably relies on an isotropic material while Hjortdal [1996] accounts for an orthotropic cornea.

19.4.6.3 Holographic interferometry

In contrast to strip extensiometry, optical interferometric techniques are non destructive, non contact and able to simultaneously record data over a whole surface.

Using holographic interferometry, Smolek [1994] measured the spatial variations of the elastic modulus. A curiously high Young's modulus of 1 GPa was obtained in the central part of human corneas subjected to a physiological IOP while the elasticity moduli across the posterior scleral hemisphere of bovine eyes ranged from 3 to 9 MPa.

Vastly different values are provided by Knox Cartwright et al. [2011]. Using radial-shearing, speckle-pattern interferometry (RSSPI), they measured the change in strain due to a small variation of the physiological IOP. They deduce that the Young's modulus of the central part of the human cornea increases from 0.27 MPa at age 20 years to 0.52 MPa at age 100 years, an age-related stiffening that they attribute to a type of naturally occurring crosslinkage different from the Riboflavin-UVA crosslinkage.

19.4.6.4 Ultrasonic techniques

Liu et al. [2007] use ultrasounds[19.2] to reconstruct, from the reflections, density, thickness and the longitudinal elastic moduli. Their method has been checked on soft contact lenses and it should be improved to be applied to fluid-saturated porous media like cornea. H. Wang et al. [1996] claim that they were able to measure the shear wave-speed and attenuation coefficient of corneal samples prepared in normal saline solutions and dextran.

19.5 Scattering of light by the corneal stroma and transmittance

The contrast of refractive indices between collagen and ground substance alters the transparency of the cornea. Before analyzing the elements that enter the phenomenon, let us recall some typical lengths of interest for human cornea: diameter of fibrils 60 nm, interfibrillar distance 60 nm, light wave-length 500 nm.

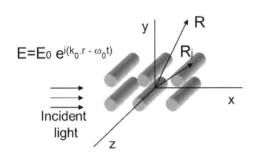

E = E₀ eⁱ⁽ᵏ₀·ʳ ⁻ ω₀ᵗ⁾

Fig. 19.5.1: Geometrical elements used in the analysis of light scattering.

Therefore fibrils appear as discrete objects at this range of wave-lengths and they *scatter* light, Fig. 19.5.1. At variance, their size is much larger than the typical wave-length of x-rays, about 1 Å, and they *diffract* x-rays. The exposition below follows Hart and Farrell [1969], Benedek [1971] and Freund et al. [1986].

A monochromatic plane-polarized incident wave,

$$\mathbf{E}^{(i)} = \mathbf{E}_0^{(i)} \exp\left(i\left(\mathbf{k}_0 \cdot \mathbf{r} - \omega_0 t\right)\right), \qquad (19.5.1)$$

impinges on the cornea, which is composed of planar lamellae, each of which contains parallel collagen fibrils embedded in a ground substance. The wave vector $\mathbf{k}_0$ has amplitude (*wave-number*) $k_0 = |\mathbf{k}_0| = 2\pi/(\lambda/n_g)$ with λ the wave-length of the incident wave in a vacuum and n_g the *refraction index* of the ground substance, that is, the wave-length in the ground substance is equal to λ/n_g. The angular frequency (pulsation) of the incident wave is equal to ω_0.

Upon impinging the cornea, the electric field induces at each point a dipole moment whose amplitude is proportional to the square of the refraction index at the point in question. In turn, these dipoles re-radiate an electric field with a *transmitted* component and a *scattered* component, Fig. 19.5.2. This scattered component decreases the incident energy and implies that the tissue is not perfectly transparent. A standard simplification consists in assuming that each fibril is submitted to the incident field only, and in neglecting the re-radiated contributions by the other fibrils, that is, the interactions between fibrils. In fact, scattering takes place because the collagen fibrils and ground substance have not identical refraction indices, and the scattered field is proportional to the difference of their squares. On

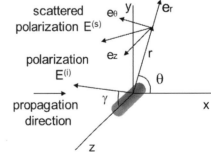

Fig. 19.5.2: Incident and scattered fields.

19.2 Normal human hearing perceives sounds in the *acoustic* range from 20 Hz to 20 kHz. The lower and upper frequency ranges are referred to as *infrasound* and *ultrasound*, respectively.

the other hand, the radiated field keeps the same wave-length and frequency as the incident field, and the scattered wave vector $\mathbf{k}$ has the same amplitude as the incident wave vector $\mathbf{k}_0$, namely $k = k_0$. Now let us denote the position of the collagen fibril j by the vector $\mathbf{R}_j$. The scattered field at a point $\mathbf{r}$ is due to the superposition of rays scattered at a reference point, say the origin of coordinates, and point j. If the point $\mathbf{r}$ is far from the origin, the difference of optical lengths between the two rays is equal to $(\mathbf{k}_0 - \mathbf{k}) \cdot \mathbf{R}_j$. The scattered field at point $\mathbf{r}$,

$$\mathbf{E}^{(s)} = \mathbf{E}_0^{(s)} \exp\left(i\left(\mathbf{k}_0 \cdot \mathbf{r} - \omega_0 t\right)\right) \exp\left(i\,\mathbf{K} \cdot \mathbf{R}_j\right), \qquad (19.5.2)$$

expresses in terms of the *scattering vector* $\mathbf{K} = \mathbf{k}_0 - \mathbf{k}$ of amplitude $K = 2\,k_0 \sin(\theta/2) = 4\,\pi/(\lambda/n_g)\,\sin(\theta/2)$, θ being the *scattering angle* between $\mathbf{k}_0$ and $\mathbf{k}$ as sketched in Fig. 19.5.3.

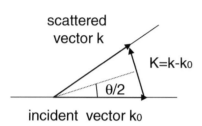

scattered vector k

K=k-k₀

θ/2

incident vector k₀

Fig. 19.5.3: Incident and scattered wave vectors $\mathbf{k}_0$ and $\mathbf{k}$, scattering vector $\mathbf{K}$.

If the fibril was at the origin, the scattered polarization vector would be $\mathbf{E}_0^{(s)}$. Since all fibrils of a lamella are identical, this quantity is the same for all fibrils. The field at point $\mathbf{r}$ scattered by N fibrils results by superposition,

$$\mathbf{E}^{(s)} = \mathbf{E}_0^{(s)} \exp\left(i\left(\mathbf{k}_0 \cdot \mathbf{r} - \omega_0 t\right)\right) J(\mathbf{K}), \quad J(\mathbf{K}) \equiv \sum_{j=1,N} \exp\left(i\,\mathbf{K} \cdot \mathbf{R}_j\right).$$
$$(19.5.3)$$

If all centers of the fibrils lie over an area A, we may rewrite the function $J(\mathbf{K})$ via the Dirac delta function as,

$$J(\mathbf{K}) = \int_A \exp\left(i\,\mathbf{K} \cdot \mathbf{R}\right) \sum_{j=1,N} \delta(\mathbf{R} - \mathbf{R}_j)\, dA(\mathbf{R}). \qquad (19.5.4)$$

The sum of the Dirac delta functions around a point $\mathbf{R}$ may be viewed as a number density $\rho(\mathbf{R})$, i.e. number of fibrils per unit area.

This expression may also be exploited in another perspective, namely when we consider the circular cross-section of a single fibril of radius r_f. Indeed, then $\rho(R) = 1/A$, with $A = \pi\,r_f^2$, is viewed as the number of fibrils per unit area. With $R = |\mathbf{R}|$, $K = |\mathbf{K}|$, the integral evaluates explicitly,

$$\int_A e^{i\,\mathbf{K} \cdot \mathbf{R}}\, dA(\mathbf{R}) = \int_0^{r_f} R\, dR \underbrace{\int_0^{2\pi} e^{i\,KR\cos\phi}\, d\phi}_{2\pi\, J_0(KR),\ \text{Eqn (7.5.67)}} = \frac{2\pi}{K^2} \underbrace{\int_0^{Kr_f} x\, J_0(x)\, dx}_{Kr_f\, J_1(Kr_f),\ \text{Eqn (7.5.74)}_1} = \frac{2\pi r_f}{K} J_1(Kr_f), \qquad (19.5.5)$$

in terms of J_0 and J_1 the Bessel functions of the first kind and orders 0 and 1, respectively, and

$$J(\mathbf{K}) = 2\,\frac{J_1(K\,r_f)}{K\,r_f}; \quad J(\mathbf{K}) \simeq \frac{1}{\pi^{\frac{1}{2}}} \left(\frac{2}{K\,r_f}\right)^{\frac{3}{2}} \cos\left(K\,r_f - \frac{3\pi}{4}\right) \quad \text{for } K\,r_f \gg 1. \qquad (19.5.6)$$

The function J is equal to 1 at $Kr_f = 0$ and dies out quickly as the radius of the fibril increases, being very small for $Kr_f \geq 3$.

The intensity of the scattered light per fibril is proportional to the square of the scattered electrical field, namely

$$\frac{1}{N}\,\frac{|\mathbf{E}^{(s)}|^2}{|\mathbf{E}^{(i)}|^2} = \Sigma(\mathbf{K})\, I(\mathbf{K}). \qquad (19.5.7)$$

It may be viewed as the product of the *interference function* $I = I(\mathbf{K})$ and of the function $\Sigma(\mathbf{K})$,

$$I(\mathbf{K}) = 1 + \frac{1}{N} \sum_{j \neq k = 1,N} \exp\left(i\,\mathbf{K} \cdot (\mathbf{R}_j - \mathbf{R}_k)\right), \quad \Sigma(\mathbf{K}) = \frac{|\mathbf{E}_0^{(s)}|^2}{|\mathbf{E}_0^{(i)}|^2}. \qquad (19.5.8)$$

When light impinges upon a single fibril, endowed with standard cylindrical coordinates (r, θ, z) and axes $\mathbf{e}_r$, $\mathbf{e}_\theta$, and $\mathbf{e}_z$,

$$\mathbf{E}_0^{(i)} = E_{0i}\left(\sin\gamma\,\sin\theta\,\mathbf{e}_r + \sin\gamma\,\cos\theta\,\mathbf{e}_\theta + \cos\gamma\,\mathbf{e}_z\right), \qquad (19.5.9)$$

the scattered polarization vector,

$$\mathbf{E}_0^{(s)} = \frac{E_{0i}}{4}\left(\frac{2\pi\,r_f}{\lambda/n_g}\right)^2 \left(\frac{\lambda/n_g}{r_\infty}\right)^{\frac{1}{2}} (m^2 - 1)\left(\sin\gamma\,\cos\theta\,\frac{2}{m^2 + 1}\,\mathbf{e}_\theta + \cos\gamma\,\mathbf{e}_z\right), \qquad (19.5.10)$$

has no radial component. It is axial if the incident polarization vector is so (polarizing angle $\gamma = 0$), and depends on the square of the fibril radius. As expected, no scattering takes place if the refraction index of the tissue is spatially uniform, that is, if the ratio $m = n_c/n_g$ of the refraction indices of collagen and ground substance is equal to 1. This expression derived by Hart and Farrell [1969] has been simplified by Benedek [1971], assuming that the wave-length λ and the distance r_∞ between the origin of coordinates and the field point $\mathbf{r} = \mathbf{r}_\infty$ are much larger than the radius of the fibril r_f.

The *transmittance* F_T, i.e., fraction of light transmitted through the tissue, is defined as

$$F_T = \exp(-\sigma_t \, \rho \, \Delta), \tag{19.5.11}$$

σ_t [unit : m] being the total scattering cross-section per fibril per unit length, ρ [unit : m^{-2}] the number of fibrils per unit area, and Δ [unit : m] the thickness of the tissue. The total scattering cross-section per fibril per unit length is obtained by integrating (19.5.7) along the circle of radius r_∞,

$$\sigma_t = \int_0^{2\pi} \sigma(\theta) \, r_\infty \, d\theta, \quad \sigma(\theta) = \frac{1}{2\pi} \int_0^{2\pi} \Sigma(\mathbf{K}(\theta,\gamma)) \, I(\mathbf{K}(\theta,\gamma)) \, d\gamma. \tag{19.5.12}$$

If the positions of the fibrils are random, then $I(\mathbf{K}) = 1$. For unpolarized incident light normal to the fibril axis, $\sigma(\theta)$ is obtained by averaging with respect to the angle γ,

$$
\begin{aligned}
\sigma_t &= \frac{\pi}{8} \, r_f^4 \left(\frac{2\pi}{\lambda}\right)^3 n_g^3 \, (m^2-1)^2 \int_0^{2\pi} \frac{1}{2\pi} \int_0^{2\pi} \left(\frac{(2\sin\gamma\cos\theta)^2}{(m^2+1)^2} + (\cos\gamma)^2\right) d\gamma \, d\theta \\
&= \frac{\pi^2}{8} \, r_f^4 \left(\frac{2\pi}{\lambda}\right)^3 n_g^3 \, (m^2-1)^2 \left(\frac{2}{(m^2+1)^2} + 1\right).
\end{aligned}
\tag{19.5.13}
$$

Note the strong dependence of σ_t with respect to the fibril radius and wave-length: smaller fibril radii and higher wave-lengths are associated with significantly higher transmittances. For $\lambda \simeq 500\,\mathrm{nm}$, $\Delta \simeq 450\,\mu\mathrm{m}$, $r_f \simeq 15\,\mathrm{nm}$, $\rho \simeq 300\,\mathrm{fibrils}/\mu\mathrm{m}^2$, $n_c = 1.47$, $n_g = 1.345$, data characteristic of rabbit cornea, a very low transmittance around 0.08 results. Accounting for the actual lattice-like spatial distribution of the fibers, Hart and Farrell [1969] obtain more realistic results: the transmittance increases sharply at the lower bound of the visible range to reach 0.80 at 400 nm and continues increasing to 0.90 at the upper bound of the visible range.

This improved analysis of Hart and Farrell [1969] is now described shortly. The first step is to replace the discrete summation over individual fibrils by an integration over the area S of the cornea. This integration displays a two center distribution function $\rho_{II}(\mathbf{R}_j, \mathbf{R}_k)$, which, when it is random, is equal to the product of the number densities $\rho(\mathbf{R}_j)\,\rho(\mathbf{R}_k)$. The key simplification consists in assuming the two center distribution function to be product of the square of the number density $\rho(\mathbf{R})$ (actually taken constant and equal to N/S) and of a function g which depends only on the distance R_{jk} between the centers (and which is equal to 1 when the two center distribution function is random). The radial distribution $g(R)$ is equal to the ratio of the number of fibrils at distance R from a given fibril over the number density. It vanishes for R smaller than the smallest fibril spacing and is equal to 1 for large R. As a further manipulation, the diffracted contribution is isolated. In summary,

$$
\begin{aligned}
I(\mathbf{K}) &= 1 + \frac{1}{N} \int_S \exp\left(i\,\mathbf{K}\cdot(\mathbf{R}_j - \mathbf{R}_k)\right) \rho_{II}(\mathbf{R}_j, \mathbf{R}_k) \, dS_j \, dS_{jk} \\
&= 1 + \frac{1}{N} \int_S \exp\left(i\,K\,R_{jk}\cos\psi_{jk}\right) \rho^2 \, g(R_{jk}) \, dS_j \, dS_{jk} \\
&\simeq 1 + 2\pi\rho \int_0^{R_c} R_{jk} \, J_0(K R_{jk}) \, (g(R_{jk}) - 1) \, dR_{jk}.
\end{aligned}
\tag{19.5.14}
$$

The last line has used the following elements: 1. the definitions of the elementary areas $dS_j = R_j \, dR_j \, d\psi_j$, $dS_{jk} = R_{jk} \, dR_{jk} \, d\psi_{jk}$; 2. the definition of the number density $\rho = N/S$; 3. the integral representation of the Bessel function J_0 as in (19.5.5); 4. the asymptotic property (19.5.6) of the Bessel function J_1. In fact, the radius of the upper bound should be large with respect to the fibrillar spacing so that $g(R_c) = 1$. With a given radial distribution $g(R)$, the integrals (19.5.14) with respect to the radius R and with respect to directions (19.5.12) may be performed remembering the dependence $K = |\mathbf{K}| = 2\,k_0 \sin(\theta/2)$.

Loosely speaking, this analysis shows that the actual ultrastructure is designed to optimize transparency: interestingly, a satisfactory transmittance requires geometric characteristics (fibril radius, fibril spacing, spatial organization of the fibrillar network) and optical contrast between collagen and ground substance to fall in strongly restricted ranges. In that respect, the influence of swelling, pathologies and surgical aggressions are worth investigating. Since the above geometrical characteristics vary along a corneal radius, transmittance is also expected to decrease from the central optical axis to limbus.

The analysis has addressed a single lamella since all fibrils have been implicitly assumed to be parallel. A more comprehensive study would account for the angular mismatch between lamellae.

19.6   Corneal surgery

Radial keratectomy (RK) consists in performing radial incisions in the cornea at some distance from the central zone. These incisions relieve the circumferential tension and result in flattening this central part which is crucial for the vision. The improvement depends on the number of incisions and their depth. This method is strongly aggressive and less used nowadays.

Phototherapeutic keratectomy (PRK) excises the corneal epithelium and ablates the stroma over a diameter of about 6 mm to a maximum depth of 160-180 µm. A smooth ablation is required so as to ensure an optimal optical correction. Checking the roughness of the ablated surface helps to qualify and compare ablation tools and techniques. Surface scanning through AFM (atomic force microscopy) consists in moving a probe on which is applied a small load (a few nN) and whose position is recorded through a laser, e.g., Lombardo et al. [2006]. An alternative technique, the scanning electron microscopy (SEM), can scan over a larger area (mm^2), but AFM can operate in air and liquid, Alonso and Goldman [2003].

The depth-dependent structure of cornea may have implications on refractive surgeries, Section 19.1.2.4. Removal of the anterior zone by photorefractive keratectomy seems therefore questionable. As for LASIK (laser in-situ keratomileusis), it leaves the anterior zone intact and removes a deeper zone. A circular flap is cut through a microkeratome, folded back as a hinge to give access to inner layers, and repositioned at the end of the procedure. The flap of about 150 µm includes epithelium (50 µm), Bowman's membrane (10 µm) and almost the whole anterior part with interwoven structure. Ablation thus concentrates on lamellae that run from limbus to limbus.

Note that, from a mechanical point of view, PRK and LASIK are essentially equivalent since the structural contribution of the anterior layers to the overall structural stiffness and stability of the cornea is small.

Partial or penetrating (total) corneal transplantation (keratoplasty) consists in removing the opaque or damaged part of cornea and replacing it by a donor clear tissue. The procedure applies to treat pseudophakic bullous keratopathy, keratoconus, and other degenerative inflammatory or non-inflammatory disorders. For partial keratoplasty, artificial patches obtained by growing keratocytes on amniotic membranes are increasingly used. Besides bio-compatibility issues, the surgery should take care to avoid to submit the tissue to excessive compression or tension when suturing the implant to the sclera by a too tight stitch leading to so-called cheesewiring.

Interestingly enough, Bourne [2003] notes that diseases that affect the endothelium may be traced to both genetics and traumas. For example, Fuchs dystrophy, in which endothelial cells secrete abnormally high amounts of Descemet membrane, is linked to a genetic mutation. Contact lens wear modifies the shape and size of endothelial cells, presumably due to hypoxia. Surgical procedures are also observed to increase the rate of cell density decrease.

19.7   Cornea engineering

The need of corneas for transplantation has increased, not only because the number of transplantations increases, but also because the corneas which have undergone surgical procedures, e.g., refractive surgery, are deemed unsuitable for transplantations. In view of the gap between donor corneas and the demand, engineering of artificial corneas is becoming an actual alternative. Typically in western countries, every year, one person out of ten thousand is in demand of corneal transplant. In Europe, the available corneas amount to less than 30,000 while the demand corresponds to about 40,000 corneas. In the United States, the number of transplants is larger than 30 000 since the year 2000. Worldwide, the number of people suffering from corneal blindness is larger than two millions. The need for transplants stems from hereditary diseases, like keratoconus, diabetic keratopathy, ulcers resulting from severe degradation of collagen associated with excessive activity of specific enzymes (matrix metalloproteinases MMP), accidental lesions, burns and failure of previous surgeries. Trachoma, an infection caused by a bacterium, is the main source of blindness in developing countries.

Naively, corneal stroma seems a likely candidate to be engineered because it is avascular. However, the spatial organization of the components of the extracellular matrix is extremely sophisticated and certainly tuned to the various physiological functions of the cornea, which include ensuring rays to enter the eyeball, refracting the rays to the retina, and protecting the optical apparatus.

During development, the epithelium forms first, the endothelium next and the space is progressively filled by the stroma. De-epithelialized transplanted stromas recover progressively their epithelium through cells that migrate from the limbus. No such cellular re-population takes place on the endothelial side. Cultures of endothelial cells and stromal cells (keratocytes/fibroblasts) are far more challenging than epithelial growth. In fact, out of the three main layers that constitute the cornea, the stroma is the one that requires the greatest efforts.

The transparency of native corneas is believed to be related to the architecture and composition of the stroma. The density of collagen fibers intermingled with proteoglycans is typical both over the corneal surface and through the depth of the stroma. Moreover, at any point, their directional distribution is strongly regulated as well.

Engineering of tissues may follow two opposite routes. On one hand, one may try to mimic nature as closely as possible, under the motivation that it has presumably optimized the use of matter. However, such a goal may be out of reach, because it may require to reproduce an excessively large numbers of tricky tools at a number of scales. A challenge worth the detective sagacity of heirs of the Princes of Serendip! As an alternative, the issue might be considered through a more heuristic approach.

Bio-compatibility, mechanical strength, permeability to gas and liquid, and chiefly transparency are the key properties that need to be optimized for *in vitro* cultures that typically last 3 to 4 weeks. In fact, current cultures so far invariably show a decrease in these properties over longer periods.

Cultures of stromal fibroblasts (keratocytes) in collagen gels, or collagen sponges, are in their infancy. Three-dimensional scaffolds are now used instead of monolayer cultures. High scaffold stiffnesses have been observed to affect favorably the mechanical properties and phenotype of the seeded cells. Mechanical compression of the cultures is thought to imply traction onto the cells deposited on the scaffolds, which in turn is thought to trigger the cell activity and production of extracellular components, mainly collagen and PGs. Still, in actual cultures, the moduli of these biodegradable scaffolds (compressive, tensile) are much lower than those of native corneas.

Among issues that are to be addressed for *in vitro* cultures and analyzed are the geometry and chemical composition of the scaffold, the density of the cells that are seeded on the scaffolds, their types, e.g., keratocytes or retinal pigment epithelial cells, and the persistence of their phenotype. Deceptively, the expression of α-smooth muscle is observed to increase with cell density.

The scaffolds in themselves represent an important technological issue. Their pore size and specific surface should be easily controlled, and they should either degrade into non toxic components in aqueous solutions, or become incorporated in the substitutes. Of course, cells should be able to adhere to their surface. Potential scaffolds used to support ocular cells before transplantation include human amniotic membranes, anterior lens capsule, fibrin gels, polyester and collagen films, silk fibroin and porous silicon membranes.

Along a step by step approach, various tools are tested during cultures to attempt to mimic the layered *in vivo* distribution of collagen fibrils, among which we may list fluid flow, gravity, magnetic alignment accounting for the small diamagnetic susceptibility of collagen molecules, embedding magnetically susceptible beads in the collagen gel, or micro- and nano-patterned polyester and crosslinked collagen films. A technique to obtain elongated cells would also be valuable, as fibroblasts are observed to produce collagen along their main axis. Mechanical loading of cultures, and cells, is expected to have positive effects: cells are known to deform and migrate along the directions of the major principal stress or strain. Ruberti and Zieske [2008] provide a few references where the influence of mechanics is highlighted. Cell growth should ideally be able to compensate for the degradation of the scaffolds. The mechanical properties of corneal stromal equivalents are considered in Borene et al. [2004].

Collagen crosslinking is necessary to ensure stability of the substitutes. Insertion of proteoglycans within collagen type I sponges is observed to increase light transmission, Orwin et al. [2003]. It is important to recall that the number and activity of keratocytes are maximum during development. In adults, the cells are quiescent, ensuring homeostasis via production of extracellular matrix (proteoglycans essentially). While collagen turnover is slow in normal tissues, it is considerably accelerated by the production of collagenases and MMPs that are elicited by injuries: the trouble is that the deposited collagen is opaque.

Finally, assuming that *in vitro* cultures can provide viable corneal substitutes, a stable connection of the transplanted corneas or patches to the existing adjacent tissue is yet another issue to be solved satisfactorily. Some current synthetic corneal devices consist of a central clear part attached to a peripheral skirt through a polymer network.

19.8   The constitutive framework

Henceforth, vector and tensor quantities are identified by boldface letters. Symbols "·" and ":" between tensors of various orders denote their inner product with single and double contraction respectively. Unless stated otherwise, the convention of summation over repeated indices does *not* apply.

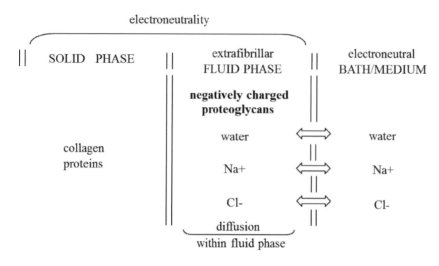

Fig. 19.8.1: The stroma is partitioned in two phases. Each geometrical point can be seen in chemical equilibrium with a fictitious bath in which the chemical and mechanical conditions are controlled.

19.8.1   Definition of the phases

The definition of the phases is mechanically motivated. A kinematical criterion on the other hand would classify components according to their velocities. Here, cornea is viewed as a two-phase, multi-component, porous medium, Fig. 19.8.1. The *solid phase* S contains the collagen fibrils denoted by the symbol c. The *(extrafibrillar) fluid phase* E contains both insoluble components, the proteoglycans (PGs), and a binary symmetric electrolyte. The latter is an aqueous solution with dissolved sodium cations Na^+ and chloride anions Cl^-. The sets of components of the solid and fluid phases are recorded below,

$$S = \{c\}, \quad E = \{w, PG, Na, Cl\}. \tag{19.8.1}$$

The main assumptions, which underly the two-phase multi-component model, follow a *strongly interacting* model, namely,

(H1) Mass balance is required for each component.

(H2) Momentum balance is required for the mixture as a whole. Water and mobile ions in the fluid phase are endowed with their own velocities so as to allow water to diffuse through the solid skeleton and ions to diffuse in their phase and satisfy their own balance of momentum.

(H3) The proteoglycans move with the velocity $\mathbf{v}_S$ of the solid phase. Thus, their balance of momentum is not required explicitly but accounted for by the balance of momentum of the mixture as a whole.

(H4) Electroneutrality holds pointwise. At physiological pH, collagen molecules can be considered electrically neutral, Li and Katz [1976], while PGs are negatively charged. Binding of chloride ions on ligands alters the *fixed* charge. Change of pH affects further the charges of collagen and PGs.

The electrical charge carried by ions is referred to as *mobile*. Cations sodium do not *bind* to PGs.

The fixed electrical charge requires a minimum concentration of ions to be present in the stromal fluid to ensure pointwise electroneutrality.

(H5) All constituents are incompressible.

19.8.2   Geometry, volumes and masses

Since we are considering a multi-phase multi-component mixture, a number of quantities that characterize the properties of components will be referred to by two indices, generically k for the component, and K for the phase. Various measures of mass and volume are used to formulate the constitutive equations. The current volume (resp. mass) of the component k of phase K is denoted by V_{kK} (resp. M_{kK}). Let the initial volume of the porous medium be V_0 and let $V = V(t)$ be its current volume. Various entities are attached to components:

- some are intrinsic like the intrinsic density ρ_k, the molar volume $\widehat{v}_k$ and molar mass $\widehat{m}_k$ linked by $\widehat{m}_k = \rho_k \widehat{v}_k$;
- some refer to the current volume, like the *volume fraction* $n^{k\mathrm{K}} = V_{k\mathrm{K}}/V$ and the *apparent density* $\rho^{k\mathrm{K}} = n^{k\mathrm{K}} \rho_k$;
- some refer to the initial volume like the *mole content* $\mathcal{N}_{k\mathrm{K}} = N_{k\mathrm{K}}/V_0$, the *volume content* $v^{k\mathrm{K}} = V_{k\mathrm{K}}/V_0 = n^{k\mathrm{K}} V/V_0$ and the *mass content* $m^{k\mathrm{K}} = M_{k\mathrm{K}}/V_0 = \rho_k v^{k\mathrm{K}}$.

The associated entities for the phase K are defined by algebraic summation of individual contributions, e.g., the current volume V_K and mass M_K, the volume fraction $n^\mathrm{K} = V_\mathrm{K}/V$ (volume fractions satisfy the closure relation $\sum_\mathrm{K} n^\mathrm{K} = 1$), the *apparent* density $\rho^\mathrm{K} = M_\mathrm{K}/V$, the volume content $v^\mathrm{K} = V_\mathrm{K}/V_0 = n^\mathrm{K} V/V_0$, and the mass content $m^\mathrm{K} = M_\mathrm{K}/V_0$.

Other entities live in their phase, e.g., the molar fractions and the concentrations. The *molar fraction* $x_{k\mathrm{K}} = N_{k\mathrm{K}}/N_\mathrm{K}$ of the component k in phase K is defined by the ratio of the mole number $N_{k\mathrm{K}}$ of that component over the total number of moles within the phase N_K. In each phase, the molar fractions satisfy the closure relation $\sum_{k\in\mathrm{K}} x_{k\mathrm{K}} = 1$, K = S, E. Since $N_{k\mathrm{K}}/V_0 = m^{k\mathrm{K}}/\widehat{m}_k$, the molar fractions can also be expressed in terms of mass contents.

The *concentration* of a fluid component is equal to its number of moles referred to the volume of fluid phase,

$$c_{k\mathrm{E}} = \frac{N_{k\mathrm{E}}}{V_\mathrm{E}} = \frac{N_{k\mathrm{E}}}{N_\mathrm{E}\widehat{v}_\mathrm{E}} = \frac{1}{\widehat{v}_k}\frac{V_{k\mathrm{E}}}{V_\mathrm{E}} = \frac{1}{\widehat{v}_k}\frac{n^{k\mathrm{E}}}{n^\mathrm{E}} = \frac{1}{\widehat{v}_k}\frac{v^{k\mathrm{E}}}{n^\mathrm{E}}\frac{V_0}{V} = \frac{x_{k\mathrm{E}}}{\widehat{v}_\mathrm{E}}, \quad k \in \mathrm{E}, \tag{19.8.2}$$

with $\widehat{v}_\mathrm{E} = \sum_{k\in\mathrm{E}} x_{k\mathrm{E}}\,\widehat{v}_k$ the molar volume of the fluid phase. For dilute concentrations, the approximation $\widehat{v}_\mathrm{E} = \widehat{v}_\mathrm{w}$ is accepted.

19.8.3 Electrochemical potentials

The incremental work per unit *reference volume* $\underline{\mathcal{W}}$ done by the (total) stress $\mathbf{T}$ in the conjugate strain $\mathbf{E}$ of the porous medium and by the *electrochemical potentials* $\mu_{k\mathrm{K}}^\mathrm{ec}$ during the change of mass contents $\delta m^{k\mathrm{K}}$ of the component k expresses as,

$$\begin{aligned}
\delta\underline{\mathcal{W}} &= \mathbf{T}:\delta\mathbf{E} + \sum_{\mathrm{K},k\in\mathrm{K}} \mu_{k\mathrm{K}}^\mathrm{ec}\,\delta m^{k\mathrm{K}}, \\
&= \mathbf{T}:\delta\mathbf{E} + \sum_{\mathrm{K},k\in\mathrm{K}} g_{k\mathrm{K}}^\mathrm{ec}\,\delta\mathcal{N}_{k\mathrm{K}}.
\end{aligned} \tag{19.8.3}$$

Here $(\mathbf{T}, \mathbf{E})$ is a generic work-conjugate stress-strain pair. Common work-conjugate pairs are listed in Section 2.8: the 1st Piola-Kirchhoff stress $\boldsymbol{\tau}$ and the deformation gradient $\mathbf{F}$, the 2nd Piola-Kirchhoff stress $\underline{\underline{\tau}} = \boldsymbol{\tau}\cdot\mathbf{F}^{-\mathrm{T}}$ and the Green strain $\mathbf{E}_\mathrm{G} = (\mathbf{F}^\mathrm{T}\cdot\mathbf{F}-\mathbf{I})/2$. Then $\mathbf{T}:\delta\mathbf{E} = \boldsymbol{\tau}:\delta\mathbf{F}^\mathrm{T} = \underline{\underline{\tau}}:\delta\mathbf{E}_\mathrm{G} = \boldsymbol{\tau}:\mathbf{D}\,\delta t$, even if there is no strain conjugated to the Kirchhoff stress $\boldsymbol{\tau}$, which is linked to the Cauchy stress $\boldsymbol{\sigma}$ and 2nd Piola-Kirchhoff stress by $\boldsymbol{\tau} = \det\mathbf{F}\,\boldsymbol{\sigma} = \mathbf{F}\cdot\underline{\underline{\tau}}\cdot\mathbf{F}^\mathrm{T} = \mathbf{F}\cdot\boldsymbol{\tau}$, and $\mathbf{D}$ is the rate of deformation.

The electrochemical potentials $\mu_{k\mathrm{K}}^\mathrm{ec}$ [unit : m^2/s^2] are mass based while the fluid-mass contents per unit initial volume of the porous medium $m^{k\mathrm{K}}$ are measured in kg/m^3. Alternatively, use can be made of the mole based electrochemical potentials $g_{k\mathrm{K}}^\mathrm{ec} = \widehat{m}_k\,\mu_{k\mathrm{K}}^\mathrm{ec}$ [unit : kg×m^2/s^2/mole].

The chemical potential of a component k in phase K identifies a mechanical contribution introduced by the intrinsic pressure $p_{k\mathrm{K}}$ and the molar volume $\widehat{v}_k$ and a chemical contribution which accounts for the molar fraction $x_{k\mathrm{K}}$. For a charged component of valence ζ_k, in presence of the electrical potential ϕ_K [unit : V], the electrochemical potential involves in addition an electrical contribution. For incompressible components,

$$g_{k\mathrm{K}}^\mathrm{ec} = \widehat{m}_k\,\mu_{k\mathrm{K}}^\mathrm{ec} = \widehat{m}_k\,\mu_{k\mathrm{K}}^0 + \widehat{v}_k\,p_{k\mathrm{K}} + RT\,\mathrm{Ln}\,x_{k\mathrm{K}} + \zeta_k\,\mathrm{F}\,\phi_\mathrm{K}, \quad k \in \mathrm{K}. \tag{19.8.4}$$

In this formula, $R = 8.31451$ J/mole/K is the universal gas constant, T the absolute temperature [unit : K], and F Faraday's constant. The term μ^0 represents the enthalpy of formation. In the context of chemical reactions, it can be viewed as an affinity, and it gives rise to an equilibrium constant different from one, Loret and Simões [2005]a. For a fluid phase without insoluble components, this term vanishes. For ions, the pressure term $p_{k\mathrm{K}}$ is dominated by the chemical contribution, although the latter decreases as the ionic strength increases. Therefore it can be neglected, at least for dilute solutions. The pressure of the fluid in phase K is denoted either by $p_{w\mathrm{K}}$ or by p_K.

19.8.4   Electroneutrality

In phase K, the *electrical density* I_{eK} [unit : coulomb/m^3] is defined as

$$I_{eK} = \frac{F}{V} \sum_{k \in K} \zeta_k N_{kK} \; [= F \, n^E \sum_{k \in E} \zeta_k \, c_{kE} \text{ for } K = E] \,. \tag{19.8.5}$$

Collagen and proteoglycans are macromolecules with a large molar mass, 285×10^3 gm for collagen and 72×10^3 gm for PGs. The molar fraction $x_{PG} = N_{PG}/N_E$ and concentration $c_{PG} = N_{PG}/V_E$ of proteoglycans are thus quite small with respect to the other components of the fluid phase. On the other hand, the valence ζ_{PG} of proteoglycans is large (and negative) at neutral pH. Moreover complexed ligands, to which chloride ions are bound, contribute to the fixed charge. Thus the *effective* molar fraction y_l and *effective* concentration e_l of collagen c, proteoglycans PGs and occupied ligands L,

$$y_l = \zeta_l \, x_l, \quad e_l = \zeta_l \, c_l, \quad l = \text{PG, c}; \quad y_{ClL} = -x_{ClL}, \quad e_{ClL} = -c_{ClL} \,, \tag{19.8.6}$$

are key parameters of the biochemical and biomechanical behaviors of the cornea. Note that even if the ligands are located within the collagen fibrils, the analysis below keeps the charges of collagen and ligands separate.

According to Assumption (H4), together the solid and fluid phases are neutral. At neutral pH, the collagen molecules are neutral, the PGs are negatively charged, as well as the complexed ligands. Consequently electroneutrality requires a minimal number of mobile cations: the charge of these cations balances the large negative charge of the proteoglycans and ligands, that is

$$N_e + \sum_{k \in E_{ions}} \zeta_k \, N_{kE} = 0 \,, \tag{19.8.7}$$

where N_e [unit : mole] represents the total fixed charge,

$$N_e = \zeta_{PG} \, N_{PG} + \zeta_c \, N_c + \zeta_{Cl} \, N_{ClL} \,. \tag{19.8.8}$$

The sum in (19.8.7) involves free ions in solution, thus excluding chloride ions bound to ligands, which contribute to the fixed charge N_e. The molar fraction y_e, effective concentration e_e, and mole content $\mathcal{N}_e$ associated with the fixed charge deduce as

$$N_e = y_e \, N_E = e_e \, V_E = \mathcal{N}_e \, V_0, \quad y_e = e_e \, \widehat{v}_E \sim e_e \, \widehat{v}_w \,. \tag{19.8.9}$$

Electroneutrality of the stroma, including the extrafibrillar fluid phase and collagen network, can be rephrased equivalently in terms of this concentration e_e, or in terms of the effective molar fraction y_e,

$$e_e + c_{NaE} - c_{ClE} = 0, \quad y_e + x_{NaE} - x_{ClE} = 0 \,. \tag{19.8.10}$$

19.9   Biochemical composition of stroma

A number of data are required to obtain the complete biochemical composition of a given volume or mass of stroma. The solid phase, or solid skeleton, contains collagen (c) and proteins (Pro), while mobile ions, PGs and water constitute the extrafibrillar fluid phase.

19.9.1   Molar masses and mass densities

Representative values of the molar masses, molar volumes and mass densities listed in Table 19.9.1 have been collected from Section 19.4.2.

19.9.2   Composition: mole numbers, masses and volumes

Hydration,

$$H_w = \frac{M_{wE}}{M_{dry}} \,, \tag{19.9.1}$$

Table 19.9.1: Representative properties of the main constituents of the stroma.

Constituent	Molar Mass (gm)	Molar Volume (cm^3)	Density (gm/cm^3)
Water	18	18	1
Proteoglycans	72×10^3	48×10^3	1.5
Collagen type I	285×10^3	200×10^3	1.425

is viewed as a parameter. The mass of PGs is the sum of the masses of GAGs and GAG-associated proteins. The relative masses of solid constituents,

$$\frac{M_{\mathrm{PG}}}{M_{\mathrm{dry}}} = 0.11; \quad \frac{M_{\mathrm{c}}}{M_{\mathrm{dry}}} = 0.71; \quad \frac{M_{\mathrm{Pro}}}{M_{\mathrm{dry}}} = 0.18 \,, \tag{19.9.2}$$

with respect to the dry mass are known from Table 19.4.1 of Section 19.4.2. The ratio of the volumes of the solid and fluid follows,

$$\frac{V_{\mathrm{dry}}}{V_{\mathrm{wE}}} = \frac{1}{H_{\mathrm{w}}} \left(\frac{\rho_{\mathrm{w}}}{\rho_{\mathrm{PG}}} \frac{M_{\mathrm{PG}}}{M_{\mathrm{dry}}} + \frac{\rho_{\mathrm{w}}}{\rho_{\mathrm{c}}} \frac{M_{\mathrm{c}}}{M_{\mathrm{dry}}} + \frac{\rho_{\mathrm{w}}}{\rho_{\mathrm{Pro}}} \frac{M_{\mathrm{Pro}}}{M_{\mathrm{dry}}} \right). \tag{19.9.3}$$

The mass density of non collagenous proteins has been taken equal to that of collagen. The number of moles of water in the specimen results as soon as either the volume V, or the mass M, of the specimen is known,

$$V_{\mathrm{wE}} = N_{\mathrm{wE}} \, \widehat{v}_{\mathrm{w}} = \frac{V}{1 + \dfrac{V_{\mathrm{dry}}}{V_{\mathrm{wE}}}}, \quad M_{\mathrm{wE}} = N_{\mathrm{wE}} \, \widehat{m}_{\mathrm{w}} = \frac{M}{1 + \dfrac{1}{H_{\mathrm{w}}}}. \tag{19.9.4}$$

As by-products, we have the numbers of moles of the solid constituents,

$$\frac{N_{\mathrm{PG}}}{N_{\mathrm{wE}}} = \frac{1}{H_{\mathrm{w}}} \frac{\widehat{m}_{\mathrm{w}}}{\widehat{m}_{\mathrm{PG}}} \frac{M_{\mathrm{PG}}}{M_{\mathrm{dry}}}, \quad \frac{N_{\mathrm{c}}}{N_{\mathrm{wE}}} = \frac{1}{H_{\mathrm{w}}} \frac{\widehat{m}_{\mathrm{w}}}{\widehat{m}_{\mathrm{c}}} \frac{M_{\mathrm{c}}}{M_{\mathrm{dry}}}, \quad \frac{N_{\mathrm{Pro}}}{N_{\mathrm{wE}}} = \frac{1}{H_{\mathrm{w}}} \frac{\widehat{m}_{\mathrm{w}}}{\widehat{m}_{\mathrm{Pro}}} \frac{M_{\mathrm{Pro}}}{M_{\mathrm{dry}}}, \tag{19.9.5}$$

and the volume fractions of extrafibrillar water and of the whole extrafibrillar phase,

$$n^{\mathrm{wE}} = \frac{V_{\mathrm{wE}}}{V} = \frac{1}{1 + \dfrac{V_{\mathrm{dry}}}{V_{\mathrm{wE}}}}, \quad \frac{n^{\mathrm{E}}}{n^{\mathrm{wE}}} = \frac{V_{\mathrm{E}}}{V_{\mathrm{wE}}} = 1 + \frac{N_{\mathrm{PG}}}{N_{\mathrm{wE}}} \frac{\widehat{v}_{\mathrm{PG}}}{\widehat{v}_{\mathrm{w}}}. \tag{19.9.6}$$

At standard hydration $H_{\mathrm{w}} = 3.2$, the resulting density of the stroma is $1.08 \, \mathrm{gm/cm}^3$. The volume fractions of water n^{wE} and of the fluid phase n^{E} are respectively equal to 0.82 and 0.84.

19.10   The fixed electrical charge: negatively charged PGs and chloride binding

In a number of soft tissues, e.g., articular cartilages, the fixed charges of PGs and collagen depend on the pH of the extrafibrillar water. Under physiological pH, PGs are negatively charged while the charge of collagen molecules is quite tiny. Sodium cations and chloride anions do not bind to the fixed charges, that is, they diffuse freely in water. On the other hand, calcium cations are known to bind reversibly and irreversibly to the PGs of articular cartilages.

The situation is different in the stroma. On equilibrating a piece of stroma in a bath of controlled chemical composition, Elliott [1980] observed total concentrations of chloride ions which were quite different from those predicted by Donnan equilibrium. This observation led him to introduce the mechanism of chloride binding. The suggestion was later confirmed by Hodson et al. [1992].

19.10.1 Chloride binding: partition of species and sites

According to Hodson et al. [1992], Elliott and Hodson [1998], stromal chloride ions bind *reversibly* on ligands L of the stroma,

$$\underbrace{\mathrm{L-Cl^-}}_{\text{occupied ligand}} \quad \rightleftharpoons \quad \underbrace{\mathrm{L}}_{\text{free ligand}} \quad + \quad \underbrace{\mathrm{Cl^-}}_{\text{fluid phase}} \ . \tag{19.10.1}$$

The ligands are unidentified, but they seem to be associated with collagen fibrils, Regini et al. [2004]. Thus the total amount of chloride ions in the stroma partitions (1) into ions which are free in water and participate to equilibrate the external solution and (2) into bound ions which contribute to the fixed charge.

The variations of the numbers of moles of the occupied and free ligands and free chloride ions, δN_{ClL}, $\delta N_{\square \mathrm{L}}$ and δN_{ClE} respectively, are linked by the stoichiometry of the reaction,

$$- \delta N_{\mathrm{ClL}} = \delta N_{\square \mathrm{L}} = \delta N_{\mathrm{ClE}} \ . \tag{19.10.2}$$

Consequently, the work done in the volume V_0 during this reaction,

$$\begin{aligned}
g_{\mathrm{ClL}}^{\mathrm{ec}} \, \delta N_{\mathrm{ClL}} &+ g_{\square \mathrm{L}} \, \delta N_{\square \mathrm{L}} + g_{\mathrm{ClE}}^{\mathrm{ec}} \, \delta N_{\mathrm{ClE}} \\[1mm]
&= \left(- g_{\mathrm{ClL}}^{\mathrm{ec}} + g_{\square \mathrm{L}} + g_{\mathrm{ClE}}^{\mathrm{ec}} \right) \delta N_{\mathrm{ClE}} \\[1mm]
&= \left(- (RT \, \mathrm{Ln}\, x_{\mathrm{ClL}} - \mathrm{F}\, \phi_{\mathrm{E}}) + RT \, \mathrm{Ln}\, x_{\square \mathrm{L}} + RT \, \mathrm{Ln}\, x_{\mathrm{ClE}} - \mathrm{F}\, \phi_{\mathrm{E}} + \mathcal{G}^0 \right) \delta N_{\mathrm{ClE}} \\[1mm]
&= \mathcal{G} \, \delta N_{\mathrm{ClE}} \ ,
\end{aligned} \tag{19.10.3}$$

can be phrased in terms of the chemical affinity $\mathcal{G}$ associated with the reaction,

$$\mathcal{G} = RT \, \mathrm{Ln}\, \frac{x_{\square \mathrm{L}}}{x_{\mathrm{ClL}}} \frac{x_{\mathrm{ClE}}}{\kappa_{\mathrm{L}}} \quad \Leftrightarrow \quad \mathcal{G} = RT \, \mathrm{Ln}\, \frac{c_{\square \mathrm{L}}}{c_{\mathrm{ClL}}} \frac{c_{\mathrm{ClE}}}{K_{\mathrm{L}}} \ . \tag{19.10.4}$$

The equilibrium of the reaction, which is defined by a vanishing affinity is characterized by an *equilibrium constant* which is expressed in various equivalent forms, namely either dimensionless κ_{L}, or with the dimension of a concentration $K_{\mathrm{L}} = \kappa_{\mathrm{L}}/\widehat{v}_{\mathrm{w}}$, or by an enthalpy of formation $\mathcal{G}^0 = -RT \, \mathrm{Ln}\, \kappa_{\mathrm{L}}$. In order to mimic the notation used to characterize the effects of pH, we shall also introduce a $\mathrm{pK_L}$. The relations between these quantities are recorded below,

$$K_{\mathrm{L}} = \frac{\kappa_{\mathrm{L}}}{\widehat{v}_{\mathrm{w}}}, \quad \mathcal{G}^0 = -RT \, \mathrm{Ln}\, \kappa_{\mathrm{L}}, \quad \underbrace{K_{\mathrm{L}}}_{\text{unit : mole/liter}} = 10^{-\mathrm{pK_L}} \ . \tag{19.10.5}$$

At equilibrium, the concentrations of the occupied ligands c_{ClL} and free ligands $c_{\square \mathrm{L}}$ deduce from the total concentration of ligands c_{L} and chloride concentration c_{ClE},

$$c_{\mathrm{L}} = c_{\square \mathrm{L}} + c_{\mathrm{ClL}}, \quad \frac{c_{\square \mathrm{L}} \times c_{\mathrm{ClE}}}{c_{\mathrm{ClL}}} = K_{\mathrm{L}} \ , \tag{19.10.6}$$

namely

$$\frac{c_{\square \mathrm{L}}}{c_{\mathrm{L}}} = \frac{K_{\mathrm{L}}}{K_{\mathrm{L}} + c_{\mathrm{ClE}}}, \quad \frac{c_{\mathrm{ClL}}}{c_{\mathrm{L}}} = \frac{c_{\mathrm{ClE}}}{K_{\mathrm{L}} + c_{\mathrm{ClE}}} \ . \tag{19.10.7}$$

Note that the equilibrium constant is the value of the concentration of chloride ions at which the ligands are equally partitioned in free and occupied sites, Figs. 19.10.1 and 19.10.2.

A measure of the total number of ligands N_{L} refers to the dry mass M_{dry} of the tissue, which includes collagen, PGs and other proteins. Since

$$\frac{N_{\mathrm{L}}}{M_{\mathrm{dry}}} = \frac{N_{\mathrm{L}}}{V_{\mathrm{E}}} \frac{V_{\mathrm{E}}}{V_{\mathrm{wE}}} \frac{H_{\mathrm{w}}}{\rho_{\mathrm{w}}} \ , \tag{19.10.8}$$

then the concentration per volume of water c_{L}, and molar fraction $x_{\mathrm{L}} = c_{\mathrm{L}}/\widehat{v}_{\mathrm{w}}$, depend on hydration $H_{\mathrm{w}} = M_{\mathrm{wE}}/M_{\mathrm{dry}}$ as defined by (19.9.1),

$$c_{\mathrm{L}} \equiv \frac{N_{\mathrm{L}}}{V_{\mathrm{E}}} = \frac{n^{\mathrm{wE}}}{n^{\mathrm{E}}} \frac{\rho_{\mathrm{w}}}{H_{\mathrm{w}}} \frac{N_{\mathrm{L}}}{M_{\mathrm{dry}}} \ . \tag{19.10.9}$$

Similarly, a measure of the effective charge of PGs is provided by its CEC,

$$\frac{e_{\mathrm{PG}}}{\mathrm{sgn}\, e_{\mathrm{PG}}} = \frac{|\zeta_{\mathrm{PG}}| \, N_{\mathrm{PG}}}{V_{\mathrm{E}}} = \frac{n^{\mathrm{wE}}}{n^{\mathrm{E}}} \frac{\rho_{\mathrm{w}}}{H_{\mathrm{w}}} \frac{\mathrm{CEC}}{\mathrm{F}} \quad \text{with} \quad \frac{\mathrm{CEC}}{\mathrm{F}} = \frac{|\zeta_{\mathrm{PG}}| \, N_{\mathrm{PG}}}{M_{\mathrm{dry}}} \ . \tag{19.10.10}$$

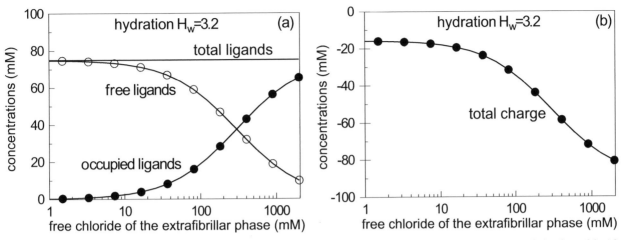

Fig. 19.10.1: (a) Concentrations of free, bound and total ligands as a function of the concentration of the free chloride ions in the extrafibrillar fluid phase; (b) total electrical charge. Equilibrium constant $K_L = 300\,\text{mM}$, hydration $H_w = 3.2$.

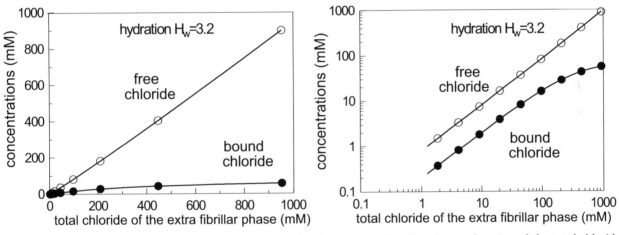

Fig. 19.10.2: Concentrations of free chloride ions and chloride ions bound to ligands as a function of the total chloride ions of the extrafibrillar fluid phase. Although the percentage of bound chloride is low, the binding process modifies significantly the fixed charge.

19.10.2 Chloride binding: the fixed electrical charge

Due to chloride binding, occupied ligands contribute to the fixed electrical charge. The concentration of this fixed charge depends, besides on hydration, on the concentration of chloride ions,

$$
\begin{aligned}
e_e &= e_{PG} + e_c - c_{ClL} \\
&= e_{PG} + e_c - \frac{c_{ClE}\, c_L}{K_L + c_{ClE}}.
\end{aligned}
\tag{19.10.11}
$$

19.10.3 Chemical equilibrium at the cornea-bath interface

A specimen of stroma is bathed in an electroneutral bath of controlled chemical composition, pressure p_{wB} and electrical potential ϕ_B, Fig. 19.10.3. The counterparts of these entities in the extrafibrillar phase are sought, assuming electro-chemomechanical equilibrium to hold.

The bath is at pH=7 so that stromal collagen molecules are not charged, namely $y_c = 0$ and $e_c = 0$. There are five unknowns, namely pressure of water, molar fractions of water, of sodium ions, of free chloride ions, and electrical potential,

$$
p_{wE},\ x_{wE},\ x_{NaE},\ x_{ClE},\ \phi_E,
\tag{19.10.12}
$$

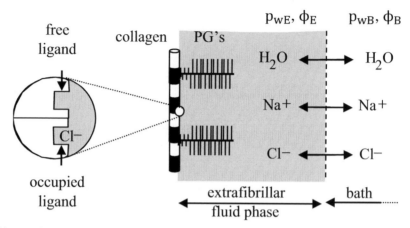

Fig. 19.10.3: Equilibrium between the extrafibrillar phase of stroma and the bath. Stromal chloride ions partition into free ions dissolved in water and ions bound onto ligands. In tests at constant hydration, the exchange of water between the bath and the stroma is zeroed.

defined by the five equations expressing the electrochemical equilibria of water, sodium ions, chloride ions, compatibility of the molar fractions and electroneutrality of the extrafibrillar phase, namely

$$g_{wE} = g_{wB}, \quad g_{NaE}^{ec} = g_{NaB}^{ec}, \quad g_{ClE}^{ec} = g_{ClB}^{ec}, \quad \sum_{l \in E} x_{lE} = 1, \quad \sum_{l \in E} \zeta_l \, x_{lE} = 0 \,. \tag{19.10.13}$$

The molar fraction of bound chloride ions deduces from that of free chloride ions by (19.10.7). The compatibility relation for the molar fractions is used to define the molar fraction of water,

$$x_{wE} = 1 - x_{NaE} - x_{ClE} - x_{PG} \,. \tag{19.10.14}$$

This relation and the equilibrium of water yields the osmotic pressure,

$$p_{wE} - p_{wB} = \frac{RT}{\widehat{v}_w} \, \mathrm{Ln} \, \frac{x_{wB}}{x_{wE}} \simeq -\frac{RT}{\widehat{v}_w} \, (x_{wE} - x_{wB}) \simeq -RT \, (c_{wE} - c_{wB}) \,. \tag{19.10.15}$$

We are now left with three unknowns,

$$x_{NaE}, \, x_{ClE}, \, \phi_E \,, \tag{19.10.16}$$

and three equations, namely

$$g_{NaE}^{ec} = g_{NaB}^{ec}, \quad g_{ClE}^{ec} = g_{ClB}^{ec}, \quad \sum_{l \in E} \zeta_l \, x_{lE} = 0 \,. \tag{19.10.17}$$

Assuming identical enthalpies of formation in the bath and in the stroma, these relations may be made more explicit,

$$\frac{F}{RT} \, (\phi_E - \phi_B) = -\mathrm{Ln} \, \frac{x_{ClB}}{x_{ClE}} = \mathrm{Ln} \, \frac{x_{NaB}}{x_{NaE}}, \quad y_e + x_{NaE} - x_{ClE} = 0 \,. \tag{19.10.18}$$

At given hydration, the fixed electrical charge is a function of the molar fraction x_{ClE} of free chloride ions. Combining (19.10.11) and the equilibrium relations (19.10.18) yields the equation for x_{ClE},

$$f = (x_{ClE})^3 + (\kappa_L - y_{PG} + x_L) \, (x_{ClE})^2 - (y_{PG} \, \kappa_L + (x_{ClB})^2) \, x_{ClE} - \kappa_L \, (x_{ClB})^2 = 0 \,, \tag{19.10.19}$$

where
  - $\kappa_L = \widehat{v}_w \, K_L$ is the dimensionless equilibrium constant;
  - $y_{PG} = \widehat{v}_w \, e_{PG}$ is the effective molar fraction of PGs;
  - $x_L = \widehat{v}_w \, c_L$ the molar fraction of ligands.

Once the molar fraction of free chloride ions is known, the molar fractions of sodium and of charge $y_e = y_e(x_{ClE})$ deduce from (19.10.18), and the difference of electrical potentials is defined by the first equality in (19.10.18).

Starting from the data of Table 19.10.1, the effective concentration e_{PG} of PGs deduces by (19.10.10), and the total concentration of ligands c_L by (19.10.9). At physiological hydration $H_w = 3.2$, these numbers are equivalent to a total concentration of ligands equal to 75 mM, to an effective concentration of PGs equal to -17 mM, and to a valence $\zeta_{PG} = -72$.

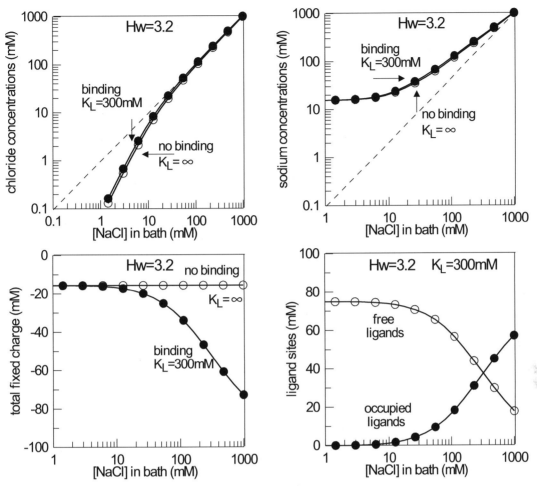

Fig. 19.10.4: Stroma in equilibrium with a bath of controlled chemical composition. Ionic concentrations in the extrafibrillar fluid phase of the cornea, total charge and ligand sites. Data: see text. The total effective fixed charge under physiological conditions is about 40 mM, in agreement with standard data, e.g., Elliott et al. [1980].

Effects of chloride binding

Combining the relations (19.10.18) implies the following inequalities,

$$x_{\text{ClB}}^2 - x_{\text{ClE}}^2 = -x_{\text{ClE}}\, y_e > 0, \quad x_{\text{NaE}}^2 - x_{\text{NaB}}^2 = -x_{\text{NaE}}\, y_e > 0, \tag{19.10.20}$$

since the total charge y_e is negative at neutral pH. Therefore, the molar fraction, or concentration, of chloride ions in the bath is larger than the molar fraction of *free* chloride ions in the stroma. Conversely, the sodium molar fraction in the stroma is always larger than in the bath, Fig. 19.10.4.

These comments hold whether or not chloride binding takes place. In absence of binding, the free chloride ions represent the total chloride ions. However, in presence of binding, it turns out that the molar fraction of total chloride ions is smaller than in the bath at low concentration, but larger at large concentration, Fig. 19.10.5. In fact, it was this observation that prompted Elliott to introduce chloride binding. The value of the molar fraction of free chloride ions, at which the two above concentrations $x_{\text{ClE}} + x_{\text{ClL}}$ and x_{ClB}, are equal, satisfies the equation,

$$(x_{\text{L}} + y_{\text{PG}})\, x_{\text{ClE}}^2 + ((\kappa_{\text{L}} + x_{\text{L}})\, x_{\text{L}} + 2\,\kappa_{\text{L}}\, y_{\text{PG}})\, x_{\text{ClE}} + y_{\text{PG}}\, \kappa_{\text{L}}^2 = 0. \tag{19.10.21}$$

This equation is obtained by inserting the expression (19.10.6) for the concentration of bound ions into (19.10.20) where the charge y_e is defined by (19.10.11). The transition value results as

$$c_{\text{ClB}} = c_{\text{ClE}} + c_{\text{ClL}} = 76\,\text{mM}. \tag{19.10.22}$$

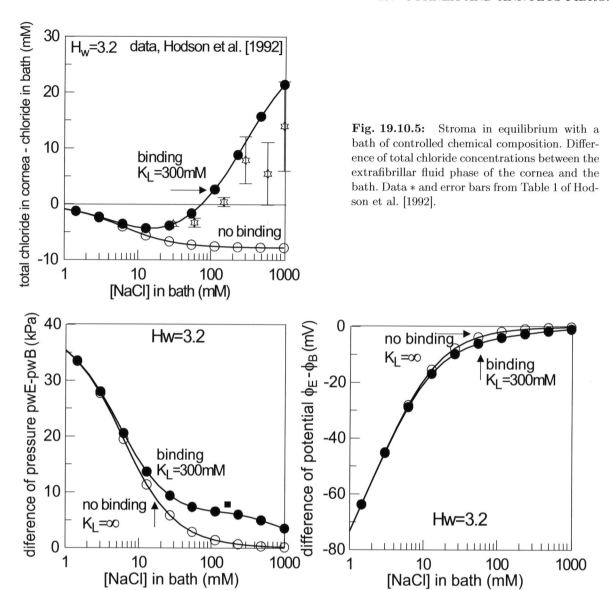

Fig. 19.10.5: Stroma in equilibrium with a bath of controlled chemical composition. Difference of total chloride concentrations between the extrafibrillar fluid phase of the cornea and the bath. Data $*$ and error bars from Table 1 of Hodson et al. [1992].

Fig. 19.10.6: Stroma in equilibrium with a bath of controlled chemical composition. Differences of pressure and of electrical potential between the extrafibrillar fluid phase of the cornea and the bath. Data: see text. The pressure under physiological ionic strength (154 mM of NaCl) and hydration ($H_w = 3.2$) measured by Hedbys and Dohlman [1963], namely about 8 kPa, and shown by the solid square, fits well with the theoretical curve accounting for chloride binding.

At physiological concentration of chloride ions, chloride binding increases the osmotic pressure but affects little the electrical potential, Fig. 19.10.6.

Particular case: no chloride binding

    Electroneutrality of the bath and stroma, and equilibrium of the ionic electrochemical potentials between the bath and stroma impose the conditions,

$$x_{\text{ClB}} = x_{\text{NaB}}, \quad x_{\text{ClE}} = x_{\text{NaE}} + y_{\text{PG}}, \quad x_{\text{NaE}}\, x_{\text{ClE}} = x_{\text{NaB}}\, x_{\text{ClB}}. \tag{19.10.23}$$

The equilibrium of the chemical potentials of water between the two compartments provides the pressure difference $p_{\text{wE}} - p_{\text{B}}$ (neglecting the concentration c_{PG} of PGs) as

$$\frac{\widehat{v}_{\text{w}}}{RT}\left(p_{\text{wE}} - p_{\text{B}}\right) \simeq -y_{\text{PG}} + 2\left(x_{\text{ClE}} - x_{\text{ClB}}\right). \tag{19.10.24}$$

The molar fractions of extrafibrillar ions express in terms of the molar fractions of the bath, namely

$$x_{\text{NaE}} = -\frac{y_{\text{PG}}}{2} + \sqrt{\left(\frac{y_{\text{PG}}}{2}\right)^2 + x_{\text{NaB}}^2}, \quad x_{\text{ClE}} = \frac{y_{\text{PG}}}{2} + \sqrt{\left(\frac{y_{\text{PG}}}{2}\right)^2 + x_{\text{NaB}}^2}. \tag{19.10.25}$$

Conversely, the ionic molar fractions in the bath express in terms of the ionic molar fractions of the fluid phase,

$$x_{NaB} = x_{ClB} = \sqrt{x_{NaE}\,x_{ClE}}\,. \tag{19.10.26}$$

19.11   Effects of pH and chloride binding on the fixed charge

It is instrumental to re-define the three sets of extrafibrillar species E, of extrafibrillar ions E_{ions}, and of extrafibrillar mobile species E_{mo}, namely

$$E = \{w, PG, Na, Cl, H, OH\}, \quad E_{ions} = \{Na, Cl, H, OH\}, \quad E_{mo} = E - \{PG\}\,. \tag{19.11.1}$$

19.11.1   Contributions to the fixed electrical charge by GAGs

The charge of proteoglycans is contributed by carboxyl and sulfonic sites on disaccharide units, as described in Section 19.1.2.2. and summarized by Fig. 19.11.1.

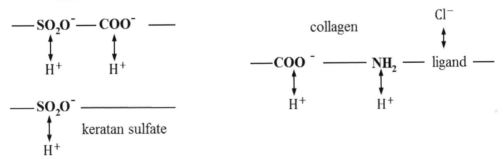

Fig. 19.11.1: Acid-base reactions at carboxyl and sulfonic sites of GAGs, at carboxyl and amino sites of the side chains of the amino acids of collagen, and chloride binding on ligands associated with collagen.

The mass of GAGs represents about 25% of the mass of a proteoglycan monomer, Axelsson and Heinegård [1978],

$$\frac{M_{GAG}}{M_{PG}} = 0.25\,. \tag{19.11.2}$$

The data for normal human stroma shown in Table 19.11.1 are extracted from Plaas et al. [2001]. The charges of the sulfonic (sg) and carboxyl (cg) groups per chain (GAG) are estimated to the following values,

$$N_{sg}^{KS} = -1 \times N_{du}^{KS1} - 2 \times N_{du}^{KS2} = -21;$$

$$N_{sg}^{CS} = -1 \times N_{du}^{CS1} = -14.4; \quad N_{cg}^{CS} = -1 \times N_{du}^{CS} = -40\,. \tag{19.11.3}$$

The cation exchange capacity CEC defined by (19.10.10), i.e., number of moles of charge per gram of dry mass,

$$\frac{CEC}{F} = \lvert N_{sg}^{KS}\rvert \times \frac{N^{KS}}{M_{dry}} + \lvert N_{sg}^{CS} + N_{cg}^{CS}\rvert \times \frac{N^{CS}}{M_{dry}}\,, \tag{19.11.4}$$

Table 19.11.1: Composition of human corneal glycosaminoglycans.

Total nb. disaccharide units per KS	$N_{\text{du}}^{\text{KS}}$	14
nbs. disaccharide units per KS		
Unsulfated	$N_{\text{du}}^{\text{KS0}}$	0.56
Monosulfated	$N_{\text{du}}^{\text{KS1}}$	5.88
Disulfated	$N_{\text{du}}^{\text{KS2}}$	7.56
Total nb. disaccharide units per CS	$N_{\text{du}}^{\text{CS}}$	40
nbs. disaccharide units per CS		
Unsulfated	$N_{\text{du}}^{\text{CS0}}$	25.6
Monosulfated	$N_{\text{du}}^{\text{CS1}}$	14.4
nb. moles of GAG of KS per g.d.m.*	$N^{\text{KS}}/M_{\text{dry}}$	2.2×10^{-6}
nb. moles of GAG of CS per g.d.m.	$N^{\text{CS}}/M_{\text{dry}}$	0.5×10^{-6}
Molar mass of GAGs of KS (gm)	$\widehat{m}_{\text{KS}}$	7.0×10^3
Molar mass of GAGs of CS (gm)	$\widehat{m}_{\text{CS}}$	17.5×10^3

* g.d.m.: gram of dry mass

takes the value 73.4×10^{-6} mole/gm dry mass for human stroma, while the CEC of ox cornea shown in Table 19.10.1 is equal to 50×10^{-6} mole/gm dry mass.

The ratios of mass of KS and CS over tissue dry mass,

$$\frac{N^{\text{KS}}\widehat{m}_{\text{KS}}}{M_{\text{dry}}} = 1.54\%, \quad \frac{N^{\text{CS}}\widehat{m}_{\text{CS}}}{M_{\text{dry}}} = 0.875\%, \tag{19.11.5}$$

yield a relative mass of GAGs of 2.415%, instead of 5% shown in Table 19.4.1. The values indicated by (19.9.2) are also slightly modified,

$$\frac{M_{\text{PG}}}{M_{\text{dry}}} = 0.0966; \quad \frac{M_{\text{c}}}{M_{\text{dry}}} = 0.71; \quad \frac{M_{\text{Pro}}}{M_{\text{dry}}} = 0.1934. \tag{19.11.6}$$

The relative numbers of moles of GAGs and PGs deduce from the mass ratios,

$$\frac{N_{\text{KS}}}{N_{\text{PG}}} = \frac{M_{\text{KS}}}{M_{\text{dry}}} \frac{M_{\text{dry}}}{M_{\text{PG}}} \frac{\widehat{m}_{\text{PG}}}{\widehat{m}_{\text{KS}}} = 1.64; \quad \frac{N_{\text{CS}}}{N_{\text{PG}}} = \frac{M_{\text{CS}}}{M_{\text{dry}}} \frac{M_{\text{dry}}}{M_{\text{PG}}} \frac{\widehat{m}_{\text{PG}}}{\widehat{m}_{\text{CS}}} = 0.37. \tag{19.11.7}$$

The charge per proteoglycan, or valence,

$$\zeta_{\text{PG}} = N_{\text{sg}}^{\text{KS}} \times \frac{N_{\text{KS}}}{N_{\text{PG}}} + (N_{\text{sg}}^{\text{CS}} + N_{\text{cg}}^{\text{CS}}) \times \frac{N_{\text{CS}}}{N_{\text{PG}}}, \tag{19.11.8}$$

is about hundred times smaller than in articular cartilages, namely -54.7, against -4000 to -6 000 in articular cartilages. The valence can be calculated equivalently from the molar mass $\widehat{m}_{\text{PG}} = 72 \times 10^3$ gm of PGs, the relative mass ratio of PGs and dry tissue, and the CEC,

$$|\zeta_{\text{PG}}| = \widehat{m}_{\text{PG}} \frac{M_{\text{dry}}}{M_{\text{PG}}} \frac{\text{CEC}}{\text{F}}. \tag{19.11.9}$$

19.11.2 Collagen contributions to the fixed electrical charge

Amino acids of the collagen molecules bear a carboxyl and an amino termination. Some of their side chains are also ionizable. The number of positively and negatively charged side chains of bovine corneal stroma is listed in Table 19.11.2. Some of these chains are turned inside the molecules and establish intramolecular links, and thus are functionally impaired. Moreover, since collagen molecules form fibrils, only the exposed chains located on molecules at the periphery of the fibrils can be considered to take part to extrafibrillar interactions, namely contribute to electroneutrality of the extrafibrillar fluid and to swelling.

Table 19.11.2: Number of charged side chains per collagen molecule (3150 amino acids) for bovine corneal stroma from Panjwani and Harding [1978], and pKs from Voet and Voet [1998].

Amino Acid	nb per Molecule	pK	Charge at Low pH	Charge at Neutral pH	Charge at High pH
Aspartic acid	151	3.90	0	-151	-151
Glutamic acid	243	4.1	0	-243	-243
Lysine	91	10.5	+91	+91	0
Hydroxylysine	26	10.5	+26	+26	0
Arginine	164	12.5	+164	+164	0
Hydroxyproline	258	10.5	0	0	-258
			+281	-113	-652

According to Regini and Elliott [2001], the collagen molecule has almost no net charge under physiological pH. On the other hand, data for tissues containing mostly type I collagen (rat tail tendon, human dermis) indicate 350 negatively charged side chains and 260 positively charged side chains at neutral pH. Table 19.11.2 matches qualitatively this trend.

A simplification has been adopted in the analysis which considers pHs of the bath varying in the range 3 to 8. The charges due to the carboxyl terminations on side chains which are negatively charged under physiological pH have been aggregated in a group of 281 charges per collagen molecule with a pK representative of carboxyl groups. Similarly, the charges due to the side chains which are positively charged under physiological pH have been aggregated in a group of 281 charges with a pK representative of amino groups.

The efficiency of these charges onto the extrafibrillar fluid is estimated through a coefficient which is defined so as to recover the isoelectric point IEP$\simeq 4$ obtained by Huang and Meek [1999] for a bath with a sodium chloride strength of $30\,\mathrm{mM}$. This efficiency can be interpreted as accounting for a partition of the collagen charges into an intrafibrillar compartment and an extrafibrillar compartment, as well as for the fact that not all charges are exposed to the outside of the molecules.

In summary, collagen contributes to electrical neutrality of the extrafibrillar compartment through the positive and negative charges of its side chains, scaled down by the efficiency coefficient.

19.11.3 Acid-base reactions, chloride binding and fixed charge

In order to compact the notation, we number the four acid-base reactions,

$$\mathrm{R}_j \; \rightleftharpoons \; \mathrm{P}_j + \mathrm{H}_{(j)}, \quad j \in [1,4], \tag{19.11.10}$$

with the following conventions for the reactants R, products P, participating hydrogen ions, and pKs:

S_1 carboxyl sites on PGs $\quad \mathrm{R}_1 = \mathrm{CO} - \mathrm{OH} \quad \mathrm{P}_1 = \mathrm{CO} - \mathrm{O}^- \quad \mathrm{H}_{(1)}^+ \quad \mathrm{pK}_1 = \mathrm{pK}_{\mathrm{cg}}$

S_2 sulfonic sites on PGs $\quad \mathrm{R}_2 = \mathrm{SOO} - \mathrm{OH} \quad \mathrm{P}_2 = \mathrm{SOO} - \mathrm{O}^- \quad \mathrm{H}_{(2)}^+ \quad \mathrm{pK}_2 = \mathrm{pK}_{\mathrm{sg}}$

S_3 amino sites on collagen $\quad \mathrm{R}_3 = \mathrm{NH}_2 - \mathrm{H}^+ \quad \mathrm{P}_3 = \mathrm{NH}_2 \quad \mathrm{H}_{(3)}^+ \quad \mathrm{pK}_3 = \mathrm{pK}_{\mathrm{ag}}$

S_4 carboxyl sites on collagen $\quad \mathrm{R}_4 = \mathrm{CO} - \mathrm{OH} \quad \mathrm{P}_4 = \mathrm{CO} - \mathrm{O}^- \quad \mathrm{H}_{(4)}^+ \quad \mathrm{pK}_4 = \mathrm{pK}_{\mathrm{cg}}$

The chemical reactions (19.11.10) for carboxyl, sulfonic and amino sites are defined through the relations,

$$c_{\mathrm{S}_j} = c_{\mathrm{R}_j} + c_{\mathrm{P}_j},$$

$$\frac{c_{\mathrm{P}_j} \times c_{\mathrm{HE}}}{c_{\mathrm{R}_j}} = 10^{-\mathrm{pK}_j} \tag{19.11.11}$$

$$\frac{c_{\mathrm{R}_j}}{c_{\mathrm{S}_j}} = \frac{\alpha_j}{1 + \alpha_j}, \quad \frac{c_{\mathrm{P}_j}}{c_{\mathrm{S}_j}} = \frac{1}{1 + \alpha_j},$$

where c_{P}, c_{R} and c_{S} are the concentrations of the free sites, occupied sites and total sites, respectively,

$$c_{\mathrm{HE}} = 10^{-\mathrm{pH}_{\mathrm{E}}}, \tag{19.11.12}$$

and

$$\alpha_j = 10^{pK_j - pH_E}, \quad j \in [1, 4].$$
(19.11.13)

Chloride binding to certain ligands contributes to the fixed charge, as indicated by (19.10.11). We have no indication concerning the possible effects of hydrogen ions on these binding sites. Therefore, the binding process is assumed to be independent of pH. The chloride binding reaction,

$$\underbrace{L-Cl^-}_{\text{occupied ligand}} \quad \rightleftharpoons \quad \underbrace{L}_{\text{free ligand}} \quad + \quad \underbrace{Cl^-}_{\text{fluid phase}},$$
(19.11.14)

is denoted by $j = L$, and according to the conventions above,

$$S_L \text{ total ligand sites L on collagen} \quad R_L \text{ occupied ligands L–Cl}^- \quad P_L \text{ free ligands L}$$

Notations similar to the acid-base reactions are introduced, namely,

$$c_{ClE} = 10^{-pCl_E},$$
(19.11.15)

with c_{ClE} expressed in mole/liter, and,

$$\alpha_L = 10^{pK_L - pCl_E},$$
(19.11.16)

and the relations (19.10.7) become,

$$\frac{c_{\square L}}{c_{S_L}} = \frac{1}{1 + \alpha_L}, \quad \frac{c_{ClL}}{c_{S_L}} = \frac{\alpha_L}{1 + \alpha_L}.$$
(19.11.17)

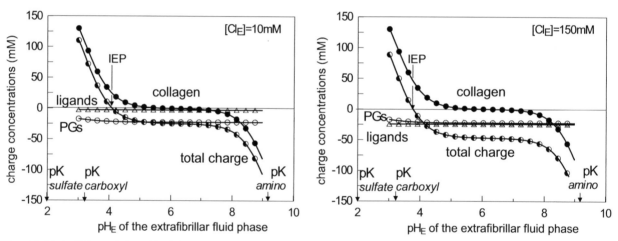

Fig. 19.11.2: Effect of the pH of the extrafibrillar fluid phase on the charges of collagen, PGs, ligands and total charge. Data: see text. A higher extrafibrillar concentration of chloride increases the negative charge of ligands and therefore shifts the isoelectric point to a somehow lower value. Still, with an efficiency coefficient of the charge of collagen equal to 1, the isoelectric point remains around 4, over a large range of chloride concentration.

The sites for acid-base reactions and chloride binding may be partitioned in the following general format,

$$\underbrace{c_{S_j}}_{\text{sites}} = \underbrace{c_{R_j}}_{\substack{\text{sites occupied} \\ \text{by hydrogen/chloride ions}}} + \underbrace{c_{P_j}}_{\text{free sites}}, \quad j \in [1, 4] \cup \{L\}.$$
(19.11.18)

For each reaction, the variations in the number of moles of reactants and products obey the relation,

$$\delta N_{XE_{(j)}} = \delta N_{P_j} = -\delta N_{R_j}, \quad j \in [1, 4] \cup \{L\}.$$
(19.11.19)

Here X stands for H if $j \in [1,4]$, and for Cl if $j = L$. The chemical affinities associated with the reactions,

$$\mathcal{G}_j = RT \operatorname{Ln} \frac{c_{P_j}}{c_{R_j}} \alpha_j, \quad j \in [1,4] \cup \{L\}, \tag{19.11.20}$$

are defined in terms of a pK=pK_j, or, equivalently by an enthalpy of formation $\mathcal{G}_j^0 = RT \operatorname{Ln} (10^{pK_j}/\widehat{v}_w)$. Since $c_{HE} = 10^{-pH_E}$ for $j \in [1,4]$, the pK_j is seen to be the pH_E at which the numbers of occupied and free sites for the acid-base reaction j are equal at equilibrium, namely for $\mathcal{G}_j = 0$. Similarly, with $c_{ClE} = 10^{-pCl_E}$, the pK_L is seen to be the pCl_E at which the numbers of occupied and free sites for chloride binding are equal at equilibrium.

The concentration of extrafibrillar electrical charge on collagen, PGs and binding ligands becomes,

$$
\begin{aligned}
e_e &= -c_{P_1} - c_{P_2} + c_{R_3} - c_{P_4} - c_{ClL}, \\
&= -\sum_{j \in [1,4]} c_{P_j} + c_{S_3} - c_{ClL}, \\
&= -\sum_{j \in [1,4]} \frac{c_{S_j}}{1 + \alpha_j} + c_{S_3} - c_{S_L} + \frac{c_{S_L}}{1 + \alpha_L}.
\end{aligned}
\tag{19.11.21}
$$

Fig. 19.11.2 displays the variations of the three sources of fixed charge, namely PGs, collagen and ligands, as the pH of the extrafibrillar fluid phase is varied. Since the charge on ligands depends on the chloride concentration, two chloride concentrations have been considered. Note that the total charge varies little as long as the extrafibrillar pH remains in the range from 5 to 8, since the pKs of interest lie outside this range. At low pH, hydrogen binding on carboxyl sites of collagen, with a pK equal to 3.2, is the main factor that turns the total charge positive.

Fig. 19.11.3 displays the effects of the variation of the pH of the bath on the corneal chemical composition, charge and osmotic pressure. The simulations have assumed the volume of the specimen to remain constant. The next section relieves this constraint by introducing chemomechanical couplings.

19.11.4 The mechanochemical effect: experimental data

Huang and Meek [1999] have submitted bovine corneas to chemical and mechanical loadings. The endothelium and epithelium were removed by scraping with a scalpel.

In a first attempt, the corneal stroma is assumed to be transverse isotropic. The collagen fibers are orthoradial (parallel to the corneal surface), and uniformly spread over all the tangential directions. This directional distribution is largely approximative, but it does not affect the simulations below where the strains are essentially radial.

The chemomechanical constitutive equations are described in Section 19.17.1. The ground substance (matrix) is endowed with isotropic elastic properties described by the nonlinear energy (19.19.70), with Lamé moduli $\lambda_e = 1.5\,\text{kPa}$ for volume increase, $\lambda_c = 6\,\text{kPa}$ for volume decrease and a shear modulus $\mu = 1.5\,\text{kPa}$. Indeed, the complex loadings to which the stromal specimens are subjected in the experiments reported by Huang and Meek [1999] include both a compression stage and a volume expansion stage. While the reported measurements do not allow to accurately backtrack the elastic properties, simulations indicate that the matrix moduli during the swelling phase are much lower than during the compression phase. Within the set of isotropic elastic materials, Section 5.4.7 indicates that such a conewise behavior can be realized by assuming the Lamé modulus λ to be different for volume expansion and for contraction.

In fact, it would be more appropriate to consider the elastic properties as displaying transversely isotropy about the radial direction, in view of the lamellar structure of the stroma. The loose connections between lamellae allow for considerable swelling. A transversely isotropic material admits two interfaces along which the elastic properties may differ, namely isochoric strain and zero radial strain, Section 5.4.6. In other words, the elastic coefficients may differ for positive volume change and negative volume change, and similarly for extension and contraction along the radial direction. The latter aspect is actually directly linked to the layered structure of the stroma.

The elastic response of collagen fibers is nonlinear as described in Section 19.17.3, with apparent modulus $K^c = 1000\,\text{kPa}$ and exponent $k_c = 100$. Let us recall that the above elastic coefficients are volume weighted coefficients, in the hypertonic state, that is, they are theoretically defined by the product of a volume fraction and an intrinsic elastic property. The volume fractions of PGs and collagen can be estimated from (19.9.5). See Section 17.12.4 for a discussion.

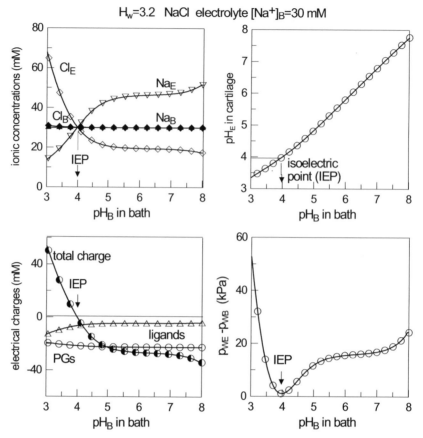

Fig. 19.11.3: Stroma at constant hydration in equilibrium with a bath of controlled chemical composition: constant concentration of sodium cations and varying pH_B. Data: see text. At the isoelectric point (IEP), the fixed charge vanishes and, since the intrinsic concentration of PGs can be neglected, the concentrations in the bath and within the stroma become identical, so that their pHs are identical as well, while the pressure differential vanishes. Note that, above the IEP, the cartilage is slightly more acid than the bath, and conversely below the IEP. Since the fixed charge becomes positive at low pH, chloride concentration increases and chloride binding becomes significant, which implies a large osmotic pressure.

The chemomechanical coupling is introduced so far only by the osmotic pressure. The chemical pressure p_{ch} in (19.17.8) is set to 0, and the coefficient $\underline{W}_{ch,2}$ is equal to 1.

The volume changes experienced during the experiments of Huang and Meek [1999] are quite large, so that a finite strain formulation is required.

The simulations mimic the experiments and proceed in four stages sketched in Fig. 19.11.4:
- in the initial state O, the bath has a hypertonic composition, with $[NaCl] \geq 2\,M$, a pH equal to 7, and, according to Table 3 of Huang and Meek [1999], hydration equal to 5.6 in absence of mechanical loading;
- stage OA: the ionic strength of NaCl in the bath is brought to $30\,mM$;
- stage AB: osmotic compression of the stromal specimens is realized through polyethylene glycol (PEG) whose concentration is brought to some target value, e.g., 5% of PEG corresponding to a pressure of 27.75 kPa;
- stage BCD: the pH of the bath is varied between 8 and 3.

Osmotic compression is realized through polyethylene glycol (PEG) with molar mass 20 kg at various concentrations. The pressure, measured in kPa, exerted by PEG can be estimated from the formula,

$$p_{PEG} = 1.28\, c_{PEG} + 0.85\, c_{PEG}^2, \tag{19.11.22}$$

the concentration c_{PEG} being measured in gm of PEG per gm of solution multiplied by 100.

As indicated in Figs. 19.11.5 to 19.11.12, a number of entities show an extremum at the IEP, e.g., specimen volume, fluid volume fraction, hydration, concentrations of PGs and osmotic pressure. When no extra load is applied, the volumes in the hypertonic state and at IEP are identical and minimal.

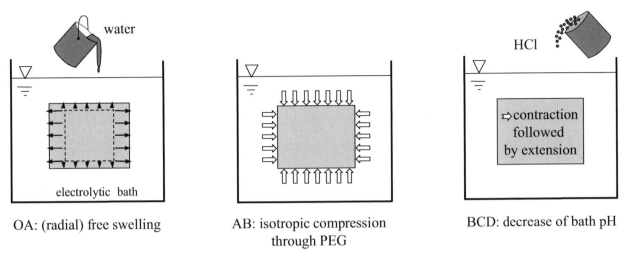

Fig. 19.11.4: Schematic of the three sequences of the chemomechanical tests on corneal stroma by Huang and Meek [1999]: OA: free swelling starting from hypertonic state; AB: isotropic compression through PEG at fixed bath chemical composition; BC: decrease of bath pH.

The shape of the curve representing the charge of PGs as a function of bath pH is the result of mechanochemical coupling. Indeed, since at the IEP, the electrical charge vanishes, the volume is minimal, as indicated in Figs. 19.11.5-19.11.7, and therefore some concentrations may increase, even if, at constant volume, they would decrease (compare the evolution of the charge of PGs in Fig. 19.11.3 at given volume, with the one here where the volume varies as shown in Figs. 19.11.5-19.11.7). The fact that the *mechanochemical coupling is strong* may also be appreciated by considering the significant increase of the concentrations of ionic and fixed charges with the PEG loading, Figs. 19.11.8 to 19.11.11. The phenomenon is to be traced to the low radial moduli which allow both large volume increase and decrease.

Below the IEP, collagen becomes positively charged. Repulsion between close charges of identical sign leads, through macroscopic electrochemical coupling, to large osmotic pressures, and to a large volume expansion along the radial direction, Fig. 19.11.12. The phenomenon can develop fully in absence of applied compression.

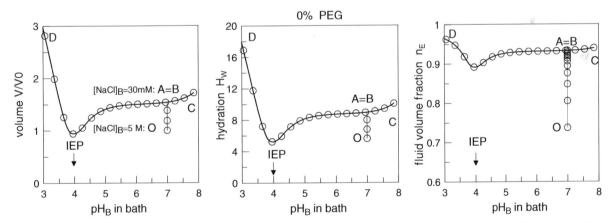

Fig. 19.11.5: Stroma in equilibrium with a bath of controlled chemical composition without mechanical confinement. Evolution of the volume, hydration and fluid volume fraction during the three loading stages. These three mechanical properties show an extremum at the IEP, around 4. Reprinted with permission from Loret and Simões [2010]b.

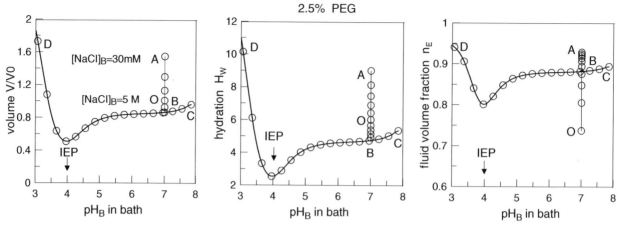

Fig. 19.11.6: Same as Fig. 19.11.5 but specimen subjected to a mechanical load through polyethylene glycol PEG at 2.5%.

19.12   Global constitutive structure: mechanics and transport

The constitutive equations are developed in a thermodynamic framework *à la Biot* where the solid skeleton is taken as reference.

19.12.1   Balance equations

The equations of mass balance are required for all mobile components in the fluid phase but water, and for the fluid phase as a whole. The balance of momentum is required for any component endowed with its own independent velocity. In practice, the balance of momentum of mobile components of the fluid phase is accounted for indirectly through the generalized diffusion relations. The balance of momentum for the mixture as a whole is required in a standard format: it includes, in particular, all the components which move with the velocity $\mathbf{v}_S$ of the solid phase.

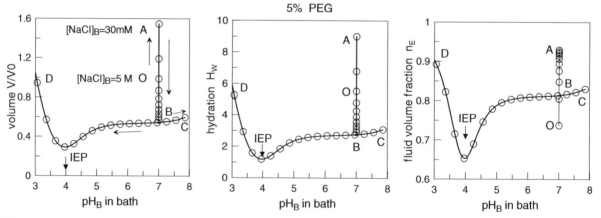

Fig. 19.11.7: Same as Fig. 19.11.5 but specimen subjected to a mechanical load through polyethylene glycol PEG at 5%.

19.12.1.1 Mass, volume fluxes and electrical current density

Balance equations are phrased in terms of several fluxes. *Absolute fluxes* are relative to the solid skeleton, while *diffusive fluxes* of a fluid component refer to the water velocity of the fluid phase,

$$
\begin{array}{llll}
\text{mass flux} & [\text{unit}: \text{kg/m}^2/\text{s}] & \mathbf{M}_{k\mathrm{K}} = \rho^{k\mathrm{K}}\,(\mathbf{v}_{k\mathrm{K}} - \mathbf{v}_{\mathrm{S}}), & k \in \mathrm{K}, \\[4pt]
\text{molar flux} & [\text{unit}: \text{mole/m}^2/\text{s}] & \mathbf{J}_{k\mathrm{K}} = n^{\mathrm{K}}\,c_{k\mathrm{K}}\,(\mathbf{v}_{k\mathrm{K}} - \mathbf{v}_{\mathrm{S}}), & k \in \mathrm{K} - \{\mathrm{w}\}, \\[4pt]
\text{diffusive molar flux} & [\text{unit}: \text{mole/m}^2/\text{s}] & \mathbf{J}_{k\mathrm{K}}^{d} = n^{\mathrm{K}}\,c_{k\mathrm{K}}\,(\mathbf{v}_{k\mathrm{K}} - \mathbf{v}_{\mathrm{wK}}), & k \in \mathrm{K} - \{\mathrm{w}\}, \\[4pt]
\text{fluid volume flux} & [\text{unit}: \text{m/s}] & \mathbf{J}_{\mathrm{wK}} = n^{\mathrm{wK}}\,(\mathbf{v}_{\mathrm{wK}} - \mathbf{v}_{\mathrm{S}})\,. &
\end{array}
\tag{19.12.1}
$$

In the analysis of transport in articular cartilages in Chapters 16 and 17, all fluxes were defined as volume fluxes. There is no fundamental difference between these definitions.

The sum of the volume fluxes in phase K defines the volume averaged flux $\mathbf{J}_{\mathrm{K}}$ of the fluid phase through the solid skeleton,

$$
\mathbf{J}_{\mathrm{K}} = \sum_{k \in \mathrm{K}} n^{k\mathrm{K}}\,(\mathbf{v}_{k\mathrm{K}} - \mathbf{v}_{\mathrm{S}})\,.
\tag{19.12.2}
$$

The *electrical current density* $\mathbf{I}_{\mathrm{eK}}$ in phase K [unit: A/m^2] is defined as the sum of constituent velocities weighted by their valences and molar densities,

$$
\mathbf{I}_{\mathrm{eK}} = \mathrm{F} \sum_{k \in \mathrm{K}} \zeta_k\,\frac{N_{k\mathrm{K}}}{V}\,\mathbf{v}_{k\mathrm{K}}\,.
\tag{19.12.3}
$$

A uniform velocity for all components of a phase satisfying electroneutrality is seen to be a sufficient condition for the electrical current density to vanish in that phase. As a further consequence of electroneutrality, $\mathbf{I}_{\mathrm{eE}}$ may be

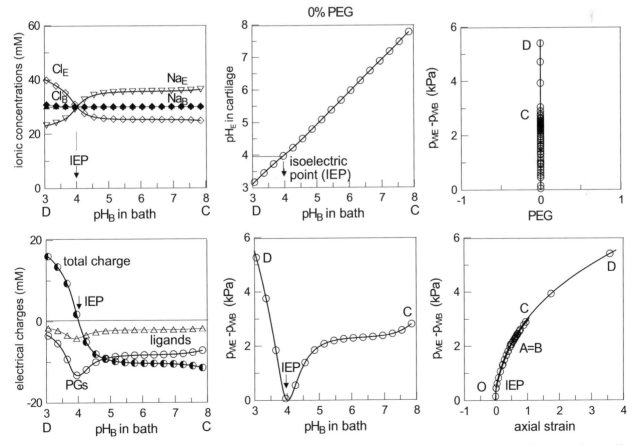

Fig. 19.11.8: Path CD in absence of isotropic compression through PEG. Chemical composition of the stromal extrafibrillar fluid, its pH, variation of the electrical charges and osmotic pressure.

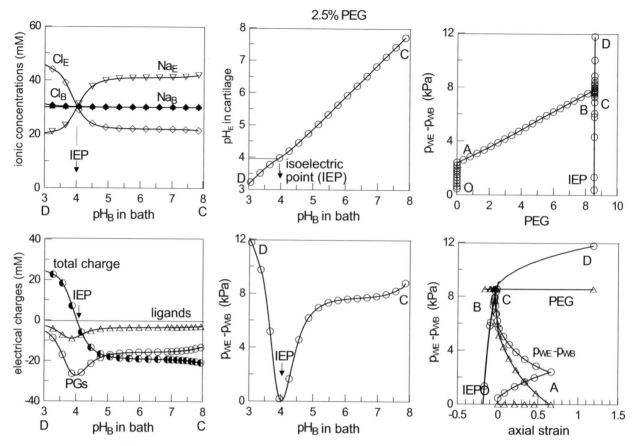

Fig. 19.11.9: Path CD with isotropic compression through PEG to 2.5%.

viewed as a sum of either absolute or diffusive fluxes, namely

$$\mathbf{I}_{eE} = F \sum_{k \in E} \zeta_k \, \mathbf{J}_{kE} = F \sum_{k \in E} \zeta_k \, \mathbf{J}^d_{kE} = F \sum_{k \in E_{ions}} \zeta_k \, \mathbf{J}^d_{kE} - F \, \tilde{e}_e \, \mathbf{J}_{wE} \,, \tag{19.12.4}$$

where $\tilde{e}_e = e_e \, n^E / n^{wE}$ can be approximated by the total charge concentration e_e.

As a side remark, note that the species that move with the velocity of the solid skeleton, e.g., PGs, have a null absolute flux. On the other hand, these species have a nonvanishing diffusive flux. Since their velocity is given, these diffusive fluxes do not come into picture in the generalized constitutive equations. On the other hand, their diffusion properties would be needed if they were not attached to the solid, and left free to diffuse.

19.12.1.2 Balances of masses

Quite generally, the change of mass of a component might be due to *transfer*, i.e., physico-chemical reaction, to *diffusion*, or to *growth*. Here, the components of the fluid phase may undergo mass change only through exchange (diffusion) with the surroundings, namely Eqn (14.6.41),

$$\frac{1}{\det \mathbf{F}} \frac{dm^{kE}}{dt} + \operatorname{div} \mathbf{M}_{kE} = 0 \,, \ k \in E. \tag{19.12.5}$$

The symbol div denotes the divergence operator, and d/dt represents the derivative following the solid phase whose velocity is $\mathbf{v}_S$.

In presence of incompressible components, the change of volume of the solid skeleton, which is the same as that of porous medium, is equal and opposite to the change of volume of the fluid phase due to diffusion, namely Eqn (14.6.49),

$$\operatorname{div} \mathbf{v}_S + \operatorname{div} \mathbf{J}_E = 0 \,. \tag{19.12.6}$$

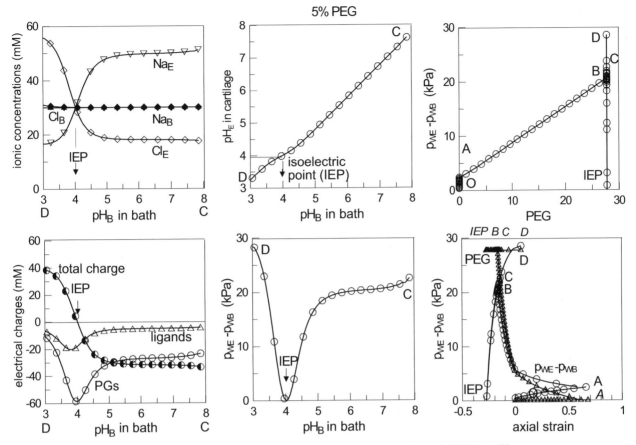

Fig. 19.11.10: Path CD with isotropic compression through PEG to 5%.

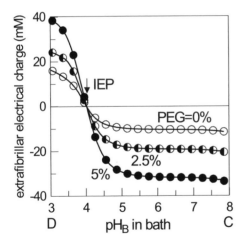

Fig. 19.11.11: Same as Fig. 19.11.5. Extrafibrillar electrical charge. Illustration of the mechanochemical coupling. The mechanical confinement influences significantly the pH, the electrical charge and therefore the electrochemical properties. Reprinted with permission from Loret and Simões [2010]b.

This form of the incompressibility condition is appropriate for transport analysis, while the mechanical version is proposed by (19.16.4).

As another consequence of (19.12.5) and of the electroneutrality in the fluid phase, the electrical current density $\mathbf{I}_{eE}$ defined by (19.12.3), or (19.12.4), may be shown to be divergence free, Eqn (14.6.46),

$$\operatorname{div} \mathbf{I}_{eE} = 0 \,. \tag{19.12.7}$$

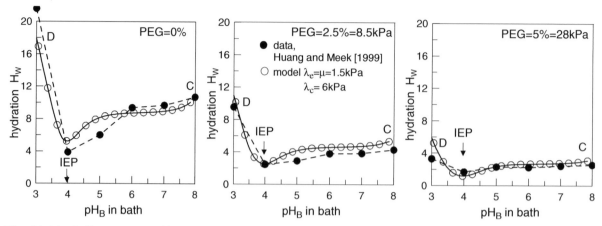

Fig. 19.11.12: Stroma in equilibrium with a bath of controlled chemical composition, and subjected to a mechanical load through polyethylene glycol PEG at 0, 2.5 and 5%. The hydration is minimum at the IEP. The large volume expansion in absence of compression is best captured via a more compliant background substance in extension. Reprinted with permission from Loret and Simões [2010]b.

19.12.1.3 Balance of momentum

The balance of momentum of the porous medium as a whole has the standard format, Eqn (14.6.47),

$$\text{div } \boldsymbol{\sigma} + \rho \, \mathbf{b} = \mathbf{0}, \tag{19.12.8}$$

where $\rho \, \mathbf{b}$ is the body force of the porous medium, built from the contributions of individual components, i.e., $\rho \, \mathbf{b} = \sum_{k,K} \rho^{kK} \, \mathbf{b}_{kK}$. Quite generally, for quasi-static analyses, the body force densities $\mathbf{b}_{kE}$ are equal to the gravity $\mathbf{g}$ and individual accelerations are neglected. For the problem at hand, the body force can be anyway neglected.

19.12.2 Clausius-Duhem inequality

A single inequality for the internal entropy is required for the porous medium as a whole. It results in an expression that contains two terms of distinct natures and which consequently are required to be positive individually, Eqn (14.7.4),

$$
\begin{aligned}
T \hat{\underline{S}}_{(i,1)} &= -\frac{d\mathcal{W}}{dt} + \mathbf{T} : \frac{d\mathbf{E}}{dt} + \sum_{k,K} \mu_{kK} \frac{dm^{kK}}{dt} \geq 0, \\
T \hat{\underline{S}}_{(i,3)} &= -\det \mathbf{F} \sum_{k \in \mathrm{E}} (\boldsymbol{\nabla} \mu_{kE}^{\mathrm{ec}} - \mathbf{b}_{kE}) \cdot \mathbf{M}_{kE} \geq 0.
\end{aligned}
\tag{19.12.9}
$$

The constitutive equations describing the chemo-hyperelastic behavior are constructed in order for the mechanical part of the entropy production $\hat{\underline{S}}_{(i,1)}$ to exactly vanish. Due to phase electroneutrality, the electrical potential does not work. This observation has lead the electrochemical potentials in the entropy production $\hat{\underline{S}}_{(i,1)}$ to be replaced by the chemical potentials. Consequently, the elastic relations do not depend directly on the electrical potential.

Satisfaction of the second inequality motivates generalized diffusion equations, described below.

19.13 Generalized diffusion in the extrafibrillar phase

Within the extrafibrillar phase, species *diffuse* and are advected by water. Their motions are governed by coupled effects including Darcy's law of seepage of water, Fick's law of diffusion of ions and Ohm's law of electrical flow. The vector of volume fluxes $\mathbf{j}$ relative to the solid phase is related to the vector $\mathbf{f}$ of electrochemical gradients,

$$
\mathbf{j} = \begin{bmatrix} \mathbf{J}_{\mathrm{wE}} \\ \mathbf{J}_{\mathrm{NaE}} \\ \mathbf{J}_{\mathrm{ClE}} \end{bmatrix}, \quad
\mathbf{f} = \begin{bmatrix} \boldsymbol{\nabla} g_{\mathrm{wE}}/\widehat{v}_{\mathrm{w}} \\ \boldsymbol{\nabla} g_{\mathrm{NaE}}^{\mathrm{ec}} \\ \boldsymbol{\nabla} g_{\mathrm{ClE}}^{\mathrm{ec}} \end{bmatrix},
\tag{19.13.1}
$$

by the generalized diffusion matrix $\boldsymbol{\kappa}$ exposed in (16.2.29), which is required to be positive semi-definite positive (PsD) in order for the dissipation inequality (19.12.9) to be satisfied,

$$\mathbf{j} = -\boldsymbol{\kappa}\,\mathbf{f}, \quad \boldsymbol{\kappa} = \boldsymbol{\kappa}^{\mathrm{T}}\,\mathrm{PsD}\,. \tag{19.13.2}$$

The matrix $\boldsymbol{\kappa}$ may be expressed in the format,

$$\boldsymbol{\kappa} = \left[\begin{array}{c|cc} \mathbf{k_{EE}} & c_{\mathrm{NaE}}\,\mathbf{k_{EE}} & c_{\mathrm{ClE}}\,\mathbf{k_{EE}} \\ \hline c_{\mathrm{NaE}}\,\mathbf{k_{EE}} & c_{\mathrm{NaE}}^{2}\,\mathbf{k_{EE}} & c_{\mathrm{NaE}}\,c_{\mathrm{ClE}}\,\mathbf{k_{EE}} \\ c_{\mathrm{ClE}}\,\mathbf{k_{EE}} & c_{\mathrm{ClE}}\,c_{\mathrm{NaE}}\,\mathbf{k_{EE}} & c_{\mathrm{ClE}}^{2}\,\mathbf{k_{EE}} \end{array}\right] + \left[\begin{array}{c|cc} \mathbf{0} & \mathbf{0} & \mathbf{0} \\ \hline \mathbf{0} & \mathbf{k}_{\mathrm{NaNa}}^{d} & \mathbf{0} \\ \mathbf{0} & \mathbf{0} & \mathbf{k}_{\mathrm{ClCl}}^{d} \end{array}\right]\,, \tag{19.13.3}$$

in terms of

$$\mathbf{k_{EE}} = \frac{\mathbf{K_{H}}}{\rho_{\mathrm{w}}\,\mathrm{g}} > 0\,, \quad \mathbf{k}_{kk}^{d} = n^{\mathrm{E}}\,c_{k\mathrm{E}}\,\frac{\mathbf{u}_{k}^{*}}{\mathrm{F}\,|\zeta_{k}|}\,, \quad k \in \mathrm{E_{ions}}\,. \tag{19.13.4}$$

Here
- $\mathbf{k_{EE}}$ is the short circuit permeability [unit : $\mathrm{m^{3}/kg{\times}s = m^{2}/Pa/s}$], $\mathbf{K_{H}}$ is referred to as hydraulic conductivity [unit : m/s], $\mathrm{g} = 9.81\,\mathrm{m/s^{2}}$ is the acceleration of gravity, and $\rho_{\mathrm{w}} = 1000\,\mathrm{kg/m^{3}}$ is the mass density of water;
- $\mathbf{u}_{k}^{*}$ [unit : $\mathrm{m^{2}/s/V}$] is the effective ionic mobility of ion k, that is, the ionic mobility times the tortuosity factor.

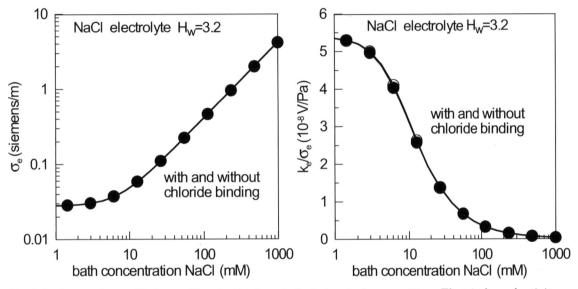

Fig. 19.13.1: Stroma in equilibrium with a bath of controlled chemical composition. Electrical conductivity σ_{e} and coefficient of streaming potential $k_{\mathrm{e}}/\sigma_{\mathrm{e}}$. Chloride binding does not affect these quantities. Data: see text in Section 19.10.3 and hydraulic short circuit permeability equal to $10^{-15}\,\mathrm{m^{2}/Pa/s}$. The coefficient of streaming potential under physiological conditions is in agreement with the value $2.5 \times 10^{-9}\,\mathrm{V/Pa}$ derived by Hodson [1974].

The constitutive diffusion equations may, alternatively but equivalently, be phrased in terms of diffusive fluxes and conjugate driving forces,

$$\mathcal{J} = \left[\begin{array}{c} \mathbf{J_{wE}} \\ \mathbf{J}_{\mathrm{NaE}}^{d} \\ \mathbf{J}_{\mathrm{ClE}}^{d} \\ \mathbf{I_{eE}} \end{array}\right]\,, \quad \mathcal{F} = \left[\begin{array}{c} \mathbf{F_{E}} \\ \mathbf{F}_{\mathrm{NaE}}^{d} \\ \mathbf{F}_{\mathrm{ClE}}^{d} \\ \mathbf{F_{eE}} \end{array}\right] = \left[\begin{array}{c} \boldsymbol{\nabla} p_{\mathrm{wE}} \\ RT\,\boldsymbol{\nabla}\mathrm{Ln}\,x_{\mathrm{NaE}} \\ RT\,\boldsymbol{\nabla}\mathrm{Ln}\,x_{\mathrm{ClE}} \\ \boldsymbol{\nabla}\phi_{\mathrm{E}} \end{array}\right]\,, \tag{19.13.5}$$

which are then related by the symmetric arrowhead diffusion matrix, $\mathcal{J} = -\mathcal{K}\,\mathcal{F}$, where

$$\mathcal{K} = \left[\begin{array}{cccc} \mathbf{k_{EE}} & \mathbf{0} & \mathbf{0} & \mathbf{k_{e}} \\ \mathbf{0} & \mathbf{k}_{\mathrm{NaNa}}^{d} & \mathbf{0} & \mathbf{k}_{\mathrm{Nae}}^{d} \\ \mathbf{0} & \mathbf{0} & \mathbf{k}_{\mathrm{ClCl}}^{d} & \mathbf{k}_{\mathrm{Cle}}^{d} \\ \mathbf{k_{e}} & \mathbf{k}_{\mathrm{eNa}}^{d} & \mathbf{k}_{\mathrm{eCl}}^{d} & \sigma_{\mathrm{e}} \end{array}\right]\,, \tag{19.13.6}$$

which features the total charge concentration e_e, the electroosmotic coefficient $\mathbf{k}_e$ [unit : $m^2/s/V$], and the electrical conductivity $\boldsymbol{\sigma}_e$ [unit : siemens/m]. Indeed,

$$\mathbf{k}_{ke}^d = n^E c_{kE} \mathbf{u}_k^* \operatorname{sgn} \zeta_k , \quad k \in \mathrm{E_{ions}},$$

$$\mathbf{k}_e = -\mathbf{k}_{EE} \operatorname{F} \tilde{e}_e, \quad \boldsymbol{\sigma}_e = n^E \operatorname{F} \sum_{k \in \mathrm{E_{ions}}} |\zeta_k| c_{kE} \mathbf{u}_k^* + \mathbf{k}_{EE} \operatorname{F}^2 \tilde{e}_e^2 . \tag{19.13.7}$$

Material data are reported in Section 19.4.5. The influence of ion concentration on the electrical conductivity and streaming potential are displayed in Fig. 19.13.1.

Remark on a reflection coefficient:

Even if not apparent, an osmotic, or reflection, coefficient is contained in the formulation, as depicted in Loret and Simões [2007], or Chapter 16.

Remark on out-of-diagonal terms in the diffusion matrices:

Note that the out-of-diagonal ion-ion interaction term in (19.13.3) is not negligible, in contrast to (19.13.6). Indeed, for a tortuosity factor of 0.5, a hydraulic permeability $k_{EE} = 10^{-15}\,m^2/Pa/s$, an ionic mobility of $50 \times 10^{-9}\,m^2/s/V$, and for concentrations of 0.1 M,

$$\frac{c_{kE}\, c_{lE}\, k_{EE}}{k_{kk}^d} \sim 0.4, \quad k,\, l \in \mathrm{E_{ions}} . \tag{19.13.8}$$

The same remark applies to the ion-water interaction terms. For $\boldsymbol{\nabla} p_{wE} \sim RT\, \boldsymbol{\nabla} c_{kE}$, the contributions to the water flux in (19.13.3) due to the gradient of pressure (diagonal term) and to the chemical (out-of-diagonal term) have comparable magnitudes.

Explicit expressions of the fluxes:

Some approximations are used routinely,
- the gradient of the enthalpies of formation is assumed to vanish;
- the pressure terms in the electrochemical potentials of ions are neglected;
- the *intrinsic* concentration of the fixed charge is neglected with respect to the ionic concentrations.

Then, from the definitions (19.8.2) of concentrations and (19.8.4) of the electrochemical potentials, the gradients driving the absolute fluxes become,

$$\boldsymbol{\nabla} g_{wE}/\widehat{v}_w = \boldsymbol{\nabla} p_{wE} - RT\, \boldsymbol{\nabla}\big(c_{NaE} + c_{ClE}\big),$$

$$c_{kE}\, \boldsymbol{\nabla} g_{kE}^{ec} = RT\, \boldsymbol{\nabla} c_{kE} + \zeta_k\, c_{kE}\, \operatorname{F}\boldsymbol{\nabla}\phi_E, \quad k \in \mathrm{E_{ions}} . \tag{19.13.9}$$

The constitutive relations in terms of either absolute or diffusive fluxes yield the volume flux of water,

$$\mathbf{J}_{wE} = -\mathbf{k}_{EE}\, \boldsymbol{\nabla} p_{wE} - \mathbf{k}_e\, \boldsymbol{\nabla}\phi_E , \tag{19.13.10}$$

the ionic molar fluxes,

$$\mathbf{J}_{kE} = -n^E\, \frac{\mathbf{u}_k^*}{|\zeta_k|}\, \frac{RT}{\operatorname{F}}\, \boldsymbol{\nabla} c_{kE} - n^E\, c_{kE}\, \mathbf{u}_k^* \operatorname{sgn}\zeta_k\, \boldsymbol{\nabla}\phi_E + \frac{n^E}{n^{wE}}\, c_{kE}\, \mathbf{J}_{wE}, \quad k \in \mathrm{E_{ions}} , \tag{19.13.11}$$

and the electrical current density,

$$\mathbf{I}_{eE} = -\mathbf{k}_e\, \boldsymbol{\nabla} p_{wE} - RT\, n^E \sum_{k \in \mathrm{E_{ions}}} \mathbf{u}_k^* \operatorname{sgn}\zeta_k\, \boldsymbol{\nabla} c_{kE} - \boldsymbol{\sigma}_e\, \boldsymbol{\nabla}\phi_E . \tag{19.13.12}$$

Explicit expressions of the fluxes for an electrically open circuit:

If the circuit is open, namely $\mathbf{I}_{eE} = \mathbf{0}$, the electrical field,

$$\boldsymbol{\nabla}\phi_E = -(\boldsymbol{\sigma}_e)^{-1} \cdot \mathbf{k}_e \cdot \boldsymbol{\nabla} p_{wE} - RT\, (\boldsymbol{\sigma}_e)^{-1} \cdot \sum_{l \in \mathrm{E_{ions}}} \mathbf{k}_{el}^d \cdot \frac{\boldsymbol{\nabla} c_{lE}}{c_{lE}} , \tag{19.13.13}$$

can be eliminated from the expressions of the fluxes of water,

$$\mathbf{J}_{wE} = -\mathbf{k}_H \cdot \boldsymbol{\nabla} p_{wE} + RT\, \mathbf{k}_e \cdot (\boldsymbol{\sigma}_e)^{-1} \cdot \sum_{l \in \mathrm{E_{ions}}} \mathbf{k}_{el}^d \cdot \frac{\boldsymbol{\nabla} c_{lE}}{c_{lE}} , \tag{19.13.14}$$

and of ions $k \in \mathrm{E_{ions}}$,

$$\begin{aligned}
\mathbf{J}_{kE} =\; & -\Big(\frac{n^E}{n^{wE}}\, c_{kE}\, \mathbf{k}_H - \mathbf{k}_{ke}^d \cdot (\boldsymbol{\sigma}_e)^{-1} \cdot \mathbf{k}_e\Big) \cdot \boldsymbol{\nabla} p_{wE} \\
& - RT \sum_{l \in \mathrm{E_{ions}}} \Big(\mathbf{k}_{kl}^d - \mathbf{k}_{ke}^d \cdot (\boldsymbol{\sigma}_e)^{-1} \cdot \mathbf{k}_{el}^d - \frac{n^E}{n^{wE}}\, c_{kE}\, \mathbf{k}_e \cdot (\boldsymbol{\sigma}_e)^{-1} \cdot \mathbf{k}_{el}^d\Big) \cdot \frac{\boldsymbol{\nabla} c_{lE}}{c_{lE}} ,
\end{aligned} \tag{19.13.15}$$

where $\mathbf{k}_H = \mathbf{k}_{EE} - \mathbf{k}_e \cdot (\boldsymbol{\sigma}_e)^{-1} \cdot \mathbf{k}_e$ is the "open circuit" hydraulic permeability.

19.14 Rest potential and active fluxes at the cell scale

We address first the difference of potential across a membrane associated with a single ion, and distinguish
- passive transport, due to diffusion, where *electrochemical equilibrium* may take place after a transient period, and
- active transport where electrochemical equilibrium does *not* take place, and instead, the *steady state* is characterized by vanishing total ionic fluxes.

Still, the above analysis is somehow paradoxical because it does not provide the reason why the concentrations are not equal across the membrane. In general, a number of ions may diffuse through cellular membranes. There may also be special channels through the membrane, which may be specific to a single ion type or to a limited number of ion types. The permeability of these channels may also vary in time, e.g., in muscle cells. The membrane may be seen as a capacitor. At rest, that is, at steady state, the difference of potential V_{rest} across the two faces of the membrane is compensated by ionic pumping. The analytical expression of this difference of potential is usually expressed through two more or less equivalent formulas, the Huxley-Hodgkin-Katz (HHK) formula and the chord conductance equation.

The results of this section rest on an *uncoupled* analysis: the ionic diffusive flux through a membrane of a given ion depends only on its own electrochemical potential. A membrane model accounting for coupled fluxes is built in Section 19.15.

19.14.1 Diffusive fluxes across a membrane

The ionic flux [unit : mole/m^2/s] for a specific ion k is assumed to be proportional to the gradient of electrochemical potential[19.3] $g_k^{ec} = RT \, \mathrm{Ln} \, (c_k/c_0) + \zeta_k \, \mathrm{F} \, \phi$,

$$J_k^{(p)} = -n \, u_k \, \mathrm{sgn}\, \zeta_k \, \frac{c_k}{\zeta_k \mathrm{F}} \, \frac{d}{dx} (RT \, \mathrm{Ln} \, \frac{c_k}{c_0} + \zeta_k \, \mathrm{F} \, \phi), \qquad (19.14.1)$$

where c_k is the concentration of the ion, ζ_k its valence, u_k its *mobility* [unit : m^2/s/V], c_0 an arbitrary concentration, n the volume fraction of water in the membrane, and x is a coordinate across the membrane. Assume the electric potential ϕ to be linear throughout the membrane, i.e., $d\phi/dx = V/h$, with $V = \phi_i - \phi_o$ and $h = x_i - x_o$ the thickness of the membrane [unit : m]. The subscripts o and i denote the **o**uter and **i**nner faces of the membrane. Then (19.14.1) may be recast in the format,

$$J_k^{(p)} = -n \, u_k \, \mathrm{sgn}\, \zeta_k \, \frac{RT}{\zeta_k \mathrm{F}} \, \exp\left(-\frac{\zeta_k \mathrm{F}}{RT} \phi\right) \frac{d}{dx} \left(c_k \exp\left(\frac{\zeta_k \mathrm{F}}{RT} \phi\right)\right), \qquad (19.14.2)$$

and integrated to

$$J_k^{(p)} = -n \, u_k \, \mathrm{sgn}\, \zeta_k \, \frac{V}{h} \, \frac{c_{ok} - c_{ik} \exp\left(\frac{\zeta_k \mathrm{F}}{RT} V\right)}{1 - \exp\left(\frac{\zeta_k \mathrm{F}}{RT} V\right)}. \qquad (19.14.3)$$

The permeability P_k [unit : m/s] to ion k of the membrane of thickness h is defined in terms of the ionic mobility u_k, or of the diffusion coefficient D_k [unit : m^2/s], linked by the Nernst-Einstein relation $u_k = D_k \, |\zeta_k| \, \mathrm{F}/RT$,

$$P_k = \frac{u_k}{h} \, \frac{RT}{|\zeta_k| \mathrm{F}} = \frac{D_k}{h}. \qquad (19.14.4)$$

The resulting expression of the ionic flux as a function of the potential difference across the membrane and ionic concentrations,

$$J_k^{(p)} = -n \, P_k \, \frac{\zeta_k \mathrm{F}}{RT} \, \frac{c_{ok} - c_{ik} \exp\left(\frac{\zeta_k \mathrm{F}}{RT} V\right)}{1 - \exp\left(\frac{\zeta_k \mathrm{F}}{RT} V\right)} \, V, \qquad (19.14.5)$$

vanishes at electrochemical equilibrium $(g_k^{ec})_i - (g_k^{ec})_o = 0$.

[19.3] The dimensionless molar fraction should enter the electrochemical potential rather than the concentration. The point is elusive at small concentrations where the ratio of the two quantities is equal to the molar volume of water.

19.14.2   Reverse Nernst potential associated with an ion

Let us assume that the electrochemical potentials of an ion k on both sides of the membrane can be equal, that is, electrochemical equilibrium can be realized. The resulting difference of electrical potential across the two faces of the membrane is the so-called reverse Nernst potential,

$$E_k = \phi_i - \phi_o = -\frac{RT}{\zeta_k \mathrm{F}} \mathrm{Ln}\, \frac{c_{ik}}{c_{ok}} . \qquad (19.14.6)$$

19.14.3   Reverse Nernst potential in presence of active transport

Instead, in the case of active transport, chemical equilibrium does not take place. Indeed, the flux J is contributed by a passive part $J^{(p)}$ and an active part $J^{(a)}$, Eqn (19.3.11).

The rest potential at steady state can be derived by considering a vanishing total flux, namely

$$J_k = J_k^{(a)} + J_k^{(p)} = J_k^{(a)} - n\, P_k\, \frac{c_{ok} - c_{ik}\, \exp\left(\dfrac{\zeta_k \mathrm{F}}{RT} V\right)}{1 - \exp\left(\dfrac{\zeta_k \mathrm{F}}{RT} V\right)}\, \frac{\zeta_k \mathrm{F}}{RT}\, V = 0 . \qquad (19.14.7)$$

This relation can be interpreted in two ways:
   - if the active flux is known, it represents an implicit equation for the rest potential, which is clearly different from the equilibrium potential given by (19.14.6);
   - conversely, if the rest potential is known, the active flux can be obtained.

19.14.4   Huxley-Hodgkin-Katz (HHK) formulas in presence of several ions

Huxley-Hodgkin-Katz formula for non-electrogenic pumps

Since fluxes J are contributed by a passive part $J^{(p)}$ and an active part $J^{(a)}$, so does the electrical current density $I_{\mathrm{e}} = \mathrm{F} \sum_k \zeta_k J_k$. For non-electrogenic exchangers and pumps,

$$I_{\mathrm{e}}^{(a)} = \mathrm{F} \sum_k \zeta_k J_k^{(a)} = 0 . \qquad (19.14.8)$$

Therefore at steady state,

$$I_{\mathrm{e}} = I_{\mathrm{e}}^{(a)} + I_{\mathrm{e}}^{(p)} = 0 \quad \Rightarrow \quad I_{\mathrm{e}}^{(p)} = \mathrm{F} \sum_k \zeta_k J_k^{(p)} = 0 . \qquad (19.14.9)$$

The HHK formula is based on approximating $\zeta_k V$ by $\epsilon_k V_{\mathrm{rest}}$ with $\epsilon_k = \zeta_k/|\zeta_k|$, namely

$$\mathrm{F}\, \zeta_k J_k^{(p)} = -n\, \mathrm{F}\, \frac{P_k\, |\zeta_k|\, c_{ok} - P_k\, |\zeta_k|\, c_{ik}\, \exp\left(\dfrac{\epsilon_k \mathrm{F}}{RT} V_{\mathrm{rest}}\right)}{1 - \exp\left(\dfrac{\epsilon_k \mathrm{F}}{RT} V_{\mathrm{rest}}\right)}\, \frac{\mathrm{F}}{RT}\, V_{\mathrm{rest}} . \qquad (19.14.10)$$

As an example, let us consider the two non-electrogenic pumps Na^+-K^+ and Na^+-Cl^-, each one using one ATP. Then

$$V_{\mathrm{rest}} = \frac{RT}{\mathrm{F}} \mathrm{Ln}\, \frac{P_{\mathrm{K}}\, [K^+]_o + P_{\mathrm{Na}}\, [Na^+]_o + P_{\mathrm{Cl}}\, [Cl^-]_i}{P_{\mathrm{K}}\, [K^+]_i + P_{\mathrm{Na}}\, [Na^+]_i + P_{\mathrm{Cl}}\, [Cl^-]_o} . \qquad (19.14.11)$$

If an ion is not pumped, it may not be involved since the rest potential has been defined as due to pumped ions. Indeed, then, at steady state, the passive flux of this ion vanishes: accordingly its inner and outer electrochemical potentials are equal, as indicated by (19.14.6). For example, if we consider only the Na^+-K^+ pump, chloride anions are not pumped, and then

$$V_{\mathrm{rest}} = -\frac{RT}{\mathrm{F}} \mathrm{Ln}\, \frac{[Cl^-]_o}{[Cl^-]_i} = \frac{RT}{\mathrm{F}} \mathrm{Ln}\, \frac{P_{\mathrm{Cl}}\, [Cl^-]_i}{P_{\mathrm{Cl}}\, [Cl^-]_o} . \qquad (19.14.12)$$

With help of elementary algebra, namely $a/b = c/d = (a+c)/(b+d)$ for any real numbers a, $b \neq 0$, c, $d \neq 0$, $d \neq -b$, the expected relation is obtained by equating (19.14.11) and (19.14.12),

$$V_{\mathrm{rest}} = \frac{RT}{\mathrm{F}} \mathrm{Ln}\, \frac{P_{\mathrm{K}}\, [K^+]_o + P_{\mathrm{Na}}\, [Na^+]_o}{P_{\mathrm{K}}\, [K^+]_i + P_{\mathrm{Na}}\, [Na^+]_i} . \qquad (19.14.13)$$

For ions of higher valences, e.g., calcium ions, the relation (19.14.9) becomes an implicit equation for V_{rest}. For example, sodium, potassium and calcium ions undergo active pumping across cardiomocytes. Still the approximate relation

$$V_{\text{rest}} = \frac{RT}{F} \, \text{Ln} \, \frac{P_K \, [K^+]_o + P_{Na} \, [Na^+]_o + 2 \, P_{Ca} \, [Ca^{2+}]_o}{P_K \, [K^+]_i + P_{Na} \, [Na^+]_i + 2 \, P_{Ca} \, [Ca^{2+}]_i} \, , \tag{19.14.14}$$

is well verified experimentally, Sperelakis [2001], p. 210. Usually, Ca^{2+} can be ignored in this relation even if its permeability was up to 10 times larger than that of Na^+ because its intracellular and extracellular concentrations are comparatively much smaller. Then, given the ionic concentrations, the rest potential is determined by the relative permeability P_{Na}/P_K of ions K^+ and Na^+, their absolute permeabilities being elusive. For $[K^+]_o = 4\,\text{mM}$, $[K^+]_i = 139\,\text{mM}$, $[Na^+]_o = 145\,\text{mM}$, $[Na^+]_i = 12\,\text{mM}$, a ratio $P_{Na}/P_K = 0.1$ gives a rest potential[19.4] of -54.1 mV and a ratio of 0.01 gives a rest potential of -86.5 mV. These numbers apply well for cardiomyocites and they are quite close to those of corneal cells, Fig. 19.3.1.

As a general trend, the ions with the largest permeability dominate the rest potential. In this context, it is worth to mention muscle cells where the permeabilities to sodium and potassium, which are controlled by external factors, vary in time. The rest potential oscillates between the reverse potential of sodium and the reverse potential of potassium.

Huxley-Hodgkin-Katz formula for electrogenic pumps

For electrogenic pumps, the electrical current contributed by active transport does not vanish,

$$I_e^{(a)} = F \sum_k \zeta_k \, J_k^{(a)} \neq 0 \, . \tag{19.14.15}$$

Therefore at steady state,

$$I_e = 0 \quad \Rightarrow \quad I_e^{(p)} = F \sum_k \zeta_k \, J_k^{(p)} = -I_e^{(a)} \neq 0 \, . \tag{19.14.16}$$

The HHK formulas above have to be modified for electrogenic pumps and exchangers.

Let us consider the Na^+-K^+-ATPase already touched upon in Section 19.3.3.3. The fluxes may be cast in the format,

$$J_{Na} = 3 \, J_{\text{Na-K-ATPase}} + J_{Na}^{(p)}, \quad J_K = -2 \, J_{\text{Na-K-ATPase}} + J_K^{(p)} \, , \tag{19.14.17}$$

in terms of the passive fluxes given by (19.14.5) and of the pump flux $J_{\text{Na-K-ATPase}}$. In a stationary state, the total fluxes of sodium and potassium vanish, so that

$$2 \, J_{Na}^{(p)} + 3 \, J_K^{(p)} = 0 \, , \tag{19.14.18}$$

and therefore,

$$V_{\text{rest}} = \frac{RT}{F} \, \text{Ln} \, \frac{3 \, P_K \, [K^+]_o + 2 \, P_{Na} \, [Na^+]_o}{3 \, P_K \, [K^+]_i + 2 \, P_{Na} \, [Na^+]_i} \, . \tag{19.14.19}$$

For the same concentrations as above, ratios $P_{Na}/P_K = 0.1$ and 0.01 yield rest potentials of -62.1 mV and -89 mV, respectively.

19.14.5   The chord conductance equation

An alternative to the HHK formula is the *chord conductance equation*. It considers ionic fluxes/currents in parallel. It involves the reverse potentials of actively pumped ions, say sodium, potassium and calcium, and their conductances σ_K, σ_{Na} and σ_{Ca}, namely

$$V_{\text{rest}} = \frac{\sigma_K \, E_K + \sigma_{Na} \, E_{Na} + \sigma_{Ca} \, E_{Ca}}{\sigma_K + \sigma_{Na} + \sigma_{Ca}} \simeq \frac{\sigma_K \, E_K + \sigma_{Na} \, E_{Na}}{\sigma_K + \sigma_{Na}} \, . \tag{19.14.20}$$

So given the ionic concentrations (and hence the reverse potentials), the rest potential depends only on the relative conductance σ_{Na}/σ_K. For example, for $E_K = -92\,\text{mV}$, $E_{Na} = -65\,\text{mV}$, V_{rest} is equal to -78 mV for $\sigma_{Na}/\sigma_K = 0.1$ and -90 mV for $\sigma_{Na}/\sigma_K = 0.01$, Sperelakis [2001]. Clearly the rest potential is dominated by the ions whose conductance is largest.

[19.4]At 37 °C, $RT/F = 26.71\,\text{mV}$.

The chord conductance equation does not apply for small concentrations of $[K^+]_o$ for which E_K becomes very large, while the HHK formula works, at least in principle, in this circumstance, Sperelakis [2001], p. 249.

The conductance [unit : siemens S/m^2] is defined as the ratio of electrical current density over electrical potential,

$$\sigma_k \equiv -\lim_{J_k \to 0} F\,\zeta_k\,\frac{dJ_k^{(p)}}{dV} = n\,P_k\,c_{ok}\left(\frac{\zeta_k F}{RT}\right)^2 \frac{\zeta_k F\,E_k}{\exp\left(\dfrac{\zeta_k F}{RT}\,E_k\right) - 1}, \tag{19.14.21}$$

where the expression of the flux (19.14.5) has been adopted and the difference of potential E_k at vanishing current is defined by (19.14.6).

For infinitesimal ionic flux/current of ion k, the cellular concentration c_{ik} may be expressed in terms of the reverse Nernst potential of ion k, Eqn (19.14.6), and then

$$F\,\zeta_k\,J_k^{(p)} = -\sigma_k\,(V - E_k). \tag{19.14.22}$$

For non-electrogenic exchangers, the passive electrical current vanishes at steady state, Eqn (19.14.9). Requiring the sum of the individual contributions (19.14.22) to vanish yields the rest potential (19.14.20).

19.15   Coupled diffusion across a membrane and active transport

19.15.1   Work-conjugate pairs for diffusion across a membrane

The continuum context

In a continuum, the contribution by species k to the dissipation inequality associated with diffusion may be expressed in terms of a mass flux $\mathbf{J}_{kM}$, a volume flux $\mathbf{J}_{kV}$, or a molar flux $\mathbf{J}_{kC}$,

$$\mathbf{J}_{kM} = \rho_k\,\frac{N_k\,\widehat{v}_k}{V}\,(\mathbf{v}_k - \mathbf{v}_S), \quad \mathbf{J}_{kV} = \frac{N_k\,\widehat{v}_k}{V}\,(\mathbf{v}_k - \mathbf{v}_S), \quad \mathbf{J}_{kC} = \frac{N_k}{V}\,(\mathbf{v}_k - \mathbf{v}_S). \tag{19.15.1}$$

The volume fraction $n^k = N_k\,\widehat{v}_k/V$ and concentration $c_k = N_k/V$ have been written in explicit form, N_k being the number of moles in the current volume V. These expressions hold for a fluid phase. In a porous medium, the volume of reference V for the volume fraction is the total volume but the concentrations refer to the fluid phase.

With μ_k and g_k mass based and mole based (electro)chemical potentials respectively,

$$g_k = \widehat{m}_k\,\mu_k, \tag{19.15.2}$$

the contribution of species k to the dissipation inequality may be cast in the three formats,

$$-\mathbf{J}_{kM} \cdot \boldsymbol{\nabla}\mu_k = -\mathbf{J}_{kV} \cdot \underbrace{\rho_k\,\boldsymbol{\nabla}\mu_k}_{=\boldsymbol{\nabla}\,g_k/\widehat{v}_k} = -\mathbf{J}_{kC} \cdot \boldsymbol{\nabla}g_k. \tag{19.15.3}$$

The work done by the (electro)chemical potential may similarly be cast in the three formats,

$$-\frac{M_k}{V}\,\mu_k = -\frac{V_k}{V}\,\frac{g_k}{\widehat{v}_k} = -\frac{N_k}{V}\,g_k. \tag{19.15.4}$$

The interface/membrane context

We leave now the continuum context, and turn to the problem of an interface/membrane. Let $\mathbf{n}$ be the unit outward normal to the membrane, which has a finite thickness. Two assumptions are adopted:

  - the motions are directed along this normal only;
  - the fluxes are spatially uniform across the interface/membrane. This simplification amounts to assume that the quantities the fluxes depend on are themselves uniform.

Let us consider an arbitrary section of the membrane, Fig. 19.15.1. Integration of the dissipation inequality along the section yields the product of a flux times the difference of the electrochemical potentials at the extremities of the section. Now, in absence of active transport, the electrochemical potentials are continuous at the boundaries between the membrane and the aqueous humor on one side, and the membrane and the stroma on the other side, unlike a number of other quantities, e.g., concentrations etc. When the integration is performed across the whole membrane,

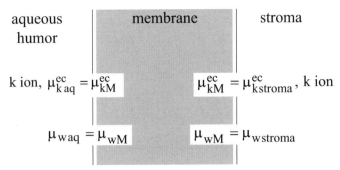

Fig. 19.15.1: In absence of active transport, electrochemical potentials are continuous at the boundaries of the endothelium (and of the epithelium).

the electrochemical potentials at the extremities are therefore the electrochemical potentials in the aqueous humor and in the stroma. In that vein, the diffusion equations developed in Section 19.15.2.1 involve driving forces which are continuous at the boundaries of the stroma, while the same can not be said for driving forces of the format used in Section 19.15.2.2.

The operator Δ denoting a spatial difference across the membrane, the contribution of this section to the dissipation inequality may be cast in the three formats, namely mass, volume or mole format,

$$ - J_{kM}\,\Delta\mu_k = -J_{kV}\,\frac{\Delta g_k}{\widehat{v}_k} = -J_{kC}\,\Delta g_k\,, \tag{19.15.5} $$

with the fluxes,

$$ J_{kM} = \rho_k\,\frac{N_k\,\widehat{v}_k}{V}\,(\mathbf{v}_k - \mathbf{v}_S)\cdot\mathbf{n}\,, \quad J_{kV} = \frac{N_k\,\widehat{v}_k}{V}\,(\mathbf{v}_k - \mathbf{v}_S)\cdot\mathbf{n}\,, \quad J_{kC} = \frac{N_k}{V}\,(\mathbf{v}_k - \mathbf{v}_S)\cdot\mathbf{n}\,. \tag{19.15.6} $$

The molar flux J_{kC} [unit: mole/m^2/s] which is preferred by, e.g., Kedem and Katchalsky [1961] can be interpreted as the number of moles crossing the membrane per unit area (of porous medium if V refers to the volume of the porous medium) in a unit time.

19.15.2 Dissipation in terms of absolute and diffusive fluxes

19.15.2.1 Dissipation in a membrane in terms of absolute fluxes

Use will be made of the volume flux for water [unit: m/s] and of the molar fluxes for ions. Placing the membrane in the context of porous media, and accounting for a metabolic source of energy, the dissipation across the whole membrane expresses through the sum,

$$ - J_{\mathrm{wM}}\,\frac{\Delta g_{\mathrm{wM}}}{\widehat{v}_{\mathrm{w}}} - \sum_{k\in\mathrm{M_{ions}}} J_{kM}\,\Delta g_{kM}^{\mathrm{ec}} - \mathcal{J}\,\mathcal{G}\,, \tag{19.15.7} $$

with $\mathrm{M_{ions}}$ the set of ions and $\mathrm{M_{mo}}$ the set of mobile species (ions and water) in the membrane, and

$$ J_{\mathrm{wM}} = n^{\mathrm{wM}}\,(\mathbf{v}_{\mathrm{wM}} - \mathbf{v}_S)\cdot\mathbf{n}\,, \quad J_{kM} = n^{\mathrm{M}}c_{kM}\,(\mathbf{v}_{kM} - \mathbf{v}_S)\cdot\mathbf{n}\,, \quad k\in\mathrm{M_{ions}}\,, \tag{19.15.8} $$

the relations (19.8.2) linking volume fractions and concentrations holding also in the membrane with M in place of E.

Close to equilibrium/or steady state, the dissipation inequality will be satisfied by assuming a linear dependence between fluxes and driving forces, in the format,

$$ \begin{bmatrix} J_{\mathrm{wM}} \\ J_{\mathrm{NaM}} \\ J_{\mathrm{ClM}} \\ \mathcal{G} \end{bmatrix} = - \begin{bmatrix} \kappa_{ww} & \kappa_{w\,\mathrm{Na}} & \kappa_{w\,\mathrm{Cl}} & \kappa_{wr} \\ \kappa_{\mathrm{Na}w} & \kappa_{\mathrm{NaNa}} & \kappa_{\mathrm{NaCl}} & \kappa_{\mathrm{Na}r} \\ \kappa_{\mathrm{Cl}w} & \kappa_{\mathrm{ClNa}} & \kappa_{\mathrm{ClCl}} & \kappa_{\mathrm{Cl}r} \\ \kappa_{rw} & \kappa_{r\mathrm{Na}} & \kappa_{r\mathrm{Cl}} & \kappa_{rr} \end{bmatrix} \begin{bmatrix} \Delta g_{\mathrm{wM}}/\widehat{v}_{\mathrm{w}} \\ \Delta g_{\mathrm{NaM}}^{\mathrm{ec}} \\ \Delta g_{\mathrm{ClM}}^{\mathrm{ec}} \\ \mathcal{J} \end{bmatrix}\,, \tag{19.15.9} $$

where the 4×4 diffusion matrix $\boldsymbol{\kappa}_{\mathrm{M}}$ is requested to be symmetric and positive semi-definite. From the definitions (19.8.4) of the electrochemical potentials,

$$
\begin{aligned}
\Delta g_{\mathrm{wM}}/\widehat{v}_{\mathrm{w}} &= \Delta p_{\mathrm{wM}} - \Delta \pi_{\mathrm{M}} & [\text{unit}: \mathrm{M/m^2}], \\
\Delta g_{k\mathrm{M}}^{\mathrm{ec}} &= RT\,\Delta\mathrm{Ln}\,c_{k\mathrm{M}} + \zeta_k\,\mathrm{F}\,\Delta\phi_{\mathrm{M}}, \quad k \in \mathrm{M_{ions}}, & [\text{unit}: \mathrm{M} \times \mathrm{m/mole}],
\end{aligned}
\tag{19.15.10}
$$

where π_{M} is the osmotic pressure[19.5],

$$
\pi_{\mathrm{M}} = RT\,(\,c_{\mathrm{NaM}} + c_{\mathrm{ClM}}\,).
\tag{19.15.14}
$$

19.15.2.2   Dissipation in a membrane in terms of diffusive fluxes

The dissipation inequality may be recast in terms of diffusive ionic fluxes, namely

$$
-J_{\mathrm{wM}}\,\Delta P_{\mathrm{wM}} - \sum_{k \in \mathrm{M_{ions}}} J_{k\mathrm{M}}^d\,RT\,\Delta\mathrm{Ln}\,c_{k\mathrm{M}} - I_{\mathrm{eM}}\,\Delta\phi_{\mathrm{M}} - \mathcal{J}\,\mathcal{G},
\tag{19.15.15}
$$

with[19.6]

$$
\begin{aligned}
\Delta P_{\mathrm{wM}} &= \Delta p_{\mathrm{wM}} - \Delta\pi_{\mathrm{M}} + \sum_{k \in \mathrm{M_{ions}}} \frac{n^{\mathrm{M}}}{n^{\mathrm{wM}}}\,c_{k\mathrm{M}}\,RT\,\Delta\mathrm{Ln}\,c_{k\mathrm{M}} \\
&\simeq \Delta p_{\mathrm{wM}} + RT \sum_{k \in \mathrm{M_{ions}}} c_{k\mathrm{M}}\,\Delta\mathrm{Ln}\,c_{k\mathrm{M}} - \Delta c_{k\mathrm{M}}.
\end{aligned}
\tag{19.15.16}
$$

The diffusive ionic fluxes are relative to water,

$$
J_{k\mathrm{M}}^d = n^{\mathrm{M}} c_{k\mathrm{M}}\,(\mathbf{v}_{k\mathrm{M}} - \mathbf{v}_{\mathrm{w}}) \cdot \mathbf{n} = J_{k\mathrm{M}} - c_{k\mathrm{M}}\,\frac{n^{\mathrm{M}}}{n^{\mathrm{wM}}}\,J_{\mathrm{wM}}, \quad k = \mathrm{Na, Cl},
\tag{19.15.17}
$$

and the electrical current density,

$$
I_{\mathrm{eM}} = \mathrm{F} \sum_{k \in \mathrm{M}} \zeta_k\,J_{k\mathrm{M}} = \mathrm{F} \sum_{k \in \mathrm{M_{ions}}} \zeta_k\,J_{k\mathrm{M}}^d - \mathrm{F}\,\frac{n^{\mathrm{M}}}{n^{\mathrm{wM}}}\,e_{\mathrm{eM}}\,J_{\mathrm{wM}},
\tag{19.15.18}
$$

expresses in terms of the diffusive fluxes of all charged species, fixed charge included. Note that the diffusive flux of the latter is proportional, and opposite, to the absolute flux of water.

The associated constitutive equations,

$$
\begin{bmatrix}
J_{\mathrm{wM}} \\
J_{\mathrm{NaM}}^d \\
J_{\mathrm{ClM}}^d \\
I_{\mathrm{eM}} \\
\mathcal{G}
\end{bmatrix}
= -
\begin{bmatrix}
k_{\mathrm{MM}} & k_{\mathrm{wNa}}^d & k_{\mathrm{wCl}}^d & k_{\mathrm{e}} & k_{\mathrm{wr}}^d \\
k_{\mathrm{Naw}}^d & k_{\mathrm{NaNa}}^d & k_{\mathrm{NaCl}}^d & k_{\mathrm{Na\,e}}^d & k_{\mathrm{Nar}}^d \\
k_{\mathrm{Clw}}^d & k_{\mathrm{ClNa}}^d & k_{\mathrm{ClCl}}^d & k_{\mathrm{Cl\,e}}^d & k_{\mathrm{Clr}}^d \\
k_{\mathrm{e}} & k_{\mathrm{eNa}}^d & k_{\mathrm{eCl}}^d & \sigma_{\mathrm{e}} & k_{\mathrm{er}}^d \\
k_{\mathrm{rw}}^d & k_{r\mathrm{Na}}^d & k_{r\mathrm{Cl}}^d & k_{r\,\mathrm{e}}^d & k_{rr}^d
\end{bmatrix}
\begin{bmatrix}
\Delta P_{\mathrm{wM}} \\
RT\,\Delta\mathrm{Ln}\,c_{\mathrm{NaM}} \\
RT\,\Delta\mathrm{Ln}\,c_{\mathrm{ClM}} \\
\Delta\phi_{\mathrm{M}} \\
\mathcal{J}
\end{bmatrix},
\tag{19.15.19}
$$

involve a 5×5 diffusion matrix $\mathcal{K}_{\mathrm{M}}$ which is also assumed to be symmetric and positive semi-definite.

[19.5]Details:

1. note that the relation below holds, to first order, in incremental form, on the basis that x_{wM} is close to 1,

$$
\frac{\Delta\mathrm{Ln}\,x_{\mathrm{wM}}}{\widehat{v}_{\mathrm{w}}} = -\sum_{k \in \mathrm{M}} \Delta c_{k\mathrm{M}}.
\tag{19.15.11}
$$

We wrote earlier in differential form,

$$
\frac{d\mathrm{Ln}\,x_{\mathrm{wM}}}{\widehat{v}_{\mathrm{w}}} = \frac{1}{x_{\mathrm{wM}}}\frac{dx_{\mathrm{wM}}}{\widehat{v}_{\mathrm{w}}} = -\frac{n^{\mathrm{M}}}{n^{\mathrm{wM}}}\sum_{k \in \mathrm{M}} dc_{k\mathrm{M}}.
\tag{19.15.12}
$$

However, to first order, if $x_1 + x_2 = 1$, with $x_1 \sim 1$,

$$
d\mathrm{Ln}\,x_1 = \frac{dx_1}{x_1} = -\frac{dx_2}{1 - x_2} \simeq -(1 + x_2 + \cdots)\,dx_2 \simeq -dx_2 + O((dx_2)^2).
\tag{19.15.13}
$$

Therefore a consistent linearization would drop the ratio $n^{\mathrm{M}}/n^{\mathrm{wM}}$ for both the differential and incremental forms.

2. The intrinsic concentration of fixed charge is neglected anyway.

3. $\Delta\mathrm{Ln}\,x_{k\mathrm{M}} = \Delta\mathrm{Ln}\,c_{k\mathrm{M}}$, using $c_{k\mathrm{M}} = x_{k\mathrm{M}}/\widehat{v}_{\mathrm{w}}$.

[19.6]Notes on line 1 below: the ratio of volume fractions comes from the definition of the diffusive fluxes, Eqn (19.15.17); on line 2 of (19.15.16): $c\,\Delta\mathrm{Ln}\,c - \Delta c \simeq (\Delta c)^2/(2\,c)$ only if $|\Delta c| \ll c$, which may not be the case for ionic concentrations. In differential form, this sum would vanish.

19.15.3 Equivalence of the formulations based on absolute and diffusive fluxes

Starting from the constitutive relations in terms of absolute fluxes, the relations in terms of diffusive fluxes deduce with the following coefficients:

$$
\begin{aligned}
&k_{\mathrm{MM}} = \kappa_{\mathrm{ww}}, \quad k_{w l}^d = \kappa_{l \mathrm{w}}^d, && k_{\mathrm{w}\,e}^d = k_e \equiv \mathrm{F} \sum_{m \in \mathrm{M_{ions}}} \zeta_m\, \kappa_{\mathrm{w}m}, \quad k_{\mathrm{wr}}^d = \kappa_{\mathrm{wr}}; \\[4pt]
&k_{k\mathrm{w}}^d = \kappa_{k\mathrm{w}}^d, \quad k_{kl}^d = \kappa_{kl}^d - \kappa_{k\mathrm{w}}^d\, c_{l\mathrm{M}}, && k_{k\,e}^d = \mathrm{F} \sum_{m \in \mathrm{M_{ions}}} \zeta_m\, \kappa_{km}^d, && k_{kr}^d = \kappa_{kr}^d; \\[4pt]
&k_{e\mathrm{w}}^d = k_e, \quad k_{el}^d = k_{le}^d, && \sigma_e \equiv \mathrm{F}^2 \sum_{m,n \in \mathrm{M_{ions}}} \zeta_m\, \kappa_{mn}\, \zeta_n, && k_{er}^d = \mathrm{F} \sum_{m \in \mathrm{M_{ions}}} \zeta_m\, \kappa_{mr}; \\[4pt]
&k_{r\mathrm{w}}^d = k_{\mathrm{w}r}^d, \quad k_{rl}^d = k_{lr}^d, && k_{re}^d = k_{er}^d, && k_{rr}^d = \kappa_{rr},
\end{aligned}
\tag{19.15.20}
$$

where the free indices k and l belong to $\mathrm{M_{ions}}$. It has been found useful to introduce the temporary coefficients,

$$
\kappa_{kl}^d \equiv \kappa_{kl} - c_{k\mathrm{M}}\, \kappa_{\mathrm{w}l}, \quad k \in \mathrm{M_{ions}}, \; l \in \mathrm{M_{mo}} \cup \{r\}.
\tag{19.15.21}
$$

Electroneutrality (for clarity, all diffusion coefficients should be marked with an M so as not to be mistaken with the stromal coefficients),

$$
\sum_{l \in \mathrm{M_{ions}}} \zeta_l\, c_{l\mathrm{M}} + e_{e\mathrm{M}} = 0,
\tag{19.15.22}
$$

and the symmetry of the coefficients κ have been used to establish (19.15.20). This symmetry is inherited by the coefficients k^d: the only point to be still checked concerns the central coefficients pertaining to ions, namely

$$
k_{kl}^d = \kappa_{kl}^d - \kappa_{k\mathrm{w}}^d\, c_{l\mathrm{M}} = \kappa_{kl} - \kappa_{k\mathrm{w}}\, c_{l\mathrm{M}} - c_{k\mathrm{M}}\, \kappa_{l\mathrm{w}} + c_{k\mathrm{M}}\, c_{l\mathrm{M}}\, \kappa_{\mathrm{ww}} = k_{lk}^d, \quad k, l \in \mathrm{M_{ions}}.
\tag{19.15.23}
$$

The coefficients corresponding to the electric current density have been deduced through the relations (19.15.19) which clearly indicate that the diffusion matrix $\mathcal{K}_{\mathrm{M}}$ is singular.

The identification procedure proposed here relies on macroscopic measurements. Alternatively, one may consider that the membrane is composed of cells separated by slits, arranged in a regular pattern. If the properties of the cells and paracellular spaces (slits) are known, the macroscopic coefficients could be obtained by spatial averaging.

The passive diffusion matrices are addressed first, formally assuming the metabolic flux $\mathcal{J}$ to vanish. Active transport is taken into account for electrically open circuits in Section 19.15.4.3, but it is left to be worked out in the general case.

The identification involves three steps:

(D1) The open circuit permeability k_{H} is known:

At uniform concentrations of ions and PGs, and at vanishing electrical current, a fluid pressure gradient necessarily gives rise to a "streaming potential",

$$
\Delta \phi_{\mathrm{M}} = -\frac{k_e}{\sigma_e} \Delta P_{\mathrm{wM}}.
\tag{19.15.24}
$$

Thus

$$
J_{\mathrm{wM}} = -k_{\mathrm{H}}\, \Delta P_{\mathrm{wM}}, \qquad \underbrace{k_{\mathrm{H}} = k_{\mathrm{MM}} - \frac{k_e^2}{\sigma_e}}_{\text{open circuit permeability}}.
\tag{19.15.25}
$$

The entity measured k_{H} [unit : $\mathrm{m}^2 \times \mathrm{s/kg}$] is thus the "open circuit" permeability. k_{H} is expected to be positive definite so that water flows against pore pressure gradient. On the other hand, water flows along the streaming potential gradient if k_e is positive, which corresponds to pHs above the isoelectric point.

Alternatively, measurement of the flow at given fluid pressure, at uniform concentrations of ions and PGs and uniform electrical potential, yields the "short circuit" permeability k_{MM} [unit : $\mathrm{m}^2 \times \mathrm{s/kg}$], which is proportional to the hydraulic conductivity K_{H} [unit : $1/\mathrm{s}$], namely

$$
J_{\mathrm{wM}} = -k_{\mathrm{MM}}\, \Delta P_{\mathrm{wM}}, \qquad \underbrace{k_{\mathrm{MM}} = \frac{K_{\mathrm{H}}}{\rho_{\mathrm{w}}\, \mathrm{g}} = \kappa_{\mathrm{ww}}}_{\text{short circuit permeability}}.
\tag{19.15.26}
$$

(D2) The effective ionic mobilities u_k^*, $k \in M_{ions}$, are known:

The velocity relative to water that the ionic component k can reach in an electrical potential ϕ_M at uniform ionic concentrations and vanishing pore pressure gradient is equal to $-\text{sgn}\,\zeta_k\,u_k^*\,\Delta\phi_M$. Hence the coefficients k_{ke}^d expresses in terms of effective ionic mobilities u_k^* [unit: m/s/V] or effective diffusion coefficients D_k^* [unit: m/s], linked via the Nernst-Einstein relation $u_k^* = D_k^*\,|\zeta_k|\,F/RT$,

$$k_{ke}^d = n^M\,c_{kM}\,u_k^*\,\text{sgn}\,\zeta_k = n^M\,c_{kM}\,\zeta_k\,D_k^*\,\frac{F}{RT}\,, \quad k \in M_{ions}\,. \tag{19.15.27}$$

The *effective* ionic mobility u^* refers to the mobility in the porous structure while the ionic mobility u refers to a blank solution. Compatibility of the above expression with the expression in (19.15.20) implies

$$\kappa_{kl}^d = n^M\,c_{kM}\,\frac{u_k^*}{F|\zeta_k|}\,I_{kl} + \sum_{m=1}^{n_{ions}-1} A_{mk}\,\zeta_{ml}^\perp\,, \quad k, l \in M_{ions}\,. \tag{19.15.28}$$

Here the $\zeta_m^\perp$'s, $m \in [1, n_{ions}-1]$, are independent pseudo-vectors orthogonal to the pseudo-vector $\zeta = (\zeta_k)$ of ionic valences. The $\mathbf{A}_m$'s, $m \in [1, n_{ions}-1]$, are arbitrary pseudo-vectors of length equal to the number of ions.

(D3) The matrix κ is enforced to be symmetric:

Use of (19.15.21) and (19.15.28) yields,

$$\kappa_{kl} = n^M\,c_{kM}\,\frac{u_k^*}{F|\zeta_k|}\,I_{kl} + \sum_{m=1}^{n_{ions}-1} A_{mk}\,\zeta_{ml}^\perp + c_{kM}\,\kappa_{wl}\,, \quad k, l \in M_{ions}\,. \tag{19.15.29}$$

Symmetry of the matrix κ implies the constraint,

$$\sum_{m=1}^{n_{ions}-1} \mathbf{A}_m \wedge \zeta_m^\perp + \mathbf{c} \wedge \kappa_w = \mathbf{0}\,. \tag{19.15.30}$$

Here $\mathbf{c} = (c_{kM})$ and $\kappa_w = (\kappa_{wk})$ are pseudo-vectors of length equal to the number of ions. The solution of this constraint depends whether the membrane is charged or not. Indeed, electroneutrality implies the constraints,

$$\zeta \cdot \mathbf{c} = -e_{eM}; \quad \zeta \cdot \zeta_m^\perp = 0, \quad m \in [1, n_{ions}-1]\,. \tag{19.15.31}$$

We envisage primarily the case of binary electrolytes, which features a single vector $\zeta^\perp$ and a single vector $\mathbf{A}$. Therefore, if the membrane is charged, then the pseudo-vectors $\zeta^\perp$ and $\mathbf{c}$ are not co-linear. If the membrane is not charged, then $\mathbf{c}$ belongs to the orthogonal set to ζ, and in fact, $\zeta^\perp$ and $\mathbf{c}$ are parallel. Ternary electrolytes will be considered as well.

19.15.4 Absolute and diffusive fluxes: charged membrane

19.15.4.1 Structure of generalized diffusion: charged membrane

The solution involves three arbitrary scalars a_i, $i \in [1, 3]$,

$$A_k = a_1\,\zeta_k^\perp - a_2\,c_{kM}\,, \quad \kappa_{wk} = a_2\,\zeta_k^\perp + a_3\,c_{kM}\,, \quad k \in M_{ions}\,. \tag{19.15.32}$$

Therefore the ionic part of the matrix κ depends on the ionic mobilities and on these three scalars,

$$\kappa_{wk} = a_2\,\zeta_k^\perp + a_3\,c_{kM}\,, \quad k \in M_{ions},$$

$$\kappa_{kl} = n^M\,c_{kM}\,\frac{u_k^*}{F|\zeta_k|}\,I_{kl} + a_1\,\zeta_k^\perp\,\zeta_l^\perp + a_3\,c_{kM}\,c_{lM}, \quad k, l \in M_{ions}\,. \tag{19.15.33}$$

Hence

$$\kappa_{kl}^d = n^M\,c_{kM}\,\frac{u_k^*}{F|\zeta_k|}\,I_{kl} + a_1\,\zeta_k^\perp\,\zeta_l^\perp - a_2\,c_{kM}\,\zeta_l^\perp\,, \quad k, l \in M_{ions}\,. \tag{19.15.34}$$

Insertion of these relations in (19.15.20) yields the diffusion coefficients,

$$
\begin{cases}
k_{\mathrm{MM}} = \dfrac{K_{\mathrm{H}}}{\rho_{\mathrm{w}}\, \mathrm{g}} = \kappa_{\mathrm{ww}} > 0, \quad k_{\mathrm{H}} = k_{\mathrm{MM}} - \dfrac{k_{\mathrm{e}}^2}{\sigma_{\mathrm{e}}}, \\[2mm]
k_{k\mathrm{e}}^d = n^{\mathrm{M}}\, c_{k\mathrm{M}}\, u_k^*\, \mathrm{sgn}\, \zeta_k, \quad k \in \mathrm{M_{ions}}, \\[2mm]
k_{\mathrm{w}k}^d = a_2\, \zeta_k^\perp + (a_3 - k_{\mathrm{MM}})\, c_{k\mathrm{M}}, \quad k \in \mathrm{M_{ions}}, \\[2mm]
k_{kl}^d = a_1\, \zeta_k^\perp \zeta_l^\perp - a_2\left(\zeta_k^\perp c_{l\mathrm{M}} + c_{k\mathrm{M}} \zeta_l^\perp\right) + (k_{\mathrm{MM}} - a_3)\, c_{k\mathrm{M}}\, c_{l\mathrm{M}} \\[2mm]
\qquad + \; n^{\mathrm{M}}\, c_{k\mathrm{M}}\, \dfrac{u_k^*}{\mathrm{F}\,|\zeta_k|}\, I_{kl}, \quad k,l \in \mathrm{M_{ions}}, \\[2mm]
k_{\mathrm{e}} = -a_3\, \mathrm{F}\, e_{\mathrm{eM}}, \quad \sigma_{\mathrm{e}} = n^{\mathrm{M}}\, \mathrm{F} \displaystyle\sum_{m \in \mathrm{M_{ions}}} |\zeta_m|\, c_{m\mathrm{M}}\, u_m^* + a_3\, \mathrm{F}^2\, e_{\mathrm{eM}}^2.
\end{cases}
\tag{19.15.35}
$$

A particular form of diffusion is derived below under the two simple following assumptions:

- *Assumption* ($\mathcal{A}1$): the diffusive flux of ions is not affected by a gradient of fluid pressure: thus $a_2 = 0$ and $a_3 = k_{\mathrm{MM}}$. Symmetry of the diffusion matrix $\mathcal{K}_{\mathrm{M}}$ implies that the ionic gradients do not affect the fluid flow;

- *Assumption* ($\mathcal{A}2$): a gradient of concentration of ion k does not affect the diffusive flux of ion $l \neq k$. Then $a_1 = 0$.

The diffusion matrix $\boldsymbol{\kappa}$ associated with the absolute fluxes has the format,

$$
\boldsymbol{\kappa} =
\begin{bmatrix}
k_{\mathrm{MM}} & c_{\mathrm{NaM}}\, k_{\mathrm{MM}} & c_{\mathrm{ClM}}\, k_{\mathrm{MM}} \\
\hline
c_{\mathrm{NaM}}\, k_{\mathrm{MM}} & c_{\mathrm{NaM}}^2\, k_{\mathrm{MM}} & c_{\mathrm{NaM}}\, c_{\mathrm{ClM}}\, k_{\mathrm{MM}} \\
c_{\mathrm{ClM}}\, k_{\mathrm{MM}} & c_{\mathrm{ClM}}\, c_{\mathrm{NaM}}\, k_{\mathrm{MM}} & c_{\mathrm{ClM}}^2\, k_{\mathrm{MM}}
\end{bmatrix}
+
\begin{bmatrix}
0 & 0 & 0 \\
\hline
0 & k_{\mathrm{NaNa}}^d & 0 \\
0 & 0 & k_{\mathrm{ClCl}}^d
\end{bmatrix},
\tag{19.15.36}
$$

while the diffusion matrix associated with diffusive fluxes becomes a symmetric arrowhead matrix,

$$
\mathcal{K} =
\begin{bmatrix}
k_{\mathrm{MM}} & 0 & 0 & k_{\mathrm{e}} \\
0 & k_{\mathrm{NaNa}}^d & 0 & k_{\mathrm{Nae}}^d \\
0 & 0 & k_{\mathrm{ClCl}}^d & k_{\mathrm{Cl\,e}}^d \\
k_{\mathrm{e}} & k_{\mathrm{eNa}}^d & k_{\mathrm{eCl}}^d & \sigma_{\mathrm{e}}
\end{bmatrix},
\tag{19.15.37}
$$

where

$$
\begin{cases}
k_{\mathrm{MM}} = \dfrac{K_{\mathrm{H}}}{\rho_{\mathrm{w}}\, \mathrm{g}} = \kappa_{\mathrm{ww}} > 0, \quad k_{\mathrm{H}} = k_{\mathrm{MM}} - \dfrac{k_{\mathrm{e}}^2}{\sigma_{\mathrm{e}}}, \\[2mm]
k_{k\mathrm{e}}^d = n^{\mathrm{M}}\, c_{k\mathrm{M}}\, u_k^*\, \mathrm{sgn}\, \zeta_k, \quad k \in \mathrm{M_{ions}}, \\[2mm]
k_{kl}^d = n^{\mathrm{M}}\, c_{k\mathrm{M}}\, \dfrac{u_k^*}{\mathrm{F}\,|\zeta_k|}\, I_{kl}, \quad k,l \in \mathrm{M_{ions}}, \\[2mm]
k_{\mathrm{e}} = -k_{\mathrm{MM}}\, \mathrm{F}\, e_{\mathrm{eM}}, \quad \sigma_{\mathrm{e}} = \sigma_{\mathrm{e}}^{\mathrm{ion}} + k_{\mathrm{MM}}\, \mathrm{F}^2\, e_{\mathrm{eM}}^2, \quad \sigma_{\mathrm{e}}^{\mathrm{ion}} = n^{\mathrm{M}}\, \mathrm{F} \displaystyle\sum_{m \in \mathrm{M_{ions}}} |\zeta_m|\, c_{m\mathrm{M}}\, u_m^*,
\end{cases}
\tag{19.15.38}
$$

which features
- the total charge concentration e_{eM} [unit: mole/l or mole/m^3];
- the electroosmotic coefficient k_{e} [unit: m/s/V],
- the electrical conductivity σ_{e} [unit: siemens/m^2],
- the short circuit permeability k_{MM} [unit: m^2/kg×s = m/Pa/s], and the hydraulic conductivity K_{H} [unit: 1/s],
- the effective ionic mobility u_k^* [unit: m/s/V] of ion k, that is, the ionic mobility times the tortuosity factor. In fact, actual data are available for the ionic permeability P_k, Eqn (19.14.4). The ionic mobility u_k deduces from the latter by $u_k = P_k |\zeta_k| \mathrm{F}/RT$.

Note the useful relation $k_{\mathrm{H}}/k_{\mathrm{MM}} = \sigma_{\mathrm{e}}^{\mathrm{ion}}/\sigma_{\mathrm{e}}$. As indicated in Section 19.13, this formulation contains a hidden osmotic coefficient. It holds formally unchanged for ternary electrolytes as shown in Chapter 16, Exercise 16.9.

19.15.4.2 Passive fluxes for an electrically open circuit in a charged membrane

If the circuit is electrically open, namely $I_{eM} = 0$, then the electrical potential

$$\Delta \phi_M = -\frac{k_e}{\sigma_e} \Delta P_{wM} - \sum_{l \in M_{ions}} \frac{k_{el}^d}{\sigma_e} RT \, \Delta \mathrm{Ln} \, c_{lM} \,, \qquad (19.15.39)$$

can be eliminated from the expressions of the water flux J_{wM}, diffusive fluxes J_{kM}^d and absolute fluxes $J_{kM} = J_{kM}^d + c_{kM} J_{wM}$, $k \in M_{ions}$, (the approximation $n^M / n^{kM} = 1$ has been used),

$$
\begin{aligned}
J_{wM}^{(p)} &= -k_H \Delta P_{wM} + \frac{k_e}{\sigma_e} \sum_{l \in M_{ions}} k_{el}^d \, RT \, \Delta \mathrm{Ln} \, c_{lM} \,, \\
J_{kM}^{(p)} &= \Big(c_{kM} - \frac{k_e}{\sigma_e} \frac{k_{ke}^d}{k_H} \Big) J_{wM} - \sum_{l \in M_{ions}} \Big(k_{kl}^d - \frac{k_{ke}^d k_{el}^d}{\sigma_e^{ion}} \Big) RT \, \Delta \mathrm{Ln} \, c_{lM} \,, \quad k \in M_{ions} \,.
\end{aligned}
\qquad (19.15.40)
$$

19.15.4.3 Passive, active and secondary active fluxes for an electrically open circuit in a charged membrane

If active transport takes place in the membrane, the above relations should be modified. Indeed the electrical current is contributed by a passive part and an active part,

$$I_{eM} = I_{eM}^{(p)} + I_{eM}^{(a)}, \quad I_{eM}^{(\alpha)} \equiv F \sum_{k \in M} \zeta_k J_{kM}^{(\alpha)}, \quad \alpha = p, a, \qquad (19.15.41)$$

with indices a and p referring to the active and passive parts of the fluxes,

$$J_{kM} = J_{kM}^{(p)} + J_{kM}^{(a)}, \quad k \in M \,. \qquad (19.15.42)$$

Starting from the passive fluxes in the format (19.15.19), the resulting absolute fluxes (19.15.17) for an electrically open circuit,

$$
\begin{aligned}
J_{wM} &= J_{wM}^{(p)} &+& J_{wM}^{(a)} &+& \frac{k_e}{\sigma_e} \big(-I_{eM}^{(a)} \big) \,, \\
J_{kM} &= \underbrace{J_{kM}^{(p)}}_{\substack{\text{passive} \\ \text{fluxes}}} &+& \underbrace{J_{kM}^{(a)}}_{\substack{\text{active} \\ \text{fluxes}}} &+& \underbrace{\big(\frac{n_M}{n_{wM}} c_{kM} \frac{k_e}{\sigma_e} + \frac{k_{ke}^d}{\sigma_e} \big) \big(-I_{eM}^{(a)} \big)}_{\text{secondary active fluxes}} \,, \quad k \in M_{ions} \,,
\end{aligned}
\qquad (19.15.43)
$$

can be cast in a format that highlights three contributions,
- passive fluxes $J_{kM}^{(p)}$, $k \in M$, given exactly by (19.15.40);
- active fluxes $J_{kM}^{(a)}$, $k \in M$;
- secondary active fluxes in presence of an electrogenic active transport.

Note that secondary active transport of water takes place via electroosmosis if
- the electroosmotic coefficient k_e does not vanish;
- the active transport is electrogenic, namely $I_{eM}^{(a)} \neq 0$.

This phenomenon is independent on whether water itself is subjected to a *primary* active transport $J_{wM}^{(a)}$. Above the isoelectric point, the electroosmotic coefficient is positive for a negatively charged membrane. Therefore the secondary flux of water takes place in a direction opposite to the electrogenic flux $I_{eM}^{(a)} \neq 0$: if the latter is from stroma to aqueous humor, the secondary active transport of water is from aqueous humor to stroma.

19.15.5   Absolute and diffusive fluxes: membrane without fixed charge

19.15.5.1   Extraction of the structure of generalized diffusion: uncharged membrane

We consider first binary electrolytes. For definiteness, the constraint (19.15.31) is satisfied by setting $\zeta^\perp = \mathbf{c}$. Then (19.15.30) becomes $(\mathbf{A} - \boldsymbol{\kappa}_\mathrm{w}) \wedge \mathbf{c} = \mathbf{0}$. Consequently, $\boldsymbol{\kappa}_\mathrm{w}$ depends on three arbitrary scalars, namely a_1 and the pseudo-vector $\mathbf{A}$,

$$\kappa_{\mathrm{w}k} = a_1 c_{k\mathrm{M}} + A_k, \quad k \in \mathrm{M_{ions}}. \tag{19.15.44}$$

Therefore the ionic part of the matrix $\boldsymbol{\kappa}$ depends on the ionic mobilities[19.7] and on these three arbitrary scalars,

$$\kappa_{\mathrm{w}k} = a_1 c_{k\mathrm{M}} + A_k, \quad k \in \mathrm{M_{ions}},$$

$$\kappa_{kl} = n^\mathrm{M} c_{k\mathrm{M}} \frac{u_k^*}{\mathrm{F}|\zeta_k|} I_{kl} + a_1 c_{k\mathrm{M}} c_{l\mathrm{M}} + A_k c_{l\mathrm{M}} + A_l c_{k\mathrm{M}}, \quad k, l \in \mathrm{M_{ions}}. \tag{19.15.45}$$

Hence

$$\kappa_{kl}^d = n^\mathrm{M} c_{k\mathrm{M}} \frac{u_k^*}{\mathrm{F}|\zeta_k|} I_{kl} + A_k c_{l\mathrm{M}}, \quad k, l \in \mathrm{M_{ions}}. \tag{19.15.46}$$

Insertion of these relations in (19.15.20) yields the diffusion coefficients,

$$\begin{cases}
k_\mathrm{MM} &= \dfrac{K_\mathrm{H}}{\rho_\mathrm{w}\, \mathrm{g}} = \kappa_\mathrm{ww} > 0, \quad k_\mathrm{H} = k_\mathrm{MM} - \dfrac{k_\mathrm{e}^2}{\sigma_\mathrm{e}}, \\[2mm]
k_{k\mathrm{e}}^d &= n^\mathrm{M} c_{k\mathrm{M}} u_k^* \operatorname{sgn} \zeta_k, \quad k \in \mathrm{M_{ions}}, \\[2mm]
k_{\mathrm{w}k}^d &= A_k + (a_1 - k_\mathrm{MM}) c_{k\mathrm{M}}, \quad k \in \mathrm{M_{ions}}, \\[2mm]
k_{kl}^d &= n^\mathrm{M} c_{k\mathrm{M}} \dfrac{u_k^*}{\mathrm{F}|\zeta_k|} I_{kl} + (k_\mathrm{MM} - a_1) c_{k\mathrm{M}} c_{l\mathrm{M}}, \quad k, l \in \mathrm{M_{ions}}, \\[2mm]
k_\mathrm{e} &= \mathrm{F} \displaystyle\sum_{m \in \mathrm{M_{ions}}} \zeta_m A_m, \quad \sigma_\mathrm{e} = n^\mathrm{M} \mathrm{F} \displaystyle\sum_{m \in \mathrm{M_{ions}}} |\zeta_m| c_{m\mathrm{M}} u_m^*.
\end{cases} \tag{19.15.47}$$

Particular diffusion equations are now developed under two assumptions.

First, assumption $(\mathcal{A}_2)$ of Section 19.15.4.1 is retained, with the consequence that $a_1 = k_\mathrm{MM}$, and therefore $A_k = k_{\mathrm{w}k}^d$, $k \in \mathrm{M_{ions}}$. Second, osmotic, or reflection, coefficients are introduced. For an electrically open circuit $I_\mathrm{eM} = 0$, and under spatially uniform concentrations, the flux of water and absolute ionic fluxes are now,

$$J_\mathrm{wM} = -k_\mathrm{H} \Delta P_\mathrm{wM}, \quad J_{k\mathrm{M}} = -\left(k_{\mathrm{w}k}^d + c_{k\mathrm{M}} k_\mathrm{H} - \frac{k_\mathrm{e}}{\sigma_\mathrm{e}} k_{k\mathrm{e}}^d\right) \Delta P_\mathrm{wM}, \quad k \in \mathrm{M_{ions}}. \tag{19.15.48}$$

The membrane is in contact with an electrolytic bath B containing only sodium chloride. A pressure does not create a diffusive ionic flux in the bath, and therefore

$$J_{k\mathrm{B}} = c_{k\mathrm{B}} J_\mathrm{wB}, \quad k \in \mathrm{B_{ions}}. \tag{19.15.49}$$

The osmotic coefficients ω are defined as the ratio between the ionic fluxes in the membrane and in the bath,

$$J_{k\mathrm{M}} = (1 - \omega_k) J_{k\mathrm{B}}, \quad k \in \mathrm{M_{ions}} = \mathrm{B_{ions}}, \tag{19.15.50}$$

at identical water flux $J_\mathrm{wM} = J_\mathrm{wB}$. Note that ω_k is a priori innate to ion k. Note also that the ionic concentrations in the bath and in the membrane are identical: this feature is in strong contrast to the case of charged medium where the fixed charge induces the discontinuity of a number of entities. Thus, for all ions, $J_{k\mathrm{B}} = c_{k\mathrm{M}} J_\mathrm{wM}$, and therefore $J_{k\mathrm{M}} = (1 - \omega_k) c_{k\mathrm{M}} J_\mathrm{wM}$. Inserting the fluxes (19.15.48) in this relation yields

$$A_k = k_{\mathrm{w}k}^d = -\omega_k c_{k\mathrm{M}} k_\mathrm{H} + \frac{k_\mathrm{e}}{\sigma_\mathrm{e}} k_{k\mathrm{e}}^d, \quad k \in \mathrm{M_{ions}}. \tag{19.15.51}$$

[19.7]These mobilities are measured at vanishing fluid pressure, and not at vanishing fluid flow.

Check of the consistency of the expression of k_e in (19.15.20), with help of (19.15.47) and (19.15.51), implies $\sum_{m \in M_{ions}} \omega_m \zeta_m c_{mM} = 0$, so that in fact the two osmotic coefficients should be equal, say $\omega_{Na} = \omega_{Cl} = \omega$.

For an electrically open circuit $I_{eM} = 0$ but non uniform concentrations, the flux of water is now,

$$
\begin{aligned}
J_{wM} &= -k_H \left(\Delta P_{wM} - RT \sum_{m \in M_{ions}} \omega \, c_{mM} \, \Delta Ln \, c_{mM} \right), \\
&= -k_H \left(\Delta p_{wM} + RT \sum_{k \in M_{ions}} c_{kM} \, \Delta Ln \, c_{kM} - \Delta c_{kM} - RT \sum_{m \in M_{ions}} \omega \, c_{mM} \, \Delta Ln \, c_{mM} \right).
\end{aligned}
\tag{19.15.52}
$$

Note that the sums in the rhs of both lines of (19.15.52) are almost the osmotic pressure: they would be so for infinitesimal variations. The same issue arises in the definition of P_{wM}, in (19.15.16).

The diffusion matrix $\boldsymbol{\kappa}$ associated with the absolute fluxes has the format,

$$
\boldsymbol{\kappa} = \left[
\begin{array}{c|cc}
k_{MM} & c_{NaM} \, k_{MM} & c_{ClM} \, k_{MM} \\
\hline
c_{NaM} \, k_{MM} & c_{NaM}^2 \, k_{MM} & c_{NaM} \, c_{ClM} \, k_{MM} \\
c_{ClM} \, k_{MM} & c_{ClM} \, c_{NaM} \, k_{MM} & c_{ClM}^2 \, k_{MM}
\end{array}
\right]
\tag{19.15.53}
$$

$$
+ \left[
\begin{array}{c|cc}
0 & k_{wNa}^d & k_{wCl}^d \\
\hline
k_{wNa}^d & k_{NaNa}^d + 2 \, k_{wNa}^d \, c_{NaM} & k_{wNa}^d \, c_{ClM} + k_{wCl}^d \, c_{NaM} \\
k_{wCl}^d & k_{wNa}^d \, c_{ClM} + k_{wCl}^d \, c_{NaM} & k_{ClCl}^d + 2 \, k_{wCl}^d \, c_{ClM}
\end{array}
\right],
$$

while the diffusion matrix associated with diffusive fluxes becomes,

$$
\mathcal{K} = \left[
\begin{array}{cccc}
k_{MM} & k_{wNa}^d & k_{wCl}^d & k_e \\
k_{wNa}^d & k_{NaNa}^d & 0 & k_{Nae}^d \\
k_{wCl}^d & 0 & k_{ClCl}^d & k_{Cle}^d \\
k_e & k_{eNa}^d & k_{eCl}^d & \sigma_e
\end{array}
\right],
\tag{19.15.54}
$$

where

$$
\begin{cases}
k_{MM} = \dfrac{K_H}{\rho_w \, g} = \kappa_{ww} > 0, \quad k_H = k_{MM} - \dfrac{k_e^2}{\sigma_e}, \\[2mm]
k_{ke}^d = n^M \, c_{kM} \, u_k^* \, \mathrm{sgn} \, \zeta_k, \quad k \in M_{ions}, \\[2mm]
k_{wk}^d = -\omega \, c_{kM} \, k_H + \dfrac{k_e}{\sigma_e} \, k_{ke}^d, \quad k \in M_{ions}, \\[2mm]
k_{kl}^d = n^M \, c_{kM} \, \dfrac{u_k^*}{F \, |\zeta_k|} \, I_{kl}, \quad k, l \in M_{ions}, \\[2mm]
\sigma_e = n^M \, F \sum_{m \in M_{ions}} |\zeta_m| \, c_{mM} \, u_m^*.
\end{cases}
\tag{19.15.55}
$$

Therefore, the electroosmotic coefficient k_e and the osmotic coefficient ω have to be provided, either as constants or as functions of the ionic concentrations. This situation is in contrast to the case of a charged membrane, where these entities result internally from the formulation.

19.15.5.2 Passive fluxes for an electrically open circuit in an uncharged membrane

If the circuit is open, namely $I_{eM} = 0$, then the electrical potential

$$
\Delta \phi_M = -\frac{k_e}{\sigma_e} \Delta P_{wM} - \sum_{l \in M_{ions}} \frac{k_{el}^d}{\sigma_e} RT \, \Delta Ln \, c_{lM},
\tag{19.15.56}
$$

can be eliminated from the expressions of the water flux J_{wM}, diffusive fluxes J_{kM}^d, and absolute fluxes $J_{kM} = J_{kM}^d + c_{kM} J_{wM}$, $k \in M_{ions}$,

$$
\begin{aligned}
J_{wM} &= -k_H \Delta P_{wM} &&+ k_H \sum_{l \in M_{ions}} \omega\, c_{lM}\, RT\, \Delta \text{Ln}\, c_{lM}, \\
J_{kM} &= (1-\omega)\, c_{kM}\, J_{wM} &&- \sum_{l \in M_{ions}} n^M D_{klM}^* c_{lM}\, \Delta \text{Ln}\, c_{lM}, \quad k \in M_{ions},
\end{aligned}
\tag{19.15.57}
$$

where

$$
n^M D_{klM}^* c_{lM} = RT \left(k_{kl}^d - \frac{k_{ke}^d\, k_{el}^d}{\sigma_e} \right) - \omega^2 c_{kM}\, c_{lM}\, RT\, k_H, \quad k,l \in M_{ions},
\tag{19.15.58}
$$

is an effective diffusion coefficient at vanishing water flux ($J_{wM} = 0$) and electrical current ($I_{eM} = 0$), associated with an absolute flux.

For sodium chloride, note that the coefficients

$$
RT \left(k_{kl}^d - \frac{k_{ke}^d\, k_{el}^d}{\sigma_e} \right) = n^M \frac{RT}{F} \frac{u_{Na}^* u_{Na}^*}{u_{Na}^* + u_{Cl}^*} c_{NaM}, \quad k,l \in M_{ions},
\tag{19.15.59}
$$

are all equal, since $c_{NaM} = c_{ClM}$ in absence of fixed charge.

If the jump of concentrations across the membrane is not too large, terms of the form $c\,\Delta \text{Ln}\, c$ can be approximated by Δc. Then the absolute fluxes of water and ions can be cast in the compact form,

$$
\begin{aligned}
J_{wM} &= -k_H \left(\Delta p_{wM} - \omega\, RT \sum_{l \in M_{ions}} \Delta c_{lM} \right), \\
J_{kM} &= (1-\omega)\, c_{kM}\, J_{wM} - \sum_{l \in M_{ions}} n^M D_{klM}^* \Delta c_{lM}.
\end{aligned}
\tag{19.15.60}
$$

These relations are similar in form to the standard equations of diffusion of a neutral solute, Eqn (19.4.17).

19.15.5.3 A ternary electrolyte in an uncharged membrane

For a ternary electrolyte, the set (19.15.28) of (pseudo-)vectors of length three which are orthogonal to the vector of valences $\boldsymbol{\zeta}$ has dimension two. It is generated by the orthogonal (pseudo-)vectors $\boldsymbol{\zeta}_1^\perp$ and $\boldsymbol{\zeta}_2^\perp$.

The solution (19.15.30) involves two (pseudo-)vectors $\mathbf{A}_1$ and $\mathbf{A}_2$ of length three. With simplified notations, the undetermined part of this solution (19.15.20) and (19.15.21) can be cast in the format,

$$
\begin{aligned}
\boldsymbol{\kappa} &= \cdots + \mathbf{A}_1 \otimes \boldsymbol{\zeta}_1^\perp + \mathbf{A}_2 \otimes \boldsymbol{\zeta}_2^\perp + \mathbf{c} \otimes \boldsymbol{\kappa}_w, \\
\boldsymbol{\kappa}^d &= \cdots + \mathbf{A}_1 \otimes \boldsymbol{\zeta}_1^\perp + \mathbf{A}_2 \otimes \boldsymbol{\zeta}_2^\perp, \\
\mathbf{k}_w^d &= \boldsymbol{\kappa}_w^d = \boldsymbol{\kappa}_w - \kappa_{ww}\, \mathbf{c}, \\
\mathbf{k}^d &= \boldsymbol{\kappa}^d - \boldsymbol{\kappa}_w^d \otimes \mathbf{c} = \boldsymbol{\kappa} - \mathbf{c} \otimes \boldsymbol{\kappa}_w - \boldsymbol{\kappa}_w \otimes \mathbf{c} + \kappa_{ww}\, \mathbf{c} \otimes \mathbf{c}.
\end{aligned}
\tag{19.15.61}
$$

Since electroneutrality implies $\boldsymbol{\zeta} \cdot \mathbf{c} = 0$, the vector $\mathbf{c}$ belongs to the set which is orthogonal to $\boldsymbol{\zeta}$. Therefore $\boldsymbol{\zeta}_1^\perp = \boldsymbol{\zeta}^\perp = \mathbf{c} \wedge \boldsymbol{\zeta}$ and $\boldsymbol{\zeta}_2^\perp = \mathbf{c}$ are legitimate solutions. Moreover $\boldsymbol{\zeta}$, $\boldsymbol{\zeta}^\perp$ and $\mathbf{c}$ constitute a basis for $\mathbf{A}_1$, $\mathbf{A}_2$ and $\boldsymbol{\kappa}_w$, say

$$
\begin{aligned}
\mathbf{A}_1 &= a_{11}\, \boldsymbol{\zeta} + a_{12}\, \boldsymbol{\zeta}^\perp + a_{13}\, \mathbf{c}, \\
\mathbf{A}_2 &= a_{21}\, \boldsymbol{\zeta} + a_{22}\, \boldsymbol{\zeta}^\perp + a_{23}\, \mathbf{c}, \\
\boldsymbol{\kappa}_w &= a_{31}\, \boldsymbol{\zeta} + a_{32}\, \boldsymbol{\zeta}^\perp + a_{33}\, \mathbf{c}.
\end{aligned}
\tag{19.15.62}
$$

The vectors of length three $\boldsymbol{\zeta}$, $\boldsymbol{\zeta}^\perp$ and $\mathbf{c}$ being known, the three vectors $\mathbf{A}_1$, $\mathbf{A}_2$ and $\boldsymbol{\kappa}_w$ are sought subject to the symmetry of $\boldsymbol{\kappa}$. They can thus be expressed in terms of $3 \times 3 - 3 = 6$ arbitrary scalars. Symmetry of $\boldsymbol{\kappa}$,

$$
\mathbf{A}_1 \wedge \boldsymbol{\zeta}^\perp + (\mathbf{A}_2 - \boldsymbol{\kappa}_w) \wedge \mathbf{c} = \mathbf{0},
\tag{19.15.63}
$$

implies

$$(-a_{13} + a_{22} - a_{32})\,\boldsymbol{\zeta} + (a_{31} - a_{21})\,\boldsymbol{\zeta}^{\perp} + a_{11}\,\mathbf{c} = \mathbf{0}\,. \tag{19.15.64}$$

Thus we are left with six unknown scalars, as expected,

$$
\begin{aligned}
\mathbf{A}_1 &= & a_{12}\,\boldsymbol{\zeta}^{\perp} &+ & a_{13}\,\mathbf{c}, \\
\mathbf{A}_2 &= a_{21}\,\boldsymbol{\zeta} + & (a_{13} + a_{32})\,\boldsymbol{\zeta}^{\perp} &+ & a_{23}\,\mathbf{c}, \\
\boldsymbol{\kappa}_{\mathrm{w}} &= a_{21}\,\boldsymbol{\zeta} + & a_{32}\,\boldsymbol{\zeta}^{\perp} &+ & a_{33}\,\mathbf{c}\,.
\end{aligned}
\tag{19.15.65}
$$

Consequently

$$\mathbf{k}_{\mathrm{w}}^{d} = \boldsymbol{\kappa}_{\mathrm{w}}^{d} = \boldsymbol{\kappa}_{\mathrm{w}} - \kappa_{\mathrm{ww}}\,\mathbf{c} = a_{21}\,\boldsymbol{\zeta} + a_{32}\,\boldsymbol{\zeta}^{\perp} + (a_{33} - \kappa_{\mathrm{ww}})\,\mathbf{c}\,, \tag{19.15.66}$$

while

$$\mathbf{k}^{d} = \cdots \quad + a_{12}\,\boldsymbol{\zeta}^{\perp} \otimes \boldsymbol{\zeta}^{\perp} + a_{13}\,(\boldsymbol{\zeta}^{\perp} \otimes \mathbf{c} + \mathbf{c} \otimes \boldsymbol{\zeta}^{\perp}) + (\kappa_{\mathrm{ww}} - a_{33} + a_{23})\,\mathbf{c} \otimes \mathbf{c}\,. \tag{19.15.67}$$

Assumption ($\mathcal{A}2$) implies $a_{12} = 0$, $a_{13} = 0$ and $a_{33} = \kappa_{\mathrm{ww}} + a_{23}$, and we are finally left with three arbitrary coefficients. Moreover, consistency of the definition of the electroosmotic coefficient k_{e} yields

$$k_{\mathrm{e}} = a_{21}\,\mathrm{F} \sum_{m \in \mathrm{M_{ions}}} \zeta_m^2\,. \tag{19.15.68}$$

In summary,

$$
\left\{
\begin{aligned}
k_{\mathrm{MM}} &= \frac{K_{\mathrm{H}}}{\rho_{\mathrm{w}}\,\mathrm{g}} = \kappa_{\mathrm{ww}} > 0, \quad k_{\mathrm{H}} = k_{\mathrm{MM}} - \frac{k_{\mathrm{e}}^2}{\sigma_{\mathrm{e}}}, \\
k_{ke}^{d} &= n^{\mathrm{M}}\,c_{k\mathrm{M}}\,u_k^{*}\,\mathrm{sgn}\,\zeta_k\,, \quad k \in \mathrm{M_{ions}}, \\
k_{\mathrm{w}k}^{d} &= a_{21}\,\zeta_k + a_{32}\,\zeta_k^{\perp} + a_{23}\,c_{k\mathrm{M}}\,, \quad k \in \mathrm{M_{ions}}, \\
k_{kl}^{d} &= n^{\mathrm{M}}\,c_{k\mathrm{M}}\,\frac{u_k^{*}}{\mathrm{F}\,|\zeta_k|}\,I_{kl}\,, \quad k,l \in \mathrm{M_{ions}}, \\
k_{\mathrm{e}} &= a_{21}\,\mathrm{F} \sum_{m \in \mathrm{M_{ions}}} \zeta_m^2, \quad \sigma_{\mathrm{e}} = n^{\mathrm{M}}\,\mathrm{F} \sum_{m \in \mathrm{M_{ions}}} |\zeta_m|\,c_{m\mathrm{M}}\,u_m^{*}\,,
\end{aligned}
\right.
\tag{19.15.69}
$$

with three arbitrary coefficients a_{21}, a_{32} and a_{23}.

The introduction of osmotic coefficients parallels exactly the analysis of a binary electrolyte. Relation (19.15.51) still holds. If the osmotic coefficients are known, so will be the three coefficients a_{21}, a_{32} and a_{23}.

Thus we have another expression for $\boldsymbol{\kappa}_{\mathrm{w}} = \mathbf{k}_{\mathrm{w}}^{d} + \kappa_{\mathrm{ww}}\,\mathbf{c}$. Electroneutrality and consistency of the expression of k_{e} in (19.15.20), with σ_{e} given by (19.15.69), yield the two relations,

$$\sum_{k \in \mathrm{M_{ions}}} \zeta_m\,c_{m\mathrm{M}} = 0, \quad \sum_{k \in \mathrm{M_{ions}}} \zeta_m\,c_{m\mathrm{M}}\,\omega_m = 0\,. \tag{19.15.70}$$

It is worth recalling that the osmotic coefficient associated with ion k is expected to be equal to 1 when the concentration of ion k vanishes, and to tend to 0 when the concentration is large. Negative values, corresponding to anomalous osmosis, may take place.

The constraints (19.15.70) imply the osmotic coefficients to have a special structure,

$$\omega_1 - \omega_3 = f\,\zeta_2\,c_{2\mathrm{M}}, \quad \omega_2 - \omega_3 = -f\,\zeta_1\,c_{1\mathrm{M}}\,, \tag{19.15.71}$$

where f is a function of the concentrations to be provided.

19.15.6   Ion-dependent osmotic coefficients

For binary electrolytes in an uncharged membrane, we have just seen that the osmotic coefficient is the same for the anion and the cation, Section 19.15.5.1. The proof holds not only for a sodium chloride electrolyte, but for any binary electrolyte. In the model of *charged* tissue developed in Loret and Simões [2007], or of charged membrane considered here, the osmotic coefficient is intrinsic to the formulation. For the two binary electrolytes studied in Loret and Simões [2007], namely sodium chloride and calcium chloride, the induced osmotic coefficient *results* to be identical for the cation and the anion. On the other hand, it is not exactly the same for sodium chloride and calcium chloride. The osmotic coefficient for ternary electrolytes in a charged tissue has not been investigated.

In summary, the osmotic coefficient is the same for the cation and anion of a binary electrolyte, whether the tissue is charged. Ion-dependent osmotic coefficients are possible only for ternary electrolytes or higher, Section 19.15.5.3. A procedure to define these coefficients remains to be described. An example of such a model for an uncharged membrane is proposed by Li [2004].

19.16   Chemomechanical framework including chloride binding

The reversible chloride binding phenomenon described in Section 19.10 is now embedded in a mechanical framework, where the cornea is viewed as a porous medium which is saturated by an electrolyte. The resulting chemo-hyperelastic equations are expressed in terms of work-conjugate generalized stresses and strains. Chemomechanical couplings imply that the varying concentrations of metallic ions and of the fixed charge affect strongly the mechanical response. The converse mechanochemical effect is significant because the radial stiffness is low.

All species in the extrafibrillar phase can exchange with the surroundings except PGs. Let

$$\mathrm{E_{mo} = E - \{PG\}}, \tag{19.16.1}$$

be the set of exchangeable extrafibrillar species. The extrafibrillar chloride ions exchange with the surroundings and react with the ligands. Let $\delta \mathcal{N}_{\mathrm{ClE}}$, $\delta \mathcal{N}_{\mathrm{ClE}}^{*}$ and $\delta \mathcal{N}_{\mathrm{ClE}_L}$ be the changes of the total mole content of chloride ions in the extrafibrillar phase, and of the contributions emanating from the surroundings and from the ligands respectively,

$$\delta \mathcal{N}_{\mathrm{ClE}} = \delta \mathcal{N}_{\mathrm{ClE}}^{*} + \delta \mathcal{N}_{\mathrm{ClE}_L}, \quad \delta \mathcal{N}_{\mathrm{ClL}} = -\delta \mathcal{N}_{\mathrm{ClE}_L}. \tag{19.16.2}$$

19.16.1   Incompressible constituents and electroneutrality

The volume change of the mixture from a reference state is equal to the sum of the volume changes of its phases,

$$\frac{\Delta V}{V_0} = \Delta v_{\mathrm{S}} + \Delta v_{\mathrm{E}}. \tag{19.16.3}$$

A particular case of interest arises when all components are incompressible. The volumes of the solvated and bound chloride ions can be considered identical. The volume change of the whole porous medium is then entirely due to the species that enter (respectively leave) the fluid phase from (respectively to) the surroundings,

$$\Delta I_{\mathrm{Inc}} \equiv \frac{\Delta V}{V_0} - \Delta v_{\mathrm{E}}^{*} = 0, \tag{19.16.4}$$

or, in rate form,

$$\begin{aligned}
\frac{\delta I_{\mathrm{Inc}}}{\delta t} &= \det \mathbf{F} \, \mathrm{div} \, \mathbf{v}_{\mathrm{S}} - \sum_{k \in \mathrm{E_{mo}}} \frac{\delta v_{*}^{k\mathrm{E}}}{\delta t} \\
&= \det \mathbf{F} \, \mathrm{div} \, \mathbf{v}_{\mathrm{S}} - \sum_{k \in \mathrm{E_{mo}}} \widehat{v}_k \frac{\delta \mathcal{N}_{k\mathrm{E}}^{*}}{\delta t}.
\end{aligned} \tag{19.16.5}$$

The constraint $\delta I_{\mathrm{Inc}} = 0$ will be satisfied by introduction of a Lagrange multiplier to be interpreted as the pressure p_{E} of the fluid phase.

Electroneutrality of the extrafibrillar phase is assumed to be satisfied by ions dissolved in the phase and the part of the fibrils which are in contact with the phase. Given that the chloride binding reaction is neutral, the variation of the electrical density I_{eE} can be phrased in terms of ions exchanging with the surroundings only,

$$\delta I_{\mathrm{eE}} = \mathrm{F} \sum_{l \in \mathrm{E_{mo}}} \zeta_l \, \delta \mathcal{N}_{l\mathrm{E}}^{*}. \tag{19.16.6}$$

19.16.2   Structure of the constitutive equations

The work done in a volume V_0 during an incremental process is contributed by work-conjugate pairs associated with mechanics, mass exchanges, chloride binding and with the electroneutrality and incompressibility constraints,

$$
\begin{aligned}
\delta \underline{\mathcal{W}} &= \mathbf{T} : \delta \mathbf{E} + \sum_{l \in \mathrm{E_{mo}}} g_{l\mathrm{E}} \, \delta \mathcal{N}_{l\mathrm{E}}^* + \mathcal{G}_L \, \delta \mathcal{N}_{\mathrm{ClE}_L} + p_\mathrm{E} \, \delta I_{\mathrm{inc}} + \phi_\mathrm{E} \, \delta I_{\mathrm{eE}} \\
&= \overline{\mathbf{T}} : \delta \mathbf{E} + \sum_{l \in \mathrm{E_{mo}}} \overline{g}_{l\mathrm{E}}^{\mathrm{ec}} \, \delta \mathcal{N}_{l\mathrm{E}}^* + \mathcal{G}_L \, \delta \mathcal{N}_{\mathrm{ClE}_L} \,,
\end{aligned}
\tag{19.16.7}
$$

where the overlined quantities are shifted through the fluid pressure,

$$
\overline{\mathbf{T}} = \mathbf{T} + p_\mathrm{E} \det \mathbf{F} \, \mathbf{F}^{-1} \cdot \mathbf{F}^{-\mathrm{T}}; \quad \overline{g}_{l\mathrm{E}}^{\mathrm{ec}} = g_{l\mathrm{E}}^{\mathrm{ec}} - \widehat{v}_l \, p_\mathrm{E}, \quad l \in \mathrm{E_{mo}} \,.
\tag{19.16.8}
$$

As detailed in Section 19.10.1, the affinity of the binding reaction $\mathcal{G}_L$ is work-conjugate with the mole content of chloride ions participating to the reaction.

The set of independent variables $\mathbb{V}_\mathrm{E}$,

$$
\mathbb{V}_\mathrm{E} = \{ \mathbf{E}, \, \mathcal{N}_\mathrm{E}^*, \, \mathcal{N}_{\mathrm{ClE}_L} \} \,,
\tag{19.16.9}
$$

consists of the strain $\mathbf{E}$ of the solid skeleton (or porous medium), of the mole contents of the exchangeable extrafibrillar species $\mathcal{N}_\mathrm{E}^* = \cup_{l \in \mathrm{E_{mo}}} \mathcal{N}_{l\mathrm{E}}^*$ and of the chloride ions participating to the binding reaction $\mathcal{N}_{\mathrm{ClE}_L}$.

The work done $\underline{\mathcal{W}}$ is used as a constitutive potential. The coupled chemo-hyperelastic constitutive equations can be cast in the format,

$$
\overline{\mathbf{T}} = \frac{\partial \underline{\mathcal{W}}}{\partial \mathbf{E}}; \quad \overline{g}_{l\mathrm{E}}^{\mathrm{ec}} = \frac{\partial \underline{\mathcal{W}}}{\partial \mathcal{N}_{l\mathrm{E}}^*}, \quad l \in \mathrm{E_{mo}}; \quad \mathcal{G}_L = \frac{\partial \underline{\mathcal{W}}}{\partial \mathcal{N}_{\mathrm{ClE}_L}} \,.
\tag{19.16.10}
$$

The potential,

$$
\underline{\mathcal{W}}(\mathbb{V}_\mathrm{E}) = \underline{\mathcal{W}}_{\mathrm{ch-mech}}(\mathbf{E}, \mathcal{N}_\mathrm{E}^*) + \underline{\mathcal{W}}_{\mathrm{ch}}(\mathcal{N}_\mathrm{E}^*, \mathcal{N}_{\mathrm{ClE}_L}) + \underline{\mathcal{W}}_{\mathrm{ef}}(\mathcal{N}_{\mathrm{ClE}_L}) \,,
\tag{19.16.11}
$$

is constitutively decomposed into
- a coupled chemomechanical contribution,

$$
\underline{\mathcal{W}}_{\mathrm{ch-mech}}(\mathbf{E}, \mathcal{N}_\mathrm{E}^*) \,;
\tag{19.16.12}
$$

- a chemical contribution,

$$
\underline{\mathcal{W}}_{\mathrm{ch}}(\mathrm{E_{mo}}, \mathcal{N}_{\mathrm{ClE}_L}) = RT \left(\sum_{l \in \mathrm{E_{mo}}} \mathcal{N}_{l\mathrm{E}} \, \mathrm{Ln} \, N_{l\mathrm{E}} - \mathcal{N}_\mathrm{E} \, \mathrm{Ln} \, N_\mathrm{E} \right.
$$
$$
\left. + \mathcal{N}_{\mathrm{P}_L} \, \mathrm{Ln} \, N_{\mathrm{P}_L} + \mathcal{N}_{\mathrm{R}_L} \, \mathrm{Ln} \, N_{\mathrm{R}_L} - \mathcal{N}_{\mathrm{S}_L} \, \mathrm{Ln} \, N_{\mathrm{S}_L} \right),
\tag{19.16.13}
$$

the indices P_L, R_L and S_L denoting the free sites, occupied sites and total sites, respectively, as defined in Section 19.11.3 and

$$
\mathcal{N}_\mathrm{E} = \sum_{l \in \mathrm{E_{mo}}} \mathcal{N}_{l\mathrm{E}} \,;
\tag{19.16.14}
$$

- a term associated with the binding phenomenon,

$$
\underline{\mathcal{W}}_{\mathrm{ef}}(\mathcal{N}_{\mathrm{ClE}_L}) = \mathcal{G}_L^0 \, \mathcal{N}_{\mathrm{ClE}_L} \quad \text{with} \quad \mathcal{G}_L^0 = -RT \, \mathrm{Ln} \left(10^{-\mathrm{pK}_L} \, \widehat{v}_\mathrm{w} \right).
\tag{19.16.15}
$$

The constitutive equations (19.16.10) can be recast in the format,

$$
\mathbf{T} = -p_\mathrm{E} \det \mathbf{F} \, \mathbf{F}^{-1} \cdot \mathbf{F}^{-\mathrm{T}} + \frac{\partial \underline{\mathcal{W}}_{\mathrm{ch-mech}}}{\partial \mathbf{E}} \,,
$$

$$
g_{l\mathrm{E}}^{\mathrm{ec}} = \widehat{v}_l \, p_\mathrm{E} + \frac{\partial \underline{\mathcal{W}}_{\mathrm{ch-mech}}}{\partial \mathcal{N}_{l\mathrm{E}}} + RT \, \mathrm{Ln} \, x_{l\mathrm{E}} + \mathrm{F} \, \zeta_l \, \phi_\mathrm{E}, \quad l \in \mathrm{E_{mo}} - \{\mathrm{Cl}\},
$$

$$
g_{\mathrm{ClE}}^{\mathrm{ec}} = \widehat{v}_{\mathrm{Cl}} \, p_\mathrm{E} + \frac{\partial \underline{\mathcal{W}}_{\mathrm{ch-mech}}}{\partial \mathcal{N}_{\mathrm{ClE}}^*} + RT \, \mathrm{Ln} \, x_{\mathrm{ClE}} - \mathrm{F} \, \phi_\mathrm{E},
$$

$$
\mathcal{G}_L = RT \, \mathrm{Ln} \, \frac{c_{\mathrm{P}_L}}{c_{\mathrm{R}_L}} \, \frac{c_{\mathrm{ClE}}}{10^{-\mathrm{pK}_L}} \,.
\tag{19.16.16}
$$

19.16.3 Equilibria for passive exchanges

At the time scale of interest here, the passive exchanges between the cornea and the surroundings, as well as chloride binding, can be considered to take place under continuing equilibrium.

To simplify the formulation, it is tacitly understood that all species in the set E_{mo} can exchange with the bath, and that the latter contains only these species. There is no difficulty to remove this simplification. Therefore, the equations that control the chemomechanical evolution of the cornea in contact with a bath of given chemical composition include

- exchange with surroundings:
$$g_{lE}^{ec} = g_{lB}^{ec}, \quad l \in E_{mo}; \tag{19.16.17}$$

- chloride binding:
$$\mathcal{G}_L = 0; \tag{19.16.18}$$

- compatibility of molar fractions,
$$\sum_{l \in K} x_{lK} = 1, \quad K = E, B; \tag{19.16.19}$$

- electroneutrality in terms of the molar fraction $y_e = \widehat{v}_w\, e_e$,
$$x_{NaK} + y_e\, I_{KE} - x_{ClK} = 0, \quad K = E, B. \tag{19.16.20}$$

The condition of mechanical equilibrium should be added to the above conditions to close the problem.

The electrochemical potentials of the species in the cornea are provided by the constitutive equations developed above while the corresponding expressions for the species in the bath are given by (19.8.4).

Remark on the kinetics of chloride binding

In this continuum thermodynamics framework, chloride binding may be described through a first order kinetics,
$$\frac{\delta N_{ClE_L}}{\delta t} = -\frac{\delta N_{ClL}}{\delta t} = -\frac{1}{t_L}\frac{\mathcal{G}_L}{RT}, \tag{19.16.21}$$

where t_L is the characteristic reaction time. When introduced in the mass balance of the ions, the above mass transfer combines with the constitutive equations of diffusion to yield partial differential equations of reaction-diffusion.

Remark on transient flows

If the chemical composition of the bath in which a piece of cornea is submerged is modified suddenly, chemical equilibrium at its boundary may possibly be considered to hold but the ensuing flow within the specimen is transiently inhomogeneous. A point inside the specimen is not in chemical equilibrium with the bath. However, the notion of *fictitious bath* allows to exhibit a bath which by definition is in equilibrium with the cornea. This fictitious bath varies in time and space.

19.17 Constitutive equations of chemo-hyperelasticity

19.17.1 Chemoelastic energy with electrical and chemical couplings

The chemomechanical part of the chemoelastic energy $\underline{\mathcal{W}}_{ch-mech}(\mathbf{E}, \mathcal{N}_E^*)$,

$$\underline{\mathcal{W}}_{ch-mech}(\mathbf{E}, \mathcal{N}_E^*) = \mathbf{T}_0 : \mathbf{E} + \underline{\mathcal{W}}_{ch,1}(\mathbf{E}, \mathcal{N}_E^*) + \underline{\mathcal{W}}_{ch,2}(\mathcal{N}_E^*)\left(\underline{\mathcal{W}}^{gs}(\mathbf{E}) + \underline{\mathcal{W}}^c(\mathbf{E})\right), \tag{19.17.1}$$

is contributed by
- the history of the deformation process encapsulated in the initial stress $\mathbf{T}_0$;
- the mechanical energy $\underline{\mathcal{W}}^{gs}(\mathbf{E})$ of the matrix, PGs and other proteins (ground substance);
- the mechanical energy $\underline{\mathcal{W}}^c(\mathbf{E})$ of the activated collagen fibrils;
- a coupled electrochemical term $\underline{\mathcal{W}}_{ch,1}(\mathbf{E}, \mathcal{N}_E^*) = -p_{ch}(\mathcal{N}_E^*)\,(\det \mathbf{F} - 1)$;
- a chemical amplification term $\underline{\mathcal{W}}_{ch,2}(\mathcal{N}_E^*)$.

For definiteness, let us consider that the work-conjugate stress-strain pair $(\mathbf{T}, \mathbf{E})$ consists of the 2nd Kirchhoff stress $\underline{\underline{\tau}}$ and of the Green strain $\mathbf{E}_G$,

$$(\mathbf{T}, \mathbf{E}) \longmapsto (\underline{\underline{\tau}}, \mathbf{E}_G).\tag{19.17.2}$$

Then the shifted stress takes the form,

$$\underline{\underline{\bar{\tau}}} = \underline{\underline{\tau}} + p_E \det \mathbf{F}\,\mathbf{F}^{-1} \cdot \mathbf{F}^{-T} = \det \mathbf{F}\,\mathbf{F}^{-1} \cdot \overbrace{(\boldsymbol{\sigma} + p_E\,\mathbf{I})}^{\bar{\boldsymbol{\sigma}}} \cdot \mathbf{F}^{-T},\tag{19.17.3}$$

and the chemical stress,

$$\underline{\underline{\tau}}_{ch} = \det \mathbf{F}\,\mathbf{F}^{-1} \cdot (-p_{ch}\,\mathbf{I})\,\mathbf{F}^{-T},\tag{19.17.4}$$

results to be isotropic in the current configuration. The mechanical constitutive equations which may now be phrased in terms of an effective stress,

$$\det \mathbf{F}\,\mathbf{F}^{-1} \cdot \big(\boldsymbol{\sigma} + \overbrace{(p_E + p_{ch})}^{p_{eff}}\,\mathbf{I}\big) \cdot \mathbf{F}^{-T} \;=\; \underline{\underline{\tau}}_0 + \underline{\mathcal{W}}_{ch,2}(\mathcal{N}_E^*) \left(\frac{\partial \underline{\mathcal{W}}^{gs}}{\partial \mathbf{E}_G} + \frac{\partial \underline{\mathcal{W}}^c}{\partial \mathbf{E}_G} \right),\tag{19.17.5}$$

express in the current configuration,

$$\left.\begin{array}{c} \boldsymbol{\sigma} + (p_E + p_{ch})\,\mathbf{I} \\[2mm] \boldsymbol{\sigma} + \tilde{p}_E\,\mathbf{I} + (\pi_{osm} + p_{ch})\,\mathbf{I} \end{array}\right\} = \boldsymbol{\sigma}_0 + \underline{\mathcal{W}}_{ch,2}(\mathcal{N}_E^*) \left(\boldsymbol{\sigma}^{gs} + \boldsymbol{\sigma}^c \right),\tag{19.17.6}$$

in terms of the Cauchy stresses,

$$\boldsymbol{\sigma}_0 = \frac{1}{\det \mathbf{F}}\,\mathbf{F} \cdot \underline{\underline{\tau}}_0 \cdot \mathbf{F}^T, \quad \boldsymbol{\sigma}^{gs} = \frac{1}{\det \mathbf{F}}\,\mathbf{F} \cdot \frac{\partial \underline{\mathcal{W}}^{gs}}{\partial \mathbf{E}_G} \cdot \mathbf{F}^T, \quad \boldsymbol{\sigma}^c = \frac{1}{\det \mathbf{F}}\,\mathbf{F} \cdot \frac{\partial \underline{\mathcal{W}}^c}{\partial \mathbf{E}_G} \cdot \mathbf{F}^T.\tag{19.17.7}$$

The pressure π_{osm} is defined as the difference between the pressures of the cornea and of the fictitious bath, namely $\pi_{osm} = p_E - \tilde{p}_E$. The notion of a fictitious bath is exposed in Section 14.5.2.

The constitutive stresses $\underline{\underline{\tau}}^{gs} = \partial \underline{\mathcal{W}}^{gs}/\partial \mathbf{E}_G$ and $\underline{\underline{\tau}}^c = \partial \underline{\mathcal{W}}^c/\partial \mathbf{E}_G$ are to be understood as volume weighted averages over the ground substance and the collagen network respectively. Both are representative of the stresses of the matrix and collagen in the hypertonic state, that is, in absence of chemical effects. Both are associated with a conewise behavior. Collagen fibrils are mechanically active only in extension, and the ground substance is stiffer when shrinking than when swelling. The purely chemical term $\underline{\mathcal{W}}_{ch,2}(\mathcal{N}_E^*)$ in (19.17.1) is intended to indicate that a smaller ionic strength amplifies the stiffness.

The constitutive functions p_{ch} and $\underline{\mathcal{W}}_{ch,2}$ depend on the (free) molar content of the fluid phase $\mathcal{N}_E^*$. On the other hand, the (fictitious) osmotic pressure may also be calculated in terms of $\mathcal{N}_E^*$. Therefore the above two constitutive functions may be expressed in terms of the (fictitious) osmotic pressure,

$$p_{ch}(\mathcal{N}_E^*) \sim p_{ch}(\pi_{osm}), \quad \underline{\mathcal{W}}_{ch,2}(\mathcal{N}_E^*) \sim \underline{\mathcal{W}}_{ch,2}(\pi_{osm}).\tag{19.17.8}$$

19.17.2   Matrix energy

The ground substance is endowed with linear isotropic elasticity, the strain energy being expressed in terms of the Green strain $\mathbf{E}_G$, or stretch $\mathbf{C} = \mathbf{I} + 2\,\mathbf{E}_G$, via the Lamé moduli λ and μ,

$$\begin{aligned} \underline{\mathcal{W}}^{gs} &= \frac{1}{2}\,\lambda\,(\operatorname{tr}\mathbf{E}_G)^2 + \mu\,\mathbf{E}_G : \mathbf{E}_G, \\[2mm] &= \frac{1}{8}\,\lambda\,(\operatorname{tr}\mathbf{C})^2 - \frac{1}{4}\,(3\lambda + 2\mu)\operatorname{tr}\mathbf{C} + \frac{1}{4}\,\mu\,\mathbf{C} : \mathbf{C} + \frac{3}{8}\,(3\lambda + 2\mu), \\[2mm] &= \frac{1}{8}\,(\lambda + 2\mu)\,I_1^2(\mathbf{C}) - \frac{1}{4}\,(3\lambda + 2\mu)\,I_1(\mathbf{C}) - \frac{1}{2}\,\mu\,I_2(\mathbf{C}) + \frac{3}{8}\,(3\lambda + 2\mu), \end{aligned}\tag{19.17.9}$$

where $I_1(\mathbf{C}) = \mathrm{tr}\,\mathbf{C}$, and $I_2(\mathbf{C}) = \frac{1}{2}\left(I_1^2(\mathbf{C}) - \mathbf{C}:\mathbf{C}\right)$. Then

$$\underline{\underline{\boldsymbol{\tau}}}^{\mathrm{gs}} = \frac{\partial \mathcal{W}^{\mathrm{gs}}}{\partial \mathbf{E}_{\mathrm{G}}} = \lambda\,\mathrm{tr}\,\mathbf{E}_{\mathrm{G}}\,\mathbf{I} + 2\,\mu\,\mathbf{E}_{\mathrm{G}}\,,$$

$$= \frac{1}{2}\lambda\,(\mathrm{tr}\,\mathbf{C} - 3)\,\mathbf{I} + \mu\,(\mathbf{C} - \mathbf{I})\,. \tag{19.17.10}$$

Alternatively, the matrix may be endowed with a nonlinear isotropic elastic behavior, as e.g., in (19.19.70).

Another nonlinear model is introduced through the function $f = f(\mathbf{C})$,

$$\mathcal{W}^{\mathrm{gs}} = \mathcal{W}_0^{\mathrm{gs}}\,f\,,$$

$$\underline{\underline{\boldsymbol{\tau}}}^{\mathrm{gs}} = \frac{\partial \mathcal{W}^{\mathrm{gs}}}{\partial \mathbf{E}_{\mathrm{G}}} = \underline{\underline{\boldsymbol{\tau}}}_0^{\mathrm{gs}}\,f + \mathcal{W}_0^{\mathrm{gs}}\,\frac{\partial f}{\partial \mathbf{E}_{\mathrm{G}}}\,, \tag{19.17.11}$$

$$\mathbb{E}^{\mathrm{gs}} = \frac{\partial^2 \mathcal{W}^{\mathrm{gs}}}{\partial \mathbf{E}_{\mathrm{G}}\,\partial \mathbf{E}_{\mathrm{G}}} = \mathbb{E}_0^{\mathrm{gs}}\,f + \frac{\partial f}{\partial \mathbf{E}_{\mathrm{G}}} \otimes \underline{\underline{\boldsymbol{\tau}}}_0^{\mathrm{gs}} + \underline{\underline{\boldsymbol{\tau}}}_0^{\mathrm{gs}} \otimes \frac{\partial f}{\partial \mathbf{E}_{\mathrm{G}}} + \mathcal{W}_0^{\mathrm{gs}}\,\frac{\partial^2 f}{\partial \mathbf{E}_{\mathrm{G}}\partial \mathbf{E}_{\mathrm{G}}}\,,$$

the index 0 denoting standard linear elasticity $f = 1$,

$$\mathcal{W}_0^{\mathrm{gs}} = \frac{1}{2}\,\lambda_0\,(\mathrm{tr}\,\mathbf{E}_{\mathrm{G}})^2 + \mu_0\,\mathbf{E}_{\mathrm{G}}:\mathbf{E}_{\mathrm{G}}\,,$$

$$\underline{\underline{\boldsymbol{\tau}}}_0^{\mathrm{gs}} = \frac{\partial \mathcal{W}_0^{\mathrm{gs}}}{\partial \mathbf{E}_{\mathrm{G}}} = \lambda_0\,\mathrm{tr}\,\mathbf{E}_{\mathrm{G}}\,\mathbf{I} + 2\,\mu_0\,\mathbf{E}_{\mathrm{G}}\,, \tag{19.17.12}$$

$$\mathbb{E}_0^{\mathrm{gs}} = \frac{\partial^2 \mathcal{W}_0^{\mathrm{gs}}}{\partial \mathbf{E}_{\mathrm{G}}\,\partial \mathbf{E}_{\mathrm{G}}} = \lambda_0\,\mathbf{I} \otimes \mathbf{I} + 2\,\mu_0\,\mathbf{I}\,\overline{\otimes}\,\mathbf{I}\,.$$

As an example of function f, consider $f = f(\det \mathbf{C})$ defined in terms of the material constant α,

$$f = \exp\big(\alpha\,(\det \mathbf{C} - 1)\big)\,,$$

$$\frac{\partial f}{\partial \mathbf{E}_{\mathrm{G}}} = 2\,\alpha\,f\,\det \mathbf{C}\,\mathbf{C}^{-1}\,, \tag{19.17.13}$$

$$\frac{\partial^2 f}{\partial \mathbf{E}_{\mathrm{G}}\,\partial \mathbf{E}_{\mathrm{G}}} = 4\,\alpha\,f\,\det \mathbf{C}\,\big((1 + \alpha\,\det \mathbf{C})\,\mathbf{C}^{-1} \otimes \mathbf{C}^{-1} - \mathbf{C}^{-1}\,\overline{\otimes}\,\mathbf{C}^{-1}\big)\,.$$

Use has been made of the differential relations (2.1.32) and (2.7.3).

19.17.3 Two families of collagen fibrils

Let $\underline{\mathbf{m}}_{\mathrm{c}}$ be a unit vector indicating the direction of a collagen fibril in a reference configuration associated with the deformation gradient $\mathbf{F} = \mathbf{I}$. The collagen fibrils are mechanically active only in extension, namely when $\langle \underline{\mathbf{m}}_{\mathrm{c}} \cdot \mathbf{C} \cdot \underline{\mathbf{m}}_{\mathrm{c}} - 1 \rangle$ is non zero, the operator $\langle \cdot \rangle$ denoting the positive part of its argument. Then the strain energy of fibrils per unit volume of collagen has the format,

$$\underline{w}_{\mathrm{c}} = \underline{w}_{\mathrm{c}}(L^2)\,, \tag{19.17.14}$$

where

$$L = \tfrac{1}{2}\,\langle \mathbf{C}:\underline{\mathbf{M}}_{\mathrm{c}} - 1 \rangle = \underbrace{\langle \mathbf{E}_{\mathrm{G}}:\underline{\mathbf{M}}_{\mathrm{c}} \rangle}_{x}\,, \tag{19.17.15}$$

and $\underline{\mathbf{M}}_{\mathrm{c}} = \underline{\mathbf{m}}_{\mathrm{c}} \otimes \underline{\mathbf{m}}_{\mathrm{c}}$. In (19.17.14), the square, actually any power strictly greater than 1, is required to ensure continuity of the stress at the extension-contraction transition $L = 0$, where, in contrast, the stiffness may, or may not, be discontinuous. Note that use of the deviatoric deformation in the above energy would lead isotropic extension, like swelling, to leave collagen fibrils mechanically inactive.

We adopt the exponential form of energy (12.2.3), defined by the modulus K_c [unit : Pa] and the dimensionless exponent k_c,

$$\underline{w}_c = \frac{K_c}{2\,k_c} \left(\exp(k_c\,L^2) - 1 \right),$$

$$\frac{\partial \underline{w}_c}{\partial \mathbf{E}_G} = K_c \exp(k_c\,L^2)\,L\,\underline{\mathbf{M}}_c,$$ (19.17.16)

$$\frac{\partial^2 \underline{w}_c}{\partial \mathbf{E}_G \partial \mathbf{E}_G} = K_c \exp(k_c\,L^2)\,\big(2\,k_c \underbrace{L^2}_{\text{continuous}} + \underbrace{\mathcal{H}(x)}_{\text{discontinuous}} \big)\,\underline{\mathbf{M}}_c \otimes \underline{\mathbf{M}}_c.$$

The energy of the network per unit volume of tissue is contributed by the two fibril families,

$$\underline{\mathcal{W}}^c(\mathbf{E}_G) = \sum_{i=1,2} \frac{K_i^c}{2\,k_{ci}} \left(\exp(k_{ci}\,L_i^2) - 1 \right) \quad \text{with} \quad L_i = \langle \mathbf{E}_G : \underline{\mathbf{M}}_{ci} \rangle,$$ (19.17.17)

with K_i^c the *apparent* modulus associated with the family i. The above summation assumes that the fibrils are mechanically independent, that is, not crosslinked. A more complete analysis would account for crosslinking of the collagen fibrils only, and possibly crosslinking of the collagen fibrils and other proteins.

19.18   Boundary value problems

19.18.1   Summary of balance equations

Under quasi-static loading and in absence of body forces, thus excluding gravity effects, the balance of momentum of the porous medium as a whole implies the divergence of the total Cauchy stress $\boldsymbol{\sigma}$ to vanish,

$$\text{div } \boldsymbol{\sigma} = \mathbf{0}.$$ (19.18.1)

All species are assumed incompressible. As a consequence, the change of volume of the porous medium is opposite to the change of volume of the extrafibrillar fluid phase due to diffusion,

$$\text{div } \mathbf{v}_S + \text{div } \mathbf{J}_E = 0,$$ (19.18.2)

where $\mathbf{v}_S = d\mathbf{u}/dt$ is the velocity of the solid skeleton, $\mathbf{J}_E = \mathbf{J}_{wE} + \sum_{k \in E_{\text{ions}}} \widehat{v}_k\,\mathbf{J}_{kE} \sim \mathbf{J}_{wE}$ the total extrafibrillar volume flux, and $\mathbf{J}_{kE}$ [unit : mole/m^2/s] the molar flux of the species k through the porous medium.

Since the species are incompressible, the balance of mass of the extrafibrillar ion k can be phrased in terms of the extrafibrillar mole content $\mathcal{N}_{kE} = N_{kE}/V_0$, namely

$$\frac{1}{\det \mathbf{F}} \frac{d\mathcal{N}_{kE}}{dt} + \text{div } \mathbf{J}_{kE} = 0, \quad k \in E_{\text{ions}}.$$ (19.18.3)

Here d/dt denotes the time derivative following the solid phase and $E_{\text{ions}} = \{\text{Na, Cl}\}$ is the subset of extrafibrillar ionic species.

The set of field equations is displayed in Table 19.18.1, together with the primary unknowns that are used in the finite element formulation. The field equations are coupled through transport and mechanics. In this Table, the unknown associated with each field equation is the one that would appear if latter was uncoupled.

19.18.2   Semi-discrete equations

In space dimension n_{sd}, there are $n_{\text{sd}} + 1 + n_{\text{ions}}$ field equations to be satisfied, namely
   - the balance of momentum of the porous medium as a whole, Eqn (19.18.1);
   - the global balance of mass, Eqn (19.18.2);
   - the balances of mass of the two extrafibrillar ions, Eqn (19.18.3).

The $n_{\text{sd}} + 1 + n_{\text{ions}}$ primary unknown fields, namely
   - the solid displacement $\mathbf{u} = \mathbf{N}_u \mathbf{u}^e$;
   - the electrochemical potentials of the extrafibrillar species $g_{kE}^{\text{ec}} = \mathbf{N}_g [\mathbf{g}_{kE}^{\text{ec}}]^e$, $k \in \{w\} \cup E_{\text{ions}}$,

Table 19.18.1: Set of field equations and list of primary unknowns. EF: extrafibrillar phase.

Nature of Equation	Field Equation	Main Unknown
• Balance of momentum of the whole mixture	$\operatorname{div} \boldsymbol{\sigma} = \mathbf{0}$	Displacement $\mathbf{u}$
• Balance of mass of the whole mixture	$\operatorname{div} \mathbf{v}_S + \operatorname{div} \mathbf{J}_E = 0$	Chemical potential of EF water g_{wE}
• Balance of mass of EF ion k	$\dfrac{1}{\det \mathbf{F}} \dfrac{d\mathcal{N}_{k\mathrm{E}}}{dt} + \operatorname{div} \mathbf{J}_{k\mathrm{E}} = 0$	Electrochemical potential $g_{k\mathrm{E}}^{\mathrm{ec}}, \ k \in \mathrm{E_{ions}}$

are interpolated, within the generic element e, in terms of nodal values through a priori distinct shape functions. Multiplying the field equations by the virtual fields $\delta \mathbf{w}$ and δg, and integrating by parts over the body V provides the weak form of the problem as:

$$\int_V \boldsymbol{\nabla}(\delta \mathbf{w}) : \boldsymbol{\sigma} \, dV = \int_{\partial V} \delta \mathbf{w} \cdot \boldsymbol{\sigma} \cdot \mathbf{n} \, dS,$$

$$\int_V \delta g \operatorname{div} \mathbf{v}_S - \boldsymbol{\nabla}(\delta g) \cdot \mathbf{J}_E \, dV = -\int_{\partial V} \delta g \, \mathbf{J}_E \cdot \mathbf{n} \, dS, \qquad (19.18.4)$$

$$\int_V \frac{\delta g}{\det \mathbf{F}} \frac{d\mathcal{N}_{k\mathrm{E}}}{dt} - \boldsymbol{\nabla}(\delta g) \cdot \mathbf{J}_{k\mathrm{E}} \, dV = -\int_{\partial V} \delta g \, \mathbf{J}_{k\mathrm{E}} \cdot \mathbf{n} \, dS, \quad k \in \mathrm{E_{ions}},$$

where $\mathbf{n}$ is the unit outward normal to the boundary ∂V. For a binary electrolyte, the global unknown vector $\mathbb{X}$,

$$\underbrace{\text{unknown vector}}_{\substack{\text{max nodal length} \\ n_{\mathrm{sd}} + 3}} \mathbb{X} = \begin{bmatrix} \mathbf{u} \\ \hline \mathbf{g}_{\mathrm{wE}} \\ \mathbf{g}_{\mathrm{NaE}}^{\mathrm{ec}} \\ \mathbf{g}_{\mathrm{ClE}}^{\mathrm{ec}} \end{bmatrix} \begin{array}{l} \left.\vphantom{\mathbf{u}}\right\} \text{displacement vector} \\[2ex] \left.\begin{array}{l} \\ \\ \\ \end{array}\right\} \begin{array}{l} \text{extrafibrillar} \\ \text{electrochemical potentials} \\ \textbf{continuous } \textit{at the boundary}\,! \end{array} \end{array} \qquad (19.18.5)$$

has therefore maximum nodal length $n_{\mathrm{sd}}+3$. The resulting nonlinear first order semi-discrete equations imply the residual $\mathbb{R}$,

$$\mathbb{R} = \mathbb{F}^{\mathrm{surf}}(\mathbb{S}, \mathbb{X}) - \mathbb{F}^{\mathrm{int}}\left(\mathbb{X}, \frac{d\mathbb{X}}{dt}\right) = \mathbb{0}, \qquad (19.18.6)$$

to vanish. Here $\mathbb{F}^{\mathrm{surf}}$ is the vector of surface loadings denoted collectively by $\mathbb{S}$. $\mathbb{F}^{\mathrm{int}}$ is the vector gathering the internal elastic and viscous forces. The contribution of a generic element e to the vector $\mathbb{F}_e^{\mathrm{surf}}$ of boundary data, and the contribution to the vector $\mathbb{F}_e^{\mathrm{int}}$ of internal forces are obtained from the weak form as,

$$\mathbb{F}_e^{\mathrm{surf}} = \begin{bmatrix} \displaystyle\int_{\partial V_e} \mathbf{N}_u^{\mathrm{T}} \, \boldsymbol{\sigma} \cdot \mathbf{n} \, dS \\[2ex] -\displaystyle\int_{\partial V_e} \mathbf{N}_g^{\mathrm{T}} \, \mathbf{J}_E \cdot \mathbf{n} \, dS \\[2ex] -\displaystyle\int_{\partial V_e} \mathbf{N}_g^{\mathrm{T}} \, \mathbf{J}_{\mathrm{NaE}} \cdot \mathbf{n} \, dS \\[2ex] -\displaystyle\int_{\partial V_e} \mathbf{N}_g^{\mathrm{T}} \, \mathbf{J}_{\mathrm{ClE}} \cdot \mathbf{n} \, dS \end{bmatrix}, \quad \mathbb{F}_e^{\mathrm{int}} = \begin{bmatrix} \displaystyle\int_{V_e} \mathbf{B}_u^{\mathrm{T}} \, \boldsymbol{\sigma} \, dV_e \\[2ex] \displaystyle\int_{V_e} \mathbf{N}_g^{\mathrm{T}} \operatorname{div} \mathbf{v}_S - \boldsymbol{\nabla}\mathbf{N}_g^{\mathrm{T}} \, \mathbf{J}_E \, dV_e \\[2ex] \displaystyle\int_{V_e} \frac{\mathbf{N}_g^{\mathrm{T}}}{\det \mathbf{F}} \frac{d\mathcal{N}_{\mathrm{NaE}}}{dt} - \boldsymbol{\nabla}\mathbf{N}_g^{\mathrm{T}} \, \mathbf{J}_{\mathrm{NaE}} \, dV_e \\[2ex] \displaystyle\int_{V_e} \frac{\mathbf{N}_g^{\mathrm{T}}}{\det \mathbf{F}} \frac{d\mathcal{N}_{\mathrm{ClE}}}{dt} - \boldsymbol{\nabla}\mathbf{N}_g^{\mathrm{T}} \, \mathbf{J}_{\mathrm{ClE}} \, dV_e \end{bmatrix}, \qquad (19.18.7)$$

where $\mathbf{B}_u$ is the strain-displacement matrix, namely $\delta \boldsymbol{\epsilon}(\mathbf{u}) = \boldsymbol{\epsilon}(\delta \mathbf{u}) = \mathbf{B}_u \, \delta \mathbf{u}$.

A generalized Galerkin procedure is adopted, in as far as the same interpolation functions are used for the primary unknowns and for the variations. The interpolation rules used for the displacements and electrochemical potentials as well as the integration rules for matrices and right hand side "forces" need to be provided.

Boundary data can be given in terms of either primary unknowns, as explained in Section 19.18.8, or in terms of the fluxes that appear in the right hand sides of the weak forms above. The expressions of the fluxes across the endothelium and epithelium are detailed in Section 19.18.4.

19.18.3   The element effective capacity matrix

Let $\mathbb{V}$ be the time rate of change of the primary variable $\mathbb{X}$. In an incremental time-marching scheme with time step Δt, e.g., the generalized trapezoidal scheme defined by the scalar $\alpha \in]0, 1]$, the effective capacity matrix $\mathbb{C}^*$ expresses in terms of the capacity matrix $\mathbb{C}$ and of the diffusion-stiffness matrix $\mathbb{K}$,

$$\mathbb{C}^* = \mathbb{C} + \alpha\,\Delta t\,\mathbb{K} \quad \text{with} \quad \mathbb{C} = \frac{\partial \mathbb{F}^{\mathrm{int}}}{\partial \mathbb{V}}, \quad \mathbb{K} = \frac{\partial \mathbb{F}^{\mathrm{int}}}{\partial \mathbb{X}}, \tag{19.18.8}$$

where it has been recognized that the derivatives of $\mathbb{F}^{\mathrm{int}}$ do not depend on $\mathbb{V}$. The element capacity matrix and element diffusion-stiffness matrix are not symmetric, nor even form symmetric,

$$\mathbb{C}^e = \begin{bmatrix} \mathbf{0} & \mathbf{0} & \mathbf{0} & \mathbf{0} \\ \mathbf{C}^e_{g_w u} & \mathbf{0} & \mathbf{0} & \mathbf{0} \\ \mathbf{C}^e_{g_{\mathrm{Na}} u} & \mathbf{0} & \mathbf{C}^e_{g_{\mathrm{Na}} g_{\mathrm{Na}}} & \mathbf{C}^e_{g_{\mathrm{Na}} g_{\mathrm{Cl}}} \\ \mathbf{C}^e_{g_{\mathrm{Cl}} u} & \mathbf{0} & \mathbf{C}^e_{g_{\mathrm{Cl}} g_{\mathrm{Na}}} & \mathbf{C}^e_{g_{\mathrm{Cl}} g_{\mathrm{Cl}}} \end{bmatrix}, \tag{19.18.9}$$

$$\mathbb{K}^e = \begin{bmatrix} \mathbf{K}^e_{uu} & \mathbf{K}^e_{u g_w} & \mathbf{K}^e_{u g_{\mathrm{Na}}} & \mathbf{K}^e_{u g_{\mathrm{Cl}}} \\ \mathbf{0} & \mathbf{K}^e_{g_w g_w} & \mathbf{K}^e_{g_w g_{\mathrm{Na}}} & \mathbf{K}^e_{g_w g_{\mathrm{Cl}}} \\ \mathbf{0} & \mathbf{K}^e_{g_{\mathrm{Na}} g_w} & \mathbf{K}^e_{g_{\mathrm{Na}} g_{\mathrm{Na}}} & \mathbf{K}^e_{g_{\mathrm{Na}} g_{\mathrm{Cl}}} \\ \mathbf{0} & \mathbf{K}^e_{g_{\mathrm{Cl}} g_w} & \mathbf{K}^e_{g_{\mathrm{Cl}} g_{\mathrm{Na}}} & \mathbf{K}^e_{g_{\mathrm{Cl}} g_{\mathrm{Cl}}} \end{bmatrix}. \tag{19.18.10}$$

The definition of the non zero components of the element matrices follow from the weak form,

- for the element capacity matrix:

$$\mathbf{C}^e_{g_w u} = \int_{V_e} \mathbf{N}_g^{\mathrm{T}} \,\mathrm{tr}\,\mathbf{B}_u \, dV_e,$$

$$\mathbf{C}^e_{g_k u} = \int_{V_e} \mathbf{N}_g^{\mathrm{T}} \frac{\mathcal{N}_{k\mathrm{E}}}{\det \mathbf{F}} \frac{Z_k}{n^{\mathrm{E}}} \,\mathrm{tr}\,\mathbf{B}_u \, dV_e, \quad k \in \mathrm{E_{ions}}, \tag{19.18.11}$$

$$\mathbf{C}^e_{g_k g_l} = \int_{V_e} \mathbf{N}_g^{\mathrm{T}} \frac{\mathcal{N}_{k\mathrm{E}}}{\det \mathbf{F}} \frac{Z_{kl}}{RT} \mathbf{N}_g \, dV_e, \quad k \in \mathrm{E_{ions}},\ l \in \mathrm{E_{ions}}.$$

- for the element diffusion-stiffness matrix:

$$\mathbf{K}^e_{uu} = \int_{V_e} \mathbf{B}_u^{\mathrm{T}} \frac{\partial \boldsymbol{\sigma}}{\partial \boldsymbol{\epsilon}} \mathbf{B}_u \, dV_e,$$

$$\mathbf{K}^e_{u g_l} = \int_{V_e} \mathbf{B}_u^{\mathrm{T}} \frac{\partial \boldsymbol{\sigma}}{\partial g_{l\mathrm{E}}^{\mathrm{ec}}} \mathbf{N}_g \, dV_e, \quad l \in \mathrm{E_{mo}}, \tag{19.18.12}$$

$$\mathbf{K}^e_{g_k g_l} = \int_{V_e} \nabla \mathbf{N}_g^{\mathrm{T}} \,\kappa_{kl}^* \,\nabla \mathbf{N}_g \, dV_e, \quad k \in \mathrm{E_{mo}},\ l \in \mathrm{E_{mo}}.$$

Two species are grouped in two sets, namely ions in $\mathrm{E_{ions}} = \{\mathrm{Na, Cl}\}$ and mobile species in $\mathrm{E_{mo}} = \{w, \mathrm{Na, Cl}\}$. The derivation of the element effective capacity matrices is detailed in Sections 19.18.5 and 19.18.6. The coefficients Z are defined by (19.18.23), the diffusion coefficients κ^* by (19.18.32)–(19.18.34), and the partial derivatives of stress by (19.18.43).

19.18.4   The contributions of the endothelial and epithelial fluxes

The external membranes are assumed to be free of fixed electrical charges. The diffusion properties for such membranes have been derived in Section 19.15.5. When the total electrical current vanishes, they further reduce to an additive format[19.8] that accounts for passive, active and secondary active fluxes, namely

$$J_{k\mathrm{M}} = J_{k\mathrm{M}}^{(p)} + J_{k\mathrm{M}}^{(a)} + J_{k\mathrm{M}}^{(sa)}, \quad k \in \mathrm{M}. \tag{19.18.13}$$

[19.8]This format has been worked out in Section 19.15.4.3 for a charged membrane.

The passive fluxes are defined by (19.15.60),

$$
\begin{aligned}
J_{\mathrm{wM}}^{(p)} &= -k_{\mathrm{hM}} \left(\Delta p_{\mathrm{wM}} - \omega_{\mathrm{M}} \, RT \sum_{l \in \mathrm{M_{ions}}} \Delta c_{l\mathrm{M}} \right), \\
J_{k\mathrm{M}}^{(p)} &= (1 - \omega_{\mathrm{M}}) \, c_{k\mathrm{M}} \, J_{\mathrm{wM}} - \sum_{l \in \mathrm{M_{ions}}} n^{\mathrm{M}} D_{kl\mathrm{M}}^{*} \, \Delta c_{l\mathrm{M}}, \quad k \in \mathrm{M_{ions}} \, .
\end{aligned}
\tag{19.18.14}
$$

The active fluxes are provided by constitutive equations or might be simply constants designed to satisfy some initial steady state conditions. Finally, the secondary active fluxes are provided by (19.15.43),

$$
\begin{aligned}
J_{\mathrm{wM}}^{(sa)} &= \frac{k_{\mathrm{e}}}{\sigma_{\mathrm{e}}} \left(-I_{\mathrm{eM}}^{(a)} \right), \\
J_{k\mathrm{M}}^{(sa)} &= \left(\frac{n_{\mathrm{M}}}{n_{\mathrm{wM}}} c_{k\mathrm{M}} \frac{k_{\mathrm{e}}}{\sigma_{\mathrm{e}}} + \frac{k_{k\mathrm{e}}^{d}}{\sigma_{\mathrm{e}}} \right) \left(-I_{\mathrm{eM}}^{(a)} \right), \quad k \in \mathrm{M_{ions}} \, .
\end{aligned}
\tag{19.18.15}
$$

The material coefficients of the membrane have been denoted by the index M so as to avoid confusion with their continuum counterparts in the stroma. Recall that their dimensions indicated in Section 19.15.4.1 are different from the latter. The scalar fluxes here are projections of the vector fluxes along the unit outward normal to the membrane. Correspondingly the symbol Δx denotes the difference $x_{\mathrm{ext}} - x_{\mathrm{stroma}}$.

Since the membrane is uncharged, the osmotic (reflection) coefficient ω_{M} and the electroosmotic coefficient k_{e} have to be defined by constitutive equations.

19.18.5 Extraction of dependent variables from primary variables

In a finite element context, the issue is to obtain the dependent variables, like
- the molar fractions x_{wE}, x_{NaE} and x_{ClE};
- the fluid pressure p_{E};
- the electrical potential ϕ_{E},

from the primary variables of the finite element procedure, namely
- displacement vector $\mathbf{u}$, its increment $\delta \mathbf{u}$, and associated strain $\delta \boldsymbol{\epsilon} = \mathbf{B}_u \, \delta \mathbf{u}$;
- electrochemical potentials g_{wE}, $g_{\mathrm{NaE}}^{\mathrm{ec}}$, $g_{\mathrm{ClE}}^{\mathrm{ec}}$.

The starting relations are the definitions of the electrochemical potentials, Eqn (19.8.4),

$$
\begin{aligned}
g_{\mathrm{wE}} &= \widehat{v}_{\mathrm{w}} \, p_{\mathrm{E}} \; + \; RT \, \mathrm{Ln} \, x_{\mathrm{wE}}, \\
g_{\mathrm{NaE}}^{\mathrm{ec}} &= RT \, \mathrm{Ln} \, x_{\mathrm{NaE}} + \mathrm{F} \, \phi_{\mathrm{E}}, \\
g_{\mathrm{ClE}}^{\mathrm{ec}} &= RT \, \mathrm{Ln} \, x_{\mathrm{ClE}} - \mathrm{F} \, \phi_{\mathrm{E}},
\end{aligned}
\tag{19.18.16}
$$

the closure relation for the molar fraction,

$$
x_{\mathrm{wE}} + x_{\mathrm{NaE}} + x_{\mathrm{ClE}} = 1 \,,
\tag{19.18.17}
$$

electroneutrality (19.8.10) with the charge given by (19.10.11),

$$
x_{\mathrm{NaE}} - x_{\mathrm{ClE}} + y_{\mathrm{e}} = 0, \quad y_{\mathrm{e}} = y_{\mathrm{PG}} - \frac{x_{\mathrm{ClE}} \, x_{\mathrm{L}}}{\kappa_{\mathrm{L}} + x_{\mathrm{ClE}}} \,.
\tag{19.18.18}
$$

Absence of chloride binding corresponds to $\kappa_{\mathrm{L}} \to \infty$.

19.18.5.1 Direct extraction

First obtain x_{ClE} as the single real positive solution of the cubic equation,

$$
\begin{aligned}
(x_{\mathrm{ClE}})^3 &+ a_2 \, (x_{\mathrm{ClE}})^2 + a_1 \, x_{\mathrm{ClE}} + a_0 = 0 \,, \\
a_2 &= \kappa_{\mathrm{L}} - y_{\mathrm{PG}} + x_{\mathrm{L}}, \\
a_1 &= -y_{\mathrm{PG}} \, \kappa_{\mathrm{L}} - \exp \left(\frac{g_{\mathrm{NaE}}^{\mathrm{ec}} + g_{\mathrm{ClE}}^{\mathrm{ec}}}{RT} \right), \\
a_0 &= -\kappa_{\mathrm{L}} \exp \left(\frac{g_{\mathrm{NaE}}^{\mathrm{ec}} + g_{\mathrm{ClE}}^{\mathrm{ec}}}{RT} \right),
\end{aligned}
\tag{19.18.19}
$$

and next, in that order,

$$y_e = y_{PG} - \frac{x_{ClE}\, x_L}{\kappa_L + x_{ClE}}$$

$$x_{NaE} = x_{ClE} - y_e,$$

$$\phi_E = \frac{1}{F}\left(g_{NaE}^{ec} - RT\,\mathrm{Ln}\,x_{NaE}\right) = \frac{1}{F}\left(-g_{ClE}^{ec} + RT\,\mathrm{Ln}\,x_{ClE}\right) \qquad (19.18.20)$$

$$p_E = \frac{1}{\widehat{v}_w}\left(g_{wE} - RT\,\mathrm{Ln}\,(1 - x_{NaE} - x_{ClE})\right).$$

19.18.5.2 Iterative extraction

The derivation below holds for any number of ions. Let us first record some useful relations, for incompressible solid constituents (V_S constant), and for infinitesimal strains (to first order),

$$V = V_E + V_S \qquad\Rightarrow\qquad \mathrm{tr}\,\delta\epsilon = \frac{\delta V}{V},$$

$$\frac{V_E}{V_E^0} = 1 + \frac{\mathrm{tr}\,\epsilon}{n_0^E} \qquad\Rightarrow\qquad \frac{\delta V_E}{V_E} = \frac{\mathrm{tr}\,\delta\epsilon}{n^E},$$

$$n^E = \frac{V_E}{V} \qquad\Rightarrow\qquad \delta n^E = (1 - n^E)\,\mathrm{tr}\,\delta\epsilon, \qquad (19.18.21)$$

$$x_{kE} = \widehat{v}_w\,\frac{N_{kE}}{V_E} \qquad\Rightarrow\qquad \frac{\delta x_{kE}}{x_{kE}} = \frac{\delta N_{kE}}{N_{kE}} - \frac{1}{n^E}\,\mathrm{tr}\,\delta\epsilon,$$

$$x_{kE} = \frac{\widehat{v}_w}{\widehat{v}_k}\,\frac{v^{kE}}{n^E}\,\frac{V_0}{V} \qquad\Rightarrow\qquad \frac{\delta m^{kE}}{m^{kE}} = \frac{\delta v^{kE}}{v^{kE}} = \frac{\delta x_{kE}}{x_{kE}} + \frac{1}{n^E}\,\mathrm{tr}\,\delta\epsilon, \quad k:\, w,\,\mathrm{ion},$$

and

$$y_{PG} = \frac{\zeta_{PG}\,N_{PG}}{V_E} \qquad\Rightarrow\qquad \delta y_{PG} = -\frac{y_{PG}}{n^E}\,\mathrm{tr}\,\delta\epsilon,$$

$$y_e = y_{PG} - \frac{x_{ClE}\,x_L}{\kappa_L + x_{ClE}} \qquad\Rightarrow\qquad \delta y_e = -\frac{y_{PG}}{n^E}\,\mathrm{tr}\,\delta\epsilon + \frac{\partial y_e}{\partial x_{ClE}}\,\delta x_{ClE}, \qquad \frac{\partial y_e}{\partial x_{ClE}} = -\frac{\kappa_L\,x_L}{(\kappa_L + x_{ClE})^2}. \qquad (19.18.22)$$

Let us now introduce some quantities that will be used repeatedly,

$$Z = \sum_{l\,\mathrm{ion}} \zeta_l^2\, x_{lE},$$

$$Z_{cb} = Z - x_{ClE}\,\frac{\partial y_e}{\partial x_{ClE}},$$

$$Z_k = 1 + \frac{y_{PG}}{Z_{cb}}\,\zeta_k, \quad k\,\mathrm{ion}, \qquad (19.18.23)$$

$$Z_{pk} = \left(1 + \frac{y_e}{Z_{cb}}\,\zeta_k\right) x_{kE} - y_e\left(1 - \frac{Z}{Z_{cb}}\right) I_{kCl}, \quad k:\, w,\,\mathrm{ion},$$

$$Z_{kl} = I_{kl} - \frac{1}{Z_{cb}}\,\zeta_k\,\zeta_l\,x_{lE} + \left(1 - \frac{Z}{Z_{cb}}\right)\zeta_k\,I_{lCl}, \quad k\,\mathrm{ion},\, l:\, w,\,\mathrm{ion}.$$

In the relations,

$$\sum_{k\in E_{mo}} x_{kE} + x_{PG} = 1, \qquad \sum_{k\in E_{ions}} \zeta_k\, x_{kE} + y_e = 0, \qquad (19.18.24)$$

the molar fraction x_{PG} of PGs can be neglected, but not their effective molar fraction $y_{PG} \equiv \zeta_{PG}\, x_{PG} < 0$. Differentiation of these relations,

$$\sum_{l:\,w,\mathrm{ion}} \delta x_{lE} = 0, \qquad \sum_{l\,\mathrm{ion}} \zeta_l\,\delta x_{lE} + \delta y_e = 0, \qquad (19.18.25)$$

and of the electrochemical potential,

$$\delta g_{kE}^{ec} = \widehat{v}_w \, \delta p_E \, I_{kw} + RT \, \frac{\delta x_{kE}}{x_{kE}} + F \, \zeta_k \, \delta \phi_E, \quad k : w, \text{ion}, \tag{19.18.26}$$

yields,

$$\begin{aligned}
\sum_{l: w, \text{ion}} x_{lE} \, \delta g_{lE}^{ec} &= x_{wE} \, \widehat{v}_w \, \delta p_E - y_e \, F \, \delta \phi_E \, ; \\
\sum_{l \text{ ion}} \zeta_l \, x_{lE} \, \delta g_{lE}^{ec} &= -RT \, \delta y_e + Z \, F \, \delta \phi_E = -RT \, \delta y_e + Z \left(-\delta g_{ClE}^{ec} + RT \, \frac{\delta x_{ClE}}{x_{ClE}} \right).
\end{aligned} \tag{19.18.27}$$

Introduction of δy_e from (19.18.22) in the last relation above delivers the increment of molar fraction of chloride ions,

$$\frac{\delta x_{ClE}}{x_{ClE}} = -\frac{1}{n^E} \frac{y_{PG}}{Z_{cb}} \operatorname{tr} \delta\epsilon + \sum_{l \text{ ion}} \left(\frac{1}{RT} \frac{1}{Z_{cb}} \zeta_l \, x_{lE} + \frac{1}{RT} \frac{Z}{Z_{cb}} I_{lCl} \right) \delta g_{lE}^{ec}. \tag{19.18.28}$$

Use of (19.18.26) with $k = Cl$ yields the increment of electrical potential,

$$F \, \delta \phi_E = -\frac{RT}{n^E} \frac{y_{PG}}{Z_{cb}} \operatorname{tr} \delta\epsilon + \sum_{l \text{ ion}} \left(\frac{\zeta_l \, x_{lE}}{Z_{cb}} - (1 - \frac{Z}{Z_{cb}}) I_{l \, Cl} \right) \delta g_{lE}^{ec}, \tag{19.18.29}$$

and further use of (19.18.26) provides the increments of molar fraction of ions other than chloride ion. The increments of mass content, volume content and mole content result from $(19.18.21)_{4,5}$,

$$\begin{aligned}
\frac{\delta x_{kE}}{x_{kE}} &= \frac{1}{n^E} (Z_k - 1) \operatorname{tr} \delta\epsilon + \sum_{l \text{ ion}} \frac{Z_{kl}}{RT} \delta g_{lE}^{ec}, \quad k \text{ ion} \neq Cl, \\
\frac{\delta m^{kE}}{m^{kE}} = \frac{\delta v^{kE}}{v^{kE}} = \frac{\delta \mathcal{N}_{kE}}{\mathcal{N}_{kE}} &= \frac{Z_k}{n^E} \operatorname{tr} \delta\epsilon + \sum_{l \text{ ion}} \frac{Z_{kl}}{RT} \delta g_{lE}^{ec}, \quad k \text{ ion}.
\end{aligned} \tag{19.18.30}$$

Finally, the fluid pressure results from $(19.18.27)_1$ and (19.18.29),

$$\delta p_E = -\frac{y_e}{x_{wE} \, \widehat{v}_w} \frac{RT}{n^E} \frac{y_{PG}}{Z_{cb}} \operatorname{tr} \delta\epsilon + \frac{1}{x_{wE} \, \widehat{v}_w} \sum_{l: w, \text{ion}} Z_{pl} \, \delta g_{lE}^{ec}. \tag{19.18.31}$$

19.18.6 Derivation of the element effective capacity matrices

Some of the algebraic manipulations necessary to obtain the element capacity and element diffusion-stiffness matrices (19.18.9) are provided below. The operator δ can be given arbitrary meaning, but, typically in an iterative process, it will a priori be interpreted as the difference between two iterates. The incremental strain $\delta\epsilon$ of the previous section derives from the incremental displacement $\delta\mathbf{u}$, namely $\delta\epsilon = \mathbf{B}_u \, \delta\mathbf{u}$.

19.18.6.1 Derivation of the terms linked to diffusion

The total flux,

$$\mathbf{J}_E = \mathbf{J}_{wE} + \sum_{k \in E_{ions}} \widehat{v}_k \, \mathbf{J}_{kE} = -\sum_{l \in E_{mo}} \kappa_{wl}^* \, \boldsymbol{\nabla} g_{lE}, \tag{19.18.32}$$

is built from the diffusion law (19.13.1)–(19.13.3),

$$\kappa_{ww}^* = \frac{\kappa_{ww}}{\widehat{v}_w} + \sum_{k \in E_{ions}} \frac{\widehat{v}_k}{\widehat{v}_w} \kappa_{kw}, \quad \kappa_{wl}^* = \kappa_{wl} + \sum_{k \in E_{ions}} \widehat{v}_k \, \kappa_{kl}, \quad l \in E_{ions}. \tag{19.18.33}$$

Line 2 of the element diffusion-stiffness matrix $\mathbb{K}^e$ follows from these expressions. The coefficients of lines 3 and 4 of the matrix $\mathbb{K}^e$,

$$\kappa_{kw}^* = \frac{\kappa_{kw}}{\widehat{v}_w}, \quad \kappa_{kl}^* = \kappa_{kl}, \quad k, l \in E_{ions}, \tag{19.18.34}$$

follow from the generalized diffusion equations for the fluxes $\mathbf{J}_{kE}$, $k \in E_{ions}$, established in Section 19.13. Lines 3 and 4 of the element capacity matrix $\mathbb{C}^e$ follow from the variation of the mole contents (19.18.30).

19.18.6.2 Derivation of the terms linked to the fictitious bath

The fictitious bath is a bath containing the same soluble species as the tissue but no insoluble species, and therefore no fixed charge. It is in chemical equilibrium with a geometrical point the tissue. If the mobile ions are the sodium cations and the chloride anions, whether chloride binding takes place or not in the tissue, the molar fractions of these species in the bath are given by (19.10.26),

$$\tilde{x}_{\mathrm{NaE}} = \tilde{x}_{\mathrm{ClE}} = \sqrt{x_{\mathrm{NaE}}\, x_{\mathrm{ClE}}}\,. \tag{19.18.35}$$

Since

$$\tilde{x}_{\mathrm{wE}} = 1 - (\tilde{x}_{\mathrm{NaE}} + \tilde{x}_{\mathrm{ClE}}), \quad x_{\mathrm{wE}} \simeq 1 - (x_{\mathrm{ClE}} + x_{\mathrm{NaE}}), \tag{19.18.36}$$

the fictitious Donnan pressure can be obtained in either exact or linearized forms,

$$\pi_{\mathrm{osm}} = p_{\mathrm{E}} - \tilde{p}_{\mathrm{E}} = \frac{RT}{\widehat{v}_{\mathrm{w}}} \mathrm{Ln}\, \frac{\tilde{x}_{\mathrm{wE}}}{x_{\mathrm{wE}}} \simeq \frac{RT}{\widehat{v}_{\mathrm{w}}} \left(\tilde{x}_{\mathrm{wE}} - x_{\mathrm{wE}} \right), \tag{19.18.37}$$

and

$$\delta\pi_{\mathrm{osm}} = \sum_{k \in \mathrm{E_{ions}}} P_k\, \delta x_{k\mathrm{E}}, \quad \text{with} \quad P_k \equiv \frac{RT}{\widehat{v}_{\mathrm{w}}} \left(\frac{1}{x_{\mathrm{wE}}} - \frac{1}{\tilde{x}_{\mathrm{wE}}} \frac{\tilde{x}_{k\mathrm{E}}}{x_{k\mathrm{E}}} \right), \text{ for } k \in \mathrm{E_{ions}}\,. \tag{19.18.38}$$

19.18.6.3 Derivation of the terms linked to chemomechanics

The Cauchy stress depends on the deformation and chemical variables as indicated by (19.17.6) and (19.17.7),

$$\boldsymbol{\sigma} = -(p_{\mathrm{E}} + p_{\mathrm{ch}})\, \mathbf{I} + \boldsymbol{\sigma}_0 + \underline{\mathcal{W}}_{\mathrm{ch},2}(\mathcal{N}_{\mathrm{E}}^*) \left(\boldsymbol{\sigma}^{\mathrm{gs}} + \boldsymbol{\sigma}^{\mathrm{c}} \right), \tag{19.18.39}$$

with the transported triplet,

$$\left(\boldsymbol{\sigma}_0,\, \boldsymbol{\sigma}^{\mathrm{gs}},\, \boldsymbol{\sigma}^{\mathrm{c}} \right) = \frac{1}{\det \mathbf{F}} \mathbf{F} \cdot \left(\underline{\underline{\tau}}_0,\, \frac{\partial \mathcal{W}^{\mathrm{gs}}}{\partial \mathbf{E}_{\mathrm{G}}},\, \frac{\partial \mathcal{W}^{\mathrm{c}}}{\partial \mathbf{E}_{\mathrm{G}}} \right) \cdot \mathbf{F}^{\mathrm{T}}. \tag{19.18.40}$$

The differential of each term on the rhs of (19.18.39) should be expressed in terms of the primary variables. First, the increment of the fluid pressure p_{E},

$$\delta p_{\mathrm{E}} = \frac{\partial p_{\mathrm{E}}}{\partial \mathrm{tr}\,\boldsymbol{\epsilon}} \delta\,\mathrm{tr}\,\boldsymbol{\epsilon} + \sum_{l:\,\mathrm{w,ion}} \frac{\partial p_{\mathrm{E}}}{\partial g_{l\mathrm{E}}^{\mathrm{ec}}} \delta g_{l\mathrm{E}}^{\mathrm{ec}}, \tag{19.18.41}$$

is provided by (19.18.31). The increments of the constitutive functions $f = p_{\mathrm{ch}}(\mathcal{N}_{\mathrm{E}}^*)$ and $f = \underline{\mathcal{W}}_{\mathrm{ch},2}(\mathcal{N}_{\mathrm{E}}^*)$ follow from (19.17.8), (19.18.30), (19.18.38),

$$\begin{aligned} \delta f &= \frac{\partial f}{\partial \pi_{\mathrm{osm}}} \delta\, \pi_{\mathrm{osm}} \\ &= \frac{\partial f}{\partial \pi_{\mathrm{osm}}} \sum_{k \in \mathrm{E_{ions}}} P_k\, \delta x_{k\mathrm{E}} \\ &= \frac{\partial f}{\partial \pi_{\mathrm{osm}}} \sum_{k \in \mathrm{E_{ions}}} P_k \left(\frac{\partial x_{k\mathrm{E}}}{\partial \mathrm{tr}\,\boldsymbol{\epsilon}} \delta\,\mathrm{tr}\,\boldsymbol{\epsilon} + \sum_{l:\,\mathrm{ion}} \frac{\partial x_{k\mathrm{E}}}{\partial g_{l\mathrm{E}}^{\mathrm{ec}}} \delta g_{l\mathrm{E}}^{\mathrm{ec}} \right). \end{aligned} \tag{19.18.42}$$

Let $\delta\mathbf{L} = \delta\mathbf{F} \cdot \mathbf{F}^{-1}$ be the incremental displacement gradient and $\delta\boldsymbol{\epsilon} = \frac{1}{2}(\delta\mathbf{L} + \delta\mathbf{L}^{\mathrm{T}})$ the incremental strain so that $\delta \det \mathbf{F} = \det \mathbf{F}\,\mathrm{tr}\,\delta\boldsymbol{\epsilon} = \det \mathbf{F}\,\mathrm{tr}\,\delta\mathbf{L}$ and $\delta\mathbf{E}_{\mathrm{G}} = \mathbf{F}^{\mathrm{T}} \cdot \delta\boldsymbol{\epsilon} \cdot \mathbf{F}$. Then the increment of Cauchy stress can be expressed as,

$$\begin{aligned} \delta\boldsymbol{\sigma} &= \left(-(\frac{\partial p_{\mathrm{E}}}{\partial \mathrm{tr}\,\boldsymbol{\epsilon}} + \frac{\partial p_{\mathrm{ch}}}{\partial \mathrm{tr}\,\boldsymbol{\epsilon}})\, \mathbf{I} \otimes \mathbf{I} + \frac{\partial \boldsymbol{\sigma}_0}{\partial \mathbf{L}} + \underline{\mathcal{W}}_{\mathrm{ch},2} \frac{\partial}{\partial \mathbf{L}} \left(\boldsymbol{\sigma}^{\mathrm{gs}} + \boldsymbol{\sigma}^{\mathrm{c}} \right) \right) : \delta\mathbf{L} \\ &\quad + \sum_{l:\,\mathrm{ion}} \left(-(\frac{\partial p_{\mathrm{E}}}{\partial g_{l\mathrm{E}}^{\mathrm{ec}}} + \frac{\partial p_{\mathrm{ch}}}{\partial g_{l\mathrm{E}}^{\mathrm{ec}}})\, \mathbf{I} + \frac{\partial \underline{\mathcal{W}}_{\mathrm{ch},2}}{\partial g_{l\mathrm{E}}^{\mathrm{ec}}} \left(\boldsymbol{\sigma}^{\mathrm{gs}} + \boldsymbol{\sigma}^{\mathrm{c}} \right) \right) \delta g_{l\mathrm{E}}^{\mathrm{ec}}. \end{aligned} \tag{19.18.43}$$

The second and third stresses in (19.18.40), $\alpha = \mathrm{gs}$ and $\alpha = \mathrm{c}$, respectively, may be expressed in the format,

$$\boldsymbol{\sigma}^\alpha = \frac{1}{\det \mathbf{F}} \, \mathbf{F} \cdot \frac{\partial \mathcal{W}_\alpha}{\partial \mathbf{E}_\mathrm{G}} \cdot \mathbf{F}^\mathrm{T} \, . \tag{19.18.44}$$

Consequently,

$$\delta \boldsymbol{\sigma}^\alpha = -\boldsymbol{\sigma}^\alpha \, (\mathbf{I} : \delta \boldsymbol{\epsilon}) + \delta \mathbf{L} \cdot \boldsymbol{\sigma}^\alpha + \boldsymbol{\sigma}^\alpha \cdot \delta \mathbf{L}^\mathrm{T} + \frac{1}{\det \mathbf{F}} \, \mathbf{F} \cdot \left(\frac{\partial^2 \mathcal{W}_\alpha}{\partial \mathbf{E}_\mathrm{G} \partial \mathbf{E}_\mathrm{G}} : \left(\mathbf{F}^\mathrm{T} \cdot \delta \boldsymbol{\epsilon} \cdot \mathbf{F} \right) \right) \cdot \mathbf{F}^\mathrm{T}, \tag{19.18.45}$$

and therefore

$$\begin{aligned}
\frac{\partial \sigma_{ij}^\alpha}{\partial L_{kl}} &= -\sigma_{ij}^\alpha \, I_{kl} + I_{ik} \sigma_{lj}^\alpha + \sigma_{il}^\alpha I_{jk} + \frac{1}{\det \mathbf{F}} \, F_{im} F_{jn} F_{kp} F_{lq} \left(\frac{\partial^2 \mathcal{W}_\alpha}{\partial \mathbf{E}_\mathrm{G} \partial \mathbf{E}_\mathrm{G}} \right)_{mnpq} ; \\
\frac{\partial \sigma_{0ij}}{\partial L_{kl}} &= -\sigma_{0ij} \, I_{kl} + I_{ik} \sigma_{0lj} + \sigma_{0il} I_{jk} \, .
\end{aligned} \tag{19.18.46}$$

19.18.7 Initial conditions

The initial concentrations, pressures, electrical potentials, active fluxes have to satisfy some conditions:
- across the membranes where there is *no pump* associated with a species, the latter should be in electrochemical equilibrium. Such is the case of water, and of chloride ion;
- when a species is *actively pumped*, the initial total flux of this species should vanish. This requirement applies to the sodium ion across epithelium and to the bicarbonate ion across endothelium. If all concentrations are known on both sides of the membrane, then the active flux is defined so as to zero the total flux. If the active flux is given, some quantity has to be adapted so that the total flux vanishes.

19.18.8 Boundary conditions

Laboratory tests aimed at measuring the time course of the electro-chemomechanical response of corneal stroma may be performed in presence or absence of the endothelium and epithelium. The time course of the pressure, ionic concentrations and electrical potential in the bath in contact with the front face of the stroma may be controlled. A few tests are sketched in Fig. 19.18.1.

19.19 The purely mechanical contribution: nonlinear elasticity

Most macroscopic models of fiber-reinforced tissues consider that the individual components of the tissue, namely fibers, matrix, etc. undergo the macroscopic deformation, that is, these components work in parallel. As an exception, Guo et al. [2006] introduce a multiplicative decomposition of the deformation gradient into a fiber extension and a shear. Homogenization techniques, like the self-consistent methods or differential methods, overcome this assumption of uniform strain, at the price of complexity and computational efforts. Discrete numerical simulations by Chandran and Barocas [2006] seem to indicate that the motion of the collagen fibrils is indeed non affine with the macroscopic motion. Nevertheless, an affine kinematics is used throughout here.

Individual lamellae display rotational symmetry about the fiber axis, Fig. 19.19.1. Still, a standard point of view is to group two adjacent lamellae and to consider a material reinforced by two families of fibers. A more refined approach consists in accounting for the directional distribution of fibers, around the two principal directions.

Consider a transversely isotropic material, with $\underline{\mathbf{m}}$ as the symmetry axis, representative of the microstructure which, as we have assumed, is convected by the deformation of the continuum characterized by the gradient $\mathbf{F}$. Then, if the reinforcement is due to fibers aligned with the axes $\underline{\mathbf{m}}$ in a reference state where $\mathbf{F} = \mathbf{I}$, their current direction will be

$$\mathbf{m} = \frac{\mathbf{F} \cdot \underline{\mathbf{m}}}{|\mathbf{F} \cdot \underline{\mathbf{m}}|} \, . \tag{19.19.1}$$

Alternatively, if the material can be seen as a deck of cards, the symmetry axis will evolve like the normal to the cards according to Nanson's rule,

$$\mathbf{m} = \frac{\mathbf{F}^{-\mathrm{T}} \cdot \underline{\mathbf{m}}}{|\mathbf{F}^{-\mathrm{T}} \cdot \underline{\mathbf{m}}|} = \frac{\underline{\mathbf{m}} \cdot \mathbf{F}^{-1}}{|\underline{\mathbf{m}} \cdot \mathbf{F}^{-1}|} \, . \tag{19.19.2}$$

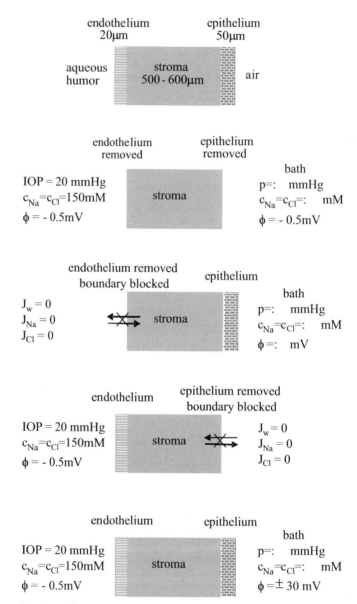

Fig. 19.18.1: Various possible boundary conditions to test the electro-chemomechanical response of the whole, de-enthotelized or de-epithelialized cornea. The values of pressure, ionic concentrations and electrical potential marked with the symbol " =: " are left to be prescribed.

When considering a small area of cornea along the normal to its surface, the structure may be seen as a deck of cards, and then it could be considered as transversely isotropic along the local normal. However, the layers themselves have a structure, which does not display rotational symmetry about this axis. In fact, layers are reinforced by parallel fibers. The fiber directions of adjacent layers are alternate. If the adjacent fibers were orthogonal, with identical properties and densities, the whole thickness of cornea could be seen as endowed with quadratic symmetry in planes parallel to the surface. However, this configuration holds approximately in the center of the cornea only. Quadratic symmetry is defined in Section 5.2.4.1.

It is appropriate to begin with the study of single layers, that is, with materials consisting of a matrix endowed with a single family of fibers. Cross-sections normal to the fibers are endowed with a special repartition in space of the fibers, Fig. 19.1.6. Strictly speaking, the layers thus do not show rotational symmetry about the fiber axis. Still, we begin by materials endowed with this rotational symmetry, where the fibers are assumed to be randomly located.

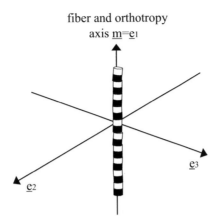

Fig. 19.19.1: Reference axes for a material reinforced by a single family of fibers.

19.19.1   A single family of fibers

Let $\underline{\mathbf{m}} = \underline{\mathbf{e}}_1$ be the unit vector that indicates the direction of the fibers in a reference configuration and

$$\mathbf{A}_1 = \mathbf{M} = \underline{\mathbf{m}} \otimes \underline{\mathbf{m}}, \tag{19.19.3}$$

be the associated second order structure tensor. The current expression of the tensor structure follows from (19.19.1),

$$\mathbf{M} = \mathbf{m} \otimes \mathbf{m} = \frac{\mathbf{F} \cdot \underline{\mathbf{M}} \cdot \mathbf{F}^{\mathrm{T}}}{|\mathbf{F} \cdot \underline{\mathbf{m}}|^2}. \tag{19.19.4}$$

The material is endowed with transverse isotropy with respect to the direction $\underline{\mathbf{m}}$ and the sense of this direction is elusive. It remains so even during deformation. Any scalar function of the second order symmetric tensor $\mathbf{C} = \mathbf{F}^{\mathrm{T}} \cdot \mathbf{F}$ that respects both invariance requirements and the material symmetries should be expressed via the scalar invariants given in Table 19.19.1.

Table 19.19.1: One family of fibers: scalar invariants of a symmetric 2nd order tensor relative to the symmetry groups of transverse isotropy (2.11.43), Boehler [1978].

Invariant I	$\partial I/\partial \mathbf{C}$	$\partial^2 I/\partial \mathbf{C}\partial \mathbf{C}$
$I_1 = \operatorname{tr} \mathbf{C}$	$\mathbf{I}$	$\mathbf{0}$
$I_2 = \frac{1}{2}(\operatorname{tr}^2 \mathbf{C} - \operatorname{tr} \mathbf{C}^2)$	$(\operatorname{tr} \mathbf{C})\mathbf{I} - \mathbf{C}$	$\mathbf{I} \otimes \mathbf{I} - \mathbf{I}\,\overline{\underline{\otimes}}\,\mathbf{I}$
$I_3 = \det \mathbf{C}$	$(\det \mathbf{C})\,\mathbf{C}^{-1}$	$\det \mathbf{C}\,(\mathbf{C}^{-1} \otimes \mathbf{C}^{-1} - \mathbf{C}^{-1}\,\overline{\underline{\otimes}}\,\mathbf{C}^{-1})$
$I_4 = \mathbf{C} : \underline{\mathbf{M}}$	$\underline{\mathbf{M}}$	$\mathbf{0}$
$I_5 = \mathbf{C}^2 : \underline{\mathbf{M}}$	$\mathbf{C} \cdot \underline{\mathbf{M}} + \underline{\mathbf{M}} \cdot \mathbf{C}$	$\mathbf{I}\,\overline{\underline{\otimes}}\,\underline{\mathbf{M}} + \underline{\mathbf{M}}\,\overline{\underline{\otimes}}\,\mathbf{I}$

For the record, let us note the first and second derivatives of a scalar $I = I(\mathbf{C})$ raised to a power $\alpha \geq 0$,

$$\frac{\partial I^\alpha}{\partial \mathbf{C}} = \alpha\, I^{\alpha-1}\,\frac{\partial I}{\partial \mathbf{C}}, \quad \frac{\partial^2 I^\alpha}{\partial \mathbf{C}\,\partial \mathbf{C}} = \alpha\,(\alpha-1)\,I^{\alpha-2}\,\frac{\partial I}{\partial \mathbf{C}} \otimes \frac{\partial I}{\partial \mathbf{C}} + \alpha\,I^{\alpha-1}\,\frac{\partial^2 I}{\partial \mathbf{C}\,\partial \mathbf{C}}. \tag{19.19.5}$$

The operator,

$$(\mathbf{A}\,\overline{\underline{\otimes}}\,\mathbf{B})_{ijkl} = \frac{1}{2}\,(A_{ik}\,B_{jl} + A_{il}\,B_{jk}), \quad (\mathbf{A}\,\overline{\underline{\otimes}}\,\mathbf{B})[\mathbf{X}] = \mathbf{A}\cdot \tfrac{1}{2}(\mathbf{X} + \mathbf{X}^{\mathrm{T}})\cdot \mathbf{B}^{\mathrm{T}}, \tag{19.19.6}$$

is instrumental to manipulate the tensor of elasticity.

Table 19.19.1 addresses compressible materials with extensible fibers. Incompressible materials are defined by the constraint,

$$I_3 - 1 = \det \mathbf{C} - 1 = 0, \tag{19.19.7}$$

and, for inextensible fibers,

$$I_4 - 1 = \mathbf{C} : \underline{\mathbf{M}} - 1 = 0. \tag{19.19.8}$$

These constraints may, or may not, be acceptable approximations of the actual behavior. Incompressibility is not appropriate in presence of chemomechanical couplings, which give rise to swelling and/or shrinking of the solid skeleton. In any case, they often facilitate the obtention of analytical solutions to boundary value problems. They are independent of one another and they may coexist in the same material. The constitutive stress is obtained by augmenting the strain energy function $\underline{W}$ via Lagrange multipliers, say p_F and T, which are defined by the boundary conditions,

$$\underline{W} \quad \Rightarrow \quad \underline{W} + \frac{1}{2} p_F \left(I_3 - 1\right) - \frac{1}{2} T \left(I_4 - 1\right), \tag{19.19.9}$$

so that the stress is modified according to the format,

$$\underline{\underline{\tau}} = 2 \frac{\partial \underline{W}}{\partial \mathbf{C}} \quad \Rightarrow \quad \underline{\underline{\tau}} \quad = \quad 2 \frac{\partial \underline{W}}{\partial \mathbf{C}} - p_F \det \mathbf{C} \, \mathbf{C}^{-1} + T \, \underline{\mathbf{M}},$$

$$\boldsymbol{\tau} \quad = \quad 2 \, \mathbf{F} \cdot \frac{\partial \underline{W}}{\partial \mathbf{C}} \cdot \mathbf{F}^{\mathrm{T}} - p_F \det \mathbf{C} \, \mathbf{I} + |\mathbf{F} \cdot \underline{\mathbf{m}}|^2 \, T \, \mathbf{M}, \tag{19.19.10}$$

$$= \quad 2 \, \mathbf{F} \cdot \frac{\partial \underline{W}}{\partial \mathbf{C}} \cdot \mathbf{F}^{\mathrm{T}} - p_F \, \mathbf{I} + T \, \mathbf{M},$$

while the augmented strain energy function is now subjected to the conditions,

$$\frac{\partial \underline{W}}{\partial \mathbf{C}} : \mathbf{C}^{-1} = 0, \quad \frac{\partial \underline{W}}{\partial \mathbf{C}} : \underline{\mathbf{M}} = 0. \tag{19.19.11}$$

The isometric matrix forms in the orthotropy axes of some tensors involved in the formulation are recorded below:

$$[\mathbf{I} \otimes \mathbf{I}] = \begin{bmatrix} 1 & 1 & 1 & 0 & 0 & 0 \\ 1 & 1 & 1 & 0 & 0 & 0 \\ 1 & 1 & 1 & 0 & 0 & 0 \\ 0 & 0 & 0 & 0 & 0 & 0 \\ 0 & 0 & 0 & 0 & 0 & 0 \\ 0 & 0 & 0 & 0 & 0 & 0 \end{bmatrix}, \quad [\mathbf{I} \,\overline{\otimes}\, \mathbf{I}] = \begin{bmatrix} 1 & 0 & 0 & 0 & 0 & 0 \\ 0 & 1 & 0 & 0 & 0 & 0 \\ 0 & 0 & 1 & 0 & 0 & 0 \\ 0 & 0 & 0 & 1 & 0 & 0 \\ 0 & 0 & 0 & 0 & 1 & 0 \\ 0 & 0 & 0 & 0 & 0 & 1 \end{bmatrix}, \tag{19.19.12}$$

$$[\underline{\mathbf{M}} \otimes \underline{\mathbf{M}}] = \begin{bmatrix} 1 & 0 & 0 & 0 & 0 & 0 \\ 0 & 0 & 0 & 0 & 0 & 0 \\ 0 & 0 & 0 & 0 & 0 & 0 \\ 0 & 0 & 0 & 0 & 0 & 0 \\ 0 & 0 & 0 & 0 & 0 & 0 \\ 0 & 0 & 0 & 0 & 0 & 0 \end{bmatrix}, \quad [\mathbf{I} \otimes \underline{\mathbf{M}} + \underline{\mathbf{M}} \otimes \mathbf{I}] = \begin{bmatrix} 2 & 1 & 1 & 0 & 0 & 0 \\ 1 & 0 & 0 & 0 & 0 & 0 \\ 1 & 0 & 0 & 0 & 0 & 0 \\ 0 & 0 & 0 & 0 & 0 & 0 \\ 0 & 0 & 0 & 0 & 0 & 0 \\ 0 & 0 & 0 & 0 & 0 & 0 \end{bmatrix}, \tag{19.19.13}$$

$$[\underline{\mathbf{M}} \,\overline{\otimes}\, \mathbf{I} + \mathbf{I} \,\overline{\otimes}\, \underline{\mathbf{M}}] = \begin{bmatrix} 2 & 0 & 0 & 0 & 0 & 0 \\ 0 & 0 & 0 & 0 & 0 & 0 \\ 0 & 0 & 0 & 0 & 0 & 0 \\ 0 & 0 & 0 & 0 & 0 & 0 \\ 0 & 0 & 0 & 0 & 1 & 0 \\ 0 & 0 & 0 & 0 & 0 & 1 \end{bmatrix}. \tag{19.19.14}$$

19.19.1.1   Interpretation of the scalar invariant I_4

The invariant I_4 represents the square of stretch along the fiber direction,

$$I_4 = \mathbf{C} : \underline{\mathbf{M}} = |\mathbf{F} \cdot \underline{\mathbf{m}}|^2. \tag{19.19.15}$$

Therefore the sign of the scalar $I_4 - 1$, or of $\sqrt{I_4} - 1$, serves to delineate whether the fiber contracts or extends. This sign is a capital information that decides of the mechanical activation of the fiber.

19.19.1.2   Insights in the strain energy function

The invariant I_5 does not have such a direct physical interpretation. Still, examples of strain energy where it is involved will be provided below.

Let $\mathbf{F} = \mathbf{R} \cdot \mathbf{U}$ be the polar decomposition of the deformation gradient $\mathbf{F}$. Then $\mathbf{C} = \mathbf{U}^2$, and $I_4 = |\mathbf{U} \cdot \mathbf{m}|^2$, $I_5 = |\mathbf{U}^2 \cdot \mathbf{m}|^2$. If the length of the fibers increases through $\mathbf{U}^2$, I_5 should increase as well through $\mathbf{U}^4$. Conversely, if the length decreases through $\mathbf{U}^4$, I_5 should decrease as well through $\mathbf{U}^2$. But $I_4 > 1$ is not equivalent to $I_5 > 1$ as indicated by Fig. 19.19.2. A proof of the above implications is provided in component form. Let $\mathbf{C} = \underline{c}_i\, \mathbf{c}_i \otimes \mathbf{c}_i$ be the spectral decomposition of $\mathbf{C}$. With $\underline{\mathbf{m}} = \underline{m}_i\, \mathbf{c}_i$, the scalar invariants become

$$I_1 = \underline{c}_1 + \underline{c}_2 + \underline{c}_3, \quad I_2 = \underline{c}_1\,\underline{c}_2 + \underline{c}_2\,\underline{c}_3 + \underline{c}_3\,\underline{c}_1, \quad I_3 = \underline{c}_1\,\underline{c}_2\,\underline{c}_3, \quad I_4 = \underline{c}_i\,\underline{m}_i^2, \quad I_5 = \underline{c}_i^2\,\underline{m}_i^2. \tag{19.19.16}$$

Note that, whether $\underline{c} > 1$ or $\underline{c} < 1$, then $\underline{c}\,(\underline{c} - 1) > \underline{c} - 1$ and of course $\underline{m}^2\,\underline{c}\,(\underline{c} - 1) > \underline{m}^2\,(\underline{c} - 1)$. Consequently $I_5 - 1 \geq 2\,(I_4 - 1)$.

Incidentally, the relations (19.19.16) can be used to check that the fifth invariant is not independent for plane strain ($\underline{c}_3 = 1$, $\underline{m}_3 = 0$), namely $I_5 = (I_1 - 1)\,I_4 - I_3$, Merodio and Ogden [2003].

The format of the scalar invariants leads naturally to the decomposition of the strain energy into the sum of an isotropic contribution and an anisotropic contribution. It is in line with the assumption that the ground substance and the fibers work in parallel, so that their deformations are identical while their contributions to the stress add algebraically, namely

$$\underline{W} = \underline{W}^{\mathrm{iso}}(I_1, I_2, I_3) + \underline{W}^{\mathrm{aniso}}(I_4, I_5). \tag{19.19.17}$$

The anisotropic contribution to the strain energy is activated only when the fibers undergo extension, namely $I_4 > 1$, even if the fibers may contribute again at large contractions once they have buckled. Thus the anisotropic contribution should involve I_4, either alone, or together with I_5, namely

$$\underline{W}^{\mathrm{aniso}}(I_4, I_5) = \underline{W}^{\mathrm{aniso}}_{(4)}(I_4) + \underline{W}^{\mathrm{aniso}}_{(4-5)}(I_4, I_5). \tag{19.19.18}$$

The term dependent on I_4 can be taken in the format,

$$\underline{W}^{\mathrm{aniso}}_{(4)}(I_4) = \underline{W}^{\mathrm{aniso}}_{(4)}(\langle I_4 - 1\rangle^\alpha), \tag{19.19.19}$$

where $\langle \cdot \rangle$ denotes the positive part of its argument and $\underline{W}^{\mathrm{aniso}}_{(4)}$ is a sufficiently smooth function, with $\alpha > 11$ a constant larger than 1, in such a way that the stress is continuous on the balance hyperplane $I_4 - 1 = 0$.

Theoretically, the term involving I_4 and I_5 is made effective only if the fibers are in extension, namely $I_4 - 1 > 0$. Along a less strict point of view, a function that smooths a step can be introduced.

19.19.1.3   The scalar invariant I_5 : *fiber-matrix interaction*

The last term in (19.19.18) influences the shear properties in a plane orthogonal to the symmetry axis, as indicated by the stiffness generated by I_5 in Table 19.19.1 and shown componentwise in Eqn (19.19.14). The invariant I_5 can be introduced through the following interpretation, Merodio and Ogden [2003], Peng et al. [2006]. Let $d\underline{\mathbf{S}}$ be an elementary surface of unit normal parallel to the symmetry axis $\mathbf{m}$. If the material undergoes a deformation gradient $\mathbf{F}$, the transformed surface becomes $d\mathbf{S} = \det\mathbf{F}\, d\underline{\mathbf{S}} \cdot \mathbf{F}^{-1}$. Thus the unit normal to the transformed surface is $\mathbf{n} = \underline{\mathbf{m}} \cdot \mathbf{F}^{-1}/|\underline{\mathbf{m}} \cdot \mathbf{F}^{-1}|$, while the transformed symmetry axis becomes $\mathbf{m} = \mathbf{F} \cdot \underline{\mathbf{m}}/|\mathbf{F} \cdot \underline{\mathbf{m}}|$, Fig. 19.19.3-(a). Therefore $\mathbf{m} \cdot \mathbf{n} = 1/\sqrt{I_4\,\underline{\mathbf{m}} \cdot \mathbf{C}^{-1} \cdot \underline{\mathbf{m}}}$. Using the Cayley-Hamilton theorem, $\mathbf{C}^3 - I_1\,\mathbf{C}^2 + I_2\,\mathbf{C} - I_3\,\mathbf{I} = \mathbf{0}$, the angle between

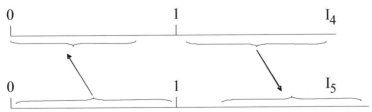

Fig. 19.19.2: The positions of the scalar invariants I_4 and I_5 with respect to 1 are not identical. Therefore, the sign of $I_5 - 1$ does not indicate whether the fibers are in extension or in contraction.

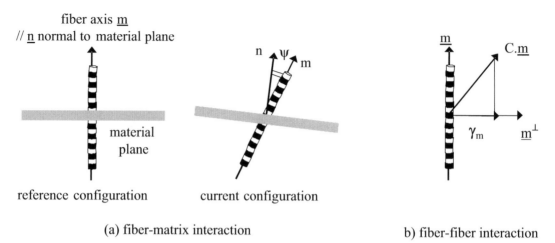

(a) fiber-matrix interaction   b) fiber-fiber interaction

Fig. 19.19.3: Two interpretations of the mechanical significance of the invariant I_5.

the deformed direction of the normal to the material surface and the deformed symmetry axis is found to express in terms of all the five scalar invariants,

$$\cos\psi = \mathbf{m}\cdot\mathbf{n} = \sqrt{\frac{I_3}{I_4\,(I_5 - I_1\,I_4 + I_2)}}\,. \tag{19.19.20}$$

For the annulus fibrosus, Peng et al. [2006] take

$$\mathcal{W}^{\text{aniso}}_{(4-5)}(I_1, I_2, I_3, I_4, I_5) = f^{(4)}(I_4)\,\tan^2\psi\,. \tag{19.19.21}$$

Thus the isotropic/anisotropic additive splitting (19.19.17) does not hold exactly. The anisotropic term will contribute both to shear and diagonal components of the stiffness since the isotropic invariants are involved in the anisotropic part. The function $f^{(4)}$ should be differentiable so as to ensure the stress to be continuous. As indicated above, it is aimed at weighting the effect of the second term $\tan^2\psi$, essentially smoothing a step function at $I_4 = 1$. Peng et al. [2006] term *fiber-matrix interaction* the contribution (19.19.21). They use a weight $f^{(4)}$ such that the interaction is fully activated only for strains larger than the toe region.

19.19.1.4   The scalar invariant I_5 : *fiber-fiber interaction* or cross-link intensity

Consider the vector $\mathbf{C}\cdot\mathbf{m}$ to be decomposed along $\underline{\mathbf{m}}$ and a unit vector $\underline{\mathbf{m}}^\perp$ orthogonal to $\underline{\mathbf{m}}$, say $\mathbf{C}\cdot\mathbf{m} = \alpha\,\underline{\mathbf{m}} + \gamma_m\,\underline{\mathbf{m}}^\perp$. Then $\alpha = \underline{\mathbf{m}}\cdot\mathbf{C}\cdot\underline{\mathbf{m}}$ and $\gamma_m = \underline{\mathbf{m}}^\perp\cdot\mathbf{C}\cdot\underline{\mathbf{m}}$ is the component of the shear strain on a plane of normal $\underline{\mathbf{m}}$ in the direction $\underline{\mathbf{m}}^\perp$ as shown in Fig. 19.19.3-(b). Consequently $|\mathbf{C}\cdot\mathbf{m}|^2 = (\underline{\mathbf{m}}\cdot\mathbf{C}\cdot\underline{\mathbf{m}})^2 + (\underline{\mathbf{m}}^\perp\cdot\mathbf{C}\cdot\underline{\mathbf{m}})^2$. The above relation can be phrased in terms of invariants,

$$(\underline{\mathbf{m}}\cdot\mathbf{C}\cdot\underline{\mathbf{m}}^\perp)^2 = \underline{\mathbf{m}}\cdot\mathbf{C}^2\cdot\underline{\mathbf{m}} - (\underline{\mathbf{m}}\cdot\mathbf{C}\cdot\underline{\mathbf{m}})^2 \quad\Leftrightarrow\quad \gamma_m^2 = I_5 - I_4^2\,. \tag{19.19.22}$$

The shear γ_m will be small if the fiber-to-fiber interaction is strong. This interaction term can also be seen as characterizing the crosslink between fibers. Said otherwise, for the fiber-to-fiber interaction to be strong, this term should be included in the strain energy with a significant coefficient, so that it requires a significant energy, or stress, to take place.

Based on this interpretation, Wagner and Lotz [2004] consider that the strain energy function of the annulus fibrosus should be a function of γ_m, in fact of $\gamma_m^2 = I_5 - I_4^2$.

19.19.2   Two families of fibers : clinotropic materials

Let $\underline{\mathbf{m}}_1$ and $\underline{\mathbf{m}}_2$ be the unit vectors that indicate the directions in a reference configuration of the two families of fibers that reinforce the lamellae of the corneal stroma and of the annulus fibrosus, Figs. 19.19.4 and 19.20.2 respectively.

In general, in addition to density, the stiffness of individual fibers may be directionally dependent, Jayasuriya et al. [2003]a [2003]b. The two families of fibers are said *mechanically equivalent* if their mechanical stiffnesses

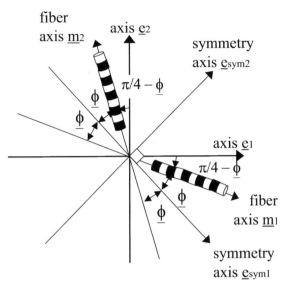

Fig. 19.19.4: The stroma is composed of lamellae which are parallel to the corneal surface. Each lamella is reinforced by parallel fibers. The structure of the stroma is similar to that of an *angle-ply composite*. The angle $(\underline{\mathbf{m}}_1, \underline{\mathbf{m}}_2)$ between the fibers of two adjacent lamellae is denoted by $\pi - 2\phi$, $0 < \underline{\phi} < \pi/4$. The value $\underline{\phi} = \pi/4$ implies the fibers to be orthogonal. For $\underline{\phi} = 0$, a transversely isotropic material is recovered.

are identical and if their densities are identical. Then the strain energy is symmetric in $\underline{\mathbf{m}}_1$ and $\underline{\mathbf{m}}_2$ and it remains unchanged by interchange of these vectors. The material displays local orthotropy with axes defined by the symmetry axes and the radial direction. Data by Boote et al. [2005] suggest that indeed the two horizontal and vertical directions are in general equally populated, although they observe that some eyes have up to 25% more collagen in one direction.

Two sets of cartesian axes are defined in Fig. 19.19.4. The orthogonal axes $\mathbf{e}$'s are chosen so that, when the fibers $\underline{\mathbf{m}}$'s are orthogonal, the two sets of axes $(\underline{\mathbf{m}}_1, \underline{\mathbf{m}}_2)$ and $(\underline{\mathbf{e}}_1, \underline{\mathbf{e}}_2)$ coincide, namely

$$\underline{\mathbf{m}}_1 = c\,\underline{\mathbf{e}}_1 - s\,\underline{\mathbf{e}}_2, \quad \underline{\mathbf{m}}_2 = -s\,\underline{\mathbf{e}}_1 + c\,\underline{\mathbf{e}}_2, \tag{19.19.23}$$

and equivalently,

$$(c^2 - s^2)\,\underline{\mathbf{e}}_1 = c\,\underline{\mathbf{m}}_1 + s\,\underline{\mathbf{m}}_2, \quad (c^2 - s^2)\,\underline{\mathbf{e}}_2 = s\,\underline{\mathbf{m}}_1 + c\,\underline{\mathbf{m}}_2, \tag{19.19.24}$$

with

$$c \equiv \cos(\frac{\pi}{4} - \underline{\phi}), \quad s \equiv \sin(\frac{\pi}{4} - \underline{\phi}). \tag{19.19.25}$$

It will be instrumental to introduce also the axes $\underline{\mathbf{e}}_1^s$, $\underline{\mathbf{e}}_2^s$, $\underline{\mathbf{e}}_3^s = \underline{\mathbf{e}}_3$ of geometrical symmetry (bisectors),

$$\underline{\mathbf{m}}_1 = \cos\underline{\phi}\,\underline{\mathbf{e}}_1^s + \sin\underline{\phi}\,\underline{\mathbf{e}}_2^s, \quad \underline{\mathbf{m}}_2 = -\cos\underline{\phi}\,\underline{\mathbf{e}}_1^s + \sin\underline{\phi}\,\underline{\mathbf{e}}_2^s, \tag{19.19.26}$$

and conversely

$$\underline{\mathbf{e}}_1^s = \frac{\underline{\mathbf{m}}_1 - \underline{\mathbf{m}}_2}{2\cos\underline{\phi}}, \quad \underline{\mathbf{e}}_2^s = \frac{\underline{\mathbf{m}}_1 + \underline{\mathbf{m}}_2}{2\sin\underline{\phi}}. \tag{19.19.27}$$

The components of the Cauchy-Green tensor in these axes are denoted by a superimposed s,

$$\mathbf{C} = C_{ij}\,\underline{\mathbf{e}}_i \otimes \underline{\mathbf{e}}_j = C_{ij}^s\,\underline{\mathbf{e}}_i^s \otimes \underline{\mathbf{e}}_j^s. \tag{19.19.28}$$

Note that the planes of geometrical symmetry are not planes of mechanical symmetry if the two families of fibers are not equivalent.

The following *structural* second order tensors will be shown to embed a good part, although not all, of the structure of the material,

$$\underline{\mathbf{M}}_1 = \underline{\mathbf{m}}_1 \otimes \underline{\mathbf{m}}_1, \quad \underline{\mathbf{M}}_2 = \underline{\mathbf{m}}_2 \otimes \underline{\mathbf{m}}_2,$$

$$\underline{\mathbf{M}}_S = \underline{\mathbf{M}}_1 \cdot \underline{\mathbf{M}}_2 + \underline{\mathbf{M}}_2 \cdot \underline{\mathbf{M}}_1 = (\underline{\mathbf{m}}_1 \cdot \underline{\mathbf{m}}_2)\,(\underline{\mathbf{m}}_1 \otimes \underline{\mathbf{m}}_2 + \underline{\mathbf{m}}_2 \otimes \underline{\mathbf{m}}_1). \tag{19.19.29}$$

19.19.2.1   Representation theorems for laminates

Let us consider a material endowed with these two families of fibers, which in general are neither orthogonal nor mechanically equivalent. The symmetry group[19.9] which characterizes its properties contains the following elements,

$$\mathbf{I}, -\mathbf{I}, \mathbf{R}, -\mathbf{R}, \tag{19.19.30}$$

where $\mathbf{R}$ is a reflection about the plane defined by the fibers and of unit normal $\underline{\mathbf{e}}_3$, namely $\mathbf{R} = \mathbf{I} - 2\,\underline{\mathbf{e}}_3 \otimes \underline{\mathbf{e}}_3$. If an element belongs to the symmetry group, its opposite also does. Thus there are no 'arrows' attached on the fibers. Consequently, the entities which will be derived are even functions of the fiber directions $\underline{\mathbf{m}}_1$ and $\underline{\mathbf{m}}_2$, that is, they introduce structure tensors like $\underline{\mathbf{M}}_1$ and $\underline{\mathbf{M}}_2$. If the angle between the fibers is involved, $\cos 2\underline{\phi} = -\underline{\mathbf{m}}_1 \cdot \underline{\mathbf{m}}_2$, then constitutive functions should be even functions of $\cos 2\underline{\phi}$ or arbitrary function of $|\cos 2\underline{\phi}|$.

Any scalar function of the second order symmetric tensor $\mathbf{C} = \mathbf{F}^{\mathrm{T}} \cdot \mathbf{F}$ that respects both invariance requirements and the material symmetries should be expressed via the scalar invariants given in Tables 19.19.2 or 19.19.3.

Table 19.19.2: Two families of fibers. List of complete and irreducible scalar invariants, $0 < 2\underline{\phi} < \pi/2$, Zheng and Boehler [1994].

Invariant I	$\partial I/\partial \mathbf{C}$	$\partial^2 I/\partial \mathbf{C}\partial \mathbf{C}$
$I_1 = \operatorname{tr}\mathbf{C}$	$\mathbf{I}$	$\mathbf{0}$
$I_2 = \frac{1}{2}\left(\operatorname{tr}^2\mathbf{C} - \operatorname{tr}\mathbf{C}^2\right)$	$(\operatorname{tr}\mathbf{C})\,\mathbf{I} - \mathbf{C}$	$\mathbf{I} \otimes \mathbf{I} - \mathbf{I}\,\overline{\otimes}\,\mathbf{I}$
$I_4 = \mathbf{C} : \underline{\mathbf{M}}_1$	$\underline{\mathbf{M}}_1$	$\mathbf{0}$
$I_5 = \mathbf{C}^2 : \underline{\mathbf{M}}_1$	$\mathbf{C} \cdot \underline{\mathbf{M}}_1 + \underline{\mathbf{M}}_1 \cdot \mathbf{C}$	$\mathbf{I}\,\overline{\otimes}\,\underline{\mathbf{M}}_1 + \underline{\mathbf{M}}_1\,\overline{\otimes}\,\mathbf{I}$
$I_6 = \mathbf{C} : \underline{\mathbf{M}}_2$	$\underline{\mathbf{M}}_2$	$\mathbf{0}$
$I_7 = \mathbf{C}^2 : \underline{\mathbf{M}}_2$	$\mathbf{C} \cdot \underline{\mathbf{M}}_2 + \underline{\mathbf{M}}_2 \cdot \mathbf{C}$	$\mathbf{I}\,\overline{\otimes}\,\underline{\mathbf{M}}_2 + \underline{\mathbf{M}}_2\,\overline{\otimes}\,\mathbf{I}$
$I_S = \mathbf{C} : \underline{\mathbf{M}}_S$	$\underline{\mathbf{M}}_S$	$\mathbf{0}$

For a material which may be compressible and with extensible fibers, the Kirchhoff stress $\boldsymbol{\tau} = 2\,\mathbf{F} \cdot \partial \underline{\mathcal{W}}/\partial \mathbf{C} \cdot \mathbf{F}^{\mathrm{T}}$ expresses as,

$$
\begin{aligned}
\boldsymbol{\tau} = 2 \quad \Big(\;\; & \frac{\partial \mathcal{W}}{\partial I_1}\mathbf{B} + \frac{\partial \mathcal{W}}{\partial I_2}\left(\operatorname{tr}\mathbf{C}\,\mathbf{B} - \mathbf{B}^2\right) \\
+ \;\; & \frac{\partial \mathcal{W}}{\partial I_4}\mathbf{M}_1 + \frac{\partial \mathcal{W}}{\partial I_6}\mathbf{M}_2 \\
+ \;\; & \frac{\partial \mathcal{W}}{\partial I_5}\left(\mathbf{M}_1 \cdot \mathbf{B} + \mathbf{B} \cdot \mathbf{M}_1\right) + \frac{\partial \mathcal{W}}{\partial I_7}\left(\mathbf{M}_2 \cdot \mathbf{B} + \mathbf{B} \cdot \mathbf{M}_2\right) + \frac{\partial \mathcal{W}}{\partial I_S}\mathbf{M}_S \Big),
\end{aligned}
\tag{19.19.31}
$$

with $\mathbf{B} = \mathbf{F} \cdot \mathbf{F}^{\mathrm{T}}$ the left Cauchy-Green tensor and $\mathbf{M}_\alpha = \mathbf{F} \cdot \underline{\mathbf{M}}_\alpha \cdot \mathbf{F}^{\mathrm{T}}$, $\alpha = 1, 2, S$, the transported structure tensors.

Table 19.19.2 splits explicitly the isotropic and anisotropic invariants. In order to respect the symmetries of the materials, the two isotropic invariants I_1 and I_2 can be replaced by the two invariants K_3 and K_7 of Table 19.19.3, in such a way that, with $\underline{\mathbf{M}}_3 = \underline{\mathbf{A}}_3 = \underline{\mathbf{e}}_3 \otimes \underline{\mathbf{e}}_3$, the three orthogonal axes play formally identical role. The property can be established using the relation $\underline{\mathbf{M}}_1 + \underline{\mathbf{M}}_2 + (c^2 - s^2)^2\,\underline{\mathbf{M}}_3 - \underline{\mathbf{M}}_S = (c^2 - s^2)^2\,\mathbf{I}$ in the first two scalar invariants of Table 19.19.2.

Table 19.19.3: Two families of fibers with arbitrary directions. Alternative and equivalent list of scalar invariants, $0 < 2\underline{\phi} < \pi/2$.

Invariant I	$\partial I/\partial \mathbf{C}$	$\partial^2 I/\partial \mathbf{C}\partial \mathbf{C}$	
$K_a = \mathbf{C} : \underline{\mathbf{M}}_a$	$\underline{\mathbf{M}}_a$	$\mathbf{0}$	$a = 1, 2, 3, S$
$K_{a+4} = \mathbf{C}^2 : \underline{\mathbf{M}}_a$	$\mathbf{C} \cdot \underline{\mathbf{M}}_a + \underline{\mathbf{M}}_a \cdot \mathbf{C}$	$\mathbf{I}\,\overline{\otimes}\,\underline{\mathbf{M}}_a + \underline{\mathbf{M}}_a\,\overline{\otimes}\,\mathbf{I}$	$a = 1, 2, 3$

If the two sets of fibers are equivalent, then the two geometrical planes of symmetry are also planes of mechanical symmetry: the material is *orthotropic* although it is usually termed *locally* orthotropic, Spencer [1992]. Note that

---

[19.9]The concept of symmetry group is exposed in Section 2.11.3.

it is tacitly assumed that the ground substance is mechanically isotropic. Then the functions accounting for this property depend on the scalar invariants through $I_4 + I_6$, $I_4 I_6$, $I_5 + I_7$, $I_5 I_7$, instead of the individual invariants. Alternatively, $I_5 I_7$ depends linearly on I_1, I_2, I_S, $II_S = \mathbf{C}^2 : \underline{\mathbf{M}}_S$, Spencer [1972], p. 87.

If the two sets of fibers are orthogonal, even if not equivalent, the material is said to possess *monoclinic symmetry* (a single plane of symmetry, 13 independent coefficients for linear elasticity). If the two sets are both orthogonal and equivalent, the material is endowed with *quadratic symmetry* (orthotropy and equivalence of two symmetry axes, 6 independent coefficients for linear elasticity). In this situation, $\underline{\mathbf{M}}_S$ vanishes and I_S is substituted by, e.g., $\mathrm{tr}\,\mathbf{C}^3$, Table 19.19.4.

Table 19.19.4: Orthotropy. List of complete and irreducible scalar invariants, Boehler [1978]. The $\underline{\mathbf{A}}_a = \underline{\mathbf{e}}_a \otimes \underline{\mathbf{e}}_a$'s are the structure tensors, with the $\underline{\mathbf{e}}_a$'s the mutually orthogonal orthotropy axes.

Invariant I	$\partial I/\partial \mathbf{C}$	$\partial^2 I/\partial \mathbf{C}\partial \mathbf{C}$	
$I_a = \mathbf{C} : \underline{\mathbf{A}}_a$	$\underline{\mathbf{A}}_a$	$\mathbf{0}$	$a = 1, 2, 3$
$II_a = \mathbf{C}^2 : \underline{\mathbf{A}}_a$	$\mathbf{C} \cdot \underline{\mathbf{A}}_a + \underline{\mathbf{A}}_a \cdot \mathbf{C}$	$\mathbf{I} \,\overline{\otimes}\, \underline{\mathbf{A}}_a + \underline{\mathbf{A}}_a \,\overline{\otimes}\, \mathbf{I}$	$a = 1, 2, 3$
$III = \mathrm{tr}\,\mathbf{C}^3$	$3\,\mathbf{C}^2$	$3\,(\mathbf{I} \,\overline{\otimes}\, \mathbf{C} + \mathbf{C} \,\overline{\otimes}\, \mathbf{I})$	

A component form of the invariants listed in Table 19.19.2 is recorded in the $\underline{\mathbf{e}}$-axes,

$$
\begin{aligned}
I_1 &= C_{11} + C_{22} + C_{33}, \\
I_2 &= C_{11}\,C_{22} + C_{22}\,C_{33} + C_{33}\,C_{11} - C_{12}^2 - C_{13}^2 - C_{23}^2, \\
I_4 &= c^2\,C_{11} + s^2\,C_{22} - 2\,c\,s\,C_{12}, \\
I_6 &= s^2\,C_{11} + c^2\,C_{22} - 2\,c\,s\,C_{12}, \\
I_5 &= c^2\,(C_{11}^2 + C_{12}^2 + C_{13}^2) + s^2\,(C_{12}^2 + C_{22}^2 + C_{23}^2) \\
&\quad - 2\,c\,s\,(C_{12}(C_{11} + C_{22}) + C_{13}C_{23}), \\
I_7 &= s^2\,(C_{11}^2 + C_{12}^2 + C_{13}^2) + c^2\,(C_{12}^2 + C_{22}^2 + C_{23}^2) \\
&\quad - 2\,c\,s\,(C_{12}(C_{11} + C_{22}) + C_{13}C_{23}), \\
I_S &= 4\,c\,s\,\Big(c\,s\,(C_{11} + C_{22}) - C_{12}\Big),
\end{aligned}
\tag{19.19.32}
$$

and in the axes of geometrical symmetry $(\underline{\mathbf{e}}_1^s, \underline{\mathbf{e}}_2^s, \underline{\mathbf{e}}_3^s)$:

$$
\begin{aligned}
I_1 &= C_{11}^s + C_{22}^s + C_{33}^s, \\
I_2 &= C_{11}^s C_{22}^s + C_{22}^s C_{33}^s + C_{33}^s C_{11}^s - (C_{12}^s)^2 - (C_{13}^s)^2 - (C_{23}^s)^2, \\
I_4 &= (\cos\underline{\phi})^2\,C_{11}^s + (\sin\underline{\phi})^2\,C_{22}^s + \sin 2\underline{\phi}\,C_{12}^s, \\
I_6 &= (\cos\underline{\phi})^2\,C_{11}^s + (\sin\underline{\phi})^2\,C_{22}^s - \sin 2\underline{\phi}\,C_{12}^s, \\
I_5 &= (\cos\underline{\phi})^2\,\Big((C_{11}^s)^2 + (C_{12}^s)^2 + (C_{13}^s)^2\Big) \\
&\quad + (\sin\underline{\phi})^2\,\Big((C_{12}^s)^2 + (C_{22}^s)^2 + (C_{23}^s)^2\Big) \\
&\quad + \sin 2\underline{\phi}\,\Big(C_{12}^s(C_{11}^s + C_{22}^s) + C_{13}^s\,C_{32}^s\Big), \\
I_7 &= (\cos\underline{\phi})^2\,\Big((C_{11}^s)^2 + (C_{12}^s)^2 + (C_{13}^s)^2\Big) \\
&\quad + (\sin\underline{\phi})^2\,\Big((C_{12}^s)^2 + (C_{22}^s)^2 + (C_{23}^s)^2\Big) \\
&\quad - \sin 2\underline{\phi}\,\Big(C_{12}^s(C_{11}^s + C_{22}^s) + C_{13}^s\,C_{32}^s\Big), \\
I_S &= \cos 2\underline{\phi}\,\Big(2\,(\cos\underline{\phi})^2\,C_{11}^s - 2\,(\sin\underline{\phi})^2\,C_{22}^s\Big).
\end{aligned}
\tag{19.19.33}
$$

Incompressible matrix and inextensible fibers:

If the tissue is incompressible and the two families of fibers are inextensible[19.10],

$$\det \mathbf{C} - 1 = 0, \quad I_4 - 1 = \underline{\mathbf{m}}_1 \cdot \mathbf{C} \cdot \underline{\mathbf{m}}_1 - 1 = 0, \quad I_6 - 1 = \underline{\mathbf{m}}_2 \cdot \mathbf{C} \cdot \underline{\mathbf{m}}_2 - 1 = 0, \tag{19.19.34}$$

the stress in the reference and actual configurations can be cast in the format, that involves an arbitrary pressure p_F and two arbitrary tensions, T_1 and T_2,

$$\begin{aligned}
\underline{\underline{\boldsymbol{\tau}}} &= 2 \frac{\partial \mathcal{W}}{\partial \mathbf{C}} - p_F \det \mathbf{C} \; \mathbf{C}^{-1} + T_1 \, \underline{\mathbf{M}}_1 + T_2 \, \underline{\mathbf{M}}_2, \\
\boldsymbol{\tau} &= 2 \, \mathbf{F} \cdot \frac{\partial \mathcal{W}}{\partial \mathbf{C}} \cdot \mathbf{F}^{\mathrm{T}} - p_F \, \mathbf{I} + T_1 \, \mathbf{M}_1 + T_2 \, \mathbf{M}_2.
\end{aligned} \tag{19.19.35}$$

The strain energy $\underline{\mathcal{W}}$ has to satisfy restrictions in the spirit of (19.19.11),

$$\frac{\partial \mathcal{W}}{\partial \mathbf{C}} : \mathbf{C}^{-1} = 0, \quad \frac{\partial \mathcal{W}}{\partial \mathbf{C}} : \underline{\mathbf{M}}_1 = 0, \quad \frac{\partial \mathcal{W}}{\partial \mathbf{C}} : \underline{\mathbf{M}}_2 = 0. \tag{19.19.36}$$

Here, the **M**'s are the current values of the structure tensors, according to (19.19.1),

$$\mathbf{m}_i = \frac{\mathbf{F} \cdot \underline{\mathbf{m}}_i}{|\mathbf{F} \cdot \underline{\mathbf{m}}_i|}, \quad i \in [1, 2]. \tag{19.19.37}$$

Let **D** be the rate of deformation. As another direct consequence, the constraints are workless,

$$\mathbf{I} : \mathbf{D} = 0, \quad \underline{\mathbf{M}}_1 : \dot{\mathbf{C}} = \mathbf{M}_1 : \mathbf{D} = 0, \quad \underline{\mathbf{M}}_2 : \dot{\mathbf{C}} = \mathbf{M}_2 : \mathbf{D} = 0, \tag{19.19.38}$$

since the power per unit reference volume expresses in the formats,

$$\frac{\delta \mathcal{W}}{\delta t} = \underline{\underline{\boldsymbol{\tau}}} : \frac{\delta \mathbf{F}}{\delta t} = \underline{\underline{\boldsymbol{\tau}}} : \frac{\delta \mathbf{E}_{\mathrm{G}}}{\delta t} = \boldsymbol{\tau} : \mathbf{D}. \tag{19.19.39}$$

19.19.2.2   Functional basis, integrity basis, irreducibility, syzygies

Regarding the list of scalar invariants of the symmetric second order tensor **C** relative to clinotropic materials, Spencer [1972] provides an *integrity basis*: a scalar-valued polynomial function of **C** is expressed as a polynomial of the integrity basis. By contrast, Zheng and Boehler [1994] work out a complete and *irreducible functional basis:* any invariant scalar function is a single-valued function of this basis. A basis is said irreducible if no subset of this basis forms a function basis.

An integrity basis forms a function basis but the elements of the integrity basis may be related by certain particular relations known as *syzygies*[19.11], Bao and Smith [1990]. In fact, the integrity basis for clinotropic materials contains an additional term with respect to the irreducible functional basis, namely $\mathrm{tr}\,\mathbf{C}^3$. Thus $\det \mathbf{C}$, or equivalently $\mathrm{tr}\,\mathbf{C}^3$, can be expressed as a function of the elements of the above tables for $0 < 2\phi < \pi/2$. In any case, in practice, $\det \mathbf{C}$ or $\mathrm{tr}\,\mathbf{C}^3$ can be substituted by a basic scalar invariant or it can be introduced as a redundant argument of the strain energy.

19.19.2.3   The scalar invariant I_S : in-plane (intralamellar) normal energy

The example below introduces the energy due to the strains normal to the fibers in the lamellar plane. Guerin and Elliott [2007] use an energy term of this type for the annulus fibrosus, Section 19.20.3.

Let $\underline{\mathbf{m}}_1^{\perp}$ and $\underline{\mathbf{m}}_2^{\perp}$ be two vectors in the fiber plane which are orthogonal to $\underline{\mathbf{m}}_1$ and $\underline{\mathbf{m}}_2$, respectively,

$$\underline{\mathbf{m}}_1^{\perp} = s \, \underline{\mathbf{e}}_1 + c \, \underline{\mathbf{e}}_2, \quad \underline{\mathbf{m}}_2^{\perp} = c \, \underline{\mathbf{e}}_1 + s \, \underline{\mathbf{e}}_2. \tag{19.19.40}$$

[19.10] Note that then $(\underline{\mathbf{m}}_1 - \underline{\mathbf{m}}_2) \cdot \mathbf{C} \cdot (\underline{\mathbf{m}}_1 + \underline{\mathbf{m}}_2) = 0$.

[19.11] As a standard example, consider the syzygy $a\,X^2 + 2\,b\,X\,Y + c\,Y^2 = 0$ between the variables X and Y, a, b and c being constants. Clearly, the variable Y can be expressed as a (non polynomial) function of X. Consequently, the two variables are independent in the polynomial sense but functionally dependent. In astronomy, a syzygy corresponds to the conjunction of unexpected events. We leave the reader to ponder on the differences between syzygy and serendipity.

where c and s are defined by (19.19.25). The normal strains to the fibers express as follows,

$$\mathbf{m}_1^\perp \cdot \mathbf{C} \cdot \mathbf{m}_1^\perp = s^2\,C_{11} + c^2\,C_{22} + 2\,c\,s\,C_{12},$$
$$\mathbf{m}_2^\perp \cdot \mathbf{C} \cdot \mathbf{m}_2^\perp = c^2\,C_{11} + s^2\,C_{22} + 2\,c\,s\,C_{12}\,. \tag{19.19.41}$$

Using the explicit expression of the invariants in the axes $\underline{\mathbf{e}}$, Eqn (19.19.32), we have the first intermediate relation,

$$4\,c\,s\,C_{12} = \frac{(\cos 2\underline{\phi})^2}{(\sin 2\underline{\phi})^2}\left(I_4 + I_6 - \frac{I_S}{(\cos 2\underline{\phi})^2}\right). \tag{19.19.42}$$

The normal strains can then be rephrased in terms of the strain invariants and of $(\cos 2\underline{\phi})^2$,

$$(\mathbf{m}_1^\perp \cdot \mathbf{C} \cdot \mathbf{m}_1^\perp)^2 + (\mathbf{m}_2^\perp \cdot \mathbf{C} \cdot \mathbf{m}_2^\perp)^2 = I_4^2 + I_6^2 + 2\,\frac{(\cos 2\underline{\phi})^2}{(\sin 2\underline{\phi})^4}\left(I_4 + I_6 - \frac{I_S}{(\cos 2\underline{\phi})^2}\right)(I_4 + I_6 - I_S)\,. \tag{19.19.43}$$

19.19.2.4 The scalar invariant I_S : out of plane (interlamellar) shear energy

In a model of annulus fibrosus, Wagner et al. [2006] introduce a term in the strain energy that contains the sum of the squares of the out of plane shear strains. The coefficients associated with this part of the strain energy are shown to correlate with increase in stiffness due to glycation, an age-related degeneration process.

Let us show that these shear strains are scalar invariants and they can be expressed in terms of the functionally irreducible basis. The proof requires some algebra that can be performed in the axes of geometrical symmetry $(\underline{\mathbf{e}}_1^s, \underline{\mathbf{e}}_2^s, \underline{\mathbf{e}}_3^s)$. From (19.19.33),

$$I_4 + I_6 = 2\,(\cos\underline{\phi})^2\,C_{11}^s + 2\,(\sin\underline{\phi})^2\,C_{22}^s, \quad \frac{I_S}{\cos 2\underline{\phi}} = 2\,(\cos\underline{\phi})^2\,C_{11}^s - 2\,(\sin\underline{\phi})^2\,C_{22}^s\,, \tag{19.19.44}$$

result the intermediate relations,

$$(\cos\underline{\phi})^2\,C_{11}^s = \frac{1}{4}(I_4 + I_6) + \frac{1}{4}\,\frac{I_S}{\cos 2\underline{\phi}}, \quad (\sin\underline{\phi})^2\,C_{22}^s = \frac{1}{4}(I_4 + I_6) - \frac{1}{4}\,\frac{I_S}{\cos 2\underline{\phi}}, \quad \sin(2\underline{\phi})\,C_{12}^s = \frac{1}{2}(I_4 - I_6)\,. \tag{19.19.45}$$

The squares of the out of plane shear strains,

$$\begin{aligned}
\underline{\mathbf{e}}_3 \cdot \mathbf{C} \cdot \mathbf{m}_1 &= \cos\underline{\phi}\,C_{13}^s + \sin\underline{\phi}\,C_{23}^s \\
\underline{\mathbf{e}}_3 \cdot \mathbf{C} \cdot \mathbf{m}_2 &= -\cos\underline{\phi}\,C_{13}^s + \sin\underline{\phi}\,C_{23}^s \\
(\underline{\mathbf{e}}_3 \cdot \mathbf{C} \cdot \mathbf{m}_1)^2 &= (\cos\underline{\phi})^2\,(C_{13}^s)^2 + (\sin\underline{\phi})^2\,(C_{23}^s)^2 + \sin(2\underline{\phi})\,C_{13}^s\,C_{23}^s \\
(\underline{\mathbf{e}}_3 \cdot \mathbf{C} \cdot \mathbf{m}_2)^2 &= (\cos\underline{\phi})^2\,(C_{13}^s)^2 + (\sin\underline{\phi})^2\,(C_{23}^s)^2 - \sin(2\underline{\phi})\,C_{13}^s\,C_{23}^s\,,
\end{aligned} \tag{19.19.46}$$

express in terms of the scalar invariants and of $\cos 2\underline{\phi}$ to the square, using again (19.19.33),

$$(\underline{\mathbf{e}}_3 \cdot \mathbf{C} \cdot \mathbf{m}_1)^2 = I_5 - \frac{1}{(\sin 2\underline{\phi})^2}\left(I_4^2 - I_4\,I_S + \frac{I_S^2}{4\,(\cos 2\underline{\phi})^2}\right),$$
$$(\underline{\mathbf{e}}_3 \cdot \mathbf{C} \cdot \mathbf{m}_2)^2 = I_7 - \frac{1}{(\sin 2\underline{\phi})^2}\left(I_6^2 - I_6\,I_S + \frac{I_S^2}{4\,(\cos 2\underline{\phi})^2}\right). \tag{19.19.47}$$

19.19.2.5 Representation theorems for materials with woven fabrics

The group of symmetries (19.19.30) describes well cross-ply and angle-ply fiber-reinforced laminates[19.12] but not *woven fabrics*, for which the sense of the fibers matters, Spencer [2001]. This situation seems to be relevant at the periphery of the stroma, Fig. 19.1.8, while the central and paracentral regions are probably better described as cross-ply laminates and angle-ply laminates, respectively.

[19.12]The cross-ply laminates run at right angles to one another while angle-ply laminates include all other fiber-reinforced laminates.

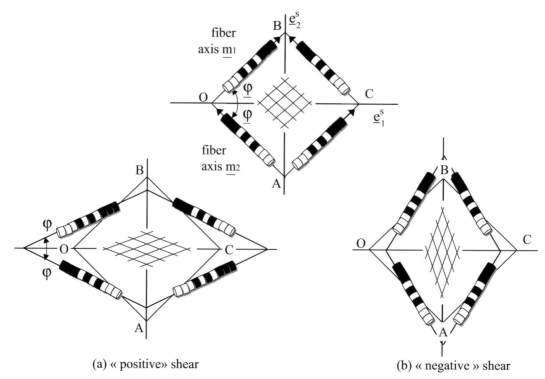

(a) « positive» shear   (b) « negative » shear

Fig. 19.19.5: In certain woven fabrics, the sense of the fibers matters. Positive and negative shears with controlled displacement of *picture frames* lead to distinct transient reaction forces, McGuinness and Ó Brádaigh [1997].

The issue is illustrated by the experiments on fabric-reinforced thermoplastic sheets set up by McGuinness and Ó Brádaigh [1997]. Sheets are deformed homogeneously by pulling along a diagonal. Pulling along the diagonal OC is termed a positive shear, and pulling along the diagonal AB a negative shear, Fig. 19.19.5. Note that the arrows leave point A and arrive at point B. The situation is different at points O and C where one arrow arrives while the other leaves. For some fabrics, positive and negative shears of *picture frames* under displacement controlled were indeed observed by McGuinness and Ó Brádaigh [1997] to produce distinct transient loads while the steady states were insensitive to the loading direction.

Spencer [2000] analyzes the elements of the constitutive equations of fabric-reinforced materials that can deliver such a direction dependent viscous response, while Spencer [2001] considers viscoplastic reinforced composites. The lines below present a short introduction to the subject in the context of lamellar tissues where we dispense of rate effects.

If the sense of the fibers matters, then the constitutive equations have to involve some terms which are odd functions of $\underline{\mathbf{m}}_1$ and $\underline{\mathbf{m}}_2$. In fact, it is appropriate to introduce the non symmetric second order tensors,

$$\underline{\mathbf{M}}_{12} = \underline{\mathbf{m}}_1 \otimes \underline{\mathbf{m}}_2, \quad \underline{\mathbf{M}}_{21} = \underline{\mathbf{m}}_2 \otimes \underline{\mathbf{m}}_1 \,. \tag{19.19.48}$$

Now observe that the two symmetric second order tensors which are linear in terms of the right Cauchy-Green tensor $\mathbf{C}$,

$$\mathbf{X} = \begin{cases} (\underline{\mathbf{M}}_{12} \,\overline{\underline{\otimes}}\, \mathbf{I} + \mathbf{I} \,\overline{\underline{\otimes}}\, \underline{\mathbf{M}}_{12})[\mathbf{C}] = \underline{\mathbf{M}}_{12} \cdot \mathbf{C} + \mathbf{C} \cdot \underline{\mathbf{M}}_{21} \,, \\ (\underline{\mathbf{M}}_{21} \,\overline{\underline{\otimes}}\, \mathbf{I} + \mathbf{I} \,\overline{\underline{\otimes}}\, \underline{\mathbf{M}}_{21})[\mathbf{C}] = \underline{\mathbf{M}}_{21} \cdot \mathbf{C} + \mathbf{C} \cdot \underline{\mathbf{M}}_{12} \,, \end{cases} \tag{19.19.49}$$

are associated with the same energy,

$$II_s = \frac{1}{2}\mathbf{X} : \mathbf{C} = \underline{\mathbf{M}}_{12} : \mathbf{C}^2 = \underline{\mathbf{M}}_{21} : \mathbf{C}^2 = \underline{\mathbf{m}}_1 \cdot \mathbf{C}^2 \cdot \underline{\mathbf{m}}_2 = \underline{\mathbf{m}}_2 \cdot \mathbf{C}^2 \cdot \underline{\mathbf{m}}_1 \,. \tag{19.19.50}$$

Conversely, with $\underline{\mathbf{M}}$ equal to $\underline{\mathbf{M}}_{12}$ or $\underline{\mathbf{M}}_{21}$ or a linear combination of these two tensors, for small $\mathbf{H} = \mathbf{H}^{\mathrm{T}}$,

$$
\begin{aligned}
II_s(\mathbf{C} + \mathbf{H}) - II_s(\mathbf{C}) &= \frac{\partial II_s}{\partial \mathbf{C}} : \mathbf{H} + O(\mathbf{H}^2), \\
&\simeq \mathbf{M} : (\mathbf{C} \cdot \mathbf{H} + \mathbf{H} \cdot \mathbf{C}) \\
&= (\mathbf{M} \cdot \mathbf{C} + \mathbf{C} \cdot \mathbf{M}) : \mathbf{H} \\
&= \left(\lambda \left(\mathbf{M} \cdot \mathbf{C} + \mathbf{C} \cdot \mathbf{M} \right) + (1 - \lambda) \left(\mathbf{M}^{\mathrm{T}} \cdot \mathbf{C} + \mathbf{C} \cdot \mathbf{M}^{\mathrm{T}} \right) \right) : \mathbf{H} \\
&= \Big(\underbrace{\left(\lambda \mathbf{M} \cdot \mathbf{C} + (1 - \lambda) \mathbf{C} \cdot \mathbf{M}^{\mathrm{T}} \right)}_{\mathbf{X}_1} + \underbrace{\left((1 - \lambda) \mathbf{M}^{\mathrm{T}} \cdot \mathbf{C} + \lambda \mathbf{C} \cdot \mathbf{M} \right)}_{\mathbf{X}_2} \Big) : \mathbf{H},
\end{aligned}
$$
(19.19.51)

where λ is an arbitrary real number and the re-ordering is aimed at producing tensors of the form shown by (19.19.49). For the tensors $\mathbf{X}_1$ and $\mathbf{X}_2$ to be symmetric, the only possibility is $\lambda = 1/2$ and then

$$
2 \frac{\partial II_s}{\partial \mathbf{C}} = a_{12} \left(\underline{\mathbf{M}}_{12} \cdot \mathbf{C} + \mathbf{C} \cdot \underline{\mathbf{M}}_{21} \right) + a_{21} \left(\underline{\mathbf{M}}_{21} \cdot \mathbf{C} + \mathbf{C} \cdot \underline{\mathbf{M}}_{12} \right),
$$
(19.19.52)

with a_{12} and a_{21} arbitrary coefficients. With the notation $\underline{\mathbf{A}}_{ij}^s = \underline{\mathbf{e}}_i^s \otimes \underline{\mathbf{e}}_j^s$, $i, j \in [1, 2]$, the component forms of these tensors in the axes of geometrical symmetry are recorded for future use,

$$
\begin{aligned}
\underline{\mathbf{M}}_{12} \cdot \mathbf{C} + \mathbf{C} \cdot \underline{\mathbf{M}}_{21} &= \left(-2 \left(\cos \underline{\phi} \right)^2 C_{11}^s + \sin 2\underline{\phi}\, C_{12}^s \right) \underline{\mathbf{A}}_{11}^s + \left(2 \left(\sin \underline{\phi} \right)^2 C_{22}^s - \sin 2\underline{\phi}\, C_{12}^s \right) \underline{\mathbf{A}}_{22}^s \\
&\quad + \left(\sin 2\underline{\phi}\, \frac{C_{22}^s - C_{11}^s}{2} - \cos 2\underline{\phi}\, C_{12}^s \right) \left(\underline{\mathbf{A}}_{12}^s + \underline{\mathbf{A}}_{21}^s \right), \\
\underline{\mathbf{M}}_{21} \cdot \mathbf{C} + \mathbf{C} \cdot \underline{\mathbf{M}}_{12} &= \left(-2 \left(\cos \underline{\phi} \right)^2 C_{11}^s - \sin 2\underline{\phi}\, C_{12}^s \right) \underline{\mathbf{A}}_{11}^s + \left(2 \left(\sin \underline{\phi} \right)^2 C_{22}^s + \sin 2\underline{\phi}\, C_{12}^s \right) \underline{\mathbf{A}}_{22}^s \\
&\quad + \left(-\sin 2\underline{\phi}\, \frac{C_{22}^s - C_{11}^s}{2} - \cos 2\underline{\phi}\, C_{12}^s \right) \left(\underline{\mathbf{A}}_{12}^s + \underline{\mathbf{A}}_{21}^s \right),
\end{aligned}
$$
(19.19.53)

and therefore

$$
\underline{\mathbf{M}}_s \cdot \mathbf{C} + \mathbf{C} \cdot \underline{\mathbf{M}}_s = \left(-4 \left(\cos \underline{\phi} \right)^2 C_{11}^s \right) \underline{\mathbf{A}}_{11}^s + \left(4 \left(\sin \underline{\phi} \right)^2 C_{22}^s \right) \underline{\mathbf{A}}_{22}^s + \left(-2 \cos 2\underline{\phi}\, C_{12}^s \right) \left(\underline{\mathbf{A}}_{12}^s + \underline{\mathbf{A}}_{21}^s \right),
$$
(19.19.54)

where

$$
\underline{\mathbf{M}}_s = \underline{\mathbf{m}}_1 \otimes \underline{\mathbf{m}}_2 + \underline{\mathbf{m}}_2 \otimes \underline{\mathbf{m}}_1, \quad \underline{\mathbf{M}}_S = (\underline{\mathbf{m}}_1 \cdot \underline{\mathbf{m}}_2) \underline{\mathbf{M}}_s = -\cos 2\underline{\phi}\, \underline{\mathbf{M}}_s.
$$
(19.19.55)

Picture frame experiments, and directional shear effects

During a *picture-frame experiment*, Fig. 19.19.5, a specimen cut along the fibers is fixed at a corner point and pulled at the opposite diagonal. The deformation is spatially homogeneous, and, during the test, symmetry with respect to the diagonal is preserved. Assuming fiber inextensibility and material incompressibility, the kinematics of the experiment is described by the following relations in the axes of geometrical symmetry,

$$
x_1^s = \underline{x}_1^s \frac{\cos \phi}{\cos \underline{\phi}}, \quad x_2^s = \underline{x}_2^s \frac{\sin \phi}{\sin \underline{\phi}}, \quad x_3^s = \underline{x}_3^s \frac{\sin 2\phi}{\sin 2\underline{\phi}}.
$$
(19.19.56)

Therefore the deformation gradient $\mathbf{F}$, right stretch $\mathbf{C}$,

$$
\begin{aligned}
F_{11}^s &= \frac{\cos \phi}{\cos \underline{\phi}}, & F_{22}^s &= \frac{\sin \phi}{\sin \underline{\phi}}, & F_{33}^s &= \frac{\sin 2\phi}{\sin 2\underline{\phi}} \\
C_{11}^s &= \frac{(\cos \phi)^2}{(\cos \underline{\phi})^2}, & C_{22}^s &= \frac{(\sin \phi)^2}{(\sin \underline{\phi})^2}, & C_{33}^s &= \frac{(\sin 2\phi)^2}{(\sin 2\underline{\phi})^2},
\end{aligned}
$$
(19.19.57)

and increment of Green strain $\delta \mathbf{E}_{\mathrm{G}}$,

$$
\delta E_{G,11}^s = -\frac{\sin \phi \cos \phi}{(\cos \underline{\phi})^2} \delta \phi, \quad \delta E_{G,22}^s = \frac{\sin \phi \cos \phi}{(\sin \underline{\phi})^2} \delta \phi, \quad \delta E_{G,33}^s = -2 \frac{(\sin 2\phi)^2 \cos 2\phi}{(\sin 2\underline{\phi})^3} \delta \phi,
$$
(19.19.58)

are principal in the loading axes (axes of geometrical symmetry).

Let us confine attention to the part of the strain energy that depends linearly on the invariant II_s (actually $\underline{W} = II_s$ shifting the coefficients in a_{12} and a_{21}) so that the effective stress (19.19.35) has the form,

$$\underline{\underline{\tau}}^{\text{eff}} \equiv \underline{\underline{\tau}} + p_F \, \mathbf{C}^{-1} + T_1 \, \underline{\mathbf{M}}_1 + T_2 \, \underline{\mathbf{M}}_2 \;\; = \;\; 2 \, \frac{\partial II_s}{\partial \mathbf{C}} \,. \tag{19.19.59}$$

Inserting (19.19.57) in (19.19.52) and (19.19.53) provides the component form of the effective stress in the axes of geometrical symmetry,

$$\underline{\underline{\tau}}^{\text{eff}} = -2 \,(a_{12} + a_{21}) \,(\cos \phi)^2 \, \underline{\mathbf{A}}_{11}^s + 2 \,(a_{12} + a_{21}) \,(\sin \phi)^2 \, \underline{\mathbf{A}}_{22}^s - (a_{12} - a_{21}) \, \frac{\cos 2\phi - \cos 2\phi}{\sin 2\phi} \, (\underline{\mathbf{A}}_{12}^s + \underline{\mathbf{A}}_{21}^s) \,, \tag{19.19.60}$$

and its change associated with a change of angle $\delta \phi$,

$$\delta \underline{\underline{\tau}}^{\text{eff}} = 2 \,(a_{12} + a_{21}) \sin 2\phi \, \delta \phi \, (\underline{\mathbf{A}}_{11}^s + \underline{\mathbf{A}}_{22}^s) + 2 \,(a_{12} - a_{21}) \, \frac{\sin 2\phi}{\sin 2\phi} \, \delta \phi \, (\underline{\mathbf{A}}_{12}^s + \underline{\mathbf{A}}_{21}^s) \,. \tag{19.19.61}$$

Let L be the length of the side BC of the frame, $\underline{h}$ be the reference thickness. The volume of the frame is $\underline{h} \, L^2$. Given that the kinematical constraints do not produce work, the incremental work (19.19.39) done in the deformation of the frame simplifies to

$$\delta \underline{\underline{\tau}} : \delta \mathbf{E}_G \times \underline{h} \, L^2 = 4 \,(a_{12} + a_{21}) \, \frac{(\sin 2\phi)^2}{(\sin 2\phi)^2} \, \cos 2\phi \, (\delta \phi)^2 \, \underline{h} \, L^2 \,. \tag{19.19.62}$$

For positive shear, the incremental displacement of point C along the diagonal is equal to

$$\delta x_C = -L \, \sin \phi \, \delta \phi \,. \tag{19.19.63}$$

Equating the work done by the opposite forces δF_C and δF_O, namely $2 \, \delta F_C \, \delta x_C$, to the work of deformation (19.19.62) yields the incremental stiffness along the first symmetry axis,

$$k_C \equiv \frac{\delta F_C}{\delta x_C} = 2 \,(a_{12} + a_{21}) \, \underline{h} \, \frac{(\sin 2\phi)^2}{(\sin 2\phi)^2} \, \frac{\cos 2\phi}{(\sin \phi)^2} \,. \tag{19.19.64}$$

For negative shear, the incremental displacement of point B along the diagonal is equal to

$$\delta y_B = L \, \cos \phi \, \delta \phi \,. \tag{19.19.65}$$

Equating the work done by the opposite forces δF_A and δF_B, namely $2 \, F_B \, \delta y_B$, to the work of deformation (19.19.62) returns the incremental stiffness along the second symmetry axis,

$$k_B \equiv \frac{\delta F_B}{\delta y_B} = 2 \,(a_{12} + a_{21}) \, \underline{h} \, \frac{(\sin 2\phi)^2}{(\sin 2\phi)^2} \, \frac{\cos 2\phi}{(\cos \phi)^2} \,. \tag{19.19.66}$$

The material responses to positive and negative shears are the same if

$$k_B \Big(\frac{\pi}{2} - \phi \Big) = k_C(\phi) \,, \tag{19.19.67}$$

which implies $a_{12} + a_{21}$ to be an odd function of $I_9 \equiv \cos 2\phi$. Otherwise, the material will be sensitive to the sense of the fibers. Table 19.19.5 encapsulates these observations.

When the fibers have a sense, one should a priori distinguish the two cases:

    1. mechanically equivalent fibers: then the strain energy should be symmetric in $\underline{\mathbf{m}}_1$ and $\underline{\mathbf{m}}_2$ and remain unchanged by interchange of these vectors;

    2. the mechanical properties display symmetry with respect to the geometrical symmetry axes $\mathbf{e}^s$'s. If the strain energy is phrased in terms of the latter axes, it should be invariant under transformations $\mathbf{e}_1^s \to -\mathbf{e}_1^s$, and $\mathbf{e}_2^s \to -\mathbf{e}_2^s$.

Table 19.19.5: For woven fabrics where the sense of the fibers matters, the last line in Table 19.19.2 should be substituted by the two lines below, $0 < 2\phi \leq \pi/2$. Directional shear effects may appear if the sum $a_{12} + a_{21}$ of the coefficients of II_s in the strain energy is not an odd function of I_9.

Invariant I	$\partial I/\partial \mathbf{C}$	$\partial^2 I/\partial \mathbf{C}\partial \mathbf{C}$
$II_s = a_{12}\,\mathbf{C}^2 : \underline{\mathbf{M}}_{12} + a_{21}\,\mathbf{C}^2 : \underline{\mathbf{M}}_{21}$	$a_{12}\,(\underline{\mathbf{M}}_{12} \cdot \mathbf{C} + \mathbf{C} \cdot \underline{\mathbf{M}}_{21})$ $+ a_{21}\,(\underline{\mathbf{M}}_{21} \cdot \mathbf{C} + \mathbf{C} \cdot \underline{\mathbf{M}}_{12})$	$a_{12}\,(\underline{\mathbf{M}}_{12}\,\overline{\underline{\otimes}}\,\mathbf{I} + \mathbf{I}\,\overline{\underline{\otimes}}\,\underline{\mathbf{M}}_{12})$ $+ a_{21}\,(\underline{\mathbf{M}}_{21}\,\overline{\underline{\otimes}}\,\mathbf{I} + \mathbf{I}\,\overline{\underline{\otimes}}\,\underline{\mathbf{M}}_{21})$
$I_9 = -\underline{\mathbf{m}}_1 \cdot \underline{\mathbf{m}}_2 = \cos 2\underline{\phi}$	$\mathbf{0}$	$\mathbf{0}$

Table 19.19.6: For woven fabrics where only the sense of the fibers matters, without directional shear effect, the last line in Table 19.19.2 should be substituted by the two lines below, $0 < 2\underline{\phi} \leq \pi/2$.

Invariant I	$\partial I/\partial \mathbf{C}$	$\partial^2 I/\partial \mathbf{C}\partial \mathbf{C}$
$I_s = \underline{\mathbf{M}}_s : \mathbf{C}$	$\underline{\mathbf{M}}_s$	$\mathbf{0}$
$I_9 = -\underline{\mathbf{m}}_1 \cdot \underline{\mathbf{m}}_2 = \cos 2\underline{\phi}$	$\mathbf{0}$	$\mathbf{0}$

The first symmetry is best scrutinized with the expression (19.19.52) based on the fiber axes and the second symmetry with the expression (19.19.53) based on the axes of geometrical symmetry. They turn out to yield the same constraint $a_{12} = a_{21}$ and the structure tensors $\underline{\mathbf{M}}_{12}$ and $\underline{\mathbf{M}}_{21}$ combine to the symmetric structure tensor $\underline{\mathbf{M}}_s$ defined by (19.19.55).

Woven fabrics where only the sense of the fibers matters

As an alternative to the above effects, one may envisage woven fabrics where only the sense of the fibers matters. The irreducible functional basis of Table 19.19.6 contains the two new invariants I_s and I_9. Unlike I_S, I_s does no longer vanish automatically when the fibers are orthogonal.

Quite generally, I_s can now be interpreted as a shear on a plane of normal $\underline{\mathbf{m}}_1$ in direction $\underline{\mathbf{m}}_2$, and conversely. In the strain energy, it can be viewed as introducing an interaction term between these two families, or between the lamellae.

For laminates, the strain energy had to be a function of $I_S = -I_9\,I_s$. By contrast, for woven fabric, the strain energy can be an arbitrary function of I_s and I_9, the odd dependence in I_s indicating the influence of the sense of the fibers.

Clearly, $\underline{\mathbf{m}}_1 \cdot \mathbf{C} \cdot \underline{\mathbf{m}}_2$ is equal to the fiber-fiber interaction term γ_m, defined by (19.19.22), when the two families of fibers are orthogonal. Therefore, one wonders if the scalar invariant $I_5 + I_7 - I_4^2 - I_6^2$, introduced in the strain energy to witness fiber-fiber interaction, may not be advantageously replaced by $I_s^2/2$. This substitution applies only for materials that qualify as woven fabrics. For the record, let us note

$$I_s = 2\,\underline{\mathbf{m}}_1 \cdot \mathbf{C} \cdot \underline{\mathbf{m}}_2 = \begin{cases} -\sin 2\underline{\phi}\,(C_{11} + C_{22}) + 2\,C_{12} & \text{in the axes } \mathbf{e}; \\ -2\,(\cos \underline{\phi})^2\,C_{11}^s + 2\,(\sin \underline{\phi})^2\,C_{22}^s & \text{in the axes } \mathbf{e}^s\,. \end{cases} \tag{19.19.68}$$

McGuinness and Ó Brádaigh [1997] provide an example, namely the 1/7 satin weave. A representative unit cell is shown in Fig. 19.19.6. The scalar invariant I_s is required, and an interpretation via the theorems of representation is presented by Spencer [2001]. Clearly the warp $\underline{\mathbf{m}}_1$ and weft $\underline{\mathbf{m}}_2$ have a sense. Symmetry holds for one diagonal only, so that the only requirement is that the strain energy should be invariant under exchanges of $\underline{\mathbf{m}}_1$ and $\underline{\mathbf{m}}_2$. It is therefore an arbitrary function of $\cos 2\underline{\phi}$ and $\underline{\mathbf{M}}_s$ via the invariant $\mathbf{C}^2 : \underline{\mathbf{M}}_s$ if directional shear effects are observed and otherwise via the invariant $\mathbf{C} : \underline{\mathbf{M}}_s$.

19.19.2.6   Glycation

Glycation is thought to improve keratoconus. Based on a qualitative biochemical analysis, Nickerson [2005], p. 155, suggests that glycation (1) glues the lamellae together and (2) stiffens the fibrils within the lamellae, in addition (3) to stiffening the fibrils themselves. When interpreted in mechanical terms, these observations imply

- a smaller change of the angle between the two families of fibers during stretch: this issue may be controlled by introduction of the invariants I_5 and I_7 as described in Section 19.19.1.4,

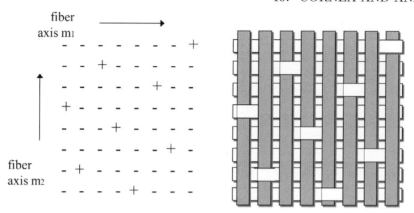

Fig. 19.19.6: A unit cell representative of the 1/7 satin weave fabric, McGuinness and Ó Brádaigh [1997], Spencer [2000] [2001]. A '+' indicates an intersection where the warp $\underline{\mathbf{m}}_1$ passes over the weft $\underline{\mathbf{m}}_2$, and conversely for the symbol '-'. On a square unit cell with eight intersections, two + in two successive rows or columns are shifted by three positions. This cell displays symmetry about one diagonal.

   - a smaller change of the interlamellar shear strain by introduction of the scalar invariants I_S or I_s as indicated in Section 19.19.2.4.

In addition, glycation is also believed to modify the charge of the amino-residues, resulting in a change of the staining pattern, a phenomenon that could be checked through electron microscopy.

   Diabetes has been correlated with structural changes on collagen, namely larger diameters and increased crosslink density, e.g., Layton and Sastry [2004]. The resulting stiffening of the mechanical behavior of peripheral nerve is suspected to lead to an increase of endoneurial fluid pressure. Excessive glycation is also pointed out as impairing the re-growth of axons in peripheral nerve tissues, therefore decreasing the long term ability of the nerves to transmit electrical signals.

19.19.3 Additive decomposition of the strain energy for the corneal stroma

Since the early work of Fung, there has been some controversy on the most appropriate format to decompose the strain energy of an anisotropic tissue. The anisotropic part is usually expressed via exponentials on the ground that the number of coefficients involved is smaller than via polynomials. Here are a few proposals for the corneal stroma.

   Pinsky et al. [2005] start from a single lamellar approach, as described in Section 19.20.2, and next accounts for a planar directional distribution of fiber directions, Section 12.3.1. The apparent strain energy

$$\underline{\mathcal{W}}^{\text{lamella}} = \underline{\mathcal{W}}^{\text{matrix}} + \underline{\mathcal{W}}^{\text{fibrils}}, \tag{19.19.69}$$

is contributed additively by the ground substance and the collagen fibrils. The ground substance that includes the extrafibrillar matrix and the fiber-matrix crosslinks is mechanically isotropic,

$$\underline{\mathcal{W}}^{\text{matrix}} = \frac{\mu}{2}\left(I_1 - 3\right) - \frac{\mu}{2}\operatorname{Ln} I_3 + \frac{\lambda}{8}\left(\operatorname{Ln} I_3\right)^2, \tag{19.19.70}$$

with $I_1 = \operatorname{tr}\mathbf{C}$, $I_3 = \det\mathbf{C}$. With help of relations (2.2.35), the moduli $K = \lambda + \frac{2}{3}\mu$ and μ can be interpreted as apparent (namely, per unit volume of tissue) bulk and shear moduli, respectively. Thanks to (2.7.3), the matrix stress and stiffness result in the form,

$$\begin{aligned}
2\frac{\partial \underline{\mathcal{W}}^{\text{matrix}}}{\partial \mathbf{C}} &= \frac{\lambda}{2}\operatorname{Ln} I_3\,\mathbf{C}^{-1} + \mu\left(\mathbf{I} - \mathbf{C}^{-1}\right), \\
4\frac{\partial^2 \underline{\mathcal{W}}^{\text{matrix}}}{\partial \mathbf{C}\partial \mathbf{C}} &= \lambda\,\mathbf{C}^{-1}\otimes\mathbf{C}^{-1} + \left(2\,\mu - \lambda\operatorname{Ln} I_3\right)\mathbf{C}^{-1}\overline{\underline{\otimes}}\,\mathbf{C}^{-1}.
\end{aligned} \tag{19.19.71}$$

The response of the matrix to an isotropic deformation gradient is shown in Fig. 19.19.7. The apparent strain energy due to fibrils is expressed in terms of the invariant $I_4 = \mathbf{C} : \underline{\mathbf{M}}$,

$$\underline{\mathcal{W}}^{\text{fibrils}} = \frac{K^{\text{c}}}{4}\left(\exp(I_4 - 1) - I_4\right), \tag{19.19.72}$$

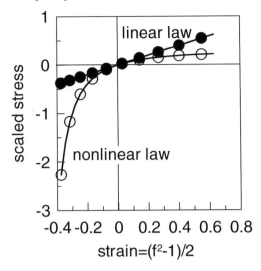

Fig. 19.19.7: Stress responses to an isotropic loading of a linear isotropic tissue $\underline{\underline{\tau}}_1 = 3\,K\,E_{G1}$ and of the nonlinear isotropic tissue introduced by Pinsky et al. [2005], namely $\underline{\underline{\tau}}_1 = 3\,\lambda\,\mathrm{Ln}\,F_1/F_1^2 + 2\,\mu\,E_{G1}/F_1^2$ for $\lambda/\mu = 1$ against the Green strain $E_{G1} = \frac{1}{2}\,(F_1^2 - 1)$. The stresses are scaled by $3\,K = 3\,\lambda + 2\,\mu$.

The nonlinearity stiffens the material response in compression, mimicking some sort of locking. It has an opposite effect in tension.

where K^c is the apparent fiber modulus, and no explicit condition to activate/deactivate the fibers is defined.

Pandolfi and Manganiello [2006] consider two families of collagen fibrils. In line with Holzapfel et al. [2005], they use the modified (isochoric) right Cauchy-Green tensor $\mathbf{C}_{\mathrm{iso}} = I_3^{-\frac{1}{3}}\,\mathbf{C}$, and adopt the additive decomposition,

$$\underline{\mathcal{W}} = \underline{\mathcal{W}}^{\mathrm{vol}}(J) + \underline{\mathcal{W}}^{\mathrm{shear}}(I_1^{\mathrm{iso}}, I_2^{\mathrm{iso}}) + \underline{\mathcal{W}}^{\mathrm{fibrils}}(I_4^{\mathrm{iso}}, I_6^{\mathrm{iso}}), \tag{19.19.73}$$

that delineates a purely volumetric component, an isotropic shear response and the directional fibril contribution via the scalar invariants, $J = \det \mathbf{F}$, $I_1^{\mathrm{iso}} = \mathrm{tr}\,\mathbf{C}_{\mathrm{iso}}$, $I_2^{\mathrm{iso}} = \frac{1}{2}\,((\mathrm{tr}\,\mathbf{C}_{\mathrm{iso}})^2 - \mathrm{tr}\,\mathbf{C}_{\mathrm{iso}}^2)$, $I_4^{\mathrm{iso}} = \mathbf{C}_{\mathrm{iso}} : \underline{\mathbf{M}}_1$ and $I_6^{\mathrm{iso}} = \mathbf{C}_{\mathrm{iso}} : \underline{\mathbf{M}}_2$. In explicit form,

$$\underline{\mathcal{W}}^{\mathrm{vol}}(J) = \frac{K}{4}\,(J^2 - 1 - 2\,\mathrm{Ln}\,J),$$

$$\underline{\mathcal{W}}^{\mathrm{shear}}(I_1^{\mathrm{iso}}, I_2^{\mathrm{iso}}) = \frac{\mu_1}{2}\,(I_1^{\mathrm{iso}} - 3) + \frac{\mu_2}{2}\,(I_2^{\mathrm{iso}} - 3), \tag{19.19.74}$$

$$\underline{\mathcal{W}}^{\mathrm{fibrils}}(I_4^{\mathrm{iso}}, I_6^{\mathrm{iso}}) = \sum_{i=4,6} \frac{K_i^c}{2k_i}\left(\exp(k_{ci}\,\langle I_i^{\mathrm{iso}} - 1\rangle^2) - 1 \right),$$

where K is the apparent bulk modulus, $\mu = \mu_1 + \mu_2$ the apparent shear modulus, k_{ci}, $i = 4, 6$, is dimensionless, and K_i^c, $i = 4, 6$, is a measure of the apparent tensile modulus of the fiber family i. The 2nd Piola-Kirchhoff stress derives from the strain energy with help of (6.6.20),

$$\underline{\underline{\tau}} = 2\,\frac{\partial \mathcal{W}}{\partial \mathbf{C}} = 2\,\frac{\partial \mathcal{W}^{\mathrm{vol}}}{\partial \mathbf{C}} + \left(2\,\frac{\partial \mathcal{W}^{\mathrm{shear}}}{\partial \mathbf{C}^{\mathrm{iso}}} + 2\,\frac{\partial \mathcal{W}^{\mathrm{fibrils}}}{\partial \mathbf{C}^{\mathrm{iso}}} \right) \cdot \frac{d\mathbf{C}^{\mathrm{iso}}}{d\mathbf{C}}. \tag{19.19.75}$$

The formula defining the activation of collagen fibers through the isochoric strain misses activation through an isotropic extension, typically swelling. Therefore, the entity to be used here is probably the strain itself, not the isochoric strain.

Fiber dispersion about the two main directions considered in a clinotropic tissue has been sometimes advocated for a more accurate description of the actual collagen fabric of corneal stroma. For each of the two families, an extended structure tensor $\alpha\,\mathbf{I} + (1 - 3\,\alpha)\,\mathbf{M}$ is introduced via the dispersion coefficient α. A value $\alpha = 1/3$ corresponds to an isotropic distribution of fibers, $\alpha = 0$ corresponds to the discrete fabric, and other values between 0 and $1/3$ describe intermediate distributions.

Li and Tighe [2006]b regard the cornea as a laminated composite shell. They start from a single lamella and use the in-plane stress-strain relation of Halpin and Tsai [1969], which is next transformed to the standard form for a laminated composite shell. The next step is to superimpose the lamellae, assuming that they are perfectly glued. The directional distribution of the fibers follows the description of Section 12.3.1. The flexural properties of laminated composites with knitted fabric reinforcement are analyzed by Huang [2004].

Note that the mechanical significance and influence of the invariants I_5, I_7 and I_S, as discussed in Sections 19.19.1.2, 19.19.1.3, 19.19.1.4, have not been explored by the above models in the context of the corneal stroma. The concept of woven fabrics sketched in Section 19.19.2.5 has not yet been explored either. The issue is more advanced for the annulus fibrosus.

19.20 Annulus fibrosus: another lamellar tissue with two families of fibers

19.20.1 Histology of the annulus fibrosus

The intervertebral disc has the shape of a biconvex lens, with thickness 10 to 15 mm. Its composition is radially heterogeneous, Fig. 19.20.1. The internal part, termed nucleus pulposus (NP), has properties which are clearly distinct from the external part, referred to as annulus fibrosus (AF), which is a fibrous cartilage[19.13]. The NP has mainly a gelatinous content. It is hydrophilic and its water content reaches up to 85% in weight or more. The AF has a lamellar structure: it is composed of about 20 approximately concentric layers, or lamellae. The thickness of the lamellae is about 0.6 mm, the inner lamellae being thinner. The structure can be discovered by a sort of *dry* peeling of the disc. Not much is known about the cement (the interlamellar septum) that links adjacent lamellae together: the presence of proteoglycan aggregates confers to the septum hydrophilic properties similar to those of the nucleus pulposus.

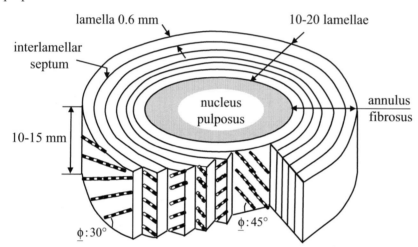

Fig. 19.20.1: The intervertebral disc consists of an inner tissue, the gelatinous nucleus pulposus (NP), surrounded and confined by the annulus fibrosus (AF). The annulus fibrosus is structured by circumferential lamellae reinforced by fibers that run on alternative directions in adjacent lamellae. The thickness of the lamellae decreases from the edge of the disc inward, while the reinforcing fibers tend to align with the disc axis.

The collagen fibrils of the nucleus pulposus are of type II, like in articular cartilages. The annulus fibrosus contains both collagen of types I and II: the relative proportion of type I varies radially from 5% in the inner lamellae to 95% in the outer lamellae, Hiltner et al. [1985]. Collagen fibrils of the AF ensure the link with the nucleus pulposus. Anchoring of the disc into the vertebral plates is ensured by further dense structures of collagen fibrils. The proteoglycan content varies in opposite direction to that of collagen, from about 2.5% of dry weight at the periphery to 10% in the nucleus pulposus, Hiltner et al. [1985]. The negative charge of proteoglycans induces the hydrophilic character of NP. Proteoglycans appear in the form of independent aggregates in the nucleus pulposus. In the annulus fibrosus, they are found in the form of monomers that interact with collagen fibrils.

The relative mass composition of the AF is only qualitatively similar to that of articular cartilage, Sun and Leong [2004]: 65 to 90% of the mass consists of water and the dry mass is shared by collagen (50-70%), PGs (10-20%) and other noncollagenous proteins (25%). In fact, the water content is higher in the NP than in the AF, with volume fractions of 93% versus 81%, so does also the proteoglycan content. By contrast, the collagen content is much higher in the AF, Jackson et al. [2009]. This biochemical composition correlates with the mechanical function of the NP which sustains axial compression, while being radially confined by the circumferential lamellae of the AF, which therefore work in tension.

The intervertebral disc is avascular and the transport of nutrients takes place through diffusion. In connective tissues, the cell density decreases with age. While cells are active in the development of the extracellular matrix in the young animal, they become sparse at maturity and are often just able to ensure maintenance, but not repair in case of injury. Still, intercellular communication, through gap junctions, are thought to help the function of the cells. From the outer layers to the inner layers, the cells change from fusiform, to wavy and finally to spherical, Bruehlmann et al. [2002]. The shape of the cells is thought to correlate with the mechanical environment.

While the NP has mainly compressive and osmotic functions, the AF undergoes compressive, bending and

---

[19.13]Two spellings are found, annulus and anulus.

torsional loads due to body weight and motion, as well as tensile loads aimed at confining the NP. The relations between structure and mechanical function are nicely addressed in Hukins and Meakin [2000].

Much like the microstructure, the mechanical properties of the AF vary radially. The elastic moduli may be twice larger in the periphery. The fibers in two adjacent lamellae run at angles $\pm\phi$ with respect to a horizontal plane. The angle ϕ varies from about 30° in the outer layers to about 45° in the inner layers, Cassidy et al. [1989]. The tensile stiffness in the periphery is thus smaller in the vertical direction than in the circumferential direction. Hukins and Meakin [2000] suggest that the alternate orientation of the lamellae is aimed at resisting torsion: one family of fibers is activated in a clockwise rotation, the other family being activated in an anticlockwise rotation. In that sense, horizontal fibers are completely inefficient, while vertical fibers would require to twist some amount first to become activated. Hiltner et al. [1985] report that the maximum tensile stresses in a transverse plane are at $\pm30°$. Correspondingly, the fracture paths of the NP are essentially aligned with the vertical axis and penetrate the vertebral bodies rather than the AF.

19.20.2 Single lamellar and multi-lamellar points of view

As far as the mechanical properties are concerned, they may be contemplated from two points of view:
- an approach considers individual lamellae, which endow the tissue with transverse isotropy about the fiber axes. The composite tensile moduli are about 50 MPa parallel to the fibers and 0.2 MPa in the transverse direction;

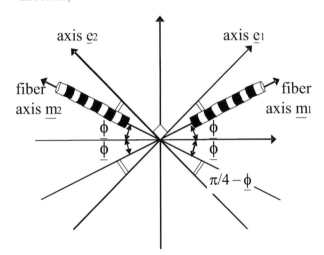

Fig. 19.20.2: Multi-lamellar model. Each lamella of the annulus fibrosus is reinforced by parallel fibers. Similar to the cornea, the fibers in two adjacent lamellae run in alternating directions. The angle $(\underline{\mathbf{m}}_1, \underline{\mathbf{m}}_2)$ is denoted by $\pi - 2\phi$, $0 < \phi < \pi/4$. The angle ϕ with the horizontal direction varies from about 30° in the outer layers to about 45° in the inner layers, Cassidy et al. [1989]. A value $\phi = \pi/4$ implies the fibers to be orthogonal.

- another approach is multi-lamellar. In fact, it groups two adjacent lamellae so as to obtain a clinotropic material as sketched in Fig. 19.20.2. The tensile circumferential moduli are reported in the ranges 15 to 20 MPa, Wu and Yao [1976]; 5 to 50 MPa, Ebara et al. [1996]; 5.5 to 17.5 MPa, Elliott and Setton [2001]. Since the fibers are oriented mainly horizontally, the tensile axial moduli are of course lower, 0.8 to 1 MPa according to Elliott and Setton [2001].

Orders of magnitude of moduli corresponding to these single-lamellar and multi-lamellar approaches are shown in Fig. 19.20.3. Given that the tensile behavior is nonlinear, the above moduli are rather representative of the physiological range with a circumferential tensile stress about 0.5 to 1 MPa associated with a stretch ratio well after the toe region but somehow before failure. Note that the axial stress over the AF resulting from daily activities ranges from 1 to 2 MPa.

The physical motivation for a single-lamellar approach is to recognize the disc as a composite material, made of layers of distinct physical properties (two adjacent lamellae have a distinct orientation). Moreover the most inner lamella is in contact with the nucleus pulposus.

In the corneal stroma, the shear stress transmitted across the lamellae is low. This low shear stiffness is witnessed by the fact that, when the cornea is isolated from the sclera and left to rest under gravity, it flattens. However, given the large number of lamellae in the cornea, the introduction of individual lamellae is somehow not realistic: there are 300 to 500 lamellae of thickness 2 μm in the corneal stroma and only 10 to 20 lamellae of thickness 0.6 mm=600 μm in the AF.

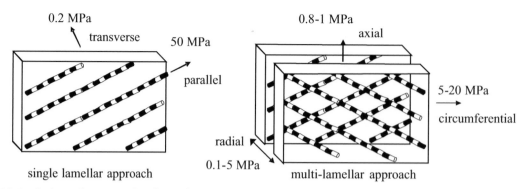

Fig. 19.20.3: Orders of magnitude of tensile experimental moduli of annulus fibrosus for a stress in the physiologic range 0.5-1 MPa associated with the stiff response at higher strains before failure. Data of moduli parallel and orthogonal to the fibers from Pezowicz et al. [2005], Holzapfel et al. [2005] for the single lamella, and for circumferential, axial and radial moduli from Wu and Yao [1976], Ebara et al. [1996], Elliott and Setton [2001] for the annulus fibrosus as a whole. The modulus of collagen fibrils for bovine Achilles tendon is reported to be 430 MPa by Sasaki and Odajima [1996].

Quite generally, the fiber-reinforced lamellar structure of cornea and annulus fibrosus may introduce several interactions in the mechanical behavior (a sketch illustrating these issues is displayed in Pezowicz et al. [2006]), namely

- an intralamellar fiber-matrix interaction, Section 19.19.1.3;
- an intralamellar interaction due to fiber crosslink, Section 19.19.1.2;
- an interlamellar interaction due to crosslinks realized by fibrils and sub-lamellae: this interaction is expected to be significant in the center of the stroma where the lamellae form parallel layers;
- an interaction between bundles of fibrils which are interwoven at the periphery of the stroma.

These four contributions may be introduced in the free energy through strain invariants. Data are needed to quantify their actual effects.

The process which consists in gathering two or several lamellae to obtain a clinotropic material should be clarified. Indeed, clinotropic materials can be thought to stem from an intrinsic microstructure, not from the gathering of individual lamellae. For example, two family of fibers could be conceived as interwoven. Clearly the degrees of interaction between interwoven and adjacent fibers, and consequently the associated strain energies, are not the same.

19.20.3   A short review of mechanical models for the annulus fibrosus

For physical consistency, the constitutive equations and strain energy should be, at given chemical content, expressed in terms of fiber angle, collagen density and PG content. This dependence of the constitutive equations with respect to composition and structure has not been reported so far in the constitutive models that view the annulus fibrosus as a fiber-reinforced hyperelastic solid. In fact, these models are assumed to be used in situ where swelling mechanisms induce the annulus fibrosus in tension. Still, this approach limits the set of calibration tests to situations where all sets of fibers are in extension.

Klisch and Lotz [1999] develop a model for the annulus fibrosus with two families of fibers, which are mechanically equivalent. They define the strain energy as the sum of exponentials of scalar invariants, rather than the exponential of the sum of scalar invariants. Bass et al. [2004] do the reverse. Klisch and Lotz [1999] try to obtain the material parameters by some optimization process. Their model seems to experience difficulties ensuring the zero lateral traction during uniaxial traction. However, they do not introduce the conewise behavior due to the fiber-reinforcement: the fibers are activated whether in extension or in contraction. These remarks apply as well to the model of Elliott and Setton [2000] which considers linear elasticity.

Bass et al. [2004] report of uniaxial and biaxial experiments. They argue that data from uniaxial loading alone or from biaxial experiments alone are not sufficient to discriminate among the various scalar invariants which should be involved and to calibrate properly the various parameters of the strain energy. In essence, the regions in strain and stress covered by uniaxial and biaxial loadings do not overlap: the strain range is larger for uniaxial loading, and the stress range is smaller. As another issue, they question the boundary conditions which maximize the homogeneity of the central part of the specimen. They conclude that keeping the bone attached to the disc has advantages over tension tests where, after complete dissection, the disc specimen is pulled through appropriately located lines of hooks. The central part of the specimen is marked with a grid defined by nine steel pins. The tested piece of AF is cut from the disc along a circumference and it has a thickness of about 1.6 mm. During biaxial experiments, the thickness may be measured through a conductance method, Mansour and Mow [1976]. However, given that Bass et al. [2004] do not measure the transverse strain, the volume change is not available.

For tendons, Guerin and Elliott [2005] start from a quadratic energy density $\underline{\mathcal{W}}$ of the strain $\mathbf{E}$ including a fiber-matrix interaction term $I_1 I_4$ that they find to be significant. In explicit form,

$$\underline{\mathcal{W}} = k_1 I_1^2 + k_2 I_2 + k_3 I_1 I_4 + k_4 I_4^2 \ (+k_5 I_3) \,, \tag{19.20.1}$$

where the k's are material constants. The toe region is defined through a transition strain $\mathbf{E}^*$. Distinct strain energy functions are provided according to the sign of $I_4(\mathbf{E}) - I_4(\mathbf{E}^*)$. Such a manipulation should take care to ensure the stress continuity at the transition strain. In the transverse direction, Guerin and Elliott [2005] note that the energy is mainly contributed by the matrix, i.e., by the two first terms in (19.20.1).

Wagner and Lotz [2004] consider a multi-lamellar approach with mechanically equivalent fibers. They introduce the strain energy associated with fiber-fiber crosslink within a lamella as defined in Section 19.19.1.4, and they sum the contributions of the two sets of fibers, $\gamma_{m_1}^2 + \gamma_{m_2}^2 = I_5 + I_7 - I_4^2 - I_6^2$, in terms of invariants of $\mathbf{C}$. Specifically they introduce the following additive decomposition of the energy $\underline{\mathcal{W}} = \underline{\mathcal{W}}(\mathbf{C})$:

$$\underline{\mathcal{W}} = \underline{\mathcal{W}}^{\text{iso}} + \underline{\mathcal{W}}^{\text{crosslink}} + \underline{\mathcal{W}}^{\text{fibrils}};$$

$$\underline{\mathcal{W}}^{\text{iso}} = \underline{\mathcal{W}}^{\text{iso-volume}}\big((I_3 - I_3^{-1})^2\big) + \underline{\mathcal{W}}^{\text{iso-deviatoric}}\big((I_1 I_3^{-1/3} - 3)^2\big);$$

$$\underline{\mathcal{W}}^{\text{crosslink}} = \underline{\mathcal{W}}^{\text{crosslink}}(I_5 + I_7 - I_4^2 - I_6^2);$$

$$\underline{\mathcal{W}}^{\text{fibrils}} = \underline{\mathcal{W}}^{\text{fibrils}}(I_4 + I_6) \,. \tag{19.20.2}$$

Thus the strain energy becomes infinite as $\det \mathbf{F} = I_3^{1/2}$ tends to 0, Section 6.1.1. Note the relations,

$$2 \frac{\partial}{\partial \mathbf{C}}(I_3 - I_3^{-1})^2 = 4\,(I_3 - I_3^{-1})\,(I_3 + I_3^{-1})\,\mathbf{C}^{-1};$$

$$2 \frac{\partial}{\partial \mathbf{C}}(I_1 I_3^{-1/3} - 3)^2 = 4\,(I_1 I_3^{-1/3} - 3)\,I_3^{-1/3}\left(\mathbf{I} - \frac{I_1}{3}\,\mathbf{C}^{-1}\right), \tag{19.20.3}$$

that justify the terminology used for the isotropic part of the energy, given that $\underline{\boldsymbol{\tau}} = 2\,\partial\underline{\mathcal{W}}/\partial\mathbf{C}$ and $\det \mathbf{F}\,\boldsymbol{\sigma} = \mathbf{F}\cdot\underline{\boldsymbol{\tau}}\cdot\mathbf{F}^{\text{T}}$. The isotropic terms are polynomials while the anisotropic terms are in exponential form. The fiber activation is not introduced in explicit form, but approximated through a slope much larger in extension than in contraction. Still, curiously, the fiber activation is addressed, not for the two families of fibers independently, but simultaneously, this coming from the fact they are equivalent and contribute additively through $I_4 + I_6$ and $I_5 + I_7$. In fact, $\langle I_4 - 1\rangle + \langle I_5 - 1\rangle \neq \langle I_4 + I_5 - 2\rangle$. Moreover, the fiber crosslink is activated through the exponential smoothing, irrespective of $\langle I_4 - 1\rangle$ and $\langle I_5 - 1\rangle$. The additive exponential decomposition in Pandolfi and Manganiello [2006] is much more reasonable in this respect. Finally note that the strain invariants I_2 and I_S are not used.

In Wagner et al. [2006], the coefficients associated with the crosslink energy are shown to correlate with increase in stiffness due to glycation.

Guerin and Elliott [2007] follow a quite similar approach to the above paper of Wagner and Lotz [2004]. They add a normal *intra*lamellar energy, Section 19.19.2.3. The normal *inter*lamellar energy is not addressed. Presumably it is small in the annulus fibrosus like in the cornea where the weak normal interaction between lamellae allows for stromal swelling along the radial direction.

Pezowicz et al. [2005] display the fine hydrated intralamellar structure of annulus fibrosus using a combined optical imaging technique. They also perform mechanical tensile tests of the lamella, parallel and perpendicularly to the fiber axis. In the reversible regime, they find a secant modulus about 50 MPa along the fibers, and 2/3 MPa in the perpendicular direction.

Holzapfel et al. [2005] test the mechanical properties of a single lamella: for a tensile stress of about 1 MPa, they report values of elastic moduli between 28 to 78 MPa along the fibers, and 0.22 MPa in the perpendicular direction.

The above experimental and theoretical models are concerned with tensile tests. Few studies address the compressive properties of annulus fibrosus. Wagner and Lotz [2004] report unconfined compression tests in the circumferential direction while Klisch and Lotz [2000] perform confined compression tests in the radial and axial directions. The compressive response in the model of Wagner and Lotz [2004] uses the same expression as in extension: a single exponential describes a continuous modulus that increases smoothly from the compression zone to the toe region, with a value of about 0.2 to 0.4 MPa at zero stretch, before reaching higher values at larger strains.

Poisson's ratios in tension are derived by Elliott and Setton [2000] [2001] while Wagner and Lotz [2004] report also data for compression.

19.20.4   A short review of chemomechanical models

The fixed charge on proteoglycans induces chemomechanical couplings giving rise to swelling and shrinkage under chemical and mechanical loadings. At neutral pH, this charge varies over the disc from 0.1 to 0.35 mole/liter resulting in osmotic pressures of 100 to 350 kPa, much like in articular cartilages, Urban and McMullin [1985], Urban and Holm [1986]. Actually, the fixed charge increases radially, peaking at the center of the NP. Measurements on bovine discs by Jackson et al. [2009] indicate average values of about 0.060 and 0.19 meq per gm of wet tissue in the annulus fibrosus and nucleus pulposus respectively. These charges correlate with the measured GAG content per dry weight of tissue, about 54 and 490 µgm per gm respectively in the AF and NP. Values for human discs do not differ much from these figures.

On the other hand, age degeneration may halve the charge of the nucleus pulposus, increasing the stress sustained by the solid matrix, Urban and Holm [1986].

The structure of the AF shows radial and regional heterogeneities. So does the biochemical composition. In order to account for these local variations in biomechanical tests, Best et al. [1994] develop a method that aims at minimizing the geometrical and biochemical perturbations during excision. They relate the compressive properties, isometric swelling pressure and hydraulic permeability to the tissue composition.

Sun and Leong [2004] introduce a comprehensive model including mechanical and chemomechanical effects. Anisotropy applies to both mechanical and transport properties. They consider *individual* lamellae: the material is thus endowed with transverse isotropy, with the symmetry axis parallel to the collagen fibers. Their strain energy is exponential in the Fung's form. Their work is presumably intended to a finite element context: the lamellae can slip, or stick, with respect to each other. Although they do not elaborate on this point, the interface conditions for entities of chemomechanical nature should be looked at as contact conditions in a porous media context.

The finite element analysis of Iatridis et al. [2003] addresses the influence of the fixed charge on the mechanical state of annulus fibrosus. The overall finite element system is split in two subsystems. The chemical-electrical subsystem is independent, while the solid-fluid subsystem involves a force term of chemical nature. The primary unknowns are the concentrations and electrical potential in the first subsystem, and the solid and fluid displacements in the second subsystem.

Drost et al. [1995] perform a combination of chemical and mechanical oedometric loadings on canine annulus fibrosus specimens excised in a radial or axial direction. The specimen which is first equilibrated with a bath at 0.6 M of NaCl swells when the bath is refreshed to 0.2 M at constant load. In a last stage, consolidation takes place when a mechanical load is applied. The axial biphasic aggregate modulus is larger in the axial direction, about 1 MPa versus 0.6 MPa. The hydraulic diffusivity and hydraulic permeability are observed to vary significantly during the three loading stages (conditioning, swelling, consolidation). During the consolidation stage, the permeabilities in the axial and radial directions are about 3.2×10^{-16} and 1.8×10^{-16} m^2/Pa/s, respectively.

Frijns et al. [1997] perform finite element simulations accounting for the presence of the fixed charge on proteoglycans and of mobile anions and cations. They simulate chemically induced swelling and mechanical consolidation in confined compression tests.

Loss of proteoglycans with age implies a decrease of osmotic pressure and therefore a decrease of hydration of the tissue at given water pressure. Indeed, according to (13.7.4) or (16.3.17), water flux is opposite to the gradient of the difference of the water and osmotic pressures. This decrease of hydration is likely to progressively transform small defects into cracks which, critically, will be hydrated, at least more hydrated than in a healthy tissue. Upon application of a sudden mechanical load, water in the crack will be pressurized, thus increasing the stress intensity factor and contributing to fracture the NP. The above argument is in fact tentative and the question of the mechanisms that trigger crack propagation as well as the crack orientation, whether radial or circumferential, is still controversial, e.g., Wognum et al. [2006] for a contribution on the subject.

Part III

Growth of biological tissues

Chapter 20

Tissue Engineering: overview of biochemical data and mechanical modeling

20.1 Biomechanical perspectives of tissue engineering

While immature articular cartilage contains vessels that circulate nutrients, adult articular cartilage is aneural, avascular and alymphatic. Nutrients from the synovial joint are transported by the extracellular fluid through diffusion and convection. Therefore the self-healing capacities of adult articular cartilage are practically nil. One method to treat damaged cartilage consists in removing the damaged part and in implanting biodegradable cell-polymer constructs. Engineered cartilages are cultured on scaffolds of agarose or alginate gels which, polymerized, display an open three-dimensional lattice structure that is prone to fix cells. Once in place, artificial cartilage proliferates reasonably well. However, the fact that its mechanical properties are lower than those of native cartilage is an issue. Recent studies have shown that the mechanical conditions to which chondrocytes, the cartilage cells, are subjected *in vivo* and *in vitro* affect the synthesis and degradation of the extracellular matrix (ECM). In fact, the chondrocytes are much more compliant than the ECM, about 1000 times according to Guilak et al. [1999]. However, they are protected by the stiff pericellular zone, in such a way that a clear picture of the transfer of the ECM deformation to the chondrocytes is still to be elucidated.

Special bioreactors have been developed to apply confined or unconfined compression on cartilage cultures. Research on the mechanosensitive effectors is still in its infancy, but certain results have been highlighted, Buschmann et al. [1992][1995] and Mauck et al. [2000]:
   - static (prolongated) compression has a detrimental influence on the synthesis of the ECM;
   - dynamic[20.1] compression on the other hand contributes to matrix synthesis. The actual efficiency of dynamic compression depends on the frequency and on the details of the loading program (amplitude, mean value, stress and strain paths, duration, etc.).

Interestingly enough, static compression seems to simply delay, not prevent, the stimulating effects of insulin-like growth factor 1 (IGF-1), suggesting that the modes of action of mechanical and biochemical factors are independent, Grodzinsky et al. [2000]. Transforming growth factor β (TGF-β) triggers also growth and differentiation but it is less efficient than IGF-1. The efficiency of these growth factors may vary depending on their *active* concentration and on the concentration of *active* receptors on cells. Indeed, a chemical whether solute or macromolecule may be present but it might be (partially) bound to some sites which prevent its action. In fact, the concentrations of IGF-1 and of IGF binding proteins are respectively lower and higher in the obese population than in the normal weight population. The resulting lower concentration in GAG has implications on osteoarthritis (OA). While mechanical interactions are likely to be significant for knee cartilage, the role of metabolism is expected to be dominant in the OA of the hip.

One possible way in which mechanics stimulates biological synthesis is by generation of *advective flow* that particularly enhances the transport of large soluble molecules (e.g., growth factors) that are thus more accessible to the chondrocytes and their immediate neighborhood, Garcia et al. [1996]. Unfortunately, advective flow seems

---

[20.1]The terminology is improper, "dynamic" actually means "time variable" by opposition to "static" which should be understood as meaning "constant in time".

also to favor the release of monomers of glycosaminoglycans out of the construct, and the loss can be considerable, Sengers et al. [2004] Kisiday et al. [2002]a [2002]b [2004].

A second mode in which mechanics might act is by *altering the metabolism*. Deformation of the chondrocytes and of their organelles modifies intracellular signaling and communication between cells. On one hand, dynamic unconfined compression affects the *gene transcription* phase, by increasing the expression of aggrecan and type II collagen mRNA, Ragan et al. [1999]. On the other hand, deformation of organelles may alter, or slow down, the post-transcription *metabolic pathways* that involve different organelles that deliver type II collagen and the various pieces of the proteoglycans, Hascall et al. [1991]. Elements that may decrease the density of electrical charges of the glycosaminoglycans are of particular relevance: indeed, the *effective* concentration of proteoglycans is a key parameter in their mechanical interactions and consequently in the overall mechanical properties of articular cartilage.

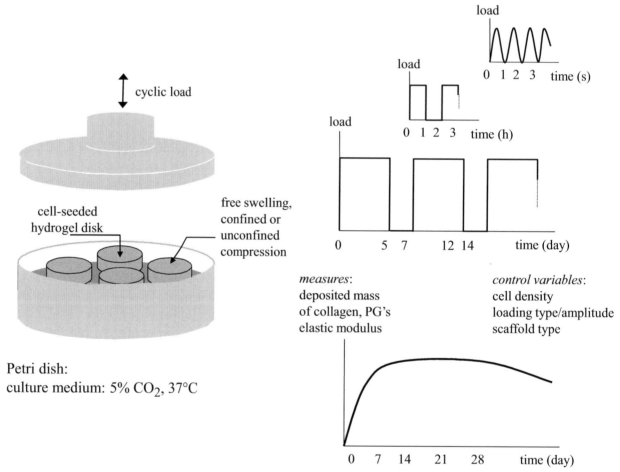

Fig. 20.1.1: Cultures of articular cartilage are submitted to various types of mechanical loading. Control parameters include cell density, loading type/amplitude/duration, scaffold type, etc. The quality of the culture is assessed in terms of the deposited mass of collagen and PGs, of the elastic stiffness and hydraulic permeability.

This chapter contains a review of the biochemical and mechanical aspects of tissue engineering with special emphasis on cartilage engineering. Basic notions of signal transduction through a cellular membrane are introduced. The main metabolic process in chondrocytes, namely glycolysis, is also sketched. The various methods of cultures are listed and compared. Data available on growth factors and cytokines are recorded for both cartilage engineering but also for osteoarthritis. Experimental data of mechanical nature and models of extracellular synthesis are given a special emphasis.

20.1.1 Issues

Since adult articular cartilage is avascular, its repair capacities to local traumas, degenerative changes (osteoathritis) and inflammation (rheumatoid arthritis) are quite small.

When damaged until the subchondral bone, the bone vasculature is recruited, either naturally or by surgery, so as to feed chondroprogenitor cells from the bone marrow. Surgery may consist in drilling or increasing the roughness of the bone. However, the resulting tissue is a fibrocartilage with low content in collagen of type II and in PGs, both constituents being essential components for a normal mechanical behavior of articular cartilages.

An alternative method consists in collecting chondrocytes, making them proliferate *in vitro*, shaping the resulting cartilage piece in an optimal form so as to ensure its adherence with the sound native cartilage in place. While it is agreed that **fetal** chondrocytes proliferate when cultured in high density layers, a real challenge is to trigger the proliferation of autologous **adult** chondrocytes. The biochemical aspects are well advanced and, in fact, grafts are now able to fill the lesions with a minimal amount of fibrous tissue. Yet the mechanical properties of the cartilage obtained are much lower than those of native cartilage, C.R. Lee et al. [2003].

Adult articular cartilage is submitted to intermittent compressive and shear loadings during walking, at a frequency of 0 to 1 Hz. It is also known that prolongated rest decreases the mass of various tissues like bone, cartilage, muscle, etc.[20.2][20.3]. In order to improve the physical properties of cultured cartilage, it seems natural to attempt to reproduce physiological stimuli, Fig. 20.1.1. Indeed, it has been observed that the metabolism of chondrocytes is sensitive to its mechanical environment and, in particular, to hydrostatic compression, its amplitude, its frequency, to streaming potential, to pH and to fluid flow in the pericellular zone. Physiological stimuli (in terms of amplitude and frequency) are expected, if not to be optimal, at least to play a special role. Permanent compression inhibits matrix synthesis in a level-dependent way while intermittent, or dynamic, compression has positive effects, Buschmann et al. [1995]. Impact, or high compressive stress, induces damage of cartilage in various forms, osteoarthritis[20.4], cell death[20.5] or fracture[20.6], and decreases the biosynthetic activity. The fact that the metabolism is depth-dependent across the cartilage layer does not ease the understanding of the individual phenomena involved.

As a biological tissue, cartilage attempts to structure in order to sustain modifications of its biochemical or mechanical environments. However, self-structuration in response to non physiological (e.g., fracture) or pathological (e.g., ostoearthritis) conditions is unsuccessful essentially due to the avascular nature of adult cartilage[20.7]. That is why cartilage needs to be engineered, *in vitro* or *in vivo*, to replace the damaged parts. Methods of cartilage engineering apply also to the intervertebral disc and to the fibrocartilage of the meniscus, e.g., Peretti et al. [2001].

Many aspects of mechanobiology encountered in the engineering of articular cartilage are common to other tissues, like bones (Nagatomi et al. [2002]), blood vessels (Ogle and Mooradian [2002], Lehoux and Tedgui [2003], Chien [2003]), ligaments, etc. These papers contain a detailed description of both biochemical and mechanical aspects of growth in the specific tissues.

Next to the difficulties in the cultures themselves, a few other issues are listed in Fig. 20.1.2. For example, autologous implantation[20.8] requires two surgical acts. In addition the implantation of the engineered tissue within the remaining native cartilage needs to occur without angiogenic response[20.9] nor immune reaction[20.10]. The culture

---

[20.2]NASA is currently studying the effects of exercising on these tissues in view of flights to Mars which are expected to long several months.

[20.3]The effects of exercising on equine joints are analyzed by van de Lest et al. [2002]. They tested several groups of foals. Their conclusions are that lack of exercising retards the normal development of cartilage but subsequent exercising re-establishes the quality of the tissue. On the other hand, excessive exercising may have irreversible detrimental effects in the long run. By extrapolation to humans, these observations seem to suggest that it is not a good idea to train young adults too intensively, see also Buckwalter and Lane [1997]. In short, light or moderate training has positive effects on cartilage while, in the long term, strenuous training has possibly irreversible degenerative effects.

[20.4]For example, it is known that about half of the compressive load on the knee is transmitted by the meniscus. Menisectomy, which thus overloads articular cartilage, has been observed to induce early osteoarthritis, Peretti et al. [2001].

[20.5]There are several methods to assess experimentally cell viability, see Kurz et al. [2001].

[20.6]Kurz et al. [2001] note that low compression rates even at large stress ($0.01 \, \text{s}^{-1}$ with a peak stress of 12 MPa) has a much less detrimental effect on cartilage biosynthesis than faster rates (0.1 to $1 \, \text{s}^{-1}$ with peak stresses of 18 MPa and 24 MPa) and that this effect may persistently damage the properties of the cartilage. Incidentally, they also note that injurious impacts lead to cartilage swelling, as a result of the disruption of the collagen network which can no longer oppose the pressure generated by GAG repulsion.

[20.7]Nutrients in mature cartilage diffuse from the synovial joint while they use vascular canals in immature cartilage.

[20.8]Autologous implantation refers to the procedure in which chondrocytes are harvested, cultured and the construct implanted back in the same person.

[20.9]A short discussion on angiogenesis is presented in Beaupré et al. [2000].

[20.10]Laboratory experiments often use nude mice (which, devoid of thymus, are immunodeficient and do not show a rejection response to grafts) as temporary hosts to grow transplants.

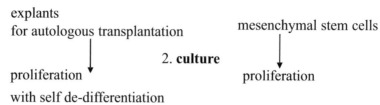

1. collecting cells

explants
for autologous transplantation   mesenchymal stem cells

2. culture

proliferation   proliferation
with self de-differentiation

⇨ chondrocytes

3. formation of cartilage constructs

cell re-differentiation: proliferation and synthesis of ECM

⎧ mono-layers
⎨ scaffolds: seeding of chondrocytes on scaffolds
⎩   or chondrocytes seeded on gels within scaffolds

⎧ favorable agents to synthesis of collagen type II and GAGs:
with ⎨ growth factors
⎩ mechanical stimuli

4. transplantation

avoid auto-immune response

improve adhesion

Fig. 20.1.2: Sketch of the steps of cartilage engineering.

of mesenchymal stem cells would alleviate part of these difficulties but methods (using growth factors) to stimulate chondrogenesis of mesenchymal stem cells need to be improved, Worster et al. [2001]. Even so, integration of engineered cartilage is an issue in itself, Obradovic et al. [2001].

Alternative surgical procedures have been used to repair damaged cartilage, e.g., modifications of the damage zone by abrasion arthroplasty, microfracture, drilling, etc. and transplantation of periosteal or perichondral autografts. They are however highly empirical and work only for small lesions and they do not seem to be sustainable.

Several reviews on cartilage engineering have been published. The review of Grodzinsky et al. [2000] is a very lucid and up-to-date account on many related biochemical and mechanical aspects. Laurencin et al. [1999] and Bronson [2002] are devoted to several qualitative aspects of orthopedic tissue engineering and target bone, cartilage, ligament, muscle. The review of Jackson et al. [2001] contains information on the composition of normal articular cartilage and presents various surgical options for the treatment of lesions. Sittinger et al. [1999] and Bentley and Minas [2000] cover the same area as Jackson et al. [2001]. Glowacki [2000] presents a brief overview with emphasis on surgical aspects. Nelson [1999] reviews the transplantation of allografts in several organs (heart, cornea, ligament, cartilage) and notes the fate of allograft cells: heart transplant is the only case where chimerism[20.11] occurs. Frenkel and di Cesare [1999] consider treatments of cartilage lesions utilizing transplantation as well and review the effects of growth factors in tissue engineering.

20.1.2   Composition of articular cartilages

Basic notions on the composition of articular cartilage have been reviewed in Chapter 14. A description of the different types of glycosaminoglycan chains, of their role in the chondrogenesis, and of many aspects of synthesis is given in Prydz and Dalen [2000]. After synthesis in the Golgi apparatus, the PGs are transported by exocytosis towards the ECM as chondroitin sulfates or dermatan sulfates. The role of the main organelles in the biosynthesis and the factors that come into picture to decide on the fate of a newly synthesized GAG chain are reviewed. The

---

[20.11]Chimerism: presence of more than one genetic line in a single individual.

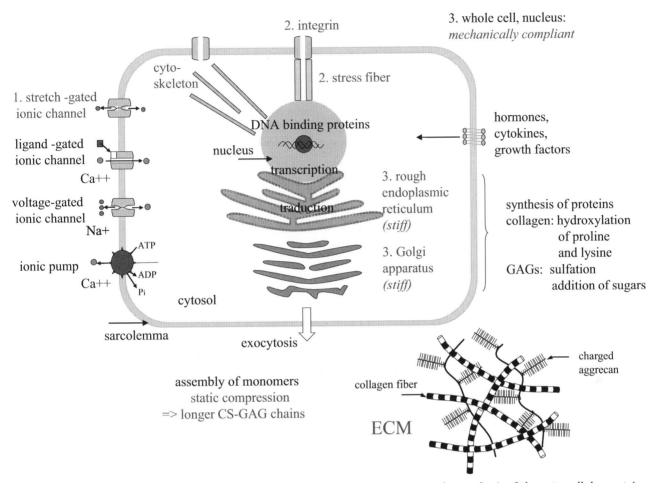

Fig. 20.1.3: Membrane and cellular mechanosensitive elements participating to the synthesis of the extracellular matrix of articular cartilages.

GAGs have a size comparable to that of the cell and they are gathered into aggrecans outside the cell. Membrane and cell elements which are involved in the synthesis of the cartilage extracellular matrix are sketched in Fig. 20.1.3. Typical biomechanical data for normal articular cartilage range as indicated in Table 20.1.1.

Table 20.1.1: Typical biomechanical data for normal articular cartilage.

Compressive Modulus	Tensile Modulus	Hydraulic Permeability	Chondrocyte Density
MPa	MPa	$m^2/Pa/s$	Cells per ml of medium
0.3 to 1	2 to 20	0.5 to 5×10^{-5}	30 to 100×10^6

A typical protocol to assess the content of cartilage consists of the following steps, Asanbaeva et al. [2008],
- weight the dry specimens;
- digest the proteins using proteinase K;
- define the DNA content, McGowan et al. [2002], and obtain the cell number with a conversion factor of 7.7 pg DNA/cell, Kim et al. [1988];
- define the GAG content, Farndale et al. [1986];
- define the hydroxyproline content, Woessner [1961], and obtain the mass of collagen as 7.25 times the mass of hydroxyproline, Herbage et al. [1977];
- define collagen-specific pyridinoline crosslinks, Uebelhart et al. [1993].

20.2   Basic notions of biochemistry

20.2.1   Macromolecules

Macromolecules are polymers built by covalent bonds between smaller subunits or monomers. A typical bacterial cell contain 70% water in weight. The 30% left are various chemicals: half of them are proteins (15%), 2% polysaccharides, 6% RNA, 1% DNA and only 2% phospholipids[20.12] and 4% ions are not macromolecules.

Polysaccharides are formed from the condensation of sugars, proteins from amino-acids and nucleic acids from nucleotides. *Condensation* consists in removing a water molecule while *hydrolysis* is the inverse process, namely using water to break a compound:

$$\text{X-OH} + \text{H-Y} \xrightleftharpoons[\text{hydrolysis}]{\text{condensation}} \text{X-Y} + H_2O$$

More details on condensation and hydrolysis are provided in Section 13.5.2.

The molar masses of macromolecules are large with respect to ions, Tables 13.2.1 and 13.2.2. Their diffusion coefficient is about inversely proportional to the square root of their molar mass, Section 15.5. It is therefore small, Tables 13.3.1 and 15.5.1, so that they move due to convection rather than diffusion.

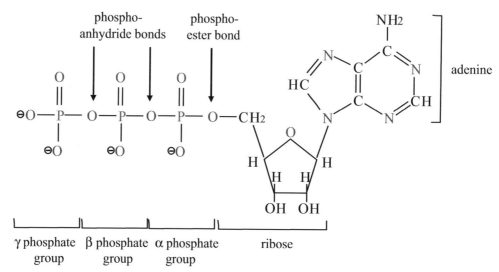

Fig. 20.2.1: Structure of ATP with chemical formula $C_{10}H_8N_4O_2NH_2(OH)_2(PO_3H)_3H$, or in compact form $C_{10}H_{16}N_5O_{13}P_3$, and molar mass of $10 \times 12.011 + 16 \times 1.008 + 5 \times 14.007 + 13 \times 15.9994 + 3 \times 30.974 \sim 507.18\,\text{g/mole}$. Breaking the phosphoanhydride bonds is highly exergonic, much more than breaking the phosphoester bond. The energetic properties of the phosphoanhydride bonds are due to the repulsion of the residues by the negatively charged oxygens. The various chemicals involved are negatively charged, namely ATP^{-4}, ADP^{-3}, AMP^{-2} and P_i^{-2}.

20.2.2   ATP and energetic coupling

The nucleotide coenzyme[20.13] adenosine triphosphate ATP^{4-} is a key element of metabolism. Its structure is sketched in Fig. 20.2.1. ATP^{4-} provides the largest storage of exchangeable chemical energy of living cells. Its hydrolysis in adenosine diphosphate ADP^{3-} and inorganic phosphate P_i^{2-} releases a chemical energy $\Delta G_{ATP}^{0'} = $ -30 to -35 kJ/mole under standard thermodynamic conditions (all concentrations equal to 1 M, temperature equal to 298.15 K and atmospheric pressure) and at pH 7. The superimposed 0 refers to the standard conditions and the prime to pH 7. In the cells, the excess of ATP^{4-} over equilibrium leads to a larger physiological energy release $\Delta G'$ of about -50 kJ/mole. This energy is used, via energetic couplings, by endergonic reactions like biosyntheses and transports.

[20.12]The phospholipids are the main components of the plasmic membranes. They are negatively charged at neutral pH.

[20.13]In most reactions which are catalyzed by enzymes, as briefly described in Section 15.12, electrons or groups of atoms are transferred from one substrate to the other. Additional substances that participate to these reactions and that bring temporarily the transferred group are called coenzymes, e.g., the nucleotide adenosine triphosphate ATP^{4-} and the nicotinamide adenine dinucleotide NAD^+.

It might be of interest to compare the energy released by the phosphoanhydride bond, namely -30 to -35 kJ/mole under standard conditions, to other energy releases in biological reactions. In fact, the energy content of this bond has an order of magnitude similar to ionic bonds in aqueous solutions. Breaking a covalent bond requires about 400 kJ/mole and a hydrogen bond amounts to about 10 kJ/mole. The importance of ATP^{4-} resides in the fact that its energy is easily exchangeable: the molecule has a relatively small molar mass, namely about 507.18 g/mole, in such a way that it can diffuse through tissues.

An example of reversible reaction will serve to explain the concept of energetic coupling. Consider transforming the chemical A to the chemical B,

$$A \quad \overset{k_+}{\underset{k_-}{\rightleftharpoons}} \quad B, \qquad 0 < \Delta G_{nc}^0, \tag{20.2.1}$$

with the subscript "nc" standing for "non coupled". However the reaction $A \to B$ is endergonic, namely the standard free enthalpy $\Delta G_{nc}^0 > 0$ is positive. Therefore at equilibrium, $\Delta G_{nc} = \Delta G_{nc}^0 + RT \operatorname{Ln} c_B/c_A = 0$,

$$\frac{c_B}{c_A} = \frac{k_+}{k_-} = K_{nc}^{eq} = \exp\left(-\frac{\Delta G_{nc}^0}{RT}\right) < 1. \tag{20.2.2}$$

Given initial concentrations c_A and c_B of A and B, the system will evolve spontaneously towards the above equilibrium. Energy is to be input to the system if we want to shift the equilibrium to the right. The hydrolysis of ATP^{4-} into adenosine diphosphate ADP^{3-} and inorganic phosphate P_i^{2-},

$$ATP^{4-} + H_2O \quad \rightleftharpoons \quad ADP^{3-} + P_i^{2-} + H^+, \tag{20.2.3}$$

is exergonic under standard conditions and pH 7,

$$\Delta G_{ATP} = \underbrace{\Delta G_{ATP}^{0'}}_{-30.5 \text{ kJ/mole}} + RT \operatorname{Ln} 10^{7-pH} + RT \operatorname{Ln} \frac{c_{ADP} \, c_{P_i}}{c_{ATP}}, \tag{20.2.4}$$

For a cell rich in energy, typically,

$$\frac{c_{ATP}}{c_{ADP} \, c_{P_i}} = 500 \text{ M}^{-1}, \tag{20.2.5}$$

the change of free enthalpy of the reaction becomes, with $RT = 2.479$ kJ/mole,

$$\Delta G_{ATP} = \Delta G_{ATP}' + RT \operatorname{Ln} 10^{7-pH}, \quad \Delta G_{ATP}' = \Delta G_{ATP}^{0'} + RT \operatorname{Ln} \frac{c_{ADP} \, c_{P_i}}{c_{ATP}} = -45.9 \text{ kJ/mole}. \tag{20.2.6}$$

Suppose now that the reaction (20.2.1) is coupled with the hydrolysis of ATP^{4-}, resulting in,

$$A + ATP^{4-} + H_2O \quad \rightleftharpoons B + ADP^{3-} + P_i^{2-} + H^+. \tag{20.2.7}$$

If the coupled reaction is totally efficient, that is, if the change of free enthalpy of the coupled reaction is equal to the sum of the changes of free enthalpies of the individual reactions, then

$$\begin{aligned} \Delta G &= \Delta G_{nc}^0 + \Delta G_{ATP}^{0'} + RT \operatorname{Ln} \frac{c_B \, c_{ADP} \, c_{P_i} \, 10^{7-pH}}{c_A \, c_{ATP}} \\ &= RT \operatorname{Ln} \frac{1}{K_{nc}^{eq}} \frac{1}{K_{ATP}^{eq'}} \frac{c_B \, c_{ADP} \, c_{P_i} \, 10^{7-pH}}{c_A \, c_{ATP}}, \end{aligned} \tag{20.2.8}$$

with

$$\Delta G_{ATP}^{0'} = -30.5 \text{ kPa}, \quad K_{ATP}^{eq'} = \exp\left(-\frac{\Delta G_{ATP}^{0'}}{RT}\right) = 2.22 \times 10^5 > 1. \tag{20.2.9}$$

At equilibrium $\Delta G = 0$,

$$\frac{c_B}{c_A} = K_c^{eq} = K_{nc}^{eq} \, K_{ATP}^{eq'} \, 10^{pH-7} \frac{c_{ATP}}{c_{ADP} \, c_{P_i}}. \tag{20.2.10}$$

For the excess of ATP defined by (20.2.5), at any pH, the initial reaction is clearly shifted to the right with respect to the uncoupled reaction,

$$\frac{K_c^{eq}}{K_{nc}^{eq}} \frac{1}{10^{pH-7}} = 2.22 \times 10^5 \times 500 = 1.11 \times 10^8. \tag{20.2.11}$$

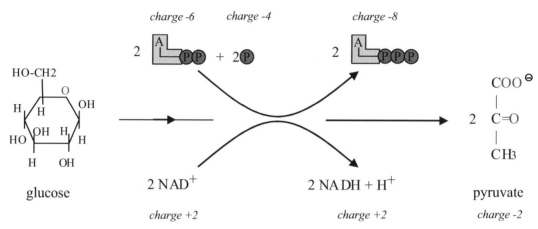

Fig. 20.2.2: During *aerobic* **glycolysis**, one mole of glucose is split in two moles of pyruvate$^-$, with a gain of two moles of ATP^{4-} and two $NADH+H^+$, adapted from APB [1999], p. 145.

20.2.3   Glycolysis

Glycolysis is a catabolic process that takes place in the cytoplasm of most cells which extract glucose from blood, Fig. 20.2.2. Glucose is the form in which sugars are transported in blood. Storage of glucose in a solid form, the glycogen, is ensured by liver and muscles.

Two main metabolites participate to glycolysis, namely the nucleotide coenzyme adenosine triphosphate ATP^{4-} and the coenzyme nicotinamide adenine dinucleotide NAD^+. NAD^+ accepts electrons and becomes reduced to NADH, namely $NAD^+ + H^+ + 2e^- \rightarrow NADH$.

Under *aerobic* conditions, one mole of glucose is split in two moles of pyruvate$^-$, together with two ATP and two $NADH+H^+$,

$$\text{glucose} + 2\,ADP^{3-} + 2\,P_i^{2-} + 2\,NAD^+ \quad \rightarrow \quad 2\,\text{pyruvate}^- + 2\,ATP^{4-} + 2\,NADH + 2\,H^+.$$

Two moles of ADP^{3-} are used for the activation and four moles of ATP^{4-} are produced, leaving a gain of two moles of ATP^{4-} per mole of glucose.

Actually, under aerobic conditions, pyruvate$^-$ and NADH enter the mitochondria where oxidation continues: the two NADH produced by glyceraldehyde 3 dehydrogenase are used in the oxidation of two pyruvate$^-$ which delivers 5 ATP^{4-}. The sole aerobic glycolysis extracts 7 moles of ATP^{4-} from each mole of glucose while the complete oxidation of a mole of glucose yields 32 moles of ATP^{4-}, Table 20.2.1.

Table 20.2.1: Summary of ATP^{4-} released by a mole of glucose during complete aerobic oxidation (glycolysis and oxidative phosphorylation OxPhos in the mitochondria) and anaerobic glycolysis under standard thermodynamic conditions and pH equal to 7, namely 1 M of each constituent and atmospheric pressure. Adapted from APB [1999], p. 140.

Chemical Reaction		Aerobic	Anaerobic
Hexokinase		-1 ATP^{4-}	-1 ATP^{4-}
Phosphofructokinase		-1 ATP^{4-}	-1 ATP^{4-}
Glyceraldehyde 3 dehydrogenase	5 $ATP^{4-} \leftarrow$	2 NADH	2 NADH
Phosphoglycerate kinase		2 ATP^{4-}	2 ATP^{4-}
Pyruvate kinase		2 ATP^{4-}	2 ATP^{4-}
Lactate dehydrogenase			-2 NADH
Balance for glycolysis		7 ATP^{4-}	2 ATP^{4-}
Complete oxidation		32 ATP^{4-}	2 ATP^{4-}

Glycolysis contains two phosphorylation steps: these steps which consist in transfering a phosphate group require two ATP^{4-} moles for each mole of glucose, Fig. 20.2.3. Aldolase hydrolyzes fructose-1,6-biphosphate in two fragments that can be converted into one another. A subsequent product, 1,3-biphosphoglycerate, contains an anhydrous bond that is much exenergetic: its hydrolysis allows for formation of ATP^{4-} in a next step. Next to oxidative phosphorylation and thiokinase in the Krebs cycle, the above 3-phosphoglycerate kinase and the final pyruvate kinase are the sole reactions that produce ATP^{4-}.

Accordingly, the energetic profile, Fig. 20.2.3, shows three steps with significant change of free enthalpy. These steps are irreversible and they are bypassed during neo-glycogenesis. Note that pyruvate kinase is exergonic while it also releases ATP^{4-}. The other steps are practically reversible and indeed they are followed in reverse direction during neo-glycogenesis.

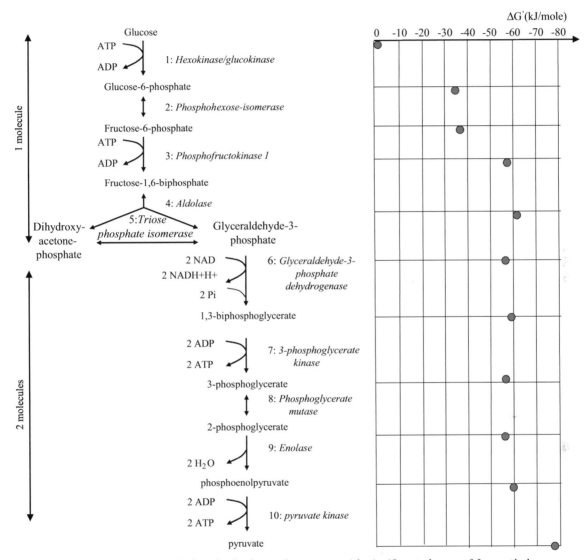

Fig. 20.2.3: The energetic profile of **glycolysis** shows three steps with significant change of free enthalpy:
- hexokinase-glucokinase that controls the rate of entrance of glucose into the pathway;
- phospho-fructokinase 1 which gives the limiting rate to the glycolysis;
- pyruvate kinase that controls the rate of production of the glycolysis.
These three steps are thus irreversible and bypassed during neo-glycogenesis. Based on APB [1999], p. 144.

Under *anaerobic* conditions, regeneration of NAD$^+$ from NADH in the mitochondria is inhibited and the reaction becomes

$$\text{glucose} + 2\,\text{ADP}^{3-} + 2\,\text{P}_i^{2-} \quad \rightarrow \quad 2\,\text{lactate}^- + 2\,\text{ATP}^{4-}.$$

Indeed, pyruvate$^-$ is also diverted from the mitochondrial pathway and it is transformed together with the NADH$^+$ produced by glyceraldehyde 3 dehydrogenase into lactate$^-$ in a reaction catalyzed by lactate dehydrogenase,

a) Cori cycle

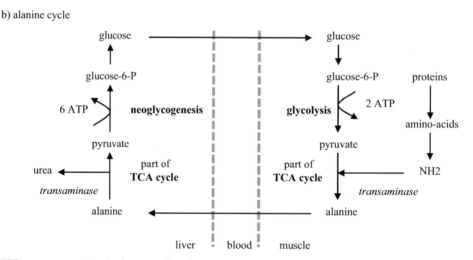

b) alanine cycle

Fig. 20.2.4: When pyruvate intake by mitochondries is not possible, there are two escape mechanisms which allow the glycolytic flow to process and ATP to be produced. One process is catalyzed by lactate dehydrogenase, the second process produces alanine, and both shift temporarily metabolism from muscle to liver. Adapted from APB [1999], p. 323.

Fig. 20.2.4. Lactate is next evacuated to blood. Under anaerobic conditions, glycolysis is the only way to form ATP^{4-}. The energy extracted from a mole of glucose then amounts to 2 moles of ATP^{4-}, Table 20.2.1.

Table 20.2.2: Plasma concentrations (mM) of metabolites involved in glycolysis.

Glucose	Lactate	Pyruvate	Alanine
3.6-6	0.5-2.5	0.02-0.1	0.3-0.6

Typical plasmic concentrations of products of glycolysis are displayed in Table 20.2.2. Concentrations in tissues and constructs are different, Section 20.6.4.

20.2.4   Hormones

The control of biological activities of living organs is ensured by a complex network of biochemical signals, internal to cells and in between cells. Intercellular signals use hormones[20.14] and neurotransmitters which specialize to the activity of synapses.

[20.14]The term hormone comes from the Greek verb meaning to arouse to activity.

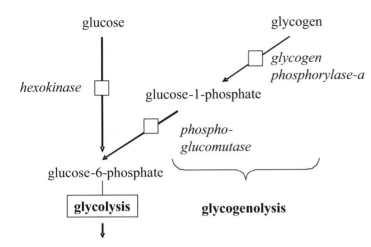

Fig. 20.2.5: The initial pathway of **glycogenolysis** involves two enzymes distinct from hexokinase and it does not require ATP.

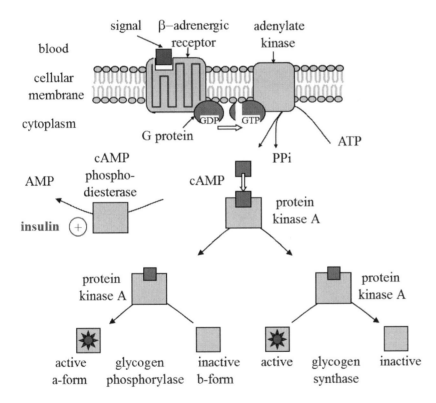

Fig. 20.2.6: Hormonal control of glycogenolysis through cAMP in a hepatic cell. Hormones fix on membrane receptors and indirectly activate cAMP which itself turns phosphorylase on (which is the first enzyme of the glycogenolysis pathway) and glycogen synthase off. Conversely, insulin diverts cAMPs from protein kinase A and reverses the above processes. Based on APB [1999], p. 365.

20.2.4.1 Hormonal control of glycogenolysis

Glycogen is the form in which animal cells store glucose. Indeed, glucose is soluble and its presence in the cytoplasm would lead to swelling by osmotic effect. Glycogen is weakly soluble and it can be quickly released in the form of glucose phosphate or glucose to feed the glycolysis pathway, Figs. 20.2.4 and 20.2.5. In case of hunger, or nervous tension, hormones such as glucagon or adrenaline are produced. They activate the release of glycogen and inhibit glycogen synthase. Indeed, as displayed in Fig. 20.2.6,

- they fix on receptors located on the cell membrane (β-adrenergic receptors for adrenaline);
- these receptors send a G protein that activates adenylate kinase;
- adenylate kinase synthesizes adenosine cyclic monophosphate cAMP from ATP^{4-};
- cAMP is an allosteric effector of protein kinase A: it binds to the regulatory units and activates protein kinase A. Protein kinase A has two distinct actions;
- on one side, protein kinase A activates phosphorylase kinase which phosphorylates the inactive glycogen

phosphorylase-b into its active form glycogen phosphorylase-a. It is the enzyme that yields glucose-1-phosphate from glycogen;

- on the other side, protein kinase A inactivates glycogen synthase;
- if the level of cAMP decreases, the reverse cascade of dephosphorylation occurs, glycogen phosphorylase turns inactive (from a form to b form) and glycogen synthase is activated;
- the level of cAMP is controlled by phosphodiesterase which transforms cAMP into AMP. Insulin activates phosphodiesterase.

There is therefore a direct relation between the activation of glycogen phosphorylase and the level of cAMP. Insulin is expected to decrease the rate of glycogenolysis and to activate glycogen synthase, Fig. 20.2.6. Glucagon and insulin are thus antagonists.

20.2.4.2 Classification of hormones

Hormones are chemical signals that are produced by specialized secretory cells and transported by the bloodstream to their target cells where they participate to the regulation of biochemical functions. They are categorized according to their modes of action, APB [1999], pp. 350-371:

- lipophilic hormones, which are weakly soluble, enter the cell and act on the nucleus. Steroids, thyroxin and retinoic acid are lipophilic hormones. They control gene transcription and the synthesis of messenger ribonucleic acid mRNA;
- hydrophilic hormones, which are easily soluble, are convected by blood. They link to specific receptors on the external side of the membrane of the target cell and act in the cell through second messengers, namely intra-cellular chemical signals. Hormones deriving from amino-acids, peptide and protein hormones are hydrophilic.

Hormones are produced by specialized cells, gathered in glands. Hypothalamus, hypophysis, thyroid, ovary, testis are endocrine glands. Depending whether they are in the cells they are created, in the neighborhood of these cells, or at distance after being transported by blood, hormones are termed autocrine, paracrine and endocrine, respectively. Sometimes, paracrine regulators, which diffuse in the extracellular fluid to the target cell are termed local mediators to contrast with endocrine hormone regulation.

Insulin for example is both paracrine and endocrine. It is hydrophilic, it favors glycogen synthase acting through second messengers, Fig. 20.2.6. It is synthesized by the β cells of pancreas[20.15] when the concentration of glucose in blood is too high. Glucagon, meaning *exciting sugar*, is synthesized by the α cells of pancreas when the concentration of glucose in blood is too low. It stimulates the secretion of glucose by liver.

20.2.5 Growth factors

A part of the stimulation of gene expression is due to various growth factors and cytokines that are located in the extracellular matrix and fix to the membrane receptors.

A growth hormone accelerates growth, prolongating growth of all organs and of bone before closure of the epiphyses, by increasing synthesis of proteins, and use of fat for energy production. The rat possesses the particularity that its bones do not close, so that growth hormones are active livelong.

Growth factors are soluble proteins of relatively small molar mass (a few hundred grams) and they are degraded rapidly. Therefore efforts have been made to develop carriers that can deliver them to cells at a more or less steady level for a long period but no optimal solution has been found so far.

Focus will be on the effects of growth factors on

- cell proliferation;
- cell differentiation to chondrocyte;
- synthesis of collagen type II and GAGs.

The main growth factors that participate to the expression of articular cartilage are, Sandell [2000]:

- insulin-like growth factor-1 IGF-1;
- transforming growth factor β TGF-β;
- bone morphogenetic protein BMP, actually a member of the TGF-β family;
- basic fibroblastic growth factor bFGF.

[20.15]These cells are also called *islets* of Langerhans, which has given the name *insulin* ("insula" island in Latin).

Growth factors IGF-1, TGF-β and BMP's increase expression of collagen type II, enhancing metabolism and preserving the phenotype[20.16]. A concentration of IGF-1 of the order of 10 ng/ml is sufficient to stimulate the metabolism.

The concentration of TGF-β is reduced in osteoarthritic cartilage with respect to normal cartilage. Worster et al. [2001] note that mesenchymal progenitor stem cells (MSCs) pretreated with TGF-β1 in monolayer cultures differentiate to chondrocytes when grown in IGF-1 supplemented cultures in 3D matrices. In absence of IGF-1, chondrogenesis was observed to be less effective. In any case, biosynthesis of GAGs and collagen was lower than in parallel fully differentiated chondrocyte cultures. On the other hand, TGF-β1 supplementation on freshly isolated chondrocytes had a negative impact on their biosynthetic activity. They attribute the difference of the effects of TGF-β1 on MSCs and on mature chondrocytes to a beneficial modification of a TGF-β1 receptor.

bFGFs are synthesized by the chondrocytes and the pituitary. They stimulate cell proliferation, but their effects on the synthesis of collagen and GAGs have been reported to be both positive and negative, or even elusive, Frenkel and di Cesare [1999].

20.2.6 Cytokines

Cytokines are peptides, proteins and glycoproteins[20.17] that are used in cellular communication[20.18] and that participate to the growth, differentiation, life and apoptosis of cells. They have a role of signal similar to hormones. Unlike hormones that are synthesized by specialized glands, cytokines are synthesized by a widespread distribution of cells of the immune system upon detection of a pathogen and by epithelial cells. Therefore their action is either autocrine or paracrine but not endocrine. Unlike hormones whose concentration is nanomolar and varies by less than an order of magnitude, the concentration of cytokines is picomolar and it may vary significantly by up to three orders of magnitude.

There are many cytokines, among which:
- interleukins IL-1 to IL-36 ;
- interferons INF-α to INF-γ;
- colony stimulating factor CSF;
- tumor necrosis factors TNF;
- macrophage migration inhibitory factor (MIF).

Cytokines have synergetic or antagonist actions: for example, IL-10 inhibits the synthesis of IL-1 and TNF-α. They participate to both anabolic and catabolic processes and the balance between these two activities determines the integrity of tissues, e.g., articular cartilage.

Cytokines act as hydrophilic signals that fix to the membranes at specific receptors. The receptors then transmit the signals through a complex path that leads to an action on the transcription of specific genes.

IL-1, interferon-γ, retinoic acid, TNF-α decrease the expression of collagen type II. IL-1 is found in higher levels in ostearthritic cartilages. It induces degenerative processes. It acts by modulating the activity of an extracellular signal-related kinase (ERK), a member of the mitogen-associated protein kinases (MAPKs). ERK transmits the stimuli from outside the cell to the nucleus. Since its activity is related to membrane receptors, it is expected to depend on the configurations, e.g., isolated chondrocytes versus chondrocytes seeded in scaffolds or treated for adhesion, Li et al. [2003].

MIF's, which are associated with rheumatoid arthritis (RA), activate cells to release harmful products into sites of inflammation.

20.2.7 Genes and gene expression

Nucleic acids play a key role in storing and translating genetic information. The whole process is governed by the very peculiar structure of DNA and by specific biochemical sequences:
- *gene structure*: a gene contains protein-coding sequences (exons) interrupted by non-coding sequences (introns).
- *gene expression* is regulated by proteins that bind to specific deoxyribonucleic acid (DNA) sequences, upstream from the first exon.

[20.16]The phenotype is understood as gathering the observable characteristics of an organism, a cell, etc. resulting from both genetics and environmental effects.
[20.17]See footnote on p. 490 for definitions.
[20.18]Etymology: "cyto" cell, "kinos" motion.

For example, the transcription is enhanced or repressed by a number of proteins, called transcription factors. Sox-9 (Sry-type HMG Box) is a transcription factor which stimulates the differentiation of mesenchymal stem cells to chondrocytes, by upregulating the genes for collagen type II and GAG, Osch et al. [2002].

Gene expression is also influenced by growth factors and cytokines located outside the cells. The transduction pathways use receptors located on the external side of the membrane of the cells which transmit the signal to the cytoplasm and next to the nucleus.

20.2.7.1   Gene structure

The genes are made up of double-stranded DNA (deoxyribonucleic acid)[20.19] which consists of the linear repetition of nucleotides: the latter are formed by a **base**, a **sugar** (deoxyribose) and a **phosphate**. The bases of the two strands are attached by hydrogen bonds according to the Watson-Crick pattern, Fig. 20.2.7: adenine pairs always with thymine, and guanine with cytosine. In humans, the 10^9 base pairs are localized in the nucleus and they are organized into 23 pairs of chromosomes. There is much more DNA than necessary to encode proteins.

DNA is the repository of the genetic information. Ribonucleic acid RNA, which differs from DNA by a base and a sugar, participates to the expression of genes and formation of proteins. Due to steric considerations, RNA can not adopt the shape of the double-stranded helix.

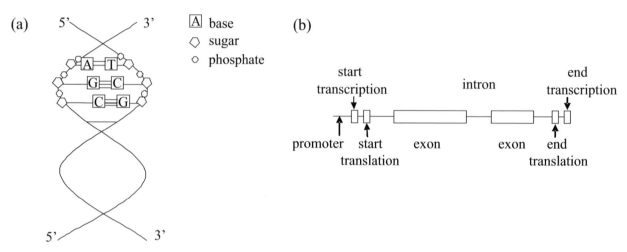

Fig. 20.2.7: Structure of DNA. (a) Watson-Crick pattern of double-stranded DNA. The strands are called 3' and 5' according to the position of the phosphate on the sugar. The bases are linked by hydrogen bonds, APB [1999], p. 83; (b) coding sequence (exons), non-coding sequences (introns), promoters, locus control region and upstream URE/dowstream DRE regulatory elements.

20.2.7.2   Gene expression

Every cell contains the same DNA but cells differentiate by the way DNA express proteins. Some parts of DNA are masked by protein binding in some cells and not in others. Alternatively, the required transcription factors are not available in some cells.

The 20 proteinogenic amino-acids are described completely by three letters (triplets) representing three successive bases. These letters are taken out of the alphabet of four letters (**codons**) of the nucleic acids, these are A, G, C, U or T[20.20]. Thy appears in DNA and Ura in RNA. For example, CAT codes for the three pairs Cyt-Gua Ade-Thy Thy-Ade of histidine. There are $4^3=64$ codons for 20 amino-acids, so that several codons are synonymous.

[20.19]ribosome: "ribo" anagram of "arabonic" acid $C_5H_{10}O_6$, "some": Greek "soma" body.

[20.20]Ade→adenine, Cyt →cytosine, Gua→guanine, Thy→thymine, Ura→uracile.

The two strands are not equivalent: the coding strand is the one which is read during transcription. Conventionally, the gene sequence is defined by the sequence of the non-coding strand in the direction 5'→3', e.g., TTC for the phenylalanine, which codes for the three pairs Thy-Ade Thy-Ade Cyt-Gua.

The expression and transmission of the genetic information include the following stages, Sandell [2000]:
- *replication*: during the phase S of the cellular cycle, each strand is used to form an associated copy;
- *transcription* or synthesis of RNA: the genetic code is translated from the language of nucleic acids (in the nucleus) to that of proteins (the ribosomes of the endoplasmic reticulum), that is, T is replaced by U, to form a heterogeneous nuclear hnRNA;
- *maturation* of the RNA: the hnRNAs are modified many times, with elimination of introns, and protection of the extremities, to form the messenger mRNA;
- *translation*: the mRNA is decoded by a ribosome to form a specific amino-acid chain. The translation is due to the transfer tRNA which fixes the appropriate amino-acid on the ribosome.

For the formation of the extracellular matrix, additional steps are needed, APB [1999], p. 327:
- *entry in the cytoplasm* and contact with the rough endoplasmic reticulum;
- *post-translational modifications*: hydroxylation (i.e., addition of hydroxyl groups -OH) of proline and lysine and glycolysation (addition of sugars) in collagen molecules, sulfation of GAGs in the *Golgi apparatus*;
- *secretion* to the extracellular space by exocytosis: secretory vesicles containing the proteins fusion with the plasmic membrane and finally release their content;
- assembly in the extracellular space of the different pieces to form functional units, a process explained in Section 14.2.3.3.

20.3 Effects of hormones and growth factors

Tissue engineering consists essentially of three steps. First, cells obtained (by biopsy) have to be multiplied and they dedifferentiate, i.e. loose their phenotype. This process of *cell expansion* is performed in monolayer cultures. Next the cells are seeded on scaffolds for some period in order to produce ECM. Finally the engineered tissue is implanted. Each of the steps might (has to) be stimulated by growth factors but the nature and levels of growth factors to be used depend on the history of the cells, Osch et al. [2002], in particular on the presence of ECM. Growth factors are soluble proteins of relatively small molar mass, they may be transported by diffusion. *In vivo*, growth factors are produced by the chondrocytes and they are present in the synovial fluid.

20.3.1 Insulin-like growth factor IGF-1

Cartilage homeostasis depends on the balance between anabolic and catabolic peptides. Insulin-like growth factor-I (IGF-1) is an important anabolic polypeptide for articular cartilage. It enhances the metabolism of cartilage, increasing the proliferation of chondrocytes and the synthesis of the components of cartilage, namely collagen, GAGs and hyaluronan, and inhibiting their degradation. In particular it acts as an antagonist of the cytokines IL-1 and TNF-α which tend to degrade cartilage[20.21]. It also contributes to the maintenance of electromechanical properties, like the elastic modulus and the electrokinetic coefficient, Sah et al. [1996].

20.3.1.1 Effects of IGF-1

IGF-1 is reported to increase the chondrogenic potential of bone-marrow cells that naturally tend to repair full-thickness damage. Fortier et al. [2001] report that endogenous expression of IGF-1 is greater inside the damaged part of the tissue than in the sound part. However, the final cartilage in the observed lesion was mainly fibrocartilaginous which is known to have poor biomechanical properties and to be prone to osteoarthritis. The authors did not analyze the effects of exogenous IGF-1.

20.3.1.2 Interactions of IGF-1 with mechanical factors

Bonassar et al. [2001] analyze the interactions of some mechanical and biochemical factors, namely dynamic compression and IGF-1. The stimuli are applied in the physiological range, i.e., 0-3% strain at 0.1 Hz and 0-300 ng/ml of

---

[20.21]Some molecules mimic a signal and bind to their natural receptors. Their efficiency to activate the receptors may be high, in which case they can rightly be called *agonists*, but it might be low, in which case they behave as *antagonists* of the signal.

IGF-1. They find that, acting independently, the mechanical program increases protein synthesis by 40% and GAG synthesis by 90%, while the corresponding figures for IGF-1 are 90% and 120%. Acting simultaneously, dynamic compression and IGF-1 yield increases of synthesis by 180% and 290%. Moreover the time constants associated with dynamic compression, with IGF-1 and with their simultaneous application are distinct. Therefore, these two sets of information convey the idea that

- the mechanical and biochemical mechanotransductions are controlled by distinct mechanisms;
- they do interact, i.e., they are not simply additive, but synergetic.

While the cultures of Bonassar et al. [2001] were relatively short, Mauck et al. [2003]a also find that, in long term cultures, TGF-β_1 and IGF-1 have synergetic effects with dynamic compression.

Tests using shearing instead of compression have come to similar results, Emkey et al. [2002]. However, the fact that shear does not induce water flow implies that there is no modification of the time characteristics by the simultaneous application of shear and IGF-1.

20.3.1.3 Transport of IGF-1

Dynamic compression increases fluid flow in the pericellular zone by enhancing convection. This flow may turn soluble growth factors more accessible to cells[20.22]: Bonassar et al. [2001] observe that the typical transport time of IGF-1, of molar mass 7.6 kg, is halved by application of the compressive 2% strain at 0.1 Hz. O'Hara et al. [1990] have already observed that the extraction out of articular cartilage of another soluble molecule, namely albumin of molar mass 65 kg, was eased by application of a cyclic load.

IGF-1 behaves as a hydrophilic hormone and binds to receptors on the cells, its signal being transmitted through a second messenger. IGF-1 is found at levels much higher than that expected by the sole presence of the cell receptors. In fact, recent studies have shown that there exist specific binding proteins (IGF-BP) whose level is particularly high in osteoarthritic cartilage, Bhakta et al. [2000]. These proteins bind IGF-1 and limit their transport through the cartilage, and therefore their accessibility to cells, Szasz et al. [2000].

Characteristic times τ are obtained by fitting data to a first order kinetics of the form $A + B \exp(-t/\tau)$. (Radial) diffusivity D is deduced by $D = r^2/\pi^2\tau$ where r is the radius of the disc. With τ=12.6 h, r =1.5 mm, Bonassar et al. [2001] obtain $D = 5 \times 10^{-8}\,\mathrm{cm^2/s}$, which falls in the range of the measurements $D = 3 - 6 \times 10^{-8}\,\mathrm{cm^2/s}$ in the absence of mechanical load, Garcia et al. [1996] [1997]. This range is an order of magnitude lower than for the free (unbound) IGF-1, Table 15.5.1.

20.3.2 Bone morphogenetic proteins BMPs

During ossification, chondrocytes undergo several specific steps of proliferation, maturation, hypertrophy, matrix calcification, apoptosis and primary bone formation.

Indian hedgehog (Ihh) and the parathyroid-hormone related protein (PTHrP) regulate the rate of chondrocyte maturation. Ihh stimulates PTHrP and, in turn, PTHrP inhibits the expression of Ihh and maturation.

On the other hand, bone morphogenetic proteins (BMPs), which are members of the TGF-β family, stimulate Ihh expression and chondrocyte maturation. But BMP-6 expression is blocked by PTHrP, Grimsrud et al. [2001].

Still BMPs induce the differentiation of mesenchymal stem cells to chondrocytes in the presence of skeletal growth factors and cartilage-derived growth factors, see Frenkel and di Cesare [1999] for references.

20.3.3 Response to injury

Response to injury is inflammation. The process involves the occurrence of a number of well regulated sequences including several pro-inflammatory and anti-inflammatory proteins.

Patwari et al. [2003] find that IL-1 and injury act synergetically to decrease the level of GAGs. On the other hand, they identify an increased level of the osteogenic protein-1 OP-1 which is chondroprotective.

Williams et al. [2003] consider injury due to injection of chymopapain which depletes PGs. They observe that hyaluronan (HA) is able to stimulate synthesis of PGs only if, after the 42 day treatment by HA, the rabbits are forced to rest for another 42 days. Thus the chondroprotective effect of HA requires rest.

[20.22]Vunjak-Novakovic et al. [2002] observe that laminar flow stimulates more the biosynthetic activities of cartilage constructs than turbulent flow. On the other hand, Chien [2003] seems to imply the converse for arteries: here however, cellular growth is detrimental, leading to the formation of atheroma associated with the proliferation of smooth muscle cells, elastin, collagen and proteoglycans.

C.R. Lee et al. [2003] compare autologous grafts made with (1) defects implanted with chondrocytes alone, (2) constructs of chondrocyte-seeded collagen scaffolds cultured for 12 h and (3) no treatment. The grafts with the scaffolds show satisfactory proliferation, with minimal fibrous tissue. However the compressive stiffness is very low, about $1/20$ the normal cartilage stiffness. The reason for the short culture of 12 h follows the rationale, that it is better that the implant be in current activity of reparation so that it can optimize its adhesion with the host cartilage. A well-formed implant with less activity is thought to be less able to develop the adhesion, Obradovic et al. [2001] and Section 20.5.3.10. It remains to see if the implant can remain in place while undergoing a progressive mechanical loading that would improve its mechanical properties.

20.4 Cell cultures

Challenges in cell cultures consist in
- conserving the phenotype;
- ensuring the proliferation, and
- obtaining mechanical properties and hydraulic permeabilities of the same level as in native cartilages.

20.4.1 Source of cells

The typical life span of a cell is a matter of weeks. For example, Baer et al. [2001] indicate that they have lost 50% of the cells 4 weeks after isolation and only 10 to 20% remains viable after 8 and 16 weeks. Cells used for *in vitro* studies are at least of three types:
- fetal animal chondrocytes are used in preliminary studies as they are known to have a high proliferation potential;
- adult chondrocytes are taken from the cartilage of the animal/person, deposited in culture to proliferate and later placed back in the damaged joint. Attention should be paid to the method of extraction of the cells. Knight et al. [2001] observe that mechanical extraction damages less the pericellular matrix than enzymatic digestion. Autologous grafts have the advantage that they prevent immunological reactions but the number of chondrocytes that can be harvested from a patient without donor-site morbidity is low[20.23]. Therefore the cells should be multiplied *in vitro* before seeding: multiplication with dedifferentiation and redifferentiation should be stimulated in the constructs[20.24];
- **mesenchymal stem cells** derive from the embryonic connective tissue. In adults, they are part of the mixture of cells found in bone marrow stroma and also in some non marrow tissues like muscle and corneal stroma. They can replicate undifferentiated, become progenitors and can be induced to differentiate to adipocytic, chondrocytic and osteocytic lineages, giving rise to specific tissues like tendon and muscle, cartilage and bone, Pittenger et al. [1999][20.25]. Their growth and differentiation are regulated in particular by soluble factors (growth factors, cytokines). Once treated for a lineage (for example by TGF-β for chondrogenic differentiation), they remarkably remain committed to the lineage at 95%.

The group of Guilak at Duke University is reported in Bronson [2002] to have isolated cells from fat stroma in liposuction waste that produced cartilage ECM: this procedure however required the cells to be rounded and to be stimulated by growth factors.

20.4.2 Cell layers and scaffolds

Isolated chondrocytes are seeded at high density ($>10^6$ cells/ml) *on* the surface of *biodegradable substrates* (calcium polyphosphate), or *within synthetic scaffolds* (poly α hydroxyacids: polylactic acid P(L)LA, polyglycolic acid P(L)GA) or *biological scaffolds* (hyaluronan, fibrin, collagen type I, agarose gel and alginate bead). They can also be *suspended* on gels which are next introduced in the scaffolds, Marijnissen et al. [2002].

[20.23]Sittinger et al. [1999] indicate that chondrocytes from the human nasal septum can be multiplied above 100-fold. However, quite generally, proliferation of human chondrocytes is more difficult than for animals.

[20.24]The chondrocytes for autologous transplantation are obtained by biopsy. About 300-500 mg of articular cartilage is harvested by arthroscopy. Only about 1% consists of chondrocytes, that is about 2000 cells/mg.

[20.25]Koike et al. [2004] report that they obtain long lasting blood vessels if they seed mesenchymal precursor cells together with human endothelial cells, the latter producing vessels that disappear after two months.

Traditional cell cultures dispose cells in high density layers (monolayer or multilayer). This method has limitations probably due to the **contact inhibition principle**, according to which non tumoral cells cease their proliferation when they come in contact with neighboring cells. Besides, the proliferation rate decreases with passage.

Three-dimensional open lattice structures are preferred: biodegradable alginate and agarose gels are used as **scaffolds** on which cells are deposited because they provide more space for the cells to proliferate while minimizing diffusion distances. The cultures are fed by serum (metabolism of cartilage cells consists essentially of glycolysis). Chondrocytes resting in these gels retain their phenotype for a long period of culture and form an extracellular matrix similar to that *in vivo*.

Alginate gels were first used for culture of chondrocytes by Guo et al. [1989]. Alginate is a highly negatively charged linear polysaccharide (mannuronic acid-guluronic acid)$_n$. Alginate forms hydrogels in presence of divalent ions like Ca^{2+} and re-solubilizes in presence of Na^+ and calcium-chelators. This property allows to recover the cells, D.A. Lee et al. [2003].

Standard alginate beads are very soft and unsuitable to mechanical loading. Wong et al. [2001] investigate the conditions (formation of the gel, concentration, etc.) that enhance the compressive properties of alginate/chondrocyte constructs. The guluronic acid is observed to increase the apparent stiffness during rapid loading. However there is a certain optimal content of guluronic acid for which the construct is mechanically stable *and* the biosynthesis is promoted. They note that the presence of the chondrocytes and pericellular matrix have a quite small contribution to the mechanical properties. This issue might be due to the fact that, even after 7 weeks of culture, the constructs do not yet form an interconnected network like PGs and collagen do in native cartilage. Wong et al. [2001] conclude that some agents should be added to render the cultures appropriate for cartilage replacement. Alternatively, the chondrons and pericellular matrix could be extracted to be used in a second biomaterial.

In contrast to cultures in layers, the fact that alginate solubilizes easily makes it a tool to study the extracellular matrix in its aggregated state. Also while cultures of layers remain phenotypically stable only for a few weeks, chondrocytes cultured on agarose and alginate beads retain their phenotype and synthesize collagen type II and collagen type IX, which plays a role in crosslinking the collagen network and aggrecans, for more than 8 months, Häuselmann et al. [1994], D.A. Lee et al. [2003].

Cultures are usually fed with Dulbecco's modified Eagle's medium and with 10% to 20% fetal bovine serum (FBS) or fetal calf serum (FCS)[20.26]. Culture media incorporate radiolabeled [^{35}S]sulfate and [^{3}H]proline. Assuming equivalent incorporation of labeled and unlabeled sulfate and proline allows the measure of synthesis of GAGs and collagen.

The influence of **cell density** in the constructs is analyzed by Mauck et al. [2002]. They find that higher densities, 20 and 60×10^6 cells/ml of construct, increase tissue properties in unloaded cultures, but that mechanical loading induces an improvement of mechanical properties on the density of 20×10^6 cells/ml, not on the larger one.

Polymer scaffolds are commercially available: their actual properties may differ depending on the provider. A comparison of the biosynthetic activities of different scaffold materials (PGA and a benzylated hyaluronan), scaffold structures (woven, nonwoven, etc.) and culture systems (Petri dish versus rotating bioreactor) is presented in Pei et al. [2002]. Galban and Locke [1997] indicate what they consider to be advantages of the PGA scaffolds: (1) histological similarity to the chondrocytes which favors adhesion of the latter; (2) able to maintain the initial growth of the chondrocytes; (3) biodegradable; (4) high porosity which allows cells to grow and nutrients to diffuse.

Once it has been seeded, the construct is submersed in a container filled with nutrients. Besides the scaffold, an important parameter for the culture is the degree of steering that can be achieved so as to maintain a steady availability of nutrients: Vunjak-Novakovic et al. [2002] compare the influence of three systems, static flask, rotating flask and rotating vessel.

Successful proliferation of cells leads to a multiplication of the cell number by 20 to 40 in two weeks.

20.4.3 Experience gained from cultures

A number of laboratory cultures have been reported in the last decades. Here are some key features:
 - Baer et al. [2001] perform alginate cultures of intervertebral disc. They observe that the phenotype of the cells of annulus fibrosus is conserved but the compressive and shear mechanical properties of the constructs are lower than those of the alginate with no cells. Therefore the question remains whether cell cultures can

---

[20.26]The use of FBS poses two problems: first, its composition is extremely variable from batch to batch. Second, there is the possibility that pathogens (prions) from the animal infect the cartilage, Sittinger et al. [1999]. As one possibility, serum from the patient himself could used throughout the culture steps. However, substitutes are sought, Chaipinyo et al. [2002].

synthesize a matrix with mechanical properties similar to that of an *in vivo* disc. Note however that Baer et al. [2001] have not submitted their cultures to mechanical stimuli during their growth;

- however, even without mechanical stimuli, Buschmann et al. [1992] report that cultures in agarose scaffolds of articular chondrocytes were able to synthesize a mechanically functional ECM: the difference might be due to either the scaffold, agarose being more appropriate than alginate, or to the tissue itself, intervertebral disc versus articular cartilage;

- a method in two steps (Alginate-Recovered-Chondrocytes) that avoids the use of a scaffold (that has to be degraded after implantation) is described in Masuda et al. [2003]. First the chondrocytes are cultured in alginate beads during a week until they form a significant PG-rich matrix. The chondrocytes are then recovered with their matrix and maintained in a tissue culture insert with a porous membrane for another week. Biochemical analyses show that the chondrocytes keep their phenotype. The optimal period of 7 days of the first step has been obtained by trial and error. DNA content is multiplied by 4 over the two weeks. The ratio of the mass contents of PGs over collagen decreases in time but remains at a level about 10 times higher than native cartilage (which is 0.25). Cultures are fed with either 10% or 20% fetal bovine serum (FBS). No capital difference in the proliferation is noted between the two concentrations but manipulations of the tissue are easier with the larger FBS concentration. The method has been tested so far on bovine chondrocytes only;

- some experiments seem to imply that the mechanotransduction mechanisms of GAG synthesis and of cell proliferation may be different. Indeed, Lee and Bader [1997] and D.A. Lee et al. [1998]b, culturing chondrocytes extracted at different depths, observed that mechanical stimuli induce different responses in terms of GAG synthesis and cell proliferation: under dynamic compression, superficial cells accelerate their proliferation more than deep cells, while the GAG synthesis was more important in deep cells;

- what motivates the choice of agarose against alginate cultures? Mauck et al. [2000] indicate that, through dynamic compression, chondrocyte-seeded agarose disks have a significant increase in elastic modulus, much higher than chondrocyte-seeded alginate disks. The biosynthetic activities show similar trends, and GAG content is higher for chondrocyte-seeded agarose disks;

- agarose can not be used for human cultures because of foreign body reactions;

- Chaipinyo et al. [2002] seeded low density chondrocytes in 3D collagen type I gels. The culture contained 3 growth factors (TGF-β1, IGF-1, bFGF) and no serum. Chondrocytes are observed to proliferate but the synthesis of GAGs is much lower than in FBS. The collagen is also more loosely arranged than in FBS;

- there are numerous attempts to optimize the hydrogel scaffold. For example, Kisiday et al. [2002]a [2002]b obtain encouraging results in the biosynthesis of both GAG and collagen with a self-assembling peptide gel consisting of the 12 amino-acid sequence KLDLKLDLKLDL at a concentration of 0.5%;

- a typical example of questionable differentiation is the expression of α-smooth actin muscle by chondrocytes, Kinner and Spector [2001]. The phenomenon has been observed both *in vitro* and *in vivo* and it raises the question of the phenotype of the cells. The role of smooth actin muscle is unclear. One possibility is that it contracts the ECM in order to deliver nutrients to the cells. Alternatively, actin muscle may contract the ECM and reduce the pore space, so restricting cell proliferation. To minimize its level, Kinner and Spector [2001] suggest to limit the number of serial passages or to use cytokines;

- the cells may be introduced in non-woven scaffolds, directly or suspended in a gel, e.g., alginate, so as to prevent them to float out. The scaffolds are either put in contact with a culture medium in a Petri dish or grafted temporarily on nude mice or rabbits. Marijnissen et al. [2002] report that the suspension in alginate gels does not influence the production rate of cartilage matrix within 8 weeks of culture, but it keeps the graft shape.

20.5 Mechanobiology: experimental data

Mechanical loading affects the biomechanical properties of cartilage. Indeed, prolonged rest has been shown to lead to cartilage atrophy as well as to bone and muscle loss. Freed et al. [1997] report of cultures of cartilage cells over seven months under microgravity 10^{-4} to 10^{-6} g in the Mir Space Station. The space experiments kept the phenotype, synthesized GAGs and type II collagen more or less as on earth but the mechanical properties of the constructs were lower compared to earth cultures and the hydraulic conductivity was larger.

Martin et al. [2000] cultivated cartilage over seven months where the fluid is either at rest or undergoes turbulent or laminar flows. For laminar flows, they observe that the period of three weeks, that other experimental data report, corresponds to a sort of local maximum. At some later stages, growth returns and, after seven months, the moduli and GAG contents have increased considerably while collagen content plateaus at 6 weeks.

Mechanical stimuli have different effects in different materials. Simmons et al. [2003] applied equibiaxial cyclic straining of 3% at 0.25 Hz on bone cell substrates (an order of magnitude supraphysiological). The loading inhibited the growth of mesenchymal stem cells into osteocytic cells. It did not induce differentiation to adipocytic or chondrocytic lineages but it increased the deposition of mineralized matrix.

A priori, laboratory experiments may be thought to circle around physiological conditions. In practice, however, there may be considerable differences in load levels and nutrient concentrations. Confined compression tests, unconfined compression tests and shear apparatus are used to mimic some of the in situ loading conditions. Since experiments use mainly a compressive mode, they would a priori be expected to upregulate the agents which contribute most to the compressive properties of cartilages, namely GAGs but also collagen.

20.5.1   The evolutions of mass and mechanical properties correlate

Laboratory data report the relative composition of components, mainly GAGs and collagen, with respect to the wet weight of the tissue. These measurements should be compared with the in vivo evolution: the dry weight per current volume of tissue increases from about 100 mg/ml in the bovine third-semester fetus to about 180-200 mg/ml in the 1-2 year old bovine adult. Values for native bovine cartilages obtained by Williamson et al. [2001] are reported in Table 20.5.1.

Table 20.5.1: Standard values for native bovine cartilages, Williamson et al. [2001].

	PG Content (g/l)	Collagen Content (g/l)	Cell Content (g/l)	Confined Compression Modulus (MPa)	Axial Modulus (MPa)
Fetus	20-25	50	15	0.08	< 1
Adult	20-25	120	2	0.7	4-5

While the evolutions of the masses of collagen and PGs in cartilage cultures are of primary significance, the evolutions of the elastic moduli, hydraulic permeability and fixed charge are also of paramount importance.

Lai et al. [1981] assume the hydraulic conductivity to be strain dependent as $k/k_0 = (1 + e)^M/(1 + e_0)^M$ where e is the void ratio. Indeed, the compressive and tensile moduli of cartilage are presumably, and actually, in direct relation with the collagen and GAG content. The hydraulic conductivity is on the other hand in inverse relation with GAG content, Maroudas [1979]. Thus, as water is expelled during compression, the relative content of GAG increases and therefore the hydraulic conductivity is expected to decrease. These correlations imply that the mechanical and transport properties should vary in depth, since the composition of cartilage does.

Here are some in vitro data:

- Mauck et al. [2000] show that the mass of GAGs decreases after day 28 of cultures of cell-seeded agarose gels at 2 percent concentration under free swelling condition. Cyclic confined compression seems to alleviate this trend, but measurements do not cover a period that allows a definite conclusion;

- Mauck et al. [2002] show that the thickness of engineered constructs is enhanced by dynamic loading. However the increase in thickness is modest and plateaus at 10-15 percent early after a week. The percentage of collagen wrt wet weight is increased both for dynamically loaded and free swelling constructs;

- Martin et al. [2000] report of previous experiments where constructs were obtained in bioreactors that ensure an improved fluid motion. After seven months of culture, the content of GAGs raises to a value largely above standard, namely 10 percent of wet weight, the double of the value at six weeks. The collagen content plateaus at six weeks at about 3 percent of wet weight. The confined compression modulus jumps from 0.2 MPa at two weeks to 1 Ma at 7 months. Thus the modulus at 7 weeks is higher than the standard data by Williamson et al. [2001] while the relative and absolute contents of GAGs and collagen are quite different;

- Williamson et al. [2001] [2003] provide data on bovine cartilages at three stages of growth, which show that the aggregate compression modulus and tensile modulus correlate with collagen and pyridinoline contents.

At any rate, most, if not all, studies show a positive correlation between equilibrium moduli and the content of GAGs and collagen, e.g., Vunjak-Novakovic et al. [2002], Williamson et al. [2001] [2003].

20.5.2 Mechanosensitive candidates

A key issue about the mechanobiological translation concerns the way in which the cells are submitted to mechanical loads. *In vivo* and in scaffolds, the issue is not trivial. *In vitro*, the cells are seeded on the scaffolds which undergo the loads. How are the loads transmitted to the cells? *In vivo*, the cells are surrounded by a pericellular matrix (PCM) which is much stiffer than the chondrocytes. Does this PCM shield the load? or does it transmit its strain to the cells? A number of studies address the mechanical properties of the chondrons. A review of cell mechanics for cartilage engineering is presented in Shieh and Athanasiou [2003]. Mechanosensitive agents potentially active in cell regulation are listed in Section 24.4.1. Here is a list of mechanosensitive candidates thought to be involved in tissue engineering:

- **mechanosensitive membrane channels**. Chondrocytes are non excitable cells. However, they have a rest membrane potential, due to the existence of channels of various natures. Ligand-gated channels have been isolated, as well as ionic channels for Na^+, K^+ and Cl^-. Their inhibition presumably affects the proliferation, Wohlrab et al. [2002]. A stretch-sensitive K^+-channel has been identified in the membrane of porcine chondrocytes by Martina et al. [1997], and in endothelial cells of the vascular system by Hoger et al. [2002]. D.A. Lee et al. [2003] indicate that the glucose transporters are stretch-sensitive as well;

- **integrins**: they are trans-membrane receptors with two units α and β, but they are also parts of fibronectins which play the role of a molecular glue between the different components of the extracellular matrix. Integrins thus link the nucleus with the ECM, Fig. 20.1.3. According to Loeser [2002], some integrins in cooperation with IL-4 may transmit mechanical signals to the chondrocytes. A similar suggestion is made by Berry et al. [2003] for dermal fibroblasts: the mechanotransduction pathways would involve G proteins, IN_2P_3 and calcium concentration. Observing that cyclic straining stimulates the expression of collagen type II and aggrecan as well as of α_2-integrin, Lahiji et al. [2002] conclude that the latter mediate the response of chondrocytes to mechanical stimuli. According to Salter et al. [2002], $\alpha 5\beta 1$ integrin is the mechanoreceptor for both normal and osteoarthritic cartilage. **Elastin-laminin receptors** play a role similar to integrins in the arterial wall: cellular proliferation is observed to be stretch- and pressure-dependent, Spofford and Chilian [2003];

- **fibril-forming collagen-cell interface**: cells bind at specific sites to collagen fibrils. Shear of the collagen network is thus transmitted to the cell membrane, Silver and Bradica [2002];

- **modification of metabolism** (glycolysis) due to altered ionic and pH levels: permanent mechanical load expels water, and so increases ionic concentrations. Glycolysis, which produces energy necessary for sulfation, is very sensitive to pH. Moreover, glucose transporters are stretch-dependent. Therefore, permanent loads act as inhibitors of glycolysis and consequently of biosynthesis, M.S. Lee et al. [2002];

- **enhanced fluid flow and transport** of convected soluble nutrients and growth factors induced by the intermittent mechanical loading, Buschmann et al. [1999]. Convection is significant for large molecules. In addition, fluid flow may induce shear stresses at the cell surface, which are known to be an important factor in the growth of endothelial cells. High frequency fluid flow and shear stresses in haversian canaliculi in cortical bones are also suggested to elicit cell activation, L. Wang et al. [2000], Mak and Zhang [2001], Swan et al. [2003]. On the other hand, cyclic fluid flow has adverse effects, by convecting some synthesized monomer elements away from the cells, preventing their aggregation.

20.5.3 Experimental observations

Mechanical tests have been performed on both cartilage *explants* and *tissue-engineered* cartilages. While there are quantitative differences between the two and also according to the culture method, the effects of mechanical stimuli are qualitatively similar. A prolongated (static) load is usually detrimental to ECM synthesis while an intermittent (dynamic) load has positive effects. A possible reason for the beneficial effects of the intermittent loads is that they trigger fluid flow, which makes nutrients available to the tissue.

Mechanical loading modifies the shape of the cells, Lahiji et al. [2002], their Fig. 4D. Although the latter indicate that the cells tested have kept their phenotype, Masuda et al. [2003] mention that the rounded shape of chondrocytes is important for the production of a functional ECM. The interpretation of experimental data should account for the possibility of non uniform strain and stress within the tested specimens.

20.5.3.1 Bioreactors

1. Basics of a bioreactor:

The cell of many microorganisms may be considered to have the following chemical composition,

$$70\% \; H_2O + 30\% \; \text{dry part} \; CH_{1.8}O_{0.5}N_{0.2} \; .$$

Bathed in a medium, cells proliferate and produce extracellular matrix,

$$\text{cells} + \text{medium} + \text{oxygen} \; O_2 \rightarrow \text{cells} + \text{ECM} + \text{carbon dioxide} \; CO_2 + \text{water} \; H_2O \; .$$

Typically the medium contains glucose $C_6H_{12}O_6$ and ammonia NH_3 (source of nitrogen N). A typical growth reaction has the form,

$$C_6H_{12}O_6 + a \, NH_3 + b \, O_2 \rightarrow \alpha \overbrace{CH_{1.8}O_{0.5}N_{0.2}}^{\text{cells}} + \beta \overbrace{CH_xO_yN_z}^{\text{ECM}} + b \, r_Q \, CO_2 + \gamma \, H_2O \; .$$

Mass balance of each element yields the coefficients α, β and γ and r_Q:
- for C: $6 = \alpha + \beta + b \, r_Q$;
- for H: $12 + 3 \, a = 1.8 \, \alpha + x \, \beta + 2 \, \gamma$;
- for N: $a = 0.2 \, \alpha + \beta \, z$;
- for O: $6 + 2 \, b = 0.5 \, \alpha + y \, \beta + 2 \, b \, r_Q + \gamma$.

The respiratory quotient r_Q is easily measurable. Aerobic synthesis is higher than anaerobic synthesis.

2. Biochemical conditions for cultures:

The testing apparatus should address several requirements:
- ensure cellular viability and metabolism: the specimens should be contained in an incubator at a given temperature of $37 \, °C$, controlled pH, controlled oxygen tension, high humidity above 95% and 5% CO_2. The culture medium should be easily changed for cultures of more than two days;
- control the mass transport of water, nutrients, growth factors, metabolites, oxygen, etc.;
- avoid contamination;
- apply prescribed loading histories in terms of stress or strain, namely confined or unconfined compression or uniaxial tension, under given frequency. Interpretation of the results is much simpler if the specimen can be considered to be uniformly stressed and strained: such is not the case of the Flexercell system where a strain is imposed at the boundary but it is not uniform inside the specimen and neither does the stress;
- allow a statistical interpretation of the results: cellular variability is large. Therefore it is helpful to test simultaneously several specimens.

3. Mechanical conditions for cultures:

In-house bioreactors have been modified so as to allow for application of mechanical loading:
- Cacou et al. [2000] developed a system where six specimens can be submitted simultaneously to independent mechanical loads, which can be un/confined compression and uniaxial traction;
- Frank et al. [2000] have built an apparatus to apply simultaneously shear and compression to 12 cartilage discs. The discs of diameter 3 mm and height 1.1 mm are located sufficiently far from the axis of the apparatus (25.4 mm) in such a way the shear can be considered uniform;
- the loading device of Mauck et al. [2000] can submit simultaneously 16 discs (1.5 mm thick, 6.76 mm in diameter) to confined or unconfined compression. Unconfined compression may be closer to physiological conditions since it entails axial compression and radial traction despite the friction between the specimen and the platens that artificially increases the actual values of the moduli;
- the fluid circulating around chondrons exerts shear stresses whose influence on growth is thought to be important. As an example, Williams et al. [2002] study, through a finite volume method, the flow in a spinflask. The constructs are attached along the sides of a central fixed cylinder while the external cup is rotating at 4 rad/s so as to induce a steady laminar flow of the fluid between the cup and the central cylinder. Periodic switch of the rotation direction improves the uniformity of the shear stress in the constructs.

Cartilage zones have different composition and mechanical properties. Ideally, chondrocytes harvested from a given zone should produce *in vivo* ECM corresponding to this zone: the experimental issue is addressed in Klein et al. [2002].

4. Cultures of substitutes other than cartilage:

Articular cartilage is one of the many tissues that are attempted to be engineered. Issues similar to the ones encountered for cartilages arise for example to engineer the anterior cruciate ligament (ACL). Altman et al. [2002] have developed a bioreactor where ACL is generated from bone marrow stromal cells submitted to multidimensional strain cycles. The bioreactor contains 24 specimens using silk fiber scaffolds, each with its own control loop of axial and torsional strain.

Other bioreactors are developed to engineer knee joint tendons, Yamamoto et al. [2002], smooth muscles, cardiac muscles, blood vessels, etc.

20.5.3.2 Chondrons

The chondrocyte is surrounded by a pericellular zone, which together form a 'chondron'. The chondron is viewed as the main functional and metabolic unit of articular cartilage. The pericellular zone is composed of an inner matrix enriched with hyaluronan and of a fibrillar capsule.

The pericellular zone is a priori stiffer than the chondrocytes, Guilak et al. [1999], Knight et al. [2001]. Therefore the pericellular matrix shields the mechanical load from the chondrocytes. Thus the actual stress and strain histories of *in vivo* cells might be quite distinct from those which are attempted in bioreactors. The issue is crucial however because the metabolism of chondrocytes is believed to depend on its mechanical environment, through possibly stretch activated ionic channels. Bader et al. [2002] compare different techniques used to characterize the mechanical properties of chondrocytes. For isolated chondrocytes[20.27], they use atomic force microscopy (AFM) and micropipette aspiration. AFM consists in testing the cell surface by a triangular indenter. The second method consists in applying negative pressure at the surface of the cell through a pipette. In both cases, the cell is considered for analytical purpose as an elastic half-plane, and therefore its heterogeneity is not accounted for (the nucleus is known to be stiffer than the cytoplasm). Bader et al. [2002] also tested chondrocytes seeded in bio-polymer scaffolds: in contrast to the above surface techniques, here the whole volume of the cell was probed.

In a subsequent work, Alexopoulos et al. [2003] obtain results on the relative stiffnesses of ECM, pericellular matrix (PCM) and chondrocytes. They report a peak of Young's modulus in the PCM, namely 0.5 to 4 kPa for chondrocytes, 67 kPa for the PCM, 20 times that value for middle to deep ECM where the outer part of the PCM forms a sort of fibrous capsule, and about 70 kPa for the surface ECM. Thus, while in any case the stiffness of the chondrocytes is much lower than the one of both PCM and ECM, the stress shielding effect by the PCM depends on the depth. For osteoarthritic cartilage, they find a lower PCM Young's modulus of about 41 kPa. These results are obtained by micropipette aspiration of the chondron. For analytical purposes, the membrane is assimilated to an infinitely long straight layer of finite thickness submitted to a pressure loading over the diameter of the pipette. As for the chondron isolation technique, a new method (different from enzymatic digestion or mechanical homogenization) is used where the chondrons are aspirated through a syringe.

In a finite element analysis of a cartilage piece surrounding a chondrocyte with electrically open circuit, Lai et al. [2002] note that the diffusion potential and streaming potential[20.28] compete, rather than collaborate. For softer tissues, the diffusion potential dominates over the streaming potential, and conversely for stiffer tissues.

In summary, these analyses convey the idea that the PCM is a mechanically significant unit that mediates the actual loading of the chondrocytes. Relations between strain, stress, pressure and composition of the chondrocytes and the ECM, are addressed in Bush and Hall [2001].

20.5.3.3 Chondrocyte metabolism: effects of mechanical stress

The metabolism in chondrocytes is reduced essentially to glycolysis. Glucose is transported in the chondrocytes by the specialized carriers Glut, M.S. Lee et al. [2002][20.29]. Glucose is transformed mainly into lactate which is released into the extracellular matrix by a specific carrier. A very small part of the pyruvate enters the Krebs cycle and oxidative phosphorylation (OxPhos) that takes place in mitochondria: in fact, it seems that the mitochondria of chondrocyte lack certain cytochromes. So the release of CO_2 by the Krebs cycle and the need of O_2 for OxPhos

---

[20.27]Attention should be paid to the method of extraction of the cells. Knight et al. [2001] observe that mechanical extraction damages less the pericellular matrix than enzymatic digestion.

[20.28]The notions of diffusion potential and streaming potentials are developed in Section 16.2.

[20.29] The negative charge of lactate mitigates the mostly acidic environment of the ECM. Low pH is due to an inward flow of hydrogen ions through the cell membrane whose inner potential is negative (it plays the role of a cathode). H^+ ions are also consumed by HCO_3^- to form CO_2 and H_2O.

to produce energy (ATP) are both low. In average, only 2 ATP are gained for each glucose mole entering the cell against 32 ATP if the full metabolic process was taking place, Table 20.2.1.

Suppression of glycolysis also suppresses biosynthesis: this implication is in agreement with the fact that incorporation of sulfates onto PGs (sulfation) requires energy. On the other hand, the level of O_2 is also important, and an optimal concentration seems to be required as both hypoxia and hyperoxia are known to strongly decrease the biosynthesis.

High permanent mechanical loads reduce the biosynthesis, inhibiting glycolysis which as seen above produces ATP necessary for sulfation, M.S. Lee et al. [2002]. As a possible mechanism, a mechanical load expels water and thus increases the concentration of ions, in particular of hydrogen ions, and decreases pH. Now it turns out that phosphosfructokinase (a reaction in the chain of glycolysis pictured in Fig. 20.2.3) is very sensitive to pH. As a second mechanism, mechanical compression may close the glucose transporters.

Conversely Wilkins et al. [2002] and Mizuno [2002] report that hydrostatic pressure turns the cell more alkaline by stimulating the Na^+-H^+ exchanger. Cellular alkalinity stimulates in turn biosynthesis, Gray et al. [1988]. Wilkins et al. [2002] report in addition that a higher cytoplasmic concentration of calcium Ca_i^{2+} increases biosynthesis. It seems that hydrostatic pressure stimulates phospholipase C which stimulates the 2nd messenger InsP3 (Inositol 1,4,5 triphosphate) which in turn triggers calcium out of the sarcoplasmic stores. Now, how hydrostatic pressure acts on phospholipase C is unknown.

Wilkins et al. [2002] note that the Golgi apparatus does not move during mechanical loading. On the contrary, Grodzinsky et al. [1999] suggest that the mechanical load may deform the intracellular organelles that participate in the metabolism of cartilage. For that purpose they study whether modifications of the shape of the organelles lead to modifications of the synthesized ECM or not. They observe that static compression decreases the volume of the nucleus, whole cell and matrix in proportion. On the other hand, the Golgi apparatus and the rough endoplasmic reticulum (RER) shrink less and undergo anisotropic reorganization, Fig. 20.1.3. These organelles are particularly important because they participate to the synthesis and sulfation of GAGs. In fact, Kim et al. [1998] have noted that static compression is accompanied by longer chondroitin sulfate CS-GAG chains.

20.5.3.4   Effects of cyclic hydrostatic pressure on constructs

Angele et al. [2003] culture mesenchymal progenitor cells in a chondrogenic medium. Hydrostatic pressure from 0 to 5 MPa is applied, with a frequency of 1 Hz, either during 4 hours on day 1 and day 3, or 4 hours daily from day 1 to day 7. The cultures are then left to rest until day 14 or day 28. The load from day 1 to day 7 is observed to increase the collagen and PG content significantly at day 28: concentrations, 35 µg/µg DNA for PG and 70 µg/µg DNA for collagen, almost double with respect to the control unloaded culture. On the other hand, no significant modification in concentrations wrt control appears at day 14. This seems to imply that the biosynthetic effect (stimulation of synthesis of ECM) due to the mechanical stimuli requires a time period to develop.

Ikenoue et al. [2003] also report that the load duration has some effect on the aggrecan mRNA levels, e.g., a load of 5 MPa during 4 h for one day increases its level by 1.5 while the same load during 4 h for four days increases its level by 1.8. There is also some effect of the load level itself, a 1 MPa load being less efficient than a 5 MPa load, being itself less efficient than a 10 MPa load. Type II collagen mRNA seems to require a minimum load level (∼5 MPa) to be upregulated. The upper bounds (load duration and load level) are not reported.

The load of 5 MPa is in the range of the stress in human articular cartilage, namely from 3 to 10 MPa according to Mow et al. [1989]. Loads below this physiological range on the other hand stimulate catabolic effects in chondrocytes. Similar comparison between different loading sequences is described by other authors, e.g., Lane Smith et al. [2000].

The mechanical properties of the cultures are not reported by Angele et al. [2003] nor by Ikenoue et al. [2003].

20.5.3.5   Static versus cyclic compression on constructs

Davisson et al. [2002] perform confined compression tests on polyglycolic acid scaffolds for a loading period of one day. They find that static compression is detrimental, but that a 50% static strain plus cyclic loading, at an appropriate frequency of 0.1 Hz and 5% strain, has positive effects on both the production of GAGs and collagen.

For the static compression tests, the axial strain is controlled and it is imposed equal to 0% (control), 10%, 30% and 50%. Compression of 50% reduces the total protein synthesis by 35% and the GAG synthesis by 57%.

For dynamic compression, a 5% strain is superimposed on an average strain of 10% or 50%. Loading frequencies from 0.001 Hz to 0.1 Hz are tested. The synthesis of proteins increases by about 50%, and that of GAG of 10% when the offset strain is 10%. At the offset strain of 50%, the synthesis increases with loading frequency, by 100% for proteins and by 180% for GAG at the frequency of 0.1 Hz.

Mauck et al. [2000] subject chondrocyte-seeded agarose gels to intermittent unconfined compression with a 10% axial strain, during 56 days, at a frequency of 1 Hz, three times one hour on, one hour off per day, five days a week. At day 28, the modulus increases six times more than in absence of mechanical loading. GAG content departs positively from its unloaded value after day 14 only. However, although both modulus and GAG content are continuously higher than in absence of mechanical loadings, their relative increase shows a maximum around day 28, while Buschmann et al. [1992] observe continued GAG accumulation throughout 70-day cultures. Since hydrogels are quite compliants, application of strain of the order of 10% produces stresses less than 8 kPa much lower than physiological (1-10 MPa). The measured moduli smaller than 100 kPa were then expectedly lower than physiological (150-800 kPa).

As already mentioned, physiological conditions might well be optimal. Elder et al. [2001] test mesenchymal cells seeded in agarose scaffolds to unconfined compression with an axial stress varying cyclically from 0.25 kPa to 9.25 kPa. The frequency is varied from 0.03 Hz to 0.33 Hz, and the duration of the tests from 12 min to 2 hours. Higher frequencies, associated with lower cumulative strains, have a more stimulating effect on biosynthesis than lower frequencies. Also a minimum duration of the loading period is observed to be necessary to elicit biosynthesis.

20.5.3.6   Confined versus unconfined compression of constructs

With respect to confined compression tests, unconfined compression tests are associated with larger strains, lower stress levels and high fluid flows. They may be expected to mimic more closely the in situ conditions of the upper zone of articular cartilages.

In fact, the mechanical and fluid flow conditions clearly vary over the thickness of the cartilage layer. Still, association of any specific loading device with the actual in situ conditions in one of the three zones is far from guaranteed. Wong and Carter [2003] do their best to place published data in this context.

20.5.3.7   Effects of cyclic shear on constructs

Frank et al. [2000] show that low amplitude shear stresses applied permanently over 24 h increase the rate of ECM synthesis in articular cartilage cultures, the synthesis of collagen being more stimulated than that of PG. Shear is not expected to induce fluid flow, except at the edges of the discs. Biochemical analysis at different locations of the discs has shown a quasi-uniform metabolic stimulation. Therefore it seems that, at least in the experiments of Frank et al. [2000], deformation rather than fluid flow is responsible of the metabolic stimulation.

Waldman et al. [2003] stimulate cultures with a 2% shear strain amplitude applied with a frequency of 1 Hz for 400 cycles (6 mn) every 2nd day. The collagen and PG contents increased both by about 20% after one week, and by about 40% after four weeks. The compressive modulus increased 6-fold after four weeks. Shear strain larger than 2% was detrimental, while lower amplitude was favorable although the minimum amplitude is not indicated in the above work. By contrast, Emkey et al. [2002] note that there is a threshold of 2% above which shear strain stimulates collagen and GAG synthesis.

In opposition to Frank et al. [2000], Beaupré et al. [2000] state that shear stress is detrimental to cartilage and stimulates ossification. They develop a one-dimensional system containing 3 layers of cartilage attached to the bone. They define a osteogenic index as $\max s + k \min \sigma$ where s is the shear stress, σ the hydrostatic stress (tension positive) and k a parameter in the range 0.35-1.0. This index is used to predict the regions where ossification occurs and the regions which remain cartilaginous. In addition, the model implies that high contact stresses (low shear stress and compression) lead to a deep cartilage zone.

Raimondi et al. [2002] have performed computer fluid dynamics (CFD) simulations in order to simulate the fluid-induced shear stress due to a fluid in which chondrocytes are bathed. The setup is intended to simulate a construct where chondrocytes are seeded on the scaffold without gel. They find stresses of the order of 3×10^{-3} Pa for a flow rate of 0.5 ml/min.

20.5.3.8   Effects of electric/magnetic fields on constructs

Under extremely low frequency electric/magnetic fields, mesenchymal progenitor cells in endochondral bone formations show an accelerated chondrogenic differentiation and an acceleration in the synthesis of the extracellular matrix, Ciombor et al. [2002].

20.5.3.9 Nitric oxide (NO) production

The level of nitric oxide (NO), an intra- and intercellular messenger, is observed to be higher in osteoarthritic cartilage as well as in presence of rheumatoid arthritis. It is the result of the process catalyzed by the NOS (NO synthase) where L-arginine is decomposed in L-citrulline and NO. There exist three isoforms, two collectively noted as cNOS and one iNOS. The activity of cNOS is controlled by the binding of calcium to calmodulin. NO is produced by articular chondrocytes, endothelial cells, and osteoblasts inter alia, at various levels, in response to biochemical and mechanical stimuli. NO half-life is only a few seconds, so that it affects only the producing cell itself and neighboring proteins. Nitrite is a stable breakdown product of NO.

The mechanical and biochemical conditions that trigger and hinder NO production are not clearly identified:

- *static versus intermittent compression.* Fermor et al. [2001] indicate that intermittent compression triggers NO production more than static compression. This conclusion is questioned in other works. Chowdhury et al. [2001] indicate that dynamic compression is able to inhibit the NO production induced by IL-1β. According to D.A. Lee et al. [1998]a, dynamic compression inhibits NO production due to shear-stress (M.S. Lee et al. [2002]) and very high stresses (Kurz et al. [2001]). M.S. Lee et al. [2002] observe that continuously applied fluid-induced shear-stress (1.64 Pa) on human chondrocytes during periods of 2 to 24 hours increases more than twofold the production of NO, and this loading reduces aggrecan and type II collagen mRNA levels by about one third;
- *biochemical versus mechanical stimulation.* Fermor et al. [2001] find that cytokine stimulation *in vitro* (IL-1, TNF-α, IL-17) releases levels of NO from chondrocytes that are several-fold higher than mechanical stimulation. On the other hand, Takahashi et al. [2001] show that hyaluronan injection in an experimental osteoarthritis model inhibits the production of NO in meniscus and synovium.

The fact that intermittent compression stimulates cartilage biosynthesis and inhibits NO production agrees with the high level of NO in osteoarthritic cartilages. On the other hand, Humphrey [2003] notes that NO is a vasodilator: it therefore dilates the arterial wall and thus allows the turnover, and mainly deposition, of constituents. In that sense, it contributes to growth.

The role of NO in *in vivo* cartilage and cultures is not fully understood. It probably affects some mediators like metalloproteases and cytokines. D.A. Lee et al. [1998]a suggest that NO is a component of the mechanotransduction pathways that affect cell proliferation but not proteoglycan synthesis. This observation leads to the conclusion that the two phenomena are governed by distinct signaling pathways. The latter authors also mention that only the cNOS isoform is involved in chondrocytes. This observation, together with the fact that the cNOS activity is controlled by the binding of calcium to calmodulin, seems to imply the presence of stretch-sensitive calcium channels on the chondrocyte surface.

20.5.3.10 Adhesion of chondrocytes and integration of the transplant

The integration of engineered cartilage within the native tissue is an issue in itself. Inflammation of the transplant seems to be cured by TGF-β growth factor, Sittinger et al. [1999]. Autologous cartilage implantation requires that the injected chondrocytes adhere to the deficient cartilage. Detachment of the donor cells, rather than death of the cells, is one possible way by which implantation can fail.

Kurtis et al. [2003] find that the adhesion strength needs a couple of hours to get fully established. The resistance to flow-induced shear stress is about 20 Pa. In alginate cultures, chondrocytes are suspended in the alginate polymer but they do not *adhere* to the ECM as they would do *in vivo*. Adhesion of chondrocytes to cartilage is mediated by integrins. Blockers of integrins β_1, $\alpha_5\beta_1$ and $\alpha_v\beta_1$ decrease this resistance, which is a proof of the involvement of the integrins in the adhesion process. Cell attachment via integrins was tested by Genes et al. [1999]. They observed that the mechanical properties were improved but, correlatively, the synthesis of GAG and collagen was decreased.

Kurtis et al. [2003] also point out that the mode of culture of chondrocytes before implantation may play a role: for implantation, chondrocytes are cultured in low-density layers that lead to de-differentiation. Implantation reinitiates the chondrogenic ligneage. The fate of integrins in these manipulations is not known.

Integration of an implant to the host cartilage has been analyzed on practical grounds in Giurea et al. [2002]. They show that, to ensure adhesion between two pieces of cartilage, devitalization of one of the two pieces is a favorable means.

The influence of the duration of the culture in polylactic acid scaffolds has been tested by Giurea et al. [2003]. They observe that the longer the culture the stronger the adhesion of transplanted perichondrial cells when submitted to a flow velocity of 0.25 to 25 mm/s. Typically a one-week culture is associated with only 7% of the chondrocytes detaching from the polylactic acid scaffold while this percentage is 57% for a one hour culture. Increased cell adhesion over time is simultaneous with increased matrix content in both GAG and collagen type II.

On the other hand, Obradovic et al. [2001] note that immature constructs, which have lower mechanical properties, integrate better than more mature constructs or cartilage explants. Indeed, cells proliferate more intensely in the former, so as to form bonds between the engineered and native cartilages. Trypsin treatment of the adjacent native cartilage improves the integration: trypsin, a proteolytic agent, solubilizes GAGs but not collagen, and GAGs are known to inhibit adhesion[20.30]. Native cartilage re-synthesizes GAGs in a second step. As a conclusion, the construct should contain active cells capable of establishing bonds to ensure a successful integration[20.31].

20.5.3.11 Biosynthesis and cell density

Mauck et al. [2002] find that dynamic loading has a positive impact on biosynthesis when the cell density is low (20×10^6 cells/ml of medium) but becomes less significant at higher cell density (60×10^6 cells/ml). Kisiday et al. [2004] did not report such a positive impact at a comparable density of 15 to 30×10^6 cells/ml. Perhaps, a clear delineation should be attempted between proliferation, diffusion, and production of cells: these processes might well require different ingredients/excitations.

After a certain typical time, the production of GAG and collagen stops due to death of cells and, more, their content decreases due to their degradation.

For collagen, Kisiday et al. [2004] find very small effects of dynamic loading with respect to free swelling. Mechanical loading excites the cell proliferation and production but it also has a negative impact, their Fig. 6, and expels GAGs, Fig. 5. Dynamic compression excites the production of GAGs (which, in articular cartilages, are the agents that are expected to resist compression) but, since many GAGs are expelled, the net aggregated number of GAGs does not differ that significantly from the unloaded case.

20.6 Mechanobiology: models of growth of mass

It might be important to precise the terminology and distinguish the modes, growth, atrophy and structuration, through which biological processes like development, maintenance, aging and healing occur. In particular, growth is defined as the creation of mass, also referred to as mass deposition, and atrophy as the loss of mass. Growth may occur by change of the number of cells (*hyperplasia*), by change of the size of the cells (*hypertrophy*), or by change of the mass of the extracellular matrix. Structuration or re-structuration is usually understood to involve changes in microstructure and material properties in order to adapt to a new environment. Morphogenesis refers to the molding and shaping of tissues and organs, and concerns the relations between forms and biological functions.

Growth of engineered cartilage involves hyperplasia and creation of mass of the extracellular matrix, namely collagen and proteoglycans.

A first issue consists in modeling the growth of mass of the different constituents of cartilage, depending on the culture conditions. A second step consists in establishing correlations between mass and mechanical and transport properties.

20.6.1 Regular turnover of cartilage

DiMicco and Sah [2003] present a model that describes maintenance of articular cartilage. For rabbit cartilage, turnover times for GAGs range from days to weeks while they are much longer for collagen, from months to years or even lifetime. The numbers are species dependent, and they are much larger for humans.

[20.30]In the same line, according to Kurtis et al. [2001], treatment of cartilage construct by chondroitinase ABC is advocated to enhance chondrocyte adhesion to cartilage.

[20.31]Peretti et al. [2001] observe that isolated chondrocytes can glue to articular cartilage and bond pieces of cartilage matrix together.

20.6.2   The model of Bonassar and coworkers: scaffold degradation

C.G. Wilson et al. [2002] take into account the synthesis of the cartilage and the scaffold degradation. Degradation of the polyester scaffold by hydrolysis is modeled by a first order kinetics with a characteristic time τ_{sc} :

$$\frac{dm_{\text{sc}}(t)}{dt} + \frac{1}{\tau_{\text{sc}}}\, m_{\text{sc}}(t) = 0 \quad \Leftrightarrow \quad \frac{m_{\text{sc}}(t)}{m_{\text{sc}}(0)} = \exp\left(-\frac{t}{\tau_{\text{sc}}}\right). \qquad (20.6.1)$$

Components i of the ECM (collagen and GAGs) grow with a first order kinetics with product-inhibition, defined by a characteristic time τ_{i} :

$$\frac{d}{dt}\bigl(m_{\text{i}}(t) - m_{\text{i}}(\infty)\bigr) + \frac{1}{\tau_{\text{i}}}\bigl(m_{\text{i}}(t) - m_{\text{i}}(\infty)\bigr) = 0 \quad \Leftrightarrow \quad \frac{m_{\text{i}}(t) - m_{\text{i}}(\infty)}{m_{\text{i}}(0) - m_{\text{i}}(\infty)} = \exp\left(-\frac{t}{\tau_{\text{i}}}\right). \qquad (20.6.2)$$

A typical characteristic time for collagen and GAGs is 20 days. The total mass is found to decrease as the scaffold degradation overweighs biosynthesis.

This model is preliminary as it does not account for mechanical or biochemical effects.

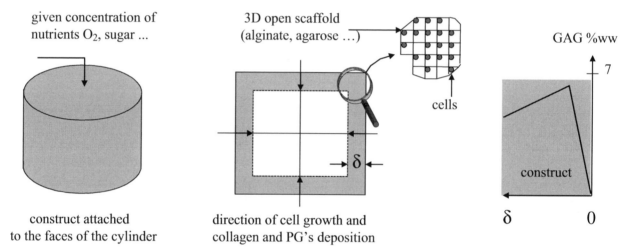

given concentration of nutrients O_2, sugar ...

3D open scaffold (alginate, agarose ...)

GAG %ww

cells

construct attached to the faces of the cylinder

direction of cell growth and collagen and PG's deposition

construct

Fig. 20.6.1: Model of cartilage culture. The scaffold is glued to the surface of a cylinder. The actual height of the cylinder is about 1 cm while the thickness δ of the scaffold varies from 1 mm to 3 mm. The pellet (cylinder + scaffold) is bathed in a medium which is steered so as to ensure a steady availability of nutrients all over the surface of the scaffold. Cell proliferation and cartilage deposition start from the outer surface in contact with the medium.

20.6.3   The chemical engineering model of Galban and Locke [1997] [1999]ab

Galban and Locke [1997] develop a mathematical model which incorporates
- the diffusion of nutrients;
- the yet unknown region where the cells grow: its surface is tracked via a moving boundary analysis. Two extreme cases are considered: in one case, the cells grow inward from the external boundary, in the second case, the cells grow parallel to the external boundary.

In a first attempt, favorable comparison with data requires that the diffusion and mass transfer coefficients be prescribed functions of the scaffold thickness δ, Fig. 20.6.1.

Galban and Locke [1999]b consider three cell kinetics involving cell death, cell density dependence, saturation and product inhibition, their Eqns (26)–(28). The generic kinetics addresses the variation of the mass of cells,

$$M_{\text{cell}}(t) = \int_{V_{\text{cell}}(t)} \rho_{\text{cell}}\, dV, \quad \frac{dM_{\text{cell}}(t)}{dt} = \int_{V_{\text{cell}}(t)} \rho_{\text{cell}}\, R_{\text{cell}}\, dV, \qquad (20.6.3)$$

through a rate of mass change which adopts a Michaelis-Menten format illustrated in Fig. 20.6.2,

$$R_{\text{cell}} = \frac{k_g\, (c_{\text{nu}})^n}{(c_h)^n + (c_{\text{nu}})^n} - k_d, \qquad (20.6.4)$$

with the nutrient concentration c_{nu} raised to some power n while the concentration c_h depends on the details on the kinetics:

$$\text{Modified Contois}: \quad c_h = c_{hc}\,(\rho^{\text{cell}})^{1/n},$$
$$\text{Moser}: \quad c_h = \text{constant},$$
$$\text{product} - \text{inhibition}: \quad c_h = c_{hc}\,(\rho^{\text{cell}})^{1/n}\left(1 + \Delta\rho^{\text{cell}}/c_{pr}\right)^{1/n}.$$

(20.6.5)

Here $\rho^{\text{cell}} \equiv n^{\text{cell}}\rho_{\text{cell}}$ is the cell apparent mass density with ρ_{cell} its intrinsic density, n^{cell} the cell volume fraction, k_g and k_d are growth and death constants, respectively. c_{hc} is a saturation constant for the Modified Contois kinetics while c_h is a constant for the Moser kinetics. The presence of the apparent density in the numerator of the Modified Contois kinetics implies density inhibition. In Galban and Locke [1999]a, product inhibition is introduced with $\Delta\rho^{\text{cell}}$ representing the change of the apparent density of the cells.

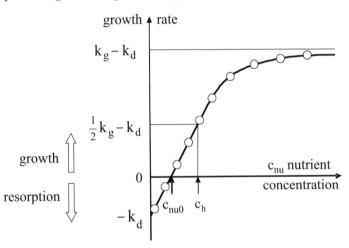

Fig. 20.6.2: Cell kinetics used by Galban and Locke [1999]b as a function of the nutrient concentration c_{nu}. The kinetics includes a constant death rate k_d, and a Michaelis-Menten growth rate. Death and growth balance for c_{nu} equal to $c_h/(k_g/k_d - 1)^{1/n}$. For $c_{nu} = c_h$, the growth rate is equal to $\frac{1}{2}k_g - k_d$.

The significance of product inhibition might be appreciated by the following observation. Nutrients will affect first the cells located close to the solution, cells located inside the construct will be reached in a time proportional to the square of their distance to the solution. Cell replication and mass deposition close to the border will further decrease the diffusion properties of the skin of the construct, and this effect will further deprive the inner part of the construct of nutrients. Product inhibition in the skin of the construct may moderate this effect. Therefore product inhibition may mitigate the heterogeneity of the engineered cartilage (at least in model simulations!).

Galban and Locke [1999]b also use an nth order *heterogeneous* kinetics

$$\text{nth order heterogeneous kinetics}: \quad R_{\text{cell}} = k_h\,\frac{(c_{nu})^n}{\rho_{\text{cell}}},$$

(20.6.6)

where k_h represents the net growth rate, including both growth and death.

No distinction is made between young and old cells, in contrast to Section 20.7.4 below. The intrinsic cell density is assumed constant in time and independent of the age of the cells, so that the rate kinetics addresses in fact the volume fraction of the cells.

20.6.4 The chemical engineering approach of the MIT and Cagliari groups

Martin et al. [1999] describe an experimental method to obtain the GAG distribution in cultured cartilage: the low magnification microscopy images of histological sections correlate with the wet weight fraction of GAG. The GAG fraction has a maximum at a typical distance from the external boundary, from 100 to 500 μm, as sketched in Fig. 20.6.1. They compare the growth of explants and engineered constructs, as well as three culture methods, namely static flasks, mixed flasks and rotating bioreactors, the latter being found more efficient and producing a continuous cartilage matrix.

Freed and coworkers stress the importance of the oxygen supply to the development of engineered cartilage: Obradovic et al. [1999] indicate that aerobic conditions result in higher amount of cartilage while anaerobic conditions suppress chondrogenesis, their Fig. 4. Oxygen is used in aerobic glycolysis to produce energy[20.32 20.33],

$$C_6H_{12}O_6 + 6\,O_2 \rightarrow 6\,CO_2 + 6\,H_2O + \text{energy}\,. \tag{20.6.7}$$

The initial conditions account for a 95% O_2 / 5% CO_2 air mixture at 1 atm and 37 °C calculated by Henry's law. Oxygen has a low solubility in water, so that it should be continuously supplied. Agitation thus creates shear stresses which may be detrimental to some cells. Carbon dioxide dissociation into carbonic acid may have detrimental effects as well because it decreases pH:

$$CO_2 + H_2O \rightarrow H^+ + H_2CO_3^-\,. \tag{20.6.8}$$

In a modeling attempt, Obradovic et al. [2000] and Pisu et al. [2003]a [2003]b use reaction-diffusion equations of the Turing type (15.4.6) where the two unknowns are the concentrations of oxygen and PGs within the tissue constructs,

$$\mathbf{X} = [c_{O_2}, c_{PG}]^T\,. \tag{20.6.9}$$

Thus collagen deposition is not accounted for. The cell kinetics is incorporated in an ad hoc way in Obradovic et al. [2000] and in a more elaborated method in Pisu et al. [2003]a [2003]b.

The reaction term in the reaction-diffusion equation for oxygen represents the *consumption* of oxygen according to a Michaelis-Menten reaction (catalyzed reaction $E + O_2 \rightarrow EO_2 \rightarrow PG$),

$$\frac{\partial c_{O_2}}{\partial t} = D_{O_2}\,\Delta c_{O_2} - \rho_{\text{cell}}\,\frac{Q_{O_2}\,c_{O_2}}{c_h + c_{O_2}}\,, \tag{20.6.10}$$

where the Michaelis-Menten coefficients Q_{O_2} and c_h (assumed to be spatially uniform) may depend on growth factors, etc. and $\rho_{\text{cell}} = 0.5 \times 10^6$ cells/ml is the cell density (the consumption of oxygen being proportional to the cell density). While here oxygen is considered as the rate-limiting substrate for cellular growth, other models consider that glucose is more appropriate. Nehring et al. [1999] use a similar equation.

The reaction term in the mass balance equation for PGs represents the deposition of PGs via a first order product-inhibited kinetics ($O_2 \rightarrow PG$),

$$\frac{\partial c_{PG}}{\partial t} = D_{PG}\,\Delta c_{PG} + \rho_{\text{cell}}\,k\left(1 - \frac{c_{PG}}{c_{PG}^{\max}}\right)c_{O_2}\,. \tag{20.6.11}$$

The model is ad hoc in the sense that the rate coefficient k is back-fitted as a function of time, following data by Obradovic et al. [2000].

To close the problem, the cell density ρ_{cell} is not constant but it is provided by an additional balance equation which includes cell division and cell death, their Eqn (8). Obradovic et al. [2000] extract an expression of the cell density from experimental data, their Eqn (11). In the analysis of Pisu et al. [2003]a [2003]b, the cells die and divide but do not differentiate, unlike in embryogenesis where differentiation is accounted for in some numerical models, see Section 15.4.6. The latter authors use the formalism of population balances. Let $\psi = \psi(\cup_{i=1}^n \xi_i, t)$ be a generic entity characterizing a population defined by n properties $\cup_{i=1}^n \xi_i$ with respective time rates of change v_i. Its evolution obeys the equation

$$\frac{\partial \psi}{\partial t} + \sum_{i=1}^n \frac{\partial(v_i \psi)}{\partial \xi_i} + D - B = 0\,, \tag{20.6.12}$$

[20.32]Some data:

- oxygen consumption by mammalian cells $\sim 140 \times 10^{-18}$ mole/cell/s, Cartwright [1994], by chondrocytes $\sim 1 \times 10^{-18}$ mole/cell/s, Obradovic et al. [2000];
- the concentration of dissolved oxygen is proportional to its partial pressure with a coefficient of solubility equal to 0.01 mmole/l/kPa=1.33 μmole/mmHg. In humans, the oxygen pressure varies between 100 mmHg (0.133 mmole/l) in the lung at oxygen intake, 40 mmHg (0.053 mmole/l) in the capillaries at oxygen release, and a few mmHg in the vena cava. At the surface of cartilage, it is about 50 mmHg and decreases to 10 mmHg at depth, Obradovic et al. [1999];
- glucose concentration in tissue 0.9 g/l, in cartilage constructs 3 g/l;
- saturation of CO_2 in medium 0.215 mmole/l, Nehring et al. [1999]. Standard cultures use a 5% CO_2/ humidified air atmosphere, i.e., 142 mmHg for O_2 and 38 mmHg for CO_2, Gooch et al. [2001];
- lactate concentrations in synovial fluid 0.13 g/l, in cartilage constructs 1.5 g/l, in bovine cartilage 0.18-0.36 g/l, Obradovic et al. [1999].

[20.33]Casciari et al. [1992] describe breakdown of glucose to lactate through glycolysis and the Krebs cycle,

$$C_6H_{12}O_6 \rightarrow 2\,C_3H_5O_3^- + 2\,H^+$$

and (20.6.7).

where $D > 0$ and $B > 0$ are the rates of death and birth respectively. If the age of the cells is not accounted for so that the population is mass-structured, then $\xi = m$ and $\psi(m, t)$ is the cell distribution function at time t,

$$\rho_{\text{cell}}(t) = \int_0^\infty \psi(m, t)\, dm\,. \tag{20.6.13}$$

Constitutive assumptions are needed for the time rates of change v_i and the death and birth rates. Since PGs are macromolecules, their transport is contributed by advection but the phenomenon is not accounted for.

The boundary value problem that is addressed considers a cylinder on the inner face of which is attached the construct. The concentration of PG is set to to vanish while the concentration of oxygen is prescribed over the construct/cylinder interface. These boundary conditions are supplemented by conditions on the axial and radial gradients of the unknowns which are set to vanish on the horizontal symmetry plane and on the symmetry axis respectively. As the medium is periodically replaced, the boundary conditions can be considered as steady.

There are two main practical methods of culture on these cylinders, Galban and Locke [1997]. The chondrocytes may be seeded on the external sides of the constructs, so that growth is directed inward. Alternatively, the chondrocytes may be spread inside the constructs so that growth is more or less isotropic.

The actual value of the initial cell seeding density matters much: a priori a large cell density should generate a large amount of tissue. Mauck et al. [2002] discuss the issue, see Section 20.4.2. The cell density ρ_{cell} duly appears in the reaction terms (20.6.10)–(20.6.11) since the chemical reactions in effect take place through the cells.

Typical profiles of the development of cells, collagen and PGs are shown in Obradovic et al. [2000], their Fig. 2: cell proliferation and concomitant deposition of collagen and PGs start at the construct's surface, which is in contact with the medium. Inner deposition comes later and is strongly dependent on the nutrient diffusion. Gentle but steady steering of the medium to ensure a steady availability of the nutrients is an important part of the efficiency of the process.

Both Obradovic et al. [2000] and Pisu et al. [2003]a [2003]b solve the boundary value problem by a finite difference method for two oxygen concentrations, 40 mm Hg and 80 mm Hg. In fact, Obradovic et al. [2000] observe that the change from 40 to 80 mmHg for O_2 results in a shift from anaerobic standstill to aerobic growth. Values of material parameters are extracted from the latter reference, their Table 1.

20.6.5 A two phase mixture model for articular cartilages

As an improvement over the above chemical engineering models, Sengers et al. [2004] introduce a framework for two-phase multi-species porous media. A number of interesting features are introduced in the computations that simulate a laboratory experiment where a specimen is subjected to specific biochemical and mechanical histories:

- the fluid is excited at the boundary: a time dependent velocity intends to mimic mixing;
- the reaction-diffusion equations for the nutrients and synthesized matter are similar to those of chemical engineering models. There are two species in the fluid: a nutrient and the synthesized matrix molecule. Only one component is synthesized. The cell density is constant;
- the solid and fluid phases do not change mass. The fluid diffuses through the solid. Thus the produced matrix molecules stay within the fluid, their mass is not aggregated to the solid;
- the solutes in the fluid undergo a reaction-diffusion process. These solutes are a nutrient 'nu' and a matrix molecule 'sm'. They diffuse with their own velocity through the fluid according to Fick's law. The rate of nutrient release by the fluid $\hat{c}_{\text{nu}}$ and the rate of matrix synthesis $\hat{c}_{\text{sm}}$ are given in the format,

$$\hat{c}_{\text{nu}} \sim -\frac{Q_{\text{nu}}\, c_{\text{nu}}}{c_{h,\text{nu}} + c_{\text{nu}}}\, \rho_{\text{cell}}, \quad \hat{c}_{\text{sm}} \sim \lambda_{\text{sm}}\, c_{\text{nu}}\, \rho_{\text{cell}}\,, \tag{20.6.14}$$

with c_{nu} and c_{sm} the concentrations of nutrient and synthesized matrix and ρ_{cell} the cell density. The material constants used are listed in Table 20.6.1. The production rate of matrix is proportional to the nutrient concentration and cell density. The release of nutrient by the fluid/uptake of nutrient by the cells is also given in an irreversible Michaelis-Menten format, once again proportional to the cell density. The steady state (no production rate) requires no cell or no nutrient;

- no influence of biosynthesized matrix on the mechanical and transport properties is introduced;
- all species are electrically neutral.

Cells produce monomers of PGs (or collagen), which are assembled in the matrix. These monomers are endowed with a relatively large diffusion coefficient. Consequently while water convection due to mechanical excitation provides more nutrients and growth factors to the cells, it also implies more monomers to diffuse out of the construct. In fact, only a small percentage of the matter produced by the cells is observed to remain in the construct after four weeks. Still, the model misses the formation of polymers, whose diffusion coefficients are smaller.

Table 20.6.1: Parameters of the Michaelis-Menten relation governing the nutrient release rate, $\hat{c}_{nu} = -\rho_{cell} \, Q_{nu} \, c_{nu}/(c_{h,nu} + c_{nu})$, with c_{nu} concentration of nutrient in mole per volume of fluid.

Nutrient	Medium	Reference	c_{nu} (mole/l)	ρ_{cell} (cells/ml)	Q_{nu} (mole/cell/s)	$c_{h,nu}$ (mole/l)
Glucose	Constructs	Sengers et al. [2004]	10^6	10^3	4×10^{-3}	0.1×10^6
BSA	"	"	"	"	4×10^{-5}	"

The release of the synthesized matter in the medium was noticed by Gooch et al. [2001]. Release is especially strong in presence of mixing (70% of GAGs) but it is strong as well under static conditions (60% of GAGs). The positive effect of mixing (more nutrients available to cells) and the negative effect (release of synthesized matter) result in a net negative effect on GAGs, but not on collagen whose mass tends still to slowly accumulate in time.

The data by Kisiday et al. [2002]a [2002]b [2004] are more optimistic: the percentage of GAGs which are lost in the medium is about one third during mixing, so that the net effect of mixing is positive in terms of mass of synthesized matter within the construct. The effect of mixing on the collagen content is less significant. Altogether, they indicate that dynamic compression results in improved synthesized mass.

20.7   Growth and structuration of the collagen network

Focus here is on the modeling of a fiber network, its orientation, the variation of its orientation as the mechanical load varies, the increase in its mass and its involvement in the mechanical properties.

Media-equivalents typically consist of type I collagen gel containing cells. The collagen fibril has a diameter of 0.05-0.5 µm which is small with respect to the cell dimension (50-100 µm). Fibroblasts are elongated prior and after compression, while chondrocytes are essentially spherical.

20.7.1   Chemical engineering aspects, the experiments by Tranquillo and co-workers

Nonlinearity and progressive collagen uncrimping and re-alignement

The initial nonlinearity shown by connective tissues in the direction of tension is believed to be due to the progressive uncrimping of the collagen. The structural elements that bear the load change progressively from the elastic fibers to the collagen network.

Tower et al. [2002] present an optical method (quantitative polarized light microscopy) to image the fiber alignment during mechanical testing. The method applies only to thin tissues. It uses the fact that collagen fibers are bi-refringent. When the fibers are aligned, the light entering the tissue undergoes a phase retardation.

Elastin

Elastin fibers are an important structural component of many biological tissues (heart valves, arteries etc.). They confer elasticity: in absence of elastin, creep leads to stenosis of arteries. Elastogenesis in tissue-engineered constructs is limited. Long and Tranquillo [2003] grow a 3D culture of smooth muscle cells in collagen gels: (1) cells are mixed with monomers of collagen I; (2) addition of NaOH entraps the cells in the hydrated collagen network; (3) the cells contract the network progressively, release crosslinked collagen fibrils and resorb the initial construct. The production of crosslinked elastic fibers and collagen is enhanced by treatment with TGF-β1 and insulin.

Glycation

Glycation is a process that creates crosslinks through a non enzymatic reaction between a reducing sugar and the lysine of collagen, without loss of cell viability. Somehow like elastin, glycation reduces creep and tends to confer elasticity to artificial constructs, Fig. 2 in Isenberg and Tranquillo [2003].

Glycated specimens have a high stiffness, but high strength is obtained only on low passage cells[20.34], Girton et al. [2000].

Cyclic loading on cultures

Isenberg and Tranquillo [2003] submit growing artificial materials to cyclic loadings (with various strain amplitudes, frequencies, etc.). In order to separate the effects of fiber alignment, they use compacted constructs with initially aligned fibers. They find that collagen based media-equivalents are stronger and stiffer when grown under cyclic distension. They mention that the amount of collagen created is not affected by cyclic loading but they observe a seldom reported significant amount of insoluble elastin.

Fibrin based cultures

The cultures using collagen fibers seem to give rise to too low stiffness and strength. One reason may be that cells are entrapped in the network and therefore exhibit little ECM synthesis. Grassl et al. [2002] and Neidert et al. [2002] report that the use of fibrin instead of collagen increases the production of collagen and the tensile strength by an order of magnitude. They cite evidence of collagen crosslinking as the source of mechanical strength and stiffness.

Stress- and strain-induced cell migration

Cell migration along an axis is observed to generate strain in the orthogonal directions, if the latter is stress-controlled. The strain, contraction or extension, depends on the cell type, according to Shreiber et al. [2003].

20.7.2 Cell and collagen fibrils alignment

Experiments by Girton et al. [2002]

Girton et al. [2002] submit a tissue-equivalent to compressive strain, the orthogonal directions being under confined compression. Cell morphology changes and cells and collagen fibrils are measured to align normal to the compression direction for a 50% strain. They attribute the alignment driving force to strain, not to stress, the latter being almost completely relaxed within the observation time interval.

In order to quantify the anisotropy, they define second order fabric tensors for the collagen fibrils and for the cells. These entities are distinct but they are related by a contact guidance parameter: cells align more than fibrils.

Experiments by Chandran and Barocas [2004]

Chandran and Barocas [2004] consider that collagen gels are endowed with a short term elasticity and long term viscosity, thus behaving like a Maxwell fluid, in their linear range.

While compression tests involve a priori both the collagen network and the fluid, extension tests probe only the collagen network. Tension in direction oblique to the main collagen orientation usually gives rise to a reorientation of the network. Glutaraldehyde crosslinking stiffens the network, decreases the viscoelasticity and prevents the reorientation.

To segregate time effects due to poroelasticity and viscoelasticity, characteristic dimensionless numbers are introduced as KHt/L^2 with K hydraulic conductivity, H representative elastic modulus, t experimental time scale and L sample length. Relaxation tests over 1800 to 2400 s are performed on samples compressed to 10% and the stresses are recorded together with a map of the orientation of the collagen network (using collagen birefringence). However the experiments are highly inhomogeneous and require an underlying constitutive model and a parallel computational approach to be cross-checked.

Micro-Macro approach of Agoram and Barocas [2001]

Agoram and Barocas [2001] develop a finite element model in which information on the collagen fibrils is contained within the elements, i.e., a multiscale approach. However, the actual efficiency of such methods is hampered by a number of practical issues, e.g., examples are 2D, etc.

Paradoxical data by Neidlinger-Wilke et al. [2001]

Not completely related to this section of collagen network, the paper of Neidlinger-Wilke et al. [2001] is nevertheless of interest. They consider a cellular network with well defined directions, the cells being either fibroblasts or osteoblasts. They plate the cells on dishes that are cyclically stretched to 4%, 8% or 12% along some direction, probably with zero stresses in the orthogonal directions. Initially the horizontal directional distribution of the cells is isotropic. At equilibrium, the main axis of the cells tend to orient away from the tensile direction: in this way, the strain in the cell axis is taking values either negative (contraction) or, anyhow, lower than the imposed stretch.

In a sense, this situation is similar to that of a deck of cards, whose axis (orthogonal to the plane of the cards) is oblique with respect to the traction axis: during a uniaxial traction, the axis is expected to become progressively

---

[20.34]Consider first a set of n cells. Grow the cells. After some typical period of time (before confluence), harvest n cells from the grown tissue. Restart the culture process. Each process is called a passage.

orthogonal to the traction axis and the plane of the cards tends to align with the traction axis. The situation is different for a fiber reinforced material. Then consider a fiber oblique with respect to the traction axis: it tends to align with the traction axis as the material is stretched.

Matrix-to-fiber and converse transformations

In a study of aortic valve, Boerboom et al. [2003] consider the re-structuration of a fiber tissue by synthesis of fibers from the extracellular matrix. They also address the converse degradation process. The rate of volume fraction of the fibers depends on the current volume fraction and stretch along the fibers: the synthesis and degradation contributions take care to keep the volume fraction between prescribed bounds.

As a follow up, Driessen et al. [2003]a consider an overall incompressible material composed

- of a matrix, with volume fraction $1 - \phi$, that sustains the stress $\boldsymbol{\tau}_m = \mu \left(\mathbf{F}^{\mathrm{T}} \cdot \mathbf{F} - \mathbf{I} \right)$, where $\mathbf{F}$ is the deformation gradient and μ the shear modulus, and

- of a fiber network, with volume fraction ϕ, that sustains a stress that vanishes if the deformation gradient is a pure rotation, their Eqn 8. The fiber stress involves the second order structural tensor $\mathbf{R} = \mathbf{F} \cdot \mathbf{S}_0 \cdot \mathbf{F}^{\mathrm{T}}$, where $\mathbf{S}_0$ is the initial value of the fabric tensor $\mathbf{S}$. The idea is to have the fiber rotating with the deformation, i.e., the fiber direction $\mathbf{e}_0$ changes to $\mathbf{e} \sim \mathbf{F} \cdot \mathbf{e}_0$. To attempt to eliminate the stretch, the structural second order tensor is defined as $\mathbf{R}/\mathrm{tr}\,\mathbf{R}$.

The total Cauchy stress is thus $\boldsymbol{\sigma} = -p\,\mathbf{I} + (1 - \phi)\,\boldsymbol{\tau}_m + \phi\,\boldsymbol{\tau}_f$. However, the fiber lattice does not evolve completely with the matter as indicated above. In fact, the rate equation for $\mathbf{S}$ would then be of the form $f(\dot{\mathbf{S}}, \mathbf{S}, \dot{\mathbf{F}}, \mathbf{F}) = \mathbf{0}$. Instead, a source term is added to the rate equation. $\mathbf{S}$ can no longer be interpreted exactly as a fabric tensor and this issue motivates a re-definition, their Eqn 20.

The volume fraction evolves according to an ad hoc first order kinetics, their Eqns 16–19, i.e., essentially according to the square of the stretch.

The two latter works probably assume an underlying growth process to take place although the phenomenon is not modeled. The modification of Eqn 14 in Driessen et al. [2003]a can be understood only in this context. Indeed, they do not address the reorientation of an existing collagen network under loading but the transformation of matrix to fibers.

Absence of viscosity in the transverse directions

Current studies seem to infer that both the elastic and viscoelastic properties of fibrillar structures are anisotropic. However, Lynch et al. [2003] report that the transverse apparent mechanical properties of tendons are rate-independent in contrast to the axial properties.

20.7.3   The partial model of Hunter and Levenston [2002]a

Hunter and Levenston [2002]a perform a finite element simulation of cylindrical discs containing an evolving gel-chondrocyte construct inside a cylindrical explant. The whole disc is submitted to a sinusoidal load and the construct/explant interface is perfect.

Previously, specific experiments, simulating biosynthesis of the construct, have been performed and the time evolutions of Young's modulus (increasing), hydraulic conductivity (decreasing) and fluid volume fraction (decreasing) have been recorded. These values are used to feed the finite element analysis: this approach dispenses with the constitutive equations of growth.

The simulations show that the fluid flow patterns are not much different in compressed and free swelling discs. The fluid pressures on the other hand are two orders of magnitude higher in compressed discs. Since there is a difference in synthesis between compressed and free swelling discs, Hunter and Levenston [2002]a conclude that pressure, rather than fluid flow, is crucial to the synthesis. This conclusion is at variance with Buschmann et al. [1999].

20.7.4   The approach of the Texas A&M group

In a biological milieu, any constituent k may have an individual growth rate as well as an individual growth tensor $\mathbf{F}_k^{\mathrm{g}}$, i.e., an individual intermediate configuration. This approach is expected to be more accurate, albeit more complex, than that adopted by Tözeren and Skalak [1988] who provide examples where the intermediate configuration is common to all constituents.

Various modeling issues are addressed in Humphrey and Rajagopal [2002] and Humphrey [2002] [2003]:

- the rates of mass changes of constituents are accounted for. Actually the rate of change is separated in a production rate and a removal rate. A first order kinetics is suggested as an attempt, that is, mass fractions are assumed to vary like in (20.6.1) for removal and like in (20.6.2) for growth, according to specific characteristic times. In fact, each constituent is endowed with two mass fractions, its initial mass fraction and an additional mass fraction, so as to allow the initial mass to be removed and the new mass to be deposited independently. The point is to account for the fact that mass removal and mass addition are usually due to different mechanisms, so that their time course evolutions are distinct. However, the way the problem is addressed seems to imply that the removal considers the mass of the initial material only, not the total current mass, as if the newly created material could not be removed. This approach leads to a question: should the old and new materials be considered as two distinct constituents?

- efforts are to be made to develop the constitutive equations for the growth rates of individual constituents depending on the mechanical and biochemical environments: these efforts concern masses but also the evolution of the mechanical properties (moduli, etc.). In fact, the first order kinetics to which the mass fractions obey, Eqn (6) of Humphrey [2002], or Eqns (20.6.1) and (20.6.2) above, are only temporary ersatz to feed the general mixture theory;

- the general approach involves two phases, a fluid phase and a solid phase. In a first attempt, the solid phase incorporates all species of interest, except the fluid of course. An alternative explanation is as follows. To simplify the formulation, momentum exchanges between these constituents are neglected[20.35], that is the velocities of all constituents are identical, Humphrey [2002], p. 535, Humphrey and Rajagopal [2002], p. 420, and a single balance of momentum is required. Therefore the theory is said "constrained". In fact, it shares similarities with "the strongly interacting model" that is developed for the intrafibrillar phase in articular cartilages in Chapter 14. A strongly interacting phase, or constrained phase, is a convenient tool when several constituents are intertwined or crosslinked, like actin filaments, intermediate filaments and microtubules inside the cells, Humphrey [2002]. The mixture theory is not completely developed in Humphrey and Rajagopal [2002] and the fluid is incorporated only through its pressure;

- the total stress is the sum of the partial stresses of the constituents: the latter are the product of the mass fraction by the intrinsic stress, Humphrey [2002];

- the deformation of all constituents are identical since their velocities are identical. On the other hand, each constituent has its own deformation decomposition and its own intermediate configuration as detailed in Section 20.7.5. The difference in the elastic deformations stems from the differences in the growth deformations. If there was no growth (parallel system), the elastic strains of all the constituents would be identical[20.36]. The actual growth rate laws are not specified. The practical question of specifying, or getting information through specific experiments on, the 'preferred' intermediate configuration of each constituent is mentioned in Humphrey and Rajagopal [2002], their Section 3.2. There are also suggestions, Humphrey [2002], that material growth might occur under optimal conditions, to be defined, e.g. "··· the cost of operation of physiological systems tends to be a minimum ···", Murray [1926]. While the principle seems appealing, the cost function remains to be defined. In essence, these authors are of the opinion that these intermediate configurations can not be measured directly, models have to be proposed, and the validity of proposals can be assessed only by their ability to represent accurately actual growth processes;

- concerning residual stresses, Humphrey and Rajagopal [2002] note that they are due to the fact that each constituent has its own intermediate configuration (and hence its own elastic strain, and hence its own stress while the strain is common to all constituent) and not to incompatible deformation as suggested by Skalak [1981];

- the stress of each constituent depends on its own elastic deformation, Eqn (2) in Humphrey [2002]. As a generalization, the total stress is given in the form of an integral of the intrinsic stress where the kernel is the mass rate of the component. The formalism summarized in Section 8.11.6 is reminiscent of a viscoelastic response with exponential kernels.

20.7.5 The approach of the UCSD group

The ideas are to be found in the papers by Hoger and coworkers, namely Rodriguez et al. [1994], Johnson and Hoger [1995], Hoger [1997], Klisch et al. [2001]a [2001]b, [2003]a [2003]b, [2005], van Dyke and Hoger [2001][2002], Lubarda and Hoger [2002] who have pioneered the use of the multiplicative decomposition of the deformation gradient in a growth context.

[20.35]Momentum exchanges are explicit if one writes the momentum balances of constituents. But they also come into picture in the generalized diffusion equations which are nothing else, in a different format, than momentum balances.

[20.36]Implicitly, with the deformation decomposition (20.7.1) is associated a mass decomposition, assuming density preserving growth. A generalization of Humphrey and Rajagopal [2002], their Eqn (9), accounting for diffusion, takes the form $\delta m = \delta m_e + \delta m_g$, with δm_e due to diffusion only and δm_g due to mass conversion (production and removal) only.

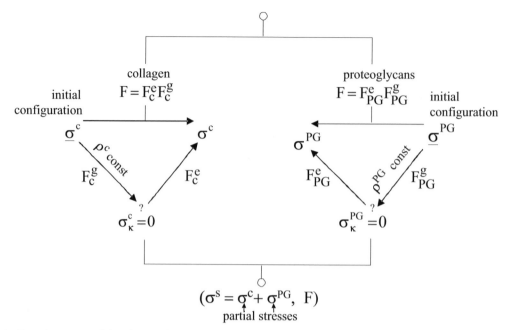

Fig. 20.7.1: In mixture models of growth, e.g., Humphrey and Rajagopal [2002], Klisch et al. [2003]a [2003]b, the constituents work in parallel. The strain is uniform over constituents and the partial stresses sum up. On the other hand, each constituent undergoes its own growth process, and the growth and elastic strains are innate to the constituents.

20.7.5.1 Gradient decomposition for mixtures

Here is a list of standard assumptions akin to mixtures:

(a1) collagen and PGs work in parallel, with the same total deformation gradient $\mathbf{F}$;

(a2) collagen and PGs are allowed to develop their own growth histories (mass deposition) but, since they live in the same cartilage (they undergo the same total deformation), they should accommodate, and the "elastic" strains which originate the individual stresses are innate to each component, Fig. 20.7.1,

$$\mathbf{F} = \mathbf{F}_k^e \cdot \mathbf{F}_k^g, \quad k = \mathrm{c}, \mathrm{PG}. \tag{20.7.1}$$

The apparent (partial) solid stress is the sum of the apparent stresses of the constituents:

$$\boldsymbol{\sigma}^s = \sum_{k \in \{\mathrm{c, PG}\}} \boldsymbol{\sigma}^k, \tag{20.7.2}$$

and each apparent stress $\boldsymbol{\sigma}^k$ is defined through the scheme (21.3.35) associated with its own transformation $\mathbf{F}_k^e$, $k = \mathrm{c}, \mathrm{PG}$;

(a3) in Rodriguez et al. [1994], the growth tensor was assumed to be symmetric. The possibility of rotations is considered. A particular case is analyzed in van Dyke and Hoger [2001];

(a4) the key point of the formulation of the growth laws in rate form may delineate mechanical and physiological influences (growth factors, nutrients ...):

- a possibility is to consider that *growth is nutrient-driven*, to use the Michaelis-Menten format in terms of a properly defined concentration, e.g., Obradovic et al. [1999], and to insert physiological influences in the parameters, as done by Galban and Locke [1999]a [1999]b for cell growth;

- another possibility is to consider that *growth is stress-driven*, see Section 21.7.1 for a review. There exists a (stable) reference stress state, called *homeostatic state*: deviation with respect to this state leads to growth or resorption. In other words, the material tries to adapt (its shape, mass) so as to retain as much as possible the homeostatic state. For example, the radial stress σ_{rr} in a tube of radius a and thickness e, loaded internally by a fluid at pressure p is equal to $p\,a/e$. Adaption to hypertension (increase of p) leads to increase of the wall thickness e;

(a5) as displayed in Fig. 4.3.2, the rate of mass supply to the species k is contributed a priori by other species of the mixture as well as by the surroundings;

(a6) a basic assumption is that the newly created material at a point is endowed with the kinematical, physical and mechanical properties of the existing material at that point, e.g., Klisch et al. [2001]a [2001]b.

(a7) still, the rate equations governing the mass evolution of the species k are *not* given an explicit form in the above publications, but examples are presented using data from their own laboratory experiments. A first assumption consists in assuming the growth transformation to be isotropic.

The above ideas are applied to model the growth of articular cartilages.

20.7.5.2 Data: PGs, collagen, crosslink density, cells, moduli from fetus to adult

Williamson et al. [2001] [2003] provide a detailed account of three growth stages of PGs, collagen and crosslinks from third-trimester fetal, 1-3 week old and adult bovine articular cartilages. They harvest cartilage layers from the patellofemoral groove and femoral condyle. They observe some spatial influences, which however do not affect the main trends. Data are reported with respect to the wet weight:

- the relative mass of water decreases little, from about 86% to about 83%;
- the apparent density of GAGs increases *little*, about $23 \, \text{mg/cm}^3$. In fact its percentage with respect to wet weight decreases from about 30 to 20 mg/g. GAG content is measured through ^{35}S-sulfate incorporation;
- the apparent density of collagen increases *much*. Its percentage with respect to wet weight increases from about $54 \, \text{mg/g}$ wet weight in the third-trimester fetus to $116 \, \text{mg/g}$ wet weight in the 1-3 week old calf to $196 \, \text{mg/g}$ wet weight in the 1-2 year old bovines. Collagen content is measured through hydroxyproline with a mass ratio of 7.25-7.30 of collagen to hydroxyproline. This ratio has been obtained for pepsin-solubilized collagen by Herbage et al. [1977];
- the pyridinoline crosslink density of collagen increases by about 730%. Two types of collagen crosslinks are delineated: lysyl-pyridinoline and hydroxylysyl-pyridinoline;
- the tensile modulus changes from less than $1 \, \text{MPa}$ to about $5 \, \text{MPa}$.
- the aggregate equilibrium modulus for confined compression increases from 0.1 to $0.7 \, \text{MPa}$;
- The aggregate moduli, both in compression and tension, correlate positively with collagen and crosslink density;
- the DNA content, a measure of the cell density, decreases considerably with maturation, from 1.01 to $0.35 \, \text{mg/g}$ of wet weight. The cell number is obtained from the DNA content through a ratio of $7.7 \, \text{pg/cell}$.

As a significant feature, the growths of PGs and collagen are not concomitant. PGs are produced more in the very early stages of development, while collagen is released progressively: the relative mass of collagen over PGs varies from about 2 in the fetus to about 4 to 5 in the adult. The release of pyridinoline is delayed with respect to that of collagen and it is thought to be a measure of cartilage maturity.

Klein et al. [2007] indicate that the compressive modulus of bovine fetal and newborn cartilage correlates positively with the GAG and collagen content, but *negatively with the cell density*: the latter decreases with depth in fetal and calf cartilage from 180×10^6 cells/ml in surface to 80×10^6 cells/ml at a depth of 0.5 mm.

The larger increase of collagen density during growth also correlates with the fact that the tensile properties increase more than the compressive properties, A.C. Chen et al. [2004]. In fact, anisotropic tensile properties are observed to develop with age, Kempson [1982], Woo et al. [1980] [1987]. By contrast, Williamson et al. [2001] and Klein et al. [2007] indicate that the compressive modulus doubles from fetal ($90 \, \text{kPa}$) to newborn ($200 \, \text{kPa}$) bovine cartilage, their Fig. 2c. The tensile properties are clearly more dependent on the collagen density, than on the density of proteoglycans. On the other hand, the compressive moduli are suggested to depend on the densities of both proteoglycans and collagen.

Finally, note that these data emphasize the relative proportions of components in the composition of a piece of cartilage at three stages of growth, but the three pieces are different so that the actual changes in mass during the growth are not reported.

20.7.5.3 Interpretation of data, Klisch et al. [2003]a [2003]b

The collagen and PGs undergo isotropic deformation $\mathbf{F}_k^e = \lambda_k^e \, \mathbf{I}$, $k = \text{c}, \text{PG}$. Therefore

$$\det \mathbf{F}_k^e = (\lambda_k^e)^3, \quad k = \text{c}, \text{PG} \, . \tag{20.7.3}$$

Since growth is assumed to preserve density, then

$$\det \mathbf{F}_k^{\mathrm{e}} = \frac{\rho^k}{\rho^k}, \quad k = \mathrm{c}, \mathrm{PG}. \tag{20.7.4}$$

The densities being measured, Table 20.7.1, the elastic stretch results, as well as the ratio between the growth stretches of collagen and PGs, since $\mathbf{F} = \mathbf{F}_{\mathrm{PG}}^{\mathrm{e}} \cdot \mathbf{F}_{\mathrm{PG}}^{\mathrm{g}} = \mathbf{F}_{\mathrm{c}}^{\mathrm{e}} \cdot \mathbf{F}_{\mathrm{c}}^{\mathrm{g}}$ or $\lambda = \lambda_{\mathrm{PG}}^{\mathrm{e}} \lambda_{\mathrm{PG}}^{\mathrm{g}} = \lambda_{\mathrm{c}}^{\mathrm{e}} \lambda_{\mathrm{c}}^{\mathrm{g}}$. Data of Table 20.7.1 yields

$$\frac{\lambda_{\mathrm{PG}}^{\mathrm{g}}}{\lambda_{\mathrm{c}}^{\mathrm{g}}} = \frac{\lambda_{\mathrm{c}}^{\mathrm{e}}}{\lambda_{\mathrm{PG}}^{\mathrm{e}}} = \underbrace{1}_{\text{fetus}}, \underbrace{0.8}_{\text{calf}}, \underbrace{0.75}_{\text{adult}}. \tag{20.7.5}$$

These numbers indicate that, taking the fetus as reference, collagen growth becomes more significant than PG growth at later stages. However, they do not provide information on the actual growth of either collagen or PGs.

Table 20.7.1: Data of apparent densities ρ^{PG} of PGs, ρ^{c} collagen and ρ^{w} water and γ_{c} crosslink density. The crosslink density is relative to the fetal configuration. Data used by Klisch et al. [2003]a [2003]b extracted from Williamson et al. [2001].

Group	ρ^{PG} (kg/m^3)	ρ^{c} (kg/m^3)	ρ^{w} (kg/m^3)	γ_{c} -
Fetal	22.2	87.6	963	1.0
Newborn	26	141	942	2.1
Adult	23.3	183	902	2.3

The partial Kirchhoff stress of collagen in the intermediate configurations obeys a linear constitutive equation,

$$\boldsymbol{\tau}_\kappa^{\mathrm{c}} = \lambda_\kappa^{\mathrm{c}} \operatorname{tr} \mathbf{E}_{G\mathrm{c}}^{\mathrm{e}} \mathbf{I} + 2 \mu_\kappa^{\mathrm{c}} \mathbf{E}_{G\mathrm{c}}^{\mathrm{e}} + \Gamma_\kappa^{\mathrm{c}} \mathbf{I} \tag{20.7.6}$$

This stress is next transported into the current configuration. The transport depends on the work-conjugate stress-strain pair. For the pair constituted of the 2nd Piola-Kirchhoff stress $\boldsymbol{\tau}_\kappa^k$ and the Green strain $\mathbf{E}_{Gk}^{\mathrm{e}}$, the Piola-Kirchhoff stress transforms to the Kirchhoff stress according to the elastic transformation,

$$\boldsymbol{\tau}_\kappa^k \xrightarrow{\mathbf{F}_k^{\mathrm{e}}} \boldsymbol{\tau}^k = \mathbf{F}_k^{\mathrm{e}} \cdot \boldsymbol{\tau}_\kappa^k \cdot (\mathbf{F}_k^{\mathrm{e}})^{\mathrm{T}}, \quad k = \mathrm{c}, \mathrm{PG}. \tag{20.7.7}$$

Confined compression tests are performed on the extracts (their duration is small such that growth does not take place during the mechanical tests). The strain applied during these tests is thus the same for PGs and collagen. The mechanical tests can be modeled through a superposition of small strain over the (possibly) large deformation due to the growth process in the animals.

The linearized moduli in the current configuration associated with the Cauchy stress are given in Section 2.9 in a particular case where the material response in the current configuration turns out to be isotropic, e.g., for collagen,

$$\boldsymbol{\sigma}^{\mathrm{c}} = \lambda^{\mathrm{c}} \operatorname{tr} \boldsymbol{\epsilon} \mathbf{I} + 2 \mu^{\mathrm{c}} \boldsymbol{\epsilon} + \Gamma^{\mathrm{c}} \mathbf{I}, \tag{20.7.8}$$

where

$$\Gamma^{\mathrm{c}} = \frac{\Gamma_\kappa^{\mathrm{c}}}{\lambda_{\mathrm{c}}^{\mathrm{e}}} + \frac{1}{\lambda_{\mathrm{c}}^{\mathrm{e}}} \left((\lambda_{\mathrm{c}}^{\mathrm{e}})^2 - 1 \right) \left(\frac{3}{2} \lambda_\kappa^{\mathrm{c}} + \mu_\kappa^{\mathrm{c}} \right), \quad \lambda^{\mathrm{c}} = \lambda_{\mathrm{c}}^{\mathrm{e}} \lambda_\kappa^{\mathrm{c}} - \Gamma^{\mathrm{c}}, \quad \mu^{\mathrm{c}} = \lambda_{\mathrm{c}}^{\mathrm{e}} \mu_\kappa^{\mathrm{c}} + \Gamma^{\mathrm{c}}. \tag{20.7.9}$$

The stress borne by the PGs is contributed by the isotropic component $\boldsymbol{\sigma}^{\mathrm{PG}} = \Gamma^{\mathrm{PG}} \mathbf{I}$. In addition, PGs are electrically charged and they give rise to an osmotic pressure π_{osm}. In this model, the osmotic pressure is viewed as exerting a fluid pressure $-\pi_{\mathrm{osm}} \mathbf{I}$ on the tissue components. The change of this contribution associated with the superposed small strain $\boldsymbol{\epsilon}$ is equal to $H_{\mathrm{osm}} \operatorname{tr} \boldsymbol{\epsilon} \mathbf{I}$. The modulus $H_{\mathrm{osm}} = -d\pi_{\mathrm{osm}}/d\operatorname{tr} \boldsymbol{\epsilon} > 0$ is deduced from the curve of Basser et al. [1998] that provides the osmotic pressure as explained in Section 14.4.1. Hence the aggregate modulus in the current configuration,

$$H_A = H^{\mathrm{c}} + H_{\mathrm{osm}}, \tag{20.7.10}$$

associated with the total stress $\boldsymbol{\sigma}^{\mathrm{c}} + \boldsymbol{\sigma}^{\mathrm{PG}} + \boldsymbol{\sigma}_{\mathrm{osm}}$.

Data are obtained from cartilages of animals at different stages of maturity. The aggregate moduli H and the stresses Γ are read from Table 20.7.2. The apparent Poisson's coefficients are taken from unconfined compression tests in Wong et al. [2000] as 0.09, 0.11 and 0.26 for fetal, newborn and adult specimens. Klisch et al. [2003]a [2003]b deduce that all collagen moduli in the current configuration correlate positively with the crosslink density.

The model is not completely satisfactory. The Lamé moduli in the intermediate configuration are not found to be positive, and the constitutive equation is phrased in terms of a partial stress, and does not include the water pressure, which is not equal to the pressure of the bath due to the presence of PGs.

Table 20.7.2: Data of stresses in the intermediate configuration and of moduli used by Klisch et al. [2003]a [2003]b deduced from Basser et al. [1998] and Williamson et al. [2001] for the aggregate moduli.

Group	Γ^{PG} (MPa)	Γ^{c} (MPa)	H^{PG} (MPa)	H^{c} (MPa)	H_A (MPa)
Fetal	-0.055	0.055	0.09	0.001	0.091
Newborn	-0.074	0.074	0.13	0.149	0.279
Adult	-0.078	0.078	0.17	0.139	0.309

20.7.5.4 Model of divergent growths of collagen and PGs

The two phase model of Klisch et al. [2001]a [2001]b, [2003]a [2003]b is developed in the following context:
(a1) they consider two phases, a fluid phase (inviscid fluid only, dissolved ions do not appear explicitly), and a solid phase defined by two growing constituents, the PGs and collagen (including noncollagenous proteins);
(a2) all solid constituents have the same velocity. However, Klisch et al. [2003]a [2003]b note that PGs are to a certain extent (20 to 40%) soluble and mobile. The stress tensor of PGs is assumed to be spherical; nevertheless, the pressure due to PGs may increase the collagen crosslinks and indirectly contribute to the shear resistance;
(a3) Klisch et al. [2003]a [2003]b impose the balance of momentum to all constituents. This requirement is probably too constraining. The point might be better treated in a theory of strongly interacting media where a single balance of momentum is enforced for a phase where all constituents have the same velocity;
(a4) growth/atrophy is defined as the mass deposition/removal while structuration involves changes in material properties, due to an evolving crosslink density or a re-alignment of the collagen fibers. The paper of Klisch et al. [2001]b describes not only *growth* but also *structuration*. The moduli of collagen are considered to change during growth as a function of a normalized crosslink density between collagen fibrils. For a numerical application, Klisch et al. [2005], the growth strains in PGs and collagen and the collagen crosslink are set a priori.

20.8 Use of light for tissue fabrication

A number of current techniques for fabricating tissue materials are reviewed in Tsang and Bhatia [2004]. They are
- using heat: the temperature of the construct is elevated above its glass value and pressure is applied;
- using light (photopolymerization, photocrosslinking, stereolithography);
- using adhesives, molding, etc.

Photoinitiation is the phenomenon by which chemical reactions are triggered at mild temperatures and physiological pH by light radiation. Therefore the concept is appropriate for tissue fabrication. It is effective for both radicals and ionic compounds. It is composed of two main steps,
- light emission and radiation to a light absorbing compound;
- ensuing chemical reactions.

The light radiations are mainly ultraviolet[20.37]. The energy absorbed is defined by the Beer-Lambert-Bouguer law (24.3.4). The entity in a photolabile compound, also called *photoinitiator*, that is responsible for light absorption is called the chromophore. Typically, there is a frequency at which absorption is maximum. Chromophores contain unsaturated groups like C=C, C=O, NO_2, etc. The excitation of chromophores involves moving a valence electron from a bonding to an antibonding orbital. On the other hand, thermal excitation involves translation and rotation motions. Once absorbed, the light energy may be used in various ways. An excited compound AB^* can, Fisher et al. [2001],
- dissociate into A+B;
- react with another compound C to ABC;
- luminesce to AB releasing the quantum energy $h f$, with h Planck constant and f frequency;
- ionize to AB^+ plus the charge of an electron e^-;
- decay without radiation to AB.

[20.37] Section 24.3.7.4 addresses photodynamic therapy where effective light radiations are mainly in the visible range.

These effects are often seen as initiating a more important sequence of reactions: a radical may be used to initiate a radical chain polymerization, and an ionic monomer to initiate an ionic chain polymerization. *Photopolymerization* concerns discrete chains of polymers while *photocrosslinking* concerns a polymer network.

Much effort has been devoted to fabricate hydrogels by photopolymerization, e.g., the hydrophilic polyethylene glycol (PEG), biodegradable scaffolds, hard tissues for dental and orthopedic applications. Light radiation of an encapsulated gel in a damaged cartilage has been shown to excite extracellular matrix production, Elisseeff et al. [1999]. The idea is to inject the compound in a soluble form and to make it insoluble by photopolymerization. The compound may be a biodegradable scaffold used for cell proliferation or for a temporary seal of leaking surface tissues after surgery.

20.9   Biochemical and mechanical factors: a synthesis

Immature articular cartilage contains vessels that circulate nutrients. On the other hand, adult articular cartilage is aneural, avascular and alymphatic. Nutrients from the synovial joint are transported by the extracellular fluid through diffusion and convection. Therefore the self-healing capacities of adult articular cartilage are practically nil. A method to treat damaged cartilage consists in removing the damaged part and in implanting biodegradable cell-polymer constructs. Engineered cartilages are cultured on scaffolds of agarose or alginate gels. Agarose and alginate gels are chosen because, polymerized, they display an open three-dimensional lattice structure that is prone to fix cells. Once in place, artificial cartilage proliferates reasonably well, but the issue is that its mechanical properties are lower than those of native cartilage. Recent studies have shown that the mechanical conditions to which chondrocytes are subjected *in vivo* and *in vitro* affect the synthesis and degradation of the extracellular matrix (ECM). In fact, the chondrocytes are much more compliant than the ECM. Still, the chondrocytes are surrounded by a stiff pericellular zone, and it is not elucidated if the ECM deformation carries over to chondrocytes.

Although experiments tend to quantify the variations in the rates of stimulation and inhibition of cartilage synthesis, the numbers should be taken with care. Indeed, the depth-distribution of strain in a cartilage layer submitted to compression at its top surface is highly inhomogeneous. Strains are much larger near the surface, and thus the percentage of tissue undergoing mechanical stimuli strongly depends on the layer thickness. These heterogeneities on the other hand may imply that there are optimal geometries for the loaded cultures. The biological environment should also be controlled, e.g., hypoxic conditions are also known to stimulate regeneration of cartilage.

Special bioreactors have been developed to apply unconfined compression on cartilage cultures. Unconfined compression, which is accompanied by lateral tension, is thought to correspond better to the *in vivo* state of stress. On the other hand, the loading programs are highly arbitrary and, in particular, the pros and cons of strain- over stress-controlled loadings are not elucidated. Therefore there are plenty degrees of freedom for optimization.

Mauck et al. [2000] loaded dynamically disks about 1 mm thick with a strain amplitude of 10%, at a frequency of 1 Hz, 3 times 1 hour on, 1 hour off, 5 days a week during 4 weeks. At the end of this period, the modulus is increased sixfold over unloaded disks. They also observed an increase in glycosaminoglycans and hydroxyproline (measurements use [^{35}S]sulfate and [^{3}H]proline radiolabel incorporation). However, the GAG content showed a peak at week 3, and further studies showed that it is later substantially reduced. At variance, Buschmann et al. [1992] observed an increase of GAG until day 70 on agarose-seeded calf chondrocytes. As another conclusion of the studies of Mauck et al. [2000], agarose gels improve the mechanical properties and matrix proliferation more efficiently than alginate gels.

The effects of shear-stress have also been tested. M.S. Lee et al. [2002] observe that fluid-induced shear-stress (1.64 Pa) application on human chondrocytes during periods of 2 to 24 hours increases more than twofold the production of nitric oxide (NO), an intra- and extracellular messenger present at increased levels in osteoarthritis, and this loading inhibits aggrecan and type II collagen mRNA levels by about one third. These observations confirm the alternative point of view, that cartilage degeneration and ossification are stimulated by shear stress but inhibited by dynamic compression, Carter and Wong [1990].

Chapter 21

Growth of soft tissues. Kinematics, formulation and examples

21.1 Natural growth and tissue engineering

The models developed here for deformation, mass transfer, generalized diffusion and growth enter the theory of thermodynamics of open mixtures. Before embarking in the details of the analysis, a short description of the physical phenomena that are intended to be addressed is in order. While the main domain of targeted applications is tissue engineering, the framework is thought to apply to the growth of biological tissues in a broad sense.

21.1.1 Mechanobiology and other points of view on biological growth

Among the many issues that the modeling of growth poses, some are rather fundamental, e.g.,
   1. What are the driving forces behind growth?
   2. What are the sensors able to detect the positive and negative signals?
   3. What are the effectors that make growth to really take place?

As far as the second and third questions are concerned, we may mention that several *mechanosensors* have been suggested, like *stretch gated ion channels*. While Wilkins et al. [2002] note that the Golgi apparatus does not move during mechanical loading, Grodzinsky et al. [1999] suggest that mechanical load may deform the intracellular organelles that participate to the metabolism of articular cartilages. In fact, they note that modifications of the shape of the organelles lead to modifications of the synthesized extracellular matrix (ECM). They observe that static compression decreases the volume of the nucleus, whole cell and matrix in proportion. On the other hand, the Golgi apparatus and the *rough endoplasmic reticulum* (RER) shrink less and undergo anisotropic reorganization. These organelles are particularly important because they participate to the synthesis and sulfation of GAGs. In fact, Kim et al. [1998] have noted that static compression is accompanied by longer CS-GAG chains.

We shall mainly concentrate on the factors that drive growth. Each discipline suggests its own favorite candidates:
1. Mechanicians advocate stress, strain, strain energy and strain rate, and insist on the importance of a side effect of growth, namely residual stresses, Carter et al. [1987], Fung [1990], Cowin et al. [1991], Omens [1998]. This approach has led to models that add or remove material at a geometric point, according to a certain growth criterion. A typical instance is the trilinear law for bone remodeling: Luo et al. [1995] describe the evolution of initial elementary forms for bone according to strain and strain rate. For soft tissues, the rate of mass deposition, or removal, is often taken proportional to a difference between certain components of the local stress and their homeostatic values, the proportionality coefficient containing information on the characteristic time scale of the process. Taber and Chabert [2002] include the influence of the deviation from the homeostatic state in terms of certain components of both stress and strain. Mechanotransduction via strain is suggested by the existence of stretch-gated ionic channels across the cell membrane in young mammals. *Integrins* are also suspected to transmit extracellular stimuli to the organelles. However, an argument against strain as a growth stimulus is that it depends on the reference configuration. For cardiovascular mechanics, Taber and Humphrey [2001] contend that growth relates better to stress than to strain, in part because isometric (active) contraction gives rise to muscle thickening. Actually, there is a real need

for substantiating the components and characteristics of the stress and strain involved in these relations. For bone mechanics, Cowin et al. [1991] and Cowin [1996] point out the role of the strain rate, which in fact is an attempt to characterize the time course of the strain, e.g. strain reversal, extremum values, average values, etc. and to account for viscous flow in channels. Specific details of the time course of strain and stress are thought to act subtly and are to be discovered. For the cardiac and aortic growth, Taber and Chabert [2002] highlight the end of diastolic and the end of systolic states of stress and strain that may affect diversely specific growth characteristics. In fact, mechanical loadings and biological processes correspond to largely different time scales, say the second and the couple of weeks $\sim 10^6$ seconds, respectively. On the other hand, even if the characteristic times of the diffusion processes (say the day) are expected to be smaller than those of growth processes, they are not separated enough so that they could be considered to act on non overlapping intervals of time;

2. Chemical engineers and biochemists insist on the availability and concentration of nutrients (glucose, oxygen), and of growth factors, Galban and Locke [1999]a [1999]b, Obradovic et al. [1999]. Standard tissue cultures use a 5% CO_2/humidified air atmosphere, that is $142\,\mathrm{mmHg}$ for O_2 and $38\,\mathrm{mmHg}$ for CO_2, Gooch et al. [2001]. Obradovic et al. [2000] observe that 40 to $80\,\mathrm{mmHg}$ for O_2 results in a shift from anaerobic standstill to aerobic growth and oxygen uptake by chondrocytes of about $1 \times 10^{-18}\,\mathrm{mole/cell/s}$;

3. Morphogeneticists suggest optimization principles, along the idea first formulated by Murray [1926], that material growth takes place under optimal conditions, "... the cost of operation of physiological systems tends to be a minimum ..." Still, such a cost function remains to be defined. The vascular network, at various scales, is supposed to develop so as to minimize the power required for the blood distribution, under a number of physiological constraints. Wolff's law suggests that bone adapts its microstructure and shape so as to be able to sustain mechanical loads with a minimal weight. Here are a few more recent attempts:

  - Taber [1998]c applies the idea to vascular morphology: the blood vessel diameter is such that the energy needed to drive the blood flow has to be minimum, see also Section 21.7.1;
  - for arterial growth, Rachev et al. [1998] assume that structuration occurs so as to keep the overall vascular compliance quasi-constant;
  - in a cylindrical model of early cardiac development, Arts et al. [1994] consider that structuration is obtained through the minimization of a cost function that involves relative and absolute variations of the sarcomere lengths and fiber orientations;
  - Lieberman et al. [2003] advance the hypothesis that Haversian structuration in cortical bone aims at maximizing strength over density.

Spatial distribution of stress might be another characteristic of arterial growth: the orthoradial stress with and without smooth muscle activation is quasi-uniform through the wall thickness, Chuong and Fung [1986].

In a restricted sense, growth is defined as the creation of mass while atrophy or resorption corresponds to mass removal, without modifications of the material properties, namely density, stress, energies. Intuitively, one might distinguish accretion on surface, Skalak et al. [1997], and volumetric growth, Skalak et al. [1996]. In fact, some growth processes occur both by surface accretion and in the volume. A typical instance is bone. Deposition of vapor to form thin films is a surface growth phenomenon. On the other hand, cartilage growth, either *in vivo* or engineered, is viewed as purely volumetric.

Growth may occur by change of the number of cells (*hyperplasia*), by change of the size of the cells (*hypertrophy*), or by change of the mass of the extracellular matrix. For example, growth of engineered cartilage involves hyperplasia and increase in extracellular matrix (collagen and proteoglycans). **Structuration** involves changes in microstructure and material properties in order to adapt to a new environment. Change of mass density is referred to as **remodeling**. A different terminology is used in the analysis of bones where remodeling is understood to include change of density and (re-)structuration mechanisms.

Morphogenesis covers the embryonic period and includes the history of the development of organs and living bodies, starting from primitive forms and ending with a functional unit. For example, Taber [1998]d reviews the mechanical influences that affect the development of the four-cavity pulsative heart from a peristaltic tube.

21.1.2   Cartilage engineering versus tumor growth

While immature articular cartilage contains vessels that circulate nutrients, adult articular cartilage is aneural, avascular and alymphatic. Nutrients from the synovial joint are transported by the extracellular fluid through diffusion and convection. The chondrocytes are not quiescent, but their activity is just sufficient for the maintenance of constituents during life. Typical turnover times are species dependent, from days-weeks for proteoglycans to months-years for collagen of rabbit cartilage, DiMicco and Sah [2003]. For humans, Muir [1983] indicates also that the turnover of proteoglycans is faster than that of collagen, but the numbers are much larger, typically a few years for proteoglycans while the half-life of collagen is about the life span. Elastin, which enters in particular the composition of arteries, is produced in the early development only and it is one of the most stable proteins of soft tissues. In fact, turnover rates may vary greatly during a lifetime. As an example, more than 80% of vascular smooth muscle cells (SMC) are replicated daily in the embryo, but this percentage decreases to 40% in the fetus, and to only a few percents in the young mammal and SMCs become quiescent in the postnatal period, with a replication rate of about only 0.06% in the adult, Stenmark and Mecham [1997].

A failing lymphatic network is thought to be the main cause of the high fluid pressure and of the limited clearance and high concentration of growth factors that exist in the tumor. This pressure hinders the advection of the drugs (macromolecules with inherently low diffusion coefficients) whose uptake by the cells is thus limited. In fact, approaches that attempt to increase the blood flow to tumors so as, in turn, to increase the drug delivery may be inefficient because they would also increase the interstitial fluid pressure, since the flow has no way out. The phenomenon is contributed as well by the tissue production which induces heterogeneous stress fields of compression and traction. These stress fields might depend strongly on the very type of tumors. Carcinomas grow on epithelia, namely skin, ovaries, etc. and constitute 90% of cancers. Sarcomas grow on mesenchymal structures, mainly fibroblasts and muscles. Carcinomas are attributed a more viscous behavior than sarcomas, Netti et al. [2000]: the incompatible strains due to cell proliferation, and the resulting residual stresses, are thus expected to have different time courses in carcinomas and sarcomas.

The nutrient transport to the center of polymer constructs is hindered by the production of the cellular matrix by the cells in contact with the medium: progresses are needed in actual cultures. A similar objective exists in tumors so as to improve the delivery of drugs (macromolecules). Unfortunately, irradiation increases the collagen content (the growth of collagen is therefore contributed in part by an exterior energy source) and decreases the permeability, Znati et al. [2003]. Thus, whether natural or engineered, biological growth typically starts over a certain surface in contact with nutrients, thus depriving the inner part of the volume of supply. Cultures of tumor spheroids larger than about 1 mm generate a central necrotic core, Sutherland et al. [1986]. Similarly, tissue engineering produces thin layers of constructs, which has prompted the development of bioreactors that enhance nutrient supply (oxygen, hormones, growth factors, pH regulation), eliminate metabolites and apply mechanical excitation, Martin et al. [2004]. While the first step of growth is mainly by volume expansion due to the proliferation of cells and their secretion, a subsequent step may involve diffusion and advection towards the surroundings. Moreover, in engineered cartilages, the stress environment is thought to have subtle effects, depending on the details of the loading (frequency, intensity). The confinement provided by gels in *in vitro* cultures of tumor cells has been observed to decrease the size of tumor spheroids with respect to spheroids grown in free suspension cultures, Helmlinger et al. [1997]a.

21.1.3 Scope

Emphasis is laid here on a thermodynamic derivation of the growth law based on the satisfaction of the internal entropy production. A distinction is made between

- **elastic-growing models** which make use of a growth strain, and
- **elastic models** which do not.

The elastic-growth decomposition is used in a mixture context, and for single solids as a particular case. Examples found in the literature are briefly reviewed in regards to this framework, namely for elastic solids,

- the models for bone remodeling of Cowin and Hegedus [1976], and Büchler et al. [2003],

and, for elastic growing solids,

- the model of MacArthur and Please [2004] for tumor growth,
- the models of Taber and co-workers for the growth and development of the cardiac muscle and of arteries, and the determination of the accompanying residual stresses,
- the model of growth of a spherical shell of Chen and Hoger [2000].

21.2 Residual stresses in elastic solids and material symmetries

In order to address residual stresses, it is useful to derive first some basic relations of general interest.

Consider a solid that occupies the volume Ω and which is assumed to be in static equilibrium with a given homogeneous traction $\mathbf{t}$ on the boundary $\partial\Omega$. The second order tensor,

$$\mathbf{\Sigma} = \frac{1}{\Omega} \int_{\partial\Omega} \mathbf{t} \otimes \mathbf{x} \, dS \, , \tag{21.2.1}$$

may be interpreted as a *macroscopic stress* in equilibrium with the traction $\mathbf{t}$ on $\partial\Omega$, namely $\mathbf{\Sigma} \cdot \mathbf{n} = \mathbf{t}$. Indeed, with help of the divergence theorem, Section 2.4.1.1,

$$\frac{1}{\Omega} \int_{\partial\Omega} t_i \, x_j \, dS = \frac{1}{\Omega} \int_{\partial\Omega} \Sigma_{ik} \, n_k \, x_j \, dS = \frac{1}{\Omega} \int_{\Omega} \left(\Sigma_{ik} \, x_j \right)_{,k} d\Omega = \Sigma_{ij} \, , \tag{21.2.2}$$

with $x_{,k}$ denoting a partial derivation with respect to the spatial coordinate x_k. The traction being given over the boundary, this macroscopic stress is also the average of the local stresses over the body,

$$\mathbf{\Sigma} = \frac{1}{\Omega} \int_{\Omega} \boldsymbol{\sigma} \, d\Omega \, , \tag{21.2.3}$$

since, again with help of the divergence theorem, $\Omega \, \Sigma_{ij} = \int_{\partial\Omega} t_i \, x_j \, dS = \int_{\Omega} \left(\sigma_{ik} \, x_j \right)_{,k} dS = \int_{\Omega} \sigma_{ij} \, dS$, provided the stress $\boldsymbol{\sigma}$ is in static equilibrium pointwise in Ω with vanishing body forces, namely $\sigma_{ik,k} = 0$, and with the traction $\mathbf{t}$ over $\partial\Omega$, namely $t_i = \sigma_{ik} \, n_k$.

The above formulas are phrased in a current configuration in terms of Cauchy stresses. Using the conservation of traction, Section 2.6, they may be written in a reference configuration in terms of 1st Piola-Kirchhoff stresses,

$$\underline{\mathbf{T}} = \frac{1}{\underline{\Omega}} \int_{\partial\underline{\Omega}} \mathbf{t} \otimes \underline{\mathbf{x}} \, d\underline{S} \, , \quad \underline{\mathbf{T}} = \frac{1}{\underline{\Omega}} \int_{\underline{\Omega}} \boldsymbol{\tau} \, d\underline{\Omega} \, . \tag{21.2.4}$$

Residual stresses correspond to a vanishing traction over the whole surface,

$$\mathbf{t} = \mathbf{0} \quad \text{on} \quad \partial\Omega \, . \tag{21.2.5}$$

Therefore, combining (21.2.1) and (21.2.3) shows that the average of the residual stresses over the body vanishes, namely

$$\int_{\Omega} \boldsymbol{\sigma} \, d\Omega = \mathbf{0} \quad \int_{\underline{\Omega}} \boldsymbol{\tau} \, d\underline{\Omega} = \mathbf{0} \, . \tag{21.2.6}$$

Hence, non zero residual stresses are spatially heterogeneous and depend on the geometry of the body.

For an elastic solid, invariance and respect of the material symmetries require the relation (2.11.26), namely $\mathbf{R} \cdot \boldsymbol{\sigma}(\mathbf{F}) \cdot \mathbf{R}^{\mathrm{T}} = \boldsymbol{\sigma}(\mathbf{R} \cdot \mathbf{F} \cdot \mathbf{R}^{\mathrm{T}})$, to be satisfied for all elements $\mathbf{R}$ of the symmetry group $\mathcal{S}$ of the solid, Coleman and Noll [1964]. Therefore in the undeformed state $\mathbf{F} = \mathbf{I}$,

$$\mathbf{R} \cdot \boldsymbol{\sigma}(\mathbf{I}) \cdot \mathbf{R}^{\mathrm{T}} = \boldsymbol{\sigma}(\mathbf{I}), \quad \forall \mathbf{R} \in \mathcal{S} \, . \tag{21.2.7}$$

For a solid which is isotropic in the residually stressed configuration, this constraint implies the stress $\boldsymbol{\sigma}(\mathbf{I})$ to be isotropic, namely $\boldsymbol{\sigma}(\mathbf{I}) = \sigma \, \mathbf{I}$. Pointwise equilibrium implies σ to be constant in space and the boundary condition (21.2.5) implies in turn the constant to vanish: thus the residual stress vanishes in an isotropic solid, whatever its form, Hoger [1985].

More generally, for an elastic solid which is orthotropic or transversely isotropic in the residually stressed body, the residual stress should be principal in the orthotropy axes which are generally position-dependent. The case of a right cylinder where one principal stress direction, say $\mathbf{e}_z$, is aligned with the axis of the cylinder and therefore independent of the coordinates is of particular interest. The decomposition of the stress $\boldsymbol{\sigma} = \boldsymbol{\sigma}_{\perp} + \sigma_{zz} \, \mathbf{e}_z \otimes \mathbf{e}_z$, is preserved by the divergence operator of static equilibrium, namely $\operatorname{div} \boldsymbol{\sigma} = \operatorname{div} \boldsymbol{\sigma}_{\perp} + \nabla \sigma_{zz} \, \mathbf{e}_z$, because $\operatorname{div} \boldsymbol{\sigma}_{\perp}$ has no component along $\mathbf{e}_z$. Then σ_{zz} should not depend on space and further, due to the boundary condition on the top and bottom surfaces, it should vanish. The remaining equilibrium equations express in terms the polar coordinates (r, θ), Eqn (2.4.84). If furthermore, the problem displays rotational symmetry about the axis $\mathbf{e}_z$, (independence with respect to θ), then the equilibrium along the tangential direction implies the stress component $\sigma_{r\theta}$ to be a constant, the latter in fact vanishing due to the lateral boundary condition. Equilibrium then boils down to the field equation $d\sigma_{rr}/dr + (\sigma_{rr} - \sigma_{\theta\theta})/r = 0$ in Ω and to $\sigma_{rr} = 0$ on the lateral boundary.

The above lines have not referred to a particular material symmetry. If the elastic solid was transverse isotropic in the residually stressed body, then, according to (21.2.7), the residual stress should be of the form $\boldsymbol{\sigma} = \sigma_{rr} (\mathbf{I} - \mathbf{e}_z \otimes \mathbf{e}_z) + \sigma_{zz} \mathbf{e}_z \otimes \mathbf{e}_z$, that is, $\sigma_{\theta\theta} = \sigma_{rr}$. Then static equilibrium implies that σ_{rr} should be spatially constant, and the boundary condition (21.2.5) on the lateral surface yields $\sigma_{rr} = 0$.

Actually, in a more general analysis, Hoger [1985] considers various material symmetries and bodies whose surface may or may not be regular. She shows that, if the axis of rotational symmetry of a transverse isotropic elastic solid is position-independent, then the residual stress should vanish irrespective of the shape of the body.

Incidentally, independence with respect to the angle θ does not mean $\sigma_{rr} = \sigma_{\theta\theta}$ while the assumption of transverse isotropy does. The distinction matters when it comes to residual stresses, Fig. 21.2.1. In fact, arteries display axial symmetry with respect to their axis but the arterial wall is traditionally modeled as a clinotropic solid which is reinforced by two families of fibers. The constitutive equations of clinotropic solids are detailed in Section 19.19.2.

The physical reasons that lead to the developments of residual stresses in bodies of engineering interest are often complex. For metallic and geological structures, they may be the result of plastic deformation. For biological organs, they are linked to the growth processes which are the main target in this and subsequent chapters. However, instead of addressing the complete problem including the generation of the residual stresses, one point of view is to develop constitutive equations starting from the (given and fixed) residual stresses and to include the latter as an additional item in the list of arguments of the standard strain energy (for hyperelastic

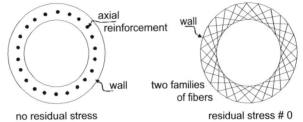

Fig. 21.2.1: Unlike an elastic clinotropic cylindrical tube, a transversely isotropic elastic cylindrical tube sustains no residual stress.

materials), Hoger [1993]. Alternatively, the constitutive stress may be contributed additively by the standard strain energy plus the residual stress.

The above discussion has considered elastic solids and the results do not apply to elastic-growing solids.

21.3 Kinematics of growth

21.3.1 Kinematics of growth in a solid and residual stresses

That mechanics is involved in the growth and structuration processes of biological tissues has been suggested for years. As early as in 1892, based on his observations as an orthopedic surgeon, Julius Wolff (1836-1902) suggested that the internal structure and external shape of bones adapt to both an increase and a decrease of the loads. As a corollary to Wolff's law of bone remodeling, Davis's law of soft tissues describes how protagonistic and antagonistic muscles thin and lengthen or thicken and shorten in response to posture. More recently has emerged the idea that the kinematics of growth of tissues is accompanied by residual stresses due to incompatible strains, Skalak [1981].

21.3.1.1 The multiplicative decomposition of the deformation gradient

The solid is used as a reference along to Biot's point of view, and the deformation gradient $\mathbf{F}$ is the deformation gradient associated with the motion of the solid. Reminiscent of the elastic-plastic decomposition of the total deformation gradient $\mathbf{F}$, e.g., Mandel [1971] [1973], a multiplicative decomposition into a growth transformation $\mathbf{F}^g$ and an elastic (accommodating) transformation $\mathbf{F}^e$ has been introduced by Rodriguez et al. [1994], as illustrated by Fig. 21.3.1-(a),

$$\mathbf{F} = \mathbf{F}^e \cdot \mathbf{F}^g . \tag{21.3.1}$$

The intermediate configurations (κ) implicitly introduced by this decomposition are defined to within a rotation. In Rodriguez et al. [1994], the growth transformation was assumed to be a pure deformation, any (left) rotation stemming from the polar decomposition being diverted to the elastic transformation. Should it be a rule, or are there definite counterexamples, as provocatively suggested in van Dyke and Hoger [2001], and what are the consequences for the rate equations that control the elastic or growth transformations? Similar to the elastic-plastic decomposition, issues of the uniqueness of intermediate configurations and invariance requirements have to be addressed. These investigations go beyond the present introduction and the reader interested by these issues is directed to, e.g., Skalak [1981], Skalak et al. [1996], Epstein and Maugin [2000], Lubarda and Hoger [2002], Cowin [2004]. The point is briefly touched upon in Section 21.3.1.3. In fact, the notions of directors, which are entities independent of material lines, and of isoclinic and invariant intermediate configurations, are central to define the orientation of the intermediate configurations, Mandel [1973].

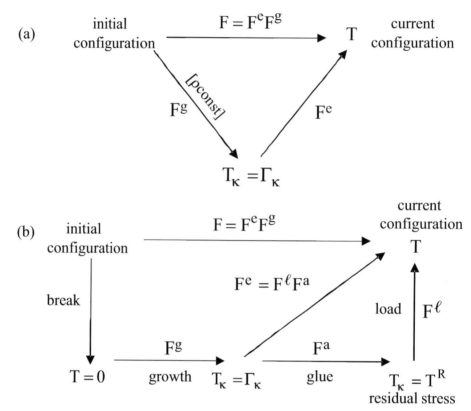

Fig. 21.3.1: Kinematics of growth according to Rodriguez et al. [1994]. The deformation gradient is decomposed in a growth part and an elastic part. The stress is due to the elastic transformation, through e.g. the Green strain $\mathbf{E}_G^e = \frac{1}{2}((\mathbf{F}^e)^T \cdot \mathbf{F}^e - \mathbf{I})$, and its value $\mathbf{\Gamma}_\kappa$ at $\mathbf{F}^e = \mathbf{I}$ may not vanish. The stress in the current configuration is obtained by a transport from the intermediate configuration via $\mathbf{F}^e$, Eqn (21.3.2).

Ideally, the intermediate configurations (κ) are stress free, although it might be difficult to reach such a situation in practice, Omens et al. [2003]. Cut the wall of an artery along a plane containing its axis, or the wall of the heart ventricles: the organ changes shape and the new configuration displays information on the residual stress, Fig. 21.3.2. Still, this new configuration might not yet be stress free. A second cut will deliver more information, and help approaching a stress free configuration, another one will do better, etc.

An understanding of the notion of residual stress is obtained by introducing the decomposition $\mathbf{F} = \mathbf{F}^\ell \cdot \mathbf{F}^a \cdot \mathbf{F}^g$, Klisch et al. [2001]a [2001]b, Taber and Humphrey [2001], where $\mathbf{F}^\ell$ is the transformation induced by the external load, and $\mathbf{F}^a$ is the accommodating transformation, that allows the material to keep its integrity, Fig. 21.3.1-(b). Indeed, in general, like plasticity, growth creates incompatibility. Still, in a simply connected region, an infinitesimal growth strain that is uniform or depends linearly on the space does not, Skalak et al. [1996] and Section 21.8.2.1. Incompatibility might be due to the strain field itself, or to the body shape, or to both. The elastic transformation arises so as to accommodate this incompatible motion, in other words, to restore compatibility of the total strain between neighboring points. The price to pay is a residual stress.

Thus the growth and elastic transformations are in general not the gradients of vector fields. Even so, they are routinely termed "gradients". Transformation, or tensor, is a more appropriate, although neutral, terminology.

Implicitly the constitutive scheme we have in mind is as follows: calculate the stress relative to the intermediate configuration and next transport it to the current configuration through some scheme via the elastic gradient:

$$\mathbf{T}_\kappa = \mathcal{T}_\kappa(\mathbf{F}^e) + \mathbf{\Gamma}_\kappa \quad \xrightarrow{\mathbf{F}^e} \quad \mathbf{T}. \tag{21.3.2}$$

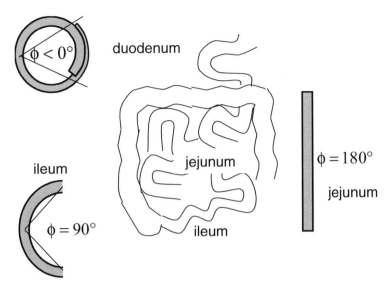

Fig. 21.3.2: Upon a radial cut along the axis, cross-sections of an unloaded intestine adopt quite distinct configurations. The inner surface (mucosal) is usually compressed while the outer (serosal) is usually under tension. The inner contractile strain is thought to protect against sudden large pulsatile flows that induce an inner extension: the initial contraction is thus expected to limit the possible damage by excessive extension. Dou et al. [2003] tested the three segments (duodenum, jejunum and ileum) of the small intestine, namely starting from the pylorus and ending at the ileo-cecal valve that communicates with the large intestine. They observe that the three parts of the small intestine undergo quite different residual strains: the jejunum returns to an almost straight configuration (ϕ=180°), the ileum has ϕ=90°, while the duodenum tends to turn inside ($\phi < 0°$). In relation with their physiological activities, the three segments have different elastic properties in both tangential and longitudinal directions.

Examples are reported at points (c3), (c4) in Chapter 23.

An interesting discussion of the kinematics of growth is presented in Cowin [1996]. Rachev et al. [1996] use this kinematics for arterial wall. Taber [1995] includes a review of modeling efforts up to 1995. A significant work on the development of heart is reported by Taber and Eggers [1996], Taber [1998]a [1998]b [2001], Taber and Chabert [2002], Zamir and Taber [2004].

21.3.1.2 Motion, mass supply and growth rate

The motion, which brings the point $\underline{\mathbf{x}} = (\underline{x}_i)$ in the reference configuration to its current position $\mathbf{x} = (x_i)$, is tracked by the deformation gradient $\mathbf{F}$ measured with respect to the reference configuration. The latter coincides with the current configuration at time $t = 0$, i.e., $\mathbf{F}(t = 0)$ is equal to the second order identity tensor $\mathbf{I}$. Entities and operators reckoned from the reference configuration are underlined, e.g., for the volume $\underline{V} = V(t = 0) \cdots$ Under the motion described by the gradient $\mathbf{F}$, an elementary volume $\underline{V}$ becomes $V = \det \mathbf{F} \, \underline{V}$.

The notion of mass content $m = M/\underline{V} = \rho \det \mathbf{F}$, which is commonly used in mixture theories, is also helpful to appreciate the effects of growth. Unlike the mass density $\rho = M/V$, it is defined as the mass M per *reference* volume. It differs from the density $\underline{\rho}$ in the reference configuration if growth takes place, and

$$M(t) = \rho(t)\, V(t) = m(t)\, \underline{V} \begin{cases} = \underline{M} = \underline{\rho}\, \underline{V} & \text{no growth;} \\[2mm] \neq \underline{M} & \text{growth.} \end{cases} \tag{21.3.3}$$

For solids, in contrast to mixtures, no diffusion is assumed to take place. The balance of mass can be cast in a format that highlights the fact that the time rate of mass content is equal to the rate of mass supply $\hat{\rho}$,

$$\frac{d\rho}{dt} + \rho \operatorname{tr} \mathbf{L} = \hat{\rho} \quad \Leftrightarrow \quad \frac{dm}{dt} = \hat{m} = \hat{\rho}\det \mathbf{F}\,, \tag{21.3.4}$$

with $\mathbf{L} = (d\mathbf{F}/dt) \cdot \mathbf{F}^{-1}$ the velocity gradient. An explicit quantification of the rate of mass supply $\hat{\rho}$ results in terms of the mass M, namely,

$$\frac{\hat{\rho}}{\rho} = \frac{d}{dt} \operatorname{Ln} M \quad \Leftrightarrow \quad \frac{M(t)}{\underline{M}} = \exp\left(\int_0^t \frac{\hat{\rho}}{\rho}\, d\tau \right). \tag{21.3.5}$$

If the mass density is constant, then

$$\frac{M(t)}{M} = \frac{V(t)}{V} = \exp\left(\int_0^t \frac{\hat{\rho}}{\rho}\, d\tau\right). \tag{21.3.6}$$

In view of the multiplicative decomposition of the deformation gradient (21.3.1), the velocity gradient is the sum of an elastic contribution and of a growth contribution,

$$\mathbf{L} \equiv \frac{d\mathbf{F}}{dt} \cdot \mathbf{F}^{-1} = \overbrace{\frac{d\mathbf{F}^{\mathrm{e}}}{dt} \cdot (\mathbf{F}^{\mathrm{e}})^{-1}}^{\mathbf{L}^{\mathrm{e}}} + \overbrace{\mathbf{F}^{\mathrm{e}} \cdot \underbrace{\frac{d\mathbf{F}^{\mathrm{g}}}{dt} \cdot (\mathbf{F}^{\mathrm{g}})^{-1}}_{\mathbf{L}^{\mathrm{g}}_\kappa} \cdot (\mathbf{F}^{\mathrm{e}})^{-1}}^{\mathbf{L}^{\mathrm{g}}}. \tag{21.3.7}$$

Note the relations,

$$\operatorname{div}\mathbf{v}_{\mathrm{s}} = \operatorname{tr}\mathbf{L} = \operatorname{tr}\mathbf{L}^{\mathrm{e}} + \operatorname{tr}\mathbf{L}^{\mathrm{g}}; \quad \operatorname{tr}\mathbf{L}^{\mathrm{g}} = \operatorname{tr}\mathbf{L}^{\mathrm{g}}_\kappa. \tag{21.3.8}$$

The constitutive equations are understood to be postulated relative to the intermediate configurations. For that purpose, the transports between the reference, intermediate and current configurations of the masses, volumes and partial densities are sketched in Fig. 22.2.1.

The transformation $\mathbf{F}^{\mathrm{g}}$ accounts for growth, so that the mass in the intermediate configuration (κ) is conserved by the elastic transformation $\mathbf{F}^{\mathrm{e}}$, i.e., $M = \rho V = M_\kappa = \rho_\kappa V_\kappa$, where V_κ is the total volume in the configuration (κ) obtained from the current volume by application of the inverse transformation $(\mathbf{F}^{\mathrm{e}})^{-1}$. Since $V = \det\mathbf{F}^{\mathrm{e}}\, V_\kappa$, then

$$\frac{\rho}{\rho_\kappa} = \frac{V_\kappa}{V} = \frac{1}{\det\mathbf{F}^{\mathrm{e}}}. \tag{21.3.9}$$

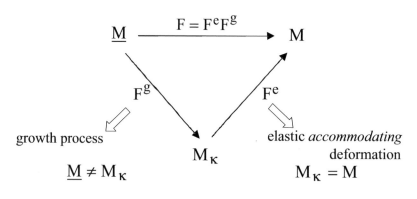

Fig. 21.3.3: Special cases of growth and elastic processes, for a single solid.

If growth may be considered to preserve density, namely $\underline{\rho} = \rho_\kappa = \rho \det\mathbf{F}^{\mathrm{e}}$, then

$$\frac{\hat{\rho}}{\rho} = \frac{d}{dt}\operatorname{Ln} M = \frac{d}{dt}\operatorname{Ln}\det\mathbf{F}^{\mathrm{g}} = \operatorname{tr}\mathbf{L}^{\mathrm{g}}_\kappa, \quad \frac{d\rho}{dt} + \rho\operatorname{tr}\mathbf{L}^{\mathrm{e}} = 0. \tag{21.3.10}$$

In other words, in these circumstances, the sole rate of growth transformation yields the aggregated mass. Otherwise, if mass density is not preserved during growth, its evolution should be, one way or the other, described by a constitutive equation, see, e.g., Section 21.3.5.3.

If the material is elastically incompressible, then

$$\rho_\kappa = \rho, \quad \det\mathbf{F}^{\mathrm{e}} = 1, \quad \operatorname{tr}\mathbf{L}^{\mathrm{e}} = 0. \tag{21.3.11}$$

Clearly, density preserving growth in an elastic incompressible solid implies ρ to remain constant, and

$$\frac{M(t)}{M} = \frac{V(t)}{V} = \frac{\det\mathbf{F}^{\mathrm{g}}(t)}{\det\mathbf{F}^{\mathrm{g}}(0)} = \exp\left(\int_0^t \operatorname{tr}\mathbf{L}^{\mathrm{g}}_\kappa(\tau)\, d\tau\right). \tag{21.3.12}$$

21.3.1.3 Intermediate configurations, isoclinic configurations and director vectors

The constitutive equations are provided in the format,

$$\frac{d\mathbf{F}^{\mathrm{g}}}{dt} \cdot (\mathbf{F}^{\mathrm{g}})^{-1} = \boldsymbol{\mathcal{L}}_{\kappa}(\mathcal{V}_{\kappa}) . \qquad (21.3.13)$$

Here the state variables relative to the intermediate configuration $\mathcal{V}_{\kappa}$ include mechanical dependence, namely stress, strain, etc. and physiological dependence, e.g., growth factors, nutrients, etc.

This format is tentative only because the derivative involved in (21.3.33) has to be objective with respect to changes of intermediate configurations. In fact, the intermediate configurations are assumed to be *isoclinic*, that is, they keep a fixed orientation. Indeed, the material which is considered is a *material with directors*: each geometrical point is endowed with a triad of director vectors. In general, *distinct* from the continuum, the latter may rotate, but also undergo in-axis and out-of-axis deformations. The motion of the body is defined by a point-to-point transformation measured by the deformation gradient, and by a transformation that describes how the directors evolve, Fig. 21.3.4. Therefore the elastic-growth decomposition applies to both transformations,

$$\mathbf{F} = \mathbf{F}^{\mathrm{e}} \cdot \mathbf{F}^{\mathrm{g}}, \quad \boldsymbol{\mathcal{F}} = \boldsymbol{\mathcal{F}}^{\mathrm{e}} \cdot \boldsymbol{\mathcal{F}}^{\mathrm{g}} . \qquad (21.3.14)$$

In the broader framework of *generalized continua*,
- the transformation of the continuum and the transformation of the director vectors are independent;
- an energetic approach would highlight two work-conjugate stress-strain pairs.

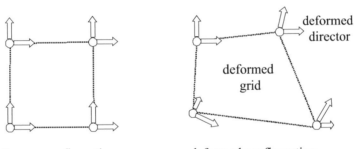

reference configuration   deformed configuration

Fig. 21.3.4: Concept of a *generalized material*: the motion of each geometrical point is defined by two a priori independent entities, the transformation of the continuum as measured by the deformation gradient, and the transformation of the director vectors. In general, the latter may both deform and rotate. In a *material with directors*, the transformation of the director vectors is distinct from the transformation of the material but it is a functional of the latter.

An exposition of the concept is to be found in Truesdell and Toupin [1960], Section 60 et seq. The *material with directors* developed below is a simplified version of a generalized material. Indeed,
- the couple stress work-conjugate to the transformation of directors is neglected;
- the time course of the director vectors is a functional of the transformation of the continuum.

A granular material may serve to illustrate the concept of a material with directors. The overall deformation of a granular material is due to the relative slips of the grains and to their own rotations. The orientation of the grains may be seen as defining the director vectors. The rotations of the micro-elements (the grains, in which the deformation is considered homogeneous) and the rotation of the macro-element (the granular medium) are distinct. As another example, consider the slip of intensity γ of a deck of cards of normal $\mathbf{e}_{\kappa 1}$ along the direction $\mathbf{e}_{\kappa 2}$,

$$
\begin{aligned}
\mathbf{L}^{\mathrm{g}}_{\kappa} &= \frac{d\mathbf{F}^{\mathrm{g}}}{dt} \cdot (\mathbf{F}^{\mathrm{g}})^{-1} = \dot{\gamma}\, \mathbf{e}_{\kappa 2} \otimes \mathbf{e}_{\kappa 1}, \\
\mathbf{L}^{\mathrm{g}} &= \mathbf{F}^{\mathrm{e}} \cdot \frac{d\mathbf{F}^{\mathrm{g}}}{dt} \cdot (\mathbf{F}^{\mathrm{g}})^{-1} \cdot (\mathbf{F}^{\mathrm{e}})^{-1} = \dot{\gamma}\, (\mathbf{F}^{\mathrm{e}} \cdot \mathbf{e}_{\kappa 2}) \otimes (\mathbf{e}_{\kappa 1} \cdot (\mathbf{F}^{\mathrm{e}})^{-1}) .
\end{aligned}
\qquad (21.3.15)
$$

Slip does not rotate the director $\mathbf{e}_{\kappa 2}$, but it does rotate the continuum. The skew-symmetric part $\mathbf{W}^{\mathrm{g}}$ of $\mathbf{L}^{\mathrm{g}} = \mathbf{F}^{\mathrm{e}} \cdot d\mathbf{F}^{\mathrm{g}}/dt \cdot (\mathbf{F}^{\mathrm{g}})^{-1} \cdot (\mathbf{F}^{\mathrm{e}})^{-1}$ represents the spin of the continuum with respect to the director vectors, Mandel [1973].

Since we assume that the intermediate configurations are isoclinic, the time derivative involved in the rate constitutive equation (21.3.13) is the material time derivative. It is worth stressing that using isoclinic configurations does *not* subsume the growth rotation to be equal to the identity. It does not prevent either the rate of growth transformation $\mathbf{L}^g$ to have a nonzero skew-symmetric part, as clearly shown by the special kinematics (21.3.15).

The next issue is to propose constitutive equations that describe how the cartesian triad of directors $\mathbf{e}_{\kappa 1}$, $\mathbf{e}_{\kappa 2}$ and $\mathbf{e}_{\kappa 3}$ transports from the isoclinic configuration to the current configuration through the elastic transformation $\mathbf{F}^e$. For example, each director may be assumed to keep a fixed unit amplitude, and, motivated by (21.3.15), lines along direction 2 are assumed to be material lines (relative to the elastic transformation $\mathbf{F}^e$), and lines along direction 1 are orthogonal to material lines, namely

$$\mathbf{e}_1 = \frac{\mathbf{e}_{\kappa 1} \cdot (\mathbf{F}^e)^{-1}}{|\mathbf{e}_{\kappa 1} \cdot (\mathbf{F}^e)^{-1}|}, \quad \mathbf{e}_2 = \frac{\mathbf{F}^e \cdot \mathbf{e}_{\kappa 2}}{|\mathbf{F}^e \cdot \mathbf{e}_{\kappa 2}|}, \quad \mathbf{e}_3 = \mathbf{e}_1 \wedge \mathbf{e}_2. \tag{21.3.16}$$

The time rates of these vectors may be expressed in the following formats,

$$\frac{D\mathbf{e}_1}{Dt} = \frac{d\mathbf{e}_1}{dt} - \left(\mathbf{W}^e - \mathbf{D}^e \cdot (\mathbf{e}_1 \otimes \mathbf{e}_1) + (\mathbf{e}_1 \otimes \mathbf{e}_1) \cdot \mathbf{D}^e\right) \cdot \mathbf{e}_1 = \mathbf{0},$$
$$\frac{D\mathbf{e}_2}{Dt} = \frac{d\mathbf{e}_2}{dt} - \left(\mathbf{W}^e + \mathbf{D}^e \cdot (\mathbf{e}_2 \otimes \mathbf{e}_2) - (\mathbf{e}_2 \otimes \mathbf{e}_2) \cdot \mathbf{D}^e\right) \cdot \mathbf{e}_2 = \mathbf{0}, \tag{21.3.17}$$

where $\mathbf{D}^e$ and $\mathbf{W}^e$ are, respectively, the symmetric and skew-symmetric parts of the elastic transformation rate $\mathbf{L}^e = \mathbf{D}^e + \mathbf{W}^e$. Actually, the spin $\mathbf{W}_{\mathrm{triad}}$ of the triad about itself,

$$\frac{D\mathbf{e}}{Dt} = \frac{d\mathbf{e}}{dt} - \mathbf{W}_{\mathrm{triad}} \cdot \mathbf{e} = \mathbf{0}, \quad \mathbf{e} = \mathbf{e}_1, \mathbf{e}_2, \mathbf{e}_3, \tag{21.3.18}$$

reads,

$$\begin{aligned}\mathbf{W}_{\mathrm{triad}} = \mathbf{W}^e \quad &+ \quad (\mathbf{e}_2 \cdot \mathbf{D}^e \cdot \mathbf{e}_1)(\mathbf{e}_1 \otimes \mathbf{e}_2 - \mathbf{e}_2 \otimes \mathbf{e}_1) \\ &+ \quad (\mathbf{e}_1 \cdot \mathbf{D}^e \cdot \mathbf{e}_3)(\mathbf{e}_1 \otimes \mathbf{e}_3 - \mathbf{e}_3 \otimes \mathbf{e}_1) \\ &+ \quad (\mathbf{e}_3 \cdot \mathbf{D}^e \cdot \mathbf{e}_2)(\mathbf{e}_3 \otimes \mathbf{e}_2 - \mathbf{e}_2 \otimes \mathbf{e}_3).\end{aligned} \tag{21.3.19}$$

To summarize this example, the director vectors are unaffected during the slip process, and they transport to the current configuration as indicated by (21.3.16), say, in essence,

$$\mathcal{F}^g = \mathbf{I}, \quad \mathcal{F}^e = \mathbf{F}^e. \tag{21.3.20}$$

Since $\mathbf{L}^e = \mathbf{L} - \mathbf{L}^g$, both the symmetric and the skew-symmetric parts of the growth rate $\mathbf{L}^g$ are required to be postulated by constitutive equations. In the deck of cards, the director vectors lie in the plane of the cards. For a continuum reinforced by a family of fibers, the director is parallel to the fibers. Both materials can be considered transversely isotropic.

Instead of the above multiplicative decomposition of the deformation, one may simply assume that particular directions in these materials (the plane of the cards, the fiber direction) transform like material lines through the complete gradient $\mathbf{F}$,

$$\mathcal{F}^g = \mathbf{F}^g, \quad \mathcal{F}^e = \mathbf{F}^e. \tag{21.3.21}$$

While the approach using multiplicative decomposition is more general, the two approaches may be shown to yield identical orientations of the directors for some pretty general kinematics when a parameter involved in the constitutive expression of the spin $\mathbf{W}^g$ is given a particular value, Dafalias [1984]. This spin is an essential ingredient of the formulation when the directors undergo significant changes of orientation (in the current configurations).

21.3.1.4 Non isoclinic configurations

The assumption of isoclinic intermediate configurations entails the possibility that the elastic transformation includes a rotation.

If, on the other hand, the intermediate configuration is defined in such a way that the elastic transformation is a pure deformation, the fact that it rotates reflects in the rate relations, if the time derivative is to be objective with respect to rotations of the intermediate configurations. Indeed, let κ_* and κ denote respectively an isoclinic configuration and a non isoclinic intermediate configuration deduced by the rotation $\boldsymbol{\beta}$ with rate $\boldsymbol{\omega} = d\boldsymbol{\beta}/dt \cdot \boldsymbol{\beta}^{\mathrm{T}}$, in symbolic form,

$$\mathbf{F} = \mathbf{F}^{\mathrm{e}}_{*} \overset{(\kappa_{*})}{\bullet} \mathbf{F}^{\mathrm{g}}_{*} \tag{21.3.22}$$

$$\mathbf{F} = \mathbf{F}^{\mathrm{e}} \circ \boldsymbol{\beta} \overset{(\kappa)}{\bullet} \boldsymbol{\beta}^{\mathrm{T}} \circ \mathbf{F}^{\mathrm{g}}$$

that is, in algebraic form,

$$\mathbf{F} = \mathbf{F}^{\mathrm{e}}_{*} \cdot \mathbf{F}^{\mathrm{g}}_{*} = \mathbf{F}^{\mathrm{e}} \cdot \mathbf{F}^{\mathrm{g}}, \quad \mathbf{F}^{\mathrm{e}} = \mathbf{F}^{\mathrm{e}}_{*} \cdot \boldsymbol{\beta}^{\mathrm{T}}, \quad \mathbf{F}^{\mathrm{g}} = \boldsymbol{\beta} \cdot \mathbf{F}^{\mathrm{g}}_{*}. \tag{21.3.23}$$

Then, the objective derivatives of the elastic and growth transformations adopt the formats,

$$\frac{D\mathbf{F}^{\mathrm{e}}}{Dt} \cdot (\mathbf{F}^{\mathrm{e}})^{-1} = \frac{d\mathbf{F}^{\mathrm{e}}}{dt} \cdot (\mathbf{F}^{\mathrm{e}})^{-1} + \mathbf{F}^{\mathrm{e}} \cdot \boldsymbol{\omega} \cdot (\mathbf{F}^{\mathrm{e}})^{-1}$$

$$\frac{D\mathbf{F}^{\mathrm{g}}}{Dt} \cdot (\mathbf{F}^{\mathrm{g}})^{-1} = \frac{d\mathbf{F}^{\mathrm{g}}}{dt} \cdot (\mathbf{F}^{\mathrm{g}})^{-1} - \boldsymbol{\omega}, \tag{21.3.24}$$

and

$$\begin{aligned}\frac{d\mathbf{F}}{dt} \cdot \mathbf{F}^{-1} &= \frac{d\mathbf{F}^{\mathrm{e}}_{*}}{dt} \cdot (\mathbf{F}^{\mathrm{e}}_{*})^{-1} + \mathbf{F}^{\mathrm{e}}_{*} \cdot \frac{d\mathbf{F}^{\mathrm{g}}_{*}}{dt} \cdot (\mathbf{F}^{\mathrm{g}}_{*})^{-1} \cdot (\mathbf{F}^{\mathrm{e}}_{*})^{-1}, \\ &= \frac{D\mathbf{F}^{\mathrm{e}}}{Dt} \cdot (\mathbf{F}^{\mathrm{e}})^{-1} + \mathbf{F}^{\mathrm{e}} \cdot \frac{D\mathbf{F}^{\mathrm{g}}}{Dt} \cdot (\mathbf{F}^{\mathrm{g}})^{-1} \cdot (\mathbf{F}^{\mathrm{e}})^{-1}. \end{aligned} \tag{21.3.25}$$

The rotations and pure deformations of the polar decomposition of the elastic and growth transformations,

$$\mathbf{F}^{\mathrm{e}} = \mathbf{R}^{\mathrm{e}} \cdot \mathbf{U}^{\mathrm{e}} = \mathbf{V}^{\mathrm{e}} \cdot \mathbf{R}^{\mathrm{e}}, \quad \mathbf{F}^{\mathrm{g}} = \mathbf{R}^{\mathrm{g}} \cdot \mathbf{U}^{\mathrm{g}} = \mathbf{V}^{\mathrm{g}} \cdot \mathbf{R}^{\mathrm{g}}$$

$$\mathbf{F}^{\mathrm{e}}_{*} = \mathbf{R}^{\mathrm{e}}_{*} \cdot \mathbf{U}^{\mathrm{e}}_{*} = \mathbf{V}^{\mathrm{e}}_{*} \cdot \mathbf{R}^{\mathrm{e}}_{*}, \quad \mathbf{F}^{\mathrm{g}}_{*} = \mathbf{R}^{\mathrm{g}}_{*} \cdot \mathbf{U}^{\mathrm{g}}_{*} = \mathbf{V}^{\mathrm{g}}_{*} \cdot \mathbf{R}^{\mathrm{g}}_{*}, \tag{21.3.26}$$

are related through the rotation $\boldsymbol{\beta}$ and their objective derivatives through the spin $\boldsymbol{\omega}$,

$$\begin{aligned} \mathbf{R}^{\mathrm{e}} &= \mathbf{R}^{\mathrm{e}}_{*} \cdot \boldsymbol{\beta}^{\mathrm{T}}, & \frac{D\mathbf{R}^{\mathrm{e}}}{Dt} &= \frac{d\mathbf{R}^{\mathrm{e}}}{dt} + \mathbf{R}^{\mathrm{e}} \cdot \boldsymbol{\omega}, \\ \mathbf{U}^{\mathrm{e}} &= \boldsymbol{\beta} \cdot \mathbf{U}^{\mathrm{e}}_{*} \cdot \boldsymbol{\beta}^{\mathrm{T}}, & \frac{D\mathbf{U}^{\mathrm{e}}}{Dt} &= \frac{d\mathbf{U}^{\mathrm{e}}}{dt} - \boldsymbol{\omega} \cdot \mathbf{U}^{\mathrm{e}} + \mathbf{U}^{\mathrm{e}} \cdot \boldsymbol{\omega}, \\ \mathbf{R}^{\mathrm{g}} &= \boldsymbol{\beta} \cdot \mathbf{R}^{\mathrm{g}}_{*}, & \frac{D\mathbf{R}^{\mathrm{g}}}{Dt} &= \frac{d\mathbf{R}^{\mathrm{g}}}{dt} - \boldsymbol{\omega} \cdot \mathbf{R}^{\mathrm{g}}, \\ \mathbf{U}^{\mathrm{g}} &= \mathbf{U}^{\mathrm{g}}_{*}, & \frac{D\mathbf{U}^{\mathrm{g}}}{Dt} &= \frac{d\mathbf{U}^{\mathrm{g}}}{dt}, \\ \mathbf{V}^{\mathrm{e}} &= \mathbf{V}^{\mathrm{e}}_{*}, & \frac{D\mathbf{V}^{\mathrm{e}}}{Dt} &= \frac{d\mathbf{V}^{\mathrm{e}}}{dt}, \\ \mathbf{V}^{\mathrm{g}} &= \boldsymbol{\beta} \cdot \mathbf{V}^{\mathrm{g}}_{*} \cdot \boldsymbol{\beta}^{\mathrm{T}}, & \frac{D\mathbf{V}^{\mathrm{g}}}{Dt} &= \frac{d\mathbf{V}^{\mathrm{g}}}{dt} - \boldsymbol{\omega} \cdot \mathbf{V}^{\mathrm{g}} + \mathbf{V}^{\mathrm{g}} \cdot \boldsymbol{\omega}. \end{aligned} \tag{21.3.27}$$

The derivatives $D(\cdot)/Dt$ are defined by convection as sketched in (2.11.18). Let $\mathbf{A}_{*}$ be a second order tensor defined in the isoclinic intermediate configurations and $\mathbf{A}$ its counterpart in non isoclinic configurations deduced from the former by the transformations $\boldsymbol{\beta}_{1}$ and $\boldsymbol{\beta}_{2}$. Then,

$$\mathbf{A} = \boldsymbol{\beta}_{1} \cdot \mathbf{A}_{*} \cdot \boldsymbol{\beta}_{2}, \quad \frac{D\mathbf{A}}{Dt} = \boldsymbol{\beta}_{1} \cdot \frac{d\mathbf{A}_{*}}{dt} \cdot \boldsymbol{\beta}_{2}, \tag{21.3.28}$$

and therefore,

$$\frac{D\mathbf{A}}{Dt} = \frac{d\mathbf{A}}{dt} - \frac{d\boldsymbol{\beta}_{1}}{dt} \cdot (\boldsymbol{\beta}_{1})^{-1} \cdot \mathbf{A} - \mathbf{A} \cdot (\boldsymbol{\beta}_{2})^{-1} \cdot \frac{d\boldsymbol{\beta}_{2}}{dt}. \tag{21.3.29}$$

These formulas also apply to a vector, setting formally one of the transformations $\boldsymbol{\beta}_{1}$ or $\boldsymbol{\beta}_{2}$ to the identity.

Like in (21.3.17) and (21.3.18), the objective derivative is set to zero if the vector $\mathbf{e}$ or the second order tensor $\mathbf{A}$ are of purely orientational nature,

$$\frac{D\mathbf{e}}{Dt} = \mathbf{0}, \quad \frac{D\mathbf{A}}{Dt} = \mathbf{0}. \tag{21.3.30}$$

Since the rotation $\boldsymbol{\beta}$ is arbitrary, it should not affect the free energy, namely

$$E_{\kappa}(\mathcal{V}_{\kappa}) = E_{\kappa_{*}}(\mathcal{V}_{\kappa_{*}}). \tag{21.3.31}$$

For materials which are isotropic in the intermediate configurations, Mandel stress results to be symmetric as indicated in Section 22.5.5. The consequences in presence of anisotropy are examined in Section 23.4.2.

21.3.1.5 Invariant intermediate configurations

The isoclinic intermediate configurations are defined as the configurations where the director vectors keep a fixed orientation.

An alternative consists in using intermediate configurations which do not rotate even if the current configuration does. With such configurations, referred to as invariant, the rate constitutive equations using the material derivative $d(\cdot)/dt$ are objective, like for the isoclinic configurations. But, at variance with the latter, the director vectors may not keep a fixed orientation, and they may be modified by the growth transformation, e.g., as indicated by (21.3.21).

21.3.2 Kinematics of growth in a mixture

The multiplicative decomposition (21.3.1) describes growth by a single kinematic entity $\mathbf{F}^g$. It does not account for the possible differential in growth between constituents of the biological milieu. A mixture theory is more appropriate. In an effort to account for this heterogeneity, Hoger and co-workers have begun to apply this kinematics to model the growth of articular cartilages, Klisch and Hoger [2003], Klisch et al. [2003]a [2003]b. In a similar mood, Humphrey and Rajagopal [2002] argue qualitatively that residual stresses in mixtures are due to the fact that each species has its own intermediate configuration, and hence its own elastic strain and its own stress, while the total strain is common to all species, along Section 21.3.3.

Henceforth, the species that grow, namely collagen and proteoglycans, are allowed to develop their own growth histories, and the "elastic" (accommodating) transformations which originate the individual stresses are innate to each component that grows[21.1],

$$\mathbf{F}_k = \mathbf{F}_k^e \cdot \mathbf{F}_k^g, \quad k \in \mathcal{S}^*. \tag{21.3.32}$$

The transformation $\mathbf{F}_k^g$ is related to the rate of mass conversion (acquired by mass transfer and/or growth), as described in Section 21.3.4. Rate equations governing the mass evolution of the species k have to be provided, say in the form

$$\frac{d\mathbf{F}_k^g}{dt} \cdot (\mathbf{F}_k^g)^{-1} = \boldsymbol{\mathcal{L}}_{\kappa k}(\mathcal{V}_{\kappa k}), \quad k \in \mathcal{S}^*. \tag{21.3.33}$$

Here $\mathcal{V}_{\kappa k}$ represents mechanical dependence, namely stress, strain, etc. and physiological dependence, e.g., growth factors, nutrients, etc. Like for a solid, the format (21.3.33) has assumed the intermediate configurations to be isoclinic. Modifications in presence of non isoclinic intermediate configurations follow identical guidelines.

The growth rate is in general assumed to be contributed by a rate independent part and a rate dependent part,

$$\frac{d\mathbf{F}_k^g}{dt} \cdot (\mathbf{F}_k^g)^{-1} = \frac{d\lambda}{dt} \boldsymbol{\mathcal{L}}_{\kappa k 1}(\mathcal{V}_{\kappa k}) + \boldsymbol{\mathcal{L}}_{\kappa k 2}(\mathcal{V}_{\kappa k}), \quad k \in \mathcal{S}^*. \tag{21.3.34}$$

The rate independent part is not accounted for, because growth processes are by nature not instantaneous. More exactly, their characteristic times are larger than the characteristic times of the other physical phenomena which are considered.

Rodriguez et al. [1994] present pros and cons of strain- and stress-driven growth, and opt for the latter, and so do Taber and Humphrey [2001] for arteries, and Lubarda and Hoger [2002] in a theoretical model. For the developing heart, Omens [1998], Taber and Chabert [2002] consider both stress and strain as mechanical stimuli of growth. Cowin [1996] highlights the difference in time scales between dynamic loading and growth. The load stimuli contain usually a variable (dynamic) component with time scale t_ℓ and a permanent component with a larger time scale, that may presumably be of the same order as the growth time scale t_g, i.e., $\mathbf{F}^\ell(t_g, t_\ell)$ and $\mathbf{F}^g(t_g)$. Small time scales will affect growth if the latter is sensitive to strain rate. In a distinct venue, Galban and Locke [1999]a [1999]b use nutrients as the main driving force of cell growth in polymer scaffolds, and so do MacArthur and Please [2004] in the analysis of tumor growth in an isotropic incompressible Maxwell fluid.

A possible method to visualize growth in an artery is to perform a cut along its axis. Still, as alluded for above, the resulting state of stress might not vanish, and many (circumferential or radial) cuts might be necessary to reach a stress free state. The developments of Klisch et al. [2003]a, based on their own cartilage cultures, are in some sense reminiscent of this issue. Schematically, they proceed as follows: they calculate the stress relative to the

---

[21.1]Warning: As a departure from the standard notation used in this work, the reference to the species k appears as a lower index, although the deformation gradient $\mathbf{F}_k$ is an apparent quantity.

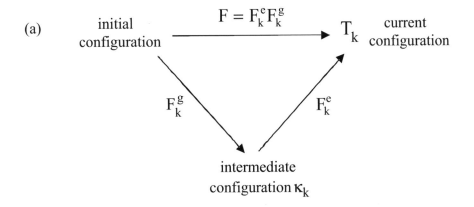

(a)

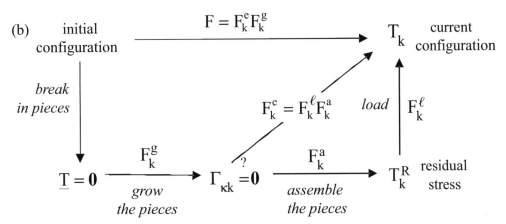

(b)

Fig. 21.3.5: Kinematics of growth in a mixture. In many mixture models of growth, the solid constituents work in parallel, the deformation gradient **F** is uniform over constituents, and the partial stresses due to each constituent contribute additively to the total stress. On the other hand, each constituent $k \in \mathcal{S}^*$ that grows undergoes its own growth process: the growth and elastic transformations $\mathbf{F}_k^g$ and $\mathbf{F}_k^e$ are innate to each growing constituent. (a) The deformation gradient **F** is decomposed multiplicatively in a growth part $\mathbf{F}_k^g$ and an elastic part $\mathbf{F}_k^e$. (b) The residual stresses $\mathbf{T}_k^R$ are the stresses that exist in the material before application of the external load, which gives rise to the transformation $\mathbf{F}_k^\ell$. The stress $\Gamma_{\kappa k}$ just after the growth step vanishes if the pieces have the proper scale.

intermediate configuration and next transport it to the current configuration through specific schemes via the elastic transformation,

$$\mathbf{T}_{\kappa k} = \mathcal{T}_{\kappa k}(\mathbf{F}_k^e) + \Gamma_{\kappa k} \xrightarrow{\mathbf{F}_k^e} \mathbf{T}_k, \quad k \in \mathcal{S}^*. \tag{21.3.35}$$

Note once again that the existence of the stresses $\Gamma_{\kappa k}$ is paradoxical in the configurations obtained after growth, and before the compatibilization process defined by the accommodating transformation $\mathbf{F}^a$, Fig. 21.3.5-(b). Klisch et al. [2003]a [2003]b observe that, as growth proceeds, collagen and proteoglycans undergo divergent growth and elastic transformations, while the stresses $\Gamma_{\kappa k}$ for both collagen $k = $ c and proteoglycans $k = $ PG increase. These stresses are thought to be due to the fact that growth can not occur without creating heterogeneities in the collagen and proteoglycans. The crosslink density in particular is increasing. Therefore, at the scale of study (which considers assemblies of monomers), growth necessary leads to internal stresses. Zero stresses after growth would probably require to decrease the scale of observation to independent material units (monomers).

As an alternative, this stress $\Gamma_{\kappa k}$, which can be transformed in a strain, may convey the idea that fibroblasts exert a tension on the collagen fibers.

One key point of this analysis consists in unveiling the restrictions the thermodynamics of irreversible processes imposes on the growth law (21.3.33): the issue is addressed in Section 21.4 and in Chapters 22 and 23.

21.3.3   The properties of the deposited mass

The equations of balance of mass, momentum, energy and entropy are detailed in Chapter 4. At the time of deposition, a mass of species $k \in \mathcal{S}^*$ is endowed with three properties, namely

  - a velocity $\tilde{\mathbf{v}}_k$,
  - an internal energy $\tilde{u}_k$,
  - an entropy $\tilde{s}_k$.

These properties have to be defined through some ad hoc assumptions, or constitutive equations:

Assumption 21.1: the components of the tissue that grow, typically collagen and proteoglycans (PGs), move with the solid, that is

$$\mathbf{v}_k = \mathbf{v}_{\mathrm{s}} \quad \Leftrightarrow \quad \mathbf{M}_k = \rho^k \left(\mathbf{v}_k - \mathbf{v}_{\mathrm{s}} \right) = \mathbf{0}, \quad k \in \mathcal{S}^* . \tag{21.3.36}$$

Note however that the fact that the velocity of the deposited mass is that of the solid does not imply necessarily that, *before* deposition (aggregation), the velocity $\tilde{\mathbf{v}}_k$ of the mass was already equal to that of the solid.

Since growing species do not diffuse with respect to the solid, then, for these species,

$$\frac{dm^k}{dt} = \hat{m}_k, \quad k \in \mathcal{S}^*; \quad \hat{m} = \sum_{k \in \mathcal{S}^*} \frac{dm^k}{dt}, \tag{21.3.37}$$

and the balance of mass in terms of mass content yields the mass of the growing species in terms of the mass supply rate in the form,

$$\frac{m^k(t)}{\underline{m}^k} = \frac{M_k(t)}{\underline{M}_k} = \exp \left(\int_0^t \frac{\hat{\rho}^k}{\rho^k} \, d\tau \right), \quad k \in \mathcal{S}^* . \tag{21.3.38}$$

In a slightly different perspective, we admit

Assumption 21.2: the deformation of the deposited mass is the same as that of the mass at the deposition site, that is in fact here, as that of the solid,

$$\mathbf{F}_k = \mathbf{F}, \quad k \in \mathcal{S}^* . \tag{21.3.39}$$

Therefore, subsequent to the deposition time, the deposited mass undergoes the same velocity gradient as the solid,

$$\mathbf{L}_k = \mathbf{L}, \quad k \in \mathcal{S}^* . \tag{21.3.40}$$

Assumption 21.2 is challenged in Section 22.6. Indeed, the deformation of the mass just before deposition could well be quite distinct from that of the existing mass at the deposition site. For example, the deposited mass could be in a natural state.

The volume fraction may be expressed in terms of mass and overall volume change, namely[21.2]

$$n^k = \frac{v^k}{\det \mathbf{F}} = \frac{1}{\det \mathbf{F}} \frac{m^k}{\rho_k}, \tag{21.3.43}$$

and its rate of change,

$$\begin{aligned} \frac{1}{n^k} \frac{dn^k}{dt} &= \frac{1}{v^k} \frac{dv^k}{dt} - \operatorname{div} \mathbf{v}_{\mathrm{s}} \\ &= \frac{1}{m^k} \frac{dm^k}{dt} - \frac{1}{\rho_k} \frac{d\rho_k}{dt} - \operatorname{div} \mathbf{v}_{\mathrm{s}}, \end{aligned} \tag{21.3.44}$$

highlights the difference of volume changes of the solid skeleton (or porous medium) and of the species.

A number of issues remain open:

[21.2] The volume fractions of the growing species, and in essence the masses, are related
  - to the intrinsic deformations (deformation gradient $\mathbf{F}_k^{\mathrm{intr}}$, infinitesimal strain ϵ_k),
  - rather than to the apparent deformations (deformation gradient $\mathbf{F}_k$, infinitesimal strain ϵ^k),

that is,

$$n^k(t) = \frac{V_k(t)}{V(t)} = \frac{V_k(0) \det \mathbf{F}_k^{\mathrm{intr}}(t)}{V(0) \det \mathbf{F}(t)} = n^k(0) \frac{\det \mathbf{F}_k^{\mathrm{intr}}(t)}{\det \mathbf{F}(t)}, \quad k \in \mathcal{S}^*, \tag{21.3.41}$$

which, for infinitesimal strains, becomes

$$n^k(t) = n^k(0) \left(\operatorname{tr} \epsilon_k(t) - \operatorname{tr} \epsilon(t) \right), \quad k \in \mathcal{S}^* . \tag{21.3.42}$$

- the influence of the deformation and energy of the deposited mass is examined in Section 22.6. The two extreme cases of deposition in a natural configuration, and of pre-deformed deposition according to assumption 21.2 will be contrasted;
- assumption 21.1 indicates that the velocity of all growing species are identical to the solid velocity. On the other hand, it is known, that upon mixing cartilage cultures, a good part of collagen is lost and convected by the fluid. Indeed, PG monomers and collagen are released by the cells, and assembled in the extracellular space. They appear as non soluble species in water. The polymerization process seems to be only partial if water is stirred;
- perhaps, the rate of mass can be considered to be contributed by N mechanisms, like deposition (aggregation), loss (resorption), e.g., Section 21.6.4.5. In the case of loss of collagen monomers, the mechanical state of water matters. The information contained in these mechanisms should be transferred into the growth transformation through a multiplicative decomposition $\mathbf{F}^{\mathrm{g}} = \mathbf{F}^{\mathrm{gN}} \cdots \mathbf{F}^{\mathrm{g2}} \cdot \mathbf{F}^{\mathrm{g1}}$. For density preserving growth, the rate of mass results from the kinematics. This conclusion applies even if the kinematics contains mechanisms other than growth, like in the case of muscle activation.

21.3.4 Consequences of mass balance for growing species

In view of the multiplicative decomposition of the deformation gradient (21.3.32), the velocity gradient is contributed by an elastic part and by a growth part,

$$\mathbf{L} \equiv \frac{d\mathbf{F}}{dt} \cdot \mathbf{F}^{-1} = \overbrace{\frac{d\mathbf{F}_k^{\mathrm{e}}}{dt} \cdot (\mathbf{F}_k^{\mathrm{e}})^{-1}}^{\mathbf{L}_k^{\mathrm{e}}} + \overbrace{\mathbf{F}_k^{\mathrm{e}} \cdot \underbrace{\frac{d\mathbf{F}_k^{\mathrm{g}}}{dt} \cdot (\mathbf{F}_k^{\mathrm{g}})^{-1}}_{\mathbf{L}_{\kappa k}^{\mathrm{g}}} \cdot (\mathbf{F}_k^{\mathrm{e}})^{-1}}^{\mathbf{L}_k^{\mathrm{g}}}, \quad k \in \mathcal{S}^* . \tag{21.3.45}$$

Note the relations,

$$\operatorname{div} \mathbf{v}_{\mathrm{s}} = \operatorname{tr} \mathbf{L} = \operatorname{tr} \mathbf{L}_k^{\mathrm{e}} + \operatorname{tr} \mathbf{L}_k^{\mathrm{g}} ; \quad \operatorname{tr} \mathbf{L}_k^{\mathrm{g}} = \operatorname{tr} \mathbf{L}_{\kappa k}^{\mathrm{g}} , \quad k \in \mathcal{S}^* . \tag{21.3.46}$$

The first equality above is due to the fact that the species in the set $\mathcal{S}^*$ which are susceptible of growth move with the velocity $\mathbf{v}_{\mathrm{s}}$ of the solid.

As a consequence of the kinematical assumption (21.3.40), the balance of mass (4.4.5) can be recast in the format,

$$\frac{d\rho^k}{dt} + \rho^k \operatorname{tr} \mathbf{L} = \hat{\rho}^k , \quad k \in \mathcal{S}^* . \tag{21.3.47}$$

Let us introduce a mass density ρ_κ^k and a rate of mass density $\hat{\rho}_\kappa^k$ relative to the intermediate configuration (κ_k),

$$\rho_\kappa^k = \rho^k \det \mathbf{F}_k^{\mathrm{e}} , \quad \hat{\rho}_\kappa^k = \hat{\rho}^k \det \mathbf{F}_k^{\mathrm{e}} , \quad k \in \mathcal{S}^* . \tag{21.3.48}$$

Then, the balance of mass during the growth step can be phrased in terms of the above entities, namely

$$\frac{d\rho_\kappa^k}{dt} + \rho_\kappa^k \operatorname{tr} \mathbf{L}_k^{\mathrm{g}} = \hat{\rho}_\kappa^k , \quad k \in \mathcal{S}^* . \tag{21.3.49}$$

The transformation $\mathbf{F}_k^{\mathrm{g}}$ is assumed to account for both mass transfer and growth, so that the mass in the intermediate configuration (κ_k) is conserved by the elastic transformation $\mathbf{F}_k^{\mathrm{e}}$, i.e.,

$$M_k = \rho^k V = M_{\kappa k} = \rho_\kappa^k V_{\kappa(k)} , \quad k \in \mathcal{S}^* , \tag{21.3.50}$$

where $V_{\kappa(k)}$ is the fictitious total volume relative to the configuration (κ_k) obtained from the current volume by application of the transformation inverse to $\mathbf{F}_k^{\mathrm{e}}$, namely $V = \det \mathbf{F}_k^{\mathrm{e}} V_{\kappa(k)}$. Then

$$\frac{\rho^k}{\rho_\kappa^k} = \frac{V_{\kappa(k)}}{V} = \frac{1}{\det \mathbf{F}_k^{\mathrm{e}}} , \quad k \in \mathcal{S}^* . \tag{21.3.51}$$

The two special cases examined by Lubarda and Hoger [2002] for single bodies can now be re-analyzed in the mixture context. Cartilage growth may be considered with reasonable approximation as preserving density. On the other hand, growth in non saturated porous media may occur by filling the internal void space and preserving the overall volume.

21.3.4.1 Density preserving growth

Let M_k be the mass of the volume V_k of species k. Let $M_{\kappa k}$ be the mass of this species in its intermediate configuration where its volume is $V_{\kappa k}$. Since $M_{\kappa k} = M_k$ and $V_k = \det \mathbf{F}_k^e \, V_{\kappa k}$, the following relations hold between the various intrinsic and apparent densities that can be defined,

$$\rho_k = \frac{M_k}{V_k}, \quad \rho_{\kappa k} = \frac{M_{\kappa k}}{V_{\kappa k}} = \rho_k \det \mathbf{F}_k^e,$$
$$\rho^k = \frac{M_k}{V}, \quad \rho_\kappa^k = \rho^k \det \mathbf{F}_k^e = n^k \rho_{\kappa k}. \tag{21.3.52}$$

Incompressibility is equivalent to a (time) constant intrinsic mass density ρ_k, and elastic incompressibly to the condition $\det \mathbf{F}_k^e = 1$. At variance, density preserving growth amounts to the apparent density in the intermediate configuration ρ_κ^k to remain constant,

$$\underline{\rho}^k = \frac{M_k}{\underline{V}} = \rho_\kappa^k = \frac{M_{\kappa k}}{V_{\kappa(k)}} = \rho^k \det \mathbf{F}_k^e, \quad k \in \mathcal{S}^*. \tag{21.3.53}$$

Then, the balance of mass (21.3.49) may be cast in the following formats,

$$\frac{\hat{\rho}^k}{\rho^k} = \operatorname{tr} \mathbf{L}_k^g = \frac{d}{dt} \operatorname{Ln} \det \mathbf{F}_k^g, \quad \frac{d\rho^k}{dt} + \rho^k \operatorname{tr} \mathbf{L}_k^e = 0, \quad k \in \mathcal{S}^*. \tag{21.3.54}$$

Integration of (21.3.54) implies,

$$\frac{\det \mathbf{F}_k^g(t)}{\det \mathbf{F}_k^g(0)} = \exp\Big(\int_0^t \frac{\hat{\rho}^k}{\rho^k} \, d\tau \Big), \quad k \in \mathcal{S}^*. \tag{21.3.55}$$

Interestingly, the relations simplify in terms of the mass contents

$$m^k = \frac{M_k}{\underline{V}} = \frac{M_k}{\underline{V}} \frac{M_k}{M_k} = \frac{M_k}{\underline{V}} \frac{V_{\kappa(k)}}{\underline{V}} = \underline{\rho}^k \det \mathbf{F}_k^g, \quad k \in \mathcal{S}^*, \tag{21.3.56}$$

or in rate form, using the time derivative following the solid skeleton,

$$\frac{1}{m^k} \frac{dm^k}{dt} = \operatorname{tr} \mathbf{L}_k^g = \operatorname{tr} \mathbf{L}_{\kappa k}^g, \quad k \in \mathcal{S}^*. \tag{21.3.57}$$

For growing constituents that move with the solid, the history of mass supplies provides the mass. Concatenation of (21.3.38) and (21.3.55) establishes a one-to-one relationship between the mass and the determinant of the growth transformation,

$$\frac{M_k(t)}{\underline{M}_k} = \exp\Big(\int_0^t \frac{\hat{\rho}^k}{\rho^k} \, d\tau \Big) = \frac{\det \mathbf{F}_k^g(t)}{\det \mathbf{F}_k^g(0)}, \quad k \in \mathcal{S}^*. \tag{21.3.58}$$

The simultaneous assumptions of density preserving growth and of elastic incompressibility imply the apparent density to remain constant,

$$\underline{\rho}^k = \frac{M_k}{\underline{V}} = \rho_\kappa^k = \frac{M_{\kappa k}}{V_{\kappa(k)}} = \rho^k = \frac{M_k}{V}, \quad k \in \mathcal{S}^*. \tag{21.3.59}$$

21.3.4.2 Volume preserving growth

In case of volume preserving growth,

$$\frac{V_{\kappa k}}{\underline{V}_k} = \det \mathbf{F}_k^g = 1, \tag{21.3.60}$$

then

$$\operatorname{tr} \mathbf{L}_k^g = 0, \quad \frac{d\rho_\kappa^k}{dt} = \hat{\rho}_\kappa^k, \quad \frac{d\rho^k}{dt} + \rho^k \operatorname{tr} \mathbf{L}_k^e = \hat{\rho}^k, \quad k \in \mathcal{S}^*. \tag{21.3.61}$$

21.3.5 Growth, remodeling, structuration and dissipation

21.3.5.1 Pure growth processes

Pure growth processes take place without change of mass density (no remodeling) and without change of the microstructure (no structuration). Therefore the masses of the growing species are excluded from the set of state variables, but the masses of the species that only transfer are included: they allow for chemoelastic couplings, as shown for articular cartilages in Chapter 14 and for cornea in Chapter 19.

The rate of supplied mass dm^k/dt to the growing species k is equal to $m^k \mathbf{I} : \mathbf{L}^{\mathrm{g}}_{\kappa k}$, in view of (21.3.46) and (21.3.57). Thus the constitutive equation for the rate of supplied mass is known, once the growth velocity "gradient" (growth rate) $\mathbf{L}^{\mathrm{g}}_{\kappa k}$ is postulated.

21.3.5.2 Bounded growth, growth versus structuration processes

Structuration embodies the microstructural re-organization of the tissues. In Chapters 22 and 23, bounded growth is modeled via an evolving homeostatic domain whose size, shape and position are controlled by *structure* variables. The structure variables in the present formulation include *inter alia* the fabric tensor describing the collagen network, the pyridinoline density that affects the overall stiffness of this network, the cell energy density, and entities attached to the homeostatic surface.

In actual models, structuration is sometimes used as an analytical tool/degree of freedom that allows to avoid unbounded growth, e.g., Section 21.7.1.1. When a linear growth law leads to unbounded growth, another ad hoc remedy consists in using nonlinear growth laws, e.g., Drozdov and Khanina [1997].

In order to decouple growth and remodeling, a remodeling variable motivated by (21.3.56),

$$\underline{r}^k = \frac{m^k}{\underline{\rho}^k} - \det \mathbf{F}^{\mathrm{g}}_k, \quad k \in \mathcal{S}^*, \tag{21.3.62}$$

is introduced. It vanishes for pure growth,

$$\frac{d\underline{r}^k}{dt} = \frac{1}{\underline{\rho}^k} \frac{dm^k}{dt} - \det \mathbf{F}^{\mathrm{g}}_k \operatorname{tr} \mathbf{L}^{\mathrm{g}}_{\kappa k}, \quad k \in \mathcal{S}^*. \tag{21.3.63}$$

21.3.5.3 Gathering the contributions: rates of growth of the growing masses

However, in general, the rate of change of the growing masses is obtained by adding the contributions due to pure growth and remodeling,

$$\frac{1}{m^k} \frac{dm^k}{dt} = \operatorname{tr} \mathbf{L}^{\mathrm{g}}_{\kappa k} + \frac{1}{\det \mathbf{F}^{\mathrm{g}}_k} \frac{d\underline{r}^k}{dt}, \quad k \in \mathcal{S}^*. \tag{21.3.64}$$

For density preserving growth, there is a one-to-one correspondence between the rate of mass supply and the rate of growth transformation, in which case the remodeling variable vanishes, Section 21.3.4.1. For volume preserving growth, the contribution due to the growth transformation vanishes, Section 21.3.4.2.

In general however, for each growing species, constitutive equations are required for two entities,
- the growth transformation (a second order tensor), and
- the remodeling variable, or the mass density.

21.4 Constitutive assumptions and restrictions: the growth law

For purely growth processes, as opposed to remodeling and structuration, we shall adopt the position in line with Hoger and coworkers that the material response in terms of density, stress, energy ... depends only on the elastic part of the deformation. The free energy $\underline{E} = \underline{E}(\underline{\mathcal{V}})$ of the mixture is then considered to be a function of the following state variables (pending invariance requirements),

$$\underline{\mathcal{V}} = \{\cup_{k \in \mathcal{S}^*} \mathbf{E}^{\mathrm{e}}_{\mathrm{G}k}, \ T, \ \cup_{k \in \mathcal{K}^*} m^k, \ \cup_{n \in \mathcal{N}} \underline{\xi}_n\}, \tag{21.4.1}$$

where $\mathbf{E}^{\mathrm{e}}_{\mathrm{G}k}$ is the elastic Green strain,

$$\mathbf{C}^{\mathrm{e}}_k = (\mathbf{F}^{\mathrm{e}}_k)^{\mathrm{T}} \cdot \mathbf{F}^{\mathrm{e}}_k, \quad \mathbf{E}^{\mathrm{e}}_{\mathrm{G}k} = \tfrac{1}{2}(\mathbf{C}^{\mathrm{e}}_k - \mathbf{I}), \quad k \in \mathcal{S}^*; \qquad (21.4.2)$$

and the $\underline{\xi}_n$'s, $n \in \mathcal{N}$, are structuration and dissipation variables. The first element of the set $\mathcal{N} = \{1, \mathcal{N}^* = \cup_{k \in \mathcal{S}^*} \mathcal{N}_k\}$ is attached to the mixture as a whole while the set $\mathcal{N}_k$ includes the structuration and dissipation variables attached to the growing constituent k. In presence of remodeling, the set of state variables of the mixture may include the mass contents of the growing species:

$$\underline{\mathcal{V}} = \{\cup_{k \in \mathcal{S}^*} \mathbf{E}^{\mathrm{e}}_{\mathrm{G}k}, \ T, \ \cup_{k \in \mathcal{K}^*} m^k, \ \cup_{k \in \mathcal{S}^*} \underline{r}^k, \ \cup_{n \in \mathcal{N}} \underline{\xi}_n\}. \qquad (21.4.3)$$

Although tentative, the present analysis takes care to completely maintain the thermodynamic framework, exposed in Section 4.7, regarding the constitutive equations of mass transfer and generalized diffusion, and to embed them in a larger framework. Henceforth we concentrate only on the contributions to the entropy production $\hat{\underline{S}}_{(\mathrm{i},1,4)}$, Eqn (4.7.28), due to mechanical (mainly viscous) effects, and to growth and structuration.

The entropy production of the mixture we are concerned with adopts the quadratic dissipation format (4.8.11), namely,

$$T\hat{\underline{S}}_{(\mathrm{i},1,4)} = -\sum_{k \in \mathcal{S}^*} \underline{\mathbf{f}}_{\kappa k} : \mathbf{L}^{\mathrm{g}}_{\kappa k} - \sum_{k \in \mathcal{S}^*} \underline{R}_k \frac{d\underline{r}^k}{dt} - \sum_{n \in \mathcal{N}} \underline{X}_n \frac{d\underline{\xi}_n}{dt}. \qquad (21.4.4)$$

In a formal presentation, the work-conjugate pairs assume all the same dimensions, namely Pa for the forces $\underline{\mathbf{f}}_\kappa$, $\underline{R}$ and $\underline{X}$, while r and ξ are dimensionless.

Constitutive equations are required for the four contributions to the external entropy production $\hat{\underline{S}}_{(\mathrm{e})}$, Eqn (4.7.29), namely for the rates of mass, momentum, energy and entropy exchanges with the surroundings, say

$$\hat{m} = \hat{m}(\mathcal{V}), \quad \hat{\underline{\mathbf{p}}} = \hat{\underline{\mathbf{p}}}(\underline{\mathcal{V}}), \quad \hat{\underline{U}} = \hat{\underline{U}}(\underline{\mathcal{V}}), \quad \hat{\underline{S}} = \hat{\underline{S}}(\underline{\mathcal{V}}). \qquad (21.4.5)$$

Constitutive assumptions of growth are motivated by the entropy production. In this chapter, the entropy production is assumed to adopt a quadratic dissipation format. Chapters 22 and 23 will embed convexity properties in the entropy production: the resulting tool turns out to be more powerful and akin to account for both dissipative processes and structuration processes to which a negative dissipation is attached.

21.4.1 Tentative assumptions

In order to illustrate the consequences of the Clausius-Duhem inequality, a number of simplifications are adopted. Note however that these assumptions are by no means essential to the goal pursued.

Assumption 21.3: the species that grow do not transfer, and conversely, the species that transfer do not grow.

Assumption 21.4: the momentum supply, internal energy and entropy supplies by the surroundings are diverted completely to the growing species.

Assumption 21.5: the velocities $\tilde{\mathbf{v}}_k$ of the masses supplied by the growth process before deposition are equal to the velocities after deposition, i.e. $\tilde{\mathbf{v}}_k = \mathbf{v}_k = \mathbf{v}_{\mathrm{s}}$, $k \in \mathcal{S}^*$. Said otherwise, the mass deposited has the same velocity as the existing mass at the point of deposition.

With assumptions 21.4 and 21.5, the external entropy production defined by (4.7.29) with supply rates (4.4.3), (4.5.2), (4.6.8) and (4.7.2) can be simplified as follows[21.3]:

$$
\begin{aligned}
-T\hat{\underline{S}}_{(\mathrm{e})} \ &= \ \tfrac{1}{2}\hat{m}\,\mathbf{v}^2_{\mathrm{s}} - \hat{\underline{\mathbf{p}}} \cdot \mathbf{v}_{\mathrm{s}} + \hat{\underline{U}} - T\hat{\underline{S}}, \\[2mm]
&= \ \sum_{k \in \mathcal{K}} \hat{m}^k \tilde{e}_k + \hat{\underline{\mathcal{E}}}^k + \tfrac{1}{2}\hat{m}^k (\tilde{\mathbf{v}}_k - \mathbf{v}_{\mathrm{s}})^2 + (\mathbf{v}_k - \mathbf{v}_{\mathrm{s}}) \cdot \hat{\boldsymbol{\pi}}^k \det \mathbf{F}, \\[2mm]
&= \ \sum_{k \in \mathcal{S}^*} \hat{m}^k \tilde{e}_k + \hat{\underline{\mathcal{E}}}^k + \\[2mm]
&\ + \sum_{k \in \mathcal{K}^*} \hat{m}^k \tilde{e}_k + \hat{\underline{\mathcal{E}}}^k + \tfrac{1}{2}\hat{m}^k (\tilde{\mathbf{v}}_k - \mathbf{v}_{\mathrm{s}})^2 + (\mathbf{v}_k - \mathbf{v}_{\mathrm{s}}) \cdot \hat{\boldsymbol{\pi}}^k \det \mathbf{F}, \\[2mm]
&= \ \sum_{k \in \mathcal{S}^*} \hat{m}^k \tilde{e}_k + \hat{\underline{\mathcal{E}}}^k,
\end{aligned}
\qquad (21.4.6)
$$

[21.3] Recall $\hat{m} = \hat{\rho} \det \mathbf{F}$, $\hat{\underline{\boldsymbol{\pi}}} = \hat{\boldsymbol{\pi}} \det \mathbf{F}$, $\hat{\underline{\mathcal{U}}}^k = \hat{\mathcal{U}}^k \det \mathbf{F}$, $\hat{\underline{S}}^k = \hat{S}^k \det \mathbf{F}$, $\hat{\underline{\mathcal{E}}}^k = \hat{\mathcal{E}}^k \det \mathbf{F}$.

with

$$
\tilde{e}_k \equiv \underbrace{\tilde{u}_k - T\,\tilde{s}_k}_{\substack{\text{specific free energy} \\ \text{of supplied mass}}} \quad, \quad \hat{\underline{\mathcal{E}}}^k \equiv \underbrace{\hat{\underline{\mathcal{U}}}^k - T\,\hat{\underline{\mathcal{S}}}^k}_{\substack{\text{supply rate of free energy} \\ \text{in other forms}}} \quad, \quad k \in \mathcal{S}^* .
$$

(21.4.7)

The interpretation of the terms $\hat{\underline{\mathcal{U}}}^k - T\,\hat{\underline{\mathcal{S}}}^k$ as supply rates of free energy assumes a constant temperature.

A number of issues are still open:

- should the energy supply be directed to the mixture as a whole, the partition between species being obtained as the result of processes to be defined (with interactions between species)?
- or can the energy supplies be specified directly for each species individually?

The first possibility would allow either competition or synergy. For example, in a certain period, a species may not need much energy, which could be used by another species which demands more.

The second line of thinking is adopted in the sequel. Then, constitutive equations are required for the supply rates by/to the surroundings, say in the format

$$
\hat{m}^k = \hat{m}^k(\underline{\mathcal{V}}), \quad \hat{\underline{\mathcal{E}}}^k = \hat{\underline{\mathcal{E}}}^k(\underline{\mathcal{V}}), \quad k \in \mathcal{S}^* ,
$$

(21.4.8)

as well as for the free energies of the supplied masses $\tilde{e}_k$. In addition, the energies in unspecified forms $\hat{\underline{\mathcal{E}}}^k$ are formally partitioned between pure growth, remodeling and structuration. The actual partition is specified by,

Assumption 21.6: the supply rate in arbitrary form $\hat{\underline{\mathcal{E}}}^k$ is linear in the growth velocity transformation, rate of remodeling variable and rates of structuration/dissipation variables of the species k only, therefore excluding couplings at this level,

$$
\hat{\underline{\mathcal{E}}}^k = \hat{\underline{\mathcal{E}}}^k_{(\mathrm{g})} + \hat{\underline{\mathcal{E}}}^k_{(\mathrm{r})} + \hat{\underline{\mathcal{E}}}^k_{(\mathrm{d})}, \quad \hat{\underline{\mathcal{E}}}^k_{(\mathrm{g})} = \underline{\beta}^{(\mathrm{g})}_{\kappa k} : \mathbf{L}^{\mathrm{g}}_{\kappa k}, \quad \hat{\underline{\mathcal{E}}}^k_{(\mathrm{r})} = \underline{\beta}^{(\mathrm{r})}_{k}\,\frac{dr^k}{dt}, \quad \hat{\underline{\mathcal{E}}}^k_{(\mathrm{d})} = \sum_{n \in \mathcal{N}_k} \underline{\beta}^{(\mathrm{d})}_{\underline{n}}\,\frac{d\xi_{\underline{n}}}{dt}, \quad k \in \mathcal{S}^* ,
$$

(21.4.9)

where the second order tensor (with symmetries to be specified) $\underline{\beta}^{(\mathrm{g})}_{\kappa k}$ [unit: Pa] and the scalars $\underline{\beta}^{(\mathrm{r})}_{k}$ [unit: Pa] and $\underline{\beta}^{(\mathrm{d})}_{\underline{n}}$ [unit: Pa] depend on the state variables.

21.4.2 A growth law consistent with the entropy production

The free energy per unit reference volume of the mixture $\underline{E} = \det \mathbf{F}\, E = m\,e$ is considered to be a function of the state variables (21.4.3). Accounting for the independence of these variables, constitutive relations are obtained in the format:

stress − strain law
$$
\underline{\underline{\tau}} = \sum_{k \in \mathcal{S}^*} \mathbf{F}^{-1} \cdot \frac{\partial \underline{E}}{\partial \mathbf{F}^{\mathrm{e}}_k} \cdot (\mathbf{F}^{\mathrm{g}}_k)^{-\mathrm{T}} + \underline{\underline{\tau}}_{\mathrm{visc}};
$$

entropy
$$
\underline{S} = -\frac{\partial \underline{E}}{\partial T};
$$

chemical potentials
$$
\mu_k = \frac{\partial \underline{E}}{\partial m^k}, \quad k \in \mathcal{K}^*;
$$
(21.4.10)

dissipative mechanisms
$$
\underline{X}_n = \frac{\partial \underline{E}}{\partial \xi_{\underline{n}}} - \underline{\beta}^{(\mathrm{d})}_{\underline{n}}, \quad n \in \mathcal{N}_k, \quad k \in \mathcal{S}^*;
$$

remodeling variables
$$
\underline{R}_k = \frac{\partial \underline{E}}{\partial r^k} - \underline{\rho}^k\,\tilde{e}_k - \underline{\beta}^{(\mathrm{r})}_{k}, \quad k \in \mathcal{S}^* .
$$

Note that the inviscid part of the Cauchy stress $\boldsymbol{\sigma} = (\det \mathbf{F})^{-1}\,\mathbf{F} \cdot \underline{\underline{\tau}} \cdot \mathbf{F}^{\mathrm{T}}$ stemming from (21.4.10)₁ may, with help of (2.7.3), be cast in the formats,

$$
\begin{aligned}
\boldsymbol{\sigma} - \boldsymbol{\sigma}_{\mathrm{visc}} &= \sum_{k \in \mathcal{S}^*} \frac{1}{\det \mathbf{F}}\,\mathbf{F}^{\mathrm{e}}_k \cdot \frac{\partial \underline{E}}{\partial \mathbf{E}^{\mathrm{e}}_{\mathrm{G}k}} \cdot (\mathbf{F}^{\mathrm{e}}_k)^{\mathrm{T}} \\
&= E\,\mathbf{I} + \sum_{k \in \mathcal{S}^*} \mathbf{F}^{\mathrm{e}}_k \cdot \frac{\partial E}{\partial \mathbf{E}^{\mathrm{e}}_{\mathrm{G}k}} \cdot (\mathbf{F}^{\mathrm{e}}_k)^{\mathrm{T}} \\
&= \sum_{k \in \mathcal{S}^*} n^k\,\mathbf{F}^{\mathrm{e}}_k \cdot \frac{\partial E_{\kappa k}}{\partial \mathbf{E}^{\mathrm{e}}_{\mathrm{G}k}} \cdot (\mathbf{F}^{\mathrm{e}}_k)^{\mathrm{T}} .
\end{aligned}
$$

(21.4.11)

The last relation based on the free energy relative to the intermediate configurations is derived in Section (23.1.1.2).

To prove (21.4.10), use the relation between stress measures (2.8.1), the multiplicative decomposition (21.3.32), and the relations (4.4.9), (21.3.37), (21.3.56), (21.4.1), (21.4.6), (21.4.9). Equating the entropy productions in (4.7.28) and (21.4.4) yields,

$$
\overbrace{\underline{\tau} : \frac{d\mathbf{F}}{dt} - \sum_{k \in \mathcal{S}^*} \frac{\partial E}{\partial \mathbf{F}_k^{\mathrm{e}}} : \left(\frac{d\mathbf{F}}{dt} \cdot (\mathbf{F}_k^{\mathrm{g}})^{-1} \right)}^{(1)}
$$

$$
\overbrace{- \left(\underline{S} + \frac{\partial E}{\partial T} \right) \frac{dT}{dt}}^{(2)}
$$

$$
+ \sum_{k \in \mathcal{K}^*} \overbrace{\left(\mu_k - \frac{\partial E}{\partial m^k} \right) \frac{dm^k}{dt}}^{(3)}
$$

$$
+ \sum_{k \in \mathcal{S}^*} \overbrace{((\mathbf{F}_k^{\mathrm{e}})^{\mathrm{T}} \cdot \frac{\partial E}{\partial \mathbf{F}_k^{\mathrm{e}}} + m^k \, \tilde{e}_k \, \mathbf{I} + \underline{\beta}_{\kappa k}^{(\mathrm{g})}) : \mathbf{L}_{\kappa k}^{\mathrm{g}}}^{(4)} \qquad (21.4.12)
$$

$$
+ \sum_{k \in \mathcal{S}^*} \overbrace{\left(\rho^k \, \tilde{e}_k - \frac{\partial E}{\partial \underline{r}^k} + \underline{\beta}_k^{(\mathrm{r})} \right) \frac{d\underline{r}^k}{dt}}^{(5)}
$$

$$
\overbrace{- \frac{\partial E}{\partial \underline{\xi}_1} \frac{d\underline{\xi}_1}{dt} + \sum_{k \in \mathcal{S}^*} \sum_{n \in \mathcal{N}_k} (\frac{\partial E}{\partial \underline{\xi}_n} + \underline{\beta}_n^{(\mathrm{d})}) \frac{d\underline{\xi}_n}{dt}}^{(6)}
$$

$$
= - \sum_{k \in \mathcal{S}^*} \overbrace{\underline{\mathbf{f}}_{\kappa k} : \mathbf{L}_{\kappa k}^{\mathrm{g}}}^{(4)} - \sum_{k \in \mathcal{S}^*} \overbrace{R_k \frac{d\underline{r}^k}{dt}}^{(5)} - \sum_{n \in \mathcal{N}} \overbrace{\underline{X}_n \frac{d\underline{\xi}_n}{dt}}^{(6)} \geq 0 \, .
$$

21.4.2.1   Stress-strain relation

The total stress in (21.4.10) can be interpreted as the sum of the partial stresses of the species. In absence of dissipative effects, the above relations induce hyperelastic constitutive equations with chemomechanical couplings contributed a priori by all species that do not grow. In actual applications however, some species might not contribute to the mechanical couplings, e.g., the ionic species of the intrafibrillar space of articular cartilages.

Mechanical dissipation might be induced via purely viscous effects, e.g. $-\underline{X}_1 = \underline{\underline{\tau}}_{\mathrm{visc}}$ and $\underline{\xi}_1 = \mathbf{E}_{\mathrm{G}}$ Green strain. Then

$$
-\underline{X}_1 \frac{d\underline{\xi}_1}{dt} = \underline{\underline{\tau}}_{\mathrm{visc}} : \frac{d\mathbf{E}_{\mathrm{G}}}{dt} \geq 0 \quad \Rightarrow \quad \underline{\underline{\tau}}_{\mathrm{visc}} = t_{\mathrm{v}} \, \underline{\underline{\mathbb{E}}}^{\mathrm{v}} : \frac{d\mathbf{E}_{\mathrm{G}}}{dt} \, , \qquad (21.4.13)
$$

with $t_{\mathrm{v}} > 0$ [unit : s] the viscous relaxation time, and $\underline{\underline{\mathbb{E}}}^{\mathrm{v}}$ [unit : Pa] a fourth order (semi-)positive definite tensor with appropriate symmetries.

21.4.2.2 Dissipative mechanisms

As for scalar dissipation mechanisms, let us consider the simplest setting where they do not couple. Then the inequality $-\underline{X}_n \, d\underline{\xi}_n/dt \geq 0$, $n \in \mathcal{N}^*$, may be satisfied in various ways, e.g., via a linear relation,

$$\frac{d\underline{\xi}_n}{dt} = -\frac{\underline{\alpha}_n^{(d)}}{t_{dn}} \underline{X}_n \,, \tag{21.4.14}$$

or via an exponential relation,

$$\frac{d\underline{\xi}_n}{dt} = \left(\exp(-\epsilon_n \frac{\underline{\alpha}_n^{(d)}}{t_{dn}} \underline{X}_n) - 1 \right) \epsilon_n \,, \quad \epsilon_n = \text{sign}(\underline{X}_n) \,, \tag{21.4.15}$$

where, for $n \in \mathcal{N}^*$, $\underline{X}_n$ is given by (21.4.10), t_{dn} [unit : s] is a characteristic time, and $\underline{\alpha}_n^{(d)}$ [unit : Pa^{-1}] a positive coefficient. This definition of the ϵ_n's bounds the rate of dissipation, Fig. 10.2.1.

Coupled dissipation mechanisms, defined by a non diagonal positive semi-definite dissipation matrix, might be envisaged as well.

21.4.2.3 Remodeling mechanisms

The terms in (21.4.12) that involve the rates of remodeling are collected,

$$-\sum_{k \in \mathcal{S}^*} \underline{R}_k \frac{d\underline{r}^k}{dt} = \sum_{k \in \mathcal{S}^*} \left(-\frac{\partial E}{\partial \underline{r}^k} + \underline{\rho}^k \, \tilde{e}_k + \underline{\beta}_k^{(r)} \right) \frac{d\underline{r}^k}{dt} \geq 0 \,. \tag{21.4.16}$$

The scalar remodeling mechanisms may be treated exactly like the scalar dissipation mechanisms. In the simplest setting where they do not couple, the inequality $-\underline{R}_k \, d\underline{r}^k/dt \geq 0$, $k \in \mathcal{S}^*$, may be satisfied in various ways, e.g., via a linear or exponential relation,

$$\frac{d\underline{r}^k}{dt} = \underbrace{-\frac{\underline{\alpha}_k^{(r)}}{t_{rk}} \underline{R}_k}_{\text{linear dissipation}} \,, \quad \text{or} \quad \frac{d\underline{r}^k}{dt} = \underbrace{\left(\exp(-\epsilon_k \frac{\underline{\alpha}_k^{(r)}}{t_{rk}} \underline{R}_k) - 1 \right) \epsilon_k}_{\text{exponential dissipation}}, \quad \epsilon_k = \text{sign}(\underline{R}_k) \,, \tag{21.4.17}$$

where, for any $k \in \mathcal{S}^*$, $\underline{R}_k$ [unit : Pa], the conjugate to the remodeling variable, is given by (21.4.10), t_{rk} [unit : s] is a characteristic time, and $\underline{\alpha}_k^{(r)}$ a positive coefficient [unit : Pa^{-1}].

21.4.2.4 The growth law

The terms in (21.4.12) that involve the rate of velocity gradients $\mathbf{L}_{\kappa k}^g$ are first collected,

$$-\sum_{k \in \mathcal{S}^*} \underline{\mathbf{f}}_{\kappa k} : \mathbf{L}_{\kappa k}^g = \sum_{k \in \mathcal{S}^*} \left(\mathbf{C}_k^e \cdot \frac{\partial E}{\partial \mathbf{E}_{Gk}^e} + m^k \, \tilde{e}_k \, \mathbf{I} + \underline{\boldsymbol{\beta}}_{\kappa k}^{(g)} \right) : \mathbf{L}_{\kappa k}^g \geq 0 \,, \tag{21.4.18}$$

and complemented by

Assumption 21.7: the relation (21.4.18) holds for arbitrary $\mathbf{L}_{\kappa k}^g$.

While this assumption is certainly more than necessary, it yields a relation that links the thermodynamic force to the sum of an internal contribution and an external contribution,

$$-\underline{\mathbf{f}}_{\kappa k} = \mathbf{C}_k^e \cdot \frac{\partial E}{\partial \mathbf{E}_{Gk}^e} + m^k \, \tilde{e}_k \, \mathbf{I} + \underline{\boldsymbol{\beta}}_{\kappa k}^{(g)}, \quad k \in \mathcal{S}^* \,. \tag{21.4.19}$$

In the simplest context of

Assumption 21.8: there is no coupling between the growth mechanisms, the growth law for the species k expresses via a characteristic time t_{gk} [unit : s] and a fourth order positive definite tensor $\boldsymbol{\alpha}_{\kappa k}^{(\mathrm{g})}$ [unit : Pa^{-1}], endowed with the major symmetry (but a priori not with minor symmetries),

$$-\underline{\mathbf{f}}_{\kappa k} = \left(\frac{\boldsymbol{\alpha}_{\kappa k}^{(\mathrm{g})}}{t_{gk}}\right)^{-1} : \mathbf{L}_{\kappa k}^{\mathrm{g}} , \quad k \in \mathcal{S}^* , \tag{21.4.20}$$

in such a way that (21.4.19) returns the growth law in the format,

$$\left(\frac{\boldsymbol{\alpha}_{\kappa k}^{(\mathrm{g})}}{t_{gk}}\right)^{-1} : \mathbf{L}_{\kappa k}^{\mathrm{g}} = \mathbf{C}_k^{\mathrm{e}} \cdot \frac{\partial E}{\partial \mathbf{E}_{\mathrm{G}k}^{\mathrm{e}}} + m^k \, \tilde{e}_k \, \mathbf{I} + \underline{\boldsymbol{\beta}}_{\kappa k}^{(\mathrm{g})} , \quad k \in \mathcal{S}^* . \tag{21.4.21}$$

We have not accounted for couples even if we might allow $\mathbf{L}_{\kappa k}^{\mathrm{g}}$ to have a non symmetric part. In fact, the phenomenon is reminiscent of the issue of couple stresses that might exist in materials (crystals, grains) in which discontinuities may occur. For example, grains in contact may undergo different rotations: while in general there exist couples at the contact points that try to oppose the relative motion, it seems sound to neglect them in a first approach, Mandel [1973]. This point of view does not negate the existence of a "microstructure" but avoids developments which currently are not substantiated by data.

The growth tensor $\boldsymbol{\alpha}_{\kappa k}^{(\mathrm{g})}$, $k \in \mathcal{S}^*$, is constrained only by its symmetry and positive definiteness, but is otherwise arbitrary: it might in particular be function of the concentrations of the nutrients and other state variables. In the formulation above, the elastic strain is a state variable, so that growth is driven by the latter. In a linear theory of elasticity, growth can equivalently be seen as driven by stress. More generally, stress-driven growth can be obtained by performing a partial Legendre transform which substitutes the stress for the elastic transformation as a state variable.

While thermodynamically consistent, the growth law is by no means too restrictive. It involves both the energy provided by, or supplied to, the surroundings and the free energy of the mixture. Therefore growth can be driven, or slowed, by properties of the individual species (strains, stresses, etc.) which have to be adequately built into the free energy, or free enthalpy, endowed with some sort of equilibrium state(s), possibly a homeostatic reference. Since the equations of growth are expected to be nonlinear, an analysis of their dynamic behavior in a phase plane, along Cowin and Hegedus [1976], is worth to be considered. The boundedness of growth is addressed in Chapters 22 and 23 by allowing the tissue structure to evolve.

Note that the models of growth we are aware of, including Lubarda and Hoger [2002], Garikipati et al. [2004], Klisch and Hoger [2003], do not ensure the positiveness of the dissipation. The growth laws, postulated by Garikipati et al. [2004], are independent of the thermodynamic derivations, and therefore they are by no means certified to be consistent with a positive dissipation. Concerned with single phase solids, Lubarda and Hoger [2002] stop short their thermodynamic analysis at (21.4.19). They further postulate growth laws, their Eqn (12.20), independently of thermodynamics. The dissipation force thus results, so that the balance of entropy is satisfied, but this point of view does not address the positiveness of the dissipation. On the other hand, here, the thermodynamics of irreversible processes is used in order to comply with the Clausius-Duhem inequality.

21.4.2.5 Growth law: the simplified small strains setting

The small deformation version of the growth equations deduces from the above developments in a straightforward manner. For each species, the multiplicative kinematics (21.3.1) implies the elastic strain $\boldsymbol{\epsilon}_k^{\mathrm{e}}$ and growth strain $\boldsymbol{\epsilon}_k^{\mathrm{g}}$ to contribute additively to the total strain $\boldsymbol{\epsilon}$,

$$\boldsymbol{\epsilon} = \boldsymbol{\epsilon}_k^{\mathrm{e}} + \boldsymbol{\epsilon}_k^{\mathrm{g}} , \quad k \in \mathcal{S}^* . \tag{21.4.22}$$

The free energy E is now a function of the elastic strains $\boldsymbol{\epsilon}_k^{\mathrm{e}}$, $k \in \mathcal{S}^*$, and the growth law (21.4.21) becomes (with a simplified notation)

$$\left(\frac{\boldsymbol{\alpha}_k^{(\mathrm{g})}}{t_{gk}}\right)^{-1} : \frac{d\boldsymbol{\epsilon}_k^{\mathrm{g}}}{dt} = \frac{\partial E}{\partial \boldsymbol{\epsilon}_k^{\mathrm{e}}} + m^k \, \tilde{e}_k \, \mathbf{I} + \boldsymbol{\beta}_k^{(\mathrm{g})} , \quad k \in \mathcal{S}^* . \tag{21.4.23}$$

21.4.3 Retrospect

The chemomechanical couplings that control much of the behavior of many soft tissues have been modeled in previous chapters, e.g., Chapters 14, 16, 17: deformation, mass transfer and generalized diffusion have been accounted for. The present chapter has consisted in aggregating the phenomenon of growth to these models. For that purpose, along the framework of Chapter 4, the mixture system has been considered thermodynamically open. Within the continuum thermodynamics of irreversible processes in a mixture context, satisfaction of the Clausius-Duhem inequality is shown here to motivate and structure the growth law.

The whole procedure is seen to have consisted in splitting the dissipation $\hat{S}_{(i,1,4)}$ into two parts associated with distinct physical phenomena,

$$
\overbrace{\hat{S}_{(i,1,4)}}^{\text{Eqn (4.7.28)}} \rightarrow \hat{S}_{(i,1)} + \hat{S}_{(i,4)}
\left\{
\begin{array}{l}
\hat{S}_{(i,1)} : \quad \overbrace{\text{mechanical dissipation}}^{\text{Sections (21.4.2.1),(21.4.2.2),(21.4.2.3)}} \quad ; \\[2em]
\hat{S}_{(i,4)} : \quad \overbrace{\text{growth dissipation}}^{\text{Section (21.4.2.4)}} .
\end{array}
\right.
\tag{21.4.24}
$$

This partition of the dissipation inequality has led to (over-)sufficient conditions, which have been exploited to obtain coupled constitutive equations governing the chemomechanical behavior of the tissue and the growth process. The procedure is qualitative at this point. It will be specialized in this and subsequent chapters.

21.5 Internal entropy production for a single solid

We consider now a solid, and drop the index k denoting a species. For simplicity, a single remodeling variable r and a single structure or dissipation variable ξ are kept in the set of state variables.

21.5.1 The relations for a single solid in the current configuration

The set of state variables reduces to

$$
\mathcal{V} = \{\mathbf{E}_{\mathrm{G}}^{\mathrm{e}}, T, r, \xi\}.
\tag{21.5.1}
$$

The remodeling variable $r = m/\rho - \det \mathbf{F}^{\mathrm{g}}$ is defined by (21.3.62). Moreover, since diffusion is absent, the rate of mass content is equal to the rate of mass supply, $dm/dt = \hat{m} = \hat{\rho} \det \mathbf{F}$.

21.5.1.1 Free energy per unit current mass

With the decomposition (21.3.45), the internal entropy production $(4.7.7)_1$,

$$
\begin{aligned}
T\hat{S}_{(i)} &= -\rho \frac{de}{dt} + \boldsymbol{\sigma} : \boldsymbol{\nabla}\mathbf{v} - \rho s \frac{dT}{dt} - \frac{\mathbf{Q}}{T} \cdot \boldsymbol{\nabla}T + \hat{\rho}\left(\tilde{e} - e + \frac{1}{2}(\tilde{\mathbf{v}} - \mathbf{v})^2\right) + \hat{\mathcal{E}}, \\[1em]
&= \left(\boldsymbol{\sigma} \cdot (\mathbf{F}^{\mathrm{e}})^{-\mathrm{T}} - \rho \frac{\partial e}{\partial \mathbf{F}^{\mathrm{e}}}\right) : \frac{d\mathbf{F}^{\mathrm{e}}}{dt} - \rho\left(s + \frac{\partial e}{\partial T}\right)\frac{dT}{dt} \\[1em]
&\quad + \boldsymbol{\sigma} : \mathbf{L}^{\mathrm{g}} - \rho \frac{\partial e}{\partial r}\frac{dr}{dt} - \rho \frac{\partial e}{\partial \xi}\frac{d\xi}{dt} - \frac{\mathbf{Q}}{T} \cdot \boldsymbol{\nabla}T + \hat{\rho}\left(\tilde{e} - e + \frac{1}{2}(\tilde{\mathbf{v}} - \mathbf{v})^2\right) + \hat{\mathcal{E}},
\end{aligned}
\tag{21.5.2}
$$

yields, with help of (2.11.13), the constitutive equations,

$$
\boldsymbol{\sigma} = \rho \frac{\partial e}{\partial \mathbf{F}^{\mathrm{e}}} \cdot (\mathbf{F}^{\mathrm{e}})^{\mathrm{T}} = \mathbf{F}^{\mathrm{e}} \cdot \rho \frac{\partial e}{\partial \mathbf{E}_{\mathrm{G}}^{\mathrm{e}}} \cdot (\mathbf{F}^{\mathrm{e}})^{\mathrm{T}}, \quad s = -\frac{\partial e}{\partial T}.
\tag{21.5.3}
$$

The reduced dissipation is cast in the format (4.8.13),

$$
T\hat{S}_{(i)} = \boldsymbol{\sigma} : \mathbf{L}^{\mathrm{g}} - \rho \frac{\partial e}{\partial r}\frac{dr}{dt} - \rho \frac{\partial e}{\partial \xi}\frac{d\xi}{dt} - \frac{\mathbf{Q}}{T} \cdot \boldsymbol{\nabla}T + \hat{\rho}\left(\tilde{e} - e + \frac{1}{2}(\tilde{\mathbf{v}} - \mathbf{v})^2\right) + \hat{\mathcal{E}} \geq 0.
\tag{21.5.4}
$$

With the constitutive equations,

$$\mathcal{X}\frac{d\xi}{dt} = -\rho\frac{\partial e}{\partial \xi} + \beta^{(d)}; \quad \boldsymbol{\kappa}\cdot\mathbf{Q} = -\frac{\boldsymbol{\nabla}T}{T}, \tag{21.5.5}$$

with $\mathcal{X} \geq 0$ a coefficient associated with the scalar dissipation phenomenon, and $\boldsymbol{\kappa}$ a positive semi-definite PsD second order tensor associated with heat conduction, we will still require,

$$\boldsymbol{\sigma} : \mathbf{L}^{\mathrm{g}} - \rho\frac{\partial e}{\partial r}\frac{dr}{dt} + \hat{\rho}\left(\tilde{e} - e + \frac{1}{2}(\tilde{\mathbf{v}} - \mathbf{v})^2\right) + \hat{\mathcal{E}}^{(\mathrm{g})} + \hat{\mathcal{E}}^{(\mathrm{r})} \geq 0, \tag{21.5.6}$$

where, with help of (21.3.63), $\hat{\rho}$ splits into a growth and a remodeling contribution,

$$\hat{\rho} = \frac{\rho}{\det \mathbf{F}^{\mathrm{e}}}\operatorname{tr}\mathbf{L}^{\mathrm{g}} + \frac{\rho}{\det \mathbf{F}}\frac{dr}{dt}. \tag{21.5.7}$$

21.5.1.2 Free energy per unit current volume

From the internal entropy production $(4.7.7)_2$, expressed in terms of the free energy per unit current volume $E = \rho\,e$,

$$T\hat{S}_{(\mathrm{i})} = -\frac{dE}{dt} - E\operatorname{div}\mathbf{v} + \boldsymbol{\sigma}:\boldsymbol{\nabla}\mathbf{v} - S\frac{dT}{dt} - \frac{\mathbf{Q}}{T}\cdot\boldsymbol{\nabla}T + \hat{\rho}\left(\tilde{e} + \frac{1}{2}(\tilde{\mathbf{v}} - \mathbf{v})^2\right) + \hat{\mathcal{E}}, \tag{21.5.8}$$

result thermoelastic constitutive equations and constitutive equations for the dissipative processes,

$$\underbrace{\boldsymbol{\sigma} - E\mathbf{I}}_{\text{Eshelby stress}} = \mathbf{F}^{\mathrm{e}}\cdot\frac{\partial E}{\partial \mathbf{E}_{\mathrm{G}}^{\mathrm{e}}}\cdot(\mathbf{F}^{\mathrm{e}})^{\mathrm{T}}, \quad S = -\frac{\partial E}{\partial T}, \quad \mathcal{X}\frac{d\xi}{dt} = -\frac{\partial E}{\partial \xi} + \beta^{(\mathrm{d})}; \quad \boldsymbol{\kappa}\cdot\mathbf{Q} = -\frac{\boldsymbol{\nabla}T}{T}, \tag{21.5.9}$$

with $\mathcal{X} \geq 0$, $\boldsymbol{\kappa}$ sPD. We are left with the inequality,

$$\left(\boldsymbol{\sigma} - E\mathbf{I}\right):\mathbf{L}^{\mathrm{g}} - \frac{\partial E}{\partial r}\frac{dr}{dt} + \hat{\rho}\left(\tilde{e} + \frac{1}{2}(\tilde{\mathbf{v}} - \mathbf{v})^2\right) + \hat{\mathcal{E}}^{(\mathrm{g})} + \hat{\mathcal{E}}^{(\mathrm{r})} \geq 0, \tag{21.5.10}$$

to be satisfied using the split (21.5.7) where $\hat{\rho}$ has the format (21.5.7).

21.5.2 The relations for a single solid in the reference configuration

For a single solid, the internal entropy productions (4.7.7) may be transformed, using the work relation (2.8.7), to,

$$\begin{aligned}
T\,\underline{\hat{S}}_{(\mathrm{i})} &= -m\frac{de}{dt} + \underline{\underline{\tau}}:\frac{d\mathbf{E}_{\mathrm{G}}}{dt} + \hat{m}\left(\tilde{e} - e + \frac{1}{2}(\tilde{\mathbf{v}} - \mathbf{v})^2\right) + \underline{\hat{\mathcal{E}}} - m\,s\frac{dT}{dt} - \frac{\mathbf{Q}}{T}\cdot\boldsymbol{\nabla}T; \\
&= -\frac{d}{dt}(m\,e) + \underline{\underline{\tau}}:\frac{d\mathbf{E}_{\mathrm{G}}}{dt} + \hat{m}\left(\tilde{e} + \frac{1}{2}(\tilde{\mathbf{v}} - \mathbf{v})^2\right) + \underline{\hat{\mathcal{E}}} - m\,s\frac{dT}{dt} - \frac{\mathbf{Q}}{T}\cdot\boldsymbol{\nabla}T.
\end{aligned} \tag{21.5.11}$$

The various forms of free energy satisfy the relations (2.6.21), namely $E\,V = \underline{E}\,\underline{V} = e\,M = \underline{e}\,\underline{M}$. The free energy per unit reference mass is such that $\underline{E} = \rho\,\underline{e}$. Therefore, using a potential per unit reference volume $\underline{E} = m\,e$ or per unit reference mass $\underline{e}$ makes no difference.

21.5.2.1 Free energy per unit current mass in the reference configuration

The potential here is the free energy per unit current mass e, but the relations are expressed in the reference configurations. The constitutive relations become

$$\underline{\underline{\tau}} = m\,(\mathbf{F}^{\mathrm{g}})^{-1}\cdot\frac{\partial e}{\partial \mathbf{E}_{\mathrm{G}}^{\mathrm{e}}}\cdot(\mathbf{F}^{\mathrm{g}})^{-\mathrm{T}}, \quad \underline{S} = -m\frac{\partial e}{\partial T}, \quad \underbrace{\underline{\mathcal{X}}}_{\geq 0}\frac{d\xi}{dt} = -m\frac{\partial e}{\partial \xi} + \underline{\beta}^{(\mathrm{d})}, \tag{21.5.12}$$

while the inequality,

$$(\mathbf{F}\cdot\underline{\underline{\tau}}\cdot\mathbf{F}^{\mathrm{T}}):\mathbf{L}^{\mathrm{g}} - m\frac{\partial e}{\partial r}\frac{dr}{dt} + \hat{m}\left(\tilde{e} - e + \frac{1}{2}(\tilde{\mathbf{v}} - \mathbf{v})^2\right) + \hat{\mathcal{E}}^{(\mathrm{g})} + \hat{\mathcal{E}}^{(\mathrm{r})} \geq 0, \tag{21.5.13}$$

remains to be satisfied.

21.5.2.2   Free energy per unit reference volume

Let $\underline{E} = m\,e = \det \mathbf{F}\,E = \det \mathbf{F}^{\mathrm{g}}\,E_\kappa$ be the free energy per unit reference volume. The thermoelastic constitutive relations result in the form,

$$\underline{\pmb{\tau}} = (\mathbf{F}^{\mathrm{g}})^{-1} \cdot \frac{\partial \underline{E}}{\partial \mathbf{E}_{\mathrm{G}}^{\mathrm{e}}} \cdot (\mathbf{F}^{\mathrm{g}})^{-\mathrm{T}}\,, \quad \underline{S} = -\frac{\partial \underline{E}}{\partial T}\,, \quad \underbrace{\underline{\chi}}_{\geq 0}\frac{d\xi}{dt} = -\frac{\partial \underline{E}}{\partial \xi} + \underline{\beta}^{(\mathrm{d})}\,. \tag{21.5.14}$$

The stress-strain relation in terms of the 2nd Piola-Kirchhoff stress may equivalently be expressed in terms of the Cauchy stress,

$$\pmb{\sigma} = \frac{1}{\det \mathbf{F}}\mathbf{F}^{\mathrm{e}} \cdot \frac{\partial \underline{E}}{\partial \mathbf{E}_{\mathrm{G}}^{\mathrm{e}}} \cdot (\mathbf{F}^{\mathrm{e}})^{\mathrm{T}} = \frac{1}{\det \mathbf{F}^{\mathrm{e}}}\mathbf{F}^{\mathrm{e}} \cdot \frac{\partial E_\kappa}{\partial \mathbf{E}_{\mathrm{G}}^{\mathrm{e}}} \cdot (\mathbf{F}^{\mathrm{e}})^{\mathrm{T}}\,. \tag{21.5.15}$$

The reduced dissipation inequality

$$(\mathbf{F} \cdot \underline{\pmb{\tau}} \cdot \mathbf{F}^{\mathrm{T}}) : \mathbf{L}^{\mathrm{g}} - \frac{\partial \underline{E}}{\partial r}\frac{dr}{dt} + \hat{m}\left(\tilde{e} + \frac{1}{2}(\tilde{\mathbf{v}} - \mathbf{v})^2\right) + \underline{\hat{\mathcal{E}}}^{(\mathrm{g})} + \underline{\hat{\mathcal{E}}}^{(\mathrm{r})} \geq 0\,, \tag{21.5.16}$$

remains to be satisfied.

21.6   Some models and their thermodynamic structure

Next to the physiological and biochemical conditions, the mechanical state that prevails during the growth of biological tissues is an important factor that may have a positive or a detrimental influence. Still, the mechanosensitive agents that transform the mechanical signals to a biochemical information are not identified. More, the actual influence of mechanics is far from being understood.

21.6.1   Homeostatic states and stimuli of growth

The models use several types of mechanical stimuli to trigger growth and resorption.

21.6.1.1   Energy, stress and strain

For cellular materials, like cancellous bones, the stimulus and homeostatic reference are phrased in terms of the free energy $\underline{E}$ [unit : Pa] or specific free energy $e = \underline{E}/m$ [unit : Pa/(kg/m^3)=m^2/s^2] divided by some power of the apparent bone density, or mass content or volume fraction of calcified tissue, Carter et al. [1987], Huiskes et al. [1987], Weinans et al. [1992]. The free energy admits a multiplicative decomposition of a function of the mass content times a function of deformation [unit : Pa],

$$\underline{E}(\mathbf{E}_{\mathrm{G}},\, m) = m\,e(\mathbf{E}_{\mathrm{G}},\, m) = \left(\frac{m}{m_h}\right)^{n_1} f(\mathbf{E}_{\mathrm{G}})\,. \tag{21.6.1}$$

The exponent n_1 takes values between 1 and 4, Carter and Hayes [1977], Gibson and Ashby [1988], p. 328. Typically, the rate of mass supply $\hat{m}$ adopts the format,

$$\frac{\hat{m}}{m_h} = \frac{\alpha}{t_{\mathrm{g}}}\left(\left(\frac{m_h}{m}\right)^{n_2} \underline{E} - \underline{E}_h\right)\,, \tag{21.6.2}$$

where α [unit : Pa^{-1}] is a constant, t_{g} a characteristic time, the subscript h refers to the homeostatic state, and the exponent $n_2 \geq n_1$ is another material constant. The latter condition is necessary, although not sufficient, to ensure stability of the growth and remodeling process. Stability is defined as 'the ability of a bone to arrive at and maintain a consistent structure' against all types of physiological perturbations, Harrigan and Hamilton [1992] [1993].

In the above formulation, the homeostatic state is defined by a single value of the free energy. In fact, for bones, the three zone law of Carter et al. [1989] introduces a "lazy zone", that separates bone resorption from bone growth, Fig. 21.6.1,

$$\frac{\hat{m}}{m_h} = \frac{\alpha}{t_{\mathrm{g}}}\begin{cases} \left(\frac{m_{\mathrm{hl}}}{m}\right)^{n_2} \underline{E} - \underline{E}_{\mathrm{hl}} & \text{if } \left(\frac{m_{\mathrm{hl}}}{m}\right)^{n_2} \underline{E} \leq \underline{E}_{\mathrm{hl}} & : \text{resorption;} \\[2mm] 0 & \text{if } \underline{E}_{\mathrm{hl}} \leq \left(\frac{m_h}{m}\right)^{n_2} \underline{E} \leq \underline{E}_{\mathrm{hu}} & : \text{homeostasis;} \\[2mm] \left(\frac{m_{\mathrm{hu}}}{m}\right)^{n_2} \underline{E} - \underline{E}_{\mathrm{hu}} & \text{if } \left(\frac{m_{\mathrm{hu}}}{m}\right)^{n_2} \underline{E} \geq \underline{E}_{\mathrm{hu}} & : \text{growth.} \end{cases} \tag{21.6.3}$$

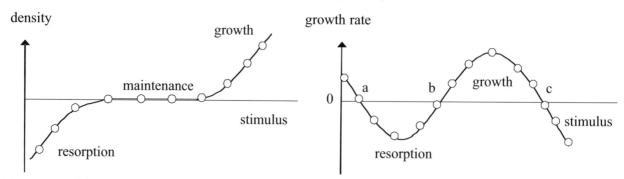

Fig. 21.6.1: (a) Typical bone growth and remodeling distinguish three zones of the stimulus. The underlying message conveyed by this behavior is that rest is detrimental to the bone mass, while activity is favorable. However, this information should be regarded with caution; (b) an alternative is Fung's proposal [1990]. Indeed, *in vitro* attempts in tissue engineering tend to show that a permanent load and a too large temporary load are detrimental to cell proliferation. Still none of these two representations captures the importance of cyclic, as opposed to permanent, loading.

where α [unit: Pa^{-1}] is a constant, t_g a characteristic time, $\underline{E}_{hl}$, $\underline{E}_{hu}$ typical free energies, m_{hl}, m_{hu} and m_h typical mass contents, and $n_2 \geq n_1$ another material constant. Homeostatic states now extend over a range of values of the stimulus, namely from $\underline{E}_{hl}$ to $\underline{E}_{hu}$, where the possible activities of agents that contribute to growth and resorption just balance, ensuring *maintenance* of the tissue.

The three zone model used for bones is subject to criticisms. Indeed, *in vitro* attempts in tissue engineering tend to show that a permanent load and a too large temporary load are detrimental to cell proliferation.

Fung's proposal [1990], p. 530, is a qualitative attempt targeting soft tissues like arteries, Fig. 21.6.1. His rule has the form,

$$\frac{\hat{m}}{m_h} \sim -(s-a)^{p_a}\,(s-b)^{p_b}\,(s-c)^{p_c}\,, \qquad (21.6.4)$$

where s is representative of the stimulus expressed in terms of the stress, a, b and c are homeostatic values, and p_a, p_b and p_c positive experimental parameters. A scaling is to be worked out because these coefficients may be non integers, and the sign of the three factors of (21.6.4) should be properly assessed. The growth rate at zero stress is considered to be positive.

Another characterization of the homeostatic state uses only strain, e.g., Eqn (21.6.12) below. The use of strain inevitably hinges on the definition of the reference state. This difficulty has led stress to be preferred to strain to define homeostatic states. Taber [2001] uses the format of Eqn (21.7.1) which involves the current stress and its value in the mature homeostatic state. Beaupré et al. [1990] define the stimulus through a weighted average, which accounts for the loading conditions, of an effective stress obtained from the free energy.

21.6.1.2 Tension and compression

In engineered cartilages, the stress environment is thought to have subtle effects, depending on the loading characteristics (frequency, intensity). In *in vitro* cultures of tumor cells, the confinement provided by gels has been observed to decrease the size of tumor spheroids with respect to spheroids grown in free suspension cultures, Helmlinger et al. [1997]a. In a model of vasculogenesis, Manoussaki et al. [1996] view traction as activator and matrix stiffness as inhibitor. These respective activator and inhibitor effects of tension and compression fit with the comments of Taber and Eggers [1996], their Section 8.3: leaving apart muscle activation, axial growth of a bar is expected if the stress is tensile, and resorption if the stress is compressive. The result is intuitively easier to guess if the growth strain is given and the motion is constrained: then growth will imply compression, and resorption tension. Under isometric conditions, growth takes place to relieve tension and resorption to relieve compression. Consider now the residual stresses due to growth/resorption in a hollow cylinder. Cutting the cylinder along an axial plane provides information on the sign of the residual stresses, Fig. 21.7.4. Growth implies the two sides of the cut to overlap. Re-establishing continuity of the matter is similar to bending that implies compression on the outer surface and tension on the inner surface. Resorption has opposite effects.

For a bar loaded along its axis 1, Rodriguez et al. [1994] take a transversal growth rate of the form $\dot{\lambda}_g = -|K|(\sigma_1 - \sigma_1^h)$, their Eqn (32), with $\sigma_1^h < 0$ the equilibrium stress: thus large axial tension implies transversal resorption, and large compression $\sigma_1 - \sigma_1^h < 0$ implies transversal growth. On the other hand, the growth model of Drozdov and Khanina [1997] implies that a bar grows in all directions under application of an axial compressive stress larger than a threshold. Laboratory tests aimed at improving the growth rate and stiffness of cartilages are mainly compressive, even if transverse extensions may appear in unconfined compression. In fact, a critical comparison of the effects of confined and unconfined compression in cartilage engineering would be welcome.

This state of affairs seems somehow paradoxical. On one hand, the observations of Helmlinger et al. [1997]a target more the *stiffness* than the *confinement*. On the other hand, the above models do not agree on the respective effects of compression and tension on the growth rate. While they are general purpose, actual applications need to account for the composition and microstructure of tissues.

21.6.1.3 Structuration and chemomechanical couplings during growth

The improvement of the elastic stiffness in *in vitro* cartilage cultures target a mechanochemical, or mechanobiological, effect. The reciprocal chemomechanical effect is active in tumors. Indeed, using elastography, McKnight et al. [2002] have observed that the shear stiffness of breast carcinoma (about 25 kPa) is 5 to 20 times higher than that of normal adipose breast tissue (about 3 kPa) or fibro-glandular breast tissue (about 8 kPa), their Fig. 9. The modifications of the material properties during growth, that we term *structuration*, are attributed to an increase of the crosslink density of collagen in engineered cartilages, Klisch et al. [2003]a. Alternatively, an increase of the binding of collagen and glycosaminoglycans is suggested for tumors by Netti et al. [2000].

21.6.2 Features of models of growth

Models of growth can be sorted out according to several characteristics, in particular,
   - whether they involve a growth strain, or not;
   - whether the mechanical behavior is elastic or viscoelastic: the distinction turns out to be an important issue for residual stresses;
   - whether they consider a single phase, or a mixture;
   - whether they can be viewed as thermodynamic systems open with respect to mass only, or to energy as well;
   - whether they include growth only, or growth and re-structuration.

21.6.3 Growth models without growth strain

We shall term these models *elastic*, so as to delineate them from the *elastic-growing* models that, in a small strain context, introduce the additive decomposition of the strain into an elastic part and a growth part. The formalism of the previous sections still applies, by setting formally the growth strain to zero.

21.6.3.1 A single solid/mixture model for bone remodeling

The model

In their pioneering continuum model of bone remodeling, Cowin and Hegedus [1976] describe a mixture, composed of two species, that is open with respect to momentum only. On the other hand, they view the bone matrix as another smaller system, open to the four types of interactions, namely mass, momentum, energy and entropy: balance and constitutive equations are of concern for this species only.

It is interesting to contrast this approach with the point of view taken here. Indeed, we do not view species as open systems with respect to other species. Species interact with the rest of the mixture via *transfer*. The whole mixture is open with respect to the surroundings, and we generically refer to these latter interactions as *growth*. The constitutive equations that control the two phenomena are quite distinct.

In the general formulation of Chapter 4, the supplied mass has been left with characteristics (velocity, internal energy, etc.) to be specified, Section 21.3.3. Cowin [2004], p. 94, presents physical arguments intended to justify that bone deposited, when interpreted in the proper time scale, can be endowed with the same internal energy as the existing bone at the deposition site.

Cowin and Hegedus [1976] neglect diffusion effects. Concentrating on the bone matrix, they postulate formal constitutive equations, in terms of the set of variables $\mathcal{V} = \{\mathbf{F}, T, \zeta\}$, with $\mathbf{F}$ deformation gradient, T temperature and $\zeta = m/\rho$ mass content of the bone matrix normalized by its reference value (in fact they include ∇T in the set of state variables, but we dispose of this dependence here). Invariance requirements are left to be applied. Equations are needed for the free energy e per unit current mass, and for the rate of energy supply (in unspecified form) $\hat{\mathcal{E}}$. The (CD) inequality then motivates hyperelastic relations for the stress (without viscous term), for the entropy and the rate of mass supply (to which we may add Fourier's law).

The constitutive equations of Hegedus and Cowin [1976]

Hegedus and Cowin [1976] deduce their constitutive equations in a thermodynamic framework, but they do not check satisfaction of the Clausius-Duhem inequality. They present arguments that motivate isothermal constitutive equations for the free energy $e = e(\mathbf{E}_{\mathrm{G}}, m)$ in terms of the Green strain and of the mass content, and for the rate of mass supply $dm/dt = dm/dt(\mathbf{E}_{\mathrm{G}}, m)$ in the form,

$$\boldsymbol{\sigma} = m\, \mathbb{E}(m) : \mathbf{E}_{\mathrm{G}}, \quad \frac{dm}{dt} = a(m) + \mathbf{A}(m) : \mathbf{E}_{\mathrm{G}} + \mathbf{E}_{\mathrm{G}} : \mathbf{C}(m) : \mathbf{E}_{\mathrm{G}}. \tag{21.6.5}$$

In essence, the multiplicative decomposition of the stiffness intends to highlight the dependence of the moduli of bone on the volume fraction of the bone matrix. The linear dependence is only demonstrative. Carter and Hayes [1977] use elastic moduli that depend on density through an exponential, while Gibson and Ashby [1988] consider a power dependence.

Hegedus and Cowin [1976] note that, while the strain is certainly small in bones, so that the stress can be considered linear in strain, clinical observations show that the remodeling rate may depend on second order terms in strain: this observation enhances the effects of periodic loading over static loading. In contrast to strain, the mass content, which is equal to intrinsic density times volume fraction times $\det\mathbf{F}$, may vary over a large range, like the volume fraction since the intrinsic density is constant. Thus they develop $\mathbb{E}(m)$ and $a(m)$ to first order and to second order in m respectively. The constitutive constant tensors involved in these expressions are endowed with the symmetry axes of the (existing) material at the deposition site, that is orthotropy. They note that some loadings are *neutral* for the remodeling, i.e. they leave the equilibrium remodeling unchanged.

In order to get an insight into the qualitative evolution of the mass during the growth process, an analysis of the dynamic behavior of the remodeling equations in a phase plane $(m, dm/dt)$ is worth investigation. Since the rate equation (21.6.5) is nonlinear, the behavior depends on the starting point. The analysis reveals the existence of a stable point and of an unstable point at either finite or infinite values of m. The second case is pathological, but the first case contains also ranges of m that are pathological, i.e., when m tends to extreme values.

Alternative constitutive equations based on the dissipation inequality

The free energy and velocity of the supplied mass are equal to their counterparts at the deposition site, namely $\tilde{e} = e$ and $\tilde{\mathbf{v}} = \mathbf{v}$. With a quadratic dissipation of the form (21.4.4), the mechanical part of the internal entropy production (21.5.2) reduces to

$$\begin{aligned}
-R\frac{dr}{dt} &= -\rho\frac{de}{dt} + \boldsymbol{\sigma} : \nabla\mathbf{v} - \rho s\frac{dT}{dt} + \hat{\mathcal{E}}, \\
&= \left(\boldsymbol{\sigma}\cdot\mathbf{F}^{-\mathrm{T}} - \rho\frac{\partial e}{\partial\mathbf{F}}\right) : \frac{d\mathbf{F}}{dt} - \rho\left(s + \frac{\partial e}{\partial T}\right) : \frac{dT}{dt} - \rho\frac{\partial e}{\partial r}\frac{dr}{dt} + \hat{\mathcal{E}} \geq 0,
\end{aligned} \tag{21.6.6}$$

so that, with use of the tensorial relation (2.11.13),

$$\boldsymbol{\sigma} = \rho\frac{\partial e}{\partial\mathbf{F}}\cdot\mathbf{F}^{\mathrm{T}} = \rho\,\mathbf{F}\cdot\frac{\partial e}{\partial\mathbf{E}_{\mathrm{G}}}\cdot\mathbf{F}^{\mathrm{T}}, \quad s = -\frac{\partial e}{\partial T}. \tag{21.6.7}$$

The rate of energy supply $\hat{\mathcal{E}}$ is contributed only by a remodeling contribution, say $\hat{\mathcal{E}}_{(\mathrm{r})} = \beta_{(\mathrm{r})}\, dr/dt$, with $\beta_{(\mathrm{r})}$ [unit: Pa] a scalar function of the state variables. The dissipation simplifies to,

$$-R\frac{dr}{dt} = \left(-\rho\frac{\partial e}{\partial r} + \beta_{(\mathrm{r})}\right)\frac{dr}{dt} \geq 0 \Rightarrow -R = -\rho\frac{\partial e}{\partial r} + \beta_{(\mathrm{r})}. \tag{21.6.8}$$

The inequality is satisfied by introducing the rate equation for the remodeling variable r,

$$\frac{dr}{dt} = -\frac{\alpha_{(\mathrm{r})}}{t_{\mathrm{r}}}\, R = \frac{\alpha_{(\mathrm{r})}}{t_{\mathrm{r}}}\left(\beta_{(\mathrm{r})} - \rho\frac{\partial e}{\partial r}\right), \tag{21.6.9}$$

and t_{r} and $\alpha_{(\mathrm{r})} \geq 0$ are respectively a characteristic time [unit: s] and a positive scalar [unit: Pa^{-1}].

21.6.3.2 A model for bone remodeling as a composite material

Büchler et al. [2003] develop a growth model where bone is viewed as a composite material constituted by a bone matrix b and fibers f. The volume fraction of the two constituents is changing, and in fact this change is considered typical of growth. The two constituents undergo the deformation gradient $\mathbf{F}$ and strain $\mathbf{E}_{\mathrm{G}}$ of the mixture. Let n^k be the volume fraction of constituent k, of course $n^b + n^f = 1$. To each species are attached also an intrinsic density ρ_k, and a specific free energy $e_k = e_k(\mathbf{F})$, actually $e_k = e_k(\mathbf{E}_{\mathrm{G}})$: this format amounts to an exponent $n_1 = 1$ in (21.6.1). To the mixture are attached a density and a mass average free energy *per unit current mass* e, namely

$$\rho = \sum_{k=b,f} n^k \rho_k \quad \Rightarrow \quad \hat{\rho} = (\rho_f - \rho_b)\frac{dn^f}{dt}, \tag{21.6.10}$$

and

$$\rho\, e = \sum_{k=b,f} n^k \rho_k e_k. \tag{21.6.11}$$

Based on observations, Büchler et al. [2003] use the following remodeling rule,

$$\frac{dn^f}{dt} = \frac{1}{t_{\mathrm{r}}}\left(M(\epsilon) - n^f\right), \tag{21.6.12}$$

where $M(\epsilon) \in [0,1]$ is an experimental function of the amplitude of the deviatoric strain $\epsilon = |\mathrm{dev}\,\mathbf{E}_{\mathrm{G}}|$. This format identifies clearly a set of homeostatic states. The function $M(\epsilon) \in [0,1]$ is monotonously increasing so that a small strain implies bone growth while a too large strain implies bone resorption and formation of fibrous material. Note that the strain is involved only through its deviatoric part, assuming that volume changes do not trigger growth/resorption.

Alternative constitutive equations based on the dissipation inequality

The (CD) inequality (4.7.7) should write,

$$T\hat{S}_{(\mathrm{i},1,4)} = -\rho\,\frac{de}{dt} + \boldsymbol{\sigma}:\boldsymbol{\nabla}\mathbf{v} + \hat{\rho}\left(\tilde{e} - e + \frac{1}{2}(\tilde{\mathbf{v}} - \mathbf{v})^2\right) + \hat{\mathcal{E}} \geq 0. \tag{21.6.13}$$

The rate of energy supply $\hat{\mathcal{E}} = \beta_{(\mathrm{r})}\,dn^f/dt$, with $\beta_{(\mathrm{r})}$ a state dependent scalar [unit : Pa], is an addition to the original equation which considered a system open to mass only. Once again, assuming the deposited mass to have the velocity and free energy of the existing mass at the deposition point, the Clausius-Duhem inequality simplifies to,

$$-\rho\,\frac{de}{dt} + \boldsymbol{\sigma}:\boldsymbol{\nabla}\mathbf{v} + \hat{\mathcal{E}} \geq 0, \tag{21.6.14}$$

and, since $e = e(\mathbf{F}, n^f)$,

$$\left(\boldsymbol{\sigma}\cdot\mathbf{F}^{-\mathrm{T}} - \rho\,\frac{\partial e}{\partial\mathbf{F}}\right):\frac{d\mathbf{F}}{dt} + \left(-\rho\,\frac{\partial e}{\partial n^f} + \beta_{(\mathrm{r})}\right)\frac{dn^f}{dt} \geq 0. \tag{21.6.15}$$

Therefore the stress is given as in (21.6.7), and the (CD) inequality is satisfied by the remodeling rate,

$$\frac{dn^f}{dt} = \frac{\alpha_{(\mathrm{r})}}{t_{\mathrm{r}}}\left(\beta_{(\mathrm{r})} + \frac{\rho_b\,\rho_f}{\rho}\,(e_b - e_f)\right), \tag{21.6.16}$$

with t_{r} a characteristic time [unit : s] and $\alpha_{(\mathrm{r})} > 0$ a positive scalar [unit : Pa^{-1}]. The derivation has used the relation $\rho\,\partial e/\partial n^f = (e_f - e_b)\,\rho_f\,\rho_b/\rho$. A more refined analysis would introduce biochemical state variables like cells, nutrients, growth factors, etc. In addition, the rate of energy supply might be seen as contributed by several physiological phenomena endowed with distinct characteristic times, for example aimed at representing the activities of the osteoclasts and osteoblasts.

There is sufficient freedom in the state dependent scalars $\alpha_{(\mathrm{r})}$ and $\beta_{(\mathrm{r})}$ to accommodate the rule (21.6.12) while satisfying the dissipation inequality.

21.6.3.3 A model for a solid including surface and volume supplies

Kuhl and Steinmann [2003] develop a single phase isothermal framework where the supplies in mass and entropy contain both volume and surface contributions as indicated in Section 4.3.3. On the other hand, the material does not display a growth strain and the velocity and free energy of the deposited mass are tacitly assumed to be equal to their counterparts at the deposition site. Perhaps, this paper is the sole work before Loret and Simões [2005]b where the dissipation inequality is actually taken care of.

The rate of mass supply in volume $\hat{m}$ adopts the format (21.6.2). The specific free energy e assumes the format (21.6.1). The rate of supply in volume $\hat{\underline{\mathcal{E}}} = 0 - T\,\hat{\underline{\mathcal{S}}}$ is chosen as $(n_1 - 1)\,e$ times the rate of mass supply $\hat{m}$, so that the associated reduced dissipation (21.5.13),

$$-m\,\frac{\partial e}{\partial m}\,\hat{m} + 0 - T\,\hat{\underline{\mathcal{S}}} = \left(-m\,\frac{\partial e}{\partial m} + (n_1 - 1)\,e\right)\hat{m}\,, \qquad (21.6.17)$$

vanishes.

Still, the dissipation inequality now includes the surface terms (4.7.30). Motivated by the Fourier's law, which states that the heat flux is proportional to the gradient of temperature, the mass flux supply $\hat{\mathbf{m}}$ is taken equal to $\boldsymbol{\nabla} m$, the internal energy flux supply $\hat{\mathbf{U}}$ is zero, and the entropy supply flux $\hat{\mathbf{S}} = -(n_1 - 1)\,e\,\boldsymbol{\nabla} m/T$ is chosen so as to satisfy the dissipation inequality,

$$-m\,\frac{\partial e}{\partial m}\,\underline{\mathrm{div}\,\hat{\mathbf{m}}} + \underline{\mathrm{div}\,\hat{\mathbf{U}}} - T\,\underline{\mathrm{div}\,\hat{\mathbf{S}}} = (n_1 - 1)^2\,\frac{e}{m}\,\boldsymbol{\nabla} m \cdot \boldsymbol{\nabla} m > 0\,. \qquad (21.6.18)$$

Therefore, in contrast to the volume supplies, the surface supplies contribute to dissipation. In fact, the flux supplies introduce diffusion and an example shows the spatial homogenization of a mass distribution which is initially heterogeneous.

Indeed, the assumption of a vanishing dissipation needs to be removed if actual growth mechanisms, which are irreversible, are addressed: anabolic and catabolic pathways are distinct.

Note 21.1 on volume fraction and specific surface area.

In essence, mass content and volume fraction play similar roles in the growth and remodeling of cellular tissues. Alternatively, the specific surface area might be thought to be the important geometrical parameter, because exchange of nutrients, proteins, etc. take place over surfaces. The concept plays a key role in energy exchanges, Section 10.8, and in fluid exchanges, Section 24.8. Quite curiously, some authors argue that the specific surface area and void ratio are related in cortical bones, Eqn (51) in Rouhi et al. [2004], so that the difference between the two approaches might be more quantitative than qualitative. Still, the latter relation is certainly not satisfied in general and, to a single volume fraction, may well correspond distinct free surface densities, and conversely.

Note however that the standard approach of the mechanics of porous media does not distinguish between surface and volume fractions, Section 4.3.2. A deeper account of the microstructure of the tissue should be introduced. A quantitative description of the evolving internal surfaces may be obtained through micro-computed tomography. Let S denote these surfaces within a volume V. Then the supply $\hat{\Psi} = \mathrm{div}\,\hat{\boldsymbol{\Psi}}$ in the volume V, of mass, momentum, internal energy and entropy due to surface supply can be obtained via the divergence theorem once the corresponding flux $\hat{\boldsymbol{\Psi}}$ and the surfaces S are known, namely $\int_V \mathrm{div}\,\hat{\boldsymbol{\Psi}}\,dV = \int_S \hat{\boldsymbol{\Psi}} \cdot d\mathbf{S}$.

21.6.4 Growth models including growth strain

21.6.4.1 A model for spherical tumors with growth and necrosis

The model

In a single phase context, MacArthur and Please [2004] represent the tissue response via an isotropic incompressible Maxwell fluid,

$$\boldsymbol{\sigma} = -p\,\mathbf{I} + \mathbf{s}\,, \quad \mathbf{s} + t_{\mathrm{v}}\,\frac{d\mathbf{s}}{dt} = 2\,\mu\,t_{\mathrm{v}}\left(\frac{d\boldsymbol{\epsilon}}{dt} - \frac{d\boldsymbol{\epsilon}^{\mathrm{g}}}{dt}\right)\,, \qquad (21.6.19)$$

with $\boldsymbol{\sigma}$ stress, p pressure, $\mathbf{s}$ stress deviator, $\boldsymbol{\epsilon}$ strain, $\boldsymbol{\epsilon}^{\mathrm{g}}$ growth strain, μ [unit : Pa] shear modulus, and t_{v} [unit : s] tissue relaxation time. The incompressibility condition applies to the elastic strain $\mathbf{I} : (\boldsymbol{\epsilon} - \boldsymbol{\epsilon}^{\mathrm{g}}) = 0$, so that the rhs of constitutive relation (21.6.19) is actually deviatoric.

Since nutrient propagation is much faster than tumor growth, the field equations are considered at steady state. Balance of the mass of the cells, whose apparent density ρ^{c} is assumed constant, yields $\partial \rho^{\mathrm{c}}/\partial t + \mathrm{div}\,(\rho^{\mathrm{c}}\mathbf{v}) = \hat{\rho}^{\mathrm{c}}$, where $\mathbf{v}$ is the cell migration velocity provided by mechanical equilibrium, $\hat{\rho}^c$ is the rate of mass supply provided by a reaction-diffusion equation. Moreover, for density preserving growth, $\mathbf{I} : d\boldsymbol{\epsilon}^g/dt = \hat{\rho}^c/\rho^c$, and, via (4.4.11),

$$\boldsymbol{\epsilon}^g = \mathrm{Ln}\, \frac{M_s(t)}{M_s(0)}\, \mathbf{I} = \mathrm{Ln}\, \frac{V_s(t)}{V_s(0)}\, \mathbf{I}. \tag{21.6.20}$$

Consequently, for simultaneous elastic incompressibility and density preserving growth

$$\mathrm{div}\, \mathbf{v} = \mathbf{I} : \frac{d\boldsymbol{\epsilon}}{dt} = \mathbf{I} : \frac{d\boldsymbol{\epsilon}^g}{dt} = \frac{\hat{\rho}^c}{\rho^c}, \tag{21.6.21}$$

and, as expected, a single information for both mass supply and growth strain rate is required.

In Landman and Please [2001], the growth strain rate is isotropic, and it depends directly on nutrient concentration c, namely $d\boldsymbol{\epsilon}^g/dt = F(c)\,\mathbf{I}/3$. The growth behavior is controlled by a critical level of nutrient concentration $c = \alpha$, and $F(c > \alpha) = a$, while $F(c < \alpha) = -a\lambda$, with a a positive constant and λ another positive constant characterizing tissue contraction (or cell death). The concentration $c = \alpha$ can be seen as homeostatic if $F(\alpha) = 0$. Since $\frac{1}{3}\,\mathrm{tr}\,\boldsymbol{\sigma} = -p$ in this model, the criteria for necrosis[21.4] can equivalently be based on the sign of the pressure or of the mean stress. The idea is to indicate that under tension cells can not communicate and/or receive nutrients[21.5].

The constitutive restrictions imposed by the dissipation inequality are considered below in two subcases which both assume a traceless elastic strain and a traceless viscous strain.

Constitutive restrictions by the dissipation inequality: the purely viscoelastic tissue

Let us consider first the thermodynamic derivation of the elastically incompressible Maxwell model without growth strain, namely $\mathrm{tr}\, d\boldsymbol{\epsilon}^e/dt = 0$. The entropy production may be recast in several formats,

$$-\rho\, \frac{de}{dt} + \boldsymbol{\sigma} : \frac{d\boldsymbol{\epsilon}}{dt} = -\mathbf{f}_v : \frac{d\boldsymbol{\xi}}{dt} \geq 0,$$
$$(\mathbf{s} - \rho\, \frac{de}{d\boldsymbol{\epsilon}^e}) : \frac{d\boldsymbol{\epsilon}^e}{dt} + \mathbf{s} : \frac{d\boldsymbol{\epsilon}^v}{dt} = -\mathbf{f}_v : \frac{d\boldsymbol{\xi}}{dt} \geq 0. \tag{21.6.22}$$

Since the free energy e is independent of the rates of elastic strain $\boldsymbol{\epsilon}^e$ and viscous strain $\boldsymbol{\epsilon}^v$, then the left parenthesis should vanish, providing the stress on the spring. The dissipative power is generated by the stress $\mathbf{s}$ that acts on the dashpot, namely $\mathbf{s} : d\boldsymbol{\epsilon}^v/dt$. Therefore

$$\mathbf{s} = \boldsymbol{\sigma} + p\mathbf{I} = \rho\, \frac{de}{d\boldsymbol{\epsilon}^e} = \underbrace{2\,\mu}_{>0}\, \boldsymbol{\epsilon}^e; \qquad \underbrace{\mathbf{s}}_{\mathbf{f}_v = -\mathbf{s}} = \underbrace{2\,\mu\, t_v}_{>0}\, \underbrace{\frac{d\boldsymbol{\epsilon}^v}{dt}}_{\boldsymbol{\xi} = \boldsymbol{\epsilon}^v}; \qquad \boldsymbol{\epsilon} = \boldsymbol{\epsilon}^e + \boldsymbol{\epsilon}^v, \tag{21.6.23}$$

which finally delivers the viscoelastic relation,

$$\mathbf{s} + t_v\, \frac{d\mathbf{s}}{dt} = 2\,\mu\, t_v\, \frac{d\boldsymbol{\epsilon}}{dt}. \tag{21.6.24}$$

Satisfaction of the dissipation inequality requires the shear modulus $\mu > 0$ [unit : Pa] and the shear viscosity $\mu\, t_v > 0$ [unit : Pa×s] to be positive.

Constitutive restrictions by the dissipation inequality: the viscoelastic-growing tissue

In presence of growth but absence of remodeling, and assuming the deposited mass to have the velocity and free energy of the existing mass at the deposition point, the thermodynamic derivation extending the purely viscoelastic model implies the inequality,

$$-\rho\, \frac{de}{dt} + \boldsymbol{\sigma} : \frac{d\boldsymbol{\epsilon}}{dt} + \hat{\mathcal{E}} = -\mathbf{f}_v : \frac{d\boldsymbol{\xi}}{dt} - \mathbf{f}_g : \frac{d\boldsymbol{\epsilon}^g}{dt} \geq 0,$$
$$(\mathbf{s} - \rho\, \frac{de}{d\boldsymbol{\epsilon}^e}) : \frac{d\boldsymbol{\epsilon}^e}{dt} + \mathbf{s} : \frac{d\boldsymbol{\epsilon}^v}{dt} + \boldsymbol{\sigma} : \frac{d\boldsymbol{\epsilon}^g}{dt} + \hat{\mathcal{E}} = -\mathbf{f}_v : \frac{d\boldsymbol{\xi}}{dt} - \mathbf{f}_g : \frac{d\boldsymbol{\epsilon}^g}{dt} \geq 0, \tag{21.6.25}$$

[21.4]*Apoptosis* is the genetically programmed cell death, where the integrity of the different parts of the cell is lost. On the other hand, *necrosis* is the death of the cells due to injuries. It gives rise to an inflammatory response and, by osmotic effect, to swelling due to an excessive ionic concentration in the cytoplasm.

[21.5]Most tumor models consider spherical regions, and nutrients are delivered at the outer boundary only. Consequently, the interior of the tumor where the level of nutrients is lower will at some step undergo a volume contraction and ultimately become necrosed. Models that account for necrosis have to track two fronts: the inner necrotic front defined as explained above and the outer boundary of the tumor. The growth of the exterior cells participates to the growth of the tumor. Moreover, since the exterior cells grow, they create, in the tissue outside the tumor, orthoradial tension and radial compression.

and therefore,

$$\mathbf{s} = \boldsymbol{\sigma} + p\,\mathbf{I} = 2\,\mu\,\boldsymbol{\epsilon}^{\mathrm{e}}; \quad \mathbf{s} = 2\,\mu\,t_{\mathrm{v}}\,\frac{d\boldsymbol{\epsilon}^{\mathrm{v}}}{dt}; \quad \boldsymbol{\epsilon} = \boldsymbol{\epsilon}^{\mathrm{e}} + \boldsymbol{\epsilon}^{\mathrm{v}} + \boldsymbol{\epsilon}^{\mathrm{g}}, \tag{21.6.26}$$

resulting in the constitutive equation (21.6.19). The reduced dissipation rate

$$\boldsymbol{\sigma} : \frac{d\boldsymbol{\epsilon}^{\mathrm{g}}}{dt} + \hat{\mathcal{E}} = -\mathbf{f}_{\mathrm{g}} : \frac{d\boldsymbol{\epsilon}^{\mathrm{g}}}{dt} \geq 0\,, \tag{21.6.27}$$

is still required to be positive. With the assumptions,

$$\hat{\mathcal{E}} = \boldsymbol{\beta} : \frac{d\boldsymbol{\epsilon}^{\mathrm{g}}}{dt}, \quad -\mathbf{f}_{\mathrm{g}} = \left(\frac{\boldsymbol{\alpha}_{\mathrm{g}}}{t_{\mathrm{g}}}\right)^{-1} : \frac{d\boldsymbol{\epsilon}^{\mathrm{g}}}{dt}, \tag{21.6.28}$$

where $\boldsymbol{\beta} = \boldsymbol{\beta}(\boldsymbol{\epsilon}^{\mathrm{e}}, p)$ [unit : Pa] is an arbitrary symmetric second order tensor, and $\boldsymbol{\alpha}_{\mathrm{g}}$ [unit : Pa^{-1}] a positive definite fourth order tensor endowed with major and minor symmetries, then

$$\frac{d\boldsymbol{\epsilon}^{\mathrm{g}}}{dt} = \frac{\boldsymbol{\alpha}_{\mathrm{g}}}{t_{\mathrm{g}}} : (\boldsymbol{\sigma} + \boldsymbol{\beta})\,. \tag{21.6.29}$$

If the set of state variables includes, besides $\boldsymbol{\epsilon}^{\mathrm{e}}$, a nutrient concentration c, then the rhs can be chosen equal to $F(c)\,\mathbf{I}/3$, in such a way that the growth law of Landman and Please [2001] can be accommodated.

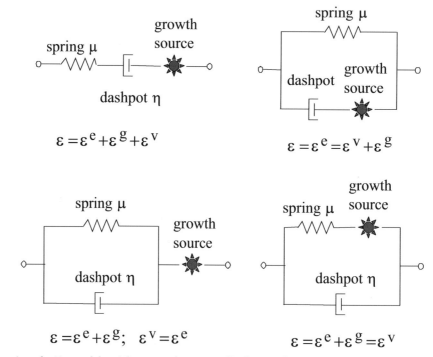

Fig. 21.6.2: Four viscoelastic models with a growth source, displaying distinct arrangements of the spring, dashpot and growth source, and therefore distinct combinations of the associated elastic strain $\boldsymbol{\epsilon}^{\mathrm{e}}$, viscous strain $\boldsymbol{\epsilon}^{\mathrm{v}}$ and growth strain $\boldsymbol{\epsilon}^{\mathrm{g}}$.

21.6.4.2   A comparison of several viscoelastic growing tissues

It is interesting to compare several viscoelastic materials endowed with a growth strain, Fig. 21.6.2. A Maxwell model where the elastic, viscous and growth strains sum is described in Section 21.6.4.1.

For a Kelvin model where the spring is in parallel with the dashpot, the dissipation inequality now has the form,

$$-\rho\,\frac{de}{dt} + \boldsymbol{\sigma} : \frac{d\boldsymbol{\epsilon}}{dt} + \hat{\mathcal{E}} = -\mathbf{f}_{\mathrm{v}} : \frac{d\boldsymbol{\xi}}{dt} - \mathbf{f}_{\mathrm{g}} : \frac{d\boldsymbol{\epsilon}^{\mathrm{g}}}{dt} \geq 0$$

$$(\mathbf{s} - \rho\,\frac{de}{d\boldsymbol{\epsilon}^{\mathrm{e}}}) : \frac{d\boldsymbol{\epsilon}^{\mathrm{e}}}{dt} + \boldsymbol{\sigma} : \frac{d\boldsymbol{\epsilon}^{\mathrm{g}}}{dt} + \boldsymbol{\beta} : \frac{d\boldsymbol{\epsilon}^{\mathrm{g}}}{dt} = -\mathbf{f}_{\mathrm{v}} : \frac{d\boldsymbol{\xi}}{dt} - \mathbf{f}_{\mathrm{g}} : \frac{d\boldsymbol{\epsilon}^{\mathrm{g}}}{dt} \geq 0\,, \tag{21.6.30}$$

and therefore,

$$\mathbf{s} = 2\,\mu\,\boldsymbol{\epsilon}^{\mathrm{e}} + 2\,\mu\,t_{\mathrm{v}}\,\frac{d\boldsymbol{\epsilon}^{\mathrm{v}}}{dt}; \quad \boldsymbol{\epsilon} = \boldsymbol{\epsilon}^{\mathrm{e}} + \boldsymbol{\epsilon}^{\mathrm{g}}, \quad \boldsymbol{\epsilon}^{\mathrm{e}} = \boldsymbol{\epsilon}^{\mathrm{v}}, \quad \frac{d\boldsymbol{\epsilon}^{\mathrm{g}}}{dt} = \frac{\boldsymbol{\alpha}_{\mathrm{g}}}{t_{\mathrm{g}}} : (\boldsymbol{\sigma} + \boldsymbol{\beta}), \tag{21.6.31}$$

the material parameters being restricted as in Section 21.6.4.1.

The viscoelastic model of Section 21.4.2.1 where the spring and growth source are in parallel with a dashpot may be given a small strain version, namely

$$(\mathbf{s} - \rho\,\frac{de}{d\boldsymbol{\epsilon}^{\mathrm{e}}}) : \frac{d\boldsymbol{\epsilon}}{dt} + \rho\,\frac{de}{d\boldsymbol{\epsilon}^{\mathrm{e}}} : \frac{d\boldsymbol{\epsilon}^{\mathrm{g}}}{dt} + \boldsymbol{\beta} : \frac{d\boldsymbol{\epsilon}^{\mathrm{g}}}{dt} = -\mathbf{f}_{\mathrm{v}} : \frac{d\boldsymbol{\xi}}{dt} - \mathbf{f}_{\mathrm{g}} : \frac{d\boldsymbol{\epsilon}^{\mathrm{g}}}{dt} \geq 0, \tag{21.6.32}$$

and therefore,

$$\mathbf{s} = 2\,\mu\,\boldsymbol{\epsilon}^{\mathrm{e}} + 2\,\mu\,t_{\mathrm{v}}\,\frac{d\boldsymbol{\epsilon}^{\mathrm{v}}}{dt}; \quad \boldsymbol{\epsilon} = \boldsymbol{\epsilon}^{\mathrm{e}} + \boldsymbol{\epsilon}^{\mathrm{g}} = \boldsymbol{\epsilon}^{\mathrm{v}}, \quad \frac{d\boldsymbol{\epsilon}^{\mathrm{g}}}{dt} = \frac{\boldsymbol{\alpha}_{\mathrm{g}}}{t_{\mathrm{g}}} : (2\,\mu\,\boldsymbol{\epsilon}^{\mathrm{e}} + \boldsymbol{\beta}). \tag{21.6.33}$$

Elastic and viscous incompressibilities imply the growth strain to be traceless as well. Therefore the second order tensor $\boldsymbol{\beta}$ is deviatoric, and the fourth order tensor $\boldsymbol{\alpha}_{\mathrm{g}}$ satisfies the restriction $\mathbf{I} : \boldsymbol{\alpha}_{\mathrm{g}} = \boldsymbol{\alpha}_{\mathrm{g}} : \mathbf{I} = \mathbf{0}$.

For a viscoelastic model where the dashpot and growth source are in parallel with a spring, the dissipation inequality,

$$(\mathbf{s} - \rho\,\frac{de}{d\boldsymbol{\epsilon}^{\mathrm{e}}}) : \frac{d\boldsymbol{\epsilon}^{\mathrm{v}}}{dt} + (\mathbf{s} - \rho\,\frac{de}{d\boldsymbol{\epsilon}^{\mathrm{e}}} + \boldsymbol{\beta}) : \frac{d\boldsymbol{\epsilon}^{\mathrm{g}}}{dt} = -\mathbf{f}_{\mathrm{v}} : \frac{d\boldsymbol{\xi}}{dt} - \mathbf{f}_{\mathrm{g}} : \frac{d\boldsymbol{\epsilon}^{\mathrm{g}}}{dt} \geq 0, \tag{21.6.34}$$

yields,

$$\mathbf{s} = 2\,\mu\,\boldsymbol{\epsilon}^{\mathrm{e}} + 2\,\mu\,t_{\mathrm{v}}\,\frac{d\boldsymbol{\epsilon}^{\mathrm{v}}}{dt}; \quad \boldsymbol{\epsilon} = \boldsymbol{\epsilon}^{\mathrm{e}} = \boldsymbol{\epsilon}^{\mathrm{v}} + \boldsymbol{\epsilon}^{\mathrm{g}}, \quad \frac{d\boldsymbol{\epsilon}^{\mathrm{g}}}{dt} = \frac{\boldsymbol{\alpha}_{\mathrm{g}}}{t_{\mathrm{g}}} : (\mathbf{s} - 2\,\mu\,\boldsymbol{\epsilon}^{\mathrm{e}} + \boldsymbol{\beta}). \tag{21.6.35}$$

The growth strain is traceless as in the previous example and the same restrictions on $\boldsymbol{\alpha}$ and $\boldsymbol{\beta}$ apply.

21.6.4.3 A single phase general purpose model

Lubarda and Hoger [2002] and Lubarda [2004]a consider that growth is triggered by tension while compression implies resorption. Actually, compression and tension are relative to the homeostatic stress.

Let $\mathbf{F}^{\mathrm{g}} = \mathbf{diag}\,[\lambda_{\mathrm{g}1}, \lambda_{\mathrm{g}2}, \lambda_{\mathrm{g}3}]$ be the spectral representation of the growth transformation. For isotropic growth, $\lambda_{\mathrm{g}i} = \lambda_{\mathrm{g}}$, $i \in [1,3]$, Lubarda and Hoger [2002] propose a rate of growth of the form

$$\frac{d\lambda_{\mathrm{g}}}{dt} = \Lambda(\lambda_{\mathrm{g}}, \mathrm{tr}\,\boldsymbol{\tau}_{\kappa}), \tag{21.6.36}$$

where $\boldsymbol{\tau}_{\kappa} = \det \mathbf{F}^{\mathrm{e}}\,(\mathbf{F}^{\mathrm{e}})^{-1} \cdot \boldsymbol{\sigma} \cdot (\mathbf{F}^{\mathrm{e}})^{-\mathrm{T}}$ is the 2nd Piola-Kirchhoff stress relative to the intermediate configurations (κ). The growth rate depends linearly on the stress and has a bounded value,

$$\frac{d\lambda_{\mathrm{g}}}{dt} = \Lambda_{0\epsilon} \times \left(\frac{\lambda_{\mathrm{g}} - \lambda_{\mathrm{g},\epsilon}^{\infty}}{1 - \lambda_{\mathrm{g},\epsilon}^{\infty}}\right)^{m_{\epsilon}} \times \mathrm{tr}\,(\boldsymbol{\tau}_{\kappa} - \boldsymbol{\tau}_{\kappa}^{h}), \tag{21.6.37}$$

with $\epsilon = \mathrm{sign}\,\mathrm{tr}\,(\boldsymbol{\tau}_{\kappa} - \boldsymbol{\tau}_{\kappa}^{h})$, $\Lambda_{0\epsilon} > 0$, $m_{\epsilon} > 0$, $\boldsymbol{\tau}_{\kappa}^{h}$ the homeostatic stress, and $\lambda_{\mathrm{g},\epsilon}^{\infty}$ the homeostatic stretch in tension or compression. Thus $d\lambda_{\mathrm{g}}/dt > 0$ and $\lambda_{\mathrm{g}} > 1$ for tension, while $d\lambda_{\mathrm{g}}/dt < 0$ and $\lambda_{\mathrm{g}} < 1$ for compression.

The above relations are extended to a transversely isotropic material in the intermediate configuration, with $\mathbf{e}_{\kappa}$ the orthotropy direction, and $\mathbf{F}^{\mathrm{g}} = \lambda_{\mathrm{g}2}\,\mathbf{I} + (\lambda_{\mathrm{g}1} - \lambda_{\mathrm{g}2})\,\mathbf{e}_{\kappa} \otimes \mathbf{e}_{\kappa}$. The rate relations for the principal stretches in the orthotropy direction $a = 1$ and orthogonally $a = 2$ take the form

$$\frac{d\lambda_{\mathrm{g}a}}{dt} = \Lambda_{0a} \times \left(\frac{\lambda_{\mathrm{g}a} - \lambda_{\mathrm{g},\epsilon_a}^{\infty}}{1 - \lambda_{\mathrm{g},\epsilon_a}^{\infty}}\right)^{m_{\epsilon_a}} \times \Sigma_a(\boldsymbol{\tau}_{\kappa}, \boldsymbol{\tau}_{\kappa}^{h}, \mathbf{e}_{\kappa} \otimes \mathbf{e}_{\kappa}), \tag{21.6.38}$$

with $\epsilon_a = \mathrm{sign}\,\Sigma_a$, and Σ_a a scalar function of the stress, homeostatic stress and orthotropy direction.

In fact, the homeostatic stress is an addition to their formulation. But the complete formulation points out that growth stops either because the stress is initially homeostatic or because the growth stretch, namely the deposited mass from a reference state, reaches progressively a homeostatic value. Thus, the definition of the homeostatic state involves both stress and strain and, of course, the issue of the reference state re-surfaces here.

21.6.4.4   A single phase model for aortic growth and cardiac development

A model and applications to the calculations of residual stresses in aortas and in the developing heart, with and without muscle activation, are reviewed in Section 21.7.1.

21.6.4.5   A mixture model for aortic growth, mass production and mass resorption

Humphrey and Rajagopal [2003] develop a two phase mixture model to study the growth of arteries and their adaption to varying blood flow. They distinguish mass production and mass resorption. The underlying idea is that these two entities are governed by distinct biochemical mechanisms and distinct characteristic times. The rates of mass production (subscript α) and mass resorption (subscript ω) are postulated in the format,

$$
\begin{aligned}
\hat{\rho}_\alpha &= \hat{\rho}_h &+& f_\alpha(\boldsymbol{\sigma} - \boldsymbol{\sigma}_h)\,\hat{\rho}_{\alpha,\text{mech}}, \\
\hat{\rho}_\omega &= -\hat{\rho}_h &+& f_\omega(\boldsymbol{\sigma} - \boldsymbol{\sigma}_h)\,\hat{\rho}_{\omega,\text{mech}},
\end{aligned}
\tag{21.6.39}
$$

where
- $\hat{\rho}_h$ denotes the rate of mass production required to ensure maintenance (homeostasis);
- $\boldsymbol{\sigma}$ is the current stress, $\boldsymbol{\sigma}_h$ the homeostatic stress;
- f_α and f_ω are dimensionless functions that vanish with their arguments. In practice, f_α is the relative distance of the orthoradial stress to its homeostatic value: higher (respectively lower) blood pressure implies higher orthoradial tension and, in turn, mass production over (respectively below) the basal rate;
- $\hat{\rho}_{\alpha,\text{mech}}$ and $\hat{\rho}_{\omega,\text{mech}}$ are the rates of growth and resorption, respectively, depending on stress conditions as well, in fact on the shear stress induced by the blood flow.

21.6.4.6   A two phase mixture model for tumor growth

The model of Roose et al. [2003] is described in Section 24.9.1.

21.7   Boundary value problems for elastic-growing solids

21.7.1   Aortic growth, cardiac growth and development

In a series of papers, Taber addresses the interactions between growth and mechanics. The existence of a homeostatic state plays a key role: growth is due to departure of the current state from the homeostatic state. In fact, he distinguishes *growth* where the homeostatic state is fixed and *development* during which the homeostatic state evolves. Therefore, modeling development up to maturity requires additional assumptions to follow the evolution in time and space of the homeostatic state.

In the early stages of its development, the heart can be assimilated to a tube, with three layers: an outer myocardium, a central layer of extracellular matrix and a thin endocardium. If the tube is cut radially as a whole, it opens to some angle, which reflects the magnitude of residual stresses. Taber [1998]b [2001] uses a two-layer model of aortic growth. He notes that the inner layer is subjected to compressive orthoradial residual stresses, due to larger growth on the inner part of the vessel. Consequently, subsequent to a radial cut, the opening angle is larger than for a single cut of the whole tube, Fig. 21.7.1. On the other hand, the outer skin is subjected to tensile residual stresses, and the opening angle of the outer ring is smaller than for a single cut of the whole tube.

In a thin wall approximation, with cylindrical coordinates (r, θ, z), the orthoradial stress in the vessel $\sigma_\theta = P\,a/e$ results from Laplace's law, with P blood pressure, a radius of the vessel, e thickness of the vessel, and the shear stress at the wall follows from the Hagen-Poiseuille relation, Eqn (15.6.14), namely $\sigma_{rz} = 4\,\eta\,Q/(\pi a^3)$, with Q flow rate along the axis z and η dynamic viscosity of blood. Typical values for the rat aorta are, Taber [1998]a [1998]c: radius 0.20 mm, thickness 0.05 mm, mature blood pressure 16 kPa, mature blood flow rate $\sim 1\,\text{mm}^3/\text{s}$, dynamic viscosity of blood 3 cP. The resulting homeostatic orthoradial and shear stresses are equal to 64 kPa and 0.5 Pa, respectively. In the human arteries, the homeostatic orthoradial and longitudinal stresses are equal to 150 kPa and 125 kPa, respectively, and the shear stress is about 1.5 kPa. A noticeable point is that, under physiological pressure, the homeostatic orthoradial stress turns out to be quasi-uniform through the thickness of the wall of a sound artery, Chuong and Fung [1986].

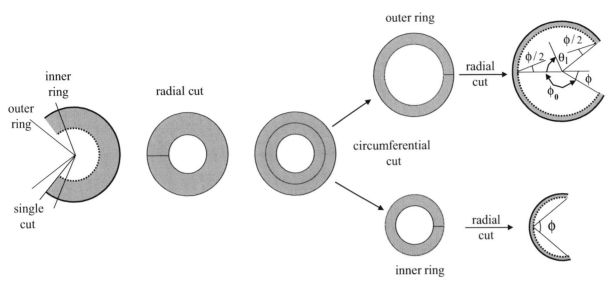

Fig. 21.7.1: Determination of the opening angle in an arterial section. A first circumferential cut is performed to form an inner ring and an outer ring. The latter are next cut transmurally and left to deform under zero external load, Taber and Humphrey [2001]. The order of the cuts, orthoradial and transmural, can be exchanged if the material behavior is elastic. The opening angle of the whole section subsequent to a single transmural cut lies between the opening angles of the inner and outer rings. A number of radial cuts do not succeed to relieve completely the stress state, Fung and Liu [1989].

The influence of the blood pressure on the shear stress and on the evolution of the above quantities during growth is analyzed in Taber [1998]c. A certain power is required to drive the blood flow, to ensure the passive and active maintenances of the vessel and to sustain metabolic processes. Optimizing this power with respect to geometry (lumen), at given blood pressure and flow, Taber finds a pressure dependent shear wall stress.

Due to hypertension (large P), the wall of the vessel e thickens, so as to maintain an approximatively constant orthoradial stress. On the other hand, increased flow rate leads to increased lumen a.

These two particular loading types occur respectively at the end of the systole ES (high pressure due to blood ejection) and at the end of the diastole ED (volume overload). The model of Lin and Taber [1995] accounts for ED wall stress while Taber [1998]a and Taber and Chabert [2002] account for both ED and ES wall stresses.

Taber [1998]a distinguishes growth of smooth muscles and of striated muscles[21.6]:
- for smooth muscles, only the total stress is involved in the growth law;
- for striated muscles, the active stress controls the transverse growth (exercise thickens the musculature), while the passive stress controls the length of the muscle fibers.

The analysis for smooth muscles is intended to arteries and that of striated muscles for the first stage of cardiac development. The subsequent stages of the morphogenesis of the heart are addressed in recent works, e.g., Ramasubramanian et al. [2006]. In these stages, the tube twists and bends to form a four cavity organ. The dorsal mesocardium which links the heart to the body is thought to play a crucial role in these processes, which are still subject to conjecture, Taber [2001].

It is interesting to note that, in all his works, Taber addresses the rate of variation of the strain, or stretches, not of mass. Of course, if the growth preserves density, the latter can be deduced from the growth strain.

Growth takes place because the stress state departs from a homeostatic state, defined in terms of stress. In other words, the rate of growth is a function of the *distance* between the current stress and the homeostatic stress denoted by an index h, say, in a format that highlights a characteristic time of growth t_{g},

$$\frac{d\mathbf{F}^{\mathrm{g}}}{dt} \quad \text{or} \quad \mathbf{L}_{\kappa}^{\mathrm{g}} = \frac{d\mathbf{F}^{\mathrm{g}}}{dt} \cdot (\mathbf{F}^{\mathrm{g}})^{-1} = \frac{\boldsymbol{\mathcal{L}}_{\kappa}}{t_{\mathrm{g}}}, \quad \boldsymbol{\mathcal{L}}_{\kappa} = \boldsymbol{\mathcal{L}}_{\kappa}\left(\boldsymbol{\sigma}^{*} - \boldsymbol{\sigma}^{h*}\right). \tag{21.7.1}$$

[21.6]Smooth muscles, skeletal muscles and cardiac muscles differ by the time course of their action potential and muscular tension. Some information is sketched in Fig. 13.11.1 for the cardiac muscle. The two latter types of muscles appear striated due to the overlapping of actin and myosin filaments. Smooth muscles are found in the walls of internal structures like blood vessels, intestines, etc. Unlike skeletal muscles, the contraction of smooth and cardiac muscles is not under conscious control.

The superimposed symbol $*$ denotes normalization via the (mature) homeostatic stress. In practice, the normalization of the stress tensor is performed component by component. According to (21.3.33), the growth law should be prescribed, not for the rate of growth transformation $d\mathbf{F}^g/dt$ (or its principal stretches $d\lambda_g/dt$), but for the growth velocity transformation $\mathbf{L}_\kappa^g$ (or for the rate of the logarithm of the stretches $d\mathrm{Ln}\,\lambda_g/dt$). The tensorial nature of the dimensionless constitutive function $\mathcal{L}_\kappa$, which vanishes in the (mature) homeostatic state, $\mathcal{L}_\kappa(\mathbf{0}) = \mathbf{0}$, embeds the evolving anisotropic properties of the growing material.

Growth stimuli: mechanics (pressure, pressure gradient), fluid flow and nutrients

There are two candidate stimuli to growth: pressure and flow, that is, pressure gradient. While one may ponder in a general approach, whether pressure or pressure gradient acts as a mechanical stimulus, here they are both involved, and their effects are distinct. Under physiological conditions, the fluid flow has two effects:

- a mechanical effect, by inducing a shear stress: since the fluid (blood) is viscous, the shear stress is proportional to the shear strain rate. This observation has given rise to the idea that, among possible mechanical stimuli, strain rate, rather than strain, is the likely stimulus of growth, Cowin [1996];
- enhanced diffusion and advection of nutrients, with potential adverse effects like advection of collagen monomers away from the cells.

More generally, a mechanical stimulation may be thought to interact in various positive and negative ways with the transport of nutrients:

- a cyclic loading is expected to trigger fluid flow;
- a permanent compressive load is certainly decreasing the porosity, and therefore the space available for fluid flow, and the permeability.

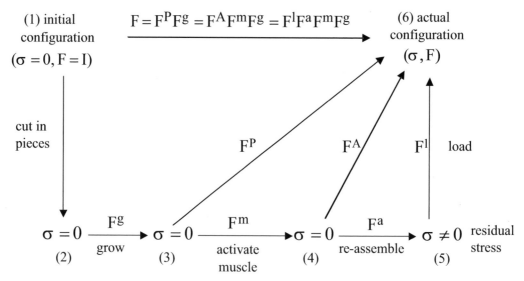

Fig. 21.7.2: Decomposition of the deformation gradient for the heart undergoing growth and muscle activation, Taber [2001], and vascular growth, Taber and Humphrey [2001]. Ramasubramanian et al. [2006] consider in addition *actin polymerization*, that is cell-shape change, defined by the transformation $\mathbf{F}^{\mathrm{ap}}$: the total deformation gradient then admits the multiplicative decomposition $\mathbf{F} = \mathbf{F}^l \cdot \mathbf{F}^a \cdot \mathbf{F}^m \cdot \mathbf{F}^{\mathrm{ap}} \cdot \mathbf{F}^g$.

21.7.1.1 Growth and structuration in a skeletal muscle tissue

The development and isometric contraction (activation) of a muscle are considered in Taber [1998]a. The fiber transversal strain is due to growth only, while the longitudinal strain is contributed by growth, muscle activation and load-induced strain, as sketched in Fig. 21.7.2,

$$\epsilon_t = \epsilon_{gt}; \quad \epsilon_l = \epsilon_{gl} + \epsilon_{Pl} = \epsilon_{gl} + \epsilon_{ml} + \epsilon_{Al}. \tag{21.7.2}$$

Stress-strain relations are needed for the passive and active behavior,

$$\sigma_P = E_P\,\epsilon_{Pl}; \quad \sigma_A = E_A\,\epsilon_{Al}. \tag{21.7.3}$$

The growth law for a striated muscle is postulated in the format,

$$\frac{d\epsilon_{gt}}{dt} = \frac{1}{t_{gt}}(\sigma_A^* - \sigma_A^{h*}); \quad \frac{d\epsilon_{gl}}{dt} = \frac{1}{t_{gl}}(\sigma_P^* - \sigma_P^{h*}). \tag{21.7.4}$$

The parameters t_g's are time constants. The justification is as follows, Taber [1998]a: during development, the muscle elongates while it is not activated. On the other hand, exercise (active stress) tends to thicken rather than to elongate muscles.

For a smooth muscle, e.g., in arteries, the wall thickens due to fluid pressure, and the muscle elongates due to fluid flow (shear stress):

$$\frac{d\epsilon_{gt}}{dt} = \frac{1}{t_{gt}}(\sigma^* - \sigma^{h*}); \quad \frac{d\epsilon_{gl}}{dt} = \frac{1}{t_{gl}}(\sigma^* - \sigma^{h*}) + \frac{1}{t_{g\tau}}(\tau^* - \tau^{h*})\,e^{-\beta\,(r/r_1-1)}, \tag{21.7.5}$$

where β is a positive constant, and r_1 the inner radius of the artery. The homeostatic reference stress is assumed fixed here. For cardiac development, the homeostatic state is varying with time, and it is either load dependent, Taber [1998]a, or strain dependent, Taber and Chabert [2002].

Development

Development is characterized by a given axial strain $\epsilon_l = \epsilon_0$, and no activation, i.e., $\epsilon_{ml} = 0$ and $\epsilon_{Pl} = \epsilon_{Al}$. For example, one may think about a muscle being elongated by a growing bone. Substituting $\epsilon_{Pl} = \epsilon_0 - \epsilon_{gl}$, Eqn (21.7.2), and $\sigma_P = E_P\,\epsilon_{Pl}$, Eqn (21.7.3), into (21.7.4)$_2$ yields a first order differential equation for ϵ_{gl},

$$t_{gl}\frac{d\epsilon_{gl}}{dt} + \chi\,\epsilon_{gl} = \chi\,(\epsilon_0 - \frac{\sigma_P^h}{E_P}), \quad \chi \equiv \frac{E_P}{\sigma_{P0}}, \tag{21.7.6}$$

where σ_{P0} is the stress normalizing the passive homeostatic stress. The solution

$$\epsilon_{gl}(t) = (\epsilon_0 - \frac{\sigma_P^h}{E_p})\left(1 - \exp(-\chi\,\frac{t}{t_{gl}})\right), \quad \sigma_P(t) = \sigma_P^h + E_P\,(\epsilon_0 - \frac{\sigma_P^h}{E_P})\exp\left(-\chi\,\frac{t}{t_{gl}}\right), \tag{21.7.7}$$

shows that the length of the muscle increases to relieve the stress which decreases in time, assuming $\epsilon_0 - \sigma_P^h/E_P > 0$.

Isometric contraction

A particular case of isometric contraction, where $\epsilon_{ml} < 0$ and $\epsilon_l = \epsilon_0$ are fixed, is considered, namely $\epsilon_0 = 0$ and $\sigma_P = \sigma_P^h$ so that no longitudinal growth occurs, Eqn (21.7.4), namely $\epsilon_{gl} = 0$. Eqn (21.7.2)$_2$ implies $\epsilon_{Pl} = 0 = \epsilon_{ml} + \epsilon_{Al}$. Thus, $\sigma_A = -E_A\,\epsilon_{ml}$ is constant, and (21.7.4)$_1$ implies that the muscle thickens unboundedly if $\sigma_A - \sigma_A^h > 0$. To avoid this pitfall, the muscle is allowed to restructure, i.e., to decrease its stiffness according to the rate law,

$$\frac{dE_A}{dt} = -\frac{1}{t_E}(\sigma_A - \sigma_A^h). \tag{21.7.8}$$

Substituting $\sigma_A = -E_A\,\epsilon_{ml}$ in (21.7.8) yields a differential equation for E_A, namely $t_E\,dE_A/dt - \epsilon_{ml}\,E_A = \sigma_A^h$, whose solution decreases in time,

$$E_A(t) = E_A(0) + \left(E_A(0) + \frac{\sigma_A^h}{\epsilon_{ml}}\right)\left(\exp\left(\epsilon_{ml}\,\frac{t}{t_E}\right) - 1\right), \tag{21.7.9}$$

while the transverse growth strain increases to a finite value, Eqn (21.7.4)$_1$,

$$\epsilon_{gt}(t) = -\frac{t_E}{t_{gt}}\frac{E_A(t) - E_A(0)}{\sigma_{A0}}, \tag{21.7.10}$$

where σ_{A0} is the stress normalizing the active homeostatic stress.

Note that the residual stress vanishes in this example since there is no spatial variation of the fields of interest.

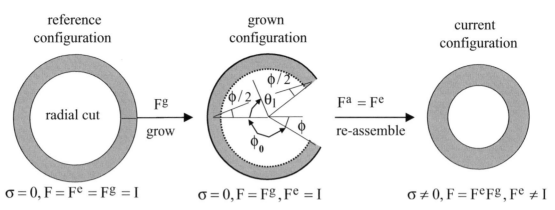

Fig. 21.7.3: An artery, cut radially, is subjected to stress free growth. For simplicity, the grown configuration is assumed to remain circular. Moreover, growth is only circumferential, so that actual growth leads to an overlap $\phi_0 > \pi$, while resorption leads to a circumferential gap $\phi_0 < \pi$ (as sketched). The integrity of the structure is next re-established, through an elastic (accommodating) transformation, that is accompanied with a change of radius, and gives rise to (residual) stresses. The coordinate systems attached to the three configurations are noted according to the convention: reference configuration $\underline{B}(\underline{r}, \underline{\theta}, \underline{z})$, grown configuration $B_e(r_e, \theta_e, z_e)$, current configuration $B(r, \theta, z)$.

21.7.1.2 Growth, accommodation and residual stresses

The actual growth process involves *simultaneously* growth and accommodation so that the tissue remains intact. Residual stresses are detected by cutting the grown organ, and observing its deformation.

An alternative method to discover residual stresses is to de-synchronize growth and accommodation: leave the material grow in absence of stress (take an artery with a radial cut), and next re-establish its integrity (the accommodating, or elastic, process). This section is devoted to this last point of view along Rodriguez et al. [1994].

In regards to Fig. 21.7.3, start from the reference configuration $\underline{B}(\underline{r}, \underline{\theta}, \underline{z})$, grow the material under free stress to $B_e(r_e, \theta_e, z_e)$, and finally re-establish the integrity through accommodation to the actual current configuration $B(r, \theta, z)$.

The motion with respect to the initial configuration, and to any configuration, is axially symmetric, and it allows for axial elongation/shortening. The cylindrical coordinates used are indicated in Fig. 21.7.3. The generic deformation gradient (2.4.81) is diagonal in the cylindrical axes linked to the artery,

$$\mathbf{F} = \mathbf{diag}\left[\lambda_r = \frac{\partial r}{\partial \underline{r}}, \ \lambda_\theta = \frac{r}{\underline{r}}, \ \lambda_z\right], \tag{21.7.11}$$

where $r = r(\underline{r}, t)$, and λ_z is the axial stretch, assumed to be spatially uniform.

As a simplification, the grown geometry is assumed to remain cylindrical, and to be free of shear stress, Fig. 21.7.3, e.g.,

$$r_e = \underline{r}, \quad \theta_e = \underline{\theta}\,\frac{\phi_0}{\pi}, \quad z_e = \lambda_{gz}\,\underline{z}. \tag{21.7.12}$$

Like λ_z, the longitudinal growth stretch λ_{gz} is considered independent of the radial position, and therefore so is the elastic stretch λ_{ez}. The growth transformation deduces as

$$\mathbf{F}^g = \mathbf{diag}\left[\lambda_{gr} = 1, \ \lambda_{g\theta} = \frac{\phi_0}{\pi}, \ \lambda_{gz}\right]. \tag{21.7.13}$$

The accommodation motion,

$$r = r(r_e), \quad \theta = \theta_e\,\frac{\pi}{\phi_0}, \quad z = \lambda_{ez}\,z_e\,, \tag{21.7.14}$$

is defined by the transformation,

$$\mathbf{F}^e = \mathbf{diag}\left[\lambda_{er} = \frac{dr}{dr_e}, \ \lambda_{e\theta} = \frac{r}{r_e}\,\frac{\pi}{\phi_0}, \ \lambda_{ez}\right], \tag{21.7.15}$$

and the elastic Green strain $\mathbf{E}_G^e$,

$$\mathbf{E}_G^e = \frac{1}{2}\left((\mathbf{F}^e)^T \cdot \mathbf{F}^e - \mathbf{I}\right) = \frac{1}{2}\,\mathbf{diag}\left[\lambda_{er}^2 - 1, \ \lambda_{e\theta}^2 - 1, \ \lambda_{ez}^2 - 1\right]. \tag{21.7.16}$$

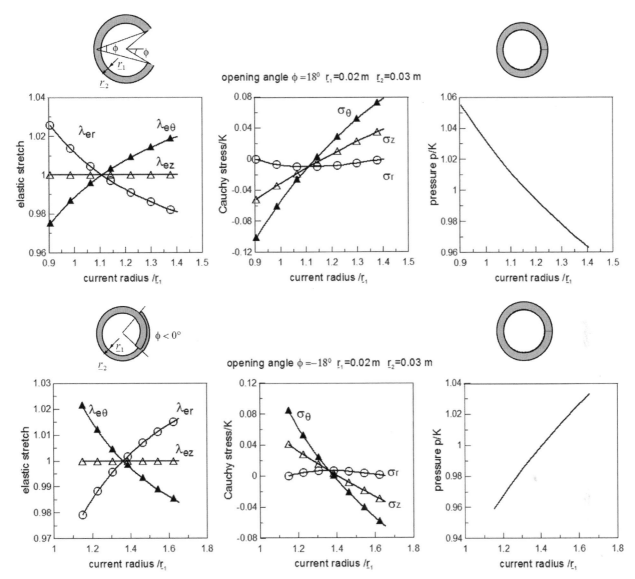

Fig. 21.7.4: Residual stresses in arteries due to the accommodation process, as described in Fig. 21.7.3, with prescribed growth transformation $\mathbf{F}^{\mathrm{g}} = \mathbf{diag}[\lambda_{\mathrm{g}r} = 1, \lambda_{\mathrm{g}\theta} = \phi_0/\pi, \lambda_{\mathrm{g}z} = 1]$ and axial stretch $\lambda_z = 1$. Note that the spatial variations of stretches and stresses are almost linear through the thickness. On the other hand, in actual arteries under physiological pressure, the transmural profile of the orthoradial stress is quasi-uniform. In the case where the opening angle is equal to $18°$, the inner radius changes from $2\,\mathrm{cm}$ to $1.81\,\mathrm{cm}$ while, for the negative opening angle $\phi = -18°$, the final inner radius end up at $2.25\,\mathrm{cm}$.

If the material is elastically incompressible, $\det \mathbf{F}^{\mathrm{e}} = 1$, the current radius expresses in terms of the grown radius or reference radius,

$$r^2 - r_1^2 = (r_{\mathrm{e}}^2 - r_{\mathrm{e}1}^2) \frac{\phi_0}{\pi} \frac{1}{\lambda_{\mathrm{e}z}} = (\underline{r}^2 - \underline{r}_1^2) \frac{\phi_0}{\pi} \frac{1}{\lambda_{\mathrm{e}z}} , \qquad (21.7.17)$$

where $r_{\mathrm{e}1} = \underline{r}_1$ is the inner reference radius. Note that the inner current radius $r_1 = r(r_{\mathrm{e}1})$ is yet unknown. Then, since $\mathbf{I} : \mathbf{L}^{\mathrm{e}} = ((\mathbf{F}^{\mathrm{e}})^{-1} \cdot (\mathbf{F}^{\mathrm{e}})^{-\mathrm{T}}) : d\mathbf{E}_{\mathrm{G}}^{\mathrm{e}}/dt$, the elastic constitutive equation yields the effective 2nd Piola-Kirchhoff stress in the relaxed configuration in terms of the elastic free energy E_κ,

$$\boldsymbol{\tau}_\kappa + p \, (\mathbf{F}^{\mathrm{e}})^{-1} \cdot (\mathbf{F}^{\mathrm{e}})^{-\mathrm{T}} = \frac{\partial E_\kappa}{\partial \mathbf{E}_{\mathrm{G}}^{\mathrm{e}}} . \qquad (21.7.18)$$

The effective constitutive Cauchy stress expresses through the transport rule $(\cdot) \to \mathbf{F}^e \cdot (\cdot) \cdot (\mathbf{F}^e)^T$, as,

$$
\begin{aligned}
\boldsymbol{\sigma} + p\,\mathbf{I} &= \mathbf{F}^e \cdot \frac{\partial E_\kappa}{\partial \mathbf{E}_G^e} \cdot (\mathbf{F}^e)^T \\
&= \mathbf{diag}\,\Big[\frac{\partial E_\kappa}{\partial E_{Gr}^e}\,\lambda_{er}^2,\ \frac{\partial E_\kappa}{\partial E_{G\theta}^e}\,\lambda_{e\theta}^2,\ \frac{\partial E_\kappa}{\partial E_{Gz}^e}\,\lambda_{ez}^2\Big].
\end{aligned}
\tag{21.7.19}
$$

Therefore the Lagrange multiplier p associated with the incompressibility constraint can be interpreted as a pressure. The Cauchy stress is thus principal in the axes of the artery, and

$$
\frac{\partial \sigma_r}{\partial r} + \frac{1}{r}\,(\sigma_r - \sigma_\theta) = 0\,,
\tag{21.7.20}
$$

is the single equilibrium equation that does not vanish identically. Since

$$
\frac{dr}{r} = \frac{\pi}{\phi_0}\,\frac{1}{\lambda_{ez}}\,\frac{1}{\lambda_{e\theta}^2}\,\frac{dr_e}{r_e}\,,
\tag{21.7.21}
$$

integration of the radial equilibrium equation can be cast in the format,

$$
\sigma_r(r) - \underbrace{\sigma_r(r_1)}_{=0} + \frac{\pi}{\phi_0}\,\frac{1}{\lambda_{ez}} \int_{r_{e1}}^{r_e} (\sigma_r - \sigma_\theta)\,\frac{1}{\lambda_{e\theta}^2}\,\frac{dr_e}{r_e} = 0
\tag{21.7.22}
$$

where $r_1 = r(r_{e1})$, $r = r(r_e)$ by (21.7.17), and $\lambda_{e\theta} = (r/r_e)\,(\pi/\phi_0)$. For $r_e = r_{e2}$ where $\sigma_r(r_2) = 0$, this relation yields the current inner radius r_1,

$$
\mathrm{Res}(r_1) \equiv \int_{r_{e1}}^{r_{e2}} (\sigma_r - \sigma_\theta)\,\frac{1}{\lambda_{e\theta}^2}\,\frac{dr_e}{r_e} = 0\,.
\tag{21.7.23}
$$

Indeed, all terms in the integrand are known functions of the radii and all radii express in terms of the radius r_e via (21.7.17).

Next, all stretches being known functions of r_e, the integrated form of the equilibrium (21.7.22) yields, via the elastic constitutive equations, the pressure $p = p(r_e)$,

$$
p(r_e) = \frac{\partial E_\kappa}{\partial E_{Gr}^e}\,\lambda_{er}^2 + \frac{\pi}{\phi_0}\,\frac{1}{\lambda_{ez}} \int_{r_{e1}}^{r_e} (\sigma_r - \sigma_\theta)\,\frac{1}{\lambda_{e\theta}^2}\,\frac{dr_e}{r_e}\,,
\tag{21.7.24}
$$

which, re-introduced in the elastic constitutive equations, provides the components of the Cauchy stress.

Rodriguez et al. [1994] consider a free energy of the Blatz-Ko type (actually neo-Hokean due to elastic incompressibility, Section 6.4),

$$
E_\kappa = \frac{K}{2\,k}\,\big(\exp(2\,k\,\mathrm{tr}\,\mathbf{E}_G^e) - 1\big) \quad \Rightarrow \quad \frac{\partial E_\kappa}{\partial \mathbf{E}_G^e} = K\,\exp(2\,k\,\mathrm{tr}\,\mathbf{E}_G^e)\,\mathbf{I}\,,
\tag{21.7.25}
$$

defined, for an isotropic material, by the modulus K and the exponent k. The stresses shown in Fig. 21.7.4, correspond to the exponent $k = 1$. For moderate strains (smaller than 0.1), the stress is quite insensitive to the value of k.

21.7.1.3 Growth and residual stresses in arteries without muscle activation

Taber and Eggers [1996] and Taber and Humphrey [2001] use a thick-walled cylinder composed of two concentric layers to describe volumetric growth and residual stresses in arteries. They note a strong coupling between material properties and residual stresses. Neither structuration nor muscle activation are accounted for, so that the sketch of Fig. 21.7.5 simplifies somehow, $\mathbf{F}^m = \mathbf{I}$ and formally $P \sim A$, which will be denoted by the common symbol e, e.g., $\mathbf{F}^P = \mathbf{F}^A = \mathbf{F}^e$. Unlike in Section 21.7.1.2 where the growth transformation was prescribed, here it is built from a growth rate law.

The motion with respect to the initial configuration is axially symmetric and it allows for axial elongation/shortening. The cylindrical coordinates used are indicated in Fig. 21.7.5. The deformation gradient is principal in the cylindrical axes and expresses as,

$$
\mathbf{F} = \mathbf{diag}\,[\lambda_r = \frac{\partial r}{\partial \underline{r}},\ \lambda_\theta = \frac{r}{\underline{r}},\ \lambda_z]\,,
\tag{21.7.26}
$$

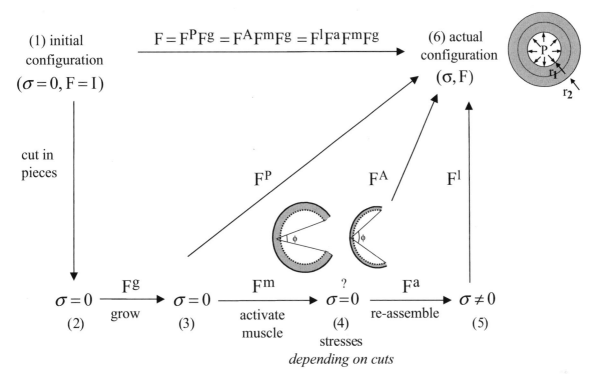

Fig. 21.7.5: Configurations for the heart undergoing growth and muscle activation to determine the opening angle obtained by one or several cuts, Taber [2001]. Coordinate systems attached to these configurations:
$(1) : \underline{B}(\underline{r}, \underline{\theta}, \underline{z})$, $(3) : B_P(r_P, \theta_P, z_P)$, $(4) : B_A(r_A, \theta_A, z_A)$, $(5) : B_l(r_l, \theta_l, z_l)$, $(6) : B(r, \theta, z)$. For vascular growth without muscle activation, Taber and Humphrey [2001], the configurations P and A collapse to e, and the coordinate systems simplify to: $(1) : \underline{B}(\underline{r}, \underline{\theta}, \underline{z})$, $(3) = (4) : B_e(r_e, \theta_e, z_e)$, $(5) : B_l(r_l, \theta_l, z_l)$, $(6) : B(r, \theta, z)$.

where $r = r(\underline{r}, t)$ is the current radius, and λ_z the axial stretch.

The elastic configuration that highlights the existence of the residual stresses displays an opening angle $\phi = \pi - \phi_0$. As a simplification, the geometries in the various configurations are assumed to remain circular, and to be free of shear stress,

$$r_l = r(r_e, t), \quad \theta_l = \theta_e \frac{\pi}{\phi_0}, \quad z_l = \lambda_{ez} z_e; \quad r = r(r_l, t), \quad \theta = \theta_l, \quad z = \lambda_{lz} z_l. \tag{21.7.27}$$

Here λ_{ez} is considered a constant, independent of the radial position. During development, the axial strain relieves the applied stress and the associated potential damage, Eqn (21.7.7). Thus the motion is defined by the mappings,

$$\mathbf{x}_e = r_e \mathbf{e}_{r_e} + z_e \mathbf{e}_{z_e} \quad \rightarrow \quad \mathbf{x}_l = r_l \mathbf{e}_{r_l} + z_l \mathbf{e}_{z_l} \quad \rightarrow \quad \mathbf{x} = r \mathbf{e}_r + z \mathbf{e}_z, \tag{21.7.28}$$

the vectors

$$\mathbf{e}_{r_e} = \cos\theta_e \mathbf{e}_1 + \sin\theta_e \mathbf{e}_2, \quad \mathbf{e}_{r_l} = \cos\theta_l \mathbf{e}_1 + \sin\theta_l \mathbf{e}_2, \quad \mathbf{e}_r = \cos\theta \mathbf{e}_1 + \sin\theta \mathbf{e}_2, \tag{21.7.29}$$

being expressed in terms of fixed unit vectors $\mathbf{e}_1$ and $\mathbf{e}_2$, so that the transformations $\mathbf{F}^a = \partial \mathbf{x}_l / \partial \mathbf{x}_e$ and $\mathbf{F}^l = \partial \mathbf{x} / \partial \mathbf{x}_l$ obtained from the differentials,

$$d\mathbf{x}_e = \mathbf{e}_{r_e} dr_e + r_e \mathbf{e}_{\theta_e} d\theta_e + \mathbf{e}_{z_e} dz_e, \quad d\mathbf{x}_l = \mathbf{e}_{r_l} dr_l + r_l \mathbf{e}_{\theta_l} d\theta_l + \mathbf{e}_{z_l} dz_l, \quad d\mathbf{x} = \mathbf{e}_r dr + r \mathbf{e}_\theta d\theta + \mathbf{e}_z dz, \tag{21.7.30}$$

are principal in the cylindrical axes,

$$\mathbf{F}^a = \mathbf{diag}\left[\lambda_{ar} = \frac{\partial r_l}{\partial r_e}, \ \lambda_{a\theta} = \frac{r_l}{r_e} \frac{\pi}{\phi_0}, \ \lambda_{az}\right], \quad \mathbf{F}^l = \mathbf{diag}\left[\lambda_{lr} = \frac{\partial r}{\partial r_l}, \ \lambda_{l\theta} = \frac{r}{r_l}, \ \lambda_{lz}\right], \tag{21.7.31}$$

and

$$\mathbf{F}^e \equiv \mathbf{F}^l \cdot \mathbf{F}^a = \mathbf{diag}\left[\lambda_{er} = \frac{\partial r}{\partial r_e}, \ \lambda_{e\theta} = \frac{r}{r_e} \frac{\pi}{\phi_0}, \ \lambda_{ez}\right]. \tag{21.7.32}$$

If the material is elastically incompressible, then

$$\det \mathbf{F}^a \, \det \mathbf{F}^l = \det \mathbf{F}^e = 1 \quad \text{but} \quad \det \mathbf{F} = \det \mathbf{F}^g \neq 1! \tag{21.7.33}$$

The equilibrium equations in the radial direction in the "elastic" (relaxed) and current configurations need to be ensured,

$$\frac{\partial \sigma_{er}}{\partial r_e} + \frac{1}{r_e}(\sigma_{er} - \sigma_{e\theta}) = 0, \quad \frac{\partial \sigma_r}{\partial r} + \frac{1}{r}(\sigma_r - \sigma_\theta) = 0. \tag{21.7.34}$$

If the cylinder is subjected to the internal pressure P, the boundary conditions on the current inner and outer radii, $r_1 = r(\underline{r}_1)$ and $r_2 = r(\underline{r}_2)$ respectively, and in the axial direction, express as,

$$\sigma_r(r_1) = -P, \quad \sigma_r(r_2) = 0, \quad 2\pi \int_{r_1}^{r_2} \sigma_z(r)\, r\, dr = \pi\, r_1^2\, P. \tag{21.7.35}$$

The boundary conditions on the "elastic" configuration which is unloaded require, besides zero lateral traction and vertical force, a vanishing circumferential moment (this condition is less stringent than the zero traction on the transmural cut), with $r_{e1} = r_e(\underline{r}_1)$, $r_{e2} = r_e(\underline{r}_2)$,

$$\sigma_{er}(r_{e1}) = 0, \quad \sigma_{er}(r_{e2}) = 0, \quad \int_{r_{e1}}^{r_{e2}} \sigma_{ez}(r_e)\, r_e\, dr_e = 0, \quad \int_{r_{e1}}^{r_{e2}} \sigma_{e\theta}(r_e)\, r_e\, dr_e = 0. \tag{21.7.36}$$

The stress $\boldsymbol{\sigma}_e = \det \mathbf{F}^e\, (\mathbf{F}^e)^{-1} \cdot \boldsymbol{\sigma} \cdot (\mathbf{F}^e)^{-T}$ in the relaxed configuration derives from the free energy E_κ,

$$\boldsymbol{\sigma}_e + p\, (\mathbf{F}^e)^{-1} \cdot (\mathbf{F}^e)^{-T} = \frac{\partial E_\kappa}{\partial \mathbf{E}_G^e}, \quad \sigma_{ej} + \frac{p}{\lambda_{ej}^2} = \frac{1}{\lambda_{ej}}\frac{\partial E_\kappa}{\partial \lambda_{ej}}, \quad j = r, \theta, z. \tag{21.7.37}$$

The pressure p is calculated by (21.7.42) below.

Growth is driven by the difference between the current stress and the homeostatic stress,

$$\frac{d\lambda_{gr}}{dt} = \frac{1}{t_{gr}}(\sigma_\theta^* - \sigma_\theta^{*h}), \quad \frac{d\lambda_{g\theta}}{dt} = \frac{1}{t_{g\theta}}(\sigma_\theta^* - \sigma_\theta^{*h}) + \frac{1}{t_{grz}}(\sigma_{rz}^* - \sigma_{rz}^{*h}), \quad \frac{d\lambda_{gz}}{dt} = 0, \tag{21.7.38}$$

where the t_g's are characteristic times (order of magnitude: 1 day for rat, 10 days for cow). Motivated by Laplace's law, the homeostatic orthoradial stress is assumed proportional to the (time varying) blood pressure. The homeostatic shear stress is estimated with help of the Hagen-Poiseuille relation. Both are assumed uniform through the wall thickness. The symbol $*$ denotes a normalization, Taber and Humphrey [2001], their Eqn (13). The normalizing stresses are identified with the normal and shear stresses at maturity. This growth rule leads to bounded growth. By contrast, the presence of the axial stress σ_z in the growth law was shown to induce unbounded growth in Taber and Eggers [1996]: structuration was shown to alleviate this issue, Eqn (21.7.8).

The solution for the opening angle

The constraints (21.7.33) provide differential relations between the radii in the initial configuration, relaxed configuration and current configuration, namely via (21.7.26) and (21.7.31),

$$\frac{\det \mathbf{F}^g}{\lambda_z}\, \underline{r}\, d\underline{r} = r\, dr = \frac{\phi_0}{\pi}\frac{1}{\lambda_{ez}}\, r_e\, dr_e, \tag{21.7.39}$$

which can be integrated between the inner radius and an arbitrary position within the artery,

$$r^2 - r_1^2 = \frac{\phi_0}{\pi}\frac{1}{\lambda_{ez}}(r_e^2 - r_e^2(\underline{r}_1)) = \frac{2}{\lambda_z}\int_{\underline{r}_1}^{\underline{r}} \det \mathbf{F}^g\, \underline{r}\, d\underline{r}. \tag{21.7.40}$$

The axial growth stretch is constant in time, Eqn (21.7.38), and it will be assumed to be constant in space. Then, since the axial stretch is independent of the radial position, so is the elastic axial stretch, which justifies the integrations (21.7.40).

Integration of the equilibrium relation (21.7.34)$_1$ between a current radius and the external radius yields,

$$\sigma_{er}(r_e(r_2)) = \sigma_{er}(r_e(r)) - \int_{r_e(r)}^{r_e(r_2)} (\sigma_{er} - \sigma_{e\theta})\frac{r_e\, dr_e}{r_e^2}. \tag{21.7.41}$$

Combined with the boundary condition (21.7.36)$_2$ and the differential relation (21.7.39), the expression (21.7.41) can be used to calculate the pressure p,

$$\underbrace{\sigma_{er}(r_e(\underline{r}_2))}_{0} = \underbrace{\sigma_{er}(r_e(\underline{r}))}_{-\dfrac{p}{\lambda_{er}^2} + \dfrac{1}{\lambda_{er}}\dfrac{\partial E_\kappa}{\partial \lambda_{er}}} - \frac{\pi}{\phi_0}\frac{\lambda_{ez}}{\lambda_z}\int_{\underline{r}}^{\underline{r}_2}(\sigma_{er} - \sigma_{e\theta})\,\det\mathbf{F}^g\,\frac{r\,dr}{r_e^2}, \tag{21.7.42}$$

where r_e is given in terms of $\underline{r}$ by (21.7.40), and the stress $\boldsymbol{\sigma}_e$ is defined in terms of $\mathbf{F}^e$ by (21.7.37). If the lower integration boundary is $\underline{r}_1$, then the boundary condition (21.7.36)$_1$ results in the condition,

$$0 = \frac{\pi}{\phi_0}\frac{\lambda_{ez}}{\lambda_z}\int_{\underline{r}_1}^{\underline{r}_2}(\sigma_{er} - \sigma_{e\theta})\,\det\mathbf{F}^g\,\frac{r\,dr}{r_e^2}. \tag{21.7.43}$$

The integral boundary conditions (21.7.36)$_{3,4}$ can be transformed with help of (21.7.39) to

$$\int_{\underline{r}_1}^{\underline{r}_2}\sigma_{ej}\,\det\mathbf{F}^g\,\underline{r}\,d\underline{r} = 0, \quad j = z, \theta. \tag{21.7.44}$$

The blood pressure and flow rate being monitored, the three equilibrium equations of the relaxed configuration (21.7.43) and (21.7.44) are solved for the inner radius r_{e1}, the axial strain λ_{ez} and opening angle ϕ_0 with help of the constitutive equations (21.7.37) and growth rate law (21.7.38). The transformations $\mathbf{F}$, $\mathbf{F}^g$ and $\mathbf{F}^e$, pressure p and stress $\boldsymbol{\sigma}$ result as by-products.

The analysis applies whether or not the relaxed configurations involve circumferential cuts.

It is useful to put into perspective the problems addressed in Section 21.7.1.2 and in this section. In Section 21.7.1.2, a partial problem was addressed, namely the growth transformation $\mathbf{F}^g$ and the axial strain were given. Here, a closed cylinder with internal pressure is considered, so that the axial strain is unknown. Moreover, the growth transformation is defined by a rate law, so that the opening angle is now an unknown. The two associated equations to be solved are the balance of axial force and the balance of bending moment (21.7.44).

The solution in the current configuration

The equations defining the current configuration deduce simply by dropping the boundary condition (21.7.36)$_4$, namely, for an internal pressure P,

$$0 = -P - \frac{1}{\lambda_z}\int_{\underline{r}_1}^{\underline{r}_2}(\sigma_r - \sigma_\theta)\,\det\mathbf{F}^g\,\frac{r\,dr}{r^2}, \quad \frac{2\pi}{\lambda_z}\int_{\underline{r}_1}^{\underline{r}_2}\sigma_z\,\det\mathbf{F}^g\,\underline{r}\,d\underline{r} = \pi\,r_1^2\,P, \tag{21.7.45}$$

where r is given in terms of $\underline{r}$ by (21.7.40), and $\det\mathbf{F}^g = \lambda_{gr}\,\lambda_{g\theta}\,\lambda_{gz}$.

The solution proceeds as above but the equilibrium equations of the current configuration (21.7.45) are solved for the inner radius r_1 and the axial strain λ_z, with help of the growth rate law (21.7.38) and of the constitutive equations,

$$\boldsymbol{\sigma} + p\,\mathbf{I} = \mathbf{F}^e \cdot \frac{\partial E_\kappa}{\partial \mathbf{E}^e_G}\cdot(\mathbf{F}^e)^{\mathrm{T}}, \quad \sigma_j + p = \lambda_{ej}\frac{\partial E_\kappa}{\partial \lambda_{ej}}, \quad j = r, \theta, z, \tag{21.7.46}$$

deduced from the relaxed configuration (21.7.37) through the transport rule $(\cdot) \to \mathbf{F}^e \cdot (\cdot) \cdot (\mathbf{F}^e)^{\mathrm{T}}$.

Determination of the opening angle in arteries: alternative versions

Methods to determine the opening angle in arteries are compared in van Dyke and Hoger [2002]. Among the differences noted are

- the difference of treatment of the axial stretch. Taber and coworkers assume a uniform axial stretch and define it through the incompressibility condition $\det\mathbf{F}^e = 1$. Chuong and Fung [1986] assume plane strain, while van Dyke and Hoger [2002] suggest that plane stress is more appropriate as the height of the arteries tested is small. In contrast to the other authors, the latter also assume a linear radial variation of the axial stretch from the open to the closed configurations;

- in order to avoid some approximations in Taber and Eggers [1996] due to the radial non uniformity of the opening angle, van Dyke and Hoger [2002] minimize the total energy which, in absence of an external load, is equal to $2\pi\int_{\underline{r}_1}^{\underline{r}_2}E_\kappa\,\det\mathbf{F}^g\,\underline{r}\,d\underline{r}$, with respect to four kinematic parameters.

All these methods assume in fact that the motion from the open to the closed configurations is a pure bending, and that the thickness of the arteries is small with respect to the radius.

A finite element analysis targeting residual stresses in arteries is presented in Raghavan et al. [2004]. Unfortunately, it does not account for growth strain.

21.7.1.4   Growth, structuration and muscle activation during heart development

Taber and Chabert [2002] use a thick-walled cylinder composed of two layers to describe volumetric growth of the developing heart of chicken. Structuration and muscle activation are accounted for, in contrast to Section 21.7.1.3. Structuration is accounted for as follows: the constitutive relations remain identical during development, but a number of constitutive elements are prescribed ad hoc functions of time so as to match experimental data.

Growth and muscle activation in unloaded and loaded states are defined by five configurations, Fig. 21.7.5, and the passive, active and muscle transformations are linked by the relation $\mathbf{F}^P = \mathbf{F}^A \cdot (\mathbf{F}^m)^{-1}$.

The stress is the sum of a passive contribution (P) and an active contribution (A), both of which derive from a potential,

$$\boldsymbol{\sigma} = \mathbf{F}_P \cdot \boldsymbol{\sigma}_P \cdot \mathbf{F}_P^{\mathrm{T}} + \mathbf{F}_A \cdot \boldsymbol{\sigma}_A \cdot \mathbf{F}_A^{\mathrm{T}},$$

$$\boldsymbol{\sigma}_i = \frac{1}{\det \mathbf{F}_i} \frac{\partial W_i}{\partial \mathbf{E}_i}(\mathbf{E}_i) = \frac{1}{\det \mathbf{F}_i} \mathbf{F}_i^{-1} \cdot \frac{\partial W_i}{\partial \mathbf{F}_i}(\mathbf{F}_i), \quad i = P, A, \tag{21.7.47}$$

with $\mathbf{E}_i = \frac{1}{2}(\mathbf{F}_i^{\mathrm{T}} \cdot \mathbf{F}_i - \mathbf{I})$, $i = P, A$, the respective Green strains. If the material is incompressible, the effective stress $\boldsymbol{\sigma} + p\mathbf{I}$ should replace the stress in the left-hand side of $(21.7.47)_2$.

The active part of the stress is triggered by calcium, Taber and Chabert [2002], Ramasubramanian et al. [2006]. Cells are considered to be circumferential with stretch $\lambda_m = \lambda_\theta$, or axial with stretch $\lambda_m = \lambda_z$. For incompressible cells, of generic axis $\mathbf{e}_1$, the muscle activation $\mathbf{F}^m$ has the form,

$$\mathbf{F}^m = \lambda_m \, \mathbf{e}_1 \otimes \underline{\mathbf{e}}_1 + \lambda_m^{-1/2} \left(\mathbf{e}_2 \otimes \underline{\mathbf{e}}_2 + \mathbf{e}_3 \otimes \underline{\mathbf{e}}_3 \right). \tag{21.7.48}$$

Taber [1998]a considers that muscle fibers are mainly parallel to the surface of the left ventricle and artery, and does not account for incompressibility. For the thick-wall analysis, Taber and Chabert [2002] consider circumferential and axial fibers with an active transformation $\mathbf{F}^A = \mathbf{F}^P \cdot (\mathbf{F}^m)^{-1}$.

The growth rate law is expressed in the format (21.3.33), i.e., addresses the velocity gradient, at variance with (21.7.38),

$$\frac{1}{\lambda_{gr}} \frac{d\lambda_{gr}}{dt} = \frac{1}{t_{gr}} (\sigma_\theta^* - \sigma_\theta^{h*})_{ES}, \quad \frac{1}{\lambda_{g\theta}} \frac{d\lambda_{g\theta}}{dt} = \frac{1}{t_{g\theta}} (\sigma_\theta^* - \sigma_\theta^{h*})_{ES} - \frac{1}{t_{g\theta_\lambda}} (\lambda_\theta - \lambda_\theta^h)_{ED}^{|r=r_1}, \quad \frac{1}{\lambda_{gz}} \frac{d\lambda_{gz}}{dt} = 0. \tag{21.7.49}$$

Deviation from an equilibrium value of the orthoradial stretch at the inner radius $r = r_1$, instead of the shear stress induced by blood flow, contributes to growth. It acts as a restoring term, and leads to bounded growth, as discussed in Taber [2001]. Moreover, the time windows where the growth engines are thought to be active are accounted for: end of systole pressure overload ES and end of diastole volume overload ED. Specifically, three characteristic times t_g are introduced.

A slightly different form, akin the striated muscle, Eqn (21.7.4), is shown in Taber [2001]. Assumptions are needed for the time course of change, during the development, of the homeostatic orthoradial stretch λ_θ^h, as well as for the ED and ES pressures.

21.7.2   Residual stresses due to the growth of a spherical shell

Consider a body which has undergone spatially heterogeneous growth. If the external load vanishes, the internal stresses are by definition the residual stresses. The purely radial motion of a spherical shell, $\underline{r} \in [\underline{r}_1, \underline{r}_2]$, described in spherical coordinates as shown in Fig. 2.4.2,

$$r = r(\underline{r}, t), \; \theta = \underline{\theta}, \; \phi = \underline{\phi}, \tag{21.7.50}$$

will serve to illustrate the purpose. The deformation gradient (2.4.64) is principal in the spherical axes,

$$\mathbf{F} = \mathbf{diag} \left[\frac{dr}{d\underline{r}}, \; \frac{r}{\underline{r}}, \; \frac{r}{\underline{r}} \right]. \tag{21.7.51}$$

In absence of body force, the sole non identically satisfied equation of the balance of momentum $\overrightarrow{\nabla} \cdot \boldsymbol{\sigma} = \mathbf{0}$, Eqn (2.4.67), expresses in terms of the radial and orthoradial stresses, σ_r and σ_θ respectively, as

$$\frac{d\sigma_r}{dr} + \frac{2}{r} (\sigma_r - \sigma_\theta) = 0. \tag{21.7.52}$$

The shell is unloaded, so that the boundary conditions in the current configuration imply vanishing radial Cauchy stress over the inner surface $r_1 = r(\underline{r}_1)$ and outer surface $r_2 = r(\underline{r}_2)$,

$$\sigma_r(r_1) = \sigma_r(r_2) = 0. \tag{21.7.53}$$

The growth tensor $\mathbf{F}^{\mathrm{g}}$ is assumed to be diagonal and of the same form as the deformation gradient,

$$\mathbf{F}^{\mathrm{g}} = \mathbf{diag}\left[g_r(\underline{r}, t),\ g_\theta(\underline{r}, t),\ g_\theta(\underline{r}, t)\right], \tag{21.7.54}$$

with g_r and g_θ representing radial and orthoradial growths respectively, and $g_r(\underline{r}, t = 0) = g_\theta(\underline{r}, t = 0) = 1$. If the material is elastically incompressible, then

$$\det \mathbf{F}^{\mathrm{e}} = 1 \quad \text{but} \quad \det \mathbf{F} = \det \mathbf{F}^{\mathrm{g}} \neq 1! \tag{21.7.55}$$

Note the polar decomposition (2.5.1) of the elastic transformation,

$$\begin{aligned} \mathbf{F}^{\mathrm{e}} &= \mathbf{U}^{\mathrm{e}} = \mathbf{V}^{\mathrm{e}} \\ &= \mathbf{F} \cdot (\mathbf{F}^{\mathrm{g}})^{-1} = \mathbf{diag}\left[v_r = \frac{1}{g_r}\frac{dr}{d\underline{r}},\ v_\theta = \frac{1}{g_\theta}\frac{r}{\underline{r}},\ v_\theta = \frac{1}{g_\theta}\frac{r}{\underline{r}}\right]. \end{aligned} \tag{21.7.56}$$

Chen and Hoger [2000] and van Dyke and Hoger [2000] consider the pros and cons of the use of the initial configuration (Lagrangian analysis) versus the current configuration (Eulerian analysis) as reference. In particular, they note that, since growth introduces new material points, an Eulerian analysis is better suited. They examine the dependence of the residual stresses with respect to the form of the *prescribed* growth tensor $\mathbf{F}^{\mathrm{g}}(\underline{r})$. For illustration, Chen and Hoger [2000] consider an elastically incompressible solid with the constitutive equation,

$$\boldsymbol{\sigma} + p\,\mathbf{I} = f_1(I_1, I_2)\,\mathbf{V}^{\mathrm{e}} + f_2(I_1, I_2)\,(\mathbf{V}^{\mathrm{e}})^2, \tag{21.7.57}$$

which, in the spherical axes, can be rewritten in the format,

$$\boldsymbol{\sigma} + p\,\mathbf{I} = \mathbf{diag}\left[f_1\,v_r + f_2\,v_r^2,\ f_1\,v_\theta + f_2\,v_\theta^2,\ f_1\,v_\theta + f_2\,v_\theta^2\right]. \tag{21.7.58}$$

The I's are the first and second invariants of the elastic stretch,

$$\begin{aligned} I_1 &= \operatorname{tr}\mathbf{V}^{\mathrm{e}} = v_r + 2\,v_\theta = \frac{1}{g_r}\frac{dr}{d\underline{r}} + \frac{2}{g_\theta}\frac{r}{\underline{r}}, \\ I_2 &= \tfrac{1}{2}\left((\operatorname{tr}\mathbf{V}^{\mathrm{e}})^2 - \operatorname{tr}(\mathbf{V}^{\mathrm{e}})^2\right) = 2\,v_r\,v_\theta + v_\theta^2 = \frac{2}{g_r\,g_\theta}\frac{r}{\underline{r}}\frac{dr}{d\underline{r}} + \frac{1}{g_\theta^2}\frac{r^2}{\underline{r}^2}. \end{aligned} \tag{21.7.59}$$

The constitutive functions f_1 and f_2 are constrained by the Baker-Ericksen inequalities (6.8.4).

A few elements of the solution to the boundary value problem are now shortly presented. The elastic incompressibility condition $(21.7.55)_2$ implies $g_r\,g_\theta^2 = (r/\underline{r})^2\,(dr/d\underline{r})$, which yields the current geometry in terms of the determinant of the growth transformation $\mathbf{F}^{\mathrm{g}}$,

$$r^3(\underline{r}, t) - r^3(\underline{r}_1, t) = 3\int_{\underline{r}_1}^{\underline{r}} s^2\,g_r(s, t)\,g_\theta^2(s, t)\,ds. \tag{21.7.60}$$

The elastic transformation deduces by (21.7.56). With help of (21.7.58), the equilibrium (21.7.52) is cast in the form,

$$\frac{d\sigma_r}{d\underline{r}} = 2\,\frac{v_\theta - v_r}{r}\frac{dr}{d\underline{r}}\left(f_1 + (v_r + v_\theta)\,f_2\right), \tag{21.7.61}$$

which, with the boundary condition (21.7.53) at the inner radius, is integrated to,

$$\sigma_r(r, t) - 0 = \int_{\underline{r}_1}^{\underline{r}} 2\left[\frac{v_\theta - v_r}{r}\frac{dr}{ds}\left(f_1 + (v_r + v_\theta)\,f_2\right)\right](s, t)\,ds, \tag{21.7.62}$$

the current radius r being given in terms of the reference geometry $\underline{r}$ by (21.7.60). Eqn (21.7.58) also provides the orthoradial stress,

$$\sigma_\theta = \sigma_r + (v_\theta - v_r)\left(f_1 + (v_r + v_\theta)\,f_2\right). \tag{21.7.63}$$

If the growth transformation $\mathbf{F}^g$ is a known function of $\underline{r}$, satisfaction of the boundary condition on the outer radius $\sigma_r(r_2) = 0$ boils down to solve

$$0 = \mathrm{Res}(r_1) \;\; = \;\; 2 \int_{\underline{r}_1}^{\underline{r}_2} \left[\frac{v_\theta - v_r}{r} \, \frac{dr}{ds} \, \left(f_1 + (v_r + v_\theta) \, f_2 \right) \right] (s,t) \, ds \,, \tag{21.7.64}$$

for the current inner radius $r_1 = r(\underline{r}_1, t)$, with $r = r(\underline{r}, t)$ given by (21.7.60). Once the latter is known, the stresses deduce by (21.7.62)-(21.7.63), and the pressure p by (21.7.58).

If the growth is uniform *and* isotropic, i.e. $g_r(\underline{r}, t) = g_\theta(\underline{r}, t) = G(t)$, or $\mathbf{F}^g(\underline{r}, t) = G(t) \, \mathbf{I}$, then $r = \underline{r} \, G(t)$, the deformation gradient is isotropic and uniform, $\mathbf{F}(\underline{r}, t) = G(t) \, \mathbf{I}$, the elastic transformation is equal to the identity $\mathbf{F}^e(\underline{r}, t) = \mathbf{I}$, and the stress, which is also the residual stress, vanishes.

For general growth, Chen and Hoger [2000] can not obtain an analytical solution to (21.7.64), but they calculate the derivatives of the stresses at time $t = 0$, and this allows them to make some comments on the sign of the residual stress, when the growth process is prescribed.

Much like in Section 21.7.1.2, the problem addressed in this section is illustrative, in the sense that the aim was to display some features of growth, like residual stresses, as independent as possible from the details of the growth itself. In actual situations, growth is not an independent process, it is expected to be influenced by, and in fact to interact with, the local mechanical and biochemical conditions. These interactions are realized by postulating constitutive equations for the growth transformation in terms of a set of independent variables, which define the mechanical and biochemical states of the material.

An attempt to discover effects of growth, somehow in the mood of Chen and Hoger [2000], is described by Klisch et al. [2001]b. Three separate cases of growth are considered, namely isotropic, radial and circumferential growths,

$$\mathbf{F}^g_{\mathrm{iso}} = \mathbf{diag} \, [\lambda_g, \lambda_g, \lambda_g], \quad \mathbf{F}^g_{\mathrm{rad}} = \mathbf{diag} \, [\lambda_{gr}, 1, 1], \quad \mathbf{F}^g_{\mathrm{cir}} = \mathbf{diag} \, [1, \lambda_{g\theta}, \lambda_{g\theta}] \,. \tag{21.7.65}$$

In order to mimic the conewise mechanical response of fiber-reinforced materials, like articular cartilages, Klisch et al. [2001]b consider two strain energy functions which exhibit a large difference of stiffnesses in contraction and extension, Eqns (6.6.14) and (6.6.15). The growth rate is assumed to be either uniform in space or linear along the radius, and vanishing on the inner surface. Of course, the magnitude and sign of the residual stress depends much of the spatial variation of the growth. Linear circumferential growth may imply a compressive orthoradial stress on the outer surface and tension on the inner surface.

21.8 Incompatible strains and residual stresses

The compatibility conditions derived here are expressed in terms of strains. For linear isotropic elasticity, these conditions are turned in terms of stresses, so-called Beltrami-Mitchell conditions, in Section 9.4 which accounts in addition for the presence of an inelastic strain. The latter may be of hydraulic, thermal or chemical nature or associated with other physical phenomena, like growth.

21.8.1 Incompatible strains

Residual stresses are the stresses that remain in the body after all external loads have been removed. Actually, the situation is difficult to imagine in a quasi-static context, unless gravity is neglected. A practical method to detect the existence of residual stresses is to make educated cut(s) and to observe the resulting motion. Fung [1990] [1991] has characterized the existence of compressive residual stresses in arteries by an opening angle. Skalak et al. [1996] attribute residual stresses to incompatible strains, which are themselves induced by growth. The total deformation is compatible by definition, since its keeps the body in a single piece. In a small strain context, it is the sum of the incompatible deformation and of an elastic strain, whose incompatibility negates the one of the growth deformation, if the latter exists. In other words, the elastic strain emerges so as to *accommodate* the incompatibility, i.e., make the total deformation compatible. Conversely, if the existing growth strain is compatible, then no additional elastic strain need to develop (in absence of external loading).

21.8.2 Compatible strains in an infinitesimal strain context

Compatibility conditions are intended to ensure that a strain field $\boldsymbol{\epsilon} = \boldsymbol{\epsilon}(\mathbf{x})$ derives from a single-valued displacement field $\mathbf{u}$, that is,

$$\boldsymbol{\epsilon} = \tfrac{1}{2}\,(\mathbf{u}\overleftarrow{\nabla} + \overrightarrow{\nabla}\,\mathbf{u}), \quad \epsilon_{ij} = \tfrac{1}{2}\,(u_{i,j} + u_{j,i})\,. \tag{21.8.1}$$

To motivate the raison d'être of the compatibility conditions, one simply should realize that the displacement vector has 3 independent components while the strain tensor has 6 components.

On the other hand, if the displacement is the primary unknown in a boundary value problem, then no compatibility condition is needed.

21.8.2.1 Saint-Venant's compatibility conditions in a simply-connected body

Let the deformation gradient be decomposed into its symmetric and skew-symmetric parts, $\boldsymbol{\epsilon}$ and $\boldsymbol{\omega}$ respectively, Eqn (2.4.40). Then[21.7]

(a) $\quad du_i = (\epsilon_{ij} + \omega_{ij})\,dx_j$

(b) $\quad \epsilon_{ij,k} + \omega_{ij,k} = \epsilon_{ik,j} + \omega_{ik,j}$ $\qquad$: 1st integrability condition $u_{i,kj} = u_{i,jk}$

(c) $\quad \omega_{ij,k} + \omega_{jk,i} + \omega_{ki,j} = 0$ $\qquad$: circular permutation of (b) and sum $\qquad\qquad$ (21.8.2)

(d) $\quad \omega_{jk,i} = -\omega_{ij,k} + \omega_{ik,j} = \epsilon_{ij,k} - \epsilon_{ik,j}$ $\quad$: by (b) and (c),

(e) $\quad \epsilon_{ij,kl} - \epsilon_{ik,jl} = \epsilon_{lj,ki} - \epsilon_{lk,ji}$ $\qquad$: 2nd integrability condition $\omega_{jk,il} = \omega_{jk,li}$.

Out of these 81 relations, some are identities and other repetitions due to the symmetry of the strain tensor. In fact, when phrased in terms of operators, the number of independent relations appear to be six, or less. It is instrumental to introduce the pseudo-vector $\tilde{\boldsymbol{\omega}}$ associated with $-\boldsymbol{\omega}$, namely $\tilde{\omega}_m = -\tfrac{1}{2}\,e_{mjk}\,\omega_{jk}$. The relation (21.8.2)-(d) may be rewritten $d\tilde{\omega}_m = \alpha_{mi}\,dx_i$ with $\alpha_{mi} = e_{mjk}\,\epsilon_{ki,j}$. Indeed,

$$\begin{aligned}
d\tilde{\omega}_m &= -\tfrac{1}{2}\,e_{mjk}\,\omega_{jk,i}\,dx_i \\
&= -\tfrac{1}{2}\,e_{mjk}\,(\epsilon_{ij,k} - \epsilon_{ik,j})\,dx_i \quad \text{by } (21.8.2) - (d) \\
&= e_{mjk}\,\epsilon_{ki,j}\,dx_i\,.
\end{aligned} \tag{21.8.3}$$

With help of (2.4.30) and (2.4.31), the second order tensor $\boldsymbol{\alpha}$ may be expressed in either form,

$$\boldsymbol{\alpha} = \overrightarrow{\nabla}\wedge\boldsymbol{\epsilon} = -(\boldsymbol{\epsilon}\wedge\overleftarrow{\nabla})^{\mathrm{T}}\,. \tag{21.8.4}$$

The integrability condition now expresses in terms of the second order *incompatibility tensor* $\boldsymbol{\eta}$,

$$\begin{aligned}
\boldsymbol{\eta} &= -\boldsymbol{\alpha}\wedge\overleftarrow{\nabla} \\
&= -(\overrightarrow{\nabla}\wedge\boldsymbol{\epsilon})\wedge\overleftarrow{\nabla} \\
&= -((\overrightarrow{\nabla}\wedge\boldsymbol{\epsilon})\wedge\overleftarrow{\nabla})^{\mathrm{T}} \\
&= \overrightarrow{\nabla}\wedge(\overrightarrow{\nabla}\wedge\boldsymbol{\epsilon})^{\mathrm{T}} \quad \text{by } (2.4.30) \\
&= -\overrightarrow{\nabla}\wedge(\boldsymbol{\epsilon}\wedge\overleftarrow{\nabla}) \quad \text{by } (2.4.30) \\
\eta_{ij} = \eta_{ji} &= e_{ikm}\,e_{jln}\,\epsilon_{mn,kl} \quad \text{by } (2.4.31)\,,
\end{aligned} \tag{21.8.5}$$

[21.7]A differential form $\mathbf{a}\cdot d\mathbf{x}$ is said to be an exact differential if it derives from a gradient, say $\mathbf{a} = \boldsymbol{\nabla}b$, or $\boldsymbol{\nabla}\wedge\mathbf{a} = \mathbf{0}$. Then it is integrable between two arbitrary points in a simply connected body, that is, its integral is independent of the path connecting these points and remaining in the body. This integrability condition is equivalent to the cross-derivatives to be equal. This property can be shown easily by performing the integration between two infinitesimally close points, and requiring that the integral be independent of the paths followed (typically the axes) to joint the endpoints.

which is required to vanish,

$$\boldsymbol{\eta} = \mathbf{0} \qquad \text{Saint-Venant's compatibility conditions}. \tag{21.8.6}$$

The incompatibility tensor $\boldsymbol{\eta}$ is symmetric. Its six components are not independent since, the divergence of a rotational vanishing due to (2.4.5), they are subjected to the three relations, termed Bianchi formulas,

$$\vec{\nabla} \cdot \boldsymbol{\eta} = \mathbf{0}, \quad \eta_{ji,j} = 0, \quad i \in [1,3]. \tag{21.8.7}$$

However, it is not sufficient to check only three equations (21.8.5). The usual procedure is to satisfy three of them in the body and three others on the boundary. Alternatively, one may consider all six equations, remembering that they are not completely independent, Malvern [1969], p. 187.

Similarly for plane displacements in the plane (x_1, x_2), the number of compatibility equations is three, out of which one is independent,

$$\epsilon_{11,22} + \epsilon_{22,11} = 2\,\epsilon_{12,12}. \tag{21.8.8}$$

In cartesian axes, the compatibility conditions (21.8.6) can be cast in the explicit format,

$$\begin{cases} \eta_{11} = \epsilon_{22,33} + \epsilon_{33,22} - 2\,\epsilon_{23,23} = 0, \\[2mm] \eta_{22} = \epsilon_{33,11} + \epsilon_{11,33} - 2\,\epsilon_{31,31} = 0, \\[2mm] \eta_{33} = \epsilon_{11,22} + \epsilon_{22,11} - 2\,\epsilon_{12,12} = 0, \\[2mm] \eta_{23} = -\epsilon_{23,11} - \epsilon_{11,23} + \epsilon_{21,31} + \epsilon_{31,21} = 0, \\[2mm] \eta_{31} = -\epsilon_{31,22} - \epsilon_{22,31} + \epsilon_{32,12} + \epsilon_{12,32} = 0, \\[2mm] \eta_{12} = -\epsilon_{12,33} - \epsilon_{33,12} + \epsilon_{13,23} + \epsilon_{23,13} = 0. \end{cases} \tag{21.8.9}$$

A basic application is proposed in Exercise 21.1. Another equivalent expression of the incompatibility tensor, Gurtin [1972],

$$\boldsymbol{\eta} = -\vec{\nabla}^2 \operatorname{tr} \boldsymbol{\epsilon} + \vec{\nabla}(\vec{\nabla} \cdot \boldsymbol{\epsilon}) + \left(\vec{\nabla}(\vec{\nabla} \cdot \boldsymbol{\epsilon})\right)^{\mathrm{T}} - \Delta \boldsymbol{\epsilon} = \mathbf{0} \tag{21.8.10}$$

$$\eta_{ij} = -\epsilon_{kk,ij} + \epsilon_{jk,ik} + \epsilon_{ik,jk} - \epsilon_{ij,kk} = 0, \quad i,j \in [1,3],$$

is proved in Exercise 21.2. In fact, the incompatibility tensor can be shown to be equal to,

$$\boldsymbol{\eta} = \left(\Delta \operatorname{tr} \boldsymbol{\epsilon} - \vec{\nabla} \cdot (\vec{\nabla} \cdot \boldsymbol{\epsilon})\right) \mathbf{I} - \left(\vec{\nabla}^2 \operatorname{tr} \boldsymbol{\epsilon} - \vec{\nabla}(\vec{\nabla} \cdot \boldsymbol{\epsilon}) - \left(\vec{\nabla}(\vec{\nabla} \cdot \boldsymbol{\epsilon})\right)^{\mathrm{T}} + \Delta \boldsymbol{\epsilon}\right), \tag{21.8.11}$$

$$\eta_{ij} = I_{ij}\left(\epsilon_{mm,kk} - \epsilon_{mk,mk}\right) - \left(\epsilon_{mm,ij} - \epsilon_{im,jm} - \epsilon_{jk,ik} + \epsilon_{ij,kk}\right),$$

and the relations

$$\Delta \operatorname{tr} \boldsymbol{\epsilon} - \vec{\nabla} \cdot (\vec{\nabla} \cdot \boldsymbol{\epsilon}) = \tfrac{1}{2} \operatorname{tr} \left(\vec{\nabla}^2 \operatorname{tr} \boldsymbol{\epsilon} - \vec{\nabla}(\vec{\nabla} \cdot \boldsymbol{\epsilon}) - \left(\vec{\nabla}(\vec{\nabla} \cdot \boldsymbol{\epsilon})\right)^{\mathrm{T}} + \Delta \boldsymbol{\epsilon}\right) = 0, \tag{21.8.12}$$

$$\vec{\nabla}^2 \operatorname{tr} \boldsymbol{\epsilon} - \vec{\nabla}(\vec{\nabla} \cdot \boldsymbol{\epsilon}) - \left(\vec{\nabla}(\vec{\nabla} \cdot \boldsymbol{\epsilon})\right)^{\mathrm{T}} + \Delta \boldsymbol{\epsilon} = \mathbf{0},$$

should hold.

Eqns (21.8.7) or (21.8.10) are *necessary*. They are *sufficient* to ensure the existence of a single-valued displacement field only when the body is *simply-connected*.

The compatibility conditions in cylindrical coordinates are given in, e.g., Malvern [1969], pp. 663 et seq.

Example: isotropic strain and thermal field

The strain $\boldsymbol{\epsilon} = a(\mathbf{x})\,\mathbf{I}$ is compatible in a simply connected domain only if it is constant or if it depends linearly on the spatial coordinates.

As examples, we may mention thermal strain and growth strain in isotropic bodies. Indeed, in absence of stress, a thermal field $\theta = \theta(\mathbf{x})$ in an isotropic body leads to the thermal strain $\boldsymbol{\epsilon} = \frac{1}{3} c_{\mathrm{T}}\,\theta\,\mathbf{I}$, with c_{T} the cubic coefficient of thermal expansion. Therefore, a given thermal field can be associated with a stress free displacement field under the above conditions only.

21.8.2.2 The reconstructed displacement in explicit form: Cesaro formula

The issue is to calculate the displacement in terms of the strain, namely from the first line of (21.8.2),

$$u_i(\mathbf{x}) = u_i(\mathbf{x}^0) + \int_{\mathbf{x}^0}^{\mathbf{x}} (\epsilon_{ij} + w_{ij}) \, dy_j \,, \tag{21.8.13}$$

where, using (21.8.2)-(d),

$$dw_{ij} = \big(\epsilon_{im,j}(\mathbf{y}) - \epsilon_{jm,i}(\mathbf{y})\big) \, dy_m \,. \tag{21.8.14}$$

The integration on a path from the arbitrary point $\mathbf{x}^0$ to the current point $\mathbf{x}$ may be worked out as follows, Sokolnikoff [1956],

$$
\begin{aligned}
\int_{\mathbf{x}^0}^{\mathbf{x}} w_{ij} \, dy_j &= \int_{\mathbf{x}^0}^{\mathbf{x}} w_{ij} \, d(y_j - x_j), \\
&= w_{ij}(\mathbf{x}^0)\,(x_j - x_j^0) - \int_{\mathbf{x}^0}^{\mathbf{x}} (y_j - x_j)\,(\epsilon_{im,j} - \epsilon_{jm,i}) \, dy_m,
\end{aligned}
\tag{21.8.15}
$$

the second line being obtained via (21.8.14). In summary, the so-called Cesaro formula gives the displacement as

$$\mathbf{u}(\mathbf{x}) = \mathbf{u}(\mathbf{x}^0) + \boldsymbol{\omega}(\mathbf{x}^0) \cdot (\mathbf{x} - \mathbf{x}^0) + \int_{\mathbf{x}^0}^{\mathbf{x}} \mathbf{U}(\mathbf{x}, \mathbf{y}) \cdot d\mathbf{y} \,, \tag{21.8.16}$$

where

$$U_{ij}(\mathbf{x}, \mathbf{y}) = \epsilon_{ij}(\mathbf{y}) + (x_k - y_k)\big(\epsilon_{ij,k}(\mathbf{y}) - \epsilon_{kj,i}(\mathbf{y})\big) \,. \tag{21.8.17}$$

In a simply connected domain, the integral is independent on the path of integration. Different choices in $\mathbf{x}^0$ result in displacement fields that differ only by an infinitesimal rigid body motion.

Example: simple shear

Consider the kinematics given in Exercise 2.2. Since the infinitesimal strain is constant, then $\mathbf{U} = \boldsymbol{\epsilon}$ from (21.8.17), and the reconstructed displacement field obtained by integration of (2.E.7) by (21.8.16) gives the second displacement field. This field differs from the original field only by an infinitesimal rotation around the x_3-axis.

21.8.2.3 Compatibility conditions in a multiply-connected body

Skalak et al. [1996], on which the subsections below are based, give an overview of the conditions for which a deformation field in a simply- or multiply-connected body is incompatible, both in a small strain context and in a finite strain context.

A body is said $n + 1$-*tuply connected* if there are n independent curves C_m, $m \in [1, n]$, such that the body can be reduced to a simply-connected body by excluding these curves, e.g., branch cuts.

For an $n+1$-*tuply connected* body, n additional conditions have to be satisfied for the strain field to be compatible, namely

$$\oint_{C_m} \mathbf{U}(\mathbf{x}^0, \mathbf{y}) \cdot d\mathbf{y} = \mathbf{0}, \quad m \in [1, n] \,, \tag{21.8.18}$$

with

$$U_{ij}(\mathbf{x}^0, \mathbf{y}) = \epsilon_{ij}(\mathbf{y}) + (x_k^0 - y_k)\big(\epsilon_{ij,k}(\mathbf{y}) - \epsilon_{kj,i}(\mathbf{y})\big) \,, \tag{21.8.19}$$

where $\mathbf{x}^0$ is a point of the curve C_m. In other words, the contribution of the integral to the displacement should be the same on each side of the branch cuts.

Example: a locally compatible but globally incompatible uniform strain, Skalak et al. [1996]

Consider the strain field in a hollow cylinder defined in cylindrical coordinates (r, θ, z) by:

$$\epsilon_{rr} = 0, \quad \epsilon_{\theta\theta} = k, \quad \epsilon_{zz} = 0 \,. \tag{21.8.20}$$

It satisfies the compatibility conditions (21.8.10), Malvern [1969], p. 669. Consequently, this strain would be compatible in a solid cylinder.

However, it does not satisfy the compatibility conditions (21.8.18). Indeed the integration of $\epsilon_{\theta\theta}$ along a circular closed path inside a plane section of the hollow cylinder gives $2\,\pi\,r\,k \neq 0$.

21.8.2.4 Incompatibility tensor and dislocations

The Burgers vector is defined by curvilinear integration of an incompatible strain field along a closed curve C,

$$\mathbf{b} = \oint_C \mathbf{t} \cdot \vec{\boldsymbol{\nabla}}\mathbf{u}\, ds, \quad b_j = \oint_C t_i\, u_{j,i}\, ds\,. \tag{21.8.21}$$

With help of the Stokes theorem (2.4.13), for any closed curve C, the Burgers vector can be expressed in terms of the *distributed dislocation density* $\boldsymbol{\alpha} = \vec{\boldsymbol{\nabla}} \wedge \vec{\boldsymbol{\nabla}}\mathbf{u}$ through an integral over the surface S of local normal $\mathbf{e}$ that the curve C encloses,

$$\mathbf{b} = \int_S \mathbf{e} \cdot \boldsymbol{\alpha}\, dS, \quad b_j = \int_S e_m\, e_{mkl}\, (u_{j,l})_{,k}\, dS\,. \tag{21.8.22}$$

Example: Volterra dislocations, Skalak et al. [1996]

Volterra dislocations exist in multiply-connected bodies only: they are displacement fields, which satisfy the compatibility conditions (21.8.10) but not the integrability condition (21.8.18) across some curve C. The displacement, as well as the skew-symmetric part of the deformation gradient, are discontinuous across the curve C, while the strain is continuous.

Volterra dislocations may be visualized by considering a cut (the above curve C) in an artery (hollow cylinder) parallel to the axis. The motion on each side of the cut differs by six rigid body degrees of freedom, which define the six Volterra dislocations. They might not be associated with a distributed dislocation density.

21.8.3 Compatible strain in a finite strain context

The motion $\underline{\mathbf{x}} \to \mathbf{x}$ is described by the deformation gradient $\mathbf{F} = \partial\mathbf{x}/\partial\underline{\mathbf{x}}$, with $\det \mathbf{F} > 0$. The deformation gradient is decomposed into a product of a proper orthogonal tensor and a pure deformation, namely $\mathbf{F} = \mathbf{Q} \cdot \mathbf{U} = \mathbf{V} \cdot \mathbf{Q}$, with $\mathbf{Q} \cdot \mathbf{Q}^{\mathrm{T}} = \mathbf{I}$, $\det \mathbf{Q} = 1$, and $\mathbf{U}$ and $\mathbf{V}$ symmetric positive definite. The left and right Cauchy-Green tensors are defined as $\mathbf{B} = \mathbf{F} \cdot \mathbf{F}^{\mathrm{T}}$ and $\mathbf{C} = \mathbf{F}^{\mathrm{T}} \cdot \mathbf{F}$ respectively, and the Green strain by $\mathbf{E}_{\mathrm{G}} = \frac{1}{2}(\mathbf{C} - \mathbf{I})$. Thus $\mathbf{U} = \mathbf{C}^{1/2}$ and $\mathbf{V} = \mathbf{B}^{1/2}$.

The compatibility conditions can be expressed in terms of the right Cauchy-Green tensor. The question can be stated as follows: given a right Cauchy-Green tensor $\mathbf{C}$, what are the conditions for the existence of a continuous displacement field $\mathbf{u}$ whose right Cauchy-Green tensor is $\mathbf{C}$?

The compatibility conditions require the curvature tensor $\mathbf{R}$ to vanish, Truesdell and Toupin [1960], Section 34,

$$\mathbf{R} = \mathbf{0}; \quad R_{ijkl} = 0, \quad (i,j,k,l) \in [1,3]\,. \tag{21.8.23}$$

The curvature tensor,

$$R_{ijkl} = \Gamma_{jli,k} - \Gamma_{jki,l} + \sum_{p,q=1,3} C_{pq}^{-1}\left(\Gamma_{jkp}\Gamma_{ilq} - \Gamma_{jlp}\Gamma_{ikq}\right), \tag{21.8.24}$$

expresses in terms of the Christoffel symbols Γ,

$$\Gamma_{ijk} = \tfrac{1}{2}\left(C_{jk,i} + C_{ik,j} - C_{ij,k}\right). \tag{21.8.25}$$

The expression above, due to Blume [1989], is actually a slight simplification of the one derived by Truesdell and Toupin [1960]. Out of the $3^4 = 81$ conditions (21.8.23), only six are distinct, which are in fact subjected to the Bianchi conditions, much like in the infinitesimal strain case, e.g., Malvern [1969], p. 195.

The condition $\mathbf{R} = \mathbf{0}$ is necessary, but it is sufficient only when the region is simply connected. Additional conditions arise for sufficiency in a multiply-connected region, defined by n curves C_m, namely

$$\oint_{C_m} (\mathbf{F} - \mathbf{I}) \cdot d\underline{\mathbf{x}}' = \mathbf{0}, \quad m \in [1,n]\,. \tag{21.8.26}$$

Indeed, the reconstructed displacement is, to within a rigid body rotation depending on $\underline{\mathbf{x}}^0$, obtained through the expression,

$$\mathbf{u}(\underline{\mathbf{x}}) = \int_{\underline{\mathbf{x}}^0}^{\mathbf{x}} (\mathbf{F} - \mathbf{I}) \cdot d\underline{\mathbf{x}}'\,. \tag{21.8.27}$$

Two practical verification expressions are given in Skalak et al. [1996], their Eqns 5.22, 5.23 and 5.25.

21.8.3.1 An example in a finite strain context

Consider the kinematics of simple shear defined in Exercise 2.3. It is compatible in a simply-connected region, since the deformation gradient is constant.

A reconstructed displacement $\mathbf{u}^r$ is based on $\mathbf{C}$, that is on $\mathbf{U}$,

$$\mathbf{u}^r(\underline{\mathbf{x}}) = \int_{\underline{\mathbf{0}}}^{\mathbf{x}} (\mathbf{U} - \mathbf{I}) \cdot d\underline{\mathbf{x}}' = \mathbf{U} \cdot \underline{\mathbf{x}} - \underline{\mathbf{x}}. \qquad (21.8.28)$$

The original displacement field was $\mathbf{u} = \mathbf{Q} \cdot \mathbf{u}^r$. Other compatible displacement fields differ from $\mathbf{u}$ by rigid body motions. The latter are not arbitrary. Indeed, in view of the relation (21.8.2)-(d), we see that, in an infinitesimal strain context, the skew-symmetric part of the displacement gradient should be spatially uniform. The result still holds in a finite strain context, Skalak et al. [1996].

21.8.3.2 Incompatible strain rate

In applications, the strain rate $d\epsilon/dt$, or rate of deformation gradient $\mathbf{D}$, may be given. Then formally, the conditions of compatibility are the same as in the infinitesimal strain context, with $\mathbf{D}$ in place of ϵ, and the velocity $\mathbf{v}$ in place of the displacement $\mathbf{u}$.

Incidentally, note that, in terms of the polar decomposition of the deformation gradient $\mathbf{F} = \mathbf{Q} \cdot \mathbf{U}$,

$$\mathbf{D} = \tfrac{1}{2} (\mathbf{v} \overleftarrow{\nabla} + \overrightarrow{\nabla} \mathbf{v}) = \tfrac{1}{2} \mathbf{Q} \cdot \left(\frac{d\mathbf{U}}{dt} \cdot \mathbf{U}^{-1} + \mathbf{U}^{-1} \cdot \frac{d\mathbf{U}}{dt}\right) \cdot \mathbf{Q}^{\mathrm{T}}. \qquad (21.8.29)$$

Let $\mathbf{T} = \mathbf{Q}^{\mathrm{T}} \cdot \mathbf{D} \cdot \mathbf{Q}$. The solution $d\mathbf{U}/dt$ to (21.8.29) is given by (2.3.17),

$$\begin{aligned}
(I_1 I_2 - I_3) &\frac{d\mathbf{U}}{dt} = \\
\mathbf{U}^2 \cdot \mathbf{T} \cdot \mathbf{U}^2 &- I_1 (\mathbf{U}^2 \cdot \mathbf{T} \cdot \mathbf{U} + \mathbf{U} \cdot \mathbf{T} \cdot \mathbf{U}^2) + (I_1^2 + I_2) \mathbf{U} \cdot \mathbf{T} \cdot \mathbf{U} - I_3 (\mathbf{T} \cdot \mathbf{U} + \mathbf{U} \cdot \mathbf{T}) + I_1 I_3 \mathbf{T},
\end{aligned} \qquad (21.8.30)$$

with I_1, I_2 and I_3 isotropic scalar invariants of $\mathbf{U}$.

Exercises on Chapter 21

Exercise 21.1 Two particular compatible kinematics

Consider the infinitesimal strain tensors,

$$\boldsymbol{\epsilon}_{(1)} = \begin{bmatrix} 0 & f & 0 \\ f & 0 & 0 \\ 0 & 0 & 0 \end{bmatrix}, \quad \boldsymbol{\epsilon}_{(2)} = \begin{bmatrix} g & 0 & 0 \\ 0 & 0 & 0 \\ 0 & 0 & 0 \end{bmatrix}, \tag{21.E.1}$$

where f and g are smooth functions of all coordinates $(x_1,\, x_2,\, x_3)$. Calculate the incompatibility tensors and find the conditions for which the strains are compatible in a simply connected domain.

Answer:

Use of (21.8.9) yields the respective incompatibility tensors,

$$\boldsymbol{\eta}_{(1)} = \begin{bmatrix} 0 & -f_{,33} & f_{,23} \\ -f_{,33} & 0 & f_{,13} \\ f_{,23} & f_{,13} & -2\,f_{,12} \end{bmatrix}, \quad \boldsymbol{\eta}_{(2)} = \begin{bmatrix} 0 & 0 & 0 \\ 0 & g_{,33} & -g_{,23} \\ 0 & -g_{,23} & g_{,22} \end{bmatrix}. \tag{21.E.2}$$

Compatibility holds for the strain $\boldsymbol{\epsilon}_{(1)}$ if $f = f(x_1,\, x_2) = f_1(x_1) + f_2(x_2)$, $f_1(x_1)$ and $f_2(x_2)$ being two arbitrary functions of their respective arguments.

The strain $\boldsymbol{\epsilon}_{(2)}$ is compatible if $g = g_1(x_1)\, x_3 + g_2(x_1)\, x_2 + g_3(x_1)$, g_1, g_2 and g_3 being arbitrary functions of x_1.

Exercise 21.2 The incompatibility tensor

Prove that the incompatibility tensor may be expressed in the form (21.8.10).

Answer:

The proof uses (2.1.11), Segel [1987], p. 158,

$$
\begin{aligned}
\eta_{ij} &= e_{ikm}\, e_{jln}\, \epsilon_{mn,kl} \\[2mm]
&= \det \begin{bmatrix} I_{ij} & I_{il} & I_{in} \\ I_{kj} & I_{kl} & I_{kn} \\ I_{mj} & I_{ml} & I_{mn} \end{bmatrix} \epsilon_{mn,kl} \\[2mm]
&= I_{ij}\left(\epsilon_{mm,kk} - \epsilon_{mk,mk}\right) - \left(\epsilon_{mm,ij} - \epsilon_{im,jm}\right) + \left(\epsilon_{jk,ik} - \epsilon_{ij,kk}\right).
\end{aligned}
\tag{21.E.3}
$$

Hence

$$\boldsymbol{\eta} = \Big(\underbrace{\Delta \operatorname{tr}\boldsymbol{\epsilon} - \vec{\nabla}\cdot(\vec{\nabla}\cdot\boldsymbol{\epsilon})}_{B}\Big)\mathbf{I} - \Big(\underbrace{\vec{\nabla}^2\operatorname{tr}\boldsymbol{\epsilon} - \vec{\nabla}\,(\vec{\nabla}\cdot\boldsymbol{\epsilon}) - \big(\vec{\nabla}\,(\vec{\nabla}\cdot\boldsymbol{\epsilon})\big)^{\mathrm{T}} + \Delta\boldsymbol{\epsilon}}_{A}\Big). \tag{21.E.4}$$

Now one easily shows the relation,

$$\Delta \operatorname{tr}\boldsymbol{\epsilon} - \vec{\nabla}\cdot(\vec{\nabla}\cdot\boldsymbol{\epsilon}) = \tfrac{1}{2}\operatorname{tr}\Big(\vec{\nabla}^2\operatorname{tr}\boldsymbol{\epsilon} - \vec{\nabla}\,(\vec{\nabla}\cdot\boldsymbol{\epsilon}) - \big(\vec{\nabla}\,(\vec{\nabla}\cdot\boldsymbol{\epsilon})\big)^{\mathrm{T}} + \Delta\boldsymbol{\epsilon}\Big). \tag{21.E.5}$$

Therefore $\boldsymbol{\eta} = \tfrac{1}{2}\operatorname{tr}\mathbf{A}\,\mathbf{I} - \mathbf{A}$. Since $\boldsymbol{\eta}$ should vanish, so should $\operatorname{tr}\mathbf{A}$, and consequently $\boldsymbol{\eta} = -\mathbf{A} = \mathbf{0}$.

Chapter 22

Elastic-growing solids. Growth laws consistent with thermodynamics

22.1 The tools: dissipation, homeostatic domain and convexity

Formal arguments have been presented in Chapter 21 in favor of a multiplicative decomposition of the total deformation into an elastic contribution and a growth contribution. The implications of this point of view are examined in this chapter through examples. Growth models that adopt this formalism are referred to as elastic-growing, in contrast to elastic growth models.

The two formalisms are compared in simple boundary value problems. The comparison addresses both the similarities and differences in the constitutive equations and the elements that come into picture to solve the boundary value problems.

Actual growth processes are triggered by a number of biochemical, chemophysical and mechanical driving engines. Emphasis is placed on the implications of mechanics on growth processes, in the framework of solids and saturated porous media. A key task to be addressed consists in devising a growth rate law that has sufficient generality to be able to encompass several families of tissues. The growth rate laws developed here are structured by the thermodynamics of irreversible processes, and are tailored to satisfy the dissipation inequality. Two directions are explored. Along an Onsagerist perspective, the dissipation is assumed to be quadratic. In a more comprehensive approach, the concept of a homeostatic domain is adopted and its convexity is exploited to produce the growth rate laws. A homeostatic domain can be conceived as an extension of the concept of homeostatic state. In fact, it can be viewed as a buffer against small variations of the loads about a homeostatic state. Growth and resorption then require some sizable amount of load deviation with respect to this homeostatic state. The model may accommodate changes of the homeostatic domain in both size and position. In fact, it has been observed that the homeostatic states certainly evolve during morphogenesis.

While the perspective is purely phenomenological, some details that account for the modes in which the mass aggregates may be incorporated in the growth models, and their actual effects quantified. Two modes of mass deposition are considered: the new mass may aggregate undeformed, its deformation is progressive, and the interactions with the existing tissue leads finally to a homogenized state. Alternatively, the mass may aggregate only if it satisfies already some conditions: for example, its deformation should fit the existing tissue. In the first mode, energy is spent by the compound while, in the second mode, it is diverted to the mass before its aggregation.

22.2 Relations relative to the intermediate configurations

Growth is considered as a dissipative mechanism. The first task is to identify the sources of dissipation. The dissipation inequality is expressed in terms of entities displayed in Fig. 22.2.1 that refer to the intermediate configurations (κ) introduced by the multiplicative decomposition of the deformation gradient. The properties of the deposited mass are a priori arbitrary, as indicated in Section 21.3.3.

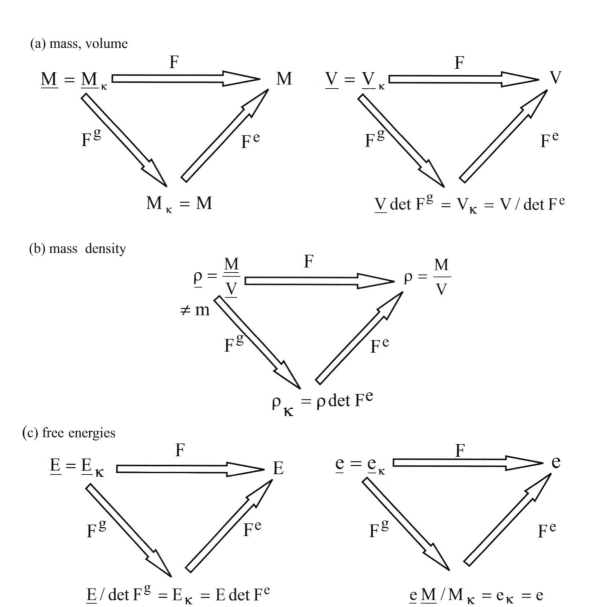

Fig. 22.2.1: Multiplicative decomposition of the deformation into a growth transformation and an elastic transformation for a solid. Transports of mass M, volume V, mass density ρ, free energy per unit volume E, free energy per unit mass e among the reference, intermediate and current configurations.

22.2.1 Entities relative to the intermediate configurations

Use will be made of the following entities, namely
- the "2nd Piola-Kirchhoff stress" τ_κ relative to the intermediate configurations,

$$\tau_\kappa = (\mathbf{F}^{\mathrm{e}})^{-1} \cdot (\det \mathbf{F}^{\mathrm{e}} \, \boldsymbol{\sigma}) \cdot (\mathbf{F}^{\mathrm{e}})^{-\mathrm{T}} \, ; \tag{22.2.1}$$

- the elastic right Cauchy-Green tensor $\mathbf{C}^{\mathrm{e}}$ and elastic Green strain $\mathbf{E}^{\mathrm{e}}_{\mathrm{G}}$,

$$\mathbf{C}^{\mathrm{e}} = (\mathbf{F}^{\mathrm{e}})^{\mathrm{T}} \cdot \mathbf{F}^{\mathrm{e}}, \quad \mathbf{E}^{\mathrm{e}}_{\mathrm{G}} = \tfrac{1}{2} \, (\mathbf{C}^{\mathrm{e}} - \mathbf{I}) \, ; \tag{22.2.2}$$

- the power relation following from (21.3.45),

$$\det \mathbf{F}^{\mathrm{e}} \, \boldsymbol{\sigma} : \boldsymbol{\nabla} \mathbf{v}_{\mathrm{s}} = \tau_\kappa : \Big(\frac{d\mathbf{E}^{\mathrm{e}}_{\mathrm{G}}}{dt} + \mathbf{C}^{\mathrm{e}} \cdot \mathbf{L}^{\mathrm{g}}_\kappa \Big) \, ; \tag{22.2.3}$$

- the free energy per unit volume E_κ at a material point, the free energy $\tilde{E}_\kappa$ of the supplied mass at that point, the entropy per unit volume S_κ, the rate of energy supply in unspecified form $\hat{\mathcal{E}}_\kappa$,

$$E_\kappa = E \det \mathbf{F}^{\mathrm{e}}, \quad \tilde{E}_\kappa = \tilde{E} \det \mathbf{F}^{\mathrm{e}} = \rho\,\tilde{e}\, \det \mathbf{F}^{\mathrm{e}}, \quad S_\kappa = S \det \mathbf{F}^{\mathrm{e}}, \quad \hat{\mathcal{E}}_\kappa = \hat{\mathcal{E}} \det \mathbf{F}^{\mathrm{e}}\,; \tag{22.2.4}$$

- the heat flux $\mathbf{Q}_\kappa$,

$$\mathbf{Q}_\kappa = \det \mathbf{F}^{\mathrm{e}}\, (\mathbf{F}^{\mathrm{e}})^{-1} \cdot \mathbf{Q}\,; \tag{22.2.5}$$

- the gradient $\boldsymbol{\nabla}_\kappa$,

$$\boldsymbol{\nabla}_\kappa(\cdot) = \boldsymbol{\nabla}(\cdot) \cdot \mathbf{F}^{\mathrm{e}}\,; \tag{22.2.6}$$

- the density ρ_κ and the rate of mass supply $\hat{\rho}_\kappa$,

$$\rho_\kappa = \rho \det \mathbf{F}^{\mathrm{e}}, \quad \hat{\rho}_\kappa = \hat{\rho} \det \mathbf{F}^{\mathrm{e}}\,; \tag{22.2.7}$$

which, using the balance of mass $d\rho/dt + \rho \operatorname{div} \mathbf{v}_{\mathrm{s}} = \hat{\rho}$, satisfy the relation,

$$\frac{d\rho_\kappa}{dt} + \rho_\kappa \operatorname{tr} \mathbf{L}_\kappa^{\mathrm{g}} = \hat{\rho}_\kappa\,; \tag{22.2.8}$$

- the remodeling variable r_κ,

$$r_\kappa = \frac{\rho_\kappa}{\rho} - 1\,, \tag{22.2.9}$$

which vanishes for density preserving growth processes.

Regarding the definition of the 2nd Piola-Kirchhoff stress $\boldsymbol{\tau}_\kappa$, it is worth stressing the relations,

$$\boldsymbol{\sigma} = \frac{1}{\det \mathbf{F}^{\mathrm{e}}}\, \mathbf{F}^{\mathrm{e}} \cdot \boldsymbol{\tau}_\kappa \cdot (\mathbf{F}^{\mathrm{e}})^{\mathrm{T}} = \frac{1}{\det \mathbf{F}}\, \mathbf{F} \cdot \underline{\underline{\boldsymbol{\tau}}} \cdot \mathbf{F}^{\mathrm{T}}\,, \tag{22.2.10}$$

and therefore

$$\mathbf{F}^{\mathrm{e}} \cdot \boldsymbol{\tau}_\kappa \cdot (\mathbf{F}^{\mathrm{e}})^{\mathrm{T}} = \det \mathbf{F}^{\mathrm{e}}\, \boldsymbol{\sigma} \neq \det \mathbf{F}\, \boldsymbol{\sigma} = \boldsymbol{\tau} = \mathbf{F} \cdot \underline{\underline{\boldsymbol{\tau}}} \cdot \mathbf{F}^{\mathrm{T}}\,, \tag{22.2.11}$$

because, as indicated by the power relations (22.2.3) and (22.2.13), $\boldsymbol{\tau}_\kappa$ is work-conjugate to the Green strain $\mathbf{E}_{\mathrm{G}}^{\mathrm{e}}$ relative to the intermediate configurations and not to the reference configurations.

In the set of state variables relative to the intermediate configurations,

$$\mathcal{V}_\kappa = \{\mathbf{E}_{\mathrm{G}}^{\mathrm{e}},\, T,\, r_\kappa,\, \cup_{n \in [1,N]} \xi_{\kappa n}\}\,, \tag{22.2.12}$$

the ξ_κ's are dimensionless *dissipation variables* of viscous or other origins. For instance, *structure variables*, of various tensor orders, will be used to characterize the mechanical state of the tissues as growth, or resorption, proceeds. If required, their transport from the intermediate to the current configurations is part of the constitutive equations.

22.2.2 Dissipation inequality in the intermediate configurations for a solid

Upon multiplication of $(4.7.7)_2$ by $\det \mathbf{F}^{\mathrm{e}}$, the dissipation expresses in terms of the state variables (22.2.12) in the form,

$$\begin{aligned}
T\hat{S}_{\kappa(\mathrm{i})} ={}& -\frac{dE_\kappa}{dt} + \boldsymbol{\tau}_\kappa : \frac{d\mathbf{E}_{\mathrm{G}}^{\mathrm{e}}}{dt} - S_\kappa \frac{dT}{dt} - \frac{\mathbf{Q}_\kappa}{T} \cdot \boldsymbol{\nabla}_\kappa T \\
&+ \left(\rho\,\tilde{e} + \tfrac{1}{2}\rho\,(\tilde{\mathbf{v}}_{\mathrm{s}} - \mathbf{v}_{\mathrm{s}})^2\right) \frac{dr_\kappa}{dt} \\
&+ \left(\boldsymbol{\Psi}_\kappa + \left(\tilde{E}_\kappa - E_\kappa + \tfrac{1}{2}\rho_\kappa(\tilde{\mathbf{v}}_{\mathrm{s}} - \mathbf{v}_{\mathrm{s}})^2\right)\mathbf{I}\right) : \mathbf{L}_\kappa^{\mathrm{g}} + \hat{\mathcal{E}}_\kappa \geq 0\,,
\end{aligned} \tag{22.2.13}$$

where the stress

$$\boldsymbol{\Psi}_\kappa \equiv \mathbf{C}^{\mathrm{e}} \cdot \boldsymbol{\tau}_\kappa\,, \tag{22.2.14}$$

with $\mathbf{C}^{\mathrm{e}} = (\mathbf{F}^{\mathrm{e}})^{\mathrm{T}} \cdot \mathbf{F}^{\mathrm{e}}$ the elastic right Cauchy-Green tensor, is sometimes referred to as Mandel stress to acknowledge the contribution of the French mechanician Jean Mandel (1907-1982) to the thermomechanical analysis of inelastic engineering materials undergoing large transformations[22.1]. In general, this stress is not symmetric, but it displays only 6 independent components since it satisfies the three independent relations,

$$\left(\boldsymbol{\Psi}_\kappa \cdot \mathbf{C}^{\mathrm{e}}\right)^{\mathrm{T}} = \boldsymbol{\Psi}_\kappa \cdot \mathbf{C}^{\mathrm{e}}. \tag{22.2.15}$$

The arbitrariness of the elastic strain and temperature yields the thermomechanical constitutive equations,

$$\boldsymbol{\tau}_\kappa = \frac{\partial E_\kappa}{\partial \mathbf{E}_{\mathrm{G}}^{\mathrm{e}}}, \quad S_\kappa = -\frac{\partial E_\kappa}{\partial T}, \tag{22.2.16}$$

and we are left with the reduced dissipation inequality,

$$
\begin{aligned}
T \hat{S}_{\kappa(\mathrm{i})} &= -\sum_{n=1}^{N} \frac{\partial E_\kappa}{\partial \xi_{\kappa n}} \frac{d\xi_{\kappa n}}{dt} - \frac{\mathbf{Q}_\kappa}{T} \cdot \boldsymbol{\nabla}_\kappa T \\
&+ \left(-\frac{\partial E_\kappa}{\partial r_\kappa} + \underline{\rho}\, \tilde{e} + \tfrac{1}{2}\, \underline{\rho}\, (\tilde{\mathbf{v}}_{\mathrm{s}} - \mathbf{v}_{\mathrm{s}})^2 \right) \frac{dr_\kappa}{dt} \\
&+ \left(\boldsymbol{\Psi}_\kappa + \left(\tilde{E}_\kappa - E_\kappa + \tfrac{1}{2}\, \rho_\kappa\, (\tilde{\mathbf{v}}_{\mathrm{s}} - \mathbf{v}_{\mathrm{s}})^2 \right) \mathbf{I} \right) : \mathbf{L}_\kappa^{\mathrm{g}} + \hat{\mathcal{E}}_\kappa \geq 0.
\end{aligned}
\tag{22.2.17}
$$

As an example of elastic constitutive equations, one may think of a material endowed with a persistent structural anisotropy characterized by a set of tensors which are fixed. Let us consider two particular cases, namely materials that remain isotropic and materials that remain transversely isotropic in their intermediate configurations. Of course, these properties are not maintained by transport to the actual configurations.

For an isotropic material, the set of mechanical state variables reduces to the elastic Green strain $\mathbf{E}_{\mathrm{G}}^{\mathrm{e}}$. Then the background elastic potential E_κ depends on the strain via the three scalar invariants of the elastic Green strain. The elastic potential adopted here for the elastic-growing material expresses in terms of the two first invariants only via the two constants a_κ and μ_κ [unit: Pa],

$$E_\kappa = a_\kappa \operatorname{tr} \mathbf{E}_{\mathrm{G}}^{\mathrm{e}} + \mu_\kappa \operatorname{tr}\left(\mathbf{E}_{\mathrm{G}}^{\mathrm{e}}\right)^2. \tag{22.2.18}$$

The coefficient a_κ in this expression should vanish if the stress is to be zero in the unstrained intermediate configuration where $\mathbf{E}_{\mathrm{G}}^{\mathrm{e}} = \mathbf{0}$. Indeed

$$\frac{\partial E_\kappa}{\partial \mathbf{E}_{\mathrm{G}}^{\mathrm{e}}} = a_\kappa \mathbf{I} + 2\,\mu_\kappa \mathbf{E}_{\mathrm{G}}^{\mathrm{e}}. \tag{22.2.19}$$

Convexity of the elastic potential in the elastic strain space $\mathbf{E}_{\mathrm{G}}^{\mathrm{e}}$ is ensured for $\mu_\kappa > 0$, irrespective of a_κ.

Let us now consider a material that displays transverse isotropy about the axis $\mathbf{m}_\kappa$. Then with $\mathbf{M}_\kappa = \mathbf{m}_\kappa \otimes \mathbf{m}_\kappa$ the structure tensor, the elastic potential can be shown to be an isotropic scalar function of the augmented set $\{\mathbf{E}_{\mathrm{G}}^{\mathrm{e}}, \mathbf{M}_\kappa\}$. The background elastic potential E_κ depends on the strain via the five scalar invariants $\operatorname{tr} \mathbf{E}_{\mathrm{G}}^{\mathrm{e}}$, $\operatorname{tr}\left(\mathbf{E}_{\mathrm{G}}^{\mathrm{e}}\right)^2$, $\operatorname{tr}\left(\mathbf{E}_{\mathrm{G}}^{\mathrm{e}}\right)^3$, $\operatorname{tr} \mathbf{M}_\kappa \cdot \mathbf{E}_{\mathrm{G}}^{\mathrm{e}}$, $\operatorname{tr} \mathbf{M}_\kappa \cdot \left(\mathbf{E}_{\mathrm{G}}^{\mathrm{e}}\right)^2$. An extension of the above isotropic potential involves four constants, namely

$$E_\kappa = a_{\kappa 1} \operatorname{tr} \mathbf{E}_{\mathrm{G}}^{\mathrm{e}} + (a_{\kappa 3} - a_{\kappa 1}) \operatorname{tr} \mathbf{M}_\kappa \cdot \mathbf{E}_{\mathrm{G}}^{\mathrm{e}} + \mu_{\kappa 1} \operatorname{tr}\left(\mathbf{E}_{\mathrm{G}}^{\mathrm{e}}\right)^2 + (\mu_{\kappa 3} - \mu_{\kappa 1}) \operatorname{tr} \mathbf{M}_\kappa \cdot \left(\mathbf{E}_{\mathrm{G}}^{\mathrm{e}}\right)^2, \tag{22.2.20}$$

and then, with help of (6.E.10),

$$\frac{\partial E_\kappa}{\partial \mathbf{E}_{\mathrm{G}}^{\mathrm{e}}} = a_{\kappa 1} \mathbf{I} + (a_{\kappa 3} - a_{\kappa 1}) \mathbf{M}_\kappa + 2\,\mu_{\kappa 1} \mathbf{E}_{\mathrm{G}}^{\mathrm{e}} + (\mu_{\kappa 3} - \mu_{\kappa 1}) \left(\mathbf{M}_\kappa \cdot \mathbf{E}_{\mathrm{G}}^{\mathrm{e}} + \mathbf{E}_{\mathrm{G}}^{\mathrm{e}} \cdot \mathbf{M}_\kappa\right). \tag{22.2.21}$$

Convexity of the elastic potential in the elastic strain space $\mathbf{E}_{\mathrm{G}}^{\mathrm{e}}$ is ensured by the positive definiteness of the Hessian,

$$\frac{\partial^2 E_\kappa}{\partial E_{\mathrm{G}ij}^{\mathrm{e}}\, \partial E_{\mathrm{G}kl}^{\mathrm{e}}} = \mu_{\kappa 1}\left(I_{ik}\, I_{jl} + I_{il}\, I_{jk}\right) + (\mu_{\kappa 3} - \mu_{\kappa 1})\left(I_{ik}\, M_{\kappa jl} + M_{\kappa il}\, I_{jk}\right), \tag{22.2.22}$$

namely for $\mu_{\kappa 3} > \mu_{\kappa 1} > 0$ irrespective of the a_κ's.

[22.1] Actually, this terminology is a bit curious, and sounds like a misunderstanding. In fact, Mandel addressed the inelastic mechanical behavior of metals where the pure elastic deformation may be neglected so that, for most practical purposes, $\boldsymbol{\Psi}_\kappa$ reduces to the symmetric 2nd Piola-Kirchhoff stress $\boldsymbol{\tau}_\kappa$. The key contribution of Mandel [1971] [1973] was to recognize the significance of the skew-symmetric part of the inelastic velocity gradient, a quantity nowadays referred to as the plastic spin. Approximating $\boldsymbol{\Psi}_\kappa$ by $\boldsymbol{\tau}_\kappa$ implies that the plastic spin does not contribute to dissipation. As he pointed out unambiguously, first this does not mean that the plastic spin vanishes, and second, the plastic spin does not emanate from the skew-symmetric part of $\boldsymbol{\Psi}_\kappa$ in actual materials.

22.3 Growth law based on quadratic dissipation

The dissipation assumptions described in Section 4.8.6 are now stated in the intermediate configurations. The entropy production associated with dissipation mechanisms, the remodeling mechanism and growth, is cast in the format,

$$T\hat{S}_{\kappa(\mathrm{i})} = -\sum_{n=1}^{N} X_{\kappa n}\frac{d\xi_{\kappa n}}{dt} - R_\kappa\frac{dr_\kappa}{dt} - \mathbf{f}_\kappa : \mathbf{L}_\kappa^{\mathrm{g}} \geq 0 \,. \qquad (22.3.1)$$

The energy in unspecified forms $\hat{\mathcal{E}}_\kappa$ is formally partitioned among structuration/dissipation, remodeling and pure growth mechanisms,

$$\hat{\mathcal{E}}_\kappa = \hat{\mathcal{E}}_\kappa^{(\mathrm{d})} + \hat{\mathcal{E}}_\kappa^{(\mathrm{r})} + \hat{\mathcal{E}}_\kappa^{(\mathrm{g})} \,, \qquad (22.3.2)$$

with

$$\hat{\mathcal{E}}_\kappa^{(\mathrm{d})} = \sum_{n=1}^{N} \beta_{\kappa n}^{(\mathrm{d})}\frac{d\xi_{\kappa n}}{dt}, \quad \hat{\mathcal{E}}_\kappa^{(\mathrm{r})} = \beta_\kappa^{(\mathrm{r})}\frac{dr_\kappa}{dt}, \quad \hat{\mathcal{E}}_\kappa^{(\mathrm{g})} = \boldsymbol{\beta}_\kappa^{(\mathrm{g})} : \mathbf{L}_\kappa^{\mathrm{g}} \,, \qquad (22.3.3)$$

where the scalars $\beta_\kappa^{(\mathrm{d})}$ [unit : Pa] and $\beta_\kappa^{(\mathrm{r})}$ [unit : Pa] and the second order tensor (with symmetries to be specified) $\boldsymbol{\beta}_\kappa^{\mathrm{g}}$ [unit : Pa] depend on the state variables.

The dissipation (22.2.17) assumes the format (22.3.1) (leaving aside the thermal diffusion). With help of the constitutive equations (22.3.3), rate equations for the dissipation, remodeling and growth variables, result, namely

$$
\begin{aligned}
-X_{\kappa n} &= \left(\frac{\alpha_{\kappa n}^{(\mathrm{d})}}{t_{\mathrm{dn}}}\right)^{-1}\frac{d\xi_{\kappa n}}{dt} &= -\frac{\partial E_\kappa}{\partial \xi_{\kappa n}} + \beta_{\kappa n}^{(\mathrm{d})}, \quad n \in [1, N] \,, \\[2mm]
-R_\kappa &= \left(\frac{\alpha_\kappa^{(\mathrm{r})}}{t_{\mathrm{r}}}\right)^{-1}\frac{dr_\kappa}{dt} &= -\frac{\partial E_\kappa}{\partial r_\kappa} + \underline{\rho}\,\tilde{e} + \tfrac{1}{2}\,\underline{\rho}\,(\tilde{\mathbf{v}}_{\mathrm{s}} - \mathbf{v}_{\mathrm{s}})^2 + \beta_\kappa^{(\mathrm{r})} \,, \\[2mm]
-\mathbf{f}_\kappa &= \left(\frac{\boldsymbol{\alpha}_\kappa^{(\mathrm{g})}}{t_{\mathrm{g}}}\right)^{-1} : \mathbf{L}_\kappa^{\mathrm{g}} &= \boldsymbol{\Psi}_\kappa + \left(\tilde{E}_\kappa - E_\kappa + \tfrac{1}{2}\,\rho_\kappa\,(\tilde{\mathbf{v}}_{\mathrm{s}} - \mathbf{v}_{\mathrm{s}})^2\right)\mathbf{I} + \boldsymbol{\beta}_\kappa^{(\mathrm{g})} \,.
\end{aligned}
\qquad (22.3.4)
$$

Here the t's are time scales, $\alpha_{\kappa n}^{(\mathrm{d})}$ and $\alpha_\kappa^{(\mathrm{r})}$ are positive scalars, and $\boldsymbol{\alpha}_\kappa^{(\mathrm{g})}$ with dimension Pa^{-1} is a fourth order positive definite tensor, endowed with the major symmetry (but a priori not with minor symmetries), all quantities depending on the state variables.

Backsubstituting (22.3.4) into (22.3.1) yields

$$T\hat{S}_{\kappa(\mathrm{i})} = \sum_{n=1}^{N} \frac{d\xi_{\kappa n}}{dt}\left(\frac{\alpha_{\kappa n}^{(\mathrm{d})}}{t_{\mathrm{dn}}}\right)^{-1}\frac{d\xi_{\kappa n}}{dt} + \frac{dr_\kappa}{dt}\left(\frac{\alpha_\kappa^{(\mathrm{r})}}{t_{\mathrm{r}}}\right)^{-1}\frac{dr_\kappa}{dt} + \mathbf{L}_\kappa^{\mathrm{g}} : \left(\frac{\boldsymbol{\alpha}_\kappa^{(\mathrm{g})}}{t_{\mathrm{g}}}\right)^{-1} : \mathbf{L}_\kappa^{\mathrm{g}} \geq 0 \,. \qquad (22.3.5)$$

If the α's are independent of the rate variables, then a homogeneous quadratic dissipation potential $\Omega_{\kappa(\mathrm{i})} \geq 0$ may be introduced,

$$T\hat{S}_{\kappa(\mathrm{i})} = 2\,\Omega_{\kappa(\mathrm{i})} = \frac{\partial\Omega_{\kappa(\mathrm{i})}}{\partial(d\xi_{\kappa n}/dt)} : \frac{d\xi_{\kappa n}}{dt} + \frac{\partial\Omega_{\kappa(\mathrm{i})}}{\partial(dr_\kappa/dt)}\frac{dr_\kappa}{dt} + \frac{\partial\Omega_{\kappa(\mathrm{i})}}{\partial\mathbf{L}_\kappa^{\mathrm{g}}} : \mathbf{L}_\kappa^{\mathrm{g}} \geq 0 \,. \qquad (22.3.6)$$

The formulation has not accounted for potential structuration mechanisms that contribute negatively to dissipation.

If the dissipation variables are needed in the current configurations, their transports from the intermediate to the current configurations should be postulated via information contained in $\mathbf{F}^{\mathrm{e}}$. These transports depend on the tensorial nature of the variables. For a scalar, the transport will be based on the scalar invariants of $\mathbf{F}^{\mathrm{e}}$, typically on $\det\mathbf{F}^{\mathrm{e}}$. For a symmetric second order tensor $\boldsymbol{\xi}_\kappa$, the transport $(\cdot)_\kappa \to (\cdot)$ may take the form,

$$\boldsymbol{\xi}_\kappa \longleftrightarrow \boldsymbol{\xi} = (\det\mathbf{F}^{\mathrm{e}})^{-p}\,(\mathbf{F}^{\mathrm{e}})^q \cdot \boldsymbol{\xi}_\kappa \cdot ((\mathbf{F}^{\mathrm{e}})^{\mathrm{T}})^q \,, \qquad (22.3.7)$$

with p and q scalars. For the stress, we have $p = q = 1$ according to Eqn (22.2.1).

The total rate of change of mass content (21.3.4) is the sum of a pure growth contribution and of a contribution associated with the change of mass density, or equivalently, a change of the remodeling variable,

$$\frac{1}{m}\frac{dm}{dt} = \mathrm{tr}\,\mathbf{L}_\kappa^{\mathrm{g}} + \frac{d}{dt}\mathrm{Ln}\,(r_\kappa + 1) \,. \qquad (22.3.8)$$

22.4   Bounded structure variables via convexity

The issue here is to investigate how a growth law and an evolution law for dissipative structure variables can be developed so as to ensure that growth and structure variables remain bounded.

Let

$$\mathcal{V} = \{\boldsymbol{\epsilon}^{\mathrm{e}}, \xi, \boldsymbol{\xi}\}, \tag{22.4.1}$$

be the set of state variables consisting of the elastic strain $\boldsymbol{\epsilon}^{\mathrm{e}}$ and of two structure variables, namely a scalar ξ [unit : 1] and a symmetric second order tensor $\boldsymbol{\xi}$ [unit : 1]. The strain assumes the additive decomposition into an elastic part and a growth part,

$$\boldsymbol{\epsilon} = \boldsymbol{\epsilon}^{\mathrm{e}} + \boldsymbol{\epsilon}^{\mathrm{g}}. \tag{22.4.2}$$

The dissipation inequality, phrased in terms of the free energy per unit volume E [unit : Pa] (to be accurate, the Cauchy stress below should be replaced by the Eshelby stress),

$$
\begin{aligned}
T\hat{S}_{(\mathrm{i},1,4)} &= -\frac{dE}{dt} + \boldsymbol{\sigma} : \frac{d\boldsymbol{\epsilon}}{dt} \geq 0 \\
&= \left(\boldsymbol{\sigma} - \frac{\partial E}{\partial \boldsymbol{\epsilon}^{\mathrm{e}}}\right) : \frac{d\boldsymbol{\epsilon}^{\mathrm{e}}}{dt} + \boldsymbol{\sigma} : \frac{d\boldsymbol{\epsilon}^{\mathrm{g}}}{dt} - \frac{\partial E}{\partial \xi}\frac{d\xi}{dt} - \frac{\partial E}{\partial \boldsymbol{\xi}} : \frac{d\boldsymbol{\xi}}{dt},
\end{aligned}
\tag{22.4.3}
$$

implies, since the elastic strain rate can be given arbitrary values, the stress-strain constitutive equation,

$$\boldsymbol{\sigma} = \frac{\partial E}{\partial \boldsymbol{\epsilon}^{\mathrm{e}}}. \tag{22.4.4}$$

The reduced dissipation,

$$T\hat{S}_{(\mathrm{i},1,4)} = \boldsymbol{\Sigma} : \frac{d\boldsymbol{\epsilon}^{\mathrm{g}}}{dt} - X\frac{d\xi}{dt} - \mathbf{X} : \frac{d\boldsymbol{\xi}}{dt} \geq 0, \tag{22.4.5}$$

with the conjugate variables $\boldsymbol{\Sigma}$, X and $\mathbf{X}$ [unit : Pa],

$$\boldsymbol{\Sigma} = \boldsymbol{\sigma}, \quad X = \frac{\partial E}{\partial \xi}, \quad \mathbf{X} = \frac{\partial E}{\partial \boldsymbol{\xi}}, \tag{22.4.6}$$

remains to be satisfied by the constitutive rate laws for the generalized strains.

22.4.1   Rates of structure variables through heuristic differential equations

The point is now to form differential equations that govern the rates of the structure variables. Several issues should be considered simultaneously:

- first, of course, the equations should ensure satisfaction of the dissipation inequality;
- second, the structure variables should remain bounded, even at large growth. Perhaps this point should be checked more carefully in examples, but boundedness seems a priori to be a nice property, referred to as *evanescent memory* property;
- third, we probably require that the structure variables vanish when growth does so. The underlying idea is that structuration takes place to accommodate growth. Still, the point might be debatable, as delayed structuration might not be excluded;
- the degree of coupling between the differential equations is an issue.

The complete dissipation inequality may be written in many ways. For example, (22.4.5) might be recast in the form,

$$T\hat{S}_{(\mathrm{i},1,4)} = (\boldsymbol{\Sigma} - \mathbf{X}) : \frac{d\boldsymbol{\epsilon}^{\mathrm{g}}}{dt} + \mathbf{X} : \left(\frac{d\boldsymbol{\epsilon}^{\mathrm{g}}}{dt} - \frac{d\boldsymbol{\xi}}{dt}\right) - X\frac{d\xi}{dt}. \tag{22.4.7}$$

Now, how do we group the terms of the rhs to form differential equations? In an Onsager approach, pairs of forces and fluxes are defined and linear, or nonlinear, relations are postulated between forces and fluxes. The definition of these pairs is by no means unique.

The present method can be viewed as an alternative, more heuristic, way of defining these pairs so as to satisfy easily the requirements listed above. For example, we may decide for,

$$
\begin{aligned}
\frac{d\boldsymbol{\epsilon}^{\mathrm{g}}}{dt} - \frac{d\boldsymbol{\xi}}{dt} &= a_1 \parallel \frac{d\boldsymbol{\epsilon}^{\mathrm{g}}}{dt} \parallel \frac{\mathbf{X}}{\mu_{\mathrm{g}}} \quad \Rightarrow \quad \frac{d\boldsymbol{\xi}}{dt} = \frac{d\boldsymbol{\epsilon}^{\mathrm{g}}}{dt} - a_1 \parallel \frac{d\boldsymbol{\epsilon}^{\mathrm{g}}}{dt} \parallel \frac{\mathbf{X}}{\mu_{\mathrm{g}}}, \\
\frac{d\xi}{dt} &= \frac{1}{t_{\mathrm{g}}} \Phi(h) \left(a_2 - a_3 \frac{X}{\mu_{\mathrm{g}}}\right) \frac{X}{\mu_{\mathrm{g}}},
\end{aligned}
\tag{22.4.8}
$$

where the three scalars a's [unit : 1] either depend on the state variables or are constants. The sign of the dissipation is ensured on a particular example of free energy and homeostatic surface.

22.4.2 Constitutive prototypes: free energy and homeostatic surface

22.4.2.1 Free energy and free enthalpy as partial Legendre duals

Free energy and free enthalpy are partial Legendre duals with respect to stress and elastic strain,

$$E(\epsilon^e, \xi, \boldsymbol{\xi}) - G(\boldsymbol{\sigma}, \xi, \boldsymbol{\xi}) = \boldsymbol{\sigma} : \epsilon^e. \tag{22.4.9}$$

Then

$$\epsilon^e = -\frac{\partial G}{\partial \boldsymbol{\sigma}}, \quad X = \frac{\partial G}{\partial \xi}, \quad \mathbf{X} = \frac{\partial G}{\partial \boldsymbol{\xi}}. \tag{22.4.10}$$

If the free energy and free enthalpy are assumed in the following additive format,

$$E(\epsilon^e, \xi, \boldsymbol{\xi}) = E_1(\epsilon^e) + E_2(\xi, \boldsymbol{\xi}), \quad G(\boldsymbol{\sigma}, \xi, \boldsymbol{\xi}) = G_1(\boldsymbol{\sigma}) + \overbrace{G_2(\xi, \boldsymbol{\xi})}^{E_2}, \tag{22.4.11}$$

then the elastic stiffness and compliance are independent of the structure variables.

22.4.2.2 An example of a free energy of growth and structuration

The free energy is decomposed additively into an elastic part and a structuration part,

$$E = \frac{1}{2}\epsilon^e : \mathbb{E} : \epsilon^e + \frac{1}{2}\xi\,\mathcal{X}\,\xi + \frac{1}{2}\boldsymbol{\xi} : \mathbb{X} : \boldsymbol{\xi}, \tag{22.4.12}$$

where $\mathbb{E}$ [unit : Pa] is the elasticity tensor, $\mathcal{X}$ [unit : Pa] is a scalar and $\mathbb{X}$ [unit : Pa] a fourth order tensor with minor and major symmetries. The conjugate variables (22.4.6) derive from (22.4.12),

$$\boldsymbol{\Sigma} = \boldsymbol{\sigma}, \quad X = \xi\,\mathcal{X}, \quad \mathbf{X} = \boldsymbol{\xi} : \mathbb{X}. \tag{22.4.13}$$

The fourth order tensors $\mathbb{E}$ and $\mathbb{X}$ are assumed positive definite, and the scalar $\mathcal{X} > 0$ is assumed positive as well. Still, the signs of the structuration contributions in the free energy are not so immediate. For example, Feigenbaum and Dafalias [2008] justify a negative contribution to the free energy due to distortion of the yield surface of metals.

With this free energy, the structure variables and their conjugate entities are simply proportional. Still, in view of comparison with another approach below, let us record the rate laws of the conjugate variables deduced from (22.4.8),

$$\begin{aligned}
\mathbb{X}^{-1} : \frac{d\mathbf{X}}{dt} &= \frac{d\boldsymbol{\xi}}{dt} = \frac{d\epsilon^g}{dt} - a_1 \left\| \frac{d\epsilon^g}{dt} \right\| \frac{\mathbf{X}}{\mu_g}, \\
\mathcal{X}^{-1}\frac{dX}{dt} &= \frac{d\xi}{dt} = \frac{1}{t_g}\,\Phi(h)\left(a_2 - a_3\frac{X}{\mu_g}\right)\frac{X}{\mu_g}.
\end{aligned} \tag{22.4.14}$$

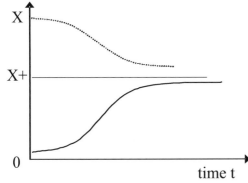

Fig. 22.4.1: Singular (limit) points of the scalar structure variable X: the value $X = X^+$ is attractive while $X = 0$ is repulsive.

We shall see that the coefficients a_2 and a_3 are positive. Then the limit point $X^+/\mu_g = a_2/a_3$ is attractive, whether the initial value of X is initially smaller or larger, Fig. 22.4.1. On the other hand, the point $X = 0$ is repulsive.

Actual satisfaction of the dissipation inequality, exposed below for a particular homeostatic surface, results in restrictions on the coefficients a_1, a_2 and a_3.

22.4.2.3   Homeostatic surface and the rate law for the growth strain

Let us assume the existence of a convex homeostatic surface in stress space [unit : 1],

$$h = h(\mathbf{\Sigma} - \mathbf{X}, X) = 0\,. \tag{22.4.15}$$

For a stress inside the homeostatic surface, namely $h \leq 0$, the rate of growth strain vanishes. For a stress outside the surface, namely $h > 0$, the growth strain rate is given in the form,

$$\frac{d\boldsymbol{\epsilon}^{\mathrm{g}}}{dt} = \frac{\mu_{\mathrm{g}}}{t_{\mathrm{g}}}\,\Phi(h)\,\frac{\partial h}{\partial \mathbf{\Sigma}}(\mathbf{\Sigma} - \mathbf{X}, X)\,, \tag{22.4.16}$$

where
- t_{g} [unit : s] is a time scale of the growth process;
- $\mu_{\mathrm{g}} > 0$ [unit : Pa] is a representative elastic modulus;
- $\Phi(x) = \Phi(\langle x \rangle)$ [unit : 1] is a sufficiently smooth, positive, dimensionless function, which vanishes for $x \leq 0$ ($\langle \cdot \rangle$ denotes the positive part of its argument). The function Φ is understood to be monotonous, i.e., the farthest is the stress from the homeostatic surface, the largest is the growth. The basic choice for Φ is a power function. Therefore, a non monotonous function to avoid unbounded growth does not fit well with the framework. Still, non monotonous functions Φ have been envisaged with a specific motivation.

In words, the rate of growth deformation is non zero only for stresses $\mathbf{\Sigma}$ outside the homeostatic convex, namely $h(\mathbf{\Sigma} - \mathbf{X}, X) > 0$. Then it has the direction of the outward normal to this surface at the current stress point $\mathbf{\Sigma}$. Convexity of the homoeostatic surface implies,

$$(\mathbf{\Sigma} - \mathbf{X}) : \frac{d\boldsymbol{\epsilon}^{\mathrm{g}}}{dt} \geq 0\,. \tag{22.4.17}$$

22.4.2.4   A prototype homeostatic surface

The homeostatic surface,

$$\mu_{\mathrm{g}}^2\,h = \frac{3}{2\,M^2}\,\mathrm{dev}\,(\mathbf{\Sigma} - \mathbf{X}) : \mathrm{dev}\,(\mathbf{\Sigma} - \mathbf{X}) + \left(\frac{1}{3}\,\mathrm{tr}\,(\mathbf{\Sigma} - \mathbf{X})\right)^2 - X^2 = 0\,, \tag{22.4.18}$$

is an ellipse whose axis is parallel to the isotropic axis, but shifted in the deviatoric space by dev $\mathbf{X}$. Its size is equal to $2\,X$ along the isotropic axis and its diameter along any deviatoric direction is equal to $2\,MX$.

The dissipation inequality (22.4.7) now becomes,

$$T\hat{S}_{(\mathrm{i},1,4)} \;=\; \frac{\mu_{\mathrm{g}}}{t_{\mathrm{g}}}\,\Phi(h)\left(2\,h + 2\,\frac{X^2}{\mu_{\mathrm{g}}^2} + a_3\,\frac{X^3}{\mu_{\mathrm{g}}^3} - a_2\,\frac{X^2}{\mu_{\mathrm{g}}^2} + a_1\,\|\frac{\partial h}{\partial \mathbf{\Sigma}}\|\,\frac{\mathbf{X}:\mathbf{X}}{\mu_{\mathrm{g}}}\right) \geq 0\,, \tag{22.4.19}$$

and it is satisfied for $a_1 \geq 0$, $2 \geq a_2 \geq 0$ and $a_3 \geq 0$.

22.4.3   The generalized normality rule

The generalized normality rule is a more formal approach to provide formal structure to the constitutive equations while ensuring the inequality dissipation.

Let us introduce a stress potential through its differential,

$$d\omega(\mathbf{\Sigma}, X, \mathbf{X}) = d\mathbf{\Sigma} : \frac{d\boldsymbol{\epsilon}^{\mathrm{g}}}{dt} - dX\,\frac{d\xi}{dt} - d\mathbf{X} : \frac{d\boldsymbol{\xi}}{dt}\,, \tag{22.4.20}$$

and a dissipation potential,

$$d\Omega\Big(\frac{d\boldsymbol{\epsilon}^{\mathrm{g}}}{dt}, \frac{d\xi}{dt}, \frac{d\boldsymbol{\xi}}{dt}\Big) = \mathbf{\Sigma} : d\Big(\frac{d\boldsymbol{\epsilon}^{\mathrm{g}}}{dt}\Big) - X\,d\Big(\frac{d\xi}{dt}\Big) - \mathbf{X} : d\Big(\frac{d\boldsymbol{\xi}}{dt}\Big)\,. \tag{22.4.21}$$

Then

$$\frac{d\boldsymbol{\epsilon}^{\mathrm{g}}}{dt} = \frac{\partial \omega}{\partial \mathbf{\Sigma}}\,, \qquad -\frac{d\xi}{dt} = \frac{\partial \omega}{\partial X}\,, \qquad -\frac{d\boldsymbol{\xi}}{dt} = \frac{\partial \omega}{\partial \mathbf{X}}\,, \tag{22.4.22}$$

and

$$\boldsymbol{\Sigma} = \frac{\partial \Omega}{\partial (d\boldsymbol{\epsilon}^g/dt)}, \quad X = -\frac{\partial \Omega}{\partial (d\xi/dt)}, \quad \mathbf{X} = -\frac{\partial \Omega}{\partial (d\boldsymbol{\xi}/dt)}. \tag{22.4.23}$$

Moreover

$$\omega(\boldsymbol{\Sigma}, X, \mathbf{X}) + \Omega(\frac{d\boldsymbol{\epsilon}^g}{dt}, \frac{d\xi}{dt}, \frac{d\boldsymbol{\xi}}{dt}) = \boldsymbol{\Sigma} : \frac{d\boldsymbol{\epsilon}^g}{dt} - X \frac{d\xi}{dt} - \mathbf{X} : \frac{d\boldsymbol{\xi}}{dt}. \tag{22.4.24}$$

This approach is referred to as generalized normality rule, Halphen and Nguyen [1975], and it has been extensively used in modeling the mechanical behavior of engineering materials, e.g., Lemaitre and Chaboche [1988].

The reduced dissipation (22.4.5) becomes with (22.4.22),

$$T\hat{S}_{(i,1,4)} \;=\; \boldsymbol{\Sigma} : \frac{\partial \omega}{\partial \boldsymbol{\Sigma}} + X \frac{\partial \omega}{\partial X} + \mathbf{X} : \frac{\partial \omega}{\partial \mathbf{X}}. \tag{22.4.25}$$

The stress potential ω is assumed to be positive at points where growth takes place. A positive dissipation results if the stress potential $\omega(\boldsymbol{\Sigma}, X, \mathbf{X})$ is
- convex with respect to all its arguments;
- negative or zero when the arguments vanish, i.e. $\omega(\boldsymbol{\Sigma} = 0, X = 0, \mathbf{X} = 0) \le 0$.

Indeed, convexity wrt to the three arguments implies convexity with respect to anyone of them, say for $\boldsymbol{\alpha}$:

$$\omega(\boldsymbol{\alpha}_1) - \omega(\boldsymbol{\alpha}_2) \le \frac{\partial \omega}{\partial \boldsymbol{\alpha}}(\boldsymbol{\alpha}_1) : (\boldsymbol{\alpha}_1 - \boldsymbol{\alpha}_2). \tag{22.4.26}$$

Therefore, for $\boldsymbol{\alpha}_1 = \boldsymbol{\alpha}$ arbitrary outside or on the convex, namely $\omega(\boldsymbol{\alpha}) \ge 0$, and $\boldsymbol{\alpha}_2 = \mathbf{0}$,

$$0 \le \underbrace{\omega(\boldsymbol{\alpha})}_{\ge 0} - \underbrace{\omega(\mathbf{0})}_{\le 0} \le \frac{\partial \omega}{\partial \boldsymbol{\alpha}}(\boldsymbol{\alpha}) : (\boldsymbol{\alpha} - \mathbf{0}) \quad \Rightarrow \quad 0 \le \frac{\partial \omega}{\partial \boldsymbol{\alpha}}(\boldsymbol{\alpha}) : \boldsymbol{\alpha}, \tag{22.4.27}$$

or, in more explicit form, at a point $(\boldsymbol{\Sigma}, X, \mathbf{X})$ outside the convex, i.e., $\omega(\boldsymbol{\Sigma}, X, \mathbf{X}) \ge 0$,

$$0 \le \underbrace{\omega(\boldsymbol{\Sigma}, X, \mathbf{X})}_{\ge 0} - \underbrace{\omega(\mathbf{0}, 0, \mathbf{0})}_{\le 0} \le \frac{\partial \omega}{\partial \boldsymbol{\Sigma}} : (\boldsymbol{\Sigma} - \mathbf{0}) + \frac{\partial \omega}{\partial X}(X - 0) + \frac{\partial \omega}{\partial \mathbf{X}} : (\mathbf{X} - \mathbf{0}), \tag{22.4.28}$$

where the derivatives are estimated at $(\boldsymbol{\Sigma}, X, \mathbf{X})$.

22.4.4   Heuristic differential equations versus generalized normality rule

The approach in Section 22.4.1 uses
- the normality rule for the growth strain rate, Eqn (22.4.16);
- differential equations for the structure variables, that ensure the existence of limit values.

A comparison with the generalized normality rule of Section 22.4.3 is of interest. The example below follows Lemaitre and Chaboche [1988]. At variance with (22.4.12), a degree of freedom is left in the scalar structuration contribution to the free energy,

$$E = \frac{1}{2}\,\boldsymbol{\epsilon}^e : \mathbb{E} : \boldsymbol{\epsilon}^e + E_\xi(\xi) + \frac{1}{2}\,\boldsymbol{\xi} : \mathbb{X} : \boldsymbol{\xi}, \tag{22.4.29}$$

from which follow the conjugate structure variables,

$$\boldsymbol{\Sigma} = \boldsymbol{\sigma}, \quad X = \frac{\partial E_\xi}{\partial \xi}, \quad \mathbf{X} = \boldsymbol{\xi} : \mathbb{X}. \tag{22.4.30}$$

The homeostatic surface is a sphere,

$$h = \frac{1}{q}\left(\frac{\|\boldsymbol{\Sigma} - \mathbf{X}\|}{\mu_g}\right)^q - \frac{1}{q}\left(\frac{X}{\mu_g}\right)^q. \tag{22.4.31}$$

X appears as the radius of the homeostatic surface, and $\mathbf{X}$ its center. $q \ge 1$ is a parameter to be fixed later: it influences, not the homeostatic surface itself, but its derivatives. Then

$$\frac{1}{q}\frac{\partial}{\partial \boldsymbol{\Sigma}}\left(\frac{\|\boldsymbol{\Sigma} - \mathbf{X}\|}{\mu_g}\right)^q = \frac{1}{\mu_g}\left(\frac{\|\boldsymbol{\Sigma} - \mathbf{X}\|}{\mu_g}\right)^{q-1}\frac{\boldsymbol{\Sigma} - \mathbf{X}}{\|\boldsymbol{\Sigma} - \mathbf{X}\|}, \quad (\boldsymbol{\Sigma} - \mathbf{X}) : \frac{\partial h}{\partial \boldsymbol{\Sigma}} = q\,h + \left(\frac{X}{\mu_g}\right)^q. \tag{22.4.32}$$

Note that Lemaitre and Chaboche [1988] focus on the viscoplastic response of metals, and therefore their *yield* surface, which depends on the deviatoric part of $\boldsymbol{\Sigma} - \mathbf{X}$ only, is a cylinder and not a sphere, in the stress space.

An additional ad hoc modification with respect to the standard approach is necessary. In fact, following Lemaitre and Chaboche [1988], p. 307, we accept that the stress potential $\omega(\boldsymbol{\Sigma}, X, \mathbf{X})$ depends on the state variables $(\xi, \boldsymbol{\xi})$ viewed as parameters. Let

$$\omega = \frac{\mu_{\mathrm{g}}}{t_{\mathrm{g}}} \, \omega_1(u), \quad u \equiv h(\boldsymbol{\Sigma}, X, \mathbf{X}) + \frac{b_1}{2 \, \mu_{\mathrm{g}}^2} \Big(\mathbf{X} : \mathbf{X} - (\boldsymbol{\xi} : \mathbb{X}) : (\boldsymbol{\xi} : \mathbb{X}) \Big), \tag{22.4.33}$$

with h given by (22.4.31), and b_1 is a dimensionless constant. Note that, in fact u and h are equal, in view of (22.4.30), while their derivatives with respect to $\mathbf{X}$ differ. Note also that the scalar structure variable ξ is not introduced in the stress potential, but the free energy is modified via the function $E_\xi(\xi)$. Then

$$\begin{aligned}
\frac{d\boldsymbol{\epsilon}^{\mathrm{g}}}{dt} &= \frac{\partial \omega}{\partial \boldsymbol{\Sigma}} = \frac{\partial \omega_1}{\partial u} \frac{\mu_{\mathrm{g}}}{t_{\mathrm{g}}} \frac{\partial u}{\partial \boldsymbol{\Sigma}} = \frac{\partial \omega_1}{\partial u} \frac{1}{t_{\mathrm{g}}} \Big(\frac{\| \boldsymbol{\Sigma} - \mathbf{X} \|}{\mu_{\mathrm{g}}} \Big)^{q-1} \frac{\boldsymbol{\Sigma} - \mathbf{X}}{\| \boldsymbol{\Sigma} - \mathbf{X} \|}, \\
-\frac{d\xi}{dt} &= \frac{\partial \omega}{\partial X} = \frac{\partial \omega_1}{\partial u} \frac{\mu_{\mathrm{g}}}{t_{\mathrm{g}}} \frac{\partial u}{\partial X} = -\frac{\partial \omega_1}{\partial u} \frac{1}{t_{\mathrm{g}}} \Big(\frac{X}{\mu_{\mathrm{g}}} \Big)^{q-1}, \\
-\frac{d\boldsymbol{\xi}}{dt} &= \frac{\partial \omega}{\partial \mathbf{X}} = -\frac{d\boldsymbol{\epsilon}^{\mathrm{g}}}{dt} + \frac{\partial \omega_1}{\partial u} \frac{b_1}{t_{\mathrm{g}}} \frac{\mathbf{X}}{\mu_{\mathrm{g}}}.
\end{aligned} \tag{22.4.34}$$

Note the relation,

$$\Big\| \frac{d\boldsymbol{\epsilon}^{\mathrm{g}}}{dt} \Big\| = \Big| \frac{\partial \omega_1}{\partial u} \Big| \frac{1}{t_{\mathrm{g}}} \Big(\frac{\| \boldsymbol{\Sigma} - \mathbf{X} \|}{\mu_{\mathrm{g}}} \Big)^{q-1}, \tag{22.4.35}$$

so that the last equation (22.4.34)$_3$ may be recast in the form,

$$\frac{d\boldsymbol{\xi}}{dt} = \frac{d\boldsymbol{\epsilon}^{\mathrm{g}}}{dt} - \Big\| \frac{d\boldsymbol{\epsilon}^{\mathrm{g}}}{dt} \Big\| \, b_1 \Big(\frac{\| \boldsymbol{\Sigma} - \mathbf{X} \|}{\mu_{\mathrm{g}}} \Big)^{1-q} \frac{\mathbf{X}}{\mu_{\mathrm{g}}}, \tag{22.4.36}$$

assuming $\partial \omega_1 / \partial u$ to be positive.

The rate equations (22.4.16) and (22.4.34)$_1$ for the growth strain, and (22.4.8) and (22.4.36) for the structure variable $\boldsymbol{\xi}$ are identical if

$$\frac{\partial \omega_1}{\partial u}(u) = \Phi(h), \quad b_1 = a_1 \Big(\frac{\| \boldsymbol{\Sigma} - \mathbf{X} \|}{\mu_{\mathrm{g}}} \Big)^{q-1}. \tag{22.4.37}$$

Since the conjugate variable $\mathbf{X}$ depends linearly on the variable $\boldsymbol{\xi}$ both in Section 22.4.1 and here, its evolution is identical under the conditions (22.4.37).

For practical purposes, the first condition above implies in fact $u = h$, which explains why, in writing u in (22.4.33), the modification to h by the conjugate variable $\mathbf{X}$ is made algebraically ineffective through incorporation of an appropriate function of the state variable $\boldsymbol{\xi}$. As indicated by Lemaitre and Chaboche [1988], building a model where both a structure variable $\boldsymbol{\xi}$ and its conjugate $\mathbf{X}$ are bounded requires some sort of non normality, that is, the derivatives of h and u with respect to $\mathbf{X}$ are distinct.

In contrast to the tensor structure variable, the evolutions of the scalar structure variable ξ, defined by Eqns (22.4.8) and (22.4.34)$_2$, can not be identical. Indeed, let us first set the exponent q to 1 so as to simplify the algebras. Then (22.4.34)$_2$ and (22.4.35) imply ξ to be the cumulated growth strain,

$$\frac{\Phi(h)}{t_{\mathrm{g}}} = \Big\| \frac{d\boldsymbol{\epsilon}^{\mathrm{g}}}{dt} \Big\| = \frac{d\xi}{dt} \quad \Rightarrow \quad \xi(t) = \xi(0) + \int_0^t \Big\| \frac{d\boldsymbol{\epsilon}^{\mathrm{g}}}{d\tau} \Big\| \, d\tau. \tag{22.4.38}$$

Still, the rate equation for the conjugate variable X can be taken in the format (22.4.14)$_2$. Combining (22.4.35), (22.4.37)$_1$ and (22.4.38) yields the differential equation $dX/d\xi = \mathcal{X} (a_2 - a_3 X/\mu_{\mathrm{g}}) (X/\mu_{\mathrm{g}})$, which can be integrated to

$$\frac{X(t)}{\mu_{\mathrm{g}}} = \frac{a_2/a_3}{1 + \big(\frac{a_2/a_3}{X(0)/\mu_{\mathrm{g}}} - 1 \big) \exp \big(-a_2 \frac{\mathcal{X}}{\mu_{\mathrm{g}}} (\xi - \xi(0)) \big)}. \tag{22.4.39}$$

Due to (22.4.30), the contribution to the free energy $E_\xi(\xi)$ is obtained by integration,

$$E_\xi(\xi) - E_\xi(\xi(0)) = \frac{\mu_g}{a_3}\frac{\mu_g}{\mathcal{X}}\,\text{Ln}\left(1 + \frac{X(0)}{\mu_g}\frac{a_3}{a_2}\left(\exp\left(a_2\frac{\mathcal{X}}{\mu_g}(\xi - \xi(0))\right) - 1\right)\right). \qquad (22.4.40)$$

We still have to check two conditions for the dissipation inequality to be satisfied. First, the generalized origin satisfies the inequality $\omega(\mathbf{\Sigma} = 0, X = 0, \mathbf{X} = 0) \leq 0$. Second,

$$\frac{\partial^2\,\|\,\mathbf{\Sigma} - \mathbf{X}\,\|}{\partial(\mathbf{\Sigma} - \mathbf{X})^2} = \frac{1}{\|\,\mathbf{\Sigma} - \mathbf{X}\,\|}\left(\mathbf{I}\,\overline{\otimes}\,\mathbf{I} - \frac{\mathbf{\Sigma} - \mathbf{X}}{\|\,\mathbf{\Sigma} - \mathbf{X}\,\|}\otimes\frac{\mathbf{\Sigma} - \mathbf{X}}{\|\,\mathbf{\Sigma} - \mathbf{X}\,\|}\right), \quad \frac{\partial^2(\mathbf{X}:\mathbf{X})}{\partial\mathbf{X}^2} = 2\,\mathbf{I}\,\overline{\otimes}\,\mathbf{I}, \qquad (22.4.41)$$

so that the convexity of u with respect to its three arguments results if $b_1 \geq 0$. Moreover, if $\omega(u)$ is a convex and monotonous increasing function of its argument, then convexity of u implies convexity of ω, since

$$\frac{d^2\omega}{d\boldsymbol{\alpha}^2} = \frac{d^2\omega}{du^2}\frac{du}{d\boldsymbol{\alpha}}\frac{du}{d\boldsymbol{\alpha}} + \frac{d\omega}{du}\frac{d^2u}{d\boldsymbol{\alpha}^2}. \qquad (22.4.42)$$

Alternatively, the dissipation inequality can be checked directly, see Lemaitre and Chaboche [1988], p. 312. Since

$$T\,\hat{S}_{(i,1,4)} = \|\,\frac{d\boldsymbol{\epsilon}^g}{dt}\,\|\,\left(\mu_g\,h + b_1\frac{\mathbf{X}:\mathbf{X}}{\mu_g}\right), \qquad (22.4.43)$$

it is satisfied for $b_1 > 0$.

Summarizing,
- in the heuristic approach of Section 22.4.1, the state variable ξ and its conjugate entity X obey both a rate law of evanescent memory type;
- with the generalized normality rule, the state variable ξ is equal to the cumulated growth strain while the conjugate variable X still obeys a rate law of evanescent memory type: the radius of the homeostatic surface remains bounded;
- the growth rate and the rate of the tensor structure variable are identical in the two formulations.

22.5 A thermodynamically consistent growth law

The growth rate law is developed in a format that satisfies automatically the internal entropy inequality. Note that the onsagerist point of view in Sections 21.4 and 22.3, based on a quadratic dissipation potential was already satisfying this inequality. Still, the approach developed below, based on Loret and Simões [2010]c, is different and more versatile. In essence, it consists in adapting to the large strain elastic-growth decomposition the ideas qualitatively described in a small strain context in Section 22.4.

A key ingredient of the constitutive equations developed here is the *homeostatic domain*, or *set*, generalizing the notion of a homeostatic state:

- there exists a whole range of stress in which no growth takes place. This feature allows load variation without growth nor resorption. This (neutral) zone may also be seen as damping the effects of load with respect to the growth response. The actual impact of this damping effect depends on the size of the homeostatic domain. A simple identification procedure is used in the examples addressed below;

- the homeostatic domain may move during the growth process, as both its size may vary and its center is attracted by the load. The fact that its size varies may not be easily quantified. On the other hand, the fact that its center may move is certainly relevant. This aspect is especially of interest during the *development* of organs where the homeostatic state is expected, and known, to evolve, e.g., Taber and Chabert [2002] for cardiac and aortic growths.

The analysis here is devoted to single phase solids. The key target is to formulate rate constitutive equations for the growth transformation and other structure variables defining the thermodynamic state of the tissue, that are motivated by and that satisfy the dissipation inequality, that is, they are consistent with a positive *internal* entropy production.

The mechanical constitutive equations are postulated in the intermediate configurations which are implicitly introduced by the multiplicative decomposition of the deformation gradient. They are derived in a thermodynamic framework where growth is viewed as a dissipative mechanism. Much like for elastic-plastic materials, the convexity of the homeostatic domain is a key ingredient of the construction. The thermodynamic state of the tissues is defined, besides the elastic transformation, by structure variables that characterize the translation and size of the homeostatic domain in stress space. The rate equations of the structure variables are built so as to satisfy the dissipation inequality. It might be worth stressing that the heuristic derivation of these rate equations appears as an alternative to the generalized normality rule developed in the early 1980s in the context of metal plasticity, e.g., Lemaitre and Chaboche [1988]. In fact, the latter approach becomes quite involved as soon as the yield, or homeostatic, surface deviates from the von Mises cylinder, and when it changes shape during the deformation process, as illustrated by Feigenbaum and Dafalias [2008] for anisotropic crystal plasticity.

22.5.1   The state of the deposited mass and the free energy of growth

When the deformation gradient is decomposed into an elastic part and a growth part, the deformation enters the set of state variables through its elastic part. The actual format of the stress-strain relation depends on the compressibility properties. If the material is elastically compressible, the stress-strain relation involves the Kirchhoff stress $\boldsymbol{\tau} = \boldsymbol{\sigma} \det \mathbf{F}$. For an incompressible solid, the stress-strain relation involves the *effective* Kirchhoff stress $\boldsymbol{\tau}' = \boldsymbol{\sigma}' \det \mathbf{F}$ where $\boldsymbol{\sigma}' = \boldsymbol{\sigma} + p_I \mathbf{I}$ is the effective Cauchy stress. The Lagrange multiplier $p_I = p_I(t)$, associated with the incompressibility constraint $\det \mathbf{F}^{\mathrm{e}} - 1 = 0$, is interpreted as a pressure and, in a boundary value problem, it is defined through boundary conditions.

The free energy of growth should take into account the mechanical state of the mass at the time where it is deposited. In essence, the history of growth, in terms of the growth velocity transformation, or of rate of mass supply, modulates the free energies. In order to simplify the exposition, the deposited mass is assumed to be endowed with the same thermodynamic state as the existing material at the deposition point. Alternatively, one might imagine that the mass could be deposited with its own state, e.g. unstrained. The present framework has to be extended to embed this aspect: the issue is addressed in Section 22.6.

In fact, the primary effort lays in the definition of the free energies relative to the intermediate configurations. The stress, and possibly other structure variables relative to these intermediate configurations, result from these free energies. Entities relative to the current configuration deduce by a transport rule.

When the mass is deposited with the same free energy as the existing material, its aggregation is by definition elusive as far as energy is concerned. The free energy of growth boils down to a background free energy per current intermediate volume E_κ. The effective 2nd Piola-Kirchhoff stress relative to the intermediate configurations (κ),

$$\boldsymbol{\tau}'_\kappa = \frac{\partial E_\kappa}{\partial \mathbf{E}_{\mathrm{G}}^{\mathrm{e}}}, \qquad (22.5.1)$$

is work-conjugate to the elastic Green strain $\mathbf{E}_{\mathrm{G}}^{\mathrm{e}} = \frac{1}{2}(\mathbf{C}^{\mathrm{e}} - \mathbf{I})$. In order to embody both elastically compressible and incompressible tissues, the symbol prime is used to denote an effective stress in case of elastic incompressibility, and otherwise the stress itself. In any case, the Cauchy stress $\boldsymbol{\sigma}$ and the 2nd Piola-Kirchhoff stress $\boldsymbol{\tau}_\kappa$, and their effective counterparts, obey the transport rule,

$$\boldsymbol{\sigma}' = \frac{1}{\det \mathbf{F}^{\mathrm{e}}} \mathbf{F}^{\mathrm{e}} \cdot \boldsymbol{\tau}'_\kappa \cdot (\mathbf{F}^{\mathrm{e}})^{\mathrm{T}} \quad \Leftrightarrow \quad \boldsymbol{\tau}'_\kappa = (\mathbf{F}^{\mathrm{e}})^{-1} \cdot (\det \mathbf{F}^{\mathrm{e}} \, \boldsymbol{\sigma}') \cdot (\mathbf{F}^{\mathrm{e}})^{-\mathrm{T}}. \qquad (22.5.2)$$

22.5.2   Dissipation inequality for the rates of growth and structure variables

The set of *dimensionless* state variables relative to the intermediate configurations includes the elastic strain, a *remodeling* variable r_κ, and two additional dissipation *structure* variables, namely a scalar variable ξ_κ and a symmetric second order tensor $\boldsymbol{\xi}_\kappa$,

$$\mathcal{V}_\kappa = \{\mathbf{E}_{\mathrm{G}}^{\mathrm{e}}, r_\kappa, \xi_\kappa, \boldsymbol{\xi}_\kappa\}. \qquad (22.5.3)$$

The scalar remodeling variable $r_\kappa = \rho_\kappa/\underline{\rho} - 1$ is defined so as to vanish for pure growth, which preserves mass density. Note that the structure variables are needed in the intermediate configurations. They are a priori not required in the current configurations: if, for some reasons, they are requested there, their transports from the intermediate to the current configurations should be defined. The physical interpretation of these structure variables will be made clear when we introduce the homeostatic surface.

Even if the analysis here addresses a single solid, it is of interest to embed the mechanical model to be developed in a more general framework, in view of an extension to a mixture context. The tissue may be supplied by the surroundings independently in mass $\hat{\rho}_\kappa = \hat{\rho} \det \mathbf{F}^e$ and in free energy $\hat{\mathcal{E}}_\kappa$ associated with light, radiations, etc. When expressed in terms of entities relative to the intermediate configurations (κ), the condition of a non negative mechanical part of the internal entropy production $\hat{S}_{\kappa(i)}$ at constant temperature T may be cast in the format, Eqn (22.2.13),

$$T\hat{S}_{\kappa(i)} = -\frac{dE_\kappa}{dt} + \boldsymbol{\tau}_\kappa : \frac{d\mathbf{E}_G^e}{dt} + \boldsymbol{\Psi}_\kappa : \mathbf{L}_\kappa^g$$

$$+ \left(\rho\,\tilde{e} + \tfrac{1}{2}\,\rho\,(\tilde{\mathbf{v}}_s - \mathbf{v}_s)^2\right)\frac{dr_\kappa}{dt} + \left(\tilde{E}_\kappa + \tfrac{1}{2}\,\rho_\kappa\,(\tilde{\mathbf{v}}_s - \mathbf{v}_s)^2 - E_\kappa\right)\mathbf{I} : \mathbf{L}_\kappa^g + \hat{\mathcal{E}}_\kappa \geq 0, \quad (22.5.4)$$

where the stress $\boldsymbol{\Psi}_\kappa \equiv \mathbf{C}^e \cdot \boldsymbol{\tau}_\kappa$ is the Mandel stress. To proceed, a number of simplifications are adopted:
- growth is preserving density, so that the remodeling variable r_κ vanishes. The rate of mass growth $\hat{\rho}/\rho = \operatorname{tr}\mathbf{L}_\kappa^g = \operatorname{tr}\mathbf{D}_\kappa^g$ *results* from the isotropic part of the rate of growth transformation as indicated by (21.3.54);
- the free energy $\tilde{E}_\kappa$ and velocity $\tilde{\mathbf{v}}_s$ of the deposited mass at the time of deposition are equal respectively to the free energy E_κ and velocity $\mathbf{v}_s$ of the existing mass at the deposition site;
- the system is thermodynamically open to mass only, in the sense that the independent source of free energy $\hat{\mathcal{E}}_\kappa$ is zero.

In case of elastic incompressibility $\det\mathbf{F}^e = 1$, the dissipation inequality can not be exploited directly since $\mathbf{F}^e$ is no longer arbitrary, unless the constraint is included in rate form in the inequality via the Lagrange multiplier p_I. Actually, since by (2.7.3),

$$\frac{\partial \det\mathbf{F}^e}{\partial\mathbf{E}_G^e} = \det\mathbf{F}^e\,(\mathbf{C}^e)^{-1}, \quad (22.5.5)$$

the terms of interest then become

$$\boldsymbol{\tau}_\kappa : \frac{d\mathbf{E}_G^e}{dt} + p_I\,\frac{d}{dt}\,(\det\mathbf{F}^e - 1) = \boldsymbol{\tau}_\kappa' : \frac{d\mathbf{E}_G^e}{dt}. \quad (22.5.6)$$

According to the convention of Section 22.5.1, the notation $\boldsymbol{\tau}_\kappa'$ denotes the 2nd Piola-Kirchhoff stress $\boldsymbol{\tau}_\kappa$ relative to the intermediate configuration if the material is elastically compressible and the effective 2nd Piola-Kirchhoff stress

$$\boldsymbol{\tau}_\kappa' = \boldsymbol{\tau}_\kappa + p_I\,(\mathbf{C}^e)^{-1}, \quad (22.5.7)$$

if the material is elastically incompressible. Then, with the *generalized stresses* [unit : Pa],

$$\boldsymbol{\tau}_\kappa' = \frac{\partial E_\kappa}{\partial\mathbf{E}_G^e}, \quad X_\kappa = \frac{\partial E_\kappa}{\partial\xi_\kappa}, \quad \mathbf{X}_\kappa = \frac{\partial E_\kappa}{\partial\boldsymbol{\xi}_\kappa}, \quad (22.5.8)$$

the reduced dissipation inequality,

$$T\hat{S}_{\kappa(i)} = \boldsymbol{\Psi}_\kappa : \mathbf{L}_\kappa^g - X_\kappa\,\frac{d\xi_\kappa}{dt} - \mathbf{X}_\kappa : \frac{d\boldsymbol{\xi}_\kappa}{dt} \geq 0, \quad (22.5.9)$$

remains to be satisfied by the constitutive equations to be developed for the growth transformation $\mathbf{F}^g$ and structure variables ξ_κ and $\boldsymbol{\xi}_\kappa$.

The material properties are expected to adapt so as to resist the load. In particular, both the magnitude and directional properties of the elastic stiffness might evolve with growth. Examples are to be found in Hegedus and Cowin [1976] for bone, and in Taber [1998]a for skeletal muscle tissue. However, as a simplification, we will consider here that the elastic properties remain isotropic and fixed in the intermediate configurations[22.2]. Adaption to growth will be embodied in the structure variables that define the homeostatic surface. A general approach should account for both aspects, namely change of elastic properties and evolution of the homeostatic state. Note that, while the elastic properties are fixed relative to the intermediate configurations, they certainly do not in the current configurations.

The free energy is *additively* contributed by a linear expression of the scalar invariants of the Green strain and by quadratic expressions of the structure variables, so that the elastic moduli do not evolve with growth,

$$E_\kappa = a_\kappa \operatorname{tr}\mathbf{E}_G^e + \mu_\kappa \operatorname{tr}(\mathbf{E}_G^e)^2 + X_\kappa(0)\,\xi_\kappa + \frac{1}{2}\,\xi_\kappa\,\mathcal{X}_\kappa\,\xi_\kappa + \mathbf{X}_\kappa^0 : \boldsymbol{\xi}_\kappa + \frac{1}{2}\,\boldsymbol{\xi}_\kappa : \mathbb{X}_\kappa : \boldsymbol{\xi}_\kappa. \quad (22.5.10)$$

The free energy adopted for the elastic-growing material expresses in terms of only the two first invariants of the Green strain via the two constants a_κ and μ_κ [unit : Pa]. The coefficient a_κ in this expression should vanish if the stress is to be zero in the unstrained intermediate configuration where $\mathbf{E}_G^e = \mathbf{0}$. Indeed

$$\boldsymbol{\tau}_\kappa' = \frac{\partial E_\kappa}{\partial\mathbf{E}_G^e} = a_\kappa\,\mathbf{I} + 2\,\mu_\kappa\,\mathbf{E}_G^e, \quad X_\kappa = X_\kappa(0) + \mathcal{X}_\kappa\,\xi_\kappa, \quad \mathbf{X}_\kappa = \mathbf{X}_\kappa^0 + \mathbb{X}_\kappa : \boldsymbol{\xi}_\kappa. \quad (22.5.11)$$

Convexity of the free energy in the elastic strain space is ensured for a positive shear modulus $\mu_\kappa > 0$, irrespective of a_κ. The constants introducing the structure variables are the scalar $\mathcal{X}_\kappa$ [unit : Pa] and the fourth order tensor with minor and major symmetries $\mathbb{X}_\kappa$ [unit : Pa].

[22.2] This restriction is alleviated in Chapter 23 where the effect of pyridinoline on collagen stiffness is accounted for.

22.5.3   A growth law based on a convex homeostatic surface

The growth transformation $\mathbf{F}^{\mathrm{g}}$ has to be provided by a constitutive equation, say in the tentative form,

$$\mathbf{L}^{\mathrm{g}}_{\kappa} \equiv \frac{d\mathbf{F}^{\mathrm{g}}}{dt} \cdot (\mathbf{F}^{\mathrm{g}})^{-1} = \mathcal{L}_{\kappa}(\mathcal{V}_{\kappa})\,. \tag{22.5.12}$$

In a general perspective, the set of state variables $\mathcal{V}_{\kappa}$, relative to the intermediate configurations, may include mechanical dependence, namely stresses or strains, and physiological dependence, e.g., cell density, growth factors, nutrients, etc. Moreover, for an open system, the surroundings contribute to the growth process in terms of mass, momentum, and free energy, Loret and Simões [2005]b.

As a further formal simplification, the (pure) elastic strain is assumed to be small, or moderate, so that $\boldsymbol{\Psi}_{\kappa} \equiv \mathbf{C}^{\mathrm{e}} \cdot \boldsymbol{\tau}_{\kappa} \simeq \boldsymbol{\tau}_{\kappa}$. Note that the dissipation inequality imposes a constraint only on the rate of growth transformation $\mathbf{D}^{\mathrm{g}}_{\kappa} = \frac{1}{2}\,(\mathbf{L}^{\mathrm{g}}_{\kappa} + (\mathbf{L}^{\mathrm{g}}_{\kappa})^{\mathrm{T}})$, since the stress $\boldsymbol{\tau}_{\kappa}$ is symmetric. That is, there might well exist a skew-symmetric rate of growth transformation $\boldsymbol{\Omega}^{\mathrm{g}}_{\kappa}$, namely

$$\mathbf{L}^{\mathrm{g}}_{\kappa} = \mathbf{D}^{\mathrm{g}}_{\kappa} + \boldsymbol{\Omega}^{\mathrm{g}}_{\kappa}\,, \tag{22.5.13}$$

but it is not limited by the dissipation inequality, because we have made the approximation $\boldsymbol{\Psi}_{\kappa} \equiv \mathbf{C}^{\mathrm{e}} \cdot \boldsymbol{\tau}_{\kappa} \simeq \boldsymbol{\tau}_{\kappa}$ and implicitly assumed that the couple stresses are negligible. To simplify the presentation, the growth spin is assumed to vanish,

$$\boldsymbol{\Omega}^{\mathrm{g}}_{\kappa} = \mathbf{0}\,. \tag{22.5.14}$$

This assumption has to be scrutinized for general anisotropic materials but, for isotropic materials, it is expected to be implied by the constitutive framework[22.3].

One might wonder of the relevance of satisfying a positive internal entropy production $\hat{S}_{(\mathrm{i})} > 0$. A brief discussion, supporting the idea that growth is a dissipative process, is presented in Section 4.8.5. Here, the stress is the engine that triggers growth. Homeostasis corresponds to $\hat{S}_{(\mathrm{i})} = 0$, and to a stress on or inside the homeostatic surface to be defined below. Mass is deposited only if the stress lays outside this surface.

Similar to the notion of yield surface for plastic materials, the key concept of *homeostatic domain* is adopted:
Assumption ($\mathcal{H}$): there exists a non empty, convex set (H) of homeostatic stress states defined by the scalar, dimensionless and smooth function $h_{\kappa}(\boldsymbol{\tau}_{\kappa} - \mathbf{X}_{\kappa}, X_{\kappa}) \le 0$.

It should be stressed that, while the elastic stress-strain relation involves the *effective* 2nd Piola-Kirchhoff stress $\boldsymbol{\tau}'_{\kappa}$ due to the elastic incompressibility, the growth law is defined in terms of the 2nd Piola-Kirchhoff stress $\boldsymbol{\tau}_{\kappa}$, as motivated by the simplified dissipation inequality. In fact, without the simplification which led the Mandel stress $\boldsymbol{\Psi}_{\kappa} \equiv \mathbf{C}^{\mathrm{e}} \cdot \boldsymbol{\tau}_{\kappa}$ to be approximated by the Kirchhoff stress $\boldsymbol{\tau}_{\kappa}$, the homeostatic surface would be phrased in terms of the Mandel stress.

The assumption of the existence of a domain, as opposed to a single value, of homeostatic states is more than purely formal. It needs to be substantiated in particular contexts, e.g., for articular cartilages, for solid tumors, etc. The underlying idea is that the amplitude of growth is a smooth function of the distance of the stress to a hypothetical homeostatic center, the close neighborhood of which is a neutral region. In other words, the stress under standard physiological conditions is allowed to vary somehow from a reference state without implying neither growth nor resorption. The homeostatic domain serves as a damper against accidental or physiological departures from the hypothetical homeostatic center. Its actual size is likely to depend on the organs.

The rate of the growth transformation is aligned with the outward normal to the homeostatic surface:

$$\mathbf{D}^{\mathrm{g}}_{\kappa} = \frac{\mu_{\kappa}}{t_{\mathrm{g}}}\,\Phi(h_{\kappa})\,\frac{\partial h_{\kappa}}{\partial \boldsymbol{\tau}_{\kappa}}\,, \tag{22.5.15}$$

where h_{κ} and its derivative are estimated at the current state $(\boldsymbol{\tau}_{\kappa} - \mathbf{X}_{\kappa}, X_{\kappa})$, and
- t_{g} [unit: s] is a scaling time of the growth process;
- μ_{κ} [unit: Pa] is a representative elastic modulus, actually the shear modulus;
- $\Phi(x) = \Phi(\langle x \rangle)$ is a continuous, convex, dimensionless function (the symbol $\langle \cdot \rangle$ denotes the positive part of its argument). Moreover, $\Phi(0) = 0$ and $\Phi(x) \ge 0$.

[22.3] The two above simplifications, namely (1). small pure elastic strain, which implies $\boldsymbol{\Psi}_{\kappa} \simeq \boldsymbol{\tau}_{\kappa}$, and (2). a vanishing growth spin $\boldsymbol{\Omega}^{\mathrm{g}}_{\kappa}$, are alleviated in Chapter 23 to model the growth of collagen in a mixture context.

(a)

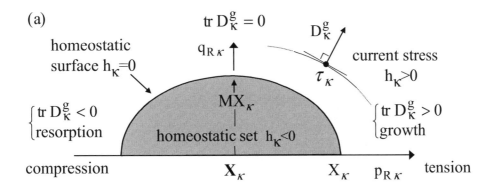

(b)

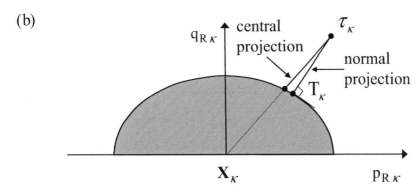

Fig. 22.5.1: Homeostatic set and geometric representation of the growth rate law in the plane of the two invariants of stress $(p_{R\kappa}, q_{R\kappa})$ relative to the tensor structure variable $\mathbf{X}_\kappa$. The direction of the rate of growth transformation is parallel to the outward normal to the homeostatic surface at the *current* point (a). In an alternative that still satisfies the dissipation inequality, the direction of the rate of growth transformation is parallel to the outward normal at the projection point (b). The latter may be obtained by normal projection, or, in some circumstances by radial projection.

In words, the rate of the growth transformation is non zero only for stresses outside the homeostatic convex, namely $h_\kappa(\boldsymbol{\tau}_\kappa - \mathbf{X}_\kappa, X_\kappa) > 0$, and then it has the direction of the outward normal to this surface at the current stress point.

In fact, the convexity of the homeostatic domain will be helpful to ensure satisfaction of the dissipation inequality. Indeed, according to (6.2.28), for two arbitrary stress states $\boldsymbol{\tau}_\kappa^{(1)} \neq \boldsymbol{\tau}_\kappa^{(2)}$, then

$$h_\kappa(\boldsymbol{\tau}_\kappa^{(1)} - \mathbf{X}_\kappa, X_\kappa) - h_\kappa(\boldsymbol{\tau}_\kappa^{(2)} - \mathbf{X}_\kappa, X_\kappa) \leq \frac{\partial h_\kappa}{\partial \boldsymbol{\tau}_\kappa}(\boldsymbol{\tau}_\kappa^{(1)} - \mathbf{X}_\kappa, X_\kappa) : (\boldsymbol{\tau}_\kappa^{(1)} - \boldsymbol{\tau}_\kappa^{(2)}). \tag{22.5.16}$$

Then for $\boldsymbol{\tau}_\kappa^{(1)} = \boldsymbol{\tau}_\kappa$, on or outside the homeostatic surface, namely $h_\kappa(\boldsymbol{\tau}_\kappa - \mathbf{X}_\kappa, X_\kappa) \geq 0$, and $\boldsymbol{\tau}_\kappa^{(2)} \in (H)$ inside the surface, e.g. $\boldsymbol{\tau}_\kappa^{(2)} - \mathbf{X}_\kappa = \mathbf{0}$, so that $h_\kappa(\boldsymbol{\tau}_\kappa^{(2)} - \mathbf{X}_\kappa, X_\kappa) < 0$, the following inequality holds,

$$0 < h_\kappa(\boldsymbol{\tau}_\kappa - \mathbf{X}_\kappa, X_\kappa) - h_\kappa(\mathbf{0}, X_\kappa) \leq \frac{\partial h_\kappa}{\partial \boldsymbol{\tau}_\kappa}(\boldsymbol{\tau}_\kappa - \mathbf{X}_\kappa, X_\kappa) : (\boldsymbol{\tau}_\kappa - \mathbf{X}_\kappa). \tag{22.5.17}$$

An alternative:
A slight modification to the above growth rate (22.5.15) can be introduced, by setting,

$$\mathbf{D}_\kappa^g = \frac{\mu_\kappa}{t_g}\, \Phi(h_\kappa)\, \frac{\partial h_\kappa}{\partial \boldsymbol{\tau}_\kappa}(\mathbf{T}_\kappa - \mathbf{X}_\kappa, X_\kappa), \tag{22.5.18}$$

where h_κ is estimated at the current point $(\boldsymbol{\tau}_\kappa - \mathbf{X}_\kappa, X_\kappa)$ but the derivative $\partial h_\kappa/\partial \boldsymbol{\tau}_\kappa$ is estimated at $(\mathbf{T}_\kappa - \mathbf{X}_\kappa, X_\kappa)$, where $\mathbf{T}_\kappa$ denotes the normal projection of the stress $\boldsymbol{\tau}_\kappa$ onto the homeostatic surface $h_\kappa = 0$. In words, the rate of growth transformation is non zero only for stresses $\boldsymbol{\tau}_\kappa$ outside the homeostatic convex, namely $h_\kappa(\boldsymbol{\tau}_\kappa - \mathbf{X}_\kappa, X_\kappa) > 0$, and then it has the direction of the outward normal to this surface at the projection point $\mathbf{T}_\kappa = \operatorname{Proj}_{h_\kappa=0} \boldsymbol{\tau}_\kappa$, as displayed in Fig. 22.5.1.

Like previously, the convexity of the homeostatic domain implies an inequality that is useful to exploit the dissipation inequality. Indeed, for $\boldsymbol{\tau}_\kappa^{(1)} = \mathbf{T}_\kappa$ on the homeostatic surface, and $\boldsymbol{\tau}_\kappa^{(2)} - \mathbf{X}_\kappa = \mathbf{0}$, (22.5.16) implies,

$$0 < 0 - h_\kappa(\mathbf{0}, X_\kappa) \le \frac{\partial h_\kappa}{\partial \boldsymbol{\tau}_\kappa}(\mathbf{T}_\kappa - \mathbf{X}_\kappa, X_\kappa) : (\mathbf{T}_\kappa - \mathbf{X}_\kappa) \,. \tag{22.5.19}$$

Now since

$$\boldsymbol{\tau}_\kappa = \mathbf{T}_\kappa + \lambda' \frac{\partial h_\kappa}{\partial \boldsymbol{\tau}_\kappa}(\mathbf{T}_\kappa - \mathbf{X}_\kappa, X_\kappa) \,, \quad \lambda' > 0 \,, \tag{22.5.20}$$

then

$$0 < \frac{\partial h_\kappa}{\partial \boldsymbol{\tau}_\kappa}(\mathbf{T}_\kappa - \mathbf{X}_\kappa, X_\kappa) : \left(\boldsymbol{\tau}_\kappa - \mathbf{X}_\kappa - \lambda' \frac{\partial h_\kappa}{\partial \boldsymbol{\tau}_\kappa}(\mathbf{T}_\kappa - \mathbf{X}_\kappa, X_\kappa) \right) \,, \tag{22.5.21}$$

and consequently,

$$0 < \frac{\partial h_\kappa}{\partial \boldsymbol{\tau}_\kappa}(\mathbf{T}_\kappa - \mathbf{X}_\kappa, X_\kappa) : (\boldsymbol{\tau}_\kappa - \mathbf{X}_\kappa) \,. \tag{22.5.22}$$

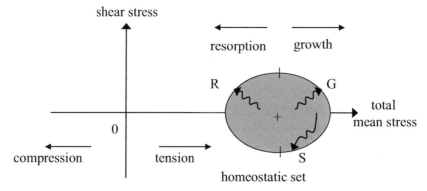

Fig. 22.5.2: Configuration where the homeostatic domain is centered around tensile stresses. Change of the stress may lead directly to growth if the stress point was already on (point G) or outside the surface with a mean stress larger than the center of the homeostatic surface. If on the other hand the stress is strictly inside the surface, the change has to be sufficiently large to induce growth. Note that, even if the stress is tensile, a stress decrease may lead to resorption, point R, or to growth, point S.

22.5.4   The ellipsoidal homeostatic surface

The mechanical effect of the stress on growth is somehow paradoxical. Growth is sometimes considered to be triggered by tension, and resorption by compression, Taber and Eggers [1996], Lubarda and Hoger [2002]. The opposite option is taken, e.g., by Rodriguez et al. [1994]. The growth rate law of Drozdov and Khanina [1997] makes no difference between tension and compression, and the rate of change of the mass can be only positive or zero, that is, resorption is not covered. On the other hand, it might be argued that the physiological mechanisms of growth and resorption are not the same. This observation would imply to develop independent constitutive equations for the two mechanisms. The macroscopic approach here is not that fine: as a first attempt, the two mechanisms use a single homeostatic surface and, in essence, obey identical rules and use the same time constant.

We adopt the former point of view, namely growth is a priori triggered by tensile states and resorption by compressive states. Actually, the model provides an explanation to the paradox relative to the load sign. In fact, if the homeostatic surface is centered about the zero stress, then growth is triggered by tensile states and resorption by compressive states. If the homeostatic surface is centered about a large tensile stress, so that the whole homeostatic surface lays in the tensile region, then growth will take place for stresses (out of the homeostatic surface) larger than the center stress and resorption for stresses smaller than the center stress, Fig. 22.5.2. The situation where the homeostatic surface is centered about a large compressive stress and completely located in the compressive region may be envisaged as well: compressive stresses algebraically larger than the center of the homeostatic surface give rise to growth.

The above description provides a general trend. Still, there might be more subtle situations. For example, consider a point located well inside the homeostatic surface. The stress path might induce a decrease in mean stress but still hit the surface at a point of mean stress larger than the center, point S in Fig. 22.5.2. Thus, we have a case where a decrease of mean stress may lead to growth, at least temporarily.

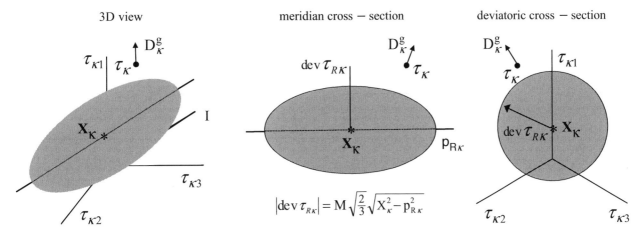

Fig. 22.5.3: Representation, in the principal stress axes, of an ellipsoidal homeostatic domain and of the growth strain rate $\mathbf{D}_\kappa^\mathrm{g}$ at a current non homeostatic stress $\boldsymbol{\tau}_\kappa$. The norm $|\mathrm{dev}\,\boldsymbol{\tau}_{R\kappa}|$ of the deviatoric part $\mathrm{dev}\,\boldsymbol{\tau}_{R\kappa}$ of the stress $\boldsymbol{\tau}_\kappa$ relative to the center $\mathbf{X}_\kappa$ is equal to $M\sqrt{X_\kappa^2 - p_{R\kappa}^2}$ times $\sqrt{2/3}$.

The set of homeostatic states is assumed ellipsoidal, with one of its axes aligned with the isotropic axis, but shifted through the variable $\mathbf{X}_\kappa$, which thus appears as the center of the homeostatic surface, Fig. 22.5.3. The homeostatic surface is thus defined by the two scalar invariants $p_{R\kappa}$ and $q_{R\kappa}$,

$$p_{R\kappa} = \frac{1}{3}\,\mathrm{tr}\,(\boldsymbol{\tau}_\kappa - \mathbf{X}_\kappa)\,, \quad q_{R\kappa}^2 = \frac{3}{2}\,\mathrm{dev}(\boldsymbol{\tau}_\kappa - \mathbf{X}_\kappa):\mathrm{dev}(\boldsymbol{\tau}_\kappa - \mathbf{X}_\kappa)\,, \tag{22.5.23}$$

of the *shifted* stress

$$\boldsymbol{\tau}_{R\kappa} = \boldsymbol{\tau}_\kappa - \mathbf{X}_\kappa\,, \tag{22.5.24}$$

with deviator $\mathbf{s}_{R\kappa} = \mathrm{dev}\,\boldsymbol{\tau}_{R\kappa} = \boldsymbol{\tau}_{R\kappa} - p_{R\kappa}\,\mathbf{I}$. The homeostatic surface is cast in the format, Fig. 22.5.1,

$$\mu_\kappa^2\,h_\kappa(\boldsymbol{\tau}_\kappa - \mathbf{X}_\kappa, X_\kappa) \;=\; \frac{q_{R\kappa}^2}{M^2} + (p_{R\kappa})^2 - X_\kappa^2 = 0\,. \tag{22.5.25}$$

The diameter of the homeostatic surface is equal to $2X_\kappa$ along the isotropic axis and to $2\,MX_\kappa$ in any direction on the deviatoric planes. For $p_{R\kappa} = 0$, then $q_{R\kappa} = M\,X_\kappa$: the ellipse can be elongated along the isotropic axis via $X_\kappa \to \infty$ while decreasing M so as to keep the product $M\,X_\kappa$ constant.

The normal to the homeostatic surface

$$\mu_\kappa^2\,\frac{\partial h_\kappa}{\partial \boldsymbol{\tau}_\kappa}(\boldsymbol{\tau}_\kappa - \mathbf{X}_\kappa, X_\kappa) \;=\; \frac{3}{M^2}\,\mathbf{s}_{R\kappa} + 2\,p_{R\kappa}\,\frac{\mathbf{I}}{3}\,, \tag{22.5.26}$$

is needed in the rate of the growth transformation (22.5.15). As expected, the outward normal can not vanish for points on or outside the homeostatic surface,

$$\mu_\kappa^4\,\|\,\frac{\partial h_\kappa}{\partial \boldsymbol{\tau}_\kappa}\,\|^2 \;=\; \frac{6}{M^4}\,q_{R\kappa}^2 + \frac{4}{3}\,p_{R\kappa}^2\,. \tag{22.5.27}$$

Note the trace of the rate of the growth transformation,

$$\mathrm{tr}\,\mathbf{D}_\kappa^\mathrm{g} = \frac{\mu_\kappa}{t_\mathrm{g}}\,\Phi(h_\kappa)\,\mathrm{tr}\,\frac{\partial h_\kappa}{\partial \boldsymbol{\tau}_\kappa} = 2\,\frac{\Phi(h_\kappa)}{t_\mathrm{g}}\,\frac{p_{R\kappa}}{\mu_\kappa}\,. \tag{22.5.28}$$

Thus, for a stress $\boldsymbol{\tau}_\kappa$ outside the homeostatic domain, with $\boldsymbol{\tau}_\kappa - \mathbf{X}_\kappa = p_{R\kappa}\,\mathbf{I} + \mathbf{s}_{R\kappa}$, with projection $\mathbf{T}_\kappa - \mathbf{X}_\kappa = P_{R\kappa}\,\mathbf{I} + \mathbf{S}_{R\kappa}$, one may have

- mass resorption, if the shifted stress is sufficiently compressive, namely $p_{R\kappa} < 0$ or $P_{R\kappa} < 0$;
- mass growth for sufficient shifted tensile stress, namely $p_{R\kappa} > 0$ or $P_{R\kappa} > 0$;
- and no mass change along the plane $p_{R\kappa} = 0$ or $P_{R\kappa} = 0$.

The resulting dissipation,

$$(\boldsymbol{\tau}_\kappa - \mathbf{X}_\kappa) : \mathbf{D}_\kappa^{\mathrm{g}} = 2 \frac{\mu_\kappa}{t_{\mathrm{g}}} \, \Phi(h_\kappa) \left(h_\kappa + \frac{X_\kappa^2}{\mu_\kappa^2} \right), \tag{22.5.29}$$

is positive as expected from the general analysis (22.5.17) since the ellipsoidal homeostatic surface is convex, and its hessian,

$$\mathbb{H}_\kappa = \mu_\kappa^2 \frac{\partial^2 h_\kappa}{\partial \boldsymbol{\tau}_\kappa^2} \;\; = \;\; \frac{2}{9} \, \mathbf{I} \otimes \mathbf{I} + \frac{3}{M^2} \, \mathbf{J}^{(4)}, \tag{22.5.30}$$

is clearly positive definite. Here $\mathbf{I}^{(4)} = \mathbf{I} \, \overline{\otimes} \, \mathbf{I}$ denotes the fourth order identity tensor over symmetric second order tensors, componentwise $I_{ijkl}^{(4)} = \frac{1}{2} (I_{ik} I_{jl} + I_{il} I_{jk})$, then $\mathbf{J}^{(4)} = \mathbf{I}^{(4)} - \frac{1}{3} \mathbf{I} \otimes \mathbf{I}$ is the fourth order identity tensor over symmetric second order traceless tensors, componentwise $J_{ijkl}^{(4)} = \frac{1}{2} (I_{ik} I_{jl} + I_{il} I_{jk}) - \frac{1}{3} I_{ij} I_{kl}$.

The homeostatic surface may be endowed with anisotropy, through the second order fabric tensor $\mathbf{B}$ introduced in Section 5.2.11 (isotropy is recovered for $\mathbf{B} = \mathbf{I}$),

$$\mu_\kappa^2 \, h_\kappa \;\; = \;\; \frac{1}{9} (\mathbf{B} : \boldsymbol{\tau}_{R\kappa})^2 + \frac{3}{2 M^2} \left((\mathbf{B} \cdot \boldsymbol{\tau}_{R\kappa}) \cdot (\boldsymbol{\tau}_{R\kappa} \cdot \mathbf{B}) - \frac{1}{3} (\mathbf{B} : \boldsymbol{\tau}_{R\kappa})^2 \right) - X_\kappa^2,$$

$$\mu_\kappa^2 \, \frac{\partial h_\kappa}{\partial \boldsymbol{\tau}_{R\kappa}} \;\; = \;\; \frac{2}{9} (\mathbf{B} : \boldsymbol{\tau}_{R\kappa}) \, \mathbf{B} + \frac{3}{M^2} \left(\mathbf{B} \cdot \boldsymbol{\tau}_{R\kappa} \cdot \mathbf{B} - \frac{1}{3} (\mathbf{B} : \boldsymbol{\tau}_{R\kappa}) \, \mathbf{B} \right), \tag{22.5.31}$$

$$\mu_\kappa^2 \, \frac{\partial^2 h_\kappa}{\partial \boldsymbol{\tau}_{R\kappa}^2} \;\; = \;\; \frac{2}{9} \, \mathbf{B} \otimes \mathbf{B} + \frac{3}{M^2} \left(\mathbf{B} \, \overline{\otimes} \, \mathbf{B} - \frac{1}{3} \mathbf{B} \otimes \mathbf{B} \right).$$

With $\mathbb{H}_\kappa = \mathbb{H}_\kappa(\mathbf{B})$ the dimensionless Hessian,

$$\mu_\kappa^2 \, h_\kappa = \frac{1}{2} \boldsymbol{\tau}_{R\kappa} : \mathbb{H}_\kappa : \boldsymbol{\tau}_{R\kappa} - X_\kappa^2, \quad \mu_\kappa^2 \, \frac{\partial h_\kappa}{\partial \boldsymbol{\tau}_{R\kappa}} = \mathbb{H}_\kappa : \boldsymbol{\tau}_{R\kappa}, \quad \mu_\kappa^2 \, \frac{\partial^2 h_\kappa}{\partial \boldsymbol{\tau}_{R\kappa}^2} = \mathbb{H}_\kappa. \tag{22.5.32}$$

Incidentally, a relation similar to (5.2.78) holds, namely

$$\boldsymbol{\tau}_{R\kappa} : \mathbb{H}_\kappa(\mathbf{B}) : \boldsymbol{\tau}_{R\kappa} = \tilde{\boldsymbol{\tau}}_{R\kappa} : \mathbb{H}_\kappa(\mathbf{I}) : \tilde{\boldsymbol{\tau}}_{R\kappa} \quad \text{with} \quad \tilde{\boldsymbol{\tau}}_{R\kappa} = \mathbf{B}^{1/2} \cdot \boldsymbol{\tau}_{R\kappa} \cdot \mathbf{B}^{1/2}. \tag{22.5.33}$$

22.5.5   Rate laws for the structure variables

With all ingredients of the framework now in place, we are in position to display the heuristic, and simple, manipulations that will lead to the rate laws for the structure variables. Irrespective of the details of the homeostatic surface, the reduced dissipation inequality (22.5.9) may be recast in the format,

$$T \, \hat{S}_{\kappa(\mathrm{i})} \;\; = \;\; (\boldsymbol{\tau}_\kappa - \mathbf{X}_\kappa) : \mathbf{D}_\kappa^{\mathrm{g}} - X_\kappa \frac{d\xi_\kappa}{dt} + \mathbf{X}_\kappa : (\mathbf{D}_\kappa^{\mathrm{g}} - \frac{d\boldsymbol{\xi}_\kappa}{dt}) \geq 0. \tag{22.5.34}$$

We build the growth law, having in mind to satisfy this inequality. For example, we may take

$$\mathbf{D}_\kappa^{\mathrm{g}} - \frac{d\boldsymbol{\xi}_\kappa}{dt} \;\; = \;\; a_1 \, \| \, \mathbf{D}_\kappa^{\mathrm{g}} \, \| \, \frac{\mathbf{X}_\kappa}{\mu_\kappa}, \tag{22.5.35}$$

and

$$\frac{d\xi_\kappa}{dt} = \frac{1}{t_{\mathrm{g}}} \, \Phi(h_\kappa) \left(a_2 - a_3 \frac{X_\kappa}{\mu_\kappa} \right) \frac{X_\kappa}{\mu_\kappa}, \tag{22.5.36}$$

where the three scalars a's [unit : 1] either depend on the state variables or are constants. More generally, the scalar a_1 could be replaced by a symmetric positive semi-definite fourth order tensor built from the second order tensor $\mathbf{X}_\kappa$, e.g. $\mathbb{A}_1 \equiv \alpha_1 \, \mathbf{X}_\kappa \otimes \mathbf{X}_\kappa + 2 \beta_1 \, \mathbf{X}_\kappa \, \overline{\otimes} \, \mathbf{X}_\kappa$, with $3 \alpha_1 + 2 \beta_1 \geq 0$ and $\beta_1 \geq 0$[22.4].

[22.4] Indeed, with the definition (2.1.25), $\mathbf{Y} : \mathbb{A}_1 : \mathbf{Y}$ is equal to $(\alpha_1 + 2 \beta_1/3) \, (\mathrm{tr} \, \mathbf{Y} \cdot \mathbf{X}_\kappa)^2 + 2 \beta_1 \, \mathrm{dev}^2(\mathbf{Y} \cdot \mathbf{X}_\kappa)$.

The dissipation becomes,

$$T\hat{S}_{\kappa(\mathrm{i})} = \frac{\mu_\kappa}{t_\mathrm{g}} \, \Phi(h_\kappa) \left(2 \, h_\kappa + (2 - a_2) \, \frac{X_\kappa^2}{\mu_\kappa^2} + a_3 \, \frac{X_\kappa^3}{\mu_\kappa^3} + a_1 \parallel \frac{\partial h_\kappa}{\partial \boldsymbol{\tau}_\kappa} \parallel \frac{\mathbf{X}_\kappa : \mathbf{X}_\kappa}{\mu_\kappa} \right). \tag{22.5.37}$$

For $h_\kappa \leq 0$, no growth, and therefore, no dissipation take place. For $h_\kappa > 0$, growth takes place, and the dissipation is ensured to be positive for

$$a_1 \geq 0, \quad 2 \geq a_2 \geq 0, \quad a_3 \geq 0. \tag{22.5.38}$$

Note that the rates of the structure variables have been assumed to be non zero only if growth takes place. Alternatively, delayed structuration might be envisaged.

In summary, the constitutive equations include
- the linear elastic equations (22.5.11) yielding the generalized stresses in terms of the generalized strains;
- the growth rate equation (22.5.15) with the homeostatic surface h_κ (22.5.25) and its derivative (22.5.26);
- the rate equations (22.5.35) and (22.5.36) for the structure variables,

$$\begin{aligned}
\frac{d\boldsymbol{\xi}_\kappa}{dt} &= \mathbf{D}_\kappa^\mathrm{g} - a_1 \parallel \mathbf{D}_\kappa^\mathrm{g} \parallel \frac{\mathbf{X}_\kappa}{\mu_\kappa}, \\
\frac{d\xi_\kappa}{dt} &= \frac{1}{t_\mathrm{g}} \, \Phi(h_\kappa) \left(a_2 - a_3 \, \frac{X_\kappa}{\mu_\kappa} \right) \frac{X_\kappa}{\mu_\kappa}.
\end{aligned} \tag{22.5.39}$$

If $a_1 = 0$ in (22.5.39) and the principal axes of the growth transformation $\mathbf{F}^\mathrm{g}$ remain fixed, the tensor structure variable $\boldsymbol{\xi}_\kappa(t)$ conveys the same information as the growth transformation $\mathbf{F}^\mathrm{g}$,

$$\boldsymbol{\xi}_\kappa(t) = \boldsymbol{\xi}_\kappa(0) + \mathrm{Ln} \, \mathbf{F}^\mathrm{g}(t) \cdot \mathbf{F}^\mathrm{g}(0)^{-1}. \tag{22.5.40}$$

Given the linear elastic equations, rate equations for the generalized stresses associated with the structure variables follow from (22.5.39), and have direct consequences:
- the radius of the homeostatic surface X_κ tends asymptotically to a constant either because X_κ/μ_κ approaches a_2/a_3, or because the homeostatic surface is approached;
- the behavior of the center of the homeostatic surface $\mathbf{X}_\kappa$ is more complex to analyze a priori. If $a_1 = 0$, and if the fourth order tensor $\mathbb{X}_\kappa$ is positive definite, then clearly $\mathbf{X}_\kappa$ moves in the direction of the growth rate and therefore is headed towards the current stress. Its evolution does not stop as long as growth takes place. For $a_1 > 0$, the rate of variation of the center is decreased and this might give rise to increased growth if the stress is fixed in time.

The use of the total derivatives to postulate the constitutive equations of the structure variables in the intermediate configurations implicitly assumes that the latter are isoclinic, namely that they have a fixed orientation. As a consequence, the elastic transformation generally includes a rotation.

In Section 22.6 where emphasis is laid on the impact of the strain energy the aggregating mass is endowed with at the time of deposition, the two last sets of constitutive equations (22.5.39) will be maintained unchanged for the generalized strains, but the generalized stresses will derive from the energy, no longer via linear algebraic relations like (22.5.11), but via integral expressions.

On the non symmetry of Mandel stress $\boldsymbol{\Psi}_\kappa \equiv \mathbf{C}^\mathrm{e} \cdot \boldsymbol{\tau}_\kappa$

Now that all the pieces of the model are in place, we can take a step backward and comment the approximation in Section 22.5.2 of the stress Mandel stress $\boldsymbol{\Psi}_\kappa = \mathbf{C}^\mathrm{e} \cdot \boldsymbol{\tau}_\kappa$ by the 2nd Piola-Kirchhoff stress $\boldsymbol{\tau}_\kappa$. The fact that the Mandel stress is in general non symmetric leads to two sorts of difficulties. First, theorems of representation of isotropic scalar functions are not available for arbitrary second order tensors: this issue pops up when considering actual expressions of the homeostatic function. Second, the fact that the dissipation would involve the complete rate of growth transformation implies that satisfaction of the dissipation inequality becomes more tricky.

The approach developed in Section 23.4.2 alleviates both difficulties and it applies in presence of arbitrary anisotropy in the intermediate configurations. Actually, in some cases, the stress $\boldsymbol{\Psi}_\kappa$ is anyway symmetric, e.g. for isotropic elasticity where $\boldsymbol{\tau}_\kappa = \partial E_\kappa/\partial \mathbf{E}_\mathrm{G}^\mathrm{e}$ and $\mathbf{C}^\mathrm{e}$ are coaxial. Then the dissipated power involves anyway only the symmetric part of the growth rate $\mathbf{L}_\kappa^\mathrm{g}$. In fact the symmetry of Mandel stress for a material which is isotropic in the intermediate configurations results from (2.11.29).

22.5.6 Application to volumetric growth

The characteristics of the constitutive equations are now explored with a tissue specimen subjected to an isotropic stress. The analytical developments are simplified for $a_1 = 0$.

It is appropriate to consider first the initial conditions. At time $t = 0$, $\boldsymbol{\xi}_\kappa$ is assumed to vanish and $\mathbf{F}^g$ is equal to the identity $\mathbf{I}$, so that $\boldsymbol{\xi}_\kappa(t) = \mathrm{Ln}\ \mathbf{F}^g(t)$ according to (22.5.40). The fourth order tensor $\mathbb{X}_\kappa$ being assumed isotropic, namely $\mathbb{X}_\kappa = \mathrm{x}_\kappa\,\mathbf{I}^{(4)}$, with $\mathrm{x}_\kappa > 0$, the center $\mathbf{X}_\kappa$ of the homeostatic surface is simply equal to $\mathrm{x}_\kappa\,\mathrm{Ln}\ \mathbf{F}^g(t)$. Since the elastic properties are isotropic, the elastic and growth transformations remain isotropic during the growth process. Cartesian axes $(\mathbf{e}_{(1)}, \mathbf{e}_{(2)}, \mathbf{e}_{(3)})$ may be chosen arbitrarily and the component notation is simplified, writing F_1 instead of F_{11}, etc.

The loading process assumes the Cauchy stress to be given, and to give rise to growth or resorption. If the homeostatic surface does not adapt by either changing its center or its size, the given stress remains non homeostatic forever. Therefore, bounded growth requires either the center or the size of the homeostatic surface to evolve.

22.5.6.1 The homeostatic surface is mobile but has fixed size

For an elastically incompressible tissue, the elastic transformation reduces to the identity, and the total deformation and growth transformation are equal. Before considering the elastically incompressible tissue, it is worth to address first the compressible tissue. While the analyses in the compressible and incompressible cases are different, a correspondence between the respective deformations and stresses can be obtained.

Compressible elastic-growing tissue

For volumetric growth, the elastic transformation $\mathbf{F}^e = F_1^e\,\mathbf{I}$, the growth transformation $\mathbf{F}^g = F_1^g\,\mathbf{I}$, and the Green strain $\mathbf{E}_G^e = \frac{1}{2}\big((F_1^e)^2 - 1\big)\,\mathbf{I}$ are isotropic. So are the stresses $\boldsymbol{\tau}_\kappa = p_\kappa\,\mathbf{I}$ and $\boldsymbol{\sigma} = p\,\mathbf{I}$, namely by (22.5.2) and (22.5.11),

$$p_\kappa = a_\kappa + \mu_\kappa\big((F_1^e)^2 - 1\big) = p\,F_1^e. \tag{22.5.41}$$

The Cauchy stress p is controlled. The above relations indicate that it is equivalent to control the elastic component F_1^e or the 2nd Piola-Kirchhoff stress p_κ. Moreover,

$$F_1^e = \frac{p}{2\,\mu_\kappa} + \sqrt{\frac{p^2}{4\,\mu_\kappa^2} + 1 - \frac{a_\kappa}{\mu_\kappa}}. \tag{22.5.42}$$

Below, the applied stress is fixed in time, namely $p(t) = p_0\,\mathcal{H}(t)$, with $\mathcal{H}(t)$ the Heaviside step function. The elastic deformation and total stress result to be discontinuous at time $t = 0$, but are constant at any later time. On the other hand, the growth strain is continuous and so are the structure variables, namely the center and size of the homeostatic surface.

For $\Phi(h_\kappa) = h_\kappa$, the rate of growth is governed by the differential equation (22.5.15),

$$\frac{d}{dt}\mathrm{Ln}\,F_1^g(t) = \frac{2}{3\,t_g}\,h_\kappa\,\frac{p_{R\kappa}}{\mu_\kappa}, \tag{22.5.43}$$

with

$$h_\kappa = \left(\frac{p_{R\kappa}}{\mu_\kappa}\right)^2 - \left(\frac{X_\kappa}{\mu_\kappa}\right)^2, \quad \frac{p_{R\kappa}}{\mu_\kappa} = \frac{p_\kappa}{\mu_\kappa} - \frac{\mathrm{x}_\kappa}{\mu_\kappa}\,\mathrm{Ln}\,F_1^g(t). \tag{22.5.44}$$

The governing differential equation,

$$d\left(\frac{p_{R\kappa}}{\mu_\kappa}\right) = -\frac{2}{3}\frac{\mathrm{x}_\kappa}{\mu_\kappa}\left(\frac{p_{R\kappa}^2}{\mu_\kappa^2} - \frac{X_\kappa^2}{\mu_\kappa^2}\right)\frac{p_{R\kappa}}{\mu_\kappa}\frac{dt}{t_g}, \tag{22.5.45}$$

integrates to

$$1 - \frac{X_\kappa^2}{p_{R\kappa}^2(t)} = \left(1 - \frac{X_\kappa^2}{p_{R\kappa}^2(0^+)}\right)\exp\left(-\frac{t}{T_g}\right), \tag{22.5.46}$$

from which $F_1^g(t)$ is easily extracted with help of (22.5.44).

The condition of growth at time $t = 0^+$, $\mu_\kappa^2\,h_\kappa(t = 0) = p_{R\kappa}^2(0^+) - X_\kappa^2 = p_\kappa^2(0^+) - X_\kappa^2 > 0$ is seen to imply continuous growth at any later time $t > 0$. Traction $p_\kappa > X_\kappa$ implies growth and compression $p_\kappa < -X_\kappa$ leads to resorption. The growth rate decreases progressively because the homeostatic surface moves along the isotropic axis so as to ultimately come in contact with the (fixed) stress point. Note that the actual relaxation time,

$$T_{\mathrm{g}} = \frac{3}{4} \frac{\mu_\kappa}{\mathbb{x}_\kappa} \frac{\mu_\kappa^2}{X_\kappa^2} t_{\mathrm{g}} , \tag{22.5.47}$$

is inversely proportional to the kinematic coefficient $\mathbb{x}_\kappa / \mu_\kappa$, as well as to the square of the radius of the homeostatic surface. The final growth amplitude depends on the initial distance $|p_\kappa(0^+)| - X_\kappa$ of the controlled stress to the homeostatic surface and on the kinematic coefficient $\mathbb{x}_\kappa$,

$$\frac{1}{3} \operatorname{Ln} \lim_{t\to\infty} \det \mathbf{F}^{\mathrm{g}}(t) = \operatorname{Ln} \lim_{t\to\infty} F_1^{\mathrm{g}}(t) = \frac{1}{\mathbb{x}_\kappa} \begin{cases} p_\kappa(0^+) - X_\kappa & \text{tension,} \\[2mm] p_\kappa(0^+) + X_\kappa & \text{compression.} \end{cases} \tag{22.5.48}$$

If the homeostatic surface (domain) was fixed, an applied non homeostatic stress would give rise to unbounded growth. The tissue reacts against this undesirable event by adapting, namely allowing the homeostatic surface to be attracted by the stress. The larger the kinematic coefficient $\mathbb{x}_\kappa / \mu_\kappa$, the shorter the growth process and the smaller the growth amplitude.

Incompressible elastic-growing tissue

Isotropic elastic incompressibility implies $F_1^{\mathrm{e}} = 1$, and therefore a vanishing elastic strain,

$$\mathbf{F}^{\mathrm{e}} = \mathbf{I}, \quad \mathbf{F}^{\mathrm{g}} = F_1^{\mathrm{g}} \mathbf{I}, \quad \mathbf{E}_{\mathrm{G}}^{\mathrm{e}} = \mathbf{0} . \tag{22.5.49}$$

The constitutive equations now yield the effective stress, rather than the total stress,

$$\boldsymbol{\tau}_\kappa' = \boldsymbol{\sigma}' = a_\kappa \mathbf{I}, \quad \boldsymbol{\sigma} = p\mathbf{I} = \boldsymbol{\tau}_\kappa = p_\kappa \mathbf{I} , \tag{22.5.50}$$

with $p_\kappa = p = a_\kappa - p_I$. In other words, the incompressibility pressure p_I is equal to $a_\kappa - p$, with $p = p_0 \mathcal{H}(t)$ the given Cauchy stress.

Growth is governed by the total 2nd Piola-Kirchhoff stress $\boldsymbol{\tau}_\kappa$. Then the growth amplitude will be identical in the compressible and incompressible cases if the respective 2nd Piola-Kirchhoff stresses are identical, or equivalently, according to (22.5.41), if the Cauchy stress p in the incompressible case is chosen equal to F_1^{e} times the Cauchy stress in the compressible case.

Influence of the saturating coefficient a_1

The differential equations for the growth transformation, shifted stress $p_{R\kappa} = p_\kappa(0^+) - X_{\kappa 1}$ and center of the homeostatic surface $X_{\kappa 1}$ are modified to,

$$d\operatorname{Ln} F_1^{\mathrm{g}} = -\frac{1}{\epsilon\, a_1 \sqrt{3}} \frac{\mu_\kappa}{\mathbb{x}_\kappa} \frac{dp_{R\kappa}/\mu_\kappa}{\dfrac{p_{R\kappa}}{\mu_\kappa} + \dfrac{1}{\epsilon\, a_1 \sqrt{3}} - \dfrac{p_\kappa}{\mu_\kappa}} = \frac{2}{3}\left(\frac{p_{R\kappa}^2}{\mu_\kappa^2} - \frac{X_\kappa^2}{\mu_\kappa^2}\right) \frac{p_{R\kappa}}{\mu_\kappa} \frac{dt}{t_{\mathrm{g}}} , \tag{22.5.51}$$

with $\epsilon = 1$ for volumetric growth and -1 for volumetric resorption.

The shifted stress and center of the homeostatic surface depend weakly on the coefficient a_1, in contrast to the growth strain, Fig. 22.5.4. In fact, the limit of the center of the homeostatic surface at large times keeps a value independent of a_1, while the growth, or resorption, depends on this coefficient:

$$\lim_{t\to\infty} X_{\kappa 1}(t) = p_\kappa(0^+) - \epsilon\, X_\kappa, \quad \operatorname{Ln} \lim_{t\to\infty} F_1^{\mathrm{g}}(t) = -\frac{\mu_\kappa}{\mathbb{x}_\kappa} \frac{1}{\epsilon\, a_1 \sqrt{3}} \operatorname{Ln}\left(1 - \epsilon\, a_1 \sqrt{3} \frac{X_{\kappa 1}(\infty)}{\mu_\kappa}\right) . \tag{22.5.52}$$

As expected, the mass growth is larger for $a_1 > 0$, Fig. 22.5.4. In fact, the coefficient a_1, which is bounded below by 0 so as to satisfy the dissipation inequality, should be bounded above so that growth remains finite. Indeed,

$$0 < a_1 \sqrt{3} < \frac{\mu_\kappa}{X_{\kappa 1}(\infty)} = \frac{\mu_\kappa}{p_\kappa(0^+) - X_\kappa} . \tag{22.5.53}$$

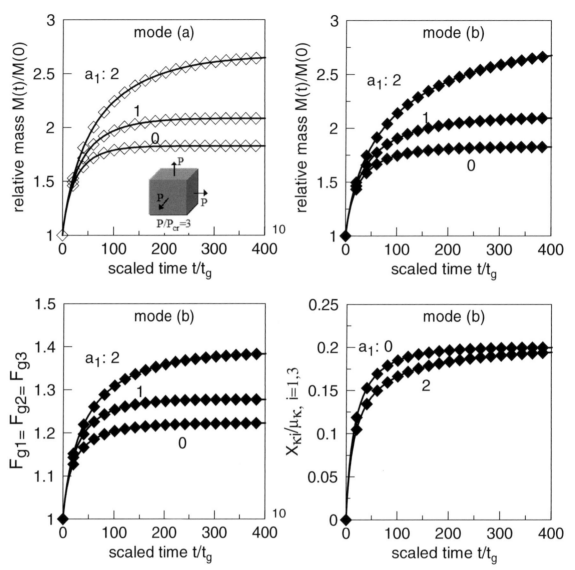

Fig. 22.5.4: Elastic-growing tissue subjected to volumetric growth. Standard parameters as given in Section 22.5.6.2, except $a_1 \neq 0$. According to (22.5.52) and (22.5.53), the value of a_1 should be bounded above by $5/\sqrt{3} \sim 2.89$. Time profiles of aggregated mass, or volume, of the growth transformation and center of homeostatic surface for several values of a_1. The position of the center of the homeostatic surface depends weakly on a_1 in contrast to the growth transformation. The modes of deposition (a) and (b) are addressed in Section 22.6.

22.5.6.2 Standard constitutive parameters

Parameters with dimension of stress are scaled by the modulus μ_κ, and parameters with dimension of time are scaled by the time t_g:

- $a_\kappa = 0 \, \text{kPa}$, elastic parameter indicating the value of the stress in the intermediate configuration at zero strain;
- $\Phi(h_\kappa)$, monotone function of its argument h_κ, typically taken as a power law, say $\Phi(h_\kappa) = h_\kappa^{n_\phi}$. Here $n_\phi = 1$;
- $M = 1.0$, relative size of the homeostatic surface in a deviatoric plane;
- a_1, parameter implying a deviation of the rate of the tensor structure variable $\boldsymbol{\xi}_\kappa$, and center of the homeostatic surface $\mathbf{X}_\kappa$, with respect to the rate of growth. Here $a_1 = 0$;
- $a_2 \in [0,2]$, parameter involved in the rate of change of the size of the scalar structure variable, and size of the homeostatic surface, $a_2 = 0.1$;
- $a_3 \geq 0$: the size X_κ/μ_κ of the homeostatic surface is asymptotic to a_2/a_3 taken equal to 0.1. Note that this value may be smaller or larger than the initial value;

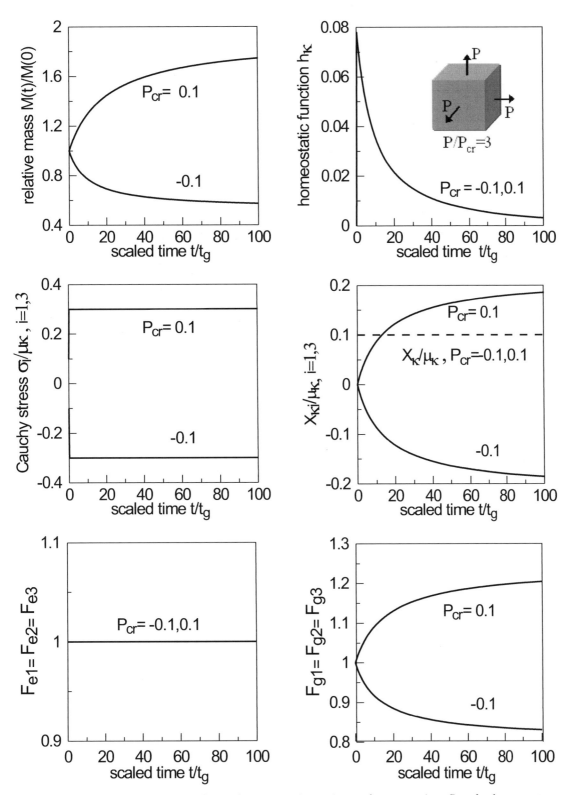

Fig. 22.5.5: Elastic-growing tissue subjected to isotropic tension and compression. Standard parameters as given in Section 22.5.6.2. The initial radius of the homeostatic surface is defined from the normalized load P_{cr}^{ext} equal to 0.1 in tension and -0.1 in compression. In both cases, the actual loads P^{ext} are equal to three times these values, namely $P^{ext}/P_{cr}^{ext} = 3$. Time profiles of aggregated mass, or volume, homeostatic function, Cauchy stress, and center and radius of the homeostatic surface. Reprinted with permission from Loret and Simões [2010]c.

- $\mathbb{X}_\kappa = \mathrm{x}_\kappa\, \mathbf{I}^{(4)}$, kinematic fourth order tensor linking the tensor structure variable and the center of the homeo-static surface taken proportional to the fourth order identity tensor, with $\mathrm{x}_\kappa/\mu_\kappa = 1$;
- $\mathcal{X}_\kappa = \mu_\kappa$, scalar linking the scalar structure variable and the size of the homeostatic surface.

22.5.6.3 Simulations of volumetric growth on elastic-growing tissues

Volumetric growth and resorption tests are shown in Figs. 22.5.5 and 22.5.6 for an elastically incompressible tissue. The standard material parameters are listed in Section 22.5.6.2. At time $t = 0$, the homeostatic surface is centered at the origin of the stress space, namely $\mathbf{X}_\kappa = \mathbf{0}$.

The normalized loads $P_{\mathrm{cr}}^{\mathrm{ext}} = p/\mu_\kappa$ which imply growth are prescribed equal to 0.1 for tension and -0.1 for compression. The scalar structure variable ξ_κ and the radius X_κ are deduced from the above datum, $X_\kappa/\mu_\kappa = \mathcal{X}_\kappa\,\xi_\kappa/\mu_\kappa = 0.1$. This value is exactly equal to the standard ratio $a_2/a_3 = 0.1$. Therefore, according to (22.5.39)$_2$, the scalar structure variable ξ_κ, and the radius X_κ of the homeostatic surface do not change during the growth process, for these standard parameters. For both tension and compression, the actual applied loads P^{ext} are equal to three times the critical values, namely $P^{\mathrm{ext}}/P_{\mathrm{cr}}^{\mathrm{ext}} = 3$.

In agreement with the analytical solution (22.5.47), the half time of growth process is observed to be about $100\,t_{\mathrm{g}}$ for the standard values of the parameters listed above. Typical times for growth of arteries are provided in Taber and Eggers [1996], their Fig. 12, for a 50% increase of the homeostatic blood pressure, namely 1 day for rats, 10 days for cows. The characteristic time for the heart of chick embryos is about 50 days after the first heart beat. These numbers do not apply directly to volumetric growth. However, analysis of the growth of pressurized cylinders with this model has indicated that the characteristic time of growth is still given by an expression similar to (22.5.47).

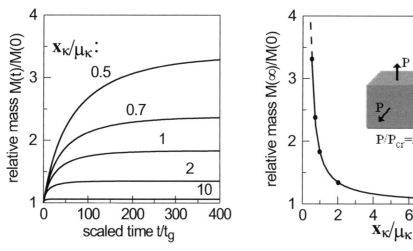

Fig. 22.5.6: Elastic-growing tissue subjected to volumetric growth. Standard parameters as given in Section 22.5.6.2, except the amplitude x_κ of the fourth order tensor $\mathbb{X}_\kappa$. Time profiles of aggregated mass, or volume, for several values of x_κ, and value at large times of the aggregated mass as a function of the kinematic coefficient x_κ. The larger this coefficient, the faster an equilibrium is reached and the less the aggregated mass. Reprinted with permission from Loret and Simões [2010]c.

22.5.6.4 The homeostatic surface is not mobile but has variable size

The coefficient tensor $\mathbb{X}_\kappa$ is now set to vanish so that, even if the tensor structure variable $\boldsymbol{\xi}_\kappa$ evolves in time, the associated center of the homeostatic surface $\mathbf{X}_\kappa = \mathbb{X}_\kappa : \boldsymbol{\xi}_\kappa$ remains equal to $\mathbf{0}$. On the other hand, the radius of the homeostatic surface varies in time. The differential equation governing its evolution simplifies to,

$$d\left(\frac{X_\kappa}{\mu_\kappa}\right) = \frac{\mathcal{X}_\kappa}{\mu_\kappa}\left(\frac{p_\kappa^2}{\mu_\kappa^2} - \frac{X_\kappa^2}{\mu_\kappa^2}\right)\left(a_2 - a_3\,\frac{X_\kappa}{\mu_\kappa}\right)\frac{X_\kappa}{\mu_\kappa}\,\frac{dt}{t_{\mathrm{g}}}\,. \qquad (22.5.54)$$

The stress being fixed, the differential equation integrates for $a_3 = 0$ to

$$1 - \frac{p_\kappa^2}{X_\kappa^2(t)} = \left(1 - \frac{p_\kappa^2}{X_\kappa^2(0)}\right) \exp\left(-\frac{t}{T_g}\right). \tag{22.5.55}$$

An analytical expression of the growth transformation deduces by comparing $(22.5.39)_2$ and $(22.5.43)$,

$$\mathrm{Ln}\, \frac{F_1^g(t)}{F_1^g(0)} = \frac{2}{3\,a_2} \frac{p_\kappa}{X_\kappa} \mathrm{Ln}\, \frac{X_\kappa(t)}{X_\kappa(0)}. \tag{22.5.56}$$

Like in the previous section, the condition of growth at time $t = 0^+$ is seen to imply continuous growth at any later time $t > 0$. The actual relaxation time,

$$T_g = \frac{1}{2\,a_2} \frac{\mu_\kappa}{X_\kappa} \frac{\mu_\kappa^2}{p_\kappa^2}\, t_g, \tag{22.5.57}$$

is inversely proportional to the coefficients X_κ/μ_κ and a_2. This feature is expected: if either of them vanishes, the radius of the homeostatic surface is fixed, and growth can not stop. The radius varies so that, once again, the homeostatic surface gets ultimately in contact with the given stress point.

22.6 Modes of mass deposition and mechanical response

With respect to the standard hyperelasticity of solids, two main types of difficulties have to be faced in the elastic-growing framework:

- the state variables are a priori defined in the intermediate configurations whose existence is implied by the multiplicative decomposition of the total deformation gradient into an elastic contribution and a growth contribution. The stress and other state variables are, in a second step, transported in the current configuration, according to transport laws to be defined by constitutive relations;
- the mixture context requires a detailed treatment. All constituents of the mixture may not grow at the same pace, and, for some of them, like ions, the growth process is irrelevant. Thus, for each constituent, the growth process and the contribution to the total stress should be defined by constitutive equations.

Constitutive equations for elastic-growing tissues have so far made use of the free energy per reference volume of the existing material, as if growth had no direct effect on this entity, e.g., Lubarda and Hoger [2002]. For thin film deposition, Jabbour and Bhattacharya [2003] proceed similarly and use the free energy per current volume. The issue here precisely consists in scrutinizing how the growth process may influence the free energy. In essence, the question addressed is the following:

How do the mechanical state of the deposited material, at the time of deposition, *and* the time course of growth affect the free energy, and consequently the material response?

22.6.1 The mechanical state of the newly deposited mass

In fact, the local (=pointwise) free energy of a growing solid depends on the mode according to which matter is deposited, or aggregated. Perhaps the two simplest modes that can be imagined are the following:

(a) the mass is deposited in a natural state;

(b) at the time of deposition, the deposited mass is endowed with the same mechanical properties as the existing material at the deposition location, e.g., with the current deformation and current state variables, and in particular with the same current free energy.

The latter mode simply produces the free energy per reference volume as if the solid was not growing. By contrast, the former mode involves the history of deformation and mass deposition in integral form.

Cowin [2004] presents *pro domo* qualitative arguments in favor of mode (b), that he used in his earlier model of bone remodeling exposed in Cowin and Hegedus [1976]. His arguments rest on a two time scale analysis. Indeed, examining the possible roles of strain and strain rates as stimuli of growth, Cowin [1996] highlights the fact that mechanical loadings and biological processes correspond to largely different time scales, say, the second and the couple of weeks $\sim 10^6$ seconds, respectively.

Qualitatively, one may consider the two above growth processes in the framework of thermodynamically open systems. In deposition mode (a), the energy supplied by the surroundings is diverted to the growing conglomerate.

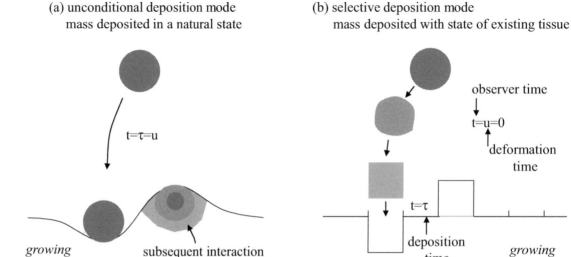

Fig. 22.6.1: Symbolic representation of two modes of deposition ($\equiv$ aggregation) of mass. Under mode (a), the mass is aggregated to the growing tissue as it is, and, once aggregated, it deforms and interacts with the existing tissue. Under mode (b), the mass has to acquire the properties dictated by the growing tissue to be able to aggregate. Thus growth under mode (a) is characterized by "unconditional deposition/interactive growth" while growth under mode (b) involves "selective deposition".

Alternatively, for mode (b), the energy supplied may be seen as diverted essentially to the supplied mass before its deposition: it aggregates when its free energy is approximately identical to that of the already existing material. The latter point of view introduces a growth process in two steps. Mode (b) seems to be accepted in most studies by convenience, e.g. Klisch et al. [2001]b, Taber and Chabert [2002]. We are aware of a single work, namely that of Drozdov and Khanina [1997], where growth is assumed to occur along mode (a). It addresses elastic and viscoelastic solids, with fixed mass density.

The conditional deposition associated with mode (b) may be compared with the optimization principles suggested by morphogeneticists: indeed, along the idea first formulated by Murray [1926], material growth takes place under optimal conditions. Still, a cost function remains to be defined:

- for example, Taber [1998]c applies the idea to vascular morphology: the blood vessel diameter is such that the energy needed to drive the blood flow has to be minimum;
- for arterial growth, Rachev et al. [1998] assume that re-structuration takes place so as to keep the vascular compliance constant;
- according to Lieberman et al. [2003], Haversian remodeling in cortical bone aims at maximizing strength over density.

An accurate description of the growth processes would require to address in detail the associated chemical reactions, energetic aspects and characteristic times. The time course of the energy of the deposited mass would emerge as an outcome of this analysis: rephrased along this axiomatic point of view, mode (b) would result from a principle that states that aggregation is possible only if the free energy of the deposited mass is equal to the free energy of the existing material at the deposition site.

Here, we are content to consider the two modes of mass deposition (a) and (b) sketched in Figs. 22.6.1 and 22.6.2. Three instants are involved in the processes:

- the time t of an observer;
- the time τ at which the mass is deposited;
- the time u at which the deposited mass begins to be deformed.

These two modes are certainly extreme, in the sense that actual growth modes are expected to follow some intermediate pathway. Moreover, two constituents of a mixture susceptible to grow may well evolve along distinct paths.

The analysis below simply explores the range of variations in material response that these two modes are able to induce.

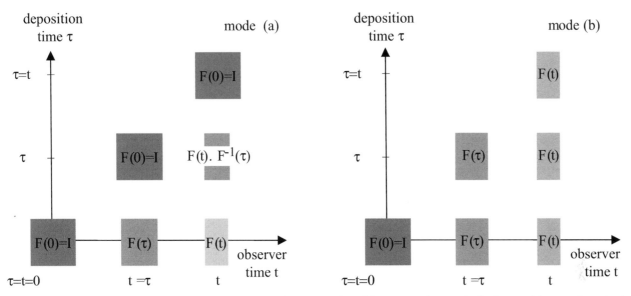

Fig. 22.6.2: Sketch of deformation history in the two distinct modes of deposition ($\equiv$creation) of mass. (a) mass deposited in a natural state: the deformation gradient $\mathbf{F}$ is the identity at the time of deposition. Then the time u at which the deformation of the deposited mass starts is the deposition time τ, namely $u = \tau$ for mode (a); (b) mass deposited with the properties of the existing material, that is, with the current state variables, and in particular current deformation, at the deposition location. Thus the deformation of the deposited material begins at time $u = 0$.

The free energies of solids and mixtures are built separately. The analysis of Section 23.2 addresses mixtures where the growing constituents are all elastic-growing. Even if the case of solids could be derived from the mixture analysis, it has been treated separately because, first, it is simpler, and, second, it is of interest in itself, Section 22.6.2. Once the free energy has been built, the constitutive equations for the generalized stresses may be derived. The approach capitalizes upon the thermodynamic analysis of Section 22.5, where the growth rate laws for the growth transformation and structure variables are motivated by, and satisfy strictly, the dissipation inequality. Still, the analysis in Section 22.5 was restricted to deposition mode (b). The quantitative influence of these deposition modes can be next appreciated via simple loading contexts in Sections 22.8 and 22.9.

22.6.2 Free energy of growth for an elastic-growing solid

The free energy of growth per intermediate volume $E_\kappa^\natural$, and per reference volume $\underline{E}_\kappa^\natural = E_\kappa^\natural \det \mathbf{F}^{\mathrm{g}}$, are built from the background free energy per intermediate volume E_κ.

The convention that $\mathbf{F}(t=0)=\mathbf{I}$, $\mathbf{F}^{\mathrm{e}}(t=0)=\mathbf{I}$, $\mathbf{F}^{\mathrm{g}}(t=0)=\mathbf{I}$ is adopted. The scalar and tensor structure variables are similarly assumed to vanish at $t = 0$, namely $\xi_\kappa(0) = 0$, $\boldsymbol{\xi}_\kappa(0) = \mathbf{0}$. With help of the sketches of Fig. 22.2.1, volume and free energy relative to the intermediate configurations are defined, first via transports from the reference configuration,

$$\underline{V}_\kappa = \underline{V} = V_\kappa(t) \left(\det \mathbf{F}^{\mathrm{g}}(t)\right)^{-1}, \quad \underline{E}_\kappa(t) = \underline{E}(t) = E(t) \det \mathbf{F}(t) = E_\kappa(t) \det \mathbf{F}^{\mathrm{g}}(t), \quad (22.6.1)$$

and, next via transports from the current configuration,

$$V_\kappa(t) = V(t) \left(\det \mathbf{F}^{\mathrm{e}}(t)\right)^{-1}, \quad E_\kappa(t) = E(t) \det \mathbf{F}^{\mathrm{e}}(t). \quad (22.6.2)$$

The free energy $E_\kappa^\natural(t)\, V_\kappa(t)$ at time t of the volume $V_\kappa(t)$ is equal to the free energy at time t of the reference volume $V_\kappa(0) = \underline{V}_\kappa$, plus the free energy of the progressively growing volume $dV_\kappa(\tau)$, $\tau \in [0, t]$,

$$E_\kappa^\natural(t)\, V_\kappa(t) = \underline{E}_\kappa^\natural(t)\, \underline{V}_\kappa = E_\kappa(t)\, \underline{V}_\kappa + \int_0^t E_\kappa(t, u)\, dV_\kappa(\tau), \quad (22.6.3)$$

where $E_\kappa(t) = E_\kappa(\mathcal{V}_\kappa(t))$, and $E_\kappa(t, u) = E_\kappa(\mathcal{V}_\kappa(u \to t))$ denotes the time dependence of the state variables from u to t,

$$\mathcal{V}_\kappa(u \to t) = \{ \mathbf{E}_G^e(t, u),\, \xi_\kappa(t - u),\, \boldsymbol{\xi}_\kappa(t - u) \} . \tag{22.6.4}$$

Here $\mathbf{E}_G^e(t, u)$ is the extended Green strain,

$$
\begin{aligned}
\mathbf{E}_G^e(t, u) &= \left(\mathbf{F}^e(u) \right)^{-T} \cdot \left(\mathbf{E}_G^e(t) - \mathbf{E}_G^e(u) \right) \cdot \left(\mathbf{F}^e(u) \right)^{-1} \\
&= \left(\mathbf{F}^e(u) \right)^{-T} \cdot \mathbf{E}_G^e(t) \cdot \left(\mathbf{F}^e(u) \right)^{-1} + \tfrac{1}{2} \left(\left(\mathbf{F}^e(u) \right)^{-T} \cdot \left(\mathbf{F}^e(u) \right)^{-1} - \mathbf{I} \right),
\end{aligned}
\tag{22.6.5}
$$

associated with the transformation $\mathbf{F}^e(t, u)$,

$$\mathbf{F}^e(t, u) = \mathbf{F}^e(t) \cdot \left(\mathbf{F}^e(u) \right)^{-1} . \tag{22.6.6}$$

22.6.2.1   The solid is deposited in a natural configuration

Under deposition mode (a), the deposition time τ and the time u at which the deformation of the deposited material begins are identical, namely $u = \tau$. Dividing both sides of (22.6.3) by $\underline{V}_\kappa$, the free energy of growth $\underline{E}_\kappa^\natural(t)$ for the growing solid results in

$$
\begin{aligned}
\underline{E}_\kappa^\natural(t) &= E_\kappa(t) + \int_0^t E_\kappa(t, \tau) \frac{V_\kappa(\tau)}{\underline{V}_\kappa} \frac{dV_\kappa(\tau)}{V_\kappa(\tau)} , \\
&= E_\kappa(t) + \int_0^t E_\kappa(t, \tau) \det \mathbf{F}^g(\tau) \frac{dV_\kappa(\tau)}{V_\kappa(\tau)} ,
\end{aligned}
\tag{22.6.7}
$$

with help of the relation $(22.6.1)_1$, which also implies $dV_\kappa(\tau)/V_\kappa(\tau) = \operatorname{tr} \mathbf{L}_\kappa^g(\tau)\, d\tau$. Thus the free energy expresses in terms of the time course of the growth transformation,

$$\underline{E}_\kappa^\natural(t) = E_\kappa(t) + \int_0^t E_\kappa(t, \tau) \frac{d}{d\tau} \det \mathbf{F}^g(\tau)\, d\tau . \tag{22.6.8}$$

Particular case: density preserving growth

For density preserving growth, $\operatorname{tr} \mathbf{L}_\kappa^g$ is explicitly linked to the rate of mass supply, namely $\operatorname{tr} \mathbf{L}_\kappa^g = \hat\rho/\rho$. The free energy thus keeps track of the time course of mass deposition. As indicated in Section 21.3.1.2, the assumption of density preserving growth is equivalent to a constant density only if the material is elastically incompressible.

22.6.2.2   The solid is deposited with the existing properties at the deposition site

For deposition mode (b), if the volume $dV(\tau)$ is deposited at time τ, its free energy at time t is $E_\kappa(t, u = 0)\, dV(\tau)$, where $E_\kappa(t, u = 0)$ is the free energy associated with the elastic transformation $\mathbf{F}^e(t, u = 0) = \mathbf{F}^e(t) \cdot (\mathbf{F}^e)^{-1}(u = 0) = \mathbf{F}^e(t)$, and therefore $E_\kappa(t, u = 0) = E_\kappa(t)$.

The free energy $E_\kappa^\natural(t)\, V_\kappa(t)$ at time t in a volume $V_\kappa(t)$ takes now the form

$$
\begin{aligned}
E_\kappa^\natural(t)\, V_\kappa(t) = \underline{E}_\kappa^\natural(t)\, \underline{V}_\kappa &= E_\kappa(t)\, \underline{V}_\kappa + \int_0^t E_\kappa(t)\, dV_\kappa(\tau) \\
&= E_\kappa(t) \left(\underline{V}_\kappa + V_\kappa(t) - \underline{V}_\kappa \right) \\
&= E_\kappa(t)\, V_\kappa(t) ,
\end{aligned}
\tag{22.6.9}
$$

so that, as expected, the free energy of growth boils down to the free energy per current volume,

$$E_\kappa^\natural = E_\kappa \quad \Leftrightarrow \quad \underline{E}_\kappa^\natural = \underline{E}_\kappa = E_\kappa \det \mathbf{F}^g , \tag{22.6.10}$$

the last relation being obtained with help of $(22.6.1)_2$.

Particular case: volume preserving growth

Then $\underline{E}_\kappa^\natural(t) = E_\kappa(t)$ holds for both modes of deposition.

22.6.2.3 Unified constitutive equations for elastic-growing solids

The rate equations for the generalized strains being in place as stated in Sections 22.5.3 and 22.5.5, the constitutive equations for the generalized stresses can now be exposed. For both mode (a) and mode (b), the generalized stresses are expressed in the format,

$$\boldsymbol{\tau}'_\kappa = \frac{\partial E^\natural_\kappa}{\partial \mathbf{E}^{\mathrm{e}}_{\mathrm{G}}}, \quad X_\kappa = \frac{\partial E^\natural_\kappa}{\partial \xi_\kappa}, \quad \mathbf{X}_\kappa = \frac{\partial E^\natural_\kappa}{\partial \boldsymbol{\xi}_\kappa}. \tag{22.6.11}$$

For mode (b), the free energy $E^\natural_\kappa$ is the background free energy E_κ defined by (22.5.10),

$$\boldsymbol{\tau}'_\kappa = \frac{\partial E_\kappa}{\partial \mathbf{E}^{\mathrm{e}}_{\mathrm{G}}} = a_\kappa \mathbf{I} + 2\,\mu_\kappa \mathbf{E}^{\mathrm{e}}_{\mathrm{G}}, \quad X_\kappa = X_\kappa(0) + \mathcal{X}_\kappa\,\xi_\kappa, \quad \mathbf{X}_\kappa = \mathbf{X}_\kappa(0) + \mathbb{X}_\kappa : \boldsymbol{\xi}_\kappa. \tag{22.6.12}$$

For deposition mode (a), the stress is still given in the format (22.6.11) but the free energy $E^\natural_\kappa = \underline{E}^\natural_\kappa / \det \mathbf{F}^{\mathrm{g}}$ results from the free energy of growth per reference volume $\underline{E}^\natural_\kappa$,

$$
\begin{aligned}
\det \mathbf{F}^{\mathrm{g}}(t)\,\boldsymbol{\tau}'_\kappa(t) &= \frac{\partial E_\kappa(t)}{\partial \mathbf{E}^{\mathrm{e}}_{\mathrm{G}}(t)} + \int_0^t \frac{\partial E_\kappa(t,\tau)}{\partial \mathbf{E}^{\mathrm{e}}_{\mathrm{G}}(t)}\,\frac{d}{d\tau}\det \mathbf{F}^{\mathrm{g}}(\tau)\,d\tau, \\
\det \mathbf{F}^{\mathrm{g}}(t)\,X_\kappa(t) &= \frac{\partial E_\kappa(t)}{\partial \xi_\kappa(t)} + \int_0^t \frac{\partial E_\kappa(t,\tau)}{\partial \xi_\kappa(t)}\,\frac{d}{d\tau}\det \mathbf{F}^{\mathrm{g}}(\tau)\,d\tau, \\
\det \mathbf{F}^{\mathrm{g}}(t)\,\mathbf{X}_\kappa(t) &= \frac{\partial E_\kappa(t)}{\partial \boldsymbol{\xi}_\kappa(t)} + \int_0^t \frac{\partial E_\kappa(t,\tau)}{\partial \boldsymbol{\xi}_\kappa(t)}\,\frac{d}{d\tau}\det \mathbf{F}^{\mathrm{g}}(\tau)\,d\tau.
\end{aligned}
\tag{22.6.13}
$$

The extended free energy,

$$
\begin{aligned}
E_\kappa(t,\tau) &= a_\kappa \operatorname{tr} \mathbf{E}^{\mathrm{e}}_{\mathrm{G}}(t,\tau) + \mu_\kappa \operatorname{tr}\left(\mathbf{E}^{\mathrm{e}}_{\mathrm{G}}(t,\tau)\right)^2 + X_\kappa(0)\,\xi_\kappa(t,\tau) + \frac{1}{2}\,\xi_\kappa(t,\tau)\,\mathcal{X}_\kappa\,\xi_\kappa(t,\tau) \\
&\quad + \mathbf{X}_\kappa(0) : \boldsymbol{\xi}_\kappa(t,\tau) + \frac{1}{2}\,\boldsymbol{\xi}_\kappa(t,\tau) : \mathbb{X}_\kappa : \boldsymbol{\xi}_\kappa(t,\tau),
\end{aligned}
\tag{22.6.14}
$$

boils down to $E_\kappa(t)$ given by (22.5.10) for $\tau = 0$, that is for deposition mode (b), where the deformation process of the aggregated mass begins simultaneously with the existing mass. For

$$\xi_\kappa(t,\tau) = \xi_\kappa(t) - \xi_\kappa(\tau), \quad \boldsymbol{\xi}_\kappa(t,\tau) = \boldsymbol{\xi}_\kappa(t) - \boldsymbol{\xi}_\kappa(\tau), \tag{22.6.15}$$

result the derivatives,

$$
\begin{aligned}
\frac{\partial E_\kappa(t,\tau)}{\partial \mathbf{E}^{\mathrm{e}}_{\mathrm{G}}(t)} &= (\mathbf{F}^{\mathrm{e}})^{-1}(\tau) \cdot \left(a_\kappa \mathbf{I} + 2\,\mu_\kappa \mathbf{E}^{\mathrm{e}}_{\mathrm{G}}(t,\tau)\right) \cdot (\mathbf{F}^{\mathrm{e}})^{-\mathrm{T}}(\tau), \\
\frac{\partial E_\kappa(t,\tau)}{\partial \xi_\kappa(t)} &= X_\kappa(0) + \mathcal{X}_\kappa\,\xi_\kappa(t,\tau), \\
\frac{\partial E_\kappa(t,\tau)}{\partial \boldsymbol{\xi}_\kappa(t)} &= \mathbf{X}_\kappa(0) + \mathbb{X}_\kappa : \boldsymbol{\xi}_\kappa(t,\tau).
\end{aligned}
\tag{22.6.16}
$$

The expressions for the stress and the conjugate variables may be recast as convolution integrals,

$$
\begin{aligned}
\det \mathbf{F}^{\mathrm{g}}(t)\,\boldsymbol{\tau}'_\kappa(t) &= a_\kappa \mathbf{I} + 2\,\mu_\kappa \mathbf{E}^{\mathrm{e}}_{\mathrm{G}}(t) \\
&\quad + \int_0^t (\mathbf{F}^{\mathrm{e}})^{-1}(\tau) \cdot \left(a_\kappa \mathbf{I} + 2\,\mu_\kappa \mathbf{E}^{\mathrm{e}}_{\mathrm{G}}(t,\tau)\right) \cdot (\mathbf{F}^{\mathrm{e}})^{-\mathrm{T}}(\tau)\,\frac{d}{d\tau}\det \mathbf{F}^{\mathrm{g}}(\tau)\,d\tau, \\
\det \mathbf{F}^{\mathrm{g}}(t)\left(X_\kappa(t) - X_\kappa(0)\right) &= \mathcal{X}_\kappa\,\xi_\kappa(t) + \int_0^t \mathcal{X}_\kappa\,\xi_\kappa(t,\tau)\,\frac{d}{d\tau}\det \mathbf{F}^{\mathrm{g}}(\tau)\,d\tau, \\
\det \mathbf{F}^{\mathrm{g}}(t)\left(\mathbf{X}_\kappa(t) - \mathbf{X}_\kappa(0)\right) &= \mathbb{X}_\kappa : \boldsymbol{\xi}_\kappa(t) + \int_0^t \mathbb{X}_\kappa : \boldsymbol{\xi}_\kappa(t,\tau)\,\frac{d}{d\tau}\det \mathbf{F}^{\mathrm{g}}(\tau)\,d\tau.
\end{aligned}
\tag{22.6.17}
$$

Therefore, if the generalized strains are constant in time, the generalized stresses are constant as well and, as expected, they retrieve exactly the same expressions as for deposition mode (b).

The above relations (22.6.16) and (22.6.17) for the structure variables display convolution integrals. Besides, as just mentioned, they pass the two criteria:

Criterion 1: replacing the argument $t - \tau$ of the structure variables by $t - 0 = t$ as if the mass was beginning to be deformed at time $\tau = 0$, the constitutive relations for deformation mode (b) are recovered;

Criterion 2: if the structure variables are constant during the deformation process, then so are their conjugate variables, again like in deformation mode (b).

22.7 Constitutive equations for purely elastic solids

A number of constitutive models of growth have been proposed before the multiplicative decomposition of the deformation gradient was formalized. As an example, in a pioneer article on the remodeling of bones, Cowin and Hegedus [1976] make use of deposition mode (b) above. We shall distinguish
   - "elastic" or "viscoelastic" tissues whose growth is characterized by a rate of mass supply only, and
   - "elastic-growing tissues" whose mechanical behavior includes both a rate of mass supply and a multiplicative decomposition of the deformation into an elastic part and a growth part, as proposed by Rodriguez et al. [1994].

   The point here is to present
   - a formulation of the constitutive equations of growth for elastic solids;
   - a qualitative comparison between the formulations of the constitutive equations of growth of elastic-growing solids and elastic solids;
   - a qualitative comparison between the formulations of boundary value problems for these two growth models;
   - a quantitative comparison between the mechanical responses of the two models to simple loading paths.

   The free energy and growth rate law of elastic solids are introduced in a format close to that developed for elastic-growing solids, so as to facilitate comparison between the formulations and the quantitative mechanical responses to the boundary value problems described in the next sections. In other words, while the constitutive equations for the elastic-growing solids are motivated by the dissipation inequality, no such a goal is pursued in building the constitutive equations for the elastic solids. Rather, we proceed by similarity. Nevertheless, the underlying frameworks are quite distinct and it is therefore necessary to inspect in detail all the steps of the construction.

   Rate constitutive equations are needed for the generalized strains that contribute to define the thermodynamic state of the tissue. These rate equations are identical for deposition modes (a) and (b). Like for elastic-growing solids in the previous section, the two modes differ in the constitutive equations for the generalized stresses.

22.7.1 The deposition modes and their free energies

The constitutive equations are phrased in terms of a work-conjugate pair, namely the 2nd Piola-Kirchhoff stress $\underline{\tau}$ relative to the reference configuration where the deformation gradient $\mathbf{F}$ is equal to the identity, and the Green strain $\mathbf{E}_{\mathrm{G}} = \frac{1}{2}(\mathbf{C} - \mathbf{I})$ or the right Cauchy-Green tensor $\mathbf{C} = \mathbf{F}^{\mathrm{T}} \cdot \mathbf{F}$. The tissue will be assumed to be elastically incompressible:

$$\det \mathbf{F} : \begin{cases} = 1 & \text{no growth}; \\ \geqslant 1 & \text{growth/resorption}. \end{cases} \tag{22.7.1}$$

   The set of state variables includes three generalized strains, namely the Green strain $\mathbf{E}_{\mathrm{G}}$, and two structure variables, namely a scalar $\underline{\xi}$ and a symmetric second order tensor $\underline{\underline{\xi}}$, all of them vanishing at time $t = 0$,

$$\mathcal{V} = \{\mathbf{E}_{\mathrm{G}}, \underline{\xi}, \underline{\underline{\xi}}\}. \tag{22.7.2}$$

The constitutive equations provide the generalized stresses,

$$\underline{\underline{\tau}}' = \frac{\partial \underline{E}^{\natural}}{\partial \mathbf{E}_{\mathrm{G}}}, \quad \underline{X} = \frac{\partial \underline{E}^{\natural}}{\partial \underline{\xi}}, \quad \underline{\underline{X}} = \frac{\partial \underline{E}^{\natural}}{\partial \underline{\underline{\xi}}}. \tag{22.7.3}$$

The free energy $\underline{E}^{\natural}$ in (22.7.3) is referred to as the "free energy of growth" per reference volume. It is related to the free energy of growth per current volume $E^{\natural}$ through the standard relation $\underline{E}^{\natural} = E^{\natural} \det \mathbf{F}$.

   These generalized stresses, which are relative to the reference configuration, transport in the current configuration according to the rules,

$$\boldsymbol{\tau}' = \det \mathbf{F}\, \boldsymbol{\sigma}' = \mathbf{F} \cdot \underline{\underline{\tau}}' \cdot \mathbf{F}^{\mathrm{T}}, \quad X = \underline{X}, \quad \mathbf{X} = \mathbf{F} \cdot \underline{\underline{X}} \cdot \mathbf{F}^{\mathrm{T}}. \tag{22.7.4}$$

A physical interpretation of the work-conjugate variables $(\underline{\xi}, \underline{X})$ and $(\underline{\underline{\xi}}, \underline{\underline{X}})$ emerges once the homeostatic surface is introduced. The superscript prime denotes an effective stress. Actually, the formulation applies to both compressible and incompressible tissues with the following convention, namely for compressible tissues,

$$\underline{\underline{\tau}}' = \underline{\underline{\tau}}; \quad \boldsymbol{\tau}' = \boldsymbol{\tau}; \quad \boldsymbol{\sigma}' = \boldsymbol{\sigma}, \tag{22.7.5}$$

and, for incompressible tissues ($\det \mathbf{F} = 1$),

$$\underline{\underline{\tau}}' = \underline{\underline{\tau}} + p_I \det \mathbf{F}\,\mathbf{C}^{-1}; \quad \tau' = \tau + p_I \det \mathbf{F}\,\mathbf{I}; \quad \sigma' = \sigma + p_I\,\mathbf{I}, \tag{22.7.6}$$

where p_I is the incompressibility pressure.

The mass density is constant, so that, with help of (21.3.58), the volume change is directly linked to the mass,

$$\det \mathbf{F}(t) = \frac{V(t)}{V} = \frac{M(t)}{M} = \exp\left(\int_0^t \frac{\hat{\rho}}{\rho}(\tau)\,d\tau\right). \tag{22.7.7}$$

Then

$$\frac{d}{d\tau}\det \mathbf{F}(\tau) = \det \mathbf{F}(\tau)\,\operatorname{tr}\mathbf{L}(\tau) = \det \mathbf{F}(\tau)\,\frac{\hat{\rho}}{\rho}(\tau). \tag{22.7.8}$$

Consequently, if the solid is deposited in an unstrained configuration, i.e., under mode (a), then, the free energy of growth expresses in terms of the time course of the velocity gradient, or of the history of the rate of mass supply,

$$
\begin{aligned}
\underline{E}^{\natural}(t) &= E(t) + \int_0^t E(t,\tau)\,\frac{V(\tau)}{V}\frac{dV(\tau)}{V(\tau)} \\
&= E(t) + \int_0^t E(t,\tau)\,\frac{d}{d\tau}\det \mathbf{F}(\tau)\,d\tau\,.
\end{aligned}
\tag{22.7.9}
$$

Here $E(t,\tau) = E(\mathcal{V}(\tau \to t))$ denotes the time dependence of the set of state variables $\mathcal{V}$ from τ to t, associated with the deformation gradient $\mathbf{F}(t,\tau) = \mathbf{F}(t)\cdot\mathbf{F}^{-1}(\tau)$ and structure variables $\underline{\xi}(t,\tau)$ and $\underline{\boldsymbol{\xi}}(t,\tau)$:

$$\mathcal{V}(\tau \to t) = \{\mathbf{E}_{\mathrm{G}}(t,\tau),\ \underline{\xi}(t,\tau),\ \underline{\boldsymbol{\xi}}(t,\tau)\}, \tag{22.7.10}$$

where $\mathbf{E}_{\mathrm{G}}(t,\tau)$ is the extended Green strain associated with the transformation $\mathbf{F}(t,\tau)$,

$$
\begin{aligned}
\mathbf{E}_{\mathrm{G}}(t,\tau) &= \mathbf{F}^{-\mathrm{T}}(\tau)\cdot(\mathbf{E}_{\mathrm{G}}(t) - \mathbf{E}_{\mathrm{G}}(\tau))\cdot\mathbf{F}^{-1}(\tau) \\
&= \mathbf{F}^{-\mathrm{T}}(\tau)\cdot\mathbf{E}_{\mathrm{G}}(t)\cdot\mathbf{F}^{-1}(\tau) + \tfrac{1}{2}\left(\mathbf{F}^{-\mathrm{T}}(\tau)\cdot\mathbf{F}^{-1}(\tau) - \mathbf{I}\right); \\
\underline{\xi}(t,\tau) &= \underline{\xi}(t) - \underline{\xi}(\tau); \\
\underline{\boldsymbol{\xi}}(t,\tau) &= \underline{\boldsymbol{\xi}}(t) - \underline{\boldsymbol{\xi}}(\tau).
\end{aligned}
\tag{22.7.11}
$$

On the other hand, if the tissue is deposited with the properties of the tissue that already exists at the deposition site, i.e., deposition mode (b), then, since aggregation of matter is elusive in terms of free energy, the free energy of growth boils down to the free energy per reference volume $\underline{E}$,

$$\underline{E}^{\natural} = \underline{E} = E \det \mathbf{F}\,. \tag{22.7.12}$$

22.7.2 The constitutive equations for the generalized stresses for mode (a)

The generalized stresses (22.7.3) express now as,

$$
\begin{aligned}
\underline{\underline{\tau}}'(t) &= \frac{\partial E(t)}{\partial \mathbf{E}_{\mathrm{G}}(t)} + \int_0^t \frac{\partial E(t,\tau)}{\partial \mathbf{E}_{\mathrm{G}}(t)}\,\frac{d}{d\tau}\det \mathbf{F}(\tau)\,d\tau, \\
\underline{X}(t) &= \frac{\partial E(t)}{\partial \underline{\xi}(t)} + \int_0^t \frac{\partial E(t,\tau)}{\partial \underline{\xi}(t)}\,\frac{d}{d\tau}\det \mathbf{F}(\tau)\,d\tau, \\
\underline{\underline{X}}(t) &= \frac{\partial E(t)}{\partial \underline{\boldsymbol{\xi}}(t)} + \int_0^t \frac{\partial E(t,\tau)}{\partial \underline{\boldsymbol{\xi}}(t)}\,\frac{d}{d\tau}\det \mathbf{F}(\tau)\,d\tau\,.
\end{aligned}
\tag{22.7.13}
$$

The free energy per reference volume $\underline{E} = \det \mathbf{F}\,E$ postulated in the quadratic form,

$$\underline{E} = a\,\operatorname{tr}\mathbf{E}_{\mathrm{G}} + \mu\,\operatorname{tr}\,(\mathbf{E}_{\mathrm{G}})^2 + \underline{X}_0\,\underline{\xi} + \frac{1}{2}\,\underline{\xi}\,\mathcal{X}\,\underline{\xi} + \underline{\underline{X}}_0 : \underline{\boldsymbol{\xi}} + \frac{1}{2}\,\underline{\boldsymbol{\xi}} : \mathcal{X} : \underline{\boldsymbol{\xi}}, \tag{22.7.14}$$

involves two constants a and $\mu > 0$ associated with an elastic potential defined by a linear term and a quadratic term, as well as two constants associated with the structure variables, namely a scalar $\mathcal{X}$ associated with the scalar variable ξ and a symmetric fourth order tensor $\mathbb{X}$ associated with the second order tensor $\underline{\boldsymbol{\xi}}$. Omens et al. [2003] use such a linear term in the elastic potential of an elastic aorta, so as to obtain a non vanishing residual stress in the unstrained configuration.

Note the relation

$$\frac{\partial E(t,\tau)}{\partial \mathbf{E}_{\mathrm{G}}(t)} \;=\; \mathbf{F}^{-1}(\tau) \cdot \frac{\partial E(t,\tau)}{\partial \mathbf{E}_{\mathrm{G}}(t,\tau)} \cdot \mathbf{F}^{-\mathrm{T}}(\tau) . \tag{22.7.15}$$

For the free energy given by (22.7.14), the integrands in (22.7.13) have the explicit expression,

$$\det \mathbf{F}(t) \frac{\partial E(t,\tau)}{\partial \mathbf{E}_{\mathrm{G}}(t)} \;=\; \mathbf{F}^{-1}(\tau) \cdot \big(a\,\mathbf{I} + 2\,\mu\,\mathbf{E}_{\mathrm{G}}(t,\tau)\big) \cdot \mathbf{F}^{-\mathrm{T}}(\tau),$$

$$\det \mathbf{F}(t) \frac{\partial E(t,\tau)}{\partial \underline{\xi}(t)} \;=\; \underline{X}_0 + \mathcal{X}\,\underline{\xi}(t,\tau), \tag{22.7.16}$$

$$\det \mathbf{F}(t) \frac{\partial E(t,\tau)}{\partial \underline{\underline{\boldsymbol{\xi}}}(t)} \;=\; \underline{\underline{\mathbf{X}}}_0 + \mathbb{X} : \underline{\underline{\boldsymbol{\xi}}}(t,\tau) .$$

The expressions for the conjugate variables may be given a more explicit form, namely

$$\det \mathbf{F}(t)\,\underline{\boldsymbol{\tau}}'(t) \quad\;=\; a\,\mathbf{I} + 2\,\mu\,\mathbf{E}_{\mathrm{G}}(t) \;+\; \int_0^t \mathbf{F}^{-1}(\tau) \cdot \big(a\,\mathbf{I} + 2\,\mu\,\mathbf{E}_{\mathrm{G}}(t,\tau)\big) \cdot \mathbf{F}^{-\mathrm{T}}(\tau) \frac{d}{d\tau}\det \mathbf{F}(\tau)\,d\tau,$$

$$\det \mathbf{F}(t)\,\big(\underline{X}(t) - \underline{X}_0\big) \;=\; \mathcal{X}\,\underline{\xi}(t) \;+\; \int_0^t \mathcal{X}\,\underline{\xi}(t,\tau) \frac{d}{d\tau}\det \mathbf{F}(\tau)\,d\tau, \tag{22.7.17}$$

$$\det \mathbf{F}(t)\,\big(\underline{\underline{\mathbf{X}}}(t) - \underline{\underline{\mathbf{X}}}_0\big) \;=\; \mathbb{X} : \underline{\underline{\boldsymbol{\xi}}}(t) \;+\; \int_0^t \mathbb{X} : \underline{\underline{\boldsymbol{\xi}}}(t,\tau) \frac{d}{d\tau}\det \mathbf{F}(\tau)\,d\tau .$$

Note that, when deriving the effective stress above, $\det \mathbf{F}$ is not differentiated: either there is no growth and then $\det \mathbf{F} = 1$ continuously or growth takes place and then $\det \mathbf{F}$ is a function of the history of the state variables. Indeed, the rate of $\det \mathbf{F}$ is provided by a growth rate law as indicated by (22.7.8), and detailed in Section 22.7.4.

22.7.3   The constitutive equations for the generalized stresses for mode (b)

The background tissue is considered to be an isotropic elastic solid whose free energy per reference volume $\underline{E}$ is given by (22.7.14). The generalized stresses deduce as,

$$\underline{\boldsymbol{\tau}}' = \frac{\partial E}{\partial \mathbf{E}_{\mathrm{G}}} = a\,\mathbf{I} + 2\,\mu\,\mathbf{E}_{\mathrm{G}}, \quad \underline{X} = \frac{\partial E}{\partial \xi} = \underline{X}_0 + \mathcal{X}\,\underline{\xi}, \quad \underline{\underline{\mathbf{X}}} = \frac{\partial E}{\partial \underline{\underline{\boldsymbol{\xi}}}} = \underline{\underline{\mathbf{X}}}_0 + \mathbb{X} : \underline{\underline{\boldsymbol{\xi}}} . \tag{22.7.18}$$

In fact, setting $\tau = 0$ in the integrands of (22.7.17) yields these expressions (22.7.18). This result is expected, since deposition mode (b) is associated with a deformation process that starts together with the one of the existing tissue.

22.7.4   The growth rate law for the mass

A constitutive equation is needed for the rate of mass supply, say in the format,

$$\frac{\hat{\rho}}{\rho} = \mathcal{L}(\underline{\mathcal{V}}) . \tag{22.7.19}$$

The set of homeostatic states $h(\boldsymbol{\sigma} - \mathbf{Y}, Y) < 0$ refers to the generalized stresses in the current configuration,

$$(\boldsymbol{\sigma}, \mathbf{Y}, Y) = \frac{1}{\det \mathbf{F}} \,(\boldsymbol{\tau}, \mathbf{X}, X) , \tag{22.7.20}$$

and expresses in terms of the shifted Cauchy stress $\boldsymbol{\sigma} - \mathbf{Y} = p_R\,\mathbf{I} + \mathbf{s}_R$ via the two scalar invariants p_R and q_R,

$$p_R = \frac{1}{3}\,\mathrm{tr}\,(\boldsymbol{\sigma} - \mathbf{Y}) , \quad q_R^2 = \frac{3}{2}\,\mathbf{s}_R : \mathbf{s}_R . \tag{22.7.21}$$

The explicit form of the homeostatic function mimics (22.5.25),

$$\mu^2 \, h(\boldsymbol{\sigma} - \mathbf{Y}, Y) = \frac{q_R^2}{M^2} + p_R^2 - Y^2 \, . \tag{22.7.22}$$

The growth rate law is provided in a stress-driven format somehow similar to the growth rate law developed for elastic-growing solids,

$$\frac{\hat{\rho}}{\rho} = \frac{\mu}{t_{\mathrm{g}}} \, \Phi(h) \, \mathrm{tr} \, \frac{\partial h}{\partial \boldsymbol{\sigma}} \, . \tag{22.7.23}$$

22.7.5 Rate equations for the structure variables

Two forms of the rate equations for the structure variables are explored. One form uses the Green strain while the other uses the logarithmic strain.

22.7.5.1 Rate equations for the structure variables using the Green strain

Mimicking (22.5.39), the rates of the structure variables assume the format,

$$
\begin{aligned}
\frac{d\underline{\underline{\boldsymbol{\xi}}}}{dt} &= \frac{d\mathbf{E}_{\mathrm{G}}}{dt} - a_1 \, \Big\| \frac{d\mathbf{E}_{\mathrm{G}}}{dt} \Big\| \, \frac{\underline{\underline{\mathbf{X}}}}{\mu} \, , \\
\frac{d\xi}{dt} &= \frac{1}{t_{\mathrm{g}}} \, \Phi(h) \, \Big(a_2 - a_3 \, \frac{\underline{\underline{X}}}{\mu} \Big) \, \frac{X}{\mu} \, .
\end{aligned} \tag{22.7.24}
$$

The generalized strains follow transport rules akin their conjugate variables (actually, the current generalized strains are not needed),

$$\mathbf{D} = \mathbf{F}^{-\mathrm{T}} \cdot \frac{d\mathbf{E}_{\mathrm{G}}}{dt} \cdot \mathbf{F}^{-1}, \quad \xi = \underline{\xi}, \quad \boldsymbol{\xi} = \mathbf{F}^{-\mathrm{T}} \cdot \underline{\underline{\boldsymbol{\xi}}} \cdot \mathbf{F}^{-1} \, . \tag{22.7.25}$$

With the transport relations (22.7.25) and with $\mathbf{C} = \mathbf{F}^{\mathrm{T}} \cdot \mathbf{F}$ and $\mathbf{B} = \mathbf{F} \cdot \mathbf{F}^{\mathrm{T}}$ the right and left Cauchy-Green tensors, respectively, then, if $a_1 = 0$, the Green strain is transported to the covariant Almansi strain,

$$\underline{\boldsymbol{\xi}} = \mathbf{E}_{\mathrm{G}} = \frac{1}{2}(\mathbf{C} - \mathbf{I}), \quad \boldsymbol{\xi} = \mathbf{E}_{\mathrm{B}} = \frac{1}{2}(\mathbf{I} - \mathbf{B}^{-1}) \, . \tag{22.7.26}$$

22.7.5.2 Rate equations for the structure variables using the logarithmic strain

It is instrumental to introduce first the standard polar decomposition $\mathbf{F} = \mathbf{R} \cdot \mathbf{U} = \mathbf{V} \cdot \mathbf{R}$ of the gradient $\mathbf{F}$ in terms of the right stretch $\mathbf{U}$, the left stretch $\mathbf{V}$ and the rotation $\mathbf{R}$, Section 2.5.1.

Again mimicking (22.5.39), the rates of the structure variables assume the format,

$$
\begin{aligned}
\frac{d\underline{\underline{\boldsymbol{\xi}}}}{dt} &= \frac{d\mathrm{Ln}\,\mathbf{U}}{dt} - a_1 \, \Big\| \frac{d\mathrm{Ln}\,\mathbf{U}}{dt} \Big\| \, \frac{\underline{\underline{\mathbf{X}}}}{\mu} \, , \\
\frac{d\underline{\xi}}{dt} &= \frac{1}{t_{\mathrm{g}}} \, \Phi(h) \, \Big(a_2 - a_3 \, \frac{\underline{X}}{\mu} \Big) \, \frac{X}{\mu} \, .
\end{aligned} \tag{22.7.27}
$$

The transports between the reference and current configurations need to be postulated, namely for the generalized stresses,

$$\boldsymbol{\tau}' = \mathbf{R} \cdot \underline{\boldsymbol{\tau}}' \cdot \mathbf{R}^{\mathrm{T}}, \quad X = \underline{X}, \quad \mathbf{X} = \mathbf{R} \cdot \underline{\underline{\mathbf{X}}} \cdot \mathbf{R}^{\mathrm{T}} \, , \tag{22.7.28}$$

and for the generalized strains,

$$\mathrm{Ln}\,\mathbf{V} = \mathbf{R} \cdot \mathrm{Ln}\,\mathbf{U} \cdot \mathbf{R}^{\mathrm{T}}, \quad \xi = \underline{\xi}, \quad \boldsymbol{\xi} = \mathbf{R} \cdot \underline{\underline{\boldsymbol{\xi}}} \cdot \mathbf{R}^{\mathrm{T}} \, . \tag{22.7.29}$$

The rate equations (22.7.27) may be expressed in the current configuration using the transport relations (22.7.28) and (22.7.29),

$$
\begin{aligned}
\frac{D\underline{\underline{\boldsymbol{\xi}}}}{Dt} &= \frac{D\mathrm{Ln}\,\mathbf{V}}{Dt} - a_1 \, \Big\| \frac{D\mathrm{Ln}\,\mathbf{V}}{Dt} \Big\| \, \frac{\mathbf{X}}{\mu} \, , \\
\frac{d\xi}{dt} &= \frac{1}{t_{\mathrm{g}}} \, \Phi(h) \, \Big(a_2 - a_3 \, \frac{X}{\mu} \Big) \, \frac{X}{\mu} \, ,
\end{aligned} \tag{22.7.30}
$$

with $D(\cdot)/Dt$ the corotational derivative based on the spin $\boldsymbol{\Omega} = d\mathbf{R}/dt \cdot \mathbf{R}^{-1}$,

$$\frac{D(\cdot)}{Dt} = \frac{d(\cdot)}{dt} + (\cdot)\cdot\boldsymbol{\Omega} - \boldsymbol{\Omega}\cdot(\cdot). \tag{22.7.31}$$

Note the relations,

$$\mathrm{Ln}\,\mathbf{U} = \mathbf{R}^{\mathrm{T}}\cdot\mathrm{Ln}\,\mathbf{V}\cdot\mathbf{R} \quad\Rightarrow\quad \|\,\mathrm{Ln}\,\mathbf{U}\,\| = \|\,\mathrm{Ln}\,\mathbf{V}\,\|,$$

$$\frac{d\mathrm{Ln}\,\mathbf{U}}{dt} = \mathbf{R}^{\mathrm{T}}\cdot\frac{D\mathrm{Ln}\,\mathbf{V}}{Dt}\cdot\mathbf{R} \quad\Rightarrow\quad \left\|\,\frac{d\mathrm{Ln}\,\mathbf{U}}{dt}\,\right\| = \left\|\,\frac{D\mathrm{Ln}\,\mathbf{V}}{Dt}\,\right\|. \tag{22.7.32}$$

Clearly, if $a_1 = 0$, then,

$$\underline{\underline{\xi}} = \mathrm{Ln}\,\mathbf{U}, \quad \boldsymbol{\xi} = \mathrm{Ln}\,\mathbf{V}. \tag{22.7.33}$$

22.8   Growth of a bar under unconfined tension/compression

We consider first a growth problem in which the tissue state is uniform in space and next the inhomogeneous growth in a pressurized hollow cylinder of fixed length. These examples highlight the qualitative and quantitative influences of the types of constitutive growth model, namely elastic-growing versus elastic tissues, and of mass deposition modes on the tissue response, in terms of stress, strain and mass.

Two types of constitutive equations are needed:
- the rate of growth equations for the generalized strains, namely the growth transformation, or mass, and the structure variables;
- the constitutive equations for the generalized stresses, which are distinct for deposition modes (a) and (b).

These examples go beyond the analysis of Drozdov and Khanina [1997] who address mass deposition of elastic solids under mode (a) only, consider a linear free energy, that is $\mu = 0$ in (22.7.14), and do not include structure variables. Here, both elastic solids and elastic-growing solids are addressed: the deformation gradient admits a multiplicative decomposition into an elastic transformation and a growth transformation, and the thermodynamic state includes, besides the stress or the elastic strain, a scalar entity and a tensor.

22.8.1   Unconfined tension/compression of the bar

22.8.1.1   Unconfined tension/compression of the elastic-growing bar

The bar is assumed elastically incompressible. Its material properties and state variables are initially uniform, and they remain so during the growth process.

Kinematics and statics

The elastic, growth and total transformation tensors are principal in the fixed axes $(\mathbf{e}_{(1)}, \mathbf{e}_{(2)}, \mathbf{e}_{(3)})$,

$$\mathbf{F}(t) = \mathbf{diag}[F_1(t),\, F_2(t),\, F_3(t)], \quad \mathbf{F}^{\mathrm{g}}(t) = \mathbf{diag}\left[G_1(t),\, G_2(t),\, G_3(t)\right], \tag{22.8.1}$$

and

$$\mathbf{F}^{\mathrm{e}}(t) = \mathbf{F}(t)\cdot\mathbf{F}^{\mathrm{g}}(t)^{-1} = \mathbf{diag}\left[v_1 = \frac{F_1}{G_1},\, v_2 = \frac{F_2}{G_2},\, v_3 = \frac{F_3}{G_3}\right](t). \tag{22.8.2}$$

Traction along the axis $\mathbf{e}_{(1)}$ is applied through a controlled axial force, the lateral sides being traction free. This force over the whole section $\mathbf{S}(t)$ can be expressed in terms of the Cauchy stress,

$$\mathbf{P}_{(1)}(t) = \int_{\mathbf{S}(t)} \boldsymbol{\sigma}(t)\cdot d\mathbf{S}(t) = \int_{\underline{\mathbf{S}}} \boldsymbol{\sigma}(t)\cdot\det\mathbf{F}(t)\,\mathbf{F}^{-\mathrm{T}}(t)\cdot d\underline{\mathbf{S}}. \tag{22.8.3}$$

The last relation is implied by the fact that an elementary area $d\underline{\mathbf{S}}$ transforms to $d\mathbf{S} = \det\mathbf{F}\,\mathbf{F}^{-\mathrm{T}}\cdot d\underline{\mathbf{S}}$ under the motion defined by the deformation gradient $\mathbf{F}$, Eqn (2.6.9). With uniform stress and strain in the bar, the Cauchy stress and incompressibility pressure are given in terms of the force $P_1(t)$ (> 0 for tension) along the axis $\mathbf{e}_{(1)}$, $\mathbf{P}_{(1)}(t) = P_1(t)\,\mathbf{e}_{(1)}$, and effective Cauchy stress as

$$\sigma_1(t) = \sigma_1'(t) - \sigma_2'(t) = \frac{P_1(t)}{\underline{S}}\,\frac{F_1(t)}{\det\mathbf{F}^{\mathrm{g}}(t)}; \quad \sigma_k(t) = \sigma_k'(t) - p_I(t) = 0, \quad k = 2, 3. \tag{22.8.4}$$

Solution method

Since the material properties respect the axial symmetry implied by the loading, the lateral directions 2 and 3 are equivalent, in both stress and deformations. Details of the derivations of the stress are provided in Appendix A. Moreover, due to elastic incompressibility, $v_1(t) \, v_2^2(t) = 1$. The equilibrium $P^{\text{ext}}(t) \equiv P_1(t)/\mu_\kappa \, \underline{S} = P^{\text{int}}(t)$ can be shown to express in terms of the elastic stretch v_1 and of the growth stretches, namely

$$P^{\text{int}}(t) = \frac{1}{G_1(t)} \left[\left(\frac{a_\kappa}{\mu_\kappa} - 1 \right) \left(v_1 \, \mathbb{I}_\kappa(\frac{1}{v_1^2}) - \frac{1}{v_1^2} \, \mathbb{I}_\kappa(v_1) \right) + v_1^3 \, \mathbb{I}_\kappa(\frac{1}{v_1^4}) - \frac{1}{v_1^3} \, \mathbb{I}_\kappa(v_1^2) \right](t) \,, \tag{22.8.5}$$

where $\mathbb{I}_\kappa$ is the mode dependent functional,

$$\mathbb{I}_\kappa(v)(t) \equiv \begin{cases} 1 + \displaystyle\int_0^t v(\tau) \frac{d}{d\tau} \det \mathbf{F}^{\text{g}}(\tau) \, d\tau & \text{mode (a)}, \\[3mm] \det \mathbf{F}^{\text{g}}(t) & \text{mode (b)}. \end{cases} \tag{22.8.6}$$

The governing equations consist of the equilibrium equation $P^{\text{int}}(t) - P^{\text{ext}}(t) = 0$, and of the integrated rate constitutive equations for the growth transformation (22.5.15) and structure variables (22.5.39). They are solved for the unknowns,

$$\mathbf{U} = \begin{bmatrix} v_1 & G_1 & G_2 & G_3 & \xi_{\kappa 1} & \xi_{\kappa 2} & \xi_{\kappa 3} & \xi_\kappa \end{bmatrix}^{\mathsf{T}}, \tag{22.8.7}$$

by a time-marching procedure, via an implicit Euler scheme, that consists in zeroing the residual, e.g., at step $n+1$,

$$\mathbf{R}_{n+1} = \begin{cases} P^{n+1}_{\text{int}} - P^{n+1}_{\text{ext}} \,, \\[2mm] G^{n+1}_k - G^n_k - L^{n+1}_{\kappa k} \, G^{n+1}_k \,, & k \in [1,3], \\[2mm] \xi^{n+1}_{\kappa k} - \xi^n_{\kappa k} - M^{n+1}_{\kappa k} \,, & k \in [1,3], \\[2mm] \xi^{n+1}_\kappa - \xi^n_\kappa - N^{n+1}_\kappa \,, \end{cases} \tag{22.8.8}$$

with

$$\mathbf{L}_\kappa = \mu_\kappa \frac{\Delta t}{t_{\text{g}}} \, \Phi(h_\kappa) \frac{\partial h_\kappa}{\partial \boldsymbol{\tau}_\kappa} \,,$$

$$\mathbf{M}_\kappa = \mathbf{L}_\kappa - a_1 \, \| \, \mathbf{L}_\kappa \, \| \, \frac{\mathbf{X}_\kappa}{\mu_\kappa} \,, \tag{22.8.9}$$

$$N_\kappa = \frac{\Delta t}{t_{\text{g}}} \, \Phi(h_\kappa) \left(a_2 - a_3 \frac{\mathbf{X}_\kappa}{\mu_\kappa} \right) \frac{\mathbf{X}_\kappa}{\mu_\kappa} \,.$$

22.8.1.2 Unconfined tension/compression of the elastic bar

The bar is incompressible. The analysis is qualitatively similar to the elastic-growing bar, but somehow different in the details.

Kinematics and statics

With the deformation gradient given by (22.8.1), the transformation $\mathbf{F}(t, \tau)$ expresses as,

$$\mathbf{F}(t, \tau) = \mathbf{F}(t) \cdot \mathbf{F}^{-1}(\tau) = \mathbf{diag}\left[\frac{F_1(t)}{F_1(\tau)}, \frac{F_2(t)}{F_2(\tau)}, \frac{F_3(t)}{F_3(\tau)} \right]. \tag{22.8.10}$$

The static relation (22.8.4) becomes,

$$\sigma_1(t) = \sigma'_1(t) - \sigma'_2(t) = \frac{P_1(t)}{\underline{S}} \frac{F_1(t)}{\det \mathbf{F}(t)} \,; \quad \sigma_k(t) = \sigma'_k(t) - p_I(t) = 0, \quad k = 2, 3 \,. \tag{22.8.11}$$

The governing integro-differential equations

The lateral directions 2 and 3 being equivalent, in both stress and strain, it suffices to obtain the time history of two unknown functions, for example F_1 and $\det \mathbf{F}$. The details of the derivations of the stress components are provided in Appendix B.

The equilibrium between the external loading force and the internal force $P^{\text{ext}}(t) \equiv P_1(t)/\mu\,\underline{S} = P^{\text{int}}(t)$ expresses in terms of the normalized internal force $P^{\text{int}}(t)$,

$$P^{\text{int}}(t) = \frac{1}{\det \mathbf{F}(t)} \left[\left(\frac{a}{\mu} - 1\right) \left(F_1\,\mathbb{I}\left(\frac{1}{F_1^2}\right) - \frac{F_2^2}{F_1}\,\mathbb{I}\left(\frac{1}{F_2^2}\right)\right) + F_1^3\,\mathbb{I}\left(\frac{1}{F_1^4}\right) - \frac{F_2^4}{F_1}\,\mathbb{I}\left(\frac{1}{F_2^4}\right) \right](t), \qquad (22.8.12)$$

where F_2 is obtained via the relation $F_2^2 = \det \mathbf{F}/F_1$. The mode dependent functional $\mathbb{I}: v \to \mathbb{I}(v)$ is defined in terms of the time history of the volume $\det \mathbf{F}$,

$$\mathbb{I}(v)(t) \equiv \begin{cases} 1 + \displaystyle\int_0^t v(\tau)\frac{d}{d\tau}\det \mathbf{F}(\tau)\,d\tau & \text{mode (a)}, \\[4pt] \det \mathbf{F}(t) & \text{mode (b)}. \end{cases} \qquad (22.8.13)$$

Governing equations

The governing equations consist of the equilibrium equation $P^{\text{int}}(t) - P^{\text{ext}}(t) = 0$, and in the integrated equations for the rate of mass growth and structure variables. They are solved for the unknowns,

$$\mathbf{U} = \begin{bmatrix} F_1 & \det \mathbf{F} & \underline{\xi}_1 & \underline{\xi}_2 & \underline{\xi}_3 & \xi \end{bmatrix}^{\mathrm{T}}, \qquad (22.8.14)$$

by a time-marching procedure, via an implicit Euler scheme, that consists in zeroing the residual, e.g., at step $n+1$,

$$\mathbf{R}_{n+1} = \begin{cases} P^{n+1}_{\text{int}} - P^{n+1}_{\text{ext}} \\[4pt] \det \mathbf{F}_{n+1} - \det \mathbf{F}_n - L_{n+1}\det \mathbf{F}_{n+1} \\[4pt] \underline{\xi}_k^{n+1} - \underline{\xi}_k^n - \underline{M}_k^{n+1}, \quad k \in [1,3] \\[4pt] \xi^{n+1} - \xi^n - N^{n+1} \end{cases} \qquad (22.8.15)$$

where

$$\begin{aligned} L &= \mu\,\frac{\Delta t}{t_{\text{g}}}\,\Phi(h)\,\text{tr}\,\frac{\partial h}{\partial \boldsymbol{\sigma}}, \\[4pt] \underline{M} &= \Delta t\,\frac{d\mathbf{E}_{\text{G}}}{dt} - a_1 \,\|\, \Delta t\,\frac{d\mathbf{E}_{\text{G}}}{dt} \,\|\, \frac{\underline{X}}{\mu}, \\[4pt] \underline{N} &= \frac{\Delta t}{t_{\text{g}}}\,\Phi(h)\left(a_2 - a_3\,\frac{X}{\mu}\right)\frac{X}{\mu}. \end{aligned} \qquad (22.8.16)$$

22.8.2   The critical load for the bar

The load process assumes that the stress state is initially homeostatic. The stress hits the homeostatic surface for some critical value of the load as sketched in Fig. 22.8.1. Below this critical value, the tissue response is elastic, and identical for the elastic-growing and elastic tissues. The analysis of this section, which uses the formalism of elastic tissues, holds as well for elastic-growing tissues.

At time $t = 0$, the load $P_1(0) = 0$ vanishes, the deformation gradient $\mathbf{F}(0)$ is equal to the identity $\mathbf{I}$ so that the effective stress $\boldsymbol{\tau}'(0) = \boldsymbol{\sigma}'(0)$ deriving from (22.7.14) is equal to $a\,\mathbf{I}$, the incompressibility pressure p_I is equal to a and therefore the total stresses vanish. For the unstrained state to be homeostatic, since $\det \mathbf{F}(0) = 1$, the structure variables should satisfy the inequality,

$$h(-\mathbf{X}, X) \leq 0. \qquad (22.8.17)$$

The condition is trivially satisfied if the tensor structure variable $\mathbf{X}(0)$ vanishes. During the load process, growth begins when the stress hits the homeostatic surface $h(\boldsymbol{\sigma} - \mathbf{0}, X) = 0$. Within the homeostatic surface, $\det \mathbf{F} = F_1\,F_2^2$ remains equal to 1, and the Cauchy and Kirchhoff stresses are, respectively, equal to,

$$\frac{\sigma_1}{\mu} = F_1^2\,\frac{\tau_{\kappa 1}}{\mu} = \left(\frac{a}{\mu} - 1\right)\left(F_1^2 - \frac{1}{F_1}\right) + F_1^4 - \frac{1}{F_1^2}. \qquad (22.8.18)$$

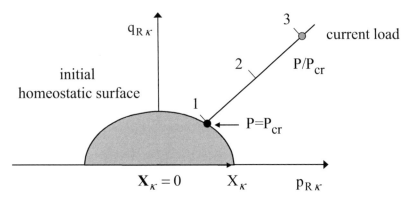

Fig. 22.8.1: The external load P^{ext} is normalized by the value $P^{\text{ext}}_{\text{cr}}$ at which it hits the initial homeostatic surface.

In the discussion below, the coefficients a and μ which define the free energy are assumed to be known, as well as the coefficient M of the homeostatic surface.

One may consider the initial size X of the homeostatic surface to be prescribed. Then the value of the strain F^{cr}_1 at which the stress hits the homeostatic surface $h = 0$ is solution of the equation

$$F_1^6 + \left(\frac{a}{\mu} - 1\right) F_1^4 - \frac{\sigma_1^{\text{cr}}}{\mu} F_1^2 - \left(\frac{a}{\mu} - 1\right) F_1 - 1 = 0, \tag{22.8.19}$$

with σ_1^{cr} the stress zeroing the homeostatic function,

$$\frac{|\sigma_1|}{\mu} = F_1^2 \frac{|\tau_{\kappa 1}|}{\mu} = \frac{X}{\mu} \frac{1}{\sqrt{\left(\frac{1}{M^2} + \frac{1}{9}\right)}}. \tag{22.8.20}$$

Then the critical normalized axial load expresses in terms of F^{cr}_1 as,

$$P^{\text{ext}}_{\text{cr}} = \left(\frac{a}{\mu} - 1\right) \left(F_1^{\text{cr}} - \frac{1}{(F_1^{\text{cr}})^2}\right) + (F_1^{\text{cr}})^3 - \frac{1}{(F_1^{\text{cr}})^3}. \tag{22.8.21}$$

Alternatively, one might consider the critical load $P^{\text{ext}}_{\text{cr}}$ at which growth begins to be prescribed. Then F^{cr}_1 is obtained as the solution of the equation

$$F_1^6 + \left(\frac{a}{\mu} - 1\right) F_1^4 - P^{\text{ext}}_{\text{cr}} F_1^3 - \left(\frac{a}{\mu} - 1\right) F_1 - 1 = 0. \tag{22.8.22}$$

The axial stress σ_1^{cr}/μ is then equal to $F_1^{\text{cr}} P^{\text{ext}}_{\text{cr}}$ or equivalently by (22.8.18), and the size X results from (22.8.20).

22.8.3   The elastic-growing bar versus the elastic bar under tension

In the simulations, the load is imposed instantaneously, $P_1(t) = P_1 \mathcal{H}(t)$ with $\mathcal{H}$ the Heaviside step function. The instantaneous responses of the elastic-growing and elastic tissues show some differences:

- for an elastic-growing tissue, the elastic deformation and the total stress result to be discontinuous at time $t = 0$. On the other hand, the growth strain is continuous and so are the structure variables and the center and the size of the homeostatic surface;

- for elastic tissues on the other hand, the strain and stress are discontinuous, and so are the tensor structure variable and the center of the homeostatic surface, while the aggregated mass and size of the homeostatic surface remain continuous.

Figs. 22.8.2 and 22.8.3 display simulations respectively of an elastic-growing bar and of an elastic bar undergoing axial tension. There are further significant differences between elastic-growing and elastic tissues:

- first, for the elastic tissue, growth may asymptotically vanish while the stress is still outside the homeostatic surface as witnessed by the fact that the values of the homeostatic function h at the current point do not tend to vanish. This happens when the latter has moved in such a way that the volumetric part of its normal in stress space estimated at the current point vanishes, because $\operatorname{tr}(\boldsymbol{\sigma} - \mathbf{Y})$ tends to 0. This phenomenon is clearly a shortcoming of the too simple structure of the elastic model. On the other hand, all components of the normal are involved for an elastic-growing tissue. Since the latter can not vanish for points outside the homeostatic surface, growth can stop only when the stress point comes in contact with the homeostatic surface;

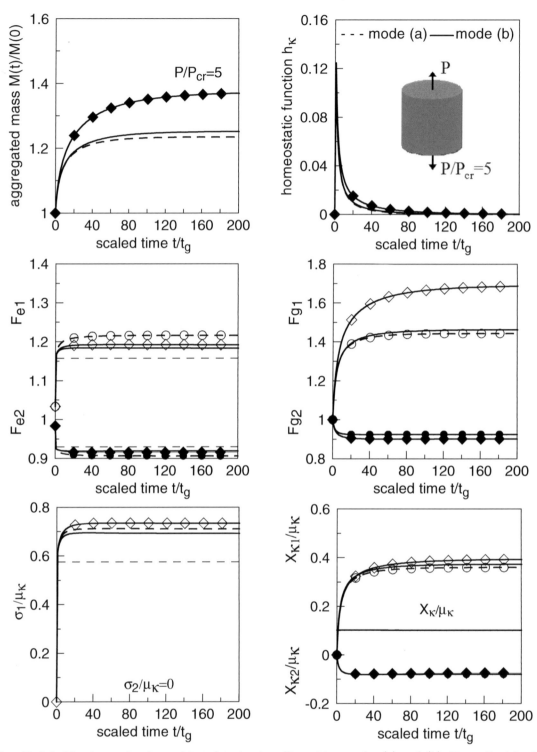

Fig. 22.8.2: Elastic-growing bar subjected to tension. Deposition modes (a) and (b). Normalized load P_{cr}^{ext} equal to 0.1 and actual load $P^{ext}/P_{cr}^{ext} = 5$. See text in Section 22.5.6.2 for tissue parameters. Time courses of aggregated mass, homeostatic function, elastic and growth transformations, Cauchy stress and center and size of the homeostatic surface. ● : radial component, ○ : axial component. The curves correspond to $a_1 = 0$, except those marked with a diamond which are associated with $a_1 = 1.0$ and deposition mode (b): ◆ : radial component, ◇ : axial component. The elastic gradients and stresses in complete absence of growth are indicated by horizontal dashed lines.

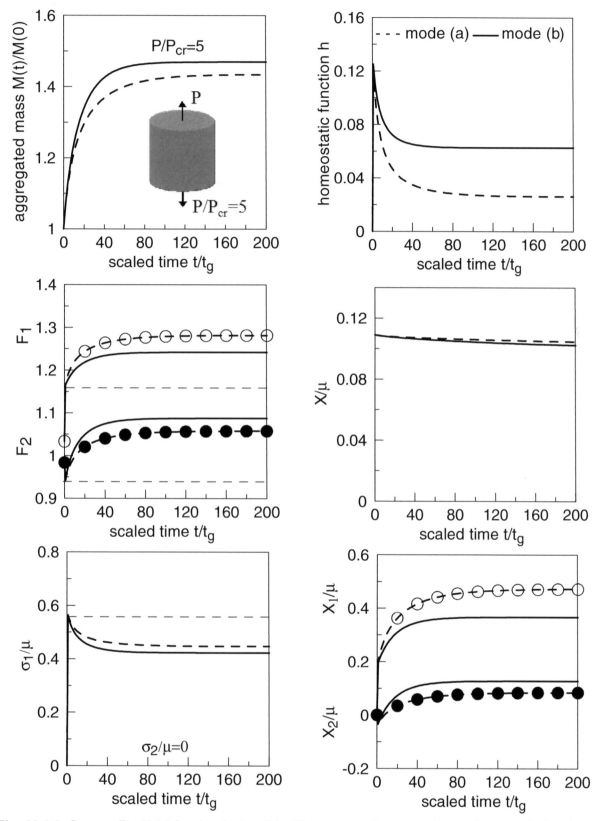

Fig. 22.8.3: Same as Fig. 22.8.2 but for elastic solids. Time courses of aggregated mass, homeostatic function, total deformation gradient, size of the homeostatic surface, Cauchy stress and center of the homeostatic surface.

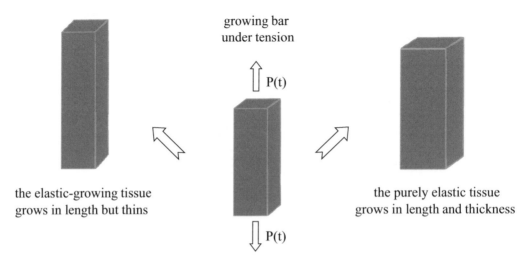

Fig. 22.8.4: An elastic bar subjected to uniaxial tension grows in length and thickness, while an elastic growing bar thins. The distinct behavior of the elastic tissue is due to the fact that the growth information is conveyed by a scalar, so that directional properties are ignored.

- another difference highlighted in Fig. 22.8.4 is that, for the bar under tension, both the axial and lateral directions undergo growth for the elastic bar. On the other hand, for the elastic-growing bar, the axial direction undergoes growth while the radial directions undergo resorption. Therefore, since the axial load is constant, and the cross-section decreases, the axial Cauchy stress increases slightly for the elastic-growing solid. Conversely, the cross-section increases for the elastic solid, so that the axial Cauchy stress decreases slightly;
- also, note that the formulation of the growth rate law for elastic-growing tissues has decided that the structure variables change only if growth takes place. Such is not the case for elastic tissues since the tensor structure variables, and therefore the center of the homeostatic surface, may change whether growth takes place or not.

The quantitative differences between deposition modes (a) and (b) are small for the material values used in the simulations reported here. However, the differences in terms of stresses, strains and aggregated masses become significant as the ability of the homeostatic surface to translate and change size decreases.

22.9 Growth of a hollow cylinder under internal pressure and with a fixed length

A hollow cylinder is submitted to the pressure $p_i(t)$ on its inner boundary $\underline{r} = \underline{r}_1$. The outer lateral boundary $\underline{r} = \underline{r}_2$ is submitted to the pressure $p_e(t)$, and the length along the axial direction $\mathbf{e}_z$ is kept constant.

If the cylinder is elastic, it is incompressible. If it is elastic-growing, it is elastically incompressible. The analysis below holds for both tissues.

22.9.1 The thin wall approximation

The strain in the cylinder is heterogeneous. The motion is purely radial and it is described by the transformation:

$$\underline{\mathbf{x}} = \underline{r}\,\mathbf{e}_r + \underline{z}\,\mathbf{e}_z \quad \rightarrow \quad \mathbf{x}(\underline{r}, t) = r(\underline{r}, t)\,\mathbf{e}_r + \underline{z}\,\mathbf{e}_z \,, \tag{22.9.1}$$

so that, in the unit cylindrical axes $(\mathbf{e}_r, \mathbf{e}_\theta, \mathbf{e}_z)$, the deformation gradient writes

$$\mathbf{F}(\underline{r}, t) = \mathbf{diag}\left[F_r(\underline{r}, t) = \frac{\partial r}{\partial \underline{r}}, \ F_\theta(\underline{r}, t) = \frac{r}{\underline{r}}, \ F_z(\underline{r}, t) = 1 \right]. \tag{22.9.2}$$

The static equilibrium equation in absence of body force expresses in terms of the radial and tangential components of the Cauchy stress,

$$\frac{\partial \sigma_r}{\partial r} + \frac{1}{r}\,(\sigma_r - \sigma_\theta) = 0 \,, \tag{22.9.3}$$

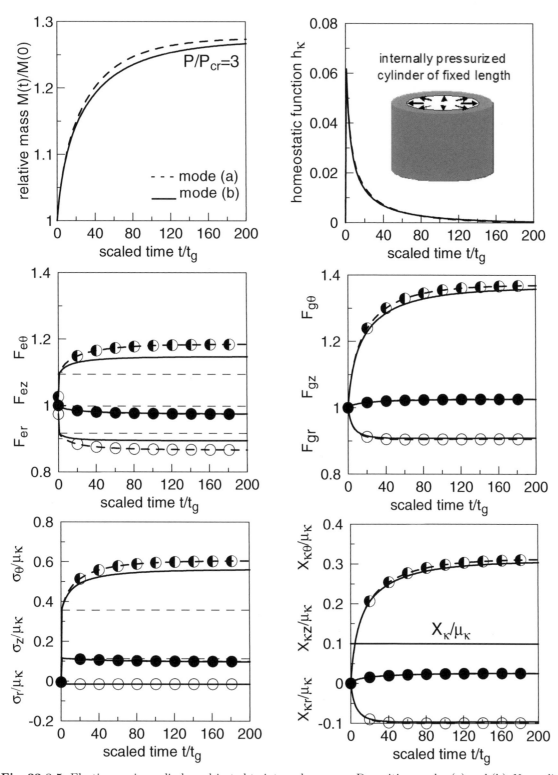

Fig. 22.8.5: Elastic-growing cylinder subjected to internal pressure. Deposition modes (a) and (b). Normalized load $P_{\mathrm{cr}}^{\mathrm{ext}}$ equal to 0.1 and actual load $P^{\mathrm{ext}}/P_{\mathrm{cr}}^{\mathrm{ext}} = 3$. See text in Section 22.5.6.2 for tissue parameters. $\bigcirc$: radial component, $\circledcirc$: tangential component, $\bullet$: axial component. Time courses of aggregated mass, homeostatic function, elastic and growth transformations, Cauchy stress and center and size of the homeostatic surface. The elastic gradients and stresses in complete absence of growth are indicated by horizontal dashed lines.

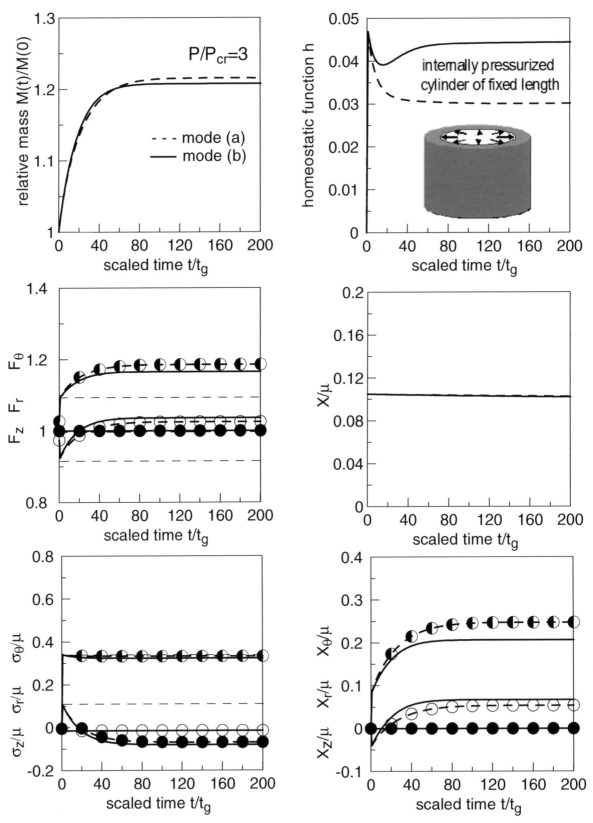

Fig. 22.8.6: Same as Fig. 22.8.5 but for elastic solids. Time courses of aggregated mass, homeostatic function, total deformation gradient, size of the homeostatic surface, Cauchy stress and center of the homeostatic surface.

and it is solved formally to provide a constraint in terms of the pressure differential,

$$
\begin{aligned}
\sigma_r(r(\underline{r},t),t) &= \sigma_r(r(\underline{r}_1,t),t) &+ \int_{r(\underline{r}_1,t)}^r \left(\sigma_\theta(u,t) - \sigma_r(u,t)\right) \frac{du}{u} \\
&= -p_i(t) &+ \int_{\underline{r}_1}^{\underline{r}} \left(\sigma_\theta(u,t) - \sigma_r(u,t)\right) \frac{1}{u} \frac{\partial u}{\partial \underline{r}} \, d\underline{r} \, .
\end{aligned}
\tag{22.9.4}
$$

Thus with help of (22.9.2), the equilibrium condition can be cast in the format,

$$
\Pi(t) = \frac{p_i(t) - p_e(t)}{\mu}, \quad \Pi(t) \equiv \int_{\underline{r}_1}^{\underline{r}_2} \left(\frac{\sigma_\theta'}{\mu}(r,t) - \frac{\sigma_r'}{\mu}(r,t)\right) \frac{F_r(\underline{r},t)}{F_\theta(\underline{r},t)} \frac{d\underline{r}}{r},
\tag{22.9.5}
$$

where $r = r(\underline{r}, t)$. In the thin wall approximation, the normalized equilibrium condition may be expressed as $P^{\mathrm{int}}(t) - P^{\mathrm{ext}}(t) = 0$, with

$$
P^{\mathrm{ext}}(t) = \frac{p_i(t) - p_e(t)}{\mu_\kappa} \frac{\underline{r}}{\Delta \underline{r}}, \quad P^{\mathrm{int}}(t) = \left(\frac{\sigma_\theta'}{\mu_\kappa}(\underline{r},t) - \frac{\sigma_r'}{\mu_\kappa}(\underline{r},t)\right) \frac{F_r(\underline{r},t)}{F_\theta(\underline{r},t)},
\tag{22.9.6}
$$

where $\underline{r} = (\underline{r}_1 + \underline{r}_2)/2$, $r = r(\underline{r}, t)$, and $\Delta \underline{r} = \underline{r}_2 - \underline{r}_1$. From now on, the space dependent entities are estimated at mid wall, and the space dependence may be omitted without ambiguity. If we admit that the mid-radial stress is equal to $-(p_i(t) + p_e(t))/2$, the incompressibility pressure p_I is equal to $\sigma_r'(r,t) + (p_i(t) + p_e(t))/2$.

22.9.2 The hollow cylinder under internal pressure

22.9.2.1 The elastic-growing hollow cylinder

The solution is sought in terms of radial, tangential and axial total stretches (F_r, F_θ, F_z), elastic stretches (v_r, v_θ, v_z), and growth stretches (G_r, G_θ, G_z). In view of the elastic incompressibility, $v_r \, v_\theta \, v_z = 1$, and of the fact that the length of the cylinder is fixed, $F_z = 1$, the axial elastic stretch expresses in terms of the axial growth stretch, namely $v_z = 1/G_z$, and the tangential elastic stretch in terms of the radial elastic stretch and of the axial growth stretch, namely $v_\theta = G_z/v_r$. Consequently, the effective Cauchy stress expresses in terms of the elastic stretch v_r and of the components of the growth transformation, and

$$
P^{\mathrm{int}}(t) = \frac{1}{G_\theta^2 \, G_z^2} \left[\left(\frac{a_\kappa}{\mu_\kappa} - 1\right) \left(G_z^2 \, \mathbb{I}_\kappa(\frac{1}{v_\theta^2}) - v_r^4 \, \mathbb{I}_\kappa(\frac{1}{v_r^2})\right) + \frac{G_z^4}{v_r^2} \, \mathbb{I}_\kappa(\frac{1}{v_\theta^4}) - v_r^6 \, \mathbb{I}_\kappa(\frac{1}{v_r^4}) \right](t),
\tag{22.9.7}
$$

where $\mathbb{I}_\kappa$ is the mode dependent functional defined by (22.8.6).

22.9.2.2 The elastic hollow cylinder

The difference of stresses, or effective stresses, is estimated from the explicit expressions provided in Appendix B. With help of the mode dependent functional $\mathbb{I}$ defined by (22.8.13), the internal force (22.9.6) can be cast in the format,

$$
P^{\mathrm{int}}(t) = \frac{1}{\det \mathbf{F}(t)} \left[\left(\frac{a}{\mu} - 1\right) \left(\mathbb{I}(\frac{1}{F_\theta^2}) - \frac{F_r^2}{F_\theta^2} \mathbb{I}(\frac{1}{F_r^2})\right) + F_\theta^2 \, \mathbb{I}(\frac{1}{F_\theta^4}) - \frac{F_r^4}{F_\theta^2} \mathbb{I}(\frac{1}{F_r^4}) \right](t),
\tag{22.9.8}
$$

with $F_\theta = \det \mathbf{F}/F_r$. The residuals, consisting of the equilibrium equation and of the integrated growth law for the mass and structure variables, are solved like in Section 22.8.1.2.

22.9.3 The critical load for the cylinder

The initial state is assumed to be homeostatic. The behavior of the tissue is then elastic, and identical whether the growth model is elastic-growing or elastic. As long as no growth takes place, the volume is constant, $\det \mathbf{F} = 1$. Since the axial length of the bar is fixed, one may introduce the intermediate quantity $Z = F_r^2 = 1/F_\theta^2$. The components of the effective stress for the two deposition modes simplify to

$$
\sigma_r' = (a - \mu) Z + \mu Z^2, \quad \sigma_\theta' = \frac{a - \mu}{Z} + \frac{\mu}{Z^2}, \quad \sigma_z' = a.
\tag{22.9.9}
$$

Let $Z = Z^{\mathrm{cr}}$ be the value of Z for which the stress hits the homeostatic surface $h = 0$. Then the normalized pressure (22.9.6) results as,

$$P_{\mathrm{cr}}^{\mathrm{ext}} = (\frac{a}{\mu} - 1)\left(1 - (Z^{\mathrm{cr}})^2\right) + \frac{1}{Z^{\mathrm{cr}}} - (Z^{\mathrm{cr}})^3 \,. \tag{22.9.10}$$

Since the pressures p_i and p_e have been assumed to be known, this expression actually provides the geometry of the cylinder, namely $r/\Delta r$.

Alternatively, one might consider the ratio $r/\Delta r$ and the critical load $P_{\mathrm{cr}}^{\mathrm{ext}}$ at which growth begins to be known. Then $Z^{\mathrm{cr}} = (F_r^{\mathrm{cr}})^2$ is obtained as the solution of the equation

$$Z^4 + (\frac{a}{\mu} - 1)Z^3 + \left(P_{\mathrm{cr}}^{\mathrm{ext}} - (\frac{a}{\mu} - 1)\right)Z - 1 = 0 \,. \tag{22.9.11}$$

The radius of the homeostatic surface X results by zeroing the homeostatic function.

22.9.4   The elastic-growing cylinder versus the elastic cylinder

The load is controlled via the internal and external pressures $p_i(t)$ and $p_e(t)$. In the simulations, the external pressure is set to vanish, $p_e(t) = 0$, while the internal pressure is raised instantaneously to a positive value. The standard thickness to radius ratio $\Delta r/r$ is equal to $1/10$, and the dimensionless tension load $P^{\mathrm{ext}} = P_{\mathrm{cr}}^{\mathrm{ext}} = 0.1$ corresponds to a point *on* the homeostatic surface.

The elastic solution in absence of growth is indicated in Figs. 22.8.5 and 22.8.6 to serve as a reference.

The instantaneous response observed for the bar at the time of application of the load takes place as well for a cylinder under internal pressure, as displayed in Figs. 22.8.5 and 22.8.6. Much similarly to the test on the bar, the quantitative differences between deposition modes (a) and (b) are small while there are main differences between elastic-growing and elastic tissues. In fact, the comments concerning the growth of the bar hold as well, namely

- for the elastic tissue, growth may asymptotically vanish while the stress is still outside the homeostatic surface;
- both the radial and circumferential directions undergo growth for the elastic cylinder, the axial length being fixed. For the elastic-growing cylinder, while the radial direction undergoes resorption, the circumferential direction, and to a lesser extent the axial direction, undergo growth.

For the elastic-growing tissue, the expression of the center of the homeostatic surface is driven by the growth transformation: actually since the principal axes are fixed, the center of the homeostatic surface varies like the logarithm of the growth transformation. For the elastic tissue, the center of the homeostatic surface is driven by the total strain. These correlations are apparent from the simulations of both the bar, Figs. 22.8.2 and 22.8.3, and the hollow cylinder, Figs. 22.8.5 and 22.8.6.

Moreover, for the standard values of the tissue parameters, the relaxation time of the growth process is about hundred times the time t_g. This estimation holds true for both elastic-growing and elastic tissues: it matches with the analytical result obtained for an elastic-growing tissue undergoing volumetric growth.

Appendix A: Details of the calculations of stress in an elastic-growing solid

The generalized strains and stresses are coaxial. The relations are expressed in the cartesian axes associated with their eigendirections.

Elastic strain associated with $\mathbf{F}^e(t)$:

$$\mathbf{F}^e(t) = \mathbf{F}(t) \cdot (\mathbf{F}^g(t))^{-1} = \mathbf{diag}\left[v_1(t) = \frac{F_1(t)}{G_1(t)},\ v_2(t) = \frac{F_2(t)}{G_2(t)},\ v_3(t) = \frac{F_3(t)}{G_3(t)}\right] \tag{22.A.1}$$

$$\mathbf{E}_G^e(t) = \frac{1}{2}((\mathbf{F}^e(t))^T \cdot \mathbf{F}^e(t) - \mathbf{I}) = \frac{1}{2}\,\mathbf{diag}\,[v_1^2(t) - 1,\ v_2^2(t) - 1,\ v_3^2(t) - 1] \tag{22.A.2}$$

$$I_1^e(t) = \mathrm{tr}\,\mathbf{E}_G^e(t) = \sum_{i=1}^{3} \frac{1}{2}\left(v_i^2(t) - 1\right),\quad I_2^e(t) = \mathrm{tr}\,(\mathbf{E}_G^e(t))^2 = \sum_{i=1}^{3} \frac{1}{4}\left(v_i^2(t) - 1\right)^2 \tag{22.A.3}$$

$$\frac{\partial}{\partial E_{Gi}^e(t)} = \frac{\partial}{\partial v_i(t)}\frac{dv_i(t)}{dE_{Gi}^e(t)} = \frac{1}{v_i(t)}\frac{\partial}{\partial v_i(t)} \tag{22.A.4}$$

$$\frac{\partial I_1^e(t)}{\partial E_{Gi}^e(t)} = 1,\quad \frac{\partial I_2^e(t)}{\partial E_{Gi}^e(t)} = v_i^2(t) - 1,\quad i \in [1,3] \tag{22.A.5}$$

Deposition mode (a): elastic strain associated with $\mathbf{F}^e(t,\tau) = \mathbf{F}^e(t) \cdot (\mathbf{F}^e(\tau))^{-1}$:

$$\mathbf{F}^e(t,\tau) = \mathbf{diag}\left[\frac{v_1(t)}{v_1(\tau)},\ \frac{v_2(t)}{v_2(\tau)},\ \frac{v_3(t)}{v_3(\tau)}\right] \tag{22.A.6}$$

$$\mathbf{E}_G^e(t,\tau) = \frac{1}{2}((\mathbf{F}^e(t,\tau))^T \cdot \mathbf{F}^e(t,\tau) - \mathbf{I}) = \frac{1}{2}\,\mathbf{diag}\left[\frac{v_1^2(t)}{v_1^2(\tau)} - 1,\ \frac{v_2^2(t)}{v_2^2(\tau)} - 1,\ \frac{v_3^2(t)}{v_3^2(\tau)} - 1\right] \tag{22.A.7}$$

$$I_1^e(t,\tau) = \mathrm{tr}\,\mathbf{E}_G^e(t,\tau) = \sum_{i=1}^{3} \frac{1}{2}\left(\frac{v_i^2(t)}{v_i^2(\tau)} - 1\right) \tag{22.A.8}$$

$$I_2^e(t,\tau) = \mathrm{tr}\,(\mathbf{E}_G^e(t,\tau))^2 = \sum_{i=1}^{3} \frac{1}{4}\left(\frac{v_i^2(t)}{v_i^2(\tau)} - 1\right)^2 \tag{22.A.9}$$

$$\frac{\partial I_1^e(t,\tau)}{\partial E_{Gi}^e(t)} = \frac{1}{v_i^2(\tau)},\quad \frac{\partial I_2^e(t,\tau)}{\partial E_{Gi}^e(t)} = \frac{1}{v_i^2(\tau)}\left(\frac{v_i^2(t)}{v_i^2(\tau)} - 1\right),\quad i \in [1,3] \tag{22.A.10}$$

$$\frac{\partial \underline{E}_\kappa^\natural(t)}{\partial E_{Gi}^e(t)} = \frac{\partial E_\kappa(t)}{\partial E_{Gi}^e(t)} + \int_0^t \frac{\partial E_\kappa(t,\tau)}{\partial E_{Gi}^e(t)}\frac{d}{d\tau}\det \mathbf{F}^g(\tau)\,d\tau,\quad i \in [1,3] \tag{22.A.11}$$

With $\det \mathbf{F}^e = 1$ due to elastic incompressibility,

$$\sigma_i'(t) = \frac{v_i^2}{\det \mathbf{F}}\frac{\partial \underline{E}_\kappa^\natural}{\partial E_{Gi}^e}(t) = \frac{v_i^2}{\det \mathbf{F}^g}\frac{\partial \underline{E}_\kappa^\natural}{\partial E_{Gi}^e}(t),\quad i \in [1,3] \tag{22.A.12}$$

$$\frac{\sigma_i'}{\mu_\kappa}(t) = \frac{1}{\det \mathbf{F}^g(t)}\left[\left(\frac{a_\kappa}{\mu_\kappa} - 1\right)v_i^2\,\mathbb{I}_\kappa\left(\frac{1}{v_i^2}\right) + v_i^4\,\mathbb{I}_\kappa\left(\frac{1}{v_i^4}\right)\right](t),\quad i \in [1,3] \tag{22.A.13}$$

Deposition mode (b):

$$\frac{\sigma_i'}{\mu_\kappa}(t) = \frac{v_i^2}{\det \mathbf{F}^e}\frac{\partial E_\kappa}{\partial E_{Gi}^e}(t) = \left(\frac{a_\kappa}{\mu_\kappa} - 1\right)v_i^2(t) + v_i^4(t),\quad i \in [1,3] \tag{22.A.14}$$

The components of the 2nd Piola-Kirchhoff stress $\boldsymbol{\tau}_\kappa'$ are linked to those of the Cauchy stress by

$$\tau_{\kappa i}' = \frac{\sigma_i'}{v_i^2},\quad i \in [1,3]. \tag{22.A.15}$$

Appendix B: Details of the calculations of stress in an elastic solid
Strain associated with $\mathbf{F}(t)$:

$$\mathbf{F}(t) = \mathbf{diag}\,[F_1(t),\ F_2(t),\ F_3(t)] \tag{22.B.1}$$

$$I_1(t) = \operatorname{tr}\mathbf{E}_\mathrm{G} = \sum_{i=1}^{3} \frac{1}{2}\,(F_i^2(t) - 1) \tag{22.B.2}$$

$$I_2(t) = \operatorname{tr}\mathbf{E}_\mathrm{G}^2 = \sum_{i=1}^{3} \frac{1}{4}\,(F_i^2(t) - 1)^2$$

$$\frac{\partial}{\partial E_{Gi}(t)} = \frac{\partial}{\partial F_i(t)}\frac{dF_i(t)}{dE_{Gi}(t)} = \frac{1}{F_i(t)}\frac{\partial}{\partial F_i(t)}, \quad i \in [1,3] \tag{22.B.3}$$

$$\frac{\partial I_1(t)}{\partial E_{Gi}(t)} = 1, \quad \frac{\partial I_2(t)}{\partial E_{Gi}(t)} = F_i^2(t) - 1, \quad i \in [1,3] \tag{22.B.4}$$

Deposition mode (a): strain associated with $\mathbf{F}(t,\tau) = \mathbf{F}(t) \cdot \mathbf{F}^{-1}(\tau)$:

$$\mathbf{F}(t,\tau) = \mathbf{diag}\,\Big[\frac{F_1(t)}{F_1(\tau)},\ \frac{F_2(t)}{F_2(\tau)},\ \frac{F_3(t)}{F_3(\tau)}\Big] \tag{22.B.5}$$

$$\mathbf{E}_\mathrm{G}(t,\tau) = \frac{1}{2}(\mathbf{F}^\mathrm{T}\cdot\mathbf{F} - \mathbf{I}) = \frac{1}{2}\,\mathbf{diag}\,\Big[\frac{F_1^2(t)}{F_1^2(\tau)} - 1,\ \frac{F_2^2(t)}{F_2^2(\tau)} - 1,\ \frac{F_3^2(t)}{F_3^2(\tau)} - 1\Big] \tag{22.B.6}$$

$$I_1(t,\tau) = \operatorname{tr}\mathbf{E}_\mathrm{G}(t,\tau) = \sum_{i=1}^{3} \frac{1}{2}\,\Big(\frac{F_i^2(t)}{F_i^2(\tau)} - 1\Big) \tag{22.B.7}$$

$$I_2(t,\tau) = \operatorname{tr}\mathbf{E}_\mathrm{G}^2(t,\tau) = \sum_{i=1}^{3} \frac{1}{4}\,\Big(\frac{F_i^2(t)}{F_i^2(\tau)} - 1\Big)^2 \tag{22.B.8}$$

$$\frac{\partial I_1(t,\tau)}{\partial E_{Gi}(t)} = \frac{1}{F_i^2(\tau)}, \quad i \in [1,3] \tag{22.B.9}$$

$$\frac{\partial I_2(t,\tau)}{\partial E_{Gi}(t)} = \frac{1}{F_i^2(\tau)}\Big(\frac{F_i^2(t)}{F_i^2(\tau)} - 1\Big), \quad i \in [1,3] \tag{22.B.10}$$

$$\frac{\partial \underline{E}^{\natural}(t)}{\partial E_{Gi}(t)} = \frac{\partial E(t)}{\partial E_{Gi}(t)} + \int_0^t \frac{\partial E(t,\tau)}{\partial E_{Gi}(t)}\frac{d}{d\tau}\det\mathbf{F}(\tau)\,d\tau, \quad i \in [1,3] \tag{22.B.11}$$

$$\sigma_i'(t) = \frac{F_i^2(t)}{\det^2\mathbf{F}(t)}\frac{\partial \underline{E}^{\natural}}{\partial E_{Gi}}(t), \quad i \in [1,3] \tag{22.B.12}$$

With $\underline{E}$ defined by (22.7.14), and with the functional $\mathbb{I}$ defined by (22.8.13),

$$\frac{\sigma_i'}{\mu}(t) = \frac{1}{\det^2\mathbf{F}(t)}\Big[(\frac{a}{\mu} - 1)\,F_i^2\,\mathbb{I}(\frac{1}{F_i^2}) + F_i^4\,\mathbb{I}(\frac{1}{F_i^4})\Big](t), \quad i \in [1,3] \tag{22.B.13}$$

Deposition mode (b):

For the energy $\underline{E}$ defined by (22.7.14), then

$$\frac{\sigma_i'}{\mu}(t) = \frac{1}{\det\mathbf{F}(t)}\Big((\frac{a}{\mu} - 1)\,F_i^2(t) + F_i^4(t)\Big), \quad i \in [1,3]\,. \tag{22.B.14}$$

Chapter 23

Elastic-growing mixtures

23.1 Thermodynamically consistent growth laws in a mixture context

Growth is an irreversible process, metabolism and catabolism being associated with distinct pathways. Energy provided by metabolism is used to grow the mass and improve the mechanical properties of the tissue constituents. Heat dissipation is in some sense minimized: the tissue recovers as much energy as possible to drive the growth and stiffening/self-healing processes. Indeed in contrast to damage, stiffening contributes negatively to the dissipation inequality. In other words, the model is built so that the dissipation can only be positive or zero. The idea is that mechanisms that require energy (active processes) are scaled down if the energy provided by passive processes is not sufficient. The energy available to the active processes is limited by satisfaction of the dissipation inequality. Thus, along Loret and Simões [2014], an effort should be made to delineate unambiguously the signs of the contributions to the dissipation inequality of all the processes involved in the growth and modification of the physical properties of the tissue. In fact, while there are myriads of analyses of passive processes, e.g., damage of engineering materials and bone tissues, stiffening and self-healing processes have been much more scarcely addressed.

23.1.1 The mixture as a whole

Along (4.2.1), the set of species $\mathcal{K}$ is partitioned in growing constituents $\mathcal{S}^*$ and non growing species $\mathcal{K}^*$. Here the growing constituents are collagen fibers and proteoglycans while non growing species include water, ionic species, and macromolecules.

As stated earlier, the velocity $\mathbf{v}_k$ of any growing constituent $k \in \mathcal{S}^*$ is the velocity $\mathbf{v}_s$ of the solid[23.1]. Thus the time derivative d^k/dt following a constituent k is equal to the time derivative following the solid skeleton d/dt. Therefore, for any growing constituent $k \in \mathcal{S}^*$, the balance of mass (21.3.4) in terms of mass content takes the form $dm^k/dt = \hat{m}^k$, and the expression (21.3.45) for the velocity gradient holds, namely

$$\mathbf{L} \equiv \frac{d\mathbf{F}}{dt} \cdot \mathbf{F}^{-1} = \overbrace{\frac{d\mathbf{F}_k^{\mathrm{e}}}{dt} \cdot (\mathbf{F}_k^{\mathrm{e}})^{-1}}^{\mathbf{L}_k^{\mathrm{e}}} + \mathbf{F}_k^{\mathrm{e}} \cdot \overbrace{\underbrace{\frac{d\mathbf{F}_k^{\mathrm{g}}}{dt} \cdot (\mathbf{F}_k^{\mathrm{g}})^{-1}}_{\mathbf{L}_{\kappa k}^{\mathrm{g}}} \cdot (\mathbf{F}_k^{\mathrm{e}})^{-1}}^{\mathbf{L}_k^{\mathrm{g}}}, \quad k \in \mathcal{S}^*. \tag{23.1.1}$$

The velocity and the energy of any constituent before and after deposition are kept distinct temporarily.

23.1.1.1 The mixture as a whole: definitions

For each growing constituent k, entities relative to the intermediate configurations are defined, namely
- the elastic right Cauchy-Green tensor $\mathbf{C}_k^{\mathrm{e}}$ and the elastic Green strain $\mathbf{E}_{\mathrm{G}k}^{\mathrm{e}}$,

$$\mathbf{C}_k^{\mathrm{e}} = (\mathbf{F}_k^{\mathrm{e}})^{\mathrm{T}} \cdot \mathbf{F}_k^{\mathrm{e}}, \quad \mathbf{E}_{\mathrm{G}k}^{\mathrm{e}} = \tfrac{1}{2}(\mathbf{C}_k^{\mathrm{e}} - \mathbf{I}), \quad k \in \mathcal{S}^*; \tag{23.1.2}$$

[23.1]So we have no diffusion of the growing constituents, neither inward nor outward. The velocity of the growing constituents should be distinct from the solid skeleton to allow for diffusion of (part of) the secreted mass via water bathing the tissue.

- the apparent mass density ρ_κ^k and the rate of mass supply $\hat{\rho}_\kappa^k$,

$$
\rho_\kappa^k = \rho^k \det \mathbf{F}_k^{\mathrm{e}} = \frac{m^k}{\det \mathbf{F}_k^{\mathrm{g}}}, \quad \hat{\rho}_\kappa^k = \hat{\rho}^k \det \mathbf{F}_k^{\mathrm{e}} = \frac{\hat{m}^k}{\det \mathbf{F}_k^{\mathrm{g}}}, \quad k \in \mathcal{S}^* ,
\tag{23.1.3}
$$

which satisfy the rate equations,

$$
\frac{dm^k}{dt} = \det \mathbf{F}_k^{\mathrm{g}} \frac{d\rho_\kappa^k}{dt} + m^k \operatorname{tr} \mathbf{L}_{\kappa k}^{\mathrm{g}} = \hat{m}^k, \quad \frac{d\rho_\kappa^k}{dt} + \rho_\kappa^k \operatorname{tr} \mathbf{L}_{\kappa k}^{\mathrm{g}} = \hat{\rho}_\kappa^k, \quad k \in \mathcal{S}^* ;
\tag{23.1.4}
$$

- the remodeling variable r_κ^k which vanishes if growth preserves density,

$$
r_\kappa^k = \frac{\rho_\kappa^k}{\rho^k} - 1, \quad k \in \mathcal{S}^* ;
\tag{23.1.5}
$$

- for the time being and for ease of writing, in the set of independent state variables relative to the intermediate configurations of the growing constituent k,

$$
\mathcal{V}_{\kappa k} = \{\mathbf{E}_{\mathrm{G}k}^{\mathrm{e}}, \ T, \ r_\kappa^k, \ \xi_{\kappa k}\} ,
\tag{23.1.6}
$$

a single additional dimensionless dissipative variable $\xi_{\kappa k}$ per growing constituent is displayed. It will be expanded later to represent the state of the homeostatic surface, the cell energy density, and, for collagen $k = \mathrm{c}$, an orientation distribution function (a set of vectors and scalars) and the pyridinoline crosslink density. The temperature T is common to all species as the mixture is in local thermal equilibrium.

Implicit in the choice of state variables are the following assumptions:
- the chemomechanical variables $\cup_{k \in \mathcal{K}^*} m^k$, typically the masses of non growing species, are independent of the kinematics of the growing constituents;
- on the other hand, for each growing constituent, the structuration/dissipation and remodeling variables are all assumed to be affected by the kinematics of the growth process.

23.1.1.2 The mixture as a whole: dissipation in the current configuration

The formalism used for a single solid can not be adopted for the mixture. Indeed, the elastic-growth multiplicative decomposition sketched in Fig. 23.1.1 endows each growing constituent with its own intermediate configuration[23.2]. The starting points are the dissipation due to deformation and growth (4.7.28) associated with the mixture as a whole, and the external entropy production in the format (21.4.6),

$$
\begin{aligned}
T \hat{S}_{(\mathrm{i},1,4)} &= -\frac{1}{\det \mathbf{F}} \frac{dE}{dt} + \boldsymbol{\sigma} : \boldsymbol{\nabla} \mathbf{v}_\mathrm{s} - S \frac{dT}{dt} + \frac{1}{\det \mathbf{F}} \sum_{k \in \mathcal{K}^*} \mu_k^{\mathrm{ec}} \frac{dm^k}{dt} \\
&+ \sum_{k \in \mathcal{S}^*} \hat{\rho}^k \left(\tilde{e}_k + \tfrac{1}{2} (\tilde{\mathbf{v}}_k - \mathbf{v}_k)^2 \right) + \hat{\mathcal{E}}^k \geq 0 .
\end{aligned}
\tag{23.1.7}
$$

The free energies of the mixture, $\underline{E}$ referred to the initial volume $\underline{V}$, and E referred to the current volume V, are obtained by summing the free energies of the growing constituents and of the non growing species. The free energy of the latter is assumed to depend only on their mass contents, namely $E_{\mathrm{ch}} = E_{\mathrm{ch}}(\cup_{k \in \mathcal{K}^*} m^k)$. The contribution to the dissipation of each constituent may be estimated *relative* to any configuration, e.g., relative to the current configuration for non growing species. Properties for growing constituents rely on the intermediate configurations. Since the mass $M_{\kappa k}$ in the intermediate configuration is equal to the current mass M_k, the free energy $V_{\kappa k} E_{\kappa k}$ with $E_{\kappa k} = E_k \det \mathbf{F}_k^{\mathrm{e}}$ and $V_{\kappa k} = V_k (\det \mathbf{F}_k^{\mathrm{e}})^{-1}$ is also equal to $V_k E_k$, with E_k the free energy per current volume of the species k. Therefore,

$$
\underline{E}\,\underline{V} = E\,V = \sum_{k \in \mathcal{S}^*} V_k\, E_k + E_{\mathrm{ch}}\, V ,
\tag{23.1.8}
$$

[23.2] Caution should be exercised when writing the dissipation inequality in the intermediate configurations. An issue of the same flavor arises for mixtures which are not in local thermal equilibrium, as explored in Section 10.6: the quantity of interest is the sum over constituents of the entropy productions $\hat{S}_{(\mathrm{i})}^k$ relative to the current configuration, and not the sum over constituents of the product T_k times $\hat{S}_{(\mathrm{i})}^k$. Here the total entropy production is also the sum of the entropy productions $\hat{S}_{(\mathrm{i})}^k$ of the constituents: the entropy $\hat{S}_{\kappa(\mathrm{i})}^k$ relative to the intermediate configuration of constituent k is simply $\hat{S}_{(\mathrm{i})}^k$ divided by $\det \mathbf{F}_k^{\mathrm{e}}$.

(a) intrinsic properties

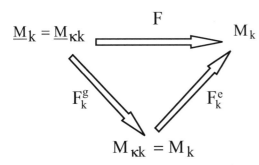

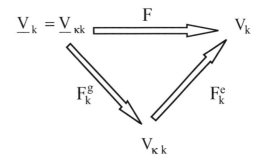

(b) partial properties

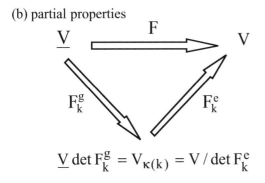

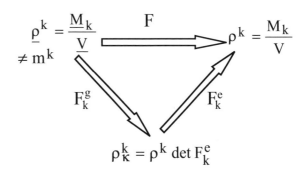

(c) free energies

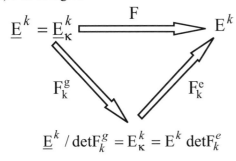

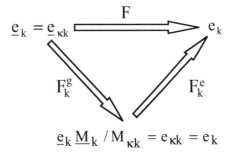

Fig. 23.1.1: Multiplicative decomposition of the deformation into a growth transformation and an elastic transformation for a growing constituent k of a mixture. Transports of mass M_k, volume of constituent V_k, apparent density ρ^k, free energy per unit volume E^k, free energy per unit mass e_k, and volume of mixture V, in the reference, intermediate and current configurations. The sketches apply to solids as well simply by removing the index k.

whence, with $n^k = V_k/V$ volume fraction, $v^k = V_k/\underline{V}$ volume content,

$$\underline{E} = \sum_{k \in \mathcal{S}^*} v^k E_k + \det \mathbf{F}\, E_{\mathrm{ch}}, \quad E = \sum_{k \in \mathcal{S}^*} n^k E_k + E_{\mathrm{ch}}. \tag{23.1.9}$$

Now accounting for the rate relations (23.1.1), (23.1.4), for the definition (23.1.5), and for the differential formula (2.5.14),

$$
\begin{aligned}
\frac{1}{\det \mathbf{F}} \frac{d(v^k E_k)}{dt} &= \left(n^k E_k\, \mathbf{I} - n^k\, \mathbf{C}_k^{\mathrm{e}} \cdot \frac{\partial E_k}{\partial \mathbf{E}_{\mathrm{G}k}^{\mathrm{e}}} \right) : \mathbf{L}_{\kappa k}^{\mathrm{g}} + n^k \left(\mathbf{F}_k^{\mathrm{e}} \cdot \frac{\partial E_k}{\partial \mathbf{E}_{\mathrm{G}k}^{\mathrm{e}}} \cdot (\mathbf{F}^{\mathrm{e}})^{\mathrm{T}} \right) : \mathbf{L} \\
&\quad + \left(\frac{\rho^k}{\rho_\kappa^k} n^k E_k + n^k \frac{\partial E_k}{\partial r_\kappa^k} \right) \frac{dr_\kappa^k}{dt} + n^k \frac{\partial E_k}{\partial \xi_{\kappa k}} \frac{d\xi_{\kappa k}}{dt} + n^k \frac{\partial E_k}{\partial T} \frac{dT}{dt},
\end{aligned}
\tag{23.1.10}
$$

where the fact that the intrinsic mass density ρ_k is constant has been anticipated. The dissipation can now be expanded in terms of the state variables, using in particular (2.5.14), (23.1.3) and (23.1.4),

$$
\begin{aligned}
T\hat{S}_{(\mathrm{i},1,4)} &= \left(\boldsymbol{\sigma} - \sum_{k\in\mathcal{S}^*} n^k\,\mathbf{F}_k^{\mathrm{e}} \cdot \frac{\partial E_k}{\partial\mathbf{E}_{\mathrm{G}k}^{\mathrm{e}}} \cdot (\mathbf{F}^{\mathrm{e}})^{\mathrm{T}}\right) : \mathbf{L} \\
&\quad - \left(S + \sum_{k\in\mathcal{S}^*} n^k\,\frac{\partial E_k}{\partial T}\right)\frac{dT}{dt} + \sum_{k\in\mathcal{K}^*}\left(\frac{\mu_k}{\det\mathbf{F}} - \frac{\partial E}{\partial m^k}\right)\frac{dm^k}{dt} \\
&\quad + \sum_{k\in\mathcal{S}^*}\left(-n^k\,\frac{\partial E_k}{\partial r_\kappa^k} + \frac{\rho^k}{\rho_\kappa^k}\left(n^k\,(\tilde{E}_k - E_k) + \tfrac12\,\rho^k\,(\tilde{\mathbf{v}}_k - \mathbf{v}_k)^2\right)\right)\frac{dr_\kappa^k}{dt} - n^k\,\frac{\partial E_k}{\partial\xi_{\kappa k}}\,\frac{d\xi_{\kappa k}}{dt} \\
&\quad + \sum_{k\in\mathcal{S}^*}\left(\mathbf{C}_k^{\mathrm{e}}\cdot n^k\,\frac{\partial E_k}{\partial\mathbf{E}_{\mathrm{G}k}^{\mathrm{e}}} + \left(n^k\,(\tilde{E}_k - E_k) + \tfrac12\,\rho^k\,(\tilde{\mathbf{v}}_k - \mathbf{v}_k)^2\right)\mathbf{I}\right) : \mathbf{L}_{\kappa k}^{\mathrm{g}} + \hat{\mathcal{E}}^k \geq 0\,,
\end{aligned}
\tag{23.1.11}
$$

where $\tilde{E}_k = \rho_k\,\tilde{e}_k$, for $k\in\mathcal{S}^*$. Using the arbitrariness of the velocity gradient $\mathbf{L}$, of the temperature T, and of the mass contents m^k of non growing species, a first set of constitutive equations deduces,

- stress-strain relation:

$$
\boldsymbol{\sigma} = \sum_{k\in\mathcal{S}^*} n^k\,\mathbf{F}_k^{\mathrm{e}} \cdot \frac{\partial E_k}{\partial\mathbf{E}_{\mathrm{G}k}^{\mathrm{e}}} \cdot (\mathbf{F}_k^{\mathrm{e}})^{\mathrm{T}}\,;
\tag{23.1.12}
$$

- entropy-temperature relation:

$$
S = -\sum_{k\in\mathcal{S}^*} n^k\,\frac{\partial E_k}{\partial T}\,;
\tag{23.1.13}
$$

- chemical potentials of non growing species:

$$
\mu_k^{\mathrm{ec}} = \det\mathbf{F}\,\frac{\partial E}{\partial m^k}\,,\quad k\in\mathcal{K}^*\,.
\tag{23.1.14}
$$

The dissipation then reduces to,

$$
\begin{aligned}
T\hat{S}_{(\mathrm{i},1,4)} &= \sum_{k\in\mathcal{S}^*}\left(-n^k\,\frac{\partial E_k}{\partial r_\kappa^k} + \frac{\rho^k}{\rho_\kappa^k}\left(n^k\,(\tilde{E}_k - E_k) + \tfrac12\,\rho^k\,(\tilde{\mathbf{v}}_k - \mathbf{v}_k)^2\right)\right)\frac{dr_\kappa^k}{dt} - n^k\,\frac{\partial E_k}{\partial\xi_{\kappa k}}\,\frac{d\xi_{\kappa k}}{dt} \\
&\quad + \sum_{k\in\mathcal{S}^*}\left(\mathbf{C}_k^{\mathrm{e}}\cdot n^k\,\frac{\partial E_k}{\partial\mathbf{E}_{\mathrm{G}k}^{\mathrm{e}}} + \left(n^k\,(\tilde{E}_k - E_k) + \tfrac12\,\rho^k\,(\tilde{\mathbf{v}}_k - \mathbf{v}_k)^2\right)\mathbf{I}\right) : \mathbf{L}_{\kappa k}^{\mathrm{g}} + \hat{\mathcal{E}}^k \geq 0\,.
\end{aligned}
\tag{23.1.15}
$$

23.1.1.3 The mixture as a whole: incompressible species

The total rate of change of mass content (23.1.4) is the sum of a pure growth contribution and a contribution associated with the change of mass density, or equivalently, a change of the remodeling variable defined by (23.1.5),

$$
\frac{1}{m^k}\frac{dm^k}{dt} = \operatorname{tr}\mathbf{L}_{\kappa k}^{\mathrm{g}} + \frac{d}{dt}\mathrm{Ln}\,(r_\kappa^k + 1),\quad k\in\mathcal{S}^*\,.
\tag{23.1.16}
$$

The incompressibility of a single species of a mixture does not introduce a constraint. A constraint arises only if all species are incompressible. The inequality (23.1.11) can not be exploited directly because the velocity gradient and rates of mass contents then satisfy the constraint $dI_{\mathrm{inc}}/dt = 0$, with

$$
\begin{aligned}
\frac{dI_{\mathrm{inc}}}{dt} &= \operatorname{tr}\mathbf{L} - \sum_{k\in\mathcal{K}}\frac{n^k}{m^k}\frac{dm^k}{dt} \\
&= \operatorname{tr}\mathbf{L} - \sum_{k\in\mathcal{S}^*} n^k\operatorname{tr}\mathbf{L}_{\kappa k}^{\mathrm{g}} + n^k\,\frac{\rho^k}{\rho_\kappa^k}\,\frac{dr_\kappa^k}{dt} - \sum_{k\in\mathcal{K}^*}\frac{n^k}{m^k}\frac{dm^k}{dt}\,.
\end{aligned}
\tag{23.1.17}
$$

The first line is obtained by summation of $(7.2.4)_4$ over all species of the mixture, assuming all species to be incompressible, i.e. their intrinsic mass density ρ_k is constant in time. The second line results from the definition of the mass density (23.1.3) and from the rate relation (23.1.4). With p_{E} the Lagrange multiplier that can be interpreted as a fluid pressure, the dissipation inequality augmented by the term $p_{\mathrm{E}}\, dI_{\mathrm{inc}}/dt$ yields the relations,

$$
\boldsymbol{\sigma}' = \boldsymbol{\sigma} + p_{\mathrm{E}}\,\mathbf{I} = \sum_{k \in \mathcal{S}^*} n^k\, \mathbf{F}_k^{\mathrm{e}} \cdot \frac{\partial E_k}{\partial \mathbf{E}_{\mathrm{G}k}^{\mathrm{e}}} \cdot (\mathbf{F}_k^{\mathrm{e}})^{\mathrm{T}};
$$
$$
\mu_k' = \mu_k - \frac{p_{\mathrm{E}}}{\rho_k} = \det \mathbf{F}\, \frac{\partial E}{\partial m^k}, \quad k \in \mathcal{K}^*.
$$
(23.1.18)

In the sequel, the superscript $'$ will be used to denote the above effective stress and chemical potentials in case where all species are incompressible, and, otherwise, the stress and chemical potentials themselves. The reduced dissipation modifies slightly to,

$$
\begin{aligned}
T\hat{S}_{(\mathrm{i},1,4)} &= \sum_{k \in \mathcal{S}^*} \mathbf{C}_k^{\mathrm{e}} \cdot \left(n^k\, \frac{\partial E_k}{\partial \mathbf{E}_{\mathrm{G}k}^{\mathrm{e}}} - p_{\mathrm{E}}\, n^k\, (\mathbf{C}_k^{\mathrm{e}})^{-1} \right) : \mathbf{L}_{\kappa k}^{\mathrm{g}} - n^k\, \frac{\partial E_k}{\partial \xi_{\kappa k}}\, \frac{d\xi_{\kappa k}}{dt} \\
&+ \sum_{k \in \mathcal{S}^*} \left(-n^k\, \frac{\partial E_k}{\partial r_\kappa^k} + \frac{\rho^k}{\rho_\kappa^k} \left(n^k\, (\tilde{E}_k - E_k) + \tfrac{1}{2}\, \rho^k\, (\tilde{\mathbf{v}}_k - \mathbf{v}_k)^2 - n^k\, p_{\mathrm{E}} \right) \right) \frac{dr_\kappa^k}{dt} \\
&+ \sum_{k \in \mathcal{S}^*} \left(n^k\, (\tilde{E}_k - E_k) + \tfrac{1}{2}\, \rho^k\, (\tilde{\mathbf{v}}_k - \mathbf{v}_k)^2 \right) (\mathbf{I} : \mathbf{L}_{\kappa k}^{\mathrm{g}}) + \hat{\mathcal{E}}^k \geq 0\,.
\end{aligned}
$$
(23.1.19)

23.1.2 Growth laws for constituents based on quadratic dissipation

The bulk of this chapter is devoted to build growth laws for the mixture constituents using the notion of convex homeostatic surfaces as an extension of the developments in Chapter 22 for a single solid. Still it is worth at this point to reconsider the other tool that has been used to deliver thermodynamically consistent growth laws, namely the quadratic dissipation assumption. Section 22.3 considers this approach for a single solid while Section 21.4.2 addresses a mixture using the free energy of the mixture as a whole. Inequalities (23.1.15), or (23.1.19) for incompressible species, make it possible to develop the growth equations for a mixture using the free energy of the constituents relative to their individual intermediate configurations.

The internal entropy production is contributed independently by dissipation mechanisms, remodeling mechanisms and growth,

$$
T\hat{S}_{(\mathrm{i},1,4)} = -\sum_{k \in \mathcal{S}^*} X_{\kappa k}\, \frac{d\xi_{\kappa k}}{dt} - \sum_{k \in \mathcal{S}^*} R_{\kappa k}\, \frac{dr_\kappa^k}{dt} - \sum_{k \in \mathcal{S}^*} \mathbf{f}_{\kappa k} : \mathbf{L}_{\kappa k}^{\mathrm{g}} \geq 0\,.
$$
(23.1.20)

To simplify the presentation, a single scalar dissipation variable is displayed per growing constituent k. In line with Section 22.3, the energy in unspecified forms $\hat{\mathcal{E}}^k$ is formally partitioned among structuration/dissipation, remodeling and pure growth mechanisms,

$$
\hat{\mathcal{E}}^k = \hat{\mathcal{E}}_{(\mathrm{d})}^k + \hat{\mathcal{E}}_{(\mathrm{r})}^k + \hat{\mathcal{E}}_{(\mathrm{g})}^k, \quad k \in \mathcal{S}^*,
$$
(23.1.21)

with

$$
\hat{\mathcal{E}}_{(\mathrm{d})}^k = \beta_{\kappa k}^{(\mathrm{d})}\, \frac{d\xi_{\kappa k}}{dt}, \quad \hat{\mathcal{E}}_{(\mathrm{r})}^k = \beta_{\kappa k}^{(\mathrm{r})}\, \frac{dr_\kappa^k}{dt}, \quad \hat{\mathcal{E}}_{(\mathrm{g})}^k = \boldsymbol{\beta}_{\kappa k}^{(\mathrm{g})} : \mathbf{L}_{\kappa k}^{\mathrm{g}},
$$
(23.1.22)

where the scalars $\beta_{\kappa k}^{(\mathrm{d})}$ [unit : Pa], and $\beta_{\kappa k}^{(\mathrm{r})}$ [unit : Pa] and the second order tensor $\boldsymbol{\beta}_{\kappa k}^{(\mathrm{g})}$ [unit : Pa] depend on the state variables.

In order to ensure positiveness of the internal entropy production, once again, the Onsager formalism is used close to equilibrium. If couplings between the various dissipation mechanisms and across constituents are disregarded, rate constitutive equations result for the dissipation, remodeling and growth variables, namely for the generic constituent $k \in \mathcal{S}^*$,

$$
\begin{aligned}
-X_{\kappa k} &= \left(\frac{\alpha_{\kappa k}^{(\mathrm{d})}}{t_{\mathrm{d}k}} \right)^{-1} \frac{d\xi_{\kappa k}}{dt} = -n^k\, \frac{\partial E_k}{\partial \xi_{\kappa k}} + \beta_{\kappa k}^{(\mathrm{d})}, \\
-R_{\kappa k} &= \left(\frac{\alpha_{\kappa k}^{(\mathrm{r})}}{t_{\mathrm{r}k}} \right)^{-1} \frac{dr_\kappa^k}{dt} = -n^k\, \frac{\partial E_k}{\partial r_\kappa^k} + \frac{\rho^k}{\rho_\kappa^k} \left(n^k\, (\tilde{E}_k - E_k) + \tfrac{1}{2}\, \rho^k\, (\tilde{\mathbf{v}}_k - \mathbf{v}_k)^2 \right) + \beta_{\kappa k}^{(\mathrm{r})}, \\
-\mathbf{f}_{\kappa k} &= \left(\frac{\boldsymbol{\alpha}_{\kappa k}^{(\mathrm{g})}}{t_{\mathrm{g}k}} \right)^{-1} : \mathbf{L}_{\kappa k}^{\mathrm{g}} = \mathbf{C}_k^{\mathrm{e}} \cdot n^k\, \frac{\partial E_k}{\partial \mathbf{E}_{\mathrm{G}k}^{\mathrm{e}}} + \left(n^k\, (\tilde{E}_k - E_k) + \tfrac{1}{2}\, \rho^k\, (\tilde{\mathbf{v}}_k - \mathbf{v}_k)^2 \right) \mathbf{I} + \boldsymbol{\beta}_{\kappa k}^{(\mathrm{g})}.
\end{aligned}
$$
(23.1.23)

Here the t's are time scales, $\alpha_{\kappa k}^{(\mathrm{d})}$ and $\alpha_{\kappa k}^{(\mathrm{r})}$ are positive scalars, and $\boldsymbol{\alpha}_{\kappa k}^{(\mathrm{g})}$ is a fourth order positive definite tensor, endowed with the major symmetry (but a priori not with minor symmetries), all quantities with dimension Pa^{-1} depending on the state variables.

As such, the formalism presents a major drawback: it does not account for mechanisms of structuration that are associated with negative dissipation.

23.1.3   The dissipation inequalities in the intermediate configurations

Returning to the analysis in intermediate configurations, a number of simplifications are adopted:
- growth is preserving density, so that the remodeling variable r_κ^k vanishes,

$$r_\kappa^k = \frac{\rho^k \det \mathbf{F}_k^{\mathrm{e}}}{\underline{\rho}^k} - 1 = 0 \,. \tag{23.1.24}$$

The rate of mass growth,

$$\frac{1}{m^k} \frac{dm^k}{dt} = \frac{\hat{\rho}^k}{\rho^k} = \operatorname{tr} \mathbf{L}_{\kappa k}^{\mathrm{g}}, \quad k \in \mathcal{S}^* \,, \tag{23.1.25}$$

results from the isotropic part of the rate of growth transformation as indicated by (21.3.54) or (23.1.4);
- the free energy $\tilde{E}_k$ of the deposited mass at the time of deposition is equal to the free energy E_k of the existing mass at the deposition site. An assumption of the same flavor applies to the velocities $\tilde{\mathbf{v}}_k$ and $\mathbf{v}_k$;
- the system is thermodynamically open to mass only, in the sense that the independent source of free energy $\hat{\mathcal{E}}^k$ is zero. Supply from external sources $\hat{\mathcal{E}}^k \neq 0$ may be considered in a general setting, but it is not envisaged here (by supply, we mean here positive supply $\hat{\mathcal{E}}^k \geq 0$, energy being *extracted* from the surroundings);
- cells are viewed as an internal energy source: this energy is transferred to the tissue by the growing process, and it therefore decreases during this process. The energy may be understood as being channeled to increase the mass and to improve the mechanical properties of the collagen network and proteoglycans;
- positive dissipation ends up to heat unless it is a source for the stiffening/self-healing processes to take place. In other words, the model is built so that the global dissipation can only be positive or zero.

The reduced dissipation contributed by growth,

$$
\begin{aligned}
T\hat{S}_{(\mathrm{i},4)} &= \sum_{k \in \mathcal{S}^*} \frac{1}{\det \mathbf{F}_k^{\mathrm{e}}} T\hat{S}_{\kappa(\mathrm{i})}^k \geq 0, \quad T\hat{S}_{\kappa(\mathrm{i})}^k = \boldsymbol{\Psi}_\kappa^k : \mathbf{L}_{\kappa k}^{\mathrm{g}} - n^k \det \mathbf{F}_k^{\mathrm{e}} \frac{\partial E_k}{\partial \xi_{\kappa k}} \frac{d\xi_{\kappa k}}{dt} \geq 0, \\
&= \sum_{k \in \mathcal{S}^*} \frac{n^k}{\det \mathbf{F}_k^{\mathrm{e}}} T\hat{S}_{\kappa k}^{(\mathrm{i})} \geq 0, \quad T\hat{S}_{\kappa k}^{(\mathrm{i})} = \boldsymbol{\Psi}_{\kappa k} : \mathbf{L}_{\kappa k}^{\mathrm{g}} - \det \mathbf{F}_k^{\mathrm{e}} \frac{\partial E_k}{\partial \xi_{\kappa k}} \frac{d\xi_{\kappa k}}{dt} \geq 0,
\end{aligned} \tag{23.1.26}
$$

will be satisfied by requiring the contribution associated with each growing constituent to be positive. This requirement is of course more than sufficient, but it has the practical advantage to restrict the analysis to each constituent. However, we still have to partition the cell energy to each growing constituent. Perhaps, a starting point is to compare the characteristic times for cell density decrease, collagen growth and proteoglycan growth, Section 20.7.5.2. Proteoglycans seem to pre-exist to the time window that is of interest in, e.g., Williamson et al. [2001]. This observation would imply that the growth windows for PGs and collagen are essentially sequential, rather than overlapping, and therefore, at the time where it grows, a constituent would benefit of the whole cell energy.

An intrinsic stress is denoted by a subscript and a partial, or apparent, stress by a superscript, e.g., respectively for the 2nd Piola-Kirchhoff stress $\boldsymbol{\tau}_\kappa$, the effective 2nd Piola-Kirchhoff stress $\boldsymbol{\tau}'_\kappa$, and the Mandel stress $\boldsymbol{\Psi}_\kappa$ of the growing constituent k,

$$\boldsymbol{\tau}_\kappa^k = n^k \boldsymbol{\tau}_{\kappa k}, \quad \boldsymbol{\tau}_\kappa^{k'} = n^k \boldsymbol{\tau}'_{\kappa k}, \quad \boldsymbol{\Psi}_\kappa^k = n^k \boldsymbol{\Psi}_{\kappa k} \,. \tag{23.1.27}$$

The Mandel stress $\boldsymbol{\Psi}_\kappa$,

$$\boldsymbol{\Psi}_\kappa^k = \mathbf{C}_k^{\mathrm{e}} \cdot \boldsymbol{\tau}_\kappa^k, \quad \boldsymbol{\Psi}_{\kappa k} = \mathbf{C}_k^{\mathrm{e}} \cdot \boldsymbol{\tau}_{\kappa k} \,, \tag{23.1.28}$$

may be expressed in terms of a partial 2nd Piola-Kirchhoff stress $\boldsymbol{\tau}_\kappa$ defined by the relations,

$$
\begin{aligned}
\boldsymbol{\tau}_\kappa^k &= \boldsymbol{\tau}_\kappa^{k'} - p_{\mathrm{E}} \, n^k \det \mathbf{F}_k^{\mathrm{e}} \, (\mathbf{C}_k^{\mathrm{e}})^{-1}, \quad \boldsymbol{\tau}_\kappa^{k'} = n^k \det \mathbf{F}_k^{\mathrm{e}} \frac{\partial E_k}{\partial \mathbf{E}_{\mathrm{G}k}^{\mathrm{e}}}, \\
\boldsymbol{\tau}_{\kappa k} &= \boldsymbol{\tau}'_{\kappa k} - p_{\mathrm{E}} \det \mathbf{F}_k^{\mathrm{e}} \, (\mathbf{C}_k^{\mathrm{e}})^{-1}, \quad \boldsymbol{\tau}'_{\kappa k} = \det \mathbf{F}_k^{\mathrm{e}} \frac{\partial E_k}{\partial \mathbf{E}_{\mathrm{G}k}^{\mathrm{e}}} \,.
\end{aligned} \tag{23.1.29}
$$

Although non symmetric in general, Mandel stress has still only 6 independent components, since Eqn (23.1.28) implies the restriction $(\boldsymbol{\Psi}_\kappa^k \cdot \mathbf{C}_k^e)^T = \boldsymbol{\Psi}_\kappa^k \cdot \mathbf{C}_k^e$.

The notation and ensuing interpretation derive from the following observation. Let us note $\boldsymbol{\sigma}^k$ the partial Cauchy stress associated with the partial 2nd Piola-Kirchhoff stress $\boldsymbol{\tau}_\kappa^k$, and similarly $\boldsymbol{\sigma}^{k'}$ the effective partial Cauchy stress associated with the effective partial 2nd Piola-Kirchhoff stress $\boldsymbol{\tau}_\kappa^{k'}$,

$$
\begin{aligned}
\boldsymbol{\sigma}^k &= (\det \mathbf{F}_k^e)^{-1} \, \mathbf{F}_k^e \cdot \boldsymbol{\tau}_\kappa^k \cdot (\mathbf{F}_k^e)^T, \\
\boldsymbol{\sigma}^{k'} &= (\det \mathbf{F}_k^e)^{-1} \, \mathbf{F}_k^e \cdot \boldsymbol{\tau}_\kappa^{k'} \cdot (\mathbf{F}_k^e)^T = \boldsymbol{\sigma}^k + p_E \, n^k \, \mathbf{I}.
\end{aligned}
\tag{23.1.30}
$$

The total stress is contributed by growing and nongrowing species, while the effective stress is contributed by growing constituents only,

$$
\boldsymbol{\sigma} = \sum_{k \in \mathcal{S}^* \cup \mathcal{K}^*} \boldsymbol{\sigma}^k, \quad \boldsymbol{\sigma}' = \sum_{k \in \mathcal{S}^*} \boldsymbol{\sigma}^{k'}.
\tag{23.1.31}
$$

The non growing species $\mathcal{K}^*$ bear the fluid pressure p_E, namely $\sum_{k \in \mathcal{K}^*} \boldsymbol{\sigma}^k = -(\sum_{k \in \mathcal{K}^*} n^k) p_E \, \mathbf{I}$. Then

$$
\boldsymbol{\sigma}' = \sum_{k \in \mathcal{S}^*} \boldsymbol{\sigma}^{k'} = \sum_{k \in \mathcal{S}^*} \boldsymbol{\sigma}^k + (\sum_{k \in \mathcal{S}^*} n^k) p_E \, \mathbf{I} = \sum_{k \in \mathcal{S}^* \cup \mathcal{K}^*} \boldsymbol{\sigma}^k + p_E \, \mathbf{I} = \boldsymbol{\sigma} + p_E \, \mathbf{I},
\tag{23.1.32}
$$

in agreement with (23.1.18).

The following remark is worth interest at this stage: since the intrinsic mass density ρ_k and apparent mass density ρ_κ^k are constant during growth, the latter being equal to the reference apparent mass density $\underline{\rho}^k = \underline{n}^k \rho_k$, the volume fraction n^k is a function of the sole elastic strain,

$$
n^k \det \mathbf{F}_k^e = \frac{\rho_\kappa^k}{\rho_k} = \underline{n}^k.
\tag{23.1.33}
$$

Its rate is given by (23.1.34)[23.3].

23.1.4 Constituents with evolving microstructure: options

The working intermediate configurations are invariant as defined in Section 21.3.1.5 so that the material derivative following the solid d/dt is objective.

23.1.4.1 Approach 1: use of a distribution density of fibers

The distribution density of fibers over the space directions varies as new fibers are deposited due to the biochemical events and mechanical loading. An issue with this approach is that the sign of the dissipation due to the change of the distribution density is not controlled.

Free energy in terms of a distribution density

The number of collagen fibers over the solid angle dS_κ is equal to $\mathcal{D}_{\kappa c}(\mathbf{m}_{\kappa c}) dS_\kappa$, with $\mathcal{D}_{\kappa c}(\mathbf{m}_{\kappa c})$ the distribution density. The sense of the fibers is elusive, so that $\mathcal{D}_{\kappa c}(-\mathbf{m}_{\kappa c}) = \mathcal{D}_{\kappa c}(\mathbf{m}_{\kappa c})$. The response of the collagen network

---

[23.3] Some relations are worth recording. From Fig. 23.1.1, since $\rho^k = n^k \rho_k$ and $\rho_\kappa^k = \rho^k \det \mathbf{F}_k^e$, then, for any $k \in \mathcal{S}^*$,

$$
\frac{1}{n^k} \frac{dn^k}{dt} = \frac{1}{\rho_\kappa^k} \frac{d\rho_\kappa^k}{dt} - \frac{1}{\rho_k} \frac{d\rho_k}{dt} - \operatorname{tr} \mathbf{L}_k^e,
\tag{23.1.34}
$$

and also,

$$
\begin{aligned}
V_k = \frac{M_k}{\rho_k} &\rightarrow \frac{1}{V_k} \frac{dV_k}{dt} = \frac{1}{M_k} \frac{dM_k}{dt} - \frac{1}{\rho_k} \frac{d\rho_k}{dt} = \frac{\hat{\rho}^k}{\rho^k} - \frac{1}{\rho_k} \frac{d\rho_k}{dt}, \\
n^k = \frac{V_k}{V} &\rightarrow \frac{1}{n^k} \frac{dn^k}{dt} = \frac{\hat{\rho}^k}{\rho^k} - \frac{1}{\rho_k} \frac{d\rho_k}{dt} - \operatorname{div} \mathbf{v}_s.
\end{aligned}
\tag{23.1.35}
$$

These relations may be compared with (7.2.18), (21.3.41) and footnote 21.2, p. 858, so as, for example, to deduce the intrinsic volumetric strain, assuming the rate of mass supply to be known.

If growth preserves apparent mass density, the first term in the rhs of (23.1.34) vanishes, and, if the tissue is intrinsically incompressible, so does the 2nd term. Then, the volume fraction changes only due to the accommodating strain, that is, the strain that allows PGs and collagen to live together (same global deformation gradient $\mathbf{F}$) while being endowed with distinct growth transformations $\mathbf{F}^g$.

is obtained by summing the directional response over the unit hemisphere as indicated by (12.1.11) (as a abuse of notation, we note $\mathcal{D}_{\kappa c}(\mathbf{m}_{\kappa c})$ or $\mathcal{D}_{\kappa c}(\theta_{\kappa c}, \phi_{\kappa c})$),

$$2 \int_{\mathbb{S}^2_+} [\cdots] \, \mathcal{D}_{\kappa c}(\theta_{\kappa c}, \phi_{\kappa c}) \, dS_{\kappa} = 2 \int_0^{2\pi} d\phi_{\kappa c} \int_0^{\pi/2} [\cdots] \, \mathcal{D}_{\kappa c}(\theta_{\kappa c}, \phi_{\kappa c}) \, \sin\theta_{\kappa c} \, d\theta_{\kappa c} \, . \tag{23.1.36}$$

Alternatively, integration may be performed over the complete unit sphere $\mathbb{S}^2$, accounting for the relation $\mathcal{D}_{\kappa c}(-\mathbf{m}_{\kappa c}) = \mathcal{D}_{\kappa c}(\mathbf{m}_{\kappa c})$. The average density of fibers, i.e., the total number of collagen fibers N_{cf} in a volume, is obtained from the distribution density,

$$\mathcal{D}_{\kappa c}^{av} = \frac{1}{4\pi} \int_{\mathbb{S}^2} \mathcal{D}_{\kappa c}(\theta_{\kappa c}, \phi_{\kappa c}) \, dS_{\kappa} \, . \tag{23.1.37}$$

Actually, the distribution density may be decomposed multiplicatively in an average factor, which depends only on the mass of collagen, or on the number of fibers which are all assumed to have identical mass, and a fluctuation orientation distribution function (ODF) in the standard sense,

$$\mathcal{D}_{\kappa c}(\theta_{\kappa c}, \phi_{\kappa c}) = \mathcal{D}_{\kappa c}^{av} \, \mathcal{D}_{\kappa c}^{fl}(\theta_{\kappa c}, \phi_{\kappa c}) \, , \tag{23.1.38}$$

with

$$\int_{\mathbb{S}^2} \mathcal{D}_{\kappa c}^{fl}(\theta_{\kappa c}, \phi_{\kappa c}) \, dS_{\kappa} = 4\pi \, , \quad \int_{\mathbb{S}^2} \frac{\partial \mathcal{D}_{\kappa c}^{fl}}{\partial t}(\theta_{\kappa c}, \phi_{\kappa c}) \, dS_{\kappa} = 0 \, . \tag{23.1.39}$$

Therefore, if the density is isotropic, then $\mathcal{D}_{\kappa c}^{fl}(\theta_{\kappa c}, \phi_{\kappa c}) = 1$.

The free energy of the collagen network $E_{\kappa}^c = E_{\kappa}^c(\mathcal{V}_{\kappa c})$ depends on the set of variables (growth is assumed to preserve density),

$$\mathcal{V}_{\kappa c} = \{\mathbf{E}_{Gc}^e, \, \cup_{\mathbb{S}^2} \mathbf{m}_{\kappa c}, \, \xi_{\kappa c}, \, \boldsymbol{\xi}_{\kappa c}\} \, , \tag{23.1.40}$$

and it is contributed additively by an elastic part and by a part associated with the structuration of the homeostatic surface, namely from (22.5.10),

$$E_{\kappa}^c = \int_{\mathbb{S}^2} n^c w_{\kappa c}(L_{\kappa c}) \, \mathcal{D}_{\kappa c}(\mathbf{m}_{\kappa c}) \, dS_{\kappa} + \frac{1}{2} \xi_{\kappa c} \, \mathcal{X}_{\kappa c} \, \xi_{\kappa c} + \frac{1}{2} \boldsymbol{\xi}_{\kappa c} : \mathbb{X}_{\kappa c} : \boldsymbol{\xi}_{\kappa c} \, . \tag{23.1.41}$$

The rate of the free energy becomes,

$$\frac{dE_{\kappa}^c}{dt} = \int_{\mathbb{S}^2} \frac{\partial(n^c w_{\kappa c})}{\partial \mathbf{E}_{Gc}^e} : \frac{d\mathbf{E}_{Gc}^e}{dt} \, \mathcal{D}_{\kappa c}(\mathbf{m}_{\kappa c}) + n^c w_{\kappa c}(L_{\kappa c}) \frac{\partial \mathcal{D}_{\kappa c}}{\partial t}(\mathbf{m}_{\kappa c}) \, dS_{\kappa} + \mathcal{X}_{\kappa c} \frac{d\xi_{\kappa c}}{dt} + \mathbb{X}_{\kappa c} : \frac{d\boldsymbol{\xi}_{\kappa c}}{dt} \, . \tag{23.1.42}$$

The time rate of change of the mass content is given by (22.3.8), namely, if growth preserves density, $dm^c/dt = m^c \, \text{tr} \, \mathbf{L}_{\kappa c}^g$. The time derivative of the distribution density is a partial derivative $\partial/\partial t$, and not a material derivative d/dt, since the integration over the unit sphere focused attention over fixed spatial directions, and not on individual fibers which may change direction.

Evolution of the distribution density of fibers

The evolution of the distribution density may be provided in the form of a differential equation involving diffusion, convection and reaction, much like the fabric associated with contact normals in particulate materials, Kuhn [2010]. The idea is as follows. Consider the conservation equation $\partial \mathcal{D}_{\kappa c}/\partial t + \text{div}_{\mathbb{S}}(\mathcal{D}_{\kappa c} \mathbf{v}) = \hat{\mathcal{D}}_{\kappa c}$ where $\mathbf{v}$ [unit : s^{-1}] is a rate associated with the distribution density, and $\hat{\mathcal{D}}_{\kappa c}$ is the supply rate of the distribution density, due to the deposition/disappearance of collagen fibers. Differential operators over the unit sphere are defined in Section 12.1.2. Then the time rate of change of the distribution density in a given direction

$$\frac{\partial \mathcal{D}_{\kappa c}}{\partial t} + \text{div}_{\mathbb{S}}(\mathcal{D}_{\kappa c} \mathbf{v}_*) + \text{div}_{\mathbb{S}}(\mathcal{D}_{\kappa c} (\mathbf{v} - \mathbf{v}_*)) = \hat{\mathcal{D}}_{\kappa c} \, , \tag{23.1.43}$$

becomes,

$$\frac{\partial \mathcal{D}_{\kappa c}}{\partial t} = \hat{\mathcal{D}}_{\kappa c} - \text{div}_{\mathbb{S}}(\mathcal{D}_{\kappa c} \mathbf{v}_*) + \text{div}_{\mathbb{S}}(D_{\mathbb{S}} \nabla_{\mathbb{S}} \mathcal{D}_{\kappa c}) \, , \tag{23.1.44}$$

assuming a diffusion relation of the form $\mathcal{D}_{\kappa c}(\mathbf{v} - \mathbf{v}_*) = -D_\mathsf{S} \boldsymbol{\nabla}_\mathsf{S} \mathcal{D}_{\kappa c}$ where D_S [unit : s^{-1}] plays the role of a diffusion coefficient and $\mathbf{v}_*$ is a convection velocity. The diffusion term is thought as inducing relative exchange between directions, and, therefore $\int_{\mathbb{S}^2} \mathrm{div}_\mathsf{S}(\mathcal{D}_{\kappa c}(\mathbf{v} - \mathbf{v}_*))\, dS = 0$, and

$$\int_{\mathbb{S}^2} \frac{\partial \mathcal{D}_{\kappa c}}{\partial t} + \mathrm{div}_\mathsf{S}(\mathcal{D}_{\kappa c}\, \mathbf{v}_*)\, dS = \int_{\mathbb{S}^2} \frac{\partial \mathcal{D}_{\kappa c}}{\partial t} + \mathrm{div}_\mathsf{S}(\mathcal{D}_{\kappa c}\, \mathbf{v})\, dS = \int_{\mathbb{S}^2} \hat{\mathcal{D}}_{\kappa c}\, dS. \tag{23.1.45}$$

With the multiplicative decompositions $\mathcal{D}_{\kappa c} = \mathcal{D}_{\kappa c}^{\mathrm{av}} \mathcal{D}_{\kappa c}^{\mathrm{fl}}(\theta_{\kappa c}, \phi_{\kappa c})$ and $\hat{\mathcal{D}}_{\kappa c} = \hat{\mathcal{D}}_{\kappa c}^{\mathrm{av}} \mathcal{D}_{\kappa c}^{\mathrm{fl}}(\theta_{\kappa c}, \phi_{\kappa c})$, Eqn (23.1.44) gives rise to two uncoupled equations for $\mathcal{D}_{\kappa c}^{\mathrm{av}}$ and $\mathcal{D}_{\kappa c}^{\mathrm{fl}}(\theta_{\kappa c}, \phi_{\kappa c})$, namely

$$\frac{\partial \mathcal{D}_{\kappa c}^{\mathrm{av}}}{\partial t} = \hat{\mathcal{D}}_{\kappa c}^{\mathrm{av}}, \quad \frac{\partial \mathcal{D}_{\kappa c}^{\mathrm{fl}}}{\partial t} = -\mathrm{div}_\mathsf{S}(\mathcal{D}_{\kappa c}^{\mathrm{fl}}\, \mathbf{v}_*) + \mathrm{div}_\mathsf{S}(D_\mathsf{S} \nabla_\mathsf{S} \mathcal{D}_{\kappa c}^{\mathrm{fl}}). \tag{23.1.46}$$

Let us examine in turn the diffusion, convection and supply contributions:

- in absence of supply, $\hat{\mathcal{D}}_{\kappa c} = 0$, and of convection, $\mathbf{v}_* = 0$, steady state $\partial \mathcal{D}_{\kappa c}/\partial t = 0$ corresponds to a distribution density $\mathcal{D}_{\kappa c}$ that is harmonic on the unit sphere, assuming the diffusion coefficient D_S to be directionally uniform as well. If the idea to convey is that the diffusion is driven by the growth rate, then the diffusion coefficient may be assumed to be a monotonous function of some scalar measure of the growth rate, namely its Eulerian norm, or the norm of its deviatoric part. In the simplest linear case,

$$D_\mathsf{S} = dS\, |\mathbf{D}_{\kappa c}^{\mathrm{g}}|, \quad \text{or} \quad D_\mathsf{S} = dS\, |\mathrm{dev}\, \mathbf{D}_{\kappa c}^{\mathrm{g}}|, \tag{23.1.47}$$

with dS dimensionless material parameter. Actually, we have in mind that $\mathbf{m}_{\kappa c} \sim \mathbf{F}_c^{\mathrm{g}} \cdot \underline{\mathbf{m}}_c$. Therefore, if the growth velocity transformation $\mathbf{L}_{\kappa c}^{\mathrm{g}}$ is isotropic, the fiber directions do not change, and therefore the distribution density $\mathcal{D}_{\kappa c}^{\mathrm{fl}}$ should not change either. Then the diffusion coefficient should be based on the deviatoric part of the growth rate. One might also consider that the diffusion process is driven, not by the growth rate, but by the growth, in the sense that the diffusion coefficient may be a function of the deviatoric part of the growth strain.

As an alternative to the last term of (23.1.44), the orientation distribution of fibers might be sought to optimize, at steady state, some specific quantity distinct from the distribution density. For example, steady state corresponds to the normal strain $\mathbf{m}_{\kappa c} \cdot (\mathbf{F}_c^{\mathrm{e}})^\mathsf{T} \cdot \mathbf{F}_c^{\mathrm{e}} \cdot \mathbf{m}_{\kappa c}$, or normal strain rate $\mathbf{m}_{\kappa c} \cdot \mathbf{D}_c^{\mathrm{e}} \cdot \mathbf{m}_{\kappa c}$, or normal Piola-Kirchhoff stress $\mathbf{m}_{\kappa c} \cdot \boldsymbol{\tau}_\kappa \cdot \mathbf{m}_{\kappa c}$, or normal Cauchy stress $\mathbf{m} \cdot \boldsymbol{\sigma} \cdot \mathbf{m}$, equivalent to $\mathbf{m}_{\kappa c} \cdot ((\mathbf{F}_c^{\mathrm{e}})^\mathsf{T} \cdot \mathbf{F}_c^{\mathrm{e}}) \cdot \boldsymbol{\tau}_\kappa \cdot ((\mathbf{F}_c^{\mathrm{e}})^\mathsf{T} \cdot \mathbf{F}_c^{\mathrm{e}})^\mathsf{T} \cdot \mathbf{m}_{\kappa c}/|(\mathbf{F}_c^{\mathrm{e}}) \cdot \mathbf{m}_{\kappa c}|^2$, with $\mathbf{m} = \mathbf{F}_c^{\mathrm{e}} \cdot \mathbf{m}_{\kappa c}/|\mathbf{F}_c^{\mathrm{e}} \cdot \mathbf{m}_{\kappa c}|$, to be uniform over the sphere. Then the Laplacian of this quantity should replace the rightmost term of (23.1.44). Still, a uniform orientation distribution of tractions is not feasible unless the stress is isotropic, see (12.1.23). At steady state, the non uniformity has to be balanced by the convection and supply contributions;

- a candidate for the convection velocity is the material rate $d\mathbf{m}_{\kappa c}/dt$. If the collagen fiber is viewed as a material line during the growth stage, then $\mathbf{m}_{\kappa c} \sim \mathbf{F}_c^{\mathrm{g}} \cdot \underline{\mathbf{m}}_c$, and then $d\mathbf{m}_{\kappa c}/dt \sim \mathbf{L}_{\kappa c}^{\mathrm{g}} \cdot \mathbf{m}_{\kappa c}$. If the vector is considered to remain of unit norm, then its rate is tangent to the unit sphere, and

$$\frac{d\mathbf{m}_{\kappa c}}{dt} = (\mathbf{I} - \mathbf{m}_{\kappa c} \otimes \mathbf{m}_{\kappa c}) \cdot \mathbf{L}_{\kappa c}^{\mathrm{g}} \cdot \mathbf{m}_{\kappa c}, \tag{23.1.48}$$

so that it can be viewed as the projection of $\mathbf{L}_{\kappa c}^{\mathrm{g}} \cdot \mathbf{m}_{\kappa c}$ on the tangent plane to the unit sphere at the current direction $\mathbf{m}_{\kappa c}$. If the growth velocity transformation $\mathbf{L}_{\kappa c}^{\mathrm{g}}$ is isotropic, the fibers do not change directions. Since $\mathbf{m}_{\kappa c}$ is aligned with the unit radial vector $\hat{\mathbf{e}}_r$, the divergence of the convection velocity $\mathbf{v}_* = d\mathbf{m}_{\kappa c}/dt$ can be calculated from (2.4.48) and (12.1.20),

$$\mathbf{v}_* = (\hat{\mathbf{e}}_\theta \cdot \mathbf{L}_{\kappa c}^{\mathrm{g}} \cdot \hat{\mathbf{e}}_r)\, \hat{\mathbf{e}}_\theta + (\hat{\mathbf{e}}_\phi \cdot \mathbf{L}_{\kappa c}^{\mathrm{g}} \cdot \hat{\mathbf{e}}_r)\, \hat{\mathbf{e}}_\phi, \quad \mathrm{div}_\mathsf{S}\, \mathbf{v}_* = -3\, \hat{\mathbf{e}}_r \cdot \mathrm{dev}\, \mathbf{D}_\kappa^{\mathrm{g}} \cdot \hat{\mathbf{e}}_r; \tag{23.1.49}$$

- the average density $\mathcal{D}_{\kappa c}^{\mathrm{av}} = N_{\mathrm{cf}}$ is deduced from the mass content $m^c = N_{\mathrm{cf}}\, m_{\mathrm{cf}}$, with m_{cf} [unit : kg/m^3] the mass content of a single collagen fiber. Along this approach, the deposited mass is split directionally following the existing distribution density $\mathcal{D}_{\kappa c}^{\mathrm{fl}}$. In other words, the fibers are secreted by the cells (fibroblasts) in agreement with the existing state of the ECM. This approach correlates with the fact that the fibroblasts are roughly aligned with the fibers, and that they secrete collagen molecules along their own morphology. In fact, Girton et al. [2002] report that muscle cell morphology and fiber orientation evolve according to the strain, and strain eigendirections. Fibers provide guidance to adherent cells, whose activity is regulated in particular by the mechanical stimuli of the fibers and ECM. Along a slightly different, and complementary, interpretation, cells secrete a soluble substance that is fibrillated outside the cell. Thus structures at higher length scales, namely fibrils, etc. are formed outside the cell by crosslinking that is responsible for their resistance to dissolution, Jackson [1958]. These structures are therefore expected to depend on the mechanical conditions of the ECM.

23.1.4.2   Approach 2: a discrete description of the collagen network

Next to the structure variables associated with the evolution of the homeostatic surface, the free energy is endowed with structure variables whose change implies an evolving elastic anisotropy. The contributions to the Clausius-Duhem inequality of the various structuration processes depend strongly of the phenomena in question:

- positive contributions to the dissipation inequality emanate from stress, through growth, and cells whose energy is progressively exhausted by the structuration processes;
- on the other hand, the development of the orientation distribution function clearly requires energy: the signs of the contributions to the dissipation associated with re-orientation of the fibers, with the change of the relative volume fractions, and with the stiffening of the fibers by pyridinoline crosslinks are negative.

The restrictions listed in Section 23.1.3 are now complemented:

- the system is thermodynamically open to mass only, in the sense that the independent source of free energy $\hat{\mathcal{E}}_\kappa^c$ is zero. Supply from external sources $\hat{\mathcal{E}}_\kappa^c \neq 0$ may be considered in a general setting, but it is not envisaged here (by supply, we mean here positive supply $\hat{\mathcal{E}}_\kappa^c \geq 0$, energy being *extracted* from the surroundings). Actually, this supply may either be unlimited, in which case the self-healing processes are guaranteed to take place. Alternatively, the external supply may be finite/limited, so that the issue whether the self-healing processes can take place or not surfaces again. The case of zero external energy supply that we consider here is simply a particular case;
- cells are viewed as an internal energy source: this energy is transferred to the tissue by the growing process, and it therefore decreases during this process. This energy may be understood to be used both to increase the mass (directional increase of mass), and to improve the mechanical properties of the collagen network through crosslinkages (the increase of pyridinoline density). Perhaps, this energy may be linked to the traction that cells exert on the ECM. This traction is presumably due to the fact that cells undergo compression as they secrete proteoglycans and collagen monomers. Since the growing constituents of the tissue work in parallel, undergoing the same strain, their contributions to the total stress add. Therefore, under a fixed external load, any change by a constituent (here compression of the cells) should be compensated by the other constituents (here tension of the ECM);
- the structuration and self-healing processes are driven by growth and the above (internal) cellular energy source. Theoretically, the mechanics can block the growth process. However, the idea is that biochemical conditions affect the homeostatic surface, so that, in presence of nutrients, the homeostatic surface is small, the stress is likely to be outside the homeostatic surface, and growth takes place. On the other hand, in quasi-absence of nutrients, the homeostatic surface is large so that the stress is more likely to be inside the homeostatic surface, and growth is blocked, and so are structuration and self-healing processes;
- positive dissipation ends up to heat unless it is a source for the self-healing processes to take place. In other words, we build the model so that the dissipation can only be positive or zero (in a broader context, one might perhaps question the consequences of a positive dissipation, that is, what will the surroundings do with this energy?).

Sufficient positive contributions to dissipation by stress and cell energy may allow the collagen network to evolve, via processes that contribute negatively to the dissipation inequality. If the positive contributions are not sufficient, the rates of change are scaled down by a factor that allows satisfaction of the dissipation. As a tenet of this approach, the sign of the dissipation contributions by the various physical phenomena involved should be controlled.

Director vectors and growth

We consider materials endowed with special directions. Typically, the collagen fibers are seen as introducing anisotropic elastic, and diffusion, properties. The fibers play the role of director vectors. Growth is considered to modify the density of collagen fibers, and their directions.

The strain energy of a set collagen fibers of direction $\mathbf{m}_{\kappa c}$ per unit volume of collagen in the intermediate configurations is expressed in terms of the elastic right stretch $\mathbf{C}_c^e = (\mathbf{F}_c^e)^\mathrm{T} \cdot \mathbf{F}_c^e$ or elastic Green strain $\mathbf{E}_{Gc}^e = (\mathbf{C}_c^e - \mathbf{I})/2$ in the format,

$$w_{\kappa c} = w_{\kappa c}(L_{\kappa c}^2), \tag{23.1.50}$$

where

$$L_{\kappa c} = \tfrac{1}{2} \langle \mathbf{C}_c^e : \mathbf{M}_{\kappa c} - 1 \rangle = \underbrace{\langle \mathbf{E}_{Gc}^e : \mathbf{M}_{\kappa c} \rangle}_{x}, \tag{23.1.51}$$

and $\mathbf{M}_{\kappa c} = \mathbf{m}_{\kappa c} \otimes \mathbf{m}_{\kappa c}$. We adopt the exponential form of energy (12.2.3), defined by the modulus K_c [unit : MPa] and the dimensionless exponent $k_{\rm c}$,

$$w_{\kappa{\rm c}} = \frac{K_{\rm c}}{2\,k_{\rm c}} \left(\exp(k_{\rm c}\,L_{\kappa{\rm c}}^2) - 1 \right),$$

$$\frac{\partial w_{\kappa{\rm c}}}{\partial \mathbf{E}_{\rm Gc}^{\rm e}} = K_{\rm c}\,\exp(k_{\rm c}\,L_{\kappa{\rm c}}^2)\,L_{\kappa{\rm c}}\,\mathbf{M}_{\kappa{\rm c}}, \tag{23.1.52}$$

$$\frac{\partial^2 w_{\kappa{\rm c}}}{\partial \mathbf{E}_{\rm Gc}^{\rm e}\partial \mathbf{E}_{\rm Gc}^{\rm e}} = K_{\rm c}\,\exp(k_{\rm c}\,L_{\kappa{\rm c}}^2)\,\big(2\,k_{\rm c}\,\underbrace{L_{\kappa{\rm c}}^2}_{\text{continuous}} + \underbrace{\mathcal{H}(x)}_{\text{discontinuous}}\big)\,\mathbf{M}_{\kappa{\rm c}} \otimes \mathbf{M}_{\kappa{\rm c}}.$$

Here $K_{\rm c} = K_{\rm cf}$ is the modulus of the collagen fibers per unit volume of collagen. In a mixture context, $w_{\kappa{\rm c}}$ would be replaced by its apparent counterpart $w_\kappa^{\rm c}$, defined by the modulus $K^{\rm c} = n^{\rm c}\,K_{\rm cf}$ equal to the modulus of collagen fibers $K_{\rm cf}$ weighted by the volume fraction $n^{\rm c}$ of the collagen fibers.

Actually, the strain energy function could be expressed in terms of five isotropic scalar invariants, the three standard isotropic invariants plus the two scalar invariants $I_4 = \mathbf{C}_{\rm c}^{\rm e} : \mathbf{M}_{\kappa{\rm c}}$ and $I_5 = (\mathbf{C}_{\rm c}^{\rm e})^2 : \mathbf{M}_{\kappa{\rm c}}$. The invariant I_4 embeds the strain energy of collagen fibers due to their extension along the fibers. The invariant I_5 can be viewed as introducing either a fiber-matrix interaction or a fiber-fiber interaction, Section 19.19.1.

23.2   Modes of mass deposition in an elastic-growing mixture

The two modes of mass deposition (a) and (b) addressed in Section 22.6 for a solid are reconsidered here for a mixture where all growing constituents are elastic-growing. The free energies are derived first for the constituents, which are all assumed to be elastic-growing. The stress is obtained next by summing the contributions of all constituents in the mixture.

With respect to a single elastic-growing solid, the analysis of individual constituents is quite similar in spirit but it displays many differences in details, so that it is worth to be re-considered in this mixture context. Attention should be paid in particular to the distinction between intrinsic and partial, or apparent, quantities.

With the underlying convention that $\mathbf{F}(t=0) = \mathbf{I}$, $\mathbf{F}_k^{\rm e}(t=0) = \mathbf{I}$, $\mathbf{F}_k^{\rm g}(t=0) = \mathbf{I}$, for any growing constituent $k \in \mathcal{S}^*$, the volume and energy relative to the intermediate configurations are transported, with help of the sketches of Fig. 23.1.1 (the relations shown on sketch (c) display apparent free energies, but they also hold for their intrinsic counterparts) first from the reference configuration,

$$\underline{V}_{\kappa k} = \underline{V}_k = V_{\kappa k}(t)\,(\det\mathbf{F}_k^{\rm g}(t))^{-1}, \quad \underline{E}_{\kappa k}(t) = \underline{E}_k(t) = E_k(t)\,\det\mathbf{F}(t) = E_{\kappa k}(t)\,\det\mathbf{F}_k^{\rm g}(t), \tag{23.2.1}$$

and next to the current configuration,

$$V_k(t) = \underline{V}_k\,\det\mathbf{F}(t), \quad V_{\kappa k}(t) = V_k(t)\,(\det\mathbf{F}_k^{\rm e}(t))^{-1}, \quad E_{\kappa k}(t) = E_k(t)\,\det\mathbf{F}_k^{\rm e}(t). \tag{23.2.2}$$

The free energy $E_{\kappa k}^\natural(t)\,V_{\kappa k}(t)$ at time t of the volume $V_{\kappa k}(t)$ is equal to the free energy at time t of the reference volume $V_{\kappa k}(0) = \underline{V}_{\kappa k}$, plus the free energy of the progressively growing volume $dV_{\kappa k}(\tau)$, $\tau \in [0, t]$,

$$E_{\kappa k}^\natural(t)\,V_{\kappa k}(t) = \underline{E}_{\kappa k}^\natural(t)\,\underline{V}_{\kappa k} = E_{\kappa k}(t)\,\underline{V}_{\kappa k} + \int_0^t E_{\kappa k}(t, u)\,dV_{\kappa k}(\tau), \tag{23.2.3}$$

where $E_{\kappa k}(t) = E_{\kappa k}(V_{\kappa k}(t))$, and $E_{\kappa k}(t, u) = E_{\kappa k}(V_{\kappa k}(u \to t))$ denotes the time dependence of the state variables from u to t, associated with the elastic deformation $\mathbf{F}_k^{\rm e}(t, u) = \mathbf{F}_k^{\rm e}(t) \cdot (\mathbf{F}_k^{\rm e}(u))^{-1}$.

23.2.1   The constituent is deposited in a natural configuration

For mode (a), set $u = \tau$ in (23.2.3), and the free energy of growth $\underline{E}_{\kappa k}^\natural(t)$ for the growing constituent results in

$$\begin{aligned} \underline{E}_{\kappa k}^\natural(t) &= E_{\kappa k}(t) + \int_0^t E_{\kappa k}(t, \tau)\,\frac{V_{\kappa k}(\tau)}{\underline{V}_{\kappa k}}\,\frac{dV_{\kappa k}(\tau)}{V_{\kappa k}(\tau)}, \\ &= E_{\kappa k}(t) + \int_0^t E_{\kappa k}(t, \tau)\,\det\mathbf{F}_k^{\rm g}(\tau)\,\frac{dV_{\kappa k}(\tau)}{V_{\kappa k}(\tau)}, \end{aligned} \tag{23.2.4}$$

with help of the relation $(23.2.1)_1$, which implies also $dV_{\kappa k}(\tau)/V_{\kappa k}(\tau) = \operatorname{tr} \mathbf{L}^{\mathrm{g}}_{\kappa k}(\tau)\, d\tau$. Thus the free energy of growth $\underline{E}^{\natural}_{\kappa k}$ expresses in terms of the time course of the growth transformation as

$$\underline{E}^{\natural}_{\kappa k}(t) = E_{\kappa k}(t) + \int_0^t E_{\kappa k}(t,\tau)\, \frac{d}{d\tau} \det \mathbf{F}^{\mathrm{g}}_k(\tau)\, d\tau, \quad k \in \mathcal{S}^*. \tag{23.2.5}$$

For density preserving growth, the rate of growth transformation is explicitly linked to the rate of mass supply by $\operatorname{tr} \mathbf{L}^{\mathrm{g}}_{\kappa k} = \hat{\rho}^k/\rho^k$, and therefore the free energy keeps track of the history of mass deposition.

23.2.2   The constituent is deposited with the existing properties at the deposition site

For mass deposition mode (b), set $u = 0$, and similar to Section 22.6.2.2,

$$
\begin{aligned}
E^{\natural}_{\kappa k}(t)\, V_{\kappa k}(t) = \underline{E}^{\natural}_{\kappa k}(t)\, \underline{V}_{\kappa k} &= E_{\kappa k}(t)\, \underline{V}_{\kappa k} + \int_0^t E_{\kappa k}(t)\, dV_{\kappa k}(\tau), \\
&= E_{\kappa k}(t)\, (\underline{V}_{\kappa k} + V_{\kappa k}(t) - \underline{V}_{\kappa k}) = E_{\kappa k}(t)\, V_{\kappa k}(t),
\end{aligned}
\tag{23.2.6}
$$

so that the free energy of growth boils down to the standard free energy,

$$E^{\natural}_{\kappa k} = E_{\kappa k} \quad \Leftrightarrow \quad \underline{E}^{\natural}_{\kappa k} = \underline{E}_{\kappa k} = E_{\kappa k}\, \det \mathbf{F}^{\mathrm{g}}_k, \quad k \in \mathcal{S}^*, \tag{23.2.7}$$

the last equality resulting from $(23.2.1)_2$.

23.2.3   The expression for the stress

The intermediate configurations are innate to each constituent. Moreover, each constituent is endowed with its own deposition mode. As an extension of the expression derived in Section 23.1.3, the effective stress for both modes (a) and (b) adopt the format,

$$\boldsymbol{\sigma}' = \sum_{k \in \mathcal{S}^*} n^k\, \mathbf{F}^{\mathrm{e}}_k \cdot \frac{\partial E^{\natural}_{\kappa k}}{\partial \mathbf{E}^{\mathrm{e}}_{\mathrm{G}k}} \cdot (\mathbf{F}^{\mathrm{e}}_k)^{\mathrm{T}}. \tag{23.2.8}$$

Remark 23.2.1 on chemomechanical couplings

   Although the point has not been made explicit, the free energy of the individual constituent $E^{\natural}_{\kappa k}$, $k \in \mathcal{S}^*$, and consequently the free energy of the porous medium $E^{\natural}_{\kappa}$, depend on the mass contents of the non growing species $\cup_{k \in \mathcal{K}^*} m_k$. This dependence introduces chemomechanical couplings, and makes the formulation chemo-hyperelastic. Chemohyperelasticity can take various forms. For example, for sensitive clays, the hyperelastic moduli have been made dependent on the content of the chemical variables, i.e., of $\cup_{k \in \mathcal{K}^*} m_k$, Gajo et al. [2002], Loret et al. [2004]a [2004]b. Alternatively, for articular cartilages, the coupling gives rise to an effective stress, Loret and Simões [2004]. The motivations for these distinct modelings stem from distinct spatial repartitions of the constituents, and in particular from distinct nanoscale configurations of the electrical *fixed* charge, which is the key engine driving the electrochemomechanical couplings.

Remark 23.2.2 on an incompressibility constraint

   The above formulation for mixtures assumes that not all the constituents are incompressible. Otherwise, a kinematical constraint arises that implies that the constitutive equations define the stress up to a pressure. Examples for chemo-hyperelastic mixtures in which such a constraint holds can be found, e.g., in Sections 17.11.1 and 23.1.1.3.

Remark 23.2.3 on surface, volume and mass averages

   Neglecting the diffusion contribution, the total stress in a mixture is equal to the sum of the partial stresses of the constituents, Eqn (4.5.7). The apparent stress $\boldsymbol{\sigma}^k$ is equal to the intrinsic stress $\boldsymbol{\sigma}_k$ weighted by the areal fraction a^k, or volume fraction n^k. Since mass is the main ingredient in growth, there are also suggestions, Humphrey and Rajagopal [2002] [2003], to use mass as a weighting factor for growing mixtures. The motivation for mass weighting stems from integral expressions of the free energy, where, for density preserving growth, the rate of mass supply over the mixture density appears in the integrand.

   So, we have the three proposals,

$$\boldsymbol{\sigma} = \sum_{k \in \mathcal{K}} a^k\, \boldsymbol{\sigma}_k, \quad \boldsymbol{\sigma} = \sum_{k \in \mathcal{K}} n^k\, \boldsymbol{\sigma}_k, \quad \boldsymbol{\sigma} = \sum_{k \in \mathcal{K}} \frac{\rho^k}{\rho}\, \boldsymbol{\sigma}_k, \tag{23.2.9}$$

with $\rho^k = n^k \rho_k$ apparent mass density of constituent k, and $\rho = \sum_{k \in \mathcal{K}} \rho^k$ mass density of the mixture. Areal and volume fractions are equal when the constituents are randomly distributed. On the other hand, volume fractions V_k/V and mass fractions M_k/M are not rigorously equal, because proteins have a larger density than water, and thus $\rho^k/\rho \neq n^k$, $k \in \mathcal{S}^*$. For normal articular cartilages, the collagen density is about $1.4\,\mathrm{gm/cm^3}$ and that of proteoglycans $1.8\,\mathrm{gm/cm^3}$. Averaging over the thickness of the cartilage, the relative mass of collagen is about $1/5$ of the cartilage, and that of proteoglycans and other proteins $1/10$, so that the density of the cartilages is about $1.11\,\mathrm{gm/cm^3}$.

To avoid confusion, it is worth recalling that the total stress in a mixture can be decomposed in at least two ways. The areal and volumetric decompositions shown in $(23.2.9)_{1,2}$ are of purely *geometric nature*, and involve all constituents. On the other hand, the expression of stress derived in $(23.2.8)$ is of *constitutive nature*, and involves only the constituents that grow. The formalism includes naturally constituents that contribute to sustain loads but do not grow, simply by setting their growth transformation equal to the identity.

For a particle material saturated by a fluid w, these decompositions take the familiar forms $\boldsymbol{\sigma} = \boldsymbol{\sigma}^{\mathrm{s}} + \boldsymbol{\sigma}^{\mathrm{w}} = \boldsymbol{\sigma}' + \boldsymbol{\sigma}_{\mathrm{w}}$ where $\boldsymbol{\sigma}^{\mathrm{s}}$ is the partial stress over the solid skeleton s, $\boldsymbol{\sigma}^{\mathrm{w}} = -n^{\mathrm{w}} p\,\mathbf{I}$ and $\boldsymbol{\sigma}_{\mathrm{w}} = -p\,\mathbf{I}$ are respectively the partial and intrinsic fluid pressures, and $\boldsymbol{\sigma}'$ is the interparticle stress, or Terzaghi's effective stress. The partial stresses can be derived by summing the forces over a plane section of the porous medium, Fig. 7.3.1-(a). The effective stress is obtained by summing the forces along a path between contact points within the fluid phase, Fig. 7.3.1-(b).

23.3 The fixed charge: electro-chemomechanical couplings

The presence of a fixed (not mobile) electrical charge on PGs gives rise to electro-chemomechanical couplings which amount to wrap a semi-permeable membrane around the cartilage. The issue is addressed in particular in Chapters 13 and 14. Therefore the determination of the incompressibility pressure p_{E} is somehow more complex than standard. In fact, it will be contributed additively by a non constitutive term and by a constitutive term.

The cartilage specimen is assumed to be in local equilibrium with a fictitious bath at pressure $\tilde{p}_{\mathrm{E}}$. The concept of a fictitious bath is exposed in Section 14.5.2. The constitutive equations $(23.1.30)$ and $(23.1.32)$ are then recast in the format,

$$\boldsymbol{\sigma}' = \boldsymbol{\sigma} + \tilde{p}_{\mathrm{E}}\,\mathbf{I} + \pi_{\mathrm{osm}}\,\mathbf{I} = \sum_{k \in \mathcal{S}^*} \frac{1}{\det \mathbf{F}_k^{\mathrm{e}}}\,\mathbf{F}_k^{\mathrm{e}} \cdot \boldsymbol{\tau}_\kappa^{k'} \cdot (\mathbf{F}_k^{\mathrm{e}})^{\mathrm{T}}\,, \tag{23.3.1}$$

that involves the osmotic pressure $\pi_{\mathrm{osm}} = p_{\mathrm{E}} - \tilde{p}_{\mathrm{E}}$. The properties of the fictitious bath are deduced from those of the cartilage. The osmotic pressure expresses in terms of molar fractions, which can themselves be expressed in terms of mole contents. Therefore, the osmotic pressure can be seen as a function of the state variables. It consequently qualifies to enter the constitutive equations.

For a general boundary value problem, the indeterminate pressure $\tilde{p}_{\mathrm{E}}$ is defined by boundary conditions. However, we consider here a local analysis, or equivalently, the special situation of a homogeneous specimen in equilibrium with a (real) bath. Therefore, the fictitious bath and the real bath can be identified, $\tilde{p}_{\mathrm{E}}$ is understood as the pressure p_{B} of the real bath and,

$$p_{\mathrm{E}} = p_{\mathrm{B}} + \pi_{\mathrm{osm}}\,. \tag{23.3.2}$$

In absence of a fixed charge, the incompressibility pressure would reduce in a standard way to the bath pressure. In a chemomechanical analysis of tissues with a fixed charge, as detailed in e.g., Chapter 17, chemical equilibrium expressed in terms of chemical potentials provides the osmotic pressure in terms of differences of chemical contents between the bath and the tissue. Alternatively, one may use experimental measurements available in the literature. The pressure π_{osm} [unit: MPa] of a solution of PGs measured by Basser et al. [1998] is reported as a function of the fixed charge density FCD [unit: mole/kg water] in Eqn (14.4.2). While the experiment of Basser et al. [1998] considers a solution of PGs, the real cartilage contains collagen as well. The fixed charge density [unit: mole/kg] expresses in terms of volumes or masses,

$$\mathrm{FCD} = \frac{|\zeta_{\mathrm{PG}}|\,N_{\mathrm{PG}}}{M_{\mathrm{w}}} = \frac{|\zeta_{\mathrm{PG}}|}{\rho_w\,\widehat{v}_{\mathrm{PG}}}\,\frac{v_{\mathrm{PG}}}{V/\underline{V} - v_{\mathrm{PG}} - v_{\mathrm{c}}}\,, \tag{23.3.3}$$

with $\widehat{v}_{\mathrm{PG}} = 1.1 \times 10^6\,\mathrm{cm^3} = 1.1\,\mathrm{m^3}$ the molar volume of PGs, $\zeta_{\mathrm{PG}} \in [-5000, -6000]$ the valence of PGs under physiological pH, and $v_{\mathrm{PG}} = V_{\mathrm{PG}}/\underline{V}$ and $v_{\mathrm{c}} = V_{\mathrm{c}}/\underline{V}$ the volume contents of PGs and collagen respectively. The volume contents of growing constituents are deduced from their masses, assuming a mass density of 1.8 for PGs and 1.4 for collagen. Hence, the experimental data of Basser et al. [1998] can be interpreted as yielding the osmotic pressure as a function of the volume change and amounts of PGs and collagen.

If growth preserves density, the differential $\delta\pi_{\text{osm}}$ expresses with help of (21.3.58) as

$$\delta\pi_{\text{osm}} = \frac{\partial\pi_{\text{osm}}}{\partial\text{FCD}} \frac{\text{FCD}}{n^{\text{w}}} \left(-\mathbf{I} : \delta\mathbf{L} + n^{\text{c}}\,\mathbf{I} : \delta\mathbf{L}_{\kappa\text{c}}^{\text{g}} + (1 - n^{\text{c}})\,\mathbf{I} : \delta\mathbf{L}_{\kappa\text{PG}}^{\text{g}} \right). \tag{23.3.4}$$

The changes of masses of growing constituents and non growing species are balanced as indicated by (23.1.17) when they are all incompressible. In the more particular case where growth preserves density and water is the sole non growing species, the change of mass of the latter can be given an explicit form in terms of the velocity gradient and rates of growth transformations,

$$\frac{1}{\rho_{\text{w}}\,V} \frac{dM_{\text{w}}}{dt} = \operatorname{tr}\mathbf{L} - \sum_{k \in \mathcal{S}^*} n^k \operatorname{tr}\mathbf{L}_{\kappa k}^{\text{g}}. \tag{23.3.5}$$

Remark 23.3.1: during growth under all around pressure, water enters the tissue

Consider a single growing constituent c and water. If density preserves growth, growth of the solid constituent implies water to enter the tissue. Indeed, let us assume the growing constituent and fluid to be incompressible. Then,

$$\begin{aligned} \operatorname{tr}\mathbf{L} &= n^{\text{c}} \operatorname{tr}\mathbf{L}_{\kappa\text{c}}^{\text{g}} + \frac{n^{\text{w}}}{m^{\text{w}}} \frac{dm^{\text{w}}}{dt} \\ &= \operatorname{tr}\mathbf{L}_{\text{c}}^{\text{e}} + \operatorname{tr}\mathbf{L}_{\kappa\text{c}}^{\text{g}}. \end{aligned} \tag{23.3.6}$$

The first line results from the incompressibility condition (23.1.17), accounting for density preserving growth. The second line results from the elastic-growth decomposition. If the tissue is bathed in a fluid so that $\boldsymbol{\sigma} + p_{\text{B}}\mathbf{I} = \mathbf{0}$, the effective stress $\boldsymbol{\sigma}' = \boldsymbol{\sigma} + p_{\text{B}}\mathbf{I} + \pi_{\text{osm}}\mathbf{I}$ is tensile because the osmotic pressure is positive: the rate of elastic volume change is positive as well. Therefore the mass of water increases,

$$\frac{n^{\text{w}}}{m^{\text{w}}} \frac{dm^{\text{w}}}{dt} = (1 - n^{\text{c}}) \operatorname{tr}\mathbf{L}_{\kappa\text{c}}^{\text{g}} + \operatorname{tr}\mathbf{L}_{\text{c}}^{\text{e}} > 0. \tag{23.3.7}$$

23.4   Growth equations with an evolving microstructure

Disposing from now on of thermal effects, the analysis of this section targets a single component, namely collagen while Section 23.5 addresses proteoglycans.

The evolution of the stiffness of the collagen network during growth is contributed by several aspects:

    1) chondrocytes secrete monomers that are assembled in the extracellular space, first by linear end-to-end aggregation, and next laterally. The resistance of a collagen fiber is estimated to increase during fibrillogenesis, not as a function of the fiber section, but essentially as a function of the length of the fibers;

    2) as indicated by (23.1.33), for density preserving growth and an incompressible constituent k, the volume fraction n^k is a function of the elastic strain;

    3) mechanical loading rotates already existing fibers. By the same token, it also rotates the chondrocytes which tend to produce collagen monomers which assemble in fibers along directions dictated by the mechanical conditions. Along an alternative point of view, only those fibers whose growth is triggered by the mechanical conditions grow, giving rise to an evolving fabric. The resulting anisotropic spatial distribution of fibers spills over on the macroscopic elastic properties of the collagen network;

    4) the stiffness of individual collagen fibers increases due to crosslinking by an increasing pyridinoline density, Williamson et al. [2003].

23.4.1   Collagen : the free energy in terms of director vectors

The set of *dimensionless* state variables relative to the intermediate configurations,

$$\mathcal{V}_{\kappa\text{c}} = \{\mathbf{E}_{\text{Gc}}^{\text{e}},\ \{\cup_{i=1}^{n_e}\mathbf{m}_{\kappa i},\ \tilde{n}_{\kappa i}\},\ c_{\kappa\text{c}},\ \xi_{\kappa\text{c}},\ \boldsymbol{\xi}_{\kappa\text{c}},\ p_{\kappa\text{c}}\}, \tag{23.4.1}$$

includes
- the elastic Green strain $\mathbf{E}_{\text{Gc}}^{\text{e}}$;

- a set of *structure*, or *director*, vectors associated with the directions $\mathbf{m}_{\kappa i}$, $i \in [1, n_e]$, of collagen fibers;
- a set of relative volume fractions $\tilde{n}_{\kappa i}$, $i \in [1, n_e]$, which are a discrete representation of an orientation distribution function, or fabric, and that satisfy the constraint,

$$\sum_{i=1, n_e} \tilde{n}_{\kappa i} = 1 \,. \qquad (23.4.2)$$

If the relative volume fractions $\tilde{n}_{\kappa i}$ are identical, then $\tilde{n}_{\kappa i} = 1/n_e$;
- the scaled dimensionless cell energy density $c_{\kappa c}$;
- two additional *structure* variables associated with the evolution of the homeostatic surface, namely a scalar variable $\xi_{\kappa c}$ controlling its size and a symmetric second order tensor $\boldsymbol{\xi}_{\kappa c}$ controlling the position of its center;
- the intrinsic pyridinoline density $p_{\kappa c}$ expressed in mole of pyridinoline per mole of collagen.

The scalar variable $c_{\kappa c}$ is interpreted as a (dimensionless) cell energy density that contributes to satisfy the dissipation inequality. The question of the partition of the total cell energy to constituents is an issue. In addition, it would perhaps be appropriate to consider the cells to be themselves anisotropic and to have the same directional distribution as the fibers they secrete, but this sophistication is not included in this model. The formulation keeps the possibility of governing the size of the homeostatic surface either by the independent scalar variable $\xi_{\kappa c}$ or by the cell energy density. Along the latter idea, the size of the homeostatic surface is inversely proportional to the energy provided by the cells. In presence of a large energy source, the homeostatic surface is small and growth is likely. Conversely, in absence of energy source, the homeostatic surface is large, and growth is unlikely.

The sense of the director vectors is elusive, and, actually, the dependence of the state variables with respect to the vectors $\mathbf{m}_{\kappa i}$ takes place via the second order tensors,

$$\mathbf{M}_{\kappa i} = \mathbf{m}_{\kappa i} \otimes \mathbf{m}_{\kappa i}, \quad i \in [1, n_e] \,. \qquad (23.4.3)$$

Then, with the *generalized stresses* [unit : Pa],

$$\begin{aligned}
\boldsymbol{\tau}_\kappa^{c'} &= \underline{n}^c \, \frac{\partial E_c}{\partial \mathbf{E}_{Gc}^e}, \quad \mathbf{A}_{\kappa i} = \underline{n}^c \, \frac{\partial E_c}{\partial \mathbf{M}_{\kappa i}}, \quad \tilde{N}_{\kappa i} = \underline{n}^c \, \frac{\partial E_c}{\partial \tilde{n}_{\kappa i}}, \quad i \in [1, n_e], \\
C_\kappa^c &= \underline{n}^c \, \frac{\partial E_c}{\partial c_{\kappa c}}, \quad X_\kappa^c = \underline{n}^c \, \frac{\partial E_c}{\partial \xi_{\kappa c}}, \quad \mathbf{X}_\kappa^c = \underline{n}^c \, \frac{\partial E_c}{\partial \boldsymbol{\xi}_{\kappa c}}, \quad P_\kappa^c = \underline{n}^c \, \frac{\partial E_c}{\partial p_{\kappa c}},
\end{aligned} \qquad (23.4.4)$$

the reduced dissipation inequality (23.1.26) ,

$$T \hat{S}_{\kappa(i)}^c = \boldsymbol{\Psi}_\kappa^c : \mathbf{L}_{\kappa c}^g - \sum_{i=1}^{n_e} \left(\mathbf{A}_{\kappa i} : \frac{d\mathbf{M}_{\kappa i}}{dt} + \tilde{N}_{\kappa i} \frac{d\tilde{n}_{\kappa i}}{dt} \right) - C_\kappa^c \frac{dc_{\kappa c}}{dt} - X_\kappa^c \frac{d\xi_{\kappa c}}{dt} - \mathbf{X}_\kappa^c : \frac{d\boldsymbol{\xi}_{\kappa c}}{dt} - P_\kappa^c \frac{dp_{\kappa c}}{dt} \geq 0, \qquad (23.4.5)$$

remains to be satisfied by the constitutive equations to be developed for the growth transformation $\mathbf{F}_c^g$, for the orientation variables $\mathbf{M}_{\kappa i}$, $i \in [1, n_e]$, for the relative volume fractions $\tilde{n}_{\kappa i}$, $i \in [1, n_e]$, for the cell energy concentration $c_{\kappa c}$, for the structure variables $\xi_{\kappa c}$ and $\boldsymbol{\xi}_{\kappa c}$, and for the pyridinoline density $p_{\kappa c}$.

The free energy is *additively* contributed by a function of the elastic Green strain, directional structure variables, pyridinoline density, and by terms involving the structure variables associated with the evolution of the homeostatic surface[23.4],

$$E_c = \boldsymbol{\Gamma}_{\kappa c} : \mathbf{E}_{Gc}^e + E_c^e(\mathbf{E}_{Gc}^e, \{\cup_{i=1}^{n_e} \mathbf{m}_{\kappa i}, \tilde{n}_{\kappa i}\}, p_{\kappa c}) + E_c^{cell}(c_{\kappa c}) + \frac{1}{2} \boldsymbol{\xi}_{\kappa c} : \mathbb{X}_{\kappa c} : \boldsymbol{\xi}_{\kappa c} + \frac{1}{2} \mathcal{X}_{\kappa c} \, \xi_{\kappa c}^2, \qquad (23.4.6)$$

with

$$E_c^e = \sum_{i=1}^{n_e} \frac{K_{ci}}{2 \, k_c(0)} \left(\exp(k_c \, L_{\kappa i}^2) - 1 \right), \quad L_{\kappa i} = \langle \mathbf{E}_{Gc}^e : \mathbf{M}_{\kappa i} \rangle \,. \qquad (23.4.7)$$

Here $K_{ci} = \tilde{n}_{\kappa i} K_{cf}$ is the modulus associated with the sector i. $\tilde{n}_{\kappa i}$ is the relative volume fraction of the set of fibers i, that altogether satisfy the constraint (23.4.2). K_{cf} is the modulus of collagen fibers per unit volume of collagen. A priori, both $K_{cf} = K_{cf}(p_{\kappa c})$ and the exponent $k_c = k_c(p_{\kappa c})$ may depend on pyridinoline density. The tensile and

---

[23.4] The pyridinoline effect has been included in the intrinsic modulus of collagen fiber since pyridinoline links collagen molecules, Fig. 14.2.5. The concentration in sound adult articular cartilage is one molecule of pyridinoline per molecule of collagen. This concentration is small in the fetus, increases during growth and decreases slightly with age.

compressive moduli of articular cartilage are observed to increase from fetus to calf, Williamson et al. [2003], p. 878. This increase seems to correlate with the increase of the concentrations of both collagen and pyridinoline, Williamson et al. [2003], their Fig. 3. The coefficients introducing the structure variables are the scalar $\mathcal{X}_{\kappa c}$ [unit : Pa] and the fourth order tensor with minor and major symmetries $\mathbb{X}_{\kappa c}$ [unit : Pa].

The generalized stresses take the explicit form,

$$
\begin{aligned}
\text{effective stress} \qquad (\underline{n}^{\mathrm{c}})^{-1}\,\boldsymbol{\tau}_{\kappa}^{\mathrm{c}'} &= \frac{\partial E_{\mathrm{c}}}{\partial \mathbf{E}_{\mathrm{Gc}}^{\mathrm{e}}} = \boldsymbol{\Gamma}_{\kappa c} + \sum_{i=1}^{n_e} K_{\mathrm{c}i}\,\frac{k_{\mathrm{c}}}{k_{\mathrm{c}}(0)}\,\exp(k_{\mathrm{c}}\,L_{\kappa i}^2)\,L_{\kappa i}\,\mathbf{M}_{\kappa i}; \\[4pt]
\text{collagen fiber} \qquad (\underline{n}^{\mathrm{c}})^{-1}\,\mathbf{A}_{\kappa i} &= \frac{\partial E_{\mathrm{c}}}{\partial \mathbf{M}_{\kappa i}} = K_{\mathrm{c}i}\,\frac{k_{\mathrm{c}}}{k_{\mathrm{c}}(0)}\,\exp(k_{\mathrm{c}}\,L_{\kappa i}^2)\,L_{\kappa i}\,\mathbf{E}_{\mathrm{Gc}}^{\mathrm{e}}, \quad i \in [1, n_e]; \\[4pt]
\text{collagen fabric} \qquad (\underline{n}^{\mathrm{c}})^{-1}\,\tilde{N}_{\kappa i} &= \frac{\partial E_{\mathrm{c}}}{\partial \tilde{n}_{\kappa i}} = \frac{K_{\mathrm{cf}}}{2\,k_{\mathrm{c}}(0)}\,\big(\exp(k_{\mathrm{c}}\,L_{\kappa i}^2)-1\big), \quad i \in [1, n_e]; \\[4pt]
\text{cell energy density} \qquad (\underline{n}^{\mathrm{c}})^{-1}\,C_{\kappa}^{\mathrm{c}} &= \frac{\partial E_{\mathrm{c}}}{\partial c_{\kappa c}} = \frac{dE_{\mathrm{c}}^{\mathrm{cell}}}{dc_{\kappa c}}; \\[4pt]
\text{homeostatic surface} \qquad (\underline{n}^{\mathrm{c}})^{-1}\,X_{\kappa}^{\mathrm{c}} &= \frac{\partial E_{\mathrm{c}}}{\partial \xi_{\kappa c}} = \mathcal{X}_{\kappa c}\,\xi_{\kappa c}; \\[4pt]
\text{homeostatic surface} \qquad (\underline{n}^{\mathrm{c}})^{-1}\,\mathbf{X}_{\kappa}^{\mathrm{c}} &= \frac{\partial E_{\mathrm{c}}}{\partial \boldsymbol{\xi}_{\kappa c}} = \mathbb{X}_{\kappa c} : \boldsymbol{\xi}_{\kappa c}; \\[4pt]
\text{pyridinoline density} \qquad (\underline{n}^{\mathrm{c}})^{-1}\,P_{\kappa}^{\mathrm{c}} &= \frac{\partial E_{\mathrm{c}}}{\partial p_{\kappa c}} = \frac{dK_{\mathrm{cf}}}{dp_{\kappa c}}\,\frac{E_{\mathrm{c}}^{\mathrm{e}}}{K_{\mathrm{cf}}} + \frac{dk_{\mathrm{c}}}{dp_{\kappa c}}\,\sum_{i=1}^{n_e}\frac{K_{\mathrm{c}i}}{2\,k_{\mathrm{c}}(0)}\,\exp(k_{\mathrm{c}}\,L_{\kappa i}^2)\,L_{\kappa i}^2\,.
\end{aligned} \tag{23.4.8}
$$

23.4.2 Collagen : rate of growth

The Mandel stress $\boldsymbol{\Psi}_{\kappa}^{k}$ is not symmetric in general and the rate of growth transformation $\mathbf{L}_{\kappa k}^{\mathrm{g}}$ has both a symmetric contribution $\mathbf{D}_{\kappa k}^{\mathrm{g}}$ and a skew-symmetric contribution $\boldsymbol{\Omega}_{\kappa k}^{\mathrm{g}}$,

$$
\mathbf{L}_{\kappa k}^{\mathrm{g}} = \mathbf{D}_{\kappa k}^{\mathrm{g}} + \boldsymbol{\Omega}_{\kappa k}^{\mathrm{g}}\,. \tag{23.4.9}
$$

Let α and β be two arbitrary scalars. The standard definition of Mandel stress is associated with $\alpha = 1$, $\beta = 0$. We will consider modified Mandel stresses with $\alpha = \beta = 1/2$, $1 \cdots$ It will appear advantageous, and costless[23.5], to introduce the rate $\mathbf{L}_{\kappa_* k}^{\mathrm{g}}$ with symmetric part $\mathbf{D}_{\kappa_* k}^{\mathrm{g}}$ and skew-symmetric part $\boldsymbol{\Omega}_{\kappa_* k}^{\mathrm{g}}$,

$$
\mathbf{L}_{\kappa_* k}^{\mathrm{g}} = (\mathbf{C}_k^{\mathrm{e}})^{1-\alpha} \cdot \mathbf{L}_{\kappa k}^{\mathrm{g}} \cdot (\mathbf{C}_k^{\mathrm{e}})^{-\beta}\,, \tag{23.4.10}
$$

and the symmetric stress,

$$
\boldsymbol{\Psi}_{\kappa_*}^{k} \equiv (\mathbf{C}_k^{\mathrm{e}})^{-1+\alpha} \cdot \boldsymbol{\Psi}_{\kappa}^{k} \cdot (\mathbf{C}_k^{\mathrm{e}})^{\beta} = (\mathbf{C}_k^{\mathrm{e}})^{\alpha} \cdot \boldsymbol{\tau}_{\kappa}^{k} \cdot (\mathbf{C}_k^{\mathrm{e}})^{\beta} = (\boldsymbol{\Psi}_{\kappa_*}^{k})^{\mathrm{T}}\,, \tag{23.4.11}
$$

with $\mathbf{C}_k^{\mathrm{e}} = (\mathbf{F}_k^{\mathrm{e}})^{\mathrm{T}} \cdot \mathbf{F}_k^{\mathrm{e}}$ the elastic right Cauchy-Green tensor, the symmetry of the modified tensor holding for $\alpha = \beta$. The decompositions (23.1.1) of the velocity gradient and of its symmetric and skew-symmetric parts adopt a more balanced form in terms of the rate $\mathbf{L}_{\kappa_* k}^{\mathrm{g}}$, namely for any growing constituent $k \in \mathcal{S}^*$,

$$
\begin{aligned}
\mathbf{L} &= \frac{d\mathbf{F}}{dt} \cdot \mathbf{F}^{-1} = \overbrace{\frac{d\mathbf{F}_k^{\mathrm{e}}}{dt} \cdot (\mathbf{F}_k^{\mathrm{e}})^{-1}}^{\mathbf{L}_k^{\mathrm{e}}} + \overbrace{(\mathbf{F}_k^{\mathrm{e}})^{-\mathrm{T}} \cdot (\mathbf{C}_k^{\mathrm{e}})^{\alpha} \cdot \mathbf{L}_{\kappa_* k}^{\mathrm{g}} \cdot (\mathbf{C}_k^{\mathrm{e}})^{\beta} \cdot (\mathbf{F}_k^{\mathrm{e}})^{-1}}^{\mathbf{L}_k^{\mathrm{g}}}\,, \\[6pt]
\mathbf{D} &= \frac{1}{2}\,(\mathbf{L}+\mathbf{L}^{\mathrm{T}}) = \mathbf{D}_k^{\mathrm{e}} + \overbrace{(\mathbf{F}_k^{\mathrm{e}})^{-\mathrm{T}} \cdot (\mathbf{C}_k^{\mathrm{e}})^{\alpha} \cdot \mathbf{D}_{\kappa_* k}^{\mathrm{g}} \cdot (\mathbf{C}_k^{\mathrm{e}})^{\beta} \cdot (\mathbf{F}_k^{\mathrm{e}})^{-1}}^{\mathbf{D}_k^{\mathrm{g}}}\,, \\[6pt]
\boldsymbol{\Omega} &= \frac{1}{2}\,(\mathbf{L}-\mathbf{L}^{\mathrm{T}}) = \boldsymbol{\Omega}_k^{\mathrm{e}} + \overbrace{(\mathbf{F}_k^{\mathrm{e}})^{-\mathrm{T}} \cdot (\mathbf{C}_k^{\mathrm{e}})^{\alpha} \cdot \boldsymbol{\Omega}_{\kappa_* k}^{\mathrm{g}} \cdot (\mathbf{C}_k^{\mathrm{e}})^{\beta} \cdot (\mathbf{F}_k^{\mathrm{e}})^{-1}}^{\boldsymbol{\Omega}_k^{\mathrm{g}}}\,.
\end{aligned} \tag{23.4.12}
$$

[23.5]Computations need to address the presence of the square root of a symmetric second order tensor and of its differential. Appropriate algebraic tools are proposed in Sections 2.3.2 or 2.3.6 to obtain the square root $\mathbf{C}^{1/2}$. Next, the differential of $\mathbf{C}^{1/2}$ satisfying the relation $d\mathbf{C} = d\mathbf{C}^{1/2} \cdot \mathbf{C}^{1/2} + \mathbf{C}^{1/2} \cdot d\mathbf{C}^{1/2}$ can be obtained by Section 2.3.4.

The dissipation,

$$\mathbf{\Psi}_\kappa^k : \mathbf{L}_{\kappa k}^g = \left((\mathbf{C}_k^e)^{1-\alpha} \cdot \mathbf{\Psi}_{\kappa_*}^k \cdot (\mathbf{C}_k^e)^{-\beta}\right) : \left((\mathbf{C}_k^e)^{-1+\alpha} \cdot \mathbf{L}_{\kappa_* k}^g \cdot (\mathbf{C}_k^e)^{\beta}\right) = \mathbf{\Psi}_{\kappa_*}^k : \mathbf{D}_{\kappa_* k}^g, \tag{23.4.13}$$

involves only the symmetric rate $\mathbf{D}_{\kappa_* k}^g$, and the growth spin $\mathbf{\Omega}_{\kappa_* k}^g$ does not contribute. While it might appear surprising at first glance, this situation is exactly similar to that of metal plasticity where the pure elastic deformation is small and the Mandel stress reduces to the symmetric 2nd Piola-Kirchhoff stress relative to the intermediate configuration. Let us insist that the skew-symmetric part of the rate of growth transformation is not driven by the non symmetry of the Mandel stress. In fact, as explained by Mandel [1973], this state of affairs comes from the fact that we have considered materials with directors but we have neglected the work-conjugate couple stresses, see also Section 21.3.1.3.

Based on the definitions (23.1.28), (23.1.29) and (23.4.11), one may also define an effective modified Mandel stress,

$$\mathbf{\Psi}_{\kappa_*}^{k'} \equiv \mathbf{\Psi}_{\kappa_*}^k + p_E\, n^k \det \mathbf{F}_k^e\, (\mathbf{C}_k^e)^{-1+\alpha+\beta}, \quad \mathbf{\Psi}_{\kappa_*}^{k'} = (\mathbf{C}_k^e)^\alpha \cdot \boldsymbol{\tau}_\kappa^{k'} \cdot (\mathbf{C}_k^e)^\beta, \quad \boldsymbol{\tau}_\kappa^{k'} = n^k \det \mathbf{F}_k^e \frac{\partial E_k}{\partial \mathbf{E}_{Gk}^e}. \tag{23.4.14}$$

Symmetric growth rate

Turning back to collagen ($k = c$), the symmetric rate $\mathbf{D}_{\kappa_* c}^g$,

$$\mathbf{D}_{\kappa_* c}^g = \frac{\mu_\kappa}{t_{gc}}\, \Phi(h_{\kappa c})\, \frac{\partial h_{\kappa c}}{\partial \mathbf{\Psi}_{\kappa_*}^c}, \tag{23.4.15}$$

is postulated exactly as in (22.5.15) but with a homeostatic surface[23.6],

$$\mu_\kappa^2\, h_{\kappa c}(\mathbf{\Psi}_{\kappa R}^c, X_\kappa^c) = \frac{(q_{\kappa R}^c)^2}{M_c^2} + (p_{\kappa R}^c)^2 - (X_\kappa^c)^2 = 0, \tag{23.4.16}$$

expressed in terms of the *shifted* stress $\mathbf{\Psi}_{\kappa R}^c = \mathbf{\Psi}_{\kappa_*}^c - \mathbf{X}_\kappa^c$, with invariants $p_{\kappa R}^c$ and $q_{\kappa R}^c$ defined by

$$p_{\kappa R}^c = \frac{1}{3} \operatorname{tr}(\mathbf{\Psi}_{\kappa_*}^c - \mathbf{X}_\kappa^c), \quad (q_{\kappa R}^c)^2 = \frac{3}{2} \operatorname{dev}(\mathbf{\Psi}_{\kappa_*}^c - \mathbf{X}_\kappa^c) : \operatorname{dev}(\mathbf{\Psi}_{\kappa_*}^c - \mathbf{X}_\kappa^c). \tag{23.4.17}$$

Here μ_κ [unit : Pa] is a reference (scaling) elastic modulus, t_{gc} [unit : s] a time scale proportional to (although potentially quite distinct from) the actual relaxation time, and M_c is a shape coefficient.

Skew-symmetric growth rate

The standard way of building the growth spin is to start from two symmetric second order tensors, and to use representation theorems, Loret [1983]. Here we have two sources of anisotropy for the collagen, namely the kinematic anisotropy associated with the center of the homeostatic surface $\mathbf{X}_\kappa^c$ and the fabric anisotropy associated with the fiber network.

If we use $\mathbf{\Psi}_{\kappa_*}^c$ and the center of the homeostatic function $\mathbf{X}_\kappa^c$, the simplest skew-symmetric tensor has the format $\mathbf{\Psi}_{\kappa_*}^c \cdot \mathbf{X}_\kappa^c - \mathbf{X}_\kappa^c \cdot \mathbf{\Psi}_{\kappa_*}^c$, or $(\mathbf{\Psi}_{\kappa_*}^c - \mathbf{X}_\kappa^c) \cdot \mathbf{\Psi}_{\kappa_*}^c - \mathbf{\Psi}_{\kappa_*}^c \cdot (\mathbf{\Psi}_{\kappa_*}^c - \mathbf{X}_\kappa^c)$, or $(\mathbf{\Psi}_{\kappa_*}^c - \mathbf{X}_\kappa^c) \cdot \mathbf{X}_\kappa^c - \mathbf{X}_\kappa^c \cdot (\mathbf{\Psi}_{\kappa_*}^c - \mathbf{X}_\kappa^c)$, resulting in either of the two forms with a dimensionless factor η_c,

$$\mathbf{\Omega}_{\kappa_* c}^g = \frac{\eta_c}{X_\kappa^c} \left(\mathbf{D}_{\kappa_* c}^g \cdot \mathbf{\Psi}_{\kappa_*}^c - \mathbf{\Psi}_{\kappa_*}^c \cdot \mathbf{D}_{\kappa_* c}^g\right), \quad \mathbf{\Omega}_{\kappa_* c}^g = \frac{\eta_c}{X_\kappa^c} \left(\mathbf{D}_{\kappa_* c}^g \cdot \mathbf{X}_\kappa^c - \mathbf{X}_\kappa^c \cdot \mathbf{D}_{\kappa_* c}^g\right). \tag{23.4.18}$$

Two other possibilities,

$$\mathbf{\Omega}_{\kappa_* c}^g = \eta_c \left(\mathbf{D}_{\kappa_* c}^g \cdot \mathbf{M}_\kappa - \mathbf{M}_\kappa \cdot \mathbf{D}_{\kappa_* c}^g\right), \quad \mathbf{\Omega}_{\kappa_* c}^g = \eta_c \frac{|\mathbf{D}_{\kappa_* c}^g|}{X_\kappa^c} \left(\mathbf{\Psi}_{\kappa_*}^c \cdot \mathbf{M}_\kappa - \mathbf{M}_\kappa \cdot \mathbf{\Psi}_{\kappa_*}^c\right), \tag{23.4.19}$$

use the fabric tensor,

$$\mathbf{M}_\kappa \equiv \sum_{i=1}^{n_e} \tilde{n}_{\kappa i}\, \mathbf{M}_{\kappa i}. \tag{23.4.20}$$

If the collagen network is isotropic, then the fabric tensor $\mathbf{M}_\kappa$ is equal to $\mathbf{I}/3$ and the growth spins based on the fabric tensor vanish, as expected. Therefore initial isotropy of the collagen network in the intermediate configurations necessarily leads the spins (23.4.19) to remain zero during arbitrary deformation processes unless the collagen network evolves in the intermediate configurations.

[23.6]The ellipsoidal shape associated with (23.4.16) makes no difference between compressive and tensile shifted stresses. One might want to open the homeostatic surface towards compressive stresses, in order to avoid growth for purely compressive shifted stresses, by replacing $p_{\kappa R}^c$ by its positive part $\langle p_{\kappa R}^c \rangle$.

23.4.3 Collagen : rate of structure variables

The rate of change of the tensor structure variable adopts the evanescent memory format (22.5.39),

$$\frac{d\boldsymbol{\xi}_{\kappa c}}{dt} = \mathbf{D}^{g}_{\kappa_* c} - a_{1c} \parallel \mathbf{D}^{g}_{\kappa_* c} \parallel \frac{\mathbf{X}^{c}_{\kappa}}{\mu_\kappa} \,. \tag{23.4.21}$$

The rate of change of the scalar structure variable (conjugated to the radius of the homeostatic surface) mimics the expression (22.5.36),

$$\frac{d\xi_{\kappa c}}{dt} = \frac{1}{t_{gc}} \, \Phi(h_{\kappa c}) \left(a_{2c} - a_{3c} \frac{X^{c}_{\kappa}}{\mu_\kappa} \right) \frac{X^{c}_{\kappa}}{\mu_\kappa} \,. \tag{23.4.22}$$

The sum of the contributions to the dissipation associated with the three above mechanisms,

$$\begin{aligned}
T\hat{S}^{c}_{\kappa(i)} &= \boldsymbol{\Psi}^{c}_{\kappa_*} : \mathbf{D}^{g}_{\kappa_* c} - X^{c}_{\kappa} \frac{d\xi_{\kappa c}}{dt} - \mathbf{X}^{c}_{\kappa} : \frac{d\boldsymbol{\xi}_{\kappa c}}{dt} \\[2mm]
&= (\boldsymbol{\Psi}^{c}_{\kappa_*} - \mathbf{X}^{c}_{\kappa}) : \mathbf{D}^{g}_{\kappa_* c} - \mathbf{X}^{c}_{\kappa} : \left(\frac{d\boldsymbol{\xi}_{\kappa c}}{dt} - \mathbf{D}^{g}_{\kappa_* c} \right) - X^{c}_{\kappa} \frac{d\xi_{\kappa c}}{dt} \\[2mm]
&= \frac{\mu_\kappa}{t_{gc}} \, \Phi(h_{\kappa c}) \, A_{(1)} \,,
\end{aligned} \tag{23.4.23}$$

where

$$A_{(1)} = 2 \, h_{\kappa c} + (2 - a_{2c}) \frac{(X^{c}_{\kappa})^2}{\mu^2_\kappa} + a_{3c} \frac{(X^{c}_{\kappa})^3}{\mu^3_\kappa} + a_{1c} \parallel \frac{\partial h_{\kappa c}}{\partial \boldsymbol{\Psi}^{c}_{\kappa_*}} \parallel \frac{\mathbf{X}^{c}_{\kappa} : \mathbf{X}^{c}_{\kappa}}{\mu_\kappa} \,, \tag{23.4.24}$$

is positive provided the dimensionless coefficients a_{1c}, a_{2c} and a_{3c} satisfy the constraints,

$$a_{1c} \geq 0, \quad 2 - a_{2c} \geq 0, \quad a_{3c} \geq 0 \,. \tag{23.4.25}$$

On the alternative determination of the size of the homeostatic surface via the cell energy density

As an alternative, the size of the homeostatic surface may be defined via the cell energy density. Then the scalar structure variable $\boldsymbol{\xi}_{\kappa c}$ is elusive and a_{2c} and a_{3c} should be set to 0, Section 23.4.7.

On the non symmetry of Mandel stress and alternative formulations

Let $\boldsymbol{\Psi}^{c}_{\kappa+} = \frac{1}{2}(\boldsymbol{\Psi}^{c}_{\kappa} + (\boldsymbol{\Psi}^{c}_{\kappa})^{T})$ and $\boldsymbol{\Psi}^{c}_{\kappa-} = \frac{1}{2}(\boldsymbol{\Psi}^{c}_{\kappa} - (\boldsymbol{\Psi}^{c}_{\kappa})^{T})$ denote respectively the symmetric and skew-symmetric parts of Mandel stress. Since the free energy E_k is an isotropic function of $\mathbf{C}^{e}_{c} = (\mathbf{F}^{e}_{c})^{T} \cdot \mathbf{F}^{e}_{c}$, of $\boldsymbol{\Gamma}_{\kappa c}$ and of the $\mathbf{M}_{\kappa i}$'s, we have the following implications from (2.1.7) and (2.11.33),

$$\begin{aligned}
&\frac{\partial E_{c}}{\partial \mathbf{C}^{e}_{c}} \cdot \mathbf{C}^{e}_{c} + \frac{\partial E_{c}}{\partial \boldsymbol{\Gamma}_{\kappa c}} \cdot \boldsymbol{\Gamma}_{\kappa c} + \sum_{i=1}^{n_e} \frac{\partial E_{c}}{\partial \mathbf{M}_{\kappa i}} \cdot \mathbf{M}_{\kappa i} \quad \text{symmetric,} \\[2mm]
\Rightarrow \quad & \mathbf{C}^{e}_{c} \cdot \frac{\partial E_{c}}{\partial \mathbf{C}^{e}_{c}} - \frac{\partial E_{c}}{\partial \boldsymbol{\Gamma}_{\kappa c}} \cdot \boldsymbol{\Gamma}_{\kappa c} - \sum_{i=1}^{n_e} \frac{\partial E_{c}}{\partial \mathbf{M}_{\kappa i}} \cdot \mathbf{M}_{\kappa i} \quad \text{symmetric,} \\[2mm]
\Rightarrow \quad & \tfrac{1}{2} \, \mathbf{C}^{e}_{c} \cdot (\boldsymbol{\tau}^{c}_{\kappa} + n^{c} \, p_{E} \, \det \mathbf{F}^{e}_{c} \, (\mathbf{C}^{e}_{c})^{-1}) - \mathbf{E}^{e}_{Gc} \cdot \underline{n}^{c} \, \boldsymbol{\Gamma}_{\kappa c} - \sum_{i=1}^{n_e} \mathbf{A}_{\kappa i} \cdot \mathbf{M}_{\kappa i} \quad \text{symmetric,} \\[2mm]
\Rightarrow \quad & \tfrac{1}{2} \, \boldsymbol{\Psi}^{c}_{\kappa} - \mathbf{E}^{e}_{Gc} \cdot \underline{n}^{c} \, \boldsymbol{\Gamma}_{\kappa c} - \sum_{i=1}^{n_e} \mathbf{A}_{\kappa i} \cdot \mathbf{M}_{\kappa i} \quad \text{symmetric,}
\end{aligned} \tag{23.4.26}$$

which provides the skew-symmetric part of $\boldsymbol{\Psi}^{c}_{\kappa}$,

$$\boldsymbol{\Psi}^{c}_{\kappa-} = \mathbf{E}^{e}_{Gc} \cdot \underline{n}^{c} \, \boldsymbol{\Gamma}_{\kappa c} - \underline{n}^{c} \, \boldsymbol{\Gamma}_{\kappa c} \cdot \mathbf{E}^{e}_{Gc} + \sum_{i=1}^{n_e} \mathbf{A}_{\kappa i} \cdot \mathbf{M}_{\kappa i} - \mathbf{M}_{\kappa i} \cdot \mathbf{A}_{\kappa i} \,. \tag{23.4.27}$$

While this relation holds for any free energy, it can be checked for the particular expression (23.4.8). Hence, while Mandel stress results to be symmetric for materials which are isotropic in the intermediate configurations, when all $\mathbf{A}_{\kappa i}$ vanish, the property does not hold in presence of an arbitrary anisotropy.

Motivated by the dissipation,

$$\boldsymbol{\Psi}^{c}_{\kappa} : \mathbf{L}^{g}_{\kappa c} = \boldsymbol{\Psi}^{c}_{\kappa+} : \mathbf{D}^{g}_{\kappa c} + \boldsymbol{\Psi}^{c}_{\kappa-} : \boldsymbol{\Omega}^{g}_{\kappa c} \,, \tag{23.4.28}$$

one might be tempted to assume $\mathbf{\Omega}^{\mathrm{g}}_{\kappa c}$ to be proportional (with a positive or negative coefficient) to $\mathbf{\Psi}^{\mathrm{c}}_{\kappa-}$. However this construction is not appropriate from a constitutive point of view, even if it provides a contribution to the dissipation inequality with a definite sign. In fact, the growth spin can not be based (exclusively) on the skew-symmetric part of Mandel stress: the plastic spin has shown its relevance in metal plasticity where the pure elastic deformation is small and therefore Mandel stress reduces to the symmetric 2nd Piola-Kirchhoff stress.

If

$$\mathbf{\Omega}^{\mathrm{g}}_{\kappa c} \simeq \mathbf{\Psi}^{\mathrm{c}}_{\kappa+} \cdot \mathbf{D}^{\mathrm{g}}_{\kappa c} - \mathbf{D}^{\mathrm{g}}_{\kappa c} \cdot \mathbf{\Psi}^{\mathrm{c}}_{\kappa+}, \tag{23.4.29}$$

the dissipation may be calculated directly,

$$\mathbf{\Psi}^{\mathrm{c}}_{\kappa-} : \mathbf{\Omega}^{\mathrm{g}}_{\kappa c} \simeq \frac{1}{2} \left((\mathbf{\Psi}^{\mathrm{c}}_{\kappa})^{\mathrm{T}} \cdot \mathbf{\Psi}^{\mathrm{c}}_{\kappa} - \mathbf{\Psi}^{\mathrm{c}}_{\kappa} \cdot (\mathbf{\Psi}^{\mathrm{c}}_{\kappa})^{\mathrm{T}} \right) : \mathbf{D}^{\mathrm{g}}_{\kappa c}. \tag{23.4.30}$$

In general, via (23.4.27),

$$\mathbf{\Psi}^{\mathrm{c}}_{\kappa-} : \mathbf{\Omega}^{\mathrm{g}}_{\kappa c} = 2 \left(\mathbf{E}^{\mathrm{e}}_{\mathrm{Gc}} \cdot \underline{n}^{\mathrm{c}} \, \mathbf{\Gamma}_{\kappa c} \right) : \mathbf{\Omega}^{\mathrm{g}}_{\kappa c} + 2 \underline{n}^{\mathrm{c}} \, \frac{k_{\mathrm{c}}}{k_{\mathrm{c}}(0)} \sum_{i=1}^{n_e} K_{ci} \exp(k_{\mathrm{c}} L^2_{\kappa i}) \, L_{\kappa i} \, (\mathbf{C}^{\mathrm{e}}_{\mathrm{c}} \cdot \mathbf{m}_{\kappa i}) \cdot \mathbf{\Omega}^{\mathrm{g}}_{\kappa c} \cdot \mathbf{m}_{\kappa i}. \tag{23.4.31}$$

Still, neither (23.4.30) nor (23.4.31) provides a definite sign of the dissipation unlike the formulation of Section 23.4.2 based on the modified Mandel stress.

23.4.4 Collagen: evolution of the fiber directions $\mathbf{m}_{\kappa i}$, for $i \in [1, n_e]$

The fiber directions in the intermediate configurations transport in the current configurations like material lines embedded in the tissue, namely, for $i \in [1, n_e]$,

$$\mathbf{m}_i = \frac{\mathbf{F}^{\mathrm{e}}_{\mathrm{c}} \cdot \mathbf{m}_{\kappa i}}{|\mathbf{F}^{\mathrm{e}}_{\mathrm{c}} \cdot \mathbf{m}_{\kappa i}|}. \tag{23.4.32}$$

The point now is to address the rate of change of the intermediate fiber directions $\mathbf{m}_{\kappa i}$. Let

$$\frac{d\mathbf{m}_{\kappa i}}{dt} = \mathbf{\mathcal{E}}_{\kappa i} \cdot \mathbf{m}_{\kappa i}. \tag{23.4.33}$$

If the vector $\mathbf{m}_{\kappa i}$ remains of unit norm, the second order tensor $\mathbf{\mathcal{E}}_{\kappa i}$ may be *skew-symmetric*, see, e.g., (21.3.17). Still that form is not unique, for example (23.1.48) uses a *projection* on the plane orthogonal to $\mathbf{m}_{\kappa i}$. Note also the two forms of the dissipation,

$$-\mathbf{A}_{\kappa i} : \frac{d\mathbf{M}_{\kappa i}}{dt} = -2 \underbrace{\mu_\kappa \mathbf{\mathcal{P}}_{\kappa i}}_{\mathbf{A}_{\kappa i} \cdot \mathbf{M}_{\kappa i}} : \mathbf{\mathcal{E}}_{\kappa i} = -2 \underbrace{\mu_\kappa \mathbf{p}_{\kappa i}}_{\mathbf{A}_{\kappa i} \cdot \mathbf{m}_{\kappa i}} \cdot \frac{d\mathbf{m}_{\kappa i}}{dt}, \tag{23.4.34}$$

and the relations,

$$\mathbf{\mathcal{P}}_{\kappa i} \equiv \frac{\mathbf{A}_{\kappa i}}{\mu_\kappa} \cdot \mathbf{M}_{\kappa i} = \mathbf{p}_{\kappa i} \otimes \mathbf{m}_{\kappa i}, \quad \mathbf{p}_{\kappa i} \equiv \frac{\mathbf{A}_{\kappa i}}{\mu_\kappa} \cdot \mathbf{m}_{\kappa i} = \frac{1}{|\mathbf{m}_{\kappa i}|^2} \mathbf{\mathcal{P}}_{\kappa i} \cdot \mathbf{m}_{\kappa i}. \tag{23.4.35}$$

In order to satisfy the dissipation inequality associated with the directional variables, the existence of a dimensionless smooth convex function $\phi_{\kappa i}(\mathbf{\mathcal{P}}_{\kappa i})$ containing the origin is postulated, and we set

$$\mathbf{\mathcal{E}}_{\kappa i} = \frac{\Phi(h_{\kappa c})}{t_{\mathrm{gc}}} \frac{\partial \phi_{\kappa i}}{\partial \mathbf{\mathcal{P}}_{\kappa i}}, \quad i \in [1, n_e]. \tag{23.4.36}$$

For the free energy (23.4.7), the entities involved in the dissipation specialize to, for any sector $i \in [1, n_e]$,

$$\mathbf{A}_{\kappa i} = \underline{n}^{\mathrm{c}} \frac{\partial E_{\mathrm{c}}}{\partial \mathbf{M}_{\kappa i}} = \underline{n}^{\mathrm{c}} K_{ci} \frac{k_{\mathrm{c}}}{k_{\mathrm{c}}(0)} \exp(k_{\mathrm{c}} L^2_{\kappa i}) \, L_{\kappa i} \, \mathbf{E}^{\mathrm{e}}_{\mathrm{Gc}};$$

$$\mathbf{\mathcal{P}}_{\kappa i} = \frac{\mathbf{A}_{\kappa i}}{\mu_\kappa} \cdot \mathbf{M}_{\kappa i} = \underline{n}^{\mathrm{c}} \frac{K_{ci}}{\mu_\kappa} \frac{k_{\mathrm{c}}}{k_{\mathrm{c}}(0)} \exp(k_{\mathrm{c}} L^2_{\kappa i}) \, L_{\kappa i} \, \mathbf{E}^{\mathrm{e}}_{\mathrm{Gc}} \cdot \mathbf{M}_{\kappa i}; \tag{23.4.37}$$

$$\mathbf{p}_{\kappa i} = \frac{\mathbf{A}_{\kappa i}}{\mu_\kappa} \cdot \mathbf{m}_{\kappa i} = \underline{n}^{\mathrm{c}} \frac{K_{ci}}{\mu_\kappa} \frac{k_{\mathrm{c}}}{k_{\mathrm{c}}(0)} \exp(k_{\mathrm{c}} L^2_{\kappa i}) \, L_{\kappa i} \, \mathbf{E}^{\mathrm{e}}_{\mathrm{Gc}} \cdot \mathbf{m}_{\kappa i}.$$

Convexity of $\phi_{\kappa i}$ ensures the dissipation to be negative. Still, here is a case where $\phi_{\kappa i}$ is not convex[23.7], but the sign of the dissipation can be controlled. Let $\alpha_{(2c)} \geq 0$ and $0 \leq \alpha_\prime \leq 1$,

$$
\begin{aligned}
\phi_{\kappa i}(\boldsymbol{\mathcal{P}}_{\kappa i}) &= \frac{\alpha_{(2c)}}{2} \left(\boldsymbol{\mathcal{P}}_{\kappa i} : \boldsymbol{\mathcal{P}}_{\kappa i} - \alpha_\prime (\operatorname{tr}\boldsymbol{\mathcal{P}}_{\kappa i})^2 \right) = \frac{\alpha_{(2c)}}{2} \left((\mathbf{p}_{\kappa i} \cdot \mathbf{p}_{\kappa i})(\mathbf{m}_{\kappa i} \cdot \mathbf{m}_{\kappa i}) - \alpha_\prime (\mathbf{p}_{\kappa i} \cdot \mathbf{m}_{\kappa i})^2 \right) \geq 0; \\
\boldsymbol{\mathcal{E}}_{\kappa i} &= \frac{\alpha_{(2c)}}{t_{\mathrm{gc}}} \Phi(h_{\kappa c}) \left(\boldsymbol{\mathcal{P}}_{\kappa i} - \alpha_\prime \operatorname{tr}\boldsymbol{\mathcal{P}}_{\kappa i}\, \mathbf{I} \right) = \frac{\alpha_{(2c)}}{t_{\mathrm{gc}}} \Phi(h_{\kappa c}) \left(\mathbf{p}_{\kappa i} \otimes \mathbf{m}_{\kappa i} - \alpha_\prime (\mathbf{p}_{\kappa i} \cdot \mathbf{m}_{\kappa i})\, \mathbf{I} \right),
\end{aligned}
\tag{23.4.38}
$$

and the contribution to the Clausius-Duhem inequality is negative,

$$
T\hat{S}^c_{\kappa(\mathrm{i})} = - \sum_{i=1}^{n_e} \mathbf{A}_{\kappa i} : \frac{d\mathbf{M}_{\kappa i}}{dt} = \frac{\mu_\kappa}{t_{\mathrm{gc}}} \Phi(h_{\kappa c})(-A_{(2)}) \leq 0,
\tag{23.4.39}
$$

with

$$
A_{(2)} = 4 \sum_{i=1, n_e} \phi_{\kappa i}(\boldsymbol{\mathcal{P}}_{\kappa i}) \geq 0.
\tag{23.4.40}
$$

In the sequel, the idea of changing the modulus of the vectors $\mathbf{m}_{\kappa i}$, is abandoned, the coefficient $\alpha_\prime$ will be set equal to 1 and the vectors $\mathbf{m}_{\kappa i}$ remain of unit norm[23.8]. One may nevertheless record a few relations for the more general case $0 < \alpha' < 1$. Let $\mathbf{m}_{\kappa i} = |\mathbf{m}_{\kappa i}| \hat{\mathbf{m}}_{\kappa i}$, with $\hat{\mathbf{m}}_{\kappa i}$ of unit norm. Then

$$
\frac{d\mathbf{m}_{\kappa i}}{dt} = \boldsymbol{\mathcal{E}}_{\kappa i} \cdot \mathbf{m}_{\kappa i} = |\mathbf{m}_{\kappa i}|^2 \frac{\alpha_{(2c)}}{t_{\mathrm{gc}}} \Phi(h_{\kappa c})(\mathbf{I} - \alpha_\prime \hat{\mathbf{m}}_{\kappa i} \otimes \hat{\mathbf{m}}_{\kappa i}) \cdot \mathbf{p}_{\kappa i}.
\tag{23.4.41}
$$

Actually, using (23.4.35), $d\hat{\mathbf{m}}_{\kappa i}/dt$ may be shown to be the projection of $\mathbf{p}_{\kappa i}$ on the plane orthogonal to $\mathbf{m}_{\kappa i}$,

$$
\begin{aligned}
\frac{d\hat{\mathbf{m}}_{\kappa i}}{dt} &= |\mathbf{m}_{\kappa i}| \frac{\alpha_{(2c)}}{t_{\mathrm{gc}}} \Phi(h_{\kappa c})(\mathbf{I} - \hat{\mathbf{m}}_{\kappa i} \otimes \hat{\mathbf{m}}_{\kappa i}) \cdot \mathbf{p}_{\kappa i}, \\
\frac{d|\mathbf{m}_{\kappa i}|}{dt} &= |\mathbf{m}_{\kappa i}| \frac{\alpha_{(2c)}}{t_{\mathrm{gc}}} \Phi(h_{\kappa c})(\hat{\mathbf{m}}_{\kappa i} \cdot \mathbf{p}_{\kappa i})(1 - \alpha_\prime).
\end{aligned}
\tag{23.4.42}
$$

Note that this rate law does not conserve right angles: indeed let $\mathbf{m}_{\kappa i}$ and $\mathbf{m}_{\kappa j}$ be two orthogonal vectors. Then $\mathbf{m}_{\kappa i} \cdot \mathbf{m}_{\kappa j} = 0$ but

$$
\frac{d}{dt}(\mathbf{m}_{\kappa i} \cdot \mathbf{m}_{\kappa j}) \simeq \frac{\alpha_{(2c)}}{t_{\mathrm{gc}}} \Phi(h_{\kappa c}) \mathbf{m}_{\kappa i} \cdot \mathbf{E}^e_{\mathrm{Gc}} \cdot \mathbf{m}_{\kappa j}.
\tag{23.4.43}
$$

23.4.5 Collagen: evolution of the relative volume fractions $\tilde{n}_{\kappa i}$, for $i \in [1, n_e]$

The effects of the orientation of the collagen fibers on the mechanical properties are less significant in the young adult bovine cartilage than in human adult cartilage, Williamson et al. [2003], p. 878. We interpret this observation as a developing orientation distribution function of collagen fibers that impacts on the mechanical properties.

We have also to address the additional problem of the sign of $L_{\kappa i} = \langle \mathbf{E}^e_{\mathrm{Gc}} : \mathbf{M}_{\kappa i} \rangle$ which should be positive for the fan of fibers i to be active. Otherwise, the fiber is not active, and the term does not appear in the free energy and dissipation inequality. Interestingly, the multiplicative decomposition of the deformation gradient may trigger elastic extension if the compression induced resorption is large enough.

In the model, the directional properties of the fiber network are described by two rate laws. The rate law in this section addresses the relative volume fractions: it conveys the idea that the *collagen network* grows in the directions that are indicated by the mechanical loading. Along another view point, we may consider the rotation of *individual*

[23.7] Indeed, with $\boldsymbol{\mathcal{P}}_{\kappa i} = \operatorname{tr}\boldsymbol{\mathcal{P}}_{\kappa i}\, \mathbf{I}/3 + \operatorname{dev}\boldsymbol{\mathcal{P}}_{\kappa i}$, then the quadratic function $2\phi_{\kappa i}(\boldsymbol{\mathcal{P}}_{\kappa i}) = (\operatorname{dev}\boldsymbol{\mathcal{P}}_{\kappa i})^2 + (1/3 - \alpha_\prime)(\operatorname{tr}\boldsymbol{\mathcal{P}}_{\kappa i})^2 = \boldsymbol{\mathcal{P}}_{\kappa i} :$ $(\mathbf{I}\overline{\otimes}\mathbf{I} - \alpha_\prime\, \mathbf{I}\otimes\mathbf{I}) : \boldsymbol{\mathcal{P}}_{\kappa i}$ is clearly nonconvex, and nonpositive, over the whole sets of second order tensors. On the other hand, it is positive over the set of dyadic second order tensors as shown by (23.4.38).

[23.8] Still microscopic measurements indicate that the ultimate energy $\int_0^l \tau\, l\, dl = \tau\, l^2/2$ required by the relative parallel sliding of two units of length l undergoing a uniform shear stress τ is proportional to the square of the length of the units. The actual length is limited by the axial strength of the units themselves. Typical tensile strengths vary from a few MPa's in the chick embryo to 50-60 MPa at a month. Measurements indicate that fibril length, rather than fibril diameter, is the key parameter in the elastic properties of collagen. Moreover, in collagen gels, the Young's modulus is reported to vary like the cubic power of the collagen concentration with concentrations varying from 0.5 to 5 mg/ml and for a strain smaller than 10%, above which the material behavior stiffens, Section 14.2.3.3.

fibers all along the growth process. Their directions change due to the accommodation strain. Their rotation during the growth process with respect to the intermediate configuration is considered in Section 23.4.4.

The contributions to the dissipation due to the change of relative volume fractions of the fibers sum to,

$$-\sum_{i=1}^{n_e} \tilde{N}_{\kappa i} \frac{d\tilde{n}_{\kappa i}}{dt} \,. \tag{23.4.44}$$

The generalized stress $\tilde{N}_{\kappa i}$ defined by (23.4.8) is positive, but the rates $d\tilde{n}_{\kappa i}/dt$ are both positive and negative. In fact, the rate equations for the relative volume fractions should satisfy the constraints,

$$\sum_{i=1}^{n_e} \tilde{n}_{\kappa i} = 1, \quad \sum_{i=1}^{n_e} \frac{d\tilde{n}_{\kappa i}}{dt} = 0, \quad \tilde{n}_{\kappa i} \in [0,1] \,. \tag{23.4.45}$$

Assume

$$\begin{aligned}
\frac{d\tilde{n}_{\kappa i}}{dt} &= \frac{\alpha_{(3c)}}{\mu_\kappa} \frac{\Phi(h_{\kappa c})}{t_{gc}} \nu_i \left(\tilde{N}_{\kappa i} + \Lambda \right) \\
&= \frac{\alpha_{(3c)}}{\mu_\kappa} \frac{\Phi(h_{\kappa c})}{t_{gc}} \frac{\nu_i}{\nu} \left(\nu \, \tilde{N}_{\kappa i} - \sum_{j=1}^{n_e} \nu_j \, \tilde{N}_{\kappa j} \right),
\end{aligned} \tag{23.4.46}$$

with $\alpha_{(3c)} \geq 0$ dimensionless, $\nu_i = \langle \tilde{n}_{\kappa i} \left(1 - \tilde{n}_{\kappa i}\right)\rangle$, $\nu = \sum_{j=1}^{n_e} \nu_j$, and $\Lambda = -\left(\sum_{i=1}^{n_e} \nu_i \, \tilde{N}_{\kappa i}\right)/\nu$. The coefficients ν_i prevent the volume fractions to exit the admissible interval $[0,1]$, and the constraints (23.4.45) are satisfied.

The sum of the contributions to the Clausius-Duhem inequality over all directions turns out to be negative. Indeed,

$$\begin{aligned}
-\sum_{i=1}^{n_e} \tilde{N}_{\kappa i} \frac{d\tilde{n}_{\kappa i}}{dt} &= \frac{\alpha_{(3c)}}{\mu_\kappa} \frac{\Phi(h_{\kappa c})}{t_{gc}} \frac{1}{\nu} \overbrace{\left(\Big(\sum_{j=1}^{n_e} \nu_j \, \tilde{N}_{\kappa j} \Big)^2 - \nu \sum_{j=1}^{n_e} \nu_j \, \tilde{N}_{\kappa j} \, \tilde{N}_{\kappa j} \right)}^{A} \\
&= \frac{\alpha_{(3c)}}{\mu_\kappa} \frac{\Phi(h_{\kappa c})}{t_{gc}} \frac{1}{\nu} \sum_{j=1}^{n_e} \nu_j \, \tilde{N}_{\kappa j} \Big(\sum_{i=1}^{n_e} \nu_i \, (\tilde{N}_{\kappa i} - \tilde{N}_{\kappa j}) \Big) \\
&= \frac{\alpha_{(3c)}}{\mu_\kappa} \frac{\Phi(h_{\kappa c})}{t_{gc}} \frac{1}{\nu} \sum_{i,j=1}^{n_e} \left(\nu_j \, (\tilde{N}_{\kappa j} - \tilde{N}_{\kappa i}) + \nu_j \, \tilde{N}_{\kappa i} \right) \nu_i \, (\tilde{N}_{\kappa i} - \tilde{N}_{\kappa j}) \\
&= \frac{\alpha_{(3c)}}{\mu_\kappa} \frac{\Phi(h_{\kappa c})}{t_{gc}} \frac{1}{\nu} \Big(\sum_{i,j=1}^{n_e} -\nu_i \, \nu_j \, (\tilde{N}_{\kappa j} - \tilde{N}_{\kappa i})^2 + \overbrace{\nu_i \, \nu_j \, \tilde{N}_{\kappa i} \, (\tilde{N}_{\kappa i} - \tilde{N}_{\kappa j})}^{-A} \Big) \\
&= -\frac{\alpha_{(3c)}}{\mu_\kappa} \frac{\Phi(h_{\kappa c})}{t_{gc}} \frac{1}{2\nu} \sum_{i,j=1}^{n_e} \nu_i \, \nu_j \, (\tilde{N}_{\kappa j} - \tilde{N}_{\kappa i})^2 \,,
\end{aligned} \tag{23.4.47}$$

or,

$$T\hat{S}^c_{\kappa(i)} = -\sum_{i=1}^{n_e} \tilde{N}_{\kappa i} \frac{d\tilde{n}_{\kappa i}}{dt} = \frac{\mu_\kappa}{t_{gc}} \Phi(h_{\kappa c}) \, (-A_{(3)}) \leq 0 \,, \tag{23.4.48}$$

with

$$A_{(3)} = \frac{\alpha_{(3c)}}{2\nu} \sum_{i,j=1}^{n_e} \nu_i \, \nu_j \left(\frac{\tilde{N}_{\kappa j} - \tilde{N}_{\kappa i}}{\mu_\kappa} \right)^2 \geq 0 \,. \tag{23.4.49}$$

Motivation and tentative alternatives:

Let us postulate, whether the fiber i is active, a rate law of the form $d\tilde{n}_{\kappa i}/dt \sim a_{\kappa i}$, with $a_{\kappa i} = \mathbf{m}_{\kappa i} \cdot \mathrm{dev}\mathbf{E}^e_{Gc} \cdot \mathbf{m}_{\kappa i}$, so that $\int_{\mathbb{S}^2} d\tilde{n}_{\kappa i} \, dS = 0$, since $\int_{\mathbb{S}^2} \mathbf{m}_{\kappa i} \otimes \mathbf{m}_{\kappa i} \, dS/(4\pi) = \mathbf{I}/3$[23.9]. The trend looks appropriate: highly activated fiber families would have a growing volume fraction and inactive fibers a decreasing volume fraction. Still one should take care to keep the volume fractions between 0 and 1. One way to satisfy this constraint is to postulate

---

[23.9]This relation assumes that the fiber directions $\mathbf{m}_{\kappa i}$ are uniformly distributed, that is, they represent the fans, and that they do not rotate, unlike what has been considered in Section 23.4.4.

$d\tilde{n}_{\kappa i}/dt \sim \tilde{n}_{\kappa i}\,(1-\tilde{n}_{\kappa i})\,a_{\kappa i}$, but then we are not sure any longer that $\int_{\mathbb{S}^2}\,d\tilde{n}_{\kappa i}\,dS=0$ (actually we should write a finite sum over i instead of the integral, but the idea is the same). Let us consider the two options:

Option 1

If

$$\frac{d\tilde{n}_{\kappa i}}{dt}=\alpha_{(3c)}\,\frac{\Phi(h_{\kappa c})}{t_{gc}}\,a_{\kappa i}\,,\quad a_{\kappa i}\equiv\mathbf{m}_{\kappa i}\cdot\operatorname{dev}\mathbf{E}_{Gc}^{e}\cdot\mathbf{m}_{\kappa i}\tag{23.4.50}$$

with $\alpha_{(3c)}\geq 0$, then the constraints $(23.4.45)_{1,2}$ are satisfied automatically, but we are not sure that the volume fractions remain between 0 and 1.

Option 2

With $\alpha_{(3c)}\geq 0$, assume

$$\frac{d\tilde{n}_{\kappa i}}{dt}=\alpha_{(3c)}\,\frac{\Phi(h_{\kappa c})}{t_{gc}}\,\nu_i\,\Big(f(a_{\kappa i})-\sum_{j=1}^{n_e}\frac{\nu_j}{\nu}\,f(a_{\kappa j})\Big)\,.\tag{23.4.51}$$

We think of $f(a_{\kappa i})$ as $a_{\kappa i}$. For fibers which undergo the largest extension, the rate of relative volume fraction is positive, and the contribution to the Clausius-Duhem inequality is negative. Conversely, the rate of relative volume fraction is negative for fibers than undergo contraction, but then the energy vanishes and there is no associated contribution with the Clausius-Duhem inequality.

Alternatively, one might define $f(a_{\kappa i})$ as $\langle a_{\kappa i}\rangle$, the positive part of $a_{\kappa i}$. Then the rate of volume fraction does not vanish even in the sets which are in contraction due to the term Λ, which is for sure negative, but the difference wrt the previous case should be more quantitative than qualitative.

Unlike in (23.4.48), the sign of the contribution to the Clausius-Duhem inequality over all directions is undecided and depends on the details of the model,

$$T\hat{S}_{\kappa(i)}^{c}=\frac{\mu_\kappa}{t_{gc}}\,\Phi(h_{\kappa c})\,(-A_{(3)})\lesseqgtr 0\,,\quad A_{(3)}=\sum_{i=1}^{n_e}\frac{\tilde{N}_{\kappa i}}{\mu_\kappa}\,\alpha_{(3c)}\,\nu_i\,\big(f(a_{\kappa i})+\Lambda\big)\gtreqless 0\,.\tag{23.4.52}$$

23.4.6 Collagen: evolution of the pyridinoline crosslink density

The contribution to the Clausius-Duhem inequality associated with the change of the density of pyridinoline,

$$-P_\kappa^c\,\frac{dp_{\kappa c}}{dt}\leq 0\,,\tag{23.4.53}$$

is negative since the pyridinoline density $p_{\kappa c}$ increases during growth and so does the collagen modulus so that the associated generalized stress P_κ^c expressed by (23.4.4) is positive, in as far as the modulus $K_{cf}(p_{\kappa c})$ and exponent $k_c(p_{\kappa c})$ are increasing functions of $p_{\kappa c}$.

Along the line of thinking of this work, the pyridinoline density increases only if growth takes place[23.10], and it saturates at one mole per mole of collagen as indicated in Table 23.4.1. The simplest assumption consists of an affine relation, namely

$$\frac{dp_{\kappa c}}{dt}=\frac{\alpha_{(4c)}}{t_{gc}}\,\Phi(h_{\kappa c})\,(1-p_{\kappa c})\geq 0\,,\tag{23.4.54}$$

with $\alpha_{(4c)}\geq 0$ dimensionless. Therefore the contribution to the Clausius-Duhem inequality associated with the change of pyridinoline density reads,

$$T\hat{S}_{\kappa(i)}^{c}=-P_\kappa^c\,\frac{dp_{\kappa c}}{dt}=\frac{\mu_\kappa}{t_{gc}}\,\Phi(h_{\kappa c})\,(-A_{(4)})\leq 0\,,\quad A_{(4)}\equiv\frac{P_\kappa^c}{\mu_\kappa}\,\alpha_{(4c)}\,(1-p_{\kappa c})\geq 0\,.\tag{23.4.55}$$

It remains to provide expressions for the functions $K_{cf}=K_{cf}(p_{\kappa c})$, and $k_c=k_c(p_{\kappa c})$. Again they can be either saturating, or affine, wrt the pyridinoline density, e.g.,

$$K_{cf}(p_{\kappa c})=K_{cf}(0)+p_{\kappa c}\,K_{cf}(1)\,,\quad k_c(p_{\kappa c})=k_c(0)+p_{\kappa c}\,k_c(1)\,.\tag{23.4.56}$$

[23.10]Observations show that the growths of PGs and collagen are sequential rather than simultaneous: PGs grow first, collagen grows next. In fact, if the load is tensile, PGs can grow only if collagen stiffness is low. Since growth and structuration/stiffening are more or less simultaneous, PGs can not but grow first. Indeed, in the model, collagen and PGs have the same total deformation. Therefore, for PGs to be able to grow in presence of another constituent, the stiffness of this other constituent should be small. Therefore the time course of the pyridinoline density seems quite important. In conclusion, in the first phase when PGs grow, the pyridinoline density and collagen stiffness should be very small.

Table 23.4.1: Evolution of the pyridinoline density for bovine patellofemoral groove, Williamson et al. [2003], and of cell density, Williamson et al. [2001].

Species	Unit	Fetal	Calf	Adult
Pyridinoline	mole/mole collagen	0.2	0.5	0.95
Cell	mg/ml tissue	16	10	2

23.4.7 Collagen: evolution of the cell energy source

Cells are viewed as providing energy to the tissue. Williamson et al. [2001] display cell densities that decrease significantly with age, Table 23.4.1. This decrease of cell density may be traced to the development of the extracellular matrix, which leads to modifications of the spatial organization of the cell network, Jadin et al. [2007]. The energy provided by a cell during growth is also expected to decrease, due to genetic programming that reduces the metabolism. Another potential reason for a reduced metabolism of chondrocytes may be of purely biochemical origin: diffusion through the extracellular matrix of nutrients and specially of macromolecules, like growth factors, may be hampered by the developing GAGs and collagen network. Cell death is also cited, although apparently not quantified, Stockwell [1979].

The above remarks intend to convey the idea that the cell energy density interacts with the growing tissue[23.11]. In other words, growth and structuration tend to be self-regulated processes. Therefore, it is appropriate to include cells in the system to be considered. An alternative would be to consider the energy provided by cells as an external input. As just argued, this approach neglects the interactions between cells and extracellular matrix[23.12].

The dimensionless cell energy density $c_{\kappa c}$, i.e., the energy stored by the cells per unit volume of the intermediate configuration available to collagen,

$$c_{\kappa c} = \gamma_c \, N_{\text{cell}} \, e_\kappa^{\text{cell}}, \tag{23.4.57}$$

involves

- the dimensionless cell energy e_κ^{cell} per unit intermediate volume and per cell;
- the cell number N_{cell} per unit intermediate volume;
- the energy partition coefficient γ_k: the total energy released by the cells is partitioned between the growing constituents, and $\sum_{k \in \mathcal{S}^*} \gamma_k = 1$.

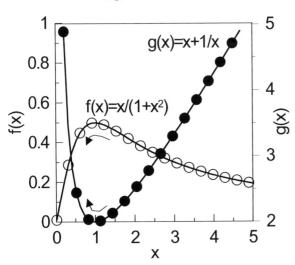

Fig. 23.4.1: During the growth process, the cell energy density x decreases, its contribution to the dissipation $f(x)$ has a maximum while the size of the homeostatic surface $g(x)$ has a minimum at $x = 1$. Reprinted with permission from Loret and Simões [2014].

Size of homeostatic surface governed by cell energy density

Let us consider first the model where the size of the homeostatic surface is a function of the cell energy density. Then a part $E_{c,1}^{\text{cell}}$ of the free energy contributed by the cells is singled out and diverted to be associated with the

---

[23.11] The main idea is that the cell energy is considered like a source of energy that is used, and that decreases, during the growth process, much like the surface energy in a cracked material that decreases as the crack surface decreases during the healing process.
[23.12] The attempt below accounts only partially for these interactions. The rate of cell energy density certainly affects the rates of the other processes involved in the growth of mass and structuration. The converse implication also holds as the cell energy density evolves only if these processes are activated.

change of the size of the homeostatic surface X_κ^c. The remaining part $E_{c,2}^{cell}$ is used for the cell evolution and a priori spread over all the other processes. The size of the homeostatic surface,

$$(\underline{n}^c)^{-1} X_\kappa^c = \frac{dE_{c,1}^{cell}}{dc_{\kappa c}} = \chi_{\kappa c} \left(c_{\kappa c} + \frac{1}{c_{\kappa c}} \right) > 0, \qquad (23.4.58)$$

with the coefficient $\chi_{\kappa c}$ [unit : Pa] to be defined in the initialization process, is minimum at $c_{\kappa c} = 1$ where consequently the activation of growth is maximum, as sketched in Fig. 23.4.1. The term associated with the cell evolution,

$$(\underline{n}^c)^{-1} \left(C_\kappa^c - X_\kappa^c \right) = \frac{dE_{c,2}^{cell}}{dc_{\kappa c}} = \mu_\kappa c_{ic} \frac{c_{\kappa c}}{1 + c_{\kappa c}^2} > 0, \qquad (23.4.59)$$

with c_{ic} a material constant (set to 1 below), is tailored such that
   - it has a maximum at the critical value $c_{\kappa c} = 1$, and
   - it vanishes in absence of cells and at large density in agreement with the phenomenon of contact inhibition.
The cell energy is used solely by the growth process, so that it changes only if growth is activated. In summary,

$$\frac{dc_{\kappa c}}{dt} = -\frac{\alpha_{(5c)}}{t_{gc}} \Phi(h_{\kappa c}) \frac{1}{\mu_\kappa} \frac{dE_{c,2}^{cell}}{dc_{\kappa c}} \leq 0, \qquad (23.4.60)$$

with $\alpha_{(5c)} \geq 0$ a dimensionless parameter.

By the assumption (23.4.60), even if the cell energy density is large, it is not used if the stress is homeostatic, that is, the stress condition can block completely the growth and resorption processes. Still the introduction of the cell energy density in the current constitutive functions should help resolve this apparent paradox.

In view of (23.4.8), the contribution to the dissipation inequality associated with cells,

$$T\hat{S}_{\kappa(i)}^c = -C_\kappa^c \frac{dc_{\kappa c}}{dt} = \frac{\mu_\kappa}{t_{gc}} \Phi(h_{\kappa c}) A_{(5)} \geq 0, \quad A_{(5)} \equiv \alpha_{(5c)} \frac{\underline{n}^c}{\mu_\kappa} \frac{dE_c^{cell}}{dc_{\kappa c}} \times \frac{1}{\mu_\kappa} \frac{dE_{c,2}^{cell}}{dc_{\kappa c}} \geq 0, \qquad (23.4.61)$$

is positive.

Size of homeostatic surface governed by the independent variable ξ_κ^c

Alternatively, if the size of the homeostatic surface is governed by the independent variable ξ_κ^c, then the only change here consists in zeroing the term $E_{c,1}^{cell}$ so that $E_{c,2}^{cell} = E_c^{cell}$.

The role of nutrients in the rate processes

The role of nutrients is not explicitly considered in the present formulation. A simple way to introduce the effects of nutrients would consist in inserting a representative nutrient concentration c_{nu} in the constitutive function Φ. Typically, the effect could be aggregated in a Michaelis-Menten format, say via the multiplicative factor $c_{nu}/(k_{nu} + c_{nu})$. For example, the typical concentration of oxygen in tissues and constructs is about 0.1 to 0.3 mole/m^3 with k_{nu} about 5×10^{-3} mole/m^3, Obradovic et al. [2000], Roose et al. [2003].

23.4.8   Collagen: gathering the contributions to the dissipation inequality

The dissipation inequality is enforced at any time. If the rate equations developed so far turn at a particular time to violate the dissipation inequality, the processes which contribute negatively to the dissipation inequality are damped. The scaling coefficient is uniform over all these processes, while a more refined analysis might introduce distinct scalings.

The complete reduced dissipation inequality may be expressed in the format,

$$T\hat{S}_{\kappa(i)}^c = \frac{\mu_\kappa}{t_{gc}} \Phi(h_{\kappa c}) \big(\underbrace{A_{(1)}}_{\geq 0} - \underbrace{A_{(2)}}_{\geq 0} - \underbrace{A_{(3)}}_{\geq 0} - \underbrace{A_{(4)}}_{\geq 0} + \underbrace{A_{(5)}}_{\geq 0} \big). \qquad (23.4.62)$$

The processes which require energy use all the energy available, but no more. In order to avoid the inequality dissipation to be violated, the rate equations of these processes as defined so far are scaled down by $\epsilon \leq 1$, say formally[23.13],

$$\frac{d\xi_{\kappa c}^{(k)}}{dt} = \frac{\mu_\kappa}{t_{gc}} \Phi(h_{\kappa c}) \cdots \times \epsilon, \quad k = 2, 3, 4, \qquad (23.4.63)$$

[23.13]Note that the growth process itself is not damped. This possibility might be envisaged.

with

$$\epsilon = \min\left(1, \frac{A_{(1)} + A_{(5)}}{A_{(2)} + A_{(3)} + A_{(4)}}\right). \tag{23.4.64}$$

The dissipation (23.4.62) becomes,

$$T\hat{S}^{c}_{\kappa(i)} = \frac{\mu_\kappa}{t_{gc}} \Phi(h_{\kappa c}) \left(A_{(1)} + A_{(5)} - (A_{(2)} + A_{(3)} + A_{(4)}) \times \epsilon\right) \geq 0. \tag{23.4.65}$$

In essence,
- if the energy available is larger than the energy required, namely $\epsilon = 1$, then the growth and structuration processes take place with a strictly positive dissipation;
- if the energy available is smaller than the energy required, namely $\epsilon < 1$, then the growth processes, associated with a positive dissipation, take place as before, but the structuration processes, which require energy, are damped as indicated by the multiplicative factor $\epsilon < 1$.

Even when the dissipation vanishes, the processes are not reversible because the constitutive equations do not allow so.

23.5 Growth equations for proteoglycans

The exposition for proteoglycans is simplified with respect to collagen, because their stress contribution in the current configuration is isotropic. No directional reorientation is involved and no improvement of the mechanical stiffness similar to that induced for collagen by pyridinoline takes place.

23.5.1 Proteoglycans: the free energy

The set of *dimensionless* state variables relative to the intermediate configurations,

$$\mathcal{V}_{\kappa PG} = \{\mathbf{E}^{e}_{GPG}, c_{\kappa PG}, \xi_{\kappa PG}, \boldsymbol{\xi}_{\kappa PG}\}, \tag{23.5.1}$$

includes
- the elastic Green strain $\mathbf{E}^{e}_{GPG}$;
- the cell energy density $c_{\kappa PG}$;
- two additional *structure* variables associated with the evolution of the homeostatic surface, namely a scalar variable $\xi_{\kappa PG}$ which controls its size and a symmetric second order tensor $\boldsymbol{\xi}_{\kappa PG}$ which defines its center.
  Then, with the *generalized stresses* [unit : Pa],

$$\boldsymbol{\tau}^{PG'}_{\kappa} = \underline{n}^{PG} \frac{\partial E_{\kappa PG}}{\partial \mathbf{E}^{e}_{GPG}}, \quad C^{PG}_{\kappa} = \underline{n}^{PG} \frac{\partial E_{\kappa PG}}{\partial c_{\kappa PG}}, \quad X^{PG}_{\kappa} = \underline{n}^{PG} \frac{\partial E_{\kappa PG}}{\partial \xi_{\kappa PG}}, \quad \mathbf{X}_{\kappa PG} = \underline{n}^{PG} \frac{\partial E_{\kappa PG}}{\partial \boldsymbol{\xi}_{\kappa PG}}, \tag{23.5.2}$$

the reduced dissipation inequality,

$$T\hat{S}^{PG}_{\kappa(i)} = \boldsymbol{\Psi}^{PG}_{\kappa} : \mathbf{L}^{g}_{\kappa PG} - C^{PG}_{\kappa} \frac{dc_{\kappa PG}}{dt} - X^{PG}_{\kappa} \frac{d\xi_{\kappa PG}}{dt} - \mathbf{X}_{\kappa PG} : \frac{d\boldsymbol{\xi}_{\kappa PG}}{dt} \geq 0, \tag{23.5.3}$$

remains to be satisfied by the constitutive equations to be developed for the growth transformation $\mathbf{F}^{g}_{PG}$, for the cellular energy concentration $c_{\kappa PG}$, for the scalar structure variable $\xi_{\kappa PG}$, and for the tensor structure variable $\boldsymbol{\xi}_{\kappa PG}$.

The free energy is *additively* contributed by a function of the elastic Green strain, terms involving cell energy and structure variables associated with the evolution of the homeostatic surface,

$$E_{\kappa PG} = \boldsymbol{\Gamma}_{\kappa PG} : \mathbf{E}^{e}_{GPG} + E^{e}_{\kappa PG}(\mathbf{E}^{e}_{GPG}) + E^{cell}_{PG}(c_{\kappa PG}) + \frac{1}{2} \boldsymbol{\xi}_{\kappa PG} : \mathbb{X}_{\kappa PG} : \boldsymbol{\xi}_{\kappa PG} + \frac{1}{2} \mathcal{X}_{\kappa PG} \xi^{2}_{\kappa PG}. \tag{23.5.4}$$

The purely mechanical part of the free energy represents the energy of the PGs and other proteins forming the ground substance. Tentatively, the PG constituent is endowed with an isotropic linear elastic response (see however remark below),

$$E^{e}_{\kappa PG} = \tfrac{1}{2} \lambda_{\kappa PG} (\mathbf{I} : \mathbf{E}^{e}_{GPG})^{2} + \mu_{\kappa PG} \mathbf{E}^{e}_{GPG} : \mathbf{E}^{e}_{GPG}, \tag{23.5.5}$$

and with an initial stress $\mathbf{\Gamma}_{\kappa\mathrm{PG}}$. Note that the moduli $\lambda_{\kappa\mathrm{PG}}$ and $\mu_{\kappa\mathrm{PG}}$ involved in (23.5.5) are the hypertonic intrinsic moduli of the ground substance. These moduli are assumed to be independent of the chemical composition of the bath (synovial fluid) the cartilage is surrounded by. They might perhaps be considered to evolve during the growth process, much like the collagen modulus increases with the pyridinoline density. The Lamé modulus $\lambda_{\kappa\mathrm{PG}}$ might also depend on the sign of the elastic volume change. Still, these two aspects would require to be quantified.

The fourth order tensor with minor and major symmetries $\mathbb{X}_{\kappa\mathrm{PG}}$ [unit : Pa] will be considered isotropic, namely $\mathbb{X}_{\kappa\mathrm{PG}} = \mathrm{x}_{\kappa\mathrm{PG}}\,\mathbf{I}^{(4)}$, with $\mathrm{x}_{\kappa\mathrm{PG}} > 0$.

Therefore, the generalized stresses take the explicit form,

$$
\begin{aligned}
\text{effective stress} \qquad & (\underline{n}^{\mathrm{PG}})^{-1}\,\boldsymbol{\tau}_{\kappa}^{\mathrm{PG}\,\prime} = \frac{\partial E_{\kappa\mathrm{PG}}}{\partial \mathbf{E}_{\mathrm{GPG}}^{\mathrm{e}}} = \mathbf{\Gamma}_{\kappa\mathrm{PG}} + \lambda_{\kappa\mathrm{PG}}\,(\mathbf{I} : \mathbf{E}_{\mathrm{GPG}}^{\mathrm{e}})\,\mathbf{I} + 2\,\mu_{\kappa\mathrm{PG}}\,\mathbf{E}_{\mathrm{GPG}}^{\mathrm{e}}; \\[4pt]
\text{cell energy density} \qquad & (\underline{n}^{\mathrm{PG}})^{-1}\,C_{\kappa}^{\mathrm{PG}} = \frac{\partial E_{\kappa\mathrm{PG}}}{\partial c_{\kappa\mathrm{PG}}} = \frac{dE_{\mathrm{PG}}^{\mathrm{cell}}}{dc_{\kappa\mathrm{PG}}}; \\[4pt]
\text{homeostatic surface} \qquad & (\underline{n}^{\mathrm{PG}})^{-1}\,X_{\kappa}^{\mathrm{PG}} = \frac{\partial E_{\kappa\mathrm{PG}}}{\partial \xi_{\kappa\mathrm{PG}}} = \mathcal{X}_{\kappa\mathrm{PG}}\,\xi_{\kappa\mathrm{PG}}; \\[4pt]
\text{homeostatic surface} \qquad & (\underline{n}^{\mathrm{PG}})^{-1}\,\mathbf{X}_{\kappa}^{\mathrm{PG}} = \frac{\partial E_{\kappa\mathrm{PG}}}{\partial \boldsymbol{\xi}_{\kappa\mathrm{PG}}} = \mathbb{X}_{\kappa\mathrm{PG}} : \boldsymbol{\xi}_{\kappa\mathrm{PG}}\,.
\end{aligned}
\tag{23.5.6}
$$

Alternatively to $(23.5.2)_3$, the radius of the homeostatic surface X_{κ}^{PG} may vary with the cell energy in the same format (23.4.58) as for collagen.

23.5.2   Proteoglycans: rate of growth and rates of structure variables

In absence of initial stress, the Mandel stress for PGs would be symmetric as implicitly shown by (23.4.27), the material being elastically isotropic in the intermediate configurations. However, due to this initial stress, the same modifications as for collagen are introduced, namely a modified rate of growth transformation $\mathbf{L}_{\kappa_* \mathrm{PG}}^{\mathrm{g}}$[23.14] and a modified Mandel stress $\mathbf{\Psi}_{\kappa_*}^{\mathrm{PG}}$,

$$
\mathbf{L}_{\kappa_* \mathrm{PG}}^{\mathrm{g}} = (\mathbf{C}_{\mathrm{PG}}^{\mathrm{e}})^{1-\alpha} \cdot \mathbf{L}_{\kappa\mathrm{PG}}^{\mathrm{g}} \cdot (\mathbf{C}_{\mathrm{PG}}^{\mathrm{e}})^{-\beta}, \quad \mathbf{\Psi}_{\kappa_*}^{\mathrm{PG}} \equiv (\mathbf{C}_{\mathrm{PG}}^{\mathrm{e}})^{-1+\alpha} \cdot \mathbf{\Psi}_{\kappa}^{\mathrm{PG}} \cdot (\mathbf{C}_{\mathrm{PG}}^{\mathrm{e}})^{\beta} = (\mathbf{\Psi}_{\kappa_*}^{\mathrm{PG}})^{\mathrm{T}}, \tag{23.5.7}
$$

with $\mathbf{C}_{\mathrm{PG}}^{\mathrm{e}} = (\mathbf{F}_{\mathrm{PG}}^{\mathrm{e}})^{\mathrm{T}} \cdot \mathbf{F}_{\mathrm{PG}}^{\mathrm{e}}$ the elastic right Cauchy-Green tensor.

The effective modified Mandel stress $\mathbf{\Psi}_{\kappa_*}^{\mathrm{PG}}$ being defined similarly to (23.4.14), the homeostatic surface (22.5.25) expresses in terms of the *shifted* stress $\mathbf{\Psi}_{\kappa R}^{\mathrm{PG}} = \mathbf{\Psi}_{\kappa_*}^{\mathrm{PG}} - \mathbf{X}_{\kappa}^{\mathrm{PG}}$, with invariants $p_{\kappa R}^{\mathrm{PG}}$ and $q_{\kappa R}^{\mathrm{PG}}$ defined as in (22.5.23),

$$
\mu_{\kappa}^2\,h_{\kappa\mathrm{PG}}(\mathbf{\Psi}_{\kappa R}^{\mathrm{PG}}, X_{\kappa}^{\mathrm{PG}}) = \frac{(q_{\kappa R}^{\mathrm{PG}})^2}{M_{\mathrm{PG}}^2} + (p_{\kappa R}^{\mathrm{PG}})^2 - (X_{\kappa}^{\mathrm{PG}})^2 = 0\,. \tag{23.5.8}
$$

The growth rate and the rate of change of the tensor structure variable adopt the form (22.5.39),

$$
\begin{aligned}
\mathbf{D}_{\kappa_* \mathrm{PG}}^{\mathrm{g}} &= \frac{\mu_{\kappa}}{t_{\mathrm{gPG}}}\,\Phi(h_{\kappa\mathrm{PG}})\,\frac{\partial h_{\kappa\mathrm{PG}}}{\partial \mathbf{\Psi}_{\kappa_*}^{\mathrm{PG}}}, \\[6pt]
\frac{d\boldsymbol{\xi}_{\kappa\mathrm{PG}}}{dt} &= \mathbf{D}_{\kappa_* \mathrm{PG}}^{\mathrm{g}} - a_{1\mathrm{PG}}\,\|\,\mathbf{D}_{\kappa_* \mathrm{PG}}^{\mathrm{g}}\,\|\,\frac{\mathbf{X}_{\kappa}^{\mathrm{PG}}}{\mu_{\kappa}}\,.
\end{aligned}
\tag{23.5.9}
$$

with t_{gPG} a characteristic time of the growth process The rate of change of the scalar structure variable (size of homeostatic surface) mimics the expression (22.5.36),

$$
\frac{d\xi_{\kappa\mathrm{PG}}}{dt} = \frac{1}{t_{\mathrm{gPG}}}\,\Phi(h_{\kappa\mathrm{PG}})\,(a_{2\mathrm{PG}} - a_{3\mathrm{PG}}\,\frac{X_{\kappa}^{\mathrm{PG}}}{\mu_{\kappa}})\,\frac{X_{\kappa}^{\mathrm{PG}}}{\mu_{\kappa}}\,. \tag{23.5.10}
$$

The sum of the contributions to the Clausius-Duhem inequality associated with these variables,

$$
T\hat{S}_{\kappa(\mathrm{i})}^{\mathrm{PG}} = \mathbf{\Psi}_{\kappa_*}^{\mathrm{PG}} : \mathbf{D}_{\kappa_* \mathrm{PG}}^{\mathrm{g}} - X_{\kappa}^{\mathrm{PG}}\,\frac{d\xi_{\kappa\mathrm{PG}}}{dt} - \mathbf{X}_{\kappa}^{\mathrm{PG}} : \frac{d\boldsymbol{\xi}_{\kappa\mathrm{PG}}}{dt} = \frac{\mu_{\kappa}}{t_{\mathrm{gPG}}}\,\Phi(h_{\kappa\mathrm{PG}})\,A_{(1)}\,, \tag{23.5.11}
$$

[23.14]The growth of PGs displays anisotropy due to the shift of the homeostatic surface by $\mathbf{X}_{\kappa}^{\mathrm{PG}}$. On the other hand, it has no structural/directional anisotropy associated with the fibers. Therefore, if the growth spin is associated with $\mathbf{X}_{\kappa}^{\mathrm{PG}}$, then it should be introduced. If the growth spin is associated with the directional properties, it is simply elusive.

where

$$A_{(1)} = 2\,h_{\kappa\text{PG}} + (2 - a_{2\text{PG}})\frac{(X_\kappa^{\text{PG}})^2}{\mu_\kappa^2} + a_{3\text{PG}}\frac{(X_\kappa^{\text{PG}})^3}{\mu_\kappa^3} + a_{1\text{PG}} \left\| \frac{\partial h_{\kappa\text{PG}}}{\partial \boldsymbol{\Psi}_{\kappa_*}^{\text{PG}}} \right\| \frac{\mathbf{X}_\kappa^{\text{PG}} : \mathbf{X}_\kappa^{\text{PG}}}{\mu_\kappa}, \qquad (23.5.12)$$

is positive for

$$a_{1\text{PG}} \geq 0, \quad 2 - a_{2\text{PG}} \geq 0, \quad a_{3\text{PG}} \geq 0. \qquad (23.5.13)$$

On the alternative determination of the size of the homeostatic surface via the cell energy density

As an alternative, the size of the homeostatic surface may be defined via the cell energy density. Then the scalar structure variable $\boldsymbol{\xi}_{\kappa\text{PG}}$ is elusive and $a_{2\text{PG}}$ and $a_{3\text{PG}}$ should be set to 0.

23.5.3 Proteoglycans: evolution of the cell energy source

Let us consider first the model where the size of the homeostatic surface is a function of the cell energy density. The equations associated with the contribution of the cell energy density to PGs mimic the developments for collagen, namely for the size of the homeostatic surface,

$$(\underline{n}^{\text{PG}})^{-1}\,X_\kappa^{\text{PG}} = \frac{dE_{\text{PG},1}^{\text{cell}}}{dc_{\kappa\text{PG}}} = \chi_{\kappa\text{PG}}\left(c_{\kappa\text{PG}} + \frac{1}{c_{\kappa\text{PG}}}\right) > 0, \qquad (23.5.14)$$

with the coefficient $\chi_{\kappa\text{PG}}$ [unit : Pa] to be defined in the initialization process, as well as for the term associated with the cell evolution,

$$(\underline{n}^{\text{PG}})^{-1}\left(C_\kappa^{\text{PG}} - X_\kappa^{\text{PG}}\right) = \frac{dE_{\text{PG},2}^{\text{cell}}}{dc_{\kappa\text{PG}}} = \mu_\kappa\,c_{i\text{PG}}\frac{c_{\kappa\text{PG}}}{1 + c_{\kappa\text{PG}}^2} > 0, \qquad (23.5.15)$$

with $c_{i\text{PG}}$ a material constant (set to 1 below).

The cell energy is used solely by the growth process, so that it changes only if growth is activated,

$$\frac{dc_{\kappa\text{PG}}}{dt} = -\frac{\alpha_{(5\text{PG})}}{t_{\text{gPG}}}\,\Phi(h_{\kappa\text{PG}})\frac{1}{\mu_\kappa}\frac{dE_{\text{PG},2}^{\text{cell}}}{dc_{\kappa\text{PG}}} \leq 0, \qquad (23.5.16)$$

with $\alpha_{(5\text{PG})} \geq 0$ a dimensionless parameter. In view of (23.5.6), the contribution to the dissipation inequality associated with cells,

$$T\hat{S}_{\kappa(i)}^{\text{PG}} = -C_\kappa^{\text{PG}}\frac{dc_{\kappa\text{PG}}}{dt} = \frac{\mu_\kappa}{t_{\text{gPG}}}\,\Phi(h_{\kappa\text{PG}})\,A_{(5)} \geq 0, \quad A_{(5)} \equiv \alpha_{(5\text{PG})}\frac{\underline{n}^{\text{PG}}}{\mu_\kappa}\frac{dE_{\text{PG}}^{\text{cell}}}{dc_{\kappa\text{PG}}} \times \frac{1}{\mu_\kappa}\frac{dE_{\text{PG},2}^{\text{cell}}}{dc_{\kappa\text{PG}}} \geq 0, \qquad (23.5.17)$$

is positive.

Alternatively, if the size of the homeostatic surface is governed by the independent variable ξ_κ^{PG}, then the only change here consists in zeroing the term $E_{\text{PG},1}^{\text{cell}}$ so that $E_{\text{PG},2}^{\text{cell}} = E_{\text{PG}}^{\text{cell}}$.

23.5.4 Proteoglycans: gathering the contributions to the Clausius-Duhem inequality

The total dissipation obtained by summing the two above contributions,

$$T\hat{S}_{\kappa(i)}^{\text{PG}} = \frac{\mu_\kappa}{t_{\text{gPG}}}\,\Phi(h_{\kappa\text{PG}})\,\big(\underbrace{A_{(1)}}_{\geq 0} + \underbrace{A_{(5)}}_{\geq 0}\big), \qquad (23.5.18)$$

is definitely positive.

23.6 Summary of constitutive equations and residual

The set of unknowns, namely

$$\mathbb{U} = \begin{bmatrix} \mathbf{F} & \mathbf{F}_{\text{c}}^{\text{g}} & c_{\kappa\text{c}} & \xi_{\kappa\text{c}} & \boldsymbol{\xi}_{\kappa\text{c}} & \cup_{i\in[1,n_e]}\,\mathbf{m}_{\kappa i} & \cup_{i\in[1,n_e]}\,\tilde{n}_{\kappa i} & p_{\kappa\text{c}} & \mathbf{F}_{\text{PG}}^{\text{g}} & c_{\kappa\text{PG}} & \xi_{\kappa\text{PG}} & \boldsymbol{\xi}_{\kappa\text{PG}} \end{bmatrix}^{\text{T}}, \qquad (23.6.1)$$

including

- $\mathbf{F}$ the total deformation gradient;
- $\mathbf{F}_c^g$ the growth transformation of collagen;
- $c_{\kappa c}$ the cell energy density for collagen;
- $\xi_{\kappa c}$ the scalar structure variable for collagen;
- $\boldsymbol{\xi}_{\kappa c}$ the tensor structure variable for collagen;
- $\cup_{i \in [1, n_e]} \mathbf{m}_{\kappa i}$ the set of fiber directions;
- $\cup_{i \in [1, n_e]} \tilde{n}_{\kappa i}$ the set of relative volume fractions;
- $p_{\kappa c}$ the pyridinoline density;
- $\mathbf{F}_{PG}^g$ the growth transformation of PGs;
- $c_{\kappa PG}$ the cell energy density for PGs;
- $\xi_{\kappa PG}$ the scalar structure variable for PGs;
- $\boldsymbol{\xi}_{\kappa PG}$ the tensor structure variable for PGs,

is obtained by a time-marching procedure, via an implicit Euler scheme, that consists in zeroing the residual, e.g., at step n with time-step $\Delta t = t_{n+1} - t_n$,

$$
\mathbb{R}_{n+1} =
\begin{cases}
\mathbf{R}_{(i)}, \; i \in [1, 3], \\[4pt]
\left((\mathbf{F}_c^g)^{n+1} - (\mathbf{F}_c^g)^n \right) \cdot \left((\mathbf{F}_c^g)^{n+1} \right)^{-1} - \Delta t \, (\mathbf{L}_{\kappa c}^g)^{n+1}, \\[4pt]
c_{\kappa c}^{n+1} - c_{\kappa c}^n - \Delta t \left(\dfrac{dc_{\kappa c}}{dt} \right)^{n+1}, \\[4pt]
\xi_{\kappa c}^{n+1} - \xi_{\kappa c}^n - \Delta t \left(\dfrac{d\xi_{\kappa c}}{dt} \right)^{n+1}, \\[4pt]
\boldsymbol{\xi}_{\kappa c}^{n+1} - \boldsymbol{\xi}_{\kappa c}^n - \Delta t \left(\dfrac{d\boldsymbol{\xi}_{\kappa c}}{dt} \right)^{n+1}, \\[4pt]
\hat{\mathbf{m}}_{\kappa i}^{n+1} - \hat{\mathbf{m}}_{\kappa i}^n - \Delta t \left(\dfrac{d\hat{\mathbf{m}}_{\kappa i}}{dt} \right)^{n+1}, \quad i \in [1, n_e], \\[4pt]
\tilde{n}_{\kappa i}^{n+1} - \tilde{n}_{\kappa i}^n - \Delta t \left(\dfrac{d\tilde{n}_{\kappa i}}{dt} \right)^{n+1}, \quad i \in [1, n_e], \\[4pt]
p_{\kappa c}^{n+1} - p_{\kappa c}^n - \Delta t \left(\dfrac{dp_{\kappa c}}{dt} \right)^{n+1}, \\[4pt]
\left((\mathbf{F}_{PG}^g)^{n+1} - (\mathbf{F}_{PG}^g)^n \right) \cdot \left((\mathbf{F}_{PG}^g)^{n+1} \right)^{-1} - \Delta t \, (\mathbf{L}_{\kappa PG}^g)^{n+1}, \\[4pt]
c_{\kappa PG}^{n+1} - c_{\kappa PG}^n - \Delta t \left(\dfrac{dc_{\kappa PG}}{dt} \right)^{n+1}, \\[4pt]
\xi_{\kappa PG}^{n+1} - \xi_{\kappa PG}^n - \Delta t \left(\dfrac{d\xi_{\kappa PG}}{dt} \right)^{n+1}, \\[4pt]
\boldsymbol{\xi}_{\kappa PG}^{n+1} - \boldsymbol{\xi}_{\kappa PG}^n - \Delta t \left(\dfrac{d\boldsymbol{\xi}_{\kappa PG}}{dt} \right)^{n+1},
\end{cases}
\tag{23.6.2}
$$

associated with
- the static equilibrium (23.7.10), namely $\mathbf{R}_{(i)} = \mathbf{0}$, $i \in [1, 3]$;
- the growth rate (23.4.10) of collagen $\mathbf{L}_{\kappa c}^g = (\mathbf{C}_c^e)^{\alpha-1} \cdot (\mathbf{D}_{\kappa_* c}^g + \boldsymbol{\Omega}_{\kappa_* c}^g) \cdot (\mathbf{C}_c^e)^\beta$ with,

$$
\mathbf{D}_{\kappa_* c}^g = \frac{\mu_\kappa}{t_{gc}} \, \Phi(h_{\kappa c}) \frac{\partial h_{\kappa c}}{\partial \boldsymbol{\Psi}_{\kappa_*}^c} : \text{Eqn } (23.4.15), \quad \boldsymbol{\Omega}_{\kappa_* c}^g : \text{Eqn } (23.4.18) \, ;
\tag{23.6.3}
$$

- the rate of change of the cell energy density for collagen (23.4.60),

$$
\frac{dc_{\kappa c}}{dt} = -\frac{\alpha_{(5c)}}{t_{gc}} \, \Phi(h_{\kappa c}) \frac{1}{\mu_\kappa} \frac{dE_{c,2}^{cell}}{dc_{\kappa c}} \, ;
\tag{23.6.4}
$$

- the rate of change of the scalar structure variable of collagen (23.4.22),

$$
\frac{d\xi_{\kappa c}}{dt} = \frac{1}{t_{gc}} \, \Phi(h_{\kappa c}) \left(a_{2c} - a_{3c} \frac{X_\kappa^c}{\mu_\kappa} \right) \frac{X_\kappa^c}{\mu_\kappa} \, ;
\tag{23.6.5}
$$

- the rate of change of the tensor structure variable of collagen (23.4.21),

$$\frac{d\boldsymbol{\xi}_{\kappa c}}{dt} = \mathbf{D}^{g}_{\kappa_* c} - a_{1c} \parallel \mathbf{D}^{g}_{\kappa_* c} \parallel \frac{\mathbf{X}^{c}_{\kappa}}{\mu_{\kappa}}; \qquad (23.6.6)$$

- the rate of change (23.4.42) of the fiber directions $\mathbf{m}_{\kappa i}$ of unit norm,

$$\frac{d\mathbf{m}_{\kappa i}}{dt} = \frac{\alpha_{(2c)}}{t_{gc}} \Phi(h_{\kappa c}) \left(\mathbf{I} - \mathbf{m}_{\kappa i} \otimes \mathbf{m}_{\kappa i}\right) \cdot \mathbf{p}_{\kappa i}, \quad i \in [1, n_e]; \qquad (23.6.7)$$

- the rate of change of the relative volume fractions (23.4.46),

$$\frac{d\tilde{n}_{\kappa i}}{dt} = \frac{\alpha_{(3c)}}{\mu_{\kappa}} \frac{\Phi(h_{\kappa c})}{t_{gc}} \nu_i \left(\tilde{N}_{\kappa i} + \Lambda\right), \quad i \in [1, n_e]; \qquad (23.6.8)$$

- the rate of change of the pyridinoline density (23.4.54),

$$\frac{dp_{\kappa c}}{dt} = \frac{\alpha_{(4c)}}{t_{gc}} \Phi(h_{\kappa c}) \left(1 - p_{\kappa c}\right) \geq 0; \qquad (23.6.9)$$

- the growth rate (23.4.10) of PGs $\mathbf{L}^{g}_{\kappa PG} = (\mathbf{C}^{e}_{PG})^{\alpha-1} \cdot (\mathbf{D}^{g}_{\kappa_* PG} + \boldsymbol{\Omega}^{g}_{\kappa_* PG}) \cdot (\mathbf{C}^{e}_{PG})^{\beta}$ with,

$$\mathbf{D}^{g}_{\kappa_* PG} = \frac{\mu_{\kappa}}{t_{gPG}} \Phi(h_{\kappa PG}) \frac{\partial h_{\kappa PG}}{\partial \boldsymbol{\Psi}^{PG}_{\kappa_*}} : \text{Eqn (23.5.9)}, \quad \boldsymbol{\Omega}^{g}_{\kappa_* PG} = \mathbf{0}; \qquad (23.6.10)$$

- the rate of change of the cell energy density for PGs (23.5.16),

$$\frac{dc_{\kappa PG}}{dt} = -\frac{\alpha_{(5PG)}}{t_{gPG}} \Phi(h_{\kappa PG}) \frac{1}{\mu_{\kappa}} \frac{dE^{cell}_{PG,2}}{dc_{\kappa PG}}; \qquad (23.6.11)$$

- the rate of change of the scalar structure variable of PGs (23.5.10),

$$\frac{d\xi_{\kappa PG}}{dt} = \frac{1}{t_{gPG}} \Phi(h_{\kappa PG}) \left(a_{2PG} - a_{3PG} \frac{X^{PG}_{\kappa}}{\mu_{\kappa}}\right) \frac{X^{PG}_{\kappa}}{\mu_{\kappa}}; \qquad (23.6.12)$$

- the rate of change of the tensor structure variable of PGs (23.5.9),

$$\frac{d\boldsymbol{\xi}_{\kappa PG}}{dt} = \mathbf{D}^{g}_{\kappa_* PG} - a_{1PG} \parallel \mathbf{D}^{g}_{\kappa_* PG} \parallel \frac{\mathbf{X}^{PG}_{\kappa}}{\mu_{\kappa}}. \qquad (23.6.13)$$

Concerning the various stresses, the points below may be worth to be stressed,
- the format (23.3.1) intends to convey the idea that the lhs is an effective stress, a quantity that applies to the whole mixture, much like Terzaghi's stress in standard porous media;
- the effective stresses in the intermediate configurations for the constituents are given by (23.4.8)$_1$ for collagen and (23.5.6)$_1$ for PGs;
- the total stresses in the actual configurations for constituents are given by (23.1.30) with $p_E = p_B + \pi_{osm}$. These are the quantities involved in the growth rates, homeostatic surfaces, etc.

23.7 Growth laws with evolving microstructure: simulations

23.7.1 Physical and constitutive parameters

The valence ζ_{PG} of PGs is about -5000 and their molar volume $\hat{v}_{PG}$ close to $1.1\,\mathrm{m}^3/\mathrm{mole}$[23.15]. The physical properties of water, collagen and PGs are listed in Table 23.7.1. All stress-like quantities are scaled by the scaling modulus $\mu_{\kappa}(=100\,\mathrm{kPa})$. Times are scaled by the scaling time t_{gPG} attached to PGs.

The constitutive equations for the growing constituent k require the parameters below to be specified:
- $\Phi(h_{\kappa k}) = (h_{\kappa k})^{n_{\phi k}}$ increasing function of its argument $h_{\kappa k}$ defining the magnitude of the growth strain;
- $\mathbb{X}_{\kappa k} = \mathrm{x}_{\kappa k}\,\mathbf{I}^{(4)}$ kinematic fourth order tensor linking the tensor structure variable and the center of the homeostatic surface taken proportional to the fourth order identity tensor.

The values of the parameters or their ranges used in the simulations are shown in Table 23.7.2. If the size of the homeostatic surface depends on the scalar structure variable, the following additional parameters should be prescribed for the two constituents k:
- $a_{2k} \in [0,2]$ and $a_{3k} \geq 0$ parameters involved in the rate of change of the size of the scalar structure variable. Typical values $a_{2k} = 0.1$, $a_{3k} = 1$;
- $\mathcal{X}_{\kappa k}$ scalar linking the scalar structure variable and the size of the homeostatic surface. Typical value $\underline{n}^k(S_+)\,\mathcal{X}_{\kappa k}(0)/\mu_{\kappa}=1$.

[23.15] The components of proteoglycans are not highlighted, even if only GAGs are charged, the mass ratio between GAGs and PGs being about 90%, see Section 17.6.3. Similarly the components other than collagen, like noncollagenous proteins and cells, are not individualized.

Table 23.7.1: Mass densities of species [unit : $\mathrm{kg/m^3}$].

Water	Collagen	Proteoglycans
ρ_{w}	ρ_{c}	ρ_{PG}
1000	1400	1800

Table 23.7.2: Growth model: constitutive parameters. The modulus μ_κ [unit : Pa] and the time t_{gc} [unit : s] are scaling coefficients.

Intrinsic elastic constants	$K_{\mathrm{cf}}(0)/\mu_\kappa \in [1,10]$	$\lambda_{\kappa\mathrm{PG}}/\mu_\kappa = 50$
	$K_{\mathrm{cf}}(1)/\mu_\kappa \in [0,10]$	$\mu_{\kappa\mathrm{PG}}/\mu_\kappa = 50$
	$k_{\mathrm{c}}(0) \in [1,10]$	-
	$k_{\mathrm{c}}(1) \in [0,10]$	-
Growth time	$t_{\mathrm{gc}} = $ -	$t_{\mathrm{gPG}}/t_{\mathrm{gc}} = 1/100$
Homeostatic surface	$M_{\mathrm{c}}=1, \quad n_{\phi\mathrm{c}}=1$	$M_{\mathrm{PG}}=1, \quad n_{\phi\mathrm{PG}}=1$
Tensor structure variable	$a_{1\mathrm{c}}=0$	$a_{1\mathrm{PG}}=0$
Tensor generalized stress	$\underline{n}^{\mathrm{c}}(S_+)\,\mathbb{x}_{\kappa\mathrm{c}}/\mu_\kappa \in [0.01,1]$	$\underline{n}^{\mathrm{PG}}(S_+)\,\mathbb{x}_{\kappa\mathrm{PG}}/\mu_\kappa = 1$
Rate factor fiber orientation	$\alpha_{(2\mathrm{c})} = 0 \in [0,10]$	-
Rate factor fiber density	$\alpha_{(3\mathrm{c})} = 10 \in [0,10]$	-
Rate factor pyridinoline	$\alpha_{(4\mathrm{c})} = 10 \in [0,10]$	-
Rate factor cell energy density	$\alpha_{(5\mathrm{c})} = 10 \in [0,10]$	$\alpha_{(5\mathrm{PG})} = 10 \in [0,10]$
Cell energy density	$c_{i\mathrm{c}}=1$	$c_{i\mathrm{PG}}=1$

23.7.2 The reference isotropic configuration

To establish an equilibrium reference configuration (S), from which the tissue growth will be analyzed, we formally define first a tentative (virtual) configuration (V) where the PGs have no fixed charge, and the pyridinoline density is minimal, say $p_{\kappa\mathrm{c}} = 0$.

Initially, the directional distribution of fibers is random, so that, with help of (12.2.9), the directional integral for the contribution of the collagen effective stress $\sum_{i=1,n_e} \tilde{n}_{\kappa i} \mathbf{M}_{\kappa i}$ can be obtained in explicit form as $(4\,\pi)^{-1} \int_{\mathbb{S}^2} \mathbf{m}_\kappa \otimes \mathbf{m}_\kappa \, dS = \mathbf{I}/3$. Note that the relative volume fraction $\tilde{n}_{\kappa i}$ is equal to $\sin\theta_i\, d\theta\, d\phi/4\pi$.

The tissue is bathed in an electrolyte with isotropic pressure p_{B}, so that $\boldsymbol{\sigma} + p_{\mathrm{B}}\,\mathbf{I} = \mathbf{0}$. As long as the tissue does not grow, the constitutive equation (23.3.1) becomes,

$$
\boldsymbol{\sigma}' = \overbrace{\boldsymbol{\sigma} + p_{\mathrm{B}}\,\mathbf{I}}^{=\mathbf{0}} + \pi_{\mathrm{osm}}\,\mathbf{I} = \frac{1}{F_1}\left(F_1^3\,\boldsymbol{\Gamma}_\kappa^{\mathrm{c}} + \tau_{\kappa 1}^{\mathrm{c}''}\,\mathbf{I} + F_1^3\,\boldsymbol{\Gamma}_\kappa^{\mathrm{PG}} + \tau_{\kappa 1}^{\mathrm{PG}''}\,\mathbf{I} \right)
$$

$$
\tau_{\kappa 1}^{\mathrm{c}''} = \underline{n}^{\mathrm{c}}\,\frac{K_{\mathrm{cf}}}{3}\,\exp(k_{\mathrm{c}}\,\langle E_{G1}\rangle^2)\,E_{G1}
\tag{23.7.1}
$$

$$
\tau_{\kappa 1}^{\mathrm{PG}''} = 3\,\underline{n}^{\mathrm{PG}}\,K_{\kappa\mathrm{PG}}\,E_{G1}\,,
$$

with $K_{\kappa\mathrm{PG}} = \lambda_{\kappa\mathrm{PG}} + 2\,\mu_{\kappa\mathrm{PG}}/3$, the axis 1 being arbitrary, actually $E_{G1} = \mathrm{tr}\,\mathbf{E}_G/3$ and $F_1 = \det \mathbf{F}^{1/3}$.

For a given PG content, (23.7.1) is seen as an equation for the deformation gradient F_1, the osmotic pressure being given by (14.4.2) and (23.3.3).

This relation may be interpreted in at least two ways. If the strain vanishes, it provides the sum of the two initial stresses $\boldsymbol{\Gamma}_\kappa^{\mathrm{c}}$ and $\boldsymbol{\Gamma}_\kappa^{\mathrm{PG}}$. An alternative point of view is adopted, according to which the absence of fixed charge corresponds to a zero strain and to vanishing initial stresses. Then (23.7.1) provides the strain. The point can be realized by observing that, at very small tensile strain, (23.7.1) linearizes to

$$
\pi_{\mathrm{osm}} = \left(n^{\mathrm{c}}\,\frac{K_{\mathrm{cf}}}{3} + 3\,n^{\mathrm{PG}}\,K_{\kappa\mathrm{PG}} \right) E_{G1}\,.
\tag{23.7.2}
$$

The procedure to establish the reference equilibrium configuration is as follows:

Table 23.7.3: Tentative initial volume fractions of species in absence of charge on PGs, point (V), and equilibrium values for the actual valence ζ_{PG}, point (S).

	Water	Collagen	PGs
Point (V)	0.7	0.25	0.05
Point (S)	0.789	0.176	0.035

- the volume fractions are fixed arbitrarily, e.g., as given by Table 23.7.3. Note that the volume fractions in Table 23.7.3 correspond to a vanishing fixed charge and a vanishing strain. The corresponding configuration (V) associated with uncharged PGs is virtual;
- the electric charge is applied and the nonlinear equation (23.7.1) is solved for the strain;
- the centers $\mathbf{X}_\kappa^c$ and $\mathbf{X}_\kappa^{\text{PG}}$ of the homeostatic surfaces are set to $\mathbf{0}$;
- the radius of the homeostatic surface of collagen or PGs is chosen so that the stress point corresponding to the configuration described above lies inside, on or outside the homeostatic surface, Section 23.7.3.

The amounts of collagen and PGs remain constant from the virtual configuration (V), with uncharged PGs, to the actual configuration (S) where PGs are charged, but the amount of water changes since the total volume does so. Therefore, at the beginning of the growth period, configuration (S), the volume fractions resulting from the above derivations and shown in Table 23.7.3 are no longer equal to the tentative volume fractions in configuration (V).

Redefinition of the reference configuration at point (S) and definition of the initial stresses $\mathbf{\Gamma}_\kappa$'s

All transformations, elastic, growth and total, may be re-initialized at point (S), which, unlike point (V), is an equilibrium point, the external load being the bath pressure $p_{\text{B}}(= 0)$, and the internal load being the osmotic pressure, both being collected in the pressure $p_{\text{E}} = p_{\text{B}} + \pi_{\text{osm}}$. The deformation gradient $\mathbf{F} = F_1\,\mathbf{I}$ at S_- has just been obtained while the deformation gradient at S_+ is set equal to the identity.

Continuity of the effective Cauchy stresses $\boldsymbol{\sigma}^{k'}$ is ensured by an appropriate redefinition of the initial stresses,

$$\boldsymbol{\sigma}^{k'}(S_-) = \Big[\frac{1}{\det \mathbf{F}_k^{\text{e}}}\,\mathbf{F}_k^{\text{e}} \cdot \boldsymbol{\tau}_\kappa^{k'} \cdot (\mathbf{F}_k^{\text{e}})^{\text{T}}\Big](S_-) = \frac{\tau_{\kappa 1}^{k''}}{F_1}\,\mathbf{I} = \boldsymbol{\sigma}^{k'}(S_+) = \mathbf{\Gamma}_\kappa^k, \quad k \in \mathcal{S}^*, \tag{23.7.3}$$

the $\tau_{\kappa 1}^{k''}(F_1)$'s being shown in explicit form in (23.7.1), with $F_1 = F_1(S_-)$ the deformation gradient at (S_-), while $F_1(S_+) = 1$. Hence

$$n^k\,\mathbf{\Gamma}_{\kappa k} = \mathbf{\Gamma}_\kappa^k = \frac{\tau_{\kappa 1}^{k''}}{F_1}\,\mathbf{I}, \quad k \in \mathcal{S}^*. \tag{23.7.4}$$

Note that the effective Kirchhoff stress is not continuous (but this discontinuity has no consequence),

$$\boldsymbol{\tau}_\kappa^{k'}(S_-) = \tau_{\kappa 1}^{k''}\,\mathbf{I} \neq \boldsymbol{\tau}_\kappa^{k'}(S_+) = \frac{\tau_{\kappa 1}^{k''}}{F_1}\,\mathbf{I}, \quad k \in \mathcal{S}^*. \tag{23.7.5}$$

Definition of the radius of the homeostatic surface at point (S)

The radius of the homeostatic surface may be defined using information at (S_+), namely

$$X_\kappa^k(0) = |\boldsymbol{\tau}_{\kappa 1}^{k'}(S_+)|. \tag{23.7.6}$$

The (isotropic) Mandel stress at (S_+) is found tensile for PGs and compressive for collagen, as explained in Section 23.7.6. If the homeostatic surface of collagen is open towards compression, the initial stress point is located inside the surface defined by the above procedure. Growth of collagen requires the initially compressive Mandel stress to move towards the tensile region before hitting the homeostatic surface.

23.7.3 Initial data

The volumes are scaled by the initial volume $\underline{V}$ of the tissue at point (S). The initial network is considered to be randomly oriented. The relative volume fraction of a fan centered around the point (θ, ϕ) and of width $\Delta\theta$ and $\Delta\phi$ is equal to $\sin\theta\,\Delta\theta\,\Delta\phi/4\pi$. Since the sense of the fibers does not matter, only the upper hemisphere $\theta \in [0, \pi/2]$ may be considered, with volume fractions equal to $\sin\theta\,\Delta\theta\,\Delta\phi/2\pi$. Clearly, a uniform size of fans is far from being optimal since the area of a ring close to the pole is small. More efficient partitions may be defined, as suggested in

Section 12.1.2. Calculations under arbitrary three-dimensional loadings have been run by discretizing the unit sphere with 36 meridians and 36 parallels.

For a collagen network and loading conditions that display symmetry about a common axis, the relative volume fraction $\tilde{n}_{\kappa i}$ corresponding to the fan $i \in [1, n_e]$ is equal to $\sin \theta_i \, \Delta \theta$, with $\theta_i = (i - 1/2) \times \Delta \theta$, and $\Delta \theta = \pi / (2 \, n_e)$ the uniform size of the n_e fans. The unit vector $\mathbf{m}_{\kappa i}$ of the fiber i has the format (12.1.9) so that $\mathbf{E}_{\mathrm{G}}^{\mathrm{e}} : \mathbf{M}_{\kappa i} = E_{\mathrm{G}1} \cos^2 \theta_i + E_{\mathrm{G}2} \sin^2 \theta_i$ accounting for the fact that the Green strain is principal in the traction axes and endowed with rotational symmetry ($E_{\mathrm{G}2} = E_{\mathrm{G}3}$). For a planar network, $\tilde{n}_{\kappa i}$ is simply equal to $1/n_e$.

When the size of the homeostatic surface is considered a function of the cell energy density, the coefficient $\chi_{\kappa k}$ is defined as follows. The procedure of Section 23.7.2 yields an estimation of the radius $X_\kappa^k(0)$ of the homeostatic surface at (S_+). The cell energy density $c_{\kappa k}(0)$ being given at the beginning of the growth process, a reference coefficient $\chi_{\kappa k}^{\mathrm{ref}}$ is estimated from (23.4.58), namely

$$X_\kappa^k(0) = \underline{n}^{\mathrm{c}} \, \chi_{\kappa k}^{\mathrm{ref}} \left(c_{\kappa k}(0) + \frac{1}{c_{\kappa k}(0)} \right). \tag{23.7.7}$$

If $\chi_{\kappa k}$ is chosen smaller than $\chi_{\kappa k}^{\mathrm{ref}}$, then the current point will lay outside the homeostatic surface, and the growth process initiates. The cell density decreases, so does the radius of the homeostatic surface until $c_{\kappa k} = 1$ as illustrated in Fig. 23.4.1, below which it will increase until the growth process stops because the cell energy density has decreased so much that the size of the yield surface has turned large as well.

23.7.4 Boundary conditions

The material properties and state variables, stresses and strains are initially uniform, and they remain so during the growth processes simulated below.

23.7.4.1 Statics

Let $(\underline{\mathbf{e}}_{(1)}, \underline{\mathbf{e}}_{(2)}, \underline{\mathbf{e}}_{(3)})$ be a cartesian triad of fixed axes. A load $\mathbf{P}_{(1)}$ is applied on a plane of initial normal $\underline{\mathbf{e}}_{(1)}$. This force over the whole section $\mathbf{S}(t)$ can be expressed in terms of the Cauchy stress,

$$\mathbf{P}_{(1)}(t) = \int_{\mathbf{S}(t)} (\boldsymbol{\sigma}(t) + p_{\mathrm{B}} \, \mathbf{I}) \cdot d\mathbf{S}(t) = \int_{\underline{\mathbf{S}}} (\boldsymbol{\sigma}(t) + p_{\mathrm{B}} \, \mathbf{I}) \cdot \det \mathbf{F}(t) \, \mathbf{F}^{\mathrm{-T}}(t) \cdot d\underline{\mathbf{S}}. \tag{23.7.8}$$

The last relation is implied by the fact that an elementary area $d\underline{\mathbf{S}}$ transforms to $d\mathbf{S} = \det \mathbf{F} \, \mathbf{F}^{\mathrm{-T}} \cdot d\underline{\mathbf{S}}$ under the motion defined by the deformation gradient $\mathbf{F}$, Eqn (2.6.9). The initial conditions and loading conditions keep the specimen pointwise homogeneous so that the above relation simplifies to,

$$\mathbf{R}_{(1)}(t) = (\boldsymbol{\sigma}(t) + p_{\mathrm{B}} \, \mathbf{I}) \cdot \det \mathbf{F}(t) \, \mathbf{F}^{\mathrm{-T}}(t) \cdot \underline{\mathbf{e}}_{(1)} - \frac{\mathbf{P}_{(1)}(t)}{\underline{S}} = \mathbf{0}. \tag{23.7.9}$$

The tissue is bathed in an electrolyte at given fixed charge, with fixed pressure p_{B}. The lateral directions $\mathbf{e} = \mathbf{e}_{(2)}$ and $\mathbf{e}_{(3)}$ can rotate in general, and the boundary conditions $\mathbf{R}_{(i)} = \mathbf{0}$, $i = 2, 3$, may be expressed in the same form as (23.7.9) with $\mathbf{P}_{(i)} = \mathbf{0}$, $i = 2, 3$. Static equilibrium is thus defined by the set of 9 scalar equations below where $\mathbf{e}_{(i)} \equiv \det \mathbf{F} \, \mathbf{F}^{\mathrm{-T}} \cdot \underline{\mathbf{e}}_{(i)}$,

$$\mathbf{R}_{(i)}(t) = (\boldsymbol{\sigma}(t) + p_{\mathrm{B}} \, \mathbf{I}) \cdot \mathbf{e}_{(i)}(t) - \frac{\mathbf{P}_{(i)}(t)}{\underline{S}} = \mathbf{0}, \quad i \in [1, 3]. \tag{23.7.10}$$

23.7.4.2 Kinematics

For uniaxial tension tests, the lateral displacements are sometimes constrained to vanish. Else the lateral stresses are equal to the bath pressure.

For shear tests, parallel to direction 2, that is with $F_{12} > 0$, on a planar network, the sides of the specimen parallel to the direction 2 and thus normal to the direction 1, are constrained to keep their original direction, namely $\underline{\mathbf{e}}_{(2)} \sim \mathbf{F}^{\mathrm{-T}}(t) \cdot \underline{\mathbf{e}}_{(2)}$, and therefore $F_{21} = F_{23} = 0$.

23.7.4.3 The special case of diagonal axisymmetric strains and stresses

Since the axis $\mathbf{e}_{(1)}$ remains fixed during the process, the deformation gradient $\mathbf{F}$ and its inverse $\mathbf{F}^{-1}$ have the form in the fixed triad,

$$\mathbf{F} = \begin{bmatrix} F_{11} & 0 & 0 \\ F_{21} & F_{22} & F_{23} \\ F_{31} & F_{32} & F_{33} \end{bmatrix}, \quad \mathbf{F}^{-1} = \frac{1}{\det \mathbf{F}} \begin{bmatrix} F_{22} F_{33} - F_{23} F_{32} & 0 & 0 \\ F_{23} F_{31} - F_{21} F_{33} & F_{11} F_{33} & -F_{11} F_{23} \\ F_{21} F_{32} - F_{31} F_{22} & -F_{11} F_{32} & F_{11} F_{22} \end{bmatrix}. \tag{23.7.11}$$

Moreover, the material is either isotropic or the microstructure displays an initial symmetry with respect to the traction axis so that all transformations, strains and stresses remain principal in the reference cartesian triad, in particular,

$$\mathbf{F}(t) = \mathbf{diag}[F_i(t)], \quad \mathbf{F}_k^{\mathrm{g}}(t) = \mathbf{diag}\left[F_{ki}^{\mathrm{g}}(t)\right], \quad \mathbf{F}_k^{\mathrm{e}}(t) = \mathbf{diag}\left[v_{ki} = \frac{F_i}{F_{ki}^{\mathrm{g}}}\right](t). \tag{23.7.12}$$

With uniform stress and strain in the bar, the Cauchy stress and incompressibility pressure are given in terms of the force $P_1(t)$ (> 0 for tension) along the axis $\mathbf{e}_{(1)}$, $\mathbf{P}_{(1)}(t) = P_1(t)\,\mathbf{e}_{(1)}$, and effective Cauchy stress as

$$\sigma_1(t) + p_{\mathrm{B}} = \sigma_1{}'(t) - \sigma_2{}'(t) = \frac{P_1(t)}{\underline{S}} \frac{F_1(t)}{\det \mathbf{F}(t)}, \quad \sigma_i(t) + p_{\mathrm{B}} = \sigma_i{}'(t) - \pi_{\mathrm{osm}}(t) = 0, \quad i = 2, 3. \tag{23.7.13}$$

Since the material properties respect the axial symmetry implied by the loading, the lateral directions 2 and 3 are equivalent, in both stress and deformation. With μ_κ a scaling modulus, the axial equilibrium may be phrased in terms of a normalized load $P^{\mathrm{ext}}(t)$, namely

$$R_1 = P_1^{\mathrm{int}}(t) - P_1^{\mathrm{ext}}(t), \quad P_1^{\mathrm{int}}(t) \equiv \frac{\det \mathbf{F}(t)}{F_1(t)} \left(\frac{\sigma_1{}'(t)}{\mu_\kappa} - \frac{\sigma_2{}'(t)}{\mu_\kappa}\right), \quad P_1^{\mathrm{ext}}(t) \equiv \frac{P_1(t)}{\mu_\kappa\,\underline{S}}. \tag{23.7.14}$$

The lateral equilibria may be formally cast in a similar format, namely $P_i^{\mathrm{ext}} = P_i^{\mathrm{int}}$, $i = 2, 3$, with $P_i^{\mathrm{ext}} = 0$ and $P_i^{\mathrm{int}} = \sigma_i(t) + p_{\mathrm{B}} = \sigma_i{}'(t) - \pi_{\mathrm{osm}}(t)$. When expressed in terms of the constitutive equations $(23.4.8)_1$ and $(23.5.6)_1$, namely

$$\begin{aligned} R_1 = P_1^{\mathrm{int}} - P_1^{\mathrm{ext}} = 0, \quad & P_1^{\mathrm{int}} = F_2^2 \sum_{k \in \mathcal{S}*} \frac{F_{k1}^{\mathrm{e}}}{(F_{k2}^{\mathrm{e}})^2} \frac{\tau_{\kappa 1}^{k'}}{\mu_\kappa} - \frac{1}{F_{k1}^{\mathrm{e}}} \frac{\tau_{\kappa 2}^{k'}}{\mu_\kappa}, \quad P_1^{\mathrm{ext}}(t) \equiv \frac{P_1(t)}{\mu_\kappa\,\underline{S}}, \\ R_i = P_i^{\mathrm{int}} - P_i^{\mathrm{ext}} = 0, \quad & P_i^{\mathrm{int}} = F_1 F_2 \sum_{k \in \mathcal{S}*} \frac{1}{F_{k1}^{\mathrm{e}}} \frac{\tau_{\kappa i}^{k'}}{\mu_\kappa} - \frac{\pi_{\mathrm{osm}}}{\mu_\kappa} = 0, \quad P_i^{\mathrm{ext}} = 0, \quad i = 2, 3, \end{aligned} \tag{23.7.15}$$

the axial and lateral equilibria may be seen as providing two independent equations for, say, the total stretches. The growth stretches and internal variables are obtained by integration of the rate equations. The total and growth stretches being known, the elastic stretches deduce readily.

23.7.4.4 Loading program

The loading program starts from the isotropic equilibrium point (S). It consists of three parts:
1. free swelling up to $t = 500\,t_{\mathrm{gPG}}$;
2. mechanical loading during $100\,t_{\mathrm{gPG}}$;
3. relaxation under constant load.

The first step is a free swelling step: no external load is applied and the tissue is subjected only to the osmotic pressure associated with the electrical charge of PGs. In a second step, a compressive or tensile load of magnitude P is applied along direction 1 while the lateral sides are subjected to a prescribed total strain (usually a zero strain). The load gives rise to a scaled nominal stress $P/(\mu_\kappa \underline{S})$, with a standard value of 2, with $\underline{S}$ the initial area of the specimen in the loading direction. The load is applied over a relatively short time interval, typically $100\,t_{\mathrm{gPG}}$, and next kept fixed.

While the tests presented below respect a rotational symmetry about direction 1, true triaxial tests with unequal prescribed strains in the lateral directions have been run to check the three-dimensional evolution of the collagen fabric.

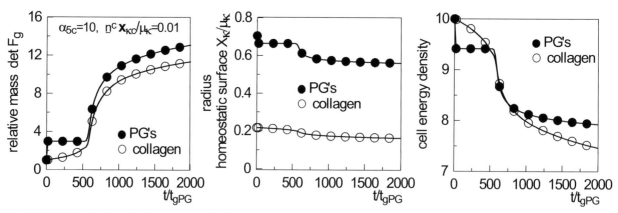

Fig. 23.7.1: Evolution of the masses of collagen and PGs, radii of the homeostatic surfaces and cell energy densities during swelling up to $500\,t_{\text{gPG}}$ and subsequent uniaxial traction with zero lateral strains using the standard parameters shown in Table 23.7.2. PGs grow first during the free swelling stage and their growth implies collagen growth albeit at a slower pace. Growth of both constituents is definitely triggered by the uniaxial loading. With the parameters used, the cell energies remain above the value 1 so that the radii of the homeostatic surfaces continue to decrease during the time window of the test. Reprinted with permission from Loret and Simões [2014].

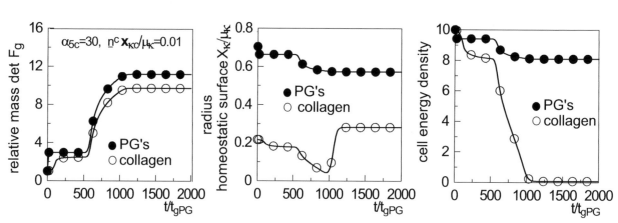

Fig. 23.7.2: Same as Fig. 23.7.1 but with a larger coefficient α_{5c} associated with the rate of change of the cell energy density. A larger coefficient implies a faster decrease of the cell energy density. As a consequence, the size of the homeostatic surface reaches its minimum faster. It later increases which impedes growth, a priori, of collagen but interactions imply that growth of PGs is also reduced significantly. Reprinted with permission from Loret and Simões [2014].

23.7.5 Asymptotic properties of the growth law

To ensure bounded growth, Taber [1998]a resorts to structuration while Drozdov and Khanina [1997] use a nonlinear relation that damps large mechanical loads: this trick appears an expedient and the physical motivation is not provided.

Qualitatively, the growth model here is endowed with two devices that allow to accommodate a stress signal by a bounded growth response. Indeed, the growth amplitude in response to a given fixed stress depends on the distance of the stress to the homeostatic surface and of the ease by which the surface moves to catch the stress. This process involves a kinematic component by allowing the center of the surface to move in stress space, and an isotropic component by allowing the size of the surface to change.

Since growth preserves density, the relative mass ($\equiv$ current mass M_k over initial mass $\underline{M}_k$) and the determinant $\det \mathbf{F}_g$ shown in Fig. 23.7.1 et seq. are, for each constituent, one and the same. As the cell energy density tends to vanish as growth proceeds, the radius of the homeostatic surface increases so that growth is self-bounded: this phenomenon is displayed in Fig. 23.7.2 where the cell energy density of collagen becomes smaller than 1 and as a consequence the size of its homeostatic surface increases. Allowing the homeostatic surface to move and catch the stress faster reduces the growth amplitude, Fig. 23.7.3.

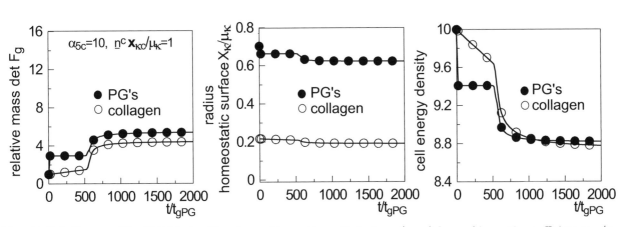

Fig. 23.7.3: Same as Fig. 23.7.1 but with a larger kinematic coefficient $x_{\kappa c}/\mu_\kappa$. A larger kinematic coefficient $x_{\kappa c}/\mu_\kappa$, allowing the homeostatic surface to move faster and catch up the applied stress, reduces the collagen growth, and, again through interactions, PG growth as well. Reprinted with permission from Loret and Simões [2014].

On another side, the actual amplitude of growth may be increased by the restoring term in the rate law (23.4.21) of the tensor structure variable if the coefficient a_1 is strictly positive.

The above features are qualitatively illustrated by the simulations shown in Figs. 23.7.1, 23.7.2 and 23.7.3. Still the actual situation in a mixture context is somehow more complex: even if the external stress is fixed, the stress applied to each growing constituent is out of control because the two constituents work in parallel, which leads to strong interactions in their growth.

Note that the time axis in the plots is scaled by the characteristic time of PGs which is smaller than that of collagen as stressed below.

23.7.6 Sequential growth of PGs and collagen

Proteoglycans seem to pre-exist to the time window that is of interest in, e.g., Williamson et al. [2001]. This observation implies that the growth windows for PGs and collagen might be more sequential, rather than overlapping.

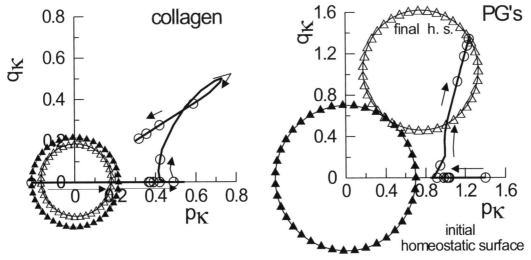

Fig. 23.7.4: Swelling step followed by uniaxial traction with standard parameters. Paths of Mandel stress (open circles) and homeostatic surfaces at initial and last time steps (solid and open symbols, respectively). The mean stress p_κ and shear stress q_κ are defined similarly to (23.4.17). The low kinematic coefficient associated with the homeostatic surface of collagen does not allow its center to move much. On the other hand, the PG stress attracts the surface towards tension. Since the axial loading rate is slow for PGs and fast for collagen, the PGs have come to steady state in few t_{gPG}'s while collagen has not yet at the end of the simulation at $1000\,t_{\mathrm{gPG}}$. Reprinted with permission from Loret and Simões [2014].

For isotropic growth of a single solid, the actual relaxation times T_g, namely

$$T_g = \frac{3}{4} \frac{\mu_\kappa}{\mathrm{x}_\kappa} \frac{\mu_\kappa^2}{X_\kappa^2} t_g, \qquad (23.7.16)$$

and amplitudes of growth associated with the catchup process are directly proportional to the time scale t_g and inversely proportional to the kinematic coefficient $\mathrm{x}_\kappa/\mu_\kappa$ and to the isotropic coefficient $\mathcal{X}_\kappa/\mu_\kappa$, Loret and Simões [2010]c and (22.5.47). The values of the parameters displayed in Table 23.7.2 and the estimations of the initial configuration adopted in Section 23.7.2 yield a ratio T_g/t_g equal to about 300 for collagen and to 0.06 for PGs. With $t_{gc}/t_{gPG} = 10$ to 100, the relaxation time of collagen is orders of magnitude larger than that of PGs, namely $T_{gc}/T_{gPG} \simeq 5 \times 10^4$ to 5×10^5. These estimations hold for a single phase solid. However, the interactions between the two constituents in the current mixture context imply that the time courses of their growth are strongly related.

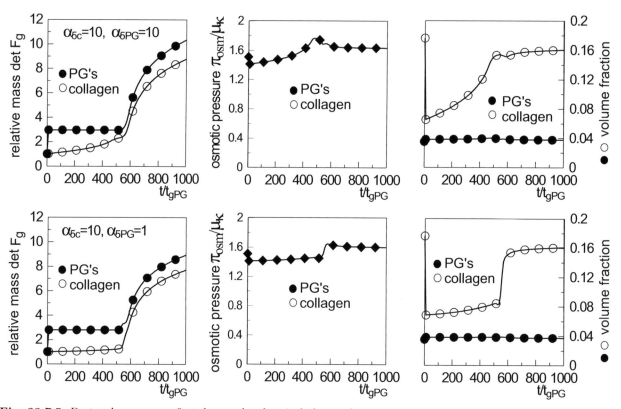

Fig. 23.7.5: Proteoglycans grow first due to the electrical charge that gives rise to an osmotic pressure and to chemomechanical couplings. Growth of collagen trails behind. The largest growth takes place when the coefficients α_{5c} and α_{5PG} that govern the rates of cell energy of collagen and PGs have close values. Dissimilar coefficients, whether α_{5c} is larger or smaller than α_{5PG}, hamper growth. The volume fraction of PGs and the osmotic pressure are only moderately affected by this change of parameters while the initial time course of the volume fraction of collagen reflects more its growth deficit with respect to PGs. Reprinted with permission from Loret and Simões [2014].

The (isotropic) Mandel stress at (S_+) just upon application of the osmotic pressure is found tensile for PGs and compressive for collagen. The explanation of this configuration is as follows. In fact, since the deformation gradients are reset to identity at (S_+), Cauchy stresses, 2nd Piola-Kirchhoff stresses and Mandel stresses are identical. Therefore, the effective partial stress (23.1.30) of constituent k, namely $\boldsymbol{\sigma}^{k'} = \boldsymbol{\sigma}^k + p_E\, n^k\, \mathbf{I}$, is equal to $\boldsymbol{\Gamma}_\kappa^k$ in view of the constitutive equations (23.4.8) and (23.5.6).

Since the elastic moduli of collagen are very low in the early stages, the term $\boldsymbol{\Gamma}_\kappa^c$ is low with respect to the osmotic pressure, and, in essence, the effective stress of collagen is low, $\boldsymbol{\sigma}^{c'} = \mathbf{0}$, while its partial stress is proportional to the osmotic pressure, $\boldsymbol{\sigma}^c = -p_{\mathrm{E}}\,n^c\,\mathbf{I}$. By contrast, the initial volume fraction of PGs is much smaller than that of collagen and its effective and partial stresses are nearly equal. The applied total stress, which is equal to the atmospheric pressure, $\boldsymbol{\sigma} = -p_{\mathrm{B}}\,\mathbf{I}(=\mathbf{0})$, is according to (23.1.31) contributed by the partial stresses of collagen and PGs and the non growing species: then $\boldsymbol{\sigma}^{\mathrm{PG}} = p_{\mathrm{E}}\,(1 - n^{\mathrm{PG}})\,\mathbf{I}$ and $\boldsymbol{\sigma}^{\mathrm{PG}'} = p_{\mathrm{E}}\,\mathbf{I}$.

If the homeostatic surface of collagen is closed towards compression, its size is defined by assuming that the stress point at (S_+) lays on its boundary. However, if the homeostatic surface of collagen is open towards compression, the stress point is located well inside the homeostatic surface defined by the above procedure. The initial size of the homeostatic surface of collagen can no longer be defined using (23.7.6) and it should be set using some other criterion.

Moreover, in order for PGs to grow first, the radius of the homeostatic surface at the equilibrium point (S) is divided arbitrarily by 2. Therefore, during the free swelling step, PGs grow and this growth is accompanied with an elastic elongation of collagen. Growth of collagen requires its initially compressive Mandel stress to move towards the tensile region before hitting the homeostatic surface. In fact, this shift is triggered by the relaxation of the PGs, which is fast (a few $t_{g\mathrm{PG}}$) at the time scale of the plots, Fig. 23.7.4.

An axial load yielding a scaled axial nominal stress of 2 is applied over $100\,t_{g\mathrm{PG}}$: the loading rate is slow for PGs but fast (dynamic) for collagen. The stress path of PGs is smooth and directed towards tension: the stress attracts the homeostatic surface and remains close to its static value. On the other hand, the high loading rate gives rise to a peak in the stress path of collagen and the relaxation is not yet complete in the time window of the simulations $(1000\,t_{g\mathrm{PG}})$.

The initial elastic extension of collagen implies by (23.1.35) an initial decrease of its volume fraction, Fig. 23.7.5. This decrease stops as soon as the growth of collagen starts. The phenomenon is amplified by application of a tensile load. The steady state values of the volume fractions of collagen and PGs, respectively $n^c \sim 0.16$ and $n^{\mathrm{PG}} \sim 0.04$, correspond to the expected values for articular cartilages.

The scaled osmotic pressure varies little in all the tests, say between 1.2 and 2, which again fits qualitatively with data on actual articular cartilages.

23.7.7 Stiffening of collagen by pyridinoline

The tensile and compressive moduli of articular cartilage are observed to increase from fetus to calf, Williamson et al. [2003], p. 878. This increase seems to correlate with the increase of the concentrations of both collagen and pyridinoline.

The main increase of stiffness of the collagen network is thought to be due to crosslinking of individual fibers associated with an increasing pyridinoline density.

As observed in Fig. 23.7.6, the time course of development of pyridinoline density is directly affected by the coefficient α_{4c}. This feature affects the time course and amplitude of growth. Indeed, Fig. 23.7.7 shows that the time course of evolution of the stiffness of the collagen network has a significant impact on the growth of both collagen and PGs. The higher the rate coefficient α_{4c}, the faster the pyridinoline crosslinks establish. Slow rate of increase of the pyridinoline density allows PGs to grow since their growth is not restrained by a stiff collagen network. By contrast, collagen develops less under this condition because its stress is lower (since its stiffness is lower) and growth is stress-driven. The large gap of growth between PGs and collagen decreases progressively as the pyridinoline density is allowed to increase faster. Most of the change takes place for α_{4c} between 0 and 2, higher values resulting in minor changes in the time course of growth of both the pyridinoline density and the masses of growing constituents.

23.7.8 Directional distribution of collagen network

The relative directional density of the collagen network is quantified through the relative volume fractions. Their rates of change governed by (23.4.46) are proportional to the coefficient α_{3c} and they depend primarily on the load direction and magnitude, Fig. 23.7.8. The change of the directional distribution of collagen fibers is clearly smaller for the compressive test. Let us reiterate that these entities are relative. The absolute density of collagen fibers in a given direction is obtained by scaling these relative entities by the (updated) mass of collagen.

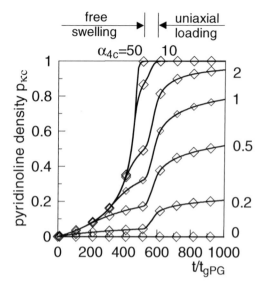

Fig. 23.7.6: Effect of the coefficient α_{4c} on the development of the pyridinoline density during the swelling step (up to $500\,t_{gPG}$) and uniaxial loading. For sufficiently large α_{4c}, the pyridinoline density has reached the value 1 during the swelling process. Reprinted with permission from Loret and Simões [2014].

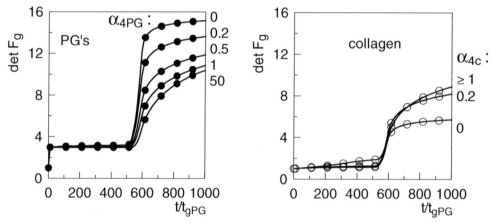

Fig. 23.7.7: Effect of the coefficient α_{4c} associated with the rate of development of the pyridinoline density and collagen stiffness on the growth during the swelling and uniaxial load steps. Higher rates of pyridinoline development tend to decrease progressively the differential of growth between PGs and collagen.

23.7.9   Growth and resorption

The shape of the homeostatic surface and the associated flow rule (23.4.15) are tailored to allow for both growth (mass increase) under tensile stresses and resorption (mass decrease) under compressive stresses. In fact, this statement should be phrased more accurately, since the stresses that govern the growth process are relative to the center of the homeostatic surface (shifted stresses). Of course, resorption of collagen requires the homeostatic surface to be closed towards compression. Still, the current scheme may be seen as a first attempt. Indeed, a more refined analysis might consider two distinct and independent rate equations for growth and resorption as the key biochemical agents in the two processes might be distinct.

Simulations illustrated by Fig. 23.7.9 show the following qualitative trends:
- resorption takes place more easily under isotropic compression than under confined compression;
- additional simulations have also shown that resorption takes place more easily for a confined compression test (zero lateral strains) than for unconfined compression (zero lateral stress);
- resorption of collagen takes place more easily than resorption of PGs: for a (scaled) isotropic compression of -5, only collagen resorbs while PGs remain elastic. As a consequence, the volume fraction of collagen decreases. At higher compression, e.g., -10, both collagen and PGs resorb;

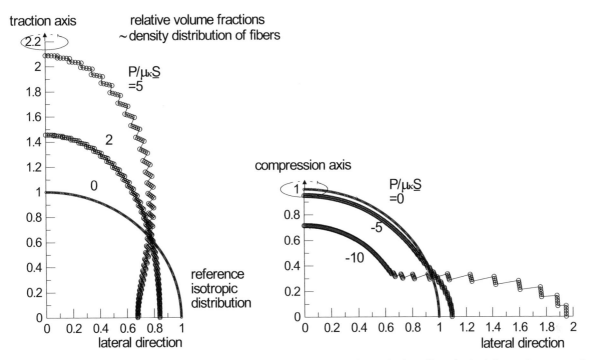

Fig. 23.7.8: Polar representation of the relative volume fractions at the end of swelling/uniaxial traction or confined compression tests with standard parameters and for two axial loads. The collagen network is initially isotropic and remains so during the swelling step. The irregularities in the curves reduce by increasing the number of angular fans n_e currently equal to 36. Reprinted with permission from Loret and Simões [2014].

- for resorption to take place, the shifted Mandel stress should be located on the contractive side of the homeostatic surface;
- for confined compression, the water content decreases so that the fixed charge density and osmotic pressure increase, which limits the changes of the stresses. For uniaxial traction, the water content increases but the amplitudes of the resulting decrease in fixed charge density and osmotic pressure are much less than for compression.

If growth/resorption had not taken place, the collagen would be mechanically inactive under isotropic compression. However, since resorption takes place, the fibers may be elastically elongated and the collagen network takes part to the mechanical response.

The resorption tests may be used to motivate the introduction in the homeostatic surface of the partial modified Mandel stress $\mathbf{\Psi}_{\kappa_*}^k = (\mathbf{C}_k^e)^\alpha \cdot \boldsymbol{\tau}_\kappa^k \cdot (\mathbf{C}_k^e)^\beta$ associated with the partial 2nd Kirchhoff stress $\boldsymbol{\tau}_\kappa^k$, as opposed to the effective modified Mandel stress $\mathbf{\Psi}_{\kappa_*}^{k'} = (\mathbf{C}_k^e)^\alpha \cdot \boldsymbol{\tau}_\kappa^{k'} \cdot (\mathbf{C}_k^e)^\beta$ associated with the effective 2nd Kirchhoff stress $\boldsymbol{\tau}_\kappa^{k'}$ defined by (23.1.29). Recalling the relation between the partial stresses and osmotic pressures in the isotropic compression case, namely for the constituent k,

$$\mathbf{\Psi}_\kappa^k = (F_k^e)^{\alpha+\beta} \left(\tau_\kappa^{k'} - n^k \, p_E \right), \tag{23.7.17}$$

is helpful for the discussion (the component indices have been deleted).

Indeed, PGs grow more than collagen during the swelling phase: collagen undergoes therefore an elastic elongation, and it is mechanically active even if the applied load is a compression. Still, as collagen growth proceeds and gets closer to that of PGs, its elastic elongation decreases so that at some time $t*$ its effective stress vanishes, Fig. 23.7.10. At this time $t*$, the modified Mandel stress is therefore equal to $-n^k \, p_E \, (F_k^e)^{\alpha+\beta}$ according to (23.7.17): it turns out that it is negative enough to induce resorption for the load $P/(\mu_\kappa \underline{S}) = -5$. Instead would the effective Mandel stress (which then vanishes) enter the homeostatic surface, resorption would not take place. In fact, further tests under various loading conditions have confirmed the fact that such a formulation would be very resistant to resorption. Subsequent to time t_*, collagen resorption requires again an elastic elongation which, leading to a positive effective stress, increases the partial Mandel stress, which finally is no longer negative enough to induce resorption. The elastic elongation being small, the difference between the Mandel and effective Mandel stresses remains approximately equal to $-n^k \, p_E$.

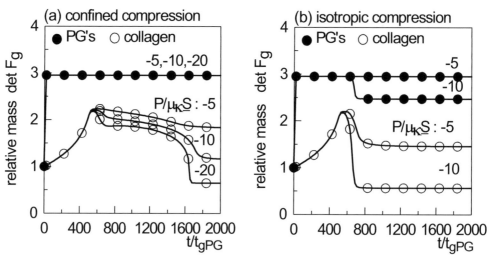

Fig. 23.7.9: Mass growth and resorption during the isotropic swelling phase up to $500\,t_{g\mathrm{PG}}$ and subsequent confined or isotropic compression for three load amplitudes. Reprinted with permission from Loret and Simões [2014].

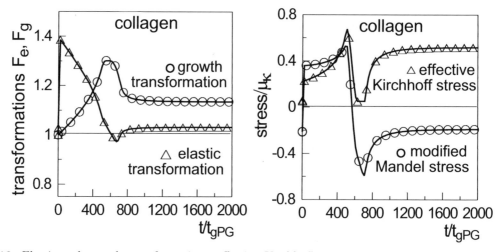

Fig. 23.7.10: Elastic and growth transformations, effective Kirchhoff stress and modified Mandel stress of collagen during free swelling followed by isotropic compression with a scaled load $P/(\mu_\kappa \underline{S}) = -5$. Reprinted with permission from Loret and Simões [2014].

The actual values of the exponent $\alpha = \beta$ may be observed to have some subtle effect by triggering or opposing resorption. The point is illustrated by Fig. 23.7.11 which pertains to PGs under a larger isotropic compression of $P/(\mu_\kappa \underline{S}) = -10$, and therefore a relatively large elastic contraction $F_{\mathrm{PG}}^{\mathrm{e}}$. Would the elastic transformation be small, the partial modified Mandel stress would be below the effective 2nd Kirchhoff stress. However the factor $(F_{\mathrm{PG}}^{\mathrm{e}})^{\alpha+\beta}$ reverses the relative position of the two stresses and, in this case, implies resorption to be less likely. The smallest the exponent $\alpha + \beta$, the more negative the partial modified Mandel stress would be and the more likely resorption could take place. Still, care should be exercised in fixing the values of these exponents since large values would favor growth in case of elastic elongation and tensile loads.

As a further remark, in the present form of the model, the structuration mechanisms of collagen take place when either growth or resorption take place. In particular, the pyridinoline density increases somehow during compression tests. However, this increase is lower than during tensile tests: in fact, the amplitude of the structuration is damped by a value of the ratio ϵ lower than 1.

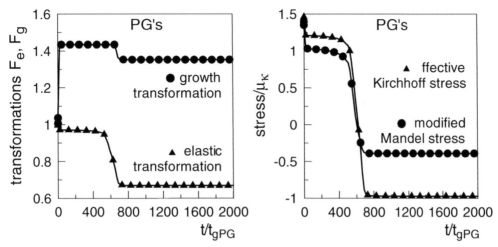

Fig. 23.7.11: Elastic and growth transformations, effective Kirchhoff stress and modified Mandel stress of PGs during free swelling followed by isotropic compression with a scaled load $P/(\mu_\kappa \underline{S}) = -10$. Reprinted with permission from Loret and Simões [2014].

23.7.10 Dissipations and structuration damping

Figs. 23.7.12 and 23.7.13 aim at quantifying the contributions of the various sources of dissipation during the swelling step and uniaxial loading process. For both collagen and PGs, the largest positive contributions are essentially associated with the rates of mass growth and cell density decrease. However, the structuration processes which contribute negatively counterweight these two positive terms during the swelling period. Pyridinoline crosslinking is seen to be the more demanding process.

In fact, the sum of all the contributions to the dissipation would be negative for most of the swelling process despite the cell input. To alleviate this issue, the mechanisms that contribute negatively to the dissipation are damped by a ratio ϵ smaller than 1, so that the total dissipation is never negative, Fig. 23.7.12. Upon application of a load, the ratio ϵ increases quickly to 1 and the structuration processes are no longer damped. For PGs, this ratio is always equal to 1 when growth takes place.

The time courses of development of pyridinoline and of its associated dissipation $-A_{(4)}$ are little affected by the uniaxial growth and, for the standard parameters, its maximum has taken place during the swelling period, Fig. 23.7.12-(c).

23.7.11 Influence of the initial cell energy density

The cell mass has been observed by Williamson et al. [2001] to decrease at increasing masses of PGs and collagen. This observation has motivated the rate equations of cell energy density. The latter has been scaled so that the value 1 corresponds to the minimal size of the homeostatic surface. Therefore maximum growth activity is expected for cell densities in the neighborhood of 1. However, as the cell energy density decreases as growth proceeds, the actual amplitude of growth depends on the path the density describes. For growth to be significant, the cell energy density should have stayed for some time in the neighborhood of 1, and therefore its initial value should be somehow larger than 1.

Fig. 23.7.14 actually shows that there is an optimum value of the initial cell energy density that maximizes growth in a given interval of time. Indeed, if the initial value is too far from 1, then the path will not have reached the neighborhood of 1 to allow significant growth to take place. Conversely, if the initial value is too close to 1 (whether smaller or larger), the cell density will decrease below 1: it will not have stayed in the neighborhood of 1 long enough to generate significant growth.

23.8 Self-healing in engineering materials

The term "self-healing" is intended to convey the idea that energy from the surroundings in the form of heat or light is not required to repair the material. Few cases of combined damage and self-healing processes have been analyzed

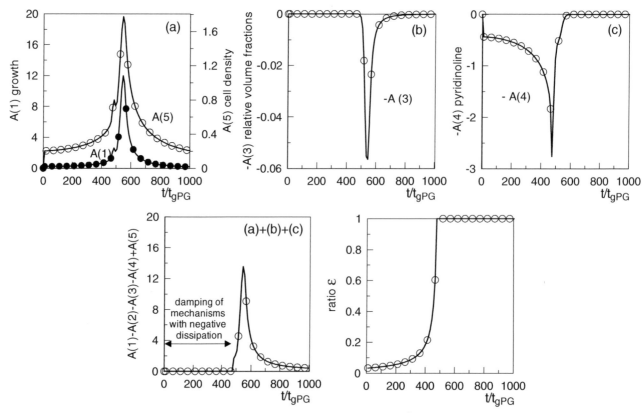

Fig. 23.7.12: Total dissipation $A_{(1)} - A_{(2)} - A_{(3)} - A_{(4)} + A_{(5)}$ (scaled by 10^3), associated, respectively, with the rates of growth, collagen fiber re-orientation ($=0$), relative density of collagen network, pyridinoline development and cell density decrease, during the steps of free swelling and uniaxial loading. Values of the ratio ϵ smaller than 1 damp the mechanisms that contribute negatively to dissipation. Reprinted with permission from Loret and Simões [2014].

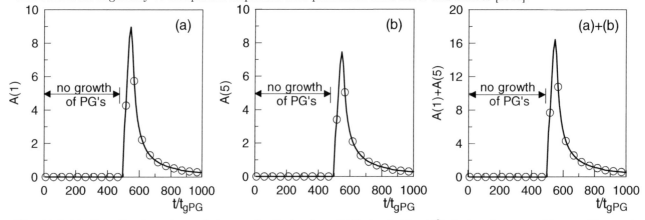

Fig. 23.7.13: Contributions $A_{(1)}$ and $A_{(5)}$ to the dissipation of PGs (scaled by 10^3) during the steps of free swelling and uniaxial loading. The very early times of the swelling are not shown so as to concentrate on the dissipation associated with the uniaxial loading. Reprinted with permission from Loret and Simões [2014].

in the literature:

- in engineering materials, like fiber-reinforced composites, concrete and ceramic materials, a chemical wrapped in micro-capsules and a catalyst are dispersed over the volume during the fabrication process. If the loading gives rise to diffuse microcracks, the micro-capsules are expected to open, releasing the chemical which ideally flows to cover the crack surfaces, and gets in contact with the catalyst dispersed in the material. The chemical and catalyst react, using the energy dissipated by the damage process, so as to glue the crack faces, Barbero and Ford [2007];

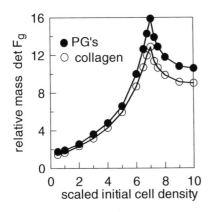

Fig. 23.7.14: Influence of the initial cell energy density on the growth volume/mass at time $1000\,t_{g\mathrm{PG}}$ due to the swelling step and uniaxial loading test. Reprinted with permission from Loret and Simões [2014].

- for rock densification, and compaction of crushed rock salt, Miao et al. [1995], the healing source is the surface energy of cracks, which is proportional to their surface.

Note that both cases correspond to self-healing processes, the energy required by the healing being extracted from the material directly. The characteristic time of the reaction may be large, especially with respect to the mechanical loading time scale. Moreover, the efficiency of self-healing is not expected to be complete.

Let us consider compaction healing in more details. Let ϵ^{e}, ϵ^{i} and $\epsilon = \epsilon^{\mathrm{e}} + \epsilon^{\mathrm{i}}$ be respectively the elastic, inelastic and total strain. The free energy,

$$ e = s(h)\,\frac{E}{2}\,\epsilon_{\mathrm{e}}^2 + \gamma\,a\,, \tag{23.8.1} $$

is contributed by the elastic strain energy and the healing energy. Indeed, the healing energy $\gamma\,a$, with $\gamma > 0$ free energy density, contributes positively to the free energy and the rate of dissipation $-\gamma\,da$ is positive because the crack area a decreases during the healing process[23.16]. This energy is transferred to the healing process. Indeed, the contribution of the elastic strain energy $s(h)\,E\,\epsilon_{\mathrm{e}}^2/2$ to the free energy, with E constant elastic modulus, is a function of the healing variable h, with $s(h)$ an increasing function, namely $ds/dh \geq 0$.

Healing is driven by the inelastic process: the rates,

$$ \frac{d\epsilon^{\mathrm{i}}}{dt} = \lambda\,\sigma; \quad \frac{da}{dt} = -\lambda\,A \leq 0, \quad A \geq 0; \quad \frac{dh}{dt} = \lambda\,H \geq 0, \quad H \geq 0\,, \tag{23.8.2} $$

where $\lambda \geq 0$ measures the advancement of the inelastic process, are provided by constitutive equations. Here $\sigma = \partial e / \partial \epsilon_{\mathrm{e}}$ is the stress. While one may think of $s(h)$ as a decreasing function of a, h increasing with a being consumed, the formalism introduces two a priori independent variables a and h.

The dissipated power,

$$
\begin{aligned}
\hat{S}_{(\mathrm{i})} &= \sigma\,\frac{d\epsilon^{\mathrm{i}}}{dt} - \frac{ds}{dh}\,\frac{E}{2}\,\epsilon_{\mathrm{e}}^2\,\frac{dh}{dt} - \gamma\,\frac{da}{dt} \\
&= \left(\sigma^2 - \frac{ds}{dh}\,\frac{E}{2}\,\epsilon_{\mathrm{e}}^2\,H + \gamma\,A\right)\lambda \geq 0\,,
\end{aligned}
\tag{23.8.3}
$$

should be positive. The last line in (23.8.3) indicates clearly that the dissipated power due to the inelastic process (first term) and decrease of the crack area (third term) are used to increase the stiffness (second term). The efficiency α of the process,

$$ \frac{ds}{dh}\,\frac{E}{2}\,\epsilon_{\mathrm{e}}^2\,H = \alpha\left(\sigma^2 + \gamma\,A\right), \tag{23.8.4} $$

is smaller than 1 since the dissipation $(1-\alpha)\left(\sigma^2 + \gamma\,A\right)$ should be positive.

The analysis of self-healing due to a chemical source spread over the material proceeds similarly, replacing formally the crack area by the chemical concentration.

If the pre-existing energy source that is used by the healing process, e.g., the chemical and catalyst concentration in the case of self-healing composite materials, does not exist, the third term in the last line of (23.8.3) vanishes. Then stiffening has to rely only on the energy provided directly by the growth process, namely the first term in the last line of (23.8.3).

[23.16]Instead of the linear term $\gamma\,a$, one may use a finite energy of the form $1 - \exp(-\gamma\,a)$, with $\gamma > 0$, which linearizes to $\gamma\,a$, and which has the sign of a, for any a.

The stiffening of the collagen network during growth of biological tissue may perhaps differ somehow from the above description:

- in fact, the self-healing process, based on the flow of a chemical from micro-capsules towards micro-cracks, is just a crude attempt to mimic bleeding (out of vessels with a finite content). A vascular network, bleeding and blood clotting make most biological tissues self-healing systems in the long range, Trask et al. [2007];
- the stiffening process during growth that we have in mind is yet slightly different. Still, in view of Fig. 4 of Williamson et al. [2001] which shows a decreasing cell mass at increasing masses of PGs and collagen, cells may be seen as an energy source, that decreases and is exhausted by the growth process.

Finally, it is of interest to contrast the phenomena of healing and *damage*, the latter corresponding to a decrease of the elastic stiffness (the variable a and related energy are elusive here). Then, one typically would set $ds/dh \leq 0$, and postulate rate equations in the format,

$$\frac{d\epsilon^i}{dt} = \lambda \sigma; \quad \frac{dh}{dt} = \lambda H, \quad \lambda \geq 0, \tag{23.8.5}$$

with free energy,

$$e = s(h) \frac{E}{2} \epsilon_e^2, \tag{23.8.6}$$

and dissipated power,

$$\hat{S}_{(i)} = \left(\sigma^2 - \frac{ds}{dh} \frac{E}{2} \epsilon_e^2 H\right) \lambda \geq 0. \tag{23.8.7}$$

The second term of (23.8.7) would be positive if $dh/dt = \lambda H \geq 0$, i.e. if damage increases continuously. This term is negative for healing where $ds/dh \geq 0$. As a key consequence, *healing is not the converse of damaging*. These considerations simply illustrate the fact that the area delimited by a closed path, that includes dissipative portions, can not be zero, e.g., the metabolic pathways, Section 4.8.5. The forward and return paths are distinct.

Chapter 24

Solid tumors: biochemical overview and mechanical modeling

24.1 Solid tumors: the recent intrusion of mechanics

Biochemists and radiologists have realized that mechanics plays a significant role in the growth of tumors. Strangely enough, as far as modeling is concerned, the field has been colonized mainly by mathematicians whose aim is to address phenomena on a qualitative point of view. The mechanical and transport aspects require specific *in vitro* experiments to be performed, e.g., to justify and quantify the constitutive decisions that have to be taken in this complex area. A few such laboratory experiments are reported. A more quantitative approach has still to be undertaken. At the time of writing these notes, several papers are announced whose title seems to indicate that they have embarked in the theory of mixtures.

In order the review to be of interest to the mechanician, the biochemical aspects are considered first. The basic notions of microcirculation summarized in Section 15.9 may be of help to interpret the perturbations of blood pressure in tumors. The interstitial fluid pressure (IFP) recorded in tumors by Jain and collaborators may be high but less than 90 mmHg against less than 3 mmHg in normal tissues: thus the dissolution of gases in blood should not change substantially. Some information on the composition of fluids (plasma, interstitium, cells) and on the transport of O_2 and CO_2 in blood may be found in Sections 3.6 and 15.11, respectively. Many biological aspects that may not be familiar to the specialist in continuum mechanics and mixture theory are briefly overviewed here. References to specialized textbooks are provided when necessary.

As for practical aspects of these notes, topics are separated to isolate issues, e.g., molecular aspects and modeling come usually under different headings. The physiological and biochemical aspects of the growth of solid tumors are sketched in the first sections. They address in turn,

- tissue composition;
- normal cellular cycle: cell cycle, differentiation, proliferation;
- growth factors and inhibitors;
- apoptosis and necrosis;
- formation of tumor cells: proto-oncogenes, products of oncogenes;
- tissue modification and capsule formation;
- angiogenesis, lymphangiogenesis;
- some therapies: anti-angiogenesis, immunotherapy, manipulation of liposomes, phototherapy.

A review of mathematical and mechanical aspects of tumor growth is presented next in a modeling context, with emphasis on the interactions between transport, mechanics and drug delivery processes. The ultimate goal of the modeling of the growth of tumors is to help in the understanding of drug delivery.

A few methods for imaging the elasticity of tissues are also briefly described. In fact, imaging the mechanical properties of tissues may be used as a tool to detect solid tumors. The issue is illustrated by the MR elastography results of Fig. 24.1.1 and by Table 24.1.1 which contrasts elastic moduli in a few normal tissues, tumor tissues and gels. This comparison aims at highlighting the fact that tumor tissues have a shear stiffness about three times larger than their normal (sound) counterparts.

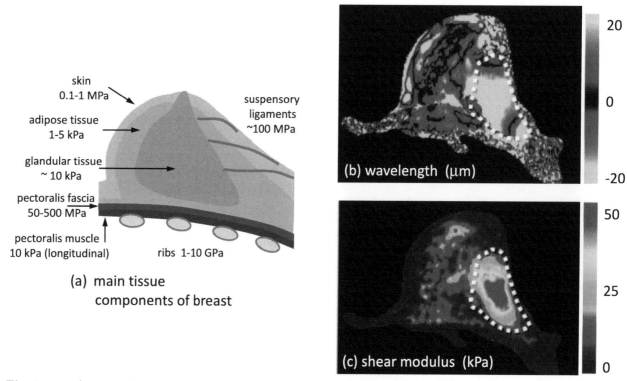

Fig. 24.1.1: Anatomical section of a mature breast (a) and range of elastic shear moduli of the main tissue components from McKnight et al. [2002] and Gefen and Dilmoney [2007]. A MR elastogram of the right breast of a volonteer patient lying in the supine position reproduced with permission from Mariappan et al. [2010] shows (b) longer wave-lengths and (c) higher shear moduli over the 5 cm long adenocarcinoma than in the adjacent normal tissue. This property has been known for a long time and checked clinically by palpation but MR elastography makes quantification possible.

As another theme, since tumor growth involves a number of species (normal cells, tumor cells, extracellular matrix, interstitial fluid, blood, vessels, drugs, etc.) and a number of coupled transport phenomena (diffusion, convection, mass transfer), emphasis is laid on a multiphase mixture framework. Still, mathematical models with a more heuristic flavor are briefly reviewed for completeness.

Therapies of tumors, especially chemotherapies, have to overcome the barrier that tumors oppose to drug uptake. This barrier is essentially due to the high interstitial pressure that exists in the tumors. The phenomenon is enhanced by the mechanical stress created by the volume expansion. The change of the structure of the tissue implies heterogeneities and associated residual stresses. Couplings between mechanics and transports are important and should be addressed. Collapse of vessels, angiogenesis and synthesis of collagen are further instances where mechanics, biochemistry and transport interact.

Textbooks of biology and physiology:

The basics of the molecular aspects of cancer are summarized in textbooks, e.g.,

- Atlas de Poche de Biochimie (1999), Koolman J. and Röhm K.H. [1999], Flammarion, Paris, referred to as APB [1999]. An English version is available under the name "Color Atlas of Biochemistry", Georg Thieme Verlag, Stuttgart, Germany, 2005.
- Atlas de Poche de Physiologie (2000), Silbernagl S. and Despopoulos A. [2000], Flammarion, Paris, referred to as APP [2000]. An English version is available under the name "Color Atlas of Physiology", Georg Thieme Verlag, Stuttgart, Germany, 2008.
- Biochimie, Voet D. and J. [1998], De Boek Université, Paris Bruxelles, pp. 1180-1189. This book is a translation of the 2nd US edition titled "Biochemistry", John Wiley & Sons, 1995.
- Principles of Physiology, Berne R.M. and Levy M. [2000], Mosby, St Louis, MO.

Various aspects of tumor growth are periodically reviewed. The reviews by Jain [1999] and Jang et al. [2003] are devoted to issues of drug delivery in tumors. Araujo and McElwain [2004]a [2004]b focus on the history of the mathematical modeling of tumor growth.

Table 24.1.1: Drained bulk modulus K, shear modulus μ and drained Poisson's ratio ν
of normal, tumor* and artificial tissues.

Tissue	Reference	K (mmHg)	K (kPa)	μ^{a} (mmHg)^{b}	μ (kPa)	ν -
Melanomas*	Roose et al. [2003]	19	2.53	14.25	1.9	0.2
Carcinomas*	Netti et al. [2000]	30-50	4-6.65	22.5-37.5	3-5	0.2
Glioblastomas*	Netti et al. [2000]	200	26.6	150	20	0.2
Sarcomas*	Netti et al. [2000]	300	40	225	30	0.2
Human lung	Roose et al. [2003]	9.76	1.3	4.2	0.56	0.2
Breast carcinomas*	McKnight et al. [2002]			187.5	25	
Adipose tissue	McKnight et al. [2002]			22.5	3	
Fibro-glandular tissue	McKnight et al. [2002]			60	8	
Rat cerebellum	Nicholson & Phillips [1981]	694	92	15	2	0.489
0.5% agarose gel	Helmlinger et al. [1997]a	0.90	0.13	0.68	0.09	0.2
1% agarose gel	Helmlinger et al. [1997]a	2.5	0.33	1.15	0.15	0.2

[a]The shear modulus μ used by the Harvard's group is obtained from the bulk modulus by assuming a Poisson's ratio equal to 0.2. This, value is somehow problematic since the shear modulus is known to be larger in tumors while the bulk modulus probably remains constant. Thus shear modulus and Poisson's ratio should evolve in time.
[b]1 mmHg=133.322 Pa.

Prerequisites:

The understanding of this chapter will be eased if the reader has beforehand become familiar with the following building blocks:
- notions on poroelasticity and phase partition in mixtures, Chapters 7, 16;
- notions of transport of ions and macromolecules in blood and interstitium, Sections 15.1 to 15.7;
- overview of blood circulation, Section 15.11.

24.2 From DNA alteration to metastatic migration

Oncology (oncos=cluster) is the branch of medicine that addresses tumors (=swollen zones), including the study of their development, diagnosis, treatment, and prevention. In French, the two terms "oncologie" and "cancérologie" are synonymous. The neoplasm or tumor tissue is the *abnormal* tissue that grows *abnormally* fast with respect to normal cells. Let us begin by clinical observations and data.

A long process:

The process from the alteration of the first cell to the formation of tumors is usually long. One distinguishes several stages whose successful treatment becomes more and more improbable:
- microscopic phase (difficult to detect, percentage of mortality: small);
- phase of latency (detection possible, mortality 5%);
- localized tumor (surgery and radiotherapy efficient, mortality 20%);
- metastases in ganglions (global treatments, mortality 50%);
- remote metastases (lethal).

A tumor is detectable when it is about 1 gm=10^9 cells, which corresponds to about 30 cell replications ($10^9 \sim 2^{30}$). Volume doubling takes about 5 days in mice and 2-3 months in humans. The total number of cells in the human body is about 10^{14}.

Tumors affect various tissues: pleura, lung, pericardium, peritoneum, thyroid, liver, nasopharynx, ocular orbit, meninges, brain, spinal cord, Nishio and Fukui [1999]. They are isolated (solitary fibrous tumors) or diffuse. To characterize tumors, e.g., mesothelioma, fibrous meningioma, fibrous histiocytomas, fibrosarcomas, melanomas, carcinomas, a number of immunohistochemical markers and ultrastructural examinations must be used.

Tumors may be **benign** (most solitary fibrous tumors) or **malignant** and invasive. In fact, there are two characteristics that have to be considered: proliferation and migration. Migration off the initial center is due to **chemotaxis**, that is, motion along the gradient of nutrient, see, e.g., Keller and Segel [1971] for a model. Usually the initial stages correspond to a central multicellular spheroid. Later on, tumor cells colonize the surrounding tissues. Highly malignant cells such as glioblastoma multiform in the brain grow in a dendritic pattern.

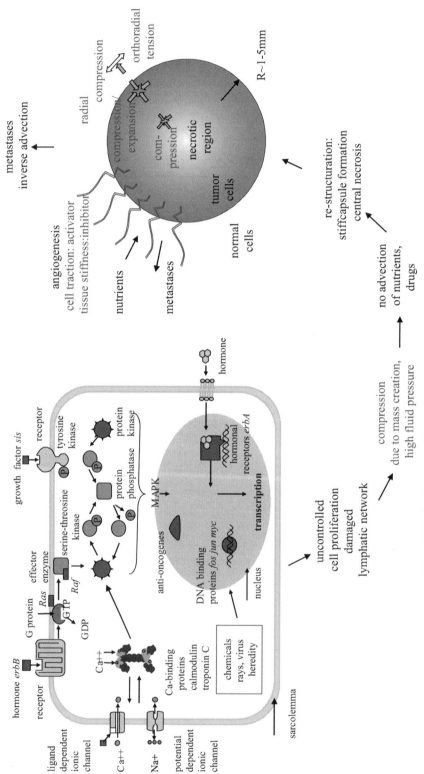

Fig. 24.2.1: Overview of molecular basis of cancer, based on APB [1999] p. 377.

Diagnosis:

Among the signs that allow a diagnosis, let us mention bleeding due to fragile neovascularization and mass syndrom due to tumor induced compression, two physiological aspects to which we will come back.

Histology:

One distinguishes **solid tumors** from cancers of hematopoietic organs (e.g., the bone marrow where undifferentiated stem cells form) like leukemia, lymphomas (e.g., Hodgkin's disease). Solid tumors are categorized in four groups:

- carcinomas[24.1] grown on epithelia, namely skin, breast, ovaries, colon, etc. constitute 90% of cancers;
- sarcomas (sarx=flesh) grown on mesenchymal structures (fibroblasts, etc.), connective and soft tissues: bones, cartilages, muscles;
- neurectoblastic[24.2] tumors of the central nervous system, cerebrospinal cavities, etc., or mesoectodermal tumors in meninges, sympathetic ganglia, Schwann's sheath, etc.;
- tumors with an embryonic structure that affect children and organs of reproduction.

The histology of tumors shows the following characteristics:

- four concentric zones: an avascular necrotic central zone, a semi-necrotic zone, a zone with steady vascularization and an expanding front, Fig. 24.2.1;
- fascicles of spindle (elongated) cells with scant cytoplasm;
- abundant eosinophilic collagen material between cells;
- vascularization: when present, the vessels are dilated, tortuous and irregularly-shaped, Jain [1990], with a fractal dimension of about 1.09 according to Baish and Jain [2000].

Etiology:

Genetics, x-rays, tobacco, benzene, virus, etc. The proportion of lethal cancers due to tobacco in France is about 40%. These cancers are localized in the following organs: lungs, larynx, nose, pharynx, oesophagus, bladder, pancreas.

Treatments:

- radiotherapy at high dose directly kills cells and at lower dose inhibits cellular division; its efficiency depends on the radiosensitivity of the cell line;
- surgery for yet avascular tumors which are still relatively isolated;
- chemotherapy, e.g., drug delivery like anti-angiogenic agents, Carmeliet and Jain [2000], Sheehan et al. [2001]. Some drugs are specific to a part of the cell cycle, some are not;
- gene therapy, e.g., delivery of genetically engineered cells, or viral particles;
- immunotherapy: delivery of antibodies.

Each approach has optimal conditions of use and it also presents its shortcomings. Radiotherapy has difficulties to target only tumor cells and may present important side effects. Surgery can be used only for localized tumors, e.g., it cannot be used for glioblastoma multiforme which is a highly malignant tumor with metastases possibly located remotely from the core of the tumor.

Tumor cells oppose several barriers to the delivery of macromolecules, among which high interstitium hydrostatic and oncotic pressures, Stohrer et al. [2000]. Monoclonal antibodies (MAB's) and adenoviruses used by gene therapy have also to overcome these barriers. Indeed, chemotherapies of cancer face several equally challenging issues. They should in turn

- first, choose a systemic (blood-borne) or topical (local) administration of the therapeutic agents,
- second, deliver the drug to the target (transport issue);
- finally, activate the therapeutic agents.

***In vivo* and *in vitro* laboratory research**:

Research on the various aspects of cancers goes from clinical observations and measurements to laboratory cultures of implanted human or animal tumors. To study cancer *in vivo* in the laboratory, small animals are often used. Two zones of the body of mice are considered, a cranial zone and a dorsal zone (dorsal skinfold chamber). Devoid of thymus, nude mice are immunodeficient, Section 24.3.7.2. This immunodeficiency allows for the stable transplantation of human tumors (xenografts). Alternatively, tumor cells are injected in a particular line of mice. However, extrapolation of observations over species is unwarranted: cell cycle may have different characteristic times, etc.

[24.1]The Roman encyclopaedist A. C. Celsus is quoted to have translated in Latin the Greek word "carcinos" meaning "crab" both as such and in the form "cancer". The suffix "-oma" has been used in modern ages to denote a "cancer lesion".

[24.2]neuro=nerve, ecto=external, blast=germ.

24.3   Oncology and molecular basis of cancer

Typically, a tumor begins with a modified DNA (deoxyribonucleic acid) due to chemicals (e.g., tobacco smoke), x-rays or viruses. DNA alteration induces more and more **de-differentiation** that in turn induces tumor cells to acquire the propensity of embryonic cells to replicate. However, unlike normal cells whose proliferation is regulated, **tumor cells proliferate without physiological control**, e.g., the telomeres do not shorten. The cells form clusters and they may also migrate via the blood vessels and create colonies (metastases) in other tissues. Metastasis is a process with many steps: cancer cells should detach from their clusters, invade the extracellular matrix (ECM), enter the blood/lymphatic vessels, flow through vessels, attach at a distant site, extravasate (cross the vessel wall), migrate, proliferate and recruit new vessels, Jain and Padera [2002].

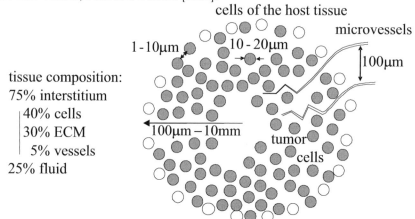

Fig. 24.3.1: Typical dimensions of tumor spheroids.

In vitro laboratory tests to study the tumor development and cell motility involve monolayers, spin flasks or 3D cultures, which are analyzed and compared with clinical data. The protocols currently in use produce spheroid models of 1 to 10 mm that display, after a few days in a spin flask at 37 °C with 5% CO_2 flow, an outer zone of proliferating cells, an intermediate hypoxic region and a central necrotic region as sketched in Fig. 24.3.1[24.3]. As an example two-photon images of MU89 melanomas are displayed by Alexandrakis et al. [2004]. Melanomas are due to the uncontrolled proliferation of pigment cells called melanocytes (=dark cells). They are rare but very serious carcinomas and they are found in the skin and the eye.

24.3.1   Tissue composition

The composition of biological tissues varies much according to their functions. The cardiac muscle and articular cartilages are extreme instances. The muscle is mainly a syncitium where cells occupy most of the volume, gap junctions facilitate the intercellular communications and rapid delivery of nutrients (O_2, etc.) by blood vessels is essential. On the other hand, adult articular cartilage is avascular, aneural and alymphatic. The cells, called chondrocytes, occupy less than 2% of the volume and are in a quiescent state. Most of the space is occupied by water and the extracellular matrix (ECM) consists of collagen fibrils, glycoproteins, glycosaminoglycans, hyaluronan and proteoglycans.

Tissue constitution may be defined by successive partitions. The decompositions,

tissue = cells + extracellular space,

and

extracellular space = vascular space + interstitial space (or interstitium),

---

[24.3]Spheroids consist of tumor cells aggregated *in vitro* with irradiated HeLa cells which can be viewed as the ECM. On the other hand, parts of tumors are maintained *in vivo* in histocultures, Jang et al. [2003]. HeLa cells are immortal cells derived from the cervical cancer cells of a person, Henrietta Lacks. "Immortal" means that, in a survivable environment, the cells divide without end: they overcome the so-called Hayflick limit to the number of cell divisions (40 to 60) associated with telomere shortening. HeLa cells have contaminated a number of cell lines which have been used under a (acknowledged or not acknowledged) false identity, e.g., Lacroix [2008] for an account of this embarrassing issue. Cross-contamination and misidentification of cell lines in research and manufacturing are far from being eradicated.

are function-dependent. A tissue analysis based on composition would lead to a quite different split. For example, there are fluids in cells, vascular space and interstitium, Table 24.3.1.

Table 24.3.1: For subcutaneous transplants of Walker carcinomas 256 on rats, Gullino et al. [1965] provide the following repartition of water per gram of tissue.

	Partition	Volume (ml)	%
Total water	a+b	0.829	100
Cellular water	a	0.453	54.6
Extracellular water	b = c+d	0.376	45.4
Vascular water	c	0.063	7.6
Interstitial water	d	0.313	37.8

The lymphatic network may be considered as part of the extracellular space and its fluid contributes (perhaps not exclusively) to the interstitial fluid.

The **extracellular matrix** is the solid part of the interstitium. It consists of a crosslinked network of collagen and elastic fibers: this network is penetrated by charged hydrophilic macromolecules of glycosaminoglycans (GAGs) structured by hyaluronate and bathed in an electrolyte containing metallic ions (mainly sodium ions). The collagen is formed by a triple helix whose exact composition affects the properties of the fibers (there exist at least 20 types of collagen). The collagen network is defined by the spatial arrangement of fibers and its properties depend both on the properties of the fibers and on the degree of crosslinking. The collagen network endows the ECM with a tensile strength while the elastic fibers provide elasticity.

Data on the content of interstitial space, in terms of water, collagen, GAGs and other macromolecules and ions for specific tumors are provided inter alii by Gullino et al. [1962],[1965], Gullino [1975], Nugent and Jain [1984], Jain [1987]a [1987]b. As a general trend, the interstitial space is larger in tumor tissues than in normal tissues, Table 24.3.2. Hindrance to diffusion is therefore expected to be smaller in tumor tissues. On the other hand, this larger space occupied by fluid is a priori paradoxical with the observation that the shear modulus of tumor tissues is larger than in normal tissues. The altered organization of the collagen network may play a role in this respect.

Table 24.3.2: Average water partition (% of total water) in liver and three tumor tissues, Gullino et al. [1965].

	Liver	Walker Carcinoma	Fibrosarcoma	Hepatoma
Cellular water	72	54.6	38	47
Vascular water	16	7.6	0.2	3
Interstitial water	12	37.8	61.8	50

Spatial heterogeneity is a characteristic of tumor tissue. There is also a large heterogeneity between different types of tumors. For example, Jain [1990] observes that the vascular space varies from 1 to 20% depending on the tumor type and size.

For rat fibrosarcomas, the relative volume n of the interstitial fluid and total tissue is reported to be 50% by Krol et al. [1999]. This relative volume is of importance in the context of a two phase mixture theory where the volume of the tissue is decomposed in a fluid compartment (the interstitial fluid) and a solid compartment which involves the rest of the tissue. More elaborate models may involve the vascular space as a third phase or attribute distinct properties to the species (constituents) of the phases. In theoretical analyses, Netti et al. [1995] adopt $n = 0.2$, and, for melanomas grown in agarose gels, Roose et al. [2003] use $n = 0.4$. Table 24.3.1 for Walker carcinomas 256 indicates a value ranging from 0.30 if only the interstitial water is considered, to more than 0.80 if the whole fluid is accounted for.

Adjacent cells communicate between each other through channels that allow the passage of small molecules. These channels, called gap junctions, are composed of proteins called connexins, and the communication process is referred to as gap junctional intercellular communication (GJC). GJC is blocked in many cancer cells. For example, the viral oncogene Src disrupts cell growth regulation. If communication is restored, the cells return to a normal state of growth (source: Walter Eckhart homepage, Salk Institute).

24.3.2   Normal cellular cycle

24.3.2.1   Cell cycle

A living body survives because its cells undergo divisions. For mammals, the cell cycle lasts from 10 h to 24 h. The duration of the cellular cycle varies much according to the type of cells (embryonic or somatic cells).

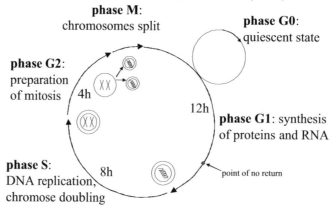

Fig. 24.3.2: The four phases of the human cellular cycle. There are several control stages which aim at checking the correct transmission of DNA (see also Fig. 24.3.4). Only the main one, termed 'point of no return' and regulated by the protein p53, is shown here.

The cycle contains four phases, Fig. 24.3.2:
- during the phase G_1, the cells begin their development;
- DNA is replicated during phase S (DNA synthesis);
- during the phase G_2, mitosis is prepared;
- during the phase M, mitosis, the cells separate.

Cells which do not undergo division, but enter a quiescent state, leave the cycle before a point of no return, to follow the phase G_0 from which they may possibly return under specific stimulations. The step from phase G_2 to phase M is controlled by a MPF, Maturing Promoting Factor, namely the serine/threosine kinase.

A short account of cell cycle may be found in Bassaglia [2001].

24.3.2.2   Cell differentiation

Any normal tissue displays many cell types. Some cells, called stem cells, are not differentiated while others are differentiated. The human genome contains about 100 000 genes but they are **not** all **expressed**. Differentiated cells express less genes and are dedicated to some special functions (cells of bone, of skin, etc.). Stem cells express the maximum of genes while differentiation is due to the repression of some genes.

A stem cell replicates in two daughters: one similar to itself, the other more specialized. The latter in turn may replicate by amplifying the differentiation process until the complete specialization has been reached. Differentiated cells finally undergo apoptosis, Fig. 24.3.3.

Every time a primary human cell divides, its telomeres get shorter, until critically short telomeres lead to arrest of the cell cycle. Cancer cells avoid replication associated with telomere shortening and therefore divide indefinitely. A fine understanding of the mechanisms of telomere shortening could mean finding a way to apply it to cancer cells, as a result restricting their uncontrolled and unlimited growth potential (source: J. Karlseder homepage, Salk Institute).

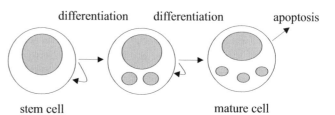

Fig. 24.3.3: The differentiation process corresponds to a specialization of the biological activities of cells.

The number of cells in a living body is the result of the competition between two adverse processes, **cell proliferation** and **apoptosis**. Both processes are strongly regulated.

24.3.2.3 Cell proliferation

Tissues proliferate at various paces:
- fast proliferation: bone marrow, mucous membrane of the liver, ovaries, testicles, skin;
- slow proliferation: lungs, kidneys, liver, vascular endothelium;
- no proliferation: muscles, bone, cartilage, nerves.

24.3.2.4 Apoptosis and necrosis

Apoptosis is the genetically programmed cell death where the integrity of the different parts of the cell is lost. The overall process is controlled so as to dispose of the remnants (macrophages, etc.) and avoid inflammation. Apoptosis is promoted by certain signal pathways, namely hypoxia, TNF-α (tumor necrosis factor), retinoic acid, ligand Fas, protein p53, as sketched in Fig. 24.3.4.

These extracellular ligands Fas are trimeres which fix on membrane receptors which transport the signal to proteases. The nuclear protein p53 is produced when certain nuclear anti-oncogenes are degraded, by x-rays for example. Other pathways inhibit apoptosis. The ICE (interleukin 1β converting enzyme) has an important regulating effect on apoptosis.

On the other hand, **necrosis** is the death of the cells due to injuries. It gives rise to an inflammatory response and to **swelling** by osmotic effect due to an excessive ionic concentration in the cytoplasm.

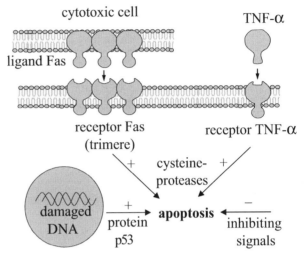

Fig. 24.3.4: The main mechanisms of regulation of apoptosis. Modified from APB [1999], p. 375.

24.3.2.5 Growth factors and inhibitors

The serum used in cellular cultures contains growth factors which are necessary for the cellular proliferation. Growth factors are polypeptides of moderate molar mass (6-30 kg/mole). Their normal plasmic concentration is low, $10^{-12} - 10^{-7}$ M.

In contrast to endocrine hormones which act remotely, growth factors affect only neighboring cells. They fix on membrane receptors: the signal crosses the sarcolemma, activates cytoplasmic pathways (phosphorylations), reaches the nucleus and finally affects the gene transcription, Fig. 24.2.1.

The main growth factors are:
- platelet derived growth factors PDGF1, PDGF2;
- vascular endothelial growth factor VEGF/vascular permeability factor VPF;
- epidermal growth factors EGF, among which transforming growth factor TGF-α;
- fibroblast growth factors FGF-1 to 6;
- insulin growth factors IGF-1, 2.

The role of factors of inhibition is in opposition to that of growth factors. The main inhibitors are:
- transforming growth factor TGF-β that fixes on a receptor whose role is to transmit the signal to a protein that uses a serine/threonine kinase, Fig. 24.2.1;
- tumor necrosis factor TNF-α, Fig. 24.3.4.

24.3.3 Formation of tumor cells

24.3.3.1 Proto-oncogenes

The normal development of cells and their differentiation are stimulated by *proto-oncogenes* which are either cellular or viral. The activities of these genes are controlled by other genes which produce control proteins, called *anti-oncogenes* or tumor suppressor genes, Fig. 24.3.5.

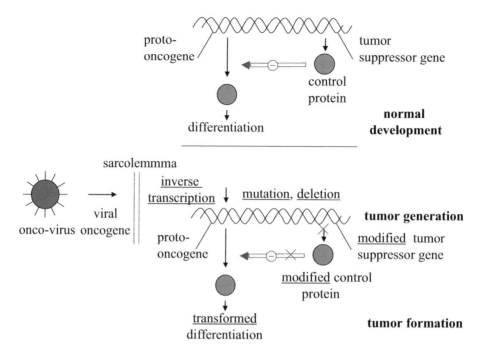

Fig. 24.3.5: Normal development, tumor generation and formation: cellular mechanisms. Simplified from APB [1999], p. 377.

Uncontrolled cell proliferation will be possible when either the activity of anti-oncogenes is reduced by deletion or the activity of proto-oncogenes is stimulated by mutation or over-expression.

The human cycle of tumor cells is reported to be about 2 to 3 days with a long S phase of about 20 h. Theoretical doubling of volume would take about 10 days. However there are a number of non-proliferating cells in tumors and a lot of cells are lost so that the actual time for doubling the tumor size is about 2-3 months. The doubling time is much smaller in small animals (mice), about 5 days.

In oncology, the time evolution of the tumors is traditionally represented by the Gompertz equation, which involves the initial and maximal sizes R_0 and $R_{\max}$ and the effective growth rate a which represents the difference between the growth rate and rate of loss,

$$R(t) = R_{\max} \exp \left(\exp(-a\,t)\,\mathrm{Ln}\frac{R_0}{R_{\max}} \right). \qquad (24.3.1)$$

This empirical relation indicates a growth slowdown with size in agreement with clinic and laboratory observations. There are other estimates that indicate that the rate of growth of tumor volume depends on the tumor type, from linear for murine tumors to cubic for murine lung and breast carcinomas cells, Liao et al. [2000], Candido et al. [2001].

The exponential or Gompertz growth rate is currently strongly criticized because it can be seen as a model where growth is driven by the nutrient availability. More recent models of global growth using fractals consider that growth is rather limited by the competition with adjacent tissue and results in linear growth rate, Brú et al. [2003]. In fact, these models advocate that tumor doubling in volume takes much more than the time of cell cycle, more than 100 cycles, because in particular the cellular activity occurs mainly on the border of the tumor only. They stress the fact that nutrients are critical for part of the growth process only (differentiation, respiration, etc.) but not in the mechanical stress exerted on the tissue. They also point out that usual biopsies collect tissue of the central part of the tumor to be sure not to be mixed with adjacent tissue: however, since the cellular activity is mainly on the periphery, such biopsies are not representative of the current activity of the tumor.

24.3.3.2 Products of oncogenes

The proteins which are affected by cellular deregulation, called onco-proteins, take part in a signal transduction process which is thus modified. Here are some onco-proteins, Fig. 24.2.1 :
- **extracellular ligands**, involved in the transmembrane transport and which present similarities with growth factors;

- **receptors with one segment** which have a tyrosine-kinase activity and that can bind growth factors and hormones that stimulate growth;
- some proteins called G proteins that bind to GTP (guanine tri-phosphate). In normal cells, the external signal associated with a 7 segment receptor uses a G protein which activates an effector protein and phosphorylates GTP in GDP. The main G proteins are the intramembrane **Ras** proteins (H-ras, K-Ras, N-Ras) that transmit the phosphorylation through the cytoplasmic **Raf**;
- many oncogenic proteins are **protein kinases**. Protein kinases stabilize the intracellular activity by phosphorylating certain proteins. Protein phosphatases dephosphorylate these proteins. Protein kinases may be activated by cAMP (they are then called PK-A), by cGMP (PK-G), by diacylglycerol (PK-C) and others are Ca^{2+}/calmodulin-dependent. According to the nature of the amino-acid residues, one distinguishes tyrosine kinases and serine-threosine kinases. Some protein kinases are soluble, others are linked to the sarcolemma;
- some factors of transcription activated by liposoluble hormones (steroids);
- anti-oncogenes;
- after the cascades of kinases and phosphatases, a family of proteins called MAP (mitogen activated proteins) enters the nucleus and phosphorylates the genes of DNA transcription.

24.3.4 Physiological consequences of the presence of tumor cells

In a normal development, cells divide (between 20 to 60 times) before senescence, or apoptosis. They cease to divide also when they come into contact (**contact inhibition**). There are however exceptions to this rule, namely embryonic cells, cells of the epithelium of the liver, bone marrow cells (which form blood cells), and tumor cells. In fact, **tumors are characterized by the proliferation of cells in absence of physiological stimulations** (provided for instance by growth factors). Tumor cells become resistant to apoptosis and the number of cellular divisions can be infinite (*immortality*). A tumor cell is also called a **transformed** cell. A distinction is made between

- a **benign tumor** which contains differentiated cells, dedicated to a specific function;
- a **malignant tumor** which is **dedifferentiated** (like embryonic cells) and which replicates at a fast pace. In fact during uncontrolled division, the cells become more and more dedifferentiated. Malignant cells have a modified boundary which is not sensitive to contact inhibition, maybe because the membrane has modified properties. Indeed, the concentration of the proteins on the membrane is different from normal cells, and the cytoskeleton is less developed, which is probably the reason of their more spherical shape. When sufficiently dedifferentiated, tumor cells produce secondary tumors (metastases). The latter leave their initial environment, degrading the tissue by secreting proteases, they next enter the blood stream, re-enter the endothelium of the vessels and recolonize new tissues where they begin a new proliferation.

The immune system is able to eliminate most of the transformed cells. Otherwise, other processes have to be used: irradiation by γ-rays, which stop the proliferation, or chemotherapy via *cytostatic* substances. However, these processes are not completely selective and they target normal cells as well as tumor cells. Moreover, the deprivation of oxygen (hypoxia) in tumor cells decreases the efficiency of radiation.

The local proliferation of cells can have strong physiological effects. For example, due to the constant cranial volume, brain tumors induce a rise of pressure that squeezes neurones and may lead to epilepsy. Lung tumors may prevent correct ventilation.

The extra-energetic demand by the presence of the tumors overloads the cardiac muscle. This demand is often met by an anaerobic glycolysis, which in turn induces hypoglycemia and **acidosis**, due to lactic acid and CO_2, Helmlinger et al. [1997]b, [2002][24.4]. More generally, tumor cells may overload the metabolism by stimulating certain tissue functions. Specific pathways are activated, e.g., the pentose phosphate pathway, and the metabolism is regulated not by the demand of the tumors but by the substrate availability, Helmlinger et al. [2002].

To summarize, the abnormal microenvironment in tumors (irregular microvessels, impaired lymphatic network) implies interstitial hypertension, acidosis and hypoxia.

24.3.5 Tissue modification and capsule formation

Collagen formation competes for matter with the tumor cells. Collagen forms a shell (capsule) with higher stiffness around the tumor. What characterizes this capsule? Besides a higher stiffness, it seems to be mainly a smaller volume

---

[24.4]Thus, the fact that pH is low in tumors may alter further physiological processes. For example, low pH regulates angiogenic molecules.

fraction of water according to Lubkin and Jackson [2002], see also Section 24.10. Its content in hyaluronan is large, Znati et al. [2003]. Gullino et al. [1962] collect data that indicate that some tumors contain more collagen than normal tissues. Netti et al. [2000] note a collagen-rich area surrounding some types of tumors, e.g., the carcinomas. The capsule around tumors is believed to serve as mechanical and chemical barrier to metastatic tumor cells, Lunevicius et al. [2001], and to reduce or even stop the growth of the tumors. However, its effectiveness is questioned. Indeed,

- on one hand, tumor cells are reported to be insensitive to the stiffness of the substrate, H.B. Wang et al. [2000]. Lubkin and Jackson [2002] suggest that the neoplastic (tumor) cells may be aggressive and their migratory motion may in fact either prevent, or be insensitive to, the capsule;

- on the other hand, Helmlinger et al. [1997]a indicate that mechanical stress inhibits the growth of tumors. Moreover, observations seem indeed to support the fact that ECM synthesis reduces tumor growth, Lunevicius et al. [2001], as then **ECM and tumor cells compete for nutrients** from the fluid. Therefore, drug that favors (controlled) ECM synthesis might be worth to consider. This conclusion is however contradictory with an efficient delivery process. Indeed, collagen hinders considerably the transport of macromolecules, Section 15.8.4.

24.3.6 Angiogenesis and lymphangiogenesis

A tumor can not grow more than $3\,\mathrm{mm}^3$ without vasculature that is necessary to provide nutrients and eliminate waste. In normal conditions, the vasculature is quiescent. In fact it seems that tumors undergo a *switch* to an angiogenic phenotype, Carmeliet and Jain [2000].

Blood vessels form by two mechanisms, *vasculogenesis* and *angiogenesis*. Vasculogenesis is mainly active during embryonic life and involves the creation of vessels from scratch, in which undifferentiated cells elongate to form a tube: on the outer part, the cells transform to smooth muscle while the inner part becomes blood cells. The differentiation of the wall of the vessels increases in time to form, from inside to outside, the intima, media and adventitia, Fung [1993], p. 322. The intima is mainly composed of endothelial cells, the media of smooth muscles and collagen fibrils and the adventitia of collagen fibers, vasa vasorum[24.5] (capillaries, arterioles, venulas), and nerves. Veins have the same layered structure as arteries although the wall is thinner and the composition slightly different in accordance with their hemodynamic function. For a detailed histology of the blood vessels and of the lymphatic network, the reader may consult Dadoune et al. [2000].

24.3.6.1 Angiogenesis

Angiogenesis corresponds to the sprouting of vessels from an existing vasculature. It is stimulated by specific conditions, e.g., hypoxia, and inhibited by others (angiostatin, endostatin). Typical profiles of oxygen and pH away from a vessel are shown in Fig. 24.3.6.

Hypoxic cells seem to send signals, TAF (tumor angiogenesis factor), TGF-β, VEGF (vascular endothelial growth factor), bFGF (basic fibroblast growth factor), that diffuse to the surrounding tissue which is then recruited to contribute to form the new vessels through complex pathways. On the other hand, cartilage produces an inhibitor of angiogenesis, Moses et al. [1990]. Another inhibitor which is found in many tissues is Heparin.

If the vasculature does not exist, inner cells are mainly deprived from nutrients which move only by diffusion. If it exists, the vasculature is perturbed by the tumors. The tumor composition is highly heterogeneous due to the uneven nutrient supply by the vasculature. High fluid pressure may give rise to a succession of passive and active reactions, in particular synthesis of collagen, Koike et al. [2004]. Correlatively, collagen, put in tension by the fluid, may compress the vasculature which may collapse, Padera et al. [2004]. Thus specific therapy that can reduce the mechanical compression and re-open the vessels, should be helpful because it will further improve the delivery of macromolecules. Griffon-Etienne et al. [1999] report of agents (taxanes) that induce apoptosis and decompress the vessels whose diameter increases apparently without angiogenesis.

Vascularized tumors are more aggressive than avascular tumors since tumoral cells can thrive due to the presence of nutrients. In fact, a relation is established between vascularization and metastases. Thus anti-angiogenic drugs are expected to decrease, if not stop, the growth of tumors, Carmeliet and Jain [2000], Sheehan et al. [2001]. There is, however, a paradox to be solved: drug has to be delivered, a priori through vessels, but the latter should not be used by the tumors. In fact, when endothelial cells are targeted, blood vessels should be destroyed and the drug delivery is compromised. Anti-angiogenic agents that do not hinder drug delivery but only *normalize* the existing vasculature are suggested, Tong et al. [2004]. The paradox is addressed in Jain [2005].

[24.5] "Vessels of the vessels" with a diameter larger than 1 mm. Smaller vessels are irrigated by diffusion.

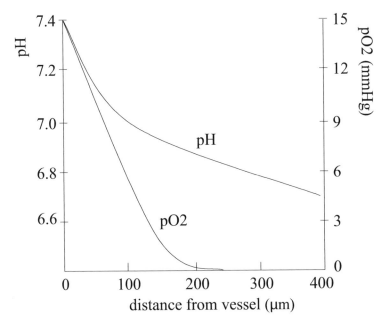

Fig. 24.3.6: Typical profiles of pH and oxygen pressure away from a blood vessel showing the acidosis and the diffusion length of oxygen, qualitatively reproduced from Helmlinger et al. [1997]b. Tumor interstitial pH~6.75 is thus significantly lower than in normal tissue, see also Martin and Jain [1994].

Angiogenesis is one of the aspects that make *in vitro* and *in vivo* tumors different. For further remarks on vasculature collapse, consult Section 24.4.2.2, and, for models of angiogenesis, turn to Section 24.6.4.1.

24.3.6.2 Lymphangiogenesis

In normal tissues, the lymphatic network drains back to the venous system the excess of fluid and cells that have been extravasated from the blood vessels to the interstitium and not re-absorbed directly by the venous circulation. Lymphatic endothelial cells express nitric oxide synthase NOS. The latter seems essentially to increase the fluid flow and does not affect the lymphatic structure, Hagendoorn et al. [2004].

Lymphangiogenic growth factors and receptors in tumor tissues are considered in Leu et al. [2000], Jain and Fenton [2002], Jain and Padera [2002], Reis-Filho and Schmitt [2003]. The lymphatic system is thought to be functionally impaired in tumors: this impairment contributes to the high hydrostatic pressure. One possible cause of damage is the collapse of the vessels due to the compressive stress generated by the proliferating cancer cells.

These days, the role of the lymphatic system in the tumor margin is advocated, besides vessels, in the dissemination of metastases, Dadoune et al. [2000], Jain and Padera [2002], Padera et al. [2003].

24.3.7 Some therapies

24.3.7.1 Antiangiogenesis

Hahnfeldt et al. [1999] generalized the Gompertz relation (24.3.1) so as to integrate semi-empirically the effects of inhibitors and stimulators of angiogenesis. They obtain the model parameters from data and compare the results of their model, their Eqns (C1),(C4), with laboratory tests on mice injected with Lewis lung carcinoma cells. The dose of anti-angiogenic drug necessary for tumor reduction is quantified, typically 20 mg/kg/day of angiostatin or endostatin. They also observe that, in late treatments corresponding to large tumors, the tumor size continues to increase for some time after initiation of the treatment.

The modes of action of inhibitors of angiogenesis are not completely understood, that is, their actual targets are sometimes unknown, e.g., for angiostatin and endostatin.

On the other hand, efforts are devoted to develop inhibitors to specific signaling pathways in the angiogenesis process, e.g., inhibitors to VEGF/VPF (vascular endothelial growth factor/vascular permeability factor). An anti-VEGF/VPF modifies strongly the morphology of the tumor vessels: their diameter becomes smaller, Fig. 1 in Yuan et al. [1996], and they are less tortuous, Tong et al. [2004], Jain [2005]. Vascular normalization by anti-VEGF induces a pressure gradient across the vasculature and thus improves drug delivery. The action of anti-VEGF/VPF needs to be prolongated to be efficient. The idea of vascular normalization is to improve the delivery of drugs and nutrients necessary to their efficiency, in a first step. Ideally, some of the drugs could act as antiangiogenic agents in a second step, so as to deprive the tumor cells from nutrients, Jain [2005].

Other inhibitors target specific adhesion molecules like $\alpha_V \beta_{3/5}$ integrins. As another example, vascular endothelial-cadherin (VEC) is an adhesion molecule that is needed to ensure the assembly of vascular tubes. Administration of an antibody to VEC has been shown to actually reduce greatly the growth of Lewis lung and human epidermoid tumors *in vitro* and *in vivo*, Liao et al. [2000]. The result is a massive apoptosis and necrosis. However the tests are not completely successful because the growth is not entirely suppressed at $50\,\mu$g per dose and higher doses create undesirable side effects.

24.3.7.2   Immunotherapy

Living bodies maintain a control on both external agents (virus, bacteria, etc.) and internal agents (e.g., tumor cells). The control involves identification, specific response (possibly deletion) and book keeping of the events. This role of self-defense against foreign bodies, so called **antigens**, is devoted to the **lymphocytes** (lymphocytes are part of the leucocytes, Fig. 3.6.1). The most important lymphocytes are the lymphocytes T, secreted in the thymus, and the lymphocytes B, secreted by the bone marrow. Lymphocytes T stimulate the secretion of immunoglobulins, called **antibodies**, by lymphocytes B. Immunoglobulins are glycoproteins located on the surface of the lymphocytes B or dissolved in blood. They bind antigens at their both ends. They involve two fragments Fab (antigen binding) and a fragment Fc (in crystal form) linked by papain. The fragment Fc binds to cells and has a major role for transport across cellular membranes. Human immunoglobulins are divided in five classes, IgA, IgD, IgE, IgG and IgM. Their concentrations are listed in Table 3.6.2. The most important ones, the IgG's, are tetrameres of 150 kg/mole, found in blood and in interstitial fluids, APB [1999], p. 280. A histologic account of the organs of the immune system is reported in Dadoune et al. [2000], pp. 147-159.

Dendritic cells can acquire efficiently foreign antigens from apoptotic cells and accelerate secondary immune responses within tumors. Candido et al. [2001] report of successful tests on murine (mouse) breast carcinoma. To enhance tumor-induced apoptosis, the tumor apoptosis-inducing agent TNF-α can be administered.

24.3.7.3   Manipulation of liposomes

Liposomes are drug delivery carriers[24.6]. They are shells, with a membrane of similar composition as the cells. Uncharged, they access easily the cytoplasm, as their membrane combines with the cell membrane.

A point seldom mentioned is the electrical charge of the drugs. Since the vascular glycocalyx (epithelium) is negatively charged, positively charged drugs should be of interest.

The physico-chemical properties of liposomes (molar mass, charge) can be modified relatively easily, Drummond et al. [1999]. Positively charged liposomes have been observed to travel far in the interstitium while, unmanipulated and negatively charged, they remain associated with the vessel wall, Campbell et al. [2002]. On the other hand, the uptake by tumors is independent of the charge. Thus cationic liposomes should be monitored to deliver their loads while within the interstitium.

24.3.7.4   Photodynamic therapy

Photoinitiation is the phenomenon by which chemical reactions are triggered at mild temperatures and physiological pH by light radiation. It is composed of two main steps:
   - production of a cytotoxic agent of lifetime sufficiently long to interact with the substrate:
      - light emission and excitation of a photolabile compound,
      - which releases energy in returning to its ground state,
      - in the form of electrons which, directly or via radicals, transfer to oxygen;
   - reactive oxygen either via superoxide anions or hydroxyl radicals (type I favored in environments with a low oxygen level) or in an excited state (type II), is highly cytotoxic (=: toxic to cells).
Photodynamic therapy acts on tumors in three ways:
   - by targeting tumor cells,
   - by damaging the vessels,

[24.6]The cell membrane is a lipid bilayer, with hydrophilic heads turned outside (in contact with the ECM and the cytoplasm) and hydrophobic tails turned inside. This conformation usually allows the diffusion of hydrophobic molecules. As for polar solutes and ions, they need special transmembrane channels to be able to access the cytoplasm.

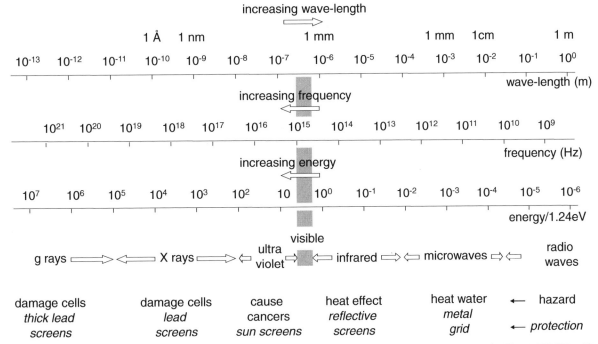

Fig. 24.3.7: The electromagnetic spectrum in terms of wave-lengths, frequencies and energies (1 eV $=1.602176 \times 10^{-19}$ J). The visible range from 380 nm to 750 nm in grey represents a tiny window of the electromagnetic spectrum. Wave-lengths out of the visible range may cause hazards to biological tissues.

- by stimulating an immune response.

The method rests on basic principles of photodynamics. The quantum theory associates with each frequency a particle of zero mass at rest (a photon) that carries an energy quantum,

$$E = h f, \tag{24.3.2}$$

where h is Planck constant. Alternatively, light may be seen to be due to the superposition of electromagnetic waves whose wave-lengths belong to a certain domain. Frequency f and wave-length λ are inverse of one another, $f \lambda = c$, with $c \sim 3 \times 10^8$ m/s the speed of light in a vacuum. Thus, the energy may be expressed in terms of either entity,

$$E = h f = \frac{h c}{\lambda}. \tag{24.3.3}$$

The intensity of light decreases along its path according to the law $I = I_0 \, e^{-A}$. The absorbance A (dimensionless) is described by the Beer-Lambert-Bouguer law,

$$A = \epsilon_\lambda \, C \, \ell, \tag{24.3.4}$$

where
- ϵ_λ is the molar absorptivity [unit : m^2/mole or mole/cm] (the extinction coefficient $\epsilon_\lambda \, C \, \lambda / 4\pi$ is a measure of the absorption of the light in the tissue). It is a property of the compound that depends on the light wave-length (or frequency);
- C is the concentration of the compound [unit : mole/m^3 or percentage];
- ℓ is the light path length [unit : m or cm].

The entity in a photolabile compound, also called *photoinitiator*, that is responsible for light absorption is called the *photosensitizer*. Typically, there is a frequency at which absorption is maximum (e.g., 630 nm for Photofrin©). Higher absorption minimizes the dose of sensitizer to be delivered.

Photosensitizers must absorb long wave-lengths which penetrate the tissue deeper. The effective light radiations are mainly visible, namely in the range 380 to 750 nm, the pure colors being attributed definite wave-lengths,

violet indigo blue green yellow orange red
440   445   475   510   570   590   650

The visible range represents a tiny window of the electromagnetic spectrum shown in Fig. 24.3.7.

Oxygen is a key actor and therefore the method seems somehow paradoxical for cancer treatment: it would not be expected to apply to tumors which are hypoxic or anoxic. This basic difficulty is partially overcome by optimizing the use of the photosensitizers. Other methods like photo*chemo*therapies do not use oxygen as an intermediate. In fact oxygen is one of the few molecules that has a triplet ground state of energy lower than the singlet (metastable) states. These states differ by the occupancy and spin of the π-orbitals, namely $\boxed{\uparrow \, . \, | \uparrow \, . \,}$ for the triplet state and $\boxed{\uparrow\downarrow \, | .. \,}$ or $\boxed{\uparrow \, . \, | \downarrow \, . \,}$ for the two singlet states.

Optimal absorption of energy by the photosensitizer is ensured for a specific wave-length. Moreover, the method capitalizes upon the fact that some photosensitizers concentrate in tumors (for reasons that are not yet completely understood). They are administered by intravenous injection or topically. Once they have reached tumors, photo-sensitizers are subjected to laser light.

Current progresses aim at specializing photosensitizers so that they localize in specialized parts of the tumors, e.g., the vasculature, mitochondria, plasma membranes, nuclei, etc. Regarding the delivery of the photosensitizers to high fluid pressure tumors, the issues to be solved are similar to those of chemotherapy. The standard photosensitizer has been Photofrin® but others like Foscan® are now tested. Next to photosensitizer localization and absorption capacity, issues related to their synthesis and transport remain to be improved. Sensitizers have to be water-soluble to be transported by blood and hydrophobic so as to cross lipid membranes: amphipathic molecules with a polar end (water-soluble) and a non polar end (not water-soluble) are likely candidates.

A brief overview of photodynamic therapies of cancer, including a historical account and clinical aspects, is presented by Dolmans et al. [2003]. An earlier, but more complete, introduction to the principles and issues regarding clinical applications is the book edited by pioneers of the method, Henderson and Dougherty [1992].

Regarding energy consumption during chemical reactions, it is worthwhile to compare photoinitiation with *photosynthesis* and with the *respiratory chain*. Photosynthesis uses light energy to introduce carbon dioxide into organic molecules: the hydrogen involved in the reaction is extracted from water with molecular oxygen as a by-product. The transport of electrons takes place against (i.e., at increasing) the electrochemical potential and requires (light) energy. On the contrary, in the respiratory chain, electrons are transported along a decreasing electrochemical potential with production of energy (ATP) and water.

24.4   Mechanobiology: molecular basis and observations

Mechanical stress interacts with the biochemical and biomechanical properties of tissues and specially with the growth of tumor cells. Indeed the latter modifies the fluid pressure which in turn interacts with the overall homeostasis and metabolism.

24.4.1   Mechanosensitive candidates

Mechanobiology rests on the existence of agents that are able to detect mechanical information and to transmit this information to biochemical effectors that ensure the *mechanotransduction*. While actual mechanobiological effects are observed, the identity of the mechanosensitive agents is far from being elucidated. Here is a list of potential candidates in tumor growth:

- **mechanosensitive membrane channels**. Many cells, although not excitable, have a rest membrane potential due to the existence of channels of various natures. There are three main types of membrane channels: ligand-gated channels, voltage-dependent channels and stretch-sensitive channels. For example glucose transporters are stretch-sensitive according to M.S. Lee et al. [2002]. Mechanosensitive channels are advocated to explain the mechanoelectric feedback in the heart[24.7]. A Ca^{2+}-mechanosensitive channel has been cloned in endothelial cells, Yao et al. [2003]. The inhibition of these channels presumably affects cell proliferation, Wohlrab et al. [2002];

[24.7]Interest in these channels has been due to their link to *commotio cordis* which is observed in young adults because they may originate arrhythmias and fibrillation. Selective activation of these stretch-sensitive channels elicits a mechanism by which sudden death is presumably originated by low-energy chest-wall impact without creating material damage.

- **mechanosensitive membrane receptors** of growth factors. For instance, Westacott et al. [2002] consider the sensitivity of the receptors of tumor necrosis factor TNF-α to pressure. TNF-α is involved in the regulation of apoptosis. Similarly, mechanical stretching contributes to the proliferation of human epidermal keratinocytes via the phosphorylation of epidermal growth factor receptors and calcium influx, Yano et al. [2004];
- **mechanosensitive protein kinases.** Arnoczky et al. [2002] study the effects of cyclic strain on stress-activated protein kinases in canine patellar tendon cells. Petermann et al. [2002] observe that cyclic strain of 5 to 20% decreases the growth of podocytes (visceral glomerular epithelial cells); this decrease is accompanied by an increase of the cyclin dependent kinase inhibitor p21. By contrast, mesangial cells (glomerular smooth muscle) proliferate under cyclic strain where specific intracellular pathways are activated;
- **mechanosensitive mitogen activated kinases** MAPK. A potent angiogenic mitogen of hypoxic tumor cells, myocardiac cells, mesangial cells, retinal pigment endothelial cells, etc. is the vascular endothelial growth factor (VEGF), Yuan et al. [1996], Seko et al. [1999]. VEGF is secreted by cyclic stretching and its expression is thought to be mediated by TGF-β (transforming growth factor-β);
- **integrins.** Integrins are trans-membrane receptors that link the ECM with the nucleus and therefore may transmit mechanical signals. During cardiac development, Nerurkar et al. [2006] suggest that integrin signaling transfers mechanical stimuli to the cytoskeleton and triggers Rho kinase which has been shown to be acting in cellular differentiation as well as in cellular proliferation;
- **enhanced fluid flow and transport** of convected soluble nutrients and growth factors induced by the intermittent mechanical loading increase the availability of nutrients to cells.

The idea above was to show that many cells are sensitive to mechanical forces and that they may change their phenotype and ECM in response to mechanical loadings. Although not concerned directly by tumors, Swartz et al. [2001] go over that stage and suggest that cells can intercommunicate not only biochemically but also mechanically. They consider a setup with two types of cells. One cell type undergoes mechanical loading which elicits gene up-regulation of transforming growth factor β. The latter reaches a second type of cells which reacts by producing collagen and fibronectin.

Mechanosensitive agents, thought to be active in the synthesis of articular cartilages, are listed in Section 20.5.2.

24.4.2   Effects of fluid pressure and mechanical stress

24.4.2.1   High hydrostatic pressures

Jain [1987]a and Boucher and Jain [1992] observe that
- the microvascular pressure (MVP) and the interstitial fluid pressure (IFP) are nearly equal: this feature is attributed to a large vascular permeability and hydraulic conductivity of the wall of tumor vessels, e.g., Fig. 2 in Eikenes et al. [2004];
- the interstitial pressure in a tumor (interstitial hypertension) is much higher (20 to 50 mm Hg) than in the normal tissue (0 mm Hg), see data in Table 1 in Jain [1987]a, Jain [1999]. One of the reasons is the absence of a functional lymphatic network in tumors, Butler et al. [1975]. The role of the lymphatic network is highlighted in Fig. 15.9.1. Another reason is the development of the neovasculature that circulates blood: in contrast, in avascular tumors, the interstitial pressure is low, Fig. 2 in Boucher et al. [1996];
- the radial distribution of IFP in tumors shows a plateau with a steep gradient at the boundary, Fig. 1 in Eikenes et al. [2004]. This picture modifies the smoother qualitative description proposed by Fig. 2 of Jain [1987]a which indicates a smooth monotonous increase from boundary to tumor center;
- the interstitial pressure increases while the perfusion rate and the pO$_2$ decrease with tumor size, Fig. 3 of Jain [1990];
- the pressure gradient of oxygen normal to the wall of vessels is large, Fig. 24.3.6.

Values of pressures and measurement techniques are discussed in Jain [1990]. The wick-in-needle technique is presented by Boucher et al. [1991] and it is detailed in Section 15.7.10.

24.4.2.2   Collapse of vasculature

The high fluid pressure (interstitial hypertension) due to cell proliferation and impaired lymphatic network is thought to have far-reaching physiological effects. As a reaction, it seems to trigger angiogenesis, Carmeliet and Jain [2000].

Vessel collapse induced by high interstitial fluid pressure has been mentioned erroneously in the literature. In fact, Boucher and Jain [1992] observe that the microvascular and interstitial fluid pressures are practically equal in tumors due to the large permeability to macromolecules. Collapse is more likely to be due to compression by proliferating cells, estimated in the range 45 to 120 mmHg, Helmlinger et al. [1997]a, Padera et al. [2004], and swelling due to necrosis.

The new vasculature resulting from angiogenesis is particularly susceptible of collapsing while the mature vessels are more resistant to buckling. Collapse of the vessels leads in turn to a necrotic center due to the absence of nutrients (oxygen and glucose): the dead cells decrease in volume, which might *in fine* re-open the vessels. Still that sequence of events seems somehow contradictory with the necrosis-induced swelling. Perhaps swelling is only an early and temporary event.

Some mechanical aspects of the collapse of veins are described in Section 15.10.2.

24.4.2.3 Effects of gel density in *in vitro* growth of tumor spheroids

An external load on laboratory cultures can be controlled in terms of stress or of strain. A number of bioreactors have been modified to accommodate controlled mechanical loadings aimed at improving the rate of proliferation and/or the mechanical quality of the engineered tissues.

This kind of *external* stress appears somehow artificial in the analysis of tumor growth. The effects of *internal* stress on the growth of multicellular tumor spheroids have been tested *in vitro* by Helmlinger et al. [1997]a. For that purpose, they embedded tumor cells in an inert matrix at various concentrations of agarose gel, namely from 0.5 to 1%. They checked that the agarose was not toxic to the tumor cells.

The tests on several tumor cell lines showed the following effects, Helmlinger et al. [1997]a, Koike et al. [2002]:
- the mechanical stress facilitates the gathering of cells into spheroids, it decreases apoptosis but does not affect proliferation significantly. This positive effect of stress might be due to improved contact and exchange between cells, Yuan [1997];
- upon removing the mechanical stress by enzymatic digestion of the gel, a growth, typical of free suspensions, is recovered in few days. The effects of internal stress are therefore reversible;
- on the other hand, increasing the gel concentration inhibits the growth of the spheroids. Helmlinger et al. [1997]a observe that the spheroids can undergo mechanical stress up to 6 kPa above which the growth becomes inhibited. The model of C.Y. Chen et al. [2001] replicates this observation.

Increase in gel concentration correlates with an increase in the *stiffness* of the composite system, and an increase of the compressive stress. Therefore, the respective roles of stress and stiffness in the above experiments are not settled. Along this line, let us note that, in the model of vasculogenesis of Manoussaki [2003], cell traction is the activator and gel stiffness the inhibitor.

Breward et al. [2002] introduce a cell-cell interaction stress that, in absence of external loading, would imply cells to get apart and the tissue to expand. Conversely, under a prescribed volume, the interaction term generates compression.

Nutrients (oxygen, glucose) diffuse through the tissue or are convected by the capillaries. In avascular tumors, nutrients reach the cells by diffusion only. Consequently the inner cells are deprived from nutrients, their metabolism is annihilated while residues accumulate and acidosis develops. Thus inner cells either remain in a quiescent state (like chondrocytes in cartilage) or starve (undergo necrosis, see Section 24.3.2). A priori one may assume that the growth rate is function solely of the available nutrient concentration. However the stress-inhibited growth as observed by Helmlinger et al. [1997]a may be explained by negative effects of compression on mechanosensitive agents and/or by the partial or total collapse of the vascular network.

24.5   Early mathematical models of tumor growth

24.5.1   Residual stresses

The multiphase approach provides a natural framework to model growth in biological tissues where balances of mass and momentum are accounted for.

24.5.1.1   Compatible and incompatible deformations

Growth might be viewed, in a single phase context, as induced by a volume increase. However since the physical source of the added volume is not considered, attention should be paid to define a consistent local material behavior and a consistent balance of momentum. Indeed, let us consider two cells with different properties or different shapes glued together along part of their boundary. When they undergo loading, the response is a priori dependent on this constraint. To check that point, let us free the cells and next let us apply an arbitrary loading. If they can be re-assembled without external stress, then the individual growths are said to be *compatible*. If, on the other hand, a remoulding stress needs to be applied for re-assemblage, the individual growths are said to be incompatible, Fig. 24.5.1. This stress is also termed *residual stress* because it remains in the material after removal of external loads. Residual stresses are observed in biological organs, Sections 21.7 and 21.8.

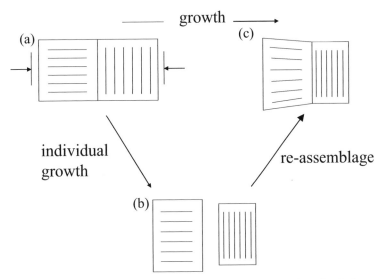

Fig. 24.5.1: Notion of residual stress. Consider two cells with different properties glued along part of their boundaries. They undergo some growth process. Consider now a thought experiment in which they undergo the growth process separately. A remoulding stress has to be applied so as to ensure the two cells to fit again.

Jones et al. [2000] use the analogy with thermal effects to derive the elastic constitutive behavior. However, as indicated in Section 24.5.3, they obtain residual stresses that grow to infinity.

24.5.1.2 Growth in a large strain framework

The influence of mechanics (stress) on growth has been put forward in the 19th century in the remodeling of bones through Wolff's law. More recently has emerged the idea that the kinematics of growth of tissues is accompanied by residual stresses, Skalak [1981], Skalak et al. [1996]. The stress free configuration is evolving during growth[24.8]. A growth tensor $\mathbf{F}^g$ has been introduced to describe these changing configurations, Rodriguez et al. [1994]. Elastic deformations associated with by the transformation $\mathbf{F}^e$ ensure the total deformation gradient $\mathbf{F} = \mathbf{F}^e \cdot \mathbf{F}^g$ to be compatible. This formalism is reminiscent of the elastic-plastic decomposition and the problem of the uniqueness of the intermediate configuration arises as well. In addition, constitutive equations should be prescribed to describe the evolution of $\mathbf{F}^g$.

This approach describes growth by a single kinematic entity $\mathbf{F}^g$. It does not account for the possible differential in growth between constituents of the biological milieu. A mixture theory is therefore more appropriate.

Growing tumors induce inhomogeneous strains. Residual stresses are a priori expected to be tensile on the interior of the tumors and compressive on the exterior, Skalak et al. [1996].

24.5.2 Spherical models

A model consists of constitutive equations defining the tissue response of the tissue (as a single phase or mixture), balance equations of mass and momentum and initial and boundary conditions.

[24.8] Cut the wall of an artery along a plane containing its axis, and observe the relaxed configuration. Perform the same operation on an artery with a thicker wall. The two relaxed configurations are not expected to be identical.

Most growth models consider spherical tumors, although the issue of loss of symmetry to a dendritic pattern is sometimes viewed as an instability.

A reaction-diffusion equation describes the transport of the nutrient and an integro-differential equation the radius of the tumors. In fact, necrosis develops in the center of the tumor when its outer radius is about 500 μm so that the active cells of the tumor form a shell. The ultimate size of the tumors is of the order of the millimeter. The unknowns in these models are thus the concentration of the nutrient and the inner and outer radii defining the tumoral shell. The main parameters are the diffusion coefficient of the nutrient, the coefficient dictating its rates of consumption and the rates of growth and necrosis of the cells.

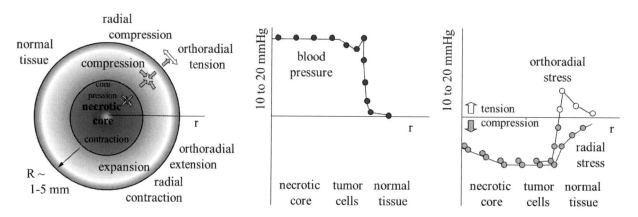

Fig. 24.5.2: Qualitative stress and strain conditions in the tumor and in the surrounding tissue. The terminology associated with stress and strain is worth to be recalled here. A positive stress (respectively strain) component is associated with tension (respectively extension). A negative stress (respectively strain) component is associated with compression (respectively contraction). Expansion characterizes a volume increase.

Models that account for necrosis have to track two fronts: the inner necrosis front defined as explained above and the outer boundary of the tumor. The growth of the exterior cells participates to the growth of the tumors. The position $X(t)$ of the *moving outer boundary* of the tumor is tracked by a so-called Stefan condition, setting that its velocity dX/dt is that of the cells, e.g., MacArthur and Please [2004], their Eqns (5),(24). Moreover, since the exterior cells grow, they induce an orthoradial tension in the surrounding tissue. This orthoradial tension in turn has to be sustained by a radial compression in the tumor, their Eqn (25), Fig. 24.5.2. The nutrient flux vanishes at the center of the tumor, their Eqn (23).

By placing beads in a gel, Gordon et al. [2003] observe that the front of advancement of the invading cells is accompanied by an inward motion. The associated tension is correlative with a re-arrangement of the collagen fibers, Guido and Tranquillo [1993]. The tensile front may be due to a proteolytic activity, which would act as predator of matter, and shift it inwards. Alternatively, it might simply be due to the enhanced tension in the vicinity of the cells which, being stiffer than the surrounding gel (that undergo radial compression and possibly orthoradial extension), act as an attractor for the beads.

24.5.3   Criterion for necrosis

The first models consider isotropic growth in a purely elastic solid. However then, stresses produced by isotropic growth become unbounded, Jones et al. [2000]. Later analyses have recourse to Maxwell viscoelasticity to relax the stresses, Please et al. [1997] [1999]. In that context, MacArthur and Please [2004] consider that necrosis forms due to mechanical conditions rather than chemical conditions. Indeed the criterion for necrosis is that the net stress becomes tensile. In addition, their single phase model shows that the unbounded stress observed in Jones et al. [2000] can be alleviated either by a viscous relaxation or by an appropriate boundary condition at the interface between the necrotic center and the active cells.

The geometrical characteristics of the vasculature, namely area and volume, are maximal at the onset of necrosis according to Jain [1990].

24.5.4 Mixture models of mathematical flavor

Most models for avascular tumors have two phases while more advanced models have three phases.

In order to proliferate, cells take up nutrients and water and the overall volume of active cells expands. Thus active cells are expected to be under compression. Where necrosis occurs, the state of stress is thought to be in tension. This idea is exploited in mixture models by Please et al. [1997] [1999]. They consider two inviscid phases, cells and ECM on one side and water and nutrients forming a second phase. Mass exchange between the two phases allows for cell proliferation. The model considers that cells can not undergo tension, in the sense that, if the intracellular pressure is smaller than the extracellular pressure, the cells get **necrotic** because isolated.

In practice, the time of growth of tumors (days) is much larger than the characteristic time of nutrient supply so that the tumor size is sometimes considered to be steady and a reaction-diffusion equation becomes a reaction equation, e.g., MacArthur and Please [2004], their Eqn (1).

Jackson and co-workers develop a model with three phases, one fluid and two solid phases, for the normal and tumor cells together with their extracellular matrix, Jackson [2003]a. They require the balance equations of the mass and momentum and necessary constitutive equations to close the set of equations. They distinguish between
   - models of *passive growth* where tumor expansion compresses the adjacent tissue;
   - models of *active growth* where the tumor triggers the growth of adjacent collagen.
In both cases, although with different intensities, the outer boundary of the tumor has higher density and stiffness: the corresponding zone (capsule) might delay or stop progression. In a different view, it might be visible only for benign tumors where it had had time to develop.

In Lubkin and Jackson [2002], the tumor grows through transfer of mass from the fluid phase but the extracellular matrix (ECM) does not grow. Jackson [2002] modifies the model to include competition for mass from the water phase as the ECM can grow and, in a further refinement, also degrade.

24.5.4.1 The multiphase models of Jackson and coworkers

The three phase model of capsule formation

The model of Lubkin and Jackson [2002] is a multiphase model. It stresses the importance of the capsule that bounds the tumor. This model is an *expansive growth* model according to the classification above. The model has three phases:
   - the fluid phase, volume fraction $1 - \theta$, velocity $\mathbf{v}$;
   - the normal cells and fibers, volume fraction $(1 - c)\,\theta$, velocity $\mathbf{w}$;
   - the tumor cells and fibers, volume fraction $c\,\theta$, velocity $\mathbf{w}$;
The following equations are used:
   - three *conservations of mass*, one for each of the three phases. The number of normal cells does not vary (no exchange). The matter that forms tumor cells is extracted from the water phase: the rate of transfer is proportional to $\theta\,c\,(1 - \theta)$, i.e., product of volume fractions of tumor cells and water, with a coefficient of proportionality k in the range $10^{-8} - 10^{-6}\,\text{s}^{-1}$, that is a time scale of transfer from 10 to 1000 days;
   - two *momentum balances*, since there are only two independent velocities, one for the fluid and one for the sum of normal and tumor cells. The format is the one of mixture theory and the momentum equations are phrased in terms of the partial stresses (pressures) with a momentum supply that has two terms, a buoyancy term $P\boldsymbol{\nabla}\theta$ where P is the *capillary pressure* and a second term $\phi^*\,\theta\,(\mathbf{w} - \mathbf{v})$ that accounts for diffusion (their $\phi^*\,\theta$ is equal to the drag coefficient k_{Sd} in Section 7.3.2). ϕ^* is taken in the format $\phi/(1 - \theta)$ which implies that the permeability which is proportional to the inverse of $\phi*$ increases with the volume fraction of fluid;
   - the incompressibility of all phases gives rise to an overall incompressibility condition, which provides a field equation for the capillary pressure P;
   - *mechanical constitutive equations* are postulated for the (tumor and normal) cells which are considered as a viscoelastic material (Lamé moduli λ_c, μ_c, pressure P_c) and the water phase whose viscosity is negligible (*interstitial pressure P_i*). Constitutive equations are proposed for these two pressures in a special format. A large positive *solvation* pressure $Y \equiv P_i - P$ is interpreted as a measure of the hydrophilic character of the tumor cell-fiber phase and Y is taken proportional to $c\,\theta$, the volume fraction of tumor cells, with a coefficient of proportionality h called *thirst* coefficient. The *contractile* pressure $\Psi \equiv P_i - P_c$ is generated by contractile or motile cells: the total contractility $\theta\,\Psi$ is assumed proportional to $c\,\theta$ times $(1 - c)\,\theta$, i.e., the product of the volume fractions of the two solid phases with a coefficient of proportionality a.

Osmotic effects (the solvation pressure) are implied by proteoglycans which are not explicitly present as a species neither through a molar fraction nor through their fixed charge.

Parameters are estimated from literature data. Special attention is paid to the drag coefficient ϕ. It is obtained as the volume average of the drag coefficients of species of the solid phases, namely cells, collagen and GAGs. Cells are considered spherical, collagen cylindrical. The following estimates are used: radius of cell $r_{\text{cell}} = 10\,\mu\text{m} = 10^{-5}\,\text{m}$, radius of collagen fibers $r_{\text{fiber}} = 10\,\text{nm} = 10^{-8}\,\text{m}$, so that the contribution of fibers dominates that of cells. The expressions of Stokes' drag coefficient for spherical and cylindrical particles are given in Section 15.7.8 and Exercise 15.1.

A boundary value problem (BVP) with spherical symmetry is considered. The initial volume fractions are specified, the fluid phase is spatially uniform unlike the tumor cells and the analytical expression of the volume fraction c introduces a length-scale r_0. The velocity vanishes at the center of the sphere and the pressure tends to an asymptotic value far away.

Simulations show that the thirst coefficient h should be different from 0 to obtain a capsule, their Fig. 1(e)-(h). Contractility highlights the capsule formation. The interstitial pressure is large within the tumor.

The model does *not* account for cell migration, wound healing inflammation, collagen synthesis, angiogenesis nor for an existing vasculature: the blood inflow and outflow may be modeled by a source term with a time constant of less than a second.

Extension of the model, Jackson and Byrne [2002]: synthesis of collagen

The model in Lubkin and Jackson [2002] is extended in Jackson and Byrne [2002] to include the "foreign body hypothesis". The extracellular matrix (ECM) stress in particular depends on both the ECM volume fraction and the volume fraction of tumor cells, their Eqn (17). The rate of tumor growth is modified to $\alpha_c\,\theta\,c\,(1-\theta) - \delta_c\,\theta\,c$ while the rate of growth of ECM which was earlier zero is now $\alpha_f\,\theta^2\,c\,(1-c)\,(1-\theta)$, which means that tumor cells and ECM compete to gain matter from the water phase in order to grow, Fig. 24.5.3. The numerical simulations indicate that the collagen deposition is not sufficient by itself to form the capsule in contrast to the passive mechanism. On the other hand, both mechanisms control the growth of the tumor.

The further issue of degradation of ECM by tumor cells which produce a protease at a rate depending on their pressure is also considered.

Thus the model does not seem to agree with observations that support the fact that ECM synthesis reduces tumor growth, Lunevicius et al. [2001]. In fact, correlation between capsule formation and tumor control is controversial, some studies supporting the idea, e.g., Wakasa et al. [1985], unlike others, e.g., Ng et al. [1992] and this model.

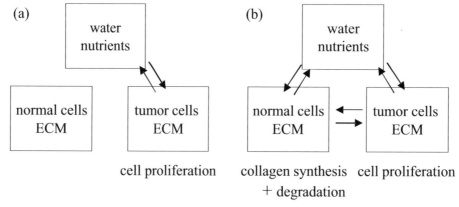

Fig. 24.5.3: Two types of mass exchange and transformation in three-phase mixtures. Typically, the nutrient intake by the tumor cells is proportional to the product of the concentrations of both the nutrient and the tumor cells. The reversibility of the exchanges should be questioned.

24.5.4.2   The models of Byrne and co-workers

Effects of the surrounding tissue

In most models, the tissue surrounding the tumor is disregarded and, instead, boundary conditions are given on the propagation front. By contrast, C.Y. Chen et al. [2001] account for interactions with the surrounding tissue. They embed the tumor in a gel which is endowed with a strain energy function. Their analysis fits with the observations of Helmlinger et al. [1997]a, namely, the higher the gel stiffness, the lower the growth rate.

Cell-cell interactions

Breward et al. [2002] [2003] introduce an interaction term between tumor cells that contribute additively to the stress. The cell-cell interaction term is motivated by the following idea. There exists a critical volume fraction α^* above which cells are closely packed, and interact mechanically. This term is equal to 0 for $\alpha < \alpha^*$, infinite for $\alpha = 1$ and it interpolates smoothly for intermediate values. The cell phase is viscous and, in a 1D context, the effective stress writes $\sigma_\alpha + p_\alpha = 2\,\mu_\alpha\dot\epsilon$. In practice this interaction term adds to the pore pressure of water, $p_\alpha = p_\beta + \Sigma_\alpha$, and therefore $\sigma_\alpha + p_\beta + \Sigma_\alpha = 2\,\mu_\alpha\dot\epsilon$, so that, at constant applied stress σ_α and pore pressure of water p_β, the interaction term $\Sigma_\alpha > 0$ induces a volume expansion.

24.5.4.3 Angiogenesis and a blood vessel phase

The model of Breward et al. [2003] has three phases:
- the tumor cells with volume fraction α, velocity $\mathbf{u}_\alpha$;
- the extracellular matrix ECM (normal cells, fibers) and liquid with volume fraction β, velocity $\mathbf{u}_\beta$;
- the blood vessel phase with volume fraction γ, velocity $\mathbf{u}_\gamma$.

The following equations are developed:
1. three conservations of mass, one for each phase, their Eqns (1)-(3);
   - tumor cells grow from matter that comes from the ECM phase and oxygen crosses the wall of the blood vessels;
   - the death (necrosis) of tumor cells is controlled by the volume fraction of blood vessels (maximal when $\gamma = 0$) while apoptosis is implicitly represented by a death rate proportional to the number of tumor cells, their Eqn (12);
   - the blood vessels grow due to angiogenesis in presence of tumor cells and they collapse when the apparent pressure in tumor cells is larger than a critical value, their Eqn (13);
   - the ECM decreases to provide matter for the tumor cells and it increases when the latter die. Part of the dead tumor cells are assumed to be evacuated by the lymphatic drainage. Liquid emanating from the blood vessels can also enter the ECM, their Eqn (14);
2. three balances of momentum with momentum supplies, their Eqns (5)-(7). The momentum supplies, their Eqns (15)-(16), include a Darcy term proportional to the phase pressure and diffusion terms as a generalization of the model of capsule formation;
3. mechanical constitutive equations: in a 1D context, the stresses are assimilated to pressures, their Eqn (8). The tumor cell pressure p_α is assumed to be the sum of the pressure in ECM p_β plus a cell-cell interaction term Σ_α, namely $p_\alpha = p_\beta + \Sigma_\alpha$, their Eqn (9). The second mechanical constitutive equations concern the pressure in the blood vessel phase: the latter is simply set to the external (constant) pressure, their Eqn (11). Thus there remains a single indeterminate pressure to be obtained from the overall mass/volume conservation.

Of course the closure relation for the rates of mass supply and for the momentum supplies should be introduced in the developments. However it is not clear enough that the rate of mass content of a species is the sum of a diffusion term and of an exchange term. In addition there should be a constraint on the mass supplies, Eqns (12)-(14).

The model introduces the following simplifications:
- oxygen does not appear as an independent species: its volume fraction is implicitly assumed to be a fixed proportion of the volume fraction of the blood vessel phase;
- necrosis and apoptosis are not distinguished. The elastic stiffness of the ECM does not appear either nor that of the collapsed vasculature;
- interactions with the surrounding tissue are not accounted for (see C.Y. Chen et al. [2001] for this issue). In fact, the surrounding tissue is elusive and it is substituted by given boundary conditions.

These models of tumor growth consider a single species per phase. In fact here, a phase results from averaging over several species which are completely elusive. Therefore, since, for example, oxygen is not present as an independent species, the model can not display unambiguously the fact that angiogenesis is triggered by hypoxia.

24.5.4.4   The models of Preziosi and coworkers

Farina and Preziosi [2001] consider the modification that mass transfer induces on Darcy's law. In fact what they are concerned with are the momentum supplies which are of course modified by the presence of the rates of mass supplies. They are not concerned with the constitutive equations of mass transfer.

Preziosi and Graziano [2003] consider a multiphase mixture framework (not in the Biot's sense) in presence of neutral species. They define the momentum supplies in terms of fluxes. The generalized diffusion laws of Chapter 16 provide hydraulic, ionic and electrical fluxes in terms of the gradients of pressure, concentrations and electrical potential. In an attempt to model chemotaxis, the body force adopts the form of a gradient of a nutrient. On that aspect, please refer to Section 24.6.4.1.

Byrne and Preziosi [2003] consider a two phase model in the line of the earlier work of Byrne. They add stress dependence in the cell proliferation rate, much like Roose et al. [2003], following in that the results of Helmlinger et al. [1997]a who observed that proliferation stops in gel cultures at sufficiently high confining stress (gel density). The cell proliferation rate is associated with a mass exchange term between the solid and fluid phases.

24.6   Models of transport of fluids, nutrients and drugs

24.6.1   Influence of the microstructure

A number of changes in the structural organization of the tissue takes place in tumors. These structural changes lead to changes of the continuum properties and highlight a correlation between tissue compliance, diffusion coefficients and hydraulic conductivity :
  - the shear modulus is multiplied by three in breast tissue, McKnight et al. [2002];
  - the extravasation coefficient is multiplied by 10 to 10,000, Sevick and Jain [1991];
  - the tissue permeability is larger by about an order of magnitude, Netti et al. [1995];
  - the volume fraction occupied by blood and lymphatic vessels is larger, say 0.06 instead of 0.02, Table 1 in Leiderman et al. [2006].

The hydraulic permeabilities of a few tumor tissues are displayed in Table 24.6.1.

Table 24.6.1: Hydraulic permeability of a few tumor tissues.
$1\,\mathrm{cm}^2/\mathrm{mmHg/s} = 0.75 \times 10^{-6}\,\mathrm{m}^2/\mathrm{Pa/s}$

Tissue	Reference	$10^{-6}\,\mathrm{cm}^2/\mathrm{mmHg/s}$	$10^{-12}\,\mathrm{m}^2/\mathrm{Pa/s}$
Carcinomas	Netti et al. [2000]	0.45-2.5	0.338-1.88
Glioblastomas	Netti et al. [2000]	0.65	0.488
Sarcomas	Netti et al. [2000]	0.092	0.069
Cervix tumors	Khosravani et al. [2004]	4.3	3.225

24.6.2   Hydraulic permeability and diffusion coefficients in tumors

The transport of a fluid across a porous medium involves a number of coupled processes:
  - in an electrically neutral tissue, seepage of distilled water through the porous medium is described by Darcy's law, Sections 7.3.2 and 15.7;
  - the presence of electrically charged species is considered in Chapter 16;
  - collagen content hinders the diffusion of macromolecules. Pluen et al. [1999] and Ramanujan et al. [2002] report diffusion coefficients in terms of the hydrodynamic radius and Darcy's permeability as a function of the collagen content. As indicated in Section 15.5, the diffusion coefficients are found to decrease at higher hydrodynamic radii of the molecules;
  - still, the hydraulic conductivity is larger in tumors than in normal tissues. Ease of transport is attributed to a deficient organization of the collagen network;
  - the decrease of permeability with GAG content is not significant at the concentrations which are typical of tumor tissues: the effect is of importance at higher GAG concentrations typical of articular cartilages and corneal stroma, e.g., Eqn (F) in Swabb et al. [1974] and Eqn (I) in Jain [1987]a;

- the hydraulic conductivity decreases as the stress state becomes more compressive, Section 15.7.7;
- for agarose gels, the measurements of Johnson and Deen [1996] indicate that there are two determinants of hydraulic permeability, namely gel concentration and compression state. The permeability is halved when the applied pressure changes from 0 to 20 kPa. Under zero pressure, the intrinsic permeability decreases from 616 nm^2 for a volume fraction of the gel n=0.019 to 22 nm^2 for n=0.072.

The above properties (hydraulic conductivity, diffusion coefficients) refer to the interstitium: the measurements take care to exclude the transport in the vessels.

24.6.3 Models that include nutrients

Data on oxygen and glucose consumptions by tumors as well as comparisons are available in Gullino et al. [1967]a [1967]b [1967]c. Eskey et al. [1993] have noted that the energetic supply to tumors is due to glucose more than to oxygen. Nevertheless, most, if not all, models that include nutrients consider oxygen rather than glucose. For instance, oxygen is accounted for in Landman and Please [2001], C.Y. Chen et al. [2001] and Breward et al. [2002].

When nutrients do not appear in the models, the growth and death rates of tumor cells and collagen are considered to depend on some appropriate available volume fractions.

In Breward et al. [2002], the growth of tumor cells is regulated by oxygen: the rate of mass exchange depends on the volume fraction of cells times a function of oxygen tension (partial pressure pO_2), their Eqn (5). Oxygen is available through a diffusion process, their Eqn (4). Its consumption is regulated by a Michaelis-Menten reaction, their Eqn (10). Its tension is assumed to be steady in time which dispenses with an equation of conservation.

Roh et al. [1991] observe that oxygen tension correlates inversely with the interstitial fluid pressure in human cervical carcinomas: this observation is in agreement with the high IFP in hypoxic tumors.

24.6.4 Models that include drug delivery aspects

As already stressed, avascular tumors are usually benign. On the other hand, vascularized tumors are aggressive and able to produce metastases. Therefore drugs like anti-angiogenic agents are expected to be helpful. But a main difficulty is precisely the transport of blood-borne drugs because the vasculature in tumors is inherently tortuous and because efficient drugs aim at deleting their own transport vehicle. In tumor tissues, at variance with normal tissues, where the velocity of red blood cells depends on the diameter of the vessels, blood flow is chaotic and spatially heterogeneous.

Whether in chemotherapy, gene therapy or immunotherapy, therapeutic agents may be molecules, particles or cells, Jain [1999]. The two main steps of drug delivery within the tumors, namely extravasation and transport through the tumor, have been described in Section 15.8. and they are sketched in Fig. 24.6.1. They both include diffusion, convection, transcytosis and possibly binding or metabolization. As stressed in that section, drug delivery has to overcome the unfavorable **tumor hypertension**.

As another issue, **resistance to drugs** manifests itself also as an increased efflux of fluid and convected drugs out of the tumors. The glycoprotein P with an excretion activity is one of the actors in this process. Other factors, like failure for the drugs to bind to the enzymes, are also advocated. In fact, the global transient concentration of a drug in the tumor is not a good indicator. First, because the drug should be homogeneously spread to act efficiently. Second, because it should not only enter the tumors but also bind to cellular macromolecules in order not to be ejected.

A number of references on drug delivery issues are listed in the review of Jain [1999].

24.6.4.1 Models of angiogenesis and inhibition of angiogenesis

Physiological aspects of angiogenesis are briefly exposed in Section 24.3.6.1.

Molecular and modeling aspects of tumor-induced angiogenesis are reported in the review by Mantzaris et al. [2004]. They distinguish continuum models that include chemical interactions between the ECM and the cells at various complexities and discrete models where cells are treated as units.

Literature indicates that, in *in vitro* tests, angiogenesis does not develop if the gel (ECM) is too stiff. Accordingly, in a mechanochemical model of vasculogenesis, Manoussaki et al. [1996] and Manoussaki [2003] view traction as

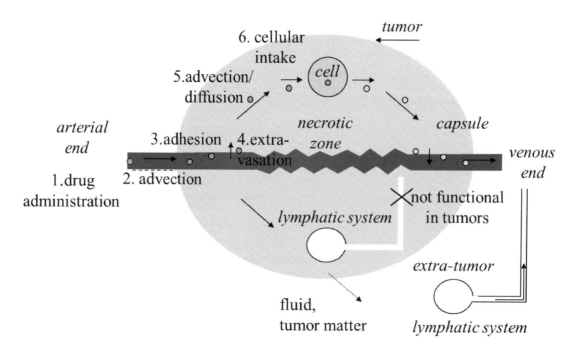

Fig. 24.6.1: The different steps of drug delivery to tumors (discs) and fluid transport (arrows). The non functional lymphatic system has important consequences: (1). the hydrostatic pressures in vessels and tissues are high, and (2). fluid and matter, which normally would have used the lymphatic system to reenter the vessels, are diverted out of the tumor, thus contributing to the tumor propagation via the lymphatic system.

activator and matrix stiffness as inhibitor. Cells move along the gradient of a chemical activator[24.9] and they exert an isotropic traction on the matrix (ECM). An initially uniform domain of cells turns out to a network of cells which is viewed as angiogenesis. The model involves the matrix which is viscoelastic, the cells and the chemical activator. Mass conservation is requested for the matrix (which is inert), the cells and the chemical activator. The diffusion coefficient of the cells depends on the strain so as to account for the special directions that represent the principal strain directions on the chemotaxis. The balance of momentum of the matrix accounts for the stress that the cells exert on it: this term can be viewed as a momentum supply term in surface form.

The model of Breward et al. [2003] briefly sketched in Section 24.5.4.3 includes angiogenesis. Stoll et al. [2003] analyze a model that takes into account endothelial cells and endothelial progenitor cells. The model indicates that angiogenesis is maximum at the tumor periphery.

Some additional references to models addressing adjacent issues might be worth to be mentioned. Vascular structuration is reviewed by Taber [2001]. In a broader perspective, let us mention the reviews of modeling of microcirculation by Schmid-Schönbein [1999], of modeling of the mechanical properties of blood vessels by Vito and Dixon [2003] and of magnetic resonance imaging of microvasculature by Neeman and Dafni [2003].

24.6.4.2 Traditional chemotherapy

The model of Jackson and Byrne [2000] involves three species, namely
   - a drug, the low molar mass doxorubicin;
   - a rapidly dividing population highly susceptible to the drug;
   - a population that is less susceptible to the drug.
The three species move with the same velocity. Their time and space evolutions are governed by reaction-diffusion equations. The rate of change of the drug concentration is taken proportional to the difference $d_b - d$, where d is the drug concentration in the tissue and d_b is the prescribed blood concentration, with a proportionality factor that defines the rate of mass transfer (vascular permeability). Besides their natural growth/death rate in absence of drug,

---

[24.9]This phenomenon is referred to as *chemotaxis*. *Haptotaxis* is the phenomenon by which cells respond to a mechanical signal.

the cell population accounts for the effect of this drug with low coupling, their Eqns (1)-(3). This effect is accounted for in the Michaelis-Menten format very much like for nutrients.

The formulation highlights two time scales: the typical time scale for drug diffusion is the minute while the time scale for tumor growth is rather the day.

In the above model, the involved concentration of drug is external to the cells. In a refined analysis, Jackson [2003]b pays attention to the intracellular levels of the drug and different cellular mechanisms used by the drug in order to simulate different cell lines. Three *compartments* are considered: the extracellular matrix, the cytosol of tumor cells and the sequestered drug in the organelles of the latter. The model considers the evolution of drug concentrations in these three compartments. The drug, doxorubicin, is brought to the ECM by extravasation and it diffuses in this compartment. A first order coupled kinetics governs the time evolution of the drug concentrations in the two cellular compartments. The tumor volume increases due to proliferation and decreases as a function of drug concentration in the organelles, their Eqns (6) or (12). Decreasing cellular permeability is observed to be the most efficient way for the tumor cells to counteract the effect of the drug.

24.6.4.3 Novel chemotherapies

Novel strategies consist in using cytotoxic agents that release ligands that attach to the tumor receptors and activate apoptosis, Fig. 24.3.4. These methods may be one-step and deliver the cytotoxic agent directly or two-step. Two two-step approaches are analyzed in Yuan et al. [1991]. Upon injection, an enzyme conjugate-antibody (ECA), or bifunctional antibody (BFA), binds to antigenic sites in the tumor and it is progressively cleared from the normal tissue. A low molar mass prodrug, or hapten (molecule recognized by an antibody), is injected next, assuming sufficient time has been left to allow the ECA to be cleared and avoid side effect.

Combined chemotherapy and radiation therapies are expected to deliver best benefit. Radiation kills cells directly while the chemotherapy targets angiogenesis, thus depriving the cells from nutrients.

24.6.4.4 Gene therapy by viral replication

Traditional gene therapy has been relatively unsuccessful because of inefficient gene transduction. Improvements and alternatives are tested. As an example, a virus that is engineered to bind to receptors of tumor cells only is delivered, it enters tumor cells by endocytosis, it replicates inside the cells and thus induces the lysis (=death) of the cells. In turn the viruses escape out of the cells and infect other tumor cells. Wu et al. [2001] analyze strategies that differ by the mode of spatial delivery of the viruses in the tumor volume, intravenous delivery being less efficient than intratumoral delivery.

24.6.5 Models that include pH influence

A distinction should be made between intracellular $pH=pH_i$ and extracellular $pH=pH_e$. The pH-gradients are reversed in normal and tumor tissues, Table 24.6.2.

Table 24.6.2: Intracellular and extracellular pH in normal and tumor tissues

Tissue	Intracellular pH	Extracellular pH
Normal	~ 7.1 - 7.2	~ 7.3
Tumor	~ 7.1 - 7.2	~ 6.75

The cells manage to maintain a constant pH irrespective of the external pH. To explain this phenomenon, Webb et al. [1999]a [1999]b examine
  - the effects of membrane channels on the intracellular pH, in particular the lactate$^-$/H$^+$ exchanger under aerobic and anaerobic conditions;
  - the role of buffer of the organelles (endoplasmic reticulum, Golgi apparatus, etc.).

They consider the possible alterations of the metabolic activities due to the low extracellular pH in tumor tissues. They conclude that low pH plays a critical role on cysteine proteinases which control apoptosis, Fig. 24.3.4.

The ratio of free PSA (prostate specific antigen) to total PSA is clinically used as an early diagnostic tool for prostate cancer. Both forms are pH-sensitive. A model that accounts for this effect is developed by K.C. Chen et al. [2004] in view of an appropriate interpretation of this ratio at physiological pH.

24.6.6   Special effects

24.6.6.1   Model for the temporal heterogeneities of tumor blood flow

Mollica et al. [2003] develop a model to describe the temporal heterogeneities of tumor blood flow. They use the membrane theory, with a virtual mass term, to obtain the differential equation in space (along the vessel) and time that governs the displacement (along the height of the capillary). The capillary stiffness accounts for both inflation (membrane tension) and compression and buckling. A second equation is obtained for the pressure by assuming a Newtonian fluid and Darcy's law to describe the fluid extravasation. The system of two coupled equations is solved by the finite element method.

The results show self-sustained oscillations with lumen closing associated with pressure increase. The model is only qualitative and suggestions to improve the tissue response are provided.

24.6.6.2   Cell migration

Tumor malignancy is often reported to correlate with the motility of the tumor cells. Cell locomotion is of importance in wound healing, regeneration and formation of tumors. It is generally believed to be triggered/hindered by chemotaxis, electrical fields and ECM architecture. Abercrombie [1979] contends that tumor cells are guided by the normal cells (fibroblasts, myoblasts, etc.) as they invade a tissue or migrate to form metastases. In 3D collagen lattices, the fibroblasts and tumor cells acquire an elongated shape about $100\,\mu$m in length and $20\,\mu$m in diameter, Friedl et al. [1998].

24.6.6.3   Electrical properties of tumor cells

Cell polarity is reported to be a criterion in the assessment of tumor grade, Klezovitch et al. [2004]. Cell polarity is altered in many cancers. In fact, genes which are responsible for cell proliferation are also responsible for cell polarity.

On another route, it has been suggested that, to defend themselves against attack by the immune system, tumor cells release positive charges, e.g., Ca^{2+}, presumably acting via negatively charged entities. The released calcium further modifies the metabolism and contributes to the proliferation of some cells and to the disappearance of others (B-cells and T-cells in case of leukemia).

24.7   Mechanical boundary value problems

Two basic boundary value problems for a hollow cylinder and a hollow sphere consisting of an isotropic solid endowed with a linear elastic mechanical response and undergoing infinitesimal strain under static conditions are recorded. They illustrate the purely mechanical aspects associated with the proliferation of tumor tissues.

24.7.1   Cylindrical problems

A hollow cylinder sketched in Fig. 24.7.1 is subjected to either a uniform axial displacement or uniform axial traction on its lower and upper bases. Several boundary conditions will be considered on the lateral boundaries. The material is endowed with a linear, isotropic elastic mechanical response, with λ and μ the first and second Lamé moduli.

The motion can then be shown to be compatible with a radial displacement $u_r = u_r(r)$ depending only on the radius r and an axial displacement $u_z = u_z(z)$ depending only on the axial coordinate z while the orthoradial displacement u_θ vanishes. Then, the strain tensor is principal in the axes of the cylinder and so does the stress tensor since the material is isotropic. The radial strain ϵ_r, orthoradial strain ϵ_θ, axial strain ϵ_z and dilatation e deduced from the symmetric part of the gradient (2.4.77) simplify to,

$$\epsilon_r = \frac{du_r}{dr}, \quad \epsilon_\theta = \frac{u_r}{r}, \quad \epsilon_z = \frac{du_z}{dz}, \quad e = \epsilon_r + \epsilon_\theta + \epsilon_z = \frac{1}{r}\frac{d}{dr}(r\,u_r) + \frac{du_z}{dz}\,. \tag{24.7.1}$$

The radial, orthoradial and axial stress components,

$$\sigma_r = \lambda\,e + 2\,\mu\,\frac{du_r}{dr}, \quad \sigma_\theta = \lambda\,e + 2\,\mu\,\frac{u_r}{r}, \quad \sigma_z = \lambda\,e + 2\,\mu\,\frac{du_z}{dz}, \tag{24.7.2}$$

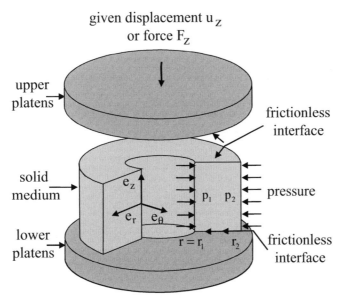

Fig. 24.7.1: A cylindrical specimen subjected along its upper and lower boundaries to a prescribed axial displacement or traction and to pressures along its lateral boundaries.

should satisfy the static equilibrium equations $\vec{\nabla}\cdot\boldsymbol{\sigma}=\mathbf{0}$ expressed in cylindrical coordinates by (2.4.79). In fact, only two equations are nontrivially satisfied,

$$\frac{d\sigma_r}{dr}+\frac{\sigma_r-\sigma_\theta}{r}=0,\qquad \frac{d\sigma_z}{dz}=0\,. \tag{24.7.3}$$

Upon insertion of (24.7.2), the equilibrium equations yield the displacements,

$$u_r=A\,r+\frac{B}{r}\,,\qquad u_z=\epsilon_z\,z+u_z(0)\,, \tag{24.7.4}$$

where A and B are constants defined by the lateral boundary conditions and ϵ_z is the constant axial strain. From the resulting strains,

$$\epsilon_r=A-\frac{B}{r^2}\,,\qquad \epsilon_\theta=A+\frac{B}{r^2}\,,\qquad e=2\,A+\epsilon_z\,, \tag{24.7.5}$$

the volume change is observed to be constant in space and therefore so is the first stress invariant $\operatorname{tr}\boldsymbol{\sigma}=3\,K\operatorname{tr}\boldsymbol{\epsilon}$ with $K\equiv\lambda+\frac{2}{3}\,\mu$ the bulk modulus.

If, instead of the axial strain, the total axial force applied on the cylinder F_z is monitored, the axial stress can be obtained as,

$$F_z=\sigma_z\,\pi\,(r_2^2-r_1^2)+p_i\,\pi\,r_1^2\,. \tag{24.7.6}$$

Several geometries and boundary conditions on the lateral faces sketched in Fig. 24.7.2 are now envisaged where the body force $\rho\,b$ vanishes.

24.7.1.1   A hollow cylinder subjected to lateral pressures

Let p_1 and p_2 be the pressures applied respectively on the internal boundary $r=r_1$ and external boundary $r=r_2$. Then

$$A=-\frac{1}{2}\frac{\lambda}{\lambda+\mu}\epsilon_z-\frac{1}{2}\frac{1}{\lambda+\mu}\frac{p_2r_2^2-p_1r_1^2}{r_2^2-r_1^2}\,,\qquad B=-\frac{p_2-p_1}{2\,\mu}\frac{r_1^2r_2^2}{r_2^2-r_1^2}\,. \tag{24.7.7}$$

Both the dilation,

$$e=\frac{\mu}{\lambda+\mu}\epsilon_z+\frac{1}{\lambda+\mu}\frac{p_1r_1^2-p_2r_2^2}{r_2^2-r_1^2}\,, \tag{24.7.8}$$

and the axial stress,

$$\sigma_z=\mu\frac{3\lambda+2\,\mu}{\lambda+\mu}\epsilon_z-\frac{\lambda}{\lambda+\mu}\frac{p_2r_2^2-p_1r_1^2}{r_2^2-r_1^2}\,, \tag{24.7.9}$$

are spatially uniform while the radial and orthoradial stresses vary radially,

$$\sigma_r = -\frac{p_2 r_2^2 - p_1 r_1^2}{r_2^2 - r_1^2} + \frac{p_2 - p_1}{r^2}\frac{r_1^2 r_2^2}{r_2^2 - r_1^2}, \quad \sigma_\theta = -\frac{p_2 r_2^2 - p_1 r_1^2}{r_2^2 - r_1^2} - \frac{p_2 - p_1}{r^2}\frac{r_1^2 r_2^2}{r_2^2 - r_1^2}. \tag{24.7.10}$$

If the two pressures are identical, $p_1 = p_2 = p$, strain and stress are uniform, and

$$\epsilon_r = \epsilon_\theta = -\frac{1}{2}\frac{\lambda\,\epsilon_z + p}{\lambda + \mu}, \quad e = \frac{\mu\,\epsilon_z - p}{\lambda + \mu}, \quad \sigma_r = \sigma_\theta = -p, \quad \sigma_z = \mu\frac{3\lambda + 2\mu}{\lambda + \mu}\epsilon_z - \frac{\lambda}{\lambda + \mu}p. \tag{24.7.11}$$

If the wall thickness h is small with respect to the average radius $\bar{r} = \frac{1}{2}(r_1 + r_2)$, then

$$\sigma_\theta(r_1) = \frac{\bar{r}}{h}(p_1 - p_2) - p_1 + O(h), \quad \sigma_\theta(r_2) = \frac{\bar{r}}{h}(p_1 - p_2) - p_2 + O(h). \tag{24.7.12}$$

24.7.1.2 A finite solid cylinder

The expressions (24.7.11) hold as well for a solid cylinder ($r_1 = 0$) subjected to the pressure $p = p_2$ at the outer boundary $r = r_2$.

24.7.1.3 A rigid cylindrical cavity in an infinite domain loaded at infinity

Let a cylindrical cavity of radius r_1 embedded in an elastic medium with Lamé moduli λ and μ be loaded at infinity by the pressure p_∞. The cavity is rigid, $u(r_1) = 0$. Then

$$A = -\frac{1}{2}\frac{\lambda\,\epsilon_z + p_\infty}{\lambda + \mu}, \quad B = -A\,r_1^2, \tag{24.7.13}$$

and therefore, for $r \geq r_1$,

$$\epsilon_r = -\frac{1}{2}\frac{\lambda\,\epsilon_z + p_\infty}{\lambda + \mu}\left(1 + \frac{r_1^2}{r^2}\right), \quad \epsilon_\theta = -\frac{1}{2}\frac{\lambda\,\epsilon_z + p_\infty}{\lambda + \mu}\left(1 - \frac{r_1^2}{r^2}\right), \quad e = \frac{\mu\,\epsilon_z - p_\infty}{\lambda + \mu}, \tag{24.7.14}$$

and

$$\sigma_r = -p_\infty - \mu\frac{\lambda\,\epsilon_z + p_\infty}{\lambda + \mu}\frac{r_1^2}{r^2}, \quad \sigma_\theta = -p_\infty + \mu\frac{\lambda\,\epsilon_z + p_\infty}{\lambda + \mu}\frac{r_1^2}{r^2}. \tag{24.7.15}$$

24.7.1.4 A cylindrical cavity with given radial displacement in an infinite domain

Let a cylindrical cavity of radius r_1 embedded in an elastic medium with Lamé moduli λ and μ to be free of load at infinity. The radius of the cylinder undergoes a small change from r_{10} to r_1, i.e., $u_r(r_1) = \Delta r_1$. Then

$$A = -\frac{1}{2}\frac{\lambda\,\epsilon_z}{\lambda + \mu}, \quad B = -A\,(\bar{r}_1)^2 + \bar{r}_1\,\Delta r_1, \tag{24.7.16}$$

with $\bar{r}_1 = \frac{1}{2}(r_{10} + r_1)$ an average value of the radius. Therefore, for $r \geq r_1$,

$$\epsilon_r = -\frac{1}{2}\frac{\lambda\,\epsilon_z}{\lambda + \mu}\left(1 + \frac{(\bar{r}_1)^2}{r^2}\right) - \frac{\bar{r}_1\,\Delta r_1}{r^2}, \quad \epsilon_\theta = -\frac{1}{2}\frac{\lambda\,\epsilon_z}{\lambda + \mu}\left(1 - \frac{(\bar{r}_1)^2}{r^2}\right) + \frac{\bar{r}_1\,\Delta r_1}{r^2}, \quad e = \frac{\mu\,\epsilon_z}{\lambda + \mu}, \tag{24.7.17}$$

and

$$\sigma_r = -\sigma_\theta = -\mu\left(\frac{\lambda\,\epsilon_z}{\lambda + \mu} + 2\frac{\Delta r_1}{\bar{r}_1}\right)\frac{(\bar{r}_1)^2}{r^2}, \quad \sigma_z = \frac{3\lambda + 2\mu}{\lambda + \mu}\mu\,\epsilon_z. \tag{24.7.18}$$

24.7.2 A sphere or spherical shell under purely radial displacement

A sphere or a spherical shell consists of a material which is linear isotropic elastic with λ and μ the first and second Lamé moduli. It is subjected to radial pressure loading and to a body force per unit volume $\rho\,\mathbf{b} = \rho\,b(r)\,\mathbf{e}_r$ which is radial as well. Therefore the displacement $\mathbf{u} = u(r)\,\mathbf{e}_r$ is radial and certainly the traction in a radial direction is radial, namely $\boldsymbol{\sigma}\cdot\mathbf{e}_r = \sigma_r\,\mathbf{e}_r$.

The radial strain ϵ_r, orthoradial strains ϵ_θ and ϵ_ϕ and dilatation e express in terms of the radial displacement u as, Section 2.4.3.1,

$$\epsilon_r = \frac{du}{dr}, \quad \epsilon_\theta = \epsilon_\phi = \frac{u}{r}, \quad e = \epsilon_r + \epsilon_\theta + \epsilon_\phi = \frac{1}{r^2}\frac{d}{dr}(r^2 u) = \frac{du}{dr} + 2\frac{u}{r}. \tag{24.7.19}$$

The stress components

$$\sigma_r = \lambda\,e + 2\,\mu\frac{du}{dr}, \quad \sigma_\theta = \sigma_\phi = \lambda\,e + 2\,\mu\frac{u}{r}, \tag{24.7.20}$$

should satisfy the static equilibrium equations $\vec{\nabla}\cdot\boldsymbol{\sigma} = \mathbf{0}$ expressed in spherical coordinates by (2.4.69). In fact, a single equation is nontrivially satisfied,

$$\frac{d\sigma_r}{dr} + 2\frac{\sigma_r - \sigma_\theta}{r} + \rho\,b = 0. \tag{24.7.21}$$

Upon insertion of (24.7.19) and (24.7.20), the equilibrium equation (24.7.21) becomes

$$(\lambda + 2\,\mu)\frac{d}{dr}\left(\frac{1}{r^2}\frac{d}{dr}(r^2\,u)\right) + \rho\,b = 0. \tag{24.7.22}$$

The solution,

$$u = A\,r + \frac{B}{r^2} - \frac{\rho\,b}{\lambda + 2\,\mu}\frac{r^2}{4}, \tag{24.7.23}$$

involves two constants A and B which are provided by boundary conditions. The resulting nonzero strain components are

$$\epsilon_r = A - 2\frac{B}{r^3} - \frac{\rho\,b}{\lambda + 2\,\mu}\frac{r}{2}, \quad \epsilon_\theta = \epsilon_\phi = A + \frac{B}{r^3} - \frac{\rho\,b}{\lambda + 2\,\mu}\frac{r}{4}, \quad \text{tr}\,\boldsymbol{\epsilon} = 3\,A - \frac{\rho\,b}{\lambda + 2\,\mu}r. \tag{24.7.24}$$

The nonzero stress components become

$$\sigma_r = 3\,K\,A - 4\,\mu\frac{B}{r^3} - \frac{\lambda + \mu}{\lambda + 2\,\mu}\rho\,b\,r, \quad \sigma_\theta = \sigma_\phi = 3\,K\,A + 2\,\mu\frac{B}{r^3} - \frac{2\,\lambda + \mu}{\lambda + 2\,\mu}\frac{\rho\,b}{2}r. \tag{24.7.25}$$

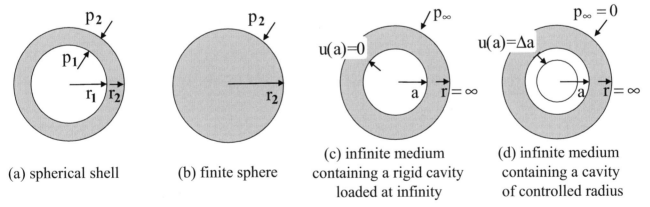

(a) spherical shell (b) finite sphere (c) infinite medium containing a rigid cavity loaded at infinity (d) infinite medium containing a cavity of controlled radius

Fig. 24.7.2: Four problems with spherical symmetry under pressure loading.

In absence of body force ($\rho\,b = 0$), the volume change is constant in space, and therefore so is the first stress invariant $\text{tr}\,\boldsymbol{\sigma} = 3\,K\,\text{tr}\,\boldsymbol{\epsilon}$, with $K \equiv \lambda + \frac{2}{3}\mu$ the bulk modulus.

24.7.2.1   A spherical shell

Let us consider now a spherical shell bounded by the surfaces of radii r_1 and $r_2 > r_1 > 0$. The inner and outer surfaces are loaded by the pressures $p_1 \geq 0$ and $p_2 \geq 0$ respectively. Therefore,

$$3\,K\,A - 4\,\mu\,\frac{B}{r_i^3} = -p_i, \quad i = 1, 2, \tag{24.7.26}$$

from which

$$3\,K\,A = -\frac{p_2\,r_2^3 - p_1\,r_1^3}{r_2^3 - r_1^3}, \quad 4\,\mu\,B = -\frac{r_1^3\,r_2^3}{r_2^3 - r_1^3}\,(p_2 - p_1), \tag{24.7.27}$$

and therefore
- the stresses are uniform only if the surface pressures are equal;
- the volume change is positive if the internal pressure is sufficiently larger than the outer pressure,

$$\mathrm{tr}\,\boldsymbol{\epsilon} = \frac{1}{K}\,\frac{p_1\,r_1^3 - p_2\,r_2^3}{r_2^3 - r_1^3} > 0 \quad \Leftrightarrow \quad \frac{p_1}{p_2} > \frac{r_2^3}{r_1^3}. \tag{24.7.28}$$

The orthoradial stresses at the boundaries become,

$$\sigma_\theta(r_1) = \frac{p_1\,r_1^3 + (p_1 - 3\,p_2)\,r_2^3/2}{r_2^3 - r_1^3}, \quad \sigma_\theta(r_2) = \frac{p_2\,r_2^3 + (p_2 - 3\,p_1)\,r_1^3/2}{r_1^3 - r_2^3}. \tag{24.7.29}$$

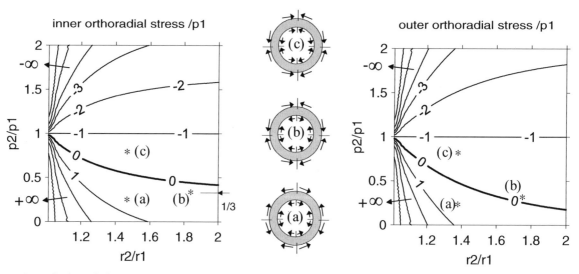

Fig. 24.7.3: Orthoradial stresses at the boundaries of the spherical shell of Fig. 24.7.2-(a) in terms of geometry and loading. Both factors combine to define the sign (compression or tension) and amplitude of the stresses. The inner and outer orthoradial stresses are both tensile, point (a), or compressive, point (c), while they are of opposite signs in a thin zone, e.g., at point (b).

Fig. 24.7.3 shows how the geometry, namely the ratio r_2/r_1, and the loading, namely the ratio p_2/p_1, combine to influence the orthoradial stresses on the boundaries of the shell:
- the stresses are uniform and compressive if the inner and outer pressures are equal;
- the stresses are independent of the elastic moduli;
- the inner orthoradial stress is positive (tension) if

$$\frac{p_2}{p_1} < \frac{1}{3} + \frac{2}{3}\,\frac{r_1^3}{r_2^3}\,. \tag{24.7.30}$$

It is negative (compression) if the inequality is reversed and it vanishes when the above inequality becomes an equality;

- similarly the outer orthoradial stress is tensile if

$$\frac{p_2}{p_1} < \frac{3}{2\,r_2^3/r_1^3 + 1}\,. \tag{24.7.31}$$

- the stresses become large for a thin shell while the dependence with respect to the geometry decreases quickly as the shell thickens;
- given a load $1/3 < p_2/p_1 < 1$, there always exists a geometry (corresponding to a sufficiently thick shell) such that the inner orthoradial stress is compressive. The proposition holds for the outer stress for $p_2/p_1 < 1$. For $p_2/p_1 < 1/3$, the inner stress is always tensile. In other words, the outer pressure has to be sufficiently large for a geometry to exist that ensures a compressive inner orthoradial stress. If $p_2/p_1 > 1$, then the two stresses are always compressive;
- under most geometries and loadings the inner and outer orthoradial stresses have the same sign except in the thin zone,

$$\frac{3}{2\,r_2^3/r_1^3 + 1} < \frac{p_2}{p_1} < \frac{1}{3} + \frac{2}{3}\frac{r_1^3}{r_2^3}\,, \tag{24.7.32}$$

where the inner stress is tensile while the outer stress is compressive.

The sign of these stresses is of particular mechanical and biological importance. Indeed biological growth is influenced differently by tension and compression.

24.7.2.2   A finite solid sphere

Under uniform external pressure, the strain and stress are uniform and isotropic, $A = -p_2/(3\,K)$, and $B = 0$.

24.7.2.3   A rigid spherical cavity in an infinite domain loaded at infinity

Let a spherical cavity of radius a embedded in an elastic medium with moduli $K = \lambda + \frac{2}{3}\mu$ and μ be loaded at infinity by the isotropic pressure p_∞. The cavity is rigid, $u(a) = 0$. Then from (24.7.23) and (24.7.25), $A = -B/a^3 = -p_\infty/(3K)$ and

$$\sigma_r(r) = -p_\infty\,(1 + \frac{4\,\mu}{3\,K}\frac{a^3}{r^3}) < 0, \quad \sigma_\theta(r) = -p_\infty\,(1 - \frac{2\,\mu}{3\,K}\frac{a^3}{r^3}) \le 0, \quad r \ge a\,. \tag{24.7.33}$$

The sign of the orthoradial stress assumes a positive Lamé modulus λ (equivalently a positive Poisson's ratio ν). Fig. 24.7.4 shows that perturbations due to the presence of the cavity extend to about two radii.

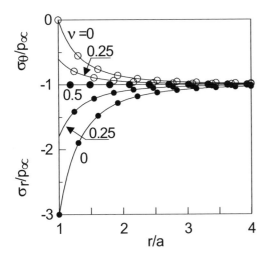

Fig. 24.7.4: The radial and orthoradial stresses outside the rigid cavity of Fig. 24.7.2-(c) loaded at infinity are both compressive for any radius r/a. Due to the inverse cubic spatial variation, the stress perturbation concerns only a skin around the sphere of about two radii. Note the strong influence of Poisson's ratio ν.

24.7.2.4 A spherical cavity with given radial displacement in an infinite domain

Let a spherical cavity of radius a embedded in an elastic medium with shear modulus μ be free of load at infinity. The radius of the sphere undergoes a small change from a_0 to a, i.e., $u(a) = A\,\bar{a} + B/\bar{a}^2 = \Delta a$, with $\bar{a}$ an average value of the radius, say $\bar{a} = \frac{1}{2}(a_0 + a)$. Then $A = 0$ from (24.7.25) and thus $B = \bar{a}^2 \Delta a$. Consequently the stress components in the medium due to the variation of radius of the cavity,

$$\sigma_r(r) = -2\,\sigma_\theta(r) = -4\,\mu\,\frac{\bar{a}^3}{r^3}\,\frac{\Delta a}{\bar{a}}, \quad r \geq \bar{a}, \tag{24.7.34}$$

have opposite signs and the means stress and dilatation vanish.

24.8 Poroelastic mixture models without growth strain

Microvessels are assumed to be uniformly spread through the interstitium. Microvessels and the interstitium exchange fluid, a phenomenon referred to as extravasation. A poroelastic mixture model accounting for extravasation may consider the microvessels to be either external or internal to the poroelastic medium.

A three phase mixture including the solid skeleton, the fluid and the microvessels is envisaged in Sections 24.8.2 and 24.8.3. On the other hand, excluding microvessels, the models developed for extravasation in Netti et al. [1995] and for a localized fluid infusion in a viscoelastic polymer gel in Netti et al. [2003] consider a two phase porous medium. In both cases emphasis is laid on the interstitium whose hydro-mechanical state is spatially heterogeneous while pressure or flow in the microvessels are controlled. By contrast, spatial homogeneity in the interstitium and heterogeneity along vessels are considered in Moran and Prato [2001].

24.8.1 A two-phase poroelastic tissue with external mass exchanges

The tumor tissue in the model of Netti et al. [1995] is poroelastic. The model involves mass exchange between the poroelastic tissue and the capillary network, or microvasculature, which is *external* to the poroelastic tissue as sketched in Fig. 24.8.1.

As a main feature of the model, the extravasated fluid appears as a source term in the balance of mass of the interstitium. Modeling the microvessel-tissue fluid exchange is motivated by the fact that the high interstitial fluid pressure (IFP) is due to a high microvascular pressure (MVP). Here only extravasation by convection is addressed while a diffusion contribution (for low molar mass macromolecules) exists as well. Some qualitative details on extravasation across the vascular wall are reported in Sections 15.8 and 15.9.3.

24.8.1.1 Overview of the extravasation model

The two phase model

  The mixture model has two phases. Stresses are represented by the solid stress and the insterstitium fluid pressure p_i (IFP). The main characteristics of the constitutive model are the following:
  - fluid from the vessels (and lymphatics) transfers to the porous medium which in fact appears as open because the vessels are not part of the mixture. The underlying assumption is that the vessels can be assumed to be homogeneously distributed in the tissue at the scale of interest. Since the typical distance between vessels is $100\,\mu$m, tumors of size 1 mm-1 cm should be considered for the homogenization scheme to be meaningful. A characteristic quantity of the exchange is the ratio S_c/V of vascular area per unit tissue volume, or *capillary specific surface area*. According to Hilmas and Gilette [1974] this surface area increases in tumor tissue due to the irregular (fractal) shape of vessels[24.10],

$$\text{capillary specific surface area } \frac{S_c}{V} = \begin{cases} 1 \times 10^4\,\text{m}^{-1} & \text{for normal tissue,} \\ \downarrow \\ 2 \times 10^4\,\text{m}^{-1} & \text{for tumor tissue.} \end{cases} \tag{24.8.1}$$

[24.10]Hilmas and Gilette [1974] use a morphologic method to quantify the changes in growing and irradiated tumors. The method consists in throwing a segment of given length in random directions and counting the number of times the endpoints fall in the tissue component of interest (tumor, necrotic tissue, etc.) and the number of times it intersects the vessels. They estimate the percent of vessel volume, the length and diameter of the vessels, and the specific surface area.

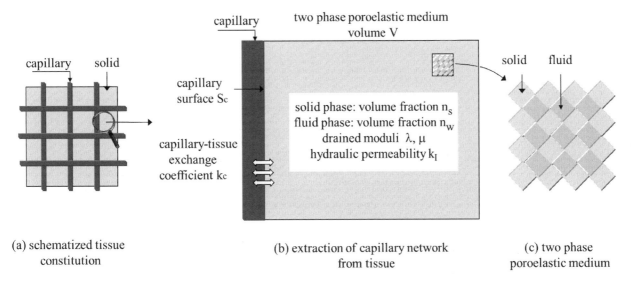

(a) schematized tissue
constitution

(b) extraction of capillary network
from tissue

(c) two phase
poroelastic medium

Fig. 24.8.1: A model for extravasation. Microvessels are homogeneously distributed, above some spatial scale. They are however considered as external to the two phase porous medium representing the tissue. In the delocalized sketch, the microvessels have been grouped together. In fact, at any geometrical point, the microvascular network is in contact with the tissue, viewed as a poroelastic medium. Fluid can be exchanged between the microvessels and the fluid phase of the tissue at a rate proportional to the vascular area and conductivity.

Measurable differences between tumors and normal tissues are listed in Jang et al. [2003]: they concern the geometry of microvessels, their higher volume fraction, their surface, their length, larger concentrations of growth factors, etc.;

- summing the balances of mass $d^k n^k/dt + n^k \operatorname{div} \mathbf{v}_k = \hat{\rho}^k$ over the two phases $k = \mathrm{s, w}$, yields the overall mass balance as $\operatorname{div}(n^s \mathbf{v}_s + n^w \mathbf{v}_w) = \sum_k \hat{\rho}^k/\rho^k = \Omega$. Extravasation may take place towards the interstitium or towards the fluid phase. Extravasation obeys Starling's law (15.9.4),

$$\Omega = \kappa \, (p_c^* - p_i),\qquad(24.8.2)$$

with κ [unit : 1/Pa/s] the coefficient of filtration between the capillaries and the pores also referred to as mass transfer coefficient,

$$\kappa = \underbrace{k_c \frac{S_c}{V}}_{\text{capillary filtration}} \quad ; \qquad(24.8.3)$$

- the scaled hydraulic permeability of the capillary wall k_c [unit : m/Pa/s] is considered to be one or two orders of magnitude larger in tumor tissues than in normal tissues,

$$\text{hydraulic permeability of capillaries } k_c = \begin{cases} 1 \times 10^{-12}\,\mathrm{m/Pa/s} & \text{for normal tissue,} \\ \downarrow & \\ 100 \times 10^{-12}\,\mathrm{m/Pa/s} & \text{for tumor tissue;} \end{cases} \qquad(24.8.4)$$

- p_i is the interstitium fluid pressure (IFP), $p_c^* = p_c - \omega\,(\pi_c - \pi_i) \geq 0$ the effective capillary pressure, with p_c the fluid pressure in the capillary, $\pi_c - \pi_i$ the oncotic pressure and $\omega \in [0, 1]$ the reflection coefficient. To simplify the analysis, p_c^* is replaced by p_c, which amounts to a vanishing reflection coefficient in Starling's law. Sarntinoranont et al. [2003] include lymphatics in this pressure differential as indicated in Section 24.9.2.

Boundary value problems and characteristic times

Two tests are simulated, Fig. 24.8.2:

- abrupt changes of the arterial pressure, actually of the microvascular pressure (MVP) p_c. The characteristic time response is the *transcapillary time*, here about 10 s,

$$t_{\mathrm{tr}} = \frac{1}{(k_c\,S_c/V)\,(\lambda + 2\,\mu)}; \qquad(24.8.5)$$

- abrupt cessation of the extravasation, i.e. $\Omega = 0$. The characteristic time response is the *percolation time*, characterizing fluid seepage through the interstitium, about $1000\,\mathrm{s}$,

$$t_{\mathrm{H}} = \frac{1}{(\pi^2\,k_{\mathrm{H}}/r_0^2)\,(\lambda + 2\,\mu)}\,, \tag{24.8.6}$$

with r_0 radius of the tumor and k_{H} [unit : $\mathrm{m}^2/\mathrm{Pa/s}$] the tissue hydraulic permeability.

Clearly, increase of tissue stiffness shortens the transient period of the IFP. Whether or not the steady state is modified by the stiffness can not be answered at this stage.

Alteration of blood pressure can be obtained by a vasoactive agent that contracts the vessels and thus increases the MVP and consequently induces extravasation[24.11]. Unfortunately transient increase of MVP leads also to transient extravasation and might actually be insufficient for macromolecular drugs to penetrate the tissue. Increase of the transcapillary time is thus sought. Repetitive changes of MVP is suggested as a more efficient alternative.

24.8.1.2 Constitutive equations and field equations

The poroelastic model is isotropic in both its transport and mechanical properties. The required parameters are the tissue hydraulic permeability k_{H} [unit : $\mathrm{m}^2/\mathrm{Pa/s}$] and the Lamé moduli of the drained tissue λ and μ [unit : Pa]. The species are all incompressible.

The following constitutive and field equations govern the problem,

$$
\begin{array}{lll}
\text{balance of mass} & \operatorname{div}\left(n^{\mathrm{s}}\,\mathbf{v}_{\mathrm{s}} + n^{\mathrm{w}}\,\mathbf{v}_{\mathrm{w}}\right) = \Omega = k_{\mathrm{c}}\,\dfrac{S_{\mathrm{c}}}{V}\,(p_{\mathrm{c}} - p_i), & \\[2mm]
\text{balance of momentum} & \operatorname{div}\boldsymbol{\sigma} = \mathbf{0}\,, & \\[2mm]
\text{Darcy's law} & n^{\mathrm{w}}\left(\mathbf{v}_{\mathrm{w}} - \mathbf{v}_{\mathrm{s}}\right) = -k_{\mathrm{H}}\,\boldsymbol{\nabla}p_i, & \\[2mm]
\text{elasticity equation} & \boldsymbol{\sigma} + p_i\,\mathbf{I} = \lambda\,e\,\mathbf{I} + 2\,\mu\,\boldsymbol{\epsilon}, & e \equiv \operatorname{tr}\boldsymbol{\epsilon}\,.
\end{array}
\tag{24.8.7}
$$

The n^k's are the volume fractions, the $\mathbf{v}_k$'s the velocities of the solid phase ($k = \mathrm{s}$) and fluid phase ($k = \mathrm{w}$) of the poroelastic medium, $\boldsymbol{\sigma}$ the total stress and $\boldsymbol{\epsilon} = \frac{1}{2}(\boldsymbol{\nabla}\mathbf{u}_{\mathrm{s}} + (\boldsymbol{\nabla}\mathbf{u}_{\mathrm{s}})^{\mathrm{T}})$ the strain of the skeleton. The net momentum supply due to the supplied mass and other potential sources which is obtained by summation over the contributions of the two phases as described in Section 4.5.2 has been assumed negligible.

24.8.1.3 Initial and boundary conditions

The analysis is restricted to radial motions about the center of the tumor with initial radius r_0. The fluid pressure is prescribed at the outer boundary of the tumor, while the fluid flow relative to solid vanishes at its center,

$$p_i(r = r_0, t) - p_{i0} = 0, \qquad \frac{dp_i}{dr}(r = 0, t) = 0\,. \tag{24.8.8}$$

The model responses to a jump in capillary pressure and to a sudden vanishing of the capillary conductivity are sought.

24.8.1.4 Control of the capillary pressure

Insertion of the poroelastic equation $(24.8.7)_4$ in the balance of momentum $(24.8.7)_2$ yields, as an extension of the linear isotropic elastic relation exposed in Section 11.2.6,

$$\boldsymbol{\nabla}p_i = (\lambda + \mu)\,\boldsymbol{\nabla}e + \mu\operatorname{div}\boldsymbol{\nabla}\mathbf{u} = (\lambda + 2\,\mu)\,\boldsymbol{\nabla}e - 2\,\mu\,\boldsymbol{\nabla}\wedge\boldsymbol{\omega}\,, \tag{24.8.9}$$

where $\boldsymbol{\omega} = \frac{1}{2}\,\boldsymbol{\nabla}\wedge\mathbf{u}$ and $\operatorname{div}\boldsymbol{\nabla} = \boldsymbol{\nabla}\operatorname{div} - \boldsymbol{\nabla}\wedge\boldsymbol{\nabla}\wedge$ according to (2.4.5).

1. The boundary value problem for radial motions

[24.11]In fact not all vasoconstrictors are efficient. Indeed extravasation should occur in tumors only, not at other locations. Angiotensin II seems to work but not epinephrine, Jang et al. [2003].

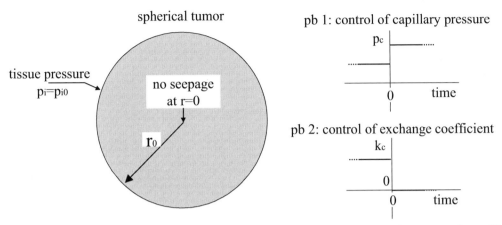

Fig. 24.8.2: A spherical tumor, with given pressure at its boundary, is subjected to either a jump in capillary pressure or to a sudden vanishing of the blood flow.

For radial motions, the rotational $\boldsymbol{\omega} = \mathbf{0}$ vanishes. Eqn (24.8.9) simplifies to

$$\boldsymbol{\nabla} p_i = (\lambda + 2\,\mu)\,\boldsymbol{\nabla} e \quad \Rightarrow \quad p_i(\mathbf{x}, t) - p_{i0} = (\lambda + 2\,\mu)\,e(\mathbf{x}, t)\,, \tag{24.8.10}$$

considering the volume change to vanish at the boundary of the tumor, namely $e(r = r_0, t) = 0$. Insertion of this relation and of Darcy's law in the balance of mass yields the reaction-diffusion equation obeyed by the volume change e

$$\frac{de}{dt} - k_{\text{H}}\,(\lambda + 2\,\mu)\,\mathrm{div}\,\boldsymbol{\nabla} e = \Omega, \quad \frac{de}{dt} - k_{\text{H}}\,(\lambda + 2\,\mu)\,\frac{1}{r^2}\frac{d}{dr}(r^2\frac{de}{dr}) = \Omega\,. \tag{24.8.11}$$

To establish this relation, the fact that the volume fractions sum up to one, $n^{\text{s}} + n^{\text{w}} = 1$, as well as the identity $\mathrm{div}\,\mathbf{v}_{\text{s}} = de/dt$ have been used.

The change of variable and unknown,

$$\hat{r} \equiv \frac{r}{r_0}, \quad E = \hat{r}\,e\,, \tag{24.8.12}$$

simplifies the expression of the Laplacian in spherical coordinates. The initial and boundary value problem is thus defined by the differential equation,

$$\frac{dE}{dt} + (\alpha^2\,E - \frac{d^2 E}{d\hat{r}^2})\,\gamma^2 = \hat{r}\,\alpha^2\,\gamma^2\,X, \quad X(t) \equiv \frac{p_{\text{c}}(t) - p_{i0}}{\lambda + 2\,\mu}\,, \tag{24.8.13}$$

where α [unit : 1] and γ [unit : $s^{-\frac{1}{2}}$] are coefficients whose interpretation will appear in the analysis,

$$\alpha^2 = r_0^2\,\frac{k_{\text{c}} S_{\text{c}}}{k_{\text{H}}\,V}, \quad \gamma^2 = \frac{k_{\text{H}}}{r_0^2}\,(\lambda + 2\,\mu)\,, \tag{24.8.14}$$

and by the boundary conditions (24.8.8) which become via (24.8.10),

$$E(\hat{r} = 1, t) = 0, \quad \frac{1}{\hat{r}^2}\Big(\hat{r}\,\frac{dE}{d\hat{r}} - E\Big)(\hat{r} = 0, t) = 0\,. \tag{24.8.15}$$

2. The nonuniform stationary solution

Before solving the differential equation, the steady state solution at time $t = 0^-$ is sought first. It satisfies the differential equation,

$$\alpha^2\,E^- - \frac{d^2 E^-}{d\hat{r}^2} = \hat{r}\,\alpha^2\,X^-, \quad X^- \equiv \frac{p_{\text{c}}(0^-) - p_{i0}}{\lambda + 2\,\mu}\,. \tag{24.8.16}$$

The solution of (24.8.16) with the boundary conditions (24.8.15) can be shown to be

$$E^-(\hat{r}) = X^-\,\Big(\hat{r} - \frac{\sinh \alpha \hat{r}}{\sinh \alpha}\Big)\,. \tag{24.8.17}$$

3. The time dependent part of the solution

The complete solution is the sum of the stationary solution and of a time dependent function $E^+(\hat{r}, t)$ which vanishes at time $t = 0^-$,

$$E(\hat{r}, t) = E^-(\hat{r}) + E^+(\hat{r}, t). \tag{24.8.18}$$

The function E^+ satisfies the homogeneous boundary conditions (24.8.15) and the differential equation,

$$\frac{dE^+}{dt} + (\alpha^2 E^+ - \frac{d^2 E^+}{d\hat{r}^2})\gamma^2 = \hat{r}\,\alpha^2\,\gamma^2\,X^+, \quad X^+(t) \equiv \frac{p_c(t) - p_c(0^-)}{\lambda + 2\,\mu}. \tag{24.8.19}$$

This equation is solved through the Laplace transform in time $E^+(\hat{r}, t) \to \overline{E^+}(\hat{r}, p)$. Let β be the dimensionless coefficient,

$$\beta^2 = \frac{p}{\gamma^2} + \alpha^2. \tag{24.8.20}$$

The Laplace transform of (24.8.19),

$$\beta^2\,\overline{E^+} - \frac{d^2\overline{E^+}}{d\hat{r}^2} = \hat{r}\,\alpha^2\,\overline{X^+}, \tag{24.8.21}$$

has the solution $\overline{E^+} = \overline{E^+}(\hat{r}, p)$,

$$\overline{E^+}(\hat{r}, p) = a(p)\,e^{\beta\,\hat{r}} + b(p)\,e^{-\beta\,\hat{r}} + \hat{r}\,\frac{\alpha^2}{\beta^2}\,\overline{X^+}. \tag{24.8.22}$$

Since $E^-(\hat{r} = 1) = 0$, the first boundary condition becomes $\overline{E^+}(\hat{r} = 1, p) = \overline{e}(\hat{r} = 1, p) = 0$ so that

$$a(p)\,e^{\beta} + b(p)\,e^{-\beta} + \frac{\alpha^2}{\beta^2}\,\overline{X^+} = 0. \tag{24.8.23}$$

The second boundary condition (24.8.15)$_2$ implies $a + b = 0$. Hence

$$a(p) = -b(p) = -\frac{\alpha^2}{2\,\beta^2}\,\frac{\overline{X^+}}{\sinh\beta}. \tag{24.8.24}$$

Therefore

$$\overline{E^+}(\hat{r}, p) = \frac{\alpha^2}{\beta^2}\,\overline{X^+}\left(\hat{r} - \frac{\sinh\beta\hat{r}}{\sinh\beta}\right). \tag{24.8.25}$$

4. Jump of the capillary pressure

To calculate the inverse Laplace transform, the function $X^+(t)$ should be specified. The capillary pressure undergoes a jump at time $t = 0^+$ and remains constant thereafter,

$$X(t) = X^- + X^+(t), \quad X^+(t) = (X^+ - X^-)\,\mathcal{H}(t). \tag{24.8.26}$$

A series expansion shows that $\beta = 0$ is not a pole of the right hand side of (24.8.25). Actually, the poles are $p = 0$ and the roots of $\sinh\beta = 0$, namely $\beta = i\,n\,\pi$, $n \neq 0$ integer, all of order 1, to which correspond by (24.8.20),

$$p_0 = 0, \quad p_n = -(\alpha^2 + n^2\pi^2)\gamma^2, \quad n \in [1, +\infty[. \tag{24.8.27}$$

Using $2\,\beta\,d\beta = dp/\gamma^2$ the inverse for $t > 0$ reads,

$$\begin{aligned}
E^+(\hat{r}, t) &= \frac{\alpha^2}{\beta^2}\,X^+(t)\left(\hat{r} - \frac{\sinh\beta\hat{r}}{\sinh\beta}\right)_{|p=0} + \sum_{n=1}^{\infty} e^{tp}\,\frac{\alpha^2}{\beta^2}\,\frac{X^+(t)}{p}\,\frac{\hat{r}\sinh\beta - \sinh\beta\hat{r}}{\dfrac{d}{dp}\sinh\beta}\bigg|_{p=p_n} \\
&= (X^+ - X^-)\left(\hat{r} - \frac{\sinh\alpha\hat{r}}{\sinh\alpha}\right) + 2\,\alpha^2\,(X^+ - X^-)\sum_{n=1}^{\infty} e^{tp_n}\,\frac{(-1)^n}{n\,\pi}\,\frac{\sin n\pi\hat{r}}{\alpha^2 + n^2\pi^2}.
\end{aligned} \tag{24.8.28}$$

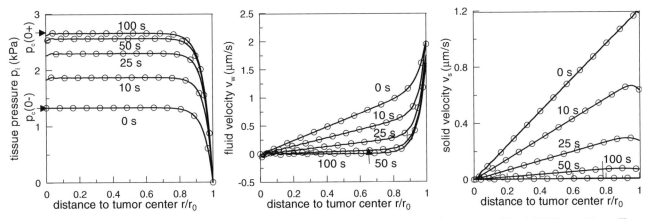

Fig. 24.8.3: The capillary pressure in the tumor is suddenly increased at $t = 0$ from $10\,\mathrm{mmHg} = 1.33\,\mathrm{kPa}$ to $20\,\mathrm{mmHg}$, while the tissue pressure is kept to a zero reference value at the boundary. The blood flow from the capillaries to the tumor increases the tissue pressure and volume and gives rise to outward motion and flow. The dependence with respect to the material properties of the transcapillary time t_{tr} necessary to reach a new steady state is commented in Section 24.8.1.6.

5. The complete solution in terms of volume change and pressure

With help of the relation,

$$\hat{r} - \frac{\sinh \alpha \hat{r}}{\sinh \alpha} = -2\,\alpha^2 \sum_{n=1}^{\infty} \frac{(-1)^n}{n\,\pi} \frac{\sin n\pi\hat{r}}{\alpha^2 + n^2\pi^2}\,, \tag{24.8.29}$$

the complete solution $p_i(\hat{r}, t) - p_{i0} = (\lambda + 2\,\mu)\,e(\hat{r}, t)$, with $e(\hat{r}, t) = E(\hat{r}, t)/\hat{r}$ may be cast in the format,

$$p_i(\hat{r}, t) - p_{i0} = (p_c(0^-) - p_{i0}) \left(1 - \frac{\sinh \alpha \hat{r}}{\hat{r}\,\sinh \alpha}\right) + 2\,\alpha^2\,(p_c(0^+) - p_c(0^-)) \sum_{n=1}^{\infty} (-1)^n \frac{e^{tp_n} - 1}{\alpha^2 + n^2\pi^2} \frac{\sin n\pi\hat{r}}{n\pi\hat{r}}\,. \tag{24.8.30}$$

6. The solid and fluid velocities

The radial component v_s of the velocity of the solid $\mathbf{v}_s$ can be calculated from $\operatorname{div} \mathbf{v}_s = r^{-2}d(r^2 v_s)/dr$ equal to de/dt, namely with $v_s(r = 0, t) = 0$,

$$\frac{v_s(\hat{r}, t)}{k_{\mathrm{H}}/r_0} = 2\,\alpha^2\,(p_c(0^+) - p_c(0^-)) \sum_{n=1}^{\infty} (-1)^n \frac{e^{tp_n}}{n^2\,\pi^2} \frac{1}{\hat{r}} \left(\cos n\pi\hat{r} - \frac{\sin n\pi\hat{r}}{n\pi\hat{r}}\right). \tag{24.8.31}$$

The tissue pressure is provided by (24.8.30). The fluid flux relative to solid $J_w = n^w\,(v_w - v_s)$ follows from Darcy's law $(24.8.7)_3$,

$$\begin{aligned}
\frac{J_w}{k_{\mathrm{H}}/r_0} = -r_0 \frac{dp_i}{dr} =\ & (p_c(0^-) - p_{i0}) \frac{\alpha}{\sinh \alpha} \frac{1}{\hat{r}} \left(\cosh \alpha\,\hat{r} - \frac{\sinh \alpha\,\hat{r}}{\alpha\,\hat{r}}\right) + \\
& - 2\,\alpha^2\,(p_c(0^+) - p_c(0^-)) \sum_{n=1}^{\infty} (-1)^n \frac{e^{tp_n} - 1}{\alpha^2 + n^2\pi^2} \frac{1}{\hat{r}} \left(\cos n\pi\hat{r} - \frac{\sin n\pi\hat{r}}{n\pi\hat{r}}\right),
\end{aligned} \tag{24.8.32}$$

and the fluid velocity $v_w = v_s + J_w/n^w$ finally results from (24.8.31) and (24.8.32).

The radial profiles of the pressure p_i and of the fluid and solid velocities at specific times are displayed in Fig. 24.8.3. The number of terms in the series defining the solutions should be larger than about 50 to reach convergence.

24.8.1.5 Control of the capillary conductivity

The tumor blood flow is suddenly stopped at time $t = 0$. Equivalently, the extravasation, or the capillary conductivity, may be set to vanish. The stationary solution remains identical to (24.8.17). On the other hand the differential equation is modified at time $t > 0$ due to the fact that $\Omega(t) = 0$ for $t > 0$.

1. The boundary value problem for radial motion

The coefficient β is redefined and a new coefficient δ is introduced as,

$$\beta^2 = \frac{p}{\gamma^2}, \quad \delta^2 = \frac{p}{\gamma^2} - \alpha^2. \tag{24.8.33}$$

The differential equation (24.8.13) becomes,

$$\frac{dE}{dt} - \frac{d^2E}{d\hat{r}^2}\gamma^2 = 0. \tag{24.8.34}$$

Using the decomposition (24.8.18), the function E^+ is seen to satisfy the boundary condition (24.8.15) and the equation,

$$\frac{dE^+}{dt} - \frac{d^2E^+}{d\hat{r}^2}\gamma^2 = \frac{d^2E^-}{d\hat{r}^2}\gamma^2\,\mathcal{H}(t). \tag{24.8.35}$$

2. The time dependent part of the solution

The Laplace transform in time of the latter equation,

$$\beta^2\,\overline{E^+} - \frac{d^2\overline{E^+}}{d\hat{r}^2} = \frac{1}{p}\frac{d^2E^-}{d\hat{r}^2} = -\frac{X^-}{p}\alpha^2\frac{\sinh\alpha\hat{r}}{\sinh\alpha}, \tag{24.8.36}$$

has the solution $\overline{E^+} = \overline{E^+}(\hat{r},p)$,

$$\overline{E^+}(\hat{r},p) = a(p)\,e^{\beta\hat{r}} + b(p)\,e^{-\beta\hat{r}} - \frac{X^-}{p}\frac{\alpha^2}{\delta^2}\frac{\sinh\alpha\hat{r}}{\sinh\alpha}. \tag{24.8.37}$$

The boundary conditions (24.8.15) define the constants and

$$\overline{E^+}(\hat{r},p) = \frac{\alpha^2}{\delta^2}\frac{X^-}{p}\left(\frac{\sinh\beta\hat{r}}{\sinh\beta} - \frac{\sinh\alpha\hat{r}}{\sinh\alpha}\right). \tag{24.8.38}$$

3. The solution

To calculate the inverse Laplace transform, the two parts of (24.8.38) are split. The pole $p_0 = 0$ of the first and second parts yields $-E^-(\hat{r})$. The poles of the first term different from 0 are the roots of $\sinh\beta = 0$, namely $\beta = i\,n\,\pi$, $n \neq 0$ integer, all of order 1, to which correspond, by $(24.8.33)_1$,

$$p_n^* = -n^2\pi^2\gamma^2, \quad n \in [1,+\infty[. \tag{24.8.39}$$

Using $2\,\beta\,d\beta = dp/\gamma^2$ the inverse reads,

$$\begin{aligned}
E^+(\hat{r},t) &= -E^-(\hat{r}) + X^-\sum_{n=1}^{\infty}e^{tp}\frac{\alpha^2}{\delta^2}\frac{\sinh\beta\hat{r}}{\cosh\beta}\frac{2\,\beta\,\gamma^2}{p}\bigg|_{p=p_n^*} \\
&= -E^-(\hat{r}) - 2\,\alpha^2\,X^-\sum_{n=1}^{\infty}e^{tp_n^*}\frac{(-1)^n}{n\,\pi}\frac{\sin n\pi\hat{r}}{\alpha^2 + n^2\pi^2}.
\end{aligned} \tag{24.8.40}$$

With help of (24.8.29), the complete solution $p_i(\hat{r},t) - p_{i0} = (\lambda + 2\,\mu)\,e(\hat{r},t)$, with $e(\hat{r},t) = E(\hat{r},t)/\hat{r}$ may be written in the alternative forms,

$$\begin{aligned}
p_i(\hat{r},t) - p_{i0} &= -2\,\alpha^2\left(p_c(0^-) - p_{i0}\right)\sum_{n=1}^{\infty}(-1)^n\frac{e^{tp_n^*}}{\alpha^2 + n^2\pi^2}\frac{\sin n\pi\hat{r}}{n\pi\hat{r}} \\
&= \left(p_c(0^-) - p_{i0}\right)\left(1 - \frac{\sinh\alpha\hat{r}}{\hat{r}\sinh\alpha}\right) - 2\,\alpha^2\left(p_c(0^-) - p_{i0}\right)\sum_{n=1}^{\infty}(-1)^n\frac{e^{tp_n^*} - 1}{\alpha^2 + n^2\pi^2}\frac{\sin n\pi\hat{r}}{n\pi\hat{r}}.
\end{aligned} \tag{24.8.41}$$

As by-products, we have the solid velocity,

$$\frac{v_s(\hat{r},t)}{k_H/r_0} = -2\,\alpha^2\left(p_c(0^-) - p_{i0}\right)\sum_{n=1}^{\infty}(-1)^n\frac{e^{tp_n^*}}{\alpha^2 + (n\,\pi)^2}\frac{1}{\hat{r}}\left(\cos n\pi\hat{r} - \frac{\sin n\pi\hat{r}}{n\pi\hat{r}}\right), \tag{24.8.42}$$

the fluid flux relative to solid $J_w = n^w\left(v_w - v_s\right)$,

$$\begin{aligned}
\frac{J_w}{k_H/r_0} = -r_0\frac{dp_i}{dr} &= \left(p_c(0^-) - p_{i0}\right)\frac{\alpha}{\sinh\alpha}\frac{1}{\hat{r}}\left(\cosh\alpha\hat{r} - \frac{\sinh\alpha\hat{r}}{\alpha\hat{r}}\right) \\
&\quad + 2\,\alpha^2\left(p_c(0^-) - p_{i0}\right)\sum_{n=1}^{\infty}(-1)^n\frac{e^{tp_n^*} - 1}{\alpha^2 + n^2\pi^2}\frac{1}{\hat{r}}\left(\cos n\pi\hat{r} - \frac{\sin n\pi\hat{r}}{n\pi\hat{r}}\right),
\end{aligned} \tag{24.8.43}$$

and the fluid velocity $v_w = v_s + J_w/n^w$. The radial profiles of the pressure p_i and of the fluid and solid velocities at specific times are displayed in Fig. 24.8.4.

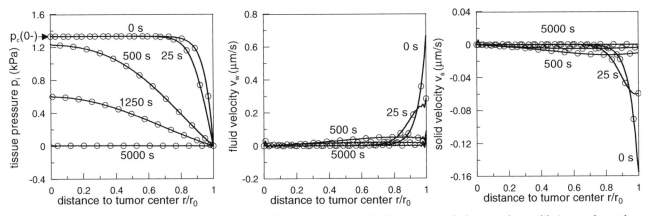

Fig. 24.8.4: The vascular conductivity is suddenly stopped at $t = 0$. The time needed to reach equilibrium, where the pressure is homogeneous and thus flow vanishes, is governed by the diffusion properties of the tumor.

24.8.1.6 Constitutive parameters and results

The material parameters used in the simulations are indicated in Table 24.8.1. The elastic moduli correspond to rat cerebellum. They are quite large with respect to the subsequent analyses in Roose et al. [2003], and Sarntinoranont et al. [2003]. The number of terms in the series defining the solutions should be larger than about 80 to reach convergence.

Table 24.8.1: Parameters used in the simulations extracted from Netti et al. [1995].

Entity	Normal Tissue	Tumor Tissue	Unit
Fluid volume fraction in tissue n^w	-	0.2	
Elastic modulus $\lambda + 2\mu$	-	95.25	kPa
Tissue hydraulic permeability k_H	-	3.1×10^{-14}	$m^2/Pa/s$
Hydraulic permeability of capillaries k_c	0.27×10^{-11}	2.7×10^{-11}	m/Pa/s
Vascular area per unit tissue volume S_c/V	7×10^3	20×10^3	m^{-1}
Tumor diameter $2r_0$		1×10^{-2}	m

The two characteristic times involved in these initial and boundary value problems, namely t_{tr} which controls the exchange, or transfer, of the fluid between capillaries and tissue and t_H which controls the seepage of the fluid within the tissue, are defined by (24.8.5) and (24.8.6), respectively. For the standard values shown in Table 24.8.1, the transfer time $t_{tr} = 20$ s is more than an order of magnitude smaller than the diffusion time $t_H \sim 860$ s.

The characteristic transfer time is of course independent of the diameter of the tumor while the diffusion process is proportional to its square. Clearly increase of the tissue stiffness accelerates the transient processes. So does an increase of capillary conductivity. Therefore the two processes analyzed are faster in tumor tissues.

The coefficients α and γ can be interpreted in terms of these characteristic times, namely

$$\gamma^2 = \frac{1}{\pi^2 t_H}, \quad \alpha^2 = \frac{\pi^2 t_H}{t_{tr}}. \tag{24.8.44}$$

Using the relations, Gradshteyn and Rhyzik [1980], p. 23,

$$\frac{1}{\pi^2} \sum_{n=1}^{\infty} \frac{1}{(\alpha/\pi)^2 + n^2} = \frac{1}{2\alpha^2}\left(\frac{\alpha \cosh \alpha}{\sinh \alpha} - 1\right), \quad \frac{1}{\pi^2} \sum_{n=1}^{\infty} \frac{1}{n^2} = \frac{1}{6}, \tag{24.8.45}$$

the initial solid velocity at the boundary of the tumor can be shown to depend only on the mass exchange properties for the pressure controlled process, Eqn (24.8.31),

$$v_s(\hat{r} = 1, t = 0) = \frac{1}{3} r_0 k_c \frac{S_c}{V}\left(p_c(0^+) - p_c(0^-)\right), \tag{24.8.46}$$

while both mass exchange and seepage properties come into picture for the capillary conductivity controlled process, Eqn (24.8.42),

$$v_s(\hat{r} = 1, t = 0) = -\left(\frac{\alpha \cosh \alpha}{\sinh \alpha} - 1\right)\frac{k_H}{r_0}\left(p_c(0^-) - p_{i0}\right). \tag{24.8.47}$$

24.8.2 A three phase poroelastic tissue with internal mass exchanges

The model presented in Section 24.8.1 involves a solid skeleton circulated by a pore fluid and has two phases. The capillaries were not part of the system. The latter was *open* to mass exchanges between the fluid of the capillaries and the pore fluid. By contrast, the model below *includes* the capillaries so that the mass exchanges are internal to the system which is *closed* with respect to external mass interactions.

24.8.2.1 Mass exchange between pores and capillaries

Let us consider a porous medium endowed with a single porosity. Further let us assume that both the solid and the fluid are incompressible. Then the mass conservation indicates that the volume change of the solid skeleton (or porous medium) should be balanced by the fluid entering/leaving the medium, namely $\operatorname{div} \mathbf{v}_s + \operatorname{div} \mathbf{J}_w = 0$, where $\mathbf{v}_s$ is the solid velocity and $\mathbf{J}_w = n^w (\mathbf{v}_w - \mathbf{v}_s)$ the flux of water relative to the solid skeleton.

Now assume that another type of porosity exists, namely capillaries. The volume of mixture V contains solid, pores and capillaries in volume fractions n^s, n^w and n^c, respectively,

$$n^s + n^w + n^c = 1 \,. \tag{24.8.48}$$

Balance of mass of the above porous medium should account for the flux across the interface between the pore fluid and the second type of porosity. The external boundary S of the volume V includes the parts S_s, S_w and S_c pertaining to solid, pores and capillaries. The flux across the boundary of unit outward normal $\mathbf{n}$ may be decomposed in several ways,

$$
\begin{aligned}
\int_S \mathbf{v} \cdot \mathbf{n} \, dS &= \int_{S_s} \mathbf{v}_s \cdot \mathbf{n} \, dS + \int_{S_w} \mathbf{v}_w \cdot \mathbf{n} \, dS + \int_{S_c} \mathbf{v}_c \cdot \mathbf{n} \, dS \\
&= \int_S n^s \mathbf{v}_s \cdot \mathbf{n} + n^w \mathbf{v}_w \cdot \mathbf{n} + n^c \mathbf{v}_c \cdot \mathbf{n} \, dS \\
&= \int_S \mathbf{v}_s \cdot \mathbf{n} + n^w (\mathbf{v}_w - \mathbf{v}_s) \cdot \mathbf{n} + n^c (\mathbf{v}_c - \mathbf{v}_s) \cdot \mathbf{n} \, dS \\
&= \int_V \operatorname{div} \mathbf{v}_s + \operatorname{div} \mathbf{J}_w \, dV + \int_{S_c} (\mathbf{v}_c - \mathbf{v}_s) \cdot \mathbf{n} \, dS \\
&\simeq (\operatorname{div} \mathbf{v}_s + \operatorname{div} \mathbf{J}_w) \, V + (\mathbf{v}_c - \mathbf{v}_s) \cdot \mathbf{n} \, S_c \,.
\end{aligned}
\tag{24.8.49}
$$

The passage from the first to the second line follows from the fact that the surface fractions S_k/S are assumed to be equal to the volume fractions $n^k = V_k/V$, $k = s, w, c$. The line before last is a consequence of Green's theorem. The last line is an approximation for a small volume V.

The outward relative flux $(\mathbf{v}_c - \mathbf{v}_s) \cdot \mathbf{n}$ is assumed to be proportional to the fluid pressure differential between the porous medium (p_i) and the capillaries (p_c),

$$(\mathbf{v}_c - \mathbf{v}_s) \cdot \mathbf{n} = k_c (p_i - p_c) \,, \tag{24.8.50}$$

with k_c [unit : m/Pa/s] permeability of the capillaries. Since the system is closed with respect to mass exchanges, the surface integral (24.8.49) vanishes, and therefore,

$$\operatorname{div} \mathbf{v}_s + \operatorname{div} \mathbf{J}_w + \kappa (p_i - p_c) = 0 \,, \tag{24.8.51}$$

with $\kappa = k_c S_c / V$ [unit : 1/Pa/s] the coefficient of filtration between the capillaries and the pores. Thus the mass balance $(24.8.7)_1$ is recovered but the interpretations in a two phase context and in a three phase context are different.

24.8.2.2 Poroelastic constitutive equations with exchanges and field equations

The constitutive equations and field equations including mass transfer can be encapsulated in the format,

$$
\begin{aligned}
&\text{balance of mass} &&\operatorname{div} \mathbf{v}_s + \operatorname{div} \mathbf{J}_w = \kappa (p_c - p_i), \\
&\text{balance of momentum} &&\operatorname{div} \boldsymbol{\sigma} = \mathbf{0}, \\
&\text{Darcy's law} &&\mathbf{J}_w = n^w (\mathbf{v}_w - \mathbf{v}_s) = -k_H \boldsymbol{\nabla} p_i, \\
&\text{elastic equation} &&\boldsymbol{\sigma} + p_i \mathbf{I} = \lambda e \mathbf{I} + 2 \mu \boldsymbol{\epsilon}, \quad e \equiv \operatorname{tr} \boldsymbol{\epsilon},
\end{aligned}
\tag{24.8.52}
$$

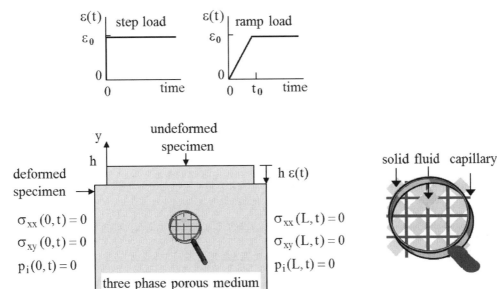

Fig. 24.8.5: A tissue specimen is submitted to unconfined compression in a plane (x, y). Platens ensure the strains, stresses and pressures to remain independent of the coordinate along the loading direction. The tissue is viewed as a three phase mixture and fluid can transfer from the pores to the capillaries and conversely.

with $\boldsymbol{\sigma}$ the total stress and $\boldsymbol{\epsilon} = \frac{1}{2}(\boldsymbol{\nabla}\mathbf{u}_s + (\boldsymbol{\nabla}\mathbf{u}_s)^{\mathrm{T}})$ the strain of the skeleton.

The poroelastic model is isotropic in both its transport and mechanical properties. Besides the filtration coefficient, the required parameters are the tissue hydraulic permeability k_{H} [unit : m^2/Pa/s], and the Lamé moduli of the drained tissue λ and μ [unit : Pa]. All species are incompressible.

24.8.2.3   Initial and boundary value problem

Leiderman et al. [2006] consider plane strain unconfined compression of a rectangular parallelepiped in the plane (x, y) as sketched in Fig. 24.8.5. The boundary conditions and loading device are such that the strains and stresses are independent of the y-coordinate, and therefore the specimen does not undergo shear strains, nor shear stresses since it is mechanically isotropic:

- the tissue is subjected in the plane (x, y) to a uniform axial strain along the axis y, either through a step loading or through a ramp loading;
- the lateral sides of the sample, $x = 0$ and $x = L = 2\,l$, are maintained at zero pore pressure, $p_i(x = 0, t) = 0$, $p_i(x = L, t) = 0$, and zero traction, $\sigma_{xx}(x = 0, t) = 0$, $\sigma_{xx}(x = L, t) = 0$;
- the lateral displacement $u_{sx}(x = l, t) = 0$ vanishes along the symmetry line $x = l$ while the axial displacement $u_{sy}(x, y = 0, t) = 0$ vanishes along the plane $y = 0$;
- the pressure in the capillaries is taken as a reference, namely $p_c(x, t) = 0$, at any time $t \geq 0$;
- the initial pressure in the tissue is equal to the pressure in the capillaries, $p_i(x, t = 0) = 0$.

24.8.2.4   Step loading

The axial strain is suddenly modified,

$$\epsilon_{yy}(\cancel{x}, \cancel{y}, t) = \epsilon_{yy}(t) = -\epsilon_0\,\mathcal{H}(t) < 0\,. \tag{24.8.53}$$

Inserting Darcy's law $(24.8.52)_3$ into the mass balance $(24.8.52)_1$ and assuming the volume fractions to remain unchanged yields a partial differential equation in space and time,

$$\frac{\partial v_{sx}}{\partial x} + \frac{\partial \epsilon_{yy}}{\partial t} - k_{\mathrm{H}}\,\frac{\partial^2 p_i}{\partial x^2} + \kappa\,p_i = 0\,, \quad t > 0\,. \tag{24.8.54}$$

Inserting the elastic equation $(24.8.52)_4$ into the momentum balance $(24.8.52)_2$ and accounting for the spatial uniformity of the strain ϵ_{yy} yields a space differential equation,

$$\frac{\partial p_i}{\partial x} - H \frac{\partial^2 u_{sx}}{\partial x^2} = 0, \tag{24.8.55}$$

with $H = \lambda + 2\mu$ the aggregate drained modulus. This relation can be integrated from $x = 0$, using the boundary conditions $p_i(x = 0, t) = 0$ and $\sigma_{xx}(x = 0, t) = 0$ in $(24.8.52)_4$,

$$p_i(x, t) - H \frac{\partial u_{sx}}{\partial x}(x, t) = \lambda \epsilon_{yy}(t). \tag{24.8.56}$$

The relation (24.8.56) is used to infer the value of the pressure p_i at time $t = 0^+$. Indeed under a step load the material behaves incompressibly[24.12] so that, with help of the elastic equation $(24.8.52)_4$,

$$\epsilon_{xx}(x, 0^+) = -\epsilon_{yy}(x, 0^+) = \epsilon_0; \quad p_i(x, 0^+) = 2\mu \epsilon_0; \quad \sigma_{xx}(x, 0^+) = 0, \quad \sigma_{yy}(x, 0^+) = -4\mu \epsilon_0. \tag{24.8.57}$$

Insertion of the time derivative of (24.8.56) into (24.8.54) yields a partial differential equation in space and time for the pressure,

$$\frac{\partial p_i}{\partial t} - \hat{k} \frac{\partial^2 p_i}{\partial x^2} + \hat{\kappa} p_i = -2\mu \frac{\partial \epsilon_{yy}}{\partial t}(t), \tag{24.8.58}$$

where

$$\hat{\kappa} = H\kappa, \quad \hat{k} = H k_{\mathrm{H}}. \tag{24.8.59}$$

It is convenient to introduce the function,

$$q(x, t) = e^{\hat{\kappa} t} p_i(x, t). \tag{24.8.60}$$

Then the initial and boundary value problem to be solved for $q(x, t)$ simplifies to,

$$\frac{\partial q}{\partial t} - \hat{k} \frac{\partial^2 q}{\partial x^2} = 0, \ x \in]0, L[, \ t > 0, \quad q(x, 0^+) = 2\mu \epsilon_0, \quad q(0, t) = q(L, t) = 0. \tag{24.8.61}$$

The solution of the Laplace transform $q(x, t) \to \bar{q}(x, s)$ takes the form, with $\alpha = \sqrt{s/\hat{k}}$,

$$\bar{q}(x, s) = 2\mu \epsilon_0 \frac{\sinh(\alpha(x - L)) - \sinh(\alpha x) + \sinh(\alpha L)}{s \sinh(\alpha L)}. \tag{24.8.62}$$

The denominator admits only single roots, namely $\sqrt{s_n/\hat{k}} L = i n \pi$, or $s_n = -n^2 \pi^2 \hat{k}/L^2$, $n \geq 1$. The contribution of the root s_n reads,

$$2\mu \epsilon_0 \frac{(\cosh(i n \pi) - 1) \sinh(i n \pi \frac{x}{L})}{\frac{1}{2} i n \pi \cosh(i n \pi)} \exp(t s_n). \tag{24.8.63}$$

The pressure can be cast in the format,

$$\frac{p_i(x, t)}{\mu \epsilon_0} = 4 \sum_{n=0}^{\infty} \frac{\sin(X_n x/l)}{X_n} \exp(-\frac{t}{t_n}), \tag{24.8.64}$$

with $l = L/2$, $X_n = (n + 1/2)\pi$ and

$$\frac{1}{t_n} = X_n^2 \frac{\hat{k}}{l^2} + \hat{\kappa}. \tag{24.8.65}$$

The strain $\epsilon_{xx}(x, t) = \partial u_{sx}/\partial x = (p_i + \lambda \epsilon_0)/H$ for $t > 0$ deduces from (24.8.56).

Characteristic times

Let us recall the units of the entities involved in the characteristic times, namely
- stiffness H [unit : Pa],
- hydraulic permeability k_{H} [unit : m^2/Pa/s],

[24.12] The reader may consider also Table 7.5.1 which provides the instantaneous response of a porous cylinder loaded by frictionless platens with traction free lateral boundary.

- diffusion coefficient D_C [unit : $\mathrm{m^2/s}$],
- mass transfer coefficient κ [unit : $1/\mathrm{Pa/s}$],
- drainage length l [unit : m].

It is worth recalling the expressions involved in the characteristic times associated with seepage, diffusion, mass transfer and parallel seepage:
- seepage : $t_H = l^2/(H\,k_H)$ involves stiffness, permeability and seepage length;
- diffusion : $t_C = l^2/D_C$ involves the diffusion coefficient and diffusion length but not the stiffness;
- mass transfer : $t_{tr} = 1/(H\,\kappa)$ involves stiffness and transfer coefficient but not length;
- seepage *and* mass transfer : $1/t_{H+tr} \sim \pi^2/(4\,t_H) + 1/t_{tr}$.

Here seepage and mass transfer take place *in parallel*, the time response is of the form $\exp(-t/t_{H+tr})$, and it is controlled by the faster of the two processes. On the other hand, for processes involving *sequential* seepage and diffusion of a chemical, the response is the sum of terms of the form $\exp(-t/t_H)$ and $\exp(-t/t_C)$: it is controlled by the slower of the two processes. An example is analyzed in Section 13.13.

24.8.2.5 Ramp loading

The axial strain is now taken in the format,

$$\epsilon_{yy}(t) = -\dot\epsilon_0 \left(t\,\mathcal{H}(t) - (t-t_0)\,\mathcal{H}(t-t_0) \right) < 0 , \tag{24.8.66}$$

with $\dot\epsilon_0$ the constant loading rate, and $t_0 > 0$ the time interval during which the load is increased.

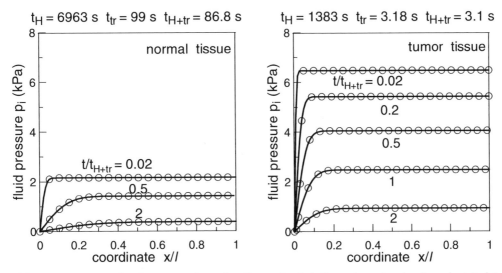

Fig. 24.8.6: Plane strain unconfined compression of a tissue, loaded through a step load as depicted in Fig. 24.8.5. Specimen width $2\,l = 1 \times 10^{-2}$ m; axial strain $\epsilon_0 = 0.10$. In comparison with a normal tissue the pore pressure is higher in the tumor tissue essentially due to the higher shear modulus and the transient process is faster.

Table 24.8.2: Representative parameters used in the simulations, Leiderman et al. [2006].

Entity	Normal Tissue	Tumor Tissue	Unit
Drained elastic modulus λ	539	517	kPa
Elastic modulus μ	11	33	kPa
Tissue hydraulic permeability k_H	0.64×10^{-14}	3.1×10^{-14}	$\mathrm{m^2/Pa/s}$
Mass transfer coefficient κ	1.8×10^{-8}	54×10^{-8}	$1/\mathrm{Pa/s}$

If $p_{i0}(x,t)$ is the response to a unit step load $\epsilon_{yy}(t) = -\epsilon_0\,\mathcal{H}(t)$, and $p_i(x,t)$ the response to the load $\epsilon_{yy}(t)$, then the Laplace transforms of the differential equation (24.8.58) in these two loading cases write,

$$\left(s - \hat{k}\frac{\partial^2}{\partial x^2} + \hat{\kappa} \right) \overline{p_{i0}}(x,s) = 2\mu\,\epsilon_0, \quad \left(s - \hat{k}\frac{\partial^2}{\partial x^2} + \hat{\kappa} \right) \overline{p_i}(x,s) = -2\mu\,\overline{\frac{d\epsilon_{yy}}{dt}}(s) , \tag{24.8.67}$$

accounting for the initial condition $p_{i0}(x, 0^-) = 0$, $p_i(x, 0^-) = 0$, and the fact that the Laplace transform of the Dirac delta function is equal to 1. Therefore,

$$\overline{p_i}(x, s) = \overline{p_{i0}}(x, s) \frac{1}{-\epsilon_0} \overline{\frac{d\epsilon_{yy}}{dt}}(s) \,, \tag{24.8.68}$$

and consequently the solution in the time domain is a convolution integral,

$$p_i(x, t) = \int_0^t p_{i0}(x, t - \tau) \frac{1}{-\epsilon_0} \frac{d\epsilon_{yy}(\tau)}{d\tau} \, d\tau \,, \tag{24.8.69}$$

which for ramp loading specializes to,

$$p_i(x, t) = \int_0^{\min(t, t_0)} p_{i0}(x, t - \tau) \frac{d\tau}{t_0} \,, \tag{24.8.70}$$

where now $\epsilon_0 = \dot{\epsilon}_0 \, t_0$. Thus the pressure due to ramp loading takes the format,

$$\frac{p_i(x, t)}{\mu \, \epsilon_0} = 4 \sum_{n=0}^{\infty} \frac{t_n}{t_0} \frac{\sin(X_n \, x/l)}{X_n} \begin{cases} 1 - \exp(-t/t_n), & t \leq t_0, \\ \exp(-(t - t_0)/t_n) - \exp(-t/t_n), & t \geq t_0. \end{cases} \tag{24.8.71}$$

24.8.2.6   Constitutive parameters and results

The transport parameters used in the simulations shown in Fig. 24.8.6 are listed in Table 24.8.2. The elastic coefficients are representative of breast tissue.

24.8.3   Infusion length in a three phase poroelastic tissue

Along with Swartz et al. [1999], we now take a close look at the neighborhood of a capillary, Fig. 24.8.7. Then the tissue appears of infinite size. In the context of this section, the capillaries are actually layers parallel to the out-of-plane axis and extravasation takes place along the x-axis. On the other hand, at variance with the previous section, the pressure of the capillary is controlled in the form $p_c(t) = p_c \, \mathcal{H}(t)$.

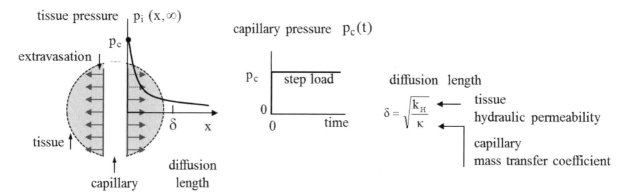

Fig. 24.8.7: Zoom in a neighborhood of a capillary inside the tissue specimen submitted to unconfined compression in a plane (x, y) and shown in Fig. 24.8.5. The capillary pressure is controlled so as to define the infusion length.

Equilibrium takes place instantaneously at time $t = 0^+$ at the capillary wall, $p_i(x = 0, 0^+) = p_c$. In view of the developments in Section 24.8.2, the initial and boundary value problem to be solved for $q(x, t) = e^{\hat{\kappa} t} \, p_i(x, t)$ expresses through the relations,

$$\frac{\partial q}{\partial t} - \hat{k} \frac{\partial^2 q}{\partial x^2} = 0, \ x > 0, \ t > 0, \quad q(x, 0) = 0, \quad q(0, t > 0) = p_c \, e^{\hat{\kappa} t}, \quad q(\infty, t) = 0 \,. \tag{24.8.72}$$

The Laplace transform $q(x,t) \to q(x,s)$ has solution,

$$\overline{q}(x,s) = p_c \frac{\exp(-x\sqrt{s/\hat{k}})}{s - \hat{\kappa}}. \tag{24.8.73}$$

Its inverse is given in Carslaw and Jaeger [1959], p. 495[24.13],

$$q(x,t) = \frac{p_c}{2} e^{\hat{\kappa} t} \left(\exp(-\frac{x}{\delta}) \operatorname{erfc}\left(\frac{x}{2\sqrt{\hat{k}\,t}} - \sqrt{\hat{\kappa}\,t}\right) + \exp(\frac{x}{\delta}) \operatorname{erfc}\left(\frac{x}{2\sqrt{\hat{k}\,t}} + \sqrt{\hat{\kappa}\,t}\right) \right), \tag{24.8.74}$$

where δ,

$$\delta = \sqrt{\frac{k_H}{\kappa}}, \tag{24.8.75}$$

has dimension of a length. Thus the pressure in the tissue,

$$\frac{p_i(x,t)}{p_c} = \frac{1}{2} \left(\exp(-\frac{x}{\delta}) \operatorname{erfc}\left(\frac{x}{2\sqrt{\hat{k}\,t}} - \sqrt{\hat{\kappa}\,t}\right) + \exp(\frac{x}{\delta}) \operatorname{erfc}\left(\frac{x}{2\sqrt{\hat{k}\,t}} + \sqrt{\hat{\kappa}\,t}\right) \right), \tag{24.8.76}$$

has a transient behavior controlled by both the characteristic times $x^2/\hat{k}$ and $1/\hat{\kappa}$.

Accounting for the properties of the error and complementary error functions recorded in Section 15.3.2.3, its steady state profile decreases exponentially from the capillary,

$$\frac{p_i(x,\infty)}{p_c} = \exp\left(-\frac{x}{\delta}\right). \tag{24.8.77}$$

Therefore the scalar δ can be interpreted as a penetration, or infusion, length. If Darcy's seepage time is defined as $t_H = \delta^2/(H\,k_H)$, then it is equal to the transfer time $t_{tr} = 1/(H\,\kappa)$.

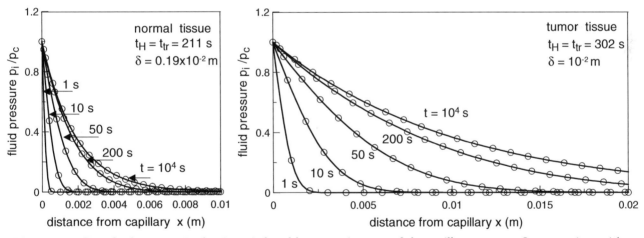

Fig. 24.8.8: Pore fluid pressure in the tissue induced by a step increase of the capillary pressure. In comparison with a normal tissue, the infusion length δ is larger and the transient process is somehow slower in a tumor tissue.

Fig. 24.8.8 shows that the infusion length is larger in tumor tissues. Indeed, Swartz et al. [1999] find the infusion length to be about 2 mm for normal tissue of mouse tail and 10 mm for edematous tissue. The measurements use labelled molecules: their transport is due to convection by the fluid as long as their diffusion can be considered negligible wrt their convection: the Péclet number $P_e = v\,x/D$ is larger than one, D being the diffusion coefficient and v a representative velocity of the molecules at position x. Swelling of tissue was associated with an increase of the hydraulic permeability, while the mass transfer coefficient changes little as indicated by Table 24.8.3.

[24.13]Exercise 15.2 constructs the inverse transform.

Table 24.8.3: Representative parameters of skin of mouse tails, Swartz et al. [1999].

Entity	Normal Tissue	Oedematous Tissue	Unit
Drained elastic modulus $\lambda + 2\mu$	14.7	–	kPa
Tissue hydraulic permeability k_H	1.13×10^{-12}	22.6×10^{-12}	m^2/Pa/s
Mass transfer coefficient κ	3.23×10^{-7}	2.25×10^{-7}	1/Pa/s

24.9 Poroelastic mixture models with a growth strain

In analyses of tumor spheroids, nutrients are provided at the interface between the tumor and the bulk tissue. Nutrients obey a transport equation involving diffusion, convection and chemical reaction. As they penetrate the spheroids, nutrients are consumed by cells. Most often, cell proliferation is controlled by the supply of one nutrient. In Roose et al. [2003], cell proliferation is made sensitive to mechanics. While oxygen supply is theoretically the rate limiting quantity, the analysis intends to suggest that the slowdown of growth may be due to a confined environment and the production of smaller daughter cells. On the other hand, while they do not address mechanical aspects, Casciari et al. [1992] account for a more complete cellular metabolism including oxygen and glucose. A qualitative illustration of these aspects is displayed in Fig. 24.9.1.

The examples below address the growth of tumor spheroids within a sound host tissue or an ersatz. The boundary value problems consider radial displacements only. If the issue is similar to a phase boundary transition in which a line moves over the matter, an Eulerian approach is appropriate. On the other hand, if the tumor only grows and pushes the tumor host, a Lagrangian approach is adequate. Even if actual growth strains may be large, the analyses below are restricted to a small strain approach.

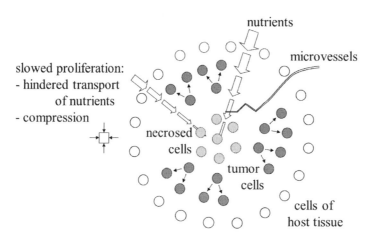

Fig. 24.9.1: Tumor cells rely on supply of nutrients which are transported mainly by convection from the host tissue towards the tumor interior. Hindered transport due to damaged vasculature and lymph and high blood pressure within the tumor as well as mechanical conditions slow proliferation of tumor cells and lead to necrosis at the center of the spheroids.

24.9.1 A two phase model including a growth strain and nutrients

The model

Roose et al. [2003] study the growth of a spheroid in an agarose gel. They consider a two phase poroelastic mixture:
- the solid phase contains the cells and the extracellular matrix gathered in a single species;
- the fluid phase contains the extracellular fluid and a nutrient whose concentration is calculated.

The model accounts for the following ingredients:
- a balance of mass of each phase, allowing for a mass supply rate, $\hat{\rho}^s$ for the solid phase, $\hat{\rho}^w$ for the fluid phase. The supply rates sum to zero, as they represent exchanges between the phases and do not include a contribution from an external source;

- seepage of water through the solid skeleton which is controlled by Darcy's law in terms of the fluid pressure p_i and hydraulic permeability of the interstitium;
- an effective stress that is a function of the elastic strain, defined as the total strain minus the growth strain in the solid phase:

$$\boldsymbol{\sigma} + p_i\,\mathbf{I} = \mathbb{E} : (\boldsymbol{\epsilon} - \boldsymbol{\epsilon}^{\mathrm{g}})\,. \tag{24.9.1}$$

For density preserving growth, the rate of mass supply provides the growth strain by Eqn (21.3.54),

$$\boldsymbol{\epsilon}^{\mathrm{g}} = \mathrm{Ln}\,\frac{M_{\mathrm{s}}(t)}{M_{\mathrm{s}}(0)}\,\mathbf{I} = \int_0^t \frac{\hat{\rho}^{\mathrm{s}}}{\rho^{\mathrm{s}}}\,d\tau\,\mathbf{I}\,; \tag{24.9.2}$$

If in addition the tissue is elastically incompressible, so that the apparent mass density ρ^{s} is constant, the above mass ratio $M_{\mathrm{s}}(t)/M_{\mathrm{s}}(0)$ is also equal to the volume ratio $V_{\mathrm{s}}(t)/V_{\mathrm{s}}(0)$;
- an equation, namely Eqn (15.4.4), for the transport of oxygen in the tissue which involves diffusion, convection by the fluid and a release term in the Michaelis-Menten form: oxygen is next used by cells as described below;
- the actual rate of mass supply to the solid,

$$\frac{\hat{\rho}^{\mathrm{s}}}{\rho_{\mathrm{s}}} = n^{\mathrm{s}}\,(1 - n^{\mathrm{s}})\,F\,\frac{\lambda_{\mathrm{s}}\,c_{O_2}}{c_h + c_{O_2}} - d_{\mathrm{s}}\,n^{\mathrm{s}} \tag{24.9.3}$$

accounts for a mass creation, which depends

- on the concentration c_{O_2} of nutrient (oxygen) in a Michaelis-Menten format,
- on the mechanical environment through a stress-dependent term F,
- on the volume fraction of the fluid $1 - n^{\mathrm{s}}$: mass supply relies on both the driving engine, namely oxygen, but also on the carrier, namely the fluid;
- on the solid volume fraction n^{s}: the supply can be effective only if the amount of cells available is sufficient: in other words, the number of cells is proportional to the mass of the solid;

and for the death of tumor cells.

Consequently, the growth of tumor tissue is controlled by oxygen supply. The coefficients involved include
- the maximum proliferation rate assuming full availability of oxygen $\lambda_{\mathrm{s}} = 1.45 - 5/\text{day}$;
- the concentration corresponding to half of the maximum proliferation rate $c_h = 8.3 \times 10^{-3}\,\text{mole/m}^3$;
- the rate of cell death $d_{\mathrm{s}} \geq 0$ [unit : $1/\text{s}$ or $1/\text{day}$].

Sections 20.6.3 and 20.6.4 present similar approaches, in the context of tissue engineering, to the rate of proliferation and to the oxygen transport, respectively.

This rate of mass supply is not phrased within a thermodynamic context, that is, it does not account for an out-of-balance of chemical potentials between the solid and the surroundings.

The coefficient F is motivated by the experiments of Helmlinger et al. [1997]a who observed that the *in vitro* growth of spheroids was inhibited by the gel density or confinement. Indeed, consider a sphere in a medium of infinite extent, of bulk modulus K and shear modulus μ, and remotely submitted to the isotropic pressure p_∞. The stress at the sphere surface is given by (24.7.33). Now, let the sphere of initial radius a_0 adopt a new radius a. The ensuing stress at the sphere surface is given by (24.7.34). The radial compressive stress on the sphere surface equivalent to these two loadings is obtained by superposition,

$$-\left(1 + \frac{4\,\mu}{3K}\right) p_\infty - 4\,\mu\,\left(\frac{a}{a_0} - 1\right)\,. \tag{24.9.4}$$

Let us consider an unconfined environment $p_\infty = 0$. A cell of radius a_0 divides into two daughter cells of same radius a_0 as the mother cell. The resulting spherical equivalent cell has radius a,

$$V_0 = \frac{4}{3}\,\pi\,a_0^3, \quad V = \frac{4}{3}\,\pi\,a^3, \quad V = 2\,V_0 \Rightarrow \frac{a}{a_0} = 2^{1/3}\,. \tag{24.9.5}$$

As a working assumption, the radial stress in the sphere is assumed to be independent of the loading conditions at infinity and at the sphere radius. Thus

$$\left(\frac{4\,\mu}{3K} + 1\right) p_\infty + 4\,\mu\,\left(\frac{a}{a_0} - 1\right) = 4\,\mu\,(2^{1/3} - 1)\,, \tag{24.9.6}$$

which yields the actual change of radius as a function of the elastic moduli and confinement,

$$\frac{a}{a_0} = 2^{1/3} - \left(\frac{1}{4\mu} + \frac{1}{3K}\right) p_\infty \, . \tag{24.9.7}$$

Thus in a confined environment, an individual cell of initial volume V_0 and radius a_0 expands to volume V and radius a, and finally splits into two cells of volume $V_{\text{new}} = V/2$ and radius $a_{\text{new}} = a/2^{1/3}$. Thus

$$\frac{a_{\text{new}}}{a_0} = 1 - \alpha\, p_\infty, \quad F = \frac{V_{\text{new}}}{V_0} = (1 - \alpha\, p_\infty)^3 \, , \tag{24.9.8}$$

with

$$\alpha = \frac{1}{2^{1/3}} \left(\frac{1}{4\mu} + \frac{1}{3K}\right) . \tag{24.9.9}$$

Thus the cells adapt to the environment by changing their size: the model interprets this volume change as a change of volume fraction which modifies (actually decreases) their proliferation rate in a confined environment.

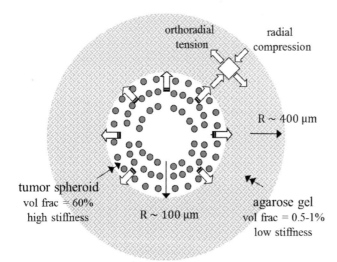

Fig. 24.9.2: Sketch of the boundary value problem analyzed by Roose et al. [2003]. The inner shell contains tumor cells that proliferate and exert compressive stresses on the outer shell consisting of agarose gel. Both the tumor spheroids and the outer shell are two phase mixtures. Tumors proliferate due to an exchange of mass between the fluid and the solid phases. The rate of tumor proliferation depends on the oxygen availability. The latter is governed by a reaction-diffusion equation.

The initial and boundary boundary value problem

The constitutive expressions above are used to solve an initial and boundary value problem. The model involves a composite sphere of radius R as sketched in Fig. 24.9.2:

- the center is a spheroid of cells of radius a at some given volume fraction. The initial radius a is taken equal to five cell diameters $5 \times 20\,\mu$m, so as to ensure a minimum homogeneity. The boundary $R = R(t)$ between the tumor spheroid and the gel evolves and its rate of change dR/dt is equal to the radial velocity of the tumor spheroid;
- the outer shell contains fibers in an agarose gel at another given volume fraction.
- water fills the voids of both the tumor spheroids and the outer shell;
- the initial fluid pressure is assumed uniform and null;
- the initial (total or effective) stress in the cell spheroid is taken as the stress required for the radius to pass from one cell diameter to 5 cell diameters, that is $\sigma_r \sim -4\mu \times 4/3$, $\sigma_\theta \sim 2\mu \times 4/3$, according to (24.7.34), with μ shear modulus of the outer shell;
- the initial stress in the outer shell is thus not uniform: it is obtained by ensuring traction continuity at the spheroid/outer shell interface;
- the initial concentration of oxygen c_{O_2} is uniform. Oxygen concentration is controlled on the outer boundary and it is considered to be continuous at this boundary $r = R$. Alternatively, Casciari et al. [1992] impose a continuity in terms of fluxes: the apparent normal flux in the bulk medium is defined via a convective boundary condition, that involves a convection coefficient k_{conv},

$$J_{O_2} = -k_{\text{conv}} \left(c_{O_2}(r = \infty) - c_{O_2}(R)\right); \tag{24.9.10}$$

- boundary conditions are given at the center of the cell spheroid (zero gradients) and at infinity (stationarity with respect to the initial state).

Results

The field equations are solved by a finite difference technique. The initial state is in mechanical equilibrium, but not in chemical equilibrium, which implies a strong discontinuity of the pore pressure, their Fig. 2, and growth to begin at time $t = 0$:

- the oxygen level is almost maximum so that nutrient availability is not the issue in the tests;
- the rate of proliferation depends strongly on the collagen content of the outer shell, which delays availability of nutrient to cells;
- the radial stress is compressive everywhere, while the orthoradial stress is compressive in the cell spheroid, tensile and large at the interface and slightly tensile in the gel;
- the pore pressure is always negative (with respect to its initial value) and tends to vanish;
- the tests involve *low* stress levels in the gel but high stress levels in the tumor and at the interface;
- however, the model does not account for the effects of *metalloproteinases* that are known to disrupt the surrounding tissue and thus lower the generated stress and the shear modulus of agarose gel which, although not accurately known, plays a critical role in the analytical results;
- the stress level in the outer shell depends linearly on the stiffness of the material. If the gel is replaced by human lung material, or tumor material, the stress could reach 100 mmHg. This stress level is able to collapse blood vessels;
- the fact that the orthoradial stress is either compressive or tensile would necessitate a conewise elastic model, Section 5.4. Moreover, tension might lead to radial fracture, but the fracture strength of agarose is not known.

24.9.2 A two phase model including the effects of lymphatics

The model is developed and illustrated in Sarntinoranont et al. [2003] by solving boundary value problems using finite elements, finite differences and the method of lines. It includes a fluid source aimed at simulating exchange between vessels, lymphatics and tissue. On the other hand, the growth strain is elementary and not associated with a mass transfer or growth process.

The model

The model has two phases. Vessels and lymphatics are external to the tissue. Stresses are represented by the solid stress and the fluid pressure of the interstitium (IFP). Here are the main characteristics of the constitutive model:

- the tumor expands due to cell proliferation, i.e., the stress is a function of the total strain minus the growth strain. A constitutive expression for the growth strain is not provided. Several spatial profiles, fixed in time, are tested: the spatial profiles aim at smoothing the growth activity in the tumor and absence of activity in the host, their Eqns 4 and 5;
- a transfer of fluid takes place from the vessels and lymphatics to the porous medium which in fact appears as open because the vessels are not part of the mixture, their Eqn 3. A similar analysis has been described in Section 24.8.1;
- the rate of mass exchange Ω is proportional to the difference between the pressure p_{cL}^* and interstitial fluid pressure p_i,

$$\Omega = \kappa \left(p_{\mathrm{cL}}^* - p_i \right), \tag{24.9.11}$$

In the normal tissue, the stress and pressures p_i and p_{cL}^* vanish so that the efflux from the vasculature equilibrates the influx into lymphatics without net exchange with the interstitium. The actual expression of p_{cL}^* is obtained by weighting the pressures of the vessels and of the lymphatics through their hydraulic permeabilities and areas,

$$p_{\mathrm{cL}}^* = \frac{k_{\mathrm{c}}\, S_{\mathrm{c}}\, p_{\mathrm{c}}^* + k_{\mathrm{L}}\, S_{\mathrm{L}}\, p_L}{k_{\mathrm{c}}\, S_{\mathrm{c}} + k_{\mathrm{L}}\, S_{\mathrm{L}}}, \quad \kappa = k_{\mathrm{c}}\, \frac{S_{\mathrm{c}}}{V} + k_{\mathrm{L}}\, \frac{S_{\mathrm{L}}}{V}, \tag{24.9.12}$$

where

- V is the volume of the mixture;
- k_{c}, k_{L} [unit : m/Pa/s] are the hydraulic permeabilities of the vessels and of the lymphatics respectively, S_{c} and S_{L} their surfaces, and p_L is the pressure in the lymphatics;
- $p_{\mathrm{c}}^* = p_{\mathrm{c}} - \omega \left(\pi_{\mathrm{c}} - \pi_i \right)$ is the driving pressure of the capillaries, p_{c} being the pressure of the capillaries, $\pi_{\mathrm{c}} - \pi_i > 0$ the oncotic pressure and $\omega \in [0, 1]$ the reflection coefficient. These quantities are assumed constant, their Table 1;

- in the tumor, the lymphatics are impaired, $S_L = 0$. In the host tissue, the lymphatics are normal and $p_{cL}^* = 0$;
    - if there are no lymphatics, the relations (24.9.11) and (24.9.12) boil down to (24.8.2).
  - a necrotic zone, characterized by a zero growth strain, develops in the center of the tumor. In this zone the mass transfer from the vessels and lymphatics vanishes. The necrotic zone is fixed arbitrarily: a constitutive criterion to decide its existence and development would be needed.

Parameters

Material coefficients are extracted from Baxter and Jain [1989] [1990] where growth and transport issues were addressed for rigid tissues. The tumor and adjacent tissue are endowed with the same elastic coefficients, because the compatibility equation (9.4.11) assumes uniform elastic properties. This requirement appears as a severe limitation, since material properties are known to change due to tumor growth. On the other hand, other parameters are modified smoothly from the tumor to the host tissue, their Eqns 4 and 5.

The boundary value problem

The boundary value problem is defined by the following ingredients:
  - the field equations consist of the balance of momentum of the mixture as a whole, in the equation of strain compatibility, and in an equation governing the fluid pressure, their Eqn 3. This equation contains a source term which is aimed at representing the exchange between the interstitium and the vessels. The displacement is purely radial;
  - the radius $a = a(t)$ of the tumor/host boundary evolves in time according to the ad hoc relation,

$$\frac{4}{3}\pi a^3(t) = V_0 \, 2^{t/T_D}, \tag{24.9.13}$$

where T_D is the time interval of volume doubling. Note incidentally that this exponential growth law is criticized by Brú et al. [2003] who find rather a linear growth law. An elementary exposition of the issue is sketched in Section 24.3.3.1. It is also interesting to compare this empirical relation with the data and simulations of Roose et al. [2003], their Fig. 1;
  - the characteristic time associated with seepage is much smaller than the growth time. Thus the partial derivatives wrt time essentially vanish. As an expedient, a number of entities are made arbitrary functions of time, e.g., size of necrotic core, tumor/host boundary, etc.

Results

The results show quasi-uniform compressive stresses within the tumor in both the radial and orthoradial directions. A growth strain increases the magnitude of the compression. On the other hand, the orthoradial stress is tensile in the adjacent tissue, their Fig. 4.

The adjacent tissue is invaded, their Fig. 6, but this feature is just the result of a tentative modeling. Actually, the surrounding tissue probably undergoes apoptosis/necrosis possibly due to (orthoradial) tension, the dead cells are either crushed or evacuated partially by macrophages and the proliferating tumor cells fill the space.

The existence of a central necrotic core (no growth strain) releases slightly (about 10%) the compressive stresses and pore pressure in the necrotic region, their Fig. 7: this result correlates with the fact mentioned above that a growth strain increases the compression in the tumor.

The pressures involved in the simulations are modest, less than 10 mmHg. The stress necessary to stop the growth of spheroids embedded in gels is in range 45-120 mmHg according to Helmlinger et al. [1997]a in the same range as the stress required to collapse vessels. In fact, compressive stress generated by the growth process, namely about 100 mmHg, are a possible engine able to collapse vasculature as suggested in Roose et al. [2003]. At steady state, the pressure difference between vessels and interstitium $p_c - p_i = \omega (\pi_c - \pi_i) > 0$ is slightly positive. Under transient conditions, this difference is not expected to become so negative as to lead to the collapse of the vessels.

As a general notice, there seems to be no direct experimental measurements of the state of stress within the tumors *in vivo*. Indirect qualitative manifestations stem from the existence of residual stresses.

24.10   Methods of imaging transport and mechanical properties

Palpation in search of abnormal tissues relies on the idea that sound and abnormal tissues possess distinct mechanical properties. Methods to image the mechanical and transport properties of soft tissues aim at improving upon palpation, first because they provide quantification and second because they can detect heterogeneities in depth.

24.10.1   Basic ideas

The basic idea to detect tumor is that their mechanical properties are distinct from those of normal tissues, Krouskop et al. [1998]. Wu et al. [2000] report also some changes in the viscoelastic properties. An understanding of the behavior of tissues including cells and extracellular matrix (ECM) should explain why there exists a correlation between the *shear* modulus and the physiological state of tissues.

Experimental procedures to date consider essentially isotropic and often quasi-incompressible materials. This simplification allows to de-correlate the longitudinal (volume) body waves (also called B-mode in ultrasound methods) from static strain patterns due to the shear modulus.

24.10.1.1   Categories of imaging methods

Greenleaf et al. [2003] propose to categorize imaging methods based on their *excitation* and *detection* approaches. Excitation is either
- *static*: a slow loading is applied, displacements and strains are measured over the tissue. Based on an a priori knowledge of the stress over the tissue, the stiffness is obtained. The method requires accurate knowledge of the boundary conditions;
- *dynamic*: dynamic methods rely on the dynamic properties of tissues, namely wave-speeds and attenuation which are local properties, and therefore boundary conditions are of less importance than for static methods.

Whether static or dynamic, methods may be further distinguished depending whether they apply the excitation on the boundary, e.g., with a plate or a vibrating device, or internally, e.g., radiation force generated in a restricted region by a beam of ultrasounds. Dynamic methods may have difficulties to transfer a well controlled excitation inside the body due to dissipation.

Detection of displacements is handled in various ways:
- dynamic ultrasound methods, also called *sonoelasticity*, measure the displacements by Doppler or pulse-echo methods[24.14];
- *acoustic* methods use tissue vibration;
- *magnetic resonance elastography* (MRE) uses phase sensitive resonance to measure the displacements and strains in the three directions of space. This method has pros and cons as the 3D wave equation needs to be inverted. It appears as an alternative to palpation for breast tumors, and to biopsy for liver fibrosis and it is used to assess pathologies of brain and cardiac and skeleletal muscles, Mariappan et al. [2010].

24.10.1.2   Standard assumptions on material properties

As a simplification, tissues have been initially considered as single phase solid, linear elastic isotropic and incompressible. The latter property subsumes that blood flow is minimized by application of a small compression. However tissue is better viewed as a porous medium, with two main contributors, the solid part and the interstitium containing pores, lymphatics and hemal vessels, see Section 24.10.7. For a length scale of 1 cm, the characteristic time of leak of fluid to vessels is fast, typically a few seconds, while seepage in the interstitium takes much longer to reach a steady state, about hundred seconds. So tissue incompressibility is an acceptable assumption only in some time windows, say from 5 to 20 seconds and above 100 seconds.

[24.14] Ultrasound techniques involve frequencies from 1 MHz to 10 MHz which are higher than human hearing which extend from 20 Hz to 20 kHz:

infrasound	_____	acoustic	_____	_____	ultrasound
	20 Hz		20 kHz	1 MHz	

Frequency and wave-length are linked by the relation $f\,\lambda = c$, where the wave-speed c is a characteristic of the medium across which the wave travels. Typically c is equal to 330 m/s in air at ambient temperature, 1500 m/s in soft tissues and 3400 m/s in compact bones. As it travels through different tissues, the ultrasound wave keeps its frequency but changes its wavelength. For a soft tissue a frequency of 1 MHz produces a wavelength of 1.5 mm. A higher frequency would yield a smaller wavelength but it would be more absorbed. This energy absorption is low and harmless in standard medical practice as tissue perfusion eliminates the absorbed energy and both the energy intensity and the time of exposure are minimized. It may however, if its intensity is large, elicit some intra- and extracellular activities like formation of bubbles.

24.10.2 Tumor tissues have distinct elastic and transport properties

The shear modulus of tumor tissues is higher

Tests using MR elastography to date seem to have addressed essentially breast tissues. McKnight et al. [2002] observe that the shear stiffness of breast carcinoma (about 25 kPa) is 5 to 20 times higher than that of normal adipose breast tissue (about 3 kPa) or fibro-glandular breast tissue (about 8 kPa), their Fig. 9. In contrast to palpation which can access only surface or depths smaller than 10 mm, elastography can detect tumors inside the tissue. It seems also that some tumors are not easily detectable through conventional techniques such as mammography or computed tomography (CT).

Tumor tissues are more nonlinear

Wellman et al. [1999] contend that the elastic nonlinearity in compression of tumor tissues becomes stronger than that of normal tissues. This conclusion was drawn from punch tests on thin specimens which had just been removed from the organ.

24.10.3 Ultrasonic techniques provide local strain

B-mode is the basic ultrasonic method by which an image of the tissue is reconstructed from the echoes received from scattering points in response to ultrasonic pulses. Displacements are obtained by sending a beam in a certain direction using correlation techniques, see Greenleaf et al. [2003] for a basic account. The strain in this direction is then easily deduced.

Currently imaging techniques aim at detecting smooth heterogeneities in mechanical moduli. The presence of fat (adipose tissue) which is much more compliant than glandular structures introduces some decorrelation (blurring): pre-compression is sometimes used to reduce the stiffness heterogeneity. Still, pre-compression may be somehow unpleasant in clinical tests.

Acoustic elastography, e.g., Ophir et al. [1991], sends ultrasounds before and after mechanical loading and use the scattered signal to produce an image of strain in the tissue sample. Initially only the strain along the beam direction was measured. The method is improved so as to record the strain in an orthogonal direction, Konofagou and Ophir [1998]. Thus, Poisson's ratio becomes a main target.

24.10.4 MR elastography (MRE) provides wave-length

Researchers of the Mayo Clinic have developed a technique based on a combination of shear waves and magnetic resonance (MR) to estimate the mechanical properties of tissues, McKnight et al. [2002]. The underlying idea is that tumors alter these properties. The method, called elastography, involves three steps:
 - generating acoustic shear waves in tissue;
 - imaging the propagation of these waves using MR;
 - processing the data to obtain information on the tissue elasticity.

For a given frequency f (75-300 Hz) of the applied waves, measurements provide the wave-length l so that the shear modulus μ of the (assumed isotropic elastic) tissue is equal to $\rho\,(f\,l)^2$, the density ρ of the tissue being assumed close to that of water.

24.10.5 Ultrasound and computer tomography (CT) for blood flow imaging

Pollard et al. [2004] and Broumas et al. [2005] compare CT and ultrasound methods to quantitatively assess the permeability and blood flow in tumors. They claim that ultrasound method has a better resolution than CT. They send a marker, namely microbubbles, in the blood flow: submitted to ultrasound waves, the bubbles oscillate with a characteristic frequency so that they can be located. The tracer remains strictly in the vessels, and does not cross the vasculature, which is why the resolution of the method is good.

The method is tentatively used to detect angiogenesis at the periphery of tumors, Figs. 4 and 11 in Broumas et al. [2005].

24.10.6 Inverse analysis based on surface measurements

Liu et al. [2003] perform a 2D analytical study of breast elastography, assuming normal and tumor tissues to be isotropic linear elastic and incompressible. They consider two problems:

- a direct problem where an inhomogeneity, of given geometry and Young's modulus, is located inside the organ. Using Eshelby analysis, they deduce the strain and stress in the organ;
- the inverse problem consists in detecting if and where a tumor exists in the organ. For that purpose, they apply some loading and measure the surface strain. The internal mechanical properties of the organ able to produce the observed surface strain pattern are sought via inverse analysis. The inverse method does not require information from inside the organ. However, it does not seem that robust at this stage.

24.10.7 Imaging poroelastic properties

24.10.7.1 Unconfined compression

Berry et al. [2006] consider unconfined compression of a cylindrical tissue specimen axially pre-compressed with a given strain. The platens are lubrificated so that the strains do not depend on the axial position. The strain and stress tensors are principal in the radial, orthoradial and axial directions. On the lateral sides, traction and pore pressure vanish so that fluid flow may take place.

The analytic solution detailed in Section 7.5 is compared with an axisymmetric finite element analysis using Abaqus. They comment in details the spatial profile and time evolution of the strain components as well as the pressure and fluid flow. A special attention is devoted to the radial-to-axial strain ratio, which represents the apparent Poisson's ratio. For a porous material, the apparent Poisson's ratio is changing in time but as the material drains, it tends to the Poisson's ratio of the solid phase. Earlier attempts to measure Poisson's ratio along similar ideas are described in Konofagou et al. [2001] and Righetti et al. [2004].

In summary, the aim of unconfined compression tests is to be able to estimate an overall value of the aggregate modulus, Poisson's ratio of the solid and permeability, Righetti et al. [2005].

24.10.7.2 Exploiting the faster extravasation in tumors

Leiderman et al. [2006] use a poroelastic model with three phases: the solid phase, including interstitial tissue and cells, the fluid phase and the microvasculature, namely blood and lymphatic vessels. Seepage takes place in the interstitium while fluid can transfer from the vessels to the interstitium, and conversely. The model is detailed in Section 24.8.2.

The effect of a local heterogeneity, representing the tumor, on the response of a 2D square sample to several loading and boundary conditions is analyzed via a plane strain finite element analysis. The versatility of the finite element method allows to test independently the influences of the diffusion through the interstitium and of the vascular permeability.

The material properties of the sound tissue and tumor are listed in Table 24.8.2. The typical extravasation time is around 3 s inside the tumor and 100 s in the sound tissue. The typical seepage time is high as 7000 s for the sound tissue and five time less for the tumor.

These observations could possibly be exploited to detect tumor tissues along the following scheme. Apply a rapid compression to the tissue. If ultrasonic measurements detect at some location a strain relaxation, presumably induced by extravasation, that is faster than standard, the underlying tissue may be suspected to be tumoral.

24.10.8 Extracting information by minimization

In order to recover material properties, analyses exploit experimental data, say displacements $\mathbf{u}$ of points located on a grid. In a complementary approach, a finite element analysis based on guessed material properties yields displacements $\mathbf{v}$. Ultimately the material properties are obtained by a trial and error so as to minimize the distance between measured and computed displacements.

For example, Harrigan and Konofagou [2004] consider an isotropic elastic linear solid. They assume a fixed Poisson's ratio but seek the Young's modulus. They proceed as follows. The target function to be minimized,

$$\Psi = (\mathbf{u} - \mathbf{v})^{\mathrm{T}} \mathbf{K} (\mathbf{u} - \mathbf{v}), \qquad (24.10.1)$$

is built from the positive definite overall stiffness $\mathbf{K} = \cup_e \mathbf{k}_e$ obtained by assembling element contributions $\mathbf{k}_e = \phi_e \, \mathbf{k}_e^0$. The subscript e denotes the element number and the superscript 0 denotes a reference Young's modulus E_e^0 and the associated element stiffness $\mathbf{k}_e^0$. Optimization is performed on the element variable ϕ_e with the understanding that the sought Young's modulus E_e is equal to $\phi_e \, E_e^0$.

The computed displacements satisfy the relation,

$$\mathbf{K}\,\mathbf{v} = \mathbf{R}\,, \tag{24.10.2}$$

where the fixed rhs represents the boundary conditions that were applied during the experimental measurements. Differentiating this relation with respect to ϕ_e yields

$$\frac{\partial \mathbf{K}}{\partial \phi_e}\,\mathbf{v} + \mathbf{K}\,\frac{\partial \mathbf{v}}{\partial \phi_e} = \mathbf{0} \quad \Rightarrow \quad \frac{\partial \mathbf{v}}{\partial \phi_e} = -\mathbf{K}^{-1}\,\mathbf{K}_e^0\,\mathbf{v}\,, \tag{24.10.3}$$

where $\mathbf{K}_e^0$ is the global matrix associated with the element matrix $\mathbf{k}_e^0$. Using (24.10.2) and the above relation, differentiation of (24.10.1) gives

$$\frac{\partial \Psi}{\partial \phi_e} = \mathbf{u}^{\mathrm{T}}\,\mathbf{K}_e^0\,\mathbf{u} - \mathbf{v}^{\mathrm{T}}\,\mathbf{K}_e^0\,\mathbf{v}\,. \tag{24.10.4}$$

Since

$$\delta \Psi = \frac{\partial \Psi}{\partial \phi_e}\,\delta \phi_e\,, \tag{24.10.5}$$

one may choose,

$$\delta \phi_e = -\frac{\partial \Psi}{\partial \phi_e} = -\left(\mathbf{u}^{\mathrm{T}}\,\mathbf{K}_e^0\,\mathbf{u} - \mathbf{v}^{\mathrm{T}}\,\mathbf{K}_e^0\,\mathbf{v}\right)\,, \tag{24.10.6}$$

so as to decrease the target function Ψ,

$$\delta \Psi = -\frac{\partial \Psi}{\partial \phi_e}\,\frac{\partial \Psi}{\partial \phi_e} \leq 0\,. \tag{24.10.7}$$

These calculations are performed for each element. The main burden lies in the recalculation of the overall stiffness matrix $\mathbf{K}$. The method rests on the assumption that the coefficients ϕ_e remain strictly positive during the iterations. The case of a Poisson's ration close to $1/2$ should also be scrutinized.

While the method is elementary, it serves as a basis for extension to more sophisticated material models where several material parameters are sought simultaneously or in a staggered approach.

Chapter 25

Units and physical constants

25.1 Units

25.1.1 Basic units

Symbol	Name (plural)	Physical Property
m	meter(s)	length
kg	kilo(s)	mass
s	second(s)	time
A	ampère(s)	electric current
K	kelvin(s)	temperature
mole	mole(s)	quantity of matter

25.1.2 Standard units

Symbol	Name (plural)	Dimension	Physical Property
Hz	hertz(-)	cycle/s	frequency
N	newton(s)	$kg \times m/s^2$	force
Pa	pascal(s)	$N/m^2 = kg/m/s^2$	stress, pressure
J	joule(s)	$N \times m = kg \times m^2/s^2$	work
W	watt(s)	$J/s = kg \times m^2/s^3$	power
°C	degree(s) Celsius	T (°C) = T(K) - 273.15	temperature
C	coulomb(s)	$A \times s$	electrical charge
V	volt(s)	$kg \times m^2/s^3/A = J/C$	electric potential
F	farad(s)	$C/V = C^2/J$	electrical capacitance
Ω	ohm(s)	V/A	electrical resistance
S	siemens(-)	$1/\Omega = A/V$	electrical conductance
		$Pa \times s = 1\ kg/m/s$	dynamic viscosity

25.1.3 Non standard units

o pressure[25.1][25.2]:

[25.1]"Hg" stands for hydrargyrum, i.e., liquid silver.

[25.2]The equivalences assume a temperature of water equal to 4°C and an atmospheric pressure: the water density is shown on Table 3.3.1.

1 bar=10^5 Pa;
dyne/cm^2=0.1 Pa
1 atm=101 325 Pa=760 mmHg=10.333 m H$_2$O;
1 mmHg=1 Torr=133.322 Pa;
10 mH$_2$O=98 064 Pa; 1 cm H$_2$O=0.7358 mmHg; 1mm Hg=1.359 cm H$_2$O

o energy:
    1 cal (calorie) = 4.185 J   energy required to increase the temperature
                                 of 1 gram of water from 13.5 to 14.5 °C

o viscosity:
    dynamic viscosity   1 poise=100 centipoise cP=0.1 Pa×s=0.1 kg/m/s
    kinematic viscosity 10^4 stokes=1 m^2/s

o hydraulic permeability: 1 darcy=$0.9869233 \times 10^{-12}$ m^2; plural: darcys

o standard conditions of temperature and pressure refer to[25.3]
    a temperature of 0°C=273.15 K and a pressure of 10^5 Pa;
    before 1990, the reference pressure was the atmospheric pressure (101 325 Pa).

o standard conditions of *ambient* temperature and pressure refer to
    a temperature of 25°C= 298.15 K and a pressure of 10^5 Pa.

25.2 Physical constants

Symbol	Definition	Value
g	acceleration of gravity at sea level	9.80665 m/s^2
N_A	Avogadro's number nb. elementary entities in 1 mole	6.02×10^{23} mole^{-1}
R	gas constant	8.314510 J/mole/K
k_B	Boltzmann constant R/N_A	1.380658×10^{-23} J/K
e^-	charge of an electron	$-1.602176 \times 10^{-19}$ C
F	Faraday's equivalent charge 1 mole of electrons $N_A \times |e^-|$	96485 C/mole
eV	electron volt	1.602176×10^{-19} J
ϵ_0	permittivity of the vacuum	8.852×10^{-12} C^2/J/m
h	Planck constant	6.626×10^{-34} J×s
		4.13567×10^{-15} eV/Hz

[25.3]Different conventions might be used by national metrological agencies, e.g., by the US National Institute of Standards and Technology, and by branches of industry.

Bibliography

* * * * A * * * *

Abercrombie M. (1979). Contact inhibition and malignancy. *Nature*, 281(5729), 259-262. (cited p. 1014).

Abousleiman Y., Cheng A.H.-D., Cui L., Detournay E. and Roegiers J.-C. (1996). Mandel's problem revisited. *Géotechnique*, 46(2), 187-195. (cited p. 214).

Abramowitz M. and Stegun I.A. (1964). *Handbook of mathematical functions with formulas, graphs and mathematical tables.* National Bureau of Standards, Applied Mathematics Series 55, New York. (cited p. 588).

Acar Y.B., Hamed J., Alshawabkeh A.N. and Gale R.J. (1994). Cadmium (II) removal from saturated kaolinite by application of electric current. *Géotechnique*, 44(2), 239-254. (cited pp. 443 and 444).

Achenbach J.D. (1975). *Wave propagation in elastic solids.* Elsevier Science Publishers B.V., Amsterdam, the Netherlands. (cited pp. 28, 361, 388, and 404).

Aghamohammadzadeh H., Newton R.H. and Meek K.M. (2004). X-ray scattering used to map the preferred collagen orientation in the human cornea and limbus. *Structure*, 12, 249-256. (cited pp. 429 and 704).

Agoram B. and Barocas V.H. (2001). Coupled macroscopic and microscopic scale modeling of fibrillar tissues and tissue equivalents. *ASME J. Biomechanical Engng.*, 123(4), 362-369. (cited p. 837).

Aifantis E. (1977). Introducing a multi-porous medium. *Developments in Mechanics*, 8, 209-211. (cited pp. 321 and 686).

Akizuki S., Mow V.C., Müller F., Pita J.C., Howell D.S. and Manicourt D.H. (1986). Tensile properties of human knee joint cartilage: I. Influence of ionic conditions, weight bearing, and fibrillation on the tensile modulus. *J. Orthopaedic Research*, 4(4), 379-392. (cited pp. 461, 462, 466, 489, 498, and 500).

Akizuki S., Mow V.C., Müller F., Pita J.C. and Howell D.S. (1987). Tensile properties of human knee joint cartilage: II. Correlations between weight bearing and tissue pathology and the kinetics of swelling. *J. Orthopaedic Research*, 5(2), 173-186. (cited pp. 462 and 498).

Alanko T. (2000). Growth factor effects on cell survival in three-dimensional cultures. Helsinki University Biomedical Dissertation n°2, Departments of Virology and Dermatalogy, Finland.

Albro M.A., Chahine N.O., Li R., Yeager K., Hung C.T. and Ateshian G.A. (2008). Dynamic loading of deformable porous media can induce active solute transport. *J. Biomechanics*, 41, 3152-3157. (cited p. 6).

Alberty R.A. (2003). The role of water in the thermodynamics of dilute aqueous solutions. *Biophysical Chemistry*, 100(1-3), 183-192. (cited p. 443).

Alexandrakis G., Brown E.B., Tong R.T., McKee T.D., Campbell R.B., Boucher Y. and Jain R.K. (2004). Two-photon fluorescence correlation microscopy reveals the two-phase nature of transport in tumors. *Nature Medicine*, 10(2), 203-207. (cited pp. 571 and 992).

Alexopoulos L.G., Haider M.A., Vail Th.P. and Guilak F. (2003). Alterations in the mechanical properties of the human chondrocyte pericellular matrix with osteoarthritis. *ASME J. Biomechanical Engng.*, 125, 323-333. (cited p. 827).

Almeida E.S. and Spilker R.L. (1998). Finite element formulations for hyperelastic transversely isotropic biphasic soft tissues. *Computer Methods in Applied Mechanics and Engineering*, 151(3-4), 513-538. (cited p. 189).

Alonso J.L. and Goldman W.H. (2003). Feeling the forces: atomic force microscopy in cell biology. *Life Sciences*, 72(23), 2553-2560. (cited p. 730).

Alshawabkeh A.N., Gale R.J., Oszu-Acar E. and Bricka R.M. (1999). Optimization of 2D-electrode configuration for electrokinetic remediation. *J. Soil Sediment Contamination*, 8(6), 617-635. (cited p. 442).

Altman G.H., Lu H.H., Horan R.L., Calabro T., Ryder D., Kaplan D.L., Stark P., Martin I., Richmond J.C. and Vunjak-Novakovic G. (2002). Advanced bioreactor with controlled application of multi-dimensional strain for tissue engineering. *ASME J. Biomechanical Engng.*, 124(6), 742-749. (cited p. 827).

Ambrosi D. and Mollica F. (2002). On the mechanics of a growing tumor. *Int. J. Engng. Science*, 40(12), 1297-1316.

Anderson D.E. and Graf D.L. (1976). Multicomponent electrolyte diffusion. *Annual Review Earth and Planetary Sciences*, 4, 95-121.

Angele P., Yoo J.U., Smith C., Mansour J., Jepsen K.J., Nerlich M. and Johnstone B. (2003). Cyclic hydrostatic pressure enhances the chondrogenic phenotype of human mesenchymal progenitor cells differentiated *in vitro*. *J. Orthopaedic Research*, 21, 451-457. (cited p. 828).

Antman S.S. (1995). *Nonlinear problems of elasticity*. Applied Mathematical Science Series, vol. 107, Springer Verlag, Berlin and New York. (cited p. 185).

Aoki T., Tan C.P. and Bamford W.E. (1993). Effects of deformation and strength anisotropy on borehole failure in saturated shales. *Int. J. Rock Mechanics Mining Sciences*, 30(7), 1031-1034. (cited pp. 222 and 223).

Appelo C.A.J. and Postma D. (1993). *Geochemistry, groundwater and pollution*. Balkema, Rotterdam, the Netherlands. (cited pp. 122, 449, and 450).

Araujo R.P. and McElwain D.L.S. (2004)a. A history of the study of solid tumour growth: the contribution of mathematical modelling. *Bulletin Mathematical Biology*, 66(5), 1039-1091. (cited p. 988).

Araujo R.P. and McElwain D.L.S. (2004)b. A linear-elastic model of anisotropic tumour growth. *European J. Applied Mathematics*, 15, 365-384. (cited p. 988).

Archer C.W. and Francis-West Ph. (2003). The chondrocyte. *Int. J. Biochemistry & Cell Biology*, 35, 401-404.

Armstrong C.G., Lai W.M. and Mow V.C. (1984). An analysis of the unconfined compression of articular cartilage. *ASME J. Biomechanical Engng.*, 106, 165-173. (cited pp. 207, 229, and 498).

Arnoczky S.P., Tian T., Lavagnino M., Gardner K., Schuler P. and Morse P. (2002). Activation of stress-activated protein kinases (SAPK) in tendon cells following cyclic strain: the effects of strain frequency, strain magnitude, and cytosolic calcium. *J. Orthopaedic Research*, 20, 947-952. (cited p. 1003).

Arts T., Prinzen F.W., Snoeckx L.H.E.H., Rijcken J.M. and Reneman R.S. (1994). Adaptation of cardiac structure by mechanical feedback in the environment of the cell: a model study. *Biophysical J.*, 66, 953-961. (cited p. 846).

Asanbaeva A., Masuda K., Thonar E.J.-M.A., Klisch S.M. and Sah R.L. (2008). Regulation of immature cartilage growth by IGF-I, TGF-α1, BMP-7, and PDGF-AB: role of metabolic balance between fixed charge and collagen network. *Biomechanics and Modeling in Mechanobiology*, 7, 263-276. (cited p. 809).

Atala A. and Koh C.J. (2004). Tissue engineering applications of therapeutic cloning. *Annual Review Biomedical Engng.*, 6, 27-40.

Ateshian G., Warden W.H., Kim J.J., Grelsamer R.P. and Mow V.C. (1997). Finite deformation biphasic material properties of bovine articular cartilage from confined compression experiments. *J. Biomechanics*, 30(11-12), 1157-1164. (cited p. 612).

Athanasiou K.A., Rosenwasser M.P., Buckwalter J.A., Malinin T.I. and Mow V.C. (1991). Interspecies comparisons of in situ intrinsic mechanical properties of distal femoral cartilage. *J. Orthopaedic Research*, 9(3), 330-340. (cited pp. 500 and 660).

Atkin R.J. and Craine R.E. (1975). Continuum theories of mixtures: basic theory and historical development. *Quarterly J. Mechanics and Applied Mathematics*, 29(2), 209-244. (cited pp. 104 and 120).

Atkinson E.R. (1979). Assessment of current hyperthermia technology. *Cancer Research*, 39, 2313-2324. (cited p. 279).

Axelsson I. and Heinegård D. (1978). Characterization of the keratan sulfate proteoglycans from bovine corneal stroma. *Biochemical J.*, 169, 517-530. (cited pp. 705 and 741).

* * * * B * * * *

Bader D.L., Kempson G.E., Barrett A.J. and Webb W. (1981). The effects of leucocyte elastase on the mechanical properties of adult human articular cartilage in tension. *Biochimica et Biophysica Acta - General Subjects*, 677, 103-108. (cited pp. 503 and 704).

Bader D.L., Ohashi T., Knight M.M., Lee D.A. and Sato M. (2002). Deformation properties of articular chondrocytes: a critique of three separate techniques. *Biorheology*, 39, 69-78. (cited p. 827).

Badulescu C., Grédiac M., Haddadi H., Mathias J.D., Balandraud X. and Tran H.S. (2011). Applying the grid method and infrared thermography to investigate plastic deformation in aluminium multicrystal. *Mechanics of Materials*, 43(1), 36-53. (cited p. 289).

Baer A.E., Wang J.Y., Kraus V.B. and Setton L.A. (2001). Collagen gene expression and mechanical properties of intervertebral disc cell-alginate cultures. *J. Orthopaedic Research*, 19, 2-10. (cited pp. 821, 822, and 823).

Baish J.W. and Jain R.K. (2000). Fractals and Cancer. *Cancer Research*, 60, 3683-3688. (cited p. 991).

Ball J. (1977). Convexity conditions and existence theorems in nonlinear elasticity. *Archives Rational Mechanics Analysis*, 63, 337-403. (cited pp. 174 and 180).

Banobre A., Alvarez T.L., Fechtner R.D., Greene R.J., Thomas G.A., Levi O. and Ciampa N. (2005). Measurement of intraocular pressure in pig's eyes using a new tonometer prototype. Proceedings 31st IEEE Bioengineering Northeast Conference, 260-261.

Contents

Bao G. and Smith G.F. (1990). Syzygies, orbits, and constitutive equations. In *Yielding, damage, and failure of anisotropic solids*, 147-154, J.P. Boehler ed., EGF Publication 5, Mechanical Engineering Publications Limited, London, UK. (cited p. 790).

Bao G. (2002). Mechanics of biomolecules. *J. Mechanics Physics Solids*, 50, 2237-2274.

Barbero E.J. and Ford K.J. (2007). Characterization of Self-Healing Fiber-Reinforced Polymer-Matrix Composite with Distributed Damage. *J. Advanced Materials*, 39(4), 20-27. (cited p. 984).

Barenblatt G.I., Zheltow I.P. and Kochina T.N. (1960). Basic concepts in the theory of seepage of homogeneous liquids in fissured rocks. *J. Applied Mathematics and Mechanics*, 24, 1286-1303. (cited p. 321).

Barocas V.H. and Tranquillo R.T. (1997)a. An anisotropic biphasic theory of tissue-equivalent mechanics: the interplay among cell traction, fibrillar network deformation, fibril alignment, and cell contact guidance. *ASME J. Biomechanical Engng.*, 119(2), 137-145.

Barocas V.H. and Tranquillo R.T. (1997)b. A finite element solution for the anisotropic biphasic theory of tissue-equivalent mechanics: the effect of contact guidance on isometric cell traction measurement. *ASME J. Biomechanical Engng.*, 119(3), 261-268.

Barron V., Lyons E., Stenson-Cox C., McHugh P.E. and Pandt A. (2003). Bioreactors for cardiovascular cell and tissue growth: a review. *Annals Biomedical Engng.*, 31(9), 1017-1030.

Bass E.C., Ashford F.A., Segal M.R. and Lotz J.C. (2004). Biaxial testing of human annulus fibrosus and its implications for a constitutive formulation. *Annals Biomedical Engng.*, 32(9), 1231-1242. (cited pp. 800 and 801).

Bassaglia Y. (2001). *Biologie Cellulaire* (in French). Librairie Maloine, Paris. (cited p. 994).

Basser P.J., Schneiderman R., Bank R.A., Wachtel E. and Maroudas A. (1998). Mechanical properties of the collagen network in human articular cartilage as measured by osmotic stress technique. *Archives Biochemistry Biophysics*, 351(2), 207-219. (cited pp. 485, 495, 496, 502, 512, 639, 713, 842, 843, 955, and 956).

Bataille J. and Kestin J. (1977). Thermodynamics of mixtures. *J. Non-Equilibrium Thermodynamics*, 2, 49-65.

Bathurst R.J. and Rothenburg L. (1990). Observations on stress-force-fabric relationships in idealized granular materials. *Mechanics of Materials*, 9(1), 65-80. (cited p. 434).

Battaglioli J.L. and Kamm R.D. (1984). Measurements of the compressive properties of scleral tissue. *Investigative Ophthalmology & Visual Science*, 25, 59-65. (cited p. 720).

Baxter L.T. and Jain R.K. (1989). Transport of fluid and macromolecules in tumors. I. Role of interstitial pressure and convection. *Microvascular Research*, 37(1), 77-104. (cited p. 1038).

Baxter L.T. and Jain R.K. (1990). Transport of fluid and macromolecules in tumors. II. Role of heterogeneous perfusion and lymphatics. *Microvascular Research*, 40(2), 246-263. (cited p. 1038).

Bayne S.C. (2005). Dental Biomaterials: Where Are We and Where Are We Going? *J. of Dental Education*, 69(5), 571-585. (cited p. 7).

Bazant M.Z. (2006). Random walks and diffusion. Lecture 11. Non-linear diffusion. MIT OpenCourseWare.

Bear J. (1972). *Dynamics of fluids in porous media*. Elsevier, Amsterdam, the Netherlands, 764 p. (cited pp. 545, 559, 561, 565, and 613).

Beaupré G.S., Orr T.E. and Carter D.R. (1990). An approach for time-dependent bone modeling and remodeling - theoretical development. *J. Orthopaedic Research*, 8(5), 651-661. (cited p. 870).

Beaupré G.S., Stevens S.S. and Carter D.R. (2000). Mechanobiology in the development, maintenance, and degeneration of articular cartilage. *J. Rehabilitation Research & Development*, 37(2), 145-151. (cited pp. 807 and 829).

Bejan A. (1993). *Heat transfer*. John Wiley & Sons, Inc., New York. (cited pp. 318, 319, and 336).

Benedek G.B. (1971). Theory of transparency of the eye. *Applied Optics*, 10(3), 459-473. (cited pp. 706, 727, and 729).

Bentley G. and Minas T. (2000). Treating joint damage in young people. *British Medical J.*, 320, 1585-1588. (cited p. 808).

Berne R.M. and Levy M.N. (2000). *Principles of physiology*. Mosby, St Louis, MO. (cited p. 988).

Berry C.C., Cacou C., Lee D.A., Bader D.L. and Shelton J.C. (2003). Dermal fibroblasts respond to mechanical conditioning in a strain profile dependent manner. *Biorheology*, 40, 337-345. (cited p. 825).

Berry G.P., Bamber J.C., Armstrong C.G., Miller N.R. and Barbone P.E. (2006). Towards an acoustic model-based poroelastic imaging method: I. Theoretical foundation. *Ultrasound in Medicine and Biology*, 32(4), 547-567. (cited pp. 231 and 1041).

Berryman J.G. (1995). Mixture theories for rock properties. *Rock Physics and Phase Relations. A handbook of physical constants. AGU Reference Shelf 3*, T.J. Ahrens ed., 205-228. (cited p. 373).

Berryman J.G. and Wang H.F. (1995). The elastic coefficients of double porosity models for fluid transport in jointed rock. *J. Geophysical Research*, 100(B12), 24611-24627. (cited p. 322).

Berryman J.G. (2006). Measures of microstructure to improve estimates and bounds on elastic constants and transport coefficients in heterogeneous media. *Mechanics of Materials*, 38(8-10), 732-747. (cited p. 561).

Beskos D.E., Papadakis C.N. and Woo H.S. (1989). Dynamics of saturated rocks. III; Rayleigh waves. *ASCE J. Engng. Mechanics*, 115(5), 1017-1034. (cited p. 388).

Best B.A., Guilak F., Setton L.A., Zhu W., Saed-Nejad F., Ratcliffe A., Weidenbaum M. and Mow V.C. (1994). Compressive mechanical properties of the human anulus fibrosus and their relationship to biochemical composition. *Spine*, 19(2), 212-221. (cited p. 802).

Bhakta N.R., Garcia A.M., Frank E.H., Grodzinsky A.J. and Morales T.I. (2000). The insulin-like growth factors (IGFs) I and II bind to articular cartilage via the IGF-binding proteins. *The J. Biological Chemistry*, 275(8), 5860-5866. (cited p. 820).

Bhattacharya A. and Mahajan R.L. (2003). Temperature dependence of thermal conductivity of biological tissues. *Physiological Measurement*, 24(3), 769-783. (cited p. 279).

Bianco P. and Robey P.G. (2001). Stem cells in tissue engineering. *Nature*, 414(6859), 118-121.

Bigoni D. and Loret B. (1999). Effects of elastic anisotropy on strain localization and flutter instability in plastic solids. *J. Mechanics Physics Solids*, 47(7), 1409-1436. (cited pp. 151, 170, and 360).

Biot, M.A. (1941). General theory of three-dimensional consolidation. *J. Applied Physics*, 12(2), 155-164.

Biot, M.A. (1955). Theory of elasticity and consolidation for a porous anisotropic solid. *J. Applied Physics*, 26(2), 182-185. (cited pp. 105 and 214).

Biot M.A. (1956). Theory of propagation of elastic waves in a fluid-saturated porous solid. *J. Acoustical Society America*, 28(2). I. Low-frequency range, 168-178; II. Higher frequency range, 179-191. (cited pp. 205, 321, 372, 373, 377, 378, 381, 388, 390, and 400).

Biot, M.A. (1962). Mechanics of deformation and acoustic propagation in porous media. *J. Applied Physics*, 33(4), 1482-1498. (cited p. 208).

Biot M.A. (1972). Theory of finite deformations of porous solids. *Indiana University Mathematics J.*, 21(7), 597-620.

Biot, M.A. (1973). Nonlinear and semilinear rheology of porous solids. *J. Geophysical Research*, 78(23), 4924-4937. (cited pp. 104 and 215).

Biot, M.A. (1977)a. Variational-lagrangian thermodynamics of nonisothermal finite strain mechanics of porous solids and thermomolecular diffusion. *Int. J. Solids Structures*, 13, 579-597.

Biot, M.A. (1977)b. Variational-lagrangian irreversible thermodynamics of initially-stressed solids with thermomolecular diffusion and chemical reactions. *J. Mechanics Physics Solids*, 25, 289-307.

Biot, M.A. (1978). Variational irreversible thermodynamics of heat and mass transfer in porous solids: new concepts and methods. *Quarterly Applied Mathematics*, 36, 19-38.

Biot, M.A. (1979). New variational-lagrangian thermodynamics of viscous fluid mixtures with thermomolecular diffusion. *Proceedings Royal Society London*, A365, 467-494.

Biot, M.A. (1984). New variational-lagrangian irreversible thermodynamics with application to viscous flow, reaction-diffusion, and solid mechanics. *Advances in Applied Mechanics*, Academic Press, Inc., Orlando, FL, 24, 1-91.

Biot M.A. and Willis D.G. (1957). The elastic coefficients of the theory of consolidation. *ASME J. Applied Mechanics*, 24, 594-601. (cited pp. 210 and 214).

Bird R.B., Stewart W.E. and Lightfoot E.N. (1960). *Transport phenomena*. John Wiley & Sons, Inc., New York. (cited pp. 69, 337, and 533).

Bischoff J.E. (2006). Continuous versus discrete (invariant) representations of fibrous structure for modeling non-linear anisotropic soft tissue behavior. *Int. J. Non-Linear Mechanics*, 41, 167-179. (cited pp. 429, 431, and 432).

Bishop A.W. and Hight D.W. (1977). The value of Poisson's ratio in saturated soils and rocks stressed under undrained conditions. *Géotechnique*, 27(3), 369-384. (cited p. 216).

Blume J.A. (1989). Compatibility conditions for a left Cauchy-Green strain field. *J. Elasticity*, 21(3), 271-308. (cited p. 894).

Boehler J.P. (1978). Lois de comportement anisotrope des milieux continus. *J. de Mécanique*, 17(2), 153-190. (cited pp. 61, 783, and 789).

Boehler J.P. (1987). Introduction to the invariant formulation of anisotropic constitutive equations. In *Applications of Tensor Functions in Solid Mechanics*. CISM Courses and Lectures n° 292, edited by J.P. Boehler, 3-65. Springer Verlag, Wien. (cited pp. 61, 62, and 63).

Boerboom R.A., Driessen N.J.B., Bouten C.V.C., Huyghe J.M. and Baaijens F.P.T. (2003). Finite element model of mechanically induced collagen fiber synthesis and degradation in the aortic valve. *Annals Biomedical Engng.*, 31, 1040-1053. (cited p. 838).

Bonanno J.A. (2003). Identity and regulation of ion transport mechanisms in the corneal endothelium. *Progress in Retinal and Eye Research*, 22, 69-94. (cited pp. 711 and 712).

Bonassar L.J., Grodzinsky A.J., Frank E.H., Davila S.G., Bhaktav N.R. and Trippel S.B. (2001). The effect of dynamic compression on the response of articular cartilage to insulin-like growth factor-I. *J. Orthopaedic Research*, 19, 11-17. (cited pp. 819 and 820).

Boote C., Dennis S., Huang Y., Quantock A.J. and Meek K.M. (2005). Lamellar orientation in human cornea in relation to mechanical properties. *J. Structural Biology*, 149, 1-6. (cited p. 787).

Boote C., Hayes S., Abahussin M. and Meek K.M. (2006). Mapping collagen organization in the human cornea: left and right eyes are structurally distinct. *Investigative Ophthalmology & Visual Science*, 47(3), 901-908. (cited p. 704).

Borene M.L., Barocas V.H. and Hubel A. (2004). Mechanical and cellular changes during compaction of a collagen-sponge-based corneal stromal equivalent. *Annals Biomedical Engng.*, 32(2), 274-283. (cited p. 731).

Boucher Y., Kirkwood J.M., Opacic D., Desantis M. and Jain R.K. (1991). Interstitial hypertension in superficial metastatic melanomas in humans. *Cancer Research*, 51(24), 6691-6694. (cited p. 1003).

Boucher Y. and Jain R.K. (1992). Microvascular pressure is the principal driving force for interstitial hypertension in solid tumors: implications for vascular collapse. *Cancer Research*, 52, 5110-5114. (cited pp. 570 and 1003).

Boucher Y., Leunig M. and Jain R.K. (1996). Tumor angiogenesis and interstitial hypertension. *Cancer Research*, 56(18), 4264-4266. (cited pp. 568 and 1003).

Bourne W.M. (2003). Biology of the corneal endothelium in health and disease. *Eye*, 17, 912-918. (cited pp. 702 and 730).

Bowen R.M. and Wiese J.C. (1969). Diffusion in mixtures of elastic materials. *Int. J. Engng. Science*, 7, 689-722.

Bowen R.M. and Garcia D.J. (1970). On the thermodynamics of mixtures with several temperatures. *Int. J. Engng. Science*, 8, 63-83. (cited pp. 287 and 323).

Bowen R.M. and Wang C.C. (1971). Thermodynamic influences on acceleration waves in inhomogeneous isotropic elastic bodies with internal state variables. *Archives Rational Mechanics Analysis*, 41(4), 287-318.

Bowen R.M. and Chen P.J. (1972)a. On the effects of diffusion and chemical reactions on the growth and decay of acceleration waves. *Archives Rational Mechanics Analysis*, 48(5), 319-351.

Bowen R.M. and Chen P.J. (1972)b. Acceleration waves in anisotropic thermoelastic materials with internal state variables. *Acta Mechanica*, 15(2), 95-104.

Bowen R.M. and Chen P.J. (1975). Waves in a binary mixture of linear elastic materials. *J. de Mécanique*, 14(2), 237-266. (cited p. 323).

Bowen R.M. (1976). Theory of mixtures. In *Continuum Physics*, Vol. III, Eringen A.C. ed., Academic Press, New York, 1-127. (cited pp. 104, 112, 115, 117, 206, 209, 355, and 366).

Bowen R.M. and Reinicke K.M. (1978). Plane progressive waves in a binary mixture of linear elastic materials. *ASME J. Applied Mechanics*, 45, 493-499.

Bowen R.M. (1980). Incompressible porous media models by use of the theory of mixtures. *Int. J. Engng. Science*, 18, 1129-1148.

Bowen R.M. (1981). Green's functions for consolidation problems. *Int. J. Engng. Science*, 19, 455-466.

Bowen R.M. (1982). Compressible porous media models by use of the theory of mixtures. *Int. J. Engng. Science*, 20(6), 697-735. (cited pp. 117 and 197).

Bowen R.M. and Lockett R.R. (1983). Inertial effects in poroelasticity. *ASME J. Applied Mechanics*, 50, 334-342.

Bowen R.M. (1989). *Introduction to continuum mechanics for engineers*. Plenum Press, New York. (cited pp. 13, 21, 26, 36, 45, 174, 285, 351, and 365).

Boyce B.L., Jones R.E., Nguyen T.D. and Grazier J.M. (2007). Stress-controlled viscoelastic tensile response of bovine cornea. *J. Biomechanics*, 40(11), 2367-2376. (cited pp. 242, 255, and 433).

Brekken Ch., Bruland Ø.S. and de Lange Davies C. (2000). Interstitial fluid pressure in human osteosarcoma xenografts: significance of implantation site and the response to intratumoral injection of hyaluronidase. *Anticancer Research*, 20, 3503-3512. (cited p. 570).

Bresler E. (1973). Anion exclusion and coupling effects in nonsteady transport through unsaturated soils. I. Theory. *Soil Science Society of America J.*, 37(5), 663-669. (cited p. 609).

Breuls R.G.M., Sengers B.G., Oomens C.W.J., Bouten C.V.C. and Baaijens F.P.T. (2002). Predicting local cell deformations in engineered tissue constructs: a multilevel finite element approach. *ASME J. Biomechanical Engng.*, 124, 198-207.

Breward Ch.J.W., Byrne H.M. and Lewis C.E. (2002). The role of cell-cell interactions in a two-phase model for avascular tumour growth. *J. Mathematical Biology*, 45, 125-152. (cited pp. 112, 1004, 1009, and 1011).

Breward Ch.J.W., Byrne H.M. and Lewis C.E. (2003). A multiphase model describing vascular tumour growth. *Bulletin Mathematical Biology*, 65, 609-640. (cited pp. 1009 and 1012).

Brinkman H.C. (1947) A calculation of the viscous force exerted by a flowing fluid on a dense swarm of particles. *Applied Science Research*, A1, 27-34. (cited pp. 208 and 565).

Brocklehurst R., Bayliss M.T., Maroudas A., Coysh H.L., Freeman M.A.R., Revell P.A. and Ali S.Y. (1984). The composition of normal and osteoarthritic articular cartilage from human knee joints. *J. Bone & Joint Surgery*, 66-A, 1, 95-106. (cited p. 488).

Bronson J.G. (2002). Bio-gardening: New frontiers. *Orthopedic Technology Review*, 4(4), 5 p. (cited pp. 808 and 821).

Brooks A.N. and Hughes T.J.R. (1982). Streamline upwind/Petrov-Galerkin formulations for convection dominated flows with particular emphasis on the incompressible Navier-Stokes equations. *Computer Methods in Applied Mechanics and Engineering*, 32(1-3), 199-259.

Brooks R.J. and Corey A.T. (1964). Hydraulic properties of porous media. Hydrology Papers, Colorado State University, Fort Collins, CO.

Broumas A.R., Pollard R.E., Bloch S.H., Wisner E.R., Griffey S. and Ferrara K.W. (2005). Contrast-enhanced computed tomography and ultrasound for the evaluation of tumor blood flow. *Investigative Radiology*, 40(3), 134-147. (cited p. 1040).

Brown E. and Dejana E. (2003). Cell-to-cell contact and extracellular matrix. Editorial overview: cell-cell and cell-matrix interactions - running, jumping, standing still. *Current Opinion in Cell Biology*, 15, 505-508.

Brown E.B., Boucher Y., Nasser S. and Jain R.K. (2004). Measurement of macromolecular diffusion coefficients in human tumors. *Microvascular Research*, 67, 231-236; erratum in *Microvascular Research*, 2004, 68(3), 313-314. (cited pp. 553 and 571).

Brú A., Albertos S., Subiza J.L., López Garcia-Asenjo J. and Brú I. (2003). The universal dynamics of tumor growth. *Biophysical J.*, 85, 2948-2961. (cited pp. 996 and 1038).

Bruehlmann S.B., Rattner J.B., Matyas J.R. and Duncan N.A. (2002). Regional variations in the cellular matrix of the annulus fibrosus of the intervertebral disc. *J. Anatomy*, 201, 159-171. (cited p. 798).

Bruhns O.T., Meyers A. and Xiao H. (2004). On non-corotational rates of Oldroyd's type and relevant issues in rate constitutive formulations. *Proceedings Royal Society London*, A460, 909-928. (cited p. 51).

Bryant M.R., Szerenyi K., Schmotzer H. and McDonnell P.J. (1994). Corneal tensile strength in fully healed radial keratotomy wounds. *Investigative Ophthalmology & Visual Science*, 35(7), 3022-3031. (cited p. 721).

Bryant M.R. and McDonnell P.J. (1998). A triphasic analysis of corneal swelling and hydration control. *ASME J. Biomechanical Engng.*, 120, 370-381. (cited pp. 713, 714, 720, 722, and 723).

Büchler P., Pioletti D.P. and Rakotomanana L.R. (2003). Biphasic constitutive laws for biological interface evolution. *Biomechanics and Modeling in Mechanobiology*, 1(4), 239-249. (cited pp. 847 and 873).

Buckwalter J.A. and Lane N.E. (1997). Athletics and osteoarthritis. *American J. Sports Medicine*, 25(6), 873-881. (cited p. 807).

Burger J., Sourieau P. and Combarnous M. (1985). Thermal methods of oil recovery. Editions Technip, Paris, France. (cited p. 207).

Bursać P., McGrath C.V., Eisenberg S.R. and Stamenović D. (2000). A microstructural model of elastostatic properties of articular cartilage in confined compression. *ASME J. Biomechanical Engng.*, 122, 347-353. (cited p. 506).

Buschmann M.D., Gluzband Y.A., Grodzinsky A.J., Kimura J.H. and Hunziker E.B. (1992). Chondrocytes in agarose culture synthesize a mechanically functional extracellular matrix. *J. Orthopaedic Research*, 10, 745-758. (cited pp. 471, 805, 823, 829, and 844).

Buschmann M.D., Gluzband Y.A., Grodzinsky A.J. and Hunziker E.B. (1995). Mechanical compression modulates matrix biosynthesis in chondrocyte/agarose culture. *J. Cell Science*, 108, 1497-1508. (cited pp. 471, 554, 805, and 807).

Buschmann M.D. and Grodzinsky A.J. (1995). A molecular model of proteoglycan-associated electrostatic forces in cartilage mechanics. *ASME J. Biomechanical Engng.*, 117, 179-192. (cited p. 506).

Buschmann M.D., Kim Y.J., Wong M., Frank E., Hunziker E.B. and Grodzinsky A.J. (1999). Stimulation of aggrecan synthesis in cartilage explants by cyclic loading is localized to regions of high interstitial fluid flow. *Archives Biochemistry Biophysics*, 366, 1-7. (cited pp. 825 and 838).

Bush P.G. and Hall A.C. (2001). The osmotic sensitivity of isolated and in situ bovine articular chondrocytes. *J. Orthopaedic Research*, 19, 768-778. (cited p. 827).

Butler T.P., Grantham F.H. and Gullino P.M. (1975). Bulk transfer of fluid in the interstitial compartment of mammary tumors. *Cancer Research*, 35(11), 3084-3088. (cited p. 1003).

Buzard K.A. (1992). Introduction to biomechanics of the cornea. *Refractive Corneal Surgery*, 8(2), 127-138.

Byrne H. and Preziosi L. (2003). Modelling solid tumour growth using the theory of mixtures. *Mathematical Medicine and Biology*, 20(4), 341-366. (cited p. 1010).

∗ ∗ ∗ ∗ C ∗ ∗ ∗ ∗

Cacou C., Palmer D., Lee D.A., Bader D.L. and Shelton J.C. (2000). A system for monitoring the response of uniaxial strain on cell seeded collagen gels. *Medical Engineering & Physics*, 22(5), 327-333. (cited p. 826).

Callen H.B. (1960). *Thermodynamics*. John Wiley & Sons, New York. (cited p. 533).

Campbell R.B., Fukumura D., Brown E.B., Mazzola L.M., Izumi Y., Jain R.K., Torchilin V.P. and Munn L.L. (2002). Cationic charge determines the distribution of liposomes between the vascular and extravascular compartments of tumors. *Cancer Research*, 62, 6831-6836. (cited p. 1000).

Candido K.A., Shimizu K., McLaughlin J.C., Kunkel R., Fuller J.A., Redman B.G., Thomas E.K., Nickoloff B.J. and Mulé J.J. (2001). Local administration of dendritic cells inhibits established breast tumor growth: implications for apoptosis-inducing agents. *Cancer Research*, 61, 228-236. (cited pp. 996 and 1000).

Cardoso L., Teboul F., Sedel L., Oddou C. and Meunier A. (2003). *In vitro* acoustic waves propagation in human and bovine cancellous bone. *J. Bone Mineral Research*, 18(10), 1803-1812. (cited p. 374).

Carmeliet P. and Jain R.K. (2000). Angiogenesis in cancer and other diseases. *Nature*, 407(6801), 249-257. (cited pp. 991, 998, and 1003).

Carroll M.M. (1979). An effective stress law for anisotropic elastic deformation. *J. Geophysical Research*, 84(B13), 7510-7512. (cited pp. 210 and 214).

Carslaw H.S. and Jaeger J.C. (1959). *Conduction of heat in solids*. Clarendon Press, Oxford, UK. (cited pp. 237, 340, 343, 538, 541, 543, 588, and 1033).

Carter D.R. and Hayes W.C. (1977). The compressive behavior of bone as a two-phase porous structure. *J. Bone & Joint Surgery*, American Volume 59, 954-962. (cited pp. 869 and 872).

Carter D.R., Fyhrie D.P. and Whalen R.T. (1987). Trabecular bone density and loading history: regulation of connective tissue biology by mechanical energy. *J. Biomechanics*, 20, 785-794. (cited pp. 845 and 869).

Carter D.R., Orr T.E. and Fyhrie D.P. (1989). Relationships between loading history and femoral cancellous bone architecture. *J. Biomechanics*, 22, 231-244. (cited p. 869).

Carter D.R. and Wong M. (1990). Mechanical stresses in joint morphogenesis and maintenance. In *Biomechanics of Diarthrodial Joints*, Mow V.C., Ratcliffe A. and Woo S.L.-Y. eds., Springer-Verlag, New York, 2, 155-174. (cited pp. 471 and 844).

Cartwright T. (1994). *Animal cells as bioreactors*. Cambridge University Press, New York. (cited pp. 550 and 834).

Casciari J.J., Sotirchos S.V. and Sutherland R.M. (1992). Mathematical modelling of microenvironment and growth in EMT6/Ro multicellular tumour spheroids. *Cell Proliferation*, 25(1), 1-22. (cited pp. 550, 553, 834, 1034, and 1036).

Casey J. (1985). Approximate kinematical relations in plasticity. *Int. J. Solids Structures*, 21(7), 671-682. (cited p. 43).

Cassidy J.J., Hiltner A. and Baer E. (1989). Hierarchical structure of the intervertebral disc. *Connecting Tissue Research*, 23(1), 75-88. (cited p. 799).

Cattaneo C. (1948,1949,1958). Sulla conduzione del calore. *Atti del Seminario matematico e fisico dell'Università di Modena*, (1948), 3, 3-21; (1949), 3, 83-101; Sur une forme de l'équation de la chaleur éliminant le paradoxe d'une propagation instantanée, *Comptes rendus hebdomadaires des séances de l'Académie des sciences*, Paris, (1958), 247, 431-433. (cited p. 285).

Cazacu O. (2002). A new hyperelastic model for transversely isotropic solids. *Zeitschrift für angewandte Mathematik und Physik*, 53, 901-911.

Chaboche J.L., Lesne P.M. and Maire J.F. (1995). Continuum Damage Mechanics, Anisotropy and Damage Deactivation for Brittle Materials Like Concrete and Ceramic Composites. *Int. J. Damage Mechanics*, 4, 5-22. (cited p. 158).

Chadwick P. (1960). Thermoelasticity: the dynamical theory. In *Progress in Solid Mechanics*, vol. 1, 263-328, Sneddon I.N. and Hill R. eds., North-Holland, Amsterdam, the Netherlands. (cited p. 281).

Chadwick P. (1989). Wave propagation in transversely isotropic elastic media. I. Homogeneous plane waves. *Proceedings Royal Society London*, A422, 23-66. (cited p. 169).

Chahine N.O., Wang C.C.B., Mason J.H., Lai W.M., Hung C.T. and Ateshian G.A. (2002). The role of osmotic swelling pressure and tension-compression nonlinearity on stress-strain responses of bovine articular cartilage. *Transactions Orthopaedic Research Society*, 48th Annual Meeting, 83.

Chaipinyo K., Oakes B.W. and van Damme M.-P.I. (2002). Effects of growth factors on cell proliferation and matrix synthesis of low-density, primary bovine chondrocytes cultured in collagen I gels. *J. Orthopaedic Research*, 20(5), 1070-1078. (cited pp. 822 and 823).

Chandran P.L. and Barocas V.H. (2004). Microstructural mechanics of collagen gels in confined compression: poroelasticity, viscoelasticity, and collapse. *ASME J. Biomechanical Engng.*, 126, 152-166. (cited p. 837).

Chandran P.L. and Barocas V.H. (2006). Affine versus non-affine fibril kinematics in collagen networks: theoretical studies of network behavior. *ASME J. Biomechanical Engng.*, 128, 259-270. (cited p. 781).

Chang D.G., Lottman L.M., Chen A.C., Shinagl R.M., Albrecht D.R., Pedowitz R.A., Brossmann J., Frank L.R. and Sah R.L. (1999). The depth-dependent, multi-axial properties of aged human patellar cartilage in tension. *Transactions Orthopaedic Research Society*, 45th Annual Meeting, Anaheim, 24, 644. (cited pp. 162 and 163).

Charlebois M., McKee M.D. and Buschmann M.D. (2004). Nonlinear tensile properties of bovine articular cartilage and their variation with age and depth. *ASME J. Biomechanical Engng.*, 126(2), 129-137. (cited pp. 241, 242, and 246).

Chen A.C., Bae W.C., Shinagl R.M. and Sah R.L. (2001). Depth- and strain-dependent mechanical and electromechanical properties of full-thickness bovine articular cartilage in confined compression. *J. Biomechanics*, 34, 1-12. (cited pp. 489 and 613).

Chen A.C., Pfister B.E., Nagakawa K., Pietryla D.W., Wong V.W., Thonar E.J., Sah R.L. and Masuda K. (2004). Effect of long-term culture on tensile properties and collagen network maturation in tissue-engineered cartilage. *Transactions Orthopaedic Research Society*, 50th annual meeting, 29, 307. (cited p. 841).

Chen C.Y., Byrne H.M. and King J.R. (2001). The influence of growth-induced stress from the surrounding medium on the development of multicell spheroids. *J. Mathematical Biology*, 43(3), 191-220. (cited pp. 1004, 1008, 1009, and 1011).

Chen K.C., Peng C.H., Wang H.E., Peng C.C. and Peng R.Y. (2004). Modeling of the pH- and the temperature-dependent deviations of the free to total PSA (prostate specific antigen) ratios for clinical predictability of prostate cancer and benign prostate hyperplasia. *Bulletin Mathematical Biology*, 66(3), 423-445. (cited p. 1013).

Chen M.M., Holmes K.R. and Rupinskas V. (1981). Pulse-decay method for measuring the thermal conductivity of living tissues. *ASME J. Biomechanical Engng.*, 103(4), 253-280. (cited p. 283).

Chen S.S., Falcovitz Y.H., Schneiderman R., Maroudas A. and Sah R.L. (2001). Depth-dependent compressive properties of normal aged human femoral head articular cartilage: relationship to fixed charge density. *Osteoarthritis and Cartilage*, 9, 561-569. (cited pp. 489 and 613).

Chen S.S. and Sah R.L. (2001). Contribution of collagen network and fixed charge to the confined compression modulus of articular cartilage. *Transactions Orthopaedic Research Society*, 47th Annual Meeting, 26, 426. (cited p. 500).

Chen Y.-C. and Hoger A. (2000). Constitutive functions of elastic materials in finite growth and deformation. *J. Elasticity*, 59(1-3), 175-193. (cited pp. 192, 847, 889, and 890).

Cheng A.H.-D. and Detournay E. (1993). Fundamentals of poroelasticity. In *Comprehensive Rock Engineering: Principles, Practice and Projects*, 2: *Analysis and Design Methods*, C. Fairhurst ed., Pergamon Press, 113-171. (cited p. 219).

Cheng A.H.-D. (1997). Material coefficients of anisotropic poroelasticity. *Int. J. Rock Mechanics Mining Sciences*, 34(2), 199-205. (cited pp. 214, 217, 223, and 224).

Cheng A.H.-D. (1998). On generalized plane strain poroelasticity. *Int. J. Rock Mechanics Mining Sciences*, 35(2), 183-193. (cited p. 214).

Chien S. (2003). Molecular and mechanical bases of focal lipid accumulation in arterial wall. *Progress in Biophysics & Molecular Biology*, 83(2), 131-151. (cited pp. 807 and 820).

Choi H.S. and Vito R.P. (1990). Two-dimensional stress-strain relationship for canine pericardium. *ASME J. Biomechanical Engng.*, 112(2), 153-159. (cited p. 190).

Chowdhury T.T., Bader D.L. and Lee D.A. (2001). Dynamic compression inhibits the synthesis of nitric oxide and PGE2 by IL-1β-stimulated chondrocytes cultured in agarose constructs. *Biochemical and Biophysical Research Communications*, 285, 1168-1174. (cited p. 830).

Christensen R.M. (1982). *Theory of viscoelasticity. An introduction.* 2nd edition. Academic Press, New York. (cited pp. 248 and 253).

Christiansen D.L., Huang E.K. and Silver F.H. (2000). Assembly of type I collagen: fusion of fibril subunits and the influence of fibril diameter on mechanical properties. *Matrix Biology*, 19(5), 409-420. (cited p. 493).

Chuong C.J. and Fung Y.C. (1986). On residual stress in arteries. *ASME J. Biomechanical Engng.*, 108(2), 189-192. (cited pp. 846, 878, and 887).

Ciarlet Ph. (1989). Introduction to numerical linear algebra and optimization. Cambridge Universiy Press, Cambridge, UK. (cited p. 14).

Ciombor D.McK., Lester G., Aaron R.K., Neame P. and Caterson B. (2002). Low frequency EMF regulates chondrocyte differentiation and expression of matrix proteins. *J. Orthopaedic Research*, 20, 40-50. (cited p. 829).

Cleary M.P. (1977). Fundamental solutions for a fluid-saturated porous solid. *Int. J. Solids Structures*, 13, 785-806.

Cleland W.-W. (1963). The kinetics of enzyme-catalyzed reactions with two or more substrates or products. *Biochimica et Biophysica Acta*, 67, I. Nomenclature and rate equations, 104-137; II. Inhibition: nomenclature and theory, 173-187; III. Prediction of initial velocity and inhibition patterns by inspection, 188-196. (cited p. 581).

Coleman B.D. and Noll W. (1960). An approximation theorem for functionals, with applications in continuum mechanics. *Archives Rational Mechanics Analysis*, 6, 355-370. (cited p. 248).

Coleman B.D. (1964). Thermodynamics of materials with memory. *Archives Rational Mechanics Analysis*, 17, 1-46; On thermodynamics, strain impulses, and viscoelasticity. *Archives Rational Mechanics Analysis*, 17, 230-254. (cited pp. 131 and 248).

Coleman B.D. and Noll W. (1964). Material symmetry and thermostatic inequalities in finite elastic deformations. *Archives Rational Mechanics Analysis*, 15, 87-111. (cited p. 848).

Colin J. and Velou S. (2003). Current surgical options for keratoconus. *J. Cataract & Refractive Surgery*, 29(2), 379-386.

Comper W.D. and Williams R.P.W. (1987). Hydrodynamics of concentrated proteoglycan solutions. *The J. Biological Chemistry*, 262(28), 13464-13471. (cited p. 553).

Comper W.D. and Lyons K.C. (1993). Non-electrostatic factors govern the hydrodynamic properties of articular cartilage proteoglycan. *Biochemical J.*, 289, 543-547. (cited p. 628).

Coon M.D. and Evans R.J. (1971). Recoverable deformation of cohensionless soils. *ASCE J. Soil Mechanics and Foundations*, 97(2), 375-391. (cited p. 134).

Cornish-Bowden A. (1979). *Fundamentals of Enzyme Kinetics*. Butterworths, London, UK. (cited p. 585).

Cosserat E. and Cosserat F. (1909). Théorie des corps déformables. Librairie Scientifique A. Hermann et Fils, Paris. (cited p. 112).

Courant R. and Hilbert D. *Methods of Mathematical Physics*. John Wiley & Sons, Inc., New York, Vol. I, Wiley-Interscience Publication, (1953); Vol. II, *Partial Differential Equations*, Interscience Publishers, (1962). (cited p. 365).

Cowin S.C. and Hegedus D.M. (1976). Bone remodeling I: theory of adaptive elasticity. *J. Elasticity*, 6(3), 313-326. (cited pp. 104, 847, 866, 871, 921, and 926).

Cowin S.C. (1985). The relationship between the elasticity tensor and the fabric tensor. *Mechanics of Materials*, 4, 137-147. (cited p. 150).

Cowin S.C., Moss-Salentijn L. and Moss M.L. (1991). Candidates for the mechanosensory system in bone. *ASME J. Biomechanical Engng.*, 113(2), 191-197. (cited pp. 845 and 846).

Cowin S.C. (1996). Strain or deformation rate dependent finite growth in soft tissues. *J. Biomechanics*, 29(5), 647-649. (cited pp. 846, 851, 856, 880, and 921).

Cowin S.C. (1999). Structural changes in living tissues. *Meccanica*, 29, 379-398.

Cowin S.C. (2000). How is a tissue built? *ASME J. Biomechanical Engng.*, 122(6), 553-569. (cited p. 551).

Cowin S.C. (2004). Tissue growth and remodeling. *Annual Review Biomedical Engng.*, 6, 77-107. (cited pp. 114, 850, 871, and 921).

Cox M.A.J., Oomens C.W.J. and Sengers B.G. (2003). Perfusion-enhanced growth of tissue-engineered cartilage in a bioreactor: A finite element study. Internal report, Department of Biomedical Engineering, Eindhoven University of Technology, the Netherlands.

Criscione J.C., Sacks M.S. and Hunter W.C. (2003). Experimentally tractable, pseudo-elastic constitutive law for biomembranes. *ASME J. Biomechanical Engng.*, 125(1); I. Theory: 94-99; II. Application: 100-105. (cited pp. 192 and 193).

Curnier A., He Q.-C. and Zysset Ph. (1994). Conewise linear elastic materials. *J. Elasticity*, 37(1), 1-38. (cited pp. 144, 146, 157, and 158).

* * * * D * * * *

Dadoune J.-P., Hadjiiski P., Siffroi J.-P. and Vendrely E. (2000). Histologie (in French). Flammarion, Paris. (cited pp. 998, 999, and 1000).

Dafalias Y.F. (1984). The plastic spin concept and a simple illustration of its role in finite plastic transformations. *Mechanics of Materials*, 3, 223-233. (cited p. 854).

Darling E.M. and Athanasiou K.A. (2003). Biomedical strategies for articular cartilage regeneration. *Annals Biomedical Engng.*, 31, 1114-1124.

Davisson T., Kunig S., Chen A., Sah R. and Ratcliffe A. (2002). Static and dynamic compressions modulate matrix metabolism in tissue engineered cartilage. *J. Orthopaedic Research*, 20, 842-848. (cited p. 828).

Daxer A. and Fratzl P. (1997). Collagen fibril orientation in the human corneal stroma and its implications in keratoconus. *Investigative Ophthalmology & Visual Science*, 38, 121-129. (cited p. 708).

De S.K. and Aluru N.R. (2004). A chemo-electro-mechanical mathematical model for simulation of pH sensitive hydrogels. *Mechanics of Materials*, 36, 5-6, 395-410. (cited p. 626).

Debes J.C. and Y.C. Fung (1995). Biaxial mechanics of excised canine pulmonary arteries. *American J. Physiology - Heart Circulation Physiology*, 269, H433-H442. (cited p. 183).

de Dear R.J., Arens E., Zhang H. and Oguro M. (1997). Convective and radiative heat transfer coefficients for individual human body segments. *Int. J. Biometeorology*, 40(3), 141-156. (cited p. 318).

de Groot S.R. and Mazur P. (1962). *Non-Equilibrium Thermodynamics*. North-Holland, Amsterdam, the Netherlands. (cited pp. 104 and 131).

de Marsily G. (1986). *Quantitative hydrogeology*. Academic Press, San Diego, CA. (cited pp. 88, 337, and 566).

Deresiewicz H. (1960). The effect of boundaries on wave propagation in a liquid filled porous solid: I. Reflection of plane waves at a free plane boundary (non-dissipative case). *Bulletin Seismological Society America*, 50(4), 599-607. (cited p. 385).

Deresiewicz H. (1962). The effect of boundaries on wave propagation in a liquid filled porous solid: IV. Surface waves in a half-space. *Bulletin Seismological Society America*, 52(3), 627-638. (cited p. 388).

Diller K.R. and Zhu L. (2009). Hypothermia therapy for brain injury. *Annual Review Biomedical Engng.*, 11, 135-162. (cited p. 279).

Di Maio C. (1996). Exposure of bentonite to salt solution: osmotic and mechanical effects. *Géotechnique*, 46(4), 695-707.

Di Maio C. and Fenelli G. (1997). Influenza delle interazioni chimico-fisiche sulla deformabilità di alcuni terreni argillosi. *Rivista Italiana Geotecnica*, 1, 695-707. (cited p. 459).

Di Maio C. (1998). Discussion on Exposure of bentonite to salt solution: osmotic and mechanical effects. *Géotechnique*, 48(3), 433-436. (cited p. 459).

Di Maio C., Santoli L. and Schiavone P. (2004). Volume change behaviour of clays: the influence of mineral composition, pore fluid composition and stress state. *Mechanics of Materials*, 36(5-6), 435-451. (cited p. 457).

DiMicco M.A. and Sah R.L. (2003). Dependence of cartilage matrix composition on biosynthesis, diffusion and reaction. *Transport in Porous Media*, 50(1-2), 57-73. (cited pp. 499, 831, and 847).

Di Silvestro M. R. and Suh J.K.F. (2001). A cross-validation of the biphasic poroviscoelastic model of articular cartilage in unconfined compression, indentation, and confined compression. *J. Biomechanics*, 34(4), 519-525. (cited p. 242).

Dohlman C.H., Hedbys B.O. and Mishima S. (1962). The swelling pressure of the corneal stroma. *Investigative Ophthalmology*, 1(2), 158-162. (cited p. 722).

Dolmans D.E., Fukumura D. and Jain R.K. (2003). Photodynamic therapy for cancer. *Nature Reviews*, 3(5), 380-387. (cited p. 1002).

Dou Y., Zhao J. and Gregersen H. (2003). Morphology and stress-strain properties along the small intestine in the rat. *ASME J. Biomechanical Engng.*, 125, 266-273. (cited p. 851).

Doughty M.J. (2001). Changes in hydration, protein and proteoglycan composition of the collagen-keratocyte matrix of the bovine corneal stroma ex vivo in a bicarbonate-mixed salts solution, compared to other solutions. *Biochimica et Biophysica Acta*, 1525(1-2), 97-107. (cited pp. 709, 710, 719, and 725).

Doughty M.J. (2004). Impact of hypotonic solutions on the stromal swelling, lactate dehydrogenase and aldehyde dehydrogenase activity of the keratocytes of the bovine cornea. *Cell Biology International*, 28(8-9), 593-607.

Douthwaite W.A. (2003). The asphericity, curvature and tilt of the human cornea measured using a videokeratoscope. *Ophthalmic and Physiological Optics*, 23(2), 141-150.

Driessen N.J.B., Boerboom R.A., Huyghe J.M., Bouten C.V.C. and Baaijens F.P.T. (2003)a. Computational analyses of mechanically induced collagen fiber remodeling in the aortic heart valve. *ASME J. Biomechanical Engng.*, 125, 549-557. (cited p. 838).

Driessen N.J.B., Peters G.W.M., Huyghe J.M., Bouten C.V.C. and Baaijens F.P.T. (2003)b. Remodelling of continuously distributed collagen fibres in soft connective tissues. *J. Biomechanics*, 36(8), 1151-1158.

Drost M.R., Willems P., Snijders H., Huyghe J.M., Janssen J.D. and Huson A. (1995). Confined compression of canine annulus fibrosus under chemical and mechanical loading. *ASME J. Biomechanical Engng.*, 117, 390-396. (cited p. 802).

Drozdov A.D. and Khanina H. (1997). A model for the volumetric growth of a soft tissue. *Mathematical and Computer Modelling*, 25(2), 11-29. (cited pp. 110, 861, 871, 912, 922, 930, and 976).

Drumheller D.S. (1998). *Introduction to wave propagation in nonlinear fluids and solids.* Cambridge University Press, Cambridge, UK. (cited p. 82).

Drummond D.C., Meyer O., Hong K., Kirpokin D.B. and Papahadjopoulos D. (1999). Optimizing liposomes for delivery of chemotherapeutic agents to solid tumors. *Pharmacological Reviews*, 51(4), 691-743. (cited pp. 569 and 1000).

Duck F.A. (1990). *Physical properties of tissues.* Academic Press, New York. (cited p. 87).

Duguid J.O. and Lee P.C.Y. (1977). Flow in fractured porous media. *Water Resources Research*, 13(3), 558-566. (cited p. 321).

Durbin F. (1974). Numerical inversion of Laplace transformation: an efficient improvement to Dubner and Abate's method. *The Computer Journal*, 17(4), 371-376. (cited p. 258).

* * * * E * * * *

Ebara S., Iatridis J.C., Setton L.A., Foster R.J., Mow V.C. and Weidenbaum M. (1996). Tensile properties of nondegenerate human lumbar anulus fibrosus. *Spine*, 21(4), 452-461. (cited pp. 799 and 800).

Edwards A. and Prausnitz M.R. (1998). Fiber matrix model of sclera and corneal stroma for drug delivery to the eye. *J. BioEngineering, Food and Natural Products, American Institute of Chemical Engineers*, 44(1), 214-224. (cited pp. 561, 718, 719, 724, and 725).

Eikenes L., Bruland Ø.S., Brekken Ch. and de Lange Davies C. (2004). Collagenase increases the transcapillary pressure gradient and improves the uptake and distribution of monoclonal antibodies in osteosarcoma xenografts. *Cancer Research*, 64, 4768-4773. (cited pp. 568, 570, and 1003).

Eisenberg S.R. and Grodzinsky A.J. (1985). Swelling of articular cartilage and other connective tissues: electromechanochemical forces. *J. Orthopaedic Research*, 3, 148-159. (cited pp. 461, 462, 485, 498, 503, 505, 626, 640, 660, and 683).

Eisenberg S.R. and Grodzinsky A.J. (1987). The kinetics of chemically induced nonequilibrium swelling of articular cartilage and corneal stroma. *ASME J. Biomechanical Engng.*, 109, 79-89. (cited p. 506).

Einstein A. (1956). Investigations on the theory of the Brownian motion, 18 p., re-publication of the 1926 English translation by Dover Publications, Inc., Mineola, NY.

Elder S.H., Goldstein S.A., Kimura J.H., Soslowsky L.J. and Spengler D.M. (2001). Chondrocyte differentiation is modulated by frequency and duration of cyclic compressive loading. *Annals Biomedical Engng.*, 29, 476-482. (cited p. 829).

El-Kareh A.W., Braunstein S.L. and Secomb T.W. (1993). Effect of cell arrangement and interstitial volume fraction on the diffusivity of monoclonal antibodies in tissue. *Biophysical J.*, 64, 1638-1646. (cited p. 438).

Elisseeff J., Anseth K., Sims D., McIntosh W., Randolph M. and Langer R. (1999). Transdermal photopolymerization for minimally invasive implantation. *Proceedings National Academy Sciences USA*, 96(6), 3104-3107. (cited pp. 101 and 844).

Elliott D.M. and Setton L.A. (2000). A linear material model for fiber-induced anisotropy of the anulus fibrosus. *ASME J. Biomechanical Engng.*, 122, 173-179. (cited pp. 800 and 802).

Elliott D.M. and Setton L.A. (2001). Anisotropic and inhomogeneous tensile behavior of the human anulus fibrosus: experimental measurement and material model predictions. *ASME J. Biomechanical Engng.*, 123, 256-263. (cited pp. 799, 800, and 802).

Elliott D.M., Narmoneva D.A. and Setton L.A. (2002). Direct measurement of the Poisson's ratio of human patella cartilage in tension. *ASME J. Biomechanical Engng.*, 124, 223-228. (cited p. 163).

Elliott G.F. (1980). Measurement of the electric charge and ion binding of the protein filaments in intact muscle and cornea, with implications for filament assembly. *Biophysical J.*, 32(1), 95-97. (cited p. 735).

Elliott G.F., Goodfellow J.M. and Woolgar A.E. (1980). Swelling studies of bovine corneal stroma without bounding membranes. *J. Physiology*, 298, 453-470. (cited pp. 718 and 739).

Elliott G.F. and Hodson S.A. (1998). Cornea, and the swelling of polyelectrolyte gels of biological interest. *Reports Progress Physics*, 61, 1325-1365. (cited pp. 702, 705, 708, 709, 710, 711, 718, 719, and 736).

Elsheikh A., Wang D., Korecha A., Brown M. and Garway-Heath D. (2006). Evaluation of Goldmann applanation tonometry using a nonlinear finite element ocular model. *Annals Biomedical Engng.*, 34(10), 1628-1640. (cited p. 726).

Emkey G., Jin M., Grodzinsky A.J. and Trippel S. (2002). Combined effects of dynamic tissue shear deformation and insulin-like growth factor-I on chondrocyte biosynthesis in cartilage explants. *Transactions Orthopaedic Research Society*, 48th Annual Meeting, 134. (cited pp. 820 and 829).

Epstein M. and Maugin G.A. (2000). Thermomechanics of volumetric growth in uniform bodies. *Int. J. Plasticity*, 16, 951-978. (cited p. 850).

Ericksen J.L. (1991). *Introduction to the thermodynamics of solids*. Chapman & Hall, London, UK. (cited p. 128).

Eringen A.C. and Ingram J.D. (1965). A continuum theory for chemically reacting media-I. *Int. J. Engng. Science*, 3, 197-212. (cited pp. 104, 107, 110, 112, 114, 115, 117, 118, 120, and 355).

Eringen A.C. (1967). *Mechanics of continua*. John Wiley & Sons, New York. (cited pp. 14 and 18).

Eskey C.J., Koretsky A.P., Domach M.M. and Jain R.K. (1993). Role of oxygen vs. glucose in energy metabolism in a mammary carcinoma perfused ex vivo: direct measurement by 31P NMR. *Proceedings National Academy Sciences USA*, 90(7), 2646-2650. (cited p. 1011).

Ethier C. Ross, Johnson M. and Ruberti J. (2004). Ocular biomechanics and biotransport. *Annual Review Biomedical Engng.*, 6, 249-273. (cited pp. 722 and 725).

* * ** F *** *

Fang H.Y. (1997). *Introduction to environmental geotechnology*. CRC Press, Boca Raton, FL. (cited p. 445).

Farina A. and Preziosi L. (2001). On Darcy's law for growing porous media. *Int. J. Non-Linear Mechanics*, 37, 485-491. (cited p. 1010).

Farndale R.W., Buttle D.J. and Barrett A.J. (1986). Improved quantitation and discrimination of sulphated glycosaminoglycans by use of dimethylmethylene blue. *Biochimica et Biophysica Acta*, 883(2), 173-177. (cited p. 809).

Farraro K.F., Kim K.E., Woo S.L-Y., Flowers J.R. and McCullough M.B. (2014). Revolutionizing orthopaedic biomaterials: The potential of biodegradable and bioresorbable magnesium-based materials for functional tissue engineering. *J. Biomechanics*, 47, 1979-1986. (cited p. 8).

Fatt I. and Goldstick T. (1965). Dynamics of water transport in swelling membranes. *J. Cell Science*, 20(9), 962-989. (cited p. 722).

Fatt I. and Hedbys B.O. (1970)a. Flow conductivity of human corneal stroma. *Experimental Eye Research*, 10, 237-242. (cited p. 724).

Fatt I. and Hedbys B.O. (1970)b. Flow of water in the sclera. *Experimental Eye Research*, 10, 243-249. (cited p. 724).

Federico S., Grillo A., La Rosa G., Giaquinta G. and Herzog W. (2005). A transversely isotropic, transversely homogeneous microstructural-statistical model of articular cartilage. *J. Biomechanics*, 38(10), 2008-2018. (cited p. 660).

Fedorov F.I. (1968). *Theory of elastic waves in crystals*. Plenum Press, New York. (cited p. 139).

Feigenbaum H.P. and Dafalias Y.F. (2008). Simple model for directional distortional hardening in metal plasticity within thermodynamics. *ASCE J. Engng. Mechanics*, 134(9), 730-738. (cited pp. 903 and 908).

Fermor B., Weinberg J.B., Pisetsky D.S., Misukonis M.A., Banes A.J. and Guilak F. (2001). The effects of static and intermittent compression on nitric oxide production in articular cartilage explants. *J. Orthopaedic Research*, 19, 729-737. (cited p. 830).

Fernandes J.C., Martel-Pelletier J. and Pelletier J.P. (2002). The role of cytokines in osteoarthritis pathophysiology. *Biorheology*, 39, 237-246. (cited p. 501).

Fischbarg J. and Diecke F.P.J. (2005). A mathematical model of electrolyte and fluid transport across corneal endothelium. *J. Membrane Biology*, 203, 41-56. (cited p. 714).

Fischbarg J., Diecke F.P.J., Iserovich P. and Rubashkin A. (2006). The role of the tight junction in paracellular fluid transport across corneal endothelium. Electro-osmosis as a driving force. *J. Membrane Biology*, 210, 117-130. (cited p. 714).

Fisher J.P., Dean D., Engel P.S. and Mikos A.G. (2001). Photoinitiated polymerization of biomaterials. *Annual Review Materials Research*, 31, 171-181. (cited p. 843).

Fletcher R. (1990). *Practical methods of optimizaton*. 2nd edition. John Wiley & Sons Ltd, Chichester. (cited pp. 178 and 195).

Flory P.J. (1953). *Principles of polymer chemistry*. Cornell University Press, Ithaca, NY. (cited p. 439).

Flory P.J. (1961). Thermodynamic relations for high elastic materials. *Transactions Faraday Society*, 57, 829-838. (cited p. 189).

Fortier L.A., Balkman C.E., Sandell L.J., Ratcliffe A. and Nixon A.J. (2001). Insulin-like growth factor-I gene expression patterns during spontaneous repair of acute articular cartilage injury. *J. Orthopaedic Research*, 19(4), 720-728. (cited p. 819).

François M., Geymonat G. and Berthaud Y. (1998). Determination of the symmetries of an experimentally determined stiffness tensor: application to acoustic measurements. *Int. J. Solids Structures*, 35(31-32), 4091-4106. (cited p. 139).

Frank E.H. and Grodzinsky A.J. (1987). Cartilage electromechanics, *J. Biomechanics*, 20(6); I. Electrokinetic transduction and the effects of electrolyte pH and ionic strength, 615-627; II. A continuum model of cartilage electrokinetics and correlation with experiments, 629-639. (cited pp. 495, 505, 612, 614, 615, 627, 639, 648, and 651).

Frank E.H., Grodzinsky A.J., Phillips S.L. and Grimshaw P.E. (1990). Physicochemical and bioelectrical determinants of cartilage material properties. In *Biomechanics of diarthrodial joints*, Mow V.C., Ratcliffe A. and Woo S. L.-Y. eds., Springer-Verlag, New York, vol I, 261-282. (cited p. 643).

Frank E.H., Jin M., Loening A.M., Levenston M.E. and Grodzinsky A.J. (2000). A versatile shear and compression apparatus for mechanical stimulation of tissue culture explants. *J. Biomechanics*, 33, 1523-1527. (cited pp. 826 and 829).

Fratzl P. (2003). Cellulose and collagen: from fibres to tissues. *Current Opinion in Colloid & Interface Science*, 8(1), 32-39. (cited p. 493).

Freed L., Vunjak-Novakovic G., Marquis J.C. and Langer R. (1994). Kinetics of chondrocyte growth in cell-polymer implants. *Biotechnology and Bioengineering*, 43, 597-604.

Freed L., Langer R., Martin I., Pellis N.R. and Vunjak-Novakovic G. (1997). Tissue engineering of cartilage in space. *Proceedings National Academy Sciences USA*, 94, 13885-13890. (cited p. 823).

Frenkel S.R. and di Cesare P.E. (1999). Degradation and repair of articular cartilage. *Frontiers in Bioscience*, 4, D671-685. (cited pp. 501, 808, 817, and 820).

Freund D.E., McCally R.L and Farrell R.A. (1986). Effects of fibril orientations on light scattering in the cornea. *J. Optical Society America A*, 3, 1970-1982. (cited pp. 703 and 727).

Freund L.B. (1990). *Dynamic fracture mechanics*. Cambridge Monographs on Mechanics and Applied Mathematics. Cambridge University Press, Cambridge, UK. (cited p. 404).

Friedl P., Zänker K.S. and Bröcker E.B. (1998). Cell migration strategies in 3D extracellular matrix: differences in morphology, cell matrix interactions, and integrin function. *Microscopy Research and Technique*, 43(5), 369-378. (cited p. 1014).

Friedman M.H. (1972). A quantitative description of equilibrium and homeostatic thickness regulation in the in vivo cornea. *Biophysical J.*, 12, 666-682.

Friedman M.H. (1973). Unsteady transport and hydration dynamics in the in vivo cornea. *Biophysical J.*, 13, 890-910. (cited p. 703).

Frijns A.J.H., Huyghe J.M. and Janssen J.D. (1997). A validation of the quadriphasic mixture theory for intervertebral disc tissue. *Int. J. Engng. Science*, 35(15), 1419-1429. (cited pp. 505, 670, and 802).

Fu B.M., Chen B. and Chen W. (2003). An electrodiffusion model for effects of surface glycocalyx layer on microvessel permeability. *American J. Physiology - Heart Circulation Physiology*, 284, H1240-H1250. (cited p. 574).

Fujita Y., Duncan N.A. and Lotz J.C. (1997). Radial tensile properties of the lumbar annulus fibrosus are site and degeneration dependent. *J. Orthopaedic Research*, 15(6), 814-819.

Fujita Y., Wagner D.R., Biviji A.A., Duncan N.A. and Lotz J.C. (2000). Anisotropic shear behavior of the lumbar annulus fibrosus: effect of harvest site and tissue prestrain. *Medical Engineering & Physics*, 22(5), 349-357.

Fullwood N.J., Tuft S.J., Malik N.S., Meek K.M., Ridgway A.E.A. and Harrison R.J. (1992). Synchrotron X-ray diffraction studies of Keratoconus corneal stroma. *Investigative Ophthalmology & Visual Science*, 33(5), 1734-1741. (cited p. 707).

Fung Y.C. (1965). *Foundations of solid mechanics*. Prentice Hall, Englewood Cliffs, NJ. (cited pp. 13, 28, and 133).

Fung Y.C. (1977). *A first course in continuum mechanics*. 2nd ed., Prentice Hall, Englewood Cliffs, NJ. (cited pp. 13, 85, 142, 143, and 364).

Fung Y.C. (1984). *Biodynamics. Circulation.* Springer-Verlag, New York.

Fung Y.C. and Liu S.Q. (1989). Change of residual strains in arteries due to hypertrophy caused by aortic constriction. *Circulation Research*, 65(5), 1340-1349. (cited p. 879).

Fung Y.C. (1990). *Biomechanics. Motion, flow, stress and growth.* Springer-Verlag, New York. (cited pp. 99, 557, 566, 571, 575, 576, 588, 845, 870, and 890).

Fung Y.C. (1991). What are the residual stresses doing in our blood vessels? *Annals Biomedical Engng.*, 19(3), 237-249. (cited p. 890).

Fung Y.C. (1993). *Biomechanics: Mechanical properties of living tissues.* Springer Verlag, New York. (cited pp. 242, 249, 252, 253, 254, 258, 493, 720, and 998).

* * * * G * * * *

Gajo A. (1996). The effects of inertial coupling in the interpretation of dynamic soil tests. *Géotechnique*, 46(2), 245-257. (cited pp. 205 and 377).

Gajo A., Loret B. and Hueckel T. (2002). Electro-chemo-mechanical couplings in saturated porous media: elastic-plastic behaviour of heteroionic expansive clays. *Int. J. Solids Structures*, 39(16), 4327-4362. (cited pp. 121 and 954).

Gajo A. and Loret B. (2003). Finite element simulations of chemo-mechanical coupling in elastic-plastic homoionic expansive clays. *Computer Methods in Applied Mechanics and Engineering*, 192(31-32), 3489-3530. (cited pp. 124, 545, and 657).

Gajo A. and Loret B. (2004). Transient analysis of ionic replacement in elastic-plastic expansive clays. *Int. J. Solids Structures*, 41(26), 7493-7531. (cited p. 124).

Gajo A. and Loret B. (2007). The mechanics of active clays circulated by salts, acids and bases. *J. Mechanics Physics Solids*, 55(8), 1762-1801. (cited p. 631).

Galban C.J. and Locke B.R. (1997). Analysis of cell growth in a polymer scaffold using a moving boundary approach. *Biotechnology and Bioengineering*, 56, 422-432. (cited pp. 822, 832, and 835).

Galban C.J. and Locke B.R. (1999)a. Effects of spatial variation of cells and nutrient and product concentrations coupled with product inhibition on cell growth in a polymer scaffold. *Biotechnology and Bioengineering*, 64(6), 633-643. (cited pp. 553, 832, 833, 840, 846, and 856).

Galban C.J. and Locke B.R. (1999)b. Analysis of cell growth kinetics and substrate diffusion in a polymer scaffold. *Biotechnology and Bioengineering*, 65(2), 121-132. (cited pp. 832, 833, 840, 846, and 856).

Gantmacher F.R. (1959). *Matrix theory.* Vols. I and II. Chelsea Publishing Company, New York. (cited pp. 40, 41, and 168).

Garcia A.M., Frank E.H., Grimshaw P.E. and Grodzinsky A.J. (1996). Contribution of fluid convection and electrical migration to transport in cartilage: relevance to loading. *Archives Biochemistry Biophysics*, 333, 317-325. (cited pp. 471, 805, and 820).

Garcia A.M., Frank E.H., Trippel S.B. and Grodzinsky A.J. (1997). IGF-I transport in cartilage: effects of binding and intratissue fluid flow. *Transactions Orthopaedic Research Society*, 43rd Annual Meeting, 410. (cited p. 820).

Garcia A.M., Szasz N., Trippel S.B., Morales T.I., Grodzinsky A.J. and Frank E.H. (2003). Transport and binding of insulin-like growth factor I through articular cartilage. *Archives Biochemistry Biophysics*, 415, 69-79. (cited p. 553).

Garikipati K., Arruda E.M., Grosh K., Narayanan H. and Calve S. (2004). A continuum treatment of growth in biological tissue: the coupling of mass transport and mechanics. *J. Mechanics Physics Solids*, 52(7), 1595-1625. (cited p. 866).

Gefen A. and Dilmoney B. (2007). Mechanics of the normal woman's breast. *Technology and Health Care*, 15, 259-271. (cited p. 988).

Gelet R. (2011). Thermo-hydro-mechanical study of deformable porous media with double porosity in local thermal non-equilibrium. PhD thesis, Institut National Polytechnique de Grenoble, France, and The University of New South Wales, Sydney, Australia. (cited p. 322).

Genes N.G., Rowley J.A., Mooney D.J and Bonassar L.J. (1999). Culture of chondrocytes in RGD-bonded alginate: effects on mechanical and biosynthetic properties. ASME Bioengineering Conference, June 16-20, 1999, Big Sky, MT. (cited p. 830).

Genes N.G. and Bonassar L.J. (2001). Regulation of chondrocyte adhesion to modified alginate surfaces. ASME Bioengineering Conference, BED-Vol. 50, 631-632.

Ghassemi A. and Diek A. (2002). Porothermoelasticity of swelling shales. *J. Petroleum Science and Engng.*, 34 123-135. (cited p. 310).

Gibson L.J. and Ashby M.F. (1988). *Cellular Solids. Structure and Properties.* Pergamon Press, Oxford, UK. (cited pp. 8, 869, and 872).

Gilbert S.F. (1996). *Biologie du Développement.* Translation of the 4th American edition, De Boeck Université, Paris, Bruxelles. (cited p. 575).

Gill B.J. and West J.L. (2014). Modeling the tumor extracellular matrix: Tissue engineering tools repurposed towards new frontiers in cancer biology. *J. Biomechanics*, 47, 1969-1978.
http://dx.doi.org/10.1016/j.jbiomech.2013.09.029 (cited p. 7).

Girton T.S., Oegema T.R., Grassl T.D., Isenberg B.C. and Tranquillo R.T. (2000). Mechanisms of stiffening and strengthening in media-equivalents fabricated using glycation. *ASME J. Biomechanical Engng.*, 122, 216-223. (cited p. 837).

Girton T.S., Barocas V.H. and Tranquillo R.T. (2002). Confined compression of a tissue-equivalent: collagen fibril and cell alignment in response to anisotropic strain. *ASME J. Biomechanical Engng.*, 124, 568-575. (cited pp. 493, 837, and 951).

Giurea A., DiMicco M.A., Akeson W.H. and Sah R.L. (2002). Development-associated differences in integrative cartilage repair: roles of biosynthesis and matrix. *J. Orthopaedic Research*, 20, 1274-1281. (cited p. 830).

Giurea A., Klein T.J., Chen A.C., Goomer R.S., Coutts D.T., Akeson W.H., Amiel D. and Sah R.L. (2003). Adhesion of perichondral cells to a polylactic acid scaffold. *J. Orthopaedic Research*, 21, 584-589. (cited p. 830).

Glansdorff P. and Prigogine I. (1971). *Structure, stabilité et fluctuations.* Masson et cie, Paris, France. (cited p. 121).

Glowacki J. (2000). *In vitro* engineering of cartilage. *J. Rehabilitation Research & Development*, 37(2), 171-177. (cited p. 808).

Goldstein H. (1980). *Classical mechanics.* Addison-Wesley, Reading, MA. (cited pp. 39 and 41).

Gooch K.J., Kwon J.H., Blunk T., Langer R., Freed L.E. and Vunjak-Novakovic G. (2001). Effects of mixing intensity on tissue-engineered cartilage. *Biotechnology and Bioengineering*, 72(4), 402-407. (cited pp. 834, 836, and 846).

Gordon V.D., Valentine M.T., Gardel M.L., Andor-Ardó D., Dennison S., Bogdanov A.A., Weitz D.A. and Deisboeck T.S. (2003). Measuring the mechanical stress induced by an expanding multicellular tumor system: a case study. *Experimental Cell Research*, 289(1), 58-66. (cited p. 1006).

Gradshteyn I.S. and Ryzhik I.M. (1980). *Tables of integrals, series and products.* 6th printing. Academic Press, Inc., San Diego. (cited p. 1027).

Graff K.F. (1991). *Wave motion in elastic solids.* Dover Publications, Inc., New York. (cited p. 388).

Grande D.A. and Nixon A.J. (2000). Orthopaedic gene therapy. Chondroprogenitors. *Clinical Orthopaedics and Related Research*, 379S, S222-S224. (cited p. 501).

Grassl E.D., Oegema T.R. and Tranquillo R.T. (2002). Fibrin as an alternative biopolymer to type-I collagen for the fabrication of a media equivalent. *J. Biomedical Materials Research*, 60(4), 607-612. (cited p. 837).

Gray M.L., Pizzanelli A.M., Grodzinsky A.J. and Lee R.C. (1988). Mechanical and physicochemical determinants of the chondrocyte biosynthetic response. *J. Orthopaedic Research*, 6(6), 777-792. (cited p. 828).

Green A.E. and Naghdi P.M. (1969). On basic equations for mixtures. *Quarterly J. Mechanics and Applied Mathematics*, 22(4), 427-438. (cited p. 366).

Green A.E. and Laws N. (1972). On the entropy production inequality. *Archives Rational Mechanics Analysis*, 45(1), 47-53. (cited p. 285).

Green A.E. and Lindsay K.A. (1972). Thermoelasticity. *J. Elasticity*, 2(1), 1-7. (cited p. 285).

Greenleaf J.F., Fatemi M. and Insana M. (2003). Selected methods for imaging elastic properties of biological tissues. *Annals Biomedical Engng.*, 5, 57-78. (cited pp. 1039 and 1040).

Griffon-Etienne G., Boucher Y., Brekken C., Suit H.D. and Jain R.K. (1999). Taxane-induced apoptosis decompresses blood vessels and lowers interstitial fluid pressure in solid tumors: clinical implications. *Cancer Research*, 59, 3776-3782. (cited p. 998).

Grimsrud Ch.D., Romano P.R., D'Souza M., Puzas J.E., Schwarz E.M., Reynolds P.R., Roiser R.N. and O'Keefe R.J. (2001). BMP signaling stimulates chondrocyte maturation and the expression of indian hedgehog. *J. Orthopaedic Research*, 19(1), 18-25. (cited p. 820).

Grodzinsky A.J. and Melcher J.R. (1976). Electromechanical transduction with charged polyelectrolyte membranes. *IEEE Transactions on Biomedical Engng.*, 23(6), 421-433. (cited p. 492).

Grodzinsky A.J., Roth V., Myers E.R., Grossman W.D. and Mow V.C. (1981). The significance of electromechanical and osmotic forces in the nonequilibrium swelling behavior of articular cartilage in tension. *ASME J. Biomechanical Engng.*, 103, 221-231. (cited pp. 468, 615, 626, 628, 631, 634, 649, 657, 659, 661, 663, and 665).

Grodzinsky A.J., Berger E., Hung H.K., Frank E.H. and Hunziker E.B. (1999). Compression of cartilage alters cartilage morphology of intracellular organelles: a potential link between mechanical stimulation and aggrecan structure. *Transactions Orthopaedic Research Society*, 45th Annual Meeting, 671. (cited pp. 828 and 845).

Grodzinsky A.J., Levenston M.E., Jin M. and Frank E.H. (2000). Cartilage tissue remodeling in response to mechanical forces. *Annual Review Biomedical Engng.*, 2, 691-713. (cited pp. 805 and 808).

Gross B. (1953). *Mathematical structure of the theories of viscoelasticity*. Hermann, Paris, France. (cited pp. 245 and 253).

Grynpas M.D, Eyre D.R. and Kirschner D.A. (1980). Collagen type II differs from type I in native molecular packing. *Biochimica et Biophysica Acta*, 626, 346-355.

Gu W.Y., Lai W.M. and Mow V.C. (1993). Transport of fluid and ions through a porous-permeable charged-hydrated tissue, and streaming potential data on normal bovine articular cartilage. *J. Biomechanics*, 26(6), 709-723.

Gu W.Y., Lai W.M. and Mow V.C. (1997). A triphasic analysis of negative osmotic flows through charged-hydrated soft tissues. *J. Biomechanics*, 30(1), 71-78. (cited pp. 512 and 609).

Gu W.Y., Lai W.M. and Mow V.C. (1998). A mixture theory for charged-hydrated soft tissues containing multi-electrolytes: passive transport and swelling behaviors. *ASME J. Biomechanical Engng.*, 120, 169-180. (cited pp. 508, 591, 592, 602, 612, and 622).

Gu W.Y., Mao X.G., Foster R.J., Weidenbaum M. and Mow V.C. (1999). The anisotropic hydraulic permeability of human lumbar anulus fibrosus. Influence of age, degeneration, direction and water content. *Spine*, 24(23), 2449-2455. (cited p. 724).

Gu W.Y., Yao H., Vega A.L. and Flager D. (2004). Diffusivity of ions in agarose gels and intervertebral disc: effect of porosity. *Annals Biomedical Engng.*, 32(12), 1710-1717. (cited p. 612).

Guccione J.M., McCulloch A. and Waldman L.K. (1991). Passive material properties of intact ventricular myocardium determined from a cylindrical model. *ASME J. Biomechanical Engng.*, 113, 42-55. (cited p. 192).

Guéguen Y. and Palciauskas V. (1992). *Introduction à la physique des roches*. Hermann, Paris, France. (cited pp. 561 and 563).

Guerin H.A.L. and Elliott D.M. (2005). The role of fiber-matrix interactions in a nonlinear fiber-reinforced strain energy model of tendon. *ASME J. Biomechanical Engng.*, 127, 345-350. (cited p. 801).

Guerin H.A.L. and Elliott D.M. (2007). Quantifying the contributions of structure to annulus fibrosus mechanical function using a nonlinear, anisotropic, hyperelastic model. *J. Orthopaedic Research*, 25(4), 508-516. (cited pp. 790 and 801).

Guido S. and Tranquillo R.T. (1993). A methodology for the systematic and quantitative study of cell contact guidance in oriented collagen gels: correlation of fibroblast orientation and gel birefringence. *J. Cell Science*, 105(2), 317-331. (cited p. 1006).

Guilak F., Jones W.R., Ting-Beall H.P. and Lee G.M. (1999). The deformation behavior and mechanical properties of chondrocytes in articular cartilage. *Osteoarthritis and Cartilage*, 7(1), 59-70. (cited pp. 805 and 827).

Gullino P.M., Grantham F.H. and Clark S.H. (1962). The collagen content of transplanted tumors. *Cancer Research*, 22, 1031-1037. (cited pp. 570, 993, and 998).

Gullino P.M., Grantham F.H. and Smith S.H. (1965). The interstitial water space of tumors. *Cancer Research*, 25, 727-731. (cited pp. 570 and 993).

Gullino P.M., Grantham F.H. and Courtney A.H. (1967)a. Utilization of oxygen by transplanted tumors in vivo. *Cancer Research*, 27, 1021-1030. (cited p. 1011).

Gullino P.M., Grantham F.H. and Courtney A.H. (1967)b. Glucose consumption by transplanted tumors in vivo. *Cancer Research*, 27, 1031-1040. (cited p. 1011).

Gullino P.M., Grantham F.H., Courtney A.H. and Losonczy I. (1967)c. Relationship between oxygen and glucose consumption by transplanted tumors in vivo. *Cancer Research*, 27, 1041-1052. (cited p. 1011).

Gullino P.M. (1975). Extracellular compartments of solid tumors. *Cancer*, 3, 327-354. (cited p. 993).

Guo J., Jourdian G. and MacCallum D. (1989). Culture and growth characteristics of chondrocytes encapsulated in alginate beads. *Connecting Tissue Research*, 19, 277-297. (cited p. 822).

Guo Z.H. (1984). Rates of stretch tensors. *J. Elasticity*, 14, 263-267. (cited p. 23).

Guo Z.Y., Peng X.Q. and Moran B. (2006). A composites-based hyperelastic constitutive model for soft tissue with application to the human annulus fibrosus. *J. Mechanics Physics Solids*, 54, 1952-1971. (cited p. 781).

Gurtin M.E. (1972). The linear theory of elasticity. In *Encyclopedia of physics*, vol. VIa/2, C. Truesdell ed., Springer-Verlag, Berlin, Germany, 1-295. (cited p. 892).

Gurtin M.E. (1981). *An introduction to continuum mechanics*. Academic Press, New York. (cited p. 59).

Gurtin M.E. and Spear K. (1983). On the relationship between the logarithmic strain rate and the stretching tensor. *Int. J. Solids Structures*, 19(5), 437-444. (cited p. 50).

* * * * H * * * *

Haase R. (1990). *Thermodynamics of irreversible processes*. Dover Publications, Inc., New York. (cited pp. 69, 104, 520, 533, and 620).

Hagendoorn J., Padera T.P., Kashiwagi S., Isaka N., Noda F., Lin M.I., Huang P.L., Sessa W.C., Fukumura D. and Jain R.K. (2004). Endothelial nitric oxide synthase regulates microlymphatic flow via collecting lymphatics. *Circulation Research*, 95(2), 204-209. (cited p. 999).

Hahnfeldt P., Panigrahy D., Folkman J. and Hlatky L. (1999). Tumor development under angiogenic signaling: a dynamic theory of tumor growth, treatment response, and post-vascular dormancy. *Cancer Research*, 59, 4770-4775. (cited p. 999).

Halphen B. and Nguyen Q.S. (1975). Sur les matériaux standards généralisés. *J. de Mécanique*, 14(1), 39-63. (cited p. 905).

Halpin J.C. and Tsai S.W. (1969). Effect of environment factors on composite materials. Air Force Technical Report AFML-TR-67-423, June 1969. (cited p. 797).

Hancock W.O., Martyn D.A. and Huntsman L.L. (1993). Ca^{2+} and segment length dependence of isometric force kinetics in intact ferret cardiac muscle. *Circulation Research*, 73(4), 603-610.

Hardin R.H. and Sloane N.J.A. (1996). McLaren's improved snub cube and other new spherical designs in three dimensions. *Discrete and Computational Geometry*, 15(4), 429-441. (cited p. 418).

Harrigan T.P. and Mann R.W. (1984). Characterization of microstructural anisotropy in orthotropic materials using a second rank tensor. *J. Materials Science*, 19(3), 761-767. (cited p. 434).

Harrigan T.P. and Hamilton J.J. (1992). An analytical and numerical study of the stability of bone remodelling theories: dependence on microstructural stimulus. *J. Biomechanics*, 25(5), 477-488. (cited p. 869).

Harrigan T.P. and Hamilton J.J. (1993). Finite element simulation of adaptive bone remodelling: a stability criterion and a time stepping method. *Int. J. Numerical Methods Engineering*, 36(5), 837-854. (cited p. 869).

Harrigan T.P. and Konofagou E.E. (2004). Estimation of material elastic moduli in elastography: a local method, and an investigation of Poisson's ratio sensitivity. *J. Biomechanics*, 37(8), 1215-1221. (cited p. 1041).

Hart R.W. and Farrell R.A. (1969). Light scattering in the cornea. *J. Optical Society America*, 59(6), 766-774. (cited pp. 706, 727, and 729).

Hartmann S. and Neff P. (2003). Polyconvexity of generalized polynomial-type hyperelastic strain energy functions for near-incompressibility. *Int. J. Solids Structures*, 40, 2767-2791. (cited p. 181).

Hascall V.C., Heinegard D.K. and Wright T.N. (1991). Proteoglycans: metabolism and pathology. In *Cell biology of extracellular Matrix*, Hay E.D. ed., Plenum Press, New York, 149-175. (cited p. 806).

Haslach H.W. (2005). Nonlinear viscoelastic, thermodynamically consistent, models for biological soft tissue. *Biomechanics and Modeling in Mechanobiology*, 3, 172-189. (cited p. 272).

Häuselmann H.J., Fernandes R.J., Mok S.S., Schmid T.M., Block J.A., Aydelotte M.B., Kuettner K.E. and Thonar E.J.-M.A. (1994). Phenotypic stability of bovine articular chondrocytes after long-term culture in alginate beads. *J. Cell Science*, 107, 17-27. (cited p. 822).

Hayes S., Boote C., Tuft S.J., Quantock A.J. and Meek K.M. (2007)a. A study of corneal thickness, shape and collagen organisation in keratoconus using videokeratography and x-ray scattering techniques. *Experimental Eye Research*, 84, 423-434. (cited pp. 704, 707, and 708).

Hayes S., Boote C., Lewis J., Sheppard J., Abahussin M., Quantock A.J., Purslow C., Votubra M. and Meek K.M. (2007)b. Comparative study of fibrillar collagen arrangement in the corneas of primates and other mammals. *The Anatomical Record*, 290, 1542-1550. (cited p. 708).

He Q.-C. and Curnier A. (1995). A more fundamental approach to damaged elastic stress-strain relations. *Int. J. Solids Structures*, 32(10), 1433-1457.

Hedbys B.O. and Mishima S. (1962). Flow of water in the corneal stroma. *Experimental Eye Research*, 1, 262-275. (cited pp. 597, 723, and 724).

Hedbys B.O. and Dohlman C. (1963). A new method for the determination of the swelling pressure of the corneal stroma *in vitro*. *Experimental Eye Research*, 2, 122-129. (cited pp. 722 and 740).

Hedbys B.O., Mishima S. and Maurice D.M. (1963). The imbibition pressure of the corneal stroma. *Experimental Eye Research*, 2, 99-111. (cited p. 722).

Hedbys B.O. and Mishima S. (1966). The thickness-hydration relationship of the cornea. *Experimental Eye Research*, 5(3), 221-228. (cited p. 718).

Hegedus D.M. and Cowin S.C. (1976). Bone remodeling II: Small strain adaptive elasticity. *J. Elasticity*, 6(4), 337-352. (cited pp. 110, 872, and 909).

Heidug W.K. and Wong S.-W. (1996). Hydration swelling of water-absorbing rocks: a constitutive model. *Int. J. Numerical Analytical Methods Geomechanics*, 20, 403-430. (cited p. 521).

Hellio Le Graverand M.-P., Eggerer J., Vignon E., Otterness I.G., Barclay L. and Hart D.A. (2002). Assessment of specific mRNA levels in cartilage regions in a lapine model of osteoarthritis. *J. Orthopaedic Research*, 20, 535-544.

Helmlinger G., Netti P.A., Lichtenbeld H.C., Melder R.J. and Jain R.K. (1997)a. Solid stress inhibits the growth of multicellular tumor spheroids. *Nature Biotechnology*, 15(8), 778-783. doi:10.1038/nbt0897-778 (cited pp. 7, 847, 870, 871, 989, 998, 1004, 1009, 1010, 1035, and 1038).

Helmlinger G., Yuan F., Dellian M. and Jain R.K. (1997)b. Interstitial pH and pO2 gradients in solid tumors in vivo: high resolution measurements reveal a lack of correlation. *Nature Medicine*, 3(2), 177-182. doi:10.1038/nm0297-177 (cited pp. 997 and 999).

Helmlinger G., Sckell A., Dellian M., Forbes N.S. and Jain R.K. (2002). Acid production in glycolysis-impaired tumors provides new insights into tumor metabolism. *Clinical Cancer Research*, 8(4), 1284-1291. (cited p. 997).

Henderson B.W. and Dougherty T.J. eds. (1992). *Photodynamic therapy. Basic principles and clinical applications*. CRC Press, Boca Raton, FL. (cited p. 1002).

Henkensiefken R., Bentz D., Nantung T. and Weiss J. (2009). Volume change and cracking in internally cured mixtures made with saturated lightweight aggregate under sealed and unsealed conditions. *Cement and Concrete Composites*, 31, 427-437. (cited p. 81).

Henson G.H. (2013). Engineering models for synthetic microvascular materials with interphase mass, momentum and energy transfer. *Int. J. Solids Structures*, 50 (14-15), 2371-2382. (cited p. 321).

Herbage D., Bouillet J. and Bernengo J.C. (1977). Biochemical and physicochemical characterization of pepsin-solubilized type-II collagen from bovine articular cartilage. *Biochemical J.*, 161, 303-312. (cited pp. 490, 809, and 841).

Hesse K., Sloan I.H. and Womersley R.S. (2010). Numerical integration on the unit sphere. In *Handbook of geomathematics*, Freeden W., Nashed Z.M. and Sonar T. eds., Springer-Verlag, 1187-1220. (cited p. 418).

Hill R. (1962). Acceleration waves in solids. *J. Mechanics Physics Solids*, 10, 1-16.

Hill R. (1968). On constitutive inequalities for simple materials-I. *J. Mechanics Physics Solids*, 16, 229-242. (cited p. 38).

Hill R. (1970). Constitutive inequalities for isotropic elastic solids under finite strain. *Proceedings Royal Society London*, A314, 457-472. (cited p. 49).

Hill R. (1978). Aspects of invariance in solid mechanics. *Advances in applied mechanics*, 18, 1-75, C.S. Yih ed., Academic Press, New York. (cited p. 39).

Hilmas D.E. and Gilette E.L. (1974). Morphometric analyses of the microvasculature of tumors during growth and after x-irradiation. *Cancer*, 33(1), 103-110. (cited p. 1020).

Hiltner A., Cassidy J.J. and Baer E. (1985). Mechanical properties of biological polymers. *Annual Review Materials Science*, 15, 455-482. (cited pp. 488, 798, and 799).

Hjortdal J.Ø. (1996). Regional elastic performance of the human cornea. *J. Biomechanics*, 29(7), 931-942. (cited p. 726).

Hodson S. (1974). The regulation of corneal hydration by a salt pump requiring the presence of sodium and bicarbonate ions. *J. Physiology*, 236, 271-302. (cited pp. 714, 718, 723, 725, and 753).

Hodson S., O'Leary D. and Watkins S. (1991). The measurement of ox corneal swelling pressure by osmometry. *J. Physiology*, 434, 399-408. (cited p. 719).

Hodson S., Kaila D., Hammond S., Rebello G. and Al-Omari Y. (1992). Transient chloride binding as a contributory factor to corneal stromal swelling in the ox. *J. Physiology*, 450, 89-103. (cited pp. 709, 718, 735, 736, 740, and 741).

Hodson S.A. (1997). Corneal stromal swelling. *Progress in Retinal and Eye Research*, 16(1), 99-116. (cited p. 723).

Hoeltzel D.A., Altman P., Buzard K. and Choe K.-I. (1992). Strip extensiometry for comparison of the mechanical response of bovine, rabbit, and human corneas. *ASME J. Biomechanical Engng.*, 114, 202-215. (cited pp. 708 and 720).

Hoffman A.H. and Grigg P. (2002). Using uniaxial pseudorandom stress stimuli to develop soft tissue constitutive equations. *Annals Biomedical Engng.*, 30(1), 44-53. (cited p. 274).

Hofmeyr J.H.S. and Cornish-Bowden A. (2000). Regulating the cellular economy of supply and demand. *Federation of European Biochemical Societies Letters*, 476, 47-51. (cited p. 579).

Hoger A. and Carlson D.E. (1984)a. Determination of the stretch and rotation in the polar decomposition of the deformation gradient. *Quarterly Applied Mathematics*, 42(1), 113-117. (cited pp. 21, 22, and 37).

Hoger A. and Carlson D.E. (1984)b. On the derivative of the square root of a tensor and Guo's rate theorems. *J. Elasticity*, 14, 329-336. (cited p. 23).

Hoger A. (1985). On the residual stress possible in an elastic body with material symmetry. *Archives Rational Mechanics Analysis*, 88, 271-290. (cited pp. 848 and 849).

Hoger A. (1987). The stress conjugate to logarithmic strain. *Int. J. Solids Structures*, 23(12), 1645-1656. (cited p. 49).

Hoger A. (1993). The constitutive equation for finite deformations of a transversely isotropic hyperelastic material with residual stress. *J. Elasticity*, 33, 107-118. (cited p. 849).

Hoger A. (1997). Virtual configurations and constitutive equations for residually stressed bodies with material symmetry. *J. Elasticity*, 48, 125-144. (cited p. 839).

Hoger J.H., Ilyin V., Forsyth S. and Hoger A. (2002). Shear stress regulates the endothelial Kir2.1 ion channel. *Proceedings National Academy Sciences USA*, 99(11), 7780-7785. (cited p. 825).

Hollenbeck K.J. (1998). INVLAP.M: A Matlab Function for Numerical Inversion of Laplace Transforms by the de Hoog Algorithm. Available online `http://www.isva.dtu.dk/staff/karl/invlap.htm` (cited p. 258).

Holmes M.H and Mow V.C. (1990). The non-linear characteristics of soft gels and hydrated connective tissues in ultrafiltration. *J. Biomechanics*, 23(11), 1145-1156. (cited p. 612).

Holzapfel G.A. and Weizsäcker H.W. (1998). Biomechanical behavior of the arterial wall and its numerical characterization. *Computers in Biology and Medicine*, 28(4), 377-392. (cited p. 190).

Holzapfel G.A. (2000). *Nonlinear solid mechanics. A continuum approach for engineering.* John Wiley & Sons, Ltd., Chichester, UK. (cited p. 174).

Holzapfel G.A., Gasser T.C. and Ogden R.W. (2000). A new constitutive framework for arterial wall mechanics and a comparative study of material models. *J. Elasticity*, 61(1-3), 1-48. (cited p. 190).

Holzapfel G.A., Gasser T.C. and Stadler M. (2002). A structural model for the viscoelastic behavior of arterial walls: continuum formulation and finite element analysis. *European J. Mechanics-A/Solids*, 21(3), 441-463. (cited pp. 173, 190, and 242).

Holzapfel G.A., Schulze-Bauer C.A.J., Feigl G. and Regitnig P. (2005). Single lamellar mechanics of the human lumbar anulus fibrosus. *Biomechanics and Modeling in Mechanobiology*, 3(3), 125-140. (cited pp. 797, 800, and 801).

Hon Y.C., Lu M.W., Xue W.M. and Zhou X. (1999). A new formulation and computation of the triphasic model for mechano-electrochemical mixtures. *Computational Mechanics*, 24, 155-165. (cited p. 670).

Hon Y.C., Lu M.W., Xue W.M. and Zhou X. (2002). Numerical algorithm for triphasic model of charged and hydrated soft tissues. *Computational Mechanics*, 29, 1-15. (cited p. 670).

Honig G. and Hirdes U. (1984). A method for the numerical inversion of the Laplace transform. *J. Computational and Applied Mathematics*, 10, 113-132. (cited p. 258).

Hoog F.R., Knight J.H. and Stokes A.N. (1982). An improved method for numerical inversion of Laplace transforms. *SIAM J. Scientific and Statistical Computing*, 3, 357-366. (cited p. 258).

Howland H.C., Rand R.H. and Lubkin S.R. (1992). A thin shell model for cornea and its application for corneal surgery. *Refractive Corneal Surgery*, 8(2), 183-186. (cited p. 720).

Hsu C.T. (1999). A closure model for transient heat conduction in porous media. *ASME J. Heat Transfer*, 121(3), 733-739. (cited p. 336).

Hu X., Adamson R.H., Liu B., Curry F.E. and Weinbaum S. (2000). Starling forces that oppose filtration after tissue oncotic pressure is increased. *American J. Physiology - Heart Circulation Physiology*, 279(4), H1724-H1736. (cited pp. 574 and 575).

Huang C.-Y., Stankiewicz A., Ateshian G.A., Bigliani L.U. and Mow V.C. (1999). Anisotropy, inhomogeneity, and tension-compression nonlinearity of human glenohumeral cartilage in finite deformation. *Transactions Orthopaedic Research Society*, 45th Annual Meeting, Anaheim, 95. (cited p. 163).

Huang C.Y., Mow V.C. and Ateshian G.A. (2001). The role of flow-independent viscoelasticity in the biphasic tensile and compressive responses of articular cartilage. *ASME J. Biomechanical Engng.*, 123, 410-417. (cited pp. 242 and 506).

Huang C.-Y., Soltz M.A., Kopacz M., Mow V.C. and Ateshian G.A. (2003). Experimental verification of the roles of intrinsic matrix visco-elasticity and tension-compression nonlinearity in the biphasic response of cartilage. *ASME J. Biomechanical Engng.*, 125, 84-93. (cited pp. 242, 254, and 255).

Huang Y. and Meek K.M. (1999). Swelling studies on the cornea and sclera: The effects of pH and ionic strength. *Biophysical J.*, 77, 1655-1665. (cited pp. 468, 626, 635, 707, 710, 711, 719, 743, 745, 746, and 747).

Huang Z.M. (2004). Progressive flexural failure analysis of laminated composites with knitted fabric reinforcement. *Mechanics of Materials*, 36, 239-260. (cited p. 797).

Hubbell J.A. (1999). Bioactive biomaterials. *Current Opinion in Biotechnology*, 10, 123-129.

Hubbert M.K. (1940). The theory of ground water motion. *J. Geology*, 48, 8, 785-944. (cited p. 561).

Hueckel T. and Drescher A. (1975). On dilatational effects of inelastic granular media. *Archiwum Mechaniki Stosowanej*, 27(1), 157-172. (cited p. 134).

Hughes T.J.R. (1987). *The finite element method. Linear static and dynamic finite element analysis* Prentice Hall, Inc., Englewood Cliffs, NJ. (cited p. 677).

Huiskes R., Weinans H., Grootenboer H.J., Dalstra M., Fudala B. and Slooff T.J. (1987). Adaptive bone remodeling theory applied to prosthetic design analysis. *J. Biomechanics*, 20(11-12), 1135-1150. (cited p. 869).

Hukins D.W.L. and Meakin J.R. (2000). Relationship between structure and mechanical function of the tissues of the intervertebral joint. *American Zoologist*, 40(1), 42-52. (cited p. 799).

Hulmes D.J.S. (2002). Building Collagen Molecules, Fibrils, and Suprafibrillar Structures. *J. Structural Biology*, 137(1-2), 2-10. (cited p. 706).

Humphrey J.D. (1995). Mechanics of the arterial wall: review and directions. *Critical Reviews in Biomedical Engineering*, 23(1-2), 1-162. (cited pp. 184 and 190).

Humphrey J.D. (1999). Remodeling of a collagenous tissue at fixed lengths. *ASME J. Biomechanical Engng.*, 121, 591-597.

Humphrey J.D. (2001). Stress, strain and mechanotransduction in cells. *ASME J. Biomechanical Engng.*, 123, 638-641.

Humphrey J.D. (2002). On mechanical modeling of dynamic changes in the structure and properties of adherent cells. *Mathematics and Mechanics of Solids*, 7, 521-539. (cited pp. 838 and 839).

Humphrey J.D. and Rajagopal K.R. (2002). A constrained mixture model for the growth and remodeling of soft tissues. *Mathematical Models and Methods in Applied Sciences*, 12(3), 407-430. (cited pp. 273, 838, 839, 840, 856, and 954).

Humphrey J.D. (2003). Continuum biomechanics of soft biological tissues. *Proceedings Royal Society London*, A459: 3-46. (cited pp. 830 and 838).

Humphrey J.D. and Rajagopal K.R. (2003). A constrained mixture model for arterial adaptations to a sustained step change in blood flow. *Biomechanics and Modeling in Mechanobiology*, 2, 109-126. (cited pp. 10, 273, 878, and 954).

Hunter Ch.J., Imler S.M., Malaviya P., Nerem R.M. and Levenston M.E. (2002). Mechanical compression alters gene expression and extracellular matrix synthesis by chondrocytes cultured in collagen I gels. *Biomaterials*, 23, 1249-1259.

Hunter Ch.J. and Levenston M.E. (2002)a. The influence of repair tissue maturation on the response to oscillatory compression in a cartilage defect repair model. *Biorheology*, 39, 79-88. (cited p. 838).

Hunter Ch.J. and Levenston M.E. (2002)b. Oscillatory compression stimulates matrix production in an *in vitro* articular cartilage repair model. *Transactions Orthopaedic Research Society*, 48th Annual Meeting, 250.

Hunter G.K. (1991). Role of proteoglycan in the provisional calcification of cartilage. A review and reinterpretation. *Clinical Orthopaedics and Related Research*, 262, 256-280. (cited pp. 596, 635, and 640).

Hunter P.J., McCulloch A.D. and ter Keurs H.E.D.J. (1998). Modelling the mechanical properties of cardiac muscle. *Progress in Biophysics & Molecular Biology*, 69(2-3), 289-331. (cited p. 470).

Hunter R.J. (1981). *Zeta potential in colloid science. Principles and applications*. Academic Press, London, UK.

Hurschler Ch., Provenzano P.P. and Vanderby R. (2003). Application of a probabilistic microstructural model to determine reference length and toe-to-linear region transition in fibrous connective tissue. *ASME J. Biomechanical Engng.*, 125, 415-422. (cited p. 432).

Huyghe J.M. and van Campen D.H. (1995). Finite deformation theory of hierarchically arranged porous solids -I. Balance of mass and momentum. *Int. J. Engng. Science*, 33, 1861-1871; -II. Constitutive behaviour. *Int. J. Engng. Science*, 33, 1873-1886.

Huyghe J.M. and Janssen J.D. (1997). Quadriphasic mechanics of swelling incompressible porous media. *Int. J. Engng. Science*, 35(8), 793-802.

Huyghe J.M. (1999). Intra-extrafibrillar mixture formulation of soft charged hydrated tissues. *J. Theoretical and Applied Mechanics*, 3, 37, 519-536. (cited pp. 508 and 591).

Huyghe J.M. and Janssen J.D. (1999). Thermo-chemo-electro-mechanical formulation of saturated charged porous solids. *Transport in Porous Media*, 34, 129-141. (cited pp. 508 and 591).

Huyghe J.M., Janssen Ch.F., Lanir Y., van Donkelaar C.C., Maroudas A. and van Campen D.H. (2002). Experimental measurement of electrical conductivity and electro-osmotic permeability of ionised porous media. In *Porous media: theoretical, experimental and numerical applications*, W. Ehlers and J. Bluhm eds., 295-313, Springer-Verlag, Berlin, Germany. (cited p. 615).

Huyghe J.M., Houben G.B., Drost M.R. and van Donkelaar C.C. (2003). An ionised/non-ionised dual porosity model of intervertebral disc tissue. *Biomechanics and Modeling in Mechanobiology*, 2(1), 3-19. (cited pp. 508 and 591).

* * * * I * * * *

IAPWS (1984). Representative equations for the viscosity of water substance. Sengers J.V. and Kamgar-Parsi B., *J. Physical Chemical Reference Data*, 13(1), 185-205. (cited pp. 87 and 96).

IAPWS (2002). The IAPWS Formulation 1995 for the Thermodynamic Properties of Ordinary Water Substance for General and Scientific Use. Wagner W. and Pruß A. eds., *J. Physical Chemical Reference Data*, 31, 387-535. (cited pp. 77 and 91).

IAPWS (2009). Revised Release on 'The IAPWS Formulation 1995 for the Thermodynamic Properties of Ordinary Water Substance for General and Scientific Use'. The International Association for the Properties of Water and Steam, Doorwerth, The Netherlands, September 2009. (cited pp. 77, 78, 87, and 91).

Iatridis J.C., Setton L.A., Foster R.J., Rawlins B.A., Weidenbaum M. and Mow V.C. (1996). Degeneration affects the anisotropic and nonlinear behaviors of human anulus fibrosus in compression. *J. Biomechanics*, 31, 535-544.

Iatridis J.C., Laible J.P. and Krag M.H. (2003). Influence of fixed charge density magnitude and distribution on the intervertebral disc: applications of a poroelastic and chemical electric (PEACE) model. *ASME J. Biomechanical Engng.*, 125, 12-24. (cited p. 802).

Ikenoue T., Trindade M.C.D., Lee M.S., Lin E.Y., Schurman D.J., Goodman S.B. and Lane Smith R. (2003). Mechanoregulation of human articular chondrocyte aggrecan and type II collagen expression by intermittent hydrostatic pressure *in vitro*. *J. Orthopaedic Research*, 21, 110-116. (cited p. 828).

Ingram J.D. and Eringen A.C. (1967). A continuum theory for chemically reacting media - II Constitutive equations of reacting fluid mixtures. *Int. J. Engng. Science*, 5, 289-322.

Isenberg B.C. and Tranquillo R.T. (2003). Long-term cyclic distension enhances the mechanical properties of collagen-based media-equivalents. *Annals Biomedical Engng.*, 31(8), 937-949. (cited pp. 836 and 837).

Itskov M. and Aksel N. (2004). A class of orthotropic and transversely isotropic hyperelastic constitutive models based on a polyconvex strain energy function. *Int. J. Solids Structures*, 41, 3833-3848. (cited p. 181).

* * * * J * * * *

Jabbour M.E. and Bhattacharya K. (2003). A continuum theory of multispecies thin solid film growth by chemical vapor deposition. *J. Elasticity*, 73, 13-74. (cited pp. 107 and 921).

Jackson A.R., Yuan T.-Y., Huang C.-Y. and Gu W.Y. (2009). A conductivity approach to measuring fixed charge density in intervertebral disc tissue. *Annals Biomedical Engng.*, 37(12), 2566-2573. (cited pp. 497, 798, and 802).

Jackson D.S. (1958). Some biochemical aspects of fibrogenesis and wound healing. *The New England J. of Medicine*, 259(17), 814-820. (cited pp. 493 and 951).

Jackson D.W., Scheer M.J. and Simon T.M. (2001). Cartilage substitutes: overview of basic science and treatment options. *J. American Academy Orthopaedic Surgeons*, 9(1), 37-52. (cited pp. 499 and 808).

Jackson G.W. and James D.F. (1986). The permeability of fibrous porous media. *Canadian J. Chemical Engng.*, 64(3), 364-374. (cited pp. 565 and 724).

Jackson T.L. and Byrne H.M. (2000). A mathematical model to study the effects of drug resistance and vasculature on the response of solid tumors to chemotherapy. *Mathematical Biosciences*, 164, 17-38. (cited p. 1012).

Jackson T.L. (2002). Vascular tumor growth and treatment: consequences of polyclonality, competition and dynamic vascular support. *J. Mathematical Biology*, 44(3), 201-226. (cited p. 1007).

Jackson T.L. and Byrne H.M. (2002). A mechanical model of tumor encapsulation and transcapsular spread. *Mathematical Biosciences*, 180, 307-328. (cited p. 1008).

Jackson T.L. (2003)a. Multiphase mechanics of tumor growth and control strategies. (cited p. 1007). http://www.math.ttu.edu.

Jackson T.L. (2003)b. Intracellular accumulation and mechanism of action of doxorubicin in a spatio-temporal tumor model. *J. Theoretical Biology*, 220, 201-213. (cited p. 1013).

Jadin K.D., Bae W.C., Schumacher B.L. and Sah R.L. (2007). 3-D imaging of chondrocytes in articular cartilage: growth-associated changes in cell organization. *Biomaterials*, 28(2), 230-239. (cited pp. 498, 499, and 965).

Jain M.K., Chernomorsky A., Silver F.H. and Berg R.A. (1988). Material properties of living soft tissue composites. *J. Biomedical Materials Research*, 22(2 Suppl.), 311-326. (cited p. 438).

Jain R.K. (1987)a. Transport of molecules in the tumor interstitium: a review. *Cancer Research*, 47(12), 3039-3051. (cited pp. 552, 553, 554, 564, 570, 571, 993, 1003, and 1010).

Jain R.K. (1987)b. Transport of molecules across tumor vasculature. *Cancer Metastasis Reviews*, 6(4), 559-593. (cited pp. 552, 569, 570, and 993).

Jain R.K. (1990). Physiological barriers to delivery of monoclonal antibodies and other macromolecules in tumors. *Cancer Research*, 50(3 Suppl), 814s-819s. (cited pp. 552, 568, 570, 991, 993, 1003, and 1006).

Jain R.K. (1999). Transport of molecules, particles, and cells in solid tumors. *Annual Review Biomedical Engng.*, 1, 241-263. (cited pp. 569, 988, 1003, and 1011).

Jain R.K. and Fenton B.T. (2002). Intratumoral lymphatic vessels: a case of mistaken identity or malfunction? *J. National Cancer Institute*, 94(6), 417-421. (cited p. 999).

Jain R.K. and Padera T.P. (2002). Prevention and treatment of lymphatic metastasis by antilymphangiogenic therapy. *J. National Cancer Institute*, 94(11), 785-787. (cited pp. 992 and 999).

Jain R.K. and Padera T.P. (2003). Lymphatics make the break. *Science*, 299, 209-210.

Jain R.K. and Stroh M. Zooming in and out with quantum dots. *Nature Biotechnology*, 22(8), 959-960.

Jain R.K. (2005). Normalization of tumor vasculature: an emerging concept in antiangiogenic therapy. *Science*, 307, 58-62. (cited pp. 998, 999, and 1000).

Jang S.H., Wientjes M.G., Lu D. and Au J.L.S. (2003). Drug delivery and transport to solid tumors. *Pharmaceutical Research*, 20(9), 1337-1350. (cited pp. 568, 988, 992, 1021, and 1022).

Jayasuriya A.C., Ghosh S., Scheinbeim J., Lubkin V., Bennett G. and Kramer Ph. (2003)a. A study of piezoelectric and mechanical anisotropies of the human cornea. *Biosensors and Bioelectronics*, 18, 381-387. (cited pp. 720 and 786).

Jayasuriya A.C., Scheinbeim J.I., Lubkin V., Bennett G. and Kramer P. (2003)b. Piezoelectric and mechanical properties in bovine cornea. *J. Biomedical Materials Research*, 66A, 260-265. (cited pp. 720 and 786).

Jeffreys H. and Jeffreys B. (1980). *Methods of mathematical physics*. Cambridge University Press, Cambridge, 3rd edition. (cited p. 142).

Jen M.C. (1995). Transport studies of component proteoglycan molecules through cartilage. Master thesis, Massachusetts Institute of Technology, September 1995. (cited p. 553).

Jin M. and Grodzinsky A.J. (2001). Effect of electrostatic interactions between glycosaminoglycans on the shear stiffness of cartilage: a molecular model and experiments. *Macromolecules*, 34(23), 8330-8339. (cited p. 505).

Johan Z. and Hughes T.J.R. (1991). A globally convergent matrix-free algorithm for implicit time-marching schemes arising in finite element analysis in fluids. *Computer Methods in Applied Mechanics and Engineering*, 87(2-3), 281-304. (cited p. 545).

Johnson B.E. and Hoger A. (1995). The use of a virtual configuration in formulating constitutive equations for residually stressed elastic materials. *J. Elasticity*, 41(3), 177-215. (cited p. 839).

Johnson D.L., Plona T.J., Scala C., Pasierb F. and Kojima H. (1982). Tortuosity and acoustic slow waves. *Physical Review Letters*, 49(25), 1840-1844. (cited pp. 205 and 374).

Johnson E.M. and Deen W.M. (1996). Hydraulic permeability of agarose gels. *American Institute Chemical Engineers J.*, 42(5), 1220-1224. (cited pp. 565 and 1011).

Jones A.F., Byrne H.M., Gibson J.S. and Dold J.W. (2000). A mathematical model of the stress induced during avascular tumor growth. *J. Mathematical Biology*, 40(6), 473-499. (cited pp. 1005 and 1006).

Jones J.P. (1961). Rayleigh waves in a porous, elastic, saturated solid. *J. Acoustical Society America*, 33(7), 959-962. (cited p. 388).

Jouzeau J.-Y., Pacquelet S., Boileau Ch., Nedelec E., Presle N., Netter P. and Terlain B. (2002). Nitric oxide (NO) and cartilage metabolism: NO effects are modulated by superoxide in response to IL-1. *Biorheology*, 39, 201-214. (cited p. 501).

Jue B. and Maurice D.M. (1986). The mechanical properties of the rabbit and human cornea. *J. Biomechanics*, 19, 847-853.

∗ ∗ ∗ ∗ K ∗ ∗ ∗ ∗

Kaliske M. (2000). A formulation of elasticity and viscoelasticity for fibre reinforced material at small and finite strains. *Computer Methods in Applied Mechanics and Engineering*, 185(2-4), 225-243.

Kambouchev N., Radovitzky R. and Fernandez J. (2006). Anisotropic materials which can be modeled by polyconvex strain energy functions. 47th AIAA/ASME/ASCE/AHS/ASC Structures, Structural Dynamics, and Materials Conference 1-4 May 2006, Newport, RI, 6 p. (cited p. 180).

Kaminski W. (1990). Hyperbolic heat conduction equation for material with nonhomogeneous inner structure. *ASME J. Heat Transfer*, 112, 555-560. (cited p. 318).

Kanatani Ken-Ichi (1984). Distribution of directional data and fabric tensors. *Int. J. Engng. Science*, 22(2), 149-164. (cited pp. 422, 435, 436, and 437).

Kanatani Ken-Ichi (1985). Procedures for stereological estimation of structural anisotropy. *Int. J. Engng. Science*, 23(5), 587-598. (cited p. 139).

Kannan K. and Rajagopal K.R. (2004). A thermomechanical framework for the transition of a viscoelastic liquid to a viscoelastic solid. *Mathematics and Mechanics of Solids*, 9, 37-59. (cited p. 173).

Katchalsky A. and Kedem O. (1962). Thermodynamics of flow processes in biological systems. *Biophysical J.*, 2, 53-78. (cited pp. 121, 715, and 716).

Katz E.P., Wachtel E.J. and Maroudas A. (1986). Extrafibrillar proteoglycans osmotically regulate the molecular packing of collagen in cartilage. *Biochimica et Biophysica Acta*, 882(1), 136-139. (cited p. 502).

Kaviany M. (1995). *Principles of heat transfer in porous media*. 2nd edition. Springer-Verlag, New York. (cited p. 336).

Kedem O. and Katchalsky A. (1961). A physical interpretation of the phenomenological coefficients of membrane permeability. *J. General Physiology*, 45(1), 143-179. (cited p. 759).

Keller E. and Segel L. (1971). Model for chemotaxis. *J. Theoretical Biology*, 30(2), 225-234. (cited p. 989).

Kemper W.D. and Quirk J.P. (1972). Ionic mobilities and electric charge of external clay surfaces inferred from potential differences and osmotic flow. *Soil Science Society of America J.*, 36(3), 426-433. (cited p. 609).

Kempson G.E., Freeman M.A.R. and Swanson S.A.V. (1968). Tensile properties of articular cartilage. *Nature*, 220(5172), 1127-1128. (cited p. 466).

Kempson G.E., Muir H., Pollard C. and Tuke M. (1973). The tensile properties of the cartilage of human femoral condyles related to the content of collagen and glycosaminoglycans. *Biochimica et Biophysica Acta*, 297, 456-472. (cited pp. 431, 466, and 489).

Kempson G.E. (1982). Relationship between the tensile properties of articular cartilage from the human knee and age. *Annals Rheumatic Diseases*, 41(5), 508-511. (cited p. 841).

Kentish J.C., ter Keurs H.E.D.J., Ricciardi L., Bucx J.J.J. and Noble M.I.M. (1986). Comparison between the sarcomere length-force relations of intact and skinned trabeculae from rat right ventricle. *Circulation Research*, 58(6), 755-768.

Kestin J. (1966). *A course in thermodynamics*, vol. I. Blaisdell Publishing Co., Waltham, MA. (cited pp. 69 and 72).

Kestin J. (1968). *A course in thermodynamics*, vol. II. Blaisdell Publishing Co., Waltham, MA. (cited pp. 69, 72, 79, 80, 81, 82, 86, 88, 89, 124, 128, 284, and 533).

Kestin J. (1978). *A course in thermodynamics*, vol. I, revised printing. Hemisphere Publishing Co., Washington. (cited p. 86).

Khaled M.Y., Beskos D.E. and Aifantis E.C. (1984). On the theory of consolidation with double porosity - III A finite element formulation. *Int. J. Numerical Analytical Methods Geomechanics*, 8, 101-123. (cited p. 321).

Khalili N. and Khabbaz M.H. (1995). On the theory of three-dimensional consolidation in unsaturated soils. *Unsaturated Soils/Sols non saturés*, E. Alonso and P. Delage eds, Presses de l'Ecole Nationale des Ponts et Chaussées, Paris, 1995, 745-750.

Khalili N. and Valliappan S. (1996). Unified theory of flow and deformation in double porous media. *European J. Mechanics-A/Solids*, 15(2), 321-336. (cited p. 321).

Khalili N. and Loret B. (2001). An elasto-plastic model for non-isothermal analysis of flow and deformation in unsaturated porous media: formulation. *Int. J. Solids Structures*, 38(46-47), 8305-8330. (cited p. 322).

Khalili N.K. (2003). Coupling effects in double porosity media with deformable matrix. *Geophysical Research Letters*, 30(22), 2153-2155. (cited pp. 322 and 326).

Khalili N.K. and Selvadurai A.P.S. (2003). A fully coupled constitutive model for thermo-hydro-mechanical analysis in elastic media with double porosity. *Geophysical Research Letters*, 30(24), 2268-2272. (cited p. 322).

Khalsa P.S. and Eisenberg S.R. (1997). Compressive behavior of articular cartilage is not completely explained by proteoglycan osmotic pressure. *J. Biomechanics*, 30(6), 589-594. (cited p. 505).

Khosravani H., Chugh B., Milosevic M.F. and Norwich K.H. (2004). Time response of intersitial fluid pressure measurements in cervix cancer. *Microvascular Research*, 68(1), 63-70. (cited pp. 566, 567, and 1010).

Kim S. and Karrila S.J. (1991). *Microhydrodynamics: Principles and selected applications*. Butterworth-Heinemann, Boston, MA. (cited p. 588).

Kim Y.J., Sah R.L., Doong J.Y.H., Grodzinsky A.J. (1988). Fluorometric assay of DNA in cartilage explants using Hoechst 33258. *Analytical Biochemistry*, 174, 168-176. (cited pp. 498 and 809).

Kim Y.J., Grodzinsky A.J. and Plaas A.H.K. (1998). Compression of cartilage results in differential effects on biosynthetic pathways for aggrecan, link protein, and hyaluronan. *Archives Biochemistry Biophysics*, 328(2), 331-340. (cited pp. 828 and 845).

Kinner B. and Spector M. (2001). Smooth muscle actin expression by human articular chondrocytes and their contraction of a collagen-glycosaminoglycan matrix *in vitro*. *J. Orthopaedic Research*, 19, 233-241. (cited p. 823).

Kisiday J.D., Jin M., Kurz B., Hung H., Semino C., Zhang S. and Grodzinsky A.J. (2002)a. Self-assembling peptide hydrogel fosters chondrocyte extracellular matrix production and cell division: Implications for cartilage tissue repair. *Proceedings National Academy Sciences USA*, 99(15), 9996-10001. (cited pp. 806, 823, and 836).

Kisiday J.D., Jin M. and Grodzinsky A.J. (2002)b. Effects of dynamic compressive loading duty cycle on *in vitro* conditioning of chondrocyte-seeded peptide and agarose scaffolds. *Transactions Orthopaedic Research Society*, 48th Annual Meeting, 216. (cited pp. 806, 823, and 836).

Kisiday J.D., Jin M., DiMicco M., Kurz B. and Grodzinsky A.J. (2004). Effects of dynamic compressive loading on chondrocyte biosynthesis in self-assembling peptide scaffolds. *J. Biomechanics*, 37(5), 595-604. (cited pp. 806, 831, and 836).

Kitano T., Ateshian G.A., Mow V.C., Kadoya Y. and Yamano Y. (2001). Constituents and pH changes in protein rich hyaluronan solution affect the biotribological properties of artificial articular joints. *J. Biomechanics*, 34, 1031-1037.

Kiviranta P., Rieppo J., Korhonen R.K., Julkunen P., Töyräs J. and Jurvelin J.S. (2006). Collagen network primarily controls Poisson's ratio of bovine articular cartilage in compression. *J. Orthopaedic Research*, 24(4), 690-699.

Klein T., Schumacher B., Li K., Voegtline M., Masuda K., Thonar E.J.-M.A. and Sah R.L. (2002). Tissue engineered articular cartilage with functional stratification: targeted delivery of chondrocytes expressing superficial zone protein. *Transactions Orthopaedic Research Society*, 48th Annual Meeting, 212. (cited p. 826).

Klein T.J., Chaudhry M., Bae W.C. and Sah R.L. (2007). Depth-dependent biomechanical and biochemical properties of fetal, newborn, and tissue-engineered articular cartilage. *J. Biomechanics*, 40(1), 182-190. (cited pp. 498 and 841).

Klezovitch O., Fernandez T.E., Tapscott S.J. and Vasioukhin V. (2004). Loss of cell polarity causes severe brain dysplasia in Lgl1 knockout mice. *Genes & Development*, 18(5), 559-571. (cited p. 1014).

Klisch S.M. and Lotz J.C. (1999). Application of a fiber-reinforced continuum theory to multiple deformations of the annulus fibrosus. *J. Biomechanics*, 32, 1027-1036. (cited pp. 189 and 800).

Klisch S.M. and Lotz J.C. (2000). A special theory of biphasic mixtures and experimental results for human annulus fibrosus tested in confined compression. *ASME J. Biomechanical Engng.*, 122, 180-188. (cited p. 802).

Klisch S.M., Chen S.S., Masuda K., Thonar E.J.-M.A., Hoger A. and Sah R.L. (2001)a. Application of a growth and re-modeling mixture theory to developing articular cartilage. *Transactions Orthopaedic Research Society*, 47th Annual Meeting, 316. (cited pp. 839, 841, 843, and 850).

Klisch S.M., van Dyke T.J. and Hoger A. (2001)b. A theory of volumetric growth for compressible elastic biological materials. *Mathematics and Mechanics of Solids*, 6, 551-575. (cited pp. 189, 839, 841, 843, 850, 890, and 922).

Klisch S.M. and Hoger A. (2003). Volumetric growth of thermoelastic materials and mixtures. *Mathematics and Mechanics of Solids*, 8(4), 377-402. (cited pp. 104, 111, 128, 856, and 866).

Klisch S.M., Chen S.S., Sah R.L. and Hoger A. (2003)a. A growth mixture theory for cartilage with application to growth-related experiments on cartilage explants. *ASME J. Biomechanical Engng.*, 125, 169-179. (cited pp. 111, 839, 840, 841, 842, 843, 856, 857, and 871).

Klisch S.M., Sah R.L. and Hoger A. (2003)b. A cartilage growth mixture model for infinitesimal strains: equilibrium solutions. Summer Bioengineering Conference, June 25-29, Key Biscane, FL. (cited pp. 111, 839, 840, 841, 842, 843, 856, and 857).

Klisch S.M., Sah R.L. and Hoger A. (2005). A cartilage growth mixture model for infinitesimal strains: solutions of boundary-value problems related to *in vitro* growth experiments. *Biomechanics and Modeling in Mechanobiology*, 3, 209-223. (cited pp. 839 and 843).

Klyce S.D. (1972). Electrical profiles in the corneal epithelium. *J. Physiology*, 226, 407-429. (cited p. 723).

Klyce S.D. and Russell S.R. (1979). Numerical solution of coupled transport equations applied to corneal hydration dynamics. *J. Physiology*, 292, 107-134. (cited p. 723).

Knackstedt M.A., Arns C.H. and Val Pinczewski W. (2005). Velocity-porosity relationships: predictive velocity model for cemented sands composed of multiple mineral phases. *Geophysical Prospecting*, 53(3), 349-372. (cited p. 212).

Knight M.M., Ross J.M., Shervin A.F., Lee D.A., Bader D.L. and Poole C.A. (2001). Chondrocyte deformation within mechanically and enzymatically extracted chondrons compressed in agarose. *Biochimica et Biophysica Acta*, 1526, 141-146. (cited pp. 821 and 827).

Knox Cartwright N.E, Tyrer J.R. and Marshall J. (2011). Age-Related Differences in the Elasticity of the Human Cornea. *Investigative Ophthalmology & Visual Science*, 52, 4324-4329. (cited p. 727).

Kohl P. and Sachs F. (2001). Mechanoelectric feedback in cardiac cells. *Philosophical Transactions Royal Society A*, 359, 1-13. (cited p. 468).

Koike C., McKee T.D., Pluen A., Ramanujan S., Burton K., Munn L.L., Boucher Y. and Jain R.K. (2002). Solid stress facilitates spheroid formation: potential involvement of hyaluronan. *British J. Cancer*, 86, 947-953. (cited p. 1004).

Koike N., Fukumura D., Gralla O., Au P., Schechner J.S. and Jain R.K. (2004). Tissue engineering: creation of long-lasting blood vessels. *Nature*, 428(6979), 138-139. (cited pp. 821 and 998).

Konofagou E.E. and Ophir J. (1998). A new elastographic method for estimation and imaging of lateral displacements, lateral strains, corrected axial strains, Poisson's ratios in tissues. *Ultrasound in Medicine and Biology*, 24(8), 1183-1199. (cited p. 1040).

Konofagou E.E., Harrigan T.P., Ophir J. and Krouskop T.A. (2001). Poroelastography: imaging the poroelastic properties of tissues. *Ultrasound in Medicine and Biology*, 27(10), 1387-1397. (cited p. 1041).

Koolman J. and Röhm K.H. (1999). *Atlas de Poche de Biochimie*. Flammarion, Paris. *Referred to as APB [1999]*. (cited pp. 99, 127, 501, 578, 580, 581, 582, 812, 813, 814, 815, 816, 818, 819, 988, 990, 995, 996, and 1000).

Krishnaswamy S. and Batra R.C. (1997). A thermomechanical theory of solid-fluid mixtures. *Mathematics and Mechanics of Solids*, 2, 143-151. (cited p. 128).

Krol A., Maresca J., Dewhirst M.W. and Yuan F. (1999). Available volume fraction of macromolecules in the extravascular space of a fibrosarcoma: Implications for drug delivery. *Cancer Research*, 59, 4136-4141. (cited pp. 568, 569, and 993).

Kronick P.L. (1988). Analysis of the effects of pH and tensile deformation on the small-deformation modulus of calf skin. *Connecting Tissue Research*, 18(2), 95-106. (cited p. 432).

Krouskop T.A., Wheeler T.M., Kallel F., Gara B.S and Hall T.J. (1998). Elastic moduli of breast and prostate tissues under compression. *Ultrasonic Imaging*, 20(4), 260-274. (cited p. 1039).

Kuang K., Li Y., Yiming M., Sanchez J.M., Iserovich P., Cragoe E.J., Diecke F.P.J. and Fischbarg J. (2004). Intracellular [Na+], Na+ pathways, and fluid transport in cultured bovine corneal endothelial cells. *Experimental Eye Research*, 79, 93-103. (cited p. 714).

Kuhl E. and Steinmann P. (2003). Theory and numerics of geometrically non-linear open system mechanics. *Int. J. Numerical Methods Engineering*, 58, 1593-1615. (cited pp. 107 and 874).

Kuhn M.R. (2010). Micro-mechanics of fabric and failure in granular materials. *Mechanics of Materials*, 42(9), 827-840. (cited p. 950).

Kurtis M.S., Tu B.P., Gaya O.A., Mollenhauer J., Knudson W., Loeser R.F., Knudson Ch.B. and Sah R.L. (2001). Mechanisms of chondrocyte adhesion to cartilage: role of $\beta 1$ integrins, CD44, and annexin V. *J. Orthopaedic Research*, 19, 1122-1130. (cited p. 831).

Kurtis M.S., Schmidt T.A., Bugbee W.D., Loeser R.F. and Sah R.L. (2003). Integrin-mediated adhesion of human articular chondrocytes to cartilage. *Arthritis and Rheumatism*, 48(1), 110-118. (cited p. 830).

Kurz B., Jin M., Patwari P., Cheng D.M., Lark M.W. and Grodzinsky A.J. (2001). Biosynthetic response and mechanical properties of articular cartilage after injurious compression. *J. Orthopaedic Research*, 19(6), 1140-1146. (cited pp. 807 and 830).

Kwok L.S. and Klyce S.D. (1990). Theoretical basis for an anomalous temperature coefficient in swelling pressure of rabbit corneal stroma. *Biophysical J.*, 57, 657-662. (cited pp. 284 and 710).

* * * *   L   * * * *

Laasanen M., Töyräs J., Korhonen R., Rieppo J., Saarakkala S., Nieminen M., Hirvonen J. and Jurvelin J.S. (2003). Biomechanical properties of knee articular cartilage. *Biorheology*, 40(1-3), 133-140.

Lachenbruch C.A. and Diller K.R. (1999). A Network Thermodynamic Model of Kidney Perfusion With a Cryoprotective Agent. *ASME J. Biomechanical Engng.*, 121, 574-583. (cited p. 5).

Lacroix M. (2008). Persistent use of 'false' cell lines. *Int. J. Cancer*, 122, 1-4. (cited p. 992).

Lahiji K., Polotsky A., Hungerford D.S. and Frondoza C.G. (2002). Cyclic strain stimulates proliferative capacity, α_2-integrin by human articular chondrocytes from osteoarthritic knee joints. *The University of Pennsylvania Orthopaedic J.*, 15, 75-81. (cited pp. 501 and 825).

Lai W.M. and Mow V.C. (1980). Drag-induced compression of articular cartilage during a permeation experiment. *Biorheology*, 17(1-2), 111-123. (cited p. 565).

Lai W.M., Mow V.C. and Roth V. (1981). Effects of non-linear stress-strain dependent permeability and rate of compression on the stress behavior of articular cartilage. *ASME J. Biomechanical Engng.*, 103(2), 61-66. (cited pp. 612 and 824).

Lai W.M., Hou J.S. and Mow V.C. (1991). A triphasic theory for the swelling and deformation behaviors of articular cartilage. *ASME J. Biomechanical Engng.*, 113(3), 245-258. (cited pp. 485, 502, 503, 504, 505, 508, 510, 591, 649, 670, and 713).

Lai W.M., Sun D.D., Ateshian G.A., Guo X.E. and Mow V.C. (2002). Electrical signals for chondrocytes in cartilage. *Biorheology*, 39, 39-45. (cited p. 827).

Laible J.P., Pflaster D., Simon B.R., Krag M.H., Pope M. and Haugh L.D. (1994). A dynamic material parameter estimation procedure for soft tissue using a poroelastic finite element model. *ASME J. Biomechanical Engng.*, 116(1), 19-29.

Lakes R., Yoon H.S. and Katz J.L. (1983). Slow compressional wave propagation in wet human and bovine cortical bone. *Science*, 220, 513-515. (cited p. 374).

Landman K.A. and Please C.P. (2001). Tumor dynamics and necrosis: surface tension and stability. *IMA J. Mathematics Applied in Medicine and Biology*, 18(2), 131-158. (cited pp. 875, 876, and 1011).

Lane Smith R., Lin J., Trindade M.C.D., Shida J., Kajiyama G., Vu T., Hoffman A.R., van der Meulen M.C.H., Goodman S.B., Shurman D.J. and Carter D.R. (2000). Time-dependent effects of intermittent hydrostatic pressure on articular chondrocyte type II collagen and aggrecan mRNA expression. *J. Rehabilitation Research & Development*, 37(2), 153-162. (cited p. 828).

Lanir Y. (1979). A structural theory for the homogeneous biaxial stress-strain relationships in flat collagenous tissues. *J. Biomechanics*, 12(6), 423-436. (cited p. 431).

Lanir Y. (1983). Constitutive equations for fibrous connective tissues. *J. Biomechanics*, 16(1), 1-12. (cited p. 431).

Lanir Y. (1987). Biorheology and fluid flux in swelling tissues. I. Bicomponent theory for small deformations, including concentration effects. *Biorheology*, 24(2), 173-187. (cited p. 670).

Laude D., Odlum K., Rudnicki S. and Bachrach N. (2000). A novel injectable collagen matrix: *in vitro* characterization and in vivo evaluation. *ASME J. Biomechanical Engng.*, 122(3), 231-235. (cited p. 565).

Laurencin C.T., Ambrosio A.M.A., Borden M.D. and Cooper Jr. J.A. (1999). Tissue Engineering: Orthopedic Applications. *Annual Review Biomedical Engng.*, 1, 19-46. (cited p. 808).

Layton B.E. and Sastry A.M. (2004). A mechanical model for collagen load sharing in peripheral nerve of diabetic and nondiabetic rats. *ASME J. Biomechanical Engng.*, 126, 803-814. (cited p. 796).

Lee C.R., Grodzinsky A.J., Hsu H.-P. and Spector M. (2003). Effects of a cultured autologous chondrocyte-seeded type II collagen scaffold on the healing of a chondral defect in a canine model. *J. Orthopaedic Research*, 21, 272-281. (cited pp. 807 and 821).

Lee D.A. and Bader D.L. (1997). Compressive strains at physiological frequencies influence the metabolism of chondrocytes seeded in agarose. *J. Orthopaedic Research*, 15(2), 181-188. (cited p. 823).

Lee D.A., Frean P.S., Lees P. and Bader D.L. (1998)a. Dynamic mechanical compression influences nitric oxide production by articular chondrocytes seeded in agarose. *Biochemical and Biophysical Research Communications*, 251, 580-585. (cited p. 830).

Lee D.A., Noguchi, T., Knight M.M., O'Donnell L., Bentley G. and Bader D.L. (1998)b. Response of chondrocyte subpopulations cultured within unloaded and loaded agarose. *J. Orthopaedic Research*, 16, 726-733. (cited p. 823).

Lee D.A., Reisler T. and Bader D.L. (2003). Expansion of chondrocytes for tissue engineering in alginate beads enhances chondrocytic phenotype compared to conventional monolayer techniques. *Acta Orthopaedica Scandinavica*, 74(1), 6-15. (cited pp. 822 and 825).

Lee K.Y., Peters M.C., Anderson K.W. and Mooney D.J. (2000). Controlled growth factor release from synthetic extracellular matrices. *Nature*, 408(6815), 998-1000. (cited p. 258).

Lee M.S., Trindade C.D., Ikenoue T., Schurman D.J., Goodman S.B. and Lane Smith R. (2002). Effects of shear stress on nitric oxide and matrix protein gene expression in human osteoarthritic chondrocytes *in vitro*. *J. Orthopaedic Research*, 20, 556-561. (cited pp. 471, 825, 827, 828, 830, 844, and 1002).

Lehmann Th., Guo Z.h. and Liang H. (1991). The conjugacy between Cauchy stress and logarithm of the left stretch tensor. *European J. Mechanics-A/Solids*, 10(4), 395-404.

Lehoux S. and Tedgui A. (2003). Cellular mechanics and gene expression in blood vessels. *J. Biomechanics*, 36, 631-643. (cited p. 807).

Lei F. and Szeri A.Z. (2007). Predicting articular cartilage behavior with non-linear microstructural model. *The Open Mechanics J.*, 1, 11-19. (cited pp. 241, 242, and 506).

Leiderman R., Barbone P., Oberai A. and Bamber J.C. (2006). Coupling between elastic strain and interstitial fluid flow: ramifications for poroelastic imaging. *Physics in Medicine and Biology*, 51(24), 6291-6313. (cited pp. 1010, 1029, 1031, and 1041).

Leikin S., Parsegian V.A. and Rau D.C. (1993). Hydration forces. *Annual Review Physical Chemistry*, 44, 369-395.

Leikin S., Rau D.C. and Parsegian V.A. (1994). Direct measurement of forces between self-assembled proteins: Temperature-dependent exponential forces between collagen triple helices. *Proceedings National Academy Sciences USA*, 91, 276-280.

Lekhnitskii S.G. (1963). *Theory of elasticity of an anisotropic elastic body*. Holden Day, Inc. San Francisco, CA. (cited pp. 139 and 150).

Lemaitre J. and Chaboche J.L. (1988). *Mécanique des matériaux solides*. Dunod, Paris, France. (cited pp. 905, 906, 907, and 908).

Leroy Ph. (2005). Transport ionique dans les argiles. Influence de la microstructure et des effets d'interface. Thèse de doctorat en Sciences, Université d'Aix-Marseille III, France. (cited p. 609).

Leu A.J., Berk D.A., Lymboussaki A., Alitalo K. and Jain R.K. (2000). Absence of functional lymphatics within a murine sarcoma: a molecular and functional evaluation. *Cancer Research*, 60, 4324-4327. (cited p. 999).

Levenston M.E., Eisenberg S.R. and Grodzinsky A.J. (1998)a. A variational formulation for coupled physicochemical flows during finite deformations of charged porous media. *Int. J. Solids Structures*, 35, 4999-5019. (cited p. 670).

Levenston M.E., Frank E.H. and Grodzinsky A.J. (1998)b. Variationally derived 3-field finite element formulations for quasistatic poroelastic analysis of hydrated biological tissues. *Computer Methods in Applied Mechanics and Engineering*, 156, 231-246. (cited p. 670).

Levenston M.E., Frank E.H. and Grodzinsky A.J. (1999). Electrokinetic and poroelastic coupling during finite deformations of charged porous media. *ASME J. Applied Mechanics*, 66, 323-333.

Levick J.R. (1987). Flow through interstitium and other fibrous matrices. *Quarterly J. Experimental Physiology*, 72, 409-437. (cited pp. 561, 571, and 724).

Li K.W., Wang A.S. and Sah R.L. (2003). Microenvironment regulation of extracellular signal-regulated kinase activity in chondrocytes: effects of culture configuration, interleukin-1, and compressive stress. *Arthritis and Rheumatism*, 48(3), 689-699. (cited p. 817).

Li L.P., Herzog W., Korhonen R.K. and Jurvelin J.S. (2005). The role of viscoelasticity of collagen fibers in articular cartilage: axial tension versus compression. *Medical Engineering & Physics*, 27(1), 51-57. (cited p. 241).

Li L.Y. (2004). Transport of multicomponent ionic solutions in membrane systems. *Philosophical Magazine Letters*, 84(9), 593-599. (cited p. 769).

Li L.Y., Tighe B.J. and Ruberti J.W. (2004). Mathematical modelling of corneal swelling. *Biomechanics and Modeling in Mechanobiology*, 3, 114-123. (cited p. 723).

Li L.Y. and Tighe B.J. (2006)a. Numerical simulation of corneal transport processes. *J. Royal Society Interface*, 3, 303-310.

Li L.Y. and Tighe B.J. (2006)b. The anisotropic material constitutive models for the human cornea. *J. Structural Biology*, 153, 223-230. (cited pp. 430 and 797).

Li S.-T. and Katz E.P. (1976). An electrostatic model for collagen fibrils. The interaction of reconstituted collagen with Ca^{++}, Na^+, and Cl^-. *Biopolymers*, 15, 1439-1460. (cited pp. 485, 496, 501, 514, 594, 629, 634, and 732).

Liao F. and 11 co-authors (2000). Monoclonal antibody to vascular endothelial-cadherin is a potent inhibitor of angiogenesis, tumor growth, and metastasis. *Cancer Research*, 60, 6805-6810. (cited pp. 996 and 1000).

Lieberman D.E., Pearson O.M., Polk J.D., Demes B. and Crompton A.W. (2003). Optimization of bone growth and re-modeling in response to loading in tapered mammalian limbs. *J. Experimental Biology*, 206, 3125-3138. (cited pp. 846 and 922).

Lim J.J. and Fischbarg J. (1981). Electrical properties of rabbit corneal endothelium as determinated from impedance measurements. *Biophysical J.*, 36, 677-695. (cited p. 257).

Limbert G. and Taylor M. (2002). On the constitutive modeling of biological soft connective tissues. *Int. J. Solids Structures*, 39(8), 2343-2358.

Limbert G. and Middleton J. (2004). A transversely isotropic viscohyperelastic material. Application to the modeling of biological soft connective tissues. *Int. J. Solids Structures*, 41(15), 4237-4260. (cited pp. 173, 191, and 273).

Lin I.E. and Taber L. (1995). A model for stress-induced growth in the developing heart. *ASME J. Biomechanical Engng.*, 117, 343-349. (cited p. 879).

Linninger A.A., Somayaji M.R., Zhang L., Hariharan M.S. and Penn R.D. (2008). Rigorous Mathematical Modeling Techniques for Optimal Delivery of Macromolecules to the Brain. *IEEE Transactions on Biomedical Engng.*, 55(9), 2303-2313. (cited p. 4).

Liu H.T., Sun L.Z., Wang G. and Vannier M.W. (2003). Analytic modeling of breast elastography. *Medical Physics*, 30(9), 2340-2349. (cited p. 1040).

Liu I-Shih (1982). On representations of anisotropic invariants. *Int. J. Engng. Science*, 20(10), 1099-1109. (cited p. 60).

Liu J., He X., Pan X. and Roberts C.J. (2007). Ultrasonic model and system for measurement of corneal biomechanical properties and validation on phantoms. *J. Biomechanics*, 40, 1177-1182. (cited p. 727).

Liu K.C. (2008). Thermal propagation analysis for living tissue with surface heating. *Int. J. Thermal Sciences*, 47, 507-513. (cited p. 318).

Liu X., Ormond A., Bartko K., Li Y. and Ortoleva P. (1997). A geochemical reaction-transport simulator for matrix acidizing analysis and design. *J. Petroleum Science and Engng.*, 17(1-2), 181-196. (cited p. 565).

Lobo V.M.M. (1990). *Handbook of Electrolyte Solutions* in two volumes. Elsevier Publishing Company, Amsterdam, the Netherlands. (cited pp. 440 and 441).

Loeser R.F. (2002). Integrins and cell signaling in chondrocytes. *Biorheology*, 39, 119-124. (cited p. 825).

Loix F., Simões F.M.F. and Loret B. (2008). Articular cartilage with intra- and extra-fibrillar waters. Simulations of mechanical and chemical loadings by the finite element method. *Computer Methods in Applied Mechanics and Engineering*, 197(51-52), 4840-4857. (cited pp. 669, 682, 683, 684, 685, 686, 687, 688, 689, 690, 691, and 693).

Lombardo M., De Santo M.P., Lombardo G., Barberi R. and Serrao S. (2006). Atomic force microscopy analysis of normal and photoablated porcine corneas. *J. Biomechanics*, 39, 2719-2724. (cited p. 730).

Long J.L. and Tranquillo R.T. (2003). Elastic fiber production in cardiovascular tissue-equivalents. *Matrix Biology*, 22(4), 339-350. (cited p. 836).

Loret B. (1983). On the effects of plastic rotation in the finite deformation of anisotropic elastoplastic materials. *Mechanics of Materials*, 2, 287-304. (cited p. 959).

Loret B. (1985). On the choice of elastic parameters for sand. *Int. J. Numerical Analytical Methods Geomechanics*, 9(3), 285-292. (cited pp. 133 and 171).

Loret B. (1990). Acceleration waves in elastic-plastic porous media: interlacing and separation properties. *Int. J. Engng. Science*, 28(12), 1315-1320. (cited p. 375).

Loret B., Prévost J.H. and Harireche O. (1990). Loss of hyperbolicity in elastic-plastic solids with deviatoric associativity. *European J. Mechanics-A/Solids*, 9(3), 225-231.

Loret B. and Harireche O. (1991). Acceleration waves, flutter instabilities and stationary discontinuities in inelastic porous media. *J. Mechanics Physics Solids*, 39(5), 569-606. (cited pp. 19 and 372).

Loret B. and Prévost J.H. (1991). Dynamic strain localization in fluid-saturated porous media. *ASCE J. Engng. Mechanics*, 117(4), 907-922. (cited pp. 209, 297, and 675).

Loret B. (1992). Does deviation from deviatoric associativity lead to the onset of flutter instability? *J. Mechanics Physics Solids*, 40(6), 1363-1375. (cited p. 19).

Loret B. and Rizzi E. (1997). Anisotropic stiffness degradation triggers onset of strain localization. *Int. J. Plasticity*, 13(5), 447-459. (cited pp. 21, 148, and 149).

Loret B., Simões F.M.F. and Martins J.A.C. (1997). Growth and decay of acceleration waves in non-associative elastic-plastic fluid-saturated porous media. *Int. J. Solids Structures*, 34(13), 1583-1608. (cited pp. 380, 381, and 382).

Loret B. and Rizzi E. (1998). On the effects of inertial coupling on the wave-speeds of elastic-plastic fluid-saturated porous media. *Material instabilities in solids*, R. de Borst and E. van der Giessen editors, John Wiley & Sons, Chichester, 41-53. (cited p. 377).

Loret B. and E. Rizzi (1999). Strain localization in fluid-saturated anisotropic elastic-plastic porous media with double porosity. *J. Mechanics Physics Solids*, 47(3), 503-530. (cited pp. 322, 327, and 328).

Loret B. and Khalili N. (2000). Thermo-mechanical potentials for unsaturated Soils. CISM Courses and Lectures *Advanced Numerical Applications and Plasticity in Geomechanics* n° 426, Udine, edited by D.V. Griffiths and G. Gioda, Springer New York, 2001, 253-276. (cited p. 322).

Loret B. and Radi E. (2001). The effects of inertia on crack growth in poro-elastic fluid-saturated media. *J. Mechanics Physics Solids*, 49(5), 995-1020. (cited p. 411).

Loret B., Rizzi E. and Zerfa Z. (2001). Relations between drained and undrained moduli in anisotropic elastic fluid-saturated porous media. *J. Mechanics Physics Solids*, 49(11), 2593-2619. (cited pp. 151, 157, 214, 219, 222, 223, 224, and 225).

Loret B., Hueckel T. and Gajo A. (2002). Chemo-mechanical coupling in saturated porous media: elastic-plastic behaviour of homoionic expansive clays. *Int. J. Solids Structures*, 39(10), 2773-2806. (cited p. 505).

Loret B., Gajo A. and Simões F.M.F. (2004)a. Chemo-mechanical couplings in clays, pp. 125-238, CISM Courses and Lectures *Chemo-Mechanical Couplings in Geomechanics and Biomechanics*, Udine, Italia, B. Loret and J.M. Huyghe eds., Springer New York, 2004. (cited pp. 101, 117, 211, 459, 460, 462, 463, 468, 469, 470, 488, 509, 599, 614, and 954).

Loret B., Gajo A. and Simões F.M.F. (2004)b. A note on the dissipation due to generalized diffusion with electro-chemo-mechanical couplings in heteroionic clays. *European J. Mechanics-A/Solids*, 23(5), 763-782. (cited pp. 124, 509, 592, 614, and 954).

Loret B. and Simões F.M.F. (2004). Articular cartilage with intra- and extrafibrillar waters. A chemo-mechanical model. *Mechanics of Materials*, 36(5-6), 515-541. (cited pp. 101, 103, 104, 111, 112, 124, 467, 485, 497, 502, 508, 510, 520, 548, 617, 621, 659, 669, 673, 680, 681, 683, and 954).

Loret B. and F.M.F. Simões (2005)a. Mechanical effects of ionic replacements in articular cartilage. *Biomechanics and Modeling in Mechanobiology*, 4(2-3). I. The constitutive model, 63-80; II. Simulations of successive substitutions of NaCl and $CaCl_2$, 81-99. (cited pp. 101, 103, 104, 111, 124, 163, 170, 464, 465, 466, 485, 508, 509, 519, 520, 617, 621, 628, 631, 668, 669, 673, 680, 681, 692, and 733).

Loret B. and Simões F.M.F. (2005)b. A framework for deformation, generalized diffusion, mass transfer and growth in multi-species multi-phase biological tissues. *European J. Mechanics-A/Solids*, 24(5), 757-781. (cited pp. 123, 874, and 910).

Loret B. and Simões F.M.F. (2007). Articular cartilage with intra- and extra-fibrillar waters. Mass transfer and generalized diffusion. *European J. Mechanics-A/Solids*, 26(5), 759-788. (cited pp. 121, 125, 597, 608, 611, 616, 617, 628, 669, 674, 754, and 769).

Loret B. and Simões F.M.F. (2010)a. Effects of pH on transport properties of articular cartilages. *Biomechanics and Modeling in Mechanobiology*, 9(1), 45-63. (cited pp. 634, 643, 648, 649, 650, 651, 652, 653, 654, and 662).

Loret B. and Simões F.M.F. (2010)b. Effect of the pH on the mechanical behavior of articular cartilage and corneal stroma. *Int. J. Solids Structures*, 47(17), 2201-2214. (cited pp. 662, 663, 664, 665, 747, 751, and 752).

Loret B. and Simões F.M.F. (2010)c. Elastic growing tissues: a growth rate law that satisfies the dissipation inequality. *Mechanics of Materials*, 42(8), 782-796, doi:10.1016/j.mechmat.2010.06.001. (cited pp. 907, 919, 920, and 978).

Loret B. and Simões F.M.F. (2014). A thermodynamically consistent growth law for collagen fiber reinforced tissues in a mixture context. *Mechanics of Materials*, 76, 45-63. (cited pp. 943, 965, 976, 977, 978, 980, 981, 982, 983, 984, and 985).

Lu M. and Connell L.D. (2007). A dual-porosity model for gas reservoir flow incorporating adsorption behaviour-part I. Theoretical development and asymptotic analyses. *Transport in Porous Media*, 2007, 68(2), 153-173. (cited p. 342).

Lubarda V.A. and Hoger A. (2002). On the mechanics of solids with a growing mass. *Int. J. Solids Structures*, 39, 4627-4664. (cited pp. 104, 839, 850, 856, 859, 866, 877, 912, and 921).

Lubarda V.A. (2004)a. Constitutive Theories Based on the Multiplicative Decomposition of Deformation Gradient: Thermoelasticity, Elastoplasticity and Biomechanics, *ASME Applied Mechanics Reviews*, 57(2), 95-108. (cited p. 877).

Lubarda V.A. (2004)b. On thermodynamic potentials in linear thermoelasticity. *Int. J. Solids Structures*, 41, 7377-7398. (cited p. 281).

Lubkin S.R. and Jackson T.L. (2002). Multiphase mechanics of capsule formation in tumors. *ASME J. Biomechanical Engng.*, 124, 237-243. (cited pp. 998, 1007, and 1008).

Luehr C.P. and Rubin M.B. (1990). The significance of projection operators in the spectral representation of symmetric second order tensors. *Computer Methods in Applied Mechanics and Engineering*, 84, 243-246. (cited p. 153).

Luenberger D.G. (1984). *Linear and nonlinear programming.* 2nd edition. Addison-Wesley Publishing Company, Reading, MA. (cited p. 178).

Lunevicius R., Nakanishi H., Ito S., Kozaki K.I., Kato T., Tatematsu M. and Yasui K. (2001). Clinicopathological significance of fibrotic capsule formation around liver metastasis from colorectal cancer. *J. Cancer Research and Clinical Oncology*, 127(3), 193-199. (cited pp. 998 and 1008).

Luo G.M., Cowin S.C., Sadegh A.M. and Arramon Y.P. (1995). Implementation of strain rate as a bone remodeling stimulus. *ASME J. Biomechanical Engng.*, 117(3), 329-338. (cited p. 845).

Lynch H.A., Johannessen W., Wu J.P., Jawa A. and Elliott D.M. (2003). Effect of fiber orientation and strain rate on the nonlinear uniaxial tensile material properties of tendon. *ASME J. Biomechanical Engng.*, 125(5), 726-731. (cited p. 838).

* * * * M * * * *

MacArthur B.D. and Please C.P. (2004). Residual stress generation and necrosis formation in multi-cell tumour spheroids. *J. Mathematical Biology*, 49(6), 537-552. (cited pp. 847, 856, 874, 1006, and 1007).

MacGregor E.A. and Bowness J.M. (1970). Interaction of proteoglycans and chondroitin sulfates with calcium or phosphate ions. *Canadian J. Biochemistry*, 49(4), 417-425. (cited pp. 635 and 641).

Mackay M. (1992). Mechanisms of injury and biomechanics: Vehicle design and crash performance. *World J. of Surgery*, 16, 421-427. (cited p. 9).

Mackie J.S. and Meares P. (1955). The diffusion of electrolytes in a cation-exchange resin membrane .I. Theoretical. *Proceedings Royal Society London*, A232, 498-509. (cited pp. 570 and 613).

Majewski M., Habelt S. and Steinbrück K. (2006). Epidemiology of athletic knee injuries: A 10-year study. *Knee*, 13(3), 184-188. (cited p. 10).

Majone M., Petrangeli Papini M. and Rolle E. (1996). Modeling lead adsorption on clays by models with and without electrostatic terms. *J. Colloid and Interface Science*, 179, 412-425.

Mak A.F.T. (1986). The apparent viscoelastic behavior of articular cartilage - The contributions from the intrinsic matrix viscoelasticity and interstitial fluid flows. *ASME J. Biomechanical Engng.*, 108, 123-130. (cited p. 241).

Mak A.F.T. and Zhang J.D. (2001). Numerical simulation of streaming potentials due to deformation-induced hierarchical flows in cortical bone. *ASME J. Biomechanical Engng.*, 123, 66-70. (cited p. 825).

Malusis M.A. and Shackelford C.D. (2002). Coupling effects during steady-state solute diffusion through a semipermeable clay membrane. *Environmental Science & Technology*, 36, 1312-1319. (cited p. 611).

Malvern L.E. (1969). *Introduction to the mechanics of a continuous medium.* Prentice Hall, Inc., Englewood Cliffs, NJ. (cited pp. 13, 32, 57, 64, 143, 164, 557, 565, 588, 892, 893, and 894).

Mandel J. (1962). Ondes plastiques dans un milieu indéfini à trois dimensions. *J. de Mécanique*, 1, 3-30. (cited p. 375).

Mandel J. (1966). *Cours de Mécanique des Milieux Continus.* Gauthier-Villars, Paris. Reprinted edition, 1994, by Editions Jacques Gabay, Paris. (cited pp. 13, 207, 248, and 588).

Mandel J. (1971). Plasticité classique et viscoplasticité. International Center for Mechanical Sciences, Courses and Lectures n° 97. Springer-Verlag, New York. (cited pp. 849 and 900).

Mandel J. (1973). Equations constitutives et directeurs dans les milieux plastiques et viscoplastiques. *Int. J. Solids Structures*, 9(6), 725-740. (cited pp. 849, 850, 853, 866, 900, and 959).

Mandel J. (1974). *Introduction à la mécanique des milieux continus déformables.* PWN Editions Scientifiques de Pologne, Varsovie, Pologne. (cited pp. 59, 66, 85, 130, 133, 177, 280, and 281).

Manoussaki D., Lubkin S.R., Vernon R.B and Murray J.D. (1996). A mechanical model for the formation of vascular networks *in vitro. Acta Biotheoretica*, 44(3-4), 271-282. (cited pp. 870 and 1011).

Manoussaki D. (2003). A mechanochemical model of angiogenesis and vasculogenesis. *Mathematical Modelling and Numerical Analysis*, 37(4), 581-599. (cited pp. 107, 1004, and 1011).

Mansour J.M. and Mow V.C. (1976). The permeability of articular cartilage under compressive strain and high pressures. *J. Bone & Joint Surgery*, 58A(4), 509-516. (cited pp. 612 and 801).

Mantzaris N.V., Webb S. and Othmer H.G. (2004). Mathematical modeling of tumor-induced angiogenesis. A review. *J. Mathematical Biology*, 49, 111-187. (cited p. 1011).

Marchand F. and Ahmed A.M. (1990). Investigation on the laminate structure of the lumbar disc anulus fibrosus. *Spine*, 15(5), 402-410.

Mariappan Y.K., Glaser K.J. and Ehmann R.L. (2010). Magnetic resonance elastography: a review. *Clinical Anatomy*, 23(5), 497-511. (cited pp. 988 and 1039).

Marijnissen W.J.C.M., van Osch G.J.V.M., Aigner J., van der Ween S.W., Hollander A.P., Verwoerd-Verhoef H.L. and Verhaar J.A.N. (2002). Alginate as a chondrocyte-delivery substance in combination with a non-woven scaffold for cartilage tissue engineering. *Biomaterials*, 23, 1511-1517. (cited pp. 821 and 823).

Maroudas A. (1968). Physico-chemical properties of cartilage in the light of ion exchange theory. *Biophysical J.*, 8, 575-595. (cited pp. 499, 612, and 681).

Maroudas A. and Thomas H. (1970). A simple physicochemical micromethod for determining fixed anionic groups in connective tissue. *Biochimica et Biophysica Acta*, 215(1), 214-216. (cited p. 497).

Maroudas A. (1975). Biophysical chemistry of cartilaginous tissues with special reference to solute and fluid transport. *Biorheology*, 12(3-4), 233-248. (cited pp. 488, 498, 499, 501, 505, 506, 512, 612, 614, 640, 649, 681, 683, and 710).

Maroudas A. (1976). Balance between swelling pressure and collagen tension in normal and degenerate cartilage. *Nature*, 260(5554), 808-809. (cited p. 488).

Maroudas A. (1979). Physico-chemical properties of articular cartilage. In *Adult articular cartilage*, M.A.R. Freeman ed., Pitman Medical, Tunbridge Wells, England, 215-290. (cited pp. 612, 643, 710, and 824).

Maroudas A., Bayliss M. and Venn M. (1980). Further studies on the composition of human femoral head cartilage. *Annals Rheumatic Diseases*, 39, 514-523. (cited p. 488).

Maroudas A., Wachtel E., Grushko G., Katz E.P. and Weinberg P. (1991). The effect of osmotic and mechanical pressures on water partitioning in articular cartilage. *Biochimica et Biophysica Acta*, 1073(2), 285-294. (cited pp. 485, 501, 502, 512, and 713).

Marsden J.E. and Hughes T.J.R. (1983). *Mathematical foundations of elasticity*. Prentice Hall, Englewood Cliffs, NJ. (cited p. 174).

Martin G.R. and Jain R.K. (1994). Noninvasive measurement of interstitial pH profiles in normal and neoplastic tissue using fluorescence ratio imaging microscopy. *Cancer Research*, 54(21), 5670-5674. (cited p. 999).

Martin I., Obradovic B., Freed L.E. and Vunjak-Novakovic G. (1999). Method for quantitative analysis of glycosaminoglycan distribution in cultured natural and engineered cartilage. *Annals Biomedical Engng.*, 27, 656-662. (cited p. 833).

Martin I., Obradovic B., Treppo S., Grodzinsky A.J., Langer R., Freed L.E. and Vunjak-Novakovic G. (2000). Modulation of the mechanical properties of tissue engineered cartilage. *Biorheology*, 37, 141-147. (cited pp. 823 and 824).

Martin I., Wendt D. and Heberer M. (2004). The role of bioreactors in tissue engineering. *Trends in Biotechnology*, 22(2), 80-86. (cited p. 847).

Martin J.A. and Buckwalter J.A. (2002). Human chondrocyte senescence and osteoarhtritis. *Biorheology*, 39, 145-152. (cited p. 500).

Martina M., Mozrzymas J.W. and Vittur F. (1997). Membrane stretch activates a potassium channel in pig articular chondrocytes. *Biochimica et Biophysica Acta - Biomembranes*, 1329(2), 205-210. (cited p. 825).

Masaro L. and Zhu X.X. (1999). Physical models of diffusion for polymer solutions, gels and solids. *Progress in Polymer Science*, 24, 731-775. (cited p. 4).

Massieu F. (1869). Sur les fonctions caractéristiques des divers fluides. *Comptes-Rendus Académie des Sciences, Paris*, 69, 858-862. Addition au précédent mémoire sur les fonctions caractéristiques. *Comptes-Rendus Académie des Sciences, Paris*, 69, 1057-1061. (cited p. 70).

Massieu F. (1876). Mémoire sur les fonctions caractéristiques des divers fluides et sur la théorie des vapeurs. *Mémoires des Savants étrangers*, XXII, 1-92. (cited p. 70).

Masuda K., Sah R.L., Hejna M.J. and Thonar E.J.-M.A. (2003). A novel two-step method for the formation of tissue-engineered cartilage by mature bovine chondrocytes: the alginate-recovered-chondrocyte (ARC) method. *J. Orthopaedic Research*, 21, 139-148. (cited pp. 823 and 825).

Mauck R.L., Soltz M.A., Wang Ch.C.B., Wong. D.D., Chao P.-H.G., Valhmu W.B., Hung C.T. and Athesian G.A. (2000). Functional tissue engineering of articular cartilage through dynamic loading of chondrocyte-seeded agarose gels. *ASME J. Biomechanical Engng.*, 122, 252-260. (cited pp. 471, 499, 805, 823, 824, 826, 829, and 844).

Mauck R.L., Seyhan S.L., Ateshian G.A. and Hung C.T. (2002). Influence of seeding density and dynamic deformational loading in the developing structure/function relationships of chondrocyte-seeded agarose hydrogels. *Annals Biomedical Engng.*, 30(8), 1046-1056. (cited pp. 822, 824, 831, and 835).

Mauck R.L., Nicoll S.B., Seyhan S.L., Ateshian G.A. and Hung C.T. (2003)a. Synergistic action of growth factors and dynamic loading for articular cartilage tissue engineering. *Tissue Engineering*, 9(4), 597-611. (cited p. 820).

Mauck R.L., Hung C.T. and Ateshian G.A. (2003)b. Modeling of neutral solute transport in a dynamically loaded porous permeable gel: Implications for articular cartilage biosynthesis and tissue engineering. *ASME J. Biomechanical Engng.*, 125, 602-614. Erratum *ASME J. Biomechanical Engng.* (2004), 126, 392.

Maurice D.M. (1951). The permeability to sodium ions of the living rabbit's cornea. *J. Physiology*, 112, 367-391. (cited pp. 713 and 723).

Maurice D.M. (1961). Use of permeability studies in the investigation of submicroscopic structure. In *Structure of the eye*, G. K. Smelser ed., Academic Press, New York, 381-391. (cited p. 725).

McGowan K.B., Kurtis M.S., Lottman L.M., Watson D. and Sah R.L. (2002). Biochemical quantification of DNA in human articular and septal cartilage using PicoGreen and Hoechst 33258. *Osteoarthritis and Cartilage*, 10(7), 580-587. (cited p. 809).

McGuinness G.B. and Ó Brádaigh C.M. (1997). Development of rheological models for forming flows and picture-frame shear testing of fabric reinforced thermoplastic sheets. *J. Non-Newtonian Fluid Mechanics*, 73, 1-28. (cited pp. 792, 795, and 796).

McKnight A.L., Kugel J.L., Rossman P.J., Manduca A., Hartmann L.C. and Ehman R.L. (2002). MR Elastography of Breast Cancer: Preliminary Results. *American J. Roentgenology*, 178(6), 1411-1417. (cited pp. 871, 988, 989, 1010, and 1040).

McLaughlin S. and Mathias R.T. (1985). Electro-osmosis and the reabsorption of fluid in renal proximal tubules. *J. General Physiology*, 85(5), 699-728. (cited pp. 558 and 712).

McTigue D.F. (1986). Thermoelastic response of fluid-saturated porous rock. *J. Geophysical Research*, 91(B9), 9533-9542. (cited pp. 199, 207, 284, 294, 300, and 309).

Meek K.M., Blamires T., Elliott G.F., Gyi T.J. and Nave C. (1987). The organization of collagen fibrils in the human corneal stroma: a synchrotron x-ray diffraction study. *Current Eye Research*, 6(7), 841-846. (cited p. 702).

Meek K.M. and Fullwood N.J. (2001). Corneal and scleral collagens - a microscopist's perspective. *Micron*, 32, 261-272. (cited pp. 705, 706, and 707).

Meek K.M. and Quantock A.J. (2001). The use of x-ray scattering techniques to determine corneal ultrastructure. *Progress in Retinal and Eye Research*, 20(1), 95-137. (cited p. 491).

Meek K.M., Tuft S., Huang Y., Gill P.S., Hayes S., Newton R.H. and Bron A.J. (2005). Changes in collagen orientation and distribution in keratoconus corneas. *Investigative Ophthalmology & Visual Science*, 46(6), 1948-1956. (cited p. 708).

Mehrabadi M.M. and Cowin S.C. (1990). Eigentensors of linear anisotropic elastic materials. *Quarterly J. Mechanics and Applied Mathematics*, 43, 15-41. (cited pp. 136 and 153).

Mei C.C. (1995). *Mathematical analysis in engineering*. Cambridge University Press, Cambridge, UK. (cited pp. 26, 474, and 589).

Mercury L. and Tardy Y. (2001). Negative pressure of stretched liquid water. Geochemistry of soil capillaries. *Geochimica et Cosmochimica Acta*, 65(20), 3391-3408. (cited p. 81).

Mergler S. and Pleyer U. (2007). The human corneal endothelium: new insights into electrophysiology and ion channels. *Progress in Retinal and Eye Research*, 26, 359-378. (cited p. 711).

Merodio J. and Ogden R. (2003). Instabilities and loss of ellipticity in fiber-reinforced compressible non-linearly elastic solids under plane deformation. *Int. J. Solids Structures*, 40, 4707-4727. (cited p. 785).

Merodio J. and Ogden R. (2005). On tensile instabilities and ellipticity loss in fiber-reinforced incompressible non-linearly elastic solids. *Mechanics Research Communications*, 32(3), 290-299.

Miao S., Wang M.L. and Schreyer H.L. (1995). Constitutive Models for Healing of Materials with Application to Compaction of Crushed Rock Salt. *J. Engng. Mathematics*, 121(10), 1122-1129. (cited p. 985).

Mishima S. and Hedbys B.O. (1967). The permeability of the corneal epithelium and endothelium to water. *Experimental Eye Research*, 6(1), 10-32. (cited p. 723).

Mitchell J.K. (1993). *Fundamentals of soil behavior*. 2nd ed., John Wiley & Sons, In., New York. (cited pp. 284, 447, 449, 479, 592, and 609).

Miyazaki H. and Hayashi K. (1999). Tensile tests of collagen fibers obtained from the rabbit patellar tendon. *Biomedical Microdevices*, 2(2), 151-157. (cited pp. 431 and 493).

Mizuno S. (2002). Hydrostatic fluid pressure enhances intracellular calcium concentration in bovine articular chondrocytes. *Transactions Orthopaedic Research Society*, 48th Annual Meeting, 377. (cited p. 828).

Mok U., Bernabé Y. and Evans B. (2002). Permeability, porosity and pore geometry of chemically altered porous silica glass. *J. Geophysical Research*, 107, B1, 2015. (cited p. 561).

Mollica F., Jain R.K. and Netti P.A. (2003). A model for temporal heterogeneities of tumor blood flow. *Microvascular Research*, 65(1), 56-60. (cited p. 1014).

Moran G.R. and Prato F.S. (2001). Modeling tissue contrast agent concentration: a solution to the tissue homogeneity model using a simulated arterial input function. *Magnetic Resonance in Medicine*, 45(1), 42-45. (cited p. 1020).

Moritz A.R. and Henriques F.C. (1947). Studies of thermal injury: II. The relative importance of time and surface temperature in the causation of cutaneous burns. *American J. Pathology*, 23(5), 695-720. (cited p. 279).

Morman K.N. (1986). The generalized strain-measure with application to nonhomogeneous deformations in rubber-like solids. *ASME J. Applied Mechanics*, 53, 726-728. (cited pp. 24 and 67).

Morrey Ch.B., Jr. (1952). Quasi-convexity and the lower semicontinuity of multiple integrals. *Pacific Journal of Mathematics*, 2(1), 25-53. (cited p. 180).

Moses M.A., Sudhalter J. and Langer R. (1990). Identification of an inhibitor of neovascularization from cartilage. *Science*, 248(4961), 1408-1410. (cited p. 998).

Mow V.C., Kuei S.C., Lai W.M. and Armstrong C.G. (1980). Biphasic creep and stress relaxation of articular cartilage in compression: Theory and experiments. *ASME J. Biomechanical Engng.*, 102, 73-84.

Mow V.C. and Schoonbeck J.M. (1984). Contribution of Donnan osmotic pressure towards the biphasic compressive modulus of articular cartilage. *Transactions Orthopaedic Research Society*, 30th Annual Meeting, Atlanta, 262.

Mow V.C., Gibbs M.C., Lai W.M., Zhu W.B. and Athanasiou K.A. (1989). Biphasic indentation of articular cartilage. Part II- A numerical algorithm and an experimental study. *J. Biomechanics*, 22(8-9), 853-861. (cited p. 828).

Mow V.C., Ateshian G.A., Lai W.M. and Gu W.Y. (1998). Effects of fixed charges on the stress-relaxation behavior of hydrated soft tissues in a confined compression problem. *Int. J. Solids Structures*, 35, 4945-4962.

Mow V.C. and Guo X.E. (2002). Mechano-electrochemical properties of articular cartilage: their inhomogeneities and anisotropies. *Annual Review Biomedical Engng.*, 4, 175-209. (cited pp. 488, 612, and 639).

Muir H., Bullough P. and Maroudas A. (1970). The distribution of collagen in human articular cartilage with some of its physiological implications. *J. Bone & Joint Surgery*, 52B, 3, 554-563. (cited p. 488).

Muir H. (1978). Proteoglycans of cartilage. *J. Clinical Pathology. Supplement (Royal College Pathologists)*, 31, Suppl (Roy. Coll. Path.), 12, 67-81. (cited pp. 488, 494, 495, 496, 627, 638, 639, and 641).

Muir H. (1983). Proteoglycans as organizers of the intercellular matrix. *Biochemical Society Transactions*, 11(6), 613-622. (cited pp. 638, 701, and 847).

Muir H. (1995). The chondrocyte, architect of cartilage. Biomechanics, structure, function and molecular biology of cartilage matrix macromolecules. *BioEssays*, 17(12), 1039-1048.

Müller I. (1967). Zum Paradoxon der Wärmeleitungstheorie. *Zeitschrift für Physik*, 198(4), 329-344. (cited p. 285).

Müller L.J., Pels E. and Vrensen G. (2001). The specific architecture of the anterior stroma accounts for maintenance of corneal curvature. *British J. Ophthalmology*, 85, 437-443. (cited p. 707).

Müller L.J., Pels E., Schurmans L. and Vrensen G. (2004). A new three-dimensional model of the organization of proteoglycans and collagen fibrils in the human corneal stroma. *Experimental Eye Research*, 78, 493-501. (cited pp. 705 and 706).

Murad M.A. (1999). Thermo-mechanical model of hydration swelling in smectitic clays. *Int. J. Numerical Analytical Methods Geomechanics*, 23(7), Part I: 673-696, Part II: 697-719.

Murnaghan F.D. (1951). *Finite deformation of an elastic solid.* Wiley, New York. (cited p. 183).

Murray C.D. (1926). The physiological principle of minimum work. I. The vascular system and the cost of blood volume. *Proceedings National Academy Sciences USA*, 12(3), 207-214. (cited pp. 839, 846, and 922).

Murray J.D. (2002). *Mathematical biolology. I. An introduction.* 3rd edition, Springer-Verlag, New York.

Murray J.D. (2003). *Mathematical biolology. II. Spatial models and biomedical applications.* 3rd edition, Springer-Verlag, New York. (cited pp. 130 and 551).

Musgrave M.J.P. (1970). *Crystal acoustics. Introduction to the study of elastic waves and vibrations in crystals.* Holden-Day, San Francisco, CA.

Myers E.R., Lai W.M. and Mow V.C. (1984). A continuum theory and an experiment for the ion-induced swelling behavior of articular cartilage. *ASME J. Biomechanical Engng.*, 106, 151-158. (cited pp. 471, 472, 477, 478, 503, and 506).

* * * * N * * * *

Nagaki S., Goya M. and Sowerby R. (1993). The influence of void distribution on the yielding of an elastic-plastic solid. *Int. J. Plasticity*, 9(2), 199-211. (cited p. 435).

Nagatomi J., Arulanandan B.P., Metzger D.W., Meunier A. and Bizios R. (2002). Effects of cyclic pressure on bone marrow cell cultures. *ASME J. Biomechanical Engng.*, 124, 308-314. (cited p. 807).

Nakayama K., Nagaki S. and Abe T. (1996). Yield stress of regularly perforated sheets and anisotropic yield function. *Transactions Japan Society Mechanical Engineers (A)* 62 (600), 1877-1882 (in Japanese). [Chaps.-12-] (cited p. 434).

Nash I.S., Greene P.R. and Foster C.S. (1982). Comparison of mechanical properties of keratoconus and normal corneas. *Experimental Eye Research*, 35(4), 413-423.

Neeman M. and Dafni H. (2003). Structural, functional and molecular MR imaging of the microvasculature. *Annals Biomedical Engng.*, 5, 29-56. (cited p. 1012).

Nehring D., Adamietz P., Meenen N.M. and Pörtner R. (1999). Perfusion cultures and modelling of oxygen uptake with three-dimensional chondrocyte pellets. *Biotechnology Techniques*, 13, 701-706. (cited p. 834).

Neidert M.R., Lee E.S., Oegema T.R. and Tranquillo R.T. (2002). Enhanced fibrin remodeling *in vitro* with TGF-β1, insulin and plasmin for improved tissue-equivalents. *Biomaterials*, 23, 3717-3731. (cited p. 837).

Neidlinger-Wilke C., Grood E.S., Wang J.H.-C., Brand R.A. and Claes L. (2001). Cell alignement is induced by cyclic changes in cell length: studies of cells grown in cyclically stretched substrates. *J. Orthopaedic Research*, 19, 286-293. (cited p. 837).

Nelson F.R. (1999). Chondrocyte engineering: in search of articular cartilage. Medical University of South Carolina's Orthopaedic Journal, Department of Orthopaedic Surgery, 2, 26-31, June 1999. (cited p. 808). http://www.musc.edu/orthosurg/research/orthojournal99.

Nemat-Nasser S. and Hori M. (1993). *Micromechanics; overall properties of heterogeneous materials.* Vol. 37, Series in Applied Mathematics and Mechanics. North-Holland, Amsterdam, the Netherlands.

Nemat-Nasser S. (2004). *Plasticity. A treatise on finite deformation of heterogeneous inelastic materials.* Cambridge Monographs on Mechanics. Cambridge University Press, Cambridge, UK.

Nerurkar N.L., Ramasubramanian A. and Taber L.A. (2006). Morphogenetic adaptation of the looping embryonic heart to altered mechanical loads. *Developmental Dynamics*, 235(7), 1822-1829. (cited p. 1003).

Netti P.A., Baxter L.T., Boucher Y., Skalak R. and Jain R.K. (1995). Time-dependent behavior of interstitial fluid pressure in solid tumors: implications for drug delivery. *Cancer Research*, 55, 5451-5458. (cited pp. 207, 570, 993, 1010, 1020, and 1027).

Netti P.A., Baxter L.T., Boucher Y., Skalak R. and Jain R.K. (1997). Macro- and microscopic fluid transport in living tissues: application to solid tumors. *American Institute Chemical Engineers J.*, 43(3), 818-834.

Netti P.A., Hamberg L.M., Babich J.W., Kierstead D., Graham W., Hunter G.J., Wolf G.L., Fischman A., Boucher Y. and Jain R.K. (1999). Enhancement of fluid filtration across tumor vessels: implication for delivery of macromolecules. *Proceedings National Academy Sciences USA*, 96, 3137-3142. (cited pp. 569 and 570).

Netti P.A., Berk D.A., Swartz M.A., Grodzinsky A.J. and Jain R.K. (2000). Role of extracellular matrix assembly in interstitial transport in solid tumors. *Cancer Research*, 60, 2497-2503. (cited pp. 258, 265, 553, 570, 571, 612, 847, 871, 989, 998, and 1010).

Netti P.A., Travascio F. and Jain R.K. (2003). Coupled macromolecular transport and gel mechanics: poroviscoelastic approach. *J. BioEngineering, Food and Natural Products, American Institute of Chemical Engineers*, 49(6), 1580-1596. (cited pp. 258, 260, 568, and 1020).

Ng I.O.L, Lai E.C.S., Ng M.M.T. and Fan S.T. (1992). Tumor encapsulation in hepatocellular carcinoma. A pathological study of 189 cases. *Cancer*, 70(1), 45-49. (cited p. 1008).

Ng L., Grodzinsky A.J., Patwari P., Sandy J., Plaas A. and Ortiz Ch. (2003). Individual cartilage aggrecan macromolecules and their constituent glysosaminoglycans visualized via atomic force microscopy. *J. Structural Biology*, 143, 242-257.

Nicholson C. and Phillips J.M. (1981). Ion diffusion modified by tortuosity and volume fraction in the extracellular microenvironment of the rat cerebellum. *J. Physiology*, 321, 225-257. (cited p. 989).

Nickerson Ch. S. (2005). Engineering the mechanical properties of ocular tissues. PhD thesis, California Institute of Technology, Pasadena, CA. (cited pp. 704 and 795).

Nishio S. and Fukui M. (1999). Solitary fibrous tumor in neurosurgical practice. *Critical Reviews in Neurosurgery*, 9(5), 319-325. (cited p. 989).

Nixon A.J., Brower-Toland B.D., Bent S.J., Saxer R.A., Wilke M.J., Robbins P.D. and Evans C.H. (2000). Insulin-like growth factor-I gene therapy applications for cartilage repair. *Clinical Orthopaedics and Related Research*, 379 Suppl, S201-S213. (cited p. 501).

Norris A.N. (1992). On the correspondence between poroelasticity and thermoelasticity. *J. Applied Physics*, 71(3), 1138-1141. (cited p. 281).

Norris A.N. (2006). Elastic moduli approximation of higher symmetry for the acoustical properties of an anisotropic material. *J. Acoustical Society America*, 119(4), 2114-2121. (cited pp. 139 and 358).

Nova R. and Wood D.M. (1978). A constitutive model for sand in triaxial compression. *Int. J. Numerical Analytical Methods Geomechanics*, 3(3), 255-278. (cited p. 135).

Nugent L.J. and Jain R.K. (1984). Extravascular diffusion in normal and neoplastic tissues. *Cancer Research*, 44(1), 238-244. (cited pp. 552, 570, 571, and 993).

Nur A. and Byerlee J.D. (1971). An exact effective stress law for elastic deformation of rock with fluids. *J. Geophysical Research*, 76(26), 6414-6419. (cited pp. 210 and 220).

Nye J.F. (1957). *Physical properties of crystals. Their representation by tensors and matrices.* Oxford University Press, Oxford, UK.

* * ** O * * **

Obradovic B., Carrier R.L., Vunjak-Novakovic G. and Freed L.E. (1999). Gas Exchange is Essential for Bioreactor Cultivation of Tissue Engineered Cartilage. *Biotechnology and Bioengineering*, 63(2), 197-205. (cited pp. 833, 834, 840, and 846).

Obradovic B., Meldon J.H., Freed L.E. and Vunjak-Novakovic G. (2000). Glycosaminoglycan deposition in engineered cartilage: experiments and mathematical model. *American Institute Chemical Engineers J.*, 46(9), 1860-1871. (cited pp. 550, 553, 834, 835, 846, and 966).

Obradovic B., Martin I., Padera R.F., Treppo S., Freed L.E. and Vunjak-Novakovic G. (2001). Integration of engineered cartilage. *J. Orthopaedic Research*, 19, 1089-1097. (cited pp. 808, 821, and 831).

Oda M., Nemat-Nasser S. and Mehrabadi M.M. (1982). A statistical study of fabric in a random assembly of spherical granules. *Int. J. Numerical Analytical Methods Geomechanics*, 6, 77-94. (cited p. 434).

Oda M., Nemat-Nasser S. and Konishi J. (1985). Stress-induced anisotropy in Granular Masses. *Soils and Foundations*, 25(3), 85-97. (cited p. 105).

Ogden R.W. (1972)a. Large deformation isotropic elasticity - on the correlation of theory and experiment for incompressible rubberlike solids. *Proceedings Royal Society London*, A326, 565-584. (cited p. 183).

Ogden R.W. (1972)b. Large deformation isotropic elasticity - on the correlation of theory and experiment for compressible rubberlike solids. *Proceedings Royal Society London*, A328, 567-583. (cited p. 183).

Ogden R.W. (1982). Elastic deformations of rubberlike solids. In *Mechanics of solids, The Rodney Hill 60th Anniversary Volume*, H.G. Hopkins and M.J. Sewell, eds., 499-537. Pergamon Press, Oxford, UK. (cited p. 183).

Ogle B.M. and Mooradian D.L. (2002). Manipulation of remodeling pathways to enhance the mechanical properties of a tissue engineered blood vessel. *ASME J. Biomechanical Engng.*, 124(6), 724-733. (cited p. 807).

O'Hara B.P., Urban J.P. and Maroudas A. (1990). Influence of cyclic loading on the nutrition of articular cartilage. *Annals Rheumatic Diseases*, 49(7), 536-539. (cited p. 820).

Olsen H.W., Yearsley E.N. and Nelson K.R. (1990). Chemico-osmosis versus diffusion-osmosis. *Transportation Research Record - Soils, Geology and Foundations*, Washington DC, 1288, 15-22. (cited p. 609).

Omens J.H. (1998). Stress and strain as regulators of myocardial growth. *Progress in Biophysics & Molecular Biology*, 69, 559-572. (cited pp. 845 and 856).

Omens J.H., McCulloch A.D. and Criscione J.C. (2003). Complex distributions of residual stress and strain in the mouse left ventricle: experimental and theoretical models. *Biomechanics and Modeling in Mechanobiology*, 1, 267-277. (cited pp. 189, 850, and 928).

Ophir J., Céspedes I., Ponnekanti H., Yazdi Y. and Li X. (1991). Elastography: a quantitative method for imaging the elasticity of biological tissues. *Ultrasonic Imaging*, 13(2), 111-134. (cited p. 1040).

Orssengo G.J. and Pye D.C. (1999). Determination of the true intraocular pressure and modulus of elasticity of the human cornea in vivo. *Bulletin Mathematical Biology*, 61, 551-572. (cited p. 726).

Ortega J.M. (1987). *Matrix theory.* Plenum Press, New York. (cited pp. 14 and 17).

Orwin E.J., Borene M.L. and Hubel A. (2003). Biomechanical and optical characteristics of a corneal stromal equivalent. *ASME J. Biomechanical Engng.*, 125, 439-444. (cited pp. 703 and 731).

Osch G.J.V.M., van Mandl E.W., Marijnissen W.J.C.M., van der Ween S.W., Verwoerd-Verhoef H.L. and Verhaar J.A.N. (2002). Growth factors in cartilage tissue engineering. *Biorheology*, 39, 215-220. (cited pp. 818 and 819).

Othmer H.G. (1981). The interactions of structure and dynamics in chemical reaction networks. In *Modelling of chemical reaction systems.* Proceedings Intl Workshop, Heidelberg, Germany, Sept. 1-5, 1980, Springer Verlag, Berlin. K.H. Ebert, P. Deuflhard and W. Jäger, eds. pp. 2-19. (cited p. 73).

Overby D., Ruberti J., Gong H., Freddo Th.F. and Johnson M. (2001). Specific hydraulic conductivity of corneal stromal as seen by quick-freeze/deep-etch. *ASME J. Biomechanical Engng.*, 123, 154-161. (cited p. 724).

Owens J.M., Lai W.M. and Mow V.C. (1991). Biomechanical effects due to Na^+-Ca^{2+} exchange in articular cartilage. *Transactions Orthopaedic Research Society*, 37th Annual Meeting, Anaheim, CA, 360. (cited pp. 461, 462, 463, 464, and 465).

* * ** P * * **

Padera T.P., Boucher Y. and Jain R.K. (2003). Correspondence re: "S. Maula et al. (2003), Intratumoral Lymphatics are Essential for the Metastatic Spread and Prognosis in Squamous Cell Carcinoma of the Head and Neck. *Cancer Research*, 63, 1920-1926". *Cancer Research*, 63, 8555-8556. (cited p. 999).

Padera T.P., Stoll B.R., Tooredman J.B., Capen D., di Tomaso E. and Jain R.K. (2004). Cancer cells compress intratumor vessels. *Nature*, 427(6976), 695. (cited pp. 998 and 1004).

Pálfi V.K. and Perczel A. (2008). The inherent stability of collagen. John von Neumann Institute for Computing, Jülich, NIC Series, 40, 349-352. (cited p. 126).

Pandolfi A. and Manganiello F. (2006). A model for the human cornea: constitutive formulation and numerical analysis. *Biomechanics and Modeling in Mechanobiology*, 5(4), 237-246. (cited pp. 797 and 801).

Panjwani N.A. and Harding J.J. (1978). Isolation and hydroxylysine glycoside content of some cyanogen bromide-cleaved fragments of collagen from bovine corneal stroma. *Biochemical J.*, 171, 687-695. (cited pp. 492 and 743).

Parker K.H., Winlowe C.P. and Maroudas A. (1988). The theoretical distributions and diffusivities of small ions in chondroitin sulphate and hyaluronate. *Biophysical Chemistry*, 32(2-3), 271-282. (cited p. 614).

Parsons J.W. and Coger R.N. (2002). A new device for measuring the viscoelastic properties of hydrated matrix gels. *ASME J. Biomechanical Engng.*, 124(2), 145-154. (cited p. 241).

Passman S.L. and McTigue D.F. (1989). Momentum Flux in Multiphase Mixtures. *J. Rheology*, 33(1), 177-182. (cited pp. 107 and 112).

Patwari P., Chubinskaya S., Hakimiyan A., Kumar B., Cole A.A., Kuettner K.E., Rueger D.C. and Grodzinsky A.J. (2003). Injurious compression of adult human donor cartilage explants: investigation of anabolic and catabolic processes. *Transactions Orthopaedic Research Society*, 49th Annual Meeting, 695. (cited p. 820).

Payton R.G. (1983). *Elastic wave propagation in transversely isotropic media*. Martinus Nijhoff Publishers, The Hague, the Netherlands. (cited p. 412).

Pei M., Solchaga L.A., Seidel J., Zeng L., Vunjak-Novakovic G., Caplan A.I. and Freed L.E. (2002). Bioreactors mediate the effectiveness of tissue engineering scaffolds. *The Faseb J., Federation of American Societies for Experimental Biology*, 16(12), 1691-1694. (cited p. 822).

Peng X.Q., Guo Z.Y. and Moran B. (2006). An anisotropic hyperelastic constitutive model with fiber-matrix shear interaction for the human annulus fibrosus. *ASME J. Applied Mechanics*, 73, 815-824. (cited pp. 785 and 786).

Pennes H.H. (1948). Analysis of tissue and arterial blood temperatures in resting human forearm. *J. Applied Physiology*, 1(2), 93-122. (cited p. 317).

Peretti G.M., Caruso E.M., Randolph M.A. and Zaleske D.J. (2001). Meniscal repair using engineered tissue. *J. Orthopaedic Research*, 19(2), 278-285. (cited pp. 807 and 831).

Pericak-Spector K.A., Sivaloganathan J. and Spector S.J. (2000). The representation theorem for linear, isotropic tensor functions in even dimensions. *J. Elasticity*, 57(2), 157-164.

Petermann A.T., Hiromura K., Blonski M., Pippin J., Monkawa T., Durvasula R., Couser W.G. and Shankland S.J. (2002). Mechanical stress reduces podocyte proliferation *in vitro*. *Kidney International*, 61(1), 40-50. (cited p. 1003).

Petrangeli Papini M. and Majone M. (2002). Modeling of heavy metal adsorption at clay surfaces. *Encyclopedia of Surface and Colloid Science*, A. Hubbard, ed., Marcel Dekker, Inc., New York, 3483-3498. (cited pp. 482, 484, and 631).

Pezowicz C.A., Robertson P.A and Broom N.D. (2005). Intralamellar relationships within the collagenous architecture of the annulus fibrosus imaged in its fully hydrated state. *J. Anatomy*, 207, 299-312. (cited pp. 660, 800, and 801).

Pezowicz C.A., Robertson P.A and Broom N.D. (2006). The structural basis of interlamellar cohesion in the intervertebral disc wall. *J. Anatomy*, 208, 317-330. (cited p. 800).

Phillips R.J. (2000). A hydrodynamic model for hindered diffusion of proteins and micelles in hydrogels. *Biophysical J.*, 79, 3350-3354. (cited p. 571).

Phillips S.L. (1984). The determination of charge density in articular cartilage via chemical titration. Master Thesis, Dept. of Electrical Engineering and Computer Science, MIT, Cambridge, MA. (cited p. 643).

Pienkowski D. and Pollack S.R. (1983). The origin of stress-generated potentials in fluid-saturated bone. *J. Orthopaedic Research*, 1(1), 30-41. (cited p. 627).

Del Piero G.P. (1998). Representation theorems for hemitropic and transversely isotropic tensor functions. *J. Elasticity*, 51(1), 43-71.

Pinsky P.M. and Datye D.V. (1991). A microstructurally-based finite element model of the incised human cornea. *J. Biomechanics*, 24(10), 907-922.

Pinsky P.M., van der Heide D. and Chernyak D. (2005). Computational modeling of the mechanical anisotropy in the cornea and sclera. *J. Cataract & Refractive Surgery*, 31, 136-145. (cited pp. 429, 796, and 797).

Pioletti D.P. and Rakotomanana L.R. (2000). Non linear viscoelastic laws for soft biological tissues. *European J. Mechanics-A/Solids*, 19, 749-759. (cited pp. 173, 187, and 273).

Pisu M., Lai N., Cincotti A., Delogu F. and Cao G. (2003)a. A simulation model for the growth of tissue engineered cartilage on polymeric scaffolds. *Chemical Engineering Transactions*, III, 130-137. (cited pp. 834 and 835).

Pisu M., Lai N., Cincotti A., Delogu F. and Cao G. (2003)b. A simulation model for the growth of engineered cartilage on polymeric scaffolds. *Int. J. Chemical Reactor Engineering*, 1, A38, 1-15. (cited pp. 834 and 835).

Pittenger M.F., Mackay A.M., Beck S.C., Jaiswal R.K., Douglas R., Mosca J.D., Moorman M.A., Simonetti D.W., Craig S. and Marshak D.R. (1999). Multilineage potential of adult human mesenchymal stem cells. *Science*, 284(5411), 143-147. (cited p. 821).

Plaas A.H., West L.A., Thonar E.J.A., Karcioglu Z.A., Smith C.J., Klintworth G.K. and Hascall V.C. (2001). Altered fine structures of corneal and skeletal keratan sulfate and chondroitin/dermatan sulfate in macular corneal dystrophy. *The J. Biological Chemistry*, 276(43), 39788-39796. (cited p. 741).

Platten J.K. (2006). The Soret effect: a review of recent experimental results. *ASME J. Applied Mechanics*, 73(1), 5-15. (cited p. 533).

Please C.P., Pettet G. and McElwain D.L.S. (1997). A new approach to modelling the formation of necrotic regions in tumors. *Applied Mathematics Letters*, 11(3), 89-94. (cited pp. 1006 and 1007).

Please C.P., Pettet G. and McElwain D.L.S. (1999). Avascular tumour dynamics and necrosis. *Mathematical Models and Methods in Applied Sciences*, 9(4), 569-579. (cited pp. 1006 and 1007).

Plona T.J. (1980). Observation of a second bulk compressional wave in a porous medium at ultrasonic frequencies. *Applied Physics Letters*, 36(4), 259-261. (cited pp. 321 and 374).

Pluen A., Netti P.A., Jain R.K. and Berk D.A. (1999). Diffusion of macromolecules in agarose gels: comparison of linear and globular configurations. *Biophysical J.*, 77(1), 542-552. (cited pp. 553 and 1010).

Pluen A., Boucher Y., Ramanujan S., McKee T.D., Gohongi T., di Tomaso E., Brown E.B., Izumi Y., Campbell R.B., Berk D.A. and Jain R.K. (2001). Role of tumor-host interactions in interstitial diffusion of macromolecules: cranial vs. subcutaneous tumors. *Proceedings National Academy Sciences USA*, 98(8), 4628-4633. (cited pp. 552 and 571).

Polge C., Smith A.U. and Parkes A.S. (1949). Revival of spermatozoa after vitrification and dehydration at low temperatures. *Nature*, 164(4172), 666. (cited p. 279).

Pollard R.E., Garcia T.C., Stieger S.M., Ferrara K.W., Sadlowski A.R. and Wisner E.R. (2004). Quantitative evaluation of perfusion and permeability of peripheral tumors using contrast-enhanced computed tomography. *Investigative Radiology*, 39(6), 340-349. (cited p. 1040).

Prange M.T. and Margulies S.S. (2002). Regional, directional and age-dependent properties of the brain undergoing large deformation. *ASME J. Biomechanical Engng.*, 124(2), 244-252. (cited p. 183).

Press W.H., Teukolsky S.A., Vetterling W.T and Flannery B.P. (1992). *Numerical recipes: The art of scientific computing.* 2nd ed., Cambridge University Press, New York. (cited pp. 234 and 419).

Prévost J.H. and Tao D. (1983). Finite element analysis of dynamic coupled thermoelasticity problems with relaxation times. *ASME J. Applied Mechanics*, 50, 817-822. (cited p. 285).

Preziosi L. and Graziano L. (2003). Multiphase models of tumor growth: general framework and particular cases. In *Mathematical modelling and computing in biology and medicine*, V. Capasso, ed., Società Editrice Esculapio, 622-628. (cited p. 1010).

Pryse K.M., Nekouzadeh A., Genin G.M., Elson E.L. and Zahalak G.I. (2003). Incremental mechanics of collagen gels: new experiments and a new viscoelastic model. *Annals Biomedical Engng.*, 31(10), 1287-1296. (cited p. 274).

Prydz K. and Dalen K.T. (2000). Synthesis and sorting of proteoglycans. *J. Cell Science*, 113(2), 193-205. (cited p. 808).

* * * * Q * * * *

Qian H., Beard D.A. and Liang S. (2003). Stoichiometric network theory for nonequilibrium biochemical systems. *European J. Biochemistry*, 270, 415-421. (cited p. 74).

Quantock A.J., Boote C., Young R.D., Hayes S., Tanioka H., Kawasaki S., Ohta N., Iida T., Yagi N., Kinishota S. and Meek K.M. (2007). Small-angle fibre diffraction studies of corneal matrix structure: a depth-profiled investigation of the human eye-bank cornea. *J. Applied Crystallography*, 40, s335-s340. (cited pp. 704 and 706).

Quiligotti S. (2002). On bulk growth mechanics of solid-fluid mixtures: kinematics and invariance requirements. *Theoretical and Applied Mechanics*, 28-29, 277-288.

Quinn T.M., Morel V. and Meister J.J. (2001). Static compression of articular cartilage can reduce solute diffusivity and partitioning: implications for the chondrocyte biological response. *J. Biomechanics*, 34, 1463-1469.

* * * * R * * * *

Rabinovitz Y.S. (1998). Keratoconus. *Survey Ophthalmology*, 42(4), 297-319. (cited p. 707).

Rachev A., Stergiopulos N. and Meister J.J. (1996). Theoretical study of dynamics of arterial wall remodeling in response to changes in blood pressure. *J. Biomechanics*, 29(5), 635-642. (cited p. 851).

Rachev A. (1997). Theoretical study of the effect of stress-dependent remodeling on arterial geometry under hypertensive conditions. *J. Biomechanics*, 30, 819-827.

Rachev A., Stergiopulos N. and Meister J.J. (1998). A model for geometric and mechanical adaptation of arteries to sustained hypertension. *ASME J. Biomechanical Engng.*, 120, 9-17. (cited pp. 846 and 922).

Radi E. and Loret B. (2007). Mode II intersonic crack propagation in poroelastic media. *Int. J. Fracture*, 147(1-4), 235-267, DOI 10.1007/s10704-007-9169-z. (cited pp. 209, 406, 407, 409, 410, and 411).

Radi E. and Loret B. (2008). Mode I intersonic crack propagation in poroelastic media. *Mechanics of Materials*, 40(6), 524-548. (cited pp. 209, 406, 409, 410, and 411).

Radner W. and Mallinger R. (2002). Interlacing of collagen lamellae in the midstroma of the human cornea. *Cornea*, 21(6), 598-601. (cited p. 707).

Rae J.L. and Watsky M.A. (1996). Ionic channels in corneal endothelium. *American J. Physiology - Cell Physiology*, 270, C975-C989. (cited p. 712).

Ragan P.M., Badger A.M., Cook M., Chin V.I., Gowen M., Grodzinsky A.J. and Lark M.W. (1999). Down-regulation of chondrocyte aggrecan and type II collagen gene expression correlates with increases in static compression magnitude and duration. *J. Orthopaedic Research*, 17(6), 836-842. (cited p. 806).

Raghavan M.L., Trivedi S., Nagaraj A., MCPherson D.D. and Chandran K.B. (2004). Three-Dimensional finite element analysis of residual stress in arteries. *Annals Biomedical Engng.*, 32(2), 257-263. (cited p. 887).

Raimondi M.T., Boschetti F., Falcone L., Fiore G.B., Remuzzi A., Marinoni E., Marazzi M. and Pietrabissa R. (2002). Mechanobiology of engineered cartilage cultured under a quantified fluid-dynamic environment. *Biomechanics and Modeling in Mechanobiology*, 1, 69-82. (cited p. 829).

Rajagopal K.R. and Srinivasa A.R. (2004). On thermomechanical restrictions of continua. *Proceedings Royal Society London*, A460, 631-651.

Ramanujan S., Pluen A., McKee T.D., Brown E.B., Y. Boucher and Jain R.K. (2002). Diffusion and convection in collagen gels: implications for transport in the tumor interstitium. *Biophysical J.*, 83, 1650-1660. (cited pp. 570, 571, and 1010).

Ramasubramanian A., Latacha K.S., Benjamin J.M., Voronov D.A., Ravi A. and Taber L.A. (2006). Computational model for early cardiac looping. *Annals Biomedical Engng.*, 34, 1655-1669. (cited pp. 879, 880, and 888).

Rambod E., Beizai M. and Rosenfeld M. (2010). An experimental and numerical study of the flow and mass transfer in a model of the wearable artificial kidney dialyzer. *BioMedical Engineering OnLine*, 9:21. doi: 10.1186/1475-925X-9-21 (cited pp. 2 and 3).

Ramirez P., Alcaraz A., Mafé S. and Pellicer J. (2002). Donnan equilibrium of ionic drugs in pH-dependent fixed charge membranes: theoretical modeling. *J. Colloid and Interface Science*, 253, 171-179. (cited p. 635).

Ramirez P., Mafé S., Aguilella V.M. and Alcaraz A. (2003). Synthetic nanopores with fixed charges: an electrodiffusion model for ionic transport. *Physical Review*, E68, 011910-1-8. (cited p. 635).

Ransom B. and Helgeson H.C. (1994). Estimation of the standard molal heat capacities, entropies and volumes of 2:1 clay minerals. *Geochimica et Cosmochimica Acta*, 58(2), 4537-4547.

Raoult A. (1986). Non-polyconvexity of the stored energy function of a Saint Venant-Kirchhoff material. *Aplikace Matematiky*, 31(6), 417-419. (cited p. 180).

Rawe I.M., Meek K.M., Leonard D.W., Takahashi T. and Cintron Ch. (1994). Structure of corneal scar tissue: an x-ray diffraction study. *Biophysical J.*, 67(4), 1743-1748. (cited p. 707).

Regini J.W. and Elliott G.F. (2001). The effect of temperature on the Donnan potentials in biological polyelectrolyte gels: cornea and striated muscle. *Int. J. Biological Macromolecules*, 28(3), 245-254. (cited pp. 710 and 743).

Regini J.W., Elliott G.F. and Hodson S.A. (2004). The ordering of corneal collagen fibrils with increasing ionic strength. *J. Molecular Biology*, 336, 179-186. (cited pp. 709, 710, and 736).

Reichel E., Miller D., Blanco E. and Mastanduno R. (1989). The elastic modulus of central and perilimbal bovine cornea. *Annals Ophthalmology*, 21, 205-208. (cited p. 726).

Reis-Filho J.S. and Schmitt F.C. (2003). Lymphangiogenesis in tumors: what do we know? *Microscopy Research and Technique*, 60(2), 171-180. (cited p. 999).

Reynaud B. and Quinn T.M. (2006). Tensorial electrokinetics in articular cartilage. *Biophysical J.*, 91(6), 2349-2355. (cited p. 597).

Rice J.R. and Cleary M.P. (1976). Some basic stress diffusion solutions for fluid-saturated elastic porous media with compressible constituents. *Reviews Geophysics and Space Physics*, 14(2), 227-241. (cited pp. 214, 217, and 220).

Righetti R., Ophir J., Srinivasan S. and Krouskop T.A. (2004). The feasibility of using elastography for imaging the Poisson's ratio in porous media. *Ultrasound in Medicine and Biology*, 30(2), 215-228. (cited p. 1041).

Righetti R., Ophir J. and Krouskop T.A. (2005). A method for generating permeability elastograms and Poisson's ratio time-constant elastograms. *Ultrasound in Medicine and Biology*, 31(6), 803-816. (cited p. 1041).

Riley M., Winkler B.S., Starnes C.A. and Peters M.I. (1997). Fluid and ion transport in corneal endothelium: insensitivity to modulators of Na(+)-K(+)-2Cl- co-transport. *American J. Physiology*, 273, C1480-C1486. (cited p. 714).

Rivlin R.S. (1955). Further remarks on the stress-deformation relations for isotropic materials. *J. Rational Mechanics and Analysis*, 4, 681-701. (cited pp. 22, 23, and 65).

Rizzi E. and Loret B. (1997). Qualitative analysis of strain localization. Part I: Transversely isotropic elasticity and isotropic plasticity. *Int. J. Plasticity*, 13(5), 461-499.

Rodriguez E.K., Hoger A. and McCulloch A.D. (1994). Stress-dependent finite growth in soft elastic tissues. *J. Biomechanics*, 27(4), 455-467. (cited pp. 189, 839, 840, 849, 850, 856, 871, 882, 884, 912, 926, and 1005).

Rodriguez-Carvajal M.A., Imberty A. and Pérez S. (2003). Conformational behavior of chondroitin and chondroitin sulfate in relation to their physical properties as inferred by molecular modeling. *Biopolymers*, 69, 15-28. (cited p. 641).

Rogers T.G. (1984). Problems for helically wound cylinders. In *Continuum theories of the mechanics of fibre-reinforced composites*. CISM course N° 282, Spencer A.J.M. ed., Springer Verlag, New York, Chapter V, 147-178. (cited p. 490).

Roh H.D., Boucher Y., Kalnicki S., Buchsbaum R., Bloomer W.D. and Jain R.K. (1991). Interstitial hypertension in carcinoma of uterine cervix in patients: possible correlation with tumor oxygenation and radiation response. *Cancer Research*, 51, 6695-6698. (cited p. 1011).

Rooney F.J., Ferrari M. and Imam A. (1996). On the pressure distribution within tumors. *J. Mathematical Modeling and Scientific Computing*, 6, 715-721.

Roose T., Netti P., Munn L.L., Boucher Y. and Jain R.K. (2003). Solid stress generated by spheroid growth estimated using a linear poroelasticity model. *Microvascular Research*, 66, 204-212. (cited pp. 550, 878, 966, 989, 993, 1010, 1027, 1034, 1036, and 1038).

Rosakis P., Rosakis A.J., Ravichandran G. and Hodowany J. (2000). A thermodynamic internal variable model for the partition of plastic work into heat and stored energy in metals. *J. Mechanics Physics Solids*, 48, 581-607. (cited pp. 288 and 289).

Rouhi G., Herzog W., Sudak L., Firoozbakhsh K. and Epstein M. (2004). Free surface density instead of volume fraction in the bone remodeling equation: theoretical considerations. *Forma*, 19(3), 165-182. (cited pp. 107 and 874).

Rovati M. and Taliercio A. (2003). On stationarity of strain energy density for some classes of anisotropic solids. *Int. J. Solids Structures*, 40, 6043-6075. (cited p. 139).

Rubashkin A., Iserovich P., Hernández J.A. and Fischbarg J. (2005). Epithelial fluid transport: protruding macromolecules and space charges can bring about electro-osmotic coupling at the tight junctions. *J. Membrane Biology*, 208(3), 251-263. (cited p. 712).

Ruberti J.W. and Klyce S.D. (2003). NaCl osmotic perturbation can modulate hydration control in rabbit cornea. *Experimental Eye Research*, 76(3), 349-359. (cited pp. 703, 710, and 725).

Ruberti J.W. and Zieske J.D. (2008). Prelude to corneal tissue engineering - Gaining control of collagen organization. *Progress in Retinal and Eye Research*, 27, 549-577. (cited p. 731).

* * ** S * * * *

Sachs J.R. and Grodzinsky A.J. (1989). An electromechanically coupled poroelastic medium driven by an applied electric current: surface detection of bulk material properties. *Physicochemical Hydrodynamics*, 11(4), 585-614. (cited p. 612).

Sacks M.S., Chuong C.J., Petroll W.M., Kwan M. and Halberstadt C. (1997). Collagen fiber architecture of a cultured dermal tissue. *ASME J. Biomechanical Engng.*, 119, 124-127.

Sacks M.S. (2000). Biaxial mechanical evaluation of planar biological materials. *J. Elasticity*, 61, 199-246. (cited p. 190).

Sacks M.S. (2003). Incorporation of experimentally-derived fiber orientation into a structural constitutive model for planar collagenous tissues. *ASME J. Biomechanical Engng.*, 125, 280-287. (cited pp. 431 and 432).

Sah R.L., Trippel S.B. and Grodzinsky A.J. (1996). Differential effects of serum, insulin-like growth factor-I, and fibroblast growth factor-2 on the maintenance of cartilage physical properties during long-term culture. *J. Orthopaedic Research*, 14(1), 44-52. (cited p. 819).

Saint-Venant, A. Barré de (1863). Mémoire sur la distribution des élasticités autour de chaque point d'un solide ou d'un milieu de contexture quelconque, particulièrement lorsqu'il est amorphe sans être isotrope. *J. de Mathématiques Pures et Appliquées (J. de Liouville)*, Série II(8), premier article 257-295, deuxième article 353-430. Available at http://portail.mathdoc.fr/JMPA. (cited p. 150).

Salter D.M., Millward-Sadler S.J., Nuki G. and Wright M.O. (2002). Differential responses of chondrocytes from normal and osteoarthritic human articular cartilage to mechanical stimulation. *Biorheology*, 39, 97-108. (cited pp. 501 and 825).

Saltzman W.M. (1997). Weaving cartilage at zero g: the reality of tissue engineering in space. *Proceedings National Academy Sciences USA*, 94(25), 13380-13382.

Contents

Sanchez J.M., Li Y., Rubashkin A., Iserovich P., Wen Q., Ruberti J.W., Smith R.W., Rittenband D., Kuang K., Diecke F.P.J. and Fischbarg J. (2002). Evidence for a central role for electro-osmosis in fluid transport by corneal endothelium. *J. Membrane Biology*, 187, 37-50. (cited p. 714).

Sandell L.J. (2000). Genes and gene expression. *Clinical Orthopaedics and Related Research*, 379S, S9-S16. (cited pp. 816 and 819).

Sarntinoranont M., Rooney F. and Ferrari M. (2003). Interstitial stress and fluid pressure within a growing tumor. *Annals Biomedical Engng.*, 31, 327-335. (cited pp. 1021, 1027, and 1037).

Sarvazyan A.P., Rudenko O.V., Swanson S.D., Fowlkes J.B. and Emelianov S.Y. (1998). Shear wave elasticity imaging: a new ultrasonic technology of medical diagnostics. *Ultrasound in Medicine and Biology*, 24(9), 1419-1435. (cited p. 8).

Sarver J.J., Robinson P.S. and Elliott D.M. (2003). Methods for quasi-linear viscoelastic modeling of soft tissues: application to incremental stress-relaxation experiments. *ASME J. Biomechanical Engng.*, 125(5), 754-758. (cited p. 254).

Sasaki N. and Odajima S. (1996). Stress-strain curve and Young's modulus of a collagen molecule as determined by the x-ray diffraction technique. *J. Biomechanics*, 29(5), 655-658. (cited pp. 493, 660, and 800).

Sawyers K. (1986). Comments on the paper Determination of the stretch and rotation in the polar decomposition of the deformation gradient by A. Hoger and D.E. Carlson. *Quarterly Applied Mathematics*, XLIV(2), 309-311. (cited p. 22).

Scanlon B.R., Nicot J.-Ph. and Massmann J.W. (2000). Soil gas movement in unsaturated systems. In *Handbook of soil science*, Chapter 8, E.M. Sumner, ed., CRC Press, Boca Raton, FA, 297-341.

Scherer G.W. and Smith D.M. (1995). Cavitation during drying of a gel. *J. of Non-Crystalline Solids*, 189, 197-211. (cited pp. 81 and 82).

Schmid-Schönbein G.W. (1999). Biomechanics of microcirculatory blood perfusion. *Annals Biomedical Engng.*, 1, 73-102. (cited p. 1012).

Scholander P.F., Hammel H.T., Hemmingsen E.A. and Bradstreet E.D. (1964). Hydrostatic pressure and osmotic potential in leaves of mangroves and some other plants. *Proceedings National Academy Sciences USA*, 52(1), 119-125. (cited p. 457).

Schröder J. and Neff P. (2003). Invariant formulation of hyperelastic transverse isotropy based on polyconvex free energy functions. *Int. J. Solids Structures*, 40, 401-445. (cited p. 181).

Schwerdtfeger H. (1961). *Introduction to linear algebra and the theory of matrices*. 2nd ed., P. Noordhoff, Groningen, the Netherlands. (cited p. 41).

Scott J.E. and Bosworth T.R. (1990). A comparative biochemical and ultrastructural study of proteoglycan-collagen interactions in corneal stroma. *Biochemical J.*, 270, 491-497. (cited pp. 705 and 708).

Secomb T.W. and El-Kareh A.W. (2001). A theoretical model for the elastic properties of very soft tissues. *Biorheology*, 38(4), 305-317. (cited pp. 283 and 438).

Seehra G.P. and Silver F.H. (2006). Viscoelastic properties of acid- and alkaline-treated human dermis: a correlation between total surface charge and elastic modulus. *Skin Research and Technology*, 12, 190-198. (cited pp. 10 and 492).

Segel L.A. (1987). *Mathematics applied to continuum mechanics*. Dover Publications, Inc., New York. (cited pp. 15 and 896).

Seko Y., Seko Y., Fujikura H., Pang J., Tokoro T. and Shimokawa H. (1999). Induction of vascular endothelial growth factor after application of mechanical stress to retinal pigment epithelium of the rat *in vitro*. *Investigative Ophthalmology & Visual Science*, 40(13), 3287-3291. (cited p. 1003).

Sen P.N., Scala C. and M.H. Cohen (1981). A self-similar model for sedimentary rocks with application to the dielectric constant of fused glass beads. *Geophysics*, 46(5), 781-795. (cited p. 205).

Sengers B.G., Oomens C.W.J. and Baaijens F.P.T. (2004). An integrated finite-element approach to mechanics, transport and biosynthesis in tissue engineering. *ASME J. Biomechanical Engng.*, 126, 82-91. (cited pp. 806, 835, and 836).

Seth B.R. (1964). Generalized strain measure with applications to physical problems. In *Second-order effects in elasticity, plasticity and fluid dynamics*, M. Reiner and D. Abir eds., published by Pergamon, Oxford, 162-182. Proceedings of the Iutam Symposium, Haifa, Israel, April 23-27, 1962. (cited p. 37).

Setton L.A, Tohyama H. and Mow V.C. (1998). Swelling and curling behaviors of articular cartilage. *ASME J. Biomechanical Engng.*, 120(3), 355-361. (cited p. 499).

Sevick E. and Jain R.K. (1991). Measurement of capillary filtration coefficient in a solid tumor. *Cancer Research*, 51(4), 1352-1355. (cited p. 1010).

Shackelford C.D., Benson C.H., Katsumi T., Edil T.B and Lin L. (2000). Evaluating the hydraulic conductivity of GCLs permeated with non-standard liquids. *Geotextiles and Geomembranes*, 18, 133-161.

Shannon M.A. and Rubinsky B. (1992). The effect of tumor growth on the stress distribution in tissue. *Advances in Biological Heat and Mass Transfer*, 231, 35-38.

Sheehan J.B., Bence A.K. and Adams V.R. (2001). Antiangiogenesis agents as cancer therapy. *J. American Pharmaceutical Association*, 41(5), 771-773. (cited pp. 991 and 998).

Sherwood J.D. (1993). Biot poroelasticity of a chemically active shale. *Proceedings Royal Society London*, A440, 365-377.

Sherwood J.D. (1994)a. Swelling of shale around a cylindrical wellbore. *Proceedings Royal Society London*, A444, 161-184.

Sherwood J.D. (1994)b. A model of hindered transport of solute in a poroelastic shale. *Proceedings Royal Society London*, A445, 679-692.

Shieh A.C. and Athanasiou K.A. (2003). Principles of cell mechanics for cartilage tissue engineering. *Annals Biomedical Engng.*, 31(1), 1-11. (cited p. 825).

Shin Th.J., Vito R.P., Johnson L.W. and McCarey B.E. (1997). The distribution of strain in the human cornea. *J. Biomechanics*, 30(5), 497-503. (cited p. 726).

Shreiber D.I., Barocas V.H. and Tranquillo R.T. (2003). Temporal variations in cell migration and traction during fibroblast-mediated gel compaction. *Biophysical J.*, 84(6), 4102-4114. (cited p. 837).

Sidoroff F. (1978). Sur l'équation tensorielle $A \cdot X + X \cdot A = H$. *Comptes-Rendus Académie des Sciences, Paris*, A286, 71-73. (cited p. 23).

Siepmann J. and Peppas N.A. (2001). Modeling of drug release from delivery systems based on hydroxypropyl methylcellulose (HPMC). *Advanced Drug Delivery Reviews*, 48, 139-157. (cited p. 4).

Silbernagl S. and Despopoulos A. (2000). Atlas de Poche de Physiologie. Flammarion, Paris. *Referred to as APP [2000]*. (cited pp. 99, 338, 572, 573, 577, 578, and 988).

Silver F.H., Freeman J.W. and DeVore D. (2001). Viscoelastic properties of human skin and processed dermis. *Skin Research and Technology*, 7, 18-23. (cited p. 493).

Silver F.H. and Bradica G. (2002). Mechanobiology of cartilage: how do internal and external stresses affect mechanochemical transduction and elastic energy storage? *Biomechanics and Modeling in Mechanobiology*, 1, 219-238. (cited pp. 431, 499, and 825).

Simões F.M.F., Martins J.A.C. and Loret B. (1999). Instabilities in elastic-plastic fluid-saturated porous media: harmonic wave versus acceleration wave analysis. *Int. J. Solids Structures*, 36(9), 1277-1295.

Simmons C.A., Matlis S., Thornton A.J., Chen S., Wang C.Y. and Mooney D.J. (2003). Cyclic strain enhances matrix mineralization by adult human mesenchymal stem cells via the extracellular signal-regulated kinase (ERK1/2) signaling pathway. *J. Biomechanics*, 36, 1087-1096. (cited p. 824).

Simon B.R., Laible J.P., Pflaster D., Yuan Y. and Krag M.H. (1996). A poroelastic finite element formulation including transport and swelling in soft tissue structures. *ASME J. Biomechanical Engng.*, 118(1), 1-9. (cited p. 670).

Simon B.R., Kaufmann M.V., Liu J. and Baldwin A.L. (1998)a. Porohyperelastic-transport-swelling theory, material properties and finite element models for large arteries. *Int. J. Solids Structures*, 35(34), 5021-5031.

Simon B.R., Kaufmann M.V., McAfee M.A., Baldwin A.L. and Wilson L.M. (1998)b. Identification and determination of material properties for porohyperelastic analysis of large arteries. *ASME J. Biomechanical Engng.*, 120(2), 188-194.

Simon B.R., Kaufmann M.V., McAfee M.A. and Baldwin A.L. (1998)c. Porohyperelastic finite element analysis of large arteries using Abaqus. *ASME J. Biomechanical Engng.*, 120, 296-298.

Sittinger M., Perka C., Schultz O., Häupl T. and Burmester G.R. (1999). Joint cartilage regeneration by tissue engineering. *Zeitschrift für Rheumatologie*, 58, 130-135. (cited pp. 808, 821, 822, and 830).

Sivan S.S., Merkher Y., Wachtel E., Urban J.P.G., Lazary A. and Maroudas A. (2013). A needle micro-osmometer for determination of glycosaminoglycan concentration in excised nucleus pulposus tissue. *European Spine J.*, 22, 1765-1773. (cited p. 497).

Sivaselvan M.V. and Reinhorn A.M. (2006). Lagrangian approach to structural collapse simulation. *ASCE J. Engng. Mechanics*, 132(8), 795-805. (cited p. 271).

Skalak R. (1981). Growth as a finite displacement field. In *Iutam symposium finite elasticity*, D.E. Carlson and R.T. Shield, eds., Martinus Nijhoff, The Hague, the Netherlands, 347-355. (cited pp. 839, 849, 850, and 1005).

Skalak R., Keller S.R. and Secomb T.W. (1981). Mechanics of blood flow. *ASME J. Biomechanical Engng.*, 103, 102-113.

Skalak R., Zargaryan S., Jain R.K., Netti P. and Hoger A. (1996). Compatibility and the genesis of residual stress by volumetric growth. *J. Mathematical Biology*, 34, 889-914. (cited pp. 846, 850, 890, 893, 894, 895, and 1005).

Skalak R., Farrow D.A. and Hoger A. (1997). Kinematics of surface growth. *J. Mathematical Biology*, 35, 869-907. (cited p. 846).

Skempton A.W. (1954). The pore pressure coefficients A and B. *Géotechnique*, 4(4), 143-147. (cited p. 217).

Smith G.F. (1970). On a fundamental error in two papers of C.C. Wang "On representations for isotropic functions, Parts I and II". *Archives Rational Mechanics Analysis*, 36, 161-165. (cited p. 61).

Smith G.F. (1971). On isotropic functions of symmetric tensors, skew-symmetric tensors and vectors. *Int. J. Engng. Science*, 9, 899-916. (cited p. 61).

Smith J.M., van Ness H.C. and Abbott M.M. (1996). *Introduction to chemical engineering thermodynamics.* 5th ed., The Mac-Graw Hill Companies, Inc., New York. (cited pp. 69, 72, 76, 83, 88, and 98).

Smolek M.K. and McCarey B.E. (1990). Interlamellar adhesive strength in human eyebank corneas. *Investigative Ophthalmology & Visual Science*, 31, 1087-1095. (cited p. 707).

Smolek M.K. (1993). Interlamellar cohesive strength in the vertical meridian of human eyebank corneas. *Investigative Ophthalmology & Visual Science*, 34(10), 2962-2969. (cited pp. 707 and 708).

Smolek M.K. (1994). Holographic interferometry of intact and radially incised human eye-bank corneas. *J. Cataract & Refractive Surgery*, 20(3), 277-286. (cited p. 727).

Smolek M.K. and Beekhuis W.H. (1997). Collagen fibril orientation in the human corneal stroma and its implication in keratoconus. *Investigative Ophthalmology & Visual Science*, 38(7), 1289-1290. (cited p. 708).

Sneddon I.N. (1946). The distribution of stress in the neighbourhood of a crack in an elastic solid. *Proceedings Royal Society London*, A187, 229-260. (cited pp. 563 and 564).

Sneddon I.N. (1980). *Special functions of mathematical physics and chemistry*, 3rd ed., Longman Publishing Group, UK. (cited p. 237).

Snijders H., Huyghe J.M. and Janssen J.D. (1995). Triphasic finite element model for swelling porous media. *Int. J. Numerical Methods Fluids*, 20, 1039-1046. (cited p. 670).

Sokolnikoff I.S. (1956). *Mathematical theory of elasticity.* Krieger Publishing Company, Malabar, FL. (cited p. 893).

Soltz M.A. and Ateshian G.A. (2000). A conewise linear elasticity mixture model for the analysis of tension-compression nonlinearity in articular cartilage. *ASME J. Biomechanical Engng.*, 122, 576-586. (cited p. 160).

Soulhat J., Buschmann M.D. and Shirazi-Adl A. (1999). A fibril-network-reinforced biphasic model of cartilage in unconfined compression. *ASME J. Biomechanical Engng.*, 121, 340-347. (cited p. 506).

Spanner D.C. (1964). *Introduction to thermodynamics.* Academic Press, Inc., London, UK. (cited p. 296).

Spencer A.J.M. (1971). Theory of invariants. In *Continuum physics*, Vol. I, Eringen A.C. ed., Academic Press, New York, 240-353. (cited pp. 60 and 61).

Spencer A.J.M. (1972). *Deformations of fibre-reinforced materials.* Oxford Science Research papers, Clarendon Press, Oxford, UK. (cited pp. 52, 789, and 790).

Spencer A.J.M. (1992). Plasticity theory for fiber-reinforced composites. *J. Engng. Mathematics*, 26(1), 107-118. (cited p. 788).

Spencer A.J.M. (2000). Theory of fabric-reinforced viscous fluids. *Composites: Part A*, 31, 1311-1321. (cited pp. 792 and 796).

Spencer A.J.M. (2001). A theory of viscoplasticity for fabric-reinforced composites. *J. Mechanics Physics Solids*, 49, 2667-2687. (cited pp. 791, 792, 795, and 796).

Sperelakis N. (2001). Origin of the cardiac resting potential. In *Handbook of Physiology - Section 2: The cardiovascular system, Volume I: The heart*, R.M. Berne, N. Sperelakis and Geiger S.R., eds., Waverly Press, Inc., Baltimore, MD, American Physiological Society, Bethesda, MD, Chapter 6, 187-267. (cited pp. 125, 757, and 758).

Spoerl E., Huhle M. and Seiler T. (1998). Induction of cross-links in corneal tissues. *Experimental Eye Research*, 66, 97-103. (cited p. 708).

Spofford C.M. and Chilian W.M. (2003). Mechanotransduction via the elastin-laminin receptor (ELR) in resistance arteries. *J. Biomechanics*, 36, 645-652. (cited p. 825).

Sposito G. (1984). *The surface chemistry of soils.* Oxford University Press, New York. (cited pp. 482, 484, and 634).

Stammers A.H., Vang S.N., Mejak B.L. and Rauch E.D. (2003). Quantification of the effect of altering hematocrit and temperature on blood viscosity. *The Journal of The American Society of Extra-Corporeal Technology*, 35, 143-151. (cited p. 207).

Stánczyk M. and Telega J.J. (2002,2003). Modeling of heat transfer in biomechanics- A review. *Acta Bioengineering Biomechanics*, I. Soft tissues, (2002), 4(1), 31-61; II. Orthopaedics, (2002), 4(2), 3-33; III. Cryosurgery, cryopreservation and cryotherapy. (2003), 5(2), 3-22. (cited p. 279).

Staroszczyk R. (1992). Rayleigh-type waves in a water-saturated porous half-space with an elastic plate on its surface. *Transport in Porous Media*, 9(1-2), 143-154. (cited p. 388).

Stenmark K.R. and Mecham R.P. (1997). Cellular and molecular mechanisms of pulmonary vascular remodeling. *Annual Review Physiology*, 59, 89-114. (cited p. 847).

Stevens A. and Lowe J.S. (1993). *Human histology.* Mosby, Inc., London, UK. (cited pp. 487 and 495).

Stiemke M.M., Roman R.J., Palmer M.L. and Edehauser H.F. (1991). Sodium activities, and concentrations of the corneal stroma, aqueous humor and plasma in the presence and absence of a transparent cornea. *Investigative Ophthalmology & Visual Science*, 32(suppl), 1065. (cited pp. 714, 718, and 722).

Stockwell R.A. (1979). *Biology of cartilage cells.* Cambridge University Press, Cambridge, UK. (cited pp. 498, 499, and 965).

Stoer J. and Bulirsch R. (1980). *Introduction to numerical analysis*, Springer-Verlag, New York. (cited pp. 381, 382, and 412).

Stohrer M., Boucher Y., Stangassinger M. and Jain R.K. (2000). Oncotic pressure in solid tumors is elevated. *Cancer Research*, 60, 4251-4255. (cited pp. 570 and 991).

Stoll B.R., Migliorini C., Kadambi A., Munn L.L. and Jain R.K. (2003). A mathematical model of the contribution of endothelial progenitor cells to angiogenesis in tumors: implications for antiangiogenic therapy. *Blood*, 102(7), 2555-2561. (cited p. 1012).

Stöverud K.H., Darcis M., Helmig R. and Hassanizadeh S.M. (2012). Modeling concentration distribution and deformation during convection-enhanced drug delivery into brain tissue. *Transport in Porous Media*, 92, 119-143. (cited p. 4).

Stumm W. (1992). *Chemistry of the solid-water interface. Processes at the mineral-water and particle-water interface in natural systems.* John Wiley & Sons, New York. (cited pp. 482 and 484).

Sun D.N., Gu W.Y., Guo X.E., Lai W.M. and Mow V.C. (1999). A mixed finite element formulation of triphasic mechano-electrochemical theory for charged, hydrated biological soft tissues. *Int. J. Numerical Methods Engineering*, 45, 1375-1402. (cited pp. 670 and 675).

Sun D.N. and Leong K.W. (2004). A nonlinear hyperelastic mixture theory model for anisotropy, transport, and swelling of annulus fibrosus. *Annals Biomedical Engng.*, 32(1), 92-102. (cited pp. 724, 798, and 802).

Sun W., Sacks M.S. and Scott M.J. (2005). Effects of the boundary conditions on the estimation of the planar biaxial mechanical properties of soft tissues. *ASME J. Biomechanical Engng.*, 127(4), 709-715.

Sutcliffe S. (1992). Spectral decomposition of the elasticity tensor. *J. Applied Physics*, 59, 762-773.

Sutherland R.M., Sordat B., Bamat J., Gabbert H., Bourrat B. and Mueller-Klieser W. (1986). Oxygenation and differentiation in multicellular spheroids of human colon carcinoma. *Cancer Research*, 46(10), 5320-5329. (cited p. 847).

Sverdlik A. and Lanir Y. (2002). Time-dependent mechanical behavior of sheep digital tendons, including the effects of preconditioning. *ASME J. Biomechanical Engng.*, 124, 78-84. (cited p. 432).

Swabb E.A., Wei J. and Gullino P.M. (1974). Diffusion and convection in normal and neoplastic tissue. *Cancer Research*, 34(10), 2814-2822. (cited pp. 552, 554, and 1010).

Swan C.C., Lakes R.S., Brand R.A. and Stewart K.J. (2003). Micromechanically based poroelastic modeling of fluid flow in haversian bone. *ASME J. Biomechanical Engng.*, 125(1), 25-37. (cited p. 825).

Swan J.S. and Hodson S.A. (2004). Rabbit corneal hydration and the bicarbonate pump. *J. Membrane Biology*, 201, 33-40. (cited p. 714).

Swartz M., Kaipainen A., Netti P., Brekken Ch., Boucher Y., Grodzinsky A.J. and Jain R.K. (1999). Mechanics of interstitial-lymphatic fluid transport: theoretical foundation and experimental verification. *J. Biomechanics*, 32, 1297-1307. (cited pp. 1032, 1033, and 1034).

Swartz M.A., Tschumperlin D.J., Kamm R.D. and Drazen J.M. (2001). Mechanical stress is communicated between different cell types to elicit matrix remodeling. *Proceedings National Academy Sciences USA*, 98(11), 6180-6185. (cited p. 1003).

Sweet M.B.E., Thonar E., Immelman A.R. and Solomon L. (1977). Biochemical changes in progressive osteoarthritis. *Annals Rheumatic Diseases*, 36, 387-398. (cited pp. 495, 496, 627, 638, and 639).

Szasz N., Buell L., Frank E.H., Grodzinsky A.J. and Morales T. (2000). IGF binding proteins specifically affect the diffusive transport of IGF-I within bovine articular cartilage. *Transactions Orthopaedic Research Society*, 46th Annual Meeting, 940. (cited p. 820).

* * ** T * * * *

Taber L. (1995). Biomechanics of growth, remodeling and morphogenesis. *ASME Applied Mechanics Reviews*, 48, 487-545. (cited p. 851).

Taber L. and Eggers D.W. (1996). Theoretical study of stress-modulated growth in the aorta. *J. Theoretical Biology*, 180, 343-357. (cited pp. 190, 851, 870, 884, 886, 887, 912, and 920).

Taber L. (1998)a. Biomechanical growth laws for muscle tissue. *J. Theoretical Biology*, 193, 201-213. (cited pp. 851, 878, 879, 880, 881, 888, 909, and 976).

Taber L. (1998)b. A model for aortic growth based on fluid shear and fiber stresses. *ASME J. Biomechanical Engng.*, 120, 348-354. (cited pp. 851 and 878).

Taber L. (1998)c. An optimization principle for vascular radius including the effects of soft muscle tone. *Biophysical J.*, 74, 109-114. (cited pp. 846, 878, 879, and 922).

Taber L. (1998)d. Mechanical aspects of cardiac development. *Progress in Biophysics & Molecular Biology*, 69, 237-255. (cited p. 846).

Taber L. (2001). Biomechanics of cardiovascular development. *Annual Review Biomedical Engng.*, 3, 1-25. (cited pp. 851, 870, 878, 879, 880, 885, 888, and 1012).

Taber L. and Humphrey J. (2001). Stress-modulated growth, residual stress, and vascular heterogeneity. *ASME J. Biomechanical Engng.*, 123, 528-535. (cited pp. 845, 850, 856, 879, 880, 884, 885, and 886).

Taber L. and Chabert S. (2002). Theoretical and experimental study of growth and remodeling in the developing heart. *Biomechanics and Modeling in Mechanobiology*, 1, 29-43. (cited pp. 191, 845, 846, 851, 856, 879, 881, 888, 907, and 922).

Takahashi K., Hashimoto S., Kubo T., Hirasawa Y., Lotz M. and Amiel D. (2001). Hyaluronan suppressed nitric oxide production in the meniscus and synovium of rabbit osteoarthritis model. *J. Orthopaedic Research*, 19(3), 500-503. (cited p. 830).

Tardy Y. and Duplay J. (1992). A method of estimating the Gibbs free energies of formation of hydrated and dehydrated clay minerals. *Geochimica et Cosmochimica Acta*, 56, 3007-3029. (cited p. 81).

Terezinsky G. (2005). Injuries of the thigh, knee, and ankle as reconstructive factors in road traffic accidents. In *Forensic science and medicine. Forensic medicine of the lower extremity: Human identification and trauma analysis of the thigh, leg, and foot.* J. Rich, D. E. Dean, and R. H. Powers, eds., The Humana Press Inc., Totowa, NJ. (cited p. 9).

Terzaghi K. (1936). The shearing resistance of saturated soils. Proc. 1st Int. Conference Soil Mechanics and Foundation Engineering. vol. I, 54-55. (cited p. 214).

Theis E.R. and Jacoby T.F. (1943). The acid-, base- and salt-binding capacity of salt-denatured collagen. *The J. Biological Chemistry*, 148, 603-609. (cited p. 635).

Thibault M., Poole A.R. and Buschmann M.D. (2002). Cyclic compression of cartilage/bone explants *in vitro* leads to physical weakening, mechanical breakdown of collagen and release of matrix fragments. *J. Orthopaedic Research*, 20, 1265-1273.

Thompson M. and Willis J.R. (1991). A reformulation of the equations of anisotropic poroelasticity. *ASME J. Applied Mechanics*, 58, 612-616. (cited pp. 210, 214, and 217).

Tong R.T., Boucher Y., Kozin S.V., Winkler F., Hicklin D.J. and Jain R.K. (2004). Vascular normalization by Vascular Endothelial Growth Factor Receptor 2 blockade induces a pressure gradient across the vasculature and improves drug penetration in tumors. *Cancer Research*, 64, 3731-3736. (cited pp. 569, 998, and 999).

Torzilli P.A. (1985). Influence of cartilage conformation on its equilibrium water partition. *J. Orthopaedic Research*, 3, 473-483. (cited pp. 485 and 502).

Tower T.T., Neidert M.R. and Tranquillo R.T. (2002). Fiber alignment imaging during mechanical testing of soft tissues. *Annals Biomedical Engng.*, 30(10), 1221-1233. (cited p. 836).

Tözeren A. and Skalak R. (1988). Interaction of stress and growth in a fibrous tissue. *J. Theoretical Biology*, 130, 337-350. (cited p. 838).

Trask R.S., Williams H.R. and Bond I.P. (2007). Self-healing polymer composites: mimicking nature to enhance performance. *Bioinspiration and Biomimetics*, 2, 1-9, doi:10.1088/1748-3182/2/1/P01. (cited p. 986).

Truesdell C. and Toupin R. (1960). The classical field theories. In *Encyclopedia of physics*, vol. III/1, 226-793, S. Flügge ed., Springer-Verlag, Berlin, Germany. (cited pp. 104, 112, 174, 355, 356, 366, 853, and 894).

Truesdell C. and Noll W. (1965). The non-linear field theories of mechanics. In *Encyclopedia of physics*, vol. III/3, S. Flügge ed., Springer-Verlag, Berlin, Germany. (cited pp. 61 and 180).

Truesdell C. (1984). *Rational thermodynamics.* 2nd ed., Springer-Verlag, New York. (cited pp. 83, 104, 111, 130, and 131).

Tsang V.L. and Bhatia S. (2004). Three-dimensional tissue fabrication. *Advanced Drug Delivery Reviews*, 56(11), 1635-1647. (cited p. 843).

Tschoegl N.W. (2000). *Fundamentals of equilibrium and steady-state thermodynamics.* Elsevier, Amsterdam, the Netherlands. (cited p. 70).

Tuncay K. and Corapcioglu Y. (1996)a. Wave propagation in fractured porous media. *Transport in Porous Media*, 23, 237-258. (cited pp. 321, 322, and 327).

Tuncay K. and Corapcioglu Y. (1996)b. Body waves in fractured porous media saturated by two Newtonian fluids. *Transport in Porous Media*, 23, 259-273. (cited pp. 321, 322, and 327).

Tuncay K. and Corapcioglu Y. (1996)c. Consolidation of elastic porous media saturated by two immiscible fluids. *ASCE J. Engng. Mechanics*, 122(11), 1077-1085.

Turing A.M. (1952). The chemical basis of morphogenesis. *Proceedings Royal Society London*, B237, 37-72. (cited p. 551).

Tyree M.T. (2003). The ascent of water. *Nature*, 423(6943), 923. (cited p. 81).

* * * * U * * * *

Uebelhart D., Thonar E.J.-M.A., Pietryla D.W. and Williams J.W. (1993). Elevation in urinary levels of pyridinium cross-links of collagen following chymopapain-induced degradation of articular cartilage in the rabbit knee provides evidence of metabolic changes in bone. *Osteoarthritis and Cartilage*, 1, 185-192. (cited p. 809).

Urban J.P.G., Maroudas A., Bayliss M.T. and Dillon J. (1979). Swelling pressure of proteoglycans at the concentration found in cartilagenous tissues. *Biorheology*, 16, 447-464. (cited pp. 495 and 638).

Urban J.P. and McMullin J.F. (1985). Swelling pressure in the intervertebral disc: influence of proteoglycan and collagen contents. *Biorheology*, 22, 145-157. (cited p. 802).

Urban J.P.G. and Holm S.H. (1986). Intervertebral disc nutrition as related to spinal movements and fusion. In *Tissue nutrition and viability*, A.R. Hargens, ed., Springer-Verlag New York, 101-119. (cited p. 802).

* * * * V * * * *

Valanis K.C. (1990). A theory of damage in brittle materials. *Engng. Fracture Mechanics*, 36, 403-416. (cited pp. 151 and 434).

van Dyke T. and Hoger A. (2000). A comparison of two methods for analyzing finite growth of soft biological materials. ICTAM Chicago, oral presentation, tvandyke@ucsd.edu. (cited p. 889).

van Dyke T.J. and Hoger A. (2001). Rotations in the theory of growth for soft biological materials. BED-Vol. 50, Bioengineering Conference ASME, June 27-July 1, 2001, Snowbird, UT, 647-648. (cited pp. 839, 840, and 849).

van Dyke T.J. and Hoger A. (2002). A new method for predicting the opening angle for soft tissues. *ASME J. Biomechanical Engng.*, 124, 347-354. (cited pp. 190, 839, and 887).

van de Lest C.H., Brama P.A. and van Weeren P.R. (2002). The influence of exercise on the composition of developing equine joints. *Biorheology*, 39(1-2), 183-191. (cited p. 807).

Vankan W.J., Huyghe J.M., Janssen J.D. and Huson A. (1996). Poroelasticity of saturated solids with an application to blood perfusion. *Int. J. Engng. Science*, 34(9), 1019-1031.

Vankan W.J., Huyghe J.M., Drost M.R., Janssen J.D. and Huson A. (1997)a. A finite element mixture model for hierarchical porous media. *Int. J. Numerical Methods Engineering*, 40, 193-210.

Vankan W.J., Huyghe J.M., Janssen J.D., Huson A., Hacking W.J.G. and Schreiner W. (1997)b. Finite element analysis of blood flow through biological tissue. *Int. J. Engng. Science*, 35(4), 375-385.

Vankan W.J., Huyghe J.M., Slaaf D.W., van Donkelaar C.C., Drost M.R., Janssen J.D. and Huson A. (1997)c. Finite-element simulation of blood perfusion in muscle tissue during compression and sustained contraction. *American J. Physiology - Heart Circulation Physiology*, 273(3), H1587-H1591.

Vermeer P.A (1978). A double hardening model for sand. *Géotechnique*, 28(4), 413-433. (cited p. 136).

de Vita R. and Slaughter W.S. (2006). A structural constitutive model for the strain rate-dependent behavior of the anterior cruciate ligaments. *Int. J. Solids Structures*, 43, 1561-1570. (cited p. 432).

Vito R.P., Shin T.J. and McCarey B.E. (1989). A mechanical model for the cornea: the effects of physiological and surgical factors on radial keratotomy surgery. *Refractive Corneal Surgery*, 5(2), 82-88. (cited p. 720).

Vito R.P. and Dixon S.A. (2003). Blood vessel constitutive models - 1995-2002. *Annals Biomedical Engng.*, 5, 413-439. (cited p. 1012).

Voet D. and Voet J.G. (1998). *Biochimie*. French translation of the 2nd American edition. De Boeck Université. Paris, Bruxelles. (cited pp. 444, 490, 491, 581, 638, 743, and 988).

Volokh K.Y. (2007). Softening hyperelasticity for modeling material failure: analysis of cavitation in hydrostatic tension. *Int. J. Solids Structures*, 44, 5043-5055. (cited p. 184).

Volokh K.Y. (2008). Fung's model of arterial wall enhanced with a failure description. *Molecular & Cellular Biomechanics*, 5(3), 207-216. (cited p. 184).

Vunjak-Novakovic G., Obradovic B., Martin I. and Freed L. (2002). Bioreactor studies of native and tissue engineered cartilage. *Biorheology*, 39(1-2), 259-268. (cited pp. 820, 822, and 824).

* * * * W * * * *

Wachtel E. and Maroudas A. (1998). The effects of pH and ionic strength on intrafibrillar hydration in articular cartilage. *Biochimica et Biophysica Acta*, 1381, 37-48. (cited pp. 492, 633, 639, 640, 643, and 650).

Wagenseil J.E., Wakatsuki T., Okamoto R., Zahalak G.I. and Elson E.L. (2003). One-dimensional viscoelastic behavior of fibroblast populated collagen matrices. *ASME J. Biomechanical Engng.*, 125(5), 719-725. (cited p. 242).

Wagner D.R. and Lotz J.C. (2004). Theoretical model and experimental results for the nonlinear elastic behavior of human annulus fibrosus. *J. Orthopaedic Research*, 22(4), 901-909. (cited pp. 786, 801, and 802).

Wagner D.R., Reiser K.M. and Lotz J.C. (2006). Glycation increases human annulus fibrosus stiffness in both experimental measurements and theoretical predictions. *J. Biomechanics*, 39(6), 1021-1029. (cited pp. 791 and 801).

Wakasa K., Sakurai M., Okamura J. and Kuroda C. (1985). Pathological study of small hepatocellular carcinomas: frequency of their invasion. *Virchows Archives*, Pathological Anatomy Histopatholy, A407(3), 259-270. (cited p. 1008).

Wakatsuki T., Kolodney M., Zahalak G.I. and Elson E.L. (2000). Cell mechanics studied by a reconstituted model tissue. *Biophysical J.*, 79(5), 2353-2368. (cited pp. 8 and 256).

Wakatsuki T., Schwab B., Thompson N.C. and Elson E.L. (2001). Effects of cytochalasin D and latrunculin B on mechanical properties of cells. *J. Cell Science*, 114(5), 1025-1036.

Wakatsuki T., Wysolmerski R.B. and Elson E.L. (2003). Mechanics of cell spreading: role of myosin II. *J. Cell Science*, 116(8), 1617-1625. (cited p. 256).

Waldman S.D., Spiteri C.G., Grynpas M.D., Pilliar R.M. and Kandel R.A. (2003). Long-term intermittent shear deformation improves the quality of cartilaginous tissue formed *in vitro*. *J. Orthopaedic Research*, 21, 590-596. (cited p. 829).

Walpole L.J. (1984). Fourth-rank tensors of the thirty-two crystal classes: multiplication tables. *Proceedings Royal Society London*, A391, 149-179. (cited pp. 136 and 171).

Wang C.C. (1970). A new representation theorem for isotropic functions: an answer to Professor Smith's criticism of my papers on representations for isotropic functions. *Archives Rational Mechanics Analysis*, 36, Part 1. Scalar-valued isotropic functions. 166-197; Part 2. Vector-valued isotropic functions, symmetric tensor-valued isotropic functions, skew-symmetric tensor-valued isotropic functions. 198-223; Corrigendum to my recent papers on "Representations for isotropic functions". *Archives Rational Mechanics Analysis*, 1971, 43, 392-395. (cited pp. 61, 62, and 63).

Wang C.C. and Truesdell C. (1973). *Introduction to rational elasticity*. Noordhoff International Publishing, Leyden, the Netherlands. (cited pp. 130 and 174).

Wang C.C.B., Guo X.E., Sun D.D., Mow V.C., Ateshian G.A. and Hung C.T. (2002). The functional environment of chondrocytes within cartilage subjected to compressive loading: a theoretical and experimental approach. *Biorheology*, 39, 11-25. (cited pp. 241 and 487).

Wang C.C.B., Chahine N.O., Hung C.T. and Ateshian G.A. (2003). Optical determination of anisotropic material properties of bovine articular cartilage in compression. *J. Biomechanics*, 36, 339-353. (cited pp. 161 and 162).

Wang H., Prendiville P.L., MacDonnell P.J. and Chang W.V. (1996). An ultrasonic technique for the measurement of the elastic moduli of human cornea. *J. Biomechanics*, 29(12), 1633-1636. (cited p. 727).

Wang H.B., Dembo M. and Wang Y.L. (2000). Substrate flexibility regulates growth and apoptosis of normal but not transformed cells. *American J. Physiology - Cell Physiology*, 279, C1345-C1350. (cited p. 998).

Wang L., Cowin S.C., Weinbaum S. and Fritton S.P. (2000). Modeling tracer transport in an osteon under cyclic loading. *Annals Biomedical Engng.*, 28, 1200-1209. Discussion *Annals Biomedical Engng.*(2001), 29, 812-816. (cited p. 825).

Warren J.B. and Root P.J. (1963). The behaviour of naturally fractured reservoirs. *Society of Petroleum Engineers Journal*, 1963, 3, 245-255. (cited pp. 342 and 343).

Wayne Broadland G. (2002). The differential interfacial tension hypothesis (DITH): a comprehensive theory for the self-rearrangement of embryonic cells and tissues. *ASME J. Biomechanical Engng.*, 124, 189-197.

Wazwaz A.M. (2005). Travelling wave solutions of generalized forms of Burgers, Burgers-KdV and Burgers-Huxley equations. *Applied Mathematics and Computation*, 169(1), 639-656.

Webb S.D., Sherratt J.A. and Fish R.G. (1999)a. Alterations in proteolytic activity at low pH and its association with invasion: a theoretical model. *Clinical and Experimental Metastasis*, 17, 397-407. (cited p. 1013).

Webb S.D., Sherratt J.A. and Fish R.G. (1999)b. Mathematical modelling of tumour acidity: regulation of intracellular pH. *J. Theoretical Biology*, 196, 237-250. (cited p. 1013).

Weinans H., Huiskes R. and Grootenboer H.J. (1992). The behavior of adaptive bone-remodelling simulation models. *J. Biomechanics*, 25(12), 1425-1441. (cited p. 869).

Wellman P.S., Howe R.D., Dalton D. and Kern K.A. (1999). Breast tissue stiffness in compression is correlated to histological diagnosis. Harvard Biorobotics Laboratory, Technical Report. (cited p. 1040).

Weng G.J. (1984). Some elastic properties of reinforced solids, with special references to isotropic ones containing spherical inclusions. *Int. J. Engng. Science*, 22, 845-856. (cited p. 438).

Werner A. and Gründer W. (1999). Calcium-induced structural changes of cartilage proteoglycans studied by ^{1}H NMR relaxometry and diffusion measurements. *Magnetic Resonance in Medicine*, 41, 43-50. (cited pp. 596 and 635).

Westacott C.I., Urban J.P.G., Goldring M.B. and Elson C.J. (2002). The effects of pressure on chondrocyte tumour necrosis factor receptor expression. *Biorheology*, 39, 125-132. (cited p. 1003).

White R.J., Bassingthwaighte J.B., Charles J.B., Kushmerick M.J. and Newman D.J. (2003). Issues of exploration: human health and wellbeing during a mission to Mars. *Advances in Space Research*, 31(1), 7-16. (cited p. 7).

Whittingham D.G., Leibo S.P. and Mazur P. (1972). Survival of mouse embryos frozen to -196 °C and -269 °C. *Science*, 178, 411-414. (cited p. 279).

Wiener N. (1958). *Nonlinear problems in random theory*. The Technology Press of Massachusetts Institute of Technology, Cambridge, MA. (cited p. 274).

Wilber J.P. and Walton J.R. (2002). The convexity properties of a class of constitutive models for biological soft tissues. *Mathematics and Mechanics of Solids*, 7, 217-236. (cited pp. 184 and 186).

Contents

Wilkins R., Browning J. and Saunders K. (2002). The effect of hydrostatic pressure on intracellular calcium concentration in a human chondrocyte cell line. *Transactions Orthopaedic Research Society*, 48th Annual Meeting, 133. (cited pp. 828 and 845).

Wilkinson J.H. (1965). *The algebraic eigenvalue problem*. Clarendon Press, Oxford, UK. (cited pp. 17, 381, and 382).

Williams K.A., Saini S. and Wick T.M. (2002). Computational Investigation of Steady-State Momentum and Mass Transfer in a Bioreactor for the Production of Tissue-Engineered Cartilage. *Biotechnology Progress*, 18, 951-963. (cited p. 826).

Williams J.M., Rayan V., Sumner D.R. and Thonar E.J.-M.A. (2003). The use of intra-articular NA-hyaluronate as a potential chondroprotective device in experimentally induced acute articular cartilage injury and repair in rabbits. *J. Orthopaedic Research*, 21, 305-311. (cited pp. 501 and 820).

Williamson A.K., Chen A.C. and Sah R.L. (2001). Compressive properties and function-composition relationships of developing bovine articular cartilage. *J. Orthopaedic Research*, 19, 1113-1121. (cited pp. 10, 824, 841, 842, 843, 948, 965, 977, and 986).

Williamson A.K., Chen A.C., Masuda K., Thonar E.J.-M.A. and Sah R.L. (2003). Tensile mechanical properties of bovine articular cartilage: variations with growth and relationships to collagen network components. *J. Orthopaedic Research*, 21, 872-880. (cited pp. 498, 824, 841, 956, 958, 962, 965, and 979).

Wilson C.G., Bonassar L.J. and Kohles S.S. (2002). Modelling the dynamic composition of engineered cartilage. *Archives Biochemistry Biophysics*, 408, 246-254. (cited p. 832).

Wilson R.K. and Aifantis E.C. (1984). A double porosity model for acoustic wave propagation in fractured-porous rock. *Int. J. Engng. Science*, 22(8-10), 1209-1217. (cited pp. 321, 322, and 326).

Wilson W., van Rietbergen B., van Donkelaar C.C. and Huiskes R. (2003). Pathways of load-induced cartilage damage causing cartilage degeneration in the knee after meniscectomy. *J. Biomechanics*, 36(3), 845-851. (cited p. 501).

Wilson W., van Donkelaar C.C., van Rietbergen B., Ito K. and Huiskes R. (2004). Stresses in the local collagen network of articular cartilage: a poroviscoelastic fibril-reinforced finite element study. *J. Biomechanics*, 37, 357-366. (cited p. 241).

Wilson W., van Donkelaar C.C., van Rietbergen B. and Huiskes R. (2005)a. A fibril-reinforced poroviscoelastic swelling model for articular cartilage. *J. Biomechanics*, 38, 1195-1204.

Wilson W., van Donkelaar C.C. and Huyghe J.M. (2005)b. A comparison between mechano-electrochemical and biphasic swelling theories for soft hydrated tissues. *ASME J. Biomechanical Engng.*, 127, 158-165. (cited p. 670).

Winslow R.L., Rice J.J., Jafri S., Marban E. and O'Rourke B. (1999). Mechanisms of altered excitation-contraction coupling in canine tachycardia-induced heart failure. II: model studies. *Circulation Research*, 84(5), 571-586. (cited p. 469).

Woessner J.F. (1961). The determination of hydroxyproline in tissue and protein samples containing small proportions of this imino acid. *Archives Biochemistry Biophysics*, 93, 440-447. (cited p. 809).

Wognum S., Huyghe J.M. and Baaijens F.P.T. (2006). Influence of osmotic pressure changes on the opening of existing cracks in 2 intervertebral disc models. *Spine*, 31(6), 1783-1788. (cited p. 802).

Wohlrab D., Lebek S., Krüger Th. and Reichel H. (2002). Influence of ion channels on the proliferation of human chondrocytes. *Biorheology*, 39, 55-61. (cited pp. 825 and 1002).

Wollensack G., Spoerl E. and Seiler T. (2003). Riboflavin/ultraviolet-A-induced collagen crosslinking for the treatment of keratoconus. *American J. Ophthalmology*, 135, 620-627. (cited pp. 708 and 720).

Wong M., Wuethrich P., Buschmann M.D., Eggli P. and Hunziker E. (1997). Chondrocyte biosynthesis correlates with local tissue strain in statically compressed adult articular cartilage. *J. Orthopaedic Research*, 15, 189-196. (cited p. 498).

Wong M., Ponticiello M., Kovanen V. and Jurvelin J.S. (2000). Volumetric changes of articular cartilage during stress relaxation in unconfined compression. *J. Biomechanics*, 33(9), 1049-1054. (cited p. 842).

Wong M., Siegrist M., Wang X. and Hunziker E. (2001). Development of mechanically stable alginate-chondrocyte constructs: effects of guluronic acid content and matrix synthesis. *J. Orthopaedic Research*, 19, 493-499. (cited p. 822).

Wong M. and Carter D. (2003). Articular cartilage functional histomorphology and mechanobiology: a research perspective. *Bone*, 33, 1-13. (cited p. 829).

Woo S.L.-Y., Simon B.R., Kuei S.C. and Akeson W.H. (1980). Quasi-linear viscoelastic properties of normal articular cartilage. *ASME J. Biomechanical Engng.*, 102, 85-90. (cited p. 841).

Woo S.L.-Y., Mow V.C. and Lai W.M. (1987). Biomechanical properties of articular cartilage. *Handbook of Bioengineering*, Skalak R. and Chien S., eds., McGraw Hill, New York, Chapter 4, 1-44. (cited p. 841).

Worster A.A., Brower-Toland B.D., Fortier L.A., Bent S.J., Williams J. and Nixon A.J. (2001). Chondrocytic differentiation of mesenchymal stem cells sequentially exposed to transforming growth factor-β1 in monolayer and insulin-like growth factor-I in a three-dimensional matrix. *J. Orthopaedic Research*, 19(4), 738-749. (cited pp. 808 and 817).

Wu H.C. and Yao R.F. (1976). Mechanical behavior of the human annulus fibrosus. *J. Biomechanics*, 9(1), 1-7. (cited pp. 799 and 800).

Wu J.T., Byrne H.M., Kyrn D.H. and Wein L.M. (2001). Modeling and analysis of a virus that replicates selectively in tumor cells. *Bulletin Mathematical Biology*, 63(4), 731-768. (cited p. 1013).

Wu Z.Z., Zhang G., Long M., Wang H.B., Song G.B. and Cai S.X. (2000). Comparison of the viscoelastic properties of normal hepatocytes and hepatocellular carcinoma cells under cytoskeletal perturbation. *Biorheology*, 37(4), 279-290. (cited p. 1039).

* * * * X * * * *

Xiao H., Bruhns O.T. and Meyers A. (1997). Logarithmic strain, logarithmic spin and logarithmic rate. *Acta Mechanica*, 124, 89-105. (cited p. 51).

Xiao H. and Chen L.-S. (2003). Hencky's logarithmic strain and dual stress-strain and strain-stress relations in isotropic finite hyperelasticity. *Int. J. Solids Structures*, 40, 1455-1463. (cited p. 50).

* * * * Y * * * *

Yamamoto E., Iwanaga W., Miyazaki H. and Hayashi K. (2002). Effects of static stress on the mechanical properties of cultured collagen fascicles from the rabbit patellar tendon. *ASME J. Biomechanical Engng.*, 124, 85-93. (cited p. 827).

Yano S., Komine M., Fujimoto M., Okochi H. and Tamaki K. (2004). Mechanical stretching *in vitro* regulates signal transduction pathways and cellular proliferation in human epidermal keratinocytes. *J. Investigative Dermatology*, 122(3), 783-790. (cited p. 1003).

Yao H. and Gu W.Y. (2004). Physical signals and solute transport in cartilage under dynamic unconfined compression: finite element analysis. *Annals Biomedical Engng.*, 32(3), 380-390. (cited p. 670).

Yao X., Kwan H.Y., Dora K.A., Garland C.J. and Huang Y. (2003). A mechanosensitive cation channel in endothelial cells and its role in vasoregulation. *Biorheology*, 40(1-3), 23-30. (cited p. 1002).

Yasuda H., Lamaze C.E. and Ikenberry L.D. (1968). Permeability of solutes through hydrated polymer membranes. Part I. Diffusion of sodium chloride. *Macromolecular Chemistry and Physics*, 118(1), 19-35. (cited p. 614).

Yeung A.T. and Datla S. (1995). Fundamental formulation of electrokinetic extraction of contaminants from soil. *Canadian Geotechnical J.*, 32, 569-583. Discussion, *Canadian Geotechnical J.*, 33, 682-684.

Yeung A.T., Hsu C.N. and Menon R.M. (1996). EDTA-enhanced electro-kinetic extraction of lead. *ASCE J. Geotechnical Engng.*, 122(8), 666-673. (cited p. 442).

Yeung A.T., Hsu C.N. and Menon R.M. (1997). Physicochemical soil-contaminant interactions during electro-kinetic extraction. *J. Hazardous Materials*, 55(1-3), 221-237.

Yu B., Herman D., Preston J., Lu W., Kirkendall D.T. and Garrett W.E. (2004). Immediate effects of a knee brace with a constraint to knee extension on knee kinematics and ground reaction forces in a stop-jump task. *American J. Sports Medicine*, 32(5), 1136-1143. (cited p. 10).

Yuan F., Baxter L.T. and Jain R.K. (1991). Pharmacokinetic analysis of two-step approaches using bifunctional and enzyme-conjugated antibodies. *Cancer Research*, 51, 3119-3130. (cited p. 1013).

Yuan F., Chen Y., Dellian M., Safabakhsh N., Ferrara N. and Jain R.K. (1996). Time-dependent vascular regression and permeability changes in established human tumor xenografts induced by an anti-vascular endothelium growth factor/vascular permeability factor antibody. *Proceedings National Academy Sciences USA*, 93, 14765-14770. (cited pp. 569, 999, and 1003).

Yuan F. (1997). Stress is good and bad for tumors. *Nature Biotechnology*, 15, 722-723. (cited p. 1004).

* * * * Z * * * *

Zahalak G.I., Wagenseil J., Wakatsuki T. and Elson E.L. (2000). A cell-based constitutive relation for bio-artificial tissues. *Biophysical J.*, 79(5), 2369-2381. (cited pp. 256, 430, and 431).

Zamir E.A. and Taber L. (2004). On the effects of residual stress in microindentation tests of soft tissue structures. *Annual Review Biomedical Engng.*, 126, 276-283. (cited p. 851).

Zeng Y., Yang J., Huang K., Lee Z. and Lee X. (2001). A comparison of biomechanical properties between human and porcine cornea. *J. Biomechanics*, 34, 533-537. (cited pp. 708 and 721).

Zerfa Z. and Loret B. (2003). Coupled dynamic elastic-plastic analysis of earth structures. *Soil Dynamics and Earthquake Engineering*, 23(6), 435-454. (cited p. 386).

Zerfa Z. and Loret B. (2004). A viscous boundary for transient analyses of saturated porous media. *Earthquake Engineering & Structural Dynamics*, 33(1), 89-110. (cited p. 386).

Zheng Q.S. and Spencer A.J.M. (1993). Tensors which characterize anisotropies. *Int. J. Engng. Science*, 31(5), 679-693.

Zheng Q.S. and Boehler J.P. (1994). Tensor function representations as applied to formulating constitutive laws for clinotropic materials. *Acta Mechanica Sinica*, 10(4), 336-348. (cited pp. 788 and 790).

Zhou J. and Fung Y.C. (1997). The degree of nonlinearity and anisotropy of blood vessel elasticity. *Proceedings National Academy Sciences USA*, 94(26), 14255-14260. (cited p. 183).

Zhou Y., Rajapakse R.K.N.D. and Graham J. (1998). A coupled thermoporoelastic model with thermo-osmosis and thermal filtration. *Int. J. Solids Structures*, 35(34-35), 4659-4683. (cited p. 310).

Zimmerman R.W., Hadgu T. and Bodvarsson G.S. (1993). Development of a dual-porosity model for vapor-dominated fractured geothermal reservoirs using a semi-analytical fracture/matrix interaction term. Proceedings, Eighteenth Workshop on Geothermal Reservoir Engineering, Stanford University, Stanford, CA, January 26-28, 1993. (cited pp. 342 and 343).

Znati C.A., Rosenstein M., McKee T.D., Brown E., Turner D., Bloomer W.D., Watkins S., Jain R.K. and Boucher Y. (2003). Irradiation reduces interstitial fluid transport and increases the collagen content in tumors. *Clinical Cancer Research*, 9, 5508-5513. (cited pp. 101, 570, 847, and 998).

Zysset P.K. and Curnier A. (1995). An alternative model for anisotropic elasticity based on fabric tensors. *Mechanics of Materials*, 21, 243-250. (cited pp. 151 and 152).

Index